ENCYCLOPEDIA OF

ENERGY TECHNOLOGY AND THE ENVIRONMENT

VOLUME 1

WILEY ENCYCLOPEDIA SERIES
IN ENVIRONMENTAL SCIENCE

ATTILIO BISIO AND SHARON BOOTS
ENCYCLOPEDIA OF ENERGY TECHNOLOGY AND THE ENVIRONMENT

ROBERT MEYERS
ENCYCLOPEDIA OF ENVIRONMENTAL ANALYSIS AND REMEDIATION

EDITORIAL BOARD

ENCYCLOPEDIA OF ENERGY TECHNOLOGY AND THE ENVIRONMENT

VOLUME 1

Attilio Bisio
Sharon Boots
Editors

A Wiley-Interscience Publication
John Wiley & Sons, Inc.
New York / Chichester / Brisbane / Toronto / Singapore

This text is printed on acid-free paper.

Library of Congress Cataloging in Publication Data:

Encyclopedia of energy technology and the environment / Attilio Bisio, Sharon Boots, editors.
p. cm.
Includes bibliographical references (p.) and index.
ISBN 0-471-54458-2
1. Power resources—Handbooks, manuals, etc. 2. Environmental protection—Handbooks, manuals, etc. I. Bisio, Attilio. II. Boots, Sharon.
TJ163.235.E53 1995
333.79'03—dc20 94-44119

Printed in the United States of America

10 9 8 7 6 5 4 3 2 1

CONTRIBUTORS

A. Warren Adam, *Consultant, Rockford, Illinois,* Organic Rankine Engines
K. A. G. Amankwah, *Syracuse University, Syracuse, New York,* Hydrogen Storage Systems
Larry Anderson, *University of Utah, Salt Lake City, Utah,* Coal Liquefaction
John Appleby, *Texas A&M University, College Station, Texas,* Fuel Cells
Emil Attanasi, *United States Geological Survey, Reston, Virginia,* Petroleum Reserve (Oil and Gas Reserve)
Jim Baggott, *Consultant, London, United Kingdom,* Lubricants, Biodegradable
William Banta, *American University, Washington, D.C.,* Oceanography
Franco Barbir, *University of Miami, Coral Gables, Florida,* Transportation Fuels, Hydrogen
Randall Barron, *Louisiana Technical University, Rusten, Louisiana,* Cryogenics
William Barry, *Co-Ag Consultants, Pittsburgh, Pennsylvania,* Coal, Transportation
Larry Baxter, *Sandia National Laboratories, Livermore, California,* Ash
Larry Beaumont, *R. W. Beck, Denver, Colorado,* Waste-to-Energy, Economics
Richard Bechtold, *EA Engineering, Silverspring, Maryland,* Methanol Vehicles
Martin Bender, *The Land Institute, Salina, Kansas,* Agriculture and Energy
Ron Berglund, *TRC, Houston, Texas,* Exhaust Control, Industrial
J. Bernaden, *Johnson Control Inc., Milwaukee, Wisconsin,* Energy Management, Principles
A. K. Bhargava, *ABB Lummus Crest, Inc., Bloomfield, New Jersey,* Coking, Delayed
Walter Bienert, *DTX Corporation, Lancaster, Pennsylvania,* Heat Pipes
Richard A. Birdsley, *USDA Forest Service, Northeastern Forest Station, Radnor, Pennsylvania,* Carbon Storage in Forests
Attilio Bisio, *Atro Associates, Mountainside, New Jersey,* District Heating; Energy Conversion Factors; Heating Value; Life Cycle Analysis; Nuclear Power; OPEC; Renewable Resources
Frank Black, *Atmospheric Research and Exposures Assessment, EPA, Research Triangle Park, North Carolina,* Automobile Emissions
George Blomgren, *Eveready Battery Co., Westlake, Ohio,* Batteries
Karl Boer, *University of Delaware, Newark, Delaware,* Global Health Index
E. R. Booser, *Consultant, Scotia, New York,* Lubricants
Gary R. Boss, *General Accounting Office, Washington, D.C.,* Energy Policy Planning
Ugo Bozzano, *UOP, Des Plains, Illinois,* Reformulated Gasoline
J. C. Bricker, *UOP, Des Plains, Illinois,* Isomerization
Pamela J. Brodowicz, *EPA, Ann Arbor, Michigan,* Air Pollution, Automobile, Toxic Emissions
F. J. Brooks, *GE Industrial and Power Systems, Schenectady, New York,* Gas Turbines
Bill Browning, *Rocky Mountain Institute, Snowmass, Colorado,* Lighting, Energy Efficient
Gregory Bryant, *Consultant, Eugene, Oregon,* Human Powered Vehicles
Bruce Bunker, *Pacific Northwest Laboratories, Richland, Washington,* Nuclear Materials, Radioactive Tank Wastes
Robert Burger, *Burger Associates, Inc., Dallas, Texas,* Cooling Towers
Andrew F. Burke, *Consultant, Idaho Falls, Idaho,* Hybrid Vehicles
Joan Bursey, *Radian Corp., Research Triangle Park, North Carolina,* Environmental Analysis, Mass Spectrometry
Richard J. Camm, *University of Manchester, Institute of Science and Technology, Manchester, United Kingdom,* Computer Applications for Energy Efficient Systems
Penny M. Carey, *EPA, Warren, Michigan,* Air Pollution: Automobile, Toxic Emissions
Lon Carlson, *Illinois State University, Normal, Illinois.* Environmental Economics
John A. Casazza, *CSA Energy Consultants, Arlington, Virginia,* Electric Power Systems and Transmission
Clarence D. Chang, *Mobil Research and Development, Princeton, New Jersey,* Fuels, Synthetic, Liquid
Scott Chaplin, *Rocky Mountain Institute, Snowmass, Colorado,* Recycling and Reuse, Energy savings; Water Efficiency
Sylvie Charpenay, *Advanced Fuel Research, Inc., East Hartford, Connecticut,* Pyrolysis
Norman Chigier, *Carnegie Mellon University, Pittsburgh, Pennsylvania,* Combustion Systems, Measurements; Liquid Fuel Spray Combustion
Robert H. Clark, *Consultant, Canada,* Tidal Power
Jerry L. Clayton, *U.S. Geological Survey, Reston, Virginia,* Coalbed Gas
Stephen Cleary, *Medical College of Virginia, Richmond, Virginia,* Electromagnetic Fields, Health Effects
Jeffrey F. Clunie, *R. W. Beck and Associates, Denver, Colorado,* Waste-to-Energy Economics
Art Cohen, *University of South Carolina, Columbia, South Carolina,* Peat: Environment and Energy Uses
Richard N. Cooper, *Harvard University, Cambridge, Massachusetts,* Global Climate Change, Mitigation
Robert M. Counce, *University of Tennessee, Knoxville, Tennessee,* Gas Cleanup Absorption
John Counsil, *Santa Rosa, California,* Geothermal Energy
Burton B. Crocker, *Chesterfield, Missouri,* Air Pollution Control Methods
Daniel A. Crowl, *Michigan Technological University, Houghton, Michigan,* Hazard Analysis of Energy Facilities
Charles Coutant, *Oak Ridge National Laboratory, Oak Ridge, Tennessee,* Thermal Pollution, Power Plants
Vincent Covello, *Columbia University, New York,* Risk Assessment; Risk Communication
N. A. Cusher, *UOP, Des Plains, Illinois,* Isomerization
T. A. Czuppon, *The M. W. Kellog Company, Houston, Texas,* Hydrogen
Bukin Danley, *University of Wollongong, Wollongong, NSW, Australia,* Coal Availability
Douglas Decker, *Johnson Control, Inc., Milwaukee, Wisconsin,* Energy Management, Principles
M. J. Lemos de Sousa, *University of Porto, Porto, Portugal,* Coal Classifications: Past and Present
V. R. Dhole, *University of Manchester Institute of Science and Technology, Manchester, UK,* Computer Applications for Energy Efficient Systems

William G. Dukek, *Summit, New Jersey,* Aircraft Fuels
K. G. Duleep, *EEA, Arlington, Virginia,* Automotive Engines, Efficiency
R. A. Durie, *CSIRO, North Ryde, NSW, Australia,* Coal Classifications: Past and Present
Jim Dyer, *Rocky Mountain Institute, Water Program, Snowmass, Colorado,* Water Efficiency
Alan D. Eastman, *Phillips Petroleum Research Center, Bartlesville, Oklahoma,* Hydrocarbons
G. Yale Eastman, *Thermacore, Inc., Lancaster, Pennsylvania,* Heat Pipes
W. W. Eckenfelder, *Eckenfelder, Inc., Nashville, Tennessee,* Water Quality Management
Eugene Ecklund, *EA Engineering, Silverspring, Maryland,* Methanol Vehicles
James A. Edmonds, *Pacific Northwest Laboratory, Washington, D.C.,* Carbond Dioxide Emissions, Fossil Fuel
Frederick Ehrich, *Consultant, Marblehead, Massachusetts,* Aircraft Engines
William D. Ehmann, *University of Kentucky, Lexington, Kentucky,* Radiation Hazards, Health Physics; Radiation Monitoring
Raymond J. Ehrig, *ARISTECH Chemical Corporation, Monroeville, Pennsylvania,* Recycling, Tertiary, Plastics
Thomas D. Ellis, *DuPont Engineering, Newark, Delaware,* Incineration
William R. Ellis, *Raytheon Constructor, Ebasco Division, New York, New York,* Fusion Energy
Donald M. Ernst, *DTX Corporation, Lancaster, Pennsylvania,* Heat Pipes
Ramon Espino, *Exxon Research and Engineering, Annandale, New Jersey,* API Engine Service Categories; Afterburning; Antiknock Compounds; Autoignition Temperature; Boundary Lubrication; Carbon Residue; Exhaust and Gas Recirculation; Flammable Limits; Flash Point; Gas Laws; Heat Balance; Heating Oil; Internal Combustion Engine; Knock; Lubricants; Additives; Middle Distillate; Octane Number; Viscosity Grades; Vapor Lock
James Fair, *University of Texas, Austin, Texas,* Distillation
Rocco Fiato, *Exxon Research and Engineering, Florham, New Jersey,* Fischer-Tropsch Process and Products
Robert Finkelman, *U.S. Geological Survey, Reston, Virginia,* Coal Quality
D. B. Firth, *University of Manchester Institute of Science and Technology, Manchester, United Kingdom, Computer Applications for Energy Efficient Systems*
Jack Fishman, *NASA, Hampton, Virginia,* Ozone, Tropospheric
Robert A. Frosch, *General Motors Research Laboratories, Warren, Michigan,* Global Climate Change, Mitigation
Conan Furber, *Consultant, Hallowell, Maine,* Railroads
Mark Gattuso, *UOP, Des Plaines, Illinois,* Gum in Gasoline
N. J. Gernert, *Thermacore, Inc., Lancaster, Pennsylvania,* Cold Fusion
Lewis M. Gibbs, *Chevron Research & Technology Company, Richmond, California,* Transportation Fuels, Automotive
Jennifer Gilitz, *Consultant, Worcester, Massachusetts,* Aluminum
J. Duncan Glover, *Failure Analysis Associates, Inc., Framingham, Massachusetts,* Electric Power Distribution
Diane R. Gold, *Channing Laboratory, Boston, Massachusetts,* Air Pollution, Indoor
Henry Gong, Jr., *Rancho Los Amigos Medical Center, Downey, California, and University of Southern California, School of Medicine,* Air Pollution, Health Effects
Ralph Gray, *Consultant, Monroeville, Pennsylvania,* Coal Coking
Michael Grub, *Royal Institute of Internal Affairs, London, UK,* Carbon Dioxide Emissions, Fossil Fuels
David Gushee, *Congressional Research Service, Library of Congress, Washington, D.C.,* Energy Taxation, Automobile Fuels
George Hagerman, *SEASUN Power Systems, Alexandria, Virginia,* Wave Power
Daniel Halacy, *Consultant, Lakewood, Colorado,* Solar Cells, Solar Cooking, Solar Heating, Solar Thermal Electric
Harold Hammershaimb, *UOP, Des Plaines, Illinois,* Alkylation
Scott Han, *Mobil Research and Development Corp., Princeton, New Jersey,* Fuels, Synthetic, Liquid Fuels
Patrick G. Hatcher, *Pennsylvania State University, University Park,* Coal Pyrolysis
Robert Heistand, *Consultant, Englewood, Colorado,* Oil Shale
John Hem, *United States Geologial Survey, Menlo Park, California,* Water Quality Issues
Byron Y. Hill, *Union Carbide Corporation, South Charleston, West Virginia,* Gas Cleanup Absorption
Parker Hirtle, *Acentech Incorporated, Cambridge, Massachusetts,* Insulation, Acoustic
Ronald Hites, *Indiana University, Bloomington,* Environmental Analysis, Mass Spectrometry Advances
Richard Houghton, *Woods Hole Research Center, Woods Hole, Massachusetts,* Carbon Cycle
James Hower, *University of Kentucky, Lexington,* Coal Availability
Thomas Hunt, *Advance Modular Power Systems, Ann Arbor, Michigan,* Sodium Heat Engines
Adrian C. Hutton, *University of Wollongong, Wollongong, NSW, Australia,* Coal Availability
Jiri Janata, *Pacific Northwest Laboratory, Richland, Washington,* Environmental Analysis, Chemical Sensor Applications
Peter A. Johnson, *United States Congress, Washington, D.C.,* Nuclear Power, Managing Nuclear Materials
Andrew Jones, *Rocky Mountain Institute, Water Program, Snowmass, Colorado,* Water Efficiency
Jack Kaye, *NASA, Washington, D.C.,* Ozone, Stratospheric
Aaron Kelly, *UOP, Des Plaines, Illinois,* Catalytic Reforming
James P. Kelly, *Pacific Northwest Laboratory, Richland, Washington,* Environmental Analysis, Optical Spectroscopy
William Kenney, *Consultant, Florham Park, New Jersey,* Energy Efficiency, Calculations; Heat Pumps; Heat Recovery; Thermodynamics; Thermodynamics, Process Analysis
Robert Kessler, *Textron Defense Systems, Everett, Massachusetts,* Magnetohydrodynamics
Naresh Khosla, *Enviro Management Research, Inc., Springfield, Virginia,* Energy Auditing
Donald L. Klass, *Entech International, Inc., Barrington, Illinois,* Fuels from Biomass
Annette Koklauner, *Gas Research Institute, Washington, D.C.,* Petroleum Markets
Paul Komor, *Office of Technology Assessment, United States Congress, Washington, D.C.,* Energy Consumption in the United States
S. A. Kruz, *The M. W. Kellog Company, Houston, Texas,* Hydrogen
Paul Kuchar, *UOP, Des Plains, Illinois,* Isomerization
Bill Kuhn, *Pacific Northwest Laboratories, Richland, Washington,* Nuclear Materials, Radioactive Tank Wastes
Karl S. Kunz, *Pennsylvania State University, University Park, Pennsylvania,* Electric Heating
James Langman, *Earth Island Institute, San Francisco, California,* Environmental and Conservation Organizations
William Lanouette, *General Accounting Office, Washington, D.C.,* Energy Policy Planning
Karen Larsen, *Office of Technology Assessment, Washington, D.C.,* Electrical Energy Efficiency, Commercial Application; Electrical Energy Efficiency, Utility Programs
Ben E. Law, *U.S. Geological Survey, Reston, Virginia,* Coalbed Gas
Salvatore Lazzari, *Congressional Research Service, Library of Congress, Washington, D.C.,* Energy Taxation, Automobile Fuels; Energy Taxation, Subsidies for Biomass; Energy Taxation; Biomass
Thomas H. Lee, *Massachusetts Institute of Technology, Cambridge, Massachusetts,* Global Climate Change, Mitigation
D. J. LeKang, *CSA Energy Consultants, Arlington, Virginia,* Electric Power Systems and Transmission
William S. Linn, *Rancho Los Amigos Medical Center, Downey, California, and University of Southern California, School of Medicine,* Air Pollution, Health Effects
W. S. Louie, *ABB Lummus Crest Inc., Bloomfield, New Jersey,* Coking, Delayed
Amory Lovins, *Rocky Mountain Institute, Snowmass, Colorado,* Agriculture and Energy; Lighting, Energy Efficient; Supercars
L. Hunter Lovins, *Rocky Mountain Institute, Snowmass, Colorado,* Agriculture and Energy

Paul Lyons, *U.S. Geological Survey, Reston, Virginia,* Coal Classification
Greg Marland, Oak Ridge National Laboratory, Oak Ridge, Tennessee, Global Climate Change, Mitigation
Amy R. Marrow, *General Accounting Office, Washington, D.C.,* Energy Policy Planning
Susan L. Mayer, *Congressional Research Service, Library of Congress, Washington, D.C.,* Clean Air Act
Loch McCabe, *Resource Recycling Systems, Inc., Ann Arbor, Michigan,* Waste Management Planning
Robin S. McDowell, *Pacific Northwest Laboratory, Richland, Virginia.* Environmental Analysis, Optical spectroscopy.
Jon McGowen, *University of Massachusetts, Amherst,* Wind Power
David E. Mears, *Unocal,* Hydrocarbons
Richard Meyer, *U.S. Geological Survey, Reston, Virginia,* Bitumen
Mark Mills, *Mills, McCarthy & Associates, Inc., Chevy Chase, Maryland,* Transportation Fuels, Electricity
John Mooney, *Englehard Corporation, Iselin, New Jersey,* Exhaust Control, Automotive
J. H. Moore, *GE Industrial and Powers Systems, Schenectady, New York,* Steam Turbines
Michael Morrison, *Caminus Energy Limited, Cambridge, United Kingdom,* Carbon Dioxide
Armand Moscovici, *Kerrite Co., Seymour, California,* Insulation, Electric; Wire and Cable Coverings
C. L. Moy, *UOP, Des Plains, Illinois,* Isomerization
Andrew Moyad, *Environmental Protection Agency, Washington, D.C.,* Energy Consumption in the United States; Nuclear Power, Decommissioning power plants
Carl Moyer, *ACUREX Environmental Corp., Mountain View, California,* Alcohol Fuels
D. S. Newsome, *The M. W. Kellog Company, Houston, Texas,* Hydrogen
Les Norford, *Massachusetts Institute of Technology, Cambridge, Massachusetts,* Building Systems
Andrew Papay, *Ethyl Petroleum Additives, Inc., St. Louis, Missouri,* Hydraulic Fluids, Synthetic
B. K. Parekh, *University of Kentucky Lexington, Kentucky,* Coal Availability
Lee Paulson, *Morgantown Energy and Technology Center, Morgantown, West Virginia,* Heat Engines, Direct-Fired Coal
Carmo J. Periera, *W. R. Grace & Co. Research Division, Columbia, Maryland,* Catalytic Cracking
Joseph J. Perona, *University of Tennessee, Knoxville,* Gas Cleanup, Absorption
Anthony Pietsch, *Chandler, Arizona,* Closed Brayton Cycle
H. J. Pinheiro, *University of Porto, Porto, Portugal,* Coal Classifications, Past and Present
Richard Pinkham, *Rocky Mountain Institue, Water Program, Snowmass, Colorado,* Water Efficiency
Edmond Piper, *Consultant, Littleton, Colorado,* Oil Shale
Thomas Perdy, *Johnson Control, Inc., Milwaukee, Wisconsin,* Energy Management, Principles
Thomas Polk, *Annandale, Virginia,* Recycling
Richard L. Powell, *R&T Centre, Runcorn, Cheshire, United Kingdom,* Refrigerant Alternatives
David E. Prinzing, *Foster Wheeler Environmental Corp., Sacramento, California,* Fuel Resources
Rod Quinn, *Pacific Northwest Laboratory, Richland, Washington,* Nuclear Materials, Radioactive tank wastes
Sam Raskin, *California Energy Commission, Sacramento, California,* Commercial Availability of Energy Technology
Raymond Regan, *Pennsylvania State University, University Park,* Activated sludge
Danny Reible, *Louisiana State University, Baton Rouge,* Chemodynamics
Dudley Rice, *U.S. Geological Survey, Denver, Colorado,* Coalbed Gas
James Robinson, *Trevose, Pennsylvania,* Water Conditioning
D. H. Root, *United States Geological Survey, Reston, Virginia,* Petroleum Reserve (Oil and Gas Reserve)
Arthur H. Rosenfeld, *Lawrence Berkeley Laboratory, University of California, Berkeley, California,* Global Climate Change, Mitigation
Marc Ross, *University of Michigan, Ann Arbor,* Energy Efficiency
Robin Roy, *United States Congress, Washington, D.C.,* Nuclear Power, Decomissioning Power Plants; Nuclear Power; Safety of Aging Power Plants
Edward S. Rubin, *Carnegie Mellon University, Pittsburgh, Pennsylvania,* Global Climate Change, Mitigation
Ted Russell, *Carnegie Mellon University, Pittsburgh, Pennsylvania,* Air Quality Modeling
Michael G. Ryan, *USDA Forest Service, Fort Collins, Colorado,* Carbon Balance Modeling
Sheppard Salon, *Rensselaer Polytechnic Institute, Troy, New York,* Electric Power Generation
Harry J. Sauer, Jr., *University of Missouri, Rolla, Missouri,* Air Conditioning; Refrigeration
Barbara Schaefer-Pederson, *UOP, Des Plaines, Illinois,* Hydrocracking
Donald Scherer, *Bowling Green State University, Bowling Green, Ohio,* Sustainable Resources, Ethics
Ron Schmitt, *Amoco Corporation, Chicago, Illinois,* Petroleum Refining, Emissions and wastes
James A. Schwartz, *Syracuse University, Syracuse, New York,* Hydrogen Storage Systems
Andrew C. Scott, *University of London, Surrey, United Kingdom,* Coal Availability
Neil Seldman, *Institute for Local Self Reliance, Washington, D.C.,* Recycling, History in the United States
Richard J. Seymour, *Texas A&M University, College Station,* Renewable Resources from the Ocean
Michael A. Serio, *Advanced Fuel Research, East Hartford, Connecticut,* Pyrolysis
Ramesh Shah, *General Motors Corporation, Harrison Divison Lockport, New York,* Heat Exchangers
Robert M. Shaubach, *Thermacore, Inc., Lancaster, Pennsylvania,* Cold Fusion
Eric Silberhorn, *Technology Sciences Group, Washington, D.C.,* PCB's
E. A. Skrabek, *Fairchild Space and Defense Company, Germantown, Maryland,* Thermoelectric Energy Conversion
B. Smith, *University of Manchester Institute of Science and Technology, Manchester, United Kingdom,* Computer Applications for Energy Efficient Systems
William Smith, *Yale University, New Haven, Connecticut,* Acid Rain
William C. Smith, *Morgantown Energy Technology Center, Morgantown, West Virginia,* Heat Engines, Direct-Fired Coal
L. Douglas Smoot, *Brigham Young University, Provo, Utah,* Coal Combustion: Coal Gasification; Combustion Modeling
Henry E. Sostmann, *Albuquerque, New Mexico,* Temperature Measurement
James Speight, *Laramie, Wyoming,* Asphalt; Asphaltenes; Caustic Washing; Cetane Number; Extra Heavy Oils; Fuels, Synthetic, Gaseous Fuels; Hydroprocessing; Kerosene; Liquefied Petroleum Gas; Manufactured Gas; Natural Gas; Petroleum Refining; Pipelines; Polymerization; Tar Sands; Underground Gasification; Visbreaking
John S. Spencer, Jr.,*USDA, St. Paul, Minnesota,* Forest Resources
Edward M. Stack, *University of South Carolina, Columbia,* Peat: Environment and Energy Use
Ronald W. Stanton, *U.S. Geological Survey, Reston, Virginia,* Coal Quality
Andrew Steer, *The World Bank, Washington, D.C.,* Global Environmental Change, Population effect
Arthur Stefani, *ABB Lummus, Bloomfield, New Jersey,* Coking, Delayed
Donald H. Stedman, *University of Denver, Denver, Colorado,* Automobile Emissions, Control
Andrew Steer, *The World Bank, Washington, D.C.,* Global Environmental Change, Population and Economic Growth
Dan Steinmeyer, *Monsanto, St. Louis, Missouri,* Energy Conservation: Energy Management, Process
J. Hugo Steven, *ICI Chemicals & Polymers Ltd., United Kingdom,* Refrigerant Alternatives
Deborah D. Stine, *National Academies of Science and Engineering, Washington, D.C.,* Global Climate Change, Mitigation
Frank Stodolsky, *Argonne National Laboratory, Washington, D.C.,* Transportation Fuels, Natural Gas

D. D. Sullivan, *UOP, Des Plains, Illinois,* Isomerization

Robert Szaro, *USDA Forest Service, Washington, D.C.,* Biodiversity Maintenance

Patrick Ten Brink, *Caminus Energy Ltd., Cambridge, UK,* Carbon Dioxide Emissions, Fossil Fuels

Clemens M. Thoennes, *GE Corporation, Schenectady, New York,* Cogeneration

Gregory Thompson, *UOP, Des Plaines, Illinois,* Hydrocracking

David Tillman, *Foster Wheeler Environmental Corporation, Sacramento, California,* Fuel Resources; Fuels from Waste

Thomas W. Tippett, *UOP, Des Plaines, Illinois,* Hydrocracking

Sergio C. Trindade, *SET International Ltd., Scarsdale, New York,* Transportation Fuels, Ethanol Fuels in Brazil

R. P. Tye, *Consultant, Surrey, United Kingdom,* Insulation, Thermal

Diane E. Vance, *Analytical Services Organization, Oak Ridge, Tennessee,* Radiation Hazards, Health Physics Radiation Monitoring

John Vandermeulen, *Bedford Institute of Oceanography, Nova Scotia, Canada,* Oil Spills

Carl Vansant, *HCI Publications, Kansas City, Missouri,* Hydropower

Luis A. Vega, *Pacific International Center for High Technology Research, Honolulu, Hawaii,* Ocean Thermal Energy Conversion

Walter Vergara, *The World Bank, Washington, D.C.,* Transportation Fuels, Ethanol Fuels in Brazil

T. Nejat Veziroglu, *University of Miami, Coral Gables, Florida,* Transportation Fuels, Hydrogen

Jud Virden, *Pacific Northwest Laboratories, Richland Washington,* Nuclear Materials, Radioactive Tank Wastes

Karl Vorres, *Argonne National Laboratory, Argonne, Illinois,* Lignite and Brown Coal

Michael P. Walsh, *Arlington, Virginia,* Air Pollution, Automobile; Clean Air Act, Mobile Sources

Richard Waring, *Oregon State University, Corvallis, Oregon,* Carbon Balance Modeling

Jeffrey Warshauer, *Foster Wheeler Environmental Corporation, Sacramento, California,* Fuel Resources

David Wear, *Forestry Sciences Laboratory, Research Triangle Park, North Carolina,* Forestry, Sustainable

Mary Wees, *HDR Engineering, Omaha, Nebraska,* Waste to Energy

Kenneth Wilund, *Argonne National Laboratory, Washington, D.C.,* Transportation Fuels, National Gas

Barry M. Wise, *Pacific Northwest Laboratory, Richland, Washington,* Environmental Analysis, Chemical Sensor Applications

Marek M. Wojtowicz, *Advanced Fuel Research, Inc., East Hartford, Connecticut,* Pyrolysis

George Wolff, *GM Research and Environmental Staff, Warren, Michigan,* air Pollution

Charles Wood, *Southwest Research Institute, San Antonio, Texas,* Automotive Engines

Markus Zahn, *MIT, Cambridge, Massachusetts,* Electromagnetic Fields

Mario Zamora, *MT Drilling Fluids Co., Houston Texas,* Drilling Fluids and Muds

John Zerbe, *Forest Products Laboratory, Madison, Wisconsin,* Wood for Heating and Cooking

Etuan Zhang, *Pennsylvania State University, University Park, Pennsylvania,* Coal Pyrolysis

PREFACE

Energy touches all aspects of life on this planet. Civilizations through the ages exploited forest and water resources. More recently we have increasingly utilized petroleum and gas, and to a lesser degree, coal to generate energy in useful forms.

Complex market forces determine how energy is generated and used. However, there is agreement that use of energy generates problems that market forces are not able to resolve satisfactorily. Indeed, the pollution associated with the production and consumption of energy involves a complex set of problems that sometimes appears almost intractable.

The processes associated with the extraction, transformation, and use of energy have great potential for damaging the environment. Considerable attention, for example, has been given over the past two decades to air pollution arising from the burning of fossil fuels. Production of coal and oil has resulted in contamination of water. More recently, considerable concern has been expressed over the contribution that combustion of fossil fuels may make to global climate change.

Concerns about the environment have had a significant impact on both energy prices and the technologies used to generate and use energy. Fears over the environmental impact of nuclear power are "credited" with bringing expansion of that option to an end. Concerns over the potential for oil spills have resulted in potentially promising areas no longer being considered for exploration and production. Energy conservation and the use of renewable forms of energy have moved to the forefront.

The nature and, to some degree, the quantity of energy used in the United States has been influenced by specific laws directed at protecting the environment, not energy policies. These laws have had a major impact on how and where energy resources are exploited and what technology can be used for the production and use of energy. In part, this is because many regulations, such as those developed under the Clean Air Act Amendments of 1990, rely on technology-based standards. For example, the Act mandated changes in the fuels used in automobiles; the changes made a significant difference in the fortunes of those companies that had the technological capability to exploit the opportunities that were created.

The goal of the editors of this *Encyclopedia* has been to provide to as wide an audience as possible information on the technology involved in energy production and use as well as information on how these technologies affect the environment and quality of life on this planet. We hope that engineers, scientists, educators, and citizens actively involved in the development of energy technology and public policy issues will benefit from this work.

The editors wish to thank the many outstanding authors who made this *Encyclopedia* possible. Their willingness to set aside time to write about the critical areas related to energy technology and the environment, made this work possible.

Attilio Bisio
Sharon Boots

CONVERSION FACTORS

Often in dealing with questions about energy, it is useful to be able to convert from one set of units to another. Most often, energy units have their origin in some historic event such as the development of steam power. Therefore, a variety of units are used throughout the world.

The following tables are provided:

Table 1. Common Energy Conversion Factors

	Joule	Quadrillion BTU	Kilogram Calorie	Metric Ton of Coal Equivalent
1 Joule	1	947.9×10^{-21}	239×10^{-6}	34.1×10^{-12}
1 Quadrillion BTU	1055×10^{15}	1	252×10^{12}	36.0×10^{6}
1 Kilogram Calorie	4184	3966×10^{-18}	1	142.9×10^{-9}
1 Metric Ton of Coal Equivalent	29.3×10^{9}	27.8×10^{-9}	7×10^{6}	1
1 Barrel of Oil Equivalent	6119×10^{6}	5.8×10^{-9}	1462×10^{3}	0.21
1 Metric Ton of Oil Equivalent	44.8×10^{9}	42.4×10^{-9}	10.7×10^{6}	1.53
1 Cubic Meter of Natural Gas	37.3×10^{6}	35.3×10^{-12}	8905	1272×10^{-6}
1 Terawatt Year	31.5×10^{18}	29.9	7537×10^{12}	1076×10^{6}

Table 2. Aggregate Energy Equivalents

1 MBDOE = 1 million barrels per day of oil equivalent
= 50 million tons of oil equivalent per year
= 76 million metric tons of oil equivalent per year
= 57 billion cubic meters of natural gas per year
= 2.2 10^{18} joules per year
= 530 × 10^{12} kilocalories per year
= 2.1 × 10^{15} Btus per year = 2.1 quads
= 620 10^{9} kwh per year

1 QUAD = 1 quadrillion Btus = 10^{15} Btus
= 500,000 petroleum barrels a day
182,500,000 barrels per year
= 40,000,000 short tons of bituminous coal = 36,363,636 metric tons
= 1 trillion (10^{12}) cubic feet of natural gas
= 100 billion (10^{11}) kwh (based on 10,000 Btu/kwh heat rate)

1 kilowatt-hour of hydropower
= 10 × 10^{3} Btus
= 0.88 lb coal
= 0.076 gallon crude oil
= 10.4 cubic feet of natural gas

1 MTCE = one million short tons of coal equivalent
= 4.48 × 10^{6} barrels of crude oil
= 67 tons of crude oil
= 25.19 × 10^{12} cubic feet of natural gas

Table 3. Weights of Typical Petroleum Products[a]

	Pounds per U.S. Gallon	Pounds per 55-gal Drum	Kilograms per Cubic Meter	Barrels (42-gal) per Short Ton	Barrels (42-gal) per Metric Ton
LP-Gas	4.52	248	541.6	10.5	11.6
Aviation gasoline	5.90	325	707.0	8.2	8.9
Motor gasoline	6.17	339	739.3	7.7	8.5
Kerosene	6.76	372	810.0	7.0	7.8
Distillate fuel oils	7.05	388	845.8	6.8	7.5
Lubricating oils	7.50	413	898.7	6.3	7.0
Residual fuel oils	7.88	434	944.2	6.0	6.7
Paraffin Wax		367	800.1	7.1	7.9
Grease		458	998.8	5.7	6.3
Asphalt		477	1039.2	5.5	6.1

[a] Source: U.S. Energy Information Administration, *Monthly Energy Review.*

Table 4. Equivalence of Mass and Energy

	1 Electron Mass	1 Atomic Mass Unit	1 Gram
Million electron volts	0.511	931.5	5.61×10^{26}
Joules	8.19×10^{-14}	1.49×10^{-10}	8.99×10^{13}
Btu	7.76×10^{-17}	1.42×10^{-13}	8.52×10^{10}
Kilowatt hours	2.27×10^{-20}	4.15×10^{-17}	2.50×10^{7}
Quads	7.76×10^{-35}	1.42×10^{-31}	8.52×10^{-8}

Table 5. Energy Factors

	1 Electron-Volt	1 Joule	1 British Thermal Unit	1 Kilocalorie	1 Kilowatt-Hour
Electron volt	1	6.24×10^{18}	6.58×10^{21}	2.61×10^{22}	2.25×10^{25}
Joule	1.60×10^{-19}	1	1054	4184	3.6×10^{6}
Calorie	3.83×10^{-20}	0.24	252	1000	860.4×10^{5}
Btu	1.52×10^{-22}	9.48×10^{-4}	1	3.97	3413
Kilocalorie	3.83×10^{-23}	2.39×10^{-4}	0.252	1	860.4
Kilowatt-hour	4.45×10^{-26}	2.78×10^{-7}	2.93×10^{-4}	1.16×10^{-3}	1
Megawatt-day	1.85×10^{-30}	1.16×10^{-11}	1.22×10^{-8}	4.84×10^{-8}	4.17×10^{-5}
Quad	1.52×10^{-40}	9.48×10^{-22}	10^{-18}	3.97×10^{-18}	3.41×10^{-15}

Table 6. Power Factors

	1 Btu per Day	1 Kilowatt-Hour per Year	1 Watt (W)	1 Kilowatt	1 Megawatt	1 Gigawatt	1 Quad per Year
Btu/day	1	9.35	81.95	8.2×10^4	8.2×10^7	8.2×10^{10}	2.74×10^{15}
Kilowatt/year	0.11	1	8.77	8766	8.8×10^6	8.77×10^9	2.93×10^{14}
Watts	0.012	0.114	1	1000	10^6	10^9	3.34×10^{13}
Kilowatts	1.22×10^{-5}	1.14×10^{-4}	0.001	1	1000	10^6	3.34×10^{10}
Megawatts	1.22×10^{-8}	1.14×10^{-7}	10^{-6}	0.001	1	1000	3.34×10^7
Gigawatts	1.22×10^{-11}	1.14×10^{-10}	10^{-9}	10^{-6}	0.001	1	3.34×10^4
Quad/yr	3.65×10^{-16}	3.41×10^{-15}	2.99×10^{-14}	2.99×10^{-11}	2.99×10^{-8}	2.99×10^{-5}	1

Table 7. Fluid Flow Factors

	1 Gallon per Day	1 Acre-Foot per Year	1 Cubic Foot per Minute	1 Cubic Meter per Second	1 Billion Gallons per Day
Gallon/day	1	892	1.077×10^4	2.28×10^7	10^9
Acre-cubic feet/yr	1.12×10^{-3}	1	12.07	2.56×10^4	1.12×10^6
Cubic feet/min	9.28×10^{-5}	8.28×10^{-2}	1	2119	9.28×10^4
Cubic feet/sec	1.55×10^{-6}	1.38×10^{-3}	1.67×10^{-2}	35.31	1547
Cubic meters/sec	4.38×10^{-8}	3.91×10^{-5}	4.72×10^{-4}	1	43.8
Billion gallons/day	10^{-9}	8.92×10^{-7}	1.08×10^{-5}	2.28×10^{-2}	1

Table 8. Useful Quantities for Global Climate Change

Quantity	Value	Ref.[a]
Solar constant	1.375 kilowatts square meters	1
Earth mass	5.976×10^{24} kilogram	2
Equatorial radius	6.378×10^6 meters	2
Polar radius	6.357×10^6 meters	2
Mean radius	6.371×10^6 meters	
Surface area	5.101×10^{14} square meters	
Land area	1.481×10^{14} square meters	3
Ocean area	3.620×10^{14} square meters	4
Ice sheets and glaciers area	0.14×10^{14} square meters	5
Mean land elevation	840 meters	4
Mean ocean depth	3730 meters	
Mean ocean volume	1.350×10^{18} cubic meters	4
Ocean mass	1.384×10^{21} kilograms	
Mass of atmosphere	5.137×10^{18} kilograms	6
Equatorial surface gravity	9.780 meters/second	2

[a] Sources

1. D. V. Hoyt, "The Smithsonian Astrophysical Observatory Solar Constant Program," *Rev. Geophys. Space Physics* **17,** 427–458 (1979).
2. F. Press and R. Siever, *Earth,* W. H. Freeman and Company, San Francisco, 1974.
3. B. K. Ridley, *The Physical Environment,* Ellis Horwood, Ltd., West Sussex, England, 1979.
4. H. W. Menard and S. M. Smith, "Hypsometry of Ocean Basin Provinces," *J. Geophys. Res.* **71,** 4305–4325 (1966), adopted as reference standard by Bolin (10).
5. M. F. Meier, ed. *Glaciers, Ice Sheets, and Sea Level: Effect of a CO_2-Induced Climatic Change.* DOE/ER-60235-1, U.S. Department of Energy, Carbon Dioxide Research Division, Office of Basic Energy Sciences, Washington, D.C., 1985.
6. K. E. Trenberth, "Seasonal Variations in Global Sea-Level Pressure and the Total Mass of the Atmosphere." *J. Geophys. Res.* **86,** 5238–5246 (1981); this supercedes value adopted as reference standard by Bolin (10).

ENCYCLOPEDIA OF

ENERGY TECHNOLOGY AND THE ENVIRONMENT

VOLUME 1

API ENGINE SERVICE CATEGORIES

RAMON ESPINO
Exxon Research and Engineering
Annandale, New Jersey

Engine Service Categories are intended to provide the consumer with recommendations concerning which engine oil should be used in a given vehicle and given climate conditions. The essence of the Service Categories is a set of tests that the engine oil must pass in order to claim a performance level. Since 1947 the tests have become increasingly more severe as well as numerous. For the gasoline engines the categories have gone from SA, alphabetically to SG, with Category SH being introduced in mid-1993. For diesel engines the highest classification level is presently CF and CG level performance expected in 1994. In 1947 the Lubrication Committee of the America Petroleum Institute (API) published the first classification and since that time, numerous revisions have been made in both the categories that define gasoline as well as diesel engines. In 1990 the American and Japanese automobile manufacturers had requested that the system be significantly revamped: The auto makers were concerned about the slow pace of introduction of higher service categories; the poor reproducibility of many of the engine tests used to measure engine oil performance; and the limited effort made by API to ensure that engine oils indeed met the classification requirements indicated in the containers. (See also AUTOMOTIVE ENGINES; LUBRICANTS; LUBRICANTS, ADDITIVES.)

Another important concern of auto makers, which is also shared by engine oil marketers as well, has been the increasing proliferation of service categories. For example, starting in 1976 a consortium of European auto makers developed common service categories. In addition, some European auto makers have had their own engine oil categories as well. The end result has been that the consumer in reading the labels of a container of engine oil can be faced with a veritable alphabet soup of classifications, SG/CD/PD2/G2, etc.

In 1992 an agreement was reached between the API and Japanese and American auto makers. The engine oil additive manufacturers also joined in the agreement under the auspices of the Chemical Manufacturers Association. The agreement has a number of critical elements:

- Definitions of base stocks and additive changes that circumscribe the testing protocol required when changes in base stocks or additives are needed to pass an engine test. The changes in question include the concentration as well as the chemical composition of the additive or base stock.
- Testing procedures are based on sound statistical concepts for the engine tests that must be passed to claim a given category. The procedures include protocols for engine stand selection and for monitoring engine test repeatability.
- A strengthen system is used to monitor the quality claims of engine oils in the market place.

A number of antiquated Service Categories will be abolished by the API, but a new label will appear in many containers: this one sponsored by the engine manufacturers in the U.S. and Japan. This label applies presently only to gasoline-powered engines; it is called GF-1. It represents the highest quality level and closely parallels the highest API classification, SH. The GF-1 label can only be obtained if the engine oil has a 2.7% improvement in fuel consumption over a reference oil.

BIBLIOGRAPHY

Reading List

Marketing Technical Bulletin 93-3, Society of Automotive Engineers Inc. (SAE).

ACID RAIN

WILLIAM SMITH
Yale University
New Haven, Connecticut

Acid rain is appropriately described as an old environmental problem with a new image. Acid rain, more than any other environmental contaminant, has focused societal concern on ecosystem toxicology. Robert Angus Smith (1) is credited with first using the term *acid rain* in 1872. During the first half of the 20th century numerous European investigators added insight to this emerging environmental challenge. Svante Odén, a Swedish scientist, is appropriately credited with detailing the regional scale nature (rain out hundreds of kilometers downwind from the source) of the acid rain problem through a series of publications and lectures between 1967 and 1977. North American interest in acid rain was significantly advanced by the pioneering efforts of Canadians Eville Gorham and Alan Gordon in their studies of the Sudbury area smelters during the 1960s (2,3). The longest continuous record of rain chemistry in the United States has been maintained at the Hubbard Brook Experimental Forest (U.S. Department of Agriculture, Forest Service) in central New Hampshire since 1963 (4). The federal governments of both Canada and the United States initiated systematic rain chemistry monitoring in 1976 (Canadian Network for Sampling Precipitation) and 1978 (National Atmospheric Deposition Program), respectively. Results of the latter program reveal that the acidity of precipitation in the eastern portion of the United States is approximately 10 times more acid than the western portion (Fig. 1). (See also AIR QUALITY MODELING; FOREST RESOURCES; FORESTRY SUSTAINABLE.)

Over the 150-yr history of acid rain research, concerns have been raised in regard to the ability of acid rain to erode and corrode buildings, sculpture, monuments, and

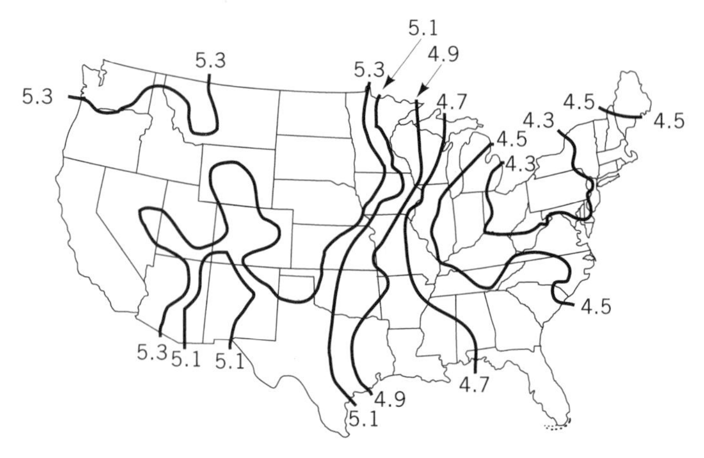

Figure 1. Average annual hydrogen ion concentrations in precipitation, expressed as pH and adjusted for amount of precipitation, for 1990. From Ref. 5.

other structures; to reduce visibility; to impair human health; and to impact adversely agricultural as well as stream, lake, and forest ecosystems. Acid rain can result in significant long-term deterioration of limestone, marble, bronze, unpainted galvanized steel, and numerous paints. In regard to visibility, it has been concluded that more than 50% of the light extinction in rural areas of the eastern United States, on an annual average basis, is due to sulfate particles in the atmosphere. A significant portion of these sulfate particles are the result of acid precursors in the atmosphere. Evidence generated from epidemiological, human clinical, and animal toxicological studies suggest that at high enough exposure acidic aerosols may produce a number of direct and indirect adverse health effects in sensitive human populations. High risk populations include the elderly and the very young and anyone with chronic lung dysfunction such as chronic bronchitis, emphysema, or asthmatic disease. Agricultural ecosystems are thought to be generally resistant to acid rain damage due to the high buffering capacity typical of agricultural soils. Known effects on less well buffered natural ecosystems such as streams, lakes, and forests, however, may range from beneficial (fertilization from nitrogen, phosphorus, calcium, potassium, and sulfur) to adverse (toxicity from hydrogen, aluminum, nitrate, sulfate, or heavy metals) (5).

In recognition of the expansive size of forest ecosystems, the multiple values of forests to people, and the potential for adverse interaction of regional-scale air pollutants on forest health, this article describes the current understanding of the relationship between forests and acid rain (6).

FORESTS

Forests cover approximately 33% of the terrestrial surface of the earth. Forests dominate the landscape in the United States by covering roughly 40% of the land area (approximately 4×10^{12} m^2, or 1 billion acres) and occurring in all 50 states. Native tree species that grow in the United States number 850. Foresters group forest stands of similar species composition and ecological development into forest types. A total of 250 distinct forest types are recognized in Canada and the United States (excluding Hawaii).

Forest ecosystems are characterized by enormous variability. Forest systems may differ in soil type, climate, aspect, elevation, age, and health as well as species composition and development stage. Forests may be uneven aged, even aged, all aged, mature, and overmature. Forests may be reproduced by seed, sprouting, and planting. Some forests have their structure completely shaped by natural forces, some may be influenced by human forces along with natural forces, and others may be completely artificial in design and establishment.

The human values associated with forest systems are almost as varied as forest types. Some values are traditional (eg, wood), long and widely appreciated, and quantifiable in standard economic terms. Other values of more recent appreciation (eg, drugs and pesticides) are developing in acceptance and sophistication and are not quantifiable in standard economic terms. For convenience, forest values may be thought of as products or services. Products include wood (lumber, particle board, plywood, paper, fuel, and mulch), wildlife (game), water (quality), forage (wildlife and livestock), food (seeds, nuts, syrup, and fruit), drugs (eg, Taxol), and pesticides (eg, neem). Services include existence value, recreation, tourism, biological diversity (genes, species, communities, and ecosystems), landscape diversity, amenity functions (microclimate amelioration, sound attenuation, visual attractiveness, and screening), runoff and erosion management, soil and nutrient conservation, pollutant sequestration, and detoxification. Global-scale or regional-scale air pollution may reduce the ability of forest ecosystems to provide, on a sustained basis, one or more forest products or services in an area with poor air quality (7).

ACID RAIN

At room temperature, pure water dissociates to produce equal amounts (0.0000001 g/L) of hydrogen (H^+) and hydroxyl (OH^-) ions. When these ionic concentrations are equal, the solution is said to be neutral. The concentration of hydrogen ions in a solution is represented by the pH scale. The pH number reflects the negative logarithm of the hydrogen ion concentration. At neutrality, when hydrogen and hydroxyl concentrations are equal, pH is 7.0. Acid solutions in water are defined as those solutions that have hydrogen ion concentrations greater than the hydroxyl concentration ($pH < 7$). Keep in mind that the pH scale is logarithmic and that a change from pH 7.0 to pH 6.0 represents a 10-fold increase in hydrogen ion concentration and, therefore, a 10-fold increase in acidity.

Natural rain, including precipitation in relatively clean or unpolluted regions, is naturally acidic, with a pH in the range of 5.0 to 6.0. This natural acidity results from the oxidation of carbon oxides and the subsequent formation of carbonic acid. Formic and acetic acids, originating primarily from natural sources, may also contribute minor amounts of acidity to precipitation.

In regions downwind from electric-generating power stations employing fossil fuels, industrial regions, or major urban centers, precipitation can be acidified below pH 5.0. Precipitation with a pH less than 5.0 is designated acid rain. This human-caused acidification of precipitation results primarily from the release of sulfur dioxide (SO_2) and nitrogen oxides (NO and NO_2) from smokestacks and tailpipes. The sulfur and nitrogen oxides are subsequently oxidized to sulfate (SO_4^{2-}) and nitrate (NO_3^-), hydrolyzed, and returned to earth as sulfuric (H_2SO_4) and nitric (HNO_3) acids. This additional acidification of precipitation by human activities can readily reduce the pH of precipitation in downwind regions to between 4.0 and 5.0 on an annual average basis. Individual precipitation events that have a pH in the 3.0 to 4.0 range are not uncommon.

The atmosphere deposits acidity onto the landscape both during and in between precipitation events. In the latter case, termed dry deposition in contrast to wet deposition, the acids are delivered in the gas phase or in association with fine particles (aerosols). *Acid deposition* is a term that includes acid delivery in the form of precipitation (rain, snow, fog, and cloud moisture) plus dry deposition. In view of the importance of both wet and dry deposition in acid transfer from the atmosphere to the biosphere, acid deposition is a much more appropriate descriptor than acid rain.

The Hubbard Brook Experimental Forest, which is part of the White Mountain National Forest in central New Hampshire, (45°56′ N, 71°45′ W), has been a focal point for North American study of acid deposition. Despite the fact that Hubbard Brook is more than 100 km from any large urban-industrial area, the average annual pH since 1963 has generally fallen in the 4.0–4.4 range. The lowest annual average pH recorded was 4.03 and occurred in 1971. Individual storm pH values have ranged from 3.0 to 5.95.

Precipitation solutions are electrically neutral, requiring the sum of cations (eg, H^+) to be balanced by the sum of anions (eg, NO_3^-). Since initiation of precipitation chemistry measurements at Hubbard Brook in 1964, precipitation cations have been dominated by hydrogen (H^+) and ammonium (NH_4^+) and anions dominated by sulfate (SO_4^{2-}) and nitrate (NO_3^-). Decreases in the concentration of sulfate in the precipitation at Hubbard Brook since 1964 have been attributed to decreases in sulfur dioxide emissions in the upwind areas of the U.S. Northeast and Midwest.

It is important to recognize that both the positive and negative effects of acid deposition on ecosystems involve the consideration of hydrogen ions, sulfate ions, nitrate ions, and heavy metal ions in acid deposition. Selected forests are judged to be at special risk to the adverse effects of acid deposition. These forests receive especially high exposure to hydrogen, sulfate, or nitrate ions due to their proximity to primary sources of sulfuric and nitric acids or because of their elevation. High elevation, or montane, forest ecosystems, eg, the high elevation coniferous forests along the Appalachian Mountain chain in the eastern United States, are exposed to high levels of acid deposition because of frequent occurrence of fog immersion and cloud water deposition. Concentrations of ionic species in fog may be 10 to 15 times that in low elevation rain. The duration of exposure to fog or cloud water events may be larger than rain events and thus result in greater deposition of ionic species to high elevation forests. Repeated alteration of high elevation precipitation events with periods of evaporation, facilitated by high elevation winds, may further act to concentrate ionic species on the surfaces of forests or forest soils. In August 1984, a wide-area cloud–fog event was simultaneously recorded in several northern U.S. forest regions. This event recorded an extremely low pH (2.80–3.09) and concentrations of sulfate and nitrate 7 to 43 times greater than previously recorded in cloud or fog water in the eastern United States (8).

ADVERSE IMPACT OF ACID DEPOSITION ON FOREST SYSTEMS

Assessment of air pollution impact on forest systems is extremely challenging for a variety of reasons; three of the most important are forest system variability, deficiency of understanding of ecosystem- and landscape-scale phenomena, and large variation in system exposure to acid deposition. In general, forest disturbance from air pollutants is exposure related, and dose–response thresholds for a specific pollutant are different among the various organisms of the ecosystem. Ecosystem response is, therefore, a complex process. In response to low exposure to air pollution, the vegetation and soils of an ecosystem function as a sink or receptor. When exposed to intermediate loads, individual plant species or individual members of a given species may be subtly and harmfully affected by nutrient stress, impaired metabolism, predisposition to entomological or pathological stress, or direct induction of disease. Exposure to high deposition may induce acute morbidity or mortality of specific plants. At the ecosystem level, the impact of these various interactions would be highly variable. In the first situation, the pollutant would be transferred from the atmosphere to the various elements of the biota and to the soil. With minimal physiological effect, the impact of this transfer on the ecosystem could be undetectable (innocuous effect) or stimulatory (fertilizing effect). If the effect of the pollutant dose on some component of the biota is harmful, then a subtle adverse response may occur. The ecosystem impact in this case could include reduced productivity or biomass, alterations in species composition or community structure, or increased morbidity. Under conditions of high dose, ecosystem impacts may include gross simplification, impaired energy flow and biogeochemical cycling, changes in hydrology and erosion, climate alteration, and major impacts on associated ecosystems.

For North American forests, it is generally concluded that acid deposition influences on forest systems are neutral, ie, no adverse effects can be discriminated from natural forest dynamics. Actual effects may be slightly stimulatory (eg, nitrogen fertilization via nitrate input) or contributory to multiple-factor forest stress (eg, in high elevation, high risk (montane) forest ecosystems). In the latter case, acid deposition is presumed to be highly interactive with other stresses, subtle in manifestation, and long-term (several decades) in development. The primary hypotheses for these subtle effects are listed below. Not all of these hypotheses are supported by equal scientific evidence. The first five hypotheses are the ones that are best understood and supported by the greatest evidence; they are summarized in the following sections.

Tree Population Interaction	Forest Ecosystem Perturbation
Increased rate of soil acidification causes altered nutrient availability and root disease.	Population dynamics, tree competition, and species composition.
Cation nutrients are leached from foliage to throughfall and stem flow.	Biogeochemical cycle rates.
Cation nutrients are leached below soil horizons of active root uptake.	Biogeochemical cycle rates.
Increased available soil aluminum results in fine-root morbidity.	Population dynamics, tree competition, and species composition.
Increased available heavy metal and hydrogen ion concentrations in soil result in enhanced root uptake or impact on soil microbiota.	Decomposer impact, biogeochemical cycle rates, and species composition.
Deposition causes alteration of carbon allocation to maintenance respiration or repair or to above-ground instead of below-ground tissues.	Productivity and energy storage.
Deposition increases or decreases phytophagous arthropod activity.	Consumer impact and insect population dynamics.
Deposition increases or decreases microbial pathogen activity.	Consumer impact and pathogen population dynamics.
Deposition increases or decreases abiotic stress influence (temperature, moisture, wind, and nutrient stresses).	Population dynamics, tree competition, and species composition.
Increased soil weathering alters soil cation availability.	Biogeochemical cycle rates.
Increased nitrogen (sulfur) deposition alters nitrogen (sulfur) cycle dynamics.	Biogeochemical cycle rates.
Deposition increases or decreases microbial symbioses.	Productivity and energy storage.
Deposition impacts one or more processes of reproductive or seedling metabolism.	Population dynamics, tree competition, and species composition.
Deposition impacts a critical metabolic process, eg, photosynthesis, respiration, water uptake, translocation, or evapotranspiration.	Population dynamics, tree competition, and species composition.

Soil Acidification

Natural processes make forest soils acid (Table 1). These processes act by controlling the chemistry of the soil cation-exchange complex. This complex consists of negative charges located on clay minerals in the soil or on soil organic matter. On clay minerals, negative charges generally result from substitution, within the mineral lattice, of a cation of lower positive charge for one of higher charge. On organic matter, negative charges result from the ionization of hydrogen ions from carboxyl, phenol, and enol groups. In acid mineral soils the cation-exchange complex is typically dominated by aluminum species, eg, Al^{3+} $Al(OH)^{2+}$, and $Al(OH)_2^+$, formed by the dissolution of soil minerals. In acid organic soils, the hydrogen ion may dominate the cation-exchange complex.

Processes that acidify forest soils include those that increase the number of negative charges, such as organic matter accumulation or clay formation, and those that remove basic cations, such as leaching of bases in association with an acid anion. Weathering by carbonic acid, organic acids (podzolization), humification, and cation uptake by roots all increase the negative charge of forest soils.

It has been noted that soil acidification is a complex of processes that cannot be quantitatively described by any single index (9). There has been an emphasis on the utility of using capacity and intensity factors. Capacity refers to the storage of hydrogen ions, aluminum ions, or base cations on the soil exchange complex or in weatherable minerals. Intensity refers to the soil solution concentration at any point in time; in the case of the hydrogen ion, it refers to the soil solution pH. In forest regions receiving acid deposition from the atmosphere, the most likely effect on the capacity will be an increase in the exchange acidity and a reduction of exchangeable bases. The former is increased directly by hydrogen ion input or, more

Table 1. Soil Acidity Designations by pH

Descriptor		pH
Extremely acid	forest soils	<4.5
Very strongly acid	forest soils	4.5–5.0
Strongly acid	forest soils	5.1–5.5
Medium acid	forest soils	5.6–6.0
Slightly acid	forest soils	6.1–6.5
Neutral		6.6–7.3
Mildly alkaline		7.4–7.8
Moderately alkaline		7.9–8.4
Strongly alkaline		8.5–9.0
Very strongly alkaline		9.1+

likely, by increasing exchangeable aluminum through the reaction of hydrogen ions with soil minerals. The latter (reduction of exchangeable bases) results via replacement of base cations on the exchange complex by aluminum species. The base cations are subsequently leached from upper soil horizons in association with strong acid anions.

The evidence for acidification (ie, lowering of pH) of forest soils at the present rates of acid deposition in North America is not great. Forest soils at greatest risk to pH reduction from acid deposition are restricted to those limited soil types characterized by no renewal by fresh soil deposits, low cation exchange capacity, low clay and organic matter content, low sulfate adsorption capacity, high input of acidic deposition without significant base cation deposition, high present pH (5.5–6.5), and deficiency of easily weatherable materials to a 1-m depth. In addition, these high risk forest systems would need to be exposed to significant levels of acid deposition for decades.

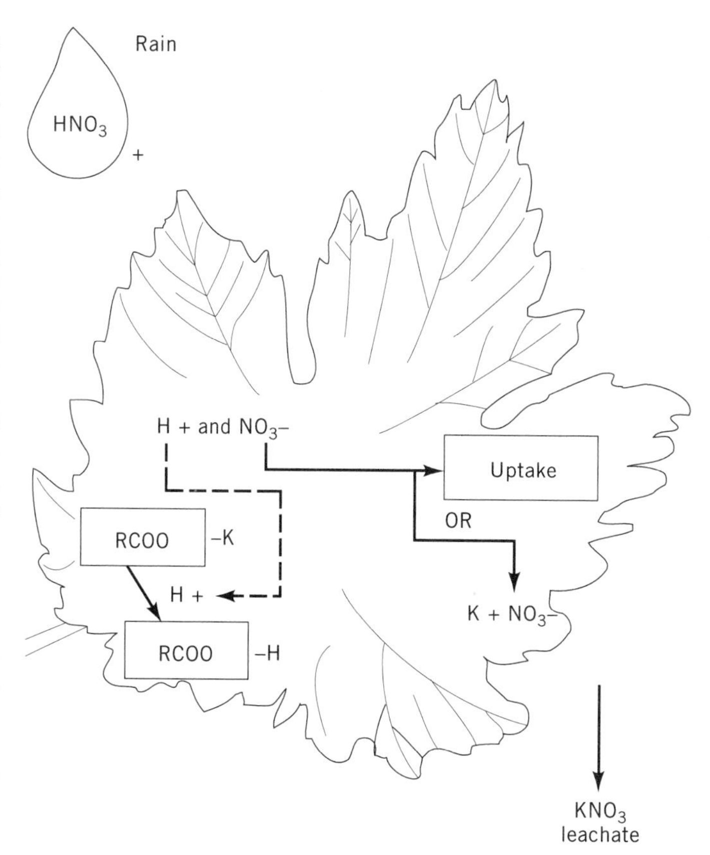

Figure 2. Foliar leaching can result from the deposition of nitric acid. Hydrogen ions from nitric acid may displace nutrient cations held on leaf cell wall exchange sites. Nitrate ions from the acid, if not taken up by the leaf or leaf microorganisms, may be available to combine with potassium and to remove this nutrient from the leaf. If potassium ions are not resupplied by the roots, forest trees may suffer from nutrient stress. From Ref. 10.

Cation Depletion from Foliage

Vegetative leaching refers to the removal of substances from plants by the action of aqueous solutions such as rain, dew, mist, and fog. Precipitation washout of chemicals from trees has been appreciated for some time. Inorganic chemicals leached from plants include all the essential macroelements. Potassium, calcium, magnesium, and manganese are typically leached in the greatest quantities. A variety of organic compounds, including sugars, amino acids, organic acids, hormones, vitamins, and pectic and phenolic substances, are also leached from vegetation. As the maturity of leaves increases, susceptibility to nutrient loss via leaching also increases and peaks at senescence. Leaves from healthy plants are more resistant to leaching than leaves that are injured, infected with microbes, infested with insects, or otherwise under stress.

Deciduous trees lose more nutrients from foliage than do coniferous species during the growing season. Conifers, however, continue to lose nutrients throughout the dormant season. The stems and branches of all woody plants lose nutrients during both the growing and dormant seasons.

The mechanism of leaching is presumed to be primarily a passive process. Cations are lost from free space areas within the plant. Under uncontaminated, natural environmental conditions, little if any cations are thought to be lost from within cells or cell walls. It has been demonstrated that leaching of cations on the leaf surface involves exchange reactions in which cations on exchange sites of the cuticle are exchanged by hydrogen from leaching solutions. Cations may move directly from the translocation stream within the leaf into the leaching solution by diffusion and mass flow through areas devoid of cuticle.

Pollutant exposure may predispose foliage to leaching loss by cuticular erosion, membrane dysfunction, or metabolic abnormality (Fig. 2). Epicuticular wax, the outermost layer of plant leaves, consists of a complex and variable mixture of long-chain alkanes, alkenes, aromatic hydrocarbons, fatty acids, ketones, aldehydes, alcohols and esters. The composition and integrity of this layer is strongly controlled by climate, foliar age, and air contaminants. Because acid precipitation may increase hydrogen ion activity by one or two orders of magnitude (pH 5.6 to 3.6) due to increasing concentrations of sulfuric and nitric acids in precipitation and because damage to cuticles and epidermal cells may also result from exposure to acid precipitation, the potential for accelerated leaching under this stress may be important.

Field investigations of foliar leaching have typically compared throughfall and stem flow chemistry to direct deposition chemistry to evaluate leaching. If cation enrichment is detected in stem flow or throughfall, relative to precipitation collected in the open, foliar leaching is presumed. Numerous studies have provided evidence for cation enrichment in throughfall collected under forest canopies. The significance of foliar cation depletion, however, in overall forest health is generally judged relatively unimportant for several reasons. First, throughfall and stem flow input to forest soil is a normal and common feature of forest nutrient cycles and is not solely due to atmospheric stress. Second, and more important, field studies of throughfall and stem flow generally do not distinguish ions actually leached from foliar surfaces from those dry deposited on canopy surfaces and subsequently washed off the foliage during precipitation events. Third, reductions of hydrogen ion input by the canopy may not be reflected in reduced hydrogen ion loading of the soil, because rhizophere acidification may result as replacement cations are taken up from the soil solution by impacted trees. Fourth, the long-term significance of foliar cation loss is regulated by the ability of forest soils to provide replacement nutrients to the available nutrient pool.

Cation Depletion from Upper Soil Horizons

Most metallic nutrient elements taken up by trees are absorbed as cations and exist in three forms in the soil: (*1*) slightly soluble components of mineral or organic material, (*2*) adsorbed into the cation exchange complex of clay and organic matter, and (*3*) small quantities in the soil solution. As water migrates through the soil profile, the movement of cations from any of the three compartments is known as cation depletion leaching.

The relative mobility of cations leached from decomposing forest litter is typically sodium > potassium > calcium > magnesium. The leaching rate largely depends on the generation of a supply of mobile anions. In forest soils, the anions are produced along with hydrogen cations and occur as acids. The hydrogen cations have a powerful substituting capability and can readily replace other cations adsorbed in the soil. The mobile anions function to move released cations through the soil. In soils of low atmospheric deposition and minimal human disturbance, carbonic and organic acids dominate the leaching process. Carbonic acid leaching dominates natural leaching processes in tropical and temperate coniferous sites; nitric acid, resulting from nitrification, dominates in nitrogen-fixing temperate deciduous sites; and organic acids dominate surface soil leaching in subalpine sites and contribute to leaching in numerous other sites. In regions with sufficient moisture, cations will be transported from the forest floor in proportion to the availability of HCO_3^- or organic acid anions or both.

It has been proposed that in forest soils that are subject to the deposition of air pollutants sulfuric and nitric acids may provide the primary, or a significant, source of H^+ for cation displacement and mobile anions for cation transport. Soil cation leaching requires the mobility of the anion associated with the leaching acid. This results from the requirement for charge balance in soil solutions, a requirement that disallows cation leaching without associated mobile anions. Nitric and sulfuric acid deposited from the atmosphere supply mobile anions in the form of nitrate (NO_3^-) and sulfate (SO_4^{2-}), respectively. In most soils, rapid biological uptake may immobilize the NO_3^- anion. As a result, risk of cation loss associated with this anion may be restricted to ecosystems rich in nitrogen where biological immobilization of NO_3^- is minimal. In a similar manner, SO_4^{2-} may be immobilized in weathered soils by adsorption to free iron and aluminum oxides. In other soils, however (especially those low in free iron or aluminum or high in organic matter, which appears to block SO_4^{2-} adsorption sites), SO_4^{2-} may readily combine with nutrient cations and leach the elements beyond the zone of fine-root uptake by forest trees (Fig. 3).

The movement of nutrient cations via leaching in forest soil profiles is an extremely important component of forest nutrient cycling. The evidence that has been provided by numerous experiments subjecting soil lysimeters to natural or artificially acidified precipitation indicates a potential for a meaningful acid precipitation influence on the soil leaching process. The threshold for significant increases in the rate of movement of calcium, potassium, and magnesium appears to require precipitation in the pH range of 3–4 for most systems examined. For certain forests, eg, subalpine balsam fir in New England, the threshold of increased leaching may be higher and in the range of pH 4.0–4.5.

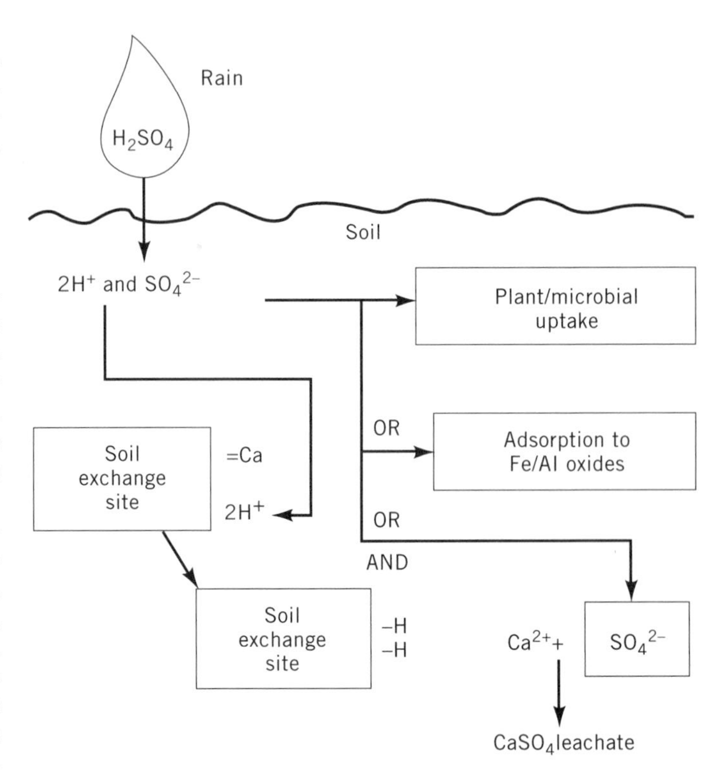

Figure 3. Leaching of forest soil can result from deposition of sulfuric acid. Hydrogen ions from the acid may displace nutrient cations held on organic or clay exchange sites. Sulfate ions from sulfuric acid, if not taken up by the soil biota or adsorbed to iron or aluminum oxides, may be available to combine with calcium and remove this nutrient from the rooting zone. If a sufficient amount of calcium ions is not resupplied by decomposition or soil weathering, forest trees may suffer from nutrient stress. From Ref. 10.

Despite the efficacy of SO_4^{2-} leaching in selected soils, it must be realized that forest soils vary greatly in their susceptibility to it (depending on chemistry, depth, and parent and bedrock material). Mineral weathering and deep rooting constantly replenish base cations for tree growth, and even for sensitive soils, leaching losses affect cation reserves only on an extended (multiple decade) time scale.

Forest soils have been judged more vulnerable to leaching influence by acid precipitation than are agricultural soils. Forest soils supporting early regeneration following harvest or severe natural disturbance events may be especially vulnerable to an adverse impact on nutrient cycling by acid rain, because the system controls on nutrient conservation are weakest at such times. The bulk of evidence, however, remains consistent with the conclusion that unless the acidity of precipitation increases substantially or the buffering capacity of forest soils declines significantly, acid rain influence will not quickly or dramatically alter the productivity of most temperate forest soils via cation depletion.

Aluminum Toxicity

Aluminum toxicity to forest trees may be expressed indirectly via soluble aluminum interference with nutrient (cation) uptake by roots or directly via soluble aluminum interference with one or more tree metabolic processes. Aluminum toxicity may influence forest tree-root health; acid deposition plus natural acidifying processes may increase soil acidity enough to increase the amount of soluble aluminum available for root uptake. Aluminum is the third most abundant element in the crust of the earth, where it occurs primarily in aluminosilicate minerals, most commonly as feldspars in metamorphic and igneous rocks and as clay minerals in weathered soils. In northern temperate zones, podzolic forest soils naturally mobilize aluminum, with organic acids resulting from the decomposition of organic matter, and leach it from upper to lower soil horizons where it precipitates. Despite high concentrations of total aluminum in forest soils, aluminum toxicity is generally presumed unimportant, because the element is present in tightly bound or insoluble form and is not available for root uptake.

Atmospheric additions of strong mineral acids to forest soils may mobilize aluminum by dissolution of aluminum-containing minerals or may remobilize aluminum previously precipitated within the soil during podzolization or held on soil exchange sites. The forms (speciation) and dynamics of aluminum equilibria in forest soils are not completely understood. It is clear that only dissolved forms of aluminum will be readily moved in the soil system and will be accessible for root uptake. In fresh water, dissolved aluminum is known to exist mainly as free ions (Al^{3+}) and as complexes with hydroxide, fluoride, sulfate, and dissolved organic matter. In aquatic systems, the formation of mononuclear (eg, $Al\,OH^{2+}$, $Al(OH)_2^+$, and $Al(OH)_4^-$) and polynuclear (eg, $Al_2\,(OH)_2^+$, $Al_6\,(OH)_5^{13+}$ and $Al_{10}\,(OH)_{22}^{8+}$) hydroxide complexes have been intensively studied. Monomeric species are thought to be especially important, and as a result, the following equilibria are assumed to be significant:

$$Al\,OH^{2+} + H^+ \rightleftharpoons H_2O + Al^{3+}$$

$$Al\,(OH)_2^+ + 2H^+ \rightleftharpoons 2H_2O + Al^{3+}$$

$$Al\,(OH)_4^- + 4H^+ \rightleftharpoons 4H_2O + Al^{3+}$$

$$Al\,F^{2+} \rightleftharpoons F^- + Al^{3+}$$

$$Al\,SO_4^+ \rightleftharpoons SO_4^{2-} + Al^{3+}$$

$$HF \rightleftharpoons H^+ + F^-$$

$$\text{Al-organic}^{(n-3)} \rightleftharpoons \text{org}^{n-} + Al^{3+}$$

Soil aluminum speciation is regulated by ligand (complexing chemical) abundance and pH. As forest soil pH falls below 5, the rate of aluminum release from silicate lattices, especially clay minerals, becomes greater. The liberated aluminum is transformed primarily into polynuclear hydroxide cations and Al^{3+} ions. The polynuclear hydroxide cations are soluble and become attached to negative charge sites but are nonexchangeable. Their residence on exchange sites results in a reduction of cation exchange capacity (CEC). Aluminum ions (Al^{3+}) also become attached to negative charges sites but are exchangeable. Due to the selective binding of Al^{3+} on exchangeable sites, Al^{3+} saturation can reach high levels at relatively low Al^{3+} concentrations in the soil solution. At $pH < 4.2$, the solubility of aluminum hydroxide compounds increases to the extent that Al^{3+} may become the dominating cation in the soil solution.

Much evidence has been gathered concerning the mechanisms of aluminum toxicity to plants. Direct root effects include reduced cell division associated with aluminum binding of DNA, reduced root growth caused by inhibition of cell elongation, and destruction of epidermal and cortical cells. Indirect effects of aluminum stress, especially in forest trees, may involve interference with the uptake, translocation, or use of required nutrients such as calcium, magnesium, phosphorus, potassium, and other essential elements.

Reductions in calcium uptake by roots have been associated with increases in aluminum uptake. As a result, an important toxic effect of aluminum on forest tree roots could be calcium deficiency. Mature forest trees have a higher calcium requirement than agricultural crops. An integrated hypothesis concerning aluminum interference with calcium uptake in red spruce in the northeastern United States has been proposed (11). It has been stressed that calcium is incorporated at a constant rate per unit volume of wood produced and is not recovered from sapwood as it matures into heartwood. As a result, when aluminum and calcium are present in approximately equimolar concentrations within the soil solution or when the aluminum: calcium ratio exceeds 1, aluminum will reduce calcium uptake by competition for binding sites in the cortical apoplast of fine roots. Reduced calcium uptake will suppress cambial growth (annual ring widths) and predispose trees to disease and injury by biotic stress agents when the functional sapwood becomes less than 25% of the cross-sectional area of the stem.

Heavy Metal Toxicity

Human activities cause increases in heavy metal deposition from the atmosphere for antimony, cadmium, chromium, copper, lead, molybdenum, nickel, silver, tin, vanadium, and zinc. Small increases also have been recorded in selected environments for cobalt, manganese, and mercury. Upper soil horizons, especially forest floors with persistent organic matter accumulation, represent important sinks for heavy metals from any source, including the atmosphere. At sufficient concentration, all heavy metals are toxic to biological systems.

Direct toxicity to trees and direct and indirect toxicity to other biotic components of the forest soil by heavy metals are exposure dependent. Exposure is a function of both deposition from the atmosphere and chemical availability. Heavy metals not available for ready exchange from binding sites or in solution are not available for root or microbial uptake. More than 90% of certain heavy metals deposited from the atmosphere may be biologically unavailable. Heavy metals may be absorbed or chelated by organic matter (humic and fulvic acids); clays; or hydrous oxides of aluminum, iron, or manganese. Heavy metals also may be complexed with soluble low molecular weight compounds. Soluble cadmium, copper, and zinc may be chelated in excess of 99%. Adsorbed heavy metals remain in equilibrium with chelated metals. Heavy metals also may be precipitated as inorganic compounds of low solubility such as oxides, phosphates, and sulfates.

Adsorption, chelation, and precipitation are strongly regulated by soil pH. As pH decreases and soils become more acid, heavy metals generally become more available for biological uptake. Natural forest soils generally become more acid as they mature. Acidification in excess of natural processes is possible especially in soils with a pH greater than 5. Under this circumstance, soil acidification, associated with acid deposition may result in increased biological availability of heavy metals in the forest floor.

The thresholds of toxicity for numerous bacterial, fungal, insect, and other components of the soil biota are in the range of 1,000–10,000 ppm metal cation on a soil dry weight basis. These concentrations are two to three orders of magnitude greater than the concentrations of heavy metals throughout most temperate forest ecosystems. In addition, fungi probably play a larger role in nutrient cycling in forests than other microbial groups, and they appear more resistant to trace metal influence than other microbes. In addition, there is considerable evidence for microbial adaptation to high trace metal exposure. While components of the forest floor soil biota may be impacted by trace metals, decomposition rates or other soil processes may remain unchanged due to adaptation or shifts in species composition unreflected in gross soil process activities.

In regard to direct toxicity of heavy metal ions to plant roots, copper, nickel, and zinc toxicities have occurred frequently in grossly polluted environments. Cadmium, cobalt, and lead toxicities have occurred less frequently. Acute direct toxicity caused by heavy metal deposition or mobilization by acidification probably does not occur in forest trees located outside of urban, roadside, or selected point source industrial and electric-generating environments.

PROGNOSIS

Across a landscape as large as the United States, some forests undoutedly benefit from fertilization by nutrients contained in acid deposition. Most forests are not detectably influenced positively or negatively by acid deposition due to low input or inability to separate acid rain responses from the background noise of forest processes, and some high risk forests, eg, those receiving cloud moisture acidity along the Appalachian Mountain chain, may be subjected to chronic and interactive stress associated with nutrient depletion and aluminum toxicity.

As the 21st century nears, it is increasingly clear that humans have the ability to influence the global environment, especially the global atmosphere. Emphasis on the atmosphere is not surprising when it is recognized that this medium integrates human activities quite independently of political boundaries. Increasing evidence emphasizes the ability of human beings to influence specific global-scale and regional-scale atmosheric processes. At the global scale, some of the most important include trace gas loading of the troposphere with the radiatively active species carbon dioxide, methane, nitrous oxide, and halocarbons, resulting in the hypothesis of global warming; chlorine loading of the stratosphere via chlorofluorocarbons, resulting in the hypothesis of stratospheric ozone depletion and potential for increased uv-b radiation at the surface of the earth; and global circulation and ultimate deposition of toxic pollutants, including chlorinated pesticides (DDT), polychlorinated biphenyls (PCBs), polynuclear aromatic hydrocarbons (PAHs), and heavy metals.

At the regional scale, the two most important air pollutants influencing terrestrial ecosystems are acid rain and oxidant (ozone) pollutants (11). Research indicates that regional air pollution is one of the significant contemporary stresses imposed on some temperate forest ecosystems. Gradual and subtle change in forest metabolism and composition over high risk areas of the temperate zone over extended time, rather than dramatic destruction of forests in the immediate vicinity of point sources over short periods, must be recognized as the primary consequence of regional air pollution stress. Global air pollution, with its associated capability to cause rapid climate change, has the potential to alter dramatically forest ecosystems in the next century. The integrity, productivity, and value of forest systems is intimately linked to air quality. Failure to give careful consideration to forest resources in societal considerations of energy technologies and in management and regulation of air resources is unthinkable.

BIBLIOGRAPHY

1. E. B. Cowling, *Environ. Sci. Technol.* **16,** 110A–123A (1982).
2. E. Gorham and A. G. Gordon, *Can. J. Botany* **38,** 307–312 (1960).
3. E. Gorham and A. G. Gordon, *Can. J. Botany* **41,** 371–378 (1963).
4. G. E. Likens, F. H. Bormann, R. S. Pierce, J. S. Eaton, and N. M. Johnson, *Biogeochemistry of a Forested Ecosystem,* Springer-Verlag, New York, 1977.
5. National Acid Precipitation Assessment Program, *1990 Integrated Assessment Report,* Washington, D.C., 1991.
6. W. H. Smith in D. C. Adriano and A. H. Johnson, eds., *Acidic Precipitation, Vol. II, Biological and Ecological Effects,* Springer-Verlag, New York, 1989, pp. 165–188.
7. W. H. Smith in J. R. Barker and D. T. Tingey, eds., *Air Pollution Effects Biodiversity,* Van Nostrand Reinhold Co., New York, 1962, pp. 234–260.
8. K. C. Weathers, G. E. Likens, F. H. Bormann, J. S. Eaton, W. B. Bowden, J. L. Andersen, D. A. Cass, J. N. Galloway, W. C. Keene, K. D. Kimball, P. Huth, and D. Smiley, *Nature,* **319,** 657–658 (1986).
9. J. O. Reuss and D. W. Johnson, *Acid Deposition and the Acidification of the Soils and Waters, Ecological Studies, Vol. 59,* Springer-Verlag, New York, 1986.
10. W. H. Smith, *Air Pollution and Forests,* Springer-Verlag, New York, 1990.
11. W. C. Shortle and K. T. Smith, Aluminum-induced calcium deficiency syndrome. *Science* **240,** 1017–1018 (1988).

ACTIVATED SLUDGE

Raymond Regan
Pennsylvania State University
University Park, Pennsylvania

Each day wastewater (sewage is the old term) is produced as a result of domestic, municipal, and industrial activities. Wastewater includes various inorganic and organic contaminants that must be removed so that the water may be safely returned to the environment. Episodes of serious pollution damage have occurred in the past when wastewaters were improperly or incompletely treated. This article will summarize some of the important features of the activated sludge process, a widely used method of wastewater treatment. Brief descriptions of the activated sludge process, and how the process has been developed over the years provides basic understandings of the microbiological and mechanical features involved. Finally, descriptions of new developments explains how the basic activated sludge process is being adapted to provide effective treatment of a range of newer pollution control needs. (See also Water Conditioning; Water Quality: Issues; Water Quality Management.)

BASIC PROCESS

This is how the activated sludge process works.

Wastewater collected from domestic and municipal sources is generally allowed to pass through a series of preliminary treatment steps basically to remove large debris (floating pieces of wood, rags, etc), materials that might damage mechanical parts of the tanks, pipes, and pumps (sand and gritty solids), and to provide conditions so that organic solids will be more readily removed from the wastewater (size reduction, odor reduction). Preliminary is followed by primary treatment, a processing step which allows organic and inorganic solids to be settled and thereby removed from the wastewater. Also during primary treatment, floating materials, such as food preparation discards and greases, are removed. When used for domestic and municipal wastewater, the treatment provided by the primary process is suitable for removing approximately 35% of the BOD_5 (biochemical oxygen demand measured after 5 days when incubated at 20°C) and 50% of the total SS (suspended solids) which are settable by gravity under the provisions provided for in the treatment plant. The discharge after primary treatment is then followed by further BOD and SS removal using the secondary treatment.

Based on the characteristics of the industrial wastewater, preliminary and primary treatment may not be needed. In this situation treatment may be initiated immediately with activated sludge.

Activated sludge is the principal process used to provide secondary levels of treatment of all types of wastewater containing biodegradable organics, which may be present in either soluble, colloidal or settable forms. The activated sludge process is suitable for removing the remainder of the BOD_5 and total SS. This process involves two main components, namely (*1*) the aeration/mixing tank and (*2*) the secondary clarifier/settling tank. Organics and other inorganic nutrients (food sources) present in the wastewater are introduced into the aeration/mixing tank where they are mixed with a suspension of microorganisms (referred to as mixed liquor suspended solids, MLSS) with atmospheric air. The air provides for the dual purposes of contacting the food sources with the MLSS by mixing, and supplying the oxygen needed by the microorganisms to metabolize the food. Microbial protoplasm is generated as part of the life cycle of the activated sludge microorganisms. The microbial population includes various species of bacteria, protozoa and rotifers.

After a period of aeration, the treated wastewater and MLSS are introduced into a clarifier. The clarifier provides the conditions needed for gravity separating of the MLSS from the treated wastewater. The clarified effluent is commonly disinfected and then discharged to a surface waterbody. The gravity settled MLSS is divided into two parts, namely (*1*) return flow to the aeration tank and (*2*) waste flow to remove the excess microbial protoplasm (waste sludge) produced during the aeration step. The activated sludge process is operated continuously. Therefore, the plant operators must understand how the growth of activated sludge microorganisms is effected by changes in the wastewater and other environmental factors controlling the life cycle of the microlife involved.

PROCESS DEVELOPMENTS

Activated sludge has changed little from the original report of Arden and Lockett in 1914 (1). Basically process changes have been made as a means of solving specific operating problems. Three general problem areas were (*1*) oxygen supply to meet the needs of microorganisms, (*2*) wastewater characteristics and (*3*) nutrient removal. Activated sludge modifications in each problem area are compared to the conventional process.

Conventional Process

The conventional process included a rectangular aeration tank with air diffusers evenly spaced along one side of the length of the tank. Wastewater would be added at one end and allowed to flow towards the other. When the aerators were operating, the wastewater containing the MLSS (1500–2000 mg/L) would be mixed in a swirling pattern. The nominal hydraulic treatment period was 6 hours. Typical food to microorganism (F/M) ratios were 0.15 to 0.40 lb BOD_5/day-lb MLSS (3). Volumetric loadings were usually less than 25 lbs BOD_5/day-1000 ft^3 of aeration volume. The sludge age (time the MLSS recirculated in the tank) was seven to 10 days. The minimum recommended operating dissolved oxygen (DO) level was 2.0 mg/L.

Oxygen Supply

As an aerobic process, activated sludge requires that a continuous supply of DO be provided in excess of the metabolic needs. Initially porous plate diffusers were used to disperse air into the aeration tank. Operational problems led to the development of non-clog coarse bubble diffusers by a number of equipment manufacturers (3). The transfer of the oxygen from the air into the wastewater was more efficient using the porous plate than the coarse bubble diffusers, but the latter type was less prone to clogging due to the growth of the microorganisms. Therefore, ease of maintenance became the controlling factor for this process modification.

Engineers and operators recognized that the amount of DO required by the activated sludge microorganisms (DO demand) was greatest at the head end of the tank and decreased while the food was removed as the wastewater and MLSS passed along the tank. However, the DO supplied by the aerators was uniform along the tank (about 20 mg/L-h). Therefore, a modification was developed whereby the wastewater flow was divided at the head end of the tank, and introduced at several locations farther down the tank. In effect, the DO demand was evened out along the tank and the capacity of the plant to treat wastewater was increased, with minimum capital expense. This modification is known as step aeration.

Wastewater Characteristics

Activated sludge plants designed to receive predominantly domestic and municipal wastewaters in larger cities worked well for many years. However, the need for efficient wastewater treatment began to change significantly during the late 1940s and early 1950s. Smaller communities and industries became regulated by environmental legislation to a greater degree. Conventional activated sludge was found to be adequate for the provision of the needed level of wastewater treatment. Process changes were made to meet these needs by equipment manufacturers (3).

The failures experienced at wastewater treatment facilities was founded primarily because of a lack of the understanding of the biochemical reactions involved with activated sludge. Original concepts for designing conventional activated sludge were empirically derived and developed on the basis of the mechanical and construction features of the physical facilities. Many novel modifications evolved with the advent of improved understanding of the fundamental principles, ie, hydraulic reactor design, biochemical mechanisms for organic utilization, and the relationship of biomass synthesis and energy transfer.

Improvements in reactor design, means of supplying DO and the treatment of industrial wastewater containing increased levels of BOD_5 were coupled in the creation of a complete mixing activated sludge (CMAS) modification. Features of CMAS included:

1. suited to small industrial and municipal applications;
2. required minimum supervision;
3. tank operated in circular or square configuration provided a complete mixing hydraulic regime;
4. surface aerators, rather than diffusers, could provide an increased DO supply (50 mg/L h and greater); and
5. the retention time of the wastewater in the reactor would be developed as a balance between the DO supply and the DO demand, rather than a predetermined retention time, as was used for the conventional activated sludge process.

Soluble, colloidal and suspended organics from industrial wastewater were excellent candidates for treatment by CMAS, provided that the proper amounts of nutrients (nitrogen and phosphorus) and trace elements needed for metabolism were present (3).

NUTRIENT REMOVAL

The need to reduce the excessive growth of algae in receiving waters has stimulated the development of biological nutrient removal systems (3). Research in South Africa has resulted in modifications of the activated sludge process that are capable of biologically reducing nitrogen and phosphorus. The biotreatment system involves a series of tanks: anaerobic (no oxygen), anoxic (no oxygen, with nitrate) and aerobic (oxygen present). Microorganisms in each of the tanks are controlled by providing the suitable oxygen regime and retention time. In the overall process, nitrate is reduced to nitrogen gas (which is allowed to escape to the atmosphere), and phosphorus is chemically precipitated or assimilated by the specialized microorganisms present in the MLSS.

CONCLUSION

An expanding number of biotreatment processes are becoming available. Our knowledge of biochemical systems continues to improve and new ways are being developed to take advantage of the biologically medicated reactions. For activated sludge microorganisms, the sky seems to be the limit.

BIBLIOGRAPHY

1. E. Arden, and W. F. Lockett, "Experiments on the Oxidation of Sewage Without the Aid of Filters," *J. Soc. Chem. Indust. London* **33,** 523–539 (1914).
2. T. D. Reynolds, *Unit Operations and Processes in Environmental Engineering,* Brooks-Cole Engineering Div, Monterrey, Calif. 1982.
3. R. E. McKinney, "Recent Developments and Future Directions in Biological Treatment," *43rd Annual Transactions Envir. Eng. Conf.,* University of Kansas, Lawrence, Kans. 1993.

AFTERBURNING

Ramon Espino
Exxon Research and Engineering
Annandale, New Jersey

In a conventional turbojet or turbofan engine a significant portion of the air entering the engine bypasses the combustion zone. The temperature in the combustion chamber and therefore the amount of fuel that can be burned in the chamber is limited by the materials available today to make turbine blades (see also Aircraft Engines). As a result, engineers must design engines with increasing amount of air bypassing the combustor. The thrust provided by this bypass air results solely from its increase in pressure and temperature provided by compression and heat exchange with the combustor zone. The bypass air exchanges heat with the combustor zone and cools it. Modern turbofan engines have air bypass ratios of 10 and greater.

Afterburning is a concept developed during the 1960s and applied in engine design since the 1970s. It takes advantage of this hot air that bypasses the combustion and burns additional fuel to provide additional engine thrust. The design of engines that have afterburners is quite complex and their operation at subsonic speeds is not very fuel efficient. Afterburners are used mainly in aircraft that operate at supersonic flow. These are mainly military aircraft and presently the Concorde supersonic transport. In the case of military aircraft, afterburners are also used at subsonic speeds since fuel economy is not a critical factor. Military jets use their afterburners to achieve better take off, higher rates of climb, and improved maneuverability.

The basic design of an afterburning turbojet or turbofan aircraft consists of a diffuser section to slow down the very hot gases exiting the combustion/turbine section and a combustion zone where additional fuel is sprayed into the air and burned. Ignition is provided by either spark ignition or by a small flame (hot streak) from the main combustor. The design of this combustion chamber is complex since one needs to maintain combustion while reducing drag. The size of a jet engine with an afterburner is double the size of a conventional engine. Finally, the afterburner unit must have the capability of expanding the gas exit area (nozzle) when the afterburners are operating in order to accommodate the increase in volumetric gas flow. The thrust of all jet engines arises from the difference in gas velocity at the exit of the engine exhaust and the velocity of the inlet air or the aircraft speed (see AIRCRAFT ENGINES).

The superiority of an afterburning engine over a conventional turbojet engine is demonstrated only at supersonic speeds. The thrust of a simple turbojet engine falls off very significantly at supersonic speeds since the temperature of the inlet air begins approaching the temperature of the combustor zone, thus limiting the temperature rise and the velocity of the exit air. Afterburning allows a significantly greater increase in the exit gas temperature and thus provides significant thrust at speeds 3–5 times the speed of sound in air.

The fuel in the afterburning section of an engine is used less efficiently. This is particularly true at subsonic speeds since the combustion in the afterburner section is carried at a lower pressure than in the primary combustor. This penalty is so large at subsonic speeds that afterburning is used only for very short periods when military jets want extra thrust to maneuver, gain altitude or take off from aircraft carriers. At supersonic speeds, afterburning engines become more efficient since the pressure in the afterburner section rises significantly. At speeds 2.5 times greater than the speed of sound, afterburning becomes highly efficient. Supersonic transports, such as the Concorde, take advantage of this fact and use their afterburners for cruising above the speed of sound.

BIBLIOGRAPHY

Reading List

W. J. Hesse and N. V. S. Mumford, Jr., *Jet Propulsion for Aerospace Applications,* 2nd ed., Pitman Publishing Corp., 1964, Chapt. 12.

J. L. Kerrebrock, *Aircraft Engines and Gas Turbines,* The MIT Press, Cambridge, Mass., 1978, Chapt. 4.

AGRICULTURE AND ENERGY

AMORY B. LOVINS
L. HUNTER LOVINS
Rocky Mountain Institute
Snowmass, Colorado
MARTY BENDER
The Land Institute
Salina, Kansas

Approximately two calories of energy are utilized per calorie of food produced in the United States. When the energy costs for processing, distribution, and preparation are added, the total energy consumption is about 10 calories of energy per calorie of food consumed in the United States. The U.S. food systems uses more than 12 quads (quadrillion Btus) of energy annually to feed 250 million Americans with a daily diet of 3,600 calories per person. (See ENERGY CONSUMPTION IN THE UNITED STATES; ENERGY TAXATION—BIOMASS, SUBSIDIES FOR BIOMASS; FUELS FROM BIOMASS).

The food systems of the rural populations in developing nations use an estimated 16.4 quads anually to feed about two billion people with a diet ranging from 1,800 to 2,400 calories daily per person. This is less than one-tenth of what the same number of people would consume were they utilizing the food system of the United States (1).

The sustainability of an agriculture depends not only on the balance between the energy it uses and produces but also on its ability to preserve its soil and water resources. In the United States, during the past 200 years, at least one-third of the cropland topsoil has been lost. Studies within the past decade indicate that from 25 to 50% more soil per acre is being lost now than in the 1930s when Hugh Hammond Bennett, founding chief of the Soil Conservation Service, was lamenting the loss of American soils (2). (See FORESTRY SUSTAINABLE; RENEWABLE RESOURCES).

Water has also become a serious resource problem since a full two-thirds of the groundwater pumped in the United States is used to irrigate crops. About one-fourth of this withdrawal is overdraft, ie, water drawn at a greater rate than it refills (3). (See WATER QUALITY ISSUES; WATER QUALITY MANAGEMENT).

Concerns over a possible future oil shortage has encouraged many people to look to agriculture to provide substantial fuels. Alcohol from biomass offers the hope of a renewable domestic source of liquid fuel to provide the mobility on which modern society depends. However, a significant biomass fuels program holds a large risk; if fuels are regarded as more important than soil and water, it would only contribute to the collapse it was meant to help forestall.

PETROFARMING

Farms in the United States have provided an ever-higher yield from the land while requiring fewer and fewer people to work on the land. Supermarkets offer a staggering

Table 1. Energy Consumption by Sector, 1974[a]

Sector	Energy Consumed (quadrillion kJ)	Percent of Total
Agriculture (on farm)	1.54	2.0
Mining	2.15	2.8
Construction	2.08	2.7
Manufacturing	22.13	28.8
Transportation	18.34	24.0
Commercial	6.01	7.8
Household	12.23	15.9
Electric utility	12.23	16.0
Total	76.70	100.0

[a] Ref. 4.

array of produce, available almost regardless of season or weather. Exports have earned thus a reputation as the breadbasket of the world. However, more food is possible only because of a temporary overabundance of subsidized fossil fuels. Thoughtful analysts for years have pointed out that highly mechanized, chemicalized, and capitalized farming cannot be sustained without cheap energy. As energy prices rose in the late 1970s, various studies were conducted to ascertain the dependence of U.S. agriculture on fossil fuels (Tables 1 and 2). Energy consumption in U.S. agriculture has not changed much since that time. As of 1974, on-farm energy use was 3.1 percent of the total U.S. energy budget, and food processing, distribution, and preparation accounted for 13.5 percent. The American public also does not see the energy that is embodied in the manufacturer of fertilizers and pesticides, in the packaging and transportation of food, etc. For example, fertilizers account for about 30% of U.S. on-farm energy use, a fact unknown to many farmers (Table 3).

The early uses of oil-derived fertilizers and pesticides yielded remarkable results. The profits made from the use of deceptively inexpensive fossil-fuel feedstocks were irresistible. Fossil fuels allowed the farmer to overwork the land while still increasing the yield. As the soil mining accelerated erosion and diminished soil quality, more and more fertilizer was needed to maintain crop production. From a basis of twelve inches of topsoil or less, corn yields are reduced annually by an average of about 4 bushels per acre for each inch of topsoil lost; oat, 2.4 bushels per acre; wheat, 1.6 bushels per acre; and soybean, 2.6 bushels per acre. The primary reasons for the reduced yields on eroded soils are low nitrogen content, impaired soil structure, deficient organic matter, and reduced availability of moisture. So, as soil quality is diminished, more and more fertilizer is needed to maintain crop production (7).

Use of pesticides also shows similarly diminishing returns. Although the chemicals were at first highly successful, many insects have quickly evolved resistance to insecticides, and similarly, a number of weeds have evolved resistance to some herbicides. Instances are now legion in which pesticides such as DDT and parathion unfavorably depressed predator and parasite populations, thus allowing insect pests to multiply so that farmers have to use even more pesticides. Around 1948, at the outset of the synthetic insecticide era, when the U.S. used 50 million pounds of insecticides, the insects destroyed 7 percent of preharvest crops. Today, using 600 million pounds, the U.S. loses 13 percent of crops before harvest (8).

Table 3. U.S. On-Farm Energy Use[a]

Component	Energy Consumed (quadrillion kJ)	Percent of Total
Direct		
Field operations	0.48	20
Transportation	0.27	12
Irrigation	0.26	11
Livestock, dairy, and poultry	0.17	7
Crop drying	0.13	5
Subtotal	1.32	55
Indirect		
Fertilizer	0.69	29
Equipment manufacture	0.32	13
Pesticides	0.07	3
Subtotal	1.07	45
Total	2.39	100

[a] Ref. 6.

Table 2. Energy Consumption with the U.S. Food System[a,b]

Sector of Food System	Energy Consumed (quadrillion kJ)	Percent U.S. Total Energy Use	Percent of U.S. Food System Energy Use
On farm			
Direct energy	1.4	1.7	10
Indirect energy	1.1	1.4	8
Food processing	3.8	4.9	30
Distribution system	1.3	1.7	10
Commercial food service	2.1	2.7	17
Home food preparation	3.3	4.2	25
Total	12.9	16.6	100

[a] There is a minor discrepancy between the on-farm numbers listed here and in Tables 1 and 3 because they were derived from different references.
[b] Ref. 5.

The addiction to petrofarming came on so gradually and insidiously that it required serious price shocks, as happened in 1974 and 1979, to bring this problem to public attention. Today, American agriculture uses more petroleum products than any other industry in the nation (9). The largest direct agricultural user of energy is farm machinery, including trucks and automobiles, which accounts for almost three-fifths of all directly used energy. In 1978, gasoline and diesel fuel used for all purposes on U.S. farms totaled 0.72 quads nationally. Energy to dry crops to prevent spoilage amounted to a further 0.12 quads. And energy to maintain livestock, dairy, and poultry totaled 0.16 quads.

Another important direct use of energy is irrigation. In 1975, to irrigate just the forty-eight million agricultural acres in the seventeen western states, ie, approximately 85 percent of all irrigation in the nation, required one-quarter of a quad. Pumping up groundwater from the San Joaquin and Imperial valleys, for example, has made the water table recede so that increasing amounts of electricity are needed to pump it; agriculture is California's biggest single users of electricity. In the Pacific Northwest, part of the justification for building several nuclear power plants under the Washington Public Power Supply System was for irrigation of the western slope desert of Washington state. WPPSS entered the largest municipal bankruptcy in history.

The energy embodied in chemicals and equipment, ie, indirect energy use, is not much less than direct use. About forty million tons of fertilizer are applied to America's fields each year, approximately 330 pounds for each person in the country. The feedstock, manufacture, and transportation of fertilizers consumed 0.65 quads of energy in 1978. Similarly, pesticides are a fossil-fuel based system of control. A full 80 percent of the one billion pounds spread annually comes out of oil wells. Pesticides required about 0.07 quads of energy in 1978 for feedstock, manufacture and transportation. The energy required to purify pesticide-contaminated water supplies should also be included. The energy embodied in manufacturing equipment is estimated to be 0.3 quads (10).

Farmers continue to depend on fossil fuels for a number of basic reasons, ie, many agronomic methods are capital intensive; farmers lack financial flexibility; perceived risk; and lack of research support on alternatives (11).

Most importantly, petroleum-powered machines have been substituted for human and animal labor. This makes it difficult for farmers to switch from machines back to human and animal labor. In 1920, there were 6,448,000 farms in the United States. The farm population at that time was 30 percent of the U.S. population, and there were twenty-five million horses and mules. By 1990 the number of farms had dropped to 2,140,420 and only 1.8 percent of the U.S. population lived on them. The number of horses and mules had dropped to about two million around 1962, and though it had risen to about eight million by 1978, almost none of these animals are used for farming (12).

Both ecology and economics are conspiring to put an end to petrofarming. The inflation caused by the escalating price of fossil fuel is giving farmers little choice. They must find new fuel-efficient farming methods or collapse under a tremendous debt burden. In 1982, farm debt was $216.3 billion, about $10 per dollar of annual farm net income.

This implied that the average farm would have lost money at annual interest rates above roughly 10 percent even if it had zero operating costs. Since 1982, net income has doubled and farm debt has declined by a fourth, so that as of 1990, the ratio of farm debt to net income was 3 to 1 (13).

ENERGY FOR FOOD PROCESSING

On-farm energy use accounts for only 18% of the energy consumption within the U.S. food system. Food processing and distribution account for 40%, and preparation accounts for the remaining 42%. Since consumer demands dictate how food is processed, distributed, and prepared, we should look at what happens beyond the farm.

The food-processing industry has become an important support industry for the farmer. In the past fifty years, canned, frozen, and other processed foods have become the principal items of our diet. During the 1960s, there was a slow but steady shift toward consumer consumption of more energy-intensive food, such as beef and highly processed foods. More than three-quarters of the food grown on farms is processed before shipment to the consumer. The food-processing industry is the fourth largest energy consumer of the Standard Industrial Classification groupings, with only primary metals, chemicals, and petroleum ahead of it. In 1974, food processing and distribution accounted for 4.8 quads or 6.6% of total U.S. energy use (Table 2) (14).

The eleven most energy-intensive food-processing industries as a whole derive about 48% of their energy from natural gas, 28% from electricity, 9% from coal, 7% from residual fuel oils, and 8% from other fuels. Most processing plants in these industires were designed and built during an era of cheap energy and thus are energy inefficient. Energy losses to the environment (such as waste heat from buildings and processing equipment and hot water) decrease the useful amount of energy available for processing. For example, only 34% of the energy put into vegetable canneries in western New York actually goes into food processing (15).

Finally, food-processing waste represents another energy loss. The energy cost of the processed food includes the energy invested in the processing and embedded in the waste. Additionally, as a consequence of the material losses, more raw food is required to obtain a given amount of processed food. The food-processing industries annually generate 14.4 million tons of solid waste, which is 9.6% of that generated by manufacturing industries in the United States every year. The 262 billion gallons of wastewater produced by the food-processing industry represent a twelve-day supply of water for U.S. urban domestic use (16).

SEEDS OF CHANGE

Since 1930, many farmers have paid the carrying charge on their debt by borrowing against inflation in land values. When land prices stabilized in 1980, farm foreclosures spread across the Midwest at a rate unseen since the Great Depression. Despite an intensive export drive, many farmers were unable to carry the debt they incurred for heavy machinery and irrigation equipment; their bankruptcies threatened the solvency of many banks. In 1982, farm production expenses of $140.1 billion had almost caught up with farm cash receipts from crops and livestock of $144.6 billion. In 1983, government farm price supports were $21.2 billion, which was almost as much as the 1982 net farm income of $22.1 billion. With the Payment-In-Kind program, one form of crop-reduction payments, and the summer drought, government subsidies exceeded net farm income in 1983, for the first time in history. Yet most farmers go deeper into debt because they are advised only on how to raise production, not on how to cut the costs of production in water, energy, chemicals, and machines (17). The heavy use of nitrogen fertilizer has resulted in dangerously high nitrate levels in the groundwater of some areas. Drinking water contaminated by nitrogen can cause severe health problems. Pesticides are also contaminating groundwater in some areas. Heavy irrigation, made economically possible by cheap oil and gas, is eroding the soil and rapidly salinizing what is left. Soil erosion threatens farm productivity, and eroded soil is filling streams and rivers with silt and the air with dust.

Either the economic or the ecological failure alone should be enough to bring about a change in agricultural methods. Together, they make an urgent need for change. But such a change need not be disruptive or demand sacrifice, if it is done sensibly and soon enough. Like energy use in general, energy efficiency on the farm can be increased a great deal without reducing productivity.

Conservation and efficiency improvements can range from weather-stripping buildings to changing the timing of irrigation systems to making better use of farm machinery. The last opportunity alone represents a saving of 0.1 quad or 21 percent over present use. Almost half of this figure could be saved solely by switching over from gasoline to diesel engines. Another potential area for immense energy saving is in irrigation, which in 1974 used 0.26 quads. Some experts report that half of the energy used in irrigation could be eliminated by improved techniques and by better pumping equipment. If farmers were to use low-temperature grain drying where climate permits, solar drying could save half the supplemental heat required (18).

The energy consumed in fertilizer applications can in many cases be much reduced, either because present usage is more than is necessary or because at least part of the nutrient additives can come from organic, rather than inorganic, sources. Only one soil test is performed for every 162 acres planted, so the nutrient levels of most cropland are not known. Poor soil testing and overfertilization are still common problems in Illinois, for example (19).

Livestock manure can also be substituted for inorganic fertilizers with considerable energy savings. U.S. livestock manure production is estimated at 1.7 billion tons per year. More than half of it is produced in feedlots and confinement rearing. If one-fifth of the manure from confinement rearing and feedlots were used as fertilizer, it could serve seventeen million acres at ten tons per acre and save 0.07 quads. At the same time, manure would add organic matter to the soil, which increases beneficial bacteria and fungi, makes plowing easier, improves soil texture, and reduces erosion.

Rotating crops with legumes can supply nitrogen to cropland in considerable quantities. Although inorganic nitrogen is commonly added to cornfields at the rate of 112 pounds per acre, it is possible to plant legumes between corn rows in late August and plow them under as green manure in early spring. In the northeastern United States, corn and winter vetch planted in this way add 150 pounds of nitrogen per acre. If this procedure were performed on fifteen million acres of corn, which is about one-fourth of U.S. corn acreage, about 0.07 quads would be saved. Using pesticides only where and when they are necessary would reduce pesticide consumption by up to one-half. This would amount to a saving of 0.03 quad, but the major benefit from decreased pesticide use is not the energy saved; rather it is the reduced contamination of the countryside.

Because of the soil erosion and energy costs associated with moldboard plowing, various forms of conservation tillage are being rapidly adopted, such as reduced-till, minimum-till, no-till. Conservation tillage can reduce erosion on many soils from 50 to 90 percent. Additional benefits of conservation tillage are lower costs for equipment, labor, and fuel, and increased soil moisture retention. In 1991, 73 million acres—20% of all U.S. cropland—were farmed with conservation tillage practices (20).

Conservation tillage has inherent problems that include increased pest populations (insects, nematodes, rodents, fungi), increased susceptibility to plant disease, herbicide carry-over that locks farmers into continuous planting of corn, evolution of weed resistance to herbicides, and shifts in weed species, to perennial weeds become a problem for which there are no fully effective chemical solutions. In addition, the farmer takes a greater economic risk.

No-till can reduce fuel use in field operations by as much as 90%; in some instances it has actually increased crop yields. However, this energy saving is partly offset because no-till requires 30 to 50% more pesticide than conventional tillage needs. In addition, no-till soil sometimes needs extra nitrogen. The result, as numerous studies have shown, is that net energy savings for no-till on individual fields range from zero to about 10%; occasionally, savings run as high as 30% (21).

However, farmers are not supported adequately by available technology or site-specific information services. If an herbicide fails to control weeds, the farmer usually cannot follow up with cultivation to salvage the crop.

Conservation tillage has become too reliant on herbicides to control weeds. Compared to research done on herbicides, little effort has been made to combine conservation tillage with integrated pest management, the system for managing insects, diseases, and weeds by combining resistant crop varieties, beneficial organisms, and crop rotations, plus other techniques. Such management uses pesticides only where necessary. Decreasing the impact on target organisms including humans. For example, the contamination of many midwestern drinking water supplies with atrazine is a cause for concern (22). Some farmers successfully use crop rotation with minimum-till without herbicides (23).

Multiple cropping (intercropping or relay cropping) can also be used to control weeds in no-till. Again, most but not all of the research on multiple cropping involves the use of herbicides to kill or weaken the grass or legume sod (annual or perennial) several weeks before the corn is planted by no-till. Herbicides are also used in research to control weeds in the interplanting of grasses or legumes with corn.

The reduction of herbicide use in conservation tillage would bring larger net energy savings and less contamination of the countryside. Too often conservation tillage with herbicides is regarded as the solution to soil erosion problems. This viewpoint discourages funding of research needed to develop alternative methods of dealing with soil erosion.

A sustainable agriculture has to result in significant energy savings. It has been estimated that the direct energy savings from such agriculture could be 0.8 to 0.9 quads. Combined with the indirect savings of oil formerly used to replace soil fertility, the total would be 1.1 to 2.2 quads, or a financial savings of $13 to $26 billion (1992 dollars) each year (24).

Clearly, it is imperative for farmers to implement techniques that remove them from the fossil-fuel treadmill. Increasing the efficiency of energy use and seeking new production techniques that minimize energy use are both necessary first steps. Ultimately, however, farmers—like all Americans—must find a replacement for fossil fuels. Thus, interest has turned to biomass fuels.

BIOMASS FUELS

Of all U.S. delivered energy, 58 percent is required in the form of heat, with three-fifths of this at temperatures below the boiling point of water. Petroleum products supply another 34 percent of this energy in the form of liquid fuels to run vehicles. Only 8 percent is needed as electricity. Heat is readily supplied by passive and active solar technologies that are commercially available already. Efficiently used electricity can be supplied just by existing hydropower, small-scale hydro, photovoltaics and windpower, without a need for any thermal power stations. For the production of both heat and electricity, the best renewable technologies now on the market actually cost less than new central power stations to deliver the same energy services in small sizes. This leaves unanswered the question of how best to provide liquid fuels for transportation (25).

A renewable liquid fuel program based on biomass feedstocks must adhere to four principles if it is to be truly sustainable.

1. *The land comes first.* All operations must be based on a concern for soil fertility and long-term environmental compatibility.
2. *Efficiency is vital.* Both the vehicle for which the fuel is intended and the means of converting the biomass into fuel must be efficient.
3. *Wastes are the source.* Use farming and forestry wastes as the principal feedstocks; no crop should be grown just to make fuels.
4. *Sustainability is a goal.* The program should be a vehicle for the reform of currently unsustainable farming and forestry practices.

PUT LAND MAINTENANCE FIRST

At present, much of American farming and forestry is little more than a mining operation. A massive biomass fuels program that simply serves to put greater pressure on overstressed land would not only risk crushing a budding energy program but could also pull down much of American agriculture.

Renewable must mean sustainable in the very long run. No biomass program can long endure unless it is based on the preservation and enhancement of soil fertility, water quality, and the biotic community on which agriculture depends. The subsidized, corn-based ethanol/Gasohol program—largely ignores these concerns. While opponents and fans argue the net energy yields, the impact of such a program on the soil is overlooked. An alcohol fuels program must take into account not only the obvious and direct costs but also such hidden requirements as soil and water. Much of modern corn production is a real soil killer. The average corn farmer who never rotates crops loses around twenty tons of soil per acre with conventionally tilled corn. This is the equivalent of 2.3 bushels of soil lost per bushel of corn harvested. Corn grown west of the hundredth meridian, which passes just west of Hays, Kansas, is also energy intensive: energy for irrigation accounts for 66% of the energy used in growing irrigated corn in Kansas (26).

Many criticisms of the use of ethanol as a fuel are really criticisms not of biomass fuels but of the agricultural system whose products are made into those fuels. Thus the first requirement of a sustainable biomass fuels program is to choose feedstocks whose production is not energy intensive and does not cause intolerable soil erosion and degradation.

Of critical importance is the proper scale of the biomass conversion techniques and their efficiency. Proper scale is essential to minimize the costs of collecting and transporting the feedstock. No definitive study has yet shown what scale is most cost effective, and there is probably no universal answer to this question, but the same diseconomies of large scale that have lately been described for electrical generating plants will probably dictate that biomass systems be relatively small in most circumstances (27).

Biomass conversion plants could be as small as mobile pyrolyzers; these devices heat a feedstock—usually a woody one—with little or no ash, producing, among other possible products, char, a low-grade fuel and gas, and a heavy oil akin to condensed woodsmoke. They could be hauled on the back of a truck to go wherever there are small collections of feedstocks. A conversion plant the size of a milk-bottling plant could serve a half-dozen towns (28).

Initially, integrated systems that use both the fuel product and its byproducts at the conversion site—the farm, the food-processing plant, or the pulp mill where the feedstock is available—make the most economic sense. For example, a pulp mill could utilize the steam from pulping processes to power a pyrolyzer plant to produce pyrolysis oil or methanol. The same steam could then be run from the pyrolysis plant through a turbine to generate electricity.

The goal in all of the considerations of scale should be to collect feedstocks with a minimum amount of transportation; the trace nutrients could be redistributed after the conversion. This transportation issue is especially important in the case of wet feedstocks, which tend to be heavy or perishable or both. Transportation can often be reduced by obtaining feedstocks from industrial plants where the wastes are already gathered into one location or, if necessary, by building the biomass conversion systems near the industrial plants. Where possible, the fuel should be sold locally, preferably through existing networks, again reducing distribution costs.

It is sometimes argued that many dispersed biomass fuel plants cannot produce enough total fuel to be nationally important. Even if this were true, a biofuels program meeting most of the needs of farms themselves would be important. Farms now get some 93 percent of their fuel from oil and gas and are often on the end of long and precarious supply lines. If the world oil trade is disrupted, farmers will suffer a double problem, ie, curtailed supplies boost fuel prices, while curtailed crop exports depress crop prices.

Conversion efficiency will decide whether a biomass fuels program will result in a net gain or a net loss of energy. The conversion technology assumed in most official studies of biomass fuels is borrowed from other processes that are inappropriate for biomass conversion. It is relatively easy to show a net energy loss if one uses old, inefficient ethanol distillation technology, or thermochemical conversion processes designed for coal, which cooks more slowly, less efficiently, and at higher temperatures than biomass. But it is also easy to find more efficient methods. Ethanol stills, for example, have been, and can still be, enormously improved. A few years ago it took 50,000–100,000 Btus to distill a gallon of ethanol to 190 proof. Today some commercial processes need only 25,000 Btus to go all the way to 200-proof ethanol, and the best demonstrated processes have reduced this to a mere 8,000–10,000 Btus. Stills can also use solar or, in some regions, geothermal heat.

PUT WASTES TO USE

Most biomass studies assume the use of special crops, notably grains, grown specifically for conversion to alcohol. The resulting potential for conflicts between food and fuel has been frequently criticized. This argument has great emotional appeal—it also largely misses the point. Most American grain feeds people only indirectly, through livestock. If our concern is to increase the amount of grain available to hungry people, the solution is not to criticize a grain-based alcohol program but to stop feeding grains to our pigs, chicken, and cattle.

Even better, however, is to run the biomass program on wastes. Many attractive feedstocks are currently a disposal problem. In California, for instance, rice straw is now burned in open fires. Used as a biomass feedstock, it would, coincidentally, solve a major air pollution problem. Each region has its example of a potential biomass feedstock, from apple pumice in Washington state to energy studies in Washington, D.C.

The diversity of waste feedstocks makes it hard to estimate their total biomass potential. The cotton-gin trash currently burned or dumped into wetlands in Texas would be enough to run every vehicle in Texas at present efficiencies. There is enough diminished-quality grain in an average Nebraska harvest to run one out of ten cars in the state at 60 mpg for a year. Walnut shells in California, peach pits in Georgia, food-processing wastes in most places—each by itself is nationally insignificant, but it might be of great importance locally. Logging waters can be used to make thermochemical methanol and, with emerging technology, perhaps ethanol directly; with other large sources of biomass, many local wastes can yield the required five to six quads per year (30).

An interesting component of municipal solid waste is urban forestry waste, such as tree trimmings, grass cuttings, and cleared brush. The Forest Service estimates that 150 million tons of urban tree wastes are generated each year. In Los Angeles County alone, 4,000—8,000 tons of separated woody material go into the landfills each day—a waste of about 1,000 megawatts thermal. Some cities deliberately manage their urban forests for wood production. Like municipal solid waste and sewage, urban forestry waste amounts to a substantial annoyance waiting to be converted to a resource. Among the technical options for converting such wood wastes are new gasifiers producing very high yields of methanol (83% in one recent design) or even making gasoline directly, as Tom Reed of the Solar Energy Research Institute stated in 1982.

There are several possible exceptions to the proscription against special crops. Among the most attractive potential biomass crops are plants of the family Euphorbiaceae, including spurges, cassavas, and poinsettias, for example. They come in hundreds of varieties, adapted to conditions ranging from deserts to rain forests. They share a sap rich in various resins, terpenes, and other fuels or fuel feedstocks. One study concluded that *Euphorbia* planted on a land area equal to that of Arizona could meet one-fourth of current U.S. petroleum consumption. *Euphorbia* grows well on otherwise marginal lands and has minimal water requirements. With careful attention to the impact of such species on the soil structure of sensitive lands, this could provide an attractive second crop in many regions (31). Other dryland crops might offer similar potential. Another possibility for special cropping is cattails. Douglas Pratt at the University of Minnesota has shown that sustainable, ecologically sensitive cropping of cattails could yield just under one quad of liquid fuels per day—up to one-fifth of ultimate national needs for vehicular liquid fuels.

A VEHICLE FOR REFORM

Agricultural practices in the 1990s are unsustainable and failure to reform those practices could leave the US short not just of fuels but also of food. Remedies are less clear: more is known about what does not work than about what does. However, a properly designed biofuels program might serve to reduce, not increase, pressure on agriculture. The residues from bioconversion should be put back on the land. Even thermochemical conversion can preserve many trace elements and wet-chemical or bacterial conversion can save everything in the feedstock except the hydrogen and much of the carbon.

Many farmers are eyeing the production of alcohol as a fuel component, with great interest. But without basic reforms to overcome the myriad problems of "modern" agriculture, such farmers have been keenly disappointed. A great deal more information about the problems inherent in agriculture, and what effect a biomass fuels program could have on land and the economics of agriculture is needed before a jump into a comprehensive biofuels campaign. More information is also needed on which feedstocks are most appropriate and which conversion processes are the most cost effective and on what scale. Most important, information on whether agriculture can be reformed so that it could sustain a biofuels program is needed.

This article is an updated version of "Energy and Agriculture," in W. Jackson, W. Berry, and B. Coleman, eds., *Meeting the Expectations of the Land: Essays in Sustainable Agriculture and Stewardship,* North Point Press, San Francisco, 1984, pp. 68–86.

BIBLIOGRAPHY

1. G. Cox and M. Atkins, *Agricultural Ecology,* W. H. Freeman, New York, 1979, p. 618; W. Vegara, "Energy Use in Processing," in R. A. Fazzolare and C. B. Smith, eds., *Changing Energy Use Futures,* Pergamon Press, New York, 1979, pp. 1862–1870; U.S. Bureau of the Census, *Statistical Abstract of the United States, 1992,* U.S. Government Printing Office, 1992.
2. D. Pimentel and co-workers, "Land Degradation: Effects on Food and Energy Resources," *Science* **194** (4261), 149–155 (1976); T. R. Hargrove, *Iowa Agricultural Home Economic Experiment Station,* Special Report no. 69, 1972; and U. S. General Accounting Office, *To Protect Tomorrow's Food Supply, Soil Conservation Needs Priority Attention,* CED-77-30 U.S. Government Printing Office, Washington, D.C., 1977.
3. U.S. Water Resources Council, *The Nation's Water Resources: 1975–2000,* U.S. Government Printing Office, Washington, D.C., 1978, p. 18.
4. W. Kolmar, "The Energy Requirements of U.S. Agriculture," *Monthly Energy Review,* Dept. of Energy, Energy Inf. Agency -0035, Vol. 83/2, p. 1.
5. Booz-Allen and Hamilton, Inc., *Energy Use in the Food System,* Office of Industrial Programs, Federal Energy Administration, FEA/D-76/083 U.S. Government Printing Office, Washington, D.C. 1976); U.S. Department of Agriculture, *Energy and U.S. Agriculture: 1974 and 1978,* Economics, Statistics, and Cooperatives Service Statistical Bulletin no. 632 (U.S. Government Printing Office, Washington, D.C. 1979), p. 64; J. S. Steinhart and C. F. Steinhart, *Science* **184** (4134), 33–42 (1974); R. A. Friedrich, *Energy Conservation for American Agriculture,* Ballinger, Cambridge, Mass, 1978, pp. 55, 69, 104.
6. USDA, *Energy and U.S. Agriculture: 1974 and 1978;* J. S. Steinhart and C. F. Steinhart, ref. 5; R. A. Friedrich, ref. 5.
7. D. Pimentel and co-workers, p. 152.
8. R. den Bosch, *The Pesticide Conspiracy,* Anchor Press/Doubleday, New York: 1980, p. 24.
9. R. A. Friedrich, *Energy Conservation for American Agriculture* Cambridge, Mass. 1978, p. 55.
10. W. Jackson, *New Roots for Agriculture,* Friends of the Earth, San Francisco 1980, pp. 26, 28.
11. H. F. Breimyer, "Outreach Programs of the Land Grant Universities: Which Publics Should They Serve?" Keynote address at Kansas State University for conference with same name and title, 1976.
12. U.S. Department of Agriculture, *Agricultural Statistics,* U.S. Government Printing Office, 1991 Washington, D.C.; W. Thomas and co-workers, *Food and Fiber for a Changing World: Third Century Challenge to American Agriculture* Interstate Printers and Publishers, Inc., Danville, Ill. 1976, p. 105.
13. U.S. Bureau of the Census, *Statistical Abstracts of the U.S.:* 1992 Washington, D.C.: U.S. Government Printing Office, 1992, and personal communication with A. Smith of the USDA Economic Research Service in Washington, D.C.
14. J. D. Buffington and J. H. Zar, "Realistic and Unrealistic Energy Conservation Potential in Agriculture," in W. Lockeretz, ed. *Agriculture and Energy* Academic Press, New York, 1977, pp. 695–711; J. S. Steinhart and C. E. Steinhart, "Energy Use in the U.S. Food System," *Science* **184** (4134), 33 (1974).
15. S. G. Unger, "Energy Utilization in the Leading Energy-Consuming Food Processing Industries," *Food Technology* **29,** (12) 33–45 (1975); W. Vegara, ref. 1, p. 1864.
16. U.S. Bureau of the Census, *Statistical Abstracts,* 206, see W. Vegara, ref. 1, p. 1864.
17. S. Tifft, "Farmers Are Taking Their PIK," *Time,* 14 (July 25, 1983).
18. Ref. 9, pp. 56, 69.
19. Buffington and Zar, ref. 14, pp. 700–701; L. Reichenberger, "Farmers Misfire on Fertilizer," *Farm Journal* **116**(4), 15 (1992).
20. W. Richards, "Restoring the Land," *Journal of Soil and Water Conservation* **46**(6), 409–410 (1991).
21. M. Hinkle, "Problems with Conservation Tillage," *J. Soil and Water Conserv.* **38**(3), 201–206 (1983). E. Axell, "The Toll of No-Till," *Soft Energy Notes* 3 (6), 14–16, 1981; W. Lockeretz, "Energy Implications of Conservation Tillage," *Journal of Soil and Water Conservation* 38(3), 207–21 (1983); D. Locker, "Problems Remain with No-Till Technology," *Prairie Sentinel* **1**(3), 8–10 (1982). Lockeretz Ref. 14, p. 211.
22. Hinkle, "Problems with Conservation Tillage," pp. 203, 204; E. Axell, ref. 21, p. 15.
23. Thompson, "No-Till Beans Without Herbicides," *Prairie Sentinel* **1**(3), 8–10 (1982); D. Looker and D. Demmel, "A Low-Cost Minimum-till Program," *Prairie Sentinel* **1**(3), II (1982).
24. W. Jackson and M. Bender, "Saving Energy and Soil," *Soft Energy Notes* **3**(6), 7 (1981).
25. A. B. Lovins, *Soft Energy Paths,* Ballinger, Cambridge, Mass. **1**(3), 1977, p. 80.
26. R. Neil Sampson, *Farmland or Wasteland: A Time to Choose,* Rodale Press, Emmaus, Pa. 1981, p. 131; D. Pimentel and D. Burgess, "Energy Inputs into Corn Production, " in *Handbook of Energy Utilization in Agriculture,* D. Pimentel, ed., CRC Press, Inc., Boca Raton, Fla., 1980, pp. 67–84.
27. A. B. Lovins and L. Hunter Lovins, *Brittle Power: Energy Strategy for National Security,* Brick House, Andover, Mass. 1982, p. 335.
28. Solar Energy Research Institute, *A Survey of Biomass Gasification,* 3 Vols. National Technical Information Service,

Springfield, Va., 1979; National Research Council, *Energy for Rural Development: Renewable Resources and Alternative Technologies for Developing Countries,* National Academy Press, Washington, D.C., 1981; National Research Council, *Producer Gas: Another Fuel for Motor Transport,* National Academy Press, Washington, D.C. 1983; and I. E. Cruz, *Producer-Gas Technology for Rural Applications,* Food and Agriculture Organization, Rome, 1985.

29. J. Harding, "Ethanol's Balance Sheet," in J. Harding and coeds. *Tools for the Soft Path,* Friends of the Earth, San Francisco, 1982, pp. 102–104.
30. C. E. Wyman, R. L. Bain, N. D. Hinman and D. J. Stevens, "Ethanol and Methanol from Cellulosic Biomass," in T. B. Johansson, H. Kelly, A. K. N. Reddy, & R. H. Williams, *Renewable Energy: Sources for Fuels and Electricity,* Island Press, Washington, D.C., 1993, pp. 865–923.

AIR CONDITIONING

HARRY J. SAUER, JR.
University of Missouri
Rolla, Missouri

The term *air conditioning* in its broadest sense implies control of any or all of the physical and chemical qualities of the air. *Heating, ventilating, and air conditioning* (HVAC) systems have as their primary function either (*1*) the generation and maintenance of comfort for occupants in a conditioned space, or (*2*) the supplying of a set of environmental conditions (high temperature and high humidity; low temperature and high humidity, etc) for a process or product within a space. Human comfort design conditions are quite different from the conditions required in textile mills or the conditions required for grain storage.

Today there is a professional engineering society devoted to promoting the art and science of air conditioning and related fields—the American Society of Heating, Refrigerating, and Air-Conditioning Engineers (ASHRAE). ASHRAE techniques and methodology are widely accepted throughout the industry and as a result will be often cited herein. In fact, much of the information contained in this article must be credited to the ASHRAE literature, and special credit is due to the Society for permission to rely so heavily on its publications. Many of the principles and their applications to air conditioning can be found in the ASHRAE Handbook Series (1–4). In addition to these reference works, there are a number of textbooks dealing with air conditioning (5–11). (See also BUILDING SYSTEMS; ENERGY CONSERVATION; ENERGY MANAGEMENT, PRINCIPLES; INSULATION, THERMAL; DISTRICT HEATING).

HISTORICAL NOTES

In order to understand the current design criteria and trends it is helpful to know something of the past. As in other fields of technology, the accomplishments and failures of the past affect the current and future design concepts.

Failure or success of comfort air conditioning depends, to a large extent, on *air movements* throughout the habited spaces. By the time of Christ, the famous Roman architect Vitruvius had already discussed the orientation and planning of buildings with regard to ventilation. The remarkable Leonardo da Vinci had built a ventilating fan at the end of the 15th Century. The Scottish physician Dr. William Cullen in 1775 pumped a vacuum in a vessel of water to make ice. And a few years later Benjamin Franklin wrote his treatise on Pennsylvania fireplaces, detailing their construction, installation, and operation with elaborate illustrations. During the 19th century the techniques of warming and ventilating were progressing well. Fans, boilers, and radiators had been invented and were in common use. In 1894 the concept of central heating was fairly well-developed, and the basic heat source took the form of either a warm air furnace or boiler. The next important innovation in the development of the furnace system was the addition of a fan to provide a mechanical means of forcing air through the duct system.

Towering above his contemporaries was the "Father of Air Conditioning," Willis H. Carrier (1876–1950), who through his brilliant analytical and practical accomplishments contributed more to the advancement of the developing industry than any other individual. In 1902, faced with a challenge to solve the lithographic industry's perennial problem of poor color register with every change in the weather, Carrier designed and installed for the Sackett-Wilhelms Lithographing & Printing Company in Brooklyn the first engineered year-round air-conditioning system, providing the four main functions of heating, *cooling, humidifying,* and *dehumidifying.* In December 1911, Carrier presented at the Annual Meeting of ASME his epoch-making paper, Rational Psychrometric Formulae, which related *dry-bulb temperature, wet-bulb temperature* and *dew-point temperatures* of the air as well as its sensible, latent and total heat, and enunciated the theory of adiabatic saturation. These formulas, together with the accompanying psychrometric chart, became the authoritative basis for all fundamental calculations in the air-conditioning industry.

Comfort air conditioning made its major breakthrough in motion picture theaters in Chicago in 1919–1920 employing CO_2 machines, and in Los Angeles in 1922 using an ammonia compressor. New centrifugal type of compressors provided the impetus for several theaters on Times Square. By the end of the 1920s, several hundred theaters has been air conditioned throughout the country. The Milam Building, in San Antonio, completed in 1928, was the first office building in the world originally planned for and provided throughout with conditioned air of a predetermined temperature and humidity. Cooling was provided by two of Carrier's centrifugal machines. Towards the end of the decade, General Electric introduced the first self-contained room air conditioner. Product technology then advanced by leaps and bounds. The air-source heat pump and the large lithium bromide water-chiller were important innovative breakthroughs right after the war. In the 1950s came automobile air conditioners and rooftop heating and cooling units. The rooftop units contained a gas- or oil-fired heating surface, a mechanical cooling system complete with air-cooled condenser, supply/exhaust fans, a filter, and an outdoor and return damper section, all assembled with the necessary controls. Only the room thermostat had to be mounted separately. Multizone packaged rooftop units were popular variations during the 1960s. However, because these units often use provided heating and cooling simultaneously, they were not energy efficient and rapidly lost favor in the 1970s.

People have now reached the moon and are seeking further space conquests. For environmental control of astronaut and spacecraft, air conditioning in its most sophisticated form has permitted people to surmount the hostile climates of moon and space, and made possible all of the significant manned missions beyond the planet Earth.

Beginning with the oil embargo of 1973, the HVAC field underwent tremendous change. It could no longer be business as usual with concern only for the initial cost of the building and the HVAC equipment. Neither could the practice of using crude rules of thumb, which greatly oversized equipment and wasted energy, be tolerated any longer. Better design of both the building envelop and the air-conditioning system began receiving concentrated attention.

As detailed in a 1989 Department of Energy Report (12), more than half of the nation's four million commercial–industrial–institutional buildings made use of energy conservation measures for heating and cooling. However, less than 4% of them had any kind of waste heat-recovery system and only 1% had a time-clock thermostat. Only 17,000 buildings featured an economizer cycle. A majority, over 80%, of the buildings had some sort of building shell conservation measures applied. The report indicates that 36% of nonresidential buildings are totally air conditioned with an additional 36% partially cooled. Two-thirds are totally heated with an additional 26% partially heated. A summary of the annual building energy use per square foot of floor space according to the year built is as follows:

Period	Energy use
pre-1900	56,000 Btu/sq ft-yr
1901–1920	78,000
1921–1945	82,000
1946–1960	80,000
1961–1973	110,000
1974–1979	90,000
1980–1983	77,000
1984–1986	78,000

The type of building and its use strongly affected the energy use as shown in the following listing:

Building type	Energy use
Assembly	55,000 Btu/sq ft-yr
Education	87,000
Food sales	212,000
Health care	217,000
Lodging	112,000
Mercantile	79,000
Office	106,000
Warehouse	55,000
Vacant	40,000

The survey also provides information on the types of heating systems used in these buildings: warm air furnaces, 1.8×10^6; individual heaters-electric baseboards, 1.1×10^6; boilers, 0.6×10^6; packaged units, 0.5×10^6; and air source heat pumps, 0.3×10^6; and on the types of cooling systems: central cooling, 1.1×10^6; individual (inc. window) units, 0.9×10^6; and packaged units, 0.7×10^6.

Another report (13) by the Energy Information Administration details how residential heating has changed over the past 40 years. The share of electricity as the main space heating fuel in new homes has grown from a 15% level in the 1960s to more than 55% today. Much of this growth is due to the *electric heat pump* with nearly one-third of the homes built between 1985 and 1987 using the *heat pump* as the primary type of heating unit. Among the nearly 18 million electrically heated homes, 4.5 million used the heat pump. Another half million homes have a heat pump as an auxiliary to the central furnace. This growth has been at the expense of natural gas, whose share of the market has declined from 70% for homes built in the sixties to 29% today.

The breakdown of residential primary fuel use and the type of heating equipment is as follows:

Primary Fuel Type

NATURAL GAS (50×10^6 HOMES)	
Central warm-air furnaces	31.6 Use, $\times 10^6$
Steam/hot water system	9.2
Floor, wall pipeless furnace	5.1
Room heater/other	4.0
ELECTRICITY (17.9×10^6 HOMES)	
Central warm air furnace	6.9
Built-in electric units	5.1
Heat pumps	4.5
Other	1.1
FUEL OIL (10.9×10^6 HOMES)	
Steam/hot water system	6.3
Central warm air furnace	4.0
Other	0.5
LPG (4.1×10^6 HOMES)	
Central warm air furnace	2.4
Room heater	0.9
Other	0.8
KEROSENE (1.3×10^6 HOMES)	

Table 1 provides very rough estimates of the relative size of heating and cooling units required for various applications.

BUILDING DESIGN LOADS

The first step in sizing and/or selecting the components for an air-conditioning system is the calculation of the design *heating load* (winter) and the design *cooling load* (summer). The design heat loss in winter is determined for the winter inside and outside design conditions of tem-

Table 1. Design Heating and Cooling Load

Building Type	Air Conditioning (ft^2/ton)	Heating [(Btu/h)/ft^3]
Apartment	450	4.9
Bank	250	3.2
Department store	250	1.1
Dormitory	450	4.9
House	700	3.2
Medical center	300	4.9
Night club	250	3.2
Office		
Interior	350	3.2
Exterior	275	3.2
Post office	250	3.2
Restaurant	250	3.2
School	275	3.2
Shopping center	250	3.2

perature and humidity; the design heat gain in summer at the summer design conditions.

For the estimation of the space heat and moisture loads, the size, construction, and use of the space must be known. The schedule of occupancy and activity of occupants must be estimated. All heat and/or moisture emitting equipment in the space and schedules of operation must be determined. The space conditions to be maintained as well as the environment external to the space must be specified. Figure 1 indicates the components of heating and cooling loads.

Standard procedures for calculating design heat losses and gains are described in the *ASHRAE Handbook 1993 Fundamental, ASHRAE Cooling and Heating Load Calculation Manual* (14), and the Air Conditioning Contractors of America *Load Calculation for Residential Winter and Summer Air Conditioning, Manual J* (15).

Indoor Design Conditions

The comfort environment is the result of simultaneous control of temperature, humidity, cleanliness and air distribution within the occupant's vicinity. This set of factors includes mean radiant temperature as well as the air temperature, odor control, and control of the proper acoustic level within the occupant's vicinity. The necessary criteria, indexes and standards for use where human occupancy is concerned are available in the ASHRAE literature (1,16) and elsewhere. Direct indexes of the sensation of comfort for the human body include the following: dry-bulb temperature; dew-point temperature; wet-bulb temperature; relative humidity; and air movement.

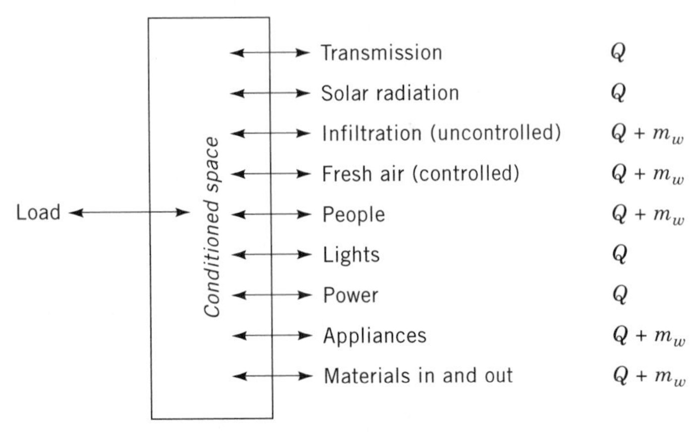

Figure 1. Components of heating and cooling loads.

Physiologists recognize that sensations of comfort and of temperature may have different physiological and physical bases, and each type should be considered separately. This dichotomy was recognized in the ANSI/ASHRAE Standard 55 (16), where *thermal comfort* is defined as "that state of mind which expresses satisfaction with the thermal environment." Unfortunately, the majority of our current predictive charts are based on comfort defined as a sensation "that is neither slightly warm or slightly cool." Figure 2 shows the winter and summer comfort zones specified in ANSI/ASHRAE Standard 55. The temperature ranges are appropriate for current seasonal clothing habits in the U.S. Summer clothing is light slacks and short sleeve shirt or comparable ensemble with an insulation value of 0.5 clo. [Traditionally, the insulating value of clothing is expressed in clo units, where 1.0 clo = 0.88 ft^2 h °F/Btu]. Winter clothing is heavy slacks, long sleeve shirt and sweater or jacket with an insulation value of 0.9 clo. The temperature ranges are for sedentary and slightly active persons. Figure 2 applies generally to altitudes from sea level to 3000 m (10,000 ft) and to the most common indoor thermal environments in which mean radiant temperature is nearly equal to the dry-bulb air temperature and where the air velocity is less than 30 fpm in winter and 50 fpm in summer. A wide range of environmental applications are covered by the ANSI/ASHRAE Standard 55. Offices, homes, schools, shops, theaters, and many others can be approximated well with these specifications.

Outdoor Design Conditions: Weather Data

Design outdoor conditions of temperature and humidity for most locations in the United States and many locations throughout the world may be found in the refs. 1 and 17. A sample of the ASHRAE weather data is given in Table 2. The winter design temperature usually selected is the outdoor air temperature that is exceeded 97.5% of the time. The temperature that is exceeded 99% of the time may be chosen when a more conservative design is desired. The summer design temperature to be used is the outdoor air temperature that is exceeded 2.5% of the time. Because humidity is also important in sizing air-conditioning equipment, the mean coincident wet-bulb temperature is tabulated along with the design (dry-bulb) temperature values.

Winter design temperatures are presented in Column 5. Two frequency levels are offered for each station representing temperatures that have been equaled or exceeded by 99% or 97.5% of the total in the months of December, January, and February (a total of 2160 hours) in the Northern Hemisphere and the months of June, July, and August (2208 hours) in the Southern Hemisphere. In a normal winter there would be approximately 22 hours at or below the 99% value and 54 hours below the 97.5% value. Column 9, the prevailing wind direction, is the

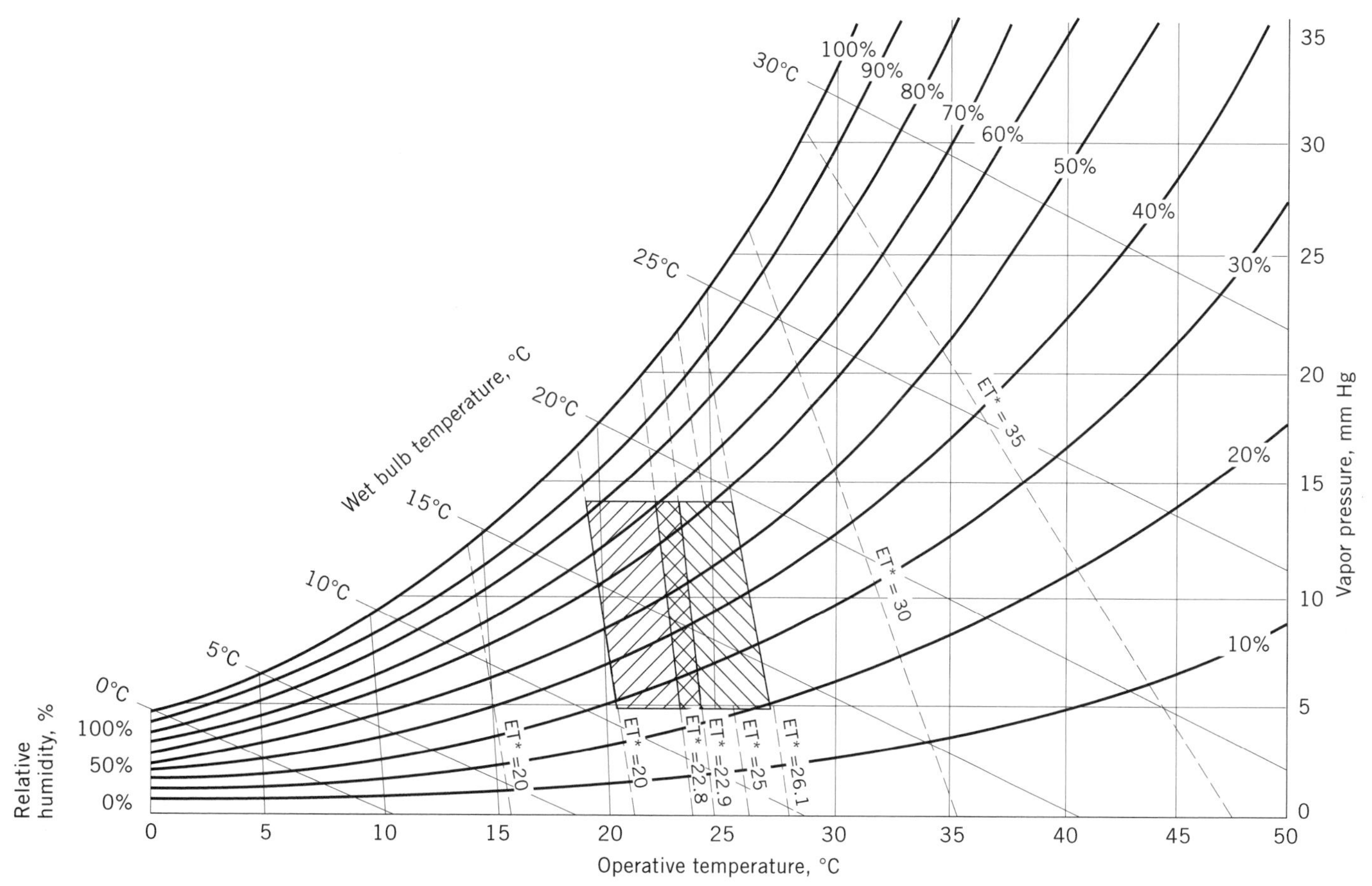

Figure 2. ASHRAE comfort zones. Courtesy of the American Society of Heating, Refrigerating, and Air-Conditioning Engineers.

wind direction occurring most frequently and the mean wind speed, which occurs coincidentally with the 97.5% dry-bulb winter design temperature. Column 10 is the median of the annual extreme minimum temperature.

For summer design, the recommended design dry-bulb and wet-bulb temperatures and mean daily range are presented in Columns 6, 7, and 8. Column 6 provides dry-bulb temperatures with their corresponding coincident wet-bulb temperatures. Wet-bulb temperatures are not available for countries other than the United States and Canada. The dry-bulb temperatures represent values that have been equaled or exceeded by 1%, 2.5%, and 5% of the total hours during the months of June through September (a total of 2928 hours) in the Northern Hemisphere and the months of December through March (2094 hours) in the Southern Hemisphere. The coincident wet-bulb temperature listed with each design dry-bulb temperature is the mean of all wet-bulb temperatures occurring at the specific dry-bulb temperature. The mean daily range shown in Column 7 is the difference between the average daily maximum and average daily minimum temperatures in the warmest month. Wet-bulb temperatures in Column 8 represent values that have been equaled or exceeded by 1%, 2.5%, and 5% of the hours during the summer months. These wet-bulb temperatures were computed independently of the dry-bulb values and are not coincident with the design dry-bulb values of Column 6. Their coincident dry-bulb values have not been determined. In Column 9, the prevailing wind direction is the wind direction occurring most frequently coincident with the 2.5% dry-bulb summer design temperature. Column 10 gives the median of the annual extreme maximum and minimum temperature. In a normal summer, approximately 30 hours would be at or above the 1% design value and 150 hours at or above the 5% design value.

Although not used in the design of building environmental control systems, the weather parameter known as the *wind chill index* (WCI) is of general interest. The WCI is an empirical index developed from cooling measurements obtained in Antarctica on a cylindrical flask partially filled with water at 33°C. The 33°C surface temperature was chosen as representative of the mean skin temperature of a resting human in comfortable surroundings. For velocities below 80 km/h, this index reliably expresses combined effects of temperature and wind on subjective discomfort. Table 3 provides equivalent wind chill temperatures for cold environments as derived from the wind chill index.

Infiltration and Ventilation

Infiltration is the air leakage through cracks and interstices, around windows and doors, and through floors and walls of a building of any type from a low one story house

Table 2. Excerpt from ASHRAE Weather Data Table[a]

				Winter,[b] °C		Summer,[c] °C							Prevailing Wind		Temp. °C	
Col. 1	Col. 2	Col. 3	Col. 4	Col. 5		Col. 6			Col. 7	Col. 8			Col. 9		Col. 10	
	Lat.	Long.	Elev.	Design Dry Bulb		Design Dry-Bulb and Mean Coincident Wet-Bulb			Mean daily	Design Wet-Bulb			Winter	Summer	Median of Annual Extr.	
State and Station[d]	° ′	° ′	m	99%	97.5%	1%	2.5%	5%	Range	1%	2.5%	5%	m/s		Max.	Min.
Arizona																
Douglas AP	31 27	109 36	1249	−3	−1	37/17	35/17	34/17	17	21	21	20			40.2	−10.0
Flagstaff AP	35 08	111 40	2136	−19	−16	29/13	28/13	27/12	17	16	16	15	NE 3	SW	32.2	−24.2
Fort Huachuca AP (S)	31 35	110 20	1422	−4	−2	35/17	33/17	32/17	15	21	20	19	SW 3	W		
Kingman AP	35 12	114 01	1079	−8	−4	39/18	38/18	36/18	17	21	21	21				
Nogales	31 21	110 55	1159	−2	0	37/18	36/18	34/18	17	22	21	21	SW 3	W		
Phoenix AP (S)	33 26	112 01	339	−1	1	43/22	42/22	41/22	15	24	24	24	E 2	W	44.9	−2.9
Prescott AP	34 39	112 26	1528	−16	−13	36/16	34/16	33/16	17	19	18	18				
Tucson AP (S)	32 07	110 56	780	−2	0	40/19	39/19	38/19	14	22	22	22	SE 3	WNW	42.7	−7.1
Winslow AP	35 01	110 44	1492	−15	−12	36/16	35/16	34/16	18	19	18	18	SW 3	WSW	39.3	−18.0
Yuma AP	32 39	114 37	65	2	4	44/22	43/22	42/22	15	26	26	25	NNE 3	WSW	46.0	−.7
District of Columbia																
Andrews AFB	38 5	76 5	85	−12	−10	33/24	32/23	31/23	10	26	24	24				
Washington, National AP	38 51	77 02	4	−10	−8	34/24	33/23	32/23	10	26	25	24	WNW 6	S	36.4	−13.7
Florida																
Belle Glade	26 39	80 39	5	5	7	33/24	33/24	32/24	9	26	26	26			34.8	−.6
Cape Kennedy AP	28 29	80 34	5	2	3	32/26	31/26	31/26	8	27	26	26				
Daytona Beach AP	29 11	81 03	9	0	2	33/26	32/25	31/25	8	27	26	26	NW 4	E		
Fort Lauderdale	26 04	80 09	3	6	8	33/26	33/26	32/26	8	27	26	26	NW 5	ESE		
Fort Myers AP	26 35	81 52	5	5	7	34/26	33/26	33/25	10	27	26	26	NNE 4	W	34.9	1.6
Minnesota																
Albert Lea	43 39	93 21	372	−27	−24	32/23	31/22	29/22	13	25	24	23				
Alexandria AP	45 52	95 23	436	−30	−27	33/22	31/22	29/21	13	24	23	22			35.1	−33.3
Bemidji AP	47 31	94 56	424	−35	−32	31/21	29/21	27/19	13	23	22	21	N 4	S	34.7	−38.3
Brainerd	46 24	94 08	374	−29	−27	32/23	31/22	29/21	13	24	23	22				
Duluth AP	46 50	92 11	435	−29	−27	29/21	28/20	26/19	12	22	21	20	WNW 6	WSW	32.7	−33.0
Fairbault	44 18	93 16	287	−27	−24	33/23	31/22	29/22	13	25	24	23			35.4	−31.3
Fergus Falls	46 16	96 04	369	−29	−27	33/22	31/22	29/21	13	24	23	22			36.1	−33.2
International Falls AP	48 34	93 23	359	−34	−32	29/20	28/20	27/19	14	22	21	20	N 5	S	34.1	−38.1
Mankato	44 09	93 59	306	−27	−24	33/22	31/22	29/21	13	25	24	23				
Minneapolis/St. Paul AP	44 53	93 13	254	−27	−24	33/24	32/23	30/22	12	25	24	23	NW 4	S	35.8	−30.0
Missouri																
Cape Girardeau	37 14	89 35	107	−13	−11	37/24	35/24	33/24	12	26	26	25				
Columbia AP (S)	38 58	92 22	237	−18	−16	36/23	34/23	33/23	12	26	25	24	WNW 5	WSW	37.5	−21.2
Farmington AP	37 46	90 24	283	−16	−13	36/24	34/24	32/23	12	26	25	24			37.7	−18.9
Hannibal	39 42	91 21	149	−19	−16	36/24	34/24	32/24	12	27	26	25	NNW 6	SSW	36.9	−22.0
Jefferson City	38 34	92 11	195	−17	−14	37/24	35/23	33/23	13	26	25	24			38.4	−21.2
Joplin AP	37 09	94 30	299	−14	−12	38/23	36/23	34/23	13	26	25	24	NNW 6	SSW		
Kansas City AP	39 07	94 35	241	−17	−14	37/24	36/23	34/23	11	26	25	24	NW 5	S	37.9	−20.2
Kirksville AP	40 06	92 33	294	−21	−18	36/23	34/23	32/23	13	26	25	24			36.8	−23.8
Mexico	39 11	91 54	236	−18	−16	36/23	34/23	33/23	12	26	25	24			38.4	−22.2
Moberly	39 24	92 26	259	−19	−16	36/23	34/23	33/23	13	26	25	24				
Poplar Bluff	36 46	90 25	116	−12	−9	37/26	35/24	33/24	12	27	26	26				
Rolla	37 59	91 43	367	−16	−13	34/25	33/24	32/23	12	26	25	24			37.4	−19.5
St. Joseph AP	39 46	94 55	252	−19	−17	36/25	34/24	33/24	13	27	26	25	NNW 5	S	38.1	−22.2
St. Louis, AP	38 45	90 23	163	−17	−14	36/24	34/24	33/23	12	26	25	24	NW 5	WSW		
St. Louis Co	38 39	90 38	141	−16	−13	37/24	34/24	33/23	10	26	25	24	NW 3	S	37.3	−19.3

[a] Reprinted with permission of the American Society of Heating, Refrigerating and Air-Conditioning Engineers.
[b] Winter design data are based on the 3-month period, December through February.
[c] Summer design data are based on the 4-month period, June through September.
[d] AP or AFB following the station name designates airport or Airforce base temperature observations. Co designates office locations within an urban area that are affected by the surrounding area. Undesignated stations are semirural and may be compared to airport data.
[e] Mean wind speeds occurring coincidentally with the 99.5% dry-bulb winter design temperature.

or commercial building to a multistory skyscraper. The magnitude of infiltration depends on the type of construction, workmanship, and condition of the building. The rate of infiltration cannot be controlled by the inhabitants of the building to any considerable extent.

Natural ventilation, on the other hand, is the intentional displacement of air through specified openings, such as windows, doors, and by ventilators.

Outside *air infiltration* may account for a significant proportion of the heating or cooling requirements for buildings. It is therefore important to be able to make an adequate estimate of its contribution with respect to both

Table 3. Equivalent Wind Chill Temperatures[a,b]

Wind Speed (in km/h)	Actual Thermometer Reading °C												
	10	5	0	−5	−10	−15	−20	−25	−30	−35	−40	−45	−50
	Equivalent Chill Temperature °C												
calm	10	5	0	−5	−10	−15	−20	−25	−30	−35	−40	−45	−50
10	8	2	−3	−9	−14	−20	−25	−31	−37	−42	−48	−53	−59
20	3	−3	−10	−16	−23	−29	−35	−42	−48	−55	−61	−68	−74
30	1	−6	−13	−20	−27	−34	−42	−49	−56	−63	−70	−77	−84
40	−1	−8	−16	−23	−31	−38	−46	−53	−60	−68	−75	−83	−90
50	−2	−10	−18	−25	−33	−41	−48	−56	−64	−71	−79	−87	−94
60	−3	−11	−19	−27	−35	−42	−50	−58	−66	−74	−82	−90	−97
70[c]	−4	−12	−20	−28	−35	−43	−51	−59	−67	−75	−83	−91	−99
	Little Danger: In less than 5 h, with dry skin. Maximum danger from false sense of security. (WCI less than 1400)				**Increasing Danger:** Danger of freezing exposed flesh within one minute. (WCI between 1400 and 2000)			**Great Danger:** Flesh may freeze within 30 s. (WCI greater than 2000)					

[a] Reprinted with permission of the American Society of Heating, Refrigerating and Air-Conditioning Engineers
[b] Cooling power of environment expressed as an equivalent temperature under calm conditions.
[c] Winds greater than 70 km/h have little added chilling effect.
Source: U.S. Army Research Institute of Environmental Medicine

design loads and seasonal energy requirements. Air infiltration is also an important factor in determining the relative humidity that will occur in buildings or, conversely, the amount of humidification or dehumidification required to maintain given humidities.

Heat gain due to infiltration must be included whenever the outdoor air introduced mechanically by the system is not sufficient to maintain a positive pressure within the enclosure to prevent any infiltration. Whenever economically feasible, it is desirable to introduce sufficient outdoor air through the air-conditioning equipment to maintain a constant outward escape of air, and thus eliminate the infiltration portion of the gain. The positive pressure maintained must be sufficient to overcome wind pressure through cracks and door openings. When this condition prevails, it is not necessary to include any infiltration component of heat gain. When the quantity of outdoor air introduced through the cooling equipment is not sufficient to build up the required pressure to eliminate infiltration, the entire infiltration load should be included in the space heat gain calculations.

There are two methods of estimating air infiltration in buildings. In one case the estimate is based on measured leakage characteristics of the building components and selected pressure differences. This is known as the *crack method,* since cracks around windows and doors are usually the major source of air leakage. The other method is known as the *air change method* and consists of estimating a certain number of air changes per hour for each room, the number of changes assumed being dependent upon the type, use, and location of the room. The crack method is generally regarded as being more accurate, provided that leakage characteristics and pressure differences can be properly evaluated. Otherwise the air change method may be justified. The accuracy of estimating infiltration for design load calculations by the crack or component method is restricted both by the limitations in information on air leakage characteristics of components and by the difficulty of estimating the pressure differences under appropriate design conditions of temperature and wind.

The introduction of some outdoor air into conditioned spaces generally is necessary. Local codes and ordinances frequently specify outdoor air ventilation requirements for public places and for industrial installations.

ASHRAE Standard 62, Ventilation for Acceptable Indoor-Air Quality, (18) is aimed at establishing better control of indoor contaminants without incurring a large energy penalty. Ventilation rates for some 91 applications are recommended for controlling indoor air quality by specifying the amount of outdoor air needed to dilute and remove contaminants. Under the standard, minimum outdoor airflow rates have been increased from 5 to 15 cfm per person, and the previous distinction between smoking and nonsmoking areas has been dropped because of that increase in ventilation rates. Studies have linked increased ventilation rates with fewer respiratory infections. Other studies have shown the need for 7.2 L/s per person (15 cfm/person) to remove occupant generated odors effectively.

Table 4 provides examples of the outdoor air requirements for ventilation specified by ASHRAE Standard 62.

Typical infiltration values in housing in North America vary by about a factor of ten, from tight housing with seasonal air change rates at 0.2 per hour to housing with infiltration rates as great as 2.0 per hour. In general, the minimum outside air requirement is 0.5 air chagnes per hour to overcome infiltration by producing a slight

Table 4. Ventilation Requirements (ASHRAE Standard 62)[a]

Application	Estimated Maximum[b] Occupancy P/1000 ft² or 100 m²	Outdoor Air Requirements cfm/person	L/s · person	cfm/ft²	L/s · m²	Comments
Offices						
Office space	7	20	10			Some office equipment may require local exhaust.
Reception areas	60	15	8			
Telecommunication centers and data entry areas	60	20	10			
Conference rooms	50	20	10			Supplementary smoke-removal equipment may be required.
Public Spaces				cfm/ft²	L/s · m²	
Corridors and utilities				0.05	0.25	
Public restrooms, cfm/wc or urinal		50	25			Mechanical exhaust with no recirculation is recommended.
Locker and dressing rooms				0.5	2.5	
Smoking lounge	70	60	30			Normally supplied by transfer air, local mechanical exhaust; with no recirculation recommended.
Elevators				1.00	5.0	Normally supplied by transfer air.
Photo studios	10	15	8			
Darkrooms	10			0.50	2.50	
Pharmacy	20	15	8			
Bank vaults	5	15	8			
Duplicating, printing				0.50	2.50	Installed equipment must incorporate positive exhaust and control (as required) of undesirable contaminants (toxic or otherwise).
Education						
Classroom	50	15	8			
Laboratories	30	20	10			Special contaminant control systems may be required for processes or functions including laboratory animal occupancy.
Training shop	30	20	10			
Music rooms	50	15	8			
Libraries	20	15	8			
Locker rooms				0.50	2.50	
Corridors				0.10	0.50	
Auditoriums	150	15	8			
Smoking lounges	70	60	30			Normally supplied by transfer air. Local mechanical exhaust with no recirculation recommended.
Hospitals, Nursing and Convalescent Homes						
Patient rooms	10	25	13			Special requirements or codes and pressure relationships may determine minimum ventilation rates and filter efficiency. Procedures generating contaminants may require higher rates.
Medical procedure	20	15	8			
Operating rooms	20	30	15			
Recovery and ICU	20	15	8			
Private Spaces						
Living areas	0.35 air changes per hour but not less than 15 cfm (7.5 L/s) per person					For calculating the air changes per hour, the volume of the living spaces shall include all areas within the conditioned space. The ventilation is normally satisfied by infiltration and natural ventilation. Dwellings with tight enclosures may require supplemental ventilation supply for fuel-burning appliances, including fireplaces and mechanically exhausted appliances. Occupant loading shall be based on the number of bedrooms as follows: first bedroom, two persons; each additional bedroom, one person. Where higher occupant loadings are known, they shall be used.
Kitchens	100 cfm (50 L/s) intermittent or 25 cfm (12 L/s) continuous or openable windows					Installed mechanical exhaust capacity. Climatic conditions may affect choice of the ventilation system.
Baths, Toilets	50 cfm (25 L/s) intermittent or 20 cfm (10 L/s) continuous or openable windows					Installed mechanical exhaust capacity.

[a] Reprinted with permission of the American Society of Heating, Refrigerating and Air-Conditioning Engineers

[b] Net occupiable space.

positive pressure within the structure. The minimum outside air standard for ventilation is (7.2 L/s) 15 cfm per person. However, local ventilation ordinances must be checked and may require greater quantities of outdoor air.

Heat Transfer Coefficients

The design of a heating, refrigerating, or air-conditioning system, including selection of building *insulation* or sizing of piping and ducts, or the evaluation of the thermal performance of parts of the system such as chillers, heat exchangers, fans, etc, is based on the principles of heat transfer. Two terms are used to indicate the relative insulating value of materials and sections of walls, floors, and ceilings. One is called U-factor and the other, R number. The U factor indicates the rate at which heat flows through a specific material or a building section, such as the one shown in Figure 3.

The smaller the U factor, the better the insulating value of the material or group of materials making up the wall, ceiling, or floor. The R number indicates the ability of one specific material, or a group of materials in a building section, to resist heat flow through them. Many of the insulating materials now have their R number stamped on the outside of the package, batt, or blanket.

Ref. 19 deals with the concepts and procedures for determining U factors and R values, and includes a brief discussion of factors that may affect the value of these coefficients and the performance of thermal insulations. Coefficients may be determined by test or they may be computed from known values of the thermal conductance of the various components. The procedures used for calculating coefficients are illustrated by examples.

Heating Load Methodology

Prior to designing a heating system, an estimate must be made of the maximum probable heat loss of each room or space to be heated, based on maintaining a selected indoor air temperature during periods of design outdoor weather conditions. The heat losses may be divided into two groups: (*1*) the transmission losses or heat transmitted through the confining walls, floor, ceiling, glass, or other surfaces, and (*2*) the infiltration losses or heat required to warm outdoor air which leaks in through cracks and crevices, around doors and windows, or through open

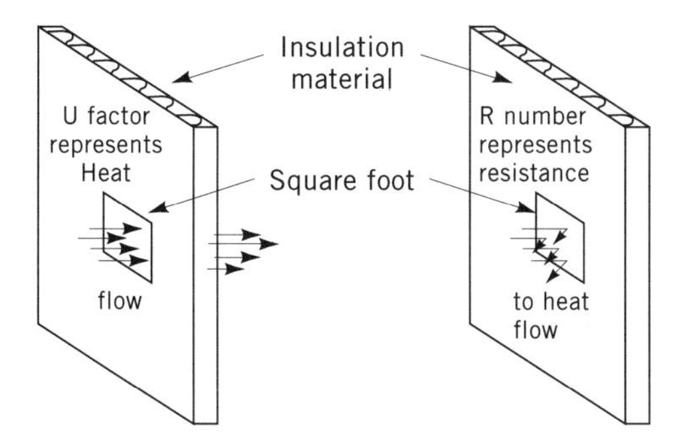

Figure 3. U factor and R number.

doors and windows, or heat required to warm outdoor air used for ventilation.

The ideal solution to the basic problem which confronts the designer of a heating system is to design a plant that has a capacity at maximum output just equal to the heating load which develops when the most severe weather conditions for the locality occur. However, where night setback is used, some excess capacity may be needed unless the owner is aware that under some conditions of operation he/she may not have the ability to elevate the temperature.

In most cases, economics interferes with the attainment of this ideal. Studies of weather records show that the most severe weather conditions do not repeat themselves every year. If heating systems were designed with adequate capacity for the maximum weather conditions on record, there would be considerable excess capacity during most of the operating life of the system.

In many cases, occasional failure of a heating plant to maintain a preselected indoor design temperature during brief periods of severe weather is not critical. However, the successful completion of some industrial or commercial processes may depend upon close regulation of indoor temperatures. These are special cases and require extra study before assigning design temperatures.

Normally the heating load is estimated for the winter design temperature usually occurring at night; therefore, no credit is taken for the heat given off by internal sources (people, lights, etc.) The heat supplied by persons, lights, motors, and machinery should always be ascertained in the case of theaters, assembly halls, industrial plants, and commercial buildings such as stores, office buildings, etc, but allowances for such heat sources must be made only after careful consideration of all local conditions. In many cases, these heat sources may materially affect the size of the heating plant and may have a marked effect on the operation and control of the system. In any evaluation, however, the night, weekend, and any other unoccupied periods must be evaluated in order to ascertain if the heating system has sufficient capacity to bring the building to the stipulated indoor temperature before the audience arrives. In industrial plants, quite a different condition may exist, and heat sources, if always available during occupancy, may be substituted for a portion of the heating requirements. In no case should the actual heating installation (exclusive of heat sources) be reduced below that required to maintain at least 4.4°C in the building.

Heat transfer through basement walls and floors to the ground depends on (*1*) the difference between the air temperature within the room and that of the ground and outside air, (*2*) the material of the walls or floor, and (*3*) conductivity of the surrounding earth. Conductivity of the earth varies with local conditions and is usually unknown. Because of the great thermal inertia of the surrounding soil, ground temperature varies with depth, and there is a substantial time lag between changes in outdoor air temperatures and corresponding changes in ground temperature. As a result, ground coupled heat transfer is less amenable to steady-state representation than is the case for above grade building elements.

Although the practice of drastically lowering the temperature when the building is unoccupied may be effective in reducing fuel consumption, additional equipment capacity is required for pickup. Because design outdoor temperatures are generally quite lower than outdoor temperatures typically experienced during operating hours, many engineers make no allowances for this additional pickup heat in most buildings. However, if optimum equipment sizing is used (minimum safety factor), the additional pickup heat should be computed and allowed for as conditions require. In the case of churches, auditoriums, and other intermittently heated buildings, additional capacity on the order of 10% of the sum of all other heating loads should be provided.

Cooling Load Principles

The design cooling load must take into account the heat gain into the space from outdoors as well as the heat being generated within the space. In calculating the heating load, the effects of the sun were ignored. In making heat gain calculations, the effect of solar radiation on glass, walls, and roofs must be included. In addition, internal loads caused by lights, people, cooking, motors, and anything that produces heat must be considered.

Peak cooling loads are more transient than heating loads. Radiative heat transfer within a space and thermal storage cause the thermal loads to lag behind the instantaneous heat gains and losses. This lag is important, especially with cooling loads, in that the peak is both reduced in magnitude and delayed in time compared to the heat gains that cause it. Early methods of calculating peak cooling loads tended to overestimate loads; this resulted in oversized cooling equipment with penalties of both first cost and part-load operating inefficiencies. Today various calculation methods are used to account for the transient nature of cooling loads. Some widely used methods of performing load analysis for building elements include the following: Transfer Function Method (TFM); Cooling Load Temperature Difference/Cooling Load Factor (CLTD/CLF) Method; Total Equivalent Temperature Differential/Time Averaging (TETD/TA) Method; Response Factor Method; and Finite Difference Method.

The primary procedure for sizing cooling equipment is the Transfer Function Method (TFM). A simplified version of this method, which applies to certain types of construction, is a one-step procedure using Cooling Load Temperature Differences and Cooling Load Factors (CLTD/CLF). Where applicable, this method should be considered as a first alternative calculation procedure. A second alternative calculation procedure uses Total Equivalent Temperature Differential values and a system of Time Averaging (TETD/TA) to calculate cooling loads. The variables affecting cooling load calculations are numerous, often difficult to define precisely, and always intricately interrelated. Many of the components of the cooling load vary in magnitude over a wide range during a 24-period. Since these cyclic changes in load components are often not in phase with each other, a detailed analysis is required to establish the resultant maximum cooling load for a building or zone.

Space *heat gain* is the rate at which heat enters into and/or is generated within a space at a given instant of time. Heat gain is classified by (*1*) the mode in which it enters the space, and (*2*) whether it is a sensible or latent gain. The first classification is necessary because different fundamental principles and equations are used to calculate different modes of energy transfer. The heat gain occurs in the form of (*1*) solar radiation through transparent surfaces; (*2*) heat conduction through exterior walls and roofs; (*3*) heat conduction through interior partitions, ceilings, and floors; (*4*) heat generated within the space by occupants, lights, and appliances; (*5*) energy transfer as a result of ventilation and infiltration of outdoor air; or (*6*) miscellaneous heat gains. The second classification, sensible or latent, is important for proper selection of cooling equipment. The heat gain is sensible when there is a direct addition of heat to the conditioned space by any or all mechanisms of conduction, convection, and radiation. The heat gain is latent when moisture is added to the space (eg, by vapor emitted by occupants). To maintain a constant humidity ratio in the enclosure, water vapor in the cooling apparatus must condense out at a rate equal to its rate of addition into the space. The amount of energy required to do this, the latent heat gain, essentially equals the product of the rate of condensation and latent heat of condensation. The distinction between sensible and latent heat gain is necessary for cooling apparatus selection. Any cooling apparatus has a maximum sensible heat removal capacity and a maximum latent heat removal capacity for particular operating conditions.

Space cooling load is the rate at which heat must be removed from the space to maintain room air temperature at a constant value. The sum of all space instantaneous heat gains at any given time does not necessarily equal the cooling load for the space at that same time. The space heat gain by radiation is partially absorbed by the surfaces and contents of the space and does not affect the room air until some time later. The radiant energy must first be absorbed by the surfaces that enclose the space (ie, walls, floor, and ceiling) and the material in the space. As soon as these surfaces and objects become warmer than the space air, some of their heat will be transferred to the air in the room by convection. Since their heat storage capacity determines the rate at which their surface temperatures increase for a given radiant input, it governs the relationship between the radiant portion of heat gain and corresponding part of the space cooling load. The thermal storage effect can be important in determining an economical cooling equipment capacity.

In general, the design cooling load determination for a building requires data for most or all of the following:

Full building description, including the construction of the walls, roof, windows, etc, and the geometry of the rooms, zones, and building. Shading geometries may also be required.

Sensible and latent internal loads because of people, lights, and equipment and their corresponding operating schedules.

Indoor and outdoor design conditions.

Geographical data such as latitude and elevation.

Ventilation requirements and amount of infiltration.

Number of zones per system and number of systems.

Commercially available computer loads programs are available for both mainframe and microcomputer which, when input with the above data, will calculate both the heating and cooling loads as well as perform a psychrometric analysis of the system. Output typically includes peak room and zone loads, supply air quantities, and total system (coil) loads.

ENERGY ESTIMATING

After the peak loads have been evaluated, the equipment must be selected with capacity sufficient to offset these loads. The air supplied to the space must be at the proper conditions to satisfy both the sensible and latent loads which were estimated. However, peak load occurs but a few times each year and partial load operation exists most of the time. Since operation is predominantly at partial load, partial load analysis is at least as important as the selection procedure.

It is often necessary to estimate the energy requirements and fuel consumption of environmental control systems for either short or long terms of operation. These quantities can be much more difficult to calculate than design loads or required system capacity. The most direct and reliable energy estimating procedure requires the hourly integrations of the various loads as a function of the weather and use schedules, and control systems. Several time-saving procedures for providing rough estimates of energy consumption are presented.

The simplest procedures assume that the energy required to maintain comfort is a function of a single parameter, the outdoor dry-bulb temperature. The more accurate methods include consideration of solar effects, internal gains, heat storage in the walls and interiors, and the effects of wind on both the building envelope heat transfer and infiltration. The most sophisticated procedures are based on hourly profiles for climatic conditions and operational characteristics for a number of typical days of the year, or better still, a full 8760 hours of operation.

Calculation of the fuel and energy required by the primary equipment to meet these loads gives consideration to efficiencies and part-load characteristics of the equipment. Often it will be necessary to keep track of the different forms of such energy, such as electrical, natural gas, or oil. In some cases where calculations are done to assure compliance with codes or standards, it is necessary to convert these energies to source energy or resource consumed, as opposed to that delivered to the building boundary.

For air conditioners and heat pumps, the American Refrigeration Institute (ARI) has established performance ratings and testing conditions and publishes directories of certified products (20,21). The *Coefficient of Performance* (COP) is the scientifically accepted measure of heating or cooling performance. The COP is defined as

$$\mathrm{COP} = \frac{\text{Heating (or cooling) provided by system}}{\text{Energy consumed by the system}}$$

The total heating output of a heat pump excludes supplemental resistance heat. The COP is a dimensionless number and thus the heating or cooling output and the energy input must be expressed in the same units. The *energy efficiency ratio (EER)* is defined in exactly the same way as the COP; however, for the EER, the heating or cooling must be expressed in Btu and the energy consumed by the system in watt-hours. The EER is numerically equal to the $3.413 \times \mathrm{COP}$.

Federal regulations specify two additional ratings for heat pumps and air conditioners: the *Heating Seasonal Performance Factor* (*HSPF*) and the *Seasonal Energy Efficiency Ratio* (*SEER*). The HSPF combines the effects of heat pump heating under a range of weather conditions applicable for a particular geographical region with performance losses due to coil frost, defrost, cycling under part-load conditions, and use of supplemental resistance heating during defrost. Standard HSPFs are determined by manufacturers for each of the six Department of Energy (DOE) regions and are provided on equipment labels. The SEER is a measure of seasonal cooling efficiency under a range of weather conditions assumed to be typical of the DOE region in question, as well as performance losses due to cycling under part-load operation. The SEER is defined as the total cooling provided during the cooling season (Btu) divided by the total energy consumed by the system (watthours). Currently, SEERS are reported for conventional heat pumps and air conditioners according to a simplified procedure that does not take weather into account explicitly.

The sophistication of the energy calculation procedure used can often be inferred from the number of separate ambient conditions and/or time increments used in the calculations. Thus, a simple procedure may use only one measure, such as annual degree days, and will be appropriate only for simple systems and applications. Here such methods will be called single-method measures. Improved accuracy may be obtained by the use of more information, such as the number of hours anticipated under particular conditions of operation. These methods, of which the bin method is the most well known, are referred to as simplified multiple-measure methods. The most elaborate methods currently in use perform energy balance calculations at each hour over some period of analysis, typically one year. These are called the detailed simulation methods, of which there are a number of variations. These methods require hourly weather data, as well as hourly estimates of internal loads, such as lighting and occupants.

Single-measure methods for estimating cooling energy are less well-established than those for heating. This is because the indoor–outdoor temperature difference under cooling conditions is typically much smaller than under heating conditions. As a result, cooling loads depend more strongly on such factors as solar gain and internal loads (besides temperature) than do heating loads. Since these loads are highly dependent on specific building features, attempts to correlate cooling energy requirements against a single climate parameter have not been highly successful.

Even when used with applications apparently as simple as residential cooling, single-measure estimates can be seriously inaccurate if they do not consider all factors that significantly affect energy use. For example, the increased use of air source heat pumps, of which the performance is extremely dependent on outdoor air temperature, has produced an awareness of some failings in using the degree day method for residences. Thus, the degree day method is restricted to small structures in which the heating and cooling loads are envelope-dominated. Other methods should be used for larger commercial or industrial buildings, where internal, cooling only zones are prevalent or where cooling loads are not linearly dependent on the outdoor-to-indoor temperature difference.

In preparing energy estimates, it should be realized that any estimating method will produce a more reliable result over a long period of operation than over a short period. Nearly all of the methods in use will give reasonable results over a full annual heating season, but estimates for shorter periods, such as a month, may produce inaccurate results. As the period of the estimate is shortened, there is more chance that some factor not directly taken into account in the estimating method will deivate from its long-term average value, and thus lead to an error in the predicted energy requirement.

Degree Day Method (Heating)

The traditional degree day procedure for estimating heating energy requirements is based on the assumption that, on a long-term average, solar and internal gains will offset heat loss when the mean daily outdoor temperature is 18.3°C, and that fuel consumption will be proportional to the difference between the mean daily temperature and 18.3°C. In other words, on a day when the mean temperature is 11.1°C below 18.3°C, twice as much fuel is consumed as on days when the mean temperature is 5.6°C below 18.3°C. This basic concept can be represented in an equation stating that energy consumption is directly proportional to the number of degree days in the estimation period.

The *degree days* equation has undergone several stages of refinement in an attempt to make it agree as closely as possible with the available measured data on an average basis. The currently recommended form is

$$E = \frac{qL \times DD \times 24}{\Delta t \times k \times V} C_D$$

where E = fuel or energy consumption for the estimate period, kW · h or Btu; qL = design heat loss, W or Btu · h; DD = number of 18.3°C (65°F) degree days for the estimate period; Δt = design temperature difference, °C; k = overall efficiency factor; V = heating value of fuel, units consistent with q_L and E; and C_D = empirical correction factor.

Table 5 lists the average number of heating and cooling degree days 18.3°C (65°F base) for various cities in the United States. Some typical heating values are natural gas, 1050 Btu/ft^3; propane, 90,000 Btu/gal; and No. 2 fuel oil, 140,000 Btu/gal.

Table 6 gives approximate values for k. In the case of heat pumps, q must be in Btuh and 1000 × HSPF is to be used for the product of k and V. The errors inherent in the established 18.3°C (65°F) based method may be adjusted by the use of an empirical factor, C_D. Table 7 gives the best presently available values for C_D.

The variable base degree day (VBDD) method is a generalization of the widely used degree day method. It retains the familiar degree day concept, but counts degree days based on the balance point temperature, defined as the average outdoor temperature at which the building requires neither heating nor cooling.

The degree-day procedure is intended to recognize that, although the energy transferred out of a building is proportional to the difference between interior space temperature and the outside temperature, the heating equip-

Table 5. Annual Degree Days, 18.3°C (65°F) Base

City, State	Heating-Degree Days	Cooling-Degree Days	City, State	Heating-Degree Days	Cooling-Degree Days
Albuquerque, N.M.	4348	1345	Memphis, Tenn.	3232	1872
Atlanta, Ga.	2961	1469	Miami, Fla.	214	4189
Birmingham, Ala.	2551	1654	Minneapolis, Minn.	8382	894
Bismark, N.D.	8851	528	New Orleans, La.	1385	2653
Boston, Mass.	5634	674	New York, N.Y.	5219	1027
Charleston, S.C.	2033	1983	Omaha, Neb.	6612	1007
Cheyenne, Wy.	7381	308	Philadelphia, Pa.	5144	1081
Chicago, Ill.	6639	713	Phoenix, Ariz.	1765	3334
Cincinnati, Ohio	4410	1147	Pittsburgh, Pa.	5897	732
Cleveland, Ohio	6351	670	Portland, Maine	7511	292
Detroit, Mich.	6293	687	Portland, Oreg.	4635	248
Fort Worth, Texas	2405	2500	Raleigh, N.C.	3393	1355
Great Falls, Mont.	7750	343	Richmond, Va.	3865	1236
Houston, Texas	1396	2745	Salt Lake City, Utah	6052	958
Indianapolis, Ind.	5699	902	San Francisco, Calif.	3015	98
Jackson, Miss.	2239	2361	St. Louis, Mo.	4900	1390
Kansas City, Kansas	4711	1475	Seattle-Tacoma, Wash.	5145	134
Los Angeles, Calif.	2061	357	Tampa, Fla.	683	3152
Louisville, Ky.	4660	1207	Tulsa, Okla.	3860	1719
Madison, Wisc.	7863	424	Washington, D.C.	4224	1491

Table 6. Efficiency Factor, k^a

Heating System	k Value
Conventional gas-fired forced-air furnace or boiler	0.60–0.70
Conventional gas-fired gravity-air furnace	0.57–0.67
Gas-fired forced-air furnace or boiler with typical energy conservation devices, intermittent ignition device, automatic vent damper	0.65–0.75
Gas-fired forced-air furnace or boiler with sealed combustion chamber, intermittent ignition device and automatic vent damper	0.70–0.78
Gas-fired pulse combustion furnace or boiler	0.80–0.90
Oil-fired furnace or boiler	0.50–0.75

[a] Reprinted with permission of the American Society of Heating, Refrigerating and Air-Conditioning Engineers.

ment needs to meet only the part not covered by free heat from internal sources such as lights, equipment, occupants, and solar gain. In other words, the free heat covers energy requirements down to the balance point temperature, below which the energy requirements of the space are proportional to the difference between the balance point temperature and the outside temperature.

Cooling Degree Day Method

Cooling degree days are widely available to a base temperature of 18.3°C, but tabulations of degree days from base 7.2°C to 18.3°C are also available from the National Climatic Center, Asheville, N.C. (*Degree Days to Selected Bases,* in microfiche or published form). Table 5 provides cooling degree day values for several U.S. cities.

Cooling energy is predicted in a similar fashion as described for heating earlier, ie,

$$E_C = \frac{q_g \times CDD \times 24}{\Delta t_d \times SEER \times 1000}$$

where E_C = energy consumed for cooling, kW · h; q_g = design cooling load, Btu · h; CDD = Fahrenheit cooling degree days; Δt = design temepratur difference, °F; $SEER$ = seasonal energy efficiency ratio, Btu · h/W; and 1000 = W/kW.

Detailed Energy and Systems Stimulation

Energy analysis programs have come to be fundamental tools used by engineers in making decisions regarding building energy use. Energy programs, along with life-cycle costing routine, are used to quantify the impact of proposed energy conservation measures in existing buildings. In new building design, energy programs are used to aid in determining the type and size of building systems and components as well as to explore the effects of design tradeoffs. The growth of the computer industry, with increases in size and speed of machines and lowering of costs, has made possible the use of detailed but time-consuming techniques, without imposing undue cost penalties. Thus, one trend has been to develop programs using

Table 7. Correction Factor, $C_D{}^a$

Quality of Construction and Relative Use of Electrical Appliances	Number of Degree Days 18.3°C (65°F)								
	1000	2000	3000	4000	5000	6000	7000	8000	9000
Well-constructed house Large quantities of insulation, tight fit on doors and windows, well sealed openings. Large use of electrical appliances. Large availability of solar energy at the house.	0.48	0.45	0.42	0.39	0.36	0.37	0.38	0.39	0.40
House of average construction Average quantities of insulation, average fit on doors and windows, partially sealed openings. Average availability of solar energy at the house. Average use of electrical appliances.	0.80	0.76	0.70	0.65	0.60	0.61	0.62	0.69	0.67
Poorly constructed house Small quantities of insulation, poor fit on doors and windows, unsealed openings. Small use of electrical appliances. Small availability of solar energy at the house.	1.12	1.04	0.98	0.90	0.82	0.85	0.88	0.90	0.92

[a] Reprinted with permission of the Electric Power Research Institute from *Heat Pump Manual,* 1985.

the resources of major computers to do very complete analyses. Programs are now available which can be used by a wide spectrum of engineers for a wide range of purposes.

BASIC HVAC SYSTEM CALCULATIONS

Application of Thermodynamics to HVAC Processes

A complete air-conditioning system is given schematically as Figure 4 and shows various space heat and moisture transfers which may be present. The symbol q_S represents a sensible heat transfer rate while the symbol m_w represents a moisture transfer rate. The symbol q_L designates the transfer of energy which accompanies the moisture transfer and is given by $\Sigma m_w h_w$ where h_w is the specific enthalpy of the added (or removed) moisture. Solar radiation and internal loads are always gains upon the space. Heat transmission through solid construction components due to a temperature difference as well as energy transfers because of infiltration may represent a gain or a loss.

Referring to the conditioner of Figure 4, it is important to note that the energy (q_c) and moisture (m_c) transfers at the conditioner cannot be determined from the space heat and moisture transfers alone. The effect of the outdoor ventilation air must also be included as well as other system load components. The designer must recognize that items such as fan energy, duct transmission, roof and ceiling transmissions, heat of lights, bypass and leakage, type of return air system, location of main fans, and actual vs. design room conditions are all related to one another, to component sizing, and to system arrangement.

The most powerful analytical tools of the air-conditioning design engineer are the first law of thermodynamics or energy balance, and the conservation of mass or mass balance. These conservation laws are the basis for the analysis of moist air processes. The following sections demonstrate the application of these laws to specific HVAC processes.

In many air-conditioning systems, air is taken from the room and returned to the air-conditioning apparatus where it is reconditioned and supplied again in the room.

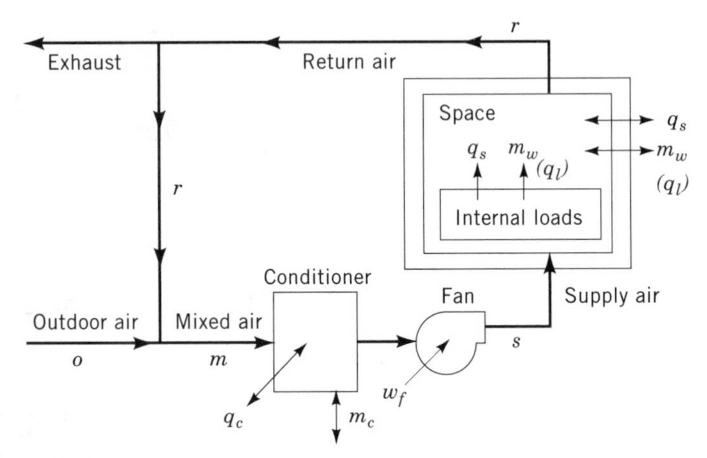

Figure 4. Schematic of air-conditioning system (7). Reprinted with permission.

In most systems, the return air from the room is mixed with outdoor air required for ventilation.

Figure 5 shows a typical air-conditioning system and the corresponding psychometric chart representation of the processes for the cooling condition. Outdoor air (o) is mixed with return air (r) from the room and enters the apparatus (m). Air flows through the conditioner and is supplied to the space (s). The air supplied to the space picks up heat q_s and moisture m_w and the cycle is repeated.

The sensible heat factor (SHF), also referred to as sensible heat ratio (SHR), is simply the ratio of the sensible heat for a process to the summation of the sensible and latent heat for the process. The sum of sensible and latent heat is also called the total heat.

As air passes through the cooling coil of an air conditioner, the sensible heat factor line represents the simultaneous cooling and dehumidifying that occurs. If the cooling process involves the removal on only sensible and no latent heat, ie, no moisture removal, the sensible heat factor line is horizontal and the sensible heat factor is 1.0. If 50% sensible and 50% latent, the SHF is 0.5.

The problem of air conditioning a space usually reduces to the determination of the quantity of moist air that must be supplied and the necessary condition which it must have in order to remove given amounts of energy and water from the space and be withdrawn at a specified condition.

The quantity q_s denotes the net sum of all rates of heat gain upon the space, arising from transfers through boundaries and from sources within the space. This heat gain involves addition of energy alone and does not include energy contributions due to addition of water (or water vapor). It is usually called the sensible heat gain. The quantity m_w denotes the net sum of all rates of moisture gain upon the space arising from transfers through boundaries and from sources within the space. Each pound of moisture injected into the space adds an amount of energy equal to its specific enthalphy h. Assuming steady-state conditions, the governing equations for the process across the space are

$$m_a h_s + m_w h_w - m_a h_r + q_s = 0$$

and

$$m_a W_s + m_w = m_a W_r$$

where m, q, h, and W are the appropriate mass flow rate, heat-transfer rate, enthalpy, and humidity ratio, respectively.

When moist air is cooled to a temperature below its dew point as it is conditioned, some of the water vapor will condense and leave the air stream. Although the actual process path will vary considerably depending on the type surface, surface temperature, and flow conditions, the heat and mass transfer can be expressed in terms of the initial and final states.

Although water may be separated at various temperatures ranging from the initial dew point to the final saturation temperature, it is assumed that condensed water is cooled to the final air temperature t_s before it drains from the system.

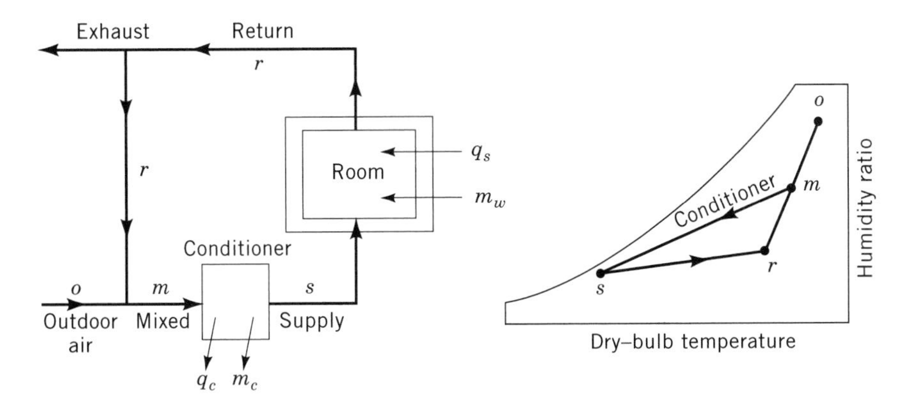

Figure 5. Typical air-conditioning system processes for cooling (7). Reprinted with permission.

For the system of Figure 5, the energy and mass balance equations for the conditioner are

$$m_c = m_a(W_m - W_S)$$
$$q_c = m_a[(h_m - h_s) - (W_m - W_s)h_c]$$

This process of heating and humidifying, generally required during the cold months of the year, follows the same mass and energy equations as given directly above.

Since the specific volume of air varies appreciably with temperature, all calculations should be made with the mass of air instead of volume. However, volume values are required for selection of coils, fans, ducts, etc. A practical method of using volume, but still including mass (weight) so that accurate results are obtained, is the use of volume values based on measurement at ASHRAE standard conditions. The value considered standard is 1.204 kg of dry air/m^3 (0.83 m^3/kg of dry air) [0.075 lb of dry air/ft^3 (13.33 ft^3/lb of dry air)], which corresponds to approximately 15.5°C at saturation and 20.6°C, dry [at 101.4 kPa (14.7 psia)].

Air-conditioning design often requires calculation of the following:

1. Sensible heat gain corresponding to the change of dry-bulb temperature for a given air flow (standard conditions). The sensible heat gain q_s, as a result of a difference in temperature Δt between the incoming air and leaving air flowing at ASHRAE standard conditions, is approximated by

$$q_s = 1.232(l/\mathrm{s})\Delta t$$

and in I-P units

$$q_s = 1.10(cfm)\Delta t$$

2. Latent heat gain corresponding to the change of humidity ratio (W) for given air flow (standard conditions). The latent heat gain in watts (Btu h), as a result of a difference in humidity ratio (ΔW) between the incoming and leaving air flowing at ASHRAE standard conditions, is

$$q_l = 3012(l/s)\Delta W$$

In I-P units

$$q_l = 4840\ (cfm)\Delta W$$

AIR-CONDITIONING SYSTEMS

The simplest form of an all-air HVAC system is a single conditioner serving a single temperature control zone. The unit may be installed within or remote from the space it serves and may operate either with or without distributing ductwork. Ideally, this can provide a system which is completely responsive to the needs of the space. Well-designed systems can maintain temperature and humidity closely and efficiently and can be shut down when desired without affecting the operation of adjacent areas.

A single-zone system responds to only one set of space conditions so its use is limited to situations where variations occur approximately uniformly throughout the zone served or where the load is stable; but when installed in multiple, it can handled a variety of conditions efficiently. A single-zone system would be applied to small department stores, small individual shops in a shopping center, individual classrooms for a small school, computer rooms, etc. A rooftop unit, for example, complete with refrigeration system, serving an individual space would be considered a single-zone system. The refrigeration system, however, may be remote and serving several single-zone units in a larger installation.

A schematic of the single-zone central unit is shown in Figure 6. The return fan is necessary if 100% outdoor air is used for cooling purposes, but may be eliminated if air may be relieved from the space with very little pressure loss through the relief system. In general, a return air fan is indicated if the resistance of the return air system (grilles and ductwork) exceeds 0.25 in. water gauge.

Control of the single-zone system can be affected by on-off operation, varying the quantity of cooling medium providing reheat, face, and bypass dampers or a combination of these. The single-duct systems with reheat satisfies variations in load by providing independent sources of heating and cooling. When a humidifier is included in the system, humidity control completely responsive to space needs is available. Since control is directly from space temperature and humidity, close regulation of the system conditions may be achieved. Single-duct systems without reheat offer cooling flexibility but cannot control summer humidity independent of temperature requirements.

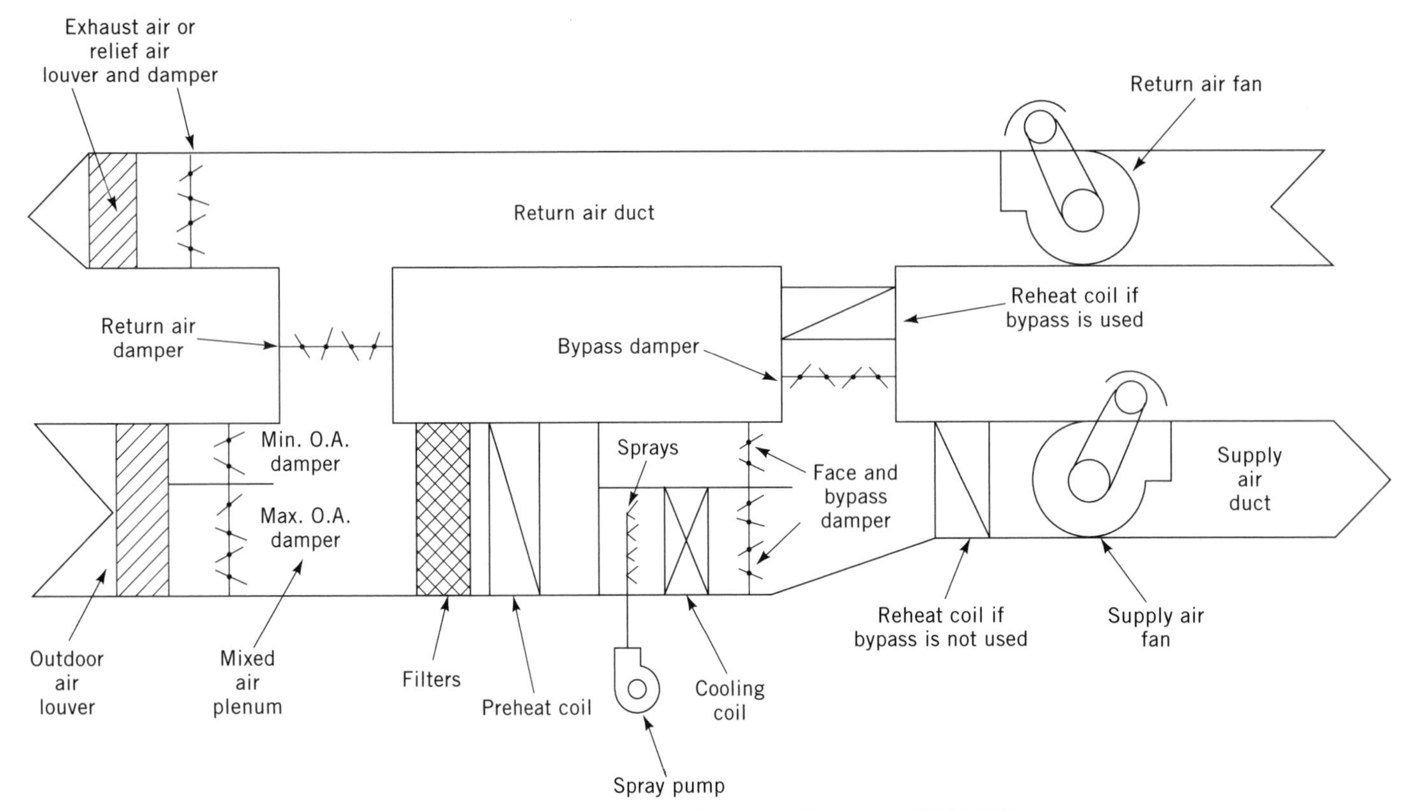

Figure 6. Single-zone central system. Courtesy of SMACNA.

System Categories

The central system is the basic system of air conditioning, originating with the forced-warm-air heating and ventilating systems, which has centrally located equipment and distributed tempered air through ducts. It was found that cooling and dehumidification equipment added to these systems produced satisfactory air conditioning in spaces where heat gains were relatively uniform throughout the conditioned area. Complication in the use of this system appeared when heat sources varied within the space served. When this occurred, the area had to be divided into sections or zones, and the central system supplemented by additional equipment and controls.

As the science of air conditioning progressed, variations in the basic design were required to meet the functional and economic demands of individual buildings. Now air-conditioning systems are categorized according to the means by which the controllable cooling is accomplished in the conditioned area. They are further segregated to accomplish specific purposes by special equipment arrangements.

There are four basic types of HVAC systems used in commercial and institutional buildings: all-air systems; air-and-water systems; all-water systems; and unitary equipment systems.

The first three types generally fall into the central system category if there is a single heating and/or cooling source for multiple areas of the building.

An *all-air* system provides through a system of ductwork all of the heating and/or cooling needed by the conditioned space. The basic all-air systems are single-zone duct systems, multizone systems, dual duct systems, terminal reheat systems, and variable air volume (VAV) systems.

The basic *air–water* system provides cooled primary air to a conditioned space from a central duct system with heating and/or additional cooling provided by a central hydronic (hot or chilled water) system to terminal units in the conditioned space. There are many variations of this system with the airside being dual duct or VAV systems and the hydronic side either two or four pipe units. Fan-coil units supplied with primary ventilation air also fall into this category. The widest use of air–water systems has been in office buildings where heating and cooling are provided simultaneously in different zones of the building.

All-water systems accomplish both sensible and latent space cooling by circulating chilled water from a central refrigeration system through cooling coils in terminal units located in the space being conditioned. Fan-coil units and unit ventilators are the most common terminal units. Heating is provided by circulating hot water through the same coil and piping system (2-pipe), through a separate coil and piping system (4-pipe), or through a separate coil but with a common return piping system (3-pipe). Alternatively, electric resistance or steam coils may be used for heating. Ventilation must be provided either through wall apertures, by an interior zone system, or by a separate ventillation air system, possibly capable of humidifying ventilation air during the winter.

Unitary air-conditioning equipment is an assembly of factory-matched components for inclusion in field assembled air-conditioning systems. Unitary equipment tends

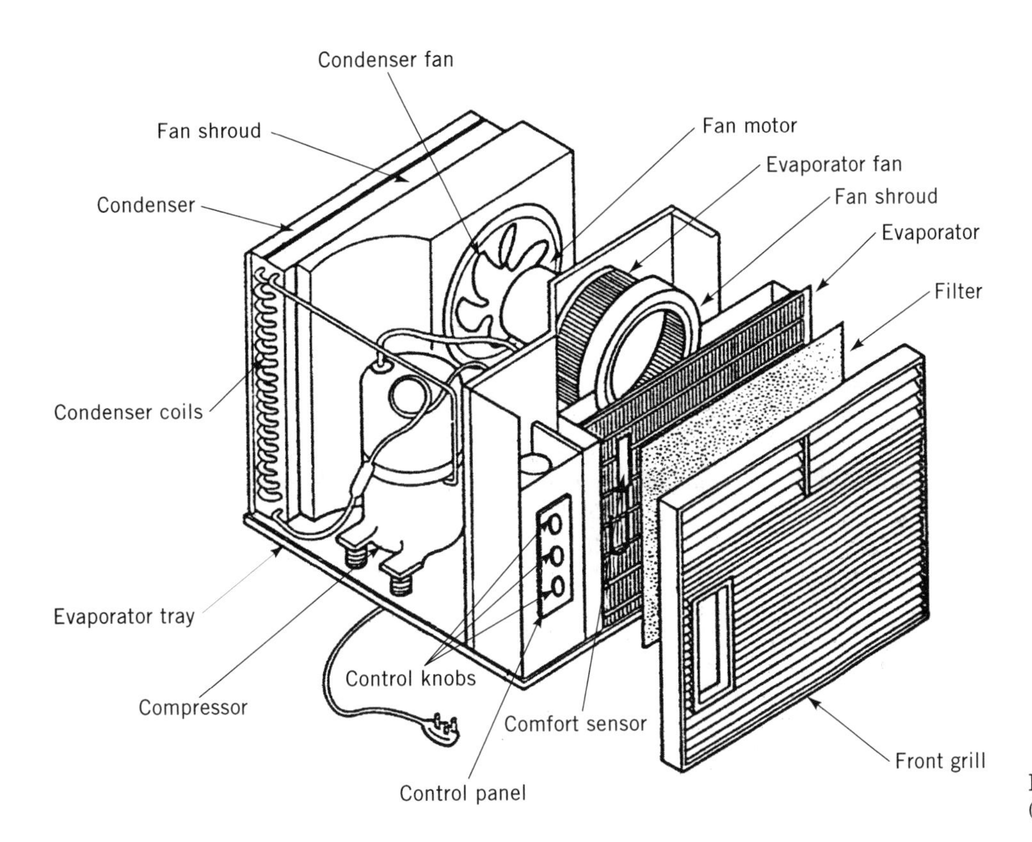

Figure 7. Window air conditioner (22). Reprinted with permission.

to serve zoned systems, with each zone served by its own unit. Although not actually in this category, the room air conditioner, depicted in Figure 7, carries this concept to relatively small rooms. Unitary air-conditioners, in contrast to room air-conditioners, include fans capable of operating ductwork.

An all-year residential air-conditioning unit, as shown in Figure 8, includes a gas furnace, a split-system cooling unit, a central system humidifier, and an air filter. The system functions as follows: Air returns to the equipment through a return air duct. It passes initially through the air filter. The circulating blower (fan) is an integral part of the gas furnace that supplies heat during the winter. The humidifier adds moisture to the heated air which is distributed throughout the house from the supply duct. When cooling is required, the circulating air passes across the evaporator coil which removes heat from the air. Refrigerant lines connect the evaporator coil to a remote condensing unit containing the refrigerant compressor, condenser, and condenser fan. Condensate from the evaporator drains away through the pipe.

Also in the unitary category is the through-the-wall conditioner designed for mounting through the wall and normally capable of providing both heating and cooling. This type of system is also referred to as a *packaged terminal air conditioner* (PTAC). Each PTAC has a self-contained, direct expansion cooling system, heating coil (electric, hot water, or steam), and packaged controls.

The Basic Central System

The basic central system is an all-air, single-zone, air-conditioning system that finds application as part of many of the systems listed previously. It can be designed for low, medium, and high pressure air distribution. Normally, the equipment is located outside the conditioned area, in a basement, penthouse, or in a service area at the core of a commercial building. It can be installed in any convenient area of a factory, particularly in the roof truss area or on the roof. The equipment can be located adjacent to the heating and refrigerating equipment, or at considerable distance from it, using circulating refrigerant, chilled water, hot water, or steam for energy transfer sources. It is most important that the heat gains and losses within the area conditioned by a central system be uniformly distributed, if a single-zone duct system is to be used. The components of a typical central HVAC system are illustrated in Figure 9.

In small commercial buildings, the packaged rooftop unit (RTU) is the standard for the HVAC system. The 3- to 20-ton rooftop unit dominates today's market for small commercial buildings. In climates with mild winters and reasonable electric rates, heat pumps have become competitive with the natural gas furnace in these units. In small buildings a single RTU can provide a comfortable environment for a single zone of concern. For larger buildings with diverse zones of heating, cooling, and ventilation demand, multiple rooftop units are installed. Each rooftop is then responsible for providing comfort for a particular zone of the building. Economizers can be installed as options on most rooftop units, saving run time and energy for applications where there is a cooling demand at the same time as the outdoor temperature is low enough to provide cooling. With high internal heat gain (eg, in restaurants) and cool climates, savings may be as much

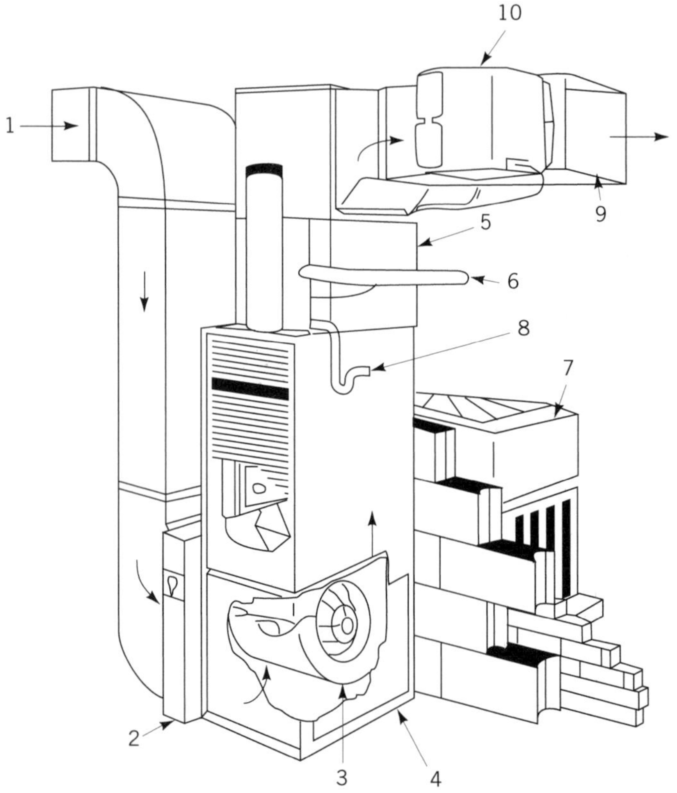

Figure 8. Residential split-system air-cooled condensing unit with upflow furnace.

1. Air duct	6. Refrigerant lines
2. Air filter	7. Remote condensing unit
3. Fan	8. Pipe
4. Gas furnace	9. Supply duct
5. Evaporator coil	10. Humidifier

Courtesy of the American Society of Heating, Refrigerating, and Air-Conditioning Engineers.

as 70% of the compressor energy used without the economizer.

Central System Mechanical Equipment

The central mechanical equipment for large building air-conditioning systems chiefly depends on economic factors, once the total required capacity has been determined. The decision as to the choice of components depends on the type of fuel available, environmental protection required, structural support, and available space, among other considerations.

In recent years the cost of energy has led to many designs based on recovering the internal heat from lights, people, and equipment to reduce the size of the heating plant. The search for energy savings has extended to total energy systems, in which on-site power generation has been added to the heating and air-conditioning project. The economics of this function is determined by gas and electric rate differentials and the electric and heat demands for the project. For these systems the waste heat from the generators is used for the input to the cooling equipment, either to drive turbines of centrifugal compressors or to serve absorption chillers.

Most large buildings have their own central equipment plant, and the choice of equipment depends on the following:

1. The required capacity and type of usage.
2. The cost and kinds of available energy.
3. The location of the equipment rooms.
4. The type of air distribution system.
5. Owning and operating costs.

Heating Equipment. Steam and hot water boilers for heating are manufactured for high or low pressure, using coal, oil, electricity, gas, and sometimes waste material for fuel. Low pressure boilers are rated for a working pressure of 103 kPa gauge (15 psig) for steam, 1100 kPa gauge (160 psig) for water, with a maximum temperature limitation of 121°C. Because of these limitations, high-rise buildings usually use high-pressure boilers. Packaged boilers with all components and controls assembled as a unit are available. Electric boilers, usually of the electrode or resistance types, are available in sizes up to 1200 kW and larger for either hot water or steam generation.

Refrigeration. The three principal types of refrigeration compressors used in large systems are reciprocating = $\frac{1}{16}$–150 hp (0.046–112 kW); helical rotary = 100–750 tons (350–2600 kW); and centrifugal = 100–10,000 tons (350–35,000 kW).

Compressors come with many types of drives including electric, gas and Diesel engines, and gas and steam turbines. The absorption chiller most commonly used for air conditioning is the indirect fired lithium bromide–water cycle unit. These are available in large tonnage units from 50 to 1500 tons (175 to 5300 kW) capacity. Their generator sections are heated with low pressure steam, hot water, or other hot liquids. Small, direct gas fired units [3.5 to 25 tons (12 to 88 kW)] are also available.

Cooling Towers. To remove the heat from the water-cooled condensers of air-conditioning systems, the water is usually cooled by contact with the atmosphere. This is accomplished by natural draft or mechanical draft cooling towers or by spray ponds. Of these, the mechanical draft tower can be designed for exacting conditions because it is independent of the wind. Large mechanical draft towers are used for electric generating stations, whereas air-conditioning systems use towers ranging from small package towers of 5–500 tons (17.5–1750 kW) or intermediate size towers of 2000–4000 tons (7000–14000 kW). Towers are usually selected in multiples so that they may be used at reduced capacity and shut down for maintenance in cool seasons. Water treatment is a definite requirement for satisfactory operation. Protection against freezing must be considered where need exists for year round operation.

Pumps. The pumps used in air-conditioning systems are usually single inlet centrifugal pumps. The pumps for the larger system and heavy duty have a horizontal split case with double suction impeller for easier maintenance and high efficiency. End suction pumps, either close coupled or flexible connected, are used for smaller tasks.

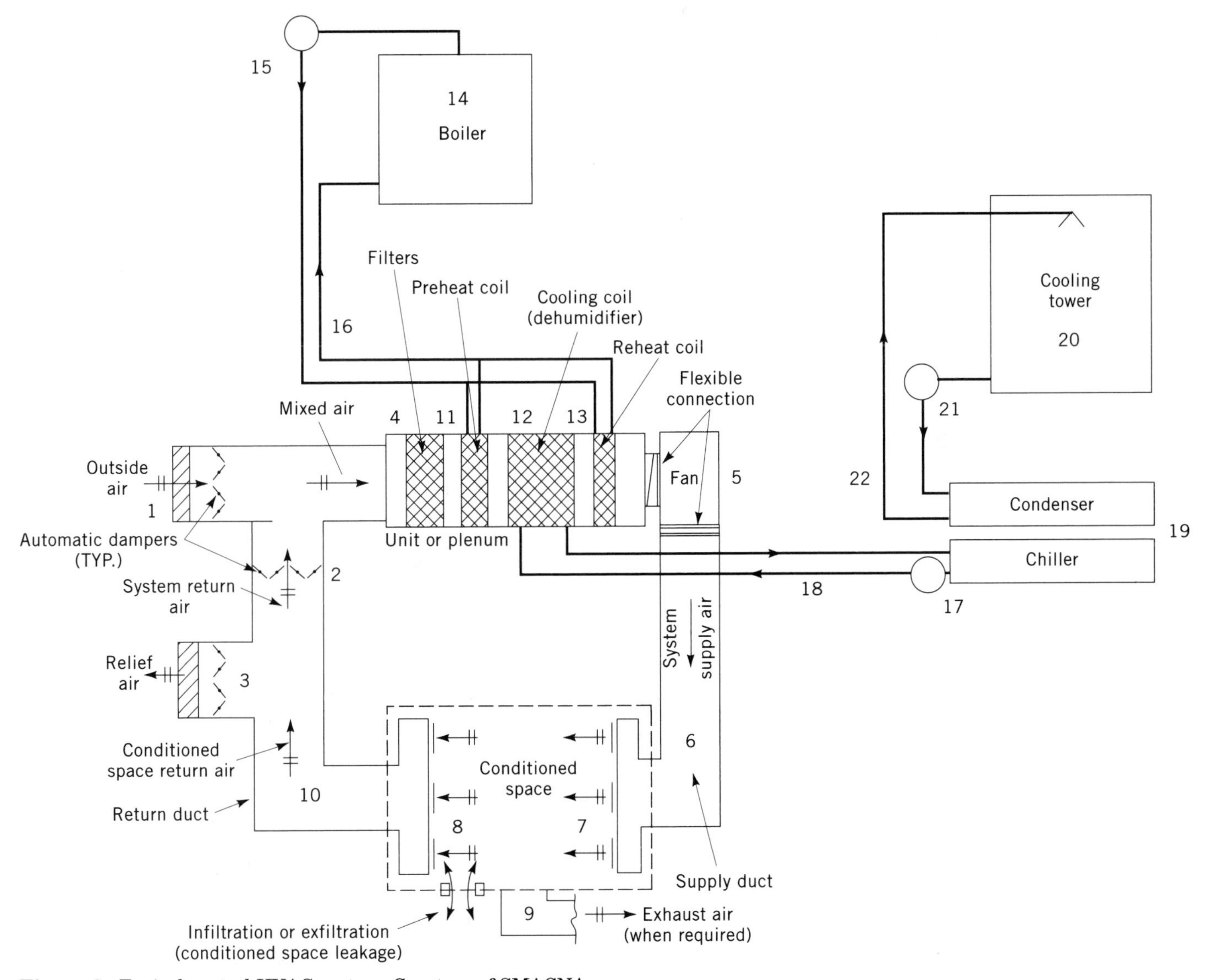

Figure 9. Typical central HVAC system. Courtesy of SMACNA.

A. Duct Systems and Plenum

1. Outdoor Air Intake	Regulation of
Louver	Air used for ventilation and/or
Screen	economizer cycle cooling.
O.A. Damper	
2. Return Air Damper	Regulation of return air to HVAC unit
3. Relief Air Damper	Regulation of relief air to outside
4. Filter Bank(s)	Removes contaminants from airstream
5. Supply Air Fan	Force to overcome system resistance
6. Supply Duct System	Path of supply air distribution
7. Air Terminal Devices	Air distribution devices to space
8. Return Air Inlets	Inlet devices from conditioned space
9. Exhaust Air System	Contaminated air removal system
10. Return Air Duct System	Path for air return to HVAC unit

B. Hydronic Systems

11. Preheat Coil	Preheats outdoor or mixed air
12. Cooling Coil	Cools and dehumidifies air
13. Reheat Coil	Heating or humidity control
14. Boiler	Source of heating hot water or steam
15. Pump (heating)	Circulating device or boiler return pump
16. Piping (heating)	Path of hot water or steam
17. Pump (chilled water)	Chilled water circulating device
18. Piping (chilled water)	Path of chilled water

C. Refrigeration System

19. Chiller	Source of chilled water for cooling
20. Cooling Tower	Disposes heat from condensing water
21. Pump (Condenser Water)	Condenser water circulating device
22. Piping (Condenser Water)	Path of condenser water

The principal applications for pumps in the equipment room are

1. Chilled water primary and secondary.
2. Hot water.
3. Condenser water.
4. Steam condensate return.
5. Boiler feed water.
6. Fuel oil.

Ducts. A fundamental requirement for an air conditioning system is the distribution of the conditioned air throughout the space. Distribution usually requires brining the conditioning fluid from a central equipment location to the individual spaces requiring environment control. For air, this means a duct system and a fan. For water, this means a piping system and a pump. For an electrical system, this means wiring. In addition to transmission of the conditioning fluid, it is vitally important to distribute the conditioned air effectively throughout the space. The distribution of conditioned air is done using air diffusers or grilles. In the design of duct systems for air transmission, the objective is to provide a system that, within prescribed limits of velocities, noise intensity, and space available for ducts, efficiently transmits the required flow rate of air to each space, while maintaining a proper balance between investment and operating cost. The size of the duct system governs how much frictional loss will occur for a given flow rate of air and thereby the size of fan and power required to operate the duct system. A duct system imposes resistance to air flow which must be overcome by the expenditure of mechanical energy; this energy is ordinarily supplied by one or more fans.

The most common methods of air duct design are (*1*) equal friction, (*2*) velocity reduction, and (*3*) static regain and variations such as total pressure. Many manuals and technical papers with detailed sample problems are referenced in the *ASHRAE Handbook 1993 Fundamentals.* The choice of design method is the designer's, and the system design with the minimum owning and operating cost depends on both the application and ingenuity of the designer. No single duct design method will automatically produce the most economical duct system for all conditions.

VEHICULAR AIR-CONDITIONING SYSTEMS

Automotive Cooling

All passenger cars sold in the United States must meet federal defroster requirements and thus ventilation systems and heaters are included in the basic vehicle design. Even though trucks are excluded from federal regulations, all U.S. manufacturers included heater/defrosters as standard equipment. Air conditioning remains an extra-cost option on nearly all vehicles. Thus, the environmental control system of modern automobiles consists of one or more of the following: heater–defroster, ventilation, and cooling and dehumidifying (air-conditioning) systems. The integration of the heater–defroster and ventilation systems is common.

For heating, outdoor air passes through a heater core which uses heater coolant as the heat source. Temperature control is achieved by either water-flow regulation or heat air bypass and subsequent mixing. A combination of ram effect from the forward velocity of the car and the electrically driven blower provides the airflow. For defrosting, some heated outside air is ducted from the heater core to defroster outlets situated at the bottom of the windshield. This air absorbs moisture from the inside surface of the windshield and raises the glass temperature above the interior dew point. Heated air also provides the heat necessary to melt ice or snow on the glass exterior surface. Some systems operate the air conditioner initially lower the dew point of the outside air. Ventilation is achieved by either of two systems: ram air or forced air. In both systems, air enters the vehicle through a screened opening in the cowl just forward of the base of the windshield. Air entering this plenum can also supply the heater and evaporator cores.

The combination evaporator–heater system in conjunction with the ventilation system dominates factory-in-

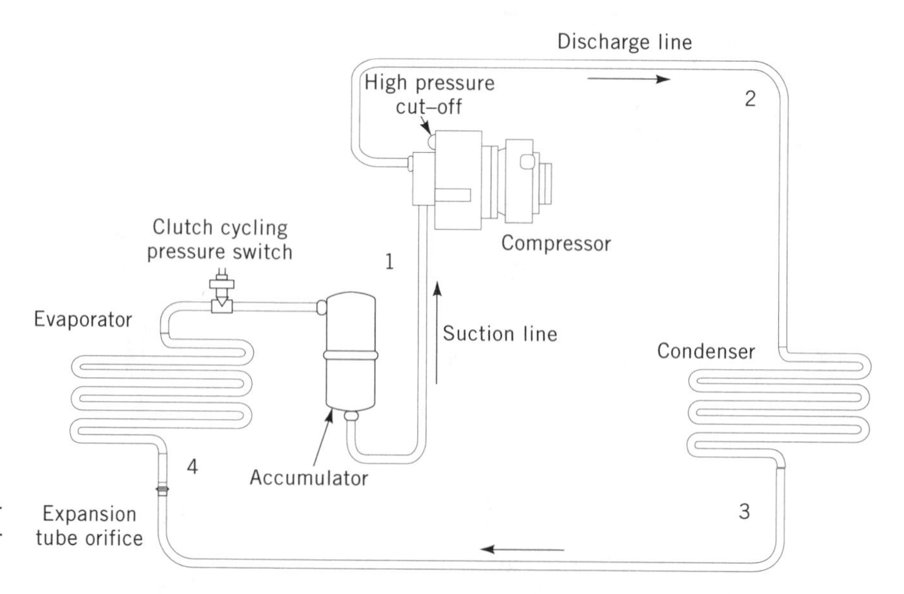

Figure 10. Automotive air conditioner. Courtesy of the American Society of Heating, Refrigerating, and Air-Conditioning Engineers.

stalled air conditioning. This system is popular because it permits dual use of components such as blower motors and outside air ducts; it generally reduces the number and complexity of control components; and it permits capacity control innovations such as automatic reheat. The conditions affecting evaporator size and design are different from residential and commercial installations in that the average operating time from a hot-soaked condition is less than a half hour. During longer periods of operation, the system is expected to cool the entire vehicle interior rather than just provide a flow of cool air. The use of an orifice tube instead of an expansion valve to control refrigerant flow through the evaporator has come into widespread use in original equipment installations. Cycling clutch systems, as illustrated in Figure 10, are common in late model cars. The clutch is controlled by a thermostat sensing evaporator temperature or by a pressure switch sensing evaporator pressure. Most units have the thermostat or pressure switch set to prevent evaporator icing. Space temperature is then controlled by blending cooled air with warm air coming through the heater core.

Aircraft Cooling

Aircraft air-conditioning equipment must meet additional requirements beyond those of building air-conditioning systems. The equipment must be more compact, lightweight, highly reliable, and unaffected by aircraft vibration and landing impact. The aircraft air-conditioning and pressurization system must maintain an aircraft environment for safety and comfort of passengers and crew and for the proper operation of onboard electronic equipment, through the wide and often rapidly changing range of ambient temperature and pressure.

Air conditioning a modern wide-body aircraft actually entails the use of a total environment control system consisting of the following subsystems:

- Bleed-air control
- Air-conditioning package
- Cabin temperature control
- Cabin pressure control

The bleed-air control system regulates the pressure, temperature, and flow of engine compressor bleed air before it is supplied to the air-conditioning system. The power source for air-cycle refrigeration is the same as that used for pressurizing the cabin. The amount of bleed air extracted from either the main engines or the auxiliary power unit (APU) by an aircraft air conditioning system depends on the cooling loads, ventilation requirements, and aircraft pressurization demands. Traditionally, the cooling load establishes the minimum requirement.

In the design for the McDonnell-Douglas DC-10, factors of performance and weight were considered and the lightest, simplest, and most maintainable system was a bootstrap air-cycle system. The center of this air-conditioning package is shown in Figure 11; it is an air-cycle machine that has three rotating components: a compressor, a turbine, and a fan. Bleed air enters the compressor section where it is compressed to a higher pressure and temperature on the order of 285°C at 460 kPa (67 psia). The air is then cooled by ram air as it passes through the heat exchanger after which it enters the 50,000 rpm turbine at a temperature of 105°C, generating power to drive the compressor and cooling the air-fan impellors. The energy removed from the air as it flows through the turbine causes a substantial temperature reduction. This cool air enters a coalescer water separator where moisture is removed. The conditioned air exiting the water separator is at 15°C and is now used for cabin and electronic compartment cooling.

LIFE-CYCLE COSTS

A properly engineered HVAC system must also be economical. The selection of any system is often based on a compromise between its performance and economic mer-

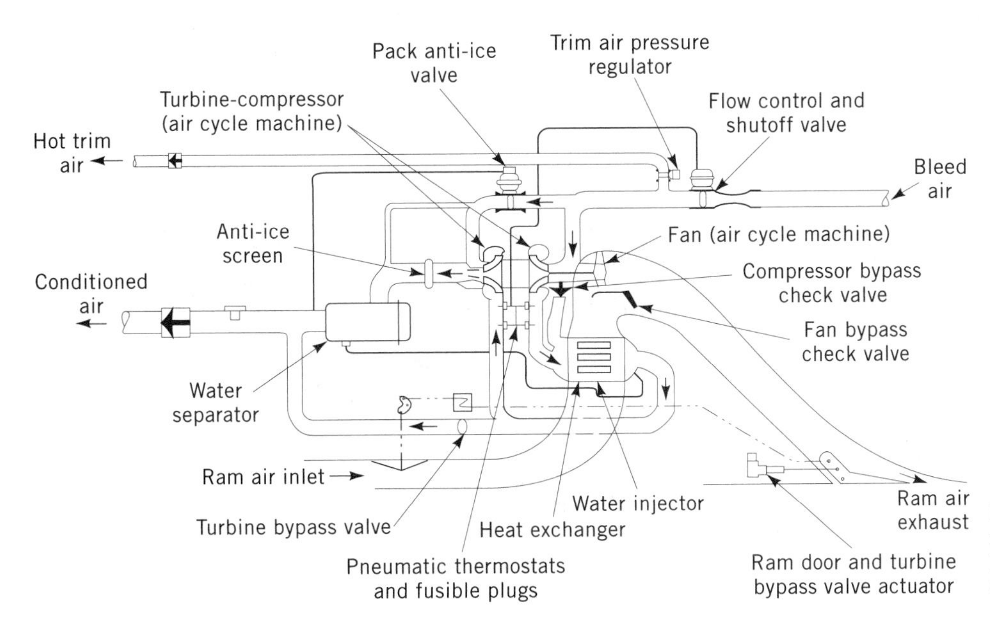

Figure 11. Aircraft air-cycle air-conditioning system. Courtesy of AiResearch Manufacturing Company, a division of The Garrett Corporation.

its. The system selected is usually determined by the user's needs, the designer's experience, local building codes, first costs, most efficient use of source energy, and the projected operating costs. Any one of the above factors may alter the choice of the system. However, when the choice is between alternatives which give apparent equal results, the system with the lowest life-cycle costs should be chosen. This does not imply the lowest initial cost system; instead, it directs a decision toward a system with the lowest long term costs. This overall cost may be divided into two main categories, owning costs and operating costs. These may be further subdivided as follows:

Owning Costs

1. Capital recovery and interest.
2. Taxes.
3. Insurance.

Operating Costs

1. Energy and fuels.
2. Operating and maintenance services.
3. Materials and supplies.

Life-cycle costs, a term which describes a broad economic philosophy, considers costs that will be experienced over an extended period of time. Many economic procedures can be used within life-cycle costing; they all produce long term costs so that comparisons can be made between systems having different initial and long-term operating costs. Any procedure, to be accurate, must include all cost factors relevant to the specific investment, such as initial costs, service life, interest, energy costs, taxes, operating expenses, and cost escalation.

Detailed procedures in handling the basic life-cycle cost techniques can be found in many good engineering economics textbooks. Techniques or procedures most commonly used include (*1*) present worth, (*2*) uniform annual owning and operating costs, (*3*) rate of return, (*4*) rate of return of investment, (*5*) benefit cost analysis, (*6*) years to payback, and (*7*) cash flow. These techniques should meet most needs. Sophisticated systems, such as *cash flow,* are used by owners and investors to meet their particular objectives. Simpler techniques, such as *benefit cost analysis* and *years to pay back* meet some needs, but can easily lead to improper decisions since important financial information is often neglected.

FUTURE DEVELOPMENTS

Although it is almost impossible to predict specifics, the general trends, on the other hand, are rather obvious. There will certainly continue to be concern about the quality of both the interior and exterior environment, about energy consumption and conservation, and, of course, about costs. Indoor air quality (IAQ) and the sick building syndrome (SBS) will continue to be of concern. Alternative refrigerants to the CFCs and HCFCs and new cooling techniques will remain under development. Microelectronics will continue to provide for better monitoring and control of HVAC systems. The new scroll compressor will experience market growth as will multi-speed compressors for year-round heat pump air-conditioning systems. The market for variable speed motors for both fans and compressors will expand. The commissioning of HVAC systems will become commonplace.

BIBLIOGRAPHY

1. *ASHRAE Handbook 1993 Fundamentals,* American Society of Heating, Refrigerating, and Air-Conditioning Engineers, Inc., Atlanta, Ga., 1993.
2. *ASHRAE Handbook 1992 HVAC Systems and Equipment,* American Society of Heating, Refrigerating, and Air-Conditioning Engineers, Inc., Atlanta, Ga., 1992.
3. *ASHRAE Handbook 1991 Refrigeration,* American Society of Heating, Refrigerating and Air-Conditioning Engineers, Inc., Altanta, Ga., 1991.
4. *ASHRAE Handbook 1990 Applications,* American Society of Heating, Refrigerating, and Air-Conditioning Engineers, Inc., Altanta, Ga., 1990.
5. J. B. Olivieri, *How to Design Heating-Cooling Comfort System,* 4th ed., Business News Publishing Company, Troy, Mich., 1987.
6. H. J. Sauer, Jr., and R. H. Howell, *Principles of Heating, Ventilating, and Air Conditioning,* American Society of Heating, Refrigerating, and Air-Conditioning Engineers, Inc., Atlanta, Ga., 1990.
7. H. J. Sauer, Jr., and R. H. Howell, *Heat Pump Systems,* John Wiley & Sons, Inc., New York, 1983.
8. W. F. Stoecker and J. D. Jones, *Refrigeration and Air Conditioning,* McGraw-Hill Book Company, New York, 1982.
9. F. C. McQuiston and J. D. Parker, *Heating, Ventilating and Air Conditioning,* John Wiley & Sons, Inc., New York, 1988.
10. G. E. Clifford, *Heating, Ventilating and Air Conditioning,* Reston Publishing Company, Reston, Va., 1984.
11. *HVAC Systems-Applications,* Sheet Metal and Air Conditioning Contractors' National Association, Vienna, Va., 1987.
12. *Commercial Buildings Consumption and Expenditures, 1986,* DOE, Superintendent of Documents, U.S. Printing Office, Washington, D. C., 1989.
13. *Housing Characteristics 1987,* DOE/EIA, Superintendent of Documents, U.S. Printing Office, Washington, D.C., 1989.
14. F. C. McQuiston and J. D. Spitler, *Cooling and Heating Load Calculation Manual,* 2nd ed., American Society of Heating, Refrigerating, and Air-Conditioning Engineers, Atlanta, Ga., 1992.
15. *Load Calculation for Residential Winter and Summer Air Conditioning, Manual J,* Air Conditioning Contractors of America, Washington, D.C., 1975.
16. *ASHRAE Standard 55-1981,* American Society of Heating, Refrigerating and Air-Conditioning Engineers, Inc., Atlanta, Ga., 1981.
17. Departments of the Air Force, Army, and Navy, *Engineering Weather Data,* AFM 88-29, #008-070-00420-8, U.S. Government Printing Office, Washington, D.C., 1978.
18. *ASHRAE Standard 62-1989* American Society of Heating, Refrigerating and Air-Conditioning Engineers, Inc., Altanta, Ga., 1989.
19. Ref. 1, Chapt. 22.
20. *Directory of Certified Unitary Air Conditioners, Unitary Air Source Heat Pumps, Sound-Rated Outdoor Unitary Equipment, Solar Collectors,* Air-Conditioning and Refrigeration Institute, Arlington, Va., semi-annually.
21. *Directory of Certified Applied Air Conditioning Products, Air Cooling and Air Heating Coils, Packaged Terminal Air Condi-*

tioners, Central Station Air-Handling Units, Packaged Terminal Heat Pumps, Room Fan-Coil Air Conditioners, Water Source Heat Pumps, Air-Conditioning and Refrigeration Institue, Arlington, Va., semi-annually.

22. D. C. Look and H. J. Sauer, Jr., *Engineering Thermodynamics,* PWS Publishers, Boston, Mass., 1986.

AIR POLLUTION

George Wolff
GM Environmental and Energy Staff
Detroit, Michigan

The concept of air pollution has changed significantly during the past several decades. Many years ago, air pollution was only associated with smoke, soot, and odors. Textbooks published in the last two decades define air pollution as any atmospheric condition in which substances are present in concentrations high enough above their normal ambient levels to produce a measurable effect on humans, animals, vegetation, or materials. Even this definition, however, is deficient today because it would not include "greenhouse" or ozone-depleting gases since the effects of these gases on humans, animals, vegetation, or materials have not been, and may never be, observed. Still, there is a need to include them because of their potential to alter the global climate and, hence, the global ecosystem. Therefore, in an attempt to be more comprehensive and timely, the following definition is offered: air pollution is the presence of any substance in the atmosphere at a concentration high enough to produce an objectionable effect on humans, animals, vegetation or materials, or to alter the natural balance of any ecosystem significantly. These substances can be solids, liquids, or gases, and can be produced by anthropogenic activities or natural sources. In this article, however, only nonbiological material will be considered. Airborne pathogens and pollens, molds, and spores will not be discussed. Airborne radioactive contaminants will not be discussed either, except for radon, which will be discussed in the context of an indoor air pollutant. (See also Global climate change; Global environmental change, population effect; Global health index.)

As the definition of air pollution changed, the breadth of air pollution also expanded. Initial perceptions of objectionable effects of air pollutants were limited to those easily detected: odors, soiling of surfaces, and smoke-belching stacks. Three rare meteorological events, however, made it clear that air pollutants can be hazardous to human health, and can even cause death at high enough concentrations. The first event was in the Meuse Valley in Belgium in 1930. Meteorological conditions produced a week-long air stagnation which caused pollutants to accumulate to presumably extremely high concentrations. As a result, 60 people died and a large number of people experienced respiratory problems. In 1948, similar meteorological conditions in Donora, Pennsylvania, resulted in nearly 7000 illnesses and 20 deaths. In 1952, 4000 deaths were attributed to a four-day "killer fog" in London. While these episodes dramatized the acute health effects of air pollution at high concentrations, it was the concern over longer term, chronic effects that led to the initiation of National Ambient Air Quality Standards (NAAQS) for six "criteria" pollutants (so named because EPA is required to summarize published information on each pollutant—these summaries are called Criteria Documents) in the United States in the early 1970s. The six criteria pollutants were: sulfur dioxide (SO_2), carbon monoxide (CO), nitrogen dioxide (NO_2), ozone (O_3), suspended particulates, and nonmethane hydrocarbons (NMHC, now referred to as volatile organic compounds or VOC). These Criteria Pollutants captured the attention of regulators because they were ubiquitous, there was substantial evidence linking them to health effects at high concentrations, three of them (O_3, SO_2, and NO_2) were also known phytotoxicants (toxic to vegetation), and they were fairly easy to measure. The NMHC were listed as Criteria Pollutants because NMHC are precursors to O_3. However, shortly after NMHC were so designated, it became obvious that the simple empirical relationship that the U.S. Environmental Protection Agency (EPA) used to relate NMHC concentrations to O_3 concentrations was not valid; consequently, NMHC were dropped from the criteria list. In the late 1970s, EPA added lead (Pb) to the list. Particulate matter with an aerodynamic diameter of less than or equal to 10 μm, PM_{10}, was added to the list in 1987.

Several developments since the establishment of the criteria pollutants greatly expanded the geographic scale of our concern. Until the 1970s, one air pollution axiom was "dilution is the solution to pollution," and the result was that tall smoke stacks were built. In the mid-1970s, however, it was shown that high concentrations of O_3 and sulfate haze could be transported hundreds of miles across state and international borders. The acid deposition studies of the 1980s clearly illustrated the international and global aspects of the issue and, finally, when stratospheric O_3 depletion and global warming became issues, air pollution became viewed in a global context. At the same time that the geographic scale of the issue was expanding, the number of pollutants of concern also increased. In the 1970s, it was realized that hundreds of potentially toxic chemicals were being released to the atmosphere. As detection capabilities improved, these chemicals were, indeed, measured in the air. This led to the establishment of a "hazardous air pollutant" category which included any potentially toxic substance in the air that was not a criteria pollutant.

AIR POLLUTION COMPONENTS

Air pollution can be considered to have three components: sources, transport and transformations in the atmosphere, and receptors. The sources are any process, device, or activity that emits airborne substances. When the substances are released, they are transported through the atmosphere. Some of the substances interact with sunlight, or other substances in the atmosphere, and are transformed into different substances. Air pollutants that are emitted directly to the atmosphere are called primary pollutants. Pollutants that are formed in the atmosphere as a result of transformations are called secondary pollutants. The reactants that undergo the transformation are referred to as precursors. An example of a secondary pollutant is O_3, and its precursors are nitrogen oxides (NO_x = nitric oxide [NO] + NO_2) and NMHC. The receptor is the person, animal, plant, material, or ecosystem affected by the emissions.

Sources

There are three types of air pollution sources: point, area, and line sources. A point source is a single facility that has one or more emissions points. An area source is a collection of smaller sources within a particular geographic area. For example, the emissions from residential heating would be treated collectively as an area source. A line source is a one-dimensional, horizontal configuration. Roadways are an example. Most, but not all, emissions emanate from a specific stack or vent. Emissions emanating from sources other than stacks, such as storage piles or unpaved lots, are classified as fugitive emissions.

EPA requires that individual states develop emissions inventories for all primary pollutants and precursors to secondary pollutants that are classified as criteria or hazardous air pollutants. The degree of geographical resolution and detail required in the inventory depends on the severity of the local air pollution problem. In clean rural areas, countrywide emissions totals for individual pollutant species may be all that is required. Emissions from large point sources are inventoried separately, however. For urban areas with severe air pollution problems, gridded emissions inventories are required. In a gridded inventory, the area is divided into squares which are typically 5 to 10 kilometers on a side; area- and line-source emissions are calculated for each grid square. Small point-sources are included in area sources, but the large point-sources are listed individually. Such detailed inventories are required as inputs to sophisticated air-quality models, which are used to develop air pollution control strategies. Details on the construction of emissions inventories have been presented by EPA (1).

Emissions rates for a specific source can be measured directly by inserting sampling probes into the stack or vent. While this has been done for most large point sources, it would be an impossible task to do for every individual source included in an area inventory. Instead, emission factors (based on measurements from similar sources or engineering mass-balance calculations) are applied to the multitude of sources. An emission factor is a statistical average or quantitative estimate of the amount of a pollutant emitted from a specific source-type as a function of the amount of raw material processed, the amount of product produced, or the amount of fuel consumed. Emission factors for most sources have been compiled (2).

To obtain emission factors for motor vehicles, the vehicles are operated using various driving patterns on a chassis dynamometer. Emissions are determined as a function of vehicle model year, speed, temperature, etc. Dynamometer-based emissions data are used in EPA's MOBILE4 model (3) to calculate total fleet emissions for a given roadway system.

Each year, EPA publishes a summary of air pollution emissions and air quality trends for the Criteria Pollutants (4). Table 1 contains the summary for 1991. United States emissions estimates for these pollutants are available back to 1940 (5).

Transport and Transformation

Once emitted into the atmosphere, air pollutants are transported and may be transformed. The fate of a particular pollutant depends upon the stability of the atmosphere and the stability of the pollutant in the atmosphere. The former will determine the concentration of the species, initially in the atmosphere, while the latter will determine the persistence of the substance in the atmosphere. The stability of the atmosphere depends on the ventilation. The stability of a pollutant depends on the presence or absence of clouds or fog, the presence or absence of precipitation, the pollutant's solubility in water, the pollutant's reactivity with other atmospheric constituents (which may be a function of temperature), the concentrations of other atmospheric constituents, the pollutant's stability in the presence of sunlight, and the deposition velocity of the pollutant.

In order to illustrate some fundamental principles of atmospheric stability, it is useful to examine a simple model which ignores transformations. The form of the Gaussian Plume Model is

$$X(x,y,z=0) = \frac{Q}{\pi\sigma_y\sigma_z u}\exp - \{(H^2/2\sigma_z^2) + (y^2/2\sigma_y^2)\}$$

where X is the concentration at the receptor located on the ground ($z = 0$); Q is the pollutant release rate; σ_y and σ_z are the crosswind and vertical plume standard deviations, which are functions of the atmospheric stability and

Table 1. Nationwide Air Pollutant Emissions Estimates for the United States in 1991[a]

	Pollutant Emissions, 10^6 t/yr[b]					
Source Category	PM_{10}	SO_x	NO_x	VOC	CO	Pb[b]
Transportation	1.5	1.0	7.3	5.1	43.5	1.6
Stationary fuel combustion	1.1	16.6	10.6	0.7	4.7	0.5
Industrial processes	1.8	3.2	0.6	7.9	4.7	2.2
Solid waste	0.3	0.02	0.1	0.7	2.1	0.7
Miscellaneous	0.7	0.01	0.2	2.6	7.1	0.0
Totals[c]	5.4	20.7	18.8	16.9	62.1	5.0

[a] Ref. 4.
[b] 10^3 t/yr.
[c] The sums of the sub-categories may not equal totals due to rounding.

the distance downwind (x); u is the mean wind speed; H is the effective stack height (which is equal to the height of release only if the plume is not buoyant); and x and y are the downwind and crosswind distances. As the ventilation increases (ie, u, σ_y and σ_z), the concentration of the pollutants decrease for a given emission rate Q. The atmospheric stability is determined by comparing the actual lapse-rate to the dry adiabatic lapse-rate. An air parcel warmer than the surrounding air will rise and cool at the dry adiabatic lapse-rate of 9.8°C/1000 m, the atmosphere is unstable, the σs become larger, and the concentrations of pollutants are lower. As the lapse rate becomes smaller, the dispersive capacity of the atmosphere declines and reaches a minimum when the lapse rate becomes positive. When the lapse rate becomes positive, a temperature inversion exists. Temperature inversions form every evening in most places as the heat from the earth's surface is radiated upward, and the air in the lower layers of the atmosphere is cooled. However, these inversions are usually destroyed the next morning as the sun heats the earth's surface, which in turn heats the air adjacent to the surface, and convective activity (mixing) is initiated. Most episodes of high pollution concentrations are associated with multiday inversions.

The stability or persistence of a pollutant in the atmosphere depends on the pollutants atmospheric residence time, which depends on the pollutant's reactivity with other atmospheric constituents, surfaces, or water. Mean atmospheric residence times and principal atmospheric sinks for a variety of species are given in Table 2. Species like SO_2, NO_x (NO and NO_2), and coarse particles have lifetimes less than a day; thus important environmental impacts from these pollutants are usually within close proximity to the emissions sources. Secondary reaction products may have a larger zone of influence, however, depending on the residence times of the products. The principal sink for the SO_2 and NO_2 is the reaction with the hydroxyl radical (OH). Sources of OH and the nature of the secondary reaction products will be discussed in subsequent sections. The residence time of O_3 varies considerably, depending on the presence of other constituents. In the presence of high concentrations of NO, the lifetime is on the order of hours to seconds. In the relatively nonpolluted environment of the free troposphere (from approximately 1500 m to the top of the troposphere [~12 km]), the maximum lifetime applies. Consequently, under the right conditions, O_3 could have important environmental impacts far downwind from the O_3 precursor sources. In fact, concentrations of O_3 near the NAAQS have been transported from the Gulf Coast of the U.S. to the Northeast over a several-day period (9).

Particles with diameters less than 2.5 μm have negligible settling velocities and, therefore, have residence times which are considerably longer than those of larger particles. As a result, observations have shown multiday transport of haze produced by fine particles over distances of more than a thousand kilometers (10). The longer lifetimes of the greenhouse gases, those listed below CO in Table 2, result in the accumulation and relatively even distribution of these gases around the globe. Chlorofluorocarbons (CFCs) and nitrous oxide (N_2O) are essentially inert in the troposphere and are only destroyed in the stratosphere by ultraviolet (uv) solar radiation. Unfortunately, the photolysis products of CFCs are the reactants which are responsible for stratospheric O_3 depletion.

To determine the fate of a pollutant after it is released, two approaches, monitoring and modeling, are available. Monitoring of the criteria pollutants is done routinely by state and local air-pollution agencies in most large urban areas and in some other areas as well. The recommended techniques for measuring the criteria and many other pollutants are found in reference 11. Monitoring is expensive and time consuming, however, and even the most extensive urban networks are insufficient to assess the geo-

Table 2. Mean Residence Time (τ) of Species in the Atmosphere

Species	CAS Registry Number	τ	Dominant Sink[a]	(Location)[b]	Refs.
SO_2	*[7446-09-5]*	0.5 days	OH	T	6
NO_x	*[10102-44-0]* *[10102-43-9]*	0.5 days	OH	T	6
Coarse particles (dia. > 2.5 μm)		<1 day	S, P	T	7
O_3 (tropospheric)	*[10028-15-6]*	90 days[c]	NO, uv, Sr, O	T	6
Fine particles (dia. < 2.5 μm)		5 days[d]	P	T	7
CO	*[630-08-0]*	100 days	OH	T	6
CO_2	*[124-38-9]*	120 yr[e]	O	T	8
CH_4	*[74-82-8]*	7–10 yr	OH	T	6
CFC-11	*[75-69-4]*	65–75 yr	uv	St	6,8
CFC-12	*[75-71-8]*	110–130 yr	uv	St	6,8
N_2O	*[10024-97-2]*	120–150 yr	uv	St	6,8
CFC-113	*[76-13-1]*	90 yr	uv	St	6

[a] Sinks: (OH), reaction with OH; S, sedimentation; P, precipitation scavenging; NO, reaction with NO; uv, photolysis by ultraviolet radiation; Sr, destruction at surfaces; O, adsorption or destruction at oceanic surface.
[b] Location of sink: *T*, troposphere; St, stratosphere.
[c] Tropospheric residence time only; shorter lifetime applies to urban areas where NO quickly destroys O_3; upper limit applies to the remote troposphere.
[d] Applies to particles released in the lower troposphere only; the most important sink is scavenging by precipitation, so in the absence of precipitation, these particles will remain suspended longer.
[e] Combined lifetime for atmosphere, biosphere, and upper ocean.

graphic distribution of pollutants accurately. Consequently, various air pollution models are employed to estimate the three-dimensional distributions of pollutants. The types of models used vary in both the scale of the area covered and in the number of processes treated. The smallest scale is the microscale which extends from the emission source to less than 10-km downwind. The Gaussian Plume Model, described above, is an example of a microscale model. From 10-km to about 100-km downwind, mesoscale or urban-scale models are applied. Such models are used to describe the pollutant patterns within and downwind of urban areas. From 100 km to about 1000 km, synoptic- or regional-scale models are employed. These models are used to estimate pollution patterns in areas the size of the eastern U.S. Above 1000 km, global-scale or general-circulation models are used, and these are the models used to calculate the distributions of species with long atmospheric residence times like the greenhouse gases. With each increase in scale, the complexity of the meteorological processes increases greatly. In addition, the complexity of the model depends on the number and kinds of transformation processes which are included (see AIR POLLUTION MODELING).

Receptors

The receptor is the person, animal, plant, material, or ecosystem affected by the pollutant emission. The criteria pollutants and the hazardous air pollutants were so designated because, at sufficient concentrations, they can cause adverse health effects to human receptors. Also, some of the criteria pollutants cause damage to plant receptors. For each criteria pollutant, an "Air Quality Criteria Document" exists and is updated periodically (see *Reading List*). These documents summarize the most current published literature concerning the effects of these pollutants on human health, animals, vegetation, and materials.

With respect to acid deposition, the receptors which have generated the most concern are certain aquatic ecosystems and certain forest ecosystems, although there is also some concern that acid deposition adversely affects some materials. For visibility-reducing air pollutants, CFCs, and greenhouse gases, the receptor is the atmosphere. Visibility-reducing species alter the optical properties of the atmosphere, and CFCs alter the natural chemical composition of the atmosphere in such a way that it becomes more transparent to potentially harmful uv solar radiation. The greenhouse gases alter the radiative properties of the atmosphere and, consequently, have the potential to alter the global heat budget.

AIR QUALITY MANAGEMENT

In the United States, the framework for air quality management was established by the 1965 Clean Air Act (CAA) and subsequent amendments in 1970 and 1977, and the comprehensive amendments of 1990. The CCA defined two categories of pollutants: criteria pollutants and hazardous air pollutants. For the criteria pollutants, the CAA required that EPA establish NAAQS and emissions standards for some large new sources and for motor vehicles, and gave the primary responsibility for designing and implementing air quality improvement programs to the states. For the hazardous air pollutants, only emissions standards for some sources are required. The NAAQS apply uniformly across the United States whereas emissions standards for criteria pollutants can vary somewhat, depending on the severity of the local air-pollution problem and whether an affected source already exists or is proposed as a new source. In addition, individual states have the right to set their own ambient air quality and emissions standards (which must be at least as stringent as the Federal standards) for all pollutants and all sources except motor vehicles. With respect to motor vehicles, the CAA allows the states to choose between two sets of emissions standards: the Federal standards or the more stringent California standards.

The two levels of NAAQS, primary and secondary, are

Table 3. National Ambient Air Quality Standards

	Primary[a]		Secondary[a]		
Pollutant	$\mu g/m^3$	ppm	$\mu g/m^3$	ppm	Averaging Time
PM_{10}	50		50		Annual arithmetic mean
	150		150		24-h[b]
SO_2	80	(0.03)			Annual arithmetic mean
	365	(0.14)			24-h[b]
			1300	(0.50)	3-h[b]
CO	(10)	9			8-h[b]
	(40)	35			1-h[b]
NO_2	(100)	0.053	(100)	0.053	Annual arithmetic mean
Pb	1.5		1.5		Maximum quarterly average
O_3	(235)	0.12	(235)	0.12	Maximum daily[c] 1-h average

[a] Parenthetical value is an approximately equivalent concentration.
[b] Not to be exceeded more than once per year.
[c] Not to be exceeded on more than three days in three years.

listed in Table 3. Primary pollutant standards were set to protect public health with an adequate margin of safety. Secondary standards, where applicable, were chosen to protect public welfare, including vegetation. The pollutant PM_{10} refers to particulate matter with an aerodynamic diameter less than or equal to 10 μm. Originally, there was a nonsize-selective NAAQS for total suspended particulates (TSP), but it was revised to the PM_{10} standard in 1987. According to the CAA, the scientific bases for the NAAQS are to be reviewed every 5 years so that the NAAQS levels reflect current knowledge. In practice, however, the review cycle takes considerably longer. In order to analyze trends in Criteria Pollutants nationwide, the EPA has established three types of monitoring systems. The first is a network of 98 National Air Monitoring sites (NAMS), located in areas with high pollutant concentrations and high population exposures. The system was established by regulations promulgated in May 1979 to provide accurate and timely data on the national air quality. In addition, EPA also regularly evaluates data from the State and Local Monitoring system (SLAMS) and from Specific Purpose Monitors (SPM). These three types of stations comprise the 274-site national monitoring system, which is required to meet rigid quality-assurance criteria. To determine if an area meets the NAAQS, the states are required to monitor the concentrations of the Criteria Pollutants in areas that are likely to be near or exceed the NAAQS. If an area exceeds a NAAQS for a given pollutant, it is designated as a nonattainment area for that pollutant, and the state is required to develop and implement a State Implementation Plan (SIP). The SIP is a strategy designed to achieve emissions reductions sufficient to meet the NAAQS within a specific deadline. The deadline is determined by the severity of the local pollution problem. Areas that receive long deadlines (six years or more) must show continuous progress by reducing emissions by a specified percentage each year. For SO_2 and NO_2, the initial SIPs were very successful in achieving the NAAQS in most areas. However, for other criteria pollutants, particularly O_3 and to a lesser extent CO, many areas are starting a third round of SIP preparations with little hope of meeting the NAAQS in the near future. If a state misses an attainment deadline, fails to revise an inadequate SIP, or fails to implement SIP requirements, EPA has the authority to enforce sanctions such as banning construction of new stationary sources and the withholding of federal grants for highways.

In nonattainment areas, the degree of control on small sources is left to the discretion of the state, and it is largely determined by the degree of required emissions reductions. Large existing sources must be retrofitted with "reasonable available control technology" (RACT) to minimize emissions. All large new sources and existing sources that undergo major modifications must meet EPA's new source performance standards at a minimum; and in nonattainment areas they must be designed with "lowest achievable emission rate" (LAER) technology, and emissions offsets must be obtained. Offsets require that emissions from existing sources within the area must be reduced below legally allowable levels so that the amount of the reduction is greater than or equal to the emissions expected from the new source. RACT usually is less stringent than LAER because it may not be feasible to retrofit certain sources with the LAER technology.

In attainment areas, new large facilities must be designed to incorporate the "best available control technology" (BACT). Generally, BACT is more stringent than RACT, and equal to or less stringent than LAER. In addition, there are also rules that specify how much deterioration in baseline air quality a new facility can cause in an attainment area, and in no situation can the facility cause a new violation in the NAAQS.

Large sources of SO_2 and NO_x may also require additional emission reductions because of the 1990 Clean Air Act Amendments. To reduce acid deposition, the amendments require that nationwide emissions of SO_2 and NO_x be reduced on an annual basis by 10 million and 2 million tons, respectively, by the year 2000.

Once a substance is designated by EPA as a Hazardous Air Pollutant (HAP), EPA has to promulgate a NESHAP (National Emission Standard for Hazardous Air Pollutants), which is designed to protect public health with an ample margin of safety. The 1990 Clean Air Act Amendments identify 189 HAPs. These are further discussed in the Air Toxics section.

AIR POLLUTION ISSUES

Photochemical Smog

Photochemical smog is a complex mixture of constituents formed when VOCs and NO_x are irradiated by sunlight. From an effects perspective, O_3 is the primary concern and it is the most abundant species formed in photochemical smog. Extensive studies have shown that O_3 is a lung irritant and a phytotoxicant. It is responsible for crop damage and it is suspected of being a contributor to forest decline in Europe and in parts of the United States. There are, however, a multitude of other photochemical smog species that have significant environmental consequences. The most important of these additional pollutant species–particles are hydrogen peroxide (H_2O_2), peroxyacetyl nitrate (PAN), aldehydes, and nitric acid.

Photochemical smog is a summertime phenomenon for most parts of the United States because temperatures are too low and sunlight is insufficient during the other seasons. In the warmer parts of the country, especially in Southern California, the smog season begins earlier and lasts into the fall. Despite almost two decades of reducing VOC emissions from stationary and mobile sources and NO_x emissions from mobile sources, progress had been slow in reducing the number of areas in the United States designated as nonattainment for O_3. For the 1985–1987 period, there were 64 areas, mostly urban, that experience O_3 concentrations above the NAAQS. This number increased to 101 areas after the anomalously hot summer of 1988, but then gradually decreased to 56 by 1991. The recent decrease is due primarily to decreased VOC emissions resulting from reducing the Reid Vapor Pressure of gasoline and the replacement of older vehicles with new, cleaner ones. An area is classified nonattainment if the O_3 design value (the design value is equal to the 4th highest maximum daily 1-h O_3 concentration within a 3-yr period) exceeds 0.12 ppm. For most of the nonattainment areas, the design value falls between 0.12 to 0.15 ppm. The three areas with the highest design values for the 1989–1991 period were the Los Angeles (0.31 ppm), Houston (0.22), and New York (0.17) metropolitan areas.

There is a significant clean air background O_3 concentration that varies with season and latitude. The clean air background is defined as the concentrations measured at pristine areas of the globe. It consists of natural sources of O_3, but it undoubtedly contains some anthropogenic contribution because it may have increased since the last century (12). In the summertime in the United States, the average background is about 0.04 ppm (13). This background O_3 has four sources: intrusions of O_3-rich stratospheric air, *in situ* O_3 production from methane (CH_4) oxidation, the photooxidation of naturally emitted VOCs from vegetation, and the long-range transport of O_3 formed from the photooxidation of anthropogenic VOCs and NO_x emissions. Although there are several mechanisms which will transport O_3-rich air from the stratosphere into the lower troposphere, the most important appears to be associated with large-scale eddy transport that occurs in the vicinity of upper air troughs of low pressure associated with the jet stream (14). This is an intermittent mechanism, so the contribution of stratospheric O_3 to surface O_3 will have a considerable temporal variation. On very rare occasions, this mechanism has produced brief ground level concentrations exceeding 0.12 ppm (15). The other three mechanisms will be described below.

In the presence of sunlight ($h\nu$), NO_2 photolyzes and produces O_3

$$NO_2 + h\nu \rightarrow NO + O \quad (1)$$

$$O + O_2 + M \rightarrow O_3 + M \quad (2)$$

$$NO + O_3 \rightarrow NO_2 + O_2 \quad (3)$$

where M is any third body molecule (most likely N_2 or O_2 in the atmosphere) that remains unchanged in the reaction. This process produces some steady-state concentration of O_3 that is a function of the initial concentrations of NO and NO_2, the solar intensity, and the temperature. Although these reactions are extremely important in the atmosphere, the steady-state O_3 produced is much lower than the concentrations usually observed, even in clean air. In order for O_3 to accumulate, there must be a mechanism that converts NO to NO_2 without consuming a molecule of O_3, as does reaction 3. Reactions among hydroxyl radicals (OH) and hydrocarbons or VOC constitute such a mechanism. In clean air, OH may be generated by:

$$O_3 + h\nu \rightarrow O_2 + O(^1D) \quad (4)$$

$$O(^1D) + H_2O \rightarrow 2\ OH \quad (5)$$

where $O(^1D)$ is an excited form of an O atom that is produced from a photon at a wavelength between 280 and 310 nm. This produces a "seed" OH which can produce the following chain reactions:

$$OH + CH_4 \rightarrow H_2O + CH_3 \quad (6)$$

$$CH_3 + O_2 + M \rightarrow CH_3O_2 + M \quad (7)$$

$$CH_3O_2 + NO \rightarrow CH_3O + NO_2 \quad (8)$$

The NO_2 will then photolyze and produce O_3 (equations 1 and 2). The CH_3O radical continues to react:

$$CH_3O + O_2 \rightarrow HCHO + HO_2 \quad (9)$$

and the HO_2 radical also forms more NO_2:

$$HO_2 + NO \rightarrow NO_2 + OH \quad (10)$$

which will result in more O_3. In addition, OH is regenerated, and it can begin the cycle again by reacting with another CH_4 molecule. Further, the formaldehyde photodissociates

$$HCHO + h\nu \xrightarrow{a} H_2 + CO \quad (11)$$

$$\xrightarrow{b} HCO + H \quad (12)$$

$$HCO + O_2 \rightarrow HO_2 + CO \quad (13)$$

$$H + O_2 \rightarrow HO_2 \quad (14)$$

and the HO_2 from both equations 13 and 14 will form additional NO_2. Furthermore, the CO is oxidized:

$$CO + OH \rightarrow CO_2 + H \quad (15)$$

and the H radical can form another NO_2 (eqs. 14 and 10). Thus, the oxidation of one CH_4 molecule is capable of producing three O_3 molecules and two OH radicals by this reaction sequence. However, this chain reaction is less than 100% efficient because there are many competing chain-terminating reactions. Two examples are

Table 4. Median Concentration of the Ten Most Abundant Ambient Air Hydrocarbons in 39 U.S. Cities and their Relative Reactivity with OH

Compound	CAS Registry Number	Median Concentration[a] ppbC	Relative Reactivity with OH[b,c]
Isopentane	[*463-82-1*]	45.3	494
n-Butane	[*106-97-8*]	40.3	351
Toluene	[*108-88-3*]	33.8	831
Propane	[*74-98-6*]	23.5	143
Ethane	[*74-84-0*]	23.3	36
n-Pentane	[*109-66-0*]	22.0	480
Ethylene	[*74-85-1*]	21.4	1013
m-, p-xylene	[*108-38-3*]	18.1	3117
p-xylene	[*106-42-3*]	18.1	1818
2-Methylpentane	[*107-83-5*]	14.9	
Isobutane	[*75-28-5*]	14.8	325
	Biogenic species		
α-Pinene	[*80-56-8*]		7792
Isoprene	[*78-79-5*]		12078

[a] Ref. 16.
[b] Ref. 17.
[c] Relative to CH_4 + OH reaction at 298°C.

$$HO_2 + HO_2 \rightarrow H_2O_2 + O_2 \quad (16)$$

$$OH + NO_2 \xrightarrow{M} HNO_3 \quad (17)$$

Both of these reactions terminate the chain by scavenging a free radical. On the average, however, the chain results in a net production of O_3 and OH.

In a polluted or urban atmosphere, O_3 formation by the CH_4 oxidation mechanism is overshadowed by the oxidation of other VOCs. The "seed" OH can be produced from equations 4 and 5, but the photodisassociation of carbonyls and nitrous acid (HNO_2, formed from the reaction of OH + NO and other reactions) are also important sources of OH in polluted environments. An imperfect, but nevertheless useful, measure of the rate of O_3 formation by VOC oxidation is the rate of the initial OH–VOC reaction. The rate of the OH-CH_4 reaction is much slower than any other OH–VOC reaction. Table 4 contains the reaction rates with OH relative to the OH–CH_4 rate for some commonly occurring VOCs and their median concentrations from 39 cities. Also shown for comparison are the relative reaction rates between OH and two VOC species emitted by vegetation: isoprene and α-pinene. It is obvious from the data in Table 4 that there is a wide range of reactivities. In general, internally bonded olefins are the most reactive, followed in decreasing order by terminally bonded olefins, multialkyl aromatics, monoalkyl aromatics, C5 and greater paraffins, C2-C4 paraffins, benzene, acetylene, and ethane. The reaction mechanisms by which the VOCs are oxidized are analogous to, but much more complex than, the CH_4 oxidation mechanism. The fastest reacting species are the natural VOCs emitted from vegetation. However, natural VOCs also react rapidly with O_3, so they are a source as well as a sink of O_3. Whether the natural VOCs are a net source or sink is determined by the natural VOCs to NO_x ratio and the sunlight intensity. At high VOC/NO_x ratios, there will be insufficient NO_2 formed to offset the O_3 loss. However, when O_3 reacts with these internally bonded olefinic compounds, carbonyls are formed and, the greater the sunshine, the better the chance the carbonyls will photolyze and produce OH, which will initiate the O_3-forming chain reactions.

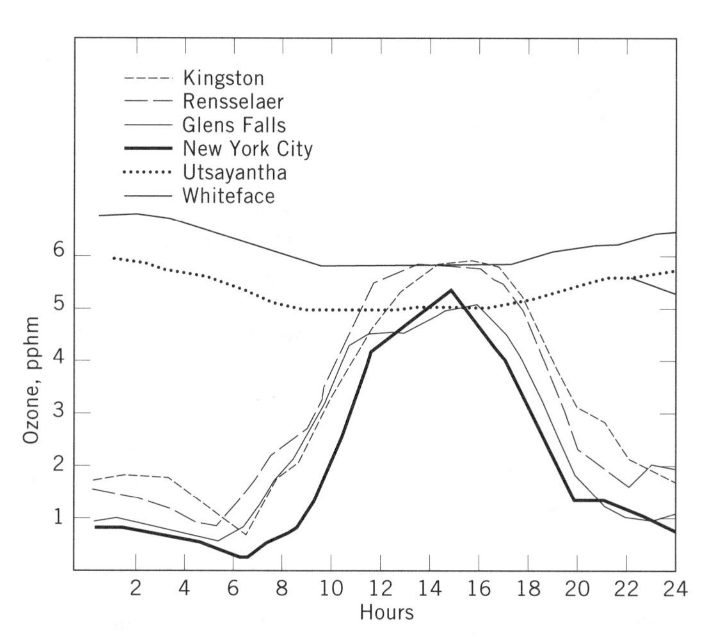

Figure 1. Hourly average ozone concentrations during August 1–17, 1973 for selected sites in New York State (18). Courtesy of Air and Waste Management Association.

Once the sun sets, O_3 formation ceases and, in an urban area, it is rapidly scavenged by freshly emitted NO by equation 3. On a typical summer night, however, a nocturnal inversion begins to form around sunset, usually below a few hundred meters. Consequently, the surface-based NO emissions are trapped below the top of the inversion. Above the inversion to the top of the mixed layer (usually about 1500 m), O_3 is depleted at a much slower rate. The next morning, the inversion dissipates and the O_3-rich air aloft is mixed down into the O_3-depleted air near the surface. This process, in combination with the onset of photochemistry as the sun rises, produces the sharp increase in surface O_3, shown in Figure 1. Notice in Figure 1 that the overnight O_3 depletion is less in the more rural areas than in a large urban area (ie, New York City). This is due to lower overnight levels of NO in rural areas. Even in the absence of NO or other O_3 scavengers (olefins, for example), O_3 will still decrease at night near the ground faster than aloft due to the destruction of O_3 at any surface (ie, the ground, buildings, trees, etc). At the remote mountaintop sites, Whiteface and Utsayantha, there is no overnight decrease in O_3 concentrations.

Although photochemical smog is a complex mixture of many primary and secondary pollutants and involves a myriad of atmospheric reactions, there are characteristic pollutant concentration versus time profiles that are generally observed within and downwind of an urban area during a photochemical smog episode. In particular, the highest O_3 concentrations are generally found 10–100 km downwind of the urban emissions areas, unless the air is completely stagnant. This fact, in conjunction with the long lifetime of O_3 in the absence of high concentrations of NO, means that O_3 is a regional problem. In the Los Angelos basin, high concentrations of O_3 are transported throughout the basin, and multiday episodes are exacerbated by the accumulation of O_3 aloft, which is mixed to the surface daily. On the East Coast, a typical O_3 episode is associated with a high pressure system anchored offshore, producing a southwesterly flow across the region. As a result, emissions from Washington, D.C., travel and mix with emissions from Baltimore and over a period of a few days will continue traveling northeastward through Philadelphia, New York City, and Boston. Under these conditions, the highest O_3 concentrations typically occur in central Connecticut (19).

It is obvious from the above discussion that in order to reduce O_3 in a polluted atmosphere, reductions in the precursors (VOC and NO_x) are required. However, the choice of whether to control VOC or NO_x or both as the optimum control strategy depends on the local VOC/NO_x ratio. At low VOC/NO_x ratios for example, O_3 formation is suppressed principally by equations 3 and 17. Consequently, reducing NO_x emissions (which are emitted mainly as NO) in this case will reduce the amount of O_3 scavenged by equation 3 and the amount of OH scavenged by equation 17. The consequence of both will be an increase in O_3 concentrations. This is illustrated in the O_3-

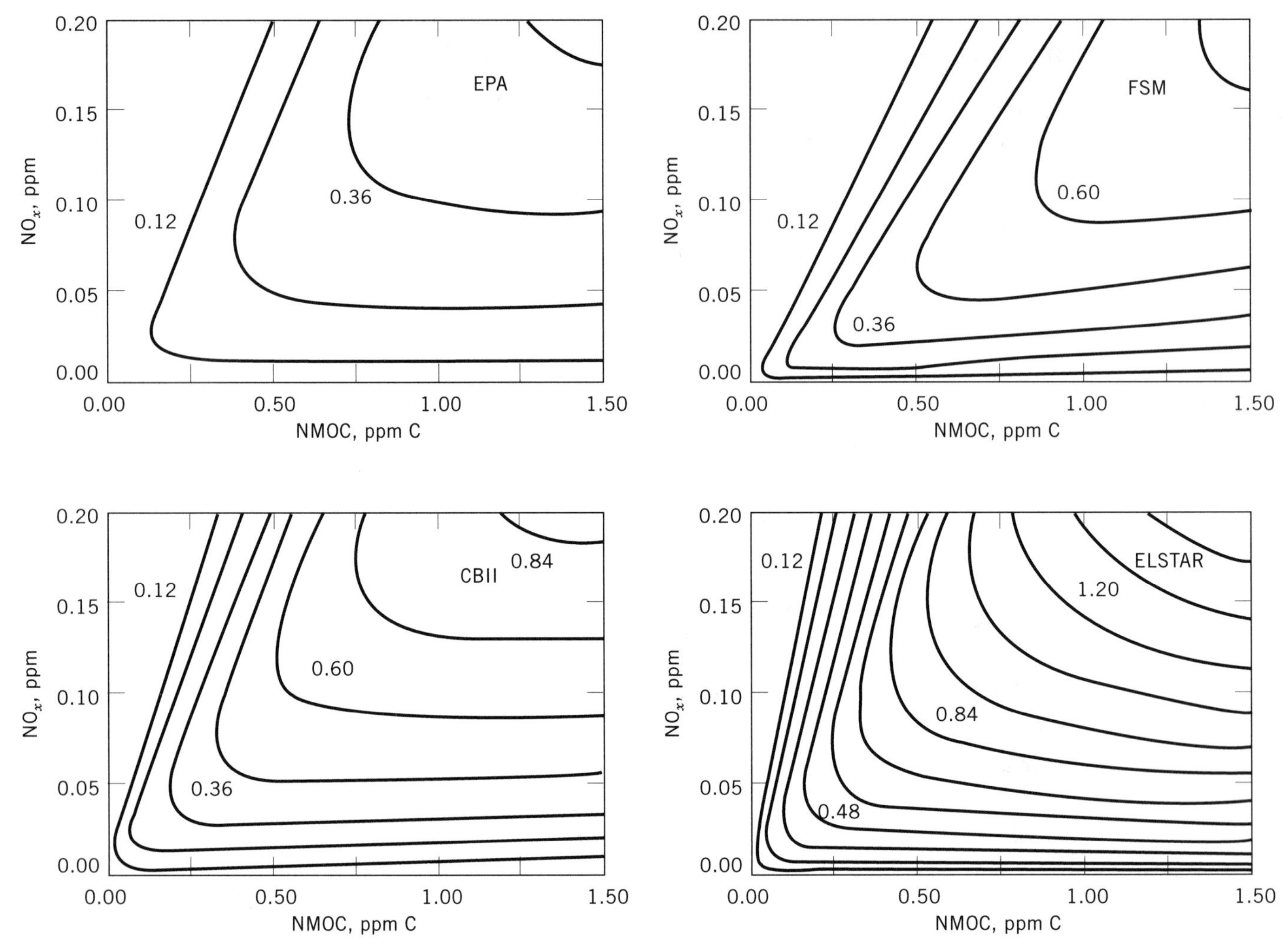

Figure 2. Examples of ozone isopleths generated using four different chemical mechanisms (20). Courtesy of Pergamon Press.

isopleth diagrams in Figure 2. Although the four different chemical mechanisms used to obtain Figure 2 give somewhat different results, the shape of the isopleths are quite similar and the key features are summarized in Figure 3. The first region in the upper left is the NO_x-inhibition region. In this region, a decrease in NO_x alone results in an increase in O_3, but a decrease in VOC decreases O_3. The region at the bottom right is the VOC or HC saturation region where reducing VOCs has no effect on the O_3. On the other hand, a reduction in NO_x in this region results in lower O_3. In the middle is the knee region, where reductions in either reduce O_3. The upper boundary of this region will vary from day to day and from place to place as its location is a function of the reactivity of the VOC mix and the sunlight intensity. As a guideline, VOC controls alone are generally the most efficient way to reduce O_3 in areas with a median 6 A.M. to 9 A.M. VOC/NO_x ratio of 10:1 or less; areas with a higher ratio may need to consider NO_x reductions as well (22). The 1990 Clean Air Act Amendments require that O_3 nonattainment areas reduce both VOC and NO_x from big stationary sources unless the air quality benefits are greater in the absence of NO_x reductions. Large cities in the Northeast tend to have ratios significantly less than 10:1, while cities in the South (Texas and east) tend to have ratios greater than 10:1. Determining a workable control strategy is further complicated by the transport issue. For example, on high O_3 days in the Northeast, the upwind air entering Philadelphia and New York City frequently contains O_3 already near or over the NAAQS as a result of emissions from

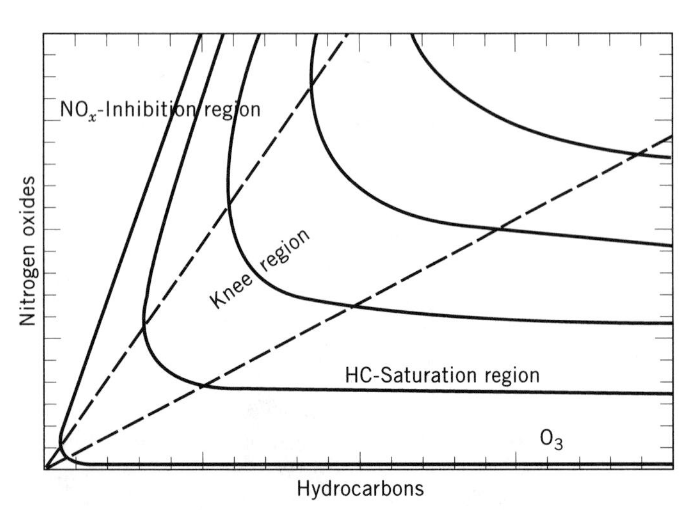

Figure 3. Typical O_3 isopleth diagram showing the three chemical regimes (21). Courtesy of Pergamon Press.

areas to the west and south (23). Consequently, control strategies must be developed on a coordinated, multistate regional basis.

Because of the complex mixture of VOCs in the atmosphere, the composition of the reaction products and intermediate species is even more complex. Some of the more important species produced in the smog process include hydrogen peroxide (H_2O_2), peroxyacetyl nitrate (PAN), aldehydes (particularly formaldehyde, HCHO [*50-00-0*]), nitric acid (HNO_3), and particles. The H_2O_2 is formed and dissolved in cloud droplets; (eq. 16) is an important oxidant, responsible for oxidizing SO_2 to H_2SO_4, the primary cause of acid precipitation. The oxidation of many VOCs will produce acetyl radicals, CH_3CO, which can react with O_2 to produce peroxyacetyl radicals, $CH_3(CO)O_2$, which in turn can react

$$CH_3(CO)O_2 + NO_2 \rightleftharpoons CH_3C(O)O_2NO_2 \text{ (PAN)} \quad (18)$$

At high enough concentrations, PAN is a potent eye irritant and a phytotoxicant. On a smoggy day in the Los Angeles area, PAN concentrations are typically 5 to 10 ppb. PAN concentrations in the rest of the United States are generally a fraction of a ppb. An important formation route for HCHO is equation 9. However, ozonolysis of olefinic compounds and some other reactions of VOCs can also produce HCHO and other aldehydes. Aldehydes are important because they are temporary reservoirs of free radicals (see equations 11 and 12), and HCHO is a known carcinogen. Nitric acid is formed by OH attack on NO_2 and by a dark-phase series of reactions initiated by O_3 + NO_2. Nitric acid is important because, next to H_2SO_4, it is the second most abundant acid in precipitation. In addition, in Southern California it is the main cause of acid fog.

Particles are the principal cause of the haze and the associated brown color that is often apparent with smog. The three most important types of particles produced in smog are composed of organics, sulfates, and nitrates. Organic particles are formed when large VOC molecules, especially aromatics and cyclic alkenes, react with each other and form condensable products. Sulfate particles (formed initially as H_2SO_4) are formed by a series of reactions initiated by the attack of OH on SO_2 in the gas phase or by liquid phase reactions. Nitrate particles are formed by

$$HNO_3(g) + NH_3(g) \rightleftharpoons NH_4NO_3(s) \quad (19)$$

or by the reactions of HNO_3 with salt (NaCl) or alkaline soil dust.

Volatile Organic Compounds (VOC). VOCs include any organic-carbon compound that exists in the gaseous state in the ambient air. In some of the older air-pollution literature the term VOCs was used interchangeably with the term nonmethane hydrocarbons (NMHC). Strictly speaking, this is incorrect organic chemistry nomenclature because hydrocarbons include only those compounds that contain carbon and hydrogen exclusively; whereas organic carbon includes any compound that contains organic carbon (ie, HCHO). Sources of VOCs include any process or activity that uses organic solvents, coatings, or fuel. Emissions of VOCs are important because some are toxic by themselves and most are precursors of O_3 and other species associated with photochemical smog. As a result of control measures designed to reduce O_3, VOC emissions are declining in the United States. Between 1979–1988, nationwide VOC emissions declined 17% (4). Trends in ambient VOC concentrations cannot be determined, however, because of the lack of measurements.

Nitrogen Oxides (NO_x). In air-pollution terminology, nitrogen oxides include the gases NO and NO_2. Most of the NO_x is emitted as NO, which is oxidized to NO_2 in the atmosphere (see eqs. 3 and 8). All combustion processes are sources of NO_x. At the high temperatures generated in the combustion process, some N_2 will be converted to NO in the presence of O_2. In general, the higher the combustion temperature, the more NO_x produced. Since NO_2 is one of the original Criteria Pollutants and it is a precursor to O_3, it has been the target of successful emissions reduction strategies for two decades in the United States. As a result, in 1987, all areas of the United States, with the exception of the Los Angeles/Long Beach area, were in compliance with the NAAQS for NO_2. From 1979 to 1988, nationwide NO_x emissions declined 8%, while ambient concentrations declined 7% (4).

Throughout the United States, however, NO_x remains an important issue because it is an essential ingredient of photochemical smog and some of the NO_x is oxidized to HNO_3, an essential ingredient of acid precipitation and fog. In addition, NO_2 is the only important gaseous species in the atmosphere that absorbs visible light, and in high enough concentrations can contribute to a brownish discoloration of the atmosphere. Additional information on NO_x appears in the sections on Acid Deposition, Particulate Matter, and Visibility.

Sulfur Oxides (SO_x). The combustion of sulfur-containing fossil fuels, especially coal, is the primary source of SO_x. Between 97 and 99% of the SO_x emitted from combustion sources is in the form of SO_2, which is a Criteria Pollutant. The remainder is mostly SO_3, which in the presence of atmospheric water vapor is immediately transformed into H_2SO_4 [*7664-93-9*], a liquid particulate. Both SO_2 and H_2SO_4 at sufficient concentrations produce deleterious effects on the respiratory system. In addition, SO_2 is a phytotoxicant. As with NO_2, control strategies designed to reduce the ambient levels of SO_2 have been highly successful. In the 1960s, most industrialized urban areas in the eastern United States had an SO_2 air-quality problem. By 1991, only Steubenville, Ohio, exceeded the annual NAAQS. Over the past 10 years, nationwide emissions declined 2%, and ambient concentrations decreased about 20% (4). However, the 1990 Clean Air Act Amendments require additional SO_2 reductions because of the role that SO_2 plays in acid deposition. Further discussion is found in the Acid Deposition, PM_{10}, and Visibility sections. In addition, there is some concern over the health effects of H_2SO_4 particles, which are not only emitted directly from some sources, but are also formed in the atmosphere from the oxidation of SO_2 (24).

Carbon Monoxide (CO). Carbon monoxide is a colorless, odorless gas emitted during the incomplete combustion of fuels. CO is emitted during any combustion process, and transportation sources account for about two-thirds of the CO emissions nationally. However, in certain areas, woodburning fireplaces and stoves contribute most of the observed CO. CO is absorbed through the lungs into the blood stream and reacts with hemoglobin to form carboxyhemoglobin, which reduces the oxygen carrying capacity of the blood.

Emissions of CO in the United States peaked in the late 1960s, but have decreased consistently since that time as transportation sector emissions significantly decreased. Between 1968 and 1983, CO emissions from new passenger cars were reduced by 96%. This has been partially offset by an increase in the number of vehicle-miles-traveled annually. Even so, there has been a steady decline in the CO concentrations across the United States, and the decline is expected to continue until the late 1990s without the implementation of any additional emissions-reduction measures. In 1991, there were still 41 urban areas in the United States that were classified as nonattainment for CO, but the number of exceedances declined by about 90% from 1982 to 1991. Over the same time period, nationwide CO emissions decreased 31%, and ambient concentrations declined by 30% (4).

Particulate Matter

In the air pollution field, the terms particulate matter, particulates, particles, and aerosols are used interchangeably and all refer to finely divided solids and liquids dispersed in the air. The original EPA primary standards for total suspended particulates (TSP) were 75 $\mu g/m^3$ as an annual geometric mean, and 260 $\mu g/m^3$ as 24-h mean. TSP is the weight of any particulate matter collected on the filter of a high volume air sampler. On the average, the high volume sampler collects particles that are less than about 30–40 μm in diameter. However, because of the unsymmetrical design of the sampler's inlet, the collection efficiencies for particles varies with both wind direction and speed. In 1987, the EPA promulgated new standards for ambient particulate matter using a new indicator, PM_{10}, rather than TSP. PM_{10} is particulate matter with an aerodynamic diameter of 10 μm or less. The 10 μm diameter was chosen because 50% of the 10 μm particles deposit in the respiratory tract below the larynx during oral breathing. The fraction deposited decreases with increasing particle diameter above 10 μm. The primary and secondary PM_{10} NAAQS are given in Table 3. Because the standard was only enacted in 1987, currently available PM_{10} data are insufficient to determine trends. However, in 1991, EPA designated 70 areas in the United States as nonattainment for PM_{10}% (4).

Atmospheric aerosols can be classified into three size modes: the nuclei-, accumulation-, and large- or coarse-particle modes. Some important characteristics of these modes are illustrated in Figure 4. The bulk of the aerosol mass usually occurs in the 0.1 to 10-μm size range which encompasses most of the accumulation mode and part of the large-particle mode. The nuclei mode is a transient mode. Nuclei, formed by combustion, nucleation, and chemical reactions, coagulate and grow into the accumulation mode. Particles in the accumulation mode are relatively stable because they exceed the size range where coagulation is important, and they are too small to have appreciable deposition velocities. Consequently, particles accumulate in this mode. Particles larger than about 2.5 μm begin to have appreciable deposition velocities, so their lifetimes in the atmosphere decrease significantly with particle size. The sources of large particles are mostly mechanical processes.

Figure 5 shows the mass size distribution of typical ambient aerosols. Note the mass peaks in the accumulation mode and between 5 and 10 μm. The minimum in the curve at about 1–2.5 μm is due to a lack of sources for these particles. Coagulation is not significant for the accumulation-mode particles, and particles produced by

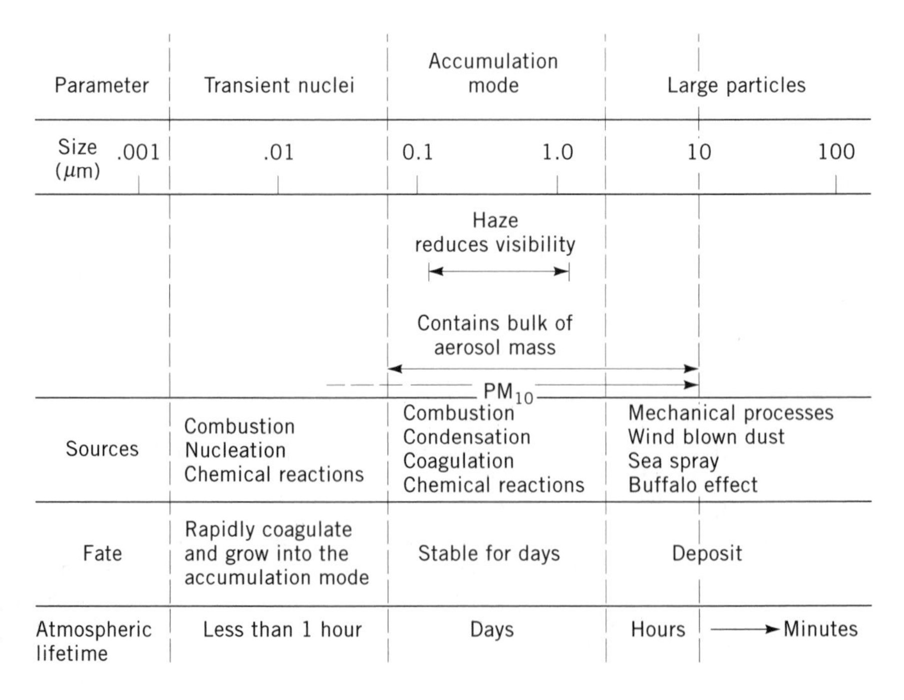

Figure 4. Some important aerosol characteristics (modified from ref. 25, which was adopted in part from ref. 26).

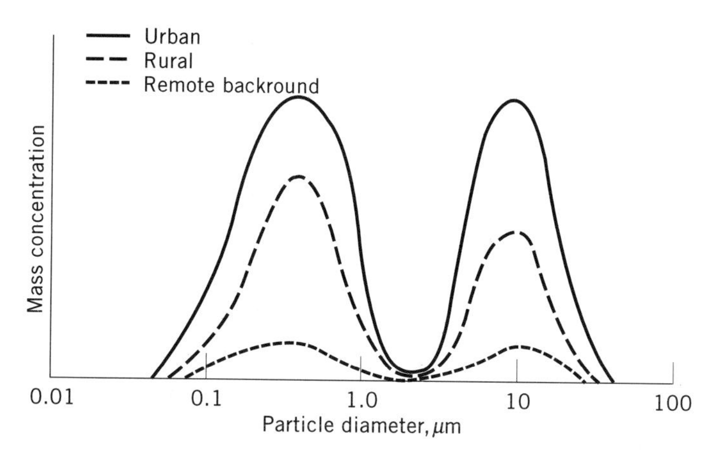

Figure 5. Typical size distributions of atmospheric particles.

mechanical process are larger than 2.5 µm. Consequently, particles less than about 2.5 µm have different sources than particles greater than 2.5 µm. Because of this, it is convenient to classify PM_{10} into a coarse-particulate-mass mode (CPM, diameter ≥ 2.5–10 µm) and a fine-particulate-mass mode (FPM < 2.5 µm). By knowing the relative amounts of CPM and FPM as well as the chemical composition of the principal species, information on the sources of the PM_{10} can be deduced. In urban areas, the CPM and FPM modes are usually comparable in mass. In rural areas, the FPM is generally lower than in urban areas, but higher than the CPM mass. The reason for this is that a significant fraction of the rural FPM is generally transported from upwind sources, whereas most of the CPM is generated locally.

Chemical composition data for CPM and FPM for a variety of locations are summarized in Table 5. The data in this table illustrate several important points. First, the distribution of the PM_{10} between CPM and FPM varies from about 0.4 to 0.7. Second, the ratio of PM_{10} to TSP varies from 0.58 to 0.79. In general, both this ratio and the ratio of FPM/PM_{10} tend to be higher at rural sites. Bermuda, because of the large influence of sea salt in the CPM, is a notable exception. The most abundant FPM species are sulfate (SO_4^{2-}), carbon (organic carbon, OC, and elemental carbon, EC), and nitrate (NO_3^-) compounds which generally account for 70–80% of the FPM. In the eastern United States, SO_4^{2-} compounds are the dominant species, but very little SO_4^{2-} is emitted directly into the atmosphere. Most of the SO_4^{2-} is a secondary aerosol formed from the oxidation of SO_2, and in the eastern United States the source of most of the SO_2 is coal-burning emissions. In the atmosphere, the principal oxidation routes include: homogeneous oxidation of SO_4^{2-} by OH, and the heterogeneous oxidation in water droplets by hydrogen peroxide (H_2O_2), O_3, or, in the presence of a catalyst, O_2. Atmospheric particles which have been identified as catalysts include many metal oxides and soot. The wa-

Table 5. Summary of Detailed Particulate Measurements at Various Urban and Rural Locations[a]

	Urban					Rural		
Species	Denver, Colo.	Detroit, Mich.	Long Beach, Calif.	Claremont, Calif.	Jacksonville, Fla.	Blue Ridge Mts., Va.	Lewes, Del.	Bermuda
				Concentrations µg / m³				
PM_{10}	nd	52.9[b]	79.1	59.7	38.8[b]	32.5	22.0	20.8
FPM	39.5	25.1	57.8	36.0	22.6	23.9	15.9	9.4
CPM	50.9[c]	27.8	23.1	22.9	16.2	8.6	6.1	11.4
FPM/PM_{10}	nd	0.48[b]	0.73	0.60	0.58	0.74[b]	0.72	0.45
TSP	104.4	90.5	nd	nd	nd	43.7	27.9	32.8
PM_{10}/TSP	nd	0.58[b]	nd	nd	nd	0.74[b]	0.79	0.63
				FPM, %				
Sulfate compounds	14	53	7	22	41	55	50	25
Nitrate compounds	20	<1	33	28	3	<1	nd	2
Organic compounds	22	27	29	29	13	24	30	15
Elemental carbon	15	4	12	5	6	5	6	<1
				CPM, %				
Crustal species	91	73	nd	nd	85	nd	48	17
Sea salt	nd	nd	nd	nd	nd	nd	43	57
References	27	28,29	30	30	[d]	31	32	33

[a] nd, not determined.
[b] PM_{10} in Detroit and Blue Ridge Mts. is PM_{15} which is ≅ to PM_{10}.
[c] CPM in Denver is diameter ≥2.5 µm to about 30 µm.
[d] Unpublished data.

ter droplets include cloud and dew droplets as well as aerosols which contain sufficient water. Under high relative humidity conditions, hygroscopic salts deliquesce and form liquid aerosols. Sulfate initially forms as sulfuric acid (H_2SO_4), which rapidly reacts with available ammonia (NH_3) to form ammonium bisulfate (NH_4HSO_4). If sufficient NH_3 is present, the final product is ammonium sulfate [$(NH_4)_2SO_4$]. In some urban areas in the western United States, NO_3^- is more abundant than SO_4^{2-}. The NO_3^- in the FPM exists primarily as NH_4NO_3, which, as discussed earlier, is formed via equation 19. However, acidic SO_4^{2-} (H_2SO_4 or NH_4HSO_4) readily reacts with NH_4NO_3 and abstracts the NH_3, leaving behind gaseous HNO_3. Consequently, unless there is sufficient NH_3 to completely convert all of the SO_4^{2-} to $(NH_4)_2SO_4$, NH_4NO_3 will not accumulate in the atmosphere. The western U.S. cities shown in Table 5 have sufficient NH_3, which is due in a large part to the presence of animal feedlots, to neutralize the SO_4^{2-} and to allow the NH_4NO_3 to accumulate. In general, in the eastern United States, there is insufficient NH_3 to neutralize the SO_4^{2-}, so NH_4NO_3 does not accumulate. OC is a principal constituent of the FPM at all sites. The main sources of OC are combustion and atmospheric reactions involving gaseous VOCs. As is the case with VOCs, there are hundreds of different OC compounds in the atmosphere. A minor but ubiquitous aerosol constituent is elemental carbon (EC). EC is the nonorganic, black constituent of soot. Its structure is similar to graphite crystals with many imperfections. Combustion and pyrolysis are the only processes that produce EC, and diesel engines and wood burning are the most significant sources. Crustal dust and water make up most of the remaining FPM mass.

Crustal dust is composed of aerosolized soil and rock from the earth's crust. Although this is natural material, human activities (traffic, which produces the buffalo effect by entraining street dust, construction activities, agricultural and land-use practices, etc) affect the rate at which crustal material is aerosolized. Since it is aerosolized by frictional processes, the diameter of most of the crustal dust is $\geq 2.5\ \mu m$. Typically, crustal dust accounts for most of the CPM and particles $> 10\ \mu m$. In global average crustal material, the principal elements contained in decreasing order are O, Si, Al, Fe, Ca, Na, K, and Mg. On the average, Si accounts for 20% of the crustal aerosol mass (34). Consequently, the crustal mass can be estimated from Si measurements alone. However, the relative amounts of the elements do vary spatially.

Lead

Lead (Pb) [*7439-92-1*] is of concern because of its tendency to be retained and accumulated by living organisms. When excessive amounts accumulate in the human body, lead can inhibit the formation of hemoglobin and produce life-threatening lead poisoning. In smaller doses, lead is also suspected of causing learning disabilities in children. From 1982 to 1991, nationwide lead emissions decreased 90%, with the primary source, transportation, showing a 97% reduction (4). The most important reason for this dramatic reduction was the removal of lead compounds from fuels, primarily gasoline. Tetraethyl lead [*78-00-2*] was added to gasoline to improve the octane rating to prevent engine knock. With the introduction of the catalytic converter, the lead compound had to be eliminated because it would poison the catalyst. Trends of lead in the ambient air have responded to the emissions reductions. In 1991, only a few isolated monitoring sites that were dominated by industrial sources of lead experienced violations of the NAAQS (4).

Air Toxics

There are thousands of commercial chemicals used in the United States. Hundreds of these substances are emitted into the atmosphere and have some potential to affect human health adversely at certain concentrations. Some are known or suspected carcinogens. Identifying all of these substances and promulgating emissions standards is beyond the present capabilities of existing air-quality management programs. Consequently, toxic air pollutants (TAPs) need to be prioritized based on risk analysis, so that those posing the greatest threats to health can be regulated. Although the criteria pollutants were so designated because they can have significant public-health impacts, the criteria pollutants are not considered TAPs because the criteria pollutants are regulated elsewhere in the CAA. A distinguishing feature between TAPs and criteria pollutants is that criteria pollutants are considered national issues while TAPs, on the other hand, are most often isolated issues, localized near the source of the TAP emissions. For example, O_3 is likely to be an issue in all large U.S. metropolitan areas, whereas TAPs usually are of concern only in areas with certain types of sources.

There are three types of TAP emissions: continuous, intermittent, and accidental. Intermittent sources can be routine emissions associated with a batch process or a continuous process operated only occasionally. An accidental release is an inadvertent emission. A dramatic example of this type was the release of methyl isocyanate in Bhopal, which was responsible for over 2000 deaths. As a result of this accident, the U.S. Congress created Title III, a free-standing statute included in the Superfund Amendments and Reauthorization Act (SARA) of 1986. Title III provides a mechanism by which the public can be informed of the existence, quantities, and releases of toxic substances, and requires the states to develop plans to respond to accidental releases of these substances. Further, it requires anyone releasing specific toxic chemicals above a certain threshold amount to annually submit a toxic-chemical-release form to EPA. At present, there are 308 specific chemicals subject to Title III regulation (35).

A valuable resource that contains a listing of many potential TAPs is the American Conference of Governmental Industrial Hygienists' *Threshold Limit Values (TLV)* (36). This booklet lists the workplace air standards for over 700 substances, many of which would be considered a TAP if present in sufficient quantity in the ambient air. The exceptions would be those listed as simple asphyxiants, or classified as Criteria Pollutants. Toxicological data for these substances can be found in reference 37.

The 1970 Clean Air Act required that EPA provide an

Table 6. Substances Listed as Hazardous Air Pollutants in the 1990 Clean Air Act Amendments

Substance	(CAS Registry Number)	Substance	(CAS Registry Number)	Substance	(CAS Registry Number)
Acetaldehyde	[75-07-0]	Dimethyl carbamoyl chloride	[79-44-7]	Parathion	[56-38-2]
Acetamide	[60-35-5]	Dimethyl formamide	[68-12-2]	Pentachloronitrobenzene	[82-68-8]
Acetonitrile	[75-05-8]	1,1-Dimethyl hydrazine	[54-14-7]	Pentachlorophenol	[87-86-5]
Acetophenone	[98-86-2]	Dimethyl phthalate	[131-11-3]	Phenol	[108-95-2]
2-Acetylaminofluorene	[53-96-3]	Dimethyl sulfate	[77-78-1]	*p*-Phenylenediamine	[106-50-3]
Acrolein	[107-02-8]	4,6-Dinitro-o-cresol, and salts	[534-52-1]	Phosgene	[75-44-5]
Acrylamide	[79-06-1]	2,4-Dinitrophenol	[51-28-5]	Phosphine	[7803-51-2]
Acrylic acid	[79-10-7]	2,4-Dinitrotoluene	[121-14-2]	Phosphorus	[7723-14-0]
Acrylonitrile	[107-13-1]	1,4-Dioxane	[123-91-1]	Phthalic anhydride	[85-44-9]
Allyl chloride	[107-05-1]	1,2-Diphenylhydrazine	[122-66-7]	Polychlorinated biphenyls	[1336-36-3]
4-Aminobiphenyl	[92-67-1]	Epichlorohydrin	[106-89-8]	1,3-Propane sultone	[1120-74-4]
Aniline	[62-53-3]	1,2-Epoxybutane	[106-88-7]	beta-Propiolactone	[57-57-8]
o-Anisidine	[90-04-0]	Ethyl acrylate	[140-88-5]	Propionaldehyde	[123-38-6]
Asbestos	[1332-21-4]	Ethyl benzene	[100-41-4]	Propoxur (Baygon)	[114-26-1]
Benzene	[71-43-2]	Ethyl carbamate	[51-79-6]	Propylene dichloride	[78-87-5]
Benzidine	[92-87-5]	Ethyl chloride	[75-00-3]	Propylene oxide	[75-56-9]
Benzotrichloride	[98-07-7]	Ethylene dibromide	[106-93-4]	1,2-Propylenimine	[75-55-8]
Benzyl chloride	[100-44-7]	Ethylene dichloride	[107-06-2]	Quinoline	[91-22-5]
Biphenyl	[92-52-4]	Ethylene glycol	[107-21-1]	Quinone	[106-51-4]
Bis(2-ethylhexyl)phthalate	[117-81-7]	Ethylene imine	[151-56-4]	Styrene	[100-42-5]
Bis(chloromethyl)ether	[542-88-1]	Ethylene oxide	[75-21-8]	Styrene oxide	[96-09-3]
Bromoform	[75-25-2]	Ethylene thiourea	[96-45-7]	2,3,7,8-Tetrachlorodibenzo-*p*-dioxin	[1746-01-6]
1,3-Butadiene	[106-99-0]	Ethylidene dichloride	[75-34-3]	1,1,2,2-Tetrachloroethane	[79-34-5]
Calcium cyanamide	[156-62-7]	Formaldehyde	[50-00-0]	Tetrachloroethylene	[127-18-4]
Caprolactam	[105-60-2]	Heptachlor	[76-44-8]	Titanium tetrachloride	[7550-45-0]
Captan	[133-06-2]	Hexachlorobenzene	[118-74-1]	Toluene	[108-88-3]
Carbaryl	[63-25-2]	Hexachlorobutadiene	[87-68-3]	2,4-Toluene diamine	[95-80-7]
Carbon disulfide	[75-15-0]	Hexachlorocyclopentadiene	[77-47-4]	2,4-Toluene diisocyanate	[584-84-9]
Carbon tetrachloride	[56-23-5]	Hexachloroethane	[67-72-1]	*o*-Toluidine	[95-53-4]
Carbonyl sulfide	[463-58-1]	Hexamethyl-1,6-diisocyanate	[822-06-0]	Toxaphene	[8001-35-2]
Catechol	[120-80-9]	Hexamethylphosphoroamide	[680-31-9]	1,2,4-trichlorobenzene	[120-82-1]
Chloramben	[133-90-4]	Hexane	[110-54-3]	1,1,2-Trichloroethane	[79-00-5]
Chlordane	[57-74-9]	Hydrazine	[302-01-2]	Trichloroethylene	[79-01-6]
Chlorine	[7782-50-5]	Hydrochloric acid	[7647-01-0]	2,4,5-Trichlorophenol	[95-95-4]
Chloroacetic acid	[79-11-8]	Hydrogen fluoride	[7664-39-3]	2,4,6-Trichlorophenol	[88-06-2]
2-Chloroacetophenone	[532-27-4]	Hydroquinone	[123-31-9]	Trimethylamine	[121-44-8]
Chlorobenzene	[108-90-7]	Isophorone	[78-59-1]	Trifluralin	[1582-09-8]
Chlorobenzilate	[510-15-6]	Lindane (all isomers)	[58-89-9]	2,2,4-Trimethylpentane	[540-84-1]
Chloroform	[67-66-3]	Maleic anhydride	[108-31-6]	Vinyl acetate	[108-05-4]
Chloromethyl methyl ether	[107-30-2]	Methanol	[67-56-1]	Vinyl bromide	[593-60-2]
Chloroprene	[126-99-8]	Methoxychlor	[72-43-5]	Vinyl chloride	[75-01-4]
Cresols/cresylic acid	[1319-77-3]	Methyl bromide	[74-83-9]	Vinylidene chloride	[75-35-4]
o-Cresol	[95-48-7]	Methyl chloride	[74-87-3]	Xylenes (isomers and mixture)	[1330-20-7]
m-Cresol	[108-39-4]	Methyl chloroform	[71-55-6]	*o*-Xylenes	[95-47-6]
p-Cresol	[106-44-5]	Methyl ethyl ketone	[78-93-3]	*m*-Xylenes	[108-38-3]
Cumene	[98-82-8]	Methyl hydrazine	[60-34-4]	*p*-Xylenes	[106-42-3]
2,4-D, salts and esters	[94-75-7]	Methyl iodide	[74-88-4]	Antimony compounds	
DDE	[3547-04-4]	Methyl isobutyl ketone	[108-10-1]	Arsenic compounds	
Diazomethane	[334-88-3]	Methyl isocyanate	[624-83-9]	Beryllium compounds	
Dibenzofurans	[132-64-9]	Methyl methacrylate	[80-62-6]	Cadmium compounds	
1,2-Dibromo-3-chloropropane	[96-12-8]	Methyl-*tert*-butyl ether	[1634-04-4]	Chromium compounds	
Dibutylphthalate	[84-74-2]	4,4-Methylene bis(2-chloroaniline)	[101-14-4]	Cobalt compounds	
1,4-Dichlorobenzene(p)	[106-46-7]	Methylene chloride	[75-09-2]	Coke oven emissions	
3,3-Dichlorobenzidene	[91-94-1]	Methylene diphenyl diisocyanate	[101-68-8]	Cyanide compounds	
Dichloroethyl ether	[111-44-4]	4,4′-Methylenedianiline	[101-77-9]	Glycol ethers	
1,3-Dichloropropene	[542-75-6]	Naphthalene	[91-20-3]	Lead compounds	
Dichlorvos	[62-73-7]	Nitrobenzene	[98-95-3]	Manganese compounds	
Diethanolamine	[111-42-2]	4-Nitrobiphenyl	[92-93-3]	Mercury compounds	
N,N-Diethyl aniline	[121-69-7]	4-Nitrophenol	[100-02-7]	Fine mineral fibers	
Diethyl sulfate	[64-67-5]	2-Nitropropane	[79-46-9]	Nickel compounds	
3,3-Dimethoxybenzidine	[119-90-4]	*N*-Nitroso-*N*-methylurea	[684-93-5]	Polycyclic organic matter	
Dimethyl aminoazobenzene	[60-11-7]	*N*-Nitrosodimethylamine	[62-75-9]	Radionuclides (including radon)	
3,3′-Dimethyl benzidine	[119-93-7]	*N*-Nitrosomorpholine	[59-89-2]	Selenium compounds	

ample margin of safety to protect against Hazardous Air Pollutants (HAPs) by establishing national emissions standards (NESHAPs) for certain sources. From 1970 to 1990, over 50 chemicals were being considered for designation as HAPs, but EPA's review process was completed for only 28 chemicals. Of the 28, NESHAPs were promulgated for only eight substances: beryllium [*7440-41-7*], mercury [*7436-97-6*], vinyl chloride [*75-01-4*], asbestos [*1332-21-4*], benzene [*71-43-2*], radionuclides, inorganic arsenic [*7440-38-2*], and coke-oven emissions. EPA decided not to list ten of the substances and intended to list the other ten substances as HAPs (38). However, in the 1990 Clean Air Act Amendments, 189 substances are listed (Table 6) that EPA must regulate by enforcing "maximum achievable control technology (MACT)." The Amendments mandate that EPA issue MACT standards for all sources of the 189 substances by the year 2000. In addition, EPA must determine the risk remaining after MACT is in place and develop health-based standards that would limit the cancer risk to one case in one million exposures. EPA may add or delete substances from the list.

Because EPA was so slow in promulgating standards for HAPs prior to the 1990 Amendments, most states developed and implemented their own TAP control programs. Such programs, as well as the pollutants they regulate, differ widely from state to state. Some states have emissions and/or ambient standards. The ambient standards for a given substance are usually selected to be some small fraction of the TLV for that substance.

Odors

The 1977 Clean Air Act Amendments directed EPA to study the effects, sources, and control feasibility of odors. Although no Federal legislation has been established to regulate odors, individual states have responded to odor complaints by enforcing common nuisance laws. Because about 50% of all citizen air pollution complaints concern odors, it is clear that a disagreeable odor is perceived as an indication of air pollution. However, many substances can be detected by the human olfactory system at concentrations well below those considered harmful. For example, hydrogen sulfide can be detected by most people at 0.0047 ppm, whereas the occupation health 8-h TLV is 10 ppm. Although exposures to such odors in low concentrations may not in itself cause apparent physical harm, the exposure can lead to nausea, loss of appetite, etc. On the other hand, the absence of an odor is no indication of healthy air. For example, CO is an odorless gas.

Odors are characterized by quality and intensity. Descriptive qualities such as sour, sweet, pungent, fishy, and spicy, to name a few, are commonly used. The strength intensity is determined by how much the concentration of the odoriferous substance exceeds its detection threshold (the concentration at which most people can detect an odor). Odor intensity is approximately proportional to the logarithm of the concentration. However, several factors affect the ability of an individual to detect an odor: the sensitivity of a subject's olfactory system, the presence of other masking odors, and olfactory fatigue (ie, reduced olfactory sensitivity during continued exposure to the odorous substance). In addition, the average person's sensitivity to odor decreases with age.

Visibility

Although there is no NAAQS designed to protect visual air quality, the 1977 Clean Air Act Amendments set as a national goal "the remedying of existing and prevention of future impairment of visibility in mandatory Class I Federal areas which impairment results from man-made pollution." Class I areas are certain national parks and wildernesses that were in existence in 1977. The 1977 Amendments also directed EPA to promulgate appropriate regulations to protect against visibility impairment in these areas. In 1981, EPA directed 36 states to amend their State Implementation Plans to develop control programs for visual impairment that could be traced to particular sources. This type of impairment is called plume blight, and it was the initial focus of EPA's effort because it involved easily identifiable sources. The 1990 Clean Air Act Amendments direct EPA to promulgate appropriate regulations to address regional haze in affected Class I areas. EPA has not dealt with a third type of visibility impairment, urban-scale haze, because the source-receptor relationships are extremely complex (39).

Visibility or visual range is the maximum distance at which a black object can be distinguished from the horizon. In other words, the ability to distinguish an object from the background horizon depends upon the contrast between the target and the background. Under certain viewing conditions, the apparent contrast, C, between a target and the horizon decreases exponentially with the distance x between the target and observer (40):

$$C = C_o e^{-b_{ext} x} \tag{20}$$

where C_o is the contrast at $x = 0$, and b_{ext} is the extinction coefficient (the proportionality constant relating the intensity of light received by an observer from a target to the intensity of light emitted by the target). The maximum distance, x_{max}, at which an observer can distinguish the target from the horizon occurs when $C = \varepsilon$, where ε is the observer's contrast threshold. Substituting in equation 20 and rearranging:

$$x_{max} = \frac{\ln C_o - \ln \varepsilon}{b_{ext}} = V \tag{21}$$

where V is the visibility. For a black target, $C_o = -1$ and experiments have shown that ε is between 0.02 and 0.05. Using the more sensitive value for ε, equation 21 reduces to (41):

$$V = 3.9/b_{ext} \tag{22}$$

This equation only applies under ideal conditions (ie, black target against a bright horizon, and a well-illuminated, homogeneous atmosphere). Actual conditions, such as a nonblack target, which will alter the value of C_o, different viewing angles, and different illumination conditions (cloud cover, different sun angles), are not described so simply, but they can have a profound effect on visual range. Nevertheless, b_{ext} is a useful indicator of the inverse of visual range, and it is widely used as an indicator of visual air quality. The total extinction can be written as the sum of a number of components:

$$b_{ext} = b_{sp} + b_R + b_{ap} + b_{ag} \tag{23}$$

Table 7. Light Extinction Budgets, Excluding b_R

	Mean Percent Contribution		
Component	Denver[a]	Detroit[b]	Blue Ridge Mts[c]
b_{ap}	29	8	3
b_{sp}	64	88	97
b_{ag}	7	4	$\ll 1$
Σ	100	100	100

[a] Ref. 42.
[b] Ref. 29.
[c] Ref. 31.

where b_{sp} is the light extinction due to light scattered by particles; b_R is the light scattered by air molecules (Rayleigh scattering) and is a function of the atmospheric pressure; b_{ap} is the light absorbed by particles; and b_{ag} is the light absorbed by gas molecules.

Rayleigh scattering accounts for only a minor part of the extinction, except on the clearest days. It is a function of atmospheric pressure alone and does not depend appreciably on the composition of the pollutant gases present. At sea level at 25°C, b_R is equal to $13.2 \times 10^{-6} m^{-1}$ which, in the absence of particles and absorbing gases, corresponds to a visual range of about 300 km. Light extinction budgets (excluding b_R) for several areas of the United States are presented in Table 7. In general, light scattering is dominated by particles, primarily fine particles. The most efficient light-scattering particles are those that are the same size as the wavelengths of visible light (0.4–0.7 μm). As shown in Figure 5, a peak in the mass distribution occurs in the size range comparable to the wavelength range of visible light. The particles in this size range, therefore, almost always dominate b_{sp}. Exceptions to this occur during fog, precipitation, and dust storms. On a per mass basis, the most efficient light-scattering fine particles are hygroscopic particles, such as sulfate, nitrate, and ammonium particles, which will sorb significant amounts of water at moderate to high relative humidities. As the particles sorb water, they become more efficient light scatterers. Light absorption by particles in the atmosphere is almost exclusively due to elemental carbon which also scatters light. The only common light-absorbing gaseous pollutant is NO_2, which usually accounts for a few percent or less of the total extinction.

As mentioned above, there are three scales of visual impairment: plume blight, urban-scale haze, and regional-scale haze. Plume blight occurs when a plume from a large point source travels into an otherwise clean area and interferes with the viewing of a particular vista. Such events can occur anywhere, but are usually most noticeable in the western U.S. and have been observed in many scenic areas. Most frequently, plume blight is associated with sulfates from a sulfur dioxide-emitting point source. Most large urban areas occasionally experience urban haze, but the public perception of the haze is highest in those cities with scenic mountain vistas like Los Angeles and Denver. Most of the light extinction in urban haze can be accounted for by sulfates (sulfuric acid and the ammonium salts), nitrate (as ammonium nitrate), organic carbon, and elemental carbon. Regional haze refers to the situation where the haze extends for hundreds of miles. Regional haze is usually dominated by sulfates. In the Southwest, occasional haze obscures scenic vistas over large portions of the area, and is attributed to a combination of sulfates from coal-fired power plant and smelter emissions, and transport of urban, Southern California haze which is composed mainly of carbonaceous particles, nitrates and, to a lesser extent, sulfates from southern California. In the East, a denser sulfate-dominated haze (ie, in the rural West, mean sulfate concentrations are $\sim$ 1 $\mu g/m^3$, while in the rural East they average $\sim$ 8 $\mu g/m^3$; also, the relative humidity is generally much higher in the East) frequently extends over much of the area during the summer. The primary source of the Eastern haze is coal-burning emissions. Natural haze, caused mainly by aerosols generated from biogenic VOC emissions from vegetation, was historically cited as the regional haze in areas such as the Blue Ridge and Smoky Mountain Regions. Except on the cleanest days, however, sulfate haze now dominates natural haze in these regions (31).

Air pollutants can also cause discolorations of the atmosphere. The most common are brownish discolorations (eg, the "brown Los Angeles haze" and the "brown clouds" observed in Denver and elsewhere). Three factors can contribute to the brown tint. The first is nitrogen dioxide which is a brownish gas. This is most commonly viewed in a plume of NO_2. In the urban hazes, the effect of NO_2 is usually overwhelmed by the effects caused by particles. Since fine particles preferentially scatter blue light in the forward direction, the light viewed through an optically thin cloud with the sun behind the observer is deficient in the blue wavelengths and appears brown. In dense haze clouds, the preferential scattering is masked by multiple-scattering effects and the haze is seen as white. However, a dense haze cloud can appear brown along its edges where it is optically thin. If the cloud is between the observer and the sun, it will appear as white. Also, a dense cloud in the distance against a bright blue sky background can appear brown through a process called chromatic adaptation. With chromatic adaptation, the blue receptors in the human eye are desensitized by the bright blue background; as a result, the white light from the haze appears to be brown.

Acid Deposition

Acid deposition is the deposition of acids from the atmosphere to the surface of the earth. The deposition can be dry or wet. Dry deposition refers to the process whereby acid gases or their precursors or acid particles come in contact with the earth's surface and are retained by the surface. The principal species associated with dry acid-deposition are SO_2(g), acid sulfate particles (H_2SO_4 and NH_4HSO_4), and HNO_3(g). Measurements of dry deposition are quite sparse, however, and usually only speciated as total SO_4^{2-} and total NO_3^-. In general, the dry acid deposition is estimated to be a small fraction of total acid deposition because most of the dry deposited SO_4^{2-} and NO_3^- has been neutralized by basic gases and particle in the atmosphere. However, the sulfate and nitrate deposited from dry deposition is estimated to be a significant fraction of the total sulfate and nitrate deposition. More specific estimates are not possible because current spatial and temporal dry deposition data are insufficient. On the other hand, there are abundant data on wet-acid deposition. Wet-acid deposition or acid precipitation is the process by which acids are deposited by the rain or snow. The principal dissolved acids are H_2SO_4 and HNO_3. Other acids, such as HCl and organic acids, usually account for only a minor part of the acidity, although organic acids can be significant contributors in remote areas.

Both acid particles and gases can be incorporated into cloud droplets. Particles are incorporated into droplets by nucleation, Brownian diffusion, impaction, diffusio-

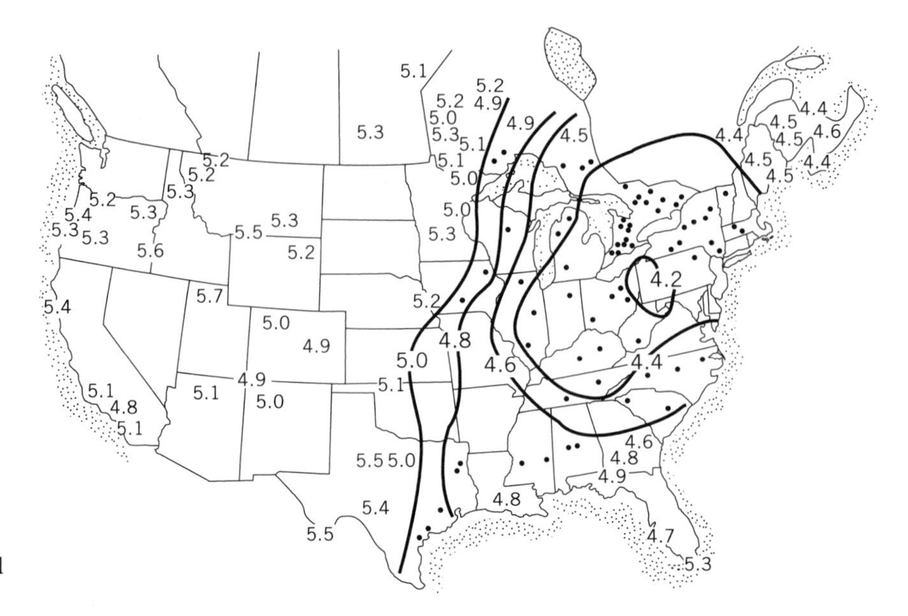

Figure 6. 1985 annual precipitation-weighted pH (43).

phoresis (transport into the droplet induced by the flux of water vapor to the same surface), thermophoresis (thermally induced transport to a cooler surface), and electrostatic transport. Advective and diffusive attachment dominate all other mechanisms for pollutant gas uptake by cloud droplets. Modeling and experimental evidence suggest that most of the H_2SO_4 is formed in cloud water droplets. SO_2 diffuses into the droplet and is oxidized to H_2SO_4 by one of several mechanisms. At pHs greater than about 5.5, oxidation of SO_2 by dissolved O_3 is the dominant reaction. At lower pHs, SO_2 oxidation is dominated by the reaction with H_2O_2. Under some conditions, oxidation by O_2, catalyzed by metals or soot, may contribute to the formation of H_2SO_4. Most of the HNO_3 in precipitation is due to the diffusion of HNO_3 into the droplet. However, there is observational evidence that some HNO_3 is formed in the droplets, but the mechanism has not been identified.

The pH of rainwater in equilibrium with atmospheric CO_2 is 5.6, a value that was frequently cited as the natural background pH of rainwater. However, in the presence of other naturally occurring species such as SO_2, SO_4^{2-}, NH_3, organic acids, sea salt, and alkaline crustal dust, the "natural" values of unpolluted rainwater will vary between 4.9 and 6.5 depending upon time and location. Across the United States, the mean annual average pH varies from 4.2 in western Pennsylvania to 5.7 in the West (see Fig. 6). Precipitation pH is generally lowest in the eastern United States within and downwind of the largest SO_2 and NO_x emissions areas. In general, the lowest pH precipitation occurs in the summer. In the East, SO_4^{2-} concentrations in precipitation are 1.5 to 2.5 times higher during the summer than winter, but the NO_3^- values are about the same year round. Consequently, the lower pHs in the summer are mostly due to the higher SO_4^{2-} concentrations. The equivalent ratio of sulfate to nitrate in precipitation is an often used (but inexact) measure of the relative contribution of these two species to the acidity. An equivalent ratio of 1.0 would mean equal contributions by both. In the East during the winter, this ratio ranges from 1.0 to 2.5, while during the summer it ranges from 2.0 to 3.0. On the average in the eastern United States, about 60% of the wet-deposited acidity can be attributed to SO_4^{2-} and 40% to NO_3^- (43).

Since SO_2 and NO_2 are criteria pollutants, their emissions are regulated. In addition, for the purposes of abating acid deposition in the United States, the 1990 Clean Air Act Amendments require that nationwide SO_2 and NO_x emissions be reduced by 10 million tons/year and 2 million tons/year, respectively, by the year 2000. The reasons for these reductions are based on concerns which include acidification of lakes and streams, acidification of poorly buffered soils, and acid damage to materials. An additional concern was that acid deposition was causing the die-back of forests at high elevations in the eastern United States and in Europe. Although a contributing role of acid deposition cannot be dismissed, the primary pollutant suspected in the forest decline issue is now O_3.

Global Warming (The Greenhouse Effect)

Solar energy, mostly in the form of visible light, is absorbed by the earth's surface and reemitted as longer wavelength infrared (ir) radiation. Certain gases in the atmosphere, primarily water vapor and to a lesser degree CO_2, have the ability to absorb the outgoing ir radiation which is translated to heat. The result is a higher atmospheric equilibrium temperature than would occur in the absence of water vapor and CO_2. This temperature enhancement is called the greenhouse effect, and gases that have the ability to absorb ir and produce this effect are called greenhouse gases. Without the naturally occurring concentrations of water vapor and CO_2, the earth's mean surface temperature would be $-18°C$ instead of the present 17°C. There is concern that increasing concentrations of CO_2 and other trace greenhouse gases due to human activities will enhance the greenhouse effect and cause global warming. Speculated scenarios based on global warming include the following: an alteration in existing precipitation patterns, an increase in the severity of storms, the dislocation of suitable land

Table 8. Summary of Important Greenhouse Gases

Gas	CAS Registry Number	Present Concentrations[a]	Concentration Increase,[a] %/Year	Warming Potential[b,c]	Atmospheric Residence Times,[b,d] Years
CO_2		350 ppm	0.3	1	120[e]
CH_4		1.68 ppm	0.8–1	11	10.5
N_2O		340 ppb	0.2	270	132
$CFCl_3$ (CFC-11)	*[75-69-4]*	226 ppt	4	3400	55
CF_2Cl_2 (CFC-12)	*[75-71-8]*	392 ppt	4	7100	116
$CHClF_2$ (HCFC-22)	*[75-46-6]*	100 ppt	7	1600	16
$C_2H_2F_4$ (HFC-134a)	*[811-97-2]*			1200	16
$C_2Cl_3F_3$ (CFC-113)	*[76-14-2]*	30–70 ppt	11	4500	110
CH_3CCl_3 (Methychloroform)	*[71-55-6]*	125 ppt	7	100	6
CCl_4 (Carbon tetrachloride)	*[56-23-5]*	75–100 ppt	1	1300	47

[a] From Ref. 44.
[b] From Ref. 45.
[c] Relative to CO_2. This is based on a 100 year time horizon (45).
[d] Residence times may be slightly different from those reported in Table 2 because the primary source is different.
[e] From Ref. 8.

for agriculture, the dislocation and possible extinction of certain biological species and ecosystems, and the flooding of many coastal areas due to rising sea levels resulting from the thermal expansion of the oceans, the melting of glaciers, and, probably less so, from the melting of polar ice caps.

Measurements of CO_2 since 1958 clearly show that CO_2 concentrations in the atmosphere are increasing at the rate of about 0.3% per year. The present concentration of ~ 350 ppm compares to the preindustrial revolution (1800) value of 285 ppm (estimated from ice cores). Projections based on the current rate of the CO_2 increase and future energy uses show that the CO_2 concentration will approach 600 ppm some time in the middle of the 21st Century. Within the past two decades, it was discovered that the concentrations of other trace greenhouse gases are also increasing. These gases include methane (CH_4), nitrous oxide (N_2O), O_3 (tropospheric), and a variety of chlorofluorocarbons (CFCs). A list of the greenhouse gases (not including water), present concentrations, current rates of increase in the atmosphere, and estimates of relative greenhouse efficiencies and atmospheric residence times are presented in Table 8. Ozone is not included in this table because there is insufficient information available for quantifying the global radiative influence of O_3 (45). CFCs are included in the table, but their warming potentials should be used with caution because there presently is a controversy over whether the warming due to the presence of CFCs is partially or completely offset by the cooling resulting from O_3 losses in the lower stratosphere (45). Three important features are evident from Table 8. First, today CO_2 is by far the most abundant anthropogenic greenhouse gas in the atmosphere. Second, all of the other trace greenhouse gases are much more efficient absorbers of ir radiation than CO_2. Third, most of the gases have very long atmospheric residence times so that even if emissions were to cease, the gases would remain in the atmosphere for decades (some for centuries). The principal sources of greenhouse gases, as they are understood today, are summarized in Table 9.

Table 9. Principal Sources of Greenhouse Gases[a]

Gases	Principal Sources
CO_2	Fossil fuel combustion, deforestation oceans, respiration
CH_4	Wetlands, rice paddies, enteric fermentation (animals), biomass burning, termites
N_2O	Natural soils, cultivated and fertilized soils, oceans, fossil fuel combustion
O_3	Photochemical reactions in the troposphere, transport from stratosphere
CFC-11	Manufacturing of foam, aerosol propellant
CFC-12	Refrigerant, aerosol propellant, manufacturing of foams
HCFC-22	Refrigerant, production of fluoropolymers
CFC-113	Electronics solvent
CH_3CCl_3	Industrial degreasing solvent
CCl_4	Intermediate in production of CFC-11, -12, solvent

[a] Ref. 44.

From the analyses of air trapped in Antarctic and Greenland ice, the concentrations of greenhouse gases (except for O_3) in the preindustrial atmosphere (averaged over ~ 1000 years) can be estimated quite accurately. The enhancement of the greenhouse effect due to current con-

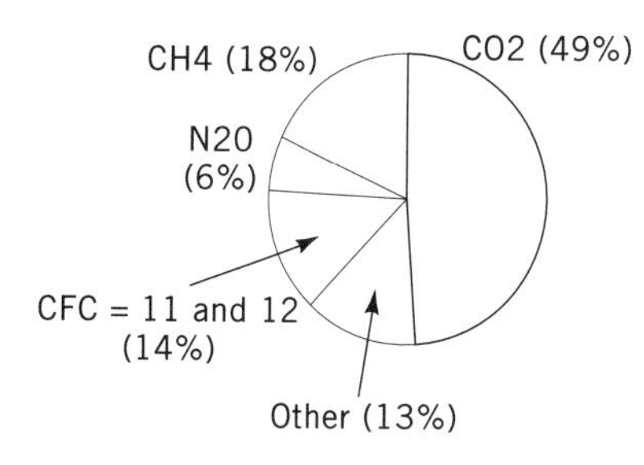

Figure 7. Estimates of greenhouse-gas contributions to global warming in the 1980s.

centrations of greenhouse gases relative to preindustrial concentrations is called the "enhanced greenhouse effect" or radiative forcing. Using a radiative convective model, the contributions from the various greenhouse gases to the radiative forcing in the 1980s can be estimated (6). Such estimates are shown in Figure 7. At present, CO_2 accounts for about half of the radiative forcing. The relative contribution from CO_2 has been shrinking and will continue to do so because other species that are much more efficient ir absorbers are increasing in concentration at a faster relative rate than CO_2.

Although there is no doubt that greenhouse gas concentrations and the radiative forcing are increasing, there is no unequivocal evidence to suggest that the forcing is actually causing a net warming of the earth. Analyses of global temperature trends since the 1860s show that the global temperature has increased about 0.5–0.7°C (47,48), but this number decreases to ~ 0.5°C when corrections for heat-island effects are considered (49). This is in reasonable agreement with modeling results which predict a temperature increase of ~ 1°C (50). However, it has been shown that most of the temperature increase occurred prior to the big increase in CO_2 (51). A detailed analysis of global temperature and CO_2 concentration time-series over the period of atmospheric measurements of CO_2 (1958–1988) shows an excellent positive correlation between the two variables, but the CO_2 changes lag temperature change by an average of 5 months (52). Thus, although there is strong evidence linking temperature and CO_2 changes, the cause and effect has not only not been demonstrated, but it is not clear which is the cause and which is the effect. The lack of a definitive relationship may also be obscured by changes in other factors that affect the earth's heat budget, such as increased atmospheric aerosols or cloud cover as well as natural climatic cycles.

Predictions of future temperature changes due to increased greenhouse gas forcing are made using global circulation models (GCMs). The GCMs are sophisticated, but incomplete models that incorporate expressions for the basic physical processes that govern the dynamics of the atmosphere and allow for some atmosphere–ocean interactions. Depending upon which model is used and the various assumptions incorporated into the model, the models predict that the average global temperature will increase from 2.0 to 5.5°C with a doubling of CO_2 concentrations from preindustrialized levels (53,54). These temperature increases will not be uniformly distributed. Although temperature change predictions from the models are in fairly good agreement, the uncertainties are considered to be large because many of the important feedback processes involving the oceans and clouds are not adequately incorporated into the models because many of them are poorly understood. Consequently, the models are an indication that global warming could occur, but the results are not definitive.

Stratospheric O_3 Depletion. In the stratosphere, O_3 is formed naturally when O_2 is dissociated by ultraviolet (uv) solar radiation in the wavelength (λ) region 180–240 nm:

$$O_2 + uv \rightarrow O + O \quad (24)$$

$$O + O_2 + M \rightarrow O_3 + M \quad (25)$$

where M is any third body molecule (most likely N_2 or O_2 in the atmosphere) that remains unchanged in the reaction. Uv radiation in the 200–300 nm λ region can also dissociate the O_3:

$$O_3 + uv \rightarrow O_2 + O \quad (26)$$

In this last reaction, O_3 is responsible for the removal of uv-B radiation (λ = 280–330nm) that would otherwise reach the earth's surface. The concern is that any process that depletes stratospheric ozone will increase the uv-B (in the 293–320 nm region) reaching the surface. Increased uv-B will lead to increased incidence of skin cancer and could have deleterious effects on certain ecosystems. The first concern over O_3 depletion was from NO_x emissions from a fleet of supersonic transport aircraft that would fly through the stratosphere and cause (55):

$$NO + O_3 \rightarrow NO_2 + O_2 \quad (27)$$

$$NO_2 + O \rightarrow NO + O_2 \quad (28)$$

The net effect of this sequence is the destruction of 2 molecules of O_3 since the O would have combined with O_2 to form O_3. In addition, the NO acts as a catalyst because it is not consumed, and therefore can participate in the reaction sequence many times.

In the mid-1970s, it was realized that the chlorofluorocarbons (CFCs) in widespread use because of their chemical inertness, would diffuse unaltered through the troposphere into the mid-stratosphere where they would be photolyzed by uv (λ < 240 nm) radiation. For example, CFC-12 would photolyze forming Cl and ClO radicals:

$$CF_2Cl_2 + uv \rightarrow CF_2Cl + Cl \quad (29)$$

$$CF_2Cl + O_2 \rightarrow CF_2O + ClO \quad (30)$$

Table 10. Ozone Depletion Potential Relative to CFC-11[a]

Compound	CAS Registry Number	Relative Ozone Depletion Potential
CFC-11	[*75-69-11*]	1.0
CFC-12	[*75-71-8*]	0.87
CFC-113	[*76-13-1*]	0.76
CFC-114		0.56
CFC-115	[*76-15-3*]	0.27
HCFC-22	[*75-45-6*]	0.043
HCFC-123	[*306-83-2*]	0.016
HFCF-124	[*2837-89-0*]	0.017
HCFC-125	[*354-33-6*]	0
HFC-134a	[*811-97-2*]	0
CCl_4	[*56-23-5*]	1.1
CH_3CCl_3	[*71-55-6*]	0.12

[a] Data from Ref. 57. Results are based on a Lawrence Livermore national laboratory 2-dimensional model.

The following reactions would then occur:

$$Cl + O_3 \rightarrow ClO + O_2 \quad (31)$$

$$ClO + O \rightarrow Cl + O_2 \quad (32)$$

In this sequence the Cl also acts as a catalyst, and two O_3 molecules are destroyed. Before the Cl is finally removed from the atmosphere in 1–2 years by precipitation, each Cl atom will have destroyed approximately 100,000 O_3 molecules (56). An estimated O_3-depletion potential of some common CFCs, HFCs (hydrofluorocarbons), and HCFCs (hydrochlorofluorocarbons) are presented in Table 10. The O_3-depletion potential is defined as the ratio of the emission rate of a compound required to produce a steady-state O_3 depletion of 1% to the amount of CFC-11 required to produce a 1% depletion. Another class of compounds, halons, are also ozone-depleting compounds. The halons are bromochlorofluorocarbons or bromofluorocarbons that are widely used in fire extinguishers. Although their emissions and thus their atmospheric concentrations are much lower than the most common CFCs, they are of concern because they are 3 to 10 times more destructive to O_3 than the CFCs.

Evidence that stratospheric O_3 depletion is occurring comes from the discovery of the Antarctic O_3 hole. In recent years during the spring, O_3 depletions of 60% integrated over all altitudes and 95% in some layers have been observed over Antarctica. During winter in the southern hemisphere, a polar vortex develops which prevents the air from outside of the vortex from mixing with air inside the vortex. The depletion begins in August, as the approaching spring sun penetrates into the polar atmosphere, and extends into October. When the hole was first observed, existing chemical models could not account for the rapid O_3 loss. Attention was soon focused on stable reservoir species for chlorine, namely HCl and $ClNO_3$. These species are formed in competing reactions involving Cl and ClO that temporarily or permanently remove Cl and ClO from participating in the O_3 destruction reactions. Two important reactions are the following:

$$Cl + CH_4 \rightarrow HCl + CH_3 \quad (33)$$

$$ClO + NO_2 + M \rightarrow ClNO_3 + M \quad (34)$$

Within the polar vortex, temperatures as low as −90°C allow the formation of polar stratospheric ice clouds. On the surfaces of the ice particles that compose these clouds, heterogeneous reactions occur which break down the reservoir species HCl and $ClNO_3$. Two important reactions are the following:

$$ClNO_3 + HCl(s) \rightarrow Cl_2 + HNO_3(s) \quad (35)$$

$$H_2O(s) + ClNO_3 \rightarrow HOCl + HNO_3(s) \quad (36)$$

During the polar winter night, Cl_2, HOCl and HNO_3(s) accumulate. When sunlight returns to the polar regions, the chlorine compounds are photolyzed, producing Cl and ClO. Nitrogen oxides remain sequestered, and without NO_2 to deplete ClO, massive O_3 destruction occurs until the polar vortex dissipates later in the spring and the mixing of air from lower latitudes occurs.

Other data show that globally, stratospheric O_3 concentrations have declined during the winter, spring, and summer in both the northern and southern hemispheres at middle and high latitudes (45). Declines were most evident during winter months (58,59).

In 1976, the United States banned the use of CFCs as aerosol propellants. No further steps were taken until 1987 when the United States and some 50 other countries adopted the Montreal Protocol which specifies a 50% reduction of fully halogenated CFCs by 1999. Since then, however, because of the Antarctic ozone hole and the observed global decreases in stratospheric ozone, there has been increased support for a faster phaseout. In 1990, an agreement was reached among 93 nations to accelerate the phaseout and completely eliminate the production of CFCs by the year 2000. The 1990 Clean Air Act Amendments contain a phaseout schedule for CFCs, halons, carbon tetrachloride, and methyl chloroform. Such steps will stop the increase of CFCs in the atmosphere but, because of their long lifetimes, they will remain in the atmosphere for centuries.

Indoor Air Pollution

Indoor air pollution is simply the presence of air pollutants in indoor air. The focus of this section is on air in residential buildings as opposed to the industrial environment which would be covered under industrial hygiene. The concentrations of indoor pollutants depend upon the strength of the indoor sources of the pollutants as well as the ventilation rate of the building and the outdoor concentrations of the pollutants. In response to the energy crisis in the early 1970s, new buildings were constructed more airtight. Unfortunately, airtight structures created a setting conducive to the accumulation of indoor air pollutants. Numerous sources and types of pollutants found indoor can be classified into seven categories: tobacco smoke, radon, emissions from building materials, combustion products from inside the building, pollutants which infiltrate from outside the building, emissions from products used within the home, and biological pollutants.

Tobacco smoke contains a variety of air pollutants. In a survey of 80 homes in an area where the outdoor TSP varied between 10–30 $\mu g/m^3$, the indoor TSP was the same, or less, in homes with no smokers. In homes with one smoker, the TSP levels were between 30–60 $\mu g/m^3$, in homes with two or more smokers, the levels were between 60–120 $\mu g/m^3$ (60). In other studies, indoor TSP levels exceeding 1000 $\mu g/m^3$ have been found in homes with numerous smokers. In addition to TSP, burning tobacco emits CO, NO_x, formaldehyde, benzopyrene, nicotine, phenols, and some metals such as cadmium and arsenic (61).

Radon-222 (Rn) is a naturally occurring, inert, radioactive gas formed from the decay of radium-226 (Ra). Because Ra is a ubiquitous, water-soluble component of the earth's crust, its daughter product, Rn, is found everywhere. Although Rn receives all of the notoriety, the principal health concern is not with Rn itself, but with its alpha (α) particle-emitting daughters (radioactive decay products). Since Rn is an inert gas, inhaled Rn will not be retained in the lungs. With a half-life of 4 days, Rn decays to polonium-218 (Po-218) with the emission of an α particle. It is Po-218, an α emitter with a half-life of 3 minutes, and Po-214, also an α-emitter with a half-life of 1.6×10^{-4} seconds, that are of most concern. Po-218 decays to lead-214 (a β-emitter with a $t_{1/2}$ = 27 minutes), which decays to bismuth-214 (a β-emitter with a $t_{1/2}$ = 20 minutes), which decays to Po-214. When inhaled, the Rn daughters, either by themselves or attached to an airborne particle, are retained in the lung and the subsequent α emission irradiate the surrounding lung tissue. Rn can enter buildings through emissions from soil, water, or construction materials. By far, the soil route is the most common, and construction material the least common source of Rn contamination (there have been isolated incidents where construction materials contained high levels of Ra). The emission rate of Rn depends on the concentration of Ra in the soil, the porosity of the soil, and the permeability of the building's foundation. For example, Rn will transport faster through cracks and sumps in the basement floor than through concrete. In the ambient air, Rn concentrations are typically 0.25–1.0 picoCurries per liter (pC/L), while the mean concentration in U.S. residences is about 1.2 pC/L (62). However, it is estimated that there are 1 million residences that have concentrations exceeding 8 pC/L, which is the action level for remedial action recommended by the National Council on Radiation Protection and Measurements (62). The highest values ever measured in U.S. homes exceeded 1000 pC/L (63). Remedial action consists of (*1*) reducing the transport of Rn into the building by sealing cracks with impervious fillers and installing plastic or other barriers that have been proven effective; (*2*) removing the daughters from the air by filtration; and (*3*) increasing the infiltration of outside air with an air-exchanger system.

Of the pollutants emitted from construction materials within the home, asbestos [*1332-21-4*] has received the most attention. Asbestos is a generic term for a number of naturally occurring fibrous, hydrated silicates. By EPA's definition, a fiber is a particle that possess a 3:1 or greater aspect ratio (length:diameter). The family of asbestos minerals is divided into two types: serpentine and amphibole. One type of serpentine, chrysotile [$Mg_6Si_4O_{10}(OH)_8$], accounts for 90% of the world's asbestos production. The balance of the world's production is accounted for by two of the amphiboles: amosite [$Fe_5Mg_2(Si_8O_{22})(OH)_2$] and crocidolite [$Na_2(Fe^{+3})_2(Fe^{+2})_2Si_8O_{22}(OH)_2$]. Three other amphiboles, anthophyllite [$(Mg,Fe)_7Si_8O_{22}(OH)_2$], tremolite [$Ca_2Mg_5Si_8O_{22}(OH)_2$], and actinolite [$Ca_2(Mg,Fe)_5Si_8O_{22}(OH)_2$], have been only rarely mined. The asbestos minerals differ in morphology, durability, range of fiber diameters, surface properties, and other attributes that determine uses and biological effects. Known by ancients as the magic mineral because of its ability to be woven into cloth, its physical strength, and its resistance to Fire, enormous heat, and chemical attack, asbestos was incorporated into many common products including roofing materials, wallboard, insulation, spray-on fireproofing and insulating material, floor tiles, pipes, filters, draperies, pot holders, brake linings, etc (64). In the 1940s and 1950s however, evidence accumulated linking exposure to the airborne fibers with asbestosis (pulmonary interstitial fibrosis), lung cancer, and mesothelioma (a rare form of cancer of the lung or abdomen). Although all forms of asbestos were implicated in the early studies, very recent studies indicate that most of the asbestos-related diseases are due to exposure to airborne amphiboles rather than the most common type, chrysotile, and to fibers greater than or equal to 5 μm in length (65). In the 1970s, the spray-on application of asbestos was banned and substitutes were found for many products. Nevertheless, asbestos was used liberally in buildings for several decades, and many of them are still standing. Asbestos in building materials does not spontaneously shed fibers, but when the materials become damaged by normal decay, renovation, or demolition, the fibers can become airborne. When such situations arise, specific procedures should be followed to contain and remove the damaged materials.

Another important pollutant emanating from building material is formaldehyde (HCHO). Formaldehyde is important because of its irritant effects and its suspected carcinogenicity. Although traces of formaldehyde can be found in the air in virtually every modern home, mobile homes and homes insulated with urea–formaldehyde foam have the highest concentrations. Higher emissions can occur in mobile homes using particle boards which are held together with an urea–formaldehyde resin. This can also be a problem in a conventional home, but it is exacerbated in a mobile home because of the usually low rate of air exchange in a mobile home. Plywood is also a source of formaldehyde as the layers of wood are held together in a similar urea–formaldehyde resin adhesive. In general, however, particle board contains more adhesive per unit mass, so the emissions are greater. Urea foam is an efficient insulation material that can be injected into the sidewalls of conventional homes. Production of the foam peaked in 1977, when about 170,000 homes were insulated. When improperly formulated or installed, the foam can emit significant amounts of formaldehyde. In 1982, the use of the foam was banned in the United States. Other sources of formaldehyde indoors are paper products, carpet backing, and some fabrics.

Whenever unvented combustion occurs indoors or when venting systems attached to combustion units malfunction, a variety of combustion products will be released to the indoor environment. Indoor combustion units include stoves and ovens (except electric), furnaces, hot-water heaters, space heaters, and wood-burning fireplaces or stoves. Products of combustion include CO, NO, NO_2, fine particles, aldehydes, polynuclear aromatics, and other organic compounds. Especially dangerous sources are unvented gas and kerosene space heaters which discharge pollutants directly into the living space. The best way to prevent the accumulation of combustion products indoors is to make sure all units are properly vented and properly maintained.

Pollutants from outdoors can also be drawn inside under certain circumstances. Incorrectly locating an air intake vent downwind of a combustion exhaust stack can cause this condition. High outdoor pollutant concentrations can infiltrate buildings. Unreactive pollutants such as CO will diffuse through any openings in the building and pass unaltered through any air-intake system. Given

Table 11. Types of Emissions of Indoor Air Pollutants Associated with Various Activities and Consumer Products

Activity or Product	Intentional Aerosol Production	Unintentional Aerosol Production	Evaporation or Sublimation	Unintentional Outgassing
Cleaning	X	X	X	X
Painting	X		X	X
Polishing	X		X	X
Stripping	X	X	X	
Refinishing	X		X	X
Hobbies, Crafts	X	X	X	X
Deodorizer	X		X	
Insecticide	X		X	
Disinfectant	X		X	
Personal grooming product	X		X	X
Plastic				X

From Ref. 67.

sufficient time, the indoor/outdoor ratio for CO will approach 1.0 if outside air is the only source of CO. For reactive species such as ozone, which is destroyed on contact with most surfaces, the indoor/outdoor ratio is usually around 0.5, but this ratio will vary considerably depending on the ventilation rate and the internal surface area within the building (66).

Air contaminants are emitted to the indoor air from a wide variety of activities and consumer products. Some of these are summarized in Table 11. It is obvious from this list that most indoor activities produce some types of pollutants. When working with these products or engaging in these activities, care should be exercised to minimize exposures by proper use of the products and by providing adequate ventilation.

Biological air pollutants found indoors include airborne bacteria, viruses, fungi, spores, molds, algae, actinomycetes, and insect and plant parts. Many of the microorganisms multiply in the presence of high humidity. The microorganisms can produce infections, disease, or allergic reactions, while the nonviable biological pollutants can produce allergic reactions. The most notable episode was the 1976 outbreak of Legionella (Legionnaires') disease in Philadelphia, where American Legion convention attendees inhaled Legionella virus from a contaminated central air-conditioning system. A similar incident in an industrial environment occurred in 1981 when more than 300 workers came down with "Pontiac fever" as a result of inhalation exposure to a similar aerosolized virus from contaminated machining fluids (68). Better preventative maintenance of air management systems and increased ventilation rates reduce the concentrations of all species, and this should reduce the incidence of adverse affects.

BIBLIOGRAPHY

1. *Procedures for Emission Inventory Preparation,* Volumes I–IV, Publication No. EPA 450/4-81-026A-E, U.S. Environmental Protection Agency, Research Triangle Park, N.C., 1981.
2. *Compilation of Air Pollution Emission Factors,* Publication No. AP-42, 5th ed., U.S. Environmental Protection Agency, Research Triangle Park, N.C., 1989.
3. *Compilation of Air Pollution Emission Factors,* Volume II, Mobile Sources, Publication No. AP-42, 5th ed., U.S. Environmental Protection Agency, Research Triangle Park, N.C., 1989.
4. *National Air Quality & Emissions Trends Report, 1991,* Publication No. EPA-450-R-92-001, U.S. Environmental Protection Agency, Research Triangle Park, N.C., 1992.
5. *National Air Pollutant Emissions Estimates 1940-1986,* Publication No. EPA-450/4-87-024, U.S. Environmental Protection Agency, Research Triangle Park, N.C., 1988.
6. V. Ramanathan and co-workers, *Review Geophys.* **25,** 1441 (1987).
7. P. Warneck, *Chemistry of Natural Atmospheres,* Academic Press, New York, 1988, p. 367.
8. Intergovernmental Panel on Climate Change, *Scientific Assessment of Climate Change, Section 2, Radiative Forcing of Climate,* United Nations, New York, 1990, p. 14.
9. G. T. Wolff and P. J. Lioy, *Environ. Sci. Technol.* **14,** 1257 (1980).
10. G. T. Wolff, N. A. Kelly, and M. A. Ferman, *Science* **311,** 703 (1981).
11. J. P. Lodge ed., *Methods of Air Sampling and Analysis,* Lewis Publishers, Chelsea, Mich., 1989, 763 pp.
12. A. M. Hough and R. G. Derwent, *Nature* **344,** 645 (1990).
13. N. A. Kelly, G. T. Wolff, M. A. Ferman, *Atmos. Environ.* **16,** 1077 (1978).
14. W. Johnson and W. Viezee, *Atmos. Environ.* **15,** 1309 (1981).
15. W. Attmannspacher and R. Hartmannsgruber, *Pure Appl. Geophysics.* **106–108,** 1091 (1973).
16. R. L. Seila, W. A. Lonneman, and S. A. Meeks, *Determination of C_2 to C_{12} Ambient Air Hydrocarbons in 39 U.S. Cities from 1984 through 1986,* Publication No. EPA/600/3-89/058, U.S. Environmental Protection Agency, Research Triangle Park, N.C., (1989).
17. P. Warneck, *Chemistry of Natural Atmospheres,* Academic Press, New York, 1988, pp. 721–729.
18. W. N. Stasiuk, Jr. and P. E. Coffey, *J. Air Pollut. Control Assoc.* **24,** 564 (1974).
19. G. T. Wolff and co-workers, *Environ. Sci. Technol.* **11,** 506 (1977).
20. A. M. Dunker, S. Kumar, and P. H. Berzins, *Atmos. Environ.* **18,** 311 (1984).

21. N. A. Kelly and R. G. Gunst, *Atmos. Environ.* **24A,** 2991 (1990).
22. *Catching Our Breath. Next Steps for Reducing Urban Ozone,* U.S. Office of Technology Assessment, Washington, D.C., 1989, pp. 101–102.
23. G. T. Wolff and co-workers, *J. Air Pollut. Control Assoc.* **27,** 460 (1977).
24. *An Acid Aerosols Issue Paper,* Publication No. EPA/600/8-88/005F, U.S. Environmental Protection Agency, Washington, D.C., 1989.
25. G. T. Wolff, *Annals NY Acad. Sci.* **338,** 379 (1980).
26. K. Willeke and K. T. Whitby, *J. Air Pollut. Control Assoc.* **25,** 529 (1975).
27. R. J. Countess, G. T. Wolff, and S. H. Cadle, *J. Air Pollut. Control Assoc.* **30,** 1195 (1980).
28. G. T. Wolff and co-workers, *Atmos. Environ.* **19,** 305 (1985).
29. G. T. Wolff and co-workers, *J. Air Pollut. Control Assoc.* **32,** 1216 (1982).
30. G. T. Wolff and co-workers, *Atmos. Environ.* **20,** 1229 (1990).
31. M. A. Ferman, G. T. Wolff, and N. A. Kelly, *J. Air Pollut. Control Assoc.,* **31,** 1074 (1981).
32. G. T. Wolff and co-workers, *J. Air Pollut. Control Assoc.* **36,** 585 (1986).
33. G. T. Wolff and co-workers, *Atmos. Environ.* **20,** 1229 (1986).
34. M. S. Miller, S. K. Friedlander, and G. M. Hidy, G. M. Hidy, *Aerosol and Atmospheric Chemistry,* Academic Press, New York, 1972, pp. 301–312.
35. P. W. Fisher, R. M. Currie, and R. J. Churchill, *J. Air Pollut. Control Assoc.* **38,** 1376 (1988).
36. *Threshold Limit Values and Biological Exposure Indices,* American Conference of Governmental and Industrial Hygienists, Cincinnati, Ohio, 1989, p. 124.
37. N. I. Sax, *Dangerous Properties of Industrial Materials,* Van Nostrand Reinhold, New York, 1979, p. 1108.
38. J. A. Cannon, *J. Air Pollut. Control Assoc.* **36,** 562 (1986).
39. J. C. Mesta, in P. S. Bhardwaja, ed., *Visibility Protection Research and Policy Aspects,* Air and Waste Management Association, Pittsburgh, Pa. 1987, pp. 1–8.
40. W. E. Middleton, *Vision Through the Atmosphere,* University of Toronto Press, Canada, 1968.
41. H. Koschmieder, *Beitr. Phys. Frein Atm.* **12,** 33 (1924).
42. P. J. Groblicki, G. T. Wolff, and R. J. Countess, *Atmos. Environ.* **15,** 2473 (1981).
43. D. Albritton and co-workers, *NAPAP Interim Assessment: Atmosphere Process and Deposition,* Vol. 2, National Acid Precipitation Assessment Program, Washington, D.C., 1987.
44. *Policy Options for Stabilizing Global Climate,* U.S. Environmental Protection Agency, Washington, D.C., 1990.
45. *Scientific Assessment of Ozone Depletion: 1991,* World Meteorological Organization/United Nations Environment Programme, 1991.
46. J. Hansen and co-workers, *J. Geophs. Res.* **93,** 9341 (1988).
47. P. D. Jones, T. M. L. Wigley, and P. B. Wright, *Nature* **322,** 430 (1986).
48. J. Hansen and S. Lebedeff, *J. Geophys. Res.* **D11,** 13, 345 (1987).
49. T. R. Karl and P. D. Jones, *Bull. Amer. Meteor Soc.* **70,** 265 (1989).
50. V. Ramanathan, *Science* **240,** 293 (1988).
51. R. S. Lindzen, *Bull. Amer. Meteorol Soc.,* **71,** 288 (1990).
52. C. Kuo, C. Lindberg and D. J. Thomson, *Nature* **343,** 709 (1990).
53. S. H. Schneider, *Scientific Amer.* **261,** 70 (1989).
54. J. F. B. Mitchell, C. A. Senior, and W. J. Ingram, *Nature* **341,** 132 (1989).
55. P. J. Crutzen, *Quart J. Royal Meteorol Soc.* **96,** 320 (1970).
56. M. J. Molina and F. S. Rowland, *Nature* **249,** 810 (1974).
57. D. A. Fisher and co-workers, *Nature* **344,** 508 (1990).
58. M. B. McElroy and R. J. Salawitch, *Science* **243,** 763 (1989).
59. F. S. Rowland, *Amer. Scientist* **77,** 36 (1989).
60. J. D. Spengler and co-workers, *Atmos. Environ.* **15,** 23 (1981).
61. California Department of Consumer Affairs, *Clean Your Room, Compendium on Indoor Air Pollution,* Sacramento, Calif. 1982, pp. III.Ei–III.E.II.
62. Mueller Associates, Inc., Syscon Corporation and Brookhaven National Laboratory, *Handbook of Radon in Buildings,* Hemisphere Publishing Corporation, New York, 1988, p. 95.
63. H. W. Alter and R. A. Oswald, *J. Air Pollut. Control Assoc.* **37,** 227 (1987).
64. P. Brodeur, *New Yorker* **44,** 117 (1968).
65. B. T. Mossman and co-workers, *Science* **247,** 294 (1990).
66. C. J. Weschler, H. C. Shields, and D. V. Naik, *J. Air Waste Manage. Assoc.* **39,** 1562 (1989).
67. *Indoor Air Pollutants,* National Academy Press, Washington, D.C., 1981, p. 101.
68. L. A. Herwaldt and co-workers, *Ann. Intern. Med.* **100,** 333 (1984).

General References

References 4, 5, 6, 7, 8, 11, 44, 45, 53, 58, 61, and 67 and the following books and reports constitute an excellent list for additional study. Reference 7 is an especially useful resource for global atmospheric chemistry.

J. H. Seinfeld, *Atmospheric Chemistry and Physics of Air Pollution,* John Wiley & Sons, Inc., New York, 1986.

B. J. Finlayson-Pitts and J. N. Pitts, Jr., *Atmospheric Chemistry Fundamentals and Experimental Techniques,* John Wiley & Sons, Inc., New York, 1986.

T. E. Graedel, D. T. Hawkins, and L. D. Claxton, *Atmospheric Chemical Compounds Sources, Occurrence and Bioassay,* Academic Press, New York, 1986.

Air Quality Criteria for Ozone and Other Photochemical Oxidants, Publication No. EPA/600/8-84-020F (5 Volumes), U.S. Environmental Protection Agency, Research Triangle Park, N.C., 1986.

EPA publishes separate Criteria Documents for all the Criteria Pollutants, and they are updated about every five years.

Atmospheric Ozone 1985, World Meteorological Organization, Geneva, Switzerland (3 volumes)_an excellent compendium on tropospheric and stratospheric processes.

G. T. Wolff, J. L. Hanisch, and K. Schere, eds., *The Scientific and Technical Issues Facing Post-1987 Ozone Control Strategies.* Air and Waste Management Association, Pittsburgh, Pa., 1988.

J. H. Seinfeld, "Urban Air Pollution: State of the Sciences," *Science* **243,** 745 (1989).

S. H. Schneider, "The Greenhouse Effect: Science and Policy," *Science,* **243,** 771 (1989).

AIR POLLUTION: AUTOMOBILE

MICHAEL P. WALSH
Arlington, Virginia

Across the entire globe, motor vehicle use has increased tremendously. In 1950, there were about 53 million cars on the world's roads; only four decades later, the global automobile fleet is over 430 million, more than an eight-fold increase. On average, the fleet has grown by about 9.5 million automobiles per year over this period. Simultaneously, the truck and bus fleet has been growing by about 3.6 million vehicles per year (1). While the growth rate has slowed in the highly industrialized countries, population growth and increased urbanization and industrialization are accelerating the use of motor vehicles elsewhere. If the approximately 100 million two-wheeled vehicles around the world is included (growing at about 4 million vehicles per year over the last decade), the global motor vehicle fleet is now approximately 675 million.

One result is that most of the major industrialized areas of the world have been experiencing serious motor vehicle pollution problems. To deal with these problems North America, Europe, and Japan have developed significant motor vehicle pollution control programs that have led to tremendous advances in petrol car control technologies. At present, similar technologies are under intensive development for diesel cars and trucks and significant breakthroughs are starting to appear with production diesel vehicles. See also AIR POLLUTION: AUTOMOBILE, TOXIC EMISSIONS.

Motor vehicle-related air pollution problems are not limited to the OECD countries. Areas of rapid industrialization are now starting to note similar air pollution problems to those of the industrialized world. Cities such as Mexico, Delhi, Seoul, Singapore, Hong Kong, Sao Paulo, Manila, Santiago, Bangkok, Taipei, and Beijing already experience unacceptable air quality or are projecting that they will in the relatively near future. See also AIR QUALITY MODELING.

The purpose of this article is to survey what is presently known about transportation related air pollution problems, to summarize the adverse impacts which result, to review actions under way or planned to address these problems, and to estimate future trends. Recent technological developments will also be summarized.

THE PROBLEM

Motor vehicles emit large quantities of carbon monoxide, hydrocarbons, nitrogen oxides, and such toxic substances as fine particles and lead. Each of these can cause adverse effects on health and the environment. Because of the growing vehicle population and the high emission rates, serious air pollution problems have been an increasingly common phenomena in modern life. Initially, these problems were most apparent in center cities but recently lakes and streams and even remote forests have experienced significant degradation as well. As more and more evidence of human impacts on the upper atmosphere accumulates, concerns are increasing that motor vehicles are contributing to global changes that could alter the climate of the planet.

Pollutants from vehicles are not an academic concern; they impair health, destroy vegetation, and inhibit the general quality of life. In regard to health especially, their effects are most pronounced in the very old, the very young, and those least able to cope.

Many sources contribute to air pollution, but the motor vehicle has been singled out as especially serious. In large part, this is due to the great number of vehicles in congested urban areas and the overall volume of pollutants which they emit. The problem is compounded by the fact that vehicles emit pollutants in close proximity to the breathing zones of people.

Until recently most of the air pollution problems were considered local in nature, but over the last two decades the evidence has been increasing that some of the most severe impacts may occur over large distances and over long periods of time with the effect far removed from the source. A comprehensive look at air pollution today must consider localized adverse impacts, regional and continental effects, and even global changes.

Urban

Carbon Monoxide. Over 90% of the carbon monoxide emitted in cities generally comes from motor vehicles. Numerous studies in humans and animals have now demonstrated that those individuals with weak hearts are placed under additional strain by the presence of excess CO in the blood. In addition, fetuses, sickle cell anemics, and young children are also especially susceptible to exposure to low levels of CO. Also, evidence indicates that CO may contribute to elevated levels of tropospheric ozone.

Oxides of Nitrogen. NO_x emissions from vehicles and other sources produce a variety of adverse health and environmental effects. They react chemically with hydrocarbons to form ozone and other highly toxic pollutants. Next to sulfur dioxide, NO_x emissions are the most prominent pollutant contributing to acidic deposition. Direct exposure to nitrogen dioxide (NO_2) leads to increased susceptibility to respiratory infection, increased airway resistance in asthmatics, and decreased pulmonary function. Short-term exposures to NO_2 have resulted in a wide ranging group of respiratory problems in school children (cough, runny nose, and sore throat are among the most common) as well as increased sensitivity to urban dust and pollen by asthmatics. Some scientists believe that NO_x is a significant contributor to the dying forests throughout central Europe.

Lead. Because lead is added to petrol, motor vehicles have been the major source of lead in the air of most cities. Several studies have now shown that children with high levels of lead accumulated in their baby teeth experience more behavioral problems, lower IQs, and decreased ability to concentrate. Most recently, in a study of 249 children from birth to 2 yr, it was found that those with prenatal umbilical cord blood lead levels at or above 10 μg/dL consistently scored lower on standard intelligence tests than those at lower levels.

Diesel Particulate. Uncontrollable diesels emit approximately 30–70 times more particulate than gasoline-fueled engines equipped with catalytic converters and burning unleaded fuel. These particles are small and respirable (less than 2.5 μm) and consist of a solid carbonaceous core on which myriad compounds adsorb. These include

- Unburned hydrocarbons.
- Oxygenated hydrocarbons.
- Polynuclear aromatic hydrocarbons.
- Inorganic species such as sulfur dioxide, nitrogen dioxide, and sulfuric acid.

It is now well established that diesel particulate represents a serious health hazard in many urban areas around the world. Cities as diverse as Santiago, Taipei, Mexico City, Manila, Bangkok, Seoul, and Jakarta stand out as experiencing particularly high levels. The World Health Organization has concluded that diesel particulate is a probable human carcinogen. Several human epidemiological studies seemed to point to this conclusion. "In the two most informative cohort studies (of railroad workers), one in the USA and one in Canada, the risk for lung cancer in those exposed to diesel engine exhaust increased significantly with duration of exposure in the first study and with increased likelihood of exposure in the second."

Other studies focusing on noncancer health effects have raised equally alarming concerns. By correlating daily weather, air pollutants, and mortality in five U.S. cities, scientists have discovered that nonaccidental death rates tend to rise and fall in near lockstep with daily levels of particulates—more than with other pollutants. Because the correlation held up even for low levels (in one city, with just 23% of the U.S. limit on particulates), these analyses suggest that as many as 60,000 U.S. residents per year may die from breathing particulates at or below legally allowed levels. Confirmation of the new findings by other researchers would make airborne particulate levels the largest known "involuntary environmental insult" to which Americans are exposed (2).

In Germany, the diesel engine has enjoyed some of its greatest success. Based on its conclusion that diesel particulate reflects an occupational health hazard, the German government has issued guidelines intended to discourage the use of diesel engines in occupational settings when possible (eg, substitute electric fork lifts for diesels) or to reduce emissions to the greatest extent feasible (eg, by using particulate traps or filters).

Other Toxics. The 1990 Clean Air Act (CAA) directed the EPA to complete a study of emissions of toxic air pollutants associated with motor vehicles and motor vehicle fuels. The study found that in 1990, the aggregate risk is 720 cancer cases. For all years, 1,3-butadiene is responsible for the majority of the cancer incidence, ranging from 58 to 72% of the total. This is due to the high unit risk of 1,3-butadiene. Gasoline and diesel particulate matter, which are considered to represent motor vehicle polycyclic organic matter (POM) in this report, are roughly equal contributors to the risk. The combined risk from gasoline and diesel particulate matter ranges from 16 to 28% of the total, depending on the year examined. Benzene is responsible for roughly 10% of the total for all years. The aldehydes, predominately formaldehyde, are responsible for roughly 4% of the total for all years.

A variety of studies have found that in individual metropolitan areas, mobile sources are one of the most important and possibly the most important source category in terms of contributions to health risks associated with air toxics. For example, according to the EPA, mobile sources may be responsible for almost 60% of the air pollution–related cancer cases in the United States per year.

Regional

Tropospheric Ozone. The most widespread air pollution problem in areas with temperate climates is ozone, one of the photochemical oxidants that results from the reaction of nitrogen oxides and hydrocarbons in the presence of sunlight. Many individuals exposed to ozone suffer eye irritation, cough and chest discomfort, headaches, upper respiratory illness, increased asthma attacks, and reduced pulmonary function. Numerous studies have also demonstrated that photochemical pollutants inflict damage on forest ecosystems and seriously impact the growth of certain crops.

Acidification. Acid deposition results from the chemical transformation and transport of sulfur dioxide and nitrogen oxides. Evidence indicates that the role of NO_x may be of increasing significance with regard to this problem.

Global

Climate Modification. A significant environmental development during the 1980s was the emergence of global warming or the greenhouse effect as a major international concern. Pollutants associated with motor vehicle use (eg, CO_2, CH_4, N_2O, etc) can increase global warming by changing the chemistry in the atmosphere to reduce the ability of the sun's reflected rays to escape. These greenhouse gases have been shown to be accumulating in recent years.

Stratospheric Ozone Depletion. The release of chlorofluorocarbons (CFCs) used in vehicle air conditioning equipment can destroy the earth's protective ozone shield. During the Antarctic spring, the ozone hole spans an area the size of North America. At certain altitudes over the Antarctic, the ozone is destroyed almost completely. By allowing more ultraviolet radiation to penetrate to the earth's surface, such a loss is expected to increase the incidence of skin cancer and impair the human immune system. Increased ultraviolet radiation might also harm the life-supporting plankton that dwell in the ocean's upper levels, thus jeopardizing marine food chains that depend on the tiny plankton.

Carbon Dioxide. Virtually the entire global motor vehicle fleet runs on fossil fuels, primarily oil. For every gallon of oil consumed by a motor vehicle, about 8.6 kg of carbon dioxide (containing about 2.4 kg of carbon) go directly into the atmosphere. In other words, for a typical fill-up at the service station (estimated at 57 L of gas), about 136 kg of carbon dioxide are eventually released into the atmosphere. Globally, motor vehicles account for about 33% of world oil consumption and about 14% of the world's carbon dioxide emissions from fossil fuel burning.

CFCs. A principal source of CFCs in the atmosphere is motor vehicle air conditioning, and in 1987 approximately 48% of all new cars, trucks, and buses manufactured worldwide were equipped with air-conditioners. (CFCs also are used as a blowing agent in the production of seating and other foam products, but this is a considerably smaller vehicular use.) Annually, about 120,000 t of CFCs are used in new vehicles and in servicing air-conditioners in older vehicles. In all, these uses account for about 28% of global demand for CFC-12. As agreed under the Montreal Protocol, CFCs are to be completely phased out of new vehicles by the turn of the century, a welcome but long overdue step. CFCs emitted over the next decade will cause damage to the planet for the next two centuries.

Motor Vehicles are a Dominant Pollution Source

Because the growth has been so great, motor vehicles are now generally recognized as responsible for more air pollution than any other single human activity. The primary pollutants of concern from motor vehicles include the precursors to ground level ozone, hydrocarbons (HC) and nitrogen oxides (NO_x), and carbon monoxide (CO).

HC, NO_x, and CO in Industrialized Countries. Motor vehicles are the dominant source of these air pollutants in Europe.

> The primary source category responsible for most NO_x emissions is road transportation roughly between 50 and 70 per cent. . . . Mobile sources, mainly road traffic, produce around 50 per cent of anthropogenic VOC emissions, therefore constituting the largest man-made VOC source category in all European OECD countries.

In calendar year 1990, the U.S. EPA estimated that transportation sources were responsible for 63% of the carbon monoxide, 38% of the NO_x, and 34% of the hydrocarbons (HC). Based on data regarding evaporative "running losses," the HC contribution from vehicles may actually be substantially higher.

Beyond the U.S. and Europe, for OECD countries as a whole, motor vehicles are the dominant source of carbon monoxide (66%), oxides of nitrogen (47%), and hydrocarbons (39%).

Selected Developing Countries. While not as well documented, it is increasingly clear that motor vehicles are also the principal source of many of the pollution problems that are plaguing the developing world. By way of examples, the air pollution problems of a few countries are summarized below.

India. The motor vehicle related air pollution problem in India is already severe and worsening. While the problem of diesel smoke and particulate is the most apparent, carbon monoxide, nitrogen dioxide, hydrocarbons, and ozone levels also exceed internationally accepted levels. High leaded petrol levels also exist. The problem continues to worsen as the vehicle population continues to grow and age; vehicles once introduced into use remain active much longer than in highly industrialized areas.

Thailand. The motor vehicle–related air pollution problem in Thailand, especially in regard to diesel particulate, is already severe and worsening. While the problems of carbon monoxide and ambient lead levels also exceed internationally accepted levels, they are apparently stabilizing. The overall vehicle population continues to increase at a rate of about 10% per year. The number of light pickup trucks, vans, and motorcycles are also growing rapidly. The adverse consequences of the pollution are especially severe because the lifestyle and climate is such that public exposure to high pollution levels is great.

Indonesia. The motor vehicle related air pollution problem in Jakarta is especially severe; a combination of densely congested traffic, poor vehicle maintenance, and large numbers of diesels and two-stroke engined (smoky) motorcycles is the cause. While air quality monitoring data are not well documented, the problems of carbon monoxide and ambient lead levels also exceed internationally accepted levels.

Philippines. A large proportion of the vehicles are diesel fueled and most of these, especially the "Jeepneys" emit excessive smoke. Due to a lack of instrumentation, actual air quality data do not exist at present. However, it is generally recognized that particulate, lead, carbon monoxide, and possibly ozone levels exceed internationally accepted levels. Based on analysis of fuel consumption, motor vehicles are estimated to be responsible for approximately 50% of the particulate, 99% of the carbon monoxide, 90% of the hydrocarbons, and 5% of the sulfur dioxide. Fuel quality is poor as reflected by the sulfur content of as much as 1.0 wt % (compared with 0.3 wt % in the United States) and a lead content of up to 1.16 g/L (compared with 0.15 g/L in the European Community).

Mexico City. The number of automobiles in Mexico City has grown dramatically in the past several decades; there were approximately 48,000 cars in 1940, 680,000 in 1970, 1.1 million in 1975, and 3 million in 1985. The Mexican government estimates that fewer than half of these cars are fitted with even modest pollution control devices. In addition, more than 40% of the cars are over 12 years old, and of these, most have engines in need of large repairs. The degree to which the existing vehicles are in need of maintenance is reflected by the results of the "voluntary" inspection and maintenance (I/M) program run by the DF during 1986 through 1988. Of the over 600,000 vehicles tested (209,638 in 1986; 313,720 in 1987; and 80,405 in the first four months of 1988), about 70% failed the gasoline vehicle standards and 85% failed the diesel standards (65HSU).

International Agreements Are Increasing Control Pressure

Historically, the thrust toward vehicle pollution controls has been directed by individual countries acting to address their local air pollution concerns, but there has evolved over the past few years an international focus. This is most evident in the Common Market, of course, where countries have joined together to prevent pollution controls from becoming a barrier to trade. But most recently, there have been a series of international protocols designed to address pollution problems of a global nature. The Montreal Protocol designed to reduce emissions of CFCs has received the most attention in this regard. However, two others are of special interest to the subject of this article.

NO_x Protocol Increases Focus on Vehicle Controls. Following the eighth session of the Working Group on Nitrogen Oxides (February 1988) and a further informal consultation of heads of delegations (April 1988), the draft *Protocol Concerning the Control of Emissions of Nitrogen Oxides or Their Transboundary Fluxes* was developed. The protocol will, as a first step, commit signatories to a "freeze" of their NO_x emissions or transboundary fluxes by 1994 on the basis of 1987 levels. At the same time, parties to the protocol must apply emission standards "based on the best available technologies which are economically feasible" to the new stationary and mobile emission sources and to introduce pollution control measures for principal existing stationary sources. They will also have to make unleaded fuel available in sufficient quantities, at least along main international transit routes, to facilitate the circulation of vehicles equipped with catalytic converters.

The second step to combat NO_x pollution, which should start not later than January 1, 1996, consists of agreed emission reductions on the basis of "critical loads" of nitrogen compounds in the environment: vegetation, soils, groundwater, and surface waters. At meetings in Sofia, Bulgaria, in November 1988 countries signed these agreements to control NO_x emissions over the next decade and presumably beyond which could also have a significant impact on both mobile and stationary source requirements in Europe. A total of 12 European countries, including France, Germany, Italy, and Spain, agreed to reduce NO_x by 30% over the next decade, an ambitious program. In summary, the protocol should increase the pressure to reduce NO_x from all sources.

VOC Protocol. The United States has joined with other European countries to form an "ad hoc workgroup," the first step toward developing an international protocol to control emissions of volatile organic compounds. After the signing of the NO_x agreement in Sofia, Bulgaria, a workgroup was formed to discuss methods for researching VOC emissions and control strategies. According to a draft report, the executive body for the Convention on Long Range Transboundary Air Pollution, recognizing damage from VOCs, which react with NO_x to form ozone, agreed to prepare "necessary substantiation for appropriate internationally agreed measures and proposals for a draft protocol to the Convention" aimed at controlling VOCs. As part of the development of the U.N. ECE protocol, a technical annex dealing with mobile sources was drafted at a meeting in Switzerland (April 6, 1990). A principal conclusion is that closed-loop three-way catalyst technology is cleaner and more efficient than either engine modifications or lean settings with open-loop catalysts.

Technology Option	Emission Level	Cost	Fuel Consumption
Uncontrolled	400		<100
Engine modifications	100	base	100
Lean setting w/oxidation catalyst	50	150–200	100
Closed-loop, three way catalyst	10	250–400	95
Advanced closed-loop catalyst	6	350–600	90

Global Warming Protocols. ***CFCs.*** As noted earlier, the U.N. IPCC process has concluded as a matter of scientific consensus that the global warming concern is real and currently under way. As a result, pressure is building at an international level to develop a collective approach. The first manifestation of this was the Montreal Protocol and its subsequent amendment, which is designed to eliminate the use of CFCs before the end of the century. Beyond this, however, and spurred in part by recent events in the Middle East, more and more countries are calling for substantial cuts in so-called CO_2 equivalent emissions. In addition to pressures for improving vehicle fuel efficiency (and therefore CO_2 emissions), this includes potential reductions in CO, HC and NO_x, which as noted earlier can increase global accumulations of methane, carbon dioxide, nitrous oxide, and ozone in direct and indirect ways; these gases are all greenhouse gases.

Carbon Dioxide. A total of 155 nations, including the European Community, signed the climate change treaty by the close of the United Nations Conference on Environment and Development, according to the U.N. Treaty Office. The treaty set guidelines for cutting emissions of greenhouse gases but stopped short of mandating firm targets and timetables for the cuts. During earlier negotiations at U.N. headquarters, the United States blocked nations from including firm targets in the treaty.

The treaty said the "return by the end of the present decade to earlier levels of anthropogenic emissions of carbon dioxide and other greenhouse gases not controlled by the Montreal Protocol would contribute" to industrialized countries "modifying longer-term trends" of emissions. It said also the aim for developed countries is to return "individually or jointly to their 1990 levels these emissions of carbon dioxide and other greenhouse gases." The EC, in signing the treaty, reaffirmed its goal of stabilizing carbon dioxide emissions by 2000.

Conclusions Regarding the Forces at Work

Air pollution problems worldwide remain serious and growing. As the principal source of HC, CO and NO_x emissions in most areas of the world, motor vehicles continue to receive priority attention from national governments and international bodies. Increasingly diesel particulate and other toxic compounds are also receiving attention.

GOVERNMENT RESPONSES TO THE PROBLEM

In 1959, California adopted legislation that called for the installation of pollution control devices as soon as three workable control devices were developed. At that time, the auto manufacturers repeatedly asserted that the technology to reduce emissions did not exist.

In 1964, the state of California was able to certify that three independent manufacturers had developed workable, add-on devices. This triggered the legal requirement that new automobiles comply with California's standards beginning with the 1966 model year. Soon afterward the large domestic manufacturers announced that they too could and would clean up their cars with technology that they had developed. Thus independently developed devices were unnecessary.

Subsequent to California's pioneering efforts, and as a reuslt of recognition of the national nature of the auto pollution problem in 1964, Congress initiated federal motor vehicle pollution control legislation. As a result of the 1965 Clean Air Act Amendments, the 1966 California auto emission standards were applied nationally in 1968.

In December 1970, the Clean Air Act was amended by Congress, "to protect and enhance the quality of the nation's air resources." Congress took particular notice of the significant role of the automobile in the nation's effort to reduce ambient pollution levels by requiring a 90% reduction in emissions models from the level previously prescribed in emissions standards for 1970 (for carbon monoxide and hydrocarbons) and 1971 (for nitrogen oxides). Congress clearly intended to aid the cause of clean air by mandating levels of automotive emissions that would essentially remove the automobile from the pollution picture.

In many ways the serious effort to control motor vehicle pollution can be considered to have begun with the passage of the landmark 1970 law. In 1977, the act was "fine tuned" by Congress, delaying and slightly relaxing the auto standards, imposing similar requirements on trucks, and specifically mandating in use directed vehicle inspection and maintenance programs in the areas with the most severe air pollution problems. Just recently, Congress passed the 1990 Clean Air Act Amendments and they were signed into law by President Bush on November 15, 1990. (A summary of these provisions is contained in Appendix A.)

Gasoline-Fueled Vehicles

Gasoline vehicle pollution control efforts reflect an approximately 30-yr effort to date. Initial crankcase HC controls were first introduced in the early 1960s followed by exhaust CO and HC standards later that decade. By the early to mid 1970s, most major industrial countries had initiated some level of vehicle pollution control program.

During the mid to late 1970s, advanced technologies were introduced on most new cars in the United States and Japan. These technologies resulted from a conscious decision to "force" the development of new approaches and were able dramatically to reduce CO, HC, and NO_x emissions beyond previous systems. As knowledge of these technological developments on cars spread, and as the adverse effects of motor vehicle pollution became more widely recognized, more and more people across the globe began demanding the use of these systems in their countries. During the mid 1980s, Austria, the Netherlands, and the FRG adopted innovative economic incentive approaches to encourage purchase of low pollution vehicles. Australia, Canada, Finland, Austria, Norway, Sweden, Denmark, and Switzerland have all decided to adopt mandatory requirements. Even rapidly industrializing, developing countries such as Brazil, Chile, Taiwan, Hong Kong, Mexico, Singapore, and South Korea have adopted stringent emissions regulations.

After years of delay, the European Economic Community has also made significant strides. As 1990 came to a close, the European Council of Environmental Ministers reached unanimous agreement to require all new models of light-duty vehicles by 1992–1993 to meet emission standards roughly equivalent to U.S. 1987 levels. Further they voted to require the commission to develop a proposal before December 31, 1992 which, taking account of technical progress, will require a further reduction in limit values (presumably the proposal should be roughly equivalent to the recently adopted U.S. standards of 0.25 NMHC, 0.4 NO and 3.4 CO).

Just as Europe was moving toward parity with U.S. standards, the United States, and to a much greater extent California, embarked on a course that could prove just as momentous to the 1990s as the 1970 Clean Air Act was to the 1970s and 1980s. Many significant changes to the federal motor vehicle emission control program will result from the passage of the Clean Air Act Amendments of 1990, including the following: the design, manufacture, and certification of new vehicles to lower tailpipe and evaporative emission standards; enhanced inspection and maintenance programs for seriously polluted areas; more durable pollution control systems; and the use of less polluting gasoline and alternative fuels.

The state of California has responded to the ozone nonattainment problem in the South Coast Air Basin with a long-term plan of new VOC and oxides of nitrogen (NO_x) control initiatives designed to achieve significant reductions in emissions of these and other criteria pollutants from current levels. To further reduce motor vehicle emissions, the California Air Resources Board (CARB) has established stringent new vehicle exhaust emission standards. Compliance with these standards will be achieved through a combination of advanced vehicle emission control technology and clean burning fuels.

The California motor vehicle emission control program is distinguished from the federal program in its general approach and philosophy as well as its more stringent emission standards and program requirements. The California Air Resources Board characterizes its program as a technology-forcing approach, which is more stringent and flexible than the federal program. Because the California program is more flexible and responsive than the federal program, in-use compliance concerns can be addressed more rapidly. The California Air Resources Board has indicated that it is committed to taking all steps necessary to ensure that California-certified vehicles meet certification standards in-use for the useful lifetime of the vehicle.

To control the evaporative emissions from vehicles in use, both EPA and CARB have developed a sequence of events designed to test vehicles for compliance with evaporative emission regulations. The California Air Resources Board was first to adopt their new procedures but EPA tightened them even more early in 1994. Although both testing programs include the events of preconditioning, diurnal heat builds and exhaust, running loss, and hot soak tests, differences exist that complicate a comparison of the emission control differences that might result from the two programs. It appears likely, however, that the EPA and CARB procedures will eventually be closely matched.

Diesel Vehicles and Engines

Not surprisingly, in view of the already serious health concerns and the continued growth in the use of diesel vehicles, many countries are pushing equally hard to reduce diesel emissions as gasoline vehicle emissions.

North America. ***United States.*** U.S. emission control requirements for smoke from engines used in heavy-duty trucks and buses were first implemented for the 1970 model year. These opacity standards were specified in terms of percent of light allowed to be blocked by the smoke in the diesel exhaust (as determined by a light extinction meter). Heavy-duty diesel engines produced during model years 1970 through 1973 were allowed a light extinction of 40% during the acceleration phase of the certification test and 20% during the lugging portion; 1974 and later model years are subject to smoke opacity standards of 20% during acceleration, 15% during lugging, and 50% at maximum power.

The first diesel exhaust particulate standards in the world were established for cars and light trucks in an EPA rule making published on March 5, 1980. Standards of 0.6 grams per mile (gpm) were set for all cars and light trucks starting with the 1982 model year dropping to 0.2 and 0.26 gpm for 1985 model year cars and light trucks, respectively. In early 1984, EPA delayed the second phase of the standards from 1985 to 1987 model year. Almost simultaneously, California decided to adopt its own diesel particulate standards: 0.4 gpm in 1985, 0.2 gpm in 1986 and 1987, and 0.08 gpm in 1989.

Subsequently, EPA revised the 0.26 gpm diesel particulate standard for certain light-duty trucks. Light-duty diesel trucks with a loaded vehicle weight of 1700 kg or greater, otherwise known as LDDT2s, were required to meet a 0.50 gpm standard for 1987 and 0.45 gpm level for 1988–1990. For the 1991 and later model years the standard was tightened to 0.13 gpm.

Particulate standards for heavy-duty diesel engines were promulgated by the EPA in March 1985. Standards of 0.60 grams per brake-horsepower-hour (G/BHP-h) (0.80 g/kW·h) were adopted for 1988 through 1990 model years, 0.25 (0.34) for 1991 through 1993 model years and 0.10 (0.13) for 1994 and later model years. Because of the special need for bus control in urban areas, the 0.10 (0.13) standard for these vehicles was to go into effect in 1991, three years earlier than for heavy-duty trucks.

Table 1 shows the standards adopted for heavy-duty diesel engines and vehicles as a result of the Clean Air Act Amendments of 1977. The U.S. Clean Air Act of 1977 required that the regulations for NO_x be based on "the maximum degree of emission reduction which can be achieved by means reasonable expected to be available." The PM standard was to require "the greatest degree of emission reduction available" according to the EPA's determination, considering cost, noise, energy, and safety. The limit was intended to be technology forcing. These regulations have an important influence on the rate at which new, less polluting technologies penetrate heavy-duty engine markets.

From 1970 to 1983, the U.S. regulations required demonstration of compliance on the U.S. 13-mode steady-state test procedure. In 1984, either the steady-state or the new U.S. transient test procedure could be used, but the latter has been required for 1985 and later engines. In addition, U.S. regulations permit no crankcase emissions of HC, except for turbo-charged engines. Compliance is required over the full useful life of 1985 and later engines, defined as 8 yr or 110,000—290,000 m, depending on the size of engine.

Changes in the 1990 Clean Air Act Amendments. As noted above, on November 15, 1990, the U.S. Clean Air Act Amendments of 1990 were signed into law. This legislation crafted a new challenge for diesel manufacturers. In summary, for cars, it requires all vehicles to meet the California particulate standard of 0.08 gpm. It initially postponed the 0.1 g/BHP-h urban bus particulate standard from 1991 until 1993. Beginning in 1994, however, the new law requires buses operating more than 70% of the time in large urban areas to cut particulate by 50% compared with conventional heavy duty vehicles (i.e., to 0.05 g/BHP-h). Furthermore, it mandates EPA to conduct testing of vehicles to determine whether urban buses are meeting the standard in use over their full useful lives. If EPA determines that 40% or more of the buses are not, it must establish a low pollution fuel requirement. Essentially, this provision allows the use of exhaust after treatment devices to reduce diesel particulate to a low level, provided they work in the field; if they fail, EPA is to mandate alternative fuels. Table 2 summarizes the heavy-duty diesel particulate standards required by current law. See also CLEAN AIR ACT, MOBILE SOURCES.

Beyond the vehicles themselves, EPA has issued a rule limiting the sulfur content of diesel fuel to 0.05 wt. % after October 1, 1993. This decision not only directly lowers particulate (sulfur) emissions but also paves the way for the use of catalytic control technology for diesel particulate since it reduces the concern over excessive sulfate emissions. The rule also allows vehicle manufacturers to certify their engines for compliance with the 1991 to 1993 emission limits with fuel of 0.10 sulfur content.

The statutory standards for HC, CO, and NO_x for heavy-duty vehicles and engines created by the 1977 Clean Air Act Amendments were eliminated in the 1990 Amendments. In their place, a general requirement that standards applicable to emissions of HC, CO, NO_x, and particulate reflect

Table 1. U.S. Emission Standards for Heavy-Duty Diesel Engines

Model Year	NO_x, g/BHP-h	HC, g/BHP-h	CO, g/BHP-h	PM, g/BHP-h	Smoke,[a] %		
					A	B	C
1988	10.7	1.3	15.5	0.60	20	15	50
1990	6.0	1.3	15.5	0.60	20	15	50
1991	5.0	1.3	15.5	0.25[b]	20	15	50
1994	5.0	1.3	15.5	0.10	20	15	50

[a] *A*, average acceleration; *B*, lug down peak; *C*, acceleration average.
[b] For urban buses, the standard is 0.10 g/BHP-h.

Table 2. Heavy-duty Vehicle Standards

Heavy-duty Trucks (3856 kg GVWR[a] or more)	
1988–1990 federal and California	0.6 g/BHP-h
1991–1993 federal and California	0.25 g/BHP-h
1994 and later model year federal and California	0.1 g/BHP-h
Urban Buses	
1988–1990 federal and California	0.6 g/BHP-h
1991–1992	
federal	0.25 g/BHP-h
California	0.1 g/BHP-h
1993 federal and California	0.1 g/BHP-h
1994 and later model year	
federal[b]	0.05 g/BHP-h
California	0.1 g/BHP-h

[a] Gross vehicle weight rating.
[b] EPA may relax to 0.07 g/BHP-h, based on technical considerations.

> the greatest degree of emission reduction achievable through the application of technology which the Administrator determines will be available for the model year to which such standards apply, giving appropriate consideration to cost, energy, and safety factors associated with the application of such technology.

Standards adopted by EPA that are currently in effect will remain in effect unless modified by EPA. EPA may revise such standards, including standards already adopted, on the basis of information concerning the effects of air pollutants from heavy-duty vehicles and other mobile sources, or the public health and welfare, taking into consideration costs.

A 4.0 g/BHP-h NO_x standard is established for 1998 and later model year gasoline- and diesel-fueled heavy-duty trucks. The useful life provisions of California apply for light and medium trucks and EPA is authorized to delay the 1998 NO_x standard for heavy trucks for 2 yr on the basis of technological feasibility. EPA is also mandated to promulgate regulations that require that urban buses in large metropolitan areas that have their engines rebuilt after January 1, 1996, shall be retrofitted to meet emissions standards that "reflect the best retrofit technology and maintenance practices."

California. Under the U.S. system, California has been allowed to adopt its own vehicle emission standards. As noted in Table 2, California retained its 0.1 particulate standard for urban buses starting wtih the 1991 model year. It also adopted new emission standards for medium-duty vehicles following a public hearing on June 14, 1990. The regulation expands the definition of medium-duty vehicles to include vehicles weighing between 6000 and 14,000 lb GVWR, establishes a chassis test procedure for those vehicles (which should greatly facilitate in use recall testing), and expands the useful life requirements to 120,000 mi.

The new emission standards are set out in Table 3. Manufacturers will be required to meet 50% certification compliance in 1995 and 100% compliance in 1996. Less stringent in-use compliance will be permitted through the 1997 model year with full 100% certification and in-use compliance required for the 1988 model year.

Demonstrating the same degree of leadership with regard to diesels as it has with gasoline fueled vehicles, the California Air Resources Board (CARB) initiated a process to explore tighter emission standards for 1996 and later model year urban buses. The staff initially was considering a 2.5 g/BHP-h NO_x standard and a 0.05 g/BHP-h particulate standard. For other classes of heavy-duty engines, the CARB staff is examining the feasibility of a 2.0 g/BHP-h NO_x standard and a 0.05 g/BHP-h particulate standard to be phased in beginning possibly as early as the 1998 model year.

Transit Buses. In light of the strong opposition of the transit industry to the proposed 1996 standards for transit buses (2.5 NO_x and 0.05 particulate) that would require the use of alternative fueled engines, the CARB staff modified its proposal as summarized below.

The staff had originally proposed that all transit buses with a gross vehicle weight rating (GVWR) of 14,001 lb and above be subject to the proposed standards. The staff has modified its proposed definition to include only transit buses that would typically use a heavy-duty engine of >33,000 lb GVWR. This change would align California's definition of transit buses with EPA's definition. Staff has determined that this revision would be beneficial to small transit agencies as they tend to operate a greater proportion of smaller buses (<33,000 lb GVWR), which would not be exempt under this new proposed definition.

The staff proposed two options for transit buses: base emission standards and optional low-emission vehicle (LEV) standards: The following base emission standards are proposed for transit buses beginning with the 1996 model year: THC, 1.3 g/BHP-h; NMHC, 1.2 g/BHP-h; CO, 15.5 g/BHP-h; NO_x, 4.0 g/BHP-h; and PM, 0.05 g/BHP-h. CARB staff believe that it is possible to meet a 4.0 g/BHP-h NO_x standard and below with heavy-duty (HD) diesel engines. In addition, with the revised proposed transit bus definition of >33,000 lb GVWR, engine manufacturers will be able to concentrate their efforts on im-

Table 3. California Emission Standards

	50,000-mi Standards, g/mi			120,000-mi Standards, g/mi			
Test Weight, lb	NMHC[a]	CO	NO_x	NMHC[a]	CO	NO_x	PM
0–3750	0.25	3.4	0.4	0.36	5.0	0.55	0.08
3751–5750	0.32	4.4	0.7	0.46	6.4	0.98	0.10
5751–8500	0.39	5.0	1.1	0.56	7.3	1.53	0.12
8501–10,000	0.46	5.5	1.3	0.66	8.1	1.81	0.15
10,001–14,000	0.60	7.0	2.0	0.86	10.3	2.77	0.18

[a] Nonmethane hydrocarbons.

proving emissions for just a few engine families. Furthermore, in 1993, diesel fuel quality regulations will limit the sulfur content to 0.05% and the aromatic HC content to 10%, which will also lower diesel engine emissions.

In addition to the base emission standards, the ARB would encourage the air quality districts to develop a program for NO_x emission credits for those transit agencies that would like to have the option of purchasing buses that are certified to a standard of 2.5 g/BHP-h NO_x. The ARB is currently developing technical guidelines for such an emission credit program, which will assist those districts interested in implementing a NO_x emission credit system. Staff proposes, beginning with the 1994 model year, that transit buses must be certified to the following optional LEV standards to be eligible for such an emission credit program: THC, 1.3 g/BHP-h; NMHC, 1.2 g/BHP-h; CO, 15.5 g/BHP-h; NO_x, 2.5 g/BHP-h; and 0.05 g/BHP-h.

Staff believes that alternative fuels will be the primary way of meeting the proposed optional LEV standards. The Detroit Diesel Corporation 1M model 6V-92TA methanol engine has been certified to levels of 1.7 NO_x and 0.03 PM and the 3M model has been certified to 2.3 NO_x and 0.06 PM. The Cummins L10 compressed natural gas engine has recently been certified to levels of 2.0 NO_x and 0.02 PM.

Currently, all HD vehicles are required to use closed crankcase emission control systems or positive crankcase ventilation (PCV) systems except for petroleum-fueled diesel-cycle engines that use turbochargers, pumps, blowers, or superchargers for air induction.

With the advent of low emission standards, crankcase emissions are becoming a significant percentage of the exhaust emissions and need to be controlled as well. In fact, EPA has indicated that their 1979 analysis, which led to their decision to allow nonnaturally aspirated petroleum-fueled diesel engines to emit crankcase gases, is outdated and represents a worst-case cost-effectiveness situation for diesels. In addition, several manufacturers have voluntarily controlled crankcase emissions of diesel engines by routing the emissions through an oil separator, and into the turbocharger. Therefore, EPA will now require that all transit bus engines have PCV systems beginning in 1996.

Heavy-duty Engines. Based on the technology feasibility assessment carried out by a contractor, CARB staff has recommended that sales-weighted average emission standards for NO_x and PM emissions be set for heavy-duty vehicles. The sales-weighted averaging scheme would allow manufacturers some time to develop low emission engines without overly impacting all engine lines. With this scheme, low emission engines can be averaged with base engines to produce a sales-weighted average.

To implement the averaged emission standards, four low emission truck classifications and three low emission bus classifications were defined. The four low emission truck classifications are the transitional low emission truck (TLET), the low emission truck (LET), the ultra low emission truck (ULET), and the zero emission truck (ZET). The three low emission bus classifications are the low emission bus (LEB), the ultra low emission bus (ULEB), and the zero emission bus (ZEB). The transitional category for buses has been eliminated as the bus standard proposed by the ARB for 1996 is lower than TLET levels.

The various low emission classifications also include a total hydrocarbon standard, a carbon monoxide standard, a reactivity-adjusted nonmethane organic gas (RANMOG) standard, a formaldehyde (HCHO) standard, and a total aldehyde standard. It was recommended that all standards be applied to all fuels in a fuel-neutral manner.

Retrofit Requirements. The rule making to establish control requirements for existing diesel heavy-duty engines is on hold pending resolution of the requirements for transit buses.

Off-the-road Vehicles or Engines. California has also taken the lead in regard to off-road emissions. At its January 10, 1992 meeting, the California Air Resources Board adopted emission standards for off-road diesel engines above 175 HP. New 1996 through 2000 model year engines must meet standards equivalent in stringency to the 1990 on-road heavy-duty diesel engine standards. Actual standards would be 6.9 g/BHP-h for NO_x and 0.4 g/BHP-h particulate, based on an eight-mode steady-state test largely culled from the ISO 8178 procedure. Beginning in the year 2001, the proposal is intended to be equivalent to 1991 on-road heavy-duty engine standards, 5.8 g/BHP-h NO_x and 0.16 g/BHP-h particulate. Beginning in 2000, CARB includes requirements for engines over 750 HP for the first time, requiring that they meet the 1996–1999 proposed levels.

CARB estimates that the 1996 standards will likely require modifications to fuel injection timing, fuel injectors, combustion chambers and jacket water after coolers. The 2000 standards will require further improvements in after cooling, improved oil control, and possibly electronic fuel injection systems. CARB noted in its proposal that alternative fuels and particulate after treatment devices may be used but no 1991 on highway engine needed them to meet similar standards.

Canada. In March 1985, in parallel with a significant tightening of gaseous emission standards, Canada adopted the U.S. particulate standards for cars and light trucks (0.2 and 0.26 gpm, respectively) to go into effect in the 1988 model year. Subsequently, Canada also decided to adopt U.S. standards for heavy-duty engines for 1988 as well. U.S. manufacturers have committed themselves to marketing 1991 and subsequent technology heavy-duty engines in Canada in the absence of specific regulations.

Mexico. In a short period of time, Mexican officials have made dramatic progress in putting an aggressive motor vehicle pollution control program in place. They have not yet adequately addressed their heavy-duty diesel particulate problem, however. As part of the refinery upgrade currently under way, diesel fuel sulfur levels will be lowered to 0.1 wt %. In addition, adoption of U.S. 1994 heavy truck standards is under consideration but has been delayed because the U.S. requirements do not address the high altitude conditions that exist in Mexico City.

Western Europe. ***Common Market.*** *Light-duty Vehicles.* The European Community Environmental ministers decided in December 1987 to adopt a particulate standard for light duty diesels of 1.1 g per test (1.4 for conformity of production). However, in response to pressure from

Germany, which has a large light-duty diesel population, and thus was pushing for much tighter standards, the ministers ordered the commission to develop a second-step proposal by the end of 1989.

As part of the consolidated directive released on February 2, 1990, the commission proposed new standards of 0.19 g/km for type approval and 0.24 g/km for conformity of production. Subsequently, following critical comments from the European Parliament, the proposed levels were reduced to 0.14 and 0.18, respectively. At the Council of Ministers meeting in October 1990 there was still some debate over the conformity of production (COP) standard, with Germany and Greece pushing for a level of 0.14 to 0.15 rather than the commission proposal of 0.18.

At the European Council of Environmental Ministers meeting on December 21, 1990, final, unanimous agreement was reached. Specifically the ministers decided to:

1. Adopt the commission proposal as amended at the October 29 meeting, to require all light-duty vehicles to meet emission standards of 2.72 g/km CO, 0.97 g/km of HC plus NO_x, 0.14 g/km of particulates for type approval and 3.16 g/km CO, 1.13 g/km for jf23HC plus NO_x, 0.18 g/km particulates for conformity of production.
2. To require the commission to develop a proposal before December 31, 1992, which, taking account of technical progress, will require a further reduction in limit values.
3. To have the council decide before December 31, 1993, on the standards proposed by the commission.
4. To allow countries before 1996 to encourage introduction of vehicles meeting the proposed requirements through "tax systems" that include pollutants and other substances in the basis for calculating motor vehicle circulation taxes.

At the local level, the German parliament agreed to include diesel vehicles in its low pollution vehicle tax incentive program. To qualify, cars must achieve a particulate standard of 0.08 g/km. (This standard is approximately halfway between the current U.S. particulate standard of 0.2 g/m and the California particulate standard of 0.08 g/m.)

Heavy-duty Vehicles. European nations have for many years negotiated harmonized standards within the U.N. Economic Commission for Europe (U.N. ECE). Once adopted in this international organization, those standards may be promulgated in law by individual countries. Regulation ECE 24 governed smoke emissions and a similar standard was required in the European Communities through EC Directive 72/306. The amount of light passing through the exhaust gas is measured with the Bosch scale and method at full load for various engine sizes.

The standards for gaseous pollutants of Regulation ECE 49, which were adopted in 1983, were no where legally required until 1986. In a first step toward reducing emissions from heavy-duty vehicles, the EEC adopted Directive 88/77/EEC, which applied as of October 1, 1990. In early 1990, after months of debate, and delayed slightly by the tremendous effort associated with the consolidated directive for light-duty vehicles, the European Commission released its proposal for cleaning up diesel trucks. The draft directive proposed compulsory EC norms to be introduced in two stages:

1. After July 1, 1992, new types of diesel engines and diesel-fuel vehicles will be required to meet the following standards:
   - carbon monoxide (CO g/kW · h = 4.5)
   - hydrocarbons (HC g/kW · h = 1.1)
   - nitrogen oxides (NO_x g/kW · h = 8)
   - particulates according to engine power: less than 85 kW (PT g/kW · h = 0.63) and more than 85 kW (PT g/kW · h = 0.36)
2. After October 1, 1996, the new diesel engines and vehicles will be required to comply with the following standards:
   - carbon monoxide (CO g/kW · h = 4)
   - hydrocarbons (HC g/kW · h = 1.1)
   - nitrogen oxide (NO_x g/kW · h = 7)
   - particulate emissions regardless of the power of the engine (PT g/kW · h = 0.3 or 0.2)

For particulate emissions, the proposal provided that the council would decide by September 30, 1994, at the latest, between the figures indicated in the proposal and the results of a report prepared by the commission before the end of 1993. The report was to indicate progress made in regard to (*1*) techniques for monitoring atmospheric pollution from diesel engines, (*2*) the availability of improved diesel fuel (sulfur content, aromatics content, cetane number) and a corresponding reference fuel to test emissions, and (*3*) a new statistical method for monitoring production compliance, which must still be adopted by the commission, needed to make a system based on a single standard operational.

In presenting the commission's proposal, Commissioner Ripa di Meana recalled the conclusions of the Task Force on Environment/Single Market of December 1989, which underlined the fact that road transport's impact on the environment could increase after 1992. From 1985 to 1992, he estimated that road transport will increase its share of goods transport from 50 to 70%. After 1992, he indicated that the opening of frontiers should increase road transport even more.

Shortly after the commission proposal, in a dramatic, surprise decision, the UK government announced on March 22, 1990 its decision to support the full U.S. heavy-duty vehicle emissions control program. This reflects British conclusions that a transient test procedure is necessary actually to achieve low in use vehicle emissions rates of nitrogen oxides and particulates.

After receiving comments from the Parliament, at the October 1, 1991, meeting of the Council of Environmental Ministers, final agreement was reached on the directive. The unanimous decision by EC environment ministers applies to both particulate and gaseous emissions (Table 4).

The directive takes effect in two stages. New vehicle models were required to conform by July 1, 1992; all new vehicles by October 1, 1993. The second stage begins October 1, 1995, for new vehicle models, and October 1, 1996, for all vehicles. The dates represent a compromise

Table 4. EC Heavy-duty Diesel Requirements

EC Step	CO	HC	NO_x	PART
1 Type Approval	4.5	1.1	8.0	0.63 (<85 kW) 0.36 (>85 kW)
COP	4.9	1.23	9.0	0.7 (<85 kW) 0.4 (>85 kW)
Step 2	4.0	1.1	7.0	0.15
Step 3 (2000) or later	new test possible			

from the original proposal by the EC Commission, in that the first stage was pushed back by nine months and the second stage was brought forward by a year.

Fuel with 0.05 sulfur content will have to be available for the second stage. Manufacturers will be allowed to certify their engines using this new fuel immediately. The sulfur content in diesel fuel will be reduced to no more than 0.2% by weight beginning October 1, 1994, and to 0.05% by weight starting October 1, 1996. As of October 1, 1995, there must be a balanced distribution of available diesel fuel, so that at least 25% of diesel fuel in each member country has a sulfur content of not more than 0.05%.

Royal Commission Report. Just as the EC was about to reach its heavy-duty vehicle decision, the Royal Commission on Environmental Pollution released a report on heavy-duty diesel particulate. Introducing the report, the chairman noted that diesel vehicles accounted for about 25% of the national emissions of oxides of nitrogen and were the principal source of smoke in urban areas. Significant points of the report are as follows:

1. Acceptable air quality levels are being exceeded regularly in London and other British cities as a result of nitrogen oxides emissions, to which diesel vehicles are major contributors. Further diesel exhaust has been classified by the World Health Organization as a probable carcinogen.
2. As proposed by the UK government, the European Community's test for new heavy-duty diesel engines should be made more demanding, along the lines of the test now used in the United States.
3. Financial incentives should be developed for lower polluting engines to accelerate the replacement of older, higher polluting engines with newer, cleaner engines.
4. The government should proceed urgently with trials of traps fitted to exhausts to catch particulates; a grant should be offered for fitting these to buses.
5. A heavy-duty engine diagnostic test should be developed to improve the actual condition of engines; new engines should be required to maintain good emissions performance over prolonged periods of operation.
6. Tighter limits for emissions from urban buses may be needed to reduce emissions in urban areas.
7. Incentives should be created for the early introduction and use of low sulfur fuel. Furthermore, subject to the outcome of a study, the government should seek to encourage bus operators to make use of alternative fuels such as petrol, LPG, CNG, or electricity.

German Field Studies Continuing. Germany has not been waiting for EC regulations to deal with the urban bus problem. For some time it has had an extensive program under way in cooperation with industry with 1500 buses distributed in urban areas across the country fitted with exhaust aftertreatment systems. Preliminary findings indicate that the after treatment technology has advanced significantly.

Non–Common Market Countries of Europe. Several other European countries have been moving toward more stringent diesel particulate requirements than those of the EC. Sweden adopted the U.S. passenger car standard in 1989 and other EFTA countries took similar actions. These countries have also been seriously considering more stringent requirements for trucks and buses. Sweden announced its intention to adopt requirements that will bring about the same degree of control as the United States by 1995. Specifically, U.S. 1990 requirements (including diesel particulate) were adopted by Sweden on a voluntary basis for 1990 light duty trucks, and on a mandatory basis in 1992. Regarding heavy trucks, the EFTA countries have been striving for U.S.-type engine technology, trying to derive the equivalent environmental benefits.

The regulatory situation in Europe is changing rapidly, however. In early 1992, the EC and EFTA countries finally reached agreement on the European Economic Area (EEA), the world's biggest (with 380 million people) and wealthiest single market. This agreement significantly alters the shape of motor vehicle pollution control in Europe. Several of the EFTA countries have also applied for membership in the EC. One can foresee that most if not all future vehicle emissions regulation in Europe will take place under the auspices of the EC. At the Edinburgh summit, the EC ministers agreed to accelerate the consideration of full community membership for Sweden, Norway, Austria, and Finland.

Asia Pacific Region

Japan. Japanese smoke standards have applied to both new and in-use vehicles since 1972 and 1975, respectively. The maximum permissible limits for both are 50% opacity; however, the new vehicle standard is the more stringent because smoke is measured at full load, while in-use vehicles are required to meet standards under the less severe no-load acceleration test.

Japan also established emission limits for heavy-duty engines as summarized in Table 5. All engines must meet the maximum value during certification testing. Also, Japanese limits are expressed in terms of parts per million (ppm), or concentrations in the exhaust gas. This differs from other countries' standards relative to engine power outputs. Without detailed information on engine power outputs and air flows in Japan, then, it is not possible to compare the Japanese with other OECD member countries' standards. In addition, the emission levels allowed in Japan for direct injection engines are higher than for indirect injection engines. Gasoline vehicles are measured for compliance on a six-mode steady-state test,

Table 5. Heavy-duty Vehicle Emission Limits in Japan, ppm

Engine	NO_x	HC	CO	Smoke	Effective Date
Diesel					
direct injection	520	670	980	50% rate of contamination	Oct. 1, 1989/Sept. 1, 1990[a]
indirect injection	350	670	980	50% rate of contamination	Oct. 1, 1989/Sept. 1, 1990[a]
Gasoline	850	520	1.6%		April 1, 1991

[a] Applies to new models/all models.

while diesel vehicles are measured on a different six-mode steady-state test.

On December 22, 1989, the Central Council for Environmental Pollution Control submitted the *Approach to Motor Vehicle Exhaust Emission Control Measures in Years Ahead* to the director-general of the Environment Agency. It includes a diesel particulate standard for the first time. In accordance with the recommendation, the Ministry of Transport revised the related laws, ordinances and regulations during 1990. Principal provisions include the following.

The emission levels of nitrogen oxides from diesel-powered motor vehicles are reduced by 38% in the case of large-size diesel-powered trucks (with direct injection-type engines) and by 56% in the case of medium-size diesel-powered passenger motor vehicles.

The NO_x emission level for the direct injection type diesel-powered motor vehicles is required to be the same as that for indirect injection type diesel-powered motor vehicles.

The emission level of particulate matter from diesel-powered motor vehicles is reduced by more than 60%. Besides the diesel smoke standards hitherto enforced, a new particulate standard was introduced, reducing the level by more than 60% from the present level by providing two-phase target values.

The emission level of diesel smoke will also be cut in half by providing two-phase target values.

The sulfur content in diesel (light) oil will be reduced from the present level of 0.4 to 0.5 wt % to 0.2 wt % and eventually to 0.05 wt %.

In Tokyo alone, the number of diesel-powered vehicles has risen from 467,000 in 1985 to 721,000. Such high growth has raised concerns about whether the government will succeed in reducing NO_x levels by the year 2000 as it has pledged to do. As a result efforts have been under way to develop additional control measures for this pollutant. Japan's Environment Agency (JEA) on June 26, 1990 announced new regulations calling for phased reductions of nitrogen oxide emissions from gasoline-powered vehicles, beginning in 1994, and from diesel-powered vehicles, beginning in 1997. Previously, the agency had planned to enforce these requirements from 1999. NO_x emissions from diesel-powered vehicles of 2.5 L or less must be reduced by 65% from 1997, the announcement said. However, it stopped short of setting targets for diesel-powered vehicles with larger engines.

An intense debate is under way within the government regarding the schedule for implementation of new truck emissions requirements, and it appears that the Environment Ministry is losing. The net result would be a significant delay. It appears that MITI and the MOT have succeeded in pushing through a grace period that would delay the implementation schedule by providing a grace period of up to 8 yr for trucks under 2.5 t and 9 yr for those over.

Also on June 26, the JEA decided to reduce the sulfur contents in light fuel oil, which is used as fuel for diesel-powered vehicles, to 0.05% from the current 0.4%. The government enacted a bill to regulate diesel powered NO_x vehicle emissions in urban areas during the Diet (parliament) session that ended June 19, 1990. The law requires vehicle operators to change vehicles that fail to achieve 1990 emission requirements to less-polluting ones. Those that failed to meet the new standards will not be issued vehicle inspection certificates and be kept off the road.

South Korea. The vehicle pollution control program in Korea continues to progress, but can still barely keep pace with the high rate of growth in the domestic vehicle population. From a current total of approximately 3 million vehicles throughout South Korea, the domestic population is expected to grow to about 15 million by the year 2000. In addition, while gasoline vehicles are gradually being controlled (with 50,000-m durability requirements for U.S. 83 standards introduced in 1990) diesels remain a serious unsolved problem. Approximately 45% of the annual kilometers in the country are accumulated by diesels, mainly trucks and buses. As reported previously, a several-stage strategy is under way to address this problem (6).

Reduce the Number of Diesel Vehicles. A program is under way to gradually discourage diesel engines in light trucks and buses. For the first stage during 1990 and 1991, 15-person mini buses and jeeps have been designated; for the second stage, 1992–1993, 35-person buses and 3-t cargo trucks have been designated as targets.

Diesel engine modifications to fumigate the diesel fuel with 30% LPG have been under investigation for several years. While this concept worked well in the laboratory, difficulties have been experienced in the field due to improper mixing. A new test is under way with electronically controlled mixing. In addition, the government is exploring the feasibility of natural gas buses. At present, about 2 million t of LNG are imported into Seoul each year from Indonesia; about 80% is used in utilities and 20% in space heaters. It would be possible to make fuel available for buses if the government concludes natural gas buses are feasible.

Increase Horsepower. The excessive smoke levels from diesel buses are at least partly caused by overloading underpowered engines. Therefore, the government has decided that new bus engines should increase their power from 185 to 235 HP.

Tighten In-use Smoke Standards. The in-use diesel smoke standard was tightened from 50% opacity to 40% in 1990. The government hopes this action will encourage good vehicle maintenance.

Develop Diesel Particulate Trap Systems. The National Environmental Protection Institute is conducting a 3-yr research program to evaluate particulate controls. Starting in August 1990, an effort was undertaken to identify the most promising systems. Laboratory evaluations are still under way. Initial results with a trap system using cerium fuel additives have been encouraging.

Emission Standards for Light and Heavy Trucks. Particulate emission standards have also been promulgated for new light and heavy trucks. By January 1, 1996 light trucks should be 0.31 g/km and heavy trucks, 0.9 g/kW · h. By January 1, 2000, light trucks should be 0.05–0.16 g/km, heavy trucks, 0.25 g/kW · h, and city buses 0.10 g/kW · h.

Singapore. The government of Singapore is moving rapidly to introduce state of the art pollution controls. As of January 1, 1991, all new diesel vehicles have been required to comply with ECE R 24.03.

Hong Kong. The government has decided to impose new car standards from January 1992 as follows:

Vehicle Type	Petrol	Standards for Diesel
Private car, taxi	U.S. 1987, FTP75; Japan 1987, 10-mode	U.S. 1987, FTP75; Japan 1990, 10-mode
Light goods vehicle, light bus not more than 1.7 t	U.S. 1990, FTP75; Japan 1988–1990, 10-mode	U.S. 1990, FTP75; Japan 1988, 10-mode
Light goods vehicle, light bus more than 1.7 t but not more than 2.5	U.S. 1990, FTP75; Japan 1988–1990, 10-mode	U.S. 1988–1990, FTP75; Japan 1988–1990, 6-mode

Taiwan (ROC). The Taiwan EPA continues to move forward with a comprehensive approach to motor vehicle pollution control. Building on its previous adoption of U.S. 1983 standards for light-duty vehicles (starting July 1, 1990) it has decided to move to U.S. 1987 requirements, which include the 0.2 gpm particulate standard as of July 1, 1995. Heavy-duty diesel particulate standards almost as stringent as U.S. 1990, 6.0 g/BHP · h NO_x and 0.7 g/BHP · h particulate, using the U.S. transient test procedure, will go into effect on July 1, 1993. While 90% of the current heavy-duty vehicles in Taiwan are of Japanese origin, none of these have yet been able to obtain approval at these levels; only Cummins and Navistar currently have been approved. It is intended that U.S. 1994 standards, 5.0 NO_x and 0.25 particulate, will be adopted soon, probably for introduction by July 1, 1997. Diesel fuel currently contains 0.5 wt % sulfur. A proposal to reduce levels to 0.3 by 1993 and 0.05 by 1997 is currently under consideration.

Latin America. ***Brazil.*** A move is under way to convert diesel buses in Sao Paulo to natural gas (CNG). Petrobras has a large surplus of natural gas. The have stopped flaring and constructed a pipeline from Rio to Sao Paulo. The municipal government in Sao Paulo has developed a 10-yr plan to convert 10,000 buses to CNG. The first CNG station went into operation in 1993, for 200 buses. Also Daimler Benz has developed a CNG bus with low HC, CO, particulate, and noise leels. NO_x levels will increase by 15% initially, because the buses will not be equipped with catalysts. DB has indicated that it intends to add catalysts at some later date.

CETESB is investigating whether or not it makes sense for the natural gas to be used in vehicles as CNG or whether it should be converted to methanol and then used. The principal advantage in their eyes is that the use of a liquid fuel would be less disruptive to the existing infrastructure. In addition, when equipped with a catalyst, the overall emissions should be lower.

Diesel fuel sulfur levels in Brazil average about 1%. CETESB is trying to get Petrobras to introduce a lower sulfur fuel, initially in the cities, starting with buses. Petrobras says the best they can do is to introduce 0.5% in the cities, several years from now. To get to 0.05% would require about $1 billion in investment, according to Petrobras, and they are expanding all available capital on increased exploration.

The motor industry is pushing the federal government to authorize the use of light-duty diesel vehicles, which is presently not allowed in Brazil. CETESB supports the present prohibition on the grounds that diesel fuel is highly subsidized and, therefore, should be used only for transport of passengers (buses) and goods. In addition, diesel oil represents the bottle neck of PETROBRAS (the consumption of diesel in Brazil defines the volume of oil which has to be refined) and, therefore, its quality has been affected over the years with the addition of both heavy and light refinery projects to meet demand. Some motor industry sources claim that Brazilian diesel use results in an increase of about one Bosch unit if compared with the European diesel.

Finally, CETESB believes that particulate and other emissions from these cars will offset any advantage due to the low CO. Besides, the introduction of these vehicles into the market means that a number of less polluting vehicles, such as the alcohol fueled, will be substituted by the diesel cars. In regard to heavy-duty diesel particulates CETESB agreed with industry to introduce EURO 1 and EURO 2 standards; however this agreement must be approved by CONAMA before it becomes effective.

Chile. Santiago, Chile, has a serious diesel particulate problem caused in large part by urban buses. To address this problem it has introduced a stringent smoke inspection program. In addition, it introduced a one day a week ban on driving with exemptions granted only to diesel buses equipped with catalysts or traps. In October 1992, the exemption program was replaced by an auction system designed to reduce the number of buses. Essentially, only 6000 buses have been granted a license to operate in the center of the city, down from approximately 9000. While the criteria for granting such licenses did not explicitly include emissions, it is intended to include particulate or smoke levels in a follow up program.

Conclusions

Diesel particulate controls are getting increasingly stringent around the world. As it has previously with gasoline-fueled vehicles, California is taking the lead in advancing diesel particulate control. CARB believes that the TLET, LET, and LEB emission levels are achievable by using current technology methanol or natural gas engines. For diesel-fueled engines to meet base 1996 bus and base 1998 truck standards, new or radically redesigned versions of current engines and a significant advancement in aftertreatment technology will be required. According to its contractor, Acurex, by using a combination of high pressure fuel injection, varaible geometry turbocharger, air-to-air aftercooler, optimized combustion chamber, electronic unit injections with minimized sac volumes, rate shaping, exhaust gas recirculation, and sophisticated electronic control of all engine systems, diesel engines could meet a 2.5 g/BPH-h NO_x standard at 0.15 g/BHP-h PM with a 5% penalty in fuel economy. Particulate traps could be used to reduce the particulate emissions to 0.05 g/BHP-h. Durability of the engine may be reduced to 80% of the 1994 counterpart. Several manufacturers and research organizations are studying these refinements in diesel technology and predict such engines may be available as early as 1998. Concentrated research and development, tied with demonstration programs, will be needed to bring these engines into reality.

Motorcycles

On February 2, 1988 the Swiss government decided to set new emission standards for motorcycles to go into effect by October 1990. These prescriptions will focus mainly on VOC control for two-stroke engines, whereas the standards for four-stroke motorcycles remain unchanged.

	Standards by Oct. 1, 1990 Two-stroke Engines, g/km	Present Standards	
		Two-stroke Engines, g/km	Four-stroke Engines, g/km
CO	8.0	8.0	13.0
VOC	3.0	7.5	3.0
NO_x	0.1	0.1	0.3

On July 1, 1991, the motorcycle standards in Taiwan were tightened to 4.5 g/km for CO (from 8.8), and 3.0 for HC and NO_x combined (from 6.5), based on the ECE R40 test procedure. With introduction of these requirements, Taiwan has the most stringent motorcycle control in the world, likely requiring use of catalytic converter controls.

GLOBAL EMISSIONS PROJECTIONS INDICATE MORE MUST BE DONE

Global Vehicle Emissions Trends by Region

Based on a continuation of the strong motor vehicle control programs in the United States and Japan and the recent tightening of requirements in EC Europe, these estimates indicate that global CO, HC, and NO_x emissions will remain fairly stable throughout the next decade. Beyond that point, however, emissions of all three pollutants will start to increase due to the projected continued growth in vehicle populations both in OECD countries but especially in other areas of the world where emissions controls are frequently minimal. This upturn in hydrocarbon and NO_x emissions will occur shortly after the turn of the century; for carbon monoxide, the downward trend is expected until about 2020 when it will also turn up.

Global Vehicle Emissions Trends, by Vehicle Type

An analysis of the trends in global emissions of CO, HC, and NO_x by vehicle type provides some startling insights. First of all, not surprisingly, cars remain the dominant source of CO for the foreseeable future. However, motorcycles, most of which are two stroke, are seen to be a significant contributor to HC emissions around the world, a fact that is largely ignored in the West due to their small contribution in that region. In regard to NO_x emissions, heavy-duty trucks are a large and rapidly growing contributor, due to the minimal NO_x control of these vehicles in most regions of the world.

TECHNOLOGICAL DEVELOPMENTS

Gasoline-Fueled Vehicles

California has demonstrated a commitment to achieving the lowest possible in use emissions levels through a comprehensive program of tighter standards, extended durability requirements, and technological (eg, onboard diagnostics) and regulatory (eg, defect reporting) innovation. Some important elements of the CARB program as it existed in early 1990 have in broad terms been mandated by the Clean Air Act Amendments of 1990 to apply to federal vehicles across the country. In regard to vehicle technology, the only comparable period of emissions and fuel technological push was following the 1970 Clean Air Act Amendments when within 5 yr unleaded fuel went from virtually zero to being the norm for all new cars, catalytic converters, which previously had virtually no real-world experience, were introduced in four out of five new cars, and the electronics revolution transformed the control mechanisms of such critical emission control components as spark and air fuel management systems. Other than controls that have already emerged from the laboratory such as preheated converters, it is difficult to anticipate fundamental technological breakthroughs that may emerge over the course of the coming decade. Therefore, in a context in which strong regulatory requirements create the most favorable environment for technological breakthroughs, another element of conservatism in the analysis that follows is that it can only be based on currently known technologies.

Technology Overview. The level of tailpipe hydrocarbon emissions from modern vehicles is primarily a function of the engine-out emissions and the overall conversion efficiency of the catalyst, both of which depend highly on proper function of the fuel and ignition systems. A fairly comprehensive system has evolved. A significant portion of the HC and CO emissions are generated during cold start, when the fuel system is operating in a rich mode and the catalyst has not yet reached its light-off temperature. There are many technological improvements, which are currently becoming widespread or are on the horizon, that make more stringent control of HC and CO feasible. These advances are expected not only to reduce the emission levels that can be achieved in the certification of new vehicles but also to reduce the deterioration of vehicle emissions in customer service.

First is the trend toward increased use of fuel injection. Fuel injection has several distinct advantages over carburetion as a fuel control system: more precise control of fuel metering, better compatibility with digital electronics, better fuel economy, and better cold-start function. Fuel metering precision is important in maintaining a stoichiometric air: fuel ratio for efficient three-way catalyst operation. Efficient catalyst operation, in turn, can reduce the need for dual-bed catalysts, air injection, and EGR. Better driveability from fuel injection has been a motivating force for the trend to convert engines from carburetion to fuel injection. In fact, it has been projected that the percentage of new California light-duty vehicles with fuel-injection will reach 95% by the early 1990s, with 70% being multipoint. Because of the inherently better fuel control provided by fuel-injection systems, this trend is highly consistent with more stringent emissions standards.

Fuel injection's compatibility with onboard electronic controls enhances fuel metering precision and also gives manufacturers the ability to integrate fuel control and emissions control systems into an overall engine management system. This permits early detection and diagnosis of malfunctions, automatic compensation for altitude, and to some degree, adjustments for normal wear. Carburetor choke valves, long considered a target for maladjustment and tampering, are replaced by more reliable cold-start enrichment systems in fuel-injected vehicles.

Closed-loop feedback systems are critical to maintain good fuel control, although when they fail, emissions can increase significantly. In fact, the CARB in-use surveillance data show that failure of components in the closed-loop system frequently has been associated with high emissions. The CARB's new requirement for onboard diagnostics will enable the system to alert the driver when something is wrong with the emission control system and will help the mechanic to identify the malfunctioning component.

Second, improvements to the fuel control and ignition systems, such as increasing the ability to maintain a stoichiometric air: fuel ratio under all operating conditions and minimizing the occurrence of spark plug misfire, will result in better overall catalyst conversion efficiency and less opportunity for catastrophic failure. These improvements, therefore, have a twofold effect: (*1*) limiting the extra engine-out emissions that would be generated by malfunctions and (*2*) helping keep the catalyst in good working condition.

Third, a trend that bodes well for catalyst deterioration rates (and therefore in-use emissions) is EPA's lead phaseout, which reduced the lead content of leaded gasoline by about 90% (to 0.1 g/gal) beginning in January 1986. The new Clean Air Act will actually lead to a complete ban on lead in gasoline in a few years. The phaseout will also reduce the small lead content of unleaded gasoline (since these amounts are due to contact with leaded gasoline facilities), which will reduce gradual low-level catalyst poisoning. Both catalyst and O_2 sensor durability will benefit from these lower gasoline lead contents.

Finally, there are alternative catalyst configurations that could and likely will be used in the future to meet lower emission standards. It is likely that dual-bed catalysts will be phased out over time, but a warmup catalyst (preceding the TWC) could be used for cold-start hydrocarbon control. To avert thermal damage and lower the catalyst deterioration rate, this small catalyst could be bypassed at all times other than during cold-start. Warmup air injection could also be used with a single-bed TWC for cold-start hydrocarbon control. As hydrocarbon standards are lowered, preheated catalysts will likely become a more important element of the pollution control system of many cars.

Other variables include engine-out emissions, which depend highly on the conditions in the combustion chamber. Over the past few years, combustion chamber geometry and turbulence levels have been optimized in an effort to minimize emissions and maximize fuel economy. Fast-burn combustion, which is being used more and more, involves changes to chamber geometry, turbulence, and the location of the spark plug. It allows greater use of EGR for NO_x control without hurting efficiency, making simultaneous control of hydrocarbons and NO_x easier.

Ignition misfire is often due to fouled or faulty spark plugs, deteriorated spark plug wires, or other ignition component malfunctions. Greater durability of the ignition system, especially spark plugs, limits misfires and resulting thermal damage to catalysts. New ignition systems currently under development (and being used experimentally by Saab and Nissan) may virtually eliminate misfires and the need for high voltage spark plug wires.

There is inevitable uncertainty associated with predicting the specific technology that manufacturers will apply to future vehicles to comply with the mandates of the LEV program. Historical evidence demonstrates that strong regulatory requirements create a favorable environment for technological breakthroughs. For example, in the 5-yr period following the adoption of the 1970 Clean Air Act Amendments, unleaded fuel use, catalytic converter technology, and electronic control mechanisms quickly became the norm. It is likely that compliance with the LEV emission standards will result in the development of new technology that is not fully accounted for in this analysis.

Virtually all vehicles used the following emission control technology to meet California 1991 emission standards:

Closed-loop, single-bed, three-way catalysts.

Fuel injection, usually multipoint.

No secondary air.

To meet the 1994 federal standards as well as to respond to competitive pressures manufacturers are likely to use the following:

- Improved catalyst formulations with higher noble metal loadings.
- Sequential multipoint fuel injection.
- Direct-fire (distributorless) ignition systems.

The latter two technologies allow for more precise control of air: fuel ratio and spark timing, relative to current technology. Improvements to catalyst formulations, coupled with reduction in gasoline contaminants such as sulfur and lead, and reduction in oil additive-based contaminants such as phosphorus will lead to reductions in catalyst deterioration, the principal cause of emissions deterioration of well-maintained cars. In addition, the electronic control and OBD systems may reduce production line variability and assist in improving in-use durability.

Evolutionary improvements to engines will also aid in meeting the 1994 federal standards. These improvements will include tuned inlet manifolds designed to reduce cylinder-to-cylinder variability, revised pistons to reduce crevice volume and oil consumption, and the incorporation of fast-burn combustion chambers. A number of engines in production today already feature such improvements.

Transitional low emission vehicles must meet a standard of 0.125 g/mi for NMHC, 3.4 g/mi for CO, and 0.4 g/mi NO_x. The California Air Resources Board projected in 1990 that for small- and medium-displacement engines, only heated fuel preparation systems and/or close-coupled catalyst completed systems will be required to meet TLEV standards. Many current gasoline engine families have certified to TLEV standards with calibration changes and evolutionary changes to hardware. The calibration changes required to meet TLEV standards will include spark retard during warmup, more careful control of air: fuel ratio after cold start, and (possibly) electronic control strategies to control precisely individual cylinder air : fuel ratio.

Low emission vehicles must meet a standard of 0.075 g/mi for CH, 3.4 g/mi for CO, and 0.2 g/mi for NO_x. Based on the most recent data, it is clear that the pace of technological progress has exceeded expectations and that the low emission vehicle standards will require less advanced technology than previously expected. Less than a year ago, to produce LEVs, manufacturers were expected to use advanced fuel control strategies, greater catalytic loading for better NO_x control for some models, and electrically heated catalysts (EHCs). In the spring of 1992, however, Ford submitted to CARB a certification application containing data that strongly indicated that LEV levels can be achieved without using EHCs. In addition to these certification materials, CARB became aware of other significant advances being made by the automotive industry to reduce emission levels. Because the rate of progress has exceeded original expectations, CARB has revised its assessment of the technologies needed in each low emission vehicle category.

At an Air Resources Board public meeting held June 11, 1992, to consider the status of implementation of the low emission vehicle program, the CARB staff outlined its current technology assessment. In summary, for TLEVs, it continues to project that only modest fuel control and catalyst improvements will be necessary. For LEVs, small- to medium-displacement engines should be able to achieve the standards with fuel control improvements and improved conventional catalysts; larger eight-cylinder engines may still require the use of poststart heated EHCs or other similarly effective technologies. For ULEVs, in addition to fuel control improvements and greater catalyst loading, CARB expects that poststart heated EHCs will be sufficient for most vehicles, although prestart heated units may be needed for some applications.

To support its new assessment of LEV technology, the CARB staff presented emission data that showed that the emission levels from a 1993 Ford Escort TLEV and a 1992 Oldsmobile Achieva were at or below the LEV standards. Just after the June 11 meeting, the CARB certified the 1993 model of the Achieva and the 1993 Honda Civic to TLEV standards. In Table 6 the certification emission levels of the Escort, Achieva, and Civic TLEVs are summarized. The Escort and Civic were certified on Indolene while the Achieva was certified on California Phase 2 gasoline.

Although certified as TLEVs, exhaust emissions of the Escort were only 77% of the LEV standard, while the emissions from the Achieva were even lower, only 65% of the LEV standard for NMOG at 50,000 miles. At 100,000 miles, the margins for compliance are even greater: 69% and 54% for the Escort and Achieva, respectively.

Advanced Engine Modifications to Achieve Low In-use Levels. The emission levels of the two vehicles were surprisingly low in view of the fact that these vehicles used little or none of the advanced technologies available to significantly reduce emissions even further. In Table 7, the advanced technologies incorporated by the Escort and Achieva are listed.

Table 6. Certification Emission Levels (g/mi) of Three 1993 TLEV Engine Families

Mileage	Model	Displacement, L	NMOG	CO	NO_x
50,000	Ford Escort	1.9	0.058	1.0	0.2
50,000	Olds Achieva	2.3	0.049	0.4	0.1
50,000	Honda Civic	1.5	0.076	0.3	0.2
50,000	LEV Standard		0.075	3.4	0.2
100,000	Ford Escort	1.9	0.062	1.3	0.3
100,000	Olds Achieva	2.3	0.049	0.4	0.1
100,000	Honda Civic	1.5	0.086	0.3	0.2
100,000	LEV Standard		0.090	4.2	0.3

Table 7. Advanced Technologies Used on the Escort and Achieva

Technology	Escort	Achieva
Heated oxygen sensor	X	
Air pump		
Sequential multiport fuel injection	X	
Aerated fuel injectors		
Heated fuel preparation		
Cylinder fueling torque control		
Dual oxygen sensor		
Adaptive transient fuel control		

As is clearly seen from Table 7, the Escort incorporates only two of the identified technologies, while the Achieva does not use any. Because both vehicle types have already shown certification emission levels (which includes a 100,000-mi durability demonstration) well below the LEV standards, even without the addition of some of these available technologies, the prospects for these and similar vehicles attaining compliance with the LEV standards in-use are excellent. With the additional technologies, virtually all but the largest vehicles should be able to be certified as LEVs without the use of EHCs.

Dual Oxygen Sensors. Toyota has been using dual oxygen sensor systems since 1988. It makes good engineering sense that dual oxygen sensors will help maintain low emission levels in-use. As oxygen sensors age, their warmup response slows considerably, and the air : fuel ratio at which the sensor switches from low to high voltage (and vice versa) can shift significantly from stoichiometric, thereby increasing emissions. Because the second sensor is placed downstream of the catalyst, it operates in a relatively low temperature environment and is better protected from poisons. It, therefore, can be used to compensate for slow response and to adjust for changes in the switch point of the front sensor. In this way, a dual oxygen sensor system helps maintain good fuel control, and consequently good emissions control, as vehicles age.

Sequential Fuel Control. Precise injection timing may be helpful in minimizing hydrocarbon emissions under steady-state conditions. An effective way to use sequential fuel injection is to optimize injection timing to occur while the intake valve is open. This can be accomplished by using aerated fuel injectors to eliminate the need to rely on evaporation, thereby allowing direct injection to significantly less fuel into the combustion chamber.

Aerated Fuel Injectors. Toyota's aerated fuel injection system demonstrates emission benefits not only over the full range of steady-state conditions but under transient and cold conditions as well. Use of aerated fuel injectors ensures good atomization during cold starts and, therefore, permits injection directly into the combustion chamber when the intake valve is open. This strategy reduces the amount of excess fuel needed to avoid driveability problems due to wall wetting, thereby minimizing emission increases and fuel economy degradation. This is especially important during transient engine operation (eg, during rapid accelerations) when excess fuel is normally added to prevent engine stumbles and sags.

Adaptive Fuel Control. CARB believes that while adaptive control of steady-state engine operating conditions has been used for some time by industry, only a few manufacturers have been successful in developing adaptive strategies for transient operating conditions. In fact, to the knowledge of the CARB staff, the first application of this technology is on a few 1992 Toyota models, and it is aware of only one other major manufacturer that is planning to employ adaptive transient fuel control in the near future.

Individual Cylinder Torque Control. CARB staff discussions with numerous manufacturers indicate that measurement of cylinder torque is being examined as a means for controlling the entire fueling strategy for their engines, enabling them to reduce the dependence on oxygen sensors for primary fuel altogether. This same technique is being used by the automotive industry for misfire detection as part of the CARB's On-Board Diagnostic II (OBD II) misfire detection requirements. It may not be necessary to monitor all engine operating conditions to determine the level of fuel compensation needed over the full range of engine operating conditions if a particular injector is underfueling or overfueling (eg, due to a partially restricted or leaking injector). In CARB's view, torque changes become more pronounced as load increases. Under more moderate load conditions and more representative vehicle driving modes, as opposed to idle conditions, torque fluctuation should be readily detectable well within a one air : fuel ratio variation.

Heated Fuel Injectors and Heated Fuel Preparation Systems. Fuel heating strategies have been used on production vehicles for some time now. Mercedes Benz has used an intake manifold heater on the 2.3-L engines in its 190 series since 1991. As with aerated fuel injectors, improved fuel vaporization allows the use of leaner air : fuel mixtures during cold engine operation, thereby providing an even greater HC and CO emission reduction.

Air Injection. To the extent that air–fuel enrichment is needed during cold engine operation to provide acceptable driveability, air injection can be utilized to provide the additional air needed fully to oxidize the excess HC and CO emissions in the catalytic converter. The need for air injection can be lessened by incorporating aerated fuel injectors, sequential fuel injection, and adaptive transient fuel compensation, either singly or in combination, to reduce the level of cold engine operating enrichment needed to achieve acceptable driveability. Air injection is a highly effective means of reducing NMOG and CO emissions but may not be needed to meet the relatively less stringent Tier I or TLEV emission levels. In fact, none of the 1993 TLEVs certified by the CARB utilize air injection. However, air pumps have been an important element in achieving reduced NMOG and CO emissions on the CARB's EHC-equipped vehicles and would also be effective in achieving LEV emission levels for vehicles without EHCs.

Heated Oxygen Sensors. Heated oxygen sensors, a technology commonly used on many of today's vehicles, will also help reduce emissions. Heating oxygen sensors is important because as oxygen sensors age, they require higher operating temperatures to maintain adequate responsiveness to changes in air : fuel ratio, particularly during cold start operation. Slow response sensors can prolong the time required to switch from open-loop to closed-loop operation or provide poor fuel control during

Table 8. In-use Compliance Test Results (g/mi) from Some 1989 Production Vehicles[a]

Manufacturer	Engine Family	NMHC	CO	NO_x
Volvo	KVV2.3V5FE8X	0.21	2.43	0.19
Ford	FKM2.2V5FXC4	0.23	4.88	0.14
Mitsubishi	KMT2.0V5FC18	0.22	2.19	0.20
Applicable standards		0.39	7.0	0.4

[a] Results shown are the average of 10 tests; mileage on the test vehicles ranged from 31,000 to 49,000.

the initial closed-loop operating period, thereby resulting in increased emissions. Oxygen sensor deterioration of this nature is most likely to occur after about 50,000 mi of driving. The deterioration can be masked, however, by electrically heating the sensor to operating temperature, thereby reducing the time needed to initiate closed-loop operation and minimizing emission increases due to improper air : fuel ratios caused by slow oxygen sensor response rates.

Additional Rhodium Loading for NO_x Control. The rhodium levels of many current vehicles are sufficient to enable achievement and maintenance of a 0.2 g/mi NO_x standard in-use. CARB in-use compliance records reveal that several 1989 models were able to achieve 0.2 g/mi NO_x levels in-use (even without the application of advanced fuel controls). The emission results of these vehicles are listed in Table 8.

Some manufacturers may need to add rhodium to match the levels used by these better performing vehicles. Adding rhodium to the catalyst is only one option for achieving low NO_x emissions. Improved control of the air : fuel ratio at stoichiometric can also improve NO_x emissions. Another option was proposed in the July 1992, issue of *Automotive Engineering,* which concluded that less expensive palladium could be a viable replacement for some of the rhodium in catalytic converters. For these reasons, the CARB considers increased rhodium loading to be an effective strategy for reducing NO_x emissions and for maintaining NO_x standards in use.

Ultra-low emission vehicles must meet a standard of 0.04 g/mi for HC, 1.7 g/mi for CO, and 0.2 g/mi for NO_x. At the very low emission levels of 0.040 HC (NMOG), it appears today that it will be necessary to have some form of additional exhaust aftertreatment. Several types of after-treatment are being investigated:

Close-coupled start catalysts.
Exhaust port catalysts.
Electrically heated catalysts.
Hydrocarbon molecular sieves.

Electrically Heated Catalysts. The electrically heated catalyst (EHC) has received the most attention and has demonstrated the potential to meet ULEV standards even in large, heavy cars. The ARB, EPA, and a catalyst manufacturer (CAMET) have collaborated in extensive testing of these types of catalysts. The main drawback of these catalysts had been that they impose a large current drain on the battery for 15–30 s before cold start. While the battery problems are not large at ambient temperatures of 21°C they may be more significant at winter temperatures of −7°C, when heating requirements are increased while battery capacity is decreased.

Innovative solutions to these problems are being researched. To support its upcoming rule making to establish reactivity adjustment factors for Phase 2 gasoline-fueled low emission vehicles, the CARB staff recently tested several late model vehicles that were retrofitted with EHCs. The warmed-up emissions performance of these vehicles are considered representative of the emission levels CARB expects to see from low emission vehicles (Table 9).

For the purpose of obtaining reactivity data to support its November rulemaking, the CARB staff has continued to conduct tests on EHC-equipped vehicles. Its most recent data using Camet's latest EHC on the Buick indicate that energy requirements range from 29 to 42 W · h (Table 10). Even then, the EHCs being used on CARB test vehicles are stand-alone units to facilitate fabrication of the new exhaust systems. Ideally, the EHCs should be close mounted to the main catalyst and contained in the same housing to minimize heat loss, thereby reducing energy

Table 9. Per Bag Emission Results (g/mi) for Late Model Year Vehicles[a]

Vehicle	Test	Bag 1	Bag 2	Bag 3	FTP Composite
1992 Lexus LS400	NMHC	0.204	0.000	0.000	0.041
	CO	1.766	0.078	0.079	0.429
	NO_x	0.313	0.131	0.248	0.201
1988 Chevrolet Corsica	NMHC	0.230	0.003	0.019	0.054
	CO	1.692	0.142	0.794	0.643
	NO_x	0.284	0.125	0.366	0.224
1990 Toyota Celica	NMHC	0.192	0.000	0.009	0.042
	CO	2.003	0.076	0.194	0.508
	NO_x	0.666	0.154	0.084	0.241
1991 VW Jetta	NMHC	0.125	0.002	0.023	0.034
	CO	2.645	0.852	1.470	1.392
	NO_x	0.357	0.038	0.184	0.144

[a] Equipped with EHCs by the CARB; vehicles operated on 1989 industry average gasoline.

Table 10. EHC Energy Requirements of Recent CARB Tests[a]

Test Date	Vehicle	Test Number	Watt · Hours	NMHC
7/10/92	Buick	28-10	33.9	0.027
7/14/92	Buick	28-11	37.2	0.025
7/16/92	Buick	28-13	38.0	0.032
7/24/92	Buick	28-15	28.8	0.027
7/29/92	Buick	28-16	33.8	0.044
7/31/92	Buick	28-17	41.7	0.022

[a] The July 1992 CARB test data involved various EHC heating and air injection strategies that contributed to the variability in the results.

requirements even further. The energy requirements of other new EHC designs arranged in this manner are much lower, as shown in Table 11. These results on a Honda Accord using an EHC developed by Corning, Inc. show NMHC emissions of 0.022 g/mi for 25 s of preheat while using only 14.8 W · h of electrical energy. For 25 s of postheat, NMHC emissions were 0.046 g/mi with an energy requirement of 14.8 W · h. The test results also indicate that heating time, electrical energy demand, and emissions should decrease even further if a 24-V energy source is used.

Energy Sources for EHCs. Recent advances made by developers of advanced power sources such as ultracapacitors and sealed bipolar lead acid (SBLA) batteries have been encouraging. Idaho National Engineering Laboratory (INEL) has provided the CARB with test data from a Panasonic carbon-based ultracapacitor and a mixed-metal oxide ultracapacitor. The data indicate that Panasonic's ultracapacitor is capable of surviving 350,000 EHC charge-discharge cycles with minimal performance losses. Tests conducted with a Camet EHC indicated that the Panasonic ultracapacitor could easily provide the energy needed to achieve a catalyst temperature well above light-off within 10 s. The ultracapacitor assembly weighs less than 4.536 kg and is similar in size to a conventional automotive battery. The mixed-metal oxide ultracapacitor appears to have the potential for even better performance than the Panasonic device and should be much smaller.

Other organizations have also recognized the potential of the EHC market and are conducting research and development on advanced energy sources for EHCs. A large battery manufacturer has been developing a deep-discharge lead-acid battery intended specifically for EHC applications. This battery has been successfully life cycle tested and is claimed to be lighter and smaller than conventional lead–acid batteries. The CARB has also been working with two other organizations to develop an SBLA battery; one of these organizations is also developing the mixed metal oxide ultracapacitor referenced above for installation on one of the CARB's test vehicles.

CARB believes that current leakage from ultracapacitors is not a problem. Improvements in contamination control have curtailed current leakage to insignificant levels. INEL has tested ultracapacitors with current leakage in the microampere to milliampere range. Also, it is well recognized in this industry that ultracapacitor current leakage can be virtually eliminated by simply maintaining a potential difference across the ultracapacitor, eg, by using the vehicle battery.

Other Strategies. While CARB has identified several technologies to achieve the low emission vehicle standards, it emphasizes that the regulations contain sufficient flexibility to accommodate nearly any strategy that can be used to reduce emissions. One viable strategy for achieving LEV and ULEV emission levels is the use of bypass start catalysts. In such a system, a small, quick light-off catalyst is mounted close to the exhaust manifold upstream of the main catalyst. During cold starts, exhaust emissions are routed to the start catalyst, thus bypassing the main catalyst. Because the start catalyst is rapidly heated to light-off temperature, it treats the exhaust gases almost immediately. The exhaust gases are routed back to the main catalyst once it has reached a sufficient operating temperature.

Bypass start catalysts should have quick light-off characteristics similar to EHCs but without the power requirements of EHCs or any fuel economy penalty. Also, because the start catalyst is used only for a short period of time during cold starts, it should be fairly durable despite its close-coupled position. Therefore, although the CARB staff believes that EHCs will continue to be more effective in the long-term, bypass start catalysts appear to be a viable strategy for manufacturers that consider EHCs to be undesirable.

In addition, there are a number of other strategies that may be used to produce a LEV or ULEV. Some of the options being explored by the automotive industry include direct injection two-stroke engines, hydrocarbon vapor traps, exhaust gas ignition, and hybrid electric vehicles. EPA has completed its initial evaluation of the direct injection two-stroke engine with promising results. One vehicle, a Ford Fiesta, had average composite FTP emission levels for all testing conducted at EPA of 0.05/g/mi nonmethane hydrocarbons, 0.2 g/mi CO, and 0.2 g/mi NO_x. In addition, average combined fuel economy was 50.2 mpg. The vehicle is equipped with a single close coupled oxidation catalyst. The CO levels and fuel economy values

Table 11. Corning EHC Test Results (June 8, 1992)

Description	Vehicle	Watt · Hours	NMHC
12-V battery, 25 s preheat	Honda Accord	14.8	0.022
24-V battery, 5 s preheat	Honda Accord	12.0	0.019
24-V battery, 25 s postheat	Honda Accord	11.6	0.027
12-V battery, 25 s postheat	Honda Accord	14.8	0.046

especially stand out relative to other clean, efficient production vehicles tested by EPA.

The hydrocarbon trap or molecular sieve is a zeolite-type material that can adsorb hydrocarbons at low temperatures, and then release them when heated to higher temperatures. It has been suggested that such materials can be used to adsorb cold-start HC before catalyst light off, and then release the HC after the catalyst is operating at high efficiency. Such a material is more energy efficient than an EHC system, and is potentially cheaper as it may not require an electrical heating and control system. Unfortunately, zeolites are temperature sensitive and little is known about their durability. In addition, research on these materials has been kept highly confidential, and no data are publicly available to gauge their efficiency or durability.

The California Air Resources Board has projected that alternative fuel (CNG and methanol) vehicles will be able to meet the LEV standards with less additional technology than gasoline-fueled vehicles. In fact, the low vapor pressure of methanol requires more cold-start enrichment and presents greater difficulty in controlling cold-start related emissions. Cold-start related formaldehyde can also be an obstacle (as it has a higher reactivity index) in meeting the reactivity weighted HC standards. However, recent tests on M85 vehicles with an electrically heated catalyst have been encouraging, as aldehyde emissions on the FTP were less than 5 mg/mi. CARB staff also stated that recent tests on a M85 vehicle at low mileage showed the potential to meet even the ULEV standards with a start catalyst–main catalyst system without electrical heat. Methanol-fueled vehicles emit a range of oxygenated compounds that are improperly measured by current systems used to measure HC emissions from gasoline vehicles. It is not clear if the measurements of HC from the M85 vehicles cited by CARB staff accounted for all of oxygenated HC emissions.

Conclusions Regarding Gasoline-fueled Technology

Starting in 1975, when HC and CO standards were tightened as a result of the 1970 Clean Air Act Amendments, the first generation oxidation catalysts were introduced on a vast majority of all new cars in the United States. By 1981, when the NO_x standard was tightened, the second-generation systems, three-way catalysts took over much of the market. As a result of the Clean Air Act Amendments of 1990 as well as the California decisions of 1989, the third-generation systems, with greater durability will begin to be introduced in 1993. As a result of the California decisions of September 1990 (coupled with Section 177 of the Clean Air Act, which allows other states to adopt the California program, a fourth generation will likely begin to be introduced by the mid to late 1990s, one based on either very rapid light off or even preheating.

FACTORS AFFECTING DIESEL EMISSIONS

Diesel engine emissions are determined by the combustion process. This process is central to the operation of the diesel engine. As opposed to Otto-cycle engines (which use a more or less homogeneous charge) all diesel engines rely on heterogeneous combustion. During the compression stroke, a diesel engine compresses only air. The process of compression heats the air to about 700° to 900°C, which is well above the self-ignition temperature of diesel fuel. Near the top of the compression stroke, liquid fuel is injected into the combustion chamber under tremendous pressure, through a number of small orifices in the tip of the injection nozzle. The quantity of fuel injected with each stroke determines the engine power output.

The high pressure injection atomizes the fuel. As the atomized fuel is injected into the chamber, the periphery of each jet mixes with the hot air already present. After a brief period known as the ignition delay, this fuel–air mixture ignites. In the premixed burning phase, the fuel–air mixture formed during the ignition delay period burns very rapidly, causing a rapid rise in cylinder pressure. The subsequent rate of burning is controlled by the rate of mixing between the remaining fuel and air, with combustion always occurring at the interface between the two. Most of the fuel injected is burned in this diffusion burning phase, except under light loads.

A mixture of fuel and exactly as much air as is required to burn the fuel completely is called a "stoichiometric mixture." The air : fuel ratio is defined as the ratio of the actual amount of air present per unit of fuel to the stoichiometric amount. In diesel engines, the fact that fuel and air must mix before burning means that a substantial amount of excess air is needed to ensure complete combustion of the fuel within the limited time allowed by the power stroke. Diesel engines, therefore, always operate with overall air : fuel ratios which are considerably lean of stoichiometric.

The air : fuel ratio in the cylinder during a given combustion cycle is determined by the engine power requirements, which govern the amount of fuel injected. Diesels operate without throttling, so that the amount of air present in the cylinder is essentially independent of power output, except in turbocharged engines. The minimum equivalence ratio for complete combustion is about 1.5. This ratio is known as the smoke limit, because smoke increases dramatically at equivalence ratios lower than this. The smoke limit establishes the maximum amount of fuel that can be burned per stroke and thus the maximum power output of the engine. This minimum air : fuel ratio explains why NO_x reduction catalysts of the type used on gasoline automobiles, which rely on stoichiometric air : fuel ratios, are not effective for diesel engines.

Pollutant Formation

The principal pollutants emitted by diesel engines are oxides of nitrogen, (NO_x), sulfur oxides (SO_x), particulate matter, and unburned hydrocarbons. Diesels are also responsible for a small amount of CO as well as visible smoke, unpleasant odors, and noise. In addition, like all engines using hydrocarbon fuel, diesels emit significant amounts of CO_2, which has been implicated in the so-called greenhouse effect. With thermal efficiencies typically in excess of 40%, however, diesels are the most fuel efficient of all common types of combustion engines. As a result, they emit less CO_2 to the atmosphere than any other type of engine doing the same work.

The NO_x, HC, and most of the particulate emissions from diesels are formed during the combustion process and can be reduced by appropriate modifications to that process, as can most of the unregulated pollutants. The sulfur oxides, in contrast, are derived directly from sulfur in the fuel, and the only feasible control technology is to reduce fuel sulfur content. Most SO_x is emitted as gaseous SO_2, but a small fraction (typically 2–4%) occurs in the form of particulate sulfates.

Diesel particulate matter consists mostly of three components: soot formed during combustion, heavy hydrocarbons condensed or adsorbed on the soot, and sulfates. In older diesels, soot was typically 40–80% of the total particulate mass. Developments in in-cylinder emissions control have reduced the soot contribution to particulate emissions from modern emission controlled engines considerably, however. Most of the remaining particulate mass consists of heavy hydrocarbons adsorbed or condensed on the soot. This is referred to as the soluble organic fraction (SOF) of the particulate matter. The SOF is derived partly from the lubricating oil, partly from unburned fuel, and partly from compounds formed during combustion. The relative importance of each of these sources varies from engine to engine.

In-cylinder emission control techniques have been most effective in reducing the soot and fuel derived SOF components of the particulate matter. As a result, the relative importance of the lube oil and sulfate components has increased. In the emission-controlled engines under development, the lubricating oil accounts for as much as 40% of the particulate matter, and the sulfates may account for another 25%. Lube oil emissions can be reduced by reducing oil consumption, but this may adversely affect engine durability. The only known way to reduce sulfate emissions is to reduce the sulfur content of diesel fuel.

The gaseous hydrocarbons and the SOF component of the particulate matter emitted by diesel engines include many known or suspected carcinogens and other toxic air contaminants. These include polynuclear aromatic compounds (PNA) and nitro-PNA, formaldehyde and other aldehydes, and other oxygenated hydrocarbons. The oxygenated hydrocarbons are also responsible for much of the characteristic diesel odor.

Oxides of nitrogen (NO_x) from diesels is primarily NO_2. This gas forms from nitrogen and free oxygen at high temperatures close to the flame front. The rate of NO_x formation in diesels is a function of oxygen availability, and is exponentially dependent on the flame temperature. In the diffusion burning phase, flame temperature depends only on the heating value of the fuel, the heat capacity of the reaction products and any inert gases presenting temperature of the initial mixture. In the premixed burning stage, the local fuel : air ratio also affects the flame temperature, but this ratio varies from place to place in the cylinder and is very hard to control.

In diesel engines, most of the NO_x emitted is formed early in the combustion process, when the piston is still near top dead center (TDC). This is when the temperature and pressure of the charge are greatest. Recent work by several researchers (3,4) indicates that most NO_x is actually formed during the premixed burning phase. It has been found that reducing the amount of fuel burned in this phase can significantly reduce NO_x emissions.

NO_x can also be reduced by actions that reduce the flame temperature during combustion. These actions include delaying combustion past TDC, cooling the air charge going into the cylinder, reducing the air : fuel mixing rate near TDC, and exhaust gas recirculation (EGR). Because combustion always occurs under near stoichiometric conditions, reducing the flame temperature by "lean burn" techniques, as in spark-ignition engines, is impractical.

Particulate matter (or soot) is formed only during the diffusion burning phase of combustion. Primary soot particles are small spheres of graphitic carbon, approximately 0.01 mm in diameter. These are formed by the rapid polymerization of acetylene at moderately high temperatures under oxygen deficient conditions. The primary particles then agglomerate to form chains and clusters of linked particles, giving the soot its characteristic fluffy appearance. During the diffusion burning phase, the local gas composition at the flame front is close to stoichiometric, with an oxygen rich region on one side and a fuel rich region on the other. The moderately high temperatures and excess fuel required for soot formation are thus always present.

Most of the soot formed during combustion is subsequently burned during the later portions of the expansion stroke. Typically, less than 10% of the soot formed in the cylinder survives to be emitted into the atmosphere. Soot oxidation is much slower than soot formation, however, and the amount of soot oxidized is heavily dependent on the availability of high temperatures and adequate oxygen during the later stages of combustion. Conditions which reduce the availability of oxygen (such as poor mixing, or operation at low air : fuel ratios), or which reduce the time available for soot oxidation for soot oxidation (such as retarding the combustion timing) tend to increase soot emissions.

The SOF component of diesel particulate matter consists of heavy hydrocarbons condensed or adsorbed on the soot. A significant part of this material is unburned lubricating oil, which is vaporized from the cylinder walls by the hot gases during the power stroke. Some of the heavier hydrocarbons in the fuel may also come through unburned, and condense on the soot particles. Finally, heavier hydrocarbons may be synthesized during combustion, possibly by the same types of processes that produce soot. Pyrosynthesis of polynuclear aromatic hydrocarbons during diesel combustion has been demonstrated in Ref. 5.

Diesel hydrocarbon emissions (as well as the unburned fuel portions of the particulate SOF) occur primarily at light loads. They are caused by excessive fuel : air mixing, which results in some volumes of air : fuel mixture that are too lean to burn. Other HC sources include fuel deposited on the combustion chamber walls or in combustion chamber crevices by the injection process, fuel retained in the orifices of the injector that vaporizes late in the combustion cycle, and partly reacted mixture that is subjected to bulk quenching by too rapid mixing with air. Aldehydes (as partially reacted hydrocarbons) and the small amount of CO produced by diesels are probably formed in the same processes as the HC emissions.

The presence of polynuclear aromatic hydrocarbons and their nitro derivatives in diesel exhaust is of special concern, because these compounds include many known mutagens and suspected carcinogens. A significant portion of these compounds (especially the smaller two- and three-ring compounds) are apparently derived directly from the fuel. Typical diesel fuel contains several percent PNA by volume. Most of the larger and more dangerous PNAs, on the other hand, appear to form during the combustion process, possibly via the same acetylene polymerization reaction that produces soot (6). Indeed, the soot particle itself can be viewed as essentially a large PNA molecule.

Visible smoke is caused primarily by the soot component of diesel particulate matter. Under most operating conditions, the exhaust plume from a properly adjusted diesel engine is normally invisible, with a total opacity (absorbance and reflectance) of 2% or less. Visible smoke emissions from heavy-duty diesels are generally the result of operating at air : fuel ratios at or below the smoke limit or to poor fuel–air mixing in the cylinder. These conditions can be prevented by proper design. The particulate reductions required to comply with the U.S. 1991 emissions standards essentially eliminate visible smoke emissions from properly functioning engines.

Noise from diesel engines is due principally to the rapid combustion (and resulting rapid pressure rise) in the cylinder during the premixed burning phase. The greater the ignition delay, and the more fuel is premixed with the air, the greater this pressure rise and the resulting noise emissions will be. Noise emissions and NO_x emissions thus tend to be related reducing the amount of fuel burned in the premixed burning phase will tend to reduce both. Other noise sources include those common to all engines, such as mechanical vibration, fan noise, and so forth. These can be minimized by appropriate mechanical design.

Odor characteristic of diesel engines is believed to be due primarily to partially oxygenated hydrocarbons (aldehydes and similar species) in the exhaust. These are believed to be due primarily to slow oxidation reactions in volumes of air : fuel mixture too lean to burn normally. Unburned aromatic hydrocarbons may also play a significant role. The most significant aldehyde species are benzaldehyde, acetaldehyde, and formaldehyde, but other aldehydes such as acrolein (a powerful irritant) are significant as well. Aldehyde and odor emissions are closely linked to total HC emissions experience has shown that modifications which reduce total HC tend to reduce aldehydes and odor as well.

Influence of Engine Variables

The engine variables having the greatest effects on diesel emission rates are the air : fuel ratio, rate of air–fuel mixing, fuel injection timing, compression ratio, and the temperature and composition of the charge in the cylinder. Most techniques for in-cylinder emission control involve manipulating one or more of these variables.

Air : Fuel Ratio. The ratio of air to fuel in the combustion chamber has an extremely important effect on emission rates for hydrocarbons and particulate matter. As discussed above, the power output of the engine is determined by the amount of fuel injected at the beginning of each power stroke. At high air : fuel ratios (corresponding to light load), the temperature in the cylinder after combustion is too low to burn out residual hydrocarbons, so emissions of gaseous HC and particulate SOF are high. At lower air : fuel ratios, less oxygen is available for soot oxidation, so soot emissions increase. As long as the equivalence ratio remains above about 1.6, this increase is relatively gradual. Soot and visible smoke emissions in a direct injection diesel engine show a strong nonlinear increase below the smoke limit at an air : fuel ratio of about 1.5, however.

In naturally aspirated engines (those without a turbocharger), the amount of air in the cylinder is independent of the power output. Maximum power output for these engines is normally smoke limited, ie, limited by the amount of fuel that can be injected without exceeding the smoke limit. Maximum fuel settings on these engines represent a compromise between smoke emissions and power output. Where diesel smoke is regulated, this compromise must result in smoke opacity below the regulated limit. Otherwise, opacity is limited by the manufacturer's judgment of commercially acceptable smoke emissions.

In turbo-charged engines, increasing the fuel injected per stroke increases the energy in the exhaust gas, causing the turbo charger to spin more rapidly and pump more air into the combustion chamber. For this reason, power output from turbo charged engines is not usually smoke limited, although this depends on the rated power of the engine. Instead, it is limited by design limits on variables such as turbocharger speed and engine mechanical stresses. However, turbo-charged engines are smoke limited at low speeds.

Turbo-charged engines do not normally experience low air : fuel ratios during steady-state operation. Low air : fuel ratios can occur during transient accelerations, because the inertia of the turbo charger rotor keeps it from responding instantly to an increase in fuel input. Thus the air supply during the first few seconds of a full power acceleration is less than the air supply in steady-state operation. To overcome this problem, turbo-charged engines in highway vehicles commonly incorporate an acceleration smoke limiter. This device limits the fuel flow to the engine until the turbocharger has time to respond. The setting on this device must compromise between acceleration performance driveability and low smoke emissions. In the United States acceleration smoke emissions are limited by regulation; elsewhere, they are limited by the manufacturer's judgment of commercial acceptability.

Air–Fuel Mixing. The rate of mixing between the compressed charge in the cylinder and the injected fuel is among the most important factors in determining diesel performance and emissions. During the ignition delay period, the mixing rate determines the amount of fuel that mixes with the air and is thus available for combustion during the premixed burning phase. The higher the mixing rate, the greater the amount of fuel burning in premixed mode, and the higher the noise and NO_x emissions will tend to be.

In the diffusion burning phase, the rate of combustion is limited by the rate at which air and fuel can mix. The more rapid and complete this mixing, the greater the amount of fuel that burns near piston top dead center, the higher the fuel efficiency, and the lower the soot emissions. Too rapid mixing, however, can increase hydrocarbon emissions, especially at light loads as small volumes of air : fuel mixture are diluted below the combustible level before they have a chance to burn. Unnecessarily intense mixing also dissipates energy through turbulence, increasing fuel consumption.

In engine design practice, it is necessary to strike a balance between the rapid and complete mixing required for low soot emissions and best fuel economy, and too rapid mixing leading to high NO_x and HC emissions. The primary factors affecting the mixing rate are the fuel injection pressure, the number and size of injection orifices, any swirling motion impared to the air as it enters the cylinder during the intake stroke, and air motions generated by combustion chamber geometry during compression. Much of the progress in in-cylinder emissions control over the last decade has come through improved understanding of the interactions between these different variables and emissions, leading to improved designs.

Air–fuel mixing rates in present emission controlled engines are the product of extensive optimization to assure rapid and complete mixing under nearly all operating conditions. Poor mixing may still occur during "lug down" high torque operation at low engine speeds. Turbocharger boost, air swirl level, and fuel injection pressure are typically poorer in these "off design" conditions. Maintenance problems such as injector tip deposits can also degrade air–fuel mixing, and result in greatly increased emissions.

Injection Timing. The timing relationship between the beginning of fuel injection and the top of the compression stroke has an important effect on diesel engine emissions and fuel economy. For best fuel economy, it is preferable that combustion begin at or somewhat before top dead center. Because there is a finite delay between the beginning of injection and the start of combustion, it is necessary to inject the fuel somewhat before this point (generally 5 to 15° of crankshaft rotation before, although this could be 10 to 25° for engines in Europe).

Because fuel is injected before the piston reaches top dead center, the charge temperature is still increasing as the charge is compressed. The earlier fuel is injected, the cooler the charge will be, and the longer the ignition delay. The longer ignition delay provides more time for air and fuel to mix, increasing the amount of fuel that burns in the premixed combustion phase. In addition, more fuel burning at or just before top dead center increases the maximum temperature and pressure attained in the cylinder. Both of these effects tend to increase NO_x emissions.

On the other hand, earlier injection timing tends to reduce particulate and light load HC emissions. Fuel burning in premixed combustion forms little soot, while the soot formed in diffusion combustion near TDC experiences a relatively long period of high temperatures and intense mixing and is thus mostly oxidized. The end of injection timing also has a major effect on soot emissions; fuel injected more than a few degrees after TDC burns more slowly, and at a lower temperature, so that less of the resulting soot has time to oxidize during the power stroke. For a fixed injection pressure, orifice size, and fuel quantity, the end of injection timing is determined by the timing of the beginning of injection.

The result of these effects is that injection timing must compromise between PM emissions and fuel economy on the one hand and noise, NO_x emissions, and maximum cylinder pressure on the other. The terms of the compromise can be improved to a considerable extent by increasing injection pressure, which increases mixing and advances the end of injection timing. Another approach under development is split injection, in which a small amount of fuel is injected early to ignite the main fuel quantity, which is injected near TDC.

Compared with uncontrolled engines, modern emission controlled engines generally exhibit moderately retarded timing to reduce NO_x in conjunction with high injection pressures to limit the effects of retarded timing of PM emissions and fuel economy. The response of fuel economy and PM emissions to retarded timing is not linear; up to a point, the effects are relatively small, but beyond that point deterioration is rapid. Great precision in injection timing is necessary; a change of one degree crank angle can have a significant impact on emissions. The optimal injection timing is a complex function of engine design, engine speed and load, and the relative stringency of emissions standards for the different pollutants. To attain the required flexibility and precision of injection timing has posed a principal challenge to fuel injection manufacturers.

Compression Ratio. Diesel engines rely on compression heating to ignite the fuel, so the engine's compression ratio has an important effect on combustion. A higher compression ratio results in a higher temperature for the compressed charge and thus in a shorter ignition delay is to reduce NO_x emissions, while the higher flame temperature would be expected to increase them. In practice, these two effects nearly cancel, so that changes in compression ratio have little effect on NO_x.

Emissions of gaseous HC and of the SOF fraction of the particulate matter are reduced at higher compression ratios, as the higher cylinder temperature increases the burnout of hydrocarbons. Soot emissions may increase at higher compression ratios, however. Because the higher compression is achieved by reducing the volume of the combustion chamber, this results in a larger fraction of the air charge being sequestered in "crevice volumes" such as the top and sides of the piston, where it is not available for combustion early in the power stroke. Thus the effective air : fuel ratio in the combustion chamber decreases, and soot emissions go up. This effect can be limited (and overall air utilization and power output improved) by reducing crevice volumes to the maximum extent possible.

Engine fuel economy, cold starting, and maximum cylinder pressures are also affected by the compression ratio. For an idealized diesel cycle, the thermodynamic efficiency is an increasing function of compression ratio. In a real engine, however, the increased thermodynamic efficiency is offset after some point by increasing friction, so that a point of maximum efficiency is reached. With most heavy-duty diesel engine designs, this optimal compression ratio is about 12 to 15. To ensure adequate starting ability under cold conditions, however, most diesel engine designs require a somewhat higher compression ratio; in the range of 15 to 20 or more. Generally, higher speed engines with smaller cylinders require higher compression ratios for adequate cold starting.

Charge Temperature. Reducing the temperature of the air charge going into the cylinder has benefits for both PM and NO_x emissions. Reducing the charge temperature directly reduces the flame temperature during combustion and thus helps to reduce NO_x emissions. In addition, the colder air is denser, so that (at the same pressure) a greater mass of air can be contained in the same fixed cylinder volume. This increases the air : fuel ratio in the cylinder and thus helps to reduce soot emissions. By increasing the air available while decreasing piston temperatures, charge air cooling can also make possible a significant increase in power output. Excessively cold charge air can reduce the burnout of hydrocarbons, and thus increase light load HC emissions, however. This can be counteracted by advancing injection timing, or by reducing charge air cooling at light loads.

Charge Composition. NO_x emissions depend heavily on flame temperature. By altering the composition of the air charge to increase its specific heat and the concentration of inert gases, it is possible to decrease the flame temperature significantly. The most common way of accomplishing this is through exhaust gas recirculation (EGR). At moderate loads, EGR has been shown to be capable of reducing NO_x emissions by a factor of two or more, with little effect on particulate emissions. Although soot emissions are increased by the reduced oxygen concentration, particulate SOF and gaseous HC emissions are reduced, due to the higher in-cylinder temperature caused by the hot exhaust gas. EGR cannot be used at high loads, however, because the displacement of air by exhaust gas would result in an air : fuel ratio below the smoke limit and thus high soot and PM emissions.

Emissions Trade-offs. It is apparent from the foregoing discussion that there is an inherent conflict between some of the most powerful diesel NO_x control techniques and particulate emissions. This is the basis for the much discussed trade-off relationship between diesel NO_x and particulate emissions. This trade-off is not absolute, various NO_x control techniques have varying effects on soot and HC emissions, and the importance of these effects varies as a function of engine speed and load. These tradeoffs do place limits on the extent to which any one of these pollutants can be reduced, however. To minimize emissions of all three pollutants simultaneously requires careful optimization of the fuel injection, fuel–air mixing, and combustion processes over the full range of engine operating conditions.

Influence of Fuel Qualities

The quality and composition of diesel fuel can have important effects on pollutant emissions. The area of fuel effects on diesel emissions has seen a great deal of study in the last few years, and a large amount of new information has become available. These data indicate that the fuel variables having the most important effects on emissions are the sulfur content and the fraction of aromatic hydrocarbons contained in the fuel. Other fuel properties may also affect emissions, but generally to a much lesser extent. A recent extensive study carried out for the Dutch Ministry of Environment for the EEC's Motor Vehicle Emissions Group (MVEG) indicates that the volatility of the diesel fuel (85 or 90% distilled temperatures) is the most important factor followed in importance by sulfur and aromatics content, in that order. The apparent discrepancy with the U.S. findings might be the result of lower volatility of European diesel fuels, compared with those in the United States. The subject is currently being studied. Finally, the use of fuel additives may have a significant impact on emissions.

Sulfur Content. Diesel fuel for highway use normally contains between 0.1 and 0.5% by weight sulfur, although some (mostly less developed) nations permit 1% or even higher sulfur concentrations. Sulfur in diesel fuel contributes to environmental deterioration both directly and indirectly. Most of the sulfur in the fuel burns to SO_2, which is emitted to the atmosphere in the diesel exhaust. Because of this, diesels are significant contributors to ambient SO_2 levels in some areas. This makes them indirect contributors to ambient particulate levels and acid deposition as well. In the United States, diesel fuel accounted for about 629,220 t of SO_2 in 1984, or about 3% of all SO_2 emissions during the same period. For Europe, diesel fuel for road traffic adds approximately 1.6% of total SO_2 emissions, a share that is growing.

Most of the fuel sulfur that is not emitted as SO_2 is converted to various metal sulfates and to sulfuric acid during the combustion process or immediately afterward. Both of these materials are emitted in particulate form. The typical rate of conversion in a heavy-duty diesel engine is about 2 to 3% of the fuel sulfur and about 3 to 5% in a light-duty engine. Even at this rate, sulfate particles typically account for 0.05 to 0.10 g/BHP-h of particulates in heavy-duty engines. In a Dutch program, a sulfur change from 0.28 to 0.07% resulted in a 10 to 15% decrease in particulate emissions. The effect of the sulfate particles is increased by their hygroscopic nature; they tend to absorb significant quantities of water from the air.

Certain precious metal catalysts can oxidize SO_2 to SO_3, which combines with water in the exhaust to form sulfuric acid. The rate of conversion with the catalyst depends on the temperature, space velocity, and oxygen content of the exhaust and on the activity of the catalyst; generally, catalyst formulations that are most effective in oxidizing hydrocarbons and CO are also most effective at oxidizing SO_2. The presence of significant quantities of sulfur in diesel fuel thus limits the potential for catalytic converters or catalytic trap-oxidizers for use in controlling PM and HC emissions.

Sulfur dioxide in the atmosphere oxidizes to form sulfate particles, in a reaction similar to that which occurs with the precious metal catalyst. Viewed in another way, the presence of the catalyst merely speeds up a reaction that would occur anyway (although this can have a significant effect on human exposure to the reaction products). According to analysis by the California Air Resources Board staff (6) roughly 0.54 kg of secondary particulate is formed per kg of SO_2 emitted in the South Coast Air Basin. For a diesel engine burning fuel of 0.29 wt% sulfur at 0.19 kg of fuel per horsepower per hour, this is equivalent to 0.85 g/BHP-h. For comparison, the average rate of primary or directly emitted particulate emissions from heavy-duty engines in use was about 0.8 g/BHP-h (7).

Quite aside from its particulate-forming tendencies, sulfur dioxide is recognized as a hazardous pollutant in its own right. The health and welfare effects of SO_2 emissions from diesel vehicles are probably much greater than those of an equivalent quantity emitted from a utility stack or industrial boiler, because diesel exhaust is emitted close to the ground level in the vicinity of roads, buildings, and concentrations of people.

Volatility. Diesel fuel consists of a mixture of hydrocarbons having different molecular weights and boiling points. As a result, as some of it boils away on heating, the boiling point of the remainder increases. This fact is used to characterize the range of hydrocarbons in the fuel in the form of a "distillation curve" specifying the temperature at which 10%, 20%, etc of the hydrocarbons have boiled away. A low 10% boiling point is associated with a significant content of relatively volatile hydrocarbons. Fuels with this characteristic tend to exhibit somewhat higher HC emissions than others. Formerly, a relatively high 90% boiling point was considered to be associated with higher particulate emissions. More recent studies (8) have shown that this effect is spurious; the apparent statistical linkage was due to the higher sulfur content of these high boiling fuels.

In a Dutch study for the EC MVEG, however, the test fuels were composed of two sets at clearly different 85 or 90% boiling points, among which sulfur content varied independently. A highly significant effect of 85 or 90% boiling point temperatures was found, in addition to a significant effect of sulfur and probably significant effect of aromatics contents. A typical effect of a 20°C change in 85% boiling point is 0.05 g/kW·h at present particulate levels. As mentioned earlier, this may be related to generally higher, 85 to 90 percentage points, that in the test fuels went up to 350 or 360°C. Commercial diesel fuels in Europe show values up to about 370°C.

Aromatic Hydrocarbon Content. Aromatic hydrocarbons are hydrocarbon compounds containing one or more "benzene-like" ring structures. They are distinguished from paraffins and naphthenes, the other principal hydrocarbon constituents of diesel fuel, which lack such structures. Compared to these other components, aromatic hydrocarbons are denser, have poorer self ignition qualities, and produce more soot in burning. Ordinarily, "straight run" diesel fuel produced by simple distillation of crude oil is fairly low in aromatic hydrocarbons. Catalytic cracking of residual oil to increase gasoline and diesel production results in increased aromatic content, however. A typical straight run diesel might contain 20 to 25% aromatics by volume, while a diesel blended from catalytically cracked stocks could have 40–50% aromatics.

Aromatic hydrocarbons have poor self ignition qualities, so that diesel fuels containing a high fraction of aromatics tend to have low cetane numbers. Typical cetane values for straight run diesel are in the range of 50–55; those for highly aromatic diesel fuels are typically 40 to 45, and may be even lower. This produces more difficulty in cold starting, and increased combustion noise, HC, and NO_x due to the increased ignition delay.

Increased aromatic content is also correlated with higher particulate emissions. Aromatic hydrocarbons have a greater tendency to form soot in burning, and the poorer combustion quality also appears to increase particulate SOF emissions. Increased aromatic content may also be correlated with increased SOF mutagenicity, possibly due to increased PNA and nitro-PNA emissions. There is also some evidence that more highly aromatic fuels have a greater tendency to form deposits on fuel injectors and other critical components. Such deposits can interfere with proper fuel/air mixing, greatly increasing PM and HC emissions.

Other Fuel Properties. Other fuel properties may also have an effect on emissions. Fuel density, for instance, may affect the mass of fuel injected into the combustion chamber at full load, and thus the air/fuel ratio. This is because fuel injection pumps meter fuel by volume, not by mass, and the denser fuel contains a greater mass in the same volume. Fuel viscosity can also affect the fuel injection characteristics, and thus the mixing rate. The corrosiveness, cleanliness, and lubricating properties of the fuel can all affect the service life of the fuel injection equipment—possibly contributing to excessive in-use emissions if the equipment is worn out prematurely.

Fuel Additives. Several generic types of diesel fuel additives can have a significant effect on emissions. These include cetane enhancers, smoke suppressants, and detergent additives. In addition, some additive research has been directed specifically at emissions reduction in recent years. Although some moderate emission reductions have been demonstrated, there is yet no consensus on the widespread applicability or desirability of such products.

Cetane enhancers are used to enhance the self ignition qualities of diesel fuel. These compounds (generally organic nitrates) are generally added to reduce the adverse impact of high aromatic fuels on cold starting and combustion noise. These compounds also appear to reduce the aromatic hydrocarbons' adverse impacts on HC and PM emissions, although PM emissions with the cetane improver are generally still somewhat higher than those from a higher quality fuel able to attain the same cetane rating without the additive. In the MVEG study cited earlier, no significant effect of ashless cetane improving additives could be detected on NO_x or particulates.

Smoke suppressing additives are organic compounds of calcium, barium, or (sometimes) magnesium. Added to diesel fuel, these compounds inhibit soot formation during the combustion process, and thus greatly reduce emissions of visible smoke. Their effects on the particulate SOF are not fully documented, but one study (9) has shown a significant increase in the PAH content and mutagenicity of the SOF with a barium additive. Particulate sulfate emissions are greatly increased with these additives, since all of them readily form stable solid metal sulfates, which are emitted in the exhaust. The overall effect of reducing soot and increasing metal sulfate emissions may be either an increase or decrease in the total particulate mass, depending on the soot emissions level at the beginning and the amount of additive used.

Detergent additives (often packaged in combination with a cetane enhancer) help to prevent and remove coke deposits on fuel injector tips and other vulnerable locations. By thus maintaining new engine injection and mixing characteristics, these deposits can help to decrease in-use PM and HC emissions. A study for the California Air Resources Board (10) estimated the increase in PM emissions due to fuel injector problems from trucks in use as being more than 50% of new-vehicle emissions levels. A significant fraction of this excess is unquestionably due to fuel injector deposits.

CONTROL TECHNOLOGIES FOR DIESEL FUELED VEHICLES

Diesel engine emissions are determined by the characteristics of the combustion process within each cylinder. Primary engine parameters affecting diesel emissions are the fuel injection system, the engine control system, the air intake port and combustion chamber design, and the air charging system. Actions to reduce lubricating oil consumption can also impact HC and PM emissions. Further, beyond the engine itself, exhaust aftertreatment systems such as trap oxidizers and catalytic converters can play a significant role. Finally, modifications to conventional fuels as well as alternative fuels can impact emissions. The following sections will review the status of each of these areas.

Engine Modifications

Air Motion and Combustion Chamber Design. The geometries of the combustion chamber and the air intake port control the air motion in the diesel combustion chamber, and thus play an important role in air/fuel mixing and emissions. A number of different combustion chamber designs, corresponding to different basic combustion systems, are in use in heavy duty diesel engines at present. This section outlines the basic consumption systems in use, their advantages and disadvantages, and the effects of changes in combustion chamber design and air motion on emissions.

Combustion Systems. Diesel engines used in heavy duty vehicles use several different types of combustion systems. The most fundamental difference is between direct injection (DI) engines and indirect injection (IDI) engines. In an indirect injection engine, fuel is injected into a separate "prechamber," where it mixes and partly burns before jetting into the main combustion chamber above the piston. In the more common direct injection engine, fuel is injected directly into a combustion chamber formed out of the top of the piston. DI engines can be further divided into high swirl and low swirl.

Fuel/air mixing in the direct injection engine is limited by the fuel injection pressure and any motion imparted to the air in the chamber as it enters. In high swirl DI engines, a strong swirling motion is imparted to the air entering the combustion chamber by the design of the intake port. These engines typically use moderate to high injection pressures, and three to five spray holes per nozzle. Low swirl engines rely primarily on the fuel injection process to supply the mixing. They typically have very high fuel injection pressures and six to nine spray holes per nozzle.

In the indirect injection engine, much of the fuel/air mixing is due to the air swirl induced in the prechamber as air is forced into it during compression, and to the turbulence induced by the expansion out of the prechamber during combustion. These engines typically have better high speed performance than direct injected engines, and can use cheaper fuel injection systems. Historically, IDI diesel engines have also exhibited lower emission levels, especially NO_x, than DI engines but, with recent developments in DI engine emission controls, this is no longer the case. The main disadvantage of the IDI engine is that the extra heat and frictional losses due to the prechamber result in a 5-10 percent reduction in fuel efficiency compared to a DI engine.

A number of advanced, low emitting fuel efficient high swirl DI engines have recently been introduced and it appears that these engines will completely displace the existing IDI designs.

DI Combustion Chamber Design Changes in the engine combustion chamber and related areas have demonstrated a major potential for emission control. Design changes to reduce the crevice volume for DI diesel cylinders increase the amount of air available in the combustion chamber. Changes in combustion chamber geometry—such as the use of a reentrant lip on the piston bowl—can markedly reduce emissions by improving air/fuel mixing and minimizing wall impingement by the fuel jet. Optimizing the intake port shape for best swirl characteristics has also yielded significant benefits. Several manufacturers are considering variable swirl intake ports, to optimize swirl characteristics across a broader range of engine speeds.

Fuel Injection. The fuel injection system, one of the most important components in a diesel engine, includes the process by which the fuel is transferred from the fuel tank to the engine, and the mechanism by which it is injected into the cylinders. The precision, characteristics and timing of the fuel injection determine the engine's power, fuel economy, and emissions characteristics.

The important areas of concentration in fuel injection system development have been on increased injection pressure, increasingly flexible control of injection timing, and more precise governing of the fuel quantity injected. Some manufacturers are also pursuing technology to vary the rate of fuel injection over the injection period, in order to reduce the amount of fuel burning in the premixed combustion phase. Reductions in NO_x and noise emissions and maximum cylinder pressures have been demonstrated using this approach. Systems offering electronic control of these quantities, as well as fuel injection rate, have been introduced. Other changes have been made to the injection nozzles themselves, to reduce or eliminate sac volume and to optimize the nozzle hole size and shape, number of holes, and spray angle for minimum emissions.

High fuel injection pressures are desirable in order to improve fuel atomization and fuel/air mixing, and to offset the effects of retarded injection timing by increasing the injection rate. It is well established that higher injection pressures reduce PM and/or smoke emissions. High injection pressures are most important in low swirl, direct injection engines, since the fuel injection system is responsible for most of the fuel/air mixing in these systems. For this reason, low swirl engines tend to use unit injector systems, which can achieve peak injection pressures in excess of 1,500 bar.

Engine Control Systems. Traditionally, diesel engine control systems have been closely integrated with the fuel injection system, and the two systems are often discussed together. These earlier control systems (still in use on most engines) are entirely mechanical. The last few years have seen the introduction of an increasing number of computerized electronic control systems for diesel engines. With the introduction of these systems, the scope of the engine control system has been greatly expanded.

The advent of computerized electronic engine control systems has greatly increased the potential flexibility and precision of fuel metering and injection timing controls. In addition, it has made possible whole new classes of control functions, such as road speed governing, alterations in control strategy during transients, synchronous idle speed control, and adaptive learning—including strategies to identify and compensate for the effects of wear and component variation in the fuel injection system.

By continuously adjusting the fuel injection timing to match a stored "map" of optimal timing vs. speed and load, an electronic timing control system can significantly improve on the NO_x/particulate and NO_x/fuel economy trade-offs possible with static or mechanically variable injection timing. Most electronic control systems also incorporate the functions of the engine governor and the transient smoke limiter. This helps to reduce excess particulate emissions due to mechanical friction and lag time during engine transients, while simultaneously improving engine performance. Potential reductions in PM emissions of up to 40% have been documented with this approach.

Turbocharging and Intercooling. A turbocharger consists of a centrifugal air compressor feeding the intake manifold, mounted on the same shaft as an exhaust gas turbine in the exhaust stream. By increasing the mass of air in the cylinder prior to compression, turbocharging correspondingly increases the amount of fuel that can be burned without excessive smoke, and thus increases the potential maximum power output. The fuel efficiency of the engine is improved as well. The process of compressing the air, however, increases its temperature, increasing the thermal load on critical engine components. By cooling the compressed air in an intercooler before it enters the cylinder, the adverse thermal effects can be reduced. This also increases the density of the air, allowing an even greater mass of air to be confined within the cylinder, and thus further increasing the maximum power potential.

Increasing the air mass in the cylinder and reducing its temperature can reduce both NO_x and particulate emissions as well as increase fuel economy and power output from a given engine displacement. Most heavy duty diesel engines are presently equipped with turbochargers, and most of these have intercoolers. In the U.S., virtually all engines will be equipped with these systems by 1991. Recent developments in air changing systems for diesel engines have been primarily concerned with increasing the turbocharger efficiency, operating range, and transient response characteristics; and with improved intercoolers to reduce the temperature of the intake charge further. Tuned intake air manifolds (including some with variable tuning) have also been developed, to maximize air intake efficiency in a given speed range.

Lubricating Oil Control. A significant fraction of diesel particulate matter consists of oil derived hydrocarbons and related solid matter; estimates range from 10 to 50%. Reduced oil consumption has been a design goal of heavy duty diesel engine manufacturers for some time, and the current generation of diesel engines already uses fairly little oil compared to their predecessors. Further reductions in oil consumption are possible through careful attention to cylinder bore roundess and surface finish, optimization of piston ring tension and shape, and attention to valve stem seals, turbocharger oil seals, and other possible sources of oil loss. Some oil consumption in the cylinder is required with present technology, however, in order for the oil to perform its lubricating and corrosion protective functions.

Aftertreatment Systems

Trap Oxidizers. A trap oxidizer system consists of a durable particulate filter (the "trap") positioned in the engine exhaust stream, along with some means for cleaning the filter by burning off ("oxidizing") the collected particulate matter. The construction of a filter capable of collecting diesel soot and other particulate matter from the exhaust stream is a straightforward task, and a number of effective trapping media have been developed and demonstrated. The most challenging problem of trap oxidizer system development has been with the process of "regenerating" the filter by burning off the accumulated particulate matter.

Diesel particulate matter consists primarily of a mixture of solid carbon coated with heavy hydrocarbons. The ignition temperature of this mixture is about 500–600°C, which is above the normal range of diesel engine exhaust temperatures. Thus, special means are needed to assure regeneration. Once ignited, however, this material burns to produce very high temperatures, which can easily melt or crack the particulate filter. Initiating and controlling the regeneration process to ensure reliable regeneration without damage to the trap is the central engineering problem of trap oxidizer development.

Traps. Presently, most of the trap oxidizer sytems under development are based on the cellular cordierite ceramic monolith trap. These traps can be formulated to be highly efficient (collecting essentially all of the soot, and a large fraction of the particulate SOF), and they are relatively compact, having a large surface area per unit of volume. They can also be coated or impregnated with catalyst material to assist regeneration.

The high concentration of soot per unit of volume with the ceramic monolith, however, makes these traps sensitive to regeneration conditions. Trap loading, temperature, and gas flow rates must be maintained within a fairly narrow window. Otherwise, the trap fails to regenerate fully, or cracks or melts due to overheating.

An alternative trap technology is provided by ceramic fiber coils. These traps are composed of a number of individual filtering elements, each of which consists of a number of thicknesses of silica fiber yarn wound on a punched metal support. A number of these filtering elements are suspended inside a large metal can to make up a trap.

Numerous other trapping media have been tested or proposed, including ceramic foams, corrugated mullite fiber felts, and catalyst coated stainless steel wire mesh.

Regeneration. Numerous techniques for regenerating particulate trap oxidizers have been proposed, and a great deal of development work has been invested in many of these. These approaches can generally be divided into two groups: passive systems and active systems. Passive systems must attain the conditions required for regeneration during normal operation of the vehicle. The most promising approaches require the use of a catalyst (either as a coating on the trap or as a fuel additive) in order to reduce the ignition temperature of the collected particulate matter. Regeneration temperatures as low as 420°C have been reported with catalytic coatings, and even lower temperatures are achievable with fuel additives.

Active systems, on the other hand, monitor the buildup of particulate matter in the trap and trigger specific actions intended to regenerate it when needed. A wide variety of approaches to triggering regeneration have been proposed, from diesel fuel burners and electric heaters to catalyst injection systems.

Passive regeneration systems face special problems on heavy duty vehicles. Exhaust temperatures from heavy duty diesel engines are normally low, and recent developments such as charge air cooling and increased turbo charger efficiency are reducing them still further. Under some conditions, therefore, it would be possible for a truck to drive for many hours without exceeding the exhaust temperature (around 400–450°C) required to trigger regeneration.

Engine and catalyst manufacturers have experimented with a wide variety of catalytic material and treatments to assist in trap regeneration. Good results have been obtained both with precious metals (platinum, palladium, rhodium, silver) and with base metal catalysts such as vanadium and copper. Precious metal catalysts are effective in oxidizing gaseous HC and CO, as well as the particulate SOF, but are relatively ineffective at promoting soot oxidation. Unfortunately, these metals also promote the oxidation of SO_2 to particulate sulfates such as sulfuric acid (H_2SO_4). The base metal catalysts, in contrast, are effective in promoting soot oxidation, but have little effect on HC, CO, NO_x or SO_2. Many experts believe that ultimately precious metal catalysis must be an important element of an effective particulate control system because it specifically attacks the "bad actors."

Catalyst coatings also have a number of advantages in active systems, however. The reduced ignition temperature and increased combustion rate due to the catalyst mean that less energy is needed from the regeneration system. Regeneration will also occur spontaneously under most duty cycles, greatly reducing the number of times the regeneration system must operate. The spontaneous regeneration capability also provides some insurance against a regeneration system failure. Finally, the use of a catalyst may make possible a simpler regeneration system.

Although normal heavy duty diesel exhaust temperatures are not high enough under all operating conditions to provide reliable regeneration for a catalyst coated trap, the exhaust temperature can readily be increased by changes in engine operating parameters. Retarding the injection timing, bypassing the intercooler, throttling the intake air (or cutting back on a variable geometry turbo charger), and/or increasing the EGR rate all markedly increase the exhaust temperature. Applying these measures all the time would seriously degrade fuel economy, engine durability, and performance. The presence of an electronic control system, however, makes it possible to apply them very selectively to regenerate the trap. Since they would be normally needed only at light loads, the effects on durability and performance would be imperceptible.

Fuel additives may play a key role in trap based systems although concerns have been raised about possible toxicity if metallic additives were widely used. Cerium based additives which do not appear to raise these concerns have been found especially promising in recent fleet studies in Athens buses; they were able to lower engine out particulate emissions as well as facilitate regeneration.

Catalytic Converters. Like a catalytic trap, a diesel catalytic converter oxidizes a large part of the hydrocarbon constituents of the SOF, as well as gaseous HC, CO, odor creating compounds, and mutagenic emissions. Unlike a catalytic trap, however, a flow through catalytic converter does not collect any of the solid particulate matter, which simply passes through in the exhaust. This eliminates the need for a regeneration system, with its attendant technical difficulties and costs. The particulate control efficiency of the catalytic converter is, of course, much less than that of a trap. However, a particulate control efficiency of even 25–35% is enough to bring many current development engines within the target range for the U.S. 1994 emissions standard.

Diesel catalytic converters have a number of advantages. First, in addition to reducing particulate emissions, the oxidation catalyst greatly reduces HC, CO, and odor emissions. The catalyst is also very efficient in reducing emissions of gaseous and particle bound toxic air contaminants such as aldehydes, PNA, and nitro-PNA. While a precious metal catalyzed particulate trap would have the same advantages, the catalytic converter is much less complex, bulky, and expensive. Unlike the trap, the catalytic converter has little impact on fuel economy or safety, and it will probably not require replacement as often.

Also, unlike the trap oxidizer, the catalytic converter is a relatively mature technology—millions of catalytic converters are in use on gasoline vehicles, and diesel catalytic converters have been used in underground mining applications for more than 20 years.

The disadvantage of the catalytic converter is the same as with the precious metal catalyzed particulate trap: sulfate emissions. The tendency of the precious metal catalyst to convert SO_2 to particulate sulfates requires the use of low sulfur fuel: otherwise, the increase in sulfate emissions would more than counterbalance the decrease in SOF. Also on the road durability has not yet been demonstrated in heavy duty applications for the required 270,000 mile useful life.

NO_x Reduction Techniques

Under appropriate conditions, NO_x can be chemically reduced to form oxygen and nitrogen gases. This process is used in modern closed-loop, three-way catalyst equipped gasoline vehicles to control NO_x emissions. However, the process of catalytic NO_x reduction used on gasoline vehicles is inapplicable to diesels. Because of their heterogeneous combustion process, diesel engines require substantial excess air, and their exhaust thus inherently contains significant excess oxygen. The three-way catalysts used on automobiles require a precise stoichiometric mixture in the exhaust in order to function, in the presence of excess oxygen, their NO_x conversion efficiency rapidly approaches zero.

A number of aftertreatment NO_x reduction techniques which will work in an oxidizing exhaust stream are currently available or under development for stationary pollution sources. These include selective catalytic reduction (SCR), selective noncatalytic reduction (Thermal Denox (tm)), and reaction with cyanuric acid (RapReNox). However, each of these systems requires a continuous supply of some reducing agent such as ammonia or cyanuric acid to react with the NO_x. Because of the need for frequent replenishment of this agent, and the difficulty of ensuring that the replenishment is performed when needed, such systems are considered impractical for vehicular use. Even if the replenishment problems could be resolved, these systems would raise serious questions about crash safety and possible emissions of toxic air contaminants.

Fuel Modifications

Modifications to diesel fuel composition have drawn considerable attention as quick and cost effective means of reducing emissions from existing vehicles. Regulations mandating low sulfur fuel (0.05 wt %) have been promulgated in both Europe and the U.S. In addition to a direct reduction in emissions of SO_2 and sulfate particles, reducing the sulfur content of diesel fuel reduces the indirect formation of sulfate particles from SO_2 in the atmosphere.

A number of well-controlled studies have demonstrated the ability of detergent additives in diesel fuel to prevent and remove injector tip deposits, thus reducing smoke levels. The reduced smoke probably results in reduced PM emissions as well, but this has not been demonstrated as clearly, due to the great expense of PM emissions tests on in use vehicles. Cetane improving additives are also likely to result in some reduction in HC and PM emissions in marginal fuels.

Alternative Fuels

The possibility of substituting cleaner burning alternative fuels for diesel fuel has drawn increasing attention during the last decade. Motivations advanced for this substitution include conservation of oil products and energy security, as well as the reduction or elimination of particulate emissions and visible smoke. Care is necessary in evaluating the air quality claims for alternative fuels, however. While many alternative fuel engines do display greatly reduced particulate and SO_2 emissions, emissions of other gaseous pollutants such as unburned hydrocarbons, CO, and in some cases NO_x and aldehydes may be much higher than from diesels.

The principal alternative fuels presently under consideration are natural gas and methanol made from natural gas, and in limited applications, LPG.

Natural Gas. Natural gas has many desirable qualities as an alternative to diesel fuel in heavy duty vehicles. Clean burning, cheap, and abundant in many parts of the world, it already plays a significant vehicular role in a number of countries. The major disadvantage of natural gas as a motor fuel is its gaseous form at normal temperatures.

Pipeline quality natural gas is a mixture of several different gases but the primary constituent is methane, which typically makes up 90–95% of the total volume. Methane is a nearly ideal fuel for Otto cycle (spark ignition) engines. As a gas under normal conditions, it mixes readily with air in any proportion, eliminating cold start problems and the need for cold start enrichment. In its purest form, it is flammable over a fairly wide range of air fuel ratios. With a research octane number of 130 (the highest of any commonly used fuel), it can be used with engine compression ratios as high as 15 : 1 (compared to 8–9 : 1 for gasoline), thus giving greater efficiency and power output. The low lean flammability limit permits operation with extremely lean air fuel ratios having as much as 60 percent excess air. On the other hand, its high flame temperature tends to result in high NO_x emissions, unless very lean mixtures are used.

Because of its gaseous form and poor self ignition qualities, methane is a poor fuel for diesel engines. Since diesels are generally somewhat more efficient than Otto cycle engines, natural gas engines are likely to use somewhat more energy than the diesels they replace. The high compression ratios achievable with natural gas limit this efficiency penalty to about 10 percent of the diesel fuel consumption, however.

Options for using natural gas in heavy duty vehicle engines are limited to the following:

Fumigation, or mixing the gas with the diesel intake air, to be ignited by diesel fuel injected in the conventional way;

Conversion of the existing diesel engine to Otto cycle operation; or

Replacement of the diesel engine with a conventional spark ignition engine.

Fumigation. Fumigation is the easiest way to use natural gas in a diesel engine. Injection and combustion of the diesel fuel ignites and burns the alternative fuel as well. Since the gas supplies much of the energy for combustion, the diesel fuel delivery for a given power level is reduced resulting in reduced smoke and particulate (PM) emissions at high load. However, incomplete combustion (especially at light loads) usually increases CO and HC emissions considerably. The increased HC emissions are of less concern with natural gas than with other hydrocarbon fuels, since the principal component, methane, is non toxic and has very low photochemical reactivity. On the other hand, methane is a very active greenhouse gas, and the other, minor components of the HC emissions include some formaldehyde as well as higher molecular weight hydrocarbons.

Spark Ignition Engines. The modifications required to convert a diesel engine to Otto cycle operation are machining the cylinder head to accept a spark plug instead of a fuel injector; redesign of the pistons to reduce the compression ratio; replacement of the fuel injection pump with an ignition system and distributor; replacement of exhaust valves and valve seats with wear resistant materials; and addition of a carburetor (or fuel injection system) and throttle assembly.

CO emissions are governed by oxygen availability, while NO_x emissions are primarily a function of flame temperature. For natural gas engines, typical NO_x emissions at stoichiometry are about 10 to 15 g/BHp–h. This can be reduced to less than 2 g/BHp–h, however, through the use of a three way catalyst and closed loop mixture controls like those on light duty passenger cars. The durability of such catalysts under the high temperatures experienced in heavy duty operation has not yet been demonstrated, however. The durability of electronic control systems under heavy duty conditions is also unproven.

An alternative approach to NO_x control is to operate very lean. Through the use of high energy ignition systems and careful optimization, homogeneous mixtures can be lean enough to lower NO_x emissions to the 2 g/BHp-h range. The driveability of such engines under transient conditions has not yet been demonstrated, however. The lean combustion limit can be extended even further by using a stratified charge strategy. Stationary engines using this technique have demonstrated NO_x emissions less than one g/BHp-h.

Liquefied Petroleum Gas (LPG). Liquefied petroleum gas is already widely used as a vehicle fuel in the U.S., Canada, the Netherlands, and elsewhere. As a fuel for spark ignition engines, it has many of the same advantages as natural gas, with the additional advantage of being easier to carry aboard the vehicle. Its major disadvantage is the limited supply, which would rule out any large scale conversion to LPG fuel.

The technologies available for LPG are the same as those available for natural gas: fumigation, or spark ignition using either stoichiometric or very lean mixtures. Due to the lower octane value of LPG, the compression ratio (and thus the thermal efficiency) possible with this fuel in spark ignition operation is lower than with natural gas, although still considerably higher than with gasoline. Aside from this, the engine technologies involved are very similar. Due to the lower octane value (and higher photochemical reactivity) of LPG, however, it is not as good a candidate for use in fumigation as natural gas.

Like natural gas, LPG in spark ignition engines is expected to produce essentially no particulate emissions (except for a small amount of lubricating oil), very little CO, and moderate HC emissions. NO_x emissions are a function of the air fuel ratio and other engine variables such as spark timing and compression ratio. LPG does not burn as well under very lean conditions as natural gas, so the NO_x levels achievable through lean burn technology are expected to be somewhat higher, probably in the range of 3 to 5 g/BHp-h. For stoichiometric LPG engines, the use of a three way catalyst and closed loop air fuel mixture control results in very low NO_x emissions, assuming that such systems can be made sufficiently durable.

Methanol. Methanol has many desirable combustion and emissions characteristics, including good lean combustion characteristics, low flame temperature (leading to low NO_x emissions) and low photochemical reactivity.

As a liquid, methanol can either be burned in an Otto cycle engine or injected into the cylinder as in a diesel. With a fairly high octane number of 112, and excellent lean combustion properties, methanol is a good fuel for lean burn spark ignition (SI). Its lean combustion limits are similar to those of natural gas, while its low energy density results in a low flame temperature compared to hydrocarbon fuels, and thus lower NO_x emissions. Methanol burns with a sootless flame and contains no heavy hydrocarbons. As a result, particulate emissions from methanol engines are very low—consisting essentially of a small amount of unburned lubricating oil.

Methanol's high octane number results in a very low cetane number, so that it is more difficult to use methanol in a diesel engine without some supplemental ignition source. Investigations to date have focused on the use of ignition improving activities, spark ignition, grow plug ignition, or dual injection with diesel fuel. Converted heavy duty diesel engines using each of these approaches have been developed and demonstrated.

The low energy density of methanol means that a large amount (about three times the mass of diesel fuel) is required to achieve for the same power output. Therefore, the diesel injection pump supplied with the engine would not be suitable in most cases; a larger volumetric capacity is required. In addition, diesel injection pumps are fuel lubricated. Since methanol is a poor lubricant, a separate oil supply to the pump is required. Other changes to the high pressure lines, injector nozzels, and so forth are required to prevent cavitation and premature wear. All of these changes are straightforward, however, and injection pumps suitable for use with methanol have been produced.

Emissions. Methanol combustion does not produce soot, so particulate emissions from methanol engines are limited to a small amount of lubricating oil. Methanol's flame temperature is also lower than that for hydrocarbon fuels, resulting in NO_x emissions which are typically 50 percent lower. CO emissions are generally comparable to or somewhat greater than those from a diesel engine; these emissions can be controlled with a catalytic converter, however.

The principal pollution problems with methanol come from emissions of unburned fuel and formaldehyde. Methanol (at least in moderate amounts) is relatively innocuous. It has low photochemical reactivity, and while acutely toxic in large doses, displays no significant chronic toxicity effects. Formaldehyde, the first oxidation product of methanol, is much less benign, however. A powerful irritant and suspected carcinogen, it also displays very high photochemical reactivity. While all combustion engines produce some formaldehyde, some early generation methanol engines exhibited greatly increased emissions compared to diesels. The potential for large increases in formaldehyde emissions with the widespread use of methanol vehicles has raised considerable concern about what would otherwise be a very benign fuel from an environmental standpoint. DDC has made major advances in formaldehyde control and levels are currently at or below the diesel engines they replace.

Formaldehyde emissions can be reduced through changes in combustion chamber and injection system design, and are also readily controllable though the use of catalytic converters, at least under warmed up conditions. In fact, the DDC engine equipped with a catalyst has attained emission levels as low as 0.06 formaldehyde, along with 0.2 HC, 0.8 CO, 2.2 NO_x and 0.04 particulate. This system is now available commercially.

ALTERNATIVE FUTURES: THE TECHNOLOGICAL STATE OF THE ART

Continued growth in emissions from the transportation sector is not inevitable. Even relatively modest steps can significantly lower emissions in the near term, and combined with slight reductions in future vehicle growth patterns, overall stability in emissions from the transport sector is possible.

In the state of the art case, emissions were estimated as if all vehicles in the world were to adopt the most stringent set of requirements considered technologically feasible today. These include currently adopted California low emission vehicle (LEV) standards for cars and light trucks, heavy truck requirements scheduled to be introduced during the 1990s in the U.S., volatility controls on petrol to reduce evaporative and refueling hydrocarbon emissions, enhanced inspection and maintenance programs to maximize the actual effectiveness of emissions standards and refueling controls.

If all vehicles in the world were to adopt these requirements, CO, HC, and NO_x emissions would go down significantly over the next twenty years. Two points are worth emphasizing:

1. Using today's state of the art technology on all vehicles, emissions are dramatically lower than they would be in the base case. For example, hydrocarbons are only a third of what they would otherwise be; in the case of NO_x, they are less than half.
2. Because of their high growth rates, even with application of state of the art controls, the non OECD countries of the world will be much more important in the future.

CONCLUSIONS

1. Motor vehicles are the largest single source of man made VOC, NO_x, and CO in the OECD as well as many rapidly industrializing countries. Controls which reduce both VOC (as well as CO) and NO_x emissions from this source to the maximum extent technologically feasible are therefore most effective at reducing ozone concentrations. In addition, motor vehicles are probably the major source of toxic pollutants as well and a significant contributor to potential climate altering emissions.
2. Depending on the ambient concentration, these emissions can cause or contribute to a wide range of adverse health and environmental effects including eye irritation, cough and chest discomfort, headaches, heart disease, upper respiratory illness, increased asthma attacks and reduced pulmonary function. The most recent studies indicate that these emissions can cause cancer and exacerbate mortality and morbidity from respiratory disease. In addition, studies indicate that air pollution seriously impairs the growth of certain crops, reduces visibility, and in sensitive aquatic systems such as small lakes and streams can destroy fish and other forms of life and damages forests.

   Whether it be localized urban problems, regional smog or global changes, it is clear that motor vehicles are a dominant source of air pollution.
3. Technologies such as closed loop three way catalysts have been developed which have the potential to reduce substantially vehicle emissions in a cost effective manner. Application of these state of the art technologies can improve vehicle performance and driveability, reduce maintenance and is consistent with improved fuel economy. Evaporative controls are also readily available and cost effective. The effectiveness of state of the art emissions controls can be improved by in use vehicle directed programs such as inspection and maintenance, recall and warranty.
4. As a result of catalysts and other controls introduced to date, CO, HC and NO_x levels are substantially below what they would otherwise be.
5. If today's state of the art emissions controls were introduced on all new vehicles around the world, it would be possible to continue to reduce vehicle emissions of CO, HC and NO_x while simultaneously absorbing the expected vehicle growth, at least until early in the next century.
6. Advances in the state of the art for vehicle technology are emerging and coupled with more modest ve-

hicle growth, they can constrain global vehicle emissions if they are widely utilized and effectively implemented.

Looking ahead to the future, it is clear that several technological challenges remain for reduced vehicle emissions. Particular areas of intense activity include:

1. Preheated or very quick light off systems which will enable manufacturers to comply with the stringent California LEV requirements.
2. Diesel flow through particulate catalysts which can lower particulate and the organics associated with diesel combustion without converting too much sulfur dioxide to sulfate.
3. Diesel trap oxidizers which can virtually eliminate diesel particulate and the associated organics.
4. Lean NO_x catalysts which would enable such technologies with inherent lean operating advantages, ie, diesels or two strokes, to take advantage of these capabilities without increasing NO_x emissions.

BIBLIOGRAPHY

1. *World Motor Vehicle Data,* Various Editions, Motor Vehicle Manufacturers Association of the United States, Inc.
2. Gardiner, Foley and Shapiro, "Priority Revision of the PM-10 NAAQS" Memo to the US EPA Administrator, July 19, 1993.
3. W. P. Cartellieri and W. F. Wachter, *Status Report on a Preliminary Survey of Strategies to Meet US-1991 HD Diesel Emission Standards Without Exhaust Gas Aftertreatment,* SAE Paper No. 870342, Society of Automotive Engineers, Warrendale, Pa., 1987.
4. W. R. Wade, C. E. Hunter, F. H. Trinker, and H. A. Cikanek, *The Reduction of NO_x and Particulate Emissions in the Diesel Combustion Process,* ASME Paper No. 87-ICE-37, American Society of Mechanical Engineers, New York, 1987.
5. P. T. Williams, G. E. Andrews, and K. D. Bartle *Diesel Particulate Emissions: The Role of Unburnt Fuel in the Organic Fraction Composition,* SAE Paper No. 870554, Society of Automotive Engineers, Warrendale, Pa., 1987.
6. California Air Resources Board staff, *Status Report: Diesel Engine Emissions Reductions through Modification of Motor Vehicle Diesel Fuel Specifications,* CARB, Sacramento, Calif., 1984.
7. Engine Manufacturer's Association "Heavy Duty Diesel Engine In-Use Emission Testing Meeting," briefing package for a presentation to the California Air Resources Board Staff, El Monte, Calif., 1985.
8. J. C. Wall and S. K. Hoekman. *Fuel Composition Effects on Heavy Duty Diesel Particulate Emissions,* SAE Paper No. 841364, Society of Automotive Engineers, Warrendale, Pa., 1984.
9. W. M. Draper, J. Phillips, and H. W. Zeller, *Impact of a Barium Fuel Additive on the Mutagenicity and Polycyclic Aromatic Hydrocarbon Content of Diesel Exhaust Particulate Emissions,* SAE paper, 1988.
10. C. S. Weaver, and R. F. Klausmeier. *Heavy Duty Diesel Vehicle Inspection and Maintenance Study: Final Report,* 4 Vol., report under ARB Contract No. A4-151-32, Radian Corporation, Sacramento, Calif., 1988.

AIR POLLUTION: AUTOMOBILE, TOXIC EMISSIONS

Pamela J. Brodowicz
Penny M. Carey
EPA
Ann Arbor, Michigan

Motor vehicle emissions are extremely complex and hundreds of compounds have been identified. This article attempts to summarize what is known about air toxic emissions from gasoline and diesel fueled motor vehicles. Specific pollutants or pollutant categories discussed include benzene, formaldehyde, 1,3-butadiene, acetaldehyde, and diesel particulate matter, all of which have been considered in previous analyses of air toxics and are also currently or in the process of being classified by the U.S. Environmental Protection Agency (EPA) as known or probable human carcinogens. For each pollutant, information is provided on emissions, atmospheric transformation, ambient levels, and health effects.

In order for a compound to be assigned an EPA classification, the experimental data must be evaluated through a process outlined in the *EPA Guidelines for Carcinogenic Risk Assessment* (1). The data from carcinogenic studies (both human and animal) are divided into five separate categories by weight of evidence collected. These five categories are defined as sufficient evidence, limited evidence, inadequate evidence, no data, and no evidence for carcinogenicity. Each compound is evaluated, giving the human and animal study criteria each its own category designation.

Human studies depend on case reports of individual cancer patients and/or epidemiological studies. These studies must be without bias, logically approached, organized, and analyzed, and the probability of chance must be ruled out in order to infer a causal relationship. Animal studies must be long-term studies, conducted in more than one species, strain, or experiment, with adequate dosage, exposure, follow-up and reporting, adequate numbers of animals, good survival rates, and the increase of tumors of the malignant type, or a combination of malignant and benign.

The categories above characterizing human and animal weight of evidence classifications can now be utilized to allow a tentative assignment to one of five categories. The tumor studies are combined with all other relative information to determine if the categorization of the weight of evidence is appropriate for a particular agent. The five categories are described below:

Group A: Human Carcinogen. Requires *sufficient evidence* from human epidemiological studies to support a causal relationship between exposure and the agent.

Group B: Probable Human Carcinogen. B1–Human evidence is *limited* no matter what is found in the animal studies. B2–If animal studies are sufficient and in humans the evidence is *inadequate, no data exists, or no evidence* is found.

Group C: Possible Human Carcinogen. There is *limited* evidence in animals in the absence of human data.

Group D: Not Classifiable as to Human Carcinogenicity. There are *inadequate* human and animal data or no data are available.

Group E: Evidence of Noncarcinogenicity For Humans. *No evidence* has been found in at least two animal studies or in both adequate human epidemiologic and animal studies. This is based on available evidence and does not rule out the possibility that this agent may be carcinogenic under other conditions.

BENZENE

Benzene is a clear, colorless, aromatic hydrocarbon which is both volatile and flammable. Benzene is present as a gas in both exhaust and evaporative emissions from motor vehicles. Benzene in the exhaust, expressed as a percentage of total organic gases (TOG), varies depending on control technology (eg, type of catalyst) and the levels of benzene and aromatics in the fuel, but is generally about 3 to 5%. Percent TOG of benzene in evaporative emissions depends on control technology (ie, fuel injector or carburetor) and fuel composition (eg, benzene level and Reid Vapor Pressure, or RVP), and is generally about 1%.

Benzene is quite stable in the atmosphere, with the only benzene reaction which is important in the lower atmosphere being the one with OH radicals (2). Yet even this reaction is relatively slow. Benzene itself will not be incorporated into clouds or rain to any large degree because of its low solubility. Benzene is not produced by atmospheric reactions. The most important source of a person's daily exposure to benzene is active tobacco smoking, accounting for roughly half of the total populations' exposure (3). Outdoor concentrations of benzene, due mainly to motor vehicles, account for roughly one-quarter of the total exposure. Of the mobile source contribution, the majority comes from the exhaust component. The annual average ambient level of benzene in urban areas ranges from 4 to 7 $\mu g/m^3$, based on 1987–1990 urban air monitoring data (4,5). Based on available data, maximum exposure levels to benzene in environments heavily impacted by motor vehicles range from 40 $\mu g/m^3$ from in-vehicle exposure (6), to 288 $\mu g/m^3$ from exposure during refueling (7).

Long-term exposure to high levels of benzene in air has been shown to cause cancer of the tissues that form white blood cells (leukemia), based on epidemiology studies with workers (8,9). Leukemias and lymphomas (lymphoma is a general term for growth of new tissue in the lymphatic system of the body), as well as other tumor types, have been observed in experimental animals that have been exposed to benzene by inhalation or oral administration (8,9). Inhalation exposure of mice to benzene (10) has also provided an animal model for the type of cancer identified most closely with occupational exposure, acute myelogenous leukemia. Exposure to benzene and/or its metabolites has also been linked with genetic changes in humans and animals (11), and increased proliferation of mouse bone marrow cells (12). Furthermore, the occurrence of certain chromosomal aberrations in individuals with known exposure to benzene may serve as a marker for those at risk for contracting leukemia (13). EPA has classified benzene as a Group A human carcinogen. The International Agency for Research on Cancer (IARC) also defines benzene as an agent that is carcinogenic to humans.

Research has been conducted on the fate of benzene in experimental animals (9,14). These studies demonstrate that species differ with respect to their ability to metabolize benzene. These differences may be important when choosing an animal model for human exposures and when extrapolating high dose exposures in animals to the low levels of exposure typically encountered in occupational situations. The recent development of a benzene physiologically based pharmacokinetic model should help in performing interspecies and route-to-route extrapolations of cancer data.

A number of adverse noncancer health effects have been associated with exposure to benzene (8,9,14). Benzene is known to cause disorders of the blood. People with long-term exposure to benzene at levels that generally exceed 50 ppm (162,500 $\mu g/m^3$) may experience harmful effects on blood-forming tissues, especially bone marrow. These effects can disrupt normal blood production and cause a decrease in important blood components, such as red blood cells and blood platelets, leading to anemia and a reduced ability to clot. Exposure to benzene at comparable or even lower levels can be harmful to the immune system, increasing the chance for infection and perhaps lowering the body's defense against tumors by altering the number and function of the body's white blood cells. Studies in humans and experimental animals indicate that benzene may be a reproductive and developmental toxicant (14).

FORMALDEHYDE

Formaldehyde, a colorless gas at normal temperatures, is the simplest member of the family of aldehydes. Formaldehyde gas is soluble in water, alcohols, and other polar solvents. It is the most prevalent aldehyde in vehicle exhaust and is formed from incomplete combustion of the fuel. Formaldehyde is emitted in the exhaust of both gasoline and diesel-fueled vehicles. It is not a component of evaporative emissions. The TOG percentage of formaldehyde in exhaust varies from roughly 1 to 4%, depending on control technology and fuel composition.

Formaldehyde exhibits extremely complex atmospheric behavior (2). It is present in emissions, but is also formed by the atmospheric oxidation of virtually all organic species. It is ubiquitous in the atmosphere because it is formed in the atmospheric oxidations of methane and biogenic (produced by a living organism) hydrocarbons. Formaldehyde is photolyzed readily, and its photolysis is an important source of photochemical radicals in urban areas. It is also destroyed by reaction with OH and NO_3. The major carbon-containing product of all gas-phase formaldehyde reactions is carbon monoxide. Because formaldehyde is often the dominant source of radicals in urban atmospheres, formaldehyde concentrations have a feedback effect on the chemical residence time of other atmospheric species. Formaldehyde is highly water soluble and participates in a complex set of chemical reactions within clouds. The product of the aqueous-phase oxidation of formaldehyde is formic acid. The mobile source contribution to ambient formaldehyde levels contains both primary (ie, direct emissions from motor vehicles) and secondary formaldehyde (ie, formed from photooxidation of volatile organic compounds, or VOCs from vehicles). The mobile source contribution is difficult to quantify, but it appears that at least 30% of formaldehyde in the ambient air may be attributable to motor vehicles (14). The annual average ambient level of formaldehyde in urban areas is roughly 5 $\mu g/m^3$, based on 1990 air monitoring data (5) which accounted for ozone interference. Based on available data, maximum exposure levels to formaldehyde in environments heavily impacted by motor vehicles range from 4.9 $\mu g/m^3$ from exhaust exposure at a service station (15), to 41.8 $\mu g/m^3$ from parking garage exposure (7).

Studies in experimental animals provide sufficient evidence that long-term inhalation exposure to formaldehyde causes an increase in the incidence of squamous cell carcinomas of the nasal cavity (9,14,16). In addition, the distribution of nasal tumors in rats has been better defined; the findings suggest that not only regional exposure, but also local tissue susceptibility may be important for the distribution of formaldehyde-induced tumors (9,14). Epidemiological studies in occupationally exposed workers suggest that long-term inhalation of formaldehyde may be associated with tumors of the nasopharyngeal cavity, nasal cavity, and sinus (9,14,16). The evidence for an association between lung cancer and occupational formaldehyde is tenuous, and collectively, the recent studies do not conclusively demonstrate a causal relationship between cancer and exposure to formaldehyde in humans. EPA has classified formaldehyde as a Group B1 probable human carcinogen. IARC also defines formaldehyde as an agent that is probably carcinogenic to humans. Both classifications are based on limited evidence for carcinogenicity in humans and sufficient evidence for carcinogenicity in animals.

Recent work on the pharmacokinetics of formaldehyde has focused on the validation of measurement of DNA-protein adducts, or cross-links (DPX), as an internal measure of dose for formaldehyde exposure (9,14). DPX is the binding of DNA to a protein to which formaldehyde is bound, thus forming a separate entity that can be quantified. This is considered a more accurate way to measure the amount of formaldehyde that is present inside a tissue. An internal dosimeter for formaldehyde exposure is desirable because the inhaled concentration of formaldehyde may not reflect actual tissue exposure levels. The difference in inhaled concentration and actual tissue exposure level is due to the action of multiple defense mechanisms that act to limit the amount of formaldehyde that reaches cellular DNA. These studies have provided more accurate data with which to quantify the level of formaldehyde in the cell.

Noncancer adverse health effects associated with exposure to formaldehyde in humans and experimental animals include irritation of the eyes, nose, throat, and lower airway at low levels (0.05–10 ppm or 123–12,300 $\mu g/m^3$) (9,14,16). There is also suggestive, but not conclusive, evidence in humans that formaldehyde can affect immune function. Studies in humans and experimental animals indicate that formaldehyde may be a reproductive and developmental toxicant (14). Adverse effects on the liver and kidney have also been noted in experimental animals exposed to higher levels of formaldehyde.

1,3-BUTADIENE

1,3-Butadiene is a colorless, flammable gas at room temperature, is insoluble in water, and its two conjugated double bonds make it highly reactive. 1,3-Butadiene is formed in vehicle exhaust by the incomplete combustion of the fuel and is not present in vehicle evaporative and refueling emissions. The TOG percentage of 1,3-butadiene in exhaust varies from roughly 0.4 to 1.0%, depending on control technology and fuel composition.

1,3-Butadiene is transformed rapidly in the atmosphere (2). There are three chemical reactions of 1,3-butadiene which are important in the ambient atmosphere: reaction with hydroxyl radical (OH), reaction with ozone (O_3), and reaction with nitrogen trioxide radical (NO_3). All three of these reactions are relatively rapid, and all produce formaldehyde and acrolein, species which are themselves toxic and/or irritants. The oxidation of 1,3-butadiene by NO_3 produces organic nitrates as well. Incorporation of 1,3-butadiene into clouds and rain will not be an important process due to the low solubility of 1,3-butadiene. 1,3-Butadiene is probably not produced by atmospheric reactions. Current estimates indicate that mobile sources account for approximately 94% of the total 1,3-butadiene emissions (17). The remaining 1,3-butadiene emissions (6%) come from stationary sources mainly related to industries producing 1,3-butadiene and those industries that use 1,3-butadiene to produce other compounds. The annual average ambient level of 1,3-butadiene in urban areas (excluding high point source areas) ranges from roughly 0.2 to 1.0 $\mu g/m^3$, based on 1987–1990 urban air monitoring data (4,5). Based on a single study (18), in-vehicle exposure to 1,3-butadiene was found to average 3.0 $\mu g/m^3$.

Long-term inhalation exposure to 1,3-butadiene has been shown to cause tumors in several organs in experimental animals (9,14,19). One recent study (9,14) demonstrates the occurrence of cancer in mice at additional sites at lower concentrations of 1,3-butadiene than those of earlier studies. Studies in humans exposed to 1,3-butadiene suggest that this chemical may cause cancer (9,14,19). These epidemiological studies of occupationally exposed workers are inconclusive with respect to the carcinogenicity of 1,3-butadiene in humans, however, because of a lack of adequate exposure information and concurrent exposure to other potentially carcinogenic substances. Studies in animals also indicate that 1,3-butadiene can alter the genetic material (9,14,19). EPA has classified 1,3-butadiene as a Group B2 probable human carcinogen. IARC has classified 1,3-butadiene as an agent that is possibly carcinogenic to humans. Both classifications are based on inadequate evidence for carcinogenicity in humans and sufficient evidence for carcinogenicity in animals.

Studies on the fate of 1,3-butadiene in the body have focused on the mechanism behind the differences in carcinogenic responses seen between species (20–21). Recent pharmacokinetic research has found marked differences among mice, rats, and human tissue preparations in their ability to metabolize 1,3-butadiene and its metabolites (22). The results suggest that the effective internal dose of DNA-reactive metabolites may be less in humans than in mice for a given level of exposure.

Exposure to 1,3-butadiene is also associated with adverse noncancer health effects (9,14,19). Exposure to high levels (on the order of hundreds to thousands ppm) of this chemical for short periods of time can cause irritation of the eyes, nose, and throat, and exposure to very high levels can cause effects on the brain leading to respiratory paralysis and death. Studies of rubber industry workers who are chronically exposed to 1,3-butadiene suggest other possible harmful effects, including heart disease, blood disease, and lung disease. Studies in animals indicate that 1,3-butadiene at exposure levels of greater than 1,000 ppm (2.2×10^6 $\mu g/m^3$) may adversely affect the blood-forming organs. Reproductive and developmental toxicity has also been demonstrated in experimental animals exposed to 1,3-butadiene.

ACETALDEHYDE

Acetaldehyde is a saturated aldehyde that is a colorless liquid and volatile at room temperature. Acetaldehyde is found in vehicle exhaust and is formed as a result of incomplete combustion of the fuel. Acetaldehyde is emitted in the exhaust of both gasoline and diesel-fueled vehicles. It is not a component of evaporative emissions. Percent TOG of acetaldehyde in exhaust varies from roughly 0.4 to 1.0 percent, depending on control technology and fuel composition.

The atmospheric chemistry of acetaldehyde is similar in many respects to that of formaldehyde (2,23). Like formaldehyde, it can be both produced and destroyed by atmospheric chemical transformation. However, there are important differences between the two. Acetaldehyde photolyzes, but much more slowly than formaldehyde. Whereas formaldehyde produces CO upon reaction or photolysis, acetaldehyde produces organic radicals that ultimately form peroxyacetyl nitrate (PAN) and formaldehyde. Acetaldehyde is also significantly less water soluble than formaldehyde. Acetaldehyde is ubiquitous in the environment and is naturally released (24). The mobile source contribution to ambient acetaldehyde levels contains both primary and secondary acetaldehyde. Data from emission inventories and atmospheric modeling indicate that roughly 40% of the acetaldehyde in ambient air may be attributable to mobile sources. The annual average ambient level of acetaldehyde in urban areas is roughly 3 $\mu g/m^3$, based on available 1990 urban air monitoring data (5) which accounted for ozone interference. Based on a single study (15), the in-vehicle exposure level of acetaldehyde was found to average 13.7 $\mu g/m^3$.

There is sufficient evidence that acetaldehyde produces cytogenetic damage in cultured mammalian cells (24,25). Although there are not many studies done with whole animals (*in vivo*) (24–26), they suggest that acetaldehyde produces similar effects *in vivo*. Thus, the available evidence indicates that acetaldehyde is mutagenic and may pose a risk for somatic cells (all body cells excluding the reproductive cells). Current knowledge, however, is inadequate with regard to germ cell (reproductive cell) mutagenicity because the available information is insufficient to support any conclusions about the ability of acetaldehyde to reach mammalian gonads and produce heritable genetic damage.

Studies in experimental animals provide sufficient evidence that long-term inhalation exposure to acetaldehyde causes an increase in the incidence of squamous cell carcinomas of the nasal cavity (24–26). One epidemiological study, in occupationally exposed workers, was insufficient to suggest that long-term inhalation of acetaldehyde may be associated with an increase in total cancers (24–26). EPA has classified acetaldehyde as a Group B2 probable human carcinogen.

Non-cancer effects in studies with rats and mice showed acetaldehyde to be moderately toxic by the inhalation, oral, and intravenous routes (24–26). Acetaldehyde is a sensory irritant that causes a depressed respiration rate in mice. In rats, acetaldehyde increased blood pressure and heart rate after exposure by inhalation. The primary acute effect of exposure to acetaldehyde vapors is irritation of the eyes, skin, and respiratory tract. At high concentrations, irritation and ciliastatic effects can occur, which could facilitate the uptake of other contaminants. Clinical effects include reddening of the skin, coughing, swelling of the pulmonary tissue, and localized tissue death. Respiratory paralysis and death have occurred at extremely high concentrations. It has been suggested that voluntary inhalation of toxic levels of acetaldehyde would be prevented by its irritant properties, since irritation occurs at levels below 200 ppm (360,000 $\mu g/m^3$). The new genotoxicity studies, which utilize lower concentrations of acetaldehyde, do not produce chromosomal aberration and/or cellular mutations.

The research into reproductive and developmental effects of acetaldehyde is based on intraperitoneal injection, intravenous, or oral administration of acetaldehyde to rats and mice, and also *in vitro* (cell culture) studies (24–26). Little research exists that addresses the effects of inhalation of acetaldehyde on reproductive and developmental effects. The *in vitro* and *in vivo* studies provide evidence to support the fact that acetaldehyde may be the causative factor in birth defects observed in fetal alcohol syndrome.

DIESEL PARTICULATE MATTER

Diesel exhaust particulate matter consists of a solid core composed mainly of carbon, a soluble organic fraction, sulfates, and trace elements (27). Light-duty diesel engines emit from 30 to 100 times more particles than comparable catalyst-equipped gasoline vehicles (27). Diesel particulate matter is mainly attributable to the incomplete combustion of fuel hydrocarbons, though some may be due to engine oil or other fuel components. The particles may also become coated with adsorbed and condensed high molecular weight organic compounds.

Diesel particulate matter has not been explicitly modeled to determine its atmospheric transformation and residence times because of its inherent complexity. In order to accomplish this modeling, consideration needs to be given to the relative abundance of the various polycyclic organic matter (POM) species in the atmosphere, the availability of emissions data, and determining a location's specific area, mobile, and point sources. Using the ambient national average total suspended particle (TSP) concentration (28,29), and the percent contribution of diesel particulate matter to TSP, the concentration of diesel particulate matter in ambient air can be estimated. For 1990, the estimated resultant concentration of diesel particulate matter is 2.5 $\mu g/m^3$.

Studies in experimental animals provide sufficient evidence that long-term inhalation exposure to high levels of diesel exhaust causes an increase in the induction of lung tumors in two strains of rats and two strains of mice (30). Studies have concentrated on the hypothesis that the carbon core of diesel particulate matter is the causative agent in the genesis of lung cancer (31,32). By exposing rats to carbon black and diesel soot and comparing the results to diesel exhaust itself, the tumor response to diesel exhaust and carbon black is qualitatively similar. Also, as a result of extensive studies, the direct-acting mutagenic activity of both particle and gaseous fractions of diesel exhaust has been shown (30). In two key epidemiological studies on railroad workers occupationally exposed to diesel exhaust (30), it was observed that long-term inhalation of diesel exhaust produced an excess risk of lung cancer. Collectively, the epidemiological studies show a positive, though limited, association between diesel exhaust exposure and lung cancer. EPA currently considers diesel particulate matter to be a Group B1 probable human carcinogen; however, a formal risk assessment is still in progress.

An understanding of the pharmacokinetics associated with pulmonary deposition of diesel exhaust particles and their adsorbed organics is critical in understanding the carcinogenic potential of diesel engine emissions. The pulmonary clearance of diesel exhaust particles has multiple phases and involves several processes, including a relatively rapid transport system and slow macrophage-mediated (a white blood cell that engulfs and digests foreign particles) processes (30). The observed dose-dependent increase in the particle burden of the lungs is due, in part, to an overloading of alveolar macrophage function. The resulting increase in particle retention has been shown to increase the bioavailability of particle adsorbed mutagenic and carcinogenic components, such as benzo[a]pyrene and 1-nitropyrene. Experimental data also indicate the ability of the alveolar macrophage to metabolize and solubilize the particle-adsorbed components. Although macromolecular binding of diesel exhaust particle-derived POM and the formation of DNA adducts following exposure to diesel exhaust have been reported, a quantitative relationship between these and increased carcinogenicity is not available.

A number of adverse noncancer health effects have also been associated with exposure to acute, subchronic, and chronic diesel exhaust at levels found in the ambient air (30). Most of the effects observed through acute and subchronic exposure are respiratory tract irritation and diminished resistance to infection. Increased cough and phlegm and slight impairments in lung function have also been documented. Animal data indicate that chronic respiratory diseases can result from long-term (chronic) exposure to diesel exhaust. It appears that normal, healthy adults are not at high risk to serious noncancer effects of diesel exhaust at levels found in the ambient air. The data base is inadequate to form conclusions about sensitive subpopulations.

BIBLIOGRAPHY

1. *Fed. Reg.,* **51**(185), 33992–34003 (Sept. 24, 1986).
2. M. P. Ligocki and co-workers, *Atmospheric Transformation of Air Toxics: Benzene, 1,3-Butadiene, and Formaldehyde,* Systems Applications International, San Rafael, Calif., SYSAPP-91/106, 1991.
3. L. A. Wallace, *Environ. Health Perspect.* **82,** 165–169, (1989).
4. Environmental Protection Agency (EPA), *AIRS User's Guide,* vol. 1-7, Office of Air Quality Planning and Standards, Research Triangle Park, N.C., 1989.
5. R. A. McAllister and co-workers, *Urban Air Toxics Monitoring Program,* EPA-450/4-91-001, 1990.
6. EPA, *The Total Exposure Assessment Methodology (TEAM) Study: Summary and Analysis:* vol. 1, EPA/600/6-87/002a, Office of Research and Development, Washington, D.C., June 1987.
7. A. Wilson and co-workers, *Air Toxics Microenvironment Exposure and Monitoring Study,* South Coast Air Quality Management District, El Monte, Calif., 1991.
8. EPA, *Interim Quantitative Cancer Unit Risk Estimates Due to Inhalation of Benzene,* Office of Health and Environmental Assessment, Carcinogen Assessment Group, Washington, D.C., 1985.
9. Clement Associates, Inc., *Motor Vehicle Air Toxics Health Information,* September 1991.
10. E. P. Cronkite and co-workers, *Environ. Health Perspect.* **82**(97-108), 1989.
11. *IARC monographs* **29,** 345–389 (1982).
12. R. D. Irons and co-workers, *Proc. Natl. Acad. Sci.* **89,** 3691–3695 (1992).
13. M. Lumley, H. Barker, and J. A. Murray, *Lancet* **336,** 1318–1319 (1990).
14. EPA, *Motor Vehicle-Related Air Toxics Study,* EPA 420-R-93-005, Office of Mobile Sources, Ann Arbor, Mich., April 1993.
15. D. C. Shikiya and co-workers, *In-Vehicle Air Toxics Characterization Study in the South Coast Air Basin,* South Coast Air Quality Management District, El Monte, Calif., May 1989.
16. EPA, *Assessment of Health Risks to Garment Workers and Certain Home Residents from Exposure to Formaldehyde,* Office of Pesticides and Toxic Substances, April 1987.
17. EPA, *Locating and Estimating Air Emissions from Sources of 1,3-Butadiene,* EPA-450/2-89-021, Office of Air Quality Planning and Standards, Research Triangle Park, N.C., 1989.
18. C. C. Chan and co-workers, *Commuter's Exposure to Volatile Organic Compounds, Ozone, Carbon Monoxide, and Nitrogen Dioxide,* Air and Waste Management Association Paper 89-34A.4, 1989.
19. EPA, *Mutagenicity and Carcinogenicity Assessment of 1,3-Butadiene,* EPA/600/8-85/004F, Office of Health and Environmental Assessment, Washington, D.C., 1985.
20. G. A. Csanády and J. A. Bond, *CIIT Activities* **11**(2), 1–8 (1991).

21. G. A. Csanády and J. A. Bond, *Toxicologist* **11,** 47, 1991.
22. L. Recio and co-workers, "Biotransformation of butadiene by hepatic and pulmonary tissues from mice, rats, and humans: relationship to butadiene carcinogenicity and *in vivo* mutagenicity," proceedings from the *Air and Waste Management Association's specialty conference "Air Toxics Pollutants from Mobile Sources: Emissions and Health Effects,"* October 16–18, 1991.
23. M. P. Ligocki and G. Z. Whitten, *Atmospheric Transformation of Air Toxics: Acetaldehyde and Polycyclic Organic Matter,* Systems Applications International, San Rafael, Calif., (SYSAPP-91/113), 1991.
24. EPA, *Health Assessment Document for Acetaldehyde,* EPA-600/8-86/015A (External Review Draft), Office of Health and Environmental Assessment, Environmental Criteria and Assessment Office, Research Triangle Park, N.C., 1987.
25. *Proposed Identification of Acetaldehyde as a Toxic Air Contaminant,* Part B Health assessment, California Air Resources Board, Stationary Source Division, August, 1992.
26. EPA, *Integrated Risk Information System (IRIS),* Office of Health and Environmental Assessment, Environmental Criteria and Assessment Office, Cincinnati, Ohio, 1992.
27. National Research Council, *Impacts of Diesel-Powered Light-duty Vehicles: Diesel Technology,* National Academy Press, Washington, D.C., 1982.
28. EPA, *National Air Pollutant Emission Estimates 1940–1990,* EPA-450/4-91-026, Office of Air Quality Planning and Standards, Research Triangle Park, NC, 1991.
29. EPA, *National Air Quality and Emissions Trends Report, 1990,* EPA-450/4-91-023, Research Triangle Park, NC, Office of Air Quality Planning and Standards, 1991.
30. EPA, *Health Assessment Document for Diesel Emissions; Workshop Review Draft,* EPA-600/8-90/057A, Office of Health and Environmental Assessment, Washington, D.C., 1990.
31. U. Heinrich and co-workers, "The carcinogenic effects of carbon black particles and tar/pitch condensation aerosol after inhalation exposure of rats," *Seventh International Symposium on Inhaled Particles,* Edinburgh, Sept. 16–20, 1991.
32. J. L. Mauderly and co-workers, "Influence of particle-associated organic compounds on carcinogenicity of diesel exhaust," *Seventh International Symposium on Inhaled Particles,* Edinburgh, Sept. 16–20, 1991.

AIR POLLUTION CONTROL METHODS

Burton B. Crocker
Chesterfield, Missouri

Air pollution has been defined as the presence in ambient air of one or more contaminants of such quantity and time duration as to be injurious to human, plant, or animal life; property; or the conduct of business (1,2) or so as to alter significantly the natural balance of an ecosystem (3). The effect of a time–dosage relationship has been clearly considered by the U.S. EPA in the establishment of the National Ambient Air Quality Standards (NAAQS) for Criteria Pollutants (Table 1). Because of more recent concerns about the effect of release of greenhouse gases and of stratospheric O_3-depleting gases to the atmosphere, the last phrase has been added to the definition above (3).

Within a local area, the definition suggests that the harmfulness of air pollutants can be reduced by reducing the downwind concentration of pollutants or the exposure time to them or both. However, atmospheric dispersion studies have indicated that high concentrations of the SO_2–sulfite–sulfate complex and of NO_x can be transported hundreds of kilometers downwind from tall stacks and deposited as acid rain, killing fish in lakes and affecting the health and growth of forests greatly removed from the source of the emission. This has been especially true of SO_x and NO_x emissions from large-scale release of fossil fuel combustion flue gases, a matter of concern to the electric power industry. Large-scale, tall stack releases of SO_x from smelting of sulfide-containing nonferrous ores can also contribute to the problem. The combination of such releases has resulted in extensive ecological damage in eastern Canada. Releases from fossil fuel combustion in western and northern Europe have been channeled by wind currents to impact the Black Forest of Germany with extensive damage to fir trees. Likewise, in the U.S. Clean Air Reauthorization Act of 1990, Title IV covers acid deposition control, or acid rain. Its purpose is to cut SO_x and NO_x emissions by 90% from 111 fossil-fueled power plants that emit major quantities of these pollutants in the central and midwestern portions of the United States beginning in 1995. An annual reduction of 9 million t of SO_2 and 1.8 million t of NO_x per year is mandated as phase 1 in 21 states, extending from Minnesota and Wisconsin to New York, New Jersey, and Pennsylvania; and from the Great Lakes to the Gulf of Mexico (Mississippi, Alabama, Georgia, and Florida). Beginning in the year 2000, the acid rain regulations will impact an additional 200 fossil-fueled power plants. The impetus for these regulations is concern that large-scale SO_2 and NO_x emissions from tall stacks of power plants in the Midwest travel northeastward for long distances and damage forests and lakes in the United States and Canada.

In regard to ozone depletion of the stratosphere, the 1990 Clean Air Reauthorization Act contains a phase-out schedule for chlorofluorocarbons (CFCs), halons, CCl_4, and methylchloroform. It includes new requirements for collecting and recovering refrigerant gases during air-conditioner maintenance such as on automobile compressors.

Selection of pollution-control methods is generally based on the need to control ambient air quality and to achieve compliance with NAAQS standards for criteria pollutants or, in the case of nonregulated contaminants, to protect human health and vegetation. There are three elements to a pollution problem: a source, a receptor affected by the pollutants, and the transport of the pollutants from source to receptor. Modification or elimination of any one of these elements can change the nature of a pollution problem. For instance, tall stacks that disperse the effluent, modifying the transport of pollutants, can reduce nearby pollution levels. Although better dispersion aloft can solve a local problem, if done from numerous sources, it can cause a regional problem, such as the acid rain problems mentioned above. Atmospheric dilution as a control measure has been discussed (4–16). A better approach is to control emissions at the source.

There are three main classes of pollutants: gases, particulates (which may be either liquid or solid or a combination), and odors (which may originate as gases or particulates). Although odors are controlled as are other

Table 1. National Ambient Air Quality Standards

	Maximum Permissible Concentration[a]			
Pollutant	Averaging Time	Primary Standard	Secondary Standard	Measurement Method
Sulfur oxides	annual arithmetic mean	80 $\mu g/m^3$ (0.3 ppmv)		West-Gaeke Pararosaniline
	24 h max	365 $\mu g/m^3$ (0.14 ppmv)		
	3 h max		1300 $\mu g/m^3$ (0.5 ppmv)	
Particulates (PM_{10})	annual arithmetic mean	50 $\mu g/m^3$	same	gravimetric 24-h high volume sample with PM_{10} classifying head
	24 h max	150 $\mu g/m^3$	same	
Carbon monoxide	8 h max	10 mg/m^3 (9 ppmv)	same	nondispersive infrared analyzer
	1 h max	40 mg/m^3 (35 ppmv)	same	
Ozone	1 h max	235 $\mu g/m^3$	same	gas-phase chemiluminescence analyzer
Hydrocarbons	3 h max (6–9 am)	160 $\mu g/m^3$ (0.24 ppmv as CH_4)	same	flame ionization detector
Nitrogen oxides	annual arithmetic mean	100 $\mu g/m^3$ (0.05 ppmv as NO_2)	same	chemiluminescence analyzer
Lead	calender quarter arithmetic mean	1.5 $\mu g/m^3$	same	lead analysis by atomic absorption spectrometry on extract from high volume sample catch

[a] Standards for periods shorter than annual average may be exceeded once per year.

pollutants, they are often discussed separately because of the different methods used for their sensing and measurement. Many effluents contain several contaminants: one or two may be present as gases; the others often exist as liquid or solid particulates of various sizes. The possibility that effluent pollutants may be present in more than one physical state must be taken into account in sampling, analysis, and control. To achieve air pollution control, reliable measurements are needed to quantify both the pollutant concentration and the contribution of individual sources. These data are necessary for designing control equipment, monitoring emissions, and maintaining acceptable ambient air quality.

PRINCIPAL ENERGY SOURCES OF AIR POLLUTANTS

Principal energy sources producing and emitting air pollutants to the atmosphere are fossil fuel combustion for electric power generation, steam production, and space heating. Another large source is fuel combustion for transportation. On a mass-emission basis, U.S. EPA estimates show the following percentage of total U.S. emissions by source: transportation, 43%; stationary fuel combustion, 29%; industrial processes, 16%; solid waste disposal (incineration), 4%; and miscellaneous, 8%. Of the transportation emissions, by far the largest component is carbon monoxide (78%). The second highest component is NO_x (10%), followed by hydrocarbons (9%), particulates (2%), and SO_x (1%). This article discusses air pollution control techniques for stationary sources only (see AUTOMOBILE EMISSIONS, AIRCRAFT ENGINES, AUTOMOTIVE; AIR POLLUTION: AUTOMOBILE TOXIC EMISSIONS).

Stationary consumption of fossil fuels takes place in boilers (about 33% of the total U.S. fossil fuel consumption), residential use, and direct and indirect heating of processes such as steel production and rolling, nonferrous metallurgical processes as well as many types of process industry operations.

Boilers can be classified industrially as utility boilers, industrial boilers, and commercial or institutional boilers. Utility boilers are generally large (averaging around 10.5×10^{12} J/h of heat release), generating steam for electric-power generation. On a combustion heat release capacity basis, the total U.S. utility boiler capacity is about 80% of the combined U.S. capacity for all other classes of boilers. More than 50% of the U.S. utility boiler capacity burns coal or residual oil. These two fuels present the most significant air pollution problems, because both contain ash and often sulfur. In the past, these boilers have also been designed for high flame temperatures, which maximizes the fixation of NO_x by reaction between N_2 and O_2 in combustion air. It is estimated that there are 4,000–5,000 utility boilers in the United States, 700,000–800,000 industrial boilers, and >1.5 million commercial or institutional boilers.

Industrial boilers average about 6.3×10^9 J/h heat release, and commercial boilers are even smaller. Approximately 33% of these boilers are natural gas or distillate oil fired. The remainder present complicated pollution-control problems because of their large number, their small size, and their proximity to population centers. In addition, many of these boilers have fluctuating load swings and unsteady operating conditions, which influence the rate and type of emissions released.

For boilers burning coal, the largest specific release, uncontrolled, is typically SO_2 (in tons released per ton of fuel fired). This can be appreciably reduced by burning low sulfur coal. NO_x is the second highest pollutant released, with particulate emissions close behind. Hydrocarbon emissions are generally quite low but important in the overall atmospheric pollution situation. Although even lower, there are small quantities of aldehydes released that are important as precursers of photochemical irritants. Emissions and their control from fossil-fuel combustion has been discussed (17), and EPA estimates of emissions from most pollution sources have been given (18). Emission release and control from sources burning unconventional fuels (19) and from municipal solid waste incinerator sources (including incineration of industrial waste) (20) have been discussed. Emissions and control in petroleum refining have been reviewed (21).

MEASUREMENT OF AIR POLLUTION

Measurement techniques are divided into two categories: ambient and source measurement. Ambient air samples often require detection and measurement in the ppmv to ppbv (parts by volume) range, whereas source concentrations can range from tenths of a volume percent to a few hundred ppmv. Federal regulations (22–23) require periodic ambient air monitoring at strategic locations in a designated air quality control region. The number of required locations and complexity of monitoring increases with region population and with the normal concentration level of pollutants. Continuous monitoring is preferable, but for particulates one 24-h sample every 6 days may be permitted. In some extensive metropolitan sampling networks, averaged results from continuous monitors are telemetered to a single data-processing center. Special problems have been investigated using portable, vehicle-carried, or airborne ambient sampling equipment. The use of remote-guided miniature aircraft has been reported as a practical, cost-effective ambient sampling method (24). Ambient sampling may fulfill one or more of the following objectives: (*1*) establishing and operating a pollution alert network, (*2*) monitoring the effect of an emission source, (*3*) predicting the effect of a proposed installation (compliance with prevention of significant deterioration (PSD) regulations requires 1 yr of background ambient air monitoring at the proposed installation site before filing an application for a construction permit for a new installation), (*4*) establishing seasonal or yearly trends, (*5*) locating the source of an undesirable pollutant, (*6*) obtaining permanent sampling records for legal action or for modifying regulations, and (*7*) correlating pollutant dispersion with meteorological, climatological, or topographic data and with changes in societal activities.

The problems of source sampling are distinct from those of ambient sampling. Source gas may have high temperature or contain high concentrations of water vapor or entrained mist, dust, or other interfering substances so that particulates or gases may be deposited on or absorbed into the grain structure of the gas-extractive sampling probes. Depending on the objective or regulations, source sampling may be infrequent, occasional, intermittent, or continuous. Typical objectives are (*1*) demonstrating compliance with regulations; (*2*) obtaining emission data; (*3*) measuring product loss or optimizing process operating variables; (*4*) obtaining data for engineering design, such as for control equipment; (*5*) determining collector efficiency or acceptance testing of purchased equipment; and (*6*) determining need for maintenance of process or control equipment.

Sampling of Gaseous Pollutants

Gaseous pollutant detection depends on the chemistry of the material involved. Reference methods for criteria (25) and hazardous (26) pollutants established by the U.S. EPA include sulfur dioxide by the West-Gaeke method, carbon monoxide by nondispersive infrared analysis, ozone and nitrogen dioxide by chemiluminescence, and hydrocarbons by gas chromatography coupled with flame-ionization detection. Gas chromatography coupled with a suitable detector can also be used to measure ambient concentrations of vinyl chloride monomer, halogenated hydrocarbons and aromatics, and polyacrylonitrile (27–28). Methods of sampling and analysis for ambient air gases have been reviewed (29).

Automated analyzers may be used for continuous monitoring of ambient pollutants and EPA has developed continuous procedures (30) as alternatives to the referenced methods. For source sampling, EPA has specified extractive sampling trains and analytical methods for pollutants such as SO_2 and SO_3, sulfuric acid mists, NO_x, mercury, beryllium, vinyl chloride, and volatile organic compounds (VOCs). Some EPA new source performance standards require continuous monitors on specified sources.

Sampling of Particulates

Ambient air suspended particulate concentration was traditionally measured gravimetrically over a 24-h period with a "Hi-Vol" sampler. However, in 1987 the EPA changed ambient particulate control to the PM_{10} reference method (31). In the PM_{10} method, a particle size classification head is attached to a Hi-Vol sampler so that only particulates finer than an aerodynamic 10 μm are collected on the filter. Although tape samplers, used for more frequent determination of suspended particulates, have been tied into the EPA Alert Warning System, it is not yet apparent how they will be correlated with PM_{10} monitoring. In the tape method, particulate quantity is measured automatically by light transmittance or β-ray attenuation and converted to an electronic signal for transmission and data processing.

Source sampling of particulates requires isokinetic removal of a composite sample from the stack or vent effluent to determine representative emission rates. Samples are collected either extractively or using an in-stack filter; EPA Method 5 is representative of extractive sampling and EPA Method 17, of in-stack filtration. Other means of source sampling have been used, but they have been largely supplanted by EPA methods. Continuous in-stack monitors of opacity use attenuation of radiation across the effluent. Opacity measurements are affected by the particle size, shape, size distribution, refractive index, and the wavelength of the radiation (32). Table 2 lists 50 specific source sampling procedures developed and promulgated by the U.S. EPA as specific regulatory compliance methods in specific situations (33). Detailed procedures for these methods have been given (34).

Table 2. EPA Reference Methods for Source Sampling[a]

Method
Method 1—Sample and velocity traverses for stationary sources
Method 2—Determination of stack gas velocity and volumetric flow rate (Type S pitot tube)
Method 2A—Direct measurement of gas volume through pipes and small ducts
Method 2B—Determination of exhaust gas volume flow rate from gasoline vapor incinerators
Method 3—Gas analysis for carbon dioxide, oxygen, excess air, and dry molecular weight
Method 3A—Determination of oxygen and carbon dioxide concentrations in emissions from stationary sources (instrumental analyzer procedure)
Method 4—Determination of moisture content in stack gases
Method 5—Determination of particulate emissions from stationary sources
Method 5A—Determination of particulate emissions from the asphalt processing and asphalt roofing industry
Method 5B—Determination of nonsulfuric acid particulate matter from stationary sources
Method 5D—Determination of particulate matter emissions from positive pressure fabric filters
Method 5E—Determination of particulate emissions from the wool fiberglass insulation manufacturing industry
Method 5F—Determination of nonsulfate particulate matter from stationary sources
Method 6—Determination of sulfur dioxide emissions from stationary sources
Method 6A—Determination of sulfur dioxide, moisture, and carbon dioxide emissions from fossil fuel combustion sources
Method 6B—Determination of sulfur dioxide and carbon dioxide daily average emissions from fossil fuel combustion sources
Method 6C—Determination of sulfur dioxide emissions from stationary sources (instrumental analyzer procedure)
Method 7—Determination of nitrogen oxide emissions from stationary sources
Method 7A—Determination of nitrogen oxide emissions from stationary sources
Method 7B—Determination of nitrogen oxide emissions from stationary sources (ultraviolet spectrophotometry)
Method 7C—Determination of nitrogen oxide emissions from stationary sources
Method 7D—Determination of nitrogen oxide emissions from stationary sources
Method 7E—Determination of nitrogen oxides emissions from stationary sources (instrumental analyzer procedure)
Method 8—Determination of sulfuric acid mist and sulfur dioxide emissions from stationary sources
Method 9—Visual determination of the opacity of emissions from stationary sources
Method 10—Determination of carbon monoxide emissions from stationary sources
Method 10A—Determination of carbon monoxide emissions in certifying continuous emission monitoring systems at petroleum refineries
Method 11—Determination of hydrogen sulfide content of fuel gas streams in petroleum refineries
Method 12—Determination of inorganic lead emissions from stationary sources
Method 13A—Determination of total fluoride emissions from stationary sources—SPADNS zirconium lake method
Method 13B—Determination of total fluoride emissions from stationary sources—specific ion electrode method
Method 14—Determination of fluoride emissions from potroom roof monitors of primary aluminum plants
Method 15—Determination of hydrogen sulfide, carbonyl sulfide, and carbon disulfide emissions from stationary sources
Method 15A—Determination of total reduced sulfur emissions from sulfur recovery plants in petroleum refineries
Method 16—Semicontinuous determination of sulfur emissions from stationary sources
Method 16A—Determination of total reduced sulfur emissions from stationary sources (impinger technique)
Method 16B—Determination of total reduced sulfur emissions from stationary sources
Method 17—Determination of particulate emissions from stationary sources (instack filtration method)
Method 18—Measurement of gaseous organic compound emissions by gas chromatography
Method 19—Determination of sulfur dioxide removal efficiency and particulate matter, sulfur dioxide and nitrogen oxides emission rates
Method 20—Determination of nitrogen oxides, sulfur dioxide, and oxygen emissions from stationary gas turbines
Method 21—Determination of volatile organic compounds leaks
Method 22—Visual determination of fugitive emissions from material sources and smoke emissions from flares
Method 24—Determination of volatile matter content, water content, density, volume solids, and weight solids of surface coating
Method 24A—Determination of volatile matter content and density of printing inks and related coatings
Method 25—Determination of total gaseous nonmethane organic emissions as carbon
Method 25A—Determination of total gaseous organic concentration using a flame ionization analyzer
Method 25B—Determination of total gaseous organic concentration using a nondispersive infrared analyzer
Method 27—Determination of vapor tightness of gasoline delivery tank using pressure-vacuum test
Appendix B—Performance Specifications
Performance Specification 1—Performance specifications and specification test procedures for transmissometer systems for continuous measurement of the opacity of stack emissions
Performance Specification 2—Specifications and test procedures for SO_2 and NO_x continuous emission monitoring systems in stationary sources
Performance Specification 3—Specifications and test procedures for O_2 and CO_2 continuous emission monitoring systems in stationary sources

[a] Ref. 33. Courtesy of McGraw-Hill.

Particle size measurements for particulates extracted by filtration, electrostatic or thermal precipitation, or impaction may be performed using microscopy, sieve analysis, gas or liquid sedimentation, centrifugal classification, or electrical or optical counters. For aerosol particulate size determination, however, questions arise such as whether the collected particles agglomerate after capture or whether they are redispersed to the same degree in the measuring media as they were originally. These problems can be avoided mainly by performing particle size measurements on the original aerosol by using devices such as cascade impactors (35), virtual impactors (36), and diffusion batteries and mobility analyzers.

AIR POLLUTION AND CONTROL REGULATIONS

There has been considerable improvement, especially in industrial areas, in U.S. air quality since the adoption of the Clean Air Act of 1972. Appreciable reductions in particulate emissions and in SO_2 levels are especially evident. In 1990, however, almost every metropolitan area was in nonattainment status on ozone air quality standards; 50 metropolitan areas exceeded the CO standard and between 50 and 100 exceeded the PM_{10} standard for particulate level (37).

The U.S. Congress adopted a new clean air act in 1990 that has three areas of emphasis: acid rain reduction in the northeastern United States; severe limitation on atmospheric emissions of 189 chemicals on the EPA hazardous or toxic substance list; and tightened regulations on vehicular exhaust, reformulated vehicular fuels, and vehicles capable of using alternative fuels (ozone compliance and smog reduction). Regulations associated with acid rain prevention emphasize reductions in sulfur oxide and NO_x emissions from combustion processes, especially coal-fired power boilers in the Midwest. The chemical process industry and their customers will be increasingly under pressure to eliminate atmospheric releases of VOCs and carcinogenic-suspect compounds.

Minimizing Pollution Control Cost

Although the first impulse for emission reduction is often to add a control device, this may not be the environmentally best or least costly approach. Process examination may reveal changes or alternatives that can eliminate or reduce pollutants, decrease the gas quantity to be treated, or render pollutants more amenable to collection. Following are principles to consider for controlling pollutants without the addition of specific treatment devices (38):

1. Eliminate the source of the pollutant.
    - Seal the system to prevent interchanges between system and atmosphere.
        - Use pressure vessels.
        - Interconnect vents on receiving and discharging containers.
        - Provide seals on rotating shafts and other necessary openings.
    - Change raw materials, fuels, etc to eliminate the pollutant from the process.
    - Change the manner of process operation to prevent or reduce formation of, or air entrainment of, a pollutant.
    - Change the type of process step to eliminate the pollutant.
    - Use a recycle gas or recycle the pollutants rather than using fresh air or venting.
2. Reduce the quantity of pollutant released or the quantity of carrier gas to be treated.
    - Minimize entrainment of pollutants into a gas stream.
    - Reduce number of points in system in which materials can become airborne.
    - Recycle a portion of process gas.
    - Design hoods to exhaust the minimum quantity of air necessary to ensure pollutant capture.
3. Use equipment for dual purposes, such as a fuel combustion furnace to serve as a pollutant incinerator.

Steps such as the substitution of low sulfur fuels or nonvolatile solvents, change of raw materials, lowering of operation temperatures to reduce NO_x formation or volatilization of process material, and installation of well-designed hoods (39–46) at emission points to reduce effectively the air quantity needed for pollutant capture are illustrations of the above principles.

Selection of Control Equipment

Engineering information (47) needed for the design and selection of pollution control equipment include knowledge of the properties of pollutants (chemical species, physical state, particle size, concentration, and quantity of conveying gas) and effects of pollutant on surrounding environment. The design must consider likely future collection requirements. Advantages of alternative collection techniques must be determined:

1. Collection efficiency.
2. Ease of reuse or disposal of recovered material.
3. Ability of collector to handle variations in gas flow and loads at required collection efficiencies.
4. Equipment reliability and freedom from operational and maintenance attention.
5. Initial investment and operating cost.
6. Possibility of recovery or conversion of contaminant into a salable product.

Known engineering principles must be applied even in areas of extremely dilute concentration.

The physical state of a pollutant is obviously important; a particulate collector cannot remove vapor. Pollutant concentration and carrier gas quantity are necessary to estimate collector size and required efficiency, and knowledge of a pollutant's chemistry may suggest alternative approaches to treatment. Emission standards may set collection efficiency, but specific regulations do not exist for many trace emissions. In such cases emission targets must be set by dose–exposure time relationships obtained from effects on vegetation, animals, and humans. With such information, a list of possible treatment methods can be made (Table 3).

Control devices that are too inefficient for a particular pollutant or too expensive can then be stricken from the list. For instance, atmospheric dispersion is often not an acceptable solution; condensation may require costly refrigeration to give adequate collection. Although both absorption and adsorption devices for contaminant gases can be designed for almost any efficiency, economics generally dictate the choice between the two. Grade-efficiency curves should be consulted in evaluating particulate collection devices, and the desirability of dry or wet particulate collection should be considered, especially with respect to material recycle or disposal.

Other factors to be evaluated are capital investment and operating cost, material reuse or alternative disposal economics, relative ruggedness and reliability of alternative control devices, and the ability to retain desired efficiency under all probable operating conditions. Control equipment needs to be both rugged and reliable, in part because managers are often reluctant to shut a process down for control equipment repair. Efficiency of control devices varies with processing conditions, flow rate, temperature, emission concentration, and particle size. Control devices should be designed to handle these variations. Combinations of gaseous and particulate pollutants can be especially troublesome as gaseous removal devices are often unsuitable for heavy loadings of insoluble solids. Concentrations of soluble particulates up to 11 g/m^3 have been handled in gas absorption equipment with some success. However, to ensure rapid particle solution, special consideration must be given to wet–dry interfaces and adequate liquid quantities.

CONTROL OF GASEOUS EMISSIONS

Five methods are available for controlling gaseous emissions: absorption, adsorption, condensation, chemical reaction, and incineration. Atmospheric dispersion from a tall stack, considered as an alternative in the past, is now less viable. Absorption is particularly attractive for pollutants in appreciable concentration; it is also applicable to dilute concentrations of gases having high solvent solubility. Adsorption is desireable for contaminant removal down to extremely low levels (less than 1 ppmv) and for handling large gas volumes that have quite dilute contaminant levels. Condensation is best for substances having rather high vapor pressures. Where refrigeration is needed for the final step, elimination of noncondensible diluents is beneficial. Incineration, suitable only for combustibles, is used to remove organic pollutants and small quantities of H_2S, CO, and NH_3. Specific problem gases such as sulfur and nitrogen oxides require combinations of methods and are discussed separately.

Table 3. Checklist of Applicable Devices for Control of Pollutants

	Pollutant Classification			
			Particulate	
Equipment Type	Gas	Odor	Liquid	Solid
Absorption	•	•		
aqueous solution				
nonaqueous				
Adsorption	•	•		
throw-away canisters				
regenerable stationary beds				
regenerable traveling beds				
chromatographic adsorption				
Air dispersion (stacks)	•	•	•	•
Condensation	•	•		
Centrifugal separation (dry)			•	•
Chemical reaction	•	•		
Coagulation and particle growth			•	•
Filtration				
fabric and felt bags				•
granular beds			•	•
fine fibers			•	•
Gravitational settling				•
Impingement (dry)				•
Incineration	•	•	•	•
Precipitation, electrical				
dry			•	•
wet	•	•	•	•
Precipitation, thermal			•	•
Wet collection[a]	•	•	•	•

[a] Includes cyclonic, dynamic, filtration, inertial impaction (wetted targets, packed towers, turbulent targets), spray chambers, and venturi.

Absorption

Absorption is a diffusional mass-transfer operation by which a soluble (or semisoluble) gaseous component can be removed from a gas stream by causing the absorbable component to dissolve in a solvent liquid through gas–liquid contact. The driving force for absorption is the difference between the partial pressure of the soluble gas in the gas mixture and the vapor pressure of the solute gas in the liquid film in contact with the gas. If the driving force is not positive, no absorption will occur. If it is negative, desorption or stripping will occur and pollution of the gas being treated will actually be enhanced.

Absorption systems can be divided into those that use water as the primary absorbing liquid and those that use a low volatility organic liquid. The system can be a simple absorption in which the liquid (usually water) is used in a single pass and then disposed of while still containing the adsorbed pollutant. Alternatively, the pollutant can be separated from the absorbing liquid and recovered in a pure, concentrated form by stripping or desorption. The absorbing liquid is then used in a closed circuit and is continuously regenerated and recycled. Regeneration alternatives to stripping are pollutant removal through precipitation and settling; chemical destruction through neutralization, oxidation, or reduction; hydrolysis; solvent extraction; pollutant liquid adsorption; and so on.

Absorption is one of the most frequently used methods for removal of water-soluble gases. Acidic gases such as HCl, HF, and SiF_4 can be absorbed in water efficiently and readily, especially if the last contact is made with water that has an alkaline pH. Less soluble acidic gases such as SO_2, Cl_2, and H_2S can be absorbed more readily in a dilute caustic solution. The scrubbing liquid may be made alkaline with dissolved soda ash or sodium bicarbonate, or with NaOH (usually no higher a concentration in the scrubbing liquid than 5–10%). Lime is a cheaper and more plentiful alkali, but its use directly in the absorber may lead to plugging or coating problems if the calcium salts produced have only limited solubility. A technique often used, such as in the two-step flue gas desulfurization process, is to have a NaOH solution inside the absorption tower, and then to lime the tower effluent externally, precipitating the absorbed component as a slightly soluble calcium salt. The precipitate may be removed by thickening, and the regenerated sodium alkali solution is recycled to the absorber. Scrubbing with an ammonium salt solution is also employed. In such cases, the gas is often first contacted with the more alkaline solution and then with the neutral or slightly acid contact to prevent stripping losses of NH_3 to the atmosphere.

When flue gases containing CO_2 are being scrubbed with an alkaline solution to remove other acidic components, the caustic consumption can be inordinately high if CO_2 is absorbed. However, if the pH of the scrubbing liquid entering the absorber is kept below 9.0, the amount of CO_2 absorbed can be kept low. Conversely, alkaline gases, such as NH_3, can be removed from the main gas stream with acidic water solutions such as dilute H_2SO_4, H_3PO_4, or HNO_3. Single pass scrubbing solutions so used can often be disposed of as fertilizer ingredients. Alternatives are to remove the absorbed component by concentration and crystallization. The absorbing gas must have adequate solubility in the scrubbing liquid at the resulting temperature of the gas–liquid system.

For pollutant gases with limited water solubility, such as SO_2 or benzene vapors, the large quantities of water that would be required are generally impractical on a single-pass basis, but may be used in unusual circumstances. Two early examples from the UK are the removal of SO_2 from flue gas at the Battersea and Bankside electric power stations, where the normally alkaline water from the Thames tidal estuary is used in large quantities on a one-pass basis (48).

Nonaqueous Systems. Although water is the most common liquid used for absorbing acidic gases, amines (monoethanol, diethanol, and triethanolamine; methyldiethanolamine; and dimethylanaline) have been used for absorbing SO_2 and H_2S from hydrocarbon gas streams. Such absorbents are generally limited to solid particulate-free systems, because solids can produce difficult-to-handle sludges as well as use up valuable organic absorbents. Furthermore, because of absorbent cost, absorbent regeneration must be practiced in almost all cases.

At first glance, an organic liquid appears to be the preferred solvent for absorbing hydrocarbon and organic vapors from a gas stream because of improved solubility and miscibility. The lower heat of vaporization of organic liquids is an energy conservation plus when solvent regeneration must occur by stripping. Many heavy oils, hexadecane (no. 2 fuel oil) or heavier, and other solvents with low vapor pressure can do extremely well in reducing organic vapor concentrations to low levels. Care must be exercised in picking a solvent that will have sufficiently low vapor pressure that the solvent itself will not become a source of VOC pollution. Obviously, the treated gas will be saturated with the absorbing solvent. An absorber-stripper system for recovery of benzene vapors has been described (49). Other aspects of organic solvent absorption requiring consideration are stability of the solvent in the gas-solvent system (such as its resistance to oxidation) and possible fire and explosion hazards.

Types and Arrangements of Absorption Equipment. Absorption requires intimate contact between a gas and a liquid. Usually means are provided to break the liquid up into small droplets or thin films (which should be constantly renewed through turbulence) to provide high liquid surface area for mass transfer and a fresh, unsaturated surface film for high driving force. The most commonly used devices are packed and plate columns, open spray chambers and towers, cyclonic spray chambers, and combinations of sprayed and packed chambers. Some of these devices are illustrated in Figures 1–4. Packed towers give excellent gas–liquid contact and efficient mass transfer. For this reason, they can generally be smaller in size than open spray towers. A countercurrent packed tower maximizes driving force, because it brings the least concentrated outlet gas into contact with fresh absorbing liquid. These features make this tower design the best choice when the inlet gas is essentially free of solid particulates. However, this design plugs rapidly when appreciable insoluble particulates are present. The

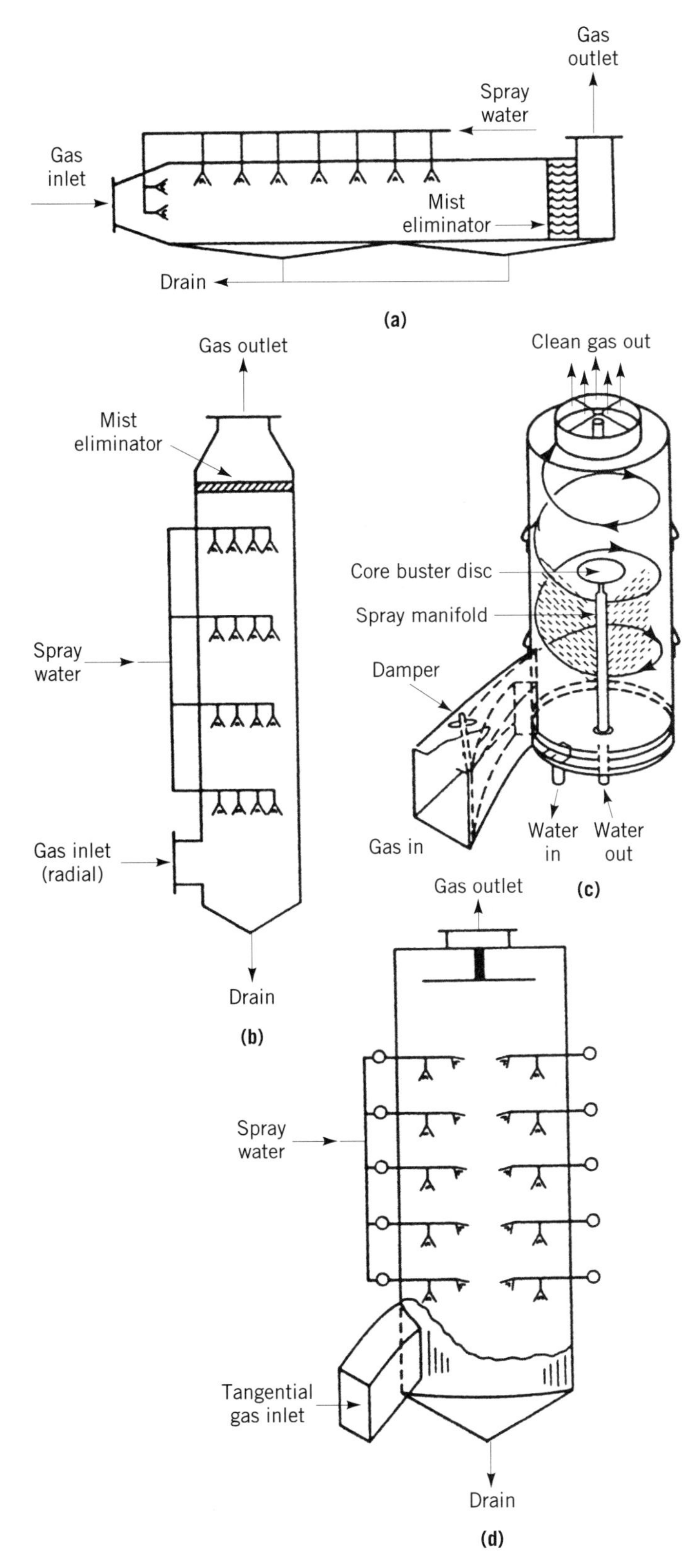

Figure 1. Types of spray towers: (**a**) horizontal spray chamber; (**b**) simple vertical spray tower; (**c**) cyclonic spray tower, Pease-Anthony type; (**d**) cyclonic spray tower, external sprays.

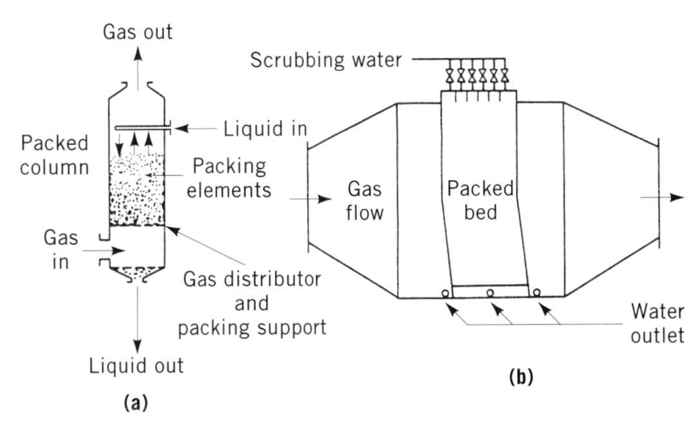

Figure 2. Two types of packed gas absorbers. (**a**) Countercurrent packed tower (54), and (**b**) cross-flow packed absorber (50).

cross-flow packed scrubber is more plug resistant when properly designed (50). In plate columns, contact between gas and liquid is obtained by forcing the gas to pass upward through small orifices, bubbling through a liquid layer flowing across a plate. The bubble cap shown is the classical contacting device. A variation is the valve tray, which permits greater variations in gas flow rate without dumping the liquid through the gas passages.

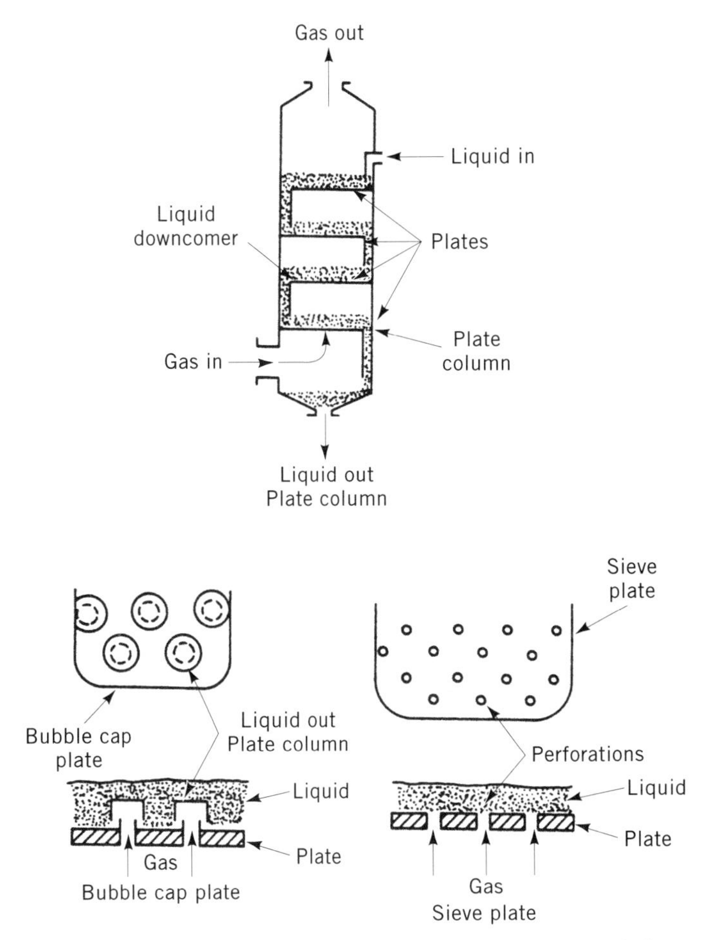

Figure 3. Typical plate column absorber and two types of plate internals (54).

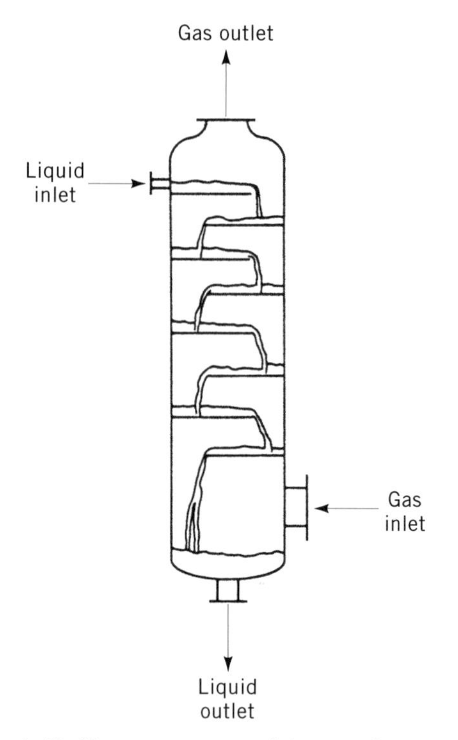

Figure 4. Baffle tray tower used for gas absorption (54).

Sieve plates are simple flat plates perforated with small holes. The advantages are low cost and high plate efficiency, but they have narrow gas flow operating ranges. Spray chambers and towers are considerably more resistant to plugging when solid particulates are present in the inlet gas, but difficulties with spray nozzle pluggage or erosion can be troublesome when the spray liquid is recycled. Particle settling followed by fine strainers or even coarse filters is beneficial.

Both horizontal and vertical spray designs have been used extensively to control gaseous emissions when particulates are present. Cyclonic spray towers may provide slightly better particulate collection as well as higher mass transfer coefficients and more transfer units per tower than other designs. Although there is theoretically no limit to the number of transfer units that can be built into a vertical countercurrent packed tower or plate column if it is made tall enough, there are definite limits to the number of transfer units that can be designed into a single vertical spray tower. As tower height and gas velocities are increased, more spray particles are entrained upward from lower levels, resulting in a loss of true countercurrency. Achievable limits have not been clearly defined in the literature, but some experimental results have been provided. There have been reports of 5.8 transfer units in a single vertical spray tower and 3.5 transfer units in horizontal spray chambers (51). Researchers (52) have attained 7 transfer units in a single commercial cyclonic spray tower. Theoretical discussion and a design equation for cyclonic spray towers of the Pease-Anthony type are available (53). Whenever more transfer units are required, spray towers in a series are used.

When heavy particulate loads must be handled or are of submicron size, it is common to use wet particulate collectors that have high particle collection efficiencies along with some capability for gas absorption. The venturi scrubber (see discussion of wet scrubbers below) is one of the more versatile such devices, but it has absorption limitations because the particles and spray liquid have parallel flow. It has been indicated that venturi scrubbers may be limited to three transfer units for gas absorption (51). The liquid-sprayed wet electrostatic precipitator is another high efficiency particulate collector with gas absorption capability. Limited research tests have indicated that the corona discharge enhances mass-transfer absorption rates, but the mechanism for this has not been established.

Another tower contacting device for absorption is the baffle tower, which has been employed occasionally when pluggage and scaling problems are expected to be severe. Gases passing up the tower must pass through sheets of downwardly cascading liquid, providing some degree of contact and liquid atomization. Baffle tower design may use alternating segmental baffles (Fig. 4) or disk and doughnut plates, in which the gas alternately flows upward through central orifices and annuli, traversing through liquid curtains with each change in direction. Mass transfer is generally poor, and information on design parameters is extremely scarce.

Theory and Design Equations. A rigorous discussion of absorption equations and models for air pollution control has been published (54). Other good references on absorption also exist (55–62). All absorption models are based on the classical Whitman two-film concept of gas–liquid mass transfer. At the gas–liquid interface, the gas is considered to be well mixed throughout the bulk of the gas by eddy diffusion. But at the interface, a stagnant gas layer or film exists, with an absorbing gas partial pressure drop across it in the direction of gas flow for absorption. The gas molecules must be transferred across this gas boundary layer by molecular diffusion—a slow process.

In the same fashion, on the liquid side of the interface, the concentration of dissolved absorbing gas in the bulk of the liquid is considered uniform due to eddy diffusion and turbulent mixing. But again at the interface, on the liquid side, is a stagnant liquid boundary layer or film, across which there is a concentration drop of dissolved gas in the liquid. Both of these reductions in absorption driving force can be viewed as resistances to the absorption process. These resistances to mass transfer can be represented by a mass-transfer coefficient for each film: Usually K_GA, the gas film mass-transfer resistance, multiplied by the area for mass transfer. Similarly, the liquid film mass-transfer resistance can be represented by K_LA, the liquid film mass-transfer resistance multiplied by its area for mass transfer. Often it is difficult or impossible to measure the area for mass transfer, so frequently the effects of the two units are treated as combined effects. For gas transfer across the interface, an equation may be written for the number of moles of gaseous component A being transferred across the interface:

$$N_A = K_G(p_A - p_A^*) = K_L(C_A^* - C_A) \quad (1)$$

The symbols are defined in "Nomenclature," below. The starred symbols represent the equilibrium values of partial pressure and concentration of component A at the actual interface. One film, either liquid or gas, may represent a larger resistance to gas flow and mass transfer than the other film. Therefore, one film may actually control the overall rate of mass transfer. But the total moles of absorbed gas traveling across the interface must be the same in both films. In such a case, one can ignore the resistance of the noncontrolling film and work with just the resistance of the controlling film. This situation is described as *gas-phase controlling* or *liquid-phase controlling.*

In the development of design equations, the operating line for the absorber is usually expressed in terms of the column material balance, involving the quantity of total gas flowing up the tower and the quantity of liquid flowing down the tower, the partial pressure of the absorbable component at the gas inlet and outlet ends of the tower, and the concentration of the absorbed gas in the entering and leaving liquid at the tower ends. For simplification, a number of assumptions are often made, which are frequently good assumptions for the dilute quantity situation of air-pollution control: (*1*) the moles of the absorbed gas in the liquid are negligible compared with the moles per hour of absorbing solvent flowing down the tower, so that the liquid flowing quantity can be represented by the flow rate of the solvent L; (*2*) the moles of absorbable gas in the gas entering is small compared to the total moles of gas entering the tower, so that the flow rate of gas in the tower can be represented by the flow rate of nonabsorbable gases G; (*3*) the pressure drop of the gas flowing through the tower is small compared to the total absolute pressure of the gas in the tower; and (*4*) if the concentration of the solute in the absorbing liquid is kept low (as is often the desired situation in air pollution control to obtain good driving force), then the vapor and liquid concentrations throughout the tower can be expressed as mole fractions.

When these assumptions are made, an equation can be written across a differential fraction of column height in terms of mass-transfer coefficients, overall driving force between bulk gas and liquid mole fractions, and the equilibrium concentrations of such gas and liquid at the interface. Integrating such an equation for tower height gives

$$Z = \frac{G_M}{K_G A P} \int_{y_2}^{y_1} \frac{dy}{y - y^*} \tag{2}$$

when gas phase is controlling, and

$$Z = \frac{L_M}{\rho_M K_L A} \int_{x_2}^{x_1} \frac{dx}{x - x^*} \tag{3}$$

when liquid phase is controlling. From these equations, the required tower height may be computed using experimental or theoretical values for $K_G A$ or $K_L A$. The value of the the integral may be obtained by analytical methods, by graphical methods, or by mathematical approximation methods (discussed below).

An alternative approach is called the transfer unit concept. In this approach, the value of the integral is defined as the number of transfer units required for the separation N_{OG} when gas phase is controlling.

$$N_{OG} = \int_{y_2}^{y_1} \frac{dy}{y - y^*} \tag{4}$$

The smaller the driving force throughout the tower, the greater the number of transfer units required to effect a given degree of pollutant removal. The height of a single transfer unit H_{OG} when the gas phase is controlling is given by the fraction in front of the integral:

$$H_{OG} = \frac{G_M}{K_G A P} \tag{5}$$

and the total column height Z is the product of these two items:

$$Z = (H_{OG})\,(N_{OG}) \tag{6}$$

Similarily, when the liquid phase is controlling, then

$$N_{OL} = \int_{x_2}^{x_1} \frac{dx}{x - x^*} \tag{7}$$

$$H_{OL} = \frac{L_M}{\rho_M K_L A} \tag{8}$$

To evaluate the number of transfer units required means having gas–liquid equilibrium data, or some means of estimating such data. To obtain an analytical solution to equation 4 or 7, one must have a mathematical relationship for the equilibrium curve. It has been suggested that to obtain good air-pollution scrubbing, the absorbing solution should contain an excess of neutralizing reactant for the pollutant being absorbed (51). Under such situations, the equilibrium vapor pressure of y^* may be 0 or extremely small. If $y^* = 0$, is a truly good assumption, then, of course, N_{OG} may be integrated as:

$$N_{OG} = \int_{y_2}^{y_1} dy\,/\,y = \ln y_1/y_2 \tag{9}$$

This is not always a satisfactory assumption. For ideal solutions, Raoult's Law and Henry's Law are valid. Even for rather strongly nonideal solutions, the relationships still apply with reasonable accuracy at dilute concentrations. In regions where Henry's Law applies, $p^* = H_x$ and the equilibrium vapor pressure $y^* = p^*/P$. When Henry's Law is linear, $y^* = (H/P)x^*$. Under such circumstances one can use the log-mean driving force existing between the top and bottom of the column to integrate the equation.

$$N_{OG} = \int_{y_2}^{y_1} \frac{dy}{y - y^*} = \frac{y_1 - y_2}{(y - y^*)_{LM}} \tag{10}$$

and

$$(y - y^*)_{LM} = \frac{(y_1 - y_1^*) - (y_2 - y_2^*)}{\ln\left(\dfrac{y_1 - y_1^*}{y_2 - y_2^*}\right)} \tag{11}$$

Adsorption

The attractive forces in a solid that exist between atoms, molecules, and ions, holding the solid together, are unsatisfied at the surface and thus are available for holding other materials such as liquids and gases. This phenomenon is known as adsorption. If the solid is produced in a highly porous form with extensive pores and microstructure, its adsorptive capacity can be greatly enhanced. Sorption can be used as a pollution-control measure to remove pollutant gases from an otherwise harmless gas desired to be released to the atmosphere. Adsorption is quite adaptable for removing such contaminants, especially VOCs, down to extremely low levels (less than 1 ppmv). Adsorption's best applications are (*1*) handling large volumes (hundred thousands of CFM) with dilute pollution levels and (*2*) removing the contamination level, regardless of gas quantity, down to only trace pollutant levels. Removal of solvent losses from large quantities of ventilation air is an example of the former, and recovery of toxic and hazardous vapors to extremely low concentrations exemplifies the latter. In any case, absorption may be used alone for the entire control requirement or in combination with other removal methods. In the latter case, adsorption typically becomes the final cleanup step because of its capability of achieving low emission concentrations.

In being adsorbed, a gas molecule travels to an adsorption site on the surface of the solid, where it is held by attractive forces and loses much of its molecular motion. This loss of kinetic energy is released as heat. The heat of adsorption is often close in magnitude to the heat of condensation for the species being adsorbed. Thus adsorption is always exothermic. Desorption is a reversal of the adsorption process and heat must be supplied to cause desorption to take place. Hence temperature rise tends to reverse the process or cause a loss in capacity of the sorbent. Cooling of the sorption bed (which is often difficult because of poor heat transfer within the bed) or precooling of the gas stream to be treated is desirable to provide a sink for the heat of adsorption being released. Some sorption processes can occur so strongly that they are irreversible, ie, the adsorbed material can only be desorbed by removal of some of the solid substrate. Such a process is referred to as chemisorption. For example, oxygen, under certain circumstances can be adsorbed so strongly on activated carbon that it can be removed from the solid only in the form of CO or CO_2. The Reinluft process (63), developed as a means for removing SO_2 from flue gas, adsorbed a portion of the SO_2 on the carbon as sulfuric acid. Desorption was achieved by heating the bed to 370°C, causing the acid and carbon to react, producing CO_2 and SO_2. Similarly, in the U.S. Bureau of Mines alkalized alumina process (64) the SO_2 became oxidized to sulfates, which could be removed only by reacting the adsorbent with hydrogen or reformed natural gas at 650°C.

The adsorbing solid is called the adsorbent or sorbent; the adsorbed material, the adsorbate or sorbate. A thorough discussion of adsorption processes for air pollution and design equations have been given (54), and other more general references are also available (65–74).

Types of Adsorbents. Commercially important adsorbents are activated carbon, other simple or complex metallic oxides, and impregnated sorbents. Activated carbon is a general adsorbent. It is composed primarily of a single species of neutral atoms with no electrical gradients between molecules. Thus there are no significant potential gradients to attract or orient polar molecules in preference to nonpolar molecules. For this reason, carbon has less selectivity than other sorbents and is one of the few that will work in absorbing organics from a humid gas stream. Because the polar water molecules attract each other as strongly as the neutral carbon, the latter tends to be slightly selective for organic molecules. However, some water is adsorbed, especially if its partial pressure is greater than that of the organic molecules. The water being adsorbed must be taken into consideration in selecting the sorptive capacity to be provided in the design. Typical sources of activated carbon are coconut and other nut shells, fruit pits, bituminous coal, hardwoods, and petroleum coke and residues.

Simple and complex metal oxides are polar and have a much greater degree of selectivity than carbon and a great preference for polar molecules. They can be useful for removal of a particular species from the gas stream but are ineffective when moisture is present, because most of these adsorbents are excellent desiccants. Siliceous adsorbents include materials such as silica gel, Fuller's and diatomaceous earth, synthetic zeolites, and molecular sieves. They are available in a wide range of capacities, with the best equaling the capacity of the best activated carbons. Synthetic zeolites can be prepared with specific and uniform pore sizes to give sorptive specificity based on the shape and size of sorbate molecules. Even this property tailored to specific organic molecules will not overcome their chemical preference for polar molecules such as water vapor. Activated alumina and other metallic oxides are even more polar than the silicas and are seldom used directly for pollution control adsorption.

Impregnated sorbents fall into three general classes: (*1*) those impregnated with a chemical reactant or reagent, (*2*) those in which the impregnant acts as a continuous catalyst for pollutant oxidation or decomposition, and (*3*) those in which the impregnant acts only intermittantly as a catalyst. Reagent impregnants chemically convert the pollutant into a harmless or adsorbable pollutant. Carbon may be impregnated with 10–20% of its mass, with bromine being used to react with olefins. Thus ethylene, which is poorly adsorbed from an air stream because of its low molecular weight, is converted at the brominated surface to 1,2-dibromoethane, which is readily adsorbed. Other impregnant reagents are iodine for collecting mercury vapor, lead acetate for collecting H_2S, and sodium silicate for collecting HF. Other applications of impregnated sorbents, continuous and intermittent catalytic sorbents have been discussed (75).

Adsorption Operation. Factors affecting the capacity of an adsorbent are given below.

1. Concentration of the adsorbate in the bed passages.
2. Total surface area of the adsorbent.

3. Relative concentration of competing adsorbable molecules.
4. Temperature of the gas and bed (the lower the better).
5. Adsorbate molecule characteristics (molecular weight, electrical polarity, chemical activity, size, and shape).
6. Sizes and shapes of adsorbent pore microstructure.
7. Chemical activity of adsorbent (polarity and chemical activity) (54).

The actual rate of adsorption is affected by the partial pressure of the sorbate, the degree of saturation of the bed, and the rate of diffusion of the sorbate into the pores of the sorbent. Principles that are typically considered in design of an adsorption system are

1. Use of low superficial gas velocities and ample bed depth to provide long bed contact time to achieve the required removal efficiency.
2. Adequate adsorptive capacity to provide for reasonable on-steam cycle time.
3. Comparatively low resistance to gas flow to conserve energy.
4. Uniform air flow distribution to prevent channeling and ensure total use of all of the sorbent.
5. Possible pretreatment to remove particulates, interfering or competing vapors, or materials easily desorbed as well as precooling and possible drying of the gas to remove moisture.
6. Provision for renewing, regenerating, or replacing the spent saturated sorbent.

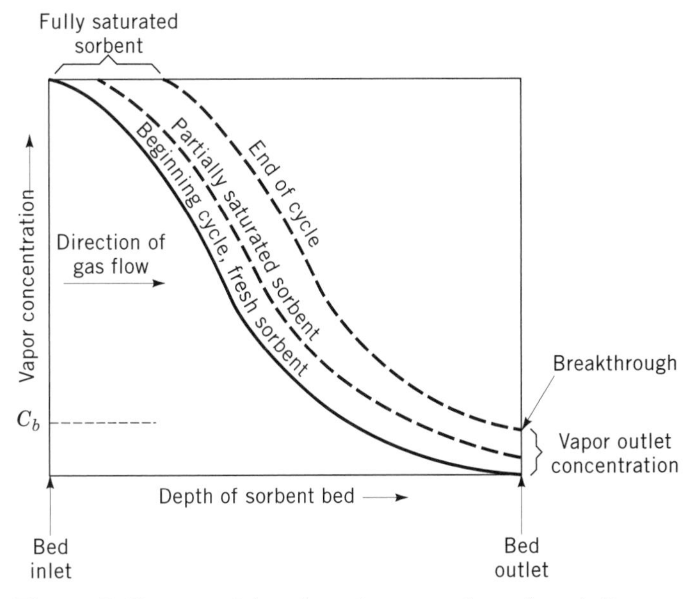

Figure 5. Passage of the adsorption wave through a stationary bed during the course of an adsorption cycle. The progressing S-shaped curves indicate the nonadsorbed vapor concentration by position in the bed at different time periods. C_b represents the maximum permissible outlet concentration for release to the atmosphere.

Adsorption of the pollutant from the carrier gas in an adsorption bed is shown in Figure 5 in the form of an adsorption wave. The solid curve shows the concentration of the pollutant in the gas as the mixture passes through a fresh bed of adsorbent. As time progresses, the sorbent becomes saturated at the inlet portion of the bed and the pollutants penetrate more deeply into the bed before being adsorbed (broken line in Fig. 5). The horizontal line C_b represents the maximum permissible pollutant concentration in the effluent. When the pollutant concentration in the outlet gas has reached this level, breakthrough has occurred and the sorbent must be regenerated. The adsorption cycle time is represented by the horizontal displacement between the sorption curves for the fresh bed and the bed at breakthrough.

Desorption or Disposal. After the pollutants have been adsorbed, they may be disposed of by discarding the saturated adsorbent or, alternatively, the sorbent may be regenerated. Disposal may be attractive when the quantity of material to be adsorbed is small, occurs infrequently, or must be recovered in an inconvenient location, such as breathing losses from a tank vent in a remote area. In these cases, the cost of fresh adsorbent is insignificant compared with the cost or inconvenience of attempting regeneration. However, before disposal can be considered, one must ascertain the nature of the chemical species that has been adsorbed. If toxic or hazardous materials have been adsorbed, the entire adsorbent cartridge and sorbent must now be considered as hazardous or toxic, which will preclude its disposal except in ways approved by applicable regulations. Even if the adsorbate is not hazardous, one must consider whether it would be leachable from the sorbent, if disposed in an ordinary landfill. When disposal is desired, often the adsorbent is contained in a paper carrier or disposable cartridge. If so, and the sorbent is carbon, incineration may be considered the best and safest method of ultimate disposal. Also, sometimes it is possible to return the spent sorbent to the manufacturer for regeneration.

When economics dictate regeneration of the sorbent, desorption may be carried out *in situ* by a number of methods: (*1*) heating the bed, (*2*) evacuating the bed, (*3*) stripping with an inert gas, (*4*) replacement of the adsorbate with a more readily adsorbed material, or (*5*) a combination of two or more of these methods. The desired form of the recovered pollutant and its subsequent handling may influence the choice of the desorption method as well as the properties of the adsorbate. Heating the bed to a temperature at which the partial pressure of the adsorbate exceeds one atmosphere will cause the pollutant to boil off in undiluted form, which can then be condensed as the pure liquid for process reuse or disposal. Absorbed heat-sensitive materials may decompose or polymerize during thermal regeneration, plugging the pores of the sorbent and not being adequately removed. A combination of heat and vacuum will also permit pollutant recovery in undiluted form, but vaporization may be made to occur at temperatures low enough so that the pollutant will not undergo chemical change. Obviously, reduced pressure vaporization will require lower con-

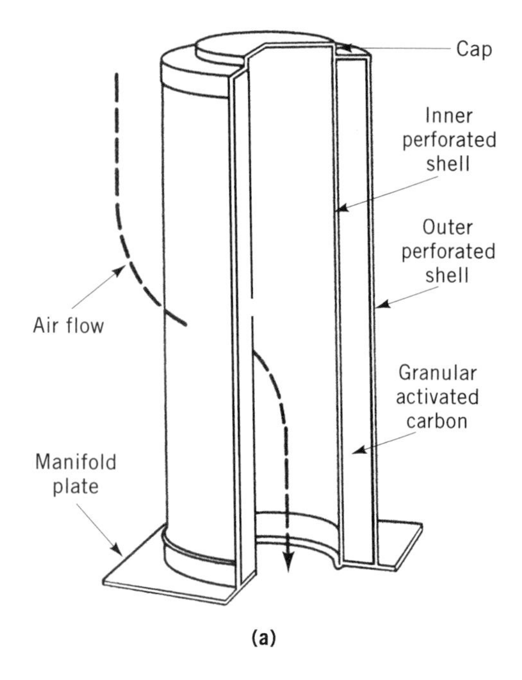

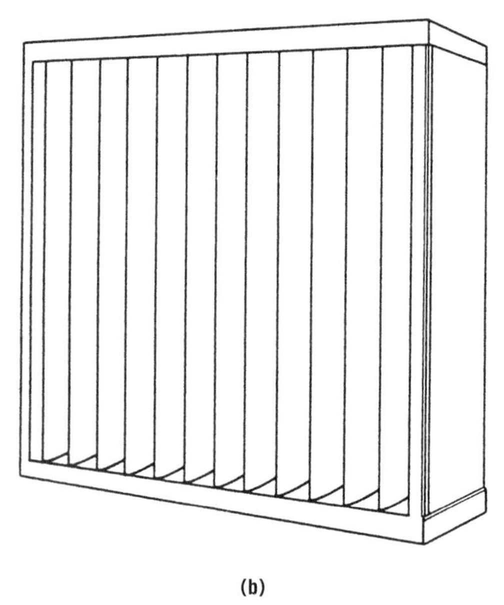

Figure 6. Typical thin bed adsorber elements. (**a**) Cylindrical thin bed canister adsorber. Courtesy of Connor, Inc., Danbury, Conn. (**b**) Pleated cell thin bed adsorber. Courtesy of Barneby-Cheney Co., Columbus, Ohio. (54,73).

denser temperatures unless the vapors are compressed before condensation.

Stripping with an inert gas often produces a more concentrated form of the pollutant than occurred in the original treated gas, but separation of the pollutant from the stripping gas is sometimes difficult. However, this pollutant concentration method may be desirable when the ultimate disposal is gaseous incineration. Up to a 40-fold increase in pollutant concentration has been achieved with adsorption followed by gas stripping (76). Pollutant concentration can result in a large heat saving in incineration by reducing the volume of inerts that must be heated to the pollutant ignition temperature. Partial condensation of the pollutant from a stripping gas, followed by reheating and recirculation, can also aid in eliminating the problem of complete separation of pollutant and stripping agent. Displacement of an adsorbed pollutant with another material is often accomplished by steaming. This requires a double regeneration cycle. Once the sorbent is saturated with water vapor, a drying cycle regeneration must be started to remove the adsorbed water. In reality, steaming is a combination of methods: adsorbate displacement, bed heating, and inert gas stripping. Regardless of the method, desorption is never entirely complete so that adsorbent capacity is always reduced over the capacity of fresh, unused sorbent.

Types of Adsorption Equipment. Five distinct types of gaseous adsorption devices are available: (1) disposable and rechargeable canisters, (2) fixed regenerable beds, (3) traveling bed adsorbers, (4) fluid bed adsorbers, and (5) chromatographic bag houses. Figures 6 and 7 illustrate some of these types.

For small flows, those of an intermittent or infrequent nature, and those with low sorbate concentration, disposable charges of carbon may be used. These can be fine carbon powder dispersed in and on an inert carrier of paper, textiles, or other support. Such elements have not received significant industrial use. More frequently, a re-

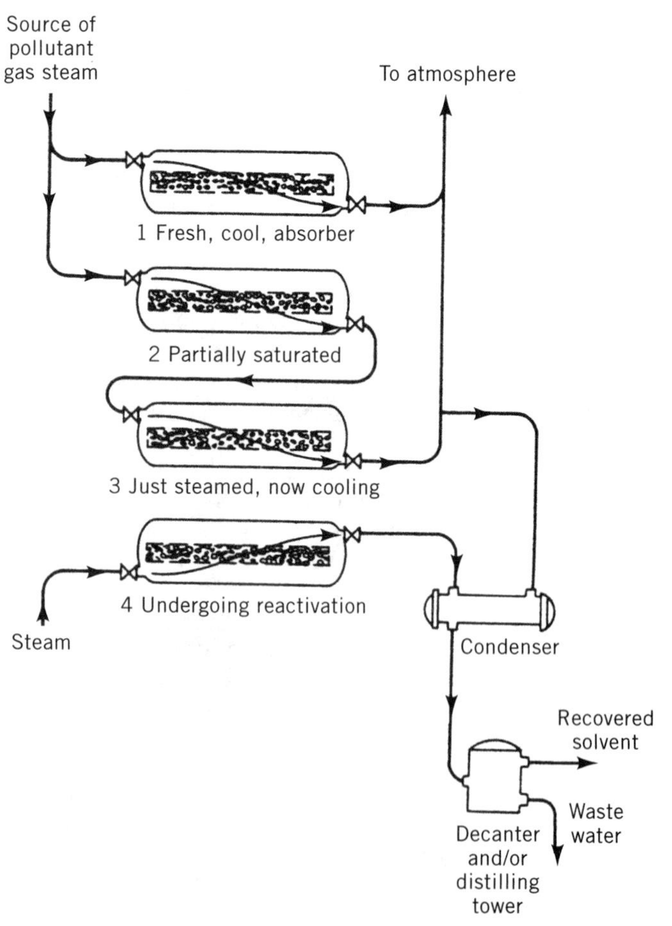

Figure 7. Four-unit deep bed adsorption train (73).

placeable or disposable canister (filled with granular sorbent and held in a permanent container) placed in a vent line is used and replaced when close to saturation. For remote location emissions, 55-gal steel drum containers of replaceable granular carbon are available from several sources.

Regeneratable beds are used when economics so dictates: the volume of gas treated or sorbate concentration is high enough to make recovery attractive, or the cost of fresh sorbate is higher than regeneration. Thin bed units (1.25–10.2 cm) are illustrated in Figure 6. Their major advantage is low resistance to gas flow. Uniformity of bed thickness in thin beds is extremely important to prevent gas channeling. The granular sorbent is usually retained between metallic screens in flat, cylindrical, or pleated shapes. The sorbent may be regenerated in place or removed for regeneration elsewhere. In thick bed adsorbers (0.3–1.0 m), the gas generally passes downward (to prevent bed lifting) through the bed and is contained in horizontal cylindrical vessels. For continuous operation, two or more such containers, suitably valved, are used to permit adsorption and regeneration. Figure 7 shows a four-bed system.

For handling larger gas flows or higher pollutant concentrations for which a static bed would be too rapidly exhausted, a fluid bed adsorber might be used. In a fluid bed, all the particles are well mixed and the typical adsorption wave concentration curve does not apply. For reducing the pollutant to low concentration levels, staging of several such beds could be practiced. An alternative is to keep the adsorbent in a single-stage bed quite fresh by removing a sizable portion of the bed continuously for regeneration in a second vessel, and then returning regenerated sorbent to the adsorber. Because the fluid bed is well mixed, all the particles are in equilibrium with desired pollutant concentration in the desired outlet gas. Thus the bed particles must be kept relatively unsaturated. This would be undesirable were it not for the ease with which the bed can be transferred between vessels. Another advantage of the fluid bed is its ability to give high heat-transfer rates between the fluidized sorbent and cooling tubes submerged in the bed to remove the heat of adsorption.

A two-fluid bed transport reactor using silica gel adsorbent for drying a gas stream has been described (77). Moisture adsorption occurs in one bed, and regeneration in the other, with continuous transport of solids between the two beds. Such a system with a less selective sorbent might be used for removal of organic vapors in a humid gas stream, where the sorbent might become saturated with water vapor before reaching the organic breakthrough concentration. Another possible application would be use of an oxcar catalytic sorbent that would both adsorb and destroy toxic organics by oxidation. A novel adaptation of transported fluid bed adsorption–desorption all in one vessel is the Linde Purasiv HR System (78), shown in Figure 8. Granular carbon is continuously recirculated within a single partitioned vessel that has two separate fluid beds.

Traveling bed adsorbers are another approach to gaseous pollution capture. They are especially applicable to large gas flows and the need to remove sizable quantities of pollutant. They have generally been used commercially for gas separation but not for pollution control, mainly because of the difference in scale. Continuous countercurrent adsorbers have been designed both as downwardly traveling packed columns and tray units. Freshly regenerated, granular sorbent is elevated and added to the top of the column at a rate to maintain a constant height of solids in the unit. Spent sorbent is continuously removed from the bottom of the column and transferred to a regeneration vessel for recycling. Attrition and loss of sorbent can be a sizable operating problem. Generally, hard and abrasion-resistant granules are required. Westvaco Pulp and Paper has considerable experience in design of both equipment and manufacture of suitable granular carbons.

A newer development, described as chromatographic adsorption, consists of injecting a cloud of adsorbent particles into the gas stream. Adsorption occurs during concurrent flow of effluent and suspended particles in pneumatic transport. The sorbent is removed from the gas stream along with the adsorbed pollutant in a conventional bag filter. Some final adsorption takes place as the gas flows through the layers of adsorbent on the surface of the filter bags.

In another new technology, called pressure-swing ad-

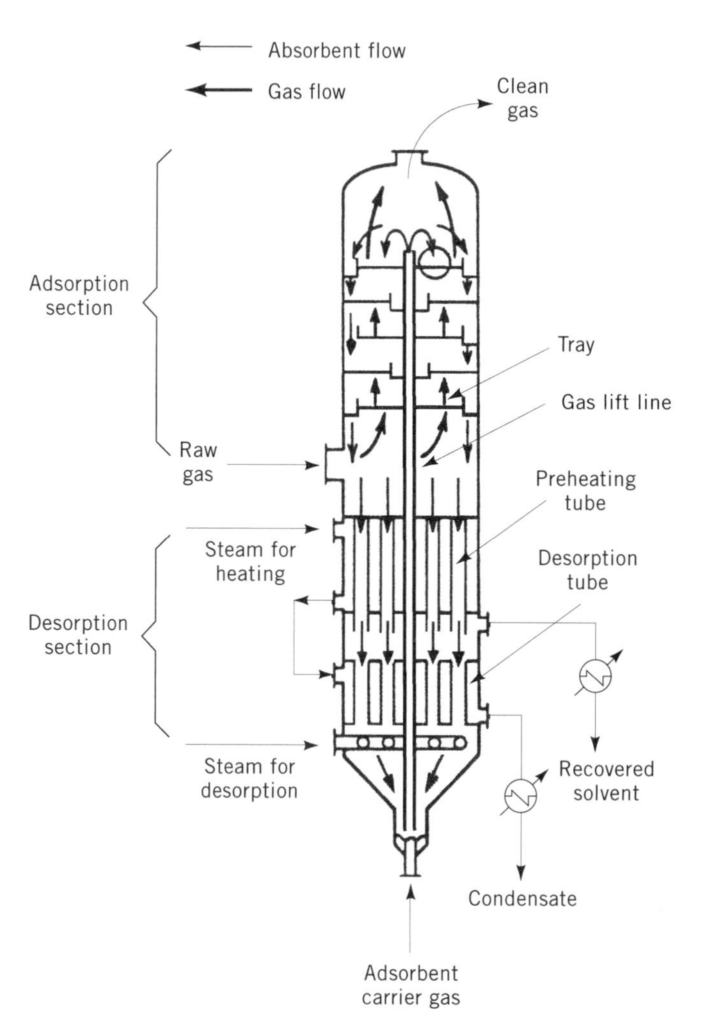

Figure 8. Linde Purasiv HR adsorption system. (63).

sorption (79), the adsorption bed is subjected to relatively short pulses of higher pressure gas containing the species to be adsorbed. The bed is then partially regenerated by reducing the pressure and allowing the adsorbed material to vaporize. By controlling the gas flow and the pressure, the pollutant can be transferred from effluent to another gas stream. This same result can be achieved by temperature fluctuation, but bed temperature changes cannot be as rapidly controlled as bed pressures.

Adsorption Applications. Typical pollution applications are

1. Odor control in food processing.
2. Solvent and odor control in chemical and manufacturing processes, including cleaning and degreasing operations; paint, coating, and printing operations; pulp and paper manufacture; and in tanneries.
3. Odor control from foundries and animal laboratories.
4. Radioactive gas control in the nuclear industry.

Small carbon-filled canisters have been mandated on U.S. automobiles in recent years to control evaporative fuel emissions from gasoline engines. Similar adsorption equipment is being used to control vapor emissions from fuel-tank filling in areas in noncompliance with NAAQS standards for ozone. Implementation of the 1990 Clean Air Reauthorization Act provisions for control of toxic and hazardous organic vapor emissions should result in increased use of adsorption pollution control applications (80).

Condensation

Condensation has been discussed as a pollutant control method (80–83). It is most applicable for vapors fairly close to their dew points. As such, it is suitable for hydrocarbons and organic compounds that have reasonably higher boiling points than ambient conditions and are present in the effluent gas in appreciable concentrations. Pollutants having reasonably low vapor pressures at ambient temperatures may be controlled satisfactorily in water-cooled or even air-cooled condensers. For more volatile solvents, two-stage condensation may be required, using cooling water in the first stage, and refrigeration in the second. Refrigeration to extremely low temperature levels is seldom attractive for pollution control alone, unless dictated by process considerations, because alternative control methods will generally be more attractive. Minimizing the presence of inert diluent gases in the condenser will reduce the need to cool to very low dewpoints, and enhance the economic attractiveness of condensation as a control method. Condensation is not a practical method of total control for reasonably volatile toxic or hazardous organics in appreciable concentrations in streams of noncondensables if the effluent concentration must be reduced to a few ppmv. Condensation may still be useful as a preliminary treatment method to recover valuable solvents or to reduce the capacity required for the final treatment method. Partial condensation can be useful when the stripped gas can be recycled to the process (rather than vented) or when it can be used as primary combustion air. (The remaining pollutants are incinerated.) Condensation can be an attractive pretreatment method when it can serve to cool the gas before final control by adsorption.

Type of Equipment. Condensation cooling can occur either by direct contact or by indirect cooling. In the former, the vapor is brought into intimate contact with a cooled or refrigerated liquid. With indirect cooling, a surface condenser having metal tubes is commonly employed. The tubes are cooled with another fluid on the other side of the wall. When appreciable noncondensibles are present, compression of the gas stream before treatment can reduce the temperature required to achieve the desired pollutant partial pressure. However, gas compression is seldom economical for pollution control unless higher pressures are needed for other process reasons. When low temperatures are required, consideration must be given to the possible presence of other materials that could solidify at the required temperature. Icing on condenser surfaces will quickly foul the heat-transfer capability of the condenser.

In direct contact condensers, intimate mixing and contact is brought about between the cold liquid and the gas to effect as close an approach to thermal and mass-transfer equilibrium as possible. The cold liquid may be sprayed into the gas in a spray tower or a jet eductor, or a tower with gas–liquid contacting internals can be used. The contacting tower can be a packed tower, sieve-plate tower, disk and doughnut or segmental baffle plate tower, or even a slat-packed chamber. Because the liquid becomes heated by both the gas and the condensing vapor, lower dewpoints can be reached with counterflow of gas and liquid, but many parallel flow devices are used. The cooling liquid is frequently recirculated through an external heat exchanger and recycled.

The recirculated cool liquid is often water, if temperatures near its freezing point are not required. (Even below the freezing point of water, a water–antifreeze mixture can be used if water itself is not being condensed from the gas, which would, of course, dilute the antifreeze mixture.) However, an appropriate low vapor pressure liquid can be used as the recirculated liquid. It may even be the substance being condensed from the gas stream. This has the advantage that no further steps are required to separate the material being condensed from the liquid being recirculated. Whatever the recirculated liquid, it should be remembered that the treated gas is going to be close to vapor–liquid equilibrium with the cooling liquid. Thus if the liquid has appreciable vapor pressure, the gas could become polluted with vapors from the cooling liquid.

When the recirculated cool liquid is a different substance from the vapor being condensed, consideration must be given to how the two materials will be separated. Use of a cool liquid with a low solubility or miscibility with the condensed vapor is often helpful, because a simple phase separator can be used.

Surface condensers are most often used when the vapors to be condensed constitute the major portion of the gas stream with only a small amount of noncondensibles present to be vented. Under such conditions, tubular condenser type heat exchangers may be used. When noncondensibles predominate, tubes finned on the gas side will give better heat transfer unless the condensing vapor will tend to scale the heat-transfer surface. In such cases, tubular condensers designed for ease of gas-side tube cleaning are used. Condensers may be either vertical or horizontal. Horizontal units are frequently pitched to provide for good drainage. When dewpoints no lower than 10°C above atmospheric air temperatures are satisfactory, air-cooled heat exchangers of the fin-fan type can be employed.

Design of either type of condenser follows heat-transfer methods used for gas dehumidifying design. Applicable heat-transfer methods for this purpose have been briefly reviewed and summarized (82). These methods consider both the rate of heat and mass transfer. Whenever the gas must be cooled more than about 40–50°C below its initial dewpoint to achieve the required condensible pollutant removal, it is possible for condensate fog to form in the bulk of the gas stream. The fog so produced will generally be 1 μm or smaller in particle size; especially difficult to collect. Fog forms whenever the rate of heat transfer appreciably exceeds the rate of mass transfer. The bulk of the gas stream subcools appreciably below the dewpoint of the condensing vapor and condensate nucleates in the bulk of the gas stream. Fog seldom forms in direct contact condensers because of the close proximity of the bulk gas to liquid droplets and films. Fog in surface condensers can be predicted by following the calculated rate of heat and mass transfer as the gas passes through the exchanger. Fog formation has been discussed (84). Fog may be avoided by switching to direct condensation either initially or at the point at which fog formation would occur in a surface condenser. Another alternative is to produce the fog and remove it from the gas stream in a suitable fine particle collector (Brownian diffusion particle filter, electrostatic precipitator, or venturi scrubber).

Chemical Reaction

Removal of an objectional or hazardous gaseous pollutant by chemical reaction has interesting possibilities. It is difficult to generalize about such means because they are so specific to the chemistry of the species of concern. In addition, the process suitability can also vary with the pollutant concentration in the gas stream as well as the temperature and composition of the carrier gas. The unit operations of absorption and wet scrubbing provide opportunities to carry on chemical reactions by adding a chemical reactant to the absorbing or scrubbing liquid, such as an alkali to enhance the absorption of an acidic gas (discussed above.) Furthermore, the chemical nature of the contacting liquid could change or destroy the pollutant vapor (if reactive in nature) by the presence of an oxidizing agent, such as potassium permanganate ($KMnO_4$), hydrogen peroxide, ozone, strong (and hot) nitric acid, hypochlorites, and chlorates. In addition, agents can be added to remove the absorbed vapors from the liquid by precipitation, forming insoluble compounds with the gaseous pollutant. Likewise, in adsorption, comment has already been made about impregnated adsorbents, such as with bromine, iodine, lead acetate, and sodium silicate. Another category is that of catalytic adsorbents, which are oxcar catalysts resulting in the oxidation of organic pollutants. More process development research has been carried out on chemical methods for removal of SO_x and NO_x, probably because of their wide-spread occurrence in flue gases. These specific chemical solutions are discussed below. Gas–solid reactions are feasible for removal of specific gaseous pollutants by injection of dry solids into the conveying gas steam. Hydrated lime injection to remove SO_2 is a prime example of such applications. Dry injection of sodium bicarbonate to remove both SO_2 and NO_x from flue gas is another. Generally, dry solid injection falls considerably short of 100% total removal, but recent research with gas humidification and the addition of activity enhancers to the injected solids have been shown to help. As would be expected, increasing solid surface area through fine particle grinding and increasing particle surface porosity are beneficial.

Some fundamental objectives to consider for chemical reactions are (*1*) convert the pollutant into a different material with a lower vapor pressure (ideally into a liquid or solid particle) or into another chemical species that is more easily collectible, (*2*) convert the pollutant molecule into a harmless or at least less harmful molecule, (*3*) destroy the pollutant. Simple examples of conversion would be to increase the molecular weight of the pollutant through gas-phase reaction, causing the particle to condense so that it can be captured in a particulate collector. Another example is modification of an organic molecule by substitution, rendering it highly water soluble for absorption. A simple example of changing a gaseous pollution problem to a particulate one is the addition of NH_3 to a gas stream containing HCl, to produce NH_4Cl smoke. An example of destruction is oxidation of a pollutant to simple harmless species such as CO_2 and H_2O.

Incineration

Incineration of gaseous contaminants is also known as thermal oxidation and fume incineration. It is primarily applicable to gaseous impurities that can be oxidized or burned to decompose the original molecules to simpler nonhazardous compounds such as CO_2 and H_2O. In a word, the process is simply controlled combustion. As such, it may be used to destroy airborne or air-mixed hydrocarbon gases, other organic vapors, and similar blowdown gases; mercaptans; and undesirable inorganic gases such as H_2S, HCN, CO, H_2, and NH_3. VOC emissions are frequently destroyed by incineration (85–86) as are gases evolving from landfills.

Substituted organic vapors can also be decomposed by incineration, but it is important to consider the complications imposed when passing the substituted radical through the combustion process. For instance, halogen-substituted organics will result in the release of the corresponding acidic hydrogen halide gas. Sulfur-containing groups will generally be oxidized to SO_2. These acidic gases generally must be scrubbed or otherwise removed from the incinerator effluent before the combustion products are released to the atmosphere. Cooling of the effluent gas in the scrubber may require adding reheat for good atmospheric dispersion. Saturating the effluent with water may also produce condensing plumes in cold weather. Recovery of the acidic products, often in dilute aqueous form, may constitute a disposal problem.

Incineration also may be used to destroy odors in those cases in which the odor substance can be decomposed by combustion. With proper design, aerosol incineration could destroy combustible airborne liquid or solid particles using a burner much like one used to burn pulverized coal or activated sludge from sewerage treatment. However, literature references for aerosol incineration are essentially nonexistent. In such an incinerator, rapid ignition of particles by quickly heating them above their kindling temperature is required as well as providing adequate residence time and flame space for complete combustion. Particulates have a considerably slower burning rate than combustible gases. (Means to reduce the particulate size to subsieve size would greatly enhance the combustion speed.) The possible presence of noncombustible residue that can produce a solid or molten ash and its subsequent handling must also be considered.

Thermal oxidizers can be designed to yield from 95 to >99% destruction of all combustible compounds. They can be designed with a capacity to handle from 0.5 to 236 m^3/s and for inlet combustible pollutant concentrations ranging from 100 to 2000 ppmv.

Combustion Principles. Combustion consists of a series of complex chemical reactions involving oxygen (usually in air) by mechanisms known as chain reactions. These numerous and rapid chain reactions, or steps, vary for each species of combustible compound present in the inlet gas stream. However, the combustion path followed by each species does not appreciably affect the overall destruction results to be achieved. Control of the combustion operation is affected more by factors such as the pollutant's initial concentration, the inlet gas temperature, the process gas throughput rate (or residence time in the combustion chamber), and the degree of component mixing or gas turbulence. In fact, most authorities refer to the three T's of combustion as a general rule affecting the completeness of the oxidation: time, temperature, and turbulence. A change in any one will change the degree of the completeness of the oxidation. However, if one variable is inadequate, it is possible to improve the overall combustion efficiency by an appropriate increase in the other controlling variables. For instance, if the residence time at the combustion temperature is inadequate to yield the desired destruction efficiency, the operation may be improved by an increase in the operating temperature. Similarly, increasing the degree of gas turbulence (degree of mixing) in the combustion zone would also improve combustion speed and increase destruction efficiency.

The actual time and temperature required for a given degree of destruction can be calculated if the chemical species and entering concentration is known along with the kinetic reaction rate as a function of temperature. The kinetic combustion rates have not been measured or reported in the literature for many compounds. Therefore, small experiments in the laboratory are often run to determine the time–temperature relationship for a specific degree of destruction. (Bear in mind that small-scale laboratory tests invariably have high degrees of gas premixing and turbulence, better than will be obtained in large-scale plant conditions). Figure 9 shows time–temperature trade-offs that can be made for 99.9% oxidation of several common hazardous chemical VOC vapors. These data were collected under laboratory conditions. Without such data, design must be started with experience from other chemical species that can be expected to have somewhat similar destruction kinetics. With data from a single point (time, temperature, and destruction efficiency) such as might be obtained from a pilot-plant off-gas incinerator, one can apply some general empirical ratios taken from Figure 10. This figure shows empirical generalities about the effect of changes in temperature, pollutant inlet concentration, and combustion residence time.

So far, specific data with respect to time and temperature have been discussed. The turbulence variable has been neglected. Degree of turbulence is hard to quantify because no good technique exists for its measurement. In general, one can be certain that as equipment size (combustion chamber, gas flow rate, etc) increases with scaleup, degree of turbulence will decrease with equal gas velocities. Thus it is common practice in scaleup to larger incinerators, to increase gas velocities and increase turbulence with swirl, with baffles and sudden gas direction changes, the introduction of jets, etc. Even so, one frequently finds that poorer destruction efficiency is obtained in a large-size incinerator scaled up from a satisfactorily operating smaller incinerator. This must be due to reduced gas mixing in the larger unit. Therefore, it is

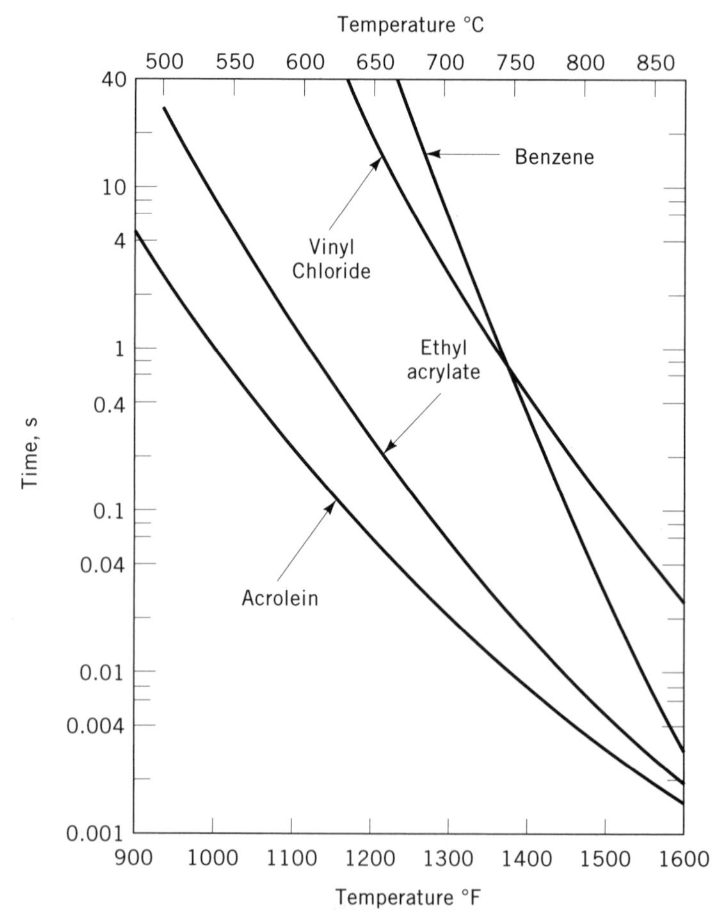

Figure 9. Relationship of time–temperature variation for incineration of some hazardous VOCs giving 99.9% destruction in laboratory tests (88).

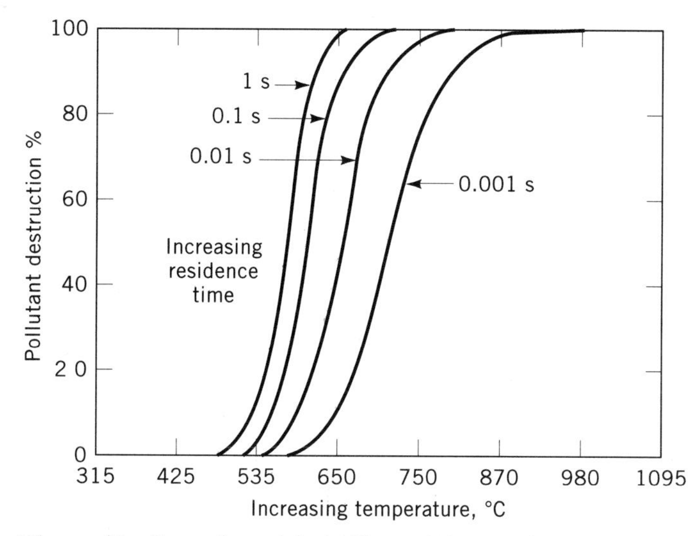

Figure 10. General empirical effects of changes in temperature and residence time in a gaseous incinerator (87). Courtesy of Van Nostrand Reinhold.

often desirable in the scaleup to provide some safety factor in planned operating temperature. In this way, the incineration temperature can be experimentally increased as needed to give the required combustion destruction efficiency. The calculation of the required time and temperature for an incinerator can also be approached with an approximate mathematical model as follows.

1. An assumption is made that the kinetic reaction rate can be approximated by a first-order reaction rate. This is a fair first approximation even if the reaction is 0 or second order. With this assumption and the further assumption that the pollutant is a hydrocarbon (HC), the differential rate equation can be written:

$$\frac{d[\mathrm{HC}]}{dt} = -k[\mathrm{HC}] \tag{12}$$

where k is the pseudo–first-order rate constant s^{-1}, which includes the oxygen concentration.

2. If the initial pollutant concentration is $[\mathrm{HC}] = C_{A,O}$ in mole fraction, and C_A represents its concentration after any desired residence time in seconds, then a solution to the equation is:

$$\ln (C_A/C_{A,O}) = -kt \tag{13}$$

where t is the residence time at temperature in seconds.

3. In the above model, the rate constant k is presumed to be of the Arrhenius form:

$$k = Ae^{-E/RT} \tag{14}$$

where A is the preexponential factor (s^{-1}), E is the activation energy (J/g mol), R is the universal gas constant (8.314 J/g mol K); and T is the absolute temperature (K). Some kinetic data reported in the literature for values of A and E are given in Table 4.

The gases entering the incinerator are the waste gas stream, which is to be incinerated (and which may contain oxygen and excess air); fuel for combustion (to raise the temperature of the mixture to the desired incineration temperature); and combustion air as may be needed by the fuel as primary air. If there is adequate oxygen in the waste gas, the incinerator may be designed to use the waste gas as secondary air only. The fuel is most often natural gas, although occasionally fuel oil is used.

The quantity of fuel required is calculated from the heat requirements needed to raise the waste stream to its incineration temperature. The heat produced from combustion of the pollutants may be included. As discussed later, the incinerator may be equipped with heat recovery devices to preheat the incoming waste stream by cooling the combusted flue gases. This recovered heat will reduce the amount of fuel and air that would otherwise be required. For maximum thermal efficiency of the incinerator, excess air should be kept to the minimum needed for obtaining complete combustion of the combustible pollutants. This is very close to the stoichiometric quantity of oxygen needed for chemical reaction. In a well-designed incinerator, an excess O_2 content needed in the exit flue gas should be no more than approximately 0.50% O_2 by volume (dry basis) by analysis. To obtain complete burning of the pollutant combustibles in the incinerator, the entire gas mixture in the combustion chamber must be kept well-mixed by turbulence to bring oxygen uniformly into contact with the combustibles in the waste gas stream. Otherwise, there are apt to be local areas in the combustion chamber where there is an excess of oxygen, and other areas where there is a deficiency of oxygen. The

Table 4. Thermal Oxidation Parameters[a]

Compound	A	E, J/g·mol
Acrolein	3.30×10^{10}	150,206
Acrylonitrile	2.13×10^{12}	217,986
Allyl alcohol	1.75×10^{6}	89,538
Allyl chloride	3.89×10^{7}	12,175
Benzene	7.43×10^{21}	401,246
Butene-1	3.74×10^{14}	243,509
Chlorobenzene	1.34×10^{17}	320,494
Cyclohexane	5.13×10^{12}	199,158
1-2, Dichloroethane	4.82×10^{11}	190,790
Ethane	5.65×10^{14}	266,102
Ethanol	5.37×10^{11}	201,250
Ethyl acrylate	2.19×10^{12}	192,464
Ethylene	1.37×10^{12}	212,547
Ethyl formate	4.39×10^{11}	187,024
Ethyl mercaptan	5.20×10^{5}	61,504
Hexane	6.02×10^{8}	143,093
Methane	1.68×10^{11}	217,986
Methyl chloride	7.34×10^{8}	171,126
Methylethyl ketone	1.45×10^{14}	244,346
Natural gas	1.65×10^{12}	206,271
Propane	5.25×10^{19}	356,477
Propylene	4.63×10^{8}	143,093
Toluene	2.28×10^{13}	236,396
Triethylamine	8.10×10^{11}	180,749
Vinyl acetate	2.54×10^{9}	150,206
Vinyl chloride	3.57×10^{14}	264,847

[a] Ref. 87. Courtesy of Van Nostrand Reinhold.

net result of poor mixing is exhaust flue gas containing both excess oxygen and unburned pollutants. (Also one should not assume that the waste gas stream entering the incinerator is itself well mixed; often it is not.)

The fuel and air requirements are calculated by conventional chemical stoichiometric and thermochemical methods. Alternatively, appropriate values may be selected from the combustion constants in ref. 87 for ease in making the required calculations.

Incinerator Precautions. Waste gases to be incinerated must be appreciably below the lower explosive limit for mixtures of the combustibles with air to prevent explosions and to protect equipment and personnel. This means only gases that are dilute in combustibles may be incinerated. The usual safety practice is to limit operation to those gases that are no more concentrated than 25% of the lower explosive limit. Table 5 gives the concentration of the lower and upper explosive limits for a number of common organic vapors in air.

Great care must be exercised to prevent feeding an explosive mixture to a gas incinerator, not only because of the possibility of blowing up the incinerator but also because of the possibility of flame flashback to the process from which the gas has come. For this reason, upstream processes must be protected against the possibility of flame flashback. If the waste gas is hydrogen rich, ordinary flame arrestors may not provide adequate flashback protection.

As an alternative to incineration, consideration should be given to collecting valuable hydrocarbons and organic solvents by other means for reuse, such as condensation, rather than destruction by combustion. If the quantities are appreciable, recovery for fuel value alone may be worthwhile. Gases containing sufficient combustibles to support combustion are not burned in an incinerator. Rather they are burned in flares, or waste heat recovery boilers, or used for process heat.

Table 5. Explosive Limits and Autoignition Temperatures for Common Hydrocarbons and VOCs

Compound	Explosive Limits in Air, Percent by Volume: Lower	Upper	Autoignition Temperature, °C
Acetone	2.15	13.0	603
Acetylene	2.5	80	335
Benzene	1.4	8	580
Butane-*n*	1.6	8.5	430
Decane-*n*	0.67	2.6	260
Diethyl ether	1.7	48.0	186
Ethane	3.12	15.0	510
Ethanol	3.28	19	426
Ethylene	3.02	34	543
Heptane-*n*	1	6	233
Hexane-*n*	1.25	6.90	247
Hydrogen	4.1	74.2	580
Methane	5.3	13.9	537
Methanol	6.0	36.5	470
Naphtha	1.2	6.0	232–260
Octane-*n*	0.84	3.2	232
Pentane-*n*	1.4	8.0	309
Propane	2.37	9.5	466
Propylene	2.0	11.1	
Toluene	1.27	7.0	811

Incinerator Types and Operation. Two general types of gaseous incinerators are in use: direct flame and catalytic. Gases also can be incinerated by indirect heating through a heat-conducting wall, but a higher temperature is generally required for their ignition and combustion than when a flame is present.

In the direct flame type incinerator, gases are heated in a fuel-fired refractory chamber to their autoignition temperature where oxidation occurs with or without a visible flame. Autoignition temperatures vary with chemical structure, but are generally in the range of 540–760°C. Use of a fuel flame aids in both combustion and mixing. The presence of the flame itself does not change the oxidation process, but it does influence the time, temperature, and turbulence factors. The fuel flame may be either short and intense (as occurs with a premix fuel burner) or luminous and diffuse. The latter accelerates ignition of entrained particles because of flame radiation. Nevertheless, the entire waste stream must be heated to the autoignition temperature to oxidize the gaseous components.

A short intense fuel flame promotes turbulence and offers an opportunity to increase the effectiveness of the holding time provided at incineration temperature. It is believed to result in more economical use of burner fuel. The residence time provided in a direct flame afterburner is generally in the range of 0.2 to 2.0 s, but this must be based on the kinetics of the compounds to be oxidized and the desired incinerator temperature. The design operating temperature chosen is generally higher than the autoignition temperature required. Incineration operation temperatures are usually in the range of 650–1100°C, but this varies with the pollutant species. A temperature of 510°C may suffice for naphtha vapors, while 870°C (1600°F) may be required for methane. Even higher temperatures are needed for some aromatic hydrocarbons and when the waste stream contains an unusually high percentage of inerts, which tend to depress oxidation. If the vapors to be incinerated are highly toxic, such as HCN, temperatures as high as 980–1100°C may be required for safety.

Length: diameter ratio of the direct flame combustion chamber is often in the range of 2 to 3. Average gas velocity can be as low as 3 m/s to as high as 16.5 m/s. (The gas velocity increases as it travels through the chamber not only due to temperature but also because the moles of reaction product gas increase with combustion). Figure 11 illustrates a simple stand-alone refractory incinerator and one built into the base of a stack. These are adequate for occasional oxidation of process blow-down gases preparatory to process shutdown for annual maintenance. However, they would have high fuel operating costs if used continuously. For such use, heat exchange or recovery is added to preheat the incoming waste gas with the hot exhaust.

Two types of thermal energy recovery systems are in common use: regenerative and recuperative. Regenerative

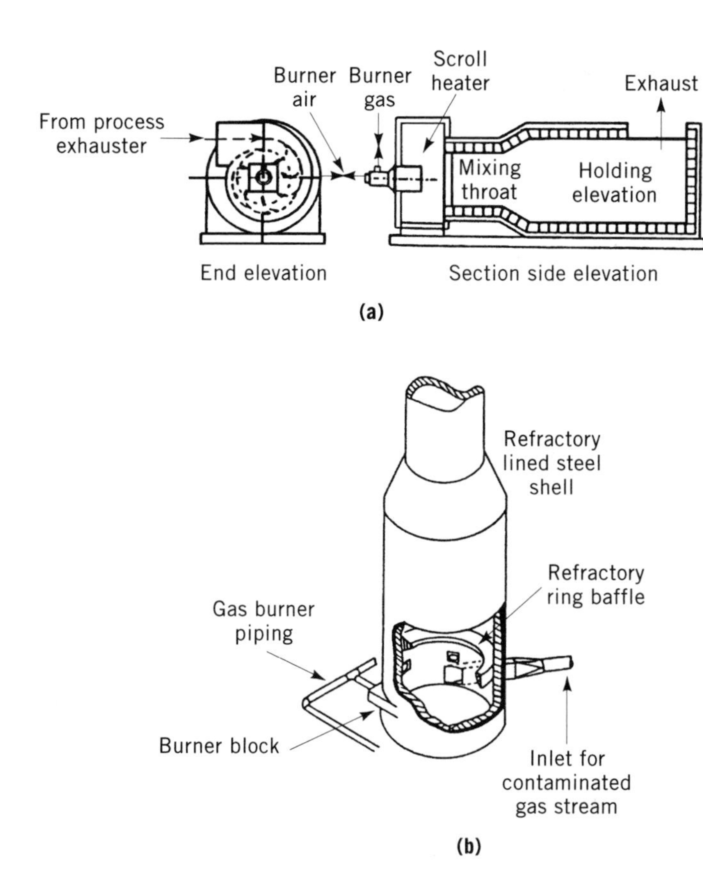

Figure 11. Simple direct flame incinerator types without heat interchange. (**a**) A modified scroll-type refractory air heater modified with a holding chamber to keep the gases at the temperature and time required for their complete oxidation. Reproduced from Ref. 90 with permission of Academic Press. (**b**) A vertical combustion chamber built into the base of a stack. The refractory ring baffle divides the chamber into a turbulent inlet section for gas and flame mixing, heat transfer, and ignition, followed by a temperature–time holding section to ensure complete destruction.

systems (Fig. 12) use ceramic (or other dense, inert material) in stationary beds to capture heat from the combustion exhaust gases. As the bed approaches the combustion temperature, the incinerator exhaust gas is switched to a lower temperature bed. The waste gas to be incinerated is passed through the hot bed to preheat the waste gas before its entry into the combustion process. When the gas flows between various beds are switched, the process is reminiscent of the ceramic-filled stoves of the steel blast furnace industry. However, by using multiple beds, regenerative systems have achieved up to 95% recovery of the energy released in the combustion process.

Thermal efficiency depends on the process operating characteristics. With relatively constant waste gas flow and combustible concentration, heat regeneration can come close to a "no external fuel requirement" (except for start-up conditions and normal safety fuel requirements for pilot flames). However, processes with cyclic operation, or highly variable waste gas total flow, and highly variable combustible content are not as compatible for heat regeneration application. During periods of low heat input waste-gas streams, much external fuel must be supplied to maintain the required operating temperature for the incinerator.

Recuperation. Recuperative systems exchange heat directly from one gas stream to the other with a heat transfer surface (Fig. 13). Typically, the interchanger may be a metallic shell and tube heat exchanger (perhaps finned to improve effective coefficients). Other alternatives are a rotating ceramic heat wheel, heated by one stream and cooled by the other, or the use of a metallic Lungstrum air preheater from steam boiler practice.

The maximum thermal recovery from a recuperative system is on the order of 70%. The advantage of the recuperative systems is the relatively short time needed for them to reach thermal equilibrium with changing conditions in incinerator operation. The large mass of regenerative systems requires long times for thermal equilibrium and relatively large initial fuel inputs to achieve steady operating conditions. Recuperative systems reach operating conditions within several minutes of startup. Their versatility yield themselves to responding well to cyclic operating conditions.

Catalytic Incineration. Catalytic incinerators oxidize substances at temperatures below those at which they would burn in air. Typical oxidation temperatures are

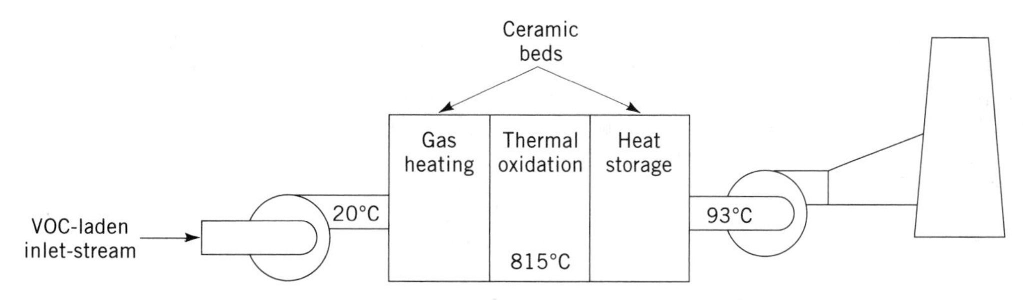

Figure 12. Schematic diagram of regenerative heat recovery for a gaseous incinerator. The hot combusted gas flows through one or more pebble-bed heaters in series, giving up its heat to the ceramic bed. The incoming waste gas to be incinerated is first passed through a ceramic bed previously heated with hot combustion products. Piping and valving provide for switching of the various beds between heating and cooling duty (86). Courtesy of the American Institute of Chemical Engineers.

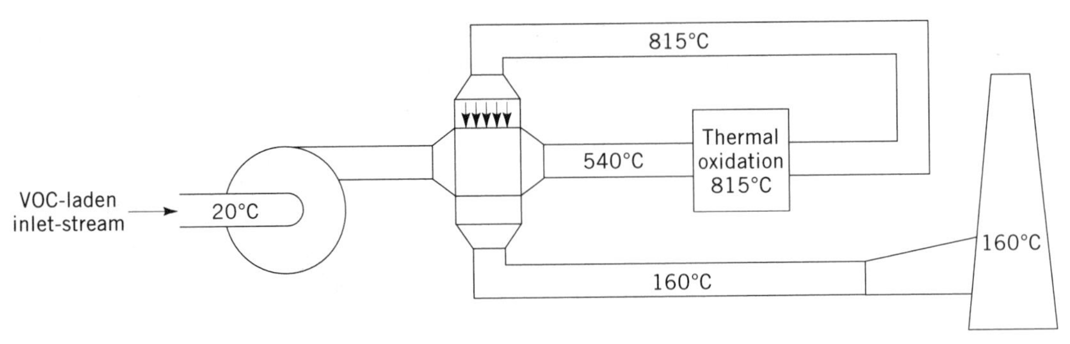

Figure 13. Schematic diagram of recuperative heat recovery for a gaseous incinerator. The hot combusted gases flow through one side of tubular heat exchanger to transfer their heat to the cool incoming waste gas to be incinerated. The waste gas is on the other side of the exchanger tubes. Alternatively, the hot and cool gases may be in separate parallel ducts. A rotating heat wheel or rotating boiler-type air preheater, rotates through the two gas ducts, absorbing heat from the hot gas and giving up heat to the cold gas (86). Courtesy of the American Institute of Chemical Engineers.

around 260°C, but vary with chemical species being incinerated. Naphtha requires a gas and catalyst temperature of 230°C to initiate catalytic oxidation. Methane requires a temperature of 400°C, and catalytic oxidation of hydrogen can be initiated at room temperature. In general, the initial oxidation temperature decreases with increase in molecular size. Chemical structure also affects the initiation temperature. Within compounds with the same number of carbon atoms, the initiation temperature is lowest for highly unsaturated aliphatic compounds. Initiation temperature increases with structure in the following order (lowest to highest): acetylenic, olefinic, normal paraffin, branched-chain paraffin, aromatic.

Metallic platinum gives the lowest oxidation initiation temperature of any material and thus is frequently used as the catalyst. Alloys of the platinum family of metals are also often used. Oxides of cobalt, nickel, manganese, chromium, and iron can be effective under specific conditions of temperature and pollutant concentration and may be used in specific situations.

Fuel must be used initially to start a cold catalytic incinerator, but once the catalyst is heated, frequently no further fuel is needed to keep the oxidation going. Thus the advantages of catalytic oxidation are lower fuel requirements, no NO_x formation (because of the low oxidation temperature), and compactness. Heat exchange is often needed to preheat the incoming waste gas with the oxidized exhaust gases. However, if the application is not carefully selected, many problems can arise. The catalyst can be poisoned by the presence of heavy metals, phosphates, and arsenic, destroying its activity. Its activity can be decreased temporarily by the presence of halogen and sulfur compounds, necessitating its off-line rejuvenation, often in a high temperature hydrogen environment. Furthermore, the surface of the catalyst may be rendered inactive with coatings of soot or fused inorganic dusts, requiring cleaning of the catalyst surfaces. Temperature control of a catalytic incinerator is critical, because the catalyst is easily damaged by overheating. Thus these units are often restricted to applications in which gas flow rate and combustible concentration is relatively constant. For flows having high variability in heat release quantity, direct flame incinerators are more rugged and can withstand short intervals of excess heat release much better and with less damage. In addition, catalytic incinerators are not well suited to situations in which the waste gas stream is cyclical or intermittent in release. The problem is one of keeping the catalyst hot during periods of no flow and of suddenly preheating the waste gas stream when its flow resumes.

When direct flame and catalytic incinerators are properly applied to suitable applications, there is essentially no difference in their efficiency of oxidation. Either can achieve >99% oxidation efficiency when properly designed and applied.

Flares. Flares burning externally in the atmosphere, without containment or shielding, are unlikely to destroy hazardous and toxic combustible gases totally. The causes include too quick flame temperature quenching by the atmosphere before combustion is complete; variations in wind turbulence and direction from minute to minute; and excessive heat losses from the flame to the surroundings by radiation. To produce efficient oxidation, the flare should be enclosed on the sides of the flame with a lightweight housing and insulated internally with a high temperature, low heat capacity, fiberous material. This shields the flame from changes in external turbulence and reduces flame heat losses to the atmosphere.

Checking Incinerator Combustion Efficiency. If the combustion in a gaseous incinerator is incomplete, compounds such as aldehydes, organic acids, carbon, and carbon monoxide will usually be present.

Therefore, one method of testing for combustion efficiency is to collect a sample of the effluent and have it laboratory screened for the presence of these incomplete combustion products. Because many of these products have a distinctive and sometimes irritating odor, another accepted test method is to form an odor panel of five or six individuals with sensitive noses and have them cautiously sniff the effluent. If all agree that they can detect no odor in the effluent, combustion is considered to be acceptable. More detailed information on design and application of gaseous incinerators is available (89–95).

Specific Problem Gases

Sulfur dioxide, nitrogen oxides, and vehicular exhaust gases are widespread gas pollutants that present specific problems. The U.S. Clean Air Reauthorization Act of 1990 requires greater control of emission of these gases. Germany and Denmark adopted acid rain regulations on sulfur and nitrogen oxides earlier. Another widely emitted gas, carbon dioxide, is predicted to become a problem emission in the not too distant future because of its green-

Table 6. Commercialized Flue Gas Desulfurization Processes

Process Name	Process Description	References
	Wet Throw-away Processes	
Dual alkali system	SO_2 absorbed in tower with $NaOH–Na_2SO_3$ recycle solution. CaOH or $CaCO_3$ added externally to precipitate $CaSO_4$, regenerate NaOH; make up NaOH or Na_2CO_3 added; process attempts to eliminate scaling and plugging problems of limestone slurry scrubbing	112,116
Limestone slurry scrubbing	Limestone slurry scrubs flue gas. SO_2 absorbed, reacted to $CaSO_3$; further air oxidized to $CaSO_4$, settled, and removed as sludge; lower cost and simpler than other processes; disadvantages: abrasive and corrosive, plugging and scaling, poor dewatering of $CaSO_4$	116–118,121
Dowa process	Similar to dual alkali except $Al_2(SO_4)_3$ solution used in scrubber; limestone addition regenerates reactant, precipitating $CaSO_4 \cdot 2H_2O$ crystals that dewater more readily; reduces plugging and scaling	
CHIYODA thoroughbred 121 process	Single vessel used to absorb SO_2 with limestone slurry and oxidize product to gypsum	
Forced oxidation	Limestone scrubbing, products air oxidized to gypsum in separate tank	
Lime spray dying	Wet–dry process; lime slurry absorbs SO_2 in vertical spray dryer forming $CaSO_3$–CaS, H_2O evaporated before droplets reach bottom or wall; dry solids collected in bag house along with fly ash	116
	Dry Throw-away Processes	
Direct injection	Pulverized lime or limestone injected into flue gas (often through burner); SO_2 absorbed on solid particles; high excess alkali required for fairly low SO_2 absorption; finer grinding, lime preheat, and flue gas humidification benefit removal; particulate collected in bag house	113,114,120
Trona sorption	Trona (natural Na_2CO_3) or Nacolite (natural $NaHCO_3$) injected into boiler; SO_2 absorbed to higher extent than with dry lime; product collected in bag house; also can capture high quantity of NO_x	113,114,119,120
	Wet Regenerative Processes	
Wellman-Lord	After flue gas pretreatment, SO_2 absorbed into Na_2SO_3 solution; solids and chloride purged, SO_2 stripped, regenerating Na_2SO_3, and SO_2 processed to S	115,116
Magnesium oxide process	SO_2 absorbed from gas with $Mg(OH)_2$ slurry, giving $MgSO_3–MgSO_4$ solids that are calcined with coke or other reducing agent, regenerating MgO and releasing SO_2	
Citrate-scrubbing	SO_2 absorbed with buffered citric acid solution; SO_2 reduced with H_2S to S; H_2S produced on site by reduction of S with steam and methane	112
Flakt-Boliden process	SO_2 absorbed with buffered citric acid solution; SO_2 stripped from solution with steam	
Aqueous carbonate	SO_2 absorbed into Na_2CO_3 solution in spray dryer, producing dry Na_2SO_3 particles	
SULF-X process	FeS slurry absorbs SO_2; product calcined producing S vapors that are condensed	
Conosox process	K_2CO_3 and K salt solutions absorb SO_2, forming K_2SO_4, which is converted to thiosulfate and $KHSO_3$, which is converted to H_2S; regenerates K_2CO_3	
	Dry Regenerative Processes	
Westvaco	SO_2 adsorbed in activated carbon fluid bed. SO_2, H_2O, and $\frac{1}{2}O_2$ react at 65–150°C forming H_2SO_4; in next vessel, $H_2SO_4 + 3H_2S$ at 150°C gives $4S + 4H_2O$; bed temperature is increased to vaporize some S; remaining S reacts with H_2 to H_2S	
Copper oxide adsorption	SO_2 adsorbed on copper oxide bed forming $CuSO_4$; bed is regenerated with H_2 or H_2–CO mixture, giving concentrated SO_2 stream; bed is reduced to Cu but reoxidized for SO_2 adsorption	113

house effect. A few exploratory research studies are beginning to be published on control techniques for CO_2, apart from avoiding its emission by ceasing to burn fossil fuels (96,97).

Major sources of SO_x are the combustion of sulfur-containing fossil fuels, the manufacture of sulfuric acid and its recycled sludge acid purification, sulfur recovery or disposal from petroleum processing, nonferrous smelting, and pulp and paper manufacture. Combustion emissions are controlled by substituting a low sulfur fuel source, by fuel desulfurization and refining, and by sulfur removal either in the combustion process or from the flue gas. Many methods of sulfur removal from flue gas have been developed and voluminous literature is available (98–122). Utility experience with more than 100 flue gas desulfurization (FGD) plants in commercial operation in the United States has been developed. Removal of 90–95% of the SO_2 in the flue gas is practical for high sulfur coal, and 70–80% removal can be achieved for low sulfur coal.

FGD systems can be classified as (*1*) throwaway vs regenerative and (*2*) wet vs dry. In throwaway processes, the removed sulfur is discarded, often as a calcium sulfate–sulfite sludge; in the regenerative processes the sulfur is recovered in a useful, but not necessarily marketable, form. Most of the earlier commercialized FGD processes are of the throwaway type, frequently using wet scrubbing and reaction with lime or limestone. Regenerative processes are attractive from a resource conversation standpoint, but frequently unattractive economically, because total recovery of sulfur constituents would exceed the world's desire for their consumption by many fold. Table 6 lists a number of FGD processes. For new boiler installations, fluid bed combustion is often used. The coal is burned in a fluid bed of limestone, which reacts with the sulfur in the fuel during combustion to produce calcium sulfate. Another advantage is that appreciable quantities of combustion-generated NO_x is also recovered in the bed. Some newer FGD processes that have been investigated or partially commercialized are available (122). Much current research is being devoted to improving dry-injection processes such as injection of fine lime or limestone particles into the combustion flame. Original work indicated that only 50% of the SO_2 could be removed, but considerably better removal has been achieved more recently with gas humidification and/or the addition of additives that have enhanced the reactivity (125–129). Also some research is being devoted to processes that can simultaneously remove both SO_2 and NO_x and even other acidic gases such as HCl. (The latter, as well as other chlorides, is a problem in flue gas emissions from municipal waste incineration.) Injection processes applicable to both pollutants have used injection of trona (natural Na_2CO_3) or nacolite (natural $NaHCO_3$), (113,119,120,130). If adequate removal efficiency can be achieved, injection throwaway processes are more attractive. They can be cheaper; there is no cooling of the flue gas with water and no scaling or corrosion problems; the residue is collected dry, perhaps in a bag filter along with the fly ash; and disposal of a dry waste solid should be easier. If one process can recover both sulfur and nitrogen compounds along with fly ash, the recovery process can be fairly simple. Still other developmental processes for removal of SO_2 or both SO_2 and NO_x are available (131–136).

SO_2 emission from sulfuric acid manufacture is controlled by interpass absorption (137) and, when necessary, by the addition of tail-gas scrubbing to reduce the SO_2 concentration in the effluent below 400–500 ppmv. In oil refining of sour crude, it is common practice to treat the separated H_2S in a Claus sulfur recovery process. In this process, 33% of the H_2S is burned to SO_2. This SO_2 is reacted with the remaining 66% of the H_2S by oxidation–reduction over a catalyst to produce elemental sulfur, which may be salable. Recovery of the sulfur values may be as high as 97%. In some areas, this is not acceptable. In such cases, the effluent from the Claus plant must be introduced into a liquid-phase reaction process such as the Stretford (Beavon) process. Three other processes for treating the Clauss tail gas (the Shell SCOT, the Wellman-Lord, and the IFP processes) have been developed (138). Emissions of SO_2 from nonferrous smelters are controlled by conversion to sulfuric acid. This is being done even in areas without a market for sulfuric acid; the acid is subsequently neutralized with limestone (139), and the gypsum produced is placed in a solid waste land disposal site. Other SO_2 emission-control methods for smelters (140,141) and pulp and paper manufacture have been discussed (142–145).

Nitrogen Oxides. Principal sources of nitrogen oxide emission are nitrogen fixation during high temperature combustion, nitric acid manufacture and concentration, organic nitrations, and vehicular emissions. During combustion in the presence of air, N_2 and O_2 react in the high temperature flame to produce NO. This reversible reaction favors NO formation as the temperature increases (Table 7). Figure 14 shows the theoretical flame temperature and thermodynamic equilibrium quantity of NO_x that would be produced at flame temperature as a function of combustion air stoichiometric ratio when burning methane in air. The kinetics for NO formation also increase rapidly with temperature. Unfortunately, the kinetic rate for the reverse (decomposition) reaction drops essentially to 0 at temperatures below 870°C. Thus, as the flue gas cools, complete reversion of the NO to N_2 and O_2 does not take place. Nitrogen present in fuel compounds tends to oxidize much more readily to NO_x than does air. This NO_x from fuel nitrogen is essentially additive in concentration to that which is produced thermally from the excess air in high temperature combustion.

The reaction kinetics are usually too slow in most high temperature furnaces to produce equilibrium amounts of

Table 7. Time to Form NO in a Gas Containing 75% N_2 and 3% O_2[a]

Temperature, °C	Time to Form 500 ppmv NO, s	NO equilibrium concentration, ppmv
1315	1370	550
1538	16.2	1380
1760	1.10	2600
1982	0.117	4150

[a] Ref. 146.

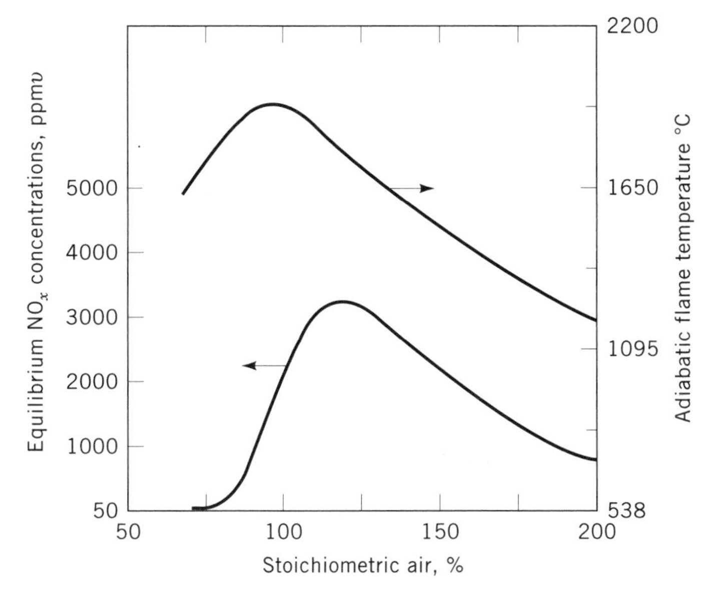

Figure 14. Thermodynamic equilibrium NO_x concentrations and flame temperatures for burning methane in air as a function of the combustion air quantity (155).

NO at the flame temperature. Another way of saying this is that residence time at flame temperature is generally much too short to produce equilibrium concentrations of NO in the combustion products. Nevertheless, in coal and oil combustion, it is fairly easy to produce NO_x concentrations of 1000–2000 ppmv with no more than 5–10% excess air.

NO reacts with O_2 in the atmosphere at a slow but steady rate such that all NO produced ends up in the atmosphere as NO_2. Combustion chemistry and NO_x formation have been reviewed (147). NO formation during combustion can be reduced by five methods: (*1*) maintaining low excess air (0.5% O_2 or less in the flue gas), (*2*) two-stage combustion, (*3*) flue gas recirculation, (*4*) lowering flame and combustion temperatures, and (*5*) burner and combustion chamber modifications. Combinations of two or more methods are also beneficial. Another method to reduce NO formation is to reduce firing capacity. This reduces the NO produced by lowering flame and gas temperatures, since the gases are chilled faster due to a larger ratio of heat sink area to heat input.

Low Excess Air. Reducing the combustion chamber excess air will reduce the amount of NO_x produced because there is less oxygen concentration to react with nitrogen and the driving force for NO_x formation is reduced. Oxygen has much more affinity for combination with carbon than with nitrogen. In a well-designed combustion chamber (with adequate turbulence and residence time), all combustibles can be completely oxidized with no more than 0.5% O_2 by volume remaining in the flue gas. Experimental data (Fig. 15) illustrates reduction in NO_x formation that can be achieved by reducing combustion excess air.

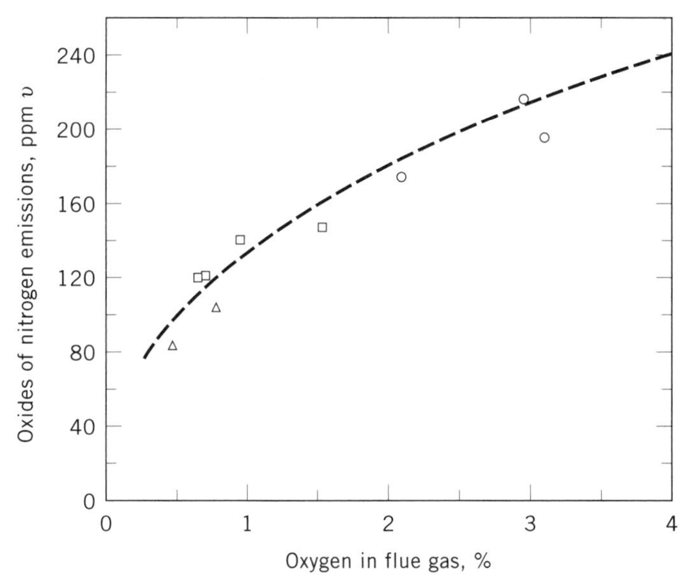

Figure 15. NO_x emissions from an oil-fired boiler as affected by quantity of excess air (146).

Two-stage Combustion. The principle of two-stage combustion can be applied to combustion chamber design with two combustion chambers in series or within the confines of individual low NO_x burners firing into a single combustion chamber. In the first stage, combustion occurs at a high temperature in a fuel-rich mixture (reducing atmosphere). With inadequate combustion air, no NO_x is produced. In the second stage, combustion is completed with the introduction of additional air, but at a low enough temperature (1000–1100°C; 1832–2012°F) that little NO_x is formed. Experimental results of two-stage combustion are shown in Figure 16. The reduction in NO formed in the test boiler is shown with both high and low excess air without combustion staging as well as the reduction with two-stage combustion, again with both high and low excess air. Furthermore, the reduction in NO_x with cut-back in boiler firing rate is also apparent for all combustion situations.

Figure 17 shows two flow schemes for two-stage low NO_x burners. The design may use either fuel-staging or air-staging. With air-staging, the combustion air is di-

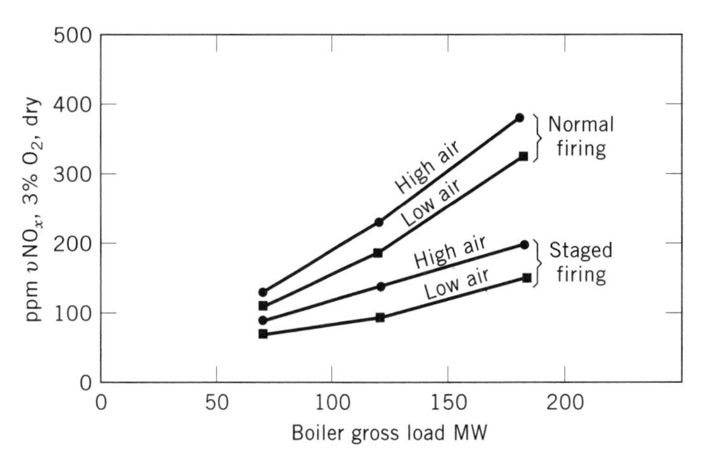

Figure 16. Average NO_x emission reduction produced by two-stage combustion in a 180 MW gas-fired boiler with front wall burners. The NO_x reduction with cut-back in boiler load is also apparent (126,152). Courtesy of the American Institute of Chemical Engineers.

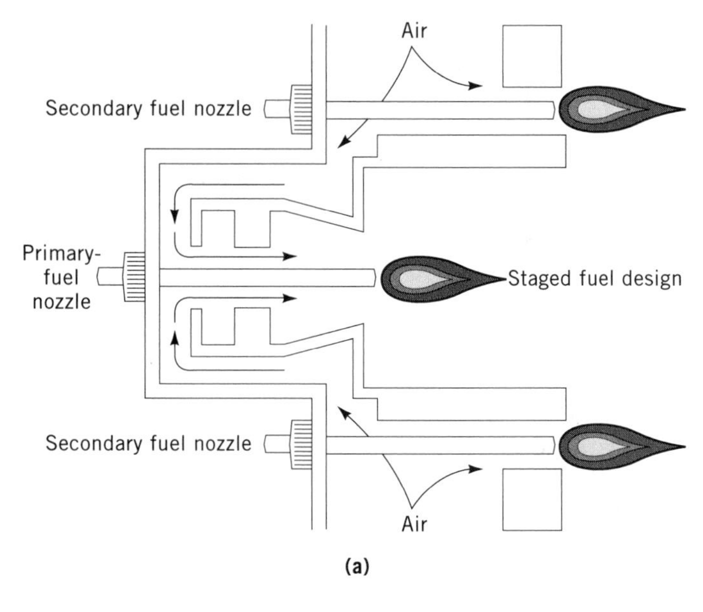

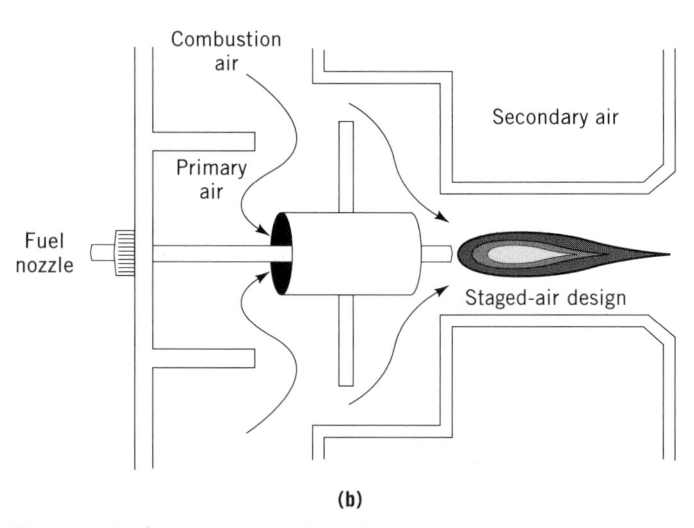

Figure 17. Two-stage fired Low NO_x burners with (**a**) fuel staging, and (**b**) with air staging (174). Courtesy of *Chemical Engineering*.

vided into primary and secondary flows. All of the fuel is injected into the burner throat, where it is combined with primary air in less than a stoichiometric amount. The fuel is only partially burned, and fuel nitrogen may be converted to ammonia, hydrogen cyanide, and NO. The latter may be reduced back to N_2, while the NH_3 and HCN remain to be oxidized in the final combustion stage. Peak flame temperature is lower because not all the fuel and the above intermediates have been oxidized. In addition, flame temperature falls rapidly by radiant heat dissipation to the boiler tubes. Some recirculation of combustion products within the burner further reduce flame temperature and oxygen concentration. Both of these factors further inhibit thermal NO_x formation. In the secondary combustion zone, injection of the additional air completes the combustion.

Staged-air low NO_x burners are relatively inexpensive and efficient for reducing NO_x. They may be used in both new boiler designs and in retrofitting existing boilers, reducing NO_x formation by 40–65%. Their main disadvantages are flames that are somewhat longer and wider than the prior conventional burners, which may require either a larger combustion chamber or reduced heat input from an individual burner. In boilers with limited combustion space, these burners may cause metal thermal stresses and fatigue in retrofit applications.

With staged-fuel burners, a portion of the fuel is injected into all the combustion air and burned in a fuel-lean (air-rich) environment. The lean combustion mixture reduces peak flame temperatures, thus lowering some of the thermal NO_x, which would otherwise be produced. The remainder of the fuel is injected in a secondary combustion zone, containing combustion products and inerts from the primary zone, further retarding NO_x formation. Staged-fuel low NO_x burners can reduce NO_x emissions from 50 to 70% and operate with a shorter flame length, minimizing problems of metal fatigue. Low NO_x burner design has been discussed (149,150).

In lieu of using either low NO_x burners or staged-combustion chambers, sometimes the combustion chamber may be divided into fuel-rich and lower temperature excess air combustion zones. A portion of the combustion air may be diverted from existing burners, producing a localized fuel-rich region, preventing NO_x formation. Then combustion is completed higher up in the boiler setting with the introduction of overfire air. (This will become clearer in a discussion of reburning, presented later.)

Flue Gas Recirculation. Flue gas recirculation operates in several ways. It introduces previously formed NO into the flame from previously consumed gases, retarding the formation of new NO by shifting the chemical equilibrium. In addition, the recycled cooler flue gas has a cooling effect on flame temperature, shifting the equilibrium in the reverse direction and retarding the kinetic reaction rate for NO formation. Finally, it dilutes the reacting concentrations, especially of oxygen, retarding the rate of NO formation. Experimental results with flue gas recirculation are shown in Figure 18.

Burner and Combustion Chamber Modifications. Burner designs can be modified to control turbulence or rate of mixing, retard combustion rate, and reduce peak flame temperatures. Combustion chambers can be retrofitted with low NO_x burners (as discussed above). Burner position and aiming can be changed and combustion chamber design modified to produce lower flame temperatures. Aiming of a number of burners to create flame impingement on a swirling fireball in the combustion chamber enhances flame temperature and NO formation. Spreading a flame out in a combustion chamber, especially along water-tube–cooled walls and other heat sinks, aids in speeding heat absorption and providing cooling for the flame. Reducing the firing rate in a fixed-size combustion chamber also reduces flame temperature and the quantity of NO produced. Combustion technology methods (148–158) are limited in the level to which NO_x can generally be reduced to around 200–300 ppmv NO_x. However, retrofitting with low NO_x burners is one of the least expensive ways to make significant reductions in NO_x emissions.

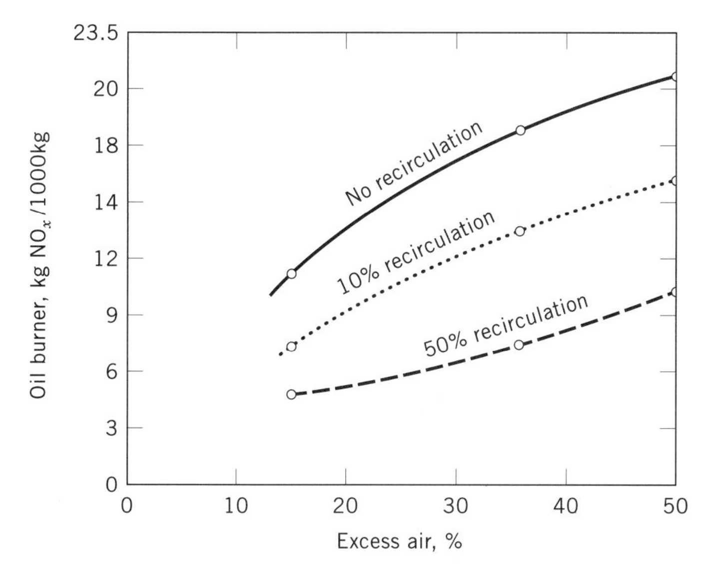

Figure 18. Combined effect of flue gas recirculation and excess air on NO_x emissions from an oil-fired domestic heating furnace (146).

Adding a portion of the combustion air (10–20%) as overfired air in boilers with tangentially fired and wall-fired burners in combination with low NO_x burners can achieve up to a 50% reduction in NO_x produced. Other procedures in use on existing power boilers to reduce NO formation are burner-off technology and reburning. Burner-off technology would appear to be applicable only when fuel-firing rate can be cut back or when excess burners have been installed. It consists of turning one burner completely off for a period of time when several burners are installed. This obviously reduces the heat input to the boiler and helps in reducing flame temperature. NO_x emissions are found to be reduced. Sometimes, after a particular burner has been off for a time, it will be restarted and another burner turned off. This change in heat release location with time may have the effect of cooling off different portions of the combustion process from time to time. Another use of burner-off technology is to reduce the air to the lower burners so that they run fuel-rich and operate the top burners with air-flow only as overfired air to complete the combustion. This is really one form of staged-air combustion. Still another technology, called biased firing, involves running certain burners fuel-rich and certain burners air-rich.

Reburning is a technique for reducing NO_x in existing flue gas by adding more fuel and air under conditions such that some of the existing NO_x will be dissociated back into N_2 and O_2. This requires a temperature and fuel : air ratio suitable to cause the NO to be reduced back to N_2. Figure 19 illustrates retrofitting an existing boiler to reduce emissions of both NO_x and SO_x. Originally totally coal-fired, 20% of the fuel heat input is replaced with natural gas, which is fired above the coal and is slightly fuel-rich. Overfire air is added above the natural gas firing, to burn the excess gas combustibles and to reduce flame temperatures. If flue gas desulfurization is desired, then still higher lime injection can be introduced, as shown in the illustrative example, to remove SO_2. If sodium bicarbonate or carbonate were also injected, some remaining NO_x would also be removed along with sulfur oxides. Reburning can employ a number of fuels. Those useful, in increasing order of effectiveness, are fine pulverized coal, fuel oil, and natural gas. Boiler retrofitting using a combination of low NO_x burners, reburning, and overfire air can approach 75% NO_x reduction. Many aspects of NO_x control retrofit have been covered (160–162).

Selective Catalytic Reduction. A number of chemical processes have been developed for the reduction of NO_x which, while costly, are capable of reducing NO_x emission levels to 80–100 ppmv (or even lower). These have generally been grouped in the literature as selective catalytic reduction (SCR) and selective noncatalytic reduction (SNCR) processes. SCR (122,163,169–172) is the most commercially developed and applied method, using post-combustion technology. It was developed in Japan and applied primarily to oil-fired boilers with capacity totaling 10,000 MW. Subsequently, it has been applied in Germany to large coal-fired boilers. In the transition to coal, difficulties were encountered with catalyst poisoning and shortened catalyst life. Successful operation has resulted from matching catalyst properties to coal impurities. The original Japanese catalysts were platinum based, alumina supported in the shape of plates, or with rectangularly cored gas passages. Alternatively, the honeycomb design was available. The honeycomb catalyst was more compact per contact surface area and preferred when space limitation was important. The plate and cored-block design had lower pressure drop and could be cleaned when used in dusty boiler areas such as before the precipitator.

Originally, the catalyst cost was quite expensive: \$9700/m^3. With its use in Germany, a German-produced catalyst has been perfected using vanadium and tungsten oxides with price reductions down to \$3300–3800/m^3. Even so, the cost of a catalyst charge for a 500-MW power boiler can run \$2–3 million, with an expected useful life

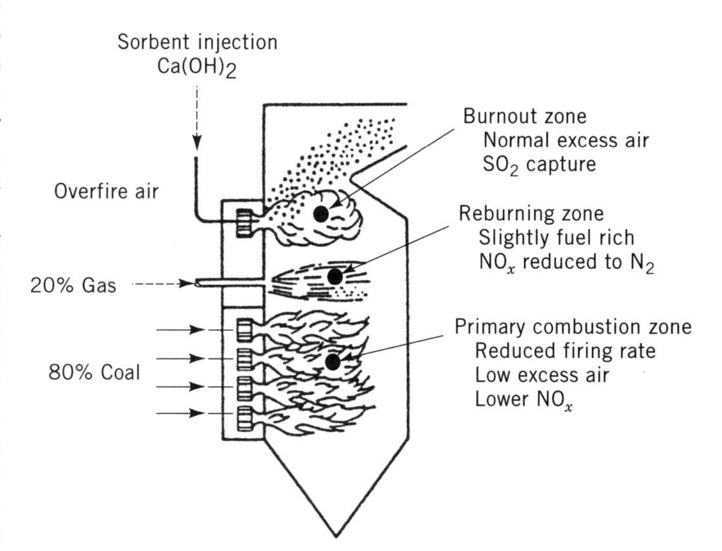

Figure 19. NO_x reduction from coal firing with gas reburning (20% of the total heat input) and sorbent injection for SO_2 removal (129). Courtesy of the American Institute of Chemical Engineers.

being 3 to possibly 5 yr. Recent literature shows that catalysts that are more resistant to poisoning or in other physical shapes, such as granular catalyst, are being developed.

All of the European applications have been with coal having alkaline ash and with coal containing less than 1% sulfur. Among substances poisoning the catalyst are arsenic, potassium, and a number of trace elements. An Austrian plant is of the opinion that water vapor, absorbed by the catalyst during plant downtime, is harmful. Therefore, they have provided an air dehumidifier to provide dried air to the unit during such periods. The effect of sulfur has not been determined specifically, but there is general concern as to whether moderate and high sulfur coals can be used successfully with the catalyst. Usually, either more catalyst than initially needed, is installed to allow for some degrading; or extra space is provided for adding more catalyst later.

Ammonia is added to the flue gas upstream of the catalyst. It must be proportioned to the NO content of the flue gas, being carefully distributed across the full width of the gas flow in the boiler. Poor distribution of the NH_3 can result in inadequate NO_x reduction if there is a local inadequacy of NH_3. An excess of NH_3 in any area will result in unreacted NH_3 being evolved in the atmospheric effluent (a pollutant). Unreacted ammonia in the effluent is referred to as ammonia slip. Designs typically achieve an ammonia slip of 1–2 ppmv in the off gas; 10 ppmv is considered the maximum that can be tolerated.

Experience has shown that the most economical way to proceed is to minimize the NO_x to be removed as much as possible ahead of the catalyst by means such as low NO_x burners, low excess air, reburning, etc. With less NO_x to be reduced, a smaller charge of SCR catalyst is required. Furthermore, less ammonia is required for operation. The catalytic reduction with ammonia can occur with temperatures in the range of 250–450°C, but installations with hotter gas have been employed (up to 520°C). Both European and Japanese experience has been mostly with hot-side, high dust catalyst installations, ie, the SCR reactor is installed between the exit of the economizer and the inlet to the air preheater (ahead of any particulate collection equipment). In this dusty atmosphere, up to 20–30% of the catalyst has become deactivated or plugged with deposited dust.

The American power industry has been exploring the use of SCR technology along with FGD. The SCR process can be installed similarly to the European installations, in a hot and dusty atmosphere, except that the catalyst would be subjected to SO_2. (Present catalysts appear to be degraded structurally in acidic environments.) An alternative is to install the SCR unit after FGD scrubbing. In this case, the catalyst would be protected from both sulfur and dust. However, scrubbing drops the gas temperature too low for present catalysts to be active and effective. Unless catalysts active at lower temperatures can be developed, reheating after scrubbing is required, which is costly from an energy standpoint. However, the catalyst charge can be smaller and catalyst life might be much longer.

Because of all these problems, the American power industry has recently been giving more attention to dry scrubbing with alkaline adsorbents. This would allow use of SCR after sulfur removal and filter collection of all particulates without dropping the gas temperature too greatly. Perhaps an even better alternative is to remove SO_2 and NO_x simultaneously with dry scrubbing, using a sorbent that can remove both simultaneously. A test installation using dry scrubbing with dry mixed sorbents has been described (130). Hydrated lime is proportioned to the SO_2 concentration, and $NaHCO_3$ is proportioned to the ratio of (SO_2 + 2NO).

Selective Noncatalytic Reduction Processes. A small number of competing processes have been developed that depend solely on the thermal energy to supply the activating energy for the reaction. Thermal DeNox is the oldest of these; others are urea injection and urea–methanol injection.

Thermal DeNox. The Exxon Thermal DeNox process introduces ammonia alone into the gas stream for NO_x reduction. The gas temperature window is rather narrow and high for the most efficient NO_x reduction. This range is 870–1000°C. In this range, 70–80% of NO_x reduction can be achieved. At still higher temperatures, the reduction is less complete. Below 870°C, ammonia is not completely reacted unless some H_2 is also injected with the NH_3. In many petroleum operations, a source of H_2 may be available, but this in unlikely to be so with many other large combustion sources. As with SCR, the ammonia must be well distributed throughout the gas stream to be treated and stoichiometrically proportioned to the NO_x concentration in the gas, or ammonia slip will also occur. Thermal DeNox has been discussed in detail (164,168,171). The process has been used primarily on gas-fired boilers and on petroleum industry still heaters. There have been some installations on oil and coal-fired utility boilers, on glass furnaces, and on some municipal solid waste incinerators. On the latter applications, removal efficiencies of 40–60% have been attained.

Urea Injection. Urea injection, developed and patented by EPRI and marketed by Fuel Tech, Inc., under the name NOxOUT process (165), has been commercially installed on two boilers in Europe. It is generally applied by making a 50% solution of urea in water, which is sprayed into the boiler flue gas to react with NO_x. When used with enhancers, it has a relatively wider temperature window than Thermal DeNox. With urea alone, it can be used at temperatures from 925 to 1040°C. The addition of enhancers widens the window from 815 to 1150°C, and with up to 7% O_2 in the flue gas. The enhancers alone can be used between 540 and 815°C.

As with other SNCR processes, the reacting solution must be uniformly distributed throughout the gas stream to be treated and be properly proportined to the NO_x present. The process has not been well demonstrated and has been deemed generally not practical for processes with large load variations or for applications to gas turbines. In boiler tests in Sweden, NO_x reduction from 35 to 70% is reported to have been demonstrated. To optimize the NO_x reduction, different urea concentrations and different enhancer solutions have been used.

Urea–Methanol Injection. Emcotek has patented a two-stage process in which urea solution is injected into the gas stream, followed somewhat further downstream with the injection of methanol. The principal purpose of the methanol injection is to reduce ammonia slip and to reduce deposits in the air preheater. NO_x reduction of 65 to 80% is reported with ammonia slip below 5 ppmv in gas temperature ranges of 815–1040°C. All of the above SNCR processes have been discussed briefly (159).

Other SNCR Research. Other SNCR research processes have involved the injection of ammonium sulfate and volatilized cyanuric acid to reduce NO_x (166–167). The cyanuric acid process appears to have good prospects for NO_x reduction from diesel engine exhaust (both large stationary engines and over-the-road engines), but cyanuric acid is a high cost reagent for large-scale use. Recent SNCR possibilities for NO_x control have been discussed (173).

Other NO_x Emission Source Controls. Control of NO_x emissions from nitric acid and nitration operations is usually achieved by NO_2 reduction to N_2 and water, using natural gas in a catalytic decomposer (175–177). NO_2 from nitric acid–nitration operations is also controlled by absorption in water to regenerate nitric acid. Modeling of such absorbers and the complexities of the NO_x–HNO_x–H_2O system have been discussed (178). Other novel control methods have also been investigated (179,180). Control of vehicular source NO_x emissions is discussed in AIR POLLUTION: AUTOMOTIVE, TOXIC EMISSIONS and elsewhere.

CONTROL OF PARTICULATE EMISSIONS

The removal of particles (liquids, solids, or mixtures) from a gas stream requires deposition and attachment to a surface. The surface may be continuous, such as the wall and cone of a cyclone or the collecting plates of an electrostatic precipitator, or it may be discontinuous, such as spray droplets in a scrubbing tower. Once deposited on a surface, the collected particles must be removed at intervals without appreciable reentrainment in the gas stream. One or more of seven physical principles (Table 8) are frequently employed to move particles from the bulk gas stream to the collecting surface. In some instances, a few other principles such as diffusiophoresis and methods of particle growth and agglomeration have also been used. The magnitude of the force developed to move a particle toward a collecting surface is influenced markedly by the size and shape of the particle. Gravity settling is efficient only for large particles ($D_p > 40$–50 μm); flow-line interception and inertial impaction are effective for particles down to 2–3 μm; diffusional deposition and thermal precipitation, increasingly efficient with a decrease in particle size, are highly efficient for particles ≤ 0.5 μm; and electrostatic forces are the strongest forces available to act on fine particles, which are loosely defined as ≤ 2–3 μm. There is a gap in collectability between 0.2 and 2.0 μm. Particles in this range are the most difficult to charge electrically.

Terminal settling velocity can be calculated for spherical particles in streamline flow (Stokes' law region) by equation 15. Small particles fall faster than predicted as the gas no longer acts as a continuum; small particles slip between gas molecules. The Cunningham-Stokes correction factor (Eq 16) must be applied. Figure 20 gives terminal settling velocities of spherical particles in air. For nonspherical particles, multiplication of equation 15 by a sphericity correction constant $K_s = 0.843 \log(\psi/0.065)$ has been recommended (196). The sphericity ψ (197) is defined as the ratio of surface area of a sphere (of volume equal to the particle) to the surface area of the particle. Further refinements when sphericity is less than 0.67 have been given (198,199).

Particles approaching targets such as a baffle, impaction element, fiber, or droplet and having appreciable velocity have sufficient momentum that they may collide with the target. Particles directly in line are collected by flow-line interception and efficiency can be predicted from equation 18. Larger particles outside the streamlines leading to the target may be collected on the sides of the target because of the particle's inertia. Target efficiency has been correlated with the inertial separation number of equation 19 (200). Figure 21 gives target efficiencies by impaction number for three target shapes. Many moderate energy collectors capture smaller particles by a combination of direct interception and inertial impaction. It has been suggested (202) that the combined efficiency for the two mechanisms can be obtained from

$$\eta_{total} = 1 - (1 - \eta_{FL})(1 - \eta_{IN}) \tag{27}$$

Efficiency of particle collection is most frequently defined on a mass basis by

$$\eta_M = \left(\frac{\text{mass of particles entering} - \text{mass of particles leaving}}{\text{mass of particles entering}}\right) \tag{28}$$

Efficiency may also be expressed in terms of the number of particles or area of particles entering and leaving the collector. An alternative terminology, used especially when collection efficiency is high, is that of penetration of particles through a collector. This focuses on the particles lost rather than those caught and penetration is defined as unity minus the fractional efficiency, $P_t = 1 - \eta$. Typical penetration of fine particles through different collectors as a function of particle size is shown in Figure 22. Measurement of the difficulty of particle collection in terms of transfer units has been suggested (204); the number of required collector transfer units is defined as $N_t = \ln[1/(1 - \eta)]$.

Equation 28 applies to the overall efficiency of a collector, but a collector's performance on a specific dust results from the integration of the collector performance for each incremental particle size fraction handled. Typical grade-efficiency curves, graphs of collector efficiency vs particle diameter, for a silica dust (specific gravity 2.3) in many different collectors have been published (205,206). For other particle densities, a correction factor based on the effect of density on the predominating collection mechanism must be applied. The concept of "cutpoint," the particle size collected with 50% efficiency, was introduced to develop generalized grade-efficiency curves (184–185). In a generalized grade-efficiency curve such as that shown in Figure 23 collection efficiency is plotted against the di-

Table 8. Physical Principles Affecting Particle Movement and Collection

Control Principle	Related Equations[a]	Conditions and Assumptions	References
Gravity settling	$U_t = \frac{D_p^2(\rho_t - \rho_g)g}{18\,\mu g} = \left[\frac{4\,D_p(\rho_t - \rho_g)g}{3\rho_g C_D}\right]^{\frac{1}{2}}$ (15)	free falling, rigid sphere; fluid continuum; for $N_{RE} < 0.1$, viscous, streamline flow, no wake formation, $C_D = 24/N_{RE}$; for $N_{RE} > 0.1$, other functions of N_{RE} must be used to calculate C_D	181–183
	$K_m \cong 1 + \frac{A\lambda}{D_p}$ (16)	Cunningham-Stokes correction factor for small particles when fluid does not behave as continuum; in air, for $D_p = 1\ \mu$m, $K_m = 1.17$; for $D_p = 0.1\ \mu$m, $K_m = 2.7$	
Centrifugal deposition	$\frac{dr}{dt} = U_{t,n} = \frac{D_p^2 \rho_t V_{cT}^2}{18\,\mu_g r}$ (17)	the tangential velocity, V_{cT}, thus the particle velocity, $U_{t,n}$, is a function of the radius r, usually $V_{cT} \propto 1/r^n$; for free gas rotation and conservation of momentum, $n = 1$; for cyclones, n varies between 0.5 and 0.7	184,185
Flow line interception	$\eta = \frac{1}{2.00 - \ln N_{RE}}\left[(1 + N_{SF})\ln(1 + N_{SF}) - \frac{N_{SF}}{2}\frac{(2 + N_{SF})}{(1 + N_{SF})}\right]$ (18)	for cylindrical target; N_{SF} is the ratio of particle diameter, D_p, to the diameter of the collector; assumes that particle will be collected if it approaches within $D_p/2$ from collector; none of the particles are reentrained	186,187
Inertial impaction	$N_{SI} = \frac{U_t U_o}{D_b} = \frac{K_m(\rho_t - \rho_g)D_p^2 U_o g}{18\,\mu_g D_b}$ (19)	viscous flow; Stokes' law region; physically, N_{SI} is the stopping distance in a quiescent fluid of a particle with initial velocity U_0/D_p; Figure 21 relates N_{SI} to collection efficiency	187,190
Diffusional deposition	$D_v = \frac{K_m kT}{3\pi\,\mu_g D_g}$ (20)	particle diffusivity from Stokes-Einstein equation; assumes Brownian motion; D_p same order of magnitude or greater than mean free path of gas molecules (0.1 μm at NTP)	
	$\eta_D = \frac{4}{N_{PE}}(2 + 0.557\,N_{SC}^{\frac{3}{8}} N_{RE}^{\frac{1}{2}})$ (21)	efficiency for spherical collector; $N_{SC} < 10^5$ or $D_p \leq 0.5$ μm in ambient air	191
	$\eta_D = \frac{\pi}{N_{PE}}\left(\frac{1}{\pi} + 0.55\,N_{SC}^{\frac{1}{3}} N_{RE}^{\frac{1}{2}}\right)$ (22)	efficiency for cylindrical collector; $1 < N_{RE} < 10^4$ and $N_{SC} < 100$	
Electrostatic precipitation	$U_e = \frac{E_o E_p D_p}{4\pi\,\mu_g K_v}$ (23)	conductive spherical particles; streamline gas flow; gas behaves as continuum; negligible particle acceleration	192
	$\eta = 1 - e^{-(U_e A_e/q_e)}$ (24)	Deutsch-Anderson equation; assumes no reentrainment from collector; well-mixed turbulent flow, turbulent eddies small compared to precipitator dimensions	
Thermal precipitation	$U_r = -\frac{1}{5(1 + \pi a/8)}\frac{k_{gTR}}{P}\frac{dT}{dX}$ (25)	D_p less than mean free path of gas; free molecular or Knudsen flow regime; a = fraction of inelastic collisions, usually taken as 0.81	193,194
(Thermophoresis)	$U_r = -\frac{3\,K_m \mu_g}{2\rho_g T(2 + k_p/k_g)}\frac{dT}{dX}$ (26)	D_p same order of magnitude or greater than mean free path of gas molecules	

[a] Terms are defined in "Nomenclature" at the end of this article.

mensionless ratio D_p/D_{pc}. An improved method for predicting collection equipment performance from particle size distribution has been suggested (207).

Gravity Settling

The gravity settling chamber is one of the oldest forms of gas–solid separation. It may be nothing more than a large room where the well-distributed gas enters at one end and leaves at the other. Such chambers were used at the turn of the century for collecting products such as lampblack. Although mechanical conveyors might minimize labor costs, gravity settlers have largely disappeared because of bulky size and low collection efficiency. They are generally impractical for particles smaller than 40–50 μm.

Centrifugal and Cyclonic Collection

A cyclonic collector is a stationary device that uses gas in vortex flow, produced either by tangential entry of the gas or by spin baffles with axial gas inlet, to collect particles.

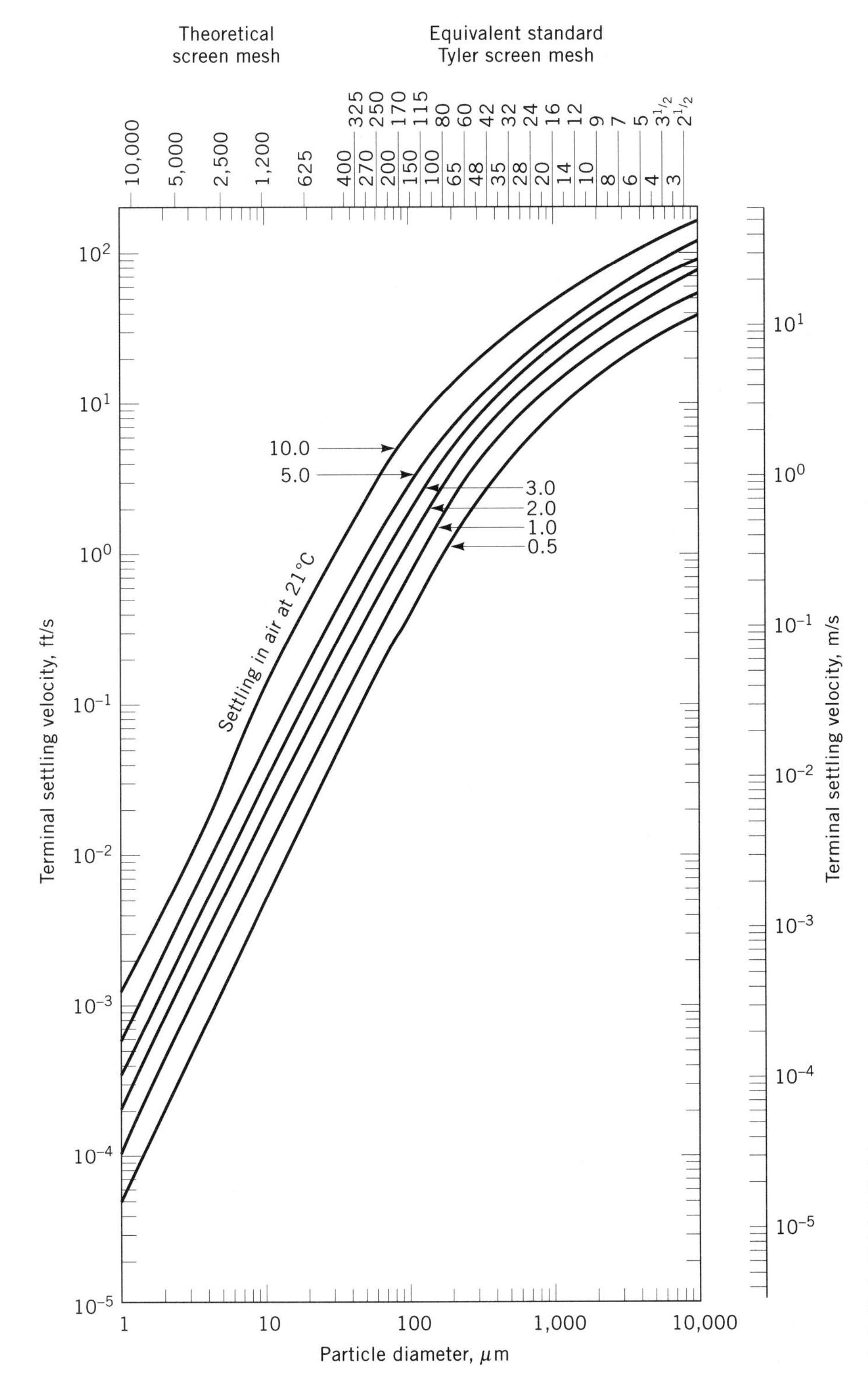

Figure 20. Terminal velocities in air of spherical particles of different densities settling at 21°C under the action of gravity. Numbers on curves represent true (not bulk or apparent) specific gravity of particles relative to water at 4°C. Stokes-Cunningham correction factor is included for settling of fine particles. The air viscosity is 0.0181 mPa·s (= cP) and density is 1.2 g/L.

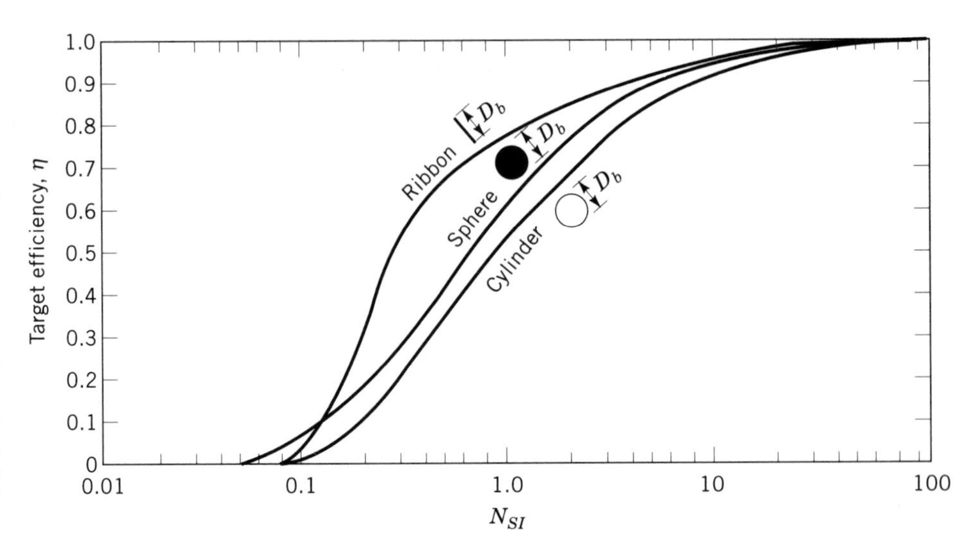

Figure 21. Target efficiency of spheres, cylinders, and ribbons. The curves apply for conditions where Stokes' law holds for the motion of the particle (see also N_{SI} in Table 8). Langmuir and Blodgett have presented similar relationships for cases where Stokes' law is not valid (200,201). Intercepts for ribbon or cylinder are $\frac{1}{16}$; for sphere, $\frac{1}{24}$. Note: If using English engineering units, calculate N_{SI} as $N_{SI} = U_t U_o / D_b g_L$ rather than by eq. 6 of Table 8 which is for SI units.

Centrifugal force acting on the particles in the gas stream causes them to migrate to the cylindrical containing wall where they are collected by inertial impaction. Because the centrifugal force developed can be many times that of gravity, small particles can be collected in a cyclone, especially in a cyclone of small diameter.

Four common types of cyclone design are illustrated in Figure 24. In the conventional cyclone (Fig. 24a), the gas enters tangentially and spins in a vortex as it proceeds down the cylinder. A conical section causes the vortex diameter to decrease until the gas reverses on itself and spins up the center to the outlet pipe or "vortex finder." The cone causes flow reversal to occur sooner and makes the cyclone more compact than if a cylinder of constant cross section were used. The dust particles collected on the wall flow down in the gas boundary layer to the cone apex and are discharged through an air lock or into a dust hopper serving a number of parallel cyclones. Collection efficiency for a given size particle decreases with increasing cyclone diameter. Conventional cyclones are usually 600–915 mm in diameter, although it is possible to build them in other sizes. The cyclones shown in Fig. 24**c** have similar gas flow but generally range from 25 to 305 mm in diameter. Because of the small diameter, these have higher collection effeciencies but lower gas capacity. Large numbers of small cyclones are mounted in a housing with a common tube sheet and dust hopper. Spin vanes in the annular gas inlets produce the vortex flow.

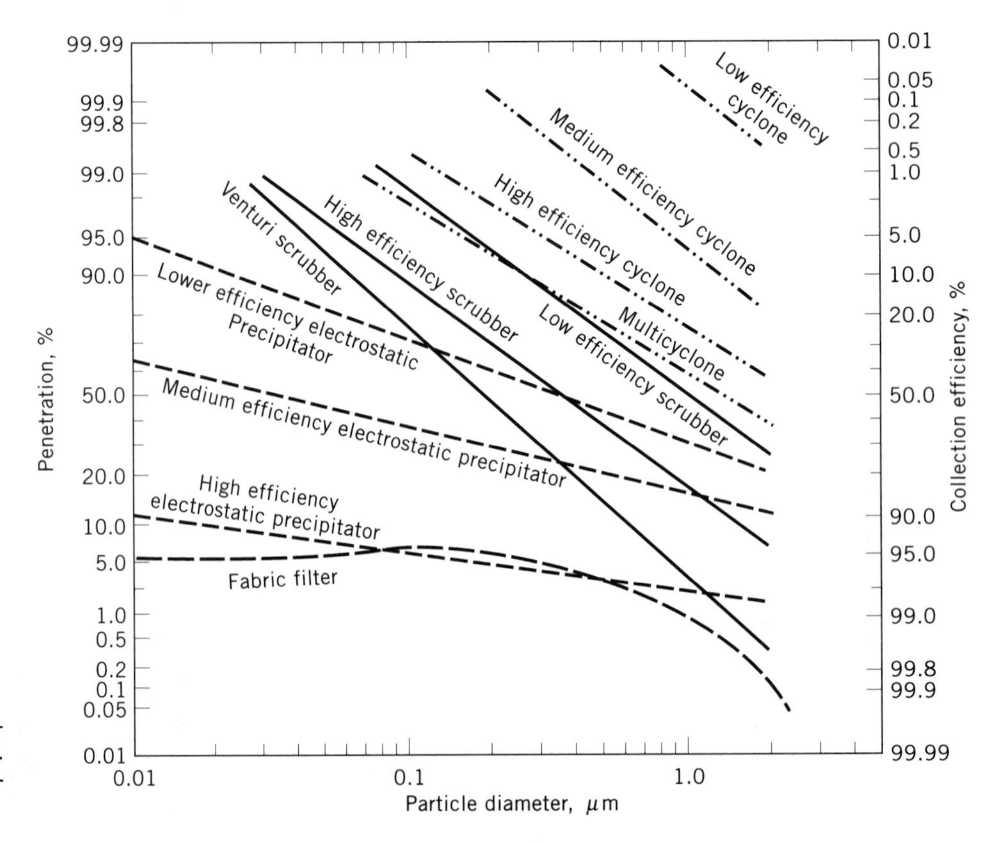

Figure 22. Penetration and fractional efficiency for fine particles. From ref. 203, courtesy of McGraw-Hill, Inc.

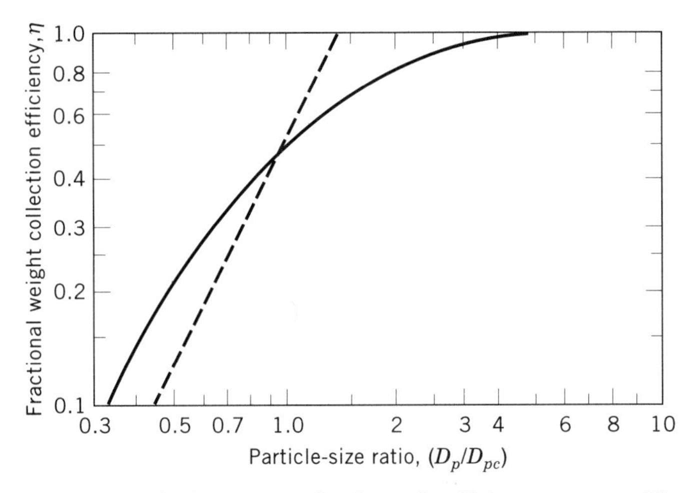

Figure 23. Cyclone generalized grade-efficiency curves. The solid line is for the Lapple cyclone dimension ratios given in Figure 25. The dotted line is theoretical efficiency based on equations of Rosin and co-workers (208). From ref. 209, courtesy of McGraw-Hill, Inc.

In other cyclone types, the dust is not completely removed from the gas stream; instead it is concentrated into about 10% of the total flow. Removal of the dust in aerosol form increases collection efficiency by reducing the dust entrainment losses that always occur at the cone apex in an ordinary cyclone. This purge dust-aerosol flow can increase overall dust collection efficiency by as much as 20–28% of that otherwise lost. Obviously, another separation device, bag filter, or cyclone, is required to complete the separation task on the purge flow. Such a cyclone dust-concentration arrangement may be especially attractive if the increased cyclone efficiency gives adequate particulate removal from a large-volume gas stream. The smaller purge stream can then be handled in a more efficient dust separation device. In the type shown in Figure 24b the gas reverses direction internally as in conventional cyclones; in that shown in Figure 24d straight-through flow, is convenient for connecting large gas volume breaching without changes in gas direction.

Cyclone Efficiency. Most cyclone manufacturers provide grade-efficiency curves to predict overall collection efficiency of a dust stream in a particular cyclone. Many investigators have attempted to develop a generalized grade-efficiency curve for cyclones (210). One problem is that a cyclone's efficiency is affected by its geometric design. Equation 29 was proposed to calculate the smallest particle size collectable in a cyclone with 100% efficiency (208).

$$D_{p\,\min} = \sqrt{\frac{9u_g W_i}{\pi N_e V(\rho_p - \rho_g)}} \tag{29}$$

For smaller particles, the theory indicates that efficiency decreases according to the dotted line in Figure 23. Experimental data (185) (solid line in Fig. 23) for a cyclone of Figure 25 dimensions show that equation 29 tends to overstate collection efficiency for moderately coarse particles and understate efficiency for the finer fraction. The concept of particle cut-size, defined as the size of particle collected with 50% mass efficiency, determined by equation 30 has been proposed (185).

$$D_{pc} = \sqrt{\frac{9u_g W_i}{2\pi N_e V_c(\rho_p - \rho_g)}} \tag{30}$$

This equation is for Figure 25 cyclone dimension ratios. The term N_e, the effective number of spirals the gas makes in the cyclone, was found to be approximately 5 for Lapple's system (185). The solid line grade-efficiency curve of Figure 23 is also used with Lapple's cyclone, which is a somewhat taller, less compact cyclone than many commercial designs.

A more compact cyclone design dimension ratio has been developed for the American Petroleum Institute (API) (213). A third set of cyclone dimension ratios has also been given (214). It is known that cyclone efficiency increases as dust loading increases, but most efficiency

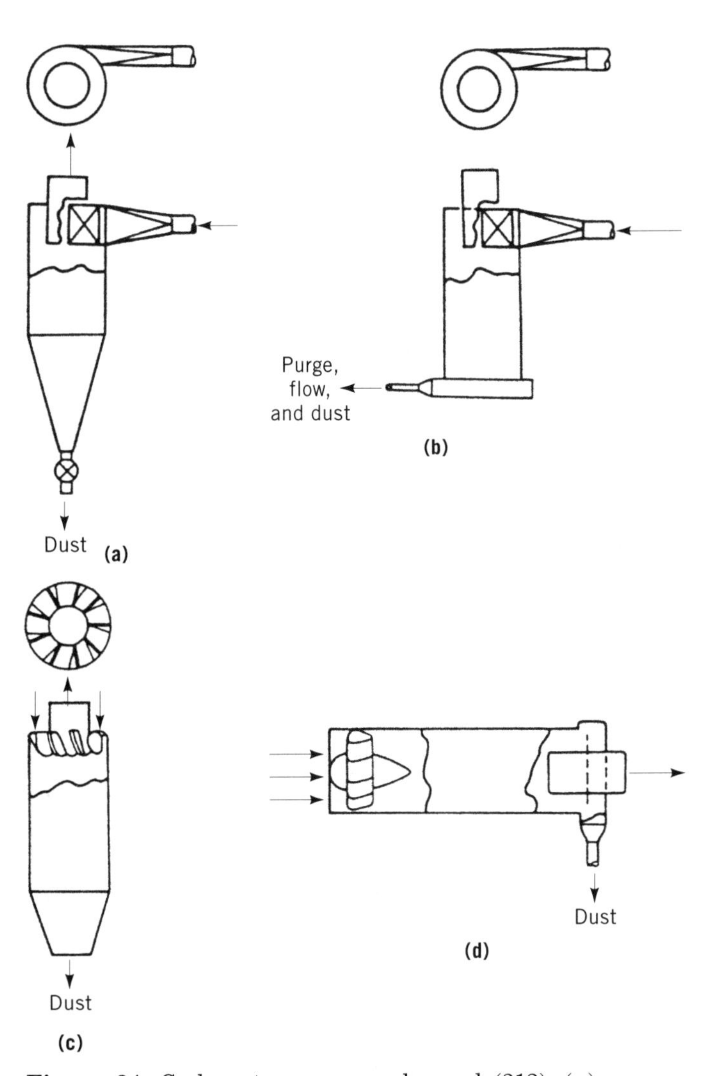

Figure 24. Cyclone types commonly used (212): (**a**) conventional, large diameter, tangential inlet, axial discharge; (**b**) smaller tube, tangential inlet, peripheral concentrated aerosol discharge; (**c**) small tube axial inlet and discharge; (**d**) smaller tube axial inlet, peripheral concentrated aerosol discharge. Courtesy of American Industrial Hygiene Association.

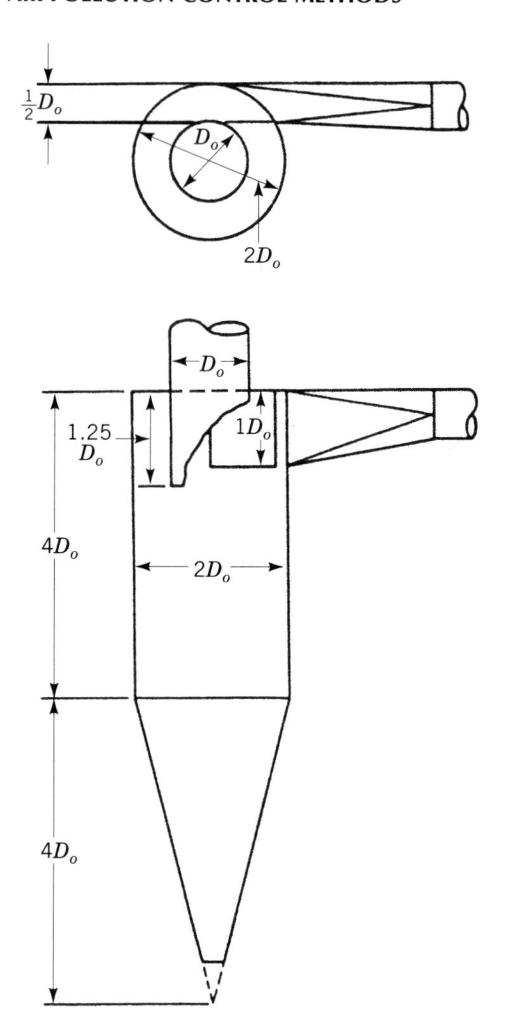

Figure 25. Dimension ratios of Lapple (185) cyclone design. From ref. 211, courtesy of Academic Press.

prediction methods other than that of the API neglect this fact. The generalized grade-efficiency curve of Figure 23 is used for a Lapple cyclone to predict collection efficiency for particle sizes other than the cut-size. Then an integrated overall collection efficiency is calculated by an interval integration. Needed are a narrow-range particle size analysis of the airborne dust as shown in Table 9 and a grade-efficiency curve for the cyclone for the dust density to be utilized. A narrow-range particle size distribution may be obtained by plotting the actual analysis on log-probability graph paper and interpolating. Each narrow particle interval is represented by its midpoint size and the corresponding collection efficiency shown in Table 9, columns 3 and 4, respectively. The product of columns 2 and 4 is then entered in column 5 as the fraction collected. The summation of column 5 gives overall collection efficiency for the cyclone.

Cyclone Design Parameters. As indicated earlier, a cyclone's geometric design affects efficiency. The outlet pipe should extend at least differentially below the bottom of the gas inlet to prevent inlet dust from short-circuiting directly to the outlet pipe. The gas inlet should be tangential to the top plate of the cyclone to eliminate eddy flows and turbulence as well as reduce top-plate erosion. Efficiency is increased with a narrower gas inlet, because the dust has a shorter distance to travel to reach the cyclone wall. However, narrower inlets mean longer, less compact cyclones so that 2:1 rectangular inlets are common. Grade efficiency is predicated on uniform distribution of gas and dust across the gas inlet. To achieve uniform distribution, duct transitions on the inlet should be gradual (no more than a 15° included angle). Elbows on cyclone inlets should not be used unless they concentrate the dust toward the outside or top of the cyclone.

Traditionally, cyclone dimensions are multiples of outlet pipe diameter D_o. Typical barrel diameters are $2D_o$ but efficiency increases at constant D_o up to a $3D_o$ barrel diameter. Efficiency also improves as barrel and cone length are increased at constant D_o up to the natural

Table 9. Technique for Calculating Cyclone Overall Efficiency for Dust Particles[a]

(1) Size Range, μm	(2) Percent by Weight in Size Range	(3) Median Diameter in Range, μm	(4) Efficiency of Median Diameter[b], Percent	(5) Fraction Collected, Percent
104–150	3	127	100.0	3.0
75–104	7	89.5	98.8	6.9
60–75	10	67.5	97.3	9.7
40–60	15	50	96.3	14.5
30–40	10	35	95.5	9.6
20–30	10	25	95.0	9.5
15–20	7	17.5	93.0	6.5
10–15	8	12.5	90.0	7.2
7.5–10	4	8.75	85.3	3.4
5–7.5	6	6.25	78.5	4.7
2.5–5	8	3.75	65.0	5.2
0–2.5	12	1.25	33.0	4.0
Total collection				*84.2*

[a] Columns 1–5 are explained in text.
[b] From a grade-efficiency curve.

length of the vortex. At constant inlet velocity, efficiency increases as outlet diameter (and all ratioed dimensions in a family of cyclones) is decreased. Improved efficiency is attained at the expense of loss in cyclone gas handling capacity. Although cut-size equations indicate that cyclone efficiency can increase without limit (subject only to available pressure drop) as inlet gas velocity is increased, this is not the case. Cyclone collection efficiency goes through a maximum as inlet velocity is increased causing separated dust to be reentrained from the cyclone wall (215). Higher efficiencies can be reached by reducing the gas outlet pipe area in relation to the gas inlet area (215).

The inside wall of a cyclone should be smooth. Bumps and projections create internal turbulence and dust layer reentrainment. Dust discharge from the cone apex is also important and a smaller apex outlet will get the dust discharge farther from the turbulent area of flow reversal. Theoretically, the apex opening should be less than 25% of the gas outlet pipe diameter. However, the outlet should not be so small that it bridges easily. Outlets of 75–150 mm are generally desireable. A good gas seal is needed for dust discharge, especially if the cyclone is under vacuum. Atmospheric air sucked in through the apex reentrains collected dust and reduces cyclone efficiency significantly. Dust hoppers are often used and a good head of dust in the hopper helps to seal against leakage through rotary valves and monitorized flap gates. It is important to keep the dust level in hoppers pulled down, however, because dust backing up into the cone of the cyclone reduces collection efficiency to zero.

Practically all cyclone performance data have been related to a preset cyclone set of geometric ratios. One model for cyclone grade-efficiency curves has been tested against reported commercial cyclone efficiencies (210). A good fit was obtained.

Cyclone Pressure Drop. Typical cyclone pressure drops range from 250 to 2000 Pa. Most data are reported for clean air flowing through the cyclone, and these data are conservative for design purposes. Many investigators have unsuccessfully attempted to relate pressure drops to inlet and outlet dimension ratios. Manufacturers' calibration curves or experimental measurements on cyclones of similar dimension should be used where possible. If a reliable experimental measurement is available, however, the pressure drop at other conditions can be estimated by first evaluating the constant K_c in equation 31.

$$\Delta P = K_c Q^2 P \rho_g / T \qquad (31)$$

Some empirical equations to predict cyclone pressure drop have been proposed (216,217). One (217) reliably predicts pressure drop under clean air flow for a cyclone having the API model dimensions. Somewhat surprisingly, pressure drop decreases with increasing dust loading. One reasonable explanation for this phenomenon is that dust particles approaching the cyclone wall break up the boundary layer film (much like spoiler knobs on an airplane wing) and reduce drag forces.

Cyclone Problems. Problems may be encountered in cyclone application because of fouling and caking or from erosion or when using multiple cyclones. Multiple cyclones are designed so that each cyclone handles a prorated share of gas and dust and the overall efficiency of the system is the same as that calculated for an individual unit. This is the case, however, only when each cyclone receives identical dust fractions (size and loading) and gas flow. Because cyclone efficiency increases with flow and dust loading and is affected by particle-size distribution, the design of the inlet gas distribution system must accomplish the proper distribution. Otherwise, those cyclones with lower gas flow and dust concentration (and perhaps finer dust) will have much poorer efficiency. When multiple cyclones share a common dust hopper, it is important that all cyclones have essentially uniform pressures at the cone apex. Wall caking, unequal gas flow or dust distribution resulting from pressure drop decreases that occur with increases in dust loading, or partial plugging of cone or cyclone inlets can all cause unequal apex pressures. Unequal pressures will cause gas from higher pressure cyclones to flow into the dust hopper and back into the cyclones having lower apex pressure. This short-circuiting can result in heavy dust reentrainment and decreased efficiency.

Fouling of cyclone walls is usually caused by sticky or hydroscopic particulates, or by moisture or other vapor condensation. To prevent particle sticking, a highly polished finish, a graniteware glass coating, or fluorocarbon plastic lining may be used on the walls or alternatively, revolving wall scrapers. Cables or chains are sometimes suspended from the center of the vortex outlet. Vortex flow causes the cable or chain to rotate and thus scrape the wall, sometimes freeing the buildup. Condensation must be prevented by decreasing the dewpoint of the gas or by heating or insulating the cyclone wall. In extreme situations, the cyclones may be enclosed in a heated chamber.

Erosion can be a severe problem even in well-designed cyclone installations when handling a heavy loading of coarse and abrasive angular particles. One answer is lining the cyclone, using protective materials. The cyclone may be made of wear-resistant plate. Linings may be hard and thick sacrificial castings; wear-resistant applied welded coatings; or of rubber, ceramic shapes, reinforced castable, or brick. Alternatively, cyclones may be used in series, increasing the velocity as dust loading is decreased. In this method, first-stage efficiency will be low, but the bulk of the coarser and more abrasive particles will be collected, permitting higher velocities in second- and even third-stage cyclones. Inlet velocities and dust loadings at which erosion may be excessive are given in Table 10 (210). Dust particles smaller than 5–10 μm do not cause appreciable erosion.

Table 10. Dust Loadings and Cyclone Inlet Velocities above Which Erosion Is Excessive[a]

Dust Load, g/m^3	Inlet Velocity, m/s
0.7	35
7.0	20
7000	2

[a] Recomputed from Ref. 218.

Other problems affecting cyclone efficiency are usually caused by abuse or poor maintenance. Problems may arise from temperature warpage, rough interior surfaces, overlapping plates and rough welds, or misalignment of parts, such as an uncentered (or cocked) vortex outlet in the barrel.

Other Centrifugal Collectors. Cyclones and modified centrifugal collectors are often used to remove entrained liquids from a gas stream. Cyclones for this purpose have been described (218–220). The rotary stream dust separator (221,222), a newer dry centrifugal collector with improved collection efficiency on particles down to 1–2 μm, is considered more expensive and hence has been found less attractive than cyclones unless improved collection in the 2–10-μm particle range is a necessity. A number of inertial centrifugal force devices as well as some others termed dynamic collectors have been described in the literature (221).

Electrostatic Precipitation

An electrical precipitator can collect either solid or liquid particles efficiently. Using a special design, it can also collect solids and liquids in combination and adsorb gases. Particles entering a precipitator are charged in an electric field and then move to a surface of opposite polarity where deposition occurs. For particles ≤ 2 μm, electrical forces are stronger than any other collectional force. Thus precipitators have the highest energy use efficiency. Advantages are low pressure drops (often 250–500 Pa), low electric power consumption, and low operating costs (mostly capital-related charges). In addition, particulates are recovered in an agglomerated form, rendering them more easily collectible in case of reentrainment. Unfortunately, precipitators are also the most capital-intensive of all control devices, and mechanical, electrical, and process problems can cause poor on-stream time and reliability.

Precipitators are currently used for high collection efficiency on fine particles. The use of electric discharge to suppress smoke was suggested in 1828. The principle was rediscovered in 1850 and independently in 1886, and attempts were made to apply it commercially at the Dee Bank Lead Works in Great Britain. The installation was not considered a success, probably because of the crude electrostatic generators of the day. No further developments occurred until 1906 when Frederick Gardiner Cottrell at the University of California revived interest (U.S. Pat. 895,729) in 1908. The first practical demonstration of a Cottrell precipitator occurred in a contact sulfuric acid plant at the Du Pont Hercules Works, Pinole, California, in about 1907. A second installation was made at Vallejo Junction, California, for the Selby Smelting and Lead Co.

Precipitators can be classed as single or two stage. In single-stage units, used for industrial gases, the particles are charged and collected in the same electrical field. Negative discharge (gas-ionizing) polarity is practically always used in single-stage precipitators, because higher voltages can be achieved without sparkover. However, negative polarity also produces O_3 from O_2-containing gases. In two-stage precipitation, the particles are charged in an ionizing section and collected or precipitated in a separate section. Two-stage units are used for air purification, air-conditioning (qv), or ventilation. They are operated at lower voltage so there is less electrical hazard, less sparking, and fewer fires. These units have also received some consideration for the collection of condensed hydrocarbon mists. A positive polarity is always used for room air-conditioning to avoid ozone formation.

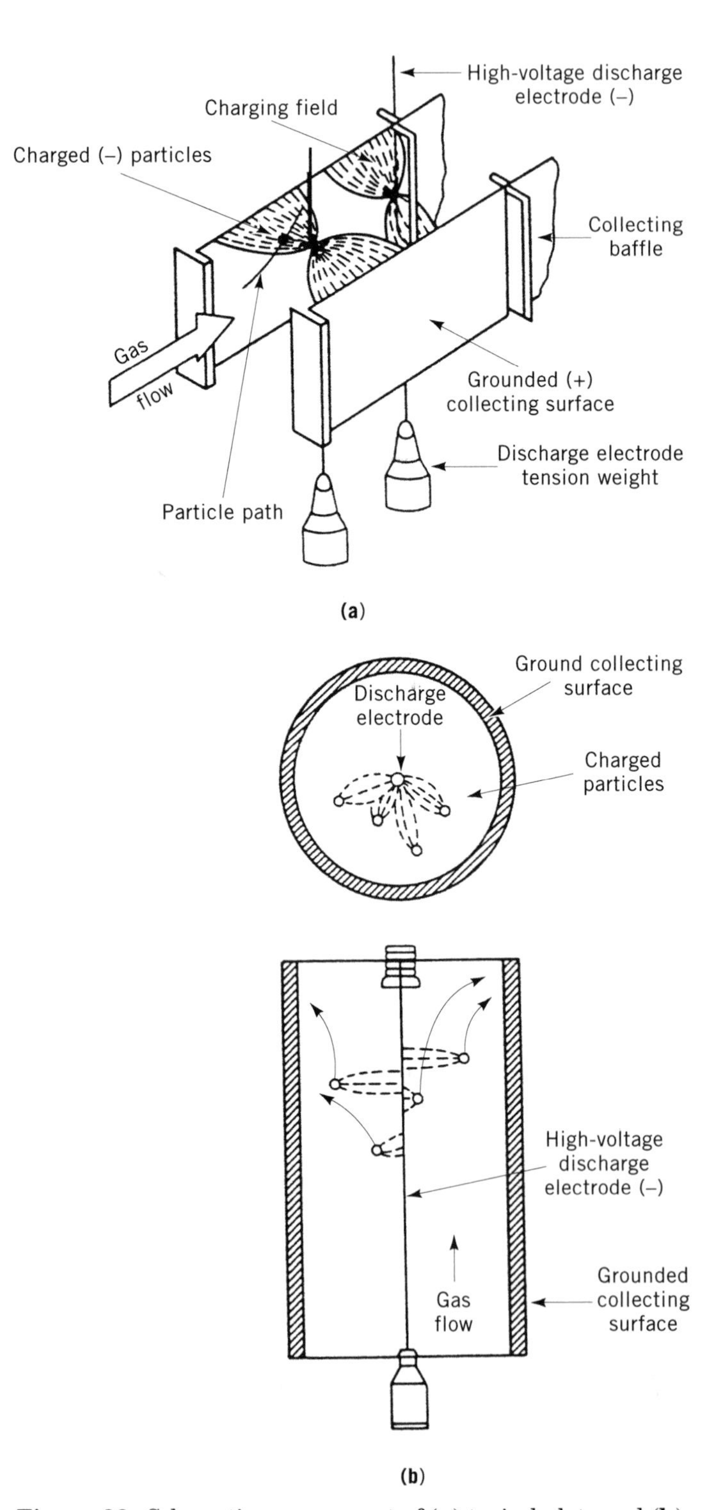

Figure 26. Schematic arrangement of (**a**) typical plate and (**b**) tube precipitators (233).

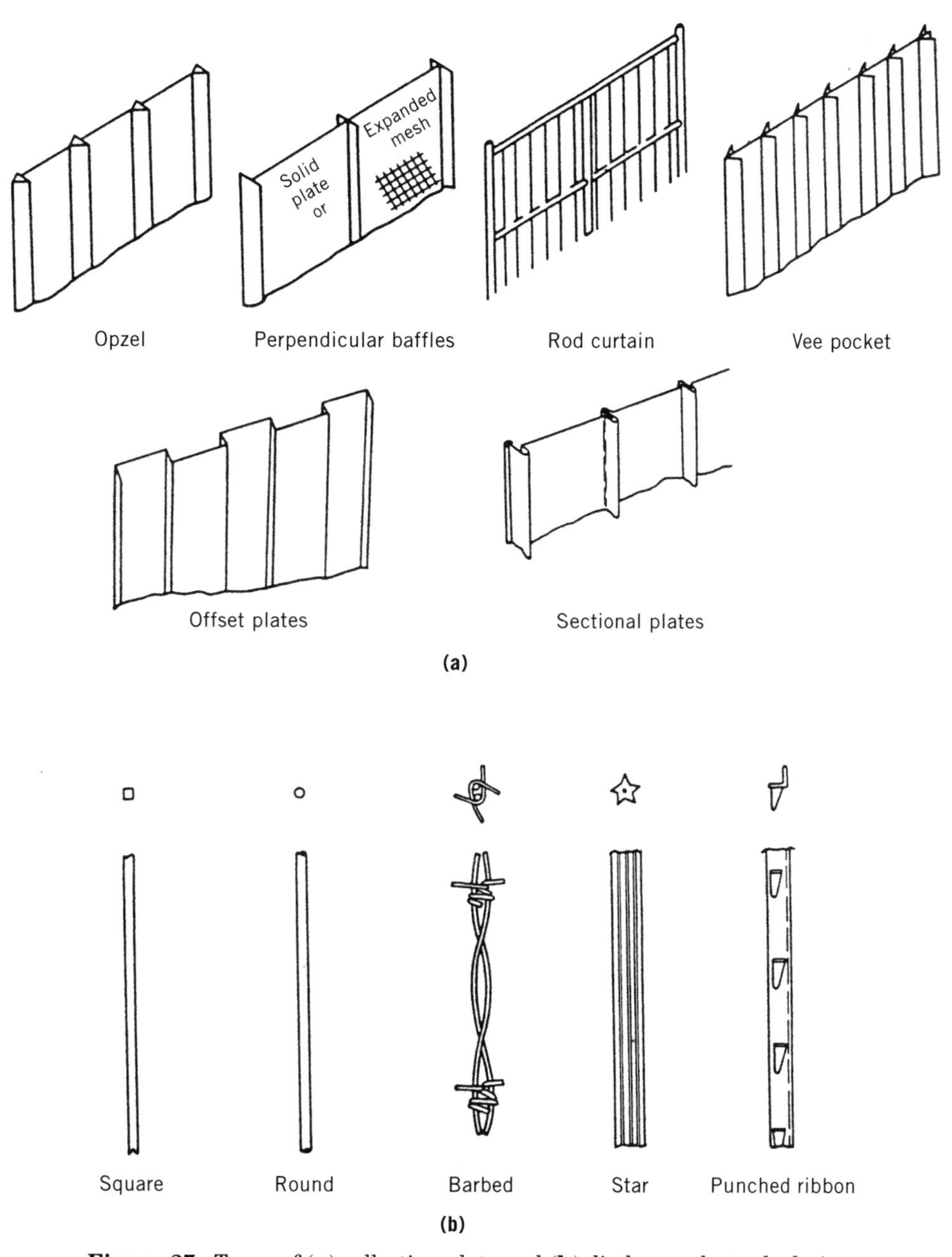

Figure 27. Types of (**a**) collecting plate and (**b**) discharge electrode designs.

Single-Stage Precipitators. Single-stage precipitators are used for control of fine particulates such as those in mists and smoke, in the 2.5–0.2-μm range. Recently, technology in the development of larger and larger units to treat flue gas from 800–1200-MW coal-fired electric power boilers has been markedly improved. This is a specialized market and these precipitators will be referred to as power boiler precipitators.

A complete precipitator consists of (*1*) discharge electrodes, (*2*) collecting surfaces (plates or tubes), (*3*) a suspension and tensioning system for discharge electrodes, (*4*) a rapping system to remove dust from tubes, (*5*) dust hoppers and dust-removal system, (*6*) gas-distribution system and precipitator housing, and (*7*) power supply and control system. Single-stage precipitators are subclassified as plate or tube type. Typical arrangements are shown in Figure 26. A plate precipitator may have tall (10–15 m), parallel, flat plates with 228–400 mm horizontal spacing and discharge electrodes (wires) that are suspended midway between the plates. The gas to be treated flows horizontally between the parallel plates. Plate length in the direction of gas flow depends on treatment time required to achieve the desired collection efficiency but is often in the range of 10–20 m in length. For power boiler applications, plate height may be increased to 15–25 m. Plate spacing would be from 300 to 400 mm. Recent designs of European origin have gone to plate spacing as much as 50–100% greater. This reduces the steel requirements and cost, but requires operating at appreciably higher voltages to increase the field intensity. Plate precipitators are used for collecting fly ash, for high gas-flow applications and for particulates that are comparatively coarser than those caught in the tube-type equipment. Plate-type precipitators are lower in cost because both sides of the plate serve as precipitating surfaces. Tube precipitators are frequently used for liquid mists and sometimes for submicrometer metallurgical fumes. Tubes vary in diameter from 150 to 400 mm; a popular size is 273 mm OD by 3.0–6.0 m long. One discharge wire is centered in each tube. Dirty gas generally enters the hopper at the bottom and travels upward through the tubes. A tube sheet may be provided top and bottom; tubes may be rolled or welded into a flat top tube sheet for a mist precipitator. For solids, however, each tube must have a round to square transition at the top end. The tubes are then welded together to form an "egg-crate" sheet. Dust, if deposited on a flat tube sheet, can buildup with a nearly 90° angle of repose, due to electrostatic charges, and short out the high tension discharge electrode support frame.

Figure 27 shows a variety of plate and discharge electrode designs. The most common collecting plate design is flat with vertical, perpendicular baffles at frequent intervals. The plate may be continuous or sectional. The Opzel design is a modification of perpendicular baffles devised to reduce gas-flow resistance. Discharge electrodes may be round wires or small (2.5–4 mm) diameter rods. The smaller the diameter, the higher the intensity of the surrounding electric field. The wire must, however, be sufficiently rugged to withstand both tensioning and thermal stresses. In the event of heavy suspended loads or long spans, wire size may become appreciable. In these cases, a 6- to 9-mm square may be a better design choice because the corners produce a higher intensity field locally. Square bars are often twisted to give one 90° turn in 25–35 mm of length, producing a high intensity field that rotates with the length of the discharge electrode. A 5-pointed star–shaped discharge electrode yields an even higher field intensity. Discharge electrodes with barbs and punched ribbons may also be used to produce higher intensity electric fields.

Discharge electrodes must be tensioned to hold the wire taut and maintain spacing. Tensioning may be accomplished by attaching a weight (5–10 kg), held in a weight-spacing frame to reduce sway, to the bottom of each wire. In some European designs, now offered in the United States, the wires are stretched between lightweight high-tension pipe frames. The wires may be heavy barbed ribbons that are unlikely to fail from arcing, an important aspect because no provision is made for wire replacement in these rigid frames. The wire supporting frames are hung from high voltage insulators in suitable enclosures on top of the precipitator. Dust, fumes, and mist must be prevented from entering these insulator compartments and coating the insulators with a conductive film. The insulator compartments may be purged with clean air and heated to prevent condensation. In a mist precipitator, liquid drains from the collecting surfaces, but a dry dust precipitator must be rapped at intervals using a weight, hammers, or vibrators. The dust, broken loose from the surface, slides down the plate into the dust hoppers. Dust is removed from the hoppers, usually batch-wise, with suitable conveyors. A precipitator's power supply is usually rectified ac.

Table 11. Sparking Potential for Small Wire Concentric in a Round Tube

Tube Diameter, mm	Peak Voltage, V[a]	Root Mean Square Voltage, V[a]
102	59,000	45,000
152	76,000	58,000
229	90,000	69,000
305	100,000	77,000

[a] Based on gases at atmospheric pressure, 38°C, containing water vapor, air, CO_2, and mist, using negative polarity electrical discharge. Recalculated from data reported in Ref. 227.

Operating Principles. No collection will occur in a precipitator, and no current will flow in the precipitator circuit, until gas ionization starts around the discharge electrode. This ionization process is known as corona formation or discharge. The starting voltage for corona in air in a tubular precipitator is given by the following dimensional equation:

$$V_s = (15.5 \times 10^5) b_r k_p D_d \left(1 + \frac{0.0436}{k_p D_d}\right) \ln\left(\frac{D_d}{D_t}\right) \tag{32}$$

If the precipitator gas is not of air composition, it is necessary to multiply equation 32 by the ratio of the precipitator gas electrical breakdown constant to that for air, both taken at 25°C and 1.013×10^5 Pa.

When corona occurs, current starts to flow in the precipitator circuit and dust particles are collected. As the voltage is increased slowly, more and more current flows and the electric field strength increases. Finally, with further potential increase, a spark jumps the gap between the discharge wire and the collecting surface. If this sparkover is permitted to occur excessively, destruction of the precipitator's internal parts can result. Precipitator efficiency increases with increase in potential and current flow; the maximum efficiency is achieved at a potential just short of heavy sparking.

Table 11 shows sparking potential in a clean tube precipitator with negative polarity. Sparking potential drops almost directly with decreasing gas density, but increases with moisture content. At elevated temperatures or reduced pressure, the corona starting voltage also drops, but sparkover voltage drops faster. Actual operating voltages are still lower because the dust layer on the collecting plate reduces the voltage at which sparkover occurs. For ease of operation and greatest efficiency, the largest possible spread between corona starting voltage and sparkover is desirable. With positive polarity, sparkover occurs at considerably lower potential and it may become nearly impossible to operate a positive polarity precipitator at elevated temperatures.

Dust particles entering the electric field become charged. Both negatively and positively charged ions exist in the small core of ionized gas surrounding the discharge electrode. In a negative polarity precipitator, the positively charged ions are quickly attracted to the discharge electrode and neutralized. Hence only electrons exist outside the corona area and travel through the gas space to the grounded collecting plate. Two mechanisms of particle

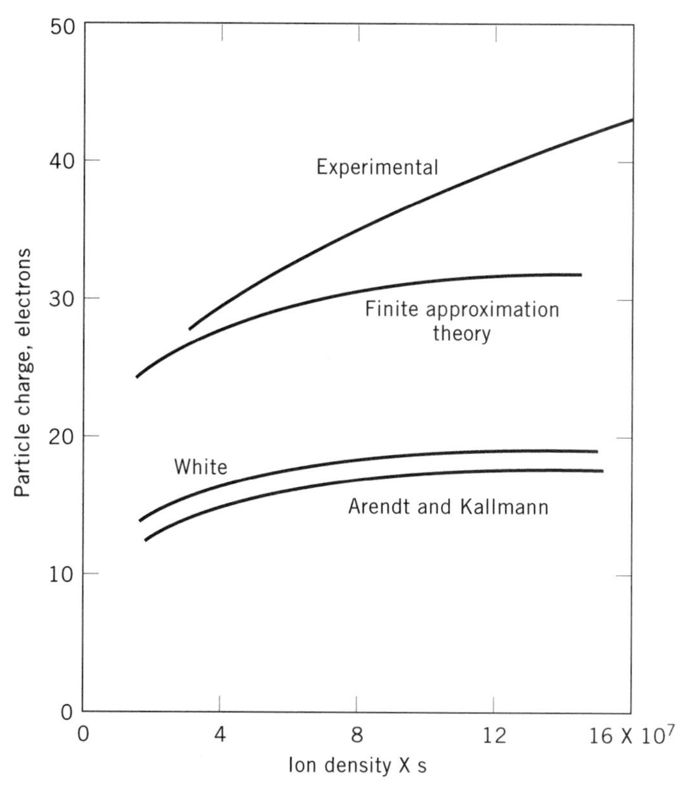

Figure 28. Comparison of actual and predicted charging rates for 0.3-μm particles in a corona field of 2.65 kV/cm (192). The finite approximation theory (224) which gives the closest approach to experimental data takes into account both field charging and diffusion charging mechanisms. The curve labeled White (192) predicts charging rate based only on field charging and that marked Arendt and Kallmann (226) shows charging rate based only on diffusion. From ref. 226, courtesy of Addison-Wesley Publishing Co.

charging exist: charging by ion bombardment (often called field charging) and charging by ion diffusion. The literature (192,226) contains a number of empirical equations to predict particle charging rate. In field charging, electrons colliding transfer charges to the particle. When charged sufficiently, the particle will repel further electron collisions, a phenomenon termed charge saturation. Field charging is rapid and most effective on larger particles. The larger the particle, the higher the charge density, and thus the easier precipitation. Coarse particles (5–30 μm) can be collected almost completely in a precipitator. The lower size range for effective field charging is on the order of 0.5–1.5 μm, depending on particle dielectric constant.

In diffusion charging, particles are too small and mobile for rapid charging by ion bombardment. They become charged by collision caused by motion of the gas molecules. Diffusion charging becomes efficient on particles smaller than 0.2 μm and has been demonstrated to be effective on particles down to 0.05 μm. These fine particles are charged rapidly and have higher charge density at saturation.

Unlike field charging, the rate of charging by diffusion is independent of electric field strength. For particles between 0.2 and 1.5–2.0 μm neither charging mechanism is highly efficient but both operate. Particles in this size range are the most difficult to collect efficiently. Figure 28 compares experimental and calculated particle charge levels for 0.3-μm particles. The field has been reviewed (228), and a particle-charging equation that has been used in computer-modeled precipitator design has been proposed (229,230).

It is possible to set up a force balance on a given size particle in a precipitator and hence calculate its trajectory. Three forces tend to move a particle toward the collecting plate: the charge on the particle, the integrated average field strength through which the particle moves, and the electric wind. This last results from the steady flow of ions from the discharge electrodes to the collecting plates, a motion resembling that of thermal convective currents. Particle movement is resisted by drag on the particle. In solving a force balance, it is convenient to define the particle migration velocity as the average velocity with which the particle moves toward the collecting surface. This velocity is a function of particle size, field strength, and such particle material properties as electrical conductivity and dielectric constant. The factors affecting migration velocity for conductive particles larger than 1 μm that are in the Stokes' law region can be expressed as (192):

$$U_e = \frac{E_o E_p D_p}{4\pi K_u \mu_g} \tag{33}$$

For conductive particles smaller than 1 μm, equation 33 should be multiplied by the Cunningham correction factor, $(1 + A\lambda/D_p)$. For air at room temperature, the molecular mean free path λ is 0.1 μm, and A is 1.72. If the particles to be collected are not electrically conductive, equation 33 should be multiplied by $[1 + 2(\delta - 1/\delta + 2)]$ to correct for the dielectric constant of the material. Equation 33 can be used to predict the relative effect of changes in particle size distribution, gas viscosity, and operating voltage on migration velocity and precipitator efficiency. An interval-integration procedure may be used to calculate average migration velocity for a given particle-size distribution, although more often an average migration velocity is measured experimentally in an existing precipitator for a similar dust. Average migration values are sometimes referred to as drift velocities to distinguish these values from migration velocities, which should be related to specific particle sizes. Table 12 lists drift velocities encountered in typical commercial precipitators.

This table represents past experience with precipitator applications based on past efficiency performance. With the demand for precipitators today having higher efficiencies, past design drift velocities must be adjusted for the required new efficiency. To achieve higher efficiency, a greater portion of the particles, difficult to charge, must be collected; consequently, a lower drift velocity must be used for design.

The theoretical development of migration velocity involves only field strength, particle charge, and electric wind force; therefore, migration velocity should be independent of forward gas velocity. However, mild turbulent flow can move particles close to the collecting surface, en-

Table 12. Precipitator Applications of Drift Velocities, U_e, cm/s

Precipitator Application	Average Value[a]	Typical Range
Pulverized coal fly ash, power boilers	13.1	3.9–20.4
Pulp and paper manufacturing particulate	7.6	6.4–9.4
Sulfuric acid mist	7.3	6.1–8.5
Cement kiln dust (wet process)	10.7	9.1–12.2
Nonferrous smelter	1.8	
Steel open hearth furnace	4.9	
Iron blast furnace	11.0	6.1–14.0
Foundry cupola	3.0	

[a] Average values are based on typical efficiencies of precipitators purchased. Drift velocities will drop if higher efficiencies are required. Recomputed from data of Ref. 226.

hancing collection, and giving improved measured migration velocity (231,232). A precipitator developed in Japan to give high migration velocities incorporates a pulse precharger and zigzag collecting electrodes (233). The zigzag collectors probably enhance particle migration through turbulent diffusion.

Collection Efficiency. The classical Deutsch and Anderson equation for predicting particle collection efficiency is

$$\eta = 1 - e^{-(U_e A_e / q_e)} \tag{34}$$

A_e / q_e can be replaced by K_e, which is defined in terms of precipitator geometric parameters and gas passage velocity. For tube precipitators, $K_e = 4L_e / D_t V_e$; for plate precipitators, $K_e = L_e B_e V_e$. Assumptions involved in applying the Deutsch and Anderson equation to an entire precipitator are that all parallel gas passages have the same gas velocity (this is not necessary if the variation in gas velocity is known because the equation can then be applied passage by passage) and that no reentrainment occurs once the particle is collected. Uniform gas distribution has been improved in some designs with inlet baffling and gas-flow modeling tests (234), and particle reentrainment can be reduced by better dust rapping. Examining K_e shows that efficiency can be improved by increasing treatment time (longer flow path in the electric field or lower gas velocity) or collection plate area, decreasing the distance from the discharge electrode to the collecting surface, or changing gas or electric field conditions to give higher migration velocity.

Precipitator Operating Problems. Dry dust precipitator operating problems may result from dust reentrainment, dust resistivity, gas distribution, or electrical arcing or lack of voltage control, and some of these problems have been discussed in detail (235). Dust reentrainment can arise from a localized high gas velocity, rapping problems, or dust resistivity. The collected dust becomes agglomerated and is easier to redeposit if entrained. There is an optimum time interval, force intensity, and direction for rapping to minimize reentrainment. A rapping force normal to the plate is most effective. The intensity of a single rap should be just enough to snap the dust cake loose from the plate and allow it to slide en masse down into the dust hopper. Too intense a blow may shatter the cake and project it out into the bulk gas stream as a cloud of small particles; an inadequate blow will require repeated raps to break the cake loose. Electrostatic forces hold the collected dust to the collection surface and the longer the dust layer is in place, the more tightly it is held. When the cake stays in place too long, greater forces are required to dislodge it and chances of reentrainment are greater. One blow of optimum intensity every 1–2 min, continuous intermittent rapping, is better than a burst of high frequency vibration as discussed in rapping parameter experiments (236,237). However, periods of 30–90 min between successive raps may be better (238). Baffles on the collecting plate tend to keep the bulk gas velocity away from the dust layer, providing a quiescent zone through which dust can slide downward during rapping.

Reentrainment can occur at either too high or too low a dust resistivity. With too low a dust resistivity (usually less than 1000 ohm·cm), the dust loses its charge to the collecting plate, randomly tumbles off, and is reentrained. Unburned carbon in fly ash is an example, A precipitator is completely unsatisfactory as a total collector for conductive and fine dusts such as carbon black. However, in such cases precipitators are used as electrostatic agglomerators and final collection occurs in a cyclone. High dust resistivity (about $2–5 \times 10^{10}$ ohm·cm) causes a phenomenon known as back corona, resulting in poor operation and reentrainment. High resistivity causes a high voltage gradient across the dust layer. The resulting weaker field potential between the discharge electrode and the dust layer surface gives lower drift velocity and collection efficiency. Moreover, the potential drop across the dust layer may be high enough to produce corona discharge, causing the particles to disperse from the collecting surface. These problems have been encountered in some coal-fired power plant precipitators when high sulfur coal is replaced with low sulfur coal. Dust resistivity varies with precipitating temperature and gas moisture content. Resistivity goes through a maximum with temperature increase, but higher moisture levels reduce resistivity at any temperature. Typical remedies are (*1*) cool the gas (a "cold side" precipitator) (239), (*2*) treat the gas at a quite high temperature ("hot side" precipitator (240) perhaps requiring alloy construction, and (*3*) condition the gas by adding steam or moisture. If the dust does not absorb moisture, chemical conditioner may be needed to make the dust hygroscopic. SO_3 is often added for alkaline dusts (241) and NH_3 for acidic ones (242). A correlation for predicting fly-ash resistivity from composition has been developed (243), and the entire dust resistivity problem has been discussed (244–246).

To achieve maximum efficiency, the precipitator should be operated at the highest voltage possible without excessive sparkover or arcing. Precipitator voltage must be a function of the conditions of the process such as gas temperature, moisture, dust resistivity, and chemical composition. A desirable option for modern precipitators is a voltage-optimizing control circuit, which usually contains current-limiting controls and spark-rate sensors. Voltage is automatically varied to maintain light sparking at 50–100 sparks per minute. Strength of sparking can be greatly affected by the pulsation frequency and wave form of the current. Full wave rectified 60–Hz a-c current permits a higher peak voltage than that obtained from a nonpulsing d-c current because the voltage is not at the peak value long enough to produce extensive gas-path electrical breakdown or a heavy arc. Higher frequency current permits the use of higher voltages resulting in higher efficiency without sparkover. The use of a 500-Hz pulse generator and special wave shape has been demonstrated as extremely beneficial (226). Unfortunately, special frequencies or wave shapes have seldom been used commercially because of the cost of power conversion equipment; however, pulse-energization equipment for precipitators (247) has become available in Denmark. Advantages of automatic voltage control systems have also been discussed (248). Precipitator on-stream reliability can be enhanced using electrical sectionalization. Fewer discharge electrodes per rectifier result in the loss of a smaller portion of the precipitator if a wire breaks and shorts out. Increasing the number of power rectifiers, however, also increases purchase cost and increases in reliability must be balanced against increases in capital cost.

Power Supply. The preferred power supply for a modern precipitator is a silicon semiconductor power rectifier submerged in oil along with a step-up transformer. Typical output voltages are 70–105 kV peak (45–67 root mean square average dc) negative polarity, producing output of either two half waves or one full wave. Lower voltages are used in some cases of smaller discharge electrode to collecting surface distances. Individual set capacities are typically 15–100 kVA, 250–1500 mA dc. Input power is usually 460 V ±5%, single-phase 60 Hz. Power supply capacity must be carefully matched to the needs of the precipitator for maximal operation. The power supply is generally chosen to operate at 70% of capacity or higher. A greatly oversize power supply can result in poor electrical control, high peak currents or excessive sparking, frequent wire burnout, and overall poor collection efficiency. Too small a power supply can result in its operation being controlled by current limitation and the inability to reach optimum voltage for maximum collection efficiency. Gas temperature and composition, dust loading and resistivity, wire alignment, adequacy of rapping system to keep thin dust layers, collecting surface per electrical section, and geometric design are all factors that can affect power package sizing and selection (249).

Precipitator Application and Cost. There were more than 4100 precipitator installations in the United States in 1970, more than 1330 of which were in electric power generating stations for fly-ash removal (250). About half this number was in each of the next three largest groups of applications: metallurgical, chemical, and fuel-gas detarring. The number of installations may well have doubled by 1992. Design approaches have been discussed (226) and the Air and Solid Waste Management Association (251) has developed an information checklist for precipitator specification and selection. General electric precipitators have been reviewed (192,226,245,246,252,254,255). A review of capital and operating costs for fly-ash precipitators at TVA steam power plants is available (253).

Wet Electric Precipitators. Tube precipitators have long been used for collection of acid mists and for removal of tar from coke oven gas. More recently, plate precipitators employing water sprays have become commercially available for the collection of dry, especially fine particulate, dusts (256). Dust resistivity problems are eliminated because the water-saturated condition renders all particles conductive. All problems with particle reentrainment are also eliminated, because the particles migrate to a flowing water film on the collecting surfaces. Conductive particles have higher migration velocities, and therefore, there also are no problems with high resistivity dust and back corona. Thus wet precipitators can be smaller for the same collection efficiency. Particle dielectric strength becomes a much more important variable, however, and materials having low dielectric constants, such as hydrocarbon mists, are much more difficult to collect than a water-wettable particle. Relationships for predicting particle charge and relative collection efficiency in a wet precipitator that take the effect of dielectric constant into account have been developed (257). A wet precipitator can also be used to absorb water-soluble gases; the pH of the spray liquid can be adjusted to enhance absorption. Comparative tests of wet precipitators with and without the electric field indicate that corona discharge also enhances absorption, but the reason has not been determined.

There are some disadvantages to wet precipitators. Water can enhance latent corrosion problems and require the use of expensive alloy construction instead of the carbon steel often used in a dry precipitator. Some wet precipitators using plastic components have been developed to lower costs in corrosive situations. In addition, the collection of pollutants in aqueous media may create water-treatment and waste-handling problems that can equal the cost and complexity of the precipitator installation itself. Spray rate and distribution can be critical in a wet precipitator and must be carefully applied so as not to produce shorting. Recirculated spray water may become supersaturated with low solubility compounds that plate out on surfaces or build up in critical parts of the precipitator. Recirculated suspended solids can erode or plug spray nozzles. Wet precipitators have been most useful in treating mixtures of gaseous pollutants and submicrometer particulates such as aluminum pot line and carbon anode baking fumes, fiberglass fume control, coke oven and metallurgical fumes, and phosphate fertilizer emissions.

Two-Stage Precipitators. In a two-stage precipitator, the particles are charged in an initial ionizing stage and then collected in a second section, consisting of closely spaced plates (Fig. 29). Basic modular units are often stacked in banks as needed for the design gas flow. The ionizing electric field is produced by fine tungsten wires, often 0.20 mm in diameter, spaced midway between large diameter (30 mm) grounded rods or flat plates at a 90-mm spacing. Corona discharge and occasional sparking to ground are produced by a 13-kV positive potential. The collecting stage usually consists of a number of parallel 20-gage plates spaced about 6–8 mm apart and separated by insulators. Alternate plates are grounded; the others are charged positively at 5–7.5 kV. Charged particles are attracted to the grounded plates while being repelled by the charged ones. The long, smooth plate surfaces prevent co-

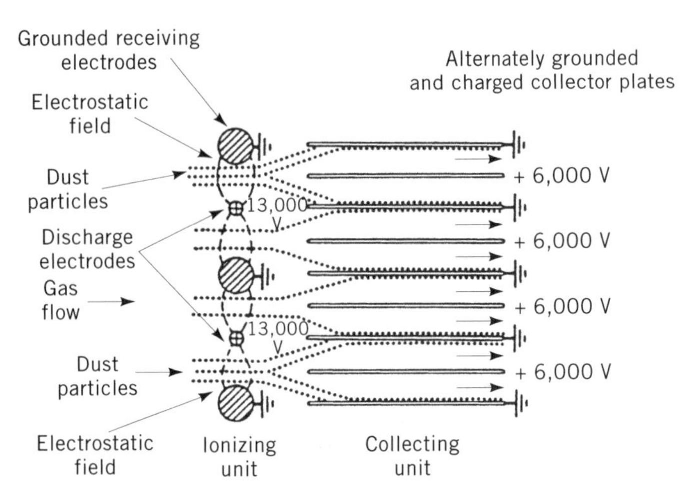

Figure 29. Operating principle of two-stage electrical precipitator. From ref. 258. Courtesy of McGraw-Hill, Inc.

rona or sparking. Thus there is no means in the collector by which entrained particles may be recharged and collected. The plates may be coated with a film of water-soluble adhesive or oil to aid collected particle retention. Collection modules, often 600–1000 mm long, are mass produced. Typical superficial gas velocities are in the range of 1.5–2.8 m/s. The two-stage precipitator has low (0.042 kW·s/m^3) power consumption and a pressure drop in the range of 25–50 Pa. Operating voltage is supplied from self-contained power packs that usually have solid state rectifiers and sufficient controls to prevent excessive arcing in the ionizer and to control shorting in the collector.

Because the collecting space for dust is small, the use of two-stage precipitators is limited to light dust loads or collection of liquid mists. The most frequent use is for air cleaning for ventilation and air-conditioning. In many of these units, the collection section must be manually removed for cleaning when dust thickness reaches about 1.5–2.0 mm. Alternatively, they can be cleaned in place with automatically controlled traveling spray cleaning and adhesive coating systems. A principal advantage of the two-stage precipitator is its low cost. Collection of liquids is possible, eliminating cleaning needs, but conductive liquids may short the spacing insulators. Two-stage units have also been used for collecting hydrocarbon mists. Single-stage Cottrell precipitators are seldom used on combustible materials because of the danger of fire or explosions from arcs. Because two-stage precipitators operate at much lower voltages and sparking is absent from the collection stage, they are considered to be inherently safer. Sparking can and does occur in the ionizer stage, however, and extreme caution should be exercised if explosive mixtures could be present.

Electrically Augmented Particle Collection. In the 1970s considerable developmental research was devoted to improved fine particle collection by using a combination of electrostatic forces and other collection mechanisms such as inertial impaction or Brownian diffusion. A number of devices were developed and some successfully applied. Little subsequent progress has been made, presumably because of the reduced availability of governmental funding. As mentioned earlier, the electrostatic attractive force is the strongest collecting force available for particles finer than 2–2.5 μm. It is, therefore, logical to couple it with other collecting mechanisms in an attempt to build more highly efficient and cheaper fine-particle collectors. The forces operating between charged and uncharged bodies decrease in magnitude in the order (259): (*1*) coulomb force, the attraction of a charged particle to an oppositely charged particle or surface; (*2*) charged particle–uncharged conducting collector; (*3*) uncharged particle–charged collector; and (*4*) charged particle in a space-charged repulsive field. These relationships have been quantified (260). The coulomb force, or migration of a particle in a gradient electric field, is, of course, employed in both single-stage and two-stage precipitators. Electrostatics is also operative for charged particles and oppositely charged water droplets. A collection efficiency of 85–95% for industrial gases in such a single-stage vertical spray tower and up to 99% collection efficiency utilizing two towers in series have been demonstrated (261,262). A number of retrofit installations have been made on existing spray towers to enhance collection of submicrometer particles.

An investigation of the collection of charged submicrometer particulates on an oppositely charged 0.8–2.0-mm fluidized bed of sand particles found up to 90% efficiency (263). The charged particle–uncharged collecting surface principle has been used and marketed in a device where particles are initially charged in a negative polarity ionizer before entering a grounded, irrigated, cross-flow, packed bed of Tellerette packing (264). The charged particles are brought close to wetted surfaces in the confines of tortuous paths through the bed, inducing a "mirror image" charge of opposite polarity in the water film that causes the particles to impact the water for collection. This device can be used for simultaneous particulate collection and gas absorption. Submicrometer particles have been collected at 85–90% efficiency in single-stage units and at 98% efficiency using two units in series. The principle of charging particles in an ionizer and collecting them dry on uncharged filter media such as cellulose fibers has been practiced for many years in some types of electronic air cleaners. This principle has been applied (265) to improved collection of fine particles in a bag dust filter. Another application of the charged particle–uncharged collector principle was its application to fine particle collection in venturi (wet) scrubbers. Precharging of particles in an ionizer permitted efficient collection of submicrometer particles in the venturi with up to a 50% reduction in venturi pressure drop, a large savings in energy. Unfortunately, a device marketed by Chemical Construction Co. disappeared after the company's bankruptcy, but a similar adaptation is available in France. A charged-droplet scrubber whose commercial design was installed on steel furnace fumes has been described (266). Severe corrosion and electrical loss problems resulted in withdrawal of the scrubber.

Research devoted to optimizing design of various electric augmentation hardware items includes exploration (267) of parameters for preionizer design and discussion (268) of detailed mathematical relations for types of charged-droplet scrubbing and means of charging spray particles. Wet scrubbing appears to have a valid place in air-pollution control because of its relatively low capital investment compared with other control techniques for fine particles. However, wet scrubbing is a marginal control method because of its poor efficiency–energy relationship on fine particles, and thus electric augmentation should be an attractive means of overcoming the weakness of wet scrubbing at moderate cost.

Particle Filtration

Filtration devices for particle collection can be divided into three categories: cloth filtration using either woven or felted fabrics in a bag or envelope, paper and mat filters, and in-depth aggregate bed filtration. The first type is used for dry particle removal from gases but cannot be employed when liquid particulates are present or condensation is imminent. Subclasses of cloth filters are dust filters using cloth in the form of a single envelope (pocket filter) and housings containing rows of stacked cannister filters. The pocket filter has low dust-handling capacity, and when pressure drop becomes too high, the element must be removed and either discarded or manually cleaned. It is used primarily for low dust loads, occasional emissions, or as a safeguard against broken bags after a normal bag-house filter. Likewise, the multiple canister filters cannot be cleaned and must be replaced once high pressure drop occurs. Filters in the form of fiber pads or pleated paper in frames are used for preparing clean air for process use or ventilation. They have limited dust-holding capacity and are seldom used for air pollution control.

Several types of aggregate-bed filters are available that provide in-depth filtration. Both gravel and particle-bed filters have been developed for removal of dry particulates but have not been used extensively. Filters have also been developed using a porous ceramic or porous metal filter surface. Mesh beds of knitted wire mesh, plastic, or glass fibers are used for the removal of liquid particulates and mist. They will also remove solid particles, but will plug rapidly unless irrigated or flushed with a particle-dissolving solvent.

Dust Filter. The cloth or bag dust filter is the oldest and often the most reliable of the many methods for removing dusts from an air stream. Among their advantages are high (often >99%) collection efficiency, moderate pressure drop and power consumption, recovery of the dust in a dry and often reusable form, and no water to saturate the exhaust gases as when a wet scrubber is used. There are also numerous disadvantages: maintenance for bag replacement can be expensive as well as a sometimes unpleasant task; these filters are suitable only for low to moderate temperature use; they cannot be used where liquid condensation may occur; they may be hazardous with combustible and explosive dusts; and they are bulky, requiring considerable installation space.

Bag filters may be woven or felted, an envelope ("pillow case") supported with an internal wire cage, or a long cylinder or stocking hung freely or containing an internal wire cage, and subject to either shaking or reverse flow for dust removal. Older filter installations employed woven cloth bags and the collected dust was generally removed by shaking. Newer bag cleaning methods such as reverse flow use nonwoven felted bags. The availability of fibers is somewhat more limited with felts and one cannot choose a type of weave as with woven fabrics. Three main types of bag filters are illustrated in Figure 30. A fourth type uses backflow of gas, thus essentially providing back-flushing of the cloth. In most bag-house designs, there is some means of slowing the entering gas and deflecting it downward so that coarse dust particles drop out into the hopper. This may be as simple as routing the gas around an inlet baffle. In passing through the filter media, the majority of the dust collects in a dust precoat built up on the bag surface and the cleaned air then flows to the gas outlet. Deposited dust is removed at intervals to prevent excessive pressure drop. In shaker-cleaning filters (envelope and stocking types), automatically controlled dampers shut off the air flow through the bag compartment when the bags are to be shaken. In filters designed for continuous use, a number of parallel compartments are provided to handle flow when one compartment is shut off for cleaning.

In the pulse-jet (reverse-blow) filter (Fig. 30c), air flow is usually not shut off during bag cleaning. Rather, one row or group of bags are cleaned while the rest of the filter remains in service. Dirty gas flows up around the outside of the bag and then through the cloth leaving the dust on the outside of the bag. An internal wire cage keeps the bag expanded and bag and cage are hung from the top tube sheet. Each tube sleeve contains a venturi casting that extends inside the bag, and the cleaned air exits through this casting. For bag cleaning, compressed air from an orifice in a manifold pipe above the casting is jetted back down through the venturi. The air pressure, usually at 620–690 kPa, is released by a solenoid valve actuated by a timer. This jet of air in the venturi induces backward flow of cleaned air from other tubes and an air bubble forms in the top of each bag being cleaned, snaps the cloth away from the cage, and displaces the dust layer. The compressed air pulse must last long enough to allow the bubble to move slowly down the length of the bag to the bottom, cleaning the entire bag surface as it goes. Frequency of cleaning is set to limit the bag pressure drop just before cleaning to the maximum desired level, often 1–1.7 kPa. Degree of cleaning can be adjusted somewhat by changing the pressure of the compressed air and the duration of the pulse. For efficient dust collection, it is desirable to leave a light dust precoat on the bag surface and for long bag life, a minimum number of cleaning cycles is preferred. Because some bags in the compartment are filtering while some are being cleaned, any reentrainment of the dust is followed by redeposition on bags remaining in the filtering operation. Dust redeposition is an appreciable problem even at low filtering velocities; it becomes considerably worse as high filtering velocities are approached (270).

Reverse or back-flow bag filters are rather similar in appearance to shaker filters except that the shaking mechanism is eliminated. The stockings are clamped to a tube sheet at the bottom and are closed at the upper end with a metal cap from which they are suspended. Dirty

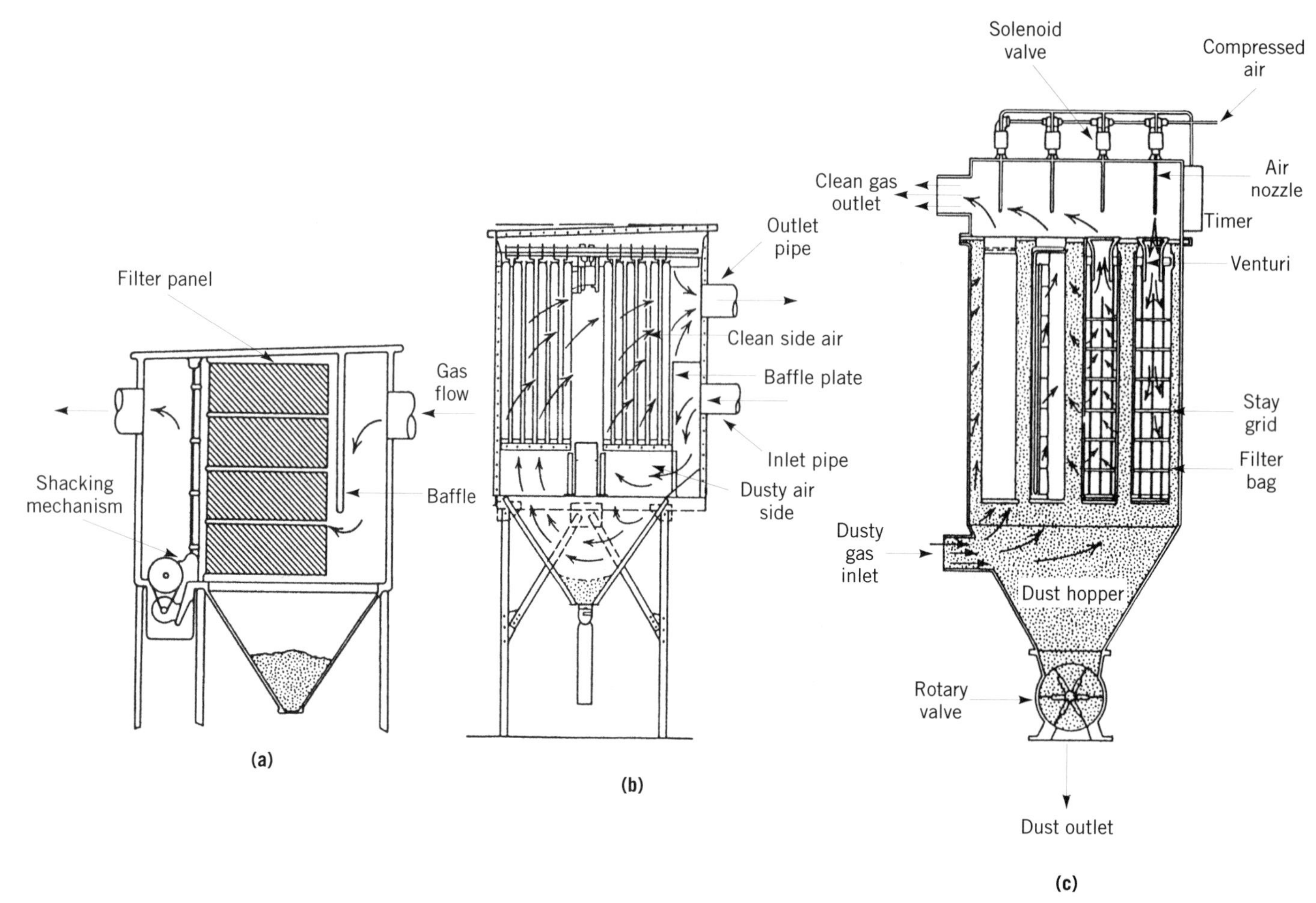

Figure 30. Types of bag filters: (**a**) panel or envelope filter; (**b**) bag or stocking filter; (**c**) pulse-jet filter (269); (**a**) and (**c**) from ref. 269, courtesy of Academic Press, (**b**) courtesy of Wheelabrator-Frye.

gas enters below the tube sheet and passes upward through the bag. When the bag cleaning cycle begins, the flow of dirty gas is shut off, and a fan forces cleaned gas backward through the bags. A series of rings, sewn into the bags at intervals, prevent the complete collapse of the bag under the reverse-flow conditions. Dust dislodged by the backflow falls down through the bag to a dust hopper below the tube sheet. The quantity of backflush gas is usually sufficient to produce a reverse-flow superficial velocity of 0.5–0.6 m/min through the bag. Woven fabrics are generally used for reverse-flow bag filters. The small reverse-flow pressures generally used would be insufficient to back flush felt bags. The principal application for reverse-flow cleaning is in bag houses using fiberglass bags that handle gas at temperatures above 150°C such as boiler flue gas containing fly ash. Bag collapse and reinflation must be sufficiently gentle that excessive stress is not applied to the fiberglass fabric.

Dust Filter Design Considerations. Separation of the dust aerosol from the carrying gas is not a sieving or simple filtration process, because filter fabric pore size is much larger than the particles collected. The efficiency of a new bag for fine dust particles is quite low until the bag fibers and interstices are coated with collected dust. A used bag always has higher collection efficiency than a new one because the entrapped dust particles reduce the effective pore size. A precoat of coarser particles serves as the filter media for finer ones and filter efficiency drops momentarily after a bag cleaning that is too thorough. Pressure drop through a cleaned, dust-impregnated bag may be as high as 10 times that of a new bag. General particle collection mechanisms, such as direct impingement, inertial impaction, gravity settling, Brownian diffusion, and electrostatic attraction, apply in initial bag coating. An understanding of the importance of these mechanisms and their effect on collector efficiency for fine particles as well as for particle release during cleaning is necessary for the development of improved filtration fabrics (271,272). Direct and inertial impingement are the most important mechanism for fabric impregnation of larger particles; Brownian diffusion is the most effective for submicrometer particles.

The effect of dust particle electrostatic attraction to fibers has been investigated (273). When air is passed through a fiber bag, it may generate an electrostatic charge on the fiber. The relative magnitude and polarity of the charge depends on fiber composition and surface configuration (eg, filament vs staple). Table 13 gives a triboelectric series for various fabrics. If fine aerosol particles develop a charge that is opposite to that of the fabric

Table 13. Triboelectric Series for Commercial Fabrics[a]

Polarity	Material
+25	
	wool felt
+20	
+15	glass, filament, heat cleaned and silicone treated
	glass, spun, heat cleaned and silicone treated
	wool, woven felt, T-2
+10	nylon-6,6, spun
	nylon-6,6, spun, heat set
	nylon-6, spun
	cotton sateen
+5	Orlon 81, filament
	Orlon 42, needled fabrics
	Arnel, filament
	Dacron, filament
	Dacron, filament, silicone treated
0	Dacron, filament, M-31
	Dacron, combination filament and spun
	Creslan, spun; Azoton, spun
	Verel, regular, spun; Orlon 81, spun
	Dynel, spun
−5	Orlon 81, spun
	Orlon 42, spun
	Dacron, needled
−10	Dacron, spun; Orlon 81, spun
	Dacron, spun and heat set
	polypropylene 01, filament
	Orlon 39B, spun
−15	Fibravyl, spun
	Darvan, needled
	Kodel
−20	polyethylene B, filament and spun

[a] Ref. 273, reprinted with permission of *J. APCA.*

used in a dust filter, initial collection should be improved. For certain fabrics, collected particles are held so tenaciously that inadequate cleaning occurs and excessive pressure drop develops very quickly. Selection of another fabric having less electrostatic attraction for the dust might well reduce this problem. The effect of charge loss may be a significant factor in cloth cleaning because dust seepage following each cleaning cycle can be a severe problem. Selection of a fabric having greater electrostatic attraction for the dust could be a remedy. Fabric surface characteristics can also affect the retention or release of fine particles. Filament fabrics tend to have slick surfaces and thus easier release than staple fabric. In woven materials, napped fabrics tend to retain fine particles and develop dust precoats better than unnapped materials. Finer and tighter weaves resulting in thicker fabrics (increased fabric weight), less vigorous cleaning, and longer intervals between cleaning cycles all help to reduce leakage of fine particles. In extreme situations, bag precoating with a filter aid after a cleaning cycle may be necessary. Fabric selection and efficiency have been discussed (274).

Pressure drop through a filter bag has two components, the drop through the dust cake and that through the cloth. The flow through both is streamline so the pressure drop varies directly with gas velocity. Pressure drop through the cloth is defined by equation 35. Values of k_c, essentially a drag coefficient for the fabric, have been tabulated (275) for various types of unused cloth. However, the pressure drop for dust-impregnated fabric can be over 10 times as great. Whenever dust cake thicknesses exceed 1.5 mm, the pressure drop across the cloth becomes insignificant. Equation 36 gives the pressure drop across the dust layer at any point in time following cleaning. If it is assumed that dust concentration, C_d and gas flow volume V_f are constant with time, then $m_d = C_d V_f \theta$, where θ is the time since the end of the last cleaning cycle, and on substitution equation 37 results. Despite the squared gas-flow term, flow is streamline, not turbulent.

$$\Delta P_c = k_c \mu_g V_f \quad (35)$$

$$\Delta P_d = k_d \mu_g m_d V_f \quad (36)$$

$$\Delta P_d = k_d \mu_g C_d V_f^2 \theta \quad (37)$$

The term k_d, essentially a drag coefficient for the dust cake particles, should be a function of the median particle size and particle size distribution, the particle shape, and the packing density. Experimental data are the only reliable source for predicting cake resistance to flow. Bag filters are often selected for some desired maximum pressure drop (500–1750 Pa) and the cleaning interval is then set to limit pressure drop to a chosen maximum value.

Bag filtering area is another variable to be determined. Selection is usually on the basis of the volume of gas to be filtered and a desired superficial gas velocity through the bag surface. The design superficial velocity chosen should depend on the dust concentration in the gas to be filtered, the type of dust, the fineness and abrasiveness, the type of cloth and weave, the cost of the cloth, and the desired frequency of bag cleaning. For woven cloth, the typical range is 0.005–0.04 m/s; although many users like to limit the velocity to 0.020–0.025 m/s: if the dust is fine or abrasive or if the dust concentration is unusually high, the velocity drops to 0.015 m/s or less. For shaking collectors, sufficient bag area is necessary that excessive velocities do not exist when one compartment is off-line for shaking. Manufacturers of pulse-jet collectors (felt bags) recommend somewhat higher velocities: 0.025–0.075 m/s. Some users, from experience, prefer to use velocities no higher than those employed for cloth bags with shaker cleaning, and experiments (270) involving redeposition appear to support keeping pulse-jet collector velocities at the lower end of the range.

Bag cleaning and control methods need to be selected. For cloth bags, shaking is well established, but pulse-jet cleaning is probably the most popular choice today whenever felted bags can be successfully used. Pulse-jet cleaning should be more desireable for applications handling heavy dust loads in continuous operation, since it is always controlled by an automatic timing device. Automatic timing can also be applied to shaker filters, but for applications of light dust loading or intermittent gas flow, remote manual control may be preferred.

Choice of bag fabric depends on chemical compatibility with the dust to be collected (Table 14) and temperature resistance (Table 15) as well as fabric cost. Additional data on fabric resistance to abrasion, temperature, and specific chemical compounds have been presented (276).

Table 14. Chemical Compatibility of Fibers in Dust Collector Bags

Resistance	In Acid Media	In Alkaline Media
Excellent	polypropylene	polypropylene
	polyethylene	polyethylene
	Saran	Dynel
	Teflon	nylon-6,6
	Orlon	Teflon
Good	Dacron	cotton
	Dynel	nylon-6
	glass	Nomex nylon
	wool	Saran
Unsuitable	cotton	wool
	nylon-6,6	glass
	nylon-6	
	Nomex nylon	

There are also other special fabric considerations. Glass fibers do not resist abrasion and flexing well. Surface coatings help, but special cleaning procedures such as reverse flow are required for these fibers, and longer intervals between cleaning are desireable. Nomex nylon has poor resistance to moisture and should be avoided whenever condensation is possible. Teflon, although highly resistant, is extremely expensive; cotton and Dacron are relatively inexpensive.

No bag fabric can withstand truly high temperature; therefore, gas cooling is often practiced. The usual methods are indirect cooling, tempering with cold air, direct water spray cooling, or a combination of any of these. Indirect cooling may take place in radiation panels or ducts exposed to the atmosphere, in waste heat boilers, or in heat-transfer devices such as finned heat exchangers and heat wheels. Tempering consists of mixing air from the atmosphere with the hot gas in a duct; good mixing must be provided to ensure temperature equilibration. Automatic temperature control can be quite precise and tempering can reduce the dewpoint of a hot, humid gas. The major disadvantages are that tempering increases the gas volume and hence the required size of the dust filter and exhaust fan, and the additional power to draw the diluted gas through the filter.

Direct water spray cooling must be carried out with care. The spray chamber must be designed to ensure complete evaporation of all liquid droplets before the gas enters the bag house. Spray impinging on the chamber walls can result in a dust mud inside the chamber, and any increase in gas dewpoint may result in bag house problems or atmospheric plume condensation. Spray nozzle wear can result in coarse or distorted spray and wetted bags, and water pressure failure can cause high temperature bag deterioration.

Bag house safety must also be taken into consideration, because the forced ventilation of filtering air can convert a bag house into a forge in short order, especially if the dust is flammable. Most bags will burn; a few will char but not support combustion. Glass is the only fireproof bag material, but it has a low softening temperature. Large installations need temperature-sensing devices to shut off air flow in the event of fire. In many installations sprinkler heads, CO_2 injection, or dry chemical systems are provided. Because bag-house explosions may also be a problem, large vents should be provided wherever there are susceptible dusts. Checklists of recommendations for safeguarding fabric filters (277) and for specifying and evaluating dust-filter purchase (278) have been prepared. Explosion venting (279) and optimization of pulse-jet filter design (280) have been discussed. The cost and performance of bag filters has been compared with other particulate control devices (281,286). Other pertinent reviews also exist (282–286).

Newer Bag Filter Applications and Developments. Bag filters have been used extensively for containing dust emissions from low temperature process operations, such as milling and screening, drying, packaging, loading and unloading, and material conveying. Efforts have been made to extend use to the control of fine particle emission from common large-volume, hot-gas streams, such as coal-fired power boilers, and ferrous and nonferrous metallurgical processes. A number of demonstration installations of bag filters on power boiler effluent have been made (287–295). New applications (296) and cost estimation procedures (297) for fabric filters have been discused. The effects of conventional fabric structure on collection efficiency have been studied (298) and nonconventional fabrics have also been investigated (299). Trilobal fibers gave improved collection efficiency and reduced pressure drop; efficiency also improved when a weave tighter (0.33 tex) than the conventional (0.67 tex = 6 den) weave was used, but at the expense of higher pressure drop. Crimped fibers also improved collection efficiency and reduced pressure drop.

Filtration by Granular Beds. Fine particle removal by filtering through a bed of granular solids has been known and used for many years, but a basis for predicting effi-

Table 15. Maximum Desirable Operating Temperature for Filter Bags[a]

Material	Temperature, °C	Material	Temperature, °C
Polyethylene	70	Arnel	120
Saran	70	Microtain	125
Cotton	80	Kodel	135
Dynel	85	Dacron	140
Polypropylene	90	Darvan	150
Wool	100	Nomex nylon	230
Nylon-6,6 and -6	105	Teflon	260
Orlon	120	Glass	290
Acrilan	120		

[a] Longer bag life is obtained if bags are not operated at their maximum temperature.

ciency and pressure drop has only been developed in the last decade or two. Granular filters have appeal because of corrosion and temperature resistance, physical robustness, and theoretical ease of cleaning. Both the Lynch filter (300), described in 1936, and the Dorfan Impingo filter (301), developed commercially in 1950 but now unavailable, used a downward-traveling gravel bed. A collection efficiency of 98% on 2- to 10-μm particles was claimed for the latter filter, but such efficiency has been found only with a fresh packing charge (302). Another cross-flow granular scrubber, developed by Combustion Power Co., (303), used 3–6-mm pea gravel. Gas face velocities were 0.5–0.75 m/s, and pressure drops were 0.5–3 kPa. Collection efficiency on submicrometer particles was low, but subsequently electric augmentation corrected this problem (304); this has been confirmed (305). The process depends on the natural charge on the dust particles passing through the charged (20–30 kV of potential is imposed) field of the gravel bed. Another type of gravel-bed filter, developed in Germany, has been described (306) in which gas, precleaned in a cyclone, is then passed to a stationary gravel bed. When the bed becomes dust saturated, forward flow is stopped and the gravel is back flushed using air to remove the dust, which is then collected in the cyclone. Tests indicated low collection efficiency on 2-μm particles (307). A cross-flow granular limestone bed was developed to remove both SO_2 and fly ash from power boilers (308). The limestone loses its reactivity because of formation of a gypsum coating. Attempts to remove the fly ash and gypsum using a combination of backflow and bed-rapping techniques were not particularly successful.

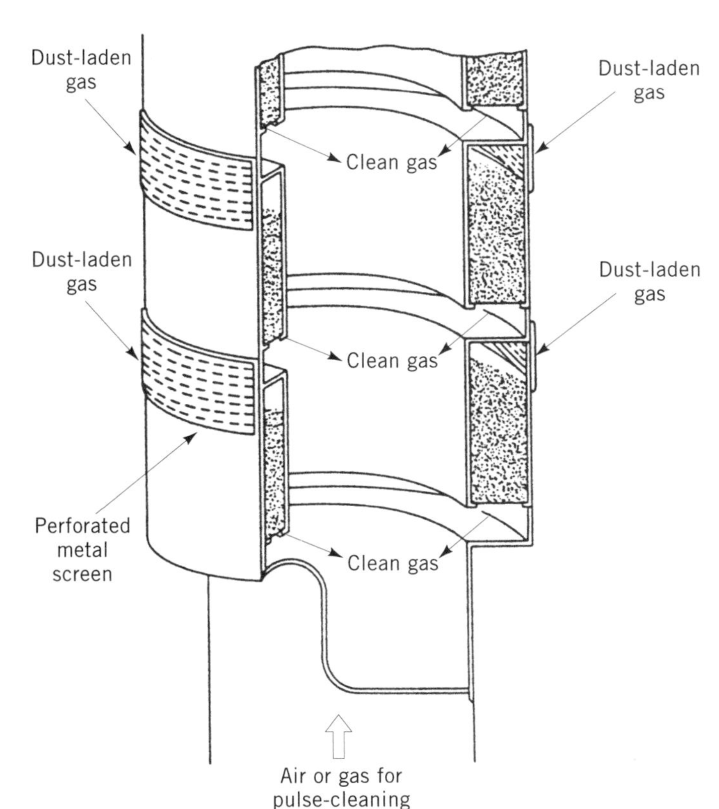

Figure 31. Zenz-Ducon expandable-bed granular filter. From ref. 311, courtesy of McGraw-Hill, Inc.

In one theoretical equation (309) for collection efficiency in granular beds important parameters are bed thickness, gravel size, and air velocity. A cleanable granular bed filter (Fig. 31) has been commercially tested (310). Filtering granules are contained in annular compartments enclosed in a housing. Dirty gas enters above the static granular bed, travels downward through the bed and emerges clean at the bottom. Collection efficiency is stated to be as high as 99% on 1–3-μm particles. When the beds become dust-laden, forward flow is shut off and a reverse flow of high pressure air in short blasts fluidizes the beds and elutriates the dust. This concentrated dust–air mixture must then be treated in another device for final dust recovery. Granules (590 μm or 30-mesh in size) of various materials have been used. One of the problems appears to be inadequate elutriation of particular dusts so that laboratory tests are conducted using the applicant dust before a unit is recommended for installation.

Particle collection in low fluidization velocity beds has been investigated (309,313); these studies indicate that collection efficiencies higher than 90% would be difficult to obtain. Using shallow fluidized beds of 25-μm particles, 99.7% collection of submicrometer particles through bed staging was obtained (314). Several conditions yielded this efficiency: three stages, each bed 76-mm thick; four stages, each bed 41-mm thick; and five stages, each bed 25-mm thick.

Fiber Bed Mist Filtration. In-depth fiber bed filters are used for the collection of liquid droplets, fogs, and mists. Horizontal pads of knitted metal wire (or plastic fibers), 100–150-mm thick, and gas upflow are used for liquid entrainment removal. Pressure drop is 250–500 Pa. Characteristics of the pads vary slightly with mesh density, but void space is typically 97–99% of total volume. Collection is by inertial impaction and direct impingement; thus efficiency will be low at low superficial velocities (usually below 2.3 m/s) and for fine particles. The desirable operating velocity is given as:

$$U_s = K_b \sqrt{\frac{\rho_L - \rho_g}{\rho_g}} \qquad (38)$$

The value for K_b is 0.11 m/s, unless liquid viscosity is high or the liquid loading is very high, when somewhat lower values of K_b are used. The same is true for materials with low liquid surface tension. At desired operating velocities, 99% efficiency will prevail for particles 5 μm and larger. For smaller particles, efficiency drops rapidly; it is about 92% for 3-μm particles. Design factors for wire-mesh mist eliminators have been discussed (315), and the use of two mesh pads in series for collection of finer mist particles at much higher pressure drops has been investigated (316,317).

For collection of fine mist particles, the use of randomly oriented fiber beds is preferable. Three types of such beds are available: chemically resistant glass, polypropylene, and fluorocarbon fibers. Beds 25–50-mm thick held between parallel screens are designated high velocity filters. Installed vertically, with horizontal gas flow and vertical

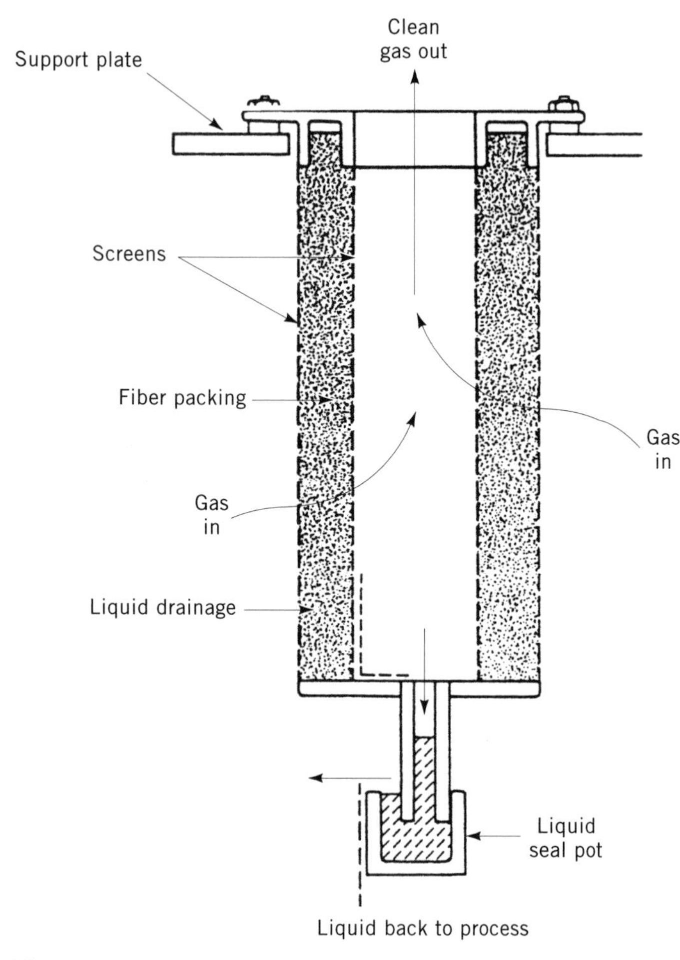

Figure 32. Brink-Monsanto Envirochem high efficiency mist eliminator element (320).

liquid drainage, such beds are sized for 2.5 m/s face velocity and pressure drops of 1–2.5 kPa. These filters collect by inertial impaction. Efficiency improves with increasing velocity and pressure drop, eg, 92–94% efficiency on 1-μm particles at 1–1.5 kPa pressure loss. The performance has been discussed (318). Large cylindrical elements similar to Figure 32, designated high efficiency mist eliminators, are custom designed for collection of submicrometer mist at efficiencies as high as 99.9%. The collection parameters of these units have been described (319). Bed thickness can vary from 75 to 101 mm, and face velocities are much lower than those for high velocity units. Pressure drop is in the range of 2.5 to 7.5 kPa. For particles smaller than 3 μm, the predominant collection mechanism is Brownian diffusion; efficiency increases with reduction in flow and no turn-down problems exist. A third type of bed, which is designated spray catcher and is lowest in efficiency, is available for 100% collection of particles 3 μm and larger. It is similar in configuration to the high velocity type but has a pressure loss of only 138–276 Pa. Fiber bed mist eliminators, designed primarily for liquid particulates, can also be efficient collectors of water-soluble solid particles if the dust loading is not high and the bed is continuously irrigated with water or another solvent.

Some effluents contain submicrometer oily or tarlike particulates that are difficult to remove from the collection surface. A disposable, dry glass fiber mat (321) that can handle such submicrometer particulates at 99% efficiency is known. A subsequent modification is the replacement of the disposable mat with a cylindrical drum or endless belt of recticulated foam that is cleaned and recycled. Cleaning procedures, using steaming, detergent washing, solvent cleaning, and ultrasonic cleaning in a variety of fluids have been developed.

Wet Scrubbing

Scrubbers can be highly effective for both particulate collection and gas absorption. Costs can be quite reasonable for the required efficiency, but the addition of water treatment for recycle or for waste disposal may make the total cost as great as any other collection method, depending on water treatment complications. Although scrubbers automatically provide cooling of hot gases, the water-saturated effluent may produce offensive plume condensation in cold weather. Many moist effluents become more corrosive than dry ones. Solids accumulation may occur at wet–dry interfaces and icing problems may occur around the stack in winter. For efficient fine-particle collection, energy consumption may also exceed that of dry collectors by an appreciable amount.

Scrubbers make use of a combination of the particulate collection mechanisms listed in Table 9. It is difficult to classify scrubbers predominantly by any one mechanism, but for some systems, inertial impaction and direct interception predominate. Semrau (204,317,323) proposed a contacting power principle for correlation of dust-scrubber efficiency: the efficiency of collection is proportional to power expended, and more energy is required to capture finer particles. This principle is applicable only when inertial impaction and direct interception are the mechanisms employed. Furthermore, the correlation is not general because different parameters are obtained for differing emissions collected by different devices. However, in many wet scrubber situations for constant particle-size distribution, Semrau's power law principle, roughly applies:

$$N_t = \alpha P_T^{\gamma} \tag{39}$$

The constants α and γ depend on the physical and chemical properties of the system, the scrubbing device, and the particle-size distribution in the entering gas stream.

Table 16 can be used as a rough guide for scrubber collection in regard to minimum particle size collected at 85% efficiency. In some cases, a higher collection efficiency can be achieved on finer particles under a higher pressure drop. For many scrubbers the particle penetration can be represented by an exponential equation of the form (326–329)

$$P_t = e^{-A(d_p)^B} \tag{40}$$

where the constants A and B depend on scrubber design: B is 0.67 for centrifugal scrubbers and essentially 2 for packed towers, sieve plates, and venturi scrubbers. Use of equation 40 permits development of generalized grade-efficiency curves of particle cut-size for a number of differ-

Table 16. Particle Size Collection Capabilities of Various Wet Scrubbers[a]

Type of Scrubber	Pressure Drop, Pa[b]	Minimum Collectible Particle Dia, μm[c]
Gravity spray towers	125–375	10
Cyclonic spray towers	500–2,500	2–6
Impingement scrubbers	500–4,000	1–5
Packed and moving bed scrubbers	500–4,000	1–5
Plate and slot scrubbers	1,200–4,000	1–3
Fiber bed scrubbers	1,000–4,000	0.8–1
Water jet scrubbers		0.8–2
Dynamic		1–3
Venturi	2,500–18,000	0.5–1

[a] Refs. 206, 324, 325.
[b] To convert Pa to psi, multiply by 1.450×10^{-4}.
[c] Minimum particle size collectible with approximately 85% efficiency.

ent types of scrubbers (220). If the entering particles have a log-normal mass distribution, Figure 33 can be used to obtain the integrated particle penetration leaving the scrubber in terms of scrubber cut-size, mass median particle diameter, exponent B, and geometric particle-size distribution. Figure 34 is simpler to use when $B = 2$. Plotting particle cut-size vs pressure drop for various scrubbers is suitable for developing generalized energy efficiency curves (326). The curves for the various devices outline an imaginary band of pressure drop vs particle size for inertial impaction (Fig. 35).

Wet scrubber collection efficiency may be unexpectedly enhanced by particle growth. Vapor condensation, high turbulence, and thermal forces acting within the confines of narrow passages can all lead to particle growth or agglomeration. Of these, vapor condensation produced by cooling is the most common. Condensation will occur preferentially on existing particles, making them larger, rather than producing new nuclei. Careful experimentation (330) has shown that the addition of small quantities of nonfoaming surfactants to the scrubbing water can enhance the collection of hydrophobic dust particles without further energy expenditure.

Scrubber Performance and Selection. A tremendous number of wet scrubbers have been marketed. General introductions are available (155,273,276). An extensive study of wet scrubbers has been done (327), and there are reports on wet scrubber selection and evaluation (333). An open spray tower, such as the countercurrent vertical one of Figure 1b, is the simplest type of scrubber. Gas velocities (typically 0.6–1.2 m/s) should be less than the termi-

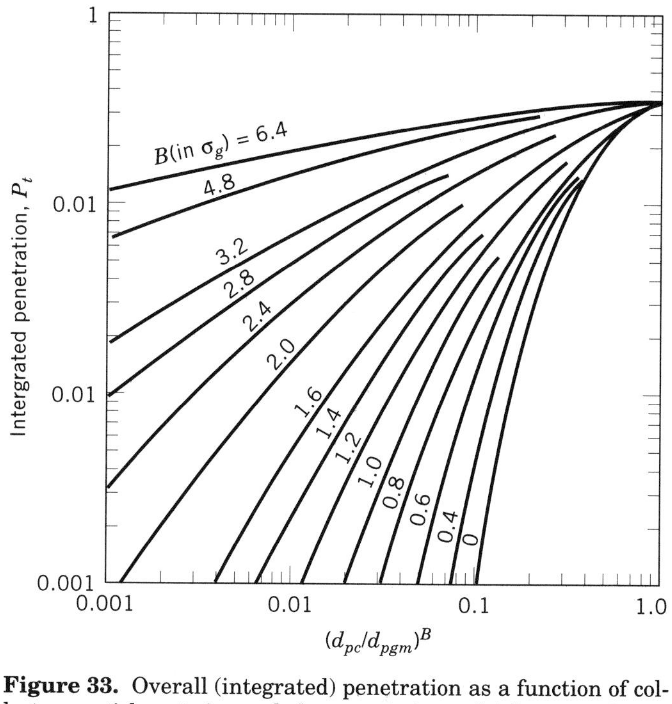

Figure 33. Overall (integrated) penetration as a function of collector particle cut-size and characteristics and inlet particle parameters for collectors that follow equation 40 (326). Courtesy of *J. APCA*.

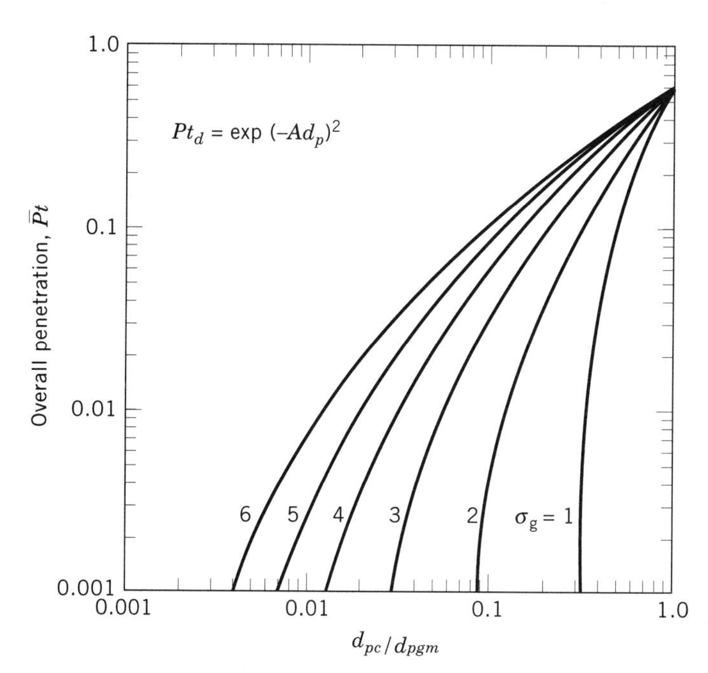

Figure 34. Integrated overall collector penetration as a function of collector cut-size for collectors following equation 40 when $B = 2$, and the entering particles are log-normally distributed, ie, d_{pgm}, is the mass median particle size of the distribution. The solid line parameters are for the standard distribution of particles, G_g, about the median (326). Reprinted from ref. 329 with permission.

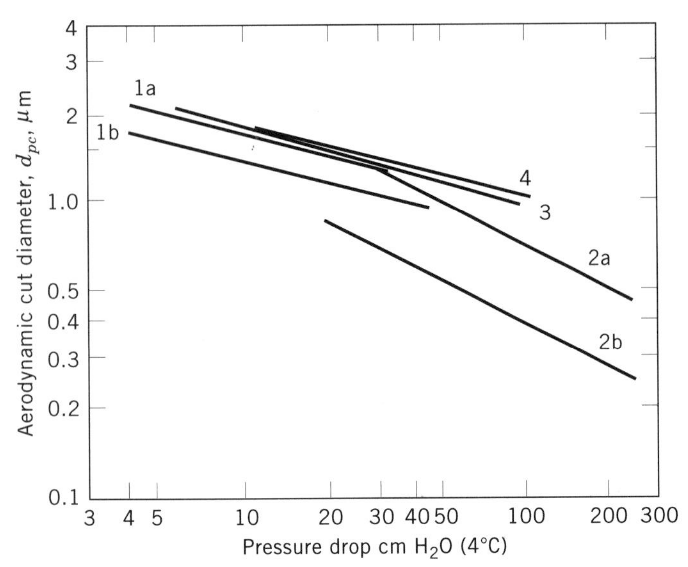

Figure 35. Generalized energy consumption vs particle cut-size relationship developed by Calvert (326). Scrubbers: 1a, sieve, $F = 0.4$, 5 mm hole dia; 1b, sieve, $F = 0.4$, 3 mm hole dia; 2a, venturi, $F = 0.25$ (hydrophobic particles); 2b, venturi, $F = 0.5$ (hydrophilic particles); 3, impingement plate; 4, packed column, 25 mm packing dia. To convert to SI pressure units of Pa, multiply by 98.07. Reprinted with permission of *J. APCA*.

nal settling velocity of the spray droplets. Optimum spray droplet size for maximum target efficiency has been calculated (334) and can be obtained from Figure 36. Figure 37**a** gives cut-size for vertical spray towers, and Figure 37**b** for horizontal spray towers. Tangential inlet of gas at the bottom of vertical spray towers (Fig. 1**c** and **d**) greatly increases the collision of aerosol particles and spray droplets as a result of centrifugal force. The optimum spray droplet size for greatest collection efficiency in centrifugal spray towers is considerably smaller than for gravity towers. Figure 38 shows calculated target efficiencies (191) for spray droplets in a centrifugal field. Cyclonic scrubbers similar to Figure 1**c** have efficiencies of 97% on 1-μm particles (338). A number of commercial scrubbers such as the type N and R Rotoclone and the Microdyne use a combination of water atomization and centrifugal force to capture particles.

Vertical columns with plates or trays are also used for particulate collection. Almost any type of perforated plate can be used: sieve plates (with or without submerged impingement targets above the holes), slot plates, valve trays, and bubble caps. Collection efficiency is good down to 1-μm particles using pressure drops of 400 Pa per plate. The best efficiency is obtained in all-wet scrubbers if the dirty gas is saturated with water vapor before entering the first contact stage. Figure 39 shows predicted cut-size (326) for a simple sieve tray. Froth density must be predicted from complex relationships for sieve plate behavior (327). Packed towers will collect particulates and cut-size is independent of packing type (Fig. 40) (326). Countercurrent towers can be used for mist collection but plug easily in the presence of solid particulates. Some particulates can be handled in concurrent downflow towers. A cross-flow packed tower can handle loads of insoluble dust up to 11 g/m³ when properly designed (339). Up to 95% collection of 3-μm particles in such a tower has been reported (340). Irrigated fiber pads commercially available are quite efficient on 3-μm particles when operated under a pressure drop of 1 kPa; they will plug easily on heavy loadings of insoluble dust. Mobile-bed packing of fluidized spheres (turbulent contactors) have both good particulate collection and good mass transfer characteristics. Collection efficiency is good on particles down to 1 μm and the constant movement and rubbing of the spherical targets prevents plugging. For greater efficiency, several stages in series can be used. Typical pressure drop per stage is 0.75–1.5 kPa.

In several commercial scrubbers, the dirty gas jets directly into a pool of water where the momentum of the larger particles allows them to penetrate. Some of the water is thus atomized into a spray aiding collision between water droplets and smaller particulate. In one such scrubber careful control of the water level is necessary for efficient particle collection (341). These scrubbers have been used to collect coarser dusts such as from metal grinding; they have good efficiencies on particles down to 3–5 μm. Dynamic collectors employ a motor-driven centrifugal device resembling a water-sprayed centrifugal fan to cause impingement of solid particles in a wetted film. Tests indicate that 1-μm particles can be collected with 50% efficiency in an ordinary water-sprayed centrifugal fan if the wheel tip speed exceeds 90 m/s.

The collection of particles larger than 1–2 μm in liquid ejector venturis has been discussed (342). High pressure water induces the flow of gas, but power costs for liquid pumping can be high because motive efficiency of jet ejectors is usually less than 10%. Improvements (343) to liquid injectors allow capture of submicrometer particles by using a superheated hot (200°C) water jet at pressures of 6,900–27,600 kPa, which flashes as it issues from the nozzle. For 99% collection, hot water rate varies from 0.4 kg/1000 m³ for 1-μm particles to 0.6 kg/1000 m³ for 0.3-μm particles.

Venturi and High Energy Scrubbers. The venturi scrubber has been studied more intensively than any other wet scrubber, perhaps because of its ability to efficiently scrub

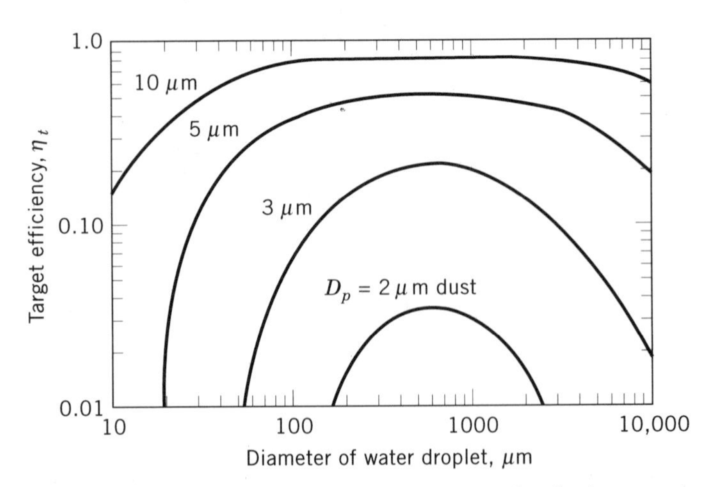

Figure 36. Target efficiency of a single water droplet in a gravitational spray tower (334,335). From ref. 356, courtesy of McGraw-Hill, Inc.

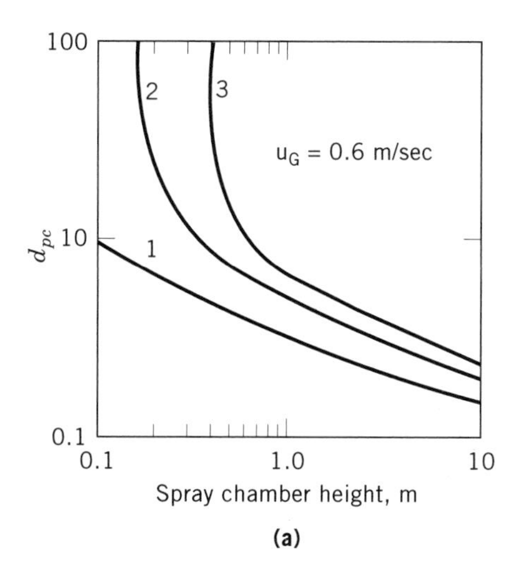

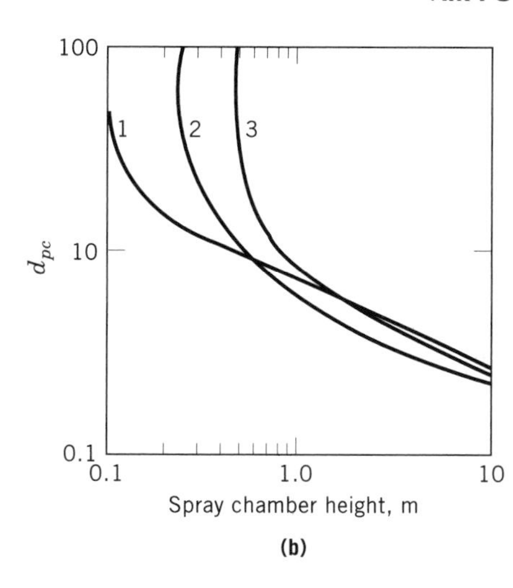

Figure 37. Predicted performance cut diameter for typical spray towers (326): (**a**) vertical countercurrent spray tower; (**b**) horizontal cross-current spray chamber. Liquid–gas ratio is 1 m^3 of liquid/1000 m^3 of gas. Drop diameter: curve 1, 200 μm; curve 2, 500 μm; curve 3, 1000 μm. U_G = 0.6 m/s. Courtesy of *J. APCA*.

any size particle by changing the pressure drop. The design readily lends itself to mathematical modeling. Gas is accelerated in the throat to velocities of 60–150 m/s where water is added either as a spray, as solid jets, or as a wall-flowing sheet. The water is atomized into small droplets by the high speed gas. Aerosol particles, moving at close to the gas speed, collide with the accelerating liquid droplets and are captured. A cyclonic collector is needed to remove the water mist produced by the venturi. Collection mechanisms have been studied (191,344–349). The liquid drop size produced has been investigated (350); the effect of water-injection method on venturi performance, examined (351); and a generalized pressure-drop prediction method has been developed (352).

Venturi scrubbers can be operated at 2.5 kPa to collect many particles coarser than 1 μm efficiently. Smaller particles often require a pressure drop of 7.5–10 kPa. When most of the particulates are smaller than 0.5 μm and are hydrophobic, venturis has been operated at pressure drops from 25 to 32.5 kPa. Water injection rate is typically 0.67–1.4 m^3 of liquid per 1000 m^3 of gas, although rates as high as 2.7 are used. Increasing water rates improves collection efficiency. Figure 41 shows the effects of throat velocity, pressure drop, and liquid–gas ratio on particle cut-size for hydrophobic particles (326). Many venturis contain louvers to vary throat cross section and pressure drop with changes in system gas flow. Venturi scrubbers can be made in various shapes with reasonably similar characteristics. Any device that causes contact of liquid and gas at high velocity and pressure drop across an accelerating orifice will act much like a venturi scrubber. A flooded-disk scrubber in which the annular orifice created by the disc is equivalent to a venturi throat has been described (353). An irrigated packed fiber bed with performance similar to a venturi scrubber was offered commercially for collection of submicrometer particles at bed pressure drops of 7.5–15 kPa.

Wet Scrubber Entrainment Separation. Fiber pads and beds to collect fine liquid entrainment have been dis-

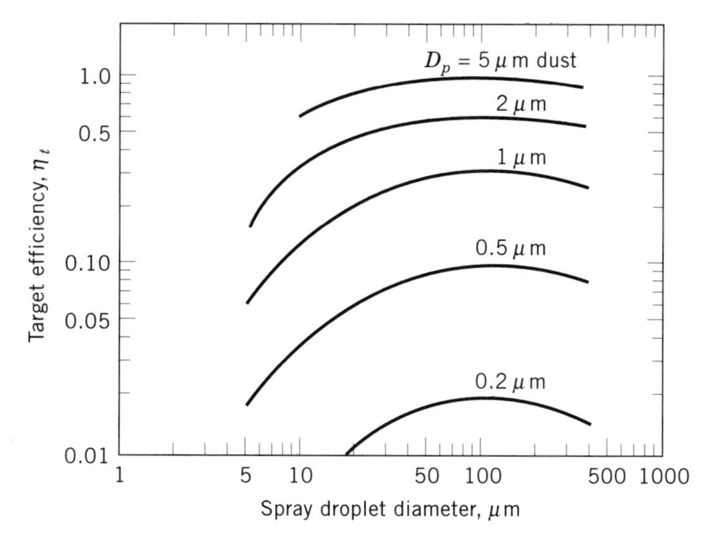

Figure 38. Spray droplet target efficiency in a centrifugal spray tower with a centrifugal field of 100 *g* (191,337).

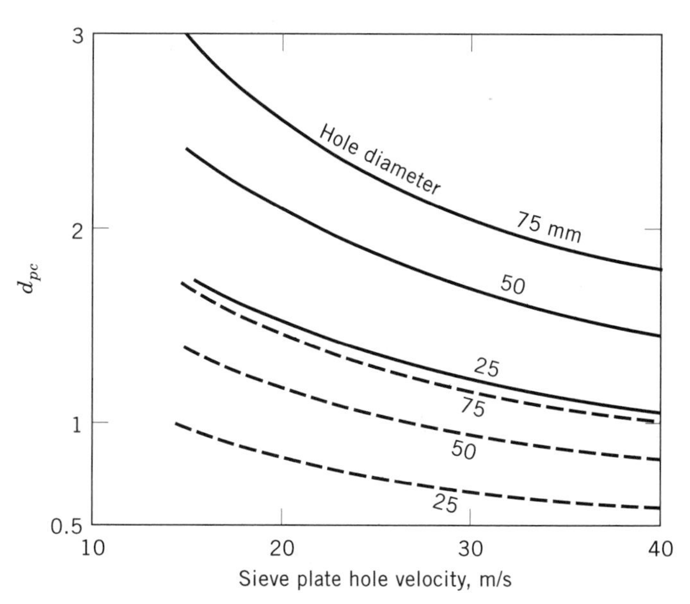

Figure 39. Performance cut diameter prediction for typical sieve plate operation on wettable particulates at foam densities *F*: solid line, *F* = 0.4 g/cm^3, dashed line, *F* = 0.65 g/cm^3 (326). Courtesy of *J. APCA*.

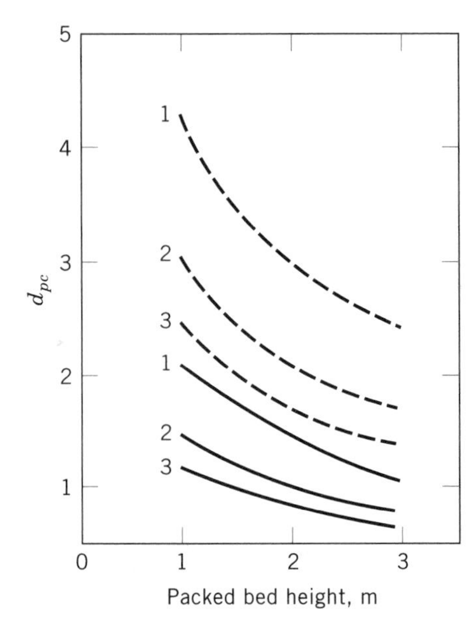

Figure 40. Performance cut diameter predictions for typical dry packed bed particle collectors as a function of bed height or depth, packing diameter D_c, and packing porosity (void area) ε. Bed irrigation increases collection efficiency or decreases cut diameter (326). Solid lines, D_c = 25 mm; dashed lines, D_c = 50 mm; ε = 0.75. Courtesy of *J. APCA.*

cussed. Unless fog is present from condensation, scrubber entrainment will be coarse and the high efficiency of fiber beds is not needed. Entrainment separators for scrubbers are usually centrifugal swirl vanes, hook and zigzag eliminators, or momentum separators using a change in direction. Efficiency and pressure drop for zigzag baffles, targets of staggered tube banks, packed beds, and knitted mesh have been compared (354), and these data have been supplemented with entrainment theory and collection efficiency in centrifugal flow and in sieve plates (359). Entrainment separator design requirements for venturi scrubbers have been discussed (356). Control of recycle water cleanliness is important to good wet scrubber performance and poor scrubber performance has been traced (357) to entrainment of dirty scrubbing liquid and also to temperature flashing (evaporation) of fine particulate contained in recycle spray water.

Developing Particulate Control Technology

Present control methods for particulates are least efficient in the size range from 0.2 to 2.5 μm; this range is the most costly to collect and energy intensive. Health studies indicate that particles in this size range are also those that penetrate most deeply into, and often become deposited in, the human respiratory system. This is the principal reason for the U.S. EPA change in the ambient air quality standard from total suspended particulate (TSP) concentration to a PM_{10} standard (ambient air particles equal to or smaller than 10-μm aerodynamic diameter). The new standard will undoubtedly place even more emphasis on the need to collect particulates in this difficult-to-control fine-particle range, and therefore, collection of this size range is most in need of improvements in technology. Improved collection requires the use of a separating force that is independent of gas velocity or of the growth of particles that can be more readily collected. Particle growth can be accomplished through coagulation (agglomeration), chemical reaction, condensation, and electrostatic attraction. Promising separation forces are the "flux forces" involving diffusiophoresis, thermophoresis, electrophoresis, and Stefan flow. Although particle growth techniques and flux-force collection theoretically can be considered independently, both phenomena are applied in many practical devices.

Thermophoresis may be considered a special form of thermal precipitation. If a hot submicrometer particle is close to a large cold particle or droplet of such relative size that it resembles a wall to the small particle, the kinetic motion of the hot gas molecules opposite the cold particle will bombard and propel the submicrometer particle toward the cold droplet. If cold droplets are introduced into a warmer saturated gas stream, water vapor will condense on the cold droplets reducing the water vapor pressure near its surface and produce a water vapor pressure gradient. The hydrodynamic flow of vapor toward the condensing surface is known as Stefan flow. If the molecular mass of the diffusing vapor is different from the molecular mass of the carrier gas, the motion of small particles is further affected by a density differential. Both of these forces influence the movement of submicrometer particles, and their algebraic sum is known as the force of diffusiophoresis. In condensation, diffusiophoresis tends to move submicrometer particles toward cold droplets; in droplet evaporation, the action is reversed, moving the particles away from the droplet. This is one of the reasons wet scrubbing collection is enhanced if the bulk

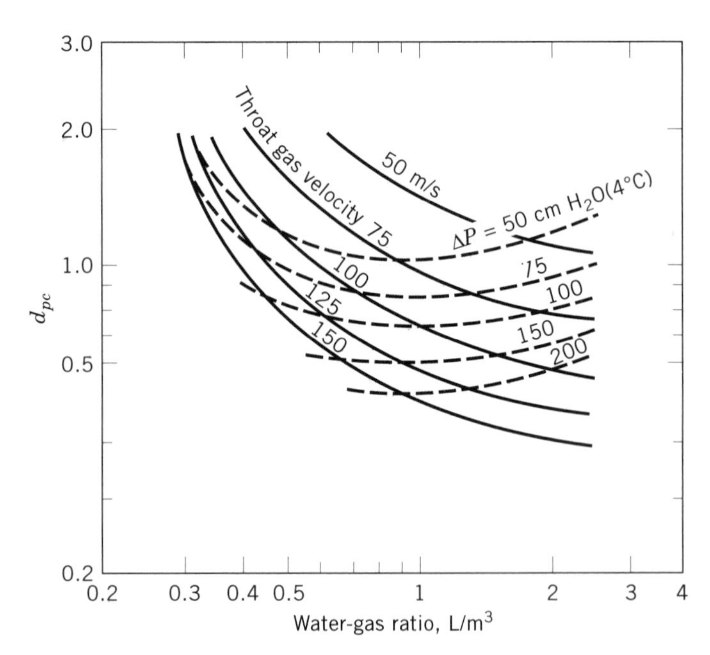

Figure 41. Performance cut diameter predictions for venturi scrubbers on hydrophobic particles (*F* parameter = 0.25). Correlating parameters are pressure drop and gas velocity through the throat, and liquid to gas ratio (320). For the empirical factor *F*, see reference 327; *F* for hydrophilic particles is 0.4–0.5. Reprinted with permission of *J. APCA.* To convert cm H_2O at 4°C to kPa multiply by 0.098.

gas stream is water saturated before its entry to the scrubber, and the scrubbing liquid is cooled below the gas dewpoint. Electrophoresis is the movement of a small particle toward a charged particle. Mathematical description of these forces has been presented (338). Diffusiophoresis has also been discussed (358).

In time, submicrometer particles will coagulate into chains or agglomerates through Brownian motion. Increasing turbulence during coagulation will increase the frequency of collisions and coagulation rate. The addition of fans to stir the gas, or gas flow motion through tortuous passages such as those of a packed bed, will aid coagulation. Sonic energy is also known as an aid to the coagulation process. Production of standing waves in the confines of long narrow tubes can bring about concentration and coagulation of aerosols in band zones in the tube. The addition of water and oil mists to the treated aerosol can improve the effectiveness of sonic agglomeration by improving the tendency of the colliding particles to stick together. Sulfuric acid mist (359) and carbon black (360) have been sufficiently agglomerated so that the coagulated product can be collected in a cyclone. Sonic agglomeration has been tested in the past for many metallurgical fumes, but has generally been found too power intensive for practical consideration. One problem is the low energy efficiency of transforming other energy sources to sonic energy in available sonic generators. The development of more energy efficient sonic generators, coupled with improved knowledge of the phenomenon of sonic coagulation, might justify further investigation. The combination of sonic agglomeration and electrostatic precipitation could result in considerable reduction of precipitator size and perhaps capital cost (361). Coagulation techniques have been discussed in mathematical terms in references 327 and 331.

Particle growth may be brought about (or the charge on particles modified) by the introduction of a gas to react with the particles. Another procedure used is chemical modification of the particle to render it hygroscopic. This can be particularly beneficial for particle growth if the particle continues to absorb moisture to form hydrates. It may also be possible to control the size of aerosols initially formed by chemical means. In studies (363) of chemical reactions producing aerosols it was found that reactions having a large chemical driving force and hence releasing large quantities of energy tended to produce very fine particles (high surface energy); conversely, reactions that occur without release of large amounts of energy tend to produce larger particles (2–6 μm). Therefore, limitation of chemical driving force may often be beneficial to prevent formation of fine aerosols. If steam or other condensable vapors can be condensed while cooling an aerosol, particle growth will occur through vapor condensation on the existing aerosol nuclei. A given mass of smaller particles will present more surface area for condensation than the same mass of larger particles. Thus the smaller particles will selectively grow faster in size than will the larger particles. Also, the addition to the system of particles opposite in electrical charge will result in mutual attraction and aid in particle growth.

Of these various possibilities, those involving electrostatic forces and condensation have received the most interest. Flux-force condensation scrubbing may be desirable for hot gases needing treatment when there is no attractive alternative available to recover the energy. Sizable amounts of low pressure waste steam are also useful. Reports have appeared on condensation scrubbing in multiple sieve-plate towers (364,365), aspirative condensation (366), and some of the parameters that affect particle growth and collection in a conventional orifice scrubber, with and without condensation (367). Fine particle collection is appreciably improved by scrubbing a hot saturated gas stream with cold water rather than using recirculated hot water. Little collection improvement resulted in cooling below a 50°C dewpoint, but much better collection was achieved when the hot gas was introduced already saturated close to its initial dewpoint (rather than admitting it unsaturated with substantial superheat). A decrease in scrubbing efficiency of an evaporative scrubber was also found. Addition of an adiabatic presaturator for hot gases ahead of a scrubber should be quite beneficial.

ODOR CONTROL

Odor is a subjective preception of the sense of smell. Its study is still in a developmental stage: information including a patent index has been compiled (368), 124 rules of odor preferences have been listed (369), detection and recognition threshold values have been given (370), and odor technology as of 1975 has been assessed (371). Odor control involves any process that gives a more acceptable perception of smell, whether as a result of dilution, removal of the offending substance, or counteraction or masking.

Odor Measurement

Both static and dynamic measurement techniques exist for odor. The objective is to measure odor intensity by determining the dilution necessary so that the odor is imperceptible or doubtful to a human test panel, ie, to reach the detection threshold, the lowest concentration at which an odor stimulus may be detected. The recognition threshold is a higher value at which the chemical entity is recognized. An odor unit, o.u., has been widely defined in terms of 0.0283 m^3 of air at the odor detection threshold. It is a dimensionless unit representing the quantity of odor that when dispersed in 28.3 L of odor-free air produces a positive response by 50% of panel members. Odor concentration is the number of cubic meters that one cubic meter of odorous gas will occupy when diluted to the odor threshold. Selection of people to participate in an odor panel should reflect the type of information or measurement required, eg, for evaluation of an alleged neighborhood odor nuisance, the test subjects should be representative of the entire neighborhood. However, threshold determinations may be done with a carefully screened panel of two or three people (372). A general population test panel of 35 people has been described (373). Odor measurement in general has been surveyed (398).

Static Dilution Methods. A known volume of odorous sample is diluted with a known amount of nonodorous air, mixed, and presented statically or quiescently to the test panel. The ASTM D1391 syringe dilution technique is the best known of these methods and involves preparation of a 100-mL glass syringe of diluted odorous air, which is allowed to stand 15 min to ensure uniformity. The test panel judge suspends breathing for a few seconds and slowly expels the 100-mL sample into one nostril. The test is made in an odor-free room with a minimum of 15 min between tests to avoid olfactory fatigue. The syringe dilution method is reviewed from time to time by the ASTM E18 Sensory Evaluation Committee, which suggests and evaluates changes. Instead of a syringe, a test chamber may be used, which can be as large as a room (374,375). A technique to make threshold determinations for 53 odorant chemicals has been described (374). The test room consisted of two chambers: an antechamber and the actual test chamber. The air in each chamber was circulated through activated carbon to provide a controlled, odor-free background for sample dilution.

Dynamic Dilution Methods. In this method, odor dilution is achieved by continuous flow. Advantages are accurate results, simplicity, reproducibility, and speed. Devices known as dynamic olfactometers control the flow of both odorous and pure diluent air, provide for ratio adjustment to give desired dilutions, and present multiple, continuous samples for test panel observers at ports beneath odor hoods. The Hemeon olfactometer (376,377) uses three ports, each designed as a face piece that surrounds the lower half of the face loosely, and allows three panelists to judge simultaneously. The Hellman odor fountain (378–380) is a similar device. An olfactometer based on forced-choice-triangle statistical design has been constructed (381). To distinguish dynamically obtained group threshold values from ASTM odor units, the ED_{50} (effective dose, 50%) designation may be used. ED_{50} is the concentration at which half the panelists begin to detect odor in a dynamic test. ED_{50} values are 5% higher at the 1000 o.u. level than ASTM odor unit values, 20% higher at 100 o.u. level, and 33% higher at the 20 o.u. level (381). Similar but greater deviations have been obtained when the Hemeon odor meter results have been compared to results of the ASTM static syringe method (381).

Another dynamic instrument, the Scentometer, is the basis for odor regulations in the states of Colorado, Illinois, Kentucky, Missouri, Nevada, and Wyoming, and in the District of Columbia (382). The portable Scentometer (Barneby-Cheney) can produce dilution ratios up to 128:1 in the field. The Scentometer blends two air streams, one of which has been deodorized with activated carbon. The dilution ratio is decreased until the odor becomes detectable (383). Improvements to dynamic methods have been recommended (384).

Odor Control Methods

Absorption, adsorption, and incineration are all typical control methods for gaseous odors; odorous particulates are controlled by the usual particulate control methods. However, carrier gas, odorized by particulates, may require gaseous odor control treatment even after the particulates have been removed. For oxidizable odors, treatment with oxidants such as hydrogen peroxide, ozone, and $KMnO_4$ may sometimes be practiced; catalytic oxidation has also been employed. Odor control as used in rendering plants (385), spent grain dryers (386), pharmaceutical plants (387–389), and cellulose pulping (390) has been reviewed (391–393); some reviews are presented in two symposium volumes (394,395) from APCA specialty conferences. The odor-control performances of activated carbon and permanganate–alumina for reducing odor level of air streams containing olefins, esters, aldehydes, ketones, amines, sulfide, mercaptan, vapor from decomposed crustacean shells, and stale tobacco smoke have been compared (396). Activated carbon produced faster deodorization in all cases. Activated carbon adsorbers have been used to concentrate odors and organic compounds from emission streams, producing fuels suitable for incineration (397). Both air pollution control and energy recovery were accomplished. A summary of odor control technology is available (399).

NOMENCLATURE

Symbol	*Definition*	*SI Units*
A	constant in Stokes-Cunningham correction factor for particle velocity; $A \cong 1.72$, but often considered a function of λ and D_p	dimensionless
A	constant in Calvert correlation for relating penetration and particle size in wet scrubbers, $P_t = \exp(-Ad_p^B)$	as required by equation 40
A	interfacial area of gas–liquid contact in absorption; often difficult to measure independently and thus often included with the mass-transfer coefficient K as K_{GA} or K_{LA}	m^2
A	gaseous component designation in absorption, such as gaseous component A; often used as a subscript such as p_A (partial pressure of component A)	
A	preexponential factor in kinetic equation (14) of Arrhenius form for combustion of a gaseous material in an incinerator	s^{-1}
A_e	collecting electrode area in an electrical precipitator	m^2
A_p	area of particle projected on plane normal to direction of flow or motion; $A_p = \pi D_p^2/4$ for spherical particles	m^2

Symbol	*Definition*	*SI Units*
a	fraction of inelastic collisions in thermophoresis, equation 11	dimensionless
B	exponential constant in Calvert correlation for relating penetration and particle size in wet scrubbers, $P_t = \exp(-Ad_p^B)$	as required by equation 37
B_e	distance between the discharge electrode and the collecting plate in a plate type electrical precipitator	m
b_r	roughness factor for discharge electrode in an electrical precipitator; increasing roughness decreases the value of b_r	dimensionless
C	concentration of absorbed gas in the bulk absorbing solvent in absorption; often used in conjunction with subscripts such as C_A (concentration of absorbed component A) or as C_1 or C_2 (concentration of the absorbed gas in the tower inlet and outlet liquid); concentration may be expressed in mass per unit volume of liquid or as mole fraction when the symbol x is used to show concentration	mg/m³
C^*	equilibrium concentration of absorbed gas in the gas–liquid interface in absorption between the liquid in contact with gas of equilibrium partial pressure p^*	mg/m³
C_A	concentration of absorbed gaseous component A in the bulk flow of the absorbing solvent at any point in the absorbing tower	mg/m³
C_A^*	same as for C^* but for absorbed component A	mg/m³
$C_{A,O}$ *	concentration of inlet combustible component A entering a gaseous incinerator in equation 13	mole fraction
C_A *	concentration of combustible component A in the combusting gas after any desired residence time t in a gaseous incinerator	mole fraction
C_D	overall drag coefficient = $2 F_d / \rho_g u^2 A_p$	dimensionless
D_b	representative dimension or diameter of impingement body or target	m
D_c	packing diameter in packed column scrubber	mm
D_d	diameter of the discharge electrode in an electrical precipitator	m
D_o	diameter of gas outlet pipe on cyclone	mm or m
D_p	diameter (or equivalent diameter) of a particle	m
D_{pc}	particle cut-size; diameter collected with 50% efficiency in a particulate collector	m
$D_{p\,min}$	diameter of smallest particle collectible with 100% collection efficiency	m
D_t	diameter of a collecting tube in a tubular type electrical precipitator	m
D_v	diffusion coefficient for particle	m²/s
d	differential operator, as in dx	dimensionless
d_p	diameter of particle (in wet scrubbing and Calvert figures)	μm
d_{pc}	particle cut-size in Calvert wet-scrubbing illustrations	μm
d_{pgm}	geometric mass mean particle diameter	μm
dt	differential change in time, equation 12	s
E	activation energy in the kinetic equation 14, used to calculate the oxidation rate of a combustible chemical species	J/g · mol
E_o	electrostatic charging field strength in an electrical precipitator	V/m or N/C
E_p	electrostatic precipitating field strength in an electrical precipitator	V/m or N/C
e	natural or Naperian logarithm base, 2.718	dimensionless
F	foam density above sieve plates	kg/L
F_d	drag or resistance to motion of a body in a fluid	N
f	Calvert correlation parameter for effect of hydrophobic (f = 0.25) or hydrophilic (f = 0.50) nature of	dimensionless

Symbol	Definition	SI Units
	particles collected in venturi scrubbers	
G	superficial gas mass velocity	$kg/(s \cdot m^2)$
G	mass flow rate of nonabsorbable gas in absorption; often the mass superficial velocity per unit cross section for gas flow is implied	kg/h or $kg/h \cdot m^2$
G	subscript in absorption equations applying to gas-phase controlling absorption	
G_M	superficial molar mass-flow gas velocity in an absorbing tower	$kg\ mol/h \cdot m^2$
g	local acceleration due to gravity	m/s^2
g_L	local acceleration due to gravity in English engineering units, ft/s^2	
H	Henry's Law constant for the absorbed solute in an absorbing solvent	Pa/mole fraction
$[HC]$	concentration of combustible hydrocarbons in a gas stream entering an incinerator	as required by equation 12
H_{OG}	height of a single mass-transfer unit in absorption, gas-phase controlling, as defined by equation 5	m
H_{OL}	height of a single mass-transfer unit in absorption, liquid-phase controlling, as defined by equation 8	m
K_b	design constant for knitted mesh mist collectors, typically 0.11 m/s	m/s
K_c	cyclone friction factor in equation 20	dimensionless
K_e	geometric design constant for an electrical precipitator; for a plate type, $K_e = L_e / B_e V_e$, and for a tube type, $K_e = 4\ L_e / D_t V_e$	s/m
K_G	mass-transfer coefficient for absorbing gas flow across the gas-film side of the gas–liquid interface	$kg\ mol/h \cdot m^2 \cdot Pa$
K_{GA}	combined mass-transfer coefficient and transfer-interfacial area for mass transfer across the gas-film side of a gas–liquid interface	$kg\ mol/h \cdot Pa$
K_L	mass-transfer coefficient for absorbed gas flow across the solvent-film side of the gas–liquid interface	$kg\ mol/h \cdot m^2 \cdot$ conc. difference
K_{LA}	combined mass-transfer coefficient and transfer-interfacial area for mass flow across the liquid-film side of a gas–liquid interface	$kg\ mol/h \cdot m^2 \cdot$ conc. difference
K_m	Stokes-Cunningham correction factor	dimensionless
K_s	sphericity correction constant	dimensionless
K_v	Coulomb's law constant, 8.987×10^9	$(N \cdot m^2)/C$
k	Boltzmann constant, 1.380×10^{-23}	J/K
k	pseudo first-order reaction rate constant for combustion of a hydrocarbon, equation 14	s^{-1}
k_c	cloth drag coefficient for a fabric bag type filter	as required by equation 35
k_d	drag coefficient for collected dust cake in a fabric bag type filter	as required by equation 36
k_g	thermal conductivity of gas	$W/(m \cdot K)$
k_{gTR}	translational thermal conductivity of gas	$W/(m \cdot k)$
k_p	thermal conductivity of particle	$W/(m \cdot k)$
k_ρ	density correction factor for corona starting potential in electrical precipitator; ratio of the precipitator gas density to that of air at 25°C and 101.3 kPa	dimensionless
L	superficial liquid mass velocity	$kg/(s \cdot m^2)$
L	liquid, mass flow rate of solvent in an absorption tower; often the mass superficial velocity per unit cross section is implied	kg/h or $kg/h \cdot m^2$
L_e	treatment path length in a charged precipitating field in an electrical precipitator	m
L_M	superficial molar mass flow gas velocity in an absorbing tower	$kg\ mol/h \cdot m^2$
LM	subscript denoting natural log-mean in absorption as in $(y - y^*)_{LM}$ and as defined by equation 11	
m_d	mass of dust solids collected per unit cloth area in a bag type filter	kg/m^2

Symbol	Definition	SI Units
N	number of moles of absorbable gas transferred across a gas–liquid interface in absorption	kg mol
N_A	number of moles of absorbing component A transferred across a gas–liquid interface.	kg mol
N_e	effective number of spirals made by the gas in a cyclonic separator	dimensionless
N_{OG}	number of transfer units required for a separation by absorption; gas-phase controlling; also value of the separation integral in equation 4	dimensionless
N_{OL}	number of transfer units required for a separation by absorption, liquid-phase controlling; also value of the separation integral in equation 7	dimensionless
N_{PE}	Peclet number $= U_o D_b / D_v$	dimensionless
N_{RE}	Reynolds number $= D_p \rho_p u / \mu_g$	dimensionless
N_{SC}	Schmidt number $= \mu_g / \rho_g D_v$	dimensionless
N_{SF}	flow-line separation number $= D_p / D_b$	dimensionless
N_{SI}	inertial separation number $= U_t U_o / D_b$	m/s^2
N_t	number of particulate collection transfer units $= \ln[1/(1 - \eta)]$	dimensionless
P	system or collector total pressure	Pa
P	total absolute pressure of the gas in the absorber at any point	Pa
P_T	total power expended in a contacting device	W
P_t	particle penetration through a collector $1 - \eta$ or $100 - \eta$ (fraction or percentage)	dimensionless
ΔP	pressure drop through a collector or control device	Pa
ΔP_c	pressure drop through the filtering cloth in a fabric bag filter	Pa
ΔP_d	pressure drop through the accumulated dust cake in a fabric bag filter	Pa
p_A	absolute partial pressure of absorbable component A at any point in the bulk gas stream in the absorber	Pa
p^*	equilibrium absolute vapor pressure of absorbable gas above a solution with solute concentration x	Pa
p_A^*	equilibrium absolute vapor pressure of absorbable gaseous component A above a solution with solute A concentration of x_A	Pa
Q	volumetric flow rate through collector	m^3/s
q_e	gas volumetric flow rate through an electrical precipitator	m^3/s
R	universal gas constant in the equation of state for a perfect gas, 8.314 J/g · mol k	J/g · mol K
r	radius, distance from center line of cyclone separator, or from center line of concentric cylinder electrical precipitator	m
T	system or collector absolute temperature	K
t	time	s
t	residence time at incineration temperature in equations 12 and 13	s
U_e	average particle migration velocity for a given size particle in an electrical precipitator; also used as the drift velocity, which is the average migration velocity for all particle sizes collected	M/s
U_G	superficial gas velocity in wet scrubber	m/s
U_o	linear gas velocity	m/s
U_s	operating superficial velocity through a knitted mesh mist collector	m/s
U_T	thermophoretic velocity of a particle	m/s
U_t	particle terminal settling velocity	m/s
$U_{t,n}$	particle terminal velocity in direction normal to gas velocity	m/s
u	velocity of particle relative to main body of fluid	m/s
V_c	average velocity of gas at cyclone inlet	m/s
V_{cT}	tangenital gas velocity component in a cyclone	m/s
V_e	average gas velocity in an individual flow passage in the precipitating field of an electrical precipitator	m/s
V_f	superficial gas velocity through the fabric in a bag filter	m/s

Symbol	Definition	SI Units
V_s	minimum potential required for the start of corona discharge in an electrical precipitator	V
W_i	width of gas inlet of cyclonic separator	m
x	local concentration of absorbed solute in the absorbing liquid at any point in the absorbing system expressed as a mole fraction with respect to solvent and solute combined; x_1 and x_2 refer to solute concentrations in the absorbing liquid at the inlet and outlet from the absorbing tower	mole fraction
x^*	equilibrium solute concentration in absorbing liquid in a contact with gas containing absorbable gas of mole fraction y^*	mole fraction
x_A	local concentration of absorbed component A in the bulk flow of the absorbing liquid expressed as a mole fraction with respect to solvent and solute combined	mole fraction
y	concentration of absorbable gas at any point in the gas stream in an absorbing tower expressed as mole fraction of total gas present at that point; y_1 and y_2 refer to mole fraction of absorbable gas present in the total gas stream at the tower gas inlets and outlets	mole fraction
y^*	concentration of absorbable gas that would be in equilibrium with the liquid at the same point in the absorption tower expressed as mole fraction of the total gas that would be present	mole fraction
$(y - y^*)_{LM}$	natural logarithmetic absorption driving force in the gas phase throughout the tower as expressed by equation 11	mole fraction
Z	total absorption tower contacting height for mass transfer	m
α	intercept constant in Semrau's power law collection principle, equation 39	as required by equation 39
γ	power exponent in Semrau's power law collection principle, equation 39	dimensionless
δ	dielectric constant for nonconductive particles	dimensionless
ε	packed bed void fraction in a packed bed wet scrubber	dimensionless
η	collection efficiency (fraction or percentage)	dimensionless
η_D	diffusional collection efficiency	dimensionless
η_{FL}	flow-line interception efficiency	dimensionless
η_{IN}	inertial deposition efficiency	dimensionless
η_M	collection efficiency by particle mass	dimensionless
η_t	target collection efficiency by inertial impact	dimensionless
λ	molecular mean free path, 0.1 μm for room temperature ambient air	m
μ_g	aerosol carrier gas viscosity	$Pa \cdot s$
π	geometric constant relating circumference and diameter of a circle, 3.14159	dimensionless
ρ_L	density of liquid or mist droplets	kg/m^3
ρ_M	molar liquid density in the absorption tower; moles contained in 1 m^3 of absorbing liquid averaged throughout the absorption tower	kg mol/m^3
ρ_g	density of aerosol carrier gas	kg/m^3
ρ_p	apparent or aerodynamic particle density	kg/m^3
ρ_t	true (not bulk) density of solids	kg/m^3
σ_g	geometric standard deviation	dimensionless
θ	time interval at any point since last bag cleaning in a fabric bag filter	s or h, as required, equation 37
Ψ	sphericity, surface area of sphere having the same volume as a particle divided by the actual surface area of the particle	dimensionless

Acknowledgment

Much of this article has been taken from J. Kroschwitz and M. Howe-Grant, eds., *Kirk-Othmer Encyclopedia of Chemical Technology,* 4th ed. (*ECT*4), Vol. 1, John Wiley & Sons, Inc., New York, 1991 and M. Grayson, ed., *Kirk-Othmer Encyclopedia of Chemical Technology,* 3rd ed. (*ECT*3), John Wiley & Sons, Inc., New York, 1978.

BIBLIOGRAPHY

1. *Guiding Principles of State Air Pollution Legislation,* U.S. Department of Health, Education, and Welfare, Washington, D.C., 1965.
2. Sect. 1420, Chapt. 111, *General Laws,* Chapt. 836, *Acts of 1969,* the Commonwealth of Massachusetts, Department of Public Health, Division of Environment, Health, Bureau of Air Use Management.
3. G. T. Wolff, in "Air Pollution," *ECT* 4, Vol. 1, p 711.
4. D. H. Slade, ed., *Meteorology and Atomic Energy 1968,* U.S. Atomic Energy Commission, July 1968; available as *TID-24190,* Clearinghouse for Federal Scientific and Technical Information National Bureau of Standards, U.S. Department of Commerce, Springfield, Va.
5. D. B. Turner, *Workbook of Atmospheric Dispersion Estimates, US EPA, OAP, Pub. AP26,* Research Triangle Park, N.C., revised 1970, U.S. Deparatment Printing Office Stock No. 5503-0015.
6. A. D. Busse and J. R. Zimmerman, *User's Guide for the Climatological Dispersion Model, US EPA Pub. No. EPA-R4-73-024,* Research Triangle Park, N.C., Dec. 1973.
7. M. Smith, ed., *Recommended Guide for the Prediction of the Dispersion of Airborne Effluents,* American Society of Mechanical Engineers, New York, 1968.
8. G. A. Briggs, "Plume Rise," *USAEC Critical Review Series TID-25075,* NTIS, Springfield, Va., 1969.
9. *Effective Stack Height: Plume Rise, US EPA Air Pollution Training Institute Pub. SI:406,* with Chapts. D, E, and G by G. A. Briggs and Chapt. H by D. B. Turner, 1974.
10. J. E. Carson and H. Moses, *J. APCA* **19,** 862 (Nov. 1969).
11. H. Moses and M. R. Kraimer, *J. APCA* **22,** 621 (Aug. 1972).
12. G. A. Briggs, *Plume Rise Predictions, Lectures on Air Pollution and Environmental Impact Analyses,* American Meteorological Society, Boston, Mass., 1975.
13. *Guideline on Air Quality Models, OAQPS Guideline Series,* U.S. Environmental Protection Agency, Research Triangle Park, N.C., 1980.
14. G. A. Schmel, *Atmos. Environ.* **14,** 983–1011 (1980).
15. N. E. Bowne, R. J. Londergan, R. J. Minott, D. R. Murray, "Preliminary Results from the EPRI Plume Model Validation Project—Plains Site," Report EPRI EA-1788, Electric Power Research Institute, Palo Alto, Calif., 1981.
16. N. E. Bowne, "Atmospheric Dispersion," in S. Calvert and H. M. Englund, eds., *Handbook of Air Pollution and Technology,* John Wiley & Sons, Inc., New York, 1984, 859–891.
17. T. W. Devitt in ref. 16, Chapt. 5, pp. 375–417.
18. *Compilation of Air Pollutant Emission Factors,* Office of Air and Waste Management, OAQPS, U.S. Environmental Protection Agency, Research Triangle Park, N.C., Publication No. AP-42, 3rd ed., April 1980 (or latest).
19. N. J. Weinstein, in Ref. 16, Chapt. 16 pp. 419–434.
20. A. J. Klee, in Ref. 16, Chapt. 19, pp. 513–550.
21. L. Short, in Ref. 16, Chapt. 25, pp. 673–695.
22. *Code of Federal Regulations* 40 (CFR 40), *Fed. Reg.,* C-50–99.
23. Ref. 22, part 58.
24. W. W. Lund and R. Starkey, *J. Air Waste Manage Assn.* **40**(6), 896–897 (June 1990).
25. Ref. 22 part 50, appendix A–G.
26. Ref. 22, part 62, appendix B, and subparts C–F, J, M, V.
27. T. A. Gosink, *Environ. Sci. Technol.* **9,** 630–634 (July 1975).
28. S. R. Heller, J. M. McGuire, and W. L. Budde, *Environ. Sci. Technol.* **9,** 210–213 (Mar. 1975).
29. H. H. Westberg, in Ref. 16, Chapt. 30, 771–783.
30. *Fed. Reg.* **40,** 46250 (Oct. 1975).
31. *Fed. Reg.* **52,** 24634–24750 (July 1, 1987); *Fed. Reg.* **54,** 41218–41232 (Oct. 5, 1989).
32. D. S. Ensor and M. J. Pilot, *J. APCA* **21,** 496–501 (Aug. 1971); B. B. Crocker, *Chem. Eng. Prog.* **71,** 83–89 (March 1975).
33. Ref. 22 part 60 as compiled in R. A. Corbett, *Standard Handbook of Environmental Engineering,* McGraw-Hill Publishing Co., New York, 1989.
34. Appendix to *Code of Federal Regulations,* CFR 40, Part 60, "New Source Performance Standards," U.S. Government Printing Office, Washington.
35. L. E. Sparks, in Ref. 16, pp. 800–818.
36. B. W. Loo, J. M. Jaklevic, and F. S. Goulding, in B. Y. H. Liu, ed., *Fine Particles—Aerosol Generation, Measurement, Sampling and Analysis,* Academic Press, Inc., New York, 1976, pp. 311–350.
37. "Hard Realities: Air and Waste Issues of the 90s" (Report of 18th Government Affairs Seminar), *J. Air and Waste Management Assn.* **40**(6), 855–860 (June 1990); also *Proceedings of the 18th Air and Waste Management Assn. Govt. Affairs Seminar,* Air and Waste Management Assn., Pittsburgh, Pa., 1990.
38. R. H. Perry, ed., *Engineering Manual,* 3rd ed., McGraw-Hill, Inc., New York, 1976. Text used with permission.
39. *Industrial Ventilation,* 15th ed., Sect. 4, American Conference of Governmental Industrial Hygienists, Committee on Industrial Ventilation, Lansing, Mich., 1978.
40. R. Jorgensen, *Fan Engineering,* 7th ed., Buffalo Forge Company, Buffalo, N.Y., 1970, pp. 471–480.
41. W. E. L. Hemeon, *Plant and Process Ventilation,* Industrial Press, Inc., New York, 1954.
42. J. M. Dalla Valle, *Exhaust Hoods,* Industrial Press Inc., New York, 1952.
43. J. L. Alden, *Design of Industrial Exhaust Systems for Dust and Fume Removal,* 3rd ed., Industrial Press Inc., New York, 1959.
44. J. A. Danielson, ed., *Air Pollution Engineering Manual, Pub. No. 999-AP-40,* U.S. Department of Health, Education, and Welfare, Cincinnati, Ohio, 1973, Chapt. 3.
45. H. D. Goodfellow, in J. A. Buonicore and T. Davis, eds., *Air Pollution Engineering Manual,* Von Nostrand Reinhard, New York, 1992, Chapt. 6, pp. 155–206.
46. B. B. Crocker, "Capture of Hazardous Emissions," in *Proceedings, Control of Specific (Toxic) Pollutants* (Conference, Feb. 1979, Gainsville, Fla.), Air Pollution Control Assn., Pittsburgh, Pa., 1979, pp. 415–433.
47. B. B. Crocker, *Chem. Eng. Prog.* **64,** 79 (Apr. 1968).
48. T. T. Frankenberg, *Mech. Eng.* **87,** 36–41 (Aug. 1965).
49. B. B. Crocker, "Removal of Hazardous Organic Vapors from Vent Gases," in *Proceedings: Control of Specific Toxic Pollutants,* APCA Specialty Conference, Gainesville, Fla, Feb. 1979, Air Pollution Control Assn., Pittsburgh, 1979, pp. 360–376.
50. A. J. Teller, *Chem. Eng. Prog.* **63,** 75 (Mar. 1967).
51. K. E. Lunde, *Ind. Eng. Chem.* **50,** 293 (Mar. 1958).
52. J. P. Jewell and B. B. Crocker, *Proceedings of the 8th Annual Sanitary and Water Resources Engineering Conference,* Vanderbilt University, Nashville, Tenn., 1969, pp. 211–228.

53. T. K. Sherwood and R. L. Pigford, *Absorption and Extraction,* 2nd ed., McGraw-Hill Book Co., New York, 1952.
54. B. B. Crocker and K. B. Schnelle, Jr., in Ref. 16, chapt. 7, pp. 135–192.
55. U. von Stockar and C. R. Wilke, "Absorption" in *ECT* 4, Vol. 1, pp. 38–93.
56. R. E. Treybal, *Mass Transfer Operations,* 3rd ed., McGraw-Hill Book Co. Inc., New York, 1980.
57. W. M. Edwards, in R. H. Perry and D. Green, eds., *Perry's Chemical Engineers' Handbook,* 6th ed., McGraw Hill Book Co., New York, 1984, Chapt. 14.
58. T. K. Sherwood, R. L. Pigford, and C. R. Wilke, *Mass Transfer,* McGraw-Hill Book Co., Inc., New York, 1975.
59. W. S. Norman, *Absorption, Distillation and Cooling Towers,* Longmans, Green & Co. Ltd., New York, 1962.
60. T. K. Sherwood and R. L. Pigford, *Absorption and Extraction,* 2nd ed., McGraw-Hill, New York, 1952.
61. P. L. Magill, F. R. Holden, and C. Ackley, eds., *Air Pollution Handbook,* McGraw-Hill Book Co. Inc., New York, 1956, pp. 13–73 through 13–81.
62. A. J. Buonicore, in Ref. 45, pp. 15–31.
63. D. Bienstock, J. H. Field, S. Katell, and K. D. Plants, *J. APCA* **15,** 459 (Oct. 1965).
64. D. Bienstock, J. H. Field, and J. G. Myers, *J. Eng. Power* **86**(3), 353 (1964).
65. D. M. Ruthven, "Adsorption" in *ECT* 4, Vol. 1, pp. 493–528.
66. J. D. Sherman and C. M. Yon, "Adsorption, Gas Separation," in *Ect* 4, Vol. 1, pp. 529–573.
67. M. Suzuki, *Adsorption Engineering,* Kodansba-Elsevier, Tokyo, 1990.
68. A. E. Rodrigues, M. D. LeVan, and D. Tondeur, *Adsorption, Science and Technology,* NATO ASI E158, Kluwer, Amsterdam, 1989.
69. R. T. Yang, *Gas Separation by Adsorption Processes,* Butterworths, Stoneham, Mass., 1987.
70. P. Wankat, *Large Scale Adsorption and Chromatography,* CRC Press, Boca Raton, Fla., 1986.
71. T. Vermeulen, M. D. LeVan, N. K. Hiester, and G. Klein, in Ref. 57, Sec. 16.
72. D. M. Ruthven, *Principles of Adsorption and Adsorption Processes,* Wiley-Interscience, New York, 1984.
73. A. Turk, in A. C. Stern, ed., *Air Pollution,* 3rd ed., Vol. 5, Academic Press, New York, 1977, pp. 329–363.
74. R. J. Buonicore, in Ref. 45, pp. 31–52.
75. Ref. 73, pp. 337–339.
76. B. Grandjacques, *Pollut. Eng.* **9,** 28–31 (Aug. 1977).
77. D. Kunii and O. Levenspiel, *Fluidization Engineering,* John Wiley & Sons, Inc., New York, 1969, pp. 31–33.
78. "Beaded Carbon Ups Solvent Recovery," *Chem. Eng.* **84,** 39–40 (Aug. 29, 1977).
79. Ref. 57, p. 16–36.
80. J. J. Spivey, *Environ. Progress* **7**(1), 31–40 (Feb. 1988).
81. S. M. Hall, *J. Air Waste Manage Assn.* **40**(3), 404–407 (Mar. 1990).
82. Ref. 54, pp. 185–189.
83. A. J. Buonicore, in Ref. 45, pp. 52–58.
84. Ref. 58, pp. 267–269.
85. V. S. Katari, W. M. Vatavuk, and A. H. Wehe, Part I, *J. APCA* **37**(1), 91 (Jan. 1987); Part II, *J. APCA* **37**(2), 198–201 (Feb. 1987); M. Kosusko and C. M. Nunez, *J. Air Waste Manage Assn.* **40**(2), 254–255 (Feb. 1990); M. A. Palazzolo and B. A. Tichenor, *Environ. Progress* **6,** 172–176 (Aug. 1987).
86. E. N. Ruddy and L. A. Carroll, *Chem. Eng. Progress* **89**(7), 28–35 (July 1993).
87. Ref. 45, p. 62.
88. K. C. Lee, H. J. Janes, and D. C. Macauley, *Preprint 78-58.6, APCA 71st Annual Meeting,* (June 25–30, 1978, Houston, Tex.) Air Pollution Control Assn., Pittsburgh, Pa.
89. J. Hirt, in Ref. 16, Chapt. 8, pp. 193–201.
90. H. J. Paulus, in A. C. Stern, ed., *Air Pollution,* 2nd ed., Vol. 3, Academic Press, New York, 1968, p. 528.
91. A. J. Buonicore, in Ref. 45, pp. 58–70.
92. D. R. van der Vaart, M. W. Vatavuk, and A. H. Wehe, *J. Air Waste Manage. Assoc.* **41,** 92–98 (Jan. 1991); 497–501 (April 1991).
93. P. Acharya, S. G. DeCicco, and R. G. Novak, *J. Air Waste Manage. Assoc.,* **41,** 1605–1615 (Dec. 1991).
94. L. Takacs and G. L. Moilanen, *J. Air Waste Manage. Assoc.* **41,** 716–722 (May 1991).
95. M. A. Palazzolo and B. A. Tichenor, *Environ. Prog.* **6,** 172–176 (Aug. 1987).
96. T. B. Simpson, *Environ. Prog.* **10,** 248–250 (Nov. 1991).
97. H. Herzog, D. Goglomb, and S. Zemba, *Environ. Prog.* **10,** 64–74 (Feb. 1991).
98. R. W. Coughlin, R. D. Siegel, and C. Rai, eds., *AIChE Symp. Ser.* **70,** (137) (1974). Contains four papers on flue gas desulfurization, four papers on coal desulfurization, and three papers on petroleum desulfurization.
99. C. Rai and R. D. Siegel, eds., *AIChE Symp. Ser.* **71,** (148) (1975). Contains seven papers on flue gas desulfurization, two on petroleum desulfurization, one on coal desulfurization, and fifteen on NO_x control.
100. J. A. Cavallaro, A. W. Dearbrouck, and A. F. Baher, *AIChE Symp. Ser.* **70,** (137); 114–122 (1974).
101. K. S. Murthy, H. S. Rosenberg, and R. B. Engdahl, *J. APCA* **26,** 851–855 (Sept. 1976).
102. A. V. Slack, *Chem. Eng. Prog.* **72,** 94–97 (Aug. 1976).
103. R. M. Jimeson and R. R. Maddocks, *Chem. Eng. Prog.* **72,** 80–88 (Aug. 1976).
104. *AIChE Symp. Ser.* **68,** (126) (1972). Contains four papers on flue gas desulfurization and two on NO_x control.
105. *Control Technology: Gases and Odors,* APCA Reprint Series, Air Pollution Control Assn., Pittsburgh (Aug. 1973). A reprint of 1970–1973 APCA journal articles: six papers on flue gas desulfurization, three on NO_x control, and two on SO_2 control from pulp and paper mills.
106. *Sulfur Dioxide Processing,* Reprints of 1972–1974 *Chem. Eng. Prog.* articles, AIChE, New York (1975). Contains thirteen papers on flue gas desulfurization, two on SO_2 control in pulp and paper, one on sulfuric acid tail gas, one on SO_2 from ore roasting, and two on NO_x from nitric acid.
107. E. L. Plyler and M. A. Maxwell, eds., *Proceedings, Flue Gas Desulfurization Symposium,* (New Orleans, La., Dec. 1973). *Pub. No. EPA-650/2-73-038,* U.S. EPA, Research Triangle Park, N.C., 1973. Contains 34 papers on flue gas desulfurization.
108. *Proceedings: Symposium on Flue Gas Desulfurization,* (Atlanta, Ga., Dec. 1974), U.S. EPA, Research Triangle Park, N.C.
109. *Proceedings: Symposium on Flue Gas Desulfurization* (New Orleans, Mar. 1976), U.S. EPA, Research Triangle Park, N.C. Contains 36 papers dealing with flue gas desulfurization.

110. *Proceedings of the Third Stationary Source Combustion Symposium, EPA 600/7-79-050a (NTIS PB 292 539)* U.S. EPA, Industrial Environmental Research Laboratory, Research Triangle Park, N.C., Feb. 1979.
111. G. M. Blythe and co-workers, *Survey of Dry SO_2 Control Systems, EPA-600/7-80-030 (NTIS PB 80 166853),* U.S. EPA, Industrial Research Laboratory, Research Triangle Park, N.C., Feb. 1980.
112. *Definitive SO_2 Control Process Evaluation: Limestone Double Alkali and Citrate FGD Process,* EPA Pub. EPA-600/7-79-177, Aug. 1979.
113. J. D. Mobley and K. J. Lim, Ref. 16, chapt. 9, pp. 193–213.
114. G. M. Blythe and co-workers, *Survey of Dry SO_2 Control Systems, EPA Pub. EPA-600/7-80-030 (NTIS PB 80-166853),* U.S. EPA, Research Triangle Park, N.C., Feb. 1980.
115. R. Pedroso, *An Update of the Wellman-Lord Flue Gas Desulfurization Process, EPA Pub. EPA-600/2-76-136a,* May 1976.
116. T. W. Devitt, in Ref. 17.
117. *Flue Gas Desulfurization Systems and SO_2 Control, Pub. GS-6121,* Electric Power Research Institute, Palo Alto, Calif., October 1988. An abstracted bibliography of EPRI Reports and Projects.
118. *Introduction to Limestone Flue Gas Desulfurization, Pub. CS-5849,* Electric Power Research Institute, Palo Alto, Calif., 1988.
119. *SO_2 Removal by Injection of Dry Sodium Compounds, Pub. RP-1682,* Electric Power Research Institute, Palo Alto, Calif., 1987.
120. C. S. Chang and C. Jorgensen, *Environ. Progress* **6**(1), 26 (Feb. 1987).
121. M. T. Melia and co-workers, "Trends in Commercial Applications of FGD," in *Proceedings: Tenth Symposium on FGD, EPRI Report CS-5167,* Electric Power Research Institute, Palo Alto, Calif. (May 1987).
122. *Proceedings of the EPA/EPRI First Combined Flue Gas Desulfurization and Dry SO_2 Control Symposium,* Oct. 25–28, 1988; St. Louis, Mo., Electric Power Research Institute, Palo Alto, Calif.
123. S. S. Tsai and co-workers, *Environ. Progress* **8**(2), 126–129 (May 1989).
124. W. Bartok and co-workers, *Environ. Progress* **9**(1), 18–23 (Feb. 1990).
125. A. Pakrasi, W. T. Davis, G. D. Reed, and T. C. Keener, *J. Air Waste Manage. Assoc.* **40,** 987–992 (July 1990).
126. P. Harriott, *J. Air Waste Manage. Assoc.* **40,** 998–1003 (July 1990).
127. F. Cunill, J. F. Izquierdo, J. C. Martinez, J. Tejero, and J. Querol, *Environ. Prog.* **10,** 273–277 (Nov. 1991).
128. J. A. Ahlbeck, *J. Air Waste Manage. Assoc.,* **40,** 345–351 (March 1990).
129. W. Bartok, B. A. Folsom, M. Elbl, F. R. Kurzynske, and H. J. Ritz, *Environ. Prog.* **9,** 18–23 (Feb. 1990).
130. D. Helfritch, S. Bortz, R. Beittel, P. Bergman, and B. Toole-O'Neil, *Environ. Prog.* **11,** 7–10, (Feb. 1992).
131. S. K. Gangwal, W. J. McMichael, and T. P. Dorchak, *Environ. Prog.* **10,** 186–191 (Aug. 1991).
132. B. W. Hall, C. Singer, W. Jozewicz, C. B. Sedman, and M. A. Maxwell, *J. Air Waste Manage. Assoc.* **42,** 103–110 (Jan. 1992).
133. S. G. Chang and G. C. Lee, *Environ. Prog.* **11,** 66–73 (Feb. 1992).
134. G. A. Kudlac, G. A. Farthing, T. Szmanski, and R. Corbett, *Environ. Prog.* **11,** 33–38 (Feb. 1992).
135. P. Chu, B. Downs, and B. Holmes, *Environ. Prog.* **9,** 149–155 (Aug. 1990).
136. K. V. Kwong, R. E. Meissner, III, S. Ahmad, and C. J. Wendt, *Environ. Prog.* **10,** 211–215 (Aug. 1991).
137. W. Moeller and K. Winkler, *J. APCA* **18,** 324 (May 1968).
138. L. Short, in Ref. 16, Chapt. 25, pp. 685–688.
139. D. S. Davies, R. A. Jiminez, and P. A. Lemke, *Chem. Eng. Prog.* **70,** 68 (June 1974).
140. R. G. Bierbower and J. H. Sciver, *Chem. Eng. Prog.* **70,** 60 (Aug. 1974).
141. K. T. Semrau, *J. APCA* **21,** 185 (Apr. 1971); its bibliography contains 140 refs.
142. J. E. Walther and H. R. Amberg, *J. APCA* **20,** 9 (Jan. 1970).
143. S. F. Galeano and B. R. Dillard, *J. APCA* **22,** 195 (Mar. 1972).
144. J. E. Walther, H. R. Amberg, and H. Hamby, III, *Chem. Eng. Prog.* **69,** 100 (June 1973).
145. J. D. Rushton, *Chem. Eng. Prog.* **69,** 39 (Dec. 1973).
146. *Control Techniques for Nitrogen Oxide Emissions from Stationary Sources,* Pub. No. AP-67, U.S. Department of Health, Education, and Welfare, Washington, D.C., 1970.
147. "Formation and Control of Nitrogen Oxides," *Catalysis Today* **2,** Elsevier Science Publishers B.V., Amsterdam, 1988, pp. 369–532; J. A. Miller and G. A. Fisk, "Combustion Chemistry," *C&EN,* 22–46 (Aug. 31, 1987).
148. W. Bartok, A. R. Crawford, and A. Skopp, *Chem. Eng. Prog.* **67,** 64 (Feb. 1974).
149. R. K. Srivastava and J. A. Mulholland, *Environ. Progress* **7**(1), 63–70 (Feb. 1988).
150. J. A. Mulholland and R. K. Srivastava, *J. APCA* **38**(9), 1162–1167 (Sept. 1988).
151. W. Bartok, V. S. Engleman, R. Goldstein, and E. G. del Valle, *AIChE Symp. Ser.* **68**(126), 30 (1972).
152. W. Bartok, A. R. Crawford, and G. J. Piegari, *AIChE Symp. Ser.* **68**(126), 66 (1972).
153. H. B. Lange, Jr., *AIChE Symp. Ser.* **68**(126), 17 (1972).
154. D. Thompson, T. D. Brown, and J. M. Beer, *Combust. Flame* **19,** 69 (1972).
155. W. Bartok and co-workers, *Systems Study of Nitrogen Oxide Control Methods for Stationary Sources, Pub. PB-192789,* NTIS, 1969.
156. J. H. Wasser and E. E. Berkau, *AIChE Symp. Ser.* **68**(126), 39 (1972).
157. G. B. Martin and E. E. Berkau, *AIChE Symp. Ser.* **68**(126), 45 (1972).
158. F. A. Bagwell and co-workers, *J. APCA* **21,** 702 (Nov. 1971).
159. T. W. Rhoads and co-workers, *Environ. Progress* **9**(2), 126–130 (May 1990).
160. M. J. Miller, *Environ. Progress* **5**(3), 171–177 (Aug. 1986).
161. J. S. Maulbetsch and co-workers, *J. APCA* **36,** 1294–1298 (Nov. 1986).
162. J. W. Jones, "Estimating Performance and Costs of Retrofit SO_2 and NO_x Controls for Acid Rain Abatement," *ACS Extended Abstract Preprint,* ACS Division of Environmental Chemistry Meeting (June 5–11, 1988, Toronto, Ontario).
163. Ref. 113, pp. 203–213.
164. W. Bartok and co-workers, *Chem. Eng. Progress* **84**(3), 54–71 (Mar. 1988).
165. *C&EN* 22 (April 18, 1988).
166. R. A. Perry and D. L. Siebers, *Nature* **324,** 657–658 (1986).

167. U.S. Pat. 4,800,068 (Jan. 24, 1989), R. A. Perry.
168. R. K. Lyon, *Environ. Sci. Technol.* **21**(3), 231–236 (1987).
169. D. Eskinazi, J. E. Cichanowicz. W. P. Linak, and R. E. Hall, *JAPCA* **39,** 1131–1139 (Aug. 1989).
170. A. Kokkinos, J. E. Cichanowicz, R. E. Hall, and C. B. Sedman, *J. Air Waste Manage. Assoc.* **41,** 1252–1259 (Sept. 1991).
171. D. Cobb, L. Glatch, J. Ruud, and S. Snyder, *Environ. Prog.* **10,** 49–59 (Feb. 1991).
172. L. C. Hardison, G. J. Nagl, and G. E. Addison, *Environ. Prog.* **10,** 314–318 (Nov. 1991).
173. S. L. Chen, R. K. Lyon, and W. R. Seeker, *Environ. Prog.* **10,** 182–185 (Aug. 1991).
174. R. McInnes and M. B. Van Wormer, *Chem. Eng.* **97**(9), 130–135 (Sept. 1990).
175. O. J. Adlhart, S. G. Hindin, and R. E. Kenson, *Chem. Eng. Prog.* **67,** 73 (Feb. 1971); D. J. Newman, *Chem. Eng. Prog.* **67,** 79 (Feb. 1971).
176. G. R. Gillespie, A. A. Boyum, and M. F. Collins, *Chem. Eng. Prog.* **68,** 72 (Apr. 1972).
177. R. M. Reed and R. L. Harvin, *Chem. Eng. Prog.* **68,** 78 (Apr. 1972).
178. R. M. Counce and co-workers, *Environ. Progress* **9**(2), 87–92 (May 1990).
179. L. L. Fornoff, *AIChE Symp. Ser.* **68**(126), 111 (1972).
180. B. J. Mayland and R. C. Heinze, *Chem. Eng. Prog.* **69,** 75 (May 1973).
181. R. Clift and W. H. Gauvin, *Can. J. Chem. Eng.* **49,** 439 (1971).
182. F. A. Zenz and D. F. Othmer, *Fluidization and Fluid Particle Systems,* Reinhold, New York, 1960, pp. 206–207.
183. C. A. Lapple and co-workers, *Fluid and Particle Mechanics,* University of Delaware, Newark, Del., 1956, p. 292.
184. C. B. Shepherd and C. E. Lapple, *Ind. Eng. Chem.* **31,** 972 (1939).
185. C. B. Shepherd and C. E. Lapple, *Ind. Eng. Chem.* **32,** 1246 (1940).
186. K. E. Lunde and C. E. Lapple, *Chem. Eng. Prog.* **53,** 385 (Aug. 1957).
187. C. Y. Shen, *Chem. Rev.* **55,** 595 (1955).
188. P. O. Rouhiainen and J. W. Stachiewicz, *J. Heat Trans.* **92,** 169 (1970).
189. W. E. Ranz, *Penn State Univ. Coll. Eng. Res. Bull.* B-66 (Dec. 1966).
190. W. E. Ranz and J. B. Wong, *Ind. Eng. Chem.* **44,** 1371 (1952).
191. H. F. Johnstone and M. H. Roberts, *Ind. Eng. Chem.* **41,** 2417 (1949).
192. H. J. White, *Industrial Electrostatic Precipitation,* Addison-Wesley Publishing Co., Reading, Mass., 1963.
193. B. Singh and R. L. Byers, *Ind. Eng. Chem. Fund.* **11,** 127 (1972).
194. L. Waldmann, *Z. Naturforsch,* **14A,** 589 (1959).
195. *Control Techniques for Particulate Air Pollutants,* AP-51, U.S. Department of Health, Education, and Welfare, Washington, D.C., 1969, p. 42.
196. E. S. Pettyjohn and E. B. Christiansen, *Chem. Eng. Prog.* **44,** 157 (1948).
197. H. Wadell, *Physics* **5,** 281 (1934).
198. H. A. Becker, *Can. J. Chem. Eng.* **37,** 85 (1959).
199. Ref. 182, pp. 209–216.
200. I. Langmuir and K. B. Blodgett, *U.S. Army Air Forces Technical Report 5418, February 19, 1946,* U.S. Department Commerce, Office Technical Services PB 27565.
201. Ref. 57, p. 20-83, Fig. 20-105.
202. J. H. Seinfeld, *Air Pollution: Physical and Chemical Fundamentals,* McGraw Hill, Inc., New York, 1975, p. 457.
203. A. E. Vandegrift, L. J. Shannon, and G. S. Gorman, *Chem. Eng. Deskbook* **80,** 109 (June 18, 1973).
204. K. T. Semrau, C. W. Margnowski, K. E. Lunde, and C. E. Lapple, *Ind. Eng. Chem.* **50,** 1615 (1958).
205. C. J. Stairmand, *J. Inst. Fuel* **29**(2), 58 (1956).
206. G. D. Sargent, *Chem. Eng.* **76,** 130 (Jan. 27, 1969).
207. C. P. Kerr, *J. APCA* **39**(12), 1585–1587 (Dec. 1989).
208. P. Rosin, E. Rammler, and W. Intelmann, *Zeit. Ver. Deut. Ind.* **76,** 433–437 (1932) (in German).
209. Ref. 200 p. 20-86, Fig. 20-109.
210. D. Leith and W. Licht, *AIChE Symp. Ser.* **68**(126), 196–206 (1972).
211. Ref. 90, p. 370.
212. Ref. 90 p. 361; *Air Pollution Manual,* Part II, American Industrial Hygiene Association, Detroit, Mich., 1968, p. 29.
213. *Manual on Disposal of Refinery Wastes, API Pub. 931,* American Petroleum Institute, Washington, D.C., 1975, Chapt. 11.
214. C. J. Stairmand, *Trans. Inst. Chem. Eng.* **29,** 356–383 (1951).
215. B. Kalen and F. A. Zenz, *AIChE Symp. Ser.* **70**(137), 388–396 (1974).
216. R. McK. Alexander, *Proc. Australas. Inst. Min Eng.* **152/3,** 202–228 (1949).
217. C. J. Stairmand, *Engineering* **16B,** 409–412 (Oct. 21, 1949).
218. A. C. Stern, K. J. Caplan, and P. D. Bush, *Cyclone Dust Collectors,* American Petroleum Institute, New York, 1955.
219. H. J. Tengbergen in K. Rietema and C. G. Verner, eds., *Cyclones in Industry,* Elsevier Publishing Co., Amsterdam, 1961 (in English).
220. B. B. Crocker in "Phase Separation" in Ref. 57, pp. 18-70–18-88.
221. W. Strauss, *Industrial Gas Cleaning,* 2nd ed., Pergamon Press, Inc., Oxford, 1975.
222. K. J. Caplan in Ref. 90, Chapt. 43.
223. Ref. 195, Figs. 4-43, 4-44.
224. A. T. Murphy, F. T. Adler, and G. W. Penney, *AIIE Trans.* Preprint, Paper 59-102 (1959).
225. Ref. 192, p. 140.
226. S. Oglesby, Jr., and G. B. Nichols, "A Manual of Electrostatic Precipitator Technology," *NTIS Report PB196380,* Southern Research Institute, Birmingham, Ala., 1970.
227. E. Anderson in R. H. Perry, ed., *Chemical Engineers' Handbook,* 2nd ed., McGraw-Hill, Inc., New York, 1941, p. 1873.
228. W. B. Smith and J. R. McDonald, *J. APCA* **25,** 168 (Feb. 1975).
229. J. P. Gooch and N. L. Francis, *J. APCA* **25,** 108 (Feb. 1975).
230. J. P. Gooch and J. R. McDonald, *AIChE Symp. Ser.* **73**(165), 146 (1977).
231. P. Cooperman, "Nondeutschian Phenomena in Electrostatic Precipitation," *Preprint 76-42.2 APCA 69th Annual Meeting, Portland, Oregon, June 27–July 1, 1976.*
232. P. L. Feldman, K. S. Kumar, and G. D. Cooperman, *AIChE Symp. Ser.* **73**(165), 120 (1977).
233. *Chem. Eng.* **83,** 51 (Aug. 30, 1976).
234. C. L. Burton and D. A. Smith, *J. APCA* **25,** 139 (Feb. 1975).

235. A. G. Hein, *J. APCA* **39,** 766–771 (May 1989).
236. W. T. Sproull, *J. APCA* **15,** 50 (Feb. 1965).
237. W. T. Sproull, *J. APCA* **22,** 181 (Mar. 1972).
238. H. W. Spencer, III, and G. B. Nichols, "A Study of Rapping Re-entrainment in a Large Pilot Electrostatic Precipitator," *Preprint 76-42.5, 68th APCA Annual Meeting, Portland, Oregon, June 27–July 1, 1976.*
239. S. Matts, *J. APCA* **25,** 146 (Feb. 1975).
240. A. B. Walker, *J. APCA* **25,** 143 (Feb. 1975).
241. R. E. Cook, *J. APCA* **25,** 156 (Feb. 1975).
242. E. B. Dismukes, *J. APCA* **25,** 152 (Feb. 1975).
243. R. E. Bickelhaupt, *J. APCA* **25,** 148 (Feb. 1975).
244. H. J. White, *J. APCA* **24,** 313 (Apr. 1974).
245. H. J. White, in Ref. 16, pp. 283–329.
246. S. Oglesby, Jr., and G. B. Nichols, *Electrostatic Precipitation,* Marcel Dekker, Inc., New York, 1978, pp. 132–155; resistivity and conditioning discussion.
247. M. R. Schioeth, *Pulse-Energized Electrostatic Precipitators,* paper at the Third Conference on Electrostatic Precipitation, Albano-Padova, Italy, October, 1987, F. L. Smidth Co., Valby, Denmark.
248. V. Reyes, *Comparison between Traditional and Modern Automatic Controllers on Fullscale Precipitators,* Seventh Symposium on Transfer and Utilization of Particulate Control Technology, Nashville, Tenn., March 1988, F. L. Smidth Co., Valby, Denmark.
249. H. J. Hall, *J. APCA* **25,** 132 (Feb. 1975).
250. H. J. White, *J. APCA* **25,** 102 (Feb. 1975).
251. *J. APCA* **25,** 362 (Apr. 1975).
252. T. T. Shen and N. C. Pereira, "Electrostatic Precipitation" in *Handbook of Environmental Engineering,* Vol. 1, The Humana Press, Clifton, N.J., 1979, Chapt. 4, pp. 103–143.
253. J. R. Benson and M. Corn, *J. APCA* **24,** 340 (Apr. 1974).
254. J. H. Turner, P. A. Lawless, T. Yamamoto, D. W. Coy, G. P. Greiner, J. D. McKenna, and W. M. Vatavuk, in Ref. 45, pp. 89–114.
255. Ref. 57, pp. 20–110 to 20–121.
256. E. Bakke, *J. APCA* **25,** 163 (Feb. 1975).
257. E. Bakke, "The Application of Wet Electrostatic Precipitators for Control of Fine Particulate Matter," *Preprint, Symposium on Control of Fine Particulate Emissions from Industrial Sources, Joint U.S.-USSR Working Group, Stationary Source Air Pollution Control Technology, San Francisco, Calif., Jan. 15–18, 1974.*
258. Ref. 57 p. 20-120, Fig. 20-152.
259. D. W. Cooper, "Fine Particle Control by Electrostatic Augmentation of Existing Methods," *Preprint 75-02.1, 68th APCA Annual Meeting, Boston, Mass., June 15–20, 1975.*
260. K. A. Nielsen and J. C. Hill, *Ind. Eng. Chem. Fundam.* **15,** 149, 157 (1976).
261. M. J. Pilot, *J. APCA* **25,** 176 (Feb. 1975).
262. M. J. Pilot and D. F. Meyer, *University of Washington Electrostatic Spray Scrubber Evaluation, NTIS PB 252653,* Apr. 1976.
263. K. Zahedi and J. R. Melcher, "Electrofluidized Beds in the Filtration of Submicron Particulate," *Preprint 75-57.8, 68th APCA Annual Meeting, Boston, Mass., June 15–20, 1975.*
264. W. L. Klugman and S. V. Sheppard, "The Ceilcote Ionizing Wet Scrubber," *Preprint 75-30.3, 68th APCA Annual Meeting, Boston, Mass., June 15–20, 1975.*
265. Helfritch, *Chem. Eng. Progress* **73,** 54–57 (Aug. 1977).
266. C. W. Lear, W. F. Krieve, and E. Cohen, *J. APCA* **25,** 184 (Feb. 1975).
267. D. H. Pontius, L. G. Felix, and W. B. Smith, "Performance Characteristics of a Pilot Scale Particle Charging Device," *Preprint 76-42.6, 69th APCA Annual Meeting, Portland, Oregon, June 27–July 1, 1976.*
268. J. R. Melcher and K. S. Sachar, "Charged Droplet Scrubbing of Submicron Particulate," *NTIS Pub. PB-241262,* Aug. 1974.
269. Ref. 90, p. 413.
270. D. Leith and M. W. First, *J. APCA* **27,** 534–539 (June 1977); **27,** 636–642 (July 1977).
271. C. E. Billings and J. Wilder, "Handbook of Fabric Filter Technology," *NTIS Pub. PB200648, PB200649, PB200651, PB200650,* 1970.
272. L. Bergmann, *J. APCA* **24,** 1187 (Dec. 1974).
273. E. R. Frederick, *J. APCA* **24,** 1164 (Dec. 1974).
274. R. Dennis, *J. APCA* **24,** 1156 (Dec. 1974).
275. R. L. Lucas in R. H. Perry and C. H. Chilton, eds., *Chemical Engineers Handbook,* 5th ed., McGraw-Hill, Inc., New York, 1973, p. 20–90.
276. V. E. Schoeck in B. E. Kester, ed., *Proceedings, Specialty Conference on Design, Operation, and Maintenance of High Efficiency Particulate Control Equipment,* Air Pollution Control Association, Pittsburgh, Pa., 1973, pp. 103–117.
277. R. A. Gross, *Proceedings of the Specialty Conference on the User and Fabric Filter Equipment II,* Air Pollution Control Association, Pittsburgh, Pa., 1975, pp. 159–163.
278. *J. APCA* **25,** 715 (July 1975). Informative Report No. 7 of the APCA TC-1 Particulate Committee.
279. G. W. Bowerman in Ref. 16, pp. 153–158.
280. E. Bakke, *J. APCA* **24,** 1150 (Dec. 1974).
281. J. D. McKenna, J. C. Mycock, and W. O. Lipscomb, *J. APCA* **24,** 1144 (Dec. 1974).
282. Ref. 57, pp. 20–97 to 20–104.
283. J. H. Turner and J. D. McKenna, "Control of Particles by Filters," in Ref. 16, Ch. 11.
284. C. Orr, *Filtration, Principles and Practice,* 2 vols., Marcel Dekker, Inc., New York, 1977.
285. J. D. McKenna and D. A. Furlong in Ref. 45, pp. 114–132.
286. J. J. McKetta, *Unit Operations Handbook,* Vol. 2, Marcel Dekker, New York, 1992.
287. M. J. Hobson, *Review of Baghouse Systems for Boiler Plants,* pp. 74–84.
288. R. A. Gross, *Proceedings of the Specialty Conference on the User and Fabric Filter Equipment II,* Air Pollution Control Association, Pittsburgh, Pa., 1975, pp. 159–163.
289. R. L. Adams, "Fabric Filters for Control of Power Plant Emissions," *Preprint 74-100, 67th APCA Annual Meeting, Denver, Col.,* June 9–13, 1974.
290. R. P. Janoso, "Baghouse Dust Collectors on a Low Sulfur Coal Fired Utility Boiler," *Preprint 74-101, 67th APCA Annual Meeting, Denver, Col., June 9–13, 1974.*
291. D. S. Ensor, R. G. Hooper, R. C. Carr, and R. W. Scheck, "Evaluation of a Fabric Filter on a Spreader Stoker Utility Boiler," *Preprint 76-27.6, 69th APCA Annual Meeting, Portland, Oregon, June 27–July 1, 1976.*
292. J. D. McKenna, J. C. Mycock, and W. O. Lipscomb, "Applying Fabric Filtration to Coal Fired Industrial Boilers: A Pilot Scale Investigation," *NTIS Pub. PB-245186,* Aug. 1975.
293. R. Dennis and co-workers, "Filtration Model for Coal Fly

Ash with Glass Fabrics," EPA Rpt. EPA-600/7-77-095a, *NTIS Pub. PB 276-489/AS,* August 1977.
294. R. L. Chang, T. R. Snyder, and P. Vann Bush, *J. APCA* **39,** Pt. I, 228; Pt. II, 361 (1989).
295. K. M. Kushing, R. L. Merritt, and R. L. Chang, *J. Air Waste Manage Assn.* **40,** 1051–1058 (July 1990).
296. J. H. Turner, *J. APCA* **24,** 1182 (Dec. 1974).
297. R. E. Jenkins and co-workers, *J. APCA* **37,** Pt. I, 749; Pt. II, 1105 (1987).
298. D. C. Drehnel, "Relationship between Fabric Structure and Filtration Performance in Dust Filtration," *NTIS Pub. PB-222237,* 1973.
299. B. Miller, G. E. R. Lamb, and P. Costanza, "Influence of Fiber Characteristics on Particulate Filtration," *NTIS Pub. PB-239997,* Jan. 1975.
300. *Fuel Econ.* **12,** 47 (Oct. 1936).
301. U.S. Pat. 2,604,187, Dorfan.
302. Taub, "Filtration Phenomena in a Packed Bed Filter," Ph.D. Thesis, Carnegie-Mellon University, Pittsburgh, Pa., 1970.
303. Reese, *TAPPI,* **60**(3), 109 (1977).
304. Parquet, *The Electroscrubber Filter: Applications and Particulate Collection Performance,* EPA-600/9-82-005c, p. 363, 1982.
305. Self and co-workers, "Electric Augmentation of Granular Bed Filters," EPA-600/9-80-039c, p. 309, 1980.
306. Schueler, *Rock Prod.,* **76**(7), 66 (1973); **77**(11), 39 (1974).
307. EPA-600/7-78-093, 1978.
308. A. M. Squires and R. Pfeffer, *J. APCA* **20,** 534 (Aug. 1970); L. Paretsky, L. Theodore, R. Pfeffer, and A. M. Squires, *J. APCA* **21,** 204 (Apr. 1971); A. M. Squires and R. A. Graff, *J. APCA* **21,** 272 (May 1971); U.S. Pat. 3,296,775, A. M. Squires.
309. S. Miyamoto and H. L. Bohn, *J. APCA* **24,** 1051 (Nov. 1974); **25,** 40 (Jan. 1975).
310. U.S. Pat. 3,410,055, F. A. Zenz.
311. Ref. 57, p. 20-89.
312. H. P. Meissner and H. S. Mickley, *Ind. Eng. Chem.* **41,** 1238 (1949).
313. D. S. Scott and D. A. Guthrie, *Can. J. Chem. Eng.* **37,** 200 (1959).
314. M. L. Jackson, *AIChE Symp. Ser.* **70**(141), 82 (1974).
315. O. H. York, *Chem. Eng. Prog.* **50,** 421 (1954); O. H. York and E. W. Poppele, *Chem. Eng. Prog.* **59,** 45 (June 1963).
316. O. D. Massey, *Chem. Eng. Prog.* **55,** 114 (May 1959); *Chem. Eng.* **66,** 143 (July 13, 1959).
317. J. W. Coykendall, E. F. Spencer, and O. Y. York, *J. APCA* **18,** 315 (May 1968).
318. J. A. Brink, W. F. Burggrabe, and J. A. Rauscher, *Chem. Eng. Prog.* **60,** 68 (Nov. 1964).
319. J. A. Brink, *Chem. Eng.* **66,** 183 (Nov. 16, 1959); *Can. J. Chem. Eng.* **41,** 134 (1963); J. A. Brink, W. F. Burggrabe, and L. E. Greenwell, *Chem. Eng. Prog.* **64,** 82 (Nov. 1968).
320. Ref. 195, p. 73.
321. J. Goldfield, V. Greco, and K. Gandhi, *J. APCA* **20,** 466 (July 1970).
322. K. T. Semrau, *J. APCA* **10,** 200 (1960).
323. K. T. Semrau, *J. APCA* **13,** 587 (1963).
324. Ref. 275, p. 20-98.
325. G. J. Celenza, *Chem. Eng. Prog.* **66,** 31 (Nov. 1970).
326. S. Calvert, *J. APCA* **24,** 929 (1974).
327. S. Calvert, J. Goldschmid, D. Leith, and D. Mehta, "Scrubber Handbook," *NTIS Pub. PB-213016 and PB-213017,* July–Aug. 1972.
328. S. Calvert, *Chem. Eng.* **84,** 54–68 (Aug. 29, 1977).
329. S. Calvert, "Particulate Control by Scrubbing," in Ref. 16, Chapt. 10, pp. 215–248.
330. H. E. Hesketh, "Atomization and Cloud Behavior in Wet Scrubbers," *U.S.-USSR Symposium on Control of Fine Particulate Emissions,* Jan. 15–18, 1974.
331. W. Strauss, "Particulate Collection by Liquid Scrubbing," in *Industrial Gas Cleaning,* 2nd ed., Pergamon Press, Oxford, 1975, Chapt. 9, pp. 367–408.
332. K. Schiffner and H. Hesketh, Ref. 45, pp. 78–88.
333. *J. APCA* **22,** 54 (June 1972).
334. C. J. Stairmand, *Trans. Inst. Chem. Eng.* **28,** 130 (1950).
335. C. J. Stairmand, *J. Inst. Fuel* **29**(2), 58 (1956).
336. Ref. 275, p. 20–97.
337. Ref. 331, p. 374.
338. R. V. Kleinschmidt and A. W. Anthony in L. C. McCabe, ed., *U.S. Technical Conference on Air Pollution,* McGraw-Hill, Inc., New York, 1952, p. 310.
339. *ECT* 4, Vol. 1, pp. 755–756.
340. S. V. Sheppard, *J. APCA* **22,** 278 (Apr. 1972).
341. H. Doyle and A. F. Brooks, *Ind. Eng. Chem.* **49**(12), 57A (1957).
342. L. S. Harris, *Chem. Eng. Prog.* **42,** 55 (Apr. 1966).
343. H. E. Gardenier, *J. APCA* **24,** 954 (Oct. 1974).
344. H. F. Johnstone, R. B. Field, and M. C. Tassler, *Ind. Eng. Chem.* **46,** 1601 (1954).
345. W. Barth, *Staub* **19,** 175 (1959).
346. S. Calvert and D. Lundgren, *J. APCA* **18,** 677 (Oct. 1968).
347. S. Calvert, D. Lundgren, and D. S. Mehta, *J. APCA* **22,** 529 (July 1972).
348. R. H. Boll, *Ind. Eng. Chem. Fundam.* **12,** 40 (Jan. 1973).
349. H. E. Hesketh, *J. APCA* **24,** 939 (Oct. 1974).
350. R. H. Boll, L. R. Flais, P. W. Maurer, and W. L. Thompson, *J. APCA* **24,** 934 (Oct. 1974).
351. S. W. Behie and J. M. Beeckmans, *J. APCA* **24,** 943 (Oct. 1974).
352. K. G. T. Hollands and K. C. Goel, *Ind. Eng. Chem. Fundam.* **14,** 16 (Jan. 1975).
353. A. B. Walker and R. M. Hall, *J. APCA* **18,** 319 (May 1968).
354. S. Calvert, I. L. Jashnani, and S. Yung, *J. APCA* **24,** 971 (Oct. 1974).
355. S. Calvert, I. L. Jashnani, S. Yung, and S. Stalberg, "Entrainment Separators for Scrubbers—Initial Report," *NTIS Pub PB-241189,* Oct. 1974.
356. E. B. Hanf, *Chem. Eng. Prog.* **67,** 54 (Nov. 1971).
357. T. R. Blackwood, *Environ. Progress* **7**(1), 71–75 (Feb. 1988).
358. L. E. Sparks and M. J. Pilat, *Atmos. Environ.* **4,** 651 (1970).
359. H. W. Danser, Jr., *Chem. Eng.* **57,** 158 (May 1950).
360. C. A. Stokes, *Chem. Eng. Prog.* **46,** 423 (1950).
361. E. P. Mednikov, *Acoustic Coagulation and Precipitation of Aerosols,* USSR Academy of Science, Moscow, 1963, translated by C. V. Larrick, Consultants Bureau, 1965.
363. G. R. Gillespie and H. F. Johnstone, *Chem. Eng. Prog.* **51,** 74F (Feb. 1955).

364. S. Calvert and N. C. Jhaveri, *J. APCA* **24,** 946 (1974).
365. S. Calvert, J. Goldshmid, D. Leith, and N. Jhaveri, "Feasibility of Flux Force/Condensation Scrubbing for Fine Particle Collection," *NTIS Pub. PB-227307,* Oct. 1973.
366. S. R. Rich and T. G. Pantazelos, *J. APCA* **24,** 952 (Oct. 1974).
367. K. T. Semrau and C. L. Witham, "Condensation and Evaporation Effects in Particulate Scrubbing," *Preprint 75-30.1, 68th APCA Annual Meeting, Boston, Mass., June 15–20, 1975.*
368. J. P. Cox, *Odor Control and Olfaction,* Pollution Sciences Publishing Company, Lynden, Washington, 1975.
369. R. W. Moncrieff, *Odour Preferences,* John Wiley & Sons, Inc., New York, 1966.
370. W. H. Stahl, ed., *Compilation of Odor and Taste Threshold Values Data,* American Society for Testing and Materials Data Series 48, Philadelphia, Pa., 1973.
371. P. N. Cheremisinoff and R. A. Young, eds., *Industrial Odor Technology Assessment,* Ann Arbor Science Publishers, Inc., Ann Arbor, Mich., 1975.
372. J. Wittes and A. Turk, *Correlation of Subjective–Objective Methods in the Study of Odors and Taste,* American Society for Testing and Materials Special Technical Publication 440, Philadelphia, Pa., 1968, pp. 49–70.
373. F. V. Wilby, *J. APCA* **19,** 96 (1969).
374. G. Leonardos, D. Kendall, and N. Barnard, *J. APCA* **19,** 91 (1969).
375. A. Turk, *Basic Principles of Sensory Evaluation,* American Society for Testing and Materials Special Technical Publication #433, 1968, pp. 79–83.
376. W. C. L. Hemeon, *J. APCA* **18,** 166 (1968).
377. W. C. L. Hemeon, *AIChE Symp. Ser.* **73**(165), 260 (1977).
378. T. M. Hellman and F. H. Small, *Chem. Eng. Prog.* **69,** 75 (1973).
379. T. M. Hellman and F. H. Small, *J. APCA* **24,** 979 (1974).
380. T. M. Hellman in Ref. 371, pp. 45–56.
381. A. Dravnieks and W. H. Prokop, *J. APCA* **25,** 28 (1975).
382. G. Leonardos, *J. APCA* **24,** 456 (1974).
383. R. A. Duffee, *J. APCA* **18,** 472 (1968).
384. S. C. Varshney, E. Poostchi, A. W. Gnyp, and C. C. St. Pierre, *Environ. Progress* **5**(4), 240–244 (Nov. 1986).
385. R. M. Bethea and co-workers, *Environ. Sci. Technol.* **7,** 504 (1973).
386. M. W. First and co-workers, *J. APCA* **24,** 653 (1974).
387. G. A. Herr, *Chem. Eng. Prog.* **70,** 65 (1974).
388. D. E. Quane, *Chem. Eng. Prog.* **70,** 51 (1974).
389. D. J. Eisenfelder and J. W. Dolen, *Chem. Eng. Prog.* **70,** 48 (1974).
390. J. E. Paul, *J. APCA* **25,** 158 (1975).
391. J. E. Yocom and R. A. Duffee, *Chem. Eng.* **77**(13), 160 (1970).
392. M. Beltran, *Chem. Eng. Prog.* **70,** 57 (1974).
393. R. M. Bethea, *Engineering Analysis and Odor Control,* Chapt. 13, pp. 203–214, ref. 371.
394. *Proceedings, State of the Art of Odor Control Technology Specialty Conference,* March 1974, Air Pollution Control Association, Pittsburgh, Pa., 1974.
395. *Proceedings, State of the Art of Odor Control Technology Specialty Conference,* March 1977, Air Pollution Control Association, Pittsburgh, Pa., 1977.
396. A. Turk, S. Mehlman, and E. Levine, *Atmos. Environ.* **7,** 1139 (1973).
397. W. D. Lovett and F. T. Cunniff, *Chem. Eng. Prog.* **70,** 43 (May 1974).
398. H. Van Langenhove and N. Schamp, in P. N. Cheremisinoff, ed., *Encyclopedia of Environmental Control Technology,* Vol. 2, Gulf Publishing Co., Houston, 1989, Chapt. 25, pp. 935–965.
399. W. H. Prokop, in Ref. 45, Chapt. 5, pp. 147–154.

Reading List

R. G. Bond and C. P. Straub, eds., *Handbook of Environmental Control,* Vol. 1, CRC Press, Cleveland, Ohio, 1972.

A. J. Buonicore and W. T. Davis, eds., *Air Pollution Engineering Manual,* Van Nostrand Reinhold, New York, 1992.

S. Calvert and H. M. Englund, eds., *Handbook of Air Pollution Technology,* John Wiley & Sons, Inc., New York, 1984.

P. N. Cheremisinoff, ed., *Encyclopedia of Environmental Control Technology,* Vol. 2, Gulf Publishing Co., Houston, 1989.

P. N. Cheremisinoff and R. A. Young, *Air Pollution Control and Design Handbook,* Parts 1 and 2, Marcel Dekker, New York, 1977.

R. A. Corbett, *Standard Handbook of Environmental Engineering,* McGraw-Hill Publishing Co., New York, 1989.

C. N. Davies, ed., *Aerosol Science,* Academic Press, Inc., New York, 1966.

H. E. Hesketh, *Air Pollution Control,* Ann Arbor Science Publishers, Ann Arbor, Mich., 1979.

L. H. Keith, *Compilation of EPA's Sampling and Analysis Methods,* Lewis Publishers, Inc. (CRC Press Inc, Boca Raton, FL), 1991.

W. Licht, *Air Pollution Control Engineering: Basic Calculations for Particulate Collection,* Marcel Dekker, Inc., New York, 1980.

B. Y. H. Liu, ed., *Fine Particles—Aerosol Generation, Measurement, Sampling and Analysis,* Academic Press, Inc., New York, 1976.

J. J. McKetta, *Unit Operations Handbook,* Vols. 1 and 2, Marcel Dekker, Inc., New York, 1992.

H. C. Perkins, *Air Pollution,* McGraw-Hill, Inc., New York, 1974.

R. H. Perry and D. Green, eds., *Perry's Chemical Engineers' Handbook,* 6th ed., McGraw-Hill Book Co., New York, 1984.

W. Ruch, ed., *Chemical Detection of Gaseous Pollutants,* Ann Arbor Science Publishers, Inc., Ann Arbor, Mich., 1966.

J. H. Seinfeld, *Air Pollution: Physical and Chemical Fundamentals,* McGraw-Hill, Inc., New York, 1975.

A. C. Stern, ed., *Air Pollution,* 3rd ed., Vols. 1–5, Academic Press, Inc., New York, (Vols. 1–3, 1976; Vols. 4–5, 1977).

W. Strauss, *Industrial Gas Cleaning,* 2nd ed., Pergamon Press, Oxford, 1975.

L. Theodore and A. J. Buonicore, *Air Pollution Control Equipment—Selection, Design, Operation and Maintenance,* Prentice-Hall, Englewood Cliffs, N.J., 1982.

L. K. Wang and N. C. Pereira, eds., *Handbook of Environmental Engineering,* Vol. 1—Air and Noise Pollution Control, The Humana Press, Clifton, N.J., 1979.

K. Wark and C. F. Warner, *Air Pollution: Its Origin and Control,* 2nd ed., Harper & Row, Publishers, New York, 1981.

P. O. Warner, *Analysis of Air Pollutants,* John Wiley & Sons, Inc., New York, 1976.

AIR POLLUTION: HEALTH EFFECTS

HENRY GONG, JR.
WILLIAM S. LINN
Rancho Los Amigos Medical Center
Downey, California and
University of Southern California School of Medicine

The recognition that air pollution can impair health is not new. Galen (born 130 AD) recognized the dangers from exposures to air contaminants in certain occupations. Treatises on occupational lung diseases and a recognition of community air pollution appeared during the Middle Ages. Governmental regulations intended to control air pollution have existed since the early 1300s. For example, harsh punishment (including death) was in store for people who violated regulations for burning coal in 14th-century England. Air pollution itself was subsequently identified as a direct cause of death during episodes in which pollutant levels were elevated well above those that occur on a daily basis, usually in association with certain meteorologic conditions. Thus, sharp increases in death rates were documented during severe pollution episodes in Belgium (Meuse River Valley) in 1930, the United States (Donora, Pennsylvania) in 1948, and England (London) in 1952 and 1962. The most notable recent example of a major episode was the London fog of December 1952 when approximately 4,000 people, out of an exposed population of 8 million, died prematurely from high levels of coal smoke. This disaster prompted serious efforts to control traditional air pollution, ie, coal smoke containing sulfur dioxide (SO_2) and a wide variety of particulate matter. The first quantitative analyses of public health statistics in Britain, along with the first report specifically associating air pollution with ill health and arguing for mitigation measures, were published during the 1660s (1).

In the United States the current system of air pollution regulations and related health research dates only from the late 1960s. Substantial progress has been made in understanding and preventing the health effects of environmental pollutants. Episodes of pollution still occur in the United States but not to the extent as in past decades. Internationally, ambient concentrations of pollutants have fallen to small fractions of their pre-1950 levels throughout the more affluent industrialized nations.

Although historical episodes of air pollution have been definitely associated with health effects, the major concern today has shifted to more subtle issues. The current emphasis is on the extent of health hazard posed by chronically low levels and intermittently high levels of outdoor and indoor pollution to which large populations are routinely exposed (2). The protection of at-risk or unusually susceptible groups is also important. Nitrogen dioxide (NO_2), ozone (O_3), particulates (including acidic aerosols), environmental tobacco smoke (ETS), asbestos, radon, and volatile organic compounds (VOCs) have become prominent health issues in the lay news media and biomedical research. Causal relationships of current air pollution to public health, such as diverse pollutant mixtures to which people are exposed, different exposure risks and heterogeneous responses within populations, and the multiplicity of environmental and lifestyle factors that can also contribute to increased morbidity and mortality, are not clear for many reasons. On the other hand, air pollution is a suspected contributor to exacerbations and premature deaths from asthma, as well as other chronic lung and heart diseases, on the basis of many statistical studies. Thus, the emphasis has shifted from avoiding unusually severe episodes with clinical diseases among highly exposed individuals to protecting large segments of the population against unacceptable risk from lower levels of pollution.

This article reviews scientific evidence concerning health effects of contemporary outdoor and indoor air pollution. Although many suggested health effects of environmental pollution remain controversial, epidemiologic, clinical, and animal toxicologic research has clearly identified changes at the molecular, cellular, organ, individual, and population levels that merit attention. This article initially reviews general concepts: common components of air pollution and methods used to measure them, health effects and their determinants, and the primary scientific approaches of assessing health effects. Respiratory effects are emphasized since the airways and lungs are the target sites for many inhaled pollutants. Known and uncertain aspects of the current database, as well as strengths and limitations of different research methodologies, will be discussed. Specific pollutants and their health effects will then be reviewed, along with possible control strategies. This review emphasizes information pertinent to the United States (U.S.), where many air pollutants have been systematically monitored, regulated, and subjected to formal health risk assessment. Substantive air quality problems in other developed countries are similar, although priorities and regulatory approaches may differ from those in the United States. The health effects of contemporary air pollution in other countries are reported elsewhere (2). It should be recognized that this review is necessarily superficial and that knowledge about health effects and risks from air pollution is continually evolving. Also, this article provides a current foundation and perspective for further reading, understanding, and questioning about this complex, multifaceted topic. See also AIR POLLUTION INDOOR; AIR POLLUTION: AUTOMOBILE; AIR POLLUTION CONTROL METHODS.

AIR POLLUTION

Definitions

Air pollution may be defined as the contamination of outdoor or indoor air by a natural or man-made agent in such a way that the air becomes less acceptable for intended uses, which, in this context, is the maintenance of human health. Although natural pollution sources are sometimes important, most pollution-related health problems result from man-made (anthropogenic) pollution involving mobile sources (eg, automobiles), outdoor stationary sources (eg, power plants, smelters, and factories), and various indoor sources (eg, building materials and combustion). Physically, air pollutants are dispersed into the atmosphere as gases, fibers, or as suspensions of liquid or solid particles in air (aerosols). Gases, in the form of discrete molecules, form true solutions within air. Fibers are arbitrarily defined as particles having a length at least three times their width. Aerosols may contain particles either of uniform size (monodisperse) or of different sizes (polydisperse or heterodisperse). The practicalities of particle monitoring dictate a focus on aerodynamic properties rather than the physical size of particles. Particles that move in an air stream, similar to one-micrometer spherical water droplets, have an aerodynamic diameter of one micrometer (μm), regardless of their physical size, shape, and density. Thus, aerosols are characterized by a mass median aerodynamic diameter (MMAD), a measure of central tendency, and a geometric standard deviation (GSD), a measure of dispersion.

Outdoor Air Pollutants

Clean air is a relative term. Chemically pristine or "pure" air is never found in nature because numerous natural sources continually contribute various agents, while meteorologic conditions mix and disperse them. For example, volcanic action and erosion inject particulate matter into the air, while decomposition of organic matter releases organic and inorganic particulates and gases. Globally, natural sources exceed man-made emissions and account for most of the atmospheric particulates and trace gases, except for SO_2 and carbon monoxide (CO). Levels of specific air pollutants at any given location depend on complex interactions of natural and anthropogenic sources which change over time (daily, seasonally, or annually) (3). Local anthropogenic sources often predominate. Thus, pollution patterns and associated health risks vary widely at different times and places because sources and meteorologic conditions continually change.

The major ambient pollutants with known or suspected adverse health effects are listed below.

Primary Pollutants	Secondary Pollutants
Carbon monoxide	Ozone
Sulfur dioxide	Nitrogen dioxide
Nitrogen oxides	Peroxyacetyl nitrate
Nitric oxide	Nitric and nitrous acid
Nitrogen dioxide	Suspended particulates
Volatile organic compounds	Sulfuric acid and sulfate salts
Suspended particulates	Nitrate salts
Metal compounds	Organic aerosols
Dusts	
Soots	

Some community air pollutants are directly released into the air. These are called primary pollutants. Those air pollutants not directly released into the atmosphere but are formed within it by chemical reactions among primary pollutants and normal constituents of air are called secondary pollutants. Secondary pollutants have been increasingly emphasized in health research. Classic urban air pollution from the combustion of coal and oil fuel, as observed in the northeastern U.S. and Europe, is characterized by primary pollutants with reducing chemical properties, ie, carbon particulates and SO_2. On the other hand, Southern California and other dry sunny regions have a characteristic oxidizing photochemical "smog" resulting from a complicated series of photochemical reactions involving primary pollutants (nitrogen oxides and hydrocarbons). Episodes of acidic sulfate particles (in the northeastern U.S.) and nitric acid vapors (in California) result from the oxidation of primary pollutants SO_2 and nitrogen oxides, respectively. Nonetheless, the distinction between primary and secondary pollutants is not always clear or practical. Both primary and secondary pollutants frequently contribute to the ambient pollutant mix in a given locale so that one cannot generally ascribe a health outcome to either alone. Certain pollutants may be primary or secondary, such as NO_2.

Sources and emission burdens of individual pollutants will not be discussed in detail here but are covered in other articles and cited references (3,4). In general, coal combustion emits more SO_2, CO, and particulates than oil combustion, whereas emission of nitrogen oxides (NO_x) is similar for both fuels. The fate of effluents from fossil fuel combustion depends on the manner in which they are released. Large modern stationary sources, such as electric power plants and smelters, typically discharge through tall stacks. This type of discharge diminishes local pollutant concentrations at ground level (except during episodic downdrafts) and enhances pollutant dispersion but increases residence and reaction times in the atmosphere, favoring the production of secondary pollutants and regional pollution problems. On the other hand, home furnaces discharge their effluents near roof level, resulting in higher, more localized ground level concentrations. The greatest source of air pollutants in the U.S. is vehicular exhaust. Fuel combustion in conventional gasoline and diesel engines is generally incomplete and results in exhaust gases containing varying amounts of CO, SO_2, NO_x, VOCs, and particulates. Industrial processes constitute a major and extremely varied and complex source of effluents. Specific industries vary greatly in the types and amounts of emissions during manufacturing, shipping, or storage of raw or processed products. The three main sources in terms of total tons emitted per year, are the chemical, primary metal, and paper industries.

Atmospheric particulates are characterized by numerous chemical species and multimodal size patterns (5). Ambient particles can be separated into two size modes—coarse and fine. Coarse particles are larger than 2.5 μm and are formed by mechanical, grinding, and other dispersive processes. Their major chemical constituents in urban areas are calcium, silicon, and aluminum. Fine particles (median size 2.5 μm or less) are generally secondary pollutants, formed by condensation processes from the vapor or gaseous state, followed by agglomeration. Sulfate and nitrate are the two largest constituents in terms of mass in outdoor urban air. From the biomedical standpoint, the aerodynamic size of particulates in air determines the potential for adverse health effects. Many coarse particles are too large to be inhaled beyond the nose and mouth (upper airways) and are not usually considered health hazards. The current federal standard regulates only particles 10 μm or less in aerodynamic diameter (PM_{10}) since these sizes are in the respirable range and potentially responsible for health effects. Thus, the upper size limit for collection of particles in contemporary air pollution studies is typically 10 μm MMAD.

Indoor Air Pollutants

Indoor air pollutants originate from both outdoor and indoor sources (Table 1) (6–9). Generally, outdoor pollutants which enter indoor environments are modified or lowered in concentration by various physical and chemical processes. For example, outdoor ozone reacts rapidly with surfaces and water vapor ("sinks") inside buildings so that its concentration is usually reduced by half or more. The concentrations of indoor air contaminants depend on many factors such as the source strength and ventilation within the indoor space. However, when indoor sources or human activities emit pollutants at rates exceeding their removal rates by ventilation or surface reactions, the indoor pollutants can accumulate to levels much higher than outdoor concentrations. Exposure to a wide range of indoor pollutants in different microenvironments (eg, residences, workplaces, schools or public buildings) is universal and virtually unavoidable. The most common exposures are to combustion emissions from appliances such as cooking ranges and furnaces and to hydrocarbon emissions from furnishing materials, hobbies, or cleaning activities. The principal pollutants generated by combustion are CO, NO_2, nitrous acid, SO_2, formaldehyde, VOCs, and particulate matter. Environmental tobacco smoke (ETS) also contributes high levels of gases and particulates. Asbestos-containing materials in thermal insulation, wallboard, and flooring release fibers into the air if disturbed or removed. Radon decay products vary in indoor concentrations and depend on the radium concentration and permeability of the soil, presence of openings through the foundations, and building ventilation. Allergens and other biological agents are present in indoor air in the form of fecal material from house dust mites and other insects, fungal spores, and dander from pets. Although biological agents are not always considered pollutants, they may impose a greater aggregate burden of illness than non-biological agents in some circumstances.

Table 1. Common Indoor Air Pollutants[a] with Known or Possible Adverse Health Effects

Pollutant	Sources
Tobacco smoke	Cigarettes, cigars, pipes
Volatile organic compounds	Solvents, cleaning chemicals, paints, glues, polishes, waxes, aerosol sprays, pesticides
Carbon monoxide	Gas and wood stoves, kerosene heaters
Nitrogen dioxide	Gas and wood stoves, kerosene heaters
Formaldehyde	Furniture stuffing, paneling, particle board, foam insulation, veneer furniture, carpets, draperies
Ozone	Outdoor ozone, photocopying machines, laser printers, electrical air cleaners
Particulates	Multiple sources, eg, wood and coal stoves, fireplaces, tobacco smoke
Asbestos	Insulation, vinyl tiles
Radon	Soil gas, water, building materials, utility natural gas

[a] Though not always considered air pollutants, bioaerosols occur both indoors and outdoors as potential infectious or antigenic agents; they include spores, pollens, molds, bacteria, viruses, animal proteins.

Other Environments with Air Pollutants

Microenvironments associated with transportation have not been well characterized but available evidence suggests that they can transiently expose occupants to high levels of different agents. For example, increased levels of CO, NO_2 and NO_x, particulates, and VOCs from automobile exhaust have been measured in automobile passenger compartments. Biologic aerosols (molds) from automobile air conditioners have also been documented. Localized high pollution levels have been measured in heavy traffic, at intersections, "street canyons" of cities, underground parking garages, parking lots, and drive-up facilities where automobile emissions are not substantially diluted before contact with people. Interestingly, some pollutants may be removed via local chemical interactions, eg, ozone is scavenged by nitric oxide in automobile exhaust, while acidic aerosols are neutralized by oral ammonia (10).

Occupational exposures to inorganic and organic dusts, VOCs, and particulates are well-documented health risks (11,12). Historically, major concerns and research have focused on inordinately high levels in the industrial setting. However, exposures with health consequences can also occur in "white-collar" offices and commercial settings where combustion products, VOCs, formaldehyde, biologic aerosols, or accumulated CO_2 are prevalent.

Air Quality Standards

Outdoor Pollutant Standards. The protection of public health has been the major driving force in air pollution control, an effort that includes legislated standards based upon judgments of scientific knowledge about health effects. The Clean Air Act of 1970, with subsequent amendments, is the basic law governing protection of air quality in the United States. The U.S. Environmental Protection Agency (EPA) is responsible for formulating and implementing the regulations governing two broad categories of air pollutants: "criteria" and "hazardous" air pollutants. The criteria pollutants are those agents identified by the U.S. EPA as airborne contaminants that "cause or contribute air pollution which may reasonably be anticipated to endanger public health" (42 USC 7408). The criteria pollutants have less alarming health effects at low exposure levels but occur widely and affect a large proportion of the population, causing a potentially great overall health impact. The Clean Air Act assumes that some level of exposure to criteria pollutants can be permitted without harm to health.

The EPA Administrator is required to 1) determine the lowest exposure level for each criteria pollutant which causes adverse health effects in the most susceptible segments of the population, and 2) set primary National Ambient Air Quality Standard (NAAQS) on the basis of this information. The NAAQS must prevent known adverse health effects and include a margin of safety to protect against health effects yet to be discovered. The Administrator must decide what constitutes an adverse health effect and how large a margin of safety is needed. Primary standards must be set without regard to cost. Cost may be considered in deciding what specific pollution control measures are required and how soon they must be implemented. Implementation is usually the responsibility of state and local governments. Thus, specific pollution control strategies and regulations vary from place to place, even though air quality standards apply nationwide.

The EPA Scientific Staff has primary responsibility for compiling and evaluating appropriate scientific evidence, and then recommending a standard to the Administrator. The staff's work is reviewed by a Clean Air Scientific Advisory Committee (CASAC) composed of independent scientists who advise the EPA on health-related issues (13,14). Other independent scientists are frequently consulted by the EPA staff or the CASAC to provide additional review. Virtually all relevant, peer-reviewed information from the scientific literature is summarized in an Air Quality Criteria Document. A shorter staff paper is then prepared to advise the EPA Administrator as to which information is most important and how it should be applied in the choice of a standard. A period of public comment is allowed when a document is issued or a standard is proposed. The Clean Air Act requires that this entire process operate on a five-year cycle. In practice, the process may take considerably longer. At the end of each cycle the EPA Administrator will either reaffirm or modify the existing Standard or set a new NAAQS on the basis of current scientific information.

Six pollutants that commonly occur in outdoor air are regulated by the U.S. EPA as criteria pollutants with individual NAAQS (Table 2): ozone, nitrogen dioxide, sulfur dioxide, carbon monoxide, particulate matter which is 10 μg or less in aerodynamic diameter (PM_{10}), and lead. The averaging times for the standards are chosen to provide a biologically relevant measure of exposure and, at the same time, a margin of safety. For example, the current NAAQS for CO is set at 9 parts per million (ppm), by volume, for an 8-hour average and 35 ppm for a one-hour average so that the blood concentrations of carboxyhemoglobin (COHb) are less than 2%, based on pharmacokinetic data about CO uptake and elimination in the body. State and local governments are required by federal law to monitor air quality in most urban areas of the U.S. In 1990, over 4,000 monitoring sites reported data to the EPA's Aerometric Information Retrieval System. Areas that do not meet the air quality standards are called "nonattainment areas." The number of nonattainment areas has decreased dramatically since the 1970s. However, many heavily populated urban and rural areas in the U.S. remain in nonattainment for one or more of the criteria pollutants. For example, 98 counties with 135 million people are presently classified as nonattainment areas for ozone (15). Seventy counties (mostly in western U.S.) are nonattainment areas for inhalable particulates (PM_{10}). Fifty counties in the Ohio River Valley and northeastern U.S. are in nonattainment for sulfur dioxide. Only metropolitan Los Angeles is in nonattainment for nitrogen dioxide. Outdoor levels of airborne lead have decreased significantly since the removal of lead from gasoline in the mid-1970s, and all monitored areas meet the lead standard.

The Clean Air Act also aims at controlling levels of other airborne chemicals for which no ambient air quality standard can be applied but "which may reasonably be anticipated to result in an increase in mortality or an increase in serious irreversible, or incapacitating reversible illness" (42 USC 7412). Title III of the 1990 Amendments (16) lists 189 substances or groups of substances as Hazardous Air Pollutants (HAP) or "air toxics" (see examples below).

	Compounds with:
Benzo[a]pyrene	Arsenic
Benzene	Beryllium
Carbon disulfide	Cadmium
Carbon tetrachloride	Chromium
Chlorine	Cobalt
Chloroform	Lead
Formaldehyde	Manganese
Hydrochloric acid	Mercury
Phenol	Nickel
Toluene	

The toxic air pollutants, which include organic compounds such as benzene and mineral compounds such as cadmium, are associated with cancer or other catastrophic health effects. Technology-based emission standards are imposed on industrial sources of HAP to protect public health. Subsequent to implementation and compliance with these standards, an analysis of the remaining "residual risks" is also required.

The EPA Administrator may also establish secondary air quality standards to prevent unfavorable effects of air

Table 2. National Ambient Air Quality Standards (NAAQS)[a] in United States, 1993

Pollutant	Maximum Allowable Concentration	Averaging Time
Ozone (O_3)	0.12 ppm (235 μg/m^3)	One hour
Nitrogen dioxide (NO_2)	0.053 ppm (100 μg/m^3)	Annual arithmetic mean
Sulfur dioxide (SO_2)	0.14 ppm (365 μg/m^3)	24 hours
	0.03 ppm (80 μg/m^3)	Annual arithmetic mean
Carbon monoxide (CO)	35 ppm (40 μg/m^3)	One hour
	9 ppm (10 μg/m^3)	8 hours
Particulates (PM_{10})[b]	150 μg/m^3	24 hours
	50 μg/m^3	Annual arithmetic mean
Lead (Pb)	1.5 μg/m^3	Quarterly arithmetic mean

[a] Primary standards designed to protect public health.
[b] Particles with aerodynamic diameter of 10 μm or less.

pollutants on public welfare, eg, damage to natural ecosystems, crop loss, or minor effects on humans that are considered nuisances rather than adverse health effects.

Indoor Pollutant Guidelines. There are no federal regulations of indoor air quality per se. Rather, the standards most widely applied to indoor air are those used to regulate worker exposures in industrial environments (17). Workplaces are regulated by the Occupational Safety and Health Administration (OSHA). The American Conference of Governmental Industrial Hygienists (ACGIH) annually publishes a list of recommended maximum acceptable workplace exposure concentrations for more than 500 toxic chemicals and dusts. These Threshold Limit Values (TLV), are defined as air concentrations of chemicals to which nearly all workers may be exposed for 40-hour work weeks over a working lifetime without ill effects. The TLV are not legal regulations but many of them have been adopted by OSHA as Permissible Exposure Limits (PEL). The TLV apply to healthy workers and should not be extended to the general population, which includes potentially more susceptible groups such as young children, the elderly, and the sick who are usually exposed for periods longer than 40 hours a week to contaminants in the home environment.

Standards for ventilation in nonresidential buildings are set by the American Society of Heating, Refrigeration, and Air Conditioning Engineers (ASHRAE). For example, ASHRAE (18) recommends a minimum indoor ventilation of 0.57 m^3 (20 ft^3) per minute per person. Indoor air quality guidelines or standards have also been recommended by other professional and governmental groups (eg, Consumer Product Safety Commission). These guidelines usually specify either ventilation requirements or acceptable indoor concentrations of specific air contaminants. They must not be interpreted as absolute values below which adverse health effects will not occur. Guidelines are developed on a pollutant-by-pollutant basis and do not consider toxic interactions that may occur in multi-contaminant indoor environments. These interactions may be antagonistic, additive, or more than additive (synergistic). Synergistically interacting pollutants might cause health effects even when exposure levels are very low. Standards or even rough guidelines are currently unavailable for most indoor pollutants derived from biologic sources (eg, allergens, microbes). When guidelines are available, one can measure contaminant levels and compare them with guidelines to judge the degrees of health risk. However, results, particularly from a single test, must be interpreted cautiously. Concentrations of indoor air contaminants vary widely depending on such factors as: the type, condition, and use of the source; indoor temperature and relative humidity; ventilation; air mixing; and the season. Also, guidelines reflect the limited state of knowledge at a given time and are subject to change; for example, the TLV for ozone has been reduced from 1.0 ppm to 0.1 ppm.

HEALTH EFFECTS

Concern about the health risks of air pollution reflects the frequent exposures to numerous pollutants in various environments, the diverse mechanisms by which these pollutants might cause health effects or diseases, and the wide range of susceptibility to pollutants in the population. Humans typically inhale 10,000 to 20,000 L (or approximately 11 to 22 kilograms or 25 to 50 pounds) of air daily, so that doses of pollutants inhaled at even low concentrations may become biologically significant with sustained exposure. The respiratory tract of humans possesses important physical, chemical, and immunologic defense mechanisms for clearing and detoxifying inhaled agents. However, the defense systems may be impaired by disease, overwhelmed by large pollutant doses, or may not be fully effective during long-term exposures.

Definition of Adverse Health Effect

There is no broadly accepted objective definition of adverse health effects caused by air pollution (19), despite widespread recognition of air pollution as a potentially large public health problem. Adversity is perceived differently by patients, practicing physicians, and scientists and depends greatly on individual circumstances. It can be defined generally as a biological change that reduces the level of well-being or functional capacity, with the implication of long-term consequences (20). Adverse respiratory health effects can be defined as medically significant changes as indicated by one or more of the following: (*1*) interference with normal activity of the affected person or persons; (*2*) episodic respiratory illness; (*3*) incapacitating illness; (*4*) permanent respiratory injury; and/or (*5*) accelerated or premature respiratory dysfunction. Examples include: (*1*) transient shortness of breath due to asthma, requiring interruption of work or school; (*2*) increased frequency of full-blown asthma attacks, requiring medical attention; (*3*) acute bronchitis or shortness of breath requiring bed rest; (*4*) pulmonary fibrosis (lung scarring) resulting from repeated lung inflammation; and (*5*) statistically lower average lung function in population groups with higher exposure levels. The final situation might not be of immediate medical significance for typical members of the affected population but portends an increased risk of disability and early death from lung disease.

Many pollution-associated effects are nonspecific (ie, are vague or transient or have possible causes other than air pollution) and go unrecognized. As investigative techniques become more sophisticated, subtle effects will be detected that may be equivocal in medical significance. For example, a small, temporary measured change in lung function may show statistical significance, eg, difference from baseline measurement with less than 5% probability of being a chance variation ($p < 0.05$). However, the change may or may not be biologically or medically relevant. Thus, not all measurable biologic, physiologic, or clinical changes are necessarily adverse for the purposes of the Clean Air Act. "Trivial" or "nuisance" effects might include low carboxyhemoglobin concentrations (<2.5%) from CO exposure or eye, nose and throat irritation during exposures to photochemical air pollution. On the other hand, some subtle, temporary physiologic changes may have medically significant long-term medical consequences. Exposure to some air pollutants may produce small but statistically significant changes in the responsiveness of airways (bronchial passages) to inhaled nonspecific bronchoconstrictors (ie, drugs that constrict the airways, such as methacholine) (21). This physiologic finding suggests not only a biological reaction to an air pollutant stimulus (ie, alteration of airway responsiveness) but also inflammation of airways, raising the possibility of acute exacerbation of asthma or even the induction of chronic airways hyperresponsiveness. The latter is believed to occur in occupational asthma caused by exposures to, for example, toluene diisocyanate, a synthetic chemical, or plicatic acid, a natural product found in cedar sawdust.

A theoretical continuum of increasing effects from a given air pollution exposure in various segments of the population is presented in Figure 1. The more intense the biological effects, the smaller is the proportion of people who experience the effects. The largest proportion of the population has some physiological or subclinical effects from air pollution, while smaller numbers of individuals develop pathophysiologic changes, morbidity (clinical illness), and death. The threshold cited above in which responses may be considered adverse and thus requires regulatory action, lies between physiologic and pathophysiologic changes. This boundary is not always well defined and varies with specific circumstances of exposure and host susceptibility. An increase in exposure concentration or duration tends to shift the middle or right portions of the curve upward, ie, to move more people into the range of adverse effects.

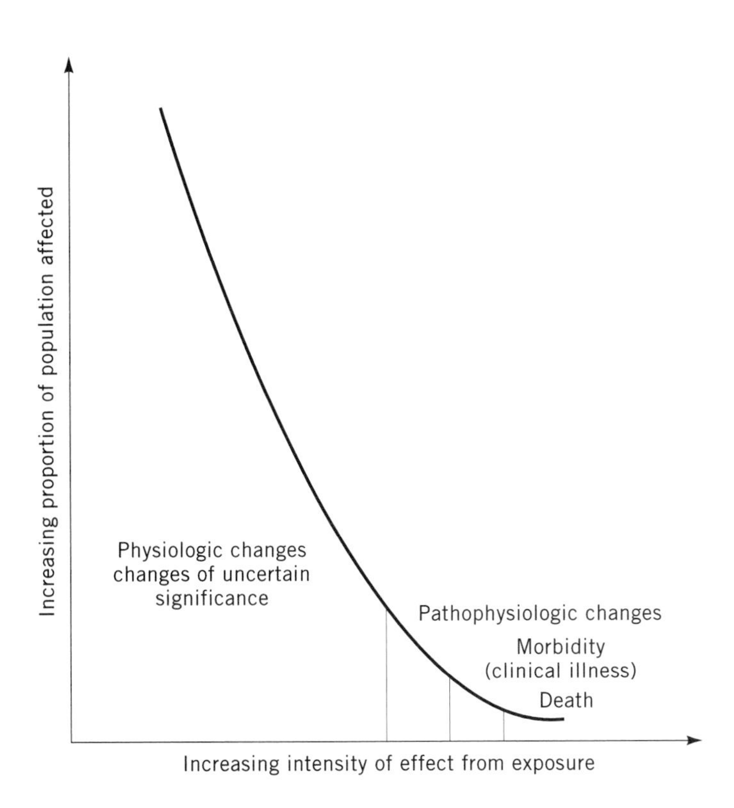

Figure 1. Relationships between increasing intensity of effect from exposure to air pollution and the proportions of the population which are affected. See text for details.

Determinants of Health Effect

The complex pathway leading to health effects begins with the generation and discharge of air pollutants from indoor and outdoor sources, continues with personal exposure to the pollutants, and, finally, may reach biologic effects of varying intensity (Fig. 2) (22). The intermediate stages in this process influence whether an exposure does or does not result in a measurable effect. Factors that affect the health outcome include the pollutant's concentration, its physical and chemical characteristics, the exposed individual's location (microenvironment) and activity pattern, the duration (contact time), and the susceptibility of the exposed individual. Exposure is defined as contact between the body and the external environment containing the agent of concern. Exposure is one (but not the only) important determinant of the dose of an agent at target sites in the human body. Important components of exposure are the concentration of the agent in inhaled air, the duration of inhalation of the contaminated air, and the ventilation rate (volume of air inhaled per unit of time) during the period. The same exposure may be achieved by various combinations of concentration, duration, and ventilation rates. However, short-term peak exposures may elicit biologic effects that are different from that observed in longer-term, lower-level exposures with the same product of concentration, time, and

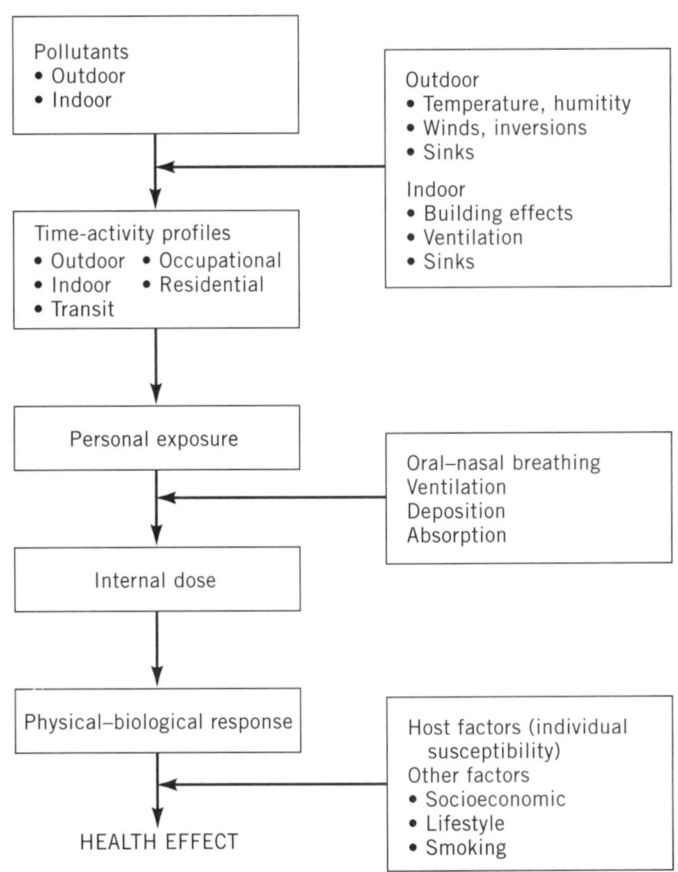

Figure 2. Factors that affect the development of adverse health effects.

ventilation rate. The respiratory tract is the most important route of exposure and thus the critical target organ system for most common inhaled pollutants, other than carbon monoxide.

Exposures can be estimated (modeled) most accurately by evaluating a person's time-activity pattern, ie, determining how much time a person spends in various locations (microenvironments) with measured or estimated air pollutant levels and how active the person is (reflecting how much air he/she breathes in) in each microenvironment (22–25). Most people are exposed to a wide range of pollutants at low levels during their normal daily activities; some individuals can be exposed to certain pollutants at levels much higher than average. The indoor environment is an important location for exposure because most people spend 80–90% of their time indoors. However, the more critical index is total exposure, which is the sum of exposures received indoors, outdoors, and in transit. The relative contributions of each setting to total exposure will vary according to age, gender, and socioeconomic, occupational, and health status. It is frequently too costly or technically unfeasible to measure air pollution exposures directly for a given individual, subgroup, or population. Thus, exposure surrogates or indices are often used to estimate exposures. These indirect measures may be pollutant concentrations in air at different but relevant times or places; measurable concentrations of the agent of interest or a related substance within the body (eg, carboxyhemoglobin [COHb] from CO inhalation); or mathematical models of exposure based on a combination of indirect measures (22–26). For example, ambient levels of O_3, NO_2, CO, and other outdoor pollutants are often used to estimate exposures for the residents of a particular community. These estimates may be reasonable for the few people outdoors near the monitoring stations. However, the accuracy of exposure estimates usually decreases with increasing distance from the stations and increasing time spent indoors and in transit. The best estimates are those based on direct measurements of personal exposure, eg, by real-time recording of CO concentrations with portable monitors, or biochemical measurements closely related to internal dose, eg, COHb in blood or nicotine/cotinine in saliva. The selection of an exposure estimator depends on the goals of the study, required level of accuracy, and costs. Generally, the cost of estimating exposure is directly proportional to the certainty of the estimate (accuracy).

The other important link in the pathway from pollutant to health effect is the internal dose, which is the quantity of agent deposited at a target site within the body where toxic effects occur. Dose and exposure are not necessarily equivalent for inhaled particles and gases. Penetration into the lung and retention at potential sites of injury depend on the physical and chemical properties of the pollutant (eg, particle size, solubility, reactivity), the exposed person's ventilation rate, the proportion of air breathed through the nose versus the mouth, airway anatomy, and efficiency of clearance mechanisms (27). Highly water-soluble gases, such as sulfur dioxide and formaldehyde, are almost completely extracted by the upper airway (nasopharynx) of a resting subject during a brief exposure, whereas less soluble gases, such as NO_2 and O_3, penetrate to the small airways and alveoli. The penetration of particles into the lung and sites of deposition within the lung depend on the aerodynamic size of the particle. Particles greater than 10 μm MMAD are effectively removed by the upper airway, whereas smaller particles penetrate and deposit in the lower airways and alveoli where local clearance mechanisms (mucociliary clearance, macrophages, permeation into blood or lymphatic vessels) are active. The response of the respiratory tract to inhaled fibers depends on fiber width, length, and susceptibility to dissolution. Regardless of the type of pollutant, the heightened ventilation during exercise increases the amount of inhaled air and the proportion of oral-to-nasal breathing, thus enhancing the dose of inhaled pollutants.

Dose measurement is more difficult and uncertain than exposure measurement. The dose of many air pollutants cannot be accurately measured in most human exposure situations. Thus, exposure–response, rather than dose–response, relationships are measured or modeled for most pollutants. In general, the likelihood of a health effect shows a strong positive correlation with exposure as well as dose. The relationship between exposure and response may have different forms, depending on the mechanisms of action (Fig. 3) (2). The shape and slope of the exposure–response relationship have important biologic and clinical implications. Response curves having a threshold that must be exceeded to produce a response or disease indicate that levels below the threshold are without risk. Thus, only individuals who exceed the threshold are at risk and a regulation could permit a safe or acceptable level of exposure. In contrast, curves without a threshold imply that any level of exposure conveys some risk. For example, the linear no-threshold relationship is widely used to assess risks of carcinogenesis for regulatory purposes. However, the assumption that a linear no-threshold model applies for carcinogenesis remains very contro-

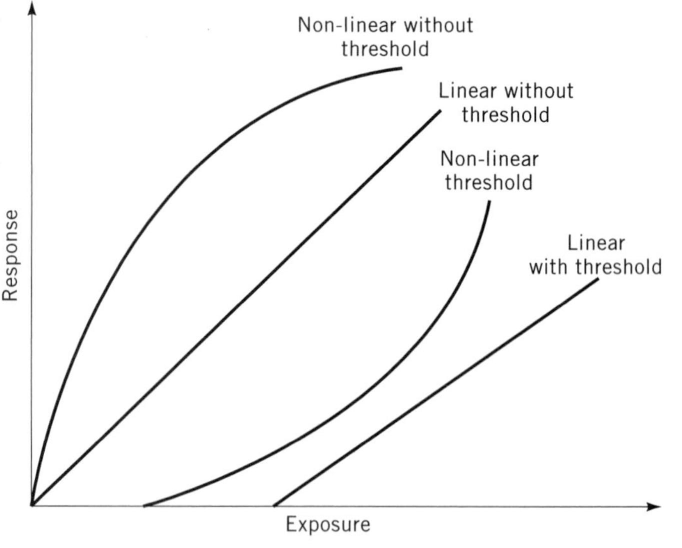

Figure 3. Theoretical exposure–response relationships for inhaled pollutants. Response indicates a biological response, eg, a change in symptoms, lung function, biochemistry, or activity of target cells, tissues, or organs.

versial. In practice, distinguishing among the different exposure–response curves may be difficult using either animal or human studies. Even when a threshold exists, it may be very low in some susceptible subgroups so that, for practical purposes, there is no safe exposure level.

The site of deposition of the inhaled material largely determines the type of clinical response (2). For example, nasal deposition of pollen may result in rhinitis in individuals with hay fever. Sulfur dioxide, aeroallergens, radon, or asbestos may enter the conducting airways (trachea, bronchi, and terminal bronchioles) where bronchoconstriction or bronchogenic carcinoma may develop. Certain fungi and inorganic dusts can affect the lung parenchyma or alveoli (where gas exchange occurs) and result in hypersensitivity pneumonitis and pneumoconiosis, respectively. The mechanisms by which deposited inhaled agents injure the airways and lungs are not fully understood but can be broadly categorized as acute irritation and inflammation, chronic inflammation accompanied by a fibrotic response (scarring), immediate and cell-mediated immune responses, and carcinogenesis (see below).

Mechanism	Pollutant
Bronchoconstriction	Sulfur dioxide, acidic aerosols
Inflammation	Ozone, environmental tobacco smoke
Fibrosis	Asbestos, silica
Carcinogenesis	Radon, asbestos, formaldehyde, active smoking, environmental tobacco smoke

The likelihood of an adverse response depends on the mass of pollutant deposited, the site of deposition, the rate of clearance or detoxification, and characteristics of the exposed person that determine susceptibility.

Certain individuals and groups may have one or more characteristics that place them above the norm or usual risk for adverse health effects (28). The Clean Air Act specifically requires protection of populations at risk, ie, groups having a significantly higher probability of developing illness or other abnormal health status than the general population (see below).

Group	Air Pollutant
Healthy young adults who exercise or work outdoors	Ozone
Children and adolescents	Ozone, lead
Pregnant women and fetuses	Carbon monoxide
Patients with asthma	Sulfur dioxide, ozone, nitrogen dioxide, particulates (sulfuric acid, sulfates)
Patients with chronic obstructive pulmonary disease (COPD)	Carbon monoxide, particulates, sulfur dioxide
Patients with ischemic heart disease	Carbon monoxide, particulates, sulfur dioxide

Inherent individual characteristics (eg, age, gender, psychologic factors, or possibly race/ethnicity) and interactions with other environmental factors (eg, nutrition, preexisting disease, medications, lifestyle or socioeconomic factors, access to health care) or agents (eg, tobacco smoking and asbestos exposure) determine a group's biological responsiveness to a specific pollutant. Particular groups may differ in their sensitivity to different pollutants. For example, most asthmatics are very sensitive to SO_2 in comparison with healthy people. On the other hand, healthy athletes develop noticeable ozone-related responses after heavy exercise, whereas asthmatics generally have little excess sensitivity to ozone in controlled exposures.

Another way to understand the potential health impact of air pollution on susceptible groups is to review the size of at-risk populations. In 1991, a total of 514 counties and 20 cities in the U.S. were designated as nonattainment areas for one or more of the six criteria pollutants. Age-specific national prevalence rates for medical conditions (National Health Interview Survey, Centers for Disease Control) and the 1990 county-specific U.S. census data can be used to derive the prevalence of populations at risk (29) (Table 3). The nonattainment sites represent an estimated 164 million persons (66% of the resident U.S. population), including 63% (approximately 31 million) of preadolescent children (13 years or less in age), 60% (19 million) of persons aged 65 years or greater, and 64% (9 million) of persons with chronic obstructive pulmonary disease (COPD). In addition, 61% of children with asthma and 65% of adults with asthma, and 33% of pregnant women and patients with coronary artery disease reside in nonattainment areas. The data do not include estimates for healthy persons who exercise extensively outdoors, an important group at risk for health effects from ozone and other outdoor pollutants. The increased ventilation associated with exercise enhances oral breathing (bypassing nasal defenses) and penetration of inhaled air into the deep lung, resulting in increased total dose. Thus, the estimates of at-risk populations in nonattainment areas may substantially underestimate the number of persons potentially exposed to unhealthy air quality.

Clinical Effects

Table 4 lists common symptoms, signs, and diagnoses associated with air pollutants. Effects in individuals can vary from trivial events to interference with normal activity to death. As discussed previously, some mild and temporary symptoms (eg, eye irritation) may be considered nuisances, while more persistent, activity-limiting symptoms such as wheezing from bronchospasm or chest pains from ischemic heart disease (ie, condition with inadequate oxygen supply to the heart muscle) would be considered adverse health effects. The table also indicates that pollutant-related effects can involve the respiratory tract at all levels (upper airways, lower airways, and alveoli) and may involve target organs outside the lungs. For example, inhaled carbon monoxide can inordinately increase myocardial oxygen needs during exercise, resulting in decreased exercise performance in healthy athletes and the early onset of chest pains (angina pectoris) in patients

Table 3. Estimated Populations at Risk[a] Residing in Communities That Have Not Attained One or More National Ambient Air Quality Standard[b], United States, 1991[c,d]

Population Subgroup at Risk	Pollutant[e]	At-risk Population Living in Nonattainment Areas No.	% [f]
Preadolescent children (aged ≤13 yrs)	PM-10, SO_2, O_3, NO_2	31,528,939	63
Elderly (aged ≥65 yrs)	PM-10, SO_2, O_3	18,846,666	60
Persons with pediatric asthma[g]	PM-10, SO_2, O_3, NO_2	2,285,061	61
Adults (aged ≥18 yrs) with asthma	PM-10, SO_2, O_3, NO_2	4,279,413	66
Persons with chronic obstructive pulmonary disease[h]	PM-10, SO_2, O_3, NO_2	8,831,970	64
Persons with coronary heart disease[i]	CO	3,493,847	33
Pregnant women[j]	CO, Pb	1,602,045	38
Children aged ≤5 yrs	Pb	74,312	3

[a] Populations at-risk estimates should be quoted individually and not added to form totals. These categories are not mutually exclusive.
[b] Only a portion of some communities are designated as nonattainment areas for some pollutants. The totals in this document are based on entire county populations and may, therefore, reflect an overrepresentation of the true populations at risk.
[c] Estimated total U.S. population living in the 514 counties and 20 cities designated as nonattainment areas = 164 million (66% of the U.S. population).
[d] Ref. 29.
[e] PM-10 = particulate matter with a diameter ≤10 μm; SO_2 = sulfur-dioxide; O_3 = ozone; NO_2 = nitrogen dioxide; CO = carbon monoxide; and Pb = lead.
[f] The proportion of each population subgroup at risk of the total population in the category.
[g] Asthma in persons aged <18 years.
[h] Includes chronic bronchitis and emphysema.
[i] Includes ischemic and cerebrovascular heart disease.
[j] The estimated number of pregnant women in each county is derived from the number of live births. Fetal losses and multiple births may have an impact on the accuracy of these estimates.

with coronary artery disease. Similarly, volatile organic compounds (VOCs) do not generally impair lung function but can produce neuropsychologic effects (30). Temporal patterns of clinical effects vary widely. Whereas asthmatics develop respiratory exacerbations within minutes of inhaling SO_2 and susceptible individuals develop hypersensitivity symptoms within hours after fungal exposure, nonsmokers or smokers exposed to ETS or asbestos may develop a lung malignancy only after a latent period of several decades. Not explicitly shown in Table 7 are subclinical effects, ie, physiologic, biochemical, or cellular measures that indicate pollutant-induced exposure or effect. There is frequently controversy as to the clinical significance of these objective but subclinical alterations, some of which are discussed later in relation to specific pollutants.

Table 4. Symptoms, Signs, and Diagnoses Associated with Exposure to Air Pollutants

Upper Respiratory Tract Effects	
Rhinitis	
Sinusitis	
Pharyngitis	
Lower Respiratory Tract Effects	
Cough, wet or dry	Worsening of asthma or COPD[a]
Chest tightness or pain	Bronchitis
Shortness of breath	Pneumonia
Wheezing	Lung cancer
Nonrespiratory Effects	
Eye irritation	Decreased exercise performance
Fatigue	Worsening of angina pectoris (cardiac chest pains)
Malaise	Malignancy
Dizziness	
Headache	
Nausea	
Skin irritation	
Fever	

[a] Chronic obstructive pulmonary disease.

The above-described responses refer to individuals. On a group or population (epidemiologic) basis, adverse health outcomes from air pollution exposure take on greater public health significance since larger numbers of persons are potentially affected. Epidemiologic evidence of adverse health outcomes can be classified according to whether the outcomes are related to acute (episodic) or chronic (long-term) exposures to air pollution (Table 5). Epidemiologic findings will be reviewed in relation to specific pollutants.

Risk Assessment

Air quality regulatory officials must compare risks and benefits of possible air pollution conditions and their control in deciding how best to protect public health. The discipline of risk assessment gives a rational structure to that process (31–33). Risk assessment is a set of decision rules widely applied in the United States for identifying and quantifying risks of chemicals and other events for adverse health effects. They are

1. Hazard identification: Is an agent causally linked to the health effect of concern?

Table 5. Sources of Evidence for Adverse Health Effects of Air Pollution in Populations[a]

Acute or Episodic Effects at More Versus Less Polluted Times
Increases in general indices of ill-health, eg, school or work absences, reduced activity days
Small temporary declines in lung function
Increased incidence of respiratory symptoms or infections
Increased physician visits or emergency room visits
Increased hospital admission rates for respiratory diseases
Increased rates of respiratory or total mortality
Chronic or Long-Term Effects in More Versus Less Polluted Regions
Increased prevalence of symptoms or respiratory disease indicators
Cross-sectional differences in lung function, eg, lower average function and/or increased percentage of population with abnormal function
Longitudinal differences in lung function, eg, slower increases in children and/or faster declines in adults
Increased mortality from respiratory disease, including long-term changes in mortality within same region coinciding with long-term changes in pollution

[a] Modified from Ref. 55.

2. Exposure assessment: What is the extent of exposure?
3. Dose-response assessment: What is the relationship between level of exposure and risk of the health effect?
4. Risk characterization: What is the risk to human health, including uncertainties?

To assure safety or a level of risk judged to be acceptable, the analysis depends on accurate and valid measurements of pollutants (34) and accurate and valid characterization of risks. In other words, the scope of risk assessment extends beyond testing for an exposure–response association to the quantification of risk at various levels of exposure and the assessment of factors that modify the exposure–response relationship. The analysis consists of evaluating the distribution of individual exposures in the population (eg, mean, variance, upper 10th percentile), the causes of high exposures (eg, sources, environmental pathways, exposure locations and settings, and time–activity patterns), and the relationship between exposure and dose. The results of risk assessment can be used to identify gaps in knowledge requiring further research, to assign priorities among environmental hazards, and to select approaches for managing risks. The strengths, limitations or costs, and methodological basis of risk assessment are controversial in both scientific and political areas (35). Risk assessment of Hazardous Air Pollutants (HAP) is especially difficult. Many noncriteria agents are classified as HAP because they are known or suspected to cause adverse, often catastrophic, health effects (36,37). Quantitative estimation of health risks is difficult for this group because of inadequate data on exposures, doses, and effects. Preliminary estimates suggest that up to 2,000 cancer cases per year may result from outdoor exposures to 45 of the 189 HAP (36). Noncancer health effects may be widespread and include nonmalignant respiratory disease, hematopoietic abnormalities, neurotoxicity, renal toxicity, and reproductive and developmental abnormalities. Approximately 50 million people live near emission sources where estimated concentrations of one or more HAP exceed "levels of concern" for noncancer health effects (37). However, a paucity of data is available to estimate actual exposures for either the general population or communities potentially at risk. Careful and extensive epidemiologic evaluations will be required to support quantitative risk assessment in these situations.

METHODS OF ASSESSING HEALTH EFFECTS

General

Research on the health effects of air pollution falls into three complementary disciplines: animal toxicologic studies, clinical (controlled exposure) studies of human subjects, and epidemiologic investigations. Research may be classified according to its approach (laboratory experiments, field observations, or review of public health statistics) or its subjects (human beings or laboratory animals). Each discipline has its particular strengths and limitations (discussed below). Results from all three disciplines are necessary to the deliberative process of setting air quality standards. Ideally, research findings should be confirmed by independent studies from other disciplines. However, generally consistent findings from different studies within one discipline are usually considered sufficient to guide most public-policy decisions.

The simplest laboratory studies of either animal or human subjects compare one defined pollutant exposure against a control (reference) condition in which the subjects breathe clean air but are otherwise treated the same way as in the pollutant exposure. In other words, the only essential difference in the exposures is the test atmosphere. The "exposed" and "control" subjects may be the same individuals studied at different times (cross-over study design) or may be different but comparable individuals (parallel-group study design). More definitive laboratory studies include a control condition (clean air) and multiple pollutant exposures at different concentrations to determine an exposure–response relationship. A statistically significant response that increases as the pollutant exposure increases gives added confidence that the pollutant is responsible and supports a causal role of the agent.

Animal (Toxicology) Studies

Animal studies provide maximal control of experimental conditions as well as a broad range of possible exposures and endpoints. Animals are typically exposed to known levels of one or more pollutants generated in laboratory exposure chambers and their responses observed (38–40). Many potential interferences such as air environment, nutrition, activity, and exposure to infection or other disease can be tightly controlled or eliminated throughout the lifespan of the animals. Results from carefully designed and performed studies can be confirmed by replication in different laboratories. Animal studies provide the important advantages of testing possible mechanisms of action of pollutants and evaluating possible interventions to prevent or limit adverse responses. On the other hand, even clear-cut findings from animal toxicology may be of uncertain relevance to humans.

Clinical Studies

Clinical studies involve temporarily exposing human volunteers to controlled concentrations of specific pollutants in a laboratory either by inhalation via a mouthpiece or mask or by unencumbered breathing in an environmentally controlled exposure chamber. Effects of acute exposure can be objectively measured with tests similar to those used by physicians in clinical practice and can be related to simultaneous subjective effects (symptoms). Lung function tests are typically used in studies of inhaled pollutants. Two very important measures of lung function are forced vital capacity (FVC), the volume of a maximum breath forced out; and forced expired volume in one second (FEV_1), the volume of the maximum breath forcibly exhaled in the first second (41). These rapid, simple, precise tests provide extensive exposure–response information that can be compared among different laboratories. Airway resistance (Raw) or specific airway resistance (SRaw = Raw/lung volume) is another common test of lung function. Increases in either Raw or SRaw indicate abnormally increased airflow obstruction because of bronchoconstriction (excessive shortening of airway smooth muscles). Other measures of pollutant–induced effects include bronchial challenges with inhaled bronchoconstricting substances (eg, methacholine, histamine, carbachol) (21) or exercise, and tests for cellular and biochemical changes in the lung, ie, by analysis of bronchoalveolar lavage (BAL) fluid obtained with a flexible fiberoptic bronchoscope (26). The wide range of research goals, study designs, and procedures in controlled clinical studies are described in more detail elsewhere (42–47). To a considerable extent, clinical studies resemble animal studies and clinical trials of new therapy in terms of their experimental designs and scientific rigor (tight control of experiments; clear demonstration of cause and effect) and relevance to human health. However, the subjects' diverse experiences and exposures away from the laboratory may introduce interferences (confounders and effect modifiers). Small groups of volunteers may not be representative of large populations exposed to community air pollution. Researchers often recruit subjects expected to be sensitive to pollutants to observe a "worst-case" response. Ethically, clinical studies are limited to exposures that cause only mild and temporary disturbances in health. Practically, the laboratory exposures are limited to small numbers of subjects and relatively low-level, short-term exposures that resemble natural conditions and do not interfere unduly with normal activities. Persistent health effects or effects that require a long time to develop (long latent period) must be investigated by animal toxicology or epidemiology.

Epidemiologic Studies

Epidemiology is the study of the distribution and determinants of disease in human populations. The primary aims of air pollution epidemiology are to identify and quantify health effects of environmental agents under ordinary exposure conditions (22). As such, epidemiologic studies potentially have the highest relevance to public health. Field studies of well-defined groups of people in their usual environment measure their health status by interviews and/or clinical tests and attempt to determine whether health changes are associated with pollution exposures. Some studies do not involve actual field observations but involve review of data routinely collected in communities with different air qualities by health or demographic agencies. Conventional epidemiologic approaches used to study the adverse effects of inhaled pollutants are the cross-sectional study (where a large group is tested at only one point in time), the cohort study (where a group of people is followed closely for a period of time), and the case–control study (where similar persons with and without a response or disease are matched and their exposures are estimated retrospectively). Each design has advantages and disadvantages for examining the effects of environmental exposures (48,49).

Epidemiologic studies are usually observational rather than experimental or interventional in nature, ie, the investigator does not usually control the exposures of the study subjects. Epidemiology has been an effective tool for investigating pollutants with either very strong or very specific effects. However, epidemiologic studies are more limited when investigating agents with weak effects (but still of public health concern). As a result, the relationship between exposure to an inhaled agent and health effect of concern may be altered by biases: selection, misclassification, and confounding (48). Bias may increase or decrease the strength of statistical association. Selection bias refers to differential patterns of subject participation depending on exposure and disease status. Misclassification bias refers to error in measuring either pollutant exposure or the health outcome. Random misclassification tends to show a lack of effect of exposure and is most frequent in epidemiologic studies of air pollution in which exposures are estimated using limited measurements or surrogates. Statistical power (ability of a study to detect exposure–effect associations) declines as the degree of random misclassification increases. Differential misclassification refers to nonvalid measures, such as from interview data in a case–control study. Confounding bias refers to another risk factor that alters the effect. Specifically, a confounder (eg, temperature) affects an outcome measure (eg, lung function) and may be correlated with the risk factor (eg, ambient concentration of ozone). An effect modifier (eg, exercise) is a variable that modifies, by antagonism or potentiation, the effect of an air pollutant (eg, ozone) on an outcome measure (eg, lung function) but does not directly affect the outcome measure. In summary, these factors can produce misleading statistical associations or mask true associations between air pollution and health outcomes. These factors may not be controllable but they should be accounted for before one can make causal inferences about the relationship between air pollution and health effects.

Accurate ascertainment of exposures and health status of the study population is usually very difficult, particularly in longitudinal studies where the cumulative health effects of pollutants are sought. For these reasons, epidemiologic studies are time-consuming, expensive, and difficult to conduct. Many older epidemiologic studies have reported results that are controversial or difficult to interpret quantitatively, in part because the studies relied on a few centrally located outdoor air monitoring stations to represent the exposure of large populations. The often invalid assumption that centralized exposure estimates for population groups will hold for individuals within those groups who have health effects is known as the "ecological fallacy." Inaccuracies in exposure assessment (misclassification) are being reduced by technological advances and methodological improvements in epidemiologic studies. Carefully developed questionnaires about indoor air quality and time-activity patterns may help to establish relative rankings of exposure for different individuals or groups (50). Questionnaires are important when monitoring is not possible and may indicate when and where monitoring is necessary. As previously discussed, individual exposures vary greatly within large populations depending on individual time–activity patterns and pollutant exposures within each microenvironment. The accuracy of individual or group exposure estimates can be improved by use of portable monitors that can record pollutant concentration when placed in various locations where individuals spend time, or when personally carried on clothing by an individual. Other epidemiologic issues remain such as the relationship between exposure and dose to the target tissue.

Interpretation and Application of Research Results

In the aggregate, a large amount of information about health effects from air pollution is available. However, much still remains unknown or uncertain, and new research results often raise more questions than they answer. Research priorities should be established based on existing knowledge and judgments about the greatest risks, since resources will never be sufficient to investigate all biomedical questions. Even when a particular health effect appears well understood in scientific terms, it may be difficult to decide on a policy to control it. Any policy decision will influence many public risks and benefits, and its overall effect may be difficult to predict (51). The limits of science and its regulatory applications are discussed elsewhere (52).

No single scientific study is completely reliable, all-inclusive, and relevant to public health. Any research data must be considered in relation to those in other studies. All reported studies require critical evaluation by an independent party (eg, journal editorial board, regulatory agency), regardless of whether the final results are alarming or reassuring, or are consistent with one's personal beliefs or not. Thus, an informed evaluation of the quality and validity of research reports can be based on the following questions:

1. What question is the research attempting to answer? All research should ideally add new, meaningful information to the scientific database. Studies should have a specific, identifiable question posed by the experimenter, ie, a hypothesis which will be tested. However, the hypothesis may or may not be highly relevant to the specific question of whether air pollution affects health. Relevance is a key but not always a direct or predictable criterion. Although some basic science research provides indirect data about subclinical effects, this research may still have important applications in the future. For example, one study may expose rats to low levels of sulfur dioxide to find the smallest dose capable of producing a certain effect. On the other hand, another study might deliberately use massive exposure to induce injury in order to investigate repair processes. The former is more immediately relevant to air pollution health risks than the latter example, although both involve the same experimental animal and pollutant. However, the latter example has scientific relevance in terms of better mechanistic understanding of pollutant injury.
2. How well does an experiment simulate real life? No experiment is perfect in this respect, but some are more realistic than others. Ideally, the concentrations of pollutants, duration of exposure, secondary stresses (temperature, humidity, exercise), the health status of the subjects, and exposure to other pollutants should closely resemble the circumstances naturally encountered by people exposed to community air pollution. The subjects should be comparable to other populations of interest when controlled exposures or epidemiologic studies are designed. For example, a study showing no effects in indoor desk workers does not rule out effects in heavily breathing outdoor laborers.
3. How do observed effects relate to pollutant exposure/dose? The 16th-century alchemist–physician, Paracelsus, stated "There is nothing which is not a poison. The dose makes the poison." This basic principle of toxicology has been consistently verified (53). Even essential substances such as oxygen become injurious at abnormally high exposure levels. Some common toxins such as CO and formaldehyde are products of normal metabolism and are present at low concentrations in the body without harmful effects. Ideally, an experiment should cover a range of concentrations to understand the health risks of air pollutants. A response pattern that consistently increases as exposure–concentration increases, generally indicates two important findings: A cause-and-effect relationship is present and a threshold (minimum dose necessary to produce a detectable effect) may exist. One can then estimate whether ambient pollution levels produce exposures exceeding a threshold.
4. Are the results significant? Variability in behavior is a characteristic of all living organisms and must be taken into account when measuring the effects of stresses such as air pollutants. Biostatistical analyses are necessary to provide confidence that the biological results are either random (reflecting chance variability) or nonrandom (pattern related to the experimental factors, eg, air pollutant exposure). Statistical analyses use laws of probability to test the (null) hypothesis, typically with an arbitrarily chosen probability (p) value of <0.05, indicating less than a 5% chance that the distribution was obtained by chance alone and was not influenced by the ex-

perimental factor. In other words, the statistically significant data have only one chance in 20 of being a "false-positive" finding (type I or alpha error). A "false-negative" finding (type II or beta error) is also a major problem. A real effect that occurs inconsistently, only in a sensitive subgroup, or is small compared to the usual variability between tests may be missed if too few subjects and/or measurements are available for analysis. In general, a "positive" finding (an effect of exposure) should be supported by evidence of statistical significance as well as precise, unbiased measurements. A "negative" finding (no effect of exposure) should be supported by evidence that the study had adequate statistical power since many studies do not provide sufficient precision to completely exclude the possibility of some increased risk of effect from the experimental agent. Interval estimation to derive confidence intervals is another useful approach to statistical hypothesis testing (48).

The other important dimension of "significance" is the biological or clinical relevance of the findings. Statistical analyses relate to numerical probabilities and not necessarily to underlying biological realities or plausibility. Test results that reflect biology can validate statistical results. For example, small, statistically significant changes in lung function or cellular function should also be considered in relation to human structure and function to decide if the alteration has possible clinical relevance. Thus, precise and unbiased experimental measurements are important, and the results must be carefully interpreted if a direct biomedical implication is possibly present.

Epidemiologic results cannot prove causality but only imply causality. Statistical methods cannot establish proof of cause and effect but can define an association with a certain probability. Careful analysis is especially important, and causal inference is especially difficult, when one attempts to differentiate the effects of a single pollutant (eg, ozone) from a complex pollutant mixture. The causal significance of an association is a matter of judgment that goes beyond any statement of statistical probability. To judge the causal significance of the association between an air pollutant and health effect, nine criteria (48,54) must be used, none of which is pathognomonic by itself: strength of association (high incidence rates); consistency (replication of findings in different populations); specificity (a cause leads to an effect); temporality of exposure and effect; biologic gradient (exposure–response relationship); biological plausibility; coherence (systematic or methodologic connectedness or interrelatedness especially when governed by logical principles); experimental evidence; and analogy. As argued by Bates (55), the epidemiologic database regarding health effects from contemporary air pollution must be actively evaluated using the central question of coherence. Coherence may exist at three different levels: within epidemiological data; between epidemiological and animal data; and between epidemiologic, controlled human exposure, and animal data. For some pollutant mixes (SO_2–particulate complex) epidemiologic coherence appears to exist over a wide range of air quality levels and human populations, but findings from other disciplines are not necessarily coherent with epidemiologic findings.

5. Has the work been peer reviewed? A peer review process for research papers submitted for publication improves the quality of scientific research and published reports. This process is the best means to assure adequate quality control of new information, but is still imperfect. Poor quality work is occasionally published and good work may have delays in publication. Preliminary data in the form of abstracts or oral presentations or work presented in news media are not peer-reviewed and cannot be assumed to be scientifically valid or accurate. Some agencies, eg, the U.S. EPA, require only peer-reviewed references to be cited and discussed in its criteria documents.

SELECTED POLLUTANTS AND THEIR HEALTH EFFECTS

Ozone, nitrogen dioxide, sulfur dioxide, particulates (including sulfuric acid), carbon monoxide, and environmental tobacco smoke have been selected for more detailed review because they represent ubiquitous pollutants with documented or strongly suspected health effects that potentially affect large proportions of the population and/or susceptible groups. Sources, ambient concentrations, biological mechanisms, and health findings from animal, clinical, and epidemiologic studies will be addressed for each pollutant. The reader is referred to other reviews (7–9,45,56–58) and EPA criteria documents for more details. It should be remembered that this selection is somewhat arbitrary and that these pollutants do not exist alone in nature but in mixtures with other compounds. Other environmental agents may have similar health effects, and still other agents may present health risks beyond those mentioned here. Lessons to be learned from the example pollutants may be generalized to other agents.

Ozone

Sources and Ambient Levels. Ozone (O_3) is the most important of the class of pollutants known as photochemical oxidants. Oxidants alter biological molecules by adding oxygen atoms with greater facility than atmospheric oxygen (O_2). Nitrogen dioxide and peroxyacetyl nitrate (PAN) are other common oxidant pollutants which appear to present less health risk than ozone.

Ozone is a secondary pollutant which forms in the troposphere by photochemical reactions of primary pollutants, ie, nitrogen oxides and nonmethane hydrocarbons (or volatile organic compounds) (59–62). These precursors originate from essentially all common combustion sources, particularly motor vehicles. The VOCs also enter the atmosphere by evaporation of volatile materials, such as gasoline, paint thinners, and cleaning solvents, and as natural emissions from vegetation.

Unlike most pollutants, ozone is a natural and beneficial constituent of the upper atmosphere (stratosphere). Stratospheric ozone absorbs potentially harmful solar radiation that would otherwise reach the earth's surface. Normal atmospheric mixing brings modest amounts of this ozone to ground levels, resulting in concentrations of 0.03–0.05 ppm in the clean air in remote areas. At one time, some medical authorities believed that ozone was a health-promoting substance because of its presence in clean air and its disinfectant properties. However, scientific observations of its substantial health risks date back as far as the 1850s. It is ironic that ozone today is spatially maldistributed in the atmosphere, ie, less than desired amounts are present in the stratosphere and greater than desired amounts are present in the troposphere. The maldistribution can be attributed to man-made ozone-destroying compounds (eg, chlorofluorocarbons) in the stratosphere and combustion emissions in the troposphere.

Significant efforts have been made to decrease photochemical oxidant air pollution, of which ozone is the most potent and common constituent. Nevertheless, the control of tropospheric ozone is a very difficult, as well as important, regulatory task. The most severe ozone pollution is found in situations with little wind (eg, a temperature inversion over a mountain-lined valley), abundant sunlight, high temperatures, and traffic congestion with large numbers of motor vehicles. Ambient ozone is characteristically highest during summer and fall months ("smog season"), although episodes can occur at other times. Ozone generally reaches its peak outdoor levels during mid-day, eg, frequently greater than 0.2 ppm and occasionally 0.30–0.35 ppm for one or more hours in Los Angeles. Multiple hours exceeding 0.12 ppm ozone can occur throughout the day and persist as episodes for 3–10 days in many urban and rural locations (59,61). Long-range transport of precursor pollutants may raise ambient ozone concentrations in areas hundreds of miles downwind from the sources. This condition commonly occurs in the northeastern U.S. during the summer. Ozone is the only known outdoor air pollutant for which permissible ambient exposures can exceed workplace permissible exposure limits (0.10 ppm, 8-h average). To some extent, nitric oxide (NO) from motor vehicular exhaust scavenges ozone in urban traffic and during the night.

Indoor ozone levels are usually much lower than concurrent outdoor concentrations since indoor sources (eg, photocopiers, laser printers) are relatively few and ozone reacts rapidly with air conditioning and building materials. Recent measurements indicate that indoor ozone values closely track outdoor levels, and, depending on the air exchange rate, are 20–80% of outdoor concentrations (63). Significant indoor/outdoor ozone ratios (>0.5) exist in homes without air conditioning and with open windows. Most people probably receive the greatest part of their personal ozone exposure (concentration × duration) while indoors because indoor levels are usually not negligible and people spend most of their time indoors.

Mechanisms. The health effects of ozone originate from its very reactive oxidizing nature and its active inflammatory reactions at all levels of the respiratory tract. The mechanisms of action of ozone in humans have been evaluated at tissue and cellular levels with various techniques. Ozone is efficiently removed (approximately 95%) by the entire respiratory tract in resting people (64). The extrathoracic airway (nose, mouth, pharynx, and larynx) removes about 40% of inhaled ozone, while the intrathoracic airways remove about 90% of the remaining ozone. Mode of breathing (oral versus nasal) has a relatively small effect on removal efficiency.

Ozone-induced pulmonary dysfunction consists of an involuntary inhibition of full inspiration, reduction of total lung capacity and vital capacity, and a concomitant decrease in maximal expiratory flow rates (65). Ozone may directly stimulate airway sensory fibers (possibly nonmyelineated C-fibers) which inhibit inspiratory muscle contraction. Either direct or indirect stimulation via biochemical mediators may explain the effects of an inhaled anticholinergic medication (atropine), which blocks ozone-induced increase in airway resistance (Raw) but only partially prevents the decrease in forced expiratory flow rates and does not prevent respiratory symptoms or decreases in vital capacity. Inhaled beta agonists (bronchodilator medications) are ineffective in preventing ozone-induced responses. Other potentially pertinent acute airway alterations include increased permeability of respiratory epithelium (ie, decreased integrity of the air–blood barrier) and increased activity of the mucociliary particle clearance system in the trachea and bronchi in nonsmokers following 2-hour exposures to 0.4 ppm ozone (64). Of note is that exposure to 0.2 ppm ozone accelerates peripheral airway clearance without significant abnormalities in routine breathing tests.

The inflammatory injury produced by inhaled ozone is evident not only from respiratory symptoms and increased methacholine responsiveness but also from results of bronchoalveolar lavage (BAL) and nasal lavage (NAL) in healthy adults undergoing exposures to 0.08 to 0.6 ppm ozone for 2.0–6.6 hours. As compared to air-exposure BALs, BAL fluid following exposure to 0.4 ppm ozone shows significantly increased polymorphonuclear leukocytes (PMN) whether sampled at 3 or 18 hour following exposure (66,67). In addition, BAL fluid shows significant ozone-induced increases in vascular permeability (total protein, albumin, immunoglobuin G), enzymes (immunoreactive neutrophil elastase, PMN elastase activity), and inflammatory markers (fibronectin, complement 3a, prostaglandin E2 and $F_{2\alpha}$, thromboxane B_2, urokinase plasminogen activator). Many of these markers, as well as other substances (lactate dehydrogenase, interleukin-6, and α_1-antitrypsin), are detected in BAL fluid when healthy adults are exposed for 6.6 hours to 0.08 and 0.10 ppm ozone (levels below the federal standard) (68). Although the BAL fluid following 0.08 ppm-ozone exposure shows smaller increases in inflammatory mediators and no major increase in protein and fibronectin, the overall results indicate that even ozone levels as low as 0.08 ppm are sufficient to initiate an inflammatory reaction in the lung. Similar increases in PMN and albumin are found in nasal lavage (NAL) performed 18 hours following exposure to 0.4 or 0.5 ppm ozone (69). Significant increases in plasma prostaglandin $F_{2\alpha}$ occurs in ozone-responsive subjects during and after exposures to ozone. Indomethacin, an inhibitor of the cyclooxygenase pathway, significantly reduces ozone-induced decrements in FVC and FEV_1 but not the increase in nonspecific bronchial reactivity, suggesting that the latter occurs via a separate, noncyclooxygenase mechanism.

Effects

Animal Studies. The effects of ozone are most pronounced in the lungs although alterations in other organs have not been ruled out. Numerous studies in different animal species have investigated various changes in the lungs after ozone exposure (14,39,40,59,62,70–74). Damage typically occurs at sites where the smallest airways (respiratory bronchioles) merge into alveoli, resulting in a respiratory bronchiolitis. The involved airways show evidence of injury and repair, ie, inflammation, edema, increased permeability (leakage) of proteins in the airways, cellular hyperplasia, replacement of normal alveolar lining cells (Type I epithelial cells) by primitive, more ozone-resistant cells (Type II epithelial cells), and scarring (fibrosis). Larger, more proximal bronchi are relatively less affected, perhaps because their surfaces are better protected by a mucus layer. Similarly, the alveoli are relatively spared perhaps because most ozone reacts with airway epithelium before entering the alveoli, which are also lined by a possibly protective surfactant layer. The nasal passages, however, can be vulnerable to inhaled ozone. The structural effects in the lower airways are most pronounced at ozone exposures of 0.5 ppm or higher (above the ambient range), but they can also occur in exposures to as little as 0.2 ppm for two hours. Exposures lasting weeks or months, either continuously or intermittently, to concentrations as low as 0.12 ppm produce persistent damage. Some studies at higher than ambient concentrations have shown airway recovery after a few days or weeks in clean air, whereas other studies indicate that ozone-induced damage persists.

Lung function changes in ozone-exposed animals typically include rapid and shallow breathing and increased resistance to air flow (Raw). These effects occur with 0.20–0.30 ppm for as little as two hours. At higher concentrations (above 0.5 ppm), the ozone-injured airways become hyperreactive and constrict in response to inhalation of non-specific bronchoconstrictors such as methacholine. Hyperreactivity presumably reflects underlying inflammation (21).

A variety of cellular and biochemical changes also occurs in the lungs at ozone concentrations of 0.2 ppm or lower. These alterations are complex and probably reflect ozone-induced inflammation. An influx of PMN and increased production of antioxidant substances (eg, vitamin E) occur in most species. Increased collagen deposition may occur with resulting fibrosis, making the lungs less compliant, ie, abnormally stiff. The implications of these findings for humans are unclear.

Ozone exposure may disturb natural defenses against microbes, as demonstrated by studies in which rodents are exposed to ozone and then to bacteria. These studies assess the proportion of ozone-exposed animals that die of respiratory infection (pneumonia) or the amount of viable bacteria remaining in the lungs. Control animals are kept in ozone-free air before exposure to bacteria. Massive amounts of bacteria are used, so that even some control animals die. Exposure concentrations as low as 0.1–0.2 ppm ozone may be sufficient to overwhelm defense mechanisms and increase death rates. The defenses which are altered by ozone include the immune system, mucociliary clearance, and alveolar macrophages.

Clinical Studies. Physiologic and symptomatic responses at current ambient concentrations of ozone have been extensively evaluated in healthy nonsmoking young adults and, to a lesser extent, in children and the elderly (14,45,59,62,64,72,75). Group responses in controlled exposures with intermittent exercise have largely correlated to the "effective dose" of ozone, which is the product of concentration $\times$ ventilation rate $\times$ duration. Most healthy adults have little response with 2–3 hours of resting exposure to ambient ozone concentrations (less than 0.3 ppm). Healthy young adults show variable but definite decreases in vital capacity and flow rates (eg, FVC, FEV_1) and increases in airway resistance (Raw) following 2-hour exposures to $\leq$0.20 ppm ozone during intermittent moderate exercise (76). As the ozone concentration increases, a strong sigmoid-shaped dose–response relationship begins at 0.12–0.16 ppm and reaches a plateau near 0.4 ppm. Exposures with heavier or more continuous exercise levels (with corresponding increases in ventilation) elicit significant respiratory responses at lower ozone levels. For example, young adults continuously exercising for 7 hours (approximately 40 L/min ventilation) in 0.08, 0.10, and 0.12 ppm ozone, showed significant progressive concentration- and duration-related reductions of FEV_1 and symptoms (especially cough and pain on deep inspiration), as well as increased sensitivity to inhaled methacholine, indicating ozone-induced airways hyperresponsiveness (77,78). Exercise performance by athletes is adversely affected by ozone exposure (79,80). Ozone-induced respiratory symptoms and lung dysfunction generally parallel each other in adults but not necessarily in children, who tend to report fewer respiratory symptoms despite comparable physiologic responses.

Several studies have evaluated whether photochemical pollutant mixture (including ambient ozone) is more toxic than ozone alone. This question can be addressed in two ways: Volunteers can be exposed to mixed pollutants of known concentrations generated in a chamber or exposed to actual oxidant air pollution in a laboratory setting. The latter has been performed in different areas in Los Angeles using a mobile exposure facility. In both cases, the most irritant responses are apparently due to ozone, and there is no clear evidence of substantial extra toxicity from other pollutants that typically accompany ozone in ambient pollution in Los Angeles.

Several potentially ozone-sensitive populations have been evaluated, but no group has yet been confirmed to show markedly greater responsiveness to inhaled ozone than observed in healthy young adults. Children, adolescents, and women appear as or slightly more sensitive than adult males in terms of lung function changes. Children show small but statistically significant decreases in FEV_1 following exposure to 0.12 ppm ozone or ambient Los Angeles air (with 0.14 ppm ozone). However, comparisons by age or gender are complicated by inherent differences in body size, lung volume, and exercise capability. Older subjects (>50 years) are less responsive to 0.20–0.45 ppm ozone and an oxidant mixture of ozone, NO_2, and peroxyacetyl nitrate (PAN) than young adults in acute 1- or 2-hour exposures. Surprisingly, healthy smokers and subjects with asthma, allergic rhinitis, COPD, or ischemic heart disease have not been conclusively shown to be more responsive than their respective controls in controlled ozone exposures. For example, adolescent asthmatics are not more responsive to 0.12 or 0.18 ppm ozone or combined ozone (0.12 ppm) and NO_2 (0.18 ppm) than healthy adolescents. Only following a 2-hour exposure to 0.4 ppm ozone with intermittent heavy exercise (ventilation 52 L/min) do adult asthmatics demonstrate greater airflow obstruction than normal subjects, but they still have similar induction of symptoms and changes in lung volumes and bronchial reactivity to methacholine. However, exposure to ozone (0.12 ppm) can increase reactivity to inhaled allergens in asthmatic subjects (81). Observed group differences (or lack thereof) may relate to multiple factors, eg, sample size, subject selection, medication usage, underlying bronchial reactivity, effective dose of ozone, and ambient exposures to inhaled irritants.

Although group differences tend to be small, individuals may show large differences in response to ozone despite being physically similar and receiving the same exposure conditions. Individual lung function response is fairly reproducible for at least 3 months in healthy young adults exposed to >0.18 ppm, although older subjects have a less consistent pattern. Individual ozone responses may be modified by repeated ambient (seasonal) exposures or by other factors which change underlying airway responsiveness. At present, an individual's sensitivity to ozone can be determined only by actual exposure, although asthma or asymptomatic nonspecific airway hyperreactivity may imply increased risk of ozone-induced lung dysfunction. Such airway-hyperreactive, ozone-responsive individuals may have difficulty developing tolerance to ozone.

Tolerance (or adaptation) results when humans (and animals) are exposed repeatedly to ozone in the laboratory. Most volunteers exposed to high concentrations (eg, 0.4 ppm) for several hours on five consecutive days show a characteristic pattern in which the worst symptoms and lung dysfunction occur during the first two exposure days, followed by less and less response on subsequent days. There is minimal response by the fourth or fifth day. The attenuated response to ozone is lost in a week or two if exposure ceases. The long-term health consequences of this phenomenon are unknown, in part, because it remains unproven that tolerance occurs naturally or is widespread in the community.

Epidemiologic Studies. Epidemiologists evaluate the short-term (acute) and long-term (chronic) health effects in people who reside in ozone-polluted areas (14,59,62,72). Short-term effects are typically evaluated as changes in reported symptoms, lung function, athletic performance, or rate of hospital admissions, in various cites such as Los Angeles and less polluted areas (Houston, Tucson, and locales in northeastern U.S. and southeastern Canada). Long-term effects are typically evaluated in terms of the respiratory health of residents in a high-pollution community, as judged by questionnaires and/or lung function tests, in comparison with the health of a matched population in a cleaner environment. Some of these studies will be reviewed.

Summer camp studies at diverse times and places have consistently shown oxidant-related changes in lung function in children. Declines in FVC, FEV_1, and peak expiratory flow (PEF) have been strongly associated with increasing ozone levels. In a 1982 study in Mendham, New Jersey, an air pollution episode occurred that involved four days of hazy weather and a peak one-hour ozone level of 0.186 ppm. Decrements in lung function were present not only during the episode but also one week following the episode.

Children participating in the Harvard Six-City Study of respiratory effects of air pollution performed lung function tests at each of six weekly sessions. Ambient ozone was monitored hourly at nearby monitoring stations. The results were in good agreement with those in camp studies despite relatively low ozone levels, ie, 0.007–0.078 ppm. A study in Tucson also showed a relationship between ambient exposure to ozone (0.12 ppm or less) and reductions in lung function.

A large-scale study in the Los Angeles area estimated the influence of pollution exposure on the rate of asthma attacks, as recorded daily by panels of asthmatic volunteers in six different communities. The best predictor of an attack on a given day was an attack on the previous day, regardless of air quality. However, increasing oxidants (primarily ozone) were associated with an increased likelihood of an attack. Similar relationships were found in Houston, Texas.

Regional ambient levels of ozone and other pollutants were compared against total hospital admissions throughout southern Ontario in Canada, for several years during the 1970s. Ambient ozone did not exceed 0.12 ppm on an hourly average. Admissions for respiratory illness during the summer tended to increase with increasing ozone, sulfur dioxide, sulfate, and temperature levels a day or two earlier. Admissions for illnesses other than respiratory illness did not show this relationship with air pollution. It remains difficult to prove which environmental factors were responsible for the increased hospital admissions, although the combination of ozone and acidic aerosols has been suggested as the most likely cause.

One large-scale study of chronic effects compared the results of lung function tests and symptom interviews among four middle-class communities in Los Angeles County with different types and levels of pollution. Some measures were worse on the average in high-ozone areas, but there was no clear overall pattern of ill health attributable to higher pollution levels. A similar association was found in another study which compared nonsmoking Seventh-Day Adventist populations between Los Angeles and other less polluted California communities.

Preliminary results from an autopsy study (82) in Los Angeles County have implicated ambient air pollution as a cause of pathologic lung effects. The lungs of 107 ostensibly healthy youths (age 15–25 years) who died by accident or violence were evaluated. Approximately 80% of the youths demonstrated some degree of respiratory bronchiolitis or other anatomic changes and 27% of the changes were considered severe and extensive. The cause of the pathologic lesions have remained unknown since the smoking and drug use histories of the youths were not determined. The causative pollutant(s) also were unknown, if, indeed, the changes are related to ambient photochemical pollution. Nonetheless, the results have suggested possible pollution-related changes in young persons. Further studies of this nature with epidemiologic data (eg, comparisons of similarly selected autopsy specimens in cities with different air qualities) will be necessary to better understand the significance of the lung changes.

Nitrogen Dioxide

Sources and Ambient Levels. Atmospheric nitrogen will thermally combine with oxygen to form nitric oxide (NO) whenever any common fuel is burned in air. Although nitric oxide is comparatively nontoxic at ambient concentrations, it can be oxidized rapidly to form nitrogen dioxide (NO_2), which has greater toxic potential. Nitrogen dioxide is the most important and ubiquitous nitrogen oxide (NO_x) from the viewpoint of human exposure. Nitrogen dioxide is a fairly water-insoluble, reddish–brown gas with moderate oxidizing capability. The major source of this pollutant in the atmosphere is the combustion of fossil fuels in stationary points (industrial heating and power generation) and motor vehicles (internal combustion engines). Indoor sources of NO_2 are unvented combustion appliances such as gas stoves and gas-fired water heaters.

Like other outdoor pollutants, urban levels of NO_2 vary according to time of day, season, meteorological conditions, and human activities (83–85). The most severe outdoor NO_2 pollution is typically found at places and times of heavy motor vehicular traffic. Typically, morning and afternoon peak NO_2 levels (following traffic emissions of NO) are superimposed on a low ambient background level. Maximum 30-minute and 24-hour outdoor values of NO_2 can be 0.45 ppm and 0.21 ppm, respectively, in metropolitan Los Angeles, which has the highest measured NO_2 concentrations in the U.S.

Indoor exposures to NO_2 may be more important and widespread than outdoor exposures for most people, ie, indoor NO_2 is the major contributor to personal exposure (7,83,85–87). Low ventilation rates ("tight" buildings), use of gas-fueled appliances, and tobacco smoking tend to produce higher indoor than outdoor concentrations of NO_2. Short-term peak concentrations in poorly ventilated kitchens with gas stoves may reach a maximum 1-hour level of 0.25–1.0 ppm and, during cooking, peak levels as high as 4 ppm in the kitchen.

Mechanisms. Potential mechanisms of airway injury by NO_2 exposure in humans have been evaluated with bronchoalveolar lavage (BAL) (45). Several laboratories have found no evidence of decreased cellular viability or increased cell numbers or inflammatory cells in BAL fluid from healthy subjects exposed continuously or intermittently for 3–3.5 hours to 0.6 to 4.0 ppm NO_2 with intermittent exercise. Alveolar macrophages collected after a 3-hour *in vivo* continuous exposure to 0.6 ppm NO_2 tend to inactivate influenza virus *in vitro* less effectively than cells collected after air exposure. However, the effect of NO_2 exposure on controlled influenza infection in humans is inconclusive. On the other hand, increased numbers of BAL lymphocytes and mast cells are found 4, 8, and 24 hours after exposure to 4 ppm NO_2 ($\times$ 20 minutes), with a return to baseline (pre-exposure) BAL cell values at 72 hours. The clinical significance of these changes is unclear since significant changes in symptoms, lung function, or airway reactivity did not occur in these subjects. The lack of a significant neutrophil response in the BAL fluid is noteworthy since NO_2 has been associated with alterations in the activity of alpha-1-proteinase inhibitor (α1-PI) and alpha-2-macroglobulin in BAL fluid. A functional or immunologic decrease in these lung antiproteases could potentially lead to a protease–antiprotease imbalance which would favor degradation of alveolar elastin and potentially cause emphysema if the condition persisted. Acute exposures to NO_2 do not significantly increase total protein or albumin in BAL fluid, suggesting that low-level NO_2 exposure (under experimental conditions investigated so far) does not alter respiratory epithelial permeability in normal subjects. Continuous exposure to NO_2 in humans evokes greater physiologic and cellular responses than intermittent peak exposures.

Effects

Animal Studies. Rats, mice, guinea pigs, and monkeys have been studied in both short- and long-term experiments with NO_2 (70,85,87–89). Nitrogen dioxide acts similarly to ozone in animals in that it injures the respiratory bronchioles, stimulates production of natural antioxidant substances, and increases susceptibility to respiratory infection. Usually a higher concentration of NO_2 than ozone is necessary to produce similar effects, and the toxic NO_2 concentration is larger than commonly occurs during ambient pollution episodes. However, enlarged air spaces due to alveolar destruction (emphysema-like changes) have been reported after six months of continuous exposure to 0.1 ppm NO_2 in conjunction with daily two-hour exposures to 1 ppm. Changes in lung collagen have been reported after less than a month of daily four-hour exposures to 0.25 ppm NO_2.

Biochemical changes may occur at NO_2 concentrations within or slightly above the ambient range. Changes in red blood cell metabolism, liver function, and kidney function in animals have been reported in exposures to 0.5 ppm or less lasting a few hours to a few days. Increased death rates after bacterial inhalation challenges have occurred in long-term (several months) exposures to as little as 0.5 ppm and short-term (three-hour) exposures to 2 ppm. The health significance of these findings for humans remains unclear.

Clinical Studies. Clinical studies using a wide range of exposure conditions have demonstrated mixed results regarding symptoms and lung function, suggesting that NO_2 exerts weak respiratory effects in the laboratory setting (45,84,85,87–91). Healthy adolescents, young adults, and older individuals have failed to consistently demonstrate acute effects of NO_2 on pulmonary function at levels ranging from 0.1 to 4.0 ppm. One study reported increased airway resistance with 0.24 ppm NO_2 but not with 0.48 ppm in resting subjects, suggesting a nonuniform dose–response pattern. Similarly, adult and adolescent subjects with primarily mild asthma have generally not responded to exposures of 0.12–4.0 ppm for 20–120 minutes. However, some asthmatics cooking on gas ranges in the home environment have demonstrated statistically significant decreases in FVC and PEF when the average NO_2 level was >0.30 ppm, although other environmental exposures could not be excluded (92). Patients with moderately severe COPD have not shown consistent lung function changes with exposures to as high as 2.0 ppm NO_2 for 1 hour. In general, subjects in the above protocols lacked significant respiratory symptoms during and following NO_2 exposure. Exposure of healthy young adults to a combination of NO_2 (0.6 ppm) and ozone (0.3 ppm) during one hour of sustained heavy exercise has not augmented symptoms or lung dysfunction above that observed with ozone alone.

Effects of NO_2 on airway reactivity have been inconsistent (45,93). Normal subjects do not change airway reactivity after exposures to as high as 0.6 ppm NO_2. However, enhanced cholinergic bronchial reactivity has been reported following exposures to 2.0 ppm ($\times$ 1 hour) at rest and 1.5 ppm ($\times$ 3 hours) during intermittent exercise. Some asthmatic subjects have not shown enhanced reactivity to inhaled methacholine, histamine, cold air, or SO_2 following exposures to 0.1 to 3.0 ppm NO_2 for two hours or less. On the other hand, NO_2 exposures have potentiated subsequent responses to the same stimuli in other asthmatic subjects. Small numbers of subjects (resulting in lack of statistical power to adequately test a hypothesis), variation in exposure protocols, and different protocols of airway challenges, have been possible sources of difficulty when comparing responses between studies. A simplified "meta-analysis" (93) combining results from 20 studies of asthmatics and 5 studies of healthy subjects exposed to NO_2 showed that there was an overall trend among both groups for airway responsiveness to increase, especially in resting exposures in asthmatics and with concentrations >1 ppm in healthy subjects.

Overall, nitrogen dioxide, at typical ambient indoor and outdoor concentrations, appears to be at least an order of magnitude weaker than ozone as an acute oxidizing agent in the laboratory setting. To date, the acute effects of NO_2 on lung health appear small, and clinically significant effects are unlikely or infrequent. However, this contaminant is probably the least understood of those discussed in this review. The inconsistent exposure–response data are currently difficult to reconcile.

Epidemiologic Studies. Epidemiologic studies of NO_2 effects are difficult because of its ubiquity indoors and outdoors and its coexistence (colinearity) with other ambient pollutants of equal or greater toxic potential (eg, ozone) (84,85,87,90). For example, studies of long-term oxidant-induced health effects in California may have some relevance to NO_2 since ambient levels of NO_2 are usually high where other oxidant levels (primarily due to ozone) are also high. Unfortunately, it is very difficult to specify effects of NO_2 apart from those of ozone, which would probably be considered the most likely cause of most health effects in high-oxidant communities.

One epidemiologic study that specifically addressed NO_2 effects was conducted near an ammunition plant in Chattanooga, Tennessee, in the late 1960s and early 1970s. Nitrogen dioxide was the predominant, although not the only, air pollutant emitted from the plant. In high-pollution areas near the plant, both schoolchildren and their parents tended to have high rates of lower-respiratory illness, when compared with residents of cleaner areas. When pollution levels improved, the illness rate appeared to decline although it remained highest in the most polluted area. Unfortunately, the standard air monitoring method then in use was eventually found to be inaccurate, leaving doubts about the reliability of NO_2 concentrations reported for the study areas.

Subsequent epidemiologic information on NO_2 has come from surveys comparing the health status of adults and children living in homes with electric stoves (with low NO_2 levels) and in homes with gas stoves (with comparatively high NO_2 levels, often exceeding current EPA air quality standards). Some studies have associated small increases in illness rates or decreases in lung function with gas stoves; other studies have reported no association. The inconsistencies may be related to numerous factors. For example, NO_2 is not the only pollutant emitted by gas stoves. Nitrogen dioxide, which was frequently not directly measured, might also be emitted by other indoor sources. Thus far, it is difficult to confirm that gas stoves in general and indoor NO_2 in particular present a meaningful health risk since most epidemiologic studies did not carry out precise exposure assessments on their study populations (86). However, a meta-analysis (94) of results from epidemiologic studies has suggested an increase of at least 20% in the odds of respiratory illness in children exposed to a long-term increase of 30 $\mu g/m^3$ NO_2.

A prospective 18-month cohort study (95), involving 1,205 healthy infants of nonsmoking parents in Albuquerque, New Mexico, evaluated whether exposure to NO_2 increases the incidence or duration, or both, of respiratory illness in infants. No association was found between NO_2 exposure and the incidence rates for any illness category or between the presence of a gas stove and illness incidence. The results cannot be generalized to older children, adults, or to more susceptible groups such as infants of smoking parents. Over 75% of the measured NO_2 concentrations in the homes were less than 0.02 ppm, whereas 5% were greater than 0.04 ppm. These levels have been similar to those in many locations in the U.S. but are much lower than those in heavily polluted cities and in poorly ventilated inner-city apartments.

Thus, the study results cannot be generalized to infants exposed to greater than 0.04 ppm NO_2. This study is considered to be one of the most carefully designed and executed assessments of the effects of NO_2 on respiratory illness conducted to date. The findings are reassuring for the segment of the very young population without uncommonly heavy exposures. Available epidemiologic evidence on the acute health effects from NO_2 exposure is insufficient to support the establishment of a short-term standard (NAAQS). However, the EPA has established a long-term standard using a maximum permissible annual average concentration (see Table 2). Apart from any direct health risk, NO_x pollution is important to health because it contributes to ozone formation.

Sulfur Dioxide

Sources and Ambient Levels. Natural fuels such as coal and oil usually contain sulfur. During combustion, the sulfur combines with oxygen to form sulfur dioxide (SO_2) gas, which is emitted into the atmosphere with other products of combustion. Additional sulfur-containing products form as small particles (eg, sulfuric acid aerosols), either during the combustion process itself or by further chemical reactions in the atmosphere. Additional particles arise from incompletely burned fuel and noncombustible mineral impurities in the fuel. Because they typically occur together in the ambient environment, SO_2 and particulate pollution must often be considered together in scientific and regulatory efforts (49,96,97). Particulate pollutants (including sulfuric acid) are discussed in the next section.

The primary outdoor source of SO_2 is the domestic, commercial, and industrial combustion of sulfur-containing fossil fuels (coal and oil). Metal smelters tend to be the largest point sources of SO_2. Power plants and other large industrial establishments are smaller point sources but constitute a larger cumulative source. Oil- or coal-heated homes and commercial buildings may also contribute substantially to SO_2 levels in urban areas. The highest ambient concentrations occur where the "plume" of emissions from a large point source reaches the ground. Many modern industries employ tall smokestacks, which reduce local ground-level pollution because the plumes are either greatly diluted before reaching ground level or dispersed elsewhere by winds. However, the long-range transport of the pollutants leads to high ambient concentrations and acid rain in downwind areas unless additional emission controls are used.

Apart from large point sources, urban SO_2 concentrations in the U.S. do not often exceed 0.2–0.3 ppm as a one-hour average. This range reflects a marked improvement relative to the 1950's, and earlier, when uncontrolled burning of coal and high-sulfur oil led to severe SO_2-particulate pollution episodes and high morbidity and mortality rates. However, down drafts near power plants may produce 5- to 10-minute peaks exceeding 1 ppm.

Indoor levels of SO_2 are dependent on outdoor concentration and the air-exchange rate. Indoor levels are generally much lower than outdoor values, owing to absorption of the gas on furniture, walls, clothing, and in ventilation systems, as well as infrequent indoor sources.

Mechanisms. Various drug pretreatment studies have provided insights into the mechanisms of SO_2-induced bronchoconstriction in asthma (45,98). Reflex parasympathetic pathways are involved since atropine inhibits the airway response to inhaled SO_2 in normal and in asthmatic subjects. However, anticholinergic drugs have not been completely protective in all asthmatics. Cromolyn sodium and nedocromil sodium provide greater protection, which is dose-dependent, than either atropine or ipratropium bromide. The combination of cromolyn and ipratropium more effectively inhibits SO_2-induced bronchoconstriction than either drug alone. The mode of action of cromolyn and nedocromil is unknown but may be related to mast cell stabilization or inhibition of mediator release. Inhaled H_1-antihistamine (clemastine) and bronchodilators (albuterol and metaproterenol) effectively prevent SO_2-induced bronchoconstriction. Pretreatment with an inhaled corticosteroid (beclomethasone dipropionate) fails to block the acute airway response to 0.75 ppm SO_2 in asthmatics.

Little information is available about the effects of inhaled SO_2 on cellular and mediator responses in the lung (45). However, dose-dependent increases in BAL lymphocytes, mast cells, macrophages, and total cells are present in healthy volunteers 24 hours following 20-minute exposures to 4, 5, and 8 ppm SO_2. BAL cell responses do not increase further after exposure to 11 ppm SO_2. Significant increases of these BAL cells occur at varying times (4, 8, 24, and 72 hours after exposure to 8 ppm SO_2) and generally peak by 24 hours. All cell counts return to preexposure baseline values by 72 hours after exposure. Neither albumin nor the ratio of helper/cytotoxic suppressor lymphocytes are changed at any time after exposure. No significant changes in lung function or increases in symptoms occur under these conditions. Although the above SO_2 exposures exceed the U.S. TLV of 5 ppm, peak indoor industrial levels of 8 ppm are not uncommon. These results indicate that acute SO_2 exposures induce cellular responses in otherwise asymptomatic (nonasthmatic) individuals with little change in lung function.

Effects

Animal Studies. Exposures to SO_2 of several minutes to hours in guinea pigs and other species have shown decreases in respiratory rate and increases in Raw and bronchial reactivity (97). These effects occur only at concentrations above 1 ppm. Airway resistance increases within a few minutes after exposure begins and recovers within a few minutes to hours after exposure ends, depending on the severity of the initial response.

Longer-term studies (ranging from a week to more than a year) have evaluated effects of SO_2 on lung injury and repair and on respiratory defense mechanisms. Evidence of airway epithelial injury has been found with light microscopy at 5 ppm or above. Chronic bronchitis (similar to that in humans) can be induced in dogs by exposure to 200 ppm SO_2 for 8 months (2 hours a day, 4–5 days a week) (99). Within the first 2–4 weeks of exposure, the dogs develop cough, mucus hypersecretion, airway epithelial thickening, an increase in the number (hyperplasia) of goblet cells, hypertrophy and hyperplasia of mucous glands, chronic airflow obstruction, and persistent lung inflammation as demonstrated by an increase in PMN in BAL fluid. On the other hand, airway responsiveness to inhaled methacholine decreases 2- to 3-fold within 8 weeks of SO_2 exposure. Daily exposures of dogs to 50 ppm for 10–11 months (with the same schedule as previously described) results in similar histologic and physiologic changes but no inflammatory cell infiltration of the airways or changes in histamine or methacholine responsiveness (100). Dogs exposed to 15 ppm SO_2 for 5 months do not demonstrate the histologic or physiologic changes noted above. These findings suggest that chronic airway inflammation induced by SO_2 may be associated with decreased responsiveness (attenuation) to inhaled bronchoconstricting agents.

Functional alterations of the immune system occur at 2 ppm SO_2 while respiratory defenses against inhaled bacteria appear unimpaired at exposure levels up to 5 ppm. Sulfur dioxide alters mucociliary clearance at ambient and higher concentrations, but the specific findings differ among species. In some species, SO_2, appears to delay clearance of inhaled tracer particles, suggesting greater retention and potential for infection or toxicity from inhaled substances. In other species, SO_2 accelerates clearance, suggesting a stress response. At present, too little is known about normal clearance mechanisms to judge the health significance of the changes attributed to SO_2.

Clinical Studies. In healthy adolescents and adults, acute exposures to high concentrations (>5 ppm) of SO_2 uniformly result in bronchoconstriction, whereas lower levels (0.15–2.0 ppm) do not elicit significant symptoms or impair lung function even with intermittent exercise and different inhalation routes (45,97,98,101,102). Older subjects (55–73 years) are not greatly more reactive to inhaled SO_2 than younger subjects. The most frequently reported symptoms are confined primarily to the upper airways, eg, an unpleasant taste or odor. The exposure combination of SO_2 (0.4 and 1.0 ppm) and ozone (0.3 and 0.4 ppm) does not produce an additive or synergistic effect.

Individuals with asthma or allergy with exercise-induced bronchospasm (EIB) represent a well-documented sensitive population, although there is some inter-individual variability. Markedly symptomatic bronchoconstriction occurs at <1 ppm SO_2 at rest and ≥0.5 ppm during moderate exercise for 10–30 minutes. At 0.25 ppm or less, some asthmatics have asymptomatic increases in specific airway resistance (SRaw). Healthy nonasthmatic individuals are usually not affected at these low concentrations. Dose-response relationships have been consistently found. Symptomatic bronchoconstriction in asthmatics can occur within 3 minutes of inhaling 0.5 or 1.0 ppm. Lengthening the exposure beyond 10 minutes does not appreciably influence the magnitude of response. Bronchoconstriction reverses spontaneously at rest within 30 minutes in clean air or within an hour with continued exposure.

Atopic (allergic) adolescents with EIB and asthmatic adolescents decrease their FEV_1 > 15% following 30-minute rest and 10-minute moderate exercise with 0.5 and 1 ppm SO_2. Atopic adolescents without EIB and healthy adolescents have similar responses to inhaled SO_2, indicating that atopy alone is insufficient for SO_2 sensitivity and that hyperreactive airways is a critical factor. Patients with moderately severe COPD do not have significant symptoms or changes in lung function or oxygen saturation with 1-hour exposures to 0.4 and 0.8 ppm SO_2 during intermittent light exercise, although it is uncertain whether low dose rates and/or airway reactivity and continued medication use, influence the results.

The bronchoconstrictor effect of SO_2 (as low as 0.10 ppm) can be potentiated in asthmatics by oral breathing, increased ventilation, subfreezing, dry air (−10° to −20°C) and prior exposure to 0.12 ppm ozone. Hot air (38°C) does not enhance the SO_2 response. The relationship between the responses to methacholine and SO_2 is unclear. Freely breathing asthmatics frequently develop significant bronchoconstriction at ≥0.5 ppm SO_2. Although the threshold is higher with unencumbered breathing than with oral breathing alone, the former does not prevent bronchoconstriction, particularly with heavy exercise. As with many inhaled pollutants, ventilation during exercise increases the effective dose to the airways and facilitates EIB. Oral and, to a lesser extent, oronasal breathing during exercise also decreases the protective SO_2-removal capability of the nose and enhances EIB (103).

Tolerance to SO_2 develops in asthmatic subjects during repeated exposures. Lung function responses are significantly attenuated in asthmatics recurrently exposed to 0.2–1.0 ppm SO_2 for ≤1 hour during exercise. The mechanism of the tolerance is unknown. Unfortunately, the attenuation is not practically protective since the response is generally modest, short-lived (<24 hours), and the initial exposure can result in a severe bronchospastic attack that halts further exercise or other activity.

Inhaled SO_2 may also affect the upper airway since as much as 99% of this water-soluble gas may be absorbed from the inspired air by the nasopharynx. Inhalation of 1.0 ppm SO_2 through a face mask has been reported to significantly increase nasal work of breathing in adolescent asthmatics. However, other investigators could not confirm either increased nasal symptoms or resistance with brief exposures to 1–4 ppm SO_2 in nonasthmatic subjects with rhinitis or in subjects with asthma and allergic rhinitis, suggesting that SO_2 responsiveness in the upper airways is not consistent among subjects and is not uniform throughout the respiratory tract.

Epidemiologic Studies. The majority of epidemiologic studies reviewed here deal with the effects of combined sulfur dioxide and particulate pollution because of their strong tendency to coexist. When health effects are found, scientists cannot be certain whether they are caused by sulfur dioxide alone, particulate pollutants alone, their combination, or an unidentified factor that tends to accompany air pollution.

Three separate international incidents between 1930 and 1952 demonstrated that several days of fog and intense sulfur dioxide-particulate pollution were associated with large increases in morbidity (illness rates) and respiratory mortality (death rates) (49,104,105). All occurred in areas with a heavy concentration of industry and uncontrolled coal burning. In the Meuse Valley of Belgium in December, 1930, about 60 premature deaths occurred in older persons with preexisting cardiac or lung disease, and complaints of respiratory irritation were widespread among all age groups. In Donora, Pennsylvania, in October, 1948, about 20 premature deaths from cardiopulmonary causes occurred in a population of 14,000. 40% of the population had some respiratory symptoms during the episode, and the incidence was higher among people with preexisting chronic respiratory illness. Although air monitoring data were not available, retrospective estimations of ambient sulfur dioxide concentrations ranged from 0.5 to 2 ppm. In London during December, 1952, 3,500–4,000 premature deaths were recorded. The maximum weekly death rate was more than twice its usual value. Sulfur dioxide levels exceeded 1 ppm as a 24-hour average. Peak particulate concentrations exceeded 4 milligrams per cubic meter (mg/m^3) or about 10 times the level considered normal at that time (but now considered as severe pollution).

Since 1952, numerous reviews have examined the relationship between London death rates and more moderate pollution (49,97,104,105). The U.S. EPA (97) drew four conclusions: (*1*) markedly increased death rates occurred, mainly among the elderly and chronically ill, at particulate concentrations above 1 mg/m^3 and SO_2 concentrations above 0.4 ppm as 24-hour averages, particularly if these concentrations persisted for several days; (*2*) during the episodes high humidity or fog was important, perhaps because it facilitated the formation of sulfuric acid or other acidic aerosols; (*3*) increased death rates were associated with particulate levels of 500 micrograms per cubic meter ($\mu g/m^3$) to 1 mg/m^3 and SO_2 levels of 0.2 to 0.4 ppm; and (*4*) small but probably real increases in death rates are likely at particulate levels below 500 mg/m^3, but not at SO_2 levels below 0.2 ppm.

A number of studies have attempted to explain the variation in death rates among different U.S. metropolitan areas in terms of their different pollution levels. Sulfur dioxide, total particulate matter, and particulate sulfate all have been suggested to cause increased mortality rates (49). The results obtained in such analyses are quite sensitive to the choice of metropolitan areas, air pollution measurements, statistical analysis, and adjustments for confounding factors. Different studies tend to yield different conclusions as to which pollutants were associated with high death rates. The U.S. EPA has concluded that full reliance cannot be placed on these analyses to assess the increased risk of mortality associated with SO_2–particulate air pollution.

Field epidemiologic studies of school-age children in North America and Europe suggest that temporary losses of lung function occur during SO_2–particulate pollution episodes. The 24-hour average concentrations measured in these episodes are 0.1 to 0.2 ppm for SO_2 and 200–400 mg/m^3 for particulate matter. Most children show only very small decreases in FVC or FEV_1. However, a few individuals have larger, perhaps medically significant losses in lung function. Test results appear to return to their pre-episode levels after several weeks.

Additional field studies have examined respiratory health in western U.S. communities where much of the SO_2 comes from smelters. Short-term exposures in these areas may be quite severe (with concentrations near 1 ppm for an hour or more), despite relatively low annual average concentrations. A comparison of four Utah towns with similar particulate levels but different SO_2 levels (annual averages ranging from near zero to above 0.4 ppm) showed that nonsmoking young adults in the most polluted community were more than twice as likely to report persistent cough, as compared to similar people in cleaner communities. An excess cough without lung function decrement was reported in Arizona school children living in smelter communities, as compared to children in cleaner areas.

Long-term studies are fewer in number. One large-scale investigation compared six northeastern and midwestern U.S. cities of differing pollution levels. More than 10,000 children were examined. Across different cities, the frequency of chronic cough was related to annual average levels of SO_2, total suspended particulates (TSP), and particulate sulfate. The frequency of lower respiratory illness was associated with total particulate levels but not SO_2. However, within cities, regional differences in pollution have not appeared to affect health measures. Most lung function differences between cities have not related significantly to pollution levels. The annual average levels of SO_2 ranged from zero to 0.07 ppm across the six cities. The annual average levels of particulate matter ranged between 30 to 160 mg/m^3 for TSP and 4 to 20 mg/m^3 for the sulfate fraction.

Particulates and Acidic Aerosols

Sources and Ambient Levels. Airborne particulate matter is a complex mixture of organic and inorganic solids-in-gas and liquid-in-gas aerosols with heterogeneous physicochemical composition, size, and biologic activity (3,5,102). Unlike other criteria pollutants, "particulates" is not a precise chemical or physical term. Particulates may be solid or liquid, chemically active or inert, soluble or insoluble in water, or may combine all these properties in different components of a particle mixture. "Smoke," "soot," "fumes," "dust," and "haze" are common terms applied to some of the diverse forms of particulate pollution. They may be emitted directly from a source (primary) or formed in the atmosphere from pollutant gases (secondary). Particulates are emitted from a wide range of natural (eg, dust storms, volcanoes, pollens, bacteria, fungi) and man-made sources (eg, coal burning, automobiles, incinerators, industrial processes, and power plants). The more widespread and clinically relevant sources are man-made and are usually concentrated in heavily populated areas. Air monitoring practices in the U.S. originally determined the concentration of total suspended particulates (TSP) by drawing a known volume of air through a highly efficient filter over a 24-hour period and measuring the gain in filter weight. Concentrations determined by this method may range from 30 $\mu g/m^3$ in clean rural air to 300 $\mu g/m^3$ or more during an urban pollution episode. Urban concentrations have been much higher in the past.

Chemical analysis of the filter samples is complex and expensive. Air quality standards relate only to the mass of particles and not to their chemical composition, except when they contain lead. A complex mixture of particles will be generated in areas with extensive coal and oil burning. Some particles have a sooty or oily nature, containing carbon or hydrocarbons from incomplete fuel combustion. Other particles will be "fly ash," ie, minerals (especially metal oxides) from noncombustible impurities in the fuel. Other particles (especially fine particles, <2.5 μm in diameter) may arise from photochemical reactions between emitted sulfur and nitrogen oxides in the atmosphere to form strong acids such as sulfuric acid (H_2SO_4), nitric acid (HNO_3). These acids can exist as airborne or aerosolized solution droplets (fog or "acid rain") with extremes in acidity that vary with meteorological factors and *in situ* interactions with other chemicals. Fog is an important factor in particulate pollution (see discussion of historical pollution episodes in "Sulfur Dioxide"). Natural fog consists of water droplets approximately 5–20 μm in diameter, which may dissolve substantial amounts of water-soluble pollutants. Clean fog droplets are slightly acidic because they dissolve some atmospheric CO_2. However, polluted fog water may contain 1,000 times the normal amount of acid. Sulfuric acid (H_2SO_4) will predominate in coal- and oil-burning areas with high levels of ambient SO_2. Most of the H_2SO_4 in ambient air results from combustion-related SO_2, which is oxidized to sulfur trioxide and hydrated to H_2SO_4 (107,108). The distribution of acidic aerosols is complex and dependent on the rates of oxidation of SO_2 and neutralization of H_2SO_4 by ammonia (NH_3) from human, animal, and fertilizer sources. With time, ammonia gas in the atmosphere or alkaline minerals in particles will neutralize the acid, forming sulfate salts. However, the partially neutralized forms of H_2SO_4, ie, ammonium bisulfate (NH_4HSO_4), letovicite ($(CNH_4)_3$ $H(SO_4)_2$) ammonium sulfate ($(NH_4)_2$ SO_4), remain strongly acidic. The acid sulfates are primarily in the fine particle size range (<2.5 μm in diameter) and can persist in the atmosphere for days and be transported downwind for long distances.

In areas with extensive motor vehicular contribution to photochemical oxidant smog, particles are likely to contain hydrocarbons and related organic substances, sulfate salts from SO_2, or nitrate salts derived from NO_2. Nitric acid (HNO_3) vapor will predominate in these areas.

Health risk from particulates depends on the effective inhaled dose (as with any pollutant) and the dose depends on both particle size and concentration. Size is expressed as aerodynamic diameter which depends on physical size, shape, and density. An airborne particle of 1 micron (micrometer, μm) in aerodynamic diameter, will settle to the ground under the influence of gravity at the same speed as a sphere that is 1 μm in physical diameter and as dense as water. Particles from any source will vary in size. Ambient air typically contains "fine" particles (largely derived from man-made pollution, with aerodynamic diameters ranging from 2.5 to less than 0.1 μm) and "coarse" particles (largely derived from soil, plants, or sea salt, with aerodynamic diameters ranging from 2.5 to 50 μm). Particles larger than 40–50 μm will not remain airborne for long and settle rapidly in windless conditions. Most particles larger than 10 μm are deposited in the upper airway during normal nasal breathing at rest and present minimal health risk.

A more relevant measurement of ambient particles, PM_{10} (particulates ≤10 μm in aerodynamic diameter), has been implemented to replace TSP measurements in order to focus on respirable particulates and their potential health effects. Particles ranging from several microns to less than one micron have presented the most health risk since they are most likely to be inhaled and deposited in the lungs and are chemically the most bioreactive. Most man-made pollutant particles fall within this respirable size range. Respirable particles that reach the tracheobronchial region tend to deposit at bifurcations of the airways; these branch points receive much higher doses of potentially toxic particles than other areas. Unless they are dissolved in the airway mucus lining, clearance of deposited insoluble particles is effected within minutes by the mucociliary defense mechanism (mucus and ciliary motion). With mouth breathing and increased ventilation (usually with exercise), more particles will penetrate deeper into the respiratory tract. During heavy exercise with mouth breathing, an estimated 10–20% of particles of 10 μm will enter the tracheobronchial tree. Very fine particles (less than 1 μm in diameter) may be exhaled or deposit in the peripheral small airways and alveoli where clearance by macrophages and lymphatics may take hours, months, or years. Some fine particles are hygroscopic, and thus attract water vapor and increase in mass during inhalation, causing them to deposit in larger, more central airways. Clearance may be impaired by some lung diseases and by some types of air pollution exposures.

The spatial and temporal distributions of ambient acidic aerosols in North America have not been monitored until recently, owing to the complex sampling systems required. Sulfate (SO_4^{2-}) has been considered to be an adequate surrogate for acidic species since it represents the predominant ambient acidic aerosol (H_2SO_4) and is easy to measure. However, the sulfate ion probably is not the toxicologically active portion of particulate sulfates, and high sulfate levels do not necessarily indicate the presence of high acid levels (107). Although the hydrogen ion (H^+) appears to be a critical marker of biological potency, most contemporary studies have expressed the concentration of H^+ as μg/m^3 of H_2SO_4. Where measured, the most frequent values of H_2SO_4 are <5 μg/m^3, and the highest levels of H_2SO_4, occurring most frequently in the summer, are 20–40 μg/m^3 (over a range of 1–12 hours) in various parts of North America. These concentrations are below the current EPA standard of 150 μg/m^3 (24-hour average) for PM_{10}. However, ambient acid aerosols are being strongly considered as a new EPA-regulated criteria pollutant owing to their distinct health risk as compared to other particulates (106). Almost 700 μg/m^3 (1-hour average) has been estimated to have occurred in London during 1962, and even higher levels were likely present in previous years. The 8-hour threshold level of H_2SO_4 for occupational exposure is 1000 μg/m^3, reflecting the observation that H_2SO_4 is not especially irritating to healthy people when inhaled.

Data for indoor levels of acidic aerosols and gases are currently scarce. A sampling of residences in the Boston area have revealed low indoor/outdoor ratios for HNO_3 and strong (free or titratable) acid (H^+) during winter and summer. However, nitrous acid (HONO), sulfate (SO_4^{2-}), nitrite (NO_2^-), nitrate ($NO3^-$), NH_3, and ammonium

(NH_4^+) have showed high indoor/outdoor ratios during both seasons. The significance of neutralization of acidic aerosols by indoor NH_3 remains to be determined.

Mechanisms. Research in acid-related mechanisms of action in humans is largely limited to characterization of the types of bronchoconstrictive stimuli present in acidic aerosols or fog (45). Titratable acidity (total available H^+) appears to be more important than pH (free H^+) alone, since oral NH_3 neutralizes inhaled H^+. The bronchoconstrictor potency of sulfate aerosols is proportional to their acidity, and inhaled buffered acids produce greater bronchoconstriction in asthmatics than unbuffered acids at the same pH, presumably by transiently decreasing airway surface pH. However, the quantitative relationship between airway response and aerosol H^+ content (dose) and the importance of differentiating the H^+ of specific acidic gases and aerosols remain uncertain. Bisulfite (HSO_3^-) without SO_2 or hypoosmolar acid aerosol can independently produce bronchoconstriction in asthmatics. Interactions of acid aerosols with co-pollutants such as oxidants may also play a role in toxicity, although combined ozone and H_2SO_4 exposures in the laboratory have not shown excess respiratory effects.

The pulmonary effect of a two-hour exposure to 1000 $\mu g/m^3$ H_2SO_4 (0.9 μm MMAD and 1.9 GSD) during intermittent exercise in healthy young adults was evaluated by BAL 18 hours following exposure (109). When compared to results from a sodium chloride exposure (control), the acid exposure did not show significant differences in alveolar cell subpopulations or alveolar macrophage function. The subjects did not have lower respiratory symptoms or changes in lung function during and after acid exposure.

Effects

Although many ambient acid aerosols and other particulate matter have potential health significance, sulfuric acid will be emphasized in this section because of its respirable nature and highest irritant potential of the particulates. Sulfate and nitrate salts have shown few effects in humans, except possibly at concentrations much higher than ambient.

Animal Studies. A substantial fraction of animal studies have dealt with particulate derivatives of SO_2, especially H_2SO_4 and certain salts that might be formed from atmospheric H_2SO_4. The latter include ammonium sulfate and ammonium bisulfate (the products of complete or partial neutralization of the acid by NH_3 gas), zinc ammonium sulfate, and sulfate salts of zinc, iron, copper, and manganese. Most of these studies have used particles of respirable sizes. Sulfuric acid appears to be the most potent respiratory irritant (102,108,110–112). In the guinea pig, dose-related increases in Raw have been observed after one-hour exposures to 100 $\mu g/m^3$ and higher. These concentrations are within the possible ambient range for total particulate matter, but probably exceed maximum contemporary ambient H_2SO_4 levels, since only a small fraction of all particulate pollution is likely to be H_2SO_4. However, ambient levels of H_2SO_4 are not well documented since only a few measurements are available and acidic particles make up a significant proportion of the fine particles. Changes in mucociliary clearance have been reported in dogs, rabbits, and donkeys during short-term and long-term exposures to H_2SO_4. Although results from different studies have been inconsistent, acute mucociliary clearance changes in humans and rabbits are similar. Continuous exposures of monkeys, dogs, and guinea pigs for a year or longer have not shown obvious damage to the lungs or blood at concentrations from less than 100 $\mu g/m^3$ to 1 mg/m^3. Thickening of airway walls occurs in monkeys at concentrations above 2 mg/m^3. Sulfuric acid and ammonium salts appear to have little effect on animal lung defenses against inhaled bacteria. Although lung defenses against infection are impaired by some metal salts, the concentrations required are much higher than ambient levels.

Sulfuric acid exposures have been studied in combination with other gaseous and particulate pollutants to compare the toxicity of mixtures and single pollutants. In some studies, ozone and H_2SO_4 together have shown excess toxic effects when ozone concentrations are near ambient levels and H_2SO_4 levels are much higher than ambient (eg, 1 mg/m^3). Coexisting sulfates and other salt aerosols appear to increase bronchoconstriction from SO_2 in guinea pigs. Exposures to mixed carbon and H_2SO_4 aerosols, both at concentrations much higher than ambient, suggest excess effects although the results have been inconsistent.

Particulate substances other than sulfates have been studied extensively to assess occupational hazards. In general, no effects have been found except at concentrations well above the ambient range.

Clinical Studies. The health effects of acidic aerosols in controlled human exposures have been summarized elsewhere (45,97,98,102,106,108,113,114). Normal subjects generally show no significant change in symptoms or lung function following 10- to 240-minute exposures to H_2SO_4 alone (10–2000 $\mu g/m^3$) or in combination with 0.3 ppm ozone with varying exercise. Nonspecific airway reactivity following brief exposures to 1000 $\mu g/m^3$ (1.1 μm diameter) of H_2SO_4, NH_4HSO_4, $(NH_4)_2SO_4$, and sodium bisulfate ($NaHSO_4$) increases according to particle acidity, with H_2SO_4 and NH_4HSO_4 significantly enhancing reactivity even though they cause only marginal changes in postexposure lung function. However, one study has indicated that bronchial reactivity to methacholine is not significantly enhanced with exposure to 2000 $\mu g/m^3$ H_2SO_4 of droplet sizes ranging between 1–20 μm.

Results of acute acid aerosol exposures in asthmatics have been mixed and less consistent than with SO_2, although it appears that many individuals with hyperreactive airways may be sensitive to inhaled H_2SO_4. In some studies, significant bronchoconstriction (adjusting for exercise effect) has not been detected in adult asthmatics exposed to as much as 2000 $\mu g/m^3$ H_2SO_4 for 10–120 minutes with intermittent exercise. However, other studies demonstrate acutely symptomatic, dose-dependent bronchoconstriction in adult asthmatics following exposures to H_2SO_4, beginning at 450 $\mu g/m^3$. A relationship between the acidity of different sulfate aerosols and degree of induced bronchoconstriction is observed. Airway reactivity to the cholinergic drug carbachol is enhanced following H_2SO_4 exposure, but reactivity to cold air inhalation is not affected. Adolescent asthmatics have been reported to be more sensitive than adult asthmatics with a 40-minute exposure to 100 $\mu g/m^3$ H_2SO_4 but this observation has not been confirmed by other investigations, suggesting that acid-sensitive and acid-insensitive subgroups of young asthmatics may exist. Most of the above studies have used fine H_2SO_4 particles less than 1 μm in aerodynamic diameter.

Inhaled H_2SO_4 acutely affects mucociliary clearance of tracer particles inhaled concurrently or after the acid exposure in healthy and asthmatic subjects. Mucociliary clearance may be more sensitive than lung function tests to detect the effect of acid. Reported clearance effects vary with dose, acid concentration, tracer particle size, and exposure duration in a complex manner. One-hour exposures to inhaled H_2SO_4 (100 and 1000 $\mu g/m^3$) show faster clearance at the low exposure concentration and slower clearance at the higher level when the inhaled tagged tracer aerosol is 7.6 μm in aerodynamic diameter. This experience suggests that lower doses stimulate clearance, whereas higher doses damage the mucociliary apparatus. However, clearance is slowed during both concentrations with a smaller tagged tracer (0.5 μm), reflecting more similar deposition sites of H_2SO_4 and tracer. Mucociliary clearance is transiently slowed in nonmedicated asthmatics following inhalation of 3.9-μm aerosol tracer and 1-hour exposure to 1000 $\mu g/m^3$ H_2SO_4. Bronchial clearance is significantly greater after 2-hour (compared to 1-hour) exposure to H_2SO_4 at 100 $\mu g/m^3$, and the reduced clearance persists as long as 3 hours following exposure. Pulmonary function does not significantly change following the above exposures. Thus, the duration of acute H_2SO_4 exposure affects mucociliary clearance both in magnitude and in persistence. The latter may correlate with some reports of delayed symptoms and bronchial hyperreactivity following H_2SO_4 exposure. Sulfuric acid fog (10.3 μm MMAD; pH 2.0; osmolarity 30 mOsm) significantly increases tracheal and outer lung zone clearance in healthy adults without affecting symptoms, lung function, or airways responsiveness to methacholine.

The inconsistent physiologic results with ambient levels of sulfuric acid warrant further clarification. The explanation may relate to biological variability of the subjects, with more sensitive subjects selected by one investigator and comparatively less sensitive volunteers by another, either by chance or by using different selection criteria. Other experimental differences may be important such as exercise intensity, times at which responses are measured, and use of mask or unencumbered breathing. Ammonia gas occurs naturally in the upper airways and can neutralize inhaled acid aerosols, presumably making them less toxic. The effectiveness of acid neutralization may vary according to the size of the acid particles, manner of breathing, amount of NH_3 present (dependent on both biological and microenvironmental characteristics), and use of acidic beverages to purposely reduce endogenous NH_3 (115–117).

Epidemiologic Studies. By necessity, most epidemiologic studies have assessed the SO_2–particulate complex rather than particulate pollution itself. Combined SO_2–particulate studies are discussed in the section on sulfur dioxide. Reviews of the possible health effects of particulate pollution have drawn different conclusions about contemporary particulate levels, pointing out the difficulties and uncertainties faced by researchers in this field (49,97,98, 102,105,106,108,118). Most early epidemiologic studies have not included measurements of acid aerosols since they were not (and still are not) part of routine aerometric monitoring networks. However, subsequent studies have included a determination of acid aerosols and have suggested that acid aerosols may aggravate symptoms in subjects with preexisting lung disease by irritation of the airways.

A potentially important finding is the association between contemporary respirable particulate matter (or a more complex pollution mix associated with respirable matter and sulfur oxides or metal ions) and daily mortality across diverse U.S. cities (119). Small but statistically significant associations of particulate pollution with all-cause daily mortality rates have been demonstrated in a number of U.S. cities or communities (Detroit, Philadelphia, Los Angeles, Steubenville, Ohio, and Santa Clara County in California) and in the Utah Valley, the site of a steel mill. This association has occurred when PM_{10} levels were below the present NAAQS (see Table 2). The increments of pollution-associated mortality have been strongest for cardiopulmonary deaths and respiratory mortality in Philadelphia. Respiratory morbidity in healthy persons and asthmatics (120) and PM_{10} pollution have shown positive correlations in the Utah Valley and Seattle, Washington. PM_{10} concentrations in Salt Lake City, Utah, have accounted for a 2–3% reduction in lung function in adult smokers with mild to moderate airflow limitation, after statistical adjustment for smoking habits (121).

All studies have used centrally located fixed-site monitors to estimate population exposures. Personal exposures to particulates have also been measured in some studies. Most studies have not been able to adequately control for cigarette smoking, temperature, season, winter-time infections, and other potential ecologic differences between communities. The particulate–mortality association has been difficult to explain mechanistically and awaits further animal and human toxicologic studies to provide an adequate interpretation. Until then, the epidemiologic data suggest that the current NAAQS for PM_{10} may not be protecting against adverse health effects with an adequate margin of safety as mandated by the Clean Air Act.

Carbon Monoxide

Sources and Ambient Levels. Carbon monoxide (CO) is formed during combustion of any carbon-containing fuel. Carbon monoxide occurs naturally as a trace constituent of the troposphere. The major source of outdoor carbon monoxide in urban areas is incompletely burned fuels (122–125). Motor vehicular exhaust accounts for two-thirds of outdoor carbon monoxide pollution, although industrial and domestic sources also contribute. In cities like Los Angeles the ambient concentrations occasionally exceed 20 ppm as a one-hour average. Before the use of catalytic converters in automobiles, community levels were substantially higher. Average background levels are higher in winter and lower in summer. Local concentrations may exceed 40 ppm on streets with very heavy traffic or in tunnels and parking garages. Thus, people who spend considerable time in such areas may receive much greater doses than would be predicted from outdoor fixed-site air monitoring stations.

Indoor microenvironments present the most important CO exposures for the majority of individuals (122–125). Indoor concentrations of carbon monoxide are a function of outdoor levels, indoor sources, infiltration, ventilation, and air mixing between and within rooms. In residences without indoor sources, average indoor CO concentrations are approximately equal to average outdoor levels. Homes with attached garages as well as homes adjacent to heavily traveled roadways, or parking garages may have high indoor CO concentrations. Personal carbon monoxide exposure studies have shown that the highest CO exposures (9–35 ppm) occur in indoor microenvironments associated with such transportation sources. Unvented gas and kerosene space heaters used indoors for prolonged periods may cause residential CO concentrations to exceed the NAAQS of an 8-hour average of 9 ppm or a one-hour average of 35 ppm. Intermittent indoor sources such as gas cooking ranges can result in peak CO levels in excess of 9 ppm and 24-hour average concentrations on the order of 1 ppm.

Of all the criteria pollutants, carbon monoxide (along with lead) is most widely recognized as a poison (126,127). Hundreds of people die every year by accident or suicide from grossly elevated indoor CO concentrations, often in the thousands of ppm. Combustion in a poorly ventilated space (eg, with a defective or improperly vented stove or heater or a car engine running in a closed garage) causes the pollution in these cases. With proper care, such serious risks can be avoided, and indoor CO levels need not be much higher than outdoor levels. Normal cooking and heating and tobacco smoking still produce some indoor CO.

Mechanism. The toxicity of carbon monoxide is due to its binding affinity for hemoglobin to form carboxyhemoglobin (COHb), a central feature of CO toxicology (123–127). The relative affinity of CO for hemoglobin is approximately 200-fold greater than that of oxygen, so that only a small concentration of CO in air can form a significant amount of COHb in blood. The COHb not only reduces the amount of hemoglobin available to carry oxygen, but also impairs the release of oxygen to body tissues. The result is reduced delivery of oxygen to tissues that require oxygen. The risk of significantly impaired oxygenation due to CO exposure increases with altitude. Oxygen requirements are most critical in the heart and brain, so most health research on CO focuses on either cardiovascular or neuropsychologic function.

A unique feature of CO exposure is that there is a biological marker of the dose the individual has received. The body burden of CO at any given time can be objectively measured by determining the percentage of COHb in a sample of blood or by measuring carbon monoxide in exhaled breath, after holding the breath for 20–30 seconds to equalize the CO between the circulating blood and alveoli. At low levels, each additional 1% of COHb in blood adds about 5 or 6 ppm of CO in the exhaled breath. Conversely, each additional 5 or 6 ppm of CO in polluted air can raise the blood COHb concentration by another 1%. The approximate COHb level resulting from any exposure can be calculated, although the actual situation is usually more complicated. In toxicologic studies, the exposure is frequently described in terms of percent COHb rather than CO concentration and exposure time. At any given concentration of CO in air, up to 10 hours of exposure will be required to "load" the hemoglobin with CO to the calculated maximum COHb concentration. Loading time is shortened by exercise, which increases ventilation and cardiac output. The body requires a slightly longer time in clean air to fully eliminate previously acquired CO because of the tight binding of CO to hemoglobin. (The conversion of COHb to oxyhemoglobin is less thermodynamically favored than the opposite reaction.) A healthy nonsmoker residing in clean air will have about 0.5% COHb which is produced by normal metabolism. Light smokers typically have 2–3% COHb, whereas heavy smokers typically have 5% or more. Cigarette smoking represents a special case since, for the smoker, smoking almost always dominates over personal exposure from other sources of carbon monoxide. Ambient CO pollution exerts little effect on smokers' preexisting COHb elevation except to prolong it during periods of smoking abstinence. Thus, the environmental exposure of nonsmokers to CO is the primary concern in terms of public health.

Animal Studies. The 1979 EPA criteria document (123) concluded that altered behavior and impaired learning might occur in laboratory animals at CO levels as low as 100 to 150 ppm. Some effects were observed in single short-term exposures (eg, 2 hours), as well as long-term exposures. However, behavioral studies in laboratory animals did not demonstrate significant effects in schedule-controlled behavior below 20–30% COHb.

Carbon monoxide exposures producing COHb concentrations of up to 39% do not produce consistent effects on lung parenchyma and vasculature or on alveolar macrophages (124). Although alveolar epithelial permeability increases during exposures to high concentrations of CO, no accumulation of edema (lung water) or significant changes in diffusing capacity of the lung have been found.

Inhaled CO (producing 2–12% COHb) can cause disturbances in cardiac rhythm and conduction in healthy as well as cardiac-impaired animals. The lowest-observable-effect level varies, depending upon the exposure regimen and species. Cardiac arrhythmias have been observed consistently after long-term intermittent exposures to CO at 100 ppm or more, producing 8–13% COHb. In a few cases, effects are reported at CO near 50 ppm, producing 4–7% COHb. Small increases in hemoglobin and hematocrit with COHb levels of 9% probably represent a compensation for reduced oxygen delivery caused by CO. Scarring or degeneration of heart muscle has been observed in some cases at the higher COHb levels. Increased myocardial ischemia was produced by COHb as low as 5% in dogs with experimentally obstructed coronary arteries. Evidence is conflicting about CO-induced enhancement of atherosclerosis and platelet aggregation.

Clinical Studies. Considerable information has been obtained on the systemic toxicity of CO, its direct effects on blood and other tissues, and the functional effects in various organs (122–129). Most laboratory-based human health studies relevant to ambient CO exposures have focused on maximal exercise performance, mental tasks, and cardiac function. A few studies have exposed volunteers to CO combined with another pollutant such as ozone; they have failed to show any additive or synergistic interaction from the combined exposure.

When the oxygen-carrying capacity of blood is reduced by the formation of COHb, the maximum rate of oxygen delivery to body tissues must decrease, potentially limiting maximal oxygen uptake by exercising muscles. Thus, CO exposure must reduce maximum exercise capability. Numerous investigators have attempted to quantify the amount of CO needed to affect exercise, ie, to determine a dose-response relationship. Decreases in maximum work rate and oxygen consumption have been consistently observed at COHb concentrations of 5% and higher and occasionally at 2–4% COHb in young, healthy nonsmoking volunteers. The desired COHb has been achieved by using higher-than-ambient CO concentrations and brief exposure times or by using several-hour exposures to 30–40 ppm CO. Submaximum exercise is not significantly affected by small increases in COHb for most sedentary healthy people.

Patients with coronary artery disease are vulnerable to myocardial ischemia which is manifested clinically by characteristic symptoms of angina pectoris (pressure–pain in chest) and by electrocardiographic (ECG) abnormalities (S-T depression). These ischemic events can be precipitated by exercise which increases myocardial oxygen requirements. The effect of CO exposure on cardiac patients can be assessed during a carefully controlled exercise test on a treadmill. The length of exercise time before the onset of angina or ECG changes should significantly decrease after CO exposure as compared to a control study with clean air, if CO (as COHb) impairs myocardial oxygenation. Carbon monoxide exposures that raise COHb to 2–4% have decreased the time to angina in some studies. A multicenter study (128) assessed a large subject population (to increase statistical power) and found that the lowest observable adverse effect level in patients with stable angina is between 3 and 4% COHb, representing a 1.5–2.2% increase in COHb from baseline values. The effect appears small compared to typical day-to-day and person-to-person variability of the test results; it is not surprising that earlier, smaller-scale studies did not show consistent results. Thus, patients with exercise-induced angina clearly are a sensitive group within the general population that is at increased risk for experiencing health effects at ambient or near-ambient CO exposure concentrations that result in COHb levels of 5% or less. Exposure sufficient to achieve 6% COHb has been shown to adversely affect exercise-related arrhythmia in patients with coronary artery disease (129).

Carbon monoxide has several effects on the nervous system. Cerebral blood flow increases, presumably due to CO-induced hypoxia, even at very low exposure levels. Studies of neurological function in CO-exposed subjects have employed numerous endpoints, including vigilance (eg, ability to respond appropriately to random signals), sensory or time discrimination (eg, visual sensitivity in the dark), complex sensory-motor performance tasks (eg, driving simulation), sleep activity, and electroencephalograms (EEG). Behaviors requiring sustained attention and performance are most sensitive to CO although the results across studies are not totally consistent. The lowest COHb level at which changes have been reliably demonstrated is 5%.

Epidemiologic Studies. Community population studies of CO in ambient air have not found significant relationships with pulmonary function, symptoms, and disease (124). Some studies have specifically addressed the relationship between CO and death rates from myocardial infarctions, cardiopulmonary symptoms, or deaths in newborn infants. Confounding factors have been abundant and results have been inconclusive. One important factor is that individual personal exposure to CO does not directly correlate with ambient CO concentrations as measured by fixed-site monitors alone because of the mobility of people, highly localized nature of CO sources, and spatial and temporal variability of CO measurements. Personal CO measurements are required to augment fixed-site monitoring data when total exposure is to be evaluated.

Environmental Tobacco Smoke

Sources, Ambient Levels, and Mechanisms. Environmental tobacco smoke (ETS) consists of a wide array of gas and particulate constituents in sidestream smoke released from the cigarette's burning end and mainstream smoke exhaled by the smoker. The inhalation of ETS is referred to as passive, involuntary, or second-hand smoking, whereas the inhalation of smoke that is drawn directly from the cigarette into the mouth or lungs is referred to as active smoking. The lower temperature in the burning cone of the smoldering cigarette releases more partial pyrolysis products in sidestream smoke as compared to mainstream smoke. As a result, sidestream smoke contains higher concentrations of some toxic and carcinogenic substances than mainstream smoke. (Table 6) Only a minority of measured substances released from a single nonfiltered cigarette have dramatically higher concentrations in sidestream than in mainstream smoke, but some known carcinogens have sidestream levels greater than 3 times that in mainstream smoke, eg, dimethylnitrosamine, pyrene, and benzo(a)pyrene. Dilution by room air can markedly reduce the concentrations inhaled by passive smokers in comparison to those inhaled by the active smoker. Thus, passive smokers are exposed to quantitatively smaller and potentially qualitatively different smoke than mainstream smokers (130–133). Substantial environmental exposure to the numerous toxic

Table 6. Selected Constituents of Cigarette Smoke[a] and Their Concentrations in Sidestream Smoke (SS) Relative to Mainstream Smoke (MS)

Constituents	MS	SS/MS Ratio
Gas Phase		
Carbon dioxide	20–60 mg	8.1
Carbon monoxide	10–20	2.5
Methane	1.3 mg	3.1
Acetylene	27 μg	0.8
Ammonia	80 μg	73.0
Hydrogen cyanide	430 μg	0.25
Methylfuran	20 μg	3.4
Acetonitrile	120 μg	3.9
Pyridine	32 μg	10.0
Dimethylnitrosamine	10–65 μg	62.0
Particulate Phase		
Tar	1–40 mg	1.3
Water	1–4 mg	2.4
Toluene	108 μg	5.6
Phenol	20–150 μg	2.6
Methylnaphthalene	2.2 μg	28.0
Pyrene	50–200 μg	3.6
Benzo[a]pyrene	20–40 μg	3.4
Aniline	360 μg	30.0
Nicotine	1.0–2.5 mg	2.7
2-Naphthylamine	2 ng	39.0

[a] Principal gas and particulate components present in mainstream smoke from a single nonfiltered cigarette.

(irritant and carcinogenic) agents from combustion of tobacco can potentially result in significant health effects.

Tobacco smoke can fill an indoor environment with high levels of respirable particles, nicotine, polycyclic aromatic hydrocarbons, CO, acrolein, and NO_2, to name a few. Their indoor concentrations depend on multiple factors such as the number of smokers, intensity of smoking, type and number of burning cigarettes, size of the room, ventilation, residence times of the individual and smoke, and use of air-cleaning devices. Monitoring studies have been conducted in public buildings and, to a lesser extent, in homes and offices. Several components of tobacco smoke have been measured as markers of the contribution of tobacco combustion to indoor air pollution. Respirable particles have been measured most frequently because they are present in large concentrations in both mainstream and sidestream smoke. Over 80% of the ETS in a room results from sidestream smoke. However, particles are a nonspecific marker of ETS because numerous other sources also discharge particles. Other components (eg, nicotine, NO_2, benzene) have also been measured as indoor markers.

Time–activity patterns strongly influence personal exposure to ETS (133). Certain groups may have greater exposures in certain locations. For example, residential exposure can be greater for infants of smoking parents than for other individuals such as nonsmoking adolescents. Nonsmoking adults may be exposed heavily at home with a smoking spouse and/or at the workplace. Concentrations of respirable particles and nicotine (contained in the vapor phase of ETS) vary widely in homes. A smoker of one pack of cigarettes daily can contribute 20 $\mu g/m^3$ to 24-hour indoor particle concentrations, and two heavy smokers can contribute more than 300 $\mu g/m^3$. Average nicotine levels in ETS at home can be 4 $\mu g/m^3$ and as high as 20 $\mu g/m^3$ in some 24-hour measurements. Indoor benzene, xylenes, ethylbenzene, and sytrene are increased in homes with smokers as compared to homes without smokers. Bars, restaurants, and other public places where smoking may be intense, show particulate concentrations as high as 700 $\mu g/m^3$. Transportation environments (trains, buses, automobiles, and airplanes) also have been documented to be contaminated by ETS. For example, NO_2, respirable particles, and nicotine in commercial aircraft are much higher with smoking than without smoking and higher in smoking than in nonsmoking sections (134). However, all passengers are exposed to nicotine (according to markers of personal exposure) regardless of seating in smoking or nonsmoking sections of the airplane cabin, especially with recirculation of cabin air (135).

Accurate exposure assessment is critical to evaluate the effects of indirect or low-dose exposure, especially when small increases in risk are present, such as with ETS. The major methodologic problem of misclassification of smoking status (ie, former smokers categorized as nonsmokers; exposed subjects as nonexposed and *vice versa*) can bias observed health effects of passive smoking. However, this problem can be mitigated with valid biological markers. A reliable and accurate biologic marker of exposure to ETS improves exposure and dose estimates, as well as validates questionnaire responses about exposure (26,133). At present, the most sensitive and specific biological markers for tobacco smoke exposure are nicotine and its metabolite, cotinine. Exposure to tobacco smoke (and perhaps ingestion of some raw vegetables) is usually the only source of nicotine or cotinine in body fluids. Nicotine concentrations in body fluids reflect recent exposures since its circulating half-life is generally less than two hours. Cotinine has a blood or plasma half-life from 10 to 40 hours, rendering it detectable for 3–6 days and capable of providing information about more subacute exposures to tobacco smoke. Cotinine measurements on the urine or saliva of passive smokers are fairly accurate and inexpensive, making them suitable for epidemiologic surveys. Cotinine levels in passive smokers range from less than 1% to 8% of the cotinine levels measured in active smokers. Thiocyanate (a metabolite of hydrogen cyanide, which is a component of tobacco smoke), exhaled CO, and COHb can distinguish active smokers from nonsmokers but are not as sensitive and specific as cotinine for assessing passive exposure to tobacco smoke. However, all the above markers are of minimal value in assessing the risk of lung cancer in children or adults since they indicate only current and not cumulative exposure. Nonetheless, the data on biological markers provide ample evidence that ETS exposure leads to absorption, circulation, and excretion of tobacco smoke components and add to the biological plausibility of associations between passive smoking and disease as documented in epidemiologic studies.

Effects

The health impact of passive smoking has yet to be fully characterized but is clearly an important public health issue since 30% of American adults are active smokers and 60% of all homes have at least one smoker. The information on the health effects of passive smoking has been largely derived from observational epidemiologic studies (130–139). In these studies ETS exposure has been primarily estimated by questionnaires and biological markers. The reliability of subjects in indicating the smoking status of spouses or parents appears high, but quantification of smoking and time spent with a smoker is less reliable. At present, both biological markers and carefully formulated questionnaires are necessary to assess acute and cumulative ETS exposures in epidemiologic studies.

Respiratory Effects in Children. In general, the association between exposure to ETS and increased respiratory disease is stronger in young children than in adults. Epidemiologic studies have linked passive smoking in children to increased occurrence of lower respiratory tract illness (presumably infectious in etiology) during infancy and childhood. Multiple investigations throughout the world have demonstrated an increased risk of bronchitis and pneumonia during the first year of life in infants with smoking parents. Maternal (rather than paternal) smoking alone can increase the incidence of lower respiratory illness in a dose-related fashion. A similar effect of passive smoking has not been readily identified after the first year of life, perhaps reflecting the change in time–activity patterns as the infant matures.

Numerous large surveys have demonstrated increased frequency of cough, phlegm, and wheeze in school-aged children of smoking parents. The frequency of reported respiratory symptoms in children increases with the number of smokers in the home, suggesting a dose-response relationship. For example, smoking by parents (largely maternal) have increased the frequency of persistent cough in their children by 30% in a Harvard study of 10,000 school children in six U.S. communities. However, the association of passive smoking with childhood asthma is less consistent despite the association with wheeze. Nonetheless, maternal smoking can adversely affect children (especially boys) with asthma, as supported by adverse effects on lung function, symptom frequency, and airway responsiveness to inhaled agonists (eg, histamine), and increased medication use and emergency room visits. Urinary cotinine concentrations are directly related to increased frequency of asthma exacerbations and reduced lung function in asthmatic children of smoking parents (140). On the other hand, the severity of children's asthma will decrease with less smoking exposure from smoking parents (141). Some studies have also found associations between passive smoking and middle-ear disease, sore throat, and school absenteeism.

Longitudinal studies indicate that passive smoking is associated with small reductions in the rate of lung function growth during childhood in children of smoking parents (especially mothers) in comparison with those of nonsmokers.

Respiratory Effects in Adults. Studies have generally evaluated the effects of ETS exposure from a smoking spouse and have found inconsistent evidence of increased chronic respiratory symptoms in adults. Similarly, epidemiologic and experimental exposure studies of asthmatic adults have yet to demonstrate consistent effects of ETS in exacerbating asthma in adults, although these studies have largely involved small numbers of subjects. Some studies, however, have indicated that a subset of ETS-sensitive adult asthmatics develop significant reductions in lung function as well as increased symptoms (142).

Lung function in adults has not been found conclusively to be adversely affected by passive smoking in cross-sectional studies. Some studies suggest that chronic exposure to ETS impairs airflow in the small peripheral airways. No data indicate that emphysema can be caused by passive smoking alone. However, further research is warranted because of the widespread exposure to ETS in workplaces and homes.

The association of passive smoking and lung cancer has generated considerable attention and controversy. The association has biologic plausibility because of the presence of respirable carcinogens in sidestream smoke and the lack of a documented threshold dose for respiratory carcinogenesis in active smokers. On the other hand, many epidemiologic studies have been affected by misclassifications of smoking or exposure status, inaccurate quantification of smoking habits, failure to assess and control for potential confounding factors (eg, diet, other exposures), and misdiagnosis of lung metastases (from cancers of other organs) as primary lung cancers. The assumption that spouse smoking alone represents ETS exposure may miss exposures outside the home or within the home. Gender-related differences may be present since women smokers generally smoke less and generate less ETS than male smokers. Inconsistencies among studies should be anticipated in view of the relatively small populations in many studies and the difficulties in estimating exposure. Thus, evidence from case-control and cohort studies does not uniformly indicate increased risk of lung cancer in persons exposed to ETS. Despite these methodologic issues, the collective weight of evidence indicates a substantial increase in the risk of lung cancer in subjects involuntarily exposed to cigarette smoke, especially in female nonsmokers married to smoking spouses. Risk appears to increase with increasing dose of ETS. Expert reviews by the National Research Council (132), U.S. Surgeon General (136), U.S. EPA (138), and the International Agency for Research on Cancer of the World Health Organization (139) have concluded that passive smoking causally increases the risk of lung cancer in nonsmokers. The U.S. EPA (138) has officially classified ETS as a known carcinogen. The magnitude of lung cancer risk remains uncertain. Simplifying assumptions about the extent and degree of exposure to ETS, exposure–response relationships, etc, must be made in risk assessments. With these limitations, involuntary smoking is estimated to be associated with 3,000 to 8,000 lung cancer deaths per year, with relative risks from 1.3 to over 2.0 versus the non-exposed population (132,143), which are small in relation to risks of typical, active smoking. In other words, assuming a mean risk of 35% (132), an average lifetime passive smoke exposure from a smoking spouse increases a nonsmoker's risk by about 35%, as compared with the risk of 1000% (or tenfold) for a lifetime of active smoking. Although the risks are small (if estimates are accurate) for individual passive smokers, their public health significance lies in the number of excess lung cancers caused by passive smoking (approximately 3000–5000 per year with this risk estimate). Regardless of the exact numbers, the calculations demonstrate that passive smoking must be considered an important cause of lung cancer deaths from the public health perspective since ETS exposure is involuntary and subject to control (ie, is avoidable).

Cardiovascular Effects. Although active smoking has been extensively established as a cause of cardiovascular disease and causes (along with lung cancer) 300,000 to 400,000 deaths annually, relatively few studies have examined passive smoking as a risk factor (133). There is no consistent data that passive smoking poses an increased risk for accelerating atherogenesis, angina pectoris/coronary artery disease, or cardiac mortality. Some epidemiologic studies indicate a two-to-threefold increase in the risk of ischemic cardiac death in passive smokers married to a male smoker, after adjustment for other known risk factors. However, cigarette smoking increases ambient CO levels and even increments of 1–2% COHb can decrease the time to the occurrence of angina during exercise. Similarly, passive smoking may adversely affect athletic performance in healthy subjects during controlled exposures to ETS. The postulated components of ETS which might affect cardiovascular function are CO and nicotine. The former (as COHb) decreases oxygen delivery to the heart muscle, whereas the latter releases epinephrine and increases oxygen demand. These studies are unlikely to be truly blinded because of the unmistakable odor of tobacco smoke. However, there is no strong evidence that cardiopulmonary responses have a psychogenic basis (144). Further research is necessary to clarify any link between cardiovascular disease and ETS.

Other Effects. Some small studies suggest that exposure to ETS is associated with health effects outside the lungs, although consistent evidence is lacking. Excess risk of malignancies at sites other than the lungs (eg, breast, cervix, nasal sinus) in nonsmoking women with smoking spouses has been reported. However, these associations show weak biologic plausibility since the cancers in question are not increased in active smokers. The associations may arise by chance or by the effect of bias (confounding factors). Several cohort studies have shown increased risk of all-cause mortality (15–40%) in passive smokers. This trend parallels cardiovascular mortality of passive smokers in some studies. Annoyance and irritation effects of ETS on the eyes, upper and lower airways are common complaints from passive smokers. ETS-sensitive individuals who are experimentally exposed to sidestream tobacco smoke report more nasal congestion, runny nose (rhinorrhea), throat irritation, headache, chest discomfort, and cough, as well as statistically significant increases in nasal resistance to airflow without changes in lung function (145).

CONTROL STRATEGIES

General

Because of the large number of potential sources of air pollutants found in outdoor and indoor locations and the wide range of health effects, different levels of preventive strategy may be considered: (*1*) Primary prevention refers to elimination or reduction of exposure to the pollutant before the onset of an adverse health effect or disease; (*2*) Secondary prevention refers to detection of asymptomatic adverse health effects, thus permitting early intervention to eliminate or reduce exposure and possibly prevent clinically evident disease; and (*3*) Tertiary prevention refers to management of a clinical effect or disease caused by the pollutant to reduce morbidity and mortality. For all pollutants, primary prevention is the most desirable and will be emphasized here. This prevention is frequently accomplished by public education and regulatory policies to limit emissions and exposures, although there are also important individual actions which may be effective.

Steps can be taken to reduce risks from environmental pollutants despite persistent controversies and points of uncertainty about their health effects. Table 7 summarizes approaches to control the health effects of air pollutants. These approaches essentially attempt to control pollutant emissions, reduce human exposure, or mitigate illness caused or exacerbated by exposure. Often multiple strategies must be pursued concomitantly to achieve substantial benefits. Each approach has inherent risk–benefit considerations which may involve scientific data, clinical experience, ethical judgments, economics, or other factors. The practicality and effectiveness of any given ap-

Table 7. Approaches to Control the Health Effects of Air Pollutants

A. *Political*
1. "Command and control" method that legally limits or forbids certain emissions; a common approach of state or federal implementation plans under the Clean Air Act
2. "Emissions trading" method that limits the quantity of emission permits distributed among sources through market mechanism; favored by economists to maximize efficiency of pollution controls
3. Direct taxes on measurable emissions
4. Taxes on goods or services associated with pollution

B. *Technological*
1. Source modification
   a. Discontinue or isolate processes that emit pollutants
   b. Substitution of less-polluting for more-polluting processes
   c. Emission controls on sources
2. Ventilation/dispersion
   a. Dilution, eg, by tall stacks (trade-off of high regional versus low local pollution)
   b. Exhaust from work areas with internal sources
3. Removal from human-occupied space
   a. Physical filtration
   b. Adsorption
   c. Catalytic destruction
   d. Electrostatic precipitation

C. *Personal*
1. Avoidance by changing activity patterns, job location
2. Medical management, eg, asthmatics inhale bronchodilator prior to exercising in SO_2
3. Education of others to practice 1 and 2

proach may vary markedly in different localities, depending on their physical and socioeconomic characteristics.

Outdoor Pollution

An ambitious national health objective for the year 2000 in the U.S. is to increase the proportion of persons who live in counties that have not exceeded any outdoor air quality standard during the previous 12 months from 49.7% in 1988 to 85% (objective 11.5) (146). The large population exposed to poor air quality and the substantial proportion who belong to high-risk subgroups underscore the importance of improved air quality for health promotion and disease prevention. How can this objective be achieved?

For some pollutants, only national policy and regulation (eg, under the Clean Air Act) can reduce risk. Pollution abatement (action intended to reduce or eliminate pollution hazards at their sources) is usually undertaken by a state or local regulatory agency, often with statutory authority delegated by the federal government (EPA). Public and private sectors contribute to the regulatory decision-making process. Community-wide or region-wide pollution abatement is a long-term process. The abatement of more limited or specific local problems may have faster progress since the sources, costs of control, and benefits of control are more easily understood or managed. For example, motor vehicular emissions are being strictly controlled in the State of California with diverse strategies, eg, catalytic converters, reformulated gasoline, “alternative” or “clean” fuels, “zero-emission” electric vehicles. Some strategies may have greater impacts and cost-effectiveness than others (51). On the other hand, reduction of acidic aerosols or ozone can be affected only by multifaceted, broadly applied regulatory strategies since the pollutants have multiple sources and spread widely in the atmosphere. In this process, there is the continual need for additional education, communication, and coordination among industries, health agencies, environmental groups, and the general public.

When outdoor pollutant exposure presents a clear risk to health and immediately effective abatement is unlikely, individual protective actions are necessary. These actions are often satisfactory for the involved individuals but are unlikely to be fully effective throughout a community. Individual protection should be considered only a temporary alternative to abatement of outdoor pollution.

For some pollutants, the individual can directly reduce risks (147). Removing the hazard from the environment is feasible in some instances (eg, hypersensitivity to a nonessential household product) but not easy with community air pollution. Conversely, removing oneself from an environment perceived or known to be hazardous (ie, personal avoidance of exposure) is frequently more effective although not necessarily as practical. Modifying time–activity patterns to limit time outside during pollution episodes often represents the most effective strategy. Staying indoors and avoiding unnecessary outdoor activity during community pollution episodes are commonly recommended by physicians and public agencies, especially for susceptible individuals or patients with lung disorders. This recommendation assumes that there are no indoor pollutant sources or hazards and that outdoor pollutant levels fall when air migrates indoors. The latter is usually true for reactive pollutants such as ozone and SO_2 but not for others such as particulates. Even cleaning shag carpeting with portable vacuum cleaners can significantly increase the number of respirable airborne particles, which may remain suspended in air for as long as one hour (148). Central vacuum cleaners generate fewer airborne particles during carpet cleaning than portable vacuum cleaners. Some medications (eg, albuterol, metaproterenol, cromolyn, nedocromil) commonly used by asthmatics and other respiratory patients can prevent acute respiratory effects (wheezing or asthma attack) induced by SO_2.

More drastic individual protective measures include changes of job and residence. These choices may involve substantial costs or risks in moving the affected person from the polluted environment, and there is no assurance of overriding benefits. They require careful evaluation of the environments in question in order to verify, insofar as possible, that the existing environment is indeed contributing to the problem and the proposed alternative is truly preferable and helpful.

Indoor Pollution

Personal modification of indoor sources is potentially effective in reducing pollutant concentrations (17,147). The available methods cover a broad range, from the control of the indoor environment where the sources are located, to modifying the sources themselves. Controlling indoor temperature and relative humidity may affect the rate at which a pollutant is generated and transported within a residence or building. For example, central, forced-air heating and cooling systems can mix pollutants within homes. Avoiding high relative humidity is important for preventing the growth of house dust mites, fungi, or various infectious pathogens which may become airborne. Low humidity may reduce emissions of formaldehyde from building materials and furnishings, as well as inhibit the growth of the aforementioned biological agents. Levels of other pollutants are not greatly affected by climate control. Some sources may be removed from the living space or a source may be replaced by a functionally equivalent product that emits less pollution. Occupant exposure may be reduced by isolating the source in a separate area and limiting access to it. Behavioral adjustments are the simplest, least expensive, and most effective means of improving indoor air quality, eg, prohibiting smoking indoors or completely segregating smoking areas; proper use and maintenance of gas appliances; proper use of humidifiers/dehumidifiers and pesticides; proper storage of cleaning materials, paints, and solvents; proper ventilation while painting or pursuing hobbies. Radon concentrations in homes can be measured inexpensively, and techniques are available for mitigating and avoiding unacceptable radon concentrations. Prevention and cessation of smoking control the hazards of both active and passive smoking, such as lung cancer, with or without concomitant asbestos or radon exposure.

Respiratory protective equipment such as masks has been developed for use in the workplace to minimize exposure to toxic gases and particles. Somewhat more sophisticated masks may include an efficient particulate filter and/or a chemical sorption agent (eg, activated carbon) to remove some pollutant gases. Many simple masks, even when effective against some particulate pollution, present a number of problems. They are uncomfortable, interfere with oral activities, do not always fit snugly, are ineffective against gases and small particles, and add to the work of breathing, which cannot be tolerated by persons with respiratory disease. For these reasons, masks are considered a last resort in air-quality protection.

Reduction of indoor air contaminants may be accomplished by natural ventilation, air purification (or more accurately, air cleaning), and mechanical ventilation systems. Personal protection by air cleaning implies remaining in a contaminated environment but cleaning the air before it is breathed. Air-cleaning devices vary greatly in size, portability, cost, and capabilities for maximum rate of cleaning and type of pollutants removed (17,147,149,150). The typical domestic air cleaner consists of a portable box containing an electrically operated blower and cleaning device. Polluted air is drawn in through the rear of the unit, passed through the purification system, and discharged through the front of the unit. The user must breathe directly in front of the clean air outlet to receive maximal benefit because of dilution by untreated room air. If the unit is to clean the air of an entire room, its flow rate (volume of clean air discharged per minute) must be a substantial fraction of the room's total volume and larger than the rate of air exchange between the room and its surroundings. Achieving these standards usually requires that the room be tightly closed, in which case there must be no significant indoor source of contaminants that the purifier cannot remove completely.

The basic mechanisms of commercially available household air-cleaning devices are as follows: high-efficiency particulate arresting (HEPA) filters, electrostatic precipitators, electret filters, negative ionizers, and ozonators. Each type has advantages and disadvantages which are reviewed elsewhere (17,147,149,150); only the most commonly used types are briefly reviewed.

Most particulate filters consist of a fibrous mat in which particles must contact and adhere to the fibers, thus removing the particles from the airstream. Common glass–fiber furnace filters are effective only against relatively large dust particles. Particles less than 1 μm are the most difficult to filter but make up a substantial fraction of the particulate matter in polluted air. Paper HEPA filters can efficiently remove particles of a wide size range and retain the same size particles with up to 99.97% efficiency. They do not control gases or sequestered allergens such as house dust mites. Increasing the filtration efficiency increases cost, both for the filter material itself and for the energy needed to move air against increased resistance of the filter. Particulate filters of any type must be discarded and replaced regularly since heavily loaded filters will obstruct airflow and may release previously captured particles.

Electrostatic precipitators use a high-voltage electric field that imparts charges to passing particles. The air then passes between pairs of charged plates to which oppositely charged particles are attracted and adhere. Although they are generally less efficient than HEPA filters, electrostatic precipitators can achieve moderate to high removal efficiencies of small particles (eg, ETS, house dust mite, and animal allergens) and larger particles (eg, pollens, mold spores) with optimal design and operating conditions. They remove radon decay products attached to particles but not unattached radon decay products. Small home systems must be deactivated regularly for cleaning the collection plates; otherwise, collection efficiency declines and some ozone may be produced by electrical arcing to the dirty collector surfaces. Large systems use automatic cleaning mechanisms. Electrostatic precipitators do not require periodic replacement and offer relatively little resistance to airflow, which can be increased to improve effectiveness.

Chemical filters are solid substances, usually in granular or pellet form. They do not actually filter but adsorb pollutant gas molecules or react with them to form more innocuous substances. An effective adsorbent must have a large surface area containing properly-sized sub-microscopic pores to capture and retain pollutant molecules. Activated carbon (charcoal) is the most common general-purpose chemical filter medium. It effectively removes organic gases of moderate to large molecular weight, including many irritating and malodorous VOCs (except tobacco smoke). Activated carbon effectively removes ozone, is only moderately effective against NO_2 and SO_2, and is ineffective against NO, CO, and organic gases of low molecular weight, such as formaldehyde. The charcoal or airstream must be chemically analyzed to determine the saturation status of the carbon. Used carbon can be either replaced or reactivated by treatment with hot air or steam to desorb (revolatize) the contaminants. Finally, activated alumina (aluminum oxide) impregnated with potassium permanganate may be more effective than activated carbon against nitrogen and sulfur oxides. Alumina-permanganate is effective against organic compounds that are readily oxidized but not against CO. Unlike activated carbon, alumina-permanganate is noncombustible but its dust is irritating and corrosive. Its functional condition may be monitored by observing the color change from purple to brown as the permanganate is consumed.

In principle, the most effective general-purpose recirculating air cleaners are those that incorporate a chemical filter and a HEPA filter. This combination should be capable of removing most particulate pollutants and most gases other than those that are relatively unreactive or of low molecular weight, such as carbon monoxide. However, the strong odor of tobacco smoke from its gas phase can be difficult to remove with either HEPA filters or electrostatic precipitators and requires separate control measures for removal (eg, ventilation).

Regular maintenance (cleaning and/or filter replacement) is absolutely essential to maintain the effectiveness of any device. In buildings equipped with central air conditioners, air-cleaning devices may be installed in the ductwork to allow the whole building's air to be treated. Particulate filtration, chemical filtration, electrostatic precipitation, or a combination may be used. The choice usually depends on economics and the nature of the pollutants to be controlled. Air-conditioning systems typically recirculate about 90% of the indoor air and draw only about 10% from outside air to minimize energy requirements for heating and cooling. Cleaning devices may be placed in the outside air inlet if there are no significant contaminant sources inside the building and maximum economy is desired. More typically, both internal and external sources of pollution are important, in which case both recirculated air and outside air should be cleaned by installing the cleaning devices in the main air-supply duct. HEPA filters can effectively remove most microbes and nonviable pollutant particles. In large buildings an engineering study is necessary to design or evaluate a system that will be satisfactory in terms of ventilation, temperature and humidity control, contaminant removal, energy use, and maintenance costs (151). Fire safety is another critically important design consideration: nonflammable materials should be used to avoid ignition and spreading smoke throughout a buildling.

The clinical perspective of the above air cleaning devices, however, is important to consider. The lack of a standard test program for these air cleaners has resulted in a scarcity of objective information regarding their performance beyond the general claims of the manufacturers. The few published evaluations of these devices indicate a wide range of performance. Similarly, the clinical utility of personal air cleaning is largely untested and anecdotal; air cleaners have not been objectively shown to improve health by reducing indoor pollutants, except in a very few small-scale studies. For example, although allergic symptoms in some patients have been reported to decrease during the domestic use of HEPA filters (152), a consensus group (153) has concluded that the lack of objective results about the health-promoting action of indoor air-cleaning devices in double-blind studies precludes any firm medical recommendations about their use by patients. The use of room air-cleaning devices in the absence of other forms of environmental control is not considered reasonable.

Acknowledgments

Parts of this chapter are based on the following sources: American Lung Association, *Health Effects of Ambient Air Pollution*, American Lung Association, New York, 1989. H. Gong, Jr., "Health Effects of Air Pollution. A Review of Clinical Studies," *Clin Chest Med* **13,** 201–230 (June 1992). American Lung Association, *Personal Protection Against Air Pollution: The Physician's Role*, American Lung Association, New York, 1981.

This publication is partly based on the authors' research which is supported by the California Air Resources Board, Electric Power Research Institute, Health Effects Institute, Southern California Edison Co., and the U.S. Environmental Protection Agency.

BIBLIOGRAPHY

1. P. Brimblecombe, *J. Air Pollut. Control Assoc.* **26,** 941–45 (Oct. 1976).
2. J. M. Samet and M. J. Utell, *J. Am. Med. Assoc.* **266,** 670–675 (Aug. 7, 1991); T. Schneider, S. D. Lee, G. J. R. Wolters, and L. D. Grant, eds., "Atmospheric Ozone Research and Its Policy Implications." *Proceedings of the 3rd US-Dutch International Symposium, Nijmegen, The Netherlands, May 9–13, 1988*, Elsevier Science Publishers B.V., Amsterdam, 1989; G. Hoek, *Acute Effects of Ambient Air Pollution Episodes on Respiratory Health of Children*, Thesis, Departments of Epidemiology and Public Health and Air Pollution, Agricultural University Wageningen, The Netherlands, Sept. 14, 1992; U. Mohr, D. V. Bates, H. Fabel, and M. J. Utell, eds., *Advances in Controlled Clinical Inhalation Studies*, Springer-Verlag, Berlin, 1993.
3. R. B. Schlesinger, "Atmospheric Pollution," *Otolaryngol Head Neck Surg.* **106,** 642–649 (June, 1992).
4. U.S. Environmental Protection Agency, Office of Air Quality Planning and Standards, *National Air Quality and Emissions Trends Report, 1989*, EPA-450/4-91-003, U.S. EPA, Research Triangle Park, N.C., 1991.
5. R. W. Shaw, *Sci. Amer.* **257,** 96–103 (Aug. 1987).
6. National Research Council, *Indoor Pollutants*, National Academy Press, Washington, D.C., 1981.
7. J. M. Samet, M. C. Marbury, and J. D. Spengler, *Am. Rev. Respir. Dis.* **136,** 1486–1508 (Dec. 1987).
8. J. M. Samet, M. C. Marbury, and J. D. Spengler, *Am. Rev. Respir. Dis.* **137,** 221–242 (Jan. 1988).
9. D. R. Gold, *Clin. Chest Med.* **13,** 215–229 (June 1992).
10. T. V. Larson, D. S. Covert, R. Frank, and R. J. Charlson, *Science* **197,** 161–163 (July 8, 1977).
11. J. B. L. Gee, W. K. C. Morgan, and S. M. Brooks, eds., *Occupational Lung Disease*, Raven Press, New York, 1984.
12. N. A. Esmen and M. A. Mehlman, eds., *Occupational and Industrial Hygiene: Concepts and Methods*, Volume 8, "Advances in Modern Environmental Toxicology," Princeton Scientific Publishers, Inc., Princeton, N.J., 1984.
13. M. Lippman, *Aerosol Sci. Technol.* **6,** 93–114 (1987).
14. M. Lippman, *J. Exposure Analysis Environ. Epidemiol.*, 103–129 (Jan.–Mar. 1993).
15. Office of Technology Assessment, *Catching Our Breath: Next Steps to Reducing Urban Ozone* (OTA-0-412), U.S. Government Printing Office, Washington D.C., July 1989.
16. Clean Air Act (CAA) Amendments, P.L. 101–549 (Nov. 15, 1990).
17. J. Samet, "Environmental Controls and Lung Disease," *Am. Rev. Respir. Dis.* **142,** 915–939 (Oct. 1990).
18. ASHRAE standard 55–1981, *The Thermal Environment Conditions for Human Occupancy*, American Society of Heating, Refrigerating, and Air Conditioning Engineers, 1989.
19. American Thoracic Society, *Am. Rev. Respir. Dis.* **131,** 666–668 (Apr. 1985).
20. I. T. T. Higgins, *J. Air Pollut. Control Assoc* **33,** 661–663 (July 1983).
21. H. A. Boushey, M. J. Holtzman, J. R. Sheller, and J. A. Nadel, *Am. Rev. Respir. Dis.* **121,** 389–413 (Feb. 1980).
22. National Research Council, Committee on the Epidemiology of Air Pollutants, *Epidemiology and Air Pollution*, National Academy Press, Washington D.C., 1985.
23. National Research Council, *Human Exposure Assessment of Airborne Pollutants: Advances and Opportunities*, National Academy Press, Washington D.C., 1991.
24. K. Sexton and P. B. Ryan, "Assessment of Human Exposure to Air Pollution: Methods, Measurements, and Models," in A. Y. Watson, R. R. Bates, and D. Kennedy, eds., *Air Pollution, the Automobile, and Public Health*, National Academy Press, Washington D.C., 1988.
25. K. Sexton, S. G. Selevan, D. K. Wagener, and J. A. Lybarger, *Arch. Environ. Health* **47,** 398–407 (Nov./Dec. 1992).
26. National Research Coouncil, Subcommittee on Pulmonary Toxicology, *Biologic Markers in Pulmonary Toxicology*, National Academy Press, Washington D.C., 1989.

27. M. Lippmann, "Introduction and Background," in M. Lippmann, ed., *Environmental Toxicants. Human Exposures and Their Health Effects*, Van Nostrand Rheinhold Co., Inc., New York, 1992, pp. 1–29.
28. J. D. Brain, B. D. Beck, A. J. Warren, and R. A. Shaikh, *Variations in Susceptibility to Inhaled Pollutants. Identification, Mechanisms, and Policy Implications*, The Johns Hopkins University Press, Baltimore, Md., 1988.
29. *MMWR Morb. Mortal Wkly. Rep.* **42,** 301–304 (Apr. 30, 1993).
30. L. Mølhave, "Volatile Organic Compounds and The Sick Building Syndrome," in M. Lippmann, ed., *Environmental Toxicants. Human Exposures and Their Health Effects*, Van Nostrand Rheinhold Co., Inc., New York, 1992, pp. 633–646.
31. National Research Council, Committee on the Institutional Means for Assessment of Risks to Public Health, *Risk Assessment in the Federal Government: Managing the Process*, National Academy Press, Washington D.C., 1983.
32. U.S. Environmental Protection Agency, *Indoor Air-Assessment. A Review of Indoor Air Quality Risk Characterization Studies,* Environmental Criteria and Assessment Office, Research Triangle Park, N.C., Mar., 1991, EPA/600/8-90/044.
33. A. C. Upton, "Perspectives on Individual and Community Risks," in M. Lippmann, ed., *Environmental Toxicants. Human Exposures and Their Health Effects*, Van Nostrand Rheinhold Co., Inc., New York, 1992, pp. 647–661.
34. C. M. Spooner, *Clin. Chest Med.* **13,** 179–192 (June 1992).
35. E. K. Silbergeld, *Environ. Health Perspect. 1993* **101,** 100–104 (June 1993).
36. U.S. Environmental Protection Agency, *Cancer Risk from Outdoor Exposures to Air Toxics*, Washington D.C., 1990, EPA-450/1-90-004A.
37. B. Hasset-Sipple and I. L. Cote, *MMWR Morb. Mortal. Wkly. Rep.* **40,** 278–279 (May 13, 1990).
38. R. F. Phalen, *Inhalation Studies: Foundations and Techniques*, CRC Press, Boca Raton, Fla., 1984.
39. D. E. Gardner, J. D. Crapo, and E. J. Massaro, eds., *Toxicology of the Lung*, Raven Press, New York, 1988.
40. J. Crapo, F. J. Miller, B. Mossman, W. A. Pryor, and J. P. Kiley, *Am. Rev. Respir. Dis.* **145,** 1506–1512 (June 1992).
41. D. V. Bates, *Respiratory Function in Disease*, 3rd ed., W.B. Saunders, Philadelphia, Pa., 1989.
42. J. D. Hackney, W. S. Linn, and E. L. Avol, *Environ. Sci. Technol.* **18,** 115A–122A (April 1984).
43. R. Frank, J. J. O'Neil, M. J. Utell, J. D. Hackney, J. Van Ryzin, and P. E. Brubaker, eds., *Inhalation Toxicology of Air Pollution: Clinical Research Considerations*, American Society for Testing and Materials, Philadelphia, Pa., 1985, ASTM no. 872.
44. L. J. Folinsbee, "Human Clinical Inhalation Exposures. Experimental Design, Methodology, and Physiological Responses," in D. E. Gardner, J. D. Crapo, E. J. Massaro, eds., *Toxicology of the Lung*, Raven Press, New York, 1988, pp. 175–199.
45. H. Gong, Jr., *Clin. Chest. Med.* **13,** 201–230 (June 1992).
46. U. Mohr, D. V. Bates, H. Fabel, M. J. Utell, eds., *Advances in Controlled Clinical Inhalation Studies*, Springer-Verlag, Berlin, 1993.
47. H. Gong, Jr., "Air Pollutants As a Provocative Challenge in Humans," in S. Spector, ed., *Provocation Testing in Clinical Allergy*, Marcel Dekker, New York, in press.
48. K. J. Rothman, *Modern Epidemiology*, Little, Brown and Company, Boston, Mass., 1986.
49. F. W. Lipfert, *Environ. Sci. Technol.* **19,** 764–770 (Sept. 1985).
50. M. D. Lebowitz, J. J. Quackenboss, M. L. Soczek, M. Kollander, and S. Colome, *J. Air Pollut. Control Assoc.* **39,** 1411–1419 (Nov. 1989).
51. J. G. Calvert, J. B. Heywood, R. F. Sawyer, and J. H. Seinfeld, *Science* **261,** 37–45 (July 2, 1993).
52. W. W. Lowrance, *Modern Science and Human Values*, Oxford University Press, New York, 1985.
53. M. A. Ottoboni, *The Dose Makes the Poison: A Plain-Language Guide to Toxicology*, 2nd ed., Van Nostrand Reinhold, New York, 1991.
54. A. B. Hill, *Proc. Royal Soc. Med.* **58,** 295–300 (May 1965).
55. D. V. Bates, *Environ. Res.* **59,** 336–349 (Dec. 1992).
56. R. A. Bethel, *Semin. Respir. Med.* **8,** 253–258 (Jan. 1987).
57. A. J. Wardlaw, *Clin. Exp. Allergy* **23,** 81–96 (Feb. 1993).
58. M. Lippmann, ed., *Environmental Toxicants. Human Exposures and Their Health Effects*, Van Nostrand Rheinhold Co., Inc., New York, 1992.
59. U.S. Environmental Protection Agency, *Air Quality Criteria for Ozone and Other Photochemical Oxidants*, Environmental Criteria and Assessment Office, Research Triangle Park, N.C., Aug., 1986, EPA/600/8-84/020eF.
60. J. H. Seinfeld, *Science* **243,** 745–752 (Feb. 10, 1989).
61. P. J. Lioy and R. V. Dyba, *Toxicol. Indust. Health* **5,** 493–505 (Mar. 1989).
62. M. Lippman, "Ozone," in M. Lippman, ed., *Environmental Toxicants, Human Exposures and Their Health Effects*, Van Nostrand Rheinhold Co., Inc., New York, 1992, pp. 465–519.
63. C. J. Weschler, C. J. Shields, and D. V. Naik, *J. Air Pollut. Control Assoc.* **39,** 1562–1568 (Dec. 1989).
64. U.S. Environmental Protection Agency, *Summary of Selected New Information on Effects of Ozone on Health and Vegetation*, Environmental Criteria and Assessment Office, Research Triangle Park, N.C., Jan., 1992, EPA/600/8-88/105F.
65. M. J. Hazucha, D. V. Bates, and P. A. Bromberg, *J. Appl. Physiol.* **67,** 1535–1541 (Apr. 1989).
66. J. Seltzer and co-workers, *J. Appl. Physiol.* **60,** 1321–1326 (Apr. 1986).
67. H. S. Koren and co-workers, *Am. Rev. Respir. Dis.* **139,** 407–415 (Feb. 1989).
68. R. B. Devlin and co-workers, *Am. J. Respir. Cell Mol. Biol.* **4,** 72–81 (Jan. 1991).
69. H. S. Koren, G. E. Hatch, and D. E. Graham, *Toxicology* **60,** 15–25 (Jan./Feb. 1990).
70. M. G. Mustafa and D. F. Tierney, *Am. Rev. Respir. Dis.* **118,** 1061–1090 (Dec. 1978).
71. M. A. Mehlman and C. Borek, *Environ. Res.* **42,** 36–53 (Feb. 1987).
72. M. Lippmann, *J. Air Pollut. Control Assoc.* **39,** 672–694 (May 1989).
73. E. S. Wright, D. M. Dziedzic, and C. S. Wheeler, *Toxicol. Lett.* **51,** 125–145 (Apr. 1990).
74. W. S. Tyler, M. D. Julian, and D. M. Hyde, *Semin. Respir. Med.* **13,** 94–113 (Mar. 1992).
75. B. E. Tilton, *Environ Sci. Technol.* **23,** 257–263 (Mar. 1989).
76. W. F. McDonnell and co-workers, *J. Appl. Physiol.* **54,** 1345–1352 (May 1983).

77. L. J. Folinsbee, W. F. McDonnell, and D. H. Horstman, *J. Air Pollut. Control Assoc.* **38,** 28–35 (Jan. 1988).
78. D. H. Horstman, L. Folinsbee, P. J. Ives, S. Abdul-Salaam, and W. F. McDonnell, *Am. Rev. Respir. Dis.* **142,** 1158–1163 (Nov. 1990).
79. W. C. Adams, *Sports Med.* **4,** 395–424 (Mar. 1987).
80. H. Gong, Jr., *J. Sports Med. Physical Fitness* **27,** 21–29 (Mar. 1987).
81. N. A. Molfino and co-workers, *Subjects Lancet* **338,** 199–203 (July 27, 1991).
82. R. P. Sherwin, *Clin. Toxicol.* **29,** 385–400 (Sept. 1991).
83. National Academy of Sciences, *Nitrogen Oxides*, National Academy Press, Washington D.C., 1977.
84. S. V. Dawson, and M. B. Schenker, *Am. Rev. Respir. Dis.* **120,** 281–292 (Aug. 1979).
85. U.S. Environmental Protection Agency, *Air Quality Criteria for Oxides of Nitrogen*, Environmental Criteria and Assessment Office, Research Triangle Park, N.C., 1982.
86. J. J. Quackenboss, J. D. Spengler, M. S. Kanarek, R. Letz, and C. P. Duffy, *Environ Sci Technol* **20,** 775–789 (Aug. 1986).
87. U.S. Environmental Protection Agency, *Air Quality Criteria for Oxides of Nitrogen*, Environmental Criteria and Assessment Office, Research Triangle Park, N.C., Aug. 1991, EPA600/8-8/049c.
88. P.E. Morrow, *J. Toxicol. Environ. Health* **13**(2–3), 205–227 (1984).
89. R. B. Schlesinger, "Nitrogen Oxides," in M. Lippmann, ed., *Environmental Toxicants. Human Exposures and Their Health Effects*, Van Nostrand Rheinhold Co., Inc., New York, 1992, pp. 412–453.
90. J. M. Samet and M. J. Utell, *Toxicol. Indust. Health* **6,** 247–262 (Mar. 1990).
91. H. Magnussen, *Eur. Respir. J.* **5,** 1040–1042 (May 1992).
92. I. F. Goldstein and co-workers, *Arch. Environ. Health* **43,** 138–142 (Mar./Apr. 1988).
93. L. J. Folinsbee, *Toxicol. Indust. Health* **8,** 273–283 (Sept./Oct. 1992).
94. V. Hasselblad, D. M. Eddy, and D. J. Kotchmar, *J. Air Waste Manag. Assoc.* **42,** 662–671 (May 1992).
95. J. M. Samet and co-workers, "Nitrogen Dioxide and Respiratory Illness in Children, *Health Effects Institute Research Report No. 58*, June, 1993.
96. National Academy of Sciences, *Sulfur Oxides*, National Academy Press, Washington D.C., 1978.
97. U.S. Environmental Protection Agency, *Air Quality Criteria for Particulate Matter and Sulfur Oxides*, Environmental Criteria and Assessment Office, Research Triangle Park, N.C., Dec., 1982, EPA-600/8-82-029aF-cF.
98. U.S. Environmental Protection Agency, *Second Addendum to Air Quality Criteria for Particulate Matter and Sulfur Oxides (1982): Assessment of Newly Available Health Effects Information*, Environmental Criteria and Assessment Office, Research Triangle Park, N.C., December 1986, EPA/600/8-86/020F.
99. S. A. Shore and co-workers, *Am. Rev. Respir. Dis.* **135,** 840–847 (Apr. 1987).
100. P. D. Scanlon, J. Seltzer, R. H. Ingram, Jr., L. Reid, and J. Drazen, *Am. Rev. Respir. Dis.* **135,** 831–839 (Apr. 1987).
101. Subcommittee on Public Health Aspects of the Energy Committee on Public Health, New York Academy of Medicine, "Environmental Effects of Sulfur Oxides and Related Particulates," *Bulletin New York Acad Med* **54,** 983–1278 (Dec. 1978).
102. M. Lippmann, "Sulfur Oxides-Acidic Aerosols and SO_2," in M. Lippmann, ed., *Environmental Toxicants. Human Exposures and Their Health Effects*, Van Nostrand Rheinhold Co., Inc., New York, 1992, pp. 543–574.
103. M. T. Kleinman, *J. Air Pollut. Control Assoc.* **34,** 32–37 (Jan. 1984).
104. J. R. Goldsmith and L. T. Friberg, "Effects of Air Pollution on Human Health," in A. C. Stern, ed., *Air Pollution*, 3rd. ed., Academic Press, New York, 1977.
105. J. R. Withey, *Toxicol Indust Health* **5,** 519–554 (May 1989).
106. F. W. Lipfert, S. C. Morris, and R. E. Wyzga, *Environ. Sci. Technol.* **23,** 1316–1322 (Nov. 1989).
107. P. J. Lioy and J. M. Waldman, *Environ Health Perspect.* **79,** 15–34 (Feb. 1989).
108. American Thoracic Society, *Am. Rev. Respir. Dis.* **144,** 464–467 (Aug. 1991).
109. M. W. Frampton, K. Z. Voter, P. E. Morrow, N. J. Roberts, Jr., D. J. Culp, C. Cox, and M. J. Utell, *Am. Rev. Respir. Dis.* **146,** 626–632 (Sept. 1992).
110. R. B. Schlesinger, *Environ. Health Perspect.* **63,** 25–38 (Nov. 1985).
111. R. B. Schlesinger, *Environ. Health Perspect.* **79,** 1212–126 (Feb. 1989).
112. J. M. Gearhart and R. B. Schlesinger, *Environ. Health Perspect.* **79,** 127–137 (Feb. 1989).
113. M. J. Utell, *Environ. Health Perspect.* **63,** 39–44 (Nov. 1985).
114. L. J. Folinsbee, *Environ. Health Perspect.* **79,** 195–199 (Feb. 1989).
115. T. V. Larson, *Environ. Health Perspect.* **79,** 7–13 (Feb. 1989).
116. M. J. Utell, J. A. Mariglio, P. E. Morrow, F. R. Gibb, and D. M. Speers, *J. Aerosol Med.* **2,** 141–147 (1989).
117. D. M. Norwood, T. Wainman, P. J. Lioy, and J. M. Waldman, *Arch. Environ. Health* **47,** 309–313 (July/Aug. 1992).
118. J. D. Spengler, M. Brauer, and P. Koutrakis, *Environ Sci. Technol.* **24,** 946–955 (July 1990).
119. M. J. Utell and J. M. Samet, *Am. Rev. Respir. Dis.* **147,** 1334–1335 (June 1993).
120. J. Schwartz, D. Slater, T. V. Larson, W. E. Pierson, and J. Q. Koenig, *Am. Rev. Respir. Dis.* **147,** 826–831 (Apr. 1993).
121. C. A. Pope III and R. E. Kanner, *Am. Rev. Respir. Dis.* **147,** 1336–1340 (June 1993).
122. National Academy of Sciences, *Carbon Monoxide*, National Academy Press, Washington D.C., 1977.
123. U.S. Environmental Protection Agency, *Air Quality Criteria for Carbon Monoxide*, Environmental Criteria and Assessment Office, Research Triangle Park, N.C., 1979, EPA-600/8-79-022.
124. U.S. Environmental Protection Agency, *Air Quality Criteria for Carbon Monoxide*, Environmental Criteria and Assessment Office, Research Triangle Park, N.C., Mar. 1990, EPA/600/8-90/045A.
125. M. T. Kleinman, "Health Effects of Carbon Monoxide," in M. Lippmann, ed., *Environmental Toxicants. Human Exposures and Their Health Effects*, Van Nostrand Rheinhold Co., Inc., New York, 1992, pp. 98–118.
126. R. J. Shepherd, *Carbon Monoxide: The Silent Killer*, Charles C. Thomas, Springfield, Ill., 1983.

127. S. R. Thom, "Carbon Monoxide Poisoning," in D. H. Simmons and D. F. Tierney, eds., *Current Pulmonology, Volume 13*, Mosby Year Book, St. Louis, Mo., 1992, pp. 289–309.
128. E. N. Allred and co-workers, *N. Engl. J. Med.* **321,** 1426–1432 (Nov. 23, 1989).
129. D. S. Sheps and co-workers, *Ann. Intern. Med.* **113,** 343–351 (Sept. 1, 1990).
130. S. T. Weiss, I. B. Tager, M. Schenker, and F. E. Speizer, *Am. Rev. Respir. Dis.* **128,** 932–942 (Nov. 1983).
131. S. T. Weiss, *J. Respir. Dis.* **9,** 46–62 (Jan. 1988).
132. National Research Council, *Environmental Tobacco Smoke: Measuring Exposures and Assessing Health Effects*, National Academy Press, Washington D.C., 1986.
133. J. M. Samet, "Environmental Tobacco Smoke," in M. Lippmann, ed., *Environmental Toxicants. Human Exposures and Their Health Effects*, Van Nostrand Rheinhold Co., Inc., New York, 1992, pp. 231–265.
134. National Research Council, Committee on Airliner Cabin Air Quality, *The Airliner Cabin Environment: Air Quality and Safety*, National Academy Press, Washington, D.C., 1986.
135. M. E. Mattson and co-workers, *J. Am. Med. Assoc.* **261,** 867–872 (Feb. 10, 1989).
136. U.S. Department of Health and Human Services, *The Health Consequences of Involuntary Smoking: A Report of the Surgeon General*, U.S. Government Printing Office, Wash., D.C., 1986, DHHS publication No. (CDC) 87-8398.
137. J. E. Fielding and K. J. Phenow, *New Engl. J. Med.* **319,** 1452–1460 (Dec. 1, 1988).
138. U.S. Environmental Protection Agency, *Respiratory Health Effects of Passive Smoking: Lung Cancer and Other Disorders*, Washington D.C., May 1992, EPA/600/6-90/006B.
139. International Agency for Research on Cancer, *IARC Monographs on the Evaluation of the Carcinogenic Risk of Chemicals to Humans: Tobacco Smoking*, Vol. 38, World Health Organization, IARC, Lyon, France, 1986.
140. B. A. Chilmonczyk and co-workers, *New Engl. J. Med.* **328,** 1665–1669 (June 10, 1993).
141. A. B. Murray and B. J. Morrison, *J. Allergy Clin. Immunol.* **91,** 102–110 (Jan. 1993).
142. B. Danuser, A. Weber, A. L. Hartmann, and H. Krueger, *Chest* **103,** 353–358 (Feb. 1993).
143. J. L. Repace and A. H. Lowrey, *Risk Analysis* **10,** 27–37 (Mar. 1990).
144. R. J. Shephard, *Arch. Environ. Health* **47,** 123–130 (Mar./Apr. 1992).
145. R. Bascom, T. Kulle, A. Kagey-Sobotka, and D. Proud, *Am. Rev. Respir. Dis.* **143,** 1304–1311 (June 1991).
146. U.S. Public Health Service, *Healthy People 2000: National Health Promotion and Disease Prevention Objectives*, U.S. Department of Health and Human Services, Public Health Service, Wash., D.C., 1991, DHHS no. (PHS) 91-50213.
147. American Lung Association, *Personal Protection Against Air Pollution: The Physician's Role*, American Lung Association, New York, 1981.
148. R. M. Sly, S. H. Josephs, and D. M. Eby, *J. Allergy* **54,** 209–212 (Mar. 1985).
149. "Air Purifiers," *Consumer Reports* **54,** 88–93 (Feb. 1989).
150. "Household Air Cleaners," *Consumer Reports* **57,** 657–662 (Oct. 1992).
151. M. J. Ellenbecker, *Clin. Chest Med.* **13,** 193–199 (June 1992).
152. R. E. Reisman, P. M. Mauriello, G. B. Davis, J. W. Georgitis, and J. M. DeMasi, *J. Allergy Clin. Immunol.* **85,** 1050–1057 (June 1990).
153. H. S. Nelson, S. R. Hirsch, J. L. Ohman, Jr., T. A. E. Platts-Mills, C. E. Reed, and W. R. Solomon, *J. Allergy Clin. Immunol.* **82,** 661–669 (Oct. 1988).

AIR POLLUTION: INDOOR

Diane R. Gold
Channing Laboratory
Department of Medicine
Brigham and Women's Hospital
Harvard Medical School
Boston, Massachusetts

Occupational pulmonary medicine began as the study of the effects of dusts and gases on the respiratory health of industrial workers. In the more economically developed countries, the proportion of adults working in the primary industries has decreased since World War II, while the proportion of adults working in offices, schools, and hospitals has increased. Investigation into indoor air pollution in these nonsmoke-stack-industrial settings was stimulated in the 1970s by an increase in health complaints that occurred after the sealing of buildings and the decreased provision of fresh air in response to the call for energy conservation during the "oil crisis." Research interest in air pollution in homes as well as offices also has reflected a growing awareness of the importance of the indoor environment, because residents of more developed countries spend most of their lives indoors (1). In poorer countries interest in indoor air pollution centers around concern for the effects of the combustion products of unvented cooking and heating fuels, and for the increasing morbidity and mortality associated with the increase of cigarette smoking.

Since the review "Health Effects and Sources of Indoor Air Pollution (2,3) was presented, additional monographs and reviews have been published (4,5). Proceedings of meetings on indoor air quality are also available (6). The American Thoracic Society has published a practical report on environmental controls and lung disease (7). This article presents an update on the health effects of each of the main indoor air pollutants and the modalities available to control them. Health effects of indoor exposure to radon and asbestos are not discussed, since they are presented elsewhere in this encyclopedia. The diagnosis and control of building-related illness that cannot be ascribed to a specific pollutant will be discussed. Finally, air quality issues related to hospitals will be reviewed. See also Air Pollution Control Methods; Air Pollution: Health Effects; Energy Management, Principles; Risk Assessment.

TOBACCO SMOKE

Active Smoking

Tobacco smoke is an aerosol containing several thousand substances that are distributed as gases (vapors) or particles. Components of the particulate phase include human carcinogens such as nickel, 2-naphthylamine and 4-aminobiphenyl, and suspected human carcinogens such as benzopyrene. Carbon monoxide, sulfur dioxide, nitrogen oxides, benzene, toluene, formaldehyde, ammonia, formic acid, acrolein, and acetic acid are some of the components of the vapor phase (8). Nicotine is partitioned between the vapor and particulate phase.

Active cigarette smoking has been well-established as a main preventable cause of bronchitis, emphysema, and lung cancer; the association with an increased risk for cardiovascular disease in men and women is also well recognized (9–11). Recent studies have focused on the association between active smoking, nonspecific bronchial reactivity, atopy, and elevated IgE levels (12).

Smoking Cessation

The beneficial effects of smoking cessation were presented in the 1990 Surgeon General's report (13). The Surgeon General's message has been generally accepted in the United States; 80% of smokers surveyed wanted to stop smoking (14). By 1987 nearly one-half of all living adults who ever smoked had quit smoking (13). The techniques that most of these individuals used in order to quit are unknown.

A doctor's advice to quit smoking has been demonstrated to increase patient cessation rates (13). Smoking cessation interventions recently reviewed by Fisher and associates include temptation management, cue extinction, aversive techniques, contingency management, and pharmacological treatment of tobacco dependence (15). By integrating these individual tactics, the American Lung Association Freedom from Smoking program has achieved a 29% smoking cessation rate at the end of 12 months of follow-up (15).

Attempts to stop initiation of smoking have met with mixed results; smoking initiation rates in the United States have actually been increasing among less educated young women (14). Since the overall decrease in smoking in the United States, cigarette companies have successfully targeted advertising to minority communities and to citizens of developing countries (14).

Passive Smoking

Environmental tobacco smoke exposes nonsmokers to many of the same toxins inhaled by active smokers (8). Salivary, plasma, and urinary cotinine have been established as markers for environmental tobacco smoke inhalation (16,2). Studies on the health effects of environmental tobacco smoke have recently been reviewed by the U.S. Environmental Protection Agency (EPA) (17), which concluded that environmental tobacco smoke is responsible for inducing new asthma cases and for increasing the number and severity of episodes among asthmatic children in the United States. In addition, environmental tobacco smoke exposure predicts acute respiratory infections (18), wheezing (18,19) and bronchial responsiveness (20) in children without a diagnosis of asthma. Additional epidemiologic studies of infants have found environmental tobacco smoke exposure associated with an increased incidence of middle-ear effusions (21) and hospitalization for chest illness (72). Published cross-sectional and longitudinal data add to the evidence for a significant effect of passive smoking on level and growth of lung function in children. Kauffman and associates studied 1160 French children between 6 and 10 years of age and found an association between maternal smoking and a decrease in FEV_1 and FEF_{25-75} (23). In a cohort of 362 children followed for an average of 9 years, Lebowitz and co-workers found slower growth of lung function in children with parents who smoke (24). A study of 80 healthy infants from East Boston suggested that maternal smoking during pregnancy may impair *in utero* airway development or alter lung elastic properties (25). Researchers are currently investigating whether the greatest effects of environmental tobacco smoke on the lung architecture and susceptibility are already established *in utero*, or whether disease associated with such exposures results from cumulative exposures during infancy and childhood.

In adults there is strong evidence for an association between environmental tobacco smoke and acute symptoms of irritation measured subjectively as eye, nose, or throat irritation, or more objectively, with eye-blink tests (8). The threshold for perception of tobacco smoke odor is tenfold lower than that for the development of irritation (4). The data for adverse effects of passive smoke on chronic respiratory symptoms or lung function in adults are less conclusive.

In 1986 the National Research Council and the Surgeon General's report concluded that involuntary smoking can cause lung cancer in nonsmokers, but that more data were necessary to estimate the magnitude of the risk in the U.S. population. Case-control studies published since 1986 have added supporting evidence for a relationship between environmental tobacco smoke and lung cancer. In a case-control study of 191 patients from the United States, Janerick and associates concluded that approximately 17% of lung cancers among nonsmokers could be attributed to high levels of exposure to cigarette smoke during childhood and adolescence (26). Their study did not find that exposure to spouses' smoke was an independent predictor of lung cancer risk. In contrast, Japanese (27,28) contribute strong evidence for an excess risk of lung cancer related to spousal smoking. This may reflect a higher exposure to spousal smoke in Japanese as compared to subjects in the Janerick study, where higher cumulative exposures may have occurred either in childhood or at work. A recent case-control study of 618 primary lung cancer patients from Missouri showed an increased lung cancer risk for lifetime nonsmokers with exposure of more than 49 pack-years from all household members or from spouses only (29).

Control of Environmental Tobacco Smoke

A totally smoke-free environment cannot be produced through presently available air-cleaning devices or through policies restricting smoking to limited areas. The commercially available modular cleaning devices can reduce the particulate load from environmental tobacco smoke but will not eliminate the constituents of tobacco smoke in the gaseous phase (7,30). Many of these gaseous elements may be irritants or carcinogens. The high ventilation rate required to eliminate the odor of the smoke is often costly and impractical from an engineering point of view. Consequently, many private and public institutions have established policies restricting or banning indoor smoking. Recent recommendations of the Joint Commission on Hospital Accreditation have specified that all hospitals are to become smoke free, and that this recommendation is to be part of the Review of Hospitals.

COMBUSTION PRODUCTS

Carbon Monoxide

Carbon monoxide results from any incomplete combustion of carbonaceous material. Water heaters, gas or coal heaters, and gas stoves are all indoor sources of carbon monoxide. Camping stoves and lanterns are sources of carbon monoxide that in the unvented space of a tent have led to accidental carbon monoxide poisoning and death. High indoor levels of carbon monoxide can also result from the entrainment of outdoor car, bus, or truck exhaust into the ventilation system of a building (31). This is a risk particularly in buildings that have delivery areas where vehicles are parked with the motors running.

Hourly carbon monoxide concentrations in homes during cooking with gas stoves range from 2 to 6 parts per million (ppm); one-hour averages may exceed 12 ppm (2). During city commuting, carbon monoxide levels in cars (6–15 ppm) may average 2 to 5 times the concentrations measured in homes (2). Cigarette smoke contains over 2% carbon monoxide; the average concentration in smoke that reaches the lungs is about 400 ppm (31).

Carbon monoxide produces toxic effects through several mechanisms (32). Because its affinity for hemoglobin is approximately 200 times greater than that of oxygen, it displaces oxygen, lowering the oxygen-carrying capacity of blood. It interferes with delivery of oxygen to the tissues by causing a leftward shift of the oxyhemoglobin dissociation curve. Carbon monoxide can also directly impair oxygen diffusion into mitochondria, interfere with intracellular oxidation, and increase platelet adhesiveness.

In a clean environment, for a nonsmoker, the normal carboxyhemoglobin saturation is 0.6%. Smoking one pack of cigarettes per day can lead to a daytime carboxyhemoglobin level of 5% to 6%; two to three packs per day can result in a carbon monoxide saturation of 7% to 9% (31). Methylene chloride, used as a paint stripper, is metabolized to carbon monoxide and after 3-hour exposure, even in a well-ventilated room, can result in a carboxyhemoglobin saturation of 16% (31).

Carbon monoxide poisoning has its most toxic acute effects on the organs with high oxygen requirements, the heart and the brain (31). The carboxyhemoglobin level is only an approximate predictor of the effect of carbon monoxide on the brain. The effects of very high levels of carbon monoxide have been well documented. Significant deterioration of judgment, calculation, and manual dexterity have been detected at levels above 20%; levels of 50% or greater can lead to coma, convulsions, or death (31). The low level and chronic effects of carbon monoxide are less well understood. Bunnell and Horvath found performance of complex tasks was impaired with increasing carboxyhemoglobin level (0–10%) at high work loads (60% maximum oxygen consumption) (32). It was recently demonstrated that exposure to inhaled carbon monoxide (200 ppm resulting in a carboxyhemoglobin of 6%) produced direct adverse effects on the heart by facilitating ventricular ectopy in patients with coronary artery disease (33,34). A multicenter study (35) suggested that exposure of patients with stable coronary disease to carbon monoxide can induce earlier subjective and objective evidence of myocardial ischemia at carboxyhemoglobin levels as low as 2% to 4%. At low levels carbon monoxide can lead to nonspecific flulike symptoms that often go unrecognized. Symptoms of headache, lethargy, nausea, or dizziness in children with carboxyhemoglobin levels between 2% and 10% have been noted (36).

Oxygen is the treatment for carbon monoxide poisoning. For an adult breathing air at 1 atmosphere, blood carboxyhemoglobin levels are reduced by half in 320 minutes; with an FIO_2 of 100%, levels are reduced by half in 80 minutes; and in a hyperbaric chamber at 3 atmospheres, levels are reduced by half in 23 minutes (37).

Nitrogen Dioxide

The National Ambient Air Quality Standard for an annual mean of NO_2 in ambient air is 53 parts per billion (ppb); on average, gas stoves add 15–25 ppb to the background concentration in a home (2), with peak levels in the kitchen of 200 to 1000 ppb. About 50% of homes in the United States have gas cooking appliances (2).

NO_2 is an oxidant that not only causes direct lung injury, but may increase susceptibility to respiratory infection through its effect on ciliary clearance and macrophage function. At high concentrations, NO_2 is known to cause diffuse pulmonary damage, not only in the animal model but in humans. Silo-fillers may develop on occupational lung disease resulting from high-dose NO_2 exposure. Freeman and co-workers demonstrated that with chronic exposure to a low level of 800 ppb of NO_2, occasional reduction or absence of bronchiolar cilia occurred in rats, but the reduction was not present in nonexposed lungs (38).

Meta-analysis utilizing the epidemiologic literature on indoor exposure to NO_2 at levels produced by gas stoves and heating in households in developed countries suggests a significant but small association (for an exposure increment of 15 ppb, odds ratio = 1:2; 95% Confidence Interval 1.14 to 1.28) (Lucas Neas, Doctoral Dissertation, 1991) between that pollutant and respiratory symptoms in children. In their study of 6,273 children aged 7 to 11 from six cities in the United States, Neas (39), and Dockery and colleagues (40) measured indoor exposure averaging 15 ppb higher in homes with gas stoves and pilot lights. They found increases in individual respiratory symptoms of 5–29%, with a combined symptom odds ratio of 1:40. In contrast, the study of 6 to 9-year-old children in the southeast region of the Netherlands did not find an effect of NO_2 on respiratory symptoms (41). A recent study found no association between nitrogen dioxide exposure and the incidence rates for respiratory illnesses in a population of healthy infants and toddlers exposed to a range of 0 to 40 ppb. They pointed out that these levels were lower than nitrogen dioxide concentrations in heavily polluted cities and in poorly ventilated inner-city apartments, but were equivalent to levels observed in many U.S. locations (42). Although recent studies of ambient NO_2 suggest an effect on pulmonary function (43), environmental chamber studies and indoor studies on NO_2 and pulmonary function are inconclusive (44). Certain asthmatics may be sensitive to acute exposure to NO_2; in a study of 11 asthmatics observed while cooking on a gas range, FVC and peak flow rates tended to drop with exposure to more than 300 ppb (45).

Biofuel

Burning of biofuel can result in production of particulates, carbon monoxide, nitrogen dioxide, sulfur dioxide (SO_2), formaldehyde, volatile organic compounds, and numerous other substances, depending on the type of fuel. In the 1970s, with the oil embargo, the shipment of woodstoves to the United States increased 10-fold (7). The newer "airtight" residential woodstoves should not leak combustion by-products into the home except during startup, stoking, and reloading. Traynor and co-workers reported indoor carbon monoxide concentrations of 0.4 to 2.8 ppm with particle concentrations only slightly above background (0 to 30 $\mu g/m^3$) with "airtight" stoves. In contrast, "nonairtight" stoves produced average levels of 1.8 to 14 ppm of carbon monoxide and particle levels of 200 to 1900 $\mu g/m^3$ (7). In developing countries, burning of wood, charcoal, crop residues, or animal dung for cooking or heating is often done without a flue or chimney and with poor ventilation. In 390 households in India, Menon found that the levels of particulates, measured during cooking, ranged between 4000 and 21,000 $\mu g/m^3$ (46). These particulate levels from biofuel combustion contrast with average particulate levels of about 40 $\mu g/m^3$ and peak concentrations of 120 $\mu g/m^3$ in homes where two packs per day of cigarettes are smoked (46).

In Shenyang, China, after adjustment for smoking, the risk for lung cancer increased in proportion to years of sleeping on brick beds heated by coal-burning stoves that were located directly under the bed (47). In Shenyang and Harbin, the number of meals cooked by deep frying and the frequency of smokiness during cooling were also associated with increased risk of lung cancer (48). In Nepal, Pandey et al. showed an association between chronic bronchitis in nonsmoking women and domestic smoke exposure as measured by the number of hours spent daily near the stove (49). Cooking in poor countries is often performed by women carrying their infants on their backs. In The Gambia, after adjustment for parental (paternal) smoking, an association between carriage on the mother's back and a child's development of an episode of difficulty in breathing was found (odds ratio of 2:80; 95% confidence interval: 1.29, 6.09) (50). Infant mortality rates due to pneumonia are higher in less-developed countries; exposure to smoke from biomass fuel may increase the vulnerability of intants to severe respiratory infection (51). In a group of 30 nonsmoking patients from Mexico, Sandoval and co-workers observed a syndrome of interstitial lung disease, with a mixed restrictive-obstructive pattern on pulmonary function testing and signs of pulmonary hypertension, which they attributed to prolonged woodsmoke inhalation (52). Studies of effects of indoor air pollution in developing countries are difficult to perform because of lack of appropriate control populations in morbidity studies, and because of the large numbers of subjects required to study mortality as an end point. The existing literature on indoor air pollution in developing countries was recently reviewed (53).

Few data are available on woodsmoke and respiratory illness in developed countries. In a study of 63 children in Michigan, an association between woodsmoke exposure and an increased prevalence of moderate to severe respiratory symptoms during the winter was found (54), yet in a similar study in Massachusetts, did not find such an association (55). In the study of 6 U.S. cities, an odds ratio of 1:32 (95% confidence interval (CI) of 0.99 to 1.76) for respiratory illness was found in households with wood-burning stoves as compared to households with other sources of heating (55).

Control of Combustion-Related Pollutants

Control options for combustion-related indoor air pollution in developed countries are outlined in the report by the American Thoracic Society (ATS) workshop on environmental controls and lung disease (7). Appliances should be ventilated, and should receive periodic service and maintenance. A visual inspection may not detect the source of a carbon monoxide leak detectable with a carbon monoxide monitor. Pilot lights should be turned off or eliminated. The cooking range should not be used for heat. Optimally, gas stoves should be exhausted to the outdoors. Woodstoves should be airtight; it should not be possible to smell the smoke indoors. The presence of soot on the inside wall above a fireplace suggests that pollutants are probably entering the house. More research is necessary on control options and interventions appropriate to the specific cultures and resources available in less-developed countries.

BIOLOGIC AGENTS

Biologic agents in buildings can cause disease through immune mechanisms, through infectious processes, or through direct toxicity (56). Of the immune-mediated respiratory diseases, allergic rhinitis and asthma are primarily associated with IgE antibody to indoor allergens, while humidifier fever and hypersensitivity pneumonitis are associated with IgG, IgA, and IgM antibody (Type-III immune complex–induced injury) and cell-mediated Type-IV hypersensitivity.

Asthma and Biologic Agents

In the United States, asthma prevalence and severity appear to be increasing. Rates of asthma and wheeze are particularly high among black peoples and inner-city inhabitants (57,58). These epidemiologic patterns of asthma prevalence may, in part, relate to changing patterns of indoor exposures to aeroallergens.

Mites, cockroaches, fungi, cats, dogs, and rodents are common sources of indoor allergens (7). Outdoor allergens such as pollens can also be entrained into the indoor ventilation system. Socioeconomic status, cultural factors, and housing conditions may be determinants of sensitization (59). In the emergency-room study of asthmatics from Charlottesville, Virginia, it was found that the prevalence of IgE antibody to one of five common inhalant allergens (dust mites, cats, cockroaches, grass pollen, and ragweed pollen) was significantly higher in asthmatics (75%) than controls (14%) (60). Whereas most asthmatics were allergic to dust mite, antibody to cats was a risk factor only among white patients, and IgE antibody to cockroaches was a risk factor only among black patients. These racial differences in allergy correlated with socioeconomic and housing conditions. The house dust mite (*Der p* I, *Der f* I) is one of the most important of the indoor aeroallergens (61). Possibly because of low humidity and temperature, homes at high altitudes in the Alps have been demonstrated to have a paucity of mites in dust (62). The use of fitted carpets, detergents effective in cool water, and tighter insulation resulting in a narrow range of temperatures may have resulted in conditions "more congenial to mite and man" (63). In a cohort of British children with a family history of allergy, they demonstrated that exposure in early childhood to house-dust-mite allergens was an important determinant of subsequent development of asthma. All but one of the children with asthma at the age of 11 had been exposed at 1 year of age to more than 10 μg of house dust mites per gram of dust. Several studies have demonstrated that levels of antigen exposure to house dust mites is associated with the degree of sensitization (64,65). Emergency-room visits and hospitalization for asthma have been related to house-dust-mite sensitization (59,66).

Molds or fungi, which thrive in conditions of home dampness and in humidifiers and vaporizers, have been associated with sensitization, wheeze symptoms, and asthma (67–69) in adults and children. Home dampness itself has been associated with increased risk of cough, phlegm, and acute respiratory illness (70,71). The high prevalence of *Alternaria* sensitivity among U.S. asthmatic children suggests the importance of this fungal antigen (72). Mold exposure may cause immediate-type hypersensitivity or may promote respiratory disease through toxic nonallergic reactions.

Sampling of Aeroallergens and Diagnosis of Type I Hypersensitivity

Refs. 62, 73–75 provide details on the sampling of aeroallergens. Ref. 76 provides a perspective on interpretation of mold/fungal measurements. While fungi and antigens from house dust mite, cockroach, cat, and dog are quantifiable in house dust, more information is needed to increase our understanding of the relationships between levels of antigen or fungi in dust and the development or exacerbation of wheeze and asthma (7). Particularly with fungi, it is not certain whether air samples rather than dust samples better represent personal exposure.

Diagnosis of sensitivity to known indoor allergens may be established by history and, for some allergens, by skin testing. Standard skin testing is not available for most fungi (7), but is available for *Alternaria*. For dust mite, cat, dog, cockroach, and few fungal allergens, specific IgE antibody can be measured through radioallergosorbent tests (RAST) (7).

Without a specific identifiable allergen, diagnosis of sensitivity to an indoor exposure can often be made only by history and by demonstration of a worsening of peak flow or peak flow variability while indoors and an improvement of symptoms and peak flows or spirometry when away from the suspected indoor setting. If the asthmatic symptoms are severe, interpretation of improvement may be confounded by the administration of more intensive asthma therapy at the same time as removal from the suspected indoor setting.

Control Measures for Aeroallergens

Control measures for indoor allergens have recently been reviewed (75). The beneficial effects of portable air filtration devices over and above those associated with central air conditioning are thought to be minimal (75,77). Although pollen entrained into a room may be effectively filtered by household air cleaners, the beneficial effect of air cleaners is considered small in removing allergens such as those from mites and cats, which exist in surface reservoirs (75). The measures can be time-consuming and costly; sensitivity to house dust mite should be documented before implementing them. Some of these measures may not be practical for people with poor housing and little money.

Mold abatement can be accomplished by prevention of the intrusion of basement air into living space, dehumidification of the basement, cleaning of moldy surfaces with 10% chlorine bleach, and cleaning of humidifiers and vaporizers with chlorine bleach before each use and by the keeping of relative humidity below 50% (7). The American Thoracic Society recommends that, because of their high rate of fungal contamination, water-spray humidifying devices should not be used in areas with asthmatics or immunocompromised persons (7).

Platts-Mills and co-workers demonstrated that over a period of months, rigorous avoidance of domestic allergens in a hospital setting substantially decreased bronchial reactivity in a group of dust mite–sensitive asthmatic patients (62). Although the majority of allergic patients lose their symptoms within days or weeks after elimination of the exposure, repeated exposure to certain allergens may lead to chronic, irreversible airway hyperresponsiveness. This has been most clearly demonstrated in occupational settings, with exposure to toluene diisocyanide (which can cause either asthma or hypersensitivity pneumonitis) and western red cedar (7).

Hypersensitivity Pneumonitis and Humidifier Fever

In hypersensitivity pneumonitis, sensitizing antigens can be microorganisms such as actinomycetes, bacteria, fungi, amoebae, animal and plant proteins including pigeon or parakeet antigen, rat urine, low-molecular-weight chemicals such as detergent enzymes, toluene diisocyanate, or pharmaceutical products (78).

Humidifier lung, a hypersensitivity pneumonitis in response to antigens such as thermophilic actinomycetes, results from indoor exposure to contaminated cooling, heating, or humidifying units. Clinically, the acute disease consists of fever, chills, cough, and dyspnea, usually with interstitial and alveolar nodular infiltrates on chest

X rays. Restrictive changes are seen on pulmonary function testing. Chronic exposure can lead to progressive dyspnea and interstitial fibrosis, with decreasing lung volumes and diffusing capacity. Acutely, on lung biopsy the tissue reaction typically consists of both an alveolar and interstitial inflammation, with prominence of lymphocytes, plasma cells, macrophages with foamy cytoplasm, granulomata, and giant cells with birefringent material. Humidifier fever is a milder form of the disease, with fewer pulmonary manifestations.

In the process of differentiating hypersensitivity pneumonitis from illnesses not due to a hypersensitivity reaction, but causing similar symptoms (eg, gram-negative toxin, *Legionella*), identification of the inciting antigen can be difficult. Three episodes of an acute, flulike illness associated with manipulations on the central air handling systems of an office building have been described (79). Symptoms were consistent with hypersensitivity pneumonitis, but no single etiologic agent could be identified. As in the Hodgson study, a search for the inciting antigen can be performed through volumetric culture plate sampling for microorganisms, though the organisms that are isolated may not be representative of personal exposure. Large volumes of air can be collected for immunologic analyses, using sera from affected patients. One disadvantage to this method is that many of the immunoassays are sensitive only to far larger amounts of antigen than would be necessary to cause symptoms in an already sensitized patient (7). Precipitating antibody testing has a use in diagnosis, but limitations include improper antigen-antibody combining ratios, the relative insensitivity of the test, and the possibility of false-positive reactions (78). Exposed subjects can have positive precipitins without the presence of disease. Skin testing is not clinically useful because many crude allergen preparations cause nonspecific irritation.

Reexposure of the patient to the suspected environment can establish the diagnosis of hypersensitivity pneumonitis, with limited risk to the patient, though not necessarily with identification of the inciting antigen. Laboratory inhalation challenge has been used clinically to identify specific etiologic agents; acute bronchospasm or severe pneumonitis are risks of the procedure (78), which should be performed in centers familiar with diagnostic challenge testing to the antigen of interest.

INFECTION

Ventilation-Related Transmission: Legionella

Legionella is a common bacterial contaminant of air conditioning and potable water systems. Since the 1976 epidemic in Philadelphia, there have been a number of reports of sporadic and nosocomial cases of legionellosis in buildings whose water systems contained the organism (3). Seven cases of nosocomial legionellosis were reported that occurred over a 7-month period in a community hospital in upstate New York (80). A case-control study demonstrated that length of hospital stay and proximity of patients' rooms to the ward shower were significant risk factors for acquiring legionellosis. The ward showers and hospital hot-water system were contaminated with *L. pneumophilia* serogroup 1. A large outbreak of Legionnaires' disease associated with Stafford District Hospital in England was reported (81). Among 68 hospitalized patients with confirmed legionellosis, 22 died. An additional 35 patients had suspected legionellosis and nearly one-third of the hospital staff had legionella antibodies. In a case-control study of the first 53 inpatient cases to be recognized, people who had visited the outpatient department were almost 100 times more likely to have developed the disease than those who had not. The cases were more likely to be smokers and more likely to have chronic illness.

Legionnella pneumophilia was identified in a chiller unit that cooled the air entering the outpatient department through the air-conditioning system. The staff with *Legionella* antibodies had not necessarily worked in the outpatient department, but were more likely to have worked in areas of the hospital ventilated by contaminated air from the cooling tower. Prior to the Stafford outbreak the water treatment company contracted to maintain the air-conditioning plants had twice isolated the epidemic strain of *Legionella* from water in the cooling tower. The outbreak ended before control measures were instituted.

Control of Legionella

The control measures utilized in Stafford included chlorinating the cooling-tower water to 50 ppm followed by continuous chlorination to 5 ppm free residual chlorine, raising the hot-water temperature in the calorifiers to 60–63°C to provide temperatures of 55–63°C at the tap, and continuous chlorination of the cold-water systems to 1–2 ppm free residual chlorine. The chilling unit for the outpatient department was replaced, the air conditioning to the outpatient department and floors above was turned off, and all spray attachments to wash-basin taps were removed.

The finding of *Legionella* in the cooling waters of air-conditioning systems is not uncommon. Better design and maintenance of hospital ventilation systems can result in disease prevention if an aerosol cannot be generated from contaminated cooling water and that bacterial counts of *Legionella* in the water are kept low. Chlorination alone may not entirely eradicate the organism; control of water temperature may also be needed. In the potable water systems, Farrell and co-workers recommend that the hot-water feed be kept at temperatures of 60°C or above and the cold-water feed temperatures do not exceed 20°C (82).

Ventilation-Related Transmission of Other Infections

Ventilation-related transmission of other infections is well reviewed in Refs. 2 and 83. Cases of nosocomial infection with *Aspergillus* have been reported primarily in immunocompromised patients. Viral infections with measles or chicken pox are easily spread to nonimmune or immunocompromised subjects. Nosocomial acquisition of tuberculosis usually requires prolonged or intimate exposure to droplet nuclei. A study of the spread of tuberculosis on a U.S. naval vessel in 1965 revealed that 140 members of the crew converted their tuberculin skin tests and 7 had evidence of active pulmonary tuberculosis after living on the same ship as a man with active pulmonary tuberculosis (84). The minimal contact between the source case and many of the tuberculin converters strongly suggested that the infection was spread by the ventilation system.

Humidity and Infection Transmission

Arundel and Sterling reviewed the literature on the effects of relative humidity on the transmission of infectious diseases for the Air Pollution Control Association (85). Of 8 epidemiologic studies, 7 suggested that the incidence of acute respiratory illness was lower in buildings with relative humidity levels between 40% and 60%, compared with buildings with lower levels of relative humidity. Very low humidity may dry protective mucous membranes and may also increase airborne or fomite transmission of organisms. Limited experimental data on the survival of viruses suggest maximum survival at low and high relative humidity and minimum survival at midrange humidity (84). More studies are needed on the risks and benefits of building and room humidification.

FORMALDEHYDE

There are numerous sources of formaldehyde in homes and offices. Because of the many health complaints by residents of homes insulated with urea–formaldehyde foam, use of this insulation has virtually ceased in the United States (3). But plywood, particle board, and pressed-wood products remain a main source of formaldehyde exposure, particularly in mobile homes. Formaldehyde is present in grocery bags, waxed paper, facial tissues, paper towels, colored newsprint, disposable sanitary products, floor coverings, carpet backings, adhesive binders, fire retardants, permanent-press cloths, household cleaning agents, cosmetics, deodorants, shampoos, fabric dyes, inks, and disinfectants. It is also present in natural gas, kerosene, and tobacco (86).

In a study of 28 residences, the range of formaldehyde levels in residences with urea–formaldehyde foam was 0.02–0.13 ppm; the range in control residences was 0.02–0.07 ppm (85). An "energy-efficient" nonurea-formaldehyde foam-insulated home evaluated by Gammage was found to have levels in the range of 0.13–0.17 ppm (85). The median level in 65 mobile homes where complaints had been noted was 0.47 ppm, with a range of less than 0.01–3.68 ppm (85).

Formaldehyde is a volatile gas that is highly water soluble, with irritant effects on the mucous membranes of the eyes and upper respiratory tract. Formaldehyde exposure can result in neuropsychologic effects that have been well-documented. Its acute human health effects are reviewed in Ref. 3. Sterling estimates that 10–20% of the general population may be sensitive to the irritant effects of formaldehyde; for these people the odor threshold can be as low as 0.01 ppm (85). Although wheeze has occasionally been reported in response to formaldehyde, immediate-type hypersensitivity to the substance has not been conclusively demonstrated because formaldehyde is a strong irritant and skin testing for hypersensitivity is not useful. An unpublished study of immunologic responses in several hundred formaldehyde-exposed individuals has demonstrated only one individual with formaldehyde-specific IgE antibody. IgM- and IgG-mediated responses were more common (G. Pier, 1991). Specific bronchial provocation with formaldehyde, though not widely available, is useful for diagnosis of formaldehyde sensitivity. Reactivity to very high concentrations of 2 ppm of formaldehyde in 12 of 230 Finnish workers referred for possible sensitivity have been reported (3). Their decrease in pulmonary function may have been effected through irritant mechanisms. Asthma and Type I reactivity related to formaldehyde exposure is probably very unusual.

The regulatory agencies recognize formaldehyde as a "probable human carcinogen" (87). In animals formaldehyde causes squamous-cell carcinoma; the dose response is nonlinear (3). This has led the Risk Estimation Panel of the Consensus Workshop to propose that formaldehyde should not be considered to have a threshold for cancer induction; a consensus has not been reached regarding a risk estimation model (3).

Evidence for carcinogenicity in humans is limited. Evidence is strongest for an association between formaldehyde and nasal and nasopharyngeal cancer (86,88). Excess risk of lung cancer in a study of 26,561 industrial workers with exposure to formaldehyde was found, but also with exposure to a variety of other substances that they believe could have been carcinogens and cocarcinogens (89,90). A clear dose–response relationship was not found. Reported also, was an excess of nasopharyngeal and oropharyngeal cancer. Significant excesses of brain cancer have been found in embalmers, anatomists, and pathologists, but not in formaldehyde-exposed industrial workers (3).

VOLATILE ORGANIC COMPOUNDS AND BUILDING-RELATED ILLNESS

In addition to formaldehyde, hundreds of other volatile organic compounds have been found in indoor air environments. Common sources of organic compounds include solvents, gasoline, printed materials, photocopying machines, chlorinated water, dry-cleaned clothes, pesticides, silicone caulk, floor adhesive, particle board, moth crystals, floor wax, wood stain, paint, furniture polish, floor finish, carpet shampoo, room freshener, and vinyl tiles (3,85,91). Even in a survey of two New Jersey communities with strong outdoor sources of volatile organic compounds, investigators from the Environmental Protection Agency found levels of most volatile organic compounds were 5 to 10 times higher indoors than outdoors (7). There is consistenty among surveys in detection of certain specific organics; 22 of the same compounds were found both in a United States building survey and in a West German survey (90).

Many of the volatile organic compounds identified in studies of indoor environments are known to be mucous irritants. Chronic exposure to high levels of some of the solvents can have effects on the central nervous system and on other organ systems such as the liver. Some known and suspected carcinogens are among these organic compounds: benzene, tetrachloroethylene, trichloroethane, and trichloroethylene. Limonene and toluene are on the EPA's priority list for carcinogenicity testing (85).

In the studies of nonindustrial indoor environments, levels of the compounds have been an order of magnitude less than the Occupational Safety and Health maximum permissible exposure levels. The individual and combined effects of low levels of volatile organic compounds have been difficult to assess in laboratory and in epidemiologic studies. In a longitudinal study of personnel from 14 Swedish primary schools, Norback, Torgen, and Edling assessed the relationship between volatile organic compounds and symptoms of eye, skin, and upper airway irritation, headache, and fatigue (92). They found a relationship between levels of VOCs and the repeated reporting of the same symptom both in 1982 and in 1986. This relationship remained after adjusting for other significant illness predictors, including "hyperreactivity," psychosocial dissatisfaction index, and wall-to-wall carpeting in the workplace. The mean indoor concentration of volatile organic compounds ranged from 70 to 180 $\mu g/m^3$.

Sterling and Sterling suggest that fluorescent lighting may generate reactions between low levels of volatile organic compounds (93). In some settings, the resultant photochemical smog, rather than the chemical antecedents, may be the most potent irritant. In their study of office workers, self-reported eye irritation decreased after replacement of fluorescent lamps with lamps emitting less ultraviolet light. No objective measurement of chemical exposure was available; the symptom improvement could have been related to a direct effect of improved lighting rather than to a reduction of photochemical smog.

Studies of health effects of volatile organic compounds sometimes have some objective measure of exposure, but they often rely solely on self-reporting of symptoms as a measure of effect. Laboratory researchers have been searching for objective markers of mucous irritation. Eyeblinking as a measure of irritation can be measured on an objective scale. Nasal lavage may be useful in evaluation of the upper respiratory tract for signs of inflammation (94). Statistically significant increases in neutrophils in nasal lavage fluid of 14 subjects exposed to a mixture of volatile organic compound has been reported (95). The mixture and quantity used (25 mg/m^3) was felt to be representative of what is found in new homes and office buildings. Although it may be impractical to use nasal lavage to evaluate large populations, the technique could be applied to subgroups among a population complaining of irritant symptoms.

Nasal lavage has not been found to be a useful tool for assessment of effects of volatile organic compounds in atopic asthmatics, who often have a high percentage of neutrophils in their lavage fluid, even without volatile organic compound exposure. With asthmatics, measurements of variability in peak flow on work days and on days off, and serial spirometric measurements may be of more use. In a chamber study of 11 subjects with bronchial reactivity to histamine and a history of asthma, Harving, Dahl, and Molhave found a decline in FEV1 with an 85-minute exposure to 25 mg/m^3 volatile organic compounds (96). The mean decline of 90.7% was statistically different from the initial value but not statistically significant from the decline to 97.4% of baseline found after sham exposure.

TIGHT BUILDING SYNDROME

Since the 1970s, there have been many outbreaks of work-related illnesses in buildings not contaminated by industrial processes. Typical building-related symptoms have included runny or stuffy nose, eye irritation, cough, chest tightness, fatigue, headache, and malaise.

Of 356 cases of building-related illnesses, the National Institute for Occupational Safety and Health found that 39% involved identifiable indoor or outdoor contaminants, and 50% were associated with inadequate provision of fresh air with no identifiable contaminant (3). The etiology of the building-related illnesses was unknown in 11% of cases. The terminology has not been uniform in the description of outbreaks of building-related illness of unknown etiology. The terms "tight building syndrome" and "sick building syndrome" have been used variably. In this article the term "tight building syndrome" is restricted to epidemics of illness in buildings with inadequate supply of fresh air and without specific contaminants associated with symptoms.

Provision and Distribution of Fresh Air

The adequacy of the supply of fresh air is often assessed by the number of cubic feet per minute provided per person, or by the level of CO_2 in a room. In the 1970s, with the pressure on energy conservation, the American Society for Heating, Refrigerating, and Air Conditioning Engineers revised downward its recommended ventilation standard for fresh air supply in the absence of smoking, recommending 5–10 cubic feet per minute (cfm) per person (3). Their guidelines recommended between 20–30 cfm for spaces where smoking was permitted. Buildings constructed according to these guidelines might not supply sufficient fresh air for the comfort of the occupants. On average, a supply of 15 cfm/occupant corresponds with a CO_2 level of approximately 1000 ppm (J. McCarthy, 1989). The newly adopted American Society for Heating, Refrigerating, and Air Conditioning Engineers standard 62-1989 (Ventilation for Indoor Air Quality) calls for maintaining CO_2 levels below 1000 ppm. Depending on other conditions in a building, its occupants may already begin to experience discomfrt and irritant symptoms at CO_2 levels of 800 to 1000 ppm. CO_2 at this level is probably a marker for poor air exchange, not a cause of symptoms.

The economic incentives for reducing the fresh air supply are evident. The annual cost of supplying and conditioning a cubic foot of air per minute may range from $2.00 to more than $4.00. Schools, office buildings, and arenas often require fresh air supply rates greater than 100,000 cfm resulting in an annual cost to the institution of more than $200,000 per year (3).

Even in a building with adequate overall supply of fresh air, the distribution of the air may be uneven, resulting in poor supply of air to some employees. This often occurs when partitions are placed in large spaces, to create small cubicle workspaces.

Researchers have not established what it is about poor supply of fresh air that causes the symptoms of tight building syndrome. The illnesses are often probably multifactorial, with retension of odors and low levels of many of the previously mentioned pollutants. A review of the literature suggests that in poorly ventilated buildings the presence of fewer negative ions may lead to decreased concentration and reaction time. Field studies are inconclusive as to whether the generation of negative ions reduces the prevalence of symptoms common in tight buildings (97).

The effect of varying levels of outdoor air supply on the symptoms of sick building syndrome was studied (98). It concluded that increasing ventilation with outdoor air above the current standard did not eliminate symptoms of sick building syndrome in the office buildings studied. An accompanying editorial points out that "this study . . . does not address the more interesting question of whether increased ventilation with outdoor air can reduce the rate of symptoms in buildings with low rates of ventilation" (99).

Temperature and Humidity

In addition to provision of adequate supply of fresh air, the actual flow of air, temperature, and humidity can effect comfort in buildings. The 1981 American Society for Heating, Refrigerating, and Air Conditioning Engineers standards define thermal environmental parameters that most—80% or more—sedentary occupants in indoor environments will find satisfactory. They recommend operative temperatures of 22.8–26°C at 50% relative humidity in summertime conditions and 20–23.6°C at 30% relative humidity in wintertime conditions. Indoor humidities below 20% are considered unacceptable because of the effect of dry air on both the mucous membranes and the skin. Cleaning and maintenance of internal ducts and filters, particularly in the setting of increased humidity, is crucial, and it is often done inadequately. Notably, most filters visible to the office workers are present for the protection of engineering structures and not for the health of the employee; they are not necessarily the filters to be concerned about. Dirty, wet, blocked systems can be the source of organisms and the reason for inadequate provision of fresh air.

Other Factors

In buildings where temperature and humidity, provision of fresh air, and distribution of fresh air are considered adequate, and where no specific contaminants are identified, building-related complaints of fatigue and malaise may, in part, relate to poor lighting and the physical and psychological effects of working in windowless environments. Psychological factors such as job dissatisfaction have been associated with physical symptoms. "Hysteria," or bias as a source of building-related illness, is a diagnosis of exclusion after what can be a long and potentially expensive investigation.

Evaluation of Building-Related Complaints

The pulmonologist/epidemiologist who wants to evaluate the relationship between conditions in a building and the health of employees as a group and as individuals needs to work hand in hand with engineers and industrial hygienists. Engineers and industrial hygienists are needed to evaluate the following:

1. Is the building "tight"? That is to say, what is the air turnover, distribution of supply of fresh air, temperature, and humidity in the work spaces in the building?
2. Are there any "significant" levels of indoor air pollutants in the building, for example, volatile organic substances from new carpets or photocopying machines, carbon monoxide from entrained exhaust produced by standing vehicles, mold from damp carpets, glass fiber or asbestos insulation?

More work is necessary to standardize questionnaires regarding building-related complaints. The pulmonologist/epidemiologist evaluates the following conditions:

1. What is the overall prevalence of irritant symptoms and symptoms of change in mental status in the building? Ideally, symptom rates are evaluated before and after changes are made to the building.
2. Are the symptoms clustered in any particular work space?
3. What is the significance of the measured levels of pollutants, and the markers of air turnover, temperature, and humidity for acute and chronic symptoms:
   a. In the entire group of building occupants?
   b. In individuals why may be "sensitive"?

If a building with a cluster of building-related illnesses is demonstrated to have poor supply of fresh air, the solution to the problem may be to provide more fresh air and adequate distribution of that air. This can be a difficult and expensive engineering task, involving extensive building maintenance as well as redesign of the building ventilation system.

Continued illness after improvement of the ventilation system may mean:

1. Continued inadequate ventilation;
2. Continued presence of a contaminant;
3. Continued disease and nonspecific reactivity following specific sensitization. It would be very unusual for this to happen in more than a few "sensitive" individuals.
4. Continued psychological concern and reporting bias.

The individual clinician is often limited to investigating whether an individual's illness symptoms are caused by building-related exposures. Decision trees for the clinician are available (4). The individual clinician can take several measures:

1. Take a careful, detailed work exposure history. Request data regarding products and materials with which the patient works. In many states Right to Know laws require that this information be provided.
2. Document disease. If the patient has respiratory complaints, provide the patient with a peak flow meter and diary to assess the level, diurnal pattern and variability of peak flow at work and on days off. For workers on the day shift, peak flows in the morning are usually lower than peak flows in the afternoon; an afternoon drop in peak flow may provide evidence of a relationship between work- or building-related exposure and lower pulmonary function.
3. Document change in disease with change in environment. If possible, try not to change medication at the same time as a change in work environment is being implemented; this may make it difficult to assess whether a change in the environment is related to improvement in symptoms. It may be appropriate to recommend that the patient stay away from the building; changes in symptoms and pulmonary function before and after leaving the building should be documented.
4. Perform specific diagnostic procedures when they are available and appropriate (eg, skin or RAST testing for specific allergens; transbronchial biopsy for suspected hypersensitivity pneumonitis).

The clinician faced with an individual patient who has building-related complaints may not be able to perform the function of either the epidemiologist or the industrial hygienist. In practice, issues that relate to job security, politics, finances, and the law often heavily influence the decision whether or not to pursue a full investigation of a building. Employers should be encouraged to recognize the importance of demonstrating an honest attempt to identify causes and institute corrective actions in terms of employer-employee relationships, employee productivity, and the avoidance of legal or regulatory action.

HOSPITAL-RELATED ILLNESS

Certain forms of indoor air pollution are more common in hospitals. Formaldehyde and ventilation-related transmission of infection have already been discussed. This section will briefly summarize recent literature regarding exposure of health personnel to antineoplastic drugs, anesthetic agents, and ethylene oxide. It will conclude with a review of literature regarding asthma in hospital workers.

Antineoplastic Agents

As Selevan points out in her study of occupational exposure of nurses to antineoplastic drugs, many antineoplastic agents have been reported to be carcinogenic, mutagenic, and teratogenic (100). In the past it has been common for nurses to prepare these agents in areas without hoods. Detectable amounts of these agents have been measured when hoods have not been used. Only vertical laminar-flow hoods reduce workers' exposure sufficiently; mutagenic agents have been found in the urine of workers using horizontal laminar-flow hoods (101,102). Not all gloves are equally protective; surgical latex gloves are less permeable to many chemotoxic drugs than the polyvinyl chloride gloves previously recommended (101).

Sotaniemi et al. reported liver damage in three consecutive head nurses after years of handling cytostatic drugs. In the first case, the 34-year-old nurse had prepared cytostatic solutions for infusion with bare hands and without a face mask in an open place (103). In her case-control study of nurses in 17 Finnish hospitals, Selevan found an association between fetal loss and occupational exposure to antineoplastic drugs during the first trimester of pregnancy (odds ratio = 2:30 [95% confidence interval 1.20 to 4.39]) (99). Although the study has been criticized for possible reporting bias in the use of self-reported fetal loss, the Selevan study does lend support to the argument that caution should be exercised in the handling of antineoplastic drugs. Work practice guidelines available for personnel dealing with cytotoxic agents include recommendations that antineoplastic agents be mixed in environmentally controlled settings by trained pharmacists. Strict protocols are also established for the administration and disposal of antineoplastic drugs, and for dealing with spills.

Anesthetic Agents

At anesthetic doses, many anesthetic agents have been demonstrated to have toxic effects on organ systems in sensitive individuals. At anesthetic doses most agents have been found to have adverse effects on reproductive events in rodents. Less well understood are the effects on hospital personnel of anesthetic gases at subanesthetic doses. In a review of the literature (104), it was concluded that there is insufficient evidence to conclude that occupational exposure to anesthetic agents causes increased rates of fetal loss or congenital anomalies. However, Rowland et al. recently reported reduced fertility among women employed as dental assistants exposed to high levels of nitrous oxide (105). Data regarding the effects of large exposures to anesthetic agents suggest the importance of maintaining an adequate anesthetic scavenging system. Studies from the 1970s have reported levels of nitrous oxide ranging from 139 to 7000 ppm, and halothane levels of 10 to 85 ppm in the breathing zone of anesthesiologists. Currently United States hospital regulations limit contamination levels in the operating room to 25 ppm of nitrous oxide and 1 to 2 ppm of halogenated agents. This is accomplished by "removal of the waste gases with a gas-scavenging trap over the escape valve of the tank and a length of tubing to provide a conduit to the outside" (103).

Ethylene Oxide

Ethylene oxide is widely used for sterilization of medical supplies and equipment within hospitals. For certain types of sterilization safer alternatives have not been identified.

In industrial accidents, short exposure to high concentrations of this agent has led to nausea, headache, respiratory irritation, bronchitis, and pulmonary edema. Direct contact with small amounts of liquid ethylene oxide may cause eye and skin irritation (106). Delayed sensitization has been reported from residues of the agent in renal dialysis equipment. With chronic exposure in hospital workers, neurological deficits have been reported, including deficits in smell, sural nerve velocity, cognition, memory, and coordination (107).

In animal models, ethylene oxide is mutagenic (108). A case-control study by Ehrenbery suggested that chronic exposure led to lymphocytosis and anemia (107). In 1986 8 cases of leukemia among 733 ethylene oxide exposed workers were reported compared with an expected 0.8 cases, and 6 cases of stomach cancer compared with 0.65 cases expected (107).

NIOSH has concluded that ethylene oxide should be treated as a cancer and reproductive hazard. In the 1977 assessment of aeration procedures in hospitals the National Institute for Occupational Safety and Health, airborne ethylene oxide concentrations of 300 to 500 ppm were noted behind some aeration cabinets (105). The survey noted a number of conditions resulting in "unnecessary, preventable exposure of hospital staff" to ethylene oxide. Federal Regulations from 1984 require an 8-hour, time-weighted, average permissible exposure limit of no more than 1 ppm (108). Short-term exposure over a 15-minute period may not exceed 10 ppm. Frequent health monitoring is required if employees are exposed above the action level of 0.5 ppm. Recommendations for modifications of workplace design and practice in hospitals should be detailed and specific.

Asthma in Hospitals

In a cross-sectional questionnaire study comparing respiratory therapists with controls, it was found that respiratory therapists in Rhode Island reported higher asthma rates (odds ratio = 3:2; 95% confidence interval, 1.9 to 5.5) (109). A similar excess of asthma was found in Massachusetts respiratory therapists (110). The mechanism for this excess of asthma has not been delineated. In the past, asthma in hospital personnel has been associated with exposure to hexachlorophene, formaldehyde, psyllium laxatives, and laboratory animals (109). Sensitization can also occur during the preparation of pharmaceutical products, particularly antibiotics.

Because of recent episodes of fatal and life-threatening anaphylaxis, recent attention has been focused on sensitization and allergy to latex, dust, and powder in the gloves of health care workers (75). A study of 28 medical center employees diagnosed with rhinitis or asthma was shown to be caused by exposure to dust from latex gloves (110). While the allergy is believed to be IgE mediated, the principal allergenic protein has not been identified.

SUMMARY

This article summarizes the health effects of indoor air pollutants and the modalities available to control them. The pollutants discussed include active and passive exposure to tobacco smoke, combustion products of carbon monoxide, nitrogen dioxide, products of biofuels including wood and coal, biologic agents leading to immune responses, such as house dust mites, cockroaches, fungi, animal dander and urine, biologic agents associated with infection such as *Legionella* and tuberculosis, formaldehyde, and volatile organic compounds. An approach to assessing building-related illness and "tight-building" syndrome is presented. Finally, the article reviews recent data on hospital-related asthma and exposures to potential respiratory hazards such as antineoplastic agents, anesthetic gases, and ethylene oxide.

Acknowledgments

This article has been reprinted from D. R. Gold, *Clinics in Chest Medicine* **13**(2), 215–229 (June 1992) Courtesy of W.B. Saunders Company.

BIBLIOGRAPHY

1. A. Szalai, ed., *The Use of Time: Daily Activities of Urban and Surburban Populations in Twelve Countries*, Mouton, The Hague, 1972.
2. J. M. Samet, M. C. Marbury, and J. D. Spengler, *Am. Rev. Respir. Dis.* **136**(6), 1486–1508 (1987).
3. J. M. Samet, M. C. Marbury, and J. D. Spengler, Part II, *Am. Rev. Respir. Dis.* **137**(1), 221–242 (1988).
4. J. E. Cone and M. J. Hodgson, ed., "Problem buildings: Building-associated Illness and the Sick Building Syndrome," *Occupational Medicine: State of the Art Reviews* **4**(4) (Oct.–Dec., 1989).
5. J. M. Samet and J. D. Spengler, ed., *Indoor Air Pollution. A Health Perspective*, The Johns Hopkins Press, Baltimore, 1991.
6. "Indoor Air '90." *Proc. of the 5th International Conference on Indoor Air Quality and Climate*, Vol. 1–4, Ottawa 1990.
7. J. Samet (chairman), *Am. Rev. Respir. Dis.* **142**(4), 915–939 (1990).
8. National Research Council (NRC), *Environmental Tobacco Smoke: Measuring Exposures and Assessing Health Effects*, National Academy Press, Washington, D.C., 1986.
9. G. A. Colditz, R. Bonita, M. J. Stampfer, and co-workers, *NEJM* **318**(15), 937–941 (1988).
10. U.S. Department of Health and Human Services. Public Health Service, Office on Smoking and Health, *Smoking and Health: A Report of the Surgeon General*. U.S. Department of Health and Human Services. U.S. Government Printing Office, Washington, D.C., 1979. DHHS 79-50066.
11. W. C. Willett, A. Green, M. J. Stampfer, and co-workers, *N. Eng. J. Med.* **317**(21), 1303–1309 (1987).
12. S. T. Weiss and D. Sparrow, *Airway Responsiveness and Atopy in the Development of Chronic Lung Disease*, Raven Press, New York, 1989.
13. U.S. Department of Health and Human Services. Public Health Service, Office on Smoking and Health. *Health Benefits of Smoking Cessation: A Report of the Surgeon General*, U.S. Department of Health and Human Services. U.S. Government Printing Office, Washington, D.C., 1990.
14. U.S. Department of Health and Human Services. Public Health Service, Office on Smoking and Health, *Reducing the Health Consequences of Smoking: Twenty-five Years of Progress. A Report of the Surgeon General* U.S. Department

of Health and Human Services. U.S. Government Printing Office, Washington, D.C. 1989, DHHS (CDC) 89-8411.

15. E. B. Fisher, D. Haire-Joshu, G. D. Morgan, and co-workers, *Am. Rev. Respir. Dis.* **142**(3), 702–720 (1990).
16. R. A. Etzel, "A Review of the Use of Saliva Cotinine as a Marker of Tobacco Smoke Exposure," *Prev. Med.* **19**(2), 190–197 (1990).
17. United States Environmental Protection Agency, *Respiratory Health Effects of Passive Smoking: Lung Cancer and Other Disorders*, United States Environmental Protection Agency. Office of Research and Development, Washington, D.C., 1992.
18. S. M. Somerville, R. J. Rona, and S. Chinn, *J. Epidemiol. Community Hlth.* **42**(2), 105–110 (1988).
19. D. R. Neuspiel, D. Rush, N. R. Butler, and co-workers, "Parental Smoking and Post-infancy Wheezing in Children: A Prospective Cohort Study," *Am. J. Public Health* **79**(2), 168–171 (1989).
20. S. M. Somerville, R. J. Rona, and S. Chinn, "Passive Smoking and Respiratory Conditions in Primary School Children," *J. Epidemiol. Community Hlth.* **42**(2), 105–110 (1988).
21. B. D. Reed and L. J. Lutz, *Fam. Med.* **20**(6), 426–430 (1988).
22. Y. Chen, W-X. Li, S. Yu, and W. Quian, *Inter. J. Epidemiol.* **17**(2), 348–355 (1988).
23. F. Kauffmann, I. B. Tager, A. Munoz, and co-workers, *Am. J. Epidemiol.* **129**(6), 1289–1299 (1989).
24. M. D. Lebowitz, C. J. Holberg, R. J. Knudson, and co-workers, *Am. Rev. Respir. Dis.* **136**(1), 69–75 (1987).
25. J. P. Hanrahan, I. B. Tager, M. R. Segal, T. D. Tosteson, R. G. Castile, H. Van Vunakis, S. T. Weiss, and F. E. Speizer, *Am. Rev. Respir. Dis.* **145**(5), 1129–1135 (1992).
26. D. T. Janerich, W. D. Thompson, L. R. Varela, and co-workers, *N. Eng. J. Med.* **323**(10), 632–636 (1990).
27. S. Akiba, H. Kato, W. J. Blot, and co-workers, *Cancer Res.* **46**(9), 4804–4807 (1986).
28. T. Hiriyama, *Prev. Med.* **13**(6), 680–690 (1984).
29. R. C. Brownson, M. C. R. Alavanja, E. T. Hock, and T. S. Loy, *Am. J. Public Health* **82**(11), 1525–1530 (1992).
30. "Household Air Cleaners," *Consumer Reports*, Consumers Union, Yonkers, NY., 1992.
31. R. R. Beard, "Inorganic compounds of oxygen nitrogen and carbon." In: Patty's *Industrial Hygiene and Toxicology, 2C*, John Wiley & Sons., Inc., New York, 1982.
32. D. E. Bunnell and S. M. Horvath, *Aviat. Space Environ. Med.* **59**(12), 1133–1138 (1988).
33. D. S. Sheps, M. C. Herbst, A. L. Hinderliter, and co-workers, *Ann. Intern. Med.* **113**(5), 343–351 (1990).
34. S. M. Walden and S. O. Gottlieb, *Ann. Intern. Med.* **113**(5), 337–338 (1990).
35. E. N. Allred, E. R. Bleecher, B. R. Chaitman, and co-workers, *N. Engl. J. Med.* **321**(21), 1426–1432 (1989).
36. M. D. Baker, F. M. Henretig, and S. Ludwig, *J. Pediatrics* **113**(3), 501–504 (1988).
37. R. P. Smith, "Toxic Responses of the Blood," in Casarett and Doull's *Toxicology*, Macmillan Publishing Company, New York, 1986, p. 231.
38. G. Freeman, S. C. Crane, N. J. Furiosi, and co-workers, *Am. Rev. Respir. Dis.* **106**(4), 563–579 (1972).
39. L. M. Neas, D. W. Dockery, J. H. Ware, J. D. Spengler, F. E. Speizer, and B. G. Ferris, Jr., *Am. J. Epidemiol.* **134**(2), 204–219 (1991).
40. D. W. Dockery, J. D. Spengler, L. M. Neas, and co-workers "An Epidemiologic Study of Respiratory Health Status and Indicators of Indoor Air Pollution from Combustion Sources. In Combustion Processes and the Quality of the Indoor Environment," *Proceedings of the AWMA Specialty Conference*, September 1988. Pittsburgh, Pa., Air and Waste Management Association, 1989, p. 262 (AWMA specialty conference no. TR-15).
41. B. Brunekreef, P. Fischer, D. Houthuijs, et al., "Health effects of indoor NO_2 pollution," In: B. Selfer, H. Esdorn, M. Rischer, and co-eds., *Indoor Air '87: Proceedings of the 4th International Conference on Indoor Air Quality and Climate; Vol. 1, volatile organic compounds, combustion gases, Particles and Fibres, Microbiological Agents*, Berlin, Federal Republic of Germany, Institute for Water, Soil and Air Hygiene, 1987, p. 304.
42. J. M. Samet, W. E. Lambert, B. J. Skipper, A. H. Cushing, W. C. Hunt, S. A. Young, L. C. McLaren, M. Schwab, and J. D. Spengler, "Nitrogen Dioxide and Respiratory Illness in Children. Part I: Health Outcomes." In: *Nitrogen Dioxide and Respiratory Illness in Children. Health Effects Institute.* Capital City Press, Montpelier, VT, 1993.
43. J. Schwartz and S. Zeger, *Am. Rev. Respir. Dis.* **141**(1), 62–67 (1990).
44. J. M. Samet and M. J. Utell, *Toxicol. Indust. Hlth.* **6**(2), 247–262 (1990).
45. I. F. Goldstein, K. Lieber, L. R. Andrews, G. Koutrakis, E. Kazembe, P. Huang, and C. Hayes, *Arch. Environ. Health* **43**(2), 138–142 (1988).
46. M. R. Pandey, K. R. Smith, J. S. M. Boleij, and co-workers, *Lancet* 427–29, February 25, 1989.
47. Z-Y. Xu, W. J. Blot, H-P. Xiao, and co-workers, *J. Natl. Cancer Inst.* **81**(23), 1800–1806 (1989).
48. A. H. Wu-Williams, X. D. Dai, W. Blot, Z. Y. Xu, X. W. Sun, H. P. Xiao, B. J. Stone, S. F. Yu, Y. P. Feng, A. G. Ershow, J. Sun, J. F. Faumeni, and B. E. Henderson, *Br. J. Cancer* **62**(6), 982–987 (1990).
49. M. R. Pandey, R. P. Neupane, and A. G. Gautam, "Domestic Smoke Pollution and Acute Respiratory Infection in Nepal." In: B. Sefert, H. Esdorn, M. Fischer, et al., eds., *Fourth International Conference on Indoor Air Quality and Climate*, Vol. 4. Berlin, Institute for Water, Soil, and Air Hygiene. 1987, p. 25.
50. H. Campbell, J. R. M. Armstrong, and P. Byass, *Lancet* 1012 (May 6, 1989).
51. N. M. H. Graham, *Epidemiol. Rev.* **12,** 149–178 (1990).
52. J. Sandoval, J. Salas, M. L. Martinez-Guerra, A. Gomez, C. Martinez, A. Portales, A. Palomar, M. Villegas, and R. Barrios, *Chest* **103**(1), 12–20 (1993).
53. B. H. Chen, C. J. Hong, M. R. Pandey, and K. R. Smith, *World Health Statistics Quart.* **43**(3), 127–138 (1990).
54. R. E. Honicky, J. S. Osborne, and C. A. Akpom, *Pediatrics* **75**(3), 587–593 (1985).
55. W. E. Pierson, J. Q. Koenig, and E. J. Bardana, *West. J. Med.* **151**(3), 339–342 (1989).
56. H. Burge, *J. Allergy Clin. Immunol.* **86**(5), 687–701 (1990).
57. J. Schwartz, D. Gold, D. W. Dockery, and co-workers, *Am. Rev. Respir. Dis.* **142**(3), 555–562 (1990).
58. D. R. Gold, A. Rotnitzky, A. I. Damokosh, J. H. Ware, F. E. Speizer, B. G. Ferris, Jr., and D. W. Dockery, *Am. Rev. Respir. Dis.* **148,** 10–18 (1993).
59. L. E. Gelber, L. H. Seltzer, J. K. Bouzoukis, S. M. Pollart, M. D. Chapman, and T. A. E. Platts-Mills, *Am. Rev. Respir. Dis.* **147**(3), 557–578 (1993).

60. T. A. E. Platts-Mills, S. Pollart, G. Fiocco, and co-workers, *J. Allergy Clin. Immunol.* **79,** 184 (abstract) (1987).

61. T. A. E. Platts-Mills, *J. Allergy Clin. Immunol.* **83**(2 Pt. 1), 416–427 (1989).

62. W. R. Solomon and K. P. Mathews, "Aerobiology and Inhalant Allergens," in *Allergy Principles and Practice.* Vol. I., CV Mosby Company, St. Louis, 1988, p. 312.

63. R. Sporik, S. T. Holgate, T. A. E. Platts-Mills, and co-workers, *N. Engl. J. Med.* **323**(8), 502–507 (1990).

64. S. Lau, G. Falkenhorst, A. Weber, I. Wethmann, and co-workers, *J. Allergy Clin. Immunol.* **84**(5 Pt. 1), 718–725 (1989).

65. M. Wickman, S. L. Nordvall, G. Pershagen, and co-workers, *J. Allergy Clin. Immunol.* **88**(1), 89–95 (1991).

66. S. M. Pollard, M. S. Chapman, G. P. Fiocco, and co-workers, *J. Allergy Clin. Immunol.* **83**(5), 875–882 (1989).

67. S. B. Lehrer, L. Aukrust, and J. E. Salvaggio, *Clin. Chest Med.* **4**(1), 23–41 (1983).

68. M. Lopez and J. E. Salvaggio, *Clin. Rev. Allergy* **3**(2), 183–196 (1985).

69. J. Salvaggio and L. Aukrust, *J. Allergy Clin. Immunol.* **68,** 327–346 (1981).

70. B. Brunekreef, D. W. Dockery, F. E. Speizer, J. H. Ware, J. D. Spengler, and B. G. Ferris, *Am. Rev. Respir. Dis.* **140,** 1363–1367 (1989).

71. R. E. Dales, R. Burnett, and H. Zwanenburg, *Am. Rev. Respir. Dis.* **143,** 505–509 (1991).

72. P. J. Gergen, P. C. Turkeltaub, and M. G. Kova, *J. Allergy Clin. Immunol.* **80,** 669–679 (1987).

73. H. Burge, *J. Allergy Clin. Immunol.* **86**(5), 687–701 (1990).

74. H. Burge, "Indoor Sources for Airborne Microbes," in R. B. Gammage and S. V. Kaye, *Indoor Air and Human Health,* Lewis, Chelsea, Michigan, 1984, p. 139.

75. A. M. Pope, R. Patterson, and H. Burge, eds., *Indoor Allergens,* National Academy Press, Washington, D.C., 1993.

76. P. P. Kozak and J. Gallup, "Endogenous Mold Exposure: Environmental Risk to Atopic and Nonatopic Patients," In R. B. Gammage and S. V. Kaye, *Indoor Air and Human Health,* Lewis, Chelsea, Michigan, 1984, p. 149.

77. H. S. Nelson, S. R. Hirsch, J. L. Ohman, T. E. Platts-Mills, C. E. Reed, and W. R. Solomon, *J. Allergy Clin. Immunol.* **82,** 661–669 (1988).

78. M. Lopez and J. E. Salvaggio, "Hypersensitivity Pneumonitis," in Murray/Nadal, *Textbook of Respiratory Medicine,* W.B. Saunders, Philadelphia, Pa., 1988, p. 1606.

79. M. J. Hodgson, P. R. Morey, J. S. Simon, et al., *Am. J. Epidemiol.* **125,** 631–637 (1987).

80. J. P. Hanrahan, D. L. Morse, V. B. Scharf, et al., *Am. J. Epidemiol.* **125**(4), 639–649 (1987).

81. M. C. O'Mahony, R. E. Stanwell-Smith, H. E. Tillett, and co-workers, *Epidemiol. infect.* **104,** 361–380 (1990).

82. I. D. Farrell, J. E. Barker, E. P. Miles, et al., *Epidemiol. Infect.* **104,** 381–387 (1990).

83. M. J. Finnegan and C. A. C. Pickering, *Clin. Allergy* **16,** 389–405 (1986).

84. V. N. Honk, D. C. Kent, J. H. Baker, and co-workers, *Arch. Environ. Hlth.* **16,** 26–35 (1968).

85. A. V. Arundel and E. M. Sterling, "Review of the Effect of Relative Humidity on the Transmission of Infectious Diseases," For presentation at the 80th Annual Meeting of APCA. Vancouver, British Columbia, Theodur D. Sterling, Ltd., 1987.

86. D. A. Sterling, "Volatile Organic Compounds in Indoor Air: An Overview of Sources, Concentrations, and Health Effects," in R. B. Gammage and S. V. Kaye, *Indoor Air and Human Health,* Lewis, Chelsea, Michigan, 1984.

87. International Agency for Research on Cancer, "IARC Monographs on the Evaluation of Carcinogenic Risks to Humans," *Overall evaluations of Carcinogenicity: an Updating of IARC Monographs Volumes 1 to 42. Supplement 7.* Lyon, France, World Health Organization, 1987.

88. D. Luce, M. Gerin, A. Leclerc, J. Morcet, J. Brugere, and M. Goldberg, *Int. J. Cancer* **53,** 224–231 (1993).

89. A. Blair, P. A. Stewart, M. O'Berg, et al., *JNCI* **76,** 1071–1084 (1986).

90. C. R. Buncher, *J. Occup. Med.* **31**(11), 885 (1989).

91. B. A. Tichenor and M. A. Mason, *JAPCA* **38,** 264–268 (1988).

92. D. Norback, M. Torgen, and C. Edling, *Br. J. Indust. Med.* **47,** 733–741 (1990).

93. E. Sterling and T. Sterling, *Canadian J. Public Hlth.* **74,** 385–392 (1983).

94. H. S. Koren and R. B. Devlin, *Ann. N.Y. Acad. Sci.* **641,** 215–224 (1992).

95. H. S. Koren, D. E. Graham, and R. B. Devlin, *Arch. of Environ. Health* **47,** 39–44 (1992).

96. H. Harving, R. Dahl, and L. Molhave, *Am. Rev. Respir. Dis.* **143,** 751–754 (1991).

97. M. J. Finnegan and C. A. C. Pickering, *Clin. Allergy* **16,** 389–405 (1986).

98. R. Menzies, R. Tamblyn, J. Farant, J. I. Hanley, F. Nunes, and R. Tamblyn, *N. Engl. J. Med.* **328,** 821–827 (1993).

99. K. Kreiss, *N. Engl. J. Med.* **328,** 877–878 (1993).

100. S. G. Selevan, M-L. Lindbohm, R. W. Hornung, and co-workers, *N. Engl. J. Med.* **313,** 1173–1178 (1985).

101. Office of Occupational Medicine, "Work Practice Guidelines for Personnel Dealing with Cytotoxic (Antineoplastic) Drugs," In: *Occupational Safety and Health Administration Instruction Publication 8-1.1,* Office of Occupational Medicine, January 29, 1986.

102. R. W. Anderson, W. H. Puckett, W. J. Dana, and co-workers, *Am. Soc. Hosp. Pharm.* **39,** 1881–1887 (1982).

103. E. A. Sotaniemi, S. Sutinen, A. J. Arranto, and co-workers, *Acto. Med. Scand.* **214,** 181–189 (1983).

104. T. N. Tannenbaum and R. J. Goldberg, *J. Occup. Med.* **27,** 659–668 (1985).

105. A. S. Rowland, D. D. Baird, C. R. Weinberg, D. L. Shore, C. M. Shy, and A. J. Wilcox, *N. Engl. J. Med.* **327,** 993–997 (1992).

106. A. R. Glaser, *Special Occupational Hazard Review with Control Recommendations for the Use of Ethylene Oxide as a Sterilant in Medical Facilities,* U.S. Department of Health, Education, and Welfare, Center for Disease Control, National Institute for Occupational Safety and Health, Division of Criteria Documentation and Standards Development, Priorities and Research Analysis Branch, Rockville, Maryland, 1977.

107. W. J. Estrin, S. A. Cavalieri, P. Wald, and co-workers, *Arch. Neurol.* **44**(12), 1283–1286 (1987).

108. V. L. Dellarco, W. M. Generoso, G. A. Sega, and co-workers, *Environ. Mol. Mutagen* **16**(2), 85–103 (1990).

109. D. G. Kern and H. Frumkin, *Ann. Intern. Med.* **110**(10), 767–773 (1989).

110. D. C. Christiani and D. G. Kern, *Am. Rev. Respir. Dis.* **143,** A441 (1991).

AIR QUALITY MODELING

ARMISTEAD G. RUSSELL
University of Colorado
Boulder, Colorado
JANA B. MILFORD
Carnegie Mellon University
Pittsburgh, Pennsylvania

One of the outcomes of humankind's activities has been to change the composition of the atmosphere. For example, the production of energy can lead to the increase in carbon dioxide levels and acid deposition. Other activities, along with energy production, have led to urban and regional smog and depletion of the stratospheric ozone layer. To better understand how particular emissions are affecting the atmosphere and thus to devise strategies to limit harmful outcomes, it is necessary to be able to describe quantitatively the source–impact relationship. This is the role of atmospheric models. Such mathematical air quality models have been developed to provide a powerful framework for understanding the dynamics of pollutants in the atmosphere and for assessing the impact emission sources have on pollutant concentrations.

Two commonly used classes of models have been developed. Empirical models provide an understanding of source impacts by statistically analyzing historical air quality data. Diagnostic models provide a comprehensive description of the detailed physics and chemistry of compounds in the atmosphere, following the evolution of pollutant emissions to their ultimate fate. At the heart of the model is a system of mathematical routines that integrate the effects of individual processes. The complexity and computational intensity of modern models have necessitated the development of algorithms for fast and accurate mathematical solution techniques. Air quality models are being applied to solving such problems as urban smog, acid deposition, regional ozone, haze in scenic regions, and the destruction of the protective stratospheric ozone layer.

Mathematical models have grown increasingly detailed in descriptions of air pollutant dynamics and are thus key tools for gaining scientific understanding of atmospheric processes. These models also are the most practical and scientifically defensible means of relating pollutant emissions to air quality. They are widely used in regulatory planning and analysis, as indicated in Figure 1. A wide variety of models is used to address problems ranging from indoor air pollution to regional acid deposition and global climate change. The models used are remarkably similar. This article focuses on the models used to understand air pollution dynamics.

Development of a mathematical air quality model proceeds through four stages. In the conceptual stage, a working set of relationships approximating the physical system is derived. Next, these relationships are expressed as mathematical equations, giving a formal description of the idealized system. The third step is the computational implementation of the model, including development of the algorithms and computer code needed to solve the equations given various inputs. The final step is the application of the model, including acquisition and processing of the necessary input data, and evaluation of model results. See also AIR POLLUTION; CARBON BALANCE MODELING; COMBUSTION MODELING; LIFE CYCLE ANALYSIS; RISK ASSESSMENT.

AIR POLLUTANTS

Air pollution problems are characterized by their scale and the types of pollutants involved. Pollutants are classified as being either primary, ie, emitted directly, or sec-

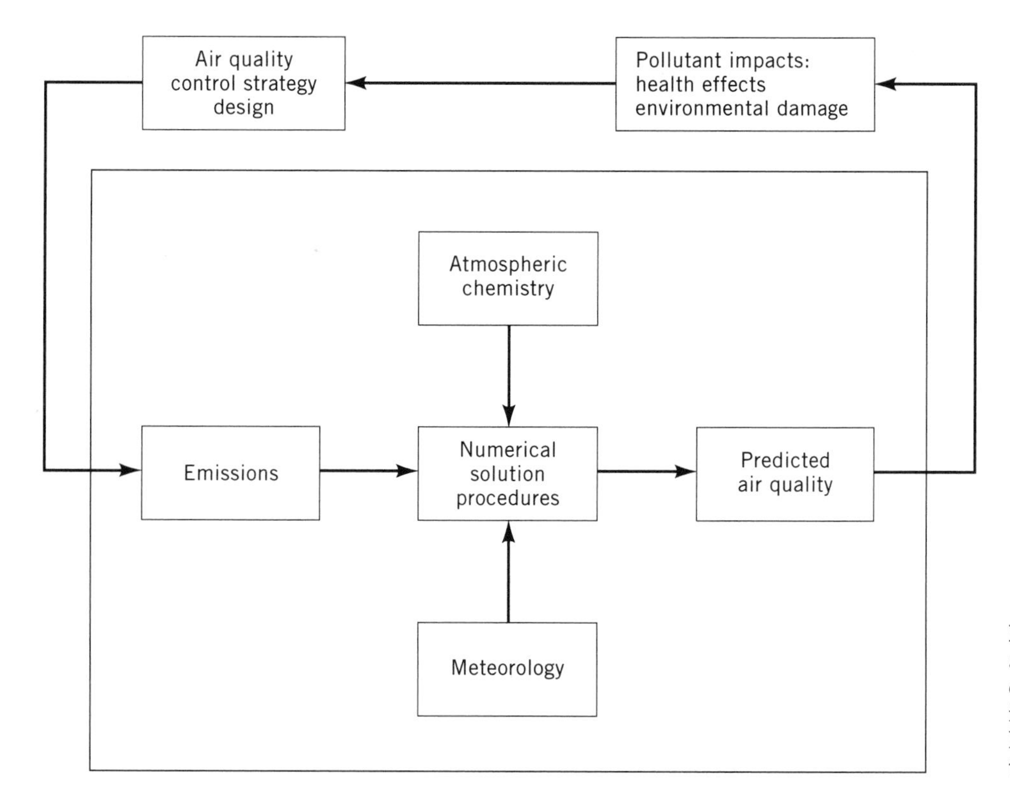

Figure 1. Schematic of the role of an air quality model, in the air quality control planning process. Some studies, eg, the prediction of indoor air pollutant concentrations, require more than one model.

ondary, ie, formed in the atmosphere through chemical or physical processes. Examples of primary pollutants are carbon monoxide [*630-08-0*], CO, lead [*7439-92-1*], Pb, chlorofluorocarbons, and many toxic compounds. Notable secondary pollutants include ozone [*10028-15-6*]; O_3, which is formed in the troposphere by reactions of nitrogen oxides (NO_x) and reactive organic gases (ROG); and sulfuric and nitric acids.

Models are used extensively to understand a wide range of problems, including indoor air pollution, such as elevated levels of radon [*10043-92-2*], Rn, and formaldehyde [*50-00-0*], CH_2O, high concentrations of carbon monoxide and particulate matter in the vicinity of congested roadway intersections, spills of volatile toxic chemicals, urban smog, acid deposition, global climate change, and stratospheric ozone depletion. Each of these problems involves pollutant transport via advection and turbulent diffusion. Several of them also involve atmospheric chemistry, and processes of wet or dry deposition may also be important. Thus the models used to describe the dynamics in each case primarily vary in temporal and spatial resolution, the chemical transformations being tracked, and the deposition processes included. If a pollutant is strictly primary in nature and the deposition processes included. If a pollutant is strictly primary in nature and is long lived in the atmosphere, the models used when dealing with carbon monoxide, many toxic chemicals, and most components of particulate matter. Following the evolution of secondary or highly reactive pollutants is generally much more complicated because chemical reactions between many compounds must be considered in addition to transport and removal. Chemically active systems include urban and regional smog, acid deposition, and stratospheric ozone depletion.

AIR QUALITY MODELS

Models used in air pollution analysis fall into two classes: empirical–statistical and deterministic, as shown in Figure 2. In the former, the model statistically relates observed air quality data to the accompanying emission patterns, and chemistry and meteorology are included only implicitly. In the latter, analytical or numerical expressions describe the complex transport and chemical processes that affect air pollutants. Pollutant concentrations are determined as explicit functions of meteorology, topography, chemical transformation, surface deposition, and source characteristics. A typical schematic is shown in Figure 3. A listing of many of the air quality models, including status, applications, and the model formulations, is available (2). Each formulation involves approximations and has certain strengths and limitations.

Empirical–Statistical Models

Empirical–statistical models are based on establishing a relationship between historically observed air quality and the corresponding emissions and other explanatory factors. The linear rollback model is simple to use and requires few data, and for these reasons has been widely applied (3,4). Linear rollback models assume that the highest measured pollutant concentration is proportional to the basinwide emission rate, plus the background value, ie,

$$c_{\max} = aE + c_b \tag{1}$$

where $c_{\max}$ is the maximum measured pollutant concentration, E is the emission rate, c_b is the background concentration resulting from sources outside the modeling region, and a is a constant of proportionality. The constant implicitly accounts for the dispersion, transport, deposition, and chemical reactions of the pollutant. The allowable emission rate E_a necessary to reach a desired ambient air quality goal c_d can be calculated from

$$\frac{E_a}{E_o} = \frac{c_d - c_b}{c_{\max} - c_b} \tag{2}$$

where E_o is the emission rate that prevailed at the time that $c_{\max}$ was observed. Presumably, pollutant concentrations at other times decrease toward background levels as emissions are reduced. The linear rollback model is a simplified approach, and its application is limited. Nonlinear processes, such as chemical reactions and spatial or temporal changes in emission patterns, are not accounted for in the rollback model.

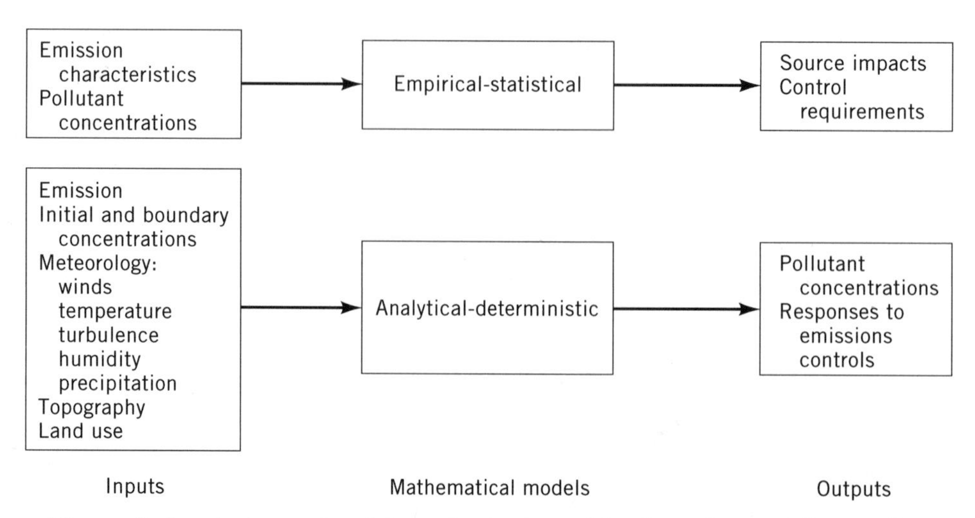

Figure 2. Inputs, types of models, and outputs used in air quality modeling studies.

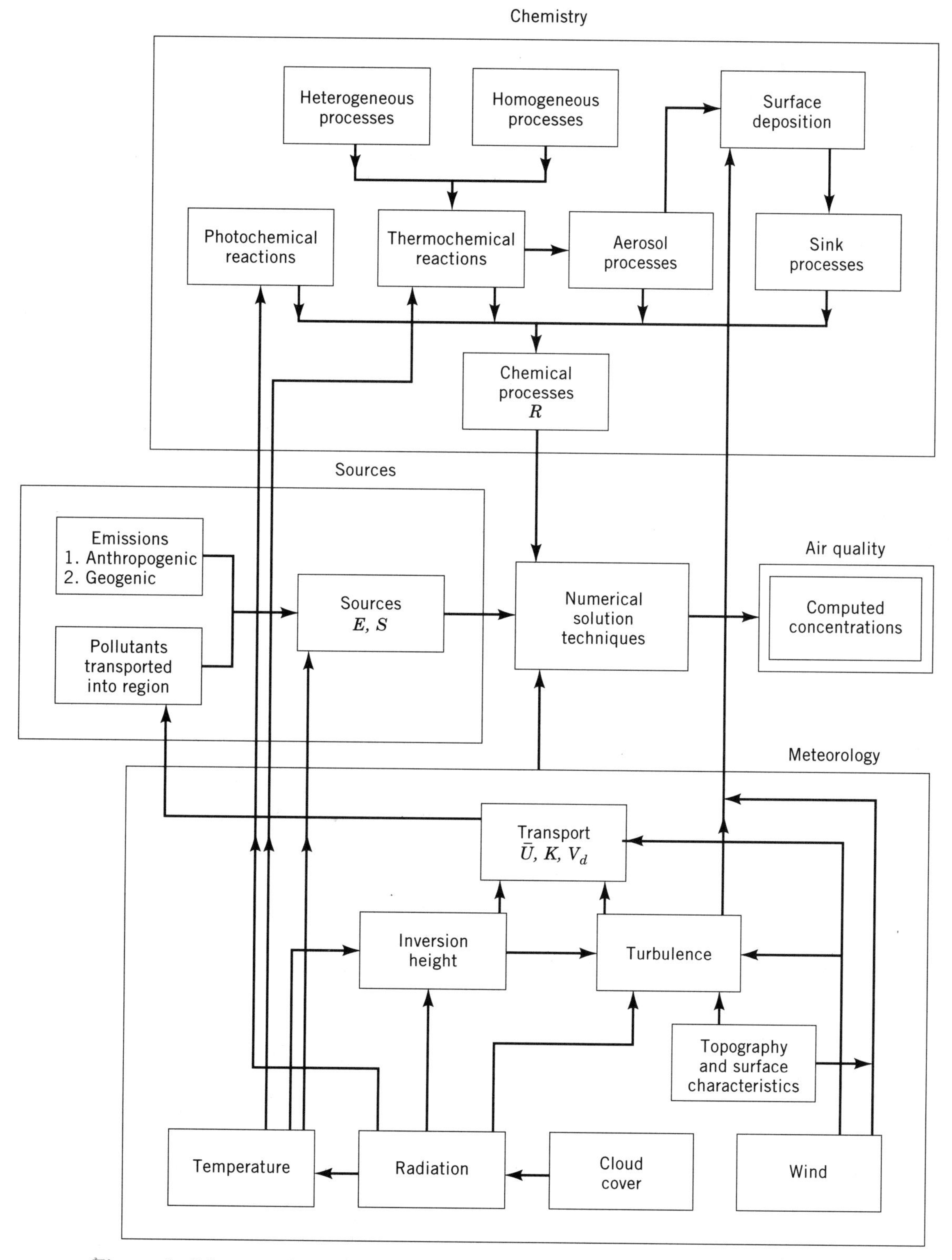

Figure 3. Schematic diagram of a deterministic air quality model, showing the model components and interactions (1); each of the boxes involves a large number of individual processes. Terms are defined in text.

Time Series and Regression Models

As in many other fields, relationships between pollutant levels and explanatory factors such as meteorological variables have been analyzed using regression models. Day-to-day variations in peak ozone concentrations, eg, can be closely reproduced by simple linear models that use temperature and relative humidity as independent variables (5). Time series models for pollutants also have been developed, accounting for temporal correlations between pollutant levels, with or without incorporation of "exogenous" explanatory variables such as temperature or emissions rates (6). Over short time horizons, time series and regression models have been used successfully to forecast pollutant levels (7,8). Such models have been used, eg, to predict whether field study crews should be mobilized in expectation of smog episodes. A clear limitation of time series and regression models is that in general they cannot provide information about how pollutant levels would respond to emissions controls.

Receptor-Oriented Model

A third class of empirical–statistical models is the receptor-oriented model, which has been used extensively for estimating the contributions that distinguishable sources such as automobiles or municipal incinerators make to particulate matter concentrations (9–18). Attempts have also been made to track nonreacting gases (19), and under special conditions, reactive organic compounds (20–22), to their sources using receptor modeling methods. Receptor models compare the measured chemical composition of particulate matter at a receptor site with the chemical composition of emissions from the primary sources to identify the source contributions at the monitoring location. There are three principal categories of receptor models: chemical mass balance, multivariate, and microscopic. Hybrid analytical and receptor (or combined source–receptor) models have been used (23), but further investigation into their capabilities is required.

Receptor models are powerful tools for source apportionment of particulates because a vast amount of particulate species characterization data have been collected at many sampling sites worldwide and because many aerosol species are primary pollutants. Most of the information available is for elemental concentrations, eg, lead, nickel, and aluminum, although more recent measurements have provided data on concentrations of ionic species and carbonaceous compounds. At a sampling (or receptor) site, the aerosol mass concentration of each species i is

$$c_i = \sum_{j=1}^{n} a_{ij} S_j \qquad i = 1, 2, \ldots m \tag{3}$$

where c_i is the mass concentration of species i at the receptor sites, S_j is the total mass concentration of all species at the receptor site that is attributable to source j, a_{ij} is the fraction that species i constitutes of the total mass concentration arriving at the sampling site from source j, m is the total number of species measured, and n is the total number of sources. The mass concentration at the receptor site and the coefficients a_{ij} that describe the chemical composition for the sources are the inputs from which S_j, the mass apportioned to source j, is determined. Because a_{ij} characterizes the source, it is referred to as the source fingerprint and should be unique to the source. When the chemical compositions of the emissions from two source categories are similar, it is extremely difficult for receptor models to distinguish between the sources. The categories of receptor models are differentiated by the techniques used to determine S_j.

If the source fingerprints, a_{ij}, for each of n sources are known and the number of sources is less than or equal to the number of measured species ($n \leq m$), an estimate for the solution to the system of equations 3 can be obtained. If $m > n$, then the set of equations is overdetermined, and least-squares or linear programming techniques are used to solve for S_j. This is the basis of the chemical mass balance (CMB) method (24,25). If each source emits a particular species unique to it, then a simple tracer technique can be used (9). Examples of commonly used tracers are lead and bromine from mobile sources, nickel from fuel oil, and sodium from sea salt. The condition that each source have a unique tracer species is not often met in practice.

Microscopic identification models are similar to the CMB methods except that additional information is used to distinguish the source of the aerosol. Such chemical or morphological data include particle size and individual particle composition and are often obtained by electron or optical microscopy.

Multivariate models, including factor analysis models (18,26–28), rely on finding the underlying structure of large sets of air quality data to determine sources. Models based on factor analysis are the most widely used. Multivariate models operate by identifying groups of elements or species, the concentrations of which fluctuate together from sampling period to sampling period, implying that these groups come from a single "source." When the composition of the hypothetical source is compared with the known composition of specific sources, it often becomes obvious what the group of cofluctuating chemical elements represents. For example, lead and bromite concentrations are usually highly correlated because they are emitted primarily by the same sources, ie, automobiles burning leaded gasoline. Multivariate techniques do not rely on a detailed knowledge of the source fingerprint a_{ij} and can be used to refine estimates of the fingerprint.

Deterministic Models

Deterministic air quality models describe in a fundamental manner the individual processes that affect the evolution of pollutant concentrations. These models are based on solving the atmospheric diffusion–reaction equation, which is in essence the conservation-of-mass principle for each pollutant species (29):

$$\frac{\delta c_i}{\delta t} + \overline{U} \cdot \nabla c_i = \nabla \cdot D_i \nabla c_i + R_i(c_1, c_2, c_3, \ldots c_n) + S_i(\overline{x}, t)$$
$$i = 1, 2, 3, \ldots n \tag{4}$$

where c_i is the concentration of species i, $\overline{U}$ is the wind velocity vector, D_i is the molecular diffusivity of species i, R_i is the net production (depletion if negative) of species i by chemical reaction, S_i is the emission rate of i, and n is the number of species. R can also be a function of the meteorological variables. Equation 4 states that the time rate of change of a pollutant depends on convective transport (term 2), diffusion (term 3), chemical reactions (term 4), and emissions (term 5), together with initial and boundary conditions. Deposition and surface level emissions enter as boundary conditions at the ground:

$$E - v_d c = -K_{zz} \frac{\delta c}{\delta z}$$

where E is the ground level emissions, v_d is the deposition velocity, and K_{zz} is the vertical diffusivity. The turbulence closure problem makes it necessary to approximate the atmospheric diffusion equation, usually by K-theory (30,31):

$$\frac{\delta c_i}{\delta t} + \langle \overline{U} \rangle \cdot \nabla \langle c_i \rangle = \nabla \cdot K \nabla \langle c_i \rangle + R_i(\langle c_1 \rangle, \langle c_2 \rangle, \ldots \langle c_n \rangle) + \langle S_i(x, t) \rangle \qquad i = 1, 2, 3, \ldots n \tag{5}$$

where the angle brackets $\langle \ \rangle$ indicate an ensemble average, and K is the turbulent (eddy) diffusivity tensor. Known analytical solutions exist only for the simplest source distributions and chemical reaction mechanisms, $\langle S_i \rangle$ and R, in equation 5.

Examination of equation 5 shows that if there are no chemical reactions, ($R = 0$) or if R is linear in $\langle c_i \rangle$ and uncoupled, then a set of linear, uncoupled differential equations are formed for determining pollutant concentrations. This is the basis of transport models, which may be transport only or transport with linear chemistry. Transport models are suitable for studying the effects of sources of CO and primary particulates on air quality, but not for studying reactive pollutants such as O_3, NO_2, HNO_3, and secondary organic species.

Lagrangian Models. There are two distinct reference frames from which to view pollutant dynamics. The most natural is the Eulerian coordinate system, which is fixed at the earth's surface and in which a succession of different air parcels are viewed as being carried by the wind past a stationary observer. The second is the Lagrangian reference frame, which moves with the flow of air, in effect maintaining the observer in contact with the same air parcel over extended periods of time. Because pollutants are carried by the wind, it is often convenient to follow pollutant evolution in a Lagrangian reference frame, and this perspective forms the basis of Lagrangian trajectory and Lagrangian marked-particle or particle-in-cell models. In a Lagrangian marked-particle model, the center of mass of parcels of emissions are followed, traveling at the local wind velocity, while diffusion about that center of mass is simulated by an additional random translation corresponding to the atmospheric diffusion rate (32,33).

Lagrangian trajectory models can be viewed as following a column of air as it is advected in the air basin at the local wind velocity. Simultaneously, the model describes the vertical diffusion of pollutants, deposition, and emissions into the air parcel as shown in Figure 4. The underlying equation being solved is a simplification of equation 5:

$$\frac{\delta c_i}{\delta t} = \frac{\delta}{\delta z} K_{zz} \frac{\delta c_i}{\delta z} + S_i(t) + R(\bar{c}, t) \tag{6}$$

Trajectory models require spatially and temporally resolved wind fields, mixing-height fields, deposition parameters, and data on the spatial distribution of emissions. Lagrangian trajectory models assume that vertical wind shear and horizontal diffusion are negligible. Other limitations of trajectory and Eulerian models have been discussed (34).

Gaussian Plume Model. One of the most basic and widely used transport models based on equation 5 is the

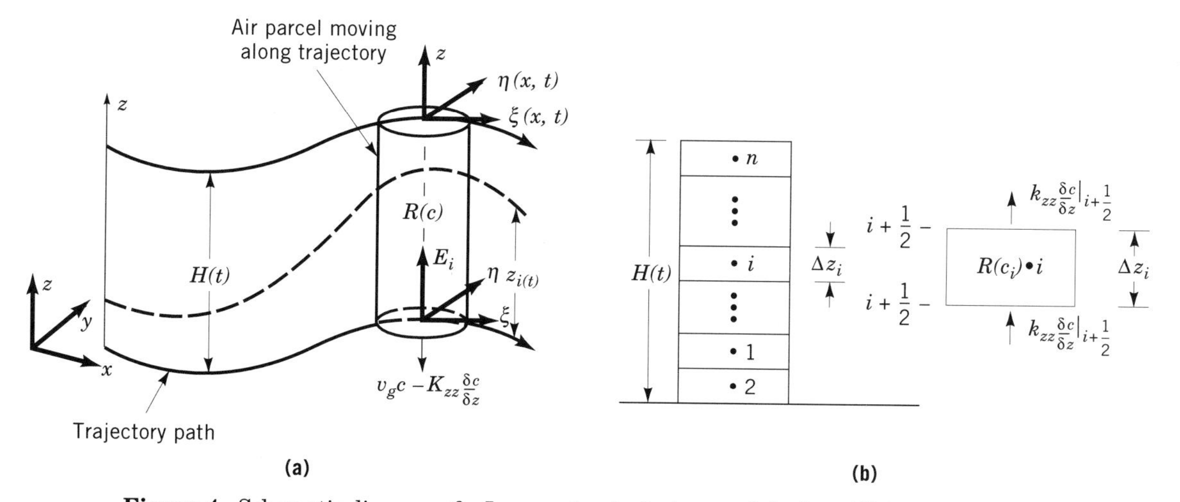

Figure 4. Schematic diagram of a Lagrangian trajectory model where $H(t)$ represents the air column height in both Eulerian (x,y,z) and Lagrangian (ξ,η,z) coordinates; c, the pollutant concentration; R, chemical reactions; v_g, deposition of pollutants; and K_{zz}, vertical diffusion. (**a**) The column of air being modeled is advected at the local wind velocity along a trajectory path across the modeling region; $z_i(t)$ represents the mixing height variation along the trajectory path; E_i, the emissions of the ith pollutant. Within the moving air parcel, the model describes the important processes affecting the pollutant evolution and concentration. (**b**) Vertical resolution is gained by dividing the column into a number of cells in the vertical direction; $R(c_i)$ represents the concentration produced by chemical reactions within the ith cell of which Δz_i is the height.

Gaussian plume model. Gaussian plume models for continuous sources can be obtained from statistical arguments or can be derived by solving:

$$\overline{U}\frac{\delta c}{\delta x} = K_{yy}\frac{\delta^2 c}{\delta y^2} + K_{zz}\frac{\delta^2 c}{\delta z^2} \tag{7}$$

where $\overline{U}$ is the temporally and vertically averaged wind velocity; x, y, and z are the distances in the downwind, crosswind, and vertical directions, respectively; and K_{yy} and K_{zz} are the horizontal and vertical turbulent diffusivities, respectively. For a source with an effective height H, emission rate Q, and a reflecting (nonabsorbing) boundary at the ground, the solution is

$$c(x, y, z) = \frac{Q}{2\pi \overline{U}\,\sigma_y(x)\sigma_z(x)} \exp\left[\frac{-y^2}{2\sigma_y^2(x)}\right]\left[\exp\frac{-(z-H)^2}{2\sigma_z^2(x)} + \exp\frac{-(z+H)^2}{2\sigma_z^2(x)}\right]$$

This solution describes a plume with a Gaussian distribution of pollutant concentrations, such as that in Figure 5, where $\sigma_y(x)$ and $\sigma_z(x)$ are the standard deviations of the mean concentration in the y and z directions. The standard deviations are the directional diffusion parameters, and are assumed to be related simply to the turbulent diffusivities, K_{yy} and K_{zz}. In practice, $\sigma_y(x)$ and $\sigma_z(x)$ are functions of x, $\overline{U}$, and atmospheric stability (2,35–37).

Gaussian plume models are easy to use and require relatively few input data. Multiple sources are treated by superimposing the calculated contributions of individual sources. It is possible to include the first-order chemical decay of pollutant species within the Gaussian plume framework. For chemically, meteorologically, or geographically complex situations, however, the Gaussian plume model fails to provide an acceptable solution.

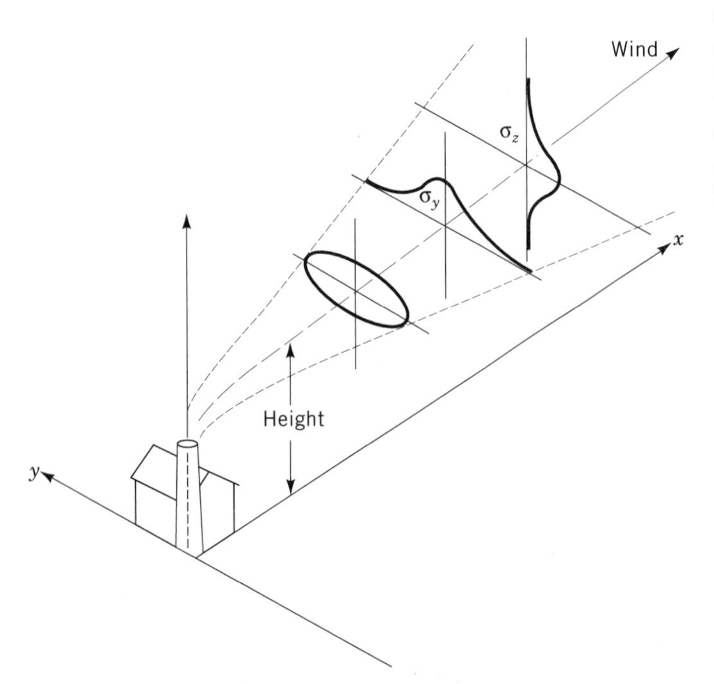

Figure 5. Diffusion of pollutants from a point source. Pollutant concentrations have separate Gaussian distributions in both the horizontal (y) and vertical (z) directions. The spread is parameterized by the standard deviations (σ), which are related to the diffusivity (K).

Eulerian Models. Of the Eulerian models, the box model is the easiest to conceptualize. The atmosphere over the modeling region is envisioned as a well-mixed box, and the evolution of pollutants in the box is calculated following conservation-of-mass principles including emissions, deposition, chemical reactions, and atmospheric mixing.

Eulerian "grid" air quality models are the most complex, but potentially the most powerful, involving the least-restrictive assumptions. They are also the most computationally intensive. Grid models solve a finite approximation to equation 5, including temporal and spatial variation of the meteorological parameters, emission sources, and surface characteristics. Grid models divide the modeling region into a large number of cells, horizontally and vertically, which interact with each other by simulating diffusion, advection, and sedimentation (for particles) of pollutant species (Fig. 6). Input data requirements for grid models are similar to those for Lagrangian trajectory models, but in addition, data on background concentrations (boundary conditions) at the edges of the grid system are required. Eulerian grid models predict pollutant concentrations throughout the entire airshed. Over successive time periods the evolution of pollutant concentrations and how they are affected by transport and chemical reaction can be tracked.

Modeling Chemically Reactive Compounds. A number of compounds are formed or destroyed in the atmosphere by a series of complex, nonlinear chemical reactions, eg, stratospheric ozone. Models that not only describe pollutant transport but also account for complex chemical transformations, $R(\bar{c},t)$ in equation 5, are necessary for these systems. Such models are also required to study the dynamics of chemically reactive primary pollutants such as nitric oxide [*10102-43-9*], NO, and pollutants that are primary as well as secondary in origin, eg, nitrogen diox-

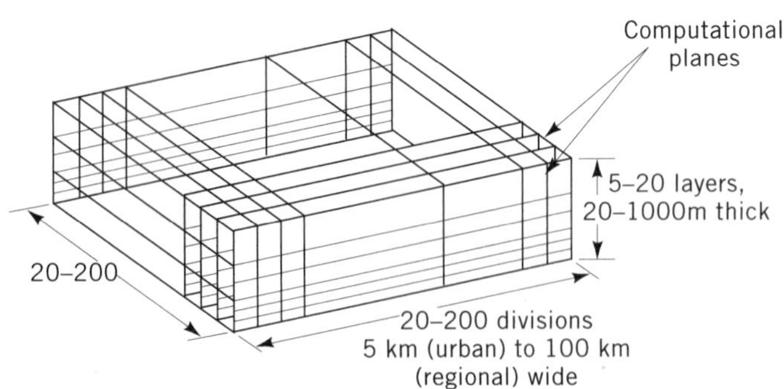

Figure 6. A typical model application would include a domain of about 60 × 100 horizontal cells, and 5 to 20 layers vertically, for a total of about 60,000 computational cells. Vertical cell spacing is usually nonuniform with layers 20- to 1000-mm thick (with the finest resolution near the ground). Horizontal resolution depends on the size of the modeling domain. Urban applications use grids of about 5 × 5 km, while regional models use grids of about 20 to 100 km per side.

ide [*10102-44-0*], NO_2, and formaldehyde. Addition of the capability to describe a series of interconnected chemical reactions greatly increases requirements for computer storage as well as computing time and input data requirements. Increased computational demands arise because the evolution of the interacting species must be followed simultaneously, leading to a system of coupled, nonlinear differential equations.

Box, Lagrangian trajectory, and Eulerian grid models have all been developed to include nonlinear chemical reactions. Box models assume that the pollutants are mixed homogeneously within the modeling region, an assumption that is often inappropriate. Trajectory and grid models resolve pollutant dynamics spatially and have been used widely and with success particularly for studying photochemical smog and acid deposition problems (30,38–49).

Temporal and Spatial Resolution

The temporal and spatial resolutions of models can vary from minutes to a year and from meters to hundreds of kilometers. The minimum meaningful resolution of a model is determined by the input data resolution and the structure of the model. Statistical models generally rely on several years worth of measurements of hourly or daily pollutant concentrations. The resolution of the input data represents the minimum resolution of a statistical model. Resolution of analytical models is limited by the spatial and temporal resolution of the emissions inventory, the meteorological fields, and the grid size chosen for model implementation. For modeling urban air basins, the size of individual grid cells is on the order of a few kilometers per side, whereas for modeling street canyons, the cell size must be reduced to a few meters on each edge. At the other extreme, regional models have horizontal resolutions varying from 20 to 100 km, and global models may have a resolution as coarse as a few thousand kilometers. The temporal resolution of models ranges from about 15 min to a few hours or days.

The information desired from modeling studies often depends on processes that occur on spatial scales much smaller than the resolution of most air quality models. The modeling of NO_x air quality in street canyons involves small-scale processes of this sort. Introduction of point-source emissions into grid-based air quality models likewise involves a mismatch between the high concentrations that exist near the source versus the lower concentrations computed by a model that immediately mixes those emissions throughout a grid cell of several kilometers on each side. Because of computational time constraints, it has often been considered impractical to describe fully the processes that take place on a scale smaller than the main model grid, ie, subgrid scale. Nevertheless, values obtained from large-scale calculations should be accurate over the spatial averaging scale adopted by the model.

MODEL COMPONENTS

A model's ability to correctly predict pollutant dynamics and to apportion source contributions depends on the accuracy of the individual process descriptions and input data, and the fidelity with which the framework reflects the interactions of the processes.

Turbulent Transport and Diffusion

There are two pollutant transport terms in equation 5: an advection term, in which pollutants are carried along with the time-averaged mean wind flow, and a dispersion term, representing transport resulting from local turbulence. The averaging time that determines the mean winds is related to the spatial scale of the system being modeled. Minutes may be appropriate for urban-scale simulations, multihour averages for the regional scale, and daily to weekly averages for determining long-term concentrations of nonreactive pollutants.

Turbulent transport is determined by complex interactions between meteorological conditions and topography. In addition to gross topographical features, surface "roughness" scales have been devised to parameterize surface characteristics according to land-use categories. Small values are assigned to smooth water or ice, and increasingly higher values to grasslands, croplands, residential areas, and urban-industrial centers. In general, the rougher the surface, the greater the local turbulence. Often, particularly during sunny days, atmospheric turbulence is generated by the heating of the earth's surface.

Turbulent fluxes are difficult to measure, and hence there are uncertainties in the various methods employed by models to describe them. As in equation 5, these fluxes are typically parameterized being proportional to the gradient of mean pollutant concentrations. Further simplifying the problem, the components of the eddy diffusivity tensor, K, are assumed to be constant along their respective axes. An obvious need when applying K-theory is some algorithm for establishing the value of K. As a result of the large variety of processes involved, there are also a number of methods to parameterize the horizontal and vertical diffusion coefficients (50). The usual limitation to the accuracy of diffusion calculations in a practical application is determined by the extent of measurements of the atmospheric structure taken during the period to be simulated. For most model applications, the number of observed factors relating to atmospheric turbulence are few and include only ground-level winds and temperatures, surface roughness, and cloud cover. At a few locations and times, the inversion base, or mixing height, wind speeds aloft, and vertical temperature gradient may also be known. As the amount and accuracy of information characterizing atmospheric structure increases, confidence in model predictions of dispersion increases.

Removal Processes

Pollutant removal processes, particularly dry deposition and scavenging by rain and clouds, are a primary factor in determining the dynamics and ultimate fate of pollutants in the atmosphere.

Dry Deposition. Dry deposition occurs in two steps: the transport of pollutants to the earth's surface, and the physical and chemical interaction between the surface and the pollutant. The first is a fluid mechanical process, the second is primarily a chemical process, and neither is completely characterized at the present time. The problem is confounded by the interaction between the pollutants and biogenic surfaces where pollutant uptake is enhanced or retarded by plant activity that varies with time (51,52). It is difficult to measure the depositional flux of pollutants from the atmosphere, though significant advances were made during the 1980s and early 1990s (53,54).

Many factors affect dry deposition, but for computational convenience air quality models resort to using a single quantity called the deposition velocity, designated v_d or v_g, to prescribe the deposition rate. The deposition velocity is defined such that the flux F_i of species i to the ground is

$$F_i = v_d c_i(z_r) \tag{9}$$

where $c_i(z_r)$ is the concentration of species i at some reference height z_r, typically from one to several meters. For a number of pollutants, v_d has been measured under various meteorological conditions and for a number of surface types (54).

Early models used a value for v_d that remained constant throughout the day. However, measurements show that the deposition velocity increases during the day as surface heating increases atmospheric turbulence and hence diffusion, and plant stomatal activity increases (54–55). More recent models take this variation of v_d into account. In one approach, the first step is to estimate the upper limit for v_d in terms of the transport processes alone. This value is then modified to account for surface interaction, because the earth's surface is not a perfect sink for all pollutants. This method has led to what is referred to as the resistance model (56,57) that represents v_d as the analogue of an electrical conductance

$$v_d = (r_a + r_b + r_s)^{-1} \tag{10}$$

where r_a is the aerodynamic resistance controlled primarily by atmospheric turbulence, r_b is the resistance to transport in the fluid sublayer near the plant surface, and r_s is the surface (or canopy) resistance. Of the three resistances, r_a is essentially the same for all species, r_b is the same for gaseous species with the same diffusivities, though it can be considerably greater for aerosols, and r_s depends greatly on the surface affinity for the diffusing species. For example, nitric acid [*7697-37-2*], HNO_3, which reacts rapidly with most surfaces, has a low surface resistance, usually taken as zero (53,58,59), whereas CO is not as reactive and has a high r_s value. Significant differences between the deposition of gases and aerosols are that aerosols have a much lower diffusivity, the rate of gravitational settling can be significant for larger particles, and the surface resistance for aerosols is not determined by species reactivity, alone. Recent models account for the variation of surface resistance and diurnal change in fluid mechanical transport. These parameterizations have been used to quantify the deposition flux of various compounds (53,60).

Scavenging by Rain, Fog, and Clouds. Wet removal, or precipitation scavenging, can be effective in cleansing the atmosphere of pollutants, and depends on the intensity and size of the raindrops (61). Fog and cloud droplets can also absorb gases, capture particles, and promote chemical reactions. Precipitation scavenging is not as important on an urban scale as on a regional scale and is not included in most urban-scale models. Fog chemistry can be important to human health on an urban scale, as evidenced in London in 1952 when thousands of persons died during an episode of excess industrial air pollution and fog (31). Cloud, rain, and fog processes are often important in regional and global scale modeling.

Representation of Atmospheric Chemistry Through Chemical Mechanisms

A complete description of atmospheric chemistry within an air quality model would require tracking the kinetics of many hundreds of compounds through thousands of chemical reactions. Fortunately, in modeling the dynamics of reactive compounds such as peroxyacetyl nitrate [*2278-22-0*] (PAN), $C_2H_3NO_5$, O_3, and NO_2, it is not necessary to follow every compound. Instead, a compact representation of the atmospheric chemistry is used. Chemical mechanisms represent a compromise between an exhaustive description of the chemistry and computational tractability. The level of chemical detail is balanced against computational time, which increases as the number of species and reactions increases. Instead of the hundreds of species present in the atmosphere, chemical mechanisms include on the order of 50 species and 100 reactions.

Three different types of chemical mechanisms have evolved as attempts to simplify organic atmospheric chemistry: surrogate (62,63), lumped (64–67), and carbon bond (68–70). These mechanisms were developed primarily to study the formation of O_3 and NO_2 in photochemical smog but can be extended to compute the concentrations of other pollutants, such as those leading to acid deposition (44,46).

Surrogate mechanisms use the chemistry of one or two compounds in each class of organics to represent the chemistry of all the species in that class. For example, the explicit chemistry of butane [*106-97-8*], C_4H_{10}, might be used to describe the chemistry of the alkanes.

Lumped mechanisms are based on the grouping of chemical compounds into classes of similar structure and reactivity. For example, all alkanes might be lumped into a single class, the reaction rates and products of which are based on a weighted average of the properties of all the alkanes present. For example, as shown in Table 1, the various alkanes, C_nH_{2n+2}, react with OH in a similar manner to form alkyl radicals, C_nH_{2n+1}. When expressed explicitly, more than 30 species and 20 reactions are involved. By lumping, the series of reactions can be reduced to 1, and the number of required organic compounds is reduced to 2. Thus lumping yields a tremendous savings in computational time, yet maintains the necessary chemical detail.

The carbon bond mechanisms (68–70), a variation of a lumped mechanism, splits each organic molecule into

Table 1. Alkane–Hydroxide Radical Reactions

Explicit reactions

$$\left.\begin{array}{c} C_2H_6 + OH \rightarrow C_2H_5^{\cdot} + H_2O \\ C_3H_8 + OH \rightarrow C_3H_7^{\cdot} + H_2O \\ \vdots \\ C_nH_{2n+2} + OH \rightarrow C_nH_{2n+1}^{\cdot} + H_2O \end{array}\right\} \text{initial reaction with OH}$$

$$\left.\begin{array}{c} C_2H_5^{\cdot} + O_2 \rightarrow C_2H_5O_2^{\cdot} \\ C_3H_7^{\cdot} + O_2 \rightarrow C_3H_7O_2^{\cdot} \\ \vdots \\ C_nH_{2n+1}^{\cdot} + O_2 \rightarrow C_nH_{2n+1}O_2^{\cdot} \end{array}\right\} \text{alkyl radical oxidation reaction}$$

Lumped representation

$$\text{alkane} + OH \rightarrow RO_2^{\cdot}$$

functional groups using the assumption that the reactivity of the molecule is dominated by the chemistry of each functional group.

Aerosol Dynamics

Inclusion of a description of aerosol dynamics within air quality models is of primary importance because of the health effects associated with fine particles in the atmosphere, visibility deterioration, and the acid deposition problem. Aerosol dynamics differ markedly from gaseous pollutant dynamics in that particles come in a continuous distribution of sizes and can coagulate, evaporate, grow in size by condensation, be formed by nucleation, or be deposited by sedimentation. Furthermore, the species mass concentration alone does not fully characterize the aerosol. The particle size distribution, which changes as a function of time, and size-dependent composition determine the fate of particulate air pollutants and their environmental and health effects. Particles of about 1 μm in diameter or smaller penetrate the lung most deeply and represent a substantial fraction of the total aerosol mass, as shown in Figure 7. The origin of these fine particles is difficult to identify because much of the fine particle mass is formed by gas-phase reaction and condensation in the atmosphere.

Simulation of aerosol processes within an air quality model begins with the fundamental equation of aerosol dynamics that describes aerosol transport (term 2), growth (term 3), coagulation (terms 4 and 5), and sedi-

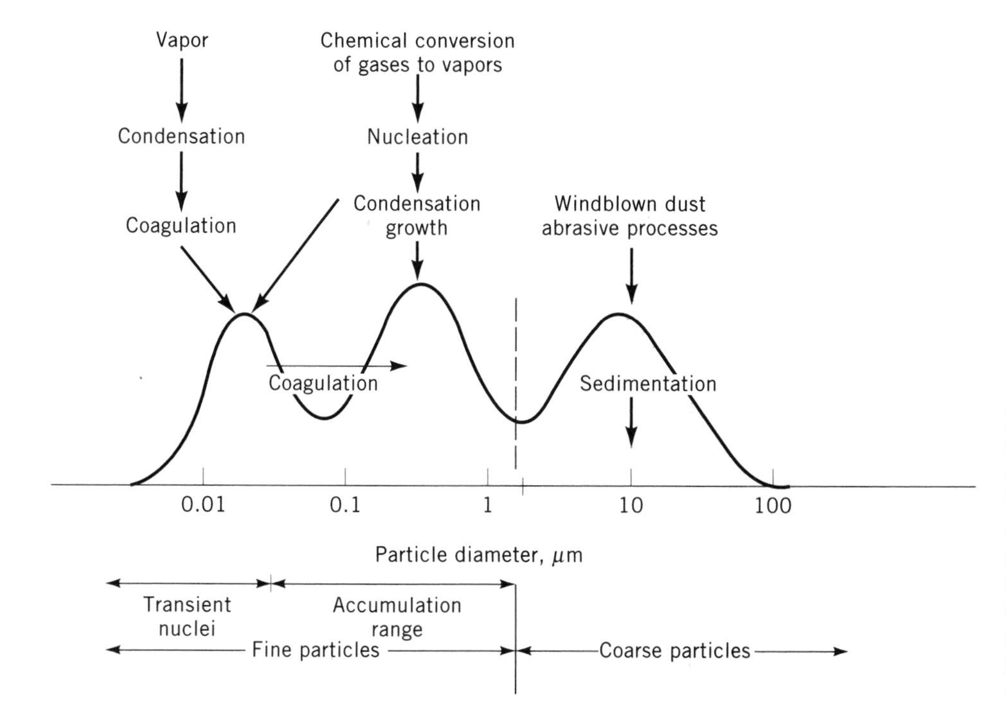

Figure 7. Size distribution of an urban aerosol showing the three modes containing much of the aerosol mass. The fine mode contains particles produced by condensation of low volatility gases. The mid-range, or accumulation mode, results from coagulation of smaller aerosols and condensation of gases on preexisting particles. Coarse particulates, the largest aerosols, are usually generated mechanically.

mentation (term 6):

$$\frac{\delta n}{\delta t} + \nabla \cdot \overline{U}n + \frac{\delta I}{\delta v} = \frac{1}{2}\int_0^v \beta(\bar{v}, v - \bar{v})n(\bar{v})n(v - \bar{v})d\bar{v} - \int_0^\infty \beta(\bar{v}, v)n(\bar{v})n(v)d\bar{v} - \nabla \cdot Cn \quad (11)$$

where n is the particle size distribution function, $\overline{U}$ is the fluid velocity, I is the droplet current that describes particle growth and nucleation resulting from gas-to-particle conversion, v is the particle volume, β is the rate of particle coagulation, and C is the sedimentation velocity. Modeling the formation and growth of aerosols is done by sectioning the size distribution n into discrete ranges. Then the size and chemical composition of an aerosol is followed as it evolves by condensation, coagulation, sedimentation, and nucleation.

Air Quality Model Inputs

Inputs to analytical air quality models can be broadly grouped as those dealing with meteorology, emissions, topography, and atmospheric concentrations. Meteorological inputs generally control the transport rate of pollutants and are used to determine reaction rates and the depositional flux of compounds. Topography influences transport and deposition. Observed compound concentrations are used to specify both initial and boundary conditions for model simulations. Especially for pollution problems involving organic compounds, emissions are a key input subject to considerable uncertainty. Although emissions from primary industrial facilities or utilities may be reasonably well known, emissions from residential or commercial facilities, mobile sources, and natural sources are often roughly estimated and difficult to verify.

The data requirements for applying models differ greatly among model types. For a Gaussian plume model, the required data could include as little as the mean wind velocity, source emission rate, atmospheric stability (and hence diffusivity), effective source height, and air quality data for comparison with model predictions (71). At the other extreme, a large grid model that incorporates chemical kinetics requires considerably more information. Ideally a comprehensive model evaluation study would incorporate spatially and temporally resolved meteorological data; eg, winds, temperature, and solar insolation; temporally varying emissions for every species in each cell of the modeling region; topographical data, eg, land use and elevation; initial species concentrations; boundary conditions for each species; and concentration data for comparison against model predictions (48).

Meteorological Inputs

Meteorological conditions determine how fast and to where pollutants are dispersed, rates of chemical reactions, and losses resulting from deposition. Inputs vary depending on the type of model being applied. Simpler models, eg, Gaussian plume models with no chemical reactions, may require only a single wind velocity and cloud cover. On the other hand, a complex, three-dimensional, chemically active model requires hourly vertically and horizontally resolved wind fields as well as hourly temperature, humidity, mixing depth, and solar insolation fields. The spatial and temporal resolution of the meteorological inputs must match the resolution of the air quality model. Much of the time the necessary inputs are determined directly from observations and relatively little mathematical modeling is used. In other cases, it is necessary to interpolate relatively sparse observations over a modeling domain using objective analysis techniques. Because of the sparseness of the data, newer air quality model applications have used dynamic, or prognostic, meteorological models. Use of dynamic models based on fundamental equations is intended to improve the accuracy of the meteorological inputs over the modeling domain.

Objective analysis techniques use a prescribed method with observations collected at a few discrete locations and times to calculate meteorological variables over the modeling domain. Usually this involves two- and three-dimensional spatial interpolation and does not involve solving for any time dependency of the flow. Particular care must be taken in developing wind fields from sparse data because the wind field should be mass consistent. Objective analysis, including diagnostic procedures, is used to reduce the divergence of interpolated wind fields and account for some topographical features (72–75). A field of input values generated by interpolation over a large geographic area from sparse data is intrinsically uncertain and leads to uncertainty in model predictions. Upper-level variables such as temperature structure (mixing depths) and wind fields are particularly susceptible to this uncertainty and have led to the wider use of dynamic models.

Dynamic meteorological models, much like air pollution models, strive to describe the physics and thermodynamics of atmospheric motions as accurately as is feasible. Besides being used in conjunction with air quality models, they are also used for weather forecasting. Like air quality models, dynamic meteorological models solve a set of partial differential equations (also called primitive equations). These equations, which are fundamental to the fluid mechanics of the atmosphere, are referred to as the Navier-Stokes equations and describe the conservation of mass and momentum. They are combined with equations describing energy conservation and thermodynamics in a moving fluid (76):

mass $$\frac{\partial \rho}{\partial t} + \nabla \cdot \rho\overline{U} = 0 \quad (12)$$

momentum $$\frac{\partial \rho\overline{U}}{\partial t} + \overline{U} \cdot \nabla\rho\overline{U} = D\nabla^2\overline{U} + \overline{F} - \nabla P \quad (13)$$

energy $$\frac{\partial \rho e}{\partial t} + \nabla \cdot (\rho e\overline{U}) = \dot{Q} - \dot{W} \quad (14)$$

thermodynamics $$\frac{p}{\rho} = RT \quad (15)$$

where ρ is the local density of the atmosphere, $\overline{U}$ is the wind velocity vector, D is the molecular diffusivity, F represents external forces such as gravity, p is pressure, e is the local internal energy of the atmosphere, and $\dot{Q}$ is the heat flux in and $\dot{W}$ is the work done by, the fluid, friction. The last equation is the Ideal Gas Law. Often it is also necessary to follow the transport of water, eg, for predicting clouds,

$$\frac{\partial q}{\partial t} + \overline{U} \cdot \nabla q = S \tag{16}$$

where q is the water concentration and S is the source or sinks of water, including rainout, etc. These equations form a set of coupled, nonlinear partial differential equations that are as formidable as those describing the chemical and physical dynamics of trace pollutants.

Experience in the solution of these governing equations is extensive, primarily because they are used in weather forecasting. This experience has led to useful simplifications and to the realization that the system is sensitive to initial conditions. The sensitivity to initial conditions leads to the temporal amplification of any errors in the initial or boundary conditions, or of computational errors. The growth of these errors seriously limits the period of time that can be simulated for use in air pollution studies unless a separate mechanism is used to dampen error growth. This same limitation makes longer range weather forecasts increasingly uncertain. A method that has been developed to reduce the growth of errors arising from initial and boundary conditions when reconstructing historical conditions is to use observations to adjust the solution back toward the actual. This is referred to as "Newtonian nudging" or four-dimensional data assimilation (FDDA) (77). In essence, the predicted solution from the dynamic model is averaged, using various weightings, and the meteorological fields developed are by objective analysis.

The equations governing air motion are generally assumed to be independent of those describing the chemical pollutant dynamics. This is because for problems such as urban smog and acid deposition, pollutant concentrations are so low that they do not significantly impact radiative transfer and hence future weather. This may not be the case for problems such as stratospheric ozone depletion and is not true for global climate change. In the latter case, the chemical evolution of the greenhouse gases, eg, carbon dioxide [*124-38-9*], CO_2, and chlorofluorocarbons (CFC), is currently assumed to be slow, and the pollutant distribution is found using little or no description of the chemistry. The radiative forcing is then found using expected concentrations.

Emissions

Especially for organic compounds and particulate matter, emissions estimates are often uncertain. Although emissions from major industrial facilities or utilities may be reasonably well known, emissions from residential or commercial facilities, mobile sources, and natural sources are often roughly estimated and difficult to verify. Moreover, model applications performed to assess the impacts of alternative control strategies often require projections of emissions out to the future. The forecasts of growth or shifts in population and economic activity that underlie emissions projections can be highly uncertain.

Emissions inventories are developed by directly measuring emissions at the point of release (for some major sources), applying emissions factors obtained from similar sources to production or operating condition estimates for local sources, and by using mass balance techniques such as inventorying compound sales or using and applying assumed evaporation rates for solvents or gasoline. In the 1985 National Acid Precipitation Assessment Program (NAPAP) inventory for the United States, less than 4% of the total VOC emissions were inventoried based on direct source tests, whereas approximately 50% were estimated using emissions factors, 25% using mass balance calculations, and 15% using "engineering judgment." Even for SO_2, direct measurements accounted for only 10% of the total emissions in the inventory (78).

In the United States, principal stationary or point sources are generally required to provide detailed information on fuel use or other measures of activity to state or local regulatory agencies. Estimates of emissions for these sources are apt to be reasonable if reliable emissions factors for sufficiently similar sources are available for combining with the activity data. By definition, area sources lie at the other end of the spectrum from point sources and are too numerous to inventory individually. For example, county-level data on vehicle registration and mileage accumulation are input into the U.S. EPA's MOBILE model (79) to generate motor vehicle exhaust emissions estimates. At the heart of the MOBILE model are exhaust emissions factors for motor vehicles in several classes (e.g., light-duty gasoline or heavy-duty diesel), which vary with vehicle model year and are based on dynamometer tests of in-use vehicles over standard driving cycles. While average speeds in a given urban area (or model grid) can be accounted for through a correction factor, the model does not explicitly account for other variables associated with local driving patterns. Estimates of natural emissions, notably biogenic emissions of VOCs, are usually derived from county-level land-use data, input to the EPA's current model, the Biogenic Emissions Inventory System (80).

Emissions projections for future years are usually traceable to forecasts of growth in energy use and shifts or growth in population and economic activity. As they are fundamental to a wide range of business and policy analyses, forecasts of these underlying trends will at least have been subjected to extensive scrutiny, even if they do not prove to be accurate. With growth projections used to adjust activity levels, what remains in projecting emissions rates is to adjust emissions factors to account for changes in fuel mix, turnover of capital equipment, improvements in control technology, and prospective regulations. The degree of uncertainty that can be associated with emissions forecasts is illustrated by recent efforts to develop emissions projections for eastern Europe to the year 2000 (81). Pre-1990 trends clearly offer little guidance to future emissions levels.

Emissions are a major source of uncertainty in air quality models for ozone and particulates. Further research is needed to develop methods for verifying emissions inventories with ambient data, using receptor modeling or other new approaches. Finally, research is also needed to develop predictive models of emissions (such as traffic simulation models and models of the impact of fuel composition, catalysts, and operating conditions on the composition of motor vehicle exhaust) and to incorporate them into emissions estimation procedures.

Mathematical and Computational Implementation

Solution of the complex systems of partial differential equations governing both the evolution of pollutant concentrations and meteorological variables, eg, winds, requires specialized mathematical techniques. Comparing the two sets of equations governing pollutant dynamics (eq. 5) and meteorology (eqs. 12–14) shows that in both cases they can be put in the form:

$$\frac{\partial A}{\partial t} + \overline{U} \cdot \nabla A = \nabla \cdot K \nabla A + S \qquad (17)$$

response convective terms source

where A is the variable being modeled, eg, pollutant concentrations, wind velocities, temperature, and water vapor; $\overline{U}$ is the wind velocity vector; and K is the turbulent diffusivity. S is a generalized source or sink, eg, chemical production–destruction and emissions for modeling pollutant concentrations; pressure gradients, gravity, and coriolis forces for momentum conservation; thermal heating for energy conservation; and evaporation for water vapor modeling. Equation 17 is the general transport equation, which can be complex and nonlinear for most atmospheric systems. The nonlinearity and coupling arise as a result of the second and fourth terms. Because the systems of equations used in pollutant dynamics and meteorological modeling have the same general form, similar mathematical techniques are used to solve them.

The generalized transport equation, equation 17, can be dissected into terms describing bulk flow (term 2), turbulent diffusion (term 3) and other processes, eg, sources or chemical reactions (term 4), each having an impact on the time evolution of the transported property. In many systems, such as urban smog, the processes have different time scales and can be viewed as being relatively independent over a short time period, allowing the equation to be "split" into separate operators. This greatly shortens solution times (82). The solution sequence is

$$\left.\frac{\partial A}{\partial t}\right|_H = -U \cdot \nabla A + \nabla \cdot K \nabla A = L_H(A) \qquad (18)$$

$$\left.\frac{\partial A}{\partial t}\right|_V = -W\frac{\partial A}{\partial z} + \frac{\partial}{\partial z} K_z \frac{\partial A}{\partial z} = L_V(A) \qquad (19)$$

$$\left.\frac{\partial A}{\partial t}\right|_S = S = L_S(A) \qquad (20)$$

$$\left.\frac{\partial A}{\partial t}\right|_{Total} = (L_H + L_V + L_S)(A) \qquad (21)$$

where L_H is the horizontal transport operator, L_V is the vertical transport operator, and L_S is the operator describing other processes, such as chemistry. Modifications of the splitting process can be used to improve accuracy and computational speed. Besides leading to smaller systems of coupled equations, splitting also allows use of solution techniques that are designed to effectively describe specific processes. For example, in a photochemical air quality model, one routine is used for horizontal transport resulting from bulk winds, another for vertical motions, which are generally diffusive processes, and a third for the chemistry.

Historically, numerical schemes used to calculate the rate of transport have been based on finite difference and, more recently, finite element techniques. Spectral methods have also shown promise, but nonlinear chemical effects tend to degrade solution accuracy in some tests (83,84). A problem with solving the set of equations is that the spatial discretization of the modeling region leads to artificial numerical dispersion, which is manifested by the formation of spurious waves and by pollutant peaks being spread out. For example, Figure 8 shows how various proposed schemes perform when following the rotation of a cone-shaped pollutant puff. In one instance, the original cone is virtually lost as a result of numerical errors. This remains a classic problem in computational fluid mechanics.

Tracking the chemical dynamics, eg, R in the atmospheric diffusion equation (eq. 5) and S (in eq. 20), is particularly difficult and time-consuming. This is because of the wide range of compound lifetimes, or more specifically, the characteristic reaction times. Some species, such as the O atom, have atmospheric lifetimes on the order of microseconds, whereas others last for weeks. It is impractical to use a standard technique, eg, Runge Kutta, having a time-step corresponding to the shortest-lived species. This would be both excessively time-consuming and lead to the accumulation of numerical errors, making the computed solution invalid. Gear methods can be used, but they are computationally slow. Instead, the chemical nature of the process is used to motivate a solution procedure. For a compound experiencing chemical production and destruction, the species concentration can be found using:

$$\frac{\partial c_i}{\partial t} = S_i = A_i - B_i c_i$$

where A_i is the production of species i from chemical reactivity and direct emissions, and $B_i c_i$ is the loss of i from chemical reactivity and deposition. In general, A_i and B_i are time varying and depend on concentrations of other compounds. As it turns out, even though species i might react rapidly, A_i and B_i are rather constant over periods of a minute or so. Thus an approximate analytical solution to the above equation is used. Such techniques have been shown to be accurate and an order of magnitude faster than others (82,85). However, solution of the chemical kinetics is still the most computationally intensive part of a chemically active air quality model, consuming about 85% of the computer time.

Historically, the computational intensity of the more complex chemically active models have limited their application. For example, modeling only the gas-phase dynamics over an area like Los Angeles requires solving about 500,000 simultaneous, nonlinear equations. Until

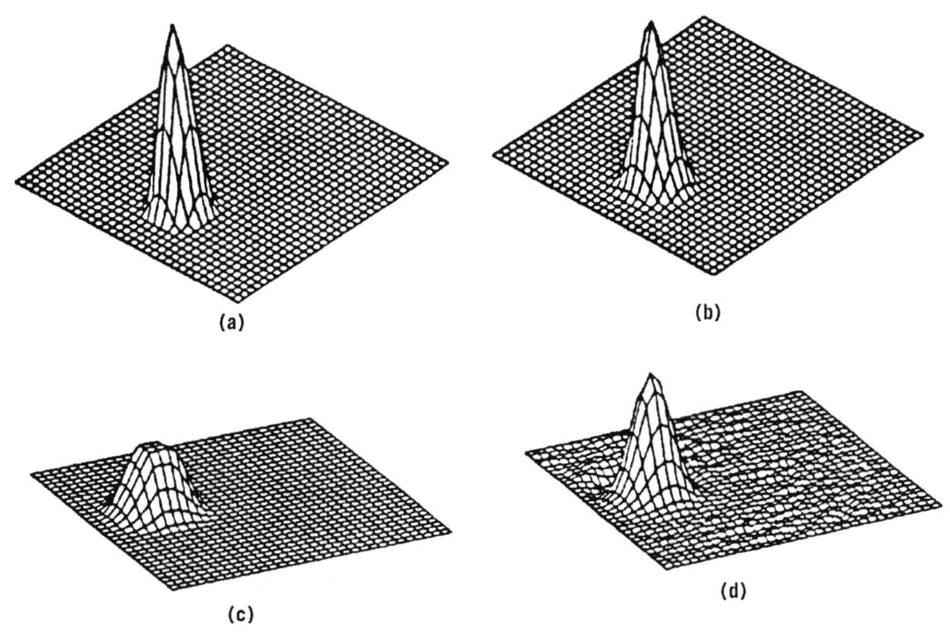

Figure 8. Comparison of various transport schemes for advecting a cone-shaped puff in a rotating windfield after one complete rotation: (**a**) the exact solution, (**b**) obtained by an accurate numerical technique, (**c**) the effect of numerical diffusion where the peak height of the cone has been severely truncated, and (**d**) where the predicted concentration field is bumpy, showing the effects of artificial dispersion. In the case of **d**, spurious waves are formed that impair the predictions, though the peak height is better maintained.

the late 1980s, computational power severely limited the use of chemically active models and particularly inhibited the development and use of regional oxidant and acid deposition models. The rapid increase in computational power is making it possible to address much larger problems, eg, regional and global pollution. One aspect of air quality models is that they are highly parallelizable. The development of parallel computers should allow for significantly more detailed modeling of large systems.

Advances in more efficient numerical algorithms as well as in computational capabilities are facilitating increased chemical and physical detail of atmospheric models and making their use and the interpretation of results more direct. One advance is the use of multiple scales in modeling urban and regional pollutant transport and chemistry simultaneously (Fig. 9). Fine scales are used over cities, and coarse scales over less populated domains. Condensed phase chemistry and dynamics are computationally intensive, though are now being included in model implementation. Visualization of air quality dynamics and synthetic construction of smog allows for quicker interpretation of emission impacts (86,87).

Model Evaluation

One question that always needs to be asked is how well an air quality model is performing, and whether the performance is adequate for the purposes of a particular application. Central to model evaluation is assessment of how closely simulated pollutant concentrations match those observed for some historical episode. As an example of guidelines for this type of performance evaluation, the California Air Resources Board (88) developed a series of tests and performance standards for evaluating the ability of photochemical models to predict ozone concentrations. Beyond model performance, however, is the question of whether the predictions are accurate for the right reason. To address this issue, separate model components such as chemical mechanisms and intermediate modeling steps such as interpolation of windfield data are tested by comparison with results from laboratory experiments or detailed field studies. To test aerosol and ozone models, a large field experiment, the Southern California Air Quality Study (SCAQS) (89) was conducted for the Los Angeles area. The SCAQS illustrated the potential of field studies of similar scope that are under way in the San Joaquin Valley of California (90), the Lake Michigan area (91), the southern United States (92), and Mexico City (93). These programs, using a variety of different models, promise a wealth of information for further developing air quality modeling systems.

Sensitivity and uncertainty analysis are valuable tools for gaining insight into the performance of complex models such as those used for ozone and acid deposition. Analyses in which uncertain parameters or inputs are varied one-at-a-time to observe the effect on the outputs have been applied to most models, if only informally during model development or in diagnosing poor performance. Taking advantage of the simple form of the differential equations that describe reaction kinetics, comprehensive functional analysis techniques that calculate local sensitivities to many parameters simultaneously have been applied to chemical mechanisms used in air quality (94–97) models and have highlighted influential reactions for which additional research might provide substantial benefits. Methods for global uncertainty analysis such as

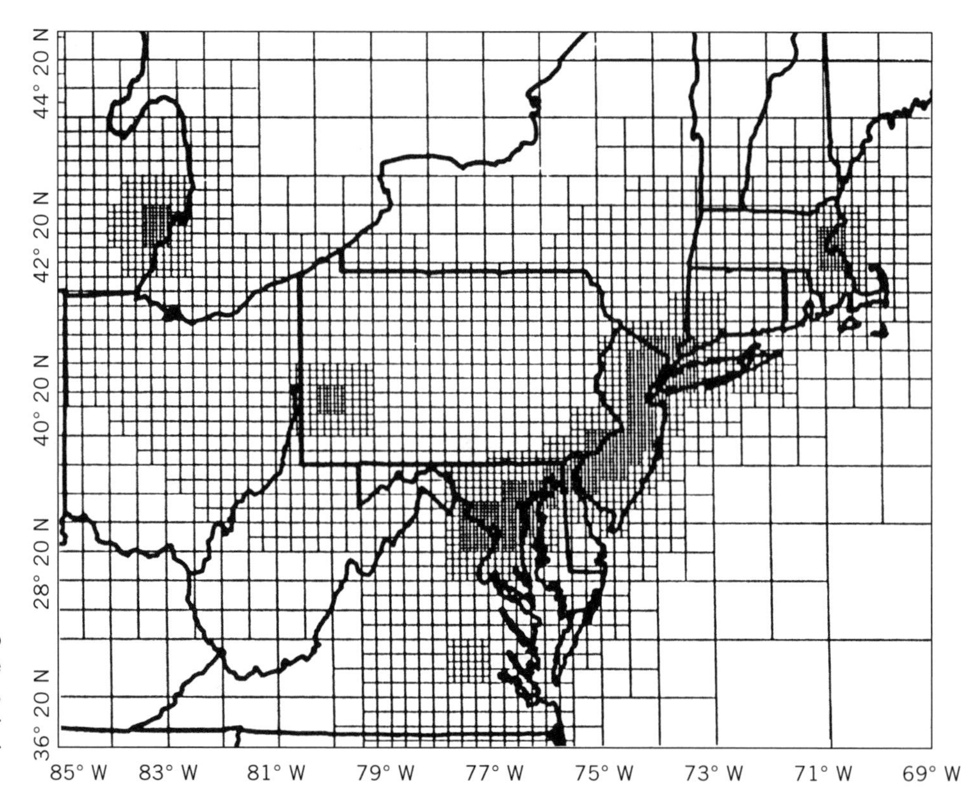

Figure 9. Multiscale grid applied to the northeastern United States. Fine grids are used in dense source regions, larger grids where less activity is present. This allows for more efficient computational implementation.

FAST (98,99) and Monte Carlo, which are essentially schemes for sampling from probability distribution functions for uncertain model parameters, have also been applied to chemical mechanisms and simplified models (100–103). Rapid expansion in computing power has lowered one of the barriers to conducting global uncertainty analyses for complex air quality models. However, development of probability distributions for the many (and potentially correlated) parameters in these models is still a formidable task.

APPLICATION OF AIR QUALITY MODELS

Both receptor and analytical air quality models have proven to be powerful tools for understanding atmospheric pollutant dynamics and for determining the impact of sources on air quality.

Receptor Models

Receptor models, by their formulation, are effective in determining the contributions of various sources to particulate matter concentrations. In classic studies, sources contributing to airborne particle loadings have been identified in Washington, D.C. (104), St. Louis (13,28), Los Angeles (11,16), Portland, Oregon (104), and Boston (105–107) as well as other areas, including the desert (108).

A receptor model (104) and a source-oriented deterministic model were combined as part of a particulate air quality control strategy analysis in Portland, Oregon. Using CMB techniques, source contributions to the ambient aerosol were identified and then dispersion modeling was used to confirm the source contributions. The results obtained with the two models were compared, and a revised particulate emissions inventory was input into the source–dispersion model. Finally, the revised emissions inventory was used in dispersion modeling of emission control strategy alternatives. This approach used the strengths of both types of models. Receptor models are suitable for predicting the outcome of perturbations in some sources but not others. They are, however, good for determining the sources of particulate matter when an accurate emissions inventory is not available. Dispersion models, on the other hand, are well-suited for modeling the impact of a wide variety of emissions changes that would result from changed emission control regulations but rely totally on an input emissions inventory, which may be uncertain or difficult to obtain.

Analytical Modeling Studies

Analytical air quality models have been used the most in modeling the dynamics of pollutants at local and urban scales. Nonreactive, mass-conservation models based on solving equation 3, including Gaussian plume models, have been used extensively for source apportionment, control strategy analysis, and source impact modeling of nonreactive pollutants, such as CO, and of carbonaceous aerosol. Models of this type have also been used to study the sources that contribute to secondary sulfate aerosol formation and they have been used to follow releases of toxic gases in local spills. Variants of the Gaussian plume model are often employed for regulatory purposes. The modest input requirements of these models are the key

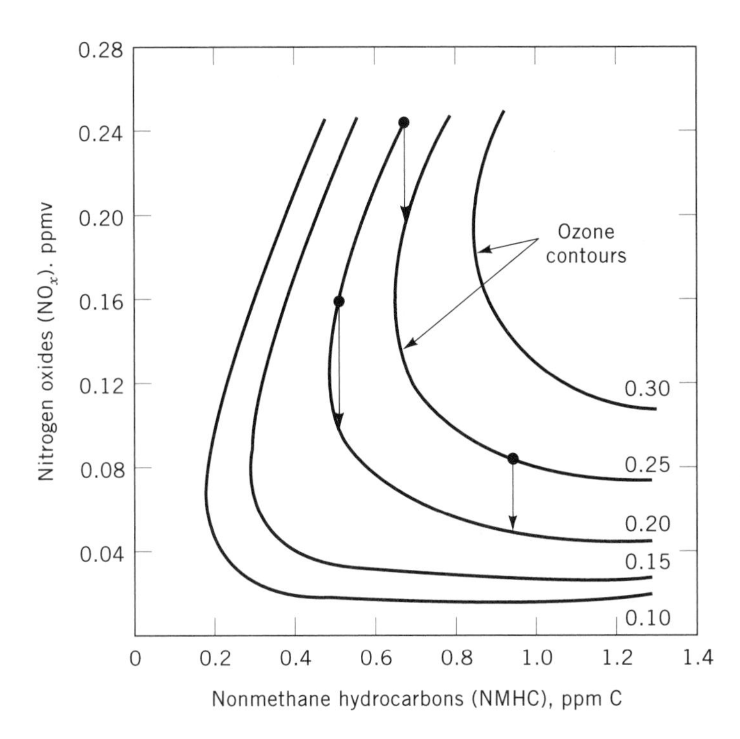

Figure 10. O_3 isopleth diagram showing the response of O_3 concentrations to changes in initial NO_x and nonmethane hydrocarbon concentrations expressed as parts per million of carbon atoms (ppm C). The response to NO_x reductions is dependent on the particular initial concentrations. At point A, the graph indicates that decreasing NO_x would increase O_3 formation, and at point C, a decrease in the NO_x concentration results in a much larger decrease in O_3 concentrations than a similar decrease at point B on the same isopleth.

advantage. Frequently, the goal is to estimate the maximum likely impact, so "worst case" meteorological conditions, ie, conditions that cause local pollutant accumulation, are modeled along with the maximum emissions rate that is allowed at the facility.

Gaussian plume models, and to a lesser extent more comprehensive models, are an integral component for risk assessment of potentially toxic emissions from industrial and commercial sources. Predicted source impacts on air quality are combined with population statistics and chemical toxicity to estimate risk to both the impacted population as a whole as well as highly exposed individuals. This information is used to determine control requirements. More detailed exposure calculations may also involve following personal activity to assess the duration of exposure in various microenvironments (eg, outdoors versus indoors versus automobiles). Important uncertainties in exposure and risk assessment include the accuracy of the predicted ambient air quality, dose–response relationship of the toxic agent, method of contact (inhalation, dermal, ingestion), and variability in personal activity and susceptibility.

Of great interest is the use of chemically active air quality models for studying and controlling urban air pollution. This is because of the high costs of controls, the complexity of the system, and the historic lack of success in reducing ozone levels in the urban areas. Interest in extending the application of chemically active models to the regional scale has heightened because the ozone problem has been recognized as extending well beyond urban areas.

The lack of success in reducing ozone levels can be partially ascribed to the nonlinearity of the system. An incremental change in the emissions of precursors to a secondary pollutant such as ozone need not lead to a proportional change in the pollutant concentration, or any change at all. For example, both NO_x and organics are precursors to the formation of O_3, but increasing the emissions of one can have a different result from increasing those of the other. In fact, decreasing NO_x emissions may increase O_3 concentrations nearby while at the same time decreasing O_3 concentrations downwind. On the other hand, reducing emissions of reactive organic gas (ROGs) can lead to significant decreases in ozone in some cities, such as New York, but have little effect in other cities and in rural areas. A common representation of the relationship between maximum O_3 concentrations and initial concentrations of NO_x and ROGs is an O_3 isopleth diagram, as shown in Figure 10. Regions of both high and low sensitivities to NO_x and ROG emissions are shown. Analysis of the effect of emission controls on O_3 air quality is further complicated by the fact that the effect of controlling two emission sources together is not necessarily equal to the sum of the incremental improvements from controlling each source alone. Thus definitively assessing the impact that a single source has on air quality is clouded in that its impact responds dynamically to changes in other sources and to varying meteorological conditions.

Because of the severity of the smog problem in Los Angeles, this location is the most studied area in regard to the role of NO_x and ROG emissions in the formation of ozone and other photochemical pollutants. The effect of NO_x and ROG emissions on ozone formation was found to vary throughout the region (109). Near downtown Los Angeles, ozone formation is inhibited by increasing NO_x emissions and is effectively reduced by lowering ROG emissions. Downwind, however, where the highest ozone levels are found, ozone is relatively insensitive to ROG emissions and is effectively reduced by lowering NO_x emissions. Modeling studies have also shown that concentrations of associated pollutants, such as nitric acid, aerosol nitrate, NO_2, and peroxyacetyl nitrate (PAN), are lowered when NO_x emissions are controlled (49). Another application of chemically active air quality models has been to study the effectiveness of using alternative automotive fuels, such as methanol, compressed natural gas, and electric power, for reducing smog (110).

Regional oxidant and acid deposition models came into use later than urban photochemical models because of the increased computational intensity, the need to describe more physical and chemical processes, and the later regulatory mandate for development and use. Regional models are similar to urban-scale photochemical models, differing primarily in their horizontal resolution, 20–100 km versus 4–5 km, and in their treatment of cloud processes and liquid-phase chemistry. Applications of regional models also cover longer periods because of the increased residence time of polluted air masses within their domain. For example, applications have followed the evolution of regional pollution episodes of more than 1 week in the northeastern United States, compared with local episodes of 2 or 3 days in Los Angeles.

Regional ozone modeling of the Northeast has shown that air pollution problems across cities within this region are linked (80). Air masses starting in Washington, D.C., can travel up the coast, impacting downwind cities such as Baltimore and New York. Emissions in New York further impact Connecticut and Massachusetts. These studies have shown that the type of controls that are most effective in one city may be counterproductive in another. For example, modeling studies have found NO_x control to be effective for controlling ozone in Boston, Washington, D.C., and Philadelphia, but lead to increased concentrations in New York (80). The horizontal resoluton of the model used for this study was about 20 km, giving some detail in urban areas but not as much as urban-scale models provide.

A variety of models have been developed to study acid deposition. Sulfuric acid is formed relatively slowly in the atmosphere, so its concentrations are believed to be more uniform than ozone, especially in and around cities. Also, the impacts are viewed as more regional in nature. This allows an even coarser horizontal resolution, on the order of 80 to 100 km, to be used in acid deposition models. Atmospheric models of acid deposition have been used to determine where reductions in sulfur dioxide emissions would be most effective. Many of the ecosystems that are most sensitive to damage from acid deposition are located in the northeastern United States and southeastern Canada. Early acid deposition models helped to establish that sulfuric acid and its precursors are transported over long distances, eg, from the Ohio River Valley to New England (111–113). Models have also been used to show that sulfuric acid deposition is nearly linear in response to changing levels of emissions of sulfur dioxide (114).

In the mid-1980s, the destruction of the ozone layer above the Antarctic was recognized to be a potential environmental disaster, and chemically active transport models were used to identify the most important processes leading to depleted stratospheric ozone levels. As with ground-level chemistry, the gas-phase chemistry that occurs in the stratosphere is relatively well understood, but the incorporation of heterogeneous chemistry is an ongoing challenge. In addition, the models that have incorporated stratospheric chemistry in the greatest detail have oversimplified the transport and dynamics of stratospheric circulations. Conversely, three-dimensional models that focus on the dynamics of the atmosphere have used limited treatments of stratospheric chemistry.

Zero- through three-dimensional models have been used to simulate the stratospheric ozone problem. Zero-dimensional models focus on radiative transfer and chemistry at a single point. One-dimensional models incorporate vertical diffusive transport as well. Two-dimensional models currently represent the best compromise between computational tractability and chemical detail (115) and are resolved by latitude as well as in the vertical dimension. Stronger flows and greater homogeneity over changes in longitude make two-dimensional treatments reasonable, except in the polar regions. Three-dimensional models are needed to study the dynamics of polar regions, including exchanges with air at lower latitudes.

Meridional circulation in two-dimensional stratospheric models has been specified based on observations or general circulation model calculations; recently, efforts have been undertaken to calculate circulations from first principles, within the stratospheric models themselves. An important limitation of using models in which circulations are specified is that these cannot be used to study the feedbacks of changing atmospheric composition and temperature on transport, factors that may be important as atmospheric composition is increasingly perturbed.

The key gas-phase reactions occurring in the stratosphere are generally known. Comprehensive reviews of kinetic data have led to general consensus on the rate parameters that should be used in stratospheric models (116). Nevertheless, discrepancies are still apparent when the chemical components of different stratospheric models are compared. Disagreeements arise from differing estimates of photolysis conditions in the stratosphere, depending in part on whether or not light scattering in the troposphere is included in the model (116).

Heterogeneous chemistry occurring on polar stratospheric cloud particles of ice and nitric acid trihydrate has been established as a dominant factor in the aggravated seasonal depletion of ozone observed to occur over Antarctica. Preliminary attempts have been made to parameterize this chemistry and incorporate it in models to study ozone depletion over the poles (116) as well as the potential role of sulfate particles throughout the stratosphere (117).

Models can be used to study human exposure to air pollutants and to identify cost-effective control strategies. In many instances, the primary limitation on the accuracy of model results is not the model formulation but the accuracy of the available input data (118). Another limitation is the inability of models to account for the alterations in the spatial distribution of emissions that occurs when controls are applied. The more detailed models are currently able to describe the dynamics of unreactive pollutants in urban areas.

Because of the expanded scale and need to describe additional physical and chemical processes, the development of acid deposition and regional oxidant models has lagged behind that of urban-scale photochemical models. An additional step up in scale and complexity, the development of analytical models of pollutant dynamics in the stratosphere is also behind that of ground-level oxidant models, in part because of the central role of heterogeneous chemistry in the stratospheric ozone depletion problem. In general, atmospheric liquid-phase chemistry and especially heterogeneous chemistry are less well understood than gas-phase reactions such as those that dominate the formation of ozone in urban areas. Development of three-dimensional models that treat both the dynamics and chemistry of the stratosphere in detail is an ongoing research problem.

BIBLIOGRAPHY

1. K. L. Demerjian, "Photochemical Diffusion Models for Air Quality Simulation: Current Status," in *Proceedings of the Conference on the State of the Art of Assessing Transportation-Related Air Quality Impacts*, Oct. 22–24, 1975, Transportation Research Board, Washington, D.C., 1975.
2. P. Zanetti, *Air Pollution Modeling*, Van Nostrand Reinhold, New York, 1990.
3. D. S. Barth, *J. Air Pollut. Control Assoc.* **20,** 519–523 (1970).
4. *Final Air Quality Management Plan*, 1982 rev., South Coast Air Quality Management District and Southern California Association of Governments, El Monte, Calif., 1982.
5. M. Das, *Multivariate Analysis of Atlanta Air Quality Data*, University of Connecticut, Mar. 1993.
6. D. P. Chock, T. R. Terrell, and S. B. Levitt, *Atmos. Environ.* **9,** 978–989 (1975).
7. G. Fronza, A. Spirito, and A. Tonielli, *Appl. Math. Model.* **3,** 409–415 (1979).
8. P. Bacci, P. Bolzern, and G. Fronza, *J. Appl. Meterol.* **20,** 121–129 (1981).
9. S. K. Friedlander, *Environ. Sci. Technol.* **7,** 235–240 (1973).
10. S. L. Heisler, S. K. Friedlander, and R. B. Husar, *Atmos. Environ.* **7,** 633–649 (1973).
11. G. Gartrell and S. K. Friedlander, *Atmos. Environ.* **9,** 279–299 (1975).
12. D. F. Gatz, *Atmos. Environ.* **9,** 1–18 (1975).
13. D. F. Gatz, *J. Appl. Meteorol.* **17,** 600–608 (1978).
14. G. E. Gordon, *Environ. Sci. Technol.* **14,** 792–800 (1980).
15. J. G. Watson, R. C. Henry, J. A. Cooper, and E. S. Macias in E. S. Macias and P. K. Hopke, eds., *Atmospheric Aerosol: Source/Air Quality Relationships*, Symp. Ser. No. 167, American Chemical Society, Washington, D.C., 1981, pp. 89–106.
16. G. R. Cass and G. J. McRae, *Environ. Sci. Technol.* **17,** 129–139 (1983).
17. J. G. Watson, *J. Air Pollut. Control Assoc.* **34,** 619–623 (1984).
18. P. K. Hopke, *Receptor Modeling in Environmental Chemistry*, John Wiley & Sons, Inc., New York, 1985.
19. R. J. Yamartino in S. C. Dattner and P. K. Hopke, eds., *Receptor Models Applied to Contemporary Pollution Problems*, Air Pollution Control Association, Pittsburgh, Pa., 1983, pp. 285–295.
20. P. F. Nelson, S. M. Quigley, and M. Y. Smith, *Atmos. Environ.* **17,** 439–449 (1983).
21. W. J. O'Shea and P. A. Scheff, *JAPCA* **38,** 1020–1026 (1988).
22. P. F. Aronian, P. A. Scheff, and R. A. Wadden, *Atmos. Environ.* **23,** 911–920 (1989).
23. J. E. Core, P. L. Hanrahan, and J. A. Cooper, in Ref. 11, pp. 107–123.
24. M. S. Miller, S. K. Friedlander, and G. M. Hidy, *J. Colloid Interface Sci.* **39,** 165–176 (1972).
25. J. A. Cooper and J. G. Watson, *J. Air Pollut. Control Assoc.* **30,** 1116–1125 (1980).
26. S. K. Friedlander, *Smoke, Dust and Haze*, Wiley-Interscience, New York, 1977.
27. R. C. Henry and G. M. Hidy, *Atmos. Environ.* **16,** 929–943 (1982).
28. P. K. Hopke in Ref. 15, pp. 21–49.
29. R. G. Lamb and J. H. Seinfeld, *Environ. Sci. Technol.* **7,** 253–261 (1973).
30. R. G. Lamb, *Atmos. Environ.* **7,** 257–263 (1973).
31. J. H. Seinfeld, *Atmospheric Chemistry and Physics of Air Pollution*, Wiley-Interscience, New York, 1986.
32. R. G. Lamb and M. Neiburger, *Atmos. Environ.* **5,** 239–264 (1971).
33. G. R. Cass, *Atmos. Environ.* **15,** 1227–1249 (1981).
34. M.-K. Liu and J. H. Seinfeld, *Atmos. Environ.* **9,** 555–574 (1975).
35. F. A. Gifford, *Nucl. Saf.* **2,** 47–57 (1961).
36. D. B. Turner, *J. Appl. Meteorol.* **3,** 83–91 (1964).
37. D. B. Turner, *A Workbook of Atmospheric Dispersion Estimates*, Public Health Service Publication No. 999-AP-26, U.S. Government Printing Office, Washington, D.C., 1967.
38. S. D. Reynolds, P. W. Roth, and J. H. Seinfeld, *Atmos. Environ.* **7,** 1033–1061 (1973).
39. A. C. Lloyd and co-workers, *Development of the ELSTAR Photochemical Air Quality Simulation Model and Its Evaluation Relative to the LARPP Data Base*, Environmental Research and Technology Report, No. P-5287-500, West Lake Village, Calif., 1979.
40. S. D. Reynolds, T. W. Tesche, and L. E. Reid, *An Introduction to the SAI Airshed Model and Its Usage*, Report No. SAI-EF79–31, Systems Applications, Inc., San Rafael, Calif., 1979.
41. J. H. Seinfeld, *J. Air Waste Management* **38,** 616–645 (1988).
42. D. P. Chock, A. M. Dunker, S. Kumar, and C. S. Sloane, *Environ. Sci. Technol.* **15,** 933–939 (1981).
43. G. R. Carmichael, T. Kitada, and L. K. Peters, *Atmos. Environ.* **20,** 173–188 (1986).
44. G. R. Carmichael, L. K. Peters, and R. D. Saylor, *Atmos. Environ.* **25A,** 2077–2090 (1991).
45. A. G. Russell and G. R. Cass, *Atmos. Environ.* **20,** 2011–2025 (1986).
46. J. G. Chang and co-workers, *J. Geophys. Res.* **92,** 14,681–14,700 (1987).
47. S. A. McKeen and co-workers, *J. Geophys. Res.* **96,** 10,809–10,845 (1991).
48. G. J. McRae and J. H. Seinfeld, *Atmos. Environ.* **17,** 501–522 (1983).
49. A. G. Russell, K. McCue, and G. R. Cass, *Environ. Sci. Technol.* **22,** 1336–1347 (1988).
50. T. W. Yu, *Mon. Weather Rev.* **105,** 57–66 (1977).
51. B. B. Hicks and M. L. Wesely, *Boundary-Layer Meteorol.* **20,** 175–185 (1981).
52. B. B. Hicks and co-workers in H. R. Pruppacher, R. G. Semonin, and W. G. N. Slinn, eds., *Precipitation Scavenging,*

Dry Deposition and Resuspension, Elsevier, New York, 1983, pp. 933–942.

53. C. J. Walcek and co-workers, *Atmos. Environ.* **20,** 949–964 (1986).
54. M. L. Wesely, D. R. Cook, R. L. Hart, and R. E. Speer, *J. Geophys. Res.* **90,** 2131–2143 (1985).
55. D. M. Whelpdale and R. W. Shaw, *Tellus* **26,** 196–205 (1974).
56. M. L. Wesely and B. B. Hicks, *J. Air Pollut. Control Assoc.* **27,** 1110–1116 (1977).
57. D. Fowler, *Atmos. Environ.* **12,** 369–373 (1978).
58. B. J. Huebert in Ref. 52, pp. 785–794.
59. B. J. Huebert and C. H. Robert, *J. Geophys. Res.* **90,** 2085–2090 (1985).
60. G. J. McRae and A. G. Russell in B. B. Hicks, ed., *Deposition Both Wet and Dry*, Acid Precipitation Series, Vol. 4, Butterworth, Boston, 1983, pp. 153–193.
61. A. Martin, *Atmos. Environ.* **18,** 1955–1961 (1984).
62. T. E. Graedel, L. A. Farrow, and T. A. Weber, *Atmos. Environ.* **10,** 1095–1116 (1976).
63. M. C. Dodge, *Combined Use of Modeling Techniques and Smog Chamber Data to Derive Ozone-Precursor Relationships*, Report No. EPA-600/3-77-001a, U.S. Environmental Protection Agency, Research Triangle Park, N.C., 1977.
64. A. H. Falls and J. H. Seinfeld, *Environ. Sci. Technol.* **12,** 1398–1406 (1978).
65. R. Atkinson, A. C. Lloyd, and L. Winges, *Atmos. Environ.* **16,** 1341–1355 (1982).
66. W. R. Stockwell, *Atmos. Environ.* **20,** 1615–1632 (1986).
67. F. W. Lurmann, W. P. L. Carter, and L. A. Coyner, *A Surrogate Species Chemical Reaction Mechanism for Urban-Scale Air Quality Simulation Models*, Report No. EPA/600/3-87-014, U.S. Environmental Protection Agency, Research Triangle Park, N.C., 1987.
68. J. P. Killus and G. Z. Whitten, *A New Carbon Bond Mechanism for Air Quality Modeling*, Report No. EPA 60013-82-041. U.S. Environmental Protection Agency, Research Triangle Park, N.C., 1982.
69. G. Z. Whitten and co-workers, *Modeling of Simulated Photochemical Smog with Kinetic Mechanism*, Vols. I and II, Report No. EPA-600/3-79-001a, U.S. Environmental Protection Agency, Research Triangle Park, N.C., 1979.
70. M. W. Gery, G. Z. Whitten, J. P. Killus, and M. C. Dodge, *J. Geophys. Res.* **94,** 12,925–12,956 (1989).
71. E. Weber, *Air Pollution, Assessment Methodology and Modeling*, Plenum Press, New York, 1982.
72. M. H. Dickerson, *J. Appl. Meteorol.* **17,** 241–253 (1978).
73. W. R. Goodin, G. J. McRae, and J. H. Seinfeld, *J. Appl. Meteorol.* **19,** 98–108 (1979).
74. G. Freuhauf, P. Halpern, and P. Lester, *Environ. Software* **3,** 72–80 (1988).
75. S. G. Douglas and R. C. Kessler, *User's Guide to te Diagnostic Wind Model (Version 1.0)*, Systems Applications, Inc., San Rafael, Calif., 1988.
76. R. A. Pielke, *Mesoscale Meteorological Modeling*, Academic Press, Orlando, Fla., 1984.
77. D. R. Stauffer and N. L. Seaman, *Mon. Weather Rev.* **118,** 1250–1277 (1990).
78. D. Zimmerman and co-workers, *Anthropogenic Emissions Data for the 1985 NAPAP Inventory*, Report No. EPA/600/7-88/022, U.S. Environmental Protection Agency, Research Triangle Park, N.C., Nov. 1988.
79. *User's Guide to MOBILE 4*, Report No. EPA-AA-TEB-89-01, Ann Arbor, Mich., U.S. Environmental Protection Agency, 1989.
80. N. C. Possiel, L. B. Milich, and B. R. Goodrich, eds., *Regional Ozone Modeling for Northeast Transport (ROMNET)*, Report No. EPA-450/4-91-002a, U.S. Environmental Protection Agency, Research Triangle Park, N.C., 1991.
81. M. Amann, *Atmos. Environ.* **24A,** 2759–2765 (1990).
82. G. J. McRae, *Mathematical Modeling of Photochemical Air Pollution*, Ph.D. dissertation, California Institute of Technology, Pasadena, Calif., 1981.
83. O. Hov and co-workers, *Atmos. Environ.* **23,** 967–983 (1990).
84. D. P. Chock, *Atmos. Environ.* **19,** 571–586 (1985).
85. M. T. Odman, N. Kumar, and A. G. Russell, *Atmos. Environ.*, in press.
86. G. J. McRae and A. G. Russell, *J. Comp. Phys.* **4,** 227–233 (1990).
87. Eldering and co-workers, *Environ. Sci. Technol.* 626–636 (1993).
88. *Technical Guidance Document: Photochemical Modeling*, Sacramento, Calif., California Air Resources Board, Aug. 1990.
89. D. R. Lawson, *J. Air Waste Manage. Assoc.* **40,** 156–165 (1990).
90. A. J. Ranzieri and R. Thullier, *San Joaquin Valley Air Quality Study (SJVAQS) and Atmospheric Utility Signatures, Predictions and Experiments (AUSPEX): A Collaborative Modeling Program*, paper presented at the 84th Annual Meeting of the Air and Waste Management Association, Vancouver, B.C., June 1991.
91. M. Koerber and N. Bowne, *Preliminary Findings of the Lake Michigan Ozone Study*, paper presented at the symposium Tropospheric Ozone and the Environment II, Atlanta, Ga., Nov. 1991.
92. W. L. Chameides, *Southern Ozone Attainment Strategies*, paper presented at the symposium Tropospheric Ozone and the Environment II, Atlanta, Ga., Nov. 1991.
93. M. D. Williams and co-workers, *Development and Testing of an Air Quality Model for Mexico City*, paper presented at the Clean Air Congress, Montreal, June 1992.
94. A. M. Dunker, *J. Chem. Phys.* **81,** 2385–2393 (1984).
95. M. Dermilap and H. Rabitz, *J. Chem. Phys.* **74,** 3362–3375 (1981).
96. S. N. Pandis and J. H. Seinfeld, *J. Geophys. Res.* **94,** 1105–1126 (1989).
97. J. B. Milford, D. Gao, A. G. Russell, and G. J. McRae, *Environ. Sci. Technol.* **26,** 1179–1189 (1992).
98. A. H. Falls, G. J. McRae, and J. H. Seinfeld, *Int. J. Chem. Kin.* **11,** 1137–1162 (1979).
99. G. J. McRae, J. W. Tilden, and J. H. Seinfeld, *Comput. Chem. Eng.* **6,** 15–25 (1982).
100. R. S. Stolarski, D. M. Butler, and R. D. Rundel, *J. Geophys. Res.* **83,** 3074–3078 (1978).
101. D. H. Ehhalt, J. S. Chang, and D. M. Butler, *J. Geophys. Res.* **84,** 7889–7894 (1979).
102. R. Derwent and O. Hov, *J. Geophys. Res.* **93,** 5185–5199 (1988).
103. A. E. Thompson and R. W. Stewart, *J. Geophys. Res.* **96,** 13089–13108 (1991).
104. G. E. Gordon, W. H. Zoller, G. S. Kowalczyk, and S. W. Rheingrover in Ref. 15, pp. 51–74.

105. P. K. Hopke and co-workers, *Atmos. Environ.* **10,** 1015–1025 (1976).
106. D. J. Alpert and P. K. Hopke, *Atmos. Environ.* **14,** 1137–1146 (1980).
107. *Ibid.*, **15,** 675–687 (1981).
108. P. D. Gaarenstroom, S. P. Perone, and J. L. Moyers, *Environ. Sci. Technol.* **11,** 795–800 (1977).
109. J. B. Milford, A. G. Russell, and G. J. McRae, *Environ. Sci. Technol.* **23,** 1290–1301 (1989).
110. T. Y. Chang, R. H. Hammerle, S. M. Japan, and I. T. Salmeen, *Environ. Sci. Technol.* **25,** 1190–1197 (1991).
111. H. Rodhe, P. Crutzen, and A. Vanderpol, *Tellus* **33,** 132–141 (1981).
112. A. Eliassen, *J. Appl. Meteor.* **19,** 231–240 (1980).
113. *U.S.–Canada Memorandum of Intent on Transboundary Air Pollution, Atmospheric Sciences and Analysis Work Group 2 Phase III Final Report*, U.S. Environmental Protection Agency, Washington, D.C., 1983.
114. J. S. Chang and co-workers, *The Regional Acid Deposition Model and Engineering Model*, State-of-Science–Technology Report 4, National Acid Precipitation Assessment Program, Washington, D.C., 1989.
115. *Scientific Assessment of Stratospheric Ozone: 1989*, United Nations Environment Program and World Meteorological Organization, New York, 1989.
116. W. B. DeMore and co-workers, *Chemical Kinetics and Photochemical Data for Use in Stratospheric Ozone Modeling*, Evaluation No. 8, JPL Publ. 87–41, Jet Propulsion Laboratory, Pasadena, Calif., 1987.
117. D. J. Hoffmann and S. Solomon, *J. Geophys. Res.* **94,** 5029 (1989).
118. W. R. Pierson, A. W. Gertler, and R. L. Bradow, *J. Air Waste Manage. Assoc.* **40,** 1495–1504 (1990).

AIRCRAFT ENGINES

FREDRIC EHRICH
DONALD W. BAHR
PHILIP R. GLIEBE
GE Aircraft Engines

AIRCRAFT ENGINE PERFORMANCE

The operation of an aircraft engine may be viewed as consisting of two related functions: (*1*) that of the prime mover, generally a gas turbine, which converts the energy derived from combustion of a liquid hydrocarbon fuel to available mechanical energy in the form of a high pressure, high temperature airstream; and (*2*) that of the propulsor, which harnesses that available energy to generate a thrust. This thrust, in turn, propels the aircraft by accelerating the air passing through the propulsor's components to an exhaust velocity higher than the aircraft's flight velocity. (See also AFTERBURNING; AIRCRAFT FUELS; GAS TURBINES; KEROSENE; PETROLEUM PRODUCTS AND USES; THERMODYNAMICS.)

Since the ultimate function of an aircraft engine is to supply the thrust for aircraft propulsion or lift, an engine's performance is most conveniently reported as the net thrust force that the engine supplies. But in certain types of engines, such as turboshaft engines and turboprop engines, the propulsor, ie, the thrust-producing device, eg, the rotor of a helicopter or the propeller of a propeller-driven aircraft, is treated as a separate system. In cases such as these, the aircraft engine's performance is reported as the power provided by the engine to drive the propulsor rather than directly, ie, as the thrust provided by the total propulsive system. In these cases, any residual thrust from the engine exhaust is either reported separately or converted to an equivalent amount of power. For turbojet engines, however, where the thrust is supplied by reaction forces in the basic power unit, or for turbofan engines, where there is a propulsor integrated into the engine, the net thrust of the engine is reported directly, without reference to the power generated within the engine.

Thrust selection is categorized by various thrust or power ratings. These may include special high ratings whose usage is limited to short durations or specified extraordinary situations. Examples are Maximum Contingency or Automatic Power Reserve (APR) or, for helicopters, One Engine Inoperative (OEI), invoked when one engine becomes inoperative during take-off in a multiengine aircraft. Use of these ratings may entail engine inspection or overhaul follow-up. Another category of high ratings, such as Take-off, Maximum Climb, and Military, Combat, or Intermediate in a military aircraft, may be limited to a specified duration or cumulative usage in order to conserve the life of the engine. Still other ratings, eg, Maximum Cruise, Normal Rated, Maximum Continuous, and Cruise, are permitted for extended use and are designed to afford long engine life and minimum fuel consumption. Additional ratings, termed Flight Idle and Ground Idle, are specified to accommodate operation where power or thrust and fuel consumption are to be minimized while maintaining the engine in a self-sustaining condition, available for rapid application of power.

Other reported operating parameters may be starting time, acceleration time, engine rotative speed (or speeds for multirotor engines), and pressures and temperatures within the engine that may be monitored. Engine operation is affected by inlet and exhaust system losses, ambient humidity, power extracted from the engine to run aircraft accessories, and compressed air extracted from the engine for use in aircraft systems.

Ambient pressure and temperature and aircraft flight speed also have significant effects on aircraft engine performance. As shown in Figure 1, average ambient pressure and temperature vary considerably with altitude. In addition to altitude variations, further extremes are experienced as a function of latitude, from the tropics to the polar regions, as a function of time of year, ie, from summer to winter, and in fluctuations from day to day. Decreasing ambient pressure at a given altitude is accompanied by proportionate reductions in air density. Since engine geometry generally accommodates a fixed volumetric flow of air, the reduced density implies a reduction in mass-flow of air and a consequent reduction in the power produced. An increase in ambient temperature also reduces air density, but more importantly, high temperature of the inlet air has a magnified effect in increasing the temperature of the compressed air, thereby limiting

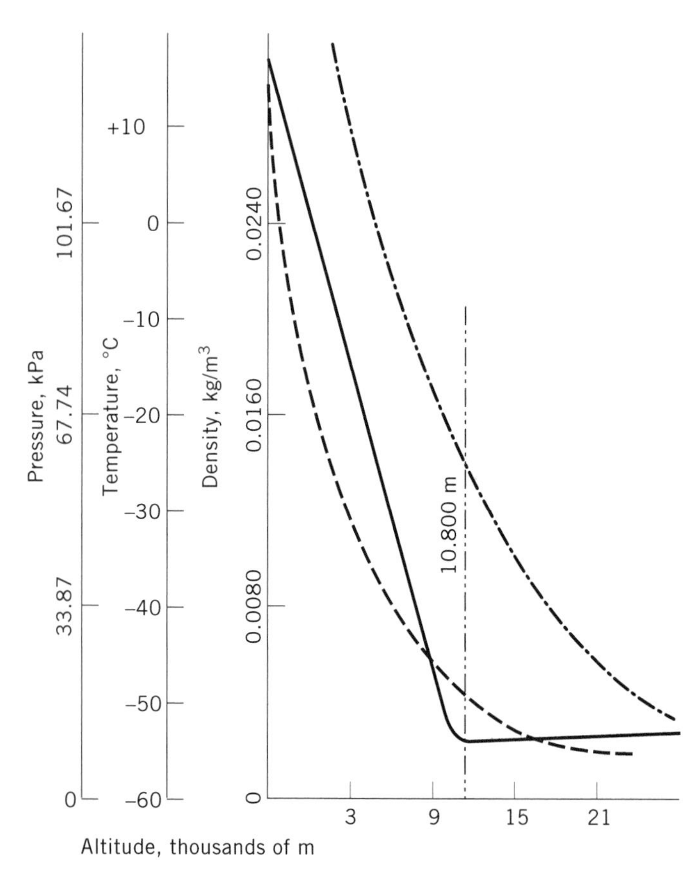

Figure 1. Variation of ambient pressure and temperature as a function of altitude (1); (− − − −), pressure; (——), temperature; and (−·−·), density. To convert kPa to in. Hg, multiply by 0.295.

the amount of fuel which may be introduced without exceeding the limiting temperature of the turbine. Consequently, often an aircraft engine size must be specified to accommodate the limiting condition of achieving the required take-off power at the hottest day anticipated at the highest altitude airport that might be encountered. Otherwise, an aircraft must have its cargo or passenger capacity restricted when these adverse conditions are encountered.

Owing to its inherent mechanical, thermodynamic, and aerodynamic limitations, an aircraft engine will generally be limited to operation over a "flight envelope," most often specified as the combinations of altitude and flight speed that are available for operation, as shown in Figure 2. For a given engine type of a given detailed design, it is then possible to assemble a comprehensive prediction of the engine's performance, ie, its power or thrust and its fuel consumption, over the entire flight envelope, using a combination of data obtained from tests of the individual engine components; tests of the engine itself in a test cell which has the facility for simulating different flight speeds and ambient conditions; data from similar components or engines which may be scaled or otherwise modified to reflect the specific design being analyzed; and theoretical or analytical simulations of engine components. An excerpt from a typical display of the performance of a turbojet engine is shown in Figure 3. Such performance summaries, more often in computerized format, are essential

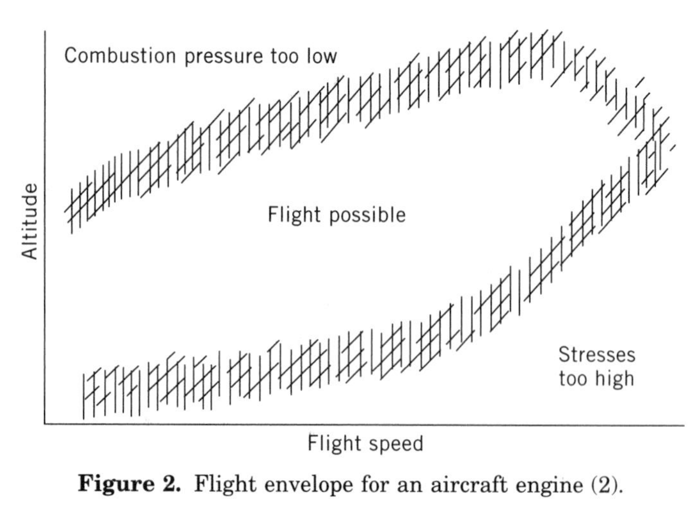

Figure 2. Flight envelope for an aircraft engine (2).

elements in the design of aircraft, the prediction of the aircraft's performance, and the specification of the aircraft's best flight plan to accomplish its specified mission. A more extensive treatment of the subject is available to the reader (4–6).

The Power Producer

The power producer or prime mover of virtually all jet engines (sometimes referred to, variously, as the core, gas

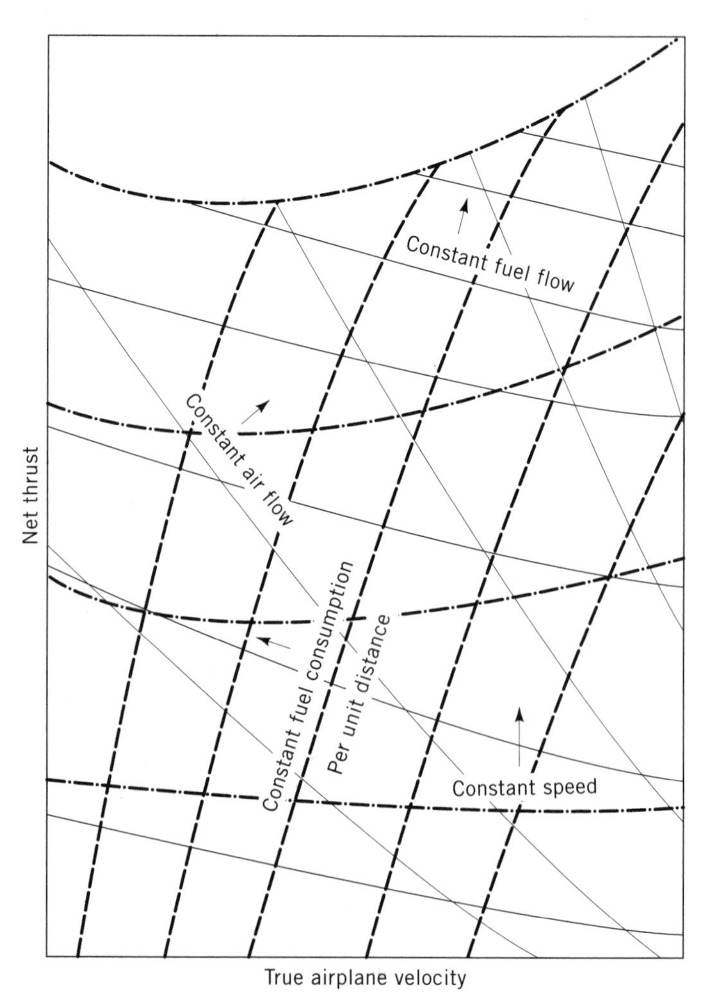

Figure 3. Typical performance chart for a turbojet engine (3).

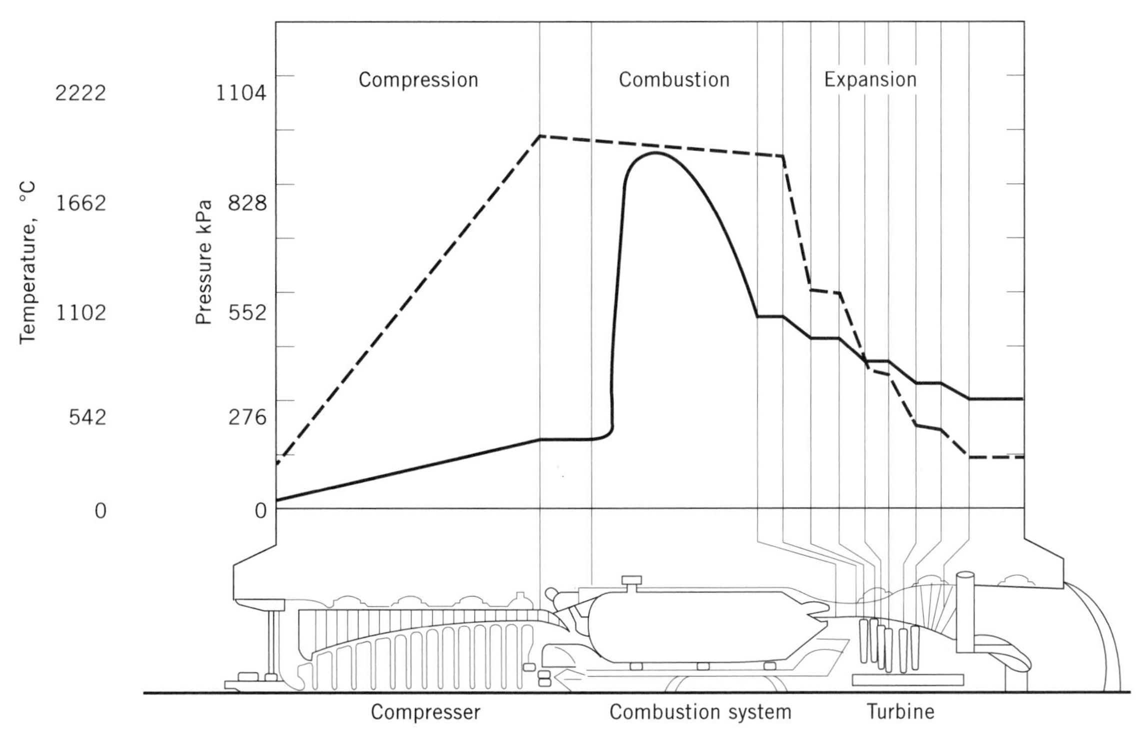

Figure 4. Cross-section of a turbojet engine and, above, graph of typical operating conditions of its working fluid (1); (– – – – –), total pressure; (——), temperature. To convert kPa to psi, multiply by 0.145.

producer, gasifier, or gas generator) is a gas turbine operating on the Brayton thermodynamic cycle, in which the working fluid is a continuous flow of air ingested into engine's inlet. As shown in Figure 4, the air is first compressed by a turbo-compressor to a pressure ratio of, typically, 10 to 40 times the pressure of the inlet air stream. It then flows into a combustion chamber where a steady stream of a hydrocarbon fuel, in liquid spray droplet or vapor form, is introduced and burned at approximately constant pressure. The continuous stream of high pressure products of combustion, whose average temperature may be typically from 980°C to 1540°C or higher, then flows through a turbine which is linked by a torque shaft to the compressor and which extracts energy from the airstream to drive the compressor. Because heat has been added to the working fluid at high pressure, the gas stream which exits from the gas generator after having been expanded through the turbine contains, by virtue of its high pressure, high temperature, and high velocity, a considerable amount of surplus energy or gas horsepower which may be harnessed for propulsion purposes.

Although this gas-turbine cycle is similar to that of a reciprocating engine which also compresses, heats, and expands the air flowing through it, the gas turbine differs in two key respects. First, the gas turbine deals with a continuous, steady flow of air, while the reciprocating engine deals with a series of discrete batches of air drawn periodically into each of the cylinders. Secondly, the gas turbine combustion process takes place at essentially constant pressure as the working fluid flows through the combustion chamber after having passed through the compressor (ie, the Brayton cycle). In contrast, combustion in the reciprocating engine takes place at essentially constant volume, while a batch of the working fluid is trapped in the cylinder after having been compressed (ie, the Otto cycle).

Thrust or power selection is usually accomplished by manipulating the fuel flow to the engine. The power producer will respond to a reduction in fuel flow by reducing its rotative speed, which reduces the airflow into the engine and the power produced. In a gas turbine engine, this reduction in rotative speed will be accompanied by a reduction in pressure ratio produced by the compressor, and typically for a turboshaft engine a considerable increase in specific fuel consumption (SFC) at low power, as shown in Figure 5. On the other hand, the compression ratio built into the cylinder and piston of a reciprocating engine does not vary with the rotative speed of the engine, and it does not experience the steep increase in SFC experienced by the gas turbine at low power.

Specific Fuel Consumption of the Power Producer

The heat released by burning a typical jet fuel in air is approximately 43,370 kJ/kg of fuel. If this process were 100% efficient, it would then produce a gas horsepower of 12.0 kW/(kg/h) for every unit of fuel flow. In actual fact, certain practical thermodynamic limitations, which are a function of the peak gas temperature achieved in the cycle, restrict the efficiency of the process to about 40% of this ideal value. The peak pressure achieved in the cycle also affects the efficiency of energy generation. This then implies that the lower bound of SFC for an engine producing gas horsepower is 0.207 (kg/h)/kW. In actual practice,

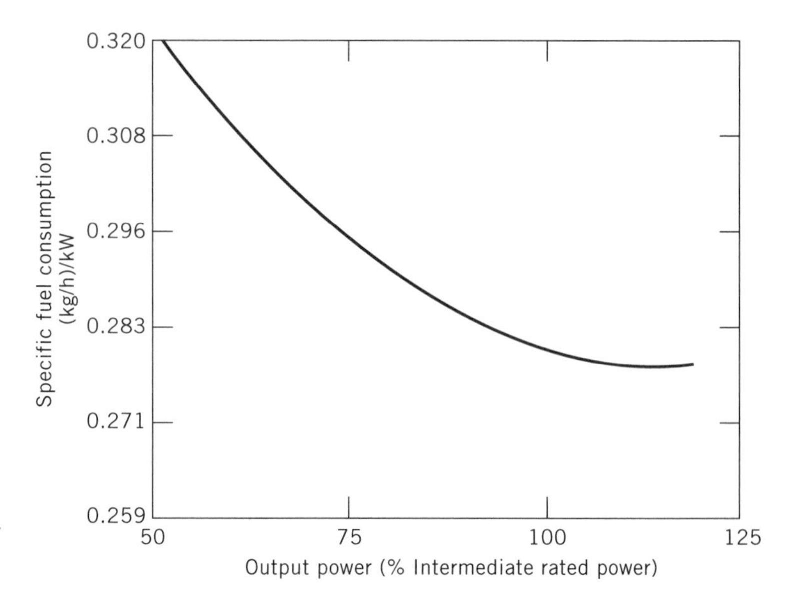

Figure 5. Part-power performance of the gas turbine power producer (7).

the specific fuel consumption will be even higher than this lower bound because of less than ideal performance in the individual components of the prime mover, including leakage of high pressure working fluid (air or products of combustion); leakage of heat from the high temperature working fluid to the lower temperature engine bay; pressure losses in the passages through which the working fluid moves from component to component; diversion of a portion of the air to cool high temperature parts; and inefficiencies in the compression, combustion, and expansion components of the engine.

As indicated by Figure 6, high cycle peak temperatures and pressure ratios are of considerable importance in achieving low SFC. Modern gas turbine power producer sections, particularly where they are used in subsonic aircraft that do not benefit from the large ram pressure ratios experienced at supersonic flight speeds, may have cycle pressure ratios of 40 : 1 or more, compared to reciprocating engines having pressure ratios of 10 : 1 or 11 : 1, necessitating the exercise of considerable ingenuity and skill in the design of the compressor.

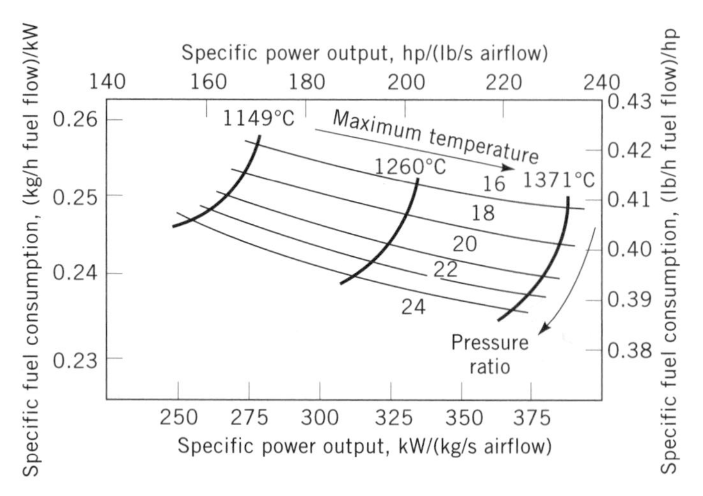

Figure 6. The influence of cycle pressure ratio and peak cycle temperature for the gas turbine power producer (7).

Specific Power

Since weight and volume are at a premium in the overall design of an aircraft, and since the powerplant represents a large fraction of any aircraft's total weight and volume, these parameters must be minimized in the course of the engine's design. The airflow that passes through an engine is a very representative measure of the engine's cross-section area, and hence its weight and volume. Therefore, an important figure of merit for the prime mover is its specific power, ie, the amount of power that it generates per unit of airflow. This quantity is a very strong function of the peak gas temperature in the core at the discharge of the combustion chamber, but increasing cycle pressure ratio has a slightly adverse effect on specific power, as shown in Figure 6. Modern engines generate from 247 to 411 kW/(kg/s).

The Propulsor

The gas horsepower generated by the prime mover is then used to drive the propulsor, which in turn produces thrust for propelling or lifting the aircraft. The principle upon which that thrust is produced is based on Newton's Second Law of Dynamics, which generalizes the observation that the force (F) required to accelerate a discrete mass (m) is proportional to the product of that mass and the acceleration (a):

$$F = ma = wa/g$$

where the mass is taken as the weight of the object divided by the acceleration due to gravity at the place where the object was weighed. In the case of a jet engine, what is generally being dealt with is the acceleration of a steady stream of air rather than a discrete mass. In this case the equivalent statement of the Second Law of Dynamics is that the force (F) required to increase the velocity of a stream of fluid is proportional to the product of the rate of mass flow (M) of the stream and the change in velocity of the stream, ie,

$$F = M(V_j - V_0) = W(V_j - V_0)/g$$

where the inlet velocity (V_0) relative to the engine is taken to be the flight velocity, and the discharge velocity (V_j) is the exhaust or jet velocity relative to the engine. W is the rate of weight flow of working fluid, ie, air or products of combustion, divided by the acceleration of gravity in the place where the weight flow is measured. (The relatively small effect of the weight flow of fuel in creating a difference between the weight flows of the inlet and exhaust streams is neglected in this example.)

It can thereby be inferred that the components of a propulsor must exert a force F on the stream of air flowing through the propulsor if that propulsor accelerates the airstream from the flight velocity V_0 to the discharge velocity V_j. The reaction to that force F is ultimately transmitted by the mounts of the propulsor to the aircraft as propulsive thrust. For engines like turbofans that have more than one exhaust stream, we must deal with the summation of the jet momentums of the several streams. For isolated propulsors like propellers or helicopter rotors, it is usual to assess thrust by focusing on the left side of the equation, ie, by summing up the pressure forces of the blades of the propulsor. For jet engines and turbofans, it is more usual and convenient to infer the forces by evaluating the right side of the equation.

The equation also illustrates another important aspect of aircraft engine performance, ie, the effect of flight speed on net thrust. The first term of the right side of the equation is referred to as gross thrust, that is, the hypothetical thrust that would be developed at zero flight speed. The second term of the right side of the equation is referred to as the ram drag or induced drag, that is, the drag on the engine induced by the ram pressure developed on the front face of the engine. For supersonic and hypersonic aircraft, the effect is considerable, with net thrust being a very small fraction of gross thrust, and the whole system's effectiveness being very sensitive to small changes in engine efficiency.

As shown in Figure 7, there are two general approaches to conversion of gas horsepower to propulsive thrust. An additional turbine (ie, a low pressure or power turbine component) may be introduced into the engine flowpath to extract additional mechanical power from the available gas horsepower. This mechanical power may then be used to drive an external propulsor such as a propeller or helicopter rotor. In cases such as these, the thrust is developed in the propulsor as it energizes and accelerates the airflow through the propulsor, that is, an airstream separate from that flowing through the prime mover. Alternatively, the high energy stream delivered by the prime mover may be fed directly to a jet nozzle which accelerates the gas stream to a very high velocity as it leaves the engine, as typified by the turbojet. In this case the thrust is developed in the components of the prime mover as they energize the gas stream. In other types of engines, such as the turbofan, thrust is generated by both approaches. A major part of the thrust is derived from the fan which is powered by a low pressure turbine and which energizes and accelerates the bypass stream. The remaining part of the total thrust is derived from the core stream which is exhausted through a jet nozzle.

Propulsive Efficiency

Just as the prime mover is an imperfect device for converting the heat of combustion of the fuel to gas horsepower, so the propulsor is an imperfect device for converting the gas horsepower to propulsive thrust. There is generally a great deal of energy left behind in the high temperature, high velocity jet stream exiting from the propulsor that is not fully exploited for propulsion. The propulsive efficiency, η_p, compares that portion of the available energy which is usefully applied in propelling the aircraft to the total energy of the jet stream. For the simple but representative case where the discharge airflow equals the inlet gas flow,

$$\eta_p = 2V_0/(V_j + V_0)$$

Figure 8 compares the variation of nondimensional thrust and propulsive efficiency as functions of jet velocity normalized by flight speed. Whereas the jet velocity V_j must be larger than the aircraft velocity V_0 in order to generate useful thrust, a large jet velocity excess over the flight speed can be very detrimental to propulsive efficiency. The maximum propulsive efficiency is approached when the jet velocity is almost equal to (but, of necessity, slightly greater than) the flight speed. This fundamental fact has given rise to a large variety of variations of jet engines, each specialized in the range of jet velocities that it generates to match the range of flight speeds of the aircraft which it must power, as indicated in Figure 7, or more generically in Figure 9.

Specific Fuel Consumption of the Engine

The net assessment of the efficiency of the engine is the measurement of its rate of fuel consumption per unit of thrust generated, eg, in terms of kg/h of fuel consumed per kg of thrust generated. There is no simple generalization of the value of SFC of the thrust engine. It is not only a strong function of the prime mover's efficiency (and hence its pressure ratio and peak cycle temperature), but also of the propulsive efficiency of the propulsor (and hence of the engine type). As has already been noted, it also is a strong function the aircraft flight speed and the ambient temperature, pressure, humidity, etc.

BASIC ENGINE TYPES

Achieving a high propulsive efficiency for a jet engine is dependent on designing it so that the exiting jet velocity is not greatly in excess over the flight speed. At the same time, the amount of thrust which is generated is proportional to that very same velocity excess which must be minimized. This set of restrictive requirements has led to the evolution of a large number of specialized variations of the basic turbojet engine, each tailored to achieve a balance of good fuel efficiency, light weight, and compactness for duty in some band of the flight speed–altitude–mis-

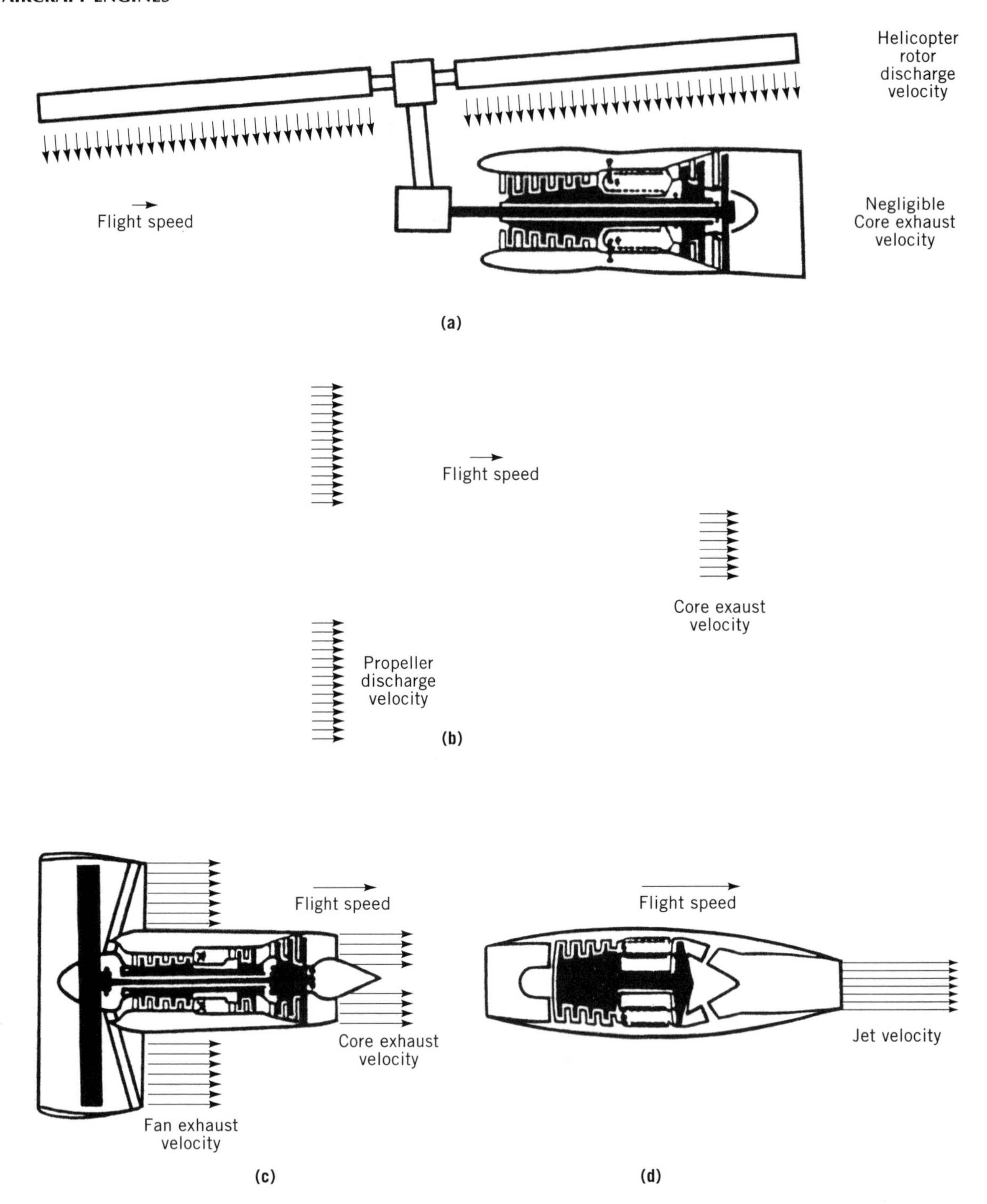

Figure 7. Variation of the jet velocity with flight speed in (**a**) the turboshaft helicopter drive; (**b**) the turboprop; (**c**) the high-bypass turbofan; and (**d**) the turbojet (7).

sion spectrum. There are, in general, two major commonalities characteristic of the different engine types. First, in order to achieve a high propulsive efficiency, the jet velocity, or the velocity of the gas stream exiting from the propulsor, is matched to the flight speed of the aircraft: slow aircraft have engines with low jet velocities, and fast aircraft have engines with high jet velocities. Secondly, as a result of designing the jet velocity to match the flight speed, the size of the propulsor varies inversely as the flight speed of the aircraft: slow aircraft have very large propulsors (eg, the helicopter rotor), and the relative size of the propulsor decreases with increasing design flight speed (turboprop propellers are relatively smaller, and turbofan fans are still smaller). A description of actual powerplants is available to the reader (9).

Although the turbojet is the simplest jet engine, and was invented and flown the earliest of all the jet engine types, it is useful to examine the spectrum of engines in the order of the flight-speed band in which they serve, starting with the slowest mission, ie, that of the helicop-

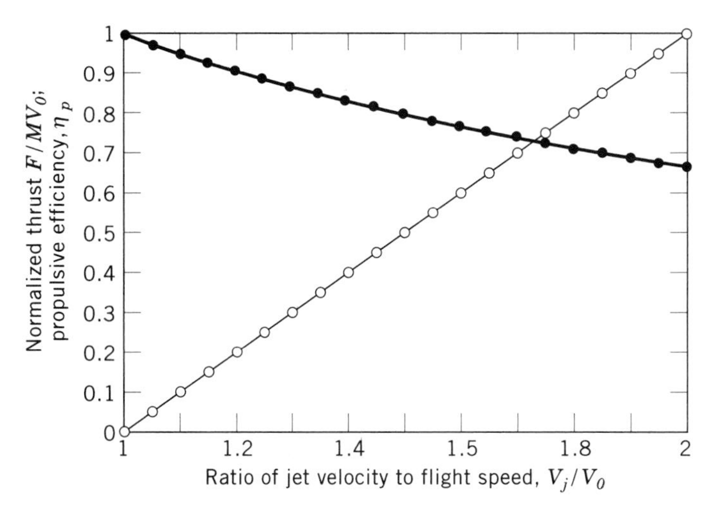

Figure 8. The propulsion dilemma; (–●–●–●–), propulsive efficiency; –○–○–○–), normalized thrust.

ter powered by the turboshaft engine, rather than dealing with the turbojet first.

Turboshaft Engines

The lowest speed aircraft in the entire spectrum of flight vehicles is the helicopter, which is designed to operate for substantial periods of time in hover, ie, at zero flight speed. Even in forward flight, helicopters rarely exceed 240 km/h or a Mach number of 0.22 (times the velocity of sound). The principal propulsor is the helicopter rotor, which is, for all modern large helicopters, driven by one or more turboshaft engines, as illustrated in Figure 7a. As we have previously noted, the propulsor is designed to give a very low discharge or jet velocity and, by the same token, is very large for a given size aircraft when compared to the propulsors of higher speed aircraft. The prime mover is a core engine whose gas horsepower is extracted by a power turbine which then drives the helicopter rotor through a speed-reducing and combining gearbox. The power turbine is usually on a spool separate from the gas generator so that its rotative speed and that of the helicopter rotor which it drives through a gearbox are independent of the rotative speed of the gas generator. This permits the helicopter rotor speed to be held constant or varied independently of the gas generator speed, which must be varied to modulate the amount of power generated.

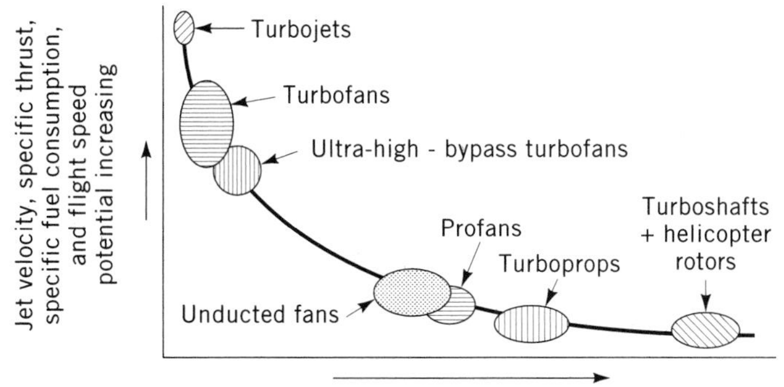

Figure 9. Effect of varying jet velocity on engine design and operation (8).

Turboprops, Propfans, and Unducted Fan Engines

The turboprop is the power plant which fills the next band of flight speeds in the flight spectrum, from a Mach number of 0.2 to 0.7. The propulsor is a propeller with somewhat higher discharge or jet velocity than that of the helicopter rotor, to match the flight speed, and with a proportionately smaller area than that of the helicopter rotor for a similarly sized aircraft. The prime mover is a turboshaft engine very similar to that which drives a helicopter rotor. In early turboprop engines, the turbine that drove the propeller was fixed to the core engine shaft and ran at the same rotative speed, as shown in Figure 10a. There are several major disadvantages to this design having to do with starting, part-power efficiency, and windmilling operation of inoperative engines. In modern turboprop engines, as in modern turboshaft engines, the turbine that drives the output (propeller) shaft is on a shaft separate from that of the core engine, ie, a "free turbine" driving at its own independent rotative speed, as shown in Figure 10b. In this context, the control mode of a turboprop is somewhat different from that of a helicopter's turboshaft engine. In the helicopter, the pilot calls for power by manipulating the pitch of the rotor blades, a greater pitch taking a bigger "bite" of the air and so demanding more power to maintain rotative speed. The engine's control responds by increasing fuel to the engine to maintain output shaft speed. In a turboprop, the pilot selects power by selecting fuel flow to the prime mover. The propeller control responds by varying propeller pitch to maintain a preselected propeller rotative speed. The turboprop also has a gearbox different from that found in helicopter. The turboprop gearbox is designed to provide a somewhat higher rotative speed for the propeller which turns faster than the much larger-diameter helicopter rotor. Many modern gearboxes involve two stages of reduction, the first being offset. This results in the propeller's centerline being offset from the engine's centerline (Fig. 10c) and is accommodated by an offset partial-arc inlet.

A recent trend in turboprop design has been the evolution of propellers for efficient operation at much higher transonic flight speeds than achieved previously—up to Mach numbers of 0.85. This usually involves a higher disk loading, ie, a higher discharge velocity from the propeller in order to achieve a smaller diameter propeller. This trend is also accompanied by an increase in the number of blades in the propeller to 6 to 12 in number instead of the more usual 2 to 4 blades in lower speed propellers. The blades are also scimitar-shaped, with swept back leading edges at the blade tips to accommodate the large Mach numbers encountered by the propeller tip at high rotative and flight speeds. Such high speed propulsors are called propfans. Another variation of the propulsor involves the application of two propellers on the same centerline, driven by the same prime mover through a gearbox which causes each propeller to rotate in a direction counter to the other, as shown in Figure 10d. Such counterrotating propellers are capable of significantly higher

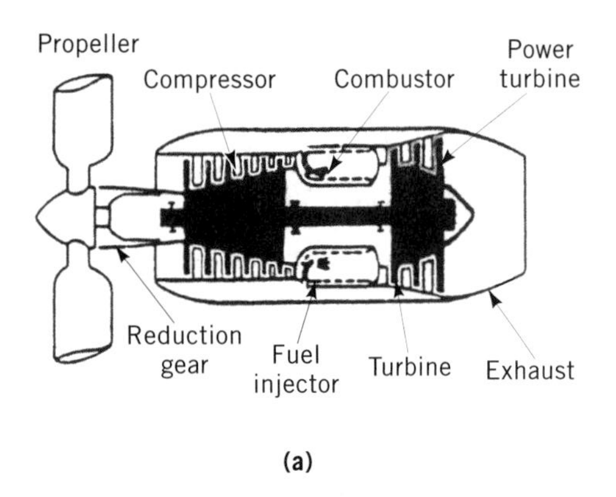

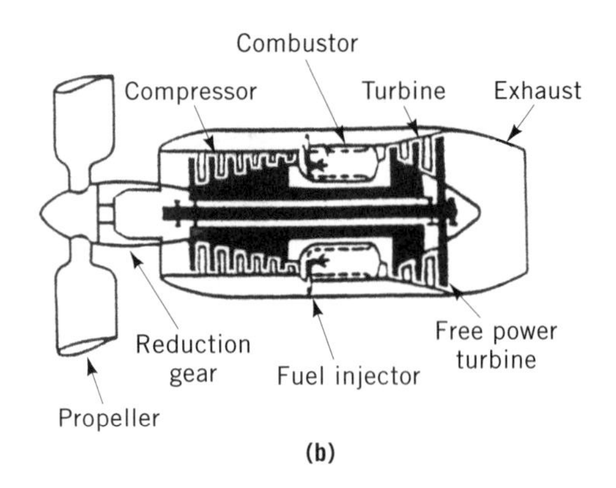

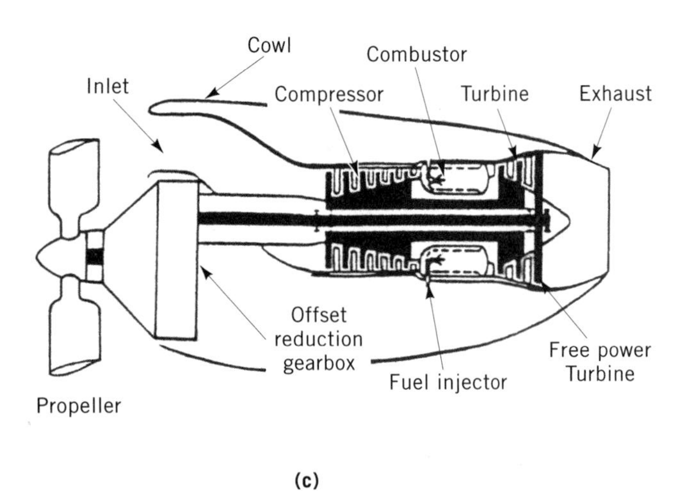

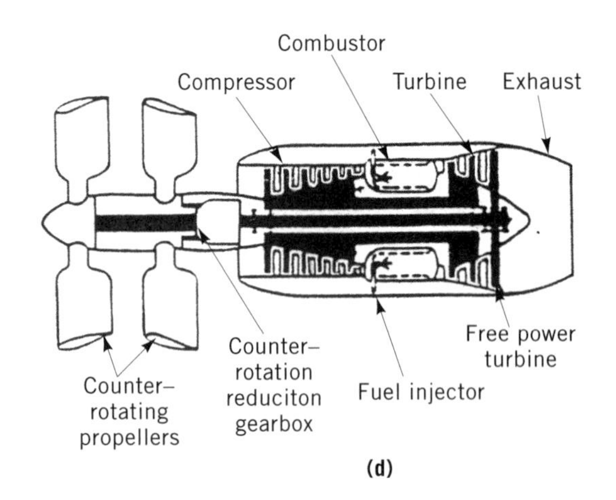

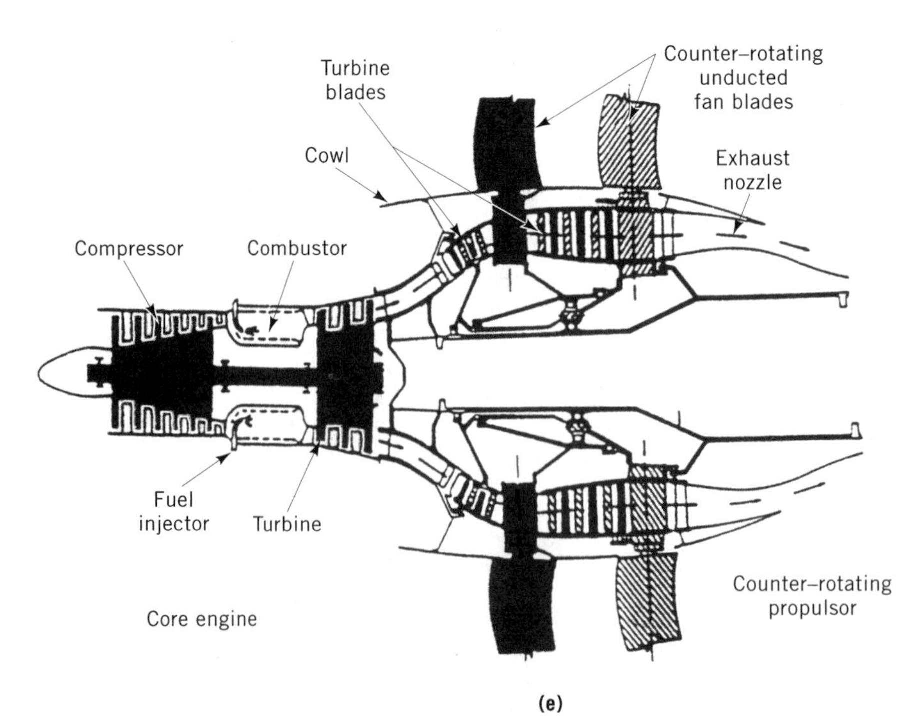

Figure 10. Turboprop engines (7). (**a**) Fixed-turbine turboprop engine driving a single rotation propeller as propulsor; tractor arrangement. (**b**) Free-turbine turboprop engine driving a single rotation propeller as propulsor; tractor arrangement. (**c**) Free-turbine turboprop engine with offset reduction gearbox driving a single rotation propeller as propulsor; tractor arrangement. (**d**) Free-turbine turboprop engine driving a counter-rotating propeller as propulsor; tractor arrangement. (**e**) Turboprop engine with counter-rotating free turbines driving ungeared counter-rotating propellers as propulsor; pusher arrangement (unducted fan) (8).

propulsive efficiency and higher disc loading than are conventional propellers. In most turboprop installations, the prime mover is mounted on the wing and the plane of the propeller is forward of the prime move (the so-called tractor layout). Modern high speed aircraft may find it more advantageous to mount the engine more toward the rear of the aircraft, with the plane of the propeller aft of the engine. These arrangements are referred to as "pusher" layouts. A recently developed engine layout, identified as the unducted fan (8), provides a set of very high efficiency, counterrotating propeller blades, with each blade mounted on one of either or two sets of counterrotating low pressure turbine stages and achieving all benefits from the arrangement without the use of a gearbox, as shown in Figure 10e.

Medium-Bypass Turbofans, High Bypass Turbofans and Ultrahigh Bypass Engines

Moving up in the spectrum of flight speeds to the transonic regime, ie, Mach numbers from 0.75 to 0.9, the most commonly found configurations of engine are turbofans such as are shown in Figure 11**a–e**. In a turbofan, only a portion of the gas horsepower generated by the core is extracted to drive a propulsor, which is usually a single low pressure-ratio shrouded turbocompression stage. Although early models of turbofans, adapted from existing jet engines, place the fan at the rear of the core as in Figure 11a, the fan is now virtually always placed in front of the core inlet, so that the air which enters the core first passes through the fan and is partially compressed by the fan. But the major portion of the air byapsses the core (hence the term bypass stream) and goes directly to an exhaust nozzle. The core stream, with some modest fraction of the gas horsepower remaining (not extracted to drive the fan), goes directly to its own exhaust nozzle.

In some early versions of the turbofan, the power turbine and the fan were directly linked to the gas generator, as in Figure 11b. But this arrangement had severe handicaps, a major one of which was the requirement for very large starter energy to energize the bypass air. This was accommodated by variable pitch fan blades which permitted the fan to be unloaded during starting. In virtually all modern turbofans the core is now configured as a separate spool which rotates independently of the spool with the fan and its driving turbine, as in Figure 11c,d. This then permits efficient starting of the engine, since it is not necessary for the starter to rotate the fan. It is also a very basic requirement in permitting the larger diameter fan spool to rotate at much lower rotative speeds than the smaller diameter core spool without the necessity of gearing between the spools.

Ideally the fan and core exhaust velocities should be nearly equal. If they are not, some benefit is derived from incorporation of a mixer between the two streams and exhausting the mixed flow (Fig. 11d) through a common nozzle rather than exhausting each stream through its own nozzle, as in the more common separated-flow configuration (Fig. 11c).

A key parameter for classifying the turbofan is its bypass ratio, defined as the ratio of the mass flow rate of the bypass stream to the mass flow rate entering the core. We would expect that the highest propulsion efficiencies are obtained by the engines with the highest bypass ratios. We actually find engines in use in this flight speed regime with a broad spectrum of bypass ratios, including medium-bypass engines with bypass ratios from 2 to 4, high bypass engines with bypass ratios from 5 to 8, and ultrahigh bypass engines, so-called UBEs, with bypass ratios from 9 to 15 or higher. Indeed, a generation of low and medium-bypass engines has completely supplanted the first generation of aircraft powered by zero-bypass turbojet engines. And that generation was itself supplanted by a third generation of medium and high bypass turbofan engines. Nevertheless, there are several reasons why we still see engines with less than the highest bypass ratios hypothetically achievable. Very high bypass ratios involve very large-diameter fans which, in turn, entail very heavy components and increase the difficulty in installing the engine on the aircraft with sufficient ground clearance. In addition, the weight and complexity of the apparatus required to reverse the direction of the bypass stream in order to achieve thrust reversal to shorten the aircraft's landing roll also increases with the bypass ratio. Thrust reversal will generally require variable pitch fan blades in such very high bypass fans, instead of the blocker-doors found in medium and high bypass engines. As shown in Figure 11e, ultrahigh bypass engines may also require a gearbox between the drive turbine and the fan in order to simplify the design of the low diameter turbine (with the attendant high rotative speed) without compromising the performance of the very large diameter fan (with the attendant low rotative speed). But the long-term trend is very definitely toward higher and higher bypass ratios.

Low Bypass Turbofans and Turbojets

The next higher regime of aircraft flight speed, the low supersonic range from Mach numbers above 1 up to Mach 2 or 3, involves the application of the simple turbojet (with no bypass stream) and low bypass turbofan engines (with bypass ratios up to 2), as shown in Figure 12a and c. Although the low bypass turbofan has the same general appearance as a turbofan with a larger bypass ratio, certain special features are unique to the lower bypass engines. The lower total flow in the fan generally involves a higher fan pressure-ratio for equivalent amounts of energy available from the drive turbine, so that a fan with more than one (ie, two or three) turbocompressor stages is usually employed.

Engines designed for operation in this flight regime are usually deficient in thrust in other flight regimes or modes where they must operate for short durations, such as acceleration through transonic speed, or in take-off at the extremes of high altitude airports with very-hot-day conditions and high gross weight, or in combat maneuvers at high supersonic-flight speed. Rather than installing a larger engine to meet these requirements, it is more effective to add an afterburner to the engine as a means of thrust augmentation, as shown in Figure 12b and d. The afterburner is a secondary combustion system that operates in the exhaust stream of the engine, before the airstream is introduced into the exhaust nozzle. Such a device is not as fuel-efficient as the main turbofan section of the engine, because heat addition takes place at a lower

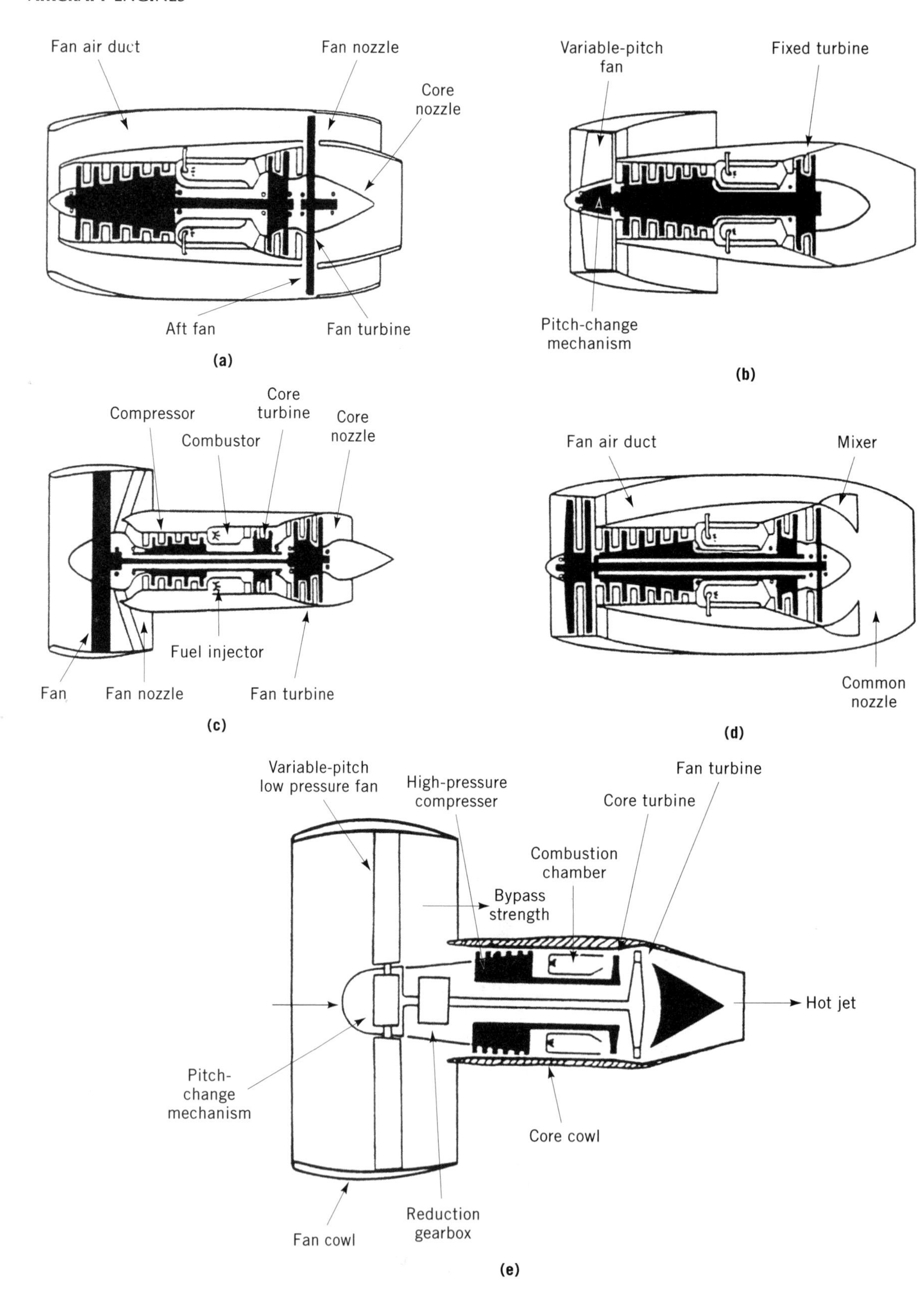

Figure 11. Medium and high bypass turbofans (7): (**a**) aft fan; (**b**) single-spool fan with variable-pitch fan blades for starting; (**c**) separated-flow turbofan; (**d**) mixed-flow turbofan; (**e**) ultrahigh bypass engine (UBE) with geared fan and variable-pitch blading for thrust reversal.

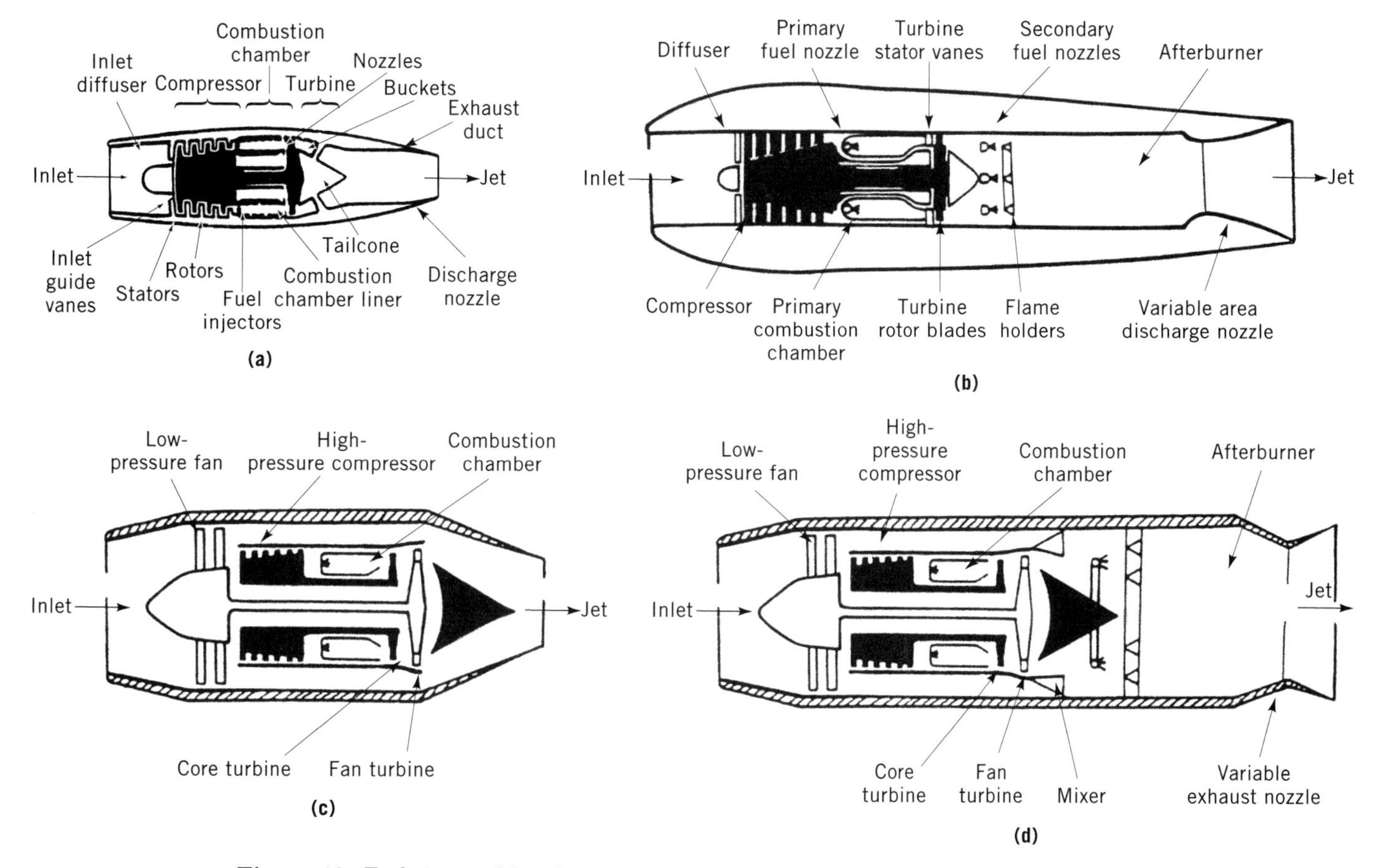

Figure 12. Turbojets and low bypass turbofans (7,10): (**a**) Turbojets; (**b**) Turboject with afterburner; (**c**) Low bypass turbofan; and (**d**) Low bypass turbofan with afterburner.

pressure than in the main burner. But the afterburner is relatively simple and lightweight, since it does not contain any rotating machinery. For the same reason, it may be operated to a much higher discharge temperature (typically 1760°C) so that it is capable of augmenting the thrust of the turbofan by as much as 50% of the thrust without afterburning. The afterburner in a turbofan usually requires a mixer to mix the relatively cool bypass air with the hot core stream, since the cooler air is otherwise very difficult to burn in a typical, low pressure afterburner environment. Additionally, in both the turbojet and the turbofan with an afterburner, the exhaust nozzle must have a variable throat area to accommodate the large variations in volumetric flow rate between the extremely hot exhaust stream from the operating afterburner, and the cooler airstream discharged from the engine when the afterburner is not in use.

Engines intended for use at supersonic flight speed will generally have a significantly lower compression pressure ratio than higher bypass machines intended for subsonic or transonic operation. A major contributor to this tendency is the additional pressure ratio developed in the engine's inlet as it slows down or diffuses the very high speed airstream, viewed from the aircraft, which is ingested as the engine's working fluid—the so called ram effect. At transonic flight speed this pressure ratio is almost 2:1, with the result that the engine's compressor may be built to provide that much less pressure where peak pressure is otherwise limiting. Early generations of jet-propelled aircraft in this low supersonic flight regime, both military combat aircraft and commercial supersonic transports, were powered by turbojet engines, but subsequent generations of aircraft have largely substituted the low bypass turbofan in these applications, in spite of the turbojet's excellent propulsive efficiency in this speed range and its extreme simplicity and relatively lower cost. This is primarily because such aircraft expend a great deal of their fuel at subsonic flight speed, such as in takeoff, climb, loiter, acceleration, approach, and landing, where the turbofan provides an advantage in propulsive efficiency and lower jet noise. Also the cooler air bypassing the core provides a cool skin to the engine, which makes it much easier to install in the aircraft.

Ramjets and Supersonic Combustion Ramjets (SCRamjets)

Ram pressure plays an increasingly important role in the thermodynamic cycle of power and thrust generation of the jet engine at supersonic flight speeds. For flight speeds above Mach 2.5 or 3, the ram pressure ratio becomes so high that a turbocompressor is no longer necessary to ensure the efficient generation of thrust. Indeed, the pressure ratio eventually rises to such high values that the associated high ram temperatures make it difficult or impossible to place high speed rotating machinery in the flowpath without provision of prohibitive amounts of cooling. This combination of circumstances naturally leads to the ramjet, an afterburning jet engine in which the pressure rise is attributable only to the ram effect of the high flight speed without any turbomachinery, and in which the main thrust producer is the afterburner, as shown in Figure 13. Ramjets are very lightweight and simple powerplants, and are ideal candidates for supersonic flight vehicles launched from other flight vehicles at very high speed. Their usage is more problematic for vehicles that must be self-sufficiently powered for subsonic take-off, climb, and acceleration to supersonic flight speed, since the subsonic ram pressure is insufficient to make possible any reasonable amount of thrust production, and alternative propulsion devices must be provided.

In the flight regime under Mach 4 or 5, it is usually efficient to decelerate the inlet airstream to subsonic velocity before it enters the combustion system. At still higher flight Mach numbers, such deceleration or diffusion becomes more difficult and more costly in terms of pressure losses, and it becomes necessary to make provision for the combustion chamber to burn its fuel in the supersonic airstream. Such specialized ramjets are called SCRamjets (for Supersonic Combustion Ramjets) and are projected to be fueled by a cryogenically liquefied gas (such as hydrogen or methane) instead of a liquid hydrocarbon. The primary reason for doing this is to exploit the greater heat release per unit-weight of fuel of the fuels having a higher ratio of hydrogen to carbon atoms than ordinary fractions of petroleum, although this gain is partially negated by those same fuels' higher volume per unit of heat release. Another incentive for using a very cold fuel is so that it may be used as a heat sink for cooling the very high speed (and hence very hot) engine and aircraft structure. A very unusual feature of SCRamjets is the fact that the inlet deceleration and the exhaust acceleration take place largely outside the enclosed engine-inlet and exhaust ducts, against external aircraft surfaces in front of and to the rear of the engine. The engine itself is little more than a sophisticated supersonic combustion chamber.

Hybrid Engine Types

As has been noted, it is possible to tailor an engine configuration so that the engine is well-suited to operation in a given band of the flight spectrum. In order to have an engine which will perform well in more than one band of the flight spectrum or more than one regime of operation, it may be necessary to configure the powerplant so that it can in effect by converted from one engine type to another by means of variable geometry built into the engine components.

Vertical and Short Take-Off and Landing (V/STOL) Propulsion. Propulsion systems that give aircraft the capability of both vertical and conventional forward flight represent a formidable challenge to the engine designer. Several major categories of engine arrangement are found in V/STOL aircraft.

As typified by the helicopter rotor in a helicopter, the propulsor is driven by one or more turboshaft engines and is installed to give vertical thrust. The entire aircraft must be tilted to give the thrust vector a forward component to achieve forward flight. There are limitations in the effectiveness of this approach, and this class of V/STOL aircraft is considerably less efficient in forward flight above a Mach number of 0.2 than fixed using aircraft without V/STOL capability.

The propulsors may be mounted on pivots so that they may be rotated from the position in which they give vertical thrust in a takeoff, hover, climb, descent, or landing maneuver, and pivoted 90° to give forward thrust for conventional forward flight (as in the tilt rotor aircraft). The prime mover that drives the propulsor may be tilted with the propulsor, or may be fixed in the wing and drive the tilting propulsor through a rotating shaft, through the pivot axis. In some configurations, the entire wing of the aircraft, carrying fixed engines and propulsors, may be tilted as an entire assembly.

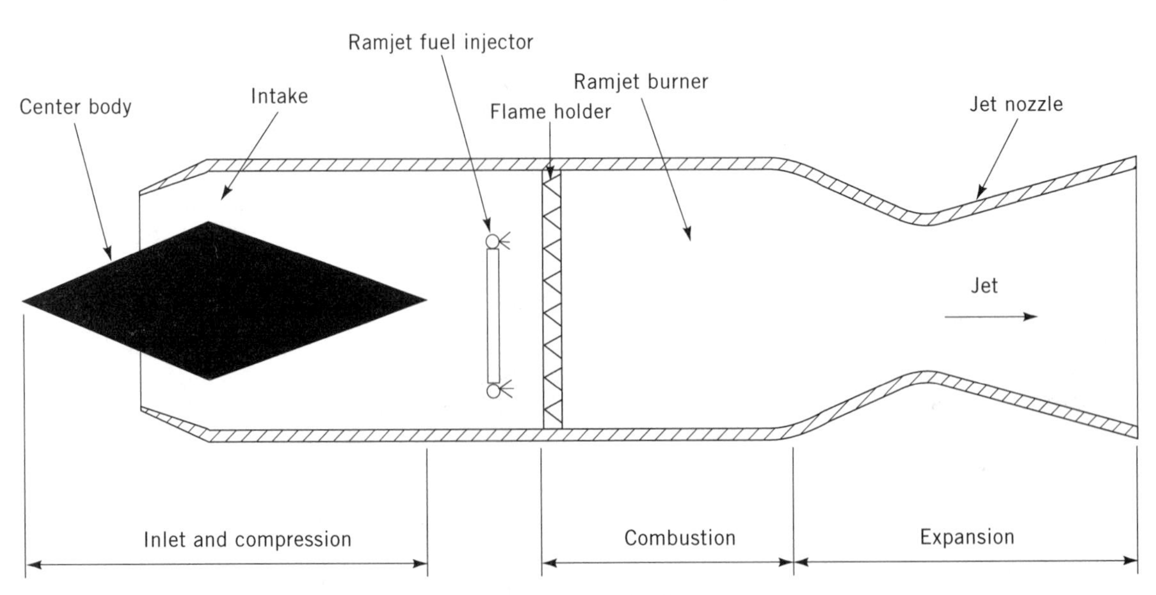

Figure 13. Ramjet (10).

The engines may be fixed in the position necessary to give thrust for forward flight. Built-in variable geometry is included in their exhaust systems to vector the exhaust nozzle(s) or divert the exhaust gases by means of valves and auxiliary ducts to nozzles mounted to give vertical thrust or lift.

The aircraft may include two different sets of engines or propulsors, both fixed in position, with one set installed for forward flight and the second set installed for vertical thrust (ie, the lift engines).

The aircraft may use a convertible engine. The convertible engine has a single prime mover which is arranged to drive a fan for efficient forward flight propulsion, or to drive a shaft which is connected to and drives the main helicopter rotor. In order to convert from vertical flight to horizontal flight, variable pitch fan blades or variable pitch stators are used to unload the fan, and the mechanical power then made available is delivered through shafting to drive a helicopter rotor for vertical flight capability.

Variable Cycle Engines. For aircraft that are designed to fly mixed missions with large amounts of fuel consumed at subsonic, transonic and supersonic flight speeds, it is desirable to have an engine with the characteristics of both a high bypass engine (for subsonic flight speed) and a low bypass engine (for supersonic flight speed). This requirement is typical of high speed commercial airliners or Supersonic Transports (SSTs) which cruise over oceans and unpopulated land areas at supersonic cruise speeds, but must also fly efficiently and quietly at subsonic flight speed for take-off, climb, cruise over populated areas, approach, and landing. This dual function is accomplished by the Variable Cycle Engine (VCE), as shown in Figure 14. If the components of an engine are designed to accommodate the extreme limits of flows, pressure ratio, etc, involved in both high bypass and low bypass operation, the engine may be operated at either extreme of bypass ratio, or at any bypass ratio between the extremes, by means of a valve or valves in the bypass stream operated in conjunction with a variable exhaust nozzle. When the valves are closed, they restrict the flow in the bypass stream to achieve the low bypass for supersonic flight. When the valves are open, the bypass is increased to its maximum value for efficient subsonic flight.

Turboramjets. As was noted in the discussion of the ramjet, that engine type provides an extremely simple and efficient means of propulsion for aircraft at relatively high supersonic flight speeds. But the ramjet is quite inefficient at transonic flight speeds, and is completely ineffective at subsonic flight velocities. The turboramjet has been developed in order to overcome this inadequacy (11,12). In effect, a turbofan engine is built into the inlet of the ramjet engine to charge the ramjet with a pressurized stream of air at subsonic flight speed where ram pressure is insufficient for effective ramjet operation. In supersonic flight, the fan blades, if they are variable pitch, may be feathered so as not to interfere with the flow of ram air to the ramjet. The separate inlet to the core engine driving the fan may be closed off so as not to expose the turbomachinery to the hostile environment of the high temperature ram air. Another variation of the turboramjet (shown in Fig. 15) does without the core inlet and the core compressor altogether. Instead, the aircraft carries a tank of an oxidizer such as liquid oxygen. The oxidizer is fed into the core combustion chamber along with the fuel to support the combustion process which generates the hot gas stream to power the turbine that drives the fan. During supersonic flight, the fan may be feathered, and a surplus of fuel may be introduced into the core combustor. The unburned fuel that passes through the fan turbine burns in the ramjet burner when it mixes with the fresh air entering by way of the bypass stream from the fan. For aircraft that are designed to proceed from high speed atmospheric flight to transatmo-

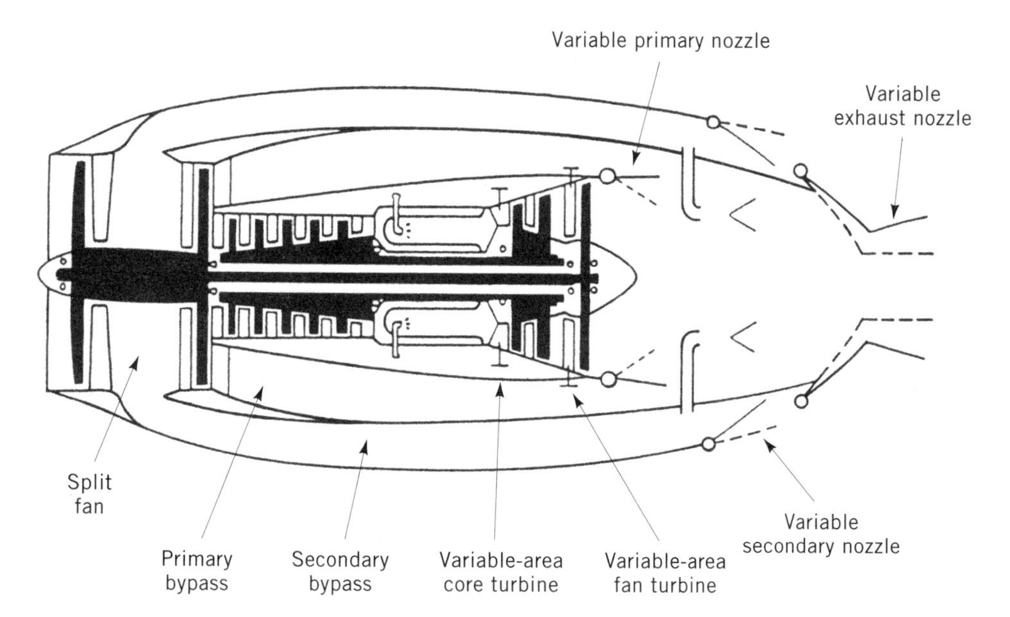

Figure 14. Variable cycle turbofan engine (7).

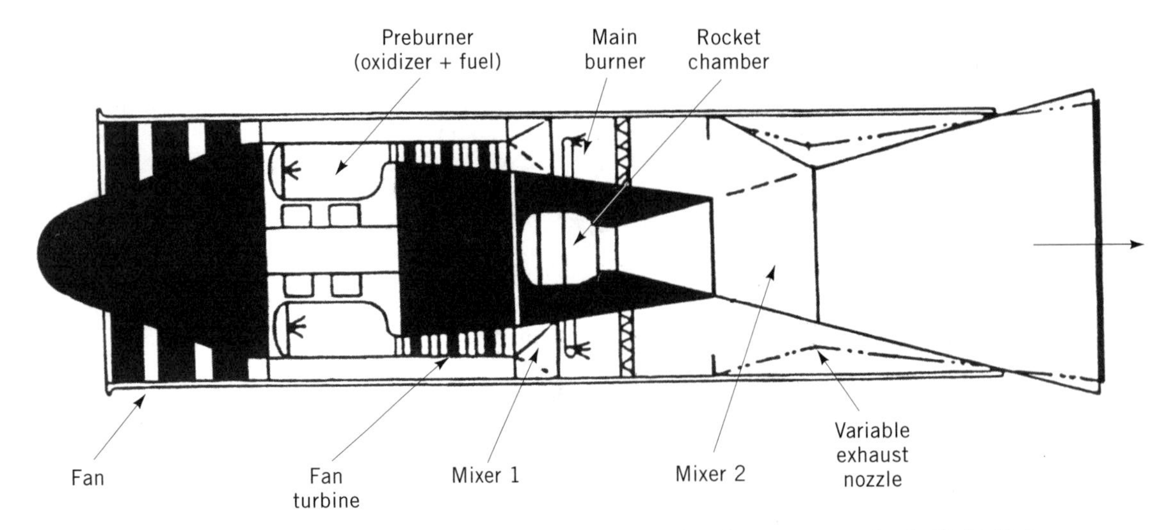

Figure 15. Turboramjet with rocket combustion chamber for exo-atmospheric flight (7,8).

spheric flight, a rocket chamber may be provided in the engine (as shown in Fig. 15), where fuel and oxidant are burned in greater quantity than is possible in the preburner, and the exhaust stream may be discharged through the thrust nozzle without having to pass through the turbine.

AIRCRAFT TURBINE ENGINE EXHAUST EMISSIONS

Since the pollutant emissions of aircraft turbine engines have been the subject of growing attention, along with the pollutant emissions from other sources. Relative to other sources, aircraft turbine engines are small contributors to global, national, and local emission burdens. On a global basis, aircraft engine-related emissions generally constitute less than 3% of all man-made emissions (13). However, some unique concerns are associated with aircraft engine emissions insofar as the bulk of these emissions are released at high altitudes.

The aircraft turbine engine exhaust emissions of key concern consist of smoke (soot), unburnt hydrocarbons (HC), carbon monoxide (CO), and oxides of nitrogen (NO_x). While not considered to be a pollutant emission in the same sense, the carbon dioxide (CO_2) emissions of aircraft engines are also of possible concern. The engine operating conditions at which these emissions are primarily generated are shown in Table 1; the most significant environmental impact issues associated with these emissions are indicated in Table 2. Some sulfur oxides (SO_x) emissions are also generated by aircraft engines; however, the quantities are very small because the sulfur contents of aircraft turbine engine fuels are extremely low.

These emission categories are typical of those of any heating or power plant operated with fossil fuels. CO_2 is a direct and major product of fossil fuel combustion. As such, the CO_2 quantities that are generated are directly proportional to the quantities of fuel that are consumed. On a global basis, the total CO_2 resulting from aircraft operations represents about 2.1% of all man-made CO_2 emissions (14). Abatement can only be achieved via the use of alternative fuels with lower carbon contents or through reduced fuel consumption. Suitable alternatives to aviation kerosene fuels do not exist at this time. For many years, however, significant emphasis has been focused by the air transportation industry on reducing fuel consumption. As a result of both engine and aircraft technology advances, fuel usage per passenger seat-mile has been dramatically reduced during the past three decades.

Table 1. Aircraft Turbine Engine Emissions

Emission Category	Engine Operation Condition(s)[a]
Smoke (soot)	Takeoff and climbout
Unburned hydrocarbons (HC)	Low power, especially ground idle
Carbon monoxide (CO)	Low power, especially ground idle
Nitrogen oxides (NO_x)	All high power, including cruise
Carbon dioxide (CO_2)	All power settings

[a] Primarily associated with the corresponding emission category.

Table 2. Key Environmental Impact Issues Associated with Aircraft Engine Emissions

Emission Category	Impact
Smoke	Visibility nuisance around airports
HC	Contribution to urban smog burdens
CO	Contribution to urban CO burdens
NO_x	
Subsonic aircraft engines	Possible contribution to global warming
Future supersonic aircraft engines	Stratospheric ozone depletion
CO_2	Contribution to global warming

Emission Standards

Limits on the allowable levels of the smoke, HC, CO and NO_x emissions of civil turbojet and turbofan engines were promulgated by the International Civil Aviation Organization (ICAO) during the 1980 to 1982 time period. These standards (15) are expressly designed to regulate the emission quantities that can be discharged in and around airports. These ICAO standards are the basis of the standards currently prescribed by several nations, including the United States. Standards are specified for both subsonic and supersonic civil aircraft engines. In the case of the supersonic engine standards, the regulatory limit values are strongly predicated on the emission characteristics of the Olympus engine, which powers the Concorde airplane.

The ICAO standard for smoke emissions is expressed in terms of a Smoke Number (SN), which is determined by means of the specific test procedure (15). This parameter is proportional to the smoke concentration of the engine exhaust. The test procedure involves smoke emission measurements over a range of sea-level static operating modes, including takeoff. This standard applies to all newly manufactured civil turbojet and turbofan engines. The maximum allowable SN is 83.6 $(F_{00})^{0.274}$ or a value of 50, whichever is lower, where F_{00} is the rated engine thrust at takeoff conditions.

The intent of this standard is to eliminate visible smoke in the engine exhaust. Visibility is dependent on the smoke concentration and on the viewing path length. As such, the allowable SN of a high thrust engine is lower than that of a low thrust engine, because of the larger exhaust diameter and associated longer viewing path length of the high thrust engine.

The ICAO standards for HC, CO, and NO_x emissions are presented in Table 3. For subsonic engines, these standards apply to newly manufactured civil turbojet and turbofan engines with thrust ratings above 26.7 kN. The standards are expressed in terms of a parameter which consists of the total mass, in grams, of the emission produced during a prescribed landing–takeoff operational (LTO) cycle per kilonewton (kN) of rated takeoff thrust at sea-level, standard day operating conditions. The prescribed LTO cycles are intended to be representative of operations that occur at or near airports at altitudes up to 0.9 km above the airport. This regulatory parameter is proportional to both emission level and engine fuel consumption characteristics.

The NO_x standards are expressed as a function of engine cycle pressure ratio at rated takeoff thrust. This feature is made use of because of the strong relationship between NO_x emission level and engine pressure ratio. In the case of subsonic engines, the NO_x standard is intended to serve as a cap, to ensure that the NO_x levels of future engines are no higher than those of existing operational engines. At the time of its promulgation in the 1980 to 1982 time period, the important elements of the rationale for this regulatory approach were the conclusions that the attainment of any significant reductions in NO_x emissions requires the use of new and more complex combustor design technology and that the contributions of aircraft engines to urban NO_x emission burdens are quite small. Based on the latter assessment, it was concluded that a mandate for the use of advanced and unproven combustor technology was not justified.

Table 3. ICAO Gaseous Emission Standards[a]

Emission, g/kN	Subsonic Turbojet/ Turbofan Engines[b]	Supersonic Turbojet/ Turbofan Engines
HC	19.6	$140(0.92)^{\pi}_{00}$
CO	118.0	$4550(\pi_{00})^{-1.03}$
NO_x	$40 + 2\ \pi_{00}$	$36 + 2.42\ \pi_{00}$

[a] π_{∞}, engine pressure ratio at rated takeoff thrust.
[b] Newly manufactured engines with rated takeoff thrust greater than 26.7 kN.

During the 1990 to 1992 time period, reassessments of the existing NO_x standard were initiated by the ICAO in view of increasing pressures to reduce NO_x emissions from all sources and in response to new concerns regarding the possible environmental impacts of NO_x emissions introduced into the upper troposphere and stratosphere. During cruise, subsonic aircraft generally operate in the upper troposphere or in the lower stratosphere. Supersonic aircraft generally operate at higher altitudes, in the stratosphere. Some preliminary findings of atmospheric modeling studies intended to quantify the environmental impacts of aircraft engine NO_x emissions indicate that the NO_x emissions of subsonic aircraft engines can be a contributing factor to global warming (13). Studies of the possible effects of a future fleet of supersonic transport aircraft indicate that the resulting NO_x emissions could result in some depletion of the ozone layer (16). As an initial result of the ongoing ICAO reassessments of the existing NO_x standards, a 20% reduction of the subsonic aircraft engine standard was proposed in 1991 by the ICAO Committee on Aviation Environmental Protection (CAEP) for adoption by the ICAO Council.

Emission Characteristics and Compliance Status

Efforts to reduce the pollutant emission levels of aircraft engines were initiated well before the promulgation of regulatory limits. During the 1965 to 1975 time period, low smoke combustors were developed and incorporated into operational engines. With these combustors, smoke visibility is virtually eliminated in most cases. During the 1970 to 1980 time period, the introduction of more fuel-efficient engines resulted in significant reductions in HC and CO levels. Later, during the 1975 to 1985 time period, combustor design features for further reducing HC and CO emission levels were developed and incorporated into operational engines. As a combined result of these efforts, significant reductions in HC and CO levels were obtained. Accordingly, significant progress in reducing the pollutant emission contributions of aircraft engines to urban air pollution burdens has been realized.

At cruise operating conditions, the emissions of modern aircraft engines consist primarily of NO_x. In the case of subsonic aircraft engines, the cruise NO_x emission indices (grams per kilogram of fuel) are always lower than the levels at takeoff conditions, because of the lower combustor inlet air temperatures and pressures that prevail at cruise operating conditions. In the case of supersonic aircraft engines, the highest indices occur at the supersonic cruise conditions because the combustor inlet air temperatures and pressures are generally highest at these conditions.

Because only NO_x emissions are generated in any significant amounts during cruise and because the HC and CO levels of modern engines have been significantly reduced at idle and other low power conditions, the emissions typically generated in a typical flight cycle of an aircraft equipped with modern engines consist primarily of NO_x.

The source of all of these emissions is, of course, the engine combustor. The emissions are mainly generated in the primary combustion zone of the combustor. The primary combustion zone consists of the forward volume of the combustor. The fuel is injected directly into this zone. In most aircraft turbine engines, annular combustors are used. A typical configuration is shown in Figure 16.

Smoke Emissions. The smoke emissions are comprised of minute agglomerates of carbon, or soot, particles. The formation of these carbon particles is known, from thermochemical considerations, to be associated with the combustion of rich fuel–air mixtures. Another probable carbon-producing mechanism is thermal cracking of liquid fuel droplets. The rates of these various formation mechanisms are dependent on the combustor pressure level, increasing rapidly as pressure is increased.

Even in older technology engine models with visible smoke emissions, the smoke concentrations are quite low, typically less than 0.003% (by weight) of the core engine exhaust gas. However, because of the extremely small and finely divided nature of these particulate emissions, even these relatively low concentrations can be very visible. Thus, these emissions can be an objectionable nuisance. In military aircraft, visible smoke emissions are also of concern, in some instances, because of tactical considerations. To eliminate such visibility concerns, very low smoke concentrations, generally less than 2 parts per million parts (by weight) of core engine exhaust gas, are required. These concentrations correspond to a SN of about 20.

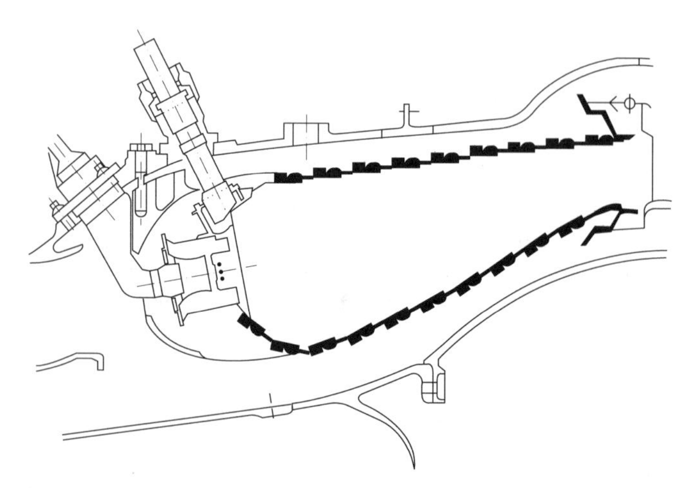

Figure 16. Cross-section drawing of a typical annular combustor.

Development efforts conducted during the 1965 to 1975 time period showed that the attainment of low smoke levels required engine combustor designs with both leaner primary combustion zone fuel–air mixtures and more effective fuel–air mixing within the primary zone, compared to those of then-existing combustors. Based on these findings, low smoke combustors were developed and incorporated into operational engines. All modern (ca 1993) aircraft engines are equipped with low smoke combustors.

HC and CO Emissions. These emissions are products of incomplete combustion. They are primarily generated at idle and other low engine power conditions. In almost all instances, both emission types are simultaneously produced under these operating conditions. The less-than-optimum combustion efficiencies, ie, the degree to which the available chemical energy of the fuel is converted to heat energy, typically obtained under these operating conditions are a result of prevailing adverse combustor operating conditions. At higher power settings, the combustion efficiency levels are generally well in excess of 99.9%, and therefore virtually all of the fuel is converted to the ideal combustion products, CO_2 and water.

The combustor design features needed to reduce HC and CO levels involve provisions for improving fuel atomization quality at low power conditions, enhancing primary zone fuel–air mixing quality at these operating conditions, and minimizing combustion quenching from fuel impingement on the combustor liners. With combinations of these design features, combustors with much-reduced HC and CO levels and low smoke levels were developed and incorporated into operational engines. Almost all modern (early 1990s) engines are equipped with combustors of this kind.

NO_x Emissions. When gases containing oxygen and nitrogen are heated to elevated temperatures, generally above 1900 K, some oxidation of the nitrogen occurs and NO_x emissions are produced. In turbine engines, these emissions consist mainly of nitric oxide (NO), together with small amounts of nitrogen dioxide (NO_2). The small amounts of NO_2 emissions result from the subsequent further oxidation of the NO that is formed in the primary combustion zone. The quantities of NO that can be generated are strongly dependent on the flame temperature levels of the combustion gases and on the availability of oxygen. Thus, the quantities increase rapidly as the primary zone fuel–air ratios approach the stoichiometric value (a stoichiometric mixture consists of the ideal proportions of air and fuel for complete combustion, with no excess air).

Within the primary zone of an engine combustor, localized regions containing stoichiometric and near-stoichiometric fuel–air mixtures generally exist at high engine power conditions and, as a result, localized high flame temperatures exist. These flame temperatures increase significantly as the combustor inlet air temperatures and pressures increase. The suppression of NO_x emissions primarily involves methods of reducing these peak flame temperatures.

A basic and well-established concept for reducing the peak flame temperatures involves operating with lean and uniform primary zone fuel–air mixtures, preferably homogeneous mixtures, at high power operating conditions. To obtain the leaner primary zone mixtures at high power, increased proportions of the total combustor airflow must be introduced into the primary zone. Doing so in a conventional combustor results in unacceptably lean primary zone mixtures at low engine power conditions, because the overall engine fuel–air ratios are quite low at these conditions. Therefore, methods are needed of obtaining both the required lean and uniform primary-zone mixtures at high power conditions and of providing adequately rich primary-zone mixtures at idle and other low power conditions. The attainment of these conflicting operating capabilities requires consideration of combustors within which the combustion process can be staged via the use of two or more combustion zones, or alternatively of combustors with variable geometry features to modulate the airflow quantities introduced into the primary zone.

Another basic concept is the use of a rich–lean, series-staged combustion process. In combustors of this kind, all of the fuel is injected into the first stage. This stage is operated with average fuel–air ratios above stoichiometric to suppress the formation of NO_x. To obtain the fuel-rich mixtures, most of the combustor airflow is bypassed around the first stage and introduced further downstream. Very rapid mixing of this bypass airflow with the first-stage combustion products is needed to complete the combustion process, with minimal NO_x formation as the rich gases are diluted. In the rich stage, the combustor liners must be cooled without the use of air film cooling, as is commonly used in current technology combustors. The use of film cooling is not acceptable because stoichiometric fuel–air mixtures would occur in regions close to the liner.

Accordingly, the abatement of aircraft turbine engine NO_x emissions requires the use of substantially more advanced combustor design technology than that used in currently operational (ca 1993) engines.

Low NO_x Combustor Design Technology

For Subsonic Aircraft Engines. To attain lower NO_x levels in the next generation of subsonic aircraft engines, efforts are underway to develop combustors with leaner primary zones and staging provisions through the use of separate combustion zones. These efforts are largely based on the results of previously conducted technology development efforts sponsored by the U.S. National Aeronautics and Space Administration (NASA) (17).

The dual annular combustor concept is generally typical of these low NO_x concepts. This concept features the use of two primary combustion zones in parallel. Both annuli, or zones, are individually fueled. One of the annuli is designed to serve as the pilot stage. The other annulus is designed to serve as the main stage. Only the pilot stage is fueled at starting, altitude relight, and idle conditions. At operating conditions above idle, both annuli are fueled and operate with lean primary-zone fuel–air ratios. An example of this design concept is shown in Figure 17.

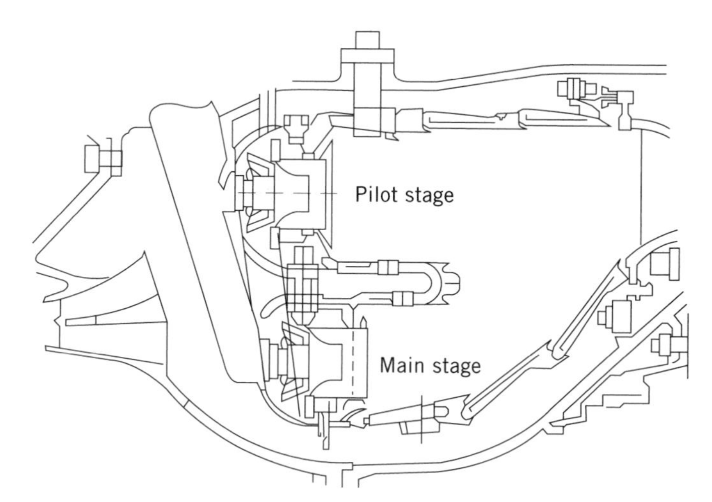

Figure 17. Cross-section drawing of a typical dual annular combustor.

Based on the results of the testing conducted to date, the NO_x abatement potential of these concepts appears to be in the range of 30–40% relative to current technology (ca 1993) combustors. Although the resulting combustors are more complex than current technology combustors, satisfactory performance and operability appear attainable with these configurations. The initial introduction of these combustors into operational engines during the 1995 to 2000 time period is a likely prospect.

For Future Supersonic Transport Aircraft Engines. Studies conducted by NASA indicate that, to prevent adverse ozone layer impacts, engines with NO_x levels up to 90% lower than those typical of current (1993) technology engines may be needed. A major thrust of the overall NASA initiative to evolve technology for future civil supersonic transport engines, therefore, involves the development of ultralow NO_x combustor technology (16). Two advanced combustor design concepts are being pursued: lean premixed–prevaporized combustion and rich–lean series-staged combustion.

Lean premixed–prevaporized combustion involves the premixing of the fuel and its combustion air upstream of the combustion zone. By premixing and prevaporizing the fuel so that lean and homogeneous mixtures result, significant NO_x reductions are possible. Combustor designs of this kind are complex because of the need for sophisticated combustion-process staging features.

As is discussed above, rich–lean series-staged combustion involves the use of a rich first stage with average fuel–air ratios above stoichiometric to suppress the formation of NO_x. A key concern associated with this concept is rich-stage liner cooling. Another concern is suppression of NO_x formation during the rapid dilution process between the rich and lean stage.

Considerable further development is needed to produce suitable versions of these two basic concepts for use in advanced supersonic transport engines. While these development efforts are specifically directed to advanced supersonic transport engines, the resulting technology is also relevant to future generations of civil subsonic aircraft engines. Hence, this technology is expected to provide a basis for reducing the NO_x levels of post-year-2000 subsonic aircraft engines to even lower levels than those of engines scheduled to enter service during the 1995 to 2000 time period.

AIRCRAFT ENGINE NOISE EMISSIONS

The design of modern aircraft propulsion systems must include consideration of the noise and acoustic fields generated by the propulsion system itself. For commercial aircraft, airport community noise is a prime consideration, and can dictate many design parameters and features in the engine. Aircraft interior noise is also a prime concern for commercial aircraft, and the engine–aircraft installation design can have a significant influence on whether or not interior noise is acceptable.

For both commercial and military aircraft, the sonic environment produced by the engine is a consideration, from the standpoint of designing the aircraft structure and nearby surfaces to withstand the generated sound levels without suffering sonic fatigue.

Sound levels generated inside the engine are also a design consideration, and care must be taken to avoid acoustic resonances inside engine ducts and cavities. Acoustically related instabilities associated with combustion can occur, and care must be taken to avoid combustor "howl" and afterburner "screech," as these phenomena can cause fatigue failures of critical engine components.

Types of Noise Emission

Various components in the aircraft engine propulsion system generate noise, and each component has its own unique characteristic sound or frequency spectrum. Two basic types of noise signature can be emitted. First, broadband or "white" noise, characterized by roaring or hissing sound, is generated when random turbulence is produced, either by flow over engine blade, vane, and duct wall surfaces, or in the mixing layers of the engine air flow streams. The frequency spectrum is typically a broad, smooth "hump" having a characteristic peak and associated peak frequency. The second type of noise produced is discrete-tone noise, characterized by a distinct whistle or whine, and is typically a sharp spike of noise occurring at a specific frequency, and often having harmonics or overtones.

In general, the fan, low pressure compressor (LPC), and low pressure turbine (LPT) produce both broadband or "white" noise and strong, discrete-frequency tones, usually at integer-multiples of blade-passing frequency of the rotor blades. The combustor and the jet exhaust mixing layers produce predominantly broadband or "white" noise. Also, fan stages that operate at supersonic tip speeds will often emit multiple-pure-tone or "buzz-saw" noise, which is characterized by discrete tones occurring at integer multiples of fan-shaft rotational frequency.

For military engine applications, the afterburner or augmentor is an additional source of noise, and can produce screech tones as a result of combustion-fed instabilities and resonances occurring in the augmentor under certain operating conditions. In addition, the exhaust jet velocities for a military engine are very high compared to a commercial engine, and so shock cells can form outside the nozzle, which interact with the turbulent eddies produced by the jet plume mixing, and this process generates both broadband and tone noise. This type of noise emission would also occur in the case of supersonic transport type aircraft propulsion systems, which are characterized by relatively low bypass ratio and high exhaust velocity (18,19).

Regulatory Issues, Current and Prospective

There are several types of prevailing rules and regulations governing commercial aircraft community noise. The primary requirement to be met is the FAR36 Stage 3 Regulation established by the Federal Aviation Administraton (FAA), and compliance demonstration is required in order to obtain FAA certification. Approximately 50% of all commercial aircraft flying today can comply with Stage 3. As older Stage 2 aircraft are phased out, the average noise levels around airports will drop until the Stage 3 saturation level is reached. Figure 18 shows a projected trend of community noise exposure vs time, with both a normal attrition rate and an imposed ban on Stage 2 aircraft.

There are other regulatory bodies which establish noise limits, and these include ICAO (International Civil Aviation Organization), the CAA (Civil Aviation Authority, Great Britain), and the JAA (Joint Aviation Authority, Europe). These agencies have regulations which are equivalent to the FAA rules, and FAA certification currently implies compliance with the others.

Local community airport rules, although not a requirement for certification, are often a primary concern for airlines, as noncompliance with these rules may restrict operations at certain airports. In general, these local airport rules are more stringent and more difficult to comply with than the FAA certification limits. Airports which have their own noise limits include Washington–National (DCA), Orange County John Wayne Airport (SNA), and London–Heathrow and London–Gatwick Airports. These

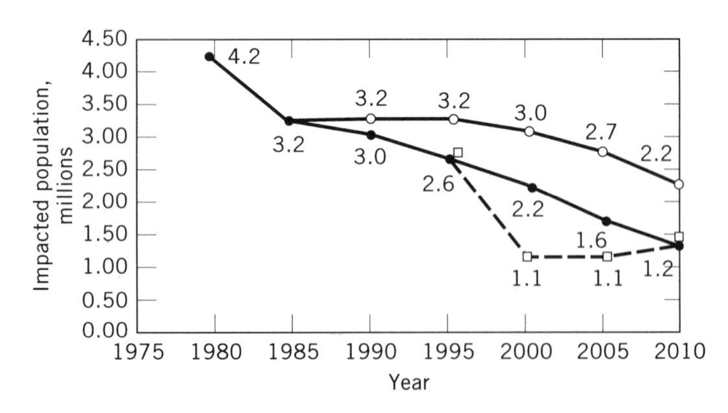

Figure 18. Trend of community noise exposure to aircraft noise from phase-out of stage 2 aircraft, given in terms of impacted population in residential areas around airports which are exposed to average day–night levels of 65 decibels or greater (18); (–○–○–), 35-year aircraft life; (–●–●–), 25-year aircraft life; (–□–□–), forced retirement.

examples each have their own noise metrics or parameters upon which their limits are based, and they are different from the FAA metrics. In addition, some airports have landing fees which are a function of the FAA/ICAO certification noise levels, and typically penalize noisier airplanes.

For a new aircraft system, aircraft community noise compliance is typically demonstrated by a controlled flight test, whereby noise measurements are made on the ground with microphones under the flight path for specific combinations of aircraft flight speed, altitude, and engine power setting. These microphones record an unsteady pressure vs time signal that forms the basis for subjective noise evaluation. The unsteady sound pressure vs time signal can be transformed to a frequency spectrum through a Fourier analysis of 0.5-second-interval time samples, as illustrated in Figure 19.

The resulting narrow-band spectrum can then be summed into frequency bands whose widths are proportional to their center frequency. When there are three frequency bands per octave (one octave corresponds to a doubling of the frequency), the spectrum is called a 1/3-octave spectrum, as shown in Figure 20. If the frequency scale is logarithmic, then the center frequencies are equally spaced. The root-mean-square pressure fluctuation amplitude (p') is expressed in terms of Sound Pressure Level (SPL) in decibels (named after Alexander Graham Bell), defined as

$$\mathrm{SPL}(f) = 20 \log (p'/p_{\mathrm{ref}})$$

The standard reference pressure (p_{ref}) is 0.0002 microbars. The concept of expressing the spectral content of sound energy on a logarithmic scale reflects the response characteristics of the human ear to sound loudness, (18,21).

The process of transforming the pressure–time history into a 1/3-octave spectrum at incremental times during the flyover can be done using 1/3-octave band analog filter analyzers, without going through the intermediate step of narrow-band (linear frequency scale) analysis. However, narrow-band analysis is often quite useful as a diagnostic tool, because it isolates more clearly the discrete tones in the spectrum.

The human ear does not respond uniformly to all sound frequencies; the response depends a good deal on the

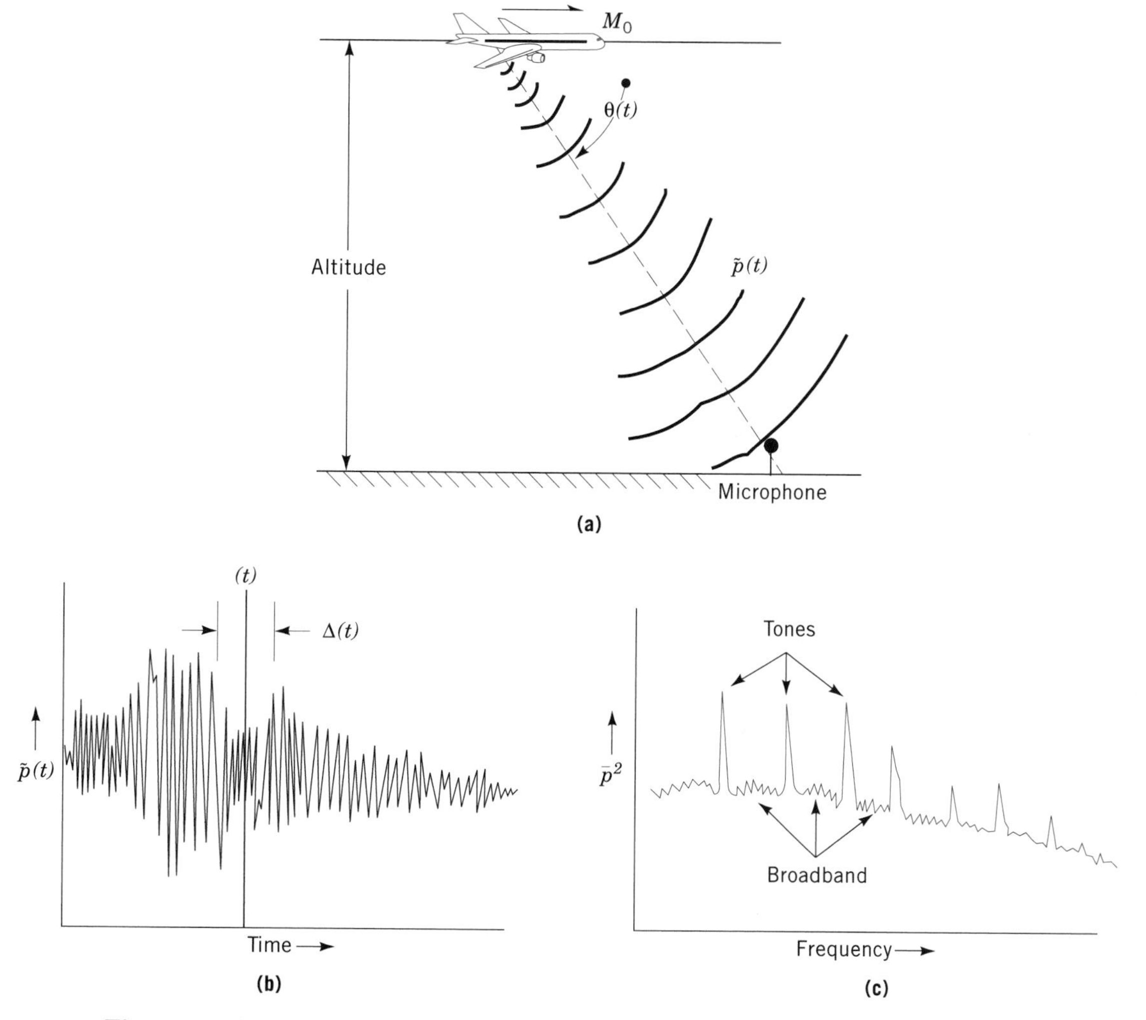

Figure 19. Aircraft flyover sound measurement (**a**) and signal processing to obtain time-history (**b**) of frequency spectrum during a flyover event. (**c**) Narrowband spectrum at time t over interval Δt.

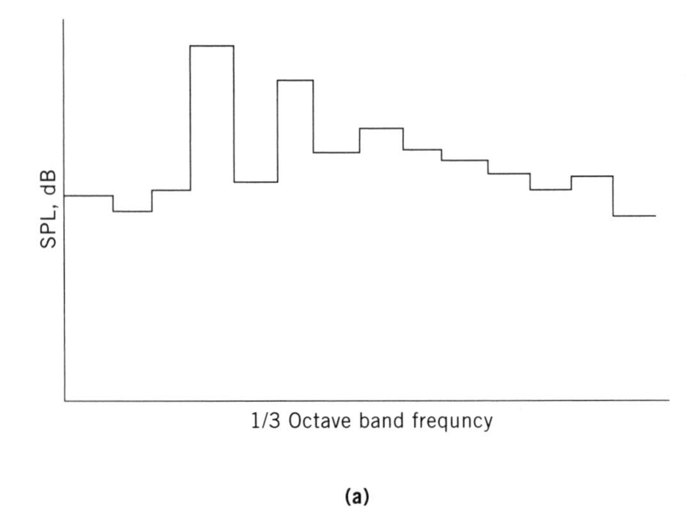

(a)

(b)

Figure 20. Noise metrics (frequency weighting) (**a**) and their time-history or directivity (**b**) (see text). (–●–●–) = OASPL (overall sound pressure level), equal weighting of all bands; (– – –) = dBA, "A" weighting, weighting factors are a function of frequency; and ( ) = PNL (perceived noise level), weighting factors are a function of both frequency and level (Noy-Weighting).

character and intensity of the sound. Based on psychoacoustic and statistical studies, different weightings are given to different sorts of noise. Some of the procedures used to evaluate noise subjectively include:

Overall Sound Pressure Level (OASPL): equal weighting for all frequencies; a measure of total sound energy.

A-Weighted Sound Level (dBA): used for evaluating industrial noise source annoyance; emphasizes sound levels in the 1000- to 5000-Hz frequency range.

Perceived Noise Level (PNL): used for evaluating aircraft community noise annoyance; emphasizes sound levels in the 2500- to 6300-Hz frequency range and applies weighting factors which are a function of the sound level itself.

Tone-Corrected Perceived Noise Level (PNLT): same as PNL, but additional correction is applied to account for a strong tone source in the spectrum.

If these metrics are calculated at all time slices during the flyover event, the resulting plot of these metrics vs aircraft location or time is called a directivity pattern, and is illustrated in Figure 20**b**. The integration of the PNLT directivity pattern over the time interval corresponding to PNLT values less than 10 dB below the peak value gives the parameter called Effective Perceived Noise Level (EPNL). EPNL is a measure of the time over which the highest noise levels must be endured, and is the metric used to demonstrate compliance with FAR36 Stage 3 Certification limits. The FAR36 Stage 3 Certification limits specify allowable limits as functions of Maximum Takeoff Gross Weight (MTOGW) and the number of engines, for three conditions, as shown in Figure 21: (*1*) sideline, typically 244–366 m (800–1200 ft) altitude; (*2*) takeoff, typically 488–792 m (1600–2600 ft) altitude; and (*3*) ap-

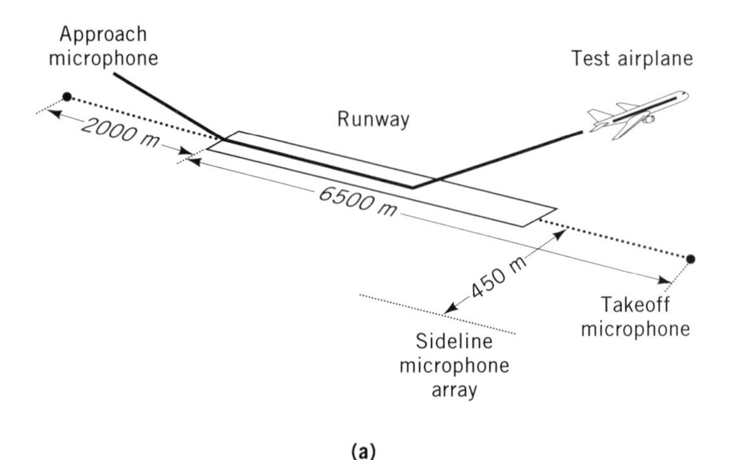

(a)

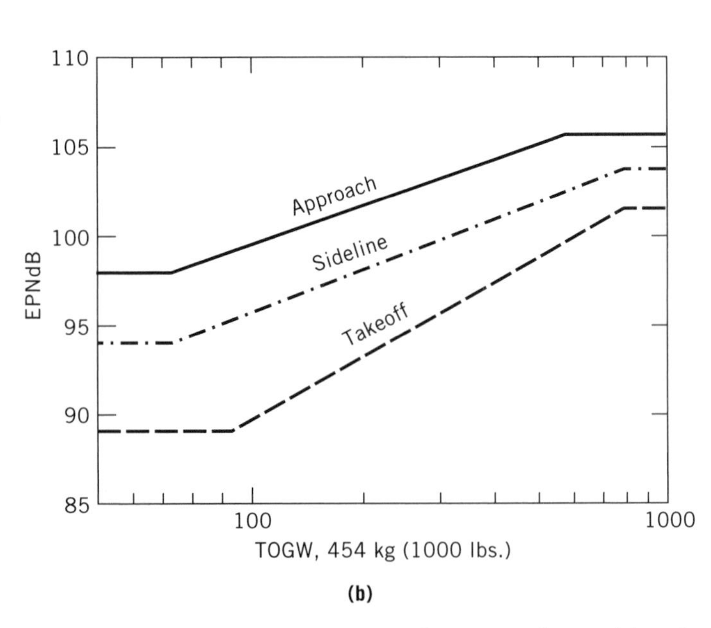

(b)

Figure 21. FAR36 Stage 3 noise certification conditions (**a**) and compliance limits (**b**) for two-engine jet-powered aircraft; ( ), approach limits; (–●–●–), sideline limits (– – –), takeoff limits.

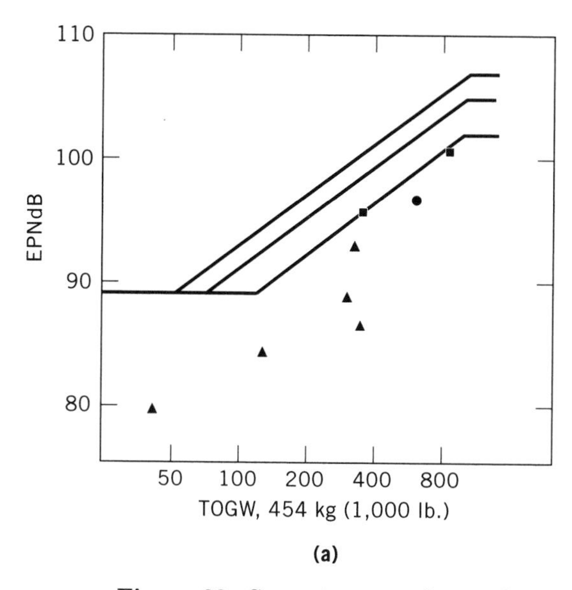

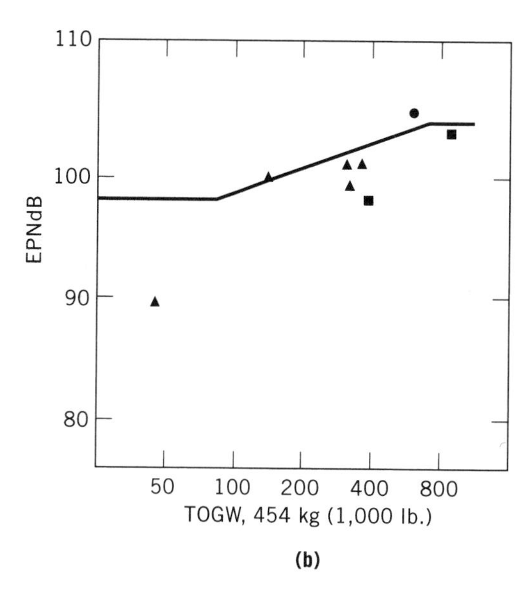

Figure 22. Current status of several typical aircraft noise levels relative to certification limits of FAR36 Stage 3; ▲, twinjet; ●, trijet; ■, quadjet. (**a**) Takeoff; (**b**), approach.

proach, 122 m (400 ft) altitude. The FAR36 Stage 3 Certification monitoring conditions are illustrated in Figure 21. The certification status of some typical aircraft are shown in Figure 22.

It can be expected that both federal and local airport noise rules and limits will become more stringent in the future, as frequency of operations and air traffic increases. As the trend in Figure 18 implies, the noise reductions gained by phase-out of the older, noisier stage 2 aircraft will eventually be mitigated by the increase in air traffic and attendant community noise exposure in the next ten to fifteen years. Future noise limit reductions will have to be implemented if noise exposure is to be contained while permitting growth in air transportation.

Design Principles and Practice for Noise Emissions Reduction

As discussed in the previous section, the major sources of commercial engine noise are the fan, LPC, LPT, the combustor, and the jet exhaust mixing. Early commercial aircraft were powered by low bypass-ratio engines, and the noise was dominated by the combustor and jet exhaust mixing, because most of the engine airflow passes through the combustor and exhausts at high jet velocity. With the advent of the high-bypass engine, more of the engine flow bypasses the combustor and exhausts at relatively low jet velocity. However, the bypass airflow is given momentum by a fan powered by a turbine, and the horsepower absorbed and produced by these two components is considerable.

The current trend in the development of advanced commercial engine propulsion systems is to employ cycles with higher and higher bypass ratios, in order to achieve high propulsive efficiency. A qualitative picture of what happens to the noise sources in an engine as a function of bypass ratio is shown in Figure 23. As the bypass ratio increases, the broadband sources (jet, combustor) decrease rapidly while the tone sources (fan, LPC, LPT) increase, until the dominant sources are the tone sources. Currently produced modern turbofan engines (ca 1993) have bypass ratios on the order of 4.5 to 6. At full-power takeoff conditions, the noise of these engines is typically dominated by jet-mixing noise, but the fan exhaust noise is only a few dB lower in level. These engines are near, but to the left of, the noise minimum in Figure 23.

The Ultra-High Bypass (UHB) engines being considered for the next generation of commercial aircraft propulsion systems have bypass ratios on the order of 8 to 12 and even higher. Figure 23 suggests that these new engine systems will have noise levels as high as or even higher than current engines, unless the technology is developed to reduce the tone sources, particularly the fan noise. This takes on even greater importance when it is

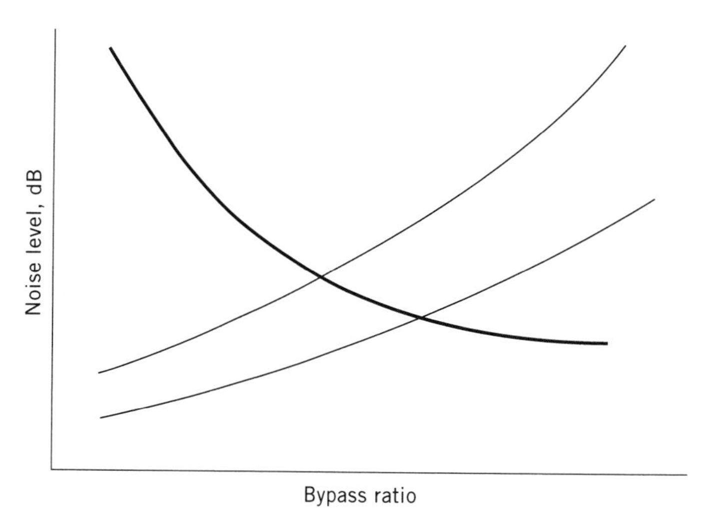

Figure 23. Effect of engine bypass ratio selection on relative contributions of engine noise sources. Curve, broadband sources (combustor and jet); hatched band, current (ca 1993) (upper limit levels) and advanced (lower limit levels) tone sources (fan, compressor, and turbine).

recognized that more stringent rules may be passed in the future, and that more local airports may follow the course set by Washington–National and Orange County Airports (22).

Figure 24 schematically shows the various sources of fan noise that we currently know about. The strongest sources of fan noise involve the unsteady aerodynamic interactions between stationary and rotating blade rows, or convected distortions and turbulence interacting with rotating blade rows. The resulting unsteady lift forces generated on the "receiving" blade row produce rotating acoustic pressure patterns in the fan duct. The efficiency of conversion of the fluctuating lift-force energy into acoustic energy is a function of the rotating blade row tip speed Mach number, the duct flow Mach numbers, and the frequency of the noise source.

Methods for reducing the noise caused by unsteady interactions include: (*1*) reducing the magnitude of the "gust" or distortion, ie, minimizing upstream blade wake defects, minimizing downstream obstacle pressure fields, reducing or eliminating inflow distortions and separations, etc; (*2*) selecting the blade number and "gust" number combination to minimize the acoustic efficiency of the resulting unsteady interaction, including using cut-off, whereby the acoustic pressure waves decay exponentially inside the duct; and (*3*) employing duct liners, ie, mufflers, to suppress the generated acoustic pressure waves inside the duct before they escape from the inlet or exhaust.

At supersonic tip speeds, fan rotors produce rotating, rotor-bound, propagating acoustic pressure waves similar to Mach waves emanating from a supersonic aircraft wing. These rotating pressure patterns produce not only discrete tones at the blade-passing frequency and harmonics thereof, but also a particularly annoying phenomenon referred to as "buzz-saw" noise, so named for its characteristic sound. This effect, also called Multiple-Pure-Tone (MPT) noise, produces tones at multiples of shaft rotation frequency, and results from the small blade-to-blade variations stemming from manufacturing irregularities, which cause the blade leading edge Mach waves to coalesce in uneven patterns as they propagate upstream in the fan duct. Duct liners are usually effective in suppressing "buzz-saw" noise, but sufficient inlet length is required for this to be so.

As has been mentioned, nacelle inlet and exhaust ducts are usually lined with sound-absorbent panels to suppress fan noise. Accommodation must be made for adequate liner surface area, panel depth and panel face sheet porosity in the nacelle design. The trend to UHB engine systems results in very large engine diameters and lower rotative speeds. The acoustic consequences are lower blade-passing frequencies, which require deeper liner designs. In addition, nacelle length:diameter ratios may be smaller to help offset the greater engine weight, which "short-changes" the liner design. Hence it becomes necessary to design liners that are more effective in suppressing fan noise, by using smaller length and depth relative to diameter.

For low bypass-ratio engines where jet-mixing noise is the dominant source, the jet-mixing noise can be reduced by using mechanical suppressors. These mechanical suppressors are basically mixer nozzles that break up the large jet into many small jets which mix and decay more rapidly, and hence reduce the mixing noise.

For current (ie, the Concorde) and future High Speed Civil Transport (HSCT) aircraft, jet mixing and shock-associated noise are the primary sources of noise at takeoff. By any estimates, it is very ambitious to expect an HSCT to meet FAR36 Stage 3 noise limits. The current Anglo-

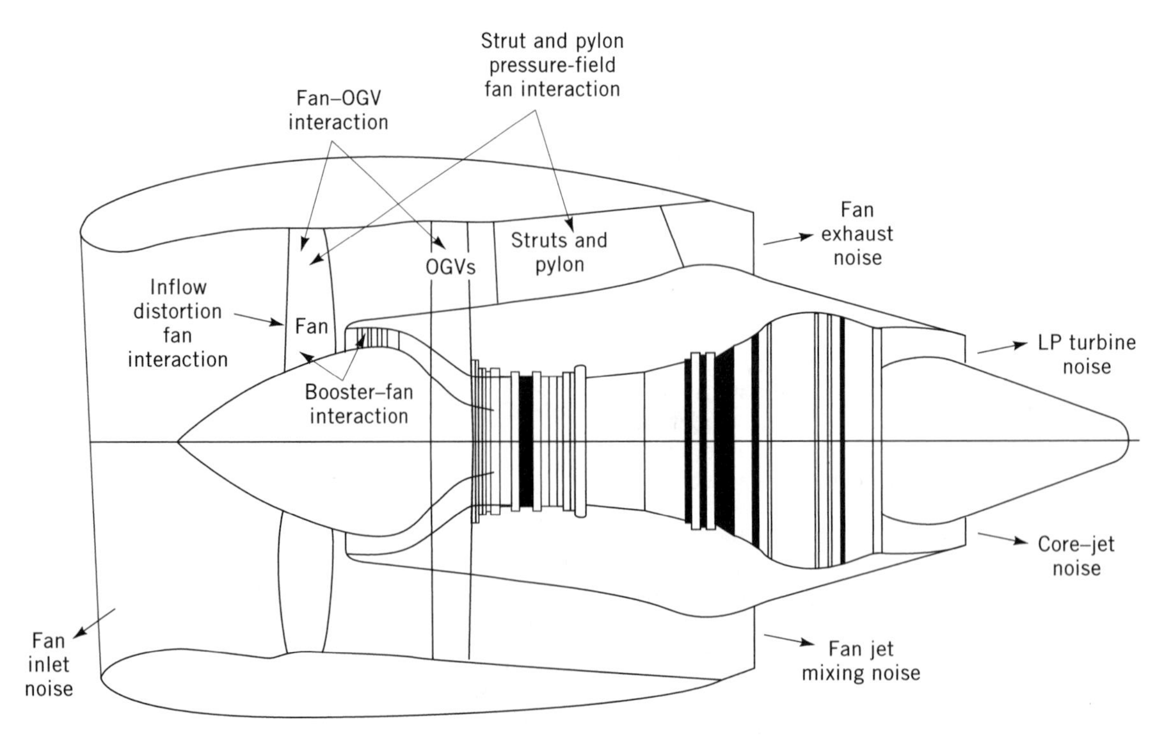

Figure 24. Noise generation mechanisms for a typical high bypass ratio turbofan engine. OGV,

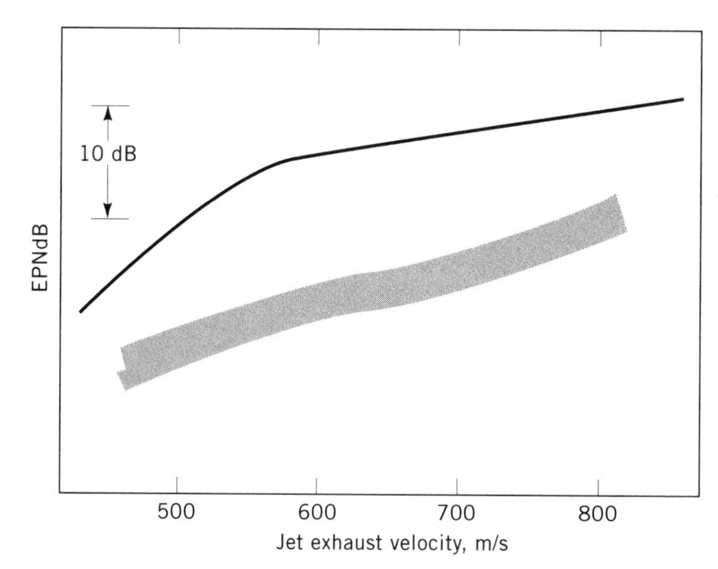

Figure 25. Jet-exhaust noise-level dependence on jet-exhaust velocity for a round, conical nozzle, and demonstrated noise reductions achieved with various types of jet exhaust suppressors. Curve, unsuppressed conic nozzle; band, range of results of various scale-model and full-scale suppressors.

French Concorde SST has an exemption from the Stage 3 rules. However, the FAA and ICAO may likely require that any new HSCT aircraft comply with at least the current Stage 3 limits. In addition, any new HSCT introduced into service must also contend with the sonic boom issue, and with the problem of emissions and ozone-layer depletion.

An HSCT engine needs to be a very high specific-thrust propulsion system. This means high jet-exhaust velocity so that that controlling HSCT noise means controlling high velocity jet noise. Figure 25 shows a typical trend curve of jet noise (including mixing noise and shock-associated noise) as a function of jet velocity for a simple conical nozzle, and a range of results for various jet-suppressor designs obtained from scale-model acoustic test programs. As much as 10 to 15 dB suppression has been demonstrated in scale-model tests, depending on the jet velocity. The best suppressors tend to be the most complex and the heaviest, and also to impose the highest performance losses.

The development of the Variable-Cycle Engine (VCE), however, makes the possibility of achieving a Stage 3 HSCT engine a little more feasible, because it allows the engine to operate at lower jet velocity at take-off while maintaining high specific thrust at supersonic cruise speeds.

BIBLIOGRAPHY

1. I. E. Treager, *Aircraft Gas Turbine Engine Technology*, 2nd ed., McGraw-Hill Book Co., Inc., New York, 1979.
2. P. J. McMahon, *Aircraft Propulsion*, Pitman, London, UK, 1971 p. 260.
3. W. Kent, *Mechanical Engineers' Handbook*, 12th ed., Vol. 2, John Wiley & Sons, Inc., New York, 1950.
4. R. D. Bent and J. L. McKinley, *Aircraft Powerplants*, McGraw-Hill Book Co., Inc., New York, 1985.
5. J. L. Kerrebrock, *Aircraft Engines and Gas Turbines*, MIT Press, Cambridge, Mass., 1978.
6. *The Jet Engine*, 4th ed., Rolls Royce plc, 1986.
7. F. F. Ehrich, "Turbine Propulsion," "Turbofan," "Turboject," "Turboprop," "Turboramjet," "Aircraft Engine Performance," *Encyclopedia of Science and Technology*, McGraw-Hill Book Co., Inc., New York, 1991; Vol. 18, pp. 618–632, Vol. 1, pp. 279–284.
8. R. C. Hawkins, *Aerospace America*, 52–55, (Oct. 1984).
9. H. Lambert *Jane's All the World's Aircraft*, Janes Info. Group, Coulsdam, UK, 1992, pp. 618–705.
10. F. F. Ehrich, "Energy Conversion: Jet Engine," *Encyclopaedia Britannica*, ed 15, 1994, Vol. 2, pp. 363–368.
11. S. W. Korthals-Altes, *Technol. Rev.* 43–51, (Jan. 1987).
12. B. Sweetman, *Interavia*, 305–308, (March 1986).
13. C. Johnson, J. Henshaw, and G. McInnes, *Nature* **355,** (1992).
14. M. Barrett, WWF Research Paper, August 1991.
15. *International Civil Aviation Organization, Annex 16*, Vol. 2, 1981.
16. H. L. Wesocky and M. J. Prather, *Proceedings of Tenth International Symposium On Air Breathing Engines*, September 1991.
17. "Aircraft Engine Emissions," NASA CP 2021, 1977.
18. E. J. Richards and D. J. Mead, eds., *Noise and Acoustic Fatigue in Aeronautics*, John Wiley & Sons, Ltd., London, 1968.
19. H. H. Hubbard, ed., *Aeroacoustics of Flight Vehicles: Theory and Practice*, NASA Reference Publication 1258, Vol. 1, WRDC Technical Report 90-3052, August 1991.
20. FAA Report to the U.S. Congress, August 1989.
21. M. J. T. Smith, *Aircraft Noise*, Cambridge University Press, Cambridge, UK, 1989.
22. P. R. Gliebe, *DGLR / AIAA 14th Aeroacoustics Conference*, paper no. DGLR/AIAA-92-02-037, Aachen, Germany, May 11–14, 1992.

AIRCRAFT FUELS

William G. Dukek
Consultant
Summit, New Jersey

Aircraft fuels have evolved since World War II from aviation gasoline for piston engines to kerosene for gas turbine jet engines. Today aviation gasoline is a shrinking specialty product used by general aviation while commercial airlines, business aircraft, and military aviation have turned jet fuel into a principal product of the petroleum industry ranking with automotive fuels and industrial oils in importance. See also Aircraft engines; Petroleum products and uses; Kerosene; Afterburning.

The gas turbine powerplant which has revolutionized aviation derives basically from the steam turbine adapted to a different working fluid. The difference is crucial with respect to fuel because steam can be generated by any heat source whereas the gas turbine requires a fuel that efficiently produces a very hot gas stream and is also compatible with the turbine itself. The hot gas stream results from converting chemical energy in fuel directly and continuously by combustion in compressed air. It is expanded in a turbine to produce useful work in the form of jet thrust or shaft power.

The jet engine combines three basic steps (Fig. 1). Air is compressed in a series of stages, fuel is burned continuously and intensively in the compressed air, and the hot gas is then expanded through a turbine which extracts energy to run the compressor and the fan and also provide shaft power. The residual exhaust mixes with fan air and exits through a nozzle as a high velocity jet which provides forward thrust to the system. The gas turbine concept was adapted to aircraft in 1930. At high speed and altitude this type of engine improved in efficiency and made jet propulsion competitive with propellor drive (1). The W-1 engine of 1937 has since proliferated into much more efficient aircraft powerplants ranging in size from small auxiliary power units to the giant turbofans which propel wide-bodied airliners. In its most advanced form, the aviation gas turbine for supersonic aircraft, a second combustion step is added to the three basic steps of compression, combustion, and turbine work extraction. Additional fuel is burned in the turbine exhaust to increase the velocity of the jet from the nozzle and gain more thrust.

Ground turbines carry out the same three basic steps of the jet engine but residual energy of the turbine exhaust is extracted by a waste heat boiler or regenerator. Industrial turbines have heavy compressors, external combustion systems, and massive turbine sections. Lightweight design for aircraft required new high temperature metals and advanced concepts for cooling compression, and combustion. Today, these lightweight types have been adapted for many ground applications in gas compression, pumping, electrical power generation, and general purpose utility units. Because the gas turbine is fundamentally a heat conversion device the combustor or gas generator is the heart of the engine. Combustion of fuel with a portion of compressed air is continuous. Early developers of gas turbines depended upon the experience with oil-fired furnaces to design fuel nozzles and combustion chambers (2).

The first fuels burned in a turbine combustor were petroleum liquids: kerosene in the prototype jet engine and diesel fuel in the industrial turbine, selected because of their availability and convenience. After fifty years of experience the gas turbine has been adapted to a wide variety of combustible gases and liquids. Although such a wide-ranging appetite for fuels capable of releasing heat energy would appear to classify the gas turbine as fuel-insensitive, clearly defined limits on fuel properties have developed to optimize its performance. The aircraft propulsion unit is most demanding, accepting only certain liquid distillates; the industrial turbine least demanding. The special requirements of gas turbine fuels that have developed relate to the individual processes that take place in the engine itself and in the support systems for handling fuels. It is evident that aviation gas turbines place more severe demands on liquid fuels than ground turbines and that gaseous fuels are much easier to utilize since they need to supply only heat energy.

A kerosene for turbine research to converse gasoline for the impending war effort was used because it had better low temperature properties than a diesel fuel. This decision tended to establish a dividing line between air and ground turbines; subsequent developments led to lighter fuels for jet engines and heavier fuels for industrial turbines. Aircraft have continued to fly on kerosene fuel or a wide-cut blend of heavy naphtha and kerosene. Supersonic transports and advanced military vehicles are also operated on kerosene fuel. Most ground turbines utilize liquid fuel systems based on standard No. 2 diesel or home burner fuel.

GAS TURBINE PRODUCTS COMPOSITION

Because the jet aircraft is a weight-limited vehicle, a high premium is assigned to hydrocarbon fuels with a maximum gravimetric heat content or hydrogen-to-carbon ra-

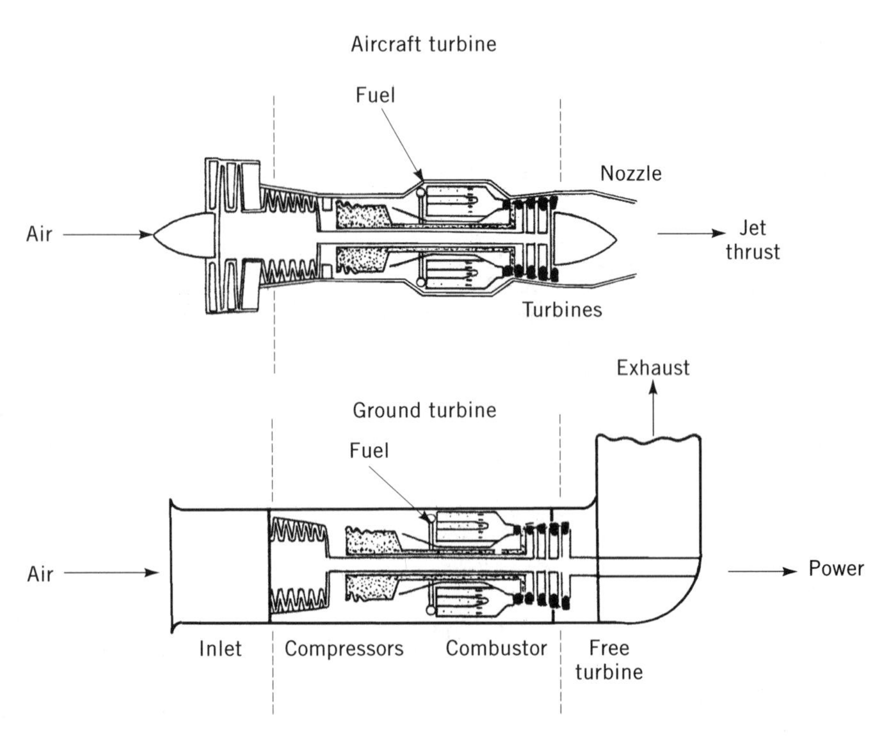

Figure 1. Schematic of aircraft and ground gas turbines.

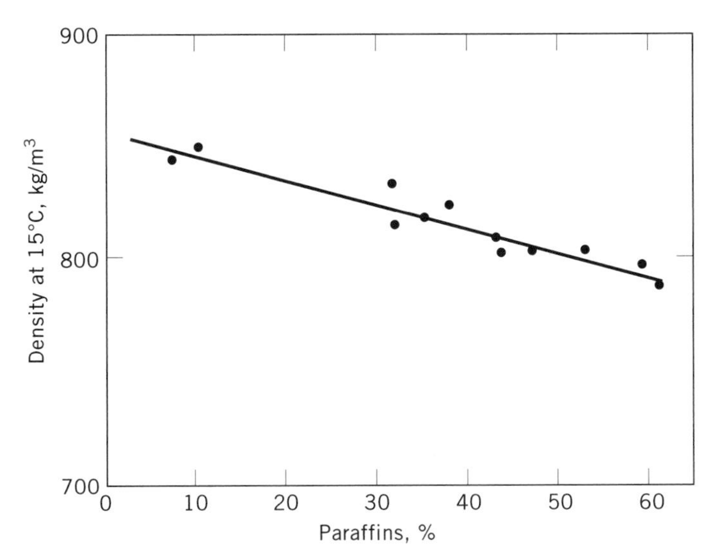

Figure 2. Density vs paraffin content of gas turbine fuels (150–288°C fractions).

tio. Paraffinic fuels have high mass heat contents (about 45 MJ/kg or 10.8 × 10^3 kcal/kg) which decrease very slowly as chain length (molecular weight) increases. However, because the density of these fuels is low, the heat content on a volumetric basis is less than that of nonparaffinic fuels. Since fuels are measured and sold on a volumetric basis, the aircraft user obtains less energy for each cubic meter of the preferred paraffinic fuel but more energy per unit mass. Ground turbine users are not concerned with weight limitations and seek maximum energy per unit of cost which means that heavier fuels are more economical for ground application.

Fuel properties are largely determined by the nature of the crude oil from which it is derived. This is illustrated by the typical density–paraffin content curve for the kerosene fractions (150–288°C) of different crudes (Fig. 2). Aromatics in this cut are of special concern because of their effects on combustion and elastomers. The mass spectrometric data in Table 1 illustrate that a wide range of single and multiring aromatics is possible even within a total aromatic concentration of 25%, the maximum level permitted by most specifications.

Specifications

The global nature of jet aircraft operations has mandated that aviation fuel quality be closely controlled in every part of the world. Specifications tend to be industry standards issued by a consensus organization like ASTM or by a government body rather than manufacturer's requirements. In Table 2 are listed some of the requirements of the principal grades of civil and military aviation turbine fuels in use throughout the world. Domestic and international airlines use fuel modeled on ASTM specification grades. Specifications issued by the British Ministry of Defence, the recommendations of the International Air Transport Association, and ASTM specifications are coordinated as a Joint System Check List to control quality at many worldwide airports where jointly operated fueling systems prevail. Specifications issued by former Soviet governments are similar to Free World aircraft fuels but not identical in all respects. International airlines that fly to East Bloc countries can utilize both types of fuels without difficulty. Except for severe Arctic areas where Jet B fuel is used, essentially all civil aviation uses kerosene fuel. About half of the demand is Jet A fuel marketed in the United States and half is Jet A1, the international fuel marketed outside the United States similar to Jet A except for freezing point. A lower temperature limit of −47°C for the (wax) freezing point of Jet A1 is required for long duration international flights. The same fuel is labeled JP-8 by the United States Air Force and used at most bases outside the United States to replace JP-4, the wide-cut fuel originally developed after World War II to assure maximum availability. JP-4 is still the primary Air Force fuel used at U.S. bases but is being replaced by JP-8 in a continuing conversion program. The other military fuel of importance is JP-5, the high flash point kerosene used by the Navy to ensure handling safety aboard carriers.

Manufacture

Aviation turbine fuels are primarily blended from straight-run distillates rather than cracked stocks because of specification limitations on olefins and aromatics. However, in refineries where heavy gas oil is hydrocracked, ie, a process that involves catalytic cracking in a hydrogen atmosphere, aviation fuels can include hydro-

Table 1. Composition[a] of 150–288°C Kerosenes

Crude Source	Middle East			North Africa	United States				South America			
Saturates, wt %	78.8	82.4	80.3	83.1	85.5	81.0	76.8	81.7	78.8	76.8	76.1	74.5
Paraffins	63.0	61.7	54.0	47.4	44.4	37.9	35.3	44.0	34.9	31.3	9.0	6.0
Cycloparaffins, single ring	10.6	12.0	14.5	23.8	24.1	22.9	27.3	26.9	27.3	29.0	33.3	31.7
Cycloparaffins, two rings	4.7	7.6	8.7	9.8	12.0	17.9	11.3	9.6	12.8	12.6	24.9	28.1
Cycloparaffins, 2 + rings	0.9	1.1	3.2	3.1	5.0	2.3	2.9	1.2	3.8	3.9	8.9	8.7
Aromatics, wt %	18.4	16.3	18.2	14.7	14.0	17.7	21.9	17.0	19.9	22.0	23.7	25.5
Single ring	16.9	14.1	14.8	11.3	11.8	13.9	19.0	15.1	14.8	16.6	16.4	16.8
Two rings	1.4	2.0	3.0	3.2	2.1		2.6	1.9	4.7	5.1	3.6	7.7
2 + rings	0.1	0.1	0.2	0.2	0.1	3.8	0.3		0.5	0.4	0.7	1.0

[a] By mass spectrometric analysis.

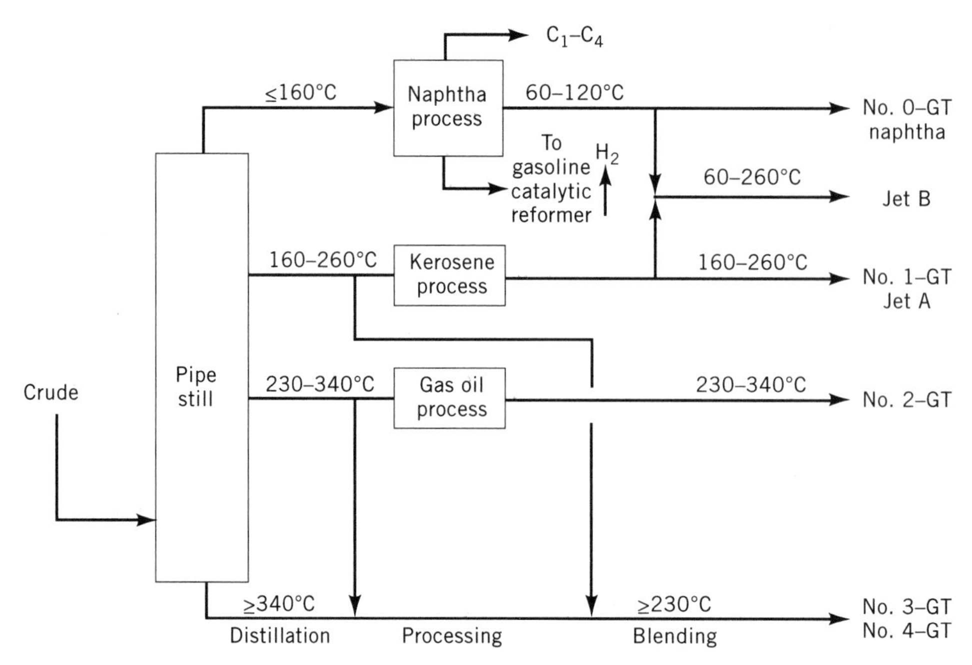

Figure 3. Refinery process for gas turbine fuels. Processing may include hydrogen treating.

cracked components since they are free of olefins, low in sulfur, and stable. Hydrocracking provides a means to extend availability of gasolines and distillates by converting most of the heavy ends in crudes to lighter products.

A schematic of a typical refinery processing sequence appears in Figure 3. In this example, the basic processes after distillation are catalytic reforming for gasoline and hydrogen processing of naphtha, kerosene, and gas oil using the hydrogen from reforming. A hydrocracking unit usually requires a separate source of hydrogen. The approximate boiling ranges of the sidestream cuts and finished gas turbine fuels are shown.

The distillation cut-points must be closely controlled to yield a product that meets the requirements of flash point on one hand and freezing point on the other. In practical terms, a jet fuel requires a fraction of about 60°C initial boiling point and a final boiling point not exceeding 300°C. The lighter portion of this fraction contains gasoline components and meets the specification for JP-4. The heavier portion above 160°C must be tailored to either the Jet A or Jet A1 freezing point, but the final boiling point is dependent on crude composition. With higher aromatic crudes, undercutting is required to meet compositional limits or a test such as smoke point. Lower boiling components will compensate for aromatic or freezing point limitations, but sometimes it is impossible to take advantage of them because a dual-purpose kerosene usually requires a higher flash point than jet fuel. The distilled fractions from the crude are apt to contain mercaptans or organic acids in excess of specification limits. A caustic wash is normally used to control acidity and remove traces of hydrogen sulfide. Removal or conversion of odorous mercaptans is carried out in a sweetening process. The most widely used chemical treatment today, Merox sweetening, utilizes dissolved air to oxidize mercaptans to disulfides over a fixed bed cobalt chelate catalyst (3). It has the advantage of simplicity and minimum waste disposal problems but does not lower the sulfur content. A modern version of the old doctor process, Bender sweetening, uses lead plumbite deposited on a fixed bed to carry out the mercaptan oxidation and spent doctor regeneration processes simultaneously, but product quality is more difficult to control and waste disposal problems are greater.

Most of the crudes available in greatest quantity today are high in sulfur and yield products that must be desulfurized to meet specifications or environmental standards. The process most widely used is mild catalytic hydrogenation. Hydrofining uses a fixed bed catalyst, usually cobalt molybdate on alumina, to convert all but ring sulfur compounds to H_2S which is then removed (and recovered) from recycle gas. At reactor temperatures below 300°C only mercaptans react. This technique is called hydrosweetening. At reactor temperatures of 400–500°C desulfurizations levels are 80% or better, and other reactive species including some multiring sulfur compounds are hydrogenated. The product of the hydrofiner is caustic washed to remove traces of dissolved H_2S and passed through sand or clay to blend tanks.

Hydrogen treating removes oxygen containing species such as phenols and naphthenic acids that are found in some crudes. The former tend to perform as natural inhibitors of hydrocarbon oxidation by trapping peroxy radicals while the latter act as natural lubricating agents. Recognition of this side-effect of hydrogen treatment led refiners to add antioxidants to hydrotreated components; this procedure is now a specification requirement.

Blending of two or more components is carried out to match as closely as practical the various specification constraints. At this point, additives are usually introduced, eg, an antioxidant to inhibit gum formation in storage or a metal deactivator to deactivate any dissolved metal ions

Table 2. Selected Specification Properties of Civil and Military Aviation Gas Turbine Fuels

	ASTM D1655[a]		Mil-T-5624-N[a]		Mil-T-83133C[a]
Characteristic	Jet A kerosene A/L and	Jet B wide-cut Gen Avn	JP-4 wide-cut USAF	JP-5 kerosene USN	JP-8 kerosene USAF
Composition					
Aromatics, vol % max	25[b]	25[b]	25	25	25
Sulfur, wt % max	0.3	0.3	0.4	0.4	0.3
Volatility					
Dist. Temp Max °C: 10% rec'd	205			205	205
Dist. Temp Max °C: 50% rec'd		190	190		
Dist. Temp Max °C: End pt	300		270	300	300
Flash pt, °C min	38			60	38
Vapor pressure at 38°C, kPa Max (psi)		21(3)	14–21 (2–3)		
Density at 15°C, kg/m^3	775–840	751–802	751–802	788–845	775–840
Fluidity					
Freezing pt, °C max	−40[c]	−50	−58	−46	−47
Viscosity at −20°C, mm^2/s (= cSt) max	8.0			8.5	8.0
Combustion					
Heat content, MJ/kg, min[d]	42.8	42.8	42.8	42.6	42.8
Smoke pt, mm, min	18[e]	18[e]	20	19	19
H_2 Content, wt % min			13.5	13.4	13.4
Stability					
Test temp[f], °C min	245	245	260	260	260

[a] Full specification requires other tests.
[b] For aromatics above 20% (22% for Jet A1) users must be notified.
[c] International airlines use Jet A1 with −47°C freeze point.
[d] To convert MJ to kcal, multiply by 239.
[e] Plus naphthalenes 3 vol % max unless smoke point exceeds 25.
[f] Thermal stability test by D3241 to meet 3.3 kPa (25 mm Hg) pressure drop and Code 3 Deposit Rating.

(see Table 2). Military fuels have made the addition of anti-icing agents and corrosion inhibitors mandatory, the former because military aircraft fuel systems do not include fuel filter heaters to prevent ice formation, the latter to reduce the tendency of fuels to pick up rust particles from pipelines and tanks. The Air Force also requires that antistatic additive be included in JP-4 to reduce the hazard of static generation and discharge in handling fuel and in operating aircraft with foam-filled tanks. It was discovered that the anti-icing additives inhibit growth of microorganisms in aircraft and storage tanks and that certain corrosion inhibitors are excellent lubricity agents for prolonging the life of engine pumps. With civil fuels, anti-icing agents are not required since air transports are equipped with fuel filter heaters to prevent ice formation. Corrosion inhibitors are not normally used in civil fuels since they tend to degrade the efficiency of filter-coalescers and would be removed in the clay filters that are commonly installed at receiving terminals and airports. However, antistatic additives are widely used in civil fuels, particularly Jet A1 at international airports. While antistatic additive is introduced into some Jet A to safeguard against static discharge in truck filling, the widespread use of clay filters in the United States to clean up fuel before airport delivery makes refinery addition impractical. Table 3 also lists a boron containing biocide used as a shock treatment additive to inhibit the tendency of organisms to form fungal mats in aircraft fuel tanks.

Distribution and Handling

Preserving the quality of gas turbine fuels between the refinery and the point of use is an important but difficult task. The difficulty arises because of the complicated distribution systems of multiproduct pipelines which move fuel and sometimes introduce contaminants. The importance is reflected by the sensitivity of gas turbine engines and fuel systems to water, corrosion products, metal salts, microorganisms, and other extraneous materials that can be introduced by the distribution system. Most products moved through pipelines contain additives, eg, detergents, deicers, antifouling dispersants, and rust inhibitors in gasoline, corrosion inhibitors and dispersants in heating oil; or ignition promotors in diesel fuel. Traces of these additives left on pipeline walls are picked up by nonadditive jet fuel which makes it difficult for pipeline operators to deliver gas turbine fuel of the same quality as that introduced into the line.

The principal means of removing contaminants are tank settling and filtration. With aviation fuels it is common practice to install several stages of cartridge type filter–coalescers between the storage tank and the aircraft delivery point. Figure 4 is a schematic of a typical airport

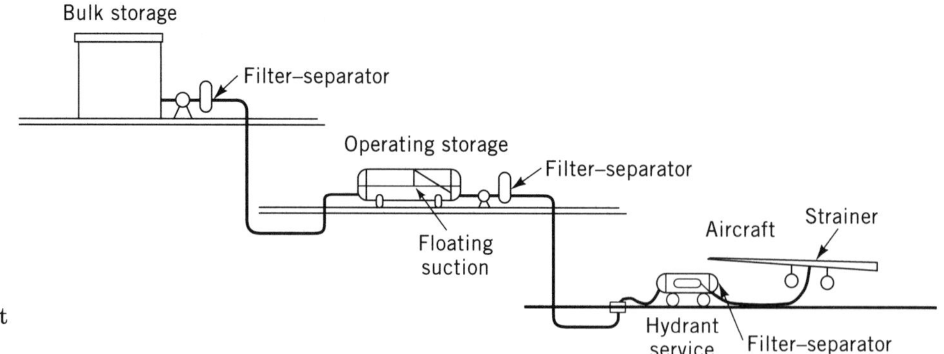

Figure 4. Typical airport hydrant system.

fueling system which transfers fuel by underground hydrant lines. Filter elements of fiberglass covered with synthetic fabric are designed to coalesce water and particulates at high flow rates. Coalesced water is prevented from passing with fuel by a hydrophobic barrier filter.

Tank settling as a means of contaminant removal is not very efficient with fuels having the viscosity of kerosene. It is common practice to design tanks with cone down drains and floating suctions to facilitate water and solids removal. Contamination control by filtration would be relatively easy if it were not for surface-active materials produced during fuel manufacture or picked up in distribution systems from pipe walls and tank bottoms. These surfactants stabilize water-in-oil emulsions, disarm filter elements, coat solid particles, and promote growth of organisms at fuel–water interfaces. When they reach the aircraft tank, the surfactants tend to trap water containing dissolved salts and particulates, causing fungal growths, coating attack, and metal corrosion.

Gas turbine fuels can contain natural surfactants if the crude fraction is high in organic acids, eg, naphthenic (cycloparaffinic) acids of 200–400 mol wt. These acids readily form salts that are water soluble and surface-active. Older treating processes for sulfur removal can leave sulfonate residues which are even more powerful surfactants. Refineries have installed processes for surfactant removal. Clay beds to adsorb these trace materials are widely used and salt towers to reduce water levels will also remove water soluble surfactants. In the field clay filters designed as cartridges mounted in vertical vessels are also used extensively to remove surfactants picked up in fuel pipelines or contaminated tankers or barges.

Specifications for gas turbine fuels prescribe test limits that must be met by the refiner who manufactures fuel; however, it is customary for fuel users to define quality control limits for fuel at the point of delivery or custody transfer. These limits must be met by third parties who distribute and handle fuels on or near the airport. Tests on receipt at airport depots include appearance, distillation, flash point (or vapor pressure), density, freezing point, smoke point, corrosion, existing gum, water reaction, and water separation. Tests on delivery to the aircraft include appearance, particulates, membrane color, free water, and electrical conductivity.

The extensive processing and cleanup steps carried out on gas turbine fuels produce a purified liquid dielectric of high resistivity which is capable of retaining electrical charges long enough for buildup of large surface voltages. Because fuel is filtered at high flow rates through filter media of extensive surface area, the ionic species that adsorb on the filter surface generate a static charge. The current carried by fuel increases sharply as fuel flows through a filter and then decreases at a rate determined by the fuel conductivity. The charge remaining in the fuel at any given point is determined by the residence time available for charges to recombine. In a hose delivering fuel to an aircraft, only a few seconds are available for charge relaxation. The result is a possible hazard: a tank filled with charged fuel that under some circumstances can discharge its energy to ground in a spark capable of igniting fuels mists or vapors (4).

This aspect of fuel handling has received much attention because of a number of accidents that have resulted in tank explosions, most often in filling tank trucks but also in fueling aircraft. Several approaches have been taken to reduce the risk of static discharge. The most common method requires introduction of an additive to increase the electrical conductivity of the fuel to a safe level, ie, to speed up charge relaxation to a fraction of a second. Charge generation can be decreased by using low-charging screens instead of filters or by reducing flow rates. The conductivity additives used in fuel are readily removed in clay filters, or lost in pipeline movement and must be monitored by conductivity tests. The additives can also degrade water separation properties and lower filter efficiency. Nevertheless the safety record since antistatic additives were introduced far outweighs the negative effects on other fuel properties.

PERFORMANCE COMBUSTION

The primary reaction carried out in the gas turbine combustion chamber is oxidation of a fuel to release its heat content at constant pressure. Atomized fuel mixed with enough air to form a close-to-stoichiometric mixture is continuously fed into a primary zone. There, its heat of formation is released at flame temperatures determined by the pressure. The heat content of the fuel is therefore a primary measure of the attainable efficiency of the overall system in terms of fuel consumed per unit of work output. Net rather than gross heat content is a more significant measure because heat of vaporization of the water formed in combustion cannot be recovered in aircraft exhaust.

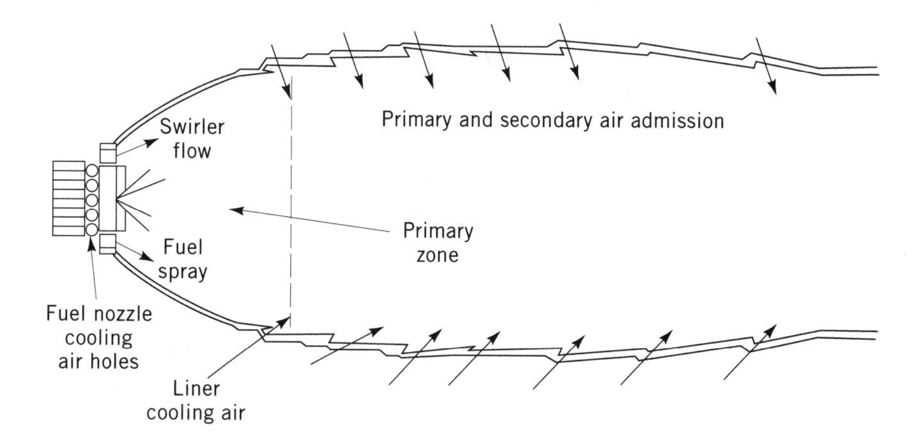

Figure 5. Gas turbine combustion chamber.

The most desirable gas turbine fuels for use in aircraft after hydrogen are hydrocarbons. Fuels that are liquid at normal atmospheric pressure and temperature are the most practical and widely used aircraft fuels; kerosene with a distillation range from 150 to 300°C is the best compromise to combine maximum mass heat content with other desirable properties.

Liquid fuel is injected through a pressure-atomizing or an airblast nozzle. This spray is sheared by air streams into laminae and droplets that vaporize and burn. Because the atomization process is so important for subsequent mixing and burning, fuel injector design is as critical as fuel properties. Figure 5 is a schematic of the processes occurring in a typical combustor.

Droplet size, particularly at high velocities, is controlled primarily by the relative velocity between liquid and air and in part by fuel viscosity and density (5). Surface tension has a minor effect. Minimum droplet size is achieved when the nozzle is designed to provide maximum physical contact between air and fuel. Hence primary air is introduced within the nozzle to provide both swirl and shearing forces. Vaporization time is characteristically related to the square of droplet diameter and is inversely proportional to pressure drop across the atomizer (5).

The vapor cloud of evaporated droplets burns like a diffusion flame in the turbulent state rather than as individual droplets. In the core of the spray, where droplets are evaporating, a rich mixture exists and soot formation occurs. Surrounding this core are a rich mixture zone where CO production is high and a flame front exists. Air entrainment completes the combustion, oxidizing CO to CO_2 and burning the soot. Soot burnup releases radiant energy and controls flame emissivity. The relatively slow rate of soot burning compared with the rate of oxidation of CO and unburned hydrocarbons leads to smoke formation. This model of a diffusion-controlled primary flame zone makes it possible to relate fuel chemistry to the behavior of fuels in combustors (5).

Aromatics readily form soot in the fuel-rich spray core as their hydrogens are stripped, leaving a carbon-rich benzenoid structure to condense into large molecular aggregates. Multiring aromatics form soot more readily than single-ring aromatics and exhibit greater smoking tendency and greater flame luminosity. At the other extreme, *n*-paraffins undergo little cyclization in the fuel-rich spray core, holding soot formation and flame radiation to a minimum. Other hydrocarbon structures exhibit smoke and radiation characteristics between the extremes of multiring aromatics and *n*-paraffins (6). Low luminosity fuels are synonymous with minimum flame radiation and clean (soot-free) combustion. In mixed fuels, two-ring aromatics have been shown to have a ten-fold greater effect luminosity rating than single-ring aromatics (7). Luminous flames in a gas turbine raise temperatures of the metal in the combustor liner and turbine inlet by the direct transmission by radiation of the heat energy in the flame core.

Air in the gas turbine combustor is carefully divided between primary and secondary uses. Primary air promotes the atomization process by adding swirl to the fuel exiting the nozzle; it must be limited to ensure that the flame core remains fuel-rich to avoid flame blow-out and facilitate reignition. Most of the air in the gas turbine combustor is directed either toward secondary reactions beyond the primary flame core or toward dilution of the hot exhaust products so that a uniform temperature gas is presented to the turbine section. The secondary reactions involve intensely turbulent mixing under stoichiometric or fuel-lean mixtures to complete the combustion to CO_2 and burn up the soot particles formed in the fuel-rich core. A combustor design that causes proper mixing is critical for ensuring smoke-free exhaust products. Nevertheless, in any given combustor there is a relationship between smoke output and the hydrogen content of fuel (8).

A gas turbine used in aircraft must be capable of handling a very wide span of fuel and air flows because the thrust output, ie, pressure, covers the range from idle to full-powered take-off. To accommodate this degree of flexibility in the combustor, fuel nozzles usually are designed with two streams, primary and secondary flow, or with alternate rows of nozzles that turn on only when secondary flow (or full thrust power) is needed. It is more difficult to vary the air streams to match the different fuel flows and as a consequence a combustor optimized for cruise conditions, ie, most of the aircraft's operation, operates less efficiently at idle and full thrust.

Exhaust emissions of CO, unburned hydrocarbons, and nitrogen oxides reflect combustion conditions rather than fuel properties. The only fuel component that degrades exhaust is sulfur; the SO_2 levels in emissions are directly proportional to the content of bound sulfur in the fuel. Sulfur levels in fuel are determined by crude type and desulfurization processes. Specifications for aircraft fuels impose limits of 3000–4000 ppm total sulfur but average levels are half of these values. Sulfur levels in heavier fuels are determined by legal limits on stack emissions.

Control of nitrogen oxides in aircraft exhaust is of increasing concern because nitrogen oxides react with ozone in the protective layer of atmosphere which exists in the altitude region where supersonic aircraft operate. Research is under way to produce a new type of combustor which minimizes NO_x formation. It is an essential component of the advanced propulsion unit needed for a successful supersonic transport fleet.

Another class of pollutant in exhaust does serious damage to the turbine itself. In this case the damage is hot corrosion of the metal blades and vanes by alkali metal (primarily sodium) salts which remove blade coatings and promote intergranular corrosion. The source of the salts can be fuel (as metal salts dissolved or entrained in water) or the air that enters the gas turbine in huge quantities. Sulfur oxides in the exhaust react with alkali to form sulfates which produce a low melting flux on the oxide coatings. The problem is especially serious with gas turbines used in ground (or marine) locations because heavier fuels are apt to contain metal contaminants. For example, some crudes contain soluble vanadium which forms V205 leading to hot corrosion.

Stability

Aviation fuel is exposed to a wide range of thermal environments and these greatly influence required fuel properties; eg, in the tank of a long-range subsonic aircraft, fuel temperatures can drop so low that ice crystals, wax formation, or viscosity increase may affect the fuel system performance. On its way to the engine the fuel then absorbs heat from airframe or engine components and in fact is used as a coolant for engine lubricant. Current high thrust engines expose fuel in the main engine control, pump, and manifold to an intense thermal environment, and fuel temperatures steadily rise as flow reaches the nozzle. Fuel stability assumes primary importance since freedom from deposits within the fuel system is essential for both performance and life.

The fuel systems of ground based turbines are far less critical since coolants other than fuel can be used and fuel lines can be well insulated. The tendency for deposit formation in fuel is not a concern in ground systems.

Deposits sometimes block fuel nozzles and distort fuel spray patterns, leading to skewed temperature distribution with the possibility of burnout of turbine parts by a hot streak exhaust. These deposits are sometimes associated with metal containing particulates but are, in general, another manifestation of fuel instability.

Oxidation of hydrocarbons by the air dissolved in fuel is catalyzed by metals and leads to polymer formation, ie, varnish and sludge deposits, by a chain reaction mechanism involving free radicals. Since it is impossible to exclude air dissolved in fuel, oxidation stability is controlled by eliminating species prone to form free radicals and by introducing antioxidants. An accelerated test that simulates the critical temperature regimes of the gas turbine engine fuel manifold and nozzle, ASTM D3241, the Jet Fuel Thermal Oxidation Test (JFTOT) has been developed to measure the oxidation stability of aircraft gas turbine fuels. In the JFTOT test deposits formed in the heated test section in 2.5 h at a specified temperature are assessed. This relative rating rates fuels that would normally be expected to operate thousands of hours in an engine without deposit formation. In those few instances when deposits have been observed in service, the fuel has been shown to fail JFTOT tests at the specification temperature.

The reactive species that initiate free radical oxidation are present in trace amounts. Extensive studies (9) of the autooxidation mechanism have clearly established that the most reactive materials are thiols and disulfides, heterocyclic nitrogen compounds, diolefins, furans, and certain aromatic–olefin compounds. Because free radical formation is accelerated by metal ions of copper, cobalt, and even iron (10) the presence of metals further complicates the control of oxidation. It is difficult to avoid some metals, particularly iron, in fuel systems.

It is possible to deactivate a metal ion by adding a compound such as disalicyclidene alkyl diamine which readily forms a chelate with most metal atoms to render them ineffective. Metal deactivator has been shown to reduce dramatically oxidation deposits in the JFTOT test and in single tube heat exchanger rigs. The role of metal deactivator in improving fuel stability is complex since quantities beyond those needed to chelate metal atoms act as passivators of metal surfaces and as antioxidants (11).

Oxidation deposits in aircraft engines are related to the thermal stresses imposed by heat soakback which results from the off/on cycles associated with many landings and takeoffs. In contrast, ground turbines tend to operate longer at constant flow and do not subject fuel to the same degree of thermal stress. Anticipating the introduction of higher performance aircraft which will place greater thermal stress on fuel, the USAF has undertaken a program to upgrade the stability of JP-8 by additive.

Volatility

The volatility of aircraft gas turbine fuel is controlled primarily by the aircraft itself and its operating environment. For example, limits of the vapor pressure of aviation gasoline were dictated by the vapor and liquid entrainment losses that could occur in a piston aircraft capable of climbing to an altitude (about 6000 m) where the vapor pressure of the warm fuel exceeds atmospheric pressure (48 kPa or 7 psi). Since early military jet aircraft could climb even higher (12000 m) and faster, it was necessary to limit further the vapor pressure of military gas turbine fuels to 21 kPa (3 psi). Originally, the United States followed the British lead and selected a kerosene of very low vapor pressure for military jet engines. Because projected production of this fuel was limited, fuel availability was the eventual determining factor for the selection of the wide-cut gasoline type fuel JP-4 as the fuel for military jet aircraft. The volatility of kerosene fuel is measured not by its vapor pressure but by its temperature at the point where its vapors just prove to be flammable, ie, the flash point.

Commercial aviation utilizes low volatility kerosene defined by a flash point minimum of 38°C. The flammability temperature has been invoked as a safety factor for handling fuels aboard aircraft carriers; Navy JP-5 is a low volatility kerosene of minimum flash point of 60°C, similar to other Navy fuels.

The dependence of vapor pressure on temperature for the fuels most commonly used in gas turbines appears in Figure 6 (12). The points on the abscissa reflect the flash point temperatures used to define the volatility of higher molecular weight fuels. When vapor pressure itself is limited, as with JP-4 or Jet B, a test temperature of 38°C is specified.

A minimum volatility is frequently specified to assure adequate vaporization under low temperature conditions. It can be defined either by a vapor pressure measurement or by initial distillation temperature limits. Vaporization promotes engine start-up. Fuel vapor pressure assumes an important role particularly at low temperature. For example, if fuel has cooled to −40°C, as at Arctic bases, the amount of vapor produced is well below the lean flammability limit.

In this case, an energetic spark igniter must vaporize enough fuel droplets to initiate combustion. Start-up under the extreme temperature conditions of the Arctic is a main constraint in converting the Air Force from volatile JP-4 to kerosene type JP-8, the military counterpart of commercial Jet A1.

The lower volatility of JP-8 is a primary factor in the U.S. Air Force conversion from JP-4 since fires and explosions under both combat and ordinary handling conditions have been attributed to the use of JP-4. In examining the safety aspects of fuel usage in aircraft, a definitive study (13) of the accident record of commercial and military jet transports concluded that kerosene type fuel is safer than wide-cut fuel with respect to survival in crashes, inflight fires, and ground fueling accidents. However, the difference in the overall accident record is small because most accidents are not fuel-related.

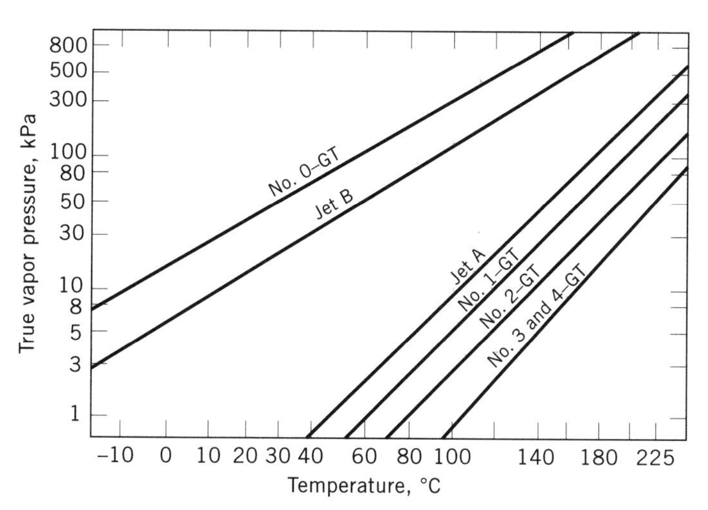

Figure 6. Variation of vapor pressure with temperature for gas turbine fuels (13). To convert kPa to psi, divide by 6.9.

Low Temperature Fuel Flow

The decrease in the temperature of fuel in the tank of an aircraft during a long duration flight produces a number of effects which can influence flight performance. Fuel viscosity increases, necessitating more pumping energy in the tank boost pump and in the engine pump. Fuel becomes saturated with water and droplets of free water form and settle. Those carried with fuel may form ice on the cold filter which protects the fuel control. For this reason, filter heaters are used in civil aircraft to avoid ice blockage and fuel starvation. Military aircraft avoid the complication of a filter heater and depend instead on an anti-icing additive in the fuel to depress the freezing point of water. At still lower temperatures, crystals of wax form in fuel. The temperature at which these wax crystals disappear is called the freezing point and distinguishes Jet A1 used by long-range international airlines from Jet A used by domestic airlines for relatively short duration flights.

The increase in fuel viscosity with temperature decrease is shown for several fuels in Figure 7. The departure from linearity as temperatures approach the pour point illustrates the non-Newtonian behavior created by wax matrices. The freezing point appears before the curves depart from linearity. It is apparent that the low temperature properties of fuel are closely related to its distillation range as well as hydrocarbon composition. Wide-cut fuels have lower viscosities and freezing points than kerosenes while heavier fuels used in ground turbines exhibit much higher viscosities and freezing points.

Low temperature viscosities have an important influence on fuel atomization and affect engine starting. Cycloparaffinic and aromatic fuels reach unacceptably high lev-

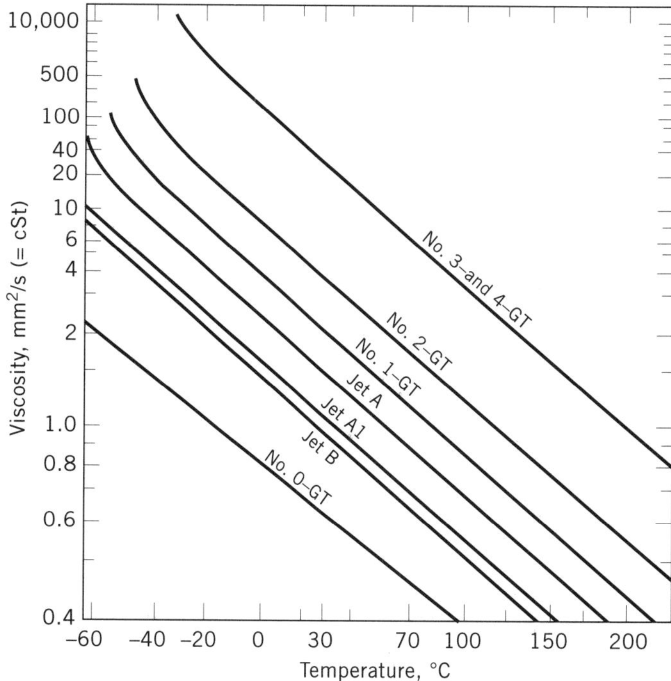

Figure 7. Variation of viscosity with temperature for gas turbine fuels (13).

els of viscosity at low temperatures. A kinematic viscosity of 35 mm2/s represents the practical upper limit for pumps on aircraft while much higher limits are acceptable for ground installations.

Water in Fuel

The solubility of water in a hydrocarbon is related to the molecular structure of the hydrocarbon and its temperature. Aromatics dissolve about five times more water than corresponding paraffins and in turn lower molecular weight paraffins dissolve more water than longer chain compounds. Figure 8 is a plot of water solubility for certain pure hydrocarbons and typical jet fuels. The water saturation values in fuel vary considerably because of composition effects. For example, wide-cut fuel normally dissolves more water than does kerosene, but because the latter may contain twice as many aromatics, the higher molecular weight fuel may in fact exhibit an equivalent water saturation curve. At a temperature of 23°C, a Coordinating Research Council program on Jet A fuel showed a range of 56–120 ppm (14).

The slope of the water solubility curves for fuels is about the same, and is constant over the 20–40°C temperature range. Each decrease of 1°C decreases water solubility about 3 ppm. The sensitivity of dissolved water to fuel temperature change is important. For example, the temperature of fuel generally drops as it is pumped into an airport underground hydrant system because subsurface temperatures are about 10°C lower than typical storage temperatures. This difference produces free water droplets, but these are removed by pumping fuel through a filter–coalescer and hydrophobic barrier before delivery into aircraft.

The amount of water actually dissolved in fuel at any given temperature is determined by equilibration with

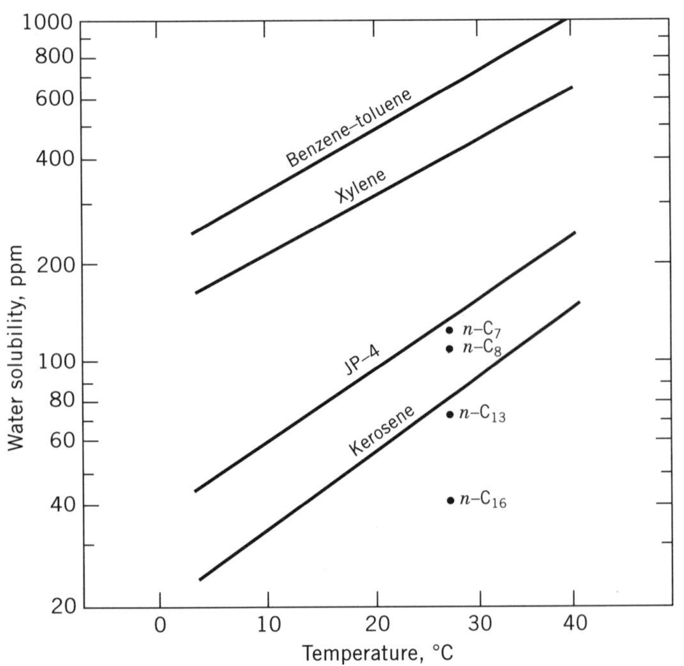

Figure 8. Variation of water solubility with temperature of gas turbine fuels (13).

the water in the atmosphere above the fuel. If a tank is vented to atmosphere, fuel may enter saturated with water but lose half its dissolved water when the relative humidity is 50%. Conversely, a tank of fuel in a humid environment will rapidly pick up water. Many storage tanks contain floating roofs which effectively eliminate any opportunity for dissolved water to be removed as vapor. The tank in the aircraft behaves like a storage tank on the ground. In the upper atmosphere, the partial pressure of water is very low and fuel tends to dry out; dissolved water leaves fuel and exits from the vent. An aircraft descending through clouds will pick up water and saturate its fuel. At the end of flight, fuel in a tank is apt to be both cold and cloudy with free water.

A stable cloud of water in fuel usually means that a surfactant is present to form a stable water-in-oil emulsion. Smaller droplets resist natural coalescence processes. A surfactant that is potent as an emulsifying agent is apt to disarm the coalescing filters allowing excess water to be delivered with fuel.

The effects of water in fuel inside a tank, particularly an aircraft tank, are important because of the demonstrated proclivity of free water to form undrainable pools where microorganisms can flourish. In ground storage tanks, an attempt is made to exclude water bottoms by proper design and regular draining practices. This is more difficult in aircraft because pools of water freeze in flight and ice may not melt when fueling turnarounds are rapid. In a large storage tank, organisms (usually fungi and yeasts) develop at fuel–water interfaces forming a cuff of growth which can plug filters. In the aircraft, the growth usually takes place on tank surfaces, forming a fungal mat under which metabolic products such as organic acids penetrate polymeric coatings to attack the aluminum skin itself. The growth may also affect the capacitance gage used to read liquid level.

It is common practice to curb growth of organisms by biocidal treatment. In storage tanks, a water soluble agent is used. Aircraft tanks are opened periodically for hand cleaning and subsequent treatment with a fuel-soluble boron-containing biocide. The anti-icing additive used by the military to lower the freezing point of water, 2 methoxyethanol, happens also to inhibit growth of organisms effectively. Therefore, aircraft tank treatment to remove fungal mats is needed only for commercial transports.

Water plays a primary role in corrosion of the metal walls of tanks (15) and pipes and increases the tendency for high speed pumps to produce wear particles and exhibit shortened life. Formation of corrosion products can be controlled by addition of corrosion inhibitors, a mandatory additive in military fuels. However, corrosion inhibitors may also degrade other fuel properties and adversely affect ground filtration equipment. Thus, they are not generally acceptable in commercial fuels where rigorous attention is given to clean and dry fuels upon aircraft fueling.

Compatibility and Corrosion

Gas turbine fuels must be compatible with the elastomeric materials and metals used in fuel systems. Elastomers are used for O-rings, seals, and hoses as well as pump parts and tank coatings. Polymers tend to swell when in contact with aromatics and improve their sealing ability, but degree of swell is a function of both elastomer type and aromatic molecular weight. Rubbers can also be attacked by peroxides that form in fuels that are not properly inhibited.

Attack on metals can be a function of fuel components as well as of water and oxygen. Organic acids react with cadmium plating and zinc coatings. Traces of H_2S and free sulfur react with silver used in older piston pumps and with copper used in bearings and brass fittings. Specification limits by copper and silver strip corrosion tests are required for fuels to forestall these reactions.

Boundary lubrication of rubbing surfaces such as those found in high speed pumps and fuel controls has been found to be related to the presence in fuel of species capable of forming a chemisorbed film that reduces friction and wear (16). The lubricity of fuel is attributed to polar materials (17), multiring aromatics (17), and a thiohydroindane species (18). Oxygen containing compounds, particularly long chain acids, tend to adsorb on metal surfaces and act as lubricity agents. Corrosion inhibitors which tend to form tenacious films on metal surfaces are generally excellent lubricity agents. Refining processes to reduce sulfur and control acidity tend to degrade a fuel's lubrication propensity. Corrosion inhibitors are mandatory in military fuels since they also improve the lubricity of fuels by extending the life of pumps and controls.

Both friction and wear measurements have been used to study boundary lubrication of fuel because sticking fuel controls and pump failures are principal field problems in gas turbine operation. An extensive research program of the Coordinating Research Council has produced a ball-on-cylinder lubricity test (BOCLE), standardized as ASTM D5001, which is used to qualify additives, investigate fuels, and assist pump manufacturers (19).

ECONOMIC ASPECTS

The exacting list of specification requirements for aviation gas turbine fuels and the constraints imposed by delivering clean fuel safely from refinery to aircraft are the factors that affect the economics. Compared with other distillates such as diesel and burner fuels, kerosene jet fuels are narrow-cut specialized product and usually command a premium price over other distillates. The prices charged for jet fuels tend to escalate with the basic price of crude, a factor which seriously undermined airline profits during the Iran revolution as crude prices increased sharply.

Availability and cost of jet fuels are also affected by the interrelationships of the market for other petroleum products. An excellent example of these interrelationships was developed for NASA in a study of how jet fuel property changes affected producibility and cost in the U.S., Canada, and Europe (20). For example the western United States is a unique market compared with the rest of the country because demand for gasoline and jet fuel is high while demand for burner and boiler fuels is low. This product distribution encourages high conversion refineries with particular emphasis on hydrocracking to make lighter products. At the same time environmental concerns over urban air quality are pressuring refineries to market reformulated gasolines and low sulfur diesel fuels. Refining these products and also producing specification jet fuel has tended to limit availability and increase the cost of aviation fuels in the western United States.

The majority of the hundreds of refineries operated by petroleum companies in different parts of the world to make local products such as gasoline and burner fuels also produce jet fuels. Even a small refinery with simple equipment can make suitable jet fuel if it has access to the right crude. However, the principal supply of both civil and military jet fuels is produced in large refineries. Many are located near large cities and airports and are frequently connected by pipeline directly to the airport. Modern airports have extensive storage and handling facilities operated by local authorities, by petroleum companies or consortia, or by the airlines themselves.

Worldwide demand for the jet fuels specified in Table 2 amounted to about 3 million B/D (477 k cubic meters per day) in 1990. About one half of this demand was kerosene Jet A sold in the United States. One third represented kerosene Jet A1 for delivery to international airlines outside the United States, while the balance comprised various military fuels used by air forces around the world.

AVIATION GASOLINE

The specifications for aviation gasoline are even more exacting than specifications for jet fuel (see selected specifications, Table 3). Antiknock requirements dictate that high grade fuel be blended mostly with iso-octane and tetraethyl lead. Table 4 summarizes the key requirements of ASTM D910 the specification used by both civil and military aircraft. Antiknock ratings, determined in a single-cylinder laboratory engine, represent the detonation limit of a spark-ignited fuel–air mixture under the same conditions as the ASTM D2700 Motor Method used to rate automotive gasoline. A second rating under rich conditions of a fuel–air mixture is required using ASTM D909 Supercharge Method to ensure adequate take-off power. Both the lean and rich ratings are related to iso-octane rated at 100 octane or performance number. Use of too low an octane grade can wreck an engine by detonation; hence, each grade is identified by a color dye. Higher powered engines require Grade 100/130 produced as either green fuel with 4mL/USG of TEL or blue fuel with half this amount of lead. Low compression engines tolerate less lead and require Grade 80/87 with only 0.5mL/USG of TEL.

The volatility requirements for all grades are determined by the need to vaporize fuel for cold starting and to control the rate of engine warm-up. Controls on distillation are also needed to exclude undesirable heavy ends while vapor pressure limits are specified to avoid excessive vapor formation at high altitude. Other exacting requirements for aviation gasoline include long-term stor-

Table 3. Selected Specification Properties of Aviation Gasoline[a]

Characteristic	Grade 80	Grade 100	Grade 100LL
Knock rating, lean min O.N.	80	100	100
Knock rating, rich min P.N.	87	130	130
Color	Red	Green	Blue
Tetraethyl lead, g Pb/Lmax	0.19	1.12	0.56
Distillation T, °C, 10% max		75	
°C, FBP max		170	
Vapor pressure, kPa[b]		38–49	
Net heat content, MJ/kg[c]		43.54	
Sulfur, % mass, max		0.05	
Freezing point, max, °C		−58	

[a] Detailed requirements appear in ASTM D910.
[b] To convert kPa to psi, multiply by 0.145.
[c] To convert MJ/kg to Btu/lb, multiply by 430.2.

age stability, noncorrosiveness, low sulfur, good low temperature performance, and minimum water reaction. The specific energy minimum tends to limit the aromatics used for blending. Antioxidants are mandatory.

The shrinking market for all grades of aviation gasoline has reduced the numbers of producers, fostered exchange arrangements, and increased the cost of distribution and handling. The resulting high price of aviation gasoline has encouraged operators of low compression engines to replace Grade 80 with unleaded premium motor gasoline, a practice supported by the Federal Aviation Administration which has issued supplemental type certificates for certain models of aircraft and engines.

FUTURE TRENDS

Aircraft Fuels

Demand for aviation gas turbine fuels has been growing more rapidly than other petroleum products since 1960, about 3–5% per year compared with 1% for all oil products. This strong demand reflects a current and predicted growth in worldwide air traffic of 4–7% annually until the end of the century. Total world oil demand will be up by 15% by the year 2000, but aviation fuel demand will increase by 50–125%. However, the fraction of the oil barrel devoted to aviation, now about 8%, will increase only slightly.

Changes in the crude supply outlook are in fact decreasing the potential yield of the straight-run kerosene fraction needed not only for aviation but also for blending into diesel and heating fuels. For example, Arabian Gulf crudes, the most abundant in the world, yield less kerosene than the primary U.S. crude. This situation will require more conversion processes at refineries such as hydrocracking of gas oil, a process that yields high quality but more expensive aviation kerosene.

The vulnerability of the aviation market was exposed in the winter of 1973–1974 when the Arab oil embargo caused actual shortages of jet fuel (21). Coupled with the dramatic rise in crude and product prices, the supply crisis precipitated a new generation of fuel-efficient aircraft

Table 4. Additives Used in Aviation Gas Turbine Fuels

Class of Additive	Chemical Types	Purpose	Specification Status[a]	
			Civil	Military
Antioxidant	Alkyl phenylene diamines, hindered alkylphenols	Improve oxidation stability–storage and high temperature	O	O
Metal deactivator	Disalicylidene alkyl diamine	Remove metal ions that catalyze oxidation	O	O
Antiicing	Glycol ether	Prevent low temperature filter icing		R
Corrosion inhibitor	Dimer acid, phosphate ester	Reduce development of corrosion products, improve lubrication		R
Antistatic	Alkyl chromium salt and calcium sulfosuccinate, olefin–sulfur dioxides and polyamines	Increase conductivity	OR	R
Biocide	Boron ester	Inhibit organic growth in fuel tanks	A	

[a] O = optional, R = required, A = airline use in maintenance.

and a relaxation of specification limits such as higher freezing point to −47°C and higher aromatics to 25%. It has been estimated that 18% of the world's crudes produce a 150–250°C fraction containing more than 20% aromatics. Two of these, Heavy Arabian and North Slope, represent the important new sources of marginal jet fuel.

Operation of aircraft gas turbines on a wider cut than the 160–260°C fraction of crude to expand availability was considered in a 1977 ASTM Symposium (22). In the case of Jet A, a reduction of flash point from 38°C to 32°C would increase yield by 17% while an increase in Jet A1 freezing point to the −40°C Jet A level used domestically would add about 25% to the kerosene pool.

Synthetic fuels derived from shale or coal will have to supplement domestic supplies from petroleum some day. Aircraft gas turbine fuels producible from these sources were assessed during the 1970s. Shale-derived fuels can meet current specifications if steps are taken to reduce the nitrogen levels. However, extracting kerogen from shale rock and denitrogenating the jet fuel are energy-intensive steps compared with petroleum refining; it has been estimated that shale jet fuel could be produced at about 70% thermal efficiency compared with 95% efficiency for petroleum (23). Such a difference would represent much higher cost for a shale product.

Synthetic jet fuel derived from coal is even more difficult and expensive since the best of the conversion processes produces a fuel very high in aromatics. With hydrogenation, overall thermal efficiency is only 50%. Without additional hydrogenation, the gas turbine fuels would contain 60–70% aromatics.

A third alternative, production of liquid jet fuel from processing of abundant natural gas, is a more promising and cheaper source of high quality product than shale or coal.

Fuels for Advanced Aircraft

The area of greatest traffic growth is predicted to be the Pacific basin where distances are great and the greatest incentive exists for an advanced aircraft that could greatly reduce travel time. Research at the National Aeronautic and Space Administration is proceeding on a high speed commercial transport (HSCT) that would satisfy this market by the year 2010. The aircraft would carry about 300 passengers nonstop over 8000 miles at three times the speed of sound and reduce travel time from 14 to 6 hours. To be economically and environmentally viable, the HSCT would have to produce less nitrogen oxides in exhaust to cause lower sonic boom and to operate on the same kerosene as subsonic aircraft.

The first commercial supersonic transport, the Concorde, operates on Jet A1 kerosene but produces unacceptable noise and exhaust emissions. Moreover, it is limited in capacity to 100 passengers and to about a 3000 mile range. At supersonic speed of Mach 2, the surfaces of the aircraft are heated by ram air. The surfaces heated by ram air can raise the temperature of fuel held in the tanks to 80°C. Since fuel is the coolant for airframe and engine subsystems, fuel to the engine can reach 150°C (24). An HSCT operated at Mach 3 would place much greater thermal stress on fuel. To minimize the formation of thermal oxidation deposits, it is likely fuel delivered to the HSCT would have to be deoxygenated.

NASA is also considering a more advanced aircraft such as Mach 5 to cut Pacific travel time to about 3 h, but in this case kerosene fuel is no longer acceptable and liquefied natural gas or liquefied hydrogen would be needed to provide the necessary cooling and stability. However, a completely new fueling system would be required at every international airport to handle these cryogenic fuels.

In the speed regime of Mach 6 and beyond, hypersonic vehicles, hydrogen is the fuel of choice because of its high energy content and combustion kinetics. For example, a first stage to orbit reusable booster for space vehicles would employ a supersonic combustor ramjet engine burning hydrogen, the only fuel capable of maintaining combustion in the shock wave. A booster that replaced 80% of the present weight of oxidizer would revolutionize space transportation by replacing the current space shuttle as a delivery vehicle.

Use of kerosene fuel rather than air as a coolant focuses attention on another fuel property (specific heat) which is a measure of its efficiency to absorb thermal energy. Heat capacity increases with temperature but decreases with density which favors a low density paraffinic fuel from petroleum or natural gas sources.

Because of tank heating, fuel volatility is also more critical in supersonic aircraft. For example, the Concorde tank is pressurized to prevent vapor losses which could be significant at high altitude where fuel vapor pressure may equal atmospheric pressure. The tank can reach 6.9 kPa at the end of flight. The need to deoxygenate fuel for thermal stability in the HSCT will doubtless require a similar pressurized system.

BIBLIOGRAPHY

1. W. G. Dukek, A. R. Ogston, and D. R. Winans, *Milestones in Aviation Fuels,* AIAA No. 69-779, American Institute of Aeronautics and Astronautics, New York, July 1969.
2. R. S. Schlaefer and S. D. Heron, *Development of Aircraft Engines and Fuels,* Harvard Business School, Boston, Mass., 1950.
3. K. M. Brown, *Commercial Results with UOP MEROX Process for Mercaptan Extraction in U.S. and Canada,* UOP Booklet 267, UOP, Inc., Des Plaines, Ill., 1960.
4. W. M. Bustin and W. G. Dukek, *Electrostatic Hazards in the Petroleum Industry,* John Wiley & Sons, Inc., Chichester, U.K., 1983.
5. N. A. Chigler, "Atomization and Burning of Liquid Fuel Sprays," in *Progress in Energy and Combustion Science,* Vol. 21 No. 4, Pergamon Press, Inc., Elmsford, N.Y., 1976.
6. *Microburner Studies of Flame Radiation as Related to Hydrocarbon Structure,* Report 3752-64R, Phillips Petroleum Co., Navy Buweps Contract NOw 63-0406d May 1964.
7. M. Smith, *Aviation Fuels,* G. T. Foulis & Co. Ltd., Henley-on-Thames, U.K., 1970.
8. H. F. Butze and R. C. Ehlers, *Effect of Fuel Properties on Performance of a Single Aircraft Turbojet Combustor,* NASA TM X-71789, National Aeronautics and Space Administration, Lewis Research Center, Cleveland, Ohio, Oct. 1975.
9. W. F. Taylor, *Development of High Stability Fuels,* contract

N00140-74-C-0618, Naval Air Propulsion Center, Trenton, N.J., Dec. 1975.
10. W. F. Taylor, *J. Appl. Chem.* **18,** 25 (1968).
11. R. H. Clark, "The Role of Metal Deactivator in Improving the Thermal Stability of Aviation Kerosenes," *3rd Intl. Conference on Stability and Handling of Liquid Fuels,* London, Sept. 1988.
12. H. C. Barnett and R. R. Hibbard, *Properties of Aircraft Fuels,* TN 3276, NASA, Lewis Research Center, Cleveland, Ohio, Aug. 1956.
13. *Aviation Fuel Safety–1975,* CRC Report no. 482, Coordinating Research Council, Inc., Atlanta, Ga., Nov. 1975.
14. *Survey of Electrical Conductivity and Charging Tendency Characteristics of Aircraft Turbine Fuels,* CRC Report no. 478, Coordinating Research Council, Inc., Atlanta, Ga., Apr. 1975.
15. H. R. Porter and co-workers, *Salt Driers Assure Really Dry Fuel to Aircraft,* SAE 710440, Society of Automotive Engineers, Warrendale, Pa.
16. J. J. Appeldoorn and W. G. Dukek, *Lubricity of Jet Fuels,* SAE 660712, Society of Automotive Engineers, Warrendale, Pa.
17. J. K. Appeldoorn and F. F. Tao, *Wear,* 12 (1968).
18. R. A. Vere, *SAE Trans.* **78,** 2237 (1969).
19. W. G. Dukek, *Ball-on-Cylinder Testing for Aviation Fuel Lubricity,* SAE 881537, Society of Automotive Engineers, Warrendale, Pa.
20. G. M. Varga and A. J. Avella, *Jet Fuel Property Changes and Their Effect on Producibility and Cost in U.S., Canada and Europe,* NASA Contract Report 174840, NASA, Cleveland, Ohio, Feb. 1985.
21. A. G. Robertson and R. E. Williams "Jet Fuel Specifications: The Need for Change," *Shell Aviation News,* Shell Oil Co. Houston, Tex., 1976, p. 435.
22. *Factors in Using Kerosine Jet Fuel of Reduced Flash Point,* ASTM STP 688, American Society for Testing and Materials, Philadelphia, Pa., 1979.
23. W. Dukek and J. P. Longwell, "Alternative Hydrocarbon Fuels for Aviation," *Esso Air World,* No. 4, Exxon International Co., Florham Park, N.J.
24. H. Strawson and A. Lewis, *Predicting Fuel Requirements for the Concorde,* SAE 689734, Society of Automotive Engineers, Warrendale, Pa., Oct. 1988.

ALCOHOL FUELS

Michael D. Jackson
Carl B. Moyer
Acurex Corporation

The use of alcohols as motor fuels gained considerable interest in the 1970s as substitutes for gasoline and diesel fuels, or, in the form of blend additives, as extenders of oil supplies. In the United States, most applications involved the use of low level ethanol [*64-17-5*] blends in gasoline [*8006-61-9*]. Brazil, however, launched a major program to substitute ethanol for gasoline in 1976, beginning with a 22% alcohol ethanol–gasoline blend and adding dedicated ethanol cars in the 1980s. By 1985 ethanol cars accounted for 95% of new car sales in Brazil. By 1988 Brazil had 3,000,000 automobiles, about 30% of the total automobile population, dedicated to ethanol. The United States has demonstration vehicles using alcohols (mostly methanol [*67-56-1*]), but otherwise has not yet passed beyond the use of limited amounts of alcohols and of ethers produced from alcohols as gasoline components. However, proposals continue to be made to implement alcohol programs similar to that of Brazil (1). Other nations such as New Zealand, Germany, and Sweden also investigated the use of alcohols as a transportation fuel.

The benefits of alcohol fuels include increased energy diversification in the transportation sector, accompanied by some energy security and balance of payments benefits, and potential air quality improvements as a result of the reduced emissions of photochemically reactive products. The Clean Air Act of 1990 and emission standards set out by the State of California may serve to encourage the substantial use of alcohol fuels, unless gasoline and diesel technologies can be developed that offer comparable advantages.

See also Air pollution: automobile; Air pollution: automobile, toxic emissions; Automobile emissions, control; Clean air act, mobile sources; Exhaust control, automotive; Methanol vehicles; Transportation fuels, automotive fuels; Transportation fuels, ethanol fuels in Brazil.

PROPERTIES OF ALCOHOL FUELS

Table 1 summarizes key properties of ethanol and methanol as compared to other fuels. Both alcohols make excellent motor fuels, although the high latent heats of vaporization and the low volatilities can make cold-starting difficult in vehicles having carburetors or fuel injectors in the intake manifold where the fuel must be vaporized prior to being introduced into the combustion chamber. This is not the case for direct injection diesel-type engines using methanol or ethanol. Both methanol and ethanol have high octane values and allow high compression ratios having increased efficiency and improved power output per cylinder. Both have wider combustion envelopes than gasoline and can be run at lean air–fuel ratios with better energy efficiency. However, ethanol and methanol have very low cetane numbers and cannot be used in compression–ignition diesel-type engines unless gas temperatures are high at the time of injection. Manufacturers of heavy-duty engines have developed several types of systems to assist autoignition of directly injected alcohols. These include glow plugs or spark plugs, reduced engine cooling, and increased amounts of exhaust gas recirculation or, in the case of two-stroke engines, reduced scavenging. Additives to improve cetane number have been effective as have dual-fuel approaches, in which a small amount of diesel fuel is used as an ignitor for the alcohol.

There are particular alcohol fuel safety considerations. Unlike gasoline or diesel fuel, the vapor of methanol or ethanol above the liquid fuel in a fuel tank is usually combustible at ambient temperatures. This poses the risk of an explosion should a spark or flame find its way to the tank such as during refueling. Additionally, a neat methanol fire has very little luminosity and, consequently, fire fighting efforts can be difficult in daylight. However, low luminosity also implies low radiative fluxes from the fire. This, combined with the high latent heat of vaporization, means that the heat release of a methanol fire is low rela-

Table 1. Properties of Fuels[a]

Properties	Methanol	Ethanol	Propane	Methane	Isoctane	Unleaded gasoline	Diesel Fuel #2
Constituents	CH_3OH	CH_3CH_2OH	C_3H_8	CH_4	C_8H_{18}	$C_nH_{1.87n}$ (C_4 to C_{12})	$C_nH_{1.8n}$ (C_8 to C_{20})
CAS Registry Number	*[67-65-1]*	*[64-17-5]*	*[74-98-6]*	*[74-82-8]*	*[540-84-1]*	*[8006-6-9]*	
Molecular weight	32.04	46.07	44.10	16.04	114.23	≈110	≈170
Element composition, wt %							
C	37.49	52.14	81.71	74.87	84.12	86.44	86.88
H	12.58	13.13	18.29	25.13	15.88	13.56	13.12
O	49.93	34.73	0	0	0	0	0
Density at 16°C and 101.3 kPa[b], kg/m³	794.6	789.8	505.9	0.6776	684.5	721–785	833–881
Boiling point at 101.3 kPa[b], °C	64.5	78.3	6.5	−161.5	99.2	38–204	163–399
Freezing point, °C	−97.7	−114.1	−188.7	−182.5	−107.4		<-7
Vapor pressure at 38°C, kPa[b]	31.9	16.0	1.297	0.5094	11.8	48–108	negligible
Heat of vaporization, ΔH_v, MJ/kg[c]	1.075	0.8399	0.4253	0.5094	0.2712	0.3044	0.270
Gross heating value, MJ/kg[c]	22.7	29.7	50.4	55.5	47.9	47.2	44.9
Net heating value							
MJ/kg[c]	20.0	27.0	46.2	50.0	44.2	43.9	42.5
MJ/m³[d]	15,800	21,200	23,400		30,600	32,000	35,600
Stoichiometric mixture net heating value, MJ/kg[c]	2.68	2.69	2.75	2.72	2.75	2.83	2.74
Autoignition temperature, °C	464	363	450	537	418	260–460	257
Flame temperature, °C	1,871	1,916	1,988	1,949	1,982	2,027	1,993
Flash point, °C	11	13			4	−43 to −39	52–96
Flame speed at stoichiometry, m/s	0.43		0.40	0.37	0.31	0.34	
Octane ratings							
Research	106	107	112	120	100	92–98	
Motor	92	89	97	120	100	80–90	
Cetane rating	0–5						>40
Flammability limits, vol % in air	6.72–36.5	3.28–18.95	2.1–9.5	5.0–15.0	1.0–6.0	1.4–7.6	1.0–5.0
Stoichiometric air–fuel mass ratio	6.46	8.98	15.65	17.21	15.10	14.6	14.5
Stoichiometric air–fuel volumetric ratio	7.15	14.29	23.82	9.53	59.55	55	85
Water solubility	complete	complete	no	no	no	no	no
Sulfur content, wt %	0	0	0	0	0	<0.06	<0.5

[a] Refs. 2–7.
[b] To convert kPa to psi, multiply by 0.145.
[c] To convert MJ/kg to Btu/lb, multiply by 430.3
[d] To convert MJ/m³ to Btu/gal, multiply by 3.59.

tive to one of gasoline or diesel fuel. Because methanol or ethanol are both water-soluble, fires can be successfully controlled by dilution with large amounts of water, a tactic that simply spreads gasoline fires. Nevertheless, fire-extinguishing foams are the preferred alcohol fire-fighting method.

Some potential problems of alcohol fuels have been addressed by adding small amounts of gasoline or specific hydrocarbons to the fuel, reducing the flammability envelope and providing luminosity in case of fire.

USES OF ALCOHOL FUELS

Early applications of internal combustion engines featured a variety of fuels, including alcohols and alcohol–hydrocarbon blends. In 1907 the U.S. Department of Agriculture investigated the use of alcohol as a motor fuel. A subsequent study by the U.S. Bureau of Mines concluded that engines could provide up to 10% higher power on alcohol fuels than on gasoline (8). Mixtures of alcohol and gasoline were used on farms in France and in the United States in the early 1900s (9). Moreover, the first Ford Model A automobiles could be run on either gasoline or ethanol using a manually adjustable carburetor (1). However, the development of low cost gasoline pushed other automobile fuels into very minor roles and the diesel engine further solidified the hold of petroleum fuels on the transportation sector. Ethanol was occasionally used, particularly in rural regions, when gasoline supplies were short or when farm prices were low.

Methanol has been used as a motor racing fuel for many decades. Its high latent heat of vaporization cools the incoming charge of air to each cylinder. Increasing the mass of air taken into each cylinder increases the power developed by each stroke, providing a turbocharger effect. Furthermore, methanol has a higher octane value than gasoline allowing higher compression ratios, greater efficiency, and higher output per unit of piston displacement volume. These power increases are advantages in racing as is the simple means by which methanol fuel quality and uniformity can be verified.

The transparency of methanol flames is usually a safety advantage in racing. In the event of fires, drivers have some visibility and the lower heat release rate of methanol provides less danger for drivers, pit crews, and spectators.

Partly for these reasons, methanol has been the required fuel of the Indianapolis 500 since 1965. Methanol is also used in many other professional and amateur races. However, transparency of the methanol flames has also been a disadvantage in some race track fires. The invisibility of the flame has confused pit crews, delayed fire detection, and caused even trained firefighters problems in locating and extinguishing fires.

Low level blends of ethanol and gasoline enjoyed some popularity in the United States in the 1970s. The interest persists into the 1990s, encouraged by the exemption of low level ethanol–gasoline blends from the Federal excise tax as well as from state excise taxes in many states.

ENERGY DIVERSIFICATION AND ENERGY SECURITY

The ethanol program in Brazil addressed that country's oil supply problems in the 1970s, at times improved the balance of payments, and served to strengthen the economy of the sugar production portion of the agricultural sector. Although the benefits are difficult to quantify because of the very high inflation rate (10) the Brazilian ethanol program generally met its goals. Ironically the inability of ethanol to substitute successfully for diesel fuel in the heavy-duty truck sector necessitated keeping refinery outputs up to provide sufficient available diesel fuel. Thus gasoline was in surplus in Brazil and oil imports were not as much affected as originally hoped. The Brazilian program demonstrated that petroleum substitution strategies need to address the "whole barrel" product slate of oil refineries.

In the 1990s world events precipitated renewed interest in energy diversification strategies for the U.S. transportation sector. However, few measures are in place to encourage fuel alternatives outside of the exemption from the Federal excise tax on motor fuel granted to ethanol blends. The Alternative Motor Fuels Act of 1988 did extend credits to automobile manufacturers in the calculation of corporate average fuel economy (CAFE) for vehicles that use methanol or ethanol or natural gas; electric vehicles had previously been granted such a credit. Under the Act's provisions, neither ethanol nor methanol is counted as fuel consumed in the calculation of fuel economy. Thus vehicles that use alcohol have very high fuel economy ratings, reflecting the value of these vehicles in reducing oil imports. The credits take effect in model-year 1993. The fuel economy calculation assumes that methanol and ethanol fuels in commerce contain 15% by volume gasoline. Vehicles that can use either alcohol fuels or gasoline receive a reduced CAFE credit. The maximum credit that can be earned by a manufacturer for selling vehicles capable of using petroleum fuels is capped because alcohol fuels usage by these vehicles is not assured.

ALCOHOL AVAILABILITY

Methanol

If methanol is to compete with conventional gasoline and diesel fuel it must be readily available and inexpensively produced. Thus methanol production from a low-cost feed stock such as natural gas [*8006-14-2*] or coal is essential. There is an abundance of natural gas (see NATURAL GAS) worldwide and reserves of coal are even greater than those of natural gas.

Natural Gas Reserves. U.S. natural gas reserves could support a significant methanol fuel program. 1990 proved, ie, well characterized amounts with access to markets and producible at current market conditions U.S. resources are 4.8 trillion cubic meters (168 trillion cubic feet = 168 TCF). U.S. consumption is about one-half trillion m^3 (18 TCF) or 18 MJ equivalents per year, but half of that is imported. Estimates of undiscovered U.S. natural gas reserves range from 14 to 16 trillion m^3 (492 to 576 TCF) or roughly a 30-year supply at current U.S. consumption rates. Additional amounts of natural gas may become available from advanced technologies.

If 10% of the U.S. gasoline consumption were replaced by methanol for a twenty year period, the required reserves of natural gas to support that methanol consumption would amount to about one trillion m^3 (36 TCF) or twice the 1990 annual consumption. Thus the United States could easily support a substantial methanol program from domestic reserves. However, the value of domestic natural gas is quite high. Almost all of the gas has access through the extensive pipeline distribution system to industrial, commercial, and domestic markets and the value of gas in these markets makes methanol produced from domestic natural gas uncompetitive with gasoline and diesel fuel, unless oil prices are very high.

It is therefore more relevant to examine world resources of natural gas in judging the supply potential for methanol. World proved reserves amount to approximately 1.1×10^{15} m^3 (40,000 TCF) (11). As seen in Figure 1, these reserves are distributed more widely than oil reserves.

Using estimates of proven reserves and commitments to energy and chemical uses of gas resources, the net surplus of natural gas in a number of different countries that might be available for major fuel methanol projects has been determined. These are more than adequate to support methanol as a motor fuel.

Coal Reserves. As indicated in Table 2, coal is more abundant than oil and gas worldwide. Moreover, the U.S. has more coal than other nations: U.S. reserves amount to about 270 billion metric tons, equivalent to about 11×10^{16} MJ (1×10^{20} BTU = 6600 quads), a large number compared to the total transportation energy use of about 3.5×10^{14} MJ (21 quads) per year (11). Methanol produced from U.S. coal would obviously provide better energy security benefits than methanol produced from imported natural gas. At present however, the costs of producing methanol from coal are far higher than the costs of producing methanol from natural gas.

Biomass. Methanol can be produced from wood and other types of biomass. The prospects for biomass reserves are noted below.

Ethanol

From the point of view of availability, ethanol is extremely attractive because it can be produced from renewable biomass. Estimates of the amount of ethanol that could be produced from biomass on a sustained basis range from about 3×10^{13} to more than 8×10^{14} MJ/yr

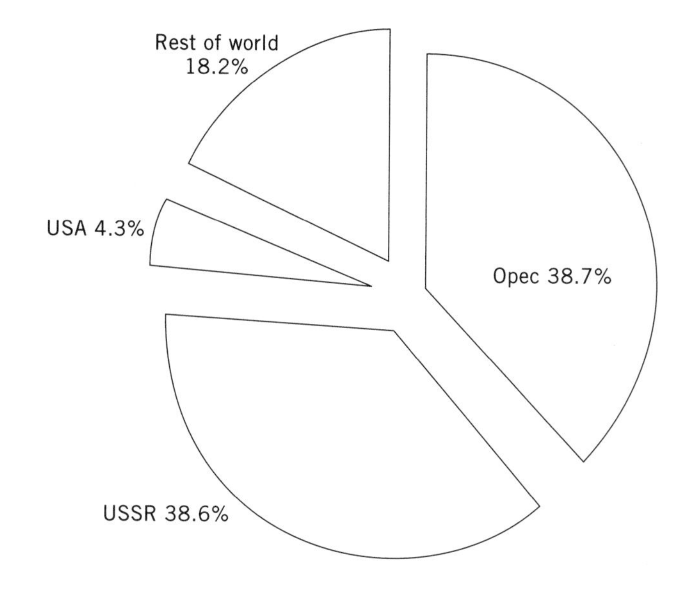

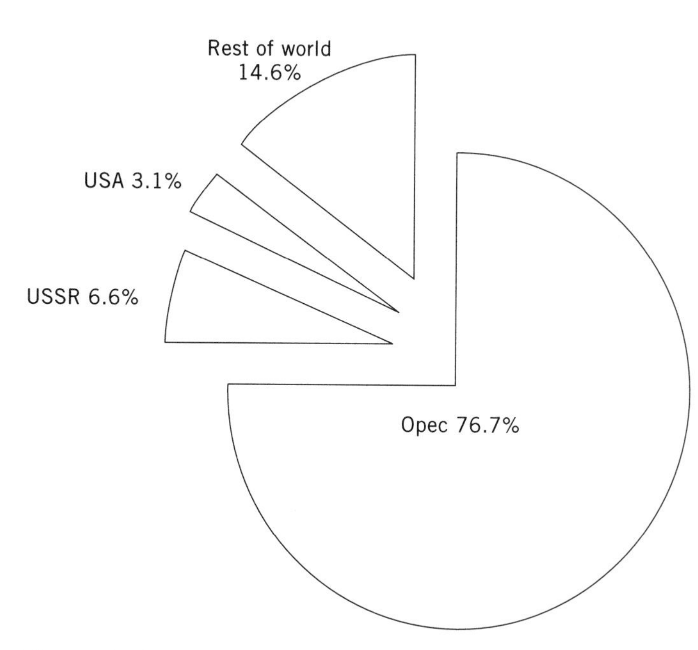

Figure 1. Distribution of the world proven gas and oil reserves as of January 1, 1988. (**a**) 107.5×10^{12} m^3 (3,800 TCF) natural gas; (**b**) 140 m^3 (890 barrels) oil.

(2–50 quad/yr), about half of which would derive from wood crops (11).

ECONOMIC ASPECTS

Alcohol Production

Studies to assess the costs of alcohol fuels and to compare the costs to those of conventional fuels contain significant uncertainties. In general, the low cost estimates indicate that methanol produced on a large scale from low cost natural gas could compete with gasoline when oil prices are around 14¢/L ($27/bbl). This comparison does not give methanol any credits for environmental or energy diversification benefits. Ethanol does not become competitive until petroleum prices are much higher.

Methanol

Produced from Natural Gas. Cost assessments of methanol produced from natural gas have been performed (13–18). Projections depend on such factors as the estimated costs of the methanol production facility, the value of the feedstock, and operating, maintenance, and shipping costs. Estimates vary for each of these factors. Costs also depend on the value of oil. Oil price not only affects the value of natural gas, it also affects the costs of plant components, labor, and shipping.

Estimates of the landed costs of methanol (the costs of methanol delivered by ship to a bulk terminal in Los Angeles), vary between 8.9 and 15.6 cents per liter (33.6 and 59.2 cents per gallon). Estimates range from 7.9 to 11.1 cents per liter (30 to 42 cents per gallon) in a large-scale established methanol market and 11.9 to 19.3 cents per liter (45 to 73 cents per gallon) during a small volume transition phase (18).

Estimated pump prices must take terminaling, distribution, and retailing costs into account as well as any differences in vehicle efficiency that methanol might offer. Estimates range from no efficiency advantage to about 30% improvement in fuel efficiency for dedicated and optimized methanol light-duty vehicles. Finally, the cost comparison must take into account the fuel specification for methanol. Most fuel methanol for light-duty vehicles is in the form of 85% methanol, 15% gasoline by volume often termed M85.

The sum of the downstream costs adds roughly 7.9 cents per liter (30¢/gal) and the adjustment of the final cost for an amount of methanol fuel equivalent in distance driven to an equal volume gasoline involves a multiplier ranging from 1.6 to 2.0, depending on fuel specification and the assumed efficiency for methanol light-duty vehicles as compared to gasoline vehicles. The California Advisory Board has undertaken such cost assessment (11).

A cost assessment must also take into account the volume of methanol use ultimately contemplated. A huge methanol program designed to replace most of the U.S. gasoline use would be likely to increase the value of even remotely located natural gas. A big program would also tend to decrease the price of oil, making it more difficult for methanol to compete as a motor fuel. Nevertheless, a balanced assessment of all the studies appears to indicate that a large scale methanol project could provide a motor fuel that competes with gasoline when oil prices are not less than about $0.17/L (1988 U.S. $27/bbl).

In small volume transition phases, methanol cannot compete directly in price with gasoline unless oil prices become very high, with the possible exception of a few scenarios in which low cost methanol is available from expansions to existing methanol plants currently serving the chemical markets for methanol. Energy diversification benefits have not been quantified but the potential air quality benefits have been studied in the work of the California Advisory Board on Air Quality and Fuels. The investment required to introduce methanol might well be

Table 2. World Estimated Recoverable Reserves of Coal in Billions of Metric Tons[a]

Location	Anthracite and Bituminous Coal[b]	Lignite[c]	Total
North America			
Canada	4.43	2.42	6.85
Mexico	1.91		1.91
United States	231.13	32.71	263.84
total	*237.47*	*35.13*	*272.60*
Central and South America			
total	*5.13*	*0.02*	*5.15*
Western Europe			
total	*32.20*	*58.23*	*90.43*
Eastern Europe and USSR			
USSR	150.19	94.53	244.72
total	*182.82*	*139.17*	*321.99*
Middle East			
total	*0.18*		*0.18*
Africa			
total	*64.67*		*64.67*
Far East and Oceania			
Australia	29.51	36.20	65.71
China	98.79		98.79
total	*130.39*	*38.61*	*169.00*
World total	*652.86*	*271.16*	*924.02*

[a] Ref. 12.
[b] Includes subanthracite and subbituminous.
[c] Includes brown coal.

justified on air quality grounds, at least in those areas such as California where air pollution programs involve substantial costs. However, the relative advantage of methanol depends on the emissions levels from future vehicles and on the costs of cleaner gasolines that might be able to offer environmental benefits that compete at least to some extent with the environmental benefits of methanol.

Produced from Coal. Estimates of the cost of producing methanol from coal have been made by the U.S. Department of Energy (DOE) (12,17) and they are more uncertain than those using natural gas. Experience in coal-to-methanol facilities of the type and size that would offer the most competitive product is limited. The projected costs of coal-derived methanol are considerably higher than those of methanol produced from natural gas. The cost of the production facility accounts for most of the increase (11). Coal-derived methanol is not expected to compete with gasoline unless oil prices exceed $0.31/L ($50/bbl). Successful development of lower cost entrained gasification technologies could reduce the cost so as to make coal-derived methanol competitive at oil prices as low as $0.25/L ($40/bbl) (17).

These cost comparisons do not assign any credit to methanol for environmental improvements or energy security. Energy security benefits could be large if methanol were produced from domestic coal.

Produced from Biomass. Estimates for methanol produced from biomass indicate (11) that these costs are higher than those of methanol produced from coal. Barring substantial technological improvements, methanol produced from biomass does not appear to be competitive.

Ethanol

Accurate projections of ethanol costs are much more difficult to make than are those for methanol. Large scale ethanol production would impact upon food costs and have important environmental consequences that are rarely cost-analyzed because of the complexity. Furthermore, for corn, the most likely large-scale feedstock, ethanol costs are strongly influenced by the credit assigned to the protein by-product remaining after the starch has been removed and converted to ethanol.

Cost estimates of producing ethanol from corn have many uncertainties (11). Most estimates fall into the range of $0.26 to 0.40 per liter ($1 to 1.50/gal), after taking credits for protein by-products, although some estimates are lower. These estimates do not make ethanol competitive with oil until oil prices are above $0.38/L ($60/bbl) (17).

For these reasons, ethanol is most likely to find use as a motor fuel in the form of a gasoline additive, either as ethanol or ethanol-based ethers. In these blend uses, ethanol can capture the high market value of gasoline components that provide high octane and reduced vapor pressure.

Impact of Incremental Vehicle Costs

The costs of alcohol fuel usage may include other costs associated with vehicles. Incremental vehicle costs have been estimated by the Ford Motor Company for the fuel-flexible vehicle that can use gasoline, methanol, ethanol, or any mixture of these, to be in the range of $300 per vehicle assuming substantial production (14). This cost may or may not be passed along to the consumer, because of incentives provided to the manufacturer by the Alternative Motor Fuels Act of 1988. There also may be incremental vehicle operating costs resulting from increased lubrication or increased frequency of oil changes.

Infrastructure Requirements

In general, infrastructure requirements resulting from the expanded use of alcohol fuels are not especially greater than those involved in the production and refining of oil. However, for a corresponding delivery of energy, the capital costs of alcohol infrastructure would presumably be larger, as capital costs of production facilities appear to be larger than corresponding oil refinery costs for an equivalent amount of energy output. Moreover, infrastructure costs for storage and distribution facilities would be higher. Facilities for hydrocarbon fuels are generally not compatible with methanol and ethanol. Thus a program to introduce substantial amounts of alcohol fuel might well require existing infrastructure modifications. These changes can be especially difficult and costly for underground pipelines and tanks, which need to be replaced or modified.

In California, the South Coast Air Quality Management District has implemented a local rule requiring that one new or replacement underground tank at each gasoline retail facility must be suitable for methanol. Replacement of an existing tank can cost $50,000 and perhaps more. But many small retail outlets are being replaced by larger more efficient ones, and many older underground tanks are being replaced to prevent possible leaks and hence underground contamination. Therefore the compatibility rule allows for the gradual development of a methanol-compatible infrastructure at low costs. The extra costs for methanol-compatible storage ranges from negligible to about $4000 per tank and dispenser, depending on the technical choices that would otherwise be made for gasoline-only facilities.

VEHICLE TECHNOLOGY AND VEHICLE EMISSIONS

One of the reasons that U.S. automobile manufacturers showed more interest in alcohols as alternative fuels in the late 1970s and early 1980s is because alcohol's energy density is closer to gasoline and diesel than other alternatives such as compressed natural gas. They reasoned that consumers would be more comfortable with liquid fuels, envisioning little change in the fuel distribution of alcohols. Most of the research in the 1970s focused on converting light-duty vehicles, to alcohol fuels. Towards the late 1970s, researchers also began to turn their attention to heavy-duty applications. In heavy-duty engines the emissions benefits of alcohols are far clearer than in light-duty vehicles. However, it is also much harder to design heavy-duty engines to use the low cetane number alcohols.

It was not until the early 1980s that the potential air quality benefits of alcohol fuels started to be investigated. It was about five years later that proponents argued that alcohols could provide significant air quality benefits in addition to energy security benefits. Low level blends of ethanol and gasoline were argued to provide lower CO emissions. The exhaust from light-duty methanol vehicles was thought to be less reactive in the formation of ozone. Uses of alcohols fuels in heavy-duty engines showed substantially reduced mass emissions in contrast to light-duty experience which showed about the same mass emissions but a reduced reactivity of the exhaust components.

LIGHT-DUTY VEHICLES

Use of Low Level Blends. The first significant U.S. use of alcohols as fuels since the 1930s was the low level 10% splash blending of ethanol in gasoline, which started after the oil crisis of the 1970s. This blend, called gasohol, is still sold in commerce although mostly by independent marketers and distributors instead of the major oil companies. EPA provided a waiver for this fuel allowing for a 6.9 kPa (1 psi) increase in the vapor pressure of gasohol over that of gasoline.

In the first years of gasohol use some starting and driveability problems were reported (19). Not all vehicles experienced these problems, however, and better fuel economy was often indicated even though the energy content of the fuel was reduced. Gasohol was exempted from the federal excise tax amounting to a $0.16/L ($0.60/gal) subsidy. Without this subsidy, ethanol would be too expensive for use even as a fuel additive.

Nearly four billion L/yr of ethanol are added to gasoline and sold as gasohol (18). The starting or driveability difficulties have been solved, in part, by the advances in vehicle technology employing fuel feedback controls.

Methanol was also considered as a gasoline additive. Table 3 summarizes some of the oxygenated compounds approved by EPA for use in unleaded gasoline. EPA waivers were granted to: Sun Oil Company in 1979 for 2.75% by volume methanol with an equal volume of tertiary butyl alcohol [*75-65-0*] (TBA) up to a blend oxygen total of 2% by weight oxygen; ARCO in 1979 for up to 7% by volume TBA; and ARCO in 1981 for the use of the blends containing a maximum of 3.5% by weight oxygen. These last blends are gasoline-grade TBA (GTBA) and OXINOL having up to 1:1 volume ratio of methanol to GTBA. Petrocoal was also granted a waiver to market up to 10% by volume methanol and cosolvents in gasoline but this waiver was revoked in 1986 after automobile manufactures complained of significant material compatibility problems and openly warned consumers against gasolines containing methanol. A waiver was issued to Du Pont in 1985 which allowed addition of up to 5% methanol to gasoline having a mixture of 2.5% cosolvents. None of these additives became very popular (20).

Vehicle Emissions. Gasohol has some automotive exhaust emissions benefits because adding oxygen to a fuel leans out the fuel mixture, producing less carbon monoxide [*630-08-2*] (CO). This is true both for carbureted vehicles and for those having electronic fuel injection.

Urban areas such as Denver, Phoenix, and others at high altitudes have problems complying with health-based carbon monoxide standards in part because of automobile emissions. Vehicles calibrated for operation at sea level that are operated at high altitudes run rich, producing more CO. Blends such as gasohol cause the engine to operate leaner because of the oxygen in the fuel. There

Table 3. EPA Approved Oxygenated Compounds for Use in Unleaded Gasoline[a]

Compound[b]	Broadest EPA Waiver	Date	Maximum Oxygen, wt %	Maximum Oxygenate, vol %
Methanol	substantially similar	1981		0.3
Propyl alcohols	substantially similar	1981	2.0	(7.1)[c]
Butyl alcohols	substantially similar	1981	2.0	(8.7)[c]
Methyl *tert*-butyl ether (MTBE)	substantially similar	1981	2.0	(11.0)[c]
Tert-amyl methyl ether (TAME)	substantially similar	1981	2.0	(12.7)[c]
Isopropyl ether	substantially similar	1981	2.0	(12.8)[c]
Methanol and butyl alcohol or higher mol wt alcohols in equal vol	substantially similar	1981	2.0	5.5
Ethanol	gasohol	1979, 1982	(3.5)[c,d]	10.0
Gasoline grade *tert*-butyl alcohol (GTBA)	ARCO	1981	3.5	(15.7)[c]
Methanol + GTBA (1:1 max ratio)	ARCO (OXINOL)	1981	3.5	(9.4)[c]
Methanol at 5 vol % max + 2.5 vol % min cosolvent	Du Pont[f]	1985	3.7	[e]
	Texas methanol (OCTAMIX)[g]	1988	3.7	[e]

[a] Ref. 21.
[b] All blends of these oxygenated compounds are subject to ASTM D 439 volatility limits except ethanol. Contact the EPA for current waivers and detailed requirements, U.S. Environmental Protection Agency, Field Operations and Support Division (EN-397F), 401 M Street, S. W., Washington, D. C. 20460.
[c] Calculated equivalent for average specific gravity gasoline (0.737 specific gravity at 16°C, NIPER Gasoline Report). Calculated equivalent depends on the specific gravity of the gasoline.
[d] Value shown is for denatured ethanol. Neat ethanol blended at 10.0 vol % produces 3.7 wt % oxygen.
[e] Varies with type of cosolvent.
[f] The cosolvents are any one or a mixture of ethanol, propyl, and butyl alcohols. Corrosion inhibitor is also required.
[g] The cosolvents are a mixture of ethanol, propyl, butyl and higher alcohols up to octyl alcohol. Corrosion inhibitor is also required.

are larger CO reductions using oxygenated blends in older, carbureted engines. But even the newer technology vehicles have lower CO emissions using gasohol and other oxygenated fuel because of periods of open loop operation, especially during cold starts.

Blended fuels increase the vapor pressure of the resulting mixture so that more hydrocarbons are evaporated into the atmosphere during operation, refueling, or periods of extended parking. Although these hydrocarbons can react with NO_x emissions in sunlight to form ozone, atmospheric modeling has indicated that ozone is probably not increased as a result of the higher fuel volatility for two reasons (see AIR QUALITY MODELING). First, CO is also an ozone precursor so reducing CO reduces ozone. Second, the hydrocarbon species are somewhat less reactive because of the lower reactivity of ethanol. Furthermore, programs for oxygenated fuel use are focused at high CO occurrences during the year. These usually occur in the wintertime, whereas most areas violate ozone standards in the summer months. Therefore, oxygenated fuel programs, as a CO control strategy, do not generally interfere with ozone attainment strategies. However, programs should be individually evaluated (20).

Ethanol blends can also have an effect on NO_x emissions. Scattered data indicate that NO_x may increase as oxygenates are added to the fuel.

Other countries have also investigated the use of low level alcohol blends as an energy substitution strategy as well as to reduce exhaust emissions of lead (22,23). Brazil implemented low level ethanol–gasoline blends throughout the twentieth century during times of oil shortages or as a hedge against international fluctuations in sugar prices. Blends ranged from 15 to 42%. In 1975 Proalcool, Brazil's ethanol fuel program, was initiated and required the blending of 20% by volume of ethanol in gasoline. This was not totally achieved throughout Brazil until about 1986 when a 22% ethanol–gasoline blend was standardized. Once the fuel was standardized, engine modifications for new vehicles were made, including higher compression and adjustments to the carburetor and timing (22).

Germany also evaluated the gasoline–alcohol blends using methanol. Early programs used 15% methanol added to gasoline (M15). This program required vehicles to be designed for this fuel. Modifications included changes to the fueling system for air–fuel control and vehicle material changes to be compatible with the higher methanol concentrations. The program ended in 1982. M15 was concluded to be feasible if higher vehicle costs could be offset by the possibility of lower fuel costs (24). Lower level blends were also investigated using up to 3% methanol with 2 or more percent of a suitable cosolvent. Unlike M15, gasoline vehicles could use this blend without any modifications (25). Germany has for several years now used low level blends of methanol in their gasoline.

Retrofits. Retrofits are vehicles designed for conventional fuels modified so that the vehicles can operate on alcohol. Generally, because both ethanol and methanol have lower energy densities, the quantity of fuel entering the engine must be increased to get the same power. Also, because the alcohols have slightly different combustion characteristics, engine parameters such as ignition timing need to be adjusted. To optimize performance and fuel economy the compression ratio of the engine should be increased. However, the economics of these conversions are such that the least amount of changes are made and adjustments to engine compression ratio is typically not done.

Retrofits were popular at the beginning of alcohol fuel programs. Kits were introduced that modified only the fuel flow rate into the engine, but material changes were also necessary because both ethanol and methanol are more corrosive to metals than gasoline. Retrofitting allowed maximum market penetration without having to wait for fleet roll over or for manufacturers to market new vehicles. There was some success in converting light-duty vehicles to methanol. Bank of America operated a fleet of 292 converted Ford and Chevrolet vehicles in the late 1970s and into the 1980s before oil prices collapsed. A conversion kit for these vehicles included hardware, material changes, and a fuel additive to help minimize corrosion (26,27).

The California Energy Commission (CEC) evaluated the conversion of 1980 Ford Pintos equipped with 2.3-L four-cylinder engines, feedback-controlled carburetors, and three-way catalysts. Four vehicles were left unmodified, four converted to methanol, and four converted to ethanol. The basic changes required to use the alcohols were: the terneplate coating in the fuel tanks was stripped; fuel level sending units and carburetors were chromated to inhibit corrosion; and the air–fuel ratio, timing, and fuel vaporization mechanisms were recalibrated for proper combustion of alcohol fuels and to comply with emission standards. Two methanol- and two ethanol-fueled engines had special pistons installed to raise compression ratios from 9:1 to 12:1 for better efficiency. These vehicles were operated for 18 months accumulating 272,000 km (169,300 miles). The methanol conversions averaged 25,000 km; ethanol conversions, 21,700 km; and gasoline controls, 22,100 km. Although both the methanol and ethanol vehicles were designed to operate on 100% alcohol, they utilized M94.5 (94.5 vol % methanol and 5.5% isopentane [*78-78-4*] added to improve cold starts and engine warmup) and CDA-20 (ethanol denatured using 2 to 5% unleaded gasoline) (28,29).

This conversion program indicated that vehicles could be converted to alcohols. Good fuel economy was obtained; methanol vehicles averaged 4.7 km/L (11.0 mpg) or 9.1 km/L (21.3 mpg) on an equivalent energy basis compared to 8.3 km/L (19.5 mpg) for the gasoline control vehicles. No driveability problems were reported and vehicles had no problems starting (lowest temperature was −1.1°C). Both the ethanol and methanol Pintos showed increased upper cylinder wear over gasoline engines. Poor lubrication from using alcohols and excess fueling because of carburetor float problems contributed to the higher wear rates. Hydrocarbon, carbon monoxide, and NO_x emissions were less for methanol, 0.14, 3.2, and 0.3 g/km, respectively, than for gasoline, 0.25, 5.6, and 0.6 g/km.

The biggest problems of the CEC Pinto fleet were that vehicle conversions were expensive and alcohol fuels were more expensive than gasoline. Changes to the fuel tank, fuel lines, and the carburetor were too labor-intensive to be done cheaply. However, these changes if designed, could be made during the vehicle manufacturing at little additional cost (30). Brazil priced ethanol at 65% the cost of gasoline (10) so that conversions could be cost-effective because of the savings on the fuel costs.

Other significant disadvantage of retrofits were the quality of the conversion kits and the ability of the conversions to meet emission regulations over the useful vehicle life. The initial phases of the Brazilian ethanol program also suffered because of poor quality vehicle retrofits. The quality was so poor that the program almost failed after a fairly substantial number of vehicles were converted and the ethanol fuel infrastructure was in place. Further incentives and automobile manufacturers introducing new vehicles designed for ethanol stabilized the program (31).

For these reasons, CEC and DOE concluded that the only cost-effective method of getting alcohol fueled vehicles would be from original equipment manufacturers (OEM). Vehicles produced on the assembly line would have lower unit costs. The OEM could design and ensure the success and durability of the emission control equipment.

Dedicated Vehicles. Only Brazil and California have continued implementing alcohols in the transportation sector. The Brazilian program, the largest alternative fuel program in the world, used about 7.5% of oil equivalent of ethanol in 1987 (equivalent to 150,000 bbl of crude oil per day). In 1987 about 4 million vehicles operated on 100% ethanol and 94% of all new vehicles purchased that year were ethanol-fueled. About 25% of Brazil's light-duty vehicle fleet (10) operate on alcohol. The leading Brazilian OEMs are Autolatina (a joint venture of Volkswagen and Ford), GM, and Fiat. Vehicles are manufactured and marketed in Brazil.

In contrast the California program has some 600 demonstration vehicles (32). Both Ford and Volkswagen participated in the dedicated vehicle phase of the California program. In 1981 Volkswagen provided the first alcohol vehicles produced on an assembly line, forty (19 methanol, 20 ethanol, and 1 gasoline) VW Rabbits and light-duty trucks were manufactured. Design incorporated continuous port fuel injection 12.5:1 compression ratio, a new ignition system calibration, and a heat exchanger for faster oil warmup. The entire fuel system was designed using materials compatible with methanol and ethanol. These vehicles operated until 1983 and logged 728,000 km of service. This fleet used the same fuel as the ethanol and methanol Pinto retrofits.

In 1981 Ford also provided CEC with 40 Escorts designed to operate on M94.5 and 15 gasoline vehicles to serve as controls. These vehicles had accumulated over 3.4 million km of service as of March 1986. The 1981 Escorts were modified to use methanol. The 1.6-L gasoline engine had a production piston used in European 1.6-L engines to raise the compression ratio to 11.4:1. Other field modifications included spark plug change, a carburetor throttleshaft material change, carburetor float redesign, and the replacement of tin-plated fuel tanks with ones of stainless-steel.

In 1983 the methanol fleet was expanded with the purchase of 506 Ford Escorts. These vehicles are equipped with engines and fuel systems redesigned from Ford's standard 1.6-L gasoline-fueled Escort. Ford also produced five advanced technology vehicles equipped with electronic fuel injection and microprocessor control, with the goal of improving fuel economy and reducing NO_x emissions to 0.25 g/km. The emissions control on these vehicles used the same technology as on the carbureted gasoline versions: standard three-way catalyst, exhaust gas recirculation, and air injection. The 1981 and 1983 Escorts have logged over 48 million km in service and some vehicles have reached gasoline equivalent fuel usage per kilometer over the lifetime of the vehicle.

The methanol fuel specification was changed for the Escorts, at the end of 1983 from M94.5 to a blend of 90% methanol and 10% unleaded gasoline (M90). In the summer of 1984 the fuel was further modified to include 15% gasoline (M85). This change from isopentane was made because gasoline was cheaper. In addition, M85 has a gasoline odor and taste, and in daylight the flame is more visible than either M94.5 or M90. Another safety benefit of the added gasoline is the increased volatility creating a richer air–vapor mixture much less likely to burn or explode in closed containers than neat or 100 percent methanol.

The results of the California fleet demonstrations indicated that fuel economy on an energy basis was equal to or better than gasoline, especially using vehicles having higher compression. None of these engines, however, was fully optimized for methanol so additional improvements are possible. Driveability was also good for the methanol vehicles. An acceleration test of two 1983 methanol-fueled vehicles and a similar 1984 model resulted in: 1983 fuel injection (EFI) methanol, 14.53 s; 1983 carbureted methanol, 15.51 s; and 1984 carbureted gasoline, 19.10 s.

Tests demonstrate that methanol vehicles can meet stringent emission standards for HC, CO, and NO_x as indicated in Figure 2. The primary benefit of methanol, however, is not the amount of hydrocarbons emitted but rather that methanol-fueled vehicles emit mainly methanol which is less reactive in the formation of ozone than the variety of complex organic molecules in gasoline exhaust. Formaldehyde [*50-00-0*] emissions from methanol vehicles are increased in comparison to gasoline vehicles. Tests of 1983 Escorts showed tailpipe levels as high as 62 mg/km, well above typical gasoline levels of 2 to 7 mg/km. The 1981 Rabbits ranged from about 6 to 14 mg/km and the 1981 Escorts had levels less than 7 mg/km. All results were obtained on relatively low mileage vehicles. Deterioration of catalyst effectiveness could increase these emissions.

The vehicles investigated were mostly adaptions of gasoline technology. For example, automobile manufacturers recommended that catalytic converters designed for gasoline automobiles be installed on vehicles in Brazil to control acetaldehyde [*75-07-0*] emissions. Brazil decided against catalysts because gasoline vehicles at that time did not have these systems. Similarly, California adopted M85 to aid in cold starting and to provide some measures of perceived safety. Research to find other additives that would assist in cold starting and provide safety characteristics at a reasonable price have been relatively unsuccessful (33). But another way to overcome the issue of cold starting is to design and optimize engines to operate on 100% methanol (M100). EPA has been pursuing M100 for

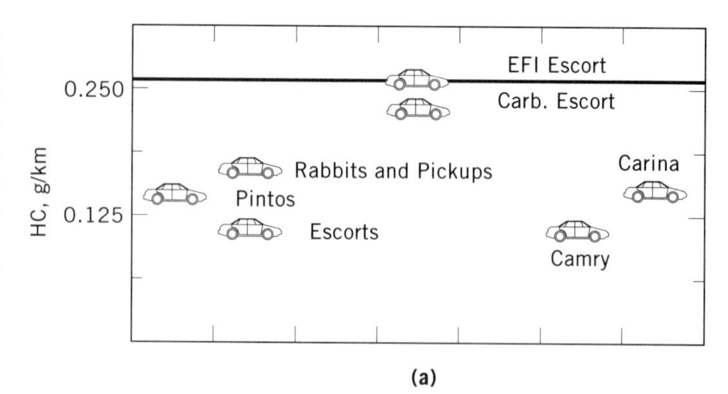

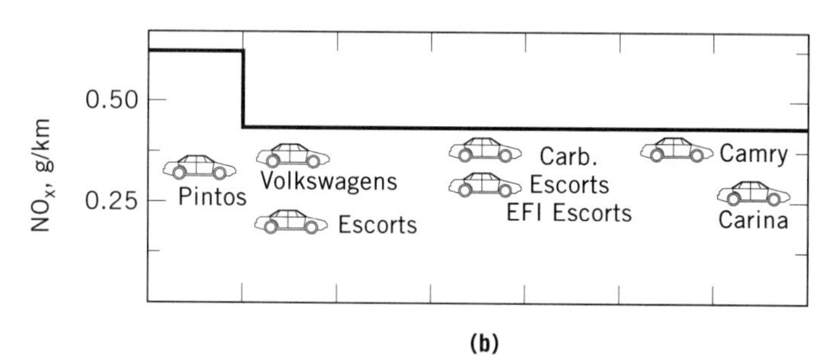

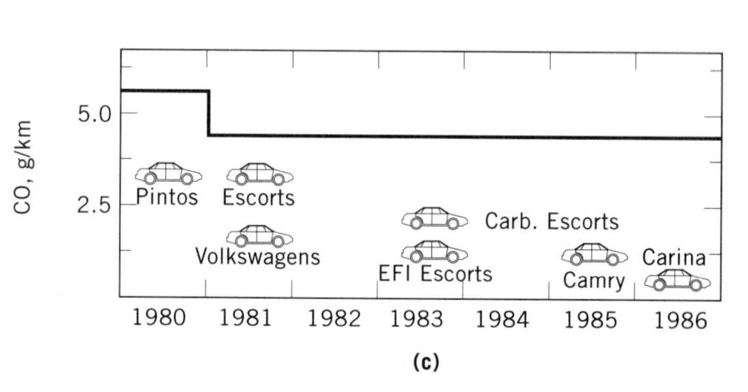

Figure 2. CEC methanol-fueled vehicle exhaust profile for (**a**) HC, hydrocarbons; (**b**) NO_x; and (**c**) CO. Solid line represents State of California standard maximum emissions. Methanol HC emissions are calculated as $CH_{1.85}$ and not corrected for flame-ionization detector response.

several years with good success (34). Cold starting is not a problem for direct injected engines where high pressure is used to atomize the fuel and results indicate that a light-duty, direct injection engine can attain very low emissions having good fuel economy and driveability. However, safety concerns of using M100 in general commerce need to be addressed (35).

Fuel Flexible Vehicles. Using dedicated alcohol fuel vehicles pointed to the importance of a wide distribution of fueling stations. Methanol-fueled vehicles require refueling more often than gasoline vehicles.

In 1981, the Dutch company TNO in cooperation with the New Zealand government converted a gasoline engine to a flexible fuel vehicle by adding a fuel sensor. The sensor determined the amount of oxygen in the fuel and then used this information to mechanically adjust the carbure-

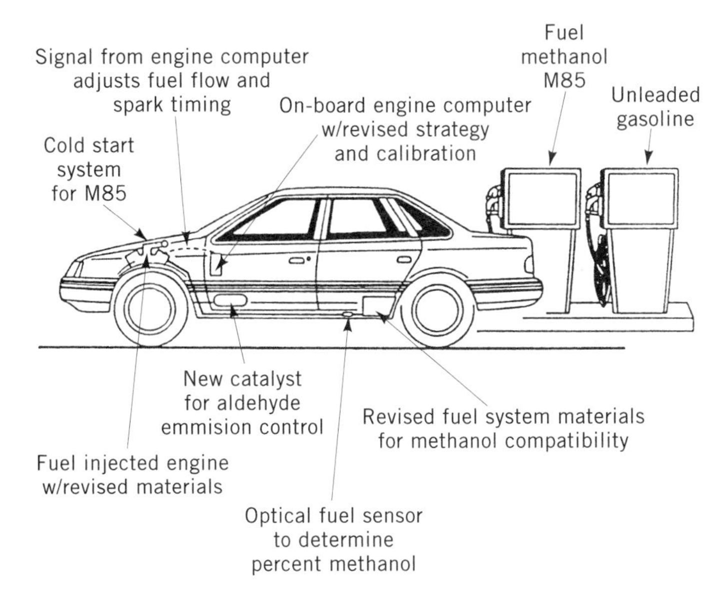

Figure 3. Components of a Ford flexible fuel vehicle (FFV).

tor jets. The initial mechanical system was crude, but the advancement of engine and emission controls, in particular the use of electronics and computers, has brought about substantial refinement (36).

Ford first tried the flexible fuel system on an Escort and called it the "flexible fueled vehicle" or FFV. As seen in Figure 3, the system included building into the electronics any necessary calibrations for gasoline and methanol fuels, adding a sensor to determine the amount of methanol in the fuel, and making necessary material changes to fuel wetted components. The sensor is one of the most critical parts of this system: its output determines parameters such as the amount of fuel to be injected and engine timing. Fuel injectors must also have a wider response range. The engine compression ratio was not changed because the vehicle is designed to operate as well on gasoline as on methanol.

Of course, FFV drivers do not have to use methanol. Emissions benefits are not obtained if methanol is not used, and fuel economy is not optimized for methanol nor are emissions. However the State of California has concluded that advantages offered by the flexibility of the FFV far outweigh the disadvantages (37).

Many U.S. and foreign automobile manufacturers are developing a fuel flexible vehicle in the 1990s. Ford has developed FFVs for 5-L engines used in Crown Victorias and for 3.0-L engines used in their Taurus car line. GM followed Ford with a variable fueled vehicle (VFV) and applied this technology first to the 2.8-L engine family used in the Corsica, and more recently to the 3.1-L engine family in the Lumina (see Fig. 4). Prototype flexible fuel vehicles are also being developed by Volkswagen, Chrysler, Toyota, Nissan, and Mitsubishi. California is in the process of obtaining an additional 5,000 of these vehicles in the next several model years (MY92 nd MY93) to be used by government and private fleets.

The experience using fuel flexible vehicles has been surprisingly successful. California is operating about 200 vehicles and driveability is excellent on whatever fuel is used. Tests performed on a Ford Crown Victoria showed slightly better fuel economy (4%) and better acceleration (6%) on methanol than gasoline (38). The fuel flexible technology is not limited to methanol but with electronic calibration changes also works for ethanol and gasoline combinations. Changes can be made by adjusting the engine maps in the computer.

EPA, the Air Resources Board (ARB), and others are investigating the possible emission benefits of alcohol fueled vehicles. EPA and ARB adopted regulations for hydrocarbon mass emissions which accounted for the oxygen components in the exhaust, so called organic mass hydrocarbon equivalent (OM-HCE). The regulations required methanol vehicles to meet the same OMHCE value as gasoline hydrocarbons, which in California is 0.155 g/km in 1993. The trend is to account also for the total mass and the reactivity of individual hydrocarbon species and the measure being proposed for total mass is nonmethane organic gases or NMOG. Reactivity of the individual species that make up NMOG are estimated (39) to give a value of ozone/km.

Vehicle emissions have been monitored over the last several years on fuel flexible vehicles and depending on when the tests were performed, reported in total hydrocarbons, OMHCE, or NMOG. The vehicle testing performed to date has shown that methanol FFVs can provide emissions benefits. Figure 5 shows the emissions from a GM Corsica for various gasoline methanol mixtures (40). This vehicle was designed to meet the California standards of 0.155 g/km hydrocarbon, 2.1 g/km CO, and 0.25 g/km NO_x. In addition California requires methanol vehicles to meet a formaldehyde standard of 9.3 mg/km. The total organic emissions decrease with increasing methanol, whereas formaldehyde increases. NO_x and CO vary but appear unaffected by methanol content. The Corsica data were taken on a green catalyst (low vehicle mileage) and some deterioration of these emissions levels can be expected with age.

Figure 6 shows data for four vehicles operated on gasoline and M85, two having an electrically heated catalyst (EHC). The two vehicles equipped with EHC both showed low values of NMOG and estimated ozone production. These data seem to indicate that methanol vehicles result in less ozone than comparable gasoline vehicles. However, the data only include exhaust emissions and not evaporative or running losses. These later sources of emissions should be lower using methanol because of the lower reactivity of the alcohols.

HEAVY-DUTY VEHICLES

The use of alcohols in heavy-duty engines developed more slowly than in light-duty engines primarily because the majority of heavy-duty engines are diesels. Diesels are unthrottled, stratified charge engines which autoignite fuel by heat generated during compression. Engine speed and load are modulated by varying the quantity of fuel injected into the cylinder rather than by throttling the fuel–air mixture as done in the Otto cycle or spark-ignition engine. Unthrottled air aspiration reduces pumping losses which in turn increases the engine's thermal

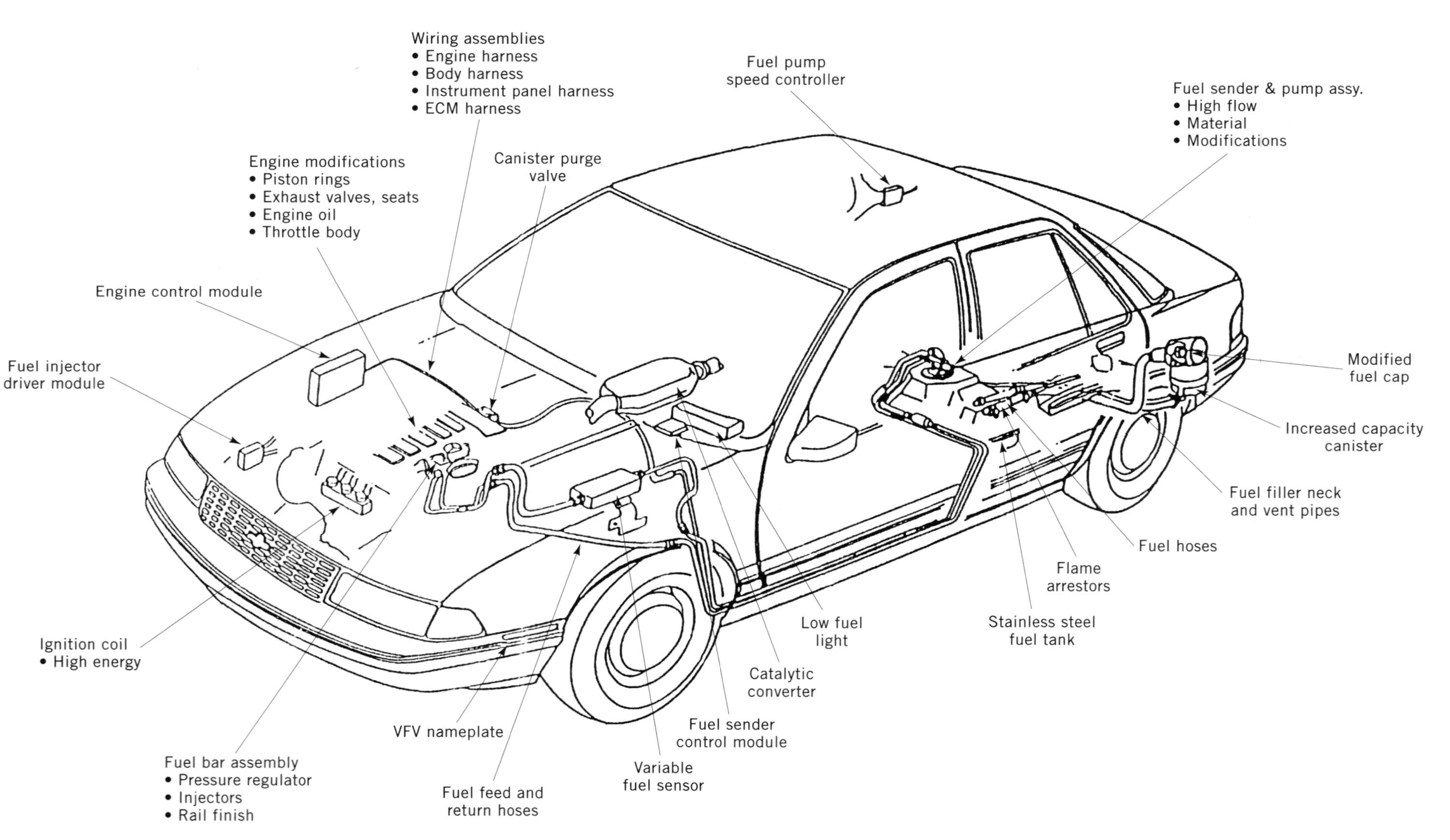

Figure 4. Components of a Lumina methanol variable fueled vehicle (VFV).

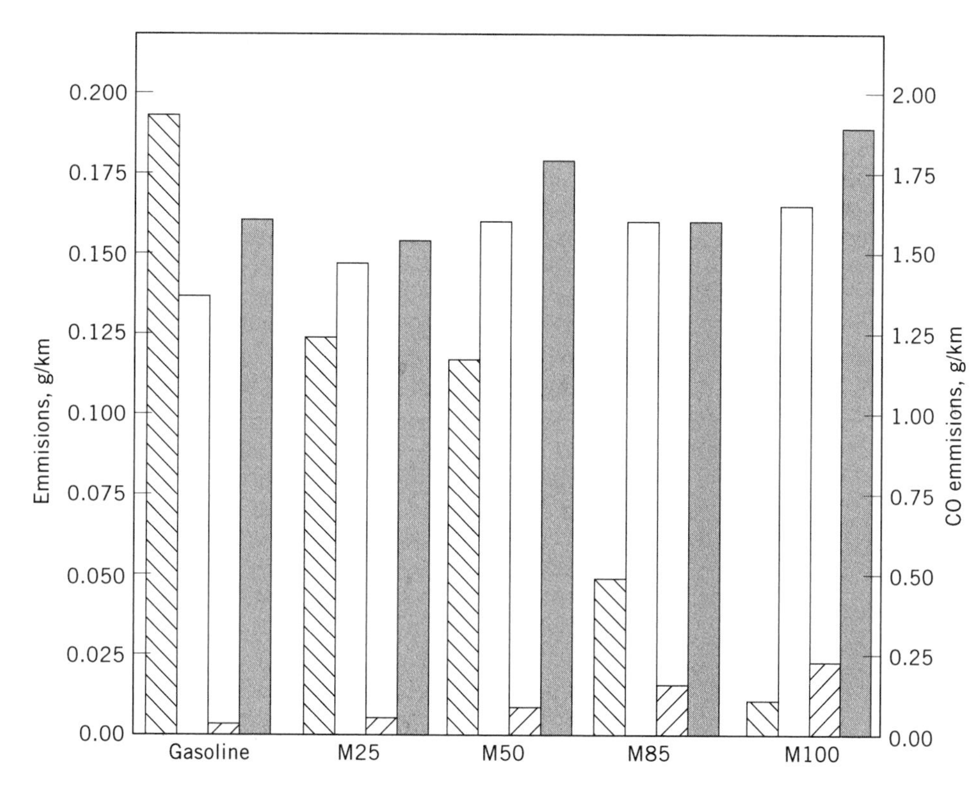

Figure 5. Emissions from a GM Corsica VFV for gasoline and gasoline–methanol mixtures where ▧ represents total organic material, including hydrocarbons, methanol, and formaldehyde; ▩ represents NO_x; ▨ formaldehyde; and ■ carbon monoxide.

efficiency. Diesel engines also are designed for higher compression ratios resulting in further efficiency improvements. The higher efficiency and excellent reliability and durability of these engines make them attractive for heavy-duty applications in trucks, buses, and off road equipment. Unfortunately, their low cetane number limited the compability of alcohol fuels with diesel engines without modifications to assist ignition.

Not until the early 1980s were prototype heavy-duty engines developed to operate on methanol and ethanol, having efficiencies equal to or better than corresponding diesel engines. These engines were quieter and had considerably cleaner exhaust emissions. Mass emissions of NO_x and particulates were substantially lower when the cleaner alcohol fuels were used, overcoming the inherent NO_x-particulate tradeoff of diesel engines. The biggest challenge facing engine manufacturers in the 1980s and 1990s was to make the alcohol engines as reliable and durable as heavy-duty diesel engines which operate in the range of 0.3 to 0.5 million kilometers without major engine maintenance and repair.

Technology Options. Because alcohols are not easily ignited in diesel-type engines changes in engine hardware or modifications to the fuel are needed. Engine modifications can include the addition of a spark ignition system or an additional fuel injection system to provide dual fuel capabilities. Fuel modifications involve the addition of cetane improvers. Low level blends do not work because methanol and diesel fuels are not miscible. Some research to emulsify diesel and alcohols was carried out, but was never successful (42). Other investigations involved adding a separate fuel system including two fuel tanks, fuel lines, injection pumps, and injectors. In this approach diesel was used to ignite the fuel mixtures, and at low speeds–low loads diesel was the primary fuel. At high speeds–high loads alcohol was the primary fuel. This dual fuel approach was both cumbersome and expensive (43).

One successful method for using alcohols was fumigation. In this technique alcohol is atomized in the engine's intake air either by carburetion or injection. Diesel is directly injected into the cylinder and the combined air–alcohol and diesel mixture is autoignited. Diesel consumption is reduced by the energy of the alcohol in the intake air. This approach, although technically feasible, also requires separate fuel systems for the diesel and alcohol fuels. Additionally, the amount of alcohol used is limited by the amount that can be vaporized into the intake air. This approach is more appropriate as an engine retro-

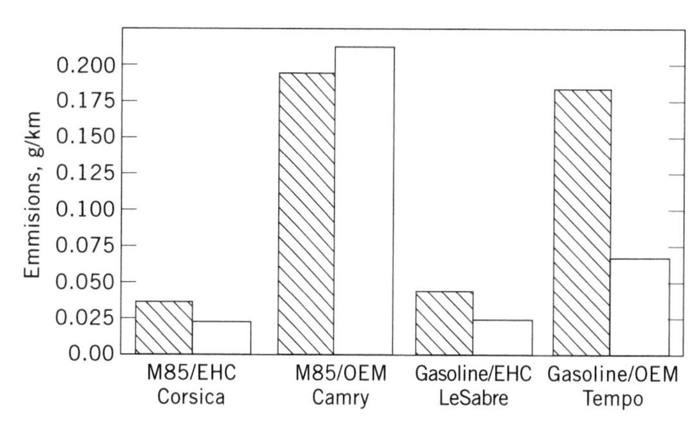

Figure 6. Comparison of M85 and gasoline emissions (including ozone) corresponding to ▧ estimated ozone (ref. 40) and ▩ NMOG for vehicles having an electrically heated catalyst (EHC) or not (OEM) (ref. 41).

fit where total energy substitution is not the primary objective (44).

Other possible technologies involve either assisted ignition or cetane improvers. Assisted ignition approaches can be divided between direct injection, stratified charge type engines, and engines converted back to Otto cycles, by throttling and lowering engine compression.

Dedicated Vehicles. As late as 1982, researchers were still arguing the worthiness of alcohols as fuels for heavy-duty engines (45). Pioneer work on multifuel engines led to modifications in diesel engines to burn neat or 100% alcohol. The German manufacturers were the first to provide prototype methanol engines. Daimler-Benz modified their four stroke M 407 series diesel engine to operate on 100% methanol, by converting the diesel version to a spark-ignited, Otto cycle engine. This required lowering the compression, adding a spark ignition system, and carburetion (throttling). To get back some of the efficiency loss caused by going to a throttling and lower compression, the Daimler-Benz design incorporated a heat exchanger to vaporize the methanol using engine cooling water. Vaporized methanol was introduced into the engine using a standard gaseous carburetor–mixing device.

The M 407 hGO methanol engine is a horizontal, water-cooled, inline six-cylinder configuration (46). Basic combustion is similar to the conventional spark-ignition Otto cycle with one significant exception. Lean combustion at part load is possible for two reasons: because of methanol's favorable flammability limits and because methanol is vaporized and introduced as a gas. Equivalence ratios (air–fuel ratio relative to stoichiometric) greater than 2 are possible without misfire, and minimum fuel consumption is obtained at an equivalence ratio of about 1.8. In the higher load range, the engine is controlled by the air–fuel ratio, rather than by intake throttling, so efficiency is increased relative to the conventional spark-ignition engine. Intake throttling is used for control in the lower load range.

The first methanol bus in the world was placed in revenue service in Auckland, New Zealand in June 1981. It was a Mercedes O 305 city bus using the M 407 hGO methanol engine. This vehicle operated in revenue service for several years with mixed results. Fuel economy on an equivalent energy basis ranged from 6 to 17% more than diesel fuel economy. Power and torque matched the diesel engine and drivers could not detect a difference. Reliability and durability of components was a problem. Additional demonstrations took place in Berlin, Germany and in Pretoria, South Africa, both in 1982.

The world's second methanol bus was introduced in Auckland shortly after the first. This was a M.A.N. bus with a M.A.N. FM multifuel combustion system utilizing 100% methanol. The FM system, more similar to a diesel engine, is a direct injection, high compression engine using a spark ignition. Fuel is injected into an open chamber combustion configuration in close proximity to the spark plugs which ignite the air–fuel mixture. Near the spark plugs the air–fuel mixture is rich and combustion proceeds to the lean fuel air mixtures in the rest of the cylinder. The air–fuel charge is thus stratified in the cylinder and these types of engines are often called lean burn, stratified charge. Engine hardware is similar to the diesel version including a high pressure injection pump and a compression ratio comparable to diesel (19:1). This technology was applied to M.A.N.'s 2566 series engines, an inline 6-cylinder engine, and for buses is configured horizontally (47). Like the Mercedes, it is a four stroke engine.

This technology was tested using diesel fuel, gasoline, methanol, and ethanol. A.M.A.N. SL 200 bus having the M.A.N. D2566 FMUH methanol engine was also demonstrated in Berlin. Results of these tests were somewhat mixed. Fuel economy was 12% less than a comparable diesel bus, but driveability was very good. Because the methanol fueled bus was not smoke limited at low speeds, higher torque was possible and bus drivers used this advantage to accelerate faster from starts. Emissions results indicated a considerable advantage in using fuels such as methanol. CO and NO_x were reduced compared to diesel engines and particulates were virtually eliminated.

The success of the New Zealand and German programs were instrumental in implementing a similar bus demonstration in California in 1982 (48). The primary objective was to assess the viability of using methanol in heavy-duty engines. The project focused on evaluating engine durability, fuel economy, driveability, and emissions characteristics. CEC also initiated a demonstration project for off-road heavy-duty vehicles using a multifuel tractor capable of operating on either neat methanol or ethanol (30).

M.A.N. and Detroit Diesel Allison, now Detroit Diesel Corporation (DDC), agreed to participate in the California bus program. DDC provided a methanol version of their 6V-92TA engine, which along with the DDC 71 series, is the most commonly used bus engine in the United States. The engine is a compression-ignited, two-stroke design having a displacement of 9.1 liters and power rating of 20,700 W (277 hp). Several design changes were incorporated for operation on methanol, including electronic unit injectors (EUI) for more precise fuel control, an increased compression ratio, a bypass blower, and glow plugs. Compression ignition is achieved by maintaining the cylinder temperatures above the autoignition temperature of methanol. Air is diverted around the blower, reducing the amount of air entering the cylinders. Glow plugs are used as a starting aid and also at low speeds and low loads to maintain the cylinder temperatures necessary for autoignition. The methanol-fueled engine is turbocharged and equipped with a blower (supercharged).

The engine was the first to incorporate compression ignition of alcohols (7). Low cetane fuels can autoignite, however, provided the in-cylinder temperature is high enough and fuel injection correctly timed. This compression ignition works for methanol as well as ethanol and gasoline.

The California bus program was run at Golden Gate Transit District (GGTD) and continued through late 1990 (49). M.A.N. supplied two European SU 240 coaches for this project, one diesel powered and one methanol powered. DDC provided a GM RTS coach powered by methanol. GGTD already had a RTS diesel powered coach. Results indicate that methanol is a viable fuel for heavy-duty engines in general and transit in particular. Drive-

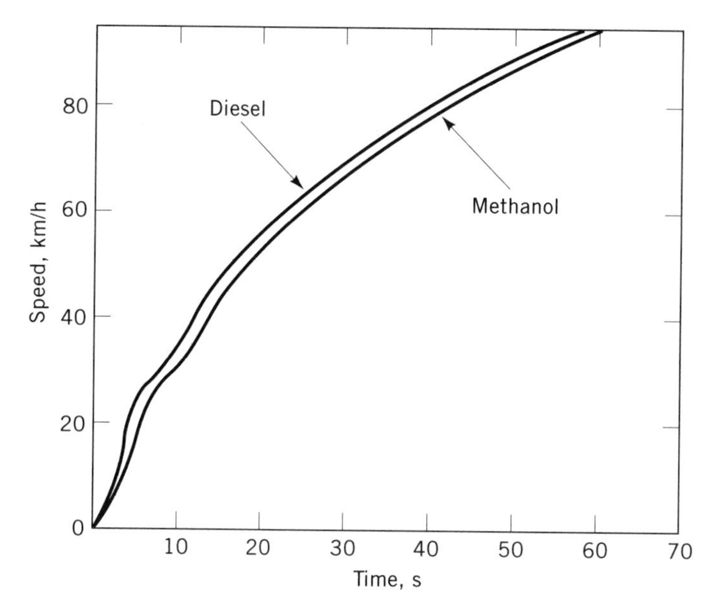

Figure 7. Full-throttle acceleration for diesel and methanol-powered GM RTS coaches having simulated full-seated passenger loads of 43 passengers.

ability including starting, full and partial throttle acceleration, and deceleration was as good or better using the methanol buses as compared to their diesel counterparts. Figure 7 illustrates the comparison of full throttle acceleration. Detailed fuel economy tests were also performed. Figures 8 and 9 compare steady-state and transient fuel consumption tests, respectively. The transient tests were performed using the Fuel Consumption Test Procedure, Type II (50). Methanol is comparable to diesel in steady-state fuel usage tests, but methanol consumption is higher at idle and during accelerations. The idle fuel consumption is higher because methanol cannot burn as lean as diesel fuel. Higher transient fuel consumption is a result of poor combustion factors resulting from poor fuel atomization, air control, and over fuelling. The methanol engine should not inherently be worse than the diesel engine during accelerations, if good combustion can be maintained.

The biggest problem of the GGTD program was engine and vehicle reliability and durability (see Fig. 10). Components needing the most frequent replacement were electronic unit fuel injectors and glow plugs, followed by the electronic control system (controlled power to the glow plugs), throttle position sensor, fuel pump, and fuel cooler fans. Other problems included increased engine deposits and ring and liner wear (51). The M.A.N. engine had similar but fewer problems; the components having the lowest lifetime were spark-plugs.

California continued the development of methanol powered vehicles primarily because of the substantial emission benefits (52). Then in 1986, the U.S. EPA promulgated technology-forcing standards for on-road, heavy-duty diesel engines which had been basically uncontrolled (53). These standards were also adopted in California by the ARB for buses in 1991 and all heavy-duty engines in 1993. Diesel engine manufacturers have made significant

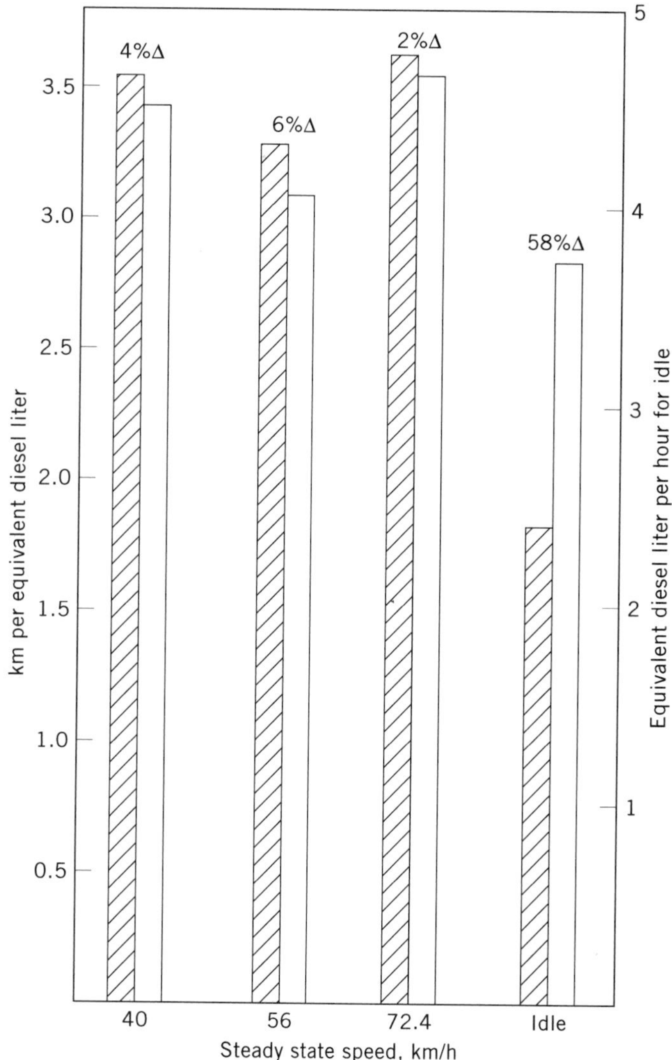

Figure 8. Fuel consumption for GM RTS coaches using ▨ diesel and □ methanol as fuel. Differences in fuel consumption are indicated by the symbol Δ. To convert km/L to mpg, multiply by 2.35.

improvements in technology and new diesel engines are projected to meet standards without a particulate trap. These improvements have made the diesel engine more competitive with alcohol fueled engines.

In 1987 Seattle Metro purchased 10 new American built M.A.N. coaches powered by methanol. Six GM buses powered by DDC methanol engines entered revenue service at Triboro Coach in Jackson Heights, New York, 2 GM buses in Medicine Hat Transit in Medicine Hat, Manitoba, and 2 Flyer coaches in Winnipeg Transit, Winnipeg, Manitoba, Canada. An additional 45 DDC powered methanol buses were introduced in California as indicated by Table 4. Figure 11 shows the distance accumulation of alternate-fueled buses in the four California transit properties.

Many of the development problems identified in the first bus programs were carried into the more recent demonstration projects. Spark-plug life continued to be an issue at Seattle Metro and the project was terminated in

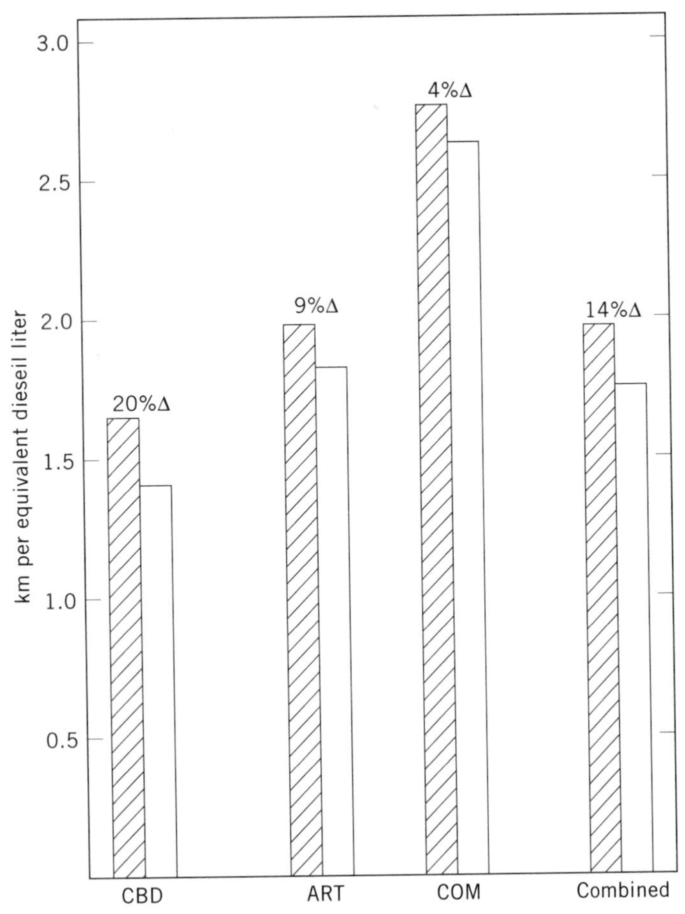

Figure 9. GM RTS coaches transient fuel consumption for ▨ diesel and ☐ methanol. SAE type road test, UMTA ADB duty cycle. Differences in fuel consumption are indicated by the symbol Δ. CBD (Central Business District), 4.3 stops per km, maximum speed 32 km/h, 18 m deceleration, 7 s dwell time, 19.3 km; ART (arterial), 1.24 stops per km, maximum speed 64 km/h, 61 m deceleration, 7 s dwell time, 13 km; COM (commuter), one stop every 6.4 km, maximum speed of 88.5 km/h, 150 m deceleration, 20 s dwell time, 13 km. To convert km/L to mpg, multiply by 2.35.

1990 because of costs of replacement parts. Costs were compounded when M.A.N. decided to discontinue manufacturing buses in the United States. DDC engines also continued to have problems with fuel injectors and glow plugs. Unit injectors were failing for a variety of reasons but the biggest problem was plugging injector tips. Injectors on some buses had to be changed at mileages as low as 1600 to 3200 km compared to diesel injectors which last up to 100 times as long. DDC and Lubrizol have since developed a fuel additive that when added to methanol at 0.06% by volume substantially reduces injector failures (see Fig. 12).

Many improvements have been made in both combustion and emissions control from the first experimental engine operating at GGTD (54). The new DDC preproduction engines have increased compression, 23:1 compared to 19:1, allowing the glow plugs to function only during starting. The rest of the time the cylinder temperatures are high enough to autoignite methanol. This revision increased the life of glow plugs from an average of 11,900 to 22,100 km between failures. Fuel economy has also improved as have exhaust emissions. Tests performed on an engine dynamometer following the Federal test procedure are many times better than California's 1991 bus standard as shown in Table 5.

Because of the success of these various alcohol fuel programs, heavy-duty demonstrations have been extended to other applications as shown in Table 4. The majority of applications are still either in transit or school buses, but California has also begun a program to utilize methanol engines in heavy-duty trucking applications (55). Domestic engine manufacturers participating in this program include Caterpillar, Cummins, DDC, Ford, and Navistar. The Caterpillar and Navistar engines are four stroke engines having glow plugs to assist in igniting methanol. These engines are very different from the two stroke DDC engine and from each other. The Navistar (56) uses a shield glow plug which neither Caterpillar (57) nor DDC do. Navistar claims these glow plugs provide both good combustion and long life. The combustion chambers and fuel control schemes differ from manufacturer to manufacturer. Ford converted a diesel engine to an Otto cycle and added electronically controlled port fuel injection, throttling, and a distributorless spark-ignition system.

Methanol has been shown to be a viable fuel for a variety of trucking applications in a local yet fairly large geographical area (58). And although the results focus primarily on dedicated methanol engines and vehicles in the U.S., the conclusions are nearly transferrable to ethanol fuels. DDC's 6V-92TA methanol engine operates on ethanol using a different engine calibration optimized for good performance and emissions. Additional hardware modifications are not anticipated.

Cetane Improvers. Compared to dedicated alcohol engines, fewer hardware changes need to be made to heavy-duty engines using cetane improvers. The early research on using cetane improvers and alcohol fuels for heavy-duty diesel engines was performed by the Germans in the late 1970s-early 1980s (59). The possibilities of using cetane improvers with ethanol for heavy-duty vehicles operating in Brazil were investigated (60). The work indicated that nitrates are the most effective cetane improvers for alcohols (61) and the Brazilian program focused on nitrates that could be manufactured from sugar cane, the feedstock for ethanol production. Additives considered included: butyl nitrate, isoamyl nitrates, 2-ethoxyethyl nitrate, and ethylene glycol nitrates. The selection of the improver depends on the method of the additive modifying the ignition delay over the entire speed and load range. Ideally ethanol should match the same ignition delay behavior as diesel.

Four buses converted to ethanol started operation in 1979 and the engine used was an OM 352, 6 cylinder, direct injection engine rated at 96,200 watt (129 hp). The ignition improver was *n*-hexyl nitrate [*20633-11-8*] which was later changed to triethylene glycol dinitrate [*111-22-8*] (TEGDN). TEGDN was mixed with ethanol at 5% or less by volume (60). The test fleet was further expanded to a variety of trucks manufactured and marketed by Mercedes-Benz in Brazil. 1,700 heavy-duty trucks were

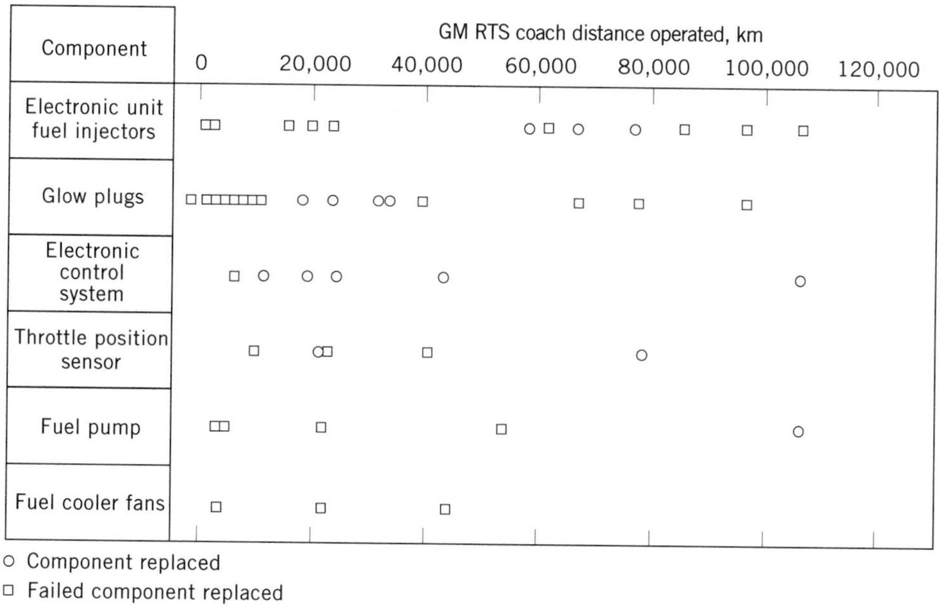

Figure 10. Components of GM RTS methanol-powered coach replaced, ○, and replaced because of component failure, □, as a function of distance operated.

converted to ethanol for use in the more gruelling sugar cane industry. Engines converted included the OM 352 O (5.7 L, 96,200 watt) and OM 355/5 O (9.7 L, 141,000 watt). Modifications to the engines to use ignition improved ethanol included increasing the fuel delivery capacity of the fuel injection pump, changing the fuel injection nozzles, and making material compatible with the TEGDN/ethanol fuel. The engines provided equal or better power and torque than the unmodified models. Some durability problems, which arose because of lack of lubrication or fuel material incompatibilities, were solved by adding lubricants to the injection pump plungers or by changing materials. Emissions were also generally lower than those from the equivalent diesel engine. Even NO_x was lower because of the lower flame temperatures compared to diesel (62).

The biggest drawback was the cost of ethanol compared to diesel fuel and the cost of the TEGDN and ethanol mixture compared to diesel. Unlike in the United States, diesel fuel in Brazil is considerably less expensive

Table 4. Distribution of California's Heavy-Duty Alternative Fuel Demonstration Vehicles

Transit District	No. of Vehicles	Fuel	OEM/Engine
South Coast Area Transit	1	methanol	DDC 6V-92TA
Riverside Transit District	3	methanol	DDC 6V-92TA
Southern California RTD	30	methanol	DDC 6V-92TA
Southern California RTD	12	methanol/Avocet[a]	DDC 6V-92TA
Southern California RTD	1	methanol	MAN D2566 MUH
Orange County Transit District	2	methanol/Avocet[a]	Cummins L10
Orange County Transit District	2	CNG[b]	Cummins L10
Orange County Transit District	2	LPG[b]	Cummins L10
		School bus demo	
Various fleets	50	methanol	DDC 6V-92TA
Various fleets	10	CNG	Bluebird/Teogen GM 454
		Trucking applications	
City of Los Angeles	1	methanol	GMC DDC 6V-92TA
City of Los Angeles	1	methanol/Avocet[a]	Peterbuilt Cummins L10
City of Glendale	1	methanol	Peterbuilt Caterpillar 3306
Golden State Foods	1	methanol	Freightliner DDC 6V-92TA
Federal Express	1	methanol	Freightliner DDC 6-71 TA
Arrowhead	1	methanol	Ford/Ford 6.61
SCE	1	methanol	Volvo/DDC 6V-92TA
South Lake Tahoe	1	methanol	International/Navistar DTG-460
Waste Management	2	methanol	Volvo/DDC 6-71TA

[a] Avocet is a cetane improver.
[b] CNG, compressed natural gas, and LPG, liquefied petroluem gas, are also used as alternative fuels.

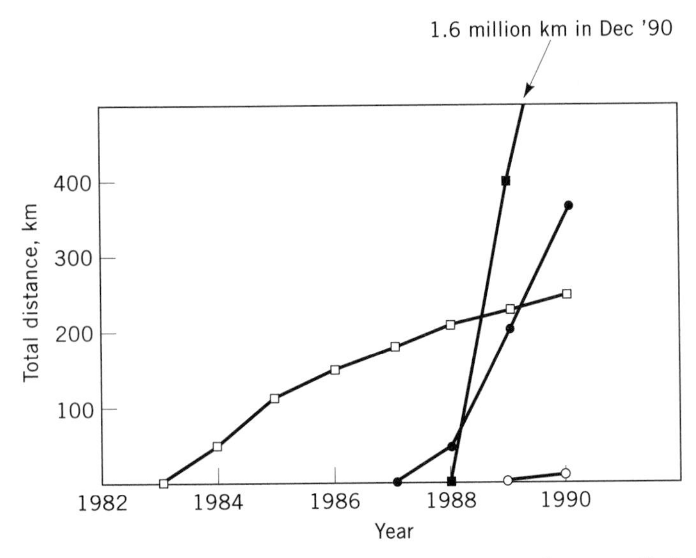

Figure 11. Distance methanol-fueled transit coaches travelled per year of operation in the □ Golden Gate Transit District (GGTD); ● Riverside Transit Agency (RTA), ■ Southern California Rapid Transit District (SCRTD); and ○ Orange County Transit District (OCTD).

on an energy basis than gasoline, because gasoline is taxed at a higher rate. This is generally the case in Europe as well. So, although technically feasible, the costs were too high for Brazil to convert many heavy-duty vehicles to ethanol.

Additional research for both ethanol and methanol showed that the amount of ignition improver could be reduced by systems increasing engine compression (63). Going from 17:1 to 21:1 reduced the amount of TEGDN required for methanol from 5% by volume to 3%. Ignition-improved methanol exhibited very low exhaust emissions compared to diesels: particulate emissions were eliminated except for small amounts associated with engine oil, NO_x was even lower with increased compression, and CO and hydrocarbons were also below diesel levels.

Auckland Regional Authority converted two M.A.N. buses to use a cetane improver and methanol and South Africa investigated the use of methanol with a proprietary cetane improver. Four Renault buses were converted in Tours, France to operate on ethanol and a cetane improver, Avocet, manufactured by Imperial Chemical Industries (ICI). The results of these demonstrations were also technically successful; slightly better fuel economy was obtained on an energy basis and durability issues were much less than the earlier tests using dedicated engines.

Cetane improvers were first investigated in the U.S. as part of a demonstration project to retrofit DDC 6V-71 two stroke engines to methanol for three transit buses operating in Jacksonville, Florida. This project started out using hardware conversion where the engine was modified to control air flow and a glow plug was added to the system. Although this was basically the same approach as used in the dedicated 6V-92TA methanol engine, the 6V-71 employed mechanical injectors as opposed to electronic injectors. These mechanical injectors were failing even with 1% castor oil [*8001-79-4*] added to the methanol. Jacksonville thus decided to use a nitrate based cetane improver containing a lubrication additive and a corrosion inhibitor and the project proved the viability of cetane-improved methanol.

The first U.S. engine manufacturer to evaluate the use of cetane improvers was Cummins Engine Company. Their methanol designed L10 engines were converted based on the knowledge gained in Brazil with ethanol and 4.5% TEGDN in their 14-L engine (64). They increased the fuel capacity of their injection pumps, for instance, and modified the combustion process to match the diesel start of combustion by increasing the compression ratio from 15.8 to 16.1:1 and adding turbocharging (both increased in-cylinder temperatures). For the methanol L10 development Avocet at 5% by volume mixed with 100%

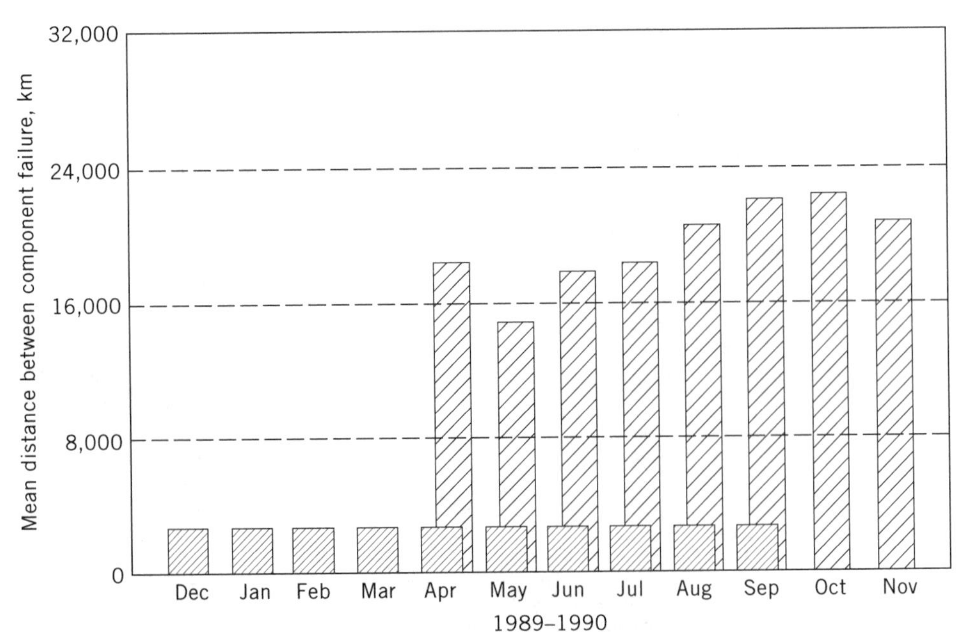

Figure 12. Methanol and coach fleet cumulative distance driven before fuel injector failure, where ▨ represents operation without, and ▨ represents operation with fuel additive.

Table 5. Exhaust Emissions and California 1991 Bus Standards, g/kW · h[a]

Exhaust Component	1991 California Standards		DC 6V-92TA (Catalyst)	
	g/(bhp · h)[a]	g/(kW · h)	g/(bhp · h)[a]	g/(kW · h)
OMHCE	1.3	1.8	0.10	0.14
CO	15.5	21.1	0.22	0.3
NO_x	5.0	6.8	2.3	3.1
particulates	0.10	0.14	0.05	0.07
aldehydes	0.10/0.05[b]	0.14/0.07[b]	0.04	0.05

[a] bhp = brake horsepower; 1 bhp = 0.735 kW.
[b] The 0.10 and 0.14 values apply from 1993 to 1995; the 0.05 and 0.07 values apply beginning in 1996.

methanol was selected (65). Engine modifications were only made to the fuel system. These included fuel pump changes for increased capacity, a different camshaft to change injection timing, and larger injectors. Changes were also made in various materials to make them compatible with methanol. The resultant L10 matched diesel power and torque, but emissions results were mixed. Data showed low emissions of particulates, but higher HC, CO, and NO_x at low speed, low load operation. These data suggest that additional optmization may be necessary for the L10 methanol engine.

The L10 methanol engines are currently being used in several demonstration fleets in California. The first use was in the City of Los Angeles Peterbilt dump truck (58), a part of the CEC truck demonstration project. Methanol L10s are also being used in the OCTD comparative alternative fuel demonstration project. In this project methanol, CNG, and LPG L10 engines are being compared to each other in a transit bus application. In the Canadian methanol in large engines (MILE) project, two L10 methanol engines were operated for several years in refuse haulers in Vancouver, British Columbia.

The largest retrofit demonstration project in the U.S. is underway at SCRTD where 12 vehicles using DDC 6V-92 with mechanical injectors are being modified to use methanol with Avocet (66). This project has evaluated the changes necessary to optimize the conversions for both fuel economy and emissions. Using the two stroke engine, changes had to be made to reduce the air into the engine, increase compression from 17 to 23:1, and use better ring packages. These changes gave both good fuel economy and reduced emissions of NO_x by 50%, considerably lowered particulates, and maintained levels of CO and hydrocarbon. New York has also modified a 8V-71 for use on Avocet-improved methanol and had similar results (67).

The success of these tests indicate that cetane-improved alcohol is technically feasible. Engines can be designed to provide equal diesel power and torque characteristics having lower NO_x and particulate emissions than diesel. However, if it is not necessary to achieve lowest possible emissions, only changes in fuel rate are required, rather than engine changes, and the commercial application of this approach depends mostly on the cost of the cetane improvers. The price of Avocet is about $4/L in small quantities and adding 3.0% in methanol nearly doubles the cost of methanol from $0.13 to $0.22/L. If Avocet were produced in larger quantities its price would drop considerably.

The biggest potential use of the cetane-improver approach may be in vehicle retrofits where for environmental reasons bus and truck fleets may be required to convert to cleaner burning fuels.

AIR QUALITY BENEFITS OF ALCOHOL FUELS

In the 1970s evaluations of alcohol fuel programs always considered environmental impacts and objectives even though the main thrust of the programs was toward energy security and diversification benefits. Assessments of performance identified these fuels as consistent with environmental goals and by the mid 1980s, the environmental benefits of the alcohol fuels had become the chief driving force for their further consideration. Detailed assessments were made of photochemical smog and air toxics reductions that might be obtained from the wide use of alcohol fuels in light-duty vehicles. Methanol received the most evaluation, because it appeared to be far more cost competitive than ethanol. The potential benefits of alcohols used in heavy-duty diesel-type engines were also studied.

The most comprehensive air quality study, supported by the California ARB and the South Coast Air Quality Management District (68), showed that if gasoline and methanol cars emitted the same amounts of carbon, an assumption that seemed reasonable based on emissions test data taken throughout the 1980s, and if methanol cars had formaldehyde emissions controlled to 9.3 mg/km (equal to the current California formaldehyde emissions standard for methanol automobiles), then substituting M85 for gasoline would produce a 9% reduction in the peak summer-day afternoon ozone level and a 19% reduction in exposure to ozone levels above the Federal standard of 0.12 ppm. These reductions constituted a substantial fraction of the reductions that would be obtained by eliminating all the emissions from vehicles. Additional assumptions were that exhaust carbon emissions were at the level of 0.15 g/km, equal to the planned certification standard for new vehicles in California beginning in 1993 (but not really expected to be characteristic of in-use vehicles) and that the distribution of hydrocarbon species in the exhaust resembled that of cars tested in the 1980s.

The results of the study, conducted at Carnegie Mellon University, are now generally accepted as the best available guides for the smog-reducing benefits of a methanol substitution strategy, at least for the conditions prevailing in the Los Angeles basin. Overall, replacement of a conventional gasoline vehicle by an equivalent M85 vehicle should provide about a 30% reduction in smog-forming potential. A 100% result would be earned by eliminating the vehicle entirely. Vehicles using M100 would provide substantially greater benefits than M85 vehicles.

Benefits depend upon location. There is reason to believe that the ratio of hydrocarbon emissions to NO_x has an influence on the degree of benefit from methanol substitution in reducing the formation of photochemical smog (69). Additionally, continued testing on methanol vehicles, particularly on vehicles which have accumulated a considerable number of miles, may show that some of the assumptions made in the Carnegie Mellon assessment are not valid. Air quality benefits of methanol also depend on good catalyst performance, especially in controlling formaldehyde, over the entire useful life of the vehicle.

Methanol substitution strategies do not appear to cause an increase in exposure to ambient formaldehyde even though the direct emissions of formaldehyde have been somewhat higher than those of comparable gasoline cars. Most ambient formaldehyde is in fact secondary formaldehyde formed by photochemical reactions of hydrocarbons emitted from gasoline vehicles and other sources. The effects of slightly higher direct formaldehyde emissions from methanol cars are offset by reduced hydrocarbon emissions (68).

Methanol use would also reduce public exposure to toxic hydrocarbons associated with gasoline and diesel fuel, including benzene, 1,3-butadiene, diesel particulates, and polynuclear aromatic hydrocarbons. Although public formaldehyde exposures might increase from methanol use in garages and tunnels, methanol use is expected to reduce overall public exposure to toxic air contaminants.

ALCOHOL FUELS USAGE AS AN AIR QUALITY STRATEGY

The cost-effectiveness of methanol substitution as an air quality strategy has been studied in some detail. Air quality planners usually rate cost-effectiveness in terms of dollars per ton of reactive hydrocarbons controlled (or removed from the inventory of emissions). Typical costs for controlling reactive hydrocarbon emissions in the U.S. are in the range of several hundreds of dollars per ton. In the Los Angeles area, the average costs of future hydrocarbon control measures average about $500 per metric ton, although some individual measures have cost-effectiveness figures above $10,000 per ton. Methanol substitution appears to be a viable and competitive control strategy. Cost-effectiveness is linked to the price of oil. Methanol appears to have a cost-effectiveness ranging from a few thousand dollars to several tens of thousands of dollars per ton (18,70–73). In heavy-duty engines, the alcohols may offer cost-effective reductions of particulates and NO_x emissions. A recent study indicates that the use of methanol was competitive with the use of cleaner diesel fuels and diesel particulate traps under some circumstances (74).

The potential air quality impacts of ethanol use have not yet been studied in detail.

Global Warming Impacts

Several studies have been made of the global warming impacts of alcohol fuels. The most useful assessments cover the entire life cycle from raw material feedstock production through processing, distribution, and fuel usage. They also consider global gases in addition to carbon dioxide (75,76). Results reflect the influence of assumptions but methanol is expected to provide slight reductions in global warming impacts compared to gasoline. Ethanol evaluations are less certain.

PUBLIC SAFETY ISSUES

Several investigators have assessed the comparative safety of methanol and conventional hydrocarbon fuels (14,77–79). The ingestion toxicity of methanol has been of some concern because of the number of gasoline ingestions associated with siphoning and in-home accidental ingestions. The use of gasoline in small engines such as those used for lawnmowers, leafblowers, and other small utility applications results in most of the siphoning ingestions and in-home accidental ingestions of gasoline. This potential problem is addressed by discouraging methanol use in small engines, by labeling and public education, and by positive siphoning prevention screens in the fill pipes of vehicles. These screens have been required in recent purchases of methanol vehicles by California agencies.

Skin contact with methanol may present a greater health threat than skin contact with gasoline and diesel fuel and is being evaluated.

The fire hazard of methanol appears to be substantially smaller than the fire hazard of gasoline, although considerably greater than the fire hazard of diesel fuel. The lack of luminosity of a methanol flame is still a concern to some, and M85 (or some other methanol fuel with an additive for flame luminosity) may become the standard fuel for this reason.

In reviewing the full range of health and safety issues associated with all alternative fuels, the California Advisory Board determined that there were no roadblocks that would prevent the near term deployment of either methanol or ethanol, assuming that adequate safety practices were followed appropriate to the specific nature of each fuel (14).

THE FUTURE OF ALCOHOL FUELS

In the late 1980s attempts were made in California to shift fuel use to methanol in order to capture the air quality benefits of the reduced photochemical reactivity of the emissions from methanol-fueled vehicles. Proposed legislation would mandate that some fraction of the sales of each vehicle manufacturer be capable of using methanol, and that fuel suppliers ensure that methanol was used in these vehicles. The legislation became a study of the California Advisory Board on Air Quality and Fuels. The report of the study recommended a broader approach to fuel quality and fuel choice that would define environmental objectives and allow the marketplace to determine which vehicle and fuel technologies were adequate to meet environmental objectives at lowest cost and maximum value to consumers. The report directed the California ARB to develop a regulatory approach that would preserve environmental objectives by using emissions standards that reflected the best potential of the cleanest fuels.

The ARB adopted a regulatory package for light-duty vehicles in 1990 that modifies the historically uniform approach to vehicle emissions, in which each and every vehicle in a regulated class must meet the same emissions standard. The new approach adopts emissions standards that apply on the average to the entire sales mix of vehicles sold by each manufacturer in each of several broad weight classes of vehicles. Thus vehicles that use fuels such as methanol and ethanol having air quality benefits in the form of lower levels of photochemical reactivity have the emissions adjusted to reflect the lower smog forming tendency of these fuels. This regulatory approach provides a powerful incentive for vehicle manufacturers to certify at least some of the sales mix of vehicles on fuels such as methanol and ethanol.

The future market response to the new form of emissions regulation is unknown. For the purpose of meeting new vehicle emissions standards, however, it is still not clear whether some combination of new emissions control approaches and reformulated gasolines can provide benefits equal to those of methanol and ethanol. It is possible that the new emissions standards will simply result in improved gasoline technologies, and that, despite the prospective air quality advantages of the alcohol fuels, the market result of the new standards will simply be cleaner gasolines. However, in 1990 the U.S. Alternative Fuels Council agreed on a goal of a 25% share of nonpetroleum transportation fuels by 2005. Although this goal may not become part of a national energy plan, it represents the first official statement of a specific goal to substitute for the use of petroleum in transportation. Alcohol fuels could capture a large part of this 25% share of nonpetroleum fuels, although vehicles powered by natural gas and electric energy will no doubt win some acceptance. For the ordinary passenger car, alcohol fuels may offer the most gasolinelike alternative in terms of range, comparable costs, and compatibility with the current gasoline/diesel storage and distribution infrastructure.

Acknowledgment

Reprinted from *Kirk-Othmer Encyclopedia of Chemical Technology*, 4th ed., Vol. 1, John Wiley & Sons, Inc., New York, 1991.

BIBLIOGRAPHY

1. J. W. Shiller, "The Automobile and the Atmosphere" in *Energy: Production, Consumption and Consequences,* National Academy Press, Washington, D.C., 1990, pp. 111–142.
2. *Engineering Data Book,* Gas Processors Suppliers Association, 1972.
3. *Reference Data for Hydrocarbons and Petro-Sulfur Compounds,* Phillips Petroleum Company, 1962.
4. J. B. Heywood, *Internal Combustion Engine Fundamentals,* McGraw-Hill, New York, 1988.
5. *Fire Hazard Properties of Flammable Liquids, Gases, and Volatile Solids,* Pub. # NFPA 325M, National Fire Protection Association, 1984.
6. P. F. Schmidt, *Fuel Oil Manual,* Industrial Press, 1969.
7. R. R. Toepel, J. E. Bennethum, R. E. Heruth, "Development of a Detroit Diesel Allison 6V-92TA Methanol-Fueled Coach Engine," *SAE Paper 831744, SAE Fuels and Lubricants Meeting* (San Francisco, Calif., Oct. 31–Nov. 3, 1983), Society of Automotive Engineers, Warrendale, Pa., 1983.
8. T. Powell, "Racing Experiences with Methanol and Ethanol-based Motor Fuel Blends," *SAE Paper 750124,* Society of Automotive Engineers, Warrendale, Pa., Feb., 1975.
9. Sypher-Mueller International, Inc., *Future Transportation Fuels: Alcohol Fuels, Energy, Mines and Resources—Canada,* Project Mile Report, A Report on the Use of Methanol in Large Engines in Canada, May 1990.
10. S. C. Trindade and A. V. de Carvalho, "Transportation Fuels Policy Issues and Options: The Case of Ethanol Fuels in Brazil" in D. Sperling, ed., *Alternative Transportation Fuels,* Quorum Books, New York, 1989, pp. 163–185.
11. *First Interim Report,* United States Alternative Fuels Council, Washington, D.C., Sept. 30, 1990.
12. Office of Policy, Planning, and Analysis, *Assessment of Costs and Benefits of Flexible and Alternative Fuel Use in the U.S. Transportation Sector,* Technical Report #3 (Methanol Production and Transportation Costs) Pub. # DOE/P/E–0093, U.S. Department of Energy, Washington, D.C., Nov. 1989.
13. Bechtel Corporation, *California Fuel Methanol Cost Study: Chevron Corporation, U.S.,* Vol. 1 (Executive Summary, Jan. 1989); Vol. 2, (Final Report, Dec. 1988), San Francisco, Calif., 1988–1989.
14. California Advisory Board on Air Quality and Fuels, Vol. 1, *Executive Summary;* Vol. 2, *Energy Security Report;* Vol. 3, *Environmental Health and Safety Report;* Vol. 4, *Economics Report;* Vol. 5, *Mandates and Incentives Report;* San Francisco, Calif., June 13, 1990.
15. California Energy Commission, *Methanol as a Motor Fuel: Review of the Issues Related to Air Quality, Demand, Supply, Cost, Consumer Acceptance and Health and Safety,* Pub. # P500–89–002, Sacramento, Calif., April 1989.
16. California Energy Commission, *Cost and Availability of Low Emission Motor Vehicles and Fuels,* Sacramento, Calif., Aug. 1989.
17. National Research Council, *Fuels to Drive Our Future,* National Academy Press, Washington, D.C., 1990.
18. Office of Technology Assessment, *Replacing Gasoline: Alternative Fuels for Light-Duty Vehicles,* Pub. # OTA–E–364, U.S. Congress, Washington, D.C., Sept. 1990.
19. J. L. Keller, *Hydrocarbon Process.* (May 1979).
20. Office of Mobile Sources, *Analysis of the Economic and Environmental Effects of Methanol as an Automotive Fuel,* U.S. Environmental Protection Agency, Ann Arbor, Mich., Sept. 1989.
21. *Alcohols and Ethers Blended with Gasoline,* Pub. # 4261, American Petroleum Institute, Washington, D.C., Chapt. 4, pp. 23–27.
22. *7th Int. Symp. on Alcohol Fuels,* Institut Francais du Petrole, Editions Technip, Paris, France, 1986.
23. *8th Int. Symp. on Alcohol Fuels,* New Energy and Industrial Technology Development Organization, Sanbi Insatsu Co., Ltd. Tokyo, Japan, Nov. 13–16, 1988.

24. H. C. Wolff, "German Field Test Results on Methanol Fuels M100 and M15," *American Petroleum Institute 48th Midyear Refining Meeting* (Session on Oxygenates and Oxygenate-Gasoline Blends as Motor Fuels) May 11, 1983.
25. H. Menrad, "Possibilities to Introduce Methanol as a Fuel, An Example from Germany," *6th Int. Symp. on Alcohol Fuels Technology,* (Ottawa, Canada, May 21–25, 1984), Vol. 2.
26. L. Schieler, M. Fischer, D. Dennler, and R. Nettell, "Bank of America's Methanol Fuel Program: An Insurance Policy that is Now a Viable Fuel," *6th Int. Symp. on Alcohol Fuels Technology,* (Ottawa, Canada, May 21–25, 1984), Vol. 2.
27. R. N. McGill and R. L. Graves, "Results from the Federal Methanol Fleet, A Progress Report," *8th Int. Symp. on Alcohol Fuels* (Tokyo, Japan, Nov. 13–16, 1988).
28. *Alcohol Energy Systems: Alcohol Fleet One Test Report,* Pub. # 500–82–058, California Energy Commission, Sacramento, Calif., Aug. 1983.
29. F. J. Wiens and co-workers, "California's Alcohol Fleet Test Program: Final Results" *6th Int. Symp. on Alcohols Fuels Technology* (Ottawa, Canada, May 21–25, 1984), Vol. 3.
30. Acurex Corporation, *California's Methanol Program: Evaluation Report,* Vol. 2 (Technical Analyses), Pub. # P500–86–012A, California Energy Commission, Sacramento, Calif., June 1987.
31. S. C. Trindade and A. V. de Carvalho, "Utilization of Alcohol Fuels in Brazil: Early Experience, Current Situation, and Future Prospects," *Int. Symp. on Introduction of Methanol-Powered Vehicles* (Tokyo, Japan, Feb. 19, 1987).
32. Acurex Corporation, *California's Methanol Program: Evaluation Report,* Vol. 1 (Executive Summary), Pub. #P500–86–012, California Energy Commission, Sacramento, Calif., Nov. 1986.
33. E. R. Fanick, L. R. Smith, J. A. Russell and W. E. Likos, "Laboratory Evaluation of Safety-Related Additives for Neat Methanol Fuel," SAE Paper 902156, (SP840), Society of Automotive Engineers, Warrendale, Pa., Oct. 1990.
34. U. Hilger, G. Jain, E. Scheid, and F. Pischinger, "Development of a Direct Injected Neat Methanol Engine for Passenger Car Applications," SAE Paper 901521, *SAE Future Transportation Technology Conf. and Expo.* (San Diego, Calif., Aug. 13–16, 1990).
35. P. A. Machiele, "A Perspective on the Flammability, Toxicity, and Environmental Safety Distinctions Between Methanol and Conventional Fuels," *AIChE 1989 Summer National Meeting* (Philadelphia, Pa., Aug. 22, 1989). American Institute of Chemical Engineers.
36. J. V. D. Weide and R. J. Wineland, "Vehicle Operation with Variable Methanol/Gasoline Mixtures," *6th Int. Symp. on Alcohol Fuels Technology* (Ottawa, Canada, May 21–25, 1984), Vol. 3.
37. T. B. Blaisdell, M. D. Jackson, and K. D. Smith, "Potential of Light-Duty Methanol Vehicles," SAE Paper 891667, *SAE Future Transportation Technology Conf. and Expo.* (Vancouver, Canada, Aug. 7–10, 1989), Society of Automotive Engineers, Warrendale, Pa.
38. Mobile Source Division, *Alcohol Fueled Vehicle Fleet Test Program: 9th Interim Report,* Pub. ARB/MS–89–09, California Air Resources Board, El Monte, Calif., Nov. 1989.
39. W. Carter, *Ozone Reactivity Analysis of Emissions from Motor Vehicles,* (Draft Report for the Western Liquid Gas Association), Statewide Air Pollution Research Center, University of California at Riverside, July 11, 1989.
40. P. A. Gabele, *J. of Air Waste Management* **40**(3), 296–304 (Mar. 1990).
41. S. Albu, "California's Regulatory Perspective on Alternate Fuels," *13th North American Motor Vehicle Emissions Control Conf.* (Tampa, Fla., Dec. 11–14, 1990), Mobile Source Division, California Air Resources Board, El Monte, Calif.
42. A. Lawson, A. J. Last, A. S. Desphande, and E. W. Simmons, "Heavy-Duty Truck Diesel Engine Operation on Unstabilized Methanol/Diesel Fuel Emulsion," SAE Paper 810346, (SP–480) *Int. Congress and Expo* (Detroit, Mich., Feb. 23–27, 1981) Society of Automotive Engineers, Warrendale, Pa.
43. B. M. Bertilsson, "Regulated and Unregulated Emissions from an Alcohol-Fueled Diesel Engine with Two Separate Fuel Injection Systems," *5th Int. Symp. on Alcohol Fuel Technology* (Auckland, New Zealand, May 13–18, 1982) Vol. 3.
44. R. A. Baranescu, "Fumigation of Alcohols in a Multicylinder Diesel Engine: Evaluation of Potential," SAE Paper 860308, *SAE Int. Congress and Expo.* (Detroit, Mich., Feb. 24–28, 1986) Society of Automotive Engineers, Warrendale, Pa.
45. R. G. Jackson, "Workshop on Diesel Fuel Substitution," *5th Int. Symp. on Alcohol Fuel Technology* (Auckland, New Zealand, May 13–18, 1982), Vol. 4.
46. H. K. Bergmann and K. D. Holloh, "Field Experience with Mercedes-Benz Methanol City Buses," *6th Int. Symp. on Alcohol Fuels Technology* (Ottawa, Canada, May 21–25, 1984), Vol. 1.
47. A. Nietz and F. Chmela, "Results of Further Development in the M.A.N. Methanol Engine," *6th Int. Symp. on Alcohol Fuels Technology* (Ottawa, Canada, May 21–25, 1984), Vol. 1.
48. M. D. Jackson, C. A. Powars, K. D. Smith, and D. W. Fong, "Methanol-Fueled Transit Bus Demonstration," *Paper 83–DGP–2,* American Society of Mechanical Engineers.
49. M. D. Jackson, S. Unnasch, C. Sullivan, and R. A. Renner, "Transit Bus Operation with Methanol Fuel," *SAE Paper 850216, SAE Int. Congress and Expo.* (Detroit, Mich., Feb. 25–Mar. 1, 1986) Society of Automotive Engineers, Warrendale, Pa.
50. *Joint TMC/SAE Fuel Consumption Test Procedure, Type 2,* SAE J1321, SAE Recommended Practice Approved October 1981, Society of Automotive Engineers, Warrendale, Pa., 1981.
51. M. D. Jackson, S. Unnasch, and D. D. Lowell, "Heavy-Duty Methanol Engines: Wear and Emissions," *8th Int. Symp. on Alcohol Fuels* (Tokyo, Japan, Nov. 13–16, 1988).
52. K. D. Smith, "California's Methanol Program," *7th Int. Symp. on Alcohol Fuels,* (Paris, France, Oct. 20–23, 1986).
53. *EPA Emissions Standards, Code of Federal Regulations,* Vol. 40, Chapt. 1, Sect. 86.088–11, 86.091–11, and 86.094–11, U.S. Environmental Protection Agency, U.S.G.P.O., Washington, D.C., 1987.
54. J. Jaye, S. Miller, and J. Bennethum, "Development of the Detroit Diesel Corporation Methanol Engine," *8th Int. Symp. Alcohol Fuels* (Tokyo, Japan, Nov. 13–16, 1988).
55. R. A. Brown, J. A. Nicholson, M. D. Jackson, and C. Sullivan, "Methanol-Fueled Heavy-Duty Truck Engine Applications: The CEC Program." *SAE Paper 890972, SAE 40th Annual Earthmoving Industry Conf.* (Peoria, Ill., April 11–13, 1989) Society of Automotive Engineers, Warrendale, Pa.
56. R. Baranescu and co-workers, "Prototype Development of a Methanol Engine for Heavy-Duty Application-Performance and Emissions," *SAE Paper 891653, SAE Future Transportation Technology Conf.* (Aug. 7–20, 1989), Society of Automotive Engineers, Warrendale, Pa.
57. R. Richards, "Methanol-Fueled Caterpillar 3406 Engine Experience in On-Highway Trucks," *SAE Paper 902160, (SP–840),* Society of Automotive Engineers, Warrendale, Pa., Oct., 1990.
58. M. D. Jackson, C. Sullivan, and P. Wuebben, "California's

Demonstration of Heavy-Duty Methanol Engines in Trucking Applications," *9th Int. Symp. on Alcohol Fuels* (Milan, Italy, April 9–12, 1991).

59. W. Bandel, "Problems in the Application of Ethanol as Fuel for Utility Vehicles," *2nd Int. Symp. on Alcohol Fuel Technology* (Wolfsburg, Germany, Nov. 21–23, 1977).
60. W. Bandel and L. M. Ventura, "Problems in Adapting Ethanol Fuels to the Requirements of Diesel Engines," *40th Int. Symp. on Alcohol Fuels* (Guaruja, Brazil, Oct. 1980).
61. A. J. Schaefer and H. O. Hardenburg, "Ignition Improvers for Ethanol Fuels," SAE Paper 810249, SP–840, *SAE Int. Congress. and Expo.* (Detroit, Mich., Feb. 23–27, 1981).
62. E. P. Fontanello, L. M. Ventura, and W. Bandel, "The Use of Ethanol with Ignition Improver as a CI-Engine Fuel in Brazilian Trucks," *7th Int. Symp. on Alcohol Fuels* (Paris, France, Oct. 20–23, 1986).
63. H. O. Hardenburg, "Comparative Study of Heavy-Duty Engine Operation with Diesel Fuel and Ignition-Improved Methanol," *SAE Paper 872093, SAE Int. Fuels and Lubricants Meeting and Expo.* (Toronto, Canada, Nov. 2–5, 1987), Society of Automotive Engineers, Warrendale, Pa.
64. E. J. Lyford-Pike, F. C. Neves, A. C. Zulino, and V. K. Duggal, "Development of a Commercial Cummins NT Series to Burn Ethanol Alcohol," *7th Int. Symp. on Alcohol Fuels* (Paris, France, Oct. 20–23, 1986).
65. A. B. Welch, W. A. Goetz, D. Elliott, J. R. MacDonald, and V. K. Duggal, "Development of the Cummins Methanol L10 Engine with an Ignition Improver," *8th Int. Symp. on Alcohol Fuels* (Tokyo, Japan, Nov. 13–16, 1988).
66. S. Unnasch and co-workers, "Transit Bus Operation with a DDC 6V–92TAC Engine Operating on Ignition-Improved Methanol," *SAE Paper 902161,* (SP–840), *SAE Int. Fuels and Lubricants Meeting and Expo.* (Tulsa, Oklahoma, Oct. 22–23, 1990).
67. C. M. Urban, T. J. Timbario, and R. L. Bechtold, "Performance and Emissions of a DDC 8V–71 Engine Fueled with Cetane Improved Methanol," *SAE Paper 892064, SAE Int. Fuels and Lubricants Meeting and Expo.* (Baltimore, Md., Sept. 25–28, 1989) Society of Automotive Engineers, Warrendale, Pa.
68. J. N. Harris, A. R. Russell, and J. B. Milford, *Air Quality Implications of Methanol Fuel Utilization, SAE Paper 881198,* Society of Automotive Engineers, Warrendale, Pa., 1988.
69. T. Y. Chang and S. Rudy, Urban Air Quality Impact of Methanol-Fueled Compared to Gasoline-Fueled Vehicles, in W. Kohl, ed., *Methanol as an Alternative Fuel Choice: An Assessment,* Johns Hopkins Foreign Policy Institute, Washington, D.C., 1990, pp. 97–120.
70. C. B. Moyer, S. Unnasch, and M. D. Jackson, *Air Quality Programs as Driving Forces for Methanol Use,* Transportation Research, Vol. 23A, No. 3, 1989, pp. 209–216.
71. T. Lareau, *The Economics of Alternative Fuel Use: Substituting Methanol for Gasoline,* American Petroleum Institute, Washington, D.C., June 1989.
72. Mobile Source Division, *Alcohol-Fueled Vehicle Fleet Test, 9th Interim Report* ARB/MS–89–09, California Air Resources Board, El Monte, Calif., Nov. 1989.
73. *Proposed Regulations for Low Emission Vehicles and Clean Fuels,* Staff Report, California Air Resources Board, Sacramento, Calif., Aug. 13, 1990.
74. S. Unnasch, C. B. Moyer, M. D. Jackson, and K. D. Smith, *Emissions Control Options for Heavy-Duty Engines, SAE 861111,* Society of Automotive Engineers, Warrendale, Pa., 1986.
75. M. A. DeLuchi, "Emissions of Greenhouse Gases from the Use of Gasoline, Methanol, and Other Alternative Transportation Fuels," in W. Kohl, ed., *Methanol as an Alternative Fuel Choice: An Assessment,* Johns Hopkins Foreign Policy Institute, Washington, D.C., 1990, pp. 167–199.
76. S. C. Unnasch, B. Moyer, D. D. Lowell, and M. D. Jackson, *Comparing the Impacts of Different Transportation Fuels on the Greenhouse Effect, Pub. # 500–89–001,* California Energy Commission, Sacramento, Calif., April, 1989.
77. *Methanol Health and Safety Workshop* (Los Angeles, Calif., Nov. 1–2, 1989) South Coast Air Quality Management District, El Monte, Calif., 1989.
78. P. A. Machiele, "A Perspective on the Flammability, Toxicity, and Environmental Safety Distinctions Between Methanol and Conventional Fuels," *AIChE 1989 Summer National Meeting* (Philadelphia, Pa., Aug. 22, 1989), American Association of Chemical Engineers.
79. P. A. Machiele, "A Health and Safety Assessment of Methanol as an Alternative Fuel," in W. Kohl, ed., *Methanol as an Alternative Fuel: An Assessment,* Johns Hopkins Foreign Policy Institute, Washington, D.C., 1990, pp. 217–239.

General References

K. Boekhaus and co-workers, "Reformulated Gasoline for Clean Air: An ARCO Assessment," *2nd Biennial U.C. Davis Conf. on Alternative Fuels* (July 12, 1990).

California's Methanol Program, Evaluation Report, Vol. 2 (Technical Analyses), Pub. # P500–86–012A, California Energy Commission, Sacramento, Calif., June 1987.

A. V. de Carvalho, "Future Scenarios of Alcohols as Fuels in Brazil," *3rd Int. Symp. on Alcohol Fuels Technology* (Asilomar, Calif., May 29–31, 1979).

C. L. Gray, Jr. and J. A. Alson, *Moving American to Methanol,* University of Michigan Press, Ann Arbor, Mich., 1985.

P. A. Lorang, Emissions from Gasoline-Fueled and Methanol Vehicles, in W. L. Kohl, ed., *Methanol as an Alternative Fuel Choice: An Assessment* Johns Hopkins Foreign Policy Institute, Washington, D.C., 1990, pp. 21–48.

J. H. Perry and C. P. Perry, *Methanol: Bridge to a Renewable Energy Future* University Press of American, Lanham, Md., 1990.

D. Sperling, *New Transportation Fuels,* University of California Press, Berkeley, Calif., 1988.

D. Sperling, *Brazil, Ethanol, and the Process of Change, Energy,* Vol. 12, No. 1, pp. 11–23, 1987.

E. Supp, *How to Produce Methanol from Coal,* Springer-Verlag, Berlin, 1990.

J. V. D. Weide and W. A. Ramackers, "Development of Methanol and Petrol Carburation Systems in the Netherlands," *2nd Int. Symp. on Alcohol Fuel Technology,* (Wolfsburg, Germany, Nov. 21–23, 1977).

1987 World Methanol Conf. (San Francisco, Calif., Dec. 1–3, 1987) Crocco and Associates, Houston, Tex.

1989 World Methanol Conf. (Houston, Tex., Dec. 5–7, 1989), Crocco and Associates, Houston, Tex.

ALKYLATION

HAROLD HAMMERSHAIMB
UOP
Des Plaines, Illinois

Alkylation as described in this article is the catalytic process whereby C_3–C_5 olefins react stoichiometrically with isobutane to produce alkylate, a highly branched, paraffinic motor fuel blendstock. Alkylate is a high-octane, clean-burning, low-volatility product that has long been one of the key components of gasoline.

Alkylation processes were originally developed in the 1930s and 1940s to produce high-octane aviation fuels. During the 1950s and 1960s, the production of alkylate grew to meet the increasing octane requirements of more efficient high-compression engines. In the 1970s and 1980s, alkylate played an important role in maintaining the octane quality of gasoline as tetraethyl lead was phased out in the United States and Western Europe. In the 1990s, alkylate is one of the key components in the low-emission reformulated gasolines (RFGs) that are being developed in response to environmental concerns.

See also AIRCRAFT FUELS; CLEAN AIR ACT, MOBILE SOURCES; PETROLEUM PRODUCTS AND USES; TRANSPORTATION FUELS, AUTOMOTIVE FUELS; PETROLEUM REFINING.

USE OF ALKYLATE

In modern refineries, significant quantities of C_3–C_5 olefins are produced by catalytic cracking and coking. In the alkylation unit, these olefins react with isobutane, which is recovered as a by-product from various refinery operations, including crude distillation, reforming, catalytic cracking, and hydrocracking. In many cases, additional isobutane derived from natural gas liquid-recovery operations is used to augment the refinery-produced isobutane supply.

The alkylate product consists primarily of C_7–C_9 branched-chain paraffins having a high octane rating (1). The carbon number of the feedstock influences alkylate yield, isobutane consumption, and alkylate octane (Table 1).

The composition of the average 1989 U.S. gasoline pool is summarized in Table 2. As the most important sources of high-octane blending components, reformate and alkylate accounted for 34 and 11% of the gasoline pool, respectively. The octane contribution of reformate is largely because of its high aromatics content. Although aromatics are a necessary part of the gasoline pool, concerns have been raised about the impact of the aromatics content in gasoline on automotive emissions. The result has been the passage of legislation that favors increased levels of nonaromatic octane boosters, for example, oxygenates and alkylate, in the gasoline used in many areas of the United States.

In 1992, the reported production capacity of alkylate in the United States was about 1.1 million barrels per stream day (BPSD). If fully used, this capacity could have supplied about 15% of the overall United States gasoline consumption.

About 350,000 BPSD of alkylation capacity was reported outside of the United States in 1992. The installation of new alkylation capacity outside of the United States is expected to be substantial because of the growth in the demand for gasoline and the worldwide trend toward cleaner burning, higher quality gasolines.

Impact of the 1990 Clean Air Act (CAA) Amendments

The 1990 Clean Air Act (CAA) Amendments established guidelines for the future composition of transportation fuel in the United States. The initial rules set by the Environmental Protection Agency (EPA) under this act set substantially lower summertime vapor pressures for gasoline starting in 1992 in most of the country. In addition, rules established under the so-called Simple Model directly or indirectly set limits on the content of oxygen, sulfur, olefins, benzene, and aromatics in the RFG sold in ozone nonattainment areas in 1995. These same parameters are also included in the more stringent Complex Model, which will govern RFG composition from 1997 onward.

To meet the 1992 vapor pressure specifications, significant quantities of normal butanes had to be removed from the gasoline pool. Refiners found that isomerization of the excess normal butane to isobutane followed by alkylation was an effective way to maintain gasoline volume at lower vapor pressure. The principal processors of natural gas liquids installed significant additional normal butane isomerization capacity targeted to meeting the increased demand for isobutane from alkylation and MTBE producers. The additional olefins required for alkylation were made available by changes in FCC operations.

Because alkylate contains essentially no sulfur, olefins, benzene, or aromatics, alkylation is one of the key techno-

Table 1. Alkylate Yield and Octane from C_3–C_5 Olefins

	Propylene	Butylene	Amylene
True alkylate yield, bbl/bbl olefin	1.75–1.78	1.70–1.78	1.9–2.1[a]
Reaction isobutane, bbl/bbl olefin	1.27–1.32	1.10–1.16	1.0–1.2
True alkylate research octane	89–92	94–98	91–93
True alkylate motor octane	88–90	92–94	88.5–90
Acid consumption, lb/gal alkylate	0.8–1.0	0.3–0.6	0.6–1.0
Based on H_2SO_4 catalyst			

[a] Ref. 2.

Table 2. Average 1989 U.S. Gasoline Pool[a]

Composition	Volume, %	Octane Range, (R + M)/2 Clear	Aromatics Range, vol %
Blending butanes	7.0	91–93	
Light straight-run	3.3	55–75	0–4
Isomerate	5.0	80–88	
Cat cracked	35.5	84–89	23–33
Hydrocracked	2.0	85–87	2–6
Coker	0.6	60–70	4–8
Alkylate	11.2	90–94	
Reformate	34.0	86–96	50–80
MTBE	1.4	106–110	
Total	100.0		
Average octane		88.4	
Average aromatics			32.1

[a] Ref. 3.

logies for meeting the Simple Model and Complex Model specifications starting in 1995. The olefins required for alkylation will be made by further changes in FCC operations and catalyst. An important trend will be the inclusion of greater quantities of FCC-derived C_5 olefins in the alkylation unit charge in response to the limitations on vapor pressure and gasoline olefin content (3).

California Air Resources Board Regulations

The California Air Resources Board (CARB) has set standards for motor fuel that are more stringent than federal standards. California Phase II RFG specifications include limits on olefins (6 vol % maximum), aromatics (25% maximum), and oxygen content (1.8–2.2 wt %). To meet these specifications, most refiners will have to convert essentially all of the light C_3 to C_5 olefins produced by the FCC unit to oxygenates or alkylate or both. If isobutane is available, alkylate production is preferred over oxygenate production because additional gasoline volume is generated.

The CARB limit on olefin concentration is related to the concerns about the atmospheric reactivity of evaporative emissions and the subsequent effect on urban ozone levels. The light olefins are of particular concern because they have both high vapor pressure and high reactivity. The vapor pressure and atmospheric reactivity (hydroxyl radical reactivity) (4) of alkylation feed olefins and several of the potential alkylation products are summarized in Table 3. The conversion of butenes and pentenes to alkylate provides a significant environmental benefit by simultaneously reducing atmospheric reactivity and vapor pressure.

PRODUCTION OF ALKYLATE

The current commercial alkylation technologies use either sulfuric acid (H_2SO_4) or hydrofluoric acid (HF) as a catalyst (5). Both types of catalyst are widely used. In 1991, alkylation catalyzed by H_2SO_4 accounted for about 53% of the alkylate produced in the United States, and HF-catalyzed alkylation accounted for 47%. Alkylation capacity in 1991 outside of the United States was about 75% HF and 25% H_2SO_4. However, most of the new installations since 1990 have used H_2SO_4 technology because of the increased cost of meeting the more stringent safety and environmental constraints being placed on plants that use HF.

Sulfuric Acid Alkylation

The modern H_2SO_4 processes are differentiated primarily by the type of reactor system that is used. The reactor must generate a high degree of mixing to ensure contacting of the largely insoluble acid and hydrocarbon liquid phases. Efficient heat removal is also important because refrigeration is required to maintain reaction temperatures in the preferred range of 5 to 10°C. Two reactor systems, the Stratco Contractor (used in the Stratco Effluent Refrigeration process) (6) and the cascade reactor (used in the Kellogg H_2SO_4 process and currently licensed as part of the Exxon Research and Engineering stirred autorefrigerated process) (7), are widely used in modern H_2SO_4 alkylation units. The Stratco Contractor is an internally circulated heat exchanger reactor that incorporates a proprietary mixer-impeller to maintain high heat and mass transfer coefficients. The reaction temperature

Table 3. Atmospheric Reactivity and Vapor Pressure

	Atmospheric Reactivity[a]	Vapor Pressure at 37°C, kPa[b]
Alkylation feedstocks		
Propene	26.3	1561
Butene-1	31.4	435
trans-Butene-2	64.0	343
Pentene-1	31.0	132
2-Methyl,1-butene	61.0	127
2-Methyl,2-butene	63.9	99
Isobutane	2.3	498
Alkylation products		
2,2-Dimethyl pentane	3.4	24
2,4-Dimethyl pentane	5.1	23
2,2,4-Trimethyl pentane	3.7	12
2,3,4-Trimethyl pentane	7.0	7

[a] Relative gas phase reaction rate with the hydroxyl (OH) (4).
[b] To convert kPa to psi, multiply by 0.145

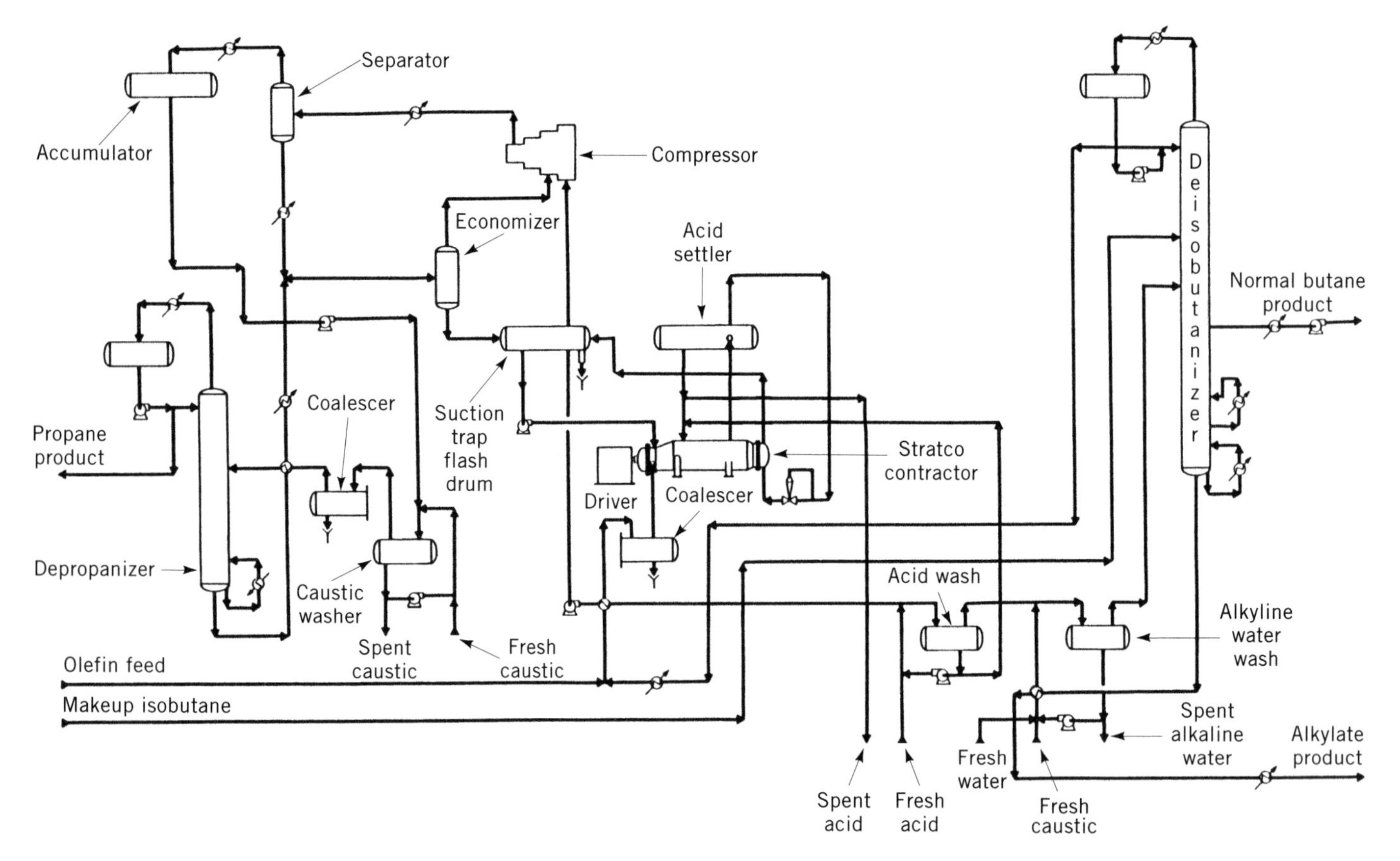

Figure 1. H_2SO_4 alkylation unit with effluent refrigeration. Courtesy of Stratco, Inc.

is controlled by effluent refrigeration, whereby flashing of the reactor effluent in the contactor tubes indirectly removes the heat of reaction. The cascade reactor employs a series of mixed stages, where increments of olefin are added to the reaction mixture through a proprietary mixing device. Direct heat removal in the cascade reactor is accomplished by partial vaporization of the reaction mixture.

A simplified flow diagram of a modern H_2SO_4 alkylation unit is shown in Figure 1. Excess isobutane is supplied as recycle to the reactor section to suppress polymerization and other undesirable side reactions. The isobutane recycle is derived from fractionation and by the return of flashed reactor effluent from the refrigeration cycle.

Propane and light ends are rejected by routing a portion of the compressor discharge to the depropanizer column. Prior to debutanization, the reactor effluent is treated to remove residual esters by means of an acid wash followed by an alkaline water wash. The deisobutanizer is designed to provide a high-purity isobutane stream for recycle to the reactor, a sidecut normal butane stream, and an alkylate product with low vapor pressure.

The H_2SO_4 concentration in the circulating acid is controlled to provide the optimum activity and selectivity. Purity is maintained by the withdrawal of system acid and replacement with fresh 98% acid. The spent acid is typically returned to an acid manufacturing plant for reprocessing, although it may also be regenerated in an onsite facility.

HF Alkylation

The modern HF alkylation processes are also differentiated primarily by the type of reactor system that is used. The Phillips process employs a gravity circulation system and a riser reactor (8). The UOP process uses a pumped acid circulation system and an exchanger reactor (9).

A simplified flow diagram of a modern HF alkylation unit is shown in Figure 2. Olefin feed and recycle isobutane are combined prior to contacting the acid catalyst in the reactor. Cooling water maintains the reactor temperature in the range of 20 to 40°C. Acid is settled from the reactor effluent and is returned to the reactor. The hydrocarbon phase is routed to the isostripper for fractionation into an isobutane recycle stream, a sidecut normal butane product, and an alkylate bottoms product. Propane is removed from the system by routing an overhead stream from the isostripper to the HF stripper. If the unit processes a significant quantity of propylene, a depropanizer is included to produce a high-purity propane product.

Before leaving the unit, the products are treated with potassium hydroxide to remove any trace acidity. In addition, any product streams that are used for liquefied petroleum gas (LPG) are processed over alumina at elevated temperatures to remove residual organic fluorides.

During operation, the HF catalyst is diluted by the production of small quantities of heavy acid-soluble hydrocarbons. The HF concentration of the circulating acid catalyst is maintained in the range of 85 to 95% by regeneration within the unit's fractionation facilities. The

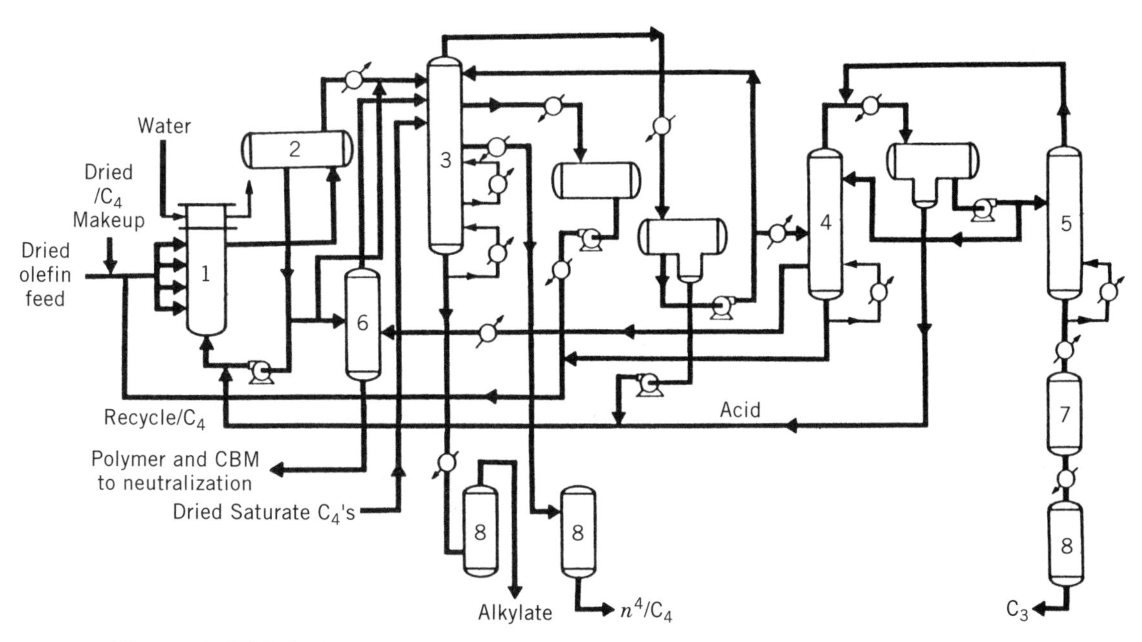

Figure 2. UOP C_3-C_4 HF alkylation process. 1, Reactor; 2, mixer settler; 3, isostripper; 4, depropanizer; 5, HF stripper; 6, acid regenerator; 7, alumina treater; and 8, KOH treater.

regeneration of acid within the unit accounts for the low consumption of fresh acid.

Solid-Catalyst Alkylation

The hazards associated with the liquid acids provided a strong incentive to develop a motor fuel alkylation process that uses a solid-acid catalyst. A number of solid-acid catalyst technologies already exist for aromatic alkylation. However, the application of solid-bed technology to paraffin alkylation has proven to be much more difficult.

Many catalyst systems have been reported in the literature as having the potential for application to a solid-catalyst alkylation process (10–16). These systems include zeolites, sulfated metal oxides, and a wide variety of catalysts prepared by incorporating normally liquid or gaseous Bronsted or Lewis acids on neutral or acidic bases.

The challenge in developing a commercially acceptable solid-catalyst alkylation process lies in finding a practical way to maintain catalyst activity. The solid acid sites are likely to be susceptible to fouling by reaction by-products in the alkylation environment. A successful catalyst must either overcome this tendency toward fouling or be easily regenerable. The regeneration must be safe and environmentally acceptable. In addition, for the process to be broadly applicable, the alkylate yield, quality, and overall production cost must be competitive with the existing liquid-acid technologies. Many laboratories throughout the world are currently working to meet this challenge, and one or more solid-catalyst alkylation technologies are expected to become commercially available before 1995.

SAFETY

Over the last 50 years, the liquid-acid alkylation technologies, both H_2SO_4 and HF, have had excellent safety records. However, in recent years, industry and the public have placed an increasing emphasis on industrial safety. One reason for this trend is the heightened awareness and concern about major accidents.

Because alkylation units process LPG, the hazards associated with a light flammable hydrocarbon are common to all H_2SO_4 and HF alkylation technologies (17). In addition, each technology has specific environmental concerns that are related to the properties of the acid and the way it is used, stored, transported, and regenerated.

Sulfuric Acid

Sulfuric acid is a strong, corrosive acid that is injurious to the skin, mucosa, and eyes (18). Because the acid has a low vapor pressure at the conditions expected in an alkylation unit, it is normally encountered as a liquid in the process or in the vicinity of any release. Consequently, safety recommendations include wearing chemical goggles, face shield, rubber gloves, and full protective clothing whenever workers are in danger of exposure.

A spill of sulfuric acid can be safely mitigated by neutralizing the acid with common chemical agents, such as bicarbonate of soda or caustic soda, by securing it with foams or dry chemicals, or by diluting it with large amounts of water. Precautions are necessary to prevent the spill from entering a public sewer system or public body of water and to take into account that H_2SO_4 gives off heat when it reacts with water.

Sulfuric acid mist in the air can present special hazards because it is a severe irritant to the respiratory tract. Exposure limits for emergency response planning are summarized in Table 4. The potential for formation of significant quantities of H_2SO_4 mist by a release from an H_2SO_4 alkylation unit is low because of the low vapor pressure of the acid.

Sulfuric acid mist can also be formed by the reaction of atmospheric moisture with sulfur trioxide, which is produced from sulfur dioxide during the regeneration or re-

Table 4. Exposure Guidelines for Sulfuric Acid, Hydrogen Fluoride, and Sulfur Dioxide[a]

Chemical Name	IDLH[b] (30 min)	ERPG-3[c] (1 h)
Sulfuric acid	20 ppm (80 mg/m^3)	7 ppm (30 mg/m^3)
Hydrogen fluoride	30 ppm (25 mg/m^3)	50 ppm (41 mg/m^3)

[a] Ref. 18, 19.

[b] IDLH (immediately dangerous to life or health): represents the maximum concentration from which one could escape within 30 minutes without any escape-impairing symptoms or any irreversible health effects. See the Pocket Guide to Chemical Hazards developed by the National Institute for Occupational Health and Safety (NIOSH).

[c] ERPG (emergency response planning guidelines): developed by the American Industrial Hygiene Association (AIHA) for a limited number of chemicals; based primarily on acute toxicity data and possible long-term effects from short-term exposure. The ERPG-3: the maximum concentration in air below which nearly all people could be exposed for one hour without life-threatening health effects.

covery of spent H_2SO_4. Spent acid from the H_2SO_4 process can contain volatile hydrocarbons. In some incidents the hydrocarbons have ignited, releasing sulfur dioxide and trioxide fumes. Because spent alkylation acid is typically regenerated in an off-site acid manufacturing plant, the acid producer is responsible for the safety of workers and the public during this operation. Modern acid-recovery plants operate at low pressures so that potential for accidental releases is minimized.

The generation of spent acid by an H_2SO_4 alkylation unit is considerable: about 100 tons per day of spent acid is generated in an alkylation unit that produces 10,000 barrels of alkylate per day. For H_2SO_4 units not having on-site recovery facilities, multiple tank cars or tank trucks are required on a daily basis to bring in fresh acid from the manufacturer and to return spent acid for recovery. When the acid plant is fairly close, acid pipelines have been used to reduce the hazard associated with surface transportation.

Hydrofluoric Acid

Pure hydrogen fluoride is a corrosive liquid that boils at 19.4°C at atmospheric conditions. It has a sharp penetrating odor that can be detected at very low concentrations in the air. Hydrogen fluoride is completely soluble in water to form HF solutions that in concentrated form will fume when exposed to moist air.

Even brief contact with HF liquid can produce serious, painful chemical burns, sometimes with delayed onset. The vapor can be extremely irritating to the eyes, skin, and respiratory tract. Short-term exposure to higher concentrations can cause serious health effects or death resulting from extensive respiratory damage. Chronic health effects, such as fluorosis, can occur from repeated exposure (20–21).

Facilities using HF must comply with the requirement of worker protection health standards and regulations applicable to the facilities' location. For example, the *Code of Federal Regulations* sets workplace exposure limits for HF (as fluorine) at 3 parts per million for an 8-hour time-weighted average. The same rule sets the short-term exposure limit at 6 parts per million for a 15-minute time-weighted average. Guidelines for emergency response planning are summarized in Table 4.

In the summer of 1986, a series of six atmospheric release tests of HF were conducted at the Department of Energy Liquified Gaseous Fuels Facility. The major conclusion from these tests was that an unmitigated release of pressurized, superheated HF can form a dense vapor cloud. An analysis of the thermodynamics of the release indicated that part of the HF was vaporized, and the remainder was present as an aerosol of very small droplets. The cloud moved downwind until the droplets vaporized and the HF vapor eventually dispersed (22).

Additional concern about HF was raised by an incident that occurred in 1987. A construction accident in a shutdown refinery alkylation unit resulted in a large unmitigated release. The result of this release was that about 1000 persons reporting to hospitals and about 100 admitted for treatment.

In 1988 further experimental investigation of atmospheric HF releases was conducted by the Industry Cooperative HF Mitigation/Assessment Program, which was supported by 20 petroleum and chemical companies. The purpose of the study was to investigate the impact of water sprays and vapor barriers on the mitigation of HF releases and to generate data for improved mathematical dispersion models that could be used to estimate the consequences of such a release. Water applied from fixed sprays or adjustable nozzles was found to be effective in reducing HF concentrations in the air. The tests showed that water sprays at high flow rates could remove more than 90% of the acid contained in a vapor cloud (23).

Under the Clean Air Act of 1990 as amended, the U.S. Congress required the EPA to carry out a study of HF to identify potential hazards to public health and the environment by considering a range of events including worst-case accidental releases. The EPA Hydrogen Fluoride Study issued in September 1993 made the general recommendation that "the EPA does not recommend legislative action from the Congress at this time to reduce the hazards associated with HF. The regulations already promulgated, and being developed, under the authorities provided to EPA in CERCLA, EPCRA, and the accidental release prevention provisions of the CAAA, provide a good framework for the prevention of accidental chemical releases and preparedness in the event that they occur" (19).

Currently, the use of HF is regulated by a number of United States federal agencies, including the EPA under several regulations, the Department of Transportation under the Hazardous Materials Transportation Act (HMTA), and the Occupational Safety and Health Administration (OHSA) under the Occupational Safety and Health Act and the Clean Air Act. In addition, the refining industry, through the American Petroleum Institute, has issued Recommended Practice 750, "Management of Process Hazards," which outlines a comprehensive program for managing all potentially hazardous processes, and Recommended Practice 751, "Safe Operation of Hydrofluoric Acid Alkylation Units," which defines specific recommended practices for HF alkylation units.

Research and development work to further improve the

safety of the HF alkylation catalyst system is continuing. Recent papers have been published describing the effectiveness of vapor-suppression additives in reducing aerosol formation when HF acid is released to the atmosphere.

BIBLIOGRAPHY

1. L. F. Albright, "Alkylation Will Be Key Process in Reformulated Gasoline Era," *Oil Gas J.* (Nov. 12, 1990).
2. P. Keefer, "Alkylation: The Cornerstone of Reformulated Gasoline," World Conference on Clean Fuels and Air Quality Control, Washington, D.C., Oct. 1993.
3. G. H. Unzelman, "Refining Options and Gasoline Composition 2000," NPRA Annual Meeting, Mar. 22–24, 1992.
4. R. Atkinson, "Gas-Phase Tropospheric Chemistry of Organic Compounds: A Review," *Atmos Environ.* **24A**(1), 1–41 (1990).
5. L. F. Albright, "H_2SO_4, HF Processes Compared and New Technologies Revealed," *Oil Gas J.* (Nov. 26, 1990).
6. L. E. Chapin, G. C. Liolios, and T. M. Robertson, *Hydrocarbon Process.* **649,** 67, (1985).
7. H. Lerner and V. A. Citarella, "Exxon Research and Engineering Sulfuric Acid Alkylation Technology," NPRA Annual Meeting, Mar. 17–19, 1991.
8. T. Hutson and W. C. McCarthy, "Phillips HF Alkylation Process," in R. A. Meyers, ed., *Handbook of Petroleum Refining Processes,* McGraw-Hill, New York, 1986.
9. B. R. Shah, "UOP HF Alkylation Process," in R. A. Meyer, ed., *Handbook of Petroleum Refining Processes,* McGraw-Hill, New York, 1986.
10. U.S. Pat. 3,893,942 (July 8, 1975), C. L. Yang (Union Carbide Corporation).
11. U.S. Pat. 5,212,136 (May 18, 1993). H. P. Angstadt, E. J. Hollstein, and C. Y. Hsu (Sun Company, Inc.).
12. T. J. Huang and S. Yurchak, in L. F. Albright and A. R. Goldsby, eds., *Industrial and Laboratory Alkylations,* American Chemical Society Symposium Series, Washington, D.C., 1977, p. 75.
13. U.S. Pat. 5,012,033 (Apr. 30, 1991), J. E. Child, A. Huss, C. R. Kennedy, D. O. Marler, and S. A. Tabak (Mobil Oil Corporation).
14. U.S. Pat. 5,157,197, (Oct. 20, 1992), M. D. Cooper, D. L. King, and W. A. Sanderson.
15. U.S. Pat. 5,190,904 (Mar. 2, 1993), C. S. Crossland, A. Johnson, J. Woods, and E. G. Pitt (Chemical Research and Licensing Company).
16. U.S. Pat. 5,220,095 (June 15, 1993), S. I. Hammeltoft and H. F. A. Topsoe (Haldor Topsoe, A/S).
17. B. Scott, "Identify Alkylation Hazards," *Hydrocarbon Process.* (Oct. 1992).
18. J. R. Donovan and J. M. Salamone in M. Grayson, ed., *Kirk Othmer Encyclopedia of Chemical Technology,* 3rd ed., vol. 22, John Wiley & Sons, New York, 1983.
19. U.S. Environmental Protection Agency, *EPA Hydrogen Fluoride Study: Final Report,* EPA 550-R-93-001, Washington, D.C., Sept. 1993.
20. D. T. Meshri, in M. Grayson, ed., *Kirk Othmer Encyclopedia of Chemical Technology,* 3rd ed., vol. 10, John Wiley and Sons, New York, 1980 pp. 732–753.
21. American Petroleum Institute (API), "Safe Operation of Hydrofluoric Acid Alkylation Units," API Recommended Practice 751, 1st ed., June 1992.
22. D. N. Blewitt, J. F. Yohn, R. P. Koopman, and T. C. Brown, "Conduct of Anhydrous Hydrofluoric Acid Spill Experiments," *International Conference on Vapor Cloud Modeling,* Cambridge, Mass., Nov. 1987.
23. K. W. Schatz and R. P. Koopman, *Effectiveness of Water Spray Mitigation Systems for Accidental Releases of Hydrogen Fluoride: Summary Report,* NTIS, No. DE9000-1181, Springfield, Va., 1989.

ALUMINUM

Jennifer S. Gitlitz
Clark University
Worcester, Massachusetts

During the last three decades, primary aluminum production has grown by an annual average of 4%: from a world production level of about 5 million metric tons (Mt) in 1960 to 19.2 Mt in 1992 (1). It has become one of the most important metals on the global market in terms of tonnage produced and dollar value: with an annual production of 721 Mt, only crude steel surpasses it, while copper lags behind at 9.3 Mt (2). The success of the industry has been accompanied by a wide range of environmental impacts, most of which are unknown to the average consumer of aluminum products. Direct impacts of the primary aluminum production process include emissions of fluorides, sulfur dioxide, and several important greenhouse gases, and the generation of large quantities of solid wastes, some of which may be hazardous. Indirect effects include the consumption of a tremendous amount of electrical energy with all of its associated environmental impacts: notably emissions of greenhouse gases and those contributing to acid rain, and the large-scale degradation and displacement of wildlife habitat and human communities by hydroelectric dams. See also Energy efficiency; Super cars; Life cycle analysis; Recycling.

PROPERTIES

Aluminum has often been called a wonder-metal. It is lightweight, and when alloyed with small quantities of other metals, is extremely strong. It is nonmagnetic, odorless and tasteless, and can be embossed, painted and printed on with ease. Because it resists corrosion by weather and by strong acids, it is ideal for use in the transportation and storage of many chemicals. It can be shaped into a wide variety of forms using heat and mechanical means. These properties combine to make it desirable in many applications, including food and drink packaging, household furniture and appliances, and cars, planes and trains. Its flexibility and resistance to corrosion give it an advantage over steel in these same applications. A good electrical conductor, aluminum is replacing heavier steel and more expensive copper in wires and cables throughout the world. Aluminum can be fabricated in many ways. Much like steel, molten aluminum can be cast into parts, and solid ingots can be rolled into sheets of various thicknesses, including extremely thin foil. It can be extruded through shaped holes into various rods and bars, drawn (or punched) into desired shapes, or

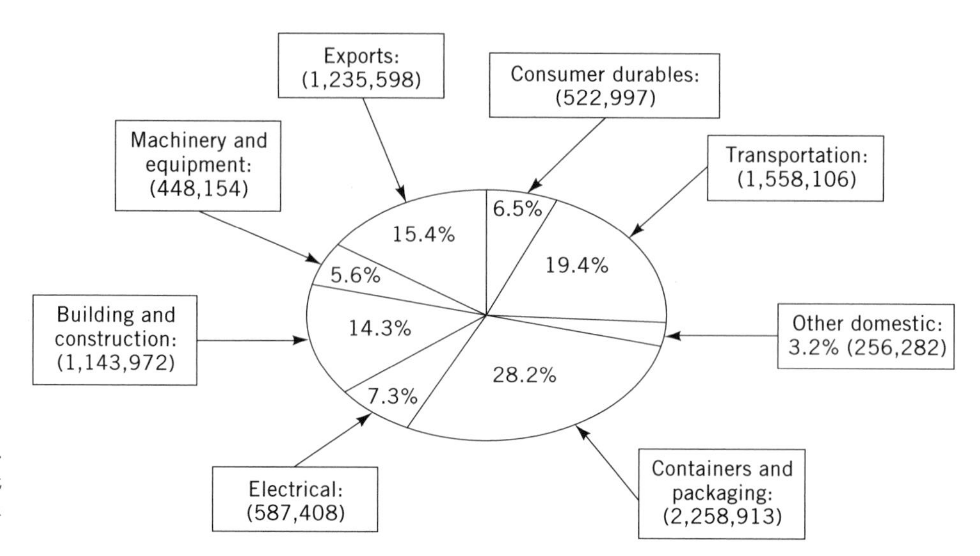

Figure 1. Distribution of net aluminum shipments by principal market in the United States, 1992 (t). From Ref. 1.

blown in powder form into complex molds and then heated, or sintered, to get its final form.

COMMON END USES AND PRODUCTS

Figure 1 shows the end use breakdown of net aluminum shipments in the United States in 1992. This includes aluminum that is fabricated in the United States and imported. Containers and packaging is the single largest principal end use of aluminum, constituting almost a third of U.S. net aluminum shipments. This category, which includes food and beverage cans, foil trays, aluminum foil, and other closures and packaging, has grown dramatically in market share during the last two decades. Production in 1992 was 2.3 Mt, 40% more than the 1982 level. Containers and packaging is the only sector of the aluminum market that is closely tracked from production through consumption through recycling, because its largest component is used beverage cans (UBCs), which are recycled in a "closed-loop," or into cans again. This sector of the market is also important because it is the only one that is specifically designed to have a short product lifetime, in many cases a few minutes between consumer purchase and disposal, as in the case of beer or pop cans, or the aluminum linings of aseptic containers (drink boxes). Aluminum foil is usually only used once, as are TV-dinner trays, pie tins, and closures for glass bottles. It is up to the consumer to recycle these items, and in areas where curbside recycling does not yet exist, the vast majority of non-can items will end up in the garbage.

Unlike packaging, other aluminum products have lifetimes of several years, as in the case of cars, to over fifty years, in the case of buildings and bridges. After containers, the transportation sector is the next largest market, with a 1992 consumption rate of 1.56 Mt. This includes parts and bodies for passenger automobiles, commercial and military planes and rockets, railroad cars, trucks, trailers, busses, etc. If recent trends continue, these uses will show more growth than all other market sectors in the coming decade. To improve gas mileage, car manufacturers in Japan, Europe, and the United States are "lightweighting" cars by substituting aluminum for steel in many applications.

Building and construction materials include window and door frames, screens, awnings, aluminum siding, girders and supports for bridges and buildings, mobile homes, and highway signs and guardrails. Electrical uses primarily consist of wires and cables. Consumer durables include aluminum furniture, appliances, cooking utensils, etc. Machinery, equipment and exports together comprised 1.7 Mt of net U.S. shipments. Unlike the other categories, exports include both primary and secondary aluminum: including prefabricated ingot, fabricated aluminum sheet, rolls, wire, etc., and scrap aluminum. The proportion of end uses in the worldwide aluminum market roughly parallels that of the United States. The major difference is that all-aluminum beverage containers have a smaller (yet rapidly-expanding) market share in other parts of the world.

BRIEF HISTORY OF PRIMARY ALUMINUM PRODUCTION

Aluminum was first produced economically in 1886 by Charles Martin Hall, an American amateur metallurgist, and Paul-Louis Héroult, a French scientist. Working independently in the United States and France, they discovered how to pass an electric current through molten cryolite to separate (or *reduce*) elemental aluminum metal from the oxygen to which it is chemically bound. Each man received a patent for the discovery in his home country. In 1888, Austrian chemist Karl Joseph Bayer received a German patent for his innovation in refining bauxite ore into alumina, the white powdery precursor to primary aluminum. In brief, he devised a method to produce alumina by first dissolving bauxite ore in sodium hydroxide to produce sodium aluminate solution and red mud, which is separated from the solution by settling; subsequently by "seeding" the sodium aluminate solution with alumina crystals. Alumina precipitates out of solution onto the crystals, which are then filtered and dried.

The first commercial aluminum reduction plants were established in 1888 by Hall in Pittsburgh, Pennsylvania, and by Héroult in Neuhausen, Switzerland. During the same period, inexpensive hydroelectricity was being developed for the first time. These three events laid the groundwork for an aluminum industry which was to burgeon from producing ten tons a year in the mid-1880's, to an annual rate of one Mt just prior to World War II, to the 19 Mt that are now produced each year (3). In 1895, Hall's company was the first large customer for the Niagara Falls hydroelectric power project in New York State. In 1907, the company's name was changed to the Aluminum Company of America, or Alcoa, now the world's largest non-public primary aluminum producer—surpassed only by the former Soviet government. After World War II, the company spawned Alcan, or the Aluminum Company of Canada. During the ensuing decades, other corporations such as Kaiser, Pechiney, Norsk-Hydro and Reynolds were established, creating the multinationally-dominated industry still in place today. With great improvements in efficiency but relatively minor technical variations, the Bayer-Hall-Héroult process is still used in producing almost 100% of the world's primary aluminum.

BAUXITE MINING AND ALUMINA PRODUCTION

Mining and Processing Bauxite Ore

The third most abundant element on Earth, aluminum comprises 8.3% of the Earth's crust by weight on average, or 15.7% alumina. Higher than average concentrations of aluminum oxide, or alumina (Al_2O_3), can be found in a wide range of rocks, including granite and basalt. Economically-recoverable alumina is extracted from bauxite, a reddish-brown ore containing 40–60% hydrated alumina by weight. The bauxite deposits mined today contain alumina which has been concentrated by surface weathering of permeable rocks under tropical conditions spanning geologic time. Although alumina can be recovered from materials other than bauxite, including clays, alunite, anorthosite, coal wastes and oil shales, these production processes are less economically feasible, and have only been employed in Eastern Bloc countries (4,5).

In some areas including parts of Australia and Jamaica, bauxite ore lies beneath a few feet of vegetation and topsoil, and can be scooped up with front-loaders and draglines with relative ease. In other areas, the deposits lie beneath up to 200 feet of overburden (unusable topsoil and rock), which is frequently vegetated. Thicknesses of ore bodies range widely: from a few feet in Arkansas, an average of 16 feet in Brazil, and 30 feet in Australia (6). Strip mining or open pit mining accounts for almost all production, because underground mining is prohibitively expensive. After vegetation is removed, the overlying rock is blasted to expose and loosen the ore. Large pieces are broken up, and are hauled to the processing site in dump trucks, rail cars or aerial trams. Strip mining has traditionally led to severe erosion and loss of topsoil. Millions of tons of overburden are generated in producing the world's aluminum. According to the U.S. Bureau of Mines, ratios of overburden to recovered ore have been as high as 13:1 in the state of Arkansas (6). Overburden has historically been left as open mounds, easily washing into lakes, streams, and the ocean, degrading the health of freshwater and marine ecosystems. However, there is an increasing industry trend in strip-mining to place the overburden back in the excavated holes, and to attempt to revegetate the site.

After mining, sizing and washing removes impurities such as sand and clay, and in the process creates about 0.7 tons of bauxite tailings in slurry form per ton alumina produced. These wet residues are usually disposed of on land, and have created huge mud lakes which have contaminated local water sources worldwide. New technologies now can reduce the moisture content of the mud to permit a more benign "dry stacking" method of disposal. The washed bauxite is then sun-dried or "calcined" (baked) in kilns to reduce its moisture content, and is then shipped to domestic alumina plants or is exported by ship for refining. Some alumina refineries are located near bauxite mines and far from smelters; some are near smelters, utilizing bauxite that is shipped from afar. In other cases, production facilities for bauxite, alumina and aluminum are integrated. Transportation is the largest single cost factor in producing alumina (7).

Alumina Production by the Bayer Process

Figure 2 is a simplified diagram of the Bayer Process, which is used in making virtually all of the world's alumina. Because the composition of bauxite ores around the world varies, alumina plants are designed to treat specific bauxites (9). First, the dried bauxite is size-reduced in a rod mill (stages 1 and 2) to particles in the 1 mm range. A recycled, concentrated solution of sodium hydroxide (2NaOH) or sodium carbonate (Na_2CO_3) is added to the ground bauxite, producing a slurry of 40–50% solids, which is fed to a series of digesters (stage 3), where more sodium hydroxide solution ("liquor") is added, and heated under pressure. The basic reaction in the digesters (stage 1) is:

$$Al_2O_3 \cdot x\, H_2O + \text{impurities} + 2\, NaOH \rightarrow 2NaAlO_2 + (x + 1)\, H_2O + \text{red mud}$$

The heat, pressure and caustic solution cause the alumina to dissolve into a sodium aluminate solution ($2NaAlO_2$). The mixture is brought through a heat-exchange "flash vessel" (stage 4) where its temperature and pressure are lowered, and the heat is recovered to pre-heat the recycled sodium hydroxide liquor. Excess water is removed, and in (stage 5), spent liquor (NaOH) is added to dilute the solution, to prevent the $NaAlO_2$ from precipitating prematurely. Stages 6–9 involve solid-liquid separation, wherein the sodium aluminate solution undergoes three types of filtration to remove insoluble impurities (red sand and mud). Gravity separators or wet cyclones are employed to remove the coarse sand. The red mud residues from the filtration are discharged into a waste lake, after being filtered to remove as much caustic liquor as possible. The resulting fluid is a clear brown sodium aluminate solution known as "new liquor" or "green liquor." It is subjected to another set of heat exchangers before it is precipitated (8,10). The equation:

$$NaAlO_2 + 2\, H_2O \xrightarrow{\text{seeding with } Al_2O_3{\cdot}3H_2O \text{ crystals}} NaOH + AL(OH)_3$$

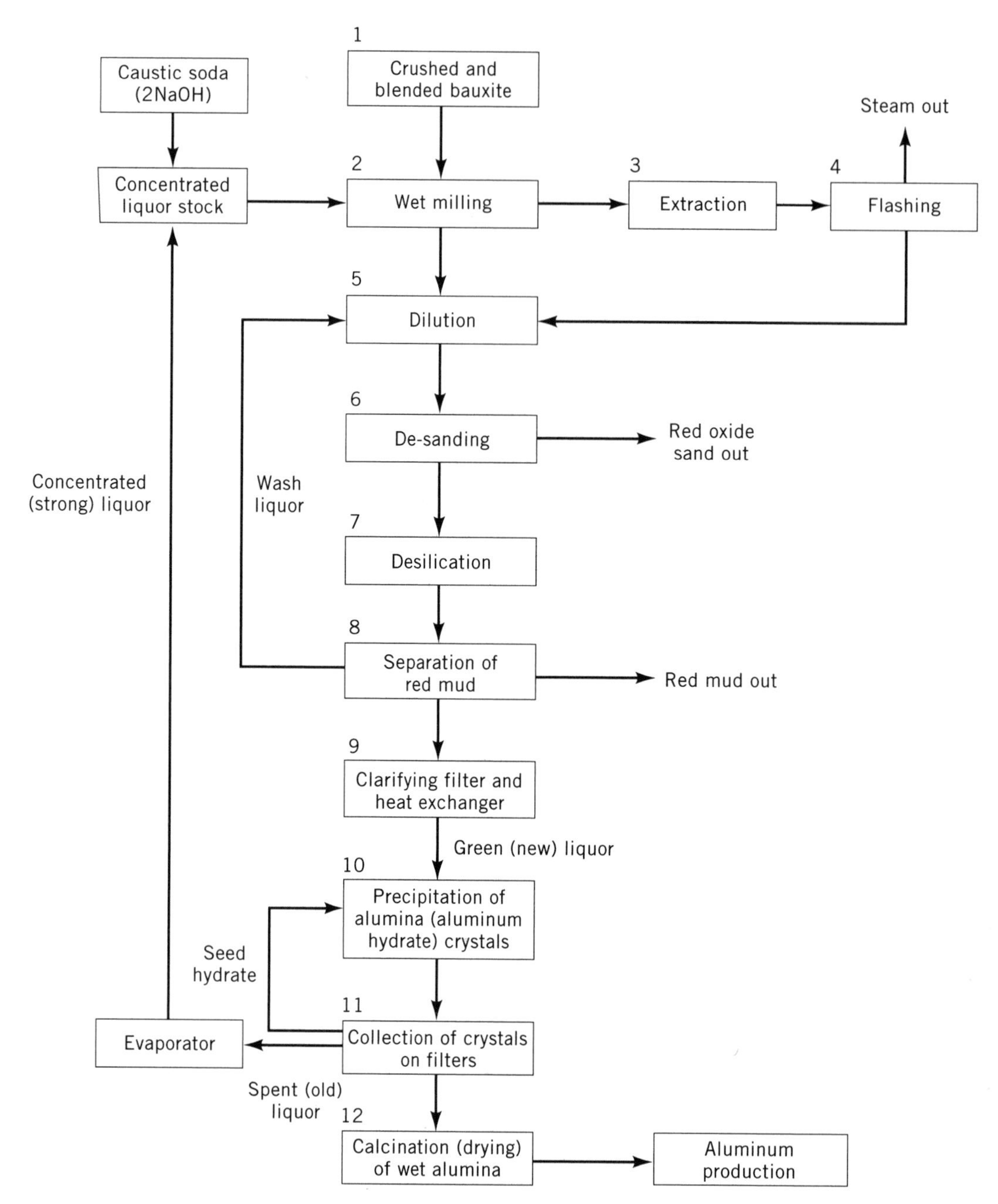

Figure 2. Simplified flow chart of the Bayer process (8). Reprinted with permission of the Controller of Her Britannic Majesty's Stationery Office.

describes the reaction wherein alumina is precipitated out from the supersaturated sodium aluminate solution by seeding the solution with aluminum trihydrate crystals (stage 10), which serve as nucleation points for the dissolved alumina. The chemical reaction is essentially the reverse of the earlier digestion reaction. The crystals grow as the reaction proceeds for up to three days in large agitated tanks. During this stage, the spent liquor is regenerated to yield caustic solution which is then recycled, along with any unprecipitated alumina, to the earlier digestion stage of the process (stage 11). The precipitated alumina is then washed and filtered for the last time to remove any residues. Flash evaporators remove excess water to attain a free moisture content of 8–10%. Some alumina is saved as "seed" for the precipitation process; the remainder is calcined in a rotary kiln at 1,200–1,300°C to further reduce the moisture content and produce anhydrous alumina powder (stage 12). Industry trends are to move from producing fine powders to producing a coarser grain product that requires less calcination. Coarse, sand-like alumina powder has the advantage of large surface area, and can be used to trap fluoride gas emissions in dry scrubbers in primary aluminum smelting operations, and can also be recycled as feedstock for the

Table 1. Material and Energy Requirements to Produce 1 Ton Primary Aluminum[a]

Alumina Production	Inputs	Outputs
Bauxite	3.5–7 tons	
Fuel oil	250–300 kg	
Soda	170–260 kg	
Lime	60–80 kg	
Steam	6.5–7.5 tons	
Alumina		1.8–2.0 tons
Residues from washing bauxite (tailings in slurry)		approx. 1.3 tons
Red mud residues from alumina processing		1.3–1.5 tons
Primary aluminum production		
Alumina	1.89–2.0 tons	
Anthracite coal	25–30 kg	
Petroleum coke	400–700 kg	
Pitch	100–150 kg	
Fuel oil	70–90 kg	
Steam	400 kg	
Cryolite	10–30 kg	
Aluminum fluoride	15–40 kg	
Calcium fluoride	3–5 kg	
Electricity	15–16 kWh/kg	
Primary aluminum		1 ton
Spent cathodes and refractory bricks		40–70 kg
Total fluorides		0.75–5 kg

[a] Ref. 14.

cell. Finished metallurgical grade alumina contains 0.3–0.8% soda, less than 0.1% iron oxide and silica, and trace amounts of other impurities (11).

Energy and Materials Requirements for Bauxite and Alumina Production

From 2.2–3.5 tons of bauxite are required for every ton of alumina produced, depending on the grade of the ore, and about 2 tons of alumina are required for each ton of aluminum. Using higher grade ores and the best technologies, the resulting ratio of bauxite ore to primary aluminum is approximately 4.5:1. (12). According to the Argonne National Laboratory, the range of energy required to mine and process bauxite is $0.26–1.06 \times 10^6$ Btu/short ton of bauxite, and the amount of fuel oil needed to ship the bauxite from producing countries to ports in the United States ranged from $0.75–2.0 \times 10^6$ Btu/short ton of bauxite. The alumina production process is much more energy-intensive, requiring 18.6×10^6 Btu/short ton of alumina (13). Added together, about 35×10^6 Btu are required per short ton of primary aluminum produced. Table 1 provides data on total inputs required and outputs generated for the entire Bayer-Hall-Héroult process for producing one metric ton of primary aluminum from bauxite ore.

Waste Products from Bauxite and Alumina Production

In the usual alumina production process described above, there are between 1.3 and 1.5 tons of red mud produced per ton of aluminum. Although these muds pose a significant disposal problem, they are still exempt from hazardous waste regulation under the Resource Conservation and Recovery Act (RCRA) in the United States. Disposal is largely unregulated in major bauxite and alumina-producing countries. Alcoa has developed a "combination process" whereby lower grade ores (with higher silica contents and thus more aluminum silicate lost in the muds) can be exploited. Their process enables the recovery of alumina from the red mud by heating the mud with calcium carbonate (limestone) and sodium carbonate (soda ash) to produce a sodium aluminate solution, which is returned to the digesters to resume the process described. The waste is called "brown mud." (15).

THE HALL-HÉROULT SMELTING PROCESS

Primary aluminum is not produced in massive fiery furnaces like those used to smelt iron; it is made electrolytically in rectangular vats (called "cells" or "pots") measuring several meters across. The smelting operation is a continuous (rather than a batch) process. Between 50 and 250 cells are connected in "potlines" in long buildings, or "potrooms." A 200,000 ton-capacity smelter might have about 700 pots in 5 separate potlines, for example. The individual cells are constructed out of rectangular steel shells lined with a refractory (non-burnable) material such as insulating brick. The refractory material on the floor of each cell is overlain with a carbon lining and a carbon cathode. This in turn is overlain with a layer of molten aluminum. The molten aluminum lies beneath a "bath" of molten synthetic cryolite, or Na_3AlF_6, at a temperature of 960–970°C. The cryolite serves as a medium to dissolve the alumina and minor additives, and to conduct electric current from the anode to the aluminum. Alumina powder is lowered from overhead hoppers onto a "crust" of cooler cryolite which forms on top of the cell. The crust serves to thermally insulate the contents of the pot, and to drive off any remaining moisture in the alumina. Periodically, the crust is mechanically broken and the alumina is stirred into the bath.

Large carbon anodes are suspended by steel bars into each cell. When a strong electric current (ranging from 50–225 kilo-Amperes) is passed through the anode into the bath, it causes the alumina to split into oxygen and molten aluminum. The oxygen combines with carbon on the anode to produce carbon dioxide (CO_2) and carbon monoxide (CO), which are vented to the atmosphere, along with fluoride compounds that are created as a result of partial evaporation of the molten cryolite.

The molten aluminum joins the pool at the bottom of the cell. The voltage at one terminal of an individual cell is 4–4.5V, while the cumulative voltage of a potline can exceed 1,000 volts. The average amount of electricity required to smelt aluminum in the 1980s was about 16.5 kWh/kg. This figure includes smelters with modern cells using about 13.5 kWh/kg, as well as older cells that are more energy-intensive. As smelters replace aging potlines with newer technology, the industry as a whole is slowly evolving to a higher average efficiency (16).

There are two main methods of smelting used today, each relying on a different type of anode. The older and more inefficient "Søderberg" process employs only one large anode per cell. The anode is comprised of a moist carbon paste that is continually baked as it descends into the cell, while fresh paste is added from above. All modern smelters use the more efficient "prebake" method, using anodes which are usually produced in on-site facilities from petroleum coke and coal tar pitch, by-products of the petroleum, coal, and steel manufacturing industries, which if not used in aluminum production, might have to be disposed of alternatively. Each prebake cell has 10 to 20 anodes. As the bottoms of the anodes are slowly consumed, they are lowered further into the bath to maintain a constant distance between the anode and the cathode (about 1.5 inches). Each anode must be replaced every few weeks, but this can be done in a staggered fashion without disrupting production. In contrast, spent potlinings need only be replaced once every five years, which is fortunate because replacement entails suspending production. Figure 3 is an illustration of a prebake cell.

In 1979, the world's annual aluminum capacity was about equally divided between Søderberg and prebake systems (17). Since then, the proportion of prebake capacity has grown to 60%, as old plants are retrofitted and new plants employing prebake are constructed (18).

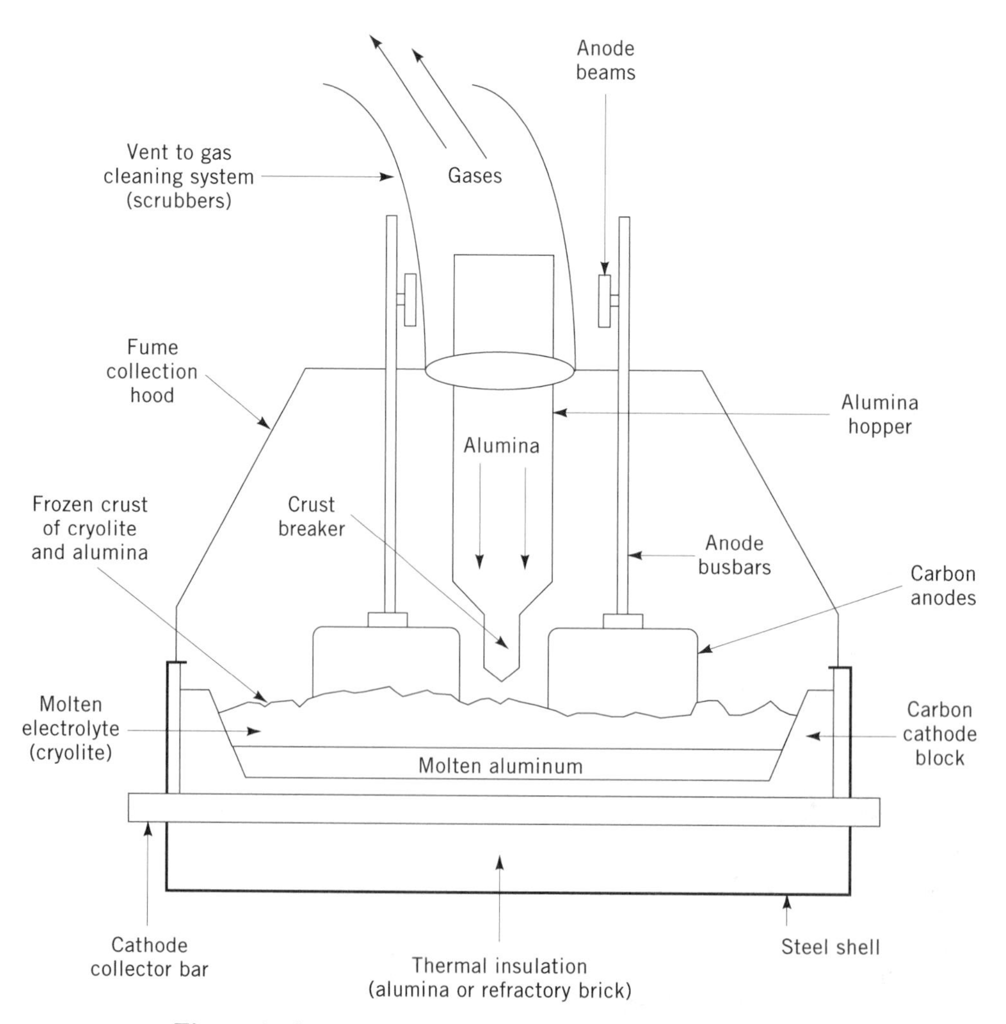

Figure 3. Cross section of a center-worked prebake anode cell.

The molten aluminum is periodically siphoned off through the tops of the cells into a holding furnace, and is then poured into "ingots" or "billets," which are large blocks or bars of solid aluminum. These in turn are either fabricated on site using the methods enumerated above, or are shipped elsewhere for fabrication, and manufacture into finished goods.

Under normal operating conditions, potlines run 24 hours a day, 365 days a year. If the electricity supply to the cells is cut off for more than three hours at a time, the molten aluminum will "freeze," or solidify in the pots, a mess that can take months and millions of dollars to clean up. This production restriction sets aluminum apart from some other large energy-intensive industries: aluminum smelters cannot purchase "interruptible power" from utilities. Their contracts usually have clauses which specify that they must take first priority in the event of power shortages.

DIRECT BY-PRODUCTS OF THE SMELTING PROCESS

The primary aluminum production process generates waste products that can impact the environment at local, regional and global levels. The following sections are concerned with solid wastes and airborne emissions from the smelting process, including fluoride compounds that can harm vegetation and animal life in the smelter locale; sulfur dioxide, a contributor to acid rain formation on a regional level; and greenhouse gases including carbon dioxide, carbon monoxide, and compounds containing carbon and fluorine. These gases trap long-wave radiation in the atmosphere, possibly leading to global warming. Solid wastes include spent carbon potlinings which can be leached by rain, contaminating groundwater. Other by-products from anode manufacture, smelting, and product forming which will not be treated here include wastewater, solid wastes other than potlinings, dusts generated by alumina and coke handling, and emissions from anode forming and baking including carbon dust, particulate and gaseous fluorides, NO_x, SO_2, and tar vapor.

Local Damages from Fluoride Emissions

During normal operations, a variety of fluorides in gaseous and particulate form evaporate from the hot cryolite bath, and are released from the anodes and the bath when the frozen crust is broken to add alumina. Particulates include cryolite (Na_3AlF_6), aluminum fluoride (AlF_3), calcium fluoride (CaF_2) and chiolite ($Na_5Al_3F_{14}$), while gases include hydrogen fluoride (HF) and silicon tetrafluoride (SiF_4) (19). The particulates are collected by bag filters. Since the mid-1970s, all new smelters have installed hoods over the cells to capture fluoride fumes and process them using dry and wet scrubbers. Dry scrubbers employ alumina which combines with the fluoride gases and is then recycled into the cell bath as fresh cryolite, thus reducing total fluorides emissions per ton aluminum (F_t/t) to 0.75–1.0 kg. The U.S. emissions standard set by the Environmental Protection Agency is 1 kg F_t/t. Some smelters, including older Søderberg plants which have not been able to afford costly equipment retrofitting, use wet scrubbers and electrostatic precipitators, enabling them to emit a net of 3–6 kg F_t/t (20). The generation of toxic wash liquors which must then be disposed of is a disadvantage of wet scrubbers. In addition to or instead of these primary gas cleaning systems, some smelters use secondary cleaning systems wherein potroom gases are processed by wet scrubbers after collection through roof vents. Due to the high cost of cleaning large volumes of potroom air with very dilute fluoride concentrations, these measures are the exception rather than the norm.

Before any control measures were taken, smelters released almost all fluorides to the atmosphere (15–25 kg F_t/t), and have been blamed for widespread environmental damage in localized areas. For example, in parts of Norway, wild and domesticated animals have contracted fluorosis of the teeth and bones from ingestion of contaminated vegetation in the vicinity of aluminum smelters (21). In Japan, fluoride emissions from the Kanbara aluminum smelter reportedly killed all vegetation within a 10-kilometer radius around the factory, leading the Japanese people to reject a proposed aluminum smelter in Okinawa in 1973, and to force the domestic aluminum industry to invest in new projects abroad, notably a large smelter in Indonesia which uses power from new dams built on the Asahan River in Sumatra (22).

When emissions lead to fluoride concentrations exceeding 40 parts per million on the surface of leaves or in plant tissue, effects can include foliar lesions, impaired growth, and reduced yield. The plants most sensitive to fluoride damage are conifers, certain berries and fruits, and grasses such as corn and sorghum. Forage plants are moderately sensitive, while most field crops, vegetables, and deciduous trees are resistant to fluoride damage. When grazing animals ingest sufficient quantities of contaminated forage, they are susceptible to bone lesions, lameness, reduced appetite, abnormal dental development, and diminished lactation. The fluoride remains in their bones, however; it is not absorbed by their flesh, milk, and eggs, and thus poses no threat to consumers. Despite these risks to animals and different types of vegetation, many smelters around the world (including most U.S. smelters) are located in agricultural areas that produce grains and vegetables for human consumption, feed and livestock, orchard products, and timber. Inhalation of ambient air around smelters is not reported to pose a further threat to animals nor humans in the vicinity, with the exception of workers in smelters with substandard fume collection units, where fluoride emissions contaminate potroom air, posing an occupational health risk (23).

Greenhouse Gases: Carbon Dioxide, Carbon Monoxide, and Fluorides

During electrolysis, 400–500 kilograms of carbon anode are consumed per ton of aluminum produced. This comes from the basic production reaction:

$$2\,Al_2O_3 + 3C \rightarrow 4Al + 3\,CO_2$$

At 100% current efficiency, the production of one ton of aluminum would thus generate 1.22 tons of carbon dioxide (CO_2), but because cells are not perfectly efficient, actual emissions are higher: between 1.5–1.8 tons CO_2 per ton of aluminum produced (21). Additional emissions from anode production bring total CO_2 generation to about 2 tons per ton of aluminum produced, yielding a global total of approximately 38 Mt of CO_2 emissions, based on 1992 world production (19.2 Mt).

When the amount of alumina in a cell falls below a certain level, an "anode effect" occurs. Until recently, technology has not allowed monitoring of alumina levels by computer; operators were required to do it manually based on experience. In addition to reducing the amount of molten aluminum produced, an anode effect causes several other greenhouse gases to be emitted, including carbon monoxide (CO), and the halocarbons dicarbon hexafluoride (C_2F_6, also known as CFC-14) and carbon tetrafluoride (CF_4, also known as CFC-116). In various texts, CF_4 is referred to as tetrafluoromethane and tetraflurocarbon, C_2F_6 is referred to as hexafluoroethane. The carbon in these compounds comes from the anodes, while the fluorine atoms come from the cryolite bath (Na_3AlF_6). Although CO is produced in significant amounts during anode effects, much of it is converted to CO_2 in burners prior to leaving plant stacks. Net CO emissions vary widely among plants, from 15 to over 1,000 kg CO/ton aluminum. The emission of halocarbons containing fluoride, however, may present a far greater threat to global warming than both CO and CO_2 emissions.

Global Implications of Halocarbon Emissions

At present, the primary aluminum industry is the only known generator of CF_4 and C_2F_6 emissions. Due to the difficulty in measuring these gases at the point of generation, and to a very limited number of reliable measurements of current atmospheric concentrations, a scientific consensus has not yet emerged as to the exact amount of CF_4 and C_2F_6 generated per ton of aluminum produced, or as to whether there are other industrial or natural sources of these compounds (24).

There are no known sinks for CF_4 and C_2F_6 molecules; they resist breakdown by combustion, sunlight, and reaction with other atmospheric gases including O_2, O_3, and NO_2. As a result, they have very long atmospheric residence times that have been estimated to be between 10,000 and 50,000 years for CF_4, and at least 10,000 years for C_2F_6. Over the course of 100 years, the global warming potential (GWP) of CF_4 has been estimated as approximately 5,000 times that of CO_2, while C_2F_6 has a GWP 10,000 times higher than CO_2. These GWP's increase with time; after 500 years, the GWP of CF_4 is estimated as 8,800, while that of C_2F_6 is 17,300 (25). The foregoing illustrates that all historical and current CF_4 and C_2F_6 emissions are essentially a permanent commitment. This situation is very unlike that of anthropogenic CO_2 emissions, which to some extent can be absorbed by terrestrial plants, soils, and oceans. If CO_2 emissions were drastically curtailed through appropriate industrial and economic policies (including improved overall energy efficiency and carbon taxes), and if CO_2 sinks were increased through the implementation of reforestation programs, atmospheric stocks of CO_2 could theoretically be stabilized and even reduced within a meaningful human timeframe (several generations). In contrast, primary aluminum producers can only reduce current emissions rates of CF_4 and C_2F_6 by reducing the frequency and severity of anode effects. Even if aluminum production were to cease today, present atmospheric stocks of CF_4 and C_2F_6 would not be measurably depleted for many thousands of years.

The formation of CF_4 and C_2F_6 is determined by a range of factors, including the type of cell used, operating temperature, and the frequency, duration, and severity of anode effects, all of which are highly variable among smelters. An anode effect can last one to several minutes, depending on how quickly an operator responds by adding alumina. Because they may occur only 0.1 times per cell per day in modern smelters using the best technology, and 2–3 times in older prebake and Søderberg plants, it is difficult to infer a worldwide average output. Early estimates of CF_4 output were as high as 1.5–2.6 kg/ton aluminum produced, while more recent estimates from Norwegian smelters show a range of 0.8 kg/ton in Søderberg plants to 0.06 kg/ton in prebake plants (21). C_2F_6 emissions are estimated to range from 5–15% of CF_4 emissions by weight. Estimates of annual generation in the United States are 6,100 tons of CF_4 and 600 tons of C_2F_6, which divided by 1992 primary aluminum production of 4.04 million tons yields an average of 1.51 kg CF_4/ton aluminum, and 0.15 kg C_2F_6/ton aluminum (26). Multiplying a conservative average emission value of 1 kg CF_4/ton aluminum produced by the 100-year GWP of 5,000 yields a result of 5 tons of CO_2 equivalent per ton, while an average emission value of 0.1 kg C_2F_6/tons times 10,000 yields 1 ton of CO_2 equivalent, for a total of 6 tons of CO_2 equivalent per ton aluminum produced. This yields an estimated world total of 115 million tons of CO_2 equivalent per year. Due to scientific uncertainties, however, this estimate may be in error by several hundred percent (Isaksen, 25).

Sulfur Dioxide

During electrolysis, organic sulfur present in the anodes reacts with alumina to form sulfur dioxide (SO_2). Since the coke and pitch used in anodes contain an average of 3% sulfur, 25–30 kilograms of SO_2 are generated per ton of aluminum produced, for a worldwide total of 480–575 thousand tons per year. Due to environmental regulations in many countries, however, some SO_2 is removed by wet scrubbers before being released to the atmosphere, so actual global emissions may be closer to 250,000 TPY, or about 1.5% of the total anthropogenic flow of SO_2 to the atmosphere as a result of metals smelting and fossil fuel combustion (20). Once airborne, SO_2 emissions can drift hundreds of miles before combining with other compounds to form acid rain. Acid rain has caused much damage to forest vegetation and lakes in temperate zones throughout the Northern Hemisphere.

Solid Wastes: Spent Potlinings

As previously stated, the life of carbon cathodes is 4–5 years, generating 40–70 kg of spent carbon potlinings and refractory materials per ton aluminum produced. Over the years, hundreds of thousands of tons these wastes have been created, much of which is stored in open piles on the grounds of aluminum production facilities worldwide. Atmospheric nitrogen combines with the carbon in the tailings to produce cyanide, which if not properly monitored and controlled by linings and leachate recovery systems, can pose a serious groundwater contamination hazard. Fluorides accumulated from the bath are also present in the wastes, accounting for 15–30% of total mass. The remainder is comprised of carbon (30–45%), alumina (15–20%), and trace amounts of calcium, silicon, iron, sulfur, and nitrogen compounds. Research is underway to process spent potlinings by incineration, intentional leaching for cryolite recovery, by reuse as fuel in other industries, and by conversion to inert substances for use in manufacturing bricks, cement, mineral wool, and steel (27).

Improvements in Smelting Technology

The aluminum industry continually strives to reduce energy consumption and improve emissions from the smelting process. Computer programs have been implemented to improve current efficiency in the cell from 70% at the turn of the century to 95% by 1990, and to reduce overall power consumption from 40 to under 13 kWh/kg aluminum during the same period (28). Design improvements have been made to lengthen anode life and thus reduce CO_2 emissions, and to reduce the frequency of anode effects and thus lower CF_4 and C_2F_6 emissions (29). Improved gas cleaning equipment has been adopted in many smelters, especially in countries with environmental regulations. Norway, for example, has reduced total fluoride emissions from 5 kg/ton in 1960 to the present 0.4 kg/ton in Søderberg plants and 1.0 kg/ton in prebake plants, for a national average of 0.76 kg/ton, an 86% reduction in three decades (21).

ELECTRICITY USE IN PRIMARY ALUMINUM SMELTING

In addition to the direct emissions and effluents enumerated above, aluminum production is responsible for significant indirect environmental impacts that stem from its heavy use of electricity. In all likelihood, the environmental damages created by electricity generation for aluminum exceed those of process-related waste products. The damages occur at all spatial scales: global effects include greenhouse gas emissions from power plants' combustion of fossil fuels; regional effects of fossil fuel combustion include acid rain; and local effects include smog, and a myriad of impacts associated with hydroelectric dams.

Worldwide Electric Demand by the Primary Aluminum Industry

The world's primary aluminum industry is a prodigious consumer of electricity. Table 2 presents data from the International Primary Aluminum Institute showing that the aluminum industry consumed 228 terawatt-hours (TWh) of electricity in 1990, and breaks this number down by fuel or energy source. Hydroelectricity is the largest source, providing 55% of all the electricity used in producing aluminum, or 126.8 TWh. This was 63% more than the next largest sector, coal, which accounted for almost 34% of the total. The other sources are less important: nuclear energy accounted for 5% of the electricity used, natural gas 4%, and oil only 1%.

More interesting is the proportion of the world's total electric generation consumed by the aluminum industry. In 1990, the aluminum industry consumed approximately 2% of the electricity generated worldwide by all sources. Almost 6% of the world's total hydroelectric generation was used in aluminum production, while 1.2% of the world's fossil fuel-generated electricity was used by smelters, both significant draws by a single manufacturing industry. The absolute meaning of these numbers is more compelling than the seemingly small percentages. In a world where electricity costs in industrialized nations are rising rapidly, and where millions of households in less-developed countries lack even basic electrification, it is worth noting the alternative uses of the electricity now devoted to aluminum production. The hydroelectricity alone used to produce aluminum every year is nearly equivalent to all the electricity consumed on the Australian continent, 132 TWh in 1987 (31). This amount could also satisfy the residential needs of about 35 million United States residents in 13 million households. All forms of electricity used to produce aluminum could power about one quarter of the 94 million homes in the United States or the entire country of India, or about 218 TWh in 1987 (32).

Table 2. Worldwide Electricity Generation, and Aluminum as a Consumer

	World Electricity Production, 1990[a]		Electricity Consumption in Primary Aluminum Production, 1990[b]		
Source	TWh Generated	% of Total	TWh Consumed	% of Aluminum Total	% of World Total, by Fuel
Coal	4645	39.3	77.9	34.1	1.7
Oil	1385	11.7	2.5	1.1	0.2
Natural gas	1578	13.3	9.5	4.2	0.6
Subtotal, fossil fuels	7608	64.3	89.9	39.4	1.2
Nuclear	2011	17.0	11.5	5.0	0.6
Hydroelectric	2142	18.1	126.8	55.6	5.9
Geothermal and other	67	0.6	0.0	0.0	0.0
Total	**11,828**	**100.0**	**228.2**	**100.0**	**1.9**

[a] Ref. 30. International Energy Agency.
[b] Ref. 30. International Primary Aluminum Institute.

The Impacts of Electric Generation Using Fossil Fuels and Nuclear Energy

The range of environmental costs associated with nuclear energy is generally well-known, as is the possibility of global climate change resulting from fossil fuel combustion. While policymakers, regulators, and the public often call on electric utilities to improve plant efficiency and institute conservation measures to reduce their emissions, they are less diligent in targeting major power consumers to determine the portion of responsibility they might bear in this regard. It is a useful exercise to look at the emissions from electric generation consumed by the world aluminum industry.

As Table 2 indicates, aluminum smelters consumed 77.9 TWh of coal-generated electricity in 1990, and 2.5 TWh of electricity generated by oil. At an average power plant conversion efficiency of 33%, the generation of this power required about 28.7 Mt of coal, and 4.7 million barrels of oil, whose combined combustion added approximately 77 Mt of CO_2 to the atmosphere. The combustion of natural gas to generate 9.5 TWh of electricity yielded an additional 3.8 Mt of CO_2 (0.4 kg CO_2 produced per kWh), for a total of about 81 Mt of CO_2 emissions generated by electricity production using fossil fuels (21). When combined with the 38 Mt of CO_2 mentioned earlier as a result of smelting, total CO_2 emissions for the smelting process is approximately 119 Mt. While this is only 0.6% of the total CO_2 emissions produced annually by fossil fuel combustion (19.4 billion tons), it is again a significant contribution by a single industry, especially since the latter figure includes *all* emissions resulting from fossil fuel use (including all coal, oil, and gas) in internal combustion engines, cement manufacturing, and in all industrial processes relying on heat rather than electricity. When 119 Mt of actual CO_2 emissions from smelting and electric generation is combined with the previously derived figure of 115 Mt of CO_2 equivalents from CF_4 and C_2F_6 emitted during smelting, the total is 234 Mt of CO_2 equivalents.

The combustion of fossil fuels is also responsible for generating oxides of nitrogen and sulfur which contribute to acid rain, especially in regions where environmental regulation is lax, and pollution control equipment is inadequate or nonexistent. Information about a variety of other environmental damages associated with coal mining and oil extraction is widely available.

Although the amount of nuclear energy consumed by smelters is slightly greater than the amount consumed by oil-generated electricity, it does not have the same emissions consequences. Because the aluminum industry consumes nuclear energy at a rate which only slightly exceeds the output of one large nuclear plant operating at full capacity (about 1,400 MW), its impacts are less significant relative to those of electric generation using fossil fuels.

The Impacts of Hydroelectric Dams Associated with Smelters

This section briefly examines some specific damages wrought by hydro projects associated with smelters in various parts of the world.

Over the last fifty years, aluminum smelters have been the sole impetus for constructing new large dams in many states and countries, including Brazil, British Columbia, Ghana, and Indonesia. Without the electric demand provided by smelters in these countries, the Tucuruí, Kenney, Volta, and Akosombo dams might never have been built. In other nations, including Chile, China, Egypt, Norway, Russia, Quebec, the United States and Venezuela, the aluminum industry is at the very least an important contributing factor in the development of existing and proposed dams. By providing large blocks of steady, long-term demand, one or more aluminum smelters can help assure an electric utility of the revenues needed to pay for dam construction. As Table 3 illustrates, the percentage of installed hydropower capacity consumed by aluminum smelters can range from 15% in Egypt, Norway and Quebec, to almost 90% in Indonesia.

The flooding of large tracts of land has had deleterious consequences for both human communities and wildlife

Table 3. Flooding, Human Relocation, and Smelter Electric Use in Selected Countries[a]

Country or Province	Main Dam(s)	Main River Affected	Area Flooded[b] (square km)	Approx. Number of People Relocated	Smelters' Consumption of Installed Power Capacity, %	Smelter(s) Capacity in 1991 (000 TPY)[e]
Brazil (Amazonia)	Tucuruí	Tocantins	2,430	24,000	32	648
British Columbia	Kenney	Nechako	800	75	67	275[d]
Chile	n/a	Cuervo	unknown, project pending			230[d]
Egypt	Aswan	Nile	4,000	100,000	15	184
Ghana	Akosombo	Volta	8,270	80,000	40	200
Indonesia	Tanga, Siguragura	Asahan	neg.	~100	88	225[d]
Norway	multiple dams	multiple rivers	n/a	neg.	15	851[d]
Quebec (James Bay)	LG1-LG4	La Grande	11,250	1,000	15	880[d]
Venezuela	Guri, Macagua	Caroní	4,250	3,600	22	680[d]
Total			31,000	208,775	18[c]	4,173[d]

[a] Ref. 34.
[b] Includes both land and water inundated (lakes, streams, ponds, etc.)
[c] Sum of all smelters' hydroelectric demands divided by the sum of the installed hydroelectric capacity linked to dams, except for Norway's contribution, which uses final domestic deliveries rather than installed capacity; and Indonesia's, which uses firm output.
[d] Indicates that existing smelter(s) may expand capacity, or that new smelters are planned in the area.
[e] Except in Chile, where all capacity is proposed.

habitat, especially among indigenous people who practice subsistence hunting, fishing and farming. In the nine countries listed in Table 3, a total of 31,000 square kilometers have already been flooded by dams built in conjunction with aluminum smelters. Worldwide, significant numbers of people have been relocated to new villages and towns, often against their wills and without more than a few weeks of warning. As Table 3 shows, the number of people relocated in selected countries varies between 75 in British Columbia to over 100,000 in Egypt, for a total of about 209,000 people that have been directly displaced throughout these eight countries alone. Many more that live in proximity to the dams or smelters have had their lives disrupted by the loss of access to and degradation of hunting or farming lands: as in James Bay, Brazil and British Columbia, and by the dams' exacerbation of diseases such as schistosomiasis, malaria, and river blindness: as in Egypt and Ghana. There are also severe secondary effects on the communities where displaced peoples are relocated, including overcrowding, competition for land and employment, incompatible cultural practices of different groups in the new community, and a variety of other urban problems caused by the large influx of construction workers and their families, as in Sumatra in Indonesia, Egyptian Nubia, and the Volta River Valley in Ghana.

The areas encompassing and surrounding these reservoirs contain unique wildlife habitats and agricultural areas which suffer irreversible changes as a result of the extensive flooding and alteration of water flow regimes. The diverse wildlife habitats include a large tract of Amazonian rainforest in the Tocantins river basin which almost definitely contained undiscovered species of plants and insects before being inundated by the mammoth Tucuruí reservoir, huge swaths of temperate forest and wetland along the La Grande, Opinaca, Caniapiscau and Eastmain rivers in James Bay, the largest remaining contiguous wilderness area in North America, and fertile riparian zones along the Nile, a river which cuts through thousands of miles of desert, and whose annual alluvial flood ecosystem has enabled the Nubians to farm for millennia. Aside from contributing to possible species extinction in tropical areas, reservoirs are a lower quality habitat than free-flowing rivers. Their water chemistry is lower in dissolved oxygen, a condition harmful to many fish. The decomposition of drowned vegetation often enables the release of toxic methylmercury which is absorbed by aquatic plants and fish, and bioaccumulates as it rises in the food chain, creating potential hazards for the fish, waterfowl and humans that feed at higher trophic levels. Reservoirs in relatively level terrain also create many square kilometers of shallow waters and a fluctuating shoreline, which leads to eutrophication and can also cause large migratory animals to get stuck in the mud. Dams also present a physical impediment to the migration of many fish species, some endangered. At this writing, few dams built in association with aluminum smelters contain fish ladders. Because large hydroelectric dams alter the seasonal patterns of freshwater flow from large rivers into estuaries, and therefore the balance of nutrient influx, ice formation, and water salinity, they may also cause long-term damage to the breeding grounds of migratory waterfowl. Adequate scientific research has yet to be conducted to determine the long-term effects of dams in various ecosystems.

Finally, two mechanisms operating in large reservoirs may also contribute to global warming. Flooding destroys forests, thus precluding the continual uptake of atmospheric CO_2 by trees and other terrestrial plants in undisturbed ecosystems. Anaerobically decaying vegetation in the inundated area also releases methane, a potent greenhouse gas. One researcher has speculated that smelters that rely on hydroelectric reservoirs with low ratios of power capacity to flooded area, such as the Akosombo in Ghana, may contribute more to global warming through methane emissions than comparable smelters using coal-fired electricity (33).

Tertiary effects of dams include damage to agricultural lands, for example the salinization of newly-irrigated lands in Egypt which had formerly relied on rainfall and annual floods. The availability of irrigation, electric power, and roads (with the concomitant access to goods and services they provide), enables the commercialization of agriculture, including the use of gasoline and pesticides in areas that had formerly used traditional farming methods without significant environmental consequences, and the possible acceleration of erosion. These changes are often accompanied by a transition from a diverse crop base for local or domestic consumption to a monocropping scheme focused on animal feeds or luxury export crops. The presence of dam and smelter access roads in previously roadless areas also opens up new opportunities for mining, logging, oil and gas drilling, illegal hunting, and human settlement in remote areas hosting sensitive wildlife communities and indigenous people.

Of course, the above paragraphs provide only a snapshot of the variety of negative effects caused by large dams. The number of displaced people and impounded area presented in Table 3 do *not* include the effects of smelter-related dams in at least 23 other areas or countries, including Argentina, China, Hungary, Iceland, Russia, the former Yugoslavia, the Pacific Northwestern United States, and the St. Lawrence River Valley that divides Quebec and New York State. Because so many of these smelters and dams are in Eastern Bloc countries and China, where accurate information has not been publicly available for decades, it is impossible to guess what the true extent of the environmental damage has been without extensive investigation. The figures also do not include flooding and relocation from future dam and smelter projects, including Quebec's Great Whale complex, Chile's Aysén Project, new dams on Venezuela's Caroní River, and any future projects in Brazil, the CIS, China, and elsewhere.

COST OF PRODUCTION

A 1988 study by the U.S. Bureau of Mines (summarized in Table 4 below) found that material inputs, such as alumina, coal, petroleum coke, soda, and lime, are the most expensive components in producing primary aluminum, accounting for about 33% of worldwide average total costs. The price of these commodities does not fluctuate

Table 4. Aluminum Production Potential and Costs for Selected Countries[a]

	Operating Costs								0% DCFROR[c]		15% DCFROR[c]		
Country	Annual Production[b] (000 tons)	Labor	Supplies/ Alumina	Energy	Overhead	Transport to Smelter	Net[d] Operating Cost	Recovery of Capital	Taxes	Total Cost	Taxes[e]	Return on Investment	Total Cost[f]
Australia	2,318	18.5	51.6	27.3	9.3	1.8	108.5	7.1	0.7	116.2	11.5	13.0	140.0
Brazil	2,407	11.0	62.4	39.7	10.8	0.9	124.8	9.7	0.9	135.4	4.4	19.6	158.5
Canada	3,057	24.7	56.9	11.9	8.6	1.3	103.4	7.3	1.1	111.8	19.2	13.0	142.9
France	639	29.5	75.4	45.2	15.4	0.9	166.4	9.9	0.7	177.0	11.0	13.4	200.8
F. R. Germany	1,518	39.5	73.0	56.0	18.5	2.4	189.4	8.2	0.9	198.4	15.7	12.8	226.0
India	610	32.2	64.6	73.0	18.5		188.3	5.3	0.2	193.8	7.9	8.8	210.3
Italy	499	41.2	75.0	46.7	17.6	0.7	181.2	9.5	0.9	191.6	10.4	12.1	213.2
Norway	1,624	38.6	67.0	24.9	14.1	2.6	147.3	6.8	0.4	154.5	12.3	11.7	178.1
Spain	681	22.7	83.8	56.7	16.8	0.2	180.1	7.9	0.2	188.3	4.9	11.5	204.4
Switzerland	126	41.0	56.9	33.1	16.3	4.9	152.1	13.7	0.7	166.4	6.4	16.3	188.5
United Kingdom	571	29.5	82.0	47.8	16.5	2.0	177.9	7.1	0.4	185.4	6.4	11.7	203.0
United States	579	28.2	47.4	45.6	14.3	1.5	137.1	3.7	1.5	142.4	5.5	8.2	154.5
Total or avg.[g]	26,840	24.7	56.0	36.4	12.3	1.5	131.0	6.6	1.3	138.9	11.5	12.3	161.4
Nonproducing operations													
Total or avg.[g]	2,410	21.8	63.1	22.9	10.4	0.9	119.0	0.4	7.1	126.5	67.9	58.9	246.3
All operations[g]													
Total or avg.,	29,250	24.5	56.4	35.3	12.3	1.5	130.1	6.2	1.8	138.0	15.7	18.7	170.6
Breakdown of costs by %		14.3	33.1	20.7	7.2	0.9	76.2	3.6	1.0	80.9	9.2	11.0	100.0

[a] Average 1988 U.S. cents per kilogram of aluminum; E. Sehnke and P. Plunkert, Ref. 4. Reprinted with the permission of the publications office of the U.S. Bureau of Mines.
[b] Based on total recovered aluminum divided by life of smelter.
[c] Discounted Cash Flow Rate Of Return.
[d] Costs may not total because of individual rounding.
[e] Taxes at 15% reflect higher revenues needed to obtain 15% return on investment.
[f] Total cost (15%) = Total cost (0%) + taxes (15%) − taxes (0%) + return on investment.
[g] Reported figures are totals or weighted averages of all operations considered in a U.S. Bureau of Mines study. Other countries included Argentina, Austria, Bahrain, Greece, Iceland, Indonesia, Iran, Mexico, the Netherlands, New Zealand, Qatar, Surinam, Turkey, the United Arab Emirates, and Venezuela.

dramatically by region or over short periods of time. Overhead, recovery of capital, transportation, profits, and taxes together accounted for 31.6% of total costs. Labor only accounted for 14% of production costs; the degree of training required limits how low wages can go, regardless of location. As the last line in Table 4 shows, energy is the second most expensive component of primary aluminum production, accounting for about 21% of total costs, although other studies have estimated this fraction to be as high as 30% (35).

The lion's share of energy costs are spent on the electricity used in smelting; a much smaller portion is fossil fuel expenditures used in mining and processing raw materials. By choosing where to build new smelters, aluminum corporations can exert some control over their electricity costs. Corporations seek regions where electricity sources are abundant and relatively cheap to develop, and where host governments have favorable industrial policies. The availability of local bauxite and other materials is of secondary concern.

PRODUCTION AND TRADE OF BAUXITE, ALUMINA AND ALUMINUM

Aluminum and its precursors, bauxite and alumina, are international commodities. A given ton of aluminum can be mined, refined, smelted, fabricated, sold, recycled and then refabricated in several different parts of the world. Because electricity expenditures comprise one fifth to one third of total production costs, while transportation costs from mine to smelter comprise less than 1% of costs, as Table 4 shows, it is standard practice to ship bauxite or alumina halfway around the world to take advantage of cheap electricity contracts, many of which are based on hydroelectricity. A Toyota Corolla, for example, may contain aluminum atoms that originated in bauxite mines on the west coast of Australia, were processed into aluminum ingot using hydroelectricity generated in the mountains of British Columbia, and were manufactured into auto parts in Japan before being shipped to the United States or Europe for sale.

The World's Bauxite Reserves

As previous sections pointed out, energy rather than materials availability is the limiting factor in the development of primary aluminum smelters. Unlike many other nonrenewable resources, the source material bauxite is not at all in short supply. In 1989, world reserves of bauxite were estimated to be 22 billion tons, up from an estimated 1 billion tons in 1945. This growth was due to increases in exploration leading to major bauxite discoveries, and to technological improvements which now make it possible to exploit lower-grade deposits previously classified as "subeconomic" reserves. Current estimates of the world's total bauxite resource base (which include subeconomic, hypothetical and speculative reserves), range from

50–70 billion tons. The world's largest known bauxite reserves are located in tropical and subtropical zones, with four countries accounting for 67% of world reserves: Guinea (5.6 billion tons), Australia (4.4 billion tons), Brazil (2.8 billion tons), and Jamaica (2 billion tons). The United States has a very small economic reserve base in comparison: about 40 million tons, all in Arkansas. Deposits in Hawaii, Washington and Oregon amount to an additional 100 Mt, but they are considered subeconomic at this time (36).

The Main Producing Countries

Bauxite is mined in 28 countries, and alumina is processed in 25 countries. About 90% of world production is refined to alumina; the rest is used in refractories and abrasives. As Table 5 shows, world bauxite production in 1992 was 103.6 million tons. The top five producers alone accounted for more than three quarters of the total output, or over 80 Mt. U.S. production is small by comparison: only half a million tons in 1988, or less than 1% of world production; but it did import 10.6 Mt of crude and dried (metallurgical grade) bauxite in 1989 (37).

Alumina production is similarly stratified; the top five producers' output was 22.8 Mt, or 59% of world production. The top five primary aluminum producers are the United States, Russia, Canada, Brazil and Australia, together producing 11 Mt, or 58% of the world total. In 1992, there were fifty aluminum-producing countries, up from forty a few years prior (the increase is due to the breakup of the Soviet Union and Yugoslavia, and the establishment of new independent republics with existing smelter capacity). The number of producing countries will grow should several smelter proposals come to fruition in countries with no existing capacity.

Since the global oil shocks of the mid-1970's, the fossil fuel fired smelters in the United States, Western Europe and Japan have suffered from price increases. Many of these smelters have been shut down entirely; others have become "swing capacity," mothballed until aluminum market prices are high. Even smelters using hydroelectricity in these countries have suffered because as fossil fuel prices rise, the demand for cheaper hydroelectricity also rises. Because the hydroelectric resources in these countries have been largely exhausted, hydroelectric prices have also been driven up. Corporations whose smelters had become uneconomical in the United States, Europe and Japan began to look abroad more favorable locations. They have collaborated with state-owned and private investor groups to build many new smelters in Australia, Brazil, and Canada—and to a lesser extent in the Middle East where natural gas abounds. Coal reserves are almost limitless in Australia, while Brazil and Canada have vast hydroelectric resources. All three countries have governmental policies (at the national and state levels) which strongly support the development of the aluminum industry.

While the bauxite-alumina-aluminum production process is vertically integrated (a few large companies control all stages of production, manufacture, and sale), it is not geographically integrated. For decades, private transnational aluminum corporations have traveled the globe to selectively exploit the cheapest resources. Certain regions of the world have been favored to mine bauxite and refine alumina inexpensively; other regions have been chosen to smelt aluminum using cheap local energy resources. A few countries have been picked to do both. "Cheap" is a relative term, however. The cost of mining or energy production is low in certain locations not only because larger or more accessible reserves exist, but because environmental regulation is often lax or non-existent, and cash-poor governments have made numerous concessions on taxes and tariffs in order to attract investment. Subsidization is common, and varied in the forms it takes. In many cases, smelters' electricity tariffs are substantially lower than rates charged to other industrial users, and are sometimes even below the cost of production. Contract periods often extend for several decades without provisions to raise rates in accordance with changing economic conditions. Other forms of government assistance include

Table 5. Principal Bauxite, Alumina, and Primary Aluminum Producing Countries[a]

Bauxite (1992)[b]	Mt	%	Alumina (1989)[c]	Mt	%	Aluminum (1992)[d]	Mt	%
1. Australia	39.95	38.6	1. Australia	10.80	27.9	1. United States	4.04	21.0
2. Guinea	13.77	13.3	2. United States	4.67	12.1	2. Russia	2.70	14.0
3. Jamaica	11.30	10.9	3. U.S.S.R.	3.50	9.0	3. Canada	1.95	10.1
4. Brazil	10.80	10.4	4. Jamaica	2.22	5.7	4. Brazil	1.22	6.3
5. India	4.48	4.3	5. Brazil	1.60	4.1	5. Australia	1.20	6.2
6. Russian Fed.[e]	4.00	3.9	6. Suriname	1.57	4.1	6. China[f]	0.95	4.9
7. Suriname	3.25	3.1	7. China[f]	1.44	3.7	7. Norway	0.81	4.2
8. China	3.00	2.9	8. India	1.42	3.7	8. Germany	0.60	3.1
9. Guyana	2.30	2.2	9. Canada	1.40	3.6	9. Venezeula	0.60	3.1
10. Greece	2.10	2.0	10. Venezuela	1.24	3.2	10. India	0.50	2.6
Subtotal, top 10	94.95	91.6	Subtotal, top 10	29.86	77.2	Subtotal, top 10	14.57	75.8
18 Other Countries	8.68	8.4	15 Other Countries	8.84	22.8	40 Other Countries	4.65	24.2
World total	103.63	100.0	World total	38.70	100.0	World total	19.2	100.0

[a] Mt = production in millions of tons; % = percent of total world production.
[b] *World Resources 1994–1995,* World Resources Institute, Oxford University Press, 1994.
[c] E. Sehnke and P. Plunkert, Ref. 4.
[d] Aluminum Statistical Review for 1992, Ref. 1.
[e] Data for the newly independent republics of the former U.S.S.R. and Yugoslavia were first collected separately in 1992.
[f] Estimate.

corporate and private income tax exemptions, free or low-cost rights to land and watercourses, and the provision of supporting infrastructure such as roads, railways, deep-water ports, and even hydroelectric dams.

By studying Table 5, it becomes apparent that countries with vast reserves of bauxite *and* cheap energy (such as Australia, Brazil and the former USSR) tend to dominate in all three production stages: bauxite, alumina and aluminum. Because these countries do not have substantial domestic markets, they are also the most prolific exporters of aluminum. Nations that possess large supplies of bauxite but have no energy sources are also major exporters, notably Guinea and Jamaica. Countries without substantial bauxite but with cheap energy and/or access to the major consuming markets tend to be in the top echelon of alumina and aluminum producers, clearly relying on imported bauxite and/or alumina. Canada and the United States are examples of the latter category, but the long-term future of the U.S. industry is threatened by rising energy costs, in part due to environmental considerations. Because the major producers of bauxite are not necessarily the major aluminum producers, it is evident that the nations with the highest per capita consumption rates, such as Japan, the U.S., and several European nations, reap the economic benefits of adding value by processing domestic and imported ingot into semifabricated and finished products. In effect, the transnationals whose ownership and management are based in the U.S., Europe and Japan are importing low-cost energy in the form of aluminum ingot, and exporting the environmental costs of primary aluminum smelting to the newly producing countries.

Production by Private and State-Owned Corporations

As Table 6 illustrates, ten principal transnational and state-run corporations account for almost 60% of world capacity. In 1990, the top five alone accounted for 43% of world capacity: the (former) Soviet Government, Alcoa, Alcan, the Chinese Government, and Reynolds. During the last decade, a number of minor producers have made significant inroads into the market, eroding the virtual supremacy of what used to be called "the Big Six," or Alcan, Alcoa, Kaiser, Reynolds, Pechiney and Alusuisse.

Although most of these corporations have their roots in one country (for example Alcoa in the United States, Pechiney in France, Norsk-Hydro in Norway, and Alcan in Canada), they all operate well beyond their countries of origin, as do smaller corporations. With the exception of the Soviet and Chinese governments, the top ten producers share access to capital. (In the future, they may also do so with aluminum companies in Russia and China as these countries seek to expand their economic opportunities by privatizing state-owned enterprises, often as joint ventures with foreign corporations and investors.) In conjunction with many smaller companies, the Big Six have convened complex webs of investment to finance smelters. It is not uncommon for competitors to cooperate to secure new projects by dividing up the equity shares in a smelter, or by divvying up different tasks, such as providing financing, technology, or the management of operations. It is also common for a main producer to team up with a host government in financing and operating smelters. In other cases, state governments are the majority owners, but clearly they depend on a major corporation for technology transfer, and will give the corporation a few shares as part of the consulting fee.

Table 6. Top Ten Aluminum Corporations (Private and State-owned), 1991[a]

Corporation	Ingot Capacity, t	% of World Capacity
1. Soviet government	2,740,000	13.7
2. Alcoa	1,867,560	9.3
3. Alcan	1,702,000	8.5
4. Chinese government	1,315,000	6.6
5. Reynolds Metals	1,023,500	5.1
6. Pechiney (French government)	851,000	4.2
7. Norsk-Hydro	623,000	3.1
8. Amax	541,630	2.7
9. Maxxam (includes Kaiser)	528,710	2.6
10. Venezuela	498,500	2.5
Total	11,690,900	58.3
World ingot capacity, 1991	20,042,000	100.0

[a] Ref. 38. World capacity figures do not correlate with production figures in Table 5 because smelters which rely on expensive fuels are used as "swing capacity," only operating when world demand is high.

COMPARATIVE RATES OF CONSUMPTION

With an apparent consumption rate of 7.1 Mt of aluminum in 1992, the United States is the largest producer and consumer in the Western Hemisphere, producing about 4 Mt and importing about 1.6 Mt of primary aluminum. (Apparent consumption is also a function of total exports and recovered scrap). This corresponds to an annual per capita consumption rate of 28 kilograms. The U.S. was surpassed only by Japan, whose 1992 per capita consumption rate was 29.4 kg. Other major per capita consumers included Austria (23.4 kg), Canada (23.2 kg), Sweden (17.6 kg), Switzerland (17 kg), and Australia (16.5 kg). These nations enjoy some of the highest standards of living in the world. Their transportation, urban and manufacturing infrastructures are well-developed, and they enjoy greater rates of personal car ownership than do less-developed nations, or nations with extensive mass transit systems. These nations also tend to have the highest consumption rates for packaging and beverage cans; the average American consumed 270 cans in 1989, while the average was 50 in Asia and 36 in Europe (1,39). Poorer countries such as El Salvador, Turkey, Argentina and Brazil had per capita aluminum consumption rates ranging from 0.3 to 3 kg in 1990–1991, while countries with moderate per capita consumption included Spain, New Zealand and Belgium, with rates of 7–12 kg.

THE FATE OF DISCARDED ALUMINUM

After aluminum consumer products have become obsolete, they are discarded, then either landfilled, burned, or recycled as "scrap" into new products. Because about 95% of the total energy used to make primary aluminum is not needed to produce secondary aluminum, including *all* the electrical energy, recycling saves the producer a tremendous amount of money. Scrap also provides a guaranteed supply of domestic aluminum which can serve as a partial buffer against periodically uncertain international market conditions. These two benefits have made scrap aluminum play an important role in total aluminum production. In 1992, for example, a full third of the total U.S. aluminum supply (primary and secondary production plus imports) was comprised of scrap—an amount equivalent to 68% of the primary aluminum produced. In Europe, about 35% of the supply was comprised of scrap in 1990. A 39-country survey conducted by the Aluminum Association, a U.S. trade organization, showed that 5.75 million tons of scrap were recovered for recycling that year, or 43.2% of the amount of primary aluminum produced in those countries. If the ratio of scrap recovery to primary aluminum production is similar in the rest of the world, then approximately 7.7 Mt of scrap may have been recovered worldwide in 1990 (1,40).

Sources of Scrap for Recycling

There are two sources of secondary aluminum: old scrap and new scrap. There are also two types of new scrap: "home scrap" is generated by the plant that produced the aluminum; it is remelted in-house without ever having entered the market. "Prompt industrial scrap" consists of cuttings, end pieces, mistakes and product overruns. It is generated in the plant of a fabricator or manufacturer of aluminum products, and is uncontaminated by debris or unwanted metals. New scrap has always been recycled. "Old scrap" or "post-consumer scrap," on the other hand, consists of products that have been purchased by consumers, and have later become old or obsolete, then discarded (41). In the United States, the ratio of old scrap to new was 1.4 to 1 in 1992, and has been rising by about 4% per year since 1982. The 39-country survey mentioned earlier did not break the scrap down into the categories of old and new scrap, which is unfortunate because only the old scrap number reveals how much aluminum is being kept out of landfills worldwide.

Landfill Rates and Replacement Production

The amount of primary aluminum produced annually worldwide has increased by 6 million tons since 1974 (1). This is roughly equivalent to the amount of aluminum landfilled every year. There is reason to believe that more than five Mt of aluminum are being landfilled annually, due to inadequate recycling opportunities, insufficient public education, and low market prices that discourage more aggressive recycling.

Precise data on the amount of aluminum landfilled worldwide do not exist, but one can try to extrapolate based on United States data. The U.S. Environmental Protection Agency has estimated that 1.54 Mt of aluminum were landfilled in 1988 in the United States alone—a quantity that surely would have astounded Karl Bayer a century earlier (42). This was equivalent to 72.6% of all the scrap recovered that year, and to 39% of the primary aluminum produced. If it is assumed that similar percentages apply to the rest of the world, then between 5.5 and 7 Mt of aluminum may be doomed to landfill burial each year, an amount equivalent to the production of about twenty-nine 250,000-ton primary smelters. Were this estimate five times too large, it would be no less appalling.

These numbers are backed up in part by data on used beverage container (UBC) recycling rates in selected countries including the United States. In 1989, beverage cans accounted for 11% of Western Europe's total aluminum consumption of about 4 Mt. In some industrialized nations with high per capita consumption rates, such as the United Kingdom, recycling programs have been widely developed only in the last five years, leaving much aluminum in the garbage. In 1989, the rate of UBC recycling was only 5% in the United Kingdom, 20% in Greece, and 22% in Switzerland. Where recycling programs have been in place longer, the 1990 UBC recovery rates are higher: 61% in the United States, 63% in Canada, and 87% in Sweden, where recycling is mandatory. In 1990 Japan initiated a Y100 million campaign to boost its UBC recycling rate from a dismal 40%. European aluminum producers kicked off similar campaigns in 1989. These "green" moves have been calculated to capitalize on the public's environmental sentiments, in an attempt to win market share from steel cans (40,43). Such a shift took place in the 1970s and early 1980s in the United States.

The used beverage container is the only aluminum product that is tracked from production through recycling, due to the distinct "closed-loop" nature of the recycling process: UBC is usually collected separately from other aluminum scrap, and is remelted, rolled into can sheet, and made into brand new cans which may be back on the shelves within a few weeks of being collected at curbside or deposited at a recycling center. Despite the fact that nine U.S. states have some form of beverage container deposit legislation, and that convenient curbside recycling programs are now available to about 100 million people in thousands of American cities, a great deal of cans still go to waste. As Table 7 illustrates, 534 thousand tons of beverage cans were landfilled in 1993. The financial losses from not recycling this aluminum exceed half a billion dollars. Because the rate of aluminum consumption has continued to grow each year, all landfilled aluminum must be replaced by a similar quantity of primary aluminum: in this case, a quantity equivalent to the maximum annual production capacity of four smelters in New York, Oregon, and Montana. The energy requirements to do this are prodigious, including approximately 8 TWh of electricity, an amount which could power more than 800,000 average U.S. households for a year. As earlier sections of this article pointed out, there are also airborne emissions and solid and liquid effluents associated with this "replacement" production.

The recycling rates of other aluminum products have not been tracked because they are mixed together at the point of collection, and as previously stated, because their useful lifetimes range from several years to many decades. It is difficult to be more precise about the amount of aluminum landfilled. Because aluminum comprises only 2.3% of all U.S. municipal solid waste by volume, and only 1.1% by weight, there is little incentive for the governmental agencies who commission waste composi-

Table 7. Landfilled Beverage Cans and Wasted Production, 1993[a]

	# of Cans	Million Pounds	Metric Tons
U.S. beverage can production (29.53 cans/lb)[a]	94.3 billion	3,193	1,448,491
Cans recycled (63.1% of total production)[a]	59.5 billion	2,015	913,998
Cans landfilled (36.9% of total production)	34.8 billion	1,178	543,493
Annual production capacity of four U.S. smelters[b]			
Alcoa Aluminum, Massena, New York			127,000
Reynolds Metals, Massena, New York			123,000
Reynolds Metals, Troutdale, Oregon			121,000
Columbia Falls Aluminum, Columbia Falls Montana			163,000
Total capacity			534,000
Foregone revenues by not recycling 534,493 tons aluminum (at $1,014/ton):[a]			$542 million
Electricity required to replace this quantity with primary aluminum (at 15 kWh/kg):			8.0 TWh
World consumption of electricity from natural gas for primary aluminum:[c]			9.5 TWh
Approximate # of U.S. homes 8.0 TWh could power for a year (at 1.127 kW/home):			812,092

[a] Ref 44.
[b] Ref. 45.
[c] International Primary Aluminum Institute, Ref. 30.

tion studies to be more accurate in examining the contribution of aluminum.

The Role of Aluminum Companies in Initiating Recycling

Without the efforts of the aluminum industry, recycling of all sorts would not have progressed as far as it has today. After World II, the intense recycling efforts that had been sustained in the national interest were abandoned. Disposal became equated with peacetime and prosperity. It was not until the growing environmental consciousness of the 1960s, coupled with a slow rise in energy prices, that aluminum recycling experienced a renaissance, at least in the United States.

Because the energy savings from recycling aluminum are so great, the aluminum industry encouraged recycling before it was environmentally popular, and long before the current landfill crisis was felt. The industry was motivated not only by energy savings, but by a scarcity in supply. As the all-aluminum beverage can gained in popularity during the 1960s and 1970s, it took over market share from glass bottles and steel or bi-metal cans. Although the shift was promoted by primary aluminum producers to keep wartime capacity in production, the growth was so rapid that they could not keep pace with the demand. Recycling was a way of ensuring adequate supply for continued can production. It became an insurance policy against an interruption in primary supplies due to geopolitical conflicts, uncertain smelter contracts, and rising energy prices.

In the United States, Reynolds, Coors and Alcoa were especially instrumental in establishing buyback centers for cans in the late 1960's and early 1970's, when no other recycling options were available. They distributed pamphlets touting the benefits of recycling, and developed programs that encouraged the participation of schools and community groups. These early recycling programs helped to re-establish recycling in the public consciousness, and paved the way for the collection of other materials, such as newspapers, steel cans and glass bottles. At first the collection efforts were spurred by volunteers with minimal cooperation from producers, but as the shortage of landfill space has intensified, local and state governments have taken on the operation of many large-scale recycling programs. This trend is also evident in Europe and Japan.

Obstacles to Recycling: Artificially Low Prices and Inadequate Public Education

It is clear from the foregoing discussion that aggressive recycling efforts are still needed to keep this valuable metal out of the trash. A 1991 survey of secondary smelter operators in the United States revealed that the single most important obstacle to increased secondary recovery is the low price of aluminum, especially that of non-UBC scrap. As the primary price rises, the competitive position of secondary aluminum improves; as the primary price drops, more aluminum is landfilled. Three quarters of survey respondents thought that the U.S. EPA estimate of 1.5 million tonnes of aluminum landfilled annually was fairly accurate. Some respondents said that market volatility discourages a number of would-be scrap collectors, and that the absence of consistent operating margins discourages scrap processing. The lack of mandatory recycling legislation, and the fact that many recycling opportunities are too far away, too inconvenient, or under-publicized, were also cited as lesser obstacles to higher aluminum recovery rates (46).

There may be yet another obstacle to increased secondary recovery: inadequate public awareness of the environmental effects of primary aluminum production. In recent years, the aluminum industry has aggressively used images of recycling to present itself as a "green" industry, without telling the public the full range of environmental costs of producing primary aluminum. Advertisements in major newspapers such as *USA Today* and London's *Financial Times,* and in numerous environmentally-oriented journals, employ abstract and oversimplified "energy savings" arguments to encourage aluminum can recycling, and to promote the use of aluminum packaging based solely on its recyclability. Using full-color pictures of butterflies and smiling little girls, they also relay anecdotes about how the revenues from various aluminum can recycling programs have been used to bolster civic organizations in small towns, to fund school tree-planting projects, and to provide medical care for marine mammals. While it is commendable for the aluminum industry to encourage the public to recycle beverage cans for all of these reasons, it may not be enough. For one, information about recycling non-UBC scrap is not widespread. These "feel-good" advertisements also convey the overall impression that aluminum consumption does not have serious environmental impacts. Even if beverage cans were recycled at rates approaching 90%, there would still be a need for primary aluminum production to replace the amount landfilled, and to contribute to many other non-beverage can end uses.

CONCLUSION

If primary aluminum production continues to grow at rates even approaching those of the late 20th century, the environmental repercussions will be severe. More rivers will be dammed, and more power plants constructed to burn fossil fuels that threaten an already-burdened global atmosphere. The problem of solid waste disposal may grow ahead of the industry's ability to control it, despite promising research and development. The extent of these problems will be determined by the rate of demand growth, which in turn will be determined by at least four factors operating at a global spatial scale:

1. Absolute population growth,
2. Growth in per capita consumption of existing products using aluminum,
3. The increasing proportion of aluminum to other materials in various uses (such as cars), and
4. The worldwide rate of aluminum recycling.

Because each of these factors is complex, it is impossible to predict future demand growth with any precision. What is known, however, is that the world population is expected to double by the year 2020, and that as historically closed socialist markets such as Russia and China open their doors to capitalist goods, their per capita consumption will grow substantially. In the interest of minimizing environmental damage in light of the first and second factors, it is useful to ask if the third and fourth factors can be controlled.

Factor 3 entails questioning the uses of aluminum. Comparative lifecycle analyses are being used to evaluate which materials are most environmentally benign for certain end uses. For example, many investigators are convinced that refillable glass or plastic bottles have lower overall environmental impacts for delivering single-serving drinks than aluminum beverage cans, even taking aggressive recycling into account. On the other hand, it is probably desirable to increase the proportion of aluminum to steel in transportation vehicles because its lighter weight enables a significant savings of fossil fuels over the life of a vehicle. These questions cannot be answered simply, but they do deserve further inquiry.

Changing factor 4 will demand that we re-examine the institutional framework in which people in society manufacture, consume and dispose of products. As previously stated, only a handful of states have deposit legislation on the books, and recycling rates are higher in these states than in those lacking such legislation. Rather than fight for new deposit laws on a state-by-state basis, many activists and policymakers are calling for a federal "bottle bill" to encourage recycling nationwide. Others are advocating "advance disposal fees" (ADF's), or taxes on the producers of manufactured goods which are tied to the amount of their products which end up in the wastestream. The present system removes manufacturers' responsibility for disposal, making it the sole fiscal burden of city and county governments. ADF's would pass the cost of disposal from the general taxpayer on to the consumer of products. Still others want to see externalities incorporated into the price of energy and primary materials production; carbon taxes and the removal of subsidies for the acquisition of land, water, and raw materials are examples. Again, these measures would pass the responsibility of paying for resource extraction from the public on to the consumer of finished goods.

Finally, many activists believe that certain areas should be declared off-limits to resource development for environmental and social reasons, including various wild rivers, coastal environments serving as wildlife refuges, and forests with importance for global biodiversity and the preservation of indigenous cultures. If more areas were protected from energy and mineral development, the price of energy and materials would increase due to competition for a more limited resource base. This in turn would hasten the transition from a global economy based on primary commodities to one that relies more heavily on secondary materials.

BIBLIOGRAPHY

1. "Aluminum Statistical Review: Historical Supplement," The Aluminum Association, Washington, D.C., 1982; "Aluminum Statistical Review for 1992," The Aluminum Association, Washington, D.C., 1993.
2. World Resources Institute, "World Resources 1993–94." Oxford University Press, 1994, Table 21.4, pp. 338–339.
3. W. Haupin, "Aluminum," *Encyclopedia of Physical Science and Technology,* Vol. 1, 2nd ed., Harcourt Brace Jovanovich, New York, 1992.
4. "The Next Phase for Aluminum," *The Mining Journal (London)* **312**(8002) (Jan. 13, 1989); E. Sehnke and P. Plunkert, "Bauxite, Alumina, and Aluminum," *Minerals Yearbook 1989,* U.S. Department of the Interior Bureau of Mines, Washington, D.C., Nov. 1990.
5. S. Y. Shen, *Energy and Materials Flows in the Production of Primary Aluminum,* Prepared by Argonne National Laboratory for the Department of Energy, (ANL-CNSV-21), Oct., 1981, pp. 2–5; I. J. Polmear, ed., *"Light Alloys: Metallurgy of the Light Metals,"* 2nd ed., Edward Arnold, London, 1989, p. 8.
6. E. Sehnke and P. Plunkert, Ref. 4, pp. 23–26.
7. B. Richardson, "Super-Companies," National Film Board of Canada, 1990; S. Y. Shen, Ref. 5, p. 2.
8. R. N. Crockett, "Bauxite, Alumina and Aluminum," Mineral

Resources Consultative Committee, Mineral Dossier No. 20, Institute of Geological Sciences, London, 1978, p. 17.

9. H. F. Kurtz and L. H. Baumgardner. "Aluminum," in *Mineral Facts and Problems,* U.S. Bureau of Mines, U.S. Government Printing Office, Washington, D.C., 1980, p. 18.
10. Ref. 9, p. 19; N. Jarrett, "Aluminum Production," *Encyclopedia of Materials Science and Engineering,* p. 166.
11. Ref. 8, p. 17; Ref. 9, p. 19; N. Jarrett, Ref. 10, p. 166; E. Sehnke and P. Plunkert, Ref. 4, p. 27; A. S. Russell, "Aluminum," in *McGraw-Hill Encyclopedia of Science and Technology,* 1987, p. 417.
12. S. Y. Shen, Ref. 5, p. 3; S. H. Patterson and J. R. Dyni, "Aluminum and Bauxite," in *United States Mineral Resources,* U.S. Geological Survey Professional Paper 820, U.S. Government Printing Office, Washington, D.C., 1973, p. 36.
13. S. Y. Shen, Ref. 5, p. 12.
14. S. Y. Shen, Ref. 5; I. J. Polmear, Ref. 5, p. 11; "The Story of Aluminum." The Aluminum Association, Washington, D.C., 1990; "Environmental Aspects of Aluminum Smelting: A Technical Review." Industry and Environment Office, United Nations Environment Programme (UNEP), 1981; "Draft: Pollution Prevention and Abatement Guidelines for the Aluminum Industry." Prepared by ICF Kaiser Engineers International, USA, for United Nations Industrial Development Organization (UNIDO) *January 1993 (Project: US/GLO/91/202-UNIDO/World Bank Industrial Pollution Guidelines).*
15. I. J. Polmear, Ref. 5, p. 11; E. Sehnke and P. Plunkert, Ref. 4, p. 1; S. Y. Shen, Ref. 5, p. 14; B. Richardson, Ref. 7.
16. S. Y. Shen, Ref. 5, p. 19.
17. *Environmental Aspects of Aluminum Smelting: A Technical Review,* Industry and Environment Office, United Nations Environment Programme (UNEP), 1981.
18. N. E. Richards, "Role of the Anode Effect in Alumina Reduction, *Proceedings, Workshop on Atmospheric Effects, Origins, and Options for Control of Two Potent Greenhouse Gases: CF_4 and C_2F_6,* sponsored by the Global Change Division of the U.S. Environmental Protection Agency Office of Air and Radiation, April 21–22, 1993, Washington, D.C.
19. *Review of New Source Performance Standards for Primary Aluminum Reduction Plants,* U.S. Environmental Protection Agency, Office of Air Quality Planning and Standards, Research Triangle Park, N.C. (EPA-450/3-86-010) Sept. 1986, pp. 3–29.
20. Y. Lallement and P. Martyn, "Industrial Sustainable Development in the Context of the Primary Aluminum Industry," *UNEP Industry and Environment,* July–Dec. 1989.
21. R. Huglen and H. Kvande, "Global Considerations of Aluminum Electrolysis on Energy and the Environment," Hydro-Aluminum a.s., Stabekk, Norway. (TMS 1994 conference paper, location unknown).
22. R. Pardy, and co-workers, "Purari Overpowering Papua New Guinea?" The International Development Action for Purari Group, IDA, Fitzroy, Victoria 3065, Australia, 1978, p. 181; "Asahan Dam: Energy for Whom?" *Environesia,* **3**(2), (June 1989), Published by WALHI, The Indonesian Environmental Forum, Jakarta, 1989.
23. *Primary Aluminum Draft Guidelines for Control of Fluoride Emissions from Existing Primary Aluminum Plants,* U.S. Environmental Protection Agency, Office of Air Quality Planning and Standards, Research Triangle Park, NC 27711 (EPA-450/2-78-049a), Feb. 1979.
24. H. I. Schiff, "The Aluminum Industry's Contribution to Atmospheric Fluorocarbons," *Proceedings, Workshop on Atmospheric Effects,* in Ref. 18.
25. A. R. Ravishankara and S. Solomon, "Atmospheric Lifetimes and Global Warming Potentials of CF_4 and C_2F_6." *Proceedings, Workshop on Atmospheric Effects,* in Ref. 18; and I. S. A. Isaksen and co-workers, "An Assessment of the Role of CF_4 and C_2F_6 as Greenhouse Gases," Center for International Climate and Energy Research, Oslo, September 1992.
26. J. Hoffman, "Introduction," *Proceedings, Workshop on Atmospheric Effects,* in Ref. 18.
27. The Aluminum Association, *Proceedings of the Workshop on Storage, Disposal and Recovery of Spent Potlining,* Washington, D.C., December 3–4, 1981; "Draft: Pollution Prevention and Abatement Guidelines for the Aluminum Industry," Prepared by ICF Kaiser Engineers International, USA, for United Nations Industrial Development Organization, (US/GLO/91/202-UNIDO/World Bank Industrial Pollution Guidelines), Jan. 1993.
28. H. A. Øye and R. Huglen, "Managing Aluminum Reduction Technology: Extracting the Most from Hall-Héroult," *JOM,* November 1990.
29. *Energy Efficiency in U.S. Industry: Accomplishments and Outlook,* A Report for the Global Climate Coalition by the EOP Group, Inc., Oct. 1993, p. 49.
30. *World Energy Outlook to the Year 2010,* International Energy Agency, Paris 1993; *Electrical Power Utilization Annual Report for 1990,* International Primary Aluminum Institute, London, 1991.
31. World Resources Institute, "World Resources 1990–1991," Oxford University Press, 1990, pp. 318–319.
32. *Statistical Yearbook of the United States: 1992,* 112th ed., U.S. Bureau of the Census. Washington, D.C., 1992, Tables 62, 937.
33. D. Abrahamson, "Greenhouse Gas Production From Aluminum Production: Some Policy Implications." *Proceedings, Workshop on Atmospheric Effects,* in Ref. 18.
34. J. Gitlitz, "The Relationship Between Primary Aluminum Production and the Damming of World Rivers." IRN Working Paper #2, International Rivers Network, Berkeley, Calif., 1993.
35. D. R. Wilburn and D. A. Buckingham, "Assessing the Availability of Bauxite, Alumina and Aluminum through the 1990's," *JOM* (Nov 1989).
36. E. Sehnke and P. Plunkert, Ref. 4, p. 25; Patterson Ref. 12, p. 39.
37. E. Sehnke and P. Plunkert, Ref. 4, pp. 1–8; R. N. Crockett, Ref. 8, p. 11.
38. *Primary Aluminum Plants Worldwide, Part 1—Detail, U.S.* Bureau of Mines, Dec. 1990.
39. "Aluminum Optimism," *The Mining Journal* **312**(8020), 393 (May 19, 1989).
40. K. Gooding, "Aluminum Producers Woo the Green Vote," *Financial Times,* London, June 16, 1989. Calculation: 43.2%. * 17.8 Mt (world production, 1990) = 7.7 Mt.
41. P. Plunkert, "Aluminum Recycling in the United States—A Historical Perspective," Presented at *TMS 2nd International Symposium-Recycling of Metals and Engineered Materials,* Williamsburg, Va., Oct. 28–31, 1990, p. 2.
42. *Characterization of Municipal Solid Waste in the United States, 1990 Update,* U.S. Environmental Protection Agency, Office of Solid Waste and Emergency Response, Washington, DC, (EPA/530-SW-90-042), June 1990.
43. "Japanese Aluminum Battle for Can Market." *Mining Journal* **314**(8074), 460 (June 8, 1990).
44. *Resource Recycling,* June 1994, Table 4, p. 26.

45. *Primary Aluminum Plants Worldwide, Part II-Summary,* U.S. Bureau of Mines, Washington, D.C., 1990.
46. J. S. Gitlitz, unpublished survey, 1991.

ANTIKNOCK COMPOUNDS

RAMON ESPINO
Exxon Research and Engineering
Annandale, New Jersey

Organic or inorganic compounds that are added to gasoline in very small amounts to reduce engine knock, are called antiknock compounds. Work done at General Motors in 1921 has identified tetraethyl lead as the most effective antiknock compound. One problem with lead compounds is that they deposit lead oxide in the combustion chamber. This limitation has been overcome with the addition of ethyl halides (bromo and chloro) which act as scavengers and volatilize the lead in the combustion chamber. Since the 1920s, lead compounds have been widely used as antiknock compounds. Other antiknock additives, both organic and inorganic, have been identified, but none approach lead alkyls in terms of cost effectiveness. See also AUTOMOBILE ENGINES; TRANSPORTATION FUELS, AUTOMOTIVE FUELS; CETANE NUMBER; OCTANE NUMBER.

Concerns surfaced regarding the toxicity of lead immediately after its introduction. In 1925 the U.S. Surgeon General undertook an investigation into the potential danger to the public health of using gasolines containing lead. In 1926 the Surgeon General decided that lead compounds could be used as long as the tetraethyl lead was handled in sealed containers that minimize inhalation and skin contact with the lead additive. Similar concerns on the health impact of lead compounds arose in Great Britain in the late 1920s, and a government inquiry cleared the product, but the acceptance of lead antiknock compounds was less rapid outside the U.S. Concerns about lead did not totally disappear and in the 1960s a number of studies showed that lead was toxic, particularly to children. Concerns about the potential health impact of breathing exhaust fumes containing volatile lead compounds started the trend to reduce the lead content of gasoline. The pressure to reduce lead was also driven by the fact that the catalyst in the exhaust emissions reduction system was poisoned by lead. Since controlling vehicle emissions was a major component of the strategy to improve air quality, States and the Federal Government legislated a program to phase lead out of gasoline. The program started in the mid 1970s and by 1990 only a very minuscule fraction of the gasoline sold in the U.S. contains lead. The rest of the world has followed this trend but at a slower pace since the use of catalytic systems to reduce auto emissions were not required to meet pollution standards. The "phasing out" of lead from gasoline has accelerated in the 1990s with stricter air pollution laws which require catalytic converters for many automobiles.

The most common lead antiknock compounds are tetraethyl lead (TEL) or tetramethyl lead (TML). The product is actually a blend of either of these lead alkyls with about 18% 1.2 dibromoethane and 1.2 dichloroethane each. TML contains about 12% of a diluent to ensure that the lead content is equal to TEL, 39.4% by weight. A mixture called CR-50 contains both TEL and TML with again the lead content fixed at 39.4%. The response of gasolines to lead alkyls varies depending on the octane of the lead-free gasoline. For example, a gasoline with RON of 72 can reach a RON of 85 when the lead content is 0.5 grams per liter of gasoline. However a gasoline with initial RON of 90 only improves 5 RONs with the same amount of lead. The selection of the appropriate alkyl lead compound depends on the gasoline type. Regular gasolines normally contain TEL while premium gasolines use mixtures of TEL and TML or CR-50. However, this commentary is a simplified picture of the selection options that a gasoline marketer has in terms of lead antiknock compounds.

The search for antiknock compounds with as much potency as alkyl leads has been very exhaustive. In 1958, the Ethyl Corporation commercialized a successful alternative: MMT-methyl cyclopentadienyl manganese tricarbonyl. MMT is in fact more effective than TEL at low concentrations. MMT's effectiveness falls rapidly at concentrations above 0.2 g of metal (manganese) per liter. Since MMT also causes buildup of deposits in the engine and also poisons the exhaust, emission catalyst is seldom used at concentrations above 0.2 g/L. During the "lead phase-down" period in the USA, many low lead gasolines used MMT as an additional antiknock compound. MMT is widely used outside the USA, but its use is decreasing for the same reason that alkyl lead compounds decreased.

Other metals have been shown to function as antiknock compounds. For example, iron as pentacarbonyl, nickel as a carbonyl derivative and tin as tetraethyl tin all suffer from the same limitations: combustion and spark plug deposits and toxicity. Since they are less effective than lead or manganese and have the same drawbacks, none have found a market.

The search for organic antiknocks has been long and strenuous and largely unsuccessful. The clear advantage of an organic antiknock compound is the elimination of all the deposit and catalyst poisoning drawbacks of metallic compounds. Some organic compounds, particularly aromatic amines, do function as antiknock compounds, but the amount required is very high. For example *n*-methylaniline (NMA) has to be used at 1% concentration to match the performance of TEL at 0.1 g/L. Aromatic amines have an additional undesirable feature: During combustion the nitrogen atom is converted to nitrogen oxides, a major air pollutant.

Oxygenated compounds have higher octane numbers than gasoline. Methanol has a blending RON of 127–136 depending on the gasoline. Ethanol ranges from 120–135 and Methyl *tertiary* butyl ether (MTBE) ranges from 115–123. These compounds are used in amounts greater than 1% oxygen content; therefore, it is not appropriate to call them antiknock compounds.

BIBLIOGRAPHY

Reading List

K. Owen and T. Coley, *Automotive Fuels Handbook,* Society of Automotive Engineers, Inc.
K. Owen, ed., *Gasoline and Diesel Fuel Additives,* John Wiley & Sons, Inc., New York.

ASH

LARRY L. BAXTER
Sandia National Laboratories
Livermore, California

The principal means of energy production from solid and heavy liquid fuels is combustion. During combustion, inorganic components of most solid and many liquid fuels lead to the formation of ash. Inorganic material typically accounts for about 10% of the mass of solid fuels, varying from over 50% to less than 1%. Coal represents the single most significant solid fuel in worldwide energy production, and statistics from coal-based electric power production in the United States accurately illustrate the magnitude of ash-related issues during combustion. About 90.7 million t of ash per year, or 363 kg per person per year, are generated in the United States from coal combustion. By comparison, Western Europe generates about 36.3 million t annually (1). Studies indicate that ash deposition, commonly referred to as fouling and slagging, causes unnecessary lost production that costs utilities between $\$5.2 \times 10^{-5}$ and $\$1.3 \times 10^{-4}$ d per kilowatt hour (kW·h) of coal-based electricity production (1985 U.S. dollars). Financial losses occurring due to other ash-related issues, such as tube leaks due to erosion or corrosion by ash, can be as much as a factor of 10 higher (2). In the United States, where coal accounts for 56% of the 2.8×10^{12} kW·h of utility power production (1992 statistics) (3), preventable ash-related production losses for U.S. utilities can be estimated to range from many hundreds of millions of dollars a year to a few billion dollars a year. These represent the bulk of unanticipated costs associated with ash-related problems. In addition, ash-related issues play a dominate role in the anticipated costs associated with scheduled shutdowns and maintenance. In general, ash-related problems can conservatively be estimated to represent a $1 to $2 billion (1990 U.S. dollars) a year problem in the U.S. electric power industry. The U.S. experience with ash during coal combustion does not differ from that of other major coal-consuming countries, nor does coal present unusually difficult ash-related problems. Energy production by combustion of essentially all solid fuels in all regions of the world is accompanied by significant ash management problems. See also COAL COMBUSTION; COMBUSTION MODELING; COMMERCIAL AVAILABILITY OF ENERGY TECHNOLOGY.

Boiler designers are keenly aware of the pronounced impact of ash on boiler performance and maintenance. Considerations of ash behavior dominate both the design and operation of combustors that burn solid and some liquid fuels. An appreciation of these dominant effects of ash-related issues on combustor design can be gained by considering designs of typical coal combustors. Figure 1 illustrates two boiler designs, both intended to produce the same amount and quality of steam electricity from coal. The first section of the boiler is called the furnace and consists of a tall narrow chamber with heat absorbing walls referred to as waterwalls. These walls are composed of steam- and water-containing tubes welded together by membranes, which are small pieces of steel between them. The second section of the boiler is referred to as the convection pass and consists of heat-exchanging tube bundles hanging in the bulk gas flow, usually with no membranes bonding them together. Coal is pneumatically transported to the burners where it is blended with combustion air as it is introduced into the bottom of the furnace. A utility power station producing 600 MW of electricity burns coal at a rate of about 181 t/h and generates about 18 t/h of ash. The combustion of this coal forms a fireball in the bottom of the furnace. Char and fly ash are carried from the fireball to the top of the furnace, through the convection pass, and eventually into cleanup systems. Dimensions for the boiler depend on its generating capacity. They are denoted as height (H), width (W), and depth (D) for a boiler designed for a typical U.S. bituminous steam coal in the design shown on the left. Utility boilers range in electrical generating capacity from about 100 to 1200 MW, with a typical 600 MW bituminous coal boiler being roughly 61 m tall and 12–15 m wide and deep. Bituminous coals typically present relatively small fouling and slagging problems relative to the lower rank subbituminous coals and lignites. The dimensions for the boiler on the right are shown relative to those for the boiler on the left. The boiler illustrated on the right has the same electrical generating capacity, but is designed for a typical U.S. lignite which poses significantly greater ash deposition (fouling and slagging) problems than bituminous coals.

The slowest step of coal (and most solid fuel) combustion is the reaction of residual char with oxygen. This reaction occurs much faster with chars generated from lignites than with those generated from bituminous coals (4). On this basis, one might expect that boilers designed for bituminous coals would be larger than those designed for lignites. The opposite trend, as is illustrated in Figure 1, attests to the dominate influence of ash on boiler design and operation. As can be seen, the lignite boiler illustrated on the right typically has 25% greater cross-sectional area and 40% greater height, for a total of 75% greater volume, than the bituminous coal boiler on the left (5). Other design features, such as tube spacing in the convection pass and maximum allowable furnace exit gas temperature, are also strongly affected by ash deposition considerations. The large change in boiler dimensions and design occur specifically to address concerns about the behavior of inorganic material in the two types of coals. This specific example applies to pulverized-coal-fired systems but is also representative of the influence ash considerations have on other technologies, such as gasifiers and fluid-bed combustors, as well as with other fuels, such as biomass and refuse. Inorganic material, from which ash is derived, has some effect on mining or collection, transportation, and preparation of fuels. As important as these effects are, the properties of inorganic material become especially important during combustion.

Principal changes in technology during the history of coal combustion are attributable in large measure to the influence of ash on combustor design and performance (6,7). These changes led to the evolution of boiler designs from moving or fixed grates to the entrained-flow, pulverized-coal boiler design currently used by most utility power stations such as those indicated in Figure 1. Advanced technology may lead to still other designs, ranging from gasifiers to fluid beds. In the opinion of many prac-

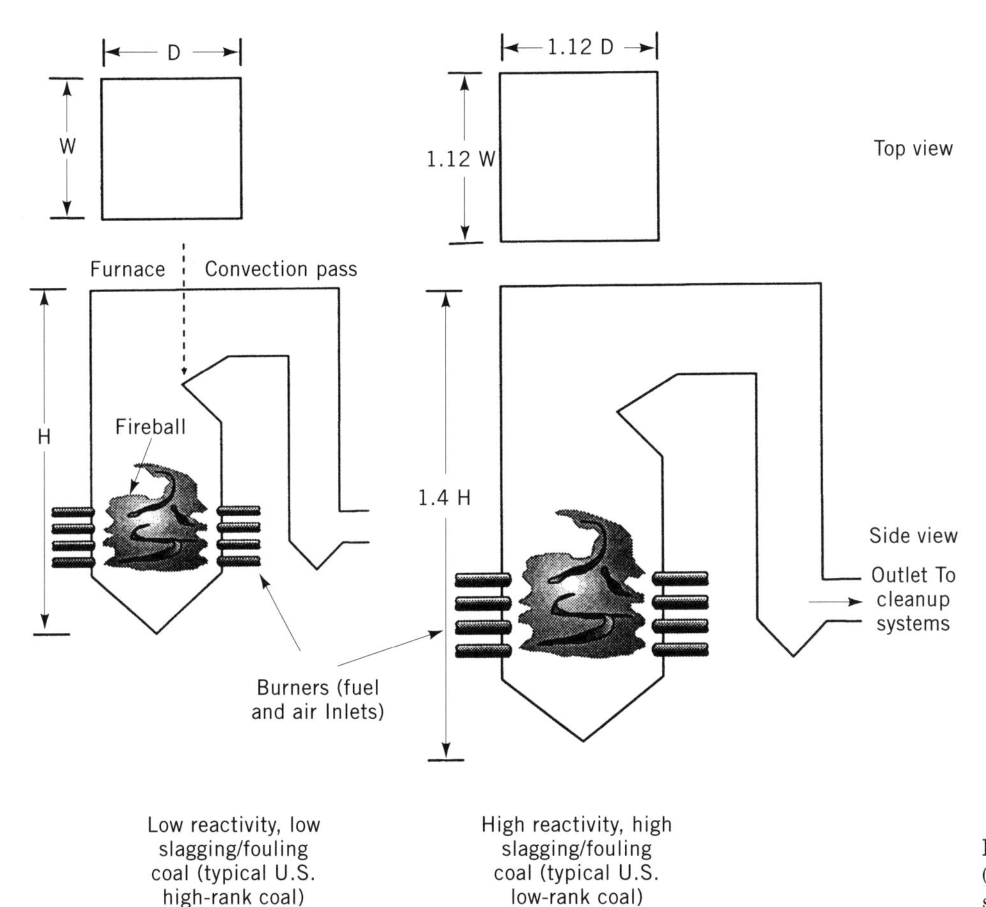

Figure 1. The effect of ash deposition (fouling and slagging) on boiler dimensions.

titioners, the behavior of inorganic material in the fuel will continue to dominate design considerations, whatever technology is employed. Liquid fuels contain far less ash than most solid fuels, the former containing rarely more than 0.2% ash. Yet even this small quantity of ash is capable of causing severe ash deposition and corrosion problems in combustors. This article defines the terminology used in discussing inorganic materials and briefly describes some of the analytical techniques used to quantify its properties. The effects of inorganic material on energy and environmental issues are then traced from the origin of the inorganics in fuel to their disposal. The discussion is intended to be general and is primarily based on ashes found in coal, biomass and refuse, and heavy oils.

DEFINITIONS

The term *ash* is often used to refer to inorganic material in a general sense. Practitioners refer to ash in the fuel, fly ash, ash deposits, bottom ash, etc. In more precise terms, ash refers to material that does not combust under a prescribed set of conditions. In this precise sense, fuels contain inorganic material, not ash. Ash is a product of combustion or gasification of the fuel. Inorganic material comprises both mineral matter and atomically distributed inorganic components. In common usage, the term *mineral matter* is sometimes used to refer to all inorganic material. More precise discussions refer to mineral matter as the portion of the inorganic material that is mineral in nature, ie, granular material that is identifiable and classifiable by common mineralogical techniques. This contrasts with atomically distributed material, which is either an integral part of the organic structure or in some ionic form absorbed on surfaces or dissolved in moisture inherent in the fuel. Common classes and examples of mineral matter found in fuels include oxides (silica, rutile), silicates (kaolinite, illite), sulfides (pyrite), carbonates (calcite), and sulfates (iron sulfate, gypsum). This material occurs in the fuel as grains that may be either intimately associated with (inherent) or extraneous to (adventitious) the fuel matrix. Minerals intimately associated with the fuel commonly occur as bands of inorganic material within a coal seam or concentrated in certain sections of a plant. This reflects their origin as detrital material (see below) or material with specific biological function. Grains vary in size from nanometers to millimeters in the raw fuel. Fuels such as coal often undergo washing, pulverization, or other fuel pretreatments, reducing maximum grain sizes to tens of microns. The variety of forms of mineral matter in fuels can be appreciated by considering the forms of silica and silicates. Grains may be dense, as is common with silica in coal, or porous, as is common for silicates in coal. Porous, imbedded grains of mineral matter are often impregnated with organic material. Grains may also be either crystalline, as is common for silica in coal, or amorphous, as is common for silica in biomass. Mineral matter may be hydrated, as is common for silicates in coal and silica in biomass, or anhydrous, as is common for silica in coal. The specific forms of minerals in most fuels can be traced to their biological and geological origin and history (see below).

Atomically distributed material includes alkali and alkaline earth elements, usually in ionic forms and often bound as cations to the organic structure. Other forms of atomically distributed inorganics include chemisorbed, complexed, and solvated material. While alkali and alkaline earth materials often dominate the atomically distributed portion of inorganic material, at least small fractions of many other elements can be found in atomically distributed forms (8). The distinctions between organic and inorganic portions of the fuel are not always clear when discussing atomically distributed material. For this discussion, only carbon, hydrogen, nitrogen, sulfur, and oxygen that is covalently bound to a hydrocarbon structure is considered organic. Thus potassium chloride in biomass, vanadium in oils, calcium cations in coal, and other similar materials with clear biological functions in living compounds are considered inorganic in this discussion. Similarly, sulfur, oxygen, and carbon in the fuel are considered inorganic when they occur in the forms of minerals.

Ash deposition in combustors is commonly referred to as fouling and slagging. There is little agreement in the precise distinction between fouling and slagging deposits. The term *slag* has connotations of molten material, whereas *foul* has connotations of solid material. The distinction between molten and solid is, in practice, one of degrees. Only a small fraction of modern combustors operate with running, fluid slag on the walls. Most operate with ash deposits that have some plastic character when stressed but do not appreciably flow. The distinction between fouling and slagging used in this discussion is that fouling deposits are confined to the convection pass and slagging deposits are confined to the furnace. This distinction is less arbitrary than it may appear. Furnace deposits occur predominantly on flat or reasonably flat surfaces such as waterwalls. Gas temperatures in the furnace are in the range of 871–1649°C. Deposits in the convection pass occur predominately on tube surfaces. Gas temperatures in the convection pass range from 149 to 1093°C. Differences in temperatures and geometries typically lead to distinct differences in deposit characteristics.

ORIGIN AND COMPOSITION OF INORGANIC MATERIAL

Figure 2 presents a useful framework for considering the relationships among the various common fuels. The lower left corner of the figure comprises a coalification diagram, which illustrates the changes in coal structure as it matures from lignite through subbituminous and bituminous coal, eventually to anthracite (9,10). Figure 2 represents an expansion of the coalification diagram and shows, eg, that biomass is a logical extension of the coalification process. Premium fuels, such as fuel oils and natural gas, are also illustrated. Liquid fuels derived from solids (pyrolysis oils, tars, etc) are slightly enriched in hydrogen and depleted in oxygen relative to the solid from which they are derived, with specific compositions dependent on processing temperature and similar parameters (11).

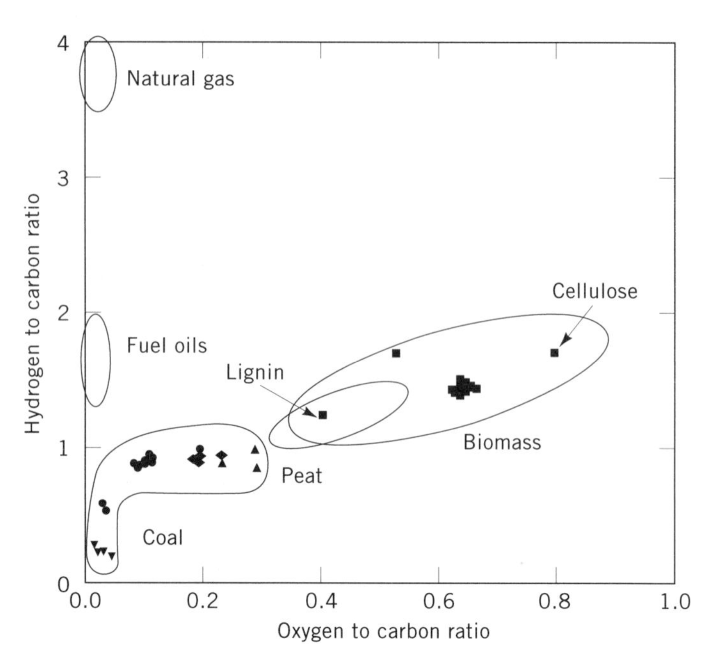

Figure 2. The relationships between various fuels based on their major atomic compositions.

Figure 2 illustrates that one distinguishing feature of solid fuels is their oxygen content. Oxygen in such fuels is commonly found in the form of functional groups such as carbonyls and hydroxyls. These forms of oxygen represent potential sites were inorganic materials, in particular alkali and alkaline earth ions, bond with the organic structure. As could be inferred from Figure 2, the amount of this organically bound inorganic material often increases as the oxygen content of the fuel increases. The other principal forms of inorganic material in fuels are not related to the fuel structure. They include salts and other loosely bound materials, often dissolved in moisture in the fuel, inherent particulate inorganic material that results from either biological or geological processes, and adventitious inorganic material that typically results from mining or fuel handling processes.

The most prevalent inorganic elements found in solid fuels are generally oxygen, chlorine, silicon, aluminum, titanium, iron, calcium, magnesium, sodium, potassium, sulfur, and phosphorus. Nitrogen is almost exclusively found as part of the organic structure. Sulfur is found in both organic and inorganic forms. The amount of the latter 10 elements in the ash is commonly reported on an oxide basis for reasons discussed below. They only occasionally occur in the forms of oxides in the fuels. In liquid fuels, an additional element, vanadium, is of considerable importance. The proportion of the various components can vary significantly from fuel to fuel, as discussed below for each principal fuel type.

Coal

Figure 3 presents typical ash compositions for several different ranks of coal in terms of the metal oxides. A primary component of all coal ashes is silicon, with significant amounts of aluminum. Subbituminous coals typically have higher calcium and lower iron contents than either bituminous coals or anthracites. Lignite ash composition varies more widely than ash for the other ranks of coal. The data shown in Figure 3 are all derived from U.S. coals. These U.S. coals are representative of a great many coals worldwide, although analyses of coal ashes from other parts of the world differ somewhat from those shown in Figure 3.

Ash compositions are usually reported on a normalized, sulfur- and phosphorus-free basis (Fig. 3). The amount of sulfur and phosphorus retained in the ash generally increases with increasing amount of alkali or alkaline earth material. Alkali and alkaline earth material that is not incorporated in silicates reacts with sulfur to form sulfates that are stable at low temperatures (below 800 to 900°C). The amount of sulfate in high rank (anthracite and bituminous coals) fuels is usually small (less than 5%). Subbituminous coals and lignites often contain as much as 25% of their ash in the form of sulfates.

The data in Figure 3 represent typical results from standard analyses of coals. They indicate little about the forms or modes of occurrence of inorganic material in coal. Geological and biological histories affect the amount and mode of occurrence in inorganic material in fossil fuels. For example, atomically distributed calcium in coals arises in part from dissolution of detrital calcite and incorporation of the dissolved material in the organic matrix. Pyrites develop to some extent by combination of organic sulfur with available iron by bacterial processes (6). The mode of occurrence of inorganic material in fuels follows reasonably consistent geographic patterns. Figures 4 and 5 indicate the partitioning of the alkali metals sodium and potassium as a function of coal rank. The total amount of sodium (Fig. 4) and potassium (Fig. 5) is divided into four forms: water soluble, ion exchangeable, acid (HCl) soluble, and residual. These four fractions are determined by a procedure known as chemical fractionation, the repeatability and precision of which have been determined, but it has not been standardized.

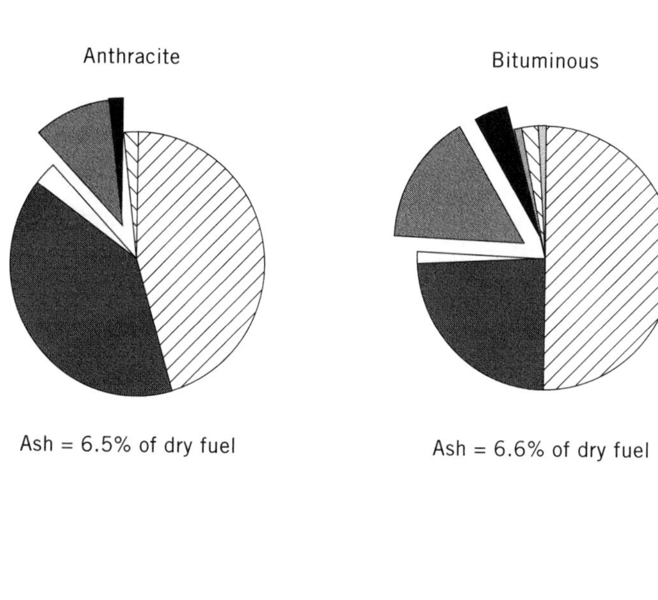

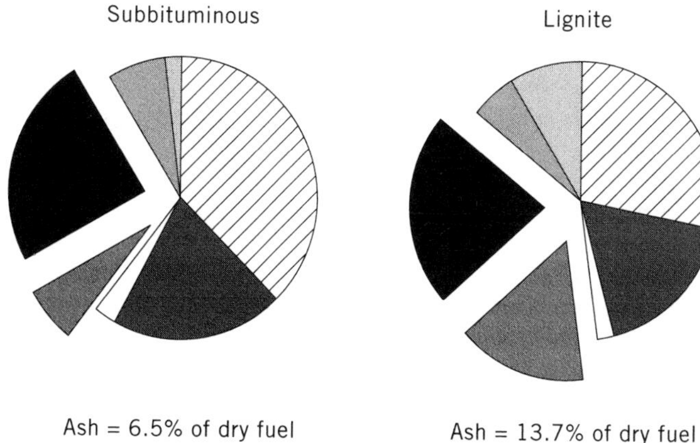

Figure 3. Representative compositions of ashes from four coals of various rank, expressed as oxides. These data are on a sulfur- and phosphorus-free basis.

Water-soluble inorganic material generally includes ions that dissolve readily in water (eg, chlorides) or chemisorbed material that is bound in the coal by weak bonds. Ion-exchangeable material includes organically associated inorganic species that complex, eg, with carboxyl or other functional groups. The sum of the water-soluble and ion-exchangeable fractions comprise the atomically dispersed (as opposed to mineralogical) fraction of the inorganic material. Certain minerals (carbonates, sulfates, etc) are soluble in hydrochloric acid and can be distinguished from other materials (many oxides, sulfides, etc) that are not.

The data indicate clear trends with respect to both rank and geography. High rank coals generally tend to have less atomically dispersed alkali material as a fraction of total alkali material. Chemically similar materials, such as sodium and potassium, occur in significantly different modes in the fuels. For example, about half of the sodium in high rank coals and nearly all of the sodium in low rank coals is atomically dispersed. By contrast, essentially none of the potassium in high rank coals and only about half of the potassium in low rank coals is atomically dispersed. Alkali material that is not atomically dispersed is typically found in clays and other impurities in the coal.

The other components show less variation as a function of rank. Aluminum occurs primarily in the form of the clays such as illite, kaolinite, and montmorillonite in all ranks of coal. Silicon also occurs in clays, with a lesser but significant fraction (1–30% of the silicon) in the form of quartzlike material (free silica). Titanium occurs primarily in the form of rutile, with a small fraction in the organic matrix. Calcium and magnesium occur as carbonates, oxides, and silicates.

The total amount of ash in coals usually varies from 5 to 15% of the dry mass of the fuel. Lignites are the most common exceptions, with ash contents sometimes reaching the 50% range. Total coal ash in some coals is reduced by washing, which is usually performed at or near the mine. Further reductions are achieved by beneficiation of several varieties. Physical beneficiation processes typically selectively remove dense, adventitious inorganic components such as pyrite. Chemical beneficiation processs can be designed either to remove some ash components selectively or to remove all components of the ash. The ash content of coals can be reduced to arbitrarily low levels by combinations of physical and chemical coal bene-

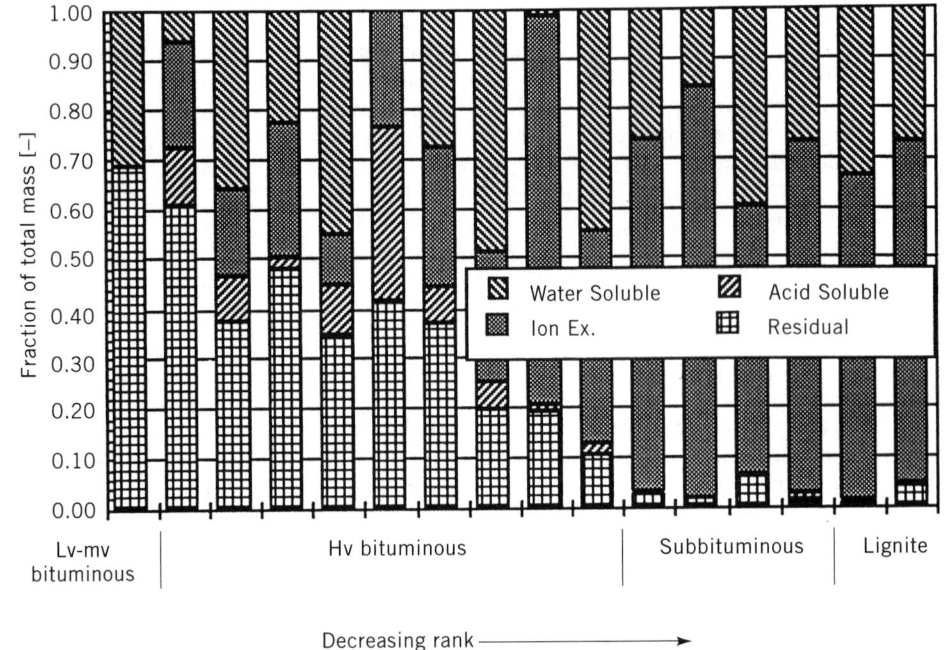

Figure 4. Mode of occurrence of sodium as a function of rank for a range of coals. The atomically dispersed sodium, represented by the sum of the water-soluble and ion-exchangeable material, increases steadily with decreasing coal rank or, equivalently, with increasing coal oxygen content (cf. Fig. 5).

ficiation and processing, but economics generally favor minimal processing for ash removal.

Biomass and Refuse

There is greater diversity in the types of ash found in various forms of biomass and refuse-derived fuels than in coal. Figure 6 illustrates typical results of ash analyses for four typical biomass fuels. In each case, the analyses are based on uncontaminated samples of biomass. Refuse-derived fuels exhibit even wider variations in their ash compositions. Handling or processing of material often increases the total amount of ash and its composition by addition of soil or other extraneous material. The figure illustrates how ash from heartwood or other woody, slowly growing, large-diameter biomass is often dominated by calcium and potassium. Leaves, stems, and other rapidly growing biomass material of small diameter have both more total ash and less calcium relative to other components. The biological activity of rapidly growing material requires greater amounts of inorganic material for both structural and metabolic reasons. Straws and grasses incorporate relatively large amounts of typically hydrated silica in the form of structural components (so-called silica cells) that give the plant rigidity to prevent lodging. Complete discussions of forms and biological functions of inorganic material in biomass are available (12).

The mode of occurrence of inherent inorganic material in biomass can usually be traced directly to its biological

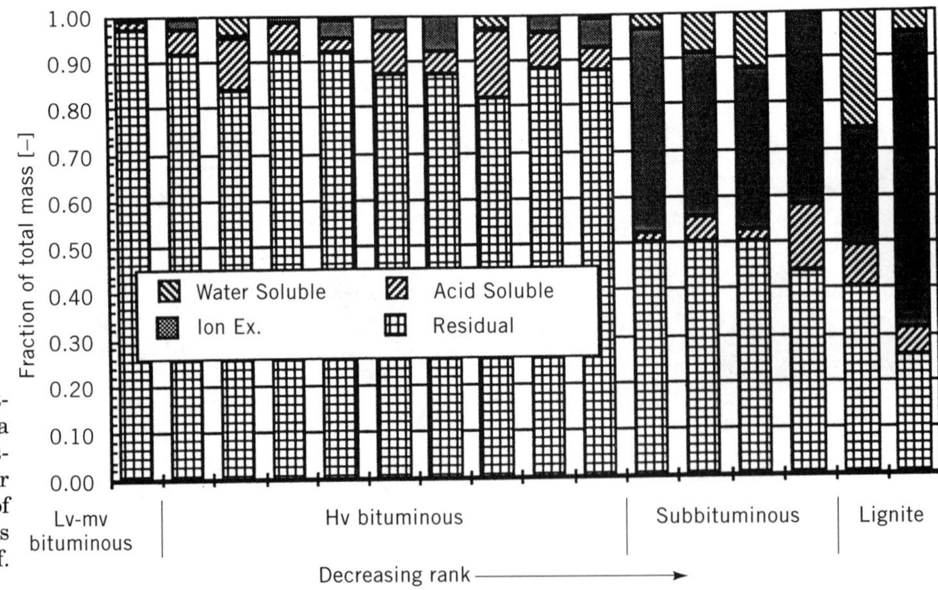

Figure 5. Mode of occurrence of potassium as a function of rank for a range of coals. The atomically dispersed potassium follows a similar trend as sodium, but the amount of atomically dispersed potassium differs significantly from that of sodium (cf. Fig. 4).

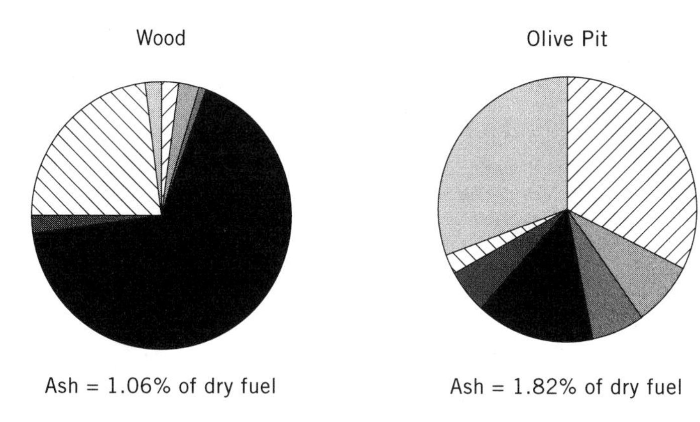

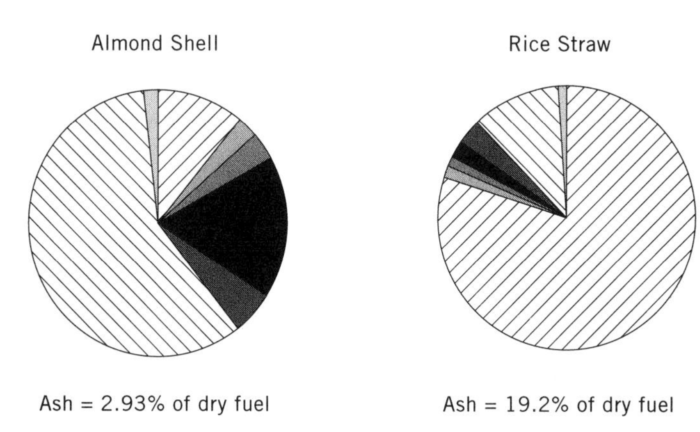

Figure 6. Typical ash analyses from four forms of biomass.

function. For example, essentially all of the potassium in essentially all forms of biomass is atomically dispersed, in strong contrast to the forms of potassium in coal. Nearly all of the silica in biomass is hydrated silica (opal), again in strong contrast to coal. Extraneous mineral matter in fuels reflects the local environment. For example, aluminum is a plant toxin and has low concentration in the inherent inorganic material of biomass fuels. Its ubiquitous nature in soils makes it useful as an indicator of soil contamination of biomass fuels. By contrast, aluminum is a major component in the inherent minerals in coal, where it occurs almost exclusively in clays (hydrated aluminosilicates).

Refuse fired directly in waste to energy plants is commonly referred to as municipal solid waste (MSW). Processed refuse is referred to as refuse-derived fuel (RDF). The amount and composition of both MSW and RDF vary to such a large extent that representative results are of little value. Typically, the amount of ash from these fuels is high, its composition is related directly to its source, and the content of noxious materials such as heavy metals and chlorine is higher than coal, biomass, or fuel oils.

Liquid Fuels

The quantity of inorganic material in fuel oils generally increases with decreasing fuel volatility. The type of inorganic material differs from that in solid fuels. The primary components of concern in liquid fuels include alkali material (primarily sodium and potassium), vanadium, and nickel. Concentrations of these inorganic materials range from a few to a few hundred ppm by weight in the fuel. Some of the material has its origin in the animal and plant matter from which the oils are formed. For example, potassium and sodium in pyrolysis oils is commonly derived from organically associated potassium or sodium in plants. Some vanadium and nickel in petroleum-based oils possibly derives from its proposed role as a porphyrin in ancient animal life. Additional inorganic material in liquid fuels derives from environmental and process sources. Inorganic material tends to concentrate in the heaviest fractions of any distillation or pyrolysis process. Therefore, most of the inorganic material is found in chars or asphalts resulting from the processes, with increasingly smaller amounts found in increasingly light fractions.

ANALYTICAL METHODS

There are both standard and developing analytical methods for determining the amount and properties of ash in fuels. The most active U.S. institute involved in developing standards for analytical procedures is the American Society for Testing and Materials (ASTM). The D and E series of ASTM standards include most of the ash-related information for coal and biomass–refuse-derived fuel, respectively. Similar standards are commonly developed on a worldwide basis by the International Organization for Standardization (ISO). Standards are also promulgated under the direction of other governments, trade and professional organizations, and research societies. These institutions publish and update their standards periodically, and reviews with more detailed discussions of the procedures are available in several reference books (13). The most relevant and common procedures are briefly reviewed below, with some indication of potential biases in the standardized procedures and, where applicable, means of avoiding them. Also described are several promising developing procedures for fuel characterization.

Standard Procedures

Total Ash. Standard methods to determine total ash in fuels involve maintaining a sample in an oxidizing environment while heating it to some prescribed temperature until its weight no longer changes. Different standards are developed for different fuels. For example, coals and cokes are typically heated to 750°C, whereas biomass is heated to 600°C. The details of the standardized techniques attempt to result in a sample that is fully oxidized, with particular emphasis on avoiding formation of carbonates and sulfates. The total ash values thus determined are typically at least 10% lower than the corresponding total inorganic material in the fuel. The difference arises because the forms of the species in the fuel (pyrites, carbonates, sulfates, chlorides, and silicates) typically weigh more than the corresponding, fully oxidized material. Furthermore, much of the inorganic material is hydrated in its pristine form. Standard methods for determining mineral matter (as opposed to ash) involve partially or completely digesting samples in acids or other strong reagents.

The principal objection to these standard methods is that they underestimate the total amount of inorganic material. In addition to converting the inorganics into different and usually lighter forms, volatile inorganics such as sodium and potassium are partially vaporized during sample heating. Alternative procedures are sometimes used, particularly by research labs, to avoid these problems. The alternative procedures have not been formally standardized, although most have been at least documented and are de facto standards for a particular lab.

Ash Elemental Composition. The elemental composition of ash is typically determined by digesting the ash sample and analyzing the resulting fluid by atomic emission or absorption spectroscopy. The high temperature source for the analysis is most commonly an inductively coupled plasma. Other common techniques for the analysis include neutron activation and x-ray fluorescence spectroscopy. The former is typically the most accurate and has the lowest detection limits, but requires equipment and facilities not widely available. The latter is rapid but is the most subject to analysis error than the other techniques. The set of 10 elements included in the analysis of most solid fuels are silicon, aluminum, titanium, iron, calcium, magnesium, sodium, potassium, sulfur, and phosphorus. Chlorine in ash is sometimes included. It is of particular importance in biomass ashes, where chlorine contents are usually higher than for most coals.

The common techniques for determining the elemental composition of ash are based on atomic composition. Results of the analysis are reported on an atomic or elemental basis or, more commonly, converted to an oxide basis. The sum of the oxides can be compared with the measured total ash content of the fuel as a consistency check. If the measured ash exceeds the sum of the oxides, as is usually the case, the difference is attributed to the presence of unmeasured species in the ash, incomplete oxidation of the ash (carbonates, sulfates, etc weigh more than their respective oxides), or experimental error. Normal procedures yield ash and sum-of-oxide numbers that agree within about 5% for many fuels.

The elemental composition of fuels ash varies in predictable ways with geographic or biological origin of the fuel. For example, coals from most western U.S. mines exhibit higher calcium concentrations and lower iron concentrations than those mined in the Midwest or eastern U.S. Herbaceous and annual plants exhibit higher concentrations of silicon and potassium and lower concentrations of calcium than the bulk of most trees and other perennial plants. However, actively growing portions of perennial plants, such as leaves and twigs, exhibit high potassium concentrations. Oils often contain soluble forms of inorganic material, including vanadium, potassium, and sodium.

Principal objections to procedures used to determine ash composition follow directly from those to total ash determination. During high temperature ashing, volatile inorganic components of the fuel are often vaporized, leading to errors in the determination of ash composition. These problems can be avoided by using lower temperature procedures in preparing the ash for analysis, as discussed above.

Fusion Temperatures. The tendency of ash to sinter on combustor walls, and the desire to predict the temperature ranges in which this occurs, has led to the development of ash fusion temperature measurements. These measurements can be performed in a completely automated fashion once the ash is prepared. Ash samples are pressed into the shape of a pyramid 19 mm high with the base forming an equilateral triangle 6.4 mm on a side and placed in a furnace. The samples are shaped like pyramids, but they are almost universally referred to as cones. Furnace temperature is increased at a rate of 8 ± 3°C/min. Changes in the shape of the cone are monitored as the furnace temperature increases. The procedure yields four critical temperatures, defined as follows: (*1*) the initial deformation temperature (tip of cone first becomes rounded), (*2*) the softening temperature (height equals width of the base), (*3*) the hemispherical temperature (height equals one-half of the width of the base), and (*4*) the fluid temperature (mass has spread to a flat layer no greater than 1.6 mm thick). These measurements are performed under both oxidizing (air or vitiated flow from a lean flame) and reducing (nominal 60% CO, 40% CO_2 gas mixture) environments.

Results of this test are sometimes several hundred degrees lower when conducted under reducing conditions than under oxidizing conditions. A primary reason for the difference is the presence of iron in the ash. Under oxidizing conditions, iron is predominately present as ferric ion (Fe^{3+}) whereas reducing conditions favor ferrous (Fe^{2+}) or, in some cases, metallic iron. Ferric iron is less likely to form eutectics or pure phases that melt at low temperatures than is ferrous iron. As the amount of iron in ash decreases, the differences between fusion temperatures measured in a reducing temperature and those measured on an oxidizing basis decrease.

Objections to ash fusion temperature measurements derive from both the technique and the interpretation of the results. The technique involves heating samples at a prescribed rate to a reasonably high temperature. The reactions that lead to the formation of liquids are often transport or kinetically controlled and occur over time periods much longer than are allowed for in the fusion temperature technique. Ash deposits in combustors have minimum lifetimes (time between soot blowing cycles, eg) of about 8 h. For this reason, the experimental technique is not too representative of actual combustor conditions.

A second objection is that the ash fusion temperature measurements are performed with fuel ash that may have a markedly different composition than the ash in deposits in a combustor fired with that fuel. Ash deposition can be selective in regard to elemental composition, sometimes highly selective. This selectivity depends on location within the combustor, combustor operating conditions, and combustor design in addition to the type of fuel fired. The reasons for this selectivity are discussed and illustrated later.

Ash Viscosity. Combustors designed to produce and remove molten slag are sensitive to ash viscosity. The most common specification of ash viscosity is the T_{250} temperature, or the temperature at which the slag viscosity is 25 Pa·s. This temperature is usually measured using a controlled-environment, high temperature, rotating-bob viscometer. Molten slag is introduced into the viscometer at temperatures of 1430°C or higher. Sample viscosity is monitored as sample temperature is reduced until the viscosity reaches a value of 25 Pa·s. T_{250} values for slags from eastern coals typically range from 1150° to more than 1500°C, with higher iron content leading to lower T_{250} values. Increasing alkali and alkaline earth concentrations often also lead to decreasing T_{250} values.

Other Analyses. The analyses discussed thus far are used primarily to address the issue of ash deposit formation during combustion. Inorganic material in fuels also affects fuel handling and preparation, potential sulfur and particulate emissions, corrosion and metal wastage, and ash salability. Analyses that give insight into these issues include the concentration of pyrite in the fuel, fly ash resistivity, concentrations of chlorine and other corrosive agents, and carbon or toxic element concentrations in the ash. Pyrite (FeS_2) is a common component of high rank coals. Pyrite is a hard mineral that is difficult to pulverize and that contributes to the overall sulfur content of the fuel. It is often the dominant form of iron in unweathered coal. Combustors using electrostatic precipitators (ESPs) as fly ash clean-up devices are sensitive to the electrical conductivity of fly ash. Specific resistivities on the order of 10^9 to 10^{10} ohm-cm yield optimal ESP performance. Higher resistivities tend to excessively insulate the collection plates, decreasing the effective strength of the electric field. High resistivities typically occur with low sulfur, low carbon, high alkali fly ash. Lower resistivities can lead to arcing in the precipitator and excessive reentrainment of collected fly ash. Corrosion and metal wastage are related to the formation of vanadium-, alkali-, sulfur-, or chlorine-laden deposits on surfaces and the abrasion of silica particles as they strike surfaces. Carbon content and other properties of ash influence its potential resale value. The impacts of ash properties on ash marketability are discussed below.

Indices of Ash Behavior. Industrial practice has led to a large number of indices of ash behavior based on the standard analyses discussed above. These are reviewed in detail elsewhere (14,15). The most commonly used are briefly indicated here.

Oxides are usually divided into acidic and basic classes. The most prevalent acidic oxides are silica, alumina, and titania (alumina and titania are actually amphoteric, but are classified as acidic in the application of ash behavior indices). Basic oxides are formed from iron and the alkali and alkaline earth elements. The most commonly cited index of ash deposition behavior is the base:acid ratio. While this is largely an empirically derived number, it has some basis in fundamental concepts. Various components of ash slag are viewed as chain formers, modifiers, or combinations. Acidic oxides are chain formers, whereas basic components are modifiers. Iron can be either a modifier or a combination, depending on its oxidation state. Aluminum is combination of former and modifier. Chain formers typically increase the length of silica-based polymers in molten silicates, increasing the viscosity of the material. Modifiers tend to destabilize the polymers, leading to lower viscosities. The empirically developed ratio of basic to acidic components of the ash relates in this way to the structure of coal slags.

There are many similar empirical indices of ash behavior. A large number of them are perturbations of the base:acid-ratio approach, but some are completely independent of it. They are, in large measure, the tools that have been used to design and operate boilers in the past. However, their successful application generally depends on some field experience with similar situations and, even then, they are perhaps only accurate in anticipating problems slightly more than half of the time. They commonly fail when extrapolated to situations new to the person using them or to situations different from those from which they were developed.

Developing Procedures

The standardized procedures discussed above are the basis for most design, operation, and fuel selection decisions in industry. There usefulness in understanding and mitigating the effects of ash on combustors and the environment is beyond question. There is general agreement, however, that these analyses and their interpretation do not provide sufficient information to address practical issues reliably, eg, to ensure trouble-free operation of combustors. The standardized techniques are even less adequate to develop detailed understanding of ash transformations and deposition. Some of the more obvious shortcomings of these analyses include (*1*) a lack of information about the mode of occurrence of the various ash components, (*2*) neglect of selective deposition of ash components, (*3*) implicit assumptions that ash is a homogeneous entity, neglecting the particle-to-particle variations, and (*4*) neglect of the influence of operating conditions and boiler design on ash deposition tendencies. The potential importance of these issues is large. For example, potassium or sodium incorporated in a clay is relatively benign with regard to ash deposition and corrosion, whereas the same amount of material in an atomically dispersed form greatly increases the deleterious effects of ash deposition and corrosion. There are several developing procedures that address the shortcomings of the standardized procedures. Some of the more significant ones are discussed here and illustrated below.

Chemical Fractionation. Figure 7 schematically illustrates a procedure called chemical fractionation which is used to determine the modes of occurrence of many inorganic components (8,16–18). A series of leaching agents selectively remove various components of fuels, with the leaching agents arranged in order of increasing aggressiveness. Interest focuses primarily on 11 inorganic components (silican, aluminum, titanium, iron, calcium, magnesium, sodium, potassium, sulfer, phosphorus, and chlorine). The modes of occurrence of each element are deduced from the fraction of material removed by each of the leaching agents. The leaching agents chosen for use most commonly include water, ammonium acetate, and hydrochloric acid and sometimes nitric acid.

Dissolved salts such as chlorides often comprise the majority of water-soluble material (stage 1). Other atomi-

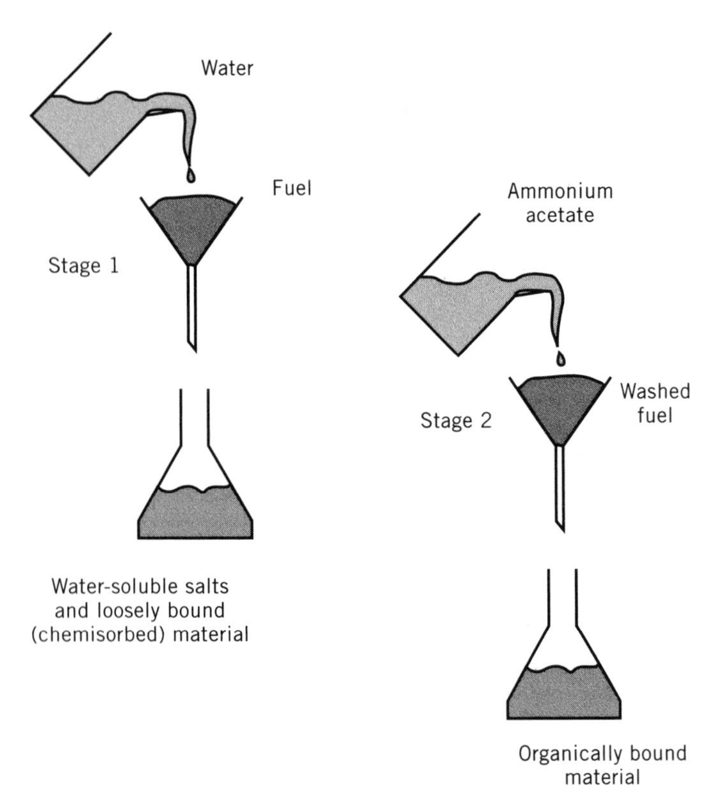

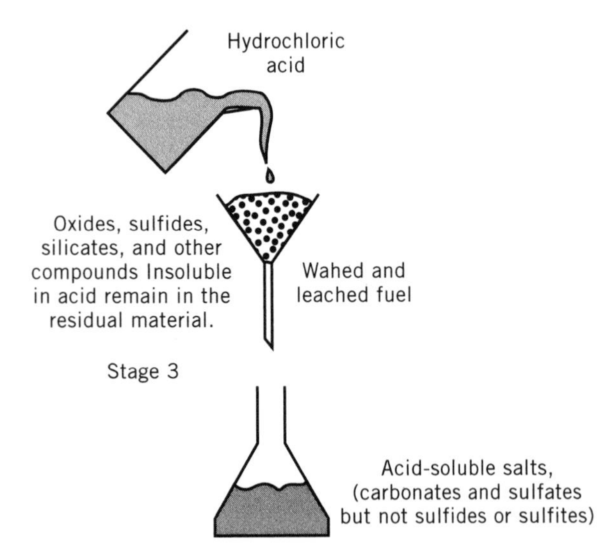

Figure 7. Chemical fractionation procedure illustrating each of the major extractions performed and examples of chemical species removed from the sample at each leaching step.

cally dispersed materials such as alkali and alkaline earth elements comprise much of the material removed by ammonium acetate (stage 2). These two categories of material are broadly classified as atomically dispersed and include the elements most likely to vaporize or otherwise be released to the gas phase during combustion. Their subsequent condensation or chemical reactions underlie many of the most severe deposition and corrosion problems in combustion systems. Hydrochloric acid dissolves carbonates, but not sulfides, and is useful for determining the fraction of, eg, calcium carbonate, in the sample. The residual material comprises primarily oxides, sulfides, or other relatively stable materials. Principal among these are oxides (silica and titania), aluminosilicates typically derived from clays found in soils (kaolinite, montmorillinite, and illite) and sulfides (pyrite). While some of these compounds are largely chemically inert at room temperature, their high temperature chemistry has profound influence on ash deposit growth and property development in combustors. Chemical fractionation results, combined with standardized coal analyses, are the essential inputs to recently developed ash deposition models (19,20).

Scanning Electron Microscope Techniques. Techniques based on scanning electron microscopy (sem) capitalize on this instruments ability to resolve the size and elemental composition of individual mineral grains in fuels. A technique that has come to be called computer-controlled scanning electron spectroscopy (ccsem) combines the automation available from computers with the elemental analysis capabilities of modern electron microscopes to produce size- and composition-classified descriptions of mineral matter in fuels. Several publications discuss and illustrate the technique in detail (21–24). The technique is not standardized, but in all of its incarnations mineral grains are identified in an sem image of the fuel and their size and elemental composition are determined. The probable chemical species or the grain is inferred from the elemental composition combined with look-up tables or logic processes. The accuracy and precision of the technique have been investigated, and recommendations for avoiding common errors are available (25). The results of ccsem analyses form the basis for some of the most recent descriptions of transformations and deposition of inorganic material (26–28). Ccsem provides information about the mineral portion of the inorganic material that is not available by alternative means.

TRANSFORMATIONS OF INORGANIC MATERIAL DURING COMBUSTION

During combustion, inorganic material transforms into ash through several mechanisms. A portion, rarely more than 30%, of the inorganic material is released from the coal particle in the form of either vapors or fumes. This occurs through vaporization, with atomically dispersed species exhibiting high vapor pressures (usually alkali metal-containing inorganic species) participating most actively (29). The partitioning of the alkali species between clays and atomically dispersed forms (see Figs. 4 and 5) is particularly important because the latter are strongly affected by vaporization whereas the former are essentially unaffected. Even when the overall combustion stoichiometry is fuel lean, reducing conditions often prevail inside individual fuel particles, significantly enhancing vaporization rates of many materials. Organic and inorganic reactions can also lead to mass release under many combustion conditions (30,31). The residual inorganic material forms a residual particle which, in turn, goes through several transformations. Mineral grains may fragment during rapid heating (23). The residual char may fragment, with the number of fragments formed increasing with increasing initial char particle size and increasing char porosity (32). Individual mineral grains and adventitious inorganic material go through a series of chemical reactions. For example, sulfides and carbonates decompose, silicates absorb available alkali and alkaline earth material, many species change phase, and grains often agglomerate to form more complex chemical mixtures (33,34).

Chlorine plays a particularly important role in the vaporization of inorganic material. Alkali chlorides are among the most stable, high temperature, alkali-containing species that form in combustion environments. Alkali chlorides also represent a favored form of chlorine in these environments (35,36). Therefore, the extent of alkali vaporization for chlorinated fuels often depends as strongly on the chlorine concentration as on the alkali concentration. The amount and effects of chlorine in coal have been carefully studied (37), but most U.S. coals and many international coals have negligible amounts of chlorine. Exceptions include some Illinois basin coals in the United States and coals mined from seaboard regions. Biomass and refuse derived fuels, on the other hand, may contain significant amounts of chlorine. Chlorine is also a significant factor in corrosion of both metallic and ceramic combustor components.

The result of these transformations is a stream of inorganic vapors and fly ash with a wide range of sizes and a wide variety of compositions. Because alkali material is preferentially vaporized, submicron-size fume particles in the fly ash that form by nucleate recondensation and coagulation are usually enriched in alkali. These alkali-laden particles and vapors generally represent a small fraction of the total fly ash, but they play an inordinate role in deposit formation and property development (38–40). Condensates on surfaces enhance the probability of particles adhering to the surface when they impact. Alkali reacts with silica and, more slowly, with silicates to form low melting point compounds that are also more likely to adhere to surfaces when they impact. Iron also plays an important role in the collection of ash particles on boiler surfaces (41). Even a thin layer of rigid particles enhances the probability of larger particles adhering to a surface when they impact (42,43). The super-micron material generally originates from agglomeration of nonvolatilized, atomically dispersed material and inorganic grains. Fly ash less than 10 μm in size (PM10) is of particular concern as a respiratory health hazard. PM10 particles are also more difficult to remove from a flue gas than larger particles. Detailed studies of fly ash formation document the development of fly ash properties and their dependence on the combustion environment and fuel properties (28,44,45).

DEPOSIT FORMATION

The most significant aspect of inorganic material during combustion is its propensity to form ash deposits on combustor surfaces. The processes that contribute to deposit formation include inertial impaction (and particle capture), thermophoresis, condensation, and heterogeneous reaction. The extent to which each of these mechanisms contributes to deposit formation depends on combustor design, combustor operating conditions, and fuel properties (46). Each of these processes is briefly reviewed below.

Inertial Impaction

Inertial impaction (Figure 8) is most often the process by which the bulk of the ash deposits. Particles depositing on a surface by inertial impaction have sufficient inertia to traverse the gas stream lines and impact on the sur-

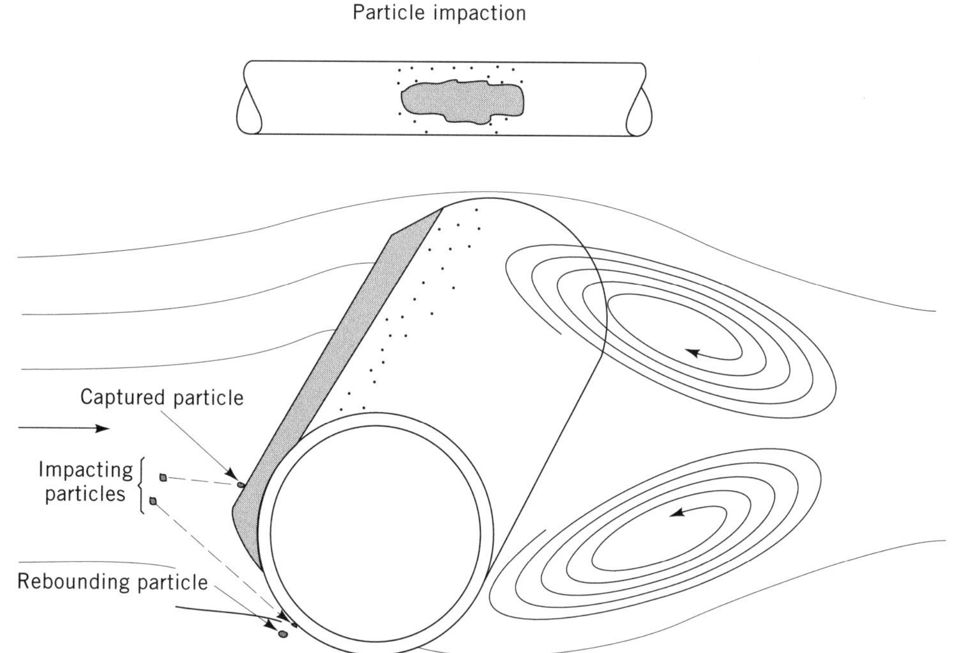

Figure 8. Inertial impaction mechanism on a cylinder in cross flow. One rebounding and one sticking particle are also illustrated.

face. The particle capture efficiency describes the propensity of these particles to stay on the surface once they impact. The rate of inertial impaction depends almost exclusively on target geometry, particle size and density, and gas flow properties. The capture efficiency depends strongly on these parameters and on particle composition and viscosity (27,47). It also depends on deposit surface composition, morphology, and viscosity (42,43). The relative magnitudes of the characteristic times and dimensions of particle and fluid relaxation processes control the rate of inertial impaction. Specifically, inertial impaction occurs when the distance a particle travels before it fully adjusts to changes in the fluid velocity is larger than the length scale of an object, or target, submerged in the fluid. The particle Stokes number is defined as the ratio of these length scales.

Inertial impaction is illustrated schematically in Figure 8 for the case of a cylinder in cross-flow. Two particles are illustrated as they approach the cylinder. Both respond to the gas flow field around the cylinder by beginning to move around the cylinder on approach. The inertia of both particles overwhelms the aerodynamic drag forces, and they impact on the cylinder. One is shown rebounding and the other sticking to the surface. Gas stream lines, including recirculation zones, are shown in light gray.

This process is most important for large particles (σ10 μm) and results in a coarse-grained deposit. The impaction rates are highest at the cylinder stagnation point, decreasing rather rapidly with angular position along the surface as measured from this stagnation point. At angular displacements larger than about 50° (as measured from the forward stagnation point), the rate of inertial impaction drops to essentially zero under conditions typical of combustor operation. Rates of inertial impaction can be correlated to particle, fluid, and target properties (20,48). The efficiency with which impacting particles adhere to surfaces has more complex dependencies on particle, surface, and gas properties (49).

Thermophoresis

Thermophoresis is a process of particle transport in a gas due to local temperature gradients. Under some circumstances, thermophoretic deposition accounts for a dominant fraction of the submicron particulate on a surface. Under most conditions relevant to coal combustion, however, the other mechanisms of deposition contribute a larger fraction of the total deposit mass than thermophoresis.

Thermophoretic forces on a particle may be induced either by the temperature gradient in the gas in which the particle is suspended or as a consequence of a temperature gradient in the particle itself. The origin of thermophoretic forces on a particle can be appreciated from the following, overly simplified argument. A particle suspended in a fluid with a strong temperature gradient interacts with molecules that have higher average kinetic energies on the side with the hot fluid than on the side with the cold fluid. The energetic collisions of the high energy molecules on the hot side of the particle create a stronger force than those of the low energy molecules on the cold side. This gives rise to a net force on the particle. In general, these forces act in the direction opposite to that of the temperature gradient, although they can act in the direction of the gradient under certain conditions of particle surface temperature.

An illustration of thermophoretic deposition is presented in Figure 9. Thermophoretic deposits are finer grained and more evenly distributed around the tube surface than deposits formed by inertial impaction, as indicated. With increasing deposit accumulation on the tube surface, there is a decrease in the temperature gradient in the thermal boundary layer, decreasing the rate of

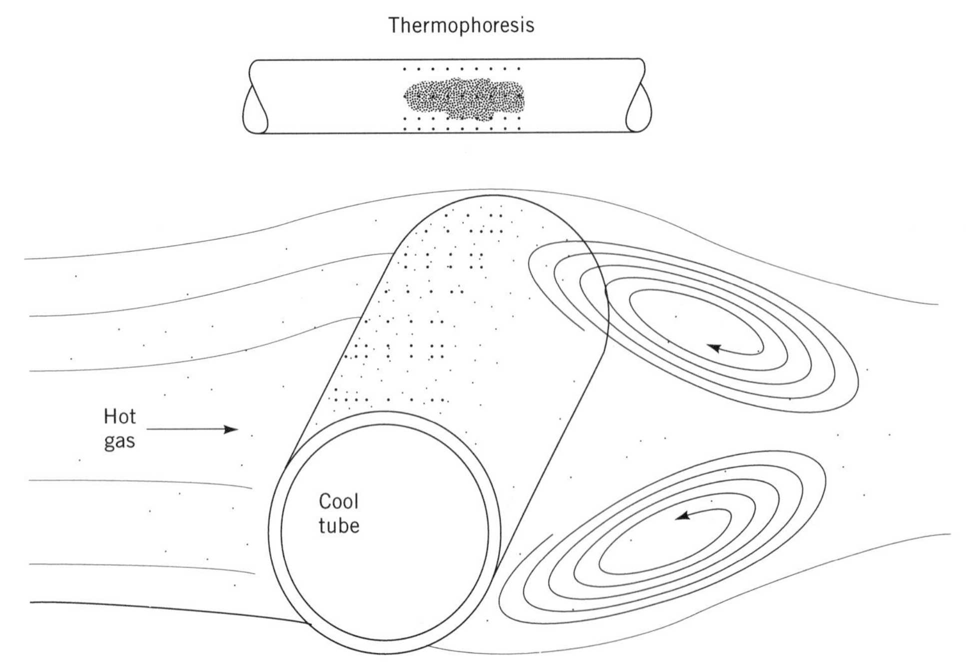

Figure 9. Thermophoretic deposition on a tube in cross flow.

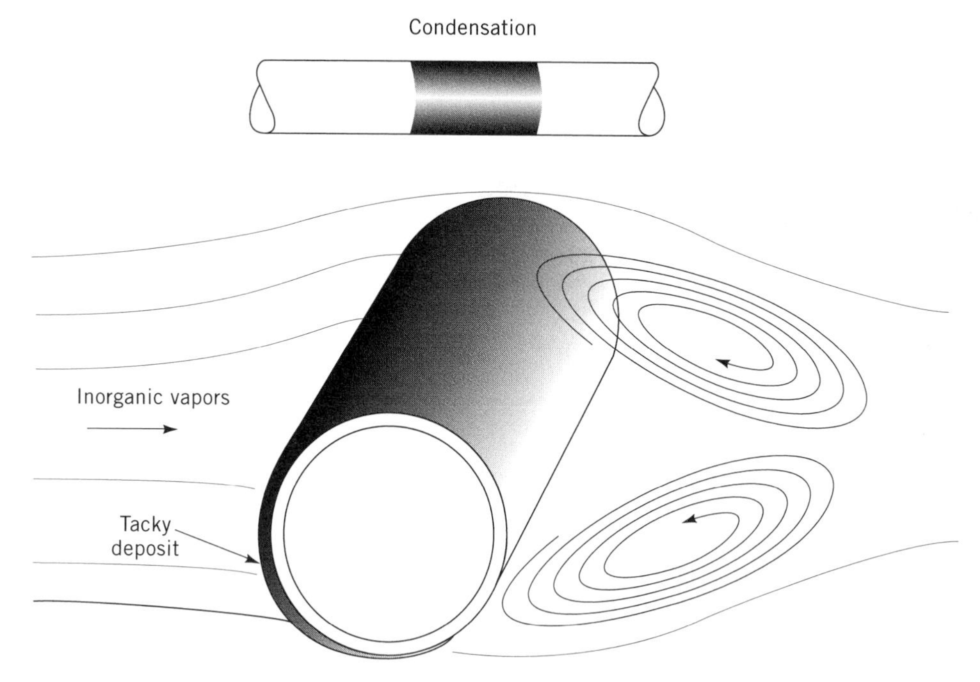

Figure 10. Condensation on a tube in cross flow.

thermophoresis. Mathematical descriptions of thermophoresis are available at several levels of complexity (17,50–52).

Condensation

Condensation is the mechanism by which vapors are collected on surfaces cooler than the local gas. An illustration of deposition by condensation on a tube in cross-flow is presented in Figure 10. The amount of condensate in a deposit depends strongly on the mode of occurrence of the inorganic material in the coal. Low rank (subbituminous) coals, lignites, and other similar fuels have the potential of producing large quantities of condensable material. Furthermore, the role of condensate in determining deposit properties can be substantially greater than the mass fraction of the condensate in the deposit might suggest. For example, condensate increases the contacting area between an otherwise granular deposit and a surface by several orders of magnitude. This increases by the difficulty of removing the deposit from the surface by a similar amount. Condensate can also increase the contacting area between particles by many orders of magnitude, having profound influences in the bulk strength, thermal conductivity, mass diffusivity, etc. of the deposit. Condensation is a relatively minor contributor to the development of deposits and their properties for most high rank coals. However, in lower grade fuels condensation becomes a significant contributor.

All vapors that enter the thermal boundary layer around a cool surface and subsequently are deposited on the surface can be thought of as condensate. Condensation occurs by at least three mechanisms: (*1*) vapors may traverse the boundary layer and heterogeneously condense on the surface or within the porous deposit, (*2*) vapors may homogeneously nucleate to form a fume and subsequently deposit by thermophoresis on the surface, and (*3*) vapors may heterogeneously condense on other particles in the boundary layer and arrive at the surface by thermophoresis (39,52,53).

Condensation deposits have no granularity at length scales larger than 0.5 μm and are more uniformly deposited on the tube than either thermophoretically or inertially deposited material. The deposits are tacky and have a strong influence on the surface capture efficiency.

Chemical Reaction

Inertial impaction and thermophoresis describe the two most significant mechanisms for transporting particulate material to a surface. Condensation involves transportation of vapors to a surface by means of a physical reaction and phase change. Chemical reactions (Fig. 11) complete the mechanisms by which mass can be accumulated in a deposit. These involve the heterogeneous reaction of gases with materials in the deposit or, less commonly, with the deposition surface itself. Some of the chemical species found in deposits are not stable at gas temperatures, alkali sulfates being typical examples. The sole source of these species is heterogeneous reactions between gas phase constituents and constituents of the lower temperature deposits.

Among the most important chemical reactions with respect to ash deposition are sulfation, alkali absorption, and oxidation. The principal sulfating species of concern are compounds containing the alkali metals, sodium and potassium. Sodium and potassium in the forms of condensed hydroxides and possibly chlorides are susceptible to sulfation (35). The formation of these materials often leads to highly reflective deposits, substantially reducing heat transfer rates in the combustor (54,55). Silica absorbs alkali material to form silicates. Silicates are less rigid and melt at lower temperatures than silica. The transformations of silica to silicates in deposits can induce sintering and significant changes in deposit properties (56,57). These reactions are slower than sulfation. Re-

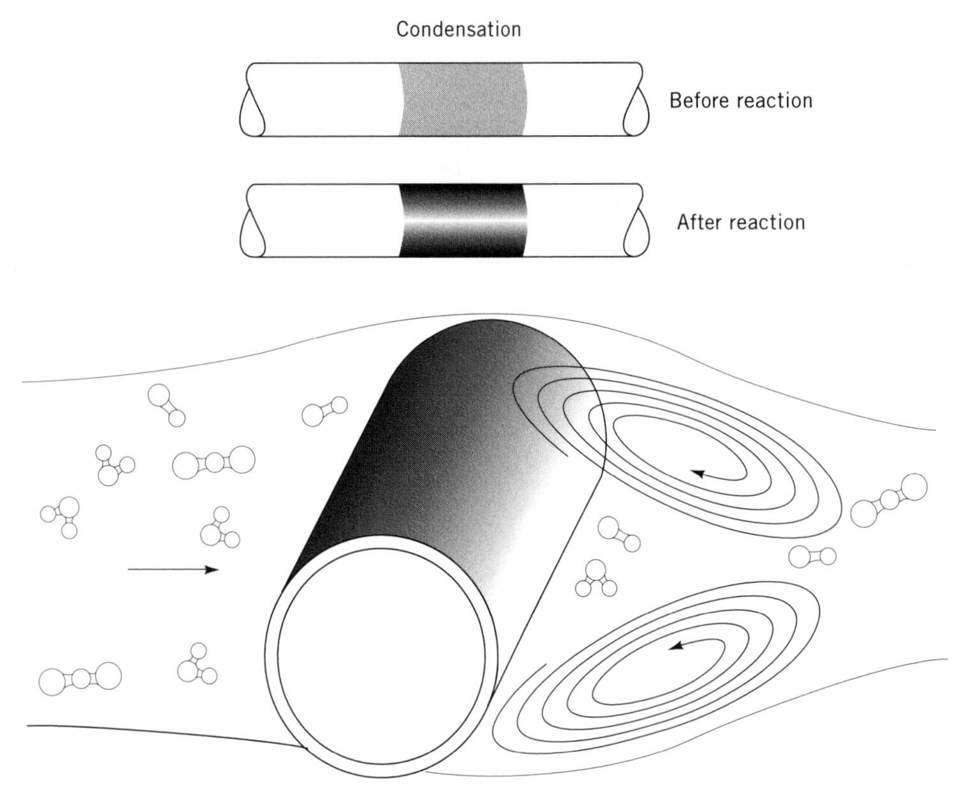

Figure 11. Chemical reaction as a mechanism of ash deposit formation. Gas-phase species such as alkali vapors and oxides of sulfur react with deposits and lead to formation of alkali silicates, alkali sulfates, and other species.

sidual char often deposits with the inorganic material on combustor surfaces. However, the char oxidizes with locally available oxygen to produce deposits with little residual carbon. In coal combustion, carbon typically accounts for less than 2% of the overall deposit mass. Chemical reactions, such as sulfation of alkali species and combustion of residual carbon in the ash, are similar to condensation in their mathematical treatment (40). Both condensation and chemical reactions are strongly temperature dependent and give rise to spatial variation in ash deposit composition. Sulfation can become particularly important when sulfur-laden fuels are blended with alkali-laden fuels (58).

EXPERIMENTAL ILLUSTRATIONS OF ASH DEPOSITION

The morphological and chemical differences in ash deposits produced by different mechanisms, as discussed above, can be demonstrated through experiment at both pilot and commercial scale (59). A deposit generated during experimental combustion of wheat straw illustrates several of the principal mechanisms of ash deposition in a single deposit. The white material on the surface of the tube in cross flow to either side of the bulk of the deposit and on the surfaces of many of the particles is composed of potassium salts that, after condensation on the surface, reacted with sulfur to form sulfates. The particles that comprise the bulk of the deposit are primarily silica, the single most prevalent element in wheat straw ash. In a commercial furnace, the deposit of this nature would extend the length of the probe, rather than form a localized mass as occurs in the experimental facility. Deposit morphologies also depend on geometry and fuel type. For example, bituminous coals in the United States typically have little atomically dispersed inorganic material and large quantities of mineral grains, whereas U.S. lignites have larger quantities of atomically dispersed material. The particulate material in the coal usually forms large fly ash particles, whereas significant fractions of the atomically dispersed material forms vapors. Therefore, the dominant mechanism for ash deposition for bituminous coals is commonly inertial impaction, whereas lignite deposits are typically initiated by condensation. This will be illustrated qualitatively by examples of deposits formed in the Sandia Multifuel Combustor. The combustor itself is described in more detail later.

The composition of deposits found in commercial systems may vary substantially from that of the fuel used to fire the system. This selective deposition occurs because different chemical species are involved in different mechanisms of ash deposit formation. For example, Figure 12 illustrates the elemental composition of a deposit collected during commercial operation of a biomass combustor burning a blend of fuels. The compositions of ash from the fuel blend and deposits found on the upper furnace walls are illustrated in the figure. While the fuel ash was composed primarily of silicon and other material, the deposit was primarily composed of alkali and sulfur. The primary mechanism of deposition in this case is condensation followed by chemical reaction. Condensation involves primarily the atomically dispersed volatile materials such as alkali metals.

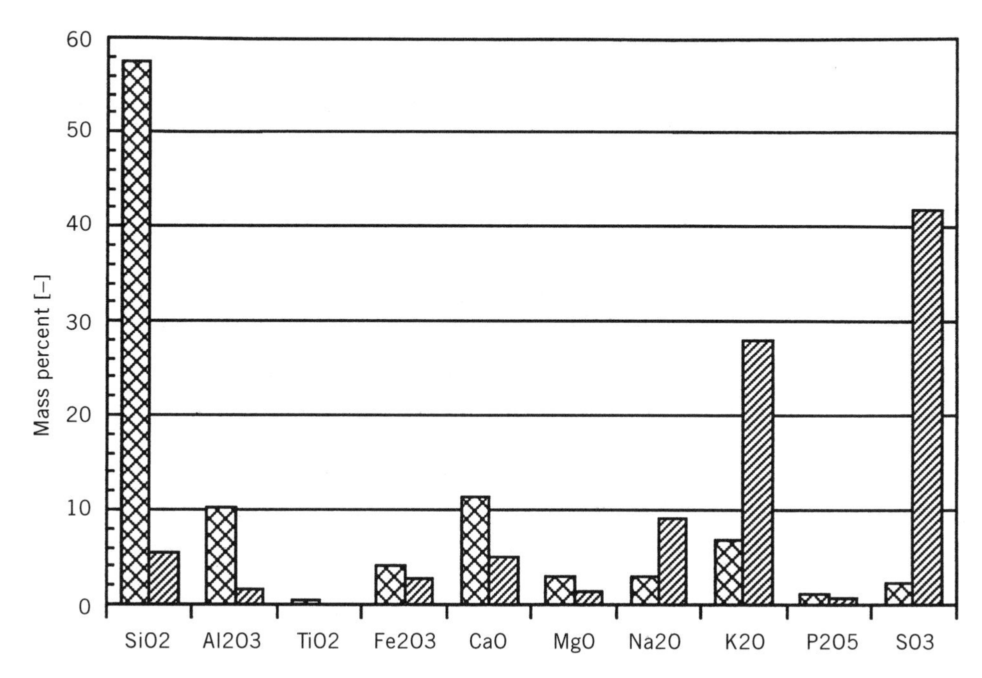

Figure 12. Elemental composition of biomass fuel and ash deposits formed on the upper walls and ceiling of the furnace. The significant differences in composition arise due to mechanisms of deposition that are selective in which elements they involve.

ASH DISPOSAL AND ENVIRONMENTAL CONCERNS

In the United States, approximately 363 kg of ash is generated per person per year from coal-fired power plants alone. The bulk of this ash is fly ash, material collected in the flue gas clean-up systems of power plants. A growing economic and environmental issue is the ultimate disposition of this material.

Coal, like any material extracted from the earth, contains low levels of inorganic contaminants that cause some concern. Landfills represent one of the most common traditional means of ash disposal. Some concern has been expressed that harmful ash components may be leached from the landfill and find their way into sensitive ecosystems or present a public health problem. If all harmful components of ash were assumed to find their way into ground water, most fly ash would be considered hazardous waste. However, most of the potentially harmful components of ash are not soluble or are only slightly soluble (60). After studying this issue, the regulatory and legal institutions in the United States determined that ash from coal-fired power plants need not be treated as hazardous waste. In several other countries, land filling of ash is not allowed, although the regulations are based more on lack of landfill space than concern about environmental or health hazards associated with the ash. In all countries, alternative markets for this ash are sought.

Two of the primary uses for coal ash and similar ashes from other fuels are as a component in Portland cement and in concrete. Some coal fly ash exhibits cementlike properties in its natural state, while others do so only when mixed with water and lime. The latter are referred to as pozzolans. Both types of ashes participate chemically in reactions that occur as concrete cures and have value as an additive during cement production. Possibly the most important measure of fly ash suitability for use in the cement industry is its carbon content. Concrete used in applications subjected to temperature changes (eg, freezing temperatures) contains small bubbles to provide relief for expansion and contraction stresses. Carbon in fly ash behaves much like activated carbon and absorbs these bubbles, producing an inferior material. Concrete also becomes increasingly discolored if the fly ash contains high carbon content or other colored components. This presents problems that are largely cosmetic but nonetheless present marketing problems for the cement industry. Commonly, the loss on ignition (weight loss during prescribed heating conditions) of fly ash for use in cement is usually limited to less than a few percent. The amount of carbon in fly ash from low rank coal is typically low (1% or less), with increasing amounts of carbon as coal rank increases and as inorganic material in the original fuel decreases. In all cases, large fly ash particles contain far greater concentrations of carbon than small fly ash particles. Carbon in fly ash also depends on the type of combustor used to generate the ash. Typically, ashes from fixed or slowly moving grates burning either coal or biomass contain higher amounts of carbon than ashes from bubbling or, especially, circulating fluid beds. Besides carbon content, specifications for fly ash to be used in cement manufacture sometimes include a combined amount of SiO_2, Al_2O_3, and Fe_2O_3 of not less than 70%, no more than 5% MgO, and no more than 4% SO_3. Its moisture content should be less than 10% and its available alkali should be less than 1.5% as Na_2O. Fly ash used in concrete (ie, not as cement but as aggregate in concrete) is less tightly constrained. Ash from low rank coals typically contains high concentrations of CaO (25–35%) and, if sulfur contents can be kept low, represents an exceptionally good cement additive. Fly ash used in concrete reduces the amount of water required, improves the strength, improves workability, lowers permeability, decreases heat generation during curing, and can lead to lower bulk densities to the extent it contains cenospheric particles. It normally should have a surface area of around 3000 cm^2/g and 88% of it should pass a 325 mesh (44 μm) (7).

Other uses for fly ash include backfill, road beds, foundry powder, fire brick (especially for cenospheric particles), and sandblasting grit. In addition to these applications, bottom ash is used as an aggregate for block manufacturing. Typically, 20% or less of the fly ash generated in the United States finds practical use due to costs associated with shipping and handling material and historically low fees for landfilling. In Europe and other densely populated areas with less available landfill volume, larger fractions of fly ash are generally used (>30% of bituminous fly ash in 1974, eg) (61). The United States will likely follow this trend as landfill volume becomes increasingly scarce.

Passage of the Clean Air Act Amendments in the United States in 1990 (62) ushered in major new legislation with regard to inorganic components of fuels in the United States. In these amendments, the EPA is directed to promulgate emission regulations for any compound that contains arsenic, beryllium, cadmium, chromium, cobalt, cyanide, lead, manganese, mercury, nickel, selenium, any radionuclide (including radon), or fine mineral fibers. Combustion systems potentially emit large quantities of some of these materials due less to their concentrations in the flue gas than to the volume of the flue gas produced. Preliminary studies indicate that mercury and possibly arsenic and selenium will be the compounds of greatest concern for most power generation facilities.

BIBLIOGRAPHY

1. J. F. Unsworth, D. J. Barratt, and P. T. Roberts in L. L. Anderson, ed., *Coal Science and Technology,* Elsevier, Amsterdam, The Netherlands, 1991.
2. R. P. Johnson, R. G. Plangemann, and M. D. Rehm, *Coal Quality Effects on Boiler Performance: An Eastern Perspective,* EPRI Report, RP2256-8, Electric Power Research Institute, 1991.
3. Energy Information Office, *Monthly Energy Review,* DOE/EIA-0035(93/09), U.S. Department of Energy, Washington, D.C., Sept. 1993.
4. R. E. Mitchell, R. H. Hurt, L. L. Baxter, and D. R. Hardesty, *Compilation of Sandia Coal Char Combustion Data and Kinetic Analyses: Milestone Report,* SAND92-8208, Sandia, 1992.
5. S. C. Stultz, and J. B. Kitto, eds., *Steam—Its Generation and Use,* Babcock & Wilcox Co., Barberton, Ohio, 1992.
6. E. Raask, *J. Inst. Energy* **57,** 231 (1984).
7. W. T. Reid in M. A. Elliot, ed., *Chemistry of Coal Utilization: Second Supplementary Volume,* John Wiley & Sons, Inc., New York, 1981, pp. 1389–1446.
8. R. B. Finkelman, Ph.D. Dissertation, University of Maryland, 1980.
9. N. Berkowitz, *The Chemistry of Coal,* Elsevier, Amsterdam, The Netherlands, 1985.
10. D. W. van Krevelen, *Coal: Typology—Chemistry—Physics—Constitution,* Elsevier, Amsterdam, The Netherlands, 1981.
11. D. C. Elliot in E. J. Soltes, T. A. Milne, eds., *Pyrolysis Oils from Biomass: Producing, Analyzing, and Upgrading,* American Chemical Society, Washington D.C., 1988, pp. 55–65.
12. H. Marschner, *Mineral Nutrition of Higher Plants,* Harcourt Brace Jovanovich, London, 1986.
13. C. Karr Jr., eds., *Analytical Methods for Coal and Coal Products,* Vol. 1 Academic Press, Inc., New York, 1978.
14. E. C. Winegartner, *Coal Fouling and Slagging,* ASME, New York, 1974.
15. R. W. Bryers, *Symposium on Slagging and Fouling in Steam Generators,* ASME, New York, 1987.
16. L. L. Baxter and D. R. Hardesty, *The Fate of Mineral Matter During Pulverized Coal Combustion: January–March,* SAND92-8210, Sandia National Laboratories, 1992.
17. L. L. Baxter and D. R. Hardesty, *The Fate of Mineral Matter During Pulverized Coal Combustion: April—June,* SAND92-8227, Sandia National Laboratories, 1992.
18. S. A. Benson and P. L. Holm, *Ind. Eng. Chem. Prod. Res. Dev.* **24,** 145–149 (1985).
19. M. F. Abbott, R. E. Douglas, C. E. Fink, N. J. Deluliis, and L. L. Baxter, paper presented at the Engineering Foundation Conference on the Impact of Ash Deposition on Coal-Fired Plants, 1993.
20. L. L. Baxter and D. R. Hardesty, *The Fate of Mineral Matter During Pulverized Coal Combustion,* SAND91-8217, Sandia National Laboratories, 1991.
21. A. Carpenter and N. Skorupska, *Coal Combustion—Analysis and Testing,* IEA Coal Research, 1993.
22. F. R. Karner, C. J. Zygarlicke, D. W. Brekke, E. N. Steadman, and S. A. Benson, *Power Eng.,* 35–38 (1994).
23. L. L. Baxter, *Prog. Energy Combustion Sci.* **16,** 261–266 (1990).
24. H. Yu, J. E. Marchek, N. L. Adair, and J. N. Harb, *The Impact of Ash Deposition on Coal Fired Power Plants.* Taylor & Francis, 1993.
25. N. Y. C. Yang and L. L. Baxter, paper presented at the Conference on Inorganic Transformations and Ash Deposition During Combustion, 1991.
26. J. M. Beér, L. S. Monroe, L. E. Barta, and A. F. Sarofim, paper presented at the Seventh International Pittsburgh Coal Conference, 1990.
27. S. Srinivasachar, C. L. Senior, J. J. Helble, and J. W. Moore, paper presented at the Twenty-Fourth Symposium (International) on Combustion. 1992.
28. C. J. Zygarlicke, M. Ramanathan, and T. A. Erickson in S. A. Benson, ed., *Inorganic Transformations and Ash Deposition During Combustion,* American Society of Mechanical Engineers, New York, 1992, pp. 525–544.
29. J. J. Helble and A. F. Sarofim, *J. Colloid Interface Sci.* **128,** 348–362 (1989).
30. L. L. Baxter and R. E. Mitchell, *Combustion Flame* **88,** 1–14 (1992).
31. L. L. Baxter, R. E. Mitchell, and T. H. Fletcher, *Combustion Flame,* in press.
32. L. L. Baxter, *Combustion Flame* **90,** 174–184 (1992).
33. S. Srinivasachar, J. J. Helble, and A. A. Boni, *Prog. Energy Combustion Sci.* **16,** 281–292 (1990).
34. S. Srinivasachar and co-workers, *Prog. Energy Combustion Sci.* **16,** 293–302 (1990).
35. S. Srinivasachar, J. J. Helble, D. O. Ham, and G. Domazetis, *Prog. Energy Combustion Sci.* **16,** 303–309 (1990).
36. J. J. Helble, S. Srinivasachar, and A. A. Boni, *Combustion Sci. Technol.* **81,** 193–205 (1992).

37. J. Stringer and D. D. Banerjee, eds., *Chlorine in Coal,* Elsevier, Amsterdam, The Netherlands, 1991.
38. L. J. Wibberly and T. F. Wall, *Fuel* **61,** 93–99 (1982).
39. J. J. Helble, M. Neville, and A. F. Sarofim, paper presented at the Twenty-First Symposium (International) on Combustion, 1986.
40. L. L. Baxter, *Biomass Bioenergy* **4,** 85–102 (1993).
41. A. T. S. Cunningham, W. H. Gibb, A. R. Jones, F. Wigley, and J. Williamson, paper presented at the Conference on Inorganic Transformations and Ash Deposition During Combustion, 1991.
42. C. L. Wagoner and X.-X. Yan, paper presented at the Conference on Inorganic Transformations and Ash Deposition During Combustion, 1991.
43. S. M. Smouse and C. L. Wagoner, paper presented at the Conference on Inorganic Transformations and Ash Deposition During Combustion, 1991.
44. R. J. Quann, M. Neville, and A. F. Sarofim, *Combustion Sci. Technol.* **74,** 245–265 (1990).
45. S.-G. Kang, J. J. Helble, A. F. Sarofim, and J. M. Beér, paper presented at the Twenty-Second Symposium (International) on Combustion, 1988.
46. L. L. Baxter and R. W. DeSollar, *Fuel* **72,** 1411–1418 (1993).
47. J. Srinivasachar, J. J. Helble, and A. A. Boni, paper presented at the Twenty-Third Symposium (International) on Combustion, 1990.
48. R. Israel and D. E. Rosner, *Aerosol Sci. Technol.* **2,** 45–51 (1983).
49. L. L. Baxter, K. R. Hencken, and N. S. Harding, paper presented at the Twenty-Third Symposium (International) on Combustion, 1990.
50. S. A. Gökoglu and D. E. Rosner, *Int. J. Heat Mass Transfer* **27,** 639–646 (1984).
51. S. A. Gökoglu and D. E. Rosner, *AIAA J.* **24,** 172–179 (1986).
52. D. E. Rosner, and M. Tassopoulos, *AIChE J.* **35,** 1497–1508 (1989).
53. J. L. Castillo and D. E. Rosner, *Int. J. Multiphase Flow* **14,** 99–120 (1988).
54. L. L. Baxter, G. H. Richards, D. K. Ottesen, and J. N. Harb, *Energy Fuels* **7,** 755–760 (1993).
55. G. H. Richards and co-workers, paper presented at the Twenty-Fifth Symposium (International) on Combustion, 1994.
56. L. L. Baxter and co-workers, paper presented at the Second International Conference on Combustion Technologies for a Clean Environment, 1993.
57. L. L. Baxter, paper presented at the EPRI Conference on The Effects of Coal Quality on Power Plants, 1992.
58. L. L. Baxter and L. Dora, Paper No. 92-JPGC-FACT-14, ASME, New York, 1992.
59. N. S. Harding and M. C. Mai, in R. W. Bryers and K. S. Vorres, eds., *Mineral Matter and Ash Deposition from Coal* Engineering Trustees, Inc., 1990, pp. 375–399.
60. C. W. Gehrs, D. S. Shriner, S. E. Herbes, E. J. Salmon, and H. Perry in M. A. Elliot, ed., *Chemistry of Coal Utilization: Second Supplementary Volume,* John Wiley & Sons, Inc., New York, 1981, pp. 2159–2223.
61. H. Jüntgen, J. Klein, K. Knoblauch, H.-J. Schröter, and J. Schulze in Ref. 60.
62. *Federal Register: Part III; Air Contaminants,* Final Rule, 29 CFR Part 1910 (1989).

ASPHALT

James G. Speight
Western Research Institute
Laramie, Wyoming

Asphalt is a dark brown to black, cementitious material in which the predominant constituents are high molecular weight hydrocarbon materials with heteroatoms (nitrogen, oxygen, and sulfur) liberally scattered throughout the molecular structures. Asphalts occur naturally (bitumens) or are obtained as end-products of petroleum processing (1). The high molecular weight polar constituents of asphalts (the asphaltenes) are soluble in carbon disulfide, pyridine, aromatic hydrocarbons, and chlorinated hydrocarbons.

There has been a tendency, although somewhat incorrect (2), to use the terms "bitumen" or "asphaltic bitumen" as a synonym for asphalt and to apply the term "asphalt" to the mixture of the organic binder (asphalt) and inorganic matter that is used for paving purposes. On the other hand, terms such as (vacuum) pitch and (vacuum) tar originated in the early days of the petroleum refining industry and have remained a part of modern refinery terminology. Pitches and tars are the products of the destructive distillation of coal, crude oils, and other organic materials (2–4).

More correctly defined, an asphalt (Table 1) is a manufactured material which is produced during petroleum processing, while a bitumen (Table 1) is defined as a naturally occurring, black, cementitious material in which the predominant constituents are high molecular weight hydrocarbons with heteroatoms (nitrogen, oxygen, and sulfur) liberally scattered throughout the molecular structures. However, the terms "asphalt" and "bitumen" are still used interchangeably in general practice.

Finally, "deasphalted oil" and "deasphaltened oil" are terms that are used interchangeably to mean (as one example) the heptane-soluble materials which are produced when the heptane-insoluble materials (asphaltenes) are separated from asphalt (Fig. 1). If low boiling liquid hydrocarbons other than heptane are employed for the separation, the names can be prefixed accordingly to alleviate any confusion; the deasphalted oil and asphaltene fractions change in composition and character with the nature of the hydrocarbon used for the separation (2). The term "petrolenes" is applied to those constituents of the deasphalted oil which can be distilled without thermal decomposition at temperatures lower than 300°C and at atmospheric pressure. On the other hand, "maltenes" are defined as that fraction of the deasphalted oil that is high boiling (>300°C) and cannot be distilled at atmospheric pressure without thermal decomposition. Thus, in terms of these definitions, asphalt prepared from a residuum contains maltenes and asphaltenes.

See also Aircraft fuels; Clean air act, mobile sources; Petroleum products and uses; Transportation fuels, automotive fuels; Petroleum refining.

NATURALLY OCCURRING MATERIALS

Native Asphalt (Bitumen)

Construction activities in ancient times depended on native asphalts (5,6). For example, there is some (but often

Table 1. Petroleum and Related Materials can be Divided into Various Class Subgroups

Natural Materials	Manufactured Materials	Derived Materials[a]
Asphaltite	Asphalt[b]	Asphaltenes
Asphaltoid	Coke	Carbenes
Bitumen (native asphalt)	Pitch	Carboids
Bituminous rock	Residuum[c]	Deasphalted oil
Bituminous sand	Synthetic crude oil	Deasphaltened oil
Heavy oil	Tar	Maltenes
Kerogen	Wax	Oils (saturates, aromatics)
Migrabitumen		Petrolenes
Mineral wax		Resins
Natural gas		
Oil sand[d]		
Petroleum		
Tar sand[e]		

[a] See Figure 1.
[b] Asphalt—a product of a refinery operation; usually made from a residuum.
[c] Residuum—the nonvolatile portion of petroleum and often further defined as "atmospheric" (bp >350°C) or vacuum (bp >565°C).
[d] Tar sands—a misnomer, tar is a product of coal processing; oil sands—also a misnomer but equivalent to usage of "oil shale": bituminous sands—more correct; bitumen is a naturally occurring asphalt.

questionable) evidence to support the concept that the bituminous product found on the surface of waters, such as the Dead Sea, was used by the Egyptians for mummification. Bitumen was used by the Sumerians as a building and ornamental mastic. There is also evidence that they used the technique of evaporation (perhaps even distillation) to remove the volatile constituents from the petroleumlike material, which they collected from various outcrops and seepages, thereby manufacturing various grades of asphalt.

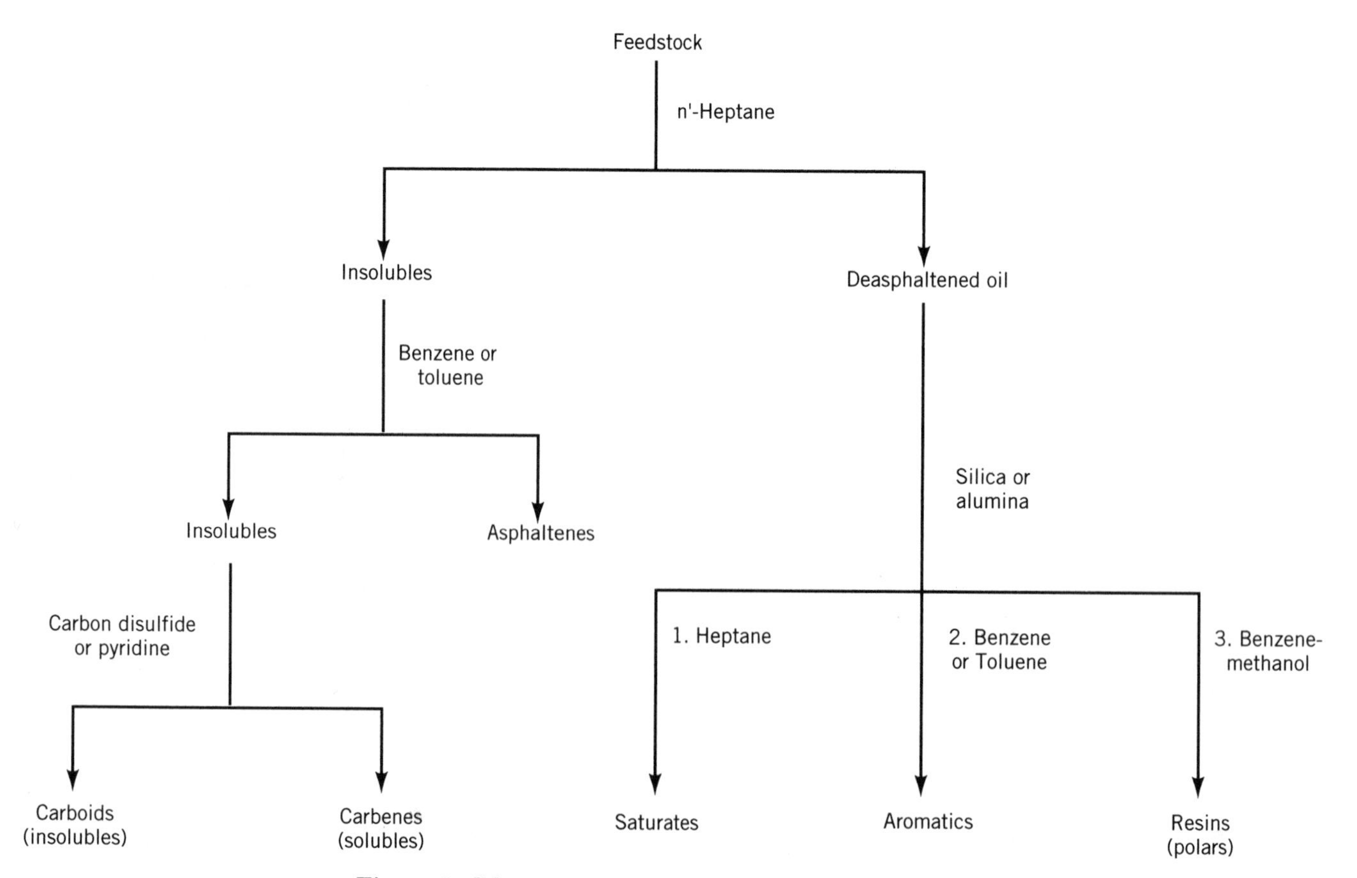

Figure 1. Schematic representation of petroleum separation.

The use of asphalt continued throughout antiquity and in more modern times, mention is made of Trinidad asphalt by Raleigh during his visit to this island in 1595 (2).

Lake Asphalt

The well-known Trinidad Lake asphalt (bitumen), which exists in a number of deposits on the island of Trinidad, near the northeast coast of Venezuela, was first used for paving in the United States in 1874. The bitumen from Bermudez Lake in Venezuela was used extensively for paving and waterproofing in the United States (7).

Trinidad asphalt has a relatively uniform composition of 29 wt % water, 39 wt % bitumen (organic material soluble in carbon disulfide), 27 wt % mineral matter (estimated as mineral ash by ignition and by combustion to the metal oxides), and 5 wt % bitumen (organic material insoluble in carbon disulfide) which remains adsorbed on the mineral matter. Refining of this lake asphalt is essentially a process of dehydration in which the crude material is heated to approximately 165°C to yield a product which contains approximately 36 wt % mineral matter.

Rock Asphalt

Rock asphalt (rock bitumen, bituminous rock) deposits contain from 5–25 wt % asphalt. When extracted, the bitumen varies from a soft, almost mobile material to a hard, immobile material. The mineral content, usually sandstone or limestone, has many gradations. Rock asphalt is found in Texas, Alabama, Oklahoma, Colorado, California, and Kentucky. The composition of Kentucky asphalt is suitable for paving in its natural condition if selected at bitumen contents of 7–9%. Some of the well-known and long-used asphalt deposits of Europe are located at Seyssel (France), Ragusa (Sicily), Val-de-Travers (Switzerland), and Vorwohle (Germany) (6,7).

Bitumen also occurs in various oil sand (also called tar sand) deposits which occur widely scattered throughout the world, and the bitumen, particularly that from the deposits in Alberta (Canada) and Utah (United States) is available by means of various extraction technologies (4,8). The properties and character of the bitumen suggest that, when used as an asphaltic binder, the bitumen compares favorably with specification-grade petroleum asphalts, may have superior aging characteristics, and may produce more water-resistant paving mixtures than the typical petroleum asphalts.

MANUFACTURED MATERIALS

Petroleum asphalts are also organic materials but, compared to bitumen, have only trace amounts of inorganic matter. They derive their characteristics from the nature of the petroleum from which they are produced, with variations in properties introduced by choice of the manufacturing process (2).

The first crude petroleum produced in the United States, from the Pennsylvania and Ohio fields, was unsuitable for manufacture of asphalt because it could not be distilled to a residuum without thermal decomposition, hence the use of asphalt from Trinidad and from Bermudez Lake (see above). However, in the 1880s, the means of preparing different grades of asphalt by conversion of residua to asphaltic products by blowing air at high temperatures became available, thus avoiding the necessity for admixture of the residua with the native asphalt. By the early 1900s, petroleum asphalt was available in appreciable quantities because of the discovery, in California, of petroleum which yielded semisolid distillation residua. Residua from the Californian crude oils were similar to the Trinidad Lake and Bermudez lake bitumens and could be used directly for paving. The advent of the first continuous tube or pipe still in 1912 (2) and the use of reduced pressure during distillation permitted the reduction of petroleum feedstocks to harder grades of residua and also permitted the use of various types of petroleum that had previously been unsuitable for asphalt manufacture.

Although there were, at one time, many refineries, or refinery units, of which the prime function was to produce asphalt, asphalt is now, primarily, a product of an integrated refinery, and is actually one of several products of the distillation sequence (Fig. 2). Crude petroleum may be selected for these refineries for a variety of other product requirements as well as asphalt content (Table 2). In fact, the asphalt (or residuum) produced may vary somewhat in characteristics depending upon the cut-point (Table 3), and may vary from one refinery-crude system to another (4).

Straight Run Asphalt

In the initial stage of refining, petroleum feedstock that has been previously heated to approximately 340°C is injected into a fractionating column at atmospheric pressure (Fig. 2). Part of the feedstock is removed as overhead (volatile) products, and the nonvolatile residuum, if it

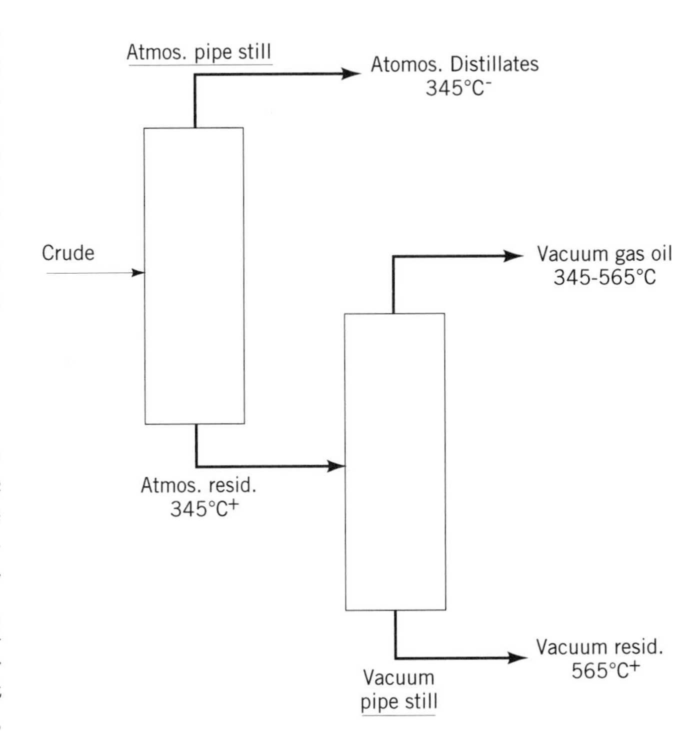

Figure 2. Simplified representation of distillation in a petroleum refinery.

Table 2. Asphalt Yields from Various Crude Oils

Crude Oil	Approximate Yield of 100 Penetration Asphalt, %
Nigerian Light	7
Arabian Light	14
North Slope	21
Arabian Heavy	31
L. A. Basin	44
California Valley	55
California Coastal	65
Boscan	79

meets the desired specifications, is termed a "straight-run" (straight-reduced) asphalt. However, most crude oils cannot be distilled to specification grade asphalts at atmospheric pressure because of the presence of high boiling, mobile (liquid) constituents which are usually referred to as "gas oils." Often, a second distillation is necessary and, as a supplement to the atmospheric distillation process, a second fractionating tower (a vacuum tower) is added (Fig. 2).

Vacuum distillation units are continuous flow-through units and include a heating unit, the vapor-liquid flash separation zone, the fractionating zone, and the vacuum producing equipment. The heated feedstock is pumped into a flash zone where the more volatile constituents vaporize; the asphalt is the nonvolatile residuum. The manufacture of asphalt by vacuum distillation does not change the chemical nature of asphalt. The temperature of the distillation unit is usually maintained below 340°C, and below the temperature (350°C) at which the onset of cracking (thermal decomposition) occurs after which the rate of thermal decomposition increases in a near-logarithmic manner. Alternatively, the high vapor velocity in a distillation unit is sufficient that the residence time at the high temperature is inadequate for cracking to occur.

Hence, distillation is a concentration process in which the nonvolatile constituents of the feedstock are retained in the asphaltic residuum, with no chemical change, although decarboxylation of organic acids can occur (9). It is generally assumed that any chemical changes of the constituents are minimal and are not sufficient to cause any major changes to the overall character of the residuum. However, the properties (eg, viscosity) of the residuum vary and depend upon the amount of the volatile constituents removed.

Propane Asphalt

Crude oils contain different quantities of nonvolatile residuum (Tables 2 and 3) and, hence, asphalt. The propane deasphalting process involves the separation of asphalt from a residuum (usually an atmospheric residuum) by treatment with liquid propane (2,4,10). Propane is more often used for the process, although propane–butane mixtures have also been employed with the necessary changes (because of the presence of butane) in process conditions, and resulting variations in product properties (eg, hardness).

Propane deasphalting is conventionally applied to low-asphalt (<15 wt % asphalt) crude oils, which are generally different in type and source from the higher asphalt crude oils. The asphalt produced from this process is normally blended with other asphaltic residua for making paving asphalt.

Propane deasphalting is a countercurrent extraction process (Fig. 3) in which the feedstock is introduced near the top of an extraction tower and the liquid propane near the bottom, using propane-to-feedstock ratios from 4:1 to

Table 3. Properties of Residua Produced from Tia Juana Light (Venezuela) Crude Oil

Property	Value								
Boiling range									
°C	whole crude	>220	>295	>345	>370	>400	>455	>510	>565
Yield on crude, vol %	100.0	70.2	57.4	48.9	44.4	39.7	31.2	23.8	17.9
Gravity, *API	31.6	22.5	19.4	17.3	16.3	15.1	12.6	9.9	7.1
Specific gravity	0.8676	0.9188	0.9377	0.9509	0.9574	0.9652	0.9820	1.007	1.0209
Sulfur, wt %	1.08	1.42	1.64	1.78	1.84	1.93	2.12	2.35	2.59
Carbon residue (Conradson), wt %		6.8	8.1	9.3	10.2	11.2	13.8	17.2	21.6
Nitrogen, wt %				0.33	0.36	0.39	0.45	0.52	0.00
Pour point, °F	−5	15	30	45	50	60	75	95	120
Viscosity:									
Kinematic, cSt									
at 38°C	10.2	83.0	315	890	1590	3100			
at 99°C		9.6	19.6	35.0	50.0	77.0	220	1010	7959
Furol (SFS)									
50°C			70.6	172	292	528			
99°C					25.2	37.6	106	484	3760
Universal (SUS) at 210°F		57.8	96.8	165	234	359	1025		
Metals									
Vanadium, ppm				185					450
Nickel, ppm				25					64
Iron, ppm				28					48

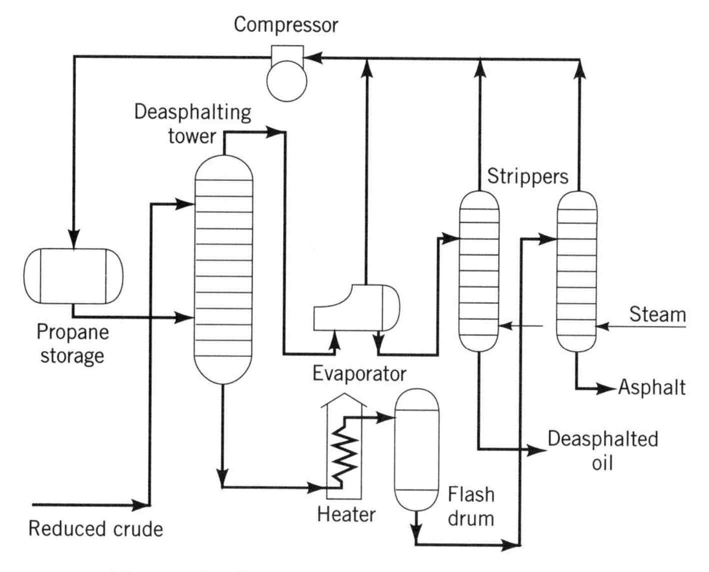

Figure 3. The propane deasphalting process.

10:1. The deasphalted oil–propane solution is withdrawn overhead, the asphalt withdrawn from the bottom, and each is subsequently stripped of propane. Temperature, solvent ratio, and pressure (Table 4) each have an effect upon the yield of the oil and asphalt components (Table 5). Contrary to the distillation (straight reduction), which is a high temperature and low pressure process, propane deasphalting is a low temperature and higher pressure process.

There are differences in the properties of asphalts prepared by propane deasphalting and those prepared by vacuum distillation from the same crude oil residuum feedstock (Fig. 4). In fact, propane deasphalting also has the ability to reduce the yield of residuum (bases on the crude oil) even further and produce an asphalt having a lower viscosity, higher ductility, and higher temperature susceptibility than other asphalts (7). Although, such properties might be anticipated to be very much dependent upon the parent crude oil.

Air-Blown Asphalt

Air blowing is an exothermic process that is dependent upon several variable parameters (2,7,10). In this process (Fig. 5), an asphalt is converted to a harder product by air contact at 200–275°C. The chemistry of the process is

Table 4. Typical Operating Conditions for a Propane Deasphalting Unit

Function	Value
Throughput, kL/d	800[a]
Charge stock	reduced crude
Gravity charge, *API[a]	20.0
Temperature of deasphalting tower, °C	65
Solvent-to-oil ratio, by volume	6:1
Pressure, kPa[c]	2170

[a] 5000 barrels/d

[b] $\text{*API gravity at } 15.6°\text{C} = \frac{141.5}{\text{sp gr at } 15.6/15.6°\text{C}} - 131.5.$

[c] To convert kPa to psi, multiply by 0.145

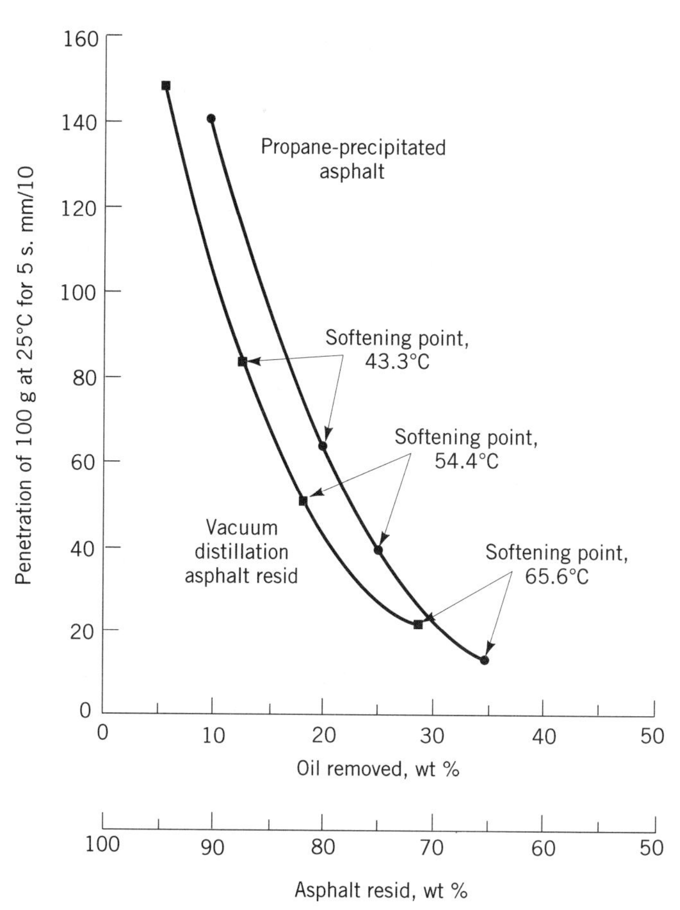

Figure 4. Comparison of product properties for the distillation and propane deasphalting processes (Lagunillas, Venezuela, crude oil).

variable and, to some extent, speculative, although dehydrogenation of some of the species (ie, hydroaromatics are converted to aromatics) are known to occur. In theory, the aerial oxygen abstracts hydrogen (to form water) from the constituents of the asphalt, thereby acting as a promoter for intermolecular condensation and cross-linking of the asphalt constituents. There is evidence for intermolecular condensation of the constituents, and molecular weight increases (of the product over the feedstock) have been noted, but not to the extent that the term "polymerization" (which is often used, incorrectly, to describe the chemistry of air blowing) suggests; oxygen is not retained in the asphalt product to any great extent (11).

However, the chemistry may be much more complex than the simple theory predicts. For example, not only are "new" asphaltenes produced during the air-blowing process, but the evidence suggests that the constituents of the asphalt, which have higher proportions of heteroatoms (ie, nitrogen, oxygen, and sulfur), will react preferentially with oxygen, leaving less polar substances unreacted (12,13). This also raises the possibility of incompatibility of the resulting products. If the more polar species (ie, the resins) are preferentially oxidized to produce more asphaltenes, the means by which the original asphaltenes are dispersed in the oily medium is removed and phase separation (asphaltene deposition, also called incompatibility) should be anticipated.

Table 5. Yields and Properties of Products Obtained By Propane and Deasphalting Various Crude Oils

Process	Crude Source				
	Pa.	Okla.	Ill.	LaDuc, Canada	Iraq
Reduced crude					
Gravity, *API[a]	25.1	19.3	13.7	12.1	7.1
Kinematic viscosity at 98.9°C, mm^2/s (Saybolt, s)	ca 47.2 (200)	ca 82.8 (385)	ca 378 (1760)	ca 441 (2050)	ca 1050 (4880)
Crude, vol %	24.0	26.1	11.3	15.2	16.4
Deasphalted oil					
Yield, vol %	83.7	75.6	50.7	47.1	34.1
gravity, °API[a]	26.6	23.6	23.4	22.7	20.2
Kinematic viscosity at 98.9°C, mm^2/s (Saybolt, s)	ca 30.4 (144)	ca 29.8 (141)	ca 40.7 (190)	ca 42.6 (199)	ca 38.2 (179)
Precipitated asphalt					
Yield, vol %	16.3	24.4	49.3	52.9	65.9
Specific gravity at 15.6°C	0.953	1.023	1.030	1.054	1.064
Softening point (ring and ball), °C		62.2	62.8	80	62.8
Penetration of 100 g at 25°C for 5 s, mm/10		23	21	0	16

[a] $°\text{API gravity at } 15.6°\text{C} = \frac{141.5}{\text{sp gr at } 15.6/15.6°\text{C}} - 131.5.$

In some cases, some of the bonds "created" during the oxidation process are unstable and, after an induction period, result in a phenomenon termed "fallback." When "fallback" occurs, usually at a time when the hardened or oxidized asphalt is stored at or near the original processing or reaction temperature, a noticeable softening of the asphalt product is observed (14).

Thermal Asphalt

Thermal or cracked asphalts differ from other asphalts in that they are products of a cracking process. Thermal asphalts have significant application as saturants for cellulosic building products, such as insulation boards, brick-finish siding, and fiber soil pipes. However, these products have a relatively high specific gravity, low viscosity, and high temperature susceptibility. They often contain insoluble constituents (carbenes; Fig. 1), as indicated by the Oliensis test (15) or other solubility tests (2), and are believed to be precursors to coke (2).

The thermal cracking process (Fig. 6) involves preheating the feedstock to 480–600°C and then discharging the heated material into a reaction vessel or chamber under pressures up to 1480 kPa (200 psi). Cracking decomposition takes place to form both lighter and heavier products, and distillation is used to separate the product stream into gas, gasoline, middle distillate, and an asphaltic residuum.

Blended Asphalt

A refinery may stock two or more grades of asphalt which vary in viscosity but cover the viscosity scale of the product grade requirements. From these, intermediate grades of asphalt can be prepared by blending (proportioning) the individual asphalts according to the requirements of the final asphalt product.

The preparation of asphalts in liquid form by blending (cutting back) an asphalt with a petroleum (distillate) fraction is customary. There are three general types of cutback asphalt which differ mainly in the diluent used. The slow-curing type is often called road oil, which is made by direct reduction, but most are commonly blended. The medium curing and rapid curing products usually have four grades within a given type; they differ mainly in the amount of diluent used and in the kinematic viscosity of the blended product.

Preparation is accomplished by blending the diluent into the hot base asphalt and is generally accomplished in tanks equipped with coils for air agitation or with a mechanical stirrer or a vortex mixer. Line blending in a batch circulation system or in a continuous fashion is used where the volume produced justifies the extra facilities. A continuous, line-blending system is applicable to the manufacture of cutback asphalts and asphalt cements (Fig. 7).

Asphalt Emulsions

An emulsion is an immiscible liquid dispersed in another liquid in the form of very fine droplets from about 1–25 μm and an average of 5 μm diameter. In the most common asphalt emulsion, the oil-in-water type, the asphalt is the dispersed (internal) phase, and water is the continuous (external) phase.

Most of the important properties of an emulsion are dependent upon the amount and type of emulsifying agent used. The more popular cationic emulsions are prepared with cation-active agents that produce emulsion droplets with positive charges. The anionic agent is usually the sodium or potassium salt of a fatty acid, such as palmitic acid $[CH_3(CH_2)_{14}CO_2H]$, stearic acid $[CH_3(CH_2)_{16}CO_2H]$, and linoleic acid $[CH_3(CH_2)_4CH=$

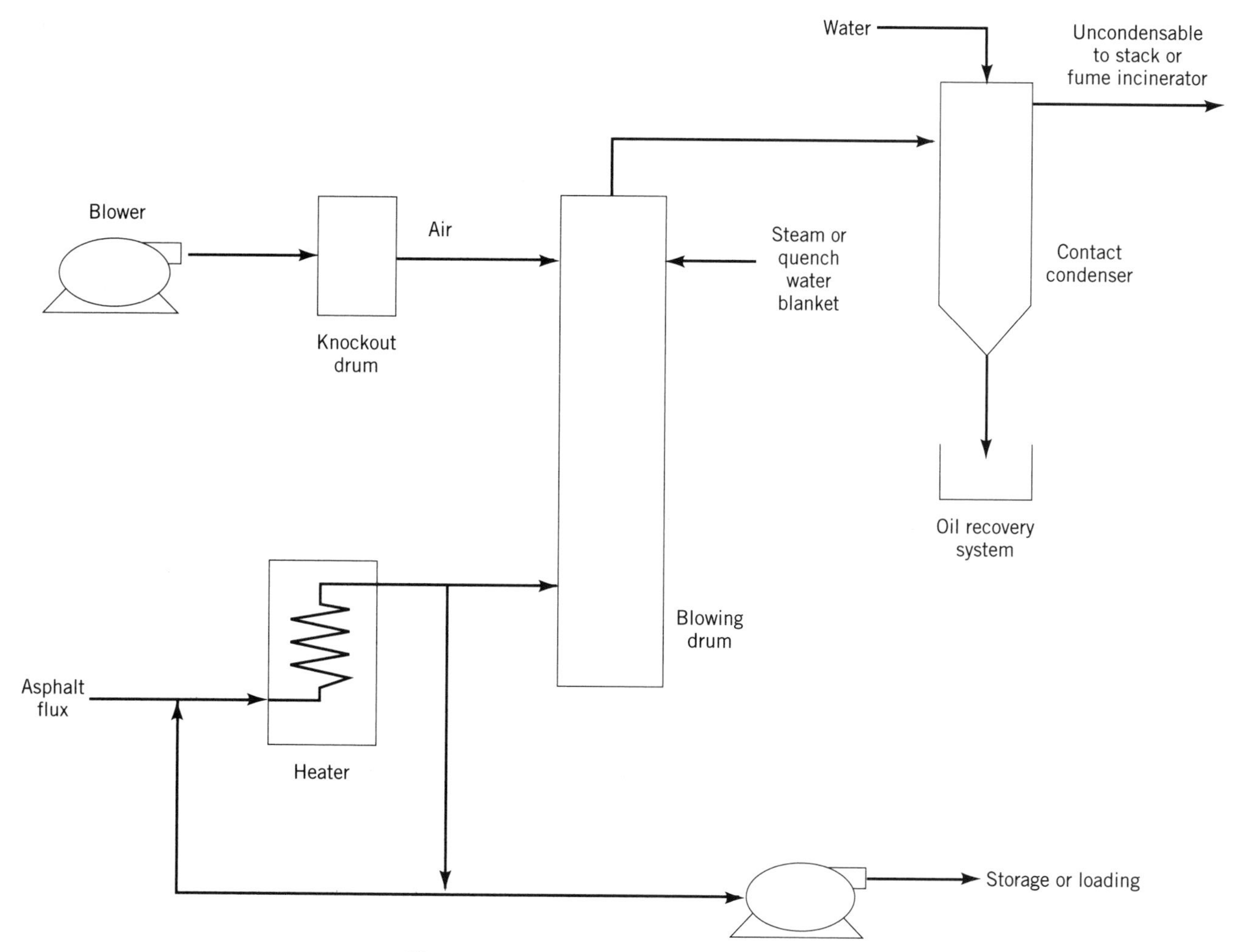

Figure 5. The air blowing (batch) process.

$CHCH_2CH{=}CH(CH_2)_7CO_2H]$ and/or high molecular weight phenols; an example of such an emulsifying agent in sodium palmitate, $CH_3(CH_2)_{14}COO^-Na^+$. Sodium lignate is often added to alkaline emulsions to effect better emulsion stability; nonionic cellulose derivatives are also used to increase the viscosity of the emulsion if needed. Diamines are frequently used as cationic agents and are prepared by the reaction of acid amines with acrylonitrile, followed by hydrogenation. An example of a cationic emulsifying agent is *N*-octadecyl-1,3-propanediamine hydrochloride, $CH_3(CH_2)_{17}NH(CH_2)_3NH_2{\cdot}HCl$.

Emulsion processes (Fig. 8) vary and the emulsions are made in rapid-, medium-, and slow-setting types for diverse application techniques in the road-building industry. Emulsions are most commonly formulated with rapid-setting properties and are designated as a spraying type, ie, for road seal-coating operations. Slow setting emulsions are usually used for soil stabilization or in mulching applications.

COMPOSITION AND PROPERTIES

The identification of the constituents of asphalts has always presented a challenge because of the complexity and high molecular weights of the molecular constituents (16,17); in fact, it is a virtually impossible task. Thus, the principle behind composition studies is to evaluate asphalts in terms of gross composition and performance. It is important to relate composition to performance since such data enable evaluations of different asphalts to be made and design of future asphalts to be formulated. If this is not the overriding goal of the work, such studies must always be suspect.

Fractional separation is the most frequently used basis for the compositional analysis of asphalt (Fig. 9) and there are four general procedures: (a) chemical precipitation, in which *n*-pentane is used for asphaltene separation and subsequent precipitation of other fractions using sulfuric acid of increasing concentration (2,18); (b) solvent fractionation, in which the separation of an insoluble "asphaltene" fraction by the use of *n*-butanol, followed by dissolution of the *n*-butanol solubles in acetone after which the acetone solution is chilled, forcing the precipitation of paraffin material from the acetone-soluble cyclic materials (2); (c) adsorption chromatography where, after removal of the asphaltenes (either by use of *n*-pentane or *n*-heptane) the remaining constituents are separated by adsorption/desorption on an adsorbent (19,20); (d) size-exclusion chromatography, in which gel-permeation chro-

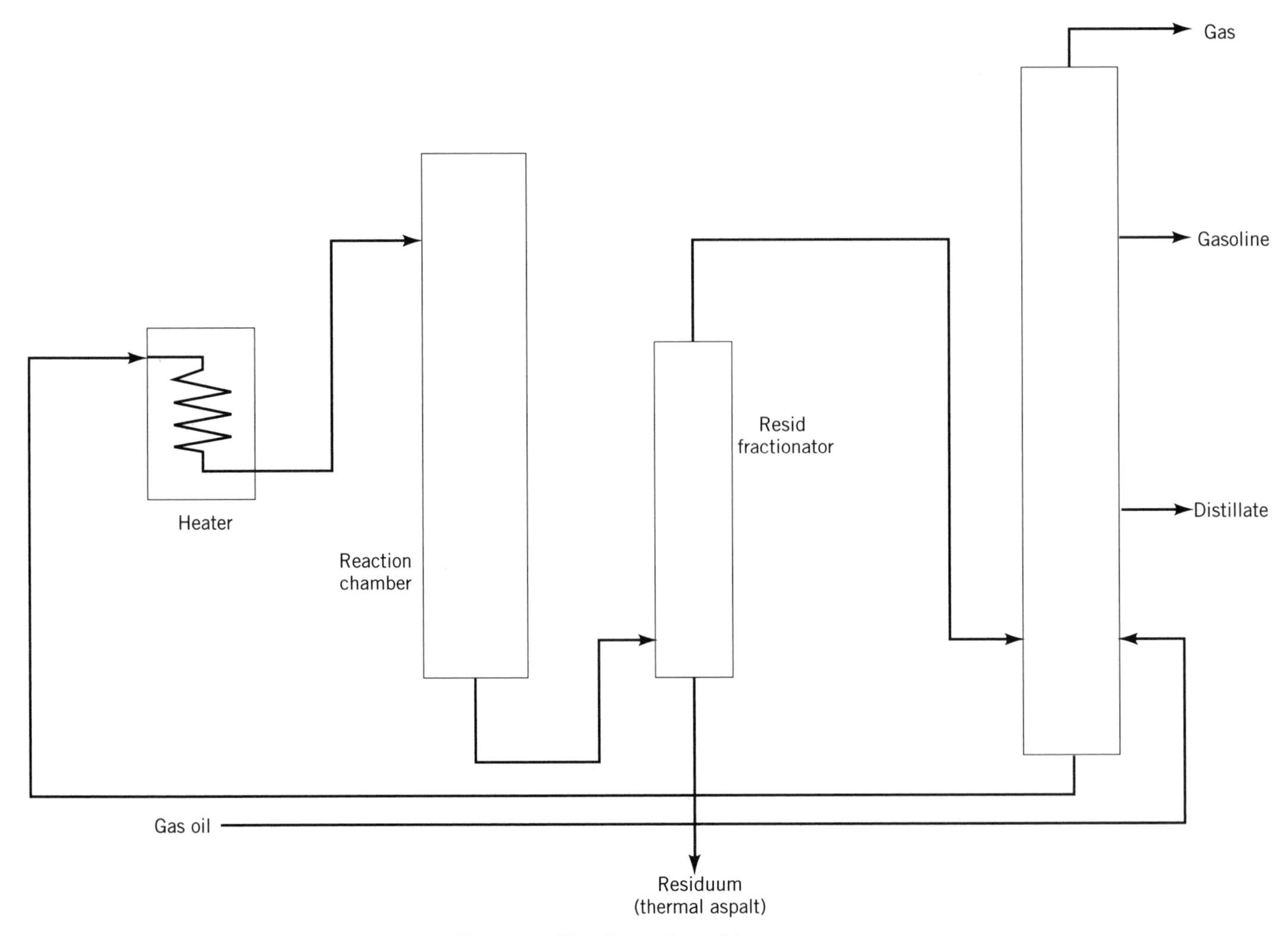

Figure 6. The thermal cracking process.

matographic (GPC) separation of asphalt constituents occurs, based on their associated sizes in dilute solutions (21).

The fractions obtained in these schemes are defined on the basis of the operations, or procedures used for the separation (2). The amount and type of asphaltenes in an asphalt are, for instance, defined by the solvent used for precipitating them (2,22) and, in fact, the fractional separation of asphalt does not yield a series of well-defined chemical components. The materials separated can only be defined in terms of the particular test procedure; in other words, the definitions are operational aids.

The component of highest carbon content is the fraction termed carboids and consists of species that are insoluble in carbon disulfide or in pyridine. The carboid fraction also contains the constituents that are highly aromatic and perhaps the most polar. The fraction that has been called carbenes contains molecular species that are soluble in carbon disulfide and in pyridine, but which are insoluble in carbon tetrachloride and in benzene (Fig. 1) (2,4,23).

Asphaltenes are relatively constant in composition, as indicated by elemental analysis, despite the source (2). The determination of asphaltenes is now a relatively standard procedure (24,25) with the yield being dependent upon the hydrocarbon liquid used for the separation (2,4,23). After the asphaltenes are removed from the asphalt, resin fractions are removed from the deasphaltened oil, usually by adsorption on activated gels or clays. Recovery of the resin fractions by desorption from the gel or clay is usually near quantitative. Continuous modifications to the separation procedures have led to a variety of fractionation methods for asphalt because of the variety of fractions possible, as well as the large number of precipitants or adsorbents (2,4).

The main outcome of many of the composition studies has been the delineation of asphalt as a colloidal system at ambient or normal service conditions. The transition from a colloid to a Newtonian liquid is dependent on temperature, hardness, shear rate, and chemical nature. At normal service temperatures asphalt is viscoelastic, and viscous at higher temperatures. The disperse phase is a micelle composed of the molecular species that make up the asphaltenes and the higher molecular weight aromatic components of the maltenes (ie, the nonasphaltene components). Complete peptization of the micelle seems probable if the system contains sufficient aromatic constituents, in relation to the concentration of asphaltenes, to allow the asphaltenes to remain in the dispersed phase.

Many investigations of relationships between composi-

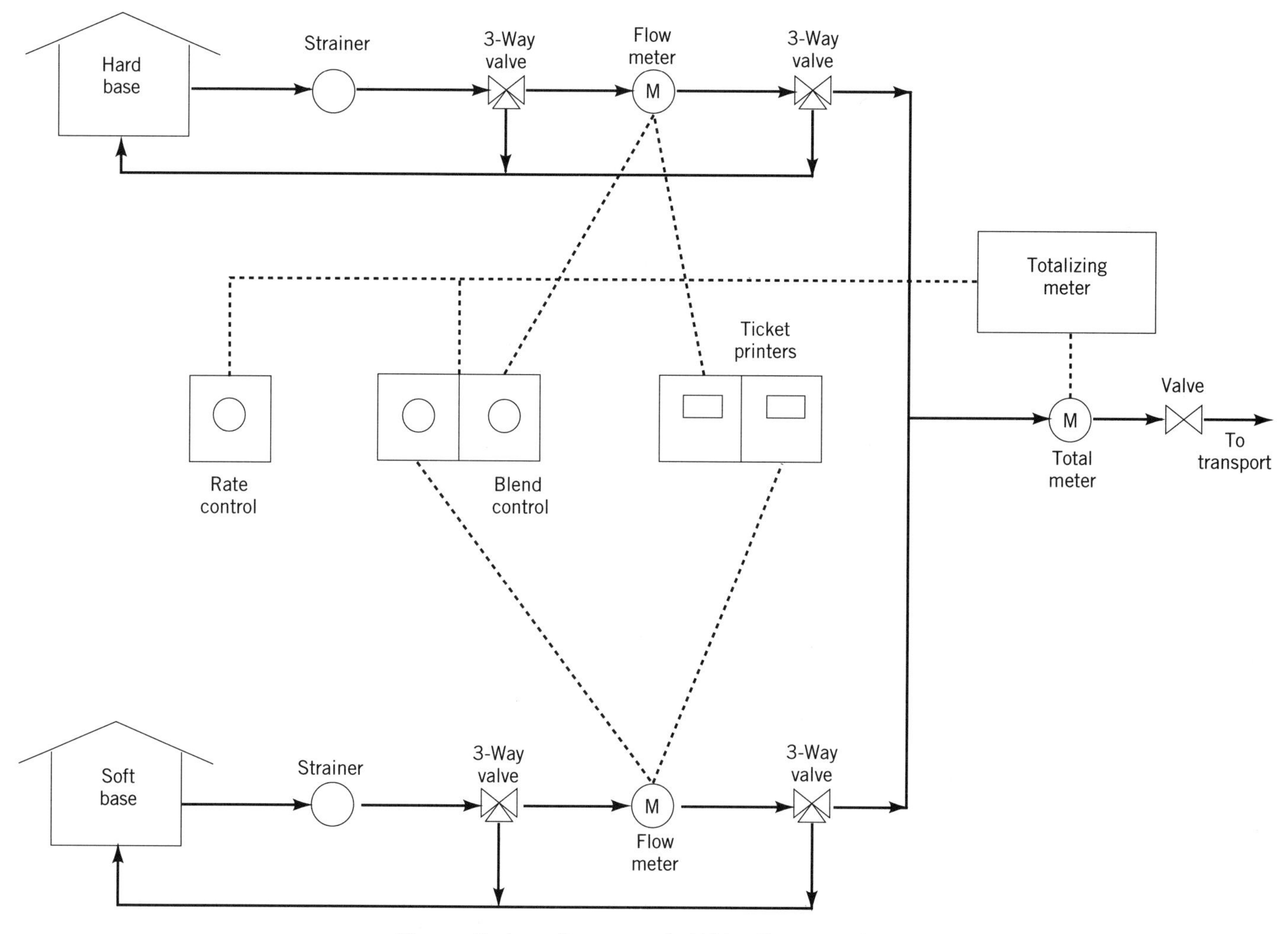

Figure 7. A continuous asphalt blending operation.

tion and properties take into account only the concentration of the asphaltenes, independently of any quality criterion (2). However, a distinction should be made between the asphaltenes which occur in straight run asphalts and those which occur in blown asphalts. Remembering that asphaltenes are a solubility class (Fig. 1) and not a distinct chemical class, means that vast differences occur in the make-up of this fraction when it is produced by different processes (2). Nevertheless, the composition data should always be applied to in-service performance (7) in order to evaluate properly the behavior of the asphaltic binder under true working conditions.

Elemental Analysis

Generally, asphalts contain carbon: 79–88 wt %, hydrogen: 7–13 wt %, nitrogen: <0.1–3 wt %, oxygen: 1–8 wt %, and sulfur: 0.1–8 wt % (Table 6). Metals, such as iron, nickel, vanadium, calcium, titanium, magnesium, sodium, cobalt, copper, tin, and zinc, occur in trace amounts by virtue of their occurrence in the parent crude oils. Vanadium and nickel are bound in organic complexes and, by virtue of the concentration (distillation) process by which asphalt is manufactured, are also found in asphalt (26). The other metals may be indigenous to the parent crude oil, but caution should be exercised in airing this conclusion, since petroleum is known to pick up metal species from pipelines and storage tanks (2).

Asphalt is not composed of a single chemical species, but is rather a complex mixture of organic molecules that vary widely in composition from nonpolar, saturated hydrocarbons to highly polar, highly condensed aromatic ring systems. Although asphalt molecules are composed predominantly of carbon and hydrogen, most molecules contain one or more heteroatoms (nitrogen, oxygen, sulfur) together with trace amounts of metals, principally vanadium and nickel. The heteroatoms, although a minor component compared to the hydrocarbon moiety, can vary in concentration over a wide range, depending on the source of the asphalt.

Because the heteroatoms often impart functionality and polarity to the molecules, their presence may make a disproportionately large contribution to the differences in physical properties among asphalts from different sources. As a result, techniques have been developed for quantitative identification of the six major types of heteroatom compounds in asphalt: carboxylic acids, 2-quinolones, phenols, pyrroles, amides, and pyridines (27,28).

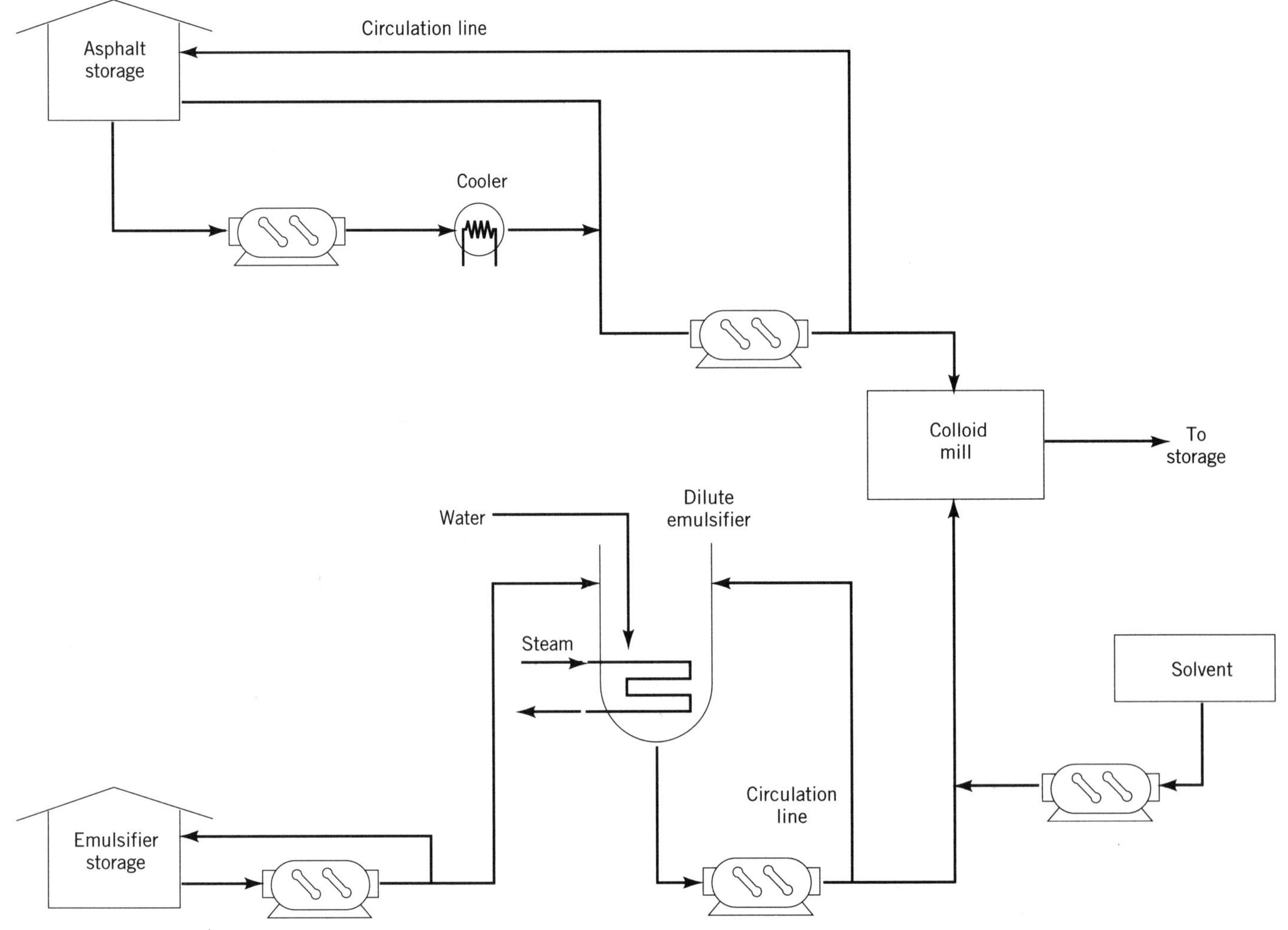

Figure 8. A plant for the manufacture of asphalt emulsions.

Molecular Weight

The molecular weights of the individual fractions of asphalt have received more attention, and have been considered to be of greater importance, than the molecular weight of the asphalt itself. The components that make up the asphalt influence the properties of the material to an extent that is dependent upon the relative amount of the component, the molecular structure of the component, and the physical structure of the component, which includes the molecular weight.

Asphaltenes have a wide range of molecular weights (from 500 to at least 2500) depending upon the method of separation, the method of determining the molecular weight, and the tendency of asphaltenes to associate in dilute solution in nonpolar solvents. This latter phenomenon results in higher molecular weights than is actually the case on an individual molecule basis (29,30). The molecular weights of the resins are somewhat lower than those of the asphaltenes and usually fall within the range 500 to 1000 (31). This is due not only to the virtual absence of any associative interactions but also to a lower absolute molecular size. The molecular weights of the oil fractions (ie, the asphalt minus the asphaltenes and resins) is usually less than 500, often 300 to 400 (2).

Acid Number

Asphalt contains a small amount of organic acids as well as other saponifiable material (7). The acid values of asphalt, which usually fall in the range 0.1–2.8 mg potassium hydroxide per gram of asphalt, are related to emulsification properties of the asphalt and the ease of dispersion. The acid content is largely determined by the percentage of naphthenic (cycloparaffinic) acids of higher molecular weight that are originally present in the crude oil. With increased hardness, asphalt from a particular crude oil normally decreases in acid number as more of the naphthenic acids are removed during the distillation process. There is also the possibility for decarboxylation of acidic functions during the distillation process (9), thereby bringing about a further decrease in the acid number.

Rheology

Asphalt is a viscoelastic material whose rheological properties reflect the nature of the parent petroleum and, to a lesser extent, the processing method used to produce the asphalt. The ability of asphalt to perform its function under many conditions depends on flow behavior. Thus, as-

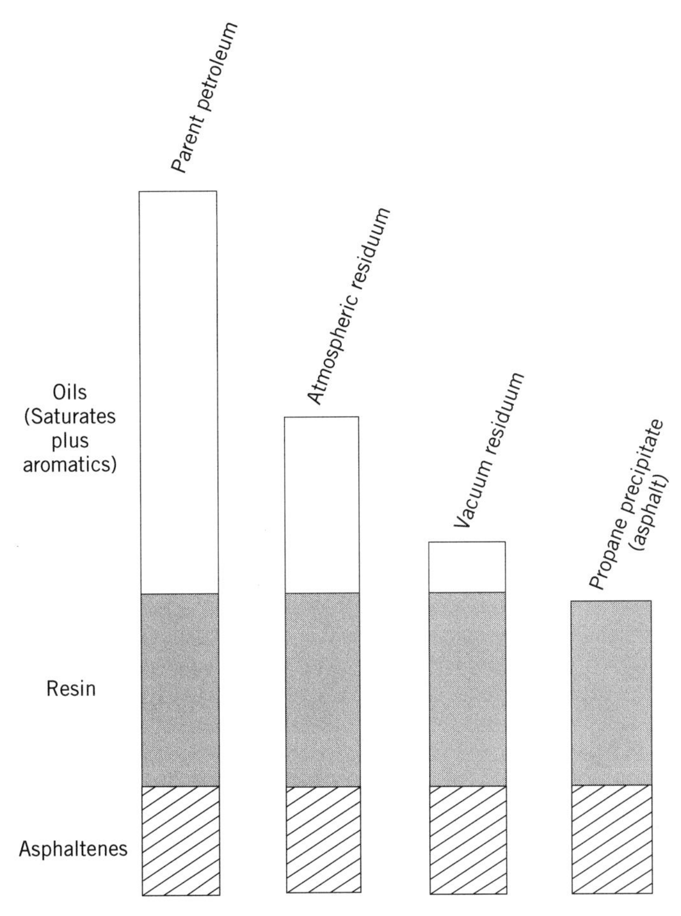

Figure 9. Relationship between a crude oil, two residua, and the propane asphalt in terms of the major fractional constituents.

phalt showing little, or no, appreciable change from original conditions is to be desired insofar as some structural movement should be allowed but without permanent deformation. Liquid asphalts, in effect, substitute diluent (or water emulsification) for heat to allow pumping, spraying, or application of the material at lower temperature.

Some procedures for measuring the viscoelastic properties of asphalt have been described (2,32). A number of viscometers have been developed for securing viscosity data at temperatures as low as 0°C (33,34). The most popular instruments in current use are the cone and plate (35), parallel plate, and capillary instruments (36,37). The cone and plate method can be used for the determination of viscosities in the range of 10 to over 10^9 Pa (10^{10} P) at temperatures of 0°C to 70°C, and at shear rates from 10^{-3} to 10^2 s^{-1}. Capillary viscometers are commonly used for the determination of viscosities at 60–135°C.

Tests developed for the measurement of viscoelastic properties are directly usable in engineering relations. Properties can be related to the inherent structure of bituminous materials. The fraction of highest molecular weight, the asphaltenes, is dispersed within the asphalt and is dependent upon the content and nature of the resin and oil fractions. Higher aromaticity of the oil fractions or higher temperatures leads to viscous (sol) conditions. A more elastic (gel) condition results from a more paraffinic nature and is indicated by large elastic moduli or, empirically, by a relatively high penetration at a given softening point.

Asphalts develop an internal structure with age, referred to as steric hardening (5), in which viscosity can increase without any loss of volatile material (38). Those with a particularly high degree of gel structure exhibit thixotropy, ie, the asphalts show an increase in flow rate with increasing duration of agitation, as well as with increased shear stress.

Durability

The term "durable" has several meanings. In the present context it is applied to an asphalt that possesses the necessary chemical and physical properties desired for the specified performance, ie, being resistant to change during the in-service conditions that are prevalent during the life of the pavement.

Asphalts are used as protective films, adhesives, and binders because of their waterproof and weather-resistant properties. One particularly valuable property, the ability of the asphalt to undergo movement without fracture, can occur because of the viscous (sol) nature. In addition, asphalts have long and continuous satisfactory service and, because of their slow rate of hardening, show resistance to heat, oxidation, fatigue, and weathering (39,40). Exposed asphalt films harden partially from a loss of volatile oils as well as from the formation of additional asphaltene material (formed from the maltene constituents) by oxidation. Such chemical change can be catalyzed by uv irradiation (7). The oxidation may, ultimately, result in the formation of water-soluble degradation products which are removed from the asphalt film. In addition, the conversion of maltene constituents to asphaltenes (13) cannot be ignored since incompatibility or uncontrolled hardening (either being responsible for asphalt failure) may be the result.

Failure of an asphalt film is reflected in the appearance of typical "crack" patterns because of a decrease in the "plasticity" of the asphalt through insufficient lower molecular weight constituents remaining in the continuous phase. Other signs of failure are loosening of the bond, such as loosening of granules in roofing shingles, to actual loss of the film, even the binder in pavement mixtures might become friable and crack. A harder asphalt, under uniform loading conditions, could reduce pavement deflection, extend fatigue life and allow less flow deformation. A softer material would normally allow longer in-service performance before any changes in composition, ie, the maltene–asphaltene ratio becomes critical. Usually the softest material allowed by initial service needs is selected.

The water resistance of asphalt films is also a manifestation of durability. Asphalts that have a low content of soluble salts show a low water absorption. The interaction, or ingestion, of water by asphalt is primarily a surface phenomenon that causes the film to soften, leading to "blistering." Even with a high rate of absorption, asphalt films show little loss of surface bonding during continued immersion in water, and continue to protect metals from corrosion for prolonged periods of time.

Mineral fillers are often added to asphalts to influence their flow properties and reduce costs, and are commonly used as stabilizers in roofing coatings at concentrations up to 60 wt %. Fillers may increase the water absorption of asphalts. Mineral-filled films show improved resistance to flow at elevated temperatures, improved impact resistance, and better flame-spread resistance. Fillers commonly used are ground limestone, slate flours, finely divided silicas, trap rocks, and mica. They are often produced as by-products in rock-crushing operations. Opaque fillers offer protection from weathering. Asbestos filler has special properties because of its fibrous structure, high resistance to flow, and toughness. It has been used in asphalt paving mixes to increase the resistance to movement under traffic (41), and in roofing materials for fire-retardant purposes.

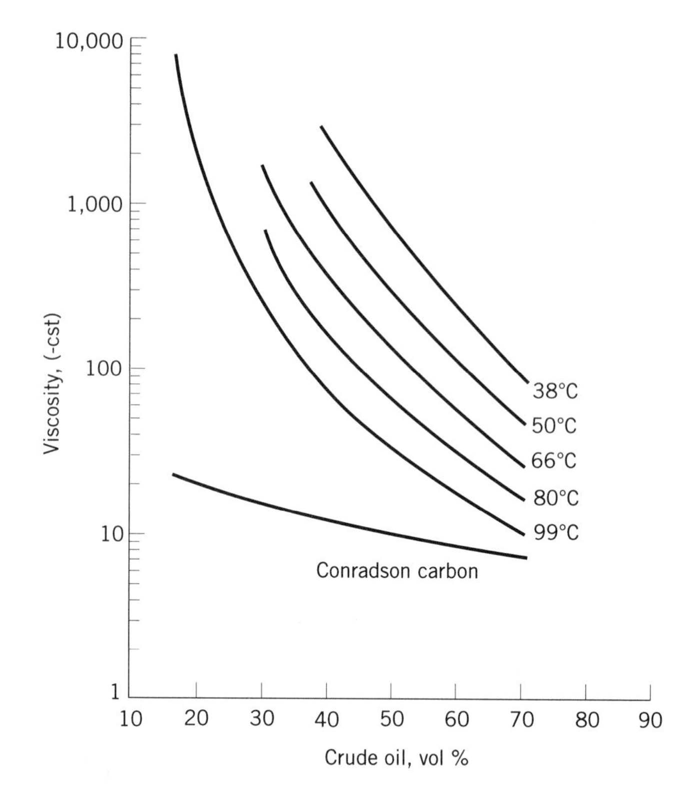

Figure 10. Relationship of selected physical properties to yield of residuum from Tia Juana light crude oil.

SPECIFICATIONS AND TESTS

In 1903 the American Society for Testing and Materials (ASTM) Committee on Road and Paving Materials was formed to develop test methods and specifications for highway materials. Test methods for volatilization and penetration were developed by the Office of Public Roads and were adopted by ASTM in 1911. For the asphalt cements produced at that time, the adoption of the volatilization and penetration tests provided some degree of control of excessive changes during plant mixing that might be reflected in more durable asphalts.

During the 1920s three national specifications for asphalt cements were first published: (a) Federal specifications, adopted in 1925; (b) the American Association of State Highway Transportation Officials (AASHTO) specifications, adopted in 1924; and (c) ASTM specifications, adopted in 1922.

The Federal (United States) specifications stipulated that only those asphalts that had been demonstrated by service tests as satisfactory for the intended use would be accepted. The specifications also indicated the type and location of construction and the relative amount of traffic for each of the grades. The AASHTO specifications indicated that the use of each grade depended on the type of road, climate, and traffic. The ASTM suggested the type of construction for which each grade would be used.

With minor exceptions, the requirements for the physical and chemical properties of asphalt were essentially the same for the three national specifications and included the following: penetration and ductility at 25°C; flash point; % loss of volatile material at 163°C; penetration of residue as a % of original; solubility in organic solvents; specific gravity at 25°C; and softening point.

The fact that the properties of residua vary with cutpoint (Table 3), ie, the volume % of the crude oil retained as asphalt, dictates that asphalt of a specific type or property be produced (Fig. 10). However, there are several properties that are usually controlled in asphalt specifications (Table 7). Asphalts are often specified in several grades for the same industry, differing only in hardness or viscosity (42). Still, the changing composition and nature of crude oil feedstocks over the past two decades dictates that supply factors, in addition to performance factors, are also an important consideration (2,7,42).

Specifications for paving asphalt cements usually include five grades differing in either viscosity or penetration level at 60°C (Table 8). The susceptibility of viscosity

Table 6. Elemental Analysis of Various Petroleum Asphalts[a]

Source	Mexican Blend	Arkansas-Louisiana	Boscan	California
carbon	83.77	85.78	82.90	86.77
hydrogen	9.91	10.19	10.45	10.94
nitrogen	0.28	0.26	0.78	1.10
sulfur	5.25	3.41	5.43	0.99
oxygen	0.77	0.36	0.29	0.20
vandium	180	7	1380	4
nickel	22	0.4	109	6

[a] Analysis in %, for vanadium and nickel in ppm.

Table 7. General Properties Used for Determining Asphalt Quality

Property	Test
Safety	Flash point
Purity	Solubility, ash, water content
Composition	Naphtha insolubles, various separation Techniques, distillation tests, homogeneity, wax content
Rheological	Penetration, viscosity, softening point, float, ductility, flow, impact or shock, break point, temperature susceptibility ratios
Durability	Thin-film oven, aging index, weathering
Density or specific gravity	Specific gravity
Special properties	Bond or adhesion, compatibility, stain, storage stability, chemical resistance, etc.

to temperature is usually controlled in asphalt cement by viscosity limits at a higher temperature, such as 135°C, and a penetration or viscosity limit at a lower temperature, such as 25°C.

Paving cutbacks are also graded at 60°C by viscosity, with usually four to five grades of each type. For asphalt cements, the newer viscosity grade designation is the midpoint of the viscosity range. The grade designation for a cutback asphalt is the minimum kinematic viscosity for the grade, with a maximum grade viscosity of twice the minimum.

Roofing and industrial asphalts are also generally specified in various hardnesses, usually with a combination of softening point and penetration to distinguish grades (43,44). Temperature susceptibility is usually controlled in these requirements by specifying penetration limits or ranges at 25°C and other temperatures, as well as softening-point ranges at higher temperatures. Asphalts for roof construction are differentiated according to application, depending primarily on pitch of the roof and, to some extent, on whether or not mineral surfacing aggregates are to be employed (43). The dampproofing grades of roof asphalts reflect whether or not a self-healing property is incorporated (44).

Most tests applied to asphalt are empirical in nature (2,7) and the ASTM tests are not the only standard procedures applicable to asphalt testing. The Institute of Petroleum (IP; England), the Deutsche Industrie Normen (DIN; Germany), and the American Association of State Highway and Transportation Officials (AASHTO) also support the standardized testing of asphalts (Table 9). The significance of a particular test is not always apparent by reading the procedure, and sometimes can only be gained through working familiarity with the test (45).

As in the petroleum industry (2), a variety of useful tests are employed, having evolved through local, or company, usage without having passed through the ASTM system of approval. Prime examples are the methods used to fractionate asphalt, of which there are a myriad in use (2,46). Those tests which have been used within companies are considered proprietary but some of these tests are now being made public and occur from time to time in literature reports. They are well worth consideration, not only because of their history of applied use, but because there is also the chance that such tests will become a part of the family of published standard tests for asphalt.

Uses

Properties, and therefore uses, depend upon the method of manufacture (47). Uses include hydraulics (dam facings, canal linings, pond linings) (48), recreation (substrate for artificial surfaces, tennis courts, running tracks) (49), transportation (railroad ballast treatment and roadbeds) (50), and metals (ore leaching pads, ore and coal briquetting, etc).

One aspect of asphalt technology that has recently received much attention involves the health and safety aspect. Steps to minimize potential safety hazards in the handling of asphalt are set forth by the American Petroleum Institute (51). Development of test methods to ensure the continued safe use of asphalt has been continu-

Table 8. Specifications (ASTM) for Asphalt Viscosity

Tests on Residue from Rolling Thin-Film Oven Test[c]	Viscosity Grade[b]				
	AR-1000	AR-2000	AR-4000	AR-8000	AR-16000
Viscosity, 60°C, Pa-s[c]	100 ± 25	200 ± 50	400 ± 100	800 ± 200	1600 ± 400
Viscosity, 135°C, min, mm²/s (= cSt)	140	200	275	400	550
Penetration of 100 g at 25°C for 5 s, min, mm/10	65	40	25	20	20
Original penetration at 25°C, min %		40	45	50	52
Ductility, 25°C, 5 cm/min, min, cm	100[d]	100[d]	75	75	75
Tests on original asphalt:					
flash point, Pensky-Martens closed tester, min, °C	205	219	227	232	238
solubility in trichloroethylene, min %	99.0	99.0	99.0	99.0	99.0

[a] Thin-film oven test may be used but the rolling thin-film oven test shall be the referee method.
[b] Grading based on residue from rolling thin-film oven test.
[c] To convert Pa · s to P, multiply by 10.
[d] If ductility is less than 100, material will be accepted if ductility at 15.5°C is 100 minimum at a rate of 5 cm/min.

Table 9. Standard Test for Asphalts

Test	Organization/Number	Description
Adsorption	ASTM D4469	Calculation of degree of adsorption of an asphalt by an aggregate.
Bond and adhesion	ASTM D1191	Used primarily to determine if an asphalt has bonding strength at low temperatures. See also ASTM D3141 and ASTM D5078.
Breaking point	IP 80/53	Indication of the temperature at which an asphalt possesses little, or no, ductility and would show brittle fracture conditions.
Compatibility	ASTM D1370	Indicates whether asphalts are likely to be incompatible and disbond under stress. See also ASTM D3407.
Distillation	ASTM D402	Determination of volatiles content; applicable to road oils and cutback asphalts.
Ductility	ASTM D113	Expressed as the distance in cm which a standard briquet can be elongated before breaking; reflects cohesion and shear susceptibility.
Emulsified asphalts	ASTM D244	Covers various tests for the composition, handling, nature and classification, storage, use, and specifications. See also ASTM D977 and ASTM D1187.
Flash point	ASTM D92	Cleveland open cup method is commonly used; Tag open cup (ASTM D3143) applicable to cutback asphalts.
Float test	ASTM D139	Normally used for asphalts that are too soft for the penetration test.
Penetration	ASTM D5	The extent to which a needle penetrates asphalt under specified conditions of load, time, and temperature; units are mm/10 measured from 0 to 300. See also ASTM D243.
Sampling	ASTM D140	Provides guidance for the sampling of asphalts.
Softening point	ASTM D36	Ring and ball method; the temperature at which an asphalt attains a particular degree of softness under specified conditions; used to classify asphalt grades. See also ASTM D2398.
Solubility in carbon disulfide	ASTM D4	Determination of the carbon amount of carboids and/or carbenes and mineral matter; trichloroethylene or 1,1,1-trichloroethane have been used for this purpose. See also ASTM D2042.
Solubility in paraffin naphtha	AASHTO T46	Designated by the American Association of State Highway and Transportation Officials; used to indicate the content asphaltene content. Other solvents such as *n*-heptane (ASTM D3279) are used instead of naphtha.
Specific gravity	ASTM D70	See also ASTM D3142.
Spot test	AASHTO T102	Distinguishes asphalts on the basis of incompatibility.
Stain	ASTM D1328	Measures the amount of stain on paper or other cellulosic materials.
Temperature–volume correction	ASTM D1250	Allows the conversion of volumes of asphalts from one temperature to another. See also ASTM D4311.
Thin film oven test	ASTM D1754	Determines the hardening effect of heat and air on the film of asphalt. See also ASTM D2872
Viscosity	ASTM D2170	A measure of resistance to flow. See also ASTM D88 (now discontinued but a useful reference), ASTM D1599, ASTM D2171, ASTM D2493, ASTM D3205, ASTM D3381, ASTM D4402, and ASTM D4957.
Water content	ASTM D95	Determines the water content by distillation using a Dean and Stark receiver.
Wax content	DIN 141	Utilizes the destructive distillation of the asphalt, followed by was separation from the distillate fractions by freezing.
Weathering	ASTM D529	Used for determining the relative weather resistance of asphalt. See also ASTM D1669 and ASTM D1670.

ing for at least (if not more than) three decades. These include test methods that examine any potential hazard from (a) sudden pressure increases from hot asphalt in contact with moisture in enclosed tanks or transports; (b) exposure to air at 150°C or above; (c) local overheating in heating coils, causing flashing of asphalt volatiles in the presence of an ignition source or possible auto-ignition; and (d) hydrogen sulfide evolution from high temperature operations.

Table 10. Analytical Techniques for Asphalt Emissions

Sample	Analytical Technique
Air	Mass spectrometry
Volatile hydrocarbons (C1–C6)	Gas chromatography
Carbon monoxide	Infrared spectrometry
Volatile sulfur compounds	Gas chromatography
Nitrogen dioxide	Spectrophotometry at 550 nm
Aldehydes	Ultraviolet spectrometry
Phenols	Ultraviolet spectrometry
Particulates	Optical microscopy
Polynuclear aromatics (PNA)	Gas chromatography

There are a variety of tests for the potential dangers of asphalt emissions (Table 10). Asphalt use, however, is not recognized as posing significant air pollution problems (52). In fact, the concentrations of gaseous substances and emissions from paving asphalt cement have been found to be in very low concentration and fall within existing United States Environmental Protection Agency (EPA) and United States Occupational Safety and Health (OSHA) standards, even when the ambient air sampling was done under confined conditions (53–55). In addition, the high molecular weight polynuclear aromatic constituents that occur in asphalt have been studied as health hazards (56–59). The general conclusion from these studies is that surveillance should be continued, although asphalt has not been shown to be a significantly hazardous material.

BIBLIOGRAPHY

1. *Terminology Relating to Materials for Roads and Pavements,* ASTM D8, *Annual Book of Standards*, American Society for Testing and Materials, Philadelphia, Pa., 1991, Sections 04.02 and 04.03.
2. J. G. Speight, *The Chemistry and Technology of Petroleum,* 2nd ed., Marcel Dekker Inc., New York, 1991.
3. J. G. Speight, *The Chemistry and Technology of Coal,* Marcel Dekker Inc., New York, 1983.
4. J. G. Speight, *Fuel Science and Technology Handbook,* Marcel Dekker Inc., New York, 1990.
5. H. Abraham, *Asphalts and Allied Substances,* 6th ed., Vol. 1, D. Van Nostrand Co., Inc., Princeton, N.J., 1960, Chapt. 2.
6. R. J. Forbes, *Bitumen and Petroleum in Antiquity,* E. J. Brill, Leiden, Netherlands, 1936.
7. J. G. Speight, "Asphalt," in J. I. Kroschwitz and M. Howe-Grant, eds., *Kirk-Othmer Encyclopedia of Chemical Technology,* 4th ed., John Wiley & Sons Inc., New York, 1992, Vol. 3, p. 689, and references cited therein.
8. N. Berkowitz and J. G. Speight, *Fuel* **54,** 138 (1975).
9. J. G. Speight and M. A. Francisco, *Rev. de l'Institut Français du Petrole,* **45J,** 733 (1990).
10. W. L. Nelson, *Petroleum Refining Engineering,* 4th ed., McGraw-Hill Book Co., Inc., New York, 1958.
11. L. W. Corbett and R. E. Swarbrick, *Proc. Assoc. Asphalt Paving Technol.* **29,** 104 (1960).
12. A. J. Hoiberg and W. E. Garris, Jr., *Ind. Eng. Chem. Anal. Ed.* **16,** 294 (1944).
13. S. E. Moschopedis and J. G. Speight, *Fuel* **57,** 235 (1978).
14. W. L. Nelson, *Oil Gas J.* **53**(16), 158 (Aug. 23, 1954).
15. G. L. Oliensis, *Proc. Assoc. Asphalt Paving Technol.* **6,** 88 (1935).
16. K. H. Altgelt, *Die Makromolekulare Chemie* **88,** 75 (1965).
17. J. B. Green, J. W. Reynolds, and S. K-T Yu, *Fuel Science and Technology Int.* **7,** 1327 (1989).
18. *Test Method for the Separation of Constituents of Asphalt,* ASTM D 2006, *Annual Book of Standards,* American Society for Testing and Materials, Philadelphia, Pa.; test discontinued but still used by some laboratories.
19. *Test Method for Characteristic Groups in Rubber Extender and Processing Oils,* ASTM D 2007, *Annual Book of Standards,* American Society for Testing and Materials, Philadelphia, Pa., 1991, section 05.02.
20. *Test Method for the Separation of Asphalt into Four Generic Fractions,* ASTM D4124, *Annual Book of Standards,* American Society for Testing and Materials, Philadelphia, Pa., 1991, section 04.03.
21. *Test Method for Molecular Weight Averages and Molecular Weight Distribution of Certain Polymers by Liquid Size-Exclusion Chromatography (Gel Permeation Chromatography GPC) Using Universal Calibration,* ASTM D 3593, *Annual Book of Standards,* American Society for Testing and Materials, Philadelphia, Pa., 1991, section 08.03.
22. D. L. Mitchell and J. G. Speight, *Fuel* **52J,** 149 (1973).
23. W. A. Gruse and D. R. Stevens, *Chemical Technology of Petroleum,* 3rd ed., McGraw-Hill Book Co., Inc., New York, 1960, p. 585.
24. J. G. Speight, R. B. Long, T. D. Trowbridge, *Preprints, Div. Fuel Chem., Am. Chem. Soc.* **27**(3/4), 268 (1984).
25. *Test Method for n-Heptane Insolubles,* ASTM D3279, *Annual Book of Standards,* American Society for Testing and Materials, Philadelphia, Pa., 1991, section 04.03.
26. R. Fish, *Anal. Chem.* **56,** 2452 (1984) and references cited therein.
27. J. C. Petersen, F. A. Barbour, and S. M. and Dorrence, *Proc. Assoc. Asphalt Paving Technol.* **43,** 162 (1974).
28. J. C. Petersen and H. Plancher, *Anal. Chem.* **47,** 112 (1975).
29. S. E. Moschopedis, J. F. Fryer, and J. G. Speight, *Fuel* **55,** 227 (1976).
30. J. G. Speight, D. L. Wernick, K. A. Gould, R. E. Overfield, B. M. L. Rao, and D. W. Savage, *Rev. Inst. Français du Petrole* **40,** 51 (1985).
31. J. A. Koots and J. G. Speight, *Fuel* **54,** 179 (1975).
32. O. E. Briscoe, ed., *Asphalt Rheology: Relationship to Mixture,* Special Publication No. 941, American Society for Testing and Materials, Race Street, Philadelphia, 1985.
33. E. C. Knowles and H. Levin, *Ind. Eng. Chem. Anal. Ed.* **13,** 314 (1941).
34. R. A. Gardner, *J. Chem. Eng. Data,* **4,** 155 (1959).
35. *Test Method for Viscosity of Asphalt with Cone and Plate,* ASTM D3205, *Annual Book of Standards,* American Society for Testing and Materials, Philadelphia, Pa., 1991, section 04.03.
36. *Test Method for Kinematic Viscosity of Asphalts,* ASTM D2170, *Annual Book of Standards,* American Society for Testing and Materials, Philadelphia, Pa., 1991, section 04.03.
37. *Test Method for Viscosity of Asphalts by Vacuum Capillary Viscometer,* ASTM D2171, *Annual Book of Standards,* American Society for Testing and Materials, Philadelphia, Pa., 1991, section 04.03.
38. C. Van der Poel in M. Reiner, ed., *Building Materials, Their Elasticity and Inelasticity,* Interscience Publishers, New York, 1954, p. 373.
39. R. N. Traxler, *Asphalt, Its Composition, Properties and Uses,* Reinhold Publishing Co., New York, 1961, Chapt. 6.
40. A. J. Hoiberg, *Bituminous Materials: Asphalts, Tars, and Pitches,* Vol. 1, 2, and 3, Interscience Publishers, a division of John Wiley & Sons, Inc., New York, 1964–1966.
41. D. A. Tamburro, A. T. Blekicki, and J. H. Kietzman, *Proc. Assoc. Asphalt Paving Technol.* **31,** 151 (1962).
42. J. L. Goodrich and H. L. Dimpfl, *Proc. Assoc. Asphalt Paving Technol.* **55,** 57 (1987).
43. *Specification for Asphalt Used in Roofing,* ASTM D312, *Annual Book of Standards,* American Society for Testing and Materials, Philadelphia, Pa., 1991, section 04.04.
44. *Specification for Asphalt Used in Dampproofing and Waterproofing,* ASTM D449, *Annual Book of Standards,* American Society for Testing and Materials, Philadelphia, Pa., 1991, section 04.04.
45. R. L. Dunning, Proc. *Assoc. Asphalt Paving Technol.* **55,** 33 (1986).
46. R. N. Traxler, *Asphalt,* Reinhold Publishing Co., New York, 1961.
47. *A Brief Introduction to Asphalt and Some of its Uses,* Manual Series No. 5, The Asphalt Institute, College Park, Md., 1975.
48. *Asphalt in Hydraulics,* Manual Series No. 12, The Asphalt Institute, College Park, Md., 197.

49. *Athletics and Recreation on Asphalt,* Information Series No. 147, The Asphalt Institute, College Park, Md. 20740.
50. *Railway Roadbeds for Tomorrow,* Information Series IS-137, The Asphalt Institute, College Park, Md. 20740.
51. *Guide for the Safe Storage and Handling of Heated Petroleum-Derived Asphalt Products and Crude Oil Residue,* Publication 2023, American Petroleum Institute, Washington, D.C., Mar. 1977.
52. Asphalt Hot-Mix Emission Study, Research Report RR-75-1, The Asphalt Institute, College Park, Maryland, 1975.
53. R. P. Hangebranck, D. J. von Lehmden, and J. E. Meaker, *Sources of Polynuclear Hydrocarbons in the Atmosphere,* U.S. Public Health Service Publication No. 999-AP-33, 1967.
54. *Particulate Polycyclic Organic Matter,* National Academy of Sciences, Washington, D.C., 1972.
55. J. M. Colucci and C. R. Begeman, *Environ. Sci. Technol.* **5,** 145 (Feb. 1971).
56. *Petroleum Asphalt and Health,* API Medical Research Rep. EA 7103, Dec. 1, 1971.
57. C. H. Baylor and N. K. Weaver, *Arch. Environ. Health* **17** (Aug. 1968).
58. W. Luinsky, *Anal. Chem.* **32,** 684 (1960).
59. W. C. Hueper, *Occupational Tumors and Allied Diseases,* Charles C Thomas, Publisher, Springfield, Ill., 1942, p. 69.

ASPHALTENES

JAMES SPEIGHT
Western Research Institute
Laramie, Wyoming

Asphaltenes are a solubility fraction that is separated from petroleum, heavy oil, and bitumen by the addition of an excess of a low-boiling liquid hydrocarbon (eg, pentane, heptane) (1). An asphaltene fraction can also be separated from synthetic fuels (coal liquids, shale oil, etc) in a similar manner; however, the character of asphaltenes from petroleum is quite different, in terms of molecular weight and polarity, to the character of asphaltenes from synthetic fuels. In other words, the definition is an operational aid (2).

See also ASPHALT; HEAVY OIL CONVERSION; PETROLEUM REFINING; PETROLEUM PRODUCTION.

SEPARATION

The procedure by which asphaltenes are separated not only dictates the yield, but also dictates the "quality" of the fraction. In summary, asphaltene separation is not only a function of the amount of diluent added, but also a function of the type and composition of the precipitant. Subsequent fractionation of petroleum, usually be adsorption chromatography (Figure 1), allows other fractions to be separated (2).

Different feedstocks contain different amounts of asphaltenes which can influence the properties of the feedstock, such as variations in the yield of coke as produced by the various carbon residue tests (3,4). Asphaltene content also causes variations in the rates of catalytic reactions, such as hydrodesulfurization, and offers several limitations in processing heavy oils and residua (3,5). Thus, the asphaltene content of a feedstock can dictate the method chosen for refining.

COMPOSITION

The elemental composition of asphaltenes isolated from various petroleum feedstocks by use of excess (greater than 30 volumes per g. of feedstock) *n*-pentane as the precipitant shows only a narrow variation in the atomic hydrogen:carbon ratio (H/C = 1.15 0.05%) (2); although values outside of this range sometimes are found. Use of a low-boiling liquid hydrocarbon, other than pentane, as the precipitant yields an asphaltene fraction that is substantially different from the pentane-insoluble material (2,6).

Asphaltenes can be fractionated by a variety of techniques indicating that asphaltenes are complex mixtures of species of varying molecular sizes and various functional types (2,7).

Data from structural studies support the hypothesis that asphaltenes contain condensed polynuclear aromatic and heterocyclic systems bearing alkyl side chains (8). However, determining the actual structures of the constituents of the asphaltene fraction has proven to be difficult. It is, no doubt, the great complexity of the asphaltene fraction that has hindered the formulation of the individual molecular structures.

Studies on the disposition of nitrogen in petroleum asphaltenes indicate that nitrogen exists as various heterocyclic types, whereas oxygen in petroleum asphaltenes has been identified in carboxylic ($—CO_2H$), phenolic (—OH), and ketonic (>C=O) locations, but is not usually found in heterocyclic systems (2).

Sulfur occurs in alicyclic and aromatic-type (eg, thiophene, benzothiophene, dibenzothiophene, and naphthenebenzothiophene systems), in alkyl–alkyl sulfides, alkyl–aryl sulfides and aryl–aryl sulfides systems, as well as in systems where it has generally replaced oxygen (2).

Metals (ie, nickel and vanadium) are much more difficult to integrate into the asphaltene system. The nickel and vanadium occur as porphyrins, but whether or not these are an integral part of the asphaltene structure is not known. Some of the porphyrins can be isolated as a separate stream from petroleum (9).

In summary, asphaltenes contain a variety of hydrocarbon and heterocyclic systems, as well as a variety of functional groups. Asphaltenes from different crude oils generally contain similar hydrocarbon and functional systems but with some variation in degree due to local and regional variations in the precursors and in the maturation conditions (2).

MOLECULAR WEIGHT

The molecular weight of petroleum asphaltenes is a contentious issue since the results are not only dependent on the nature of the solvent, but also on the temperature at

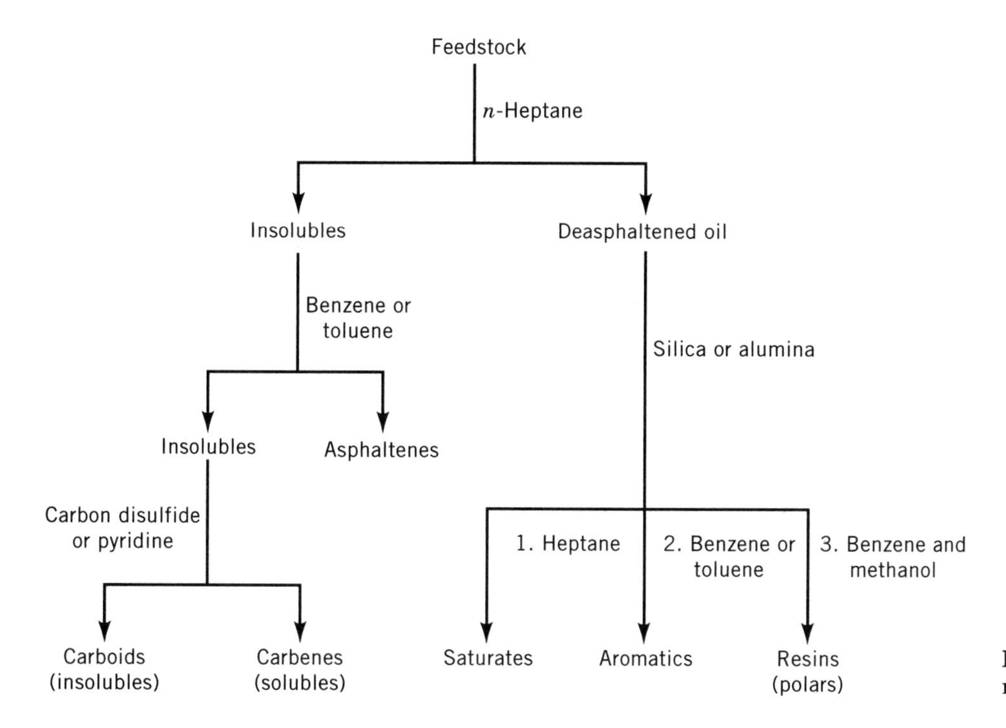

Figure 1. Separation and general nomenclature of petroleum fractions.

which the determinations are performed. In fact, the molecular complexity of the asphaltene fraction is not conducive to determination of absolute molecular weights by any one method (2).

Asphaltenes form molecular aggregates, even in dilute solution, and this association is influenced by solvent polarity, asphaltene concentration, and the temperature of the determination (10). In fact, it is recommended that in order to diminish concentration and temperature effects, the molecular weight determination should be carried out at three different concentrations at three different temperatures. The data for each temperature are then extrapolated to zero concentration, and the zero concentration data at each temperature are then extrapolated to room temperature (Fig. 2).

DISPERSION IN PETROLEUM

The asphaltene fraction, being ubiquitous to all petroleum, has received considerable attention after the realization that it is the constituents of this fraction which play a prominent, usually adverse, role in such operations as reservoir chemistry and physics (which includes oil-rock interactions and gas deasphalting effects), recovery operations (including reservoir plugging and pipeline flow), process operations (including coke formation and

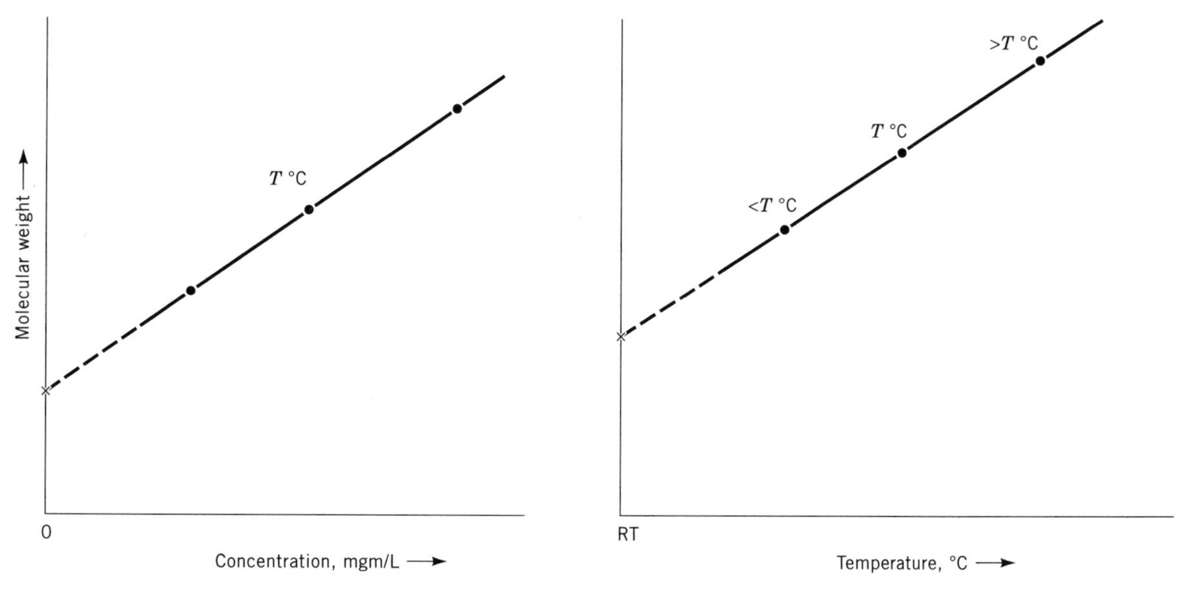

Figure 2. General representation of the method for determination of asphaltene molecular weights.

catalyst deterioration), and asphalt production, as well as asphalt-aggregate interactions.

Asphaltenes are insoluble in the oil fractions (saturates plus aromatics; Figure 1), and the means by which the asphaltenes remain dispersed in petroleum has been the subject of several studies and much speculation. It is generally believed, and there is evidence for this, that asphaltene dispersion in petroleum is attributable to the resins (11).

It is the asphaltenes which affect the ability of the asphalt to adhere to the binder (12) and effect the properties of air-blown asphalts. For example, not only are "new" asphaltenes produced during the air blowing process, but the evidence suggests that the constituents of the asphalt which have higher proportions of heteroatoms (ie, nitrogen, oxygen, and sulfur) will react preferentially with oxygen leaving less polar substances unreacted. This raises the chances of the incompatibility of the resulting products. If the more polar species (ie, the resins) are preferentially oxidized to produce more asphaltenes, the means by which the original asphaltenes are dispersed in the oily medium is removed, and phase separation (asphaltene deposition, also called incompatibility) should be anticipated.

BIBLIOGRAPHY

1. J. G. Speight, R. B. Long, T. D. Trowbridge, Preprints, *Div. Fuel Chem.*, **27**(3/4), 268 (1984).
2. J. G. Speight, *The Chemistry and Technology of Petroleum*, Second Edition, Marcel Dekker Inc., New York, 1991.
3. J. G. Speight, *The Desulfurization of Heavy Oils and Residua*, Marcel Dekker Inc., New York, 1981.
4. Annual Book of ASTM Standards, Section 5.01, *Petroleum Products Lubricants and Fossil Fuels*, American Society for Testing and Materials, Philadelphia, Pa., 1991.
5. J. G. Speight, *In Fuel Science and Technology Handbook*, Marcel Dekker Inc., New York, 1990.
6. J. G. Speight and S. E. Moschopedis, in J. W. Bunger and N. C. Li, eds. *Chemistry of Asphaltenes*, American Chemical Society, Washington, D.C., 1981.
7. R. B. Long, Preprints, *Div. Petrol. Chem.*, **24**(4), 891 (1979).
8. T. F. Yen, *Energy Sources*, **1**, 447 (1974).
9. J. F. Branthaver, in J. G. Speight, ed. *Fuel Science and Technology Handbook*, Marcel Dekker Inc., New York, 1990.
10. J. G. Speight, D. L. Wernick, K. A. Gould, R. E. Overfield, B. M. L. Rao, and D. W. Savage, *Rev. Inst. Francais du Petrole*, **40**, 51 (1985).
11. J. A. Koots and J. G. Speight, *Fuel*, **54**, 179 (1975).
12. J. G. Speight, "Asphalt," in Kirk-Othmer's *Encyclopedia of Chemical Technology*, John Wiley & Sons, Inc., New York, 1992, Vol. 3, p. 689.

AUTOIGNITION TEMPERATURE

RAMON ESPINO
Exxon Research and Engineering
Annandale, New Jersey

There is a temperature at which a combustible material can begin to burn even in the absence of a flame or an electrical spark. This is the autoignition temperature; it is significantly higher than the flash point temperature which requires a flame or spark to initiate the rapid combustion process associated with a burning material (see also COMBUSTION MODELING).

The autoignition temperature of a compound decreases if the oxygen content is above the 21% content of the earth's atmosphere. As a general rule, increased care is essential when operating in oxygen-enriched atmospheres since combustion is favored leading to lower autoignition and flash temperatures, broader flammability limits and more violent fires.

In Table 1 below the autoignition temperature of some commonly used materials are given in both air and in pure oxygen.

Table 1. Autoignition Temperatures of Commonly Used Materials

	Autoignition Temperature, °C	
Compound	In Air	In Oxygen
Methane	537	Less than 532
n-Butane	288	278
Ethylene	490	485
Gasoline	440	316
Kerosene	227	216
Ethyl ether	193	182
Chloroform	Nonflammable	
Methanol	385	Less than 385
Ethanol	365	Less than 365
Acetone	465	Less than 465
Benzene	560	Less than 560
Ammonia	651	Less than 551
Copper and brass	1107	829
Iron	927	621

[a] Ref. 1.

Most households do not experience temperatures in the range required for autoignition to occur (except cooking ovens). However, if a fire occurs, it is not uncommon for temperatures to reach the autoignition temperature of fabrics, resulting in the fire spreading throughout the house at a dangerously rapid rate. Automobile engine blocks can get extremely hot after high load road conditions or high speed. If this happens when a gasoline leak occurs, the autoignition temperature of the gasoline can be reached with a very serious fire ensuing.

In manufacturing operations the danger of fires resulting from combustibles being in contact with surfaces hot enough for autoignition to occur is greater. In steel operations, for example, fires that have resulted in deaths were caused by grease and lubricating oils spilling onto very hot surfaces.

BIBLIOGRAPHY

1. *Fire Protection Handbook*, 15th ed., National Fire Protection Association.

AUTOMOBILE EMISSIONS

F. M. Black
Mobile Source Emissions Research Branch
U.S. Environmental Protection Agency
Research Triangle Park, North Carolina

A relationship between automobile emissions and air pollution was first suggested in the United States by studies that began during the 1940s in California. Air pollution characterized by plant damage, eye and throat irritation, stressed rubber cracking, and decreased visibility was detected in Los Angeles as early as 1943. By 1948, the California legislature had established air pollution control districts empowered to curb emission sources. Initial efforts were aimed at reducing industrial particulate emissions and resulted in improved visibility, but eye irritation and other smog symptoms remained.

See also Clean air act, mobile sources; Air pollution: automobile, toxic emissions; Exhaust control, automotive; Automobile engines; Methanol vehicles; Transportation fuels—automotive fuels; Reformulated gasoline; Alcohol fuels; Transportation fuels—ethanol fuels in Brazil.

Research directed by A. J. Haagen-Smit at the California Institute of Technology, published in 1952, demonstrated that in the presence of sunlight and nitrogen dioxide (NO_2), hydrocarbon (HC) compounds react to form a variety of oxidants, including ozone, that could account for many of the observed environmental effects (1). Because the automobile was thought to be a significant source of oxidant-precursors, the Automobile Manufacturers Association established the Vehicle Combustion Products Committee in 1953 to further define the problem and to search for solutions. Simultaneously, California began an effort to reduce stationary-source HC emissions in the Los Angeles area. However, smog remained a serious problem, and in 1958, the decision was made to require some form of automotive HC emissions control. In 1959, the California legislature enacted air quality standards for oxidants and carbon monoxide (CO), and in 1960, the California Motor Vehicle Pollution Control Board was created to implement these standards. Its function was to set specifications for vehicle exhaust and evaporative emissions and to certify that vehicles sold in California met the specifications. It was estimated that achieving the air quality goals for 1970 would require an 80% reduction of motor vehicle HC emissions and a 60% reduction of motor vehicle CO emissions (2). As illustrated in Figure 1 for cars without emission controls, engine exhaust accounted for about 60% of polluting emissions, crankcase vapors for about 20%, and fuel evaporation for about 20% (3). California's first emission control requirement addressed crankcase blowby emissions. In 1963, devices to eliminate crankcase emissions were required on all cars sold in California and were installed on all new automobiles nationally. Tailpipe HC and CO emissions were first regulated by California in 1966.

The federal government became involved with air pollution in 1955 by empowering the Department of Health, Education, and Welfare to provide technical assistance for resolution of problems related to air pollution. This activity was enhanced by the Clean Air Act of 1963, which was structured to stimulate state and local air pollution control efforts. A 1965 amendment to the Clean Air Act specifically authorized the writing of national standards for emissions from all new motor vehicles sold in the United States Control of motor vehicle exhaust emissions was required nationally beginning in 1968 (4). Gasoline vehicles were required to be equipped with closed crankcase ventilation systems, and tailpipe concentrations of HC and CO emissions were lmited to prescribed concentrations (275 ppm HC, 1.5% CO) over a defined dynamometer driving schedule. Beginning in 1970, a preproduction certification practice controlled the mass emission rates (eg, grams emitted per vehicle mile traveled, g/mi) of regulated compounds, as summarized in Table 1 (5).

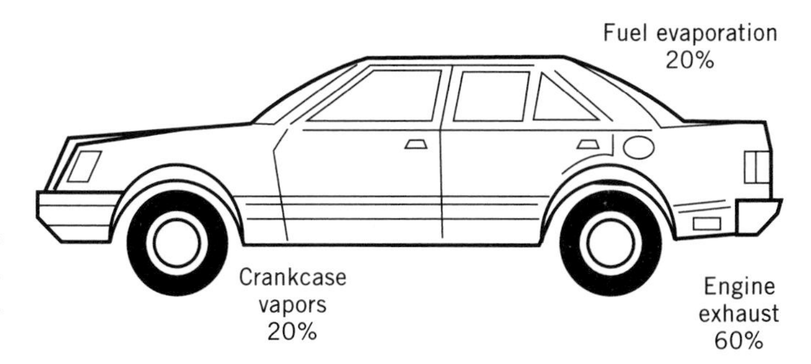

Figure 1. Sources of automobile pollutants prior to emissions control (3).

Since motor vehicle emissions control was instituted in the United States, standards have been progressively reduced and test procedures have been improved. For example, 1971 light-duty vehicle tailpipe emissions, certified at 2.2 g/mi total hydrocarbon (THC) and 23 g/mi CO, were determined using a seven-mode steady-state driving schedule. Since 1972, transient driving schedules have been used. Current tailpipe emission standards are 0.41 g/mi THC, 3.4 g/mi CO, 1.0 g/mi oxides of nitrogen [NO_x, the sum of nitric oxide (NO) and NO_2], and 0.2 g/mi particulate matter. In 1971, motor vehicle evaporative emissions, certified at 6 g per test, were determined with a gravimetric carbon trap technique by collecting emissions from selected point sources on the vehicle. Beginning in 1978, evaporative emissions were determined with a sealed housing for evaporative determination (SHED) technique, in which the entire vehicle is enclosed so that emissions are collected from all evaporative sources. The current evaporative emissions standard is 2 g per test. Starting in 1974, tailpipe emissions from heavy-duty vehicles [gross vehicle weight (GVW) 3856 kg (8500 lb) or more] were regulated using a 13-mode steady-state driving schedule. Since 1985, transient driving schedules have been used. The tailpipe emission standards have ranged from 40 g per brake-horsepower-hour (g/bhph) CO in 1974 to current gasoline truck and bus standards of 1.1 g/bhph THC, 14.4 g/bhph CO, and 5.0 g/bhph NO_x [standards are higher for GVW greater than 6350 kg (14,000 lb)]. The evaporative emission standard for heavy-duty gasoline trucks and buses is 3 g per test. Current diesel

Table 1. U.S. Motor Vehicle Emissions and Fuel Economy Standards

Model Year	Vehicle Category	Tailpipe Hydrocarbon	Carbon Monoxide	Oxides of Nitrogen	Evaporative Hydrocarbon	Particulate	Fuel Economy
1970[a]	LDV, LDT	2.2 g/mi	23 g/mi	—	—		
1971	LDV, LDT	2.2 g/mi	23 g/mi	—	6 g/test[b]		
1972[a]	LDV, LDT	3.4 g/mi	39 g/mi	—	2 g/test		
1973	LDV, LDT	3.4 g/mi	39 g/mi	3.0 g/mi	2 g/test		
1974[d]	HDGV, HDDV	—	40 g/bhphr	—	—		
1975[e]	LDV	1.5 g/mi	15 g/mi	3.1 g/mi	2 g/test		
	LDT	2.0 g/mi	20 g/mi	3.1 g/mi	2 g/test		
1977	LDV	1.5 g/mi	15 g/mi	2.0 g/mi	2 g/test		
1978	LDV	1.5 g/mi	15 g/mi	2.0 g/mi	6 g/test[f]	—	18 mpg
	LDT	2.0 g/mi	20 g/mi	3.1 g/mi	6 g/test		
1979	LDV	1.5 g/mi	15 g/mi	2.0 g/mi	6 g/test	—	19 mpg
	LDT	1.7 g/mi	18 g/mi	2.3 g/mi	6 g/test		
	HDGV, HDDV	1.5 g/bhphr	25 g/bhphr	—	—		
1980	LDV	0.41 g/mi	7.0 g/mi	2.0 g/mi	6 g/test	—	20 mpg
1981	LDV	0.41 g/mi	3.4 g/mi	1.0 g/mi	2 g/test	—	22 mpg
	LDT	1.7 g/mi	18 g/mi	2.0 g/mi	2 g/test		
1982	LDV	0.41 g/mi	3.4 g/mi	1.0 g/mi	2 g/test	—	24 mpg
1983	LDV	0.41 g/mi	3.4 g/mi	1.0 g/mi	2 g/test	—	26 mpg
1984	LDV	0.41 g/mi	3.4 g/mi	1.0 g/mi	2 g/test	—	27 mpg
	LDT	0.8 g/mi	10 g/mi	2.3 g/mi	2 g/test		
1985	LDV	0.41 g/mi	3.4 g/mi	1.0 g/mi	2 g/test	0.6 g/mi	27.5 mpg
	LDT	0.8 g/mi	10 g/mi	2.3 g/mi	2 g/test	0.6 g/mi	
[g]	HDGV (<14k)	2.5 g/bhphr	40.1 g/bhphr	—	3 g/test		
[g]	(> 14K)				4 g/test		
[g]	HDDV	1.3 g/bhphr	15.5 g/bhphr	—	—		
1987	LDV	0.41 g/mi	3.4 g/mi	1.0 g/mi	2 g/test	0.2 g/mi	27.5 mpg
	LDT	0.8 g/mi	10 g/mi	1.2 g/mi	2 g/test	0.26 g/mi	
	HDGV (<14K)	1.3 g/bhphr	15.5 g/bhphr	6.0 g/bhphr	3 g/test		
	(>14K)	2.5 g/bhphr	40 g/bhphr	6.0 g/bhphr	4 g/test		
	HDDV	1.3 g/bhphr	15.5 g/bhphr	6.0 g/bhphr	—		
1988	HDGV (<14k)	1.1 g/bhphr	14.4 g/bhphr	6.0 g/bhphr	3 g/test		
	(>14K)	1.9 g/bhphr	37.1 g/bhphr	6.0 g/bhphr	4 g/test		
	HDDV	1.3 g/bhphr	15.5 g/bhphr	6.0 g/bhphr	—	0.6 g/bhphr	
1991	HDGV (<14K)	1.1 g/bhphr	14.4 g/bhphr	5.0 g/bhphr	3 g/test		
	HDDV	1.3 g/bhphr	15.5 g/bhphr	5.0 g/bhphr	—	0.25 g/bhphr	
1993	HDDV (bus)	1.3 g/bhphr	15.5 g/bhphr	5.0 g/bhphr	—	0.1 g/bhphr	
1994[h]	LDV	0.25 g/mi	3.4 g/mi	0.4 g/mi	2 g/test	0.08 g/mi	27.5 mpg
[h]	LDT	0.32 g/mi	4.4 g/mi	0.7 g/mi	2 g/test	0.26 g/mi	
	HDDV	1.3 g/bhphr	15.5 g/bhphr	5.0 g/bhphr	—	0.1 g/bhphr	
	urban bus					0.05 g/bhphr	
1995[i]	LDT	0.32 g/mi	4.4 g/mi	0.7 g/mi	2 g/test	0.08 g/mi	
1998	HDDV	1.3 g/bhphr	15.5 g/bhphr	4.0 g/bhphr	—	0.1 g/bhphr	

Notes: LDV = light-duty passenger car, LDT = light-duty truck, HDGV = heavy-duty gasoline truck/bus, HDDV = heavy-duty diesel truck/bus, g/mi = grams per mile, g/bhphr = grams per brake horsepower hour, mpg = miles per gallon.

[a] Seven-mode steady-state driving procedure.
[b] Carbon trap technique.
[c] CVS-72 transient driving procedure.
[d] Thirteen-mode steady-state driving procedure.
[e] CVS-75 transient driving procedure.
[f] SHED technique.
[g] Transient driving procedure.
[h] Phased in from 1994 to 1996.
[i] PM phased in from 1995 to 1997.

truck and bus standards are 1.3 g/bhph THC, 15.5 g/bhph CO, 5.0 g/bhph NO_x, and 0.1 g/bhph particulate (0.05 g/bhph for urban buses). Summarizing, tailpipe HC, CO, NO_x, and particulate matter emissions, evaporative HC emissions, and fuel economy currently are regulated with light-duty passenger cars and trucks; similarly, tailpipe and evaporative emissions (from vehicles using gasoline fuel) are regulated with heavy-duty trucks and buses, but not fuel economy.

The emission standards are enforced initially by a program of preproduction design review, testing, and certification. Production prototypes are certified to emit at or below standard levels for 80,450 km (50,000 miles). The federal government also administers assembly-line audits and surveillance-recall programs designed to ensure that production motor vehicles perform as well as the certification prototypes. Thus, emission controls are examined at three stages: initially, to certify that cars are designed to meet federal emission standards before they are manufactured (preproduction); subsequently, to ensure that assembly-line cars actually meet the emission standards they were designed and certified to meet (production); and finally, to ensure that the emission control systems on cars that have been in normal everyday consumer use for several years, having accumulated substantial mileage, perform at an acceptable level (postproduction). If a substantial number of properly maintained and operated vehicles of a specific "family" fail an emission standard during in-use surveillance, the manufacturer can be required to recall the entire vehicle family and restore each vehicle to satisfactory operating condition at no cost to the owner.

Studies during and since the 1970s have revealed that large percentages of vehicles on roadways are exceeding the emission standards for which they were designed. This results primarily from inadequate maintenance by vehicle owners and, in some instances, from owners or their mechanics intentionally disabling emission control systems. Inspection and maintenance (I&M) and antitampering (ATP) programs have been instituted to identify such vehicles and to require their repair at the owners' expense. The 1977 Clean Air Act Amendments required state or local governments to administer I&M programs in all cities exceeding carbon monoxide or ozone air quality standards. A U.S. Environmental Protection Agency (EPA) 1988 tampering survey (7259 vehicles, 15 cities) suggested that 23% of vehicles not covered by I&M and ATP programs had emissions control system tampering, compared with 16% in areas with I&M and ATP (6). The 1990 Clean Air Act amendments require enhancements of programs to identify and repair inoperative vehicles (7).

The U.S. EPA periodically publishes National Air Pollutant Emission Estimates that assess the relative importance of transportation and other anthropogenic sources of air pollution (8). As suggested in California during the 1940s, transportation sources do contribute significantly to air pollution; in 1985, transportation sources contributed about 36% of volatile organic carbon (VOC) emissions, 72% of CO emissions, and 45% of NO_x emissions, according to estimates based on national annual average emission rates [calculated at the vehicle-miles-traveled (VMT) weighted annual average temperature of 14°C]. Figure 2 illustrates the trends in these emissions from 1940 to 1985. Mobile source emission rates are sensitive to ambient temperature, as are air pollution problems. As indicated in Figure 3, ozone air quality violations occur most frequently during the hot summer months, when automobile HC emission rates are elevated (relative to emissions at 14°C); and CO air quality violations occur most frequently during the cold winter months, when automobile CO emission rates are elevated (relative to emissions at 14°C) (9,10). Thus, the contribution of transportation sources to the atmospheric VOC burden is greater than 36% during periods when ozone is an air quality problem, and their contribution to the atmospheric CO burden is greater than 72% during periods when CO is an air quality problem.

Motor vehicles emission estimates generally are based on estimates of fleet average HC, CO, and NO_x emission rates (g/mi) obtained with the U.S. EPA computer model MOBILE and estimates of fleet VMT (11). MOBILE calculates emission rates for eight individual vehicle types in two altitude ranges of the United States. The emission estimates depend on various conditions such as ambient temperature, speed, and vehicle model year mileage accrual rates. The computer model permits user specification of many of the variables affecting average vehicle emission rates. The model can estimate emission rates for any calendar year between 1960 and 2020. The 25 most recent vehicle model years are considered to be in operation in each calendar year.

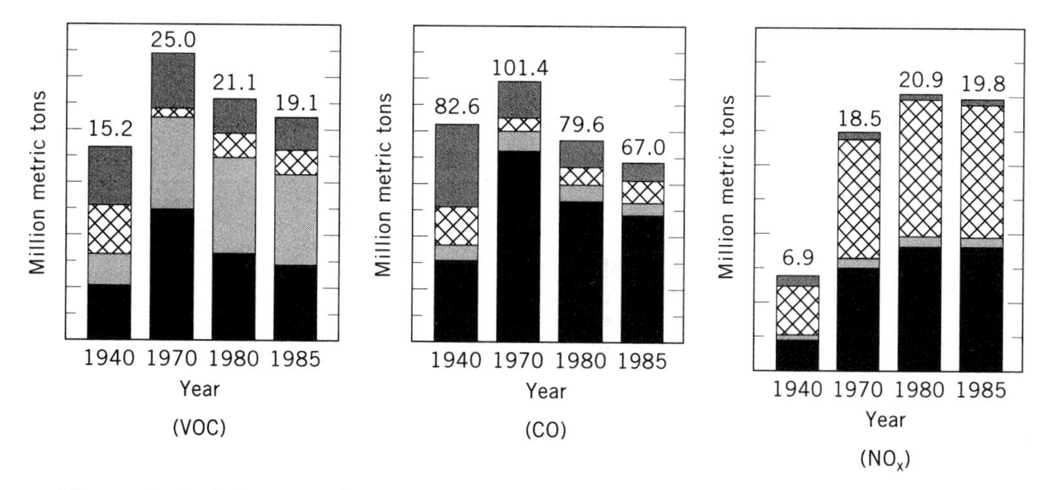

Figure 2. Emissions trend, 1940–1985 (8); (■) solid waste and miscellaneous; (⊠) stationary combustion; (▦) industrial processes; (■) transportation.

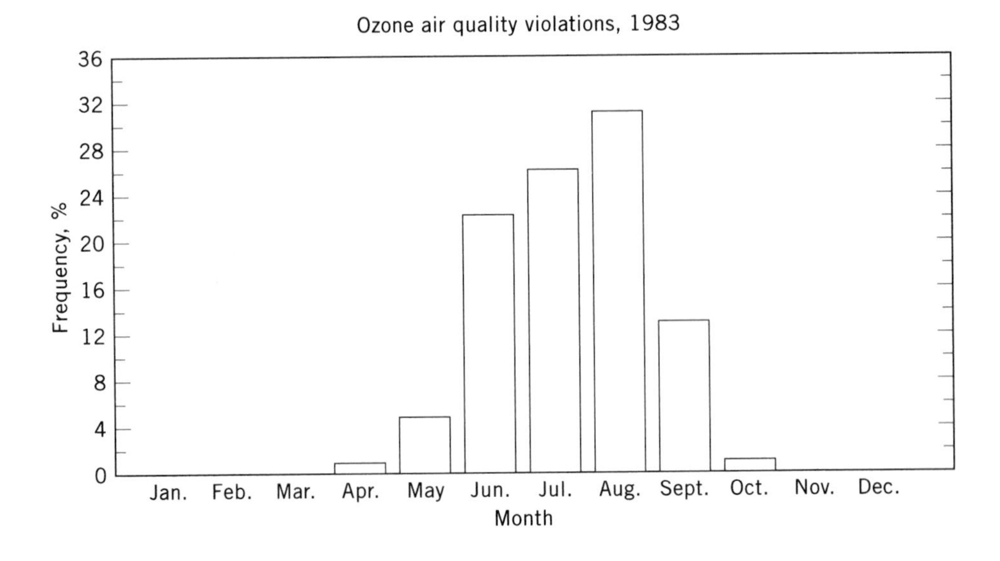

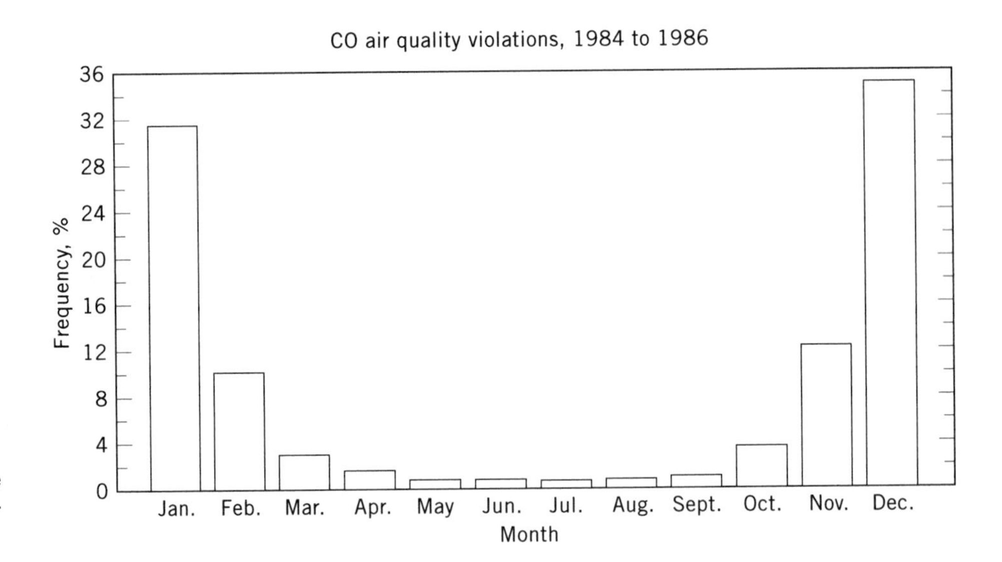

Figure 3. Monthly frequency of ozone and carbon monoxide air quality violations.

Recent roadway studies have suggested that MOBILE may underestimate the HC and CO emission rates of actual roadway motor vehicles (12). The computer model estimates are based on motor vehicle emissions data obtained with laboratory dynamometer simulations of roadway driving conditions. The dynamometer simulations may not represent actual roadway conditions, and the vehicles tested may not represent actual roadway motor vehicles. The laboratory dynamometer driving simulations used to estimate emission rates were developed during the late 1960s and early 1970s and may not adequately represent today's driving patterns (13), and poorly maintained or tampered-with vehicles may not be appropriately represented. Current research efforts are examining both the characteristics of roadway vehicles (eg, their emissions control performance) and the adequacy of laboratory simulations for determining both tailpipe and evaporative emission rates.

MOTOR VEHICLE EMISSIONS CONTROL

Beginning in the early 1950s, Americans tended to move from cities to surrounding suburbs. Because the majority of the suburbs were not served by mass transit, this population shift created an attractive market for automobile manufacturers. The number of vehicles produced and sold increased dramatically, as did construction of highways between cities and suburbs. Families began to own more than one vehicle, and it became more common for each family member to have a vehicle. As vehicle ownership and use increased, so did the pollutant levels in and around cities, as the suburbanites commuted daily between their businesses and employment in the city and their homes in the suburbs (3). As indicated in Figure 4, the automobile remains the favored form of transportation in the United States (14). If the current rate of VMT growth continues, the contribution of motor vehicle emis-

Railroad
1970 – 11 billion
1987 – 12 billion

Bus
1970 – 25 billion
1987 – 23 billion

Domestic air
1970 – 119 billion
1987 – 341 billion

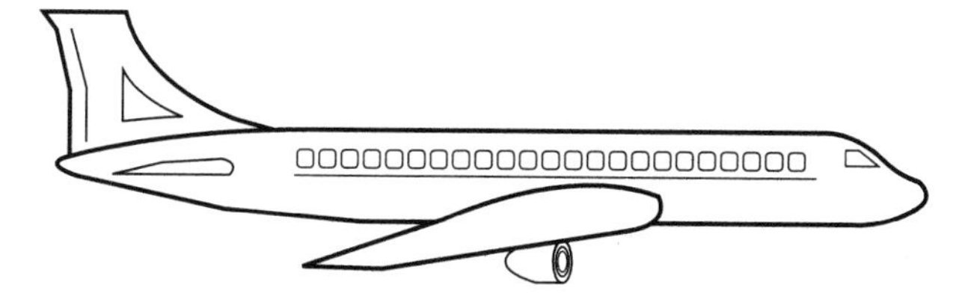

Private auto
1970 – 1.03 trillion
1987 – 1.49 trillion

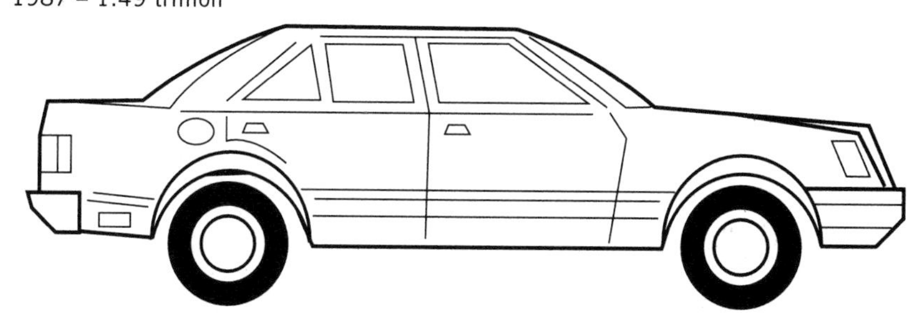

Figure 4. How Americans travel (passenger miles).

sions to air pollution is projected to bottom out in the mid to late 1990s and begin to grow in magnitude, unless further emission control improvements are realized (15). Changes in both vehicle technology and fuel formulations are expected to be used to achieve further emissions reductions.

Efforts by the U.S. government and motor vehicle industries to improve air quality in U.S. cities have resulted in a dramatic reduction of emission rates from roadway motor vehicles. Estimates obtained with MOBILE 4 suggest that from 1975 to 1990, fleet average nonmethane hydrocarbon (NMHC) and CO emission rates were reduced by about 70% and NO_x emission rates by about 55% (11), as shown in Figure 5. During that same period, however, VMT increased about 45%, substantially offsetting the emission rate reductions. Most of the emission reductions achieved to date have resulted from the changes in vehicle technology responding to new federal motor vehicle emission standards. The emissions certification practice involves measurement of the emission rates of regulated compounds (or groups of compounds) with laboratory simulations of roadway driving conditions. Figure 6 schematically illustrates a typical light-duty chassis dynamometer test cell and a speed-versus-time display of the CVS-75 driving schedule used for laboratory simulation of roadway driving conditions. Figure 7 provides a flowchart overview of light-duty motor vehicle tailpipe and evaporative emissions certification test procedures. The heavy-duty vehicle certification procedure involves a transient engine dynamometer test wherein engine speed and load are varied according to a prescribed schedule. The engine dynamometer test is used for heavy-duty vehicles because in the United States heavy-duty truck engines and chassis often are manufactured by different companies. The adequacy of current practice for laboratory simulation of roadway driving conditions is being examined, and both tailpipe and evaporative emission test procedures may be modified in the near future.

The three major categories of emissions from motor vehicles without emissions control are crankcase ventila-

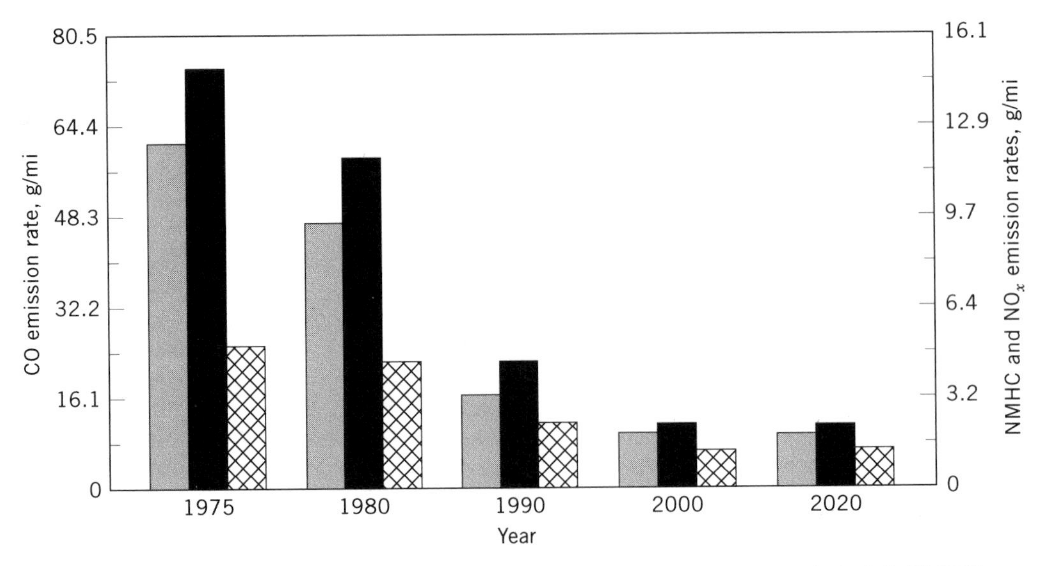

Figure 5. Fleet average NMHC, CO, NO_x, emission rates—1975–2020 (11); (▤) NMHC; (■) CO; (▧) NO_x. 1975–1990 reductions: NMHC, 73%; CO, 71%; NO_x, 55%. Note: 16–29°C diurnal range; 79.3 × 10^3 Pa (11.5 psi), 72.4 × 10^3 Pa (10.5 psi) (89) RVP fuel; 31.5 km/h (19.6 mph); I&M, antitampering.

tion, fuel evaporation, and engine exhaust. Control technologies have been developed to reduce emission rates from each of these sources.

Crankcase Emissions

Crankcase emissions, responsible for approximately 20% of all harmful automotive pollutants before emission controls, result from compression gases being forced past the piston rings on both the compression and power strokes, resulting in an accumulation of "blowby" gases in the crankcase. These gases mix with vapors from the agitated lubricating oil and must be vented from the crankcase area to prevent damaging pressures from developing. Before the 1960s, a road draft tube was used to ventilate the crankcase, emitting the pollutants into the atmosphere. By 1968, all vehicles manufactured in the United States were equipped with a closed crankcase ventilation system that prevented the escape of blowby gases and oil vapor to the atmosphere. Figure 8 is a schematic diagram of a typical positive crankcase ventilation (PCV) system. During idle and cruise periods, the emissions are vented to the intake manifold where they are mixed with the cylinder fuel–air charge for subsequent combustion. In the closed crankcase ventilation system, the fresh air intake, located in the air cleaner, has a dual role. It not only provides fresh air for crankcase ventilation, but it also allows overload release of blowby gases into the carburetor airstream should the PCV valve fail to control the buildup of blowby gases; this prevents excess gases from escaping into the atmosphere. With this system functioning properly, crankcase emissions are essentially eliminated as a source of air pollution from motor vehicles (3).

Evaporative Emissions

Annual average fuel evaporation was responsible for about 20% of harmful automobile pollutants prior to emissions control, the severity increasing with elevated ambient temperature and fuel volatility. The primary sources of evaporative hydrocarbon emissions were the fuel tank and carburetor bowl, both of which were vented to the atmosphere. Another problem was gasoline spillage when fuel tanks were overfilled. The evaporative emissions control system is designed to prevent gasoline leakage and to trap vapors. Figure 9 is a schematic diagram of a typical system. The key design features include a sealed fuel tank cap, a fuel tank fill limiter, a liquid–vapor separator, and a carbon canister. The vented fuel tank filler cap is replaced by a sealed cap with the fuel tank vented through a carbon canister designed to store the fuel vapors. A pressure/vacuum relief valve is incorporated into the sealed cap to prevent damage to the tank should excessive internal or external pressures exist. The fuel tank has been redesigned to provide approximately 10% of the total tank volume for expansion space should the fuel expand because of temperature changes. Overfill protection is provided by a filler neck that assures the expansion space is maintained when the tank is filled. Tank vapor ventilation is controlled by having a vent tap at or near the expansion chamber dome. A foam-type filter or a vapor separator allows the vapors to pass to a storage point, while preventing the passage of liquid which is returned to the fuel tank. For most vehicles, the liquid–vapor separator assembly is located within the vent line. Vehicles may be equipped with separate expansion chambers, with separate evaporation chambers, and with one or more check valves in the lines. These components usually are located near the fuel tank.

An important component of the fuel evaporative control system is the canister of activated charcoal (porous carbon). Activated charcoal can absorb and store fuel vapors, and it can be cleaned by purging with fresh air, allowing it to be reused many times. When the engine is not running, the fuel vapors are routed to the canister, where

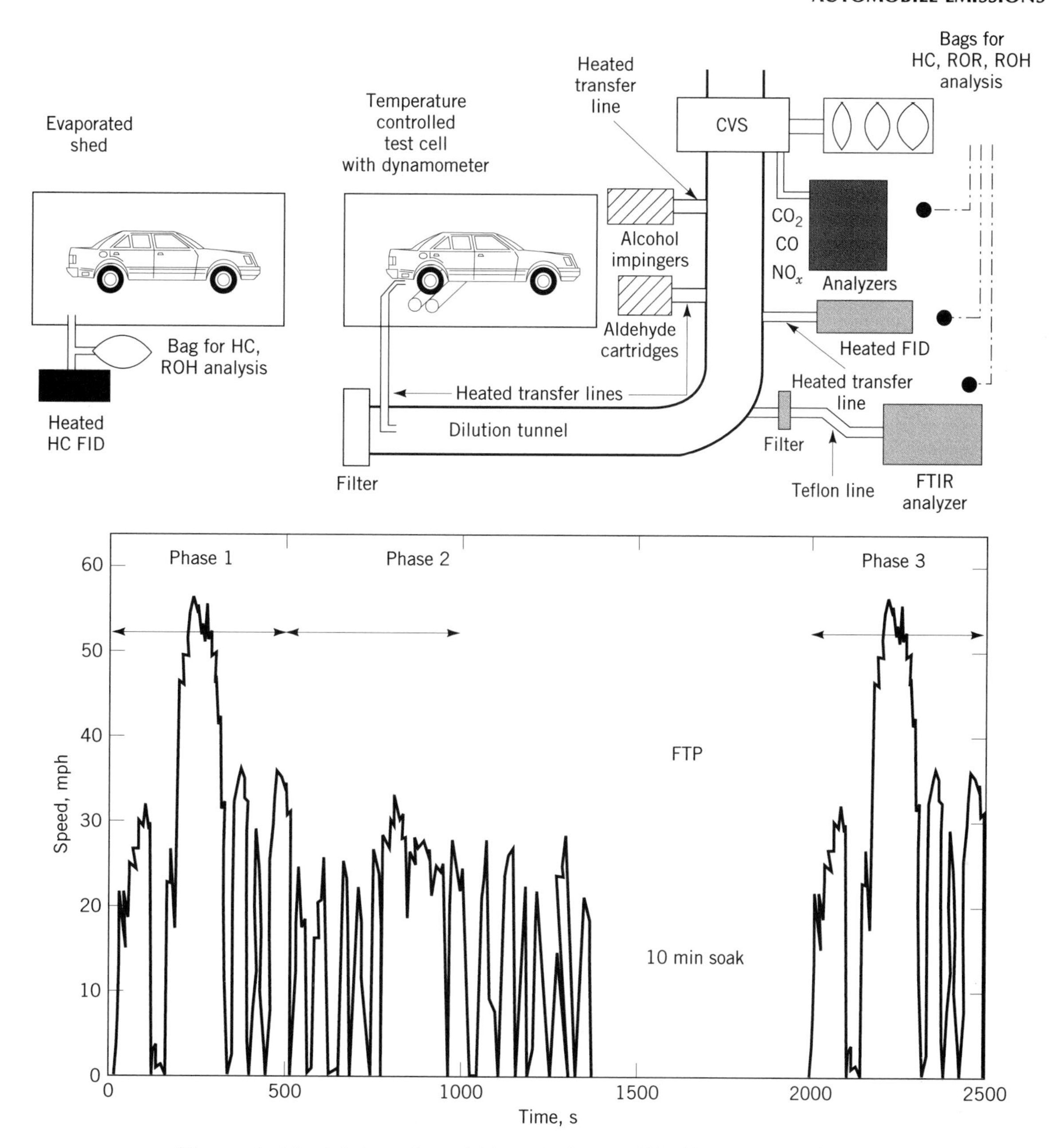

Figure 6. Light-duty motor vehicle emissions certification test procedures.

they are absorbed and stored on the activated charcoal. When the engine is started, either engine vacuum or the airflow through the carburetor air cleaner causes a metered amount of air to pass over and through the charcoal, purging the vapors. The purge air is routed into the engine's induction system. Various types of purge control systems have been used. The evaporative emissions control system schematically illustrated in Figure 9 uses a combination of constant and demand purging. Constant purging is provided by a continuous restricted flow through the canister to the intake manifold, whereas the demand purging is provided by flow only when the throttle plate is open (not during idle operation) (3).

Another category of evaporative emissions receiving recent attention is "running-loss" emissions, which occur when the vapor generated while the vehicle is being operated on the roadway exceeds the system's canister storage and purging capacity. This category of emissions can be significant during hot summer days, when ozone (O_3) problems prevail.

Fuel volatility, often expressed in terms of Reid vapor pressure (vapor pressure at 38°C; RVP), strongly affects the magnitude of motor vehicle evaporative emissions. Higher evaporative emission rates are associated with higher gasoline volatilities. Historically, the volatility of marketed gasolines has increased as vehicle designs have tolerated more volatile fuels (i.e., have become more resistant to fuel delivery system vapor lock). National average summer gasoline volatilities increased 14% from 1974 [65.5×10^3 Pa (9.5 psi) RVP] to 1985 [74.5×10^3 Pa (10.8 psi) RVP]; the increase varied regionally from 10 to 20% (16). The volatility of federal emissions certification gaso-

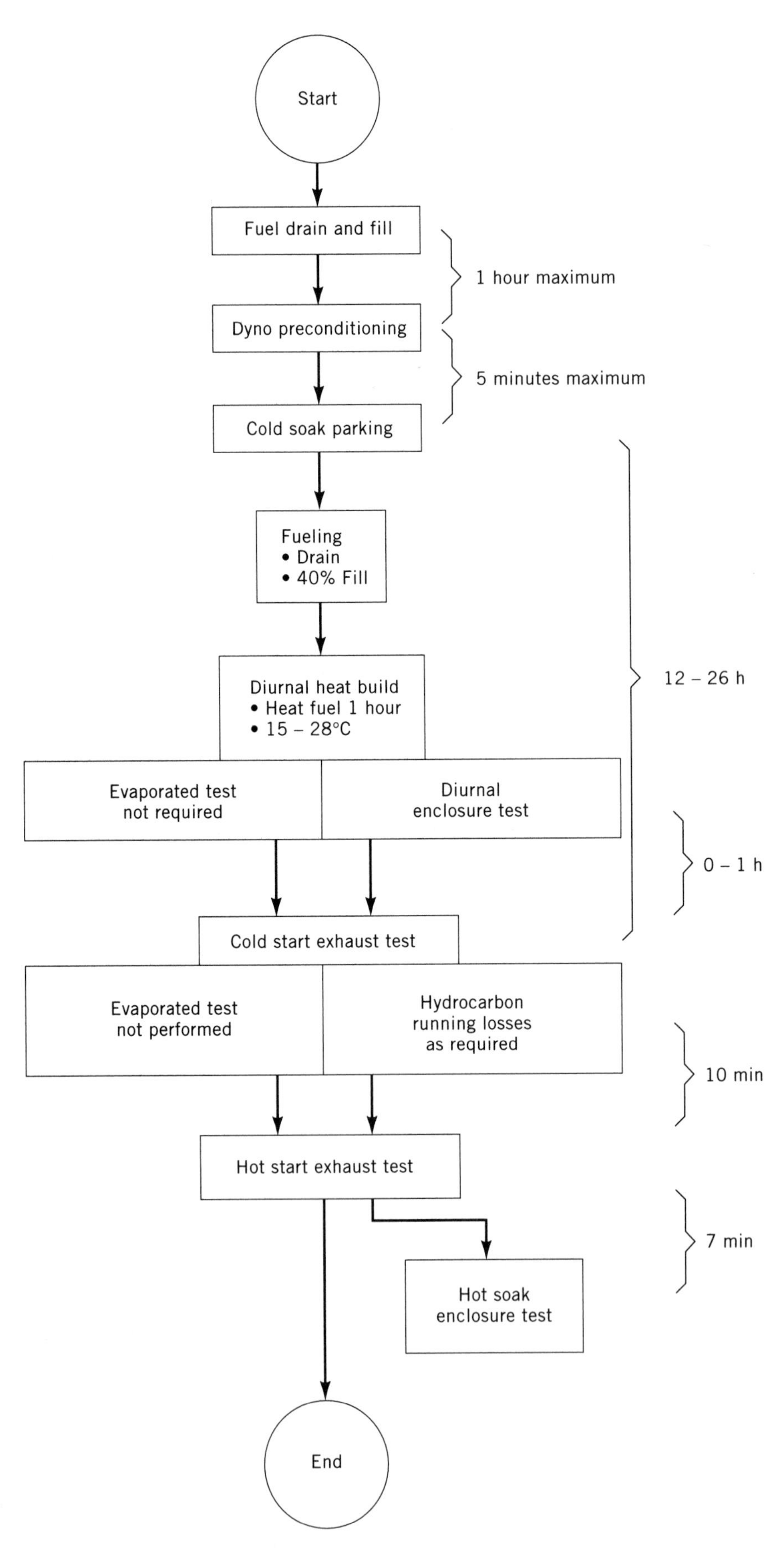

Figure 7. Light-duty motor vehicle emissions certification test sequence.

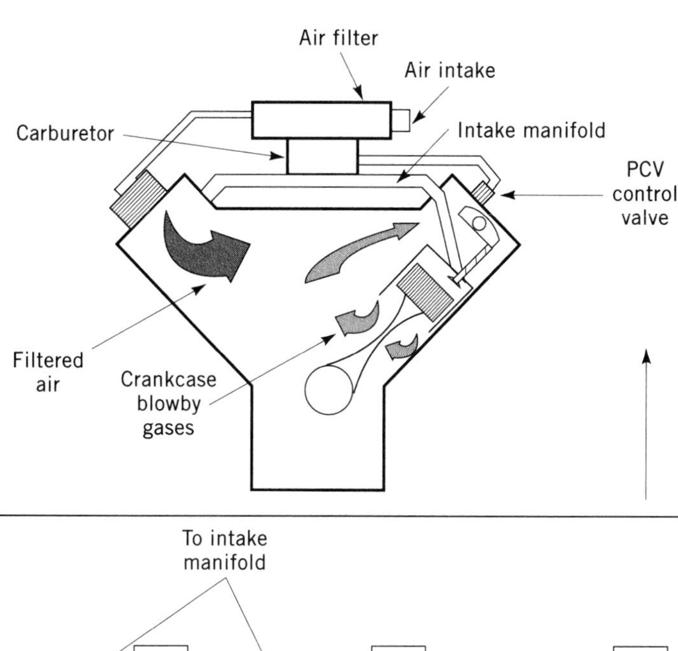

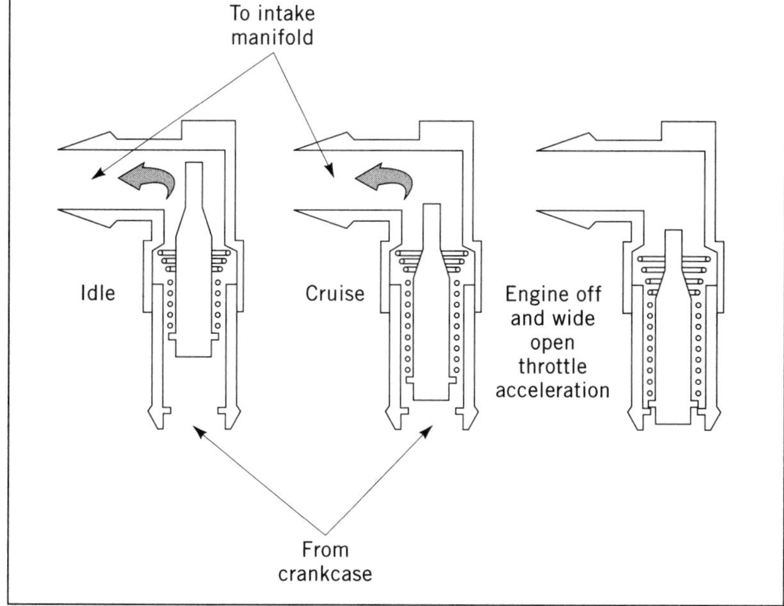

Figure 8. Crankcase emissions control (3).

line is 60 × 10^3–63.4 × 10^3 Pa (8.7–9.2 psi) RVP. Because of the disparity between certification and marketed gasoline volatilities, certified evaporative emissions control systems performed inadequately under roadway conditions. Since 1989, the U.S. EPA has regulated marketed gasoline volatility (9). Summer (May to September) gasoline volatilities currently vary regionally from 53.8 × 10^3 to 62.1 × 10^3 Pa (7.8–9.0 psi) RVP (17).

Engine Exhaust Emissions

Engine exhaust was estimated to be responsible for about 60% of pollution from automobiles prior to emissions control. The technology that has evolved to reduce engine exhaust emissions is complex and varies from vehicle to vehicle. The control system may include many different elements or subsystems, such as the following:

- thermostatically controlled air cleaner,
- air injection systems,
- ignition timing controls,
- increased temperature control,
- transmission or speed-controlled spark system,
- exhaust gas recirculation system,
- catalytic converter system,
- engine feedback systems,
- carburetor and choke modifications,
- fuel injection (throttle body and port) systems, and
- computer-based engine controls.

Figure 10 is a schematic diagram of several elements of a typical exhaust emissions control system. A key factor in simultaneous reduction of emissions of HC, CO, and NO_x is accurate control of the air–fuel ratio; many of the control system components serve that purpose. A common practice is to operate the engine slightly fuel rich (ie, at an air–fuel ratio lower than the stoichiometric ratio), optimizing NO_x conversion in the front half of the catalyst, and then to inject air midway, optimizing conversion of excess HC and CO in the rear half of the catalyst, as illustrated in Figure 11. Figure 12 shows the impact of the

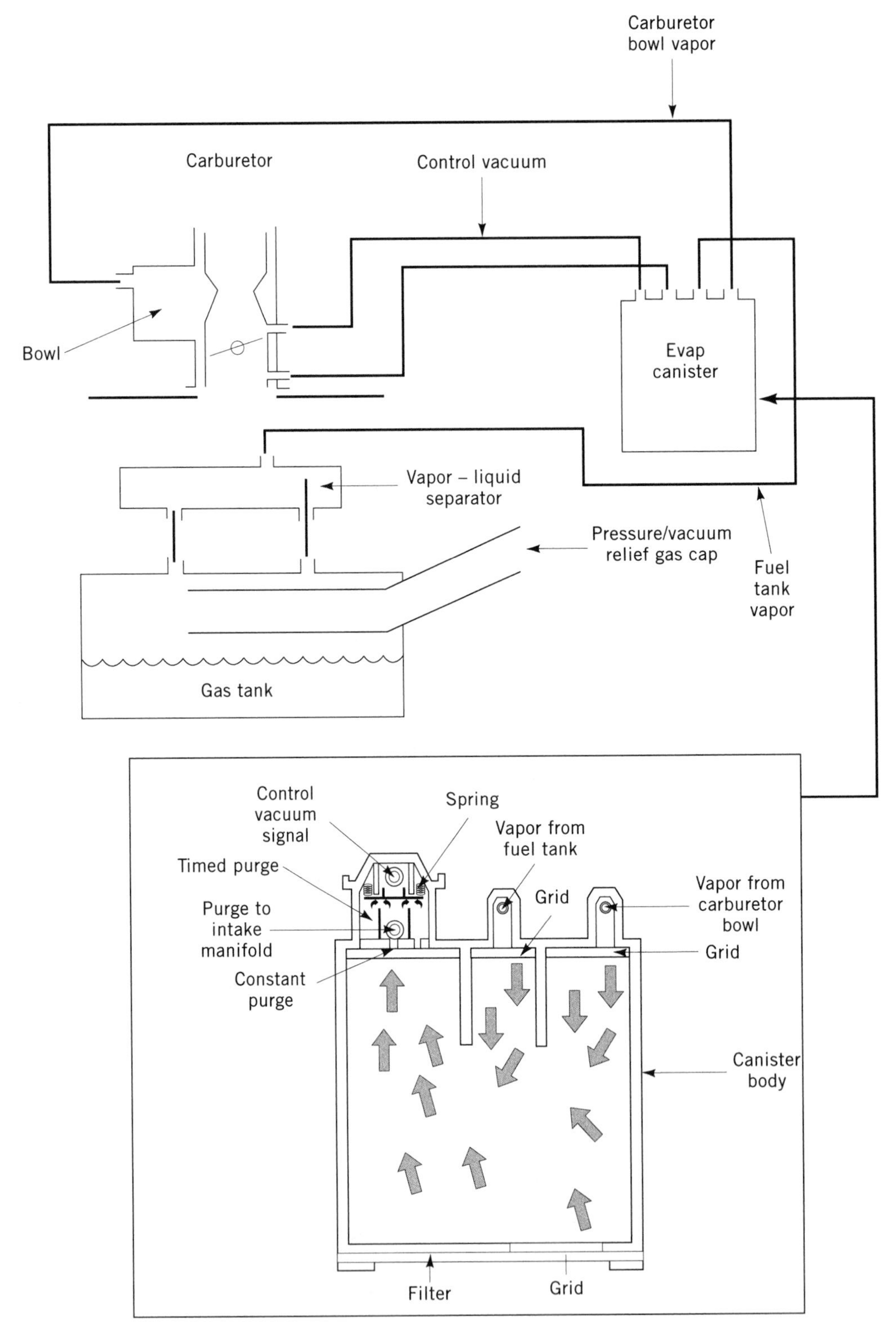

Figure 9. Evaporative emissions control (3).

air–fuel ratio on engine exhaust emissions and on catalyst conversion efficiency. This system is commonly referred to as a "three-way" catalyst, because of its combined reduction of HC, CO, and NO_x. Fuel injection and air injection are controlled by an electronic control module (ECM) in association with an exhaust oxygen sensor. The ECM is the keystone of the entire emissions control system, acquiring information from numerous sensors and controlling numerous elements and devices of the engine and the emissions control system, as summarized in Figure 13 (3).

In addition to catalytic reduction, exhaust gas recirculation (EGR) often is used to reduce NO_x emissions. Because formation of NO_x is sensitive to combustion-zone peak temperature, systems that lower the peak temperature will reduce NO_x emissions. Peak temperature can be

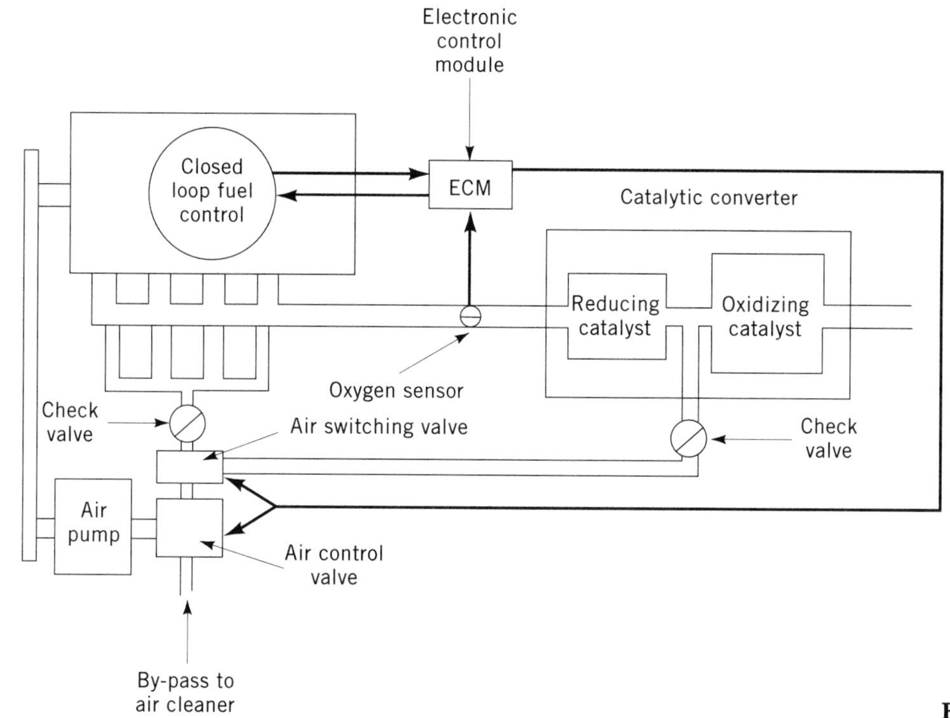

Figure 10. Tailpipe emissions control (3).

reduced by recirculating some exhaust gas into the intake air–fuel mixture. Figure 14 schematically illustrates a typical EGR system. The amount of recirculated exhaust gas must be carefully controlled. If too much exhaust gas is supplied at the wrong time (which may occur if the EGR valve sticks in the open position), the engine may stall or idle rough. If not enough exhaust gas is recirculated (which may occur if the vacuum activation system fails), NO_x will not be reduced (3).

Tampering

Emission control technology can significantly reduce the contribution of motor vehicles to atmospheric pollution, but only if the equipment is properly maintained. During 1988, the emissions control systems of 23% of U.S. vehicles not covered by I&M and ATP were purposely rendered inoperative, compared with 16% of vehicles in areas with I&M and ATP (6). As shown in Figure 15, of 7259 vehicles surveyed, 19% were clearly tampered with and an additional 12% may have been tampered with (i.e., components became inoperative either through tampering or as a result of poor maintenance). Table 2 shows the frequency of tampering with various components. Tampering has serious implications for emissions because a disconnected air pump can increase HC emissions 200% and CO emissions 800%, a disconnected EGR system can increase NO_x emissions 175 %, a missing or damaged ca-

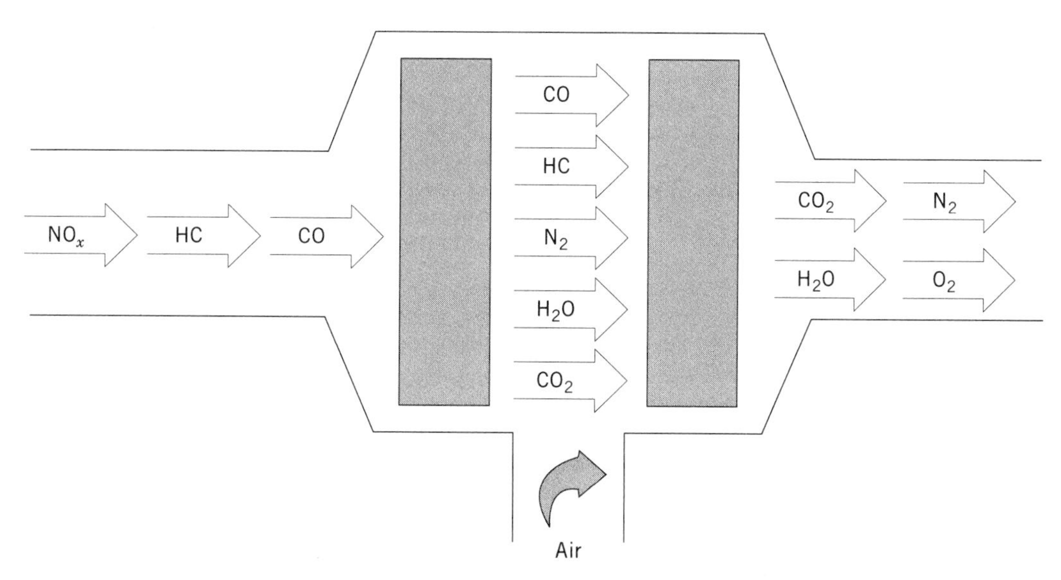

Figure 11. Three-way catalyst control (3).

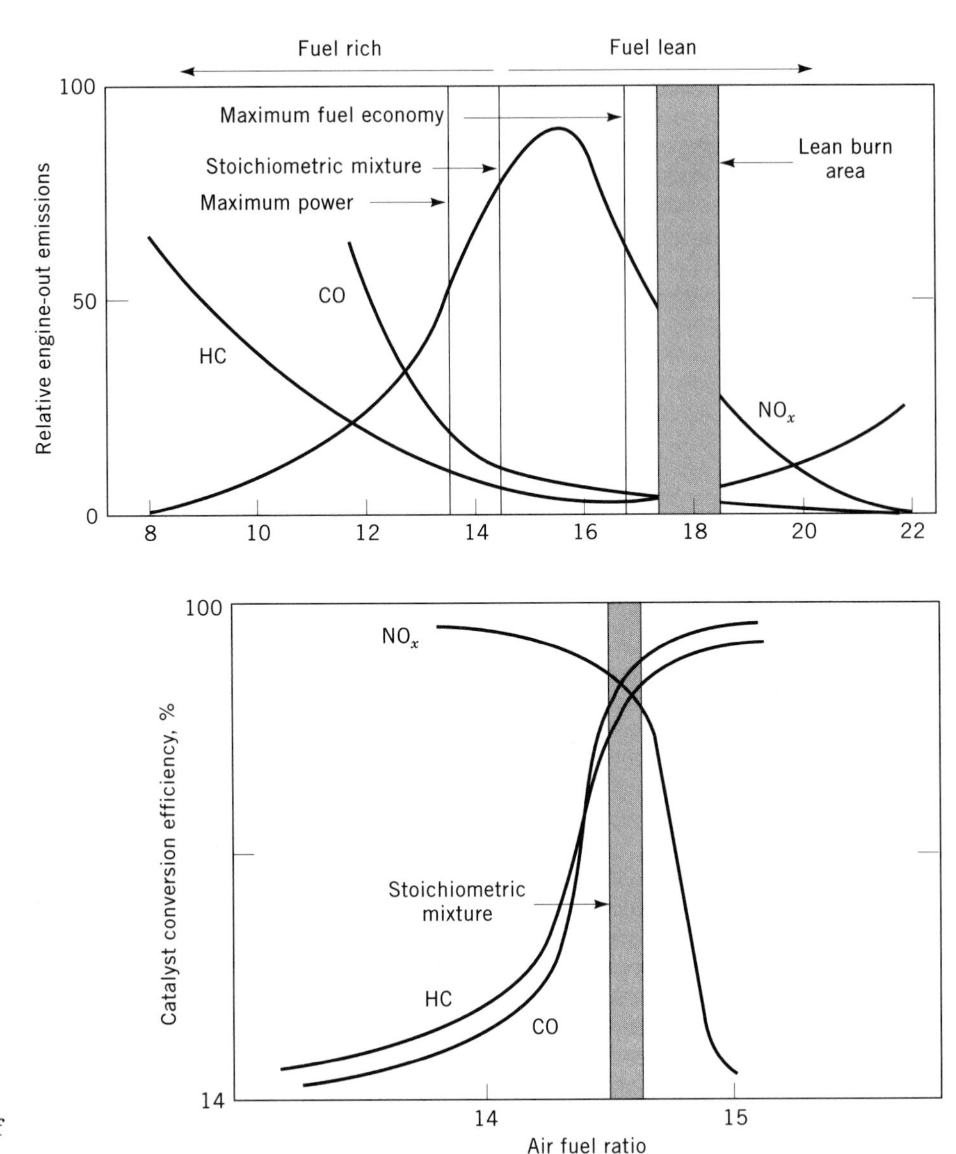

Figure 12. Emissions as a function of air–fuel ratio (18,19).

talytic converter can increase HC emissions 475% and CO emissions 425%, and a disabled oxygen sensor can increase HC emissions 445% and CO emissions 1242% (20).

Inspection and maintenance and ATP reduce tampering somewhat, but rates continue to be excessive, and the U.S. Congress has amended the Clean Air Act to address this problem. Measures for improved, more effective identification of vehicles requiring repair include enhanced I&M programs providing more comprehensive and accurate evaluation of emission control system performance, onboard diagnostic equipment providing the operator with continuous information on the performance of the vehicle's emission control systems, and enforcement programs using remote monitoring of emissions from vehicles on roadways.

MOTOR VEHICLE EMISSIONS CHARACTERISTICS

Certification procedures, emissions regulations, and emissions characteristics are different for light-duty [less than 3856 kg (8500 lb) GVW] and heavy-duty (3856 kg GVW or more) motor vehicles. Light-duty gasoline passenger cars and trucks are responsible for more than 90% of the motor vehicle NMHC and CO emissions in the United States and about 67% of motor vehicle NO_x emissions, and heavy-duty trucks and buses are responsible for more than 90% of fine particulate emissions (8). This distribution is due in part to differences in emissions characteristics (most light-duty vehicles have catalyst-equipped gasoline engines, and most heavy-duty vehicles have diesel engines) and in part to the distribution of VMT. In 1989, light-duty gasoline-fueled vehicles accounted for about 92% motor vehicle miles traveled and heavy-duty diesel-fueled vehicles for about 4% (11). Therefore, efforts to characterize motor vehicle NMHC and CO emissions have focused on light-duty gasoline vehicles, particulate matter emissions on heavy-duty diesel vehicles, and NO_x emissions on both categories of vehicles.

Motor vehicle emissions (especially those of gasoline-fueled vehicles) are sensitive to changes in operating conditions such as ambient temperature, average speed, alti-

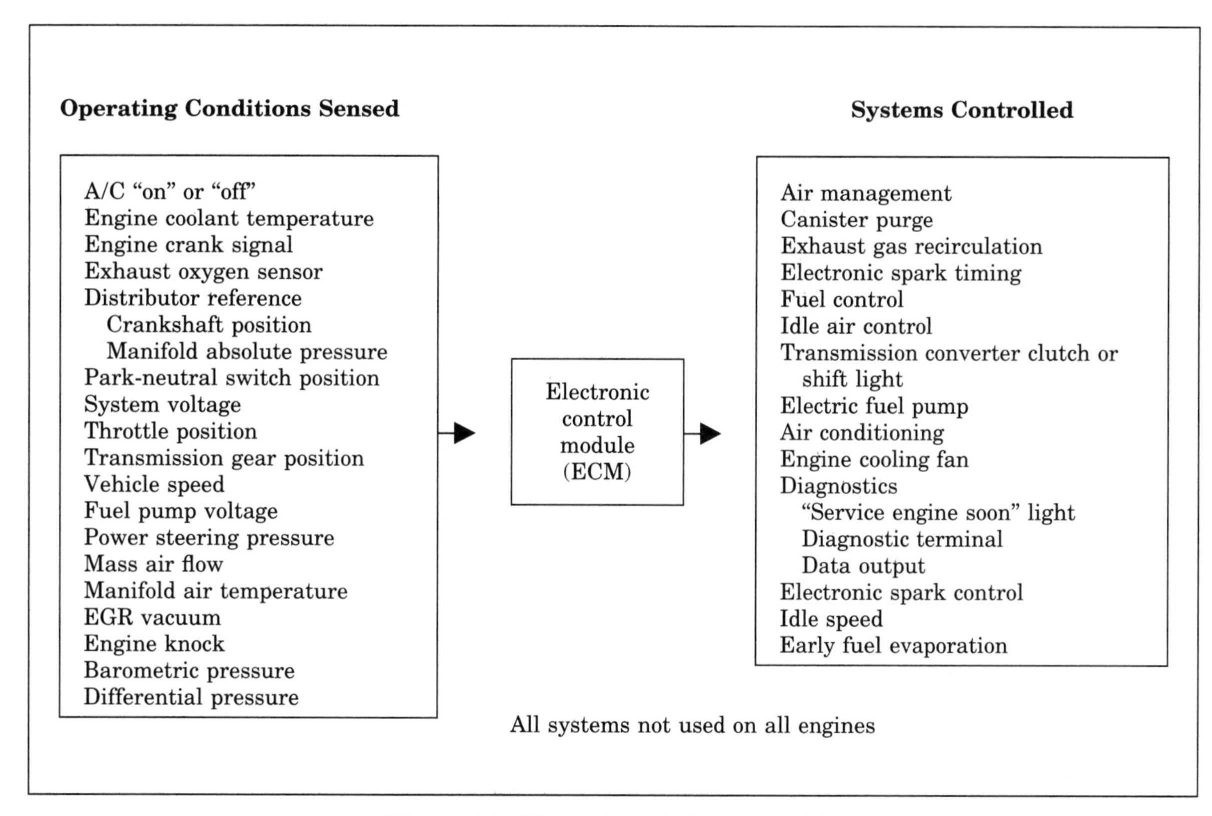

Figure 13. Electronic emission control (3).

tude, and age (because of malfunctions). Figure 16 shows how NMHC, CO, and NO_x emissions vary with ambient temperature and speed, as estimated by MOBILE 4 (11). Federal emissions certification practice involves a narrow window within this range of conditions: 31.5 km/h (19.6 mi/h) average speed and 21.1°C average temperature, with a diurnal cycle of 15.6°–28.9°C. Hydrocarbons are emitted from the tailpipe and from evaporative sources (running losses) when vehicles are operated on roadways, from evaporative sources (diurnal and hot soak) when vehicles are parked, and from the gasoline tanks when vehicles are refueled. Figure 17 illustrates the relative contribution of each of the sources at various average vehicle speeds and ambient temperatures for 11.5- and 9.0-RVP gasoline volatilities. The relative contribution of evaporative sources is much greater at higher temperatures. It

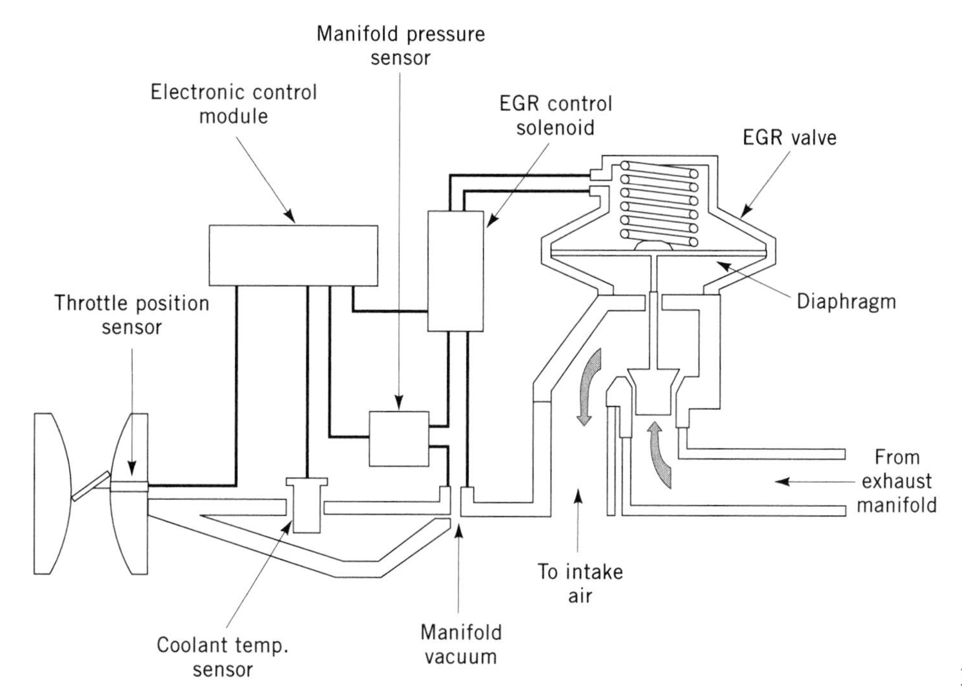

Figure 14. Exhaust gas recirculation (3).

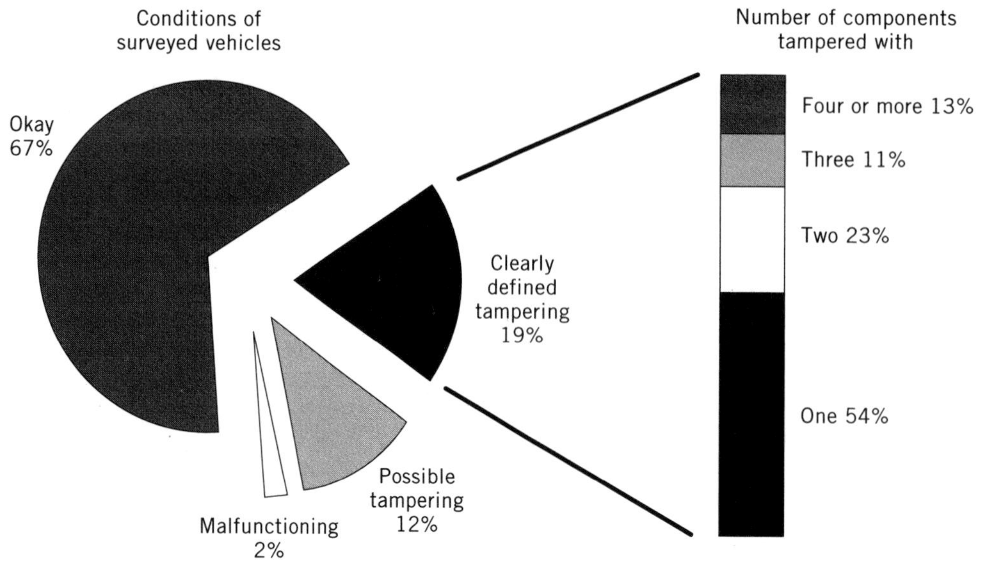

Figure 15. Frequency of emissions control system tampering (6).

should be noted that in the MOBILE 4 model, vehicle evaporative emissions are not affected by changes in vehicle average speed, nor are refueling emissions affected by fuel volatility or changes in vehicle average speed. Available data are not adequate to define the mathematical algorithms necessary to incorporate these variables in the model. Since the control technology for evaporative emissions involves a regenerative carbon canister that generally is purged more rapidly at higher average speeds and engine loads, one would expect the emission rates to vary with speed.

Historically, the primary incentive for regulation of motor vehicle emissions has been to improve air quality by reducing O_3, CO, and particulate matter concentrations. The impact of motor vehicle emissions on atmospheric O_3 is sensitive to many variables, including the relative importance of other O_3 precursor sources, the organic emissions composition (and thus their reactivity), and local HC–NO_x ratios. The relative contributions of motor vehicles and other sources (anthropogenic and biogenic) vary from one region to another, depending on ambient temperature, traffic congestion (average speed), and many other manageable and unmanageable variables. The emissions composition depends on many of these same variables, the composition of the fuel, and the characteristics of the vehicle fleet (e.g., the age of the vehicles, distribution of trucks, buses, and cars). Table 3 illustrates how U.S. commercial gasoline and diesel fuels can vary in composition. As indicated in Table 4, emissions from tailpipe, evaporative, and refueling sources are expected to vary in composition; the aggregate HC emissions composition for a vehicle depends on the relative contribution of each category, which varies with speed, temperature, and other factors (22). Since HC and NO_x emission rates differ in sensitivity to operating variables (as shown in Fig. 16), the HC–NO_x ratio of emissions also varies with operating conditions, as shown in Figure 18.

Table 2. Control System Tampering

Component/System	Tampering Rate (%)
Catalytic converter	5
Filler neck restrictor	6
Air pump system	11
Air pump belt	8
Air pump/valve	7
Aspirator	2
Positive crankcase ventilation	6
Evaporative control system	6
Exhaust gas recirculation	7
Heated air intake	3
Oxygen sensor and computer system	1

Note: From ref. 6. Vehicles with aspirated air systems are not equipped with other listed air injection components and conventional systems do not include aspirators.

In addition to O_3 precursor, CO, and particulate matter emissions, recent attention has been directed to toxic automobile emissions such as benzene, 1,3-butadiene, formaldehyde, and gasoline vapors. The emission rates of these hazardous substances are sensitive to many of the variables discussed above. For example, benzene emission rates are sensitive to fuel benzene level and to higher molecular weight fuel aromatics (eg, toluene and xylenes), which can be a source of tailpipe benzene through the dealkylation processes that occur during combustion (23–25). The limited data available suggest that tailpipe benzene emission rates increase as temperature decreases but that the benzene fraction of THC emissions remains relatively constant (26). Few data are available relating the composition of evaporative emissions to temperature; however, distillation theory suggests that aromatics constitute a larger fraction of the emissions at higher temperatures.

Estimates of cancer risk from motor vehicle emissions are presented in Table 5 (27). The largest potential risk is associated with diesel particulate emissions, which have been the subject of much study. The U.S. EPA is scheduled to publish a comprehensive diesel emissions hazard assessment during 1994. Diesel particles include

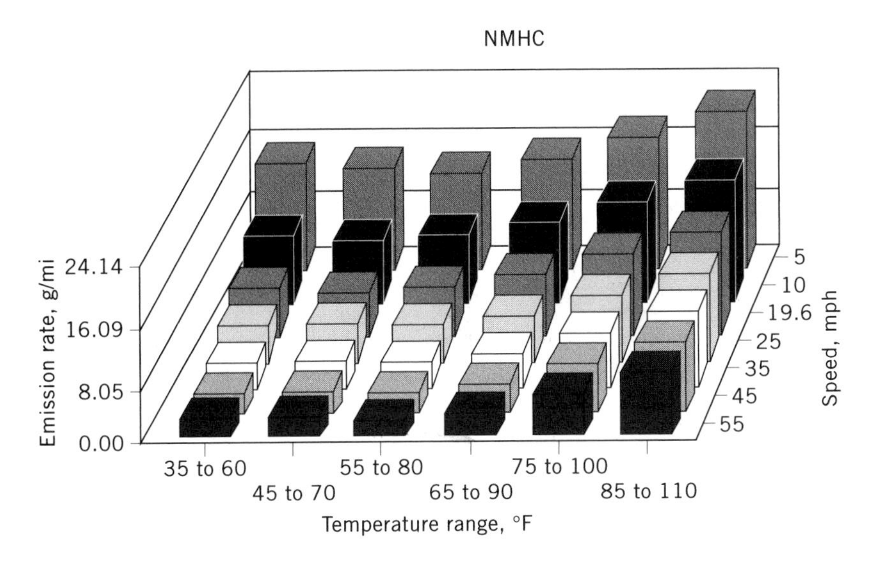

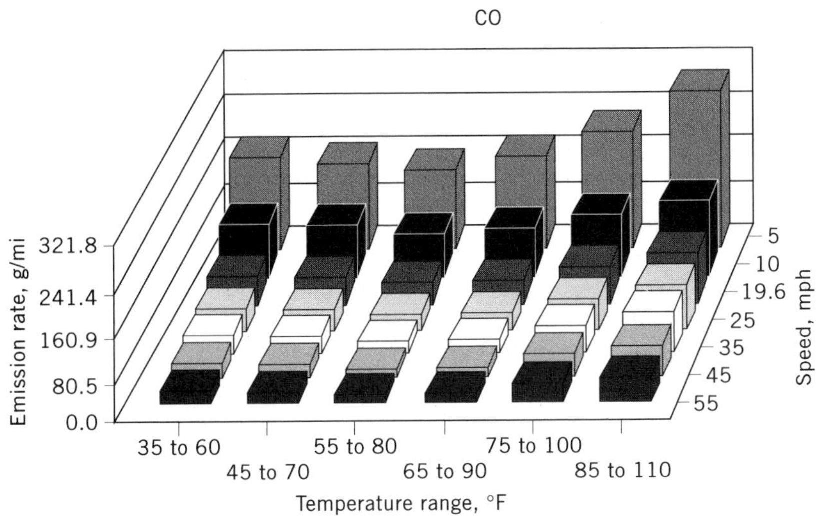

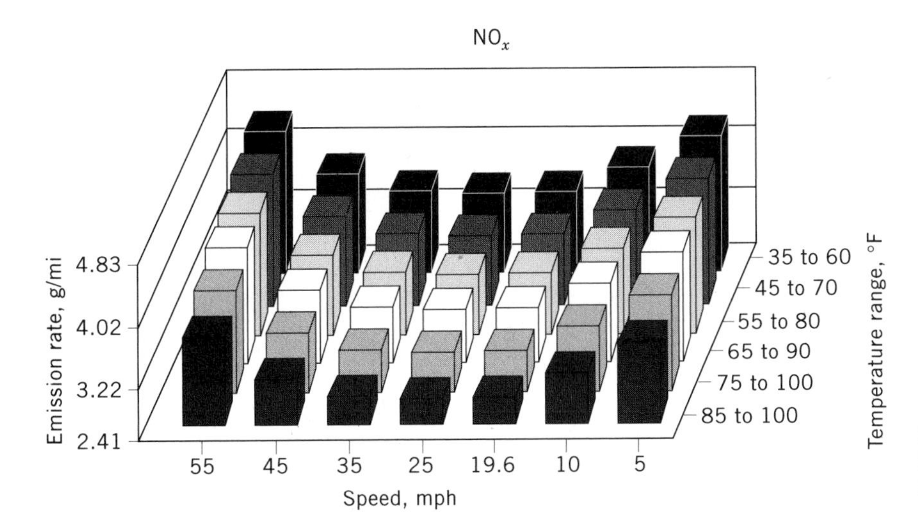

Figure 16. Motor vehicle emissions as a function of speed and ambient temperature (11).

elemental and organic carbon, sulfate, and small amounts of trace metals (28,29). The cancer risk from these particles is associated primarily with adsorbed and condensed organic substances; nitropolycyclic aromatic hydrocarbon compounds are an important subgroup (30). Technology-forcing particulate emissions standards have resulted in improved engine designs and fuel formulations (31–33). Diesel fuel sulfur content was reduced to 0.05 wt % in 1993 (the average was 0.32 wt % in the 1989 national fuel composition survey).

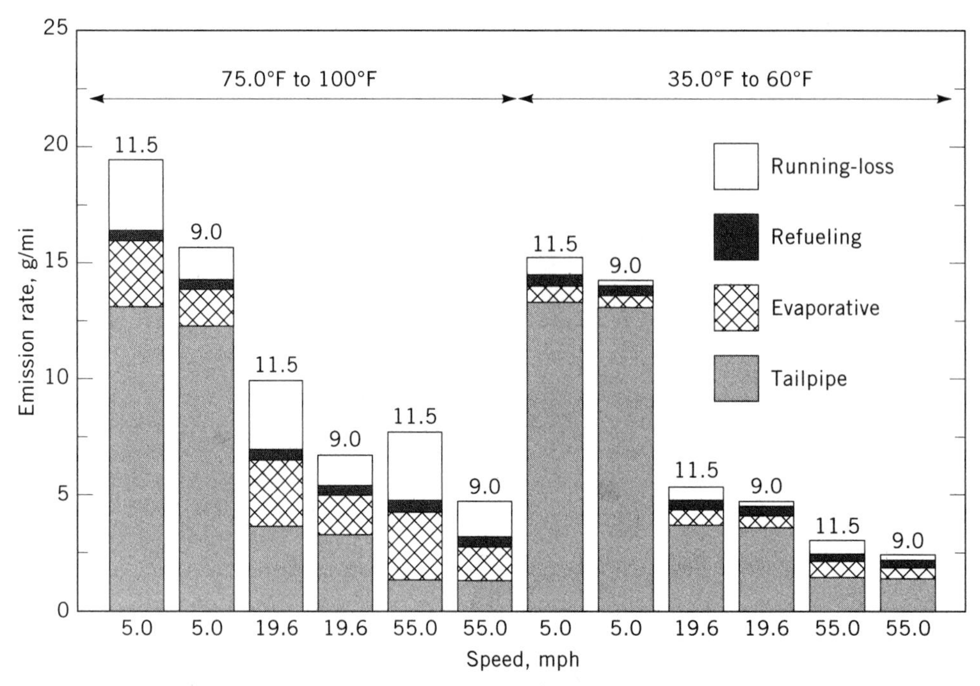

Figure 17. NMHC motor vehicle emissions distribution (1985): tailpipe, evaporative, refueling, and running losses (11).

Motor vehicles also contribute to the atmospheric burden of the radiatively important trace gases that influence global climate (34,35). Motor vehicle engine emissions of greatest importance in this regard are CO_2, methane (CH_4), and nitrous oxide (N_2O). When emissions associated with fuel production, distribution, and use in motor vehicles are considered, CO_2 is responsible for about 77% of the global warming effect from motor vehicle fuel combustion with conventional gasoline fuels and about 91% of the global warming effect with conventional diesel fuels (34). Carbon dioxide emission rates are directly related to fuel economy, and fleet average fuel economy has improved in response to federal standards and consumer market demands, as shown in Figure 19. Table 6 presents CO_2 emission rates for 1975, 1985, and 1995 (22). From 1975 to 1985, fleet average CO_2 emission rates decreased about 22%, but VMT increased about 30%, thus increasing the atmospheric CO_2 burden from motor vehicle emissions.

Chlorofluorocarbon (CFC) emissions from automobile

Table 3. National Fuel Composition Survey: Summer 1989[a]

	Aromatic (%)	Olefins (%)	Saturates (%)	MTBE (%)	RVP (psi)	Sulfur (wt %)	Octane, (R + M)/2	Cetane No.
Premium unleaded								
Maximum	67.0	17.8	82.5	10.3	10.9	0.071	97.3	
Minimum	11.0	1.1	30.6	<0.1	5.9	<0.001	87.0	
Average	37.1	6.2	56.6	<1.9	8.9	<0.016	92.3	
Intermediate								
Maximum	48.6	23.1	66.6	8.3	10.3	0.111	91.1	
Minimum	25.1	2.9	37.6	0.1	8.2	0.005	87.2	
Average	32.5	11.4	55.9	<1.6	9.1	0.036	89.3	
Regular unleaded								
Maximum	42.2	42.8	73.2	5.4	10.9	0.187	94.9	
Minimum	17.7	1.0	33.9	<0.1	7.7	<0.001	84.4	
Average	29.9	11.8	58.1	<0.2	8.9	<0.037	87.3	
No. 2 diesel								
Maximum	43.3	2.3	86.1	—	—	0.60	—	56.4
Minimum	13.3	0.4	55.3	—	—	0.03	—	39.2
Average	32.7	0.9	66.2	—	—	0.32	—	45.4

[a] From ref. 21. MTBE = methyltertiarybutyl ether; RVP = Reid vapor pressure.

[b] Research octane + motor octane)/2.

Table 4. Tailpipe, Evaporative, and Refueling Emissions Hydrocarbon Composition

Organic Classification	Tailpipe Emissions	Evaporative Emissions	Refueling Emissions
Paraffinic, %	55	72	85
n-Butane, %	5	23	32
Isopentane, %	4	15	19
Olefinic, %	18	10	11
Aromatic, %	25	18	4
Acetylenic, %	2	0	0

air conditioners also are important greenhouse gases. Amann estimated that vehicle lifetime diflurodichloromethane (CFC-12) emissions could have a greater global climate impact than exhaust CO_2 emissions with gasoline fuel, assuming 27.5 mpg fuel economy and a 10,000 CFC-12 to CO_2 mass-based relative global warming effect (36). Replacements for CFC-12 that have less greenhouse effect are being evaluated, with 1,1,1,2-tetrafluoroethane (HFC-134a) a current leading candidate. The greenhouse effect of HFC-134a is estimated to be about 10% that of CFC-12 (36). Further, air conditioner servicing practices are changing to include capture and reuse of CFCs, rather than venting to the atmosphere.

Fleet average methane emission rates are estimated at about 0.1 g/mi by MOBILE 4 at 23°C, 9.0-RVP fuel, for 1989. Catalyst emission control systems do not reduce methane as effectively as other hydrocarbon compounds, so methane emission rates have not been reduced as much as NMHC emission rates. Typically, methane constitutes as increasingly larger fraction of the THC emission rate as THC is reduced. In an examination of consumer passenger cars, methane accounted for 7.2% (0.33 g/mi) of 4.58 g/mi THC for 1975 model year cars and 24% (0.14 g/mi) of 0.57 g/mi THC for 1982 model year cars (37).

Few data on motor vehicle N_2O emissions are available. Table 7 provides an analysis of available data (22). The catalysts used to reduce other emissions may increase N_2O emission rates by about an order of magnitude.

Table 5. Motor Vehicle Emissions Risk Estimates, Cancer Incidents Per Year[a]

	1986	1995
Diesel particles	178–860	106–662
Formaldehyde	46–86	24–43
Benzene	100–155	60–107
Gasoline vapors	17–68	24–95
1,3-Butadiene	236–269	139–172
Gasoline particles	1–176	1–156
Asbestos	5–33	Not determined
Cadmium	<1	<1
Ethylene dibromide	1	<1
Acetaldehyde	2	1
Total	586–1650	355–1236

[a] From ref. 27.

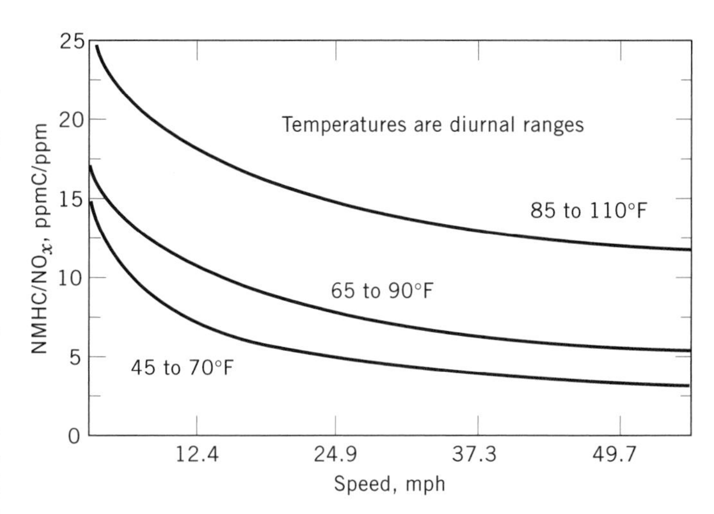

Figure 18. The variation of NMHC–NO_x ratio with speed and ambient temperature (11).

FUTURE DIRECTIONS

Although significant progress has been made in reducing motor vehicle emission rates, more than 80 million Americans still live in areas exceeding U.S. national ambient air quality standards (38). As indicated in Figure 20, excessive levels of O_3, CO, and fine particulate matter are the prevalent air quality problems, each with significant contribution from motor vehicle emissions. A growing data base also suggests that motor vehicles contribute significantly to the population's exposure to other hazardous compounds such as benzene, formaldehyde, gasoline vapors, and 1,3-butadiene. For these reasons and others, the U.S. Congress has amended the Clean Air Act to reduce the impact of motor vehicles on risk to public health and welfare. In an effort to render the total transportation system as environmentally benign as possible, both motor

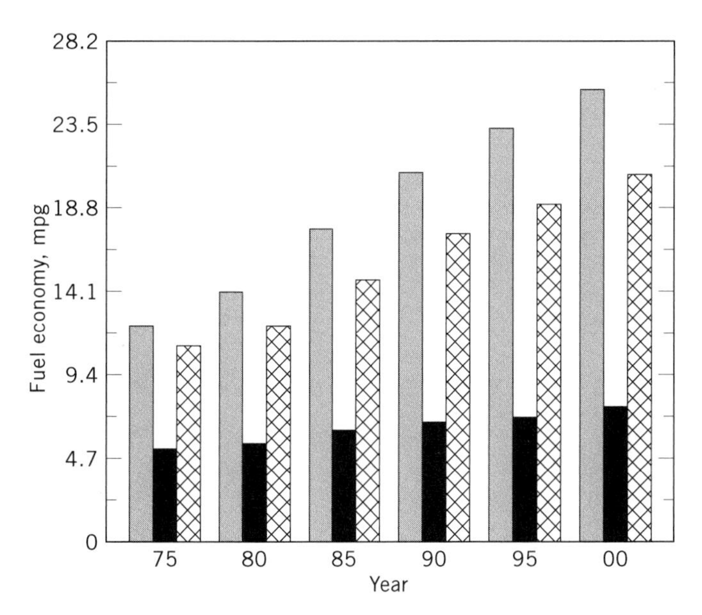

Figure 19. Fuel economy trend: (▒) light duty; (■) heavy duty; (⊠) all vehicles.

Table 6. Motor Vehicle CO_2 Emission Rates (g/mi)

Year	LDGV	LDGT	HDGV	LDDV	LDDT	HDDV	Types
1975	567.0	636.8	988.6	455.8	328.6	2081.6	685.8
1985	434.4	532.3	856.8	379.9	436.0	1731.4	537.6
1995	336.0	458.2	893.5	321.0	405.5	1461.1	431.5

Note: LDGV = light-duty gasoline vehicle; LDGT = light-duty gasoline truck; HDGV = heavy-duty gasoline vehicle; LDDV = light-duty diesel vehicle; LDDT = light-duty diesel truck; HDDV = heavy-duty diesel truck.

vehicle emissions and fuel composition are being regulated.

In the 1990 Clean Air Act amendments, Congress required a number of changes to current mobile source emissions regulations. Target dates have been established for a two-phase reduction of tailpipe emission standards. Phase I will reduce light-duty motor vehicle emission rates to 0.25 g/mi for NMHC, to 0.40 g/mi for NO_x, and to 0.08 g/mi for particulate matter (from current standards of 0.4, 1.0, and 0.2 g/mi, respectively) by 1994. Phase II, if enacted (depends on air quality in 1997), will require further reductions of emission rates to 0.125 g/mi for NMHC, to 0.2 g/mi for NO_x, and to 1.7 g/mi for CO (from the current standard of 3.4 g/mi) by about 2003.

The amendments also require the use of "clean fuel" motor vehicles in centrally fueled fleets in the nation's most severe nonattainment areas. Additionally, a demonstration program in California will provide for the introduction of 150,000 clean fuel cars per year beginning in 1996 and 300,000 per year beginning in 1999. Management of the emission characteristics of the clean fuel vehicles will be similar to the California Low-Emission Vehicle and Clean Fuel Program (discussed below).

The amended Clean Air Act also calls for regulation of the composition of gasoline in specified nonattainment areas by limiting the volumetric fractions of benzene, requiring a specific amount of oxygen in the fuel (2% during ozone nonattainment periods and 2.7% during CO nonattainment periods, probably by addition of ethanol, methyl-*tert*-butyl ether, or ethyl-*tert*-butyl ether), and requiring reduction of volatility. By 1995, ozone-forming volatile organic compounds and toxic compounds (the aggregate of benzene, 1,3-butadiene, formaldehyde, acetaldehyde, and polycyclic organic matter) will be reduced 15% by gasoline reformulation from the emissions characteristic of 1990 model year vehicles using national average unleaded gasoline, without concurrent increase in NO_x emissions. By 2000, the required reductions in volatile organic compounds and toxic compounds will be increased to 25%. The Auto/Oil Air Quality Improvement Research Program has demonstrated that fuel aromatic and olefinic hydrocarbon fractions, oxygen level, sulfur level, and distillation T_{90} are among the fuel specifications that can be used to manipulate motor vehicle emissions characteristics (39).

Additional standards reducing CO emissions at low ambient temperatures (−6.7°C) and controlling refueling emissions are required by the amended Clean Air Act. The amendments also include a variety of requirements addressing emissions from malfunctioning in-use vehicles, including on-board emission control system diagnostics, enhanced inspection and maintenance programs, and remote sensing of emissions from vehicles on roadways as an enforcement tool.

In addition to these federal programs, California approved new state vehicle and fuel regulations in September 1990, requiring even greater reduction of motor vehicle emissions (40). Compliance with California's more stringent emission standards can be achieved with advanced vehicle emission control technology, cleaner burning fuels, or a combination of the two (encouraging cooperation between the vehicle and fuel industries). The program requires a phased introduction of low-emission vehicles and the clean fuels used by those vehicles. The low-emission vehicles are defined according to four sets of emissions criteria, described in Table 8. The vehicle–fuel categories include transitional-low-emissions vehicles (TLEVs), low-emissions vehicles (LEVs), ultra-low-emissions vehicles (ULEVs), and zero-emissions vehicles

Table 7. Motor Vehicle N_2O Emission Rates

Vehicle Category[a]	Emission Factor (mg/mi)	Emission Factor Range (mg/mi)
LDGV (no-catalyst)	6.0	3.1–15.9
LDGV (catalyst)	61.0	3.1–234.0
LDDV	—	10.9–48.0
HDGV	73.1	48.0–97.0
HDDV	47.0	31.1–74.0

Note: LDGV = light-duty gasoline vehicle; LDDV = light-duty diesel vehicle; HDGV = heavy-duty gasoline vehicle; HDDV = heavy-duty diesel vehicle.

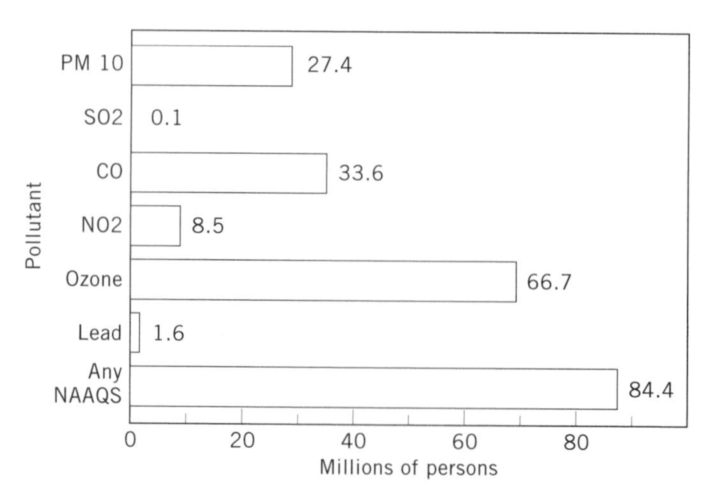

Figure 20. Persons living in counties with air quality violating NAAQS, 1989 (38).

Table 8. California Low-Emissions Motor Vehicle Standards

Category	NMOG[a] (g/mi)	CO (g/mi)	NO_x (g/mi)
Transition low-emission vehicle	0.125	3.4	0.4
Low-emission vehicle	0.075	3.4	0.2
Ultra-low-emission vehicle	0.040	1.7	0.2
Zero-emission vehicle	0.000	0.0	0.0

[a] Nonmethane organic gases before reactivity adjustment.

(ZEVs). Beginning in the mid-1990s, vehicle manufacturers will be required to certify portions of their new motor vehicles to TLEV, LEV, ULEV, or ZEV standards. New California motor vehicles under 1701 kg (3750 lb) must certify to fleet average nonmethane organic gas (NMOG) emission standards of 0.25 g/mi by 1994 and 0.062 g/mi by 2003 (other standards apply to larger vehicles). The clean-fuel element of the regulation requires that the fuels used to certify the low-emissions vehicles be readily available for consumer use. Other states may opt to require California motor vehicles if considered necessary to achieve the air quality required by the 1990 Clean Air Act amendments.

Clearly, alternatives to conventional petroleum-based fuels will be important to the United States in the future, for both environmental and economical–political reasons. The nation needs to improve its energy security by decreasing its dependence on imported oil. That is, it needs to diversify its sources of transportation energy and develop motor vehicle technologies compatible with fuels produced from natural gas, coal, and renewable biomass such as agricultural crops (corn, wheat, and sugar cane). About 63% of U.S. petroleum consumption is associated with transportation (41). In addition to the Clean Air Act amendments requiring regulation of fuel composition, the Alternative Motor Fuels Act of 1988 encourages development and widespread use of such alternative fuels as methanol, ethanol, and compressed or liquified natural gas (CNG or LNG) (42).

Alternative fuels often are classified as fuels to replace conventional gasoline and diesel fuels or fuels to extend conventional fuels with environmentally favorable reformulations. Table 9 lists several replacement fuels and fuel reformulations currently under consideration, and Figure 21 illustrates some of the feedstocks and production processes being evaluated (41). Table 10 lists several of the emissions impacts associated with the replacement fuels being studied. Potential for reduced O_3, CO, particulate matter, and other toxic compounds in urban air has been demonstrated, but several important uncertainties have been identified requiring further study, such as the significance of formaldehyde emissions from methanol-fueled motor vehicles.

The primary gasoline specifications that are being studied for reformulation are volume percentages of aro-

Table 9. Alternative Fuel Possibilities

Replacements for gasoline and/or diesel fuel
- Methanol
- Ethanol
- Compressed natural gas
- Liquefied petroleum gas
- Electricity

Gasoline and diesel fuel reformulation
- Varied gasoline volatility, T90, aromatic and olefinic hydrocarbon, sulfur, and oxygen content
- Varied diesel sulfur and aromatic hydrocarbon content, natural cetane number, and cetane improver

Table 10. Potential Emissions Impact of Replacement Fuels

Ozone precursor organic emissions reactivity reduced: methanol, ethanol, CNG (methane) less photochemically reactive than gasoline hydrocarbon mixtures

Evaporative hydrocarbon emission rate reduced

Aldehyde emissions elevated
- Methanol: formaldehyde
- Ethanol: acetaldehyde

Ozone benefit sensitive to many variables
- Formaldehyde (and possibly methyl nitrite) emission rate
- Local VOC/NO_x ratio
- Relative significance of motor vehicle and other sources of local VOC and NO_x burden

CO and NO_x impact depends on engine design
- Those designed for stoichiometric combustion emit less NO_x but provide little CO improvement (lower flame temperature with methanol, ethanol)
- Those designed for fuel lean combustion emit less CO but provide little NO_x improvement (versus closed-loop 3-way catalyst gasoline engines)

Particulate emissions from heavy-duty diesel engines significantly reduced with methanol and CNG fuels

NO_x emissions from heavy-duty diesel engines reduced with methanol fuels

Toxic organic (benzene and other gasoline hydrocarbons) emissions reduced

Table 11. Potential Emissions Impact of Fuel Reformulations

Increased fuel oxygen reduces CO and HC emissions (technology sensitive, less impact with advanced "adaptive learning" closed-loop 3-way catalysts) and increases NO_x.

Reduced fuel volatility reduces evaporative (diurnal, hot soak, running-loss) emissions, and summer CO.

Reduced fuel aromatic and olefinic hydrocarbons reduce ozone precursor reactivity (important at lower vocal VOC/NO_x ratios).

Reduced fuel aromatic hydrocarbons reduce toxic benzene emissions.

Increased fuel MTBE (ETBE) increases isobutylene emissions.

Increased fuel oxygenates (ethanol, MTBE, ETBE) increase aldehyde emissions.

Reduced fuel T90 reduces HC emissions with current technology vehicles and increases CO with old technology vehicles.

Reduced fuel sulfur reduces HC, CO, and NO_x with current technology vehicles.

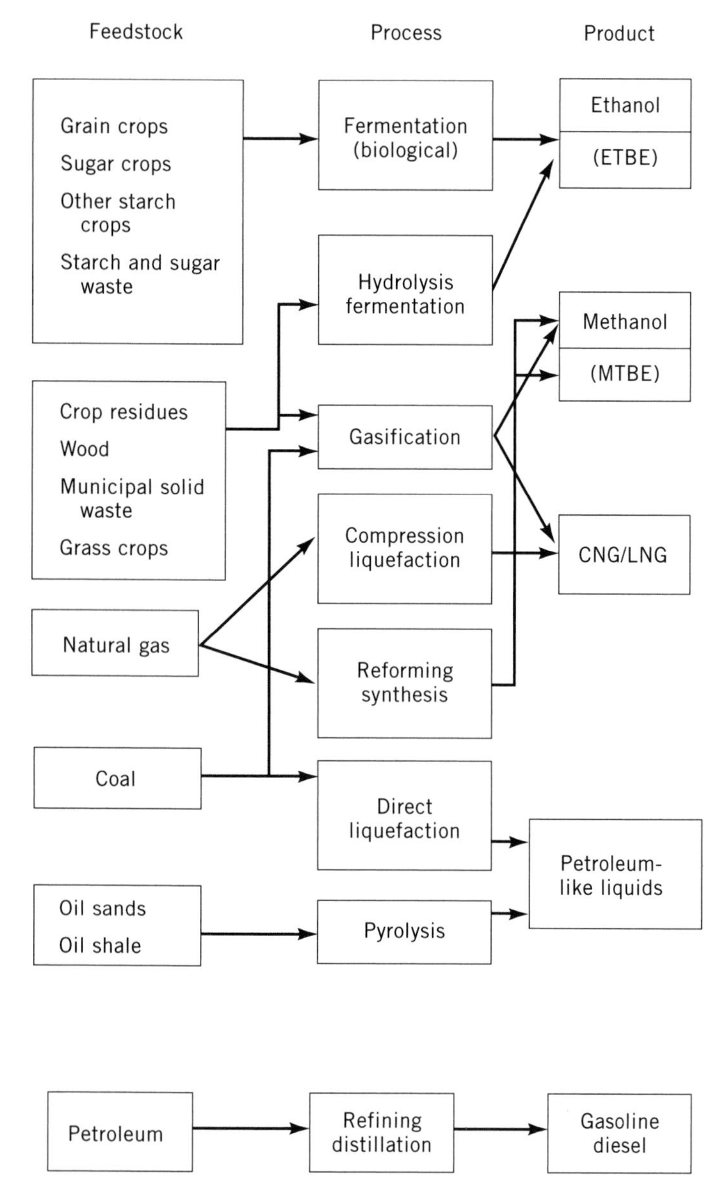

Figure 21. Alternative fuel feedstocks and processes (41).

matics and olefinics, sulfur level, T_{90}, volatility, and oxygen level. Some of the potential changes in emissions associated with changes in these specifications are given in Table 11.

SUMMARY

Many emissions improvements have been realized with U.S. motor vehicle fleets since the late 1960s as a result of cooperation between government and industry. But further improvements will be required for many major cities to achieve the air quality required by the Clean Air Act. Miles traveled by the motoring public continue to increase, and vehicle emissions control system tampering is excessive, both factors tending to compromise some of the potential of mobile source emissions control. The U.S. Congress continues to legislate further improvements, preserving the freedom of travel that Americans desire. New laws requiring additional control of motor vehicle emissions and improved fuel formulations are being enacted. At some point in the future, it may become necessary to limit the number of motor vehicles on urban roadways.

BIBLIOGRAPHY

1. A. J. Haagen-Smit, "Chemistry and Physiology of Los Angeles Smog," *Ind. Eng. Chem.* **44**(6), 1342 (1952).
2. J. E. Krier, *Pollution and Policy, A Case Essay on California and Federal Experience with Motor Vehicle Air Pollution 1940–1975,* University of California Press, Los Angeles, 1977, chap. 2.
3. K. A. Freeman, D. H. Lee, R. J. Rivele, and W. C. Settle (eds.), *Chilton's Emission Diagnosis and Tune-Up Manual,* Chilton Book, Radnor, 1988, chap. 1.
4. "Control of Air Pollution from New Motor Vehicles and New Motor Vehicle Engines," *Federal Register,* 31 CFR, Part 61, March 30, 1966.
5. Control of Air Pollution from New Motor Vehicles and New Motor Vehicle Engines, *Federal Register,* 33 CFR, Part 108, June 4, 1968.
6. U.S. Environmental Protection Agency, *1988 Motor Vehicle Tampering Survey,* Office of Mobile Sources, Washington, D.C., 1989.
7. 101st Congress, *Clean Air Act Amendments Of 1990,* Report 101-952, U.S. Government Printing Office, Washington, D.C., October 26, 1990.
8. U.S. Environmental Protection Agency, *National Air Pollutant Emission Estimates, 1940–1988,* EPA report no. EPA-450/4-90-001, Office of Air Quality Planning and Standards, Research Triangle Park, N.C., 1990.
9. "Volatility Regulations for Gasoline and Alcohol Blends Sold in Calendar Years 1989 and Beyond; Final Rule," *Federal Register,* 40 CFR, Part 80, March, 1989.
10. U.S. Environmental Protection Agency, *Regulatory Support Document-Interim Cold Ambient Temperature CO Vehicle Emission Standards,* Office of Mobile Sources, Ann Arbor, Mich., September 1989.
11. U.S. Environmental Protection Agency, *User's Guide to MOBILE 4.0 (Mobile Source Emission Factor Model),* EPA report no. EPA-AA-TEB-89-01, and *User's Guide to MOBILE 4.1 (Mobile Source Emission Factor Model),* EPA report no. EPA-AA-TEB-91-01 Office of Mobile Sources, Ann Arbor, Mich., 1991.
12. W. R. Pierson and A. W. Gertler, "Comparison of the SCAQS Tunnel Study with Other Onroad Vehicle Emissions Data," *JAWMA* **40**(11), 1495 (1990).
13. Scott Research Laboratories, *Vehicle Operations Survey,* CRC APRAC project no. CAPE-10-68 (1-70), for Coordinating Research Council, Atlanta, Ga., and Environmental Protection Agency, Ann Arbor, Mich., December 1971.
14. Motor Vehicle Manufacturers Association (MVMA), *MVMA Motor Vehicle Facts & Figures '88,* MVMA, Detroit, Mich., 1988.
15. M. P. Walsh, *Pollution on Wheels II: The Car of the Future,* American Lung Association, Office of Government Relations, Washington, D.C., 1990.
16. C. L. Dickson and P. W. Woodward, *Motor Gasolines, Summer 1985,* National Institute for Petroleum and Energy Research (NIPER), for American Petroleum Institute (API), June 1986.

17. "Volatility Regulations for Gasoline and Alcohol Blends Sold in Calendar Years 1992 and Beyond, Notice of Final Rulemaking," *Federal Register,* 40 CFR, Part 80, June 1990).
18. C. M. Heinen, "We've Done the Job—What's it Worth?" Society of Automotive Engineers paper no. 801357, Warrendale, Pa., 1980.
19. C. A. Amann, "The Powertrain, Fuel Economy and the Environment," General Motors Corp., Research paper no. 4949, Warren, Mich., 1985.
20. U.S. Environmental Protection Agency, *Anti-Tampering and Anti-Misfueling Programs to Reduce In-Use Emissions from Motor Vehicles,* EPA report no. EPA-AA-TTS-83-10, Office of Mobile Sources, Ann Arbor, Mich., 1983.
21. Motor Vehicle Manufacturers Association (MVNA), *National Fuel Survey-Gasoline and Diesel Fuel, Summer 1989,* MVMA, Detroit, Mich., 1989.
22. F. M. Black, "Motor Vehicles as Sources of Compounds Important to Tropospheric and Strospheric Ozone," in T. Schneider, S. D. Lee, G. J. R. Woters, and L. D. Grant, eds., *Atmospheric Ozone Research and Its Policy Implications,* Elsevier Science, Amsterdam, The Netherlands, 1988, p. 85.
23. F. M. Black, L. E. High, and J. M. Lang, "Composition of Automobile Evaporative and Tailpipe Hydrocarbon Emissions," *J. Air Pollut. Control Assoc.* **30,** 1216 (1980).
24. D. L. Raley, P. W. McCallum, and W. J. Shadis, "Analysis of Benzene Emissions from Vehicles and Vehicle Refueling," Society of Automotive Engineers paper no. 841397, Warrendale, Pa., 1984.
25. D. E. Seizinger, W. F. Marshall, F. W. Cox, and M. W. Boyd, *Vehicle Evaporative and Exhaust Emissions as Influenced by Benzene Content of Gasoline,* Coordinating Research Council, Atlanta, Ga., 1986.
26. F. Stump, S. Tejada, W. Ray, D. Dropkin, F. M. Black, W. Crews, R. Snow, P. Siudak, C. O. Davis, L. Baker, and N. Perry, "The Influence of Ambient Temperature on Tailpipe Emissions from 1984–1987 Model Year Light-Duty Gasoline Motor Vehicles," *Atmos. Environ.* **23,** 307 (1989).
27. J. M. Adler and P. M. Carey, "Air Toxics Emissions and Health Risks from Mobile Sources," Air and Waste Management Association paper no. 89-34A.6, Pittsburgh, Pa., 1989.
28. C. T. Hare and F. M. Black, "Motor Vehicle Particulate Emission Factors," Air Pollution Control Association paper no. 81-56.5, Pittsburgh, Pa., 1981.
29. F. M. Black, J. Braddock, R. Bradow, and M. Ingalls, "Highway Motor Vehicles as Sources of Atmospheric Particles: Projected Trends 1977 to 2000," *Environ. Int.* **11,** 205 (1985).
30. J. Lewtas, "Evaluation of the Mutagenicity and Carcinogenicity of Motor Vehicle Emissions in Short-Term Bioassays," *Environ. Health Perspect.* **47,** 141 (1983).
31. J. A. Alson, J. M. Adler, and T. M. Baines, "Motor Vehicle Emission Characteristics and Air Quality Impacts of Methanol and Compressed Natural Gas," in D. Sperling, ed., *Alternative Transportation Fuels: An Environment and Energy Solution,* Quorum/Greenwood Press, Westport, Conn., 1989.
32. T. L. Ullman, *Investigation of the Effects of Fuel Composition, and Injection and Combustion System Type on Heavy-Duty Diesel Emissions,* Coordinating Research Council, Atlanta, Ga., 1989.
33. U.S. Environmental Protection Agency, *Regulation of Fuels and Fuel Additives: Fuel Quality Regulations for Highway Diesel Fuel Sold in 1993 and Later Calendar Years [draft],* Office of Mobile Sources, Ann Arbor, Mich., 1990.
34. S. Unnasch, C. B. Moyer, D. D. Lowell, and M. D. Jackson, *Comparing the Impact of Different Transportation Fuels on the Greenhouse Effect,* California Energy Commission, Sacramento, Calif., 1989.
35. M. A. DeLuchi, R. A. Johnston, and D. Sperling, *Transportation Fuels and the Greenhouse Effect,* Universitywide Energy Research Group, UERG California Energy Studies report no. UER-182, University of California, Davis, Calif., 1987.
36. C. A. Amann, "The Passenger Car and the Greenhouse Effect," Society of Automotive Engineers paper no. 902099, Warrendale, Pa., 1990.
37. J. E. Sigsby, S. Tejada, W. Ray, J. M. Lang, and J. W. Duncan, "Volatile Organic Compound Emissions from 46 In-Use Passenger Cars," *Environ. Sci. Technol.* **21,** 466 (1987).
38. U.S. Environmental Protection Agency, *National Air Quality and Emissions Trends Report, 1989,* report no. EPA 450/4-91-003, Office of Air Quality Planning and Standards, Research Triangle Park, N.C., February 1991.
39. Coordinating Research Council, Auto/Oil Air Quality Improvement Research Program technical bulletin no. 1, "Initial Mass Exhaust Emissions Results from Reformulated Gasolines," December 1990, and technical bulletin no. 2, "Effects of Sulfur Levels on Mass Exhaust Emissions," February 1991, Atlanta, Ga., 1991.
40. D. Donohoue, L. Kao, C. Marvin, B. Takemoto, and R. Vincent, "California's Low-Emission and Clean Fuels Program," Air and Waste Management Association paper no. 91-106.7, Pittsburgh, Pa., 1991.
41. D. Sperling, *New Transportation Fuels-A Strategic Approach to Technological Change,* University of California Press, Berkeley, 1988, chaps. 1 and 3.
42. U.S. Congress, *Alternative Motor Fuels Act of 1988,* Public Law 100-494 100th Congress, 102 STAT.2441, October 14, 1988.

Disclaimer
The information in this document has been funded in part by the United States Environmental Protection Agency. It has been subjected to the Agency's peer and administrative review, and it has been approved for publication. Mention of trade names or commercial products does not constitute endorsement or recommendation for use.

AUTOMOBILE EMISSIONS—CONTROL

GARY A. BISHOP
DONALD H. STEDMAN
University of Denver
Denver, Colorado

Picture for a moment any major industrialized city in the world on a hot summer day. Perhaps Los Angeles, Mexico City, or London come to mind. Few residents can ignore the ever-present signs of humankind that linger in the air, many of which are directly attributable to the internal-combustion engine and the automobile. Carbon monoxide (CO), hydrocarbons (HC), nitrogen oxides (NO_x), fine particles, and lead are some of the compounds emitted by automobiles that foul the air.

See also AIR QUALITY MODELING; CLEAN AIR ACT, MOBILE SOURCES; ENVIRONMENTAL ANALYSIS.

In the United States, air pollution control measures to mitigate mobile emisisons in nonattainment areas (as defined by the Clean Air Act) include inspection and maintenance (I/M) programs, oxygenated fuel mandates, and transportation control measures. Nonetheless, many areas remain in nonattainment past the 1987 deadline for compliance with federal standards, and some are projected to remain in nonattainment for many more years despite the measures currently undertaken. A major emphasis in controlling mobile source emissions has centered around periodic testing for identifying vehicles with excessive emissions.

Current federal and state governments employ a variety of testing methods and protocols for certifying new and in-use vehicle emissions. These range from the extremely comprehensive new car certification test called the Federal Test Procedure, which all newly manufactured vehicles sold in the United States must comply with, to the more common one- or two-speed idle measurements performed in many state I/M programs (1–4). The former requires at least 12 h to complete and costs in excess of $700. The later requires less than 15 min and is priced around $10. Both tests are designed as surrogates for evaluating the emissions of vehicles under actual operating conditions. The limitations of each type of testing have been hotly debated. The following criteria have been outlined (5) for an "ideal" test: it should evaluate the vehicle under real-life conditons and be reproducible, accurate, quick, inexpensive, and measure all pollutants of concern.

In 1987, with support from the Colorado Office of Energy Conservation at the University of Denver developed an infrared (ir) remote-monitoring system for automobile carbon monoxide exhaust emissions (6). Significant fuel economy improvements result if rich-burning (high CO and HC emissions) or misfiring (high HC emissions) vehicles are tuned to a more stoichiometric and more efficient air : fuel (A : F) ratio. Therefore, the University of Denver CO–HC remote sensor is named the Fuel Efficiency Automobile Test (FEAT). The basic instrument measures the carbon monoxide : carbon dioxide ratio ($CO:CO_2$) and the hydrocarbon : carbon dioxide ratio ($HC:CO_2$) in the exhaust of any vehicle passing through an ir light beam that is transmitted across a single lane of roadway. Figure 1 shows a schematic diagram of the instrument (7).

THEORY OF OPERATION

The FEAT instrument was designed to emulate the results one would obtain using a conventional nondispersive infrared (NDIR) exhaust gas analyzer. Thus FEAT is also based on NDIR principles. An ir source sends a horizontal beam of radiation across a single traffic lane, approximately 0.25 m above the road surface. This beam is directed into the detector on the opposite side and divided between four individual detectors: CO, CO_2, HC, and a reference. An optical filter that transmits infrared light of a wavelength known to be uniquely absorbed by the molecule of interest is placed in front of each detector, determining its specificity. Reduction in the signal caused by absorption of light by the molecules of interest reduces the voltage output. One way of conceptualizing the instrument is to imagine a typical garage-type NDIR instrument in which the separation of the ir source and detector is increased from 0.08 to 6–12 m. Instead of pumping exhaust gas through a flow cell, a car now drives between the source and the detector.

Because the effective plume path length and amount of plume seen depends on turbulence and wind, the FEAT can only directly measure ratios of $CO:CO_2$ or $HC:CO_2$. These ratios are constant for a given exhaust plume and by themselves are useful parameters to describe the combustion system. However, with a fundamental knowledge of combustion chemistry, it is possible to determine many other parameters of the vehicle's operating characteristics from these ratios, including the instantaneous air : fuel

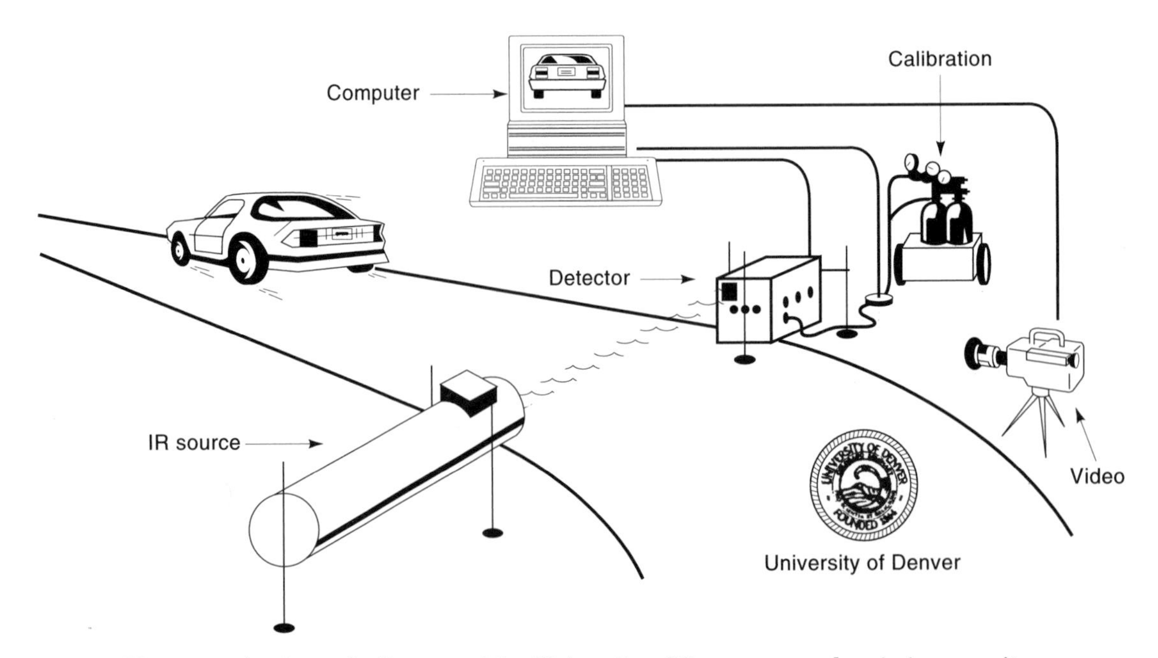

Figure 1. A schematic diagram of the University of Denver on-road emissions monitor. It is capable of monitoring emissions at vehicle speeds betwee 4.0 and 241 km/h in under 1 s/vehicle.

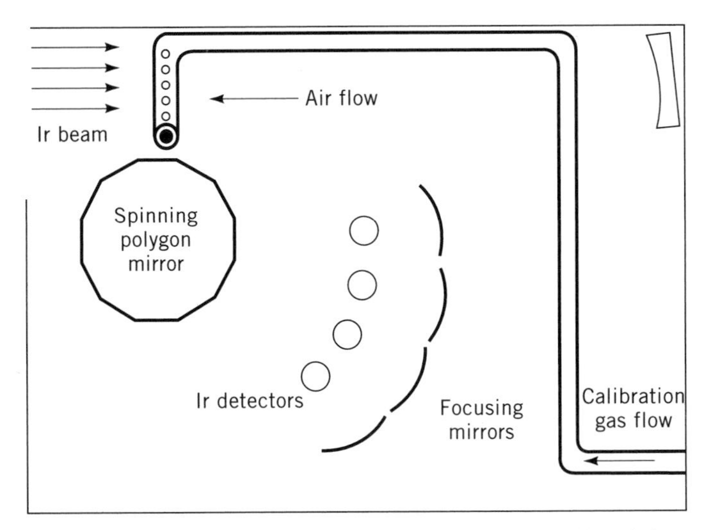

Figure 2. Schematic view (not to scale) of the optical bench layout for a University of Denver remote sensor. In-coming radiation is received by a pickup mirror in the upper right and focused off of the spinning polygon mirror and a detector-focusing mirror onto each of the four detectors for the reference, CO_2, HC, and CO. The daily gas calibrations are performed by releasing a small amount of gas directly into the beam path via the delivery tube shown. Note that the delivery tube continues up toward the viewer for an additional 6 cm before ending.

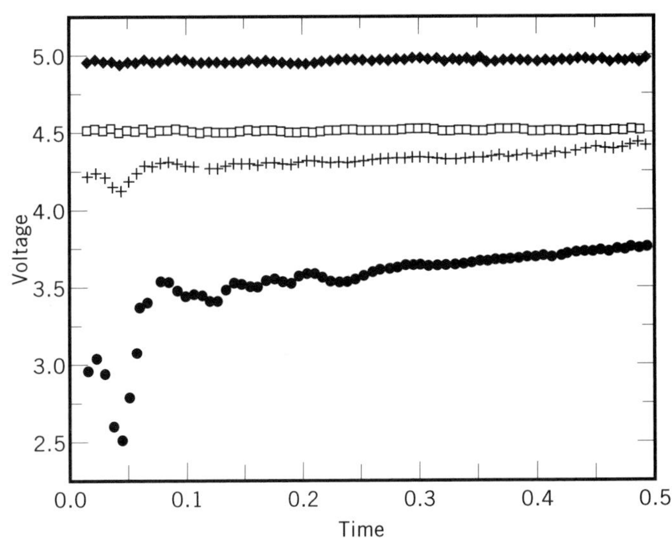

Figure 3. For each of the four detectors, 0.5 s of data. HC, ◆; reference, □; CO, +; and CO_2, ●. As the tailpipe exhaust rolls through the beam, the CO_2 voltages rise and fall until the plume finally dissipates.

ratio, grams of CO or HC emitted per liter of gasoline (g CO/L or g HC/L) burned, and the percent CO or percent HC in the exhaust gas. Most vehicles show ratios of zero, because they emit little to no CO or HC. To observe a $CO:CO_2$ ratio greater than zero, the engine must have a fuel-rich air : fuel ratio and the emission control system, if present, must not be fully operational. A high $HC:CO_2$ ratio can be associated with either fuel-rich or fuel-lean air : fuel ratios coupled with a missing or malfunctioning emission-control system. A lean air : fuel ratio, while impairing driveability, does not produce CO in the engine. If the air : fuel ratio is lean enough to induce misfire then a large amount of unburned fuel is present in the exhaust manifold. If the catalyst is absent or nonfunctional, then high HC will be observed in the exhaust without the presence of high CO. To the extent that the exhaust system of this misfiring vehicle contains some residual catalytic activity, the HC may be partially or totally converted to a $CO:CO_2$ mixture.

INSTRUMENT DETAILS

The present design of University of Denver FEAT instruments incorporates CO (4.6 μm), CO_2 (4.3 μm), HC (3.4 μm), and background (3.9 μm) channels using interference filters built into Peltier-cooled lead selenide detectors. The instrument uses a mirror to collect the light and focus it onto a spinning 12-faceted polygon mirror that provides a periodic detector illumination of 2400 Hz. The reflected light from each facet of the rotating mirror sweeps across a series of four focusing mirrors that in turn direct the light onto the four detectors as shown in Figure 2. Each detector thus gets a burst of full signal from the source in a sequential fashion for each measurement mode.

Each detector provides a pulse train at 2400 Hz equivalent to the intensity of the ir radiation detected at its specific wavelength. Electronic circuitry averages 24 of these pulses, subtracts any background signal, and provides the averaged DC level to four signal ports. These are connected to a computer through an analogue-to-digital converter. Figure 3 shows 0.5 s of zero-corrected exhaust data voltages from the HC, reference, CO, and CO_2 detectors. Voltage levels are monitored in front of and behind each passing vehicle to eliminate effects of variable background concentrations.

All data from the CO, CO_2, and HC channels (I) are corrected by ratio to the reference channel (I_0). This procedure eliminates other sources of opacity such as soot, turbulence, spray, license plates, etc from providing data that could be incorrectly identified as CO or HC. These reference-corrected values are then apportioned according to the detector-specific response factors and the resulting values for CO and HC are compared with CO_2 as shown in Figure 4. The resulting slope for $CO:CO_2$ (and $HC:CO_2$) is the path-independent measurement for that vehicle.

Software written for these instruments computes percent CO, percent CO_2, and percent HC on a dry, excess air-corrected basis from the measured $CO:CO_2$ and $HC:CO_2$ ratios. These reported values are designed to be equivalent to the percentages a standard garage analyzer would measure if one could run along behind the vehicle with a tailpipe probe. The percent HC is reported as an equivalent concentration of propane. This procedure is different from the reported HC measurements in most analyzers used in I/M programs. Most I/M instruments are tested for a single propane : hexane response ratio. All subsequent calibrations are performed with propane, yet the data are reported as "hexane equivalent" by dividing the measured number by the propane : hexane response ratio (a divisor usually close to two). The response factor for the FEAT is also close to two. Nevertheless, HC data

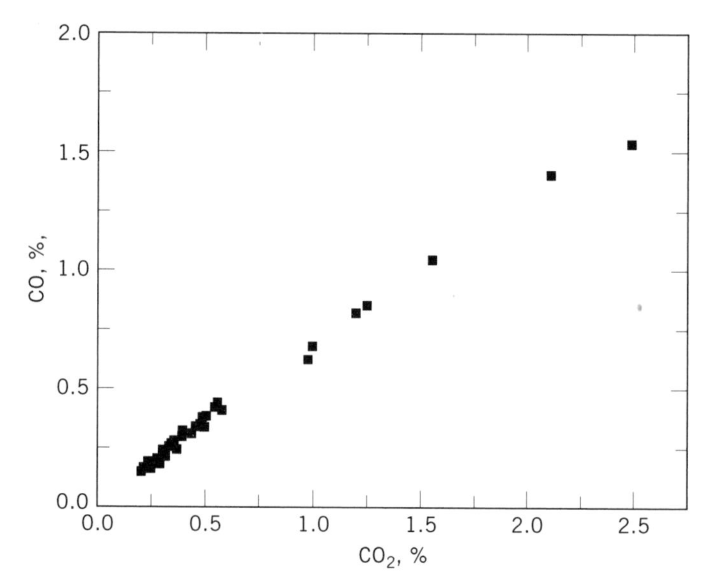

Figure 4. The final CO : CO_2 correlation graph of the data in Figure 3 used to obtain a unitless slope. The HC : CO_2 ratio is determined via the same process.

are reported in propane units because the device is, in fact, calibrated daily with propane not hexane.

The FEAT can be accompanied by a video system to record license plates. The video camera is coupled directly into the data analysis computer so that the image of the back of each passing vehicle is frozen onto the video screen. The computer writes the date, time, and the calculated exhaust CO, HC, and CO_2 percentage concentrations at the bottom of the image. These images are stored on videotape or digital storage media.

Before each day's operation in the field, a quality assurance calibration is performed on the instrument with the system set up in the field. A puff of gas designed to simulate all measured components of the exhaust is released into the instrument's path from a cylinder containing industry-certified amounts of CO, CO_2, and propane. The ratio readings from the instrument are compared with those certified by the cylinder manufacturer and used as a daily adjustment factor.

The FEAT is effective across traffic lanes of up to 15 m in width. It can be operated across double lanes of traffic with additional video hardware; however, the normal operating mode is on single-lane traffic (8). The FEAT operates most effectively on dry pavement, as rain, snow, and tire spray from a wet pavement scatter the ir beam. At suitable locations exhaust from more than 2,000 vehicles per hour can be monitored. The unit has been tested successfully for vehicle speeds exceeding 241 km/h. The FEAT has been used to measure the emissions of more than 750,000 vehicles worldwide. Table 1 compares the results for many of the locations that have been sampled (9–16).

The FEAT has been shown to give accurate readings for CO and HC in double-blind studies of vehicles both on the road and on dynamometers (17–19). A fully instrumented vehicle with tailpipe emissions controllable by the driver or passenger has been used in a series of drive-by experiments with the vehicles emissions set for CO between 0% and 10% and between 0% and 0.35% (propane) for HC to confirm the accuracy of the on-road readings

Table 1. Worldwide Data Summary

Location	Date	Number of Measurements	Mean Percent CO	Median Percent CO	Mean Percent HC	Median Percent HC
United States						
Denver	1991–1992	35,945	0.74	0.11	0.057	0.033
Denver	1993	58,894	0.58	0.13	0.022	0.013
Chicago	1990	13,640	1.10	0.37	0.139	0.087
Chicago	1992	8,733	1.04	0.25	0.088	0.064
California	1991	91,679	0.82	0.14	0.076	0.042
Provo, Utah	1991–1992	12,066	1.17	0.45	0.220	0.127
El Paso, Tex.	1993	15,986	1.22	0.37	0.073	0.044
Mexico						
Juarez	1993	7,640	2.96	2.18	0.170	0.091
Mexico City	1991	31,838	4.30	3.81	0.214	0.113
Europe						
Göteborg, Sweden	1991	10,285	0.71	0.14	0.058	0.046
Denmark	1992	9,038	1.71	0.67	0.177	0.058
Thessaloniki, Greece	1992	10,536	1.40	0.55	0.155	0.082
London, UK	1992	11,666	0.96	0.17	0.136	0.071
Leicester, UK	1992	4,992	2.32	1.61	0.212	0.131
Edinburgh, UK	1992	4,524	1.48	0.69	0.129	0.084
Asia						
Bangkok, Thailand	1993	5,260	3.04	2.54	0.948	0.567
Hong Kong	1993	5,891	0.96	0.18	0.054	0.037
Kathmandu, Nepal	1993	11,227	3.85	3.69	0.757	0.363
Seoul, Korea	1993	3,104	0.82	0.26	0.044	0.019
Taipei, Taiwan	1993	12,062	1.49	0.88	0.062	0.050
Australia						
Melbourne	1992	15,908	1.42	0.57	0.107	0.058

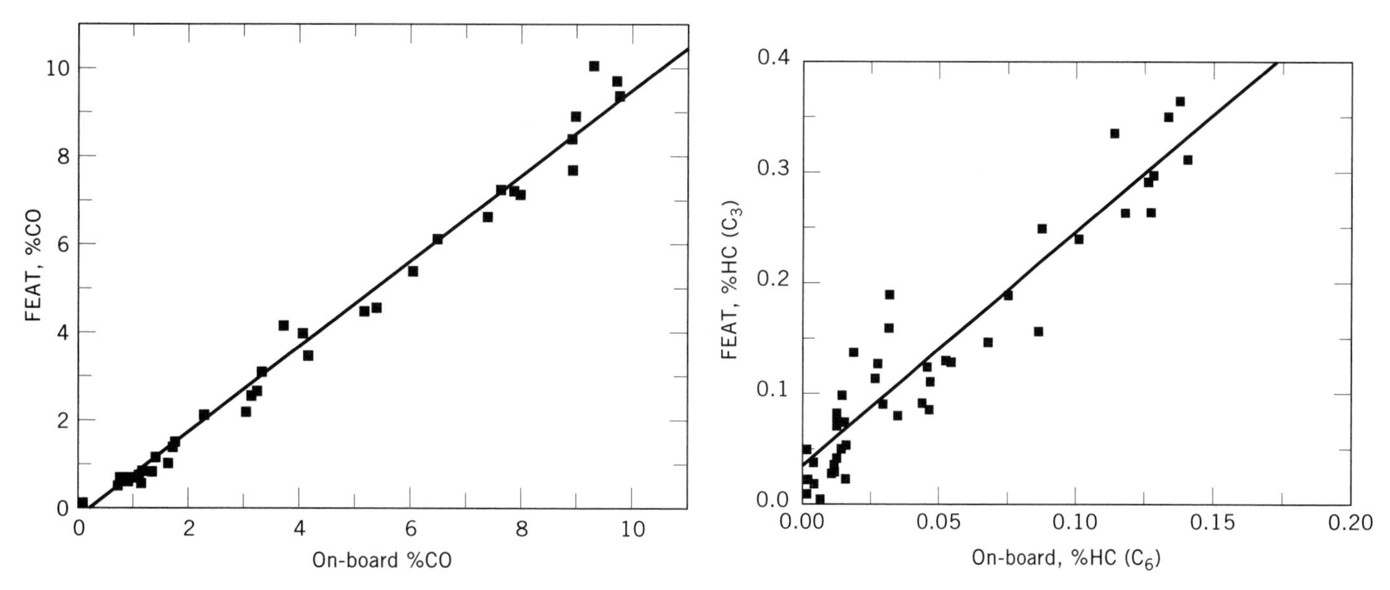

Figure 5. (**a**) Comparison of FEAT measurements to on-board measurements for carbon monoxide. The equation for the line is FEAT percent CO = 0.98 · on-board percent CO − 0.23 with an $r^2 = 0.99$. (**b**) Comparison of FEAT measurements to on-board measurements for hydrocarbons. The equation for the line is FEAT percent HC (C_3) = 2.16 · on-board percent HC (C_6) + 0.028 with an $r^2 = 0.85$. Note that because the FEAT reports HC as propane, the correct slope for the comparison with the on-board instrument will be 2.

(20). The comparison between the on-board measurements and the FEAT measurements is shown in Figure 5. The results have an accuracy of ± 5% for CO and ± 15% for HC. Recently, the abilities to measure nitric oxide and exhaust opacity have been added. NO is measured by ultraviolet absorption, whereas opacity is determined by means of ir absorption at 3.9 μm. Diesel soot (black) and so called "steam" from cold cars both cause observable opacity. Current units measure visible steam as HC. A system to eliminate this interference is under development.

REAL-WORLD VEHICLE EMISSION CHARACTERISTICS

Through the use of FEAT it is now possible to collect data quickly and easily on a large number of vehicles. This enables the study of important questions concerning the automobile's contribution to urban air quality and what mitigation actions might be taken. Following is a brief overview of general characteristics about automobile emissions that have been determined so far through the use of FEAT. Several interesting data sets have been collected that highlight the usefulness of large real-world databases.

Not all cars have equal emissions. Data show that a small fraction of the passing vehicles is responsible for half or more of the emissions in any given area. In Denver half the emissions come from only 7% of the vehicles. In Kathmandu half the emissions come from 25% of the vehicles. Figure 6 shows data collected at a single site in the Los Angeles area. At this location half of the CO is emitted by 7% of the vehicles and half of the HC is emitted by 11% of the vehicles. The few vehicles emitting half of the CO and HC are referred to as *gross polluters*. For automobile emissions the old adage the "tail wags the dog" holds true.

The overall characteristics of these fleets are similar regardless of age, location, or the presence or absence of I/M programs and can be mathematically described by a γ distribution (21). Most vehicles show mean emissions of 1% CO and 0.1% HC (as propane) or less in the exhaust.

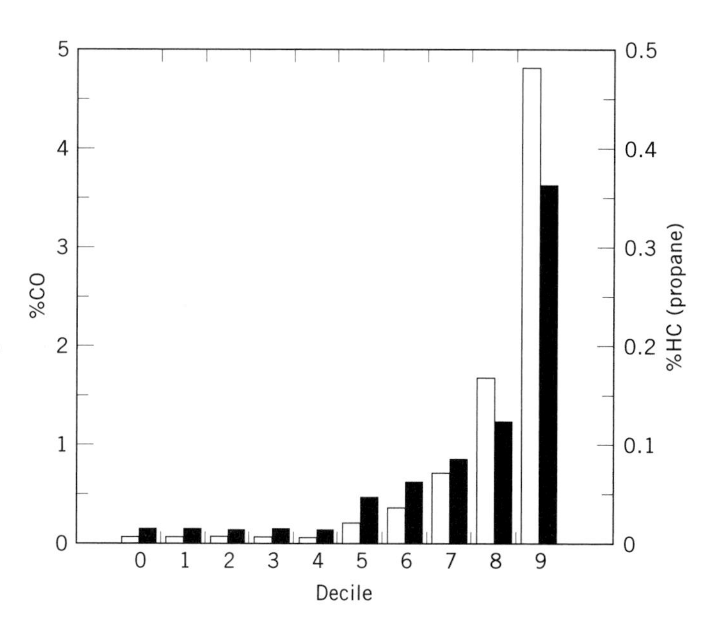

Figure 6. FEAT data collected in 1991 in El Monte, California (Rosemead Boulevard site). The fleet is rank ordered and divided into deciles; the average percent CO and percent HC for each decile is plotted. The solid bars denote CO while the empty bars are HC data. The first five deciles are displayed as an average of all five. The graph illustrates that automobile emissions are not normally distributed.

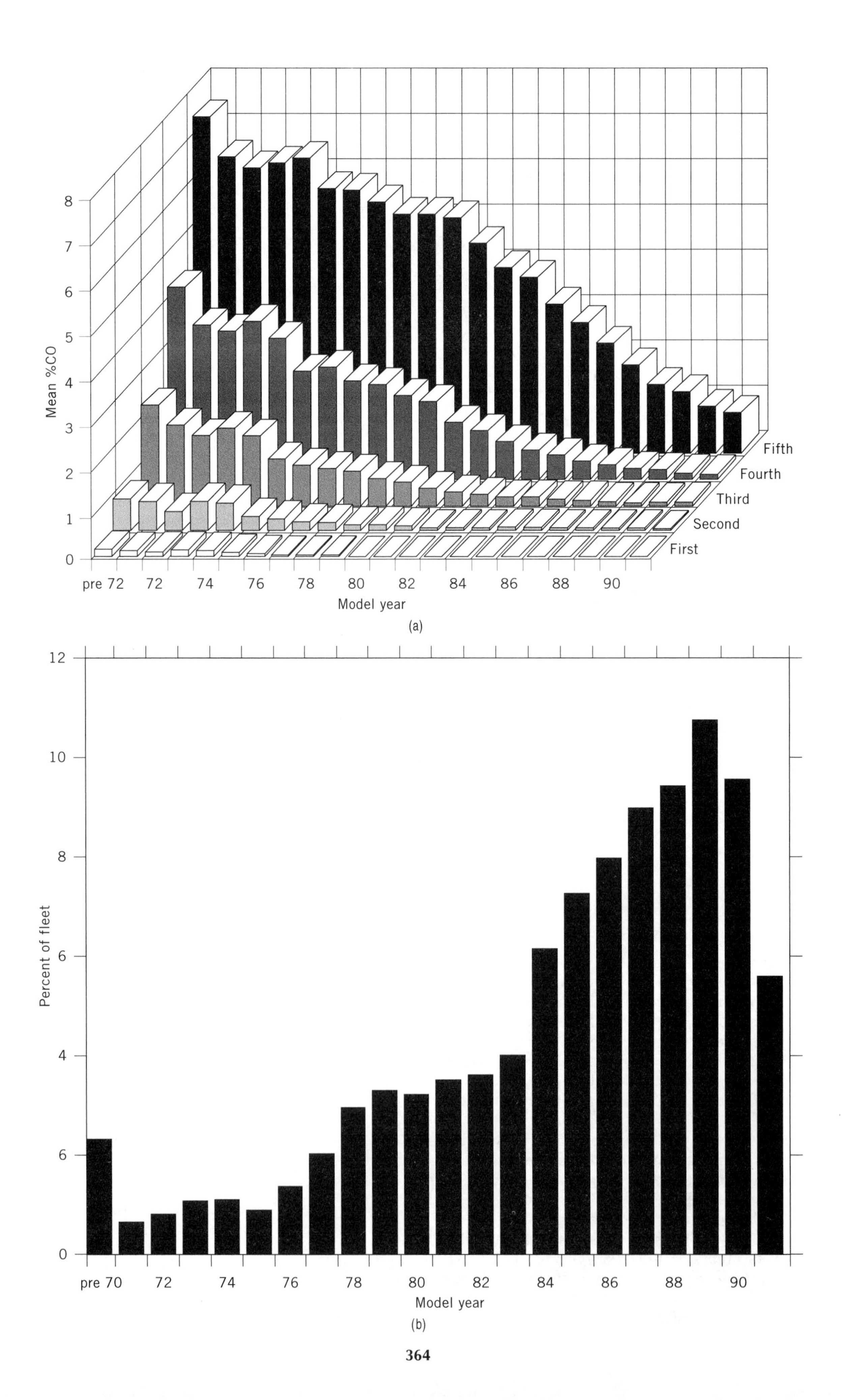

Mean %CO
0
1
2
3
4
5
6
7
8
Fifth
Fourth
Third
Second
First
pre 72
72
74
76
78
80
82
84
86
88
90
Model year
(a)
Percent of fleet
0
6
4
6
8
10
12
pre 70
72
74
76
78
80
82
84
86
88
90
Model year
(b)

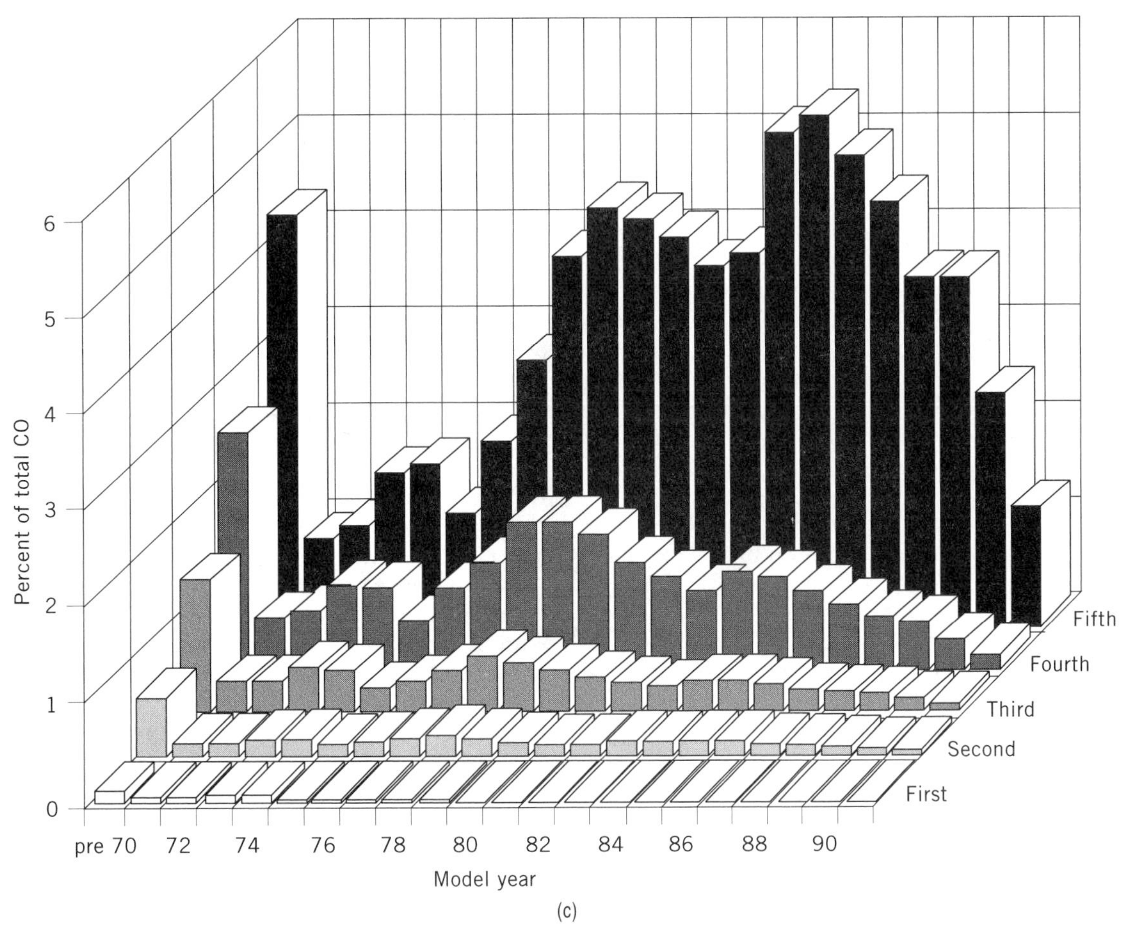

(c)

Figure 7. (**a**) The 1991 California data for CO presented as emission factors by model year for each of the data's quintiles. (**b**) The distribution of vehicles by model year for the data shown in part **a**. (**c**) The product of parts **b** and **c**, showing the average contribution of CO emissions by model year. Note that the majority of emissions is contributed by vehicles in the fifth quintile that are post-1980 models.

The newer the fleet the more skewed the emissions. This is because more of the vehicles have near zero emissions, and thus a smaller number of gross polluters dominates the total emissions.

The good news is that for the U.S. fleet 50% of the vehicles produce only 4% of the CO emissions and 16% of HC, using current gasoline formulas as fuel. This shows that alternative and reformulated fuels are not likely to solve a problem that apparently arises due to a lack of maintenance.

Not all gross polluters are old vehicles (only about 25% of pre-1975 vehicles in the United States). In fact, even the majority of precatalyst vehicles emit relatively low levels. There is a strong correlation between fleet age and fleet emissions (Fig. 7**a**). However, this correlation has less to do with emissions-control technology than it does with vehicle maintenance. Any well-maintained vehicle regardless of age can be relatively low emitting, as seen by the mean emissions for all model years in the lowest-emitting quintile. It can also be seen that the most rapid deterioration in emissions occurs during the first 11 yr of ownership. It is interesting to note that the vehicles with the most rapidly deteriorating emissions are all newer technology vehicles with computer-controlled fuel delivery systems and three-way catalysts (installed in the United States after 1982). All of the studies note that these vehicles have negligible emissions when first purchased, emphasizing the need for proper maintenance.

If the problem is not the old clunker, then which vehicles are responsible? The product of the average emissions (Fig. 7a) and the fleet distribution (Fig. 7b) shown in Figure 7c, represents the overall contribution for each model year. What is obvious is that although older vehicles (pre-1980) have higher average emissions, it is the newer and more prevalent (post-1980) vehicles that actually contribute the most to the total. This has important implications for regulatory agencies that design programs around vehicle age only.

For HC, the fleet emissions tend to be less skewed than for CO, with a larger percentage of vehicles responsible for the majority of the fleet emissions. Only four cities (Bangkok, Hong Kong, Kathmandu, and Taipei) have half of the HC emissions produced by more than 15% of the fleet. Most of the same conclusions that are drawn regarding CO emissions and fleet characteristics hold true for HC emissions, because HC emissions increase as engine combustion gets richer and produces more CO. Two cities in Asia (Bangkok and Kathmandu) stand out for HC emissions, in large part due to the high percentage of two-cycle engines (necessarily high HC emitters by design), many of which also appear to be poorly tuned and maintained.

REMOTE SENSING APPLICATIONS

The ability quickly and unobtrusively to monitor real vehicles under real conditions opens up the possibility for many new applications. The ideas have run the gamut from pure research on emissions and control equipment performance to using the measurements as a basis for emission pricing schemes (by which the toll or vehicle registration is prorated according to the emission level). In the following section, data from several research programs that have used FEAT and the potential implications and logical extensions of that data to emission-control programs are addressed.

Roadside Pullover Studies

The ability of FEAT to identify gross polluters has been tested in several pilot studies. Two such programs in California have sought out explanations for vehicles with excessive emissions. In both studies, FEAT was used to identify gross polluting vehicles. A police officer stops the vehicle and performs a roadside inspection of the vehicle's tailpipe emissions and emission-control equipment status. In California, this type of inspection is called a smog check and is required once every 2 yr for gasoline-powered vehicles registered in most parts of the state.

The first California study involved the nonrandom inspection of 60 vehicles, 50 of which were stopped for having FEAT readings of greater than 2% CO (17). What was discovered was that of the 50 vehicles stopped for elevated CO emissions 45 failed the roadside inspection. One of the chief reasons for the high failure rate was that 12 of the 45 vehicles were found to have tampered emissions-control equipment. This means that emission-control equipment that was originally on the vehicle had been removed or otherwise disabled. The most glaring example was a 1984 GMC pickup that was originally purchaed equipped with a diesel engine but was found to have a gasoline engine without any required emissions-control equipment. Because this vehicle was still registered as a diesel-powered vehicle, it was exempt from the biannual state inspection program.

To verify these results on a larger sampling of vehicles, an expanded pullover program was conducted during June 1991 in a suburb of Los Angeles (13). Following a similar protocol, vehicles were selected nonrandomly based on a 4% CO or 0.4% HC on-road emissions level. Two inspection teams conducted roadside inspections of the vehicles that were stopped. In 2 weeks of measurements 60,487 measurements on 58,063 vehicles were performed. In all 3,271 gross polluting vehicles were identified, from which 307 vehicles were stopped and inspected. In the roadside inspection, 92% failed, with tampered emissions control equipment again a leading reason. A total of 41% of the vehicles were found out of compliance and an additional 25% had defective emissions-control equipment that may not have been caused by intentional tampering. The tailpipe portion of the test had an 85% failure rate.

These data strongly suggest that excessive tailpipe emissions are often the result of vehicles with broken or defective emissions-control equipment. This undoubtly has been influenced by the willful tampering that has taken place in a large number of the examples. It is also apparent that the scheduled testing program that California has depended on to find these tampered vehicles is being circumvented by the public. The classic example of this is the case of the diesel to gasoline engine switch mentioned previously (an additional case was found in the second pullover study). Aggressive antitampering programs using FEAT as probable cause for a pullover inspection could begin quickly and effectively to address another segment of the excessive mobile source tailpipe emissions.

Provo Pollution Prevention Program

In a study conducted in Provo, Utah, gross polluting vehicles were recruited from the public through FEAT measurements to investigate repair costs and effectiveness (22). Provo was chosen because it is an area that regularly violates the national ambient CO standard during the winter months. Measurements were made at two off-ramps (one northbound and one southbound) from I15 at the entrance to Provo during November 1991 and January 1992 in search of vehicles that exceeded a 4% CO emission level on at least two occasions. Two ramps were used to impose a control on the study, ie, vehicles were only recruited from one of the two ramps. In this way it was possible convincingly to compare emission changes at one ramp with those at the companion ramp as a judge of overall repair effectiveness. At these locations it was observed that half of the CO emissions were being produced by only 9% of the vehicles. For the entire program 17,000 measurements of more than 10,000 individual vehicles were made.

With the help of the city and a local community college, 47 out of 114 identified vehicles were recruited with the promise of a rental vehicle and free emission-related repairs. All of the work was carried out by local repair shops with the community college overseeing that the work paid for was in fact performed. Repairs ranged from the simple, such as fixing automatic chokes, to the extensive repair of replacing a cam shaft, lifters, and timing chain. In all, repairs for the 47 vehicles averaged $195 with an additional $43/vehicle spent on providing rental cars. In April 1992, through more on-road emission measurements, the effectiveness of the diagnosis and repairs at reducing CO emissions from these 47 vehicles was evaluated.

Table 2 shows, by model year, the 28 vehicles that were successfully remeasured. On average, the repairs resulted in a 50% reduction in the observed on-road emissions, with 25 of the remeasured vehicles showing statistically significant CO emission reductions. The remaining three vehicles were measured to have statistically the same or increased CO emission when compared with the before-repair measurements. These repair reductions were also found to be statistically significant when compared with the control group of vehicles that were identified from the control ramp but not included in the repair program.

In terms of the cost effectiveness of the repairs, it was estimated that CO was reduced from the 47 vehicles for approximately $181/t. This compares favorably with an

Table 2. Repair Data Summary According to Vehicle Emissions Technology Grouping for the 28 Vehicles that Were Successfully Remeasured after Repairs

Model Year Grouping	Number of Vehicles	Average g CO/L before Repairs	Average g CO/L after Repairs	Average g CO/L Reduction	Average g CO/L for Provo Fleet
post-1982	6	5931	2540	3391	867
1981–1982	1	5821	2305	3516	1911
1975–1980	16	5314	2702	2612	2589
pre-1975	5	6662	4285	2377	4012

estimate for I/M programs of \$708/t (23), oxygenated fuel programs of \$0 to \$1147/t (9,23,24), and of \$930/t for an old vehicle scrappage program (25). A major advantage of this approach is the ability to measure directly the program's effectiveness, allowing alterations and adjustments to be made to improve the results further.

Swedish Vehicle Study

In the 1991 data from Los Angeles, the European nameplate vehicles tend to have the lowest emissions. It was speculated that this is because they are well maintained. Data from the California Air Resources Board (CARB) listing manufacturer-specific failure rates for smog check reinforces this perception (26). The CARB data show Saab and Volvo with the lowest and third lowest smog check failure rates, respectively.

In September 1991, a study was conducted in Göteborg, Sweden (14). The location was a freeway interchange ramp (Gullbergsmotet) just across the river from the Volvo factory and downriver from the Saab manufacturing facility. In the Swedish study, emissions from 4011 Saabs and Volvos were measured. Sweden has a stringent I/M program (fail badly and the vehicle is towed to a repair shop). Sweden mandated closed-loop catalytically controlled systems in 1988. They were phased in during the 1987 model year, with about 50% of the vehicles. The 1986 and older Saabs and Volvos in Sweden are not equipped with any type of catalytic convertor.

The data from Sweden (4011 vehicles) and Los Angeles (536 vehicles) were used to examine the effects of technology and maintenance on vehicle emissions. By comparing these presumably well-maintained high technology vehicles to the well-maintained lower technology Swedish vehicles, the effects of technology ought to be readily observable. Figure 8 shows the emission data for CO and HC. For 1978–1986 model years, the CO and HC emissions of the Los Angeles vehicles average about 0.4% and 0.04%, respectively. For the Swedish vehicles of the same model years, the CO and HC emissions average about 1.5% and 0.08%, respectively. The improved technology of the Los Angeles fleet of Saabs and Volvos has clearly resulted in lower emissions, even for older vehicles. For 1988 model years and newer, when both fleets incorporated the same technology, the Swedish vehicles in Los Angeles and Göteborg are indistinguishable.

The dramatic drop in average vehicle emissions in Sweden following the 1987–1988 introduction of catalysts is only barely decernable in the first three quintiles of the California database (Fig. 7**a**), because catalysts were introduced longer ago. In Melbourne, Australia, catalysts were introduced in 1986. The dramatic improvement shown in Swedish vehicles also is not observed in the 15,908 vehicle Australian database. It was suspected that

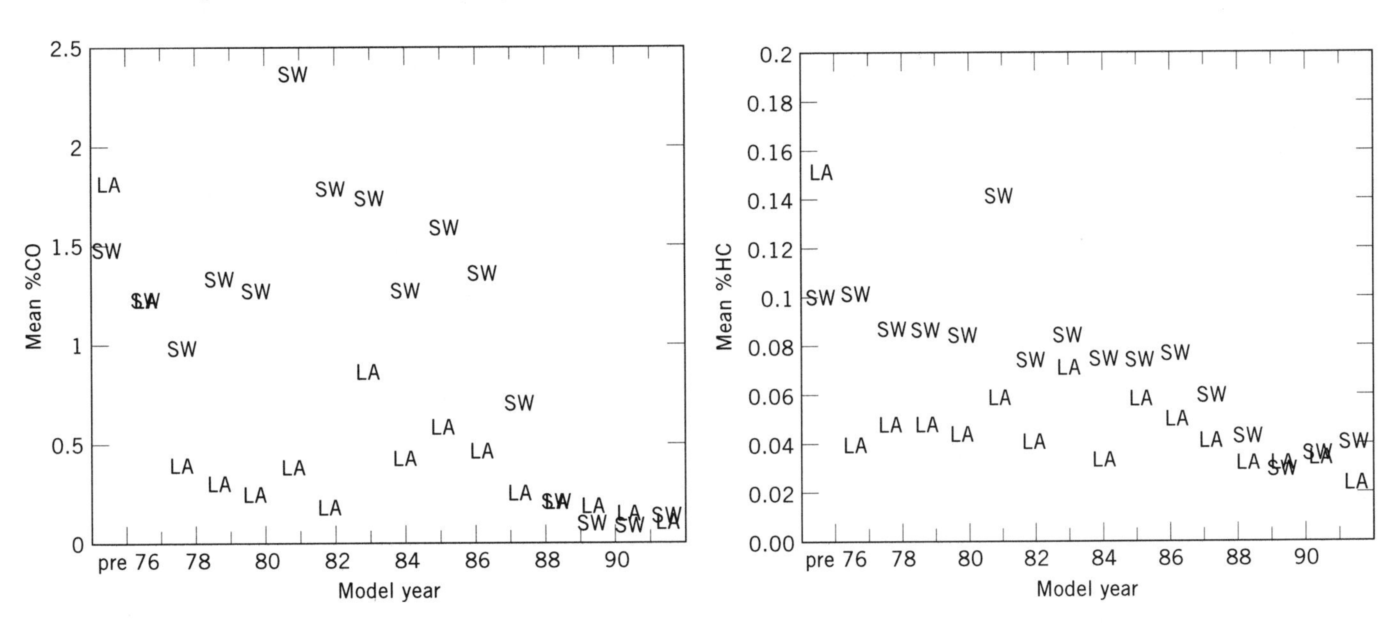

Figure 8. (**a**) Average percent CO for Saabs and Volvos measured in Los Angeles (LA) during the summer of 1991 compared with the same model year vehicles measured in Göteborg, Sweden (SW), during September of the same year. (**b**) Average percent HC for Saabs and Volvos measured in the same study.

Australian maintenance is more like California and less like Sweden.

To examine further the effect of maintenance on emissions, emissions from the Los Angeles fleet of 1978–1986 model year U.S. vehicles were compared with the same model year noncatalyst vehicles in Sweden. The Swedish vehicles averaged 1.5% CO and 0.08% HC. The emissions of U.S. vehicles were slightly lower for CO and comparable for HC. In other words, the well-maintained Swedish noncatalyst vehicles emit nearly the same CO and HC as the overall (less well-maintained) U.S. fleet in Los Angeles. This demonstrates that a high level of maintenance is as important as technology to the higher emitting (on average) older model year vehicles.

Finally, emissions from Saabs and Volvos in Los Angeles are higher in the pre-1976 fleet than in the Swedish fleet. Because vehicles rust faster in Sweden, the pre-1976 fleet is much older, on average, in Los Angeles. The older Saabs have two-stroke engines, which are notorious for HC emissions and often tuned to produce high CO; thus it is not surprising that the older fleet in Los Angeles has higher average emissions.

Swedish-manufactured vehicles appear to be well maintained in both Sweden and Los Angeles. In both locations, they have used computer-controlled port fuel injection for more than 20 yr. In Los Angeles, these vehicles also have used catalysts since 1980, whereas in Sweden catalysts were not introduced until 1987. These data have been used to conduct two thought experiments in which the citizens of Los Angeles are imagined to all drive Swedish nameplate vehicles. The first assumes that all vehicles are constructed, operated, and maintained as in Los Angeles (ie, their emissions match the entire California fleet for all makes). The second assumes they are constructed, operated, and maintained as in Sweden. The overall emissions of the vehicle fleet measured in California in the 1991 study averaged 0.79% CO and 0.076% HC. The same age distribution as this overall fleet but with the emissions distribution of the Swedish-manufactured vehicles currently in use in Los Angeles gives average CO and HC emissions of 0.49% and 0.056%, respectively. The same age distribution but with the emissions of the Swedish-manufactured vehicles currently in use in Sweden gives average CO and HC emissions of 0.9% and 0.066%, respectively. The better maintenance with catalytic control provides a reduction of 38% and 26% for CO and HC, respectively. The better maintenance alone provides an increase of 14% for CO and a reduction of 13% for HC. It was concluded that better maintenance of the current fleet in Los Angeles could provide on-road emissions reductions greater than 25% for both CO and HC. To the extent that remote sensing offers lower cost emissions testing, the money saved could be spent on more important diagnosis and repair functions.

CONCLUSIONS

On-road remote sensing can now be used to analyze the emissions of a large fleet of vehicles in a cost-effective manner without inconveniencing the driving public. The statistics of the data can be used to plan and evaluate emission-control programs. The identification of gross polluters can be used as a component of a program designed to ensure that those vehicles receive effective repair (an inspection and maintenance program). Excessive on-road readings have been used as evidence for pulling vehicles over to the side of the road to check for tampering with the emissions-control system finding a high incidence of such behavior.

Remote sensing can be applied at the same site under the same conditions periodically to determine how well efforts are working to reduce on-road emissions. Because the correlation between low emissions and proper maintenance is high, most people operate vehicles that do not contribute significantly to pollution. These people could be rewarded for their socially responsible behavior based on low on-road emissions readings. On-road remote sensing of motor vehicle emissions is a new tool that has progressed from a university prototype to a commercially available system, which is either portable or mounted in a mobile van. Everywhere the system has been tried, local air pollution officials have suggested new ways in which this tool could be applied to solving their mobile source emissions problems.

BIBLIOGRAPHY

1. *Federal Register*, Part II, **21**(60) (1966).
2. *Federal Register*, Part II, **33**(108) (1968).
3. *Federal Register*, Part II, **35**(214) (1970).
4. *Federal Register*, Part II, **36**(128) (1971).
5. M. P. Walsh, *Critical Analysis of the Federal Motor Vehicle Control Program*, NESCAUM, Albany, N.Y., 1988.
6. G. A. Bishop, J. R. Starkey, A. Ihlenfeldt, W. J. Williams, and D. H. Stedman, *Anal. Chem.* **61,** 671A–676A (1989).
7. U.S. Pat. 5,210,702 (May 11, 1993), G. Bishop and D. H. Stedman (to Colorado Seminary).
8. G. A. Bishop, Y. Zhang, S. E. Mclaren, P. L. Guenther, S. P. Beaton, J. E. Peterson, D. H. Stedman, W. R. Pierson, K. T. Knapp, R. B. Zweidinger, J. W. Duncan, A. Q. McArver, P. J. Groblicki, and J. F. Day, *J. Air Waste Manage. Assoc.* **44,** 169–175 (1994).
9. PRC Environmental Management, Inc., *Performance Audit of Colorado's Oxygenated Fuels Program, Final Report to the Colorado State Auditor Legislative Services*, Denver, Dec. 1992.
10. G. A. Bishop and D. H. Stedman, *Environ. Sci. Technol.* **24,** 843–847 (1990).
11. D. H. Stedman, G. A. Bishop, J. E. Peterson, P. L. Guenther, I. F. McVey, and S. P. Beaton, *On-Road Carbon Monoxide and Hydrocarbon Remote Sensing in the Chicago Area*, ILENR/RE-AQ-91/14, Illinois Department of Energy and Natural Resources, Springfield, Ill., 1991.
12. D. H. Stedman, G. Bishop, J. E. Peterson, and P. L. Guenther, *On-Road CO Remote Sensing in the Los Angeles Basin*, Contract No. A932-189, California Air Resources Board, Sacramento, Calif., 1991.
13. D. H. Stedman, G. A. Bishop, S. P. Beaton, J. E. Peterson, P. L. Guenther, I. F. McVey, and Y. Zhang, *On-Road Remote Sensing of CO and HC Emissions in California*, Contract No. A032-093, California Air Resources Board, Sacramento, Calif., 1994.
14. J. E. Peterson, D. H. Stedman, and G. A. Bishop, *J. Air Waste Manage. Assoc.* (1991).
15. Å. Sjödin, *Rena Och Smutsiga Bilar: En pilotstudie av avgasutsläpp från svenska fordon i verklig trafik*, Swedish Environmental Research Institute, Göteberg, 1991.

16. S. P. Beaton, G. A. Bishop, and D. H. Stedman, *J. Air Waste Manage. Assoc.* **42,** 1424–1429 (1992).
17. D. R. Lawson, P. J. Groblicki, D. H. Stedman, G. A. Bishop, and P. L. Guenther, *J. Air Waste Manage. Assoc.* **40,** 1096–1105 (1990).
18. D. H. Stedman and G. A. Bishop, *Remote Sensing for Mobile Source CO Emission Reduction*, EPA 600/4-90/032, U.S. Environmental Protection Agency, Las Vegas, Nev., 1991.
19. D. Elliott, C. Kaskavaltizis, and T. Topaloglu, *Soc. Automotive Eng.* **922314** (1992).
20. L. L. Ashbaugh, D. R. Lawson, G. A. Bishop, P. L. Guenther, D. H. Stedman, R. D. Stephens, P. J. Groblicki, B. J. Johnson, and S. C. Huang, paper presented at the A&WMA International Specialty Conference on PM10 Standards and Non-traditional Source Controls, Phoenix, 1992.
21. Y. Zhang, G. A. Bishop, and D. H. Stedman, *Environ. Sci. Tech.*, in press.
22. G. A. Bishop, D. H. Stedman, J. E. Peterson, T. J. Hosick, and P. L. Guenther, *J. Air Waste Manage. Assoc.* **43,** 978–990 (1993).
23. A. M. Michelsen, J. B. Epel and R. D. Rowe, *Economic Evaluation of Colorado's 1988–1989 Oxygenated Fuels Program*, RCG/Hagler, Bailly, Inc., Boulder, Colo., 1989.
24. J. B. Epel, M. Skumanich and R. Rowe, *The Colorado Oxygenated Fuels Program: Evaluation of Program Costs for 1989–1990 and for Alternative Program Designs*, RCG/Hagler, Bailly, Inc., Boulder, Colo., 1990.
25. *SCRAP A Clean-Air Initiative from UNOCAL*, UNOCAL Corp., Los Angeles, 1991.
26. *CVS News*, Jan. 1992.

AUTOMOTIVE ENGINES

CHARLES WOOD
Southwest Research Institute
San Antonio, Texas

Automotive engines are those power plants used to propel both highway and off-highway vehicles. Essentially all these power plants are reciprocating internal combustion engines of one of two types: spark-ignition engines or diesel engines. In either type, a piston reciprocates within a close-fitting cylinder. This reciprocating linear motion is transformed into continuous rotation of a crankshaft by means of a connecting rod that is pinned to the piston and crankshaft. Once each engine cycle, the piston compresses a quantity of air (or air-fuel mixture). The fuel in the gas is ignited and burned when the piston has compressed the gas to its minimum volume, and the heat produced by combustion raises the temperature and pressure of the mixture. The subsequent motion of the piston expands the hot, high-pressure gases and extracts work from the gas mixture, which is used to rotate the crankshaft and through various types of power transmission devices, propel the vehicle. The expanded gas mixture is then released to the surrounding atmosphere and the cylinder is filled with a fresh charge. Fuels most commonly used for automotive engines are mixtures of various hydrocarbon compounds refined from crude petroleum. Gasolines having a relatively high vapor pressure and a high autoignition temperature are used in spark-ignition engines; diesel fuels having lower vapor pressures and autoignition temperatures are used in diesel engines. The products of combustion from such fuels are principally nitrogen (from the air), carbon dioxide, and water vapor. The exhaust gases may also contain smaller percentages of carbon monoxide, unburned or partially-burned hydrocarbons, nitrogen oxides, and particulate matter consisting mostly of carbon (soot). The quantity of these latter materials are regulated by law in most of the nations of the world, since these emissions contribute significantly to atmospheric pollution. In addition, it is hypothesized that carbon dioxide in the atmosphere creates a "greenhouse" effect that may eventually result in global warming, although no legal restrictions have yet been placed on the quantity of carbon dioxide produced. The development of methods to reduce harmful emissions from automotive engines is undoubtedly the most pressing concern of engine designers and manufacturers.

See also METHANOL VEHICLES; SUPERCARS; EXHAUST CONTROL, AUTOMOTIVE; TRANSPORTATION FUELS, AUTOMOTIVE FUELS; KNOCK; GUM IN GASOLINE; INTERNAL COMBUSTION ENGINE.

ENGINE TYPES AND ARRANGEMENTS

The Four-Stroke Cycle

The four-stroke cycle, as its name implies, requires four strokes of the piston to complete one cycle. An engine using this cycle has intake and exhaust valves in the cylinder head (see Fig. 1). During the intake stroke, the intake valve is open and the piston moves downward, drawing into the cylinder a charge of air and fuel. As the piston reaches the end of its downward stroke, the intake valve closes. During the subsequent upward stroke, the fuel–air mixture is compressed (without significant heat transfer to or from the charge, or "isentropically"). At the end of this compression stroke, the fuel–air mixture is ignited (by a spark, for instance) and burnt. The exothermic reaction produced by oxidation of the fuel increases the pressure and temperature of the gases in the cylinder. During the next downward stroke the hot, high pressure gases are expanded. At the end of the expansion stroke, the exhaust valve opens and during the next upward stroke the gases are exhausted to the atmosphere.

The four strokes of the cycle (intake, compression, expansion, exhaust) are repeated continuously to produce a continuous rotation of the crankshaft. The piston produces a quantity of work during the expansion stroke that is greater than that required for the other three strokes, the friction of the engine, and for the operation of accessories (eg, alternators, fuel pumps, water pumps, fans). The net work is applied to the crankshaft and provides the energy needed to propel the vehicle.

Figure 1 also shows the various events of the cycle plotted on a pressure–volume chart. The compression stroke traces out the curve 1–2. Combustion (2–3) raises the pressure at constant volume. Expansion (3–4) is terminated by a sharp drop in pressure (4–5 as the exhaust valve opens. At the end of the exhaust stroke (5–6), the exhaust valve closes and the intake valve opens. As the piston begins to move downward, the pressure in the cylinder falls to the intake pressure (6–7), followed by the intake stroke (7–1).

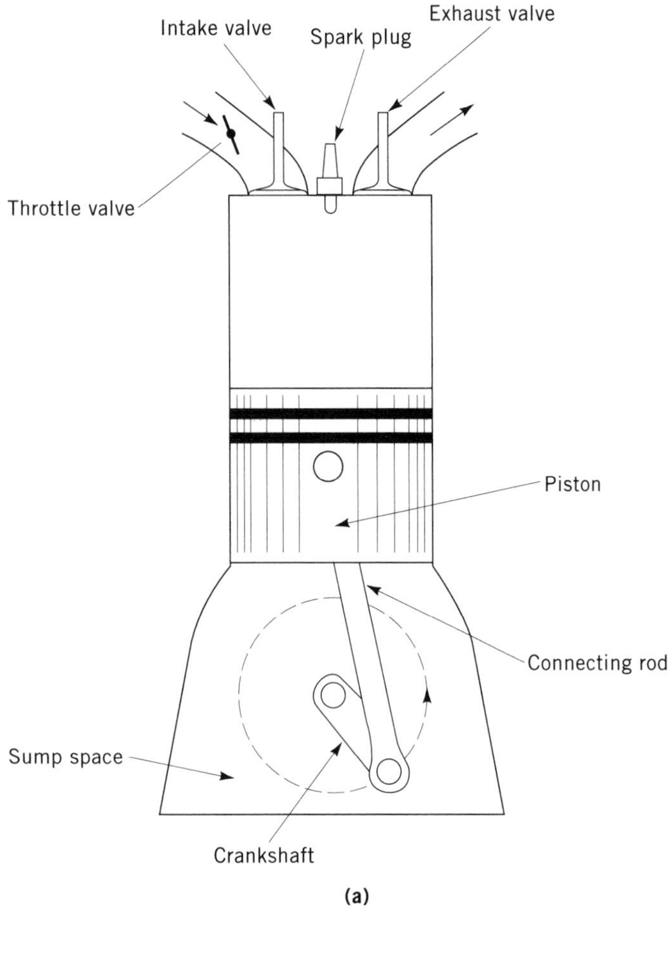

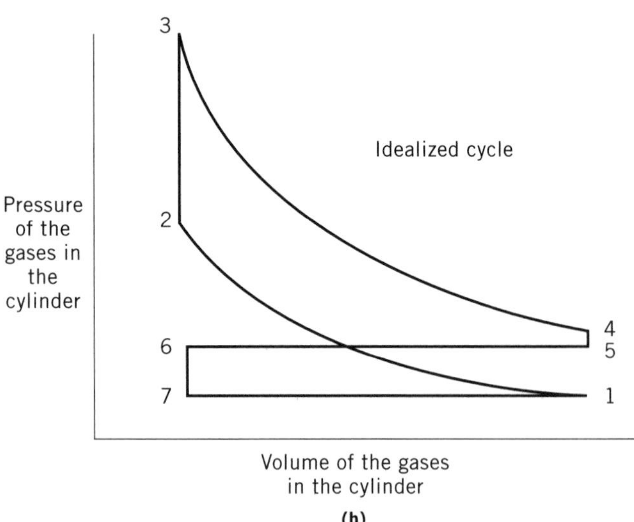

Figure 1. Four-stroke engine and its pressure–volume diagram.

The amount of net work produced by a spark-ignited engine is changed by changing the gas pressure during the intake stroke. A throttle valve in the intake pipe, when partially closed as shown in Figure 1, reduces the pressure during the intake stroke (7–1) below that of the exhaust stroke (5–6). This reduces the amount of fuel-air mixture inducted by the engine. Since the power produced by the engine is approximately proportional to the rate of fuel supplied to the engine, the power output is reduced. By this means, the engine output can be matched to the current load requirement. (Load control in diesel engines is different, and is discussed below.)

A complete engine design requires consideration of many factors, a few of which will be mentioned here. The gases above the cylinder must be contained in the cylinder, and the piston is fitted with seals (piston rings) to inhibit gas leakage from the gas volume to the sump space. Piston rings are usually made of an alloy of iron and are so designed that the high gas pressure above the piston forces the rings outward against the wall of the cylinder and downward against the bottom face of a rectangular groove cut into the piston.

Means to operate the intake and exhaust valves must be provided. This is accomplished by a cam shaft and associated valve gear. Two of many possible arrangements are shown in Figure 2.

Piston rings, bearings supporting the crankshaft and connecting rod, and valve gear must be lubricated. Lubricating oil is stored in the enclosed volume containing the crankshaft (the sump) and is distributed to the various components under pressure produced by an oil pump. The oil is recirculated and filtered.

The engine must be cooled. Critical components are the valves, cylinder head, spark plug, and piston. Cooling is most often accomplished by circulating a water-based liquid through cooling passages surrounding the cylinder head and cylinder, which is effective in maintaining valves, cylinder head, spark plug, and the cylinder at acceptable temperatures. In other engines, the cooling medium is air that is forced over the finned outside surface of the engine to carry away the heat. The piston is cooled both by conduction through the piston rings to the cylinder wall, and by the lubricating oil that is used to lubricate the upper connecting rod bearing. In high-output engines, lubricating oil is often sprayed against the bottom side of the piston to maintain an acceptable piston temperature. In such cases, it is often necessary to provide a heat exchanger to cool the oil in addition to the heat exchanger (radiator) used to cool the water-based coolant.

The Two-Stroke Cycle

In the two-stroke cycle the processes of intake, compression, expansion, and exhaust are accomplished within two piston strokes instead of four. A common two stroke cycle engine arrangement is shown in Figure 3.

The compression stroke follows the path 1–2, followed by combustion (2–3) and expansion (3–4). During the expansion stroke, fuel-air mixture in the sump is compressed by the downward motion of the piston. When the piston reaches point 4, the exhaust port begins to be uncovered by the piston and cylinder pressure falls. Shortly thereafter, the transfer port begins to open and compressed mixture from the sump enters the cylinder. Ideally, the fresh mixture displaces the remaining exhaust gases out the exhaust without mixing between fresh mixture and exhaust gas. Actually, there is a significant amount of mixing between the fresh mixture and the products of combustion, and the gases leaving the exhaust unavoidably contain fresh mixture. For the same reason, combustion products remain in the cylinder mixed with the fresh change. This process in which the fresh mixture "scavenges" the exhaust gases from the cylinder lasts until the piston reverses direction and closes the transfer port. Cylinder pressure then begins to rise due to compression by the piston even though the exhaust port is still open because of the decreasing exhaust port area as well as the rapidity of piston movement. Near the end of the compression stroke (1–2), the bottom of the piston uncovers the inlet port, allowing fresh mixture to enter the sump in time to be compressed during the expansion stroke.

In the two-stroke engine arrangement of Figure 3, lubricating oil cannot be stored in the engine sump as oil vapors and droplets would be carried into the cylinder with the fresh mixture, and in such engines it is conventional to mix lubricating oil with the fuel to lubricate bearings and piston rings. Other arrangements include the use of a separate air compressor driven by the crankshaft which supplies fresh mixture to the cylinder during the scavenging event. In such an arrangement, since the sump is not used as a compression chamber it may be used for lubricating oil storage and the lubrication system is similar to that of the four-stroke engine.

Two-stroke engines may also have poppet valves like the four-stroke engine. One arrangement uses a piston-controlled port for introducing the fresh mixture to the cylinder and two or more poppet valves (cam-operated) in

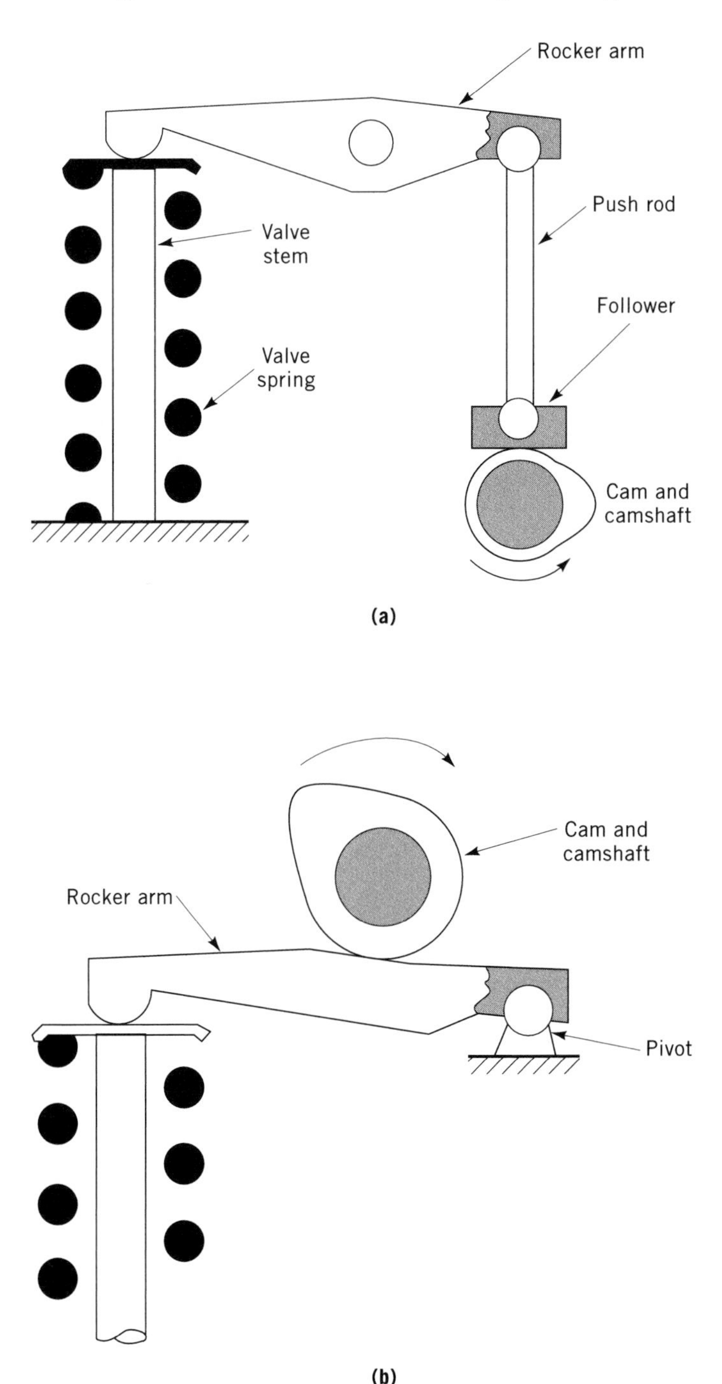

Figure 2. Two types of engine valve gear.

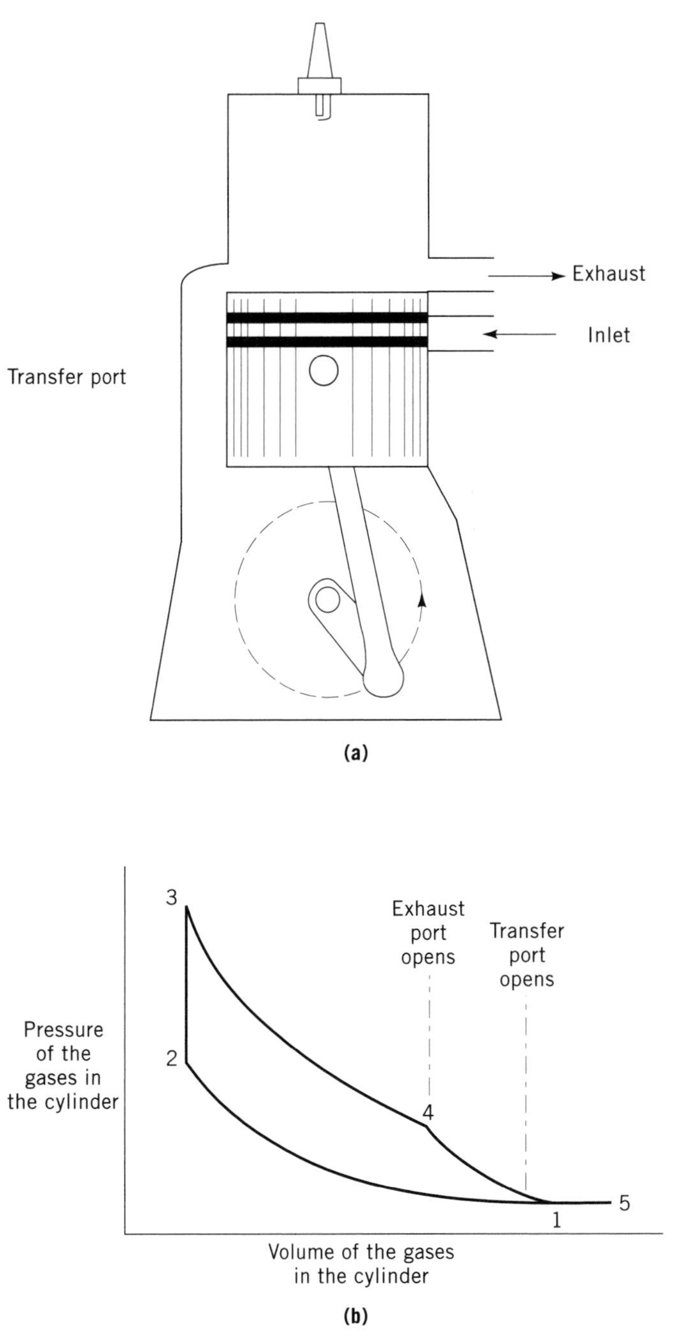

Figure 3. Two-stroke engine and its pressure–volume diagram.

the cylinder head for exhausting the products of combustion.

It has already been noted that some fresh mixture is carried out the exhaust port of two-stroke engines, and this is detrimental to good fuel economy and produces high hydrocarbon emissions as well. This problem can be remedied by injecting the fuel directly into the cylinder after the exhaust ports close, but the relatively high cost of a fuel injection system makes this impractical for the smaller two-stroke engines.

Other differences between two-stroke and four-stroke engines are

- The power density (power produced per unit volume of piston displacement) is higher with the two-stroke engine because there are twice as many expansion strokes per unit time, as compared to a four-stroke engine operating at the same crankshaft rotation speed.
- However, the power density of the two-stroke engine is less than twice that of the four-stroke engine because of the truncated expansion stroke (see Fig. 3) as well as the presence of relatively large amounts of combustion products in the cylinder during the compression stroke. These inert gases displace fresh mixture and contribute to reduced power.
- The two-stroke engine is very sensitive to the details of the scavenging process which in turn are affected by the dynamics of gas flow through the engine. This makes it more difficult to optimize two-stroke engine performance and emissions over wide ranges of speed and load.

Spark-Ignition Engines

In addition to the classification of reciprocating internal combustion engines by the number of piston strokes required for a complete cycle, they are further classified by the two methods of igniting the fuel-air mixture in the cylinder. These are ignition by spark and ignition by compression (diesel ignition).

Spark-ignition engines use a timed high-voltage spark produced across two electrodes in the combustion chamber to initiate the combustion process. The spark produces a kernel of flame between the electrodes that spreads throughout the combustion chamber by initiating exothermic reactions in the unburned fuel-air mixture at the boundary between the hot products of combustion and the unburned mixture.

The rate of spread of the flame front (the flame speed) must be fast enough so that combustion is complete relatively early in the expansion stroke. Flame speeds in fuel–air mixtures are influenced by the turbulence of the mixture and the fuel–air ratio (the weight of fuel in the mixture divided by the weight of air). In engines, sufficient turbulence is produced by the flow of mixture past the intake valve to produce acceptable flame speed so long as the fuel-air ratio of the mixture is near the stoichiometric value. (The stoichiometric fuel–air ratio is the mixture ratio that, when burned, theoretically just consumes all the oxygen in the air leaving no oxygen, fuel, or partially-reacted products in the exhaust.) However, as the fuel–air ratio is reduced below the stoichiometric value the flame speed decreases and eventually (at a fuel–air ratio of about 80 percent of stoichiometric in most engines) becomes too slow to travel across the combustion chamber in the time available. Since the use of fuel-air ratios greater than about 110% of stoichiometric results in reduced efficiency, the permissible range of fuel-air ratios for good engine efficiency is from about 80–110% of the stoichiometric value.

Spark-ignition engines for automobiles have strict legal limits on the exhaust emissions of carbon monoxide (CO), hydrocarbons, and nitrogen oxides (NO and NO_2, called NO_x for short). To attain these limits, the manufacturers have chosen to use an exhaust catalyst that oxidizes CO and hydrocarbons while simultaneously reducing NO_x. This "three-way catalyst" requires enough oxygen in the exhaust to oxidize the CO and hydrocarbons, and enough CO and hydrogen to reduce the NO_x. These conditions are fulfilled only in a very narrow range of fuel-air ratios near the stoichiometric value. The resulting need to control the fuel-air ratio very closely has led to the use of fuel injectors (supplanting carburetors), an oxygen sensor in the exhaust gas downstream of the catalyst to measure the fuel-air ratio, and a closed-loop computer-based control system to maintain close limits on fuel–air ratio.

In other applications, lean-burn engines (engines operating at fuel-air ratios less than stoichiometric) can be used. One such application is the use of natural gas fuel in stationary engines. These engines operate at relatively constant speed and load and optimization of lean-burn combustion is feasible. Since efficiency increases as fuel-air ratio is decreased, these engines have higher efficiency than engines operated stoichiometrically and can better compete with high-efficiency diesel engines. Emissions of NO_x are low because lower combustion temperatures (corresponding to lower fuel-air ratios) reduce the rate of NO_x formation in the combustion process. The emissions of CO are also low because combustion of mixtures at low fuel-air ratio leaves an excess of oxygen that oxidizes the CO. Hydrocarbon emissions are below the current legal limit.

The increase of engine efficiency with decreasing fuel-air ratio has led to many concepts to permit reliable spark-ignited combustion at low fuel-air ratios. For instance, "stratified-charge" engine designs attempt to place a fuel-rich mixture near the spark plug so that combustion can begin under favorable conditions, while the remaining mixture at locations more remote from the spark plug are fuel-lean. The "stratification" of the fuel mixtures allows the fuel-lean mixture to be heated by the prior combustion of the fuel–rich mixture, thereby increasing flame speed in the former. In one scheme, the spark plug and the fuel-rich mixture are brought together in a small combustion volume in the cylinder head connected to the cylinder volume by an orifice, while the fuel–lean mixture is placed in the cylinder above the piston. Ignition and combustion of the fuel–rich mixture creates a jet of hot products of combustion that ignites and accelerates the combustion of the fuel-lean mixture. While this concept has been used in automobile engines, it has been supplanted by stoichiometric mixture engines with catalysts because of increasingly severe emission standards.

Spark-ignition engines are subject to a combustion abnormality called "knock", which describes the sound produced by the engine so affected. Knock occurs when a portion of the fuel–air mixture in the cylinder explodes spontaneously. The exploding portion of the mixture is always the unburned portion remaining after most of the normal combustion is complete. After ignition by the spark, and during the normal spread of the flame through the mixture, the temperature and pressure of the burned gases increase. Since the pressure throughout the cylinder is uniform, the pressure of the unburned gases also increases. As the unburned gases are compressed by the rise in cylinder pressure, their temperature increases due to this compression process. High temperature and pressure maintained over a sufficient period of time causes the remaining unburned mixture to explode spontaneously. The propensity for knock is increased by high engine metal temperatures, long combustion duration, fuel–air ratio near the stoichiometric value, low engine speed, high compression pressure, high inlet mixture temperature, an earlier spark event, and a fuel with low octane number.

Knock is a severe limitation to the performance of spark-ignition engines. Engine efficiency increases with higher compression ratios (the ratio of maximum to minimum cylinder volume), but the occurrence of knock limits compressions ratios to between 8 and 12 (approximately) for spark ignition engines operating on the range of fuels normally available for such engines. Providing fuels with sufficiently high octane number is therefore a primary requisite for spark-ignition engines.

Compression–Ignition Engines

Compression–ignition (or diesel) engines produce ignition of the fuel–air mixture in the cylinder by compressing the charge of air in the cylinder to high pressures and temperatures, under which conditions the fuel-air mixture ignites spontaneously. This ignition method is identical in principle with the phenomenon of knock in spark-ignition engines, and should the fuel-air mixture in the diesel engine be homogeneous throughout the cylinder the explosion that would occur upon ignition would be extremely violent. Therefore, the engine ingests only air during the intake stroke. Near the end of the compression stroke, when the air is hot and under high pressure, the fuel is injected into the cylinder. The high-pressure fuel jet immediately begins to mix with the air, and ignition occurs in those portions of the fuel jet where the local fuel-air ratio is near the stoichiometric value. Combustion of the fuel continues as the fuel in the jet mixes with the air, so the rate of combustion is dependent on the rate of mixing between the fuel jet and the air in the combustion chamber. Unlike spark-ignition combustion, there is no orderly progression of flame through a homogeneous mixture of fuel and air, but rather a very complex combustion event that involves not only fuel jet mixing but fuel droplet evaporation, molecular changes in the as-yet unmixed fuel due to heating, and interactions of the fuel jet with the walls of the combustion chamber.

There are a number of important consequences of this type of combustion. First, regardless of the amount of fuel injected, ignition is reliable. Therefore, the engine load can be controlled by changing the amount of fuel injected and there is no need to throttle the air supply. As a result, the overall fuel-air ratio (weight of fuel injected divided by weight of air in the cylinder) can be very low without encountering a fuel-lean mixture ratio at which combustion fails. The ability to operate with fuel-lean mixtures increases diesel efficiency, and the absence of air throttling eliminates the losses associated with this process as well.

Second, the diesel must operate at high compression ratios (16 to 22, approximately) to increase the air temperature and pressure sufficiently for reliable ignition, particularly when the engine is started at low ambient temperatures. High compression ratios also increase efficiency, but the high pressures require strong structures and increase engine weight.

Third, the diesel engine exhaust contains particulate matter which is mostly carbon soot and which is sometimes visible as black smoke. Soot is produced when combustion occurs in or near a fuel-air mixture that is well above the stoichiometric value, and it is difficult to avoid fuel-rich combustion in the highly nonhomogeneous mixtures occurring in the diesel engine. The quantity of allowable diesel exhaust particulate matter is regulated by federal law in the U.S. and many other countries. Particulate emissions tend to increase with increasing overall fuel-air ratio, and for this reason diesel engines seldom operate at an overall fuel-air ratio greater than about 60% stoichiometric.

Fourth, diesel fuel must be easy to ignite spontaneously. The degree to which a fuel satisfies this requirement is measured by its "cetane number". With respect to ignition qualities, diesel fuel must readily ignite spontaneously (have a high cetane number) while spark-ignition engine fuel must resist spontaneous ignition to prevent knock (have a high octane number).

Fifth, the combustion process of the diesel readily accepts high levels of turbocharging. A turbocharger consists of a turbine and a compressor connected by a shaft. Exhaust gases drive the turbine which in turn drives the compressor. The compressor receives air from the atmosphere, compresses it, and delivers it to the engine cylinder. By this means, the weight of air delivered to the engine each cycle is increased. More fuel, in proportion to the increased weight of air, can be injected and the engine power is thereby increased. Turbocharging is a powerful means for increasing engine horsepower and greatly enhances the economic feasibility of diesel engines, since more power can be delivered with a modest increase in engine cost. Engine efficiency is improved as well. Since higher cylinder pressures, resulting from turbocharging, make spontaneous ignition of the fuel occur more readily, the use of turbocharging can allow the use of fuels with poorer ignition quality, or alternatively can allow the engine designer to reduce the compression ratio to reduce structural stress. In spark-ignition engines turbocharging is also possible but increases the propensity of the engine to knock, because the last fuel to burn is subjected to higher pressure and temperatures. The use of turbocharging in spark-ignition engines is limited to engines operating on high-octane fuels, or is accompanied by control features that invoke knock-preventive measures during conditions of high turbocharging (eg, ignition spark retard).

Diesel exhaust emissions of concern are NO_x, hydrocarbons, CO, and particulates. NO_x is produced in high temperature combustion zones. Hydrocarbons originate from either very fuel-lean zones where combustion is incomplete, or from combustion chamber walls that are wetted by liquid fuel from the fuel jet. CO is normally quite low in the diesel because the overall fuel-air ratio is fuel-lean and excess oxygen is plentiful. The source of particulates has been discussed. The option of using exhaust catalysts is currently quite restricted for diesel engines. Three-way catalysts, described earlier in the context of spark-ignition engines emission control, require stoichiometric fuel-air mixtures which are not feasible in diesel combustion. Even oxidation catalysts, which are capable of oxidizing hydrocarbons and CO (and can also reduce the weight of particulate emissions), are less than completely effective in diesel engines due to the relatively low exhaust temperatures in the diesel engine which are below the threshold needed for high catalyst efficiency.

Research work is underway towards the development of NO_x reducing catalysts that can operate in oxygen-rich atmospheres and that therefore would be applicable to diesel engines. Concurrently, particulate traps are being developed to filter the particulate matter from the exhaust stream and then periodically burn the collected soot on the filter surface to prevent filter plugging.

A comparison of the major features of diesel engines and spark-ignition engines is shown in Table 1. As a result of these, the diesel engine produces very high thermal efficiency (the net engine power divided by the rate of fuel energy). On the other hand, the diesel engine is heavier than the spark-ignition engine for the same power output. Because of the higher pressures in the diesel engine resulting in heavier structures as well as greater emphasis on piston sealing and structural stiffness, the diesel engine is more expensive. The cost of the diesel engine is further increased by the need for a high-pressure fuel injection system. The exhaust emissions of the diesel engine are more difficult to control to the levels now believed necessary for automobiles.

Since the combustion process of the diesel is relatively insensitive to the engine size (or more accurately, the diameter of the cylinder bore) while the spark ignition engine combustion process is sensitive to bore diameter (increasing bore diameter encourages knock due to longer flame travel time), one finds diesel engines in a very wide range of sizes. The smaller automobile diesels have a bore diameter of about 90 mm, while marine diesel propulsion engines have a bore diameter of about ten times this value. These largest engines currently have the highest thermal efficiency (exceeding 50%) of any internal combustion engine.

Table 1. Comparison Between Diesel and Spark-Ignition Engines

Diesel Engine	Spark-Ignition Engine
Nonhomogeneous fuel-air mixture	Homogeneous fuel–air mixture
Needs fuel with high cetane number	Needs fuel with high octane number
Maximum compression ratio limited by structural considerations	Maximum compression ratio limited by knock
Load is controlled by fuel quantity per cycle-no throttle	Load is controlled by inlet mixture density using a throttle
Fuel–air ratio is variable with load	Fuel–air ratio is constant
Must operate under fuel—lean conditions to limit exhaust particles	Cannot operate very fuel—lean because of flame speed degradation and/or exhaust catalyst requirements
Readily accepts turbocharging	Uses turbocharging only with very high octane number or with knock-preventive measures
Produces exhaust soot, particularly at high fuel–air ratios	Produces little soot
Emissions more difficult to control	Emissions can be controlled by three-way catalyst technology

ENGINE PERFORMANCE

The performance of an engine refers to its power output as well as to its efficiency, and to the factors that affect these attributes. Both spark-ignition and compression-ignition engines in their various configurations are subject to some common physical principles.

Engine efficiency is expressed in different ways. "Brake thermal efficiency" is the engine power output at the crankshaft divided by the rate of fuel energy supplied to the engine. This is the overall measure of engine efficiency. "Indicated thermal efficiency" is the net power produced by the compression and expansion of the gases above the piston divided by the rate of fuel energy supplied to the engine. "Mechanical efficiency" is the ratio of the output power at the crankshaft to the net power produced by the compression and expansion of the cylinder gases. By definition,

$$\eta_b = \eta_i * \eta_m$$

where

η_b = brake thermal efficiency

η_i = indicated thermal efficiency

η_m = mechanical efficiency

Mechanical efficiency is almost always less than one because of the power lost due to friction of rotating and reciprocating components, the difference in pressure on the piston during the intake and exhaust strokes (in four-stroke engines), and the power required to drive engine auxiliaries such as fans and pumps. In turbocharged engines, the pressure in the cylinder during the intake stroke can be higher than that during the exhaust stroke, in which case the mechanical efficiency can be greater than one. On the other hand, at idle conditions (where the engine produces no output power), the mechanical efficiency is zero. Therefore, the mechanical efficiency is not a constant but varies with both engine power output and speed.

By definition,

$$P = \bar{M}_f * Q_{\mathrm{HV}} * \eta_b$$

where P is engine power output, $\bar{M}_f$ is the mass rate of fuel supplied to the engine, and Q_{HV} is the heating value of the fuel. The fuel heating value Q_{HV} is the quantity of heat energy produced by completely burning a unit mass of fuel. This equation may be expanded as follows:

$$P = F * \bar{M}_a * Q_{\mathrm{HV}} * \eta_i * \eta_m$$

where F is the fuel-air ratio (the mass ratio of fuel to air in the engine cylinder during combustion) and $\bar{M}_a$ is the mass rate of air engaged in the combustion process. The mass rate of air $\bar{M}_a$ is dependent not only on the density of the air supplied to the engine, but also on the cylinder volume and the engine speed. The effectiveness of the engine in ingesting air is quantified by the volumetric efficiency η_v, which is the ratio of the mass rate of air ingested to the theoretical mass rate.

$$\eta_v = \frac{\dot{M}_a}{\rho V \frac{N}{n}}$$

where ρ is the density of the ingested air, V is a characteristic volume (the volume swept by the piston in four-stroke engines; the maximum volume of the cylinder above the piston in two-stroke engines), N is the rotational speed of the crankshaft, and n is the number of crankshaft revolutions each complete cycle (2 for four-stroke engines, 1 for two-stroke engines). Using this definition, the engine power is

$$P = (F \rho V \frac{N}{d} Q_{\mathrm{HV}})\,(\eta_v\, \eta_i\, \eta_m)$$

This equation shows that power output can be controlled by changing the supplied air density, and this is the control method used in spark-ignition engines where a throttle is used to reduce inlet air pressure (hence density). Power control in the diesel is accomplished by changing fuel-air ratio at constant inlet air density. Engine power is strongly affected by speed and volume, i.e., larger engines and faster-running engines produce more power. Increases in any of the efficiency terms increase engine power. A two-stroke engine usually develops more power than a four-stroke engine of the same volume operating at the same speed and fuel-air ratio, given a reasonably high value of volumetric efficiency.

Consider the interactions between the terms in the preceding equation. The terms within the first bracket of this equation can be independently controlled by the engine designer. The efficiency terms in the second bracket are affected by the designer's choices, as well as by limitations imposed by physics.

Volumetric efficiency η_v is increased by reducing flow losses in the inlet ports of the engine. It is also influenced, particularly in two-stroke engines, by pressure waves in the inlet and exhaust pipes that coincide with port opening or closing in such a way as to temporarily increase inlet pressure or decrease exhaust pressure. In two-stroke engines, the scavenging process is critical. A scavenging process in which the fresh mixture thoroughly sweeps out the residual combustion products with minimal loss of fresh mixture out the exhaust port improves volumetric efficiency, and this process is affected by details of the engine design.

Mechanical efficiency is increased by reducing friction of mechanical components and the power required by auxiliaries. Mechanical efficiency increases with higher engine power output because friction and auxiliary power losses tend to remain constant at constant speed. Also (in four-stroke spark-ignition engines), the losses due to the exhaust pressure exceeding the inlet pressure decrease as power (and inlet pressure) are increased. For this reason, reducing the size of the engine in a given vehicle improves fuel economy because the smaller engine must operate closer to its maximum power output at all the vehicle operating conditions than a larger engine and the smaller engine, as a result, has a higher mechanical efficiency. Mechanical friction increases with increased mean piston speed. (Mean piston speed is twice the product of the stroke length and the number of crankshaft revolutions per unit time.) As a result of these effects on mechanical efficiency, it is usually found that engines produce their highest brake thermal efficiency at low speeds and maximum power. As a corollary to the effect of mean piston speed on mechanical efficiency, all engines regardless of size are limited to a maximum piston speed of about 600 to 900 meters/second because of increased friction losses as well as increased wear rates and stress levels. Hence, large engines (with long piston stroke lengths) must operate at lower rotational speeds than small engines in about the inverse proportion of their respective stroke lengths.

Indicated thermal efficiency η_i is dependent on fuel–air ratio, compression ratio, and heat losses (or gains) from the gases in the cylinder to the walls of the cylinder. The limits on indicated thermal efficiency are based on thermodynamic principles. A simple analysis of the indicated thermal efficiency can be done using a so-called "air cycle". For such an analysis, it is assumed that the gases in the cylinder have a constant specific heat equal to that of air at room temperature and a molecular weight equal to that of air, that there is no heat transfer between gases and cylinder walls, that heat is added to the gases instantaneously when the piston is at its uppermost position (top dead center), and that the exhaust and inlet processes require no energy. With these assumptions, an energy balance can be accomplished for each event in the cycle and the indicated thermal efficiency is

$$\eta_i = 1 - \left(\frac{1}{r}\right)^{k-1}$$

where r is the compression ratio of the engine (the maximum volume of the gases in the cylinder divided by the minimum volume), and k is the ratio of the constant pressure specific heat of the gases to their constant volume specific heat. The value of k (for air at room temperature) is 1.4.

When the situation is made more realistic by taking into account the actual properties of the fuel-air mixture and the products of combustion, it is found that the indicated thermal efficiency is also strongly affected by fuel-air ratio, in addition to the dependence on compression ratio discovered in the analysis of the air cycle. Efficiency decreases as fuel–air ratio increases, because increasing fuel-air ratio raises the temperature of the gases during combustion which decreases the average value of k during the expansion stroke. The inclusion of the real properties of the gases reduces the predicted efficiency of the engine.

In an actual engine, the combustion process is not instantaneous, there is heat transfer between the gas and the cylinder wall, and combustion of the fuel is not complete. Actual engine efficiency is therefore reduced even further.

These effects are shown in Figure 4. The air cycle efficiency is a theoretical efficiency that illustrates the basic thermodynamic limits on efficiency. The "fuel–air cycle" efficiency accounts for real gas properties, and is the realistic limit to the efficiency of actual engines. The approximate ranges within which current engines operate are also shown.

It will be recalled that spark-ignition engines operate with a stoichiometric fuel-air mixture, and that their compression ratio is limited by knock. Comparisons should be made between the fuel-air cycle with a stoichiometric mixture and actual spark-ignition engines. On the other hand, diesel engines are limited to a fuel-air ratio of about 60% of stoichiometric because of particulate emissions and require a compression ratio of about 16 or greater. Actual diesel engine efficiency should be compared to the fuel-air cycle efficiency with a mixture 60 percent of stoichiometric.

Figure 4 illustrates the fact that, while gains in indicated engine efficiency are possible, they will not exceed the fuel-air cycle efficiencies shown.

The efficiency of diesel engines declines slightly at the highest compression ratios. High compression ratios are used in smaller, high-speed automotive diesel engines which usually employ a small pre-combustion chamber in the cylinder head into which the fuel is injected. Partial combustion of the fuel produces a jet of hot gases from the precombustion chamber that quickly mixes with the air in the cylinder, accelerates the combustion process, and permits high speed operation. However, heat losses to the walls of the precombustion chamber are high and indicated efficiency suffers, as shown in Figure 4.

APPLICATIONS

Spark-ignition engines have their most visible application in automobiles. These engines almost exclusively use the four-stroke cycle, and are made in a wide variety of sizes ranging from 3 to 12 cylinders and with cylinder bore diameters between about 65 and 100 mm. In most nations of the world, automobile engine emissions are limited by law and various systems are used to enforce these laws. In the U.S., for instance, the automobile manufacturers must certify each vehicle model made using procedures provided by the Environmental Protection Agency. As a result of such laws, U.S. automobiles use exhaust catalysts for control of NO_x, hydrocarbons, and carbon monoxide, and sophisticated electronic systems to control fuel–air ratio and the timing of the ignition event.

The spark-ignition engine is favored for automobile use for a number of reasons. Compared to a diesel engine the spark-ignition engine is easier to start, smaller, lighter, less noisy at low engine speeds, and less expensive to manufacture. While the fuel consumption of the diesel engine is lower than the spark ignition engine, the cost of fuel in the U.S. is currently (1993) too low to justify the additional expense of the diesel engine unless the vehicle

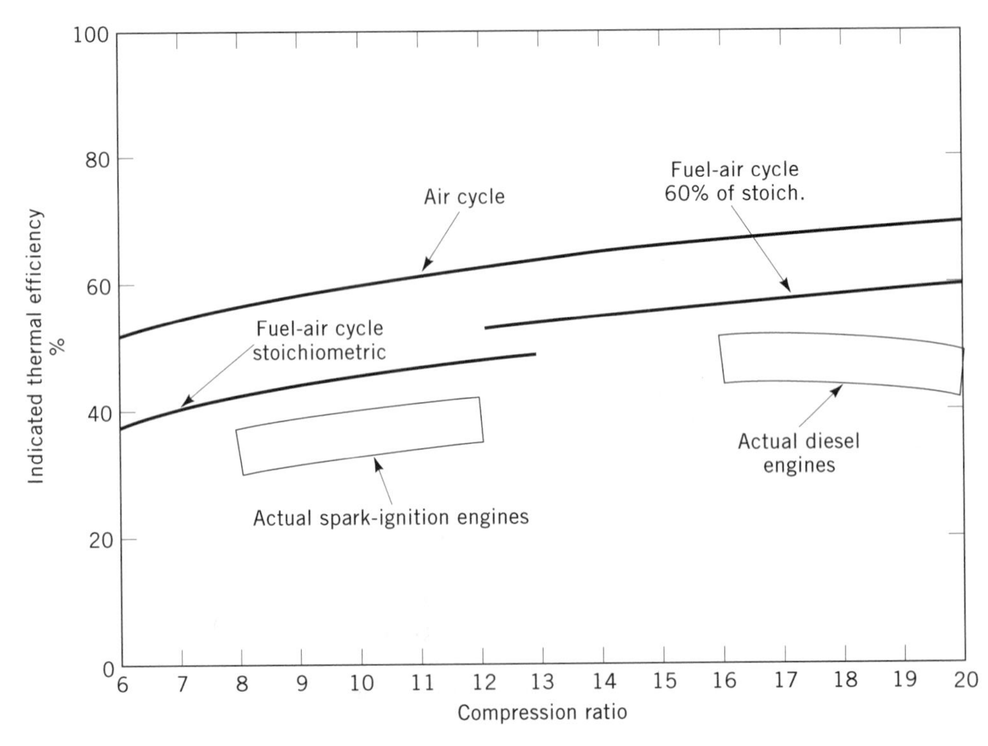

Figure 4. Comparison of indicated thermal efficiency of engines at full load.

is very heavily utilized. This is seldom the case with automobile engines.

Small spark-ignition engines are also for utility purposes (eg, lawn and garden equipment, small pumps and generators). These engines use both four and two-stroke cycles and are often made with only one cylinder. The main attributes of these engines are low manufacturing cost and convenience of operation. In many applications, these engines are not subject to emission controls and therefore have a very simple control system and no exhaust aftertreatment.

In other applications, the spark-ignition engine is dominant because of its low weight and small size (high power density), as in outboard motors and motorcycles. Since size and weight are of primary importance in outboard motors, the two-cycle engine is used in a very compact multicylinder arrangement with water-cooling. Motorcycles use both two and four-stroke multicylinder engines, the majority of which are air-cooled.

The spark-ignition engine is used, to a limited extent, in large, low-speed engines and this is the case usually when a low-cost spark-ignition engine fuel is available. The best example is engines used for natural gas compression service, where it is economical and convenient to burn natural gas, which is not suitable for diesel engines because of its high autoignition temperature. This property of natural gas also gives it a high octane number and thereby allows the use of large bore diameters which would produce knock with gasoline fuel with its lower octane number.

In general, diesel engines are used wherever fuel economy is the primary requirement, or where an engine with very large horsepower is required. Because of its good fuel economy, the diesel engine is favored in heavy trucks, buses, and earthmoving equipment. These vehicles typically have a high utilization and fuel costs are therefore a major operating expense. The higher cost of the diesel engine is therefore easily justified, and the disadvantages of increased weight, size, and noise assume less importance. Because diesel engines are often used in these heavy-duty applications in which many engine operating hours are accumulated quickly, the engines are designed to operate for more engine hours between overhauls than are spark-ignition engines used in light-duty automobile service. Thus, the longer life observed in diesel engines is the result of consumer demand rather than any inherent feature of the diesel cycle.

Diesel engines used in heavy trucks, buses, and off-highway vehicles are multicylinder (usually six cylinders in line), four-stroke cycle, liquid-cooled engines. In the U.S., these engines are turbocharged because turbocharging provides more power for heavier vehicles and higher road speeds. In Japan, for instance, trucks and buses are smaller and are operated more in urban areas under stop-and-go conditions and turbocharging is less attractive.

Since diesel engines can be made with very large bore diameters without fear of knock, diesel engines are used in very large engines. The largest of these are used for the main propulsion plant in ships, and can have a piston diameter of a meter or so. When turbocharged to high inlet pressures, these engines produce over 5000 horsepower per cylinder. The two stroke cycle is used in such engines and their maximum speed is about 100 revolutions per minute. Very high efficiency is realized, for several reasons. The heat loss from the hot combustion gases to the cylinder, piston, and cylinder head relative to the total internal energy of the gases is inversely proportional to the bore diameter and speed, so larger and slower engines suffer less relative heat loss. Because of the low speed, the combustion event occupies fewer degress of crankshaft rotation and burning is therefore completed early in the expansion stroke, which also improves efficiency.

FUTURE DEVELOPMENTS

The trends of future automotive engine design will be largely dictated by the increasingly stringent requirements for low exhaust emissions, and the decreasing reserves of hydrocarbon fuels. Lower emissions and higher fuel economy are therefore the major future goals.

Theoretically at least, the emissions of NO_x, hydrocarbons, and carbon monoxide can be reduced essentially to zero (but not carbon dioxide, as discussed later in this article). For instance, very large and efficient catalyst beds at a controlled high temperature can now accomplish this on spark-ignition engines burning a closely-controlled fuel–air ratio. However, in practice there are limits to the maximum size of the catalyst, and the catalyst temperature is not always high, particularly when starting a cold engine.

The theoretical limit to efficiency is determined by thermodynamic principles and fluid and mechanical friction, and is always well less than 100%. Current engines approach the theoretical limits, and could approach them much closer by using leaner fuel-air mixtures to increase indicated thermal efficiency. However, leaner mixtures in spark-ignition engines upset the delicate balance currently required for efficient three-way catalyst operation for emissions control.

These few examples illustrate the kinds of problems facing engine designers. Some of the current development efforts are

- Development of new catalyst materials that will reduce NO_x produced by the combustion of fuel–lean mixtures. Success in this development will permit the use of leaner mixtures in spark-ignition engines for better fuel economy and will provide a NO_x exhaust aftertreatment method for diesel engines.
- Development of methods to pre-heat the catalyst used in the exhaust of spark-ignition engines, so that the exhaust emissions during a cold start are reduced. Currently, exhaust emissions are high until the exhaust stream heats the catalyst up to its operating temperature.
- Development of better systems to control fuel-air ratio in spark-ignition engines. Currently, the exhaust oxygen sensor provides a feedback signal to the fuel economy system to maintain a constant fuel-air ratio, but during transient changes in speed and load the response of this sensor is too slow to provide exact control.
- Development of exhaust filters to remove the particulate matter from diesel exhaust. Since the size of such

filters is limited for vehicular use, means must be developed to periodically remove the collected particulate matter from the filter surface to prevent excessive pressure loss of the exhaust stream. Various methods to burn the collected particulates on the filter surface automatically are under development.

- Development of improved diesel fuel injection systems. In general, higher fuel injection pressures produce lower exhaust emissions, but for significant effects fuel pressures need to exceed about 150 MPa. Accurate control of the time of the start of injection ($\pm 3 \times 10^{-5}$ s) and the quantity of fuel injected (roughly $\pm 1 \times 10^{-3}$ g) are now achieved with mechanical systems, but electronically controlled injection systems (now being introduced in truck and bus engines) permit more flexible control at high injection pressures.
- Evolution of engine design. In diesel engines, changes in combustion chamber shape and inlet port design (as influenced by computational fluid dynamics) will continue to improve the rate of mixing between air and fuel to reduce particulate emissions and NO_x. In all engines, better analysis procedures will permit reduction of oil consumption, which contributes to increased emissions both directly (by partial combustion of the oil) and by reducing catalyst efficiency because of "poisoning" of the catalyst. Engine design continues to change because of the greater control possibilities offered by computers. Strategies such as exhaust gas recirculation to the combustion chamber (to reduce NO_x), variable inlet valve timing (to improve volumetric efficiency over the load/speed range), variable injection timing (to improve indicated thermal efficiency), and the use of variable geometry turbochargers (to more closely tailor the enging torque over the speed range to the torque requirements) become more effective given the ability for precise control.
- Development of new materials. Cast iron, alloy steels, and aluminum have historically been the structural materials in engines. New materials, material surface coatings or material surface modifications will offer opportunities for improved design. Ceramic materials, not yet widely used, maintain their strength at high temperature, are lightweight and stiff, and resist corrosion and wear, and therefore are candidates for various engine components such as valve gear and combustion chamber elements. Ceramic coatings may be useful for corrosion resistance (on diesel engine piston crowns, for instance). Various methods to alter the crystalline structure of a ceramic substrate material by inserting metallic ions into the structure are under investigation, as this surface modification can reduce mechanical friction and may be a suitable process for piston rings or cylinder walls.
- Reductions in carbon dioxide emissions. While carbon dioxide (CO_2) is not a regulated exhaust emittant, there is increasing worldwide concern that CO_2 in the atmosphere may have future environmental effects. For a given power requirement, the CO_2 emissions from an engine are a function of the molecular structure of the fuel, the fuel heating value, and the engine efficiency. The weight of CO_2 produced per unit weight of fuel burned. The weight of fuel burned per unit time for a particular level of engine power output is inversely proportional to the product of fuel heating value and engine brake thermal efficiency. The possible strategies for CO_2 emission reduction are therefore clear: choose fuels with a high hydrogen/carbon ratio, increase engine brake thermal efficiencies, and reduce power requirements. Since the choice of fuel types is restricted and efficiency is near its upper limits, large reductions in CO_2 emissions imply a main reduction in the amount of power used.

OTHER POWERPLANT TYPES

Since the advent of gas turbines in the 1940s, this engine type has periodically been considered for automotive use. In its basic form, the gas turbine consists of a compressor and a turbine connected to the same shaft. Air is compressed by the compressor, fuel is injected into this pressurized air and burnt, and the hot products of combustion expand through the turbine to power the compressor and to produce excess power used to drive the load attached to the shaft. More complex turbines employ preheating of the compressed air by exhaust gas, an additional turbine on a separate shaft, and other features to improve efficiency and the ability of the engine to match specific load-speed requirements. Gas turbines, at maximum power and speed, can approach the efficiency of a diesel engine but have higher fuel consumption at part loads. The performance of a gas turbine is strongly linked to the temperature of the gas entering the turbine; the higher the gas temperature the better the efficiency and the higher the power output. However, high turbine inlet temperatures are expensive to achieve as the turbine blades are continuously exposed to the hot gases and must be cooled by air flowing through internal passages in the turbine blades (for instance). As a result of these things, gas turbines are best used in applications where power demand is continuously high and where initial engine cost is not a major factor. Gas turbines therefore do not appear to be a strong competitor for automobile use, but may have future application in heavy cross-country trucks and locomotives.

Fuel cells are now under development for both vehicular and stationary applications. Fuel cells are essentially electrical storage batteries using the chemical energy of a fuel (hydrogen) and oxygen stored outside the cell. Hydrocarbon fuels can be used providing a reformer process that converts the hydrocarbon to hydrogen and oxides of carbon. Theoretical efficiency is high (up to 70%); actual projected efficiencies using natural gas fuel and air are 40–50%. Since this development is in its early stages it is difficult to predict its eventual success, but a major benefit of a successful fuel cell would be very low emissions of undesirable compounds.

The use of alternative fuel sources for automotive power plants is under intense study, primarily as a way to reduce exhaust emissions. Alternative fuels include natural gas, alcohols, fuels produced from vegetation (so-called "bio-fuels"), and hydrogen. Emissions from hydrogen combustion, for instance, are limited to water vapor and NO_x. Natural gas and alcohols produce reduced amounts of carbon monoxide and carbon dioxide, and the hydrocarbon emissions of these fuels, particularly natural gas, have low reactivity and hence reduce detrimental environmental effects. Natural gas is produced in large quantities in the U.S. for other purposes but is difficult to store on-board a vehicle in sufficient quantities to achieve the vehicle range obtained with liquid fuels. Hydrogen has an even greater problem in this regard, and has no established source of supply. Similarly, alcohols are now produced in quantities too small for extensive transportation needs, as are bio-fuels. In sum, the use of alternative fuels depends largely on establishing an infrastructure for their distribution for vehicular use, and this infrastructure is lacking for any of the alternative fuels. At this time, natural gas appears to have the advantage in the U.S. because of our large reserves of this resource, but large investments are required to bring about its widespread use in vehicles.

When considering various alternative powerplants and fuels, one must understand that the current engines used with petroleum-based fuels do their required functions exceedingly well: produce high power within small weight and volume envelopes, can be built and sold at low cost, and have high reliability and durability. For consumer acceptance, alternatives must match these attributes or have other very strong advantages that overcome their deficiencies. In the future the need for lower emissions, a domestically-produced fuel, or a renewable fuel source could change these currently-demanded powerplant characteristics.

BIBLIOGRAPHY

Reading List

C. F. Taylor, *The Internal-Combustion Engine in Theory and Practice*, Volumes 1 and 2, M.I.T. Press, Cambridge, Mass., 1985.

J. B. Heywood, *Internal Combustion Engine Fundamentals*, McGraw-Hill, New York, 1988.

L. C. Lichty, *Combustion Engine Processes*, McGraw-Hill, New York, 1967.

R. S. Benson and N. D. Whitehouse, *Internal Combustion Engines*, Pergamon Press, 1979.

E. M. Goodger, *Hydrocarbon Fuels*, Macmillan, London, 1975.

M. H. Edson and C. F. Taylor, "The Limits of Engine Performance—Comparison of Actual and Theoretical Cycles," *Digital Calculations of Engine Cycles, SAE Prog. in Technology*, **7**, pp. 65–81 (1964).

W. Traupel, "Reciprocating Engine and Turbine in Internal Combustion Engineering," in *Proc. CIMAC Int. Congr. on Combustion Engines*, Zurich, pp. 39–54, 1957.

A. A. Quader, "Lean Combustion and the Misfire Limit in Spark Ignition Engines," *SAE Trans.*, **83,** paper 741055 (1974).

R. Maly, "Spark Ignition: Its Physics and Effect on the Internal Combustion Process," in J. C. Hilliard and G. S. Springer, eds., *Fuel Economy in Road Vehicles Powered by Spark Ignition Engines*, Plenum Press, New York, 1984, chapt. 3.

J. R. Smith, R. M. Green, C. K. Westbrook, and W. J. Pitz, "An Experimental and Modeling Study of Engine Knock," *Proceedings of Twentieth International Symposium on Combustion*, The Combustion Institute, 1984, pp. 91–100.

A. M. Douaud and P. Eyzat, "Four-Octane-Number Method for Predicting the Anti-Knock Behavior of Fuels and Engines," *SAE Trans.*, **87,** paper 780080 (1978).

W. T. Lyn, "Study of Burning Rate and Nature of Combustion in Diesel Engines," in *Proceedings of Ninth International Symposium on Combustion*, The Combustion Institute, 1962, pp. 1069–1082.

M. Alperstein and R. L. Bradow, "Exhaust Emissions Related to Engine Combustion Reactions," *SAE Trans.*, **75,** paper 660781 (1966).

E. Keller, "International Experience with Clean Fuels," *SAE paper 931831*, August 1993.

R. L. Bechtold and G. J. Wilcox, "Assessment of Ethanol Transportation and Marketing Infrastructure," *SAE paper 932827*, Oct. 1993.

M. P. Walsh and R. Bradow, "Diesel Particulate Control Around the World," *SAE paper 910130*, Feb. 1991.

E. F. Obert, *Internal Combustion Engines and Air Pollution*, Intext Educational Publishers, 1973 edition.

R. Bosch, *Automotive Handbook*, 1st Ed., Robert Bosch GmbH, 1976.

H. Hiroyasu, "Diesel Engine Combustion and Its Modeling," in *Diagnostics and Modeling of Combustion in Reciprocating Engines, COMODIA 85, Proceedings of Symposium, Tokyo*, Sept. 4–6, 1985, pp. 53–75.

D. J. Patterson and N. A. Henein, *Emissions from Combustion Engines and Their Control*, Ann Arbor Science Publishers, Ann Arbor, Mich., 1972.

W. A. Daniel and J. T. Wentworth, "Exhaust Gas Hydrocarbons—Genesis and Exodus," *SAE paper 486B*, March 1962; also *SAE Technical Progress Series*, vol. 6, p. 192, 1964.

AUTOMOTIVE ENGINES–EFFICIENCY

K. G. Duleep
EEA
Arlington, Virginia

Automotive engine efficiency can be defined in terms of the percent of fuel energy that is converted to useful work, but should not be confused with vehicle fuel efficiency, or fuel economy. Automotive fuel economy is broadly understood by the public as a measure of the distance traveled per unit volume of fuel consumed (because most automobiles use liquid fuel), and is usually expressed in the United States in miles per gallon, or MPG. In other developed countries, fuel consumption is preferred to fuel economy and is usually expressed in liters per 100 km or the volume of fuel consumed per unit distance. Fuel economy (or consumption) of vehicles usually varies by the size, as defined by the physical exterior dimensions, and the weight. The larger or heavier the vehicle, the lower the MPG values typically obtained. However, this need not imply that the engines in larger cars have proportionally lower efficiency. Because larger cars demand more power to be driven a specific distance, fuel economy will be lower than for a smaller car with an engine of the same efficiency as that in the larger car.

See also Thermodynamics; Energy efficiency; Energy efficiency, calculations; Knock; Octane number.

Almost all currently produced automobiles and trucks worldwide use the Otto cycle engine or the Diesel cycle engine for motive power. Most of the Otto cycle engines use gasoline and are of the "four-stroke" type, because the "two-stroke" type has been too inefficient and polluting to be widely used in automobiles. The Otto cycle two-stroke engine is still used in small motorcycles and other two-wheelers, especially in developing countries, and new versions in the prototype stage hold the promise of low emisisons and high efficiency. Diesel cycle engines use diesel fuel and are of both the two- and four-stroke types, but their primary use has been in commercial heavy trucks because they are significantly more fuel efficient and much heavier than Otto cycle engines of the same power. Other engines, such as the gas turbine, have been installed in prototypes or research vehicles but have never progressed to production status because of various real and perceived drawbacks. Electrical batteries and fuel cells, which are not heat engines, have also been tried for automotive use and could replace heat engines in the future, but are not considered in the following discussion on engine efficiency of conventional heat engines.

Heat engine efficiency can be stated in several ways. One intuitively appealing method is to express the useful energy produced by an engine as a percent of the total heat energy that is theoretically liberated by combusting the fuel. This is sometimes referred to as "the first law" efficiency, implying that its basis is the first law of thermodynamics, the law of conservation of energy. Another potential but less widely used measure is based on the second law of thermodynamics, which governs how much of that heat can be converted to work. Given a maximum combustion temperature (usually limited by engine material considerations and by emission considerations), the second law postulates a maximum efficiency based on an idealized heat engine cycle called the Carnot cycle. The ratio of the "first law efficiency" to the Carnot cycle efficiency can be used as a measure of how efficiently a particular engine is operating with reference to the theoretical maximum based on the second law of thermodynamics. However, the most common measure of efficiency used by automotive engineers is termed *brake specific fuel consumption* (bsfc), which is the amount of fuel consumed per unit time per unit of power produced. In the United States, the bsfc of engines is usually stated in pounds of fuel per brake horsepower-hour, whereas the more common metric system measurement units are in grams per kilowatt-hour (g/kwh). The term 'brake' here refers to a common method historically used to measure engine shaft power. Of course, all three measures of efficiency are related to each other.

The efficiency of Otto and Diesel cycle engines are not constant but depend on the operating point of the engine as specified by its torque output and shaft speed (revolutions per minute, or RPM). Engine design considerations, frictional losses, and heat losses result in a single operating point where efficiency is highest. This maximum efficiency usually occurs at relatively high torque and at low to mid-RPM within the operating RPM range of the engine. At idle, the efficiency is zero because the engine is consuming fuel but not producing any useful work.

When considering automotive engine efficiency, the maximum efficiency need not, by itself, be an indicator of the average efficiency under normal driving conditions, because engine speed and torque vary widely under normal driving. The maximum efficiency of an engine is of interest to automotive designers, but a more practical measure of efficiency is its average efficiency during "normal" driving. In most developed countries, vehicle fuel economy figures provided to consumers are measured over a government-prescribed driving cycle, or cycles, which represent "normal" driving. In the United States, two cycles are used, one for city driving, at an average speed of about 32 km/h (20 mph) and one for highway driving, at an average speed of about 80 km/h (50 mph). These driving cycles provide convenient reference measures over which engine efficiency can be evaluated. A composite fuel economy, which is a weighted sum of city and highway fuel economy, is often used, and fuel efficiency benefits of technology improvements are with reference to the composite, unless otherwise stated. The engine operating points when the vehicle is driven on these reference cycles depends on the weight and size of the car, and how the engine is geared to the wheels. Hence, the same engine can display different average efficiencies over the driving cycle depending on the gear ratios employed and the power demanded by the vehicle. These considerations are important to the discussion on engine efficiencies that follows.

THEORETICAL ENGINE EFFICIENCY

The characteristics features common to all piston internal combustion engines are (a) intake and compression of the air or air-fuel mixture; (b) raising the temperature (and hence, the pressure) of the compressed air by combustion of fuel; (c) the extraction of work from the high-pressure products of combustion by expansion: (d) exhaust of the products of combustion. The four-stroke cycle requires two complete revolutions of the crankshaft or four up-and-down motions of the piston, to complete the entire cycle from intake to exhaust. The first engine to use this approach successfully was built by N.A. Otto in 1876. The two-stroke cycle, which requires only one revolution of the crankshaft to complete the cycle, was developed in 1878 by Sir Dugald Clerk. The Otto cycle engine is sometimes called the spark ignition (s.i.) because the combustion is initiated by an externally powered spark. It is also called the gasoline engine, because most spark ignition engines use gasoline, although many other liquid and gaseous hydrocarbon fuels can be used in this engine. Combustion of the homogenous air-fuel mixture takes place very quickly relative to piston motion, and is represented in idealized calculations as an event occurring at constant volume.

According to classical thermodynamic theory (1), the thermal efficiency (η) of an idealized Otto cycle, starting with intake air-fuel mixture drawn in at atmospheric pressure, is given by

$$\eta = 1 - 1/r^{n-1}$$

where r is the compression (and expansion) ratio and n is the ratio of specific heat at constant pressure to that at constant volume for the mixture. The equation shows that efficiency increases with increasing compression ratio.

Compression ratios are limited by the octane number of gasoline, which is a measure of its resistance to preignition, or "knock." At high compression ratios, the heat of compression of the air-fuel mixture becomes high enough to induce spontaneous combustion of small pockets of the mixture, usually those in contact with the hottest parts of the combustion chamber. These spontaneous combustion events are like small explosions that can damage the engine and reduce efficiency depending on when they occur during the cycle. Higher-octane-number gasolines prevent these events, but also cost more and require greater energy expenditure for manufacture at the refinery. The octane number is measured using two different procedures, resulting in two different ratings for a given fuel, called "motor octane" and "research octane" number. Octane numbers displayed at the pump are an average of research and motor octane numbers, and most engines sold in the United States require regular gasoline with a pump octane number of 87.

Using an n value of 1.4 for air, the equation predicts an efficiency of 58.47% at a compression ratio of 9 : 1, common in today's engines. A value of $n = 1.26$ is more correct for products of combustion of a stoichiometric mixture of air and gasoline. A stoichiometric mixture corresponds to an air-fuel ratio of 14.7 : 1, and this air-fuel ratio is commonly used in most cars sold in the United States today. At this air-fuel ratio, calculated efficiency is about 43.5%. Actual engines yield still lower efficiencies even in the absence of mechanical friction, because of heat transfer to the walls of the cylinder and the inaccuracy associated with assuming combustion to be instantaneous. Figure 1 shows the pressure-volume cycle of a typical spark ignition engine (1) and its departure from the ideal relationships.

The Diesel engine was developed by Rudolf Diesel about 20 years after Otto cycle engine was invented. The Diesel engine differs from the spark ignition engine in that only air, rather than the air-fuel mixture, is compressed. The diesel fuel is sprayed into the combustion chamber at the end of compression in a fine mist of droplets, and the diesel fuel ignites spontaneously upon contact with the compressed air because of the heat of compression. As a result, this engine is also referred to as a compression ignition (c.i.) engine. The sequence of pro-

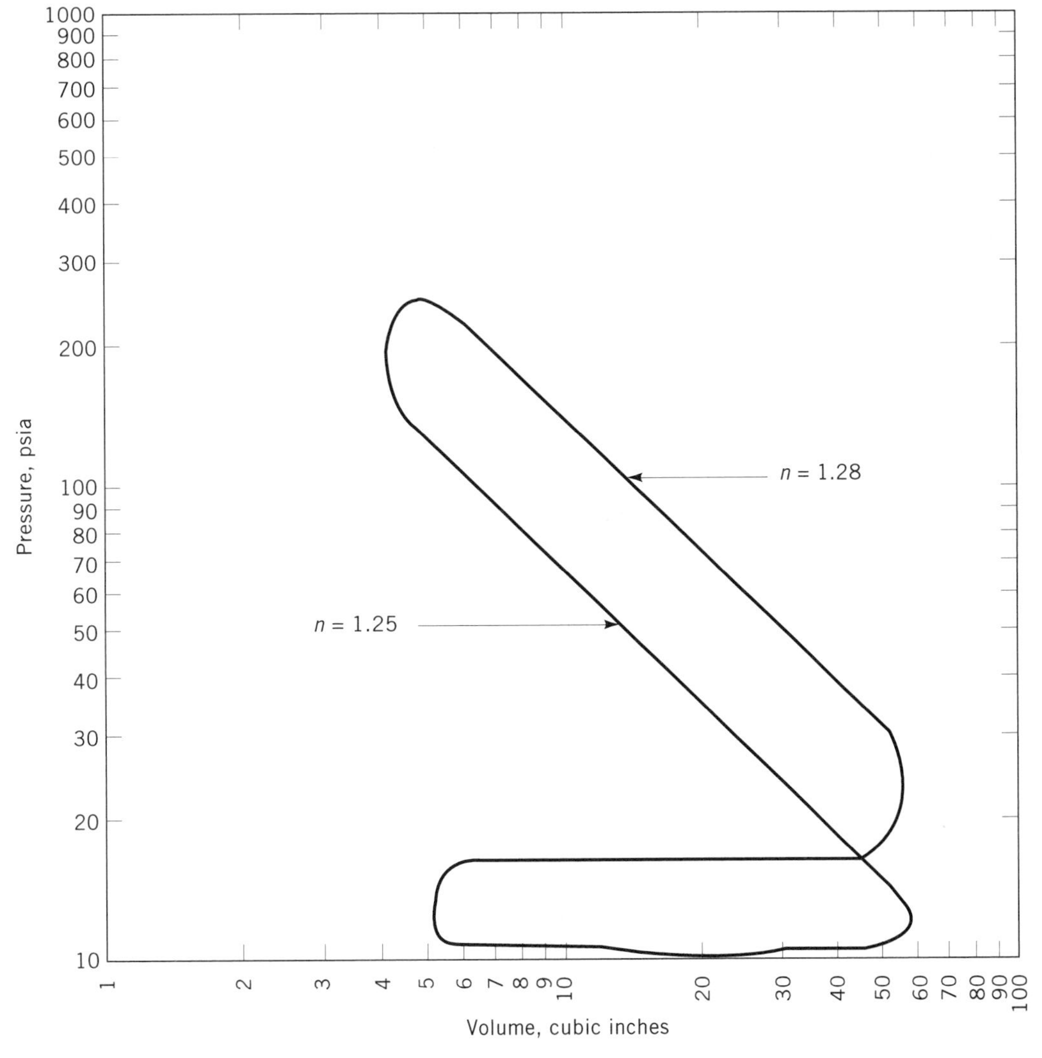

Figure 1. Pressure–volume diagram for a gasoline engine. Compression ratio = 8.7.

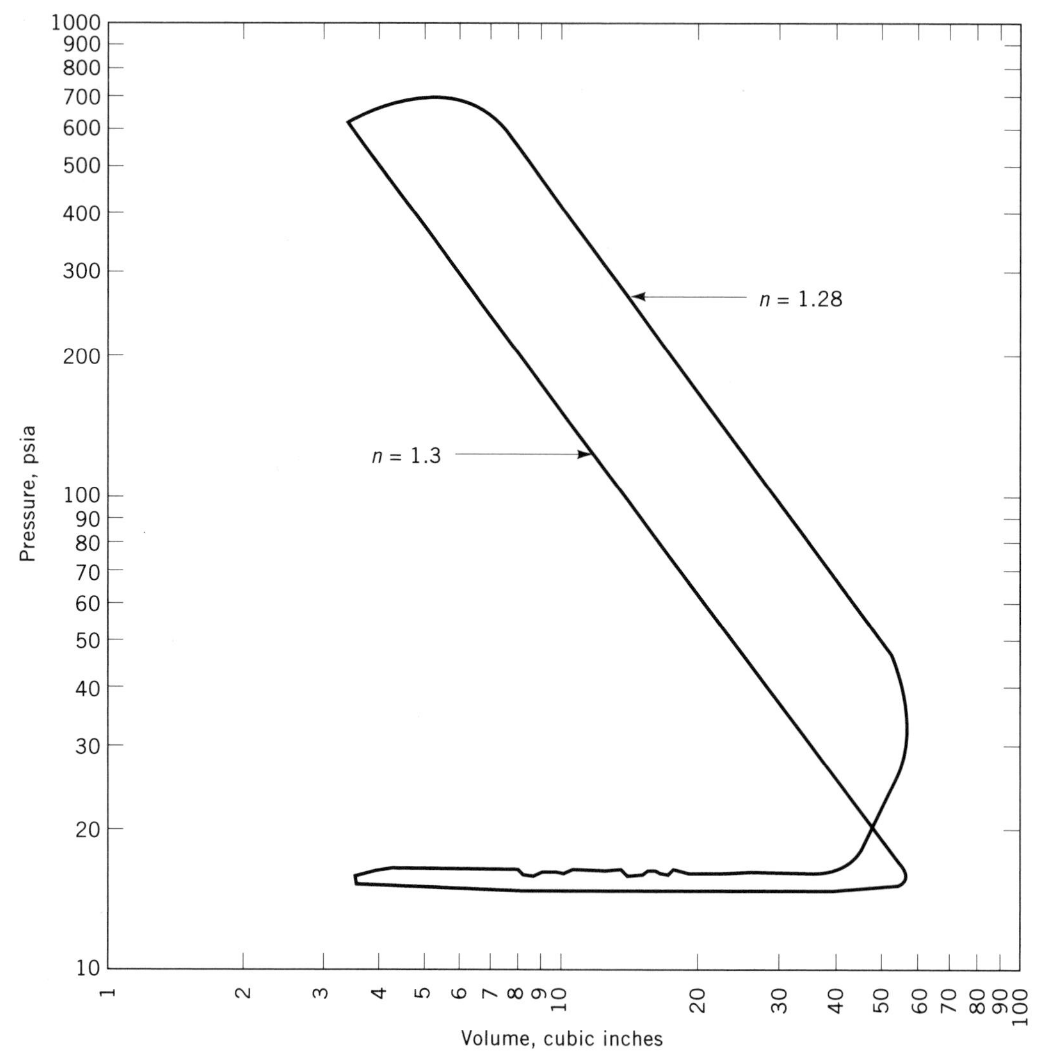

Figure 2. Pressure–volume diagram for a diesel engine. Compression ratio = 17.0.

cesses (ie, intake, compression, combustion, expansion, and exhaust) are similar to those of a Otto cycle. However, the combustion process occurs over a relatively long period and is represented in idealized calculations as an event occurring at constant pressure, that is, combustion occurs as the piston moves downward to increase volume and decrease pressure at a rate offsetting the pressure rise due to heat release. Figure 2 shows the pressure-volume cycles for a typical diesel engine and its relationship to the ideal Diesel cycle (2). If the ratio of volume at the end of the combustion period to the volume at the beginning of the period is r_c, or the "cut-off ratio," the thermodynamic efficiency of the idealized constant-pressure combustion cycle (3) is given by

$$\eta = \frac{1}{r^{n-1}}\left[\frac{r_c^n - 1}{n(r_c - 1)}\right].$$

It can be seen that for $r_c = 1$, the combustion occurs at constant volume and the efficiency of the Diesel and Otto cycle are equivalent.

The term r_c also measures the interval during which fuel is injected, and increases as the power output is increased. The efficiency equation shows that as r_c is increased, efficiency falls so that the idealized Diesel cycle is less efficient at high loads. The combustion process also is responsible for a major difference between Diesel and Otto cycle engines: in an Otto cycle engine, intake air is throttled to control power while maintaining near-constant air-fuel ratio; in a Diesel engine, power control is achieved by varying the amount of fuel injected while keeping the air mass inducted per cycle at near-constant levels. In most operating modes, combustion occurs with considerable excess air in a c.i. engine, whereas combustion occurs at or near stoichiometric air-fuel ratios in modern s.i. engines.

At the same compression ratio, the Otto cycle has the higher efficiency. However, Diesel cycle engines normally operate at much higher compression ratios, because there are no octane limitations associated with this cycle. In fact, spontaneous combustion of the fuel is required in such engines, and the ease of spontaneous combustion is measured by a fuel properly called cetane number. Most current c.i. engines require diesel fuels with a cetane number over 40.

In practice, there are two kinds of c.i. engine, the direct injection type (DI) and the indirect injection type (IDI). The DI type uses a system in which fuel is sprayed directly into the combustion chamber. The fuel spray is premixed and partially combusted with air in a prechamber in the IDI engine, before the complete burning of the fuel in the main combustion chamber occurs. DI engines generally operate at compression ratios of 15 to 20 : 1, whereas IDI engines operate at 18 to 23 : 1. For illustration, the theoretical efficiency of a c.i. engine with a compression ratio of 20 : 1, operating at a cutoff ratio of 2, is about 54% (for combustion with excess air, n is approximately 1.3). In practice, these high efficiencies are not attained, for reasons similar to those outlined for s.i. engines.

ACTUAL AND ON-ROAD EFFICIENCY

Four major factors affect the efficiency of s.i. and c.i. engines. First, the ideal cycle cannot be replicated because of thermodynamic and kinetic limitations of the combustion process, and the heat transfer that occurs from the cylinder walls and combustion chamber. Second, mechanical friction associated with the motion of the piston, crankshaft, and valves consumes a significant fraction of total power. Because friction is a stronger function of engine speed than torque, efficiency is degraded considerably at light-load and high-RPM conditions. Third, aerodynamic frictional losses associated with air flow through the air cleaner, intake manifold and valves, exhaust manifold, silencer, and catalyst are significant, especially at high air-flow rates through the engine. Fourth, pumping losses associated with throttling the air flow to achieve part-load conditions in spark ignition engines are very high at light loads. C.i. engines do not usually have throttling loss, and their part-load efficiencies are superior to those of s.i. engines. Efficiency varies with both speed and load for both engine types.

Hence, production spark ignition or compression ignition engines do not attain the theoretical values of efficiency, even at their most efficient operating point. In general, for both types of engines, the maximum efficiency point occurs at an RPM that is intermediate to idle and maximum RPM, and at a level that is 70% to 85% of maximum torque. "On-road" average efficiencies of engines used in cars and light trucks are much lower than peak efficiency, because the engines generally operate at very light loads during city driving and steady-state cruise on the highway. High power is used only during strong accelerations, at very high speeds or when climbing steep gradients. The high-load conditions are relatively infrequent, and the engine operates at light loads much of the time during normal driving.

Figures **3a** and **b** illustrate the brake specific fuel consumption (bsfc) maps for an s.i. and an IDI c.i. engine of late-1970s vintage (4). Lines of constant bsfc resemble "islands," with the lowest bsfc indicated in the central island. The spark ignition engine using gasoline is of 1.6 liters' displacement, and the diesel is of 2.1 liters' displacement. Both engines produce approximately the same power, 48 kw, and also have very similar torque ratings over the 1000 to 4500 rpm range. The c.i. produces less specific power (power per unit displacement) because it always operates with excess air; the brake specific fuel consumption of 260 g/kw-h corresponds to an efficiency of about 33%. In contrast, the lowest bsfc of 270 g/kw-h for the s.i. engine corresponds to an efficiency of about 32%. Although the peak efficiency is comparable, the part-load efficiency of the s.i. engine is significantly worse than that

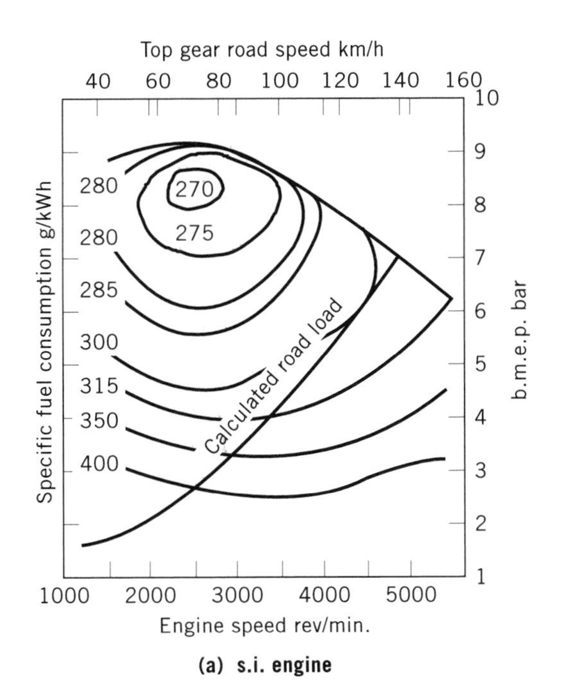

(a) s.i. engine

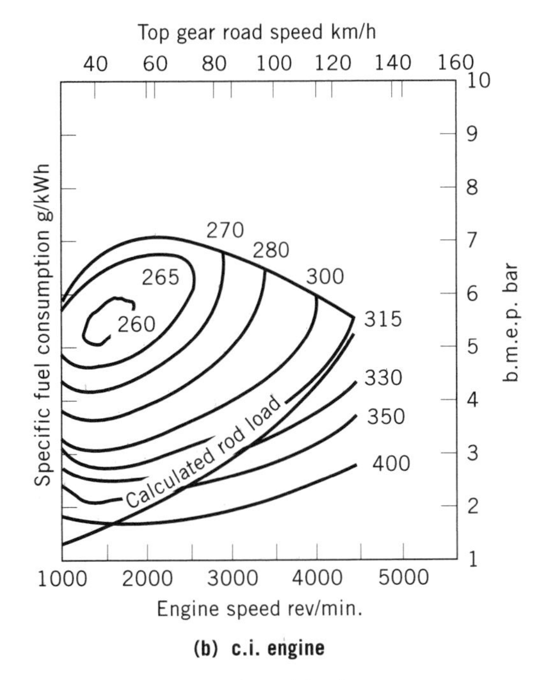

(b) c.i. engine

Figure 3. Fuel consumption map of engines. With calculated road load.

of the c.i. engine of equal power. The difference at part load is largely due to the effect of pumping loss associated with throttling in an s.i. engine, which more than compensates for the higher frictional loss in a c.i. engine.

THE SOURCES OF ENERGY LOSS

During normal driving, the heat of fuel combustion is lost to a variety of sources and only a small fraction is converted to useful output, resulting in the low values for on-road efficiency. Figure 4 provides an overview of the heat balance for a typical modern small car with a spark ignition engine under a low-speed (40 km/h) and a high-speed (100 km/h) condition (7). At very low driving speeds typical of city driving, most of the heat energy is lost to the engine coolant. Losses associated with "other waste heat" include radiant and convection losses from the hot engine block, and heat losses to the engine oil. A similar heat loss diagram for a Diesel c.i. would indicate lower heat loss to the exhaust and coolant and an increased fraction of heat converted to work, especially at the low-speed condition. During stop-and-go driving conditions typical of city driving, efficiencies are even lower than those indicated in Figure 4, because of the time spent at idle, where efficiency is zero. Under the prescribed U.S. city cycle conditions, typical modern spark ignition engines have an efficiency of about 18%, and modern IDI c.i. engines have an efficiency of about 21%.

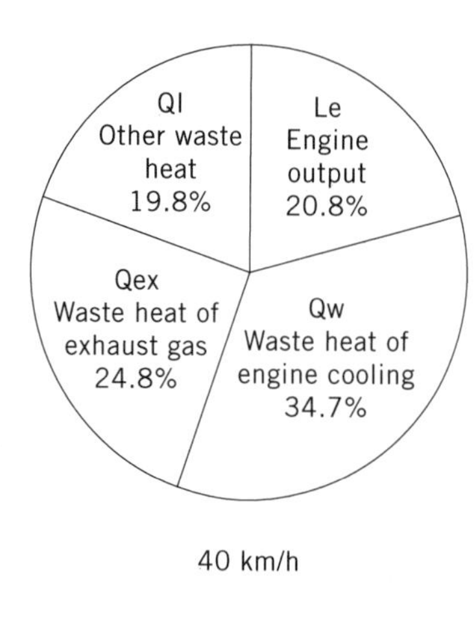

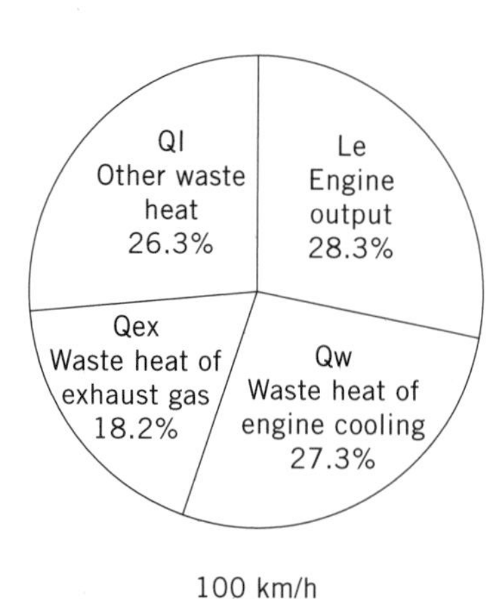

Figure 4. Heat balance of a passenger car equipped with 1500cc engine.

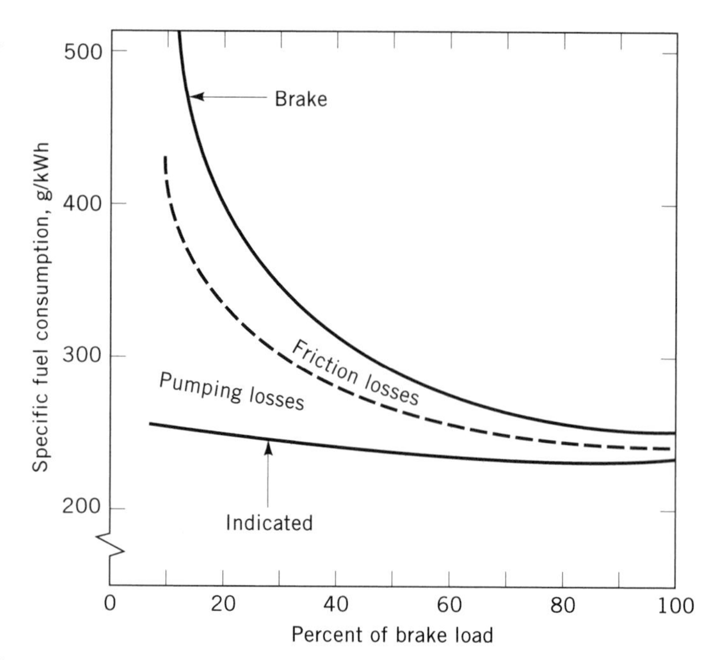

Figure 5. Specific fuel consumption vs engine load.

Another method of examining the energy losses is by allocating the power losses starting from the power developed within the cylinder. The useful work corresponds to the area that falls between the compression and expansion curve depicted in Figures 1 and 2.

The pumping work that is subtracted from this useful work, referred to as indicated work, is a function of how widely the throttle is open, and to a lesser extent, the speed of the engine. Figure 5 shows the dependence of specific fuel consumption (or fuel consumption per unit of work) with load, at constant (low) engine RPM (6). Pumping work represents only 5% of indicated work at full-load, low-RPM conditions, but increases to over 50% at light loads of less than two-tenths of maximum power.

Mechanical friction and accessory drive power on the other hand, increase non-linearly with engine speed, but do not change much with the throttle setting. Figure 6 shows the contribution of the various engine components as well as the alternator, water pump and oil pump to total friction, expressed in terms of mean effective pressure, as a function of RPM (7). The mean effective pressure is a measure of specific torque, or torque per unit of displacement; typical engine brake mean effective pressure (bmep) of spark ignition engines that are not supercharged range from 8.5 to 10 bar. Hence, friction accounts for about 25% of total indicated power at high RPM (~6000), but only for about 10% of indicated power at low RPM (~2000) in spark ignition engines. Friction in c.i. engines is higher because of the need to maintain an effective pressure seal at high compression ratios, and the friction mean effective pressure is 30% to 40% higher than that for a dimensionally similar s.i. engine at the same RPM. Because the brake mean effective pressure of a diesel is also lower than that of a gasoline engine, friction accounts for 15% to 16% of indicated maximum

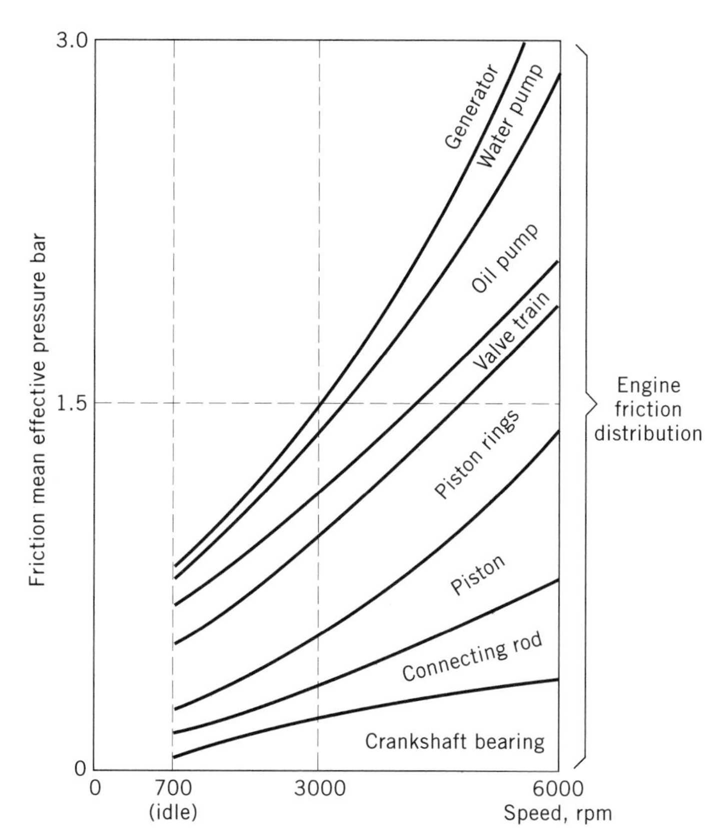

Figure 6. Friction distribution as a function of engine speed.

power even at 2000 RPM. Typical bmep of naturally aspirated c.i. engines range from 6.5 to 7.5 bar.

EVOLUTION OF SPARK IGNITION ENGINES

During the 1980s, most automotive engine manufacturers have improved engine technology to improve thermodynamic efficiency, reduce pumping loss, and decrease mechanical friction and accessory drive losses.

Design Parameters

Engine valvetrain design is a widely used method to classify spark ignition engines. The first spark ignition engines were of the side-valve type, but such engines have not been used in automobiles for several decades, although some engines used in off-highway applications, such as lawn mowers or forklifts, continue to use this design. The overhead valve (OHV) design supplanted the side-valve engine by the early 1950s, and continues to be used in many U.S. engines in a much-improved form. The overhead cam engine (OHC) is the dominant design used in the rest of the developed world. The placement of the camshaft in the cylinder heads allows the use of a simple, lighter valvetrain, and valves can be opened and closed more quickly as a result of the reduced inertia. This permits better flow of intake and exhaust gases, especially at high RPM, with the result that an OHC design can produce greater power at high RPM than an OHV design of the same displacement.

A more sophisticated version of the OHC engine is the double overhead cam (DOHC) engine, in which two separate camshafts are used to activate the intake and exhaust valves. The DOHC design permits a very light valvetrain, as the camshaft can actuate the valves directly without any intervening mechanical linkages. The DOHC design also allows some layout simplification, especially in engines that feature two intake valves and two exhaust valves (four-valve). The four-valve engine has become popular since the mid-1980s, Japanese manufacturers, in particular, have embraced the DOHC four-valve design. The DOHC design permits higher specific output than an OHC design, with the four-valve DOHC design achieving the highest ratings.

Typical specific output values for the different designs (8) in the early 1990s are as follows:

- OHV 40 to 45 BHP/liter
- OHC 50 to 55 BHP/liter
- DOHC 55 to 60 BHP/liter
- DOHC 4-valve 60 to 70 BHP/liter

A DOHC four-valve engine (9) has a maximum BMEP of up to 12 bar, and is expected to be the dominant engine design in the near future.

Thermodynamic Efficiency

Increases in thermodynamic efficiency within the limitations of the Otto cycle are obviously possible by increasing the compression ratio. However, compression ratio is also fuel octane limited, and increases in compression ratio depend on how the characteristics of the combustion chamber and the timing of the spark can be tailored to prevent knock while maximizing efficiency.

Spark timing is associated with the delay in initiating and propagating combustion of the air-fuel mixture. To complete combustion before the piston starts its expansion stroke, the spark must be initiated a few crank angle degrees ("advance") before the piston reaches top dead center. For a particular combustion chamber, compression ratio, and air-fuel mixture, there is an optimum level of spark advance for maximizing combustion chamber pressure and hence, fuel efficiency. This level of spark advance is called MBT for "maximum for best torque." However, MBT spark advance can result in knock if fuel octane is insufficient to resist preignition at the high pressures achieved with this timing. Hence, there is an interplay between spark timing and compression ratio in determining the onset of knock. Retarding timing from MBT reduces the tendency to knock but decreases fuel efficiency. Emissions of hydrocarbons and oxides of nitrogen (NO_x) are also dependent on spark timing and compression ratio, so that emission-constrained engines require careful analysis of the knock, fuel efficiency, and emission tradeoffs before the appropriate value of compression ratio and spark advance can be selected. In conventional two-valve spark ignition engines with a compression ratio of 8.5 to 9 : 1, spark timing close to MBT appears to be possible with regular gasoline having an octane rating of 87, even under the stringent 1993 emission regulations in force in the United States (8).

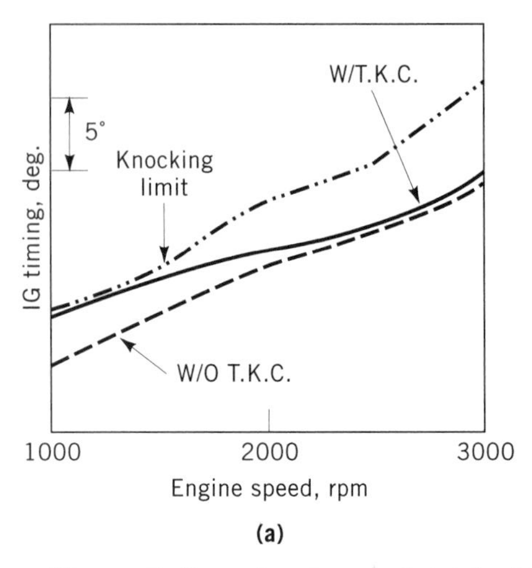

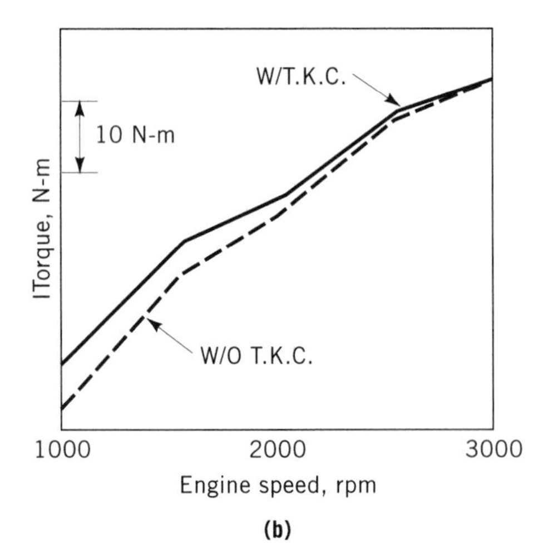

Figure 7. Trace knock control. (**a**) Ignition timing with/without trace knock control. (**b**) Effect of trace knock control on torque.

Electronic control of spark timing has made it possible to set spark timing closer to MBT relative to engines with mechanical controls. Because of production variability and inherent timing errors in a mechanical ignition timing system, the average value of timing in mechanically controlled engines had to be retarded significantly from the MBT timing so that the fraction of engines with higher-than-average advance due to production variability would be protected from knock. The use of electronic controls coupled with magnetic or optical sensors of crankshaft position has reduced the variability of timing between production engines and has also allowed better control during transient engine operation. More recently, engines have been equipped with knock sensors which are essentially vibration sensors tuned to the frequency of knock. These sensors allow for advancing ignition timing to the point where trace knock occurs, so that timing is optimal for each engine produced regardless of production variability. Figure 7 shows the ignition timing advance possible and the increase in torque with a trace knock controller in an engine of recent design. Electronic spark timing control has improved engine efficiency and fuel economy by 2% to 3% in the 1980–1990 period (10).

High-swirl, fast-burn combustion chambers have been developed during the 1980s to reduce the time taken for the air-fuel mixture to be fully combusted. The shorter the burn time, the more closely the cycle approximates the theoretical Otto cycle with constant volume combustion, and the greater the thermodynamic efficiency. Reduction in burn time can be achieved by having a turbulent vortex within the combustion chamber that promotes flame propagation and mixing. The circular motion of the air-fuel mixture is known as swirl, and turbulence is also enhanced by shaping the piston so that gases near the cylinder wall are pushed rapidly toward the center in a motion known as "squish." Recent improvements in flow visualization and computational fluid dynamics have allowed the optimization of intake valve, inlet port, and combustion chamber geometry to achieve desired flow characteristics. Typically, these designs have resulted in a 2% to 3% improvement in thermodynamic efficiency and fuel economy. The high-swirl chambers also allow higher compression ratios and reduced "spark advance" at the same fuel octane number. The use of these types of combustion chambers has allowed the compression ratio for two-valve engines to increase from 8:1 in the early 1980s to 9:1 in the early 1990s, and further improvements are likely.

Compression ratios in the future are likely to increase further with improved electronic control of spark timing and improvements in combustion chamber design. In newer engines of the four-valve DOHC type, the spark plug is placed at the center of the combustion chamber, and the chamber can be made very compact by having a nearly hemispherical shape. Engines incorporating these designs have compression ratios up to 10:1 while still allowing the use of regular 87 octane gasoline. Increases beyond 10:1 are expected to have diminishing benefits in efficiency and fuel economy (13) as shown in Figure 8, and compression ratios beyond 12:1 are not beneficial unless fuel octane is raised simultaneously.

Reduction in Mechanical Friction

Mechanical friction losses are being reduced by converting sliding metal contacts to rolling contacts, reducing the weight of moving parts, reducing production tolerances to improve the fit between pistons and bore, and improving the lubrication between sliding or rolling parts. Friction reduction has focused on the valvetrain, pistons, rings, crankshaft, crankpin bearings, and oil pump.

Valvetrain friction accounts for a larger fraction of total friction losses at low engine RPM than at high RPM. The sliding contract between the cam that activates the valve mechanism through a pushrod in an OHV design, or a rocker arm in an OHV design, can be substituted with a rolling contact by means of a roller cam follower, as illustrated in Figure 9. Roller cam followers have been found to reduce fuel consumption by 3% to 4% during city driving and 1.5% to 2.5% in highway driving (12). The use of lightweight valves made of ceramics or titanium is another possibility for the future. The lightweight valves reduce valve train inertia and also permit the use of lighter springs with lower tension. Titanium alloys are

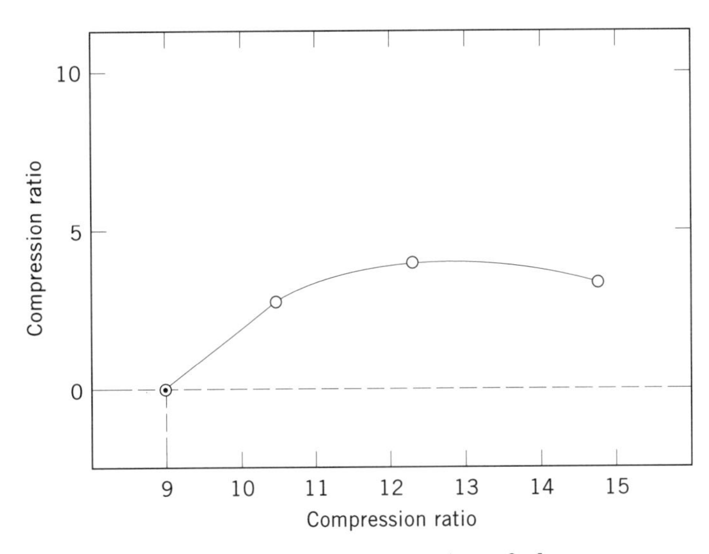

Figure 8. Effect of compression ratio on fuel economy.

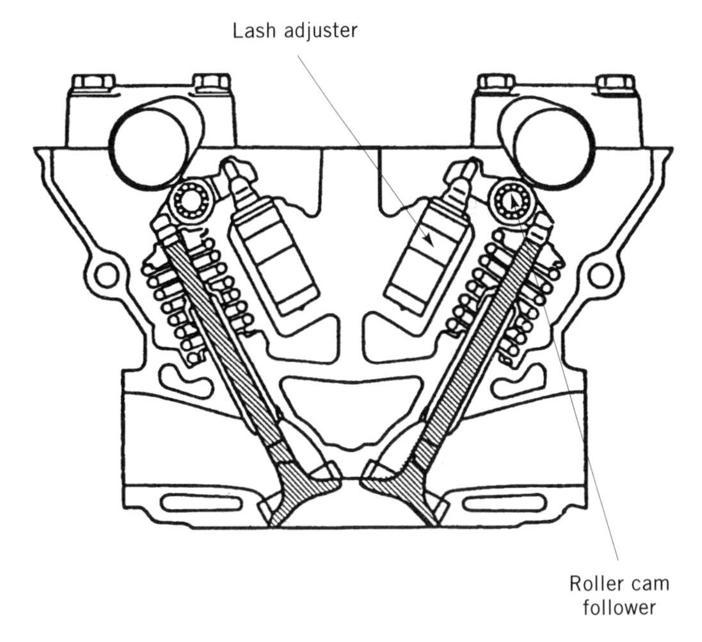

also being considered for valve springs that operate under heavy loads. There alloys have only half the shear modules of steel, and fewer coils are needed to obtain the same spring constant. A secondary benefit associated with lighter valves and springs is that the erratic valve motion at high RPM is reduced, allowing increased engine RPM range and power output (13).

The pistons and rings contribute to approximately half of total friction. The primary function of the rings is to minimize leakage of the air-fuel mixture from the combustion chamber to the crankcase, and oil leakage from the crankcase to the combustion chamber. The ring pack for most current engines is composed of two compression rings and an oil ring. The rings have been shown to operate hydrodynamically over the cycle, but metal-to-metal contact occurs often at the top and bottom of the stroke (14). The outward radial force of the rings are a result of installed ring tension, and contribute to effective sealing and friction. A wide variety of low-tension ring designs have been introduced in the 1980s especially since the need to conform to axial diameter variations or bore distortions has been reduced by improved cylinder manufacturing techniques. Reduced-tension rings have yielded friction reduction in the range of 5% to 10%, with fuel economy improvements of 1% to 2%. Elimination of one of the two compression rings has also been tried on some engines, and two-ring pistons may be the low-friction concept for the 1990s (15).

Pistons have also been redesigned to decrease friction. Before the 1980s, piston had large "skirts" to absorb side forces associated with side-to-side piston motion due to engine manufacturing inaccuracies. Pistons with reduced skirts diminish friction by having lower surface area in contact with the cylinder wall, but this effect is quite small. A larger effect is obtained for the mass reduction of a piston with smaller skirts, and piston skirt size has seen continuous reduction in the 1980s (Fig. 10). Reducing the reciprocating mass reduces the piston-to-bore loading; analytical results indicate that 25% mass reduction reduces friction mean effective pressure by 0.7 kPa at 1500 RPM (6). Secondary benefits include reduced engine weight and reduced vibration. Use of advanced materials also result in piston weight reduction. Current light-

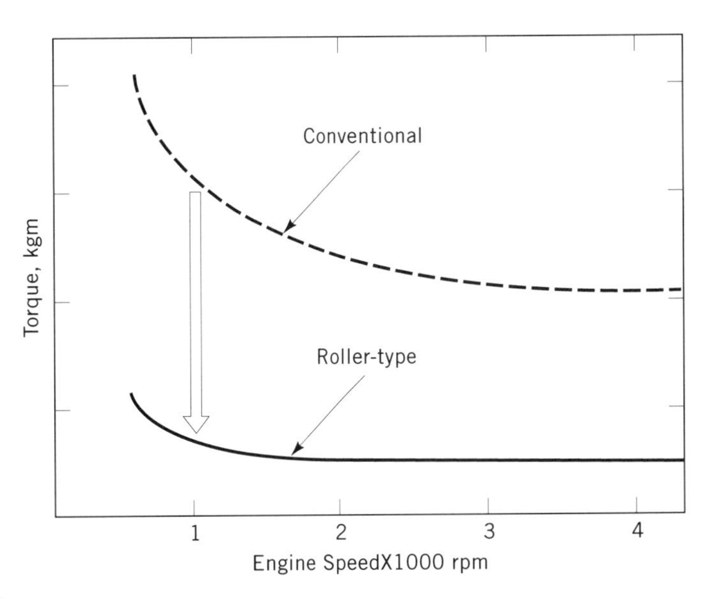

Figure 9. Roller cam follower.

weight pistons use hypereutectic aluminum alloys, and future pistons could use composite materials such as fiber-reinforced plastics. Advanced materials can also reduce the weight of the connecting rod, which also contributes to the side force on a piston.

The crankshaft bearings include the main bearings that support the crankshaft and the crankpin bearings, and are of the journal bearing type. These bearings contribute to about 25% of total friction while supporting the stresses transferred from the piston. The bearings run on a film of oil, and detailed studies of lubrication requirements have led to optimization of bearing width and clearances to minimize engine friction. Studies on the use of roller bearings rather than journal bearings in this application have shown that further reduction in friction is possible (16). Crankshaft roller bearings are currently

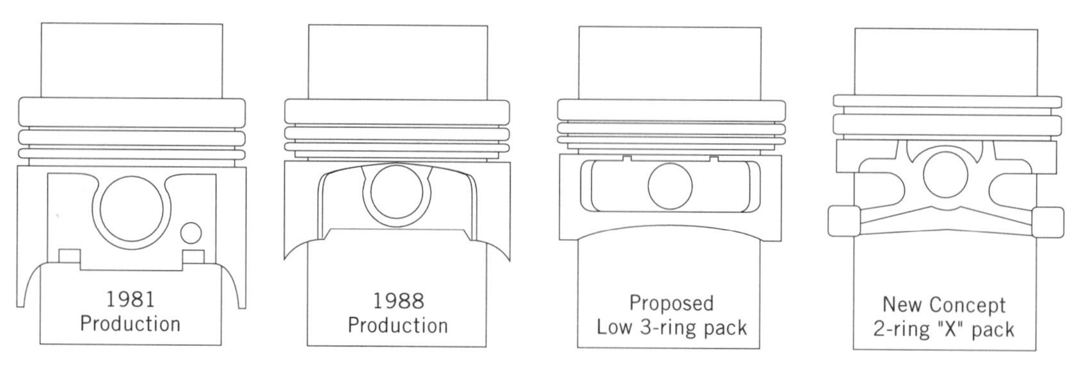

Figure 10. Evolution of piston design.

used only in some two-stroke engines, such as outboard motors for boat propulsion, but their durability in automotive applications has not been established. The use of roller bearings may contribute to a 2% to 3% improvement in fuel economy.

Coatings of the piston and ring surfaces with materials to reduce wear also contribute to friction reduction. The top ring, for example, is normally coated with molybdenum, and new proprietary coating materials with lower friction are being introduced. Piston coatings of advanced high-temperature plastics or resin have recently entered the market and are claimed to reduce friction by 5% and fuel consumption by 1% (17).

The oil pumps generally used in most engines are of the gear pump type. Optimization of oil-flow rates and reduction of the tolerances for the axial georotor clearance have led to improved efficiency, which translates to reduced drive power. Friction can be reduced by 2% to 3% with improved oil pump designs for a gain in fuel economy of about 0.5% (14).

Improvements to lubricants used in the engine also contribute to reduced friction and improved fuel economy. There is a relationship between oil viscosity, oil volatility, and engine oil consumption. Reduced viscosity oils traditionally resulted in increased oil consumption, but the development of viscosity index (VI) improvers had made it possible to tailor the viscosity with temperatures to formulate "multigrade" oils such as 10W-40 (these numbers refer to the range of viscosity covered by a multigrade oil). These multigrade oils act like low-viscosity oils during cold starts of the engine, reducing fuel consumption, but retain the lubricating properties of higher-viscosity oils after the engine warms up to normal operating temperature. The recent development of 5W-30 oils and 5W-40 oils can contribute to a fuel economy improvement by further viscosity reduction of 0.5% to 1% relative to the higher-viscosity 10W-40 oil. Friction modifiers containing molybdenum compounds have also reduced friction without affecting wear or oil consumption. Future synthetic oils combining reduced viscosity and friction modifiers can offer good wear protection, low oil consumption, and extended drain capability along with small improvements to fuel economy in the range of 1% to 1.5% over current 10W-40 oils (18).

Reduction in Pumping Loss

Reductions in flow pressure loss can be achieved by reducing the pressure drop that occurs in the flow of air (air-fuel mixture) into the cylinder, and the combusted mixture through the exhaust system. However, the largest part of pumping loss during normal driving is due to throttling, and strategies to reduce throttling loss have included variable valve timing and "lean-burn" systems.

The pressure losses associated with the intake system and exhaust system have been typically defined in terms of volumetric efficiency, which is a ratio of the actual airflow through an engine to the air flow associated with filling the cylinder completely. The volumetric efficiency can be improved by making the intake air-flow path as free of flow restrictions as possible through the air filters, intake manifolds, and valve ports. The shaping of valve ports to increase swirl in the combustion chamber can lead to reduced volumetric efficiency, leading to a tradeoff between combustion and volumetric efficiency.

More important, the intake and exhaust processes are transient, as they occur only over approximately half a revolution of the crankshaft. The momentum effects of these flow oscillations can be exploited by keeping the valves open for durations greater than half a crankshaft revolution. During the intake stroke, the intake valve can be kept open beyond the end of the intake stroke, because the momentum of the intake flow results in a dynamic pressure that sustains the intake flow even when the piston begins the compression stroke. A similar effect is observed in the exhaust process, and the exhaust valve can be held open during the initial part of the intake stroke. These flow momentum effects depend on the velocity of the flow, which is directly proportional to engine RPM. Increasing the valve opening duration helps volumetric efficiency at high RPM, but hurts it at low RPM. Increasing the exhaust valve opening time during part of the intake stroke when the intake valve is also open ("valve overlap") results in good high-RPM performance, but poor low-RPM performance when the exhaust gases lack sufficient momentum and are drawn back into the cylinder, diluting the intake change with burned gases, which can lead to rough combustion. Valve timing and overlap are selected to optimize the tradeoff between high and low RPM performance characteristics.

Improving Volumetric Efficiency

The oscillatory intake and exhaust flows can allow volumetric efficiency to be increased by exploiting resonance effects associated with pressure waves similar to those in organ pipes. The intake manifolds can be designed with pipe lengths that resonate, so that a high-pressure wave

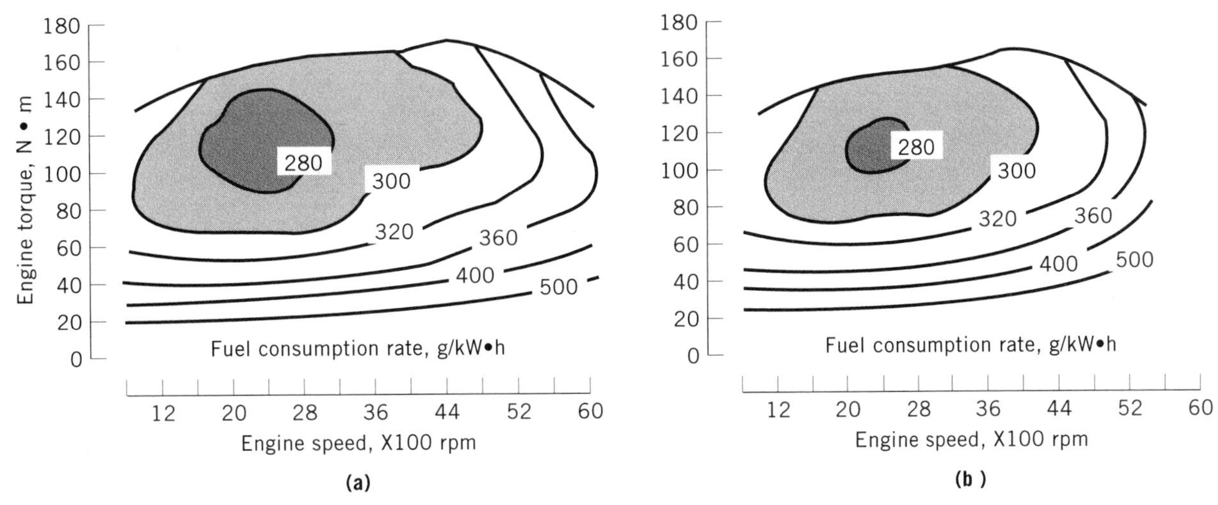

Figure 11. Fuel consumption map: Comparison of 4-valve **(a)** engine with 2-valve engine **(b)**.

is generated at the intake valve as it is about to close, to cause a supercharging effect. Exhaust manifolds can be designed to resonate to achieve the opposite pressure effect to purge exhaust gases from this cylinder. For a given pipe length, resonance occurs only at a certain specific frequency, and its integer multiples so that, historically, "tuned" intake and exhaust manifolds could help performance only in certain narrow RPM ranges. The incorporation of resonance tanks using the Helmholtz resonator principle, in addition to tuned length intake pipes, has led to improved intake manifold design that provide benefits over broader RPM ranges. More recently, variable resonance systems have been introduced, in which the intake tube lengths are changed at different RPM by opening and closing switching valves to realize smooth and high torque across virtually the entire engine speed range. Typically, the volumetric efficiency improvement is in the range of 4% to 5% over fixed resonance systems (10).

Another method to increase efficiency is by increasing valve area. A two-valve design is limited in valve size by the need to accommodate the valves and spark plug in the circle defined by the cylinder bore. The active flow area is defined by the product of valve circumference and lift. Increasing the number of valves is an obvious way to increase total valve area and flow area, and the four-valve system which increases flow area by 25% to 30% over two-valve layouts, has gained broad acceptance. The valves can be arranged around the cylinder bore and the spark plug placed in the center of the bore to improve combustion. While the peak efficiency or bsfc of a four-valve engine may not be significantly different from a two-valve engine, there is a broader range of operating conditions where low bsfc values are realized, as shown in Figure 11 (13). Analysis of additional valve layout designs that take into account the minimum required clearance between valve seats and the spark plug location suggests that five valve designs (three intake, two exhaust) can provide an additional 20% increase in flow area, at the expense of increased valvetrain complexity (19). Additional valves do not provide further increases in flow area either because of noncentral plug locations or valve-to-valve interference.

Efficiency improvements can be realized by changing the valve overlap period to provide less overlap at idle and low engine speeds, and greater overlap at high RPM. In DOHC engine, where separate crankshafts actuate the intake and exhaust valves, the valve overlap period can be changed by rotating the camshafts relative to each other. Such mechanisms have been commercialized in 1990–1991 by Nissan and Mercedes; these engines show low-RPM torque improvements of 7% to 10% with no sacrifice in maximum horsepower attained in the 5500 to 6000 RPM range. Variable valve overlap period is just one aspect of a more comprehensive variable valve timing system, as described in the following section.

Reduction in Throttling Loss

Under most normal driving conditions the throttling loss is the single largest contributor to reduction in engine efficiency. In s.i. engines, the air is throttled ahead of the intake manifold by means of a butterfly valve that is connected to the accelerator pedal. The vehicle's driver demands a power level by depressing or releasing the accelerator pedal, which in turn opens or closes the butterfly valve. The presence of the butterfly valve in the intake air stream creates a vacuum in the intake manifold at part throttle conditions, and the intake stroke draws in air at reduced pressure, resulting in pumping losses. These losses are proportional to the intake vacuum and disappear at wide-open throttle.

Measures to reduce throttling loss are varied. The horsepower demand by the driver can be satisfied by any combination of torque and RPM because

$$Power = Torque \times RPM$$

The higher the torque, the lower the RPM to satisfy a given power demand. Higher torque implies less throttling, and the lower RPM also reduces friction loss so that the optimum theoretical fuel efficiency at a given level of horsepower demand occurs at the highest torque level the engine is capable of. In practice, the highest level is never chosen because of the need to maintain a large reserve of torque for immediate acceleration, and also because engine vibrations are a problem at low RPM, especially near or below engine speeds refererd to as "lugging" RPM. Nevertheless, this simple concept can be exploited to the maximum by using a small displacement, high specific output engine in combination with a multispeed transmission with five or more forward gears. The larger number of gears allows selection of the highest torque/lowest RPM combination for fuel economy at any speed and load, while maintaining sufficient reserve torque for instantaneous changes in power demand. A specific torque increase of 10% can be used to provide a fuel economy benefit of 3% to 3.5% if the engine is downsized by 8% to 10%.

"Lean-burn" is another method to reduce pumping loss. Rather than throttling the air, the fuel flow is reduced so that the air-fuel ratio increases, or becomes "leaner." (In this context, the c.i. engine is a lean-burn engine.) Most s.i. engines, however, do not run well at air-fuel ratios leaner than 18 : 1, as the combustion quality deteriorates under lean conditions. Engines constructed with high swirl and turbulence in the intake charge can run well at air-fuel ratios up to 21 : 1. In a vehicle, lean-burn engines are calibrated lean only at light loads to reduce throttling loss, but run at stoichiometric or rich air-fuel ratios at high loads to maximize power. The excess air combustion at light loads has the added advantage of having a favorable effect on the polytropic coefficient (n) in the efficiency equation. Modern lean-burn engines do not eliminate throttling loss, but the reduction is sufficient to improve vehicle fuel economy by 8% to 10% (21). The disadvantage of lean-burn is that such engines cannot yet use catalytic controls to reduce emissions of oxide of nitrogen (NO_x), and the in-cylinder NO_x emission control from running lean is sometimes insufficient to meet stringent NO_x emissions standards. However, these are developments in "lean NO_x catalysts" that could allow lean-burn engines to meet the most stringent NO_x standards proposed in the future.

Another type of lean-burn s.i. engine is the stratified charge engine. Current research is focused on direct-injection stratified charge (DISC) engines in which the fuel is sprayed into the combustion chamber, rather than into or ahead of the intake valve. Typically, this enables the air-fuel ratio to vary axially or radially in the cylinder, with the richest air-fuel ratios present near the spark plug or at the top of the cylinder. Stratification requires very careful design of the combustion chamber shape and intake swirl, and of the fuel injection system. Laboratory tests have indicated the potential to maintain stable combustion at total air-fuel ratios as high as 40 : 1 (21). Maintaining stratification over a wide range of loads and speeds has been problematic in practice. As a result, DISC engines have not been commercialized, but remain an interesting possibility for the future.

Variable valve timing is another method of reducing throttling loss. By closing the intake valve early, the intake process occurs over a smaller fraction of the cycle, resulting in a lower vacuum in the intake manifold. It is possible to completely eliminate the butterfly valve that throttles air and achieve all part-load settings by varying the intake-valve opening duration. However, at very light load, the intake valve is open for a very short duration, and this leads to weaker in-cylinder gas motion and reduced combustion stability. At high RPM, the throttling loss benefits are not realized fully. Throttling occurs at the valve when the valve closing time increases raltive to the intake stroke duration at high speeds, because of the valvetrain inertia. Hence, throttling losses can be decreased by 80% at light-load, low-RPM conditions, but by only 40% to 50% at high RPM, even with fully variable valve timing (22).

Variable valve timing can also provide a number of other benefits, such as reduced valve overlap at light loads/low speeds (discussed earlier) and maximized output over the entire range of engine RPM. Fully variable valve timing can result in engine output levels of up to 100 BHP/liter at high RPM with little or no effect on low speed torque. In comparison to an engine with fixed valve timing that offers equal performance, fuel efficiency improvements of 7% to 10% are possible. The principal drawback has historically been the lack of a durable and low-cost mechanism to implement valve timing changes. Recently, Honda has commercialized a two-stage system in its four-valve/cylinder engines in which, depending on engine speed and load, one of two valve timing and lift schedules are realized for the intake valves. This type of engine has been combined with lean-burn to achieve remarkable efficiency in a small car (23). Fuel economy benefits of up to 15% are possible with such strategies.

Another simpler version of the system of variable valve timing simply shuts off individual cylinders by deactivating the valves. For example, an eight-cylinder engine can operate at light load as a four-cylinder engine (by deactivating the valves for four of the cylinders) and as a six-cylinder engine at moderate load. Such systems have also been tried on four-cylinder engines in Japan with up to two cylinders deactivated at light load. At idle, such systems have shown a 40% to 45% decrease in fuel consumption, and composite fuel economy has improved by 10% to 12% because both pumping and frictional losses are reduced by cylinder deactivation (24). However, the system has problems associated with noise, vibration, and emissions that have resulted in reduced acceptance in the marketplace.

EVOLUTION OF THE COMPRESSION IGNITION (DIESEL) ENGINE

Fuel Efficiency Relative to S.I. Engines

Compression ignition engines, commonly referred to as Diesel engines, are in widespread use. Most c.i. engines in light-duty vehicle applications are of the indirect-injection type (IDI), whereas most c.i. engines in heavy-duty vehicles are of the direct-injection type. In comparison to s.i. engines, c.i. engines operate at much lower brake mean effective pressures of (typically) about 7 to 8 bar at full load. Maximum power output of a c.i. engine is limited by the rate of mixing between the injected fuel spray and hot air. At high fueling levels, inadequate mixing leads to high black smoke, and the maximum horsepower is usually smoke limited for most current c.i. engines. Naturally aspirated diesel engines for light-duty vehicle use have specific power outputs of 25 to 35 BHP per liter, which is about half the specific output of a modern s.i. engine. However, fuel consumption is significantly better, and c.i.

engines are preferred over s.i. engines where fuel economy is important.

Because of the combustion process and the high internal friction of a c.i. engine, maximum speed is typically limited to less than 4500 RPM, which partially explains the lower specific output of c.i. engines. In light-duty vehicle use, an IDI engine can display between 20% and 40% better fuel economy depending on whether the comparison is based on engines of equal displacement or of equal power output in the same RPM range (4). The improvement is largely due to the superior part-load efficiency of the c.i. engine, as there is no throttling loss. At high vehicle speeds (>120 km/h), the higher internal friction of the c.i. engine offsets the reduced throttling loss, and the fuel efficiency difference between s.i. and c.i. engines narrows considerably.

Most of the evolutionary improvements for compression ignition engines in friction and pumping loss reduction are conceptually similar to those described for s.i. engines, and this section focuses on the unique aspects of c.i. engine improvements.

Design Parameters

C.i. engines have also adopted some of the same valvetrain designs as those found in s.i. engines. While most c.i. engines are of the OHV type, many recent European c.i. engines for passenger car use are of the OHC type. The c.i. engine is not normally run at high RPM, so the difference in specific output between an OHV and an OHC design is small. The OHC design does permit a simpler and lighter cylinder block casting, which is beneficial for overcoming some of the inherent weight liabilities. OHC designs also permits the camshaft to activate the fuel injector directly in "unit injector" designs, which are capable of high injection pressure and very fine atomization of the fuel spray. DOHC designs are not used as of 1993 although they are possible in future four-valve engines of the DI type.

Thermodynamic Efficiency

The peak efficiency of an IDI engine is comparable to or only slightly better than the peak efficiency of an s.i. engine, based on average values for engines in production. The contrast between theoretical and actual efficiency is notable; part of the reason is that the prechamber in the IDI diesel is a source of energy loss. The design of the prechamber is optimized to promote swirl and mixing of the fuel spray with air, but the prechamber increases total combustion time. Its extra surface area also results in more heat transfer into the cylinder head. As can be seen from Figure 12, the prechamber is connected to the main combustion chamber by a small passage (or passages), and the flow of hot, partially combustsed gas through these passages at high velocity promotes further mixing but also results in pressure losses. The main advantage of the prechamber is that it promotes smoother and more complete combustion and does not require a very-high-pressure fuel-injection system (4).

Direct-injection (DI) systems avoid the heat and flow losses from the prechamber by injecting the fuel into the

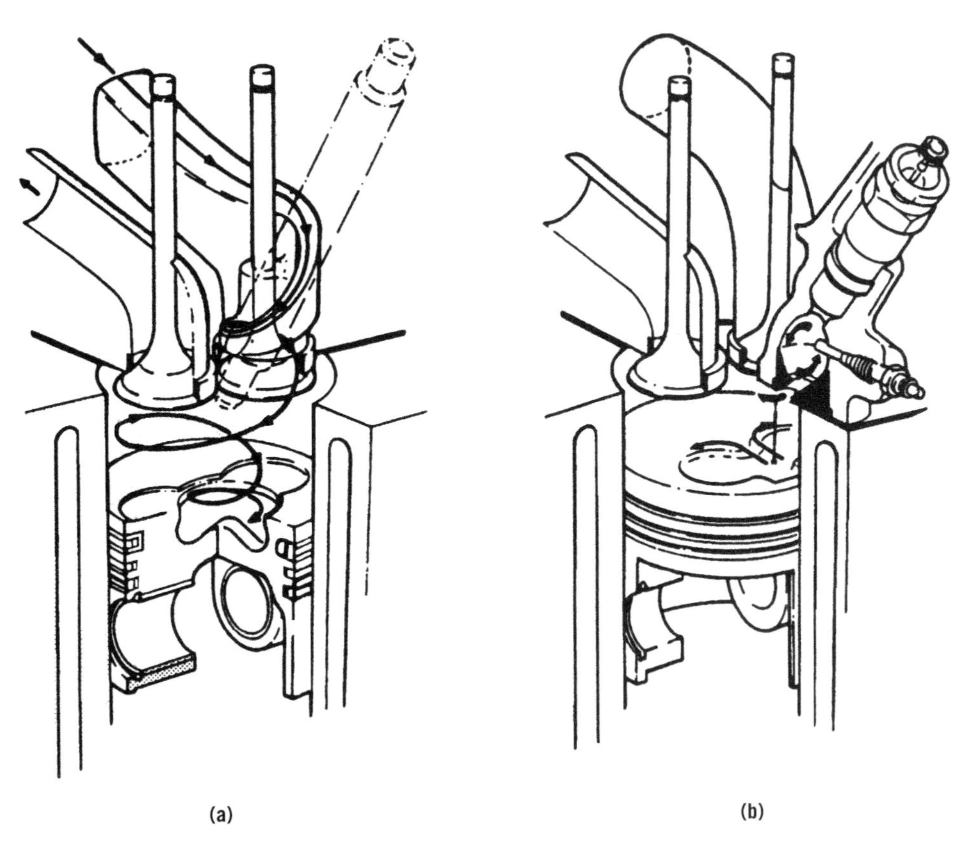

Figure 12. Comparison of combustion chambers of a DI **(a)** and IDI **(b)** engine.

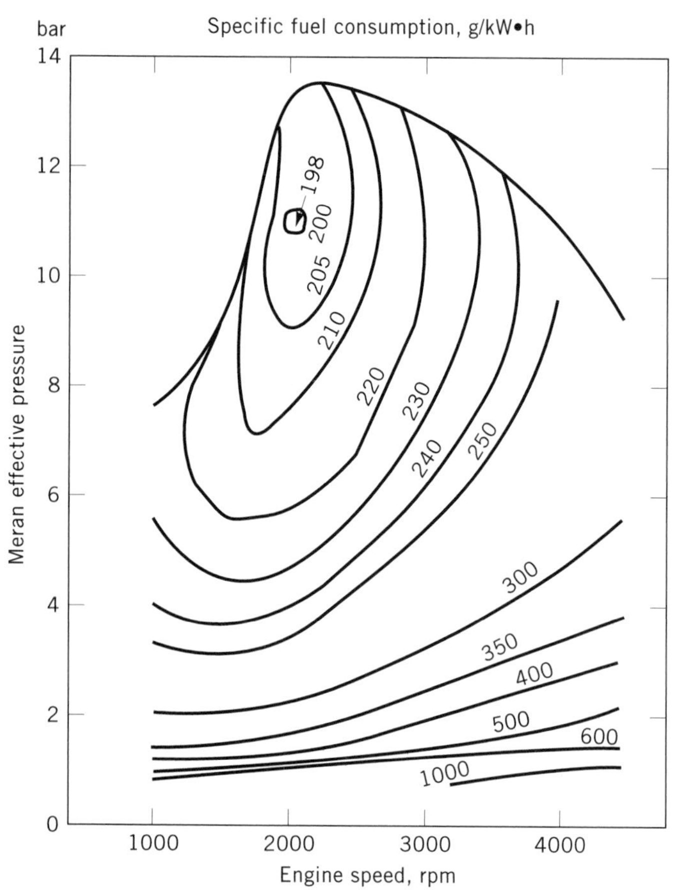

Figure 13. Fuel consumption map for a modern turbocharged DI diesel passenger car engine.

combustion chamber. The combustion process in DI diesels occurs in two phases. The first phase consists of an ignition delay period followed by spontaneous ignition of the fuel droplets. The second phase is characterized by diffusion burning of the droplets. The fuel-injection system must be capable of injecting very little fuel during the first phase, and providing highly atomized fuel and promoting intensive mixing during the second. Historically, the mixing process has been aided by creating high swirl in the combustion chamber to promote turbulence. However, high swirl and turbulence also lead to flow losses and heat losses, thus reducing efficiency. The newest concept is the "quiescent" chamber in which all the mixing is achieved by injecting fuel at very high pressures to promote fine atomization and complete penetration of the air in the combustion chamber. New fuel injection systems using "unit injectors" can achieve pressures in excess of 1000 bar, almost twice as high as injection pressures used previously. Quiescent combustion chamber designs with high-pressure fuel-injection systems have proved to be very fuel efficient and are coming into widespread use in heavy-duty truck engines. These systems have the added advantage of reducing particulate and smoke emissions (2).

DI engines have only recently entered the light-duty vehicle market, but these engines still use swirl-type combustion chambers. In combination with turbocharging, the new DI engines have attained peak efficiencies of 43%. Fuel economy improvements in the composite cycle relative to IDI engines are in the 12% to 15% range, and are up to 40% higher than naturally aspirated s.i. engines with similar torque characteristics (25). It is not clear if quiescent combustion chambers will ever be used in DI engines for cars, because the size of the chamber is quite small and fuel impingement on cylinder walls is a concern. The bsfc map of a modern high-speed automotive DI engine is shown in Figure 13.

Although the efficiency equation shows that increasing compression ratio has a positive effect on efficiency, practical limitations preclude any significant efficiency gain through this method. At high compression ratios, the size of the combustion chamber is reduced, and the regions of "dead" air trapped between the cylinder and piston edges and crevices became relatively large, leading to poor air utilization, reduced specific output, and, potentially, more smoke. Moreover, the stresses on the engine increase with increasing compression ratio, making the engine heavy and bulky. Currently, the compression ratios are already somewhat higher than optimal to provide enough heat of compression so that a cold start at low temperature is possible.

The "adiabatic diesel" has also received much attention. In this c.i. engine, the combustion chamber walls are insulated to prevent heat loss; the availability of low-thermal-conductivity ceramic materials has provided the impetus for this research. In such an insulated engine, the reduction in heat lost to the walls simply increases heat lost to exhaust gas, because the expansion ratio of a piston engine is fixed. Recovery of the waste heat of exhaust by a separate power turbine can provide significant gains in total efficiency. Unfortunately, the characteristics of turbomachinery are such that they recover energy only over a restricted range of operating conditions, typically conditions close to the maximum torque and RPM of the engine. Engines incorporating a waste heat recovery turbine, known as turbocompound engines, are best suited to heavy-duty vehicles where full power operation is common, but are not well suited to light-duty vehicles (26).

Insulated engines have other disadvantages that may prevent their commercialization. The hot cylinder walls heat the intake gases, reducing volumetric efficiency and specific power output from the piston engine. The high temperatures also lead to NO_x emission problems and problems with fuel ignition upon wall contact. Hence, theoretical efficiency advantages have not been attained in practice. However, some manufacturers have reported modest successes in insulating the prechamber in IDI diesels, although efficiency gains have been small (27).

Friction and Pumping Loss

Most of the friction-reducing technologies that can be adopted in s.i. engines are conceptually similar to those that can be adopted for diesels. There are limitations to the extent of reduction of ring tension and piston size owing to the high compression ratio of c.i. engines, but roller cam followers, optimized crankshaft bearings, and multigrade lubricants have also been adopted for c.i. engine use. Because friction is a larger fraction of total loss, a 10% reduction in fraction in a c.i. engine can lead to a 3% to 4% improvement in fuel economy.

Pumping losses are not as significant a contributor to overall energy loss in a c.i. engine, but tuned intake manifolds and improved valve port shapes and valve designs have also improved the volumetric efficiency of modern c.i. engines. Four-valve designs, in widespread use in the heavy truck market, have only recently appeared in passenger cars, but their benefits are smaller in c.i. engine use because of the low maximum RPM relative to s.i. engines. Nevertheless, the four-valve head with a centrally mounted injector is particularly useful in DI engines because it allows for symmetry in the fuel spray with resultant good air utilization.

Variable valve timing or any form of valve control holds little benefit for c.i. engines because of the lack of throttling loss or the need for high RPM performance. Valve timing can be varied to reduce the effective compression ratio so that a very high ratio can be used for cold starts, and a lower, more optimal, ratio can be used for fully warmed up operation. In future, improvements to volumetric efficiency and reduction in throttling loss are expected to contribute a 3% to 5% improvement in fuel economy.

INTAKE CHARGE BOOSTING

Most c.i. and s.i. engines for light vehicles use intake air at atmospheric pressure. One method to increase maximum power at wide-open throttle is to increase the density of air supplied to the intake by precompression. This permits a smaller displacement engine to be substituted without loss of power and acceleration performance. The use of a smaller displacement engine reduces pumping loss at part load and friction loss. However, intake charge compression has its own drawbacks. Its effect is similar in some ways to raising the compression ratio as regards peak cylinder pressure, but maximum cylinder pressure is limited in s.i. engines by the fuel octane. Charge-

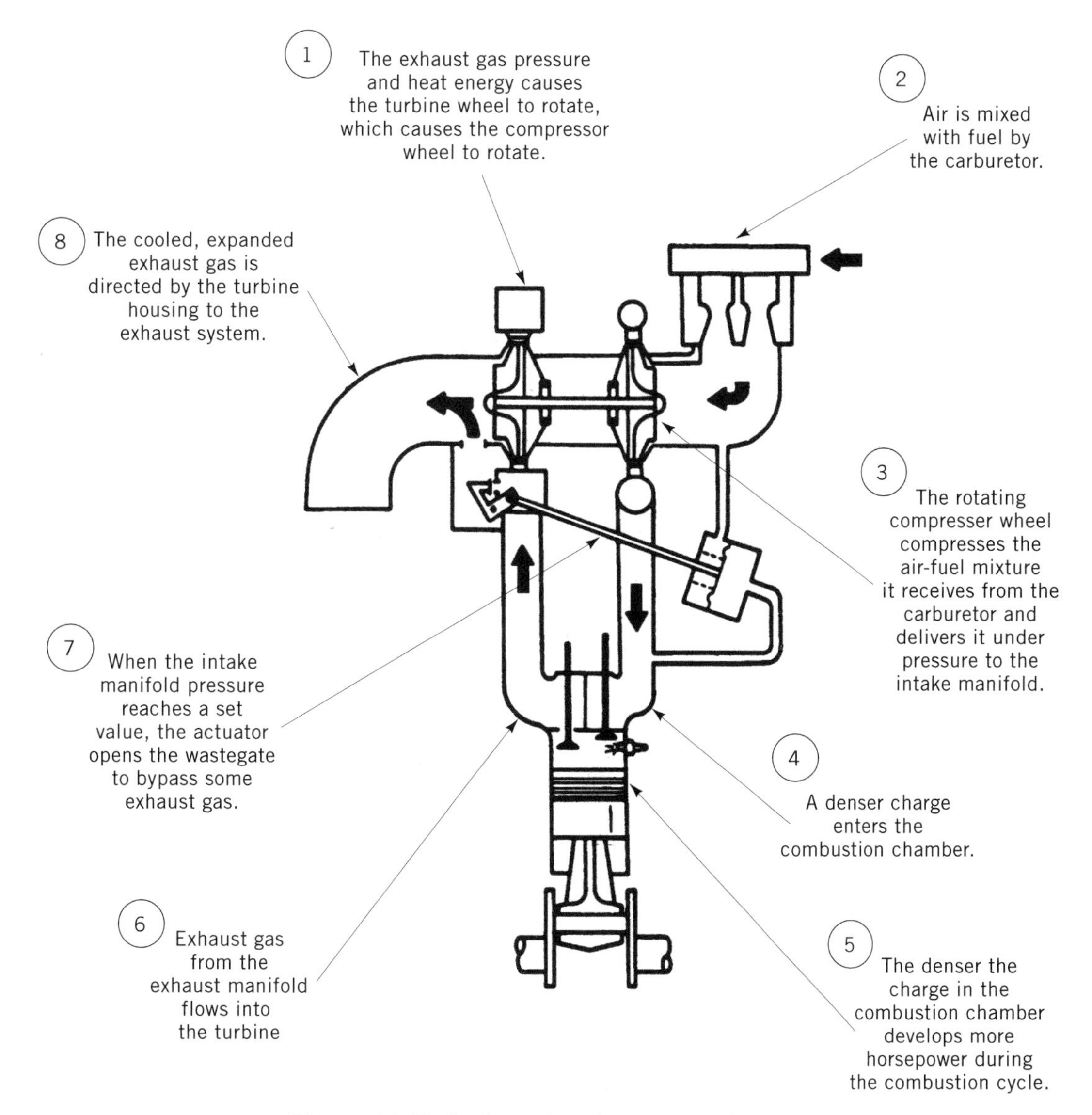

Figure 14. Turbocharged engine system schematic.

boosted engines generally require premium gasolines with higher octane number if the charge-boost levels are more than 0.2 to 0.3 bar over atmospheric in vehicles for street use. Racing cars use boost levels up to 1.5 bar in conjunction with a very high octane fuel such as methanol. This limitation is not present in a c.i. engine, and charge boosting is much more common in c.i. engine applications.

Intake charge boosting is normally achieved by the use of turbochargers or superchargers. Turbochargers recover the wasted heat and pressure in the exhaust through a turbine, which in turn drives a compressor to boost intake pressure. Superchargers are generally driven by the engine itself and are theoretically less efficient than a turbocharger. Many engines that use either device also use an aftercooler that cools the compressed air as it exits from the supercharger or turbocharger before it enters the s.i. engine. The aftercooler increases engine specific power output by providing the engine with a denser intake charge, and the lower temperature also helps in preventing detonation, or knock. Charge boosting is useful only under wide-open throttle conditions in s.i. engines, which occur rarely in normal driving, so that such devices are usually used in high-performance vehicles. In c.i. engines, charge boosting is effective at all speeds and levels.

Turbochargers

Turbochargers in automotive applications are of the radial flow turbine type. The turbine extracts pressure energy from the exhaust stream and drives a compressor that increases the pressure of the intake air (27). The process is shown schematically in Figure 14. There are a number of issues that affect the performance of turbomachinery, some of which are a result of natural laws governing the interrelationship between pressure, airflow, and turbocharger speed. Turbochargers do not function at light load because there is very little energy in the exhaust stream. At high load, the turbocharger's ability to provide boost is a nonlinear function of exhaust flow. At low engine speed and high load, the turbocharger provides little boost, but boost increases very rapidly beyond a certain flow rate that is dependent on the turbocharger size. The turbocharger also has a maximum flow rate, and the matching of a turbochargers' flow characteristics to a piston engines' flow requirements involves a number of tradeoffs (28). If the turbocharger is sized to provide adequate charge boost at moderate engine speeds of 2000 to 3000 RPM, high-RPM boost is limited and there is a sacrifice in maximum power. A larger turbocharger capable of maximizing power at high RPM sacrifices the ability to provide boost at normal driving conditions. At very low RPM (for example, when accelerating from a stopped condition), no practical design provides boost immediately. Moreover, the addition of a turbocharger requires the engine compression ratio to be decreased by 1.5 to 2 points (or 1 to 1.5 with an aftercooler) to prevent detonation. The net result is that turbocharged engines have *lower* brake specific fuel efficiencies than engines of equal size, but can provide some efficiency benefit when compared to engines of equal midrange or top-end power. During sudden acceleration, the turbocharger does not provide boost instantaneously because of its inertia, and turbocharged vehicles have noticeably different acceleration characteristics than naturally aspirated vehicles. This factor, coupled with the lack of low speed boost, has limited consumer acceptance of turbocharged s.i. engines to those primarily interested in very high performance.

Maximum boost pressure in an s.i. engine is limited by the fuel octane and engine compression ratio, and boost is capped at this level by bleeding the exhaust gas energy before it reaches the turbine. A turbocharged engine that emphasizes midrange boost can be substituted for a naturally aspirated engine that is 30% to 35% larger, and a net fuel economy gain of about 7% to 10% is possible because of pumping loss reduction. Turbocharged engines that maximize high-RPM power are not considered to be fuel efficiency enhancing because the lack of low-speed boost inherently limits that potential to downsize the engine without a large sacrifice in low-speed performance. Such designs can increase specific power by 40% to 50% at high RPM.

Turbochargers are much better suited to c.i. engines because these engines are unthrottled and the combustion process is not knock limited. Airflow at a given engine load/speed setting is always higher for a c.i. engine relative to an s.i. engine, and this provides a less restricted operating regime for the turbocharger. The lack of a knock limit also allows increased boost and removes the need to cap boost pressure under most operating conditions. In general, turbocharged c.i. engines offer up to 50% higher specific power and torque, and about 10% better fuel economy than naturally aspirated c.i. engines of approximately equal torque capability (26).

Superchargers

Most s.i. engine superchargers are driven off the crankshaft and are of the Roots blower or positive displacement pump type. In comparison to turbochargers, these superchargers are bulky and weigh considerably more. In addition, the superchargers are driven off the crankshaft, absorbing 3% to 5% of the engine power output depending on pressure boost and engine speed.

The supercharger, however, does not have the low-RPM boost problems associated with turbochargers, and also can be designed to nearly eliminate any time lag in delivering the full boost level. As a result, the superchargers are more acceptable to consumers from a drivability viewpoint. The need to reduce engine compression ratio and the supercharger's drive power requirement detract from overall efficiency. In automotive applications, a supercharged engine can replace a naturally aspirated engine that is 30% to 35% larger in displacement, with a net pumping loss reduction (30). Overall, fuel economy improves by about 8% or less, if the added weight effects are included.

Superchargers are less efficient in combination with c.i. engines, because these engines run lean even at full load, and the power required for compressing air is proportionally greater. Supercharged c.i. engines in passenger car applications are not commercially available, because the turbocharger appears far more suitable in these applications.

ALTERNATIVE HEAT ENGINES

A number of alternative engines types have been researched for use in passenger cars but have not yet proved successful in the marketplace. A brief discussion of the suitability of four engines for automotive power plants follows.

Wankel Engines

The Wankel engine is the most successful of the four engines in that it has been in commercial production in limited volume for two decades. The thermodynamic cycle is identical to that of a four-stroke engine, but the engine does not use a reciprocating piston in a cylinder. Rather, a triangular rotor spins eccentrically inside a figure eight–shaped casing. The volume trapped between the two rotor edges and the casing varies with rotor position, so that the intake, comparison, expansion, and exhaust stroke occur as the rotor spins through one revolution. The engine is very compact relative to a piston s.i. engine of equal power, and the lack of reciprocating parts provides very smooth operation. However, the friction associated with the rotor seals is high, and the engine also suffers from more heat losses than an s.i. engine. For these reasons, the Wankel engine's efficiency has always been below that of a modern s.i. piston engine (31).

Two-Stroke Engines

The two-stroke engine is widely used in small motorcycles but was considered too inefficient and polluting for use in passenger cars. A more recent development is the use of direct-injection stratified charge (DISC) combustion with this type of engine. One of the major problems with the two-stroke engine is that the intake stroke overlaps with the exhaust stroke, resulting in some intake mixture passing uncombusted into the exhaust. The use of a DISC design avoids this problem because only air is inducted during intake. Advanced fuel-injection systems have been developed to provide a finely atomized mist of fuel just before spark initiation and sustain combustion at light loads. The two-stroke engines of this type are thermodynamically less efficient than four-stroke DISC engines, but the internal friction loss and weight of a two-stroke engine are much lower than those of a four-stroke engine of equal power (32). As a result, the engine may provide fuel economy equal or superior to that of a DISC (four-stroke) engine when installed in a vehicle. Experimental prototypes have achieved good results, but the durability and emissions performance of advanced two-strokes is still not established.

Gas Turbine Engines

These engines are widely used to power aircraft, and considerable research has been completed to assess their use in automobiles. Such engines use continuous combustion of fuel, which holds the potential for low emissions and multifuel capability. The efficiency of the engine is directly proportional to the combustion temperature of the fuel, which has been constrained to 1200°C by the metals used to fabricate turbine blades. The use of high-temperature ceramic materials for turbine blades and the use of regenerative exhaust waste heat recovery were expected to increase the efficiency of gas turbine engines to levels significantly higher than the efficiency of s.i. engines (33).

In reality, such goals have not yet been attained, partly because the gas turbine components become less aerodynamically efficient at the small engine sizes suitable for passenger car use. Part-load efficiency is a major problem for gas turbines because of the nonlinear efficiency changes with air-flow rates in turbomachinery. In addition, the inertia of the gas turbine makes it poorly suited to passenger car applications, where speed and load fluctuations are rapid in city driving. As a result, there is little optimism that the gas turbine–powered car will be a reality in the foreseeable future.

Stirling Engines

These engines have held a particular fascination for researchers because the cycle closely approximates the ideal Carnot cycle, which extracts the maximum amount of work theoretically possible from a heat source. This engine is also a continuous combustion engine like the gas turbine engine. While the engine uses a piston to convert heat energy to work, the working fluid is enclosed and heat is conducted in and out of the working fluid by heat exchangers. To maximize efficiency, the working fluid is a gas of low molecular weight, such as hydrogen or helium. Prototye designs of Stirling engine have not yet attained efficiency goals and have had other problems, such as the containment of the working fluid (34). The Stirling engine is, like the gas turbine, not well suited to applications where the load and speed change rapidly, and much of the interest in this engine has faded in recent years.

BIBLIOGRAPHY

1. C. Lichty, *Combustion Engine Processes*, 7th ed., 1967.
2. M. J. Hower and co-workers, *The New Navistar T444E Direct Injector Turbocharged Diesel Engine*, SAE Paper 930269, March 1993.
3. H. R. Ricardo, *The High Speed Internal Combustion Engine*, 4th ed., 1953.
4. H. W. Barnes Moss, and W. M. Scott, *The High Speed Engine for Passenger Cars*, Institute of Mechanical Engineers Paper C15/75, 1975.
5. H. Omori and S. Ogino, *Waste Heat Recovery of Passenger Car Using a Combination of Rankine Bottoming Cycle and Evaporative Engine Cooling System*, SAE Paper 930880, March 1993.
6. J. H. Tuttle, *Controlling Engine Load by Means of Early Intake Valve Closing*, SAE Paper 820408, 1982.
7. FEV of America, *Spark Ignition Engine Development*, 1992.
8. Martin Marietta Energy Systems, *Documentation of the Attributes of Technologies to Improve Automotive Fuel Economy*, Contractor Report, 1993.
9. J. Abthoss and co-workers, *Diamler Benz 2.3 Litre, 16-valve High Performance Engine*, SAE Paper 841226, 1984.
10. T. Sakono and co-workers, *Mazda New Lightweight and Compact V6 Engines*, SAE Paper 920677, 1992.
11. K. Suzuki and M. Takimoto, "Fuel Economy Potential and Prospects," Presentation to the National Academy of Sciences, July 1991.

12. Automotive Engineering, *Technical Highlights of the 1987 Automobiles*, **94**(10), Volume (Oct. 1987).
13. Toyota Motors, *Toyota Engine Technology*, 1990.
14. J. T. Kovach, E. A. Tsakiris, and L. T. Wong, *Engine Friction Reduction for Improved Fuel Economy*, SAE Paper 820085, 1982.
15. Nissan Research and Development, *The SR18DI Engine*, Nissan Press Release, 1990.
16. Automotive Engineering, *Roller Bearings for I.C. Engines?* **95**(4) (1987).
17. A. Tanake, T. Sugiyama, and A. Kotani, *Development of Toyota JZ Type Engine*, SAE Paper 930881, 1993.
18. T. J. Cousineau, T. F. McDonnell, and D. G. Witt, *Second-Generation SAW 5W-30 Passenger Car Oils*, SAE Paper 8615, 1986.
19. K. Aoi, K. Nomura, and H. Matsuzaka, *Optimization of Multi-Valve, Four Cycle Engine Design*, SAE Paper 860032, 1986.
20. Honda Motors, *The VTEC-E Engine*, Honda Press Information, July 30, 1991.
21. M. Misumi, R. Thring, and S. Ariga, *An Experimental Study of a Low Pressure DISC Engine Concept*, SAE Paper 900653, 1990.
22. Y. Urata and co-workers, *A Study of Vehicle Equipped with Non-Throttling SI Engine with Early Intake Valve Closing Mechanism*, SAE Paper 930820, 1993.
23. *EPA Test Car List*, Honda Civic VX, 1994.
24. K. Hatano and co-workers, *Development of a New Multi-Mode Variable Valve Timing Engine*, SAE Paper 93078, 1993.
25. D. Stock and R. Bauder, *The New Audi 5-Cylinder Turbo Diesel Engine*, SAE Paper 900648, 1990.
26. C. Amman, "Promises and Challenges of the Low-Heat Rejection Diesel," GM Research Publication GMR-6188, 1988.
27. H. Kawamura, *Development Status of Isuzu Ceramic Engine*, SAE Paper 880011, 1988.
28. H. H. Dertian and co-workers, *Turbochanging Ford's 2.3 Liter Spark Ignition Engine*, SAE Paper 790312, 1979.
29. H. Hiereth and G. Withelm, *Some Special Features of the Turbocharged Gasoline Engine*, SAE Paper 790207, 1979.
30. *EPA Test Car List*, Vehicle Specifications, 1992.
31. H. A. Kuck, V. Fleischer, and W. Schnorbus, *VW's New 1.3L High Performance Supercharged Engine*, SAE Paper 860102, 1986.
32. C. Amman, *The Automotive Engine—A Future Perspective*, GM Research Publication GMR-6653, 1989.
33. K. Hellman and co-workers, *Evaluation of Research Prototype Vehicle Equipped with DISC Two Stroke Engines*, EPA Technical Report EPA/AA/CTAB 92-01, 1992.
34. C. Amman, *Why Not A New Engine?* SAE Paper 801428, 1980.

B

BATTERIES

George E. Blomgren
Eveready Battery Co., Inc.

Batteries are storehouses for electrical energy on demand. Common sizes range from large house-sized batteries for utility storage to several liter-sized batteries for starting, lighting, and ignition of vehicles to tiny coin- and button-sized cells for electronic applications that require only small amounts of capacity. The most important aspect of battery technology is that the chemicals involved in the battery reaction are converted directly into electrical energy in contrast to heat engines, which involve an intermediate thermal or combustion process to convert the chemical energy into electrical energy. The result of this direct conversion is that all the free energy of the chemical system is available for the conversion rather than being limited by the Carnot cycle, as are all heat engines. Furthermore, the conversion of electrical to mechanical energy is quite efficient because of the inherent simplicity and low friction of electrical motors. This has led to many applications for batteries, although the practical large-scale use of batteries in vehicle motive power has remained elusive. Conversion of electricity to light or sound is also efficient and easily controlled. As a result, many applications for batteries in lighting and the creation and reproduction of sound have also been developed.

There are, of course, losses in batteries as in any other energy conversion device. As a result, much of the history of battery technology has been concerned with finding ways to reduce these losses to as low a level as possible. All types of losses increase with current, so they become particularly important as the application demands higher current (see also FUEL CELLS).

The three main types of batteries are primary, secondary, and reserve. A primary battery is designed to be discharged only one time and then discarded. A secondary battery is rechargeable and can be used like the primary battery, then recharged and used again, repeating the cycle until the capacity fades or sudden loss of capacity occurs, due usually to an internal short circuit. A reserve battery involves a special construction that normally keeps active materials well separated until the time for use occurs; then an activation device readies the battery for use. This type of battery is designed for long storage before use. This chapter discusses the history of batteries, fundamental principles of battery science, and all three types of batteries, but will not include lead acid or other batteries of interest to vehicular and standby power applications. Fuel cells also will not be discussed here.

The term *battery* is usually used to mean one or more electrically connected galvanic cells, although some authors restrict the definition to more than one cell. Some other useful definitions follow.

Anode is the negative electrode of a primary cell and is always associated with oxidation or the release of electrons into the external circuit. In a rechargeable cell, the anode is the negative pole during discharge and the positive pole during charge.

Cathode is the positive electrode of a primary cell and is always associated with reduction or taking of electrons from the external circuit. In a rechargeable cell, the cathode is the positive pole during discharge and the negative pole during charge.

Electrolyte is a material that provides ionic conductivity between the positive and negative electrodes of the cell.

Separator is a physical barrier between the positive and negative electrodes to prevent direct shorting of the electrodes. Separators must be permeable to ions, but must not conduct electrons. They must be inert in the total environment.

Open-circuit voltage is the voltage measured across the terminals of the cell or battery when no external current is flowing. When measured on a single cell, it is usually close to the thermodynamic electromotive force (emf).

Closed-circuit voltage is the voltage measured across the terminals of the cell or battery when current is flowing into the external circuit.

Discharge is the operation of a cell when current flows spontaneously from the battery into an external circuit.

Charge is the operation of a cell when an external source of current reverses the electrochemical reactions of the cell to restore the battery to its original state of charge.

Internal resistance or *impedance* of the battery is the resistance or impedance of the battery to the flow of current. This resistance operates in addition to the resistance of the external load.

Electrochemical couple is the combination of the electrode reactions of the anode and cathode to form the complete galvanic cell. The number of electrons given up by the anode to the external circuit must be identical with the number of electrons withdrawn from the external circuit by the cathode.

Batteries have had both a positive and a negative impact on the environment. Lead, cadmium, and mercury are highly toxic elements that have frequently been discarded to land fills or incinerators. As individual sections of this chapter will show, this practice has changed greatly, and the use of some of these elements has also declined sharply. The positive aspect of their use has been as well-contained energy conversion devices that contribute much to the needs of mankind. This aspect is made emphatic by the requirement in the near future that zero emission vehicles be produced in the United States and the fact that only battery-powered electric vehicles qualify for this application. Also, many nonpolluting energy conversion devices such as photovoltaic systems require the concomitant use of rechargeable batteries for energy storage so the devices can be used at night or in cloudy conditions.

HISTORY OF BATTERIES

An excellent account of early battery history is given in the book by Cahoon and Heise (1), to which the reader is referred for details. Developments of battery science in the twentieth century are discussed in the Proceedings Volume of The Electrochemical Society (2). A brief discussion of the most important events, however, follows.

Galvani (3) concluded in 1786 that the effect of dissimilar metals on the nerves of frogs was electrical, and even though he mistakenly believed that the animal tissue was the source of the electricity, his name has been used to denote the galvanic current of the battery, and a unit cell of a battery is generally called a galvanic cell.

Allessandro Volta (4) discovered that a stack of alternating dissimilar metals with a paper layer between the metal bilayers gave rise to a source of electricity. He mistakenly believed that the simple fact of contact of dissimilar metals was the source of current and in 1800 announced what is now called a voltaic pile as the source of unfailing charge, or perpetual power. Volta's name has been applied to the intensive property describing the ability of the battery to do work (ie, the voltage of the battery). In addition, a series assembly of cells is sometimes called a voltaic pile.

Michael Faraday (5), in a long series of studies beginning in the 1830s, discovered that the chemical reaction at each electrode was the source of the electricity produced by a battery, and as a result, his name is used to describe electrode operation as a faradaic process. Faraday also coined the terms "electrode" as general term for a battery pole, "anode" for the negative pole of the battery, "cathode" for the positive pole of the battery, "anion" for the charge that migrates to the anode, "cation" for the charge that migrates to the cathode, "electrolyte" for the solution containing anion and cation, and "electrolysis" for the process occurring during passage of electrical current.

Batteries were nearly always used as the source of electricity for experiments in physics and chemistry for most of the remainder of the nineteenth century. Early batteries were voltaic piles, which consisted of a series of zinc and copper plates fastened back-to-back to form a high-voltage, low-current source. Between the plates was initially only a moistened pasteboard that must have had a very low conductivity because of adventitious ions. A great many other batteries were invented and developed in that century, but the inventions of the lead acid battery by Planté in 1859 (6) and the zinc-manganese dioxide cell by Leclanché in 1866 (7) were landmarks because of their importance as battery systems even to the present day.

The first half of the twentieth century saw the rise of the "dry cell," the invention of which is generally credited to Gassner (8). In this cell, the electrolyte is immobilized by starch or some other gelling agent. Because of the much greater ease of sealing such a cell, it permitted the advance to portable power; one of the major thrusts of present-day batteries. Rechargeable batteries were dominated by the lead acid system, which saw continuous development in this period, although the nickel-cadmium and nickel-iron cells were invented and developed as storage cells in the same period (see discussion under Rechargeable Batteries).

The small rechargeable battery did not really become a factor until the 1950s, when the applications of the German development of the nickel-cadmium system for wartime needs was translated into a battery for consumer applications (9). Also in the 1950s, the switch of zinc–manganese dioxide cells to alkaline electrolyte became a reality. The Ruben cell (a zinc-mercuric oxide battery with alkaline electrolyte) was also developed during World War II for electronic circuits used in military applications (10).

The second half of the twentieth century has seen the rise of many new battery systems. Several batteries are now based on the use of lithium as an anode material. Prominent new rechargeable batteries are based on metal hydride negative electrodes. A special new rechargeable battery that is touted by some as the herald of the future is based on carbon intercalated by lithium as the negative electrode. These batteries will be discussed further in the appropriate section of the article.

BASIC PRINCIPLES

Faraday discovered the relationship between current (I) and chemical reactions in that the amount of chemical change was found to be directly proportional to the quantity of charge passed (It), where t is the time, and to the equivalent weight of the active material:

$$w = MIt/nF \tag{1}$$

where w is the mass of the material (g) that reacts during the electrochemical reaction, M is the molecular weight of the material (g/mol), n the number of electrons involved in the electrode reaction (equiv/mol), and F is Faraday's constant (26.8 Ah/g-equiv).

In a very important relation that is derived from thermodynamics and Faraday's law (Eq. 1), the Gibbs free energy (ΔG) for the cell reaction is found to be directly related to the electromotive force (emf) or voltage (E) of the system:

$$\Delta G = -nFE = \Delta H - T\Delta S \tag{2}$$

where n is the number of electrons involved in the cell reaction as it is written to compute the Gibbs free energy. ΔH and ΔS are the enthalpy change and entropy change, respectively, for the reaction, and T is the absolute temperature in degrees Kelvin. A standard emf is defined by eq. 2 if the standard free energy at 25°C and unit activities for all reactants are used. This quantity is widely tabulated for electrode reactions, and the standard potential of combinations of couples can be calculated from these tabulations (11). The standard potentials are referred to the standard states for the reaction of hydrogen gas to form hydronium ion as the zero of potential. Details of the methods for using these tables can be found in (12). From eq. 2, the more negative the value of ΔG, the more useful is the work obtained from the reaction (from the definition of the free energy) and the larger is the voltage of the cell. Further development of the thermodynamic relationships gives the important equations

$$\Delta S = nF(\partial E / \partial T)_P \quad (3)$$

and

$$\Delta H = nF[E - T(\partial E / \partial T)_P] \quad (4)$$

which allows the determination of the reaction enthalpy, ΔH, and reaction entropy, ΔS, from the emf and its temperature dependence. Conversely, if the entropy and enthalpy are known (eg, from thermodynamic tables), the temperature behavior of the emf can be precisely predicted. The Nernst equation results from combination of the Van't Hoff isotherm with the relationship for the standard and total emf's, equation 2:

$$E = E^o - (RT/nF)\ln[\Pi A^s(products)/\Pi A^s(reactants)] \quad (5)$$

where ΠA^s(products) is the product of the activities of the products, each raised to its stoichiometric coefficient, and ΠA^s(reactants) is the corresponding product of activities for the reactants, and R is the gas constant. This equation allows the determination of activities for reactants and products or correspondingly to determine the cell emf when the activities are known. The Nernst equation is often used with concentrations, rather than activities, when the activity coefficients are unknown or are expected to be close to unity.

Battery efficiency is an important consideration in comparing various battery types and couples. The most fundamental efficiency is the thermodynamic efficiency given by

$$\varepsilon = \Delta G^o/\Delta H^o = 1 - T\Delta S^o/\Delta H^o \quad (6)$$

This is the highest efficiency that can be obtained from a given electrochemical couple and can be determined from thermodynamic tabulations of enthalpy and free energy. Many couples used in commercial batteries have thermodynamic efficiencies of 90% or higher whereas the heat engine efficiencies based on the Carnot cycle are seldom higher than 40%.

In actual operation, a cell performs with less than the thermodynamic efficiency because of irreversible losses involved in the electrode reactions (polarization). The overall electrochemical efficiency is defined as

$$Eff_{electrochem} = \int_{t=0}^{t=t} E'\, Idt / \Delta G \quad (7)$$

where E' is the closed-circuit terminal voltage when the net current i is flowing at time t. Depending on the type of discharge occurring, either or both E' and I may be functions of the time, t.

A battery is usually designed with one of the electrodes of smaller capacity than the other in order to achieve special properties, such as overcharge protection, that provide safe operation. Therefore, it is important to know the coulombic efficiency of the active material. This quantity is defined by

$$Eff_{coulombic} = \int_{t=0}^{t=t} Idt / Q \quad (8)$$

where Q is the quantity of coulombs (It) calculated from Faraday's law (Eq. 1). This efficiency depends on the current, design of the electrode, cut-off voltage, and other properties.

Heat analysis is an important aspect of battery design as well. The difference between the enthalpy and the free energy (T times the entropy) of the cell reaction is the reversible heat of the reaction. As noted previously in discussing the thermodynamic efficiency, this term is often quite small for practical batteries. An irreversible heat due to the polarization of the electrodes is always present. The total amount of heat released during operation of the battery is given by

$$q = T\Delta S + \int I(E - E')dt \quad (9)$$

where again I and E' may be functions of time. The heat is released inside the battery at the reaction site. Heat release can be a serious problem for high rate applications, and the battery designer must make provisions for heat dissipation in this situation. Thermal runaways and other catastrophic situations may result from a failure to accommodate battery heat.

The electrolyte phase plays a critical role in batteries. Ions created at one electrode must either precipitate, in which case a counter-ion of opposite charge must be provided in the reaction zone, or migrate to the opposite electrode where combination and/or precipitation with ions created at that electrode will occur. In aqueous batteries, an ionogenic salt, a Brönsted acid, or a Brönsted base is dissolved in water to create the electrolyte solution. In nonaqueous electrolytes, a strong electrolyte salt is usually dissolved in the nonaqueous liquid to form the electrolyte. The neutral acid, base, or salt dissociates to a greater or lesser extent into ions depending on several factors, such as the dielectric constant of the medium, specific solvation of the ions by solvent or additive molecules, and the strength and type of bonding in the neutral entity. The ions then carry the current through the electrolyte phase from one electrode to the other.

In solid electrolytes (which are ionic solids), the current may be carried by ions or by vacancies, depending on the nature of the solid. It is very important that no electronic current is carried by the electrolyte. This is normally not a problem for liquid electrolytes, because only a few unusual liquid solutions are electronic conductors, but many solids conduct both ions and electrons. Electrons that flow internally in this way are not available for the external circuit to do work, and represent a loss of efficiency of the cell. In time, during storage, all the battery capacity can be lost through this kind of internal short circuit.

Ionic conduction in the electrolyte obeys Ohm's law:

$$I = \kappa V \quad (10)$$

where I is the current in amperes and V is the voltage drop in volts across the electrolyte of conductivity, κ in S/cm. The conductivity is determined by the degree of ionic dissociation and the mobility of each kind of ion present, which is closely related to the viscosity of the medium and the availability of special conduction mechanisms. Conductivity ranges from about 1 S/cm in concentrated sulfuric acid in lead-acid batteries and 9 N KOH in alkaline batteries to as low as 1 mS/cm for organic solvent–based electrolytes in lithium batteries. Conductivity may be as low as 1 μS/cm for some solid electrolyte batteries.

The very high conductivity of aqueous acids and bases is due to a special conductance mechanism that allows the rapid transfer of protons from water molecule to water molecule or to hydroxyl ion in the case of bases. Water also solvates both anions and cations strongly, which tends to keep them apart as individual ions. In organic media, the cations are generally strongly solvated; the anions are not. The ions can approach each other closely and form ion pairs that act as neutral entities that do not move in the electric field of the electrolyte. Only a very few solids are capable of conducting ions sufficiently well for a battery operating at or near room temperature. Even these special solids must be used in very thin layers to keep the ohmic loss to a minimum. The time response to the conductance of the electrolyte is almost instantaneous. This is also true of other ohmic contributions to loss (or polarization), such as resistance of the electrodes and connectors, resistance of leads to the external circuit, and so on. Also, because the change in ohmic polarization depends on the total current, the effect increases steadily as the current increases. An even greater increase in ohmic polarization is observed if the resistance increases because of precipitation of product to block the electrolyte conduction or the formation of insulating products from conducting ones.

Another source of loss that has a slower response than the ohmic loss described previously is the so-called activation polarization. The source of this loss is the kinetic limitation of transferring charge across the electrode interface with the electrolyte. Because of the complexity of the origin and description of the activation polarization, reference is made to (13), which describes the effect. However, a simple relationship derived from the fundamental equations is the Tafel equation:

$$\eta = a - b \log I \tag{11}$$

where η is the polarization (defined as the departure from the equilibrium or open-circuit potential, $E' - E$) and i is the current density. While the origin of the equation was empirical, later theory has related the parameters a and b with two fundamental parameters i_0, the exchange current, and α, the transfer coefficient. The exchange current is a measure of the forward and reverse current at equilibrium (in which condition they are equal in value). The transfer coefficient is related to the transmission coefficient of absolute rate theory and represents the fraction of the potential drop that operates on the activated complex in the forward direction of the reaction. In particular, the coefficients a and b are related to these quantities by

$$a = (2.303RT/\alpha nF) \log i_o;\ b = 2.303RT/\alpha nF. \tag{12}$$

From a plot of the voltage versus the logarithm of the current, the linear region is selected to determine the coefficients a and b, which in turn yield the transfer coefficient and the exchange current. Because there are two electrodes, each one has characteristic values of these two parameters. Experiments in special cells called half cells, in which the test electrode dominates the behavior of the cell, must be carried out in order to analyze the activation polarization in detail. Activation polarization has an intermediate time of response to a change in the current and is usually in the range of microseconds to milliseconds. Because the variation of activation polarization with current is logarithmic, the effect is most pronounced at low- to intermediate-current levels. The activation polarization, however, is dependent on surface area (the current density, not the total current, is the independent variable). Therefore, near the end of discharge at a constant current or resistance, the activation polarization can become important even at higher current, because the surface area of the active material drops to low levels as it is exhausted by the discharge reaction.

The final type of polarization is related to the transport properties of the cell and is generally called concentration polarization. This behavior is the most difficult to analyze, because a complete solution of the hydrodynamic equations of diffusion and fluid flow is extremely complex. Because batteries have restricted fluid flow in most designs, however, the problem is most often treated only in terms of the diffusion of the ions within the electrolyte or in the solid phases of the active materials. In the simplest case, the ion in the electrolyte phase, which is necessary for the electrode reaction to occur, is brought to the electrode by electromigration (motion caused by the charge of the ion in the electric field within the electrolyte) and diffusion.

When the electrode is a sink for the ion, a depletion of the ion in the vicinity of the electrode occurs. When the electrode produces the ion in question, the electrode acts as a source of the ion. In either case, the polarization caused by the limiting transport process can be analyzed through the Nernst equation (eq. 13) to give

$$\eta_c = (RT/nF) \ln (C_e/C) \tag{13}$$

where C_e is the concentration of the ion at the electrode surface and C is the concentration in the bulk electrolyte. When the electrode reaction is a sink for the ion, the concentration at the electrode surface can tend to zero as the current is increased. This causes a limiting current situation in which the polarization tends to infinity as the concentration tends to zero, and is evidenced by a sharp drop in the cell potential. The current density (A/cm^2) at which the sharp drop occurs is called the limiting current density, i_l. In terms of this measurable current, the concentration polarization is given by

$$\eta_c = (RT/nF) \ln (1 - i/i_l). \tag{14}$$

When the electrode is a source of the ion, according to eq. 13, the concentration polarization increases gradually in time until a steady state is reached. The steady-state polarization also increases monotonically with an increase in current in this case, according to eq. 14, unless precipitation or some other process occurs. This behavior is difficult to see in most batteries because the other electrode polarization increases so dramatically when it begins to approach the limiting current situation. This situation is usually the ultimate limitation to the passage of current in most batteries, although the situation is quite complicated, as the internal temperature increases with the increasing irreversibility and the transport properties are thereby made more facile until some other event intervenes. As mentioned previously, the onset of concentration polarization is the slowest of the three types of polarization. In an ideal stirred electrolyte, a steady state can occur within milliseconds, but in a battery with a separator and porous electrodes, the time to steady state can be many minutes or even hours.

Theoretical analysis of the type discussed previously is complicated also by the fact that at least one of the electrodes in the batteries is porous and contains an additional conductor phase to minimize ohmic polarization and a binder phase (usually an electrically insulating polymer or cementation additive) to minimize delamination or shedding of the active material. Often, both electrodes are of this type. The unraveling of the complex voltage-current-time behavior of batteries is often facilitated experimentally by the inclusion of a reference electrode within the cell. This is an electrode that has a stable potential and is allowed to remain at its open-circuit value by the use of a very-high-impedance electrometer. Each operating electrode is then independently measured against the reference electrode while current passes through the cell, and many of the individual electrode effects can be sorted out. To determine the effect of ion transport, the placement of the reference electrode is sometimes varied in order to sample the electrolyte at different points, for example on either side of the separator, and sometimes deep within the porous electrode. Another diagnostic tool of considerable use has been variable frequency impedance measurements. This technique is valuable because of the differing time constants of the various types of polarization. Sometimes a reference electrode is included in impedance measurements to help sort out the contributions of the two electrodes.

In recent years, the development of semiempirical modeling methods has permitted the simulation of porous electrodes of the type used in batteries. The models require a number of physical assumptions and simplifications. However, the capability of high-speed digital computers to solve the nonlinear simultaneous differential equations derived for the types of polarization described previously for the models proposed has permitted the detailed simulations to be carried out successfully. This information has expanded the view of the operation of batteries and increased the awareness of battery designers to the importance of porous electrode parameters, such as porosity, pore size, and tortuosity, and how they interact with the local impedance of the electrodes. The models have shown that the current is not produced uniformly throughout the electrode except at very low current, but has a complex distribution that depends on time, current, and local properties such as electrolyte concentration, resistance, and so on. Reference (14) reviews the techniques of this kind of study.

Battery comparisons have undergone development in the face of the introduction of new systems with very different voltages from standard aqueous batteries. Battery scientists used to compare battery performance in terms of hours of service. However, it is now the practice to compare energies per unit of volume, called the energy density and measured in Wh/L, and per unit of weight, called specific energy and measured in Wh/kg, rather than simply the capacities. A full comparison is often made of the variation of specific energy with specific power (W/kg) and of the variation of energy density with power density (W/L). Plots of this type are sometimes called Ragone plots and reveal the power at which the energy begins to fall off rapidly, usually due to the approach of a limiting current or other abrupt failure modes.

There are other important criteria. Some are market-driven, such as cost per unit of energy or power, capital cost of rechargeable cell and charger versus the cost of a primary cell, and suitability of the capabilities of the cell to the application. The active materials and cell components in primary batteries must be stable over time (five years or more) in the operating environment to provide adequate value and safety. Rechargeable cells may suffer self-discharge during storage and are returned to their starting state by the charging process. Some criteria are driven by customer preference or convenience, such as primary versus rechargeable, difficulty and sophistication for a given charging regime, or primary cell capacity (generally large) versus the necessity of frequent recharging but overall longer life of a rechargeable cell. Another criterion is legislative: some battery materials are regarded by some legislative bodies as too undesirable to be used without collection and special disposal and recycling techniques, such as the lead in lead-acid batteries.

Temperature range of operation and storage life is important for the particular climatic condition for use of the batteries. For consumers, the range is generally from $-10°$ to $+50°C$, and military requirements may range from $-50°$ to $+75°C$. For rechargeable systems, some special requirements apply, such as the cycle life and how well the capacity is maintained from cycle to cycle, overcharge ability, cell reversal, and special charger requirements. Finally, safety is the most important single criterion for a battery system and must be carefully evaluated for any new battery. The battery should not leak, vent hazardous materials, or undergo violent reactions under any condition of use or even under abuse conditions likely to be encountered by an uninformed user.

PRIMARY CELLS

As noted previously, primary cells are galvanic cells designed to be discharged only once. Manufacturers of primary cells strongly recommend that consumers should not attempt to recharge them because of possible safety hazards, such as leakage or gas generation causing venting. The cells are designed to have the maximum possible energy in each cell size because of the single discharge, so that comparisons between battery types on the basis of the energy density in Wh/L, and the specific energy, in Wh/kg, are especially apt.

The main categories of primary cells are carbon-zinc cells (heavy duty and general purpose), alkaline cells (cylindrical and miniature), and lithium cells.

The breakdown in sales of primary cells in the United States in 1992 by category was alkaline, 80.7%; carbon-zinc, 14.9% (12.3% heavy duty, 2.6% general purpose); others (including specialty and lithium), 4.4% (15). The total primary cell market in that year was $3,084 million (an annual growth rate of 5.9% averaged over five years) with a unit volume of 2,811 million (an average annual growth rate of 4.5% over the same period), including rechargeable small sealed cells. The growth of sales has been tied mostly to growth in the electronics industry, although the sales in battery-operated toys has also shown substantial growth. The market in other parts of the world reflects the historical dominance of carbon-zinc cells. For example, in 1991 in Japan, about 40% of all sales were carbon-zinc cells, about 25% were alkaline, about 26% were lithium cells, and 9% were silver oxide cells (16). The high proportion of lithium cells reflects the important photographic market in Japan. Western European sales are similar to the Japanese sales, whereas the sales in less developed countries are almost totally dominated by carbon-zinc batteries.

A number of multinational suppliers of batteries have manufacturing facilities in many countries and a broad line of products. These companies account for a major share of the batteries manufactured in the world. Table 1 lists the companies that have sales of primary batteries of greater than $100 million, the regions of their headquarters, and their product lines.

Table 1. Battery Companies Manufacturing Primary Cells

Company	Cell Type		
	Carbon–Zinc	Alkaline	Lithium
North America			
Duracell International		x	x
Eastman Kodak		x	
Eveready Battery Co.	x	x	x
Rayovac Corporation	x	x	x
European Economic Community			
SAFT			x
Varta Batterie AG	x	x	x
Far East			
Gold Peak	x		
Matsushita	x	x	x
Sanyo	x	x	x
Toshiba	x	x	x
Yuasa	x	x	x

Carbon–Zinc Cells

Carbon–zinc batteries are the most commonly found primary cells worldwide and are produced in every major country. They traditionally have a carbon rod (for cylindrical cells) or a carbon-coated plate (for flat cells) to collect the current at the cathode with a zinc anode, which is also used as the primary container of cylindrical cells, and it is this combination that has led to their name. There are two basic verions, the Leclanché cell and the zinc chloride or heavy-duty cell. Both types have zinc anodes and manganese dioxide cathodes and include zinc chloride in the electrolyte. The Leclanché cell also has an electrolyte saturated with ammonium chloride (additional undissolved ammonium chloride is usually added to the cathode) whereas the zinc chloride cell has at most a small amount of ammonium chloride added to the electrolyte. Both types are dry cells in the sense that there is no excess liquid electrolyte in the system. The zinc chloride cell is often made with synthetic manganese dioxide and gives higher capacity than the Leclanché cell, which uses inexpensive natural manganese dioxide for the active cathode material. Because MnO_2 is only a modest conductor, the cathodes in both types of cell contain 10% to 30% carbon black in order to distribute the current. Because of the ease of manufacture and the long history of the cell, the system can be found in many sizes and shapes.

Leclanché cells are usually made with paste separators, in which electrolyte solution and various types of cereal are cooked until thick, although some designs use a cold set paste. The paste is metered into the cell and the prepressed cathode body is inserted, forcing the paste into a separating layer. Figure 1 shows a cutaway view of a typical pasted cylindrical cell. Paper-lined cells have a starch- (or modified starch-) coated paper separator, which is much thinner and more conductive than the starch paste separator. This type of separator is typically used in premium Leclanché cells and zinc chloride cells. In these cells, the separator is inserted into the zinc can and the cathode mix is extruded into the can, followed by insertion of the carbon rod into the cathode mix. Figure 1 also shows a cutaway of a typical paper-lined cell. Flat cells are used to stack in multicell batteries, most commonly for 6- or 9-volt batteries. Figure 2 shows a cutaway of a typical flat cell, as used in a 9-volt battery. Note that in this cell, the zinc plate is carbon coated to serve as the collector for the cathode of the cell beneath when the cells are stacked.

Recent work on the mechanism of carbon-zinc cell reactions has been summarized by Tye (17). Considerable insight has been gained, particularly for the equilibrium behavior, although there are still unresolved questions regarding the mechanism and the dynamics of this long-studied system.

The electrochemical behavior of electrolytic (EMD), chemical (CMD), or natural (NMD) manganese dioxide are slightly different because of structural, composition, and surface differences among the materials. Battery-grade NMD is most commonly in the form of the mineral nsutite, which is a structural intergrowth of the minerals ramsdellite and pyrolusite (18). Occasionally, the pyrolusite or cryptomelane form is used in batteries; the pure ramsdellite form is quite rare and expensive so is not

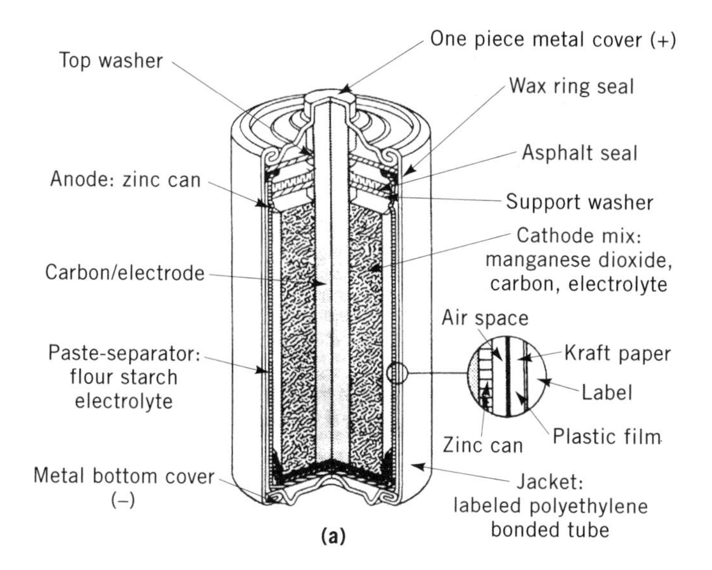

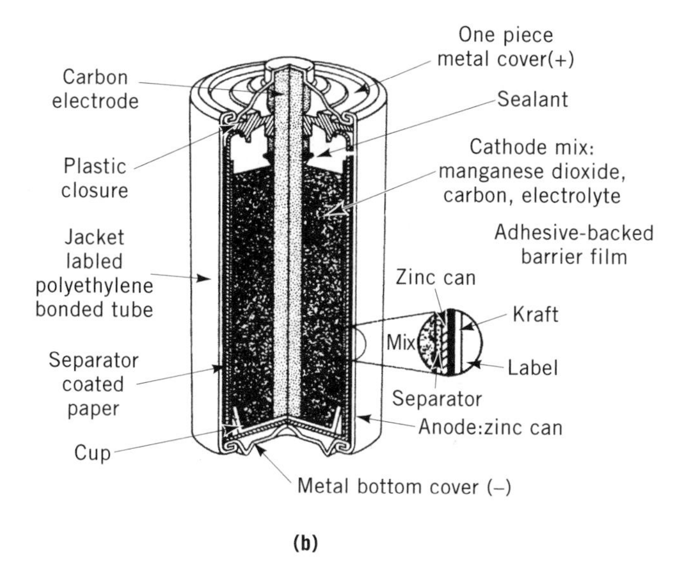

Figure 1. Cutaway view of (a) a "D"-size pasted Leclanché cell, and (b) a "D"-size paper-lined carbon–zinc cell (1). Courtesy of Eveready Battery Co.

used for this purpose. The mineral types and the synthetic types have manganese oxidation states less than 4 and contain bound protons (generally considered to be present as hydroxyl ions) as well as internal and adsorbed water. NMD normally contains higher levels of impurities than the other types, so that the overall activity (or theoretical capacity) is considerably lower. However, the cost of NMD is considerably less because the mineral is relatively abundant and no chemical or electrochemical processing is required. Thus, NMD is the material of choice for low-priced carbon-zinc cells. Premium cells, usually with zinc chloride electrolyte and paper separators, are made with some proportion of the higher-priced, higher-activity CMD or EMD in the cathode. CMD (which is made by a rather complex chemical process) often has comparable activity to EMD on a weight basis and has a lower cost, but its high surface area and poor packing capability make it less favorable in some applications. The most active form is EMD, which is also the most expensive. It is made by electrolytic deposition of MnO_2 from a bath of $MnSO_4$ in sulfuric acid.

The mechanism of the cathode reaction for all three types of MnO_2 can best be described by the following two approximately one-electron steps:

$$MnO_2 + H_3O^+(NH_4^+) \rightarrow MnOOH + H_2O(NH_3) \quad (15)$$

and

$$MnOOH + H_3O^+(NH_4^+) \rightarrow Mn(OH)_2 + H_2O(NH_3) \quad (16)$$

where NH_4^+ and NH_3 are appropriate to Leclanché cells and H_3O^+ and H_2O to zinc chloride cells. The reaction in eq. 15 is believed to occur in a single overall phase with the gradual incorporation of protons and electrons. It appears that the end product of reaction (eq. 15) is an intergrowth material in which the manganese ions occupy the same octahedral sites as in MnO_2 and the remaining octahedral sites are filled with protons, although the hexagonal close-packed oxide ion sublattice is somewhat distorted. This kind of homogeneous reaction is similar to a reaction in solution, and the potential is related to the proton activity (expressed in mole fraction) within the MnO_2.

As the reduction of the cathode proceeds through reaction (eq. 15) and the potential decreases, the reduction reaction (eq. 16) becomes favored through a complex series of steps. First the divalent $Mn(OH)_2$ is solubilized to some extent in the mildly acidic electrolyte.

$$Mn(OH)_2 + 2H_3O^+ \rightarrow Mn^{2+} + 4H_2O \quad (17)$$

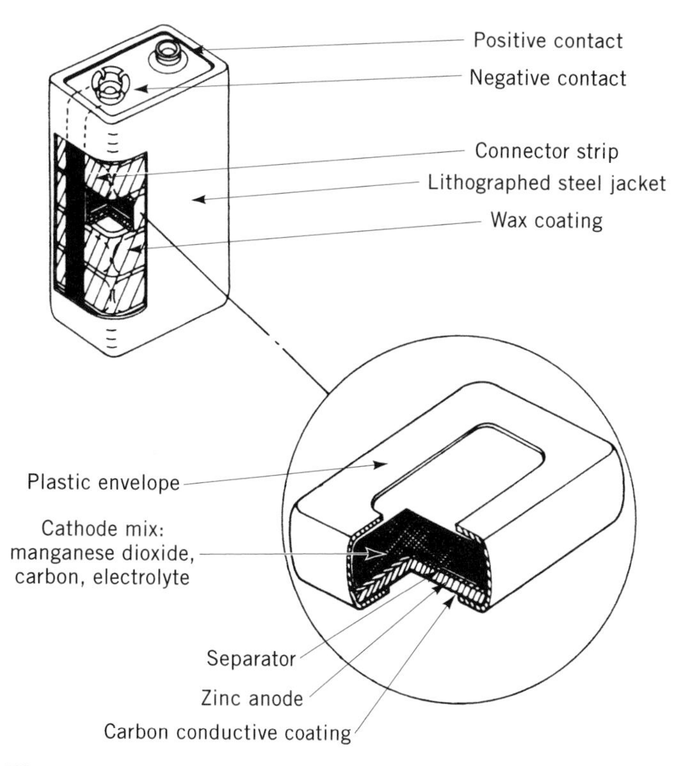

Figure 2. Cutaway view of a Leclanché flat cell used in multicell batteries (1). Courtesy of Eveready Battery Co.

The manganous ion in solution then reacts with higher valent manganese oxide and zinc ions in solution to form a new phase called hetaerolite ($ZnMn_2O_4$) or hydrohetaerolite. This reaction is usually written as

$$MnO_2 + Mn^{2+} + Zn^{2+} + 2H_2O \rightarrow ZnMn_2O_4 + 4H^+. \quad (18)$$

Hetaerolite is the most stable form of trivalent manganese at low cathode potentials; this sequence of reactions causes in part the familiar recovery of potential on open circuits, giving rise to higher cell capacity on intermittent discharge compared to continuous discharge. A second mode of potential recovery is the equilibration of the pH, due to the slow kinetics of zinc complex formation in solution and the dependence of the potential of the MnO_2 electrode on the pH.

In most cylindrical carbon–zinc cells, the zinc anode also serves as the container for the cell. The zinc can is made by drawing or extrusion. Mercury has been traditionally incorporated in the cell to improve the corrosion resistance of the anode, but the industry is in the process of removing this material because of environmental concerns. For example, Eveready Battery Co. has now removed added mercury completely from its carbon-zinc cells in the U.S. market.

The anode reaction depends on the electrolyte used, but for both types the charge transfer step is

$$Zn \rightarrow Zn^{2+} + 2e^-. \quad (19)$$

In Leclanché cells, the high concentration of ammonium chloride leads to the formation of insoluble diaminozinc chloride through the reaction in the electrolyte:

$$Zn^{2+} + 2NH_3 + 2Cl^- \rightarrow Zn(NH_3)_2Cl_2 \quad (20)$$

where the sources of the ammonia are the cathode reactions (eq. 15) and (eq. 16). The initial form of the zinc ions is predominantly $ZnCl_4^{2-}$ because of the high concentration of NH_4Cl. As the reaction proceeds and NH_4Cl is consumed, however, lower chlorocomplexes, $ZnCl_3^-$, $ZnCl_2$, and $ZnCl^+$, are formed. The final stage in the anode reaction is the precipitation of basic zinc chloride.

$$5Zn^{2+} + 2Cl^{2-}17H_2O \rightarrow ZnCl_2 \cdot 4Zn(OH)_2 \cdot H_2O + 8H_3O^+ \quad (21)$$

In the zinc chloride cell, precipitated basic zinc chloride is the major anode product because of the low concentration of ammonium chloride in the cell. Water and zinc chloride are consumed in reactions (eq. 15) and (eq. 21) and must be provided in adequate amounts for the cell to discharge efficiently. Usually, more carbon is used in zinc chloride cells than in Leclanché cells in order to increase the electrolyte absorptivity of the cathode and thus allow the use of a larger volume of electrolyte. Also, the use of a thin paper separator, which decreases internal resistance, allows less space for water storage than the thick, pasted separator construction traditionally used in Leclanché cells.

Corrosion and other unwanted reactions cause difficulties in the design of carbon-zinc cells. Fortunately, the overpotential for hydrogen evolution on zinc is quite high in mildly acidic or alkaline solutions. Sometimes, basic ZnO is added to zinc chloride electrolyte to raise the pH and lower the corrosion rate. Also, good control of heavy metal impurities in the cathode is important, because they may be solubilized by the electrolyte and deposit on the zinc to become low overpotential sites for hydrogen evolution. Another type of corrosion is the direct reaction of oxygen from air ingress to the cell. To protect against this loss, the cell is sealed and access of air is restricted in all ways possible. The cell must not be hermetically sealed, however, as hydrogen and other gases, mainly CO_2, must be able to escape from the cell as they are formed. The source of CO_2 is the cathode and results from the excellent oxidative properties of MnO_2. Thus, if a starch paste separator is used, some glucose becomes available to the cathode from starch hydrolysis and readily reacts with MnO_2. The carbon used as conductor in the cathode also reacts with MnO_2, but more slowly. The higher the voltage of the MnO_2, the more easily it reacts with these materials, so premium cells have special separator materials such as methyl cellulose–coated paper to minimize this reaction.

The porous carbon rod is often the main pathway that allows escape of the gases from the cell, but of course this pathway also allows ingress of oxygen to the cell. This compromise situation in carbon-zinc cells limits the shelf life of the system and has also prohibited the construction of a truly leakproof cell. The use of shrink tube outer wrapping and other devices, however, have improved the leakage property dramatically over older-generation cells.

As noted previously, carbon-zinc cells perform best under intermittent use. Thus many standardized tests have been devised that are appropriate to these applications, such as light and heavy flashlight tests, radio tests, cas-

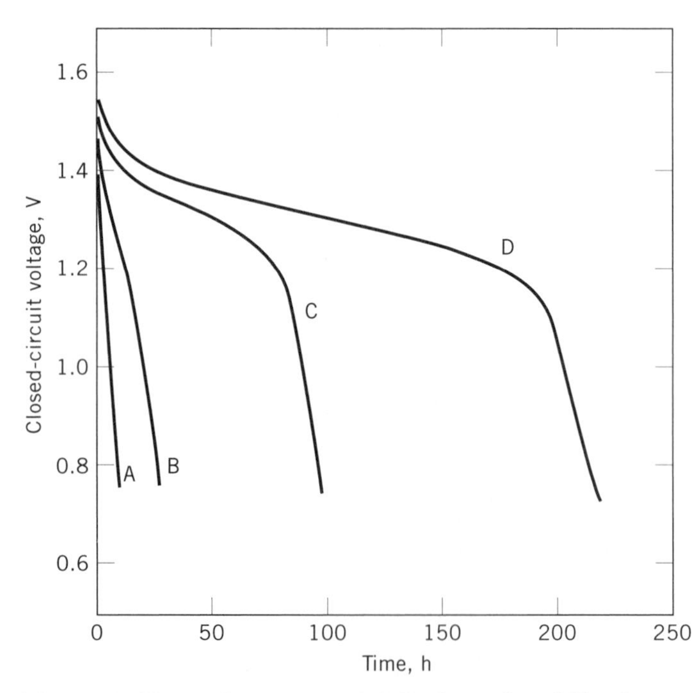

Figure 3. Hours of service on 40-Ω discharge for 4 h/d radio test at 21°C for A, RO3 "AAA"; B, R6 "AA"; C, R14 "C"; and D, R20 "D" paper-lined, heavy-duty zinc chloride cells.

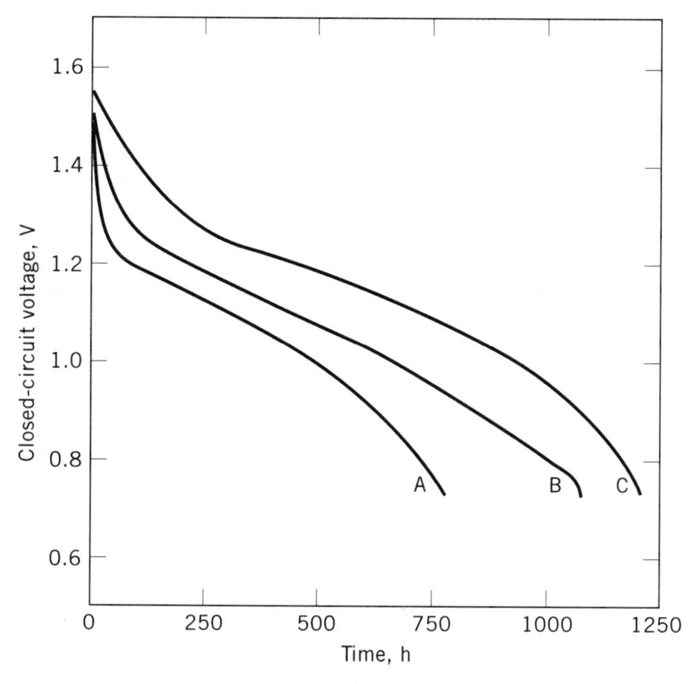

Figure 4. Hours of service on 4-Ω discharge, 15 min/h, 8 h/d, LIF test (see Table 1) at 21°C for A, general purpose (pasted Leclanché) "D"-size; B, premium Leclanché (paper-lined Leclanché); and C, heavy-duty (paper-lined $ZnCl_2$) cells.

sette tests, and motor (toy) tests. The most frequently used tests are detailed in reference (19), the American National Standards Institute (ANSI) tests. The tests are carried out at constant resistance and the results reported in minutes or hours of service. Figure 3 shows typical results under a light load for different size cells, and Figure 4 shows results for different types of R20 (D) size cells under heavy intermittent load. Typical computed values of the energy density and specific energy for R20 (D) and R6 (AA) size cells of the Leclanché and zinc chloride battery types are given in Table 2.

Note that, for heavy-duty cells, the specific energies (Wh/kg) are higher for the larger R20 cells than for the R6 cells, whereas the energy densities (Wh/L) are higher for the R6 cells. This is due to the fact that the can weight makes a much greater contribution to the weight of the system for small cells, whereas the smaller cathode of the R6 cell allows a more efficient design of the cell. These relationships are not exactly preserved in the general-purpose cells in Table 2, because the R6 cell shown here is now made in the more efficient zinc chloride cell design. The effect of the discharge rate is especially pronounced for the general-purpose cells. On intermittent tests, the heavy-duty cell operates at high efficiency even at high rate. On continuous test at high rate, heavy duty cells provide 60% to 70% of the intermittent service, whereas general purpose cells give only 30% to 50% of the intermittent service values.

Table 2. Specific Energies and Energy Densities of Carbon Zinc Cells[a]

Battery Designation and Type	Test, Ω[b,c]	Wh/L	Wh/kg
R20 (D) size cells:			
General purpose Leclanché	40Ω Radio	120	73
	LIF	70	41
Heavy duty zinc chloride	40Ω Radio	160	100
	LIF	150	90
R6 (AA) Size Cells:			
General purpose zinc chloride	75Ω Radio	130	72
	LIF	100	51
Heavy duty zinc chloride	75Ω Radio	170	81
	LIF	150	68

[a] Calculations based on data from reference 1.
[b] The radio test is one 4 h continuous discharge daily.
[c] Each discharge performed at 2.25Ω on the light industrial flashlight (LIF) test: one 4 min discharge at 1 h intervals for eight consecutive hours each day, with 16-h rest periods; ie, 32 min of discharge per day.

Alkaline Cells

Primary alkaline cells use sodium hydroxide or potassium hydroxide as the electrolyte. Early alkaline cells were of the wet cell type, but the alkaline cells of the 1990s are mostly of the limited electrolyte, dry cell type. Most primary alkaline cells are made using zinc as the anode material; a variety of cathode materials can be used. Primary alkaline cells are commonly divided into two classes, based on type of construction: the larger, cylindrical shaped batteries, and the miniature, button-type cells. Cylindrical alkaline batteries are mainly produced using zinc-manganese dioxide chemistry (manufacture of cylindrical zinc-mercury oxide cells has now been discontinued, because of environmental concerns). Miniature cells are produced with a much larger number of chemical systems, to meet the needs of particular applications.

Cylindrical alkaline cells are zinc-manganese dioxide cells having an alkaline electrolyte, which are constructed in the standard cylindrical sizes, R20 "D," R14 "C," R6 "AA," R03 "AAA," and a few other less common sizes. They can be used in the same types of devices as ordinary Leclanché and zinc chloride cells. Moreover, their high level of performance makes them ideally suited for applications such as toys, audio devices, and cameras.

Figure 5 shows a cross-sectional diagram of a typical R20-size cylindrical alkaline cell. The battery is contained in a steel can, which also serves as the current collector for the cathode. Inside the can is a dense, compacted cathode mass containing manganese dioxide, carbon, and sometimes a binder. The cathode has a hollow center lined with a separator to isolate the cathode from the anode. Within the separator-lined cavity is placed the anode mix, consisting of zinc powder and alkaline electrolyte, together with a small amount of gelling material to immobilize the electrolyte and suspend the zinc powder. Contact to the anode is made by a metal pin or leaf inserted into the anode mix. The cell has a plastic seal assembly, to keep electrolyte from leaking out of the cell and keep air from getting in. Contact to the anode collector is made through this seal. The seal contains a fail-safe vent, which is activated when the battery internal pressure exceeds a certain level.

Alkaline batteries having high output capacity and high current-carrying ability are now made by many manufacturers throughout the world. There is continuing

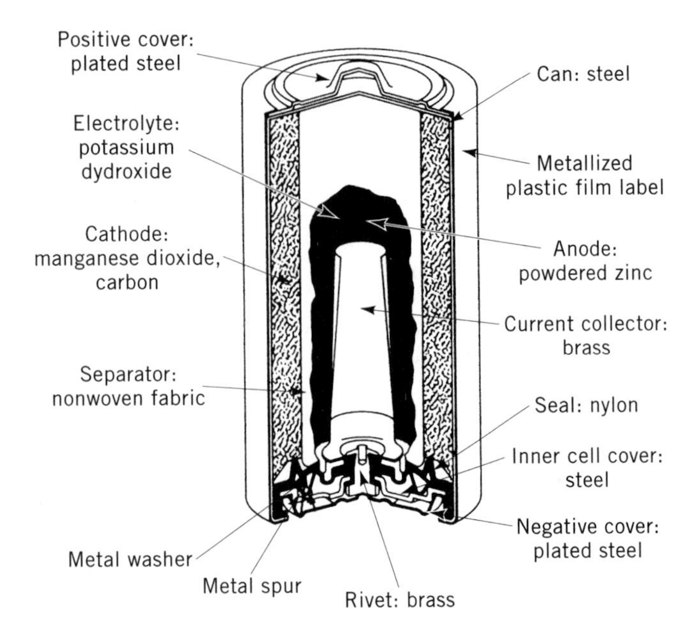

Figure 5. Cutaway view of typical R20 or "D"-size alkaline manganese battery showing components and the corresponding materials of construction (8). Courtesy of Eveready Battery Co.

competition among manufacturers to improve the performance of cylindrical alkaline batteries.

The alkaline cell derives its power from the reduction of the manganese dioxide cathode and the oxidation of the zinc anode. The anode reaction is

$$Zn + 2OH^- \rightarrow ZnO + H_2O + 2e^- \quad (22)$$

and the cathode reaction is

$$2e^- + 2\,MnO_2 + 2H_2O \rightarrow 2\,MnOOH + 2OH^-. \quad (23)$$

Both the anode (eq. 22) and cathode (eq. 23) reactions are more complicated than is indicated by the reactions shown above. Details of the individual electrode reactions are discussed below.

Zinc might at first appear to be an unusual choice for battery anode material, because it is thermodynamically unstable in contact with water. However, the reaction with water can be made to be extremely slow. Because the alkaline electrolyte is corrosive of human tissue as well as materials in devices, it is more important to have a good seal against electrolyte leakage in an alkaline battery than in a carbon-zinc cell. The formation of a good seal is incompatible with the formation of a noncondensable gas such as hydrogen, so it is important to maintain a low level of gas.

The reaction of zinc with water is not a simple homogeneous one. Rather, it is a heterogeneous electrochemical reaction, involving a mechanism similar to that of a battery. There are two steps to the reaction. Zinc dissolves at some locations as shown in equation 22 and hydrogen gas is generated at other sites.

$$2e^- + 2H_2O \rightarrow H_2 + 2OH^- \quad (24)$$

In the corrosion reaction, the anode and cathode reactions (eq. 22) and (eq. 24) occur at different locations within the anode. Because the anode is a single, electrically conductive mass, the electrons produced in the anode reaction travel easily to the site of the cathode reaction and the zinc acts like a battery with the positive and negative terminals shorted together.

Certain impurities on zinc can act as catalysts for the generation of hydrogen, thereby greatly increasing the corrosion rates. For this reason, zinc in alkaline cells must be of high purity and careful control must be exercised over the level of the harmful impurities. Moreover, other components of the cell must not contain harmful levels of impurities that might dissolve and migrate to the zinc anode.

Alternatively, there are also inhibitors that can decrease the rate of hydrogen generation and thus decrease the corrosion. Mercury, effective at inhibiting zinc corrosion, has long been used as an additive to zinc anodes. More recently, however, because of increased interest in environmental issues, the amount of mercury in alkaline cells has been sharply reduced. For example, Eveready Battery Co. no longer adds mercury to alkaline batteries.

The oxidation reaction for the zinc anode (eq. 22) takes place in several steps, ultimately resulting in zincate ions ($Zn(OH)_4^{2-}$) that dissolve in the electrolyte. As a battery discharges, the concentration of zincate in the electrolyte increases. Eventually, the concentration exceeds the solubility of zinc oxide, and zinc oxide can precipitate from the solution. γ-MnO_2 is useful in batteries because of the unusual nature of its discharge reaction. The synthetic form, EMD, is reduced in a homogeneous single-phase reaction. Thus, when MnOOH is formed at the surface of MnO_2, there is a rapid diffusion of protons from the surface MnOOH into the interior of the particle, coupled with hopping of electrons, with the net result that the lower valent material initially formed on the surface in effect diffuses into the particle. In this way the entire particle is homogeneously reduced, and the surface tends to have the same composition as the overall particle (21). The cathode material starts with a composition containing nearly all manganese(IV) and ends with a composition containing manganese(III).

This homogeneous phase reduction of EMD results in a discharge curve that differs from that of many other battery systems. In any battery system there can be a voltage drop from the open-circuit value due to polarization effects. But for a typical heterogeneous (two-phase) reaction, the open-circuit recovery voltage is substantially constant throughout the discharge. In the homogeneous phase reduction of EMD, however, the open-circuit voltage itself changes as the cathode is reduced (Figure 6). Thus the operating voltage of the alkaline zinc-manganese dioxide battery starts around 1.5 volts and drops to below 1 volt by the time a one-electron reduction of the MnO_2 is complete.

Alkaline manganese-dioxide batteries have relatively high energy density, as can be seen from Table 3. This is due in part to the use of highly pure materials, formed into electrodes of near optimum density. Moreover, the cells are able to function well with a rather small amount

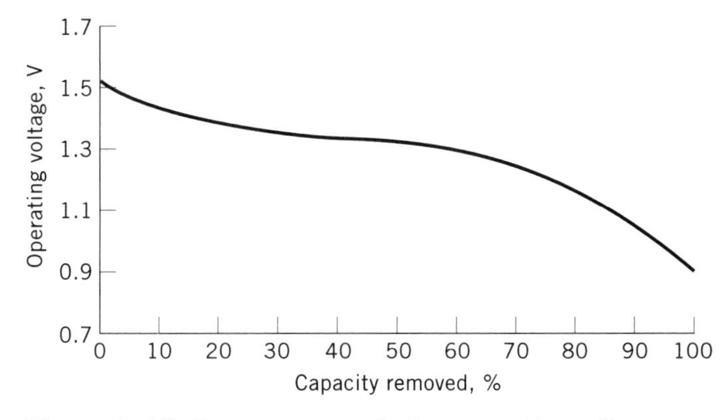

Figure 6. Alkaline–manganese battery operating voltage as a function of remaining capacity (21). Courtesy of Eveready Battery Co.

of electrolyte. The result is a cell with relatively high capacity, at a fairly reasonable cost.

The performance of the cells is influenced not only by the relatively high theoretical capacity, but also by the fact that the cells can provide good efficiency at various currents over a wide variety of conditions. The high conductivity and low polarization of the alkaline electrolyte, combined with the good electronic and ionic conductivity in the electrodes, leads to high performance even under heavy drain conditions. Alkaline zinc-manganese dioxide batteries show less variation in output capacity with variation in discharge rate than do Leclanché or zinc chloride batteries. Thus, alkaline cells have a moderate advantage over Leclanché cells in capacity on light drains (Fig. 7**a**), but they have a much greater advantage on heavy drains (Fig. 7**b**). Because alkaline cells perform nearly as well on continuous as on intermittent discharge, they have an even greater advantage on high-rate continuous tests.

Batteries tend to perform more poorly as the operating temperature is decreased because of decreased conductivity of the electrolyte and slower electrode kinetics. Ultimately, freezing of the electrolyte occurs and the battery fails. Batteries tend to perform better at higher temperature only up to the point that loss of performance occurs because of cell venting and drying out or parasitic reactions in the cell. Overall, alkaline cells have less performance loss at low and high temperatures than do Leclanché cells, as can be seen in Figure 8.

Alkaline zinc-manganese dioxide cells have good capacity retention on long-term storage. The use of high-purity materials ensures very low rates of parasitic reactions in the cells. Moreover, the alkaline cells are well sealed so there is minimal vapor transmission into or out of the cell. Figure 9 shows the capacity retention of alkaline cells versus Leclanché cells after prolonged storage at different temperatures.

Miniature alkaline cells are small, button-shaped cells, which use alkaline NaOH or KOH electrolyte and generally have zinc anodes, but may have a variety of cathode materials. They are used in watches, calculators, cameras, hearing aids, and other miniature devices. The chemistry of the anode is essentially the same as in the larger cylindrical alkaline cells. The cathode chemistry varies with the type of cathode material used.

Figure 10 shows the construction of a typical miniature alkaline cell. The cathode mix is placed into the can, followed by the insertion of a separator layer. The type of separator materials used and the number of layers of material vary according to the type of cathode used in the cell. The anode mix is then placed in contact with the anode cup. An insulating, sealing gasket electrically separates the anode cup from the can and prevents leakage of electrolyte from the cell. Miniature alkaline batteries are highly specialized, with the cathode material, separator type, and electrolyte all chosen to match the particular application.

The combination of a zinc anode and manganese-dioxide cathode is used in some miniature alkaline cells and in the cylindrical cells discussed previously. This type of battery is used where an economical power source is desired and where the application can tolerate the sloping discharge curve (Figure 11). The chemistry is the same as in the cylindrical batteries. The construction features are typical of miniature alkaline batteries.

Miniature zinc-mercuric oxide batteries have a zinc anode and a cathode containing mercuric oxide, HgO. The cathode reaction is

$$HgO + H_2O + 2e^- \rightarrow Hg + 2OH^-. \quad (25)$$

Water is not used in the overall reaction. Therefore, these cells have very high capacity, exceeding that of zinc-manganese dioxide batteries (Table 3).

The cathode reaction is rather simple: mercuric oxide is reduced to mercury metal, which is a liquid and does

Table 3. Characteristics of Aqueous Primary Batteries

Parameter	Carbon Zinc (Zn/MnO_2)	Alkaline Manganese Dioxide (Zn/MnO_2)	Mercuric Oxide (Zn/HgO)	Silver Oxide (Zn/Ag_2O)	Zinc Air (Zn/O_2)
Nominal voltage, V	1.5	1.5	1.35	1.5	1.25
Working voltage, V	1.2	1.2	1.3	1.55	1.25
Specific energy, Wh/kg	40–100	80–95	100	130	230–400
Energy density, Wh/L	70–170	150–250	400–600	490–520	700–800
Temperature range, °C	−40 to 50	−40 to 50	−40 to 60	−40 to 60	−40 to 50
Storage Operating	−5 to 55	−20 to 55	−10 to 55	−10 to 55	−10 to 55

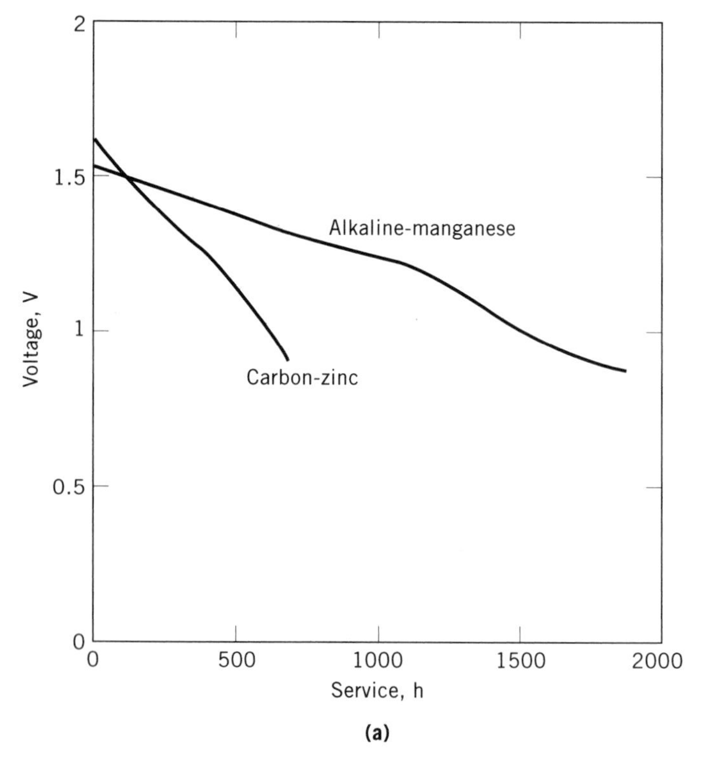

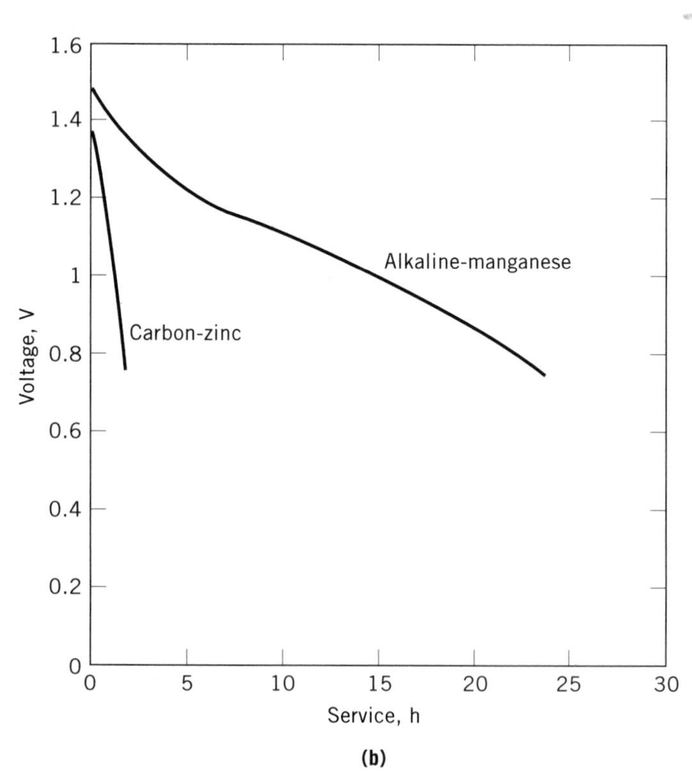

Figure 7. Performance comparison of "D"-size alkaline–manganese vs carbon–zinc batteries at 21°C on (a) a light drain 150-Ω continuous test at 21°C, and (b) a heavy drain 2.2-Ω continuous test (8). Courtesy of Eveready Battery Co.

not block off the surface of the cathode. Therefore, the reaction is a normal heterogeneous (two-phase) reaction, producing a substantially constant voltage during discharging, in contrast to the sloping voltage of a cell containing a manganese-dioxide cathode. Thus, the zinc-mercuric oxide cell has high capacity and a flat discharge curve as shown in Figure 12. Thick separators are used to keep the liquid mercury from bridging to the anode and short-circuiting the cell.

Some cells are made with small amounts of manganese dioxide added to the cathode. These cells have somewhat more sloping discharge curves (Figure 12) and are used in some devices that do not require a constant voltage. The added manganese dioxide provides for improved high-rate capability for the cells. In addition, such cells have a slightly more gradual voltage drop at the end of battery life, giving the consumer some advance warning that the battery should be replaced.

Although the zinc–mercuric oxide battery has good performance properties, a serious disposal problem has led to a deemphasis in the use of this system. Both the mercuric oxide in the fresh cell and the mercury reduction product in the used cell have long-term toxic effects, and landfill disposal has been made illegal in several states.

Miniature zinc–silver oxide batteries have a zinc anode and a cathode containing silver oxide, Ag_2O. The cathode reaction is

$$Ag_2O + H_2O + 2e^- \rightarrow 2Ag + 2OH^-. \tag{26}$$

The overall reaction neither consumes or produces water. As in the case of mercury cells, these batteries have very high energy (Table 3), and because of the good environmental properties of the silver materials compared to mercury, much of the industry has converted to this system. The cathode reaction involves reduction of silver oxide to metallic silver. The reaction is two-phase and heterogeneous, producing a substantially constant voltage during discharge. Some manganese dioxide may be added to the cathode, as in the case of mercury oxide cells.

Unlike some other cathode materials, such as manganese dioxide, which are quite insoluble, silver oxide has a fair degree of solubility in alkaline electrolyte. If the soluble silver species were allowed to be transported to the zinc anode it would react directly with the zinc, and as a result the cell would self-discharge. In order to prevent this from happening, zinc–silver oxide cells use special separator materials such as cellophane that are designed to inhibit migration of soluble silver to the anode.

Zinc–silver oxide cells have high energy density, nearly as high as that of mercury cells. They operate at higher voltages than mercury cells (1.5 versus 1.3 volts), but for somewhat less time as shown in Figure 13 (comparison based on cells of equal volume).

As with mercury batteries, miniature zinc–silver oxide batteries may be made with KOH or with NaOH electrolyte. Again, KOH-containing cells operate more efficiently than NaOH ones at high current drains, whereas batteries with NaOH are easier to seal and are more resistant to leakage.

Miniature zinc–silver oxide batteries are commonly used in electronic watches and in other applications where high energy density, a flat discharge profile, and a higher operating voltage than a mercury cell are needed. These batteries function efficiently over a wide range of temperatures and are comparable to mercury batteries in this respect. Miniature zinc–silver oxide batteries also

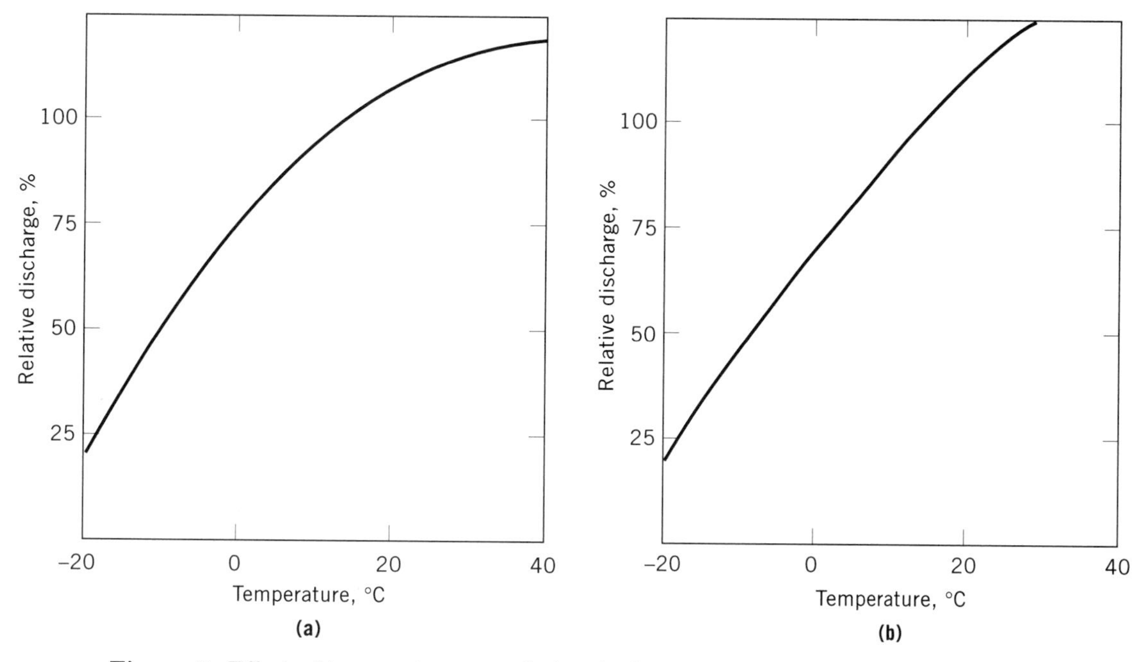

Figure 8. Effect of temperature on relative discharge performance of a fresh "D"-size battery for service on simulated radio use, 25-Ω 4-h/d test for (a) an alkaline–manganese battery undergoing 260 h of service, and (b) a carbon–zinc battery undergoing 70 h of service (22). Courtesy of Eveready Battery Co.

have good storage life. However, silver cells stored at elevated temperature lose capacity at a slightly higher rate than mercury cells because of separator degradation.

It is possible to produce a silver oxide in which the silver has a higher oxidation state, approaching the composition of AgO. This material can provide higher capacity and energy density than Ag_2O. Alternatively, a battery can be made with the same capacity as a monovalent silver cell, but with cost savings. However, some difficulties with regard to material stability and voltage regulation must be addressed.

The cathode reaction is

$$AgO + H_2O + 2e^- \rightarrow Ag + 2OH^-. \quad (27)$$

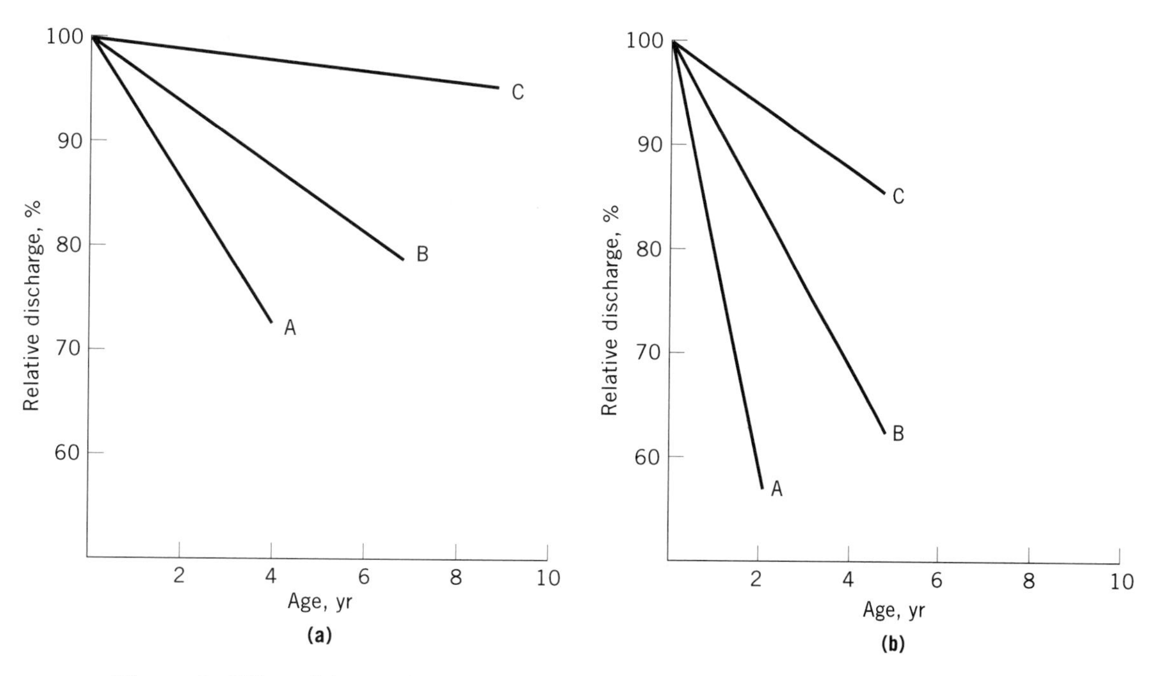

Figure 9. Effect of time and storage temperatures. A, 40°C; B, 20°C; and C, 0°C, on relative discharge performance of fresh and aged "D"-size cells on simulated radio use, 25-Ω 4-h/d test for (**a**) alkaline–manganese, and (**b**) carbon–zinc batteries (22). Courtesy of Eveready Battery Co.

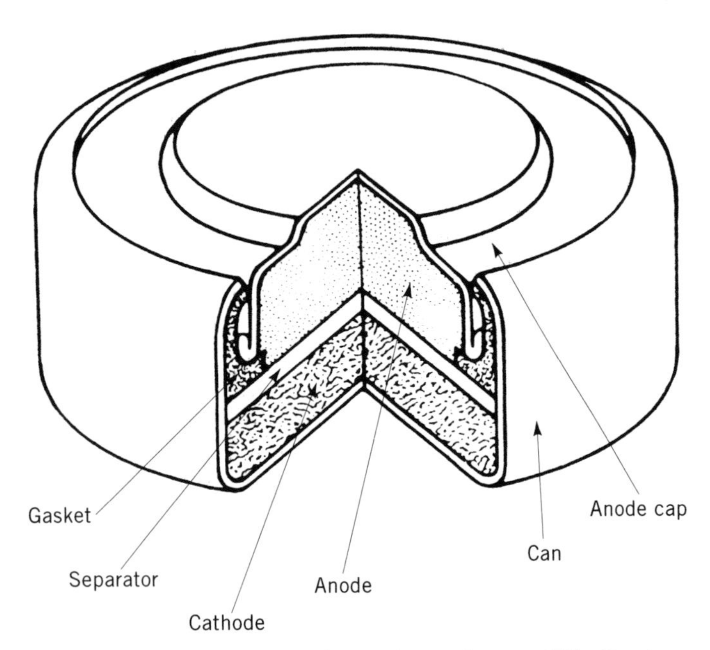

Figure 10. Cutaway view of a miniature battery (21). Courtesy of Eveready Battery Co.

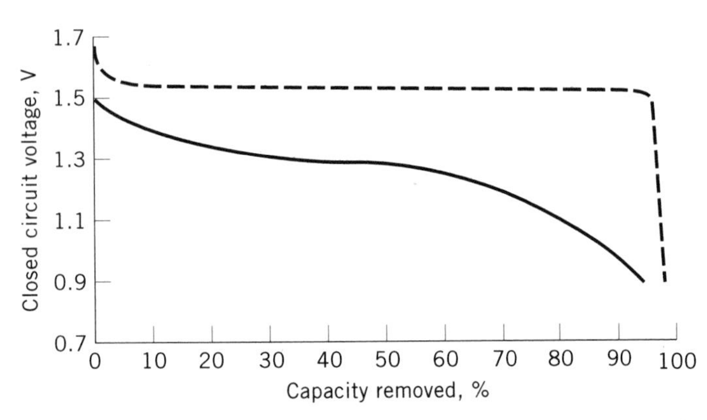

Figure 11. Discharge curves for miniature zinc–silver oxide batteries (. . .), and zinc–manganese dioxide batteries (—) (21). Courtesy of Eveready Battery Co.

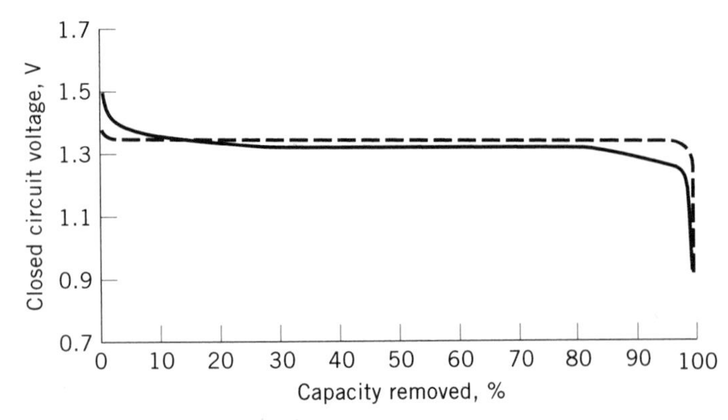

Figure 12. Typical discharge curve comparison for zinc–mercuric oxide batteries: (— — —) model 325, HgO, and (———) model E312E, HgO–MnO_2, cathodes (21). Courtesy of Eveready Battery Co.

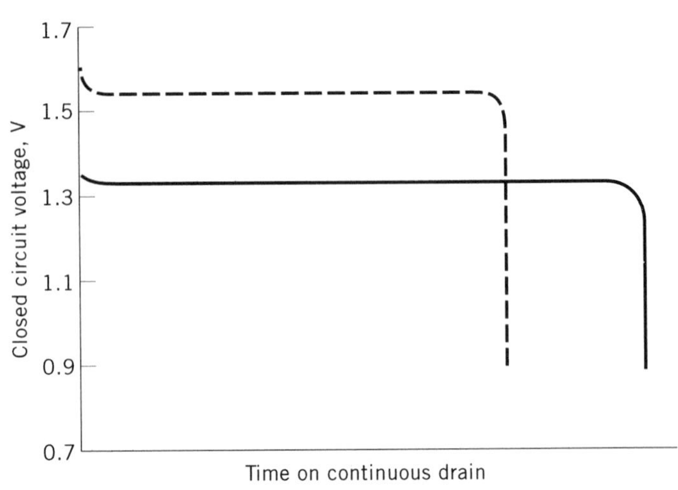

Figure 13. Relative discharge curves for (. . .) zinc–silver oxide, and (—) zinc–mercuric oxide batteries. Cells are of equal volume (21). Courtesy of Eveready Battery Co.

The reaction of divalent silver oxide would normally tend to occur in two steps. In the first step, divalent silver oxide would be reduced to monovalent silver oxide. Then, in the second step, the monovalent silver oxide would be further reduced to silver metal. These two reactions occur at different voltages, so the battery would normally operate at a higher voltage during the first part of the discharge than during the second part. Such discharge behavior, however, is unacceptable for many devices using the battery, and it is desirable to make the cell discharge at a constant voltage. The material at the cathode reaction surface is caused to be all Ag_2O, and the interior AgO is protected from contact with the electrolyte. The voltage level of the monovalent silver oxide is thus obtained, but with the enhanced coulombic capacity of the divalent material. Several schemes have been developed to accomplish this behavior. Construction features of divalent silver oxide batteries are similar to those of monovalent silver oxide batteries.

Zinc-air batteries offer the possibility of obtaining extremely high energy densities. Instead of having a cathode material placed in the battery when manufactured, oxygen from the atmosphere is used as cathode material, allowing for a much more efficient design. The construction of a miniature air cell is shown in Figure 14. From the outside, the cell looks like a typical miniature cell, except for the air access holes in the can. On the inside, however, the anode occupies much more of the internal volume of the cell. Rather than the thick cathode pellet, there is a thin layer containing the cathode catalyst and air distribution passages. Air enters the cell through the holes in the can and the oxygen reacts at the surface of the cathode catalyst. The air access holes are often covered with a protective tape, which is removed when the cell is placed in service.

The cathode reaction is

$$O_2 + 2H_2O + 4e^- \rightarrow 4OH^-. \quad (28)$$

The oxygen reaction is quite complex. Complete reduction from oxygen gas to hydroxide ion involves four electrons

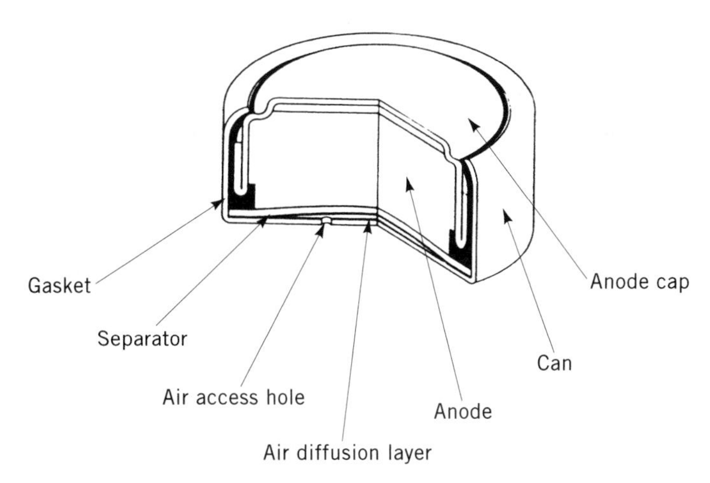

Figure 14. Cutaway view of a miniature air cell battery (21). Courtesy of Eveready Battery Co.

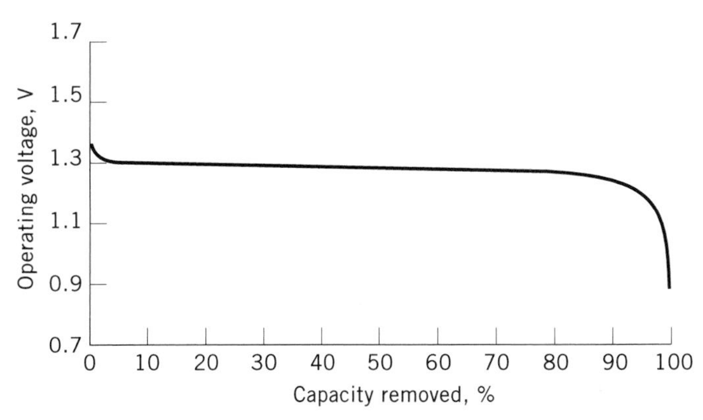

Figure 15. Discharge curve for miniature zinc–air battery (21). Courtesy of Eveready Battery Co.

and requires several steps. Initially, oxygen is reduced to peroxyl ion. The peroxyl may be further reduced or it may be catalytically decomposed, producing hydroxide and oxygen. In a typical miniature air cell, the reaction goes through the catalytic decomposition step. The performance of the air cathode will depend largely on its ability to catalyze this reaction. If the reaction is too slow, then during discharge large amounts of peroxide will build up in the cathode, resulting in large polarization and low operating voltage. If the reaction is well catalyzed, the battery can operate at a higher, more useful voltage. The miniature air cell cathode typically contains carbon to provide a surface for the initial reduction of oxygen (eq. 28). Some types of carbon are also particularly effective at catalyzing the peroxyl decomposition and are used in air cells. Alternatively, small amounts of metal oxides can be used as catalysts.

Figure 15 shows a typical discharge curve for a miniature air cell. It discharges at 1.2 to 1.3 volts, with a flat discharge profile. Its discharge voltage is slightly lower than that of a mercury cell, but its capacity is far greater than for any of the other miniature cells, including lithium cells (22).

Air cells are good choices for use in some special applications, but they are not general-purpose cells. They must be used in applications where the usage is largely continuous, and where the discharge level is relatively constant and well defined. The reason for these limitations lies in the fact that the cell must be open to the atmosphere to allow oxygen into the cell. The holes that allow oxygen into the cell can allow other gases to enter or leave the cell. Carbon dioxide from the air also enters the cell and reacts with the electrolyte to form potassium carbonate. This results in decreased cell performance. Water vapor enters the cell (if the outside humidity is high) or leaves the cell (if outside humidity is low). In the former case, the cell fills up with water and eventually bulges or leaks. In the latter case, the cell dries out, so that it no longer can function. Cretzmeyer and co-workers (23) have shown how both the rate of diffusion of water vapor into or out of the cell, and the rate of oxygen diffusion into the cell, are related, because both must diffuse through the same openings in the cell. Therefore, an air cell must be carefully designed for a particular use. The air access passages are designed to be just big enough to handle the required oxygen flux. Miniature air cells are mainly used in hearing aids, where they are required to produce a relatively high current for a relatively short period of time (a few weeks). In this application they provide exceptional performance compared to other batteries.

Air cells are packaged with sealing tape over the air holes, or in sealed containers. In the sealed state, they maintain their capacity well during storage (Figure 16). Once the sealing tape is removed, or the cell is taken from its sealed package, it must be put into use.

Miniature air cells can be used for continuous service over a temperature range of $-10°C$ to $+55°C$. However, performance at high temperatures suffers because of drying out of the cells.

Lithium Cells

Cells with lithium anodes are generally called lithium cells regardless of the cathode. They can be conveniently separated into two different types: a) cells with solid cathodes, and b) cells with liquid cathodes. Cells with liquid cathodes have liquid electrolytes. At least one component of the electrolyte solvent and the cathode active material

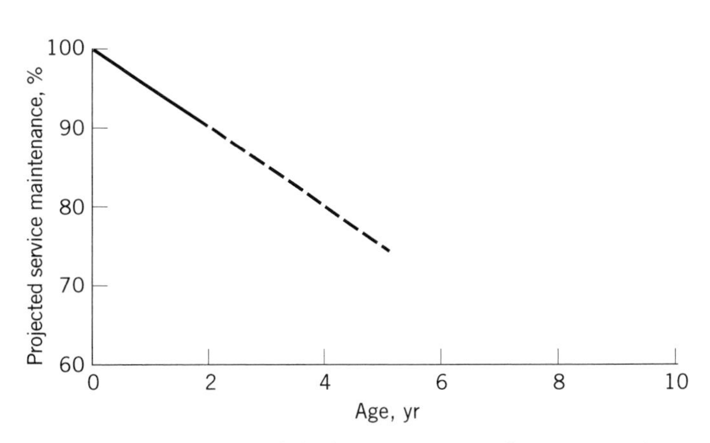

Figure 16. Retention of discharge capacity of miniature zinc–air battery having an unopened sealed cell after storage at 20°C (. . .) projected data (21). Courtesy of Eveready Battery Co.

are the same. Cells with solid cathodes may have liquid or solid electrolytes but, except for the lithium-iodine system (see the following), solid electrolyte systems have not yet matured to commercial status.

Many systems of all types involving lithium have been investigated, but only those being manufactured at the present time are discussed in this chapter. All the cells take advantage of the inherently high energy of lithium metal and its unusual film-forming property. This property, discussed in detail in reference 24, allows for lithium compatibility with solvents and other materials with which it is thermodynamically unstable, yet permits high electrochemical activity when the external circuit is closed. This is possible because the film formed is conductive to lithium ions, but not to electrons. Thus, corrosion reactions occur only to the extent of forming thin, electronically insulating films on lithium that are both coherent and adherent to the base metal. The thinness of the film is important in order to allow reasonable rates of transport of lithium ions through the film when the circuit is closed. Many materials, such as water and alcohols, which are thermodynamically unstable with lithium, do not form this kind of passivating film.

Much detailed analytical study has been required to establish materials for use as solvents and solutes in lithium batteries (see references 25 and 26 for discussions of electrolytes). Among the best organic solvents that have been found are the cyclic esters, such as propylene carbonate (PC), ethylene carbonate (EC), and butyrolactone, and the ethers, such as dimethoxyethane (DME), the glymes, tetrahydrofuran (THF), and dioxolane. Among the most useful electrolyte salts are lithium perchlorate, lithium trifluoromethanesulfonate, lithium tetrafluoroborate, and, more recently, lithium hexafluorophoshate. Lithium hexafluoroarsenate has good electrolyte properties, but because of possible environmental problems with arsenic-containing materials, it is now used mostly in military applications.

A limitation of these so-called organic electrolytes is their relatively low conductivity, compared to aqueous electrolytes. This limitation, combined with the generally slow kinetics of the cathode reactions, has forced the use of certain designs such as thin electrodes and very thin separators for all lithium batteries. This usage led to the development of coin cells rather than button cells for miniature batteries and jelly-roll or spiral-wound designs rather than bobbin designs for cylindrical cells. Figure 17**a** shows a cutaway view of a typical coin cell and Figure 17**b** a typical cylindrical jelly-roll lithium–manganese dioxide battery. Many of the cylindrical cells have glass-to-metal hermetic seals, although this is becoming less common because of the high cost associated with this type of seal. Alternatively, cylindrical cells have compression seals carefully designed to minimize the ingress of water and oxygen and the egress of volatile solvent. These construction changes are costly, and the high price of the lithium cell has limited its use thus far. However, the energy densities are superior and, in some applications, there is a definite economic benefit compared to aqueous systems.

The Li/MnO_2 cell is becoming the most widely used 3-volt solid cathode lithium primary battery. The critical

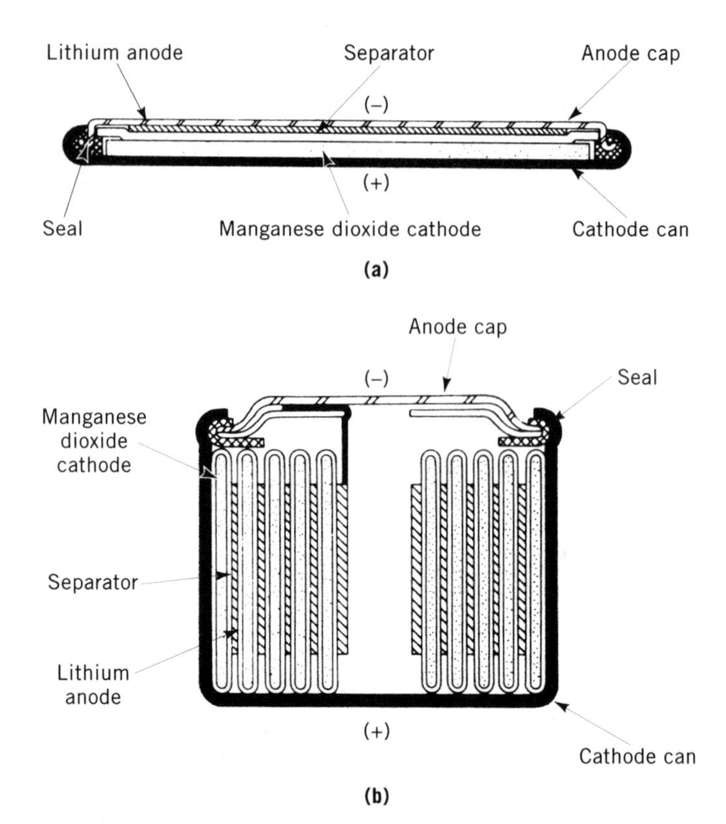

Figure 17. Cutaway view of (a) 2016-size Eveready lithium manganese dioxide coin cell, and (b) jelly roll cylindrical lithium manganese dioxide cell (28). Courtesy of Eveready Battery Co.

step in obtaining good performance is the heat treatment of the MnO_2 to at least 300°C before incorporation in the cathode (27). This removes from the EMD any excess water, which can otherwise be dissolved in the electrolyte during storage or discharge and affect the operation of the anode, as well as activate the cathode for the discharge reaction.

While there is still some controversy about the exact nature of the reduction of the cathode, it seems agreed that the reduction occurs in two steps (28). The first step, in close analogy with the aqueous reduction, is a homogeneous process that concludes with the formation of a partially lithiated material:

$$x\,\mathrm{Li}^+ + xe^- + \mathrm{MnO_2} \rightarrow \mathrm{Li}_x\mathrm{MnO_2} \qquad (29)$$

where $x \approx 0.4$. This step is followed by a heterogeneous process to a new phase, the structure of which has not yet been determined:

$$(1 - x)\mathrm{Li}^+ + (1 - x)\mathrm{e}^- + \mathrm{Li}_x\mathrm{MnO_2} \rightarrow \mathrm{LiMnO_2} \qquad (30)$$

The anodic reaction is conventionally written as

$$\mathrm{Li} \rightarrow \mathrm{Li}^+ + e^- \qquad (31)$$

where it is understood that the Li^+ ion is solvated predominantly by the strongest Lewis base among the solvent components and that, before the dissolution/solvation step, the lithium ion must diffuse through the passivating film. The most common electrolytes used with the system are $LiClO_4$ or $LiCF_3SO_3$ in a mixture of PC and DME (usually 1 : 1).

The kinetics are not very sensitive to the electrolyte, so the choice is largely dependent on safety, toxicity, and cost. The relatively slow kinetics of the system have necessitated the use of thin electrodes in order to obtain sufficient current-carrying capability; these cells are designed as coin cells or as jelly rolls with alternating anode, separator, and cathode and another separator layer. These 3-V cells are made in sizes not used for aqueous 1.5-V cells to help prevent the use of these 3-V cells in circuits designed for 1.5 V.

The energy density of the system depends on the type of cell and the current drain. Table 4 gives the specification of lithium batteries and shows that Li-MnO_2 is very competitive with the other systems. The coin cells are widely used in electronic devices such as calculators, watches, and so on whereas the cylindrical cells have found their main application in fully automatic cameras.

The 3-V lithium-carbon monofluoride (Li-CF) cell (29) was the first lithium battery to be manufactured in quantity for consumer applications. However, it is becoming less popular now because of the relatively high cost of the cathode material and the ability of the less expensive Li/MnO_2 cells to perform most of the same functions. CF is an extremely interesting material in its own right. The major use is as a chemically inert, high-temperature lubricant. It has a laminar structure that is similar to graphite. When graphite is reacted with fluorine to make CF, the flat planes of graphite become puckered as layers of fluorine atoms are interspersed among the carbon layers.

It is thought that the CF cathode reaction is a heterogeneous process, initiated by the insertion of lithium ions between the CF planes. It is completed by the extrusion of LiF and the collapse of the structure to carbon:

$$Li^+ + CF + e^- \rightarrow LiF + C \qquad (32)$$

Various lithium salts with butyrolactone or PC-DME mixtures are usually used as electrolytes. Table 4 shows the close competitive situation in performance of CF and MnO_2 cathodes. The construction of cells is also very similar for the two systems. The uses are generally the same as mentioned for the lithium–manganese dioxide system, except for automatic cameras.

The lithium–iron disulfide battery is unusual among lithium cells in that the voltage is a nominal 1.5 V. The cells were first manufactured in button-cell sizes, and because of the voltage similarity, they may be used as direct replacements for Zn/Ag_2O cells (30). Due to the lower conductivity of the organic electrolyte compared to KOH electrolyte in the Zn/Ag_2O cell, they cannot be used in the highest drain applications, but such uses are relatively few. Recently, AA-size, cylindrical lithium–iron disulfide batteries were introduced (31). Figure 18 shows the high-surface-area jelly-roll construction and a plastic gasket compression seal. The high area construction permits the use of these cells in all applications for which alkaline Zn/MnO_2 cells can be used and gives higher energy than that system, especially at high power levels. The compression seal helps to keep the cost at reasonable levels.

The chemistry of the system is quite complex. There are at least two steps to the reaction at low discharge rates. The first reaction is an approximately two-electron reduction to a new phase, which is a lithiated FeS_2 compound:

$$2Li^+ + FeS_2 + 2e^- \rightarrow Li_2FeS_2 \qquad (33)$$

The second step is also heterogeneous and involves the breakdown of the intermediate compound with further lithiation into lithium sulfide and very finely divided iron particles:

$$2Li^+ + Li_2FeS_2 + 2e^- \rightarrow 2Li_2S + Fe \qquad (34)$$

The crystal structure of the intermediate is not well understood. The final iron phase is termed superparamagnetic because the particle size is too small to support ferromagnetic domains. At low rates, the discharge occurs in two steps separated by a small voltage difference. At high rates, however, the two steps become one, indicating that the first step is rate limiting and the second step, (eq. 34), occurs immediately after formation of the intermediate, (eq. 33).

The performance of the cylindrical cell is superior to alkaline cells at all rates of discharge, but because of the high-area electrode construction, the performance advantage is magnified during high rate discharge. Table 4 shows the energy density and comparison to Tables 2 and 3 shows the superiority of the cells to aqueous alkaline and Leclanché cells and the parity for Zn/Ag_2O button cells. The cell is widely used as a power source for electronic camera flash guns; three to five times more flashes are obtained than with use of alkaline cells. Automatic

Table 4. Characteristics of Lithium Primary Batteries

Cathode: Cell System	Nominal Voltage, V	Working Voltage, V	Specific Energy, Wh/kg	Energy Density, Wh/L	Operating Temperature, Range, °C
Solid Cathodes					
Carbon monofluoride (Li/CF)	3	2.5–2.7	200	400	−20 to 60
Manganese dioxide (Li/MnO_2)	3	2.7–2.9	200	400	−20 to 55
Iron disulfide (Li/FeS_2)	1.5	1.3–1.7	250	460	−20 to 60
Iodine (Li/I_2)	3	2.4–2.8	250	920	37
Liquid Cathodes					
Sulfur dioxide (Li/SO_2)	3	2.7–2.9	250	440	−55 to 70
Thionyl chloride (Li/$SOCl_2$)	3.6	3.3–3.5	300	950	−50 to 70

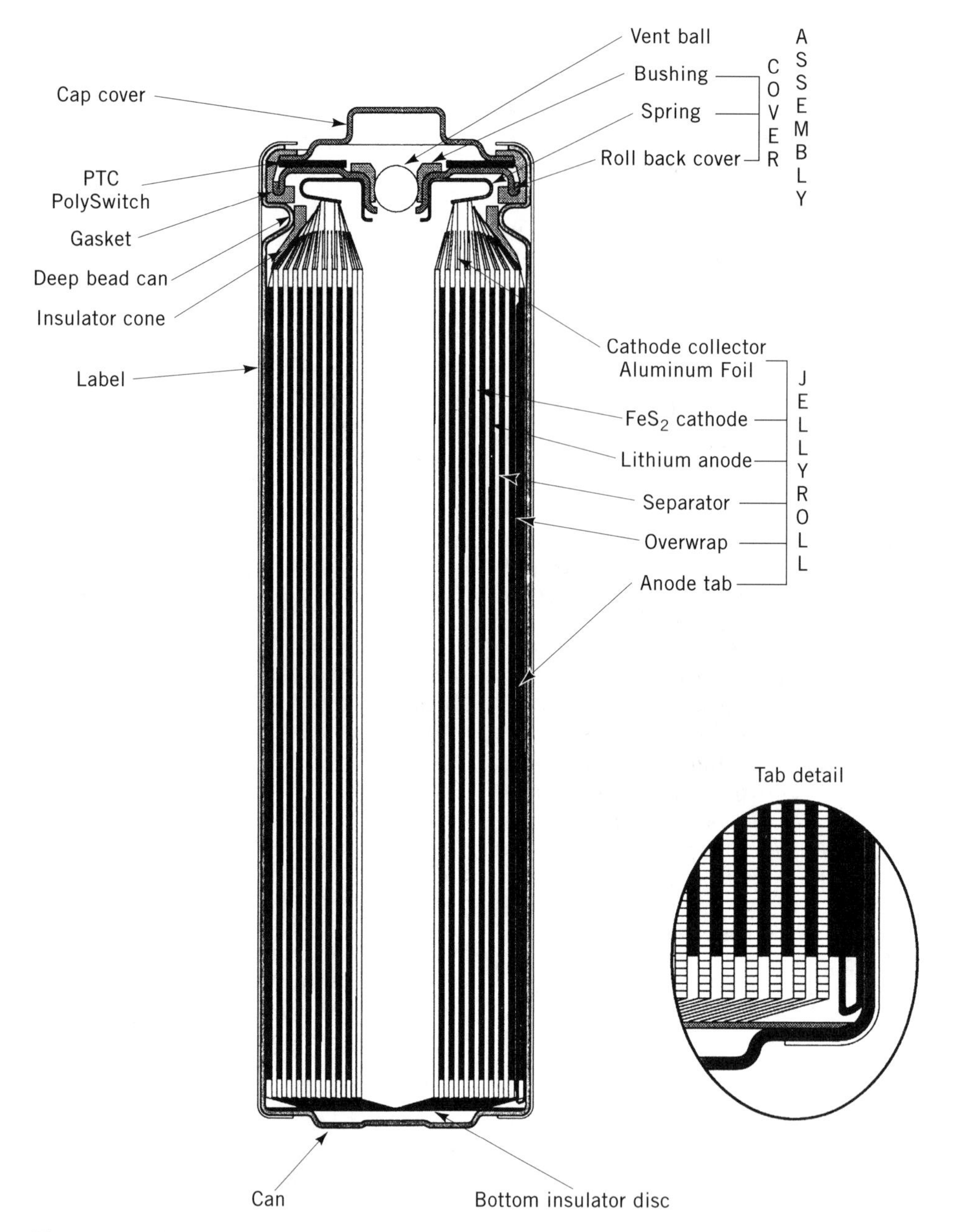

Figure 18. Cut-away view of R6 (AA) size lithium–iron disulfide high rate battery. Courtesy of Eveready Battery Co.

cameras, camcorders, and many other constant power electronic applications show similar gains over the alkaline cell, as a range of four to eight times the energy from alkaline cells is obtained (31).

Lithium-iodine cells are widely used in medical applications (32). Iodine forms a charge transfer complex with poly(2-vinylpyridine) (PVP), which makes up the cathode for these cells. In one process for making the cell, the two materials (with excess iodine) are heated together to form the molten charge transfer complex. Then, the liquid is poured into the waiting cell, which already contains the lithium anode and a cathode collector screen. After cooling and solidification of the cathode, the cell is hermetically sealed. In another process, the iodine-PVP mixture is reacted and then pelletized. The pellets are then pressed into the lithium to form a sandwich structure, which is then inserted into the waiting cell and hermetic sealing is done. The direct contact of the excess iodine and the lithium results in the immediate reaction of the two to form an *in situ* LiI film, which then protects the lithium from further reaction. The cathode reaction is the reduction of iodine to form LiI at the carbon collector sites as lithium ions diffuse to the reaction site. The anode reaction is lithium ion formation and diffusion through the thin LiI electrolyte layer. If the anode is corrugated and coated with PVP before adding the cathode fluid, the impedance of the cell is lower and remains lower until late in the discharge. As can be imagined, the cell eventually fails because of high resistance, even though the drain rate is low.

The $Li\text{-}I_2$ battery system has allowed simple heart pacers to operate for as long as 10 years compared to the 1.5

to 2 years of operation for alkaline batteries. This is mostly because of the very high stability of the lithium batteries allowing for trouble-free operation. Many other lithium batteries have been tried in this application, but the Li-I_2 system is now almost universally used, and the gain in patient health from fewer surgeries is substantial. Table 4 gives the energy density of the system, and Figure 19 shows a cutaway view of the letter *D*–shaped cell that is used to fit the electronics pacer package. These cells are extremely expensive because of the great care taken in manufacturing and the careful testing, labeling, and monitoring of each cell. The system is very sensitive to moisture because corrosion of both the lithium anode and the stainless steel container occur in the presence of moist iodine vapor. However, a completely dry cell does not form any gas and has no fluids to leak; it is this stability that has made the Li-I_2 cell so useful for medical batteries.

The limitation to low current densities seems intrinsic with the Li-I_2 system, however, so medical electronics developers are seeking equally stable batteries with higher current capabilities. The lithium–silver vanadium oxide battery shows promise for the defibrillator application (32).

Less expensive coin-type cells have also been developed for consumer electronics applications, but the severe current limitations have restricted the use of the cells.

The second class of lithium batteries is the liquid cathode group. The first successful lithium cell of the category was the lithium–sulfur dioxide cell (33). Lithium–sulfur dioxide cells (34) generally use either acetonitrile (AN) or PC or a mixture of the two as cosolvents with the SO_2, usually in amounts as high as 50 vol. %. This has the advantage of lowering the vapor pressure of the sulfur dioxide which at 25°C is about 300 kPa (3 atm).

The liquids are usually chilled, however, to make management of the toxic SO_2 gas easier during cell filling and sealing, which are the last steps in the cell construction. AN can be used as a cosolvent only because of the excellent film-forming properties of SO_2. Lithium is known to

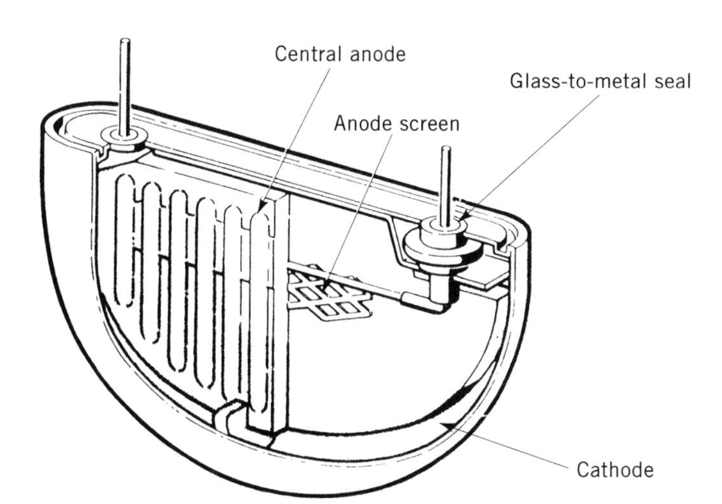

Figure 19. Cutaway view of lithium–iodine pacemaker cell in case-grounded, central anode configuration (36). Courtesy of Plenum Press.

react extensively with AN in the absence of sulfur dioxide. The other properties of AN, relatively high dielectric constant and very high fluidity, enhance the electrolyte conductivity and permit construction of high rate cells. The cell reaction proceeds according to

$$2Li + SO_2 \rightarrow Li_2S_2O_4 \quad (35)$$

where the cathode reaction, reduction of SO_2, takes place on a highly porous, >80%, carbon electrode. The lithium dithionite product is insoluble in the electrolyte and precipitates in place in the cathode as it is formed. The reaction involves only one electron per SO_2 molecule, which, along with the high volume of cosolvent in the cell, limits the capacity and energy density. Lithium bromide is the most commonly used salt, although sometimes $LiAsF_6$ is used, especially in reserve cell configurations. Because of the high rate capability, the cell has been widely used for military applications.

For both this system and the thionyl chloride system, safety has been a major focus, partly because of the high toxicity of the materials. Also, the potentially high rate of reaction during an accidental short circuit or other incident has led to some runaway reactions of high energy. However, tight control has allowed usage and they have become important batteries for military communications. In addition, the careful engineering of electrodes and vent mechanisms has controlled incidents. The cells have been subjected to extensive abuse testing.

All sulfur dioxide batteries have hermetic glass-to-metal seals to prevent loss of the highly volatile sulfur dioxide. They also have pressure-operated vents to accomplish emergency evacuation of the cell under abuse to prevent explosions. Safety has also been improved by using a balanced cell design (ie, an equal capacity of lithium and sulfur dioxide), so that low-rate discharge uses up all the lithium. The performance of these cells is generally optimized for high-current-density operation, and the battery is usually limited by blockage of the porous cathode structure with the insoluble lithium dithionite reaction product. Figure 20 gives a comparison of the energy output for lithium–sulfur dioxide cells with that for lithium-thionyl chloride liquid cathode cells. Both types of batteries have high energy density and specific energy (Table 4), because of the efficient use of space characteristic of a liquid cathode cell. Many cell types and sizes have been manufactured and many of these cells have been combined in series-parallel networks in batteries. They have been mostly made under contract for government use or, in some cases, for original equipment use for electronic device manufacturers. It is difficult for a consumer to purchase the cells separately.

Lithium-thionyl chloride cells, discovered shortly after Li-SO_2 batteries (35), have very high energy density. One of the main reasons is the nature of the cell reaction

$$4Li + 2SOCl_2 \rightarrow 4LiCl + S + SO_2. \quad (36)$$

The reaction involves two electrons per thionyl chloride molecule. Also, one of the products, SO_2, is a liquid under the internal pressure of the cell, facilitating a more complete use of the reactant. Finally, no cosolvent is required

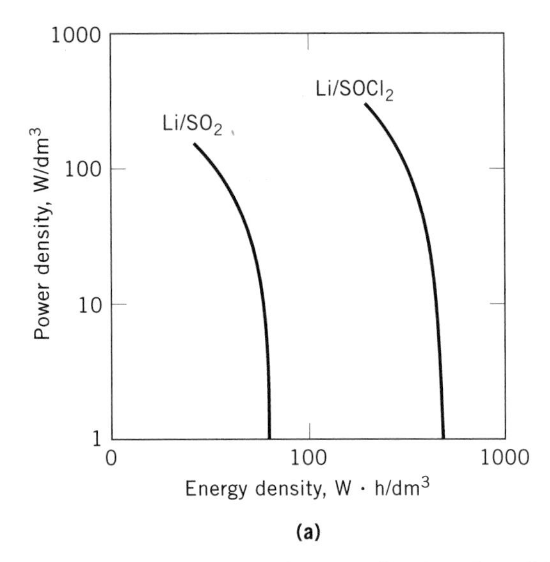

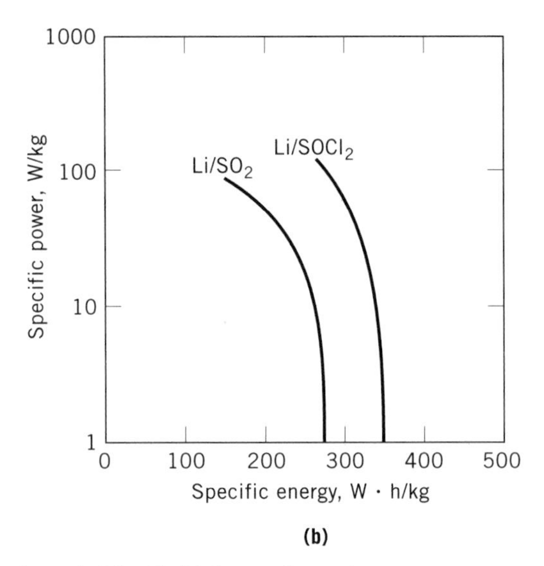

Figure 20. Lithium–sulfur dioxide and lithium–thionyl chloride high rate batteries profile with (a) power density vs energy density, and (b) specific power vs specific energy.

for the solution, because thionyl chloride is a liquid having only a modest vapor pressure at room temperature. The electrolyte salt most commonly used in $LiAlCl_4$. Initially, the sulfur product is also soluble in the electrolyte, but as the composition changes to higher SO_2 composition and the sulfur concentration builds up, a saturation point is reached and the sulfur precipitates. The end of discharge occurs as the anode is used up because the cells are usually anode limited. However, if very high discharge rates are used, the collector, which is usually made of a porous carbon such as that in the SO_2 cell, becomes blocked with the insoluble LiCl. In either case, the discharge curve is flat over time, then declines abruptly at the end of life. The system has an open-circuit voltage of 3.6 V. Because of the high activity of the cathode and the good mass transfer of the liquid catholyte, the cell is capable of very-high-rate discharges, as seen in Figure 20. The anode is also very active except for a tendency to form thick films of LiCl during open-circuit stand, much like the lithium-iodine system. The film can cause delay effects during startup of the cell. These effects can be largely eliminated by treating the anode with various types of polymeric films, which decrease the rate of corrosion and LiCl formation markedly. One of the best films for this purpose is polyvinyl chloride. The wide liquid range of the catholyte solutions gives the system a very good operating range, from very low (−50°C) to unusually high (>100°C) temperatures. The cells should not be stored at too high temperatures, however, because corrosion reactions can occur that reduce the cell life or cause an enhanced delay effect.

Several applications are developing for the lithium–thionyl chloride system. Because of the excellent voltage control, the battery is finding increasing use on electronic circuit boards to supply a fixed voltage for memory protection and other standby functions. These cells are designed in low-rate configuration, which maximizes the energy density and cell stability. Military and space applications are also developing because of the wide range of temperature/performance capability along with the high specific energy. They are also used in medical devices such as neurostimulators and drug delivery systems (32).

SECONDARY BATTERIES

Secondary or rechargeable batteries can be conveniently divided into the categories of lead acid, alkaline, and others including lithium/lithium-ion batteries. While many other batteries have been proposed and used in the past, these are the systems that are currently produced. Lead acid and iron air batteries are treated in another section; only the actively produced cells among the others are discussed in this section.

The estimated total sales (35) in the United States of secondary batteries in 1989 was $4,250 million. Of this total, over 85% was accounted for by the lead acid battery. The sales can be broken down further in terms of automotive batteries—$3,000 million, industrial; $450 million, sealed (or portable); $450 million; and $350 million, stationary. While large batteries are dominated by secondary cells, the small portable rechargeable cells have always been only a small fraction of the primary cell market. Another estimate puts the sales of small sealed rechargeables in 1992 at $312 million (15). However, as energy densities of rechargeable cells have improved with the introduction of nickel metal hydride cells and lithium ion cells, as well as improvements in nickel-cadmium cells, the markets for these batteries have also improved—and at a faster rate of growth than the primary cell market. Global production of all types of nickel-cadmium batteries was 1,100 million units, with Japanese manufacturers responsible for about 600 million units, according to another source (36).

Several other systems are in various stages of research and development, but have not yet reached production

Table 5. Rechargeable Battery Systems in Research and Development

Common Name	Nominal Voltage	Comments
Nickel–zinc	1.65	Limited cycle life, high rate capability
Zinc–bromine	1.85	Bromine complex in circulating liquid phase, containment problems
Lithium aluminum–FeS	1.33	High temperature fused salt electrolyte, corrosion problems
Sodium–sulfur	2.1	High temperature liquid electrodes, solid β-Al_2O_3 separator, containment problems
Sodium–nickel chloride (or iron chloride)	2.5 (or 2.3)	Solid β-Al_2O_3 separator, solid positive electrode, with high temperature molten salt as catholyte, molten sodium negative electrode
Aluminum–air	1.6	Replaceable negative electrode, circulating electrolyte, low efficiency negative electrode
Lithium–TiS_2	2.3	Room temperature operation, air sensitive positive electrode material, lithium plating problems
Lithium–MnO_2	3.0	Room temperature operation, somewhat air sensitive positive, lithium plating problems
Lithium–V_6O_{13}	2.5	Solid polymer electrolyte based on radiation cross linking, must maintain electrode-polymer interfaces through long term cycling

status. These systems are summarized in Table 5. The lithium and lithium ion cells are now on the market, but are in limited production and sales figures are not yet available.

Rechargeable Alkaline Batteries

Many types of rechargeable alkaline batteries have been manufactured over a long period. The early history is documented in reference (9), to which the reader is referred for details. F. de Lalande and G. Chaperon recognized the benefits of the alkaline electrolyte (concentrated KOH or NaOH solutions in water) in a zinc–copper oxide cell as early as 1881 and the system was subsequently developed as a secondary battery. Both W. Jungner and T. A. Edison recognized the strong deficiencies of the copper oxide (which had too high a solubility and migrated to the zinc electrode). With the discovery by Jungner of nickel oxyhydroxide and its high positive potential both he and Edison independently developed nickel-cadmium and nickel-iron batteries. The nickel-iron battery is no longer an important commercial battery, although it was widely used in the early part of the twentieth century as a traction battery and is again under study for this application. The nickel-cadmium system is still an important battery with many uses. Other alkaline batteries have a more recent history, but are gaining importance compared with the nickel-cadmium system. The different types of alkaline cells are summarized in Table 6 along with their principal applications and a list of advantages and disadvantages. Table 7 gives some of the electrochemical properties, the open-circuit voltages, the capacities of typical cells, the energy densities and specific energies at the usual rates of discharge for the cells.

Because nickel-cadmium cells have been produced for so many years, the design principles and basic electrochemistry are well understood. The battery is normally manufactured in the discharged state and the active materials are composed principally of $Ni(OH)_2$ (the nickel electrode) and cadmium hydroxide or oxide (the cadmium electrode). The system has been widely used in the sealed version as the workhorse for high-current applications such as power tools and for high-cycle-life applications such as computer power supplies.

A typical design for the sealed, jelly-roll cell is shown in Figure 20. The latter application is also fostered by the capability of the system for high-rate charging (as short a time as 15 minutes is sufficient when using special chargers). The cell is also manufactured as a wet cell in two versions; the sintered cell, which is mainly used for standby power and aircraft starting batteries, and the pocket cell (developed originally by Jungner in 1896), which is used for diesel engine starting, emergency lighting, and other applications where the extra expense compared to lead acid batteries is exchanged for the excellent performance capability. Large sealed cells have also been widely used in space applications where the excellent reliability of the system and high cycle life has been utilized for synchronous low earth orbit applications in conjunction with solar cells. The battery supplies power on the dark side of each orbit and is recharged on the light side by photovoltaic solar cells.

The chemistry is the same for all three types of batteries. The cadmium electrode is charged and discharged in the overall reaction:

$$Cd + 2OH^- \rightleftharpoons Cd(OH)_2 + 2e^- \quad (37)$$

where the upper arrow shows the direction of the discharge reaction and the lower arrow the charge reaction. The nickel reaction is complicated by the existence of two additional phases that are transformed under various conditions into the original phases. Thus, the normal reaction is

$$\beta - NiOOH + H_2O + e^- \rightleftharpoons \beta - Ni(OH)_2 + OH^- \quad (38)$$

and the extra phases undergo the electrochemical reactions

$$\gamma - NiOOH + (1 - x)H_2O + e^- \rightleftharpoons \alpha - Ni(OH)_2 \cdot xH_2O + OH^-. \quad (39)$$

Table 6. Alkaline Rechargeable Batteries

System	Battery Types	Applications	Advantages	Disadvantages
Nickel–Cadmium	1. Prismatic cells with pocket electrodes 2. Vented prismatic cells with sintered electrodes 3. Sealed cylindrical cells with jelly roll electrodes and prismatic cells with stacked electrodes. 4. Button cells with pressed powder electrodes	1. Stationary power-Remote power uses, load leveling, etc. 2. Aircraft batteries for starting and standby applications. 3. Portable high energy, high power applications. 4. Portable low power applications.	1. Easy to manufacture; long history 2. High rate capability, high electrode efficiencies, high cycle life. 3. Capable of mass production, no maintenance design. 4. Mass production capable, small size.	1. Low energy per volume and weight, relatively low efficiency. 2. Suitable for large size only, high maintenance. 3. Relatively low electrode efficiencies. 4. Relatively low current capability. All-cadmium toxicity.
Nickel–Hydrogen	Pressure vessel-thin electrode stacks.	Satellite power systems.	Very high cycle life and high electrode efficiency. Easily charged and abuse resistant.	Poor energy density, difficult to manufacture, poor charge retention, very expensive.
Nickel–metal hydride	Sealed cylindrical, jelly roll electrodes	Portable high energy, high power applications.	High energy density, high negative electrode efficiency. High cycle life capability. No cadmium or lead.	Poor charge retention compared to Ni-Cd. Cost and availability of rare earth and transition metal elements.
Manganese dioxide–zinc	Sealed cylindrical cells with concentric cylinder electrodes	Portable low to moderate power applications.	Inexpensive, direct replacement for primary alkaline batteries, environmental advantages.	Very poor cycle life, rapid capacity fade. Mercury contamination in some designs.
Silver–zinc	Prismatic cells with flat plate electrodes	Military aerospace, submarine and communications.	Very high rate and energy density.	Very expensive due to silver component, poor cycle life.

The γ-NiOOH phase appears in the system as a result of overcharging of the β-NiOOH electrode. This material is then reduced during discharge by eq. 48 to the α-form, which slowly converts in the strongly alkaline electrolyte to the β-phase. These reactions are disruptive to the smooth functioning of the nickel electrode because of the additional water demand to form the γ oxidized phase and the corresponding volume change of this solid material.

The cell is typically designed in a positive limited configuration. That is, the capacity of the negative electrode is greater than that of the positive electrode. The result of this is that as the system approaches full charge, the positive electrode begins to undergo oxidation of the hydroxyl ion to form oxygen:

$$2OH^- \rightarrow 1/2(O_2) + H_2O + 2e^- \quad (40)$$

The oxygen thus generated migrates to the negative electrode where part of the current continues to be the reduction of cadmium hydroxide to cadmium (eq. 37), but now also includes the reduction of the oxygen to hydroxyl ion, the reverse of eq. 40. This cycle of oxygen generation and

Table 7. Electrochemical Properties of Rechargeable Alkaline Batteries

Cell Type	Open Circuit Voltage Cell Capacity	Energy Density, Wh/L	Specific Energy, Wh/kg
Nickel–cadmium			
1. Stationary	[1.3 Volts]		
2. Aircraft	1. 315 Ah	1. 36	1. 21
3. Sealed cylindrical or prismatic	2. 40 Ah	2. 72	2. 31
	3. 4 Ah	3. 90	3. 34
4. Button	4. 60 mAh	4. 63	4. 21
Nickel–hydrogen	[1.3 Volts] 55 Ah	60	50
Nickel–metal hydride	[1.3 Volts] 1.5 Ah	184	55
Manganese dioxide–zinc	[1.5 Volts] 1.5 Ah	240 (cycle 1)	80 (Cycle 1)
		125 (cycle 10)	42 (Cycle 10)
Silver–zinc	[1.6 Volts] 100 Ah	190	90

reduction occurs smoothly as long as there is sufficient porosity within the electrodes and separator to allow adequate oxygen gas transport. The electrolyte is carefully metered into the cell to ensure that the required porosity is present. Because the reaction does no useful work, however, considerable heat is generated because of the reaction. The design must accommodate this heat because the reaction is essential to safe operation of the cell. If the cell is accidentally made negative limiting, a hydrogen cycle occurs, but the reduction of hydrogen gas is not very effective. Therefore, pressure can build up in the cell and venting can occur with corresponding loss of electrolyte. This situation is undesirable, and cells are carefully controlled to avoid it. However, because excess pressure is always to be avoided in batteries, sealed nickel-cadmium batteries include a vent that is activated before deformation of the can so that pressure is relieved safely in the cell (37). If the vent is the resealable type (38), as in the most advanced designs, the cell can go on functioning for some time after such an event.

The electrodes in sealed nickel-cadmium batteries have undergone considerable development in recent years in a search for higher energy density. Originally, sintered nickel strips of relatively high porosity were impregnated by solutions of nickel or cadmium salts and the corresponding hydroxides precipitated in place by addition of a sodium or potassium hydroxide solution. Alternatively, a water electrolysis reaction at the sintered surface gives a local change in pH to alkaline conditions, resulting in precipitation of nickel hydroxide on the sintered surface. Details are given in reference (36) and see also Fig. 21. These sintered electrodes are capable of high currents during both charge and discharge and, because of the good mechanical and electrochemical stability of the nickel sinter, give very high cycle life. The construction is expensive, however, and is not capable of high loading of active material because of the difficulty of impregnation and the relatively low porosity of the sintered substrate (about 75%). Negative electrodes utilizing polymeric binders allow an increase in energy density and a lowering of cost (37). There is a sacrifice in cycle life with this approach, however, as the cadmium has more freedom of motion than in a sintered body. Another direction used to increase the energy density of the negative electrode is through the use of a fibrous nickel mat of very fine fibers and high porosity (37,39). A recent development for the positive electrode involves pasting a blend of nickel hydroxide and conductor into a very high porosity (exceeding 95%) nickel metal foam structure (37).

Figure 21. Partial cutaway of a coiled, sintered-plate, "D"-size nickel cadmium cell (39).

Special precipitation processes can be used to make a spherical high-density nickel hydroxide for these electrodes that allows an additional increase in energy density. These electrodes and other cell components are still undergoing development, and substantial improvements in energy density are likely, at least for moderate and low-rate cells. Additives such as cobalt oxide, cadmium, and zinc have also been shown to be important in reducing swelling of the positive electrode, evolution of oxygen, and minimizing the formation of the α-phase (37). Cells of over 50 Wh/kg and 140 Wh/l have been demonstrated (40) under special conditions.

Nickel–hydrogen batteries have gained prominence mainly by their use in high-profile applications in space missions. They are currently widely used in satellites for both civilian and military purposes. The principles are simple, as they represent a fusion of the gaseous hydrogen electrode technology developed many years ago in fuel cell work, usually utilizing platinum metal catalysts, and the nickel electrodes from aerospace nickel-cadmium batteries. Because hydrogen electrodes are quite reversible when properly catalyzed, the same electrodes are used to generate the hydrogen from the water in the cell on charge and to reduce the hydrogen on discharge:

$$H_2 + 2OH^- \rightleftharpoons H_2O + 2e^- \quad (41)$$

The nickel electrode operates in the same way as in the nickel-cadmium cell (eq. 38) except that the negative electrode is now the limiting electrode so that oxygen is not evolved in the charging process. Thus a potentially dangerous situation in which oxygen and hydrogen could be present in an explosive mixture is avoided. The hydrogen is allowed to increase in pressure to a safe design level (eg, 50 bar), with a burst pressure of the vessel at 200 bar (41). The pressure vessel is carefully constructed of Inconel (40) or stainless steel (42) and operation is generally designed at about 10°C to facilitate heat dissipation. The nickel electrode reacts only slowly with the gaseous hydrogen, so the cell is well suited to the short delays between charging and discharging and the short discharge times of the low earth orbit. A finite and continuous self-discharge does occur, however, that necessitates the dissipation of heat and makes the cell unattractive for more varied terrestrial applications. The specific energy is quite high for the system (Table 7), which is an important feature for space applications, and the cycle life behavior is extraordinarily good, with over 40,000 cycles obtained. Thus over 10 years of low earth orbit can be obtained from such a battery with 10 charge/discharge cycles occurring per day. Present developments attempt to increase the flexibility of the system by enclosing multiple cell stacks in a common pressure vessel (42).

The most recently developed rechargeable alkaline cell is the nickel–metal hydride cell. There are actually two competing versions of the cell involving two different alloy types to store the hydrogen as a metal hydride. One alloy is an AB_2 type that uses a group VIII metal such as Ni for the B element and a group IV metal such as titanium for the A element. The other type is an AB_5 type that uses a rare earth element such as lanthanum for the A element and a Group VIII metal such as nickel for the B component. Many additional alloying elements have been added to both base formulations to improve the hydride formation properties (37). As can be seen from Table 7, the voltage of the nickel–metal hydride cell is virtually the same as that of the nickel–cadmium system. The energy density is superior to that of the nickel–cadmium battery.

Of course, because the nickel–metal hydride cell is quite new, recent improvements in the nickel electrode have been incorporated. Thus, the energy density should be compared with the values of 110 to 150 Wh/l quoted for the most advanced types of nickel–cadmium. The losses on shelf are much less than that of the nickel–hydrogen system because the alloy is designed to reduce the vapor pressure of hydrogen gas to only a few bars compared with the 50 or so bars of the nickel-hydrogen cell. Also, because the negative electrode reaction is based on the hydrogen contained in the alloy, the gas does not need to have access to the electrode stack, and efforts are made in the cell design to limit the access of the atmospheric hydrogen within the cell to the positive electrode. Again, the negative electrode is the limiting electrode in order to forestall the evolution of oxygen from the positive electrode during charge and prevent the presence of an explosive mixture of hydrogen and oxygen.

Experimentation continues on alloys for the negative electrodes to further reduce the hydrogen equilibrium pressure and the cost, the separators to reduce the flux of hydrogen between the electrodes, the positive electrode structure to continue to densify the electrode, and additives for the positive electrode to reduce electrode swelling (43). If the shelf properties and cycle life can be improved together, the system shows much promise for critical needs in powering portable computers, camcorders, portable telephones, and so on.

The most recent rechargeable alkaline battery in the marketplace is the well-known primary battery system, the manganese dioxide–zinc battery. The system is in a second generation as a previous attempt to market the battery was made in the 1960s by Union Carbide in the form of a two-cell battery for portable television sets and other consumer electronics of the time (9). Because of shape change and dendrite growth of the zinc in the negative electrode, special requirements are placed on the separator. Generally a laminated cellophane separator is used to restrict the growth of the dendrites, which otherwise short out the battery in a few cycles. In recent years, some improvements in these zinc containment properties and in battery design have led to a resurgence of interest in the system. The main advantage of the system compared with other rechargeable alkaline batteries is the low cost, partly because of low materials cost and partly because of low manufacturing cost. A relatively few cycles could justify the purchase of such batteries if the properties are acceptable and the charger costs are not too high.

Rayovac Corp. has introduced such a battery based on technology developed at Battery Technology, Inc. (44). The use of proprietary separator materials, graphite in the negative electrode as a conductive agent, and various additives to act as corrosion inhibitors have been advanced as improvements for system function. The system still has a severe fade in capacity, however. The capacity after about 10 cycles at 300 mA drain in an AA cell is about half of the initial capacity (44). Special chargers are required to control the upper limit of voltage and current on charge. The cell must be carefully made to maintain an anode limited configuration to prevent a deep discharge of the manganese dioxide, which becomes highly irreversible in this condition.

The silver-zinc alkaline battery is of importance for its military applications. The system no longer receives much research and development effort, however, so it is likely that it will be surpassed by other systems for many of these applications (9,36). The electrode reactions are the same as for the primary cell (eq. 26–27) except that the high voltage plateau of divalent silver oxide is purposely sought in order to raise the energy density of the system. For torpedo propulsion and many other applications, the two-step voltage characteristic is not damaging, but the resulting extra energy is very important. When the system is discharged fully, only about 10 to 25 cycles, or 80%, of the initial capacity is obtained. The problems are shape change and dendrite growth at the zinc electrode on cycling, such as the manganese dioxide–zinc cell, and attack of the separator with very highly oxidizing silver ions in solution. The exceptional energy density of the system, however, makes up for this deficiency in some special applications. The high cost of the system noted in Table 6 owing to the expensive silver cathode material is a further detriment of the system. If the cells are stored at room temperature or lower, the charge retention is quite good, which is another advantage for military applications. It is unlikely that this system will be developed for nonmilitary applications because of the deficiencies noted.

Other Secondary Batteries

As in the field of rechargeable alkaline batteries, a great many other electrochemical systems have been proposed, developed and produced for some period of time. This section deals with currently produced batteries and briefly discusses the systems that are under development.

A group of battery systems called lithium-ion, rocking chair, or swing batteries is of great current interest. The concept of the batteries is deceptively simple. A lithiated transition metal oxide such as $LiCoO_2$, $LiNiO_2$, or $LiMn_2O_4$ is mixed with a conductor and binder and coated on a metal foil for use as a positive electrode. Only the $LiCoO_2$ is used at present in commercial cells. The material is in the discharged state. The negative electrode is composed of one of several different kinds of carbon or graphite and a binder and coated on another metal foil and is also in the discharged state. The cell is charged after production. The nature of the cell reaction on charge is that the positive electrode is oxidized in a homogeneous reaction to form a lithium-deficient transition metal oxide with the transition metal in a higher oxidation state (45):

$$LiCoO_2 \leftrightharpoons Li_xCoO_2 + (1 - x)Li^+ + (1 - x)e^- \quad (42)$$

The negative electrode reaction depends to a certain extent on the nature of the carbon (46). An amorphous carbon appears to undergo a homogeneous reaction with the lithium to form a lithium insertion compound of variable stoichiometry. A graphitic carbon undergoes a staged reaction with the lithium whereby definite stoichiometric compounds are formed sequentially as the electrode is charged. The final state has the stoichiometry C_6Li. The reaction can be written generically as

$$C + yLi^+ + ye^- \leftrightharpoons CLi_y \quad (43)$$

with the understanding that the reaction may go through fixed stages of lithiation. Batteries of this type have been on the market for only about two years in special applications such as portable telephones, camcorders, and computers (47). The midlife voltage is about 3.6 V and the charged open-circuit voltage is about 4.1 V. The unit cells are cylindrical, jelly-roll designs, often of the same size as nickel-cadmium batteries. Because of the substantial voltage difference from aqueous systems, however, they are sold only in battery packs. The energy density of the cells is as high as 250 Wh/l and the specific energy 115 Wh/kg. Sony Energytech quotes a cycle life at 1200 cycles, reasonably high current capability and a self-discharge rate of 10% per month (47). A & T Battery Corp. of Japan claims similar performance with their battery (48) as does Japan Storage Battery (49). In order to ensure safety, a complicated and expensive charging system that monitors the voltage of each cell in the battery has been developed. Many researchers are studying variations of the rocking chair battery and further developments are assured. A national goal in Japan is to develop this battery for use in electric vehicles and for energy storage services (49).

Rechargeable lithium batteries have also undergone rigorous testing at many laboratories. A commercial venture in the field was withdrawn, however, when several cells vented in the portable telephones in which they were installed (50). The cells had a lithium foil negative electrode and an MoS_2 positive electrode that had been specially treated. Considerable work continues on rechargeable lithium batteries with liquid electrolytes because of the high energy densities and specific energies possible with this technology (51). Typical positive electrode materials under study include MnO_2, V_2O_5, and TiS_2. Many of the problems with the liquid electrolytes for rechargeable lithium cells have been discussed (52).

Another approach to the rechargeable lithium and lithium-ion cells is to use an immobilized polymer electrolyte as the ionically conductive medium rather than a liquid electrolyte with a polymer separator. Some lithium salts dissolve to a high degree in certain polymers such as polyethylene oxide; the conductivity is sufficiently high that battery operation can occur at somewhat elevated temperatures such as 50°C. However, when the polymer salt combination is enhanced by incorporation of a stable solvent to plasticize the polymer, the conductivity rises and room temperature operation is possible. These plasticized polymers are currently under intensive study and show promise of enhanced battery safety and high energy at moderate to low current drains (52). Some positive electrode materials under study for polymer electrolytes include V_6O_{13}, MnO_2, TiS_2, and $LiMn_2O_4$.

Two elevated-temperature batteries have received extensive development efforts because of the possibility that they could become the favored power source for electric vehicles (51). The U.S. Department of Energy and many private companies and other governments have contributed greatly to the effort, although the batteries are not yet available for market testing. The two systems are the iron sulfide–lithium aluminum battery with a molten chloride electrolyte and the sodium polysulfide-sodium battery with a solid beta alumina separator. The iron sulfide battery operates at about 400°C, below the melting point of the lithium aluminum alloy. The sodium battery operates at about 300°C, well above the melting point of sodium and the sodium polysulfide. Thus, in the first case, the electrodes are solids and the electrolyte is a highly conductive molten salt. In the second case, the electrodes are both in the liquid phase, but the separator is a sodium-ion conductive ceramic material that keeps the two liquids completely separate. A variation of these cell designs is the sodium-nickel (or iron) chloride battery, which uses a molten salt electrolyte in the positive electrode compartment and molten sodium in the negative compartment; the two compartments are separated by β-Al_2O_3 tubes (55). The temperature of operation is about 250°C. The problems of sealing the batteries for long-term operation, containing the highly reactive constituents, and providing for safety under conditions of severe abuse such as would occur in an automobile accident have proved to be formidable. The need for such power sources is so great, however, that the efforts with these systems continue.

RESERVE BATTERIES

This type of battery has been developed for applications that require an intense discharge of high energy and power, and sometimes operation at low ambient temperature. The batteries must also have the capability of a long inactive shelf period and rapid activation to the ready state. These batteries are usually classified by the mechanism of activation employed. The types are a) water-activated batteries, using fresh or sea water, b) electrolyte-activated batteries, using the complete electrolyte or only the solvent, c) gas-activated batteries, using the gas as an active cathode material or as part of the electrolyte, and d) heat-activated batteries or thermal batteries, using a solid salt electrolyte that is activated by melting on application of heat. Activation involves adding the missing component just before use, which can be done in a very simple way such as pouring the water into an opening in the cell for water-activated cells or in a more complicated way by using pistons, valves, or heat pellets activated by gravitational or electric signals for the electrolyte- or thermal-activation types. Such batteries may be stored for 10 to 20 years while awaiting use. Reserve batteries are usually manufactured under contract for various government agencies (mostly defense departments) although occasional industrial or safety uses have been found. Because of the many electrochemical systems involved in these types of batteries, this type of cell is reviewed only briefly here, and the reader is referred to reference (54) for details on these systems.

Many battery types are used in reserve cells, including several discussed under Primary Batteries. For example, the lithium–thionyl chloride and the lithium–sulfur dioxide systems are often used in reserve configurations, in which electrolyte, stored in a sealed compartment, is released on activation and may be forced by a piston or inertial forces into the interelectrode space. These high-energy-density systems have taken over some of the applications of the older liquid ammonia reserve batteries, which usually use a magnesium anode and a metadinitrobenzene cathode. Much of the engineering done on flowing liquid ammonia into the battery was taken over by the lithium cells with suitable material changes. Another new reserve battery is the lithium–vanadium pentoxide battery, which performs at high rates and high energy. Most applications for such batteries are in mines and fuzes used in military ordnance.

An interesting variant of the liquid cathode reserve battery is the lithium water battery in which water is the liquid cathode and is also used as the electrolyte. A certain amount of corrosion of the lithium occurs, but sufficient lithium to compensate for the loss is provided. The reaction product is soluble lithium hydroxide. In some cases, a solid cathode material, such as silver oxide, or another liquid reactant, such as hydrogen peroxide, is used in combination with the lithium anode and aqueous electrolyte to improve the rate or decrease the amount of gas given off by the system. These cells are mostly used in the marine environment where water is available or compatible with the cell reaction product. Common applications are for torpedo propulsion and powering sonobuoys and submersibles. An older system still used for these applications is the magnesium–silver chloride seawater-activated battery. This cell is much heavier than the corresponding lithium cell, but the buoyancy of the sea makes this less of a detriment. The magnesium-silver chloride cell is also useful for powering emergency communication devices for airplane crews whose planes have come down in the sea.

The last type of reserve cell is the thermally activated cell. The older types use calcium or magnesium anodes; newer types use lithium alloys as anodes. Lithium forms many high melting alloys such as those with aluminum, silicon, and boron. Furthermore, lithium can diffuse rapidly within the alloy phase, permitting high currents to flow. The electrolyte for both the older and newer chemistries is usually the eutectic composition of lithium chloride and potassium chloride which melts at 352°C. This electrolyte has temperature-dependent conductivities an order of magnitude higher than the best aqueous electrolytes. The high conductivity and the enhanced kinetics and mass transport allow the battery to be discharged at a very high rate of several A/cm^2 with complete discharge in 0.5 s. The cathodes for the older calcium anode cells are typically metal chromates such as calcium chromate. The anode reaction product is calcium chloride, and the cathode product is a mixed calcium chromium oxide of uncertain composition. One of the best cathodes for lithium alloy cells is FeS_2, which forms a system very similar in reaction mechanism to the lithium–iron disulfide cell.

The heat pellet used for activation is usually a mixture of a reactive metal such as iron or zirconium and an oxidant such as potassium perchlorate. An electrical or mechanical signal ignites a primer, which then ignites the heat pellet, which melts the electrolyte. Sufficient heat is given off by the high current to sustain the necessary temperature during the lifetime of the application. Many millions of these batteries have been manufactured for military ordnance—rockets, bombs, missiles, and so on.

BIBLIOGRAPHY

1. G. W. Heise and N. C. Cahoon, eds., *The Primary Battery*, Vol. 1, John Wiley & Sons, New York, 1971.
2. A. J. Salkind, ed., *History of Battery Technology*, Proc. Vol. 87–14, The Electrochemical Society, Pennington, NJ, 1987.
3. A. Galvani, quoted in Ref. 1.
4. A. Volta, quoted in Ref. 1.
5. M. Faraday, *Experimental Researches in Electricity, 1831–55*, Vol. I, Richard and John Edward Taylor, London, 1849; Vol. II, Bernard Quaritch, London, 1844; Vol. III, Taylor and Francis, London, 1855.
6. G. Planté, *Compt. rend.* **50,** 640 (1860).
7. Fr. Pat. 71,865 (1866), G. Leclanché.
8. U.S. Pat. 373,064 (1887), C. Gassner.
9. S. U. Falk and A. J. Salkind, *Alkaline Storage Batteries*, John Wiley & Sons, New York, 1969.
10. U.S. Pats. 2,422,045 (1947); 2,422,046 (1947) S. Ruben.
11. A. J. Bard, R. Parsons, and J. Jordan, eds., *Standard Potentials in Aqueous Solution*, Marcel Dekker, Inc., New York, 1985.
12. R. Brodd, "Batteries" (Introduction) in *Kirk-Othmer Encyclopedia of Chemical Technology—4th ed.*, John Wiley & Sons, Inc., New York, 1992, p. 963.
13. K. J. Vetter, *Electrochemical Kinetics*, Academic Press, New York, 1967.
14. J. Newman, *Electrochemical Systems*, Prentice-Hall, Englewood Cliffs, NJ, 1973.
15. Anon., *The Dry Cell Battery Market, Packaged Facts*, The International/Research Co., New York, 1993.

16. I. Kuribayashi, *JEE* **30**:322, 51 (Oct. 1993).
17. F. L. Tye, in T. Trans and M. Skyllas-Kazeos, eds., *Proceedings of the 7th Australian Electrochemistry Conference*, The Electrochemistry Division, The Royal Australian Chemical Institute, 1988, pp. 37–48.
18. R. G. Burns and V. M. Burns, in A. Kozawa and R. J. Brodd, eds., *Proceedings of the Manganese Dioxide Symposium*, Vol. 1, Cleveland, I. C. Sample Office, Cleveland, Ohio, 1975, p. 306.
19. *American National Standard Specification for Dry Cells and Batteries*, ANSI C18.1-1979, American National Standards Institute, New York, May, 1979.
20. Anon., *Eveready Battery Engineering Data*, Vol. 2A, Eveready Battery Co., St. Louis, MO, 1990.
21. A. Kozawa and J. F. Yeager, *J. Electrochem. Soc.* **112,** 959 (1965); A. Kozawa and R. A. Powers, *J. Electrochem. Soc.* **113,** 870 (1966).
22. R. A. Putt and A. I. Attia, *Proceedings of the 31st Power Sources Symposium*, The Electrochemical Society, Pennington, NJ, 1984, p. 339.
23. J. W. Cretzmeyer, H. R. Espig, and R. S. Melrose, in D. H. Collins, ed., *Power Sources 6*, Academic Press, New York, 1966, p. 269.
24. E. Peled in J. P. Gabano, ed., *Lithium Batteries*, Academic Press, New York, 1983, p. 43.
25. G. E. Blomgren in J. P. Gabano, ed., *Lithium Batteries,* Academic Press, New York, 1983, p. 13.
26. H. V. Venkatasetty, ed., *Lithium Battery Technology*, John Wiley & Sons, Inc. New York, 1984, pp. 11 and 13.
27. H. Ikeda, T. Saito, and H. Tamura, in Ref. 25, p. 384.
28. J. C. Nardi, *J. Electrochem. Soc.* **132,** 1787 (1985).
29. U.S. Patent. 3,536,532 (1970) and 3,700,502 (1972), N. Watanabe and M. Fukuda (to Matsushita Electric Co.).
30. M. B. Clark in Ref. 25, p. 115.
31. R. A. Langan, *Wescon/92, Conference Record*, Western Periodicals Co., Ventura, CA, 1992, p. 551.
32. C. F. Holmes in *Lithium Batteries: New materials, Developments and Perspectives*, Elsevier, New York, 1994, p. 377.
33. U.S. Pat. 3,567,515 (1971), D. L. Maricle and J. P. Mohns (to American Cyanamid).
34. C. R. Walk in Ref. 25, p. 281.
35. Ger. Pat. 2,262,256 (1972), G. E. Blomgren and M. L. Kronenberg (to Union Carbide Corp.).
36. A. J. Salkind and M. Klein, "Batteries (Secondary Cells)" in *Kirk-Othmer Encyclopedia of Chemical Technology—4th ed.*, John Wiley & Sons, Inc., 1992, p. 1028.
37. D. Berndt, *Maintenance Free Batteries*, John Wiley & Sons, New York, 1993.
38. Anon., *Rechargeable Batteries Applications Handbook*, Butterworth-Heinemann, Boston, 1992.
39. M. Oshitani, K. Takashina, and Y. Matsumara, in D. A. Corrigan and A. H. Zimmerman, eds., *Nickel Hydroxide Electrodes*, Proceedings Vol. 90–49, The Electrochemical Society, Pennington, NJ, 1990, p. 197.
40. U. Kohler and C. Claus in G. Kreysa and K. Juttner, eds., *Elecktrochemische Energiegewinnung*, DECHEMA-Monographien; Bd 128, Deutsche Gesellschaft für Chemisches Apparatewesen, Chemische Tecknik und Biotechnologie e.V., Frankfurt am Main, 1993, p. 213.
41. M. R. H. Hill, J. Lumsdon, T. L. Markin, and D. N. E. Smith in T. Keily and B. W. Baxter, eds., *Power Sources 12*, International Power Sources Symposium Committee, Leatherhead, Surrey, UK, 1989, p. 379.
42. J. R. Wheeler and W. D. Cook, *Proc. 34th Internat. Power Sources Symp.*, Institute of Electrical and Electronics Engineers, New York, 1990, p. 239.
43. See *New Sealed Rechargeable Batteries and Supercapacitors*, B. M. Barnett, E. Dowgiallo, G. Halpert, Y. Matsuda and Z-i. Takehara, eds., *Proceedings* Vol. 93–23, The Electrochemical Society, Pennington, NJ, 1993.
44. K. Kordesch, L. Binder, W. Taucher, C. Faistauer, and J. Daniel-Ivad in A. Attewell and T. Keilly, eds., *Power Sources 14*, International Power Sources Symposium Committee, London, 1993, p. 193.
45. C. Delmas in G. Pistoia, ed., *Lithium Batteries: New Materials, Developments and Perspectives*, Elsevier, New York, 1994, p. 457.
46. J. R. Dahn, A. K. Sleigh, H. Shi, B. M. Way, W. J. Weydanz, J. N. Reimers, Q. Zhong, and U. von Sacken, ibid. Ref. 41, p. 1.
47. *J. Electronics Industry*, **40:**2, Series No. 462, 28 (Feb. 1993).
48. I. Kuribayashi, *JEE*, **30:**322, 51 (Oct. 1993).
49. *J. Electronics Industry*, **41:**2, Series No. 474, 18 (Feb. 1994).
50. G. Stix, *Scientific American*, 108 (October 1993).
51. P. R. Gifford, "Batteries (Secondary Cells)" in *Kirk-Othmer Encyclopedia of Chemical Technology*—4th ed., John Wiley & Sons, Inc., New York, 1992, pp. 1107–78.
52. L. A. Dominey, in Ref. 51, Ref. 41, p. 137.
53. J. Coetzer, *J. Power Sources*, **18,** 377 (1980).
54. D. Linden, ed., *Handbook of Batteries and Fuel Cells*, McGraw-Hill, New York, 1984.

BIODIVERSITY MAINTENANCE

ROBERT C. SZARO
USDA Forest Service, Research
Washington, D.C.

One of today's most pressing environmental issues is the conservation of biodiversity. Many factors threaten the world's biological heritage. The challenge is for nations, government agencies, organizations, and individuals to protect and enhance biodiversity while continuing to meet people's needs for natural resources. This challenge exists from local to global scales. If not met, future generations will live in a biologically impoverished world and perhaps one that is less capable of producing desired resources as well. (See also ENVIRONMENTAL AND CONSERVATION ORGANIZATION; FOREST RESOURCES; FORESTRY, SUSTAINABLE.)

Biodiversity is a multi-faceted issue that crosses all traditional resource boundaries and requires a true interdisciplinary approach to develop potential solutions. Many of the concepts and theories on biodiversity are still in the formulation stages but this situation does not prevent the implementation of a variety of projects designed for its conservation. Many pieces of the puzzle have been under study for years and provide an adequate base for conservation decisions. Threatened, endangered, and sensitive species programs have long been the focal point of biodiversity concerns on the species level but they represent only the tip of the iceberg. The magnitude of the problem (more than 500 threatened and endangered species in the U.S. alone with at least another 1000 proposed for listing) makes it clear that we do not have the resources for intense management programs for all species if only because of the enormity of the problem. There is even less hope in tropical environments where the potential numbers of species, particularly insect species, number in the millions (1). Perhaps because organic diversity is so much larger than previously imagined, it has proved difficult to express as a coherent subject of scientific inquiry (2).

Clearly every effort should be made to conserve biodiversity (3–5). The conservation of biodiversity encompasses genetic diversity of species populations, richness of species in biological communities, processes whereby species interact with one another and with physical attributes within ecological systems, and the abundance of species, communities, and ecosystems at large geographic scales (6). In recent years, traditional uses of forested lands in North America have become increasingly controversial (7). The demands and expectations placed on these resources are high and widely varied, calling for new approaches that go beyond merely reacting to resource crises and concerns (8).

The greatest challenge for managing biodiversity lies in preparing for the disturbances and changes that are a fundamental feature of natural ecosystems. Many of these disturbances are the direct result of human activities. Expanses of pristine forest that support an enormous diversity of wildlife and plants and a richness of human cultures are being rapidly converted into vast wastelands that support a few tough, fire-resistant weeds and perhaps some cattle, while people scrounge for food and fuelwood from the newly–degraded soils and sparse shrubbery (9). Yet, we cannot conserve biodiversity simply by preserving areas and trying to prevent all changes, whether naturally occurring or human-caused. Nor can we conserve biodiversity by trying to maximize diversity on any particular site.

How can land managers react to the oftentimes painful dilemmas they face on an almost daily basis when making management decisions that can have potentially devastating impacts on ecosystem stability? The discipline of Conservation Biology has been described as a "crisis discipline, where limited information is applied in an uncertain environment to make urgent decisions with sometimes irrevocable consequences" (10). This really speaks to the heart of all land managers. They find themselves trying to find the balance between maintaining and sustaining forest systems while still providing the forest products needed by people. Trade-offs will be inevitable and will necessitate formulating and using alternative land management strategies to provide an acceptable mix of commodity production, amenity use, protection of environmental and ecological values, and biodiversity. Conserving biodiversity now is likely to alter immediate access to resources currently in demand in exchange for increasing the likelihood that long-term productivity, availability, and access are assured.

But is this dilemma something new? Are we the first to wrestle with these kinds of decisions? With massive simplification of landscapes? Plato in approximately 2350 B.C. describes an area in ancient Greece that was stripped of its soil following clearing and grazing (11). In fact, since the development of agriculture, there have been extensive modifications to the natural vegetation cover of every continent except Antarctica (12). Yet, never before have there been so many humans on earth taking advantage of its resources.

It is hardly surprising then, that global awareness and concerns for conserving biodiversity are continually increasing. When we have concerns for biodiversity we are saying we have a concern for all life and its relationships (4). As arguably the most intelligent species on earth we have a responsibility to try as much as possible for the continuance of all forms of life. But how can we go about this? The first step of trying to determine the amount, variety, and distribution of species, ecosystems, and landscapes will require more comprehensive inventories which must be followed by monitoring efforts to determine the impacts of management activities. Next, strategies must be developed and implemented for the preservation, maintenance, and restoration of forest ecosystems. These efforts should also incorporate strategies for the sustainable use of forest resources including more efficient utilization, recycling programs, and forest plantations in order to meet human needs.

DEFINITION OF BIODIVERSITY

Simply stated, biodiversity is the variety of life on earth and its myriad of processes (13). It includes all lifeforms, from the unicellular fungi, protozoa, and bacteria to complex multicellular organisms such as plants, fishes, and mammals. The recently negotiated Global Convention on Biological Diversity (1992) defined it as follows:

> "Biological diversity" means the variability among living organisms from all sources including, inter alia, terrestrial, marine and other aquatic ecosystems and the ecological complexes of which they are a part; this includes diversity within species and of ecosystems (14,15).

Biodiversity can be divided into four levels: genetic, species, ecosystem, and landscape diversity. Most attention is often given to *Species Diversity*, the number of different kinds of organisms found at a particular locale, and how it varies from place to place and even seasonally at the same place. A less obvious aspect of biodiversity, *Genetic Diversity*, is the variety of building blocks found within individuals of a species. It allows populations of a species to adapt to environmental changes. *Ecosystem Diversity* is the distinctive assemblage of species that live together in the same area and interact with the physical environment in unique ways. The interaction of organisms within the ecosystem is often more than the simple sum of its parts. On a broad regional scale, *Landscape Diversity* refers to the placement and size of various ecosystems across the land surface and the linkages between them.

IMPORTANCE OF BIODIVERSITY

Why should people care about all this variety of life, about sustaining and enhancing genetic resources, recovering endangered species, restoring riparian areas, maintaining old-growth forests, or conserving trees, insects, and marshes? The answer is both aesthetic and practical. A diversity of living things provides subtle needs. People enjoy picnicking, visiting seashores, and a variety of other recreational activities. Our homes, air, livestock, vegetables, fruits, and grains all derive from the products of diverse and healthy ecosystems. Diverse communities of plants, animals, and microorganisms also provide indispensable ecological services. They recycle wastes, maintain the chemical composition of the atmosphere, and play a major role in determining the world's climate. Moreover, in many cultures, diversity and the maintenance of mountains or other landforms are important because of their religious significance. Many people also feel that we must maintain biodiversity because our role as the dominant species on earth confers upon us the responsibility for the wise and careful stewardship of life.

But the value of biodiversity goes far beyond the aesthetic aspects (16). We hardly know a fraction of the species on this planet, especially in tropical ecosystems, despite years of intensive scientific effort. Every year species are lost before we have a chance to know anything about them. Who can guess what potentially valuable foods, medicines, and commercial products are forever lost with each extinction? Wild plants, animals, and microorganisms have provided essential products since humans first walked the earth, including virtually everything we eat. They continue to provide a basis for human society to respond to future changes.

Waiting in the wings, are tens of thousands of species that are edible, and many are demonstrably superior to those already in use. The vast insect faunas contain large numbers of species that are potentially superior crop pollinators, control agents for weeds, and parasites of insect pests (17). A case in point is in the area of natural sweeteners where a plant found in West Africa, the katemfe (*Thaumatococcus danielli*), produces proteins that are 1,600 times sweeter than sucrose (17). One in ten plant species contains anticancer substances of variable potency, but relatively few have been bioassayed. Similarly, few species have been screened for other potential medicinal values even though about a third of all modern prescription drugs contain compounds that plants have evolved to defend themselves against their enemies (18).

Probably most important, however, is that interacting communities of plants, animals, and microorganisms functioning as ecosystems also provide us with many indispensable services. Balanced systems recycle wastes, maintain the chemical composition of the atmosphere, and play a major role in determining the world's climate. The "ecosystem services" provided to us gratis also include supplying us with fresh water, generating soils, supplying plant pollinators, and maintaining a giant genetic "library" (18).

LOSSES OF BIODIVERSITY

Over the past few decades, the rate of global biotic impoverishment has increased dramatically. Exponential growth in human populations and even faster growth in consumption of the world's natural resources, have led to high rates of loss of species and habitats. Current rates of species loss exceed anything found in the past 65 million years (19). If the trend continues, by 2050 we may see the loss of up to one quarter of the world's species (20) and potentially dramatic changes in the climate and hydrology of entire regions such as Amazonia (21). The biotic resources we stand to lose are of immediate future value to humanity and essential for the maintenance of productive ecosystems.

Many of our most serious problems are centered in the tropics, where biodiversity is highest and losses of species and whole ecosystems is being lost most rapidly (22). In developing countries, the issues are most intense, because for hundreds of millions of people, the struggle is for survival (23). The destruction of Third World forests amounts to more than 11 million hectares annually (7.5 million ha closed forest and 3.8 million ha open forest) (23). Between 1950 and 1983, forest and woodland areas dropped 38% in Central America and 24% in Africa (23). Four factors are of special importance: (1) the explosive growth of human populations; (2) widespread and extreme poverty; (3) an ignorance of the ways in which to carry out productive agriculture and forestry; and (4) government policies that encourage the wasteful uses of forest resources (22,23).

Archaeological evidence points to the interaction between humans and tropical forests that extends far into the past when population densities were actually higher than they are today (24,25). In Mexico, studies clearly document the existence of ancient civilizations with high population densities integrated within tropical forest ecosystems. Examples are both the Olmec and Maya civilizations of southeastern Mexico that existed in that region for a combined period of at least 3000 years (26). Population densities in the rural Mayan area today is only about 5 people per km^2 compared to the peak of 400–500 people per km^2 during the height of the Olmec and Maya civilizations (26). These findings indicate the current extensive areas of tropical forests in Mexico that have been cut over the last 50 years were not untouched primeval forest but the result of regeneration since the last cycle of abandonment (24).

Recent tropical deforestation is associated with a pervasive cycle of initial timber extraction followed by shifting cultivation, land acquisition, and subsequent conversion to pasture (27) which leads to loss of forest resources, reduction of biodiversity, and impoverishment of rural people (24). The effect of past civilizations on the structure and composition of today's forests is more than just an intriguing question but is important in determining those practices used by those civilizations to maintain the tropical biodiversity left by previous generations. In fact, one of the primary causes of tropical deforestation in Mexico is due to the neglect of traditional people's vast experience with resource management. The persistence of forest resources and ecosystems following widespread human intervention indicates that a knowledge of management techniques practiced by ancient civilizations, such as the Olmec and Maya, could help in reverting current processes of landscape degradation in the tropics (24).

INVENTORY AND ASSESSMENT

Implementing biodiversity goals will require resources and knowledge. The role of science in the conservation of biodiversity is critical. More research to improve methodologies, distributional and status information, and strategies based on sound information will ultimately provide the basis for all sound policy and management decisions. Current scientific understanding of ecological processes is far from perfect. Existing resources could be reallocated, but additional resources will be necessary for improving efforts in inventory, monitoring, and basic research. Better inventories and assessments are needed of current conditions, abundances, distributions, and management direction for genetic resources, species populations, biological communities, and ecological systems. Even in the United States with the intensity of scientific effort focused on natural resources, it is extremely difficult to assess the overall status of biodiversity. Rarely have studies been done examining all vascular plants and vertebrates and their relationships in any given ecosystem, let alone the thousands of other species found within it: Particularly as Maini (1992) point out that "forests are a rich repository of planet earths' genetic heritage" (28). Forests are usually delineated by the presence of a few dominant species but this delineation barely touches the surface of their species richness. For example, in two pine systems in the southeastern United States, tree species make up less than 10% of the plant and vertebrate animal species. And this arrangement does not take into account all the thousands of other species likely to be found. R. M. May (1992) states "If we speak of total number of species, then to a good approximation everything is a terrestrial insect" (29). In fact, if one uses an estimate of 3–5 million (29) as total species on earth, then vascular plants (5.4–13.3%) and vertebrates (0.9–1.5%) only represent 6.3 to 14.8%. These percentages may be off by an order of magnitude if Erwin's (1988) hypothesis of 30 to 50 million species is closer to the true number of species (30). This example illustrates the extent of the problem as the traditional approaches to forest management have not incorporated a consideration of the vast majority of species (30). Even on an area the size and geographic scope of a typical National Forest in the United States, native biodiversity can easily encompass thousands of species of plants and animals, dozens to hundreds of identifiable biological communities, and an incomprehensible number of pathways, processes, and cycles through which all that life is interconnected. Obviously, it is not possible to address each and every aspect of this complexity. Therefore, identification is of specific aspects of diversity, such as distinct species, biological communities, or ecological processes that warrant special consideration (31).

Biodiversity Inventories

Biodiversity at the species, community and ecosystem levels can be assessed by surveys (analyses of the geographic distribution of the biota) and inventories (catalogues of the elements of biodiversity present at a site or in a region). Surveys and inventories are critical in helping determine which areas are in greatest need of preservation or management, and in identifying threatened species and habitats. A wide range of such activities needs to be supported, including:

- Rapid Assessment Programs, which give a "snapshot" of species richness in an area.
- Surveys, to map out the distribution of the Earth's ecosystems. Such analyses would be greatly aided by the development of a common and consistent scheme to classify communities and ecosystems. There is currently no widely accepted vegetation/ecological classification system.
- Broad-based inventory efforts, focused on poorly known habitats (soil communities, tropical and marine systems) that take advantage of existing protected sites, such as UNESCO-MAB's Biosphere Reserves and analogous national networks, wherever possible. However, degraded and multiple-use habitats should also be included, as well as better-understood taxa, such as vertebrates and vascular plants, that can be used as "benchmarks" for habitat quality.
- Intensive inventories, which attempt to determine all of the species present, from microbes to vertebrates, should be undertaken at two or three sites around the world. Such analyses will require intense effort and a long time-frame (a decade or more), but they are key to answering such questions as how biodiversity at the species level affects diversity at the ecosystem and community levels) (32).

It is a challenging task to formulate an integrated, concise and relevant approach to inventory. However, there are some recommended broad principles outlined in the Keystone Biodiversity Report (1991) on the components of an inventory program (13). These include:

- The inventory should be hierarchical and "top-down" in the sense that landscape level assessments such as the Fish and Wildlife Service's "gap analysis" are used to identify priorities for inventory at the local level, and local assessments are used to identify priorities at the site level.
- The inventory should make maximum use of existing data management systems.
- The inventory should be landscape based in the sense that abundance and distribution of plant and animal species are correlated with soils, vegetation, plant and animal community characteristics, and landscape features.
- The inventory at a minimum should include natural vegetation, all vertebrate and vascular plant species and at least some indicator species of non-vascular plants and invertebrates, and some indicators of other elements of biological diversity, such as sensitive communities or human-influenced processes and elements of structural diversity.
- Provision should be made for systematic inventories of all candidate, threatened, endangered, and sensitive species and for all other elements that are imperiled due to human activities or natural events.
- Inventories should be guided by an inter-agency master plan that coordinates acquisition of aerial photography, soil survey, vegetation survey, and vertebrate

inventory that ensures compatibility of data within and among agencies.

- The above mentioned master plan should be implemented for all regional ecosystems and vegetation mapping and inventory of vertebrates should be completed within the next 10 years.
- The inventory should be compatible with, and feed information directly into, development and implementation of Geographic Information System (GIS) methodology, monitoring programs, and research activities.
- The inventory should provide the basis for determining species (including genetic level assessment), species groups, population guilds, habitats, landscapes or processes that require more intensive studies.
- Inventories should be coordinated with and make maximum use of the fifty state Heritage Program databases, procedures and technology.
- The inventory process should identify levels or intensities of inventory that are appropriate for each level of planning, type of management activity or impact, type of land classification or degree of rarity or sensitivity of the element being inventoried.
- The inventory should have a strong element of quality control and assurance, including setting specific standards of accuracy and precision, timing the inventory to encompass the life-cycles of the target elements, standardizing methods and databases to the extent possible, and using trained personnel to conduct the inventories.

The need for more specific data and more efficient ways for collecting and managing data will lead to significant changes in inventory processes including the use of methods and technology that will: (1) provide resource estimates for specific geographic units and evaluate the reliability of such estimates; (2) display estimates and units spatially; (3) make maximum use of existing information and new technology, such as remote sensing and geographic information systems; (4) provide a baseline for monitoring changes in the extent and condition of the resource; (5) eliminate redundant data collection, develop common terminology, and promote data sharing through corporate data bases; (6) utilize information management systems to provide maximum flexibility for data integration, manipulation, sharing, and responding to routine and special requests; and (7) provide up-to-date data bases using modeling techniques, accounting procedures, and re-inventories.

Monitoring Effects of Land Management and Conservation Programs

Present concepts of monitoring vary depending on who is expressing them, their background, and the objectives of the monitoring being considered. There is a need for greater coordination, with considerable direction and standardization set at both National and Regional levels. Monitoring means different things to different people. Just what it is depends solely on monitoring objectives. This definition should emphasize the need to exercise great care in defining the objectives for monitoring. Monitoring should provide sufficient information about the abundance of animals or plants targeted for monitoring to assure that current management practices are not threatening the long-term viability of their populations (33). But viability concerns have added an additional layer of complexity to monitoring programs particularly when we try to derive a number for minimum viable population size. This theoretical concept espoused by Michael Soule (34) is useful from the standpoint that there probably is some minimum size population threshold that, when crossed, will lead to the demise of a species' population. The level of this threshold clearly varies with the species concerned; for passenger pigeons it probably was in the hundreds of thousands while for desert pupfish the number may be as low as 200. But as useful a concept this may seem, in the real world its close to impossible to determine minimum viable population size with any degree of certainty. There are too many variables involved; demographic variables such as immigration emigration, recruitment, birth rates, survivorship, dispersal mechanisms, etc, catastrophic events, habitat loss and even changing climatic conditions. It is difficult to envision that we can determine minimum viable population size for every species. Perhaps a more subjective approach, that being that larger numbers of a species are obviously better then fewer numbers of the same species, should be considered. Maybe we should try to maintain species with as large a population size as practical.

Monitoring efforts are often severely hampered by the lack of prior planning and thought given to the desired results from any given monitoring effort. It is not enough to select a management indicator species, guild, or other monitoring target with the idea that this will allow us to assess the impact of any given management activity (35,36). Ideally, the results from monitoring should feedback into the system to correct or fine tune management activities.

Effort should not be wasted on monitoring systems that fail to give the level of confidence needed to deduce the most likely effects of management activities on wildlife, fish or sensitive plant resources. In an era when humankind's activities are the dominant force influencing biological communities, proper management requires understanding of pattern and process in biological systems and the development of assessment and evaluation procedures that assure protection of biological resources (37,38). It is essential that appraisals of these resources give us the ability to forecast the consequences of human-induced environmental changes accurately (39). But we have a long way to go in this process.

We must first have a clear understanding of our goals and objectives. The next step is to assess risk and assign priorities. We need to perform a kind of environmental triage. We must be able to say when enough is enough. With limited financial and physical resources at our disposal we may have to make the highly undesirable decision that we no longer will try to prevent the extinction of a particular species. Once those species have been determined and not forgetting about the importance of ecological processes, we need to formulate the types of questions that need to be answered to determine that we are meeting our goals and objectives. It is absolutely critical to ask the right question in the first place. Why monitor shade cover over perennial streams in order to maintain water temperature for trout when directly monitoring stream temperature is a more appropriate measure. Thus our efforts will be geared to maintaining as much shade cover as needed to maintain water temperature.

Monitoring should also have a strong element of quality control and assurance, including setting specific levels of accuracy and precision, timing the inventory to encompass the life-cycles of the target species, as much as possible, and standardizing methods and databases for all management units, especially when monitoring the same species. However, whatever is done must be as cost-effective as possible. Some possibilities are risk analysis, increasing the scope of monitoring efforts, and determining the needs of monitoring objectives. We might want to limit direct monitoring only on high risk species, on a priority basis, while relying on habitat relationships for most other species. Monitoring efforts should be spread over as large a geographical base as possible (and feasible) to reduce the cost per forest and to increase the scope of applicability. Whenever possible, monitoring should only be asked to detect declining trends because of the potential cost savings (almost 90%) (33).

Along with this monitoring is the need to develop a quality control and assurance program that ensures: (1) objectives are measurable (and thus monitorable); (2) appropriate measurement techniques and procedures are being used; and (3) management thresholds are clearly identified and incorporated into the planning process so that, if crossed, they automatically trigger a reanalysis of the planned activities.

The advent of geographic information systems holds future promise for the development of a comprehensive biological information system (40,41). Much work and coordination will be involved to bring this information system to pass including building in mechanisms to continually check and update the information in the database. It is not enough to set up the system and then use it without regard to the dynamics of ecological systems.

MAINTAINING BIODIVERSITY

The responsibility for biodiversity belongs to all people and institutions: both public and private (31). The repeated association of biodiversity with preservationist approaches leads to the perception that biodiversity requires wilderness and preserves and cannot be sustained where human activities are prevalent. This view has disastrous consequences. The World Resources Institute in 1986 reported that only 3% of the surface of the earth had major protection status as a national park, nature preserve, wilderness, or sanctuary. How much diversity can be sustained on that land base, even if it were doubled or tripled, politically unlikely events (31)? It is clear that the major accomplishments on behalf of biodiversity must occur in conjunction with human activities.

To maintain biodiversity, we must ensure that a sufficient amount of each ecosystem is conserved and managed through a variety of actions that address different and related concerns. And because these actions must occur on lands under a variety of ownerships, goals, and uses, considerations for biodiversity must be blended into a myriad of management approaches (31). We must strive to understand the functions and processes of natural ecosystems, and make the wise, tough decisions that are necessary to maintain and enhance the productivity of those systems for all purposes and uses. This means that biodiversity, and an understanding of ecosystems, should be the underlying basis for the management of all lands.

Restoration of Biodiversity

While research and management are urgently needed to slow continuing losses of biodiversity, the remediation of past losses can help offset unavoidable future losses. Restoration of ecosystems and biological communities is one important means of maintaining biodiversity, or at least of slowing its net loss. Biodiversity is threatened not only by a reduction of habitat acreage and of connections between habitats, but also a degradation of quality of the remaining habitats. Restoration activities respond to these problems by restoring eliminated habitat types (eg, native prairies and wetlands) and enhancing the condition of remaining habitat fragments. By restoring both the extent and quality of important habitats, restoration programs provide refuges for species and genetic resources that might otherwise be lost. Moreover, surrounding landscapes are habitats that disperse into these disturbed areas, and so restoration programs can also affect the recovery and renewed diversity of their biota.

Restoration is not a substitute for the preservation or good management and is both time-consuming and expensive. In tropical forests, the incredible diversity and complexity of the ecosystem make restoration of the original vegetation and ecosystems particularly difficult (9,43). But even though it addresses the symptom of deforestation rather than the causes, restoration ecology is worth serious consideration. It can speed regeneration in managed systems, make non-productive land productive again, relieve pressure on natural forest resources, and protect closed-canopy forest. It is a strategy that focuses on areas of severe erosion and soil compaction, where quick action is desperately needed (9).

Many techniques are used to restore ecosystems, depending on the ecosystem and impact type being addressed. These techniques include vegetation planting to control erosion, fertilization of existing vegetation to encourage growth, removal of contaminated soils, fencing to exclude cattle, reintroduction of extirpated species, restoration of hydrologic connections to wetlands, and others. However, not all of these restorations strategies are compatible with the goal of maintaining biodiversity. For example, eroding lands can be rapidly restored by introducing some types of aggressively spreading plants, but the same plants can spread beyond the site and imperil the diversity of flora in adjacent areas. Thus, restoration actions can be either a savior or a nemesis for regional biodiversity, depending on their design, application methods, and existing conditions of the landscape.

Just as there are many restoration techniques, there are a very large number of species that exist in habitats that are candidates for restoration. Each has particular environmental requirements, minimum viable population size, and expected recovery rate and pattern–knowledge which is essential to evaluating restoration potential. Although responses of some species to specific restoration techniques is known, the theoretical basis is weak for grouping species so that results can be extrapolated to other combinations of techniques and species. Ecosystem restoration does not always require intervention. Left to natural processes, many ecosystems will return to something like their pre-disturbance condition if populations of original species still exist nearby (44). For example, a temperate climate and productive soils promote natural re-establishment of forests in most regions of the United States. However, restoration technologies can speed the recovery of communities and ecosystems after disturbance and can enhance in-situ conservation (44).

Ecosystem Management

What exactly does "ecosystem management" mean? An ecosystem is a community of organisms and its environment that functions as an integrated unit. Forests are ecosystems. So are ponds, rivers, rotting logs, rangelands, whole mountain ranges, and the planet. Ecosystems occur at many different scales, from micro sites to the biosphere. They constantly change in their species composition, structure, and function over time. And they do not have natural boundaries. All ecosystems grade into others; all are nested within a matrix of larger ecosystems. We arbitrarily describe the boundaries of ecosystems for specific purposes.

Management means using skill or care in treating or handling something. Thus, "ecosystem management" means the use of skill and care in handling integrated units of organisms and their environments (45). It implies that the system, or integrated ecological unit, is the context for management rather than just its individual parts. The pivotal question is, what should we manage ecosystems for? Goals are not obvious from definitions or from merely stating that one manages ecosystems. Ecosystem management is the means to an end. It is not an end in itself. Ecosystems are usually managed for specific purposes such as: producing, restoring, or sustaining certain ecological conditions; desired resource uses and products; vital environmental services; and aesthetic, cultural, or spiritual values. This idea is an important point. There is much discussion these days about whether ecosystem management means to preserve the intrinsic values or natural conditions of ecosystems, with resource uses and products as secondary byproducts, or whether it means to produce desired resource values, uses, products, and services in ways that do not impair the long-term sustainability, diversity, and productivity of the ecosystems. It is unlikely that a consensus will ever be reached on this point; however, this debate does not mean that the conservation of biodiversity should not be a common concern of all humankind whatever our individual desires or objectives. In some places the emphasis will be on ecological conditions and environmental services. In others it will be on resource products and uses. Overall, the mandate should be to protect environmental quality while also producing resources that people need. Therefore, ecosystem management cannot be reduced to a simple matter of choosing to have either resource products or natural conditions. It must chart a prudent course to attain both of these goals together. This policy can only happen in areas large enough to allow for compatible patterns of different uses and values in a landscape or region.

A thorough understanding of vegetation dynamics is critical for sound vegetation management and the maintenance of landscape diversity. The role of landscape ecology in this regard is to focus attention on hierarchy theory that considers vegetation patterns at different scales (46). Natural areas should also be a part of any long-term landscape management strategy designed to conserve landscape diversity (46). The degree of predictability decreases as the scale decreases to specific sites where most management strategies are formulated (47). Successful management strategies will therefore be more likely when planning is done on a broader scale. Planning also needs to consider the dynamics of landscape scale forest patterns, both natural and managed, and their effects on hydrology, wildlife, and other resources. The results need to be geared toward long-term design of forest management to minimize cumulative effects while maintaining biodiversity. Geographic information systems (GIS) can be used to effectively integrate land uses for forestry, agriculture, wildlife, tourism and other non-commodity values. Graphic displays of resource information allow for an easier conceptualization of the land management patterns and their interrelationships. GIS allows many different scenarios to be tried and evaluated for the possible long term effects of proposed management activities.

Future Strategies for Conserving Biodiversity

Biodiversity provides a focus for a global conservation program (48–50). But even the best-conceived conservation program could not have anticipated the enormous changes of the 20th century, with its resultant pressures on natural ecosystems. Current global concerns about worldwide losses of massive expanses of forest ecosystems and their associated species provide opportunities for resource management agencies to build on past accomplishments and chart new courses in conservation for multiple benefits. Conserving biodiversity involves restoring, protecting, conserving, or enhancing the variety of life in an area so that the abundances and distributions of species and communities provide for continued existence and normal ecological functioning, including adaptation and extinction. This definition does not mean all things must occur in all areas, but that all things must be cared for at some appropriate geographic scale.

In recent years, traditional uses of public lands in the United States and around the world have become increasingly controversial (7). The demands and expectations placed on these resources are high and widely varied, a situation calling for new approaches that go beyond merely reacting to resource crises and concerns (3,8). These new approaches must incorporate fundamental shifts in the scale and scope of conservation practice (49). These approaches include the shift of focus from the more traditional single-species and stand level management approach, to management of communities and ecosystems (50). The greatest challenge for conserving biodiversity is developing an integrative ecosystem approach to land management on a broader landscape scale. Spatial scale, be it local, regional, or global, greatly influences our perceptions of biodiversity (51). An understanding of the importance of scale is critical to accurately assess the impact of land management practices on biodiversity since many significant biological responses and cumulative management effects develop at the landscape level. Planners and managers are increasingly aware that adequate decisions cannot be made solely at the stand level. If context is ignored in conservation decisions and the surrounding landscape changes radically in pattern and structure, patch content also will be altered by edge effects and other external influences (52). Particularly when, land-use patterns characteristic of a human-dominated landscape are ones in which large, continuous tracts of natural habitat become increasingly fragmented and isolated by a network of developed lands.

The protection and maintenance of biodiversity is a long-term issue which will create problems in political systems that deal primarily with the short-term goals and objectives. The most obvious conflicts will be political and financial. There are inherent biases in a market economy that tend to result in environmental degradation. The environmental costs are generally passed to the public at large. Attempts to internalize such costs are resisted strongly by private industry. For example, installation of scrubbers in smokestacks and catalytic converters on cars in the United States were not welcomed by industry. Similar results are likely in actions to protect biodiversity. The Tellico Dam in the United States was chosen over possible extinction of a small fish, the snail darter. Today the timber industry is against setting aside old growth forests on the basis of the impacts on local jobs and economies. The actual monetary costs may not be as large as the political costs. For every action taken, there will be winners and losers. The loss of diversity will have long-term effects as we lose what are essentially building blocks for human survival. Using a financial analogy, we should manage our biological resources so that we can live off the interest. Living off the principal eventually leads to bankruptcy.

Moreover, ecosystems by their very nature are in a constant state of flux, always shifting and changing from one condition to the next (52). Reproductive strategies and other ecological characteristics of many ecosystems are strikingly adjusted to disturbance regimes resulting from catastrophes such as fire, insect outbreaks, and flooding. Change per se is not necessarily something to be avoided. It ultimately may be the underlying motivating factor in management decisions. Land-managers may wish to alter vegetation structure or composition to emphasize rarer or endangered species. Old management paradigms are difficult to shed, but only new, dynamic efforts are likely to succeed in conserving the biodiversity of all our ecosystems (3). Ecosystem-level management is going to require new approaches in planning, monitoring, coordinating, and administering. We are being challenged by shifts in public attitudes to manage our public as well as private lands for uses other than commodity production. A new paradigm is needed, one that balances all uses in the management process and looks beyond the immediate benefits. Maintaining biodiversity requires attention to a wider array of components in determining management options as well as the management of larger landscape units. There will be trade-offs, commodity production may decline in the short-term, but in the long-term these trade-offs will result in gains in sustained productivity while maintaining biodiversity with its complete range of ecological processes.

Threatened and endangered species have long been the focus of biodiversity concerns at the level of plant and animal species, but they represent only one aspect of a larger issue: conservation of the full variety of life, from genetic variation in species populations to the richness of ecosystems in the biosphere (31). And for certain species and biological communities, the pressing concern is perpetuation or enhancement of the genetic variation that provides for long-term productivity, resistance to stress, and adaptability to change. A biologically diverse forest holds a greater variety of potential resource options for a longer period of time than a less diverse forest. It is more likely to be able to respond to environmental stresses and adapt to a rapidly changing climate. And it may be far less costly in the long run to sustain a rich variety of species and biological communities operating under largely natural ecological processes than to resort to the heroic efforts now being employed to recover California condors (*Gymnogyps californianus*), peregrine falcons (*Falco peregrinus*), and grizzly bears (*Ursus horribilis*). Resource managers know from experience that access to resources is greater and less costly when forests and rangelands are sufficiently healthy and diverse.

It is easy to understand why endangered species have received the focus of attention. Many are large, easily observable, and oftentimes aesthetically pleasing. Thereby, most efforts at restoration and rehabilitation have been directed towards endangered species and harvested species. However, endangered species are fundamental indicators of environmental disturbance. Since extinction is a process, not a simple event, the recognition that a species is endangered is little more than a snapshot of a moving vehicle. Attempts at therapy most often address symptoms rather than causes. We have failed to communicate successfully why rehabilitation and restoration beyond the narrow focus of the endangered and harvested, are essential. Processes from which endangered species arise are primarily degradation at the landscape scale: fragmentation, physiographic alteration, vegetation removal or replacement, pollution, etc (54). The environmental variables which affect the health and welfare of all the flora and fauna also affect people: water and air quality, recycling of organic and inorganic substances, microclimate, etc. Loss of biodiversity means loss of ecological services and options for the future. The cost of replacing ecological services, already great, will increase to staggering proportions. The real and potential wealth represented by conserved biodiversity cannot be replaced (54). So by shifting the focus of conservation activities from the species and population level to the landscape level we may start to make progress (54).

The tough choices posed in the spotted owl (*Strix occidentalis*) case in the Pacific Northwest of the United States typify many future issues as the conservation of biodiversity becomes a higher social priority. Regardless of the eventual outcome of this issue, there is an important lesson to be learned: Conserving biodiversity will not be cheap or noncontroversial. Federal land management agencies in the United States have increasingly come under fire over management decisions that appear to decrease biodiversity. The USDA Forest Service faces numerous appeals and lawsuits on forest plans for insufficient and sometimes conflicting consideration of forest biodiversity in management decisions. The dispute over the spotted owl and old growth forests is the most visible example of how tough it is to blend the conservation of biodiversity with other uses and values of public resources. It illustrates the reality of "no free lunch" in resource allocations. Even though parks, reserves, set-asides, and easements are critical components in the mix for the conservation of biodiversity, they will become more difficult to come by and ultimately will require an expansion beyond the "reserve mentality" (55). Multiple-use of public lands is deeply ingrained. Somehow we have to come up with management prescriptions for our public lands that will allow both consumptive and nonconsumptive uses but will do so in such a way that no net loss of native species will occur. Such a prescription will require that the livestock, timber, and mining industries take their turns at the trough instead of always going to the head of the line. This prescription will require encouraging resource conservation and recycling programs that reduce the need for raw materials from public lands (55). The scope of conservation has been too constrained and steps must be taken to incorporate the benefits of biodiversity and the use of biological resources into local, regional, national and international economies (42,56).

Future conservation at larger scales will always be confounded by the potentially large number of political authorities that conduct land management practices on watershed, basin, or even landscape scales (57). Imposing coarse scale structure on the landscape, like protecting large blocks or habitat or the creation of regional-scale corridors, can preserve connectivity in the face of some acceptable amount of habitat destruction (58). Management of larger parcels for permanent and temporary openings has the potential to reduce edge impacts and allow for the regeneration of larger stands of mature forest which can benefit area-sensitive species. The juxtaposition of habitat patches within a landscape context and its importance to the maintenance of wildlife diversity is essential, but only beginning to be appreciated (51). Unfortunately, many management practices used today often ignore the natural dynamics of ecosystems (3). Managers need to be open-minded about the use of a suite of disturbance-creating tools to manipulate flora and fauna, particularly in systems that have evolved under a disturbance regime.

The scale and scope of conservation has been too restricted (42). Spatial scale be it local, regional, or global, greatly influences our perceptions of biodiversity (51). Understanding the importance of scale is critical to accurately assessing the impact of land management practices on biodiversity. The scale of a conservation endeavor affects the strategy involved, the determination of realistic goals, and the probability of success (57). For example, a strategy to maximize species diversity at the local level does not necessarily add to regional diversity. In fact, oftentimes in our hast to "enhance" habitats for wildlife we have emphasized "edge" preferring species at the expense of "area" sensitive ones and consequently may have even decreased regional diversity. It is important to realize that principles that apply at smaller scales of time and space do not necessarily apply to longer time periods and larger spatial scales (51). Long-term maintenance of species and their genetic variation will require cooperative efforts across entire landscapes (42). This maintenance is consistent with the growing scientific sentiment that biodiversity should be dealt with at the scale of habitats or ecosystems rather than species (58). However, how to deal with this technically is still in the developmental stages.

The key to maintaining global biodiversity is ecosystem and landscape protection and management, not crisis management of an increasing number of endangered species. The time to save a species is when it is still common. To meet the challenge we must continually develop new programs as information becomes available. Adaptive management which depends on monitoring the consequences of various management treatments, should be used to adjust and refine management practices (52). Much remains to be learned about managing wildlands for biodiversity, and only through extensive cooperative efforts can we provide and protect essential diversity for ourselves and future generations.

BIBLIOGRAPHY

1. T. L. Erwin, "The Tropical Forest Canopy: The Heart of Biodiversity," in E. O. Wilson and E. M. Peter, eds., *Biodiversity*. National Academy Press, Washington, D.C., 1988, pp. 123–129.
2. R. C. Szaro, "Biodiversity and Biological Realities," in D. Murphy, ed., *Proceedings of the National Silvicultural Workshop, Cedar City, Utah*, May, 1991, USDA Forest Service, Washington, D.C., 1993.
3. R. C. Szaro, "Conserving Biodiversity in the United States: Opportunities and Challenges," in D. Bandu, H. Singh and A. K. Maitra, eds., *Environmental Education and Sustainable Development (Proceedings of the 3rd International Conference on Environmental Education*, Panjim, Goa, India, October 3–7, 1989, Indian Environmental Society, New Delhi, India, 1990, pp. 173–180.
4. R. C. Szaro, "Biodiversity Inventory and Assessment," in A. Evendon, Compiler, *Proceedings—Northern Region Biodiversity Workshop*, USDA Forest Service, Northern Region, Missoula, Mont., 1992, pp. 57–62.
5. K. E. Evans and R. C. Szaro, "Biodiversity: Challenges and Opportunities for Hardwood Utilization, *Diversity* **6,** 27–29 (1990).
6. C. Harrington, D. Debell, M. Raphael, K. Aubry, A. Carey, R. Curtis, J. Lehmkuhl and R. Miller, *Stand-level Information Needs Related to New Perspectives*, Pacific Forest and Range Experiment Station, Olympia, Wash., 1990.
7. R. Szaro and B. Shapiro, *Conserving Our Heritage: America's Biodiversity*, The Nature Conservancy, Arlington, Va., 1990.
8. R. C. Szaro and H. Salwasser, "The Management Context for Conserving Biological Diversity," 10th World Forestry Congress, Paris, France, Sept., 1991, *Revue Forestiere Francaise, Actes Proceedings* **2,** 530–535 (1991).
9. J. Gradwohl and R. Greenberg, *Saving the Tropical Forests*, Island Press, Washington, D.C., 1988.
10. L. A. Maquire, "Risk Analysis for Conservation Biologists," *Conservation Biology* **5,** 123–125 (1991).
11. R. T. T. Formann, "The Ethics of Isolation, the Spread of Disturbance, and Landscape Ecology," in M. G. Turner, ed., *Landscape Heterogeneity and Disturbance*, Springer-Verlag, New York, 1987, pp. 213–229.

12. D. A. Saunders, R. J. Hobbs, and C. R. Margules, "Biological Consequences of Ecosystem Fragmentation: A Review," *Conservation Biology* **5,** 18–32 (1991).

13. Keystone Center, *Final Consensus Report of the Keystone Policy Dialogue on Biological Diversity on Federal Lands*, The Keystone Center, Keystone, Colorado, 1991.

14. United Nations Environment Programme, *Convention on Biological Diversity*. Na. 92-7807, Nairobi, Kenya, 1992.

15. United Nations Environment Programme, *Conference for The Adoption of The Agreed Text of the Convention on Biological Diversity*, Na. 92-7825. Nairobi, Kenya, 1992.

16. D. Ehrenfeld, "Why Put a Value on Biodiversity?" in E. O. Wilson and F. M. Peter, eds., *Biodiversity*, National Academy Press, Washington, D.C., 1988, pp. 212–216.

17. E. O. Wilson, "The Biological Diversity Crisis," *Bioscience* **35,** 703–704 (1985).

18. P. R. Ehrlich, "Extinctions and Ecosystem Functions: Implication for Humankind," in R. J. Hoague, ed., *Animal Extinction: What Everyone Should Know*, Smithsonian Institution Press, Washington, D.C., 1985, p. 162.

19. E. O. Wilson, "The Current State of Biological Diversity," in E. O. Wilson and F. M. Peter, eds., *Biodiversity*, National Academy Press, Washington, D.C., 1988, pp. 3–18.

20. J. A. McNeely, K. R. Miller, W. V. Reid, R. A. Mittermeier, and T. B. Werner, *Conserving the World's Biological Diversity*, World Conservation Union (IUCN), Gland, Switzerland; World Resources Institute, Conservation Internationa, World Wildlife Fund–U.S. and the World Bank, Washington, D.C., 1990, p. 193.

21. E. Salati and P. B. Vose, "Depletion of Tropical Rain Forests," *Ambio* **12,** 67–71 (1983).

22. P. H. Raven, "We're Killing Our World: Preservation of Biological Diversity," *Vital Speeches of the Day*, May 15, 1987, pp. 472–478.

23. R. Repetto, *The Forest for the Trees? Government Policies and the Misuse of Forest Resources*, World Resources Institute, Washington, D.C, 1988.

24. A. Gomez–Pompa and A. Kaus, "Traditional Management of Tropical Forests in Mexico," in A. B. Anderson, ed., *Alternatives to Deforestation: Steps Toward Sustainable Use of the Amazon Rain Forest*, Columbia University Press, New York, 1990, pp. 45–64.

25. J. R. Parsons, "The Changing Nature of New World Tropical Forests Since European Civilization," in *The Use of Ecological Guidelines for Development in the American Humid Tropics*, 1975, IUCN Publication New Series No. 31, pp. 22–38.

26. B. L. Turner, II, "Population Density in the Classic Maya lowlands: New Evidence for Old Approaches," *Geographical Review* **66,** 73–82 (1976).

27. W. L. Partridge, "The Humid Tropics Cattle Ranching Complex: Cases from Panama Reviewed," *Human Organization* **43,** 76–80 (1984).

28. J. S. Maini, "Sustainable Development of Forests," *Unasylva* **43**(169), 3–8 (1992).

29. R. M. May, "Past Efforts and Future Prospects Towards Understanding How Many Species There Are," in O. T. Solbrig, H. M. van Emden and P. G. W. J. van Oordt, eds., *Biodiversity and Global Change*, International Union of Biological Sciences, Paris, Monograph No. 8, 1992.

30. T. L. Erwin, "The Tropical Forest Canopy: the Heart of Biotic Diversity," in E. O. Wilson and F. M. Peter, eds., *Biodiversity*, National Academy Press, Washington, D.C, 1988, pp. 123–129.

31. H. Salwasser, "Conserving Biological Diversity: A Perspective on Scope and Approaches," *Forest Ecology and Management* **35,** 79–90 (1990).

32. O. T. Solbrig, ed., *From Genes to Ecosystems: A Research Agenda for Biodiversity*, International Union of Biological Scientists, Cambridge, Mass., 1991.

33. J. Verner, "Future Trends in Management of Nongame Wildlife: A Researchers Viewpoint," in J. B. Hale, L. B. Best and R. L. Clawson, eds., *Management of Nongame Birds in the Midwest—A Developing Art*, North Central Section, Wildlife Society, Madison, Wis., 1986, pp. 149–171.

34. M. E. Soule, *Viable Populations for Conservation*, Cambridge University Press, N.Y., 1987.

35. R. C. Szaro, "Guild Management: An Evaluation of Avian Guilds as a Predictive Tool," *Environmental Management* **10,** 681–688 (1986).

36. R. C. Szaro and R. P. Balda, *Selection and Monitoring of Avian Indicator Species: An Example from a Ponerosa Pine Forest in the Southwest*, Rocky Mountain Forest and Range Experiment Station, Fort Collins, Colo., 1982, USDA Forest Service General Technical Report RM-89.

37. J. R. Karr, "Biological Monitoring and Environmental Assessment: A Conceptual Framework," *Environmental Management* **11,** 249–256 (1987).

38. F. B. Goldsmith, ed., *Monitoring for Conservation and Ecology*, Chapman and Hall, New York, 1991.

39. T. W. Hoekstra and C. H. Flather, "Theoretical Basis for Integrating Wildlife in Renewable Resource Inventories," *Journal of Environmental Management* **24,** 95–110 (1986).

40. Council of Environmental Quality, "Linking Ecosystems and Biodiversity," in *Environmental Quality: 21st Annual Report*, Washington, D.C., 1990, pp. 135–187.

41. F. W. Davis, D. M. Stoms, J. E. Estes, J. Scepan, and J. M. Scott, "An Information Systems Approach to the Preservation of Biological Diversity," *International J. of Geographic Information Systems* **4,** 55–78 (1990).

42. Global Biodiversity Strategy, *Guidelines for Action to Save, Study, and Use Earth's Biotic Wealth Sustainably and Equitably*, World Resources Institute, Washington, D.C.; The World Conservation Union (IUCN), Gland, Switzerland; United Nations Environment Programme (UNEP), Nairobi, Kenya, 1992.

43. R. L. Peters and T. E. Lovejoy, eds., *Global Warming and Biological Diversity*, Yale University Press, New Haven, Conn., 1992.

44. W. V. Reid and K. R. Miller, *Keeping Options Alive: The Scientific Basis for Conserving Biodiversity*, World Resources Institute, Washington, D.C., 1989.

45. S. Woodley, J. Kay, and G. Francis, *Ecological Integrity and the Management of Ecosystems*, St. Lucie Press, Ottawa, Ont., 1993.

46. W. A. Niering, "Vegetation Dynamics (Succession and Climax) in Relation to Plant Community Management," *Conservation Biology* **1,** 287–295 (1987).

47. D. L. Urban, R. V. O'Neill, and H. H. Shugart, "Landscape Ecology," *Bioscience* **37,** 119–127 (1987).

48. O. T. Solbrig, H. M. van Emden, and P. G. W. J. van Oordt, eds., *Biodiversity and Global Change, Monograph No. 8*, International Union of Biological Sciences, Paris, 1992.

49. IUCN/UNEP/WWF, *Caring for the Earth. A Strategy for Sustainable Living*, The World Conservation Union (IUCN), Gland, Switzerland; United Nations Environment Programme (UNEP); World Wide Fund for Nature (WWF), 1991.

50. J. Lubchenco, A. M. Olson, L. B. Brubaker, S. R. Carpenter, M. M. Holland, S. P. Hubbell, S. A. Levin, J. A. MacMahon, P. A. Matson, J. M. Melillo, H. A. Mooney, C. H. Peterson, H. R. Pulliam, I. A. Real, P. J. Regal, and P. G. Risser, "The Sustainable Biosphere Initiative: An Ecological Research Agenda," *Ecology* **72,** 371–412 (1991).
51. T. R. Crow, "Biological Diversity and Silvicultural Systems," in *Proceedings of the National Silvicultural Workshop: Silvicultural Challenges and Opportunities in the 1990's*, USDA Forest Service, Timber Management, Washington, D.C. (1991).
52. R. F. Noss, "Biodiversity Conservation at the Landscape Scale," in R. C. Szaro, ed., *Biodiversity in Managed Landscapes: Theory and Practice*, Oxford University Press (In Press).
53. E. D. Schulze and H. A. Mooney, *Biodiversity and Ecosystem Function*, Springer–Verlag, New York, 1993.
54. P. Bridgewater, D. W. Walton, and J. R. Busby, "Creating Policy on Landscape Diversity," in R. C. Szaro, ed., *Biodiversity in Managed Landscapes: Theory and Practice*, Oxford University Press (In Press).
55. P. F. Brussard, D. D. Murphy, and R. F. Noss, "Strategy and Tactics for Conserving Biological Diversity in the United States," *Conservation Biology* **6,** 157–159 (1992).
56. K. R. Miller, "Conserving Biodiversity in Managed Landscapes," in R. C. Szaro, ed., *Biodiversity in Managed Landscapes: Theory and Practice*, Oxford University Press (In Press).
57. S. M. Pearson, M. G. Turner, R. H. Gardner, and R. V. O'Neill, "Scaling Issues for Biodiversity Protection," in R. C. Szaro, ed., *Biodiversity in Managed Landscapes: Theory and Practice*, Oxford University Press, in press.
58. M. L. Hunter, *Wildlife, Forests, and Forestry: Principles of Managing Forests for Biodiversity*, Prentice-Hall, Inc., Englewood Cliffs, N.J., 1990.

BITUMEN

RICHARD MEYER
U.S. Geological Survey
Reston, Virginia

Numerous oil seeps and asphalt deposits are found in the Middle East, along the Tigris and Euphrates rivers and in and beside the Dead Sea, occurrences which have been exploited for millennia. Initially used as an adhesive, the bitumen soon found use as a binder in bricks, as a waterproofing agent for boats and water courses, and as a building cement. Finally, the use of bitumen led to the construction of stills for decanting the liquid petroleum as a fuel. The art of distilling petroleum products was lost after the 12th century (1) but commenced again in the middle part of the 19th century, following the successful drilling for oil, when oil became available for the first time in large quantities. See also ASPHALT; HEAVY OIL CONVERSION; TAR SANDS; COAL CLASSIFICATION.

Occurrences of natural bitumen are now known throughout the world, in a number of varieties. All have been exploited at various times, most notably in the 19th and early 20th centuries, when the most abundant variety, natural asphalt, was widely used as paving material. This application has since been mostly displaced by manufactured asphalt, which makes possible a much more accurately controlled mixture of the bitumen and stone. However, the quantities of natural asphalt worldwide are exceedingly large, representing nearly three trillion barrels of oil in place, and the technology for its recovery and conversion to petroleum products is available. The present constraints on the exploitation of natural bitumen are economic and environmental, not technological.

DEFINITION

Natural asphalt is commonly referred to as natural bitumen or as bitumen, even though the term bitumen technically includes the other, minor species. Natural asphalt describes one species of natural bitumen. The term has been used for any altered crude oil (2). A brief description of this terminology is found in ref. 3.

Such terms as tar sand, oil sand, hydrocarbonaceous rock, and oil-impregnated rock refer to natural asphalt plus the rock which contains it. This rock is most frequently sandstone or a friable sand cemented by the natural asphalt and contains a significant proportion of fine material, predominantly or entirely clay. The rock may also be limestone or dolomite. In this discussion, such rock is simply referred to as asphalt rock, suitably modified according to the lithology. In the subsurface, the asphalt-containing rock constitutes a reservoir.

Natural bitumen is a black, highly complex mixture of molecules, some of which, the asphaltenes in particular, are extremely large and, therefore, heavy (high molecular weight). These large molecules act like colloids and thus affect the ability of the fluid to flow.

The choice of criteria for distinguishing bitumen from crude oil must be arbitrary. A gas-free dynamic viscosity measured at reservoir temperature of 10,000 centipoises (cP) in customary units or its metric equivalent, 10,000 millipascal seconds (mPa·s), appears to satisfy most requirements for such a standard: (*1*) Viscosity can be determined by any of several widely applied laboratory methodologies, such as the rotational viscometer. (*2*) 10,000 cP closely approximates the maximum viscosity at which most oils can be recovered from a reservoir without thermally enhanced oil recovery methods. (*3*) 10,000 mPa·s is a value reached by many scientists and engineers (3) or falls within the range of values suggested by others (4,5).

CLASSIFICATION OF NATURAL HYDROCARBONS

The better-known varieties of natural bitumens and their relationship to crude oil and coal are depicted in Figure 1. Coals are mostly autochthonous, form in place on land, are readily combustible, and are comprised of at least 50 wt % organic matter. Two coal groups are distinguished: sapropelic (unbanded) and humic (banded). The sapropelic coals are relatively rare and are formed in organic mud under mostly anaerobic conditions. The banded coals form from woody plant material under both anaerobic and aerobic conditions, first being transformed to peat and then to higher-ranked coals through metamorphism. The coals may be grouped diagrammatically according to their hydrogen to carbon (H/C) and nitrogen

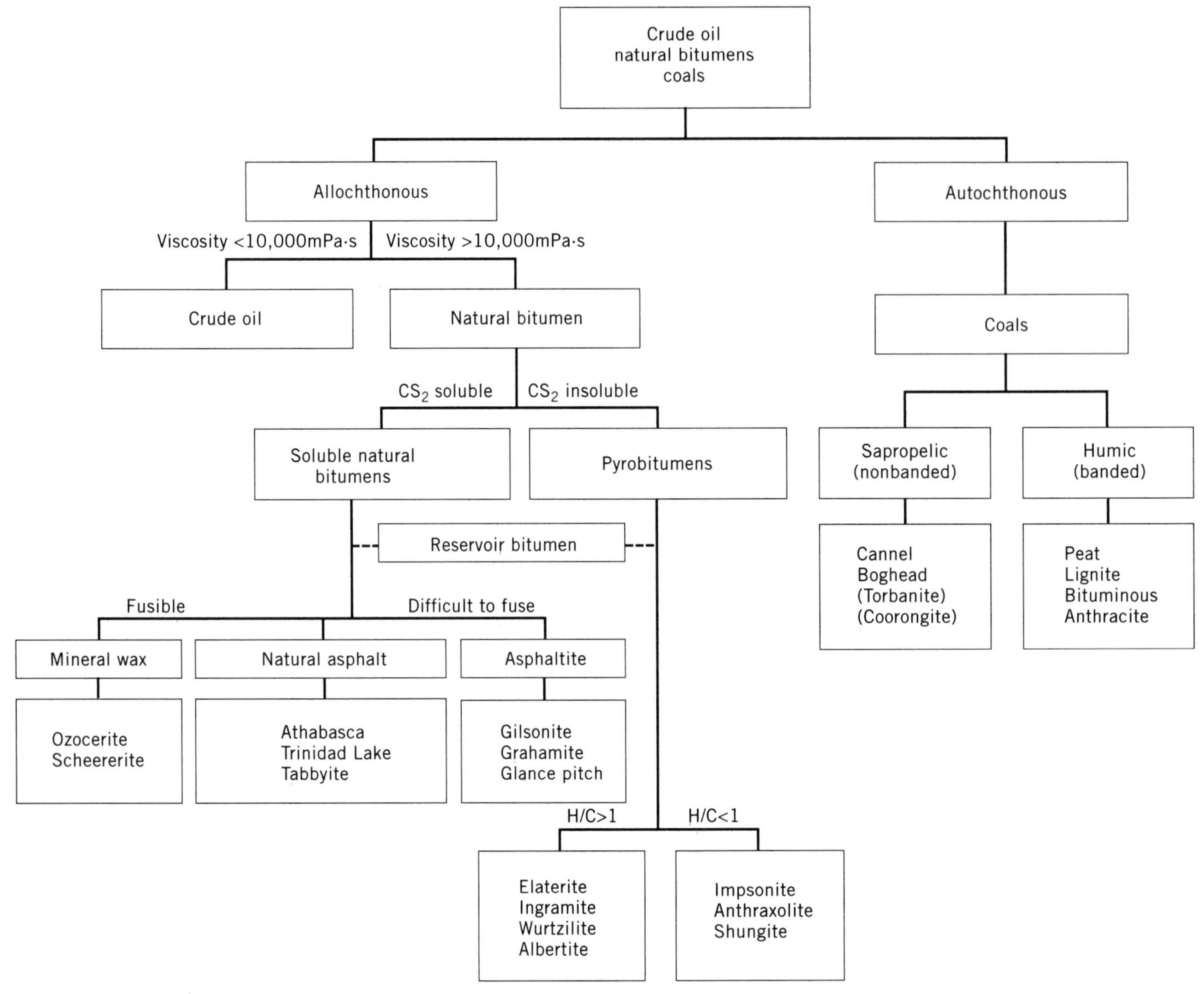

Figure 1. A classification of natural bitumens, crude oil, and coals. H/C, hydrogen–carbon ratio (3), Modified from ref. 6 and ref. 8. Courtesy of American Association of Petroleum Geologists. Modified from ref. 7, courtesy of W. H. Freeman and Company.

plus sulfur to oxygen ((N+S)/O) ratios and thus compared with the natural bitumens (Fig. 2).

Crude oil and natural bitumens are allochthonous in origin, that is, they have migrated from source rock to place of entrapment. The natural bitumens may be divided into two main groups on the basis of their solubility in carbon disulfide (3). One group, the pyrobitumens, is dark in color, hard, and insoluble in carbon disulfide. These can be divided into two subgroups, based on their H/C ratios. The impsonite–anthraxolite–shungite subgroup consists of metamorphosed bitumen whose H/C ratios are less than 1.0, usually in the range of 0.1 to 0.8, and are insoluble in chloroform. The elaterite–albertite subgroup has H/C ratios as great as 1.6 and these varieties are chloroform-soluble. Elaterite and wurtzilite contain 3–4 wt % sulfur and have undergone varying degrees of vulcanization, the sulfur imparting elasticity. Albertite and ingramite are low in sulfur, have H/C ratios only slightly greater than 1.0, and appear to be highly indurated forms of natural asphalt (7).

The natural bitumens that are soluble in carbon disulfide are composed of hydrocarbons containing minor amounts of nitrogen, oxygen, sulfur, and trace metals, notably nickel and vanadium, incorporated into the molecular structures. The carbon disulfide-soluble natural bitumens generally contain varying amounts of admixed mineral matter from their host rocks and are separated from each other most easily by their relative fusibilities, that is, the ease with which they melt.

The mineral waxes are readily fusible, have H/C ratios greater than 1.0, are soft, and are generally yellowish in color. Ozocerite is common but scheerite is rare. They differ from other bitumens in composition, being made up of paraffins and cycloparaffins rather than mostly asphaltic constituents, as are the pyrobitumens, or aromatic and asphaltic constituents, as are the others.

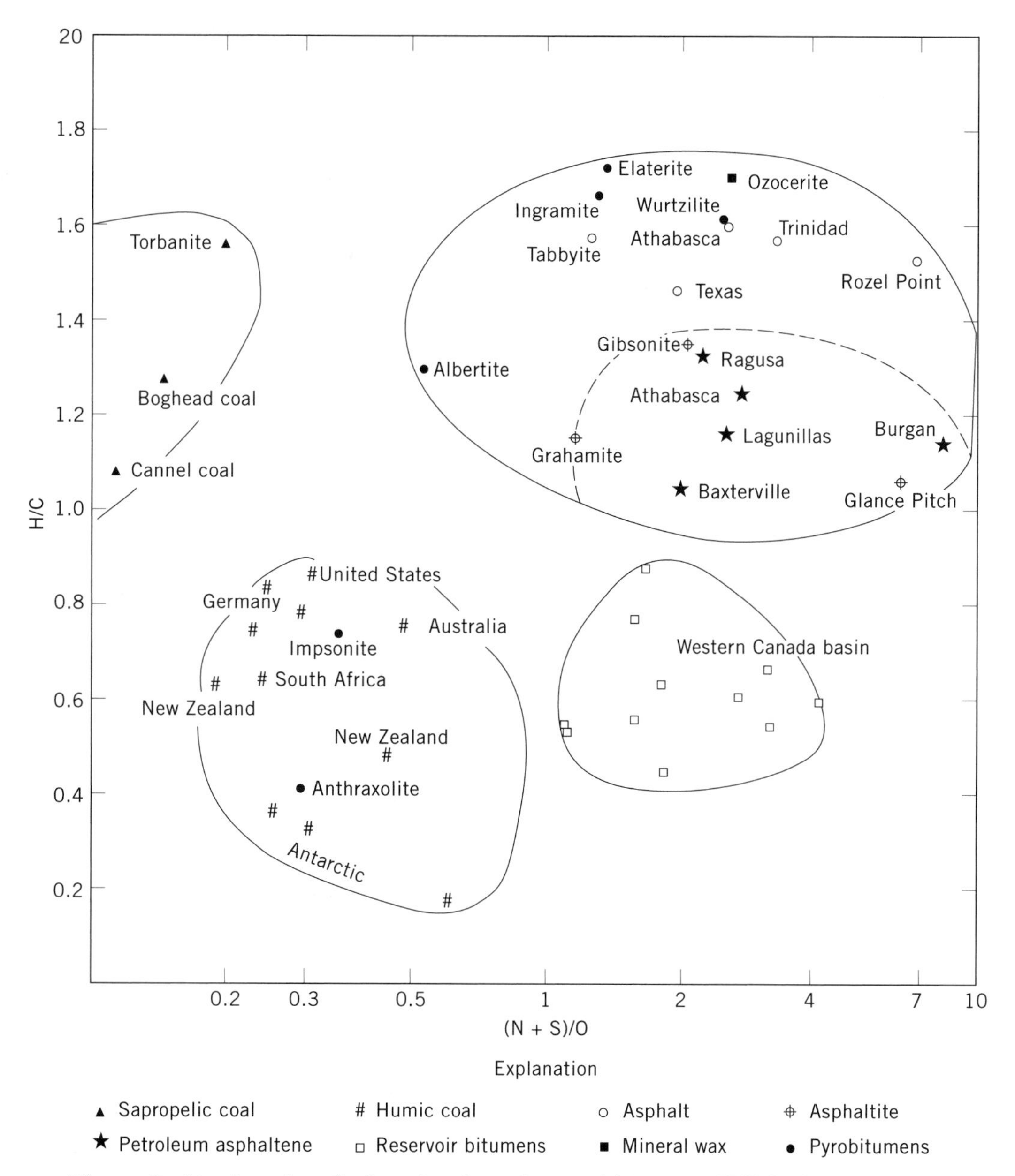

Figure 2. Atomic ratios of selected coals and natural bitumens. H/C, hydrogen–carbon ratio; (N + S)/O, (nitrogen + sulfur)–oxygen ratio (3), modified from ref. 7. Courtesy of W. H. Freeman and Company.

The asphaltites are very dark colored, are almost completely soluble in carbon disulfide, but will fuse only at temperatures of about 110°C.

Reservoir bitumens consist of asphaltene particles scattered through the reservoir rock and are not recoverable (Fig. 2) (6).

The bitumens described above are extremely limited in occurrence and of only slight present or future economic significance. Natural asphalt, on the other hand, represents an enormous resource worldwide. Almost 1.5 billion barrels of crude oil were produced from natural asphalt through 1991.

The organic matter from which the bitumens evolved, and the mineral matter with which it is admixed, were transported to places of accumulation where the diagenetic processes leading to the formation of bituminous rock began. The organic matter contained in the source rock, mostly reproductive tissue of simple plant life, such as spores and algae, is termed kerogen, which is not soluble in organic solvents (9). The principal types of kerogen and their maturation paths are shown in Figure 3.

Type I kerogen is particularly characteristic of lacustrine deposits, such as the Green River oil shales, and is relatively rare. It evolves mainly from algal and microbially reduced organic matter and is characterized by high H/C ratios and low O/C ratios. The source material is largely autochthonous.

Type II kerogen most commonly evolves from algae, diatoms, and other plankton deposited in marine environments. The commonest of source material, it has yielded most crude oil and natural gas. These source beds feature relatively high H/C and low O/C ratios and are largely autochthonous, the organic matter having been deposited under reducing conditions.

Type III kerogen derives mainly from woody and other terrestrial plant tissues accumulating in thick near-shore sediments. The organic matter therefore is mostly allochthonous. Type III kerogen displays low H/C and high O/C ratios and, as a result, forms uncommon source beds which yield little oil but do yield gas.

The alteration of the raw organic matter to kerogen takes place at low temperature during the diagenesis of the sediment. Once the stage of kerogen formation has been reached then further alteration to petroleum requires an increase in temperature. This increase results from increasing depth of burial within the basin of deposition.

The three main types of kerogen and their fates are most easily depicted on a van Krevelen diagram (Fig. 3) (9). Such a diagram relates changes in the atomic ratios of hydrogen to carbon (H/C) and oxygen to carbon (O/C) with increasing burial and, hence, increasing temperature. The direction of thermal maturation is inexorably toward carbonization, with loss of hydrogen and oxygen. The conversion of the kerogen to oil and gas results in a disproportionation of the hydrogen, the kerogen becoming deficient in hydrogen and progressively more aromatic and the maturation products rich in hydrogen (Fig. 4) (10). The temperature range of the maturation process is shown in Figure 5; obviously, the depth of burial required for the onset and conclusion of oil generation depends on the geothermal gradient within individual basins. The

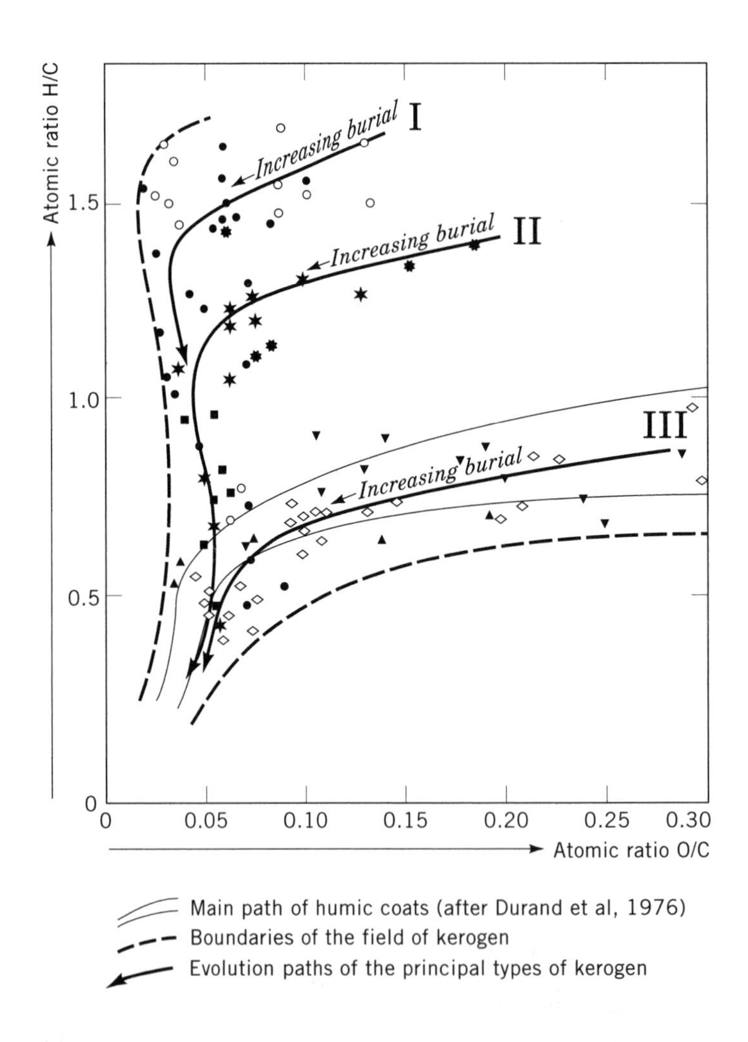

Type	Age and/or formation	Basin, country	
I	Green River shales (Paleocene-Eocene)	Uinta, Utah, U.S.A	●
	Algal kerogens (Batryococcus, etc...) Various oil shales		○
II	Lower Toarcian shales	Paris, France, W. Germany	✶
	Silurian shales	Sahara, Algeria and Libya	■
	Various oil shales		✸
III	Upper Cretaceous	Douaia, Cameroon	◇
	Lower Mannville shales	Alberta, Canada	▲
	Lower Mannville shales (McIver, 1967)	Alberta, Canada	▼

Figure 3. Principal types and paths of evolution of kerogen (9). Courtesy of Springer-Verlag.

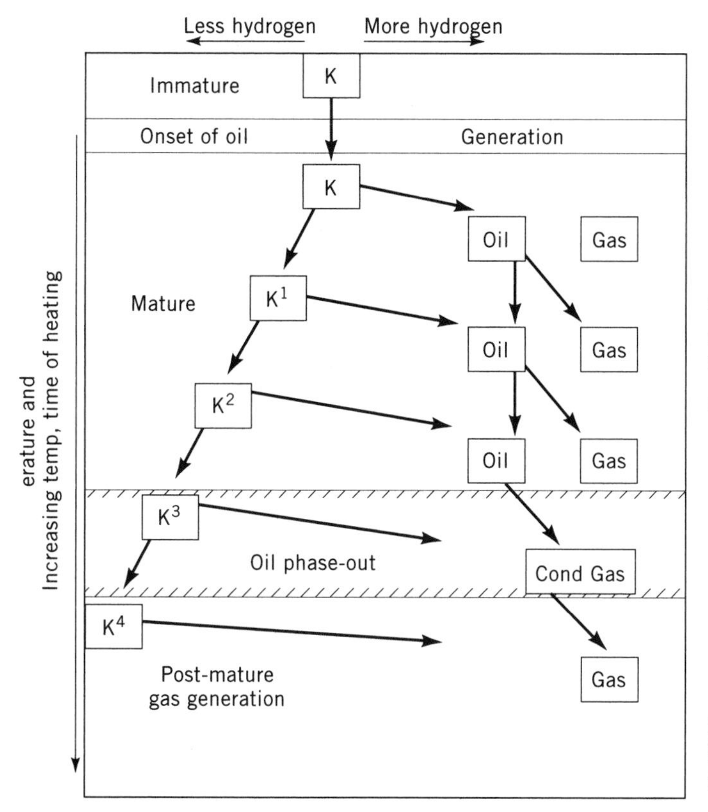

Figure 4. Disproportionation of hydrogen during maturation; K, kerogen in advancing stages of metamorphism (10). Courtesy of Academic Press, Inc. (London).

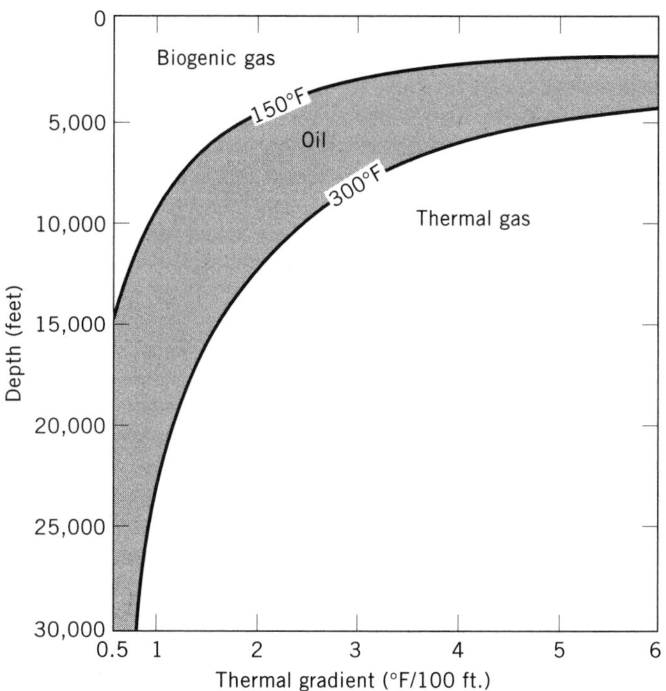

Figure 5. Oil and gas maturation as a function of depth of burial and thermal gradient (11). Courtesy of American Association of Petroleum Geologists.

bounding temperatures define an oil-generation window. Although time and natural catalysis must play a role in the maturation process, it is exceedingly difficult to quantify the extent to which they affect the process. The earliest formed, immature oil most closely resembles in chemical composition the original organic matter. It is constituted of large and complex molecules and includes many heterocompounds. Some oil can be found today in an immature form as, for example, oil from the Miocene-Pliocene Monterey Formation of California (12). However, most of the world's oil matured into mixtures of varying proportions of saturated and aromatic hydrocarbons, resins, and asphaltenes.

Following generation, the oil is expressed from the source rock as a result of compaction, moving into water-wet porous and permeable carrier beds. It moves through these beds as a result of its buoyancy in a hydrodynamic environment until reaching a position of equilibrium in a reservoir capped by an impermeable seal.

If conditions remain stable the trapped oil may remain essentially unchanged over time. If, during migration or entrapment, the oil is exposed to an excess of natural gas, the oil may be deasphaltened, the large asphaltene molecules precipitating out of the oil. These asphaltenes remain scattered in small particles through the rock as reservoir bitumens (Fig. 2) (6).

If, however, the oil is exposed to the flow of meteoric waters following entrapment, due to erosion of overlying sediments, then the oil is likely to be altered to heavy oil and natural asphalt. Water washing and bacterial degradation are the two principal agencies by which this alteration takes place.

Water washing is probably of minor significance but could result in the removal of the more water-soluble hydrocarbons. These will be of low molecular weight, such as naphthenes and lighter saturated hydrocarbons.

Bacterial degradation is very significant in the alteration of medium and light oil to natural asphalt. Barker (11) has described the requirements for biodegradation as follows: (*1*) an active water supply to carry the bacteria, inorganic nutrients, and oxygen, and to remove toxic by-products, such as hydrogen sulfide; (*2*) contact with the reservoir, in which the contained oil provides the food for the bacteria; and (*3*) temperatures up to 60°C for bacterial activity, although the bacteria can survive to about 88°C. Degraded oils show an obvious decrease in normal paraffins, the preferred bacterial food, increases in sulfur and nitrogen contents because compounds containing these elements resist biodegradation, and decreases in API gravity (increased density). This degradation being largely the result of bacterial action, asphalt deposits characteristically form near the earth's surface at the margins of geologic basins. However, asphalt may be found at depths of thousands of feet as a result of subsidence of the basin in which it evolved.

It is instructive to compare the results of biodegradation of crude oil with its earlier thermal alteration during maturation (Table 1). The thermal alteration follows a path analogous to a crude stream in a refinery distillation tower; the product becomes lighter through rejection of the heavy carbon compounds and heterocompounds. In both cases the original material is separated by molecular weight, leaving behind as a residuum the large, heavy asphaltene and small, medium molecular weight resin mole-

Table 1. Comparison of Thermal and Microbial Alteration of Crude Oil[a]

Property	Thermal	Microbial
Paraffins	Increase	Decrease
Organic sulfur	Decrease	Increase
Organic nitrogen	Decrease	Increase
Light aromatics	Increase	Decrease
API gravity	Increase	Decrease
Asphaltenes	Decrease	Increase
Optical activity	Decrease	Increase

[a] Ref. 11.

cules. The resins are highly polar because of their content of nitrogen, oxygen, and sulfur atoms, leading to their characteristic electron asymmetry. Microbial alteration, on the other hand, results in destruction of the light molecules but appears to have little if any effect on the polar molecules. The resulting heavy, asphaltene- and polar-rich natural bitumen characterizes natural asphalt. These changes may be seen graphically in ternary compositional diagrams of altered and unaltered oils (Fig. 6).

The processes involved in the alteration of a crude oil following entrapment are summarized in Figure 7. This diagram depicts the changes in a crude oil leading to the formation of a natural asphalt deposit. Both shallow meteoric water and formation brine circulating upward from depth may result in water washing of light, readily-soluble constituents. Near the earth's surface, oxidation and evaporation of the hydrocarbons occurs and the meteoric water introduces bacteria to degrade the oil. These processes lead to a heavy oil residue. Natural gas precipitates heavy, asphaltic constituents from the oil, leading to reservoir bitumen deposition and light oil. At greater depths, and therefore higher temperatures, disproportionation leads to the carbon-rich bitumens, such as pyrobitumen, with the release of natural gas.

NATURAL ASPHALT COMPOSITION AND PROPERTIES

The foregoing physical attributes of natural asphalt reflect its chemical composition and are summarized in ref. 3. Natural asphalt consists predominantly of hydrocarbons but with the saturates severely depleted. The hydrocarbons also are comprised of heterocyclic compounds, resulting in the asphalt containing as much as 5 wt % sulfur, 500 parts per million (ppm) vanadium, and 200 ppm nickel. Asphalt also contains numerous nitrogen and oxygen substituents and many other trace elements.

Speight (14) indicates the principal components of a crude oil residuum or natural asphalt by means of a fractionation diagram (Fig. 8). The asphaltenes are precipitated by *n*-pentane and make up as much as 50% of the asphalt. These include nitrogen, sulfur, and oxygen heterocompounds, termed N-S-O compounds, which are very polar, formed from the complexing of N-S-O heteroatoms and trace metals in the asphaltene structure.

The resins, or nonasphaltic polar compounds, are left in the balance of the solution, termed the maltene fraction. The resins may be precipitated from the maltene with propane. They are strongly polar compounds, significant because of their attraction to water and clay. Because of this attraction, polar compounds adversely affect the rheological properties of the asphalt, making recovery from the reservoir rock difficult. The resins can be subdivided into hard and soft resins.

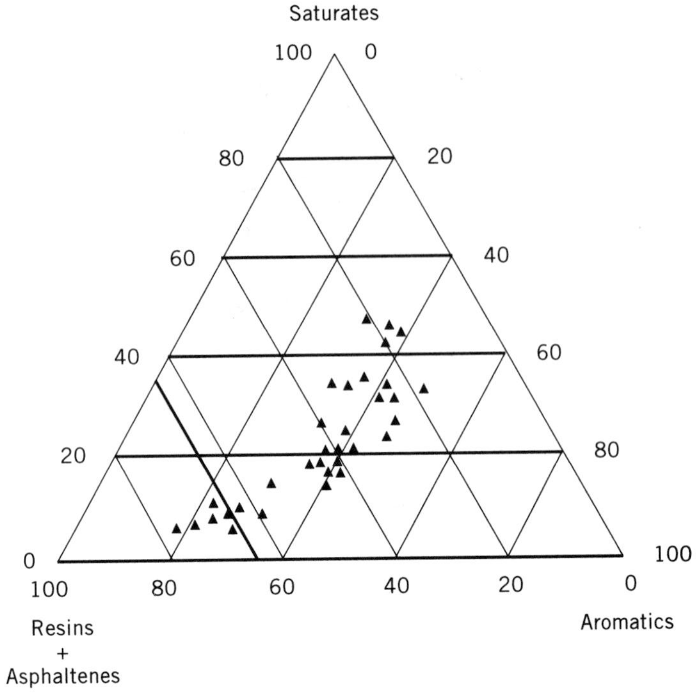

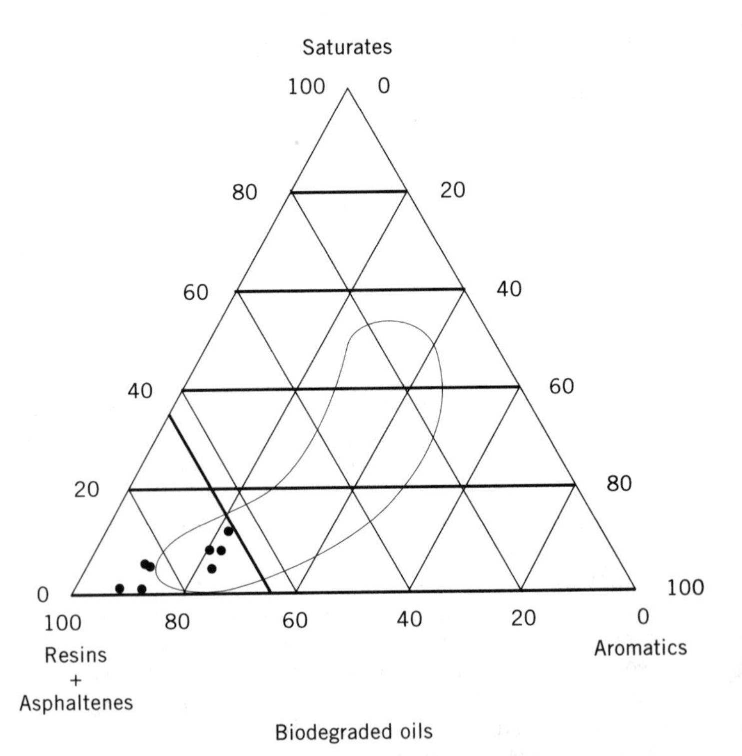

Figure 6. Gross composition of unaltered and biodegraded oils from the Aquitaine basin, France (13). Courtesy of Academic Press, Inc. (London).

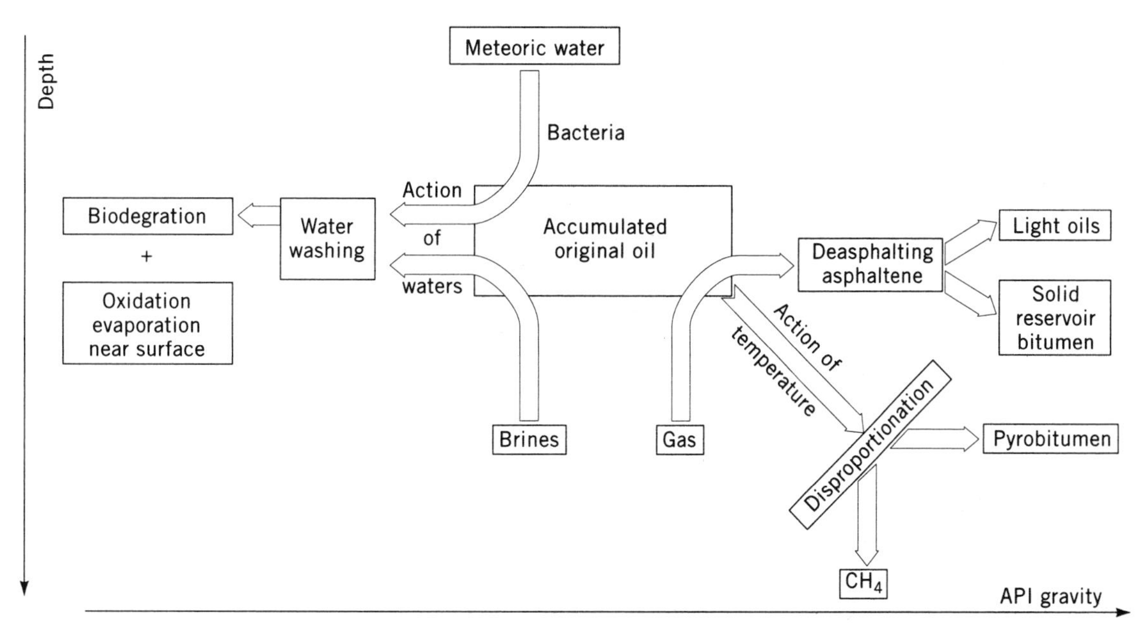

Figure 7. Alteration paths of crude oil following entrapment (13). Courtesy of Academic Press, Inc. (London).

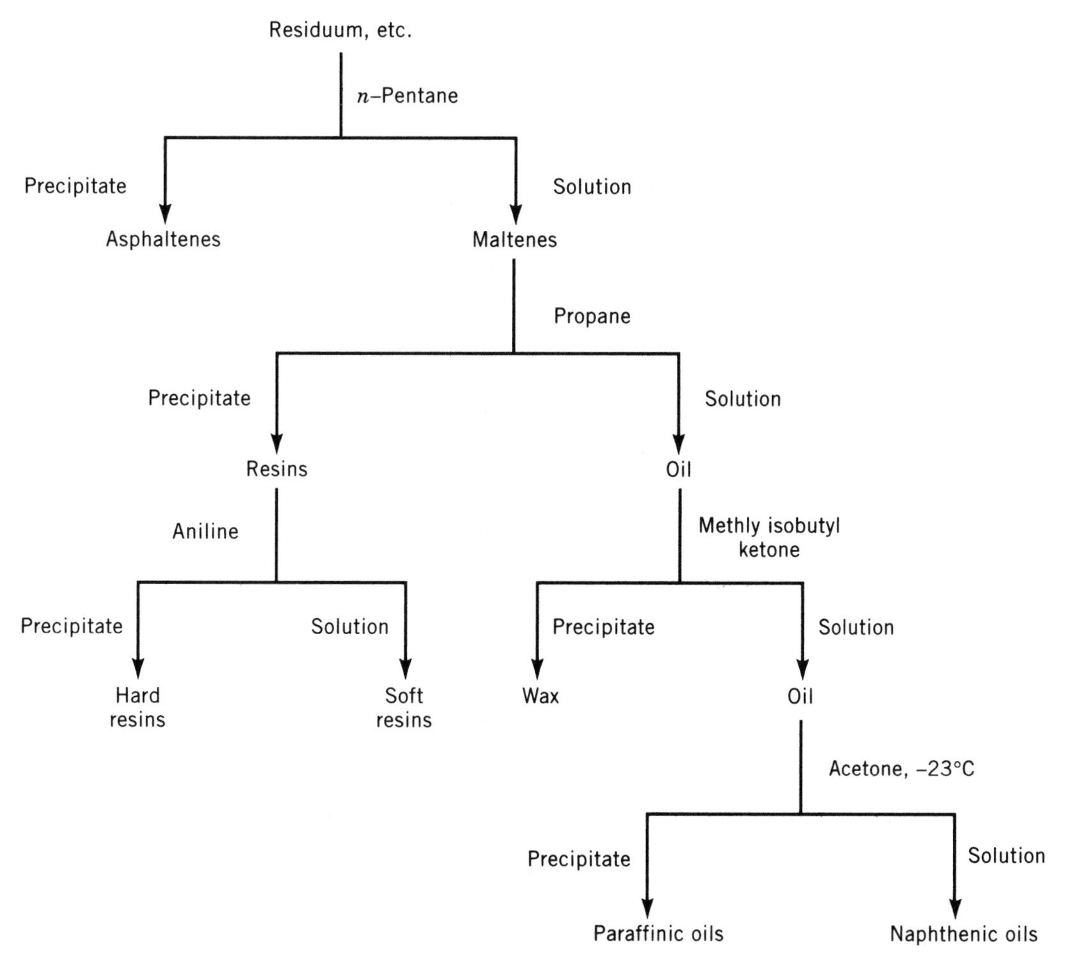

Figure 8. Procedure to derive principal hydrocarbon components from natural asphalt or refinery residuum by solvent fractionation (14). Courtesy of Marcel Dekker Inc.

The maltene, following removal of the resins, then comprises the oil fraction of the asphalt and is comprised mostly of aromatic and the heavier saturated hydrocarbons.

Unpublished data from analyses of 47 bitumen samples show 38 trace elements (Table 2). One of these elements is a nonmetal, sulfur; the balance are metals. Analyses were not made for nitrogen, oxygen, gold, silver, or platinum. In the table, the single largest value for each element was excluded from the average but this value appears in the table. The individual element maxima are seemingly randomly scattered among the samples, with more than three maxima occurring in only one sample, from the Sunnyside deposit in Utah. This single sample contained nine of the maximum values: sodium, scandium, and iron show no relationship in the periodic table, but the other six, cerium, lanthanum, samarium, europium, terbium, and ytterbium, are all rare earths. Nickel, vanadium, and sulfur are found to considerably exceed their crustal abundances in reservoir rocks of all lithologies, joined by molybdenum in carbonate rocks. It appears likely that most of the studied trace elements present in the asphalt, other than a large share of the nickel, vanadium, sulfur, and molybdenum, are derived from the reservoir rock in which the asphalt occurs or from the carrier beds through which the original oil migrated prior to entrapment and degradation.

The criteria most commonly employed for distinguishing between natural bitumen and crude oil depend upon two physical properties of these hydrocarbons, viscosity and API gravity (density). There is a definite and important correlation between viscosity and temperature (Fig. 9). This is of practical significance because the introduction of heat to a reservoir commonly can lower the vis-

Table 2. Statistical Summary of Trace Elements Found in 47 Bitumen Samples[a]

Element	Count	Maximum	Minimum	Average	Sample Maximum
Na	47	1210	4.65	171.2	1270
K	44	2053	8.6	332.9	7505
Sc	47	2.27	0.0004	0.432	3.06
Cr	47	114	0.127	10.065	206
Fe	47	5678	2.6	999.623	7120
Co	47	10.5	0.091	1.791	11.6
Ni	47	193	12.2	85.466	228
Zn	47	1343	0.15	84.0	3409
As	47	8.66	0.038	1.632	22
Se	46	40	0.124	4.043	40.5
Br	47	29.5	0.295	3.673	40.7
Rb	16	7.14	0.668	2.416	8
Sr	20	290	3.2	53.525	296
Sb	42	1.34	0.012	0.155	1.4
Cs	44	0.434	0.006	0.106	0.498
Ce	43	26.4	0.26	4.776	28.8
Ba	34	142	2.7	29.371	310
La	43	14	0.092	2.385	14.8
Nd	15	7.8	1.21	3.951	9.81
Sn	43	1.51	0.019	.342	1.91
Eu	47	0.332	0.005	0.077	0.529
Tb	39	0.219	0.005	0.051	0.346
Yb	35	0.55	0.014	0.155	1.32
Lu	32	45.6	0.003	1.469	61.1
Hf	41	0.675	0.014	0.165	1.01
Ta	30	0.402	0.015	0.069	0.445
W	33	1.45	0.02	0.238	11.4
Th	44	3.38	0.014	0.658	3.57
U	43	7.54	0.075	1.423	24.6
V	42	860	25	269.286	920
Cu	40	85	1.6	13.963	86
Hg	19	0.15	0.01	0.058	0.6
Pb	36	15	1.1	5.603	96
Li	20	29	4.5	11.72	41
Y	12	20	4.4	9.983	22
Nb	20	21	1	4.5	23
Mo	31	58	1.4	10.087	110
S	43	120,000	6,700	44,688.37	130,000

[a] Source: unpublished U.S. Geological Survey data; all values in parts per million; maximum sample value excluded from average.

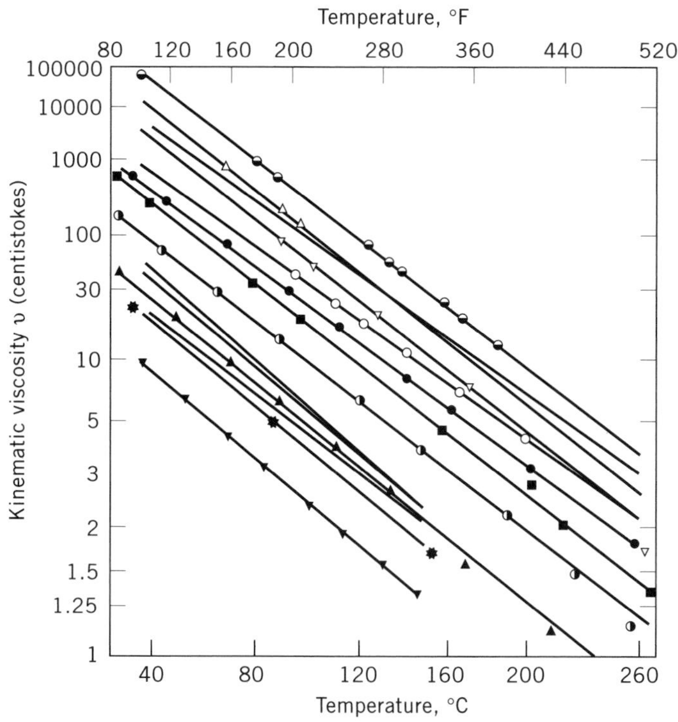

Figure 9. Relationship of kinematic viscosity in centistokes (cP/specific gravity) to temperature for gas-free oil from selected oilfields (15). Courtesy of Editions Technip.

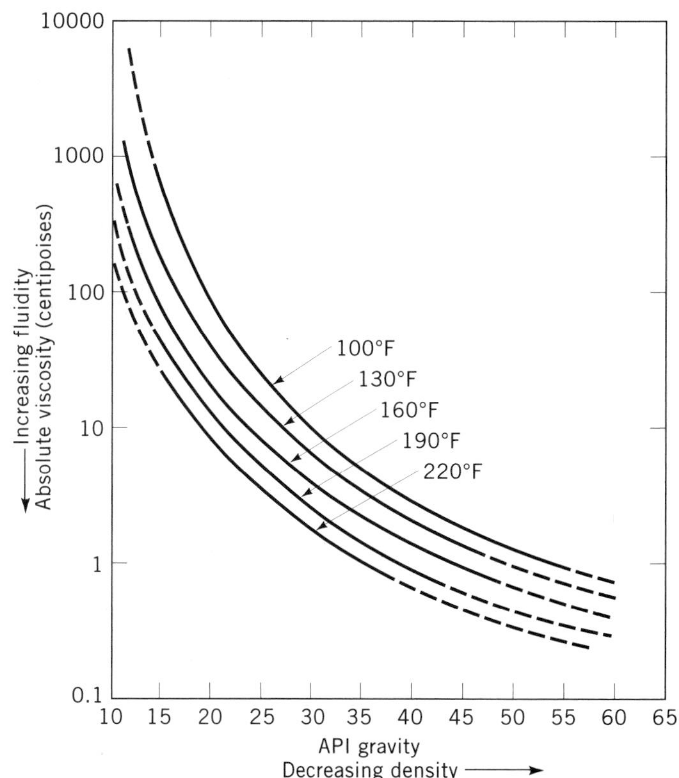

Figure 10. Correlation of viscosity with API gravity for gas-free crude oil at various temperatures (3). Modified from ref. 17. Courtesy of American Institute of Mining and Metallurgical Engineers.

cosity sufficiently for bitumen to flow to a well bore from which it can be pumped. A viscosity of 10,000 mPa·s is close to the limit at which an oil may flow to the well bore without the application of heat. Thus, the limit of 10,000 mPa·s is generally set as the demarcation between crude oil and natural bitumen. However, petroleum is a continuum of hydrocarbon compounds containing varying amounts of heteroatoms, such as sulfur, oxygen, nitrogen, and trace metals, such as vanadium and nickel, incorporated into the molecular structure.

The precise value of viscosity, 10,000 mPa·s, is thus arbitrary, but approximates a physical limit and fills a need for a value upon which to base judgments, recovery costs, and statistical series. Although viscosity can be measured with precision by a number of methods, rotational viscometry has the advantage of providing continuous measurements at a given rate of shear or shear stress for extended periods of time. Subsequently, other measurements may be made at other conditions in the same sample and with the same instrument. Time dependency may be determined. These attributes are not characteristic of most other viscometers so that rotational viscometers have become the most widely used class of instruments for rheological determinations (16).

An approximate relationship exist between viscosity and API gravity as well as temperature (Fig. 10). At a viscosity of 10,000 mPa·s bitumen has a density of about 1000 grams per cubic meter, that is, a specific gravity of 1.0 (that of water) or an API gravity of 10°. However, this relationship should be viewed as a guide and not as a substitute for the more precisely determined value of viscosity for the very heavy oils and bitumen.

Methods for classifying crude oil and bitumen, based on viscometry, have been reviewed (18) and suggest the need for standardization. (*1*) Choose a method for sample capture, preservation, and handling that minimizes sample alteration and preserves reservoir conditions. The method may involve extraction from core, cold bailing by down-hole sampler, or sampling from a produced-oil stream. (*2*) Prepare samples for viscometry to minimize alteration of oil properties, especially with respect to solvents, which will change viscosity values. (*3*) Select a method of viscosity determination, such as use of a rotational viscometer, to provide a standard of measurement. (*4*) Be aware that not all heavy oils and bitumens consistently behave as Newtonian fluids, so that viscosity may vary as a function of the magnitude and possibly duration of the stress applied at a fixed temperature. Viscosity decreases of as much of 10% have been reported because of measurements made at different temperatures.

With respect to sampling, it must be considered that reservoirs are not homogeneous; the viscosity, density, and other attributes of the oil or bitumen contained within them may vary within the reservoir, both laterally and vertically. A point-source sample may or may not be representative of the fluid in the entire reservoir. To maximize reservoir-fluid characterization, several determinations are required from separate wells or surface locations within the reservoir or outcrop area.

Reservoirs containing light oil of low molecular weight may also contain, at their base, a zone of bitumen called a tar mat. This zone results from the precipitation of asphaltic constituents from the oil (deasphaltening) by an excess (more than 40 volumes) of very light hydrocarbons, such as propane or natural gas. This action is the same as that leading to the formation of reservoir bitumens, except that the precipitate is concentrated rather than scattered. A tar mat can also result from water washing, by solution of water-soluble hydrocarbons at the contact of moving water with the base of a reservoir of oil. The water-soluble hydrocarbons are of low molecular weight, particularly aromatics, thus altering the composition toward heavy oil (11).

NATURAL BITUMEN RESOURCES

Ref. 19 reports reserves of pyrobitumens in countries of the former Soviet Union (FSU) as follows (millions of tons): Tajikistan, Uzbekistan, and Kirgizstan together, greater than 1,000; Georgia, 2.1; European Russia, 7; and Asian Russia, 141. Ozocerite reserves are reported to be as follows (millions of tons): Tajikistan, Uzbekistan, and Kirgizstan together, 3; Georgia, 0.5; Turkmenistan, 1.5; Ukraine, 2; and Asian Russia, 0.5. Similar deposits are present in many other countries, including the United States but quantities have not been reported or else, as is frequently the case, the deposits have been exhausted.

Estimated natural asphalt resources for the world are given in Table 3. The data are derived principally from ref. 20, with principal changes for the FSU from ref. 19, and for Alberta, Canada from ref. 21. About 86% of the world's demonstrated resources are found in the Province of Alberta, Canada. These resources have been carefully delineated. The next largest resources are assigned to Russia, 9.5%, about equally divided between European Russia and Asian Russia.

The inferred resources amount to about 32% of the total resource. Alberta possesses about 90% of the inferred resources, those less accurately known because of fewer drill holes and outcrops. The inferred resources of Canada, the United States, Nigeria, and Madagascar are estimated from as yet incomplete geological surveys. In many other countries, additional studies may add resources of this category to the inventory and may switch some resources from inferred to demonstrated, based on the acquisition of additional data.

Rarely do the country totals represent a single deposit; instead, they comprise the total of a few to many individual occurrences. Generally, each of these is sufficiently large to be exploitable with existing technology but seldom under extant economic conditions.

Table 3. World Natural Asphalt Resources, by Area, in Measured Deposits (million barrels)[a]

Area	Demonstrated	Inferred
North America		
Canada		
Alberta	1,697,359	831,000
Melville Island		100
Canada total	1,697,359	831,100
United States		
Alaska		11,000
Alabama		7,970
California	6,210	2,540
Kentucky	1,720	1,690
Missouri–Kansas–Oklahoma	220	2,730
New Mexico	91	
Oklahoma	42	22
Texas	2,670	300
Utah	16,856	6,429–12,362
Wyoming	120	70
United States total	27,929	32,751–38,684
Area total	*1,725,288*	*863,851–869,784*
South America		
Trinidad	60	
Venezuela	62	
Area total	*122*	
Europe		
Albania	371	
Azerbaijan	82	
Georgia	636	
Germany	315	
Italy	1,260	
Romania	25	
Russia	94,516	
Ukraine	599	
Area total	*97,804*	
Asia		
Kazakstan	2,112	
Kirgizstan, Tajikistan, Uzbekistan	48	
People's Republic of China	10,050	
Russia	91,779	
Turkmenistan	3	
Area total	*103,992*	
Africa		
Madagascar	1,600	3,400
Nigeria	42,740	42,000
Zaire	30	
Area total	*44,370*	*45,400*
Middle East		
Syria	13	
Area total	*13*	
Southeast Asia		
Indonesia	8	
Phillippines	3	
Area total	*11*	
World Totals	1,971,600	909,251–915,184
World Demonstrated Plus Inferred	2,880,851–2,886,784	

[a] Ref. 20, modified from ref. 19 and ref. 21.

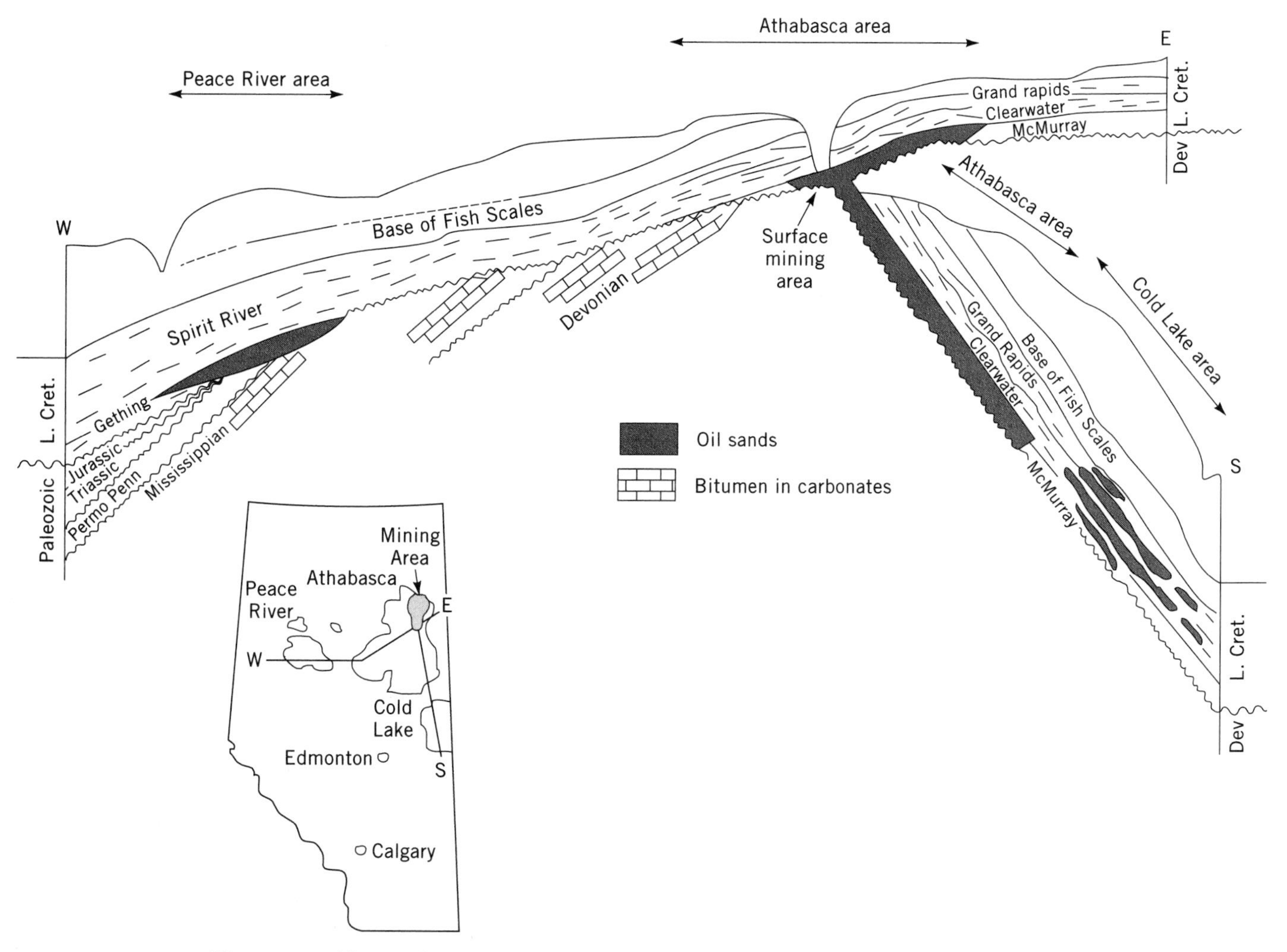

Figure 11. Alberta, Canada location map and geologic cross sections showing depth and stratigraphic relations of bitumen to oil sands and underlying carbonate rocks (29). Courtesy of AOSTRA.

Table 4. Basic Data on Alberta Natural Asphalt Deposits[a]

Parameter	Athabasca	Cold Lake	Peace River
	Sandstone reservoirs		
Asphalt in place (10^9 bbl)	950	220	76
Area (acres)	11564	3961	2439
Depth (m)	750	274–525	460–760
Thickness (m)	134	62	46
Porosity (%)	28–30	30	24
API gravity (°)	8–10	9–15	9
Viscosity mPa·s at 24°C	6,900–26,500	20,600–17,600	12,500–15,500
	Carbonate reservoir		
Asphalt in place (10^9 bbl)	384		44
Area (acres)	11530		563
Depth (m)	260–350		500–800
Thickness (m)	10		20
Porosity, %	16		18–20
API gravity (°)	7		8–10
Viscosity MPa·s, @ 650°C	5,100–8,300		

[a] Refs. 29, 30.

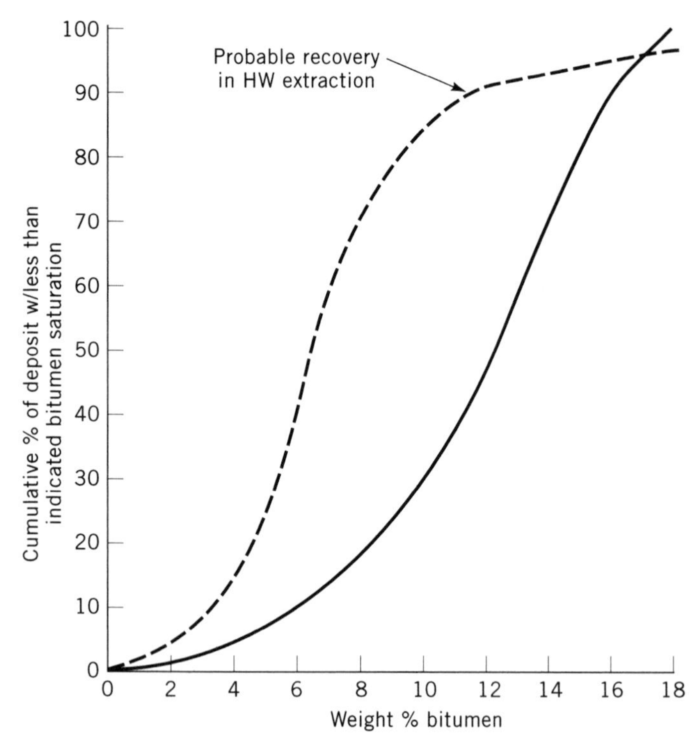

Figure 12. Bitumen grade in wt% bitumen, estimated hot water extraction recovery in %, and amount of resource in cumulative % at each grade in the Athabasca deposit (31).

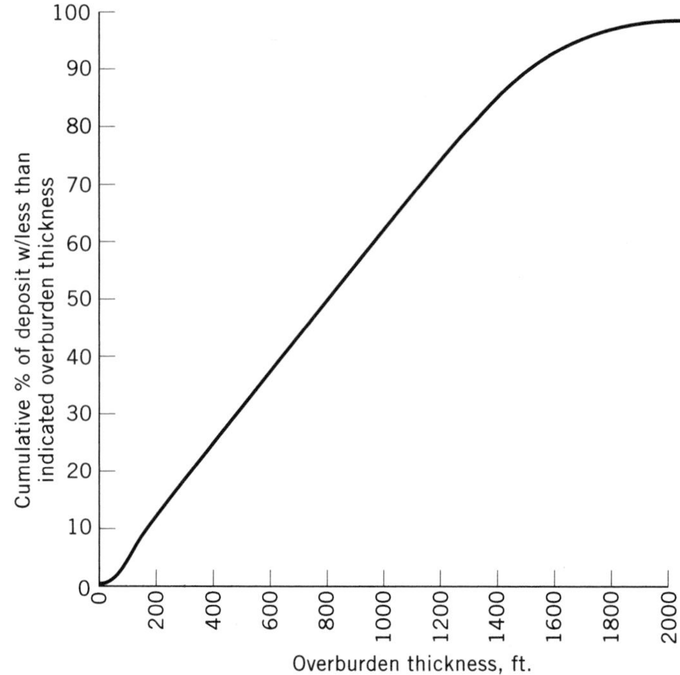

Figure 13. Proportion of Athabasca deposit at various overburden thicknesses (31).

NATURAL BITUMEN PRODUCTION

The production of natural bitumens other than asphalt is extremely limited, although many of the deposits have been exploited at some time in the past. Some ozocerite and pyrobitumen are still mined in countries of the FSU (22). Asphaltites and some asphalt are mined in Turkey (23). Ref. 24 reports 50,000 tons of asphalt rock produced annually in Kazakhstan for use as road-paving material. According to ref. 25, asphalt mines were in operation in 1962 in France in Gard (Avejan and Saint-Jean-de-Maruejols), Auvergne (Puy-de-Dome), and Ain and Haute-Savoie (Seyssel, Bourbogne, and Gardebois and Montrottier); Germany (Vorwohle); Angola (Caxito); Spain (Maestu); Italy (Abruzzi and Ragusa); Switzerland (Val de Travers); Syria (Latakia); United States in Texas (Uvalde); and Trinidad and Tobago (Pitch Lake). Other asphalt mines have included those in India (Bombay Island); Albania (Selenitza); Philippines (Leyte Island); and Indonesia (Buton Island). Cumulative production of natural bitumens for 1901–1935 is given as 21,354,000 tons of asphalt rock, 4,959,000 tons of asphalt (Trinidad), 32,000 tons of glance pitch (manjak) (Trinidad and Barbados), and 42,000 tones of ozocerite (Poland) (26). The 1970 production was 379,000 tons of asphalt rock (France, Germany, Italy, and Angola), 166,000 tons of asphalt (Trinidad, Spain, and Turkey), and about 4500 tons of grahamite and impsonite from Argentina (27). The U.S. Bureau of Mines no longer carries individual listings for bitumens in its *Minerals Yearbook* because production in the United States is limited to minor amounts of asphalt and gilsonite.

Table 5. Commercial Surface-mining Projects in the Athabasca Deposit, 1988[a]

	Suncor	Syncrude
Start date	1967	1978
Production (bbl/day)		
Capacity	56,700	163,800
Actual	43,218	136,710
Mining	Bucketwheel excavator	Dragline and bucket-wheel reclaimer
Extraction	Hot water	Hot water
Upgrading	Delayed coking	Fluid coking and hydrocracking

[a] Ref. 34.

Today, almost all rock asphalt mining is for the purpose of bitumen extraction, the bitumen then being upgraded to synthetic crude oil. In the late 19th century the rock asphalt was used widely for paving, in such cities as New York and Philadelphia. However, as early as 1906 the production of natural asphalt rock was exceeded in amount by asphalt produced from crude oil. By 1950, in the United States, less than 7% of the asphalt produced was natural.

In contrast to production for road material, Syncrude Canada Ltd. had produced 600 million barrels of oil by 1992 (28), all processed from asphalt rock recovered by open-cast mining.

RECOVERY

Three methods are employed for the recovery of natural asphalt from the earth for the purpose of converting it to petroleum products. Each of these methods either is in use or else has been tested for use in Alberta. These methods are (*1*) surface mining, (*2*) in situ thermal, and (*3*) combined mining and thermal.

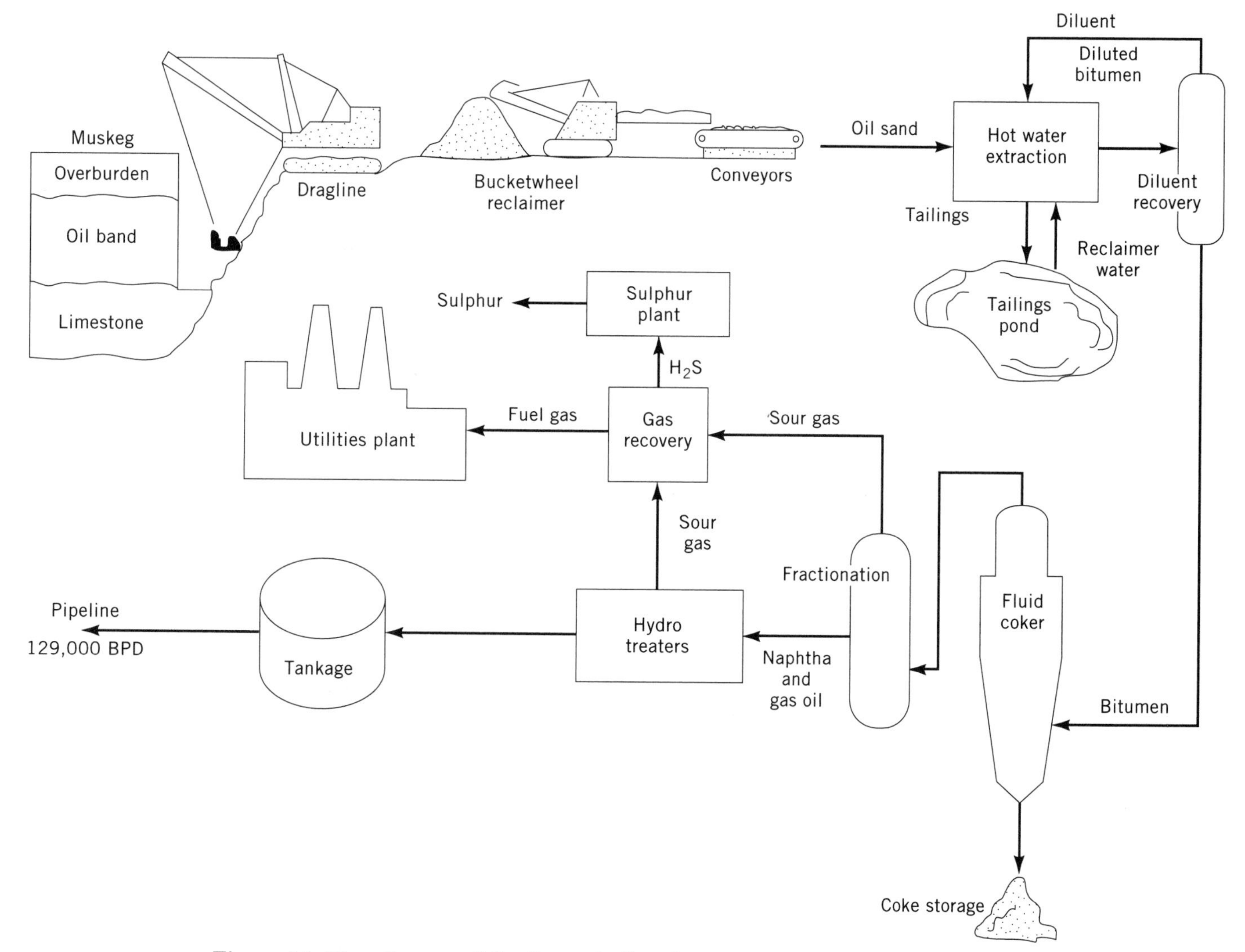

Figure 14. Flow diagram of the Syncrude Canada operation at the Athabasca deposit, showing the mining, bitumen extraction, bitumen upgrading, and utilities areas (35).

Figure 11 shows the general location of the three great asphalt (oil sand) deposits of Alberta, their geologic relations, and their relative amounts of overburden. Basic data on the reservoirs are given in Table 4, in which the presently productive sandstone reservoirs are separated from the underlying carbonate rocks. In all three areas, present recovery efforts are directed to the Lower Cretaceous sandstone deposits. However, large resources are also contained in the unconformably underlying carbonate rocks. In all cases, recovery is very sensitive to ore grade. This is particularly so in the case of surface mining (Fig. 12), which entails separation of the bitumen from the sand by the hot water process.

The Athabasca oil sands deposit is the largest of the three vast Albertan natural asphalt deposits. About 10% of the area is considered to be surface minable, the overburden above the asphalt rock being less than about 45 m (Fig. 13). Conventional underground mining of the oil sand is not economical because one cubic yard of oil sand which, at a saturation of 14.3% weighs 1.413 tons, must be extracted for each 0.8784 barrels of oil produced (32–33). Deeper sands require recovery by an *in situ* method. About 90% of the mined oil sand can be recovered in surface mining. There are two commercial surface-mining projects in the Athabasca deposit at present (Table 5). A flow diagram for the largest surface mining operation in the Athabasca deposit, that of Syncrude Canada, is given in Figure 14.

The muskeg on the surface is stripped in the winter, when frozen, and set aside for later use in land reclamation. The rock overburden is then removed with hydraulic shovels and trucked to areas mined out in earlier operations. Draglines then pile the oil sand (asphalt-impregnated rock) in windrows, from which it is placed on conveyor belts by bucketwheel reclaimers for delivery to the extraction plant. Material handling is thus a principal part of the operation, especially in light of the severe weather conditions.

The mechanical problems involved in the surface mining of the Alberta natural asphalt deposits are summarized in ref. 33. Some of these problems will be encountered wherever such deposits are mined. (*1*) The clay shale overburden swells when exposed to fresh water. This is usually not a problem when the overburden is removed by shovel and truck. However, when the overburden is removed by the technique of dredging and hydraulic transport, the result is a greater than *in situ* volume, causing a handling problem. The sand, even though unconsolidated, also may swell after extraction. This is the result of the oil sand having formerly been buried under younger formations, subsequently eroded, and thick glacial ice, subsequently melted, which led to a preferred orientation of the sand grains; this orientation is lost during mining and processing, resulting in volume increase. (*2*) Pit wall slope sloughing is encouraged by the high gas pressures in the pores of the sand. This problem is reduced by preblasting of the oil sand, as is done at the Suncor facility, and by keeping the heavy excavating equipment away from the slope crests. (*3*) The mining equipment is subjected to a high rate of wear by the abrasive sand, especially when the sand is frozen in the winter. As well, the oil sand is very strong and tough in such weather, placing heavy demands on the drag lines, bucketwheel extractors, and conveyor belt systems. (*4*) Very large quantities of tailings from the hot water extraction process—over 350,000 tons daily—must be disposed in huge settling ponds enclosed by sand dikes, which are built on swelling clay-shale foundations. The settling, or sludge pond, at Syncrude covers more than 11×10^6 m^2.

At the plant the bulk of the bitumen is separated from the sand by the hot water process. The raw bitumen next is upgraded in a fluid coker and then in a catalytic hydrotreater to produce a refinery feedstock of 32° API gravity.

In the hot-water extraction process described by the Alberta Oil Sands Technology and Research Authority (AOSTRA) (36), the crushed sand is fed into a rotating drum, with the addition of air to aid frothing of the bitumen and caustic to enhance the liberation of the bitumen. The process works and is commercially successful but has these drawbacks: (*1*) recovery is sharply reduced with low grades of ore; (*2*) the caustic disperses the fine material in the wastewater sludge, preventing its settling in the tailings ponds; (*3*) the process is energy inefficient; and (*4*) bitumen is lost to tailings and oversize rejected oil sand.

To upgrade the extracted bitumen the Syncrude operation employs a fluid coker (Fig. 15). In this process, small spheres of coke are suspended in a flow of steam into which the extracted bitumen is blown. The large molecules of asphaltic components are cracked at temperatures of about 529°C, the small molecules being carried off to a condenser and the residual coke agglomerating on the small coke particles and settling out of the steam-coke fluid. This type of process, by rejecting carbon, yields only about 86 barrels of synthetic crude per 100 barrels of processed raw bitumen (36).

The upgraded synthetic fluid, after condensing, is distilled in a fractionating tower to separate naphtha, gas oil, and sour gas. Finally, the fluids are desulfurized in the hydrotreater and the gas is itself desulfurized, to be used in the plant as fuel.

Other upgrading schemes involve hydrocracking, that is, cracking the bitumen in a hydrogen-rich environment in order to consume all the carbon rather than rejecting a large part of it. These hydrogen-addition processes yield 104 barrels of synthetic fuel per 100 barrels of raw bitumen as a result of the hydrogen uptake. The higher yields have resulted in improved economics for the processes, despite the higher pressures at which they operate, as reflected in the choice of hydrogen addition for oil sands expansion at the Bi-Provincial Upgrader and for the proposed OSLO project (which would utilize VEBA Combi-Cracking).

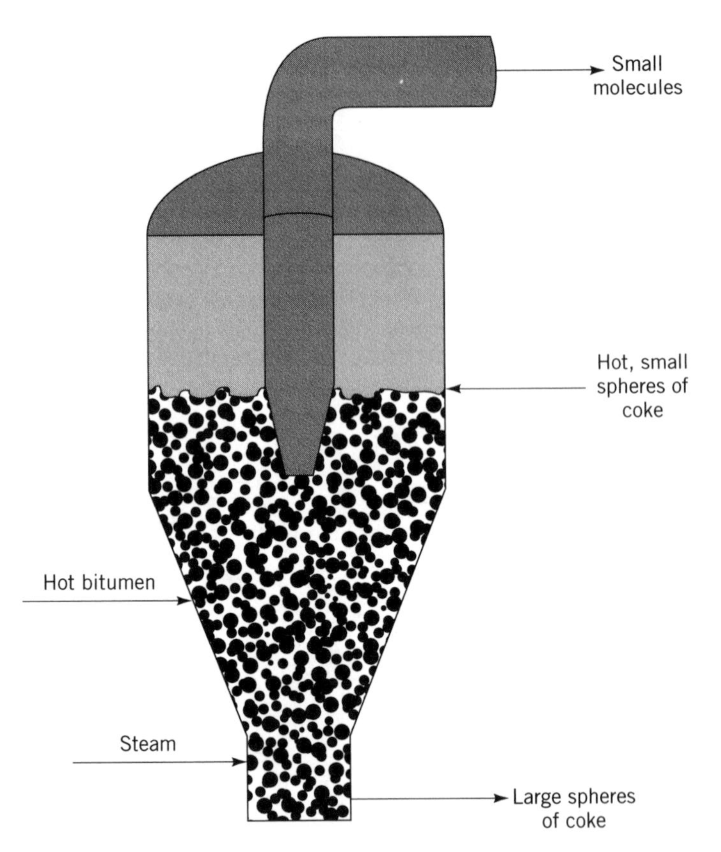

Figure 15. Fluid coker upgrading unit (36). Courtesy of AOSTRA.

Since 1976 AOSTRA has been developing a means for simultaneously extracting the bitumen from the sand and upgrading it to a residue-free, distillable oil (36). The result of this effort is the AOSTRA Taciuk Processor (Fig. 16), which uses heat alone, much of it derived from spent, processed sand, to separate the bitumen from the sand, crack it to drive off the hydrocarbons as vapor, and utilize the coked sand product for heat. Following this, the coke burns off and the spent sand then is cycled for preheat and finally for disposal. This process has distinct advantages over the extraction-upgrading scheme shown in Figure 14 (36): (*1*) product yield is increased; (*2*) plant flow from input sand to product is simplified; (*3*) bitumen losses to tailings and rejects, and the need for tailings ponds, are eliminated; (*4*) the process is more energy efficient than the hot water scheme; (*5*) no chemicals or additives are required; and (*6*) capital and operating costs may be reduced.

Because surface mining operations are only cost-effective where overburden thickness is no more than about 45 m and conventional, high-pressure steam-assisted in situ recovery requires a caprock, that is, overburden of about 152 m, a window of asphalt-impregnated rock as

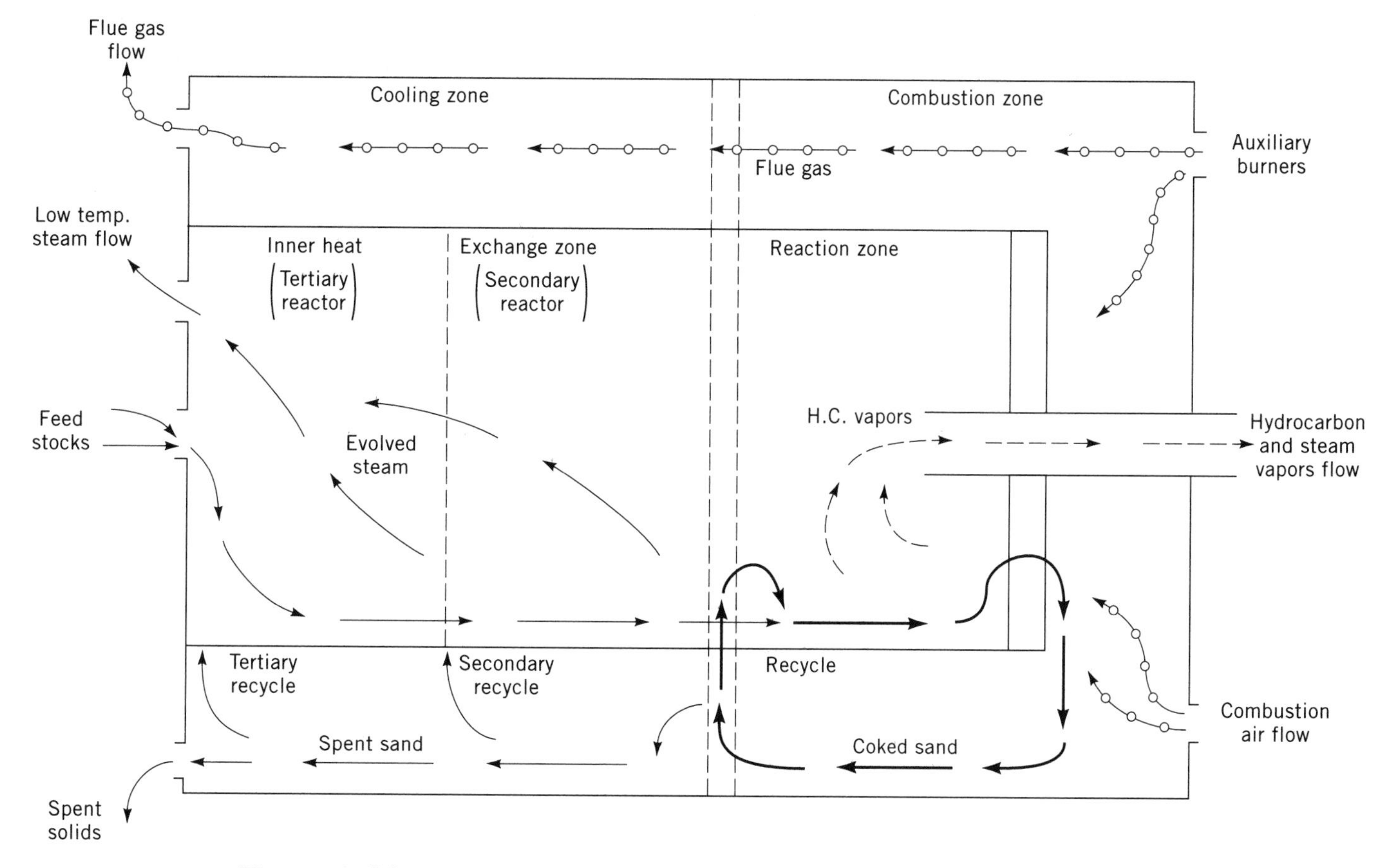

Figure 16. Schematic diagram of AOSTRA Taciuk Processor, showing major zones and flow streams (37). Courtesy of AOSTRA.

thick as 106 m or more has not been exploitable until recently. The high pressures may lead to fracturing and inter-well communication with the producing wells, resulting in premature entry of water and the bypassing of bitumen. One solution to this is underground mining above or below the oil-rich formation. From the mine tunnels, wells are drilled at close spacing to permit low-pressure steam injection into the reservoir. Oil-mining schemes have been employed many times in the past and some, such as the Pechelbronn Field, France, Wietze Field, Germany, and Yarega Field, Russia are still in operation. In Yarega Field, tunnels are drilled above the reservoir, vertical or slant wells are drilled from the tunnel into the reservoir, and steam is injected into the reservoir to improve the oil viscosity. A producing gallery is developed below the reservoir, from which wells are drilled upward into the reservoir at an inclination and the oil drains by gravity through these to the producing gallery (38). At Yarega the bitumen is similar to that at Athabasca but is less viscous (15,300 mPa · s) and contains less sulfur (1%). Reservoir parameters are comparable but with an important difference: it is necessary to fracture the reservoir to achieve production and this necessitates development by production units.

In the Athabasca deposit research is being conducted on experimental recovery of shallow bitumen at an underground test facility (UTF) (39). A schematic diagram of this facility is given in Figure 17. Two shafts, each about 3 m in diameter penetrate about 131 m of overburden, 18 m of reservoir, and 18 m of underlying limestone bedrock. Connecting tunnels are then drilled for the production operations. From the production tunnels, pairs of horizontal wells are drilled (Fig. 18). Steam is circulated through both wells until the bitumen is heated sufficiently to flow (about two months); although the system is open, little of the steam enters the formation. The steam chamber continues to grow through continuous steam injection in the upper well while the lower well is converted to production, receiving the bitumen by gravity drainage. This two-well system is called the Steam Assisted Gravity Drainage (SAGD) process, a term also applied to other gravity-drainage systems. The twin horizontal well concept is unique in that it operates at low pressures, even in reservoirs such as Athabasca, which contain bitumen with viscosities measuring in millions of cP. The close spacing of the wells (3–5 m) permits hot communication, and after communication is achieved, the pressure drop is small. SAGD with twin wells can be operated at any depth but this is the only process, because of its low working pressure, that is applicable to shallow in situ operations. This process may have other applications to supplement or replace surface mining. Here, the need for tailings ponds and the reclamation of mined out open pits would no longer be necessary.

For deeper asphalt reservoirs, below about 152 m, a number of projects illustrate ways by which the bitumen may be recovered. There were a total of 18 in situ projects in 1988 (34). Most of these are based on the injection into

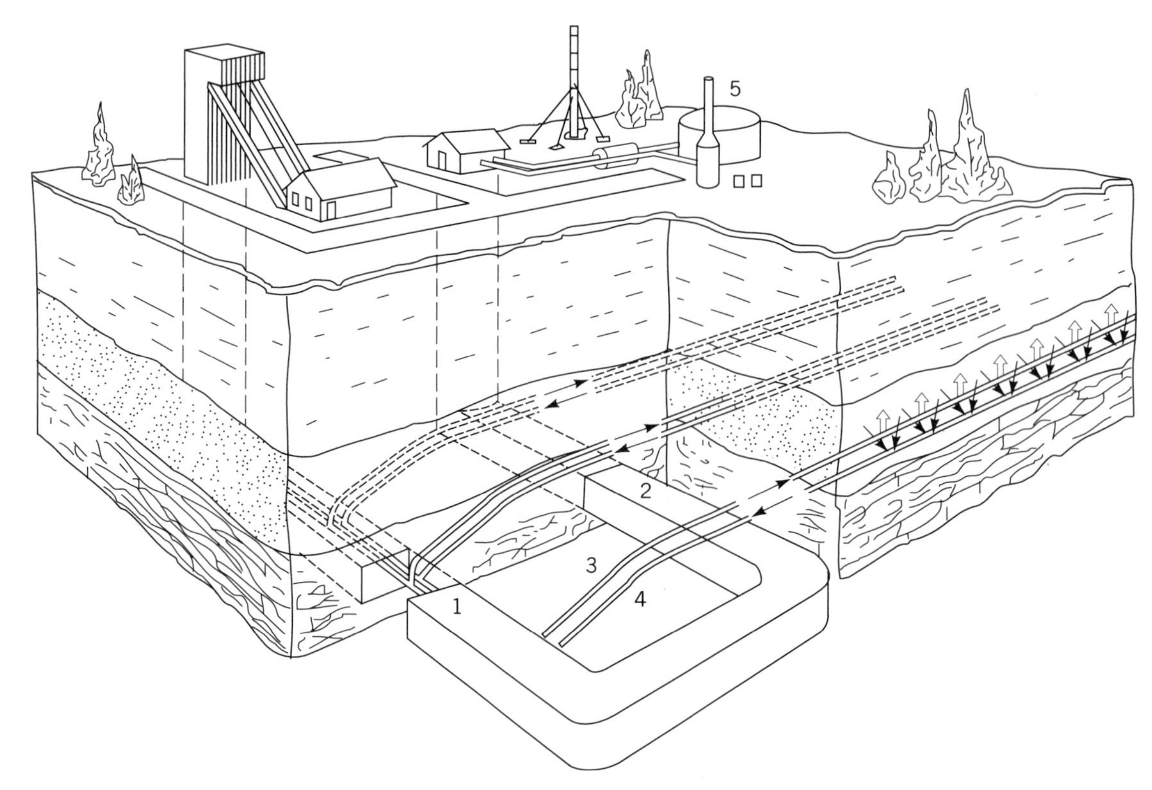

Figure 17. Schematic diagram of the Underground Test Facility (UTF); 1, tunnel for drilling horizontal wells and for steam and oil-collection pipes; 2, tunnel for access, ventilation, and oil and produced-water pipes; 3, steam-injection pipes in reservoir; 4, oil-collection pipe in reservoir; 5, surface facilities for produced water and oil (39). Courtesy of AOSTRA.

the reservoir of steam as the heat carrier, either as cyclic steam stimulation or as a steamflood. Steam is a very effective heat carrier but generally most effective if the steam can be injected below formation fracture pressure. The problem is that if the oil sand is fractured then heat is concentrated along the fractures and does not invade the entire body of the sand. However, commercial success has been achieved in such deposits as Cold Lake, in which steam injection has resulted in fracturing.

If the asphalt is extremely viscous it may reduce injectivity of the steam to, or close to zero. This is usually overcome by fracturing. Another approach is by application of the HASDrive (heated annulus steam drive) process (40), which has not yet proven to be commercial. Here, a horizontal conduit contains a high-pressure steam injection pipe and a return pipe, forming a closed system. Steam is circulated through this nonperforated conduit, termed the HAS pipe, above fracture pressure, heating but not fracturing an annulus of the formation around the HAS pipe. The viscosity of the bitumen in the annulus is thus substantially lowered. Steam below fracture pressure then is injected along the path of the annulus as a steam drive from an injector well, pushing the asphalt to a production well. This process is being tested in the parts of the Athabasca deposit too deep for surface mining.

If the asphalt zone has an overlying water leg, this water zone will dissipate heat from the injected steam, which reaches it by gravity override. The same thing is true of a basal water leg. In either case, the water legs may assist the steam in heating the asphalt if the water can be controlled at the edges of the steam project, so that the heat is focussed within the project pattern and not simply lost to the formation beyond it.

A depth of 914 m is frequently set as the depth limit for steam projects (41). This is primarily the result of heat loss while the steam is being moved through the injector pipe. However, the successful application of a steam stimulation project at approx 2500 m in the Boscan Field, Venezuela has been reported (42). A 38% production increase was achieved by injecting the steam through preinsulated tubing, with only a 5% heat loss. This was heavy oil with a viscosity of 300 mPa·s at reservoir temperature.

The oil sands in the Peace River deposit are encountered at a depth of about 502 m and are about 21 m thick at the Shell Canada pilot (43). The viscosity of the asphalt in the reservoir is 90,000–150,000 mPa·s. At this site, Shell found, when attempting steam stimulation, that most of the steam by-passed the formation in the highly permeable bottom water leg. To overcome this, Shell Canada developed a production scheme whereby 100% quality steam was injected below fracture pressure at pattern injectors (7-spot patterns). Upon steam breakthrough, the producing wells are back-pressured to build up reservoir pressure. Finally, production is maximized by blowing down the producing wells, completing the injection-production cycle and resulting in a pressure cycle steam drive. Through four such cycles a bitumen recovery of

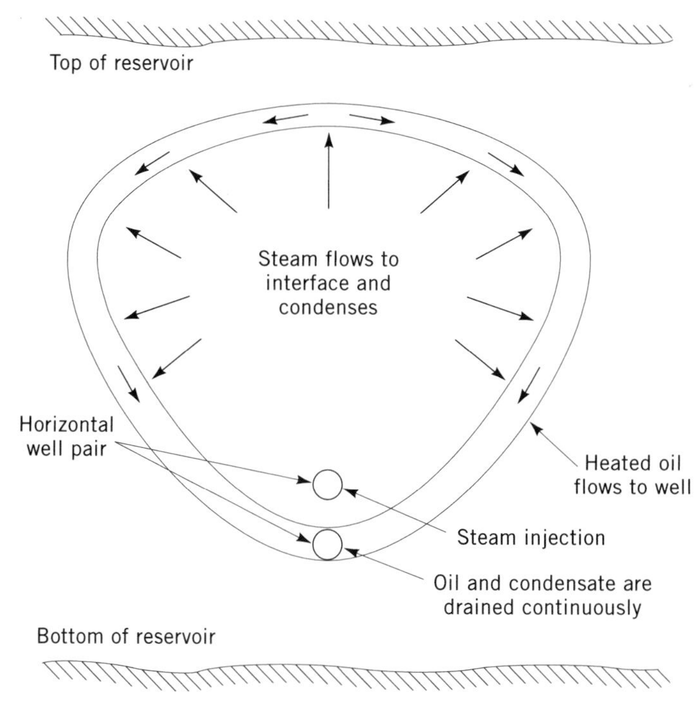

Figure 18. Schematic diagram of cross section of the UTF steam-oil production arrangement (39). Courtesy of AOSTRA.

59% was achieved, with an oil–steam ratio of 0.29. Research continues in order to find ways to steam the entire reservoir volume more completely, reaching those wells which still remain as blind spots. Shell Canada also is attempting to upgrade the bitumen in the reservoir by adding transition metals to the formation to catalyze a high-temperature bond-breaking reaction in the asphaltene molecules. This would have the salutary effect of reducing the molecular size, reducing viscosity, and thereby improving fluid flow as well as reducing the needed steam temperatures.

The Kearl Lake pilot area in the Athabasca deposit illustrates success in meeting two problems, an overlying mobile water zone above the middle oil sand and a need for fracturing the immobile, water-free lower oil sand (44). Here the reservoir is shallow, less than 213 m, and the asphalt is very viscous (3×10^6 mPa·s in the reservoir). The asphalt in the middle zone was rendered mobile by injecting steam in the water zone, which acted as the heat carrier and heated the oil sand. At first, the steam–oil ratio was unacceptably high due to heat loss beyond the pilot well limits but this was overcome by blocking the water zone outside the pilot limits, using a patented process. The thinner lower zone had an extremely low injectivity until it was successfully fractured horizontally. This permitted the steam heat to reach the entire oil sand zone with consequent reduction of the viscosity. This production scheme is illustrated in Figure 19.

Another steam stimulation project in the Athabasca deposit deals with the oil sand at about 274 m in depth and with a reservoir viscosity of more than 1×10^6 mPa·s (45). Although the formation has good injectivity for the steam, it was reduced to zero by the highly immobile asphalt. This required steam injection above fracture pressure. The long fractures reduce heat effectiveness because they appear to start in a vertical plane but then assume a horizontal aspect. Vertical fractures dissipate heat into the overlying shaley zone. Therefore, small steam slugs were found to be more effective in initial cycles, with the slug sizes increased somewhat for later cycles. In spite of tests indicating that the induced fractures become horizontal, temperature data from monitor wells still show steam rising to the top of the formation, a problem that must be overcome to achieve heating of the entire reservoir.

Successful production has been obtained from the Cold Lake deposit by use of in situ thermal methods at depths

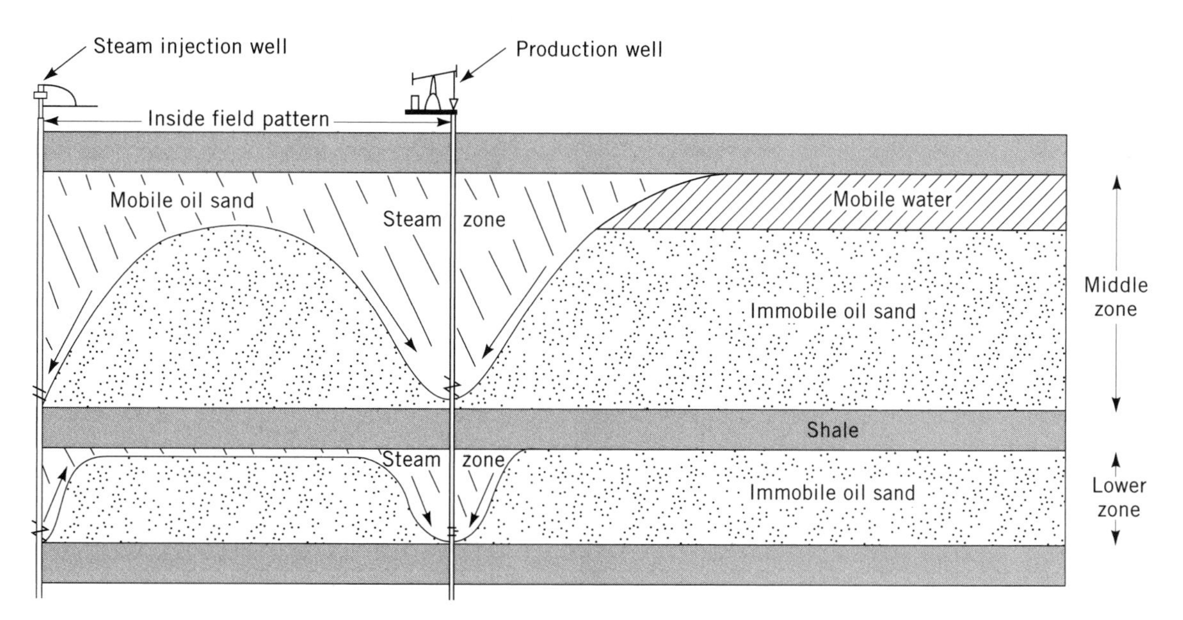

Figure 19. Schematic diagram of Kearl Lake pilot recovery process (44). Courtesy of AOSTRA.

of 396 m. Here, BP Exploration Canada has investigated what is called pressure cycle oxygen fire flooding (36) at the Marguerite Lake and Wolf Lake pilot projects. The heat sequence involves steam stimulation followed by oxygen and water injection to make a wet combustion phase. This leads to a building up of pressure. When the injection is stopped the bitumen, mobilized by the viscosity reduction, is produced during the blow-down phase. This is an example of the use of combustion (fireflood) to provide heat, with the accompanying water creating steam. Fireflooding theoretically should be the most effective of thermal processes but in practice it has never achieved the effectiveness of steam stimulation or steamflooding.

Steam stimulation tests have been conducted at Cold Lake in an area where the viscosity at reservoir conditions is 100,000 mPa·s rather than 1×10^6 mPa·s (46). The tests have been commercially successful, the main efforts being concentrated on improving recovery efficiency. Injection exceeds fracture pressure for at least five steam cycles and the fractures appear to be vertical, at least initially. In later cycles the fractures become horizontal or else yield a preferred horizontal mobility. There is a bottom water leg present but it is shielded from the rich oil sand by a shaley layer.

The cost of producing the upgraded synthetic oil feedstock reflects the elements of the operation: 40% mining, 30% upgrading, 15% bitumen extraction from the mined sand, and 15% utilities (35). Because of the large fixed costs, with the plant alone requiring an investment of about $5 billion, profitability is closely related to production rates.

These illustrations of recovery are taken from experience with the Alberta asphalt deposits, the only ones now being produced for the recovery of bitumen for conversion to crude oil. This experience, however, will be applicable in most places in the world at such time as conventional crude oil supplies are reduced sufficiently to create demand. Projects are in progress or planned in many places, including Madagascar (47), Nigeria (48), Russia (24), Trinidad (49), and Utah in the United States (50).

ENVIRONMENTAL CONSIDERATIONS

Many of the environmental considerations required in the process of converting a natural asphalt deposit to a suite of useful products are no different from those involved with conventional crude oil. The objective is to minimize each step that leads to environmentally undesirable consequences and to maximize an upgraded product as rich as possible in hydrogen and as lean as possible in carbon and nonhydrocarbon contaminants. A major difference between natural asphalt and conventional crude oil is that much of it is, and in the future probably will be produced by surface mining rather from oil wells.

Exploration

The natural bitumens, except for natural asphalt and reservoir bitumen, occur in small deposits and those that have been found occur on the surface of the earth. The reservoir bitumens are disseminated in the subsurface and hold no economic interest. Therefore, these species of bitumen have no environmental impacts in the exploration process or in their later exploitation and need not be considered further.

Natural asphalt is found in the subsurface but is discovered only by serendipity in the search for crude oil. Surface deposits of natural asphalt are known and did not incur particular environmental impacts in their discovery.

Recovery and Extraction of Bitumen

Subsurface in situ production of asphalt requires heat to reduce the viscosity and fresh water steam as the heat carrier has proved to be most effective. Usually, 3–7 barrels of steam as water are needed to produce a barrel of bitumen. Recycling of produced water reduces the requirement for fresh water and eases the problem of produced water disposal. Produced water may contain some or all of dispersed or soluble oil, heavy metals, radionuclides, treatment chemicals, salt, and dissolved oxygen (51). Many of the most toxic compounds in the dispersed oil droplets are volatile and escape, unless trapped in sediment somewhere in the production system. Of the soluble compounds, phenols are biodegradable. Levels of polynuclear aromatic compounds are below toxic levels in produced water. Heavy metals in produced water, such as cadmium and lead, may be at toxic levels and must be treated for removal or else properly disposed. The problem of radionuclides is strongly dependent on geographic location and thus is site specific. Risk-assessment studies suggest that predicted excess human cancers from this source would be comparable to background concentrations of radium. Treatment chemicals are generally present below toxic limits, with the exception of those injected as slugs. Slugs containing such chemicals, including scale and corrosion inhibitors, may be captured separately to permit proper disposal of the concentrate. If the produced water is saline it can be treated at the surface facility. Usually, the produced water is low in oxygen since that is the nature of formation water; its low oxygen content, inimical to life, must be considered if the water is discharged at the surface.

Recycled produced water can be used directly as makeup water for steam if total dissolved solids (TDS) are less than 8000 ppm, as is the case at Peace River. At higher TDS levels, some treatment is required. Figure 20 illustrates a conventional treatment facility, with oil separated by gravity, although this becomes difficult in the case of bitumen, which approaches or exceeds the specific gravity of water, and suspended solids removed by gravity settling or gas or granular media flotation (36).

Use of the hot water bitumen-extraction process leads to very large tailings in the form of sludge; at Syncrude, this amounts to about 3.5×10^5 million cubic m per day (33) in a pond covering 4.25 square miles. If left alone, this material may take centuries to consolidate. Clay minerals become thoroughly dispersed in the sludge, which also contains residual bitumen and fine-grained non-clay minerals (52). A number of solutions to the sludge problem have been proposed but at present the material is still segregated in ponds; it cannot be used or recycled.

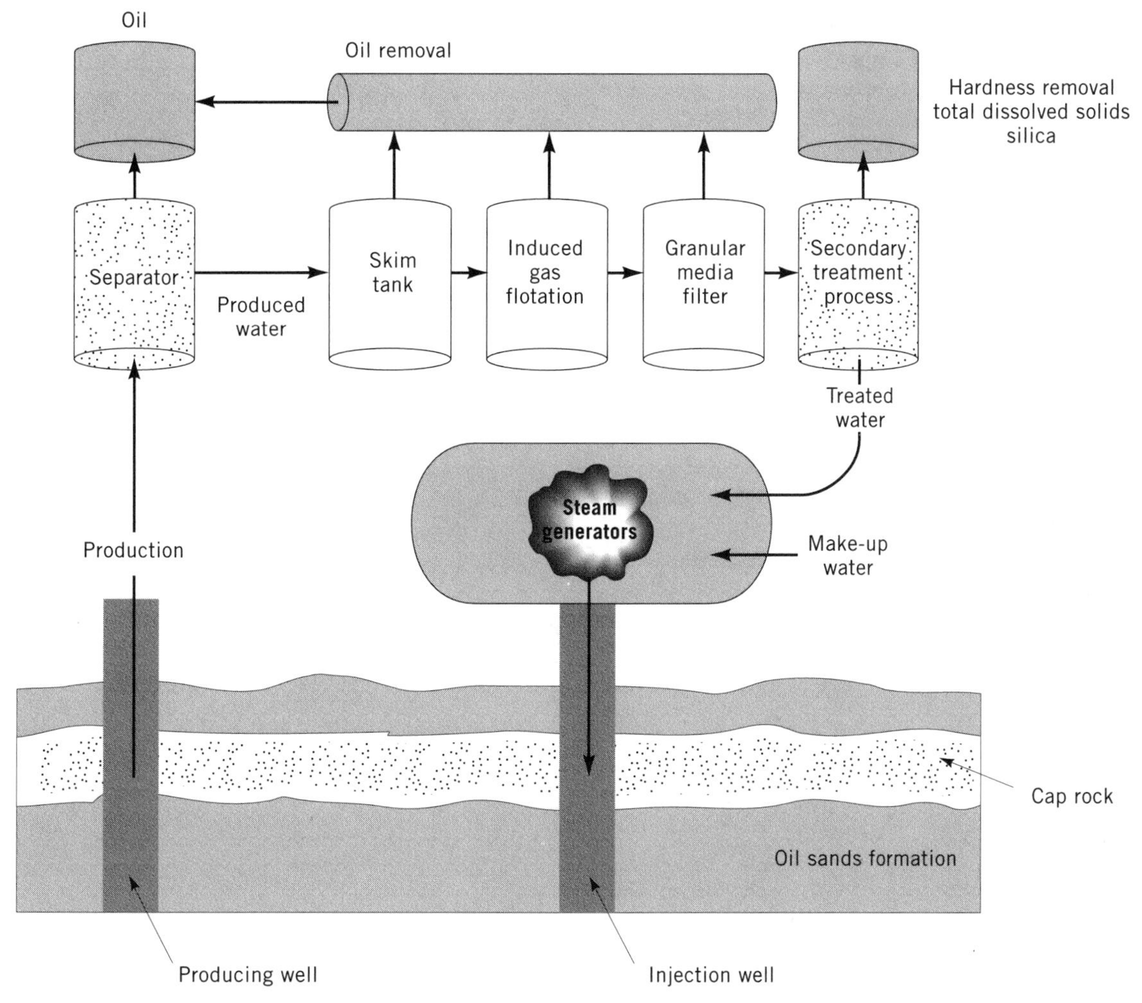

Figure 20. Schematic diagram of a conventional oil well produced-water treatment facility; water is recycled to supplement make-up water for steam production (36). Courtesy of AOSTRA.

Until recently oily-sand waste was simply used for surfacing country roads (34). This procedure must now give way to methods yielding clean sand and separated oil. The Taciuk processor, previously described as a recovery process to replace hot water extraction and the attendant tailings ponds, also may be used to clean oily wastes (Fig. 20). The result is clean, disposable sand, hydrocarbon liquids acceptable to a refinery, and fuel gas for plant use (36). The effluent solids from the ATP are comprised of 93 wt % bulk tailings, 4.9 wt % flue-gas cyclone fines, 1.7 wt % baghouse fines, and 0.4 wt % hydrocarbon cyclone fines (37). The solids mix is a dry silty sand, free of hydrocarbons, with only 21 ppm organic carbon, the latter being a function of the degree of coke burn-off in the combustion zone. Chlorides and sulfates are contained primarily in the flue gas particulates and could be separately disposed. Analysis of the leachate (37) found the presence of phenols and heavy metals low or not detectable, oil and grease not detectable, total carbon low, and concentrations of chlorides and sulfate below permissible levels for potable water. The effluent water from the preheat zone and the sour water from the oil fractionator showed high

Table 6. Summary of Net Upgrader Inputs and Outputs[a]

	VCC	LC-Fining/ Delayed Coker	Flexicoker
Inputs			
Crude, bbl/day	60,000	60,000	60,000
Natural gas, 10^9 J/day[b]	49,899	36,725	0
Electric power, MW·h	25	33	14
Additive, t/day	42	0	0
Catalyst kg/day	0	5443	0
Outputs			
Synthetic crude, bbl/day	63,910	60,920	51,950
Sulfur, t/day	361	352	359
Coke, t/day	0	576	29
Low btu gas (10^9 J/day)[b]	0	0	18,502
Residue, t/day	279	0	0
Spent catalyst, kg/day	0	5443	0

[a] Ref. 53.
[b] To convert J to Btu, divide by 1.054×10^3.

Table 7. Emissions from Selected Upgraders[a]

	VCC	LC-Fining/ Delayed Coker	Flexicoker
SO_2, (kg/h)	8.00	8.20	9.90
NO_x, (kg/h)	1.66	1.60	9.80
CO_2, (t/h)	0.88	0.84	5.20

[a] Ref. 53

phenol concentrations, biological and chemical oxygen demand, and sulfur, and are thus similar to refinery process water generated in thermal cracking of high-sulfur heavy oil and residue. These waters would have to be treated or disposed in deep wells. Off-gas could be flared or used as fuel. The flue gas from combustion of the coke on the spent solids is low in SO_x, NO_x, H_2S, and solids and can be safely discharged from the stacks.

Upgrading

With respect to upgrading, the goal is to minimize the production of residue and, particularly, sulfur. This can be done most effectively by use of high-conversion, hydrogen-addition processes to replace the current carbon-rejection reactors. High-pressure reactors will sharply reduce residue and in fact increase yield. Ref. 53 compares examples of different upgrading schemes with respect to their environmental impacts. For example, Flexicoking for carbon-rejection is used, giving moderate conversion; LC-Fining plus delayed coking for hydrogen addition plus catalysis, giving medium conversion; and VEBA-COMBI-Cracking (VCC) for hydrogen addition giving high conversion. Assuming each is treating 60,000 bbl/day of Cold Lake undiluted bitumen, the inputs to, and outputs from each process are presented in Table 6. Flexicoking yields a large quantity of low-Btu gas at the expense of liquids product, the gas being used for fuel, reducing the electric power requirement. The two hydrogen-addition processes increase yield by converting gas to liquids with hydrogen. The synthetic crude oil from the three processes is very similar in quality. Table 7 shows the emissions from the upgraders, based on production of 1000 bbl/day of synthetic crude. All meet legal emission limits but the two hydrogen-addition processes are much superior to Flexicoking carbon rejection. The carbon-rejection yield of SO_2 is higher because of the very large amount of coke gasified; as well, the NO_x and CO_2 are much higher due to the combustion of the low-Btu gas, which also adds SO_2. Particulate emissions depend upon the degree of stack gas protection but will be highest for the delayed coker, less for the Flexicoker, and none for the VCC, which produces no coke.

The LC-Finer/delayed coker will produce about three times as much gypsum as the VCC, the gypsum being a by-product of flue-gas desulfurization. Such gypsum is generally disposed in landfills.

VCC catalyst residue may be used as binder in coke manufacturing but the amount of catalyst residue from LC-Fining is greater and must be disposed in landfill after leaching for vanadium. Significant amounts of vanadium may be produced from the spent catalysts. Processes which yield such useful byproducts as vanadium are to be preferred because they lead to higher synthetic oil production without increasing emissions.

Ref. 53 concludes that hydrogen-addition processes, especially those of high conversion like VCC, have the least impact because of gaseous and solid emissions. They also have the greatest liquid yields.

Upgrading is the final step in natural asphalt utilization prior to transportation to the refinery. The high-conversion hydrogen-addition upgrading processes such as VCC are most advantageous with respect to product output and environmental consequences in terms of emission of gases and particulates. Such advantages will increase in merit as demand for the products increases and environmental constraints become more stringent. At present, however, the process used will be the one that is most cost effective. However, it also must meet demand for high-quality products that can themselves meet stringent regulatory requirements and for processes that will lead to the capture and productive use of carbon dioxide from combustion and process sources (54).

Land Use

The sludge waste resulting from use of the hot-water extraction process impacts the environment in several ways. (*1*) Areally, it removes land from other uses. (*2*) It may trap wildlife, most notably wildfowl, although protective steps have been instituted. (*3*) It would despoil any stream into which it leaked, which has resulted in stringent disposal regulations.

The land required for surface mining operations cannot be utilized for other purposes until mining stops and reclamation is completed.

With respect to in situ recovery, a number of environmental advantages accrue to the drilling of slant wells from central pads rather than from pattern spacing. In terms of surface disturbance, this means fewer roads and location sites. The pads concentrate oily and produced water wastes in one place. Drilling pads make it feasible to collect vent gases and utilize them at the site, rather than flare them. Both economically and environmentally, the use of pads is essential in soft ground areas of tidelands, swamp, muskeg, and permafrost.

Acknowledgments

The technical assistance and critical reviews of C. J. Schenk and Kathy Varnes, U.S. Geological Survey, Denver, Colorado, R. W. Stanton, U.S. Geological Survey, Reston, Virginia, and D. E. Towson, Consulting Engineer, Calgary, Alberta, are gratefully acknowledged.

BIBLIOGRAPHY

1. Z. Bilkady, "Bitumen—a History," *Aramco World Magazine,* 2–9 (Nov.–Dec., 1984).
2. J. Connan and B. M. van der Weide, in G. V. Chilingarian and T. F. Yen, eds., *Bitumens, Asphalts and Tar Sands,* Elsevier, Amsterdam, 1978, pp. 27–55.
3. R. F. Meyer and Wallace de Witt, *U.S. Geological Survey Bulletin 1944,* 14 pp., 1990.
4. A. M. Danyluk, B. Galbraith, and R. A. Omana, "Toward Definitions of Heavy Crude and Tar Sands," in R. F. Meyer, J. C. Wynn, and J. C. Olson, eds., *The Future of Heavy Crude and Tar Sands, Second International Conference,* McGraw-Hill, New York, 1984, pp. 3–6.

5. E. M. Khalimov, and co-workers, "Geological Problems of Natural Bitumen," *World Petroleum Congress, 11th, Proceedings,* vol. 2, pp. 57–70, 1983.
6. M. A. Rogers, J. D. McAlary, and N. J. L. Bailey, *American Association of Petroleum Geologists Bulletin, 58*(9), 1806–1824 (1974).
7. J. M. Hunt, *Petroleum Geochemistry and Geology,* Freeman, San Francisco, 1979, 617 pp.
8. C. D. Cornelius, "Classification of Natural Bitumens—A Physical and Chemical Approach," in R. F. Meyer, ed., *Exploration for Heavy Crude Oil and Natural Bitumen,* American Association of Petroleum Geologists Studies in Geology no. 25, pp. 165–174, 1987.
9. B. P. Tissot, and D. H. Welte, *Petroleum Formation and Occurrence,* 1st ed., Springer-Verlag, New York, 1978, 538 pp.
10. B. Horsfield, in J. Brooks and D. Welte, eds., *Advances in Petroleum Geochemistry,* v. 1, Academic Press, London, 1984, pp. 247–298.
11. Colin Barker, *Organic Geochemistry in Petroleum Exploration,* American Association of Petroleum Geologists Education Course Note Series #10, 159 pp., 1979.
12. N. F. Petersen, and P. J. Hickey, "California Plio-Miocene Oils: Evidence of Early Generation," in R. F. Meyer, ed., *Exploration for Heavy Crude Oil and Natural Bitumen,* American Association of Petroleum Geologists Studies in Geology no. 25, pp. 351–360, 1987.
13. J. Connan, "Biodegradation of Crude Oils in Reservoirs," in ref. 10, 1984, pp. 299–335.
14. J. G. Speight, *The Chemistry and Technology of Petroleum,* Marcel Dekker, New York, 1991, 760 pp.
15. J. Burger, P. Sorieau, and M. Combarnous, *Thermal Methods of Oil Recovery,* Éditions Technip, Paris, 1985, 430 pp.
16. J. R. Van Wazer, and co-workers, *Viscosity and Flow Measurement,* Wiley-Interscience, New York, 1963, 399 pp.
17. C. Beal, *American Institute of Mining and Metallurgical Engineers Transactions,* **165,** 94–115, (1946).
18. B. J. Gibson, "Methods of Classifying Heavy Crude Oils Using the UNITAR Viscosity-based Definition," in R. F. Meyer, J. C. Wynn, and J. C. Olson, eds., *The Future of Heavy Crude and Tar Sands, Second International Conference,* McGraw-Hill, New York, 1984, pp. 17–21.
19. I. S. Goldberg, ed., "Prirodnye Bitumy SSSR [National Bitumens of the USSR]," *Nedra, Leningrad,* **195,** [223] (1981).
20. R. F. Meyer, and J. M. Duford, "Resources of Heavy Oil and Natural Bitumen Worldwide," in R. F. Meyer, and E. J. Wiggins, eds., *The Fourth UNITAR / UNDP International Conference on Heavy Crude and Tar Sands Proceedings,* vol. 2, 1988, pp. 277–307.
21. Energy Resources Conservation Board [Alberta], "Alberta's Reserves of Crude Oil, Oil Sands, Gas, Natural Gas Liquids and Sulphur," *ERCB ST 92-18,* pp. 3-1–3-6, 1992.
22. N. S. Beskrovnyi, and co-workers, [Natural Solid Bitumens in the USSR—An Important Raw Material Reserve for the National Economy]," *Geologiia Nefti i Gaza,* (4), 14–20, (1985).
23. "Heavy Oiler," *Turkey,* (29) 6–7, (1989).
24. G. G. Vakhitov, and L. Ferdman, "Current Production Problems of Natural Bitumens and Bituminous Rocks in the Soviet Union," in R. F. Meyer, ed., *Heavy Crude and Tar Sands—Hydrocarbons for the 21st Century, UNITAR International Conference on Heavy Crude and Tar Sands,* 5th, Edmonton, Canada, Alberta Oil Sands Technology and Research Authority, 1991, pp. 423–426.
25. J. Armanet, *Les Mines d'Asphalte dans le Monde,* Le Syndicat des Producteurs & Entrepreneurs d'Asphalte and l'Office des Asphaltes, Paris, 1962, 48 pp.
26. G. Sell, in A. E. Dunstan, ed., *The Science of Petroleum,* vol. 1, Oxford University Press, 1938, p. 29.
27. Institute of Geological Sciences [United Kingdom], *Statistical Summary of the Mineral Industry,* Mineral Resources Division, p. 245, 1970.
28. *Oil & Gas Journal,* **90,** (349), 37, (1992).
29. D. Wightman and co-workers, in L. G. Hepler, and Chu Hsi, eds., *AOSTRA Technical Handbook on Oil Sands, Bitumens and Heavy Oil,* AOSTRA Technical Publications Series #6, pp. 1–9, 1989.
30. F. A. Seyer, and C. W. Gyte, in L. G. Hepler, and Chu Hsi, eds., *AOSTRA Technical Handbook on Oil Sands, Bitumens and Heavy Oil,* AOSTRA Technical Publications Series #6, pp. 153–184, 1989.
31. A. R. Allen, in L. C. Ruedisili, and M. W. Firebaugh, eds., *Perspectives on Energy,* Oxford University, New York, 1978, pp. 472–482.
32. F. K. Spragins, in G. V. Chilingarian, and T. F. Yen, eds., *Bitumens, Asphalts and Tar Sands,* Elsevier, Amsterdam, 1978, pp. 93–121.
33. J. A. Franklin, and Dusseault, *Rock Engineering Applications,* New York, McGraw-Hill, 1991, 431 pp.
34. R. N. Houlihan, and R. G. Evans, "Development of Alberta's Oil Sands," in R. F. Meyer, and E. J. Wiggins, eds., *The Fourth UNITAR / UNDP International Conference on Heavy Crude and Tar Sands Proceedings,* vol. 1, pp. 95–110, 1989.
35. R. W. Smith, "Progress in Surface Mining Production from Canada's Oil Sands," *World Petroleum Congress, 12th, Podium* Discussion Topic 17, Preprint, 1987, 9 pp.
36. Alberta Oil Sands Technology and Research Authority, *AOSTRA, a 15 Year Portfolio of Achievement,* Edmonton, Alberta, Canada, 176 pp., [1990].
37. L. R. Turner, "Treatment of Oil Sands and Heavy Oil Production Wastes Using the AOSTRA Taciuk Process," in R. F. Meyer, ed., *Heavy Crude and Tar Sands—Hydrocarbons for the 21st Century, UNITAR International Conference on Heavy Crude and Tar Sands,* 5th, Alberta Oil Sands Technology and Research Authority, Edmonton, Canada, v. 4, pp. 347–356, 1991.
38. V. P. Tabakov, and co-workers, "Thermal Mining Technology as an Effective EOR Method," in R. F. Meyer, eds., *Heavy Crude and Tar Sands—Hydrocarbons for the 21st Century, UNITAR International Conference on Heavy Crude and Tar Sands,* 5th, Alberta Oil Sands Technology and Research Authority, Edmonton, Canada, v. 1, pp. 427–436, 1991.
39. J. C. O'Rourke, and co-workers, "AOSTRA's Underground Test Facility (UTF) in the Athabasca Oil Sands of Alberta," in R. F. Meyer, ed., *Heavy Crude and Tar Sands—Hydrocarbons for the 21st Century, UNITAR International Conference on Heavy Crude and Tar Sands,* 5th, Alberta Oil Sands Technology and Research Authority, Edmonton, Canada, **1,** 487–508 (1991).
40. D. J. Anderson, "The Heated Annulus Steam Drive Process for Immobile Tar Sands," in R. F. Meyer, and E. J. Wiggins, eds., *The Fourth UNITAR / UNDP International Conference on Heavy Crude and Tar Sands Proceedings,* **4,** 127–145.
41. National Petroleum Council, 1984, *Enhanced Oil Recovery,* Washington, D.C., 96 p., 1984.
42. E. Chacin and co-workers, "Feasibility of Steam Injection at 8200 Feet," in R. F. Meyer, and E. J. Wiggins, eds., *The Fourth UNITAR / UNDP International Conference on Heavy Crude and Tar Sands Proceedings,* **4,** 5–11 (1989).
43. Les Parsons, "Current Performance and Future Direction of the Peace River in Situ Pilot and Expansion Projects, Alberta, Canada," in R. F. Meyer, ed., *Heavy Crude and Tar Sands—Hydrocarbons for the 21st Century, UNITAR International Conference on Heavy Crude and Tar Sands,* 5th, Ed-

monton, Canada, Alberta Oil Sands Technology and Research Authority, **2,** 59–68 (1991).

44. K. E. Kisman, and co-workers, "Steamflooding for in Situ Bitumen Recovery at Kearl Lake Pilot in Athabasca Oil Sands," in R. F. Meyer, ed., *Heavy Crude and Tar Sands—Hydrocarbons for the 21st Century, UNITAR International Conference on Heavy Crude and Tar Sands,* 5th, Alberta Oil Sands Technology and Research Authority, Edmonton, Canada, **2,** 71–83 (1991).
45. D. Towson and co-workers, "The PCEJ Steam Stimulation Project," in R. F. Meyer, ed., *Heavy Crude and Tar Sands—Hydrocarbons for the 21st Century, UNITAR International Conference on Heavy Crude and Tar Sands, 5th,* Alberta Oil Sands Technology and Research Authority, Edmonton, Canada, **2,** 113–120 (1991).
46. R. J. Gallant, and A. G. Dawson, "Evolution of Technology for Commercial Bitumen Recovery at Cold Lake," in R. F. Meyer, and E. J. Wiggins, eds., *The Fourth UNITAR/UNDP International Conference on Heavy Crude and Tar Sands Proceedings,* **4,** 577–589 (1989).
47. H. Andrianasolo-Ralaimiza, "Heavy-oil and Bitumen Projects in Madagascar," in R. F. Meyer, ed., *Exploration for Heavy Crude Oil and Natural Bitumen,* American Association of Petroleum Geologists Studies in Geology no. 25, pp. 137–141, 1987.
48. R. D. Adelu, and E. A. Fayose, "Development Prospects for the Bituminous Deposits in Nigeria," in R. F. Meyer, ed., *Heavy Crude and Tar Sands—Hydrocarbons for the 21st Century, UNITAR International Conference on Heavy Crude and Tar Sands, 5th,* Alberta Oil Sands Technology and Research Authority, Edmonton, Canada, **1,** pp. 509–515 (1991).
49. W. T. Rajpaulsingh, and L. N. S. Kunar, "An Evaluation of the Mineable Tar Sands, Trinidad, West Indies," in R. F. Meyer, ed., *Heavy Crude and Tar Sands—Hydrocarbons for the 21st Century, UNITAR International Conference on Heavy Crude and Tar Sands, 5th,* Edmonton, Canada, Alberta Oil Sands Technology and Research Authority, vol. 1, pp. 437–442, 1991.
50. J. W. Bunger, "Extraction and Production of Asphalt from Utah Tar Sands," in R. F. Meyer, ed., *Heavy Crude and Tar Sands—Hydrocarbons for the 21st Century, UNITAR International Conference on Heavy Crude and Tar Sands, 5th,* Alberta Oil Sands Technology and Research Authority, Edmonton, Canada, vol. 1, pp. 481–486, 1991.
51. M. T. Stephenson, *JPT,* **44**(5), 548–550, 602–603 (1992).
52. M. B. Dusseault, D. W. Scafe, and J. D. Scott, *AOSTRA Journal of Research,* **5**(4), 303–320 (1989).
53. F. W. Wenzel, "Comparison of Different Upgrading Schemes with Respect to their Environmental Impacts," in R. F. Meyer, ed., *Heavy Crude and Tar Sands—Hydrocarbons for the 21st Century, UNITAR International Conference on Heavy Crude and Tar Sands, 5th,* Alberta Oil Sands Technology and Research Authority, Edmonton, Canada, **4,** 357–370 (1991).
54. Alberta Oil Sands Technology and Research Authority, *1992: Annual Report,* 99 p.

BOUNDARY LUBRICATION

RAMON ESPINO
Exxon Research and Engineering
Annandale, New Jersey

A key function of a lubricant is to keep the friction between two sliding surfaces low. If the sliding surfaces are completely separated by the lubricant film, the fluid dynamic properties of the lubricant dominate the frictional properties of the system. There are lubricating regimes where the load between the surfaces is so high and the speed so low that we no longer have a continuous lubricant film. In this case the composition and properties of the boundary between the sliding surfaces dominate.

See also LUBRICANTS; LUBRICANTS, ADDITIVES.

In the boundary lubrication regime, the key to reducing friction and thus wear, is to modify the surface properties of the sliding materials. Surface layers can be formed from the lubricant by adsorption, chemisorption and tribochemical processes. These layers affect lubrication by separating the sliding elements. The resulting surface layers should have a lower coefficient of friction than the sliding elements and also show some resistance to shear. These surface layers can range from one to dozens of molecule thick, with physical adsorption layers being a few molecules thick and chemical reaction layers significantly thicker.

If the boundary lubrication is provided by physical adsorption, the concentration of the substance to be adsorbed in the lubricant and the temperature are important. The molecules to be adsorbed should be polar and their effectiveness as friction reducing agents improves with chain length too. For example, fatty acids are better than fatty acid esters and alcohols. The ability of physically adsorbed films to reduce friction is markedly decreased if the lubrication temperature is above the melting point of the adsorbed substance.

When boundary lubrication is provided by chemical adsorption or tribochemical reactions, the reactive surface layer must have a favorable friction coefficient as well as good temperature stability and mechanical strength. Most molecules that provide boundary lubrication by reaction are compounds of sulfur, halogens or phosphorous. Halides have the lowest friction coefficients but are not effective about 250°C; phosphorous compounds have higher friction coefficients, but their resistance to shear is very high.

Typical examples of boundary lubrication are the lubrication of slideways of machine tools, the lubrication of shafts with frequent start-ups and shut-downs, the lubrication of low speed/high load gears and high pressure hydraulic system pumps. On an automobile engine, most parts are lubricated by hydrodynamic lubrication, but the valve train and the bearings can operate in the boundary lubrication regime, particularly under high load and rapid acceleration conditions.

In the design of lubricants, the formulators try to balance two contrasting needs: The need to ensure that the lubricant is viscous enough to provide a good lubricating film, but keeping in mind the need that higher viscosity means more friction losses. Often, the formulator uses a boundary lubrication additive to reduce friction in the event of surface to surface contact.

BIBLIOGRAPHY

Reading List

"Friction, Lubrication and Wear Technology" *ASM Handbook,* ASM International.

E. R. Booser, ed., *CRC Handbook of Lubrication,* CRC Press.

BRAYTON CYCLE (CLOSED)

ANTHONY PIETSCH
Chandler, Arizona

The closed Brayton cycle bears the name of George Brayton (1839–1892), a New England inventor who produced a continuous flow internal combustion reciprocating engine that incorporated many unique design features to control combustion and burn a variety of fuel (1). It was supplanted by the Otto engine that at that time had a higher efficiency.

Today, the cycle utilizes high speed gas turbine machinery and perhaps should have been called the Closed Gas Turbine Cycle. An investigation by J. H. Potter of the Steven's Institute of Technology found an early patent (1791) in Great Britain by John Barber (2). Joule, also of Great Britain, proposed an indirect fired gas turbine about 1851.

In 1935 Drs. Ackeret and Keller of Escher Wyss, Switzerland, patented the closed Brayton gas turbine cycle, which is the forerunner of today's system (3). In 1939 materials and technology was sufficiently advanced to allow Escher Wyss to construct a 2MW demonstration plant which operated at the remarkably high efficiency of 32% for such a small plant. The turbine inlet temperature was a modest 973K (1292F). Over 6000 hours of operation was achieved.

See also THERMODYNAMICS; THERMODYNAMICS, PROCESS ANALYSIS; HEAT ENGINES, DIRECT-FIRED COAL.

Cycle Description

The closed Brayton gas turbine cycle, like the open cycle gas turbine used extensively for aircraft propulsion and utility power, is a continuous flow process. The open cycle gas turbine uses ambient air as the working fluid. The closed Brayton gas turbine is modified to use heat exchangers to add and reject the heat in the cycle. The benefits of closing the cycle are many, including:

- A wide selection of working fluids can be chosen
- A wide variety of energy sources can be utilized such as coal, oil, gas, solar and nuclear
- The system can be pressurized which results in dramatic size and weight reductions as well as performance gains
- The system can be hermetically sealed to operate in hostile environments such as undersea and in space

Figure 1 is a schematic of the closed Brayton cycle and Figure 2 shows the temperature–entropy diagram. The compressor raises the pressure and temperature of the gas. The gas temperature is raised further by heat transfer from the waste heat of the turbine exhaust via the recuperator. The gas temperature is further increased in the heat source heat exchanger. The gas is then expanded through the turbine which provides power to drive the compressor and the load. The remaining waste heat is rejected by means of a low temperature heat exchanger to complete the cycle.

The addition of the input heat exchanger, recuperator and waste heat exchanger significantly increases the size and weight of the closed Brayton system when it is compared to the open cycle gas turbine. These disadvantages exclude it only from being used for aircraft propulsion. For terrestrial, marine, undersea and space applications, it is very competitive.

The closed Brayton cycle can achieve efficiencies from mid 30% values for solar powered space systems to in excess of 50% for nuclear powered terrestrial systems. This high range of efficiencies can be achieved at power levels as low as a few kilowatts to more than 350 megawatts using only one turbomachinery set at each power level.

The recuperator plays a large role in the accomplishment of these efficiencies because it recycles as much as two-thirds of the heat in the cycle, thus minimizing the heat added and rejected. When a high degree of recuperation is utilized, the cycle optimizes at low pressure ratios (less than 3:1 for diatomic gases such as air or nitrogen and 2:1 for monatomic gases such as helium, neon, argon, krypton and xenon and mixtures of these). The reason is to minimize the addition of heat during compression and the drop in temperature when expanded through the turbine. This optimization applies whether the gas turbine is closed or open cycle. This system is counter to the optimization of unrecuperated gas turbines where high pressure and expansion ratios are utilized to improve efficiency.

With the loop closed it is possible to raise the pressure level in the system to 50 to 75 atmospheres (5.07 to 7.6 MPa). This level dramatically increases the mass flow per

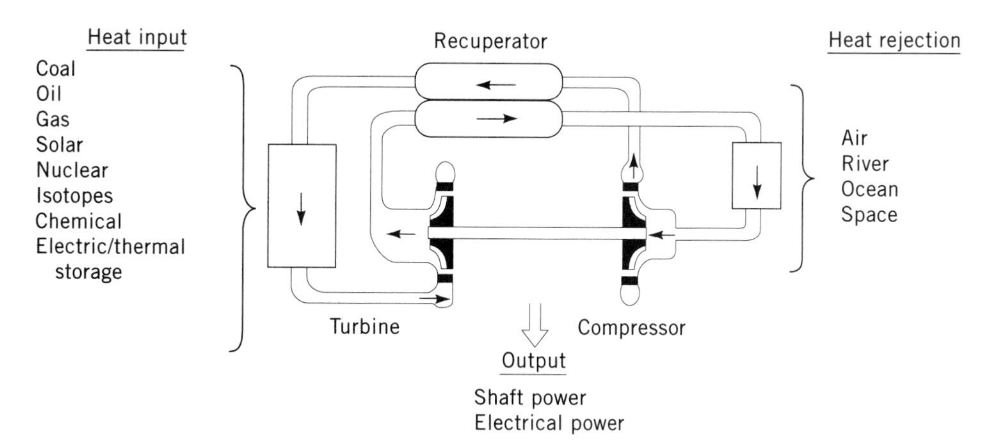

Figure 1. Closed Brayton cycle schematic.

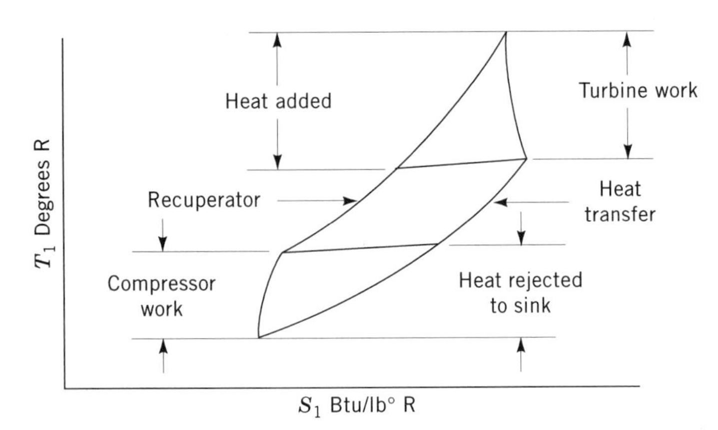

Figure 2. Temperature–entropy diagram.

unit of volume through the system resulting in significant improvement in the power density of the components (ie, the size and weight of the turbomachinery and the heat exchangers). This result is true for the megawatt size system outputs. At low power levels such as a few kilowatts, the pressure may be subatmospheric in the loop to keep the size of the turbomachinery from becoming minuscule in size and inefficient due to clearance losses.

The part load efficiency of the system is excellent. Power output is adjusted by pumping gas into and out of the loop to change the mass flow. Power output can be reduced 85 to 90% with no loss in efficiency. The explanation for this is that the heat exchangers, which are sized for the full load condition, become effectively oversized for part load. This circumstance results in lower fractional pressure losses and improved heat transfer performance. At the low power end, the parasitic losses in the system cause the efficiency to rapidly fall off.

The use of helium as the working fluid, greatly increases the heat transfer in the heat exchangers, particularly the recuperator. Helium is the ideal working fluid for systems that are directly coupled with a gas cooled nuclear reactor.

In some systems it is optimum to tailor the working fluid such as mixing xenon with helium to achieve a desired molecular weight of the gas. This tailoring is usually done because the sonic velocity and low molecular weight of helium adversely impacts the design of the turbomachinery (ie, too many turbomachinery stages are needed or their size becomes too large). Fortunately, there is a synergistic relationship when helium is mixed with other inert gases that results in higher heat transfer coefficients as illustrated in Figure 3.

Another feature of using a closed loop is the avoidance of foreign object damage or having combustion products foul the loop, both of which cause a loss of performance as well as maintenance costs.

Early Applications

During the 1950s more than a dozen closed Brayton power plants were constructed in Germany, Russia, England, France, Japan and Austria. Some were experimental units but most, particularly those in Germany, were for utility and industrial use (3,4). They burned a multitude of fuels including bituminous coal, natural gas, oil, blast furnace gas, mine gas, lignite and mixtures of these. The units were relatively small ranging from 2 to 50 megawatts of output. They were used to power small communities, mines and steel plants. Not only did they provide electrical and shaft power but the waste heat was of high enough quality to provide heating for homes, factories and mines. This combined use of power and heat utilized more than 80% of the fuel's thermal content.

These plants were very successful. They are not being replicated due to: environmental problems in burning coal; the high cost of manufacturing the heater system and; oil has become cheaper than coal. Today, solutions to these problems are being investigated using atmospheric and pressurized fluidized bed combustion systems.

In the United States a large number of companies have investigated the use of the closed Brayton system for various applications (5,6). A partial listing includes Westinghouse, Allied Signal, Rockwell International, United Technologies, General Atomics, LaFleur Corp., Babcock and Wilcox, Grumman and General Electric.

The government agencies that have provided funding are the National Aeronautics and Space Agency (NASA), Department of Energy (DOE), Department of Defense (DOD), and the Maritime Administration (MARAD). Some of these agencies (NASA, DOE, DOD) have assembled systems and performed tests at their respective sites. Research and study funding has also been provided by a consortia of utility power companies, the Electric Power Research Institute (EPRI).

Examples of work performed by and for the industrial companies, government agencies and research institutions are discussed in the following paragraphs.

Nuclear Utility Power Plant

G.A. Technologies developed the helium cooled High Temperature Gas Reactor (HTGR). During the 1970s an advanced direct cycle high temperature gas cooled nuclear reactor option using closed cycle gas turbines (HTGR-GT) was extensively studied (6). The motivation for this effort included:

1. further exploitation of increased reactor outlet temperature above the 700°C (1292°F) value to achieve high efficiency

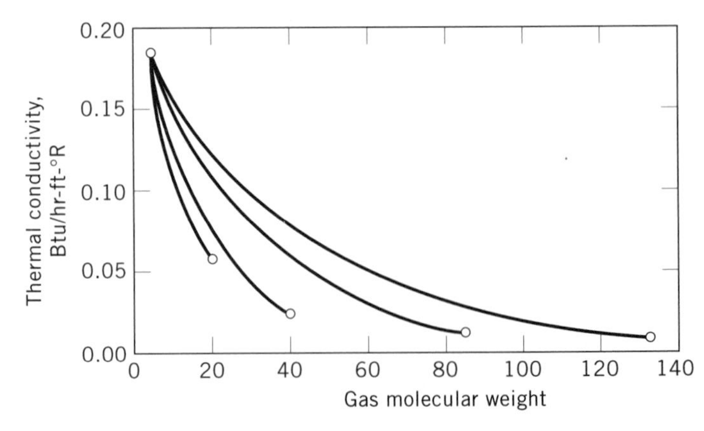

Figure 3. Monatomic gas mixture thermal conductivity.

2. to take full advantage of the favorable Brayton cycle characteristics related to economic dry-cooling capability
3. potential of very high efficiency when operated in a combined cycle mode, and
4. projected plant simplification with attendant economic benefits.

In late 1979 a reference plant was established based on a non-intercooled cycle, 850°C (1562°F) turbine inlet temperature, and dry cooling resulting in an overall plant efficiency of 40%. Figure 4 shows the plant design consisting of the nuclear reactor that powers two 400 MWe power conversion loops, each consisting of turbomachinery, electrical generator, recuperator, cooler and control valves. All of the primary reactor and power conversion equipment is installed in the prestressed concrete reactor vessel.

A motivating factor for study of the dry-cooled HTGR-GT plant was the siting flexibility (ie, lack of cooling water in certain U.S. regions) together with ever increasing environmental demands. The implementation plan for this system was targeted for the year 2000. The moratorium on new nuclear utility power plant construction has placed this unit as well as others on an indefinite hold.

Coal Fired Utility Power Plant

Allied Signal (Garrett Engine Division) under DOE sponsorship studied large coal fired closed Brayton utility power plants (7). The study was comprehensive, including power plant design, siting, environmental concerns and cost analysis. The combustion system selected after careful analysis and trade-offs was the atmospheric fluidized bed.

Figure 5 shows a schematic of the complete power plant including provision for utilization of the high quality waste heat for process applications if wanted. Since particulates and the oxides of sulfur and nitrogen are controlled to federal standards during the combustion process by the addition of sorbents and an electrostatic precipitator, costly and inefficient scrubber systems are not needed.

Figure 6 is the schematic for a 350 MWe closed Brayton cycle gas turbine power conversion loop. The working fluid is air. The turbine inlet temperature selected was 844°C, an optimum sorbent reaction temperature for the maximum removal in the atmospheric fluidized bed of the oxides of sulfur.

The overall coal pile to bus bar efficiency of the plant, including all combustion and parasitic losses is above 38%. Note the waste heat leaving the recuperator is 174°C. This minimizes the cooling tower requirements or, if desired, can be used in conducting many manufacturing processes. A unique feature of the system is the capability of increasing this waste heat temperature by bypassing part of the recuperator flow. The penalty for doing this is a reduction in overall power generation efficiency. If waste heat is used for process work, the utilization of the heating value in the fuel will exceed 80%!

Figure 7 shows the atmospheric fluidized bed. Combustion air is forced into the bottom of the unit through nozzles to fluidize (levitate) a bed of small grain-sized inert material. Crushed coal and sorbent (usually limestone) are injected into the bed where the combustion takes place and the sorbent reacts with the gaseous oxides of sulfur to produce a solid which is removed during recycling of the bed material. The heat in the bed is absorbed by heat exchangers immersed in the bed. The formation of oxides of nitrogen during combustion is very low due to

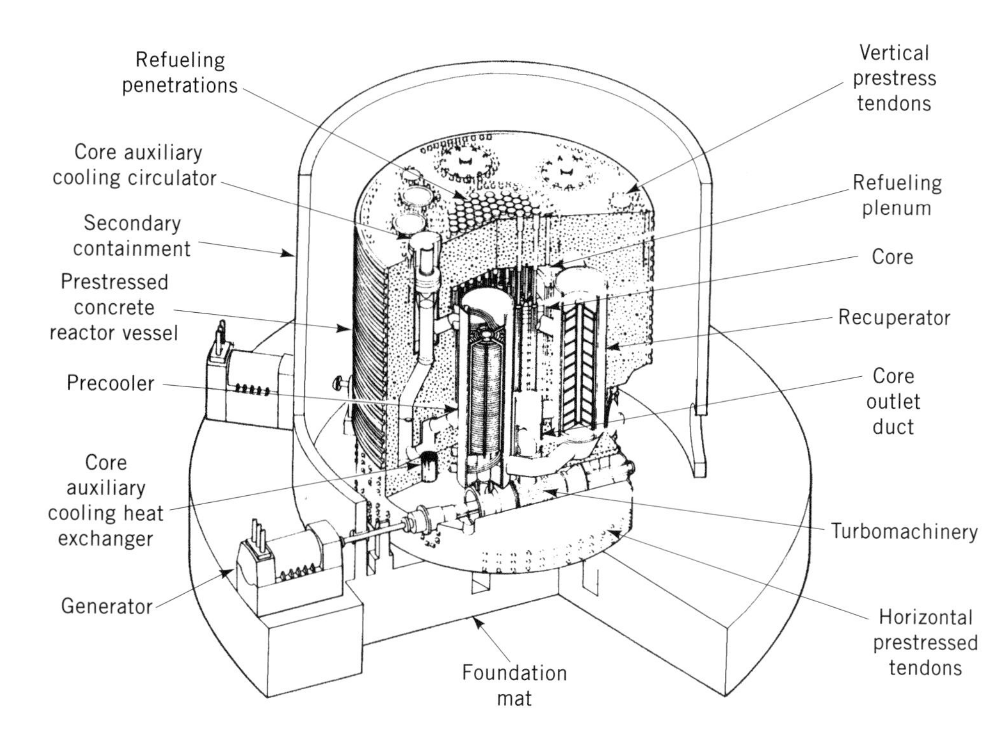

Figure 4. Nuclear closed-cycle gas turbine power plant concept, 2-Loop, 800MWe, (HTGR-GT).

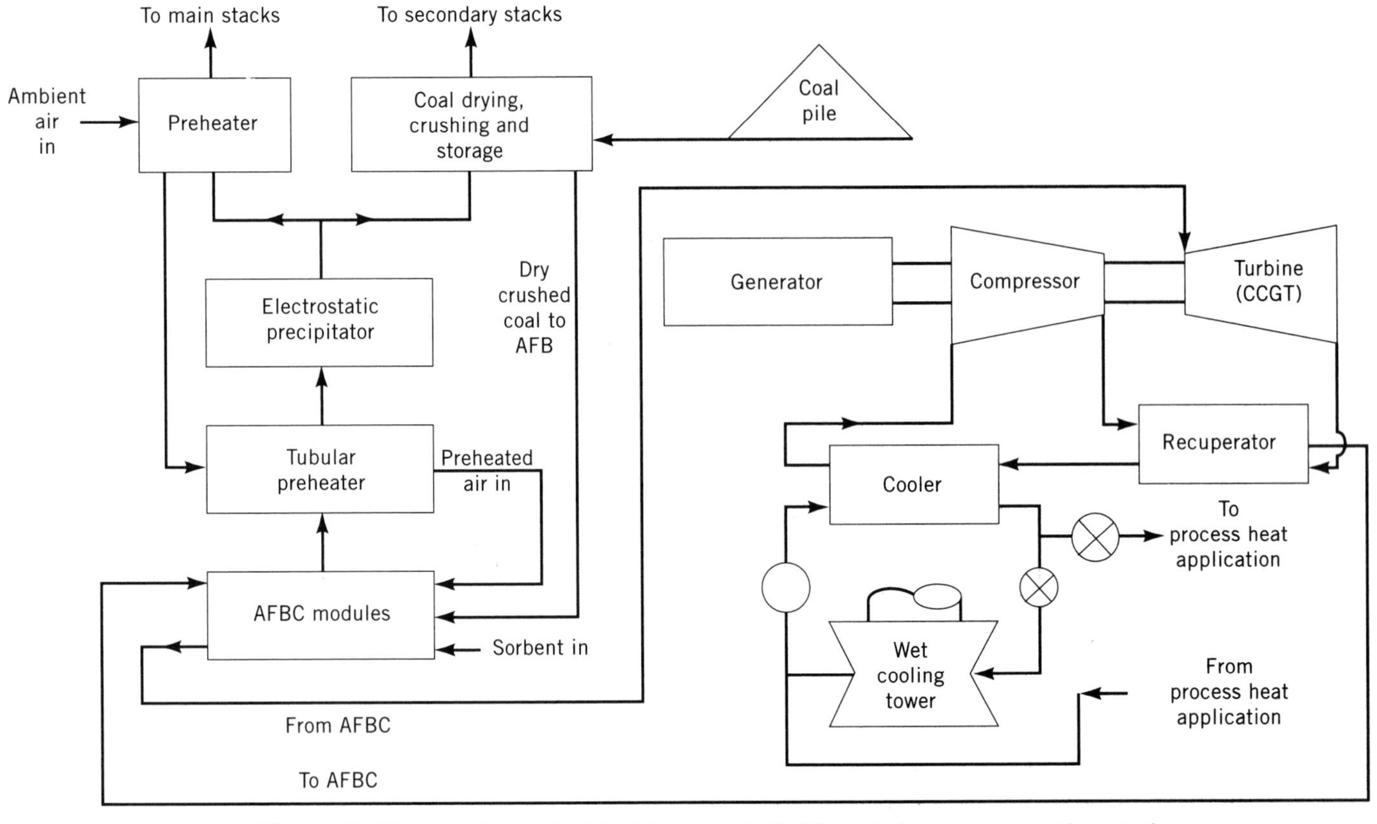

Figure 5. Closed-cycle gas turbine/atmospheric fluidized bed power generating station schematic.

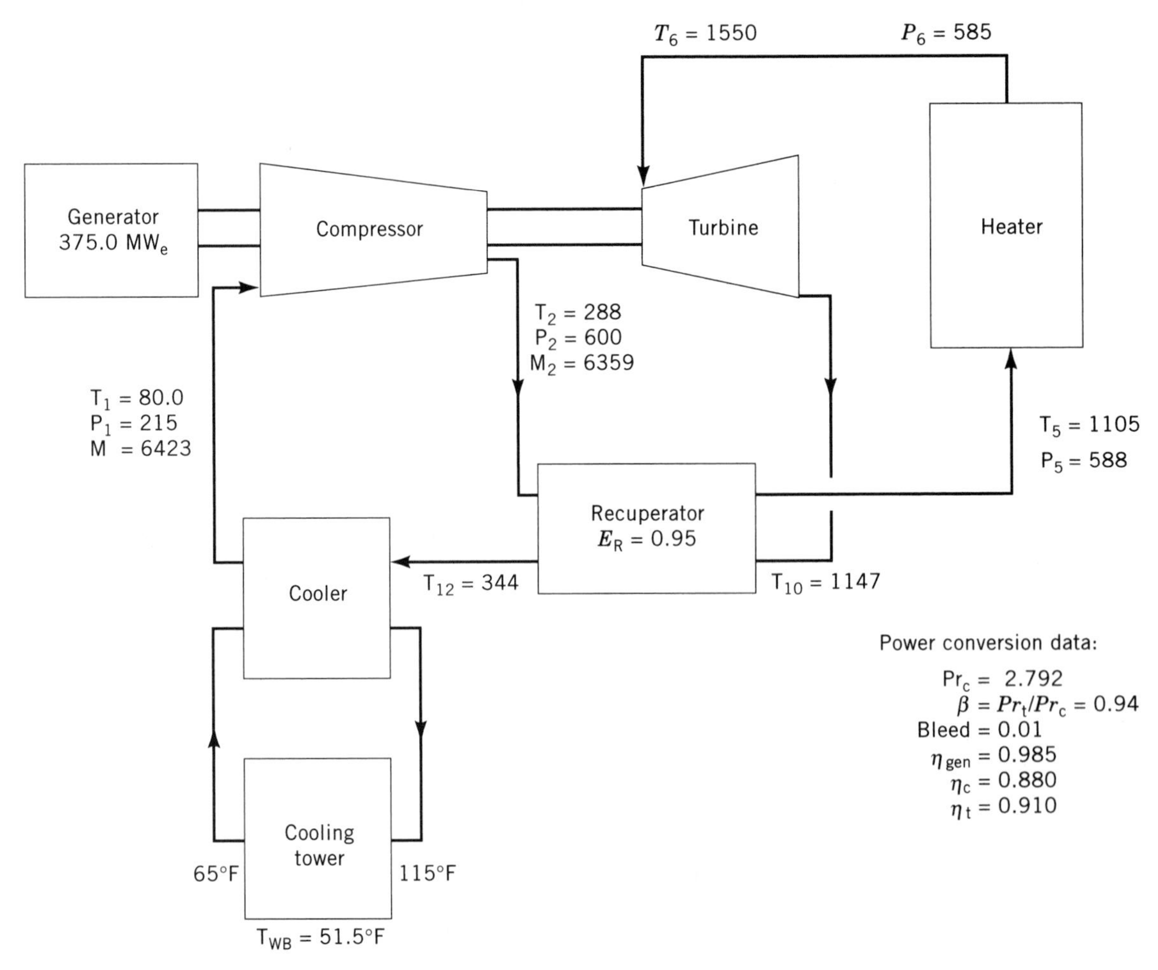

Figure 6. Closed cycle gas turbine power conversion loop schematic, 350 MWe.

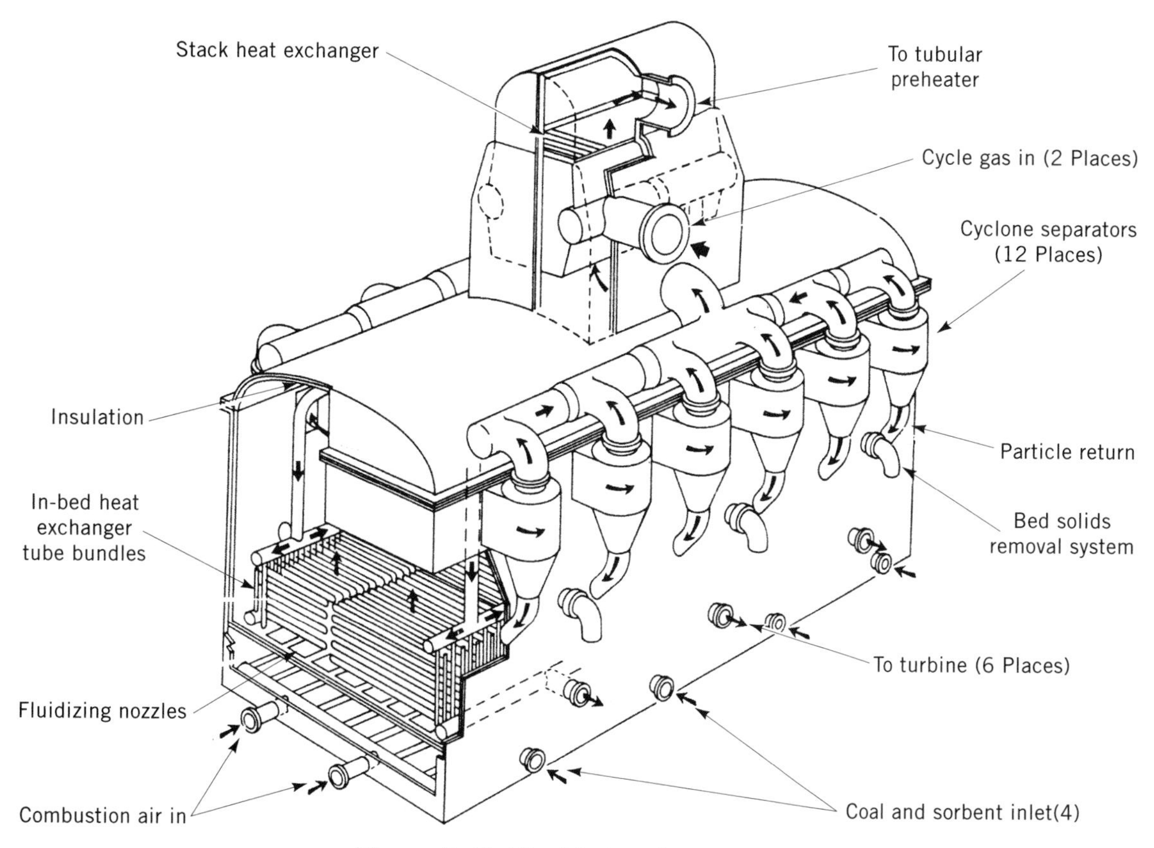

Figure 7. Fluidized bed combustor heater.

the low bed temperature (844°C). Unfortunately, many fuels have nitrogen bound up in them that is released during combustion. The hot gases leaving the bed are mixed with a sorbent to reduce the oxides of nitrogen. The gas then passes through cyclone separators to remove the heavier particulates before it is passed through a heat exchanger in the stack that preheats the gas entering the in-bed heat exchangers.

Figure 8 shows the layout of the turbomachinery, electrical generator, recuperators and waste heat exchangers. Encasing the recuperators and waste heat exchangers in the same pressure vessel maximizes the integration of these to minimize heat losses and to reduce costs.

Figure 9 is a cross section of the closed cycle gas turbine. The pressure ratio is low, 2.792:1, resulting in only nine compressor stages and three turbine stages being needed. The pressure entering the compressor is 14.63 atmospheres (1.48 MPa), thus increasing the mass flow through the machine by this factor when compared to an open cycle machine. Although this coal fired system has many superior advantages, it languishes because of the few new large plants being constructed.

Merchant Ship Power Plant

The Maritime Administration (MARAD) in 1974 had funded a study with Allied-Signal to study an oil fired 40,000 horsepower closed Brayton cycle for merchant ship propulsion (8). Figure 10 shows the installation of the power plant in the stern of a ship. An outstanding feature of the installation was the small amount of space needed, both in the engine room and the ducting required in the upper spaces for air intake and exhaust. Many of the features described for the 350 MWe system are incorporated in this layout, particularly the packaging of the recuperator and waste heat exchanger. The overall efficiency of this power plant is more than 38%. Coal as a fuel and nuclear power has been considered as heat source options for the power plant.

Submarine Power Plants

Under Navy auspices during the 1970s, Westinghouse exhaustively studied a closed Brayton cycle system coupled with a helium cooled nuclear reactor power plant for submarines (9). Figure 11 is a section view of the results of this Light Weight Nuclear Propulsion (LWNP) study. The design requirements for this power plant

- Life: 10,000 full power hours
- Output shaft power: 140,000 horsepower
- Maximum specific weight shielding: 6.8 kg per HP 4π (10 MR/H at 6 m, from reactor centerline at full power
- Coolant: Helium

Two 70,000 horsepower turbomachines with attendant heat exchangers comprise the power conversion equipment. The planned turbine inlet temperature is 927°C (1700°F).

The high power level of this unit was chosen to provide larger submarines with the speed necessary to meet the evolving mission requirements envisioned at that time.

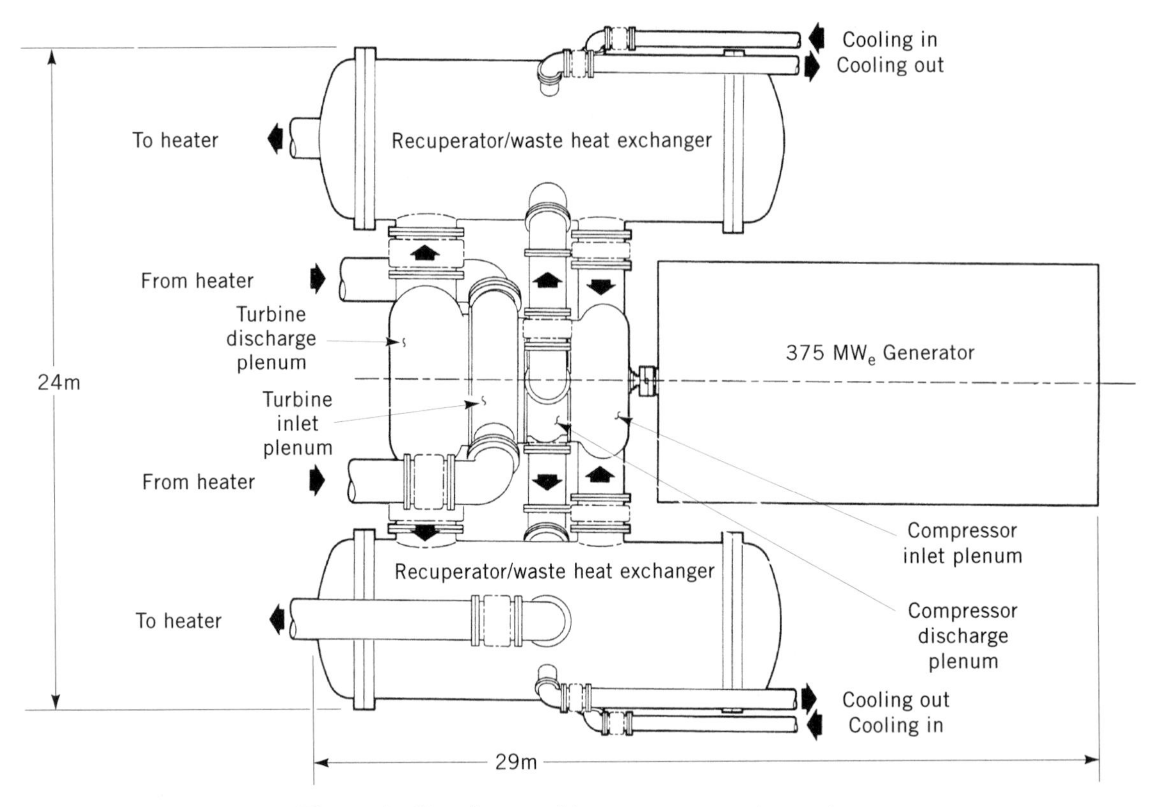

Figure 8. Closed gas turbine power conversion system.

The power plant was vibration free and extremely quiet because it was an electric drive system and the directly coupled compressor–turbine–generator assembly was supported by gas bearings.

Also under Navy auspices, Allied-Signal studied submersible power plants using chemical reactions (such as lithium plus sulfurhexafluoride), phase change latent heat storage salts (lithium fluoride and others) and sensible heat storage (carbon blocks) as energy sources to power the closed Brayton cycle (6).

Figure 12 is a cutaway isometric drawing of a 746 kilowatt (1000 horsepower) unit powered by a carbon block heat storage energy supply. Prior to submersing, the carbon block is electrically heated (by electrical resistance or

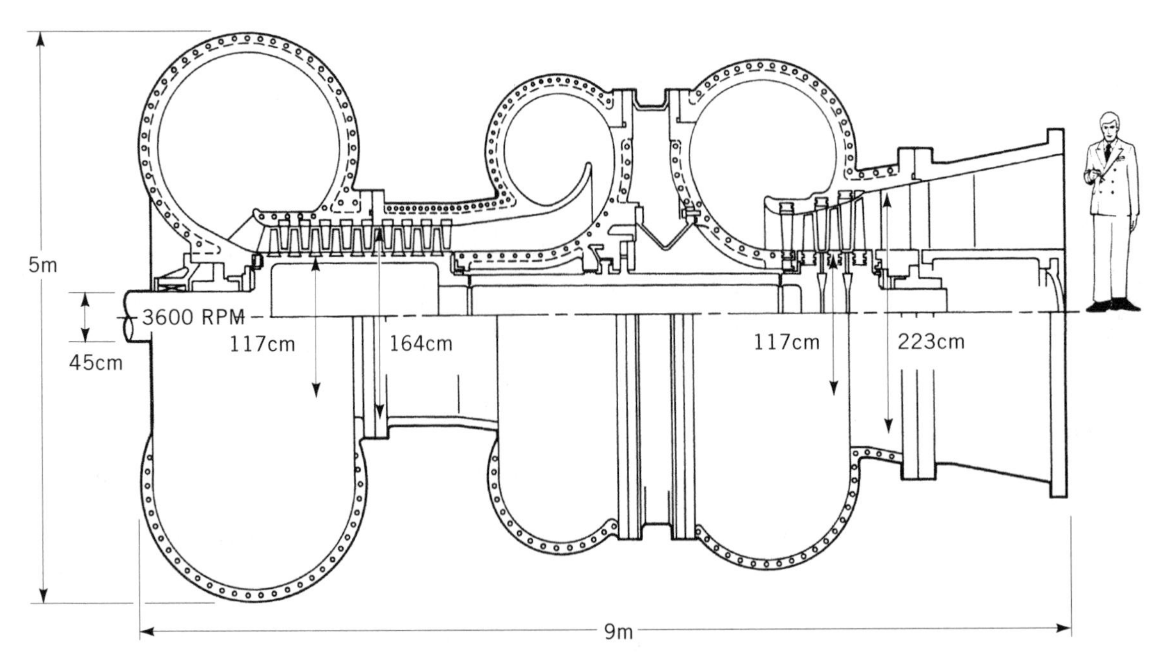

Figure 9. Closed cycle gas turbine, 350 MWe.

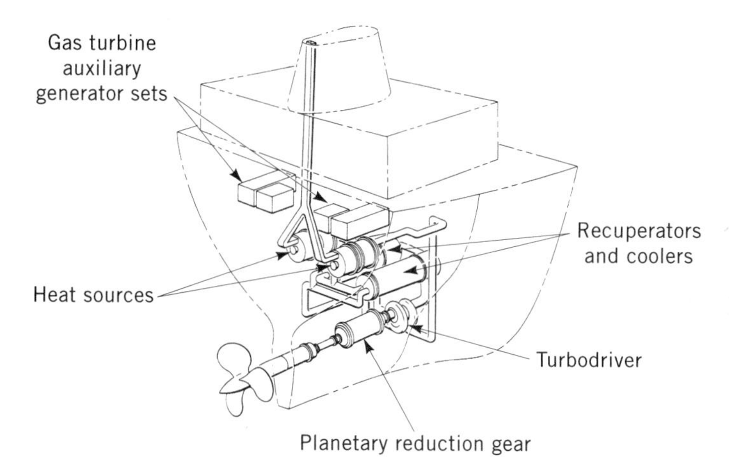

Figure 10. Closed gas turbine installation.

induction) to a temperature in excess of 2205°C (4000°F). The inert gas working fluid is convectively heated as it is passed through the carbon block. It is mixed with bypass flow to lower the turbine inlet temperature to 954°C (1750°F). The energy stored in the carbon is a function of the size of the block and the temperature to which it is raised. The mission duration of the submersible is a function of the efficiency of the engine and the mission profile. The power demand increases as the cube of the speed; therefore, low cruising speeds will prolong the mission. For smaller submarines (the size of the World War II boats) the duration for remaining submerged could last twenty or more days.

Recharging of the thermal storage system could take place while surfaced by using conventional diesel generator sets. Again, like the larger submersible power systems, the closed Brayton cycle driving electrical generator is very quiet and about 47% efficient at cruising power.

Space Power

NASA, since the early 1960s, has been interested in the development of the closed Brayton cycle for higher powered space missions. These missions included a few kilowatts of power for satellites, several hundred kilowatts for space stations, and megawatts for lunar bases and electric propulsion. The heat sources investigated were radioisotopes, solar and nuclear reactors. Extensive development was conducted on solar and radioisotope power systems ranging from 2.5 to 15 kilowatts (10).

During the mid 1980s NASA chose a solar powered closed Brayton cycle to power the growth of the Space Station Freedom to 300 kilowatts (11). Rockwell International has been the prime contractor on the complete Space Station Freedom power and distribution system. Allied Signal has supplied the closed Brayton components.

Figure 13 shows the closed Brayton solar dynamic power module. Each module produces a minimum of 25.75 kilowatts of power. The original Space Station Freedom utilized a hybrid power system consisting of 25 kilowatts of solar cell panels and two dynamic power systems with

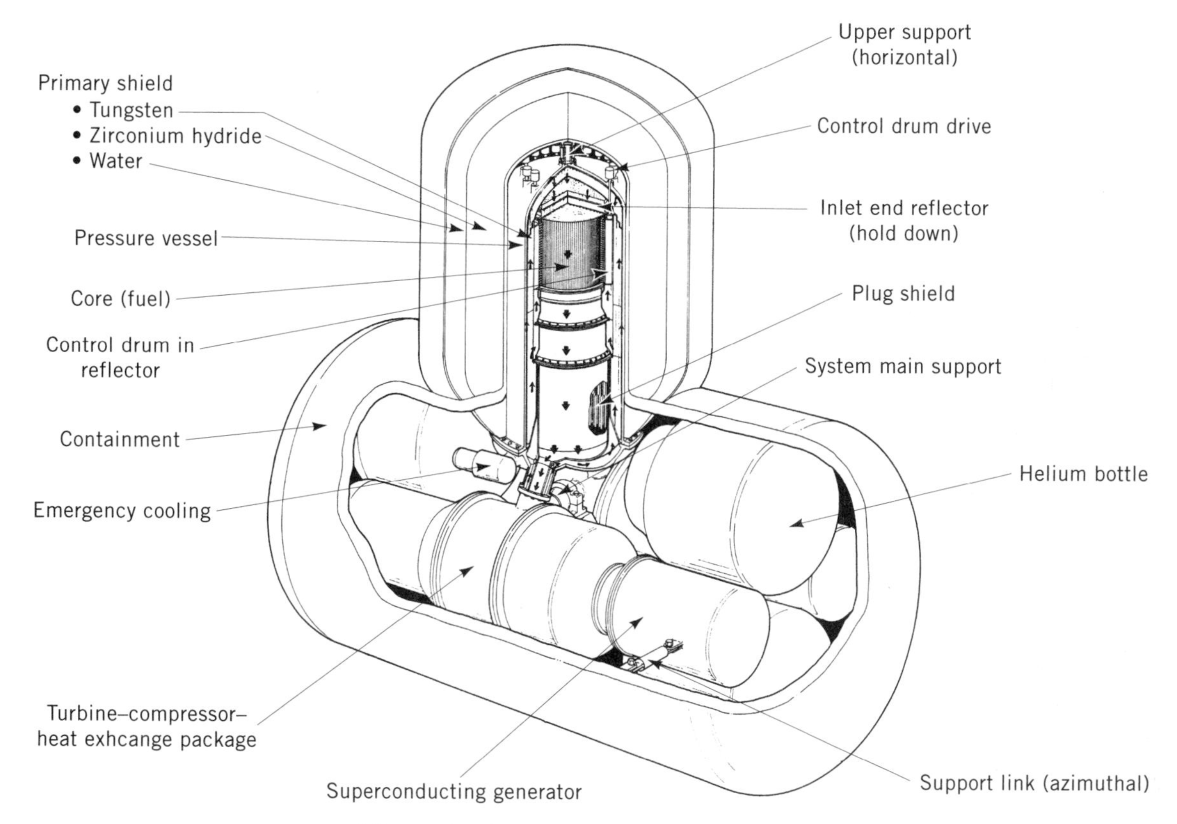

Figure 11. Low specific weight power plant.

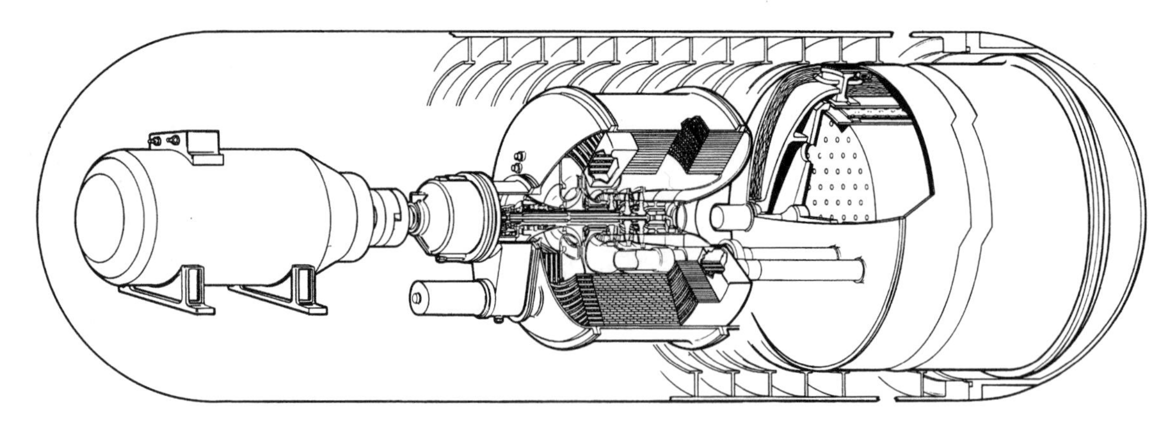

Figure 12. Engine system, sensible heat source.

the growth to be supplied by additional dynamic power units.

The solar collector is an offset Newtonian configuration to allow pivoting of the power system to take place at the junction of the power conversion system, the closest position to the center of the system mass. The solar receiver, turbomachinery, heat exchangers, radiator and controls are packaged in a single unit prior to launch and are self deployable in space. The solar collector is assembled in space and attached to the power unit.

Figure 14 is a schematic of the power unit. Operation is complicated by the need to operate during both the sunlight and shadow periods of the orbit, variations in the orbital period (ie, ratio of sunlight to shadow times) caused by the decay in altitude due to atmospheric drag and subsequent reboosting, the change in solar insolation during the revolution of the earth around the sun and the decay of the solar collector reflectivity with time. These values are listed below:

Minimum Insolation Orbit

- Orbital period = 91.02 min (180 nautical miles)
- Solar insolation = 1.323 kW/m^2
- Concentrator reflectivity = 0.90 (end of life)
- Eclipse period = 36.33 min

Maximum Insolation Orbit

- Orbital period = 93.67 minutes (250 nautical miles)
- Solar insolation = 1.419 kW/m^2
- Concentrator reflectivity = 0.93 (start of life)
- Eclipse period = 28.13 minutes

Due to the above variations there is no steady state performance condition. The worst case is the minimum solar insolation orbit when the space station is leaving the eclipse period (ie, sunrise). In order to produce a net power to the space station of 25.75 kWe per power module (including all the space station power and distribution losses) each module must generate 31.7 kWe. Under these conditions, the net cycle efficiency is 36.4%. To accommodate the maximum solar insolation orbit the module generates 40.39 kWe which dictates the alternator size.

The intercepted solar energy during the sunlight period simultaneously powers the Brayton cycle engine and stores heat in the eutectic mixture of lithium. The solar-powered closed Brayton cycle is ideally suited to the Space Station application. The high efficiency of the system results in reduced aerodynamic drag, thus saving reboost propellants. The conservative design of the power unit exceeds the postulated 30-year life of the Space Station. This longevity is accomplished by design principles supported by extensive experience. The use of inert noble gas as the working fluid in a hermetically sealed loop con-

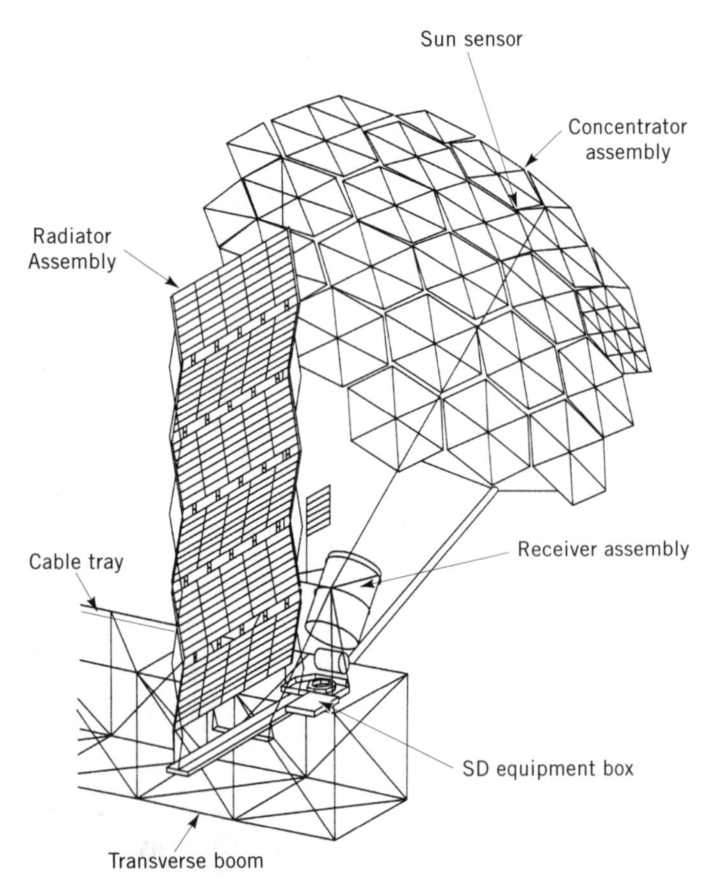

Figure 13. Space station solar dynamic power module.

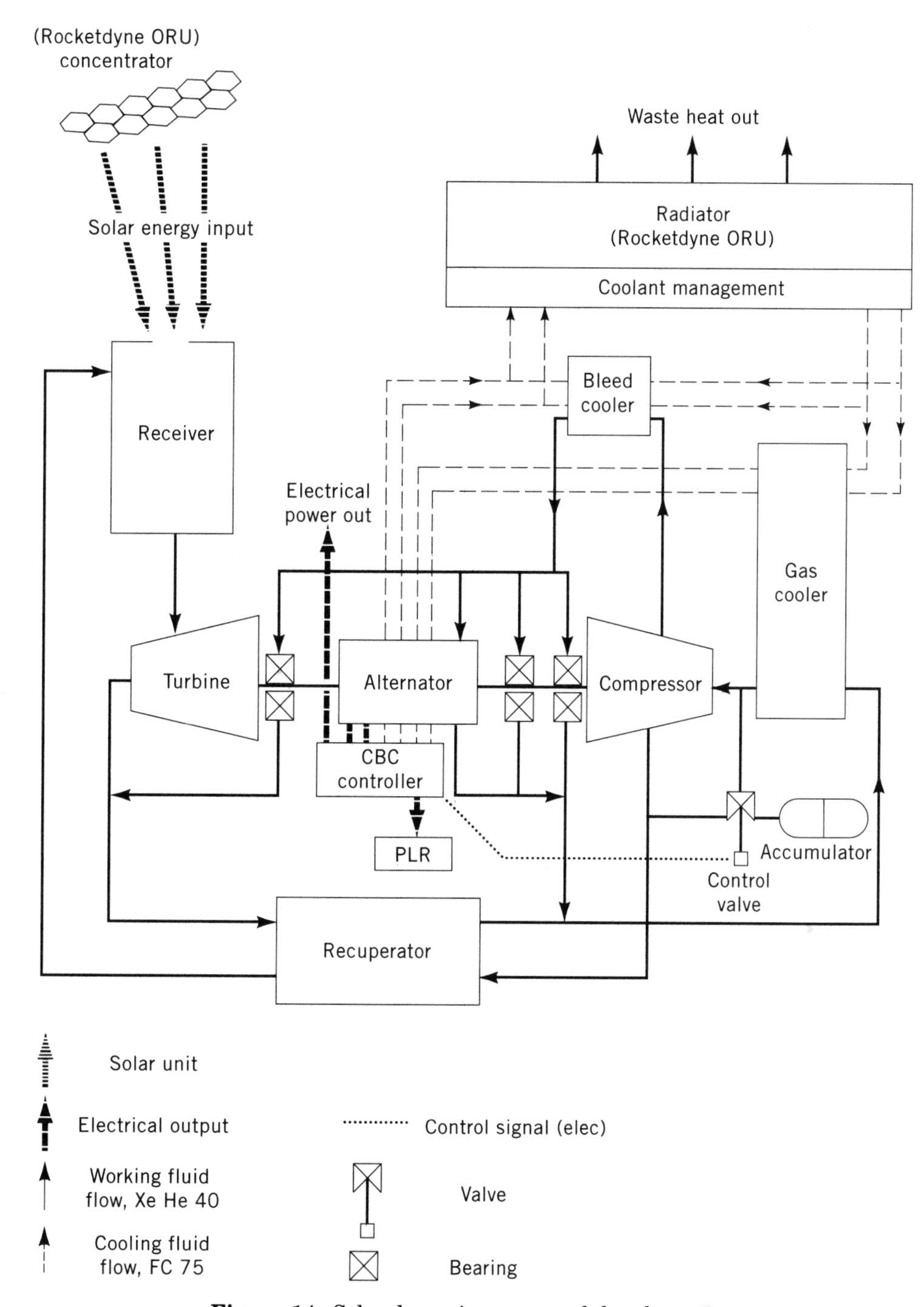

Figure 14. Solar dynamic power module schematic.

tributes to the long life, high reliability and safety of the system since it is extremely stable, single phase and nontoxic. The constant speed turbomachine operates at modest temperatures and low tip speeds.

Environmental Aspects

Environmental aspects of the closed Brayton cycle have been touched on in the foregoing discussions. The Clean Air Act of 1970 was a strong stimulus to conduct many of the studies that were made to adapt the closed Brayton cycle to vehicular, rail, ship, refrigeration, and utility applications. This attempt was due to the continuous combustion process that permitted better emission control augmented by the high cycle efficiencies that resulted in less fuel being burned. The oil embargo of 1973 further accelerated these studies to find ways to reduce fuel consumption and use indigenous fuels such as coal.

The possibility of using the high grade waste heat rejected from the closed Brayton cycle for heating, cogeneration, or industrial processes further decreased the fuel consumption with the result of further reducing unwanted pollution. The higher heat rejection temperature also made dry cooling very effective. This ability eliminates the heating of river and lake water normally used to cool power plants.

SUMMARY

The closed Brayton cycle is adaptable to a wide range of applications. It can be heated by a wide range of combustion fuels, nuclear power, radioisotope decay, chemical reactions and the release of stored latent and sensible heat. The cycle is very efficient both at full and part load. It is adaptable to a wide range of environments including undersea and space as well as terrestrial and marine.

The power plant has very long life and high reliability due to a sealed working fluid environment, constant speed operation and the use of gas bearings to suspend the rotating assembly.

Fortunately, all of the technology needed to produce and advance the closed Brayton cycle exists in the design and manufacture of open cycle gas turbines.

BIBLIOGRAPHY

1. C. L. Cummins, Jr., *Internal Fire,* Carnot Press, 1976, pp. 184–187.
2. J. H. Potter, *The Gas Turbine Cycle,* ASME Winter Annual Meeting, 1972, New York.
3. C. Keller, "Forty Years of Experience on Closed Cycle Gas Turbines," *Annals of Nuclear Energy* **5,** 405–422 (1978).
4. K. Bammert, and G. Graschup, "Status Report on Closed-Cycle Power Plants in the Federal Republic of Germany," *ASME Conference 76-GT-54,* 1976.
5. C. F. McDonald, "Large Closed-Cycle Gas Turbine Power Plants," in *Sawyers Gas Turbine Engineering Handbook 3rd ed.,* vol. 2, Chapter 9, 1985.
6. A. Pietsch, "Closed-Cycle Gas Turbines 50MW and Smaller," in *Sawyers Gas Turbine Engineering Handbook, 3rd ed.,* vol. 2, Chapter 8, 1985.
7. "Closed Cycle Gas Turbine Heater Program," *summary report,* 1977, U.S. Department of Energy Contract EF-77-C-01-2611.
8. F. Critelli, A. Pietsch, and N. Spicer, "Closed Gas Turbine Engines for Engineers," *Marine Technology* **13**(3), 309–315 (1975).
9. R. E. Thompson, "Lightweight Nuclear Powerplant Applications of a Very High Temperature Reactor (VHTR)," *1975 Intersociety Energy Conversion Engineering Conference,* paper no. 759164.
10. W. B. Harper, Jr., A. Pietsch, W. G. Baggenstoss, "The Future of Closed Brayton Cycle Space Power Systems," *Space Power,* **8**(1,2), (1989).
11. A. Pietsch, S. W. Trimble, "Closed Brayton Solar Dynamic Power for the Space Station," *Thirty-Seventh International Astronautical Congress,* Innsbruck, Austria, Oct. 1986.

BUILDING SYSTEMS

LES NORFORD
Massachusetts Institute of Technology
Cambridge, Massachusetts

Buildings are lighted and conditioned primarily for the benefit of their occupants. Conditioning requirements also exist for many industrial buildings, ranging from very stringent regulation of temperature, humidity and air quality in semiconductor manufacturing plants to relaxed criteria for warehouses. This article will focus on the needs of people and not processes, because processes are difficult to consider in general and because the indoor environment is becoming increasingly important as people spend more time within buildings, now as much as 90% (1). After occupant needs are established, this article will review how these needs can be met with low-energy, so-called passive techniques, and finally consider active mechanical and electrical systems.

See also AIR CONDITIONING; DISTRICT HEATING; AIR POLLUTION: INDOOR; LIGHTING, ENERGY EFFICIENT.

Occupant Needs

Thermal. The body is a source of heat, converting approximately 20% of the energy released by the metabolism of food into work and the remaining 80% into heat. Heat production depends on activity level, which ranges from 40 W per m^2 of body surface area while sleeping to 10 times this level, or 400 W/m^2, for strenuous sports activities (2). To maintain the near-constant deep-body temperature necessary for survival, heat associated with the body's metabolic rate must be transferred to the environment.

The body exchanges heat with its surroundings, via conduction, convection, radiation and evaporation of water. In common situations, conduction of heat through direct contact with hot or cold surfaces is typically very small, convection and radiation each account for about 40% of the net heat transfer and evaporation of water vapor providing the remaining 20% (3). Convective heat transfer depends on the surrounding air temperature, the surface temperature of exposed parts of the body and of clothing, which in turn is governed by the thermal resistance of the clothing, and the speed at which the air moves past the body. Radiation depends on the surface temperature of clothing and the body and of surrounding surfaces. Evaporation of water vapor, from the lungs or as sweat from the skin, is influenced by the amount of water in the air. The heat transfer processes have been modeled in detail (4).

People regulate heat flow from the body in a number of involuntary and voluntary ways. Sweating, shivering, and dilation and contraction of the blood vessels near the skin to control skin temperature are involuntary mechanisms (2,5). Choice of clothing, activity level, curling up or stretching out, and migration within rooms, buildings or across continents are voluntary means. Voluntary activities are motivated not only by survival but by more stringent thermal comfort criteria. That is, a feeling of satisfaction with the thermal environment occurs within a narrow range of conditions that permit the body to regulate its temperature.

Buildings are designed and operated to ensure the thermal comfort of the occupants. Temperatures that heating and cooling systems are sized and controlled to maintain are based on tests of large numbers of people who are subjected to different conditions and asked to identify what is comfortable. These results can be adjusted for wall surface temperatures that differ from the room air temperature, as would be the case for people working near large, poorly insulated glass in winter. Relative humidity limits and bounds on airspeed have been similarly derived.

Thermal comfort tests have shown that a single indoor thermal environment will probably not satisfy all of the occupants. People in buildings dress in a variety of styles. Tests have shown different comfort preferences in different geographical regions and between men and women (3,4). Even for a uniform population there is a range of preferences. The abundant data on the subject have been distilled into equations that predict the mean vote of occupants concerning their thermal comfort, on a seven-point scale, and the percent of occupants who will be dissatisfied for a given predicted mean vote (4). Guidelines for dry-bulb temperature, mean-radiant temperature, humidity and air-speed control in buildings typically provide a range of conditions within which 80% of test subjects were satisfied (2,6).

Most space-conditioning systems in commercial buildings provide little or no choice to individuals. Choice can come from local control within an individual work space or, for a small but growing percentage of the work force, from freedom to move within a building or outside it, to the point of staying home and telecommuting. Workers in buildings may be more productive with local controls that allow individually tuned conditions. Energy-consumption simulations of one local-control system, for which productivity changes have not been documented, showed the system capable of saving 7% of the energy of a more typical system or requiring 15% more energy, depending on configuration (7). Small changes in productivity are difficult to reliably measure and attribute to single factors but, if established, can justify considerable expense in capital and operating costs of conditioning systems. Labor costs about 100–200 times more than the energy to condition a building, although strictly economic arguments fail to capture environmental externalities that have not been embedded in fuel costs, a societal dilemma that confronts industry and transportation as well as buildings.

Air Quality. Recently surveyed occupants of large office buildings have claimed that air quality and air movement affect their comfort (8). Comfort is influenced by pollutants that are offensive but not harmful, such as body odors, as well as those pollutants that have been shown to have adverse health effects, such as many volatile organic compounds. In addition, there are pollutants such as radon and carbon monoxide that cannot be detected by the human senses but can be present in quantities considered to be dangerous. Table 1 lists several pollutants and, where specified, maximum concentration limits.

Table 1. Maximum Acceptable Concentrations of Pollutants Found in Buildings

Pollutant	Limit	Ref.
Radon	4 picocuries/liter (EPA guideline)	9
	0.027 working level	10
Asbestos	0.1 fibers longer than 5 μm/cm^3 abatement level (NIOSH)	
Tobacco smoke	dilution with smoke-free air	10
Formaldehyde	0.4 ppm (state standard)	10
Other volatile organics	occupational standards	11
	0.1 occupational standards for general public	10
Carbon dioxide	0.10%	10
Microorganisms and allergens	No standards	
Nitrogen dioxide	100 μg/m^3 ambient air long-term limit	12
Sulfur dioxide	80 μg/m^3 ambient long-term average	12
Carbon monoxide	40 mg/m^3 ambient one-hour limit	12
Total particulates	75 μg/m^3 ambient long-term average	12
Oxidants	235 μg/m^3 ambient one-hour limit	12

Indoor air is considered to be acceptable if outdoor air brought into the building meets the ambient limits shown in Table 1 and if there is sufficient outdoor air or active filtration to keep into pollutant levels below the mandated maximum levels. For CO_2, the maximum level of 0.1% does not represent a threshold for health effects but instead is a surrogate for body odors, which influence comfort.

Lighting. The amount of light admitted to or generated in a building should meet the needs of the occupants, which vary with tasks that range from walking down stairs to precision machining. Light can be characterized by an objective measurement of the luminance of such luminous surfaces as lamps, reflectors, a sheet of paper, windows, walls, and the sky, or by the illuminance, the amount of luminance flux falling onto a unit surface area. The eye detects luminance, which can be related to illuminance through the reflectivity of a surface. The ability to discern the detail of an object, for example the difference between the letters 'C'and 'O,' depends on the visual or radial size of the object, the amount of allowed time, the contrast between the object and its background, the luminance of the background, and the required accuracy of the visual task. Visibility studies have attempted to quantify the importance of each of these factors, with the goal of identifying the required background luminance given values for the other factors. That is, for a specified time and accuracy to identify an object of given size and contrast with its background, background luminance can be selected. However, the calculated luminance varied over orders of magnitude for modest changes in contrast, a sensitivity that raised doubts about the accuracy and applicability of the supporting and placed no reasonable bound on luminance and the associated electric lighting capable of producing the desired luminance (13).

Visual performance studies have attempted to directly show how speed and accuracy in performing a task vary with lighting conditions and therefore answer questions about the impact of lighting levels on productivity. Early studies required participants to match split rings with similar orientation (14) while more recent work scored the ability of participants to identify mistakes in lists of similar numbers (15). The relative visual performance from the latter tests shows a sharp increase as contrast or luminance increase from very low values, followed quickly by very small increase in performance as lighting conditions further improve.

As a potential basis for lighting standards, visual performance studies suffer from being tied to particular tasks that may not accurately represent a given business. Also, lighting must be provided in circulation and recreation spaces where visual performance criteria may not be appropriate. Accordingly, visual satisfaction studies have placed participants in controlled environments and asked them to express whether or not they are satisfied. These studies are similar to thermal comfort studies that have the same goal and have similar results: a Gaussian curve of percent satisfied as a function of horizontal illuminance. In one study the curve peaked at 80 percent satisfaction at 2000 lumens per square meter, with 10% finding that light level too dark and the remaining 10 percent considering it too bright (14). For circulation areas, the appropriate measure is vertical illuminance required to reveal facial features. Visual preference studies have also identified desirable luminances for room surfaces. Surfaces which have excessively high luminances or represent a sharp change in luminance from neighboring surfaces may be a source of direct or reflected glare, which can reduce visual performance in a direct, measurable way or cause discomfort that can indirectly affect well being and productivity.

Quantitative lighting criteria attempt to strike a balance between visual performance, visual preferences and prevailing economic conditions. In 1979 the Illuminating Engineering Society provided a simplified set of illuminance levels for lighting designers (16). Illuminance varied from 22 to 22,000 lumens/m^2 (2–2000 footcandles), covering eight tasks that ranged from walking in public areas to sewing and surgery. Within each category are a triplet of slightly different illuminances, with guidelines for selecting a single value on the basis of the age of the occupants and the required speed and accuracy of the work to be performed.

Illuminance levels as the sole criteria for lighting design can lead to undue preoccupation with providing a uniform, wall-to-wall lighting that does not recognize spatial and temporal variability in occupants' needs. Surface luminance guidelines, based on absolute luminance levels or ratios of luminances from different surfaces within a room, may prohibit bright light sources that cause glare, such as bare incandescent bulbs, but do not account for luminance as a source of both visual variety and stimulation, as would be produced by a brilliant chandelier. In response to the failure of quantitative standards to discriminate desirable and uncomfortable lighting with similar numerical values, another approach to lighting design concentrates on the information content of light. Occupants of buildings require visual information to satisfy biological needs as well as task needs (17). Biological needs include security, orientation, definition of personal space, and relation to nature. Features of a room that help to fulfill these needs should be emphasized while visual distractions should be minimized.

Understanding the process by which humans interpret visual information has led to fresh insights about lighting. Light levels should be higher during the day, to prevent the inside of a building from appearing unduly gloomy compared to outdoors, and lower at night. Humans are insensitive to constant spatial gradients in luminances, relieving the need for uniform lighting levels and making possible the extensive use of daylighting. Changes in the gradient are acceptable if justified by the building structure, making coffered ceilings acceptable but raising cautions about scallops of light and shadow on walls for no apparent reason. Luminances can be high if occupants understand and appreciate the reason, which might be the brilliance of a chandelier or a patch of sunshine. Illuminance levels can be very low in such circulation areas as subway stations, with sufficient light to guide riders to the platforms.

Bioclimatic Chart

Thermal comfort preferences can be summarized graphically in charts that use temperature and moisture content as axes. One of the most useful graphs is a bioclimatic chart, Figure 1, which defines a comfort region as a function of dry-bulb temperature and relative humidity (3).

The thermal resistance of clothing and the body's metabolic rate are implicit variables. Radiation, from the sun or from surfaces that affect the mean radiant temperature, and air speed alter the boundaries of the comfort region, as indicated. That is, wind can compensate for temperatures above the comfort region, while solar radiation or proximity to heated surfaces that increase the mean radiant temperature can balance temperatures below the comfort region.

The chart also shows evaporation as a cooling mechanism. Just as the body loses heat by evaporation of water vapor, the air surrounding a person will be cooled if it gives up energy to evaporate water. Evaporative cooling clearly works better in dry climates, as indicated on the chart, and requires that air flow over the water to be evaporated. If the air is stagnant, it will reach saturation and evaporative cooling will cease.

It is important to note that the chart is defined for outdoor conditions. For use within buildings, it must be modified to account for the sources of heat that raise the indoor temperature above outside conditions—lights, appliances, solar radiation that is absorbed by structure and furnishings which in turn heat the air. The impact of these heat sources can be quantified with the aid of an energy balance for the building.

BUILDING ENERGY BALANCE

A thermal balance equation for a building can be formed on the basis of conservation of energy. This equation relates energy stored within the building structure to heat flows into or out of the building:

$$\left(\sum_i m_i \cdot C_i\right) \cdot \frac{dT_{in}}{dt} = q_{\text{int}} + q_{solar} + q_{aux} + q_{cond} + q_{airflow}$$

Here, the mass of materials within the building is combined with the specific heats. The rate at which energy is stored is proportional to the time rate of change of the inside-air temperature, in the simple approximation that

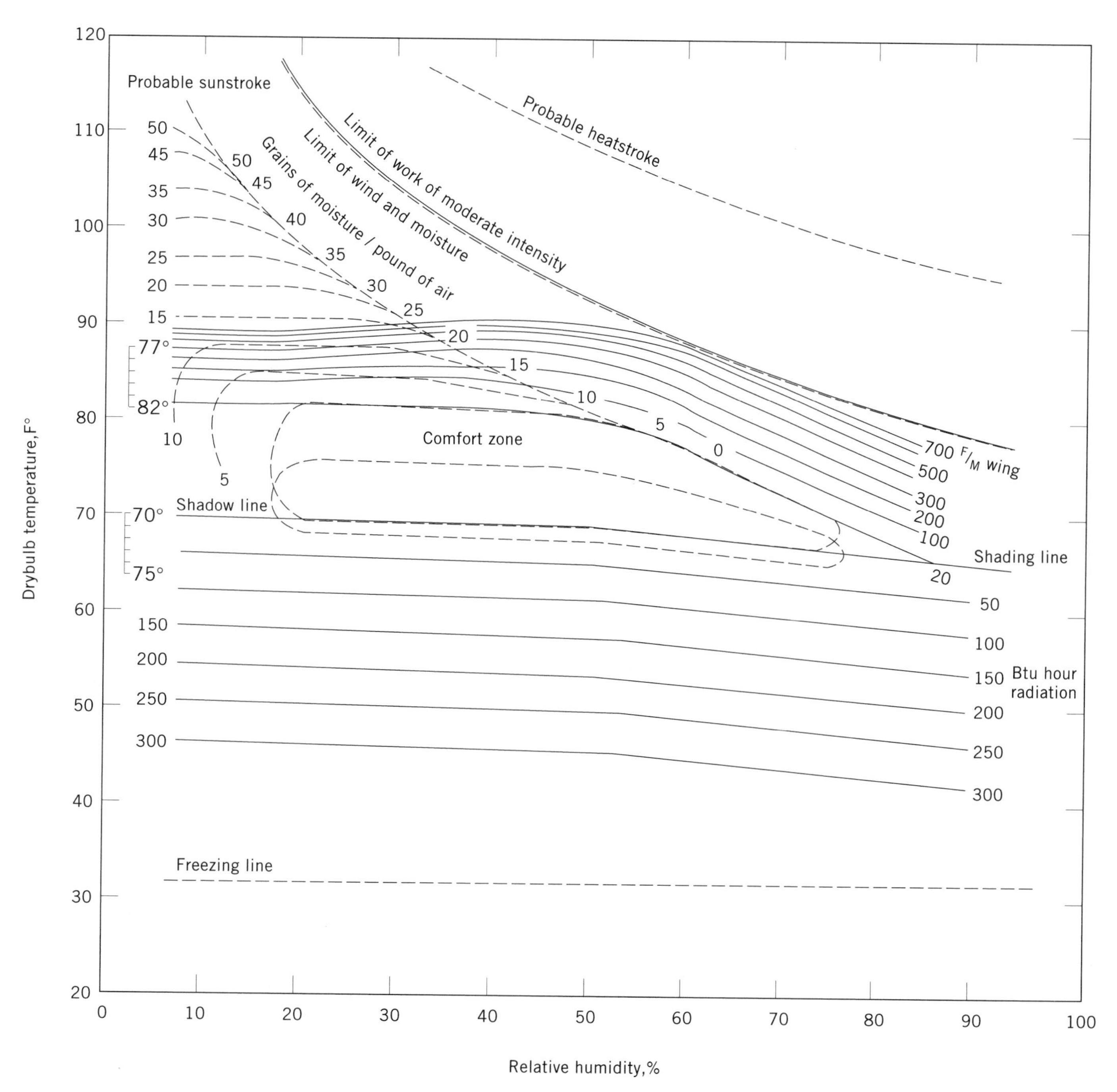

Figure 1. Bioclimatic chart for U.S. moderate zone inhabitants (3).

the air and the masses within the building are all at nearly the same temperature, T_{in}. The inside temperature rises in response to internal heat flows, from lights or office equipment, and from solar energy that enters through windows and to a lesser extent is absorbed by outside walls. Auxiliary heating or cooling comes from mechanical equipment. Conduction through the building walls includes surface heat transfer due to both convection and radiation while the airflow term embraces both forced ventilation and infiltration. Both conduction and airflow are proportional to the indoor-outdoor temperature difference (18) and the sum of the two terms can be expressed as $K(T_{in} - T_{out})$.

At steady state, there is no net flow of energy into or out of the building mass. With no mechanical heating or cooling, the equilibrium indoor-air temperature can be related to the outside temperature and sources of heat gain; its value can be used instead of the outdoor temperature to apply the bioclimatic chart to indoor conditions:

$$T_{in\text{-}equilibrium} = \frac{q_{int} + q_{solar}}{K} + T_{out}$$

The steady-state equation can also be used to determine the magnitude of ventilative cooling required to maintain the indoor temperature at or below a specified upper bound, or the amount of auxiliary heating or cooling needed when ventilation alone is insufficient:

$$q_{aux} = K \cdot (T_{in\text{-}setpo\,\mathrm{int}} - T_{out}) - q_{\mathrm{int}} - q_{solar}$$

The building mass can absorb heat during the day, reducing the need for cooling during the day and heating at night, if its temperature is allowed to increase. The air temperature must increase as well. The more mass that can effectively store heat, the lower the rise in temperature.

PASSIVE CONDITIONING AND LIGHTING

The bioclimatic chart identifies a number of strategies for providing comfortable conditions when the dry-bulb temperature and relative humidity fall outside the thermal comfort region. The building structure, its windows and openings can effectively stretch the comfort region, by promoting ventilation and controlling solar radiation, or shrink it, by passing heat and air inside when not needed. Solar heat can be used to increase the radiant temperature felt by an occupant or to increase the equilibrium temperature, via absorption of radiant energy by building mass and subsequent convective transfer to room air. Passive conditioning and lighting can reduce energy bills and environmental degradation, save space within a building otherwise required by mechanical systems, and reduce building capital costs.

Ventilation

Air flows across openings in a building when there is a pressure drop created by wind or by differences between indoor and outdoor temperatures. The pressure due to air movement can be derived from conservation of momentum, expressed as Bernoulli's equation for steady, irrotational flow of incompressible frictionless fluids (19):

$$p_{stagnation} = \frac{\rho \cdot v^2}{2}$$

where ρ is the density of air and v is the mean air velocity.

What follows are a number of empirical relationships. The mean air velocity, increases from zero at ground level up to its free-stream value, varying with height raised to a power of 0.1–0.3, depending on terrain. When wind strikes a building, the increase in pressure p_w on the windward surface is about 80% of what would be achieved if the wind completely stagnated at an obstruction of infinite size. On the leeward side, the pressure drop has a slightly lower magnitude and the opposite sign (20). Finally, the turbulent volumetric flow through an opening varies with area, pressure drop across the opening and an empirical discharge coefficient, according to an orifice equation (19):

$$\dot{V} = C_d \cdot A \cdot \sqrt{\frac{2 \cdot \Delta p}{\rho}}$$

With a number of openings in the building, the indoor pressure will adjust to conserve mass: the mass flow into the building will, in steady state, equal the flow going out. Wind-driven ventilation can be enhanced by building designs that provide ample operable windows or other openings on facades that are windward and leeward relative to prevailing winds in warmer months. The surface-to-volume ratio of the building should be large, which may lead to floor areas less than what building codes might permit as the maximum on a given site. Older office buildings, lacking mechanical cooling, often had larger interior courtyards and double-loaded corridors, rather than expansive open-plan office bays and fully used core spaces, with no intent of using natural ventilation. In relatively benign climates, concerns about energy consumption have prompted a renewed interest in natural ventilation, particularly in smaller commercial buildings. In very hot climates, vernacular architecture has made use of roof-mounted wind scoops to capture a breeze and direct it into a dwelling.

Buoyancy-driven airflow within buildings works exactly as the flow of hot air within a chimney. The pressure of a column of air decreases with height, according to the equation of hydrostatics (19):

$$p = p_o - \rho \cdot g \cdot h$$

where p_o is the pressure where the height is taken to be zero and g is the gravitational constant. Density is inversely related to temperature through the ideal gas law. When indoor temperatures are warmer than outside, the pressure inside will decrease more gradually with height. Outside air will be drawn into the building at low levels, where inside pressure is lower than out, and expelled at higher levels, where inside pressure exceeds outside pressure. The relation between volumetric flow and pressure drop across an open window governs buoyancy-driven as well as wind-driven airflows. In climates and building types where buoyancy-driven flows are desirable, window openings should be of generous size and at maximum vertical spacing. Within the building there should be ample paths for air to flow upward and out. Examples of architectural features that promote buoyancy-driven ventilation and examples of calculation of airflows are found in references 21, 22 and 23.

The cooling effect of buoyancy-driven flows comes from the velocity of the air and the enhanced removal of body heat by convection. While the volumetric flow can remove unwanted heat from the building, indoor temperatures must exceed outdoor levels for airflow to be generated. Wind-driven flows can provide additional cooling by reducing indoor temperatures to the outside value, as shown in the steady-state energy equation when the conductivity term K becomes very large.

Both wind-driven and buoyancy-driven airflows can create unwanted heat flow across the building envelope. These unintended flows can be drawn through small cracks around window and door frames and through other openings in the building structure. Passive heating strategies work better with well-sealed buildings that minimize heat flow due to infiltration. However, even when airflow is not needed on the basis of thermal criteria, fresh air is still required to maintain satisfactory air quality.

Heating, Cooling and Lighting with the Sun

Solar energy can serve as the energy source for a passive heating system, particularly on the scale of a house. Solar radiation can be brought directly into living spaces, can be confined to a sunspace attached to the house, or can be absorbed in a glass-covered, dark colored wall named after its original developer, Felix Trombe of France (23,24). The design of a few low-rise office buildings has also made use of solar energy, particularly by incorporating atria to collect heat that can maintain the temperature of the atria at near-comfortable conditions and potentially provide additional heat for storage and distribution through the building (25).

Solar energy enters the building at the time of day when it is least needed; outside temperatures are at their warmest and, for commercial buildings, internal loads often at their largest. To offset the need for auxiliary heat at night, solar energy must be stored and later released. Storage can be provided by mass inherent in the structure or by mass provided specifically for the solar heating system. For example, drums of water or masonry walls exposed to the sun store heat and release it naturally, when the temperature of the mass exceeds the indoor-air temperature. Less frequently, the thermal mass may be a phase-change material that can store substantial heat without large increases in temperature. Isolating the storage medium from the remainder of the house, as with a sunspace, allows the diurnal temperature cycle in the storage area to exceed that which full-time occupants would likely label as comfortable.

To complete the system, stored heat must be transferred to the building. A Trombe wall can transfer heat by conduction through the wall and then by convection and radiation to the indoor air, or by convection from the side of the wall facing the sun to air that flows to occupied spaces through holes in the upper part of the wall, as shown in Figure 2. Alternatively, heat collected in a sunspace or atrium can be drawn by a fan through ducting to underground storage in a rock bed, then withdrawn by fan to heat occupied portions of the building.

Although solar heating systems are often mechanically quite simple, they nevertheless require careful design to specify the size of windows and type of glazing, the size and configuration of the storage area and the means of heat transfer to the building. Design goals may include maximum reduction of the annual heating needs, where the benchmark is a building of comparable size and design that does not include a passive-heating system, as well as minimal amplitude of the diurnal temperature oscillations. System design is complicated by variations in the design of the building itself, which is often an integral part of the system: passive heating systems are not a mass-produced item. Design techniques include rational analysis of individual buildings using heat-transfer calculations and correlations derived from previous simulations (21–23,25).

In warmer weather, use of solar energy depends on whether there is a need for heat at night even though daytime temperatures are within or exceed the comfort region of the bioclimatic chart. The mass of the building structure can reduce the thermal impact of solar radiation that penetrates opaque walls by conduction or is transmitted and conducted through windows. Heat stored in adobe walls or tile floors can be vented through open windows at night or be used to keep indoor temperatures above cool outdoor conditions. Windows in occupied spaces should be shaded, either with structure or vegetation, to reduce solar heat gain at times of day and year when the bioclimatic chart indicates that no solar radiation is needed, day or, through storage, night, to ensure thermal comfort. Fixed shading devices typically are a compromise design, because the position of the Sun is symmetric about the summer solstice while outdoor temperatures are higher after the solstice, due to the thermal mass of the Earth. Vegetation, which grows in response to light and warmth, can in some cases more closely match admitted solar radiation to seasonal changes in outdoor temperatures. The sun can be used to draw air through a building by the buoyancy effect, without directly heating the building interior to create the driving force, by means of a solar chimney admits solar radiation above the occupied regions (24). Design strategies for shading and passive cooling are found in references 21, 22, 23 and 25.

The Sun is also the source of natural lighting. Daylighting is difficult to control due to its intensity, variability, and attendant heat. The Sun is a more efficient light source than many man-made devices, producing a luminous flux at the Earth of 90 lumens for each Watt of radiant energy (13). Incandescent lights typically produce 15 lm/W, the addition of a halogen gas improves the efficiency to about 23 lm/W, fluorescent lamps vary from 55–90 lm/W and high-intensity discharge lamps with metal halides also operate at about 90 lm/W (26). Light in excess

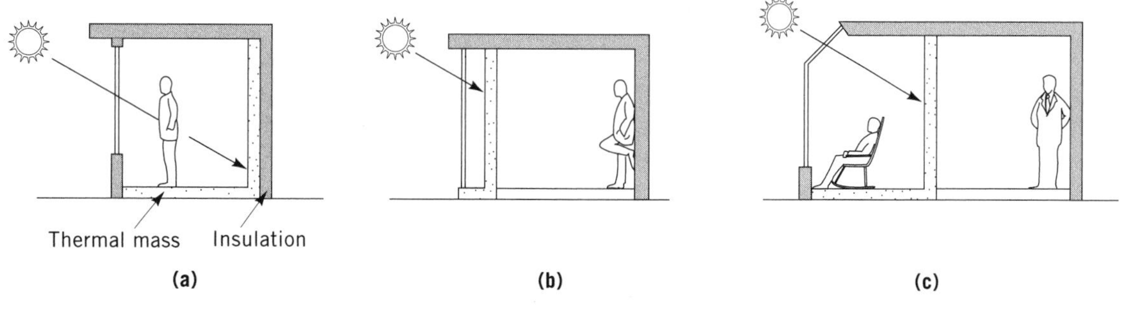

Figure 2. Three main types of passive solar space heating systems. (**a**) Direct gain; (**b**) Trombe wall; (**c**), sunspace (23).

of that required for visual tasks can be distracting or fatiguing, as the eye muscles adjust the size of the pupil to regulate the luminous energy admitted to the retina. Illuminating Engineering Society guidelines (16) specify a 3:1 maximum allowable ratio between the luminance, or brightness, of a visual task and the luminance of the adjacent background and a 10:1 variation between the task and distant backgrounds within a room.

Daylighting design in buildings centers on bringing light in above eye level and diffusing it, to evenly illuminate large areas. Windows are a good source of daylight when separated by a horizontal lightshelf into a lower portion for a view of the outdoors and an upper, lighting portion for reflecting light off the shelf, onto the ceiling, and then down into the room. The lightshelf may be solely on the outside of the glass or on the inside as well, and are sized to reduce glare and shade the lower or vision part of the window. On the roof, vertically mounted glass will admit light into "scoops" that direct the light downward into occupied spaces, corridors or atria, as shown in Figure 3. The light can be directed to a wall, producing a luminous surface, or to work surfaces, through diffusing baffles as needed to control glare. Skylights, mounted horizontally, also admit light but are not as easily shaded at times of year when the heat associated with the light is not desirable. Light scoops are similarly designed although glare is usually less of a problem due to their position in a building. Light scoops benefit from indirect radiation bouncing off the roof, particularly if the roof is covered with a highly reflective material such as marble chips in lieu of the gravel typically used to protect the waterproof roofing membrane (27).

Methods for calculating horizontal illuminance in daylit spaces (13) start with estimation of solar illuminance and sky luminance under clear and cloudy conditions. The lumen method then estimates horizontal illuminance from the luminance at windows or skylights and correlations based on the geometry and surface properties of the room, which acts as an optical cavity. The daylight-factor method, an alternative procedure appropriate for predominantly cloudy conditions, excludes direct sunlight and defines analytic expressions for the contribution of luminance directly from the sky and also reflected light from the ground and external surfaces. Designers frequently

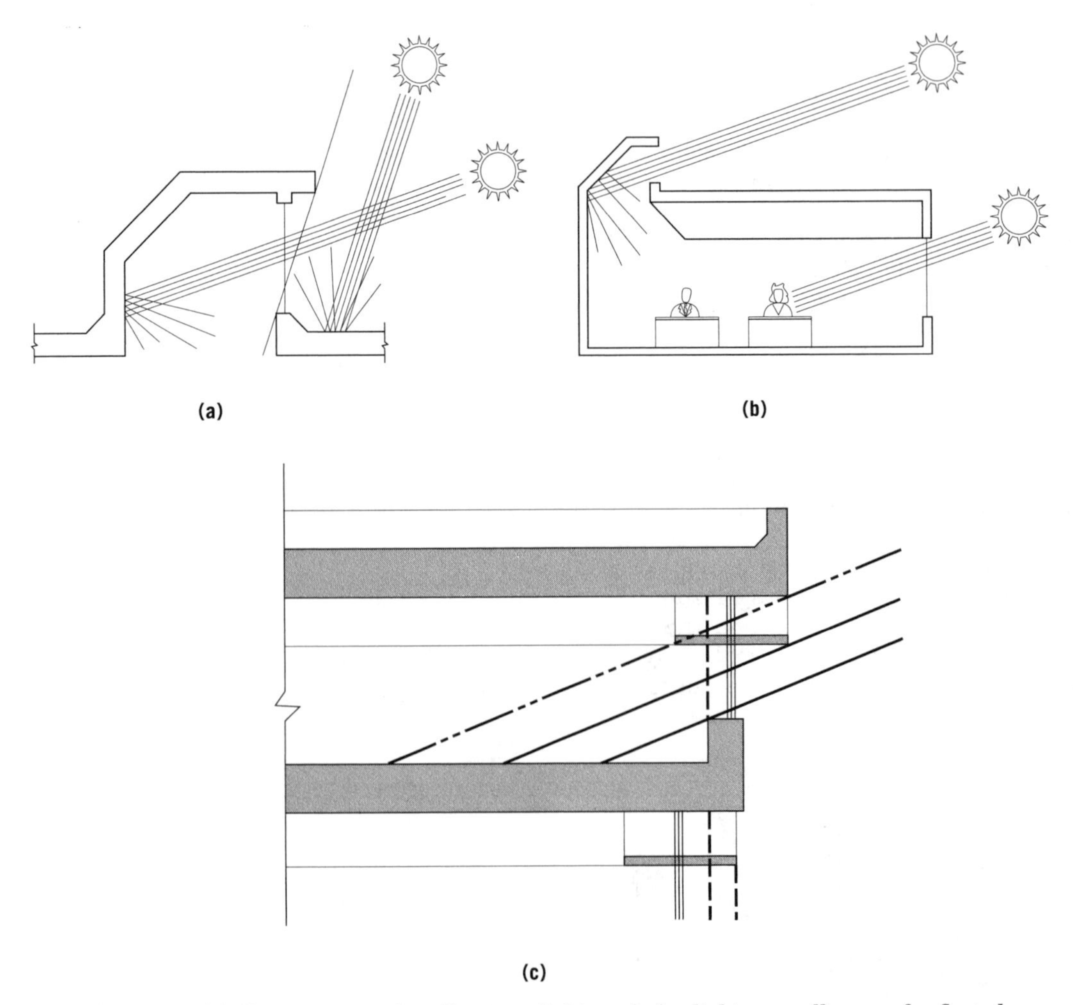

Figure 3. (**a**) Sunscoops receive direct sunlight and sky light as well as roof-reflected light. (**b**) Sunscoops are better than windows for controlling potential glare of low-angle sunlight. (**c**) South wall section diagram at 2PM on December 15 shows how interior lightshelf as designed controls glare at eye level. From ref. 17.

use scale models for complex geometries that can be tested outdoors (27) and simulation programs to reduce computation time.

Evaporative Cooling

The temperature of a stream of air will be reduced if the air transfers some of its energy to evaporate water. Passive evaporative cooling strategies require wind or buoyancy forces to generate the flow of air. While the water can be in a pool, spraying the water into the air increases the surface area between air and water and therefore the evaporation rate. Evaporation can also cool wetted building surfaces. Examples of evaporative cooling include courtyard fountains, pools upwind of plazas, wood scoops where air flows over dampened material, roof sprays, and water walls bounding occupied areas (23,28). The visual and acoustic appeal of splashing water augment the cooling that can be achieved by evaporation. Evaporative cooling also plays a role in active conditioning systems, where it can be more precisely designed. In passive applications, uncertainty about its efficacy, due to lack of experience or inadequate modeling methods, has inhibited its use by architects as a prominent feature in high-profile projects where there is an expectation of substantial energy benefit.

HEATING, VENTILATING AND AIR CONDITIONING SYSTEMS

The Need for Mechanical Systems

For many of the world's people, living in tropical and subtropical areas, the shells of residential and smaller commercial buildings have been designed to provide shelter from rain and modest thermal protection. Entrances to hotel lobbies in Hawaii may have large overhangs and supporting structure but no completely sealed walls. The airport terminal in Jeddah, Saudi Arabia used by Muslim pilgrims (29) has a fabric, tent-like roof but open sides, to screen the sun but bring in breezes. Homes in India require ample airflow for inefficient wood stoves used for cooking.

But there are many other situations where a combination of internal sources of heat, occupant expectations, and harsh climate make mechanical heating and cooling a necessity. Heating is required in homes in latitudes well removed from the Equator. Cooling brings indoor temperatures and humidity within the comfort region identified on the bioclimatic chart. While one person's luxury may often be another's necessity, mechanical systems are increasingly prevalent.

When climatic conditions become a foe rather than a friend, at least during a portion of the year, the building shell becomes much more a barrier than a sieve. Houses in Scandinavian countries are tightly sealed to minimize unwanted air leakage that increases winter heat loss. Wind-driven and buoyancy-driven ventilation is minimized. Even if the thermal balance in the house can be maintained with little or no heat from a mechanical system, equipment is still required to bring fresh air into the house and to exhaust this air. The complexity of the mechanical system is balanced by the control it offers: an even amount of airflow, regardless of changes in weather, that can be carefully directed through a house. Fresh air typically is brought into all rooms of the house and is exhausted in bathrooms and the kitchen (30).

In commercial buildings, mechanical ventilation systems make it possible to increase the floor area of a structure on a given site. No longer must office areas be sufficiently shallow for everyone to be close to an operable window. But the advantages of carefully controlled ventilation in harsh weather are often countered by occupants' inability to open up the building during mild weather.

Most buildings incorporate some passive heating, cooling and lighting features: a simple window admits solar energy and light. Most commercial buildings worldwide and most residential buildings in industrialized countries require active thermal conditioning and lighting as well. Active systems that work well in conjunction with passive strategies are preferable from an environmental perspective and offer occupants more contact with the outdoors. Also preferable are active systems that can efficiently and simultaneously serve regions of a building that may have very different thermal and lighting requirements.

Thermal Zones

Most space-conditioning systems must provide heating and cooling that varies both spatially and temporally. Late-afternoon solar radiation pours through windows on the West facade of a building on a chilly day, while the East side needs heat. Rooms in the core of a building, buffered by conditioned spaces on all sides, have no heat loss. With heat gain throughout the year due to lights and office equipment, these spaces require constant cooling, even when perimeter spaces need heating.

The space-conditioning system is quite simple for a building with a single thermal zone. Because the space is thermally uniform, a single thermostat is sufficient to measure slight deviations in temperature with respect to the set point and trigger either heating or cooling, which can be delivered with either conditioned air transported to the thermal zone via ducts or with hot or cold water. Single-zone systems typically have a constant flow of air. Whether air or water is the transport media, the thermostat controls the flow of chilled or heating water through heat exchangers in the duct or in the room itself. From the thermal balance relationship, it is expected that the amount of heating will increase as outside temperatures drop, while cooling increases with outside temperature above the cooling balance point. In most houses, energy consumption strongly correlates with outside temperature (31).

Larger buildings offer behave as multiple thermal zones. Zonal boundaries may match physical partitions or may be fuzzy borders between perimeter and core areas in an open office plan. Each thermal region requires its own thermostat. Different types of heating and cooling systems can be judged in part on the basis of how efficiently they can condition numerous thermal zones. Some older systems cool all the air centrally and then reheat it locally in those zones that require heat, a wasteful practice similar to opening all the windows in a house when the kitchen is warm and expecting the heating system to compensate. With such systems, the relationship between heating or cooling and outside temperature is different than in a single-zone building. More heat is required at mild temperatures, for example, because some of that heat is tempering air that has been deliberately cooled.

Thermal loads in buildings vary over time as well as from zone to zone at a single moment. Passive heating and cooling strategies seek to smooth out the diurnal or even seasonal variation in cooling and heating loads. Some mechanical systems use water or other substances for energy storage to accomplish the same task. Thermal storage can also serve as repository of heat from zones that require cooling and a source of heat for other zones, eliminating the need for simultaneous heating and cooling.

Ventilation

A standard issued by the American Society of Heating, Refrigerating and Air-Conditioning Engineers (ASHRAE) governs building ventilation rates for acceptable indoor air quality (32). Ventilation systems may also serve a number of other purposes, distributing heating or cooling and maintaining a specified pressure within a building, but all systems must maintain pollutant levels below the maximum values shown in Table 1. Typically, most pollutants in an office environment are controlled by careful selection of building materials and the ventilation system removes CO_2 and other bioeffluents. An amount of outdoor air proportional to the maximum expected number of occupants is usually set as a fixed minimum air intake; the current standard prescribes 20 L/s-occupant. An alternative, performance-based methodology is also permitted in the standard, whereby the ventilation system controls airflow to maintain indoor CO_2 concentrations below 1000 ppm. This alternative will reduce outdoor-air intakes when occupancy is below the design level and when CO_2 levels are building up to equilibrium values after periods of reduced occupancy (33), saving the energy required to condition the outdoor air at times of year when that is needed. Control of airflow to regulate CO_2 requires CO_2 sensors and additional control logic.

Air Movement with Fans. Consider a system that provides heated or cooled air, as shown in Figure 4. First, there are one or more central fans to circulate the air. This circulation process provides for filtration and tempering via heating or cooling, to maintain the building in thermal equilibrium. The supply fan also draws into the building the required amount of outdoor air for occupants. The supply fan raises the pressure of the air an amount sufficient to push the air through the ducts and into the occupied areas of the building. Often there is a return fan to bring most of the air the back. The remaining air leaves the building via exhausts in toilet or electrical rooms, and by exfiltrating through cracks in the building shell. Laboratory and industrial buildings may require large amounts of outside air, in excess of the needs of occupants, to remove pollutants and have large exhaust fans to remove all of the air, with no recirculation. All of this air must be tempered, for the benefit of occupants and the experimental or industrial process.

The mass flow of air out of the building must equal, in steady state, the amount of air coming in. Designers typically assume that the outdoor air intake is the only source of fresh air when the ventilation system is operating, because the ventilation system is designed to pressurize the building, but recent measurements have indicated that there may be locally depressurized areas and some infiltration as well. While pressurizing a building offers enhanced comfort to occupants working near outside walls by eliminating cold drafts of infiltrating winter air, it also drives moist interior air into the walls, where it can condense without proper attention to vapor barriers.

Displacement Ventilation. Typically, ventilation is provided to occupied spaces via diffusers that mix the incoming air with air already in the room. This is intended to ensure that outdoor air is distributed throughout the room and to prevent occupants from directly feeling a strong stream of heated or cooled air that may be thermally uncomfortable. An alternative approach, implemented in some European buildings in recent years, introduces ventilation air near the floor and avoids mixing. The goal is to produce a nearly uniform, upward-moving flow that sweeps heat and pollutants, particularly bioeffluents, to the ceiling, where they are exhausted. This approach, known as displacement ventilation (34,35), offers a higher ventilation efficiency than well-mixed ventila-

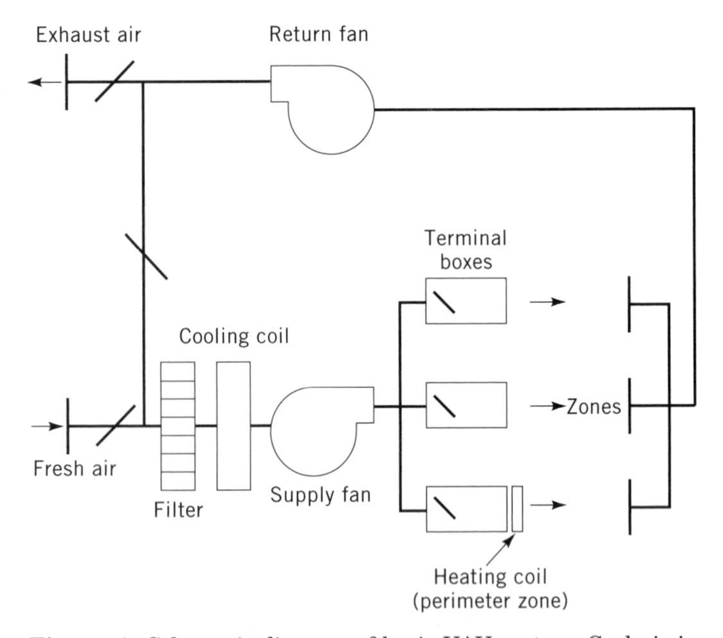

Figure 4. Schematic diagram of basic VAV system. Cool air is supplied to the terminal boxes at a constant temperature; terminal boxes vary the flow as necessary to cool individual spaces, or zones. During the heating season, the cooling coil is off, and terminal boxes add heat locally as required. Terminal boxes may use built-in fans to recirculate room air and provide a constant outlet volume, or the outlet volume may vary, with no recirculation. The return fan tracks the flow of the supply fan to maintain a fixed fresh air flow rate. A large building may have one or more VAV systems serving many zones.

tion, by reducing pollutant concentrations in the breathing zone. It can be successful if the thermal gradient is not so large as to be uncomfortable and if stratification is not unduly compromised by thermal plumes by local heat sources. Cooled ceilings are the subject of current investigation as a means of providing additional cooling to zones with large heat sources.

Air and Water Systems for Heating and Cooling

Systems for conditioning buildings can be classified by the transport medium used for cooling: air or water (36). All-air systems employ central cooling equipment and deliver cooled air to end users but may transport hot water for local heating; mixed air-water systems provide a portion of the required cooling centrally and the remainder locally, via cooling coils; and all-water systems provide all cooling locally but still need to bring in outdoor air, either centrally or locally.

All-Air Systems. ***Single-Duct Systems.*** Removing heat from a building involves both sensible and latent heat transfer. Air passing over a cooling coil will leave at a lower temperature and, if the exit temperature is below the dew point, water vapor will condense on the surface of the coils. Latent cooling is often about one third of the total, seasonal cooling load in commercial buildings, a proportion that varies with local climate.

Cooling coils can be located in a central mechanical room or locally, in individual occupied spaces. If air is cooled centrally, as shown in Figure 4, it not only provides the amount of outdoor air required to maintain indoor air quality but also preserves the thermal balance at the required temperature. Centrally located cooling offers the advantages of centralized maintenance and removal of condensed water vapor, at the cost of constraints on efficiently serving multiple thermal zones, as will be discussed, and relatively inefficient transport of thermal energy. Cooling water can be more cheaply pumped than can cool air of the same energy content be pushed through ducts by fans. In addition, cool air requires ducts larger than water pipes, occupying valuable space within the building shell (36).

The simplest system maintains a constant flow of air to a building that behaves as a single thermal zone. For multiple zones, the most basic extension of this system still keeps the airflow constant to each zone. The temperature of the air can be kept constant or can be controlled to maintain thermal equilibrium in the zone with the largest heat gains. In the former case, all zones need a source of heat at all times except during the hottest weather for which the system was designed. In the latter, the controlling zone alone requires no source of heat. Such systems are holdovers from an era of cheap energy and seemingly infinite, secure energy supplies.

Single-Duct Systems with Variable Airflow. The waste associated with simultaneous heating and cooling can be partially eliminated by varying the flow of air to each thermal zone, while still maintaining a fixed temperature. Thermostats in each zone throttle the airflow to maintain the desired space temperature, down to a minimal airflow required for air quality. At this point the thermostat controls a source of heat in the zone.

Varying the airflow has the secondary advantage of reducing the energy consumed by fans. Energy savings depend on how much energy is dissipated across dampers in the system. Zone-level thermostats drive local dampers. As they close, these dampers tend to raise the air pressure in the supply-air duct, causing dampers at the fan itself to throttle. The combined action of the dampers results in modest energy savings if the central dampers are located immediately downstream of the fan and higher savings if the dampers are placed at the inlet to centrifugal fans and impart a swirl to the incoming air that reduces fan power. Further, if the fan blades are inclined at a forward angle, savings due to flow control approach those due to replacing the central dampers with controllers that vary the speed of the fan motor. These variable-speed motor drives (VSDs) have been measured to save 35–60% of the energy used by fans with inlet dampers (37). Larger savings are typically achieved when a fan is not constrained to maintain pressure in the duct at a specified level, as is true for single-zone ventilation systems and most return or exhaust fans. In this case, fan power varies with the third power of airflow. Supply fans which serve multiple zones by maintaining the duct at a constant pressure show a weaker dependence of power on flow and reduced savings. For multi-zone supply fans, further savings can be achieved by connecting the *local* dampers and thermostats in each zone of a building to a central supervisory controller which slows the central fan until at least one local damper is fully open. This has been shown (38) to save 20–40% of the energy needed when a supply fan has a VSD but local thermostat action cannot be centrally monitored.

Dual-Duct Systems. Single-duct systems serving multiple zones provide heat locally, through hot water or electric resistance coils. Dual-duct systems, in contrast, provide warm as well as cool air and are truly all-air systems, as seen from the occupied spaces. In its simplest form, air from both ducts is tapped in each thermal zone and regulated by the thermostat, which mixes the two streams in the proportion needed to maintain the desired temperature. This offers effective control with no possibility for water leakage in occupied spaces but involves simultaneous production of both heated and cooled airstreams. Improvements focus on controlling the temperature of the air in the two ducts to reduce the energy penalty of coincident heating and cooling.

Combined Air and Water Systems. ***Induction Systems.*** Reducing the volume of cooled air distributed to occupied spaces requires that air temperature be lowered or that some cooling be provided locally. So-called induction cooling systems cool the air centrally to remove latent heat, confining condensate to a central location. The dry air is distributed via fans to office areas, where it is forced out through nozzles at high speed, inducing room air into the airstream. The warmer, higher volume of mixed air flows over secondary coils and into the room. Less space and smaller fans are needed for air distribution, but chilled water must be piped and pumped to office areas, where it is a potential source of leakage, and the systems are considered to be noisier than all-air systems by some mechanical engineers.

All-Water Systems. Performing all cooling locally will reduce airflow back to the minimal level required for air quality. Local cooling can be accomplished with cooling coils, now equipped with drip pans and drains for the condensate, or local heat pumps. Chilled water lines must be carefully insulated to prevent drippage in occupied areas. Heat pumps can be coupled with large water storage tanks and are particularly effective in serving buildings where there is simultaneous need for heating and cooling, or where there is substantial cooling needed during the day and heating at night.

Use of Outdoor Air for Cooling. In houses, the indoor equilibrium temperature, defined earlier, is typically about 3–5°C above the outside temperature. Passive-solar houses will have a much larger temperature difference, with indoor temperatures in the comfort range even when outdoor temperatures are much lower. This is achieved with large solar gains and small thermal conductivity. The indoor equilibrium temperature is also well above outdoor temperatures, by 10–15°C, for many commercial buildings, where internal heat gains are large. This temperature difference reduces the need for heat in the winter but necessarily yields a longer cooling season. With outdoor air intakes at the minimum amount needed to meet ventilation standards, this leads to situations when cooling equipment is running even though the outdoor-air temperature is lower than the indoor temperature.

All-air systems are often designed with outdoor-air intakes large enough to bring in more outdoor air than in needed for ventilation, to provide "free" cooling in mild weather. Such a so-called economizer strategy (39) requires an exhaust-air duct to remove the extra airflow from the building, and additional controls to vary the amount of outdoor air required to maintain the supply-air temperature at a set point. Only if outdoor air alone is not sufficient will the cooling equipment be turned on. Outdoor airflow is returned to its minimum value when its energy content exceeds that of the indoor air returned to the supply fan, as measured by enthalpy or approximated by temperature alone.

Economizers that rely on additional outdoor air require more space within a building for air-intake ducting. This is acceptable in small buildings, where the fans are in a basement or roof-top equipment room, but may be undesirable in high rise office buildings, where there are separate fan systems for each floor, outdoor air is ducted to each floor from a small number of intakes, and larger ducts reduce space available for occupants. In addition, mixed air-water and all-water systems cannot take advantage of free cooling from outdoor air. In these cases, it is still possible to take advantage of outdoor air in mild weather by water-side economizers that remove heat from the cooling water not with chillers but with outdoor air, via a cooling tower. The tower is normally used to reject heat from a chiller: the chiller removes heat from the water, transfers it to a separate condenser-water loop, and the condenser water is cooled in the tower by evaporation and convection. For free cooling, the condenser water loop is coupled to the building chilled-water loop via a heat exchanger, rather than a chiller. Less frequently, the cooling water is sent straight to the tower and then strained to remove impurities before use within the building. Water-side economizers are effective as long as the outdoor wetbulb temperature (the equilibrium temperature of a wetted thermometer in a moving stream of air) is below the desired chilled-water temperature.

Motive Power. As has been noted, the electrical power required by fan motors in variable-air-volume systems can be reduced when airflows are below design values by employing VSDs in lieu of dampers or vanes at the fan. The VSDs regulate pressure or airflow. VSDs, which are based on electronic components similar in principle to the transistor so ubiquitous in home and office electronics but capable of handling large amounts of current, are also used in a variety of other applications. In buildings, these applications include chilled-water and hot-water pumping. Here, the control is very similar to that for supply fans: the pump speed is regulated to maintain a constant pressure in the piping loop and individual throttle valves vary in position to control the flow of water to such end uses as radiators and convectors. VSDs are also used in condenser-water piping loops, to vary the flow of condenser water to cooling towers, in cooling-tower fans, and in chiller compressor motors. The energy savings depends on the distribution of required air or water flows throughout a year and on the type of control that would have been provided without VSDs. For example, cooling towers with multiple cells and multiple fans for each cell can closely match fan power to the thermal load from the condenser water and VSDs may offer little improvement.

High-efficiency motors also offer energy savings. Motors which convert electrical power to mechanical power at an efficiency higher than those at the low end offered by manufacturers, and at higher cost, are effective when operating hours are long, loads are large and the added investment is quickly paid back. While efficiency improvements are just a few percentage points, from 90% to 94% as an example, the savings can be substantial for a large motor in constant use.

Selection and Design of Systems. System selection depends on the individual experience of the design engineer and a host of issues that concern the building owner and occupants, ranging from first cost and operating cost to space for equipment, maintenance, noise, accuracy of control of temperature and humidity and cleanliness. Packaged, roof-top units that provide ventilation and conditioning are overwhelmingly popular in small commercial buildings. These units may be constant volume or, for multiple zones, variable volume. All-air, variable airflow systems are a common choice in larger buildings, where fans, filters and cooling equipment are not packaged but are assembled into a system on-site. Once system selection is made, an engineer employs design procedures, available in the form of manuals (39) or computerized algorithms for component sizing and selection available from equipment manufacturers.

Sources of Heating and Cooling

Equipment. Heat released from the combustion of fossil fuels can be used to directly condition buildings, either individually or via a district heating system. Fossil or nuclear fuels that are burned to generate electricity can also be a source of heat for buildings, using a heat-delivery network. Power plants typically operate at efficiencies less than 40%, due to mechanical losses and limitations embodied in the Second Law of Thermodynamics. The Second Law requires that some portion of the heat extracted from a high-temperature source be delivered to a low-temperature sink. Typically, power plants use a body or water or a cooling tower to dispose of the unusable heat. Ideally, buildings can serve as a heat sink, making use of waste heat via a districting heating system. Practically, the cost and thermal losses of distribution pipe reduce the thermal advantages. In addition, a combined heat-and-power plant must extract steam at the end of the turbine before it reaches as low a temperature as can be achieved with an electricity-only plant, a further loss in efficiency.

Heat in buildings can also be provided by the drop of electrical potential across resistive wire. The primary energy required at the power plant is larger by a factor of 2.5–3.0 (the reciprocal of the overall efficiency of generating, transmitting and distributing electricity). Hydroelectric power eliminates the thermodynamic inefficiency of generating electricity via a heat source. The overall inefficiency of electric resistance heat, computed at the power plant, is usually reflected in the higher cost of electricity compared to fossil fuels, when normalized by delivered energy. However, electric-resistance heat has lower capital and installation costs, requires less space, and is more easily controlled to match thermal loads in different thermal zones.

Heat pumps are a popular type of heating equipment in some areas not served by natural gas utilities and where cooling loads are high enough that mechanical cooling is needed in summer. Like mechanical cooling equipment, heat pumps are heat engines essentially run in reverse (40). Instead of taking heat from a high temperature source, producing work, and rejecting the remaining heat to a low temperature sink, a heat pump or chiller takes heat out of a low temperature source by means of mechanical work and adds this heat to a high-temperature sink. For a heat pump, the low-temperature source is outdoors; for the chiller, it is inside the building.

Both a heat pump and a chiller use the same four basic components, shown in Figure 5. Heat flows from the low-temperature source across the heat-exchange surfaces of the evaporator to the colder low-pressure liquid refrigerant, which is evaporated. The slightly superheated gas is compressed to a higher pressure and temperature. The hot gas then transfers heat, primarily due to a phase change back to the liquid state, across the condenser to outdoor air or condenser water. Finally, the pressure is dropped across an expansion valve, which is regulated to maintain a constant superheat temperature at the evaporator outlet (41).

The heat pump can be run as a chiller in the summer by directing the compressor discharge to the outdoor

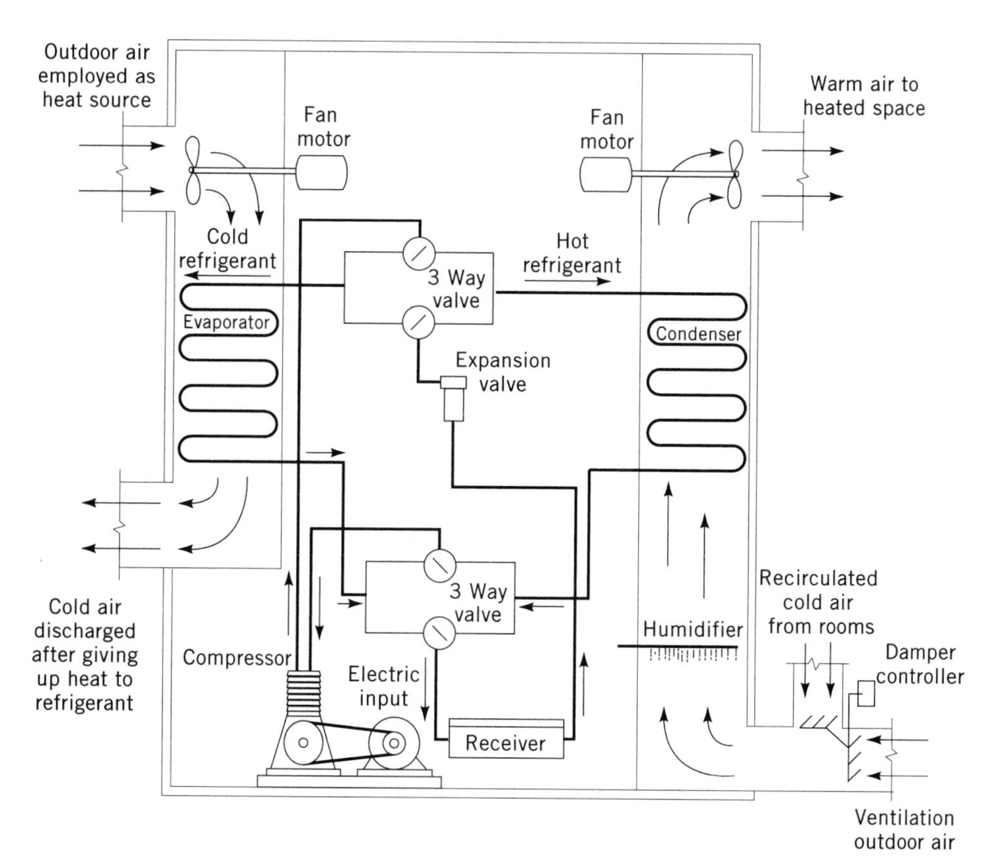

Figure 5. Diagram of heat pump used to heat residence (41).

rather than indoor heat exchanger and uses the same distribution system. Like the chiller, its efficiency is highest when the indoor-outdoor temperature difference or thermal hill is small. Water-cooled condensers offer higher efficiencies than air-cooled units, because the water can be cooled via evaporation to a temperature that approaches the outdoor wet-bulb temperature. The cost of this improved performance is consumption of water to replace what is evaporated. However, higher efficiency translates to less make-up water at those power plants that use cooling towers.

Thermal Storage. Thermal storage is applied in commercial buildings for two major and different reasons: to make the building as a whole behave as a single thermal zone, thereby minimizing simultaneous heating and cooling during occupied periods as well as daytime cooling followed by nighttime heating; and to shift cooling and less frequently heating to nighttime periods, when electricity rates are lower. In the first case, thermal storage is provided by large water tanks, which serve as a heat source or sink for individual heat pumps. One example of this technology, a 150,000 square foot office building in Boston, Massachusetts, uses 750,000 gallons of water.

In the second case, chillers are operated at night to lower the temperature of tanks of water or to freeze the water. Ice storage is a more complex technology but has a higher energy storage density. Ice storage requires chillers to operate at lower evaporator temperatures than chilled-water storage. For ice-buildup systems, the chillers cool a glycol solution to below-freezing temperatures and the glycol, passing through pipes or coiled tubes in a storage tank, freezes the stored water. During the day, the glycol solution is circulated to cooling coils in the ventilation system, extracting heat from the building that is then transferred to the storage tanks, melting the ice. In ice-harvesting systems, the refrigerant in the chiller directly freezes a thin layer of ice on plates above the storage tank. Hot refrigerant gas is periodically pulsed to the plates to melt the boundary layer of ice, allowing the remaining sheet to fall into the tank. Recent experimental work has produced a "slippery" ice slurry that falls off the plates without need for the hot-gas cycle, which, even when properly tuned, reduces the efficiency of the ice-making system by about 10%.

The overall efficiency of ice-storage systems, measured in delivered cooling per unit of fuel consumed at the power plant, suffers from low evaporator temperatures in the chiller, a penalty partially eliminated with chilled-water storage. Both types of systems benefit from lower nighttime outdoor temperatures for the condenser. Ventilation systems that take advantage of melted ice or glycol solutions near the freezing point can provide the same cooling to the building with lower airflows by reducing the temperature of the supply air. The ventilation system can use smaller fans and ducts, which take up less space and are less expensive to purchase and operate. Reduce space for ducting yields more usable space within a given building volume, a substantial economic advantage. Less appreciated but of major importance is the higher marginal efficiency of electricity production during off-peak periods, which contributes to lower electricity prices.

The mass inherent in the building structure may also be used for thermal storage, to shift loads to off-peak utility periods. This strategy has been rarely practiced because it requires more a more capable control system than is often installed in buildings. Before a building is occupied in the morning, the air temperature in a building is cooled to a point at or near the lower bound of the comfort region established by the building tenants. This sub-cooling requires more energy than conventional cooling strategies or dedicated thermal storage, because the building mass at lower temperature absorbs more heat from its environment. However, it is done at times of lower energy costs. In the day, the building air temperature is allowed to rise, letting heat flow into the cooler mass. If the thermostat is simply set back up to its normal position after the sub-cooling period, the benefits of sub-cooling will last for a relatively short period of time and the required chiller power will then be as high as if no sub-cooling were provided. Many electric utility rates for commercial buildings include a charge for demand as well as energy consumption, and smoothing out electricity demand is a primary goal of thermal storage. Electronic controls make it possible to carefully and gradually raise the thermostat set point over the day, to minimize the cost of running the chiller.

CFCs. Modern air-conditioning and refrigeration equipment uses chlorofluorcarbon (CFC) compounds as refrigerants; as a class these compounds are non-flammable, low in toxicity, and very stable. Fully halogenated CFCs have all hydrogen atoms replaced with chlorine or fluorine atoms. Examples of fully halogenated CFCs (42) include R-11, trichlorofluormethane (CCl_3F), and R-12, dichlorodifluoromethane (CCl_2F_2), as well as those derived from hexachloroethane (C_2Cl_6): R-113 (CCl_2FCClF_2), R-114 ($CClF_2CClF_2$), and R-115 ($CClF_2CF_3$). Examples of CFCs that are not fully halogenated include R-21, dichlorofluoromethane ($CHCl_2F$), and R-22, chlorodifluoromethane ($CHClF_2$). In many cases, CFCs that are used as refrigerants eventually leak or are discharged to the atmosphere. Those that are fully halogenated are more stable and a higher percentage of chlorine atoms reach the stratosphere and deplete, via catalytic reaction, the concentration of ozone molecules that screens the Earth's surface from ultraviolet radiation from the Sun. The relative ozone-depletion potential for the fully halogenated compounds varies inversely with the order listed above and ranges from 1.0 for R-11 to 0.32 for R-115. By contrast, R-22 has an ozone depletion potential of 0.05 (43).

R-11 and R-12 are used in refrigeration equipment and in foam insulation materials. Both are used in centrifugal chillers in commercial buildings, while R-11 is used in refrigerators, freezers, automobile air conditioners, water coolers and dehumidifiers. R-114 is used in shipboard chillers. The non-fully halogenated refrigerant R-22 is used in larger centrifugal compressors and in reciprocating and screw compressors. R-22 dominated 1985 refrigerant purchases by companies that are members of the American Refrigeration Institute, with 77% of the total, followed by R-11 with 11% and R-12 with 10% (43). The fully halogenated CFCs are being phased out as part of an international agreement known as the Montreal Protocol.

Replacement refrigerants may be near drop-in substitutes (R-123 or R-141b for R-11 and R-134a for R-12) or may require substantial equipment modification to accommodate higher operating pressures and temperatures (R-22 for R-12). In the future, R-22 and all other CFCs containing any chlorine may be banned, leading to a need for entirely new refrigerants or such alternatives as ammonia or propane. National energy impacts of changes in the use of fully halogenated refrigerants range from a penalty of about 3% of national energy consumption to a gain of about 1%, as a depending on refrigerant substitutes and advanced technologies for insulations and refrigeration cycles (44). In all cases, the dominant component of the overall energy impact is refrigerators and freezers, not centrifugal compressors.

Improved Control Strategies

Building systems rarely if ever operate at optimal efficiency. Mathematical techniques for optimization are well known but can only be applied when these criteria are met: there is a model for the dynamic system being controlled, which in this case is the thermal environment within the building, as affected by the thermal dynamics of the building structure and the space conditioning system; the model captures all of the dynamics that affect the variable to be minimized, which might be cost to maintain a given environment; and the parameters of the model can be accurately determined via prior knowledge or suitable experiments. An accurate model makes the most demands on parameter estimation and most models have therefore considered a limited number of variables, in some cases at time intervals as long as an hour. Improved computational speeds and more use of distributed intelligence in building control systems have made it possible to gather and process more information, encouraging recent efforts to perform on-line optimization calculations (45).

Optimization of equipment operation within an individual building is more reasonably based on cost rather than energy consumption, because the cost of electricity often varies with time of day. Thermal storage systems may be financially attractive to building owners and customers for this reason. More efficient base-load power plants are one factor that may contribute to lower electricity rates at night: during the day, these plants may be augmented with less-efficient plants to serve peak loads. In this case, thermal storage may lead to less fuel consumption at the power plant than would be the case with day-time operation of chillers even though, at the building, the chiller used to make ice at the building may operate at lower efficiency than conventional equipment due to lower evaporator temperatures.

Electricity rates that show a simple step change between on-peak and off-peak periods do not capture hourly fluctuations in the marginal costs that a utility incurs in generating and distributing electricity to its customers. Several electric utilities have computed and transmitted hourly prices to customers capable or responding to them, and tests have shown that careful choice of the least costly hours to charge a thermal storage system can nearly double the benefit of storage in reducing utility costs, relative to no storage (46).

Optimization serves as one basis for fault detection, where a fault is defined as monitored equipment performance below the calculated optimum. Faults can also be detected by identifying departures from historic data and abnormal hours of equipment operation. The same trend toward electronic control of building systems that is enhancing optimization also promotes on-line fault detection.

LIGHTING

Light Sources

Lighting systems can be designed on the basis of illuminance levels, luminance ratios, and direct attention to biological needs, or some combinations of all three. In any case, lighting starts with lamps and fixtures. The principal types of lamps can be classified as incandescent, in which a filament is heated to the point where it emits light in the visible as well as infrared spectra and discharge lamps, in which gas under low or high pressure emits light in response to electrical excitation. Low-pressure discharge lamps include those based on sodium vapor, which can also operate at high pressure, and fluorescent lamps. High-pressure discharge lamps, which offer a concentrated source of light from a small discharge tube (14), include sodium, mercury and metal halide lamps, in which halide salts improve the color rendition of ordinary mercury lamps. Metal halide lamps, long used outdoors in such applications as sports fields and parking lots, are also available for circulation spaces and indirect lighting in office areas, where the lamp points toward the ceiling. Discharge lamps require a device to ignite the arc and a device to limit the current of the arc; high-pressure discharge lamps need an outer glass tube to protect the very hot discharge tube.

Lamps vary in their efficiency, as already noted, and in color rendition. The most efficient low-pressure sodium lamps are more often used in outdoor applications because the visible output is dominated by the yellow part of the spectrum. Fluorescent lamps come with a wide variety of visible spectra, some more toward the blue and others the red. Color output can be controlled to some extent by varying the mix of phosphors that coat the inside of the fluorescent tube and are excited by colliding ultraviolet photons from the mercury vapor produced by the electric arc between the two ends of the tube.

The efficiency at which light is provided to work spaces or circulation depends on the light fixture as well as the lamp itself. The amount of light delivered per Watt of electrical power will be higher for fixtures that direct light downward rather than toward the ceiling. Direct light has a higher potential for glare than indirect and may not be acceptable in rooms with computer displays unless the light fixture includes a carefully designed diffuser.

Small light sources are more easily controlled to distribute light where needed than large ones. Incandescent spot lights are often used in retail stores to spotlight merchandise. Task lighting in the home and office is directed to illuminate small areas. Compact fluorescent lamps adapted to incandescent light electrical sockets bring increased efficiency to local lighting. Fixture manufacturers have recognized the benefits of small fluorescent lamps and have increased the variety of fixtures that are sized to accept such lamps, which are larger than most incandescent lamps.

In addition to size, color and efficiency, lamps can be classified on the basis of luminous output, luminance of the lamp surface, and lifetime. Incandescent lamps, because they are smaller than most discharge lamps, will have higher luminance for a given luminous flux. They also have lifetimes less than 10% that of discharge lamps. Lamps, regardless of type, deteriorate over time and the luminous output decreases; as a result, illuminance in a room when lamps are new should exceed the minimum acceptable value, which will be attained just prior to relamping.

Office Lighting

Lighting for open-plan offices usually involves a regular array of ceiling-mounted lamps, designed to provide a uniform illumination at desk height. The number and spacing of a given lamp fixture can be determined on the basis of room geometry and surface reflectances, for standard lamp designs and configurations. A uniform illumination is most immune to changes in office layout and necessarily the most wasteful, with light provided to unused or infrequently used areas. However, office furnishings can sufficiently reduce desk-top illumination to the point where task lighting is required. In this case, reduced general-area lighting combined with user-controlled task lighting offers a system that is potentially very efficient and that gives office workers the opportunity to vary the lighting in their individual work environment. In office areas near a source of natural light, photosensors that dim or switch the general-area lighting can achieve further energy savings. The control must be sufficiently smooth to maintain occupant visual comfort, which favors the relatively new and expensive dimmable ballasts over switchable ballasts for fluorescent lamps. Case studies of newly constructed or renovated buildings have shown that lighting power of about 15–20 W/m^2, as specified by standards, can be reduced to about 5-7 W/m^2 by use of daylighting sensors, reduced general-area lighting and task lights (47).

Electric lighting design, especially for simple systems with uniform levels of light, can be done with procedures based on manuals that describe a room as three optical cavities: from the ceiling to the height of the lamps, from the lamps to the work plane, and from the work plane to the floor. The size and surface properties of each cavity alone with the amount and distribution of luminance from the lamps affects the illuminance at the work plane (16). More complex designs, including those involving task lighting and complex wall and ceiling surfaces, may be designed with computerized methods or mock-ups.

PERFORMANCE CRITERIA

Performance criteria include such factors as lifetime, noise, required service and pollutant emission as well as peak-power demand and annual energy consumption. Energy performance of building systems as well as individual pieces of equipment is subject to labeling as well as regulation. Labels with energy consumption and operating costs (common on many residential appliances), manufacturers' specification sheets, and information-dissemination campaigns conducted by government agencies, environmental groups and energy utilities give consumers opportunity to make rational purchase decisions. Regulation is intended to remove from the market the most inefficient equipment, protecting bill-payers from purchase decisions made in ignorance or by those not paying for operating costs. Many regulations originate in non-binding, building-energy standards (32) that are incorporated into state building codes and therefore made into enforceable limits. The standards are formed in a consensus-building process that involves manufacturers as well as environmental advocates. For example, the standard for energy-efficient design of new buildings except small houses published by the American Society of Heating, Refrigerating and Air-Conditioning Engineers and the Illuminating Engineering Society of North America specifies a number of performance criteria, some of which are summarized in Table 2.

The criteria prescribe minimal acceptable performance. The standard also includes a path to compliance based on performance; for example, a building envelope will meet the standard if the *overall* thermal conductivity does not exceed a specified limit, even if some individual elements do not meet the prescriptive criteria. Further, the standard includes a third path to compliance, the energy-cost budget approach, which provides designers the most flexibility but requires that a design energy cost be computed and compared with a target. Such an approach allows for extensive use of glazing, for example, if it can be shown that the solar heat gain and daylighting benefits offset the increased thermal conductivity.

The National Energy Policy Act adopted in the U.S. in 1992 mandates that states review commercial building codes and certify that energy performance criteria are at least as stringent as those included in the ASHRAE standard. Similarly, state-level residential codes must meet or exceed the model energy code published by the Council of American Building Officials (48). The Energy Policy Act also requires energy-efficiency information programs for lights, office equipment and windows, mandates minimum efficiency criteria for heating and cooling equipment manufactured after January 1, 1994, and specifies minimum efficiencies for electric motors.

Table 2. Performance Criteria for Energy-Efficient Building Design

Component	Performance Criteria
Lighting	Electrical power per floor area
Building envelope	Thermal conductivity of individual elements
Systems	Limited oversizing
	Zone controls to limit simultaneous heating and cooling
	Economizer controls to reduce cooling costs in mild weather
	Fan power per unit flow rate
	Frictional losses in pumping systems
Equipment	Efficiencies of air conditioners, chillers, heat pumps, boilers and furnaces

BIBLIOGRAPHY

1. P. E. McNall, "Indoor Air Quality Status Report," in *CLIMA 2000,* VVS Kongres-VVS Messe, Copenhagen, 1985, Vol. 1, pp. 13–26.
2. American Society of Heating, Refrigerating and Air-Conditioning Engineers (ASHRAE), *Handbook of Fundamentals,* ASHRAE, Atlanta, 1989, pp. 8.1–8.32.
3. V. Olgyay and A. Olgyay, *Design with Climate,* Princeton University Press, Princeton, New Jersey, 1963, pp. 16.
4. P. O. Fanger, *Thermal Comfort: Analysis and Applications in Environmental Engineering,* McGraw-Hill, New York, pp. 19–64.
5. B. Givoni, *Man, Climate & Architecture,* Von Nostrand Reinhold, New York, 1981, pp. 30–59.
6. *Thermal Environmental Conditions for Human Occupancy: ANSI/ASHRAE Standard 55-1981,* American Society of Heating, Refrigerating and Air-Conditioning Engineers, Atlanta, 1981.
7. J. E. Seem and J. E. Braun, "The Impact of Personal Environmental Control on Building Energy Use," in *ASHRAE Transactions,* American Society of Heating, Refrigerating and Air-Conditioning Engineers, Atlanta, Vol. 98, Part 1, 1992.
8. F. S. Baumann, R. S. Helms, D. Faulkner, E. A. Arens, and W. J. Fisk, *ASHRAE Journal,* **35**(3), (1993).
9. *Radon Reduction Methods, a Homeowner's Guide,* U.S. Environmental Protection Agency, Washington, D.C. 1986.
10. *Ventilation for Acceptable Indoor Air Quality: ASHRAE Standard 62-1989,* American Society of Heating, Refrigerating and Air-Conditioning Engineers, Atlanta, 1989.
11. National Institute of Occupational Safety and Health, "Criteria for Recommended Standards," U.S. Department of Health and Human Services, various dates.
12. U.S. Environmental Protection Agency, *National Primary and Secondary Ambient Air Quality Standards,* Code of Federal Regulations, Title 40, Part 50.
13. J. P. Murdoch, *Illuminating Engineering from Edison to the Laser,* MacMillan, New York, 1985.
14. J. B. deBoer and D. Fischer. *Interior Lighting.* Philips Technical Library, Kluwer, Antwerp. 1981, p. 19.
15. M. S. Rea, *Journal of the Illuminating Engineering Society,* **15,** (2) 41–57 (1986).
16. *IES Lighting Handbook Reference Volume,* Illuminating Engineering Society of North America, New York, 1981.
17. W. M. C. Lam, *Perception and Lighting as Formgivers for Architecture,* Von Nostrand Reinhold, New York, 1992.
18. F. C. McQuiston and J. D. Parker, *Heating, Ventilating, and Air Conditioning Analysis and Design,* John Wiley & Sons, Inc., New York, 1988.
19. W-H Li and S-H Lam, *Principles of Fluid Mechanics,* Addison-Wesley, Reading, Mass. 1976.
20. E. Simiu and R. H. Scanlan, *Wind Effects on Structures: An Introduction to Wind Engineering,* John Wiley & Sons, Inc., New York, 1986.
21. D. Watson and K. Labs, *Climatic Building Design,* McGraw-Hill, New York, 1983.
22. G. Z. Brown and co-workers, *Insideout: Design Procedures for Passive Environmental Technologies,* John Wiley & Sons, Inc., New York, 1992.
23. N. Lechner, *Heating, Cooling, Lighting Design Methods for Architects,* Wiley Interscience, New York, 1991.
24. F. Moore, *Environmental Control Systems Heating Cooling Lighting,* McGraw-Hill, New York, 1993.
25. "Enerplex: Office Complex Exploring Sophisticated Energy Solutions," *Architectural Record,* **170**(6) (May).
26. *Sylvania GTE Light Sources Product Information/Engineering Bulletins,* Sylvania Lighting Center, Danvers, Mass.
27. W. M. C. Lam, *Sunlighting as Formgiver for Architecture,* Von Nostrand Reinhold, New York, 1986.
28. *Structure, Space, and Skin: The Works of Nicholas Grimshaw & Partners,* Phaidon Press, London, 1993.
29. *Skidmore, Owings & Merrill: Architecture and Urbanism 1973–1983,* Van Nostrand Reinhold, New York, 1983, p. 382.
30. A. Blomsterberg, *Ventilation and Airtightness in Low-Rise Residential Buildings: Analysis and Full-Scale Measurements,* Swedish Council for Building Research D10:1990.
31. M. F. Fels, "PRISM: an Introduction," in *Energy and Buildings,* Elsevier Sequoia S. A., Lausanne, Switzerland, Volume 9, Numbers 1&2, 1986.
32. American Society of Heating, Refrigerating and Air-Conditioning Engineers (ASHRAE), *ASHRAE Standard 90.1-1989,* "Energy Efficient Design of New Buildings Except New Low-Rise Residential Buildings," Atlanta, 1989.
33. A. K. Persily and W. S. Dols, "The Relation of CO_2 Concentration to Office Building Ventilation."
34. M. Sandberg and C. Blonquist, *ASHRAE Transactions,* **95,** Pt. 2 (1989).
35. A. K. Melikov, G. Langkilde, and B. Derbiszewski, *ASHRAE Transactions,* **96,** Pt. 1 (1990).
36. American Society of Heating, Refrigerating and Air-Conditioning Engineers (ASHRAE), *Handbook of HVAC Systems and Equipment,* Atlanta, 1989, pp. 1.1–4.4.
37. D. Lorenzetti, and L. Norford, *ASHRAE Transactions,* **98,** Pt. 2 (1992).
38. D. M. Lorenzetti and L. K. Norford, "Pressure Reset Control of Variable Air Volume Ventilation Systems," *Proceedings of the ASME International Solar Energy Conference, Washington, D.C.,* 1993.
39. *Air-Conditioning Systems Design Manual,* American Society of Heating, Refrigerating and Air-Conditioning Engineers, Atlanta, 1993.
40. J. B. Fenn, *Engines, Energy, and Entropy,* W. H. Freeman, New York, 1982.
41. B. H. Jennings, *Environmental Engineering Analysis and Practice,* International Textbook Co., New York, 1970.
42. R. Downing, "Development of Chlorofluorocarbon Refrigerants," in *CFCs: Time of Transition,* American Society of Heating, Refrigerating and Air-Conditioning Engineers, Atlanta, 1989.
43. R. J. Denny, 1989. "The CFC Footprint," in *CFCs: Time of Transition,* American Society of Heating, Refrigerating and Air-Conditioning Engineers, Atlanta, 1989.
44. S. K. Fischer and F. A. Creswick, 1989. "How Will CFC Bans Affect Energy Use?" in *CFCs: Time of Transition,* American Society of Heating, Refrigerating and Air-Conditioning Engineers, Atlanta.
45. Z. Cumali, *ASHRAE Transactions,* **94,** Pt. 1, (1988).
46. B. Daryanian, R. E. Bohn, and R. D. Tabors, *IEEE Transactions on Power Systems* **6**(4), 1356–1365 (1991).
47. R. K. Watson and co-workers, "Office Building Retrofit to Produce Energy Savings of 75%," in *Proceedings of the ACEEE 1992 Summer Study on Energy Efficiency in Buildings,* American Council for an Energy-Efficient Economy, Washington, D.C., Vol. 1, p. 251.
48. Energy Policy Act of 1992, *Public Law 102-486-October 24, 1992.*

C

CARBON BALANCE MODELING

RICHARD H. WARING
Oregon State University
Corvallis, Oregon
MICHAEL G. RYAN
U.S.D.A. Forest Service
Fort Collins, Colorado

Carbon is a constituent of all terrestrial life; the energy locked in its chemical bonds in living and dead plant material provides sustenance for animals and microbes. Plants extract energy from sunlight to combine carbon dioxide (CO_2) from the atmosphere and water to form carbohydrates, which serve as a source of energy and the substrate from which leaves, wood, and roots are built (see Fig. 1). Dead plant material is consumed by soil arthropods and microbes, and in the process releases nutrients. Most of the carbon is quickly returned as CO_2 to the atmosphere; however, a small fraction of the organic debris is converted by microbes into complex organic molecules that remain in the soil for many years.

Modeling the carbon balance requires the development of mathematical expressions to describe the rate that carbon flows through the cycle (from the atmosphere into biota and the soil, and out again). There is a plethora of models that describe some or all of the carbon cycle. Most are designed to answer specific management or research questions; a few are more general. On a small scale, models may estimate the effects of management on crop, range, or forest yields (2). At larger scales, models forecast commodity production, examine interactions between biota and the atmosphere, and predict the effects of global change.

This review of carbon balance modeling concentrates on the principles underlying the equations in quantitative treatments and processes described by those equations, in the context of global change. The mathematical expressions, which are not presented here, vary, depending on the purpose and form of particular models, and can be obtained from the literature cited. In addition, the focus is on forests rather than crops or grasslands, as forests represent the most complicated types of vegetation and store more than 70% of the total terrestrial carbon in their biomass and soil organic matter. Forests demonstrate the principles by which most vegetation captures and exchanges carbon with the atmosphere.

See also AIR QUALITY MODELING; CARBON CYCLE; CARBON STORAGE IN FORESTS; FOREST RESOURCES; GLOBAL CLIMATE CHANGE, MITIGATION.

CARBON DIOXIDE AND THE CARBON CYCLE

The sum total of carbon on the Earth and in its atmosphere is constant. What varies is the fraction present in the atmosphere, in terrestrial vegetation and soils, in the oceans, and locked in sediments of limestone and dolomite or in fossil fuels (see Fig. 2). Carbon moves between the atmosphere and the biota in the form of carbon dioxide, a gas that in the atmosphere is transparent to short-wave radiation from the sun but traps long-wave radiation from the relatively cool surface of the Earth. As a result, atmospheric CO_2 has the potential, along with other gases, to cause atmospheric heating and affect weather patterns globally, ie, the "greenhouse effect."

As industrial societies have recaptured and utilized the energy in fossil fuels during the last 150 years, the atmospheric pool of carbon has increased by more than 160 billion tons. In the last 30 years, the increase in CO_2 has exceeded the rate of increase at any time previously. Deforestation has played a part in increasing the atmospheric CO_2 level, but the main contributor is combustion of fossil fuel. Because the amount of carbon released in the combustion of fossil fuel is now about 5.4×10^9 annually, whereas the increase in atmospheric content is only about 3.4×10^9 annually (Fig. 2), there is an imbalance in the carbon source–sink relationships between the Earth and atmosphere. The difference between carbon released in fossil-fuel consumption and the increase in atmospheric carbon content indicates that either the oceans or the land, or both, must be increasing their efficiency in capturing CO_2.

At the scale of a forest or agricultural field, increased carbon dioxide concentrations can affect growth and decay processes, and consequently the rate of CO_2 accumulation in the atmosphere. To understand any given response, the processes involved must be analyzed. Although understanding of many key processes is still limited, as well as the ability to make accurate measurements of some variables, considerable progress can be made if the carbon budget for each system can be closed, ie, accounting for the fluxes in and out of the system, and hence the net gains, losses, and changes in carbon storage.

TYPES OF MECHANISTIC MODELS

An array of models has been developed to estimate components of the forest carbon balance. In a recent review (4), models have been classified into four types: physiological, ecosystem, plant succession, and regional or global. Physiological models simulate carbon, water, and occasionally nutrient transfers (fluxes) for single plants or small areas of perhaps a hectare (10^4 m^2) with a homogeneous cover of a single species. The time-step applied in physiological models generally ranges from an hour to a year. These models usually describe leaf-level processes, such as photosynthesis, in great detail because they are well-understood. Less understood processes, such as carbon allocation within plants, are usually modelled in much less detail (5–7).

Ecosystem models determine the same fluxes for vegetation as physiological models but generalize responses for groups of species with similar characteristics. These models may apply to areas of several hectares and resolve time at daily, monthly, and yearly intervals for periods

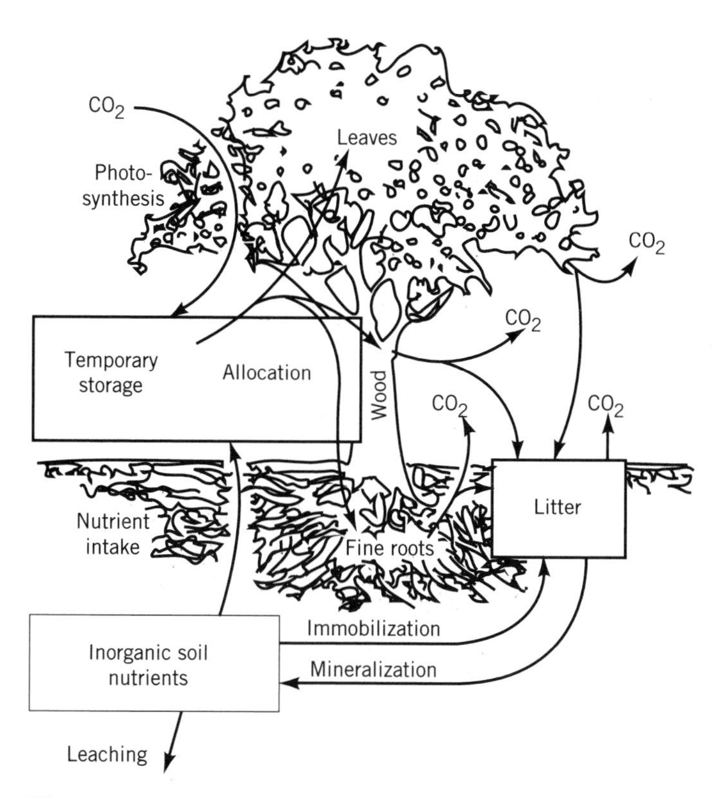

Figure 1. The carbon balance of an ecosystem represents a set of counterbalancing processes (1).

up to several decades. Most ecosystem models assume that the vegetative cover does not significantly change over the selected time period. They simplify the leaf-level detail of many physiological models and tend to focus on larger-scale processes such as carbon storage and decomposition in the soil and the interactions of nitrogen and other essential nutrient with the carbon cycle (8,9).

Modeling Compositional Changes in Vegetation Over Time

Most carbon balance models that predict exchanges between the land and the atmosphere do not incorporate changes in the composition of vegetation. Yet the composition of vegetation will change as those species that may have established on disturbed areas grow and create environments favorable for other species adapted to more shaded conditions. Disturbances, which include insect and disease outbreaks, fire, wind damage, and logging, may occur at different scales. They may kill individual plants, selectively exclude certain species, or, in the case of fire, consume all aboveground biomass. The consequences of such disturbances may be modelled and predicted in stochiastic terms (ie, as probabilities), as a function of climatic conditions, the presence of native and introduced pests and pathogens, and soil conditions associated with fertility, aeration, and slope stability.

A series of "plant succession" and forest "gap" models predict compositional changes and the growth and death of species for a wide range of vegetation. After disturbances, these succession models use simple equations to estimate tree growth and probability functions to predict when trees die and new seedlings take their place (see Fig. 3). Because of the probability functions, model predictions vary widely from simulation to simulation. Mean estimates must be derived, therefore, from the average of many simulations. Examples of plant succession models have been described (10,11). Later versions of these models incorporate nutrient cycling (12,13) and considerable physiological detail (14).

Predictions from plant succession models have favorably compared with the record determined from pollen cores for patterns of vegetation change that followed the last full-glacial period in the eastern United States (10). With the rapid changes that are projected in regional climates and land use, however, rates of disturbances are

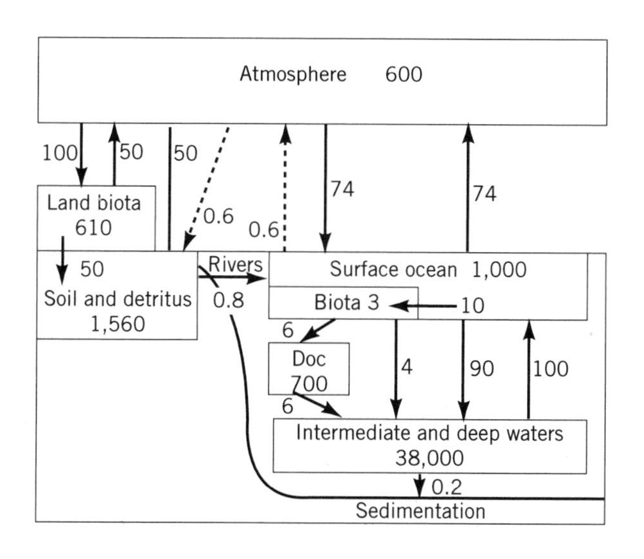

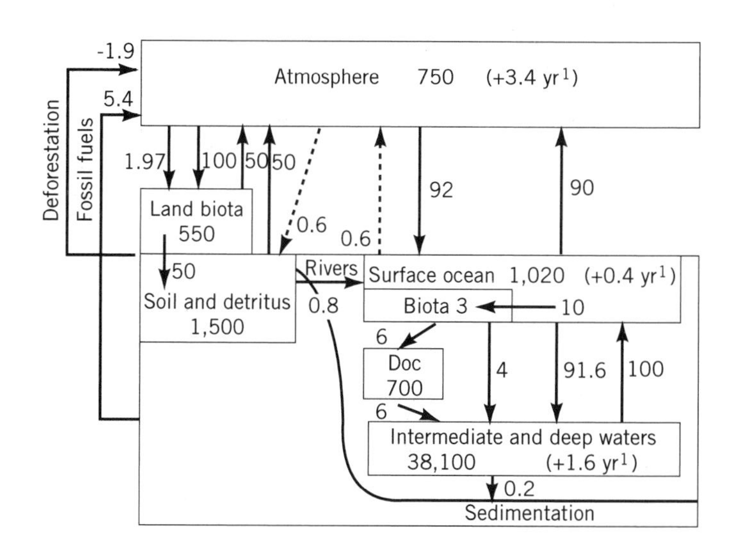

Figure 2. The global carbon cycle reservoirs and fluxes are contrasted between (**a**) the pre-industrial carbon cycle and (**b**) the carbon cycle averaged for 1980–1989. Differences caused by human activities are indicated in bold numbers in (**b**). Arrows indicate estimated annual fluxes in or out of various reservoirs; units are in billions of tons (Gt) (3).

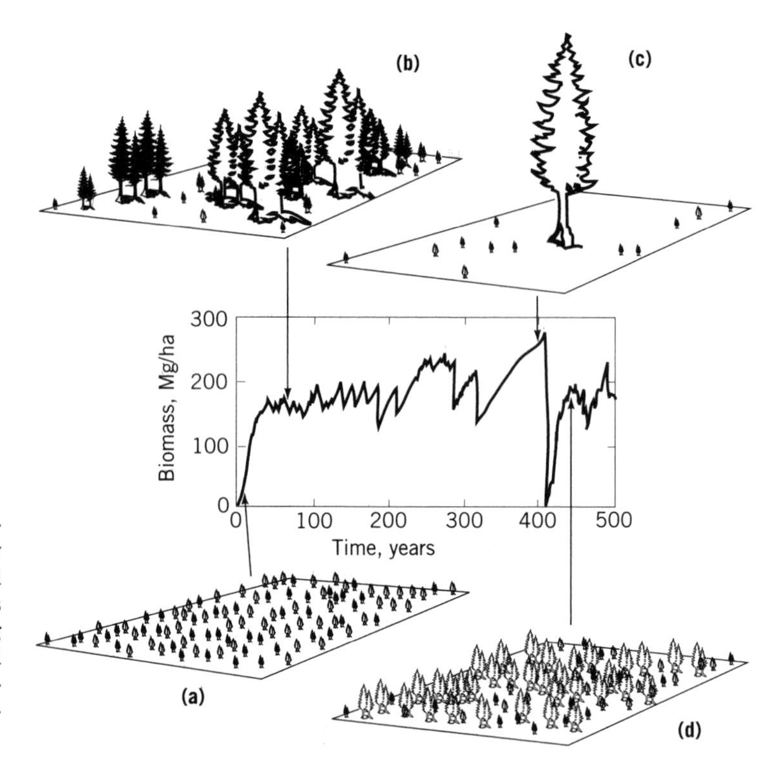

Figure 3. The changing structure of a forest can be modelled to take into account natural succession, probability of disturbance from fire, insects, disease, windstorms, and harvesting activities by humans. In this diagram, changes are depicted for a cycle that starts with (**a**), regenerating forest with many small trees, progresses to (**b**), a mixed forests with fewer trees, (**c**), harvesting of all but one dominant tree, and (**d**) regeneration of a mixture of species following death of the dominant tree (10).

likely to increase; some components of regional flora may begin to migrate and others may become extinct. These types of changes are difficult to model, even with greater knowledge of the behavior of individual species and populations. Simplification of succession models, not greater complexity, is needed to extend the models to regional or global scales.

Several approaches to scaling have proven useful. The simplest approach to estimating the effects of climatic change on primary production is to extrapolate correlations obtained between current climate and production rates associated with various types of vegetation (15,16). More sophisticated approaches combine ecosystem models with succession models (13,17); at the global scale, the precise distribution of the main vegetation types becomes critical. Some imaginative approaches have been outlined but await implementation and critical testing (18,19).

COMPONENTS OF TERRESTRIAL CARBON MODELS

Most terrestrial carbon models contain terms that represent reservoirs or pools where carbon accumulates, and include a set of equations that describe the rates at which carbon is transferred from one structural component to another. The next section introduces the components of a typical ecosystem level model.

Structural Components

Carbon balance models of terrestrial ecosystems that calculate fluxes of CO_2 to the atmosphere contain at least three structural components: living plant material, dead plant material, and carbon stored in the soil. Because humans remove a significant fraction of carbon from forests by making wood products that last for several to many decades, some models track the type and amount of materials harvested each year and the rates at which various products breakdown and are recycled into the atmosphere (20). Although a simple balance sheet may be constructed with only three structural components, reasonable predictions of the rates at which carbon cycles through a system demand a more detailed knowledge of structure. Most ecosystem models recognize leaves and woody material in stems and roots (as in Fig. 1); as these structures die, the carbon in them is released to the atmosphere or incorporated into soils at different rates. Many ecosystem models, therefore, recognize a number of separate soil reservoirs (4,8).

ECOSYSTEM PROCESSES

In undisturbed ecosystems, the rates of carbon accumulation and loss over long periods are mostly in balance. Through photosynthesis, CO_2 is transformed into simple carbohydrates that provide energy and building blocks for the construction of all structural elements. Respiration metabolically degrades large organic molecules into CO_2, which is released back to the atmosphere. The basic functioning of ecosystems, however, depends not only on the cycling of carbon, but also on water and nutrients. Currently, there are just a handful of ecosystem models that combine carbon, water, and nutrients cycles (4,5,8,21,22).

The following section outlines the primary processes involved in plant community and ecosystem carbon balances, and discusses the state and stage of development of the models being used to simulate those processes.

Photosynthesis

Photosynthesis is the process by which green plants convert CO_2 into carbohydrates, using the energy of sunlight. About 50% of the sun's radiant radiation is in the visible part of the spectrum (400-700 nm), which is the range that provides the energy for photosynthesis. The intensity and duration of sunlight varies with climate, season, and latitude, and can be modelled, but is more accurately measured directly. In addition to sunlight, chlorophyll and associated photosynthetic enzymes in leaves require nitrogen for their synthesis, and limitations in the availability of nitrogen and certain other elements reduce the capacity of leaves to photosynthesize. The strong correlation between photosynthesis and leaf nitrogen content illustrates how carbon and nutrient cycles are linked.

The light energy absorbed by chlorophyll and other leaf pigments is used by enzymes embedded in the chloroplasts of leaf tissue to split water into hydrogen and oxygen, and to combine CO_2 into simple sugars. The rate of CO_2 diffusion into leaves is restricted by pores, called stomata, that open and close, depending on environmental conditions. Stomata are especially sensitive to air humidity and dry soil, but also respond to light, ambient CO_2 concentrations, and temperature. When the rate of water loss by diffusion through stomata is greater than the rate of the movement through the plant into the leaves, the leaf water content falls and stomata close. The wet mesophyll cells behind stomatal pores are essential to trap CO_2, but necessarily involve the loss of the water, a process called transpiration.

Transpiration is a physical process driven by the energy balance of leaves. Radiant energy falling on the leaves must be disposed of as sensible heat, driven by temperature gradients, or latent heat, ie, the energy required to evaporate water. If the rate of evaporation is too slow, leaf temperatures rise and the humidity deficit at the leaf surfaces increases; stomata then tend to close. The opening and closing of stomata represent a compromise between limiting the loss of water while maximizing the rate that CO_2 enters the leaf for photosynthesis. As a result, stomatal opening is closely related to net photosynthesis.

The horizontal distribution of leaves in plant canopies affects the efficiency with which light is intercepted. These effects are recognized in most models, some of which use complex procedures to account for differences in leaf arrangement (23). Large, flat leaves that form umbrella-like canopies around individual trees capture light more effectively than small clumps of dispersed needles. In general, the amount of light intercepted by plant communities is calculated in terms of amounts intercepted by cumulative layers of leaves. Forests composed of broadleaved trees intercept nearly all solar radiation with the equivalent of about six layers of leaves projected onto an equal area of ground surface (referred to as leaf area index). In coniferous forests, made up of trees with narrow crowns, light penetrates more easily through the canopy. For this reason, coniferous forests may display up to 12 layers of projected leaf area (24).

Transpiration and Evaporation

Because the movement of water through ecosystems affects the carbon cycle in many ways, water and carbon balance models share many features, as illustrated in Figure 4. Canopy leaf area is critical in both cases; in addition to absorbing solar energy for photosynthesis, leaves intercept precipitation and evaporate water from their surface or, as noted earlier, through their stomata in the process of transpiration. Radiant energy can raise leaf temperatures and increase the humidity deficit of the air at leaf surfaces.

The size of vegetation affects the transfer of heat, water vapor, and CO_2. The foliage in tall vegetation is usually near ambient air temperatures, because air flow is turbulent around such vegetation and transfer processes are more effective. Shorter vegetation, particularly that with broad leaves, can be substantially heated above air temperature if wind speeds are low and transpiration is constrained, preventing evaporative cooling (25). This interdependence of carbon and water fluxes requires, in most cases, that additional environmental variables be measured or modelled.

Consequently, one of the variables required for water balance models is total radiant energy; clearly, another is precipitation. Water that is not evaporated from the surfaces of vegetation enters the litter or soil, or in some cases runs off the surface. The water storage capacity of the rooting zone must be known in order to model the water balance of plant communities. When between two-thirds to three-quarters of the available water has been extracted from the soil, the imbalance between the rate of loss from leaves and the rate of uptake by the generally sparse network of roots in the deeper part of the soil profile causes water stress (drought) that may induce partial or complete stomatal closure. Extreme or prolonged drought may also lead to the shedding of leaves. Given values of the appropriate variables, ie, leaf area index, atmospheric factors, and a stomatal response function, models can provide good estimates of changes in root zone water storage (see Fig. 5).

Respiration by Plants

Green plants expend metabolic energy and respire CO_2 to the atmosphere during photosynthesis (photorespiration), in producing new tissue (construction respiration), and in maintaining live tissue (maintenance respiration). Photorespiration only occurs when enough light exists for photosynthesis; maintenance and construction respiration can occur in the light and dark. All three types of respiration respond positively to increasing temperature for different reasons.

Photorespiration is a product of the same enzyme that converts CO_2 into simple sugars. This enzyme will bind with O_2 as well as with CO_2; when it does, there is a breakdown of organic molecules and production of CO_2. As temperature increases, the CO_2 concentration within leaves normally decreases, because photosynthesis continues while stomata begin to close and restrict CO_2 diffusion into the leaf.

Currently, few ecosystem models treat photorespiration explicitly. With projected increases in atmospheric CO_2 and temperature, however, the process of photorespiration becomes more significant: the compensation point where photosynthesis and respiration are in balance should be at a lower light level than today's environment permits (26). Photorespiration does not occur in some plants, such as corn and other related tropical grasses, so CO_2 rise may affect competitive interactions between species regardless of a temperature change.

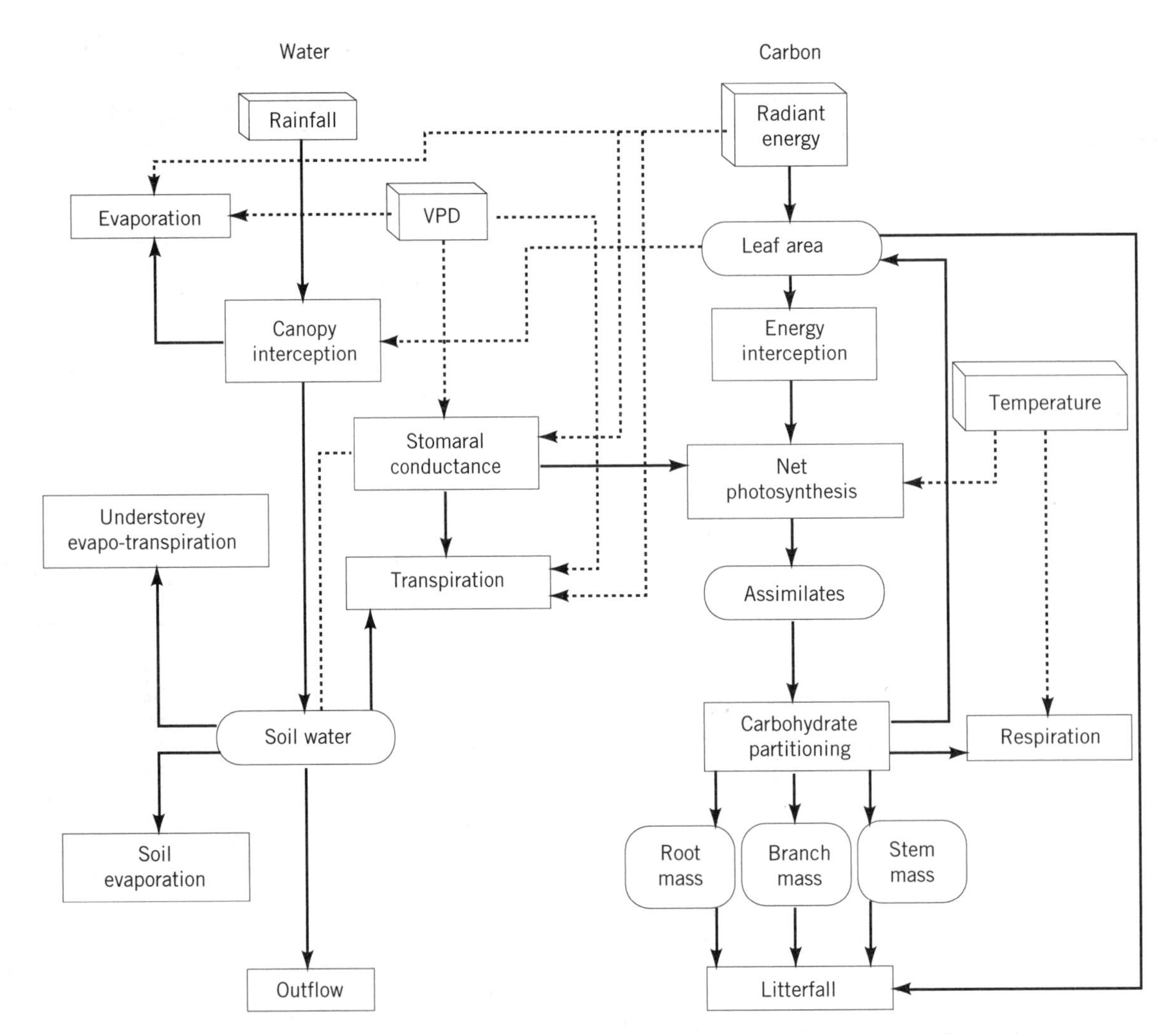

Figure 4. Coupled plant carbon and hydrologic models share many features. Transpiration and photosynthesis are both limited when stomatal conductance and leaf area are reduced. Increasing temperature potentially allows greater evaporation and transpiration, and also affects photosynthesis and respiration. As soils dry, carbon allocated to root growth usually increases until water stores in the soil are fully exhausted. Reprinted with permission of R. E. McMurtrie, University of New South Wales, Australia.

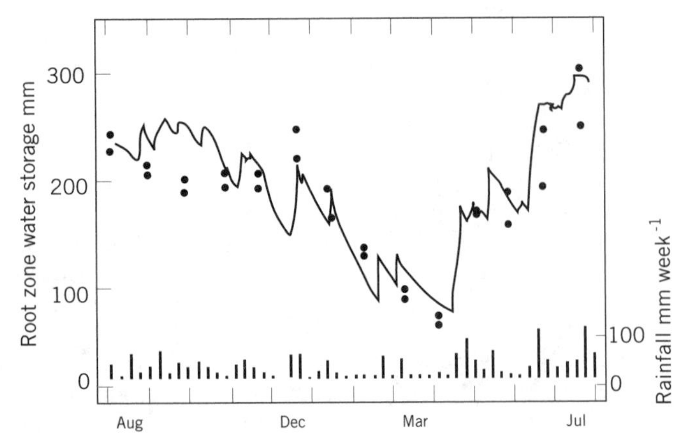

Figure 5. A process-based hydrologic model, which integrates leaf area and controls on stomatal conductance, can closely match predicted changes in root zone water storage with those observed (●). Water storage was measured in the top 3.25 m of the root zone of a 9-year-old plantation of pine growing in Australia (5).

Construction of plant tissue is metabolically expensive. During the production of organic compounds and their assembly into structure, construction respiration costs are incurred, equivalent to about 15–50% of the carbon in the material produced. Carbon incorporated into organic tissues comprises about 50% of the dry weight of plants. If growth increases with temperature, respiration associated with construction cost must therefore increase proportionally. The growth of aboveground plant material is relatively easy to measure but root production is exceedingly difficult, particularly the growth of small-diameter roots that are produced and die within a year or less. Increase in knowledge about root growth will occur upon the development of reliable models of photosynthesis, aboveground growth, and respiration; the carbon not accounted for will then be allocated belowground.

Maintenance respiration occurs in all living cells in the process of restoring enzymes, maintaining gradients of ions, and converting inorganic nutrients into biochemical compounds. Leaves are usually the most active respiratory organs, with the highest concentration of enzymatic machinery and living cells. Small-diameter roots are also active in acquiring inorganic nutrients and in converting these into biochemical compounds. In ecosystems with tall vegetation, the stems and large-diameter roots also significantly contribute to the total maintenance respiration. In general, maintenance respiration can be modelled as a linear function of the increasing nitrogen content in the living cells of plants and as an exponential function of increasing temperature (27). Because annual net photosynthesis also tends to increase with rising temperatures and an extended growing season, the relative cost of maintenance respiration as a fraction of net annual CO_2 uptake increases linearly, not exponentially, as shown in Figure 6. Although the maintenance respiration required to support the vascular system is small compared with leaves, it increases substantially with increasing temperature.

Microbial Respiration

Most of the organic matter produced by plants, whether eaten by animals or directly transferred into the litter and soil, is eventually consumed by microbes. Microbial respiration produces CO_2 while releasing nutrients from the organic matter. Decomposition of organic material depends on the local environment, where microbial activity occurs. Thus, the decomposition rates differ substantially between dead standing trees, leaf litter, and soil.

Drying and freezing conditions essentially halt microbial activity everywhere. In soils, however, conditions are often sufficiently favorable, even under snow cover and during extended drought, to allow some decomposition to proceed at reduced rates. Flooding reduces the amount of O_2 available and decreases the rate of decomposition, but it can stimulate anaerobic production of methane (CH_4) by specialized microbes. Methane production has been estimated at about 3% of the total net primary production in rice paddies, sedge meadows, and bogs (28). The environments supporting types of vegetation that vent large amounts of methane to the atmosphere are easily recognized and become important when considering whole landscapes.

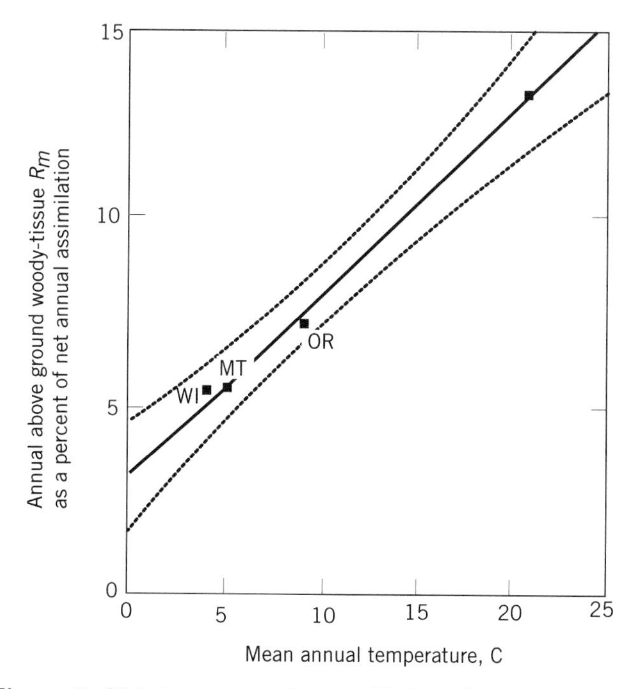

Figure 6. Maintenance respiration (R_m) of aboveground woody tissue in evergreen trees growing in Wisconsin (WI), Montana (MT), Oregon (OR), and Florida (FL) increases linearly as a fraction of the total estimated net annual assimilation of carbon. Reprinted with permission of M. G. Ryan, Rocky Mountain Forest and Range Experiment Station, Fort Collins, Colorado.

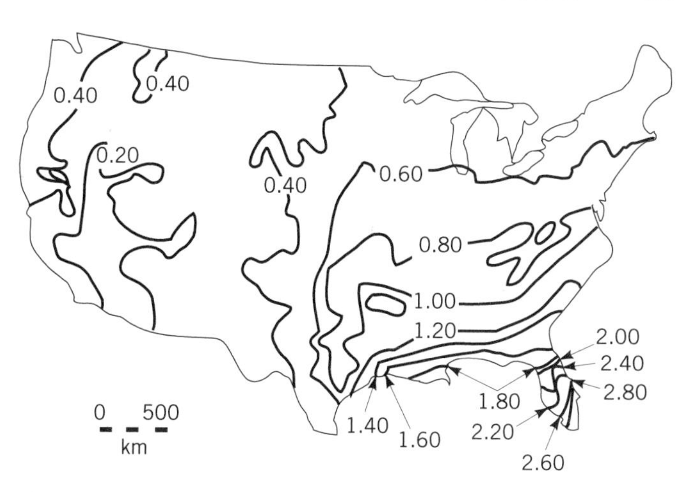

Figure 7. Rates of decomposition of fresh litter in the United States predicted by a simulation model using actual evapotranspiration as a predictive variable. Isopleth values are the fractional loss rate of mass from fresh litter during the first year of decay (29).

In general, microbial respiration associated with the decay of surface leaf litter is closely correlated with surface moisture and temperature conditions. Evaporation is also a function of these same variables, so it is not surprising that the litter decay rates (turnover) and annual evaporation from soil and transpiration from plants (evapotranspiration) are related. The relation with evapotranspiration provides a means of estimating decomposition rates across regions (see Fig. 7). The chemical composition of litter also affects its rate of decay. Deciduous leaves generally decay more rapidly than evergreen leaves, because they contain a more favorable balance of nitrogen-rich proteins to hard-to-decay organic molecules, such as the lignin, that binds together cell walls in plants. In modeling decomposition, therefore, the lignin–nitrogen ratio, or some other index of organic matter quality, is often taken into account to predict the annual weight loss of leaf litter (see Fig. 8).

Soil Carbon Accumulation

A small fraction of the total carbon entering the detrital pool of dead organic material, in the form of leaves, stems, roots, and root exudates, is converted to organic acids and other products of decomposition that become part of the soil, or are dissolved in water and eventually enter streams. Soils carbon decays more slowly than that in litter because some is bound to clay particles in a form that resists further decomposition. In some cold, anaerobic en-

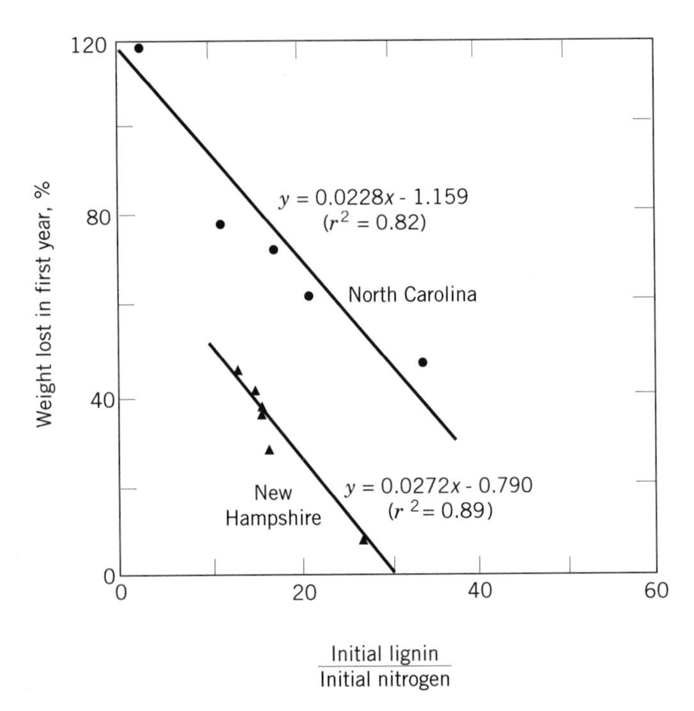

Figure 8. Decomposition of fresh leaves as a function of the lignin–nitrogen ratio of various species in New Hamsphire and North Carolina forests shows parallel slopes associated with differences in mean annual temperature. Reprinted with permission of the authors and Academic Press (30).

vironments, pure organic material may also accumulate and form peat. Even following drastic disturbance such as clear-cutting of forests, little change occurs in the total storage of soil carbon if forest cover is allowed to return. Conversion to permanent pasture or agricultural use will, however, commonly reduce soil carbon levels. With climate change and changing land-use patterns, a new equilibrium in soil carbon content is likely to result as vegetation patterns shift.

The fraction of carbon transferred from the litter to the soil is modelled in relation to decomposition rates of a single large reservoir, or of several reservoirs consisting of different materials that decay at different rates. In the soil, separate pools may receive the breakdown products. A rapidly decaying pool of organic material provides nutrients to plants, whereas a more recalcitrant carbon pool turns over carbon and releases nutrients much more slowly.

Ecosystem models generally couple the transfer of carbon to soil with the release of organically bound nitrogen and other essential minerals to plant roots, as illustrated in Figure 9. Decomposition rates are usually calculated at monthly time steps based on soil temperature, water content, and the carbon–nitrogen ratio of the decaying organic matter. When conditions are favorable, nutrients in their elemental form are released and made available to plants.

Models usually assume that all soil carbon is confined to the upper meter of soil or less, although this is not the case. One of the most difficult problems in modeling soil carbon is obtaining accurate estimates of carbon distribution throughout the entire rooting zone that may, in some ecosystems, extend to a depth of 10 m or more.

Nutrient Uptake and Carbon Allocation by Plants

The pool of nutrients released in decomposition of organic matter by microbial activity is supplemented from other sources, which include inputs from precipitation and chemical exchanges with minerals on the exposed surfaces of clays and organic residues. Most fertile soils hold calcium, magnesium, and potassium ions readily available on mineral exchange sites. Soil acidity, texture, organic matter content, and the type of clay minerals present strongly affect the availability of these essential bases. Uptake of all elements essential for plant growth must annually balance what is incorporated into new tissues, minus that lost in the shedding of leaves and other nutrient-rich organic residues.

Carbon allocation to roots generally increases over shoot growth when nutrients, particularly nitrogen and phosphorus, are not readily available. Including other nutrients in models would be desirable because the total complex affects carbon allocation and, through competition, the composition of plant communities (21). A more mechanistic approach to modeling interactions between carbon, water, and nutrients has recently been published (31). This model links transport of water, carbohydrates, and nutrients to movement through a plant's vascular system, which includes xylem, which transfers water and nutrients from roots to shoots, and phloem, which transfers carbohydrates and amino acids from leaves to roots and other organs.

A close linkage exists between carbon allocation in plants and the decomposition process because reduced nutrient availability lowers foliar nutrient levels, which in turn limits decomposition. Some carbon balance models adjust the carbon–nitrogen ratio in foliage to mirror that in litter, and use the ratio as a control on modeling carbon allocation (22).

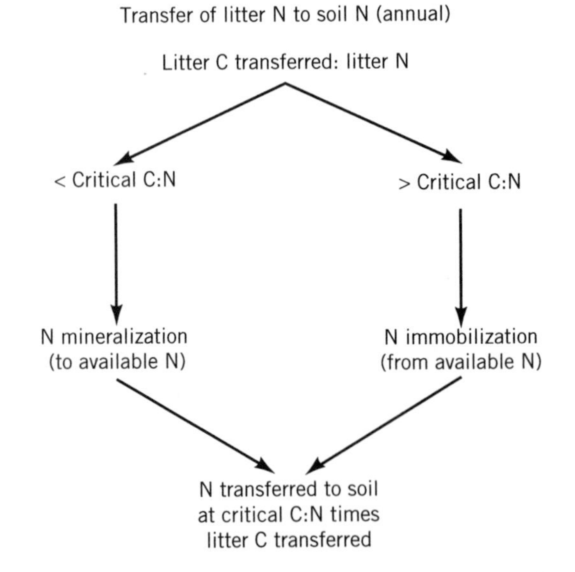

Figure 9. The fraction of nitrogen (N) and carbon (C) transferred annually to soil is partly a function of critical C–N ratios in most ecosystem models. The total amount of litter entering the soil depends on the expected death of leaves, woody material, and roots. When the system is disturbed, additional material may be added or removed. Reprinted with permission of Raymond Hunt, University of Montana, Missoula, Montana.

Some checks on the reliability of how models allocate carbon can be obtained by comparing independent estimates of the amount of carbon required to grow and maintain foliage, stems, and roots against that provided through photosynthesis (e.g., see Fig. 10). Where the forest is in equilibrium, an independent estimate of carbon allocation to roots for growth and respiration is obtained by measuring the annual CO_2 flux from the soil, minus the carbon in aboveground litterfall (33). Under varying conditions, changes in how the carbon is allocated to above- and belowground components can thus be estimated and empirically modelled.

Most models assume plant roots are evenly distributed throughout the soil or that all plants growing in a particular environment have similar rooting characteristics. Water and nutrients are not equally available to plant species; the roots of some are restricted to surface horizons, whereas others extend more deeply. When plants with deep roots are absent from a community, mineral cycling from the lower profile stops. As a result, the surface organic and mineral layers may become impoverished. Such vertically defined interactions in the soil are absent in most current ecosystem models, although not all (34).

TESTING AND SIMPLIFYING CARBON BALANCE MODELS

Testing Models

The reliability of carbon balance models may be assessed by:

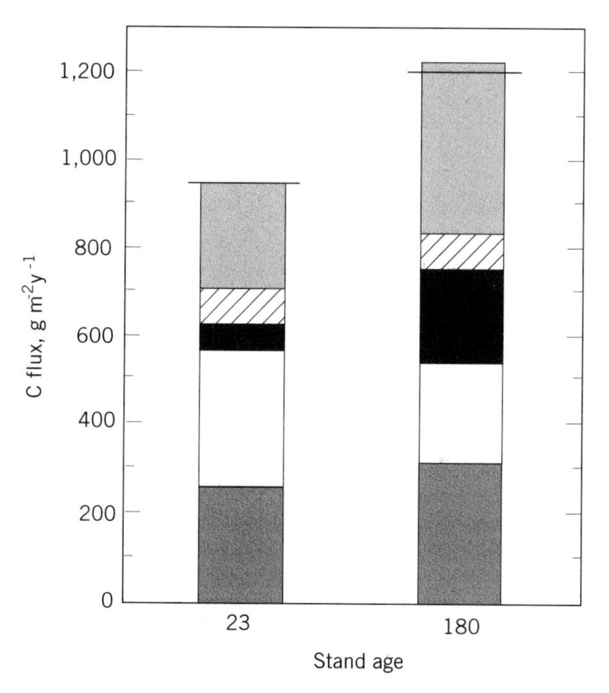

Figure 10. Annual carbon flux estimates for a 23- and 180-year-old forest of fir trees show good agreement between the amount of photosynthate available estimated by a simulation model (— —) and the sum of production (p), construction (c), and maintenance respiration (m) by foliage (F); stems (S); and roots (R) estimated independently. Reprinted with permission of M. G. Ryan and *Tree Physiology* (32).

F_m S_{p+c}
F_{p+c} R_{m+p+c}
S_m — Photosynthesis

- Direct comparisons with field measurements
- Projecting simulations over long periods
- Predicting changes in hydrologic and nutrient components
- Applying sensitivity analysis to identify critical relationships

A variety of approaches is desirable because some questions are not subject to direct experimentation (eg, changing climate), whereas others require simultaneous measurements of many variables for long periods.

Comparison of model predictions with field measurements is valuable but may be difficult to accomplish. The amount of carbon in aboveground biomass is easily measured, whereas that in roots is more difficult to estimate accurately. Carbon stocks in soil are large and vary from site to site, particularly where rooting depth is affected by rocks. Net fluxes of CO_2 from the soil upward through the vegetation can be measured, and various transfer equations tested on a short-term basis (35). Although improvements in micrometeorological instrumentation can provide accurate net exchange measurements of carbon dioxide from forest ecosystems over short periods (see Fig. 11), difficulties in making accurate measurements of growth, respiration, and photosynthesis over extended periods limit the data sets available to compare model predictions.

Using model simulations to make predictions over many years can illustrate whether the transfer rates estimated for a single year are reasonable. If such simulations resulted in litter accumulations on the ground to unrealistic depths, for example, clearly the estimated rates of annual litterfall must be too high or the rate of decomposition too low. Because of the close link between the carbon, water, and nutrient cycles, comparing simulations of nutrient and water export from monitored streams draining closed basins can provide another type of model validation (37).

Sensitivity analysis can identify critical assumptions and interactions. For example, in simulating the response of a pine forest to a doubling of CO_2 in the atmosphere, the net annual growth should increase less than 8% because decomposition rates should limit nitrogen availability (38). Sensitivity analyses can also indicate the relative need for more accurate measurements of certain variables. This might be accomplished, for example, by comparing changes in predicted growth, litter accumulation, transpiration, and photosynthesis in response to a 10% change in precipitation, carbon–nitrogen ratios, soil water storage, and leaf area index.

Simplifying Models

Carbon balance models must be simplified to make them applicable to larger spatial and temporal scales. Sensitivity analysis, discussed earlier, is one way of identifying those variables that can be estimated less precisely without losing much accuracy. An alternative approach is to compare predictions made by refined models with those from simpler models. Thus, hourly predictions of canopy photosynthesis derived from a model with separate layers of leaves can be compared with a whole canopy model that generates estimates of photosynthesis from meteorologi-

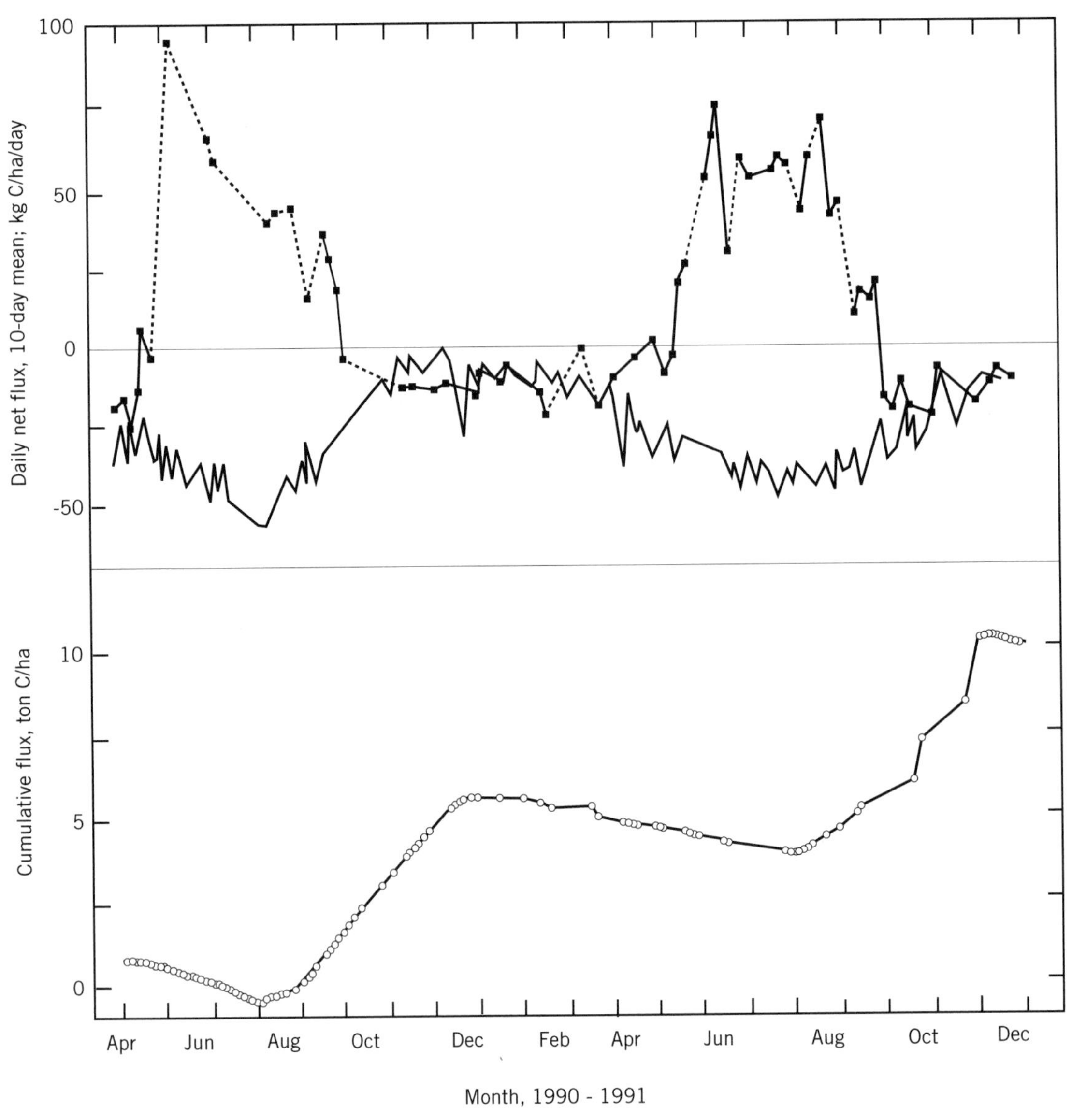

Figure 11. Net ecosystem exchanges measured with eddy correlation methods from a tower over a northern deciduous hardwood forest in Massachusetts show large seasonal differences. In the upper diagram are 10-day means for daily net CO_2 flux from the atmosphere to the ecosystem, the middle diagram presents respiration on a similar time scale, and the lower graph shows cumulative flux (36).

cal data averaged for the day (38,39). If the comparisons indicate that the simpler models provide acceptable results, they can serve as the basis for further simplifications that predict photosynthesis, respiration, evapotranspiration, decomposition, and growth on a monthly basis. Finally, monthly balance sheets can be compared with annual runoff from gauged watersheds, annual litterfall, and annual aboveground production. Sequential comparisons, from detailed to less complex models, help identify the extent to which simplifications are warranted and the minimum detail needed to provide a certain reliability of estimates from coarser-scale models.

SCALING CARBON BALANCE MODELS SPATIALLY

Although the composition of vegetation differs over time and space, some ecosystem models have been simplified to assume that generalizations of process rates can be made across landscapes by recognizing only the broadest categories of vegetation (19,40).

An even simpler model of energy utilization efficiency (41) has proven applicable to a wide range of forests in the western United States, as illustrated in Figure 12. In this model, the upper limits to production are set by the ability of vegetation to absorb solar radiation. At extremes in latitude and where leaf area development is restricted, the growth of all vegetation is limited. Even at mid-latitudes where forests with dense canopies dominate, production is severely constrained if frost, soil drought, or excessive atmospheric humidity deficits occur over extended periods (see Fig. 13).

Satellite coverage of the Earth now provides a means of estimating seasonal changes in leaf area index (43) and cloud cover (44). This capability, combined with a network

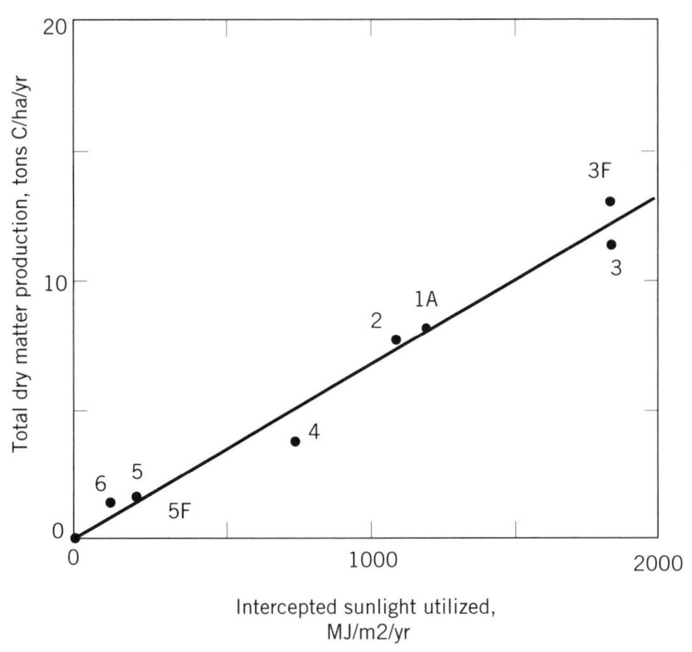

Figure 12. The net amount of carbon that is transferred into dry matter production increases proportionally with the amount of visible radiation that can be intercepted and converted by photosynthesis into carbohydrates when stomata are open. The numbered sites refer to ecosystems dominated by (1A) deciduous alder, (2) mixed forests of conifers with deciduous oak, (3) fertilized (F) and unfertilized fast-growing conifers on the western slopes of the Cascade Mountains, (4) subalpine forests, (5) fertilized (F) and unfertilized open pine forest, and (6) juniper woodland (42).

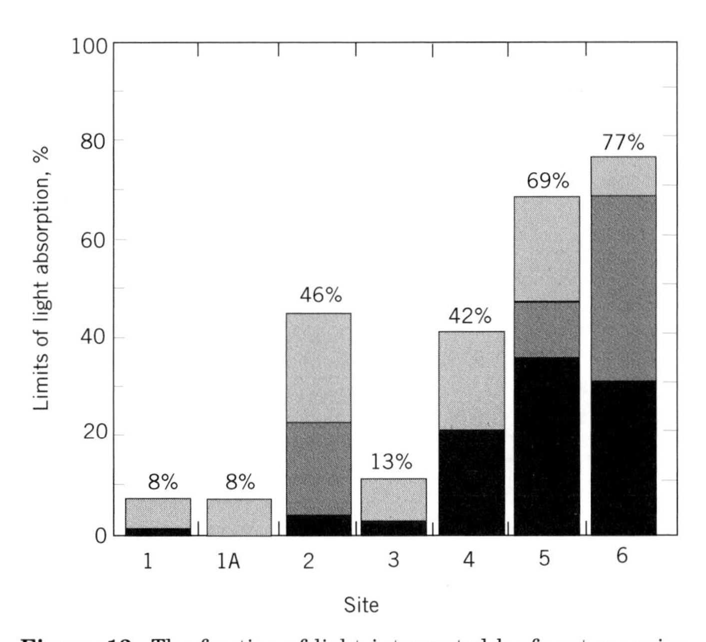

Figure 13. The fraction of light intercepted by forest canopies varies from about 20 to nearly 100% across the environmental gradient, corresponding to the forest types described in Figure 12. As environments become harsher, the percentage of light that is intercepted and can be used in photosynthesis is progressively reduced from > 90% by the two coastal forests (1 and 1A) to < 25% in the juniper woodland (42). ▢, humidity deficit; ▩, drought, ■, freezing

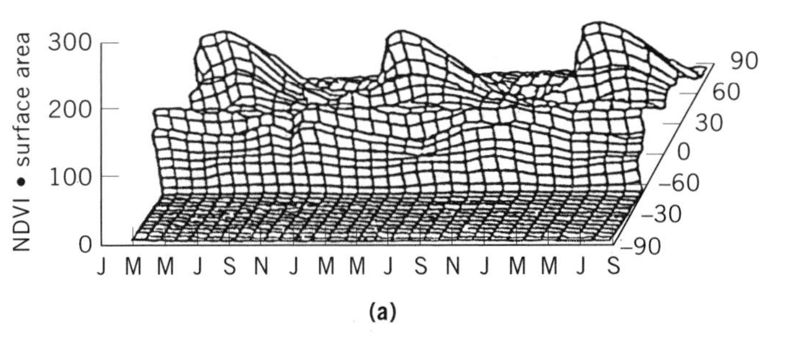

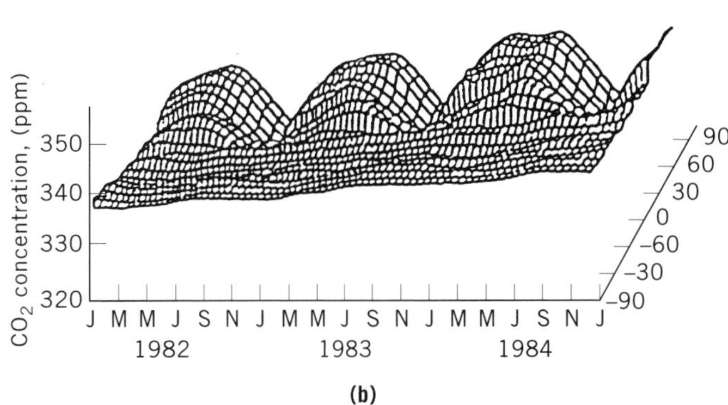

Figure 14. (**a**) A satellite-derived index of greenness (NDVI), related to reflectance in the near-ir and red parts of the spectrum, varies most at mid- and far northern latitudes and least at the equator. (**b**) Seasonal variation in the concentration of CO_2 in the atmosphere parallels changes in vegetation greenness (47).

of climatic stations, provides the essential information for modeling seasonal changes in photosynthesis, transpiration, growth, and other ecosystem processes (45). The first general global-scale ecosystem model using satellite-derived estimates of monthly vegetative cover has recently appeared (46). Previously, as illustrated in Figure 14, seasonal variations in atmospheric CO_2 concentrations were shown to vary with this same satellite-derived index of vegetative greenness associated with photosynthetic activity (47).

These models are being improved to incorporate annual variations in climate and atmospheric CO_2. A few global models include atmospheric circulation and provide a basis of comparison with measured variations in atmospheric CO_2 (46,48–50). These models will be improved further as more is learned about the coupling between land, atmosphere, and oceans (51,52), and satellite-derived data is extended to estimate climatic variation as well as vegetative cover (53).

CONCLUSION

Carbon balance modeling is still in its infancy, but a sufficient number of models is now available to allow comparisons. Leaf-level processes are well-understood and have been the first to be generalized and scaled in time and space. The implications of changes in species composition are recognized as important but continue to present difficulties to modellers. Increased understanding of how nutrients and water interact with carbon cycling is leading to more realistic, but more complicated, models. At the same time, simplified models are providing new insights at regional and global scales. Improved monitoring techniques in micrometeorology, atmospheric chemistry, and remote sensing are available to help assess the reliability of larger-scale predictions. Field experiments, model comparisons, and sensitivity analyses will continue to advance understanding and improve model predictive capacities.

Acknowledgments

We appreciate comments made on earlier drafts of this article by Joe Landsberg, Warwick Silvester, Beverley Law, Jeanne Panek, and Lise Waring.

BIBLIOGRAPHY

1. *The Ecosystems Center Annual Report,* Marine Biological Laboratory, Woods Hole, Mass., 1987.
2. L. A. Joyce and R. N. Kickert, "Applied Plant Growth Models for Grazing Lands, Forests, and Crops," in *Plant Growth Modelling for Resource Management,* Vol. 1, CRC Press, Boca Raton, Fl., 1987, pp. 17–55.
3. U. Siegenthaler and J. L. Sarmiento, *Nature* **365,** 119–125 (1993).
4. G. I. Ågren, R. E. McMurtrie, W. J. Parton, J. Pator, and H. H. Shugart, *Ecological Applications* **1,** 118–138 (1991).
5. R. E. McMurtrie, D. A. Rook, and F. M. Kelliher, *Forest Ecology and Management* **39,** 381–413 (1990).
6. Y.-P. Wang and P. G. Jarvis, *Agricultural and Forest Meteorology* **51,** 257–280 (1990).
7. G. B. Bonan, *Journal of Geophysiocal Research* **96,** 7301–7312 (1991).
8. E. B. Rastetter, M. G. Ryan, G. R. Shaver, J. M. Melillo, K. J. Nadelhoffer, J. E. Hobbie, and J. D. Aber, *Tree Physiology* **9,** 101–126 (1991).
9. R. L. Sanford, Jr., W. J. Parton, D. S. Ojima, and D. J. Lodge, *Biotropica* **23,** 364–372 (1991).
10. H. H. Shugart, *A Theory of Forest Dynamics,* Springer-Verlag, New York, 1984.
11. D. B. Botkin, J. F. Janak, and J. R. Wallis, *Journal of Ecology* **60,** 849–872 (1972).
12. J. D. Aber and J. M. Melillo, FORTNITE: A Computer Model of Organic Nitrogen Dynamics in Forest Ecosystems," *University of Wisconsin Research Bulletin R3130,* 1982.
13. J. Pastor and W. M. Post, *Nature* **334,** 55–58 (1988).
14. A. D. Friend, H. H. Shugart, and S. W. Running, *Ecology* **74,** 792–797 (1993).
15. H. Lieth, *Nature and Resources* **8,** 5–10 (1972).
16. E. Box, *Macroclimate and Plant Forms: An Introduction to Predictive Modeling in Phytogeography,* Dr. W. Junk Publishers, The Hague, the Netherlands. 1981.
17. I. C. Burke, T. G. F. Kittel, W. K. Lauenroth, P. Snook, C. M. Yonker, and W. J. Parton, *Bioscience* **41,** 685–692 (1991).
18. F. I. Woodward, *Climate and Plant Distribution,* Cambridge University Press, Cambridge, U.K., 1987.
19. I. C. Prentice, W. Cramer, S. P. Harrison, R. Leemans, R. A. Monserud, and A. M. Solomon, *Journal of Biogeography* **19,** 117–134 (1992).
20. M. E. Harmon, W. K. Ferrell, and J. F. Franklin, *Science* **247,** 699–702 (1990).
21. W. J. Parton, J. W. B. Stewart, and C. V. Cole, *Biogeochemistry* **5,** 109–131 (1988).
22. S. W. Running and S. T. Gower, *Tree Physiology* **9,** 147–160 (1991).
23. J. M. Norman, "Scaling Processes Between Leaf and Canopy Levels," in *Scaling Physiological Processes: Leaf to Globe,* Academic Press, San Diego, Calif., 1993, pp. 41–76.
24. J. D. Marshall and R. H. Waring, *Ecology* **67,** 975–979 (1986).
25. J. Grace, "Some Effects of Wind on Plants," in *Plants and Their Atmospheric Environment,* Blackwell Scientific Publications, London, 1981, pp. 31–56.
26. S. P. Long and P. R. Hutchin, *Ecological Applications* **1,** 139–156 (1991).
27. M. G. Ryan, *Ecological Applications* **1,** 157–167 (1991).
28. G. J. Whiting and J. P. Chanton, *Nature* **364,** 794–795 (1993).
29. V. Meentemeyer, "Climatic Regulation on Decomposition Rates of Organic Matter in Terrestrial Ecosystems," in *Environmental Chemistry and Cycling Processes,* National Technical Information Service, Springfield, Va., 1978, pp. 779–789.
30. J. M. Melillo, A. D. McGuire, D. W. Kicklighter, B. Moore, III, C. J. Vorosmarty, and A. L. Schloss, *Nature* **363,** 234–240 (1993).
31. R. C. Dewar, *Functional Ecology* **7,** 356–368 (1993).
32. M. G. Ryan, *Tree Physiology* **9,** 255–266 (1991).
33. J. W. Raich and K. J. Nadelhoffer, *Ecology* **70,** 1346–1354 (1989).
34. P. J. H. Sharpe, J. Walker, L. K. Penridge, and H.-I. Wu, *Ecological Modelling* **29,** 189–213 (1985).
35. D. D. Baldocchi, "Scaling Water Vapor and Carbon Dioxide Exchange from Leaves to Canopy: Rules and Tools," in *Scaling Physiological Processes: Leaf to Globe,* Academic Press, San Diego, Calif., 1993, pp. 77–114.
36. S. C. Wofsy, M. L. Gouldon, J. W. Munger, S.-M. Fan, P. S. Bakwin, B. C. Duabe, S. L. Bassow, and F. A. Bazzaz, *Science* **260,** 1314–1317 (1993).
37. F. H. Bormann, G. E. Likens, T. G. Siccama, R. S. Pierce, and J. S. Eaton, *Ecological Monographs* **44,** 255–277 (1974).
38. R. E. McMurtrie, H. N. Comins, M. U. F. Kirschbaum, and Y.-P. Wang, *Australian Journal of Botany* **40,** 657–677 (1992).
39. Y.-P. Wang, R. E. McMurtrie, and J. J. Landsberg, "Modelling Canopy Photosynthetic Productivity," in *Crop Photosynthesis: Spatial and Temporal Determinants,* Elsvier Science Publishers, Amsterdam, the Netherlands, 1992, pp. 43–67.
40. J. M. Melillo, J. D. Aber, and J. F. Muratore, *Ecology* **63,** 621–626 (1982).
41. J. L. Monteith, *Philosophical Transactions of the Royal Society, Series B* **281,** 277–294 (1977).
42. J. Runyon, R. H. Waring, S. N. Goward, and J. M. Welles, *Ecological Applications* **4,** 226–237 (1994).
43. M. Spanner, L. Johnson, J. Miller, R. McCreight, J. Runyon, P. Gong, and R. Pu, *Ecological Applications* **4,** 258–271 (1994).
44. T. Eck and D. Dye, *Remote Sensing of the Environment* **38,** 135–146 (1991).

45. P. J. Sellers, J. A. Berry, G. J. Collatz, C. B. Field, and F. G. Hall, *Remote Sensing of the Environment* **42,** 187–216 (1992).
46. C. S. Potter, J. T. Randerson, C. B. Field, P. A. Matson, P. M. Vitousek, H. A. Mooney, and S. A. Klooster, *Global Biogeochemical Cycles* **7,** 811–841 (1993).
47. C. J. Tucker, I. Y. Fung, C. D. Keeling, and R. H. Gammon, *Nature* **319,** 195–199 (1986).
48. P. P. Tans, T. J. Conway, and T. Nakazawa, *Journal of Geophysical Research* **94,** 5151–5172 (1989).
49. P. P. Tans, I. Y. Fung, and T. Takahashi, *Science* **247,** 1431–1438 (1990).
50. P. J. Sellers, S. O. Los, C. J. Tucker, C. O. Justice, D. A. Dazlich, G. J. Collatz, and D. A. Randall, *Journal of Climate* (in press).
51. R. F. Keeling, *Nature* **363,** 399–400 (1993).
52. G. D. Farquhar, J. Lloyd, J. A. Taylor, L. B. Flanagan, J. P. Syvertsen, K. T. Hubick, S. C. Wong, and J. R. Ehleringer, *Nature* **363,** 439–443 (1993).
53. S. N. Goward, R. H. Waring, D. G. Dye, and J. Yang, *Ecological Applications* **4,** 322–343 (1994).

CARBON CYCLE

RICHARD HOUGHTON
Woods Hole Research Center
Woods Hole, Massachusetts

Concern about the effects of a rapid global warming has focused attention on the global carbon cycle, first because carbon dioxide (CO_2) has been the most important human-induced greenhouse gas in the past and is expected to remain so in the future (1) and second because an understanding of the carbon cycle is required for predicting how future emissions will affect the rate of warming. The rate and extent of a global warming depend on the concentration of CO_2 and other greenhouse gases in the atmosphere. The concentration of CO_2 in the atmosphere, in turn, is a function of the emissions of CO_2 to the atmosphere from the combustion of fossil fuels and deforestation (sources) and the removal of CO_2 from the atmosphere by the oceans and terrestrial ecosystems (sinks). To constrain the rate of warming within acceptable limits, the concentration of CO_2 must be stabilized. Stabilization of the concentration requires a reduction in current emission rates. The decision of how much to reduce emissions requires an understanding of the relationship between emissions and atmospheric concentrations. If scientists do not know the relationship, there is no accurate way to determine policies regarding energy use and land use, the major determinants of CO_2 emissions. The distinction between emissions and concentration is an important one. Slowing the rate of warming will require stabilization of atmospheric concentrations, and stabilization of concentrations will require a reduction in emissions.

See also AIR QUALITY MODELING; CARBON BALANCE MODELING; CARBON STORAGE IN FORESTS; FOREST RESOURCES; GLOBAL CLIMATE CHANGE, MITIGATION.

Stabilization of emissions at present rates will not prevent climatic change. At present rates of emission, concentrations of greenhouse gases in the atmosphere will increase to the point at which global warming will occur many times more quickly than it did 10,000 yr ago, with the retreat of the glaciers. Carbon dioxide is estimated to have accounted for more than half of the increased radiative forcing in the 1980s and is projected to account for more than half in the future. Because the concentration of CO_2 in the atmosphere is determined not only by its human-induced emissions but also by natural processes that remove the gas from the atmosphere, understanding the global carbon cycle is essential for predicting future concentrations of CO_2 and calculating the relationship between emissions and the rate of warming.

The processes important in the global carbon cycle vary depending on the time frame of interest. Over millions of years, the processes that govern the distribution of carbon are largely geological: weathering of the land surface, sedimentation of organic and inorganic material on the ocean floor, vulcanism, and sea floor spreading. At the other end of the spectrum are processes such as photosynthesis and hydration of a molecule of CO_2 at the sea surface that occur over seconds. The time scale of immediate interest to the current inhabitants of the earth and to the next few generations is years to a century. The geological processes are too slow to be important in the next decades; the rapid processes are integrated into other processes more appropriately studied over years. This discussion emphasizes processes that are important over years to a century.

THE NATURAL CYCLE

In its most simple abstraction, the global carbon cycle has only four major compartments or reservoirs: atmosphere, oceans, land, and fossil fuels. The reservoirs and the exchanges of carbon shown in Figure 1 pertain to the natural world before human interference. The actual date is probably 6000 yr ago, although human disturbance affected the storage of carbon in vegetation and soils with

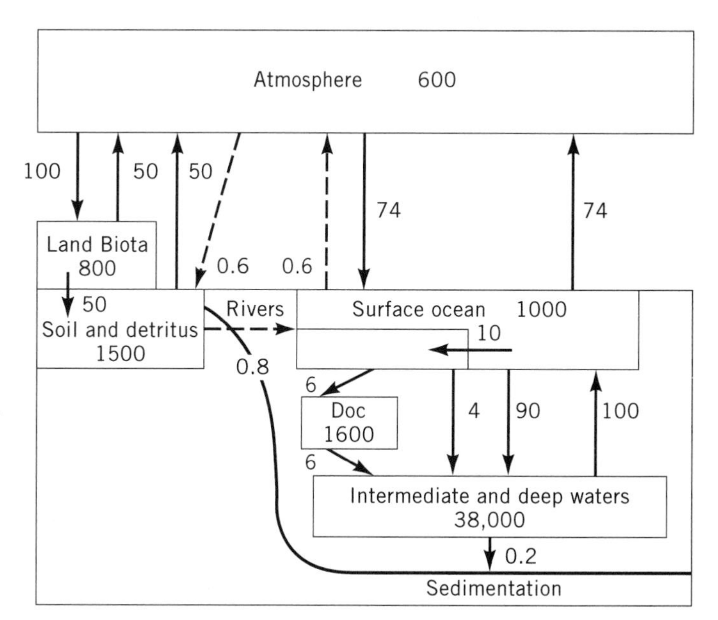

Figure 1. The global carbon cycle before human disturbance (about 6000 BP). Units are petagrams carbon and petagrams of carbon per year. Modified from Ref. 2.

the development of settled agriculture about 10,000 yr ago. Well before that, fire may have been used for hunting. But 10,000 yr ago the earth's climate was different enough from today's to have affected the distribution of forests and hence the amount of carbon held on land. Thus the natural cycle shown in Figure 1 is an attempt to represent a world with a climate similar to today's but before major human effects. The important point is that the amount of carbon in the atmosphere (600 Pg) was about equal to the amounts in the surface layers of the ocean (1000 Pg) and in the terrestrial biota (800 Pg), a condition still true today. Because the exchanges of carbon between the atmosphere and these two reservoirs are rapid, changes in either reservoir has the potential to cause large and rapid changes in the concentration of atmospheric CO_2.

Most of the carbon in the atmosphere occurs as CO_2, but its concentration is low. Nitrogen and oxygen gases comprise more than 99% of the atmosphere; CO_2 accounts for about 0.03%. If 0.03% were the accuracy of measurement today, there would be no evidence for its increase; however CO_2 is presently measured within a few tenths of a part per million by volume (ppmv), or 0.00003%. In 1990, the concentration of CO_2 was 353 ppmv. For the 1000 yr before the Industrial Revolution, the concentration was about 280 ppmv and varied by less than 10 ppmv (Fig. 2). This preindustrial concentration of CO_2 was equivalent to about 600 Pg carbon (1 Pg = 1 billion t).

The concentration of CO_2 in the atmosphere has varied over most of the earth's history. Analyses of CO_2 in the bubbles of air trapped in glacial ice from Vostok, Antarctica, show that the concentrations varied over the last 160,000 yr, in phase with the advance and retreat of glaciers (7). The concentration was about 180 ppmv during glacial periods and about 280 ppmv during interglacial periods (Fig. 3). The concentration of CO_2 increased as global temperature increased and decreased as temperature fell. The coupling is evidence for a positive feedback between the carbon cycle and global climate. It suggests not only a greenhouse effect, in which atmospheric CO_2 warms the earth, but another effect by which warming raises the CO_2 concentration of the atmosphere. The greenhouse effect was advanced almost a century ago (8). Only recently have scientists been concerned about the positive feedback to a warming. The glacial–interglacial difference of 100 ppmv corresponds to a temperature difference of about 5°C, globally (10°C at Vostok). For comparison, the warming projected for the year 2100 under the business-as-usual scenario of the Intergovernmental Panel on Climatic Change (IPCC) is 2° to 5°C (1).

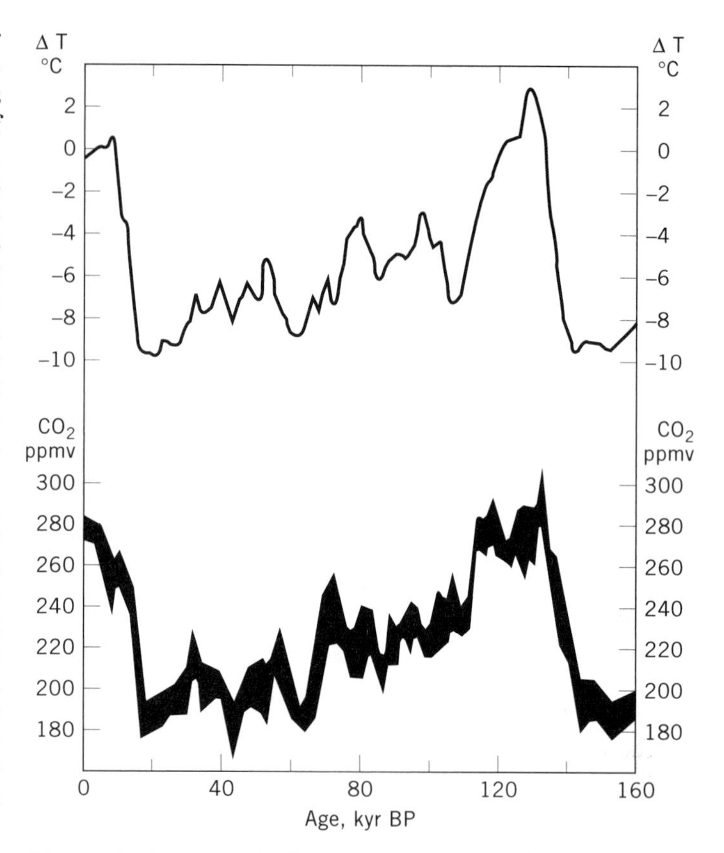

Figure 3. Atmospheric CO_2 concentrations and changes in surface temperatures over the last 160,000 yr determined from analysis of glacial ice at Vostok, Antarctica. Temperature changes were estimated based on measured deuterium concentrations. From Ref. 7.

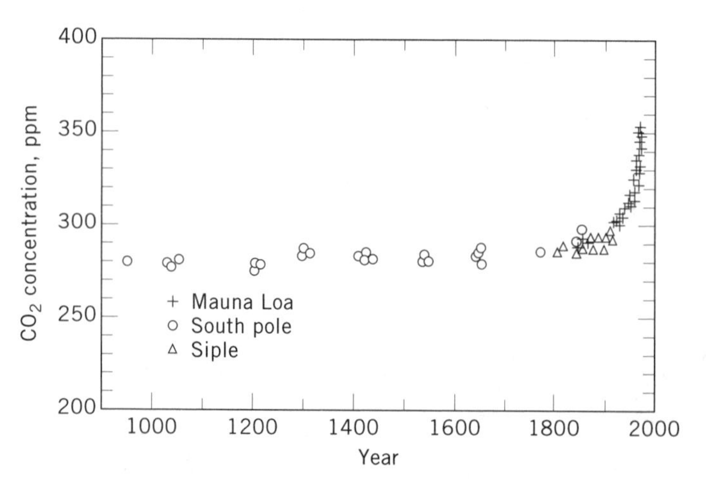

Figure 2. Atmospheric CO_2 concentrations for the last 1000 yrs determined from direct analysis of air at Mauna Loa, Hawaii (3), and from analysis of Antarctic ice cores from Siple station (4,5) and the South Pole (6).

A small amount of the carbon in the atmosphere exists as methane (CH_4) and as carbon monoxide (CO). Methane, another greenhouse gas, is at about 1/100 the concentration of CO_2 but is about 25 times more effective as a greenhouse gas than CO_2 per kilogram of emissions. The concentration of CH_4 was about 0.7 ppmv preindustrially. Methane, like CO_2, is produced as a result of respiration and decay, but under anaerobic conditions. Thus its major sources include wetlands and peatlands and the digestive processes of animals, notably ruminants. Unlike CO_2, CH_4 is chemically reactive, and a major sink for CH_4 is its destruction in the atmosphere by the hydroxyl radical. Carbon monoxide is not a greenhouse gas, but its reactivity with the hydroxyl radical means that it indirectly affects the concentrations of other greenhouse gases, principally CH_4.

Terrestrial ecosystems held about four times more carbon than the atmosphere before human disturbance. This terrestrial carbon is largely organic and exists in many forms, including living leaves and roots, animals, microbes, wood, decaying leaves, and humus. The turnover of these materials varies from less than 1 yr to more than 1000 yr. The living component of terrestrial ecosystems held approximately 800 Pg carbon, almost all of it in vegetation. Animals, including humans, account for less than 0.1% of the carbon in living organisms. About half the mass of dry plant material is carbon. The soils of the earth are estimated to have contained about 1500 Pg carbon, about twice the amount in the living biota. The carbon in soil is derived from the decomposition of plant material. Some reduction in the storage of terrestrial carbon had already occurred by 6000 yr ago; the values given in Figure 1 are approximate. Most of the carbon in terrestrial ecosystems was in forests. Forests once covered about 50% of the land surface and held more than 90% of the live organic carbon. When soils are included in the inventory, forests held about half of the terrestrial carbon. Grasslands, tundra, and wetlands store large amounts of organic carbon in their soils.

The amount of carbon in the surface ocean (ca. 1000 Pg) was also similar to the amount in the preindustrial atmosphere. The intermediate and deep waters of the ocean, not directly in contact with the atmosphere, held about 38 times this amount. At equilibrium, the concentration of CO_2 in the atmosphere is proportional to the concentration in the ocean, and hence, the oceans control the atmospheric concentration of CO_2. The equilibrium may require hundreds of years to reach; however, because the rate of circulation of oceanic waters is on the order of 1000 yr. The dramatic increase in atmospheric CO_2 over the last 30 yr demonstrates the extent of the present disequilibrium.

The form of carbon in the oceans, as in the atmosphere, is mostly inorganic carbon. The ocean's chemistry is such that less than 1% of this carbon is dissolved CO_2. Almost 90% exists in the form of the bicarbonate ion and the rest as carbonate ion. The three forms are in chemical equilibrium. About 1600 Pg carbon in the oceans (out of the total of more than 38,000 Pg carbon total) is dissolved organic carbon. Carbon in living organisms amounts to less than 2 Pg in the sea, compared with about 800 Pg on land. The mass of animal life in the oceans is almost the same as on land, however, pointing to the different trophic structures in the two environments. The ocean's plants are microscopic. They have a high productivity, but the production of plant material does not accumulate; it is either grazed or sinks to deeper layers where it decays. In contrast, terrestrial plants accumulate large amounts of carbon in long-lasting structures (ie, trees).

The amount of carbon stored in recoverable reserves of coal, oil, and gas is estimated to be about 5000 Pg, larger than any other reservoir except the deep sea, and about 10 times the carbon content of the atmosphere. The form of this carbon is organic, ie, it is the remains of terrestrial and marine vegetation once alive on the earth's surface. Until 100 to 150 yr ago, this reservoir of fossil carbon was not a significant part of the short-term cycle of carbon; the rate of burial of organic matter and its diagenesis to fossil deposits is several orders of magnitude lower than the cycling of carbon between land, sea, and air.

Each year the atmosphere exchanges about 100 Pg of carbon (about 15%) with terrestrial ecosystems and somewhat less with the surface ocean (Fig. 1). The exchanges with terrestrial systems are biotic flows of carbon. Terrestrial photosynthesis withdraws approximately 100 Pg carbon from the atmosphere annually in the production of organic matter. At least half of this photosynthetic production is respired by the plants, themselves, and the rest is consumed by animals or respired by decomposer organisms in the soil. This transfer of plant material to the consumer and decomposer communities is represented by the arrow from vegetation to soil in Figure 1 (50 Pg carbon/yr).

The most striking feature of CO_2 concentrations in the atmosphere is the regular sawtooth pattern (Fig. 4). This pattern repeats itself annually. The cause of the oscillation is the metabolism of temperate and boreal terrestrial ecosystems. Highest concentrations occur at the end of winter, following the season in which respiration has exceeded photosynthesis and caused a net release of CO_2 to the atmosphere. Lowest concentrations occur at the end of summer, following the season in which photosynthesis has exceeded respiration and drawn CO_2 out of the atmosphere. The latitudinal variability in the amplitude of this oscillation is consistent with the spatial and seasonal distribution of terrestrial metabolism: the highest amplitudes (up to about 16 ppmv) are in the Northern Hemisphere with the largest land area; the phase of the amplitude is reversed in the Southern Hemisphere, corresponding to seasonal metabolism there.

The annual rate of photosynthesis in the world's ocean is estimated to be about 50 Pg carbon. Marine photosynthesis does not affect the seasonal oscillation of atmospheric CO_2 greatly, because the chemical properties of CO_2 in seawater are such that little of the carbon involved in either photosynthesis or respiration in the sea is exchanged with the atmosphere. On the time scale of 1000 yr, however, marine photosynthesis is important in controlling the CO_2 concentration of the atmosphere. Marine photosynthesis and the sinking of organic matter out of the surface water (the so-called biological pump) are estimated to keep the concentration of CO_2 in air about 50% of what it would be in their absence.

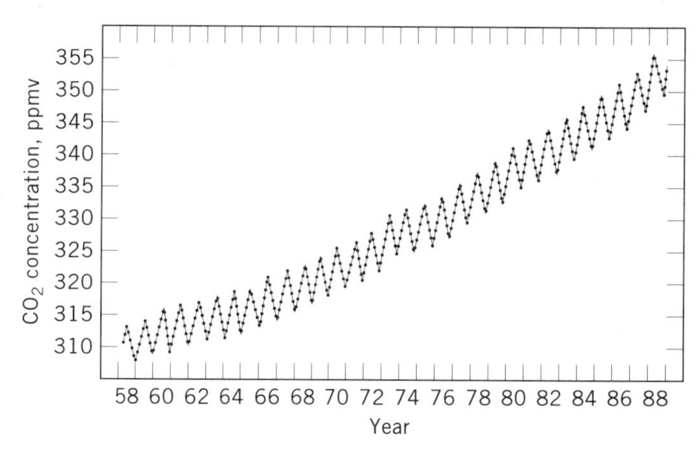

Figure 4. Atmospheric CO_2 concentrations as measured directly in air at Mauna Loa, Hawaii, since 1958. From Ref. 3.

The gross fluxes of carbon between the surface ocean and the atmosphere each year are largely the result of physicochemical processes. The rate of transfer of CO_2 across the sea surface is estimated to have been 74 Pg carbon annually before human disturbance. The flux is in both directions. The gross flows of carbon between the surface ocean and the intermediate and deep ocean are estimated to have been about 100 Pg per year, 10 Pg through the cycling of dissolved organic carbon (DOC) and the settling of particulate organic carbon (POC), and 90 Pg through physical processes of upwelling and downwelling.

CHANGES INDUCED BY HUMANS

Changes in the Atmosphere

For the 1000 years before 1800, or so, the concentration of CO_2 in the atmosphere remained between 275 and 285 ppmv (6). Since 1700, the concentration of CO_2 has increased from about 280 ppmv to about 353 ppmv (1990), an increase of more than 25% (Fig. 2). This increase is believed to have been due to the widespread replacement of forests with cleared land and emissions of CO_2 from the combustion of fossil fuels.

Numerous measurements of atmospheric CO_2 concentrations were made in the 19th century, but no reliable measure of the rate of increase was possible until after 1957 when the first continuous monitoring of CO_2 began at Mauna Loa, Hawaii, and at the South Pole (3). Measurements of the concentration of CO_2 in air made before 1957 had neither the continuous record nor the precision required to show the increase. In 1958, the average concentration of CO_2 in air at Mauna Loa was about 315 ppmv, and in 1990, it was 353 ppmv (Fig. 4); thus the average rate of increase was about 1 ppmv per year. The rate of increase has itself been increasing; the increase in recent years has been about 1.5 ppmv per year.

The atmosphere is completely mixed in about a year, so any monitoring station free of local contamination will show approximately the same year-to-year increase in CO_2. Today there are more than 30 monitoring stations, worldwide. They generally show the same year-to-year increase but vary with respect to absolute concentration, seasonal variability, and other characteristics useful for investigating the global circulation of carbon.

The concentration of methane (CH_4) in the atmosphere has increased by more than 100% in the past 100 yr, from background levels of less than 0.8 ppmv to a current value of about 1.7 ppmv. The pattern of the increase is similar to that of CO_2, with annual increases of 1.5 ppbv (parts per billion by volume) between 1700 and 1900, accelerating to 17 ppbv per year during the last decade. For the 1000 yr before 1700 there was no apparent trend in the concentration, but over the glacial–interglacial time frame, methane, like CO_2, was positively correlated with the mean global surface temperature.

Methane is released from anaerobic environments, environments with low oxygen concentrations, such as the sediments of marshes, peatlands, rice paddies and the guts of cattle and termites. The major sources of the increased methane are uncertain, but are thought to include the expansion of paddy rice, the increase in the world's population of cattle, and the warming of high latitude wetland ecosystems. Raising the temperature of peatlands stimulates anaerobic and aerobic respiration and may increase the release of both CH_4 and CO_2 during the last century. Atmospheric CH_4 budgets are more difficult to construct than CO_2 budgets because increased concentrations of CH_4 occur not only from increased sources from the earth's surface but from decreased destruction (by hydroxyl radicals) in the atmosphere itself. The increase in atmospheric CH_4 has been more significant for the greenhouse effect than it has for the carbon budget. The doubling of CH_4 since 1700 has contributed less than 1 ppmv, compared with the CO_2 increase of 70 ppmv. On the other hand, CH_4 is 25 times more effective than CO_2 as a greenhouse gas per kilogram released to the atmosphere.

Changes in Land Use

Changes in land use either release carbon to or withdraw it from the atmosphere. Releases occur when forests are replaced with agricultural crops, because the mass of carbon in vegetation per unit area is 20 to 50 times greater in forests than in croplands. The carbon in trees is released to the atmosphere when the wood is oxidized through burning or decay. Some of the organic matter of soils is also oxidized following disturbance and cultivation, and it contributes to the net release of carbon. Carbon is withdrawn from the atmosphere when agricultural land is abandoned and allowed to return to forest. Since about 6000 BP, the replacement of forests and grasslands with cultivated lands has released an estimated 250 Pg carbon to the atmosphere. The release is not known well because the amount of carbon held in vegetation and soils 6000 yr ago is uncertain. Over the last 150 yr the net loss of carbon to the atmosphere has been about 120 Pg as forests have been cleared and soils cultivated (9). This value is more reliable, but probably has an uncertainty of ± 30%. Estimates of the area and carbon content of major terrestrial ecosystems in 1980 are shown in Table 1.

The flux of carbon since 1850 is determined using historical records of land-use change and ecological data pertaining to the changes in vegetation and soil that accompany land-use change. The amount of carbon released to the atmosphere or accumulated on land depends not only on the magnitude and types of changes in land use but on the amounts of carbon held in different ecosystems. For example, the conversion of grasslands to pasture causes a smaller release of carbon to the atmosphere than the conversion of forests to pasture. The net release or accumulation also depends on the time lags introduced by the rates of decay of organic matter, the rates of oxidation of wood products, and the rates of regrowth of forests following harvest or following abandonment of agriculture land. The calculations include the accumulation of carbon in buildings, furniture, and land fills.

The long-term (1850–1990) flux of carbon to the atmosphere from global changes in land use is estimated to have been 120 Pg (Fig. 5). The annual flux of carbon increased from about 0.4 Pg/yr in 1850 to about 1.7 Pg/yr in 1990. The rate of the release has increased. It took 100 yr

Table 1. Area, Total Carbon, and Mean Carbon Content of Vegetation and Soils in Major Ecosystems of the Earth in 1980

Ecosystem	Area (10^6 ha)	Carbon in Vegetation (10^{15} g)	Carbon in Soil (10^{15} g)	Mean Carbon Content	
				Vegetation (t/ha)	Soil (t/ha)
Tropical evergreen forest	562	107	62	190	111
Tropical seasonal forest	913	116	84	119	89
Temperate evergreen forest	508	81	68	161	134
Temperate deciduous forest	368	48	49	131	134
Boreal forest	1,168	105	241	90	206
Tropical fallows (shifting cultivation)	227	8	19	36	83
Tropical woodland	776	28	53	39	68
Tropical grass and pasture	1,021	17	49	16	48
Temperate woodland	264	7	18	27	69
Temperate grassland and pasture	1,235	9	233	7	189
Tundra and alpine meadow	800	2	163	3	204
Desert scrub	1,800	5	104	3	58
Rock, ice, and sand	2,400	0.2	4	0.1	2
Temperate cultivated land	751	3	96	4	128
Tropical cultivated land	655	4	35	7	53
Swamp and marsh	200	14	145	68	725
Total	*13,684*	*554*	*1,423*		

for the first increase of 0.6 Pg/yr (from 0.4 to 1.0 Pg/yr) and only 35 yr (1950–1985) for the next increase of 0.6 Pg/yr. Until about 1940, the region with the greatest flux was the temperate zone. Since 1950, the tropics have been increasingly important. Current emissions from northern temperate and boreal zones are thought to be close to zero, with releases from decaying wood products approximately balanced by annual accumulations of carbon in regrowing forests (10). Globally, the annual flux of carbon from changes in land use is estimated to have increased during the 1980s from about 1.5 Pg in 1980 to about 1.7 Pg in 1990. Tropical Asia, America, and Africa are estimated to have contributed about 0.7, 0.6, and 0.3 Pg/yr, respectively.

About 66% of the total long-term flux, or 80 Pg carbon, was from oxidation of plant material, either burned or decayed. About 33%, or 40 Pg carbon, was from oxidation of soil carbon, largely from cultivation. Relative to the stocks

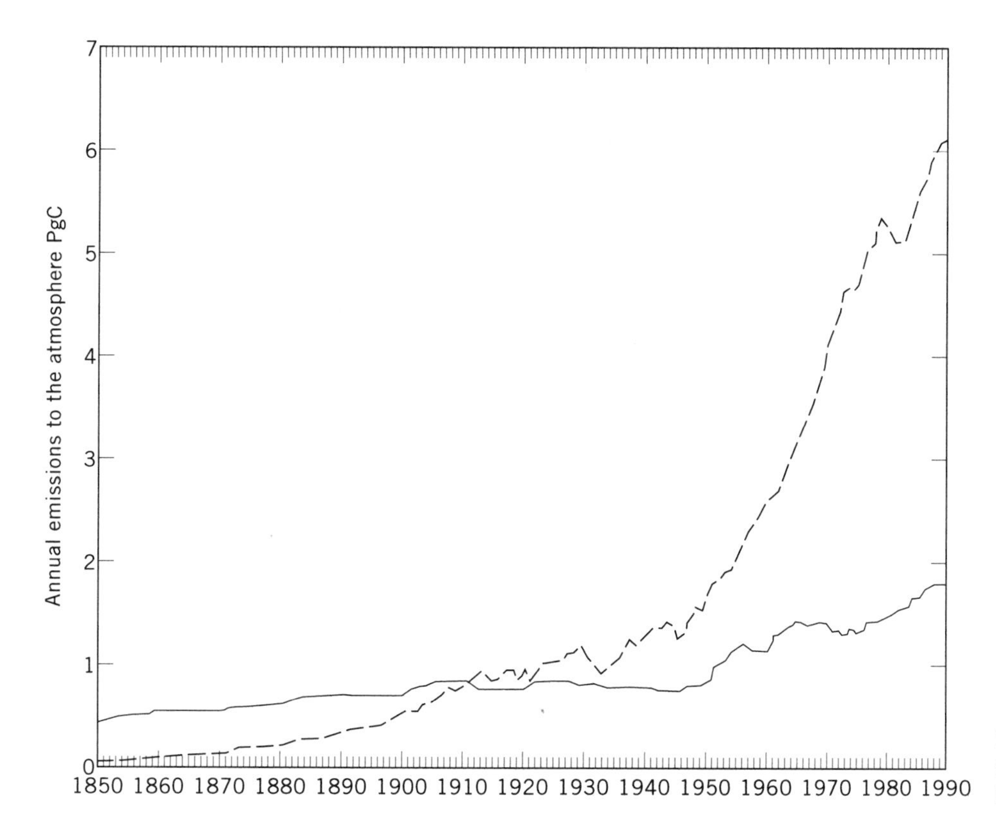

Figure 5. Annual sources of carbon from fossil fuels (dashed line) and land-use change (solid line).

of carbon in 1850, carbon in vegetation was reduced by about 12% over this 130-yr interval, and organic carbon in soil was reduced by about 4% worldwide.

The ratio of soil loss to biomass loss (1:2 in units of carbon) is somewhat surprising given the ratio of total stocks (1450:600, or approximately 2.4:1). The difference is explained by the fact that, on average, only 25 to 30% of the soil carbon is lost with cultivation, and a smaller fraction is usually lost when forests are converted to pastures or to shifting cultivation. In contrast, about 100% of the biomass is oxidized and released as CO_2 to the atmosphere with clearing. Assuming that, on average, 25% of soil carbon is lost, the effective ratio of soil carbon to biomass carbon becomes about 350:600, or approximately 1:2, in agreement with the ratio observed above.

Combustion of Fossil Fuels

The major fossil fuels are coal, oil, and natural gas. The CO_2 released annually from the combustion of these fuels is calculated from records of fuel production published by the United Nations since 1950 (11). Emissions before 1950 were also estimated (12). The rate of combustion has generally increased exponentially; the exceptions are interruptions caused by the two world wars and by the increase in oil prices during the 1970s. Despite the increase in price, the use of fossil fuels and the annual emission of CO_2 to the atmosphere increased yearly to 1980. From 1980 through 1983 the annual emissions decreased each year, but the downward trend was temporary. Starting in 1984, the annual emissions increased again; by 1985 they were similar to 1979. Since 1984, emissions have continued to increase. Annual emission of carbon from combustion of all fossil fuels was 6.0 Pg in 1991. Between 1850 and 1990, the total emission of carbon to the atmosphere from fossil fuels is estimated to have been about 225 Pg.

Total Emissions of Carbon from Human Activities

The net annual fluxes of carbon to the atmosphere from changes in land use (largely deforestation) and from fossil fuels are shown in Fig. 5. The net annual flux of carbon to the atmosphere from changes in land use before 1800 was probably less than 0.5 Pg and probably less than 1 Pg carbon per year until 1950. The combined emissions from both biotic and fossil fuel sources first exceeded 1 Pg at about the start of the 20th century. It was not until the middle of this century that the annual emission of carbon from combustion of fossil fuels exceeded the net biotic flux. Since then the fossil fuel contribution has predominated, although both fluxes have accelerated in recent decades with the intensification of industrial activity and the expansion of agricultural area. For the 1980s, the contemporary net flux averaged about 7.0 Pg carbon per year, 5.4 Pg from fossil fuels and 1.6 Pg from biotic sources (Fig. 6).

Although the industrial flows of carbon now dominate the net biotic flows, the gross natural flows of carbon still greatly exceed the anthropogenic flows. The natural flows of carbon in terrestrial photosynthesis and respiration are about 100 Pg annually, as are the gross flows between oceans and atmosphere (Figs. 1 and 6).

Changes in the Oceans

Direct measurement of changes in the amount of carbon contained in the world's oceans has not been possible for two reasons. First, the oceans are not mixed as rapidly as the atmosphere, so that spatial heterogeneity is large. Second, the background concentration of dissolved carbon in seawater is large relative to the change, so measurement of the change requires accurate methods that have

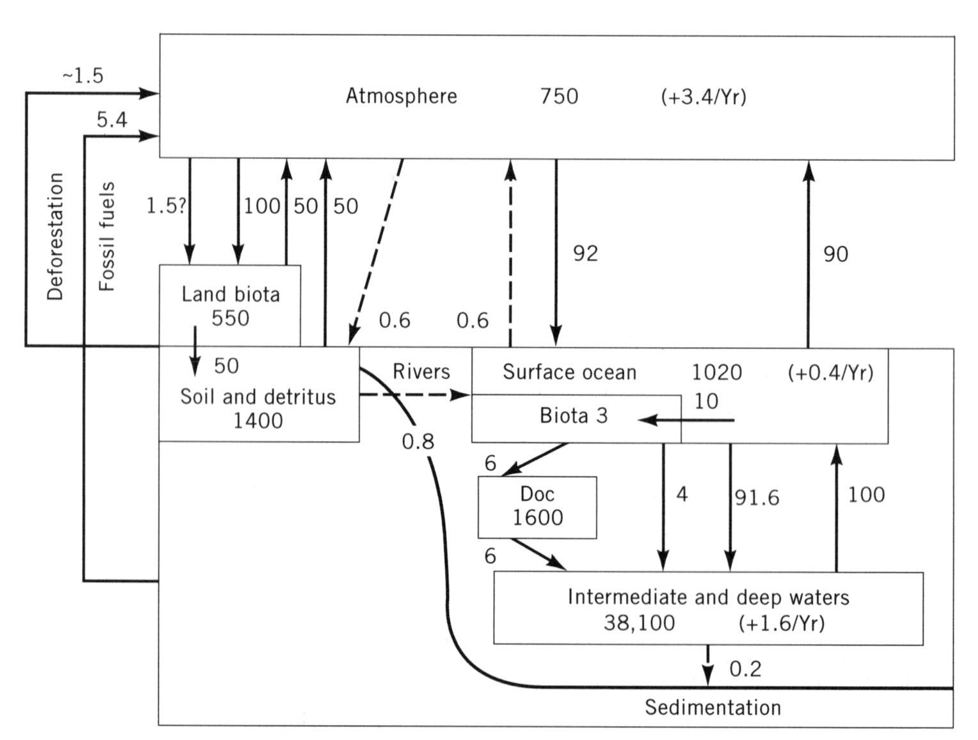

Figure 6. The global carbon cycle for the 1980s. Units are petagrams of carbon and petagrams of carbon per year. The atmosphere and oceans hold more carbon than they did before disturbance (Fig. 1), and the terrestrial biota and soils hold less. From Ref. 2.

not existed until the past few years. Most estimates of the uptake of terrestrial and fossil fuel CO_2 by the oceans have, therefore, been with models.

The world's oceans contain about 60 times more carbon than either the atmosphere or the world's terrestrial vegetation. Thus, at equilibrium, the ocean might be expected to have absorbed about 60 times more of the released carbon than the atmosphere, or 98% of total emissions. The capacity of the oceans to take up CO_2 decreases, however, as CO_2 is added to the atmosphere–ocean system. In preindustrial times, the distribution of CO_2 between atmosphere and ocean was 1.5 and 98.5%, respectively. If 150 Pg carbon are added to the atmosphere, that distribution becomes 16 and 84% (13). To date, the oceans have absorbed considerably less than 84%. As of 1980, the oceans are thought to have absorbed only about 40% of the emissions (20 to 47%, depending on the model used) (14), so there are clearly other mechanisms slowing the oceanic absorption of CO_2.

Two other mechanisms are the transfer of CO_2 across the air–sea interface and the mixing of water masses within the sea (15). The rate of transfer of CO_2 across the sea surface has been estimated by several different methods. One method is based on the fact that the transfer rate of naturally produced ^{14}C into the oceans should balance the decay of ^{14}C within the oceans. Both the production rate and the inventory of natural ^{14}C in the oceans are known, and the rate of transfer estimated by this method is about 100 Pg/yr (Fig. 1). A similar estimate is obtained by comparing the inventory of bomb ^{14}C in the oceans with the input of bomb ^{14}C from the atmosphere (16).

A third method for estimating the CO_2 gas transfer coefficient is based on measurement of radon gas in the surface ocean. Radon gas is generated by the decay of radon 226. The concentration of the parent radon 226 and its half-life allow calculation of the expected radon gas concentration in the surface water. The observed concentration is about 70% of expected, so 30% of the radon must be transferred to the atmosphere during its mean lifetime of 6 days. Correcting for differences in the diffusivity of radon and CO_2 allows an estimation of the transfer rate for CO_2. The transfer rates given by the ^{14}C methods and the radon method agree within about 10%. The rate of transfer of CO_2 across the air–sea interface is believed to have reduced the oceanic absorption of CO_2 by about 10% (15).

The most important process in slowing the oceanic uptake of CO_2 is the rate of vertical mixing. The surface layer of the ocean is approximately in equilibrium with atmospheric CO_2, but the bottleneck in oceanic uptake is in transporting surface waters to the intermediate and deeper layers.

The earliest attempts to model the uptake of carbon by oceans recognized that the ocean was not homogeneous but made up of at least two layers: a surface layer in contact with the atmosphere and a deep layer that can accumulate carbon only in exchange with the surface layer (17). Later versions of a one-dimensional ocean model consisted of a shallow surface ocean and a deep ocean composed of many boxes, stacked vertically, that exchanged carbon with those boxes immediately above or below (18). This type of model is commonly referred to as the box diffusion model.

More realistic models of the ocean are three-dimensional and include the mixing of tracers among water masses with similar densities (19). In addition to the surface and deep layers, these models include the thermocline, that portion of the ocean in which the temperature gradient is steepest. It is largely the steepness of this gradient in density that creates the bottleneck for deep-ocean absorption of CO_2.

All of the different ocean models are calibrated with tracers of some kind, either natural ^{14}C, bomb-produced ^{14}C, bomb-produced tritium, or other tracers. The profiles of these tracers were obtained during two extensive oceanographic surveys: geochemical ocean sections (GEOSECS), carried out between 1972 and 1978, and transient tracers in the ocean (TTO), carried out in the early 1980s. Both surveys measured profiles of carbon, radioisotopes, and other tracers along transects in the Atlantic and Pacific oceans. Current estimates of oceanic uptake of carbon based on all of these models range between 1.5 and 2.5 Pg/yr for the 1980s (2).

THE IMBALANCE

The global carbon cycle can be summarized by a simplified equation showing the atmospheric increase to be the sum of sources (combustion of fossil fuels and land-use change) and sinks (oceanic uptake) of carbon:

$$\text{Atmospheric increase} = \text{fossil source} + \text{land-use source} - \text{oceanic sink} - \text{residual sink}$$

For 1980 to 1989 the increase (in petagrams carbon per year) can be calculated as

$$3.2 \pm 0.2 = 5.4 \pm 0.5 + 1.6 \pm 1.0 - 2.0 \pm 0.5 - 1.8 \pm 1.2$$

and for 1850 to 1990:

$$140 = 225 + 120 - 100 - 105$$

As the equations show, an additional sink is needed to balance the equations for the last decade as well as for the longer term (Fig. 7). This unidentified sink (sometimes called the "missing carbon" sink) has received considerable attention over the last decade. The unidentified carbon sink makes projections of future atmospheric concentrations difficult. Should the sink be assumed to continue in the future? Will it increase in proportion to atmospheric concentrations of CO_2? Will it decrease? Will it saturate? When? What and where is the sink?

Two or three independent lines of geophysical evidence suggest that the unidentified sink is in terrestrial ecosystems. On the basis of a model of atmospheric transport, it was determined that the gradient in CO_2 concentrations between the Northern and Southern hemispheres was less steep than would be expected on the basis of the distribution of fossil fuel sources of CO_2 (20). Most of the CO_2 released from combustion of fossil fuels is released in the midlatitudes of the Northern Hemisphere and should

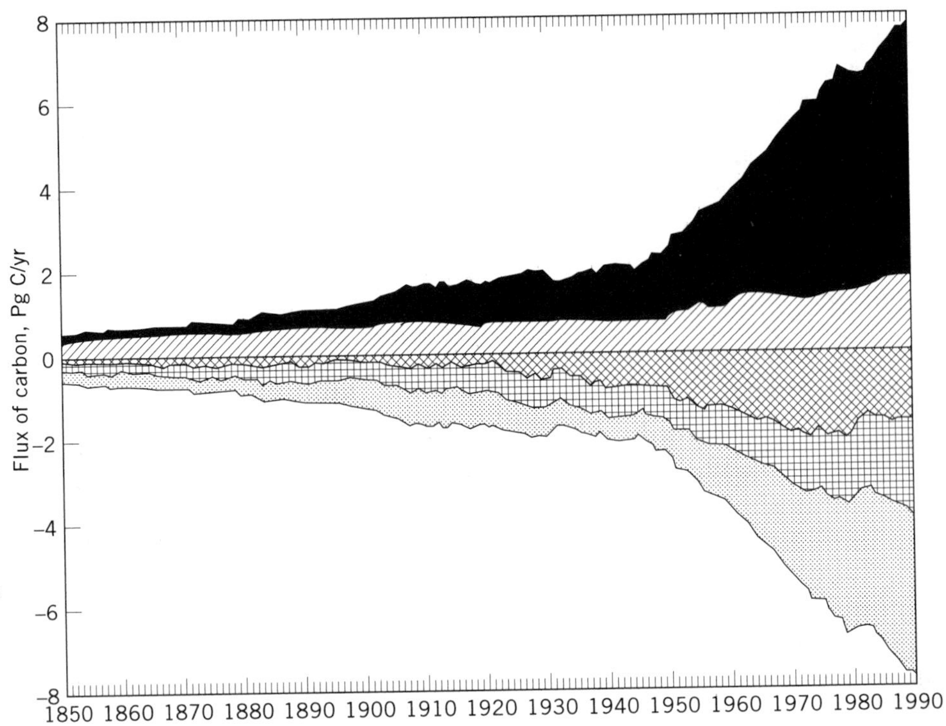

Figure 7. Annual sources and sinks of carbon in the global carbon cycle since 1850. In 1990, eg, the total source of 7.7 Pg (6.0 Pg from fossil fuels and 1.7 Pg from land-use change) must be equivalent to the total sink.
Emissions from fossil fuel
Net release from land-use change
Unidentified sink
Oceanic uptake
Atmospheric accumulation

cause a bulge in the concentrations at those latitudes. The bulge is there, but not as large as expected, indicating that CO_2 must be removed from the Northern Hemisphere before crossing the equator. Measurements of the partial pressure of CO_2 (P_{CO_2}) in surface waters suggest that the difference between atmospheric and surface ocean concentrations is not great enough to force CO_2 into the oceans. Thus it was concluded that the unidentified sink was on land in the northern midlatitudes (20).

Qualitatively, the inferred terrestrial sink is similar to that obtained with models of oceanic uptake, described earlier. However, the P_{CO_2} differences described above led the researchers to infer an oceanic uptake of carbon of only 0.3 to 0.8 Pg/yr, much less than the rate found by the more traditional ocean models (2.0 ± 0.5 Pg/yr). Could the oceanic models be that wrong?

Three factors have been proposed to reconcile the difference: carbon in rivers, the skin temperature of the ocean, and carbon monoxide (21). The earlier estimate (20) concerned the air–sea exchange but not necessarily the net uptake of carbon by the oceans. As part of the natural carbon cycle, the oceans are estimated to release about 0.6 Pg/yr to the atmosphere to balance the input from rivers. Rivers discharge about 0.8 Pg/yr into the ocean, of which about 0.2 Pg of carbon is buried in ocean sediment (Figs. 1 and 6). Thus the background or steady-state flux is not 0 but an exchange of 0.6 Pg/yr of carbon from ocean to atmosphere.

The air–sea exchange is sensitive to temperature at the surface, and measurements of temperature (20) were the traditional bulk temperature measurements of waters near the surface. Because of evaporation, however, the skin temperature at the very surface of the ocean may be 0.1° to 0.2°C cooler than the surface waters. The correction is equivalent to about 0.4 Pg/yr of carbon.

Finally, there is a correction required to account for the fact that some of the carbon emitted from fossil fuel combustion is released as CO rather than as CO_2. The CO is eventually oxidized to CO_2 in the atmosphere, but not until some of it has been redistributed in the atmosphere. The net effect of the CO emissions and redistribution is to reduce the bulge in the north–south gradient of CO_2. Correction for the effect raises the estimate of oceanic uptake of carbon by about 0.2 Pg/yr.

Adding all three of these corrections to the earlier estimate (20) gives an oceanic uptake of about 1.8 Pg/yr, within the range determined independently by ocean models. If all of these corrections applied only to the northern oceans, the need for a large terrestrial sink in midlatitudes would vanish. As it is, the northern sink is only partially reduced. Other analyses, however, show the midlatitude terrestrial sink to be much less. The sink was inferred to be terrestrial on the basis of the difference in CO_2 concentrations between the surface ocean and air (20). The gradient in concentrations was not steep enough to make the oceans a significant sink. If, instead of using the air–sea gradient in CO_2 concentrations, one uses the observed ^{13}C:^{12}C ratio in atmospheric CO_2, the data support an oceanic sink rather than a terrestrial one (3). ^{13}C is a heavier isotope of carbon than the more common ^{12}C. The heavier isotope is discriminated against in photosynthesis, so plants and soils (and fossil fuels) are depleted in ^{13}C. Trees have a lighter isotopic ratio (−22 to −27 parts per thousand) than does air (−7 ppt); the ratios are expressed in relation to a standard. If there were a large terrestrial sink at midlatitudes, preferentially removing ^{12}C from the atmosphere, the ^{13}C:^{12}C ratios of atmospheric CO_2 would be enriched in ^{13}C. But the observed isotopic ratios of atmospheric CO_2 from northern midlatitudes do not show such an enrichment. On the contrary, they show a small depletion in ^{13}C, suggesting that terrestrial ecosystems at that latitude are a source rather than a sink.

The magnitude of a terrestrial midlatitude sink is in question. P_{CO_2} data from the oceans suggest a large sink; ^{13}C:^{12}C ratios in air do not. Nevertheless, these geophysical analyses, based on atmospheric and oceanic data and models as well as other analyses based on short-term records of ^{13}C:^{12}C in the ocean all require a terrestrial sink of approximately the same magnitude as the terrestrial source from tropical deforestation. The location of that sink, however, is unclear. Another inversion approach, similar to the one discussed earlier (20), found the same northern midlatitude sink but also found a small net source in the tropics. The tropical source is of the magnitude expected from the outgassing of CO_2 from the tropical oceans, indicating that the large source of carbon from tropical deforestation must be balanced by an equally large accumulation somewhere in the terrestrial ecosystems of the tropics.

Nowhere, either in the tropics or in northern midlatitude forests, has an accumulation of carbon been observed in trees or soil. Some scientists argue that the accumulation of 1 to 2 Pg/yr is so small relative to the background levels of carbon in these ecosystems and relative to the natural exchanges of carbon with the atmosphere, that it could not be detected. But is this the case? Recent analyses of forest growth in northern regions, indeed, show an accumulation of carbon, but the rate of accumulation is that expected from past histories of logging, ie, recovering forests accumulate carbon. What does this imply for the missing carbon? If additional carbon were accumulating in the northern temperate zone, could it be measured? The global carbon imbalance is approximately 1.5 Pg/yr. If all of that carbon were accumulating in the trees of the northern temperate zone (area of temperate zone forests is about 600×10^6 ha), the accumulation would be approximately 2.5 Mg/(ha · yr). In comparison, the observed accumulation of carbon in European forests was 0.4 to 0.6 Mg/(ha · yr) (22). Thus the accumulation of missing carbon would be approximately five times higher than observed. It would be difficult to miss in the data on growth rates. Including boreal forests in the calculation lowers the accumulation rate of carbon to 1.0 Mg/(ha · yr), but it is still high relative to observed growth rates.

If the unidentified carbon sink were accumulating throughout the temperate zone, not just in forests but in woodlands, grasslands, and cultivated lands, the average accumulation rate of carbon would be about 0.5 Mg/(ha · yr), about the rate observed in European forests. This rate of accumulation in the vegetation of grasslands and cultivated lands is high relative to the standing stocks of carbon there and would be obvious. Finally, if the missing carbon is accumulating in trees throughout the world, the rate, again, would average about 0.5 Mg/(ha · yr). However, this is the rate observed for regrowing European forests, which have a long history of intensive use. That an additional accumulation of this magnitude, representing the missing sink, could be in the forests of the world, many of which have not been logged and, hence, are not regrowing (eg, large regions of Canada, central and eastern Russia, and Brazil), seems unlikely. Regrowing forests are young; mortality is low, and as a result, little carbon is lost. Even if undisturbed forests around the world were growing larger in response to elevated concentrations of CO_2 or to increased deposition of nitrogen, they would be unlikely to be accumulating carbon at rates that approximate rates of regrowth immediately after logging. By these arguments, it seems unlikely that carbon is accumulating in vegetation anywhere at the rate required to balance the global carbon budget (23). The same is true for soils. Observed rates of carbon accumulation in soils of undisturbed ecosystems are too low to account for the unidentified sink (24).

In summary, most analyses of the contemporary carbon cycle based on geophysical models point to terrestrial ecosystems as the unidentified sink. Direct measurements of tree growth do not show such a sink, however. Uncertainties in these models leave the location of the sink ambiguous. Tropical as well as northern temperate zone lands are candidates, and the ocean itself cannot be ruled out as accounting for some portion of the unidentified sink. If the sink is spread throughout the terrestrial ecosystems of the earth, and in both vegetation and soil, it may be undetectable.

Further insights as to the mechanism or location of the unidentified sink may be obtained from a consideration of its change through time. The temporal pattern of the sink has been determined by comparing geophysical estimates of a terrestrial flux of carbon with estimates based on changes in land use (9). Models of oceanic uptake are traditionally used to estimate the concentration of CO_2 in the atmosphere that will result from different trends in emissions. The models can also be run in an inverse mode. In this mode, past concentrations of CO_2 are used as input to the models, and the models calculate what the sources and sinks of carbon must have been to yield the observed concentrations. In this way, the oceanic models have been used to infer the terrestrial flux of carbon over the last 150 yr or so. The historic record of CO_2 concentrations is available from ice cores. Bubbles of air trapped in glacial ice contain samples of air that was present at the time of ice formation. The ice is crushed under carefully controlled conditions, and the air is analyzed for CO_2. This record of CO_2 concentrations is then used, together with the direct atmospheric measurements since 1958, by models of oceanic uptake to calculate the annual sources of carbon necessary to reproduce the CO_2 record. Subtracting the sources of CO_2 from combustion of fossil fuels since about 1850 leaves a residual pattern of sources and sinks that is generally assumed to represent the net exchanges of carbon with terrestrial ecosystems. The assumption is consistent with the ratio of ^{13}C to ^{12}C in the air over this period.

The cumulative flux of this residual for the period 1850–1990 is calculated to have been a release to the atmosphere of between 25 and 50 Pg of carbon according to these inverse calculations (3,19,25). The temporal patterns of the estimates are almost identical. According to the most recent of these analyses, the residual flux (presumably terrestrial) was a net release of carbon of about 0.4 Pg/yr in 1800, rising gradually to 0.6 Pg/yr by 1900, remaining at 0.6 Pg/yr until about 1920, and then falling to 0 by the late 1930s (Fig. 8). The flux continued to fall, indicating a net accumulation of carbon in terrestrial eco-

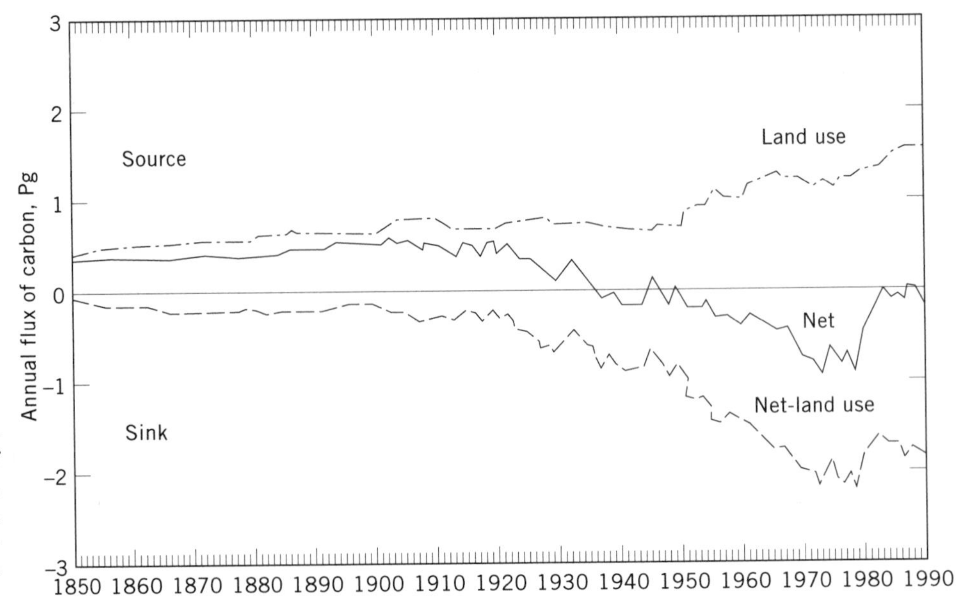

Figure 8. The residual flux of carbon determined by inverse methods (assumed to represent the total net terrestrial flux), the annual net flux of carbon from changes in land use, and the difference between them (assumed to represent changes in terrestrial ecosystems caused by factors other than changes in land use). From Ref. 19.

systems after 1930, reaching a maximum sink of 0.8 Pg/yr in the 1970s, and then abruptly returning to 0 by the early 1980s. It has remained near 0 since ca. 1982.

If this residual flux is from terrestrial ecosystems, it bears little resemblance to that calculated from changes in land use, which was about 120 Pg of carbon over the period since 1850 and increased almost steadily (Fig. 5). The disagreement, however, may not be the result of errors in one or the other of the approaches. The difference between these two estimates may represent a flux of carbon between terrestrial ecosystems and the atmosphere that is not related to changes in land use but to changes in the global environment.

Analyses of flux based on changes in land use assume that ecosystems not directly affected by land-use change are in a steady state with respect to carbon. This assumption is probably not valid. The amount of carbon stored in vegetation and soils is changed not only by deliberate human activity but also from inadvertent changes in climate, CO_2 concentrations, nutrient deposition, and pollution. Analyses of the flux of carbon due to land-use changes generally take into consideration deforestation (with the land turned into crops or pasture or degraded), reforestation (either purposefully or as a result of the abandonment of agriculture), and changes in the biomass within the forests (due to logging and regrowth, shifting cultivation and degration). Such studies tend to omit from consideration unmanaged, undisturbed areas (because the carbon cycle is assumed to be in a steady state); forest decline or enhancement; eutrophication (including CO_2 fertilization); frequency of fires (including fire surpression); climatic change, and desertification.

It is useful to distinguish between two causes or types of change: deliberate and inadvertent. Deliberate changes result directly from management, eg, cultivation. Inadvertent changes occur through changes in the metabolism of ecosystems, more specifically through a change in the ratio of the rates of primary production and respiration (respiration includes autotrophic and heterotrophic respiration, or plant and microbial respiration). Clearly, management practices affect metabolism, and the distinction between deliberate and inadvertent effects is blurred. Nevertheless, the distinction is a useful one because changes in carbon stocks that result from deliberate human activity are more easily documented than changes resulting from metabolic changes alone. Deliberate changes usually involve a well-defined area, eg, the area in agriculture or the area reforested. Furthermore, the changes in the density of carbon (t/ha) associated with land-use change are generally large: forests hold 20 to 50 times more carbon in their vegetation than the ecosystems that replace them following deforestation. In contrast to these large and well-defined changes associated with management, changes that occur as a result of metabolic changes are generally subtle. They occur slowly over poorly defined areas. They are difficult to measure against the large background levels of carbon in vegetation and soils and against the natural variability of diurnal, seasonal, and annual metabolic rates.

Assuming that the flux of carbon obtained from inversion of ocean models is a terrestrial flux, it includes both the flux from land-use change and a flux caused by other changes in terrestrial ecosystems. The flux of carbon from land-use change, on the other hand, includes only those lands directly and deliberately managed. Thus the difference between the two estimates of flux may define the flux of carbon to or from terrestrial ecosystems not directly modified by humans. This flux is the unidentified sink.

The difference is shown in Figure 8 (dashed line). The flux has always been negative and is interpreted to mean that some terrestrial ecosystems were accumulating carbon independent of land-use change. Positive differences, if they occurred, would indicate that some terrestrial ecosystems were releasing carbon to the atmosphere in addition to that released from changes in land use.

The long-term difference in these two estimates is a net sink of 70–95 Pg of carbon (a net release of 25 to 50

Pg minus the release of izo Pg from land use). As with the shorter-term sink, if this sink has been in mid- and high latitude forests, the increase in storage should be observed. Even if 66% of the sink has been in trees, and 33% in soil, the long-term sink in trees is equivalent to about a 20% increase in biomass or ring increment. Studies of tree rings do not consistently find an increased growth. It is important to recall here that the unidentified sink is over and above the sink attributable to regrowth of forests following abandonment of agricultural land or logging, ie, to changes in land use.

FEEDBACK BETWEEN THE CARBON CYCLE AND GLOBAL CLIMATE

One of the questions important to both climate modelers and ecologists is how the present-day dynamics of the carbon cycle will change if climate changes. If the earth warms, will the oceans take up carbon more rapidly or more slowly? Will terrestrial ecosystems release additional carbon with an increase in temperature, or will rising concentrations of CO_2 enhance plant productivity and store more carbon on land? How will changes in precipitation affect this terrestrial storage? Answers to these questions about climatic change are probably related to the unidentified sink for carbon. The processes responsible for the unidentified sink, if they could be identified, might shed light on how the carbon cycle will respond to a warmer earth with higher CO_2 concentrations. Feedback between the global carbon and global climate systems is defined as a process that either enhances (positive feedback) or reduces (negative feedback) the warming. Feedback (26) and biotic feedback (27) have been reviewed.

Oceans

In general, the oceans tend to buffer changes in atmospheric CO_2 and thus serve as a negative feedback for any change. With respect to feedback between temperature and carbon, however, it is useful to consider physical, chemical, and biological processes separately. Physically, as the oceans warm, they hold less carbon because of the lower solubility of gases at higher temperatures. Furthermore, a warming may lead to increased stratification of the water column and, thereby, reduced circulation and reduced uptake of carbon. The greatest warming is expected to occur in higher latitudes; the least warming, at the equator. Such a pattern of warming would reduce the north–south thermal gradient and might, thereby, reduce the intensity of circulation. In both of these physical respects, the oceans will serve as a positive feedback.

Chemically, the ocean's buffering capacity tends to keep atmospheric concentrations of CO_2 from changing. As the oceans take up more carbon, they become more acidic and thus have a diminished capacity to absorb additional carbon. The net effect of the buffering is still a negative feedback, however. Furthermore, as additional carbon is taken up by the oceans, the increasing acidity of the waters will erode calcium carbonate sediments, thereby increasing alkalinity and decreasing atmospheric CO_2, again a negative feedback. This process is important over hundreds and thousands of years but is too slow to be significant over the next 100 yr.

There is another potential feedback between the oceanic carbon cycle and global warming that is more difficult to assess: the role of the marine biota. The *biological pump* refers to photosynthetic production of organic matter in the surface ocean, its sinking or export to depth, and its subsequent decomposition or decay, which releases nutrients and CO_2 back into solution at depth. In preindustrial times, the biological pump was responsible for maintaining the concentration of CO_2 in the atmosphere at about half of what it would have been in the absence of the biological pump. Might the effectiveness of the biological pump change with a warming? First, it is unlikely that marine phytoplankton will respond to an increased concentration of CO_2 in seawater because the supply from the other chemical forms of inorganic carbon (bicarbonate and carbonate ions) is almost inexhaustible. Not carbon but nutrients and light are the major factors the limit marine productivity. So the question becomes, How might a global warming affect the concentrations of nutrients in surface waters?

According to one argument, a global warming will reduce the quantity of nutrients delivered to the surface through a reduction in upwelling. Upwelling is related to the intensity of oceanic mixing, and that intensity might be diminished if the north–south gradient in temperature were reduced through global warming. But upwelling waters bring high CO_2 concentrations to the surface as well as nutrients. The two are related: they both result from decomposition of organic matter produced at the surface. Thus it is not clear that a reduction in the biological pump will result in higher atmospheric CO_2 concentrations. Reduced upwelling will bring fewer nutrients to the surface and thus reduce productivity, but it will also bring less high CO_2 water to the surface and thus reduce the release of CO_2 to the atmosphere.

Because of this link between nutrients and carbon, there are only two ways that the biological pump might change the net removal of carbon from the euphotic zone, according to this argument. One way is to change the total amount of nutrient in the oceans; the other way is to allow the carbon:nutrient ratios to change. There is little evidence that carbon–nutrient ratios are flexible, but a change in sea level might increase the nutrient inventory of the ocean through inundation of nutrient-rich soils. If a changing climate increased the intensity of winds or increased aridity, it might also increase the effectiveness of the biological pump through increased delivery of terrestrially eroded iron to the ocean. Iron is thought to limit production in many parts of today's ocean. Under this circumstance, the feedback would be negative.

The counterargument is that reduced upwelling will reduce the effectiveness of the biological pump and will reduce the rate of oceanic uptake of carbon, at least in the short term. The argument distinguishes between the preindustrial fluxes of carbon and the perturbational fluxes, or those due to human activity. Nutrients and CO_2 in upwelled water were presumably in balance preindustrially. Today, however, the atmosphere and deep ocean are not in equilibrium. Anything that enhances circulation or mixing thus enhances the equilibration of atmosphere and ocean. Although upwelled waters contain CO_2 at absolutely higher concentrations than the current atmospheric CO_2, the concentration in upwelled waters is low relative to an equilibrated ocean. Hence, decreased upwelling reduces the biological pump and thereby reduces oceanic uptake of carbon. The mechanism is largely a physical one (described above). In the oceans, positive feedback seems more likely to prevail with global warming.

Terrestrial Ecosystems

CO_2 Fertilization (Negative Feedback). There is also considerable uncertainty surrounding the role of terrestrial ecosystems in carbon–climate feedbacks. The two factors receiving most attention are the effects of temperature and CO_2 on carbon storage. The increase in the concentration of CO_2 over the last centuries is thought by many researchers to have increased the storage of carbon on land. Based on short-term experiments in greenhouses, it is clear that many crops, annual plants, and tree seedlings grow more rapidly under elevated levels of CO_2. It is less clear that such an increased growth will occur in natural ecosystems, that the growth response found in annual crops will apply to natural systems, that it will apply to periods longer than a growing season, or that any enhanced growth that occurs will lead to an increase in the amount of carbon held in ecosystems.

There have been two moderately long-term experiments in which natural ecosystems were exposed to elevated concentrations of CO_2, and the results were not consistent (28). The net uptake of CO_2 by a tundra community showed an increased storage of carbon, but the effect lasted only 3 to 5 yr, or until some other factor became limiting to the growth of plants and the storage of carbon in the system. The results suggested a limited response to elevated levels of CO_2. In a continuing study in a tidal marsh, the increased storage of carbon has continued for 6 yr. Rates of photosynthesis increased and rates of decomposition seemed to have decreased. In the 6th yr, the experimental systems still showed an annual accumulation of carbon greater than the control and as large a response as in the first 5 yr. Whether an increased storage will or has already occurred in forests is unclear. The expense and logistical problems of enclosing a forest in chambers have so far prevented such an experiment.

Tree rings might provide an answer. Do trees show an increased rate of growth over the last decades? If the answer were yes, then perhaps the unidentified sink would have been identified. The next question becomes whether the increase in growth can be attributed to elevated CO_2, to changes in climate, or to other factors. Studies of tree rings have been equivocal so far, with about as many showing decreasing rates as showing increasing rates of growth. An exception appears in trees selected for temperature sensitivity. A sytematic investigation of trees from high latitudes and high elevations in both hemispheres indicated increased rates of growth, apparently related to the warming over the last century (29). The authors of that study caution, however, that the trees selected were not necessarily representative of the earth's forested regions. Indeed, studies in midlatitude regions indicate that rates of forest growth there are consistent with the logging histories in those regions (23); there is no indication of an enhanced growth. Nevertheless, a systematic survey relating global changes in tree rings to carbon storage has not been carried out.

The increase in amplitude of the winter–summer oscillation of CO_2 concentrations is often interpreted to indicate a trend of increased photosynthesis. But increased photosynthesis is not the same as increased storage of carbon. An increase in photosynthesis might be balanced by an equivalent increase in respiration. Furthermore, the increasing amplitude is also consistent with an increased winter respiration associated with warmer winters (30). If the cause of the increase in the seasonal amplitude of CO_2 concentrations has been increased respiration in winter, it would indicate a decreased storage of carbon on land rather than an increased storage. The increasing amplitude is interesting, but its significance is not yet clear.

The difference (evaluated above) between the total net terrestrial flux of carbon (obtained from inverse models of oceanic uptake) and the flux from changes in land use implies an increased storage of carbon over the last 70 yr (Fig. 8). The difference exhibits three features: (1) a period before 1920 that showed a small terrestrial sink (not different from 0) with little variation, (2) a period between ca. 1920 and 1975 when some of the world's terrestrial ecosystems were apparently accumulating carbon at an increasing rate, and (3) a period since the mid-1970s in which the rate of accumulation has decreased. Over the period 1850 to 1990, the generally increasing sink was proportional to increasing industrial activity. This suggests that CO_2 fertilization or increased nitrogen deposition may have been responsible for an accumulation of carbon in undisturbed ecosystems. Over the shorter period since 1940, however, the difference is poorly correlated with increasing CO_2 concentrations. Temperature seems to explain more of the variation and suggests a positive feedback.

Temperature. Temperature is likely to cause a net release of carbon to the atmosphere in addition to that released from changes in land use. Warmer temperatures are known to increase the rates of respiration and decay, which convert organic matter to CO_2 and, under anaerobic conditions, to CH_4. Warmer temperatures may also increase rates of photosynthesis and growth, especially in cold regions, but respiration is generally more sensitive to temperature than photosynthesis is. Furthermore, the amount of carbon held in the world's soils is about three times greater than the amount held in plants and two times greater than that in the atmosphere. Soil carbon is thus a large reservoir of terrestrial carbon that might be released to the atmosphere. Only a fraction of the organic carbon in soils is labile, however. Cultivation, eg, causes a loss of carbon from soils, but typically about 25% of the amount held in the top meter of soil. The larger fraction of soil carbon is more refractory and less likely to be mobilized with an increase in temperature. Although a warming can be expected to enhance decomposition of organic matter in soil, decomposition of organic matter releases nutrients as well as CO_2. The release of additional nutrients may, in turn, allow productivity to increase and may lead to an increased storage of carbon in wood. Based solely on the difference in carbon:nitrogen ratios of soil (about 8) and wood (about 200), carbon released from soil might be more than compensated for by carbon accumulated in wood. The calculation assumes that the nitrogen mineralized will be incorporated into an enhanced production of wood. If little of the nitrogen is lost, the net effect of warming may be to store carbon.

The evidence for positive or negative feedback between carbon and temperature seems to show that positive feedback dominates at most time scales. Over tens of thousands of years, the CO_2 concentration of the atmosphere is related to mean global temperature. During the last glacial period, CO_2 concentrations were about 180 ppmv; during the interglacial, they were 280 ppmv. The temporal resolution of the Vostok core is not sufficient to determine whether the warming preceded the change in CO_2 or vice versa, but during cooling periods, changes in temperature preceded changes in CO_2. Whatever, the initiating factors, the correlation is interpreted as evidence for a positive feedback.

Over the last 30 yr, there is also a positive correlation between CO_2 concentration and mean global temperature (3). The change in CO_2 again followed changes in temperature, in this case with a lag of a few months. Based on atmospheric ^{13}C:^{12}C ratios in CO_2, it was suggested that the source of the carbon related to warming is terrestrial (3).

The unidentified flux of carbon shown in Figure 8 is also correlated with temperature over the last 50 yr but not in the preceding 100 yr (9). Over the shorter period since 1940, the difference is correlated more with temperature than with atmospheric concentrations of CO_2. Larger sinks were associated with cooler years; reduced sinks were associated with warmer years. The fact that a similar relationship was not apparent before 1940 suggests either that the relationship is accidental or that different mechanisms were dominating in the two periods. Perhaps before 1940, ecosystems were largely rebounding from the little Ice Age, while, after 1940, they were responding to a more rapid warming.

Although most evidence seems to support positive feedback, the magnitudes of the various mechanisms cannot be predicted at present. The effects may cancel each other or alternate in time. A warmer earth might support a larger area of forests with a greater storage of carbon. However, unless deliberate steps are taken to reduce emissions of greenhouse gases, the immediate prospect is not a new climate but a continuously changing one. The ability of forests to keep up with such changes through migration will be difficult. It is more likely that a continuously changing climate will result in increased areas affected by forest dieback or fires or both, releasing additional amounts of carbon to the atmosphere.

Other Factors. Finally, there are a number of factors that do not strictly feed back to the global climate system but that can be expected to affect the ability of other feedback systems to operate. If carbon has been accumulating in undisturbed forests, balancing the global carbon equation over the last decades, there are reasons to expect that continuation of this accumulation is limited. The area in forests has decreased by about 15%, globally, over the last 140 yr, and rates of deforestation in the tropics are higher now than ever before. If the world's population doubles in 30 yr, the need for additional agricultural land will only increase the pressure on the remaining forests. At the same time, decreasing concentrations of stratospheric ozone may increase the amount of ultraviolet radiation reaching the earth's surface. The effect is likely to reduce both marine and terrestrial productivity and, as a result, carbon storage. The effects of acid rain, which may have increased the productivity of some European forests through increased deposition of nitrate, may eventually saturate the system and decrease productivity. Most of these effects will not only enhance a global warming but reduce the capacity of the earth to support the human enterprise. The future need not be bleak, however. The steps for stabilizing concentrations of CO_2 and other greenhouse gases are clear. They include large reductions in emissions of greenhouse gases, ie, large reductions in the use of fossil fuels and in rates of deforestation. Such reductions would be good for a number of reasons besides slowing the rate of global warming.

BIBLIOGRAPHY

1. J. T. Houghton, G. J. Jenkins, and J. J. Ephraums, eds., *Climatic Change. The IPCC Scientific Assessment,* Cambridge University Press, Cambridge, UK, 1990.
2. U. Siegenthaler and J. L. Sarmiento, *Nature* **365,** 119–125 (1993).
3. C. D. Keeling, R. B. Bacastow, A. F. Carter, S. C. Piper, T. P. Whorf, M. Heimann, W. G. Mook, and H. Roeloffzen in D. H. Peterson, ed., *Aspects of Climate Variability in the Pacific and the Western Americas, Geophysical Monograph* **55,** American Geophysical Union, Washington, D.C., 1989, pp. 165–236.
4. A. Neftel, E. Moor, H. Oeschger, and B. Stauffer, *Nature* **315,** 45–47 (1985).
5. H. Friedli, H. Lotscher, H. Oeschger, U. Siegenthaler, and B. Stauffer, *Nature* **324,** 237–238 (1986).
6. U. Siegenthaler, H. Friedli, H. Loetscher, E. Moor, A. Neftel, H. Oeschger, and B. Stauffer, *Ann. Glaciol.* **10,** 151–156 (1988).
7. J. M. Barnola, D. Raynaud, Y. S. Korotkevich, and C. Lorius, *Nature* **329,** 408–414 (1987).
8. S. Arrhenius, *Philos. Mag.* **41,** 237 (1896).
9. R. A. Houghton in G. M. Woodwell and F. T. Mackenzie, eds., *Biotic Feedbacks in the Global Climate System: Will the Warming Feed the Warming?,* Oxford University Press, Oxford, UK, in press.
10. R. A. Houghton, R. D. Boone, J. R. Fruci, J. E. Hobbie, J. M. Melillo, C. A. Palm, B. J. Peterson, G. R. Shaver, G. M. Woodwell, B. Moore, D. L. Skole and N. Myers, *Tellus* **39B,** 122–139 (1987).
11. G. Marland and R. M. Rotty, *Tellus* **36B,** 232–261 (1984).
12. C. D. Keeling, *Tellus* **25,** 174–198 (1973).
13. J. L. Sarmiento, *Chem. Eng. News* **71,** 30–43 (1993).
14. B. Bolin in B. Bolin, B. R. Doos, J. Jager, and R. A. Warrick, eds., *The Greenhouse Effect, Climatic Change, and Ecosystems, SCOPE* **29,** John Wiley & Sons, Inc., New York, 1986, pp. 93–155.

15. W. S. Broecker, T. Takahashi, H. H. Simpson, and T.-H. Peng, *Science* **206,** 409–418 (1979).
16. W. S. Broecker, T.-H. Peng, G. Ostlund, and M. Stuiver, *J. Geophys. Res.* **90,** 6953–6970 (1988).
17. R. Bacastow and C. D. Keeling in G. M. Woodwell and E. V. Pecan, eds., *Carbon and the Biosphere, U.S. Atomic Energy Commission Symposium Series* **30,** National Technical Information Service, Springfield, Va., 1973, pp. 86–135.
18. H. Oeschger, U. Siegenthaler, U. Schotterer, and A. Gugelmann, *Tellus* **27,** 168–192 (1975).
19. J. L. Sarmiento, J. C. Orr, and U. Siegenthaler, *J. Geophys. Res.* **97,** 3621–3645 (1992).
20. P. P. Tans, I. Y. Fung, and T. Takahashi, *Science* **247,** 1431–1438 (1990).
21. J. L. Sarmiento and E. T. Sundquist, *Nature* **356,** 589–593 (1992).
22. P. E. Kauppi, K. Mielikainen, and K. Kuusela, *Science* **256,** 70–74 (1992).
23. R. A. Houghton, *Global Biogeochem. Cycles* **7,** 611–617 (1993).
24. W. H. Schlesinger, *Nature* **348,** 232–234 (1990).
25. U. Siegenthaler and H. Oeschger, *Tellus* **39B,** 140–154 (1987).
26. D. A. Lashof, *Climatic Change* **14,** 213–242 (1989).
27. G. M. Woodwell and F. T. Mackenzie, eds., *Biotic Feedbacks in the Global Climate System: Will the Warming Speed the Warming?* Oxford University Press, Oxford, UK, in press.
28. H. A. Mooney, B. G. Drake, R. J. Luxmoore, W. C. Oechel, and L. F. Pitelka, *BioScience* **41,** 96–104 (1991).
29. G. C. Jacoby and R. D. D'Arrigo in Ref. 27.
30. R. A. Houghton, *J. Geophys. Res.* **92,** 4223–4230 (1987).

CARBON DIOXIDE EMISSIONS, FOSSIL FUELS

JAE EDMONDS
Pacific Northwest Laboratory, Washington, D.C.
MICHAEL GRUBB
Royal Institute of International Affairs, London
PATRICK TEN BRINK
MICHAEL MORRISON
Caminus Energy Limited, Cambridge, UK

In the late 1980s, interest flourished in the issue of global climate change. Many studies focused on the options for limiting anthropogenic emissions of greenhouse-related gases and managing the consequences of global warming and climate change. Making appropriate policy choices requires information on both the costs and benefits, as they occur over time, of policy interventions, and an increasing number of studies have sought to quantify the costs especially of limiting CO_2 emissions, as the dominant anthropogenic source. Such analyses now form an important part of overall policy assessments and influence international negotiations on policy responses.

See also COAL COMBUSTION; COMBUSTION MODELING; COMMERCIAL AVAILABILITY OF ENERGY TECHNOLOGY.

The majority of work in estimating the costs of reducing greenhouse gas emissions has occurred since 1988, but interest in the issue of costing emissions reductions began more than a decade earlier with the work of Nordhaus (1,2). Nordhaus's early work focused on the issue of reducing fossil-fuel CO_2 emissions, as did that of others (3–13). Only references 6, 11, and 13 contain data on non-CO_2 emissions, and these studies treated them separately and in an ad hoc manner; none of the studies took land-use change into account explicitly.

Although not the primary focus of their analysis, some of the studies conducted prior to 1988 analyzed the cost of emissions reductions. The results of these studies foreshadow the current debate. Reference 14 contains an assessment that explores the subsequent development, deepening, and broadening of research veins, focusing on the past five years of research on the costs of limiting CO_2 emissions. While this article focuses on the cost of reducing fossil fuel CO_2 emissions, it should be recalled that it is only one, albeit important, element in the full analysis of costs and benefits of policy intervention. A full analysis requires consideration of all greenhouse related gases from all related human activities, relevant biogeochemical processes, and damages to human and natural systems.

MODELING AND COSTING DEFINITIONS AND PARADIGMS

The cost of emissions reductions is always calculated as a difference in a given measure of performance between a reference scenario and a scenario that involves lower emissions. By far the most commonly used measures of performance are the net direct financial costs to the energy sector assessed at a specified discount rate, and the estimated impact on gross national product (GNP) or its close cousin GDP. GNP is the monetary value of new final goods and services produced in a given year, and it provides a measure of the scale of human activities that pass through markets, plus imputed values of some nonmarket activities. It is generally assumed that financial costs in the energy sector can be closely related to impacts on GNP, although this is not always the case.

Neither direct financial costs nor GNP provide direct measures of human welfare. One factor is that human welfare does not necessarily increase linearly with the degree of consumption; a given loss of income will likely matter far more to poor people, or poor countries, than to richer ones, for example. Some studies attempt to capture this through "equivalent welfare" measures, but these still rely centrally on a GNP basis. Some of the models discussed use the Hicks equivalent welfare variation, which assumes a national utility function to estimate the overall GNP increase that would be required to maintain the original level of utility.

GNP does not incorporate many nonmarket factors that affect welfare. Some studies have sought to examine explicitly the impact of abatement on various non-market costs, such as example for local air quality, and concluded that these can be very significant. However, in general, studies focus on financial costs or GNP impact. In the broader literature, other welfare indexes have been attempted (such as the United Nations Development Program's (UNDP) Human Development Index), but data are rarely adequate to quantify impacts in such terms in abatement costing studies. At present, for quantifying results there is little practical alternative to working with monetary cost and GNP impacts.

Economic studies use top-down models which analyze aggregated behavior based on economic indexes of prices and elasticities, and focus implicitly or explicitly on the use of carbon taxes to limit emissions. These studies have mostly concluded that relatively large carbon taxes (eg, that could much more than double the mine–mouth cost of coal) would be required to achieve goals such as the stabilization of fossil-fuel carbon emissions. Technology-oriented studies use bottom-up engineering models, which focus on the integration of technology cost and performance data. Many such studies have concluded conversely that emission reductions could be achieved with net cost savings. The division between the economic and engineering paradigm is closely related, but not identical, to the division between top-down and bottom-up models, as it has emerged in the literature.

There are also many modeling differences within each category. Most notably, since 1990 an important general division has become evident between the application of top-down models developed for long-run equilibrium analysis of energy and abatement costs (reflecting an idealized economy with optimal allocation of resources), and conventional macro-economic models designed for shorter-run analysis of the dynamic responses of economies which reflect many existing imperfections. Long-run equilibrium models generally estimate the costs of reducing emissions to be positive and high by the standards of most environmental measures implemented to date. Macro-economic models indicate a far more complex pattern of responses and cost indicators which may move in different directions and vary over time.

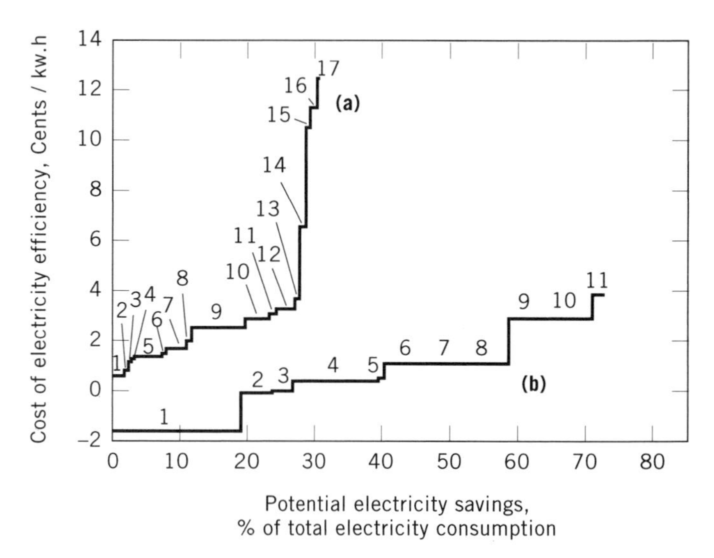

Figure 1. Discrete energy efficient technology cost curves for U.S. electricity (**a**) 1. Industrial process heating; 2. Residential lighting; 3. Residential water heating; 4. Commercial water heating; 5. Commercial lighting; 6. Commercial cooking; 7. Commercial cooking; 8. Commercial refrigeration; 9. Industrial motor drives; 10. Residential appliances; 11. Electrolytics; 12. Residential space heating; 13. Commercial and industrial space heating; 14. Commercial ventilation; 15. Commercial water heating (heat pump or solar); 16. Residential cooking; and 17. Residential water heating (heat pump or solar). (**b**) 1. Lighting; 2. Lighting's effect on heating and cooking; 3. Water heating; 4. Drive power; 5. Electronics; 6. Cooling; 7. Industrial process heat; 8. Electrolysis; 9. Residential process heat; 10. Space heating; 11. Water heating (solar).

Technology Cost–Curve Results

Clearly a principal determinant of CO_2 abatement costs will be the costs and adequacy of technologies that can reduce emissions. Many studies of the technologies that could help to limit greenhouse gas emissions have been conducted; important reviews are given in Refs. 15–20. In addition, many international databases with information on energy technologies have now been established; the UNEP study (21) lists no less than 13 technology databases available.

Technology cost curves provide a useful way of summarizing the technical potential for limiting emissions as identified in such studies. The simplest approach is to stack up different technologies in order of the cost of emission reductions or energy displaced, though cost curves can be used to represent the output for almost any degree of sophistication in modeling. There are various ways of generating cost curves of successively greater sophistication and consistency, as discussed in UNEP (21).

Discrete technology cost curves for various developed countries are presented in references 22–25. An EPRI (26) study examined potential savings in the U.S. electricity sector and concluded that "if by the year 2000 the entire stock of electrical end-use stock were to be replaced with the most efficient end-use technologies (nearly all of them estimated to be cheaper than equivalent supply), the maximum savings could range from 24% to 44% of electricity supply." References 22 and 26 suggest much higher potential savings still.

The cost curves associated with the EPRI and references 22 and 26 are shown in Figure 1 as a percentage of system demand available by using various more efficient technologies, in order of increasing cost per unit saved. These costs compare with typical U.S. electricity prices of 6–8¢/kW·h; all costs below this thus involve net economic savings for the user at the discount rates employed for the analysis. The upper curve is an estimate by the U.S. Electric Power Research Institute; the lower curve is by the Rocky Mountain Institute (22,26). Figure 1 demonstrates that a good database does not necessarily ensure comparable results; the potential estimated in the upper curve is clearly much smaller than that estimated in the lower curve. Compared against typical U.S. electricity prices of at least ~7¢/kW·h, however, both illustrate that substantial emission reductions appear to be available at net cost savings.

Technology cost curves have by no means been confined to developed countries. A number are presented in studies of the TATA Energy Research Institute (27).

Limitations and Interpretation of Technology Cost Curves

Abatement cost curves reflect the weaknesses and strengths of the procedures used to produce them. The simplest technology curves usefully summarize technical data, but may have substantial limitations as guides to actual abatement costs. In part this is because, unless they are developed iteratively using quite sophisticated system models, they may neglect interdependencies among abatement options and thus "double count" some emissions savings; eg, the CO_2 savings from reducing electricity demand may be much reduced if nonfossil sources are introduced later to displace coal power generation. They may also neglect interactions between various end uses, for example that between heat and lighting in widely diverse building environments. Frequently they also do not reflect adequately the timescales involved in bringing the technologies into place and the underlying growth in demand that may occur in the interim. Even after such issues are carefully incorporated, technology assessments and cost curves (particularly for end-use technologies) demonstrate a large potential for emission savings apparently at a negative cost, technologies that would both reduce emissions and yield net financial savings. Typically, the technologies suggest a potential to reduce emissions by well over 20% at net cost savings. The uncertainties, however, are very large, and to be meaningful the numbers have to be defined carefully in terms of scope and timescale. Based on an extensive review of technologies and related cost–curve studies, it is concluded that "substituting identified and cost-effective technologies on OECD countries could in principle increase the efficiency of electricity use by up to 50%, and of other applications by 15–40%, over the next two decades. Fully optimizing energy systems would yield larger savings, but it is far from clear how much of this potential can be tapped" (18).

Technology studies and cost curves show that a large energy efficiency gap exists between the apparent technical potential for cost-effective improvements and what is currently taken up in energy markets. In well-functioning markets, cost-effective options should be exploited anyway, because someone should profit by doing so. Some of the cost-effective potential may be taken up over time, but if such technologies are not being exploited, this may indicate that other important factors are not captured in technology analysis. For example, there may be hidden costs, people may be unaware of the options, or there may be other obstacles to uptake. The acceptability of different options may also vary, for least cost is by no means the only criterion that matters to people. This illustrates the fact that the apparent technical potential in fact comprises a number of different components. As shown in Figure 2, realizable gains consist of

- Those that are economically attractive in their own right and that will be installed without policy changes.
- Those that will not be obtained unless institutional constraints and barriers are removed, and/or other micro-economic policies are implemented to increase the take-up of cost-effective options.
- Those that are justified on the basis of nongreenhouse external benefits (eg reduced other environmental impacts, increased energy security).

In addition, some apparently cost-effective savings cannot be realized. The real economic potential can differ from apparent potential due to

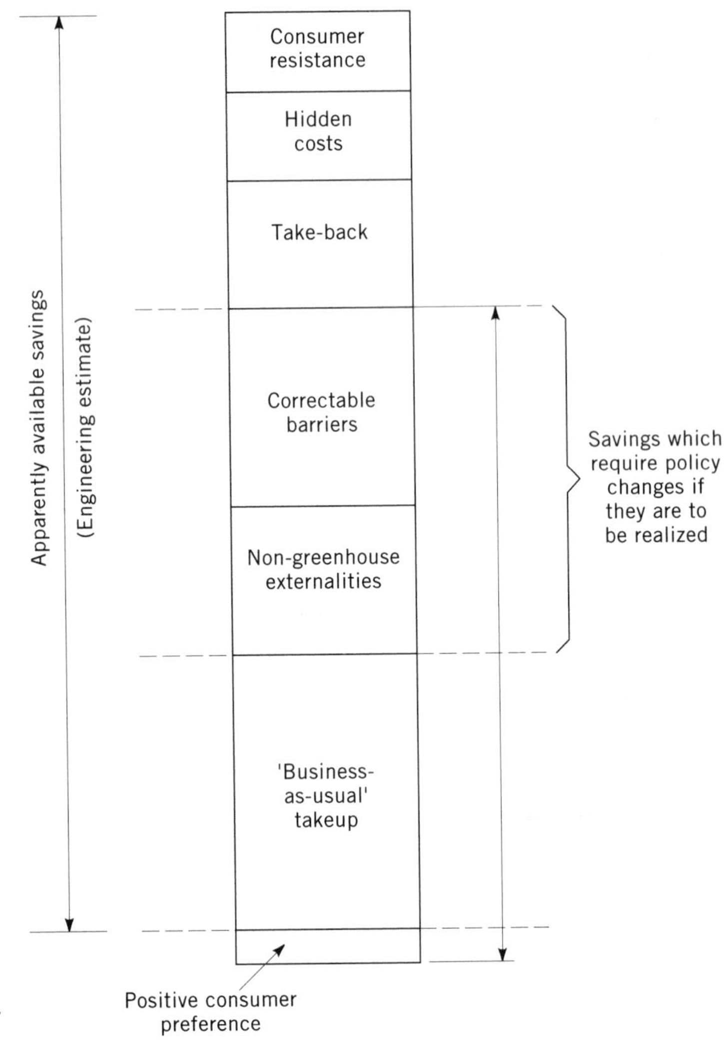

Figure 2. Energy efficiency: engineering potential and realizable gains, a classification. Hidden cost = unavoidable hidden costs and other barriers.

- Take back or rebound of savings (improved efficiency reduces the cost of the associated energy service and therefore stimulates increased demand for the energy service).
- Unavoidal hidden costs (there may be costs associated with the use of a technology or policies to stimulate its uptake that are not revealed in a simplified analysis).
- Consumer preference (the technologies may not be a perfect substitute in the provision of the energy service, which may either increase or decrease consumer readiness to take up more efficient technologies, depending on their characteristics). For example, it would be technically possible and highly cost-effective to mandate a doubling of car efficiency in most countries, but it would probably make the vehicles available smaller and less powerful, which people may consider unacceptable (often this can also be considered an aspect of hidden costs).

These different components need to be understood before drawing conclusions about the scope for cost-free re-

Table 1. Global CO_2 Abatement Cost Modeling Studies

Author (reference)	Key	CO_2 Reduction from Baseline % Reduction	CO_2 Reduction from Reference Year % Reduction[a]	GNP Impact/Cost from Baseline % Reduction
Anderson and Bird (28)	AB (2050)	68%	−17%	2.8%
Burnlaux and co-workers (29)	B (2020)	37%	17%	1.8%
Burnlaux and co-workers (30)	B (2050)	64%, 64%, 66%	−18%, −18%, −11%	2.1%, 1.0%, 0.3%[b]
Edmonds and Barns (31)	EB (2025)	14%, 36%, 47%, 70%		0.1%, 0.5%, 0.7%, 2.2%[c]
Edmonds and Barns (32)	EB (2020, 2050, 2095)	45%, 70%, 88%	22%, 41%, 53%	1.9%, 2.7%, 5.7%
Manne and Richels (33)	MR (2100)	75%	−16%	4.0%
Manne (34)	M (2020, 2050, 2100)	45%, 70%, 88%	13%, 25%, 21%	2.9%, 2.7%, 4.7%
Mintzer (13)	Mi (2075)	88%	67%	3.0%
Oliveira Martins and co-workers (35)	OM (2020, 2050)	45%, 70%	−2%, 2%	1.9%, 2.6%
Perroni and Rutherford (36)	PR (2010)	23%		1.0%
Rutherford (37)	R (2020, 2050, 2100)	45%, 70%, 88%	15%, 28%, 43%	1.5%, 2.4%, 3.6%
Whalley and Wigle (38)	WW (2030)	50%		4.4%, 4.4%, 4.2%[d]
Goldemberg and co-workers (19)	G (2020)	50%	0%	0%[e]

[a] Negative values imply an increase in CO_2.
[b] Toronto-type agreement in all three cases, with tradeable permits in the second and third cases and energy subsidy in the third case.
[c] Costs as estimated from consumer + producer surplus (see text).
[d] The three numbers refer to three different tax forms: a national producer tax, a national consumer tax, and a global tax with per capita redistribution of revenues.
[e] The cost value is indicative of the estimate by this study's authors that their scenario would incur no additional costs; it is not a modeling result.

ductions. Cost curves may identify a technical potential, but there is no expectation that this can all be realized. Part of the key is to consider specific policies for introducing better technologies set against explicit baseline scenarios that incorporate some "business as usual" uptake.

SYSTEM-WIDE ABATEMENT COST ESTIMATES

A wide variety of abatement costing studies at the global and national levels have been carried out. In addition, many local studies, which focus on policy-based studies for particular cities or utilities, have been conducted. These generally emphasize implementing the cost-free potential identified in bottom-up studies, but are too varied and specific to cover here.

Global Abatement

Table 1 reports to estimates of global costs of limiting CO_2 emissions. Four of the six models employed by these studies are fundamentally different. One uses an economic growth modeling framework oriented toward technology development, and one is a global bottom-up study. Many other global models have been developed. They have not been considered because they either are based on models already covered or they do not generate results in a relevant, comparable form.

The principal published results from these modeling studies are summarized in Figure 3; plotting the degree of abatement against the cost measured directly (for energy sector models) or as GNP loss, relative to the projected GNP. Figure 3**a** shows the results in terms of reductions from the baseline projection generated by the model; simple calculation shows that the average rate of abatement in almost all these studies, despite their apparent diversity, is 1.3–2.0% per year below the baseline projection. Figure 3**b** illustrates the same results, but in terms of the level relative to the base year (usually 1990).

Comparing the two curves in Figure 3 indicates that emission changes relative to a base year show less of a pattern, due primarily to the wide variation in the baseline emission projections. Therefore, one should focus on reductions relative to the reference projection without considering the base year.

In Figure 4**a** the relationship between the relative CO_2 reduction and required carbon tax (marginal abatement cost as reponed by the model in the target year) is shown. Figure 4**b**, shows the relationship between the reported carbon tax and the average GNP loss in the target year. For a fixed analysis the marginal cost must always be greater than the average cost, this plot shows no such relationship. For some models this may reflect anomalies in the way that marginal cost impacts are translated into GNP impacts; another source of the difference may be that in some models the CO_2 constraints are introduced to impose a much higher marginal cost over the first few decades than in the very long run. All these global studies impose a fixed path of emission constraint; only the simpler CETA (39) and DICE (40) models optimize the path of abatement to reach a given concentration level.

Even when normalized relative to the differing baseline projections, however, the results still show great variation. The Whalley-Wigle results define the high end of the cost spectrum. The relatively high costs probably reflect the limitations imposed by only having one generic carbon fuel, the lack of technology representation, and an extraordinarily high baseline projection, which scales up the global energy system 10-fold over the century because there is no allowance for autonomous efficiency improvements. The Goldemberg and co-workers (19) bottom-up

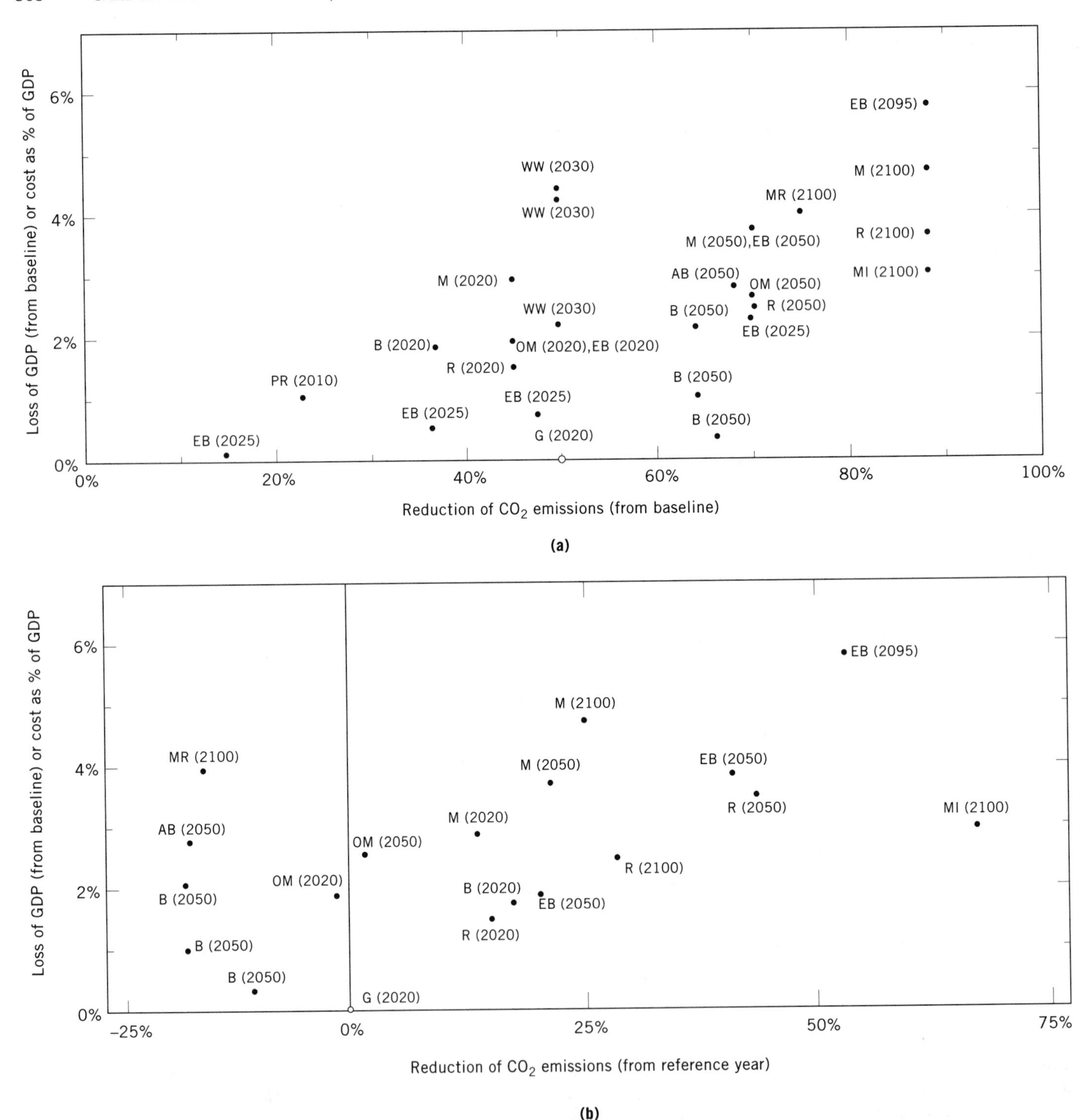

Figure 3. Global CO_2 emissions reductions cost estimates. (**a**) Reduction from baseline. Loss of GDP (from baseline) or cost as % of GDP. • Top-down; economic model, estimate, ○ Bottom-up, technology based estimate. For key see Table 1. (**b**) Reduction from reference year. Note that a negative reduction indicates an increase in emission.

results define a lower bound, with emissions reductions perhaps up to 50% estimated at no cost (7). However, some of the sensitivity runs of the global top-down models, when given more optimistic estimates of energy efficiency improvements and supply technology development, can also produce very low costs.

Excluding the Whalley-Wigle outlier, the spread of results roughly indicates that the costs of a long-run 50% reduction in global CO_2 emissions could range from negligible to a loss of about 3% in global GNP. Reductions of 80–90% depress GNP by 2–6%; at the other end of the curve, the global economic models, as well as the engi-

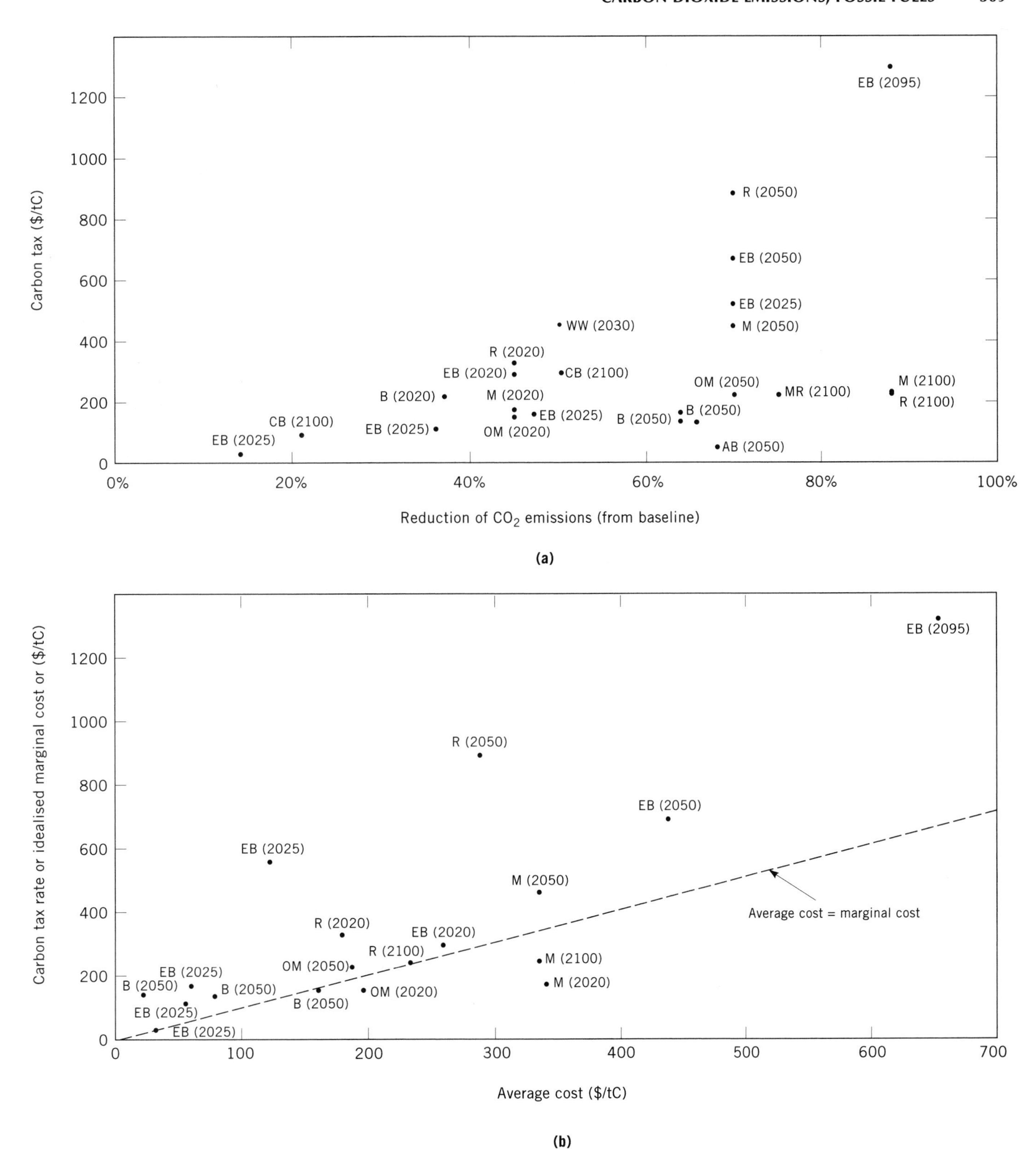

Figure 4. (**a**) Carbon taxes and CO_2 emissions reductions (from baseline), global studies. Some of the plotted carbon tax values are averages of different regional taxes. Where taxes vary over the study period, the end year tax is shown. For key see Table 1; (CB(2100), not shown in Table 1, refers to the CBO study). (**b**) Idealized marginal cost or carbon tax and the average cost, global studies. Some of the plotted carbon tax values are averages of different regional taxes. Where taxes vary over the study period, the end year tax is shown. For Key see Table 1.

neering models, suggest that emission reductions of at least 10–15% can be obtained at very low cost.

These global models are highly aggregated and capture technical issues, macro-economic issues, regional differences, and trade effects with varying degrees of detail. They inevitably sacrifice local and technological detail in order to represent important regional differences and (in some) incorporate trade.

Abatement Costing Studies for the United States

A wide range of national studies has been carried out; of these, those of the U.S. have received the most attention, as shown in Table 2. The range of estimates of the impact on GNP of CO_2 emissions reductions relative to the projected baseline emissions is shown in Figure 5.

U.S. studies are predominantly top-down economic studies. Many bottom-up/engineering studies have also been carried out, but most have been confined to cost–curve or subsectoral studies. In addition to the earlier examples, important recent studies include those of the National Academy of Sciences (NAS) (49) and the Office of Technology Assessment of the U.S. Congress (OTA) (50). These, like most engineering-based studies, maintain that main efficiency improvements (and hence CO_2 reductions) can be obtained at little or no cost. These estimates have been included in Figure 5 as indicative bottom-up cost estimates.

As with the global studies, a wide range of cost estimates is observable. For example, for the same loss of GNP of about 2%, CO_2 emission reductions can range from 20 to 80% below baseline (52,53). One big reason for such differences is that the reductions are sought on different timescales; the CBO study seeks a 20% reduction by 2000, while the 80% reduction in the reference 53 study is achieved at the end of the next century.

Ignoring the rapid reductions imposed in the CBO studies for the year 2000, early studies form a high cost outlier (45). Bottom-up studies and those that examine the recycling of tax revenues have produced some very low and negative abatement cost estimates. With these exceptions, the spread of results is pretty consistent with the global results, with 50% emission reductions from baseline yielding losses up to a little more than 2% of GNP. The coincidence between U.S. and global results may partially reflect the importance of U.S. emissions and costs, but also the dominance of U.S.-based modeling approaches in global studies.

Non-U.S. OECD Studies

A number of emissions reduction studies of non-U.S. OECD countries are listed in Table 3. One striking fea-

Table 2. U.S. CO_2 Abatement Cost Modeling Studies

Author (reference)	Key[a]	CO_2 Reduction from Baseline % Reduction	GNP Impact/Cost from Baseline % Reduction
Barns and coworkers (41)	BG (2020)	26%, 45%, 60%	0.6%, 2.0%, 3.2%
Barns and co-workers (41)	BG (2050)	45%, 70%, 84%	1.9%, 4.9%, 7.5%
Barns and co-workers (41)	BG (2095)	67%, 88%, 96%	4.3%, 8%, 10.9%
Brinner and co-workers (42)	B (2020)	37%	1.8%
CBO-PCAEO, DRI (43	CB1, CB2 (2000)	8%, 16%	1.9%, 2%
CBO-DGEM (43)	CB3 (2000)	36%	0.6%
CBO-IEA-ORAU (43)	CBG (2100)	11%, 36%, 50%, 75%	1.1%, 2.2%, 0.9%, 3.0%[b]
Edmonds and Barns (44)	EB (2020), EB (2100)	35%, 59%	1.3%, 2.3%
Goulder (45)	G (2050)	13%, 18%, 27%	1.0%, 2.2%, 4.5%
Jackson (24)	J (2005)	34%, 40%, 46%	−0.2%, −0.1%, 0.1%
Jorgenson and Wilcoxen (45)	JW (2060)	20%, 36%	0.5%, 1.1%
Jorgenson and Wilcoxen (46)	JW (2100)	10%, 20%, 30%	0.2%, 0.5%, 1.1%
Jorgenson and Wilcoxen (47)	JW (2020)	8%, 14%, 32%	0.3%, 0.5%, 1.6%
Manne and Richels (48)	MR (2020)	45%	2.2%
Manne and Richels (48)	MR (2100)	50%, 77%, 88%	0.8%, 2.5%, 4.0%[c]
Manne (34)	MRG (2020)	26%, 45%, 60%	0.8%, 2.2%, 4.2%
Manne (34)	MRG (2050)	45%, 70%, 84%	1.4%, 2.7%, 3.3%
Manne (34)	MRG (2100)	67%, 88%, 96%	2.3%, 3.1%, 3.4%
Mills and co-workers (23)	M (2010)	21%	−1.2%[d]
NAS (49)	N	24%, 40%	0%, 0.8%
Oliveira Martins and co-workers (35)	OMG (2020)	26%, 45%, 60%	0.2%, 1.1%, 2.4%
Oliveira Martins and co-workers (35)	OMG (2050)	45%, 75%, 84%	0.4%, 1.3%, 2.4%
OTA (50)	O (2015)	23%, 53%, 53%	−0.2%[e], −0.2%[e], 1.8%
Rutherford (37)	RG (2020)	26%, 45%, 60%	0.5%, 1.3%, 2.5%
Rutherford (37)	RG (2050)	84%, 45%, 70%	2.4%, 1.2%, 2.5%
Rutherford (37)	RG (2100)	67%, 88%, 96%	1.8%, 2.6%, 2.8%
Shackleton and co-workers (51)	SJW (2010), SLINK (2010)	22%, 2%	−0.6%, 0.1%
Shackleton and co-workers (51)	SDR1 (2010), SG (2010)	5%, 28%	−0.4%, 0.2%

[a] If the model used is a global model, the Key includes the letter "G" before the date.
[b] The first two results use multilateral taxes, others use unilateral taxes; taxes are flat only in the 1st and 3rd estimates, rising in other estimates.
[c] Values represent different assumptions in technological developments: an optimistic, an intermediate, and a pessimistic view.
[d] Arising from 11 specified regulatory changes: estimated from claimed savings of $85 bn/yr.
[e] The benefit shown in the OTA cost estimates is only an indicative value; no explicit modeling value has been calculated.

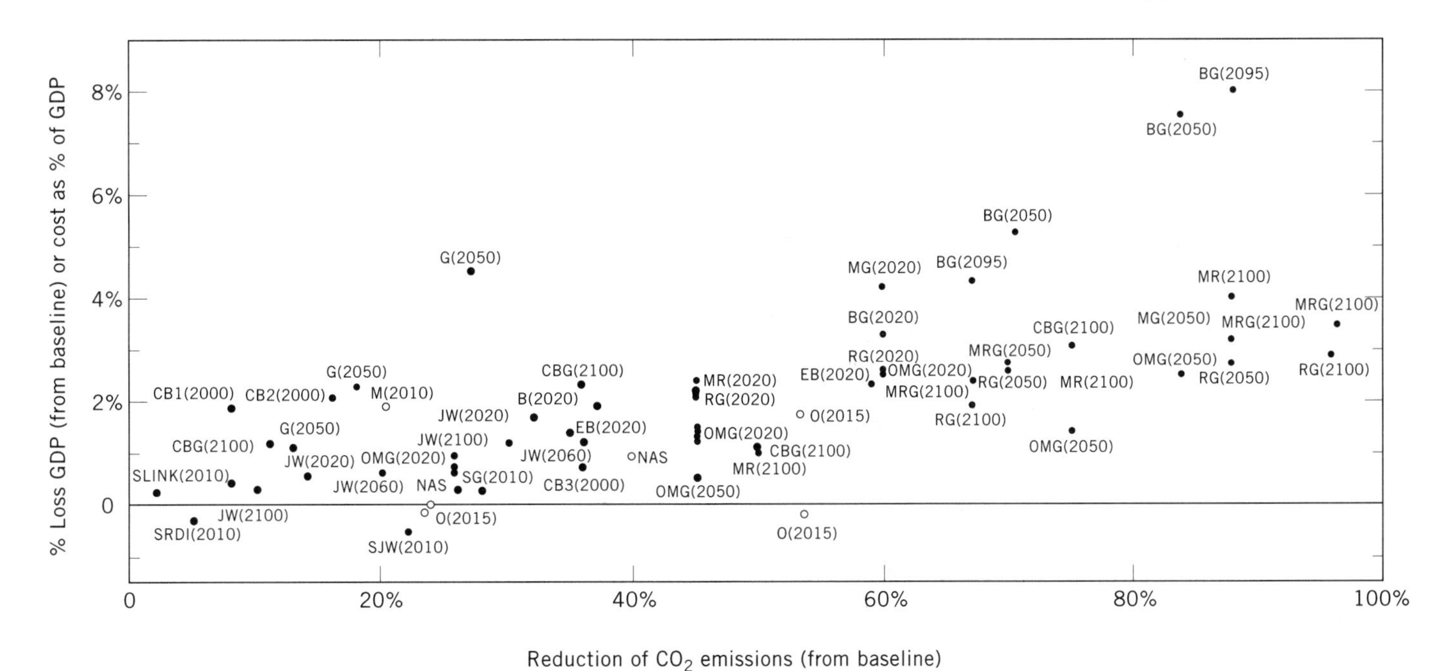

Figure 5. CO_2 emissions reductions cost estimates, United States, reduction from baseline. • Top-down; economic model estimate, ○ Bottom-up, technology based estimate. Labels are not attached to all points given danger of crowding. For key see Table 2.

ture is that nearly all these studies have a much shorter term focus than the global studies or most of the U.S. studies; most in fact are focused either on emissions stabilization by 2000 or on the Toronto target of 20% reduction by 2005. The results are displayed in Figure 6. Immense variation is once again apparent, even in terms of reductions from the baseline projection. No clear pattern emerges, other than the fact that the bottom-up studies again give much lower costs.

In general, the cost range is broader than in the global studies, perhaps reflecting the impact of transitional costs arising from the relatively rapid reductions required in some of these studies, captured by the short-run models, an issue discussed further below. The detailed EC studies of the macro-economic impacts of carbon taxation, using short-run macro-economic models, produce a wide variety of results; these results depend heavily on how the tax revenues are used, as they reflect rather the use of a tax to shift resources from one kind of economic activity to another. The Finnish study (54), with very high losses for modest reductions by 2010, is a similar outlier on the high side, and two of the Japanese studies also yield exceptionally high costs (55); this is because the carbon tax revenues are removed from the economy. The variations make it hard to discern any pattern, but even with these outliers excluded, the relative costs for this short-run national reductions are mostly larger than for equivalent long-run reductions in the U.S. and global studies, with several exceeding 3% GNP losses.

The Transitional Economics

Studies of abatement costs for the former centrally planned economies of Eastern Europe have also used both top-down and bottom-up approaches, but data are insufficient to summarize usefully as a scatter diagram. An energy technology cost curve estimated for Poland has been shown (70,71) (Fig. 7). All of the bottom-up studies indicate a large potential for reducing CO_2 emissions with net economic savings; specifically, the marginal cost of savings only rises above that of new supply for savings well in excess of 10 EJ, which is more than 20% of Soviet primary energy demand in 1989. This potential arises from the history of highly subsidized energy prices in these regions, and other cumulative inefficiencies in the structure of incentives. The economic savings potential is around 10% greater if the Soviet Union can access Western technologies.

It was estimated in 1991 that increasing prices to world market levels in Poland, Hungary, and Czechoslovakia would reduce emissions by 30% (72). Of course, the realizable potential may be a very different matter and depends in part on the progress of economic restructuring. In fact, CO_2 emissions in the former East Germany have collapsed by at least 30% as uncompetitive heavy industry has more or less shut down in the process of unification. In Poland, provisional trend/technology results also indicate significant CO_2 reductions as a by-product of the economic restructuring process (73).

Manne and Richels include the Soviet Union as an independent region in their Global 2100 model (48), and calculate much higher GNP losses (5% in (33), reduced to 3% in (53)) there than for the rest of the world in the first half of the next century. This striking contrast with bottom-up studies reflects the difficulties top-down studies have with economies undergoing restructure. They rely on the existence of a market mechanism that in many instances is

Table 3. Non-U.S. OECD CO_2 Abatement Cost Modeling Studies

Country	Author (reference)	Key[a]	CO_2 Reduction from Baseline % Reduction	GNP Impact/Cost from Baseline % Reduction
Australia	Dixon and co-workers (56)	AD (2005)	47%	2.4%
Australia	Industry Commission (57)	AICG (2005)	40%	0.8%
Australia	Marks and co-workers (58)	AM (2005)	44%	1.5%[b]
Belgium	Proost and van Regemorter (59)	BP (2010)	28%	1.8%
EC	DRI (42)	ECDRI (2005)	12%	0.8%
Finland	Christenson (54)	FC (2010)	23%, 21%	6.9%, 4.8%[c]
France	Hermes-Midas (60)	FHM (2005)	11%	0.7%
Germany	Hermes-Midas (60)	GHM (2005)	13%	1.3%
Italy	Hermes-Midas (60)	IHM (2005)	13%	1.9%
Japan	Ban (61)	FB (2000)	18%, 18%	0.4%, 1.7%[d]
Japan	Goto (62)	JG (2000, 2010, 2030)	23%, 41%, 66%	0.2%, 0.8%, 1%
Japan	Nagata and co-workers (63)	JN (2005)	26%	4.9%
Japan	Yamaji and co-workers (55)	JYC (2005)	36%	6%[e]
Netherlands	NEPP (64)	NEN (2010)	25%, 25%	4.2%, 0.6%[f]
Norway	Bye, Bye, and Lorentson (65)	NB (2000)	16%	1.5%[g]
Norway	Glomsrod and co-workers (66)	NG (2010)	26%	2.7%
Sweden	Bergman (67)	SWB (2000)	10%, 20%, 30%, 40%, 51%	0%, 1.4%, 2.6%, 3.9%, 5.6%
Sweden	Mills and co-workers (23)	SWM (2010)	44%, 87%	−0.3%, −0.2%[h]
UK	Barker and Lewney (68)	UKBL (2005)	32%	0%
UK	Sondheimer (69)	UKS (2000)	4%	−0.5%
UK	Hermes-Midas (60)	UKHM (2005)	7%	1.9%

[a] The letters in the Key refer to the country and author; in cases where a global model is used a "G" is added before the date of the estimate.
[b] Study combines technology and macro-economic assessment of GDP impact.
[c] Unilateral and global action.
[d] Tax case and regulation case.
[e] Values of both 5 and 6% are given.
[f] National and global policy scenario.
[g] GDP costs for OECD range of 1–2%.
[h] Estimated from average value of saved emissions ($143/tC and $41/tC, respectively).

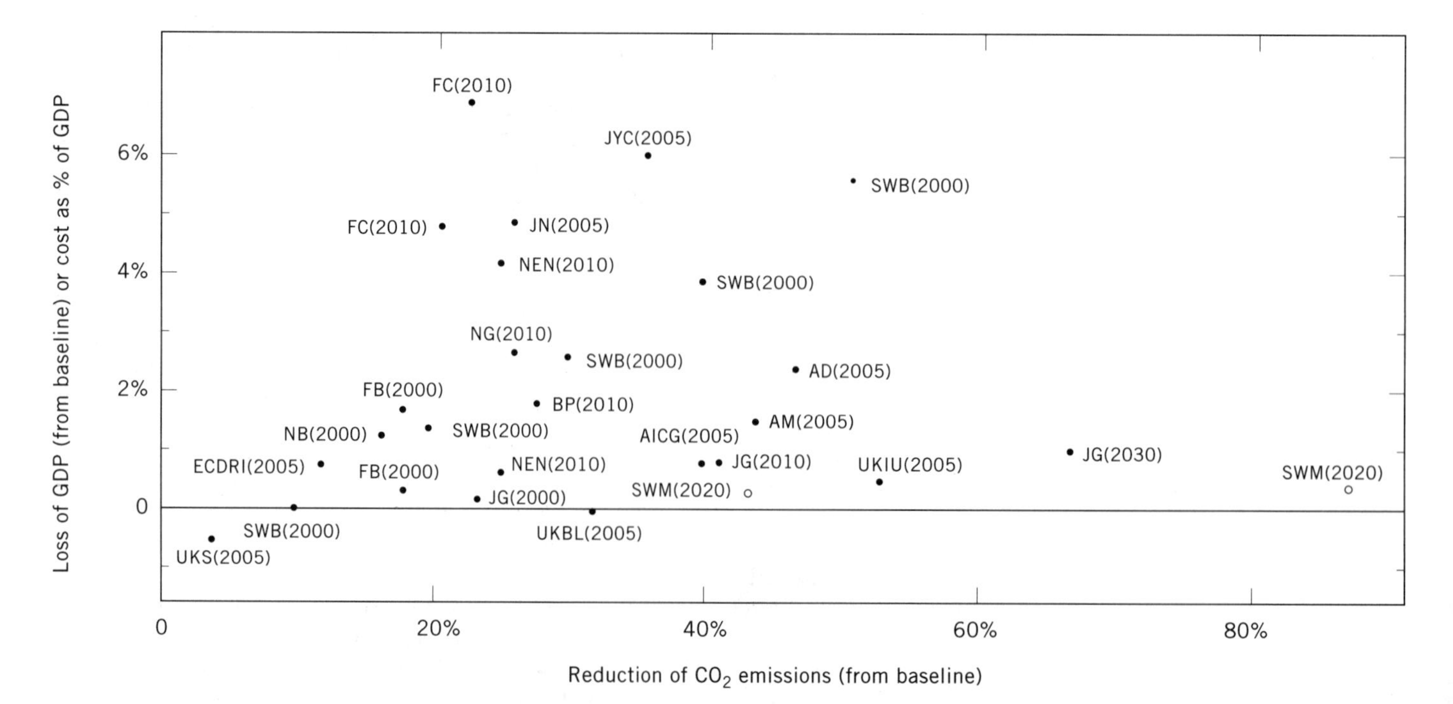

Figure 6. CO_2 emissions reductions cost estimates, industrialized countries, Non United States, reduction from baseline. • Top-down; economic model estimate, ○ Bottom-up, technology based estimate. For key see Table 3.

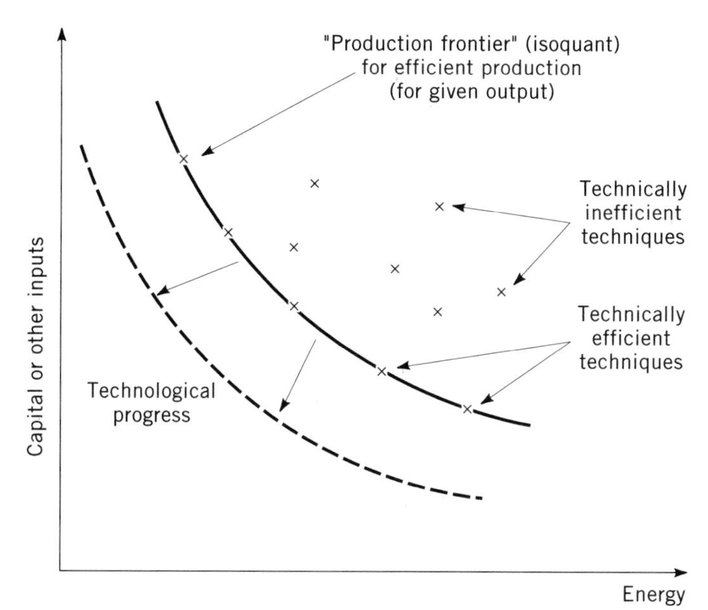

Figure 7. Technological efficiency and the "production frontier" UNEP, (21).

barely functioning, and such studies will likely miss many of the important features of these economies, such as current structural inefficiencies. It may take some time before a market-based modeling approach becomes appropriate. For the present, bottom-up engineering assessments appear much more relevant.

Developing Countries Including China. Some of these issues also apply to developing countries. Although concerns are frequently expressed about the limited data available concerning developing countries, considerable data exist concerning the situation in most principal developing countries, especially with respect to commercial energy supply. Data on detailed end uses, agriculture, and noncommercial energy sectors is sparser and less reliable, though usable estimates exist.

The past few years have also seen a number of studies of the potential for abating greenhouse gases in developing countries. Country reports to the IPCC Energy and Industries subgroup (74) include a number of studies from developing countries, some drawing on more extensive internal work. In addition, initial results from a series of nine developing-country studies coordinated by the U.S. Lawrence Berkeley Laboratory were reported (75). However, these studies focused on scenarios and did not attempt to estimate abatement costs. Some of the detailed studies for the Asian Energy Institute network reported in the UNEP study (21) attempt bottom-up estimates of short-term abatement costs or investment requirements.

Some of the global models do separate developing-country and oil-exporting regions. The global studies have also highlighted the growing importance of China in contributing to greenhouse gas emissions during the next century. The early results (33) suggested that Chinese GNP in the year 2050 could be depressed by up to 8% below the (greatly increased) reference level, if emissions are restricted to no more than double current levels, partly because of the apparent lack of alternatives to coal and modeling inability to import energy (though this was reduced to a 2% GNP loss, rising to 5% by 2100, in revised analysis (53)). The GREEN model suggests much lower abatement costs for China (76).

In estimates derived from global studies that model trade (eg, references 77 and 78), GNP impacts for some of the smaller developing countries especially can be dominated by trade and price effects arising from the action of other countries. These studies emphasize the large potential GNP loss for energy exporting developing countries; the loss could arise from abatement efforts elsewhere that depress the market for traded fuels. Conversely, energy-importing countries (which include the poorest countries) would gain from such effects.

Concerning domestic abatement efforts, a number of cost curves have been estimated which indicate substantial technical potential for savings with net economic benefits. However, the only integrated system-wide cost estimates found, excepting those from global models for China, are those of Blitzer and co-workers (79,80) for Egypt and an unpublished study of Zimbabwe by the UK consultants Touche Ross, reported in UNEP (21).

The Blitzer studies estimate a very large potential GNP impact from stabilizing CO_2 emissions in Egypt, with losses in some cases of more than 10% of GNP. This is based on a short-run macro-economic model of the Egyptian economy that is modeled with very limited capital mobility between sectors, and with oil and gas as the only future energy supply options. It recognizes none of the technical inefficiencies in the economy (ie, assumes zero scope for cost-free energy-efficiency improvements). Excepting the modest abatement available from switching from oil to gas, emissions savings can only be achieved by reducing energy consumption, which, given the constraints on capital mobility, can only be achieved by reducing economic activity or changing its structure. Consequently, the costs reported are clearly excessive. However, separate runs of the model that placed emission constraints on each sector of the economy individually did emphasize that GNP losses would be greater still if these additional restrictions were imposed.

The Touche Ross study of Zimbabwe reached precisely opposite conclusions. Using an engineering approach, widespread cost-effective options were identified that could both limit emissions growth and improve overall economic performance. However, these assessments neglect a variety of hidden costs and fundamental institutional obstacles; they also include some elements that are expected to be achieved anyway as part of current structural adjustments in the Zimbabwean economy. These and other limitations, which suggest that abatement costs in this case may be substantially underestimated, are summarized in the Zimbabwean case study of the UNEP report (21).

The limited range and appropriateness of studies for the transitional and developing economies make the use of scatter diagrams, as were used for presenting OECD results, not, in the authors' judgment, very meaningful in this case. One striking observation of note: the gap between top-down and bottom-up approaches is larger even than observed for OECD countries. Top-down models mostly report restricting developing-country emissions, even relative to projected increases, to be more expensive than equivalent relative constraints in OECD countries (37). It is not clear why this should be the case. Bottom-up studies, conversely, identify a potential for improving energy-efficiency in these regions at a net economic benefit that is even larger than that identified for OECD countries.

Despite this, it seems possible to draw two firm observations from the existing developing; country emissions abatement studies: many cost-effective technology options exist for improving energy-efficiency, but such potential will be swamped by the pressure for emissions growth in such rapidly expanding economies, so that actually stabilizing developing-country emissions at current levels is nevertheless likely to be very costly. More sophisticated and quantitative system-wide analysis of abatement impacts is, however, only just beginning, and as outlined in UNEP (21), the complexities are such that it may take many years to mature toward consensus even on very rough cost estimates and understanding of the key issues.

MODEL SCOPE AND TYPE

Cost estimates clearly vary greatly among models. It is difficult to disentangle the various reasons for these differences, due to the nonlinearity of the relationships involved, the diversity of the tools used to develop emissions reductions cost estimates, the many and varied assumptions employed, and the enormity of the task required to obtain all of the models, establish a protocol for analysis, and systematically unravel the relative contributions. Steps in that direction have been taken by teams at the Energy Modelling Forum (29,81) and OECD (45,82). In all these studies, participating modelers were asked to adopt standard assumptions to the degree possible and to provide standardized model results. Both sets of comparisons have focused on top-down models (the Energy Modelling Forum conducted a similar exercise with bottom-up models). It is clear from these activities that great variation in results can be generated through the use of different models. This variation is greatly reduced through the use of standardized assumptions.

All models share certain unavoidable limitations:

- Models are necessarily a simplified representation of reality, in terms of what the concerned researchers feel are important aspects that should be captured. A given model may not capture all important economic relationships.
- Despite their simplifications, all models are still rather complex and must necessarily rely on a large quantity of data and numerous parameter estimates. Robust estimation of these is itself significant research undertaking and serious doubts may arise about the validity of many of the actual numerical values employed. Studies of model sensitivities, and structured uncertainty studies in which the values of key parameters are varied over plausible ranges, are required to examine how much these uncertainties may affect model results. Such studies have not often been adequately performed.
- The timescales involved require assumptions to be made about changes in technology and lifestyles. Conjectures about such changes are inevitably uncertain and cannot be formally validated.

These limitations may be exacerbated by the fact that most studies employ models that were not initially designed to shed light on the cost of emissions reductions (the exceptions are the OECD's GREEN model (30,83) the ERB (3,4), and the Second Generation Model (84).

Consequently, all modeling results need to be treated with some caution, depending in part on the timescale of application, the care with which the model has been developed, the extent to which it is appropriate to the application used, and the care with which inputs have been formulated. The ultimate argument for such modeling efforts is not that they give precise and certain answers, but rather that they are the only consistent way of estimating abatement costs at all and of identifying the important factors that affect them.

Top-Down and Bottom-Up Models

The principal distinction between top-down economic models and bottom-up engineering/technology-based models have been noted. High positive abatement costs are frequently associated with top-down economic approaches and low and negative costs are frequently associated with bottom-up engineering approaches.

The underlying theoretical distinction lies between the economic and engineering paradigms is illustrated in Figure 8. In economics, technology is features as the set of techniques by which inputs, such as capital, labor, and energy, can be transformed into useful outputs.

Figure 8 is a graph of energy versus other (eg, capital) inputs. Each cross represents an individual technique or technology. The best techniques define the production frontier, as illustrated. In principle, efficient markets should result in investment only in the technically efficient techniques on this frontier (after allowing for lags associated with old stock), because such investments can reduce all costs compared with other technologies.

Economic models all assume that markets work efficiently, therefore all new investments (after allowing for hidden costs) define the production frontier. This is assumed to be consistent with cost-minimizing (or utility maximizing) behavior in response to the observed price signals: various models can encompass other inefficiencies, such as externalities and fiscal imperfections arising from taxes and subsidies, but still share this assumption. Observed behavior (historical data) combined with the optimizing assumption defines an observed production frontier. The models assume that no investments are available that lie beyond this frontier, though future technical change may move it. To the extent that real-world inefficiencies exist, they are implicitly incorporated in the inferred frontier. Relative price changes move investments along this frontier (eg, substitute labor for energy) as defined by the estimated elasticities; a purely economic model has no explicit technologies, which are simply implicit in the elasticities used.

Studies using engineering models have often focused on identifying potential least-cost abatement opportunities by assessing directly the costs of all the technological options. Such an assessment is independent of observed market behavior and also defines a production frontier. If markets are technically efficient, the frontier revealed by market behavior should correspond to that calculated by engineering studies. However, this is not the case; engineering studies reveal widespread potential for investments beyond the limit of the production frontier suggested by market behavior and built into economic models. The explanation can be considered in terms of the contrasting limitations of the economic and engineering paradigms. Economic models are slave to the assumption of cost-minimizing behavior noted above. Limitations of purely engineering models include

- The cost concept is based on an idealized evaluation of technologies and options. The existence of hidden costs is typically ignored.
- The cost of implementation measures (eg, information campaigns, standard setting, and compliance processes) is not included.
- Market imperfections and other economic barriers mean that the technical potential can never be fully realized.
- Macro-economic relationships (multiplier effects, structural effects, price effects) and indicators (GNP, employment, etc) are not included in the models.

Top-down and bottom-up are very imprecise terms. Although models generally known as top-down, all determine energy demand through aggregate, economically driven indexes (GNP and/or productivity growth, and price elasticities), they can vary greatly in the modeling of energy supply. Some of the top-down models are purely economic, with supply changes being driven only by substitution elasticities. Others are primarily economic, but incorporate a "backstop technology"—a technology that can come in, in unlimited quantities, once a certain price threshold is reached. Yet in other top-down models (such as the ERB), supply is driven largely on an engineering basis of supply technology costs, chosen from a database of supply technology cost curves.

Nearly all bottom-up models contain extensive representation of supply technologies, but the key practical distinction, as it has emerged in the CO_2 costing literature, centers on the modeling of energy end use and the introduction of end-use technologies. Bottom-up end-use studies indicate a large potential for reducing both emissions and costs relative to a traditional top-down extrapolation of energy demand. In other words, they show that the top-down projections are not optimal interms of the technologies available, and the main savings come by contrasting this with a scenario that is an engineering optimum.

Why does this create such a large difference between bottom-up and top-down studies? The primary reason is to be found in Figure 2. Neglecting the segment referring to externalities, which may or may not be reflected in either top-down or bottom-up studies, top-down projections of energy demand incorporate only efficiency improvements corresponding to the bottom segment of the column—the "business as usual" takeup. Bottom-up models, on the other hand, include all the available technologies, without distinction as to under which category in the column they fall. Consequently, neglecting externalities:

- Top-down modeling studies tend to underestimate the potential for low-cost efficiency improvements (and overestimate abatement costs) because they ignore a whole category of gains that could be tapped by nonprice policy changes; whereas
- Bottom-up end-use modeling studies overestimate the potential (and underestimate abatement costs) because they neglect various hidden costs and constraints that limit the uptake of apparently cost-effective technologies.

Which is more realistic depends on the relative size of different segments in Figure 2, something that cannot be determined without separate study of specific implementation policies. But the near-term potential for limiting CO_2 emissions at low or negative costs lies somewhere between the optimism of bottom-up studies, and the relative pessimism of many top-down studies.

The systematically differing results are largely a reflection of the non-optimality of the baseline implicit in such bottom-up studies, and the questions this raises about the assumptions built into top-down model baseline and abatement projections. This is perhaps the key difference between the modeling approaches. If the baseline in bottom-up studies used optimal technologies, the baseline emission projections would be much lower. This was illustrated clearly in a study in reference 85, which included end-use technologies in the MARKAL engineering model and found that obtaining base-case emissions anything near as high as official or macro-economic forecasts proved almost impossible; the model chose more energy-efficient end-use technologies, and more renewable energy technologies, irrespective of CO_2 constraints. Further reductions were, however, relatively expensive, relying more on supply substitution, as the stock of more efficient end-use technologies was largely selected already in the optimal baseline.

Thus there is no inherent reason why top-down studies should yield positive costs or bottom-up models should yield negative costs. The sign of the cost hinges critically on the approach applied to computing costs, in particular, assumptions regarding optimality of the baseline. For example, references 86 and 87 recognize a non-optimal baseline and illustrate negative abatement costs within a top-down energy-economy approach; whereas the study in reference 85 uses a wholly engineering model and obtains positive costs for any reductions beyond those captured in the (optimal, and much lower) baseline.

Some attempts have been made to integrate top-down and bottom-up models explicitly. Most notably, the Global 2100 model has been linked with the MARKAL engineering model (by replacing the energy technology submodelin Global 2100 that formed the energy component previously) in a bid to combine the best features of both into a single computational framework (88). However, this still does not resolve the dilemma about whether projections, for baseline or abatement scenarios, adopt the engineering optimum or econometric extrapolation of energy demand, and this linked model has been criticized on the grounds that one still dominates the other (81).

Time Horizons and Adjustment Processes: Short-term Transitional versus Long-term Equilibrium

Different models are designed for application over different timescales. There are no standard definitions, but in many relevant branches of economic analysis the short term is taken to be less than 5 years, the medium term is between 3 and 15 years, and the long term is more than 10 years. The timescale is a distinction of significant importance, particularly for economic models, because different economic processes are important on different timescales, and thus the timescale for which models are designed fundamentally affects their structures and objectives. Models for relatively long-run analysis may to a reasonable approximation assume an economic equilibrium in which resources are fully allocated. Short-run models focus on transitional and disequilibrium effects such as transitional market responses, capital constraints, unemployment, and inflation. This distinction parallels the structural distinction drawn by Boero and co-workers (89) between resource allocation models and macro-economic models.

MEDIUM TO LONG TERM: EQUILIBRIUM/RESOURCE ALLOCATION MODELS

These models focus on the allocation of available resources, within the energy sector or the broader economy. This category includes both optimizing bottom-up models (which seek to optimize resource allocation within the bounds set by available technology), and all the main long-run and global energy/CO_2 models. The latter are generally termed equilibrium models.

At one extreme of the long-term modeling dimension are models that can only consider the energy/investment mix for a "snapshot" year and compare this to another, without any information on the transition between them; these are comparative static models (29,77,78). Such models can enable detail in representing the system, but at the expense of modeling developments over time. In contrast, dynamic models cover medium- and long-term phenomena, extending across several time periods.

At the opposite of this extreme within the equilibrium models, some are designed to run in annual steps over a period of a few decades. These can include considerable detail on different sectors whose use of different resources in response to price changes is estimated econometrically from data over previous years. The main examples are the Jorgensen/Wilcoxen (90) and Goulder (45) models for the United States.

Equilibrium models such as Global 2100 and GREEN (in its more recent versions (30,83)) and the ERB model lie between these extremes. They are designed to operate in steps of 5–15 years, to look at the changing allocation of resources under different constraints over periods of many decades, and the way this may change under CO_2 constraints.

The treatment of capital stocks in these models can have important implications for costs. "Putty–putty" models represent capital stocks as perfectly interchangeable between sectors and over time. Nuclear power plants can be transformed into solar photovoltaic arrays instantaneously and without cost. "Putty–clay" models, on the other hand, allow no transfer of capital between applications. Once an investment has occurred, the technology cannot be altered. Resource allocation/equilibrium models cannot, however, model other aspects of transitional costs arising from disequilibria. In this respect, and in their assumption of optimal investments subject to constraints, they have been criticized for underestimating likely abatement costs (though the same caveats apply to the reference projection as well, which is similarly optimal within constraints and free from disequilibria).

SHORT TO MEDIUM RUN: MARKET SIMULATION/ MACROECONOMIC MODELS

Short-run models by contrast focus primarily or exclusively on the dynamics of transition, rather than the long-term equilibrium allocation of resources. One class of short-run models are sectoral market simulation models, such as detailed models of electricity or oil markets and pricing, or of industrial sector energy demand. The diversity of such models reflects the range of specific markets that they have attempted to model. Few such models as yet appear to have been applied to assessing abatement costs; one notable exception is the application of separate market models for the industrial, domestic, and transport sectors in the United Kingdom (82). However, such models may come to assume much greater importance as governments move closer to considering detailed policy measures tailored to specific sectors. Frequently they focus on sector responses to carbon taxes rather than costs (82).

Of more general interest for costing is the recent application of models known usually simply as macro-economic models. This is the name usually (if imprecisely) given to the models developed over many years for studying the short-run dynamic behavior of national economies. Typically these models contain explicit representation of investment and consumption in different sectors, and markets do not necessarily clear; there can be unemployment, idle production capacity, or capital shortage. Such models generally contain a strong Keynesian component, though many other aspects of macro-economic theory have also been brought to bear in them. Recent applications to CO_2 abatement are discussed below.

Such models can generate a wide range of macro-economic indexes such as GNP, inflation, employments, etc. For this reason they are of particular interest for assessing the short-term macro-economic impact of CO_2 abatement. However, such models may contain very limited representation of the energy sector, and some may not even model energy as a separate good within the economy. Few contain a representative set of energy-technology options. Such short-run macro-economic models are thus highly country and model-specific, and vary greatly in the extent to which they can be applied to assessing

CO_2 abatement. The models are best at representing transitional costs; results may become unstable and questionable when the models are run too far ahead, because the economic feedbacks that keep economies from straying too far from economic equilibrium are generally not well represented.

No study has focused primarily on a comparison of short-run macroeconomic with general equilibrium (GE) models, but such a comparison is implicit in the studies of Shackleton and co-workers (51). This took two macroeconomic models (DRI and LINK) and two general equilibrium models (DGEM and Goulder), all of which the model authors considered appropriate to run over a period of 2–3 decades. Each was subject to the same carbon tax ($40/ton C). The models behaved in very different ways. Concerning the impact on CO_2, the macro models suggest a reduction of 0–8% depending on the way in which carbon tax revenues are used; even a CO_2 increase is observed from a tax recycling that boosts economic growth; the equilibrium models suggest 20–30% reductions depending only on which model is used. Even more striking, the macro models show a wide variety of GNP responses, varying greatly over time, and including substantial increases from some tax recycling options; the GE models show a much smoother response, with more modest impacts.

All this corresponds to the theoretical differences. The short-run macro models reflect the resistance of the economy to change, but also the cumulative impact of greater investment unconstrained by equilibrium requirements. The GE models allow capital to move easily across the economy to respond to the price changes, giving a much stronger CO_2 response. It seems reasonable to suggest that the longer-term results from the short-run macroeconomic models are questionable (and highly sensitive to various assumed macro-economic investment and other responses), as are the short-term GE results; the most realistic outcome may be to assume a progression over time from the macro toward the GE results, but even this is speculative. Beyond this we cannot generalize, but we emphasize the importance of resolving such great variation based purely on model structure. McKibben and Wilcoxen (91) are the only researchers to have yet applied a model that can simultaneously address unemployment and GE class problems.

Finally we note that between the short- and long-run models, some progress has been made in developing medium-run models that combine elements from both short-run macro-economic and resource allocation/equilibrium models (see discussion in Reference 89). These may be seen as short-run models that are expanded to include adjustment to long-term equilibria. That is, medium-run models are mainly demand-determined and allow for market disequilibria; however, a central part of the models aims to describe the adjustment process from short-term market disequilibria to long-term market equilibria. Finally, there are some top-down modeling approaches that do not fit into these categories at all, such as the growth models employed in References 28 and 40.

Sectoral Coverage: Energy versus General Economy. Another important distinction is that between models that address only a limited part of the economy (in this case, the energy sector) and those that encompass the whole economy, which usually have a much more simplified representation of any particular sector. Among equilibrium models, the distinction is known as that between partial equilibrium and general equilibrium models; the parallel distinction for short-run models is that between energy market and macro-economic models; a more general terminology would be that of [energy] sector and general economy model.

Sectoral models focus heavily or exclusively on particular economic sectors (in this case, usually just the energy sector or parts thereof); insofar as the rest of the economy is represented, it is in a highly simplified way. They address the problem of describing behavior in a single area of the economy, for example energy, but ignore or treat summarily all other economic functions. An energy market model would have no description of the labor or capital markets.

Energy sector models come in many varieties. Bottom-up technology models are all sectoral; so are many equilibrium models, such as the ERB (3,4). Investment planning models, such as the electricity expansion models widely used for assessing power sector investments, are more focused sectoral models. Models of the international oil market—a very important but little analyzed issue in assessing abatement costs—are sectoral but global models.

Sectoral models yield an estimate of costs in a particular sector, but cannot take account of the macro-economic linkages of that sector with the rest of the economy. They cannot, for example, estimate how the labor or investment requirements in that sector may affect the resources available to other sectors.

By contrast, general economy models encompass all principal economic sectors simultaneously. They recognize feedbacks and interrelationships between sectors. In principle, this enables them to estimate the full long-run GNP impacts of restrictions (such as CO_2 constraints). In practice, this is obtained at the expense of considerable simplification of the energy sector.

General economy models are the only kind that can reflect important features in the rest of the economy. This may include, for example, economic distortions outside the energy sector. In particular, existing taxes impose varying burdens on economies. If the revenues from a carbon tax are used to reduce such taxes, the gains may in principle offset the losses, completely altering assumptions about the net impact on the GNP of such taxes. This seems to have attracted attention only recently (92); relatively early discussions of this are given by the CBO (43) and Grubb (93), and subsequent modeling studies are reviewed.

The past few years have seen the development of linked models that seek to integrate the detail of energy sector models with the economic consistency of general economy models. In Europe, the macro-economic HERMES model has been linked with the MIDAS energy supply model and applied to analyzing abatement strategies in the four largest EC countries, including interactions between these economies (60), to generate the results illustrated in Figure 6. The Manne and Richels

GLOBAL 2100 model (33) contains considerably more energy supply detail than other general equilibrium models by integrating an energy technology model (ETA) into a simple general equilibrium model (MACRO, which clears markets for all goods and services treated as a single aggregate). The SGM (84), extends this approach. It is a general equilibrium model that was designed to address greenhouse-related issues, and thus desegregates economic activities on the basis of importance to the greenhouse issue.

The ERB model is an energy market rather than general economy model, but it contains a GNP feedback parameter, which was incorporated to ensure that the impact of large changes in energy sector costs on the scale of economic activity would be reflected in terms of its impact back on energy demand. The parameter is not intended to provide consistent estimates of the costs of emissions reductions, though many studies have used it as such. As noted above, costs must be developed using consumer plus producer surplus techniques (31,32). Other sectoral models, such as Fossil (94) require similar approaches to make consistent cost estimates.

Optimization and Simulation Techniques. Models adopt different approaches to optimization. Some models optimize energy investments over time by minimizing explicitly the total discounted costs (or per capita consumption), using linear or nonlinear techniques. Several engineering models use linear programming, most notably the EFOM model (31) used extensively within the EC, and MARKAL (32), promoted internationally by the ETSAP program of the International Energy Agency. The top-down Global 2100 model uses nonlinear dynamic optimization, as does its derivative CETA (39). Such optimization approaches in effect assume perfect investment (within the confines of the model), with perfect foresight. It is debatable whether this is a drawback or advantage.

Modeling partial foresight is, however, difficult, and the main alternative approach is to simulate investment decisions on the basis of static expectations, ie, static projection of conditions at the time of investment. This myopic assumption is used in ERB, GREEN, and the CRTM trading derivative of Global 2100. Moreover, no software yet exists for solving such large dynamic general equilibrium models under the assumption of perfect foresight. The SGM model incorporates a variety of options for determining investments on the basis of future expectations (including a formulation of partial foresight).

The mechanism for selecting investments is fundamentally different between investment simulation and optimization models, and this might be expected to have a significant impact on results. In fact, this does not appear to be the case. A recent comparison by the Energy Modelling Forum (unpublished) of results from Global 2100 (dynamic optimization) with those from the ERB model (investment simulation) shows that standardizing for key assumptions leads to remarkably similar energy and emission results. Assuming a competitive economy, variations in key input assumptions (such as those discussed below) thus appear far more important than the approach used for selecting investments in the model.

Short-run macro-economic models are concerned primarily with the simulation of aggregated investment responses in terms of labor, capital, etc, but do not seek to optimize investments; they do not represent specific technologies at all. Some other models are purely for simulation of system operation, without automated investment modeling, and they report on the implications of an investment strategy that is specified externally. This can enable much greater detail in representing the system, and avoids the limitations of linear optimization especially, though there are inevitably drawbacks from having to specify investments manually and check their consistency (21).

Level of Aggregation

Models differ greatly in their degree of disaggregation. To some extent this is the obverse of the model scope. Models that, for example, focus on household electricity demand can represent this and the options for improving household electricity efficiency in great detail. Global, economy-wide models have to be highly aggregated.

At one end of the spectrum are models such as LEAP (95) and the BRUS model (96) used in the Danish Energy 2000 study (97). Their demand sectors are generally disaggregated with respect to specific industrial subsectors and processes, residential and service categories, transport modes, etc, with the aim of achieving homogeneous entities whose long-term behavior can be defined through consistent scenario projections. Similarly, energy conversion and supply technologies are represented at the plant-type and device levels. This allows detailed modeling of the alternatives for technical innovation, fuel switching, etc.

With regard to emissions of pollutants and CO_2, this type of disaggregation into specific technologies makes it possible to take account of the different characteristics of energy technologies. A detailed analysis of abatement options can thus be carried out. This includes energy savings at the end-use level, changes in the conversion system, and fuel substitution. At the other end of the spectrum are models such as GREEN and Global 2100, which treat energy and the world in a highly aggregated manner.

In general, the level of aggregation is closely related to the other aspects; for example, a multiperiod, global, general-economy model by necessity has a highly aggregated representation of energy demand and supply, with little if any technological detail. Great detail in representing energy supply, conversion, and end-use markets and technologies is only possible in models that are specific to the energy sector, and focus on simulation rather than full system optimization. The benefits need to be weighed against these limitations.

Modeling Classifications

An understanding of the differences that are incorporated into models is required because each have different strengths and weaknesses. Models for studying CO_2 abatement costs have been developed in different ways, often by adapting existing models. They are able to handle some issues (or sectors) better than others; different models are thus suited to different purposes. Models cannot, or at least should not, be interpreted as giving complete and accurate answers, but rather used for the insights they offer when the results are combined with an understanding of the model structure and limitations.

There is no universal or accepted way of classifying models. Not all the attributes of model scope and type have been reviewed here. Attributes of particular importance are

- Time horizons and adjustment processes: short-term transitional versus long-term equilibrium.
- Sectoral coverage: energy versus general economy.
- Optimization and simulation techniques.
- Level of aggregation.
- Geographic coverage, trade, and "leakage" (20).

The division between top-down economic and bottom-up engineering is of great practical importance, despite its occasional ambiguity. Within top-down models, the distinction between long-run equilibrium and shorter-run macro-economic models is central, as is that between partial (sectoral) and general economy models. Neither of these latter distinctions is relevant to bottom-up models, for which important distinctions are those between partial models (generating a cost curve of savings related to a top-down or unspecified baseline), and full system representation, and within the latter, the choice between optimization and simulation. Table 4 classifies the primary models discussed in this paper according to this scheme, which has a pragmatic focus on the factors of greatest importance to abatement costing.

NUMERICAL ASSUMPTIONS AND SENSITIVITIES

Assumptions drive model results. Critical parameters can be usefully, though not exclusively, divided into those that govern the overall scale of the system and reference missions, and those that directly affect the relative cost of emissions reductions. GNP and population growth rates, income elasticities, and the rate of "autonomous end-use energy-intensity improvement" primarily affect the baseline scale; background fossil-fuel prices strongly affect both the baseline emissions and abatement costs; the cost of low carbon technologies and price elasticities largely drive abatement costs though they also affect baseline emissions.

Population and GNP Growth Rates

The demand for energy is driven by population and per capita energy demand, and all economic models at least assume that the latter is driven by per capita GNP. It is consequently much more difficult to restrict the growth in emissions for a developing country with a high population and economic growth rate, such as India, than for a more slowly growing developed economy, for example, Germany, which has a static or declining population.

Uncertainties in future global population are reflected in abatement studies; estimates for the year 2025 for example include 9.5 billion for the NAS (49) and 8.2 billion for Edmonds and Barns (31). However, the latter study found that there was little impact from a reduction in population growth for approximately 15 years, that is until labor force and therefore GNP was affected. Projections suggest that the increase in global GNP will be much greater and more uncertain than population growth, and so differences in GNP projections account for a greater part of variation in baseline emissions.

Different baseline GNP and energy demand assumptions across studies complicate comparisons of the GNP loss associated with a target CO_2 reduction. In models that derive GNP from labor productivity, this is correspondingly critical. The difference that baseline GNP makes to energy demand is more complex, as growth in any economy inevitably varies across sectors over time;

Table 4. Model Classification and Examples Applied to CO_2 Abatement Costing

	Bottom-up	Top-down						
		Equilibrium (resource allocation)			Growth	Energy market/macro-economic		
Sectoral division Models/authors	Partial		Linked	General		Partial	Linked	General
National	OTA NAS Mills (and many others)	MARKAL	ETA-MACRO MARKAL-MACRO	DGEM (Jorgensen and Wilcoxen) Goulder Bergman Glomsrod Proost and Van Regemorter		eg, electricity market models Ingham and Ulph	GDMEEM (N. Goto)	LINK MDM (Barker) ORANI (Marks)
Regional	EFOM	MIDAS					MIDAS-HERMES	QUEST HERMES G-CUBED (Mckubben and Wilcoxen) DRI
Global	Goldemburg	ERB	Global 2100 CRTM CETA	SGM GREEN Whalley and Wigle WEDGE	(Optimising) DICE (Nordhaus) (Non-optimising) Anderson and Bird	eg, international oil market models		

greater GNP growth in reality would not necessarily imply that the growth within all sectors of the economy is increased proportionately.

Energy/GNP Relationships

Whereas GNP is a principal determinant of energy demand, many factors can affect the relationship between them. A few models incorporate explicitly a non-unitary energy-income elasticity, which implies a changing energy/GNP ratio as GNP grows. Most models, however, express such a change, if any, in terms of an exogenous parameter that defines the rate at which the energy/GNP ratio would change in the absence of price changes. This rate of exogenous (or autonomous) end-use energy-intensity improvement (AEEI) then becomes a determinant of baseline energy demand for long-term projection; the higher the rate of energy-intensity improvement, the lower the baseline CO_2 emissions, and the lower the costs of reducing relative to a given base year. The parameter has been widely described as a measure of technical progress, but it compounds many different elements including for example composition of energy using activities and the influence of non-price policy interventions.

Future Energy Prices, Resource Modeling, and Supply Elasticities

High background fossil-fuel prices lower energy demand and CO_2 emissions, and reduce the relative costs of moving to lower carbon fuels. Limited resources (eg, of oil) have the same effect, implicitly or explicitly raising the prices as resources are depleted. Resources are, however, uncertain, and the course of fuel prices is even more uncertain.

Various approaches may be taken toward estimating the future cost and availability of different fuels. National studies may define national production costs but define exogenous global prices with great variation.

Global models reflect resource/supply cost issues explicitly, through direct estimation of the resource base and supply elasticities. High supply elasticities mean lower fuel price rises as supply increases. GREEN assumes zero supply elasticity for oil outside OPEC (ie, volume set by fixed production constraints), but price is determined by OPEC supply elasticities varying from 1 to 3; supply elasticities are higher for gas and coal and much lower for nonfossil sources until backstop technologies become relevant.

Price/Substitution Elasticity of Demand for Energy

The impact of price changes on energy demand is determined by the price elasticity of energy demand, or in general economy models, the substitution elasticity between energy and other factors of production. The lower the relevant elasticity, the less energy demand is curtailed by higher energy prices, and the greater the tax that is required to reduce energy demand and consequently CO_2 emissions.

In the global models, assumed long-run elasticities range from −0.3 to −0.4 for the Manne and Richels U.S. and Global 2100 studies to −0.6 to −1.0 for the OECD's GREEN model; short-run elasticities are about one-tenth this value.

The notion of elasticities assumes a symmetric response to price changes; if prices rise and then fall back to previous levels, the energy-intensity (after allowing for lags) returns to former levels. This basic assumption, and the values assumed for modeling, are disputed and have been particularly called into question by recent trends and studies.

Technology Developments and Costs

Assumptions concerning the cost and rate of implementation of more efficient or lower carbon technologies affect both baseline emissions and relative abatement costs. The initial results for the United States (33) were strongly criticized (98) as being based on unreasonably pessimistic assumptions for efficiency improvements and the costs of alternative supply technologies. In response, Manne and Richels (99) examined three different background scenarios, which they termed technology optimistic, technology intermediate, and technology pessimistic. The last of these corresponds to their initial famous estimate that CO_2 abatement could cost the U.S. $3.6 trillion over the next century, and yields some of the higher cost points on Figures 3 and 5 above.

The assumptions for the technology optimistic scenario reduced these total costs by a factor of 20 for the (fixed) abatement target set, because of the combined impact on baseline emissions (primarily from the higher AEEI) and the halved costs for backstop low carbon supply technologies. As a result, Manne and Richels noted that "the direct economic losses are quite sensitive to assumptions about both demand and supply . . . for the losses (from carbon constraints) to approach zero; however, the most optimistic combination of supply and demand assumptions must be adopted." This confirms that results are sensitive to the assumptions concerning technology costs, a result again noted in the sensitivity study in reference 31. Almost all studies fix the costs of supply technologies as exogenous data; Reference 28 contains the only study in which technology costs decline with increasing investment.

Energy Sector Impact on GNP

The impact of changes in energy demand and energy sector costs on GNP is complex. General economy models capture the relationship consistently, but the resulting elasticity can still vary considerably. For example, the Global 2100 modeling approach (and by implication, the CRTM and CETA models also) contains a nested CES (constant elasticity of substitution) production function to relate energy input to economic output. (A CES function has the form $Q = (\sum_{i=1}^{N} a; x_i^{\rho})^{1/\rho}$ where Q is output, x_i for $i =$ 1. . . . N are inputs, and a; for $i = 1, \ldots, N$ and ρ are parameters.) Cline (100) criticizes the parameters chosen, claiming that they yield an excessive impact of energy sector changes on GNP, a claim disputed in reference 99.

For sectoral or partial equilibrium models, the impact of energy sector costs or carbon taxes on GNP may be estimated, by a direct elasticity (GNP feedback) parameter, as available in the ERB model. This approach has been criticized for overestimating the feedback and consequently overestimating the cost of reducing CO_2; Reference 31 recognizes this limitation and suggests the use of consumer plus producer surplus changes as the best method of computing cost for the ERB model.

Interfuel Substitution Elasticities

Substantial CO_2 reduction can be achieved by switching toward less carbon-intensive fuels. In models having a purely engineering approach to supply this is captured by supply technology costs; for models with econometric supply modeling, it is governed by the interfuel substitution elasticity. GREEN assumes a long-run interfuel elasticity in production of 2.0 and a short-run value (reflecting existing supply infrastructure) of 0.5. Halving these values lowered global baseline emissions by 13% in 2050; the impact on abatement costs may be expected to be much larger unless the bulk of substitution is governed by backstop technologies.

KEY DETERMINANTS

The range of reported results on the costs of limiting CO_2, and the technical modeling and data issues that affect such estimates, have been examined in a recent review (20). The gulf between "technology-engineering" and "economic" perspectives, in terms of assumptions concerning energy demand and policy, structural, and technological assumptions as well as the abatement strategy and scope of analysis affect the reported results.

It would be easy to suppose that the differences between economic and engineering views are confined to a few modeling parameters. They reflect very different perspectives, almost paradigms, about driving forces in the energy economy. A report from a UNEP workshop that sought to bridge the divide (101) observed that:

> To economists, energy is a factor of production: it is an input into economic growth, and one which can substitute for labor or capital, depending on relative prices. While energy-efficiency may improve due to technical development, so does that of other factor inputs; and efficiency improvements lower the relative price of energy, increasing the extent to which it may be applied. Also, fossil fuels dominate because of demonstrable convenience and low cost. Thus whilst recognizing the potential importance of technical improvements, and even market imperfections which prevent optimal energy use, to most economists there is every reason to expect energy consumption to grow with expanding GNP, and no particular reason to expect technical developments to reduce CO_2 emissions relative to the business-as-usual case without incurring substantial costs. It is a compelling case, with much long-term historical weight behind it.
>
> To scientists and engineers, on the other hand, energy is not an abstract input but a physical means to particular ends. The applications which consume much energy are those of heating, heavy construction, metals, etc; basic infrastructure and comfort; and travel. In developed economies, most infrastructure needs have been met, travel may be approaching limits of congestion and time budgets, and much new economic activity is in areas which consume trivial amounts of energy, such as information technology, general entertainments, etc. Thus to most scientists and engineers economic growth is becoming less and less relevant to energy needs (in developed economies). In addition, very large technical improvements in efficiency, which need not incur much extra cost, are readily demonstrable; technology is powerful and adaptable to changing conditions (such as requirement for nonfossil sources); and it is hard to believe that human society is incapable of finding ways of putting such options into effect (this applies primarily to developed economies, but may also be of great relevance to developing ones if they can move directly to advanced technologies). This too appears to be a strong case, with at least partial support from recent trends, but it leads to a very different outlook from the economist's outlook sketched above.

This shows that the different perspectives lead to widely different assumptions concerning several factors: the relationship between energy demand and future economic growth in the absence of any abatement measures; the scope for exploiting more efficient technologies; and the scope for developing new technologies as needs (such as CO_2 reduction) require. The complex issues involved in each of these have been reviewed recently (20).

Baseline Energy Demand and the AEEI

Energy demand in the absence of abatement measures depends on GNP (population times per capita GNP) growth rates; energy prices, and the response of demand to these; and the parameter usually translated as the rate of autonomous end-use energy-intensity improvement (AEEI). The impact of GNP and energy price changes is recognized by all analysts, as are the uncertainties. The main contrast in views arises from the differing assumptions about the AEEI which has an enormous impact. Reference 83 states that "unfortunately there is relatively little backing in the economic literature for specific values of the AEEI . . . the inability to tie [it] down to a much narrower range . . . is a severe handicap, an uncertainty which needs to be recognized."

Among the principal global and U.S. studies, adopt the lowest values have been adopted for AEII (0 and around 0.5, respectively) (33,48,77,78). Williams (98) strongly criticized such values as too low; Manne and Richels (102) defend their value of AEEI on the grounds that it appears consistent with observed trends in the post world war two United States. However, it is difficult to separate the various factors in their analysis (eg, price, income distribution, and time sampling effects), and Wilson and Swisher (81) strongly dispute their interpretation, concluding that "one can produce an experiment that justifies whatever AEEI one likes within very broad ranges."

Whatever the value for AEEI, it is important to understand what it is. The parameter is badly misnamed: it is a measure of all nonprice-induced changes in gross energy-intensity, which may be neither autonomous, nor concern energy-efficiency alone. It is not simply a measure of technical progress, for it conflates at least three different factors: one is technical developments that increase energy productivity, but another is structural change, ie, shifts in the mix of economic activities (which may require widely different amounts of energy per unit value added). The third is policy-driven uptake of more efficient technologies, due to regulatory (as opposed to price) changes, greater than would occur without those changes.

Technical change is indeed hard to predict. Studies by (103,104) suggest that in manufacturing alone, technical change has increased energy productivity in OECD countries by about 2% per year for at least the past two decades; this includes price effects, but in fact there is no clear change in the trend that correlates with the energy price shocks (perhaps because of the lags in manufacturing equipment). Technical change appears more closely related to the price shocks in other sectors.

Structural change incorporates the phenomena of saturation in energy intensive activities (such as home heating and primary heavy industries), and shifts toward less energy-intensive activities. A range of studies have noted that structural change, both between sectors and within manufacturing industry, has played an important part in restraining energy demand in the OECD in the past 20 years (104). Williams and co-workers (105) provide considerable evidence for expecting the observed trend in manufacturing to continue. On the basis of this and other relevant literature, and various saturation effects, reference 93 argues that an autonomous structural trend toward lower energy-intensities (ie, rising AEEI) is to be expected as countries develop and as economic growth moves toward increasingly refined products and services.

Low values of AEEI in long-term studies (significantly below 1.0, especially for OECD countries) are dubious because of saturation and structural change effects. Low values make any fixed target much harder to reach, also increasing the scale and hence relative costs of reductions. This is, however, tentative; there are substantial uncertainties and a clear need for greater understanding of technical and structural trends, and integration of such studies in abatement cost modeling.

The AEEI is not the only controversial issue surrounding the relationship between GNP and energy demand. Considerable uncertainty also surrounds estimates of the price elasticity of energy demand. Most studies have sought to estimate elasticities from the response of demand to the energy price rises of the 1970s and early 1980s. More recent trends, if anything, increase the uncertainties. Although energy demand has started to rise again in OECD countries following the price falls since the mid-1980s, the response has not been nearly as great as predicted by the simple reversible statistics of elasticity. Engineers have long maintained that the efficiency gains would not be lost, because they are embodied in better knowledge and techniques that will not be abandoned even if energy prices fall.

Regulatory Instruments and the Energy-Efficiency Gap

Most top-down models assume the optimal operation of markets in response to observed price signals, in which case economic theory suggests taxes to be the optimal means of abatement. The observation of the large efficiency gap demonstrated by engineering studies calls this into question. Many economic discussions reject the relevance of this data.

The efficiency gap comprises many different components. If it can be wholly explained in terms of lags in take-up, unavoidable hidden costs, etc, then it is a phenomenon of little direct relevance to the actual costs of limiting CO_2 emissions. However, many barriers to the uptake of more efficient technologies have been identified; an extensive and clear analysis of the different kinds of barriers has been given (106).

Some things can be done to improve the uptake of efficient techniques, for example, with government campaigns to improve information and the awareness of consumers of energy-efficiency and conservation. Most top-down studies in effect assume such measures to be enacted irrespective of CO_2 abatement, and ignore or dispute the scope for other more direct cost-effective actions. However, experience and modeling studies of regulatory policies demonstrate that such measures can and have been effective.

Regulatory changes to encourage utility demand-side management programs have been widely advanced as a way of capturing the "free lunch" by getting utilities to invest in end-use efficiency. The hidden costs in such policies are debated. Joskow and Marron (107) conducted a survey of experience in U.S. programs and concluded that "reported costs exceed those of the technology potential analysis because program costs are higher and energy savings are lower than these studies assume . . . although many of the programs still appear cost effective."

The important fact remains that modeling studies of specific regulatory options frequently yield lower, rather than higher, estimates of abatement costs than those derived from carbon taxation designed to achieve similar abatement. Such studies of regulatory measures thus contrast with the common economic assumption that regulatory options are more expensive than using economic incentives. This is because such policies address areas of significant market failure.

So how large is the potential "free lunch" of zero-cost energy-efficiency improvements captured by regulatory change? End-use technology studies frequently suggest (104) a long-run technical potential to reduce energy demand without extra costs by 20–50%. Schipper and coworkers discuss the implementation and potential of energy-efficiency programs across the OECD in detail. They too conclude that there is large potential, but emphasize that the achievable potential will always be substantially smaller than the apparent technical potential, and that exploiting it will depend on more aggressive and sophisticated policies. Taking account of the various implementation issues points at best to the lower end of the 20–50% range available.

A different view is given by other comparisons of top-down and bottom-up results. A critique and comparison of data from economic modeling is presented (81), and indicates that bottom-up studies suggest cheaper abatement right across the range, up to abatement of 70% or more. These data suggest that the whole cost curve from the top-down studies reviewed by Nordhaus has to be shifted by 20–40% to reflect the technical opportunities identified by bottom-up studies, which would greatly alter the pattern shown in the graphs of Section 4. In effect this means that if the baseline in top-down studies is equated with a "business as usual" rather than a "least cost" path, all the points derived from top-down models in the scatter diagrams of Section 4 should be shifted perhaps 20 percentage points to the right so that emissions reductions of 20% are costless. This excepts the shorter-run studies run over 10–15 years, when a "free lunch" potential of perhaps 1% per year (ie, an increase in the AEEI by 1 percentage point) below baseline might be more appropriate.

The data from studies are too scattered to reach a definitive conclusion; the actual situation is likely to lie between these extremes, implying a need for some reduction in abatement costs from top-down models across the range to allow for the potential of regulatory-driven improvements in energy-efficiency. A further factor is that, with expanded markets for more energy-efficient technologies, these technologies might develop faster, thus permanently raising the AEEI.

Technology Assumptions

Ultimately, the costs of limiting CO_2 emissions will depend heavily on the technologies available, not only technologies for more efficient use of energy, but also for the production, conversion, and utilization of lower carbon energy sources. The importance of technology, as well as assumptions concerning its developments and costs, has been widely acknowledged: it forms the central element of Williams's (98) critique of the Manne and Richels (78) conclusions, and sensitivity studies by Manne and Richels (102) and Edmonds and Barns (31) have demonstrated that estimates of abatement costs depend crucially on technology assumptions.

Yet the care with which such assumptions have been developed varies widely, and the models employed to date have limited representation of technology issues. Models that do incorporate some explicit representation of technology include the Global 2100 model and its derivatives (CRTM and CETA), more recent variants of the OECD GREEN model, the SGM model and the ERB model, which has a fuller representation of technology in compensation for its weaker macro-economic linkages. Sensitivity studies with all these models illustrate the crucial importance of technological assumptions.

To establish reasonable assumptions, it is pertinent to start with current data, and visible trends and options. Supply-side options span technologies that are proven and largely mature (such as combined-cycle gas turbines (CCGTs) proven but still developing (such as wind energy and higher efficiency clean coal conversion), confidently predicted (such as integrated biomass gasification), to a wide variety of lesser or more speculative options. Most recently, an immense study of the prospects for modernized renewable energy technologies (108) argued that these could meet about half global energy demand by the middle of the next century at little if any additional cost; the studies of wind energy in this volume estimated that costs of modern wind turbines were already almost competitive against coal power stations for large-scale exploitation in countries such as the United States with extensive wind energy resources. None of the CO_2 abatement models contain explicit representation of wind energy technology, and for large countries such as the United States and the former Soviet Union, this alone might substantially lower abatement cost estimates.

No models can capture the full range of options available; by inevitably excluding some, there can be a builtin tendency to overestimate abatement costs (unless this is offset by using over-optimistic assumptions concerning those that are included). Some abatement studies use data already being rendered obsolete, as options already identified for cost reductions are exploited. Williams (98) provides one of the most detailed critiques of the technological assumptions used in the Manne and Richels base case assumptions and their derivatives, and argues that many of the assumed costs are higher than can already be predicted with confidence.

This naturally leads to the issue of technology development and cost reductions. This is uncertain terrain, but not a complete black box. Technologies do not arise, improve, and penetrate markets at random, especially for large and complex technologies such as those involved in energy provision. Technology development follows market demand, with the associated public and private R&D investment and learning processes as technologies develop toward market maturity. Yet despite this well-attested and understood fact, almost all the abatement costing studies to date model technology development as exogenous—the costs of abatement technologies are defined as input data and do not vary explicitly with the level of investment, incentives, and market penetration in the model. That alone must be considered as a severe limitation.

A different approach to the issue of endogenous technological development is that by Hogan and Jorgenson (109). This econometric study related changes in productivity trends (which are equated with technical progress), in different sectors of the U.S. economy, to price changes in the different inputs. Although energy productivity did improve when energy prices rose, this was more than offset by reduced productivity growth in other factor inputs at the same time. They found that overall, "technology change has been negatively correlated with energy prices . . . if energy prices increase, the rate of productivity growth will decrease." However, such results may be sensitive to the model specification, and as argued in Grubb (93), the transition from "has been" to "will" in this excerpt conceals the importance of innumerable extraneous factors in the years analyzed, most notably the macroeconomic impact of the sudden and externally imposed oil price shocks. It is highly debatable whether the data reveal anything useful about the economy-wide and long-term technological impact of smoother price changes arising from domestic policies such as carbon taxes and other abatement policies.

The potential for substantial cost reductions associated with larger-scale deployment of low CO_2 technologies, combined with the observations above about possible irreversibilities in the impact of price changes, points to the possibility that there may be various choices of technological trajectories differing little in cost. One is to continue along a carbon-intensive path. Another is to invest enough to alter the course of new investments over the next decade or two toward more efficient and low carbon technologies. As investment patterns and institutions and infrastructures adapt to these new technologies, their costs will fall, perhaps until they become the naturally preferred options. The world would be on a different technological trajectory. Although the transition may be costly, especially if it is forced rapidly, given the nature of technology development and economies of scale it cannot be assumed that this would be a much more costly long-term path (110,111).

The prospects for technology development, production-scale economies, and exploitation of bifurcations to lower emissions, suggest that the cost estimates in many economic studies are implausibly high. Indeed, some use data that appear to indicate costs higher than those of some currently identified technologies, and make little or no allowance for future technical improvements, especially in nonfossil sources. However, there are considerable constraints on the rate at which such technologies could be developed and deployed, as modeled most clearly by Anderson and Bird (28), the deployment of significant new supply technologies will take many decades. On these ground, the higher long-term (beyond ca 2025) costs illustrated in Figures 3–7 may be considered relatively implausible.

Subsidies, Tax Forms, and the Use of Tax Revenues

Various forms of taxes and subsidies can be used to limit emissions. Different types of taxes lead to different reductions in CO_2 and to different impacts on the economy. Most economic models assume abatement to be achieved by a carbon tax, imposed on the carbon content of primary fuels. However, taxes could be applied to subsets of fuels, downstream on derived fuel products, or otherwise not in proportion to carbon content of fuels, eg, on gasoline only or on the energy content of fuels. This generally results in greater economic costs (lower tax efficiency). Thus a gasoline tax is less efficient than a carbon tax at reducing carbon emissions (112), and taxes on electricity production are much less efficient than a tax on input fuels; the latter do not encourage fuel switching, only reducing CO_2 by depressing demand for electricity (113).

Of greater relevance to the assessment of total abatement costs is that energy production in most countries is subject to a complex set of taxes and subsidies, and abatement costs inevitably depend on the existing tax structure. Where heavy taxes are already imposed (as with oil products in many OECD countries), the macro-economic impact of additional taxation is likely to be greater than in the absence of existing taxation; conversely, where energy is subsidized, removal of subsidies (or equivalent taxation) often yields macro-economic benefits. Many models ignore initial subsidies and taxes.

If the tax revenues are taken out of the economy (eg, unaccounted for or spent abroad), all impacts on the national economy are negative. For other uses of the tax revenues, different macro-economic indexes frequently move in different directions. For the carbon tax levels considered (up to about $80/t C), nearly all these studies find some ways in which the net effect is to boost GNP, as compared to a projection in which existing tax structures are unchanged.

A carbon tax raises money largely from consumption; this may be transferred to qualitatively different economic activities. If the revenues are used to stimulate investment directly, this increases savings and reduces consumption with a short-term negative impact (though obviously, other benefits are associated with deficit reduction). Conversely, if the revenues are used to boost investment, the resulting GNP growth in part reflects simply the transfer of resources from consumption to investment.

Carbon taxes at the levels considered may be genuinely less distortionary than the most distortionary existing taxes. There is still some debate as to how valid it is to count such changes as a credit for carbon taxation. As with the efficiency "free lunch" debate, should not governments make the current tax structures optimal irrespective of abatement efforts, with gains that should not be credited to carbon taxation? There are no easy answers to this. Certainly it is hoped that tax structures become more optimal over time. But there are also real constraints on taxation policy, and objectives other than just efficiency; Hourcade (114) notes growing political and trade constraints on traditional taxes and concludes that "it is timely to consider the taxation of 'bads' (such as pollution) as an answer to the general problem of raising revenues." In any case, if carbon taxes represent an efficient way of limiting emissions, it would be perverse to say that the revenues should not be used in the most desirable way, or alternatively, to say that economists should not model reality because the starting point is not optimal.

For longer-term assessment, current quirks of taxation systems may be less significant, but it is still important to recognize that governments need to raise revenues; carbon taxes will reduce the revenue required from other taxes and models need to reflect this reality. In general, the distortionary impact of a tax increases nonlinearly with its level, so it is efficient to spread the tax base as broadly as possible. Consequently, even in the absence of a CO_2 problem, the optimal level of energy/carbon taxation would not be zero.

Modeling studies that neglect such distortions in the rest of the tax system—and the consequent potential gains from carbon tax recycling—tend to overestimate the GNP impact of carbon taxation. This includes all the long-run (post 2010) points in Figures 3–7 and many of the shorter-term studies as well, where the practical GNP gains may be more significant.

Scope of Abatement

Climate change is a global problem, and the more countries that take part in abatement, the lower the costs of limiting global emissions. Studies emphasize that action by industrialized countries may have a significant short-term impact, but that the costs of restraining global emissions rise rapidly if developing countries are not soon included. Furthermore, even if all the principal countries participate, the costs are greater if each is bound to fixed emission targets, because some regions may then incur much higher costs than others as noted by, eg, Edmonds and co-workers (87).

Table 5 shows one estimate of the impact of allowing countries freedom to choose where to invest to limit emissions (ie, trade in emission commitments). Despite a no-trade case that involves similar rate of emissions reduction below baseline (2% per year), there are still significant differences in marginal abatement costs (reflected in

Table 5. Cost Savings from Emissions Trading[a]

Target-Date	Trade	ERB		GREEN	
		Tax ($/tC)[b]	GDP loss, %[c]	Tax ($/tC)[a]	GDP loss, %
2020	No trade	283	1.9	149	1.9
	Trade	238	1.6	106	1.0
2050	No trade	680	3.7	230	2.6
	Trade	498	3.3	182	1.9

[a] Ref. 83.
[b] Global tax or mean of taxes required to achieve CO_2 reduction at 2%/yr below baseline trend in each region.
[c] GDP loss reported by model, not net cost.

the tax rate) between regions, and thus savings (10–45%) to be made from trading. Less uniform (relative to baseline) initial commitments would increase the benefits of trading accordingly; an extreme case was a special run of the Whalley-Wigle model (77,78) that showed the costs of obtaining the same degree of abatement, but with equal per capita emissions globally, would be roughly twice the costs of achieving the same reductions if emissions trading is allowed. Clearly models that do not allow such trade report costs greater than necessary.

Rate and Pattern of Emissions Abatement

Most studies, including those reviewed, have focused on measuring costs for a given degree of emissions abatement. It is clear that the costs must also depend on the rate of abatement. This is partly because relatively slow abatement can be achieved by introducing lower carbon technologies as older vintages are naturally retired. The modeling improvement has little effect on the costs of CO_2 abatement at a rate of 1% per year below the baseline trend, but for abatement at a rate of 3% per year below the baseline trend, the costs are much higher when the stock effect is included.

In reality, many other factors also make more rapid abatement more costly. A variety of macro-economic disequilibria come into play; capital or labor cannot move rapidly from one sector to another, in response to relative price and other changes. Since resource allocation/equilibrium models do not capture these effects, they tend (other things being equal) to underestimate the costs of rapid abatement; unlike most other factors noted, this suggests that most of the points in Figures 3–5, most of which are derived from resource-allocation models, underestimate costs in this respect, especially for more rapid (shorter-term and more extreme) abatement.

The observation that short-run macro-economic models indicate a much smaller response of emissions to a given level of carbon tax is another indication that such models show the economy to be more resistant to change than do equilibrium models such as GREEN. A plot of the data from modeling studies against the rate of abatement (not reproduced here) does not, however, reveal a particularly strong relationship. This may be because the results are mostly from equilibrium models, which themselves have a wide variety of capital stock modeling (if any). Few studies with short-run macro-economic models have been carried out, and these focus on different ways of recycling carbon tax revenues; the issue of how costs vary with the rate of abatement does not seem to have been examined methodically with macro models.

A third issue relating directly to the rate of abatement is that of technology development and diffusion. This is inherently a process that takes time, particularly in energy supply. Marchetti and others at IIASA (115) proposed a general principle that it takes approximately 50 years for a new supply technology to become dominant. More detailed technology studies reveal a more complex picture, with the rates of diffusion dependent on the nature of the technology and existing infrastructure; Grubb & Walker (116) argue that electricity supply could change much more rapidly than the “50 years” rule implies, but that fundamental changes to noncarbon transport fuels could be still slower.

Clearly, long timescales are involved, and these estimates do not include timescales for basic research and development. Consequently, rapid reductions will have to use less developed technologies than will slower reductions. This observation has been interpreted as a way of saying that delayed reductions are cheaper, but the implication rather is that costs will be minimized by setting the process of transition in motion as early as possible, because that will give the maximum time in which to develop, deploy, and refine lower carbon technologies. For a given total emissions over the next century, for example, simply delaying initial abatement efforts would both delay the start of such a transition, and increase the rate and degree of abatement that would ultimately have to be pursued.

In this context, the relevance of the whole pattern of abatement imposed in abatement costing studies is noted. The ultimate objective is to reduce accumulation of CO_2 in the atmosphere. Studies that impose a set time path of emissions make no attempt to explore possible less costly emission paths to the same end. The original Manne-Richels studies, for example, show carbon taxes rising to more than $650/t C, settling back down to a level of $250/t C set by the backstop technology. Peck and Tiesberg (39) point out that this is clearly not efficient; greater long-run abatement can be achieved with less long-run GNP impact by a time path that involves steadily rising carbon taxes. A more efficient time path of abatement would thus lower the GNP impact of studies shown in Figures 3–6. By how much is difficult to say, indeed, there are so many issues inadequately addressed in this context (eg, concerning technology development and diffusion) that firm conclusions would be premature, but it is an important area for further exploration.

Externalities and Multicriteria Assessment

CO_2 is but one of many external impacts associated with energy. Other pollutants are produced, and other issues, ranging from energy security to road congestion, are important but not included in traditional economic studies. Attempts to quantify other externalities (eg, Hohmeyer (117); Ottinger and co-workers (118)) suggest these can be significant. A study by Glomsrod and co-workers for Norway (66) appears to be the only one that includes the reduction of these other impacts as

CONCLUSIONS

Our survey and analysis of the literature on fossil-fuel CO_2 abatement costing shows a wide spread of reported results, and there are many different features that explain these differences. The principal factors that affect estimates are as follows:

- The choice of modeling approach and focus, notably
  - top-down or bottom-up
  - associated regulatory modeling
  - short-run macro-economic or long-run equilibrium
  - the linkage between energy sector in impacts and GNP
  - modeling of technology development in response to incentives
- The numerical assumptions concerning
  - energy-GNP trends as governed by income elasticities and autonomous energy-intensity improvements
  - fossil-fuel reserves and associated future prices and supply elasticities
  - the degree (elasticity) and reversibility of energy demand responses to price changes
  - technology development and costs (including production scale economies), or equivalent substitution elasticities
- The nature of the abatement strategy and the scope of analysis:
  - the reflection of existing tax and subsidy distortions and associated use of carbon tax revenues
  - the scope of abatement and allowance for emissions trade among participants, the rate and pattern of emissions abatement
  - the extent to which external costs and benefits associated with energy supply are reflected

Bottom-up engineering models tend to underestimate costs by neglecting issues of implementation and other hidden costs. Top-down economic models tend to overestimate costs by neglecting the potential for enhancing structural change and energy-efficiency gains through regulatory policy. We suggest that real cost-free reductions of up to 20% below the baseline projection over a couple of decades is a reasonable estimate of this potential, but realizing such reductions requires extensive implementation of policy instruments to improve market efficiencies.

Within the range of abatement issues and options there are at least four kinds of options that reduce CO_2 emissions but may incur macro-economic gains: regulatory policies to reduce the efficiency gap as noted above; removal of subsidies for carbon fuels; use of carbon tax revenues to reduce existing tax distortions; and the reduction of high non-CO_2 externalities. In each case the possibilities raise similar questions about the extent to which the benefits of associated policy reform should be credited to CO_2 abatement. In each case it is concluded that some credit is justified, but not all the potential should be credited as benefits of CO_2 abatement.

For comparative cost purposes, it is useful to express abatement relative to a baseline projection of expected emissions in the absence of specific abatement efforts. More attention needs to be devoted to definition of the baseline case. However, in general, the following is concluded.

Short-run Abatement Costs. The results for regional and shorter-term studies of abatement costs vary widely; the 15 reported costs of emission reductions of 15 to 40% below baseline (not base year) over the next 15–20 years, for example, range from GNP gains of more than 1% to a several percent loss in GNP. However, the higher losses in this range reflect the fiscal effect of removing carbon tax revenues from the economy, and also ignore the potential for energy-efficiency programs, while the larger GNP gains either reflect the use of carbon taxes to transfer resources from consumption to investment or reflect pure technology potentials. These extremes are of questionable relevance as measures of abatement costs, which in reality may be expected to lie well within these outlier values.

Long-run Abatement Costs. The range in long-run cost estimates is not quite so broad. With few exceptions, long-run modeling results portray that reducing CO_2 emissions at a rate averaging up to 2% per year below baseline, leading to a halving of relative emissions by the middle of the next century, may reduce the associated long-run global GNP by up to 3%, with the lower bound of costs being almost negligible.

For most but not all of the issues listed above, using more realistic or plausible assumptions reduces abatement cost estimates as compared with the most widely reported top-down economic modeling studies. The evidence presented from sensitivity studies and other modeling studies suggests that reasonable allowance for technology development, removal of subsidies and recycling of carbon tax revenues, and the reflection of avoided externalities alone might halve the more pessimistic estimates of abatement costs. Consequently, the realistic range for the GNP loss from halving long-run CO_2 emissions (relative to the emissions in most baseline projections) is 0 to 1.5%.

The upper bound of this range in absolute terms (for example it represents $600 billion out of a projected global GNP of $40,000 billion), but it small compared with other uncertainties and influences on GNP; it is equivalent to reducing average GNP growth rates over 50 years from 3% per year to 2.97% per year. Efficient distribution of abatement efforts between regions and over time could further lower these.

Costs will depend heavily on the rate at which abatement is imposed. The conclusions cited above refer to moderate rates of abatement. Study of this issue is seriously inadequate, but the costs may start to rise sharply for abatement at rates exceeding 1–2% per year below the baseline trend.

Projections by the World Energy Council (119) and many others suggest that global CO_2 emissions, in the absence of CO_2 abatement policies or other policy changes that significantly restrain CO_2 emissions, may grow at a rate averaging around 1.6% per year ($\pm$0.4% per year) for many decades. Given all the uncertainties and possibilities for lower abatement costs identified in this paper, the survey and analysis therefore suggests that keeping long-term global CO_2 emissions to about current levels, which is much more severe than current proposals to stabilize emissions in industrialized countries, may if carried out in an efficient manner be expected to reduce global GNP toward the middle of the next century by no more than 1–1.5%, and average GNP growth rates over the period by less than 0.02–0.03% per year.

The real difficulties of abatement lie in the design and implementation of nationally and globally efficient policies, and the geographical, sectoral, and social distribution of abatement impacts. Many analytic issues remain to be resolved, but for constraining CO_2 emissions, the key problems appear not to be massive, macro-economic losses, but rather the politics of implementation, winners, and losers.

Acknowledgments

The work for this paper was supported in part by the Office of Energy Research of the U.S. Department of Energy under contract DE-AC06-76RLO 1830. The authors wish to express their appreciation to those who helped make the paper possible, including Ari Patrinos, John Houghton, Andrew Dean, Robert Shackleton, Wendy Nelson, and Liz Malone.

BIBLIOGRAPHY

1. W. D. Nordhaus, *Strategies for the Control of Carbon Diolide*, Cowles Foundation, Discussion Paper, No. 443, Yale University, New Haven, Conn., 1977.
2. W. D. Nordhaus, *The Efficient Use of Energy Resources*, Yale University Press, New Haven, Conn., 1979.
3. J. A. Edmonds and J. Reilly, *Energy J.* **4**(3), 21–47 (1983).
4. J. A. Edmonds and J. Reilly, *Global Energy: Assessing the Future*, Oxford University Press, New York, 1985.
5. R. F. Kosobud, T. A. Daly, and Y. I. Chang, "Long Run Energy Technology Choices," CO_2 abatement policies and CO_2 shadow, prepared for the *Atlantic Economic Society Meeting*, Paris, France, 1983.
6. S. Seidel and D. Keyes, *Can We Delay a Greenhouse Warming? I˜S* Environmental Protection Agency, Washington, D.C., 1983.
7. D. J. Rose, M. M. Miller, and C. Agnew, *Global Energy Futures and CO_2-Induced Climate Change*, MITEL 83 015, MIT Energy Laboratory; also as MITNE-259, MIT Dept. of Nuclear Engineering Cambridge, Mass., 1983.
8. A. B. Lovins, L. H. Lovins, F. Krause, and W. Bach, "Energy Strategies for Low Climate Risk," (R&D No. 104 02 513), report for the German Federal Environmental Agency, International Project for Soh-Energy Paths, San Francisco, Calif., 1981.
9. R. H. Williams, J. Goldemberg, T. B. Johansson, A. K. N. Reddy, and E. Larson, "Overview of an End-Use-Oriented Global Energy Strategy," presented at the Hubert Humphrey Institute of Public Affairs Symposium Greenhouse Problems: Policy Options, May 29–31, 1984, University of Minnesota, Minneapolis, 1984.
10. A. S. Manne, ETA-MACRO projections, in *Global Carbon Dioxide Emissions—A Comparison of Two Models*, Electrical Power Research Institute, Palo Alto, Calif., 1984, Chapt. 2.
11. A. M. Perry, K. J. Araj, W. Fulkerson, D. J. Rose, M. M. Miller, and R. M. Rotty, *Energy* **7**(12), 991–1004 (1982).
12. W. D. Nordhaus and G. W. Yohe, in *Changing Climate*, National Academic Press, Washington, D.C., 1983, pp. 87–153.
13. I. Mintzer, *A Matter of Degrees: The Potential for Controlling the Greenhouse Effect*, research report #5, World Resource Institute, Washington, D.C, 1987.
14. J. Edmonds and J. Reilly, "Future Global Energy and Carbon Emissions," in J. Trabalka, ed., *Atmospheric Carbon Dioxide and the Global Carbon Cycle, DOE/ER-0239*, National Technical Information Service, U.S. Dept. of Commerce, Springfield, Va., 1985.
15. IPCC EIS (Energy and Industry Subgroup), report to the Intergovernmental Panel on Climate Change, UNEP/World Meteorological Organization (WMO), Geneva, 1990.
16. W. Fulkerson, S. I. Auerbach, A. T. Crane, D. E. Kash, A. M. Perry, D. B. Reister, *Energy Technology R&D: What Could Make a Difference?* ORNL-6541, Oak Ridge National Laboratory, Oak Ridge, Tenn., 1989.
17. International Energy Agency, *Energy Conservation in IEA Countries*, OECD, Paris, 1987.
18. M. J. Grubb, *Energy Policies and the Greenhouse Effect*, Vol. 11, Country Studies and Technical Options, Dartmouth, Brookfield, Vt., 1991.
19. J. Goldemberg, T. B. Johansson, A. Reddy, and R. Williams, *Energy for a Sustainable World*, World Resources Institute, Washington, D.C., Wiley-Eastern, New Delhi, India, 1987.
20. M. Grubb, J. Edmonds, P. ten Brink, and M. Morrison, *Ann. Rev. Energ. Environ.* **18,** 397–470 (1993).
21. United Nations Environmental Programme, *UNEP Greenhouse Gas Abatement Costing Studies*, Phase I report, UNEP Collab. Cent. Energy Environ. Riso Natl. Lab., Denmark, 1992.
22. A. B. Lovins and L. H. Lovins, *Ann. Rev. Energ. Environ.* **16,** 433–531 (1991).
23. E. Mills, D. Wilson and T. B. Johansson, *Energy Policy* **19**(6), 526–542 (1991).
24. T. Jackson, "Least-cost Greenhouse Planning: Supply Curves for Global Warming Abatement," *Energy Policy* (Jan/Feb. 1991).
25. F. Krause, W. Bach, and J. Kooney, *Energy Policies in the Greenhouse*, Int. Proj. Sustain. Energy Paths, Palo Alto, Calif., 1991.

26. Electronic Power Research Institute (EPRI), *Efficient Electricity Use: Estimates of Maximum Savings*, EPRI, CU-6746, Palo Alto, Calif., 1990.
27. Tata Energy Research Institute *UNEP National Greenhouse Costing Study—Indian Country Report* (Phase I Report), TATA Energy Research Institute (TERI), New Delhi, 1992.
28. D. Anderson and C. D. Bird, "Carbon Accumulations and Technical Progress—A Simulation Study of Costs," *Oxford Bull. Econ. Stat.* **42**(1) (1992).
29. J.-M. Burniaux, J. P. Martin, G. Nicoletti, and J. O. Martins, *The Costs of Policies to Reduce Global Emissions of CO_2: Initial Simulation Results with GREEN*, OECD Dept. Economic Statistics, Working Paper No. 103, OECD/GD, Resource Allocation Division, Paris, 1991, p. 115.
30. J.-M. Burniaux, G. Nicoletti, and J. O. Martins, "GREEN: A Global Model for Quantifying the Costs of Policies to Curb CO_2 Emissions," *OECD Economic Studies 19*, Winter 1992.
31. J. Edmonds and D. W. Barnes, *Estimating the Marginal Cost of Reducing Global Fossil Fuel CO_2 Emissions*, PNL-SA- 18361, Washington, D.C., Pacific Northwest Laboratory, 1990.
32. J. A. Edmonds and D. W. Barns, *Int. J. Glob. Energy Issues* **4**(3), 140–166 (1992).
33. A. S. Manne and R. G. Richels, *Energy J.* **12**(1), 87–102 (1990).
34. A. S. Manne, *Global 2100: Alternative Scenarios for Reducing Emissions*, OECD Working Paper 111, OECD, Paris, 1992.
35. J. O. Martins, J.-M., Bumiaux, J. P. Martin, and G. Nicoletti, *The Cost of Reducing CO_2 Emissions: A Comparison of Carbon Tax Curves with Green*, OECD Economic Working Paper No. 118, OECD, Paris, 1992.
36. C. Perroni and T. Rutherford, *International Trade in Carbon Emissions Rights and Basic Materials: General Equilibrium Calculations for 2020*, Dept. of Economics, Wilfred Laurier University of Waterloo, Ontario and Dept. of Economics, University of Western Ontario, London, Ontario, 1991.
37. T. Rutherford, *The Welfare Effects of Fossil Carbon Reductions: Results from a Recursively Dynamic Trade Model*, Working Papers, No. 112, OECD/GD Paris, 1992, p. 89.
38. J. Whalley and R. Wigle, *Energy J.* **12**(1), 109–124 (1990).
39. S. C. Peck and T. J. Tiesberg, *Energy J.* **13**(1) (1992).
40. W. D. Nordhaus, *Energy Econ.* **15**(1), 27–50 (1993).
41. D. W. Barns, J. A. Edmonds, and J. M. Reilly, *The Use of the Edmonds-Reilly Models to Model Energy Related Greenhouse Gas Emissions*, Economic Department Working Paper No. 113, OECD, Paris, 1992.
42. Data Resources, Inc. (DRi), *The Economic Effects of Using Carbon Taxes to Reduce Carbon Dioxide Emissions in Major OECD Countries*, prepared for the U.S. Dept. Commerce DRI/McGraw-Hill, Lexington, Mass., 1993; structural comparison of the models in EMF-12; Special issue on policy modeling for climate change, *Energy Policy*, **21**(3) (1992).
43. The Congressional Budget Office, *Carbon Charges as a Response to Global Warming: The Effects of Taxing Fossil Fuel*, Congress of the United States, Washington, D.C., 1990.
44. J. A. Edmonds and D. W. Barns, *Use of the Edmonds-Reilly Model to Model Energy-Sector Impacts of Greenhouse Emissions Control Strategies*, prepared for the U.S. Dept. Energy under Contract DE-AC06-76RLO 1830, Washington, D.C., 1991.
45. L. H. Goulder, *Effects of Carbon Taxes in an Economy with Prior Tax Distortion: An Intertmporal General Equilibrium Analysis for the U.S.* Working paper, Stanford University, Stanford, Calif., 1991.
46. D. W. Jorgenson and P. J. Wilcoxen, "The Cost of Controlling U.S. Carbon Dioxide Emissions," presented at the *Workshop on Econ.l Energy/Environ. Model. Clim. Policy Anal., Washington, D.C.*, Harvard University, Cambridge, Mass., 1990.
47. D. W. Jorgenson and P. J. Wilcoxen, *Reducing U.S. Carbon Dioxide Emissions: The Cost of Different Goals*, Harvard Institute Economic Research, Harvard University, Cambridge, Mass., 1990.
48. A. S. Manne and R. G. Richels, *Energy J.* **22**(2), 51–74 (1990); *Emissions from the Energy Sector: Cost and Policy Options*. Stanford Press, Stanford, Calif., 1990.
49. D. Gaskins and J. Weyant, eds. National Academy of Science (NAS), *Policy Implications of Greenhouse Warming*, National Academy Press, Washington, D.C., 1991.
50. U.S. Congress, Office of Technological Assessment (OTA), *Changing by Degrees: Steps to Reduce Greenhouse Gases: Summary*, Washington, D.C., 1991.
51. R. Shackleton and co-workers, *The efficiency value of carbon tax revenues*, 1993; *Proc. Semin. PRISTE-CNRS*, Paris, Oct. 1992.
52. Congressional Budget Office, *Carbon Charges as a Response to Global Warming: The Effects of Taxing Fossil Fuel*, U.S. Congress, Washington, D.C., 1990.
53. A. S. Manne and R. G. Richels, *Buying Greenhouse Insurance: The Economic Costs of CO_2 Emission Limits*, MIT Press, Cambridge, Mass., 1992. This book uses revised assumptions that lower costs, especially for the global studies.
54. A. Christensen, *Stabilization of CO_2 Emissions—Economic Effects for Finland*, Ministry of Finance Discussion Paper No. 29, Helsinki, Mar. 1991.
55. K. Yamaji, See W. U. Chandler, ed. *Carbon Emissions Control Strategies: Case Studies in International Cooperation.* World Wildlife Fund & The Conservation Foundation, Baltimore, 1990, pp. 155–172.
56. P. B. Dixon, R. E. Marks, P. McLennan, R. Schodde, and P. L. Swan, *The Feasibility and Implications for Australia of the Adoption of the Toronto Proposal for CO_2 Emissions*, report to Charles River & Assoc., Sydney, Australia, 1989.
57. Ind. Comm. *Costs and Benefits of Reducing Greenhouse Gas Emissions*, Vol. I, report (draft), Australia, 1991.
58. R. E. Marks and co-workers, *Energy J.* **12**(2), 135–152 (1990).
59. S. Proost and D. van Regemorter, *Economic Effects of a Carbon Tax—With a General Equilibrium Illustration for Belgium*, Public Economic Research Paper No. 11, Katholieke University Leuven, 1990.
60. P. Karadeloglou, Energy Tax versus Carbon Tax: A Quantitative Macro-economic Analysis with the Hermes/Midas models, *Eur. Econ.* Special Ed. No. 1, CEC DG-II, Brussels, 1992.
61. K. Ban, *Energy Conservation and Economic Performance in Japan—An Economic Approach to CO_2 Emissions*, Faculty Econ., Osaka University, Discussion Paper Series No. 112, Oct. 1991.
62. U. Goto, "A study of the impacts of carbon taxes using the Edmonds-Reilly model and related sub-models," in A. Amano, ed., *Global Warming and Economic Growth*, Cent. Glob. Environ. Res., 1991.

63. Y. Nagata, K. Yamaji, and N. Sakarai, *CO_2 Reduction by Carbon Taxation and its Economic Impact*, Economic Research Institute Report No. 491002, Central Research Institute of the Electronic Power Industry, 1991.
64. National Environment Policy Plan (NEPP), *To Choose or to Lose*, Second Chamber of the States General Gravenhage, the Netherlands, 1989.
65. B. Bye, T. Bye, and L. Lorentson, *SIMEN: Studies in Industry, Environment and Energy Towards 2000*, Cent. Bur. Stat. Discuss. Paper No. 44, Oslo, 1989.
66. S. Glomsrod, H. Vennemo, and T. Johnson, *Stabilization of Emissions of CO_2: A Computable General Equilibrium Assessment*, Cent. Bur. Stat., Discussion paper No. 48, Oslo, 1990.
67. L. Bergman, *General Equilibrium Effects of Environmental Policy: A CGE-Modelling Approach*, Research Paper No. 6415, Handelshgskolan, Stockholm, 1990.
68. T. Barker and R. Lewney, A green scenario for the UK economy. In *Green Futures for Economic Growth: Britain in 2010*, Cambridge Econometrics, Cambridge, UK, 1991, Chapt. 2.
69. J. Sondheimer, Energy policy and the environment: Energy policy and the Green 1990s—macroeconomic effect of a carbon tax, paper at *Semin. sponsored by Cambridge Econometrics and UKCEED: The Economy and the Green 1990s, 11–12 July, Ribonson College*, Cambridge, UK, 1990.
70. S. Sitnicki, K. Budzinski, J. Juda, J. Michna, and A. Szpilewica, Poland in Ref. 55, pp. 55–80.
71. S. Sitnicki, K. Budzinski, J. Juda, J. Michna, and A. Szpilewica, *Energy Policy* **19**(10) (1991).
72. E. Unterwurzacher and F. Wirl, Wirtschaftliche und politcsche Neuordnung in Mitteleuropa: Auswirkungen auf die Energiewirtschaft und Umweltsituation in Polen, der CSFR und Ungarn. *Z. Energie-wirtschaft* (in German) (1991).
73. G. Leach and Z. Nowak, *Energy Policy* **19**(10) (1991).
74. IPCC–EIS (Energy and Industry Subgroup), *Report to the Intergovernmental Panel on Climate Change*, UNEP/World Meteorological Organization (WMO), Geneva, 1990.
75. J. Sathaye, ed. *Energy Policy*, Special issue: Climate Change—Country Case Studies, **19**(10), 1991.
76. J. P. Martin, J. M. Burniaux, G. Nicoletti, and J. Oliveira-Martins, The costs of international agreements to reduce CO_2 emissions: Evidence from GREEN. *OECD Econ. Stud.* 19—Winter (formerly Dept. Econ. Stat. Working Paper No. 115).
77. J. Whalley and R. Wigle, *Energy J.* **12**(1), 109–124 (1990).
78. J. Whalley and R. Wigle, in R. Dornbusch and J. M. Poterba, eds., *Global Warming: Economic Policy Responses*, MIT Press, Cambridge, Mass., 1990.
79. C. R. Blitzer, R. S. Ekhaus, S. Lahiri, and A. Meeraus, 1990. A general equilibrium analysis of the effects of carbon emission restrictions on economic growth in developing country. In *Proc. Workshop Econ. / Energy / Environ. Model. Clim. Policy Anal.* World Bank, Washington, D.C.
80. C. R. Blitzer, R. S. Ekhaus, S. Lahiri, and A. Meeraus, 1990. The potential for reducing carbon emissions from increased efficiency: a general equilibrium methodology. See Ref. 57.
81. D. Wilson and J. Swisher, *Energy Policy* **21**(3), 48 (1993). Exploring the gap: top-down versus bottom-up analyses of the cost of mitigating global warming.
82. A. Ingham, J. Maw, and A. Ulph, *Oxford Rev. Econ. Policy* **7**(2) (1991). Empirical measures of carbon taxes.
83. A. Dean and P. Hoeller, *OECD Econ. Stud.* **19** (Winter, 1992) (formerly published as OECD Econ. Dept. Working Papers No. 122). Costs of reducing CO_2 emissions: Evidence from six global models.
84. J. A. Edmonds, H. M. Pitcher, D. W. Bams, R. Baron, and M. A. Wise, in *Proc. United Nations Univ. Conf. "Global Change and Modelling,"* Tokyo, 1992. Modeling future greenhouse gas emissions: The second generation model description.
85. S. C. Morris, B. D. Solomon, D. Hill, J. Lee, and G. Goldstein, in J. W. Tester and N. Ferrari, eds., *Energy and Environment in the 21st Century*, 1990. A least cost energy analysis of U.S. CO_2 reduction options.
86. R. A. Bradley, E. C. Watts, and E. R. Williams, eds. *Limiting Net Greenhouse Gas Emissions in the United States*, DOE/pElolol. U.S. Dept. Energy, Washington, D.C. See especially Chapt. 7.
87. J. Edmonds, D. Barnes, M. Wise, and M. Ton, *Carbon Coalitions: The Cost and Effectiveness of Energy Agreements to Alter Trajectories of Atmospheric Carbon Dioxide Emissions*, prepared for the U.S. Office of Technological Assessment and the PNL Global Studies Program, 1992.
88. C.-O. Wene, Top down—bottom up: A system engineers's view. Paper presented to *Int. Workshop Costs, Impacts, and Possible Benefits of CO_2 Mitigation*, Sept. 28–30, IIASA, Laxenburg, Austria; Energy Systems Technology, Dept. Energy Conversion, Chalmers University of Technology, S-41296 Goteborg, Sweden, 1992.
89. G. Boero, R. Clarke, and L. A. Winters, *The Macroeconomic Consequences of Controlling Greenhouse Gases: A Survey*. Dept. Environ., Environ. Econ. Res. Ser., London, Her Majesty's Stationery Office, 1991.
90. D. W. Jorgenson and P. J. Wilcoxen, *Intertemporal General Equilibrium Modelling of U.S. Environmental Regulation*, Harvard Inst. Econ. Res. Cambridge, Mass., 1989.
91. W. J. McKibben and P. J. Wilcoxen, *The Global Costs of Policies to Reduce Greenhouse Gas Emissions*, Brookings Discussion papers no. 97, Brookings Institute, Washington, D.C., 1992.
92. R. C. Dower and M. B. Zimmemman, *The Right Climate for Carbon Taxes*, World Resources Institute, Washington, D.C., 1992.
93. M. J. Grubb, *Energy Policies and the Greenhouse Effect, Vol. I: Policy Appraisal*, Dartmouth Brookfield, Vt., 1990.
94. AES Corp., *An Overview of the Fossil2 Model*, prepared for the U.S. DOE Office Policy Eval., Arlington, Va., 1990.
95. SEI-B, LEAP: *Long-Range Energy Alternative Planning System, Vol. 1: Overview, Vol. 2: User Guide*, Stockholm Environmental Institute, Boston, 1992.
96. P. E. Morthorst, and J. Fenhann, *Brundtland Scenarienmodelien—BRUS (The Brundtland scenario model)*, Danish Ministry of Energy, Denmark, 1990.
97. Danish Ministry of Energy, *Energy 2000: A Plan of Action of Sustainable Development*, Danish Ministry of Energy, Denmark, 1990.
98. R. Williams, *Energy J.* **11**(4) (1990).
99. A. Manne and R. G. Richels, *Energy J.* **11**(4) (1990).
100. W. R. Cline, *The Economics of Global Warming*, Institute of International Economics, Washington, D.C., 1992.
101. *Report of UNEP Workshop on the Costs of Limiting Fossil Fuel CO_2 Emissions*, London, Jan. 7–8, 1991, UNEP Environmental Economics Series paper no. 5, UNEP, Nairobi, 1991.
102. A. S. Manne and R. G. Richels, *Buying Greenhouse Insur-*

ance: The Economic Costs of CO_2 Emission Limits, MIT Press, Cambridge, Mass., 1992. This book uses revised assumptions that lower costs, especially for the global studies.

103. S. Meyers, and L. Schipper, *Annu. Rev. Energy Environ.* **17,** 463–505 (1992). World energy use in the 1970s and 1980s: Exploring the changes.
104. L. Schipper, S. Meyers, with R. Howarth and R. Steiner, *Energy Efficiency and Human Activity: Past Trends, Future Prospects*, Cambridge University Press, Cambridge, UK, 1992.
105. R. H. Williams, E. D. Larson, and M. Ross, *Annu. Rev. Energy* **12,** 99–144 (1987).
106. A. K. N. Reddy, *Energy Policy* **19**(10) (1991).
107. P. L. Joskow and D. B. Marron, *Science* **260.** (1993).
108. T. B. Johansson, H. Kelly, A. K. Reddy, R. H. Williams, and L. Bumham, *Renewable Energy: Sources for Fuels and Electricity*, Island Washington, D.C., 1993.
109. W. W. Hogan and D. W. Jorgenson, *Energy J.* **12**(1), 67–86 (1991). Productivity trends and the cost of reducing CO_2 emissions.
110. S. Boyle and co-workers, *A Fossil-free Energy Future*, technical appendices, Greenpeace International, London, 1993.
111. T. B. Johansson, H. Kelly, A. K. Reddy, R. H. Williams, and L. Burnham, *Renewable Energy: Sources for Fuels and Electricity*, Island, Washington, D.C., 1993.
112. W. U. Chandler and A. K. Nicholls, *Assessing Carbon Emissions Control Strategies: A Carbon Tax or a Gasoline Tax?* American Council on Energy Efficiency and Economics, 1990.
113. Capros and co-workers, *The Impact of Energy and Carbon Tax on CO_2 Emissions*, report to the CED, DGXII, Brussels, 1991.
114. J. C. Hourcade, *Energy Policy* **21**(3) (1993). Modeling long-run scenarios: methodology lessons from a prospective study on a low CO_2 intensive country.
115. J. Andere, W. Hafele, N. Nakicenovic, and A. McDonald, *Energy in a Finite World*, Ballinger, Cambridge, Mass., 1981.
116. M. J. Grubb and J. Walker, eds, *Emerging Energy Technologies: Impacts and Policy Implications*, Dartmouth, Brookfield, Vt., 1992. Especially Chapt. I.
117. O. Hohmeyer, *Social Costs of Energy Consumption*, Springer-Verlag, Berlin, 1988.
118. R. Ottinger, D. Wooley, N. Robinson, D. Hodas, and S. Babb, *The Environmental Costs of Electricity*, Oceana, New York, 1990.
119. World Energy Council, *Energy for Tomorrow's World*, London, 1992.

CARBON RESIDUE

Ramon Espino
Exxon Research and Engineering
Annandale, New Jersey

There is not a universally-accepted definition of the term carbon residue. Anyone involved in energy technology is bound to encounter as many variants of this term as there are applications. Generally, it refers to the fraction of a hydrocarbon-containing product or byproduct that is largely composed of carbon atoms. The residue in most cases is not pure carbon since it can contain measurable amounts of other elements, particularly hydrogen. Sulfur, nitrogen, oxygen and metals are also present in many carbon residues.

See Coal coking; Coal pyrolysis; Petroleum production; Pyrolysis.

The fraction of a hydrocarbon that is converted to a "carbon residue" is strongly dependent on temperature. While very small amounts are produced at temperatures below 200°C, significantly greater amounts are generated at higher temperatures. Carbon hydrogen bonds begin to break, at a significant rate, at temperatures above 400°C, and most hydrocarbons are thermally unstable above 500°C, resulting in the evolution of hydrogen and the combination of the carbon atoms into larger molecular structures comprised mainly of carbon with small amounts of hydrogen. The conversion of hydrocarbons into a carbonaceous residue and hydrogen can be partially controlled by the presence of hydrogen at high pressure, and further suppressed if a catalyst that favors hydrogenation is present.

The thermal stability of the carbon-hydrogen bond varies. The hydrogens in methane and benzene are much more stable than hydrogens in paraffinic or olefinic hydrocarbons, particularly branched olefins and paraffins. In the latter, there are carbon atoms with only one hydrogen atom. These carbon-hydrogen bonds are the weakest and tend to break at lower temperatures. The thermal and catalytic chemistry of carbon-hydrogen molecules is well-advanced with a significant amount of theory and experimental data that allows the scientist and engineer to predict the yield of carbonaceous residue depending on the process conditions. Molecules containing carbon-oxygen, carbon-sulfur, and carbon-nitrogen can also produce carbonaceous residues, and the amount is also determined by the bond strength of these systems. The level of scientific understanding is not as advanced as for carbon-hydrogen systems but is still possible to estimate the amount of carbonaceous residue as a fraction of temperature, pressure, and reaction time.

The thermal decomposition of hydrocarbons at high temperatures (500°C and above) is the basis for the production of coke from petroleum residua and coal. Byproducts of these thermal reactions are volatile hydrocarbons that have a higher ratio of hydrogen to carbon than the original feed (residua or coal). Smaller amounts of molecular hydrogens, methane, water, carbon oxides, hydrogen sulfide, and ammonia are also produced if the feed contains oxygen, sulfur and nitrogen. These thermal processes are optimized for the production of maximum amounts of hydrogen-rich volatile compounds and a solid "residue" that is enriched in its carbon content. In these processes, the thermal decomposition behavior of hydrocarbons is used to produce a useful carbon residue in the form of coke.

However, in many other circumstances carbonaceous residues are not the preferred products, but rather undesirable byproducts. For example, during the combustion of hydrocarbons, the desired products are those of the complete combustion of carbon to carbon dioxide and carbon monoxide, and of hydrogen to water. However, in most combustion processes a fraction of the hydrocarbon is converted in a carbonaceous residue. This residue can be either small carbon particles called soot, or carbonaceous residues that deposit on the surfaces of the combustion chambers. A well-controlled combustion system produces very small amounts of these two byproducts, but their total elimination has not been achieved in practical applications. After all, at the high temperatures of most combustion processes the carbon-hydrogen bonds are unstable and both carbon–carbon and carbon–oxygen reactions are favored. The key to "clean combustion" is to maximize the latter at the expense of the former. In combustion processes carbon residues are undesirable. Soot particles are not only visually unattractive, but if inhaled can deposit in the lungs. Since soot particles can adsorb cancer-causing polynuclear aromatic hydrocarbons (PAH), they can also represent a health hazard. Solid carbonaceous residues can plug hydrocarbon feed nozzles (fuel injectors) and deposit on valves, combustion chambers and heat transfer equipment. Carbonaceous residues are carefully monitored in most combustion processes. For example, soot particles are analyzed for their content of soluble hydrocarbons (where PAHs are concentrated), ash (sulfates, nitrates and metals), and insoluble hydrocarbons (mainly coke-type molecules). The amount of carbonaceous residues is measured on fuel injector nozzles, inlet valves, and combustion chambers since these deposits can interfere with the operation of internal combustion engines.

Carbonaceous residues are observed in many refinery and chemical processes. These residues can deposit on reactor walls, flow systems, catalyst particles and many other types of process equipment. In many of these processes, the carbon residue is measured and monitored. For example, in the most widely used refining process, fluid catalytic cracking, the carbon residue that is deposited on the cracking catalyst rapidly kills its catalytic activity. In a fluid catalytic cracker, the catalyst flows through pipes and reactors like a fluid. The carbon containing catalyst is transferred to a combustion zone where the carbon is burned off the catalyst with air and thus regenerated. A large portion of the heat generated during the burning of carbon is retained by the hot catalyst particles which flow back to the catalytic reactor and provide the energy required for the cracking reaction to occur. The elegant simplicity of this two-step process is the reason for its worldwide applicability.

When hydrocarbons are used as lubricants, carbon residues often appear and interfere with the lubrication process. Therefore the tendency of lubricants to form carbon residues is evaluated in laboratory tests as well as in their actual application. There are a myriad of tests since the temperature and reaction time as well as the environment (catalyst, presence of hydrogen, oxygen, air, combustion gases) can vary. The carbon residue that is produced can also vary in composition, including carbonaceous deposits on surfaces, sludge and varnish on surfaces, and soot or carbonaceous particles dispersed in the liquid lubricant. The composition of these carbon residues can be highly variable, ranging from very carbon rich solids to solids containing significant amounts of metal salts and rich in nitrogen, oxygen and hydrogen.

BIBLIOGRAPHY

Reading List

Harry Marsh (Ed), *Introduction to Carbon Science*, Butterworth, Boston, 1989.

Martin A. Elliott, *Chemistry of Coal Utilization*, John Wiley and Sons, Inc. New York, 1981, 2nd suppl. vol.

E. F. Obert, *Internal Combustion Engines and Air Pollution*, Harper and Row, Publishers.

G. D. Hobson and W. Pohl, *Modern Petroleum Technology*, Applied Science Publishers, Ltd., Chapt. 17.

CARBON STORAGE IN U.S. FORESTS

RICHARD A. BIRDSEY
LINDA S. HEATH
U.S.D.A.—Forest Service
Radnor, Pennsylvania

Global concern about increasing atmospheric concentrations of greenhouse gases, particularly carbon dioxide (CO_2), and the possible consequences of future climate changes have generated interest in evaluating the role of terrestrial ecosystems in the global carbon cycle. Recent efforts to estimate changes in the global carbon budget by accounting for CO_2 emissions from burning fossil fuels, deforestation in the tropics and CO_2 uptake by the atmosphere and oceans have revealed an imbalance between sources and sinks of CO_2. Additional carbon sinks of approximately 1.8 Pg/yr are suspected in the terrestrial biosphere or the oceans (1). Identification of all sources and sinks, which are small relative to the magnitude of global carbon reservoirs, has been difficult because of inability to measure and monitor accurately the effects of natural and human disturbances on ecosystems over large regions.

See also AIR QUALITY MODELING; CARBON BALANCE MODELING; CARBON CYCLE; FOREST RESOURCES; GLOBAL CLIMATE CHANGE, MITIGATION.

Carbon compounds have a central role in plant processes that build structures and maintain physiological functions. Rates of carbon uptake through photosynthesis and release through respiration are a function of light, atmospheric CO_2, temperature, water, nutrients, and other factors such as air pollution. Loss of leaves and woody material and death of fine roots are sources of organic matter for soil processes that in turn support uptake of nutrients and plant growth. Forest ecosystems are capable of storing large quantities of carbon in biomass and soils. Forest disturbances such as fire and timber harvest may reduce biomass and cause oxidation of soil carbon, adding to the pool of CO_2 in the atmosphere. Growing forests may reduce atmospheric CO_2 through increases in biomass and soil carbon. Carbon in wood products may be effectively stored for long periods of time, depending on the end use of the wood. With a complete accounting for all of the forest changes and effects on carbon in each of the components of the system, it would be possible to determine whether a land area containing forests is a net source or sink of CO_2.

Increasing inventories of timber in European forests, and corresponding increases in biomass and soil carbon, could account for 5–9% of the "missing" carbon (2). Similar estimates for U.S. forests could account for 12–21% of the unexplained flux since 1952 (3). It is likely that Canadian forests have been a carbon sink in recent decades (4), and estimates for the territory of the former Soviet Union indicate a carbon sink there as well (5). Together, the world's temperate and boreal forests have been adding carbon in biomass and soil at a rate of 0.7–1.2 Pg/yr (6). This accounts for only some of the suspected imbalance in the global carbon cycle.

ESTIMATING CARBON IN FOREST ECOSYSTEMS

Carbon storage is usually estimated for different forest ecosystem components: biomass in trees, soil organic matter, litter on the forest floor, coarse woody debris, and understory vegetation. Live-tree biomass includes above-ground and below-ground portions of all trees: the merchantable stem, limbs, stump, foliage, bark and rootbark, and coarse tree roots. The soil component includes organic and inorganic carbon in mineral soil horizons to a specified depth (usually 1 m), excluding coarse tree roots. The forest floor includes all dead organic matter (litter and humus) above the mineral soil horizons except standing dead trees. Coarse woody debris includes standing and fallen dead trees and branches. Understory vegetation includes all live vegetation besides live trees.

Estimates of carbon storage and flux in trees over large regions are typically based on periodic forest inventories designed to provide statistically valid estimates of timber volume, growth, removals, and mortality (7). Timber volume includes only the merchantable portions of live trees that meet specified criteria for size and quality. Estimation of carbon from timber volume involves several steps to account for nonmerchantable tree biomass and to convert from volume units to weight units. Total tree biomass, if not estimated directly, may be estimated by multiplying timber volume times conversion factors derived from representative biomass inventories (8). Depending on species and location, total biomass in all trees averages from 1.7 to 2.6 times the biomass in merchantable timber (9). Approximately 50% of tree biomass is carbon.

Less direct methods are used to estimate carbon storage in other ecosystem components over large regions. Simple models can be devised to estimate carbon storage in the soil, forest floor, coarse woody debris, and understory vegetation, using data from compilations of intensive ecosystem studies (10). Soil carbon is related to mean annual temperature, precipitation, soil texture, and stand age and is sometimes estimated with a regression model (11).

Because carbon cannot be directly measured for any but the smallest of areas, estimates using the preceding methods include some uncertainty. Regional forest inventories are based on a statistical sample designed to represent the broad range of forest conditions actually present in the landscape. Therefore, estimates of carbon storage in forest trees are representative of the true average values, subject to sampling errors, estimation errors, and errors in converting data from one reporting unit to another. The estimated error for tree carbon is usually quite small if the forest inventories used to derive the estimates have small sampling errors over large areas. Accurate regional forest inventories are available for some countries with advanced forest inventory systems such as the United States.

Most regional estimates of carbon storage in the soil and forest floor are not based on statistical samples but on compilations of the results of many separate ecological studies of specific ecosystems. Published estimates for soil carbon show wide variation for terrestrial ecosystems (12).

CARBON STORAGE IN THE UNITED STATES

Forest ecosystems in the United States contain approximately 54.6 Pg organic carbon above and below the ground (Table 1). This is about 4% of all the carbon stored in the world's forests (13). The area of U.S. forests is 298×10^{10} m^2, or 5% of the world's forest area. The average forest in the United States contains 18.3 kg/m^2 of organic carbon. Trees, including tree roots, account for 29% of all forest ecosystem carbon (Fig. 1). Live and standing dead trees contain 15.9 Pg carbon, or an average of 5.3 kg/m^2. Of this total, 50% is in live tree sections classified as timber, 30% is in other solid wood above the ground, 6% is in standing dead trees, and 3% is in the foliage.

The largest proportion of carbon in the average U.S. forest is found in the soil, which contains 61% of the carbon in the forest ecosystem, or approximately 11.2 kg/m^2.

Table 1. Area of Forest Land[a] and Carbon Storage by Region and Forest Ecosystem Component, 1992

		Forest Ecosystem Component, pg				
Region	Area, m^2	Soil	Forest Floor	Understory	Trees	Total
Northeast	$34{,}554 \times 10^4$	4.73	0.62	0.06	2.24	7.64
North Central	$33{,}634 \times 10^4$	3.50	0.51	0.05	1.68	5.73
Southeast	$35{,}645 \times 10^4$	2.52	0.21	0.11	2.09	4.92
South Central	$50{,}085 \times 10^4$	3.29	0.25	0.17	2.83	6.54
Rocky Mountains	$56{,}549 \times 10^4$	3.89	0.72	0.06	2.71	7.39
Pacific Coast	$35{,}408 \times 10^4$	3.41	0.63	0.15	2.24	6.41
Alaska	$52{,}259 \times 10^4$	12.01	1.63	0.20	2.16	16.00
United States	$298{,}134 \times 10^4$	33.33	4.56	0.80	15.94	54.64

[a] Land at least 10% stocked by forest trees of any size, including land that formerly had such tree cover and that will be naturally or artificially regenerated.

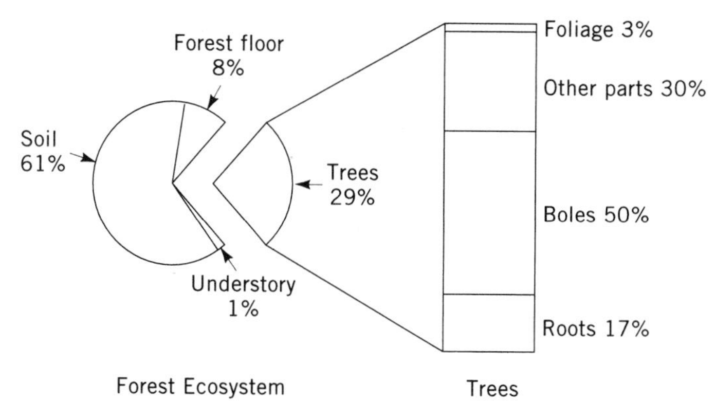

Figure 1. Allocation of carbon in forest ecosystems and trees for U.S. forests in 1992. Total storage in the United States is 54.6 billion t.

Most of the soil carbon is found in the top 30 cm of the soil profile, the location of most fine roots and adjacent to the litter inputs to the soil. About 8% of all carbon is found in litter, humus, and coarse woody debris on the forest floor, and about 1% is found in the understory vegetation. By adding carbon in tree roots to the carbon in the soil, the average proportion of carbon below the ground in the United States is estimated to be 66%.

Carbon storage and accumulation rates in a particular region or forest are influenced by many factors such as climate, solar radiation, disturbance, land-use history, age of forest, species composition, and site and soil characteristics. Even though all trees have similar physiological processes, there are significant differences in growth rates and wood density between species and individual organisms. The combination of species and site differences produces a wide variety of carbon densities across a landscape. Historical land-use patterns and landscape attributes produce characteristic regional profiles of carbon storage.

The proportion of carbon in the different ecosystem components varies considerably between regions (Fig. 2). Alaska has the highest estimated amount of carbon in the soil, about 75% of the total carbon. The southeastern and south-central states each have about 50% of total carbon in the soil but have a higher percentage in the trees. Soil carbon is closely related to temperature and precipitation, with higher amounts of soil carbon being found in regions with cooler temperatures and higher precipitation. Cooler temperatures slow the oxidation of soil carbon, while higher rainfall tends to produce greater growth of vegetation, fine roots, and litter, which are the main sources of organic soil carbon. Carbon in the forest floor varies by region in a way similar to carbon in the soil. Western and northern states contain the most carbon on the forest floor, and southern states contain the least.

Significant differences in carbon storage are apparent for different forest types (Fig. 3). The differences are related both to biological and human influences on the landscape. For example, loblolly pine plantations are younger on average than natural loblolly pine, so there is less carbon in the trees. Likewise, because loblolly pine is located in the South, there is less carbon in the soil than in forest types such as Douglas fir or spruce–fir that grow in cooler, wetter climates. Of the examples shown, piñon–juniper has the lowest amount of carbon because it occurs in dry climates that support lower vegetation densities.

The age of the forest and degree of disturbance or management intensity have significant effects on carbon storage. Generally, older forests contain more carbon than younger forests, but younger forests accumulate carbon faster than older forests. These patterns are evident in plots of the relationship between carbon storage and age of forest for representative forest ecosystems (Fig. 4). For loblolly pine, a type in which trees represent a large proportion of total ecosystem carbon, a fourfold increase in total carbon is possible over an 80-yr period. Spruce–fir, which grows in the northeastern and north-central United States, shows less total variation over the same period of time, because the most dynamic ecosystem component, trees, represents a smaller proportion of the total carbon. The effects of management and natural disturbance on carbon are less well known than the effects of age but can be approximated by substituting “time since disturbance” for “stand age” and adjusting the initial estimates for each of the ecosystem components.

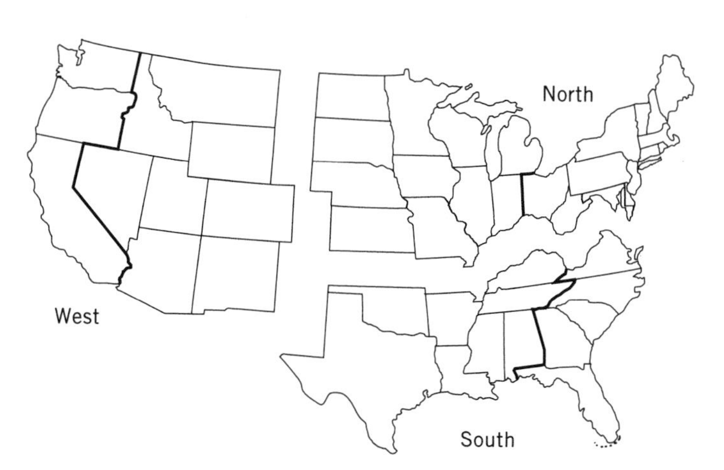

Figure 2. Regions of the United States used to estimate trends in carbon storage.

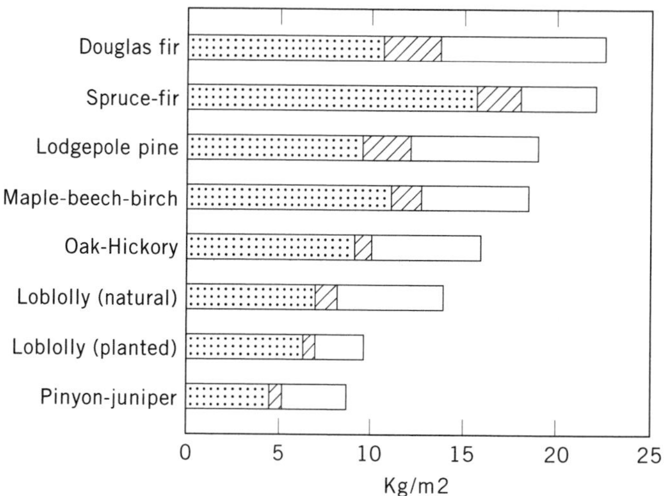

Figure 3. Average carbon storage in the soil, forest floor, and trees for selected forest types. ⊡, soil; ▨, forest floor; □, trees.

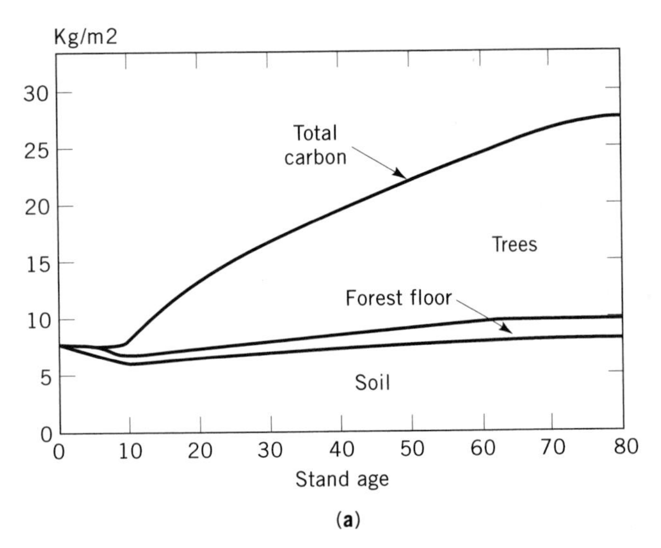

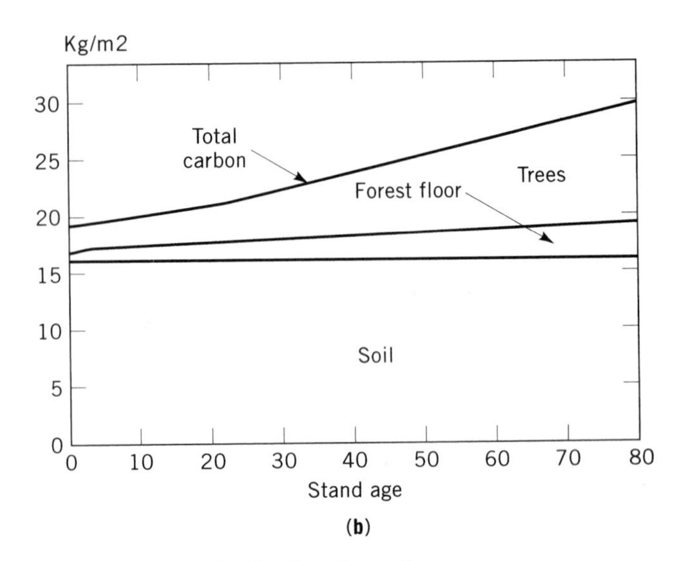

Figure 4. Carbon storage by forest ecosystem component for (**a**) a loblolly pine plantation in the Southeast and (**b**) a spruce–fir forest in the Northeast. Both forests have been clearcut and regenerated.

RECENT TRENDS IN CARBON STORAGE

U.S. forests are constantly changing. The total area of forest land declined by 1.6×10^{10} m^2 between 1977 and 1987 (7). Most of the net loss was removal of trees from forest clearing for urban and suburban development, highways, and other rights-of-way. A much larger area was cleared for agricultural use, but this loss was roughly balanced by agricultural land that was planted with trees or allowed to revert naturally to forest. In addition to land-use changes, each year about 1.6×10^{10} m^2 of timberland are harvested for timber products and regenerated to forests, 1.6×10^{10} m^2 are damaged by wildfire, and 1×10^{10} m^2 are damaged by insects and diseases. And of course, all forests change continually as individual trees and other vegetation germinate, grow, and die.

The sum effect of all these changes is evident in periodic carbon estimates for the last 40 yr. Between 1952 and 1992, carbon stored on forest land in the conterminous United States increased by an estimated 11.3 Pg (Table 2). This is an average of 0.281 Pg of carbon sequestered each year over the 40-yr period, an amount that has offset about 25% of U.S. emissions of CO_2 to the atmosphere (14). Most of the increase occurred in the eastern and central regions of the United States, offsetting a much smaller decline in the western region. Over the past 100 yrs or more, large areas of the East have reverted from agricultural use to forest. As these reverted forests have grown, biomass and soil organic matter have increased substantially. Recently in the South, increased harvesting and intensive forest management have slowed the rate of increase in carbon storage. Northeastern and north-central forests have continued to accumulate carbon at a rapid rate, because the mixed hardwood forests are less intensively used or managed for wood products. Although the West has not had the major land-use shifts characteristic of the East, forest disturbance nonetheless has dominated the landscape as the original forests have been harvested and converted to second-growth forests. Declining carbon storage in the Pacific Coast region reflects the smaller amount of carbon that younger forests contain.

Past changes in forest carbon storage vary significantly by ownership group (Table 3). Most of the recent historical increase in carbon storage has been on privately owned land in the East, the regrowing forests that were cleared by the early 1900s. As these lands approach full stocking of relatively large trees with low rates of biomass accumulation or are more intensively used for timber products, accumulation of carbon may decline in the future. Carbon storage on forest industry lands has increased slightly in the past, but with intensive manage-

Table 2. Summary of Estimates (in petagrams) of Carbon Storage by Geographic Region, Conterminous U.S. Forest Land, 1952–1992

Region	1952	1962	1970	1977	1987	1992
Northeast	3.92	4.76	5.48	6.27	7.02	7.64
North Central	2.70	3.41	3.84	4.41	5.27	5.73
Southeast	3.10	3.51	3.96	4.50	4.91	4.92
South Central	3.76	4.47	4.99	5.60	6.21	6.54
Rocky Mountains	6.91	6.68	6.68	7.06	7.41	7.39
Pacific Coast	6.08	5.91	5.71	5.63	5.53	5.50
Totals	*26.47*	*28.74*	*30.67*	*33.47*	*36.35*	*37.72*

Table 3. Summary of Estimates (in petagrams) of Carbon Storage by Ownership Group, U.S. Timberland,[a] 1952–1992

Owner	1952	1962	1970	1977	1987	1992
National forest	8.42	8.93	8.99	8.96	8.16	7.90
Other public	3.11	3.46	3.75	4.00	3.98	3.85
Forest industry	3.57	3.93	4.21	4.45	4.50	4.44
Other private	10.15	11.58	12.85	14.20	16.69	18.13
Totals	*25.25*	*27.90*	*29.80*	*31.61*	*33.33*	*34.32*

[a] Forest land that is producing or is capable of producing crops of industrial wood and not withdrawn from timber use by statute or administrative regulation.

ment for a steady supply of timber, carbon is expected to remain relatively constant. Carbon storage in national forests has declined in the recent past, a consequence of a reduction in the area of western old-growth forests, which contain large quantities of carbon, and an increase in the area of younger, faster-growing forests. With recent reductions in the amount of allowable harvest in western national forests, less carbon will be released, producing a net gain in carbon storage.

Projections show additional increases in carbon storage in U.S. forests, but at a slower rate than the past 50 yr. This projected trend reflects (*1*) a slow down in the rate of accumulation in the North as the average forest has reached an age of slower growth than in the past and increases in soil carbon on reverted land are less, (*2*) increasingly intensive use of forests for wood products in the South so that accumulation is balanced by removal, and (*3*) reduced harvest of public forests in the West coupled with a large area of younger, more vigorous and intensively managed forests on former old-growth forest land.

The effects of increasing atmospheric CO_2 and prospective climate change on productivity could have a significant impact on carbon storage in forests, but there is little agreement on the direction of such changes for temperate forests. Boreal forests, such as found in Alaska, are expected to become a net source of atmospheric CO_2 under global warming scenarios because respiration of soil carbon would likely increase more than biomass growth under higher temperatures.

CARBON IN HARVESTED WOOD

There is a substantial amount of carbon removed from forests for wood products. Some ends up in long-term storage in durable wood products and landfills, and some is burned for energy or decomposes and is released to the atmosphere (15). Wood products are goods manufactured or processed from wood, including lumber and plywood for housing and furniture and paper for packaging and newsprint. A large portion of discarded wood products is stored in landfills. This carbon eventually decomposes and is counted as emissions on release to the atmosphere. Emissions also include carbon from wood burned without generation of usable energy and from decomposing wood residues and byproducts not stored in landfills. Wood used for energy is usually considered as a separate category from emissions, because substitution of renewable for nonrenewable sources of energy has an effect on carbon flux.

Harvested wood can represent a substantial carbon sink. An estimated 0.04 Pg of additional carbon may be added to physical storage in wood products and landfills each year. This estimate is sensitive to assumptions about recycling, size of harvested timber, other factors that affect the amount of wood entering the various pools and the average retention periods.

USE OF BIOMASS FOR ENERGY CONSERVATION

The use of biomass energy from wood or wood products and byproducts and the conservation of energy by using trees for shading, shelter belts, and windbreaks are considered important factors in reducing CO_2 emissions (16). Biomass is converted to energy using a variety of technologies that, depending on the efficiency of the process, can offset the combustion of a substantial amount of nonrenewable fossil carbon. Use of biomass can effectively reduce CO_2 emissions when substituted for fossil fuels at a favorable conversion efficiency. Because biomass is renewable, the amount of carbon released through combustion can be replaced with new biomass.

Living trees have a major role in energy conservation. Windbreaks, shelter belts, and properly placed urban trees can reduce the use of energy to heat and cool buildings and to farm affected croplands. Such trees also increase carbon storage in woody plants and soil. Globally, the potential savings are estimated to be substantial.

MITIGATION OPTIONS

As a consequence of expected increases in emissions of greenhouse gases and an acceleration of global change effects on terrestrial and oceanic systems, analysts have proposed various strategies to reduce emissions of CO_2 to the atmosphere or to offset emissions by storing additional carbon in forests or other terrestrial carbon sinks (17). Carbon sinks are a component of the U.S. strategy for limiting national contributions to greenhouse gas concentrations in the atmosphere. Of particular interest in the United States are options for increased tree planting, increased recycling, changes in harvesting and ecosystem management practices, and carbon offset programs.

One mitigation option favored for increasing carbon sinks is an increase in reforestation of marginal cropland and pasture. Investigation of several scenarios shows the possibility of medium- to long-term gains in carbon storage, on the order of 5–10 Tg/yr for a relatively low cost program treating about 8.094×10^{10} m^2 of timberland that would also produce an economic return on investment (9). A more ambitious program that treated more of the biologically suitable land in the United States could achieve substantially higher gains.

Increased recycling of paper and other wood products would keep more of the harvested carbon in use instead of decomposing and reduce the area of forest that needs to be harvested for wood products. Harvesting and forest management practices that are less disturbing to forest sites could increase carbon storage in forest areas used for timber production. Although the savings from increased recycling and different management practices are likely to be substantial, few reliable estimates of the potential carbon storage changes are available yet.

Biomass plantations, energy conservation measures, and energy offset programs have substantial potential for increasing carbon storage or offsetting carbon emissions. Short-rotation woody crops specifically established for energy production could be established on marginal cropland and pasture land unsuitable for economic production of other crops. Energy offset programs have been proposed as a way to increase the storage of carbon in the terrestrial biosphere by establishing tree plantations to offset emissions of CO_2.

All together, the various carbon sink options can be a significant part of the U.S. program to reduce net greenhouse gas emissions. With the participation of other nations in a global emissions-reduction program, it may be possible to have an impact on atmospheric chemistry that would lessen the possible effects of global change.

BIBLIOGRAPHY

1. E. T. Sundquist, *Science* **259,** 934–937 (1993).
2. P. E. Kauppi, K. Mielikäinen, and K. Kuusela, *Science* **256,** 70–74 (1992).
3. R. A. Birdsey, A. J. Plantinga, and L. S. Heath, *Forest Ecol. Manage.* **58,** 33–40 (1993).
4. M. J. Apps and W. A. Kurz, *World Resources Rev.* **3,** 333–344 (1991).
5. T. P. Kolchugina and T. S. Vinson, *Can. J. Forest Res.* **23,** 81–88 (1993).
6. R. N. Sampson and co-workers, *Water Air Soil Pollution* **70,** 3–15 (1993).
7. K. Waddell, D. D. Oswald, and D. S. Powell, *Forest Statistics of the United States*, Resource Bulletin PNW-RB-168, U.S. Department of Agriculture, Forest Service, Pacific Northwest Research Station, Portland, Ore., 1989.
8. N. D. Cost, J. Howard, B. Mead, W. H. McWilliams, W. B. Smith, D. D. Van Hooser, E. H. Wharton, *The Biomass Resource of the United States*, General Tech. Rep. WO-57, U.S. Department of Agriculture, Forest Service, Washington, D.C., 1990.
9. R. A. Birdsey in R. N. Sampson and D. Hair, eds., *Forests and Global Change, Vol. 1, Opportunities for Increasing Forest Cover, American Forests*, Washington, D.C., 1992.
10. K. A. Vogt, C. C. Grier, and D. J. Vogt, *Adv. Ecolog. Res.* **15,** 303–377 (1986).
11. W. M. Post and co-workers, *Nature* **298,** 156–159 (1982).
12. R. A. Houghton and co-workers in J. R. Trabalka, ed., *Atmospheric Carbon Dioxide and the Global Carbon Cycle*, DOE/ER-0239, U.S. Department of Energy, Washington, D.C., 1985.
13. L. L. Ajtay, P. Ketner, and P. Duvigneaud in B. Bolin and co-workers, eds., *The Global Carbon Cycle*, SCOPE Rep. No. 13, John Wiley & Sons, Inc., New York, 1979, pp. 129–181.
14. T. A. Boden, P. Kanciruk, M. P. Farrell, *Trends '90—A Compendium of Data on Global Change*, ORNL/CDIAC-36, Oak Ridge National Laboratory, Oak Ridge, Tenn., 1990.
15. C. Row and R. B. Phelps in *Agriculture in a World of Change, Proceedings of Outlook '91, 67th Annual Outlook Conference*, U.S. Department of Agriculture, Washington, D.C., 1991.
16. R. N. Sampson and co-workers, *Water Air Soil Pollution* **70,** 139–159 (1993).
17. Intergovernmental Panel on Climate Change, *Climate Change—The IPCC Response Strategies*, Island Press, Washington, D.C., 1991.

CATALYTIC CRACKING

K. Rajagopalan
Grace Davison, W. R. Grace & Co.-Conn.
Columbia, Maryland

Carmo J. Pereira
Research Division, W. R. Grace & Co.-Conn.
Columbia, Maryland

HISTORY

Thermal cracking of crude oil was used to produce gasoline by early refiners. At the beginning of the 20th century, the demand for gasoline exceeded the supply produced by the thermal cracking process, thus motivating the development of catalytic cracking. Early cracking processes used pelletized catalysts of a few millimeters in diameter. These catalysts were eventually replaced by a powdered catalyst of about 70-μm diameter, which is used in the Fluidized Catalytic Cracking (FCC) process.

See also Petroleum markets; Petroleum production; Petroleum refining; Transportation fuels, automotive fuels; Reformulated gasoline.

Processes

In the early 1900s, Gray and McAfee carried out catalytic cracking of oil with aluminum chloride. A commercially feasible catalytic cracking process was developed by E. J. Houdry in the 1920s. The first commercial catalytic cracking unit employed a fixed bed process, with a design capacity of 2,000 barrels per day; it came on stream in 1936 at Socony Vacuum Oil Co. Improvement of this process was pursued by both Socony Vacuum and Houdry. These efforts resulted in the development of a moving bed process called Thermofor Catalytic Cracking (TCC) (1). The first commercial TCC unit was begun by Magnolia Oil Co. in 1943. The onset of World War II and the increased demand for high octane gasoline motivated the development of a process by Standard Oil Development Co. that used a fluidized solid catalyst. This process, later termed Fluid Catalytic Cracking (FCC), provided the potential for high capacity and gained rapid commercial acceptance. The first commercial FCC unit was started up by Standard Oil Co. in 1942 at their Baton Rouge, Louisiana refinery. The FCC process soon became the method

Table 1. Key Developments Related to Cracking Technology[a]

Year of Innovation or Commercialization	Event	Innovator
1913	Thermal cracking process.	W. M. Burton
1915	Aluminum chloride catalyst for oil cracking.	A. M. McAfee
1922	Tetraethyl lead additive for gasoline octane enhancement.	Standard Oil Co. of New Jersey
1928	Acid activated clays as catalysts.	E. Houdry
1936	Start-up of first 2000 b/d commercial fixed bed catalytic cracking unit.	Socony-Vacuum Oil Co.
1940	First synthetic silica–alumina catalyst produced commercially.	Houdry & Socony-Vacuum Oil Co.
1942	Start-up of first commercial FCC unit (Model I).	Standard Oil Co. of New Jersey
1943	Start-up of first commercial thermofor catalytic cracking (TCC) unit.	Magnolia Oil Co.
1948	Commercial production of microspheroidal FCC catalyst.	Davison Chemical Co.
1954	Synthesis and commercialization of zeolite X.	Union Carbide Co.
1956	Riser cracking.	Shell Oil Co.
1959	Synthesis and commercialization of zeolite Y.	Union Carbide Co.
1962	Zeolite cracking catalysts.	Mobil Oil Co.
1964	Development of USY and REY FCC catalysts.	Davison Chemical Co.
1971	Short contact time riser FCC.	Kellogg Co.
1974	Platinum-based CO combustion promoter.	Mobil Oil Co.
1974	SO_x reducing catalysts.	Amoco
1975	Ni passivation with antimony compounds.	Phillips Petroleum Co.
1978	V passivation with tin compounds.	Gulf Oil Co.
1983	Synthesis of high silica Y zeolites using $(NH_4)_2SiF_6$.	Union Carbide Co.
1986	ZSM-5 octane additive.	Mobil Oil Co.

[a] Ref. 2.

of choice to conduct catalytic cracking on a commercial scale. Key developments related to catalytic cracking are summarized in Table 1.

Catalysts

The first commercial synthetic fluid cracking catalyst (powdered silica—alumina containing 13 wt % Al_2O_3) was manufactured by the Davison Chemical Co. in 1942, using a grinding process. In the commercial unit, the ground catalyst attrited into fine (< 20 μm) particles, which were lost to the atmosphere. The use of spray drying to produce microspheroidal particles was later discovered to improve attrition resistance dramatically and to reduce catalyst usage. The first microspheroidal FCC catalyst was made commercially available by the Davison Chemical Co. in 1948. The early FCC process involved cracking in a fluidized bed reactor. A transfer line carried the catalyst from the regenerator to the reactor. As catalyst activity improved, the short residence time in the transfer line became sufficient to convert the feed oil. Selectivity could be improved by minimizing cracking in the reactor bed. Accordingly, FCC unit design was modified to conduct most of the cracking in a transfer line called the riser. This design was commercially implemented by Shell Oil in 1956.

A major catalyst development in FCC was the invention of zeolite-containing, high activity cracking catalysts by Mobil in 1962 (3), followed by the commercial development of catalysts containing ultrastable Y (USY) and rare earth-exchanged Y (REY) zeolite by the Davison Chemical Co. in 1964. *In situ* zeolitization, a process that converts a portion of the calcined clay microspheres into zeolite resulting in catalyst particles that contain both zeolite and matrix, was developed and commercialized by Engelhard. Early FCC catalysts contained about 5–10 wt % zeolite dispersed in amorphous SiO_2—Al_2O_3, referred to as the matrix. Increases in zeolite content lowered the attrition resistance of the catalysts. However, developments in matrix technology resulted in large improvements in catalyst attrition resistance and enabled production of attrition-resistant catalysts with high (40–45 wt %) zeolite content (4). The developments in matrix technology are summarized in Table 2. These matrix technology developments also provided catalyst manufacturers with the ability to vary the ratio of zeolite to matrix over a wide range, in order to control catalyst selectivity. Using improved matrix technology, a high activity FCC catalyst that was selective for producing high octane gasoline was introduced in 1980 (5).

Hardware

Over the years, process hardware has changed to accommodate catalyst improvements and to increase desired product yields. The Model I FCC reactor was an upflow pipe with an expanded diameter section designed to increase residence time. Model II was a downflow reactor in which the expanded reactor section was elongated, a dense fluid bed was formed, and solids were withdrawn for the bottom of the bed. The invention of zeolite cracking catalysts eliminated the need for the dense bed to achieve the desired conversion. FCC reactors were modified to provide all upflow riser cracking, usually by extending the riser through the reactor vessel (6). The hydrothermal stability of early amorphous catalysts limited

Table 2. Properties of FCC Matrices[a]

Description	Typical Catalyst Properties			
	Microactivity of Steamed Matrix[b]	DI[c]	ABD[d] (g/cm^3)	Commercialization of Matrix Technology
SiO_2–Al_2O_3 gel (13% Al_2O_3)	35	40	0.50	1963
SiO–Al_2O_3 gel (25% Al_2O_3 and clay; semisynthetic)	35	35	0.52	1965
Silica hydrosol and clay	10	6	0.77	1972
Silica hydrosol with increased matrix activity	25	6	0.76	1978
Alumina hydrosol and clay	25	5	0.80	1980
Clay-based, XP	58	4	0.73	1986
Silica hydrosol with Ni trap	25	6	0.76	1990

[a] Ref. 4.
[b] Steamed six hours at 760°C, 100% steam, 5 psig; microactivity test (MAT) conditions, 527°C, WHSV = 30, C—O ratio = 4, sour imported heavy gas oil feed.
[c] Attrition, Davison Index (DI).
[d] Apparent Bulk Density (ABD).

the operation of early FCC regenerators to temperatures of below 593°C. Temperature control was primarily obtained by controlling the air rate to limit excess oxygen to less than 0.2% in the flue gas. Additionally, spray water systems were used to counter the runaway afterburning of carbon monoxide in the dilute phase. The improved hydrothermal stability of the zeolite catalysts has resulted in increased operating temperatures of up to 732°C with complete carbon monoxide burning. Regenerator efficiency has been improved by operating the fluid bed at higher linear velocities. Examples of other hardware developments (e.g., nozzle design for gasoil—catalyst contacting, cracked product—catalyst separation, spent catalyst stripping, regenerator air distribution grids, and catalyst transfer lines) have been summarized (7).

Additives

Several solid and liquid additives have been used to enhance the flexibility of the FCC process. Invention of Pt-based combustion promoter (added in small quantities with the catalyst) in 1974 enabled control of coke burning in the FCC regenerator. Solid additives to control SO_2 emissions from the regenerator have also been discovered. Early SO_x control additives were developed by Amoco; subsequently, a high activity SO_x control additive based on magnesium aluminate spinel was invented by Arco, and developed and commercialized by Katalistiks. Heavier oil feeds contain contaminant metals like nickel and vanadium. Phillips Petroleum Co. discovered the passivation of Ni with antimony compounds in 1975. A liquid additive containing antimony is added to commercial FCC units processing high Ni feeds. The use of tin compounds to passivate vanadium was discovered by Gulf in 1978. These passivation additives enabled FCC units to process metals-containing feeds. The use of ZSM-5 in small quantities with the catalyst to increase the octane number of product gasoline was discovered and commercially demonstrated by Mobil in 1986 (8). The use of matrix technology to provide a fourfold increase in ZSM-5 activity was reported by the Davison Chemical Co. in 1990 (9). A more detailed discussion of additives used in FCC is provided in Reference 10.

PROCESS DESCRIPTION

The FCC process consists of a riser, in which all the feed cracking reactions occur, and a regenerator, which reactivates the catalyst by burning off the coke deposited in the riser. A more detailed discussion of the subject is available in Reference 11.

A simplified illustration of the process is shown in Figure 1. Preheated gasoil (which may also contain recycle streams) is injected into the bottom of the riser and contacts the hot active catalyst. The gasoil is vaporized and flows upward through the riser, together with the catalyst. The gasoil cracks, and, in the process, the catalyst becomes deactivated by coke and tramp metals contained in the feed. Coked catalyst is referred to as spent catalyst. Since cracking reactions are endothermic, reaction temperature decreases up the riser. The residence time of catalyst in the riser is in the range of 2–10 s.

The cracked product is separated from the spent catalyst by cyclones at the top of the riser and sent to a fractionation column. The spent catalyst contains coke, which consists of heavy hydrocarbons adsorbed on the catalyst surface, and product gases entrapped in the catalyst pores. Spent catalyst is stripped of entrapped and other volatile hydrocarbons using counter currently flowing steam. The catalyst then passes through a slide valve into the regenerator.

Spent catalyst flow into the regenerator may be tangential as well as downward. Compressed combustion air enters the bottom of the regenerator through a distribution grid, generating a hot, dense fluidized bed in which the coke is burnt off. Combustion products and entrained catalyst are conveyed upward, out of the dense fluidized bed, into a dilute-phase zone where cyclones separate combustion gases from catalyst, which is then returned to the bed. A typical FCC regenerator operates in turbulent fluidization at a gas velocity of approximately 1 m/s.

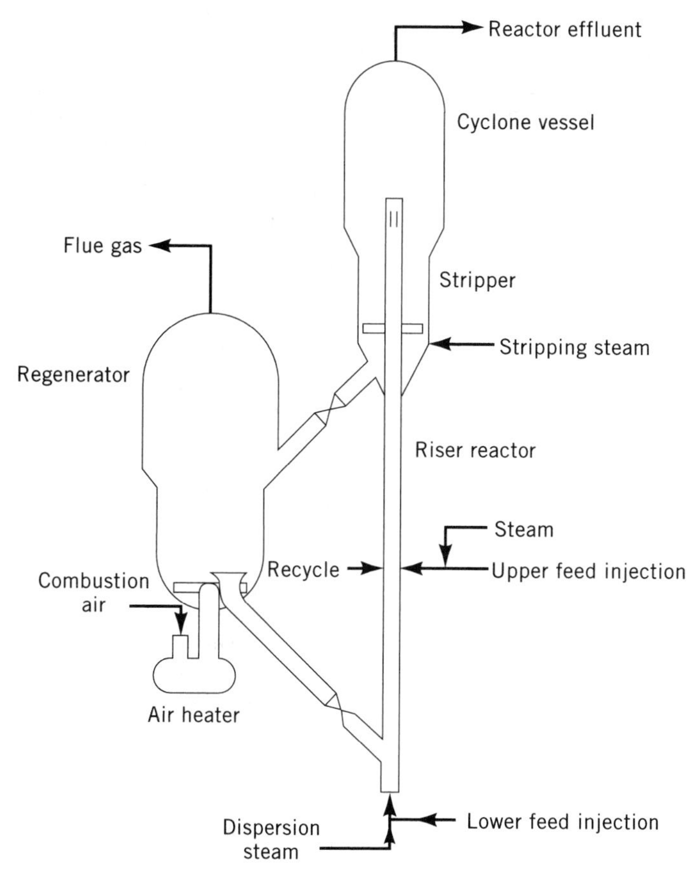

Figure 1. Typical FCC unit incorporating GULF FCC process design (12).

Thus, there is considerable carryover of catalyst to the cyclones; in fact, the whole bed circulates through the cyclones every 5 min (6). Regenerator exhaust gases contain carbon dioxide, carbon monoxide and water (from coke combustion), sulfur oxides (from the oxidation of adsorbed sulfur-containing hydrocarbons), nitrogen oxides (from adsorbed nitrogen-containing hydrocarbons), and some catalyst fines.

The hot, nearly coke-free regenerated catalyst passes through a slide valve (which controls temperature at the riser top) and contacts the gasoil feed, thereby completing the cycle. Since regenerator oxidation is exothermic and riser reactions are endothermic, the FCC unit operates in heat balance. However, as will be discussed later, process flexibility can be improved by decoupling the riser from regenerator by installing a catalyst cooler.

Operating Variables

For a given catalyst type, the key independent operating variables are the riser top temperature, the feed rate and feed preheat temperature, and the reactor pressure. Dependent variables include the catalyst circulation rate, the regenerator temperature, and conversion (defined as the percent weight or volume fraction of feed converted to gasoline and lighter products and coke). The ratio of catalyst circulation rate to the feed rate is referred to as the catalyst—oil ratio.

The complex relationship between the riser and the regenerator can be illustrated by considering the effect of increasing the feed preheat temperature. As feed preheat temperature rises, the catalyst circulation rate drops in order to maintain a constant riser top temperature. Lower catalyst—oil ratio will result in lower conversion and in lower coke make. Although there is less coke to burn, the regenerator temperature increases because less heat is transferred from the regenerator to the riser at the lower catalyst—oil ratio.

Operating variables (see Table 3) influence the heat balance of the FCC unit. Changes in these variables affect not only the coke make (or heat output) but also the coke that must be produced to maintain heat balance (or heat load). Coke yield can only be changed at altering the variables that affect heat load. Any change that disrupts the

Table 3. Classification of FCC Unit Variables[a,b]

Independent Variables that Affect Heat Load	Independent Variables that Affect Coke Make But Not Heat Load
Feed preheat	Feed Properties except enthalpy
Reactor temperature	Catalyst activity
Fresh feed rate	Catalyst selectivity
Recycle rates	Pressure
Light catalytic gas oil	
Heavy catalytic gas oil	
Slurry oil	
Steam rates	
Dispersion steam	
Regenerator steam or water	
Air temperature	
	Independent Variables that Affect Coke Make or Heat Output and Heat Load
	Reactor temperature
	Recycle rates
	Light catalytic gas oil
	Heavy catalytic gas oil
	Slurry oil
	Steam rates
	Dispersion (pressure effect)
	Stripping steam
Dependent Variables that Affect Heat Load	**Dependent Variables that Affect Coke Make or Heat Output**
Recycle temperature	Catalyst—oil ratio
Heavy catalytic gas oil	
Slurry oil	
Conversion (heat of reaction)	
	Dependent Variable that Affects Coke Make and Heat Load
	Regenerator temperature

[a] Ref. 13.
[b] According to effect on heat balance.

overall heat balance will cause the dependent variables (such as catalyst—oil ratio and regenerator temperature) to readjust to return the unit to heat balance. An increase in severity of an independent variable that increases coke make or heat output, but which does not affect heat load (eg, increasing catalyst activity or Conradson Carbon Ratio (CCR)), will cause a simultaneous change in dependent variables (eg, lower catalyst—oil ratio and higher regenerator temperature) and return the unit to heat balance (with a slightly higher coke yield). Changes in independent variables that affect both heat load and heat output (eg, riser top temperature and recycle rate) will also cause the dependent variables to readjust at a slightly different coke yield. Changes in variables that increase heat load, but which of themselves do not cause a change in coke selectivity and heat output, will cause a readjustment in dependent variables. For example, an increase in steam rate or catalyst cooling will usually increase the coke yield, decrease the regenerator temperature, and increase the conversion (13).

Factors Affecting Performance

Key factors that affect the performance of an FCC unit include feed properties, catalyst properties, and process hardware. Feed properties include the type, boiling range, and composition of the feed. General feedstock quality effects can be determined by a characterization factor, referred to as the UOP K factor (1):

$$K = \frac{(\text{MABP})^{1/3}}{\text{Specific gravity @ 15°C/15°C}}$$

where MABP = Mean Average Boiling Point in °R. K values for many commercial FCC feeds fall in the range of 11.2 to 12.2. Higher K values indicate a paraffinic feedstock.

Some components (such as CCR and nitrogen) cause a reversible loss in catalyst activity, whereas others (such as organometallic feed components and rust) cause an irreversible loss in catalyst activity and/or selectivity. Effects are summarized in Table 4. Catalyst properties include the type and concentration of zeolite, and the composition of the matrix. Catalyst properties affect the conversion and product yields from the unit, deactivation behavior on exposure to metals, and loss of catalyst as fines from the regenerator. Modifications in unit hardware have led to feed processing flexibility and to improvements in FCC performance. Some of these modifications will be discussed in the section on Developments.

Table 4. Effect of Feed Properties on FCC Performance

Element	Source	Effects
Ni	Organometallic feed components	Increased yields of gas and coke; loss of gasoline selectivity.
V	Organometallic feed components	Destruction of zeolite in the catalyst; loss in conversion and gasoline yield.
Cu, Fe	Contamination	Similar to Ni but less significant effects.
N	Feed component	Loss in catalyst activity and unit conversion; temporary poison; catalyst activity is restored when N components in the feed are removed; increased NO_x emissions from regenerator.
Na	Contamination	Destruction of zeolite; loss in conversion
Ca	Contamination	Similar to Ni but less significant effects.
S	Feed component	Increased SO_x emissions from the regenerator.
CCR	Feed component	Reversible loss of activity in the riser due to coke laydown.

CHEMICAL PATHWAYS AND CATALYSTS

Chemical Reactions

FCC feeds contain long chain (greater than 20 carbon atoms) paraffinic, naphthenic, and aromatic molecules. The reactions that occur during the catalytic cracking process are generally classified as primary and secondary reactions. The extent to which these reactions take place determines the yield and quality (chemical composition) of the different products from a catalytic cracking unit. Primary reactions involve molecules present in the feed, examples of which follow (1,14).

1. Cracking of a paraffin to produce a paraffin and an olefin:

$$C_nH_{2n+2} \rightarrow C_mH_{2m} + C_pH_{2p+2}, \text{ where } n = m + p$$

2. Dealkylation of an alkyl aromatic compound to produce an olefin:

$$\underset{}{ArC_n\,H_{2n+1}} \rightarrow \underset{\textbf{(aromatic hydrocarbon)}}{ArH + C_nH_{2n}}$$

3. Cracking of an aromatic side chain to produce a paraffin:

$$ArC_n\,H_{2n+1} \rightarrow \underset{\textbf{(aromatic with olefinic side chain)}}{ArC_m\,H_{2m-1}} + \underset{\textbf{(paraffin)}}{C_pH_{2p+2}}$$

4. Cracking of a naphthene to produce olefins:

$$C_nH_{2n} \rightarrow \underset{\textbf{(olefin)}}{C_mH_{2m}} + \underset{\textbf{(olefin)}}{C_pH_{2p}} \quad \text{where } n = m + p$$

5. Cracking of an alkylnaphthene to produce an olefin:

$$N_pC_nH_{2n+1} \rightarrow \underset{\textbf{(naphthene)}}{N_pH} + \underset{\textbf{(olefin)}}{C_nH_{2n}}$$

Secondary reactions involve products from the primary reaction. Examples of secondary reactions are

1. Cracking of olefins:

$$C_nH_{2n} \rightarrow C_mH_{2m} + C_pH_{2p}, \text{ where } n = m + p$$

2. Isomerization:

$$n\text{-olefin} \rightarrow i\text{-olefin}$$

3. Hydrogen transfer (15):

$$\text{naphthene} + \text{olefin} \rightarrow \text{aromatic} + \text{paraffin}$$

4. Alkyl group transfer:

$$C_6H_4(CH_3)_2 + C_6H_6 \rightarrow 2\ C_6H_5CH_3$$

In addition, other secondary reactions, such as polymerization of olefins, dehydrocyclization of olefins to aromatics, alkylation of aromatics, and condensation of alkyl aromatics into multiring aromatics, can occur during catalytic cracking. The polymerization, condensation, and cyclization reactions involving olefins and aromatics can produce high molecular weight species that are highly hydrogen-deficient. These condensed species do not desorb from the catalytic surface and are referred to as coke.

The pathway for most of the primary and secondary reactions described above involve a tricoordinated carbocation (also referred to as a carbenium ion). An acid site on the catalyst surface can cause the formation of a carbenium ion. Both Brönsted (proton donor denoted as HX) and Lewis (electron acceptor denoted as L) acid sites are present on a catalyst surface. Carbenium ions can be formed by the abstraction of a hydride ion (H^-) from a paraffin (RH). The following reaction illustrates the formation of a tertiary carbenium ion:

$$RH + L \rightleftharpoons LH^- + R^+$$

Carbenium ions can also be formed by donation of proton from an acid site to an olefin:

$$H_2C = CHCH_3 + HX \rightleftharpoons H_3C - \overset{\displaystyle H}{\underset{+}{\overset{|}{C}}} - CH_3 + X^-$$

In this reaction, a secondary rather than a primary carbenium ion is formed since it is more stable. The stability of carbenium ions decreases in the order:

$$\text{tertiary} > \text{secondary} > \text{primary} > \text{methyl}$$

Cracking typically occurs at the bond located β to the carbenium ion. Cracking of a straight-chain secondary carbenium ion results in the formation of a primary carbenium ion. This cracking reaction is referred to as β-scission:

$$RCH_2\ \underset{+}{C}HCH_2 - CH_2CH_2R' \rightleftharpoons R\ CH_2CH = CH_2 + \underset{+}{C}H_2CH_2R'$$

The primary ion can undergo a rapid hydrogen shift to give a more stable secondary carbenium ion:

$$\underset{+}{C}H_2CH_2R \rightleftharpoons CH_3\ \underset{+}{C}HR$$

A continued pattern of β-scission leads to formation of propylene and butylenes in high yields. In a commercial catalytic cracker, secondary reactions result in formation of stable products of higher molecular weight in the gasoline range. Carbenium ion intermediates act in all the primary and secondary cracking reactions described earlier. In addition, formation of a pentacoordinated carbocation, referred to as a carbonium ion, by protonation of a paraffin by a Brönsted acid site has also been postulated (14):

$$R - \underset{H}{\overset{H}{C}} - CH_3 + HX \rightleftharpoons \left\{ R - \overset{H}{\underset{H\ \ H}{C}} - CH_3 \right\}^+ \rightleftharpoons R - \overset{H}{\underset{+}{C}} - CH_3 + H_2$$

The above reaction leads to the formation of molecular hydrogen, one of the products of catalytic cracking.

Catalysts

Commercial fluid cracking catalysts contain faujasite or zeolite Y dispersed in an inorganic oxide matrix (11), as shown in Figure 2. Zeolites are crystalline aluminosilicates containing tetrahedrally coordinated silicon and

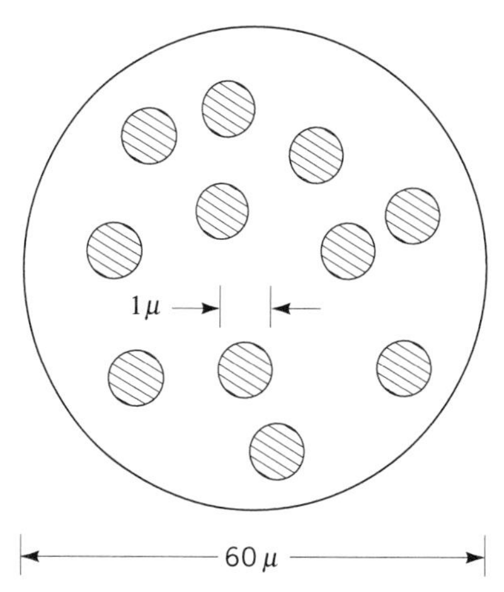

Figure 2. FCC catalyst; zeolite particulates are dispersed in a porous amorphous matrix.

aluminium with high surface area (> 400 m^2/g) and fine (< 15 Å) pores. A faujasite zeolite possesses a cubic unit cell containing 192 tetrahedral (Si or Al) atoms with an approximate unit cell dimension of 25 Å. The unit cell dimension, a_o, which can be measured by x-ray diffraction using ASTM method D3942, depends on the Si—Al ratio, R, in the zeolitic framework. This relationship has been determined as (16):

$$a_o = \frac{1.708}{1+R} + 24.233$$

The structure of faujasite is presented as Figure 3. A typical unit cell formula for NaY faujasite is Na_{55} $[(AlO_2)_{55}(SiO_2)_{137}]260H_2O$. The alumino—silicate framework of faujasite is negatively charged and can be balanced by a cation (eg, Na^+). Ion-exchange of Na^+ in a zeolite with NH_4^+, followed by calcination at 352°C, can result in formation of a Brönsted acid site associated with a framework Al atom:

```
  Na+                          H+
O   O   O                  O   O   O
 \ / \ /     NH4+,Δ         \ / \ /
 Si   Al-    ------>        Si   Al-
 / \ / \                    / \ / \
O   O O  O                 O   O O  O
                            Brönsted site
```

Dehydroxylation of the zeolite at temperatures exceeding 452°C can lead to the formation of a Lewis acid site (14):

```
O           O
 \         /
  Al    +Si
 / \    / \
O   O  O   O
  Lewis acid
```

Thus, each Al atom in a faujasite has potential to form an acid site, which catalyzes the cracking reaction. Synthesized faujasite is a microporous material, with most of the pore volume in the < 20 Å range. Postsynthesis modification of the zeolite can create a secondary mesopore (~ 50 Å) structure. Catalytic performance (activity and selectivity) of faujasite zeolites can be influenced by controlling its hydrothermal stability properties, acid site density during commercial use, pore size distribution, and crystallite size.

Na can catalyze the sintering of faujasite in the high (> 602°C) temperature environment of an FCC regenerator. Thermal and hydrothermal stabilities of faujasites are improved by lowering the Na content and increasing its Si—Al ratio. The Na content of faujasite is reduced from about 55 atoms per unit cell as synthesized, to the range of 3–8 atoms per unit cell in most commercial FCC catalysts. This can be accomplished by ion exchanging Na^+ with either NH_4^+ or with mixed, rare earth cations. La, Ce, and Nd are rare earth elements commonly found in commercial catalysts.

Economically viable processes for faujasite synthesis produce zeolites with Si—Al ratio ranging from 1.5 to 3.

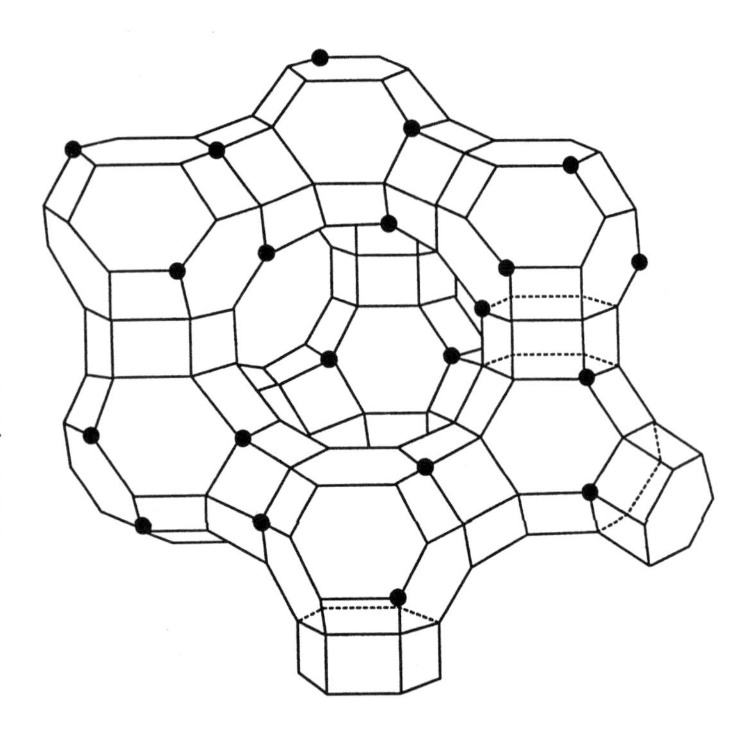

Figure 3. A faujasite zeolite.

An increase in this Si—Al ratio from 2.5 to 3.8 can be achieved either by chemical dealumination or by a combination of ion-exchange and high temperature treatment (527–727°C) in the presence of steam. The high Si—Al ratio, low Na zeolites are referred to as "ultrastable" zeolites (17).

During use of a catalyst in a commercial FCC unit, further dealumination of the zeolite component (or an increase in the framework Si—Al ratio) and a reduction in unit cell size a_o occurs due to the high temperature (727°C) steam environment in an FCC regenerator. The unit cell size of an FCC catalyst during use (a measure of the extent of dealumination) determines the acid site density of the catalysts. Hence, it is an important parameter in predicting activity and selectivity characteristics (18–20). As an example, the influence of unit cell size on gasoline composition is described in Table 5. Higher unit cell size or higher acid site density facilitates bimolecular hydrogen transfer reactions, resulting in a more paraffinic

Table 5. Gasoline Molecular Type Shifts Due to Unit Cell Size[a]

	10% REY	20% USY
Unit cell size, Å	24.5	24.3
Gasoline composition, wt %		
Paraffins	37.5	29.3
Olefins	21.9	34.2
Naphthenes	12.6	12.1
Aromatics	22.4	20.2
Research Octane Number	88.6	92.6
Motor Octane Number	77.4	80.0

[a] Ref. 21.

and less olefinic gasoline. The unit cell size during use is primarily controlled by the degree of mixed rare earth ion exchange during preparation (19,22). The bulky cations in mixed rare earth catalysts retard dealumination during use. Unit cell size of commercial equilibrium catalysts can range from 24.24 to 24.40 Å.

Though the zeolite provides most of the cracking activity of the FCC catalyst, the matrix fulfills both physical and catalytic functions. A major function of the matrix is to bind the zeolite crystallites together in a microspheroidal particle, with appropriate physical requirements to survive interparticle and reactor wall collisions in the FCC unit. Important physical characteristics of FCC catalysts are particle size distribution and average particle size, attrition resistance, and apparent bulk density. Average particle size of most commercial catalysts range from 60 to 80 μm, whereas apparent bulk densities range from 0.65 to 0.9 g/cm^3. Attrition resistance can be measured by a variety of methods developed by different catalyst manufacturers, for example, the Davison Index (DI) (23). Harder catalysts used commercially typically exhibit DI values less than 7. Attrition resistance of the catalysts generally decreases with increasing zeolite content (see Fig. 4).

The matrix facilitates heat transfer during cracking and regeneration, thus protecting the zeolite from structural damage, and also stabilizes the zeolite from structural damage by sodium (24). Acid sites in the matrix can also significantly contribute to the catalyst activity. Catalysts with an active, as well as with a nearly inert, matrix are commercially available. Properties of two of these catalysts are summarized in Table 6. An active matrix cracks the feed components with high boiling point (> 477°C) more effectively than zeolite (25). In general, zeolite exhibits better product selectivity than an active matrix. The benefits and drawbacks of the use of an active matrix are summarized in Table 7. The choice of an active or inert matrix for an FCC operation depends on the type of feedstock processed and on the objectives and constraints of the FCC unit.

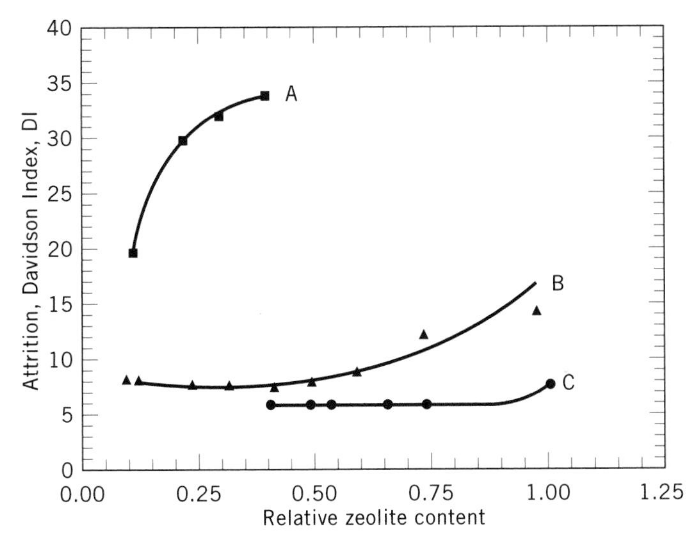

Figure 4. Effect of zeolite content on attrition resistance: (**a**) silica—alumina gel; (**b**) silica hydrosol, 1982 process; (**c**) silica hydrosol, new process (4).

Table 6. Properties of Commercial FCC Catalysts

Description	Low Matrix Activity Catalyst	High Matrix Activity Catalyst
Na_2O, wt %	0.42	0.30
Al_2O_3, wt %	28	43
RE_2O_3, wt %	0	0
Fe_2O_3, wt %	0.4	0.6
TiO_2, wt %	1.0	1.6
SiO_2, wt %	69	54
BET surface area, m^2/g	260	271
Zeolite surface area, m^2/g	224	135
Matrix surface area, m^2/g	36	136
Apparent bulk density, g/cm^3	0.68	0.70
Average particle size, μm	70	70
Attrition (Davison Index)	5	2

Matrices are multicomponent systems. One of the components is the binder that provides attrition resistance. Binders are inorganic polymers (eg, alumina or alumino—silicates) that can exhibit high or low cracking activity depending on their composition. With nearly inert binders, additional amorphous solid acid components can be added to increase the activity of the matrix. Matrices also contain nearly inert fillers such as kaolin clay.

In addition to the zeolite cracking catalyst, solid additives are used to facilitate operation of FCC units and to alter selectivity to desired products. Four types of additives are commonly used: (1) combustion promoter to facilitate complete combustion of coke to CO_2, (2) SO_x transfer additive to reduce SO_2 and SO_3 emissions from the regenerator, (3) octane and light olefin-enhancing additive, and (4) passivation additive. Combustion promoters contain low concentration (100–1,000 ppm) of Pt group metals supported on alumina or other supports. The noble metal catalyzes oxidation of CO to CO_2. SO_x transfer additives capture SO_x in the regenerator and release it as H_2S on the reactor side, with the use of a solid base with an optimum strength. (SO_x transfer will be discussed further in the Environmental section.) Octane and olefin-enhancing additives are based on the zeolite ZSM-5, which has a smaller pore opening than faujasite. The mechanism by which ZSM-5 increases C_3, C_4 olefins, and octane number of the gasoline has been described (8).

Table 7. Reported Advantages of Low and High Activity Matrices[a]

Low Activity Matrix	High Activity Matrix
High gasoline selectivity	Better bottoms cracking
Low coke and dry gas yields at constant conversion	Improved activity in the presence of V
Low coke and H_2 yields in the presence of Ni	Synergistic interaction with zeolite resulting in improved stability and activity
	Improved light cycle oil (LCO) quality
	Improved MAT activity with high N_2 feeds

[a] Ref. 4.

Table 8. Ingredients Used in FCC Catalysts

Function	Required Properties	Selected Ingredients
Gasoline selective cracking	Acidity (Brönsted, Lewis), hydrothermal stability	USY; Rare earth-exchanged Y
Large pore cracking	Acidity, hydrothermal stability	Pseudoboehmite Al_2O_3; NaOH leached, calcined kaolin; HCl leached, calcined kaolin
Binding	Inorganic polymer with optimized chain length distribution	$Al_2(OH)_5Cl$-based polymer; SiO_2, SiO_2—Al_2O_3-based polymers
Filler	Ability to provide macroporosity, no adverse effects on zeolite, active matrix	kaolin clay
SO_x control	Solid base, surface area stability, large pore diameter	Alkaline earth oxides, $Mg_2Al_2O_5$ spinel
V tolerance	Solid base, surface area stability, large pore diameter, resistance to S poisoning	MgO, La_2O_3, CaO
Combustion promotion	Ability to catalyze CO oxidation	Pt—Al_2O_3

Processing of heavy crudes results in deposition of contaminant metals like Ni and V, which poison the activity and selectivity of the catalysts. V destroys the zeolite due to formation of volatile H_3VO_4 species (26). Alkaline earth oxides (eg, MgO and CaO) have been proposed as ingredients to combat V poisoning. These oxides can also capture sulfur oxides in a commercial FCC regenerator, which may limit their ability to interact with V (26). Ni deposited on cracking catalyst catalyzes dehydrogenation reactions, which lead to the formation of undesirable products such as coke and H_2. Addition of Sb compounds to the feed has been proposed as a method of mitigating the undesirable effects of Ni. A listing of the ingredients used in FCC catalysts and catalyst additives is provided in Table 8.

USES

As discussed previously, the FCC process is used to convert gasoils in the 343–510°C boiling range into valuable gasoline and distillate fuel products. The process, which is at the heart of a modern refinery, can operate at high capacity to produce either gasoline or distillate fuel, depending on market needs.

A primary product of the refinery is gasoline; contribution of the FCC process to the total gasoline pool is shown in Table 9. FCC naphtha accounts for approximately 35% of the U.S. gasoline pool, and other major processes contributing to the pool are reforming (32%) and alkylation (11%). Light gases from the FCC are used as feedstock for alkylation, further increasing the gasoline pool. Environmental regulations restricting the concentration of benzene and aromatics in reformulated gasoline are expected to lower both reformer severity and the fraction of reformate in the gasoline pool, and to expand the role of the FCC process.

A simplified illustration of an FCC unit in a typical refinery is shown in Figure 5. The FCC process is affected by upstream units, and products from the FCC are utilized in downstream processes to produce fuels and chemicals. The role of the FCC process in a refinery is discussed in the work by Venuto and Habib (11).

Table 9. Contributions to the Gasoline Pool[a]

	Average U.S. Pool (Liquid Volume %)			Typical Properties		
	1979	1987	1989	$R + M/2$	Aromatics	Olefins
Blending butanes	6.0	7.0	7.0	91–93		
FCC naphtha	35.0	35.5	35.5	84–88	20–35	
Reformate	32.0	34.0	34.0	86–96	50–80	
Hydrocrackate	3.0	2.5	2.0	85–87	<10	
Alkylate	10.0	11.0	11.2	90–94		
Coker gasoline	2.0	1.0	0.6	60–70	<15	40–60
Light straight run	12.0	4.5	3.3	55–75	<10	
Isomerate		3.5	5.0	80–88		
Ethers		1.0	1.4	106–110		
Totals	**100.0**	**100.0**	**100.0**			
Average ($R + M/2$)	82.9	87.8	88.4			
Estimated aromatics[b]	27	31.5	32.1			
Estimated olefins	8	12	12			
Estimated 90% Pt			360			

[a] Refs. 27 and 28.
[b] Davison estimates.

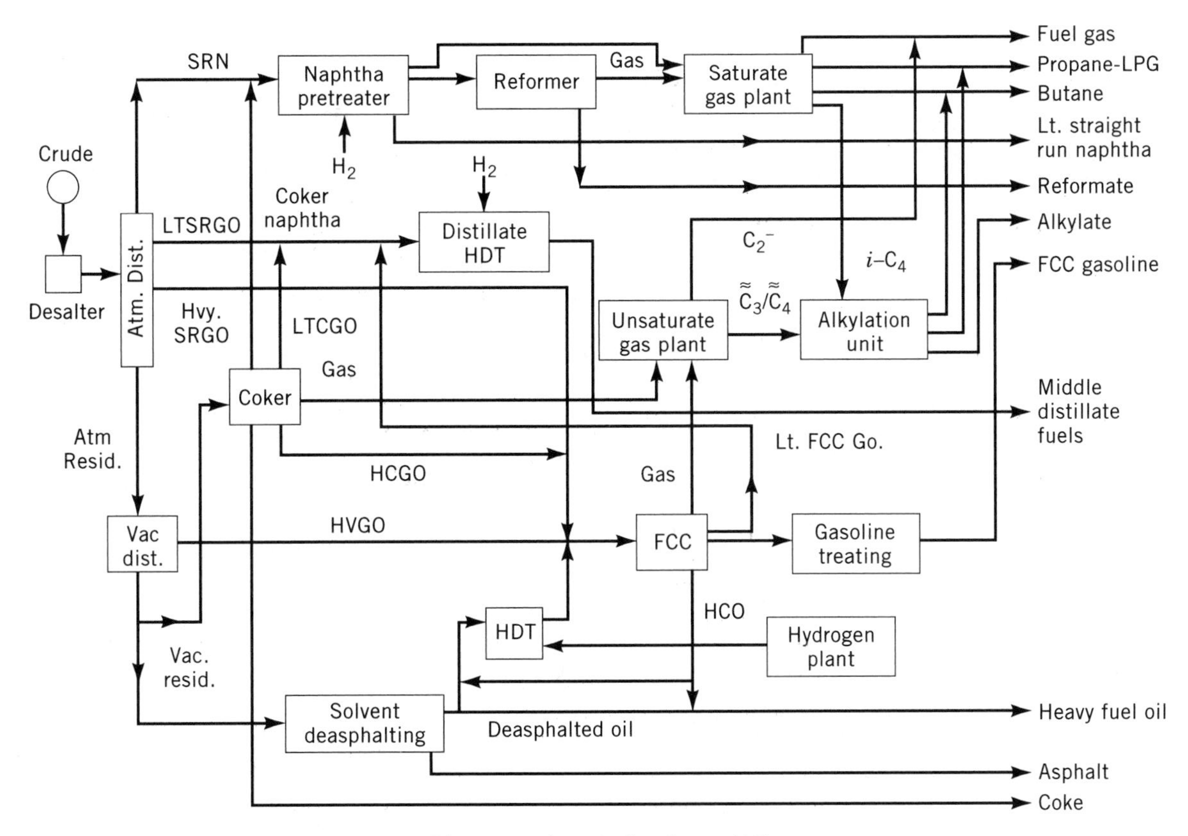

Figure 5. A typical refinery (11).

Upstream Units

Upstream units generate the gasoils in the 343–510°C boiling range that are used as FCC feeds. In these units, crude oil is separated into various boiling fractions, which are processed to achieve the required properties for downstream refining.

Atmospheric distillation separates the incoming crude into four fractions: straight-run naphtha (in the gasoline boiling range), light straight-run gas oil (in the distillate boiling range), heavy straight-run gasoil (in the FCC feed boiling range), and atmospheric resid. Atmospheric resid can be processed further by vacuum distillation to yield a lighter, heavy vacuum gasoil (which can be sent to the FCC unit) and vacuum resid. Vacuum resid is characterized by high boiling materials (typically greater than 621°C) and a high carbon—hydrogen content. The carbon content may be decreased in a coker (by the rejection of carbon as coke) or in a hydrotreater (by the addition of hydrogen). In addition to modifying the carbon—hydrogen ratio, some thermal and/or catalytic cracking also occurs in these processes, resulting in upgraded products having a lower boiling range than the vacuum resid feed. In some cases, vacuum resid is treated in a solvent deasphalter to obtain heavy fuel oil and by-product asphalt. Heavy fuel oil can be further hydroprocessed and routed to the FCC unit.

FCC Complex

The FCC complex can be divided into three major sections: riser—regenerator, regenerator gas handling and riser product fractionation. The riser—regenerator, as discussed in the Process Description section, cracks the gasoil into desirable fuel products.

Exhaust gases from the regenerator may be routed to a CO boiler. Here, CO may be completely converted to CO_2 for heat recovery and to meet emission standards. The heat recovery unit uses the heat of CO combustion and the sensible heat of the flue gas to produce high pressure steam, which may be used for power generation. Before venting through the stack, the flue gas may be passed through an electrostatic precipitator to control particulate emissions.

Product gases from the riser are fed to the base of the main fractionation column. Heat is removed at various elevations of the fractionation column to yield product streams having the desired boiling point ranges. Products in the boiling range of gasoline and lower are removed at the top of the main column and sent to gas concentration units, which consist of the main column overhead accumulator, wet gas compressors, and the debutanizer complex. Gas concentration units further separate the fraction into a light gas stream (mainly C_2^-, used as refinery fuel or sent to the gas plant), a C_3—C_4 cut rich in propylenes and butenes (fed to the alkylation plant), and a C_5^+ (or debutanized) gasoline cut (blended into the gasoline pool). Three side-cut streams are produced from the mail column: a heavy naphtha, a light cycle fuel oil, and a heavy cycle fuel oil. The heavy cycle fuel oil can be recycled to the riser and further cracked into gasoline and distillate fuel. Heat is removed from the column to achieve the precise gasoline endpoint and the boiling range of light and heavy fuel oils, as required by product specifications.

A portion of the main column bottoms passes to the slurry settler and is separated by gravity into a recyclable bottoms fraction (containing catalyst fines) and an overhead stream, known as clarified slurry oil. This latter stream, which contains small amounts of catalyst, is suitable for use as fuel or for conversion into carbon black.

The gasoline and lighter products from the top of the main column are routed to a gas plant that further separates the stream into heavy gasoline, light gasoline, C_4—$C_4^=$, C_3—$C_3^=$, and a light gas fraction containing C_2—$C_2^=$, C_1, and H_2. The latter stream, known as refinery gas, is used as a fuel.

Table 10. Description of FCC Products

Name	Description
Dry gas	H_2, CH_4, C_2H_6, C_2H_4, H_2S
Wet gas	C3, C4 compounds.
Gasoline	Liquid product containing C5 to C12 hydrocarbons; typical end boiling point = 215°C.
Light cycle oil	Liquid product containing C_{12}–C_{20} hydrocarbons; typical boiling point range = 215–338°C.
Slurry oil	Liquid product containing C_{20^+} hydrocarbons.
Coke	Solid carbonaceous deposit on the catalyst; typical C—H ratio = 1.

Downstream Units

The gasoline from an FCC unit is treated and blended into the gasoline pool. Distillate and heavy fractions may be hydrotreated and sold as fuel oils. Light gases from the FCC are separated into saturates and unsaturates. $C_3^=$—$C_4^=$ and isobutane are often routed to an alkylation unit to produce alkylate, which is blended into the gasoline pool. Regulations on reformulated gasoline are requiring changes in downstream refinery processing, as will be discussed in the Products section.

PRODUCTS

The FCC process results in formation of solid, liquid, and gaseous products. The solid product is a carbonaceous deposit that forms on the catalyst. This deposit, which can be considered to be a polymerized hydrogen-deficient hydrocarbon, is referred to as "coke." As discussed in the Process section, the FCC process itself consumes the coke. The exothermic conversion of coke to CO_2 and H_2O provides the heat necessary to drive the endothermic cracking reactions.

The liquid and gaseous products are separated from the catalyst in the reactor and stripper. These products are separated into different fractions in the main fractionation column. The gaseous products are referred to as "dry" or "wet" gas, depending on the composition. The major liquid products are gasoline, light cycle oil, and slurry oil. A description of these products are provided at Table 10. Note that the boiling point range for a product varies among refineries and depends on the properties desired for the product.

Liquid Products

Gasoline from fluid catalytic cracking is a liquid product containing C_5–C_{12} molecules. Gasoline is a mixture of thousands of compounds, which can be classified into four types: (1) paraffins or saturated hydrocarbons; (2) olefins or unsaturated hydrocarbons; (3) naphthenes, which include cyclic paraffins and cyclic olefins; and (4) aromatics, which are primarily alkyl benzenes and benzene. A 1990 survey by the National Petroleum Refiners Association (NPRA) determined that the average gasoline from the commercial FCC units in the United States contained 30 vol % olefins and 29 vol % aromatics. Table 11 lists the key properties for gasoline and other FCC products.

The relative concentration of different types of hydrocarbons in the gasoline determines its road octane. The road octane, which measures the quality of gasoline as an automotive fuel, is an arithmetic average of Research (RON) and Motor (MON) Octane Numbers. Road octanes of commercially sold gasolines range from 87 to 93. RON and MON values can be determined by methods published by the American Society of Testing Materials (ASTM). The octane number of a hydrocarbon decreases with carbon number and increases with the degree of branching (29). In general, olefins and aromatics exhibit higher octane number than paraffins of the same carbon number.

The degree of volatility of the gasoline can be measured by its Reid Vapor Pressure (RVP), usually expressed in Pascal or pounds per square inch (psi). The U.S. industry average is reported to be 4.83×10^4 Pa (7 psi) for FCC gasolines. Higher RVP values result in significant evaporative losses of gasoline to the atmosphere during fueling of automobiles. As a result, the refining industry and environmental regulations limit the RVP of gasoline.

Typical concentration of benzene in FCC gasoline is about 1%. Due to its carcinogenic properties, benzene concentration may be regulated. Sulfur (S) compounds are present in the feed to the FCC unit. As a result, gasoline can contain S compounds like mercaptans and alkyl thiophenes (30), which can adversely affect automotive emissions and hence is an important property. The 1990 industry average reported by the NPRA survey is 626 ppm S in FCC gasoline.

Table 11. Important FCC Product Properties

Product	Properties
Dry gas	H_2 content
Wet gas	Olefinicity
Gasoline	Boiling point distribution (T_{50}, T_{90}); RON, MON; S content; aromatics, benzene content; bromine number—olefin content
Light cycle oil	Cetane Index; aromatics content; smoke point; pour point; S content
Slurry oil	API gravity; viscosity; Bureau of Mines Correlation Index; ash content

Certain refiners split the gasoline from FCC into two fractions: light cut naphtha (LCN) and heavy cut naphtha (HCN). A typical boiling range for HCN is 177–221°C, whereas LCN represents the lighter fraction of the gasoline. Aviation gasolines have narrower boiling ranges (38–171°C) than conventional automobile gasoline. The narrower boiling range ensures better distribution of vaporized fuel through the induction systems of aircraft engines (31). Aircrafts operate at altitudes where the prevailing pressure is significantly less than 1 atm. The vapor pressure of aviation gasoline must therefore be limited to reduce boiling in tanks and fuel lines.

Light cycle oil from the FCC typically contains C_{12}–C_{20} molecules. High molecular weight olefins (carbon number > 12) that are formed due to cracking are readily consumed due to their high reactivity. Thus, light cycle oil is comprised primarily of paraffins, aromatics, and naphthenes. Light cycle oil quality is measured by the Cetane Index, an indicator of diesel fuel performance. The Cetane Index of light cycle oil depends on the type of feed used and conversion obtained during catalytic cracking. Lower conversion and higher feed K value result in higher Cetane Index (ie, better performance) for light cycle oil (see Fig. 6). In addition, hydrodesulfurization of the FCC feed or the light cycle oil product from FCC can also increase the Cetane Index, as well as other factors (32).

Light cycle oil can be used either as diesel fuel or for home heating; however, the sulfur content of diesel fuel is being regulated. In some areas of the United States, regulations limit S content to below 0.05 wt %. To lower the S content, hydrotreating of FCC feed or the diesel product occurs, whereby S compounds in the feed are converted to H_2S. Other significant properties of light cycle oil are pour point and smoke point.

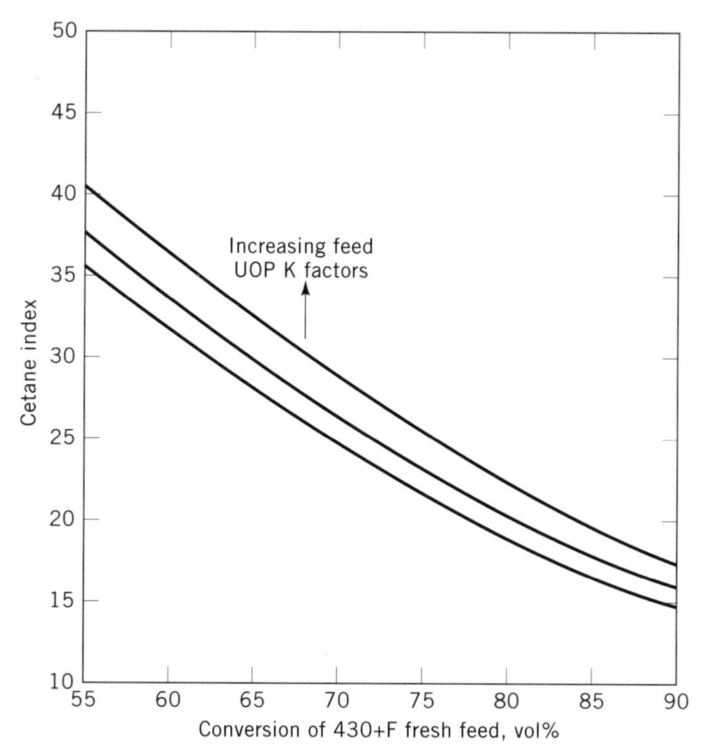

Figure 6. Predicted LCO Cetane Index from conversion level and feed quality; feed = 21.0°API (32).

The heaviest fraction of the liquid product from catalytic cracking contains C_{20^+} and higher hydrocarbons. A small portion of the catalyst (especially the fine particles) cannot be completely separated from the hydrocarbon product and become part of the heaviest liquid product fraction. Due to the presence of the catalyst fines, the heavy cut is referred to as "slurry oil." The catalyst content of slurry oil, usually referred to as "ash content," is typically less than 1 wt %. Density (usually measured as API gravity) and viscosity are important properties of slurry oil. Slurry oil can be used as a fuel in boilers.

Gaseous Products

Gaseous products from catalytic cracking can be divided into dry gas and wet gas; the predominant components in these fractions are presented in Table 10. Hydrogen, one of the components in dry gas, is also needed in the refinery processes of hydrocracking and hydrotreating. Due to the low concentration of H_2 in dry gas, it is usually too costly to recover it. However, processes to separate H_2 from FCC dry gas are under development. Dry gas can be used as refinery fuel.

Components in the wet gas are used in certain refineries to augment the gasoline yield. Olefins in the wet gas (propylene and butenes) can be alkylated using *i*-butane in the wet gas to form gasoline range components called alkylate. Normal butane from FCC can also be added to the gasoline pool. The amount of *n*-butane added is limited by the RVP specification of the gasoline, as addition of *n*-butane increases RVP. In certain refineries, components of wet gas (especially the saturates) are used as liquified petroleum gas (LPG).

An important characteristic of wet gas is its olefinicity; propylene from wet gas can be used as a petrochemical feedstock to make polypropylene. In some refineries, highly olefinic wet gas is desired, whereas at other locations, higher concentration of isobutane is needed. One of the wet gas constituents, *i*-butene, is the feed to produce methyl *t*-butyl ether (MTBE), which is added to gasoline to obtain the legally mandated oxygen content. High concentration of *i*-butene in the wet gas can facilitate increased capacity to produce MTBE.

ENVIRONMENTAL ASPECTS

Pollution control relevant to FCC can be classified into two categories: (1) during the operation of a FCC unit in a refinery, and (2) during the use of products from FCC (eg, gasoline and diesel).

Catalyst Fines

The microspheroidal catalyst particles undergo attrition during use in a commercial FCC unit. The fines (< 20 μm) produced during this attrition can be lost from both the reactor and regenerator sides of the FCC unit and emitted into the atmosphere. To minimize this loss, cyclones are used in the reactor and regenerator; electrostatic precipitators are also employed in some commercial units to capture the fine particles. In addition, the use of an attrition-resistant catalyst helps to minimize emission of catalyst fines. Particulate emissions are usually monitored by measuring the opacity of the flue gas. Limits are subject to local laws and vary among refineries.

Carbon Monoxide

The combustion of coke in an FCC regenerator produces a mixture of CO and CO_2, termed a "partial burn" operation. In certain commercial units, this mixture leaves the regenerator, and a CO boiler is used to convert CO in the flue gas to CO_2, and to produce steam for use in the refinery. Conversion in commercial CO boilers is nearly 100%; as a result, CO emissions to the atmosphere is negligible.

Certain commercial regenerators operate under "full burn" conditions, where the coke is almost entirely converted to CO_2. A noble metal catalyst, termed a "combustion promoter," is used to catalyze the conversion of CO to CO_2. Combustion promoters contain 100–1,000 ppm Pt on Al_2O_3 or other mixed oxide supports (33). These combustion promoters can be manufactured with physical characteristics (particle size distribution, attrition resistance, bulk density) similar to the FCC catalyst and be added along with the cracking catalyst to the FCC unit. Typically, only a few kilograms of combustion promoter is needed per metric ton of catalyst.

Sulfur Oxides

Many of the feedstocks processed in FCC units contain significant concentration of S (0.5–2.0 wt %), which is then distributed in all of the FCC products. About 4–8% of this S becomes part of the coke (34). During regeneration of the catalyst, sulfur oxides (a mixture of SO_2 and SO_3, referred to as SO_x) are produced. Limits on SO_x emissions (expressed as kilograms of SO_x emitted per barrel of feed processed) vary among refineries at different locations.

The factors influencing SO_x emissions include feed sulfur, oxygen and carbon monoxide levels, carbon monoxide promotor, coke make, temperature, carbon on catalyst, simultaneous change in major parameters, and catalyst addition rate (35). Four methods are used to control SO_x emissions (36): (1) hydrodesulfurization of the feed to the FCC unit reduces the concentration of S in the coke, (2) flue gas from the FCC regenerator can be scrubbed to remove SO_x, (3) regenerator operating conditions can be modified to minimize SO_x, and (4) a SO_x transfer additive can be used to control regenerator SO_x emissions. The last two methods will be discussed further.

Methods to minimize SO_x emissions include operation of the regenerator in the full burn mode and using sufficient air to have excess O_2 present in the flue gas. The above regenerator operating conditions favor oxidation of SO_2 to SO_3. SO_3 is adsorbed on some of the cracking catalyst components (eg, Al_2O_3 and La_2O_3).

SO_x transfer additives are solid bases and can be manufactured with physical characteristics (particle size, bulk density, attrition resistance) similar to the FCC catalyst. An example of an effective SO_x transfer additive is $MgAl_2O_4$ spinel containing CeO_2 (37). Usage rate of SO_x transfer additives is typically 100–200 kilograms per metric ton of catalyst.

The mechanism (2) by which a SO_x transfer additive controls SO_x emissions is described in Figure 7. SO_x is captured by the additive in the regenerator and is released as H_2S along with product gases from the reactor. This release regenerates the sites in the additive for additional SO_x capture. H_2S released on the reactor side can be handled using the Claus process for S recovery (38). The SO_x transfer additive needs to have the ability to oxi-

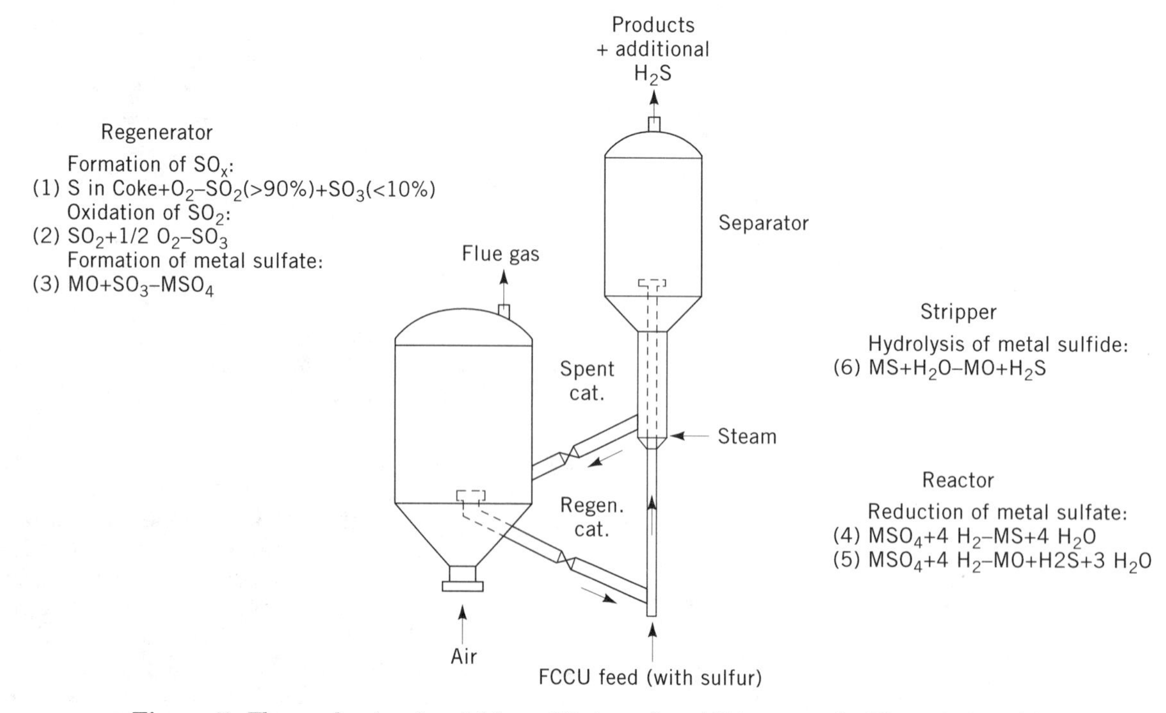

Figure 7. The mechanism by which an SO_x transfer additive controls SO_x emissions (2).

dize SO_2 to SO_3, capture SO_3 under FCC regeneration conditions, and release S under FCC reactor—stripper conditions. A strong base may capture SO_x but not release it; a weak base may not be able to capture SO_x in the regenerator. Thus, the basicity needs to be optimized.

Certain FCC units may not have the air blower capacity to operate the regenerator with excess O_2. Full burn operation may not be feasible due to temperature restrictions imposed by regenerator metallurgy. Under these circumstances, use of an SO_x transfer additive may be the most cost-effective way of meeting emission standards. The additive usage rate can be increased as feed S increases; the additive can be withdrawn during periods of processing low S feeds, thus providing flexibility. However, the use of a SO_x transfer additive has certain limitations. These additives are generally not as effective in a partial burn operation as they are in a full burn operation. Hence, higher usage rates are recommended with a partial burn operation to achieve the same degree of SO_x reduction. In addition, SO_x transfer additives can also be poisoned by SiO_2 present in the cracking catalyst (39). Despite these limitations, use of an SO_x transfer additive is a commercially proven approach for SO_x emission control.

Nitrogen Oxides

Nitrogen present in the FCC feed can become part of the coke during cracking. During regeneration, this results in formation of a mixture of nitrogen oxides (NO and NO_2), which is referred to as NO_x. NO_x emission limits (expressed as weight of NO_x emitted per barrel of feed processed) vary among refineries.

NO_x formed during catalyst regeneration can react with either CO (which is also formed in the FCC regenerator) or carbon in the coke. Thus, NO_x emissions in a partial burn regenerator, which produces a mix of CO and CO_2, can be lower than emissions from a full burn regenerator, which produces mostly CO_2. The use of a combustion promoter that converrts CO to CO_2 has thus been observed to have an adverse effect on NO_x emissions.

Regenerator operating variables that cause a reduction in SO_x emission (high CO_2—CO, high excess O_2, use of combustion promoter) generally result in an increase in NO_x emissions. Burning of coke in the regenerator with minimum oxygen, without the benefit of combustion promoter, helps to minimize NO_x. Injection of NH_3 and hydrocarbons in the regenerator (40,41) and adsorbent compositions to capture NO_x in the regenerator (42) have also been suggested. These adsorbents are expected to be used as additives to the FCC catalyst.

Reformulated Gasoline

The composition of FCC gasoline can impact the environment during its use as a motor fuel. Conventional gasoline is predominantly a mixture of hydrocarbons. The presence of oxygen-containing compounds (alcohols and ethers) in the fuel has been observed to reduce CO emissions from automobiles. The Clean Air Act of 1990 mandated the use of oxygenated fuels (oxygen content of 2.7 wt %) in 42 CO nonattainment areas in the United States, effective November 1, 1992. This and other legal requirements are expected to significantly change the chemical composition of gasoline produced at the refinery. The modified gasoline meeting the new legal requirements is referred to as reformulated gasoline.

Oxygenated compounds added to the gasoline are typically methyl-*t*-butyl ether (MTBE) and *t*-amyl methyl ether (TAME). The ethers have to be either produced at the refinery or purchased for blending with the gasoline. The ethers can be produced by reacting iso-olefins (*i*-butene for MTBE and *i*-pentene for TAME) with methanol; the FCC unit can be a source of these iso-olefins. Faujasites that exhibit low unit cell size values (< 24.25 Å) during use in a commercial FCC unit exhibit higher selectivity for *i*-olefins. Research to find catalysts with improved selectivity has intensified.

The Clean Air Act also mandated special gasoline formulation requirements for ozone nonattainment areas effective January 1, 1995. The ozone nonattainment areas include nine metropolitan areas in the United States (Baltimore, Chicago, Greater Connecticut, Houston, Los Angeles, Milwaukee, New York, Philadelphia, and San Diego). The goal of these requirements is to reduce the emission of volatile organic compounds (VOCs) and toxics during the use of gasoline as a motor fuel by 15%. A reformulated fuel meeting these requirements will be certified by the Environmental Protection Agency (EPA) using what is termed a "simple model." The fuel specifications (see Table 12) include limits on benzene and total aromatics concentration, and Reid Vapor Pressure (RVP) of the gasoline. Refiners are required to meet the more stringent of the following two standards: (1) fuel formula standard that limits aromatics to 25 vol % maximum, along with other specifications provided in Table 12, or (2) fuel performance standard which requires the 15% reduction in VOCs and toxics. By early 1994, 12 states and the District of Columbia have opted into this program and require gasoline to meet these specifications.

Typical unleaded gasoline used prior to the availability of reformulated fuels contained 2.2 wt % benzene, 30 wt % aromatics, and exhibited RVP ranging from 6.21 to 72.40 × 10^4 Pa (9–10.5 psi). Gasoline from the FCC unit

Table 12. Reformulated Gasoline Specifications Using the Simple Model

Duration	Jan. 1 1995–Mar. 1, 1997
Target	15% reduction in volatile organic compound emission; 15% reduction in toxic emissions
Summer Reid Vapor Pressure	7.2 maximum (Baltimore, Houston, Los Angeles, San Diego); 8.1 maximum (Chicago, Greater Connecticut, Milwaukee, New York, Philadelphia)
Benzene, vol %	1.0 maximum
Oxygen, vol %	2.0 minimum
Heavy metals (eg, Pb, Mn)	No
Simple model parameters	Benzene, total aromatics, RVP, O_2 content
Other significant properties	Sulfur, T90, olefins, aromatics cannot exceed refiner's 1990 average

can contribute significantly to the benzene and aromatics in the pool gasoline, and FCC catalysts that reduce concentration of aromatics in the FCC gasoline have been reported (43). Aromatics are generally concentrated in the heavy end of gasoline (C_8–C_{12} compounds). Thus, undercutting the gasoline by removing the heavier compounds for inclusion in other fuels is expected to lower the concentration of aromatics. Because RVP of the pool gasoline is controlled by the amount of *n*-butane added to the gasoline pool, RVP cannot generally be controlled by modifications of the FCC unit operation or catalyst.

After March 1, 1997, a "complex model" must be used to certify gasoline as a reformulated fuel; the EPA is still finalizing the details of this model. The State of California is considering a 40-ppm limit on the concentration of S in gasoline by March 1996. (Leaded gasoline prior to reformulation typically contained 1500 ppm S.) If the proposed California requirement passes, a substantial reduction in S in gasoline needs to be achieved. Since virtually all the S in the pool gasoline comes from the FCC gasoline, the options for a refiner to lower S in gasoline are hydrotreating the FCC feed to lower S, hydrotreating FCC gasoline, and use of a specially formulated FCC catalyst to produce low S gasoline. Hydrotreating FCC gasoline usually results in a reduction in octane number of the gasoline, which can be undesirable, and changes in catalyst composition have only a small effect on S in gasoline (44). As a result, research toward developing new FCC catalysts for S reduction has intensified.

DEVELOPMENTS

The compositional requirements of reformulated gasoline are forcing the integration of the FCC unit with a petrochemical plant. Environmentally driven refinery needs will include increased isobutylene and isoamylene production as a precursor to oxygenated compounds (such as MTBE, ETBE, and TAME), lower gasoline aromatic levels, reduced gasoline sulfur levels, and decreased SO_x, NO_x, and particulate emissions from the FCC unit. In addition, there is a trend toward processing heavier feeds with higher levels of contaminant metals (Ni and V) in an FCC unit. For example, many of the new FCC units built in Asia are designed to process heavier feedstock. Technology developments, both in the areas of improved FCC catalysts and process hardware, are allowing refiners to meet product demand and environmental compliance requirements. Trends in catalyst and hardware developments have recently been reviewed (45).

FCC Catalyst

The trend toward higher zeolite content of the FCC catalyst is likely for processing of atmospheric resids. In addition to increasing catalyst activity, increased zeolite content will increase resistance to deactivation by vanadium contained in the feed. An additional factor in favor of increasing the concentration of zeolite Y will be the increased use of additive catalysts that will lower the concentration of zeolite Y in the FCC catalyst inventory.

Research on new zeolites to complement or replace zeolite Y-based catalysts has increased. As discussed in the Environmental section, the lower benzene—aromatic content and the mandated levels of oxygen-containing compounds in reformulated gasoline will require the development of new zeolitic materials with a higher selectivity light olefins. These olefins can then be converted to oxygenates or alkylated into gasoline range fractions. An example of such a developmental zeolite is shown in Table 13. The catalyst, called RFG-5, which is being developed by Grace Davison, shows comparable light gas yields to a high ZSM-5 loaded zeolite catalyst without any loss in gasoline yield. Similar developments are also being reported by other catalyst manufacturers.

Table 13. Product Yields for Developmental Catalyst RFG-5 Circulating Riser Product Distribution Data[a,b]

Catalyst Identification	90 USY Catalyst 10 ADD O—HS	RFG-5
C—O	7.10	5.20
Conversion, wt %	68.0	68.0
Yields, wt %		
H_2	0.105	0.055
Total C1 + C2	3.23	2.86
C3=	8.7	6.3
Total C3s	10.3	7.5
1-butene	1.8	1.8
Isobutylene	3.6	3.8
t-2-butene	2.3	2.3
cis-2-butene	1.9	1.9
Total C4=	9.6	9.8
i-C4	2.9	2.1
n-C4	0.7	0.6
Total C4s	13.2	12.5
C5 + gasoline, 221°C	37.4	41.3
G-CON RON	96.4	96.4
G-CON MON	82.3	81.5
LCO, 221–366°C	16.7	15.7
Bottoms, 338°C or higher	15.3	16.3
Coke	3.58	3.74

[a] Ref. 46.
[b] At 521°C.

Catalysts with advanced matrix technology are likely to be deployed for processing atmospheric resids. Catalyst improvements will include increased metals tolerance, selective bottoms cracking, and improved physical properties. Amorphous alumina has been used to increase nickel tolerance of the catalyst and bottoms conversion; the alumina type and its pore size distribution are both important for the latter. Mesopores in the 30–100 Å range are believed to be directly responsible for reducing bottoms yield with aromatic or naphthenic feedstocks, whereas the smaller pores and the zeolite are more important for paraffinic feeds. Thus, the primary cracking of the larger molecules in resids is achieved by the matrix (25).

Increased demand for light olefins and alkylate should increase the use of ZSM-5 additives. Two recent developments in ZSM-5 additive technology include (1) higher activity additives that have cut catalyst makeup by at least a factor of two, and (2) high selectivity additives that crack less gasoline and produce less C_3 and C_4 hydrocarbons to achieve the same octane increase. The improved

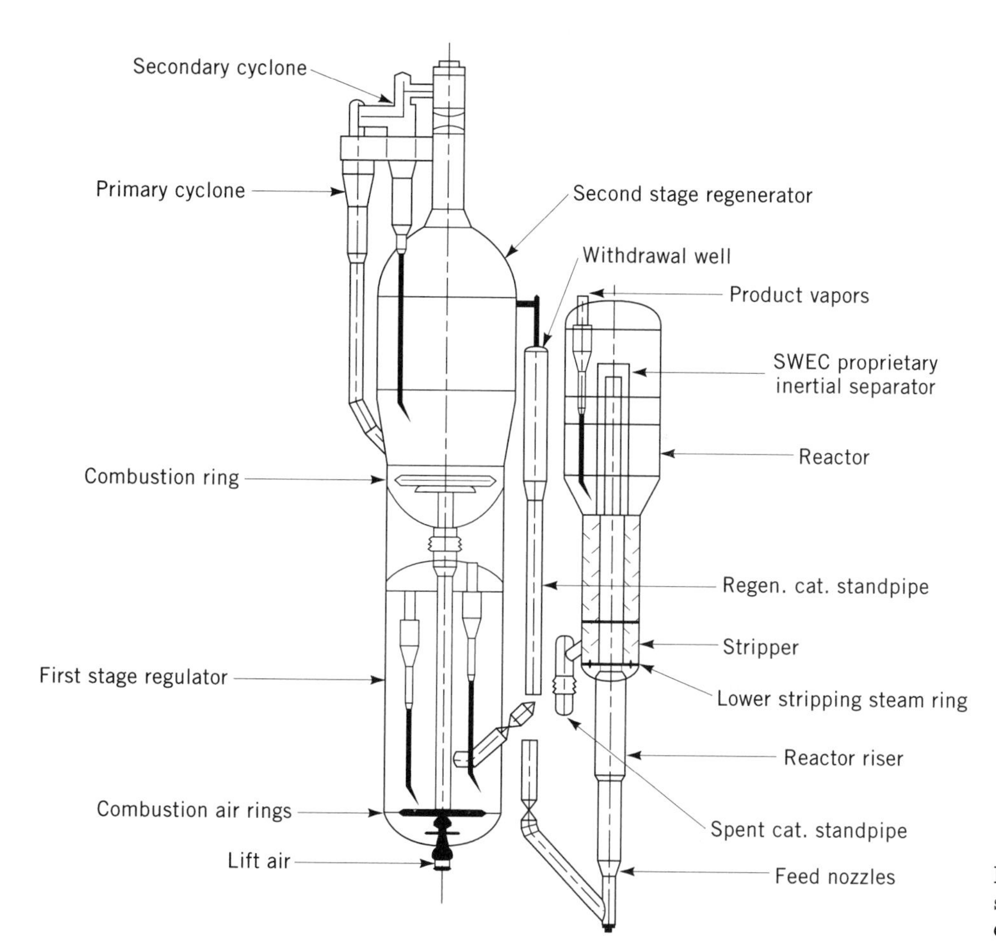

Figure 8. Stone & Webster two-stage stacked FCC unit without a catalyst cooler (51).

selectivity is achieved by using ZSM-5 of high (> 500) silica—alumina ratio (47,48). The increase in octane with the high selectivity additive is believed to be mainly due to increased isomerization activity.

The use of combustion promoters and SO_x additives for controlling FCC unit emissions is likely to increase. There are two reasons for developing FCC catalyst compositions that reduce the sulfur in gasoline: (1) sulfur-containing organics are oxidized to sulfur oxides in an automobile engine and are exhausted to the atmosphere, and (2) sulfur oxides in automobile exhausts inhibit the performance of the catalytic converter; thus, reducing sulfur oxides will reduce the carbon monoxide and hydrocarbon concentration in the exhaust, which are ozone precursors. Research in this area continues; additives that are effective for reducing nitrogen oxide emissions from the regenerator are also being developed.

The feasibility of installing Selective Catalytic Reduction (SCR) for reduction of NO_x from the regenerator is being explored. In SCR, NO_x is reacted with injected ammonia in the presence of a catalyst to form primarily nitrogen and water. The most widely used SCR catalysts are vanadia supported on anatase titania. The operating temperature window for this catalyst is between 315 and 399°C (49). Uncertainties in application of SCR include the presence of particulates, low oxygen levels, and catalyst deactivation.

Disposal costs of "equilibrium" catalysts containing nickel and vanadium may drive the development of processes for the reclamation or reprocessing of FCC catalysts.

Hardware

Developments in hardware are being driven by product demands, processing objectives, catalyst improvements, and the need for reliability, operating flexibility, and environmental compliance. Hardware developments have been recently reviewed (50,52); several of these developments are illustrated in Figures 8–10.

Improved feed—catalyst contacting is generally believed to minimize coking and thermal cracking at the bottom of the riser, and thereby to increase yields. To improve mixing between feed and catalyst, multinozzle feed injection and contacting arrangements have been developed, and low and high pressure feed atomizers used. A pre-acceleration zone, in which a recycle stream of light gases contacts the regenerated catalyst and provides fluidization at the base of the riser, has also been developed. In this configuration, light hydrocarbons are claimed to passivate active metal sites prior to contact with heavy oil.

Secondary (or segregated) feed injection is suitable in cases where the refiner has to treat two separate feeds: one more refractory with a higher nitrogen content, and the other easier to crack, having low nitrogen. The high nitrogen feed is introduced at the riser bottom, where it contacts hot regenerated catalyst. The low nitrogen feed

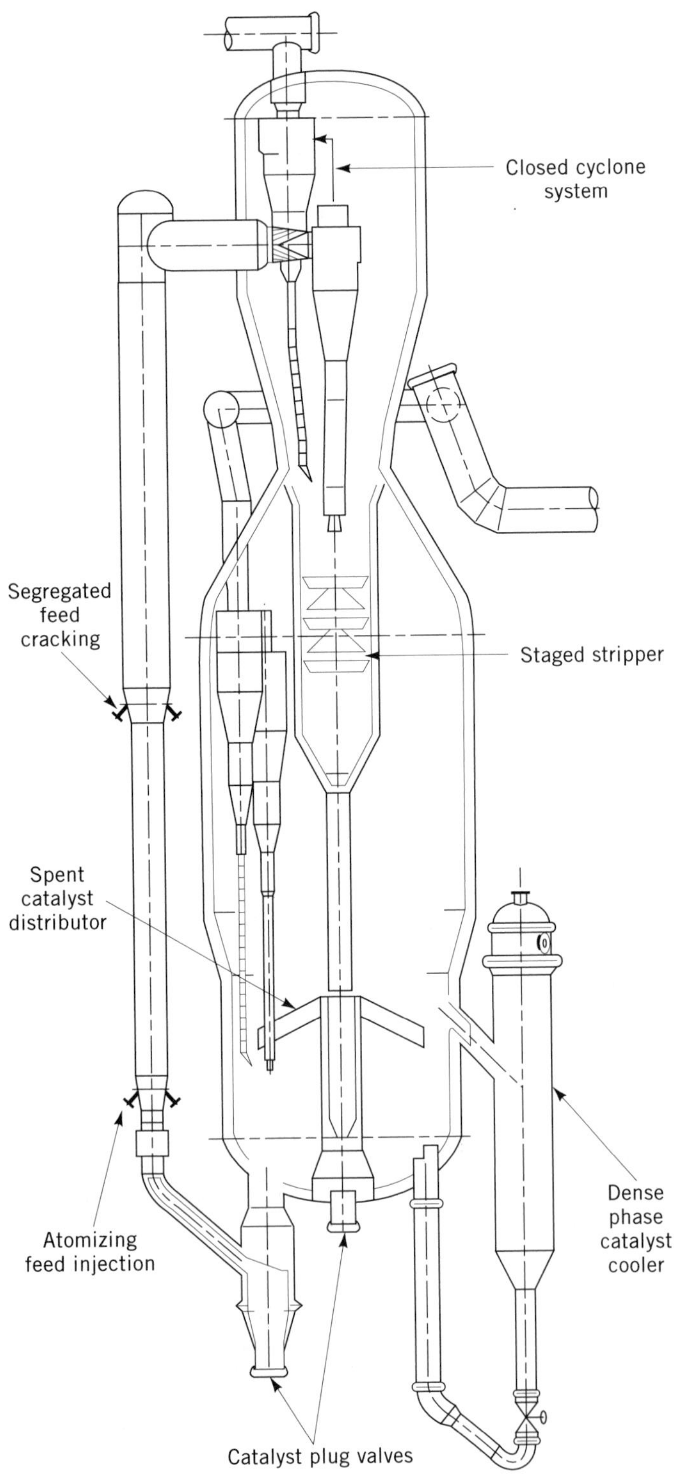

Figure 9. Orthoflow FCC converter (52).

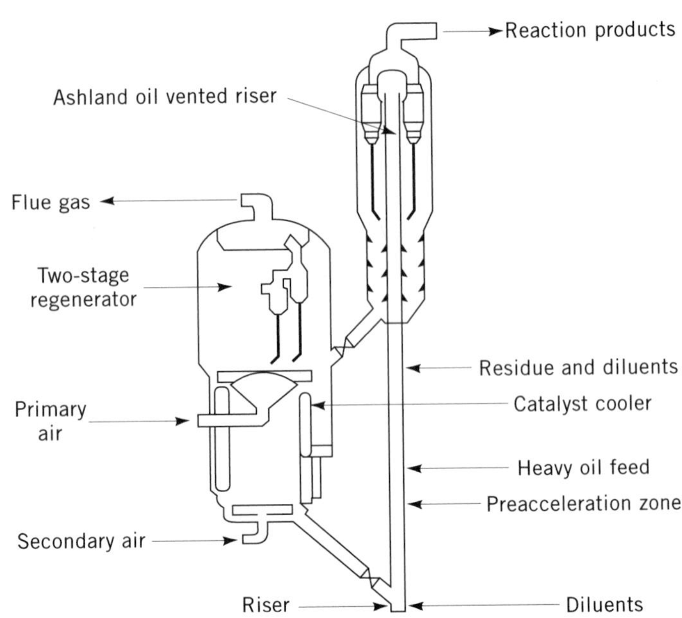

Figure 10. UOP two-stage FCC unit with catalyst cooler (50).

is injected at a point some distance from the base of the riser, permitting for the cracking of the more refractory feed at more severe conditions. Such contacting methods provide the refiner with additional flexibility to increase conversion of the heavy ends.

Systems for rapid separation of spent catalyst from gasoil vapors to reduce nonselective thermal cracking are being offered. Thermal cracking may be reduced by connecting cyclones directly to the riser; closed cyclones are claimed to reduce dry gas components by more than 40% and to increase gasoline and distillate yields and gasoline octane (53). The benefits of closed cyclone systems are expected to increase with the trend toward short contact times and higher riser top temperatures. Efficient catalyst—oil separation also results in low catalyst carryover to the main fractionator.

Staged stripping improves stripping efficiency, thereby increasing product yields and reducing heat generation in the regenerator.

Two-stage regenerator designs burn most of the coke in the partial combustion mode; the remainder of the coke is burnt under complete combustion. Excess air from the second stage passes through the first stage before leaving the regenerator. The result is a more efficient use of oxygen and a reduction in the air blower requirements to burn a given amount of coke from the catalyst. Advantages claimed for this design include increased flexibility for processing heavy feeds, lower hydrothermal deactivation, increased metals tolerance of the catalyst, and lower operating expenses.

Improved catalyst—air contacting, using advanced catalyst and air distribution systems to the regenerator, results in more uniform regenerator temperatures that increase coke burning efficiency and reduce afterburning and nitrogen oxide emissions.

The increase in regenerator temperature in treating high Conradson Carbon feeds reduces catalyst circulation, which in turn reduces conversion. It can also increase dry gas yield due to thermal cracking at the point of feed injection. Dense-phase catalyst coolers are being used to improve operating flexibility and to increase conversion and throughput in treating resids. Cooler duty may be independently controlled by the refiner to adjust the unit heat balance and coke yield. Catalyst coolers can replace or be used together with CO boilers, which also remove heat from the FCC.

BIBLIOGRAPHY

1. E. C. Luckenbach, A. C. Worley, A. D. Reichel, and E. M. Gladrow, "Catalytic Cracking," in J. J. McKetta, ed., *Encyclopedia of Chemical Processing and Design*, Vol. 13, Marcel Dekker, Inc., New York, 1981.
2. J. Scherzer, *Octane-Enhancing Zeolitic FCC Catalysts: Scientific and Technical Aspects*, Marcel Dekker, Inc., New York, 1990.
3. C. J. Plank, E. J. Rosinski, and H. P. Hawthorne, *Ind. Eng. Chem. Prod. Res. Dev.* **3,** 165–169 (1964).
4. K. Rajagopalan and E. T. Habib, *Hydrocarbon Processing*, 43–46 (Sept. 1992).
5. J. S. Magee and R. E. Ritter, "Formation of High Octane Gasoline by Zeolite Cracking Catalysts," paper presented at the ACS Meeting, Miami Beach, Fla., 1978.
6. A. A. Avidan and R. Shinnar, *Ind. Eng. Chem. Res.* **29,** 931–942 (1990).
7. J. R. Murphy, *Oil & Gas Journal*, 59–67 (May 18, 1992).
8. P. H. Schipper, F. G. Dwyer, P. T. Sparrell, S. Mizrahi, and J. A. Herbst, "Zeolite ZSM-5 in Fluid Catalytic Cracking: Performance, Benefits, and Applications," in M. L. Occelli, ed., *Fluid Catalytic Cracking—Role in Modern Refining, ACS Symp. Ser. 375*, American Chemical Society, Washington, D.C., 1988.
9. A. W. Peters, L. Rheaume, T. G. Roberie, and P. G. Thiel, "Effect of Catalyst Base Composition on the Performance of FCC Additives," paper presented at the AIChE Spring National Meeting, Orlando, Fla., Mar. 1990.
10. A. S. Krishna, C. R. Hsieh, A. R. English, T. A. Pecoraro, and C. W. Kuehler, *Hydrocarbon Processing*, 59–66 (Nov. 1991).
11. P. B. Venuto and E. T. Habib, *Fluid Catalytic Cracking with Zeolite Catalysts*, Marcel Dekker, Inc., New York, 1979.
12. J. J. Blazek, *Catalagram* **42,** 3–13 (1973).
13. J. R. Murphy and Y. L. Cheng, *Oil & Gas Journal*, 89 (Sept. 3, 1984).
14. B. C. Gates, J. R. Katzer, and G. C. A. Schuit, *Chemistry of Catalytic Processes*, McGraw-Hill, Inc., New York, 1979.
15. C. L. Thomas and D. S. Barmby, *J. Catal.* **12,** 341–346 (1968).
16. H. Fichtner-Schmittler, V. Lohse, G. Engelhardt, and V. Patzelova, *Crystal. Res. Tech.* **19,** K1–K3 (1984).
17. C. V. McDaniel and P. K. Maher, "Zeolite Stability and Ultrastable Zeolites," in J. A. Rabo, ed., *Zeolite Chemistry and Catalysis*, American Chemical Society, Washington, D.C., 1976.
18. L. A. Pine, P. J. Maher, and W. A. Wachter, *J. Catal.* **85,** 466–476 (1984).
19. K. Rajagopalan and A. W. Peters, *J. Catal.* **106,** 410–416 (1987).
20. R. E. Ritter, J. E. Creighton, T. G. Roberie, D. S. Chin, and C. C. Wear, *Paper No. AM-86-45*, presented at the NPRA Annual Meeting, Los Angeles, Calif., Mar. 1986.
21. Edwards and co-workers, "Strategies for Catalytic Octane Enhancement in a Fluid Catalytic Cracking Unit," in M. L. Occelli, ed., *Fluid Catalytic Cracking—Role in Modern Refining, ACS Symp. Ser.* **375**, American Chemical Society, Washington, D.C., 1988.
22. J. S. Magee, W. E. Cormier, and G. M. Wolterman, *Proc. Refin. Dep. Am. Pet. Inst.* **64,** 29–35 (1985).
23. J. S. Magee and J. J. Blazek, "Preparation and Performance of Zeolite Cracking Catalysts," in J. A. Rabo, ed., *Zeolite Chemistry and Catalysis*, American Chemical Society, Washington, D.C., 1976.
24. C. J. Plank, "The Invention of Zeolite Cracking Catalysts—A Personal Viewpoint," in B. H. Davis and W. P. Hettinger, Jr., eds., *Heterogeneous Catalysis—Selected American Histories, ACS Symp. Ser.* **222,** American Chemical Society, Washington, D.C., 1983.
25. G. W. Young and K. Rajagopalan, *Ind. Eng. Chem. Proc. Des. Dev.* **24,** 995–999 (1985).
26. R. F. Wormsbecher, A. W. Peters, and J. M. Maselli, *J. Catal.* **100,** 130–137 (1986).
27. W. E. Morris, *Oil & Gas Journal* **83**(11), 99–106 (Mar. 18, 1985).
28. G. H. Unzelman, *Paper No. AM-88-25*, presented at the NPRA Annual Meeting, San Antonio, Tex., Mar. 1988.
29. American Petroleum Institute, *Research Project 45*, 1957.
30. R. F. Wormsbecher, G. D. Weatherbee, G. Kim, and T. S. Dougan, *Paper No. AM-93-55*, presented at the NPRA Annual Meeting, San Antonio, Tex., Mar. 1993.
31. J. G. Speight, *Fuel Science and Technology Handbook*, Marcel Dekker, Inc., New York, 1990.
32. J. M. Collins and G. H. Unzelman, *Oil & Gas Journal* **80**(22), 87–94 (May 31, 1982).
33. A. W. Chester, A. B. Schwartz, W. A. Stover, and J. P. McWilliams, *Chemtech* **11,** 50–58 (1981).
34. G. P. Huling, J. D. McKinney, and T. C. Readal, *Oil & Gas Journal* **73**(20), 73–79 (May 19, 1975).
35. K. Baron, A. W. Wu, and L. D. Krenzke, *Prepr. ACS Div. Pet. Chem.* **28**(4), 934–943 (1983).
36. J. A. Sigan, W. A. Kelly, P. A. Lane, W. S. Letzsch, and J. W. Powell, "Reducing FCC Emissions with No Capital Cost," paper presented at the AIChE Spring National Meeting, Orlando, Fla., Mar. 1990.
37. A. A. Bhattacharyya, G. M. Woltermann, J. S. Yoo, J. A Karch, and W. E. Cormier, *Ind. Eng. Chem. Res.* **27,** 1356–1360 (1988).
38. G. P. Masologites and L. H. Beckberger, *Oil & Gas Journal* **71**(47), 49–53 (1973).
39. L. Rheaume and R. E. Ritter, "Use of Catalysts to Reduce SO_x Emissions from Fluid Catalytic Cracking Units," in M. L. Occelli, ed., *Fluid Catalytic Cracking—Role in Modern Refining, ACS Symp. Ser. 375*, American Chemical Society, Washington, D.C., 1988.
40. U. S. Patent 4,434,147 (Feg. 28, 1984), W. A. Blanton and W. L. Dimpfl (to Chevron Research and Technology Co.).
41. U. S. Patent 4,309,309 (Jan. 5, 1982), W. A. Blanton (to Chevron Research and Technology Co.).
42. U. S. Patent 5,002,654 (Mar. 26, 1991), A. A. Chin (to Mobil Corp.).
43. D. W. Cotterman, "Fluid Cracking Catalysts for Fuel Reformulation," paper presented at the National Conference on Octane Quality and Reformulated Gasoline, New Orleans, La. (Mar. 25, 1992).
44. R. F. Wormsbecher, D. S. Chin, R. R. Gatte, R. H. Harding, and T. G. Albro, *Paper No. AM-92-15*, presented at the NPRA Annual Meeting, New Orleans, La., Mar. 1992.
45. A. A. Avidan, *Oil & Gas Journal*, 59–67 (May 18, 1992).
46. T. G. Roberie, G. W. Young, D. S. Chin, R. F. Wormsbecher, and E. T. Habib, Jr., *Paper No. AM-92-43*, presented at the NPRA Annual Meeting, New Orleans, La., Mar. 1992.
47. U. S. Patent 4,309,276 (1982), S. J. Miller (to Chevron Research and Technology Co.).

48. S. J. Miller and C. R. Hsieh, "Octane Enhancement in Catalytic Cracking Using High Silica Zeolites," in M. L. Occelli, *Fluid Catalytic Cracking II—Concepts in Catalyst Design, ACS Symp. Ser. 452*, American Chemical Society, Washington, D.C., 1991.
49. F. P. Boer, L. L. Hegedus, T. R. Gouker, and K. P. Zak, *Chemtech* **20,** 312–319 (1990).
50. *Hydrocarbon Processing*, 169–175 (Nov. 1992).
51. S. L. Long, A. R. Johnson, and D. J. Dharia, *Paper No. AM-93-50*, presented at the NPRA Annual Meeting, San Antonio, Tex., Mar. 1993.
52. T. E. Johnson and A. A. Avidan, *Paper No. AM-93-51*, presented at the NPRA Annual Meeting, San Antonio, Tex., Mar. 1993.
53. A. A. Avidan, F. J. Krambeck, H. Owen, and P. H. Schipper, *Oil & Gas Journal* 56–62 (Mar. 26, 1990).

CATALYTIC REFORMING

Aaron P. Kelly
UOP
Des Plaines, Illinois

Catalytic reforming has played an important role over the years in a petroleum refiner's ability to produce increasing amounts of high quality motor fuels and petrochemical feedstocks. Initially developed as a way to upgrade low octane straight-run and natural gasoline feedstocks, catalytic reforming has become the primary provider of high octane blending components to the gasoline pool. With the increased demand in the 1990s to produce more environmentally benign products, catalytic reforming operations have been adjusted to produce cleaner-burning fuels and a greater volume of hydrogen for use in the production of other "clean" fuels in the refinery.

In catalytic reforming, gasoline-range molecules are altered, or reformed, into higher octane molecules of the same carbon number. This article includes a brief history of the development of the reforming process, a description of the commercial process, and a detailed explanation of reforming chemistry and catalysts. In addition, the article describes the current product market served by the process, identifies some of the environmental issues affecting the process, and outlines future developments expected in the continuing evolution of the process.

See also Petroleum Markets; Petroleum Production; Petroleum Refining; Transportation fuels—automotive fuels; Reformulated gasoline.

HISTORY

Up until the 1940s, the dominant refinery process for gasoline production was thermal cracking of naphthas, gas oils, and residuum streams. Although the use of the thermal-cracking process was widespread in the industry, its limitations in producing high yields of high octane motor fuels were well known and the basis for industrial research into new processing methods. The desired processes would ultimately be catalytic and, to the extent possible, free from undesired thermal reactions. In the absence of a catalyst, high temperature, high pressure processes give rise to thermal reactions that are difficult to control. As a result, a large percentage of the feedstock is converted into fuel gas. Thus the search was on for a catalytic system that would produce the desired motor fuels at conditions that were not conducive to thermal reactions.

The development of an efficient process for producing high octane gasoline began with the identification of the appropriate crude oil fraction to be used as the feedstock for the process. Straight-run naphtha cuts were chosen as the feedstock because their distillations, or boiling-point ranges, approximated those of finished gasolines and they contained many naturally occurring high octane components. The desired catalytic system for upgrading the naphtha stream had to be capable of promoting a number of catalytic reactions because of the range of hydrocarbon classes that make up natural gasolines and straight-run naphthas. Each hydrocarbon class (paraffins, naphthenes, and aromatics) requires a different reaction to convert it to more valuable products. A significant amount of conversion is required by the process because no easy way has been found to separate the high octane reaction products from the unreacted low octane components of the feedstock.

The first catalytic-reforming unit was placed on-stream at the Pan American Refining Corp. (now American Oil Co.) in Texas City, Texas, in November 1940. The Hydroforming process was a fixed-bed cyclic unit that used two reactors in the process while two additional reactors were undergoing regeneration. The initial catalyst was composed of 9 wt % molybdenum oxide on an activated alumina base.

The development of the Platforming process was announced by Universal Oil Products Co. (now UOP) in early 1949. This new process was the first catalytic-reforming process to use a catalyst containing platinum. The first commercial application of the process took place at the Old Dutch Refining Co. refinery in Muskegon, Michigan, in October 1949. Unlike the Hydroforming process, the initial Platforming unit did not use swing reactors, nor was the catalyst regenerated. The unit processed feedstock over the catalyst for a period of 9 to 18 months, at which point the catalyst was dumped and replaced with a fresh loading. In the early 1950s, improvements in catalyst preparations led to the regeneration of spent catalysts. A single batch of catalyst could thus be used in the process, and when catalytic activity declined sufficiently, the catalyst could be regenerated to near fresh catalyst condition. Multiple operating and regenerating cycles were now possible, and this development led to improved process economics.

Following the introduction of the Platforming process in 1949, nine new catalytic-reforming processes were introduced in the petroleum-refining industry. Among the new processes were examples of semiregenerative, fluid-bed, and moving-bed technologies. The 1950s were characterized by a refining industry growing in complexity and striving to produce increasing qualities and quantities of gasoline.

The 1960s saw the introduction of bimetallic catalysts by Chevron and UOP. These catalysts significantly improved catalyst stability, which allowed further reductions in operating pressure and led to improved liquid product and hydrogen yields. The reformer solidified its position as the key process unit for meeting the demand for premium high octane gasoline.

The 1970s saw the introduction of the first truly continuous catalyst regeneration process (CCR Platforming) by UOP, and another large step down in reactor pressure was achieved, again leading to further increases in the yield of a key gasoline-blending component. Lead phase-out from gasoline in the United States began during this period, and refiners relied even more on the reformer to provide the necessary high octane blending components to the gasoline pool.

In the 1980s, improvements were made to bimetallic catalyst stabilities, and process improvements were made to the continuous catalyst regeneration reforming process to allow for operations at the ultralow operating pressures of 345 kPa. Low pressure continuous regeneration reforming units played an ever-increasing role in the production of feedstocks for downstream aromatics complexes.

The 1990s have started off as the decade of increased concern for the environment, and many regions of the world are addressing the issue of lead phaseout from gasoline. In the United States, the refining industry is facing the issue of tightened environmental regulations in the production of transportation fuels. Improvements in reforming catalyst and process development continue to be announced, and the role of the reforming unit in the refinery continues to evolve.

PROCESS DESCRIPTION

The basic steps in catalytic reforming are

Feed preheat → Reaction
→ Product separation and recovery

Within the reaction step, three distinct unit configurations (semiregenerative, cyclic, and continuous catalyst regeneration) have been developed in an evolution of continuous process efficiency improvement. The basic unit flow scheme is described first followed by a description of the primary unit configurations. A typical Platforming flow diagram is shown in Figure 1.

Catalytic Reforming Flow Scheme

Fresh charge to the unit is first hydrotreated to remove sulfur, nitrogen, and metal catalyst poisons before being sent to the Platforming unit. The end point of the feedstock is normally set by the cut point in the crude tower, and the initial boiling point is typically set by fraction-

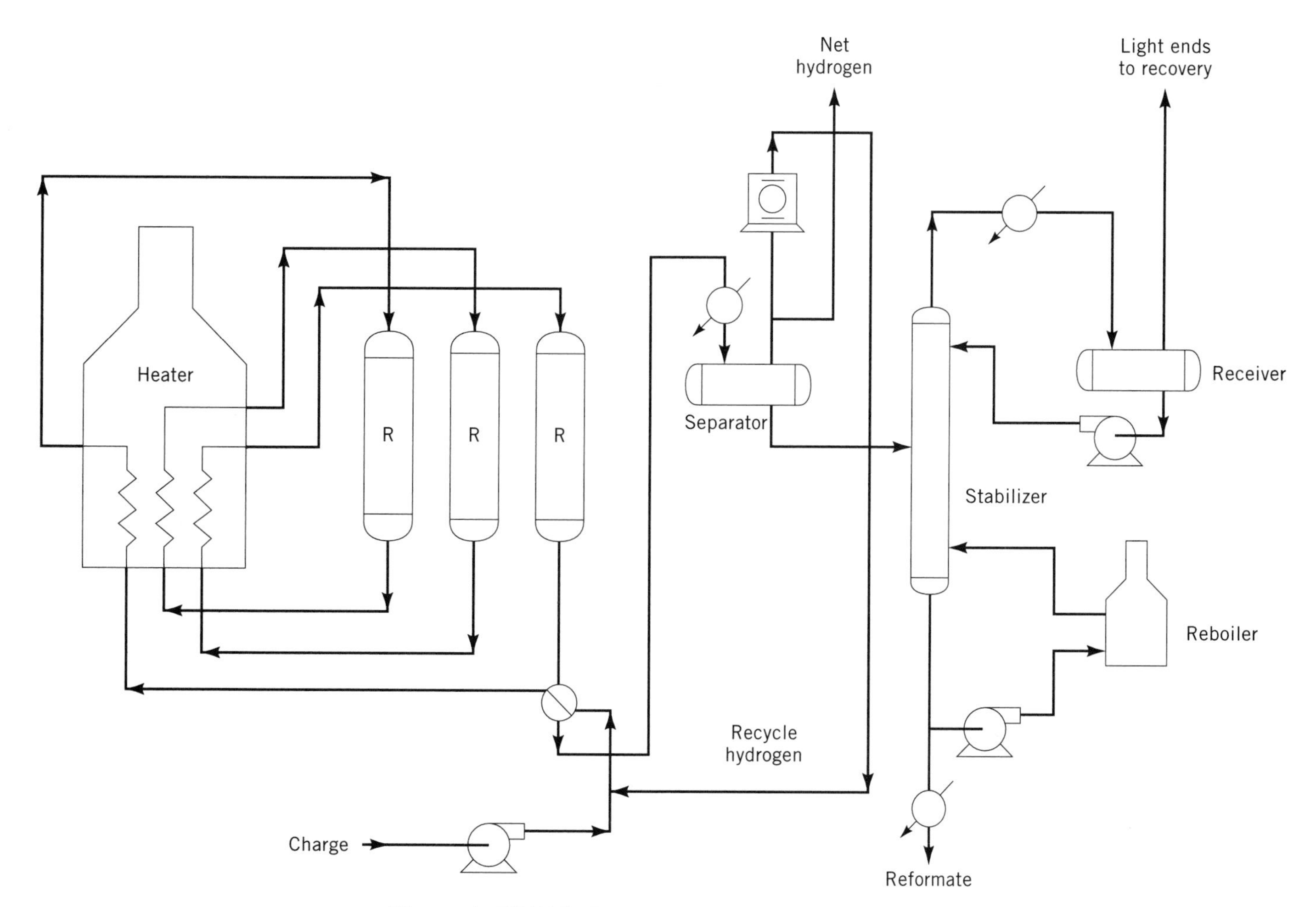

Figure 1. UOP Platforming process. Courtesy of UOP.

ation before the hydrotreater unit. The choice of proper feed end point is influenced by the relative demand for gasoline versus kerosine. The end point of the naphtha charge stock must not be too high (typically less than 204°C), or excessive coking and rapid deactivation of the catalyst occurs. The initial boiling point of the feed is often set by the desire to produce benzene. If benzene production is not desired, a large portion of the C_6 paraffin and naphthene components is removed by fractionation from the unit charge and sent either to an isomerization unit or straight to gasoline blending.

The hydrotreated naphtha charge is mixed with recycle hydrogen and heated to the reaction temperature in two steps: the combined feed is heated by exchange with effluent from the last reactor before being brought to the inlet temperature of the first reactor in the charge heater. The combined feed exchanger can be a bank of horizontal exchangers in series, vertical exchangers in parallel, or a vertical welded plate exchanger. As design unit operating pressure decreases, the last two options become more advantageous because of the lower pressure drop achieved. The charge heater is normally erected in a bank with the reactor interheaters. The heaters often use a convection section so that waste heat can be used for steam generation or column reboiling duty.

The reactor section of the Platforming unit is typically made up of three to five reactors. Multiple reactors are necessary because the majority of the desired reforming reactions are highly endothermic. In the first reactor, which typically contains 10 to 20% of the total catalyst volume, the temperature drops so quickly that the rate of reaction becomes too slow for commercial purposes. Interheaters are installed between the reactors to maintain the desired catalyst temperature. As the reactor charge moves through the reactor series, the reactions become less endothermic, and the reactor temperature differential decreases with each succeeding reactor.

Most modern reactor designs are radial flow because of the low pressure drops experienced with that design versus a downflow design. The reactors may be erected in either a side-by-side or stacked reactor configuration. Design operating conditions on Platforming units cover pressures of 345–4150 kPa and temperatures of 470–550°C.

Reactor effluent is cooled first by heat exchange with the combined feed to obtain maximum heat recovery and then with air or water to near-ambient conditions. The cooled effluent is then fed to the separation section, where the product gas is separated from the product liquid. A portion of the gas product is recycled by using a compressor, combined with fresh charge, and sent to the combined feed exchanger. The remaining hydrogen-rich net gas stream is sent to other users of hydrogen in the refinery. The separator liquid product, which contains gases and light ends as well as the desired reformate product, is sent to a product stabilizer. Reformate product from the bottom of the stabilizer is sent directly to gasoline blending when the reformer is used for motor fuel production or to an aromatics-extraction unit when the reformer is feeding a petrochemicals aromatics complex.

Reformer Configurations

The reformer diagram shown in Figure 1 is greatly simplified. One of the major classifications of reforming units is by the method of catalyst regeneration. The main reformer classifications are semiregenerative, cyclic, and moving bed or continuous catalyst regeneration.

Semiregenerative. In a semiregenerative unit, the reformer processes feedstock for a substantial time and then shuts down for regeneration of the catalyst. During catalyst regeneration, the catalyst is exposed to a number of different environments: carbon burn, by which the coke is burned from the catalyst; oxidation, by which the metal promoters are redispersed on the catalyst surface and the catalyst base is adjusted to the proper chloride level; drying; and reduction of the metal promoters. The operating time between regenerations is called a *cycle*, which is generally described in terms of months and typically will fall between 3 months and 2 yr. The length of a cycle is greatly influenced by the operating conditions (severity), feedstock quality, and the composition of the catalyst.

Cycles may be terminated as part of a normal refinery turnaround or when evidence of decreasing catalyst performance is witnessed. Typically the falloff in performance is evidenced by decreasing liquid reformate yield and increased reactor inlet temperatures. A normal temperature cycle can be estimated for a given catalyst, feed composition, operating conditions, and desired product quality. This temperature cycle recognizes the fact that as feed is processed, coke is being laid down on the catalyst. The coke reduces catalytic activity and must be compensated for by increased reactor inlet temperatures. Following regeneration, the catalyst in a semiregenerative unit is expected to regain fresh catalyst performance following regeneration for multiple cycles. Thus ultimate catalyst life can be 7 to 12 yr. To maintain reasonable cycle lengths at the reformer severities required by markets in the 1990s, operating pressures for semiregenerative units are generally not below 1300 kPa. Older reformer units may be operating at design pressures of 3450–4150 kPa.

Cyclic. Cyclic units are characterized by a special valve-and-manifold system that allows each of the reactors in the unit to be isolated and regenerated while the remaining reactors remain on stream in the reforming process. This system is in contrast to the semiregenerative unit, which must undergo a complete shutdown to regenerate the catalyst. Cyclic systems have higher onstream efficiency because the unit can operate for longer periods of time between shutdowns.

As mentioned in the description of the semiregenerative unit, the purpose of regeneration is to remove the carbon (coke) that has been laid down on the catalyst. A Cyclic unit can operate at higher severities than could a semiregenerative unit because the increased coke laydown rate is compensated for by the increased frequency of regeneration. Typical operating pressures for cyclic reformers are 1035–2070 kPa.

Cycles for the individual reactors within a cyclic unit vary from a few days to a few months, depending mainly on the reactor's position in the unit and on operating severity. Reactors toward the end of the series perform the more difficult reforming reactions and must be regenerated more frequently than reactors in the lead position.

Moving-Bed, or Continuous-Catalyst, Regeneration. The moving-bed process unit design continuously circulates catalyst through the reactors, withdraws spent catalyst from the last reactor, regenerates the catalyst in a continuous regeneration section, and then returns regenerated catalyst in a fresh state back to the inlet of the first reactor. A simplified flow scheme of a continuous reforming process unit is shown in Figure 2. Reforming reactors in a continuous-catalyst regeneration unit must be radial flow to allow for catalyst movement through the reactors. Reactors may be stacked on top of one another (Fig. 2) or may be erected in a side-by-side configuration. The stacked configuration requires only two catalyst lifting steps. The side-by-side configuration requires lifting steps between each reactor in addition to catalyst transfer to and from the regenerator.

Because the catalyst is continuously being regenerated, the continuous reformer allows for the greatest operating severity. Typical operating pressures for a continuous unit are between 345 and 1380 kPa. The ability to operate efficiently at lower pressures leads to several advantages for the continuous regeneration process over the semiregenerative process. Greater yields of C_5^+ liquid reformate and hydrogen are possible. The reformate stream is of a more consistent quantity and quality because the unit is, in effect, operating continuously at start-of-run conditions. The unit produces the maximum possible operating efficiency because continuous uninterrupted processing runs lead to reduced maintenance and operating costs.

With the reduction in operating pressure associated with the continuous regeneration units, the hydrogen–hydrocarbon flash across the product separator becomes less effective, and the net hydrogen stream contains a greater percentage of hydrocarbon impurities. To improve the recovery of C_5^+ hydrocarbons into the liquid product stream, a high pressure separator is added downstream of the low pressure separator. This recontact step improves recovery of C_3^+ material from the net hydrogen stream and boosts the pressure of the net gas stream so that it may be used by downstream units such as hydrotreaters.

PROCESS CHEMISTRY AND CATALYSTS

Composition of Reformer Feed and Product

Feed naphtha to a reformer typically contains C_6 through C_{11} paraffins, naphthenes, and aromatics. The purpose of the reforming process is to produce aromatics from the naphthene or paraffin feed components. The aromatics-rich product stream can be used as a high octane gasoline-blending component (because of the high octane values for the aromatic components) or as a feedstock to a petrochemical complex that wants certain aromatic compounds (benzene, toluene, and xylene). In motor fuel applications, the feedstock generally contains the full range of C_6 through C_{11} components to maximize gasoline production from the associated crude run. In petrochemical applications, the feedstock may be adjusted to contain a more select range (C_6 to C_7, C_6 to C_8, C_7 to C_8, and so forth) of hydrocarbons to tailor the composition of the reformate product to the desired aromatic components. For either case, the basic reforming reactions are the same; however, for aromatics processing the C_6–C_7 reactions, which are slower and more difficult to promote, are often emphasized.

Naphthas from different crude sources vary greatly in their hydrocarbon composition and thus in their ease of reforming, which is generally determined by the mix of

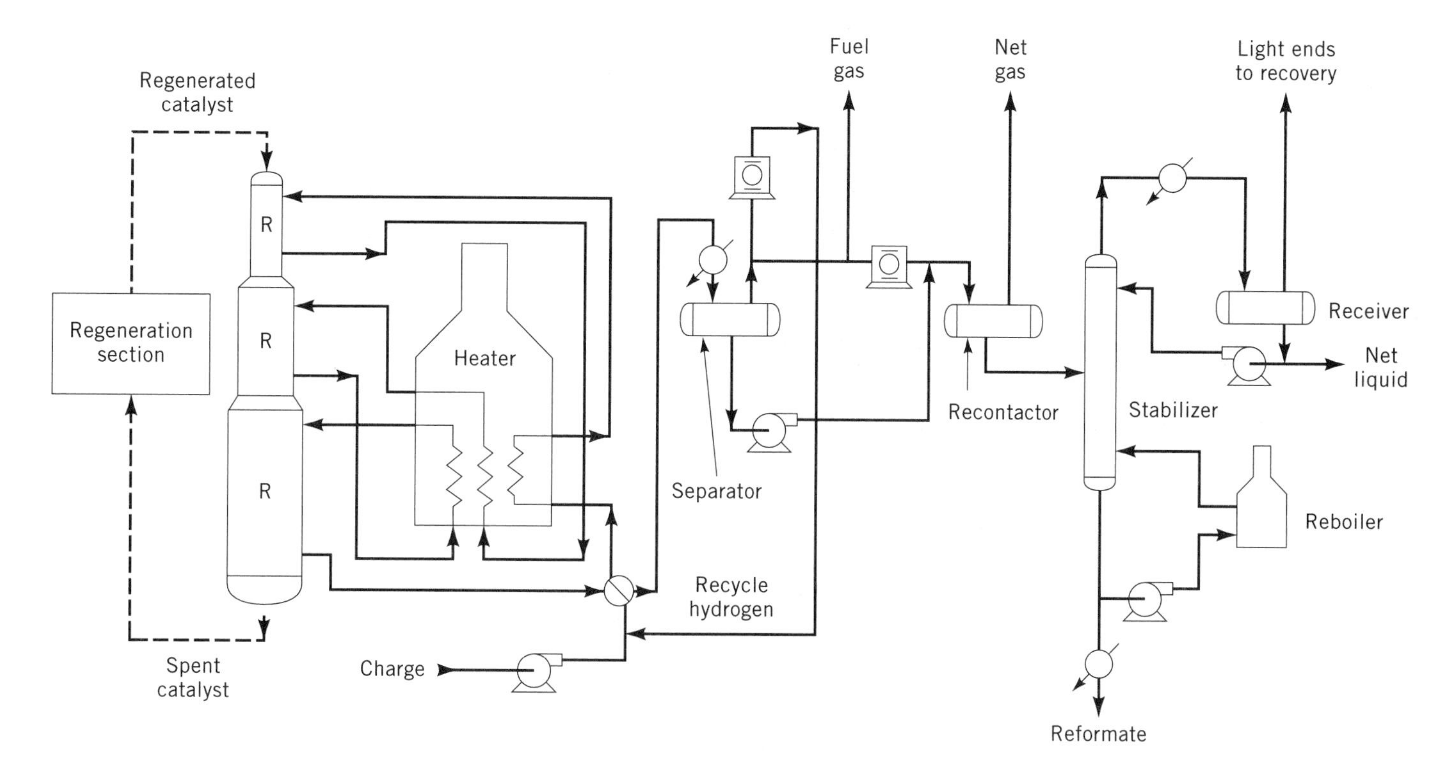

Figure 2. UOP CCR Platforming process. Courtesy of UOP.

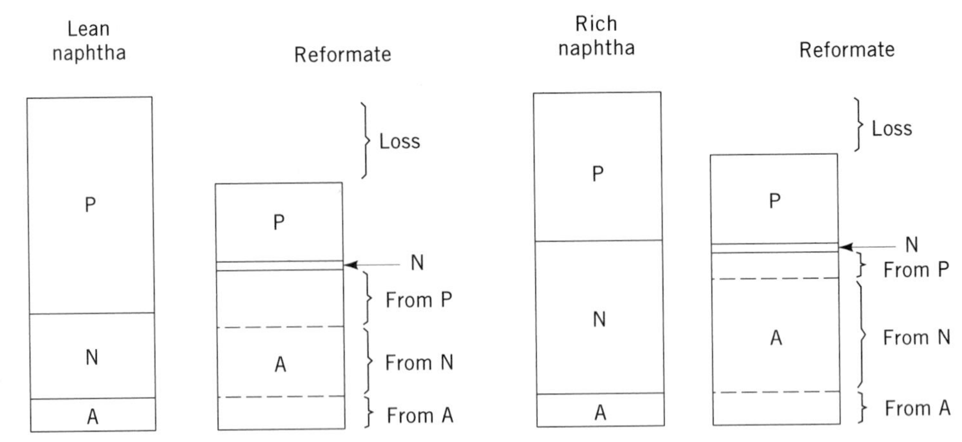

Figure 3. Conversion by hydrocarbon type. P, paraffins; N, naphthenes; A, aromatics. Courtesy of UOP.

paraffins, naphthenes, and aromatics in the feed. Aromatic hydrocarbons pass through the unit essentially unchanged. Naphthenes generally react quickly and are highly selective to aromatic compounds. Paraffin compounds are the most difficult to convert. The conversion of paraffins to aromatics is a multistep process that yields a slower and less selective reaction mechanism than for naphthene conversion. In lower severity operations, relatively little paraffin conversion is required, but higher severity operations require a significant degree of conversion.

Naphthas can be characterized as lean (high paraffin content) or rich (low paraffin content). The rich naphthas are easier to process in a reformer. Figure 3 illustrates the effect of the naphtha composition on the relative conversion of the feedstock under constant operating conditions in a reformer. The rich naphthenic feedstock produces a greater volumetric yield of reformate than does the lean feedstock.

Reforming Reactions

Reforming reactions can be broadly classified under four categories: dehydrogenation, isomerization, dehydrocyclization, and cracking (including dealkylation and demethylation). The extent to which each of the reactions occurs for a given reforming operation depends on the feedstock quality, operating conditions, and catalyst type.

Because of the many paraffin and naphthene isomers occurring in the reformer feed, multiple reforming reactions are taking place simultaneously in the reforming reactor. The rates of reaction vary considerably with the carbon number of the reactant, and so these multiple reactions occur in series and in parallel to each other:

$$\text{S}\text{—R} \rightleftharpoons \text{—R} + 3\text{H}_2$$

Dehydrogenation of Naphthenes. The principal reforming reaction in producing an aromatic from a naphthene (either a alkylcyclopentane or a alkylcyclohexane) is the dehydrogenation of an alkylcyclohexane:

$$\text{S}\text{—R} \rightleftharpoons \text{S}\text{—R}'$$

This reaction takes place rapidly and proceeds essentially to completion. Because of this reaction, which produces hydrogen as well as aromatics, naphthenes are the most desirable component in a reformer feedstock. The reaction is highly endothermic, is favored by high reaction temperature and low pressure, and is promoted by the metal catalyst function.

Isomerization of Paraffins and Naphthenes. The isomerization of an alkylcyclopentane to an alkylcyclohexane must take place before an alkylcyclopentane can be converted to an aromatic:

$$\text{R—C—C—C—C} \rightleftharpoons \text{R—C—}\underset{}{\overset{\text{C}}{\overset{|}{\text{C}}}}\text{—C}$$

The reaction involves ring rearrangement and thus the possibility for ring opening (to form a paraffin) exists. Paraffin isomerization occurs rapidly at commercial operating temperatures:

$$\text{R—C—C—C—C} \rightleftharpoons \text{S}\text{—R}' + \text{H}_2$$
$$\text{R—C—C—C—C} \rightleftharpoons \text{S}\text{—R}'' + \text{H}_2$$

Thermodynamic equilibrium slightly favors the more highly branched isomers. Thus this reaction makes a contribution to octane improvement. Isomerization reactions are promoted by the acid-catalyst function.

Dehydrocyclization of Paraffins. The dehydrocyclization of paraffins is the most difficult of the reforming reactions to promote:

$$\mathrm{R{-}C{-}\underset{}{\overset{\displaystyle C}{\overset{|}{C}}}{-}C + H_2 \rightarrow RH + C{-}\underset{H}{\overset{\displaystyle C}{\overset{|}{C}}}{-}C}$$

This reaction consists of the difficult molecular rearrangement of a paraffin to a naphthene. The paraffin cyclization reaction becomes easier with the increasing molecular weight of the paraffin because the probability of ring formation increases. Somewhat mitigating this effect is the greater likelihood of the heavy paraffins to hydrocrack. Dehydrocyclization reactions are favored by low pressure and high temperatures and require both metal and acid catalyst functions.

Cracking. Because of the difficulty of naphthene isomerization and paraffin cyclization reactions and the fact that a catalyst needs an acid function, the possibility of paraffin hydrocracking is significant:

$$\mathrm{R{-}C{-}C{-}C{-}C + H_2 \rightarrow R{-}C{-}C{-}CH + CH_4}$$

and

$$\mathrm{C_6H_5{-}R{-}C + H_2 \longrightarrow C_6H_5{-}RH + CH_4}$$

Paraffin hydrocracking is favored by high temperature and high pressure. When paraffins crack and disappear from the gasoline boiling range, the remaining aromatics become concentrated in the product, and hence the octane is increased. However, the hydrocracking reaction consumes hydrogen and reduces net liquid product yield.

Demethylation reactions generally occur only in severe reforming operations (high temperature and pressure):

$$\mathrm{C_6H_5{-}R + H_2 \longrightarrow C_6H_5{-}R' + R''}$$

This reaction is metal catalyzed and is favored by high temperature and pressure. The addition of a second metal (as in some bimetallic catalysts) or sulfur can help to inhibit this reaction by attenuation of the metal catalyst function.

The dealkylation of aromatics (Fig. 4) is similar to aromatic demethylation. The only difference is the size of the fragment removed from the ring. If the alkyl side chain is large enough, the reaction is similar to paraffin cracking. The dealkylation reaction is favored by high temperature and high pressure.

Relative Reaction Rate

The primary reactions for the C_6 and C_7 paraffins occur at vastly different rates. The hydrocracking rate for hexane is at least three times that for dehydrocyclization so that only a small fraction of normal hexane is converted into aromatics. Normal heptane hydrocracks at about the same rate as normal hexane so that the faster rate of dehydrocyclization (about four times that of hexane) means that a substantially greater conversion of normal heptane to aromatics occurs than for hexane.

Reactions of naphthenes in the feedstock show significant differences between the alkylcyclopentanes and the alkylcyclohexanes. The alkylcyclohexanes dehydrogenate rapidly and nearly completely to aromatics. The alkylcyclopentanes react more slowly and follow two competing reaction paths: isomerization to an alkylcyclohexane followed by dehydrogenation or decyclization to form paraffins.

The ratio of alkylcyclopentane isomerization rate to total alkylcyclopentane reaction rate illustrates the expected selectivity to form aromatics. At low pressure, this ratio is 0.67 for methylcyclopentane and 0.81 for dimethylcyclopentane. The relative ease of isomerization increases with increasing carbon number.

The conversion of hydrocarbon types as a function of position in the catalyst bed is shown in Figures 5 to 7 for a moderate-severity reforming operation. The feedstock is a rich BTX (benzene, toluene, and xylene) naphtha with a paraffin, naphthene, and aromatics (PNA) content of 42, 34, and 24 wt %, respectively. Total aromatics concentration increases while the concentration of naphthenes and paraffins decrease as they undergo conversion (Fig. 5). The high rate of conversion of cyclohexanes is shown by the rapidly decreasing concentration of naphthenes in the first 30% of the catalyst bed. The remaining naphthene

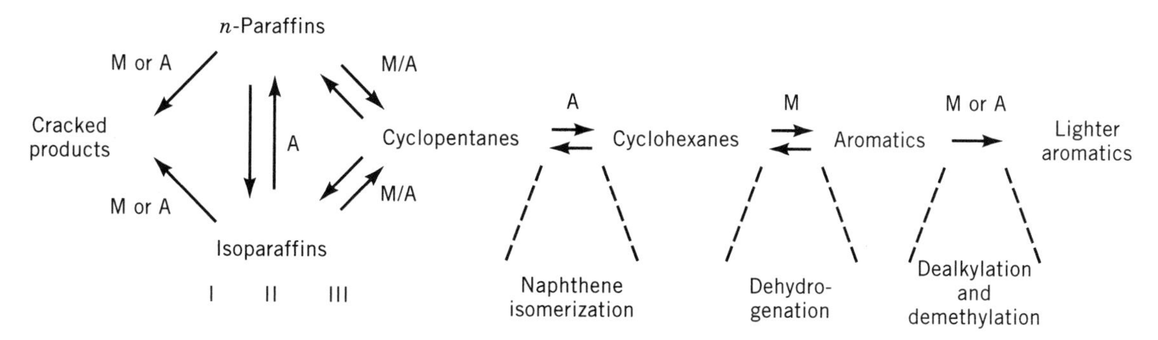

Figure 4. Generalized Platforming reaction scheme. (I), hydrocracking and demethylation (M); (II), paraffin isomerization; (III), dehydrocyclization; predominant active sites: A, acid; M, metal.

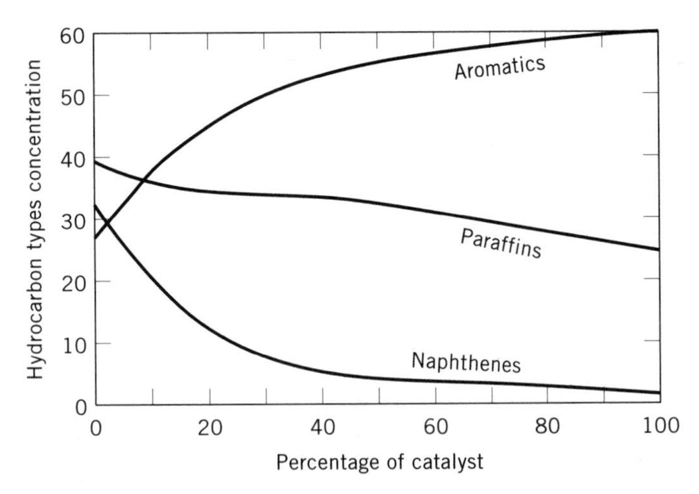

Figure 5. Hydrocarbon-type profiles. Courtesy of UOP.

conversion takes place at a slower rate and is indicative of cyclopentane conversion and dehydrocyclization of paraffins through a naphthene intermediate. At the reactor outlet, naphthene concentration approaches a low steady-state value, representing the naphthene intermediary present in paraffin dehydrocyclization. Paraffin conversion is nearly linear across the reactor bed.

Figure 6 illustrates the conversion of the three reactive species in the reforming feedstock. The relative rates of conversion are markedly different. In the first 20% of the catalyst, 90% of the cyclohexanes are converted, but conversion is only 15% for cyclopentanes and 10% for paraffins. Cyclopentanes are much less reactive than cyclohexanes.

Figure 7**a** shows the relative reaction rate of cyclopentanes by carbon number. The heavier components, which have a greater probability of isomerizing from a five- to a six-carbon ring, convert more readily than do the lighter components. The most difficult reaction, the conversion of paraffins, is characterized by carbon number in Figure 7b. Again, as with the alkylcyclopentanes, the heavier components convert more readily than do the lighter paraffins. The relative ease for conversion associated with increasing carbon number for alkylcyclopentanes and paraffins helps to explain why higher boiling range feedstocks are easier to process.

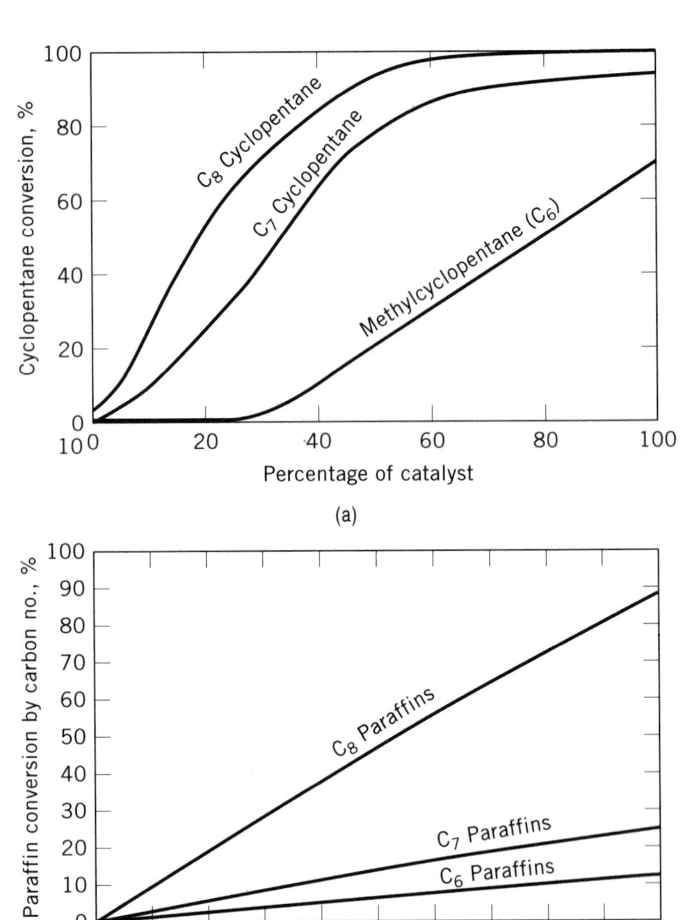

Figure 7. (**a**) Cyclopentane conversion by carbon number. (**b**) Paraffin conversion by carbon number. Courtesy of UOP.

In summary, paraffins have the lowest reactivity and selectivity to aromatics. Alkylcyclopentanes are more reactive and selective than are the paraffins but still yield a significant loss to nonaromatic products. Alkylcyclohexanes are converted rapidly and quantitatively to aromatics and make the best reforming feedstock. In addition, the removal of the lighter hydrocarbons (C_6 paraffins and alkylcyclopentanes) from the feedstock and their replacement with easier-to-convert C_9 components improves the ease of reforming operations.

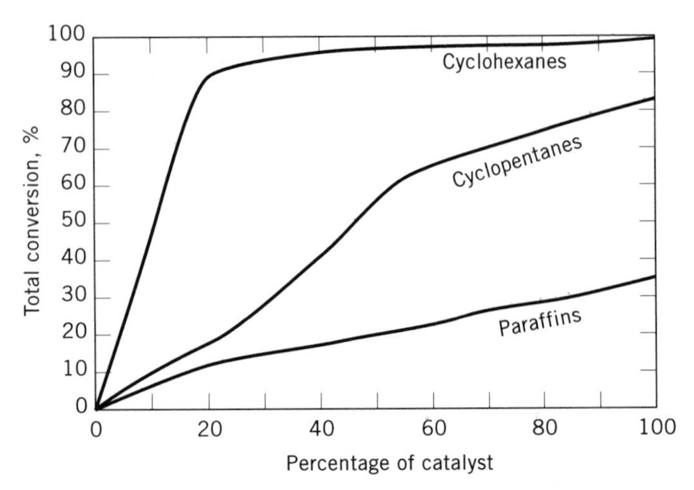

Figure 6. Reactant-type conversion profiles. Courtesy of UOP.

Heats of Reaction

Typical heats of reaction for the three broad classes of reforming reactions are

Reaction	ΔH, kJ/mol H_2
Paraffin to naphthene	+43.93 (endothermic)
Naphthene to aromatic	+70.71 (endothermic)
Hydrocracking	−56.48 (exothermic)

The dehydrocyclizaiton of paraffins and the dehydrogenation of naphthenes are endothermic. Evidence of this fact is found in commercial reforming units, where the reactor temperature differentials are large and negative across the first two reactors, which is where the majority of these reactions take place. In the final reactor, where a combination of paraffin dehydrocyclization and hydrocracking takes place, the net heat effect in the reactor may be slightly endothermic or exothermic, depending on processing conditions, feed characteristics, and catalyst.

Reforming Catalysts

Catalytic reforming implies that at the heart of the process are the catalytically controlled and promoted chemical reactions for aromatics production. Reforming catalysts are evaluated on three broad criteria: selectivity, activity, and stability. The history of catalyst development for the last 50 yr has been one of improvement in all three areas.

Selectivity. Catalyst selectivity can be easily described as how much of the desired product can be yielded from a given feedstock. Usually, the selectivity of one catalyst is compared with that of another. At constant operating conditions and feedstock properties, the catalyst that can yield the greatest amount of reformate at a given octane in motor fuel applications or the greatest amount of aromatics in a BTX operation has the greatest selectivity.

Activity. Activity is the ability of a catalyst to promote a desired reaction with respect to reaction rate, space velocity, or temperature. Again, activity is usually expressed in a relative sense in that one catalyst is more active than another. In motor fuel applications, activity is generally expressed as the temperature required to produce reformate at a given octane, space velocity, and pressure. A more active catalyst can produce the desired octane reformate at a lower temperature.

Stability. The ability of a catalyst to maintain its activity and selectivity over a reasonable time frame is a measure of its stability. If a catalyst has superior activity and selectivity but little stability, it has no use in the modern reforming industry.

Catalyst stability is expressed in vastly different terms, depending on the type of reformer operation. For semiregenerative units, the refiner often desires a catalyst that will complete a stable cycle in excess of 12 months before needing regeneration. For continuous regeneration reformers, the catalyst needs to be stable for much shorter periods of time, expressed in days. Consequently, different catalyst formulations are developed for different types of reforming operations.

Catalyst Development

Historically, reforming catalyst development has followed the path of maximizing the efficiency and balance found in a dual-function catalyst system. As shown earlier, some reforming reactions are acid catalyzed, and others are promoted by a metallic hydrogenation–dehydrogenation function.

Reforming catalysts are heterogeneous catalysts composed of a base support material (usually Al_2O_3) on which catalytically active metals are added. Early reforming catalysts were monometallic (platinum); and although these catalysts are able to produce a high octane, they are quickly deactivated by the formation of coke on the catalyst. New methods for regenerating the catalyst burn off the coke in an inert atmosphere with a low level of oxygen. The regeneration procedure essentially restored fresh catalyst performance.

During the 1960s, the need to produce higher octane reformate at increased liquid volume yields led to the development of bimetallic catalysts. The metals, eg, platinum, iridium, rhenium, and germanium, are typically added at levels of less than 1 wt % of the catalyst by using techniques that ensure a high level of dispersion of the metals over the surface of the catalyst. To ensure the dual functionality of the catalyst, a promoter such as chloride or fluoride is added. As outlined above, the metal function catalyzes dehydrogenation of paraffins and naphthenes and contributes to dehydrocyclization and isomerization. The acid function, provided by the chloride-support interaction, catalyzes isomerization, cyclization, and hydrocracking.

With the introduction of the UOP CCR Platforming process in 1971 (the first truly continuous catalyst regeneration reforming process) reforming catalyst development began a second parallel track to address the specific needs of the continuous process. The first UOP CCR unit used a conventional platinum-rhenium catalyst, but UOP quickly developed the R-30 series catalyst to provide higher yields of gasoline and hydrogen. Catalyst development for the continuous regeneration reformers has focused on the following broad areas:

- Lower coke make to reduce regenerator investment.
- Higher tolerance to multiple regeneration cycles to maximize catalyst life and minimize catalyst costs; this area includes improvement (reduction) in the rate of surface area decline, important because reduced catalyst surface area makes evenly dispersing the metals on the catalyst surface more difficult.
- Improved catalyst strength to reduce attrition in the unit as a result of catalyst transfer.
- Metals optimization developments to reduce the metal content of the catalyst and thus reduce the refinery working capital requirement.

PROCESS VARIABLES

This section describes the major process variables and their effect on unit operations. The variables addressed are catalyst type, reactor temperature, space velocity, reactor pressure, hydrogen : hydrocarbon ratio (H_2 : HC), and charge stock properties. The relationship between the variables and process performance is generally applicable to both semiregenerative and continuous regeneration modes of operation.

Catalyst Type

Catalyst selection is usually tailored to the refiner's individual needs. A particular catalyst is typically chosen to meet the yield, activity, and stability requirements of the refiner. Some basic differences in catalyst types can effect other process variables. For instance, the required temperature to produce a given octane is directly related to the type of catalyst.

Reactor Temperature

The primary control for product quality in a reforming process unit is the temperature at which the catalyst beds are held. Reforming catalysts are capable of operating over a wide range of temperatures, and a refiner can change the octane of the reformate by adjusting the heater outlet temperatures.

The reactor temperature can be expressed as the weighted average inlet temperature (WAIT), which is the summation of the product for the fraction of catalyst in each reactor multipled by the inlet temperature of the reactor, or as the weighted average bed temperature (WABT), which is the summation of the product for the fraction of catalyst in each reactor multiplied by the average of its inlet and outlet temperatures. Both measurements have a practical use in industry although neither perfectly describes the average catalyst temperature. Temperatures in this article are the WAIT calculation. Typically, semiregenerative reformers operate over temperature cycles of −1.1 to 4.4°C and have a WAIT range of 490 to 525°C. Continuous regeneration units operate at temperatures (WAIT) of 525 to 540°C.

Space Velocity

Space velocity is a measurement of an amount of naphtha processed over a given amount of catalyst over a set length of time. Space velocity is an indication of the residence time of contact between reactants and catalyst. When the hourly volume charge rate of naphtha is divided by the volume of catalyst in the reactors, the resulting quotient, expressed in units of h^{-1}, is known as the liquid hourly space velocity (LHSV). Alternatively, if the weight charge rate of naphtha is divided by the weight of catalyst, the resulting quotient, again expressed in units of h^{-1}, is the weighted hourly space velocity (WHSV). Although both terms are expressed in the same units, the calculations yield different values. The choice is using LHSV or WHSV is usually based on the custom of expressing feed rates in a given location. If charge rates are normally expressed in barrels per stream day (BPSD) then LHSV is typically used. If the rates are expressed in terms of metric ton per day (MTD), then WHSV is preferred.

Space velocity determines the limits of product quality. The greater the space velocity, the lower the maximum product octane achievable. Space velocity works directly with reactor temperature to set the octane of the product. The greater the space velocity, the higher the temperature required to produce a given product octane. If refiners wish to increase the severity of a reformer operation, they can either increase the reactor temperature or lower the space velocity, which would require cutting back on the reactor charge rate.

Reactor Pressure

Average reactor operating pressure is generally referred to as reactor pressure. For most purposes, a close approximation of the average can be obtained by using the last reactor inlet pressure. The separator pressure is generally not used because it can be greatly affected by what type of heat exchange is present between the outlet of the last reactor and the separator. Even for a given unit with constant exchange equipment, the pressure can vary substantially with feed rate, recycle gas rate, and so forth.

The reactor pressure affects reformer yields, reactor temperature requirement, and catalyst stability. Reactor pressure has no theoretical limitations although practical operating constraints have led to an historical range of operating pressures from 50 to 700 psig. Decreasing the reactor pressure increases hydrogen and reformate yields, decreases the required temperature to achieve product quality, and shortens the catalyst cycle (increases the catalyst coking rate). Because of the high coking rates associated with operating pressures as low as 50 psig, continuous catalyst regeneration is required when operating at these low pressures.

Hydrogen : Hydrocarbon Ratio

The hydrogen : hydrocarbon (H_2 : HC) ratio is the ratio of moles of hydrogen in the recycle gas to moles of naphtha charged to the unit. Recycle hydrogen is necessary to maintain catalyst stability and helps to sweep reaction products from the catalyst. An increase in the H_2 : HC ratio increases the linear velocity of the combined feed and supplies a greater heat sink for the endothermic heat of reaction. Increasing the ratio also increases the hydrogen partial pressure and reduces the coking rate, thereby increasing the stability with little or no effect on product quality or yields.

Charge Stock Properties

Charge stock characterization by hydrocarbon type (paraffin, naphthene, and aromatics) and a distillation analysis are required to predict reformer unit performance in terms of total liquid reformate and hydrogen yield as well as catalytic activity and stability. Charge stocks with a low initial boiling point (IBP) generally contain a significant amount of C_5 material, which does not react in the reformer and only serves to dilute the final product. For this reason, feedstocks are generally C_6^+ naphthas. The end point of the charge stock is usually set by the gasoline specifications for the refinery with the realization that a significant rise (typically 30 to −1.1° to 15.6°C) in end point takes place between naphtha charge and reformate product.

The effect of hydrocarbon types in the charge stock on aromatics yield has been discussed earlier and can be further illustrated by examining a broad range of charge stock compositions. Licensors typically develop a large database of feedstocks that have been analyzed and tested under controlled conditions to characterize expected reforming yields over a range of octanes. This database allows yields to be predicted for future charge stocks of known composition. Four charge stocks of widely varying compositions were chosen from such a database and are summarized in Figure 8.

The charge stock range chosen covers lean through rich feeds. The aromatics plus cyclohexanes content is a measure of their ease of conversion, and the paraffins plus cyclopentanes content indicates the difficulty of reforming reactions. The effect of feedstock composition on aromatics yield is shown in Figure 9. Increasing conversion leads to an increase in total aromatics yield for each of the feedstocks, and feeds that are easier to process produce the highest yield of aromatics at any level of conversion.

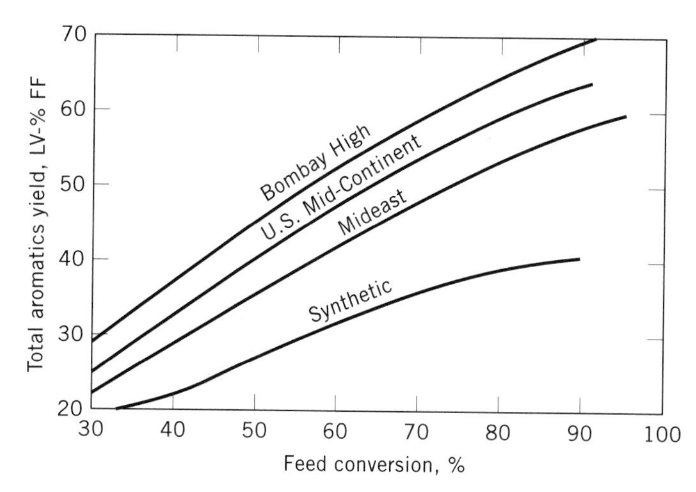

Figure 9. Feedstock conversion and aromatics yield. Courtesy of UOP.

PRODUCTS

Modern reforming operations focus on two different intermediate product markets: (*1*) the gasoline market, which requires the production of a high octane blending component to be used in the preparation of final gasoline blends, and (*2*) the aromatics market, which requires the production of a petrochemical feedstock containing a large percentage of aromatics (chiefly benzene, toluene, and xylene).

Gasoline Market

The impact of the reforming process on the gasoline market has been substantial. The evolution of the reforming process to produce a higher quality, higher yield of reformate over time has mirrored the growing pool octane demand for the U.S. refining industry. Figure 10 shows the pool octane history of the U.S. refining industry on a research octane number (RON) clear basis. With the introduction of the modern reforming process using platinum-containing catalysts, the refinery of the 1950s was able to meet the demand for higher octane.

During the 1950s, the lead susceptibility of the pool was high, and a number of unrefined components were included in the pool. By the 1960s, leaded premium gasoline was marketed at 99 RON, and any unreformed naphtha at 60 RON used in blending was a clear octane drag on the pool. During this period, higher compression engines and a growing demand for higher octane gasoline caused the refinery gasoline pool to increase from 83 to 86 RON clear. The demand for increased product quality occurred at the same time as the demand for increased volume. Taking the lead susceptibility into account, required reformate RON clear through the 1950s and 1960s rose from 80 to approximately 95 RON clear.

During the 1970s, the U.S. refining industry adjusted to the mandated phaseout of lead in gasoline. Figure 11 shows the transition of gasoline grades. With the phaseout of lead use in gasoline, the RON clear pool requirement again rose, this time from 86 to 90 between 1970 and 1985. As in the previous periods, premium-grade octane controlled the reformer-octane requirements. Figure

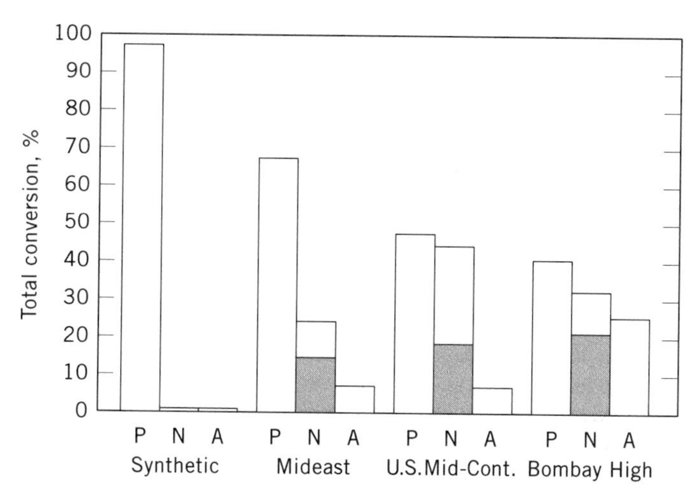

Figure 8. Naphtha characterization by hydrocarbon types. □, cyclopentanes; ▨, cyclohexanes. Courtesy of UOP.

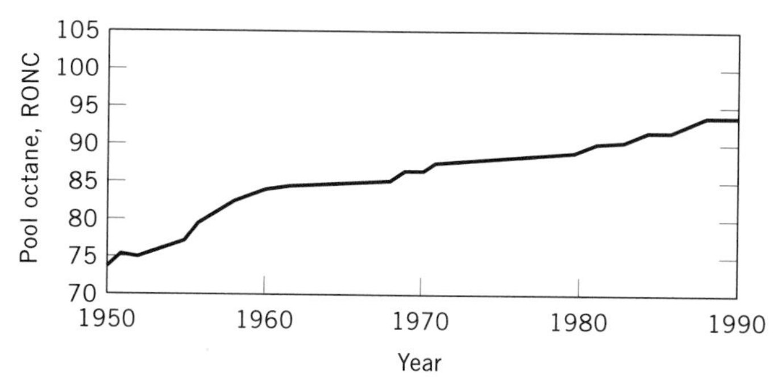

Figure 10. Pool octane history of U.S. refining industry. Courtesy of UOP.

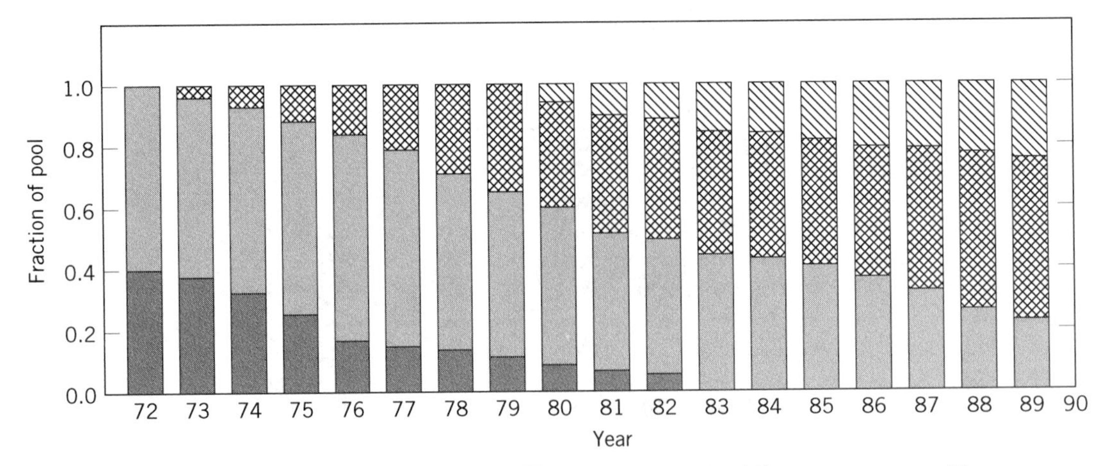

Figure 11. Pool transition to unleaded. ▨, leaded premium; □, leaded regular; ⊠, unleaded regular; ▧, unleaded premium. Courtesy of UOP.

12 shows trends in gasoline pool octane requirements as well as the required octane from the reforming unit during the leaded and unleaded gasoline eras.

As evidenced by the preceding discussion, reforming operations have been required to improve consistently product yields and quality to keep up with the changes in demand and quality for blended gasolines. As an example, four different UOP Platforming units (two semiregenerative and two continuous regeneration units) processing the same feedstock show the progression of Platforming technology. The operating conditions and yields for the four units are shown in Table 1. The reformate yield in these units has remained largely unchanged even though the product octane demands have increased by 12 numbers.

Demands on the modern-day reformer have again shifted and are markedly different for the domestic U.S. refiner than for the foreign refiner. Outside the United States much of the world is undergoing a lead phaseout in gasoline that is similar to what occurred in the United States in the 1970s through early 1980s. As in the United States, the refiners want increased octane from the reformer to replace the contribution from lead addition.

Inside the United States, the 1990 Clean Air Act (CAA) Amendments target a substantial reduction in emissions of toxics and volatile organic compounds (VOCs) from vehicles and their fuels. These regulations, which have a direct impact on reforming operations, are summarized as follows:

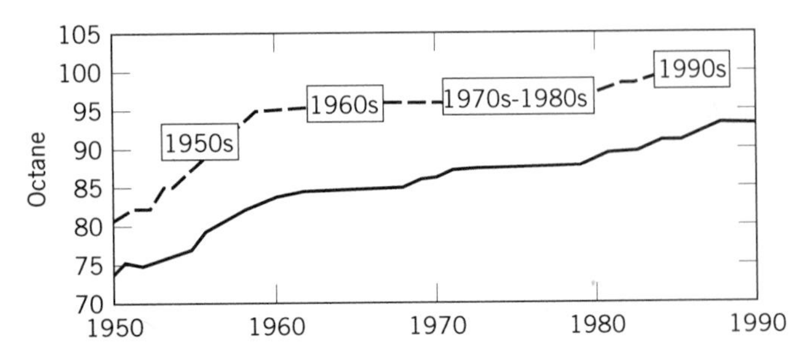

Figure 12. Required reformer octane. Courtesy of UOP.

- *Reduction in Reformer Octane Demand.* The CAA Amendments require the use of oxygenates in gasoline blends to provide 2 wt % oxygen in gasoline marketed in ozone nonattainment areas. If, eg, methyl *tertiary*-butyl ether (MTBE) is chosen as the blended oxygenate, a gasoline pool containing 11 LV % MTBE (blending RON = 117) can reduce the required reformer severity from the 98 to 100 range to the 92 to 94 C_5^+ RON clear range when producing gasoline solely for use in an ozone nonattainment area.
- *Hydrogen Production.* Hydrogen production and reformer operating severity are directly related. A drop in required reformer operating severity leads to a reduction in hydrogen production that occurs at the same time that hydrogen demand is increasing in other areas of the refinery. The new environmental regulations mandating the production of cleaner burning transportation fuels increases the demand for fuel hydrotreating processes that increase hydrogen consumption.
- *Benzene Reduction.* The CAA Amendments specify a maximum level of 1 LV % benzene in the gasoline pool in 1995. Typical reformate streams contain 3–5 LV % benzene and contribute 50–75% of the gasoline pool benzene. Benzene production in a reformer is a function of feed composition, operating severity, and operating pressure. The reduction in reformer severity from 100 to 92 C_5^+ RON clear can significantly decrease the yield of benzene but may not be enough to meet the new gasoline specifications.

Clearly, domestic reforming operations are going to face some new challenges in the reformulated gasoline environment. Some of the potential changes in reformer operations are outlined elsewhere in this article.

Petrochemical Market

The demand for BTX intermediates has grown steadily over the more than 40-yr life of the modern reforming process and is expected to continue through the 1990s, as

Table 1. Operating Conditions and Yields of the Historic Units

Operating Conditions and Yields	1950s (Semiregenerative)	1960s (Semiregenerative)	1970s (Continuous)	1990s (Continuous)
Operating conditions				
Pressure, kPa[a]	3450	2070	860	345
Capacity, BPSD	10,000	15,000	20,000	40,000
LHSV, h^{-1}	0.9	1.7	2.0	2.0
H_2 : HC, mol : mol	7.0	6.0	2.5	2.5
RON clear	90	94	98	102
Yields				
C_5^+, LV %	80.8	81.9	83.1	82.9
H_2, SCFB	387	679	1,175	1,622
Total aromatics, LV %	38.0	45.0	53.7	61.6
Catalyst life, months	12.9	12.5		
Regeneration, kg/h			317.5	2,041.2

[a] To convert kPa to psi, multiply by 0.145.

the result of new grassroots aromatics complexes being forecast for several developing nations. Aromatics have been produced by catalytic reforming since the introduction of the process, when semiregenerative units were operated at pressures as high as 700 psig. Aromatics productivity increased as the reactor operating pressures were reduced to 300 psig with the introduction of the more stable bimetallic catalysts in the 1960s. As continuous regeneration reformers were brought onstream in the 1970s, aromatics productivity again increased as reactor operating pressure was initially reduced to 690 kPa and then further to the ultralow level of 345 kPa in 1993.

Ultralow pressure reforming offers petrochemical producers the highest possible yields of aromatics per ton of naphtha processed. The benefit of ultralow pressure reforming is a result of:

- Superior selectivity in the conversion of paraffins and naphthenes to aromatics at 50 psig.
- High overall conversion made possible by the use of continuous catalyst regeneration.
- Higher yields from catalysts designed for the continuous regeneration process than from typical semiregenerative platinum–rhenium catalyst systems.

Figure 13 illustrates the difference in aromatic yields for three reforming operations processing the same lean (P/N/A = 58.9/29.3/11.8 wt %) BTX naphtha. The three operations are a 200 kPa semiregenerative unit using a platinum–rhenium catalyst and two continuous catalyst regeneration units operating at pressures of 860 and 345 kPa, respectively. As can be seen from the figures, lowering the reactor pressure from 207 to 345 psig leads to a 20 wt % increase in aromatics yield from 52 to 72 wt % of fresh feed. The reduction in pressure leads to increases in aromatics yield for each carbon number, although the increase is greater for the heavier components. This increase is directly related to less dealkylation of heavy aromatics occurring at the reduced operating pressures.

ENVIRONMENTAL ASPECTS

The reforming process produces aromatic compounds. The most toxic hydrocarbon present in the unit is benzene, whose health effects from exposure are well documented. Toluene, xylene, and heavier aromatics are also continu-

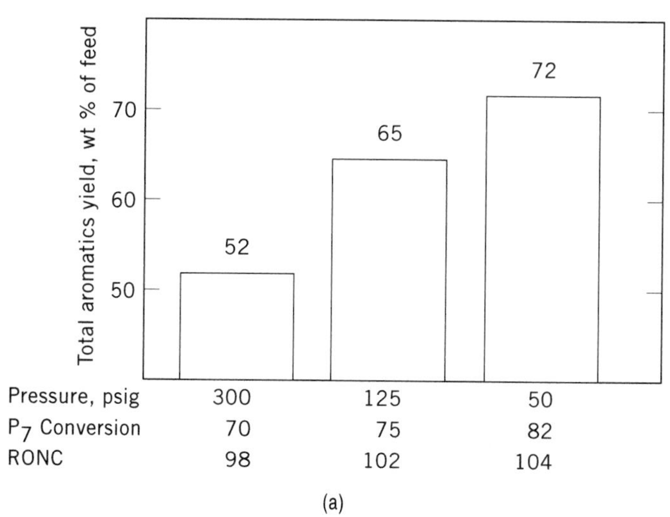

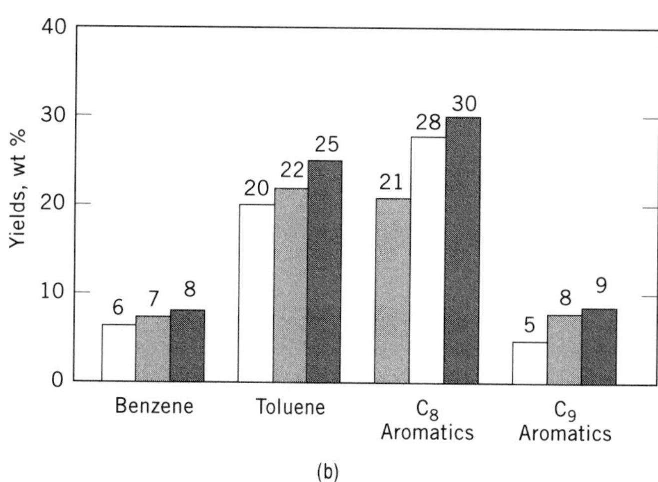

Figure 13. (**a**) Total aromatics production. (**b**) Total aromatics breakdown. Courtesy of UOP.

ously present in the process during normal operations. These compounds are moderately toxic and are believed to not have the destructive effect on health that benzene does.

The aromatic compounds contained in reformate are fundamental octane components of gasoline blends and essential building blocks for petrochemical manufacture. The refining industry has a long history of handling such compounds safely. The ability to produce aromatic compound efficiently has allowed the United States to phase out the use of lead as an antiknock additive in commercial gasolines. Much of the rest of the world is also making the transition to a lead-free transportation-fuel base and, like the United States, is looking to increased use of high octane reformate in gasoline to compensate for lost octane from the removal of lead additives.

The U.S. CAA Amendments of 1990, described in the previous section, will have an effect on future reforming operations and the production of aromatic compounds for motor fuel use. The legislated use of oxygenates in the gasoline pool can lead to a reduced octaine demand from the reformer, and this reduced demand would cause a reduction in benzene-aromatic compounds contributed to the pool by the reformer. Even with the reduction in benzene production accompanying any reduction in reformer operating severity, most refiners will need to make additional operating changes to comply with the new benzene specification. The refiner has several options to reduce reformer benzene production further.

The refiner can prefractionate the feed to remove the benzene and benzene precursors from the naphtha charge or postfractionate the reformate to remove benzene in a reformate splitter. If prefractionation is used, the benzene in the light fraction may be saturated in either an isomerization or saturation unit. However, a small amount of benzene is still present in the reformate product as a result of dealkylation reactions.

If postfractionation is employed; the benzene-rich light reformate needs to be processed to remove or saturate the benzene before blending the stream back into the gasoline pool. If a market is available, the benzene-rich stream can be sent to an extraction unit to recover petrochemical-grade benzene.

The loss of hydrogen production from the reformer as the result of decreased operating severity hurts the refinery hydrogen balance at a time when demand for hydrogen production is increasing because of other environmental regulations. Chief among these regulations are the increased quality specifications for low sulfur diesel fuel and the possible saturation of benzene to achieve specified product quality.

The refiner has several options to address the hydrogen management issue in the reformulated gasoline refinery. The easiest option to evaluate is the addition of incremental, or grassroots, hydrogen plant capacity. However, these solutions may not always be the most cost-effective for meeting the shortfall in hydrogen production. Obtaining the required environmental approval for new refinery process unit construction may also be difficult. Depending on the current reforming unit configuration, various options exist to increase the hydrogen production within the unit. The four main alternatives for semiregenerative and cyclic units are the following:

- Change to the most selective catalyst system.
- Change catalyst and reduce operating pressure (minimal revamp).
- Revamp to the lowest possible pressure (maximum hydrogen production).
- Build a new low pressure, continuous regeneration unit to consolidate existing fixed-bed capacity.

Choosing the right option depends highly on the individual refinery situation. For all cases, operating the reformer at lower pressure decreases benzene production and increases hydrogen production. This result counters the adverse effects of prefractionation and postfractionation on the refinery hydrogen balance. The reduced benzene content of the reformate is due to increased selectivity, reduced dealkylation, and reduced hydrocracking reactions at the lower operating pressure.

Organic chlorides are added to the process to maintain good metals dispersion on the catalyst. Ongoing data collection and assessment of the toxicity and health risk involved in the handling and use of the various chloriding agents may result in further legislation limiting the use of some or all of these compounds. All effluent streams from the reformer are treated to remove chlorides and prevent their discharge into the surrounding environment.

FUTURE DEVELOPMENTS

Much research is devoted to continuing the improvement in reforming catalyst performance. The following highlights a few of the efforts:

- *Improvements in Aromatic Selectivity.* For several years, patents and publications have documented the use of nonacidic zeolitic catalysts. These catalysts offer superior selectivity for paraffin conversion to aromatics, especially in the BTX range.
- *Surface Area Stability.* Catalysts are being designed to maintain surface area stability and thus extend the ultimate life of catalysts for continuous catalyst regeneration systems.
- *Higher Yields.* Although much of the "easy" work has been done over the more than 40-yr life of the reforming process, research continues into improving the overall yield of the process and improving the yields of desired components (eg, hydrogen, benzene, and xylene).

Improvements in the reforming process focus on improving both normal operating procedures and regeneration procedures. Operating design and procedure improvements seek to improve hydrogen and liquid-product recoveries, energy efficiency, and ease of unit operations. Improvements in regeneration procedures seek to reduce the severity of regeneration operations (lower temperature) and thus extend catalyst life for all types of reforming operations.

CAUSTIC WASHING

JAMES SPEIGHT
Western Research Institute
Laramie, Wyoming

Caustic washing is the treatment of materials, usually products from petroleum refining, with solutions of caustic soda (see also PETROLEUM REFINING).

The treating of petroleum products by washing with solutions of alkali (caustic or lye) is almost as old as the petroleum industry itself (1). The industry discovered early that product odor and color could be improved by removing organic acids (naphthenic acids and phenols) and sulfur compounds (mercaptans and hydrogen sulfide) through the use of a caustic wash.

Thus it is not surprising that caustic soda washing (lye treatment) has been used widely on many petroleum fractions (1–4). In fact, it is sometimes used as a pretreatment for sweetening and other processes. A caustic wash consists of mixing a water solution of lye (sodium hydroxide or caustic soda) with a petroleum fraction. The treatment is carried out as soon as possible after the petroleum fraction is distilled, since contact with air forms free sulfur, which is very corrosive and difficult to remove. The lye reacts with any hydrogen sulfide present to form sodium sulfide, which is soluble in water:

$$H_2S + 2\,NaOH \rightarrow Na_2S + 2\,H_2O$$

or with mercaptans followed by oxidation to the less nocuous disulfides:

$$RSH + NaOH \rightarrow NaSR + H_2O$$

$$4\,NaSR + O_2 + 2\,H_2O \rightarrow 2\,RSSR + 4\,NaOH$$

In the case of liquids from coal, which often contain high percentages of phenolic species (5), caustic washing can be employed to remove the phenols as phenolate salts:

$$C_6H_5 + NaOH \rightarrow C_6H_5O^-Na^+$$

which are water soluble and can be recovered at a later stage of the process.

GENERAL PROCESSING CONCEPTS

Caustic treating requires the control of the process through monitoring several process parameters. These parameters are provided below:

Treating Temperature

The usual range of temperatures is 30–50°C and the extraction of mercaptans is generally increased at the lower temperatures, whereas air-inhibitor sweetening is more rapid at higher temperatures.

Degree of Mixing

Treating effectiveness improves with mixing intensity but there is also the tendency for caustic to carry out of the settler at higher mixing rates. For treating units already in operation, the maximum acceptable mixing level should be determined by observing the amount of caustic entrainment at different recycle caustic rates and mix valve pressure drop settings.

Effect of Added Air

Efficient utilization of the air is not usually achieved; excess air is required. The amount of air should be at least 30 scf/lb of mercaptan instead of the stoichiometric amount (14.8 ft^3/lb of mercaptan sulfur converted.

Sweetening Inhibitor

Phenylenediamine ($H_2N.C_6H_4.NH_2$) inhibitors are generally used and the sweetening rate is usually increased by increasing the inhibitor addition rate.

Caustic Strength

Stronger caustic solutions generally provide more efficient extraction of sulfur species. However, to prevent severe caustic entrainment with conventional gravity settlers, it is recommended that the strength of the caustic be limited to a 25 wt %.

Number of Stages

Increasing the number of treating stages, especially in the form of countercurrent operation, augments the treating efficiency by adding mixing, reaction time, and the opportunity for caustic containing less mercaptides to work on the cracked gasoline.

PROCESSES

Caustic washing processes involves many configurations. For example, lye treatment is carried out in continuous treaters (Fig. 1) that consist of a pipe containing baffles or other mixing devices into which the oil and lye solution are both pumped. The pipe discharges into a horizontal tank where the lye solution and oil separate. Treated oil is withdrawn from near the top of the tank; lye solution is withdrawn from the bottom and recirculated to mix with incoming untreated oil. A lye-treating unit may be incorporated as part of a processing unit. For example, the overhead from a bubble tower may be condensed, cooled, and passed immediately through a lye-treating unit. Such a unit is often referred to as a worm-end treater, since the treater is attached to the particular unit as a point beyond the cooling coil or cooling worm.

Caustic solutions ranging from 5 to 20 wt % are used at 20–40°C and 34–276 kPa (5–40 psi). High temperatures and strong caustic are usually avoided because of the risk of color body formation and stability loss. Caustic-product treatment ratios vary from 1:1 to 1:10.

Spent lye is the term given to a lye solution in which about 65% of the sodium hydroxide content has been used by reaction with hydrogen sulfide (H_2S), light mercaptans (R-SH), organic acids ($R\text{-}CO_2H$), or mineral acids (H^+X^- or $H_2^{2+}X^{2-}$). A lye that is spent, as far as hydrogen sulfide

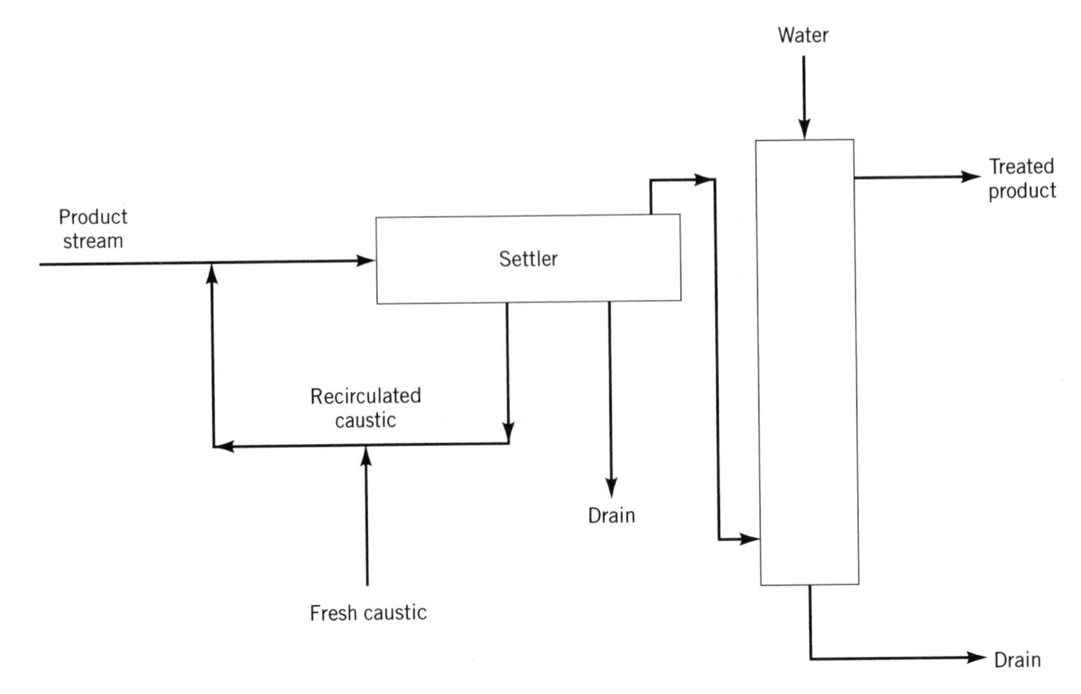

Figure 1. A caustic (lye) treating unit.

is concerned, may still be used to remove mineral or organic acids from petroleum fractions. Lye spent by hydrogen sulfide is not regenerated, but lye spent by mercaptans can be regenerated by blowing with steam, which reforms sodium hydroxide and mercaptans from the spent lye. The mercaptans separate as a vapor and are normally burned in a furnace to *destroy them*. Spent lye can also be regenerated in a stripper tower with steam, and the overhead consists of steam and mercaptans, as well as the small amount of oil picked up by the lye solution during treatment. Condensing the overhead allows the mercaptans to separate from the water.

On the other hand, a continuous two-stage, 25 wt % caustic treating and sweetening system (Fig. 2) involves extraction (first stage), followed by a combined extraction/air-inhibitor-sweetening step (second stage). The fresh, make-up, caustic solution is added to the second stage of this countercurrent treater and the rate of addition is set to give an acid oil content of at least 15 vol % in the second stage recirculating caustic to provide adequate solutizer. Caustic is transferred from the second stage to the first stage and is withdrawn from the first stage. Air and a sweetening inhibitor are added to the hydrocarbon stream between stages.

Nonregenerative caustic treatment is generally economically applied when the contaminating materials are low in concentration and waste disposal is not a problem. However, the use of nonregenerative systems is on the decline because of the frequently-occurring waste disposal problems that arise from environmental considerations, and because of the availability of numerous other processes that can effect more complete removal of contaminating materials.

Steam-regenerative caustic treatment (Fig. 3) is directed toward removal of mercaptans from such products as gasoline and low-boiling solvents (naphthas). The caustic is regenerated by steam blowing in a stripping tower. The nature and concentration of the mercaptans to be removed dictate the quantity and temperature of the process. However, the caustic gradually deteriorates because of the accumulation of material that cannot be removed by stripping: the caustic quality must be maintained by either continuous or intermittent discard, and/or by replacement of a minimum amount of the operating solution.

There are several other processes which have been given specific names and are noteworthy, not only because of their historical usage, but because of their continuous usage to the present time.

In the Sodasol process, the treating solution is composed of lye solution and for alkyl phenols (acid oils), which occur in cracked naphthas and cracked gas oils and are obtained by washing cracked naphtha or cracked gas oil with the lye solution. The lye solution, with solutizers incorporated, is then ready to treat product streams, such as naphthas and gasolines. The process is carried out by pumping a sour stream up a treating tower countercurrent to a stream of Sodasol solution that flows down the tower. As the two streams mix and pass, the solution removes mercaptans and other impurities, such as phenols and acids as well as some nitrogen compounds.

The treated stream leaves the top of the tower; the spent Sodasol solution leaves the bottom of the tower to be pumped to the top of a regeneration tower, where mercaptans are removed from the solution by steam. The regenerated Sodasol solution is then pumped to the top of the treatment tower to treat more material. A variation of the Sodasol process is the Potasol process, which uses potassium hydroxide instead of lye (sodium hydroxide).

Polysulfide treatment is a nonregenerative chemical treatment process used to remove elemental sulfur from refinery liquids. The treatment solution is prepared by

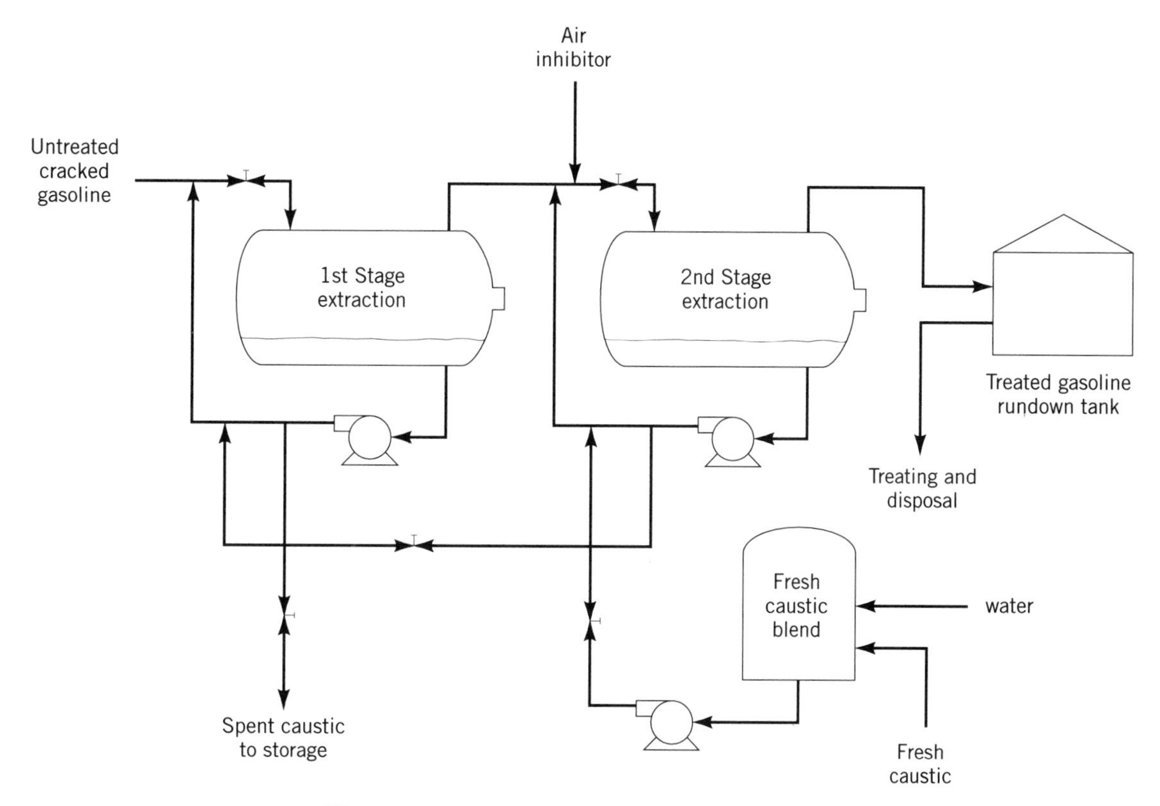

Figure 2. A two-stage caustic (lye) treating unit.

dissolving 1 pound of sodium sulfide (Na_2S) and 0.045 kgm (0.1 lb) of elemental sulfur in each gallon (3.785 L) of caustic. The sodium sulfide can actually be prepared in the refinery by passing hydrogen sulfide, an obnoxious refinery by-product gas, through caustic solution.

The solution is most active when the composition approximates Na_2S_2 to Na_2S_3 (or $Na_2S_{2.5}$), but activity decreases rapidly when the composition approaches Na_2S_4. When the solution is discarded, a portion (ca. 20%) is retained and mixed with fresh caustic-sulfide solution, which eliminates the need to add free sulfur. Indeed, if the material to be treated contains hydrogen sulfide in addition to free sulfur, it is often necessary simply to add fresh caustic.

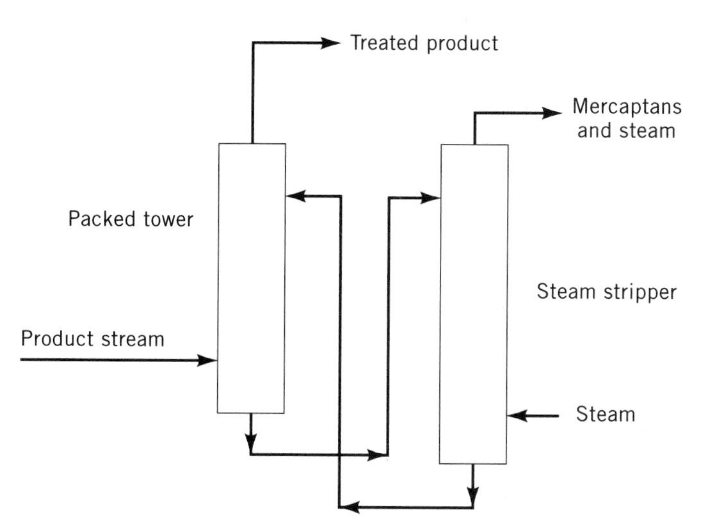

Figure 3. A steam-regenerative caustic treating unit.

The Dualayer distillate process (Fig. 4) uses caustic solution and cresylic acid (cresol, methylphenol). The process extracts organic acids, as well as mercaptans, from distillate fuels. In a typical operation, the Dualayer reagent is mixed with the distillate at about 55°C and passed to the settler, where three layers separate with the aid of electrical coagulation. The product is withdrawn from the top layer; the Dualayer reagent is withdrawn from the bottom layer, relieved of excess water, fortified with additional caustic, and recycled. The Dualayer gasoline process is a modification of the Dualayer distillate process in that it is used to extract mercaptans from liquefied petroleum gas (LPG), gasoline, and naphtha using the Dualayer reagents.

The ferrocyanide process (Fig. 5) is a regenerative chemical treatment for removing mercaptans from gasolines and from naphthas using caustic–sodium ferrocyanide reagent.

For example, gasoline is prewashed with caustic to remove hydrogen sulfide, and then washed countercurrently in a tower with the treating agent. The spent solution is mixed with fresh solution containing ferricyanide; the mercaptans are converted to insoluble disulfides and are removed by a countercurrent hydrocarbon wash. The solution is then recycled, and part of the ferrocyanide is converted to ferricyanide by an electrolyzer.

The Mercapsol process (Fig. 6) is another regenerative process for extracting mercaptans by means of sodium (or potassium) hydroxide, together with cresols, naphthenic

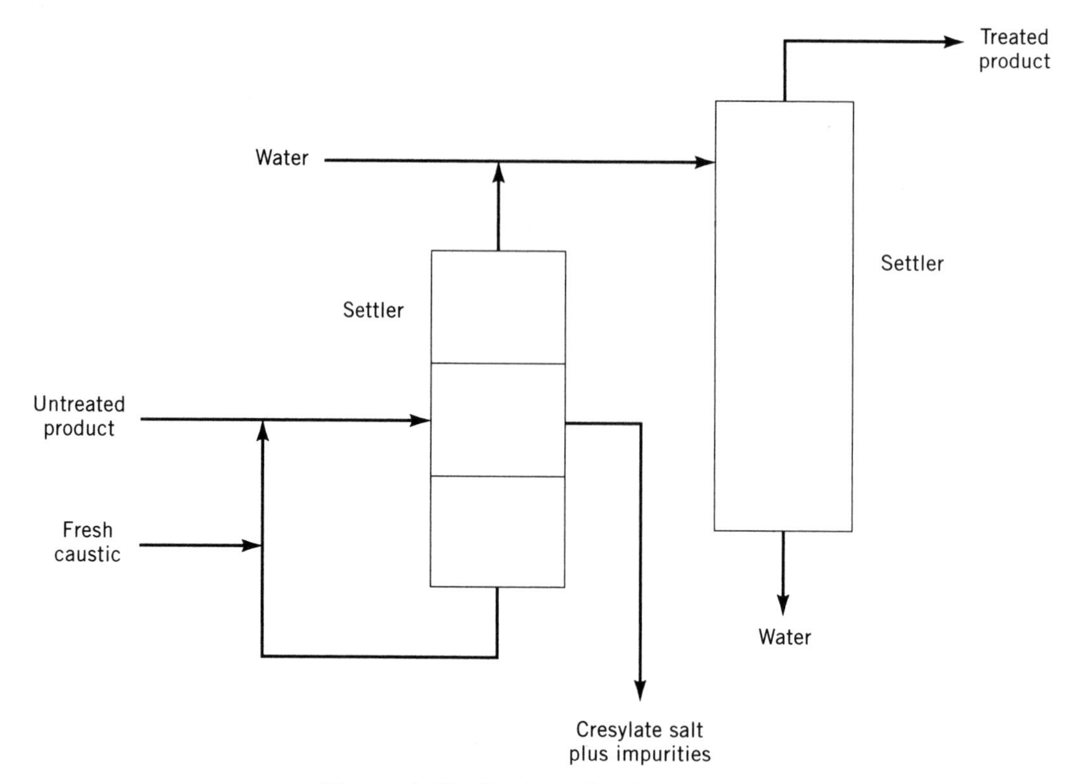

Figure 4. The Dualayer distillate process.

acids, and phenol. Gasoline is contacted countercurrently with the mercapsol solution, and the treated product is removed from the top of the tower. Spent solution is stripped to remove gasoline, and the mercaptans are then removed by steam stripping.

The Unisol process (Fig. 7) is a regenerative method for extracting not only mercaptans, but also certain nitrogen compounds from sour gasolines or distillates. The gasoline, free of hydrogen sulfide, is washed countercurrently with aqueous caustic–methanol solution at about 40°C.

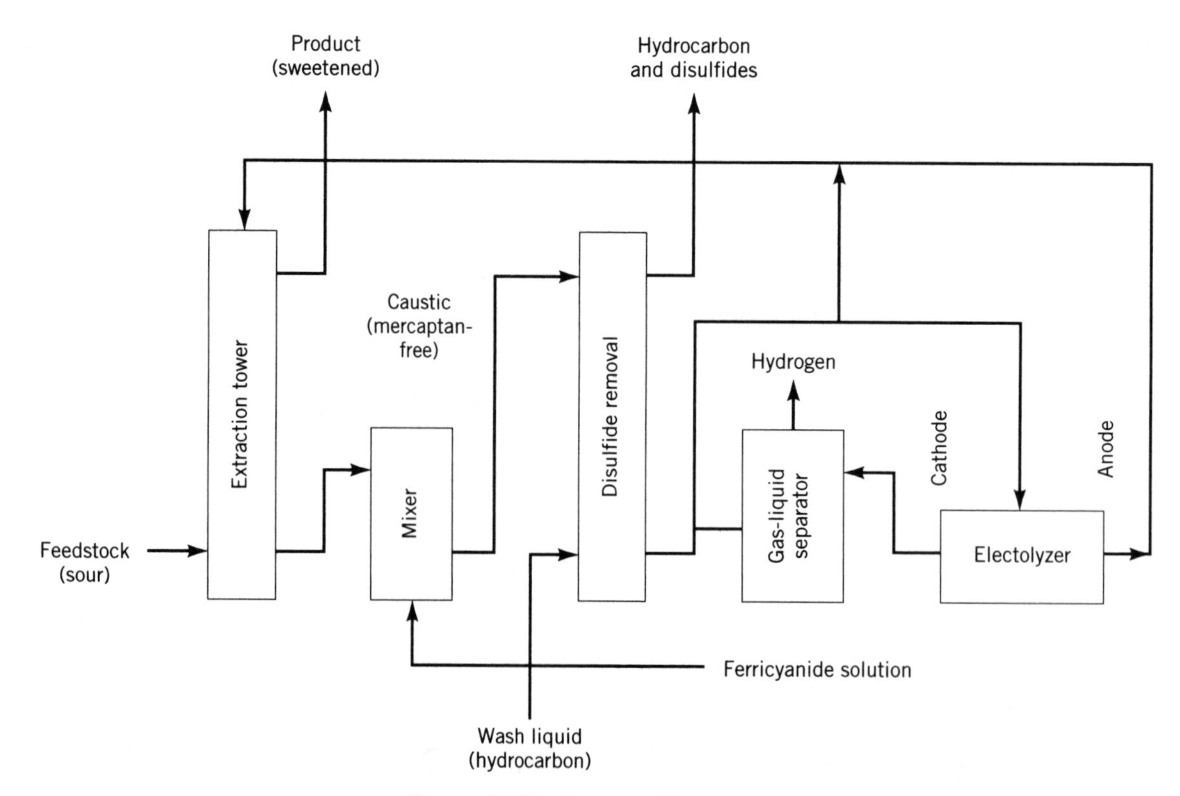

Figure 5. The ferrocyanide process.

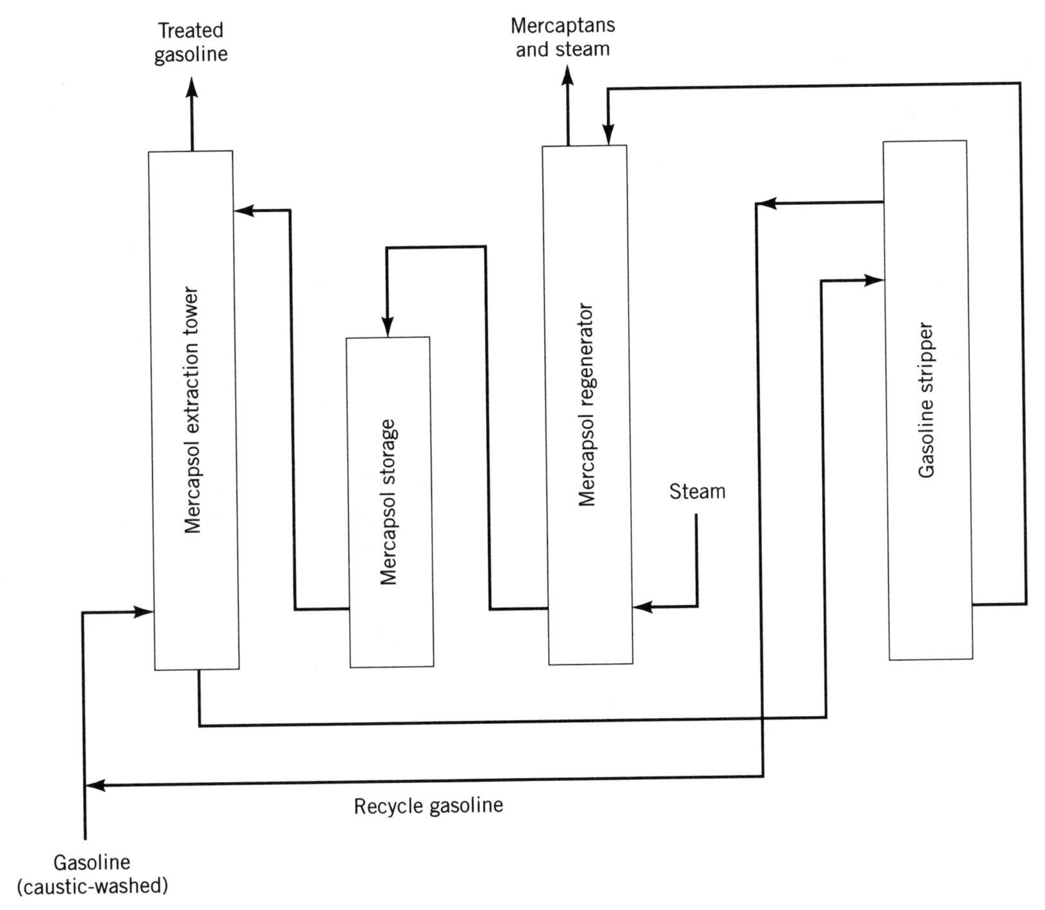

Figure 6. The Mercapsol process.

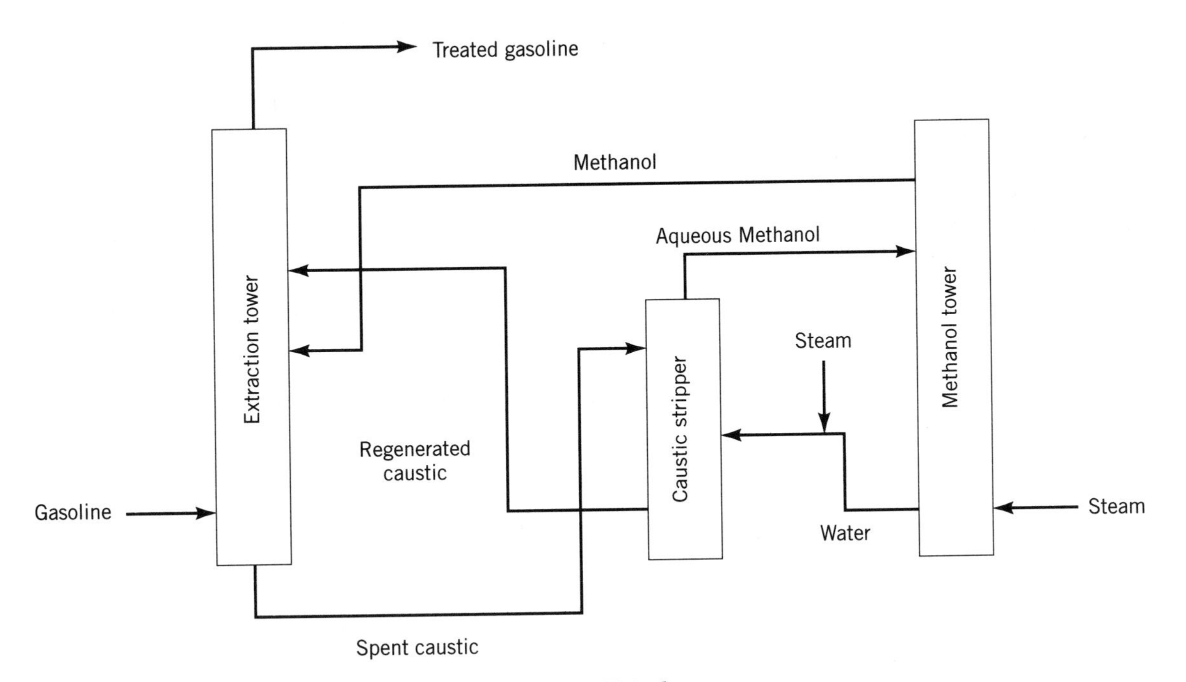

Figure 7. The Unisol process.

The spent caustic is regenerated in a stripping tower (145–150°C), where methanol, water, and mercaptans are removed.

With the widespread use of caustic treating of gasolines, numerous improvements have been introduced. For example, the introduction of electrostatic precipitation has been employed to minimize carryover of caustic in treaters. In addition, with the introduction of "static mixers," the size distribution of the dispersed droplets can be relatively narrow. Thus, for a given extraction efficiency, the carryover of caustic is relatively smaller than with mix valves.

BIBLIOGRAPHY

1. J. G. Speight, *The Chemistry and Technology of Petroleum*, 2nd ed., Marcel Dekker Inc., New York, 1991.
2. V. de Nora and J. A. King, in W. F. Bland and R. F. Davidson, eds., *Petroleum Processing Handbook*, McGraw-Hill Book Co. Inc., New York, 1967, p. 6-1.
3. G. H. Unzelman and C. J. Wolf, in W. F. Bland and R. F. Davidson, eds., *Petroleum Processing Handbook*, McGraw-Hill Book Co. Inc., New York, 1967, p. 3-110.
4. K. E. Clonts and R. E. Maple, in J. J. McKetta, ed., *Petroleum Processing Handbook*, Marcel Dekker Inc., New York, 1992, p. 727.
5. J. G. Speight, *The Chemistry and Technology of Coal*, Marcel Dekker Inc., New York, 1983, Chapt. 13.

CETANE NUMBER

JAMES SPEIGHT
Western Research Institute
Laramie, Wyoming

The cetane number of a diesel fuel is a number that indicates the ability of a diesel engine fuel to ignite quickly and burn smoothly, after being injected into the cylinder. In high-speed diesel engines, a fuel with a long ignition delay tends to produce "rough" operation.

See also AIR QUALITY MODELING; WATER QUALITY ISSUES; WATER QUALITY MANAGEMENT; GLOBAL HEALTH INDEX.

The cetane number should not be confused with the "cetene number," which is an obsolete designation for the starting and running quality of diesel fuel, using cetene ($C_{16}H_{30}$) as the reference fuel. The cetene number has been replaced by the cetane number.

Diesel fuel oil is essentially the same as furnace fuel oil, but the proportion of cracked gas oil is usually less since the high aromatic content of the cracked gas oil reduces the cetane value of the diesel fuel. Cetane number is a measure of the tendency of a diesel fuel to "knock" in a diesel engine. The scale is based upon the ignition characteristics of two well-defined hydrocarbons: cetane (n-hexadecane) $CH_3(CH_2)_{14}CH_3$; and heptamethylnonane

$$\begin{array}{c} CH_3CH_3CH_3CH_3CH_3CH_3CH_3 \\ | \quad | \quad | \quad | \quad | \quad | \quad | \\ CH_3CH\ CH\ CH\ CH\ CH\ CH\ CH\ CH_3 \end{array}$$

Cetane has a short delay period during ignition and is assigned a cetane number of 100; heptamethylnonane has a long delay period and has been assigned a cetane number of 15. Just as the octane number is meaningful for automobile fuels, the cetane number is a means of determining the ignition quality of diesel fuels and is equivalent to the percentage by volume of cetane in the blend with heptamethylnonane, which matches the ignition quality of the test fuel (1).

To determine cetane number of a fuel sample, a specially designed diesel engine is operated under specified conditions with the given fuel. The fuel is injected into the engine cylinder each cycle at 33 cm before top center. The compression ratio is adjusted until ignition takes place at top center (33 cm delay). Without changing the compression ratio, the engine is next operated on blends of cetane (n-hexadecane), a short-delay fuel, and heptamethylnonane, which has a long delay. When a blend is found that also has a 33 cm delay under these conditions, the cetane number of the fuel sample may be calculated from the quantity of cetane required in the blend.

Thus, the cetane number of a diesel fuel is determined by comparison and corresponds to the number of parts by volume of cetane in a cetane–heptamethylnonane blend, which has the same ignition quality as the fuel. The cetane number of diesel fuel usually falls into the range 30–60; a high cetane number is an indication of the potential for easy starting and smooth operation of the engine.

For general comparison, the cetane number of paraffinic hydrocarbons, particularly those with straight rather than branched chains, is high, whereas the cetane numbers of aromatic and hydroaromatic (naphthenic) hydrocarbons is low. A fuel having a high octane number tends to have a low cetane number, and vice-versa. Diesel fuels usually contain 60–80 vol % paraffinic materials and 20–40 vol % aromatic-naphthenic materials. Diesel fuels having low cetane numbers (that is, around 60 vol % paraffinics) can be improved by the addition of chemicals, such as amyl nitrate ($C_5H_{11}NO_3$) or hexyl nitrate ($C_6H_{13}NO_3$). Such compounds are referred to as "cetane number *improvers*."

Other methods are also available for the estimation of diesel fuel quality. For example, the diesel index (DI) is defined by the following relation:

$$DI = (A^0F \times {}^0API)/100$$

where A^0F = aniline point in degrees Fahrenheit and ${}^0API$ is the American Petroleum Institute gravity. A high aniline point corresponds to a high proportion of paraffins in the diesel fuel; such a fuel will have a high diesel index and, therefore, a high cetane number. The cetane number can also be estimated from the API gravity and the mid-boiling point (2,3) or by spectroscopic methods and other physical property methods (4).

BIBLIOGRAPHY

1. *ASTM D 613, Test Method for the Ignition Quality of Diesel Fuels by the Cetane Method*, Annual Book of ASTM Standards,

Section 05.04, American Society for Testing and Materials, Philadelphia, Pa., 1991.

2. *ASTM D 975, Specification for Diesel Fuel Oils, Annual Book of ASTM Standards*, Section 05.01, American Society for Testing and Materials, Philadelphia, Pa., 1991.
3. C. G. A. Rosen, In V. B. Guthrie, ed., *Petroleum Products Handbook*, McGraw-Hill Book Co. Inc., New York, 1960, p. 6–1.
4. C. T. O'Connor, R. D. Forrester, and M. S. Scurrell, *Fuel* **71**, 1323 (1992).

CHEMODYNAMICS

DANNY D. REIBLE
Louisiana State University
Baton Rouge, Louisiana
University of Sydney
Sydney, Australia

The assessment of the ambient concentration and thus the harm resulting from the release of chemicals is dependent on the dynamic transport, dilution and fate processes operative in the environment. Environmental chemodynamics is a term coined by Virgil Freed and popularized by Louis Thibodeaux (1) to describe the study of the dynamics of pollutants in the natural environment. The evaluation of the fate and transport of contaminants in the environment has often been limited to consideration of the equilibrium partitioning of these contaminants between environmental phases. This approach is one that is rarely likely to be valid and even then only locally. Instead the environment is in a state of disequilibrium and fate and transport cannot be separated from the concept of time. It is recognition of this dynamic that characterizes the study of environmental chemodynamics.

Environmental chemodynamics involves the application of fundamental energy and mass conservation principles to model and quantify pollutant concentrations and fluxes in the environment. Often, due to the relatively crude nature of our understanding of the natural environment and its inherent variability, the quantitative models are necessarily crude. Generally, these models attempt to capture only the key mechanisms of an environmental transport process and thus cannot hope to describe the full variance of field observations. The models are often lumped parameter models which assume limited spatially heterogeneity or quasi-steady conditions, while at other times, sufficient information may exist to allow spatially distributed and fully time dependent models of environmental transport processes that can be developed and applied.

Applications of environmental chemodynamics arise in any problem involving the assessment of the fate of chemicals in the environment and the resulting human or ecological exposure. The tools that are generated by chemodynamics, for example, allow one to estimate the mobility of contaminants and the exposure of potential receptors in the ecosystem. In addition, the formalism allows the evaluation of potential treatment processes by defining contaminant losses during in-situ treatment or during the collection and transportation of contaminants for conventional treatment processes. Environmental chemodynamics is thus the key to exposure assessment and the selection of remedial processes on the basis of minimum exposure. The Superfund Exposure Assessment Manual (2) illustrates the application of a variety of simple models to the hazardous waste site exposure assessment.

The purpose of the present discussion is to identify the fundamental basis for environmental chemodynamics and present some of the tools that allow estimation of the environmental dynamics of pollutants. The article is not intended to be comprehensive but instead designed to provide engineers and scientists new to the field, a starting point for the collection of information necessary to solve their own problems. In particular, the article will focus on the physico–chemical processes that influence the fate and transport of hydrophobic organic compounds. Metals and polar organic compounds are not considered although they often exhibit similar or analogous behavior. It should be emphasized that this discussion will include processes in all four phases commonly encountered in the environment: air, water, soil and nonaqueous liquid phases (eg, an oily phase). To avoid confusion, media specific parameters will be subscripted with a number to indicate air (phase 1), water (phase 2), soil (phase 3) and nonaqueous liquid phases (phase 4).

See also AIR QUALITY MODELING; WATER QUALITY ISSUES; WATER QUALITY MANAGEMENT; GLOBAL HEALTH INDEX.

THERMODYNAMICS AND TRANSPORT PHENOMENA

Fundamental Tools

The fundamental tools of the environmental chemodynamicist are no different than the tools used by the chemical process designer; that is, mass and energy conservation equations. This approach applies regardless of whether the mobile phase is air, water, a nonaqueous phase liquid or the pore fluid in soil. The mass conservation equation relates the mobile phase concentration of a pollutant A (C_A) to position (z) and time (t). One form of this equation can be written

$$\frac{\partial C_A^T}{\partial t} + U\frac{\partial C_A}{\partial z} = (D_A + D_{eff})\frac{\partial^2 C_A}{\partial z^2} - k_{fate} f(C_A) \qquad (1)$$

where C_A denotes the concentration of contaminant A in moles per unit volume of mobile phase, ρ_A will be used to denote mass concentration.

The first term on the left in equation 1 represents the time rate of change of total concentration of pollutant A (C_A^T). The total concentration is equal to the sum of the moles of contaminant in each phase that constitutes the system divided by the total volume of the system. The total concentration is equal to the mobile phase concentration, C_A, if the system does not contain any of the pollutant adsorbed onto immobile phases. Often the accumulation of contaminant occurs only in the mobile phase or is dependent only on the mobile phase concentration so that equation 1 can be written in terms of the single dependent variable C_A.

The second term in equation 1 represents the advection of contaminant as a result of the fluid velocity (U). The first term on the right-hand side represents transport by molecular diffusion (D_A) or by an effective diffusion process (D_{eff}). Molecular diffusion is the process of material transport by the random motion and collisions of molecules. D_A is dependent on the medium, the diffusing chemical, and the ambient temperature. The effective diffusion coefficient, however, is dependent primarily on the motion of the mobile phase (air or water). In the atmosphere or in surface bodies of water, the magnitude of D_{eff} is determined by the intensity of turbulence. In soil vapor or water transport, the D_{eff} is determined by the interrelationship of the mean flow with the soil heterogeneities. In some situations, however, other effective diffusion coefficients may be applicable. For example bioturbation, or the mixing of sediments by the normal activities of benthic organisms, gives rise to contaminant transport that is often modeled with an effective diffusion coefficient related to the density and activity of the benthic organisms.

The final term on the right represents chemical fate processes such as degradation reactions. The term is written in the equation as a loss term. Generally the magnitude of the loss term is a function of the local chemical concentration (ie, $f(C_A)$ as shown in equation 1). Often biological and hydrolysis reactions in soils, solution phase reactions in surface waters and reactions and deposition in the atmosphere are modeled as first order processes and the loss term is written as $k_{fate}C_A$. Irreversible fate processes will generally not be considered herein.

The relationship between the total system concentration, C_A^T, and the mobile phase concentration, C_A will be further examined. In the atmosphere or in bodies of water, C_A^T is generally taken equal to C_A. In groundwater systems, however, a contaminant can often accumulate on the immobile solid phase. Consider a linear partition coefficient, α_{32}, between the immobile solid (phase 3) and water (phase 2), that is, $\alpha_{32} = C_{A3}/C_{A2}$, where C_{A3} is the contaminant concentration on the solid phase (eg, in moles/kg) and C_{A2} is the contaminant concentration in the water phase (eg, in moles/L). Representing the water fraction or void volume by ε and the solid phase bulk density by ρ_s, then $C_A^T = C_{A2}(\varepsilon + \rho_s\alpha_{32})$. The ratio of the total concentration of the pollutant in the system to the concentration in the mobile phase (here, $\varepsilon + \rho_s\alpha_{32}$) is often termed a retardation factor, R_f, since it slows the response of the mobile phase concentration to transport via advection and diffusion. This relationship can be most readily seen by rewriting equation 1 with $C_A^T = C_AR_f$ and assuming R_f constant. Then equation 1 can be written:

$$\frac{\partial C_A}{\partial t} + \frac{U}{R_f}\frac{\partial C_A}{\partial z} = \left(\frac{D_A + D_{eff}}{R_f}\right)\frac{\partial^2 C_A}{\partial z^2} - \frac{k_{fate}}{R_f}f(C_A) \qquad (2)$$

Thus the ratio of the total concentration to the mobile phase concentration effectively produces a proportionate reduction in the contaminant velocity, diffusivity and reaction rate. Note that under steady state conditions, there is no transient accumulation in an immobile phase and therefore there is no retardation of the mobile phase processes.

A form of equation 1 (or 2) is widely used to predict dynamic contaminant concentrations and fluxes in the environment. In general, application of equation 1 requires generation of a numerical or analytical solution for concentration as a function of space and time. Carslaw and Jaeger (3) and Crank (4) present a wide variety of solutions to Equation 1 subject to various initial and boundary conditions. The vast majority of analytical solutions are limited to constant effective diffusion coefficient, linear or no reaction, and constant or zero velocity. These limitations have not hindered their usefulness in a variety of important situations when the available data does not support more sophisticated models.

Useful Solutions

Two analytical solutions have proven especially useful for environmental applications. In the first of these solutions, depicted in Figure **1a**, it is assumed that a uniform and constant concentration, C_1, exists at a plane perpendicular to the direction of transport (designated the origin). That is, $C_A = C_1$ at $z = 0$. The environment above this plane is assumed quiescent ($U = 0$) and is initially uniform at a different concentration, C_0. Transport is assumed to occur only via a diffusive process (D_A or $D_{eff} \neq 0$) without reaction ($k_{fate} = 0$). The solution to Equation 1, subject to these conditions is given by the equation after substituting R_fC_A for C_A^T.

$$C_A = C_1 + (C_0 - C_1)\,erf\left(\sqrt{\frac{z^2R_f}{4t(D_A + D_{eff})}}\right) \qquad (3)$$

The erf (η) is the error function which is related to the cumulative normal distribution function and is tabulated

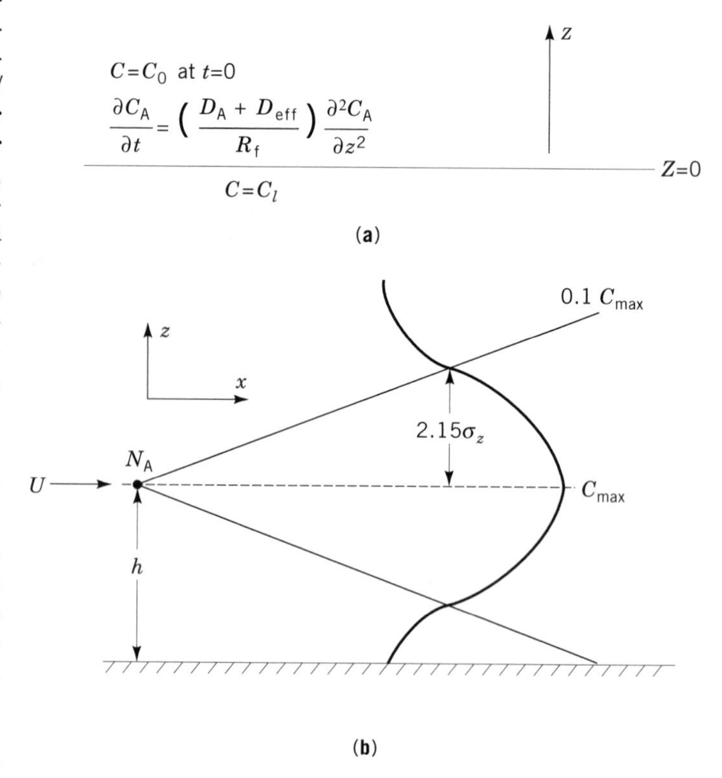

Figure 1. (**a**) Depiction of system described by equations 3 and 4. (**b**) Depiction of system described by equations 6 and 7.

Table 1. Values of the Error Function

x	erf(x)
0	0.0
0.2	0.227
0.4	0.4284
0.6	0.6039
0.8	0.7421
1.0	0.8427
1.2	0.9103
1.4	0.9523
1.8	0.9891
2.2	0.9981
2.8	0.9999

(eg, CRC Standard Math Tables, 1974). Some values of the error function are included in Table 1. Note that the solution requires a constant ratio of total contaminant concentration to mobile phase concentration, ie, constant R_f. In the case of diffusion in soil groundwaters, this configuration is equivalent to assuming local equilibrium with linear partitioning between the soil and porewater. While this is a very common assumption, it can be quite restrictive.

The flux of contaminant through the interface ($z = 0$) by this model is given by

$$\dot{n}_A = -(D_A + D_{eff})\frac{\partial C_A}{\partial z}\bigg|_0 = \left[\frac{(D_A + D_{eff})R_f}{\pi t}\right]^{1/2}(C_1 - C_0) \quad (4)$$

where use is made of the symbol convention of lower case letters representing a flux with the dot overstrike presenting a rate. An upper case symbol for mass or moles would represent an area integrated value. Assuming a constant total concentration in the contaminated region, C_1^T and recognizing $C_1 = C_1^T/R_f$, this model suggests that sorption reduces the flux by the factor $R_f^{1/2}$.

A second commonly encountered solution of equation 1, depicted in Figure 1b, is the concentration about a point source of strength $\dot{N}_A$ (eg, moles/s) released over the period δt. Since the emissions spread in all directions, the three-dimensional form of equation 1 is required. Advection ($U \neq 0$) is assumed in the x-direction and effective diffusion processes are assumed to occur in all three directions. If D_x, D_y, and D_z are the effective diffusion coefficients in the x, y and z directions, respectively, then the concentration field away from the source is given by

$$C_A = \frac{\dot{N}_A \delta t}{8(\pi\tau)^{3/2}\sqrt{D_x D_y D_z}} \exp\left[\frac{-(x - U_\tau)^2}{4D_x\tau}\right] \exp\left[-\frac{y^2}{4D_y\tau}\right]\left(\exp\left[-\frac{(z-h)^2}{4D_z\tau}\right] + \exp\left[\frac{-(z+h)^2}{4D_z\tau}\right]\right) \quad (5)$$

where τ is t/R_f. h is the location of the source relative to a plane of reflection $z = 0$. This model is only valid for $\delta t < \tau$. For $\delta t \gg \tau$, the source appears essentially continuous and the model can be integrated over time to give the steady solution. Under these conditions, transient accumulation in an immobile phase does not occur ($R_f \rightarrow 1$). Taking $\tau = x/U$, equation 5 becomes

$$C_A = \frac{\dot{N}_A}{4\pi\sqrt{D_y D_z}x} \exp\left[-\frac{y^2}{4D_y\frac{x}{U}}\right]\left(\exp\left[-\frac{(z-h)^2}{4D_z\frac{x}{U}}\right] + \exp\left[\frac{-(z+h)^2}{4D_z\frac{x}{U}}\right]\right) \quad (6)$$

This model is useful for estimating transport and dispersion from

1. an elevated emission source (stack) in the atmosphere at height h above the ground,
2. an outfall at depth h below the water surface of a lake, or,
3. a subsurface source at a height h above an impermeable layer in an aquifer.

In cases 1) and 2) the effective diffusion coefficients scale with the size of the turbulent eddies. In case 3) the effective diffusion coefficient scales with the size of the soil heterogeneities.

Equation 5 or 6 is often rewritten by setting $\sigma_i^2 = 2D_i\tau = 2D_i x / U$, where σ_i represents the standard deviation in concentration in direction i, a measure of the width of the concentration distribution. Equation 6, for example, becomes

$$C_A = \frac{\dot{N}_A}{2\pi\sigma_y\sigma_z U} \exp\left[-\frac{y^2}{2\sigma_y^2}\right]\left(\exp\left[-\frac{(z-h)^2}{2\sigma_z^2}\right] + \exp\left[\frac{-(z+h)^2}{2\sigma_z^2}\right]\right) \quad (7)$$

This form of the equation demonstrates that the model predicts profiles that are identical in shape to the Gaussian probability distribution function. The definition of the standard deviations in concentration suggest that they should grow with the square root of time or distance. Experiments in environmental media suggest, however, that linear rates of growth occur, at least initially. Empirical relationships relating σ_y and σ_z to turbulence and time or distance traveled are normally used to overcome this problem.

Note that the maximum concentration is directly downstream of the point source ($y = 0$) and at the same height or depth as the source ($z = h$). The edge of the dispersing plume can be arbitrarily selected as the location where the predicted concentration is 10% of this maximum. From the properties of the Gaussian probability distribution, this location is given in the y direction by $y = \pm 2.15\ \sigma_y$. In excess of 98% of the area in the Gaussian concentration distribution is located within these limits.

Interface Processes

The above solutions to the mass conservation equation (equation 1) describe intraphase transport processes. They assume simple forms of the concentrations and fluxes at the interface, either constant concentration or constant emission rate. Many attempts to assess transport and fate of chemicals in the environment, however, require prediction of the flux at the interface or more complicated models of its behavior. That is, in general, we must explicitly model the interface processes as well.

An environmental interface (eg, the air–water, air–soil or sediment–water) is typically assumed at equilibrium while the bulk phases are assumed to be in a state of disequilibrium. The interface is modeled as a two-dimensional surface which has no volume and in which accumulation (and thus disequilibrium) cannot occur. The rate of approach to equilibrium in the bulk phase as a result of transport across the interface is generally assumed to be proportional to the concentration difference between the interface and bulk phase, that is,

$$\dot{n}_A = k(C_A - C_{Ai}) \tag{8}$$

where $\dot{n}_A$ represents the chemical flux of A (eg, moles · area^{-1}·time^{-1}) and C_{Ai} represents the interfacial concentration of A in the same phase as the bulk phase. The coefficient of proportionality, or mass transfer coefficient, is a function of the processes operative in the particular phase for which the flux is being estimated. The concentration difference represents a deviation from equilibrium within the phase since rapid mass transfer and equilibration would result in the bulk phase attaining the concentration present at the interface.

A relationship equivalent to equation 8 could be written in an adjacent phase with a separate mass transfer coefficient describing processes within that phase and the concentration difference representing the concentration difference in that phase. As indicated above, the fluxes and concentrations are generally assumed in equilibrium at the interface, i, and thus the flux of contaminant, $\dot{n}_A$, out of phase j is the same as that entering phase k.

$$\dot{n}_A = k_j(C_A - C_{Ai})_j = -k_k(C_A - C_{Ai})_k \tag{9}$$

Here the subscript on the concentration differences refer to the phase (j or k).

The single phase mass transfer coefficients are primarily dependent on the flow conditions and are only weakly chemical dependent as a result of the influence of molecular diffusivity. The effect of molecular diffusivity can be approximated by

$$\frac{k_A}{k_B} = \left(\frac{D_A}{D_B}\right)^n \qquad \frac{1}{2} \leq n \leq 1 \tag{10}$$

where 1/2 is the power predicted by the penetration and surface renewal theories, 2/3 is predicted by boundary layer theory and 1 is predicted by film theory (5). There is very little variation in molecular diffusion coefficients between compounds in the same phase. Forty-six gases, organic and salt compounds displayed an average diffusivity in water of $1.1 \pm 0.48 \times 10^{-5}$ cm^2/s at 20–25°C (6). Similarly, the average molecular diffusivity in air at 20–25°C for forty-two compounds was 0.11 ± 0.07 cm^2/s. The specified ranges are standard deviations. This range suggests that the molecular diffusivity of most compounds in air or water at 25°C is well approximated by 0.1 and 10^{-5} cm^2/s, respectively. Thus the individual phase mass transfer coefficients are a strong function of only the flow conditions within the phase, and measurements for a particular compound are approximately applicable to other compounds either directly or through use of equation 10.

Use of equation 9, however, is hindered by the use of interfacial concentrations, C_{Ai}. Generally, only the bulk phase concentrations are of interest or are directly measurable. Thus it is often convenient to rewrite the flux relationship in terms of a global deviation from equilibrium,

$$\dot{n}_A = K_{jk}(C_{Aj} - C^*_{Aj}) = -K_{kj}(C_{Ak} - C^*_{Ak}) \tag{11}$$

Here K_{jk} represents the mass transfer coefficient for transport from phase j to phase k and C^*_{Aj} represents the concentration in phase j that would be in equilibrium with phase k. If α_{jk} represents the equilibrium ratio of the concentrations in phases j and k,

$$C^*_{Aj} = \alpha_{jk}C_{Ak} \tag{12}$$

Similarly, K_{kj} represents the mass transfer coefficient for transport from phase k to phase j and C^*_{Ak} represents the concentration in phase k that would be in equilibrium with phase j ($C^*_{Ak} = \alpha_{kj}C_{Aj} = C_{Aj}/\alpha_{jk}$). With equation 11, the chemical flux can be determined from a bulk concentration in each phase, thermodynamic equilibrium relationships and the "overall" mass transfer coefficients K_{jk} or K_{kj}. The overall mass transfer coefficients in equation 11 are dependent upon both the chemical (specifically the equilibrium partitioning of a chemical at the interface) and the flow conditions.

The more fundamental definition of the individual phase mass transfer coefficients (eq. 9) can be related to the lumped coefficients (eq. 11) for which the driving force is more easily measured. Equality of fluxes regardless of the form of the equation suggests that:

$$\frac{1}{K_{jk}} = \frac{1}{k_j} + \frac{1}{k_k}\left(\frac{C_{Aj}}{C_{Ak}}\right)_{eq} \tag{13}$$

The ratio of concentrations in the two phases (ie, the group in parentheses in eq. 13) is simply the equilibrium distribution coefficient, α_{jk}, since the interface is at equilibrium. Note that for large values of α_{jk}, the last term in eq. 13 dominates and the overall mass transfer coefficient is controlled by the mass transfer coefficient in phase k. Similarly for small values of α_{jk} (or large values of α_{kj}), the overall mass transfer coefficient is controlled by the mass transfer coefficient in phase j.

Quantitative determination of the mass transfer rate thus requires determination of the equilibrium partitioning of the contaminant between the two adjacent phases. Only linear partitioning will be considered here and is discussed below for each of the most common environmental interfaces.

Fluid–Fluid Interfaces. The distribution coefficient for a hydrophobic organic compound at the air–water interface is the ratio of the concentration in air to that in water,

$$\alpha_{12} = \frac{C_{A1}}{C_{A2}} = \frac{p_A^*}{RT\,C_{A2}^*} \tag{14}$$

where p_A^*/RT is the molar air concentration of the contaminant at its pure compound vapor pressure and C_{A2}^* is the molar solubility in water. This relationship reflects the fact that any compound exerts its pure compound vapor pressure, p_A^*, when present at its molar solubility. Thus a water body containing, for example, benzene at its water solubility of 1780 g/m^3 exerts the same benzene vapor pressure as pure benzene. For hydrophobic compounds (HOC) in water, this distribution ratio, α_{12}, remains essentially constant at any concentration below the solubility limit and is termed a Henry's Law constant. The Henry's Law constant is tabulated in a variety of different units which include the ratio of air and water concentrations or mole fractions (both dimensionless) and pressure over solubility units (eg, atm-m^3/mole). Note that Equation 13 requires use of an α_{jk} defined by the same units of concentration used in the definition of the fluxes, in this case the ratio of molar concentrations in the air and water phases. Because of their low solubility, α_{12} tends to be large for hydrophobic organic compounds even if the pure component vapor pressure is low. The PCB mixture Aroclor 1260, for example, exhibits a Henry's Law constant that is slightly higher than that of benzene despite having a vapor pressure that is more than a million times smaller (7). As a result, the evaporation of contaminants from water tends to be controlled by water-side mass transfer resistances since the water phase term dominates in equation 13. Vapor pressures, water solubilities and air–water distribution coefficients are given for selected compounds in Table 2.

The distribution coefficient for a hydrophobic organic compound at an air–nonaqueous phase liquid (NAPL) interface is governed by Raoult's Law if the compound is assumed to form an ideal solution in the NAPL phase. This can be written

$$\alpha_{14} = \frac{C_{A1}}{C_{A4}} = \frac{p_A^*}{RT\,C_4} \tag{15}$$

where C_4 is the molar density of the NAPL phase. Note that C_4 is, at high contaminant fractions (eg, >5%), a function of the concentration of the contaminant. At an air–oil interface, α_{14} tends to be small, especially for low vapor pressure compounds. The evaporation of an organic compound will then tend to be controlled by air side resistances.

At the interface between water and an oily phase, the distribution coefficient, α_{24}, is simply the ratio α_{14}/α_{12}.

$$\alpha_{24} = \frac{C_{A2}}{C_{A4}} = \frac{C_{A2}^*}{C_4} \tag{16}$$

Fluid–Solid Interfaces. At the sediment–water interface, the mobile fluid on both sides of the interface is water (ie, pore-water and overlying water) and the distribution coefficient between these "phases" is unity. Similarly, the thermodynamic distribution coefficient between soil, air and the atmosphere is unity. The partitioning between the soil or sediment phase and the adjacent pore fluid, however, is much more complicated. Partitioning between a solid and fluid phase is generally assumed to follow one

Table 2. Physical Properties of Selected Compounds[a]

Compound	MW	P_v mm Hg	Water Solubility mg/L	α_{12}	Log α_{oc}
Acenaphthene	154	0.005	3.47	0.011934	1.3
Anthracene	178	2×10^{-4}	0.045	0.042547	4.3
Benzene	78	95	1800	0.221397	1.9
Benzo[a]pyrene	252	5.5×10^{-9}	0.004	1.86E-05	6.0
Chlordane	410	1×10^{-5}	0.056	0.003938	5.2
p,p′-DDT	354	1.9×10^{-7}	0.003	0.001206	5.4
1,2-Dichloroethane	99	233	5060	0.24517	1.8
Dieldrin	381	3×10^{-6}	0.2	0.000307	4.6
Fluoranthene	202	5×10^{-6}	0.26	0.000209	4.6
Hexachlorobenzene	285	1×10^{-5}	0.005	0.030655	3.6
Methylene chloride	85	439	19000	0.105623	0.94
Naphthalene	128	0.233	30	0.053465	3.1
PCB-1242 (avg)	261	4×10^{-4}	0.24	0.023395	3.7
PCB-1254 (avg)	327	7.7×10^{-5}	0.057	0.023757	5.6
Pentachlorophenol	266	1.7×10^{-4}	20	0.000122	3.0
Phenanthrene	178	6.8×10^{-4}	1	0.00651	3.7
Pyrene	202	6.9×10^{-7}	0.135	5.55E-05	4.7
Tetrachloroethylene	166	20	150	1.190348	2.4
Trichloroethylene	131	74	1100	0.473955	2.0

[a] From Ref. 7. All data are estimates at 25°C.

of the following equilibrium relationships:

- Linear partitioning
- Langmuir or monolayer partitioning
- Fruendlich or empirical power-law partitioning
- Brunauer-Emmett-Teller or multiple layer partitioning

The latter three relationships are nonlinear and are necessary to describe partitioning in many systems. Space precludes formal presentation of the nonlinear partitioning relationships but the linear partitioning relationship is commonly used and directly analogous to Henry's Law for fluid phase equilibrium. Hydrophobic organic compounds in pore water tend to follow this linear partitioning relationship if present at low concentrations (concentrations small relative to the water solubility limit). Under such conditions, the soil or sediment–water partition coefficient is constant and can be written

$$\alpha_{32} = \frac{C_{A3}}{C_{A2}} \tag{17}$$

Since the concentration of pollutant in the solid phase is normally measured in units of mass or moles per mass of solid (eg, mg/kg) and that in the pore water is mass or moles per volume of fluid (eg, mg/L), the distribution coefficient has units of volume fluid per mass of solid (eg, L/kg). For hydrophobic organics, the sorbed phase concentration also tends to be directly proportional to the organic carbon fraction of the solid phase, f_{oc}. The constant of proportionality is the organic carbon based-partition coefficient, α_{oc}, which is a tabulated chemical-specific property, as listed in Table 2. This quantity is a measure of the hydrophobicity of a compound and therefore has been found to correlate well with the octanol-water partition coefficient (8), given α_{oc}, $\alpha_{32} = f_{oc}\alpha_{oc}$. The ratio of the total contaminant concentration in the soil–water system to the concentration in the water phase is given by

$$R_f = (\varepsilon + \rho_s f_{oc}\alpha_{oc}) \tag{18}$$

Partitioning between soil and an adjacent vapor phase is often described by the Brunauer-Emmett-Taylor isotherm (9). Linear partitioning is sometimes assumed, however, for simplicity. In addition, the unsaturated soil zone generally contains significant amounts of water as well as air and soil. Thus accumulation in each of these phases must often be considered. Since the water tends to wet the soil surface more than the air, chemical partitioning in such a system seems to be well-represented by separate partitioning between the air and water phases and between the water and soil phases. Thus the retardation factor or the ratio of the total concentration of the contaminant in the soil–water–air system to its vapor phase concentration is given by

$$R_f = \left(\varepsilon_1 + \frac{\varepsilon_2}{\alpha_{12}} + \rho_s f_{oc}\frac{\alpha_{oc}}{\alpha_{12}}\right) \tag{19}$$

where ε_1 is the air-filled porosity and ε_2 is the water-filled porosity.

The picture of the air–soil equilibrium provided by equation 19 is valid as long as linear partitioning can be assumed and as long as the volumetric water content in the soil exceeds 1–3%. At lower volumetric water contents, portions of the solid surface may be exposed to the vapor space and direct sorption of vapors onto the surface may occur. This disposition can lead to significantly greater sorption of the vapors onto the solid phase and sorption that can rarely be modeled as a linear process. The process of sorbed vapor release, as a solid is wetted, has been explained as simply the reverse of this competitive sorption process. This behavior has important practical implications; for example, rapid release of pesticide or herbicide vapors can occur upon wetting an initially dry field surface.

The previous sections have introduced the dynamic, spatially distributed mass balance as the basic tool for the evaluation of contaminant transport and fate in the environment. These sections have also identified the equilibrium relationships at environmental interfaces that determine the boundary conditions for the intraphase transport processes as well as define the concentration differences that serve as driving forces for pollutant transport. The subsequent sections will offer selected examples of the application of these principles to problems in each of the environmental media: air, soil and water.

APPLICATIONS TO ATMOSPHERIC TRANSPORT

To illustrate the application of chemodynamic principles to atmospheric transport of pollutants let us consider the estimate of exposure of a receptor (eg, a nearby home) downwind of a surface impoundment containing a volatile pollutant as shown in Figure 2. The exposure assessment requires estimation of the emission rate from the impoundment and estimation of the transport and dispersion of the pollutant in the atmosphere. Each of these portions of the analysis will be discussed in turn.

Emission Estimation

A typical situation at a surface impoundment is the presence of a volatile hydrophobic organic in an aqueous solution with a negligible background air concentration. The transport of a chemical across the air–water interface is controlled by the turbulence levels on each side of the interface. On the air side, friction with the earth's surface

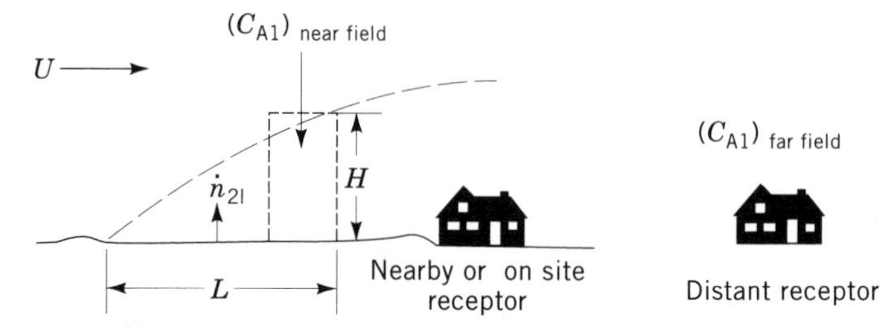

Figure 2. Depiction of setting for estimating near and far-field air concentrations emitting from a surface impoundment.

drives the mechanical generation of turbulence, and heating (cooling) of the air adjacent to the earth drives generation (consumption) of turbulence by buoyancy. On the water-side, friction at the air–water interface drives circulations and turbulence within the surface impoundment.

Experiments have suggested that the following correlations provide estimates of the mass transfer coefficients on each side of the air–water interface. On the air side, with k_1 in cm/hr,

$$k_1 = 1150(U_1 + U_2) \quad (20)$$

where U_1 and U_2 represents the wind speed and water current speed, respectively, in m/s. On the water-side, with k_2 in cm/h, where again the wind and water current speeds are in m/s. Z is the depth of the water column in meters. If there is no water current (eg, in an enclosed surface impoundment) wind driven

$$\begin{aligned} k_2 &= 23.4\, U_2\, Z^{-2/3} && \text{if } U_1 < 1.9 \text{ m/s} \\ k_2 &= 23.4\, U_2\, Z^{-2/3}\, e^{0.526(U_1 - 1.9)} && \text{if } 1.9 \text{ m/s} < U_1 < 5 \text{ m/s} \end{aligned} \quad (21)$$

circulations would be of the order of 3% of the wind speed ($U_2 \approx 0.03\ U_1$) assuming continuity of shear stresses across the air–water interface. Equations 20 and 21 are slight modifications of those presented by (10) and represent only single examples of a wide variety of correlations that have been generated for this purpose. Springer and co-workers review a number of empirical correlations applicable to surface impoundments (11).

The numerical value of the coefficients on each side of the air–water interface often differ significantly. A typical air–side mass transfer coefficient is of the order of 3000 cm/hr while the water–side coefficient is typically 3–30 cm/h. Their relative importance, however, is largely defined by the relative magnitudes of the respective terms in equation 13. For large values of Henry's Law constant (α_{12}), water–side mass transfer resistances control and the flux across the air–water interface can be written

$$\dot{n}_{21} = K_{21}(C_{A2} - C^*_{A2}) = k_2\left(C_{A2} - \frac{C_{A1}}{\alpha_{12}}\right) = k_2 = C_{A2} \quad (22)$$

where $\dot{n}_{21}$ represents the molar flux from water to air. Note that since the background air concentration of the pollutant is zero ($C_{A1} = 0$) in this example, thus the evaporative flux is *independent of the vapor pressure or the Henry's Law constant of the compound.* Hydrophobic organic compounds present at the same water concentration will often evaporate at the same rate despite vastly different pure component vapor pressures or Henry's Law constants. Note, however, that the Henry's Law constant still appears indirectly in that it must be sufficiently large to result in control by water-side resistances.

Two other common scenarios are of interest. In the first, the surface impoundment may be filled with a nonaqueous phase liquid, or oil, which contains the evaporating contaminant. A hydrophobic organic species in an oily phase is more likely to act in an ideal fashion and the air–liquid equilibrium is defined by Raoult's Law. If the contaminant exhibits a low pure component vapor pressure, the magnitude of the air–liquid partition coefficient tends to be small and the air–side mass transfer resistances tend to control evaporation.

In the second additional scenario, the surface impoundment may be filled with soil. Thus the evaporation of the pollutant involves evaporation from the covered liquid and migration through the overlying soil cap into the atmosphere. In this situation, the primary resistance to mass transfer is likely to be diffusional resistances in the soil cap. Quasi-steady conditions would develop rapidly within the soil cover suggesting that the flux can be estimated by

$$\dot{n}_{21} = \frac{D_{eff}}{L} C^*_{A1} = \frac{D_{A1}}{L} \frac{\varepsilon_1^{10/3}}{\varepsilon_t^2} \alpha_{12}\, C_{A2} \quad (23)$$

where use is made of the effective diffusion coefficient model of Millington and Quirk (12). C^*_{A1} is the vapor phase concentration at the base of the soil cover which would be in equilibrium with the liquid phase ($\alpha_{12}\, C_{A2}$). This disposition again assumes that the background air concentration of the contaminant is negligible.

Estimation of Downwind Concentration

The flux estimates outlined above can be used to estimate concentrations downwind of the surface impoundment. Atmospheric concentrations may be of interest at points downwind or immediately adjacent to the source area. Simple models for each of these cases will be considered. A review atmospheric dispersion modeling including more sophisticated approaches can be found in Hanna and coworkers (13) or Zanneti (14).

Estimation of the atmospheric concentration on or near the source area may be of interest to estimate exposure to on-site workers or at immediately adjacent receptor locations as shown in Figure 2. Often a long-term average concentration may be of interest rather than a concentration under a specific set of meteorological conditions and at a particular location. In such situations, it may be sufficient to employ average meteorological conditions and a "box" model which averages the pollutant concentration over depth. A mass balance over the box shown in Figure 2 suggests that the concentration at any position x downwind from the upwind edge of the source area (the impoundment) is given by

$$C_{A1} = \frac{\dot{n}_{21} LW}{HW\langle U\rangle} \quad (24)$$

Here, $\dot{n}_{21}$ is the emission flux (eg, moles/m^2) from a source of length L and width W. H is the height over which the contaminant is approximately well-mixed and $\langle U\rangle$ is the height averaged wind velocity. Under neutral conditions, or conditions under which surface heating or cooling does not significantly influence atmospheric turbulence, the wind velocity as a function of height can be written,

$$u = \frac{u_*}{\kappa} ln\left(\frac{z}{z_0}\right) \quad (25)$$

where z_0 is the effective aerodynamic roughness height and κ is Von Karman's constant (0.4). Also under neutral conditions, Pasquill (15) has shown that the vertical extent of the dispersing pollutant is defined by the implicit formula,

$$\frac{L}{z_0} = 6.25 \left[\frac{H}{z_0} \ln \frac{H}{z_0} - 1.58 \frac{H}{z_0} + 1.58 \right] \tag{26}$$

The neutral atmospheric condition under which equations 25 and 26 are valid typically occur in mid-morning and late afternoon, under overcast conditions day or night, and during high wind conditions. It is also intermediate between the conditions observed in the middle of the night or afternoon when surface heating or cooling most influence atmospheric turbulence. Thus neutral conditions represent, in this sense, an "average" condition over the course of the day. Use of these equations requires specification of the aerodynamic roughness height which is typically 10–30 times smaller than the physical roughness height. In an area with open terrain and occasional buildings, often typical of a landfill or impoundment area, a reasonable estimate of aerodynamic roughness might be about 0.1 m. Using this estimate of the aerodynamic roughness, the definition of the plume height given by equation 26 and integrating the velocity profile (eq. 25) over this height, and substituting into equation 24 gives

$$C_{A1} = \left[\frac{L}{0.22\, H \ln (2.5\, H)} \right] \frac{\dot{n}_{21}}{U} \tag{27}$$

where U is the wind velocity measured at the standard meteorological measurement height of 10 m. Subject to the assumption of neutral conditions and a roughness of 0.1m, the bracketed term is essentially constant and equal to 27. The air concentration is thus independent of distance from the upwind edge of the source, ie, independent of L. Since this situation occurs under neutral atmospheric conditions, which represents an average condition in the sense outlined above, the estimate of $C = 27\, \dot{n}_{21}/U$ represents a useful long-term average concentration on or immediately adjacent to an impoundment or landfill source. Accurate short-term predictions of downwind concentrations, however, requires an alternative approach, for example that of Chitgopekar and co-workers (16).

Estimating exposure for receptor locations further downwind (ie, not located on or adjacent to the source, the far-field receptor in Fig. 2) also requires a different approach than that outlined above. Far away from the impoundment, the source appears as a point and the use of equation 7, the Gaussian plume model becomes appropriate. The Gaussian plume model is the basis for most of the regulatory air dispersion models in the United States (17). Use of this model requires definition of the standard deviation in concentrations as a function of downwind distance. In general these parameters are a function of both downwind distance and the turbulence structure of the atmosphere. The turbulent intensity is a function of the balance between generation or consumption of turbulent energy by buoyancy forces (largely driven by surface heating or cooling) and mechanical generation of turbulence by velocity shear. The Richardson's number provides a convenient characterization of this balance and is given by

$$Ri = \frac{g}{T} \frac{\partial \theta / \partial z}{(\partial u / \partial z)^2} = \frac{f_{B-V}}{(\partial u / \partial z)^2} \tag{28}$$

where $\partial\theta/\partial z$ is the potential temperature gradient and f_{B-V} is the Brunt-Vaisala frequency, a measure of the static stability of the atmosphere. In a dry atmosphere, the potential temperature gradient is related to the actual temperature gradient by

$$\frac{\partial \theta}{\partial z} \approx \frac{\partial T}{\partial z} + \frac{g}{C_p} \approx \frac{\partial T}{\partial z} + 0.0098 \text{ K/m} \tag{29}$$

This relationship indicates that in a dry, statically neutral atmosphere (vanishing potential temperature gradient), the actual temperature decreases at a rate of about 9.8 K/km. If the temperature decreases less rapidly with height, the atmosphere is said to be statically stable and turbulence and contaminant mixing (dilution) would be expected to be damped. The Richardson's number suggests, however, that this static stability can be offset by the mechanical generation of turbulence by velocity shear.

The Richardson's number is large and negative under convective conditions; that is, when surface heating gives rise to significant enhancement of turbulence and turbulent mixing. The Richardson's number is large and positive under cool surface conditions that minimize turbulence and turbulent mixing. The Richardson's number is zero when buoyancy forces lead neither to consumption or generation of turbulence and neutral conditions prevail. This neutrality can occur when the potential temperature gradient is near zero or when the wind speeds and velocity gradients are sufficiently high to negate the influence of buoyancy. Although the crosswind and vertical dispersion parameters (the standard deviations in concentration) can be most appropriately related to the distance downwind and the Richardson number, practical applications of the model generally employ more qualitative criteria for characterizing the atmospheric turbulence. The Pasquill–Gifford stability classes divide the atmosphere into 6 stability categories according to the amount of solar insolation or surface radiative cooling (characterizing the influence of buoyancy forces) and wind speed (characterizing mechanical turbulence). Tables 3 and 4 summarizes the Pasquill–Gifford dispersion parameters and empirical correlations of the dispersion parameters as a function of distance downwind (13,18). Substitution of the dispersion parameters, wind speed and emission rate into equation 7 then allows estimation of concentration as a function of distance downwind, height and crosswind distance.

APPLICATIONS TO TRANSPORT IN SURFACE WATERS

The application of chemodynamic principles to transport of pollutants in surface waters is quite similar to that outlined above for the atmosphere. In general, fate and

Table 3. Meteorological Conditions Defining Pasquil Turbulence Types

A: Extremely unstable conditions	D: Neutral conditions
B: Moderately unstable conditions	E: Slightly stable conditions
C: Slightly unstable conditions	F: Moderately stable conditions

Surface Wind Speed, m/s	Daytime insolation			Nighttime conditions	
	Strong	Moderate	Slight	Thin Overcast or >4/8 Low Cloud	<3/8 Cloudiness
<2	A	A–B	B		
2–3	A–B	B	C	E	F
3–4	B	B–C	C	D	E
4–6	C	C–D	D	D	D
>6	C	D	D	D	D

transport modeling entails estimation of the emission rate into the surface water body and analysis of its subsequent fate. If the emission rate is known, for example, as a result of measurements at a specific industrial outfall, the dispersion modeling essentially follows that described for the atmosphere using dispersion parameters appropriate for the body of water. Again correlations with the local Richardson number and distance from the source would be expected unless mixing is limited by the boundaries of the water body. Correlations for both standard deviation in concentration and effective diffusion coefficients have been determined (eg, see ref. 6). In order to introduce as many different types of chemodynamic problems as possible in the limited space available, however, let us consider here the fate of a contaminant in a lake such as that shown in Figure 3. Water-borne contaminants in such a system would rapidly become well-mixed. Let us consider a contaminant that initially inhabits the sediments and which is migrating into the overlying water as a result of sediment processes. The concentration within the lake would be assumed to be affected by the flushing rate of the lake and the evaporation of the chemical from the surface of the lake. The solution of such a problem proceeds by solving an overall mass balance for the chemical in the lake. The steady state form of this mass balance can be written

$$C_{A2} = \frac{\eta K_{32} C_{A3} A}{\dot{Q}(1 + \alpha_{oc}\rho_{oc}) + K_{21} A} \quad (30)$$

Note that although this model assumes steady state conditions, the overlying water is not assumed in chemical equilibrium with the sediment unless the flushing and evaporation rates from the water column are assumed zero. Use of this model requires definition of the contaminant loss rates from the sediment to the water column ($\eta K_{32} C_{A3} A$) and from the water column to either the overlying air (evaporation-$K_{21} C_{A2} A$) or the adjacent water body (flushing-$Q\ C_{A2}^{T} = Q[1 + \alpha_{oc}\rho_{oc}]C_{A2}$). Each of these will be examined in turn.

Contaminant losses from the sediment to the water column are controlled by the mass transfer processes at the sediment–water interface. A review of the processes influencing benthic exchange of contaminants can be found in Gschwend and co-workers (19) or Reible and co-workers (20). For hydrophobic organic chemicals that are strongly sorbed to the sediment particles, particle suspension processes are very important. However, consider the case of a stable sediment bed in which there is no active scouring of the sediment bed. Under such conditions, bio-

Table 4. Formulas Recommended for $\sigma_y(x)$ and $\sigma_2(x)$ ($10^2 < x < 10^4$ m)[a]

Pasquill type	σ_y, m	σz, m
	Open-Country Conditions	
A	$0.22 \times (1 + 0.0001\,x)^{-\frac{1}{2}}$	$0.20 \times$
B	$0.16 \times (1 + 0.0001\,x)^{-\frac{1}{2}}$	$0.12 \times$
C	$0.11 \times (1 + 0.0001\,x)^{-\frac{1}{2}}$	$0.08 \times (1 + 0.0002\,x)^{-\frac{1}{2}}$
D	$0.08 \times (1 + 0.0001\,x)^{-\frac{1}{2}}$	$0.06 \times (1 + 0.0015\,x)^{-\frac{1}{2}}$
E	$0.06 \times (1 + 0.0001\,x)^{-\frac{1}{2}}$	$0.03 \times (1 + 0.0003\,x)^{-1}$
F	$0.04 \times (1 + 0.0001\,x)^{-\frac{1}{2}}$	$0.016 \times (1 + 0.0003\,x)^{-1}$
	Urban Conditions	
A–B	$0.32 \times (1 + 0.0004\,x)^{-\frac{1}{2}}$	$0.24 \times (1 + 0.001\,x)^{-\frac{1}{2}}$
C	$0.22 \times (1 + 0.0004\,x)^{-\frac{1}{2}}$	$0.20 \times$
D	$0.16 \times (1 + 0.0004\,x)^{-\frac{1}{2}}$	$0.14 \times (1 + 0.0003\,x)^{-\frac{1}{2}}$
E–F	$0.11 \times (1 + 0.0004\,x)^{-\frac{1}{2}}$	$0.08 \times (1 + 0.00015\,x)^{-\frac{1}{2}}$

[a] Ref. 18.

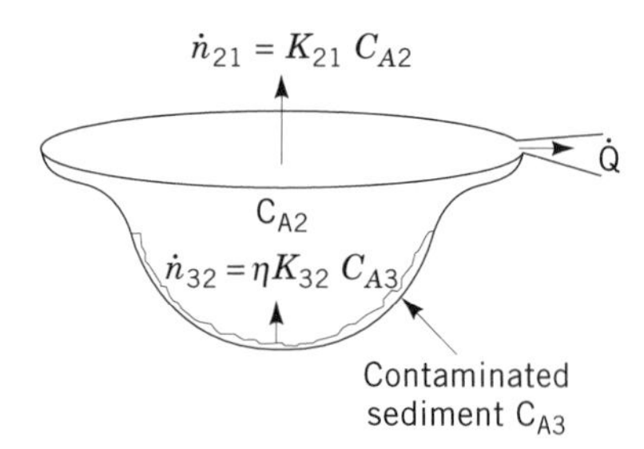

Figure 3. Depiction of setting for estimating water concentrations as a result of bioturbation induced transport from contaminated sediments.

turbation is likely to be the most significant means of moving contaminants to the sediment-water interface in the absence of active scouring of the sediment bed (20). Bioturbation is the movement of sediment and its associated contaminants as a result of the normal activities of benthic organisms. A variety of organisms contribute to the bioturbation of the sediment and their mechanism of action is also highly variable. Perhaps the most important class of organisms are conveyor belt feeders that process sediments for food and defecate at the surface. These organisms are very effective at moving sediment containing a sorbed hydrophobic organic contaminant to the surface. Initially, the organisms are likely to accumulate the contaminants in their bodies. Feeding of higher animals on these organisms can result in bioaccumulation in the food chain. The organism will also pass contaminants with their feces, to the surface. There the contaminant can desorb into the overlying water column before being covered by subsequent defecation. Among the more important conveyor-belt feeders is the worm, *Tubifex tubifex*, which tend to dominant benthic populations in contaminated freshwater sediments. Densities of 10,000–100,000 worms/m^2 have been observed in Great Lakes harbor sediments and sediment processing rates are of the order of 1–10 mg of sediment per worm per day (21). Only a fraction (η in Equation 30) of the contaminants brought to the surface by the organisms is released before being buried by fresh sediment. Measurements (22) with *Tubifex tubifex* indicate that about 20% of the contaminants brought to the surface desorb before being reburied as a result of water-side and intraparticle mass transfer resistances. It has been observed that the percentage of contaminants released at the surface was smaller for more hydrophobic compounds and for conditions (such as high dissolved oxygen content in the water) which led to the organisms penetrating deeper into the sediment (21).

The rate of particle movement to the surface has been estimated by various researchers through use of an effective diffusion coefficient, an effective mass transfer coefficient, or an effective overturning rate of the upper layer of the sediment. Ref. 23 gives estimates an effective diffusion coefficient of about 10 cm^2/yr for PCBs over the upper 10 cm of sediment in New Bedford Harbor estuarine sediment. This coefficient corresponds to an effective mass transfer coefficient of about 1 cm/yr or an effective overturning rate of the upper 10 cm of sediment of 0.1 yr^{-1}. Reible and co-workers (21) also report transport rates equivalent to an effective mass transfer coefficient of about 1–2 cm/yr for a laboratory experiment with *Tubifex tubifex* at organism densities of about 25,000/m^2. A wide variation in measured effective bioturbation diffusion coefficients depending on type of organism, organism density, and temperature has been reported (24). Values were most commonly in the range of 0.3–30 cm^2/yr for inland and near-coastal water conditions, however, or equivalent to a mass transfer coefficient of 0.03–3 cm/yr if biodiffusion were active over a layer of the order of 10 cm. The flux of contaminants via bioturbation is then estimated by

$$\dot{n}_{32} = \eta K_{32} C_{A3} \qquad (31)$$

where η is the fraction of contaminants brought to the surface that are released and K_{32} is the effective bioturbation mass transfer coefficient.

The release of contaminants to the water column by bioturbation of the contaminated sediment is balanced by evaporation and flushing. Only the truly dissolved portion of a hydrophobic organic contaminant is volatile. The total water column burden of contaminant also includes portions sorbed to particulate and colloidal matter. Operationally, particulate matter is generally defined as the mass collected on a 0.45 μm filter and everything smaller is considered dissolved. In reality, however, natural organic colloids formed by the natural decomposition of plants and animals exist within the "dissolved" fraction and can significantly influence the "solubility" of a hydrophobic organic compound in the water. This colloidal material is typically high molecular weight humic and fulvic acids. A surrogate for direct measurement of this natural organic matter is dissolved organic carbon. Its concentration (ρ_{oc}) is often used to indicate the distribution of a contaminant between the dissolved and colloidally-sorbed state. It is also often assumed that the dissolved organic carbon is as effective at sorbing a hydrophobic organic species as solid phase associated carbon. That is, the partition coefficient between water and the dissolved organic carbon is assumed to be equal to α_{oc}. The total "dissolved" concentration in the water is then the sum of the dissolved and colloidally sorbed fraction and can be written

$$C_{A2}^{T} = (1 + \alpha_{oc}\rho_{oc})C_{A2} \qquad (32)$$

where C_{A2} represents just the contaminant in the truly dissolved state. The total concentration in water of a strongly sorbing compound (large α_{oc}) can be significantly larger than the dissolved concentration. As little as 10 mg/L dissolved organic carbon in water would mean that about half of the moderately hydrophobic compound pyrene ($\alpha_{oc} \sim 10^5$) would be present in the colloidal fraction and about half in the truly dissolved fraction. Only the truly dissolved fraction is directly available for evaporation or partitioning with the underlying sediment. Since the operational definition of "dissolved" is typically that which passes a 0.45 μm filter, ie, the sum of the truly dissolved and colloidal concentration, this concentration would overestimate the evaporative flux and the amount sorbed on suspended particulate carbon. The latter problem has been identified as a likely cause of the apparent constancy of the effective distribution coefficient between water and suspended particulates (eg, see ref. 25). The correct evaporative flux, for example, is given by

$$\dot{n}_{21} = K_{21} C_{A2} = K_{21} \frac{C_{A2}^T}{1 + \alpha_{oc}\rho_{oc}} \quad (33)$$

Flushing of the water column by the flow of water through the system, however, removes contaminants in dissolved, colloidal and particulate forms. If ρ_{oc} represents the water column density of organic matter (either colloidal or particulate), the contaminant removal rate by flushing is given by

$$\dot{N}_{22} = QC_{A2}^T = Q(1 + \alpha_{oc}\rho_{oc})C_{A2} \quad (34)$$

Here the upper case symbol $\dot{N}_{22}$ represents total molar flushing rate and not a flux. Equations 31, 33 and 34 can be combined through an overall molar balance on the lake to give equation 30.

Capped Sediment

One means of controlling the water column concentration is by reduction of the flux from the sediments by capping with clean sediments. Since contaminated sediments are generally of interest only in low energy environments, the physical integrity of such a cap is likely to be maintained for long periods of time. It is also possible to armor a cap with large diameter stone or gravel to more effectively maintain its physical integrity. The role of a cap in containing the chemical contaminants, however, also requires assessment of its chemical integrity. The effects of a capping layer include elimination of the bioturbating population at the original contaminated sediment–water interface and extension of the path length through which the chemical contaminants must diffuse. Under steady conditions and assuming that only diffusion occurs in the cap, the capping layer will reduce the contaminant flux over that expected from the uncapped sediment (assumed controlled by bioturbation) by

$$\frac{(\dot{n}_{32})_{Capped}}{(\dot{n}_{32})_{Uncapped}} = D_{A2} \frac{\varepsilon_2^{4/3}}{H K_{32}\,\alpha_{32}} \quad (35)$$

where H is the effective capping layer height and use is once again made of the Millington and Quirk model (12) for the effective diffusion coefficient in the porous cap. Given a typical water diffusivity of the order of 10^{-5} cm^2/s and an effective bioturbation mass transfer coefficient of about 1 cm/hr, the capped flux will be about 5000 times smaller for a strongly sorbing compound ($\alpha_{32} > 10^4$) with a cap of the order of 50 cm. The primary effect is elimination of bioturbation, which effectively moves both sediment particles and porewater for a purely porewater driven transport process (diffusion).

In addition to reducing the steady contaminant flux, the presence of a cap will significantly delay the movement of the contaminant to the overlying water. The model of equation 3 can be used to estimate the time before contaminants begin to break through the capping layer. At a value of the error function of about 1.8, the model predicts a concentration at the top of the cap equal to about 1% of that at the bottom. Thus the time required for this to occur is given by

$$\tau = \frac{H^2 R_f}{13\, D_{A2}\, \varepsilon^{4/3}} \quad (36)$$

This time period is about 20,000 years for a 40% porosity cap with an effective depth of 50 cm over a sediment containing a sorbing compound with an R_f of 10^4 (effectively $\alpha_{32} \sim 10^4$). Although the physical integrity of the cap cannot be guaranteed over this period, it is clear that a well-designed and maintained cap can be a very effective means of isolating contaminants from the water column.

APPLICATIONS OF CHEMODYNAMICS TO SOIL CONTAMINATION

Contaminant migration in soils involves many of the same processes that influence contaminant migration in sediments. Commonly, however, contaminant migration in soils is complicated by multiple phases. The contaminant may be present as a separate nonaqueous phase or one of its components. If the nonaqueous phase is present as a discontinuous, immobile residual, the contaminant would be essentially immobile except for that which might partition into a mobile phase such as air or water. Transport in the mobile phase is further complicated by sorption onto the immobile soil. See Ref. 26 for a description of the processes influencing contaminant transport in soils. A shorter discussion including chemodynamic models can be found in Ref. 27. General hydrogeological concepts are discussed in Refs. 28 and 29.

To illustrate the application of chemodynamic principles to contaminant transport in soils, let us consider a gasoline or similar nonaqueous phase liquid (NAPL) that has spilled or leaked in the unsaturated zone of the soil. This phase would tend to migrate downward as a result of gravity. A residual that might fill 10–20% of the soil pore space would remain, however, trapped by capillary forces. The NAPL is generally ineffective at displacing the residual water that typically wets the media and is also trapped by capillary forces at a volumetric content of also about 10–20%. Air, however, is easily displaced by the nonaqueous phase.

If an unsaturated soil contains low permeability regions, the displacement of air results in entrapment of the nonaqueous phase in regions with high capillary suctions. The discontinuity in capillary suctions between a fine media and an underlying coarse media can result in a pore space essentially saturated with the NAPL perched at the base of the fine material. The physical situation at long times after the spill or leak is then an essentially homogeneous residual NAPL in coarse zones with essentially saturated pockets of NAPL in fine grained material. If the fine-grained material is initially saturated with water, however, the inability of the nonaqueous phase to displace the water under natural hydraulic gradients results in the nonaqueous phase remaining perched on top of the fine-grained material. Thus in an unsaturated soil, a residual pool of nonaqueous phase material is likely to be

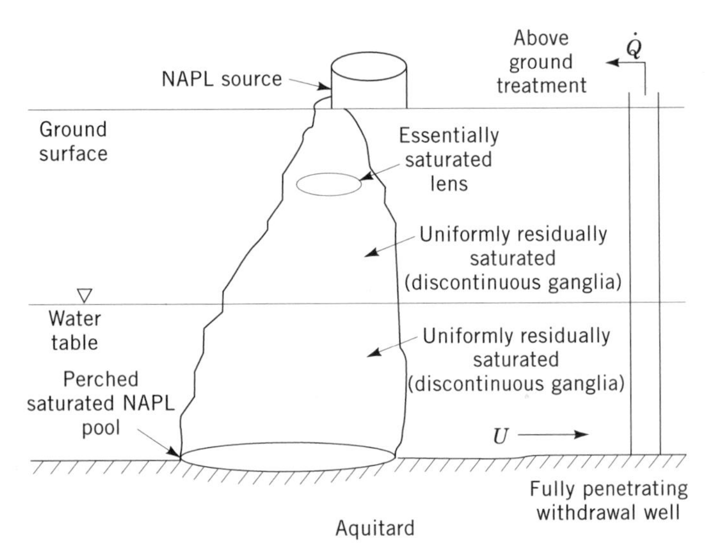

Figure 4. Soil and groundwater contamination scenario. For purposes of illustration, the withdrawal well is not assumed to penetrate the contaminated region.

found *within* zones of low permeability, while in saturated soil, a residual pool of nonaqueous phase liquid is likely to be found *on the surface* of the low permeability media. This configuration is depicted in Figure 4.

If sufficient nonaqueous phase liquid is released, the contamination may reach the underlying water table before being completely trapped by capillary forces. If the nonaqueous phase liquid were less dense than water, the phase would tend to remain at the surface of the water table. Draw-down of the water table, for example by seasonal variations or by depression by the weight of the nonaqueous phase, however, would allow deeper penetration. A subsequent rise in the water table could result in the burial of this phase within the water table. A nonaqueous phase liquid more dense than water would tend to penetrate the water table and continue to move downward until hindered by a low permeability formation. Because the water would tend to be more wetting, the nonaqueous phase would tend to come to rest on top of the lens without significant penetration as indicated above.

Any residual contaminant would be subject to dissolution by infiltrating water and, if in contact with the underlying aquifer, dissolution into the ground waters. Any residual contaminant in the unsaturated zone would also be subject to vaporization. For purposes of illustration, let us consider a groundwater aquifer contaminated by a chlorinated solvent, eg, tetrachloroethylene, in an essentially water-immiscible heavy oil. If the system is more dense than water, penetration of the water table may occur as indicated previously. Let us assume that the region below the water table is uniformly contaminated except in a pool immediately above a low permeability clay lens. It is desired to remediate this contamination via in-situ means. In-situ remediation of soil contamination is a desirable goal in that soil removal with the incumbent expense and potential for losses (by evaporation, during transportation, etc) is eliminated. In addition, for contaminants below the water table, in-situ remediation may represent the only viable remediation option. In this case, in-situ remediation entails the pumping and aboveground treatment of the contaminated ground water. Water injection or recharge, addition of surfactants and other means, are often used to increase water flowrate or the solubility of the contaminant in the water. The processes of interest here are the dissolution of the contaminant and transport of the contaminated water to a withdrawal well. Initially, the contaminant levels in the withdrawn water are controlled by the sorption/desorption processes in the zone between the contaminated region and the withdrawal well. The contaminant removal rates are then controlled by mass transfer from the essentially uniform residual that fills much of the aquifer. Ultimately, however, this residual will be largely removed and the pool of contaminant perched on the low permeability layer at the bottom of the aquifer remains. This zone will pose a long-term source of contaminants and represents the primary reason that pumping and treating of contaminated ground water is often only effective as a containment technology and ultimate cleanup is extremely slow. The analysis of all three stages of the cleanup will be considered separately.

Mobilization Stage

Passage of the water through the essentially uniformly contaminated zone in the water column results in dissolution of the chemical into the water. The rate of the transfer is dependent upon the local velocity of the water and the surface area of the residual nonaqueous phase liquid ganglia or droplets in the water. Some correlations for the estimation of the transfer rate between the nonaqueous phase and the water are found in Ref. 38. If the length of the nonaqueous phase contaminated region is sufficiently large and the residual spread sufficiently uniform, however, the water exiting the contaminated region might be assumed in equilibrium with the nonaqueous phase as given by equation 16. This situation also represents an upper bound to the effectiveness of a pump and aboveground treat system in that the effluent concentration can be no greater than that at equilibrium with the nonaqueous phase. Note that the equilibrium is not the pure component solubility but, as given by equation 16, is directly proportional to the molar concentration of the partitioning contaminant in the nonaqueous phase.

The contaminants picked up by the water in the contaminated region are transported to the withdrawal well. The zone between the contaminated region and the withdrawal well is initially devoid of contaminants and thus a sorbing contaminant will initially accumulate in this zone, delaying transport to the withdrawal well. During this transient period the velocity of the chemical front toward the withdrawal well is given by:

$$u_A = \frac{U}{R_f} \tag{37}$$

Here u_A is the contaminant front velocity and U is the D'Arcy velocity; that is, a superficial velocity given by the product of the hydraulic gradient and the conductivity of the soil. There will also be dispersion about this front which leads to more rapid transport at lower concentrations. One source of dispersion is stratification or layering of different permeability materials. For a given average hydraulic gradient and hydraulic conductivity, the velocity in the more permeable layers will be higher than in low permeability layers, spreading the contaminant front horizontally. Spreading also occurs perpendicular to the direction of flow but at a slower rate. The dispersion coefficient is typically written as the product of a dispersivity and the effective contaminant velocity plus a contribution for molecular diffusion,

$$D_x = \frac{d}{R_f} U + D_{A2}\, \varepsilon^{4/3} \qquad D_y D_z \approx 0.1 \frac{d}{R_f} U + D_{A2}\, \varepsilon^{4/3} \quad (38)$$

The dispersivity, d, has units of length and scales with the size of the heterogeneities encountered by the contaminant plume. Since many different scales of soil heterogeneities exist, the dispersivity tends to change with distance as these different scales are encountered. Ref. 31 suggests

$$d \approx 0.017\, L^{1.5} \quad (39)$$

where L is the travel distance in m. The relative magnitude of advective and dispersive transport rates are described by the Peclet number, which for Neuman's model, becomes,

$$Pe = \frac{u_A L}{D_x} \approx 59\, L^{-0.5} \quad (40)$$

Equation 40 also assumes that molecular diffusion can be neglected. Although dispersivities and dispersion coefficients vary a great deal under field conditions, equation 40 provides at least a rough indication that advective transport dominates over the travel distances and conditions of interest; that is, 10–100 m distances with sufficient groundwater withdrawal such that molecular diffusion is negligible. The time required for the contaminant to reach the withdrawal well is then the transport distance ($x_w - x_s$, where x_w and x_s refer to the withdrawal well and source area locations, respectively) divided by the interstitial contaminant velocity, equation 37.

Uniform Residual Dissolution Stage

Ultimately, however, the soil will no longer accumulate contaminants and the mass flow rate of contaminants at the withdrawal well is then simply equal to the rate of release of contaminants from the source area. Under the ideal condition that the water passing through the source area becomes equilibrated with the contaminant and that the withdrawal well removes only water that had passed through the contaminated zone, the contaminant removal rate from the residual nonaqueous phase liquid to the withdrawn water is given by

$$\dot{N}_{42} = \dot{Q}\, C^*_{A2} \qquad \text{for } t > \frac{x_w - x_s}{u_A} \quad (41)$$

where $\dot{Q}$ is the volumetric production rate of the withdrawal well. The use of a saturated concentration during this period assumes that the uniformly contaminated region continues to feed contaminates at this concentration into the system. In reality, flow nonuniformities will tend to result in a reduction of concentration and thus removal efficiency with time.

Pool Dissolution Stage

Ultimately, however, the pool floating on the low permeability lens will be the primary source of contamination. This pool is unlikely to be displaced by the water movement to the well because of the low water velocities in the vicinity of the low permeability region and because water tends to wet the medium making it difficult for a nonaqueous phase to displace water from water-filled pore spaces. Thus the nonaqueous phase pool represents a significant quantity of contaminants that must be remediated during a pump and above-ground treatment process by dissolution into the water flowing to the withdrawal well. Because only a limited volume of withdrawn water passes near the contaminant pool, the concentration in the withdrawal well is much reduced over that defined by equation 41. The volume of contaminants in the water being convected horizontally toward the well is largely a function of the vertical rate of diffusion or dispersion processes upward from the essentially saturated pool. This process can be modeled with the error function diffusion model, equation 3, where time is taken as the contact time for water flowing over the nonaqueous phase pool (L/U), and the effective diffusion coefficient is taken as the dispersion coefficient in the direction perpendicular to flow. The water concentration well above the pool is assumed to be zero since the uniformly contaminated region has been presumed cleaned of contaminant before significant dissolution of the pool would occur. The concentration in water at the pool–groundwater interface is the equilibrium concentration, $\alpha_{24} C_{A4}$, defined by Raoult's Law. The contaminant release rate to the withdrawn water by this model can be found by integrating equation 4 over the length of the pool to give

$$\dot{N}_{42} = W L\, \alpha_{24}\, C_{A4} \sqrt{\frac{4 D_z R_f U}{\pi L}} \quad (42)$$

This contaminant release rate is equal to the rate of removal at the withdrawal well.

The contaminant removal rates during each of the three stages of mobilization, uniform residual dissolution and pool residual dissolution are depicted in Figure 5. Unfortunately, the long "tail" in the removal rate associated with the dissolution of the pooled residual at the base of the aquifer can last for centuries. As a result, conventional pump and treat technology is mostly used to contain the contamination by controlling the hydraulic gradient in a contaminated area and is less effective as a final cleanup process.

SUMMARY AND CONCLUSIONS

The key concepts of chemodynamics as applied to exposure assessment and the evaluation of in-situ remediation and containment technologies have been presented. The

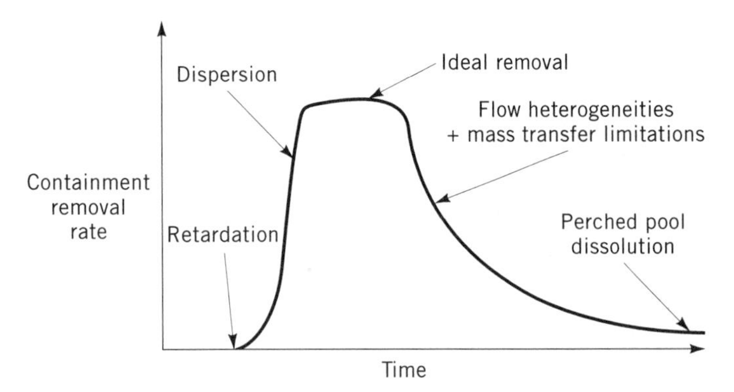

Figure 5. Hypothetical removal versus time during groundwater pumping and surface treating of contaminants in scenario shown in Figure 4.

concepts have also been illustrated by consideration of examples in each of the environmental media: air, water and soil. The discussion of the concepts and the examples, emphasize the similarity of approaches to analyze contaminant migration problems in these media.

The key feature of many chemodynamic models is the capture of just the most fundamental physiochemical processes. By such an approach it is often not possible to describe all of the behavior of an environmental system, but the models presented herein do describe at least semi-quantitatively the systems for which they were designed. In some cases more sophisticated models are not yet supported by experiment despite the obvious limitations of the simpler models. In addition, the sparse dataset typically available in any particular field application often limits the ability to use more sophisticated models regardless of their level of validation by theory and laboratory experiments. The level of models presented herein often represent an excellent compromise between the ability to describe a physical system and the data input required to drive the model. The ultimate test of such models, of course, is a comparison to field observations. In this work, the field observations have been presented only in the form of empirical correlations for some of the model parameters. The reader is referred to the original works for the data and motivation for the form of the models.

Acknowledgments

The author would like to acknowledge partial support for this work by the U.S. Environmental Protection Agency Risk Reduction Engineering Laboratory (supporting studies of capping contaminated sediments); the U.S. Environmental Protection Agency Office of Exploratory Research and Louisiana State University Hazardous Substances Research Center (supporting studies of bioturbation); the Louisiana Board of Regents; and the United States National Science Foundation (supporting studies of nonaqueous phase liquid dynamics and colloidal facilitation of transport).

BIBLIOGRAPHY

1. L. J. Thibodeaux, *Chemodynamics Movement of Chemicals in Air, Water and Soil*, Wiley-Interscience, New York, 1979.
2. Environmental Protection Agency. *Superfund Exposure Assessment Manual*, 1988, EPA/540/188/001.
3. H. S. Carslaw, J. C. Jaeger, *Conduction of Heat in Solids,* 2nd ed., Oxford Science Publications, Oxford, UK, 1959.
4. J. Crank, *The Mathematics of Diffusion*, Clarendon Press, Oxford, UK, 1975.
5. Treybal, *Mass Transfer Operations*, McGraw-Hill, New York, 1974.
6. L. J. Thibodeaux, *Chemodynamics*, Wiley-Interscience, New York, 1979.
7. J.H. Montgomery and L. M. Welkom. *Groundwater Chemicals Desk Reference*, Lewis Publishers, Chelsea, Mich., 1990.
8. W. J. Lyman, W. F. Reehl, and D. H. Rosenblatt, *Handbook of Chemical Property Estimation Methods*, American Chemical Society, Washington, D.C., 1990.
9. K. T. Valsaraj and L. J. Thibodeaux. "Equilibrium Adsorption of Chemical Vapors on Surface Soils, Landfills and Landfarms—A review," *J. of Hazardous Materials,* **19,** 79–99 (1988).
10. R. G. Thomas, "Volatization from Water," in W. J. Lyman, W. F. Reehl, and D. H. Rosenblatt, eds., *Handbook of Chemical Property Estimation Methods*, American Chemical Society, New York, 1990.
11. C. Springer, P. D. Lunney, K. T. Valsaraj, and L. J. Thibodeaux, *Emission of Hazardous Chemicals from Surface and Near Surface Impoundments to Air. Part A. Surface Impoundments*., Final Report to U.S. EPA, Hazardous Waste Engineering Research Laboratory, Cincinnati, Ohio, 1986.
12. R. J. Millington and J. P. Quirk, "Permeability of Porous Solids," *Trans. Faraday Soc.,* **57,** 1200 (1961).
13. S. R. Hanna, G. A. Briggs, and R. P. Hosker, *Handbook on Atmospheric Diffusion*, Technical Information Center, U.S. Department of Energy, 1982.
14. P. Zanneti, *Air Pollution Modeling*, Van Nostrand Reinhold Co., Inc., New York, 1990.
15. F. Pasquill, "The Dispersion of Material in The Atmospheric Boundary Layer—The basis for Generalization," in *Lectures on Air Pollution and Environmental Analysis*, American Meteorological Society, Boston, Mass., 1975.
16. N. P. Chitgopekar, D. D. Reible, and L. J. Thibodeaux, "Modeling Short Range Air Dispersion From Area Sources of Non-Buoyant Toxics," *J. Air Waste Manage. Assoc.* **40,** 1121–1128 (1990).
17. D. B. Turner, "Atmospheric Dispersion Modeling: A Critical Review," *J. Air Pollut. Control Assoc.* **29,** 502–519 (1979).
18. G. A. Briggs, *Diffusion Estimates for Small Emissions*, ATDL Contribution File No. 79, Atmospheric Turbulence and Diffusion Laboratory, 1973.
19. P. M. Gschwend, S. Wu, O. S. Madsen, J. C. Wilkin, R. B. Ambrose, and S. C. McCutcheon, *Modeling the Benthos-Water Column Exchange of Hydrophobic Chemicals*, Final Report to U.S. EPA Environmental Research Laboratory, Athens, Ga., 1987.
20. D. D. Reible, K. T. Valsaraj, and L. J. Thibodeaux, "Chemodynamic Models for Transport of Contaminants From Sediment Beds," in O. Hutzinger, ed., *Handbook of Environmental Chemistry*, Vol. 2 part F, 1991.
21. D. D. Reible, L. J. Thibodeaux, K. T. Valsaraj, F. Lin, M. Dikshit, V. Popov, M. A. Todaro, and J. W. Fleeger, "Pollutant Fluxes to Aquatic Systems Via Coupled Biological and Physicochemical Bed-Sediment Processes," *IAWQ Conference on Contaminated Sediment*, Milwaukee, Wis., June, 1993.

22. K. R. Morris and S. M. Karickhoff. *Environ. Toxico. Chem.* **4,** 469 (1985).
23. L. J. Thibodeaux, D. D. Reible, W. S. Bosworth, and L. C. Sarapas, *A Theoretical Evaluation of The Effectiveness of Capping PCB Contaminated New Bedford Harbor Sediment*, Final Report to the LSU Hazardous Waste Research Center, Baton Rouge, La., 1990.
24. G. Matisoff, "Mathematical Models of Bioturbation," in R. L. McCall and M. I. S. Tevasz, eds., *Animal Sediment Relations*, Plenum Press, N.Y., 1982.
25. J. E. Baker, S. J. Eisenreich, and D. L. Swackhamer, "Field Measured Associations Between Polychlorinated Biphenyls and Suspended Solids in Natural Waters: An Evaluation of the Partitioning Paradigm," chapter 4 in R. A. Baker, ed., *Organic Substances and Sediments in Water*, Lewis Publishers, Boca Raton, Fla., 1991.
26. R. C. Knox, D. A. Sabatini, and L. W. Canter, *Subsurface Transport and Fate Processes*, Lewis Publishers, Boca Raton, Fla., 1993.
27. J. H. Clarke, D. D. Reible, and R. D. Mutch. "Contaminant Transport and Behavior in the Subsurface, in D. J. Wilson and A. Clarke, eds., *Soil Clean-Up: Technology and Applications*, Marcel-Dekker, Inc., New York, 1993.
28. R. A. Freeze and J. A. Cherry, *Groundwater*, Prentice-Hall, Inc., Englewood Cliffs, N.J., 1979.
29. C. W. Fetter, *Applied Hydrogeology, 2nd ed.* Merrill Publishing, Columbus, Ohio, 1988.
30. H. O. Phannkuch, "Determination of The Contaminant Source Strength From Mass Exchange Processes at the Petroleum Ground-Water Inteface in Shallow Aquifer Systems," *Petroleum Hydrocarbons and Organic Chemicals in Ground Water*, National Water Well Association, Worthington, Ohio, 1984, pp. 111–129.
31. S. P. Neuman, "Universal Scaling of Hydraulic Conductivities and Dispersivities in Geologic Media," *Water Resources Research* **26**(8), 1749–1758 (1990).

THE CLEAN AIR ACT

SUSAN MAYER
Congressional Research Service
Washington, D.C.

The Clean Air Act (CAA), codified as 42 U.S.C. 7401 *et seq.*, seeks to protect human health and the environment from emissions that pollute ambient, or outdoor, air. To ensure that air quality in all areas of the United States meets certain federally mandated minimum standards, it assigns primary responsibility to States to assure adequate air quality. The Act deems areas not meeting standards as nonattainment areas, and requires them to implement specified air pollution controls. In addition to provisions relating to ozone nonattainment, the Act addresses mobile sources, air toxics, and the special problem of acid rain. It establishes a comprehensive permit system for all sources. Other provisions address ozone depleting substances, enforcement, clean air research, disadvantaged business concerns, and employment transition and assistance.

BACKGROUND

Prior to 1955, air pollution was controlled at the State and local level. Federal legislation controlling air pollution was first passed in 1955. The Federal role was strengthened in subsequent amendments and changed significantly with the passage of the Clean Air Act Amendments of 1990 (P.L. 101-549) which was signed into law on November 15, 1990.

Significant new provisions added to the Act by the 1990 amendments included (1) the acid rain control program, including the use of marketable allowances for introducing flexibility into the sulfur oxides reduction program; (2) a State-run program requiring permits for the operation of many sources of air pollutants, including a requirement that fees be imposed to cover administrative costs; and (3) the authorization for a 5-year, $250 million program for retraining and unemployment benefits for workers displaced by requirements of the CAA.

Changes to the Act by the 1990 amendments included, in part, provisions to (1) classify areas in nonattainment according to the extent to which they exceed the standard and to tailor deadlines, planning, and controls according to each area's status and problem; (2) tighten automobile emission standards and to require reformulated and alternative fuels in the most-polluted areas; (3) revise the air toxics section, establishing a new program of technology-based standards and addressing the problem of sudden, catastrophic releases of toxics; (4) change the stratospheric ozone protection provision to a phase-out of the most ozone-depleting chemicals; and (5) update the enforcement provisions so they parallel those in other pollution control acts, including authority for EPA to assess administrative penalties.

See also COAL COMBUSTION; COMBUSTION MODELING; ELECTRIC POWER GENERATION; ENERGY EFFICIENCY-ELECTRIC UTILITIES; ENVIRONMENTAL ECONOMICS.

The following sections describe each of these new and revised programs required by the Act.

NATIONAL AMBIENT AIR QUALITY STANDARDS (NAAQS) (SECTION 109)

The Act requires EPA to establish NAAQS for several types of air pollutants. The NAAQS must be designed to protect public health and welfare with an adequate margin of safety. The NAAQS must be attained as expeditiously as possible, and in no case more than five years after EPA determines that an area does not meet the standards, unless EPA provides an extension of the deadline. For areas not in attainment with NAAQS, the Clean Air Act Amendments of 1990 established special compliance schedules, staggered according to each area's air pollution problem and status, as described below.

EPA has promulgated NAAQS for six air pollutants: sulfur dioxide (SO_2), particulate matter (PM_{10}), nitrogen dioxide (NO_2), carbon monoxide (CO), ozone (O_8), and lead (Pb). The Act requires EPA to review the scientific data upon which the standards are based, and revise the standards, if necessary, every five years. More often than not,

Table 1. Clean Air Act and Principal Amendments[a]

Year	Act	Public Law Number
1955	Air Pollution Control Act	P.L. 84-159
1959	Reauthorization	P.L. 86-353
1960	Motor vehicle exhaust study	P.L. 86-493
1963	Clean Air Act Amendments	P.L. 88-206
1965	Motor Vehicle Air Pollution Control Act	P.L. 89-272, title I
1966	Clean Air Act Amendments of 1966	P.L. 89-675
1967	Air Quality Act of 1967	P.L. 90-148
	National Air Emission Standards Act	P.L. 90-148, title II
1970	Clean Air Act Amendments of 1970	P.L. 91-604
1973	Reauthorization	P.L. 93-13
1974	Energy Supply and Environmental Coordination Act of 1974	P.L. 93-319
1977	Clean Air Act Amendments of 1977	P.L. 95-95
1980	Acid Precipitation Act of 1980	P.L. 96-294, title VII
1981	Steel Industry Compliance Extension Act of 1981	P.L. 97-23
1987	Clean Air Act 8-month Extension	P.L. 100-202
1990	Clean Air Act Amendments of 1990	P.L. 101-549

[a] Codified generally as 42 U.S.C. 7401-7671.

however, EPA has taken more than five years in reviewing and revising the standards.

STATE IMPLEMENTATION PLANS (SIPS) (SECTION 110)

Although the Act authorizes the EPA to set NAAQS, the States are responsible for establishing procedures to attain and maintain the NAAQS. The States adopt plans, known as State Implementation Plans (SIPs), and submit them to EPA to ensure that they are adequate to meet the statutory requirements. Each State is responsible for achieving the NAAQS within its jurisdiction. Areas not meeting NAAQS are called "nonattainment" areas.

SIPs are based on emission inventories and computer models to determine whether air quality violations will occur. If these data show that standards would be exceeded, the State imposes necessary controls on existing sources to ensure that emissions do not cause "exceedences" of the standards. Proposed new and modified sources must obtain State construction permits in which the applicant shows how the anticipated emissions will not exceed allowable limits. In "nonattainment" areas, emissions from new or modified sources must be offset by reductions in emissions from existing sources.

The 1990 amendments *require* EPA to impose one of two sanctions in areas which fail to submit an SIP, fail to submit an adequate SIP, or fail to implement an SIP: either a 2-to-1 emissions offset for the construction of new polluting sources, or a ban on Federal highway grants (an additional ban on air quality grants is discretionary).

NONATTAINMENT REQUIREMENTS (PART D OF TITLE I)

In a great departure from the prior law, the 1990 Clean Air Act Amendments group nonattainment areas into classifications based on the extent to which the NAAQS is exceeded, and establish specific pollution controls and attainment dates for each classification.

Nonattainment areas are classified on the basis of a "design value," which is derived from the pollutant concentration (in parts per million) recorded by air quality monitoring devices. The design value for ozone is the fourth highest reading measured over a 3-year period. The Act creates 5 classes of ozone nonattainment, as shown in Table 4. Only Los Angeles falls into the "extreme" class. A simpler classification system establishes moderate and serious nonattainment areas for carbon monoxide and particulate matter with correspondingly more stringent control requirements for the more polluted class.

For ozone nonattainment areas, the deadlines are established for each classification. For carbon monoxide, the attainment date for moderate areas is December 31, 1995, and for serious areas is December 31, 2000. For particulate matter, the deadline for areas currently designated moderate nonattainment areas is December 31, 1994; for those areas subsequently designated as moderate, the

Table 2. Ozone Nonattainment Classifications

Class	Marginal	Moderate	Serious	Severe	Extreme
Deadline	3 years	6 years	9 years	15 years[a]	20 years
Areas	42 areas	31 areas	14 areas	9 areas	1 area
Design Value	0.121 ppm– 0.188 ppm	0.138 ppm– 0.160 ppm	0.160 ppm– 0.180 ppm	0.180 ppm– 0.280 ppm	>0280 ppm

[a] Areas with a 1988 design value between 0.190 and 0.280 ppm have 17 years to attain, rather than 15 years.

deadline is 6 years after designation. For serious areas, the respective deadlines are December 31, 2001 or 10 years after designation.

Although areas with more severe air pollution problems have a longer time to meet the standards, more stringent control requirements are imposed through the SIP. Each category of nonattainment must have specified air pollution control measures imposed. Marginal ozone nonattainment areas are required to have few controls imposed. Each area must meet the requirements of lower nonattainment categories in addition to the requirements of its own category. A summary of the primary ozone control requirements for each nonattainment category follows.

Marginal Areas. Inventory emissions sources to be updated every 3 years. Require 1.1 to 1 offsets, ie, industries must reduce emissions from existing facilities by 10 percent more than emissions of any new facility opened in the area. Impose reasonable available control technology (RACT) on all major sources emitting more than 100 tons per year for the nine industrial categories where EPA had already issued control technique guidelines describing RACT prior to 1990.

Moderate Areas. Meet all requirements for marginal areas. Impose a 15 percent reduction in volatile organic compounds (VOCs) in 6 years. Adopt a basic vehicle inspection and maintenance program. Impose RACT on all major sources emitting more than 100 tons per year for all additional industrial categories where EPA will issue control technique guidelines describing RACT. Require vapor recovery at gas stations selling more than 10,000 gallons per month. Require 1.15 to 1 offsets.

Serious Areas. Meet all requirements for moderate areas. Reduce definition of a major source of VOCs from emissions of 100 tons per year to 50 tons per year for the purpose of imposing RACT. Reduce VOCs 3 percent annually for years 7–9 after the 15 percent reduction already required by year 6. Improve monitoring. Adopt an enhanced vehicle inspection and maintenance program. Require fleet vehicles to use clean alternative fuels. Adopt transportation control measures if the number of vehicle miles travelled in the area is greater than expected. Require 1.2 to 1 offsets. Adopt contingency measures if the area does not meet required VOC reductions.

Severe Areas. Meet all requirements for serious areas. Reduce definition of a major source of VOCs from emissions of 50 tons per year to 25 tons per year for the purpose of imposing RACT. Adopt specified transportation control measures. Implement a reformulated gasoline program for all vehicles in the area. Require 1.3 to 1 offsets. Impose $5,000 per ton penalties on major sources if the area does not meet required reductions.

Extreme Areas. Meet all requirements for severe areas. Reduce definition of a major source of VOCs from emissions of 25 tons per year to 10 tons per year for the purpose of imposing RACT. Require clean fuels or advanced control technology for boilers emitting more than 25 tons per year of NO_x. Require 1.5 to 1 offsets.

As with ozone nonattainment areas, CO nonattainment areas are subjected to specified control requirements. A summary of the primary CO control requirements for each nonattainment category follows.

Moderate Areas. Conduct an inventory of emissions sources. Forecast total vehicle miles travelled in the area. Adopt an enhanced vehicle inspection and maintenance program, and demonstrate annual improvements sufficient to attain the standard.

Serious Areas. Adopt specified transportation control measures. Implement a reformulated gasoline program for all vehicles in the area, and reduce definition of a major source of CO from emissions of 100 tons per year to 50 tons per year if stationary sources contribute significantly to the CO problem.

Serious areas failing to attain the standard by the deadline have to revise their SIP and demonstrate reductions of 5 percent per year until the standard is attained.

EMISSION STANDARDS FOR MOBILE SOURCES (TITLE II)

The 1990 amendments tightened automobile emission standards. (For cars, the hydrocarbon standard is reduced by 40%, the nitrogen oxides (NO_x) standard by 50%). These standards would be phased in over the 1994–1996 model years. If a study due in 1997 shows both the technological feasibility and the need for air quality improvement, yet another round of tightened standards would be imposed in 2004.

The 1990 amendments also require that a cleaner, "reformulated" gasoline be sold, starting in 1995, in the nine worst ozone nonattainment areas (Los Angeles, San Diego, Houston, Baltimore, Philadelphia, New York, Hartford, Chicago, and Milwaukee). Other ozone nonattainment areas will be able to opt in to the reformulated gasoline program at a later date.

Use of alternative fuels other than gasoline will be stimulated by two programs. First, California is required to develop a program requiring low emission vehicles (LEVs) and ultralow emission vehicles (ULEVs) for use in the State to be introduced starting in 1996 at 150,000 vehicles per year and rising to 300,000 vehicles per year in 1999. Second, in all of the most seriously polluted ozone nonattainment areas and the worst CO nonattainment areas, centrally fueled fleets of 10 or more vehicles must purchase some "clean fuel vehicles" when they add new vehicles to existing fleets starting in 1998. The percentage of any new vehicles purchased that must be clean fuel vehicles is 30 percent in 1998, 50 percent in 1999, and 70 percent in 2000. The "clean fuel" vehicles must be the same as those sold in California.

The 1990 amendments also imposed tighter requirements on certification (an auto's useful life is defined as 100,000 miles instead of the earlier 50,000 miles), on emissions allowed during refueling, on low temperature CO emissions, on in-use performance over time, and on warranties for the most expensive emission control components (8 years/80,000 miles for the catalytic converter, electronic emissions control unit, and onboard emissions diagnostic unit). Regulations were also extended to include nonroad fuels and engines.

HAZARDOUS AIR POLLUTANTS (SECTION 112)

Completely rewritten by the Clean Air Act Amendments of 1990, section 112 of the Act establishes programs for protecting the public health and environment from exposure to toxic air pollutants. Section 112, as revised by the 1990 amendments, contains four major provisions including, Maximum Achievable Control Technology (MACT) requirements, health-based standards, standards for stationary area sources, and requirements for the prevention of catastrophic releases.

First, EPA is to establish technology-based emission standards, called MACT standards, for sources of 189 pollutants listed in the legislation, and to specify categories of sources subject to the emission standards. (Public Law 102-187, enacted on December 4, 1991, deleted hydrogen sulfide from the list of toxic pollutants leaving only 188.)

EPA is to revise the standards periodically (at least every eight years). EPA can, on its initiative or in response to a petition, add or delete substances or source categories from the lists. Section 112 establishes a presumption in favor of regulation for the designated chemicals; it requires regulation of a designated pollutant unless EPA or a petitioner is able to show "that there is adequate data on the health and environmental effects of the substance to determine that emissions, ambient concentrations, bioaccumulation or deposition of the substance may not reasonably be anticipated to cause any adverse effects to human health or adverse environmental effects."

Section 112 requires EPA to set standards for sources of the listed pollutants that achieve "the maximum degree of reduction in emissions" taking into account cost and other non-air-quality factors. The standards for new sources "shall not be less stringent than the most stringent emissions level that is achieved in practice by the best controlled similar source." The standards for existing sources may be less stringent than those for new sources, but must be more stringent than the emission limitations achieved by either the best performing 12 percent of existing sources (if there are more than 30 such sources in the category or subcategory) or the best performing 5 similar sources (if there are fewer than 30). Existing sources are given 3 years following promulgation of standards to achieve compliance, with a possible 1-year extension; additional extensions may be available for special circumstances or for certain categories of sources. Existing sources that achieve voluntary early emissions reductions will receive a 6-year extension for compliance with MACT.

For solid waste incinerators, EPA is to establish performance standards for compliance with both hazardous air pollutant provisions (section 112) and the new source performance requirements (section 111) of the Act. However, rules for both small and large incinerators are overdue.

The second major provision of section 112 sets health-based standards to address situations in which a significant residual risk of adverse health effects or a threat of adverse environmental effects remains after installation of MACT. This provision requires that EPA, after consultation with the Surgeon General of the United States, submit a report to Congress on the public health significance of residual risks, and make recommendations as to legislation regarding such risks. If Congress does not legislate in response to EPA's recommendations, then EPA is required to issue standards for categories of sources of hazardous air pollutants as necessary to protect the public health with an ample margin of safety or to prevent an adverse environmental effect. A residual risk standard is required for any source emitting a cancer-causing pollutant that poses an added risk to the most exposed person of more than 1-in-a-million. Residual risk standards would be due 8 years after promulgation of MACT for the affected source category. Existing sources have 90 days to comply with a residual risk standard, with a possible 2-year extension. In general, residual risk standards do not apply to area sources.

This provision also directs EPA to contract with the National Academy of Sciences for a study of risk assessment methodology, and creates a Risk Assessment and Management Commission to investigate policy implications and appropriate uses of risk assessment and risk management. These studies are designed to ensure that regulatory decisions are well based technically.

Third, in addition to the technology-based and health-based programs for major sources of hazardous air pollution, EPA is to establish standards for stationary "area sources" determined to present a threat of adverse effects to human health or the environment. (Area sources are numerous, small sources such as gas stations or dry cleaners that may cumulatively produce significant quantities of a pollutant.) The provision requires EPA to regulate the stationary area sources responsible for 90 percent of the emissions of the 30 hazardous air pollutants that present the greatest risk to public health in the largest number of urban areas. EPA is to list the sources and pollutants within 5 years of enactment, and promulgate regulations within 10 years. In setting the standard, EPA can impose less stringent "generally available" control technologies, rather than MACT.

Finally, section 112 addresses prevention of sudden, catastrophic releases of air toxics by establishing an independent Chemical Safety and Hazard Investigation Board. The Board will investigate accidents involving releases of hazardous substances, and conduct studies and prepare reports on the handling of toxic materials, as well as measures to reduce the risk of accidents.

EPA is also authorized to issue prevention, detection, and correction requirements for catastrophic releases of air toxics, which shall require owners and operators to prepare risk management plans for the listed chemicals, including a hazard assessment, measures to prevent releases, and a response program. EPA is to issue these regulations by November 1993, which become effective 3 years after promulgation (or after a substance is listed, whichever is later).

NEW SOURCE PERFORMANCE STANDARDS (NSPS) (SECTION 111)

NSPS establish nationally uniform, technology-based standards for categories of new industrial facilities. These standards accomplish two goals: first, they establish a consistent baseline for pollution control that competing firms must meet, and thereby remove any incentive for States or communities to weaken air pollution standards in order to attract polluting industry; and second, they preserve clean air to accommodate future growth, as well as for its own benefits.

NSPS establish maximum emission levels for new or extensively modified principal stationary sources—powerplants, steel mills, and smelters—with the emission levels determined by the best "adequately demonstrated" continuous control technology available, taking costs into account. EPA must regularly revise and update NSPS applicable to designated sources as new technology becomes available, since the goal is to prevent new pollution problems from developing and to force the installation of new control technology.

PREVENTION OF SIGNIFICANT DETERIORATION (PSD) (TITLE I, PART C)

Prevention of Significant Deterioration reflects the principle that areas where the air quality is better than required by NAAQS should be protected from significant new air pollution even if NAAQS would not be violated. The Act divides clean air areas in three classes, and specifies the increments of SO_2 and particulate pollution allowed in each. Class I areas include international and national parks, wilderness areas, or other such pristine areas; allowable increments of new pollution are very small. Class II areas include all attainment and not classifiable areas, not designated as Class I; allowable increments of new pollution are modest. Class III represents selected areas that States designate for development; allowable increments of new pollution are large (but not exceeding NAAQS).

Polluting sources in PSD areas must install best available control technology (BACT) that may be more strict than that required by NSPS. The justifications of the policy are that it protects air quality, provides an added margin of health protection, preserves clean air for future development, and prevents firms from gaining a competitive edge by "shopping" for "clean air" to pollute.

ACID DEPOSITION CONTROL (TITLE IV)

The Clean Air Act Amendments of 1990 added an acid deposition control program to the Act. It sets goals for the year 2000 of reducing annual SO_2 emissions by 10 million tons from 1980 levels and reducing annual NO_x emissions by 2 million tons, also from 1980 levels.

The SO_2 reductions are imposed in two steps. Under Phase 1, owners/operators of 111 facilities listed in the law that are larger than 100 megawatts must meet tonnage emission limitations by January 1, 1995. This would reduce SO_2 emission by about 3.5 million tons. Phase 2 would include facilities larger than 75 megawatts.

To introduce some flexibility in the distribution and timing of reductions, the Act creates a comprehensive permit and emissions allowance system. An allowance is a limited authorization to emit a ton of SO_2. Issued by EPA, the allowances would be allocated to Phase 1 and Phase 2 units in accordance with baseline emissions estimates. Powerplants which commence operation after November 15, 1990 would not receive any allowances. These new units would have to obtain allowances (offsets) from holders of existing allowances. Allowances may be traded nationally during either phase. The law also permits industrial sources and powerplants to sell allowances to utility systems under regulations to be developed by EPA. Allowances may be banked by a utility for future use or sale.

Beginning in 1993, the Act provides for two sales to improve the liquidity of the allowance system and ensure the availability of allowances for utilities and independent power producers who need them. A special reserve fund consisting of 2.8% of Phase 1 and Phase 2 allowance allocations is set aside for sale. Allowances from this fund (25,000 annually from 1993–1999 and 50,000 thereafter) are sold at a fixed price of $1,500 an allowance. Independent power producers have guaranteed rights to these allowances under certain conditions. An annual auction of allowances (150,000 from 1993–1995, and 250,000 from 1996–1999) is an open auction with no minimum price. Utilities with excess allowances may have them auctioned off at this auction, and any person may buy allowances.

The Act essentially caps SO_2 emissions at individual existing sources through a tonnage limitation, and at future plants through the allowance system. First, emissions from most existing sources are capped at a specified emission rate times an historic baseline level. Second, for plants commencing operation after November 15, 1990, emission must be completely offset with additional reductions at existing facilities beginning after Phase 2 compliance. However, as noted above, the law provides some allowances to future powerplants which meet certain criteria. The utility SO_2 emission cap is set at 8.9 million tons, with some exceptions.

The Act provides that if an affected unit does not have sufficient allowances to cover its emissions, it is subject to an excess emission penalty of $2,000 per ton of SO_2 and required to reduce an additional ton of SO_2 of pollution the next year for each ton of excess pollutant emitted.

The Act also requires EPA to inventory industrial emissions of SO_2 and to report every 5 years, beginning in 1995. If the inventory shows that industrial emissions may reach levels above 5.60 million tons per year, then EPA is to take action under the Act to ensure that the 5.60 million ton cap is not exceeded.

The Act requires EPA to set specific NO_x emission rate limitations—0.45 lb. per million Btu for tangentially fired boilers and 0.50 lb. per million Btu for wall-fired boilers—unless those rates can not be achieved by low-NO_x burner technology. Tangentially and wall-fired boilers affected by Phase 1 SO_2 controls must also meet NO_x requirements. EPA is to set emission limitations for other types of boilers by 1997 based on low-NO_x burner costs. In addition, EPA is to propose and promulgate a revised new source performance standard for NO_x from fossil fuel steam generating units by January 1, 1994.

PERMITS (TITLE V)

The Clean Air Act Amendments of 1990 added a Title V to the Act which requires States to administer a compre-

Table 3. Principal U.S. Code Sections of the Clean Air Act

42 U.S.C.	Section Title	Clean Air Act, as Amended
Subchapter I	Programs and Activities	
Part A	Air Quality Emissions and Limitations	
7401	Findings, Purpose	sec. 101[a]
7402	Cooperative activities	sec. 102[a]
7403	Research, investigation, training	sec. 103[a]
7404	Research relating to fuels and vehicles	sec. 104[b]
7405	Grants for support of air pollution and control programs	sec. 105[a]
7406	Interstate air quality agencies; program cost limitations	sec. 106
7407	Air quality control regions	sec. 107[a]
7408	Air quality criteria and control techniques	sec. 108[c]
7409	National primary and secondary air quality standards	sec. 109[c]
7410	SIPs for national primary and secondary air quality standards	sec. 110[c]
7411	Standards of performance for new stationary sources	sec. 111[c]
7412	National emission standards for hazardous air pollutants	sec. 112[c]
7413	Federal enforcement procedures	sec. 113[c]
7414	Recordkeeping inspections, monitoring and entry	sec. 114[c]
7415	International air pollution	sec. 115[a]
7416	Retention of State authority	sec. 116[c]
7417	Advisory committees	sec. 117[a]
7418	Control of pollution from Federal facilities	sec. 118[a]
7419	Primary nonferrous smelter orders	sec. 119[d]
7420	Noncompliance penalty	sec. 120[d]
7421	Consultation	sec. 121[d]
7422	Listing of certain unregulated pollutants	sec. 122[d]
7423	Stack heights	sec. 123[d]
7424	Assurance of adequacy of State plans	sec. 124[d]
7425	Measures to prevent economic disruption	sec. 125[d]
7426	Interstate pollution abatement	sec. 126[d]
7427	Public notification	sec. 127[d]
7428	State boards	sec. 128[d]
Part B	Ozone Protection (repealed)	
Part C	Prevention of Significant Deterioration of Air Quality	
Subpart I	Clean Air	
7470	Congressional declaration of purpose	sec. 160[d]
7471	Plan requirements	sec. 161[d]
7472	Initial classifications	sec. 162[d]
7473	Increments and ceilings	sec. 163[d]
7474	Area designations	sec. 164[d]
7475	Preconstruction requirements	sec. 165[d]
7476	Other pollutants	sec. 166[d]
7477	Enforcement	sec. 167[d]
7478	Period before plan approval	sec. 168[d]
7479	Definition	sec. 169[d]
Subpart II	Visibility Protection	
7491	Visibility protection of Federal class I areas	sec. 169a[d]
7492	Visibility	(New sec. 816)
Part D	Plan Requirements for Nonattainment Areas	
7501	Definitions	sec. 171[d]
7502	Nonattainment plan provisions	sec. 172[d]
7503	Permit requirements	sec. 173[d]
7504	Planning procedures	sec. 174[d]
7505	Environmental Protection Agency grants	sec. 175[d]
7506	Limitations on certain Federal assistance	sec. 176[d]
7507	New motor vehicle emission standards in nonattainment areas	sec. 177[d]
7508	Guidance documents	sec. 178[d]
7509	Sanctions (NEW)	(NEW, sec. 102(g)
7511	Additional Provisions for Ozone Nonattainment Areas (NEW)	(NEW, sec. 103, 819)
7512	Additional Provisions for Carbon Monoxide Areas (NEW)	(NEW, sec. 104)
7513	Particulate Matter (PM_{10})	(sec. 105(a))
7514	Additional Provisions for Areas Designated Nonattainment for Sulfur Oxides, Nitrogen Dioxide, and Lead (NEW)	(NEW, sec. 106)
7415	Savings Provisions (NEW)	(NEW, sec. 108(1))

Table 3. *(Continued)*

42 U.S.C.	Section Title	Clean Air Act, as Amended
Subchapter II	Emission Standards for Moving Sources	
Part A	Motor Vehicle Emission and Fuel Standards	
7521	Emission standards for new motor vehicles or engines	sec. 202[e]
7522	Prohibited acts	sec. 203[e]
7523	Actions to restrain violations	sec. 204[e]
7524	Penalties	sec. 205[e]
7525	Motor vehicle and engines testing and certification	sec. 206[e]
7541	Compliance by vehicles and engines in actual use	sec. 207[e]
7542	Records and reports	sec. 208[e]
7543	State standards	sec. 209[b]
7544	State grants	sec. 210[b]
7545	Regulation of fuels	sec. 211[b]
7546	Low emission vehicles (repealed)	sec. 212[c]
7547	Fuel economy improvement for new vehicles	sec. 213[c]
7548	Study of particulate emissions from motor vehicles	sec. 214[d]
7549	High altitude performance adjustments	sec. 215[d]
7550	Definitions	sec. 216[e]
7551	Study and report on fuel consumption	sec. [d]
7552	Compliance Program Fees (NEW)	(NEW, sec. 225)
7553	Prohibition on Production of Engines Requiring Leaded Gasoline (NEW)	(NEW, sec. 226)
7554	Urban Buses (NEW)	(NEW, sec. 227(a))
Part B	Aircraft Emissions Standards	
7571	Establishment of standards	sec. 231[c]
7572	Enforcement of standards	sec. 232[c]
7573	State standards and control	sec. 233[c]
7574	Definitions	sec. 234[c]
Part C	Clean Fuel Vehicles (NEW)	
7581	Definitions (NEW)	(NEW, sec. 229(a))
7582	Requirements Applicable to Clean Fuel Vehicles (NEW)	(NEW, sec. 229(a))
7583	Standards for Light-Duty Clean Fuel Vehicles (NEW)	(NEW, sec. 229(a))
7584	Administration and Enforcement as Per California (NEW)	(NEW, sec. 229(a))
7585	Standards for Heavy Duty Clean Fuel Vehicles (NEW)	(NEW, sec. 229(a))
7586	Centrally Fueled Fleets (NEW)	(NEW, sec. 229(a))
7587	Vehicle Conversions (NEW)	(NEW, sec. 229(a))
7588	Federal Agency Fleets	(NEW, sec. 229(a))
7589	California Pilot Test Program	(NEW, sec. 229(a))
7590	General Provisions	(NEW, sec. 229(a))
Subchapter III	General Provisions	
7601	Administration	sec. 301[a]
7602	Definitions	sec. 302[a]
7603	Emergency powers	sec. 303[c]
7604	Citizen suits	sec. 304[c]
7605	Representation in litigation	sec. 305[c]
7606	Federal procurement	sec. 306[c]
7607	Administrative proceedings and judicial review	sec. 307[c]
7608	Mandatory licensing	sec. 308[c]
7609	Policy review	sec. 309[c]
7610	Other authority	sec. 310[a]
7611	Records and audits	sec. 311[a]
7612	Cost studies	sec. 312[b]
7613	Additional reports to Congress (repealed)	sec. 313[b]
7614	Labor standards	sec. 314[b]
7615	Separability of provisions	sec. 315[a]
7616	Sewage treatment plants	sec. 316[d]
7617	Economic impact assessment	sec. 317[d]
7618	Financial disclosure (repealed)	sec. 318[d]
7619	Air quality monitoring	sec. 319[d]
7620	Standardized air quality modeling	sec. 320[d]

Table 3. *(Continued)*

42 U.S.C.	Section Title	Clean Air Act, as Amended
7621	Employment effects	sec. 321[d]
7622	Employee protection	sec. 322[d]
7623	Repealed	sec. 323[d]
7624	Cost of vapor recovery equipment	sec. 324[d]
7625	Vapor recovery for small business marketers of petroleum products	sec. 324[d]
7625a	Statutory construction	sec. 325[d]
7626	Authorization of Appropriations	sec. 326[d]
7651	Acid Disposition Control (NEW)	(NEW, sec. 401, 404, 406, 407, 408)
7661	Permits (NEW)	(NEW, sec. 501)
Title VI	Stratospheric Ozone Protection (New)	
7671	Stratospheric Ozone Protection	(NEW, sec. 602(a))
[29 U.S.C. 655]	Chemical Process Safety Management (NEW)	(NEW, sec. 304)
[29 U.S.C. 1502]	Grants for Clean Air Employment Transition Assistance	(NEW, sec. 1101(b)(2))

Codified generally as 42 U.S.C. 7401-7671. Shows original sections renumbered by P.L. 88-206, and indicates when added, although similar sections may have existed earlier. "Added" usually indicates a general amendment or complete revision. For a detailed history, see the notes in the U.S. [a] Code. P.L. 88-206, general amendment which renumbered sections. [b] added by P.L. 90-148, Nov. 21, 1967. [c] added by P.L. 91-604, Dec. 21, 1970. [d] added by P.L. 95-95, August 7, 1977; (New) added by P.L. 101-548. [e] added by P.L. 89-272, Oct. 20, 1965.

hensive permit program for the operation of sources emitting air pollutants. These requirements are modeled after similar provisions in the Clean Water Act. Previously, the Clean Air Act contained limited provision for permits, requiring only new or modified major stationary sources to obtain construction permits (Title I of the CAA).

Sources subject to the new permit requirements would generally include major sources that emit or have the potential to emit 100 tons per year of any regulated pollutant, plus stationary and area sources that emit or have potential to emit lesser specified amounts of hazardous air pollutants. However, in nonattainment areas, the permit requirements also include sources which emit as little as 10 tons per year of VOCs, depending on the severity of the region's nonattainment status (serious, severe, or extreme).

States are required to develop permit programs and to submit those programs for EPA approval by November 15, 1993. If a State does not have an approved plan within 5 years, EPA can impose and administer its own plan in the State. States are to collect annual fees from sources sufficient to cover the "reasonable costs" of administering the permit program, with revenues to be used to support the agency's air pollution control program. The fee would be not less than $25 per ton of regulated pollutants (excluding carbon monoxide). Permitting authorities will have discretion to not collect fees on emissions in excess of 4,000 tons per year and may collect other fee amounts, if appropriate.

The permit states which air pollutants a source is allowed to emit. As a part of the permit process, a source must prepare a compliance plan and certify compliance. The term of permits is limited to no more than 5 years; sources are required to renew permits at that time. State permit authorities must notify contiguous States of permit applications that may affect them; the application and any comments of contiguous States must be forwarded to EPA for review. EPA can veto a permit; however, this authority is essentially limited to major permit changes. EPA review need not include permits which simply codify elements of a State's overall clean air plan, and EPA has discretion to not review permits for small sources. Holding a permit to some extent shields a source from enforcement actions: the Act provides that a source cannot be held in violation if it is complying with explicit requirements addressed in a permit, or if the State finds that certain provisions do not apply to that source.

ENFORCEMENT (SECTION 113)

In general, this section establishes Federal authority to impose penalties for violations of the requirements of the Act. The 1990 Amendments elevated penalties for some knowing violations from misdemeanors to felonies; removed the ability of a source to avoid an enforcement order by ceasing a violation within 30 days of notice; gave authority to EPA to assess administrative penalties; and authorized $10,000 awards to persons supplying information leading to convictions under the Act. The Act authorizes EPA to require sources to submit reports, to monitor emissions, and to certify compliance with the Act's requirements.

STRATOSPHERIC OZONE PROTECTION (TITLE VI)

As amended in 1990, the Act establishes a national policy of phasing out production and use of "Class I" ozone-depleting substances (ie, all fully halogenated chlorine or bromine containing compounds and carbon tetrachloride) by the year 2000, and methyl chloroform by 2002. "Class II" substances (ie, other halogenated chlorine and bromine compounds, and others designated by EPA) will have their production frozen in 2015, and production ended by 2030. EPA can accelerate the phase-out schedules. Limited exemptions are allowed (eg, for medical, aviation, and fire-suppression uses). EPA is to issue regulations on use and disposal of these chemicals that will minimize production and foster recycling and reuse. This title's authorities are expressly linked to the Montreal Protocol (an international accord phasing out chlorofluorocarbons), specifying that the more stringent requirement of either is effective.

In the omnibus budget reconciliation act of 1989 (P.L. 101-239), Congress imposed a tax on CFCs and halons covered by the Montreal Protocol to accelerate the transition to substitutes. The tax starts at $1.37/lb. in 1990, escalates to $2.65/lb in 1993, and will increase an additional 45 cents/lb/year after 1994.

BIBLIOGRAPHY

Reading List

U.S. Office of Technology Assessment, *Catching our Breath: Next Steps for Reducing Urban Ozone, Summary,* U.S. Office of Technology Assessment, Washington, D.C. 1989, 24 p.

U.S. Environmental Protection Agency, *The Clean Air Act Amendments of 1990 Detailed Summary of Titles,* Environmental Protection Agency, Washington, D.C. 1990.

CLEAN AIR ACT AMENDMENT OF 1990—MOBILE SOURCES

MICHAEL WALSH
Arlington, Virginia

As the dominant source of hydrocarbons, carbon monoxide, and nitrogen oxides, motor vehicles were singled out for special attention in the 1970 Clean Air Act Amendments. Twenty years later, in spite of significant progress, this situation has not changed. Because more than 50 million additional cars are on U.S. highways than in 1970, motor vehicles remain the dominant emissions source and therefore received special attention and focus in the Clean Air Act Amendments of 1990. In addition to more stringent standards for cars, trucks, and buses, the Amendments will require substantial modification to conventional fuels, provide greater opportunity for the introduction of alternative fuels (but without mandating them), and extend the manufacturers' responsibility for compliance with auto standards in use to 10 years or 160,900 km (100,000 miles).

See also EXHAUST CONTROL, AUTOMOTIVE; REFORMULATED GASOLINE; TRANSPORTATION FUELS—AUTOMOTIVE FUELS; METHANOL VEHICLES; AUTOMOBILE EMISSIONS; AUTOMOBILE EMISSIONS, CONTROL; ENERGY TAXATION.

Major provisions dealing with mobile sources are summarized as follows.

CONVENTIONAL VEHICLES

Light-Duty Vehicle Tailpipe Standards—Tier 1

Emissions Standards. The Tier 1 tailpipe standards for light-duty vehicles are summarized in Tables 1 and 2.

Intermediate In-Use Standards. For the first two years that passenger cars and light-duty trucks are subject to the Tier I certification standards shown in Table 1, less stringent, intermediate in-use emission standards apply for purposes of recall liability, and the useful life period is only five years or 80,450 km (50,000 miles), whichever occurs first.

The intermediate in-use standards for passenger cars and light-duty trucks up to (6000 lbs) GVWR are shown in Table 3 and for light-duty trucks above (6000 lbs) GVWR in Table 4.

Implementation Schedule. For purposes of certification, the standards in Table 1 will be phased in over a three-

Table 1. Emission Standards for Light-Duty Vehicles (Passenger Cars) and Light-Duty Trucks of up to 2721.6 kg (6000 lbs) GVWR

	Column A (5 yrs/80,450 km)				Column B (10 yrs/160,900 km)			
Vehicle Type	NMHC	CO	NO_x	Part.	NMHC	CO	NO_x	Part.
Non diesel								
LTDs 0–1701 kg (0–3, 750 lbs) LVW and light-duty vehicles	0.25	3.4	0.4		0.31	4.2	0.6	
LDTs 1701.4–2608.2 kg (3,751–5,750 lbs) LVW	0.32	4.4	0.7		0.40	5.5	0.97	
Diesel								
LTDs 0–1701 kg (0–3, 750 lbs) LVW and light-duty vehicles	0.25	3.4	1.0	0.08	0.31	4.2	1.25	0.10
LDTs 1701.4–2608.2 kg (3,751–5,750 lbs) LVW	0.32	4.4		0.08	0.40	5.0	0.97	0.10

Table 2. Emission Standards for Light-Duty Trucks of More Than 2721.6 kg (6000 lbs) GVWR[a]

	Column A (5 yrs/80,450 kg)			Column B (11 yrs/193,080 kg)			
LDT Test Weight	NMHC	CO	NO_x	NMHC	CO	NO_x	PM
1701.4–2608.2 kg (3,751–5,750 lbs)	0.32	4.4	0.7[b]	0.46	6.4	0.98	0.10
Over 2608.2 kg (5,750 lbs)	0.39	5.0	1.1[b]	0.56	7.3	1.53	0.12

[a] Standards are expressed in grams per mile (gpm).
[b] Not applicable to diesel-fueled LDTs.

year period (applicable to 40% of MY 1994 vehicles, 80% of MY 1995 vehicles, and 100% of MY 1996 vehicles, as illustrated in Table 5). In use, starting in 1994, the standards will start to be phased in—40% the first year, 80% the second and 100% the third year (1996) for NO_x. For HC, in the first year, 40% will be required to meet the *intermediate* in use standard (0.32 grams per mile NMHC) with the remainder achieving the current standard. This will rise to 80% in the second year. By the third year only 60% will meet the intermediate standard with the other 40% meeting the *final* (0.25 NMHC) standard. In the fourth year, 80% will be required to meet the 0.25 level, rising to 100% by 1998.

Starting in 1996, new in-use standards start to be phased in for 160,900 kg (100,000 miles) (120,675 kg [75,000 miles] for Recall testing) that allow for 25% higher emissions levels between 80,450 and 160,900 kg (50,000 and 100,000 miles) for HC and CO, and 50% higher for NO_x. Diesel vehicles are also allowed to comply with a relaxed 1.0 gpm NO_x standard. Light trucks under 1701 kg (3750 lbs) loaded vehicle weight (LVW) will be required to achieve the same standards as cars.

Recall Requirements. For the purpose of Recall, if a vehicle is tested in-use before five years or 80,450 kg (50,000 miles), the 80,450-kg (50,000-mile) certification standards apply. If the vehicle is tested after five years or 80,450 kg (50,000 miles), the 160,900-kg (100,000-mile) certification standards apply in the case of passenger cars and light-duty trucks up to 2721.4 kg (6000 lbs) GVWR and 193,080-kg (120,000-mile) certification standard for light-duty trucks over 2721.4 kg (6000 lbs) GVWR.

While passenger cars and light-duty trucks are expected to meet standards in-use for their full useful lives, EPA may not conduct recall testing on passenger cars and light-duty trucks up to 2721.4 kg (6000 lbs) that are over seven years old or have been driven more than 120,675 kg (75,000 miles) or on light-duty trucks over 2721 kg (6000 lbs) GVWR that are over seven years old or been driven more than 144,810 kg (90,000 miles).

Table 3. Intermediate In-use Standards for Passenger Cars and Light-duty Trucks up to 2721.6 kg (6000 lbs) GVWR

Vehicle Type	NMHC	CO	NO_x
Passenger cars	0.32	3.4	0.4[a]
LDTs 0–1701 kg (0–3750 lbs) LVW	0.32	5.2	0.4[a]
LDTs 1701.4–2608.2 kg (3751–5750 lbs) LVW	0.41	6.7	0.7[a]

[a] Not applicable to diesel-fueled vehicles.

Tier 2 Standards

EPA is required to study and report to Congress no later than June 1, 1997, on the technological feasibility, need for, and cost-effectiveness of the standards shown in Table 6 for passenger cars and light-duty trucks of 1701 kg (3750 lbs) LVW or less.

Within three years of reporting to Congress, but not later than December 31, 1999, EPA is required to (a) promulgate the Tier 2 standards for model years commencing not earlier than model year 2003 and not later than model year 2006, (b) promulgate standards different from those shown in Table 6 provided they are more stringent than the Tier 1 standards for model years commencing not earlier than January 1, 2003, or later than the 2006 model year, or (c) determine that standards more stringent than the Tier 1 levels are not technologically feasible, needed, or cost-effective. If the administrator fails to act, the Tier 2 standards become effective with the model years commencing after January 1, 2003.

Potential Revision of Standards

EPA retains the authority to revise emission standards for all classes of motor vehicles and engines based on the need to protect the public welfare *except that* the administrator may not revise the specific emission standards established in the Act for passenger cars, light-duty trucks, and heavy-duty trucks before the 2004 model year.

Cold Temperature CO Standards. EPA is required to establish a cold-temperature CO standard (−6.66°C) beginning with the 1994 model year of 10.0 gpm for passenger

Table 4. Intermediate In-use Standards for LDTs More than 2721.6 kg (6000 lbs) GVWR

Vehicle Type	NMHC	CO	NO_x
LDTs 1701.4–2608.2 kg (3751–5750 lbs) LVW	0.40	5.5	0.88[a]
LDTs 2608.2 kg (over 5750 lbs) LVW	0.49	6.2	1.38[a]

[a] Not applicable to diesel-fueled vehicles.

Table 5. Implementation Schedule for Tier 1 Standards for Light-duty Vehicles (Passenger Cars) and Light-duty Trucks of up to 2721.4 kg (6000 lbs) GVWR

Model Year	Certification in Use	Percentage of Manufacturers Sales Volume %
1994	1996	40
1995	1997	80
1996	1998	100

Implementation Schedule for Standards for Light-duty Trucks of More Than 2721.4 kg (6000 lbs) GVWR

Model Year	Certification in Use	Percentage of Manufacturers Sales Volume, %
1996	1998	50
1997	1999	100

cars and a level comparable in stringency for light-duty trucks according to the phase-in schedule listed in Table 7.

If, as of June 1, 1997, six or more nonattainment areas (not including Stubenville and Winebago, where the CO problem is primarily caused by stationary sources) have a carbon monoxide design value of 9.5 ppm or greater, EPA is required to establish second-phase cold-temperature (−6.66°C) CO standards beginning with the 2002 model year of 3.4 gpm for passenger cars, 4.4 for light-duty trucks up to (6000 lbs) and a level comparable in stringency for light-duty trucks (6000 lbs) LVWR and above.

The useful life for the cold-temperature CO standards is 5 years/80,450 km (50,000 miles), but EPA may extend this period, if it determines such requirements to be technologically feasible.

Trucks and Buses

New Heavy-Duty Vehicles. The statutory standards for HC, CO, and NO_x for heavy-duty vehicles and engines created by the 1977 Amendments are eliminated. In their place, a general requirement that standards applicable to emissions of HC, CO, NO_x, and particulate reflect "the greatest degree of emission reduction achievable through the application of technology which the Administrator determines will be available for the model year to which such standards apply, giving appropriate consideration to cost, energy, and safety factors associated with the application of such technology." Standards adopted by EPA that are currently in effect will remain in effect unless modified by EPA.

EPA may revise (relax) such standards, including standards already adopted, on the basis of information concerning the effects of air pollutants from heavy-duty vehicles and other mobile sources, on the public health and welfare, taking into consideration costs.

Table 6. Tier 2 Emission Standards for Gasoline and Diesel-Fueled Passenger Car and Light-duty Trucks 1701 kg (3750) lbs LVW or Less

Pollutant	Emission Level
NMHC	0.125 gpm
NO_x	0.2 gpm
CO	1.7 gpm

Table 7. Phase-in Schedule for Cold Start Standards

Model Year	Percentage of Manufacturer's Sales Volume, %
1994	40
1995	80
1996 and after	100

A 4.0 g/BHP-h NO_x standard is established for 1998 and later model year gasoline- and diesel-fueled heavy-duty trucks.

Standards adopted for heavy-duty vehicles must provide at least four years lead time before they go into effect and then must remain in effect for at least three years before they are changed.

Rebuilt Engines. The EPA is required to study the practice of rebuilding heavy-duty engines and the impact rebuilding has on engine emissions. EPA *may* establish requirements to control rebuilding practices, including emission standards. No deadlines are established.

Cold Temperature CO. EPA *may* establish cold temperature CO standards for heavy-duty vehicles and engines.

Urban Buses (New). The existing 1991 model year 0.1 g/BHP-h particulate standard is relaxed to 0.25 g/BHP-h for the 1991 and 1992 model years. For the 1993 model year, the particulate standard is 0.1 g/BHP-h.

Beginning with the 1994 model year, EPA is required to establish separate emission standards for urban buses including a particulate standard 50% more stringent than the 0.1 g/BHP-h (eg, 0.05 g/BHP-h) standard. If EPA determines the 50% level is technologically infeasible, EPA is required to increase the allowable level of particulate to no greater than 70% of the 0.1 g/BHP-h level (eg, 0.07 g/BHP-h).

EPA is required to conduct annual tests on a representative sample of operating urban buses to determine whether such buses meet the particulate standard over their full useful life. If EPA determines that buses are not meeting the particulate standard, EPA must require buses sold in areas with populations of 750,000 or more to operate on low-polluting fuels (methanol, ethanol, propane natural gas, or any comparably low polluting fuel). EPA may extend this requirement to buses sold in other areas if it determines there will be a significant benefit to the public health. The low-polluting fuel requirement would be phased in over five model years commencing three years after EPA's determination.

Retrofit Requirement for Buses. Not later than November 15, 1991, EPA is required to promulgate emission standards or emission-control technology requirements reflecting the best retrofit technology and maintenance practices reasonably achievable. Such standards are to apply to engines replaced or rebuilt after January 1, 1995, for buses operating in areas with populations of 750,000 or more.

Onboard Refueling Control Systems

By November 15, 1991, and after consultation with the Department of Transportation on safety issues, EPA is required to issue regulations mandating that passenger cars be equipped with vehicle-based ("onboard") systems to control 95% of the HC emissions emitted during refueling. The regulations take effect beginning in the fourth model year after the regulations are adopted and are to be phased in over three years (40% of total sales in the first year, 80% in the second, and 100% in the third and subsequent years).

Warranties

The warranty period for 1995 and later model year passenger cars and light-duty trucks for the catalytic converter, electronic emission control unit, onboard diagnostic device, and other emission-control equipment designated by EPA as a "specific major emission control component" will be eight years or 128,720 km (80,000 miles), whichever occurs first. To be designated a "specific major emission control component," the device or component could not have been in general use on vehicles and engines before the 1990 model year and must have a retail cost (exclusive of installation costs) that exceeds $200.

The Warranty for all remaining emission control components is two years or 38,616 km (24,000 miles), whichever occurs first.

EPA is given the authority to establish the warranty period for other classes of motor vehicles and engines.

Evaporative Controls

By June 15, 1991, EPA is required to establish evaporative hydrocarbon emission standards for all classes of gasoline-fueled motor vehicles. The standards, which are to take effect as expeditiously as possible, must require the greatest degree of reduction reasonably achievable of evaporative HC emissions during operation ("running losses") and over two or more days of nonuse under ozone-prone summertime conditions.

Toxics

EPA is required to complete a report by June 15, 1992, on the need for and feasibility of controlling unregulated motor vehicle toxic air pollutants including benzene, formaldehyde, and 1, 3-butadiene.

By June 15, 1995, EPA is required to issue regulations to control hazardous air pollutants to the greatest degree achievable through technology that will be available considering cost, noise, safety, and necessary lead time. These regulations must apply, at a minimum, to benzene and formaldehyde. No effective date for the regulations is specified in the Amendments.

Onboard Diagnostics

EPA must issue regulations requiring all passenger cars and light-duty trucks to be equipped with onboard diagnostic systems capable of

1. accurately identifying emission-related system deterioration or malfunctions including, at a minimum, the catalytic converter and oxygen sensor;
2. alerting vehicle owners of the need for emission-related component or system maintenance or repair; and
3. storing and retrieving diagnostic fault codes that are readily accessible.

The regulations are to be phased in starting in 1994 (covering 40% of sales the first year, 80% the second, and 100% the third) but EPA may delay them for up to two years for any class or category of motor vehicles based on technological feasibility considerations.

EPA *may* also establish onboard diagnostic control requirements for heavy-duty vehicles.

States are required to establish programs for inspecting onboard diagnostic systems as part of their periodic inspection and maintenance program requirements.

Compliance Program Fees

EPA may establish fees to recover the costs from manufacturers of (a) new vehicle or engine certification tests, (b) new vehicle or engine compliance monitoring (eg, assembly line testing), and (c) in-use vehicle or engine compliance monitoring (eg, recall). This could be extremely critical as the budget deficit squeezes harder.

Motor Vehicle Testing and Certification

By November 15, 1991, EPA must revise the certification test procedures to add a test to determine whether 1994 and later model year passenger cars and light-duty trucks are capable of passing state inspection emission (I/M) tests.

By June 15, 1992, EPA must review and revise, *as necessary,* the certification test procedures to ensure that motor vehicles are tested under conditions that reflect actual current driving conditions, including fuel, temperature, acceleration, and altitude.

Manufacturers with projected U.S. sales of vehicles or engines of no more than 300 are subject to less stringent durability demonstration requirements.

Information Collection

EPA is given authority to require manufacturers to maintain records and perform emission tests and to report testing results to the Agency. EPA is also given expanded authority to inspect manufacturers' facilities and records.

High-Altitude Testing

EPA is required to establish at least one high-altitude emission testing center to evaluate in-use emissions of vehicles at high altitudes. As part of the testing center, EPA must establish a research facility to evaluate the effects of high altitude on aftermarket emission-control components, dual-fueled vehicles, conversion kits, and alternative fuels. The center must also offer training courses designed to maximize the effectiveness of inspection and maintenance (I/M) programs at high altitude.

Antitampering

The Amendments extend the prohibition against tampering with emission controls to individuals. Also, the manufacture, sale, or installation of an emission-control defeat device is prohibited.

CONVENTIONAL FUELS

Reformulated Gasoline

Conventional gasoline will undergo significant modification during the 1990s as a result of the Amendments. There are two separate sets of requirements: one for ozone problems and one for CO problems.

Severe and Serious Ozone Nonattainment Areas. By November 15, 1991, EPA is required to establish requirements for reformulated gasoline requiring the greatest reduction of ozone-forming VOCs and toxic air pollutants achievable considering costs and technological feasibility.

Beginning January 1, 1995, cleaner, "reformulated" gasoline must be sold in the nine worst ozone nonattainment areas ("severe" and "serious") with populations over 250,000. Other ozone nonattainment areas ("marginal," "moderate," or "serious") are permitted to "opt-in," but EPA may delay on a limited basis requirements for reformulated gasoline in these areas if it determines the fuel will not be available in adequate quantities.

Emission Performance Requirements. At a minimum, reformulated gasoline must (a) not cause NO_x emissions to increase (EPA may modify other requirements discussed below if necessary to prevent an increase in NO_x emissions), (b) have an oxygen content of at least 2.0% by weight (EPA may waive this requirement if it would interfere with attaining an air quality standard), (c) have a benzene content no greater than 1.0% by volume, and (d) contain no heavy metals, including lead or manganese (EPA may waive the prohibition against heavy metals other than lead if it determines that the metal will not increase on an aggregate mass or cancer-risk basis, toxic air-pollution emissions from motor vehicles).

In addition, VOC and toxic emissions must be reduced by 15% over baseline levels beginning in 1995 and by 25% beginning in the year 2000. EPA may adjust the 25% requirement up or down based on technological feasibility and cost considerations, but in no event may the percent reduction beginning in the year 2000 be less than 20%. Toxic air pollutants are defined by the Amendments in terms of the aggregate emissions of benzene, 1,3 butadiene, polycyclic organic matter (POM), acetaldehyde, and formaldehyde.

Credits. Fuel refiners and suppliers may accumulate and trade credits if they produce reformulated gasoline that exceeds the minimum requirements specified by EPA.

Anti Dumping Provisions. By November 15, 1991, EPA must establish regulations that prohibit gasoline from being introduced into commerce that on average results in emissions of VOC, NO_x, or toxics greater than gasoline sold by that refiner, blender, or importer in 1990. These regulations must go into effect by January 1, 1995.

Carbon Monoxide Nonattainment Areas. Areas with a carbon monoxide design target of 9.5 ppm or above for 1988 and 1989 must have as part of their state implementation plan (SIP) a requirement that during that portion of the year in which the area is prone to high ambient concentrations of CO (winter months), all gasoline sold must contain not less than 2.7% oxygen by weight. Such requirements are to take effect no later than November 1, 1992 (or such other date in 1992 as determined by the Administrator). For areas that exceed the 9.5 CO design target for any two-year period after 1989, the 2.7% oxygen requirement must go into effect in that area no later than three years after the end of the two-year period.

Areas classified as serious CO nonattainment areas that have not achieved attainment by the date specified in the Act must require that the oxygen level in gasoline be 3.1% by weight.

Waivers of this requirement for up to two years can be granted by EPA on an area-by-area basis if it is petitioned and if it determines a) the use of oxygenated gasoline would present or interfere with the attainment by a given area with a federal or state ambient air quality standard; b) mobile sources do not contribute significantly to the CO levels in the area; or c) there is an inadequate domestic supply of, or distribution capacity for, oxygenated gasoline meeting the applicable requirements. EPA must act on any petition within six months of its filing.

Fuel Volatility

By June 15, 1991, the EPA must promulgate regulations limiting the volatility of gasoline to no greater than (9.0 pounds per square inch—(PSI) Reid Vapor Pressure (RVP) during the high ozone season. EPA may establish a lower RVP in an individual nonattainment area if it determines that a lower level is necessary to achieve comparable evaporative emissions (on a per-vehicle basis) in nonattainment areas. The fuel volatility requirements are to take effect not later than the high ozone season for 1992.

For fuel blends containing 10% ethanol, the applicable RVP limitation may be one pound per square inch greater than for conventional gasoline.

Detergent Requirements

Beginning January 1, 1995, any gasoline sold nationwide must contain additives to prevent the accumulation of deposits in the engine and fuel supply systems.

Lead Phasedown

After December 31, 1985, it will be unlawful to sell, supply, dispense, transport, introduce into commerce, or use gasoline that contains lead or lead additives for highway use.

Lead Substitute Gasoline Additives

EPA is required to develop a test procedure for evaluating the effectiveness of lead substitute gasoline additives and to arrange for independent testing and evaluation of each additive proposed to be registered as a lead substitute.

EPA may impose a user fee to cover the cost of testing any lead-substitute fuel additive.

Prohibition on Engines Requiring Leaded Gasoline

EPA is required to prohibit the manufacture, sale, or introduction into commerce of any motor vehicle or nonroad engine that can only be operated on leaded gasoline and is manufactured after the 1992 model year.

Diesel Fuel Sulfur Content

Effective October 1, 1993, the sulfur content of motor vehicle diesel fuel may not exceed 0.05% by weight and must meet a cetane index minimum of 40.

EPA may exempt Alaska and Hawaii from the diesel fuel quality requirements.

The sulfur content of fuel used in certification of 1991–1993 model year heavy-duty vehicles and engines must be 0.10% by weight.

EPA may require diesel fuel manufacturers and importers to dye fuels to differentiate highway and nonhighway diesel fuels.

Ethanol Substitute for Diesel Fuel

EPA is required to commission a study and report to Congress within three years of enactment of the Amendments on the feasibility, engine performance, emissions, and production capability associated with an alternative diesel fuel composed of ethanol and high-erucic rapeseed oil.

Fuel and Fuel Additive Importers

The statutory requirements applicable to fuel and fuel additive manufacturers are expressly made applicable to importers.

Misfueling

The Amendments extend to individuals the prohibition against misfueling with leaded gasoline. After October 1, 1993, there will be a prohibition against fueling a diesel-powered vehicle with fuel containing a sulfur content greater than 0.05% by weight or that fails to meet a cetane index minimum of 40 (or such equivalent aromatic level as prescribed by the Administrator).

CLEAN ALTERNATIVE FUELS

The Amendments define "clean alternative fuel" as any fuel including methanol, ethanol, or other alcohols including any mixture thereof containing 85% or more by volume of such alcohol with gasoline or other fuels (reformulated gasoline, diesel, natural gas, liquified petroleum gas, and hydrogen) or power source (including electricity) used in a clean-fuel vehicle that complies with the Amendments' performance requirements.

Fleet Program

By November 15, 1992, EPA is required to issue regulations implementing the Clean Fuel Fleet Vehicle Program.

The fleet program applies in serious nonattainment areas to fleets of 10 or more vehicles that are centrally refueled or capable of being so (but not including vehicles garaged at personal residences each night under normal circumstances).

The program agreed upon will mandate California's Low Emission Vehicle (LEV) standards for light-duty vehicles (0.075 grams per mile nonmethane organic material, 3.4 CO and 0.2 NO_x) and light trucks below 2721.6 kg (6000 lbs) (0.1 grams per mile NMOG, 0.4 NO_x, 4.4 CO) by 1998, provided these vehicles are offered for sale in California. By 2001, these vehicles will be required without regard to availability in California.

EPA is mandated to establish an equivalent wraparound standard (exhaust, evaporative, and refueling emissions combined) for LEVs below 3855.6 kg (8500 lbs) GVWR. This wraparound standard is to be based on LEV vehicles using reformulated gasoline meeting the reformulated gasoline standards for the applicable time period. It will be left to the manufacturers to decide which standard to use—the LEV tailpipe standards or the wrap around standards.

Congress followed California's lead in substituting a nonmethane organic gas (NMOG) standard in place of the total or nonmethane hydrocarbon standards currently used. A nonmethane organic gas is defined as the sum of nonoxygenated and oxygenated organic gases containing 5 or fewer carbon atoms and all known alkanes, alkenes, alkynes, and aromatics containing 12 or fewer carbon atoms. The test procedure for measuring NMOG is the recently adopted California NMOG Test Procedure. NMOG is a more appropriate substance to measure when evaluating the emission performance of alternative fueled vehicles.

Covered Fleets. ***Centrally Fueled Fleets.*** Centrally fueled fleets with 10 or more vehicles that are owned or operated by a single person and operate in a covered area are subject to the clean vehicle requirements. A number of vehicle fleets are exempted including rental fleets, emergency vehicles, enforcement vehicles, and nonroad vehicles. Also, vehicles garaged at personal residences each night under normal circumstances would not be covered.

Covered areas include any ozone nonattainment area with a population of 250,000 or more classified as Serious, Severe, or Extreme (approximately 26 areas), or any carbon monoxide nonattainment area with 250,000 or more population and a CO design value at or above 16.0 ppm.

States are required to implement clean-fuel vehicle phase-in as shown in Table 8.

In complying with this section, passenger cars and light-duty trucks up to 2721.6 kg (6000 lbs) GVWR must meet the Phase II emission standards shown in Table 9. Also, if passenger cars and light-duty trucks meeting the Phase II standards are sold in California in model years before 1998, then that model year becomes the applicable model year for phasing in clean vehicle fleets.

Federal Agency Fleets. The fleet requirements apply to federally owned fleets except vehicles exempted by the Secretary of Defense for national security reasons.

Conversions. The requirements for purchase of clean

Table 8. Clean Fuel Vehicle Phase-in Requirements for Fleets

Vehicle Type	Applicable Model Year 1998	1999	2000
Light-duty trucks up to 2721.6 Kg (6000 lbs) GVWR and light-duty vehicles	30%	50%	70%
Heavy-duty trucks above 3855.6 Kg (8500 lbs) GVWR	50%	50%	50%

vehicles may be met through vehicle conversions of existing gasoline- or diesel-powered vehicles to clean fuel vehicles. EPA is required to establish regulations defining criteria for conversions.

Dedicated Clean-Fuel Light-Duty Vehicles. The emission standards applicable to clean-fuel vehicle passenger cars and light-duty (trucks weighing up to 2721.6 kg (6000 lbs) GWVR but not more than 1701 kg (3750 lbs) LVW are shown in Table 9.

Table 9. Clean Fuel Vehicle Emission Standards for Light-Duty Vehicles and Light-duty Trucks of up to 6000 lb GVWR[a]

Pollutant	NMOG CO NO_x PM HCHO (formaldehyde)
Light-Duty Vehicles and Light-Duty Trucks of up to 1701 kg (3750 lbs) Loaded Vehicle Weight	
Phase I—1996 Model Year	
80,450-km (50,000-mile) standard	0.1253.40.40.015
160,900-km (100,000-mile) standard	0.1564.20.60.8[b]0.018
Phase II—2001 Model Year	
80,450-km (50,000-mile) standard	0.0753.40.20.015
160,900-km (100,000-mile) standard	0.904.20.30.080.018
Light-Duty Trucks From 1701.4 to 2608.2 kg (3751 to 5750 lb) Loaded Vehicle Weight	
Phase I—1996 Model Year	
80,450-km (50,000-mile) standard	0.1604.40.70.018
160,900-km (100,000-mile) standard	0.2005.50.90.08.0.023
Phase II—2001 Model Year	
80,450-km (50,000-mile) standard	0.1004.40.40.018
160,900-km (100,000-mile) standard	0.1305.50.50.080.023

[a] Standards are expressed in grams per mile.
[b] Standards for particulates (PM) shall apply only to diesel-fueled vehicles.

Table 10. Clean Fuel Vehicle Emission Standards for Light-duty Trucks Greater Than 2721.6 kg (6000 lbs) GVWR[a]

Pollutant	NMOG CO NO_x PM[b] HCHO (formaldehyde)
Test Weight Category: Up to 1701 kg (3750 lb)	
80,450-km (50,000-mile) standard	0.1253.40.4[c]0.015
193,080-km (120,000-mile) standard	0.1805.00.60.080.022
Test Weight Category: Above 1701 (3750) but not above 2608.2 kg (5750 lb)	
80,450-km (50,000-mile) standard	0.1604.40.7[c]0.018
193,080-km (120,000-mile) standard	0.2306.41.00.100.027
Test Weight Category: Above 2608.2 (5750) but not above 3855.6 kg (8500 lb) GVWR	
80,450-kg (50,000-mile) standard	0.1955.01.1[c]0.022
54,432-kg (120,000-mile) standard	0.2807.31.50.120.032

[a] Standards are expressed in grams per mile.
[b] Standards for particulates (PM) shall apply only to diesel-fueled vehicles.
[c] Standard not applicable to diesel-fueled vehicles.

The emission standards for light-duty trucks greater than 2721.6 kg (6000 lb) GVWR are shown in Table 10.

Flexible-Fueled Light-Duty Vehicles. The emission standards for flexible-fueled passenger cars and light-duty trucks up to 2721.6 kg (6000 lbs) GVWR are shown in Table 11 (when operating on clean alternative fuel) and Table 12 (when operating on conventional fuel).

Possible Modification to Requirement. The clean-vehicle and flexible-fueled vehicle standards in the Amendments are based on standards recently adopted by California as part of its Low Emission Vehicles and Clean Fuel Program. The Amendments provide that if California adopts less stringent standards applicable to clean fuel and dual-fueled passenger cars and light-duty trucks, the standards shown in Tables 11 and 12 are to be relaxed as well.

Enforcement of Standards. Where the numerical clean-fuel vehicle standards applying to vehicles of not more than 3855.6 kg (8500 lbs) GVWR are the same as the numerical standards applicable in California under the Low Emission Vehicle and Clean Fuels Regulations of the California Air Resources Board (CARB); the standards are to be administered and enforced by EPA in the same manner, and with the same flexibility, subject to the same requirements and using the same interpretations and policy judgments as applied by CARB including those applying for Certification, Assembly Line Testing, and In-Use Compliance.

Table 11. NMOG Standards for Flexible- and Dual-fueled Vehicles When Operating on Clean Alternative Fuel

Light-duty Trucks up to 2721.6 kg (6000 lbs) GVWR and Light-duty Vehicles		
Vehicle Type	Column A 80,450-km (50,000-mi) Standard (gpm)	Column B 160,900 km (100,000-mi) Standard (gpm)
Beginning MY 1996: LTDs 0–1701 kg (0–3750 lb) LVW and light-duty vehicles	0.125	0.156
LTDs 1701.4–2608.2 kg (3751–5750 lbs) LVW	0.160	0.200
Beginning MY 2001: LTDs 0–1701 kg (0–3750 lb) LVW and light-duty vehicles	0.075	0.090
LTDs 1701.4–2608.2 kg (3751–5750 lb) LVW	0.100	0.130
Light-duty Trucks More than 2721.6 kg (6000 lb) GVWR		
	Column A 80,450-kg (50,000-mi)	Column B (120,000-mi)
Beginning MY 1998: LTDs 0–1701 kg (0–3750 lbs) tw.	0.125	0.180
LTDs 1701.4–2608.2 kg (3751–5750 lbs) tw.	0.160	0.230
LTDs above 2608.2 kg (5750 lbs) tw.	0.195	0.280

Heavy-Duty Clean Fuel Vehicles. Model years 1998 and later heavy-duty vehicles and engines greater than 3855.6 kg (8500 lbs) GVWR and up to 11,793.6 kg (26,000 lbs) GVWR are required to meet a combined NO_x and non–methane hydrocarbon (NMHC) standard of 3.15 g/BHP-hr (equivalent to 50% of the combined HC and NO_x emission standards applicable to conventional 1994 model year heavy-duty diesel fueled vehicles or engines). EPA may relax this standard upon a finding that it is technologically infeasible for clean diesel-fueled vehicles. EPA must, however, require at least a 30% reduction from conventional-fueled vehicle and engine 1994 NO_x and HC standards (combined).

California Pilot Program

The Amendments also contain a clean fuels pilot program for California. Beginning in 1996, 150,000 Clean Fuel vehicles must be produced for sale in California; by 1999, this number must rise to 300,000. These vehicles must meet California's TLEV standards (0.125 NMOG, 0.4 NO_x, 3.4 CO, and 0.015 formaldehyde) until the year 2000 when they must meet the LEV requirements (0.075 NMOG, 0.2 NO_x, 3.4 CO, and 0.015 formaldehyde). California is required to develop an implementation plan revision within one year to ensure that sufficient clean fuels are produced, distributed, and made available so that all clean-fuel vehicles required can operate to the maximum extent practicable exclusively on such fuels in the covered area. If California fails to adopt a fuels program that meets the requirements, EPA is required to establish such a program within three years.

Other states with serious, severe, or extreme ozone nonattainment areas are authorized to "voluntarily" opt in to the program in whole or in part. This voluntary opt-in cannot include any production or sales mandate for vehicles or fuels but must rely on incentives to encourage their sale and use.

Table 12. NMOG Standards for Flexible- and Dual-fueled Vehicles When Operating on Conventional Fuel

Light-duty Trucks of up to 2721.6 kg (6000 lbs) GVWR and Light-duty Vehicles		
Vehicle Type	Column A 80,450-km (50,000-mi) Standard (gpm)	Column B 160,900 km (100,000-mi) Standard (gpm)
Beginning MY 1996: LTDs 0–1701 kg (0–3750 lbs) LVW and light-duty vehicles	0.25	0.31
LTDs 1701.4–2608.2 kg (3751–5750 lbs) LVW	0.32	0.40
Beginning MY 2001: LTDs 0–1701 kg (0–3750 lbs) LVW and light-duty vehicles	0.125	0.156
LTDs 1701.4–2608.2 kg (3751–5750 lbs) LVW	0.160	0.200
Light-duty Trucks More than 2721.6 kg (6000 lbs) GVWR		
Vehicle Type	Column A 80,450-km (50,000-mi) Standard	Column B 160,900-km (120,000-mi) Standard
Beginning MY 1998: LTDs 0–1701 kg (0–3750 lbs) tw.	0.25	0.36
LTDs 1701.4–2608.2 kg (3751–5750 lbs) tw.	0.32	0.46
LTDs (above 2608.2 kg (5750 lbs) tw.	0.39	0.56

Urban Buses

The Amendments also set a performance criteria which mandates that beginning in 1994, buses operating more than 70% of the time in large urban areas using any fuel must reduce particulate by 50% compared to conventional heavy-duty vehicles (ie, 0.05 grams per brake horsepower-hour particulates). EPA is authorized to relax the control requirements to 30% on the basis of technological feasibility. Beginning in 1994, EPA is to do yearly testing to determine whether buses subject to the standard are meeting the standard *in use over their full useful life.* If EPA determines that 40% or more of the buses are not, it *must* establish a low pollution fuel requirement. Essentially, this provision allows the use of exhaust aftertreatment devices to reduce diesel particulates to a very low level provided they work in the field; if they fail, EPA will mandate alternative fuels.

OFF-HIGHWAY ENGINES

All Non-Road Engines/Vehicles Except Locomotives

By November 15, 1991, EPA must complete a study of the health and welfare effects of nonroad engines and vehicles (except locomotives). Within 12 months of completing the study, EPA must determine if HC, CO, or NO_x emissions from new or existing nonroad engines and vehicles significantly contribute to ozone or CO concentrations in more than one area in nonattainment for ozone or CO.

If EPA makes an affirmative finding, it must regulate those classes or categories of nonroad engines or vehicles by requiring the "greatest degree of emission reduction achievable considering technological feasibility cost, noise, energy, safety and lead time factors." EPA, in setting standards, must consider standards equivalent in stringency to onroad vehicle standards. No deadline for establishing standards has been set.

EPA *may* also regulate other pollutants from nonroad engines and vehicles (eg, diesel particulate) if it determines such standards are needed to protect the public health and welfare.

Locomotives

By November 15, 1995, EPA must establish standards for locomotive emissions that require the use of the best technology that will be available considering cost, energy, and safety. The standards are to take effect at the earliest possible date considering the lead time needed to develop the control technology.

State Standards

States, including California, are prohibited from setting emission standards for a) new engines smaller than 175 horsepower used in construction equipment, vehicles, or farm equipment, and b) new locomotives or new engines used in locomotives.

SECTION 177 OPT IN TO THE CALIFORNIA STANDARDS

One of the most contentious issues, which reared its head very late in the process, was whether other states should continue to be allowed to adopt California emissions standards for motor vehicles, as is allowed by current law. As New York approached final adoption of the 1993 California requirements in the late summer of 1990, paving the way for other northeastern states to do the same, vehicle manufacturers persuaded some in Congress that such programs would create a patchwork of cars across the country and a production nightmare. Others raised concerns that the existing law did not provide sufficient flexibility for states to adopt California cars when these vehicles require new fuels because Section 211 C 4 could preempt state authority in this area. States were concerned that efforts to ensure that only vehicles meeting identical standards as applied in California and clarifying fuel authority were actually creating major new procedural hurdles that would make it more difficult for states to adopt these programs or make it easier for car manufacturers to challenge state enforcement decisions. Some proposals would have removed states entirely from the enforcement process and left it entirely to California and EPA.

The basic thrust of the provision is that states cannot carry out requirements that would force manufacturers to produce a third car.

COAL: AVAILABILITY, MINING, AND PREPARATION

James C. Hower
University of Kentucky
Lexington, Kentucky

Andrew C. Scott
Royal University of London
Halloway, Egham, Surrey
United Kingdom

Adrian C. Hutton
University of Wollongong
Wollongong, New South Wales
Australia

B. K. Parekh
University of Kentucky
Lexington, Kentucky

Bukin Dauley
University of Wollongong
Wollongong, New South Wales
Australia
(Co-Author on Indonesia Section)

RESOURCES AND RESERVES

Discussion of the quantity of coal in any coal field requires some understanding of the concept of resources and reserves. Regretably, there is no single practice for the estimation of those quantities. Indeed, in scanning some literature the reader cannot always be certain that resources and reserves are not being used interchangeably. A concise treatment of the concept may be found in Ref. 1. Generally, *resources* refers to the total amount of coal, discovered and undiscovered, given restrictions of coal thickness and depth. *Reserves* are that portion of the

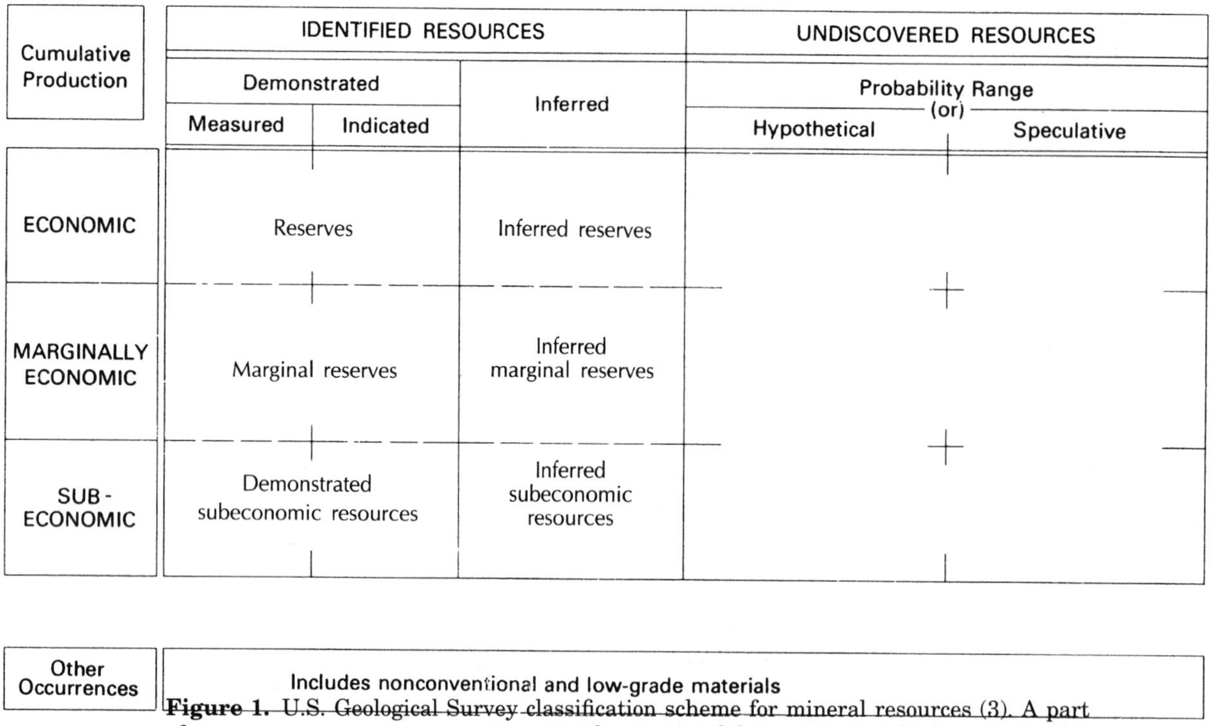

Cumulative Production	IDENTIFIED RESOURCES			UNDISCOVERED RESOURCES	
	Demonstrated		Inferred	Probability Range (or)	
	Measured	Indicated		Hypothetical	Speculative
ECONOMIC	Reserves		Inferred reserves		
MARGINALLY ECONOMIC	Marginal reserves		Inferred marginal reserves		
SUB-ECONOMIC	Demonstrated subeconomic resources		Inferred subeconomic resources		

Other Occurrences	Includes nonconventional and low-grade materials

Author: Date:

Figure 1. U.S. Geological Survey classification scheme for mineral resources (3). A part of reserves or any resource category may be restricted from extraction by laws or regulations (see text). Area: mine district, field, state, etc. Units: tons, barrels, ounces, etc.

resources which are known through relatively tight exploration and which are considered to be economic and mineable within a certain time frame. For example, accessible reserves by the International Energy Agency definition is the amount of coal that could be mined to meet demand over the next 20 years using current mining technology (2). At the other end of the reporting scale is categories such as identified-hypothetical coal-in-place and speculative coal-in-place, which encompasses undiscovered potential resources in formations which are coal bearing in explored areas (3; 23 cited with discussion of North America below). A good example would be the known, but relatively unexplored, extensions of the Appalachian coal field beneath the Cretaceous Gulf Coast cover in the southeastern United States. The U.S. Geological Survey System is illustrated on Figure 1 (3).

World estimated recoverable reserves in tons is listed on Table 1. Production and consumption trends of high and low rank coals is listed on Table 2. The world coal flow (Table 3) provides an interesting corollary to the latter table. The United States and Australia contribute about half of the energy in the total coal flow. Coal involved in coal flow amounts to over 10% of the world consumption of coal-derived energy (4).

SOUTH AMERICA

Chile and Argentina

The Tertiary Magallanes coal field in Chile and the Río Turbio coalfield in Argentina are the most important coal-bearing basins in southern South America. Summaries of the coal fields are after Refs. 5–7.

The Magallanes deposits of Tierra del Fuego were originally developed to supply fuel to ships passing through the Strait of Magellan. Low- to moderate-ash, low- to high-sulfur subbituminous coal, averaging about 19.8 MJ/kg, occurs in northwestward-trending synclines containing the Oligocene Loreto Formation. Coal quality increases to the northwest. Coal at Natales at the northern end of the deposit, with 8.5% moisture and 45.2% volatile matter (daf) could be transitional to high volatile C bituminous rank. The number of coals varies between the five sub-basins. The greatest cummulative thickness of ca 15 m was reported for up to 10 coals on Isla Riesco. Accessible reserves total 5 Gt, about 20% of its surface is mineable (6).

The Río Turbio coal field of Argentina lies within 100 km of the Natales, Chile, deposits at about 51°S latitude. The most important coals are in the Paleocene to Oligocene Sierra Dorotea Group. Coals also occur in the Miocene Río Turbio Group. The coals are high volatile C bituminous/subbituminous with heating values in the 16.8–29.4 MJ/kg range (as received?), 11 > 40% ash, and low sulfur. Ref. 6 indicates that reserves totaled 479 Mt. IEA (2) accessible deep reserves total 224 Mt.

Brazil

Brazil has five coal-bearing regions: four in the Amazon basin and the commercial deposits in the Paraná Basin (Fig. 2). Thin (<1.5 m) Miocene-Pliocene lignites occur in the Pébas Formation in the upper Amazon basin. Thin

Figure 2. Coal fields of Brazil (14). I, Upper Amazonas region; II, Rio Fresco region; III, Tocantims-Araguaia region; IV, Western Piauí region; V, Southern Brazil region.

(<1 m) anthracite-rank "coals" of Cambro-Ordovician age are found in the Rio Fresco region in the state of Pará. The Parnaiba basin in the Tocantins-Arguaia region contains lenses of Westphalian-Stephanian age coals. Thin high volatile C bituminous Mississippian coals occur in the eastern Parnaiba basin, western Piauí state. None of the latter four basins are economic at the present time although reserves of 10 Gt for the Amazon region have been estimated (8).

The Paraná Basin in Rio Grande do Sol, Santa Catarina, and Paraná states contains Permo-carboniferous coals in the Rio Bonito Formation of the Tubarão Group. The following summary is after Refs. 8–15. The coals are associated with the *Glossopteris-Gangamopteris* flora typical of Gondwana coals found elsewhere in the southern hemisphere continents and India. Coal quality is typical of other Gondawana coals, with a high inertinite percentage and ash percentages generally exceeding 20% with some samples >50%. Sulfur ranges up to 5% (8) with some Rio Grande do Sol coals having <1% S (dry) (albeit with 40% ash). Rank increases to the north with Rio Grande do Sol having subbituminous rank with vitrinite reflectances from 0.4–0.5% (12) and the northern states having high volatile bituminous coal. Santa Catarina coals are caking. Reserves, largely in Rio Grande do Sol, total 23 Gt out of resources of 88 Gt (8). IEA (2) listed accessible reserves of 1 Gt, 10% of which is surface mineable.

Colombia

Colombia has a number of coal basins ranging in age from Maestrichtian–Paleocene in the Eastern Cordillera to middle Oligocene in the Western Cordillera (Fig. 3). Coal rank is generally high to medium volatile with some areas up to meta-anthracite where intruded by andesite.

The Eastern Cordillera deposits in the Bogata area occur in the Maastrichtian-Paleocene Guaduas Fm. The lower Guaduas Fm. has scarce low volatile bituminous coals. The 600-m thick middle Guaduas Fm. contains a

Table 1. World Estimated Recoverable Reserves of Coal[a]

	Recoverable Anthracite and Bituminous		Recoverable Lignite		Total Recoverable Coal	
Region Country	World Energy Council[b]	British Petroleum[c]	World Energy Council[c]	British Petroleum	World Energy Council	British Petroleum
North America						
Canada	4,905	3,716	1,999	3,044	6,965	6,760
Mexico	1,885	1,222		634	1,885	1,856
United States[d]	208,600	112,670	31,963	127,894	240,920	
Total	215,450	117,608	33,963	131,572	249,413	249,182
Central and South America						
Argentina	130				130	
Brazil	1,245	1,933		2,323	1,245	4,256
Chile	1,180				1,180	
Colombia	4,663	9,612			9,663	9,611
Peru	960		100		1,060	
Venezuela	47	416			417	416
Other		983	23	1,403	23	2,389
Total	13,595	12,948	123	3,726	13,719	16,673
Western Europe						
Belgium	410				410	
France	258	177		43	258	220
Germany, West	23,913	23,694	34,233	34,825	59,053	58,523
Greece			2,999	2,862	2,999	2,862
Netherlands	497				497	
Spain	535	321	235	342	770	663
Turkey	175	158	5,928	5,778	6,102	5,936
United Kingdom	3,294	9,482	500		3,794	9,102
Yugoslavia	1,569		14,996	500	16,565	
Other	55	860	110	145	165	1,048
Total	30,710	33,859	59,909	44,405	90,619	78,355
Eastern Europe and U.S.S.R.						
Bulgaria	30		3,699		3,729	
Czechoslovakia	1,870		3,494		5,369	
Germany, East			20,994	20,139	20,994	20,139
Hungary	1,578		2,882		4,460	
Poland	28,692	28,182		11,487	40,390	39,670
U.S.S.R.	140,962	102,491	11,697	136,521	40,936	239,020
Other		2,221		26,792		29,013
Total	173,132	132,901	142,745	194,941	315,877	327,842
Middle East						
Iran	193	189			193	189
Total	193	189			193	189
Africa						
Botswana	3,499				3,499	
Mozambique	240				240	
Nigeria	190				190	
South Africa	55,318	54,812			55,318	54,812
Swaziland	1,820				1,820	
Tanzania	200				200	
Zaire	600				600	
Zimbabwe	734	722			734	722
Other	278	6,553		279	278	6,832
Total	62,878	62,087		279	62,878	62,366
Far East and Oceania						
Australia	49,027	44,894	41,890	45,461	90,916	90,355
China	610,537	152,833	119,965	13,292	730,505	166,125
India	60,632	60,099	1,900	1,874	62,531	61,973
Indonesia	1,400	986	1,599	2,000	2,999	2,986
Japan	855	826	17	17	873	843
Korea, North	600				600	
Korea, South	158	593			158	93
New Zealand	108	20	9	89	117	108

Table 1. *(Continued)*

Region Country	Recoverable Anthracite and Bituminous		Recoverable Lignite		Total Recoverable Coal	
	World Energy Council[b]	British Petroleum[c]	World Energy Council[c]	British Petroleum	World Energy Council	British Petroleum
Pakistan	1,020				102	
Taiwan	200	97		100	200	194
Thailand	14		890		914	
Vietnam	50				150	
Other	150	387	9	1,341	165	1,728
Total	723,939	260,234	166,290	64,175	890,230	324,410
World Total	1,219,900	619,827	403,030	439,189	1,622,930	105,902

[a] Amount in Place that can be recovered (extracted from the earth in raw form) under present and expected local economic conditions with existing available technology. British Petroleum definitions of "Proved Reserves": Proved reserves of coal are generally taken to be those quantities which geological and engineering information indicate with reasonable certainty can be recovered in the future from known deposits under existing economic and operating conditions. The Energy Information Administration does not certify the international reserves data but reproduces the information as a matter of convenience for the reader. Comparisons between sources for both the "Recoverable Anthracite and Bituminous" category and the "Recoverable Lignite" category require careful interpretation because of the different definitional groupings for subbituminous coal. Sum of components may not equal total due to independent rounding.
[b] Recoverable anthracite and bituminous data for the World Energy Council include subbituminous.
[c] Recoverable lignite data for British Petroleum includes subbituminous.
[d] U.S. data are calculated from ETA file information. Excludes certain resource data current under review: 7,315 million short tons of anthracite in 5 States: 1,407 short tons of subbituminous coal in Alaska, and a total of 164 million short tons of coal resources in noncoal-producing States. Data represent both measured and indicated tonnage, as of January 1, 1990. Those data have been combined prior to depletion adjustments and cannot be recaptured as "measured alone."

number of mineable coals of medium and high volatile bituminous rank. The Paleocene upper unit is generally devoid of coals. "Reserves" (1968 figures) of 6.28 Gt in five Bogata area basins with total coal thickness ranging from <7 to nearly 18 m for the individual basins have been noted (17). Total "resources" for all of Colombia were listed as 10 Gt, with half of that total being in the "possible additional" category.

The coal reserves of the Cerrejón basin were poorly known at the time of Suescún-Gomez's work, with indicated reserves of 312 Mt according to the 1968 figures. The Cerrejón deposit has since been recognized as a main resource, with growing importance in world coal trade, with measured reserves of 3.6 Gt in the north and central blocks and in situ resources of 3 Gt in the south block (18). Located in the Guajura department, the deposit lies in the Cesar-Rancheria syncline and extends for 50 km north-south and is up to 5 km wide. The basin is bounded by the Oca fault to the north and the Cerrejón fault to the east. The 900 m-thick Tertiary Cerrejón Fm. consists of a lower unit with thin coals, a ca 300 m thick middle section with 0.6–6 m-thick coals, and a ca 400-m thick upper section with extensive 1.4–26 m thick coals. Coal rank spans the high volatile bituminous range with rank increasing with stratigraphic depth and from southwest to northeast within the basin. Coal quality is generally high with ash in the 5–10% range and sulfur less than 1%.

As noted above, the basin has been divided into three blocks. The North Block, with an area of 380 km^2, contains 20–26 mineable seams (of 55 total) with an average 3-m thickness. The block contains 3 Gt of measured reserves and supports the single (as of 1992) operating mine in the basin. The Central Block, with an area of 100 km^2, contains 28 mineable seams (of 38) from 1–12 m thick. The 200 km^2 South Block contains 15 seams from 1–11 m thick.

Venezuela

Tertiary coals occur in several small coal fields (Fig. 4), three of which were identified as significant by IEA (2).

The Naricual coal field, Anzoátegui state, contains perhaps 20 coal beds from 1–2.5 m thick in the lower Miocene Naricual Formation. The coals are weakly caking high volatile bituminous rank with low sulfur content. The coals are faulted and folded, with dips up to 85° (7). IEA (2) listed accessible reserves of 53 Mt in the 50 km^2 coal field.

The Lobatera coal field, Táchira state, contains a few 0.5–2.5 m, high ash, low sulfur lignites in the upper Paleocene/lower Eocene Los Cuervos and the upper Eocene/lower Oligocene Carbonera Formations (2,19). IEA (2) listed accessible reserves of 20 Mt in the 25 km^2 coal field.

The Guasare coal field, Zulia state, contains 25–30 coals, some up to 13-m thick, in the upper Paleocene/lower Eocene Marcelina Formation. The Marcelina Formation is correlative with the coal-bearing Cerrejón Formation in Colombia. The coals are noncaking, high volatile bituminous rank with variable ash and low sulfur content. Coal quality for the Paso Diablo mine is 7% moisture, 7.5% ash, 0.6% sulfur, and 29.4 MJ/kg heating value (20). Resources were estimated at 4.6 Gt (21). Latest IEA estimates (9) are 353 Mt of proven reserves with an additional 2.1 Gt of indicated resources.

CENTRAL AMERICA

Owing to its relatively recent age and active tectonic setting, Central America does not have extensive coal deposits. The oldest coals are in the Triassic-Jurassic El Plan Formation of Honduras. Analyses published (22) for Mesozoic coals in Honduras and Guatemala show a wide range of coal quality, most of it poor with high ash con-

Table 2. World Coal[a] Production and Consumption, 1989–1990, 10^6 t

Region Country	Production		Consumption	
	1989	1990[b]	1989	1990[b]
North America				
Canada	71	68	56.7	51.43
Mexico	10	11	10.1	10.56
United States[c]	890	934	807.9	811.54
Total	970	1,012	874.7	873.54
Central and South America				
Brazil	6	7	17.3	18.21
Chile	2	2	3.59	3.77
Colombia	20	20	8.31	8.72
Other	3	3	4.1	4.32
Total	30	32	33.35	35.02
Western Europe				
Austria	3	3	6.7	7.20
Belgium and Luxembourg	3	2	16.32	17.74
France	15	14	31.40	30.62
Germany, West	187	184	140.2	188.21
Greece	52	50	53.28	51.08
Italy	2	2	21.47	21.57
Norway	[d]	[d]	1.44	1.56
Spain	42	36	53.66	46.45
Turkey	40	39	41.77	43.56
United Kingdom	100	89	115.0	101.68
Yugoslavia	74	76	79.02	79.75
Other	[d]	[d]	21.17	18.67
Total	518	494	652.5	632.11
Eastern Europe and USSR				
Albania	3	3	2.65	2.79
Bulgaria	34	30	40.58	36.37
Czechoslovakia	118	107	117.93	107.70
Germany, East	301	256	306.93	257.74
Hungary	20	18	22.54	20.37
Poland	250	259	218.27	209.19
Romania	62	38	71.51	46.26
USSR	690	630	661.59	599.60
Total	1,477	1,343	1,442.99	1,300.00
Middle East				
Iran	1	1	1.56	1.64
Total	1	1	3.83	4.02
			5.39	5.66
Africa				
Morocco	1	1	1.87	1.97
South Africa	175	168	131.81	124.21
Zambia	[d]	[d]	0.39	0.38
Zimbabwe	5	5	5.17	5.43
Other	1	2	4.15	4.47
Total	183	177	143.39	136.47
Far East and Oceania				
Australia	196	210	95.28	107.04
China	1,054	1,107	1,040.77	1,092.80
India	209	219	212.49	223.12
Indonesia	4	4		
Japan	10	10	109.40	110.85
Korea, North	54	56	56.23	59.05
Korea, South	21	22	46.39	48.70
Mongolia	8	8	7.35	7.7
New Zealand	3	3	2.13	2.22
Pakistan	3	3	3.65	2.73
Philippines	2	2	2.48	2.61
Thailand	17	18	12.10	12.44
Vietnam	1	5	4.92	5.16

Table 2. *(Continued)*

Region Country	Production		Consumption	
	1989	1990[b]	1989	1990[b]
Other	1	1	32.49	3.41
Total	1,584	1,667	1,631.68	1,708.54
World Total	4,766	4,727	4,783.05	4691.33

[a] Coal includes anthracite, subanthracite, bituminous, subbituminous, lignite, and brown coal. Sum of components may not equal total due to independent rounding.
[b] Preliminary.
[c] Consumption data do not reflect net exports of coke.
[d] Denotes less than 450 thousand tons.

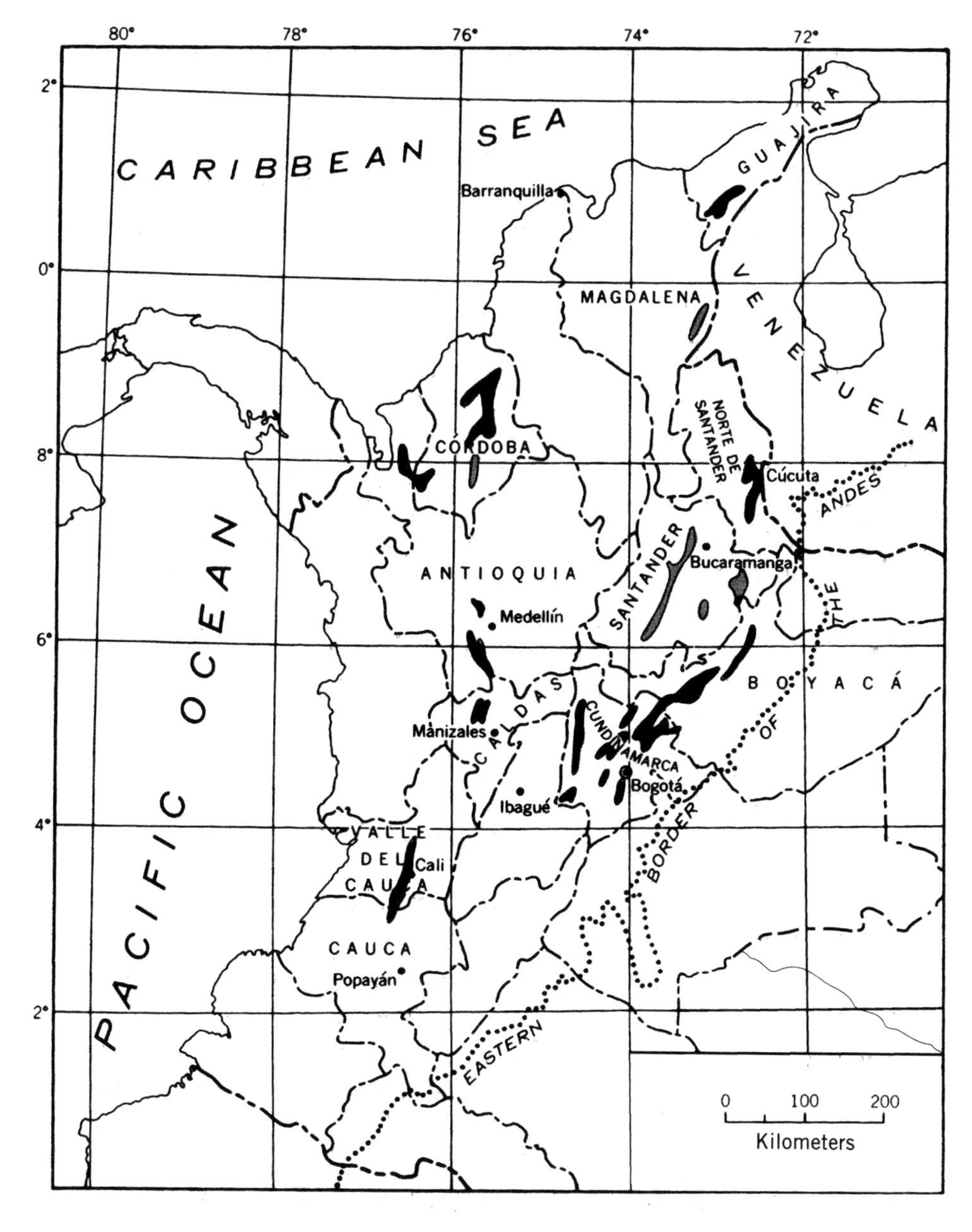

Figure 3. Coal fields of Colombia (7). Age of coal deposits ■, Tertiary; ▤ late Cretaceous.

Table 3. World Coal Flow[a], 1989, P (10^{15}) J

	Exporting Region and Country															
	North, Central and South America			Western Europe				Eastern Europe			Africa	Far East				
Importing Region and Country	Canada	United States	Colombia	Belgium	Germany, West	Netherlands	United Kingdom	Czechoslovakia	Poland	USSR	South Africa	Australia	China	Japan	World Other	World Total
North America																
Canada	0	448	0	0	4	0	0	0	0	0	0	2	3	0	1	458
United States	37	0	28	2	73	1	1	0	3	0	0	24	3	24	0	137
Other	1	4	0	[b]	[b]	0	0	0	0	0	0	[b]	0	0	0	5
Central and South America																
Brazil	0	160	0	0	0	0	0	0	0	0	0	35	1	0	83	280
Chile	0	24	2	0	0	0	0	0	0	0	0	6	0	0	13	46
Other	0	11	2	0	79	0	[b]	33	0	5	0	8	2	0	0	81
Western Europe																
Austria	0	0	0	0	14	0	0	31	42	25	[b]	0	0	0	0	113
Belgium and Luxembourg	3	173	2	0	62	0	3	0	7	2	86	45	12	0	22	417
Denmark	16	76	61	0	[b]	0	12	[b]	21	19	0	74	4	0	0	282
Finland	0	24	4	0	[b]	0	5	0	57	64	0	0	1	0	0	156
France	19	163	38	8	44	0	8	0	0	64	18	79	30	0	40	493
Germany, West	[b]	55	8	2	0	[b]	2	5	44	22	80	9	4	0	0	233
Ireland	0	30	21	[b]	1	1	9	[b]	17	0	3	4	[b]	0	1	88
Italy	1	293	16	0	33	0	0	0	21	15	130	30	8	0	0	547
Netherlands	15	139	42	5	20	0	0	0	21	4	31	82	8	0	[b]	367
Portugal	0	34	0	0	0	0	6	0	2	44	0	0	0	1	8	96
Spain	0	90	13	1	9	0	3	0	0	7	130	26	1	0	0	280
Sweden	17	34	2	0	[b]	[b]	[b]	[b]	16	15	0	25	[b]	0	0	111
Turkey	0	57	0	0	0	0	0	0	0	13	0	30	0	0	0	99
United Kingdom	22	141	7	15	8	46	0	0	0	0	9	83	7	0	0	341
Yugoslavia	0	43	1	1	2	0	0	0	0	53	0	11	0	0	0	111
Other	1	27	2	15	14	1	6	4	17	[b]	1	3	1	1	0	95
Eastern Europe and U.S.S.R.																
Bulgaria	0	2	0	0	0	0	0	0	0	139	0	0	0	0	0	141
Czechoslovakia	0	0	0	0	0	0	0	0	35	67	0	0	0	0	16	119
Germany, East	0	0	0	0	0	0	0	22	22	83	0	1	1	0	14	145
Hungary	0	0	0	0	0	0	0	0	0	57	0	2	0	0	8	61
Romania	0	36	0	0	0	0	0	0	194	27	0	0	0	0	0	257
USSR	0	0	0	0	0	0	0	0	206	0	0	60	[b]	0	[b]	265
Other	0	12	0	0	0	0	0	2	3	5	0	5	0	0	0	26
Middle East																
Israel	0	12	12	0	0	0	0	0	0	0	60	15	0	0	0	97
Other	2	5	0	0	0	0	3	0	0	0	0	0	0	0	3	14
Africa																
Algeria	0	27	0	0	0	0	0	0	0	0	0	0	0	0	3	31
Egypt	0	14	0	0	0	0	0	0	0	12	0	0	0	0	0	26
Other	0	20	0	0	0	0	[b]	0	0	0	0	34	0	0	4	58
Far East and Oceania																
Hong Kong	0	0	9	0	0	0	0	0	0	0	103	100	44	0	0	257
India	0	5	0	0	0	0	0	0	0	0	0	95	0	0	15	118
Japan	527	341	3	0	2	0	0	0	0	72	287	1,443	115	0	0	2,792
Korea, South	264	140	0	0	0	0	0	0	0	40	0	220	42	0	4	708
Taiwan	26	131	0	0	0	0	0	0	0	0	137	158	5	1	0	406
Other	21	41	3	0	44	0	5	0	6	[b]	31	87	30	43	0	311
World Total	910	2,809	276	50	291	50	67	99	728	830	1,108	2,800	327	79	235	10,716

[a] Includes coke. Sum of components may not equal total due to statistical discrepancies, losses, unaccounted for coal and coal trade not in national accounts (eg, U.S. shipment to U.S. Armed Forces in Europe).
[b] Less that 527 billion kJ.

tents, and a rank range from lignite to anthracite. Speculative Carboniferous deposits in Guatemala totaling 641 Mt have been noted (23).

Tertiary deposits are generally confined to the Miocene Gatún Formation, primarily in Panama and Costa Rica, and various Pliocene and Pleistocene lignites throughout the region. The lignites and subbituminous coals of Costa Rica, in particular those in Limon state on the Caribbean coast, have some of the greatest reserves and highest quality of the Gatún deposits. Analyses published (22) in-

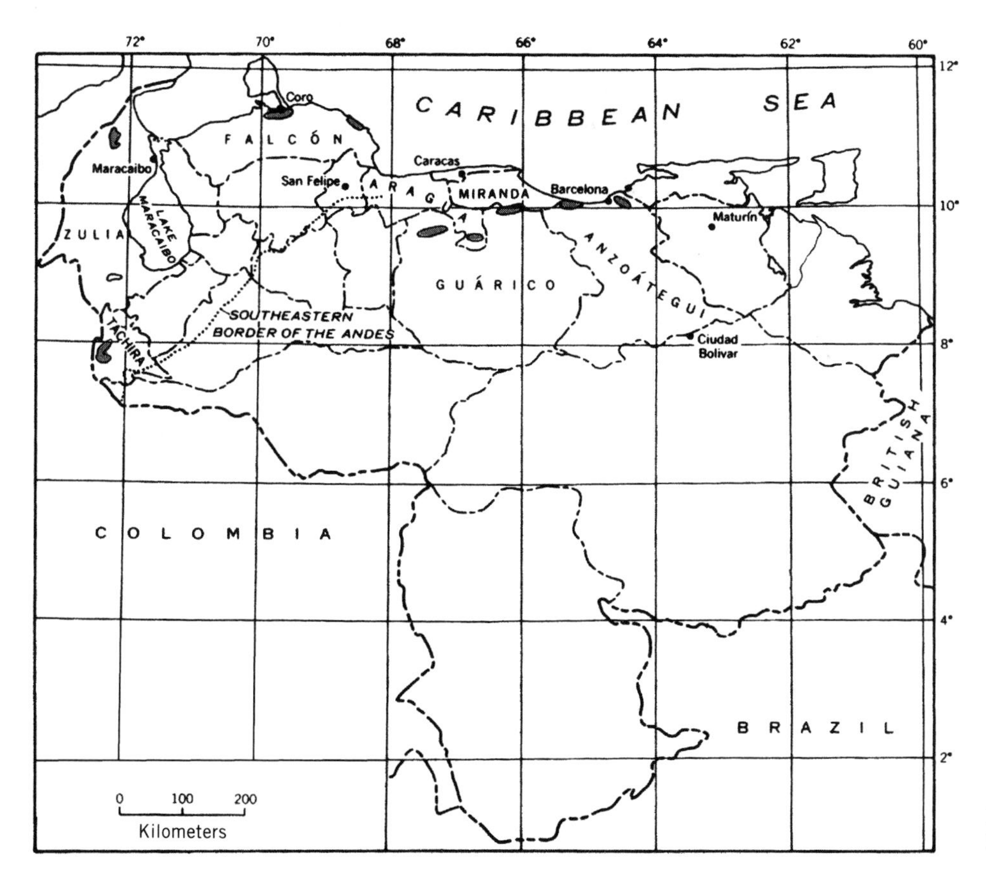

Figure 4. Coal fields of Venezuela (7). ■, Coal deposits of Tertiary age.

dicated a range of ash from 3.71–15.58%, sulfur from 0.42–8.93%, and heating values from 20.3–25.4 MJ/kg (no indicated basis) for Zent coals. Similar ranges have been published (24–26). Limited petrographic analyses (24) indicated that the Baja Talamanca coals had vitrinite contents of 64–79%. Vitrinite maximum reflectances average 0.4%. Reserves are limited, with about 13 Mt measured and indicated reserves in the Baja Talamanca field (25). The latter field has 54.6 Mt identified-hypothetical coal-in-place (23), half of Costa Rica's total resources. Mesozoic and Cenozoic deposits in Central America total 320 Mt, most of it in small, poorly explored deposits.

NORTH AMERICA

Mexico

Economic coal deposits have been found in three regions of Mexico. General summaries are in Ref. 27.

Coal occurs in the Oaxaca region in the south in the early to middle Jurassic Rosario, Zorrillo, and Simón Formations. Coal beds up to 3-m thick are complexly folded and have been influenced by post-Mesozoic intrusions. Rosario Formation coals are low-volatile bituminous and semianthracite and medium-volatile bituminous coals occur in the Zorrillo Formation. Analyses published (28) indicate that the coal is low sulfur but high ash, generally >30% (dry). The most recent resource figures quoted (27) were 24 Mt inferred and 26 Mt possible resources. Ref. 23 lists 48 Mt of identified-hypothetical resources and 1.8 Gt of additional speculative resources.

Coal in Sonora in northwest Mexico occurs in the upper Triassic Barranca Formation (27,29). Two separate small basins contain seven to nine 1-m coals of anthracite and semianthracite rank. In places, intrusions have coked the coal. There are 247 Mt of identified-hypothetical resources and an additional 1.06 Gt of speculative resources (23).

The Sabinas basin of the Coahuila region in northeast Mexico has a 2-m thick split seam in the upper Cretaceous Olmos Formation which has supported mining since the 1880s (27,30–33). In contrast to the latter two regions, the geologic structure is relatively simple, with a series of northwest-trending anticlines and synclines. The coal is medium-volatile bituminous with low sulfur and variable ash content. Much of the coal has been used for coking. IEA (2) considered the region to contain 1 Gt of accessible coal reserves out of 8.56 Gt identified-hypothetical and 12.5 Gt speculative resources (23).

Canada

Coal in Canada occurs in the western and Maritime provinces (Fig. 5) in Carboniferous through Tertiary formations (10). Resources and reserves are given in Tables 4 and 5.

Western Provinces. Coal reserves in western Canada occur from the Pacific islands in British Columbia; through to the Cordillera, foothills, and plains coal fields of Alberta; and to the lignite deposits of Saskatchewan. The summaries which follow are after Ref. 9, the provincial summaries in Refs. 34 and 35.

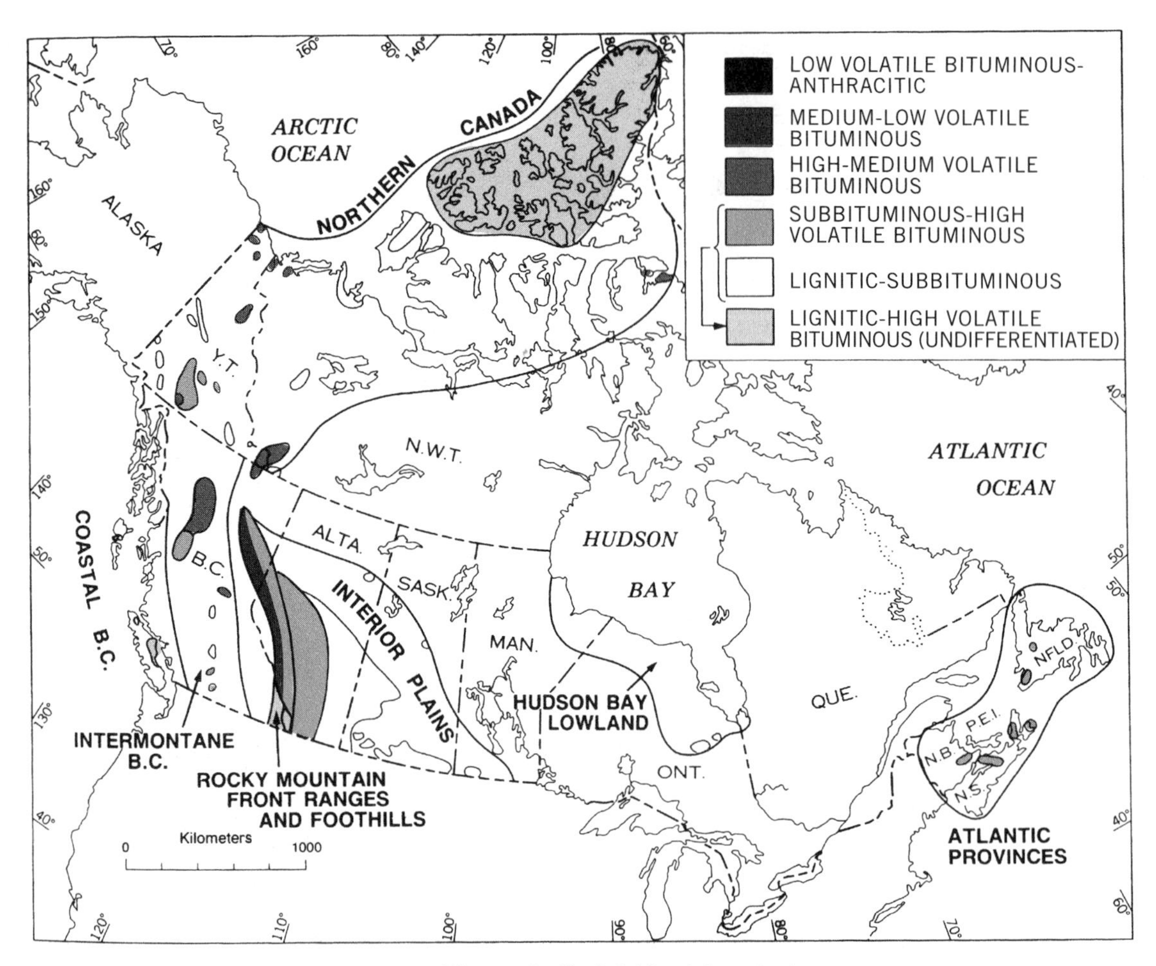

Figure 5. Coal fields of Canada (35).

Coal-bearing strata were deposited from the middle Jurassic to the Tertiary. The Jurassic-early Cretaceous phase is represented by the coal-bearing Kootenay Group in southeastern British Columbia and southwestern Alberta and the partly correlative Minnes Group in northeastern British Columbia and northwestern Alberta. The Kootenay-Minnes sedimentary wedge reaches 2.7 km and contains at least 14 mineable 1–15 m coals.

The lower to upper Cretaceous Bullhead and Ft. St. John Groups of northeastern British Columbia–northwestern Alberta and the correlative Luscar Group of north central Alberta represent the second sedimentary wedge. The Gething Formation of the Bullhead Group ranges from 120-m thick with two 2–5 m coals in the south to 1 km with about 100 coals, some to 4 m-thick. The Gates Formation of the Luscar Group has seven <1–13-m coals and the correlative Ft. St. John Group has 11 coals with thicknesses up to 10 m.

The upper Cretaceous to Tertiary Edmonton and Wapiti Groups, northeastern British Columbia–northwestern Alberta; the Belly River Group of southeastern British Columbia–southwestern Alberta; and the Saunders Group, north central Alberta, represent the third sedimentary wedge. The Belly River Formation contains three coal zones with a total of 14 seams from <1–18-m thick. The Saunders Group Coalspur Formation's Coal Valley zone and the correlative Edmonton Group Scollard Formation contain a number of coals up to 22-m thick, including the coals in the 60 m Ardley coal zone in the latter formation. The upper Cretaceous Horseshoe Canyon Formation in the Drumheller, Alberta, region contains up to 13 coal beds.

The Saskatchewan deposits are in the Paleocene Ravenscrag Formation, correlative with the Fort Union Group in the contiguous coal fields of North Dakota and Montana.

Coal rank trends in western Canada have been reviewed (36–40).

Insular Coal Fields, British Columbia. Deposits of low-ash, low-sulfur Tertiary lignites and Mesozoic high volatile bituminous and anthracite rank coal, the latter being low ash, occur on Graham Island. Three Cretaceous high volatile bituminous deposits are found on Vancouver Island. The Comox and Nanaimo coal fields supported mining up to 1953. Identified-hypothetical coal-in-place totals 1.47 Gt (22).

Interior Coal Fields, British Columbia. High-ash, low-volatile bituminous to anthracite rank coals up to 5-m thick occur in the Klappan-Groundhog coal field in north central British Columbia. Coals belong to the Middle to Late Jurassic Bowser Lake Group and the Early Cretaceous Skeena Group.

High volatile to meta-anthracite rank coals are found in the Telkwa-Red Rose coal field.

Table 4. Summary of Canada's Coal Resources, 10^6 t[a]

Coal Region	Coal Rank	Immediate Interest			Future Interest		
		Measured	Indicated	Inferred	Indicated	Inferred	Speculative
Coastal British Columbia							
Vancouver Island	h-mvb	35	80	200		300	
Queen Charlotte Islands	lvb-an			10			
	h-mvb		15	10			
	lig-sub			50			500
Intermontane British Columbia							
Northern District	lvb-an	100	500	1,000			4,000
	h-mvb	30	50	100			100
Southern District	sub-hvb	40	120	340			
	lig-sub	450	320	270			
Rocky Mountains and Foothills							
Front Ranges							
East Kootenay	h-mvb	1,390	1,320	4,040	2,700		
Crowsnest	m-lvb	265	140	510	200		
	h-mvb	330	170	630			
Cascade	lvb-an	240	120	455	210		
Panther River-Clearwater	lvb-an				15	700	
Inner Foothills							
Southern District	m-lvb	635	320	1,145	245		
	h-mvb	150	75	275			
Northern District	m-lvb	1,115	2,385	6,270	100		
Outer Foothills	sub-hvb	830	740	1,955	200		
Plains							
Mannville Group	lig-sub		35	100		30	
Belly River/Edmonton/Wapiti	sub-hvb	1,240	585	1,860	820		
	lig-sub	11,860	4,935	16,575	14,115		
Paskapoo	sub-hvb	120	60	175	25		
Ravenscrag	lig-sub	1,445	2,680	3,440	3,910	23,510	
Deep coal	sub-hvb				4,000	50,000	85,000
Hudson Bay Lowland, Onakawana	lig-sub	170	10				
Atlantic Provinces	h-mvb	345	365	770	1,500	215	
Northern Canada							
Yukon Territory and District of Mackenzie	lvb-an			90	(no available estimates of resources of future interest for this region)		
	h-mvb			150			
	sub-hvb			350			
	lig-sub			2,290			
Arctic Archipelago	sub-hvb				500	550	4,500
	lig-sub				7,000	7,500	31,000
Totals	lvb-an	340	620	1,555	225	700	4,000
	m-lvb	2,015	2,845	7,925	545		
	h-mvb	2,280	2,075	6,175	4,200	515	100
	sub-hvb	2,230	1,505	4,680	5,545	50,550	89,500
	lig-sub	13,925	7,980	22,725	25,025	31,040	31,500

[a] Ref. 35

Tertiary high volatile coals are found in the Bowron River, Similkameen, and Merritt coal fields and lignite–subbituminous coals occur in the Princeton coal field. The latter field has about 70 m of coal within 500 m of section.

The Tertiary Hat Creek coal field in south central British Columbia occupies a 16 × 2.5 km graben. Two separate deposits, each with four 200–430-m coal zones, constitute the 200 Mt of accessible reserves (2) of the 5 Gt of resources. The deposit is low-sulfur, high-ash subbituminous rank coal.

Mountains, British Columbia and Alberta. The East Kootenay-Crowsnest Pass area coalfields contain high volatile A to low volatile bituminous Kootenay Group Mist Mountain Formation coals. Up to 15 coals occur in portions of the coal field. Severe tectonism has influenced mining conditions. Tectonic thickening has increased coal thickness from 10 to nearly 80 m in places. Coal quality ranges from 12 to >30% ash, generally less than 1% sulfur, and 20–29 MJ/kg heating value (moist basis). Measured resources are 12.7 Gt with an additional 13 Gt indicated resources in the Flathead, Crowsnest, and Elk Valley coal fields.

Table 5. Estimated Coal Reserves in Canada[a]

Province / Region / Coalfield	Predominant Rank	Coal in Mineable Seams (megatonnes)	Recoverable Coal (megatonnes)
British Columbia			
Vancouver Island			
Nanaimo	hvb	3	2
Comox	hvb	16	15
Southern Intermontane			
Hat Creek	lig-sub	739	566
Rocky Mountain Front Ranges			
Fernie/Flathead	m-lvb	540	436
Elk Valley	m-lvb	1133	1069
Northern Inner Foothills			
Peace River	mvb	710	476
(Sub-totals-British Columbia)	m-lvb	2383	1981
	hvb	19	17
	lig-sub	739	566
Alberta			
Inner Foothills			
Cadomin-Luscar	mvb	161	77
Smokey River	lvb	420	163
Outer Foothills			
Coalspur	hvb	449	316
McLeod River	hvb	387	330
Plains			
Obed Mountain	hvb	197	154
Alix	sub	4	3
Ardley	sub	2	1
Battle River	sub	136	76
Drumbeller	sub	42	11
Mayerthorpe	sub	3	2
Morinville	sub	13	8
Sheerness	sub	192	134
Thorhild-Abee	lig	2	1
Tofield-Dodds	sub	1	1
Wabamun	sub	642	480
Wetaskiwin	sub	250	155
(Sub-totals-Alberta)	m-lvb	581	240
	hvb	1033	800
	lig-sub	1287	872
Saskatchewan			
Cypress	lig	(published data on separate coalfields are not available)	
Willow Bunch	lig		
Wood Mountain	lig		
Estevan	lig		
(Sub-totals-Saskatchewan)	lig	2088	1670
New Brunswick			
Minto	hvb	10	9
Beersville	hvb	13	12
(Sub-totals-New Brunswick)	hvb	23	21
Nova Scotia			
Inverness	hvb	3	1
Sydney	hvb	567	413
(Minor reserves in Inverness and Springhill coalfields)			
(Sub-totals-Nova Scotia)	hvb	570	415
Totals, Canada	*m-lvb*	*2964*	*2221*
	hvb	*1645*	*1253*
	lig-sub	*4114*	*3108*

[a] Ref. 35, Dec. 1985.

The Cadomin-Luscar area, Alberta, contains one main Luscar Formation coal. The high volatile A to medium volatile coal (27–30 MJ/kg, moist basis) has low sulfur content and 15–25% ash in a 10 m, with thickening to 60 m, coal bed. The Smoky River area, Alberta, has medium to low volatile rank Luscar Formation coals in 11 seams of up to 15-m thick. The Cadomin-Luscar area has 110 Mt measured and 55 Mt indicated resources and the Smoky River area has 265 Mt measured and 130 Mt indicated resources.

The Peace River area, British Columbia, contains six seams in the Bullhead Group Gething Formation and the Ft. St. John Group Gates Formation. The medium to low volatile bituminous, 20–35% ash coals reach 10-m thickness. The Gates Formation coals are better coking coals than the underlying Gething coals. Measured resources total 1.1 Gt with additional 2.4 Gt indicated resources.

IEA (2) listed 700 Mt accessible coal for all of the mountain coal fields.

Foothills, Alberta. The Coalspur, McLeod River, and Obed Mtn. coal fields contain a total of 800 Mt of measured and 700 Mt of indicated resources in several 3–20-m Cretaceous-Tertiary high volatile C bituminous coals. All deposits are low sulfur with the McLeod River field containing lower ash than the Coalspur deposit. IEA (2) listed 700 Mt of accessible coal.

Plains, Alberta. Other than high volatile bituminous coals in the Upper Cretaceous Oldman Formation in the Lethbridge area, the Plains coal fields contain numerous low ash, low sulfur subbituminous coals with heating values in the 15–21 MJ/kg range (moist basis). The coal zones are in the Horseshoe Canyon and Paskapoo Formations. Measured resources total 10 Gt and indicated resources total 4.1 Gt. IEA (1983) placed accessible coal at 2 Gt.

The Cretaceous and Tertiary deposits which are coal bearing in the Alberta plains continue to the north into the Northwest Territories. Coals are known from the Arctic islands, including extensive Devonian and Mississippian deposits, but are subeconomic.

Ravenscrag, Saskatchewan. The lignite-bearing Paleocene Ravenscrag Formation extends along the U.S.–Canada boundary from about 102° to 107° W longitude with discontinuous fields to Alberta and ranges from 80–300 m thick. Up to six economic coal zones with up to eight coal beds each comprise the resources. Maximum thickness for the low sulfur, low ash lignites is about 7 m. Ash content increases to the west.

Resources of immediate interest include:

- Coal ≥1.5 m with overburden ≤45 m
- Coal 0.9–1.5 m-thick with <15:1 stripping ratio or coal underlain by coal in the first category
- Coal 0.6–0.9 m underlain by coal in either of the above categories or overlain by an acceptable seam with rock intervals less than twice the thickness of the lower seam

Resources of future interest include all other coals ≥0.9 m down to 460 m depth. Resources of immediate interest total 8.1 Gt and resources of future interest total 27.6 Gt. IEA (2) considered 1.3 Gt to be accessible.

Maritime Provinces. Pennsylvanian subbituminous to high volatile A bituminous, high ash, high sulfur reserves totaling 20 Mt occur in the Minto-Chipman area of central New Brunswick (Fig. 6). More important reserves occur in Nova Scotia in the Sydney coal field; in several coalfields on the west side of Cape Breton Island; in the Pictou, Stellarton, and several small mainland coal fields; and in the Springhill and Joggins coal fields in Cumberland County, western Nova Scotia. Ref. 41 includes the Minto and western Cape Breton Island deposits together in the largely submarine 46,600 km^2 Gulf of St. Lawrence basin.

Eleven Westphalian C to Stephanian coal beds have been mined in the Sydney coal field. The Harbour, Hub, and Phalen coal beds support all current mining. The basin extends under the Cabot Strait to Newfoundland for a total extent of 36,000 km^2. Quality of the high volatile A bituminous coal is variable with high sulfur coal common. Coal reserves in the Sydney coal field are 1 Gt (42). Resources in the submarine coal field are defined as coal greater than 1.3-m thick at a depth of less than 1220 m within 13 km of the coast with at least 55 m of solid cover.

High volatile coals of variable ash and sulfur content occur in the other coal fields. Westphalian A Riversdale Group coals occur in the Chimney Corner-St. Rose and Port Hood coal fields on the west coast of Cape Breton Island. Westphalian B Cumberland Group coals occur in the Joggins, Springhill, and Saltsprings coal fields. Westphalian C-D Pictou Group coals occur in the Inverness and Mabou coal fields on the west coast of Cape Breton Island and the Pictou and Debbert-Kemptown coal fields on the mainland. Total resources in the province are 3 Gt with a recoverable reserve of 1 Gt.

United States

The United States is one of the top three coal producers in the world and, indeed, the states of Wyoming, West Virginia, and Kentucky would individually rank with the top ten producing countries. The United States has an estimated 3.37 Tt of total resources, over 26% of the world total. Discussion of U.S. coal fields will proceed along the lines of the coal provinces: Appalachians, Eastern and Western Interior, Gulf Coast (including Cretaceous coals in Texas), Great Plains and Rocky Mountains, Pacific Northwest, and Alaska.

An overview of coal rank trends throughout the U.S. is presented in Ref. 43.

Appalachians. The U.S. Appalachian coal fields extend from Pennsylvania to Alabama (Fig. 6). With the exception of two breaks in Tennessee and Alabama, there is a contiguous region of coal producing counties from north-central Pennsylvania to Tuscaloosa County, Alabama, a distance of about 1300 km with a maximum width of about 340 km through Ohio and West Virginia. The demonstrated reserve base is 101 Gt (identified hypothetical resources on Table 6). In terms of producing regions: the Pennsylvania Anthracite fields produced 3 Mt, the Northern Appalachians (Pennsylvania, Ohio, Maryland, portions of Kentucky and West Virginia) produced 143 Mt, the Central Appalachians (southern West Virginia, most of eastern Kentucky, Virginia, and nothern Tennessee)

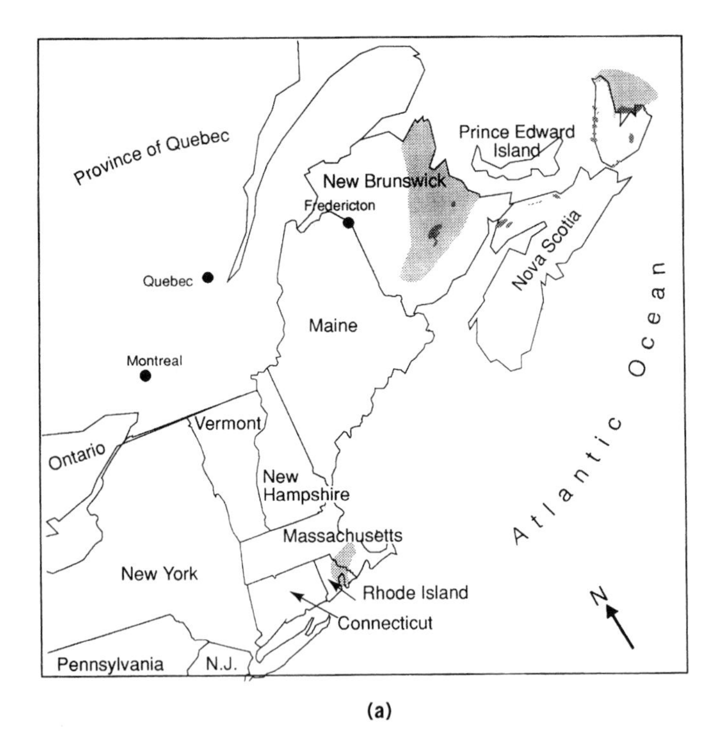

(a)

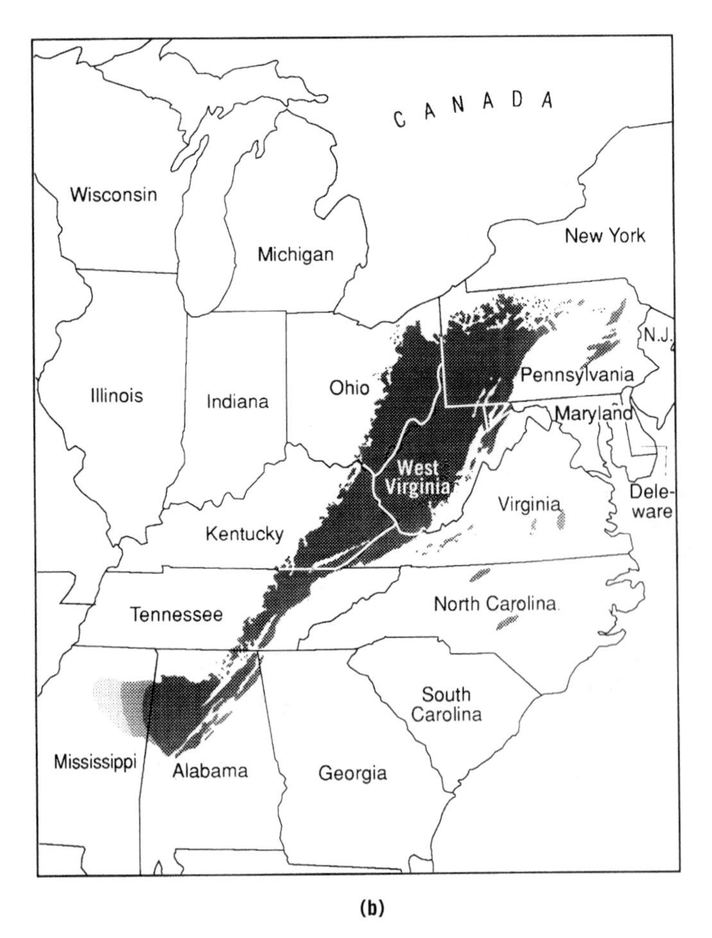

(b)

Figure 6. Coal fields in the (**a**) Maritime provinces, New England, and (**b**) Appalachian coal fields.

produced 249 Mt, and the Southern Appalachians (southern Tennessee, Georgia, and Alabama) produced 25 Mt (1989 Figures).

Economic coals range in age from Tournaisian (Mississippian) in the Valley fields, Virginia, to Autunian (Permian) in the Dunkard basin and possibly in the Pennsylvania Anthracite fields.

The summaries which follow are adapted from summaries refs. 9,13,44,45.

Anthracite Fields, Pennsylvania. Semianthracite to meta-anthracite occurs in several basins in eastern Pennsylvania totaling about 720 km^2 (Fig. 7). Structural complexity decrease to the north and northwest, passing from thrusted, overturned folds in the Southern field to gentle folds in the Bernice basin on the Allegheny Plateau. Coal rank trends are not coincident with structural trends but rather rank decreases from the eastern end of the Southern field ($R_{max} > 6\%$) to the western end of the Southern and Western Middle fields and the Bernice basin ($R_{max} < 3\%$) (46). Summaries of the structure and stratigraphy of the Anthracite fields are provided in Ref. 47.

Coal measures range in age from Namurian B-C of the Pottsville Formation's Tumbling Run member at least to the top of the Stephanian (Llewellyn Formation) (Fig. 8). The uppermost Llewellyn Formation contains floras found elsewhere in the Permian of the Dunkard basin, western Pennsylvania–northern West Virginia. The Southern field contains the thickest section with the greatest age span. The maximum thickness of the Pennsylvanian section is at least 1.5 km. The Pottsville Formation (Namurian B-C to Westphalian D) contains 10 persistent coals and the Llewellyn Formation (Westphalian D to Autunian (?)) contains at least 40 persistent coals. The coals average less than 3-m in thickness but may be up to 15-m thick with tectonic thickening to 29 m. Sulfur content is generally low, averaging 0.8%, increasing to the west and north.

The demonstrated resource base is estimated to be 7.3 Gt. Structural complexity and the associated mining costs inhibit production. Production peaked at nearly 90 Mt in both 1917 and 1918.

Northern Appalachians. The Northern Appalachian coal field, also known in part as the Dunkard basin and the Pittsburgh-Huntington synclinorium, extends from the North-central fields in Pennsylvania (the semianthracite-bearing Bernice field is also one of the North-central fields) to the Princess Reserve District in northeastern Kentucky and includes the outlying Broad Top and Georges Creek basins. Coals range in age from the Westphalian A-C Pottsville Group (not strictly correlative with the Pottsville Formation in the Anthracite fields) through the Westphalian D Allegheny Formation, the Stephanian Conemaugh and Monogahela Groups, to Autunian within the Dunkard Group. The entire section is about 1-km thick and contains perhaps 40 coal beds or zones, of which at least 12 are commonly mined. The best known, most extensively mined coals are the Allegheny Formation's Lower Kittanning and Upper Freeport coal beds and the Monongahela Group's Pittsburgh coal bed.

Coal quality varies between coal beds and localities but, in general, sulfur content is higher than in Central Appalachian coals. Petrographic composition is also

Table 6. Coal Resources 10^9 t of Carboniferous Coals in the Eastern U.S.[a]

	Identified-Hypothetical Coal-in-Place	Onshore Speculative Coal-in-Place	Total Coal-in-Place
New Brunswick	0.11	9.96	10.06
Newfoundland	0.14	1.53	1.67
Nova Scotia[b]	1.56		3.25
MARITIMES[b]	1.81	11.49	14.98
Rhode Island/ Massachusetts	0.27		0.27
Pennsylvania			
(anthracite)	21.76		21.76
(bituminous)	66.59		66.59
Maryland	4.06		4.06
Ohio	47.59		47.49
West Virginia	96.86		96.86
Virginia			
(Valley fields)	0.40		0.40
(Bituminous)	16.25		16.25
Kentucky (east)	49.96		49.96
Tennessee	5.62		5.62
Georgia	0.15		0.15
Alabama	14.46	2.53	16.99
Mississippi		8.92	8.92
APPALACHIANS	323.97	11.45	335.42
MICHIGAN BASIN	2.23		2.23
Illinois	256.80		256.80
Indiana	34.07		34.07
Kentucky (west)	58.60		58.60
Missouri	0.15		0.15
EASTERN INTERIOR	349.62		349.62
Iowa	21.19	7.33	28.53
Missouri	44.53		44.53
Nebraska	8.13		8.13
Kansas	18.69	20.82	39.51
Oklahoma	25.37	29.00	54.37
Arkansas	3.99	14.34	18.33
Texas	14.13	35.54	49.67
Mississippi		0.25	0.25
WESTERN INTERIOR	136.03	107.28	243.31

[a] Ref. 23.
[b] Off shore speculative coal-in-place, 1.69 × 10^9 t.

highly variable, as with all Appalachian coals. Rank varies with respect to position relative to the Allegheny Front on the southeastern edge of the Allegheny Plateau (50,51). Low volatile bituminous rank coal is found in the Barclay, Ralston, and Cogan Station fields among the north central fields, in the Broad Top coal field in south central Pennsylvania, and in Cambria and Somerset counties, Pennsylvania. Rank decreases to high volatile C bituminous to the northwest.

Most coals average about 1–2-m thick. The Pittsburgh coal bed is nearly 7-m thick in the Georges Creek basin in Maryland.

Remaining resources are estimated at 277 Gt (48). The U.S. Geological Survey is in the process of revising resource estimates for the Appalachians on the basis of a wide range of land-use restrictions (49).

Central Appalachians. The Central Appalachian coal field, which includes the Pocahontas basin, and, for the purpose of this discussion, includes mineable coals in Virginia's Valley fields, extends from the southern side of the Northern Appalachians to central Tennessee. Coals range in age from the two Tournaisian Price Formation coals in the Valley fields, to the Namurian Pocahontas Formation coals, the Westphalian A New River formation (West Virginia)/Lee (Virginia and Kentucky), to the Westphalian B-D Breathitt (Kentucky)/Kanawha (West Virginia)/Norton-Wise (Virginia) Formation. The uppermost mineable coals are correlative with lower portions of the Allegheny Formation. The older portions of the basin are to the southeast, the Valley fields being in the thrust-faulted Valley and Ridge province. Pocahontas and older New River coals are in the portion of the basin to the northwest of the St. Clair fault (Allegheny Front) and are generally absent beneath the coal measures of Kentucky. The coal field in its entirety is a succession of coal-bearing units topped by marine units and, at least for units through to Westphalian A, bounded on the northwest by sandstone units. Each successive unit extended further northwest

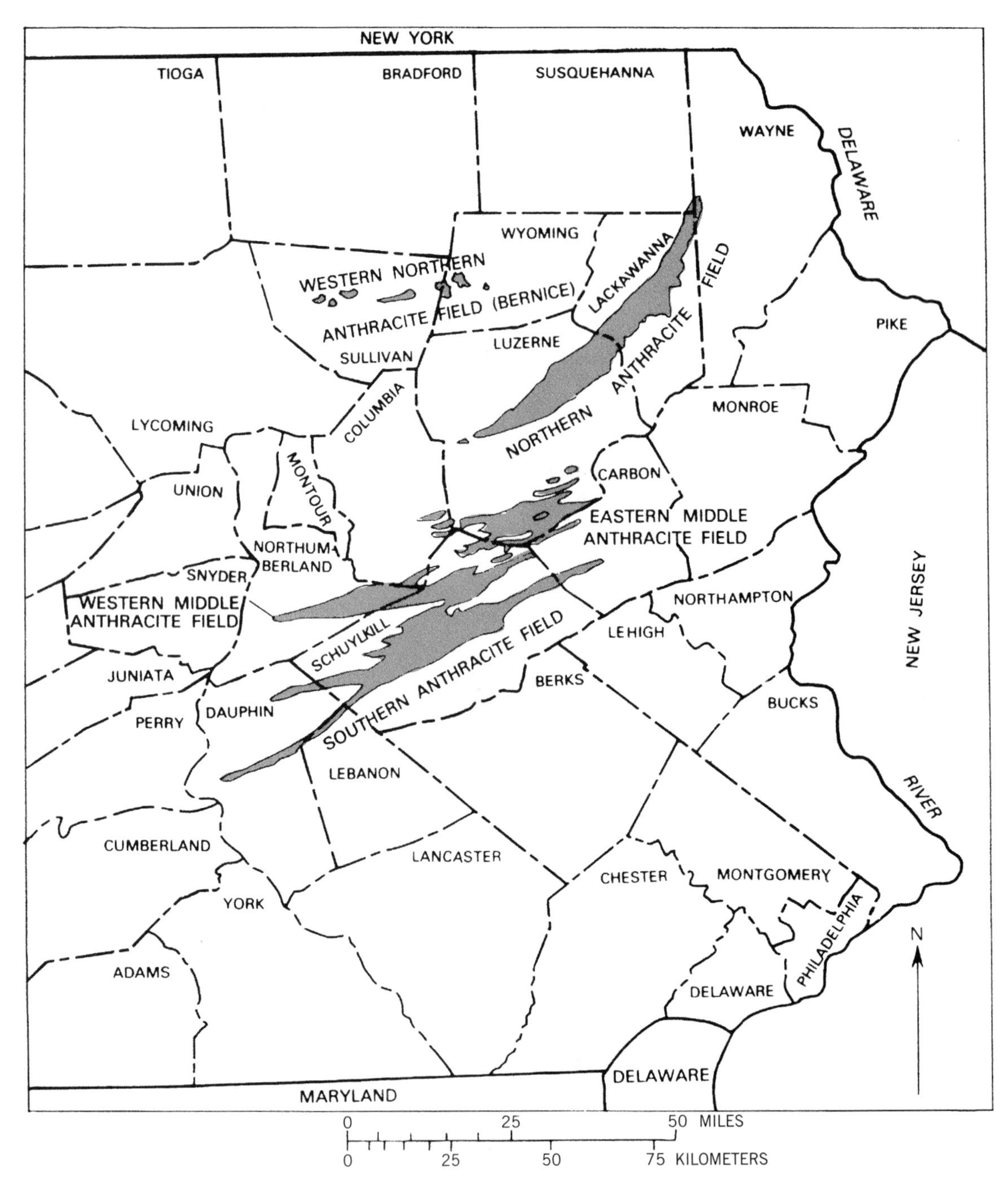

Figure 7. Anthracite coal fields, eastern Pennsylvania (47).

than the underlying unit (52). Well over 60 principal coal beds, along with perhaps 200 splits and riders, are recognized in the region. Mined thicknesses range from <0.5–>4 m but is generally in the 1–2-m range.

Coal quality varies but most coals where mined are low sulfur, low-to-moderate ash, and have 60–80% vitrinite. The Blue Gem coal bed in southeastern Kentucky has <1% ash at some mines. With 1995 and 2000 Clean Air Act deadlines for SO_2 reductions, the Central Appalachians, along with the Powder River basin, should see an increase in production to meet demands for low sulfur coal. The field continues to be a producer of high volatile A to low volatile bituminous metallurgical coal. Coal rank ranges from semianthracite and low volatile bituminous for the Valley field coals; to low volatile bituminous for Pocahontas Formation coals in the structurally deeper basins in Buchanan County, Virginia, and McDowell County, West Virginia; to high volatile C bituminous in coals near the northwest boundary of the coalfield. Refs. 52 and 53 discuss coal rank trends in the Valley and Plateau coal fields.

Remaining resources estimated at 283 Gt (48).

Southern Appalachians. The Southern Appalachian coal field includes the Lookout Mountain (Georgia and Alabama), Soddy-Daisy (Tennessee), Sand Mtn. (Georgia), Plateau (Alabama) fields, the Eastland-Clifty/Smartt Mountain/Tracy City (Tennessee), Warrior/Plateau (west fields) (Alabama) fields, and the Cahaba and Coosa (Ala-

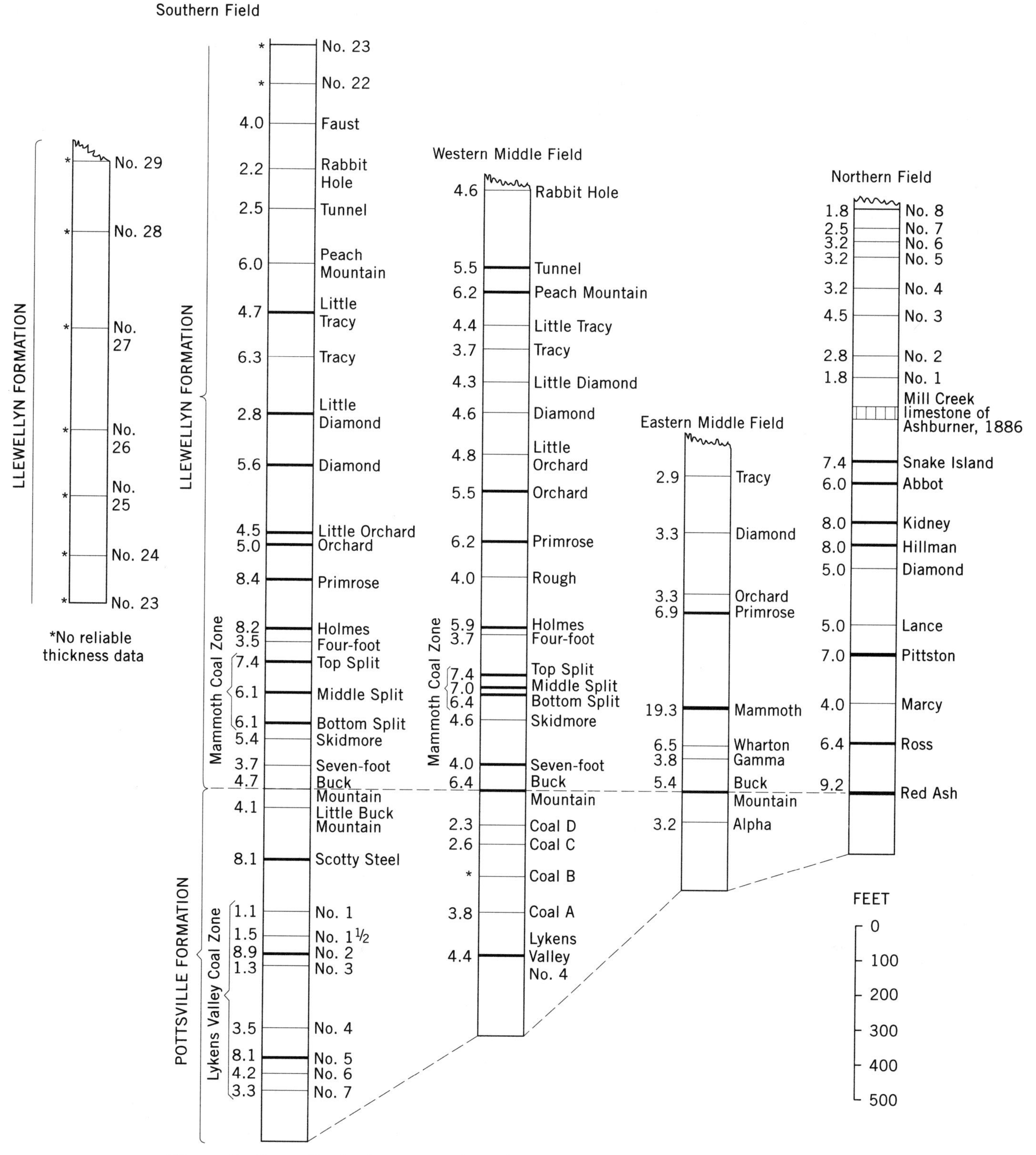

Figure 8. Generalized columnar sections showing names, average thickness, and intervals between coals. Thickness in feet, 1 ft = 0.305 m (47). *No reliable data.

bama) fields. The coal fields extend into Mississippi in the subsurface. The southern termination of the coal fields is beneath the Cretaceous (and, in the subsurface, Jurassic) sediments of the Gulf coastal plain. The Pottsville coal measures are Namurian C through Westphalian A, equivalent to the Pocahontas and New River Formations in the Central Appalachians. There are over 20 mineable coals or coal zones with coal thicknesses averaging 1–2 m. The

Pottsville Formation is up to 2.7-km thick with about 60 coal beds in the Cahaba coal field.

Coal quality varies widely between beds with sulfur ranging from <1–>6% (generally <2%). Coal rank varies from low/medium volatile bituminous in portions of Tennessee and in the Warrior-Plateau fields to high volatile A bituminous in the Cahaba and Coosa fields and in the northwestern portions of the main coal fields (55). Remaining resources are estimated to be 149 Gt (48).

Eastern Interior. The Eastern Interior, or Illinois Basin, includes the coal fields in Illinois, Indiana, and western Kentucky (Fig. 9). Descriptions of the fields are after Refs. 9,34,44,56. Pennsylvanian coals also occur in Michigan but are not economic at the present time. Coals range in age from the Westphalian A Caseyville Formation Gentry coal bed in Kentucky and Illinois through the major Westphalian D Carbondale Formation coals such as the Springfield (Illinois 5/Western Kentucky 9/Indiana V) and the Herrin (Illinois 6/Western Kentucky 11) to late Stephanian coals in Kentucky's Mattoon Formation. Permian strata, without coal, overlie the Mattoon in a graben in the Rough Creek fault zone, Union County, Kentucky. The entire section exceeds 1200 m in Kentucky and contains over 20 mineable coals and in excess of 60 additional thinner splits and coal beds. Michigan's Saginaw Formation contains up to 13 coals (57).

Most of the principal coals are high sulfur throughout most of their extent. Notable exceptions are the "quality circle" of Herrin coal in central Illinois where sulfur is as low as 0.5% in contrast to >4% elsewhere. Certain other coals are known to be <1% sulfur, such as the Amos coal bed in Butler County, Kentucky (58). Coal rank is generally high volatile C bituminous but high volatile A bituminous coal is found in southwestern Kentucky and southern Illinois and medium volatile bituminous rank is known from the Hick's Dome area in southern Illinois where the coal was influenced by igneous intrusions. Mineable thicknesses range from <25 cm for near-surface low-sulfur coals to >4 m. Typical thicknesses are in the 1–2 m range.

The demonstrated reserve base is 91 Gt, about 70% of it in Illinois.

Figure 9. Eastern and Western Interior coal fields and the Michigan basin.

Western Interior. The Western Interior coal fields include fields from Iowa to the Arkoma basin, Oklahoma and Arkansas, and in north central Texas. Coals range in age from Westphalian C/Atokan (Midcontinent series), which includes the Hartshorne coal bed in Oklahoma and Arkansas, through the Westphalian D/Desmoinesian, which includes the Croweburg and Bevier coal beds, to the Stephanian/Missourian and Virgilian. The uppermost Cisco Group, Texas, coal beds may be lower Permian. Up to 20 mineable coals occur in any of the producing states.

Coal quality is variable but in general the northern coals are high sulfur while Oklahoma has some low-sulfur, metallurgical-grade coal, notably the Hartshorne coal bed. Texas coals are generally high sulfur with the Strawn Group Thurber coal bed having the highest quality. Coal rank through the various fields ranges from semianthracite/anthracite in Arkansas, to low volatile bituminous in portions of Arkansas and Oklahoma, medium volatile bituminous in Oklahoma, and high volatile bituminous in the remainder of the fields. Subbituminous coal is reported from Texas (59).

The demonstrated reserve base is 10.2 Gt.

Gulf Coast. The Gulf Coast province stretches from Mexico north to Kentucky and east at least to Georgia. Commercial deposits are in Texas and Louisiana with Arkansas having some potential. Strippable low-rank coals occur in long belts along the strike of the Tertiary units. Lignites occur in the lower Paleocene Midway Group in Louisiana to Alabama, the Paleocene–lower Eocene Wilcox Group in Texas, the middle Eocene Yegua Formation in Texas, and the middle upper Eocene Jackson Group in Texas (Fig. 10). Lignites in Kentucky, Tennessee, and Mississippi are in the middle Eocene Claiborne Formation. Upper Cretaceous Olmos Formation coals occur at Eagle Pass, Maverick County, Texas, and Piedras Negras, Coahuila, Mexico. The Tertiary formations thicken downdip to the Gulf of Mexico. The Wilcox thickens from its erosional edge to over 3 km. Coals up to 12-m thick are known with most in the 1–4 m range.

The coals, where mined, are moderate ash with low sulfur content. Wilcox lignites in Texas are lower sulfur than the Jackson lignites. Rank ranges from lignite to subbituminous in east Texas. The Eagle Pass coals are high volatile bituminous. Petrography of the Texas lig-

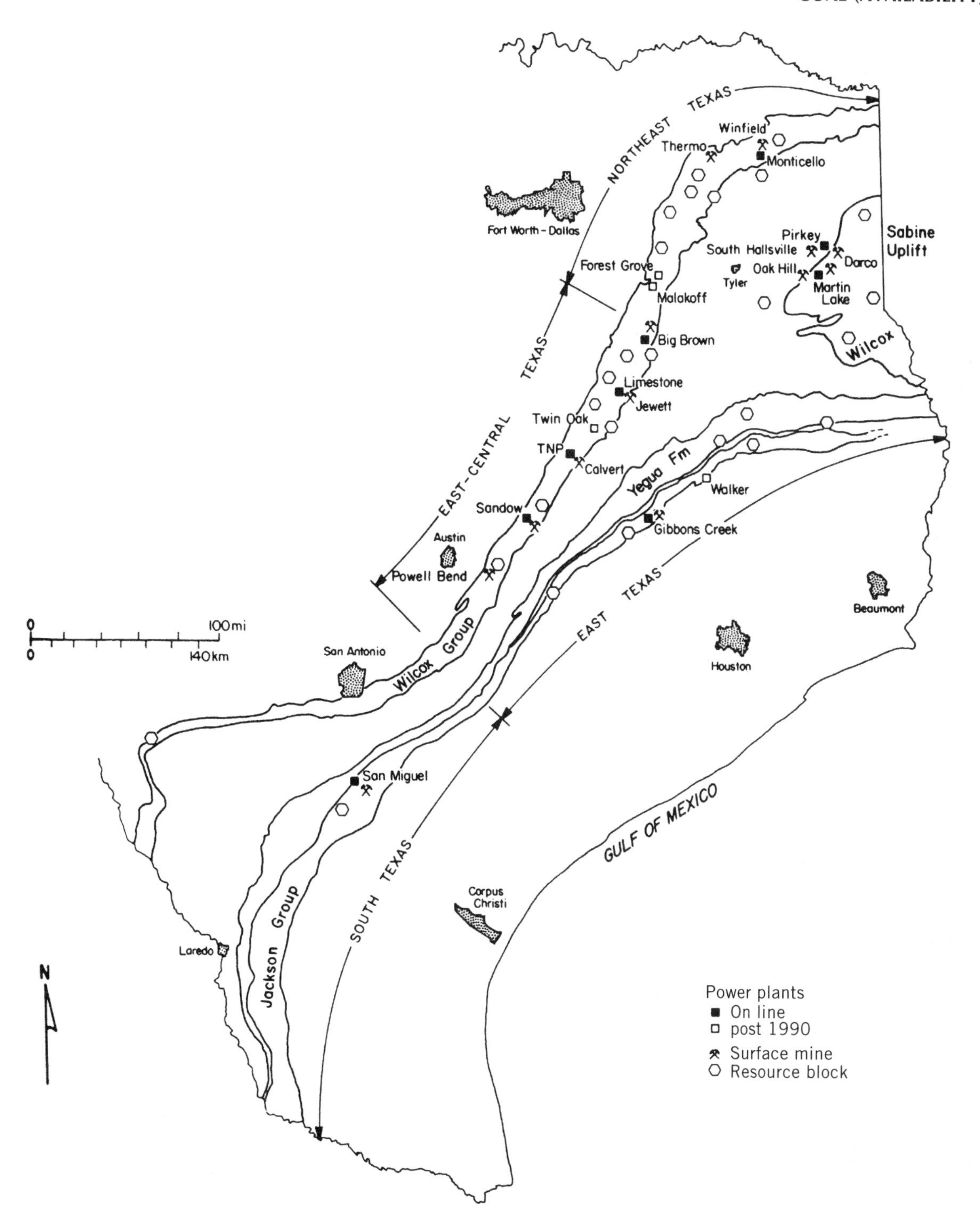

Figure 10. Lignite belts in east and south Texas (34). W. R. Kaiser, 1991.

nites was summarized (60). Most of the lignites are dominated by humic macerals. The Claiborne Group in Webb County, Texas, does contain two mineable cannels.

Reserves total 20.5 Gt (61) (Table 7). The demonstrated reserve base is 13.3 Gt.

Great Plains/Rocky Mountains. The coal fields of the Great Plains and Rocky Mountain provinces include some of the most important producing districts in the U.S. The discussions are linked, following the treatment by Flores and Cross (62), due to the continuum of Cretaceous-Tertiary coal-bearing strata. Coal fields in the U.S. extend from central New Mexico to the U.S.–Canada boundary (Fig. 11). Further discussions are found in Refs. 9 and 34. Resources for main basins are given on Table 7 (23).

The Cretaceous coal-bearing formations are thicker in the southern and central Rocky Mtn. region than in the north, up to 4.9 km vs 3 km. Important coals are early to middle Campanian age in the south and late Campanian to Maastrichtian in the central and northern Rockies.

Table 7. Coal Resources 10^9 t of Cretaceous and Tertiary Coals in the Gulf Coast Province, Great Plains, Rocky Mountain Provinces, Washington, and Alaska[a]

	Identified-Hypothetical Coal-in-Place	Onshore Speculative Coal-in-Place	Offshore speculative Coal-in-Place	Total Coal-in-Place
Alabama/Florida	26.85	16.47		43.32
Mississippi	53.20	27.21		80.41
Kentucky/Tennessee	2.78			2.78
Missouri	1.09			1.09
Arkansas	58.32	14.34		72.66
Louisiana	30.93	20.57		51.50
Texas				
(Tertiary)	94.97	68.28		162.75
(Cretaceous)	1.20	2.91		4.11
Nuevo Leon, Mexico	5.17			5.17
Tamaulipas, Mexico		1.90		1.90
Gulf Coast	274.01	151.68		425.69
North Central	28.70	38.58		67.28
Fort Union/Powder River				
North Dakota	507.63	28.21		535.84
South Dakota	11.07			11.07
Montana	400.73			400.73
Wyoming	941.39	3.64		945.03
Total	1860.82	31.85		1892.67
Denver Basin	40.91			40.91
Raton Basin				
Colorado	27.61			27.61
New Mexico	14.02			14.02
Total	41.63			41.63
Bighorn Basin				
Montana	4.08			4.08
Wyoming	21.39			21.39
Total	25.47			25.47
Wind River/Jackson Hole	79.49			79.49
Hanna	3.09			3.09
Green River				
Wyoming	215.77			215.77
Utah	0.80			0.80
Colorado	58.91			58.91
Total	275.48			275.48
Uinta				
Colorado	153.44			153.44
Utah	27.29	35.07		62.36
Total	180.73	35.07		215.80
North and Middle Park	3.81			3.81
Hams Fork				
Wyoming	5.00	25.05		30.05
Utah	0.02	0.20		0.22
Total	5.02	25.25		30.27
San Juan River				
Colorado	149.68	3.17		152.85
Utah	1.29			1.29
New Mexico	264.93	30.66		295.59
Arizona	0.54			0.54
Total	416.44	33.83		450.27
Henry Mountains	0.96			0.96
Kaiparowits	11.70	3.93		15.63
Black Mesa	19.95			19.95

Table 7. *(Continued)*

	Identified-Hypothetical Coal-in-Place	Onshore Speculative Coal-in-Place	Offshore speculative Coal-in-Place	Total Coal-in-Place
Washington				
Bellingham basin	7.08			7.08
Puget basin	9.17			9.17
Chehalis basin	16.97			16.97
Cowlitz basin	0.08			0.08
Roslyn basin	11.74			11.74
Alaska				
Cook Inlet-Susitna[b]	17.28	2.20	2.04	21.52
Nenana	6.41			6.41
Northern Alaska[c]	227.68	22.75	22.75	273.18

[a] Ref. 23.
[b] Offshore speculative coal-in-place, 2.04×10^9t.
[c] Offshore speculative coal-in-place, 22.75×10^9t.

Coals of Turonian to Santonian age occur in the southwest San Juan basin and Kaiparowits Plateau fields. The coal-bearing formations were deposited in foreland basins to the east of the Sevier orogenic belt. The basin migrated to the east through the Cretaceous until about 80 Ma. The second phase of basin development from 76 Ma into the Eocene corresponds to a period of low-angle subduction.

The thickness of the Paleocene and Eocene varies between basins, exceeding 3 km in the Hanna Basin.

Great Plains Williston Basin.
The Williston Basin supports mining in the Paleocene Fort Union Formation of North Dakota. The basin extends into South Dakota and Montana but production is centered in west-central North Dakota.

The deposits are lignite, with R_o of about 0.31%, and are low ash and low sulfur. The two petrographic analyses published by Flores and Cross (62) shows 67–83% huminite, 1–12% liptinite, and 16–21% inertinite. The Beulah-Zap coal bed varies from about 3–7 m-thick. Strippable reserves total 14.6 Gt in North Dakota and 12.9 Gt in Montana.

Powder River Basin. The Powder River Basin, encompassing about 50,000 km^2 in Wyoming and Montana, is an asymetric syncline bounded on the west by the Big Horn Mountains and on the east by the Black Hills. It is underlain in part by the upper Cretaceous Mesaverde Group which contains thin coals on the southern end of the basin. The thickest coals are in the 250–550-m thick Paleocene Tongue River Member of the Fort Union Formation and the 300–600 m Eocene Wasatch Formation. Production in Wyoming is along the eastern flank of the basin where the Tongue River Member has 8–12 coals, including the Anderson and Canyon coal beds which merge to form the Wyodak coal bed, up to 53-m thick with an average 30-m thickness (Fig. 12). Production in Montana is centered on the northwest side of the basin. The importance of the basin is underscored by the fact that Campbell County, Wyoming, accounted for over 14% of U.S. coal production in 1989 and that 3 of the top 7 producing counties were in the basin (the other four are in the Central Appalachians).

Coal quality is generally high with sulfur contents below 1% and the ash content low. Rank of the Tertiary coals is subbituminous. The huminite percentage tends to exceed 90% (62–65). The remaining strippable reserve base is 54.8 Gt.

Denver Basin. The basin occupies 19,425 km^2 in northeastern Colorado. Coal-bearing formations include the upper Cretaceous Laramie Formation and the 100–150-m Paleocene portion of the Denver Formation. The northern portion of the basin contains 3–9-m lignites with a maximum thickness of 16.8 m. The southern portion of the basin contains 1.5–3-m lignites with a maximum of about 9 m.

Rank varies from subbituminous in the Laramie coals near the Front Range (western basin) to lignite elsewhere. Ash varies from <4–>40% and sulfur content is generally low. Total Laramie and Denver Formation resources are 65.5 Gt.

Raton Mesa Region. The basin lies to the east of the Front Range in south central Colorado and north central New Mexico. Coal-bearing formations are the Maastrictian Vermejo Formation and the Maastrictian–Paleocene Raton Formation. The Vermejo Formation is up to about 170 m thick and the Raton Formation is up to 630-m thick.

Coal quality is generally high with low sulfur and low to moderate ash content. Coal rank varies from subbituminous to low volatile bituminous and, where influenced by igneous intrusions, anthracite and natural coke. Metallurgical grade coal has been produced from the coal field.

Rocky Mountains. Coal occurs in a number of structural basins, some too small to be considered, throughout the Rockies. Some of the more important basins will be discussed below.

Red Lodge–Bighorn. Important coals occur in the upper Cretaceous Mesaverde and Meeteetse formations and the Paleocene Fort Union Formation. Mesaverde coals are commonly 1–2-m thick with thicknesses up to nearly 4 m. Meeteetse coals are in a similar thickness range. Fort Union coals are thickest in the southern end of the field

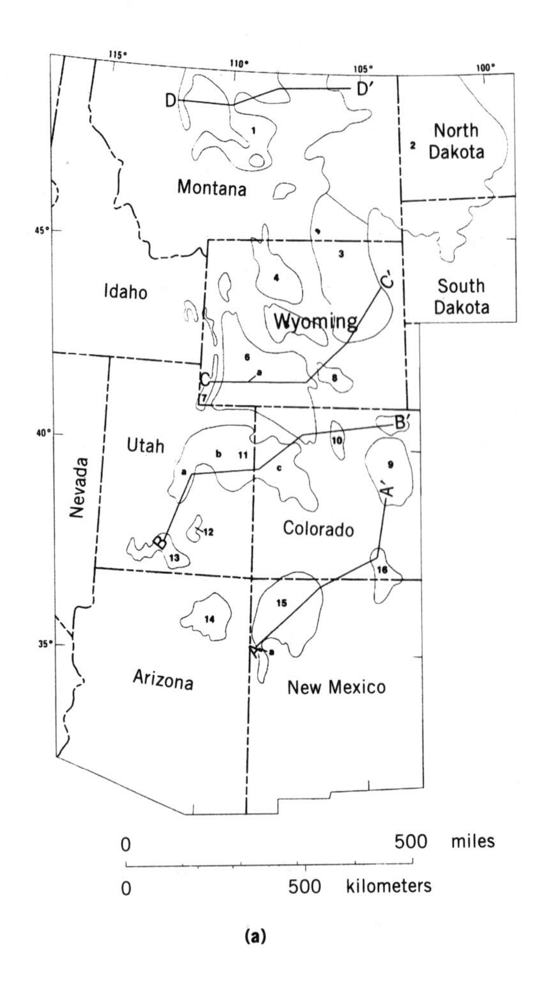

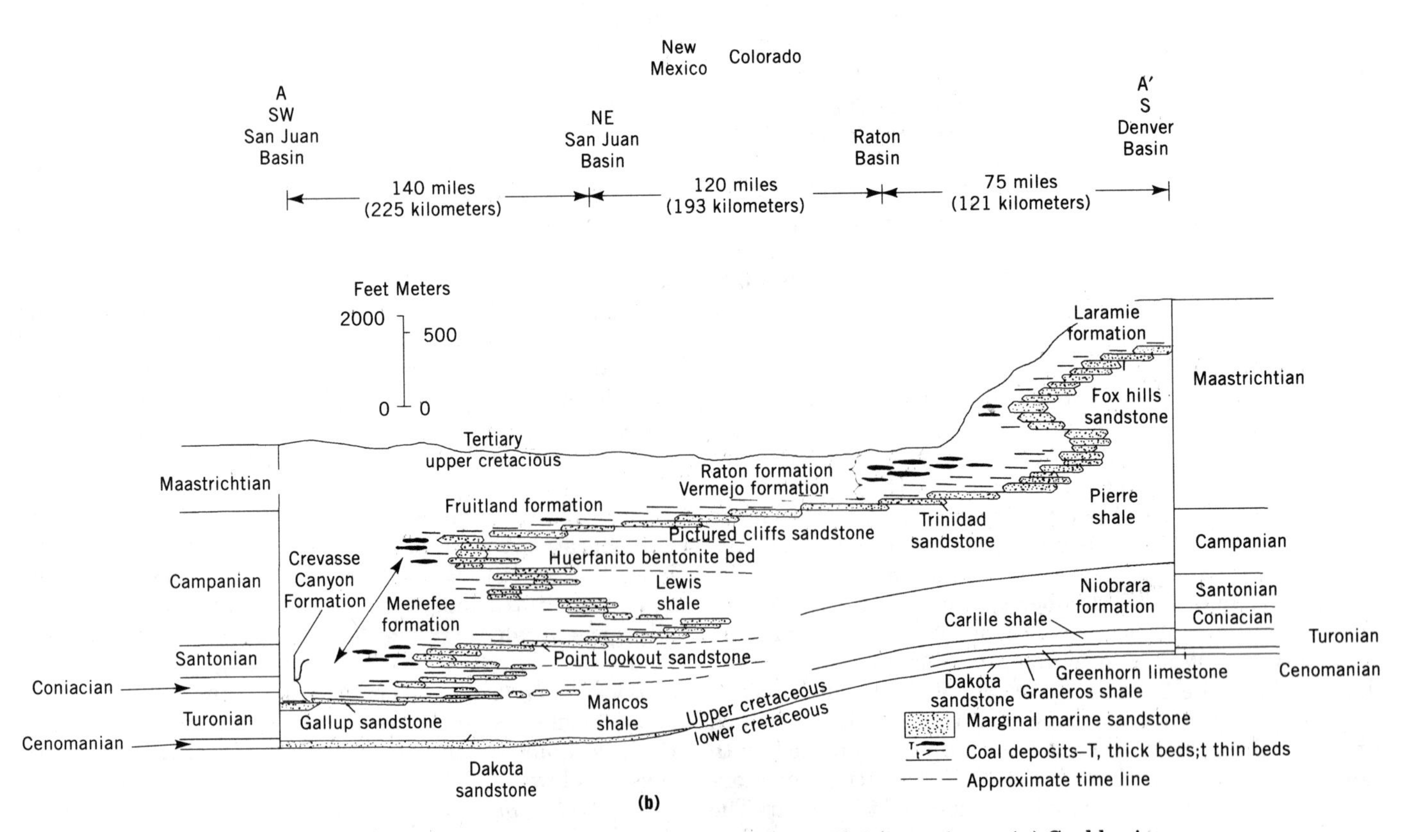

Figure 11. Coal fields of the Rocky Mountain and Great Plains provinces. (**a**) Coal basins and fields. 1, North Central field; 2, Williston basin; 3, Powder River basin; 4, Bighorn basin; 5, Wind River basin; 6, Green River basin (a. Rock Springs uplift); 7, Hams Fork field; 8, Hanna basin; 9, Denver basin; 10, North Park Field; 11, Unita-Piceance basin (a. Wasatch plateau; B. Uinta basin; c. Piceance basin); 12, Henry Mountain field; 13, Kaiparowits Plateau field; 14, Black Mesa field; 15, San Juan basin (a. Gallup Sag); 16. Raton basin. (b) Illustrated section.

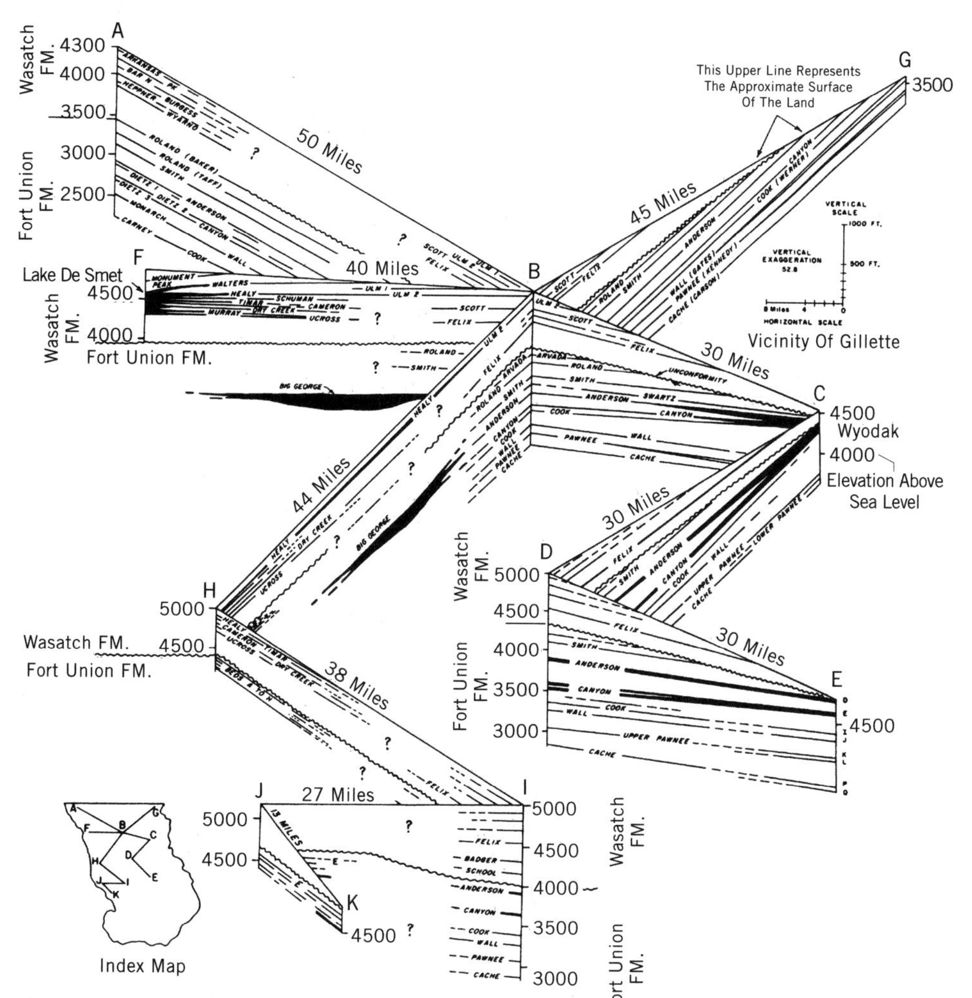

Figure 12. Correlation of coals in the Powder river basin (34). Compiled and modified from published reports and open-file reports of the U.S. Geological Survey.

with thicknesses over 12 m reported. Coals are low ash, low sulfur with rank of high volatile C bituminous.

Wind River. Subbituminous Meeteetse Formation coals and subbituminous to high volatile Mesaverde Formation coals are the important reserves in the Wind River basin. In-place reserves are estimated at 73.7 Gt.

Hanna Basin. Coals occur in a 1940 km^2 basin in Carbon County, Wyoming. Coal-bearing units include the upper Cretaceous Mesaverde and Medicine Bow formations, the upper Cretaceous-Paleocene Ferris Formation, and the Paleocene-Eocene Hanna Formation. The combined thickness of the Hanna and Ferris formations is up to 3.2 km. In excess of 75 coals occur in the Tertiary formations which together account for most of the reserves. Coal thicknesses range up to about 9.5 m.

Most mined coals are low ash and low sulfur. Coal rank ranges from subbituminous to high volatile C bituminous. In-place resources are estimated to be 21 Gt with a strippable reserve base of 590 Mt.

Green River. The Green River region comprises about 60,000 km^2 in southwestern Wyoming and northwestern Colorado. The Sandwash basin in Colorado contains resources to a depth of 3 km. Coal-bearing strata are the Mesaverde Group Rock Springs (400-m thick) and the overlying Almond (150-m thick) formations; the upper Cretaceous Lance Formation; and the Tertiary Fort Union and Wasatch formations (300- and 1000-m thick, respectively). Up to 10 coals occur in each of the formations. The Fort Union Deadman coal is up to 9 m where unsplit.

Coal quality is generally high, with low to moderate ash and low sulfur contents. Wasatch Formation coals are higher sulfur, up to 5.4% for the Latham No. 3 coal bed in the northwestern part of the coal field. Coal rank ranges from subbituminous to high volatile C bituminous. Inplace resources to a depth of 1.8 km may be 270 Gt with about 2.5 Gt of strippable resources.

Uinta Region. The Uinta region covers 60,000 km^2 in west-central Colorado and east-central Utah. Productive coals are in the Mesaverde Group which includes the Iles and Williams Fork formations in the north and the Mount Garfield Formation in western Colorado and the Blackhawk Formation in Utah. Coal-bearing rocks occur to at least 3 km depth in the Piceance basin, Colorado. Thickness of coals varies widely with thicknesses of up to 17 m.

Coal quality varies with coals having from 2–>20% ash and <0.5–2.0% sulfur. Coal rank varies from subbituminous to high volatile bituminous over most of the field. The eastern and southeastern portions of the field have low volatile bituminous to anthracite ranks where igneous intrusions have influenced the coal, for example, in the Crested Butte field in Gunnison County, Colorado. Coking quality coal is found in several of the basins.

Identified resources to 1.8 km depth total about 52 Gt.

Henry Mountains. The coal field covers about 1165 km^2 in south central Utah. Coal-bearing units are the upper Cretaceous Emery Sandstone, Ferron Sandstone, and Dakota Formation. The Emery Sandstone contains the bulk of the 127 Mt of surface-mineable coal. Emery coals are up to 4-m thick and have sulfur contents <1%. Ferron coals are up to about 2.5-m thick and have nearly 2% sulfur. Coal rank is high volatile bituminous.

Kaiparowits Plateau. The coal field in southern Utah contains reserves in four coal zones in the upper Cretaceous Straight Cliffs Formation. Coal thicknesses up to 9 m are known but are generally less than half of that. Coal rank is subbituminous to high volatile C bitmuninous and sulfur contents are generally below 1%. The field contains 7.2 Gt of measured, indicated, and inferred reserves with resources several times that.

San Juan River Region. The San Juan River coal fields cover about 70,000 km^2 in northwestern New Mexico and southwestern Colorado with an extension into Utah. The bulk of the production comes from San Juan and McKinley counties, New Mexico. Coal-bearing units include the upper Cretaceous Mesaverde Group, which includes the Gallup, Crevasse Canyon, and Menefee formations, and the Fruitland Formation. Menefee coals are 3–5.5-m thick and Fruitland coals are up to 20-m thick.

Coal rank ranges from subbituminous in the south to high volatile bituminous in Colorado. Coking quality coal is produced in the Colorado portion of the field. The sulfur content is generally below 1% but ash is variable in the <5–>25% range. Further details of the quality of the New Mexico coals can be found in Ref. 66.

The coal field contains 3.8 Gt of strippable reserves.

Black Mesa. The Black Mesa coal field covers 8300 km^2 in northeastern Arizona. The mesa is an erosional remnant of upper Cretaceous strata. Coal-bearing units are the Toreva and Wepo formations of the Mesaverde Group. Coal is produced from eight zones in the Wepo Formation. Coal thicknesses are generally 2–2.5 m with thicknesses up to 8.5 m.

Coal rank is subbituminous to high volatile C bituminous. Wepo coals have <10% ash and <1% sulfur. Toreva and Dakota Sandstone coals have wider ranges of ash and sulfur.

Strippable reserves to 40 m are 910 Mt with 19 Gt total resource for the entire 520 m upper Cretaceous section.

Pacific Northwest. Washington state contains the most important reserves in the western states exclusive of Alaska. Discussion of other western fields may be found in Refs. 34, 67, and 68. Coal occurs in several fields in four sedimentary basins (Fig. 13). The Bellingham sedimentary basin includes fields in Whatcom and Skagit counties on the east side of Puget Sound south of British Columbia. The Puget sedimentary basin includes the Issaquah-Grand Ridge, Green River, Wilkeson-Carbonado, and Fairfax-Ashford fields. The Chehalis basin to the west includes the Morton and the Centralia-Chehalis fields. The Cowlitz basin to the south of the latter basin includes the Kelso-Castle Rock coal field. The latter field does include coal-bearing Oligocene strata. Otherwise the coal-bearing units are Eocene age. Total reserves for the state total 6.3 Gt for ranks from lignite to anthracite (Table 7).

Production in the Centralia-Chehalis area comes from the late Eocene Skookumchuck Formation. Seven coal beds have been mined since production started in 1971. Thicknesses range from 1 m to 6 m for the Big Dirty coal bed. Coal rank is subbituminous. Sulfur content for the mined coals ranges from 0.6–<2% and ash content is generally less than 12%. Reserves for the area total 3.36 Gt.

Production in the Issaquah-Grand Ridge area is from the high volatile bituminous Franklin No. 7, 8, 9, and 10 coal beds. The coal has a sulfur content of 0.5–1% and ash content of about 14%.

Alaska. Alaska has an estimated 5.1 Tt of hypothetical resources compared to the 3.37 Tt of hypothetical resources for the conterminous United States. Much of the resource is north of the Arctic Circle. Summaries of Alaska coal deposits are from Refs. 34, 69, and 70. Three-Gt-resource provinces will be discussed below (Fig. 14).

The Cook Inlet-Susitna fields in south-central Alaska contains the state's second largest and most accessible resource. Coal-bearing units are the Miocene-Pliocene Kenai Group Tyonek, Beluga, and Sterling formations and the Paleocene Chickaloon Formation. Individual coal fields include the Beluga, Yentna, Susitna, Matanuska, Kenai, and Broad Pass. Coal rank varies from subbituminous for the younger coals to anthracite for intensely folded Chickaloon formation coals in the Mananuska field. Sulfur content is generally less than 1%. Identified resources total 9.67 Gt.

The Nenana province is on the north slope of the Alaska range, north of the Cook Inlet-Susitna fields. Coals are in the Miocene Usibelli Group Healy Creek, Suntrana, and Lignite Creek formations. The Suntrana Formation contains the most continuous coals, with coal thicknesses in the 3–18 m range. Coal production comes from the Hosanna Creek (Lignite Creek) field in the western portion of the region. The subbituminous coals are low sulfur. The Nenana basin contains 6.37 Gt of identified resources with an additional 1.4 Gt in outlying fields.

The Northern Alaska province extends for 650 km along the Arctic Ocean from Corwin Bluff to Prudhoe Bay north of the Brooks Range. The field is defined by the outcrop belt of the Cretaceous Nanushuk Group. The latter group contains the better quality coals. The overlying Colville Group contains thin, bony coals. The Cape Lisburne area, to the west of the province, contains Mississippian Kapaloak Formation low volatile bituminous and semianthracite rank coals. Coal rank of the Cretaceous coals increases from subbituminous to high volatile bituminous towards the Brooks Range front. A 1.11% R_{max} (high volatile A/medium volatile bituminous) for a high ash coal in one test well has been reported (71). The coal field contains 227.5 Mt of measured resources, 136.5 Gt of identified (indicated and inferred) resources, and 3.64 Tt of hypothetical resources. Development of the field will be hindered by the remote location, climate, and difficulty of mining in the tundra.

EUROPE

The coal industry in Europe has undergone significant decline over the past ten years. Coal production is likely to

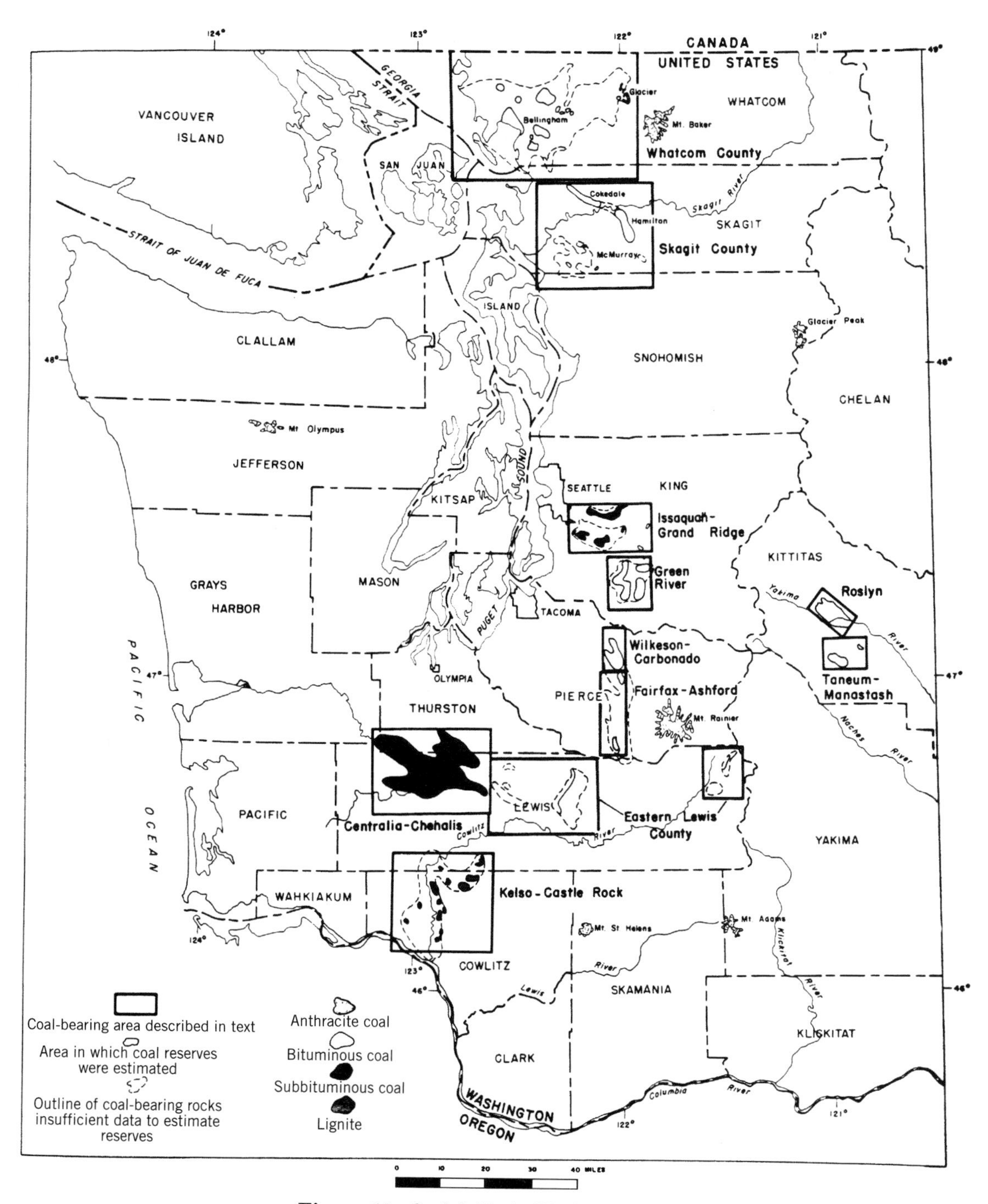

Figure 13. Coal fields in Washington (68).

continue to be cut in many European countries and any shortfall made up by imports (72). A significant coal industry is likely to remain in Germany and Poland but the prospects for the British Coal Industry looks bleak.

Great Britain

The British coal industry has undergone severe decline over the past ten years and future prospects are uncertain. Currently there are three major coal basins (Midland Valley, South Wales, and Pennine) (Fig. 15) two of which now have very few operating mines. Proved recoverable reserves are 3.8 Gt of bituminous coal and anthracite (73).

Midland Valley Basin, Scotland. In Scotland, the Midland Valley Basin is a rift basin which underwent big subsidence in the late Palaeozoic. Overlying thick Devonian Molasse sediments of Old Red Sandstone facies are Lower Carboniferous sediments, predominantly nonmarine but with extensive volcanics (74). Within the basin there are several coalfields which have been structurally separated. The earliest coals are late Visean age but they have little significance. Reserves of coal are difficult to estimate, but probably over 3 Gt (75), with only a small percentage of that is likely to be worked by opencast methods, most of the deep mines having been closed. The oldest worked coals are from the Limestone Coal Group of Namurian age worked mainly by opencast methods, both in the East

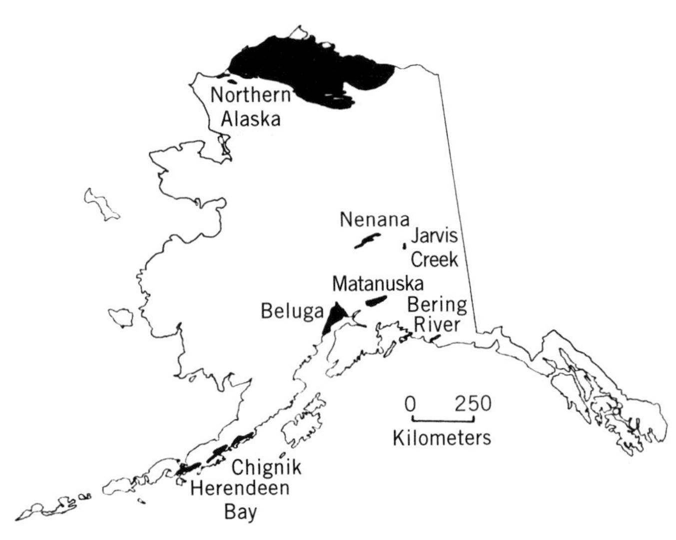

Figure 14. Coal fields in Alaska (70).

Lothian region and in the Douglas coalfield. These coals are high volatile bituminous to anthracites and range in ash from 3–7% and sulfur 0.5–0.9% (76). In the Douglas coalfield, coals up to 2-m thick are worked in opencast mines. Reserve estimates are uncertain. These coals were deposited in typical coastal plain/deltaic settings.

The other big phase of coal deposition was during the late Namurian Passage Group and the Westphalian Productive Coal Measures. These occur as thin seams often less than 1 m thick and have been worked by both deep mining and opencast techniques in several coalfields, although most deep mines are now closed and mining appears for the future to be restricted to a few opencast operations. Coals of this age are typically high to medium volatile bituminous coals with ash ranging 3–7% and sulfur 0.4–1.3% (76).

Pennine Basin, England. In central and northern England the Pennine basin is a broad sag basin which overlies an earlier series of blocks and grabens of Lower Carboniferous age (77). The main coals, belonging to the Coal Measures, are of Upper Carboniferous (Westphalian A-C)

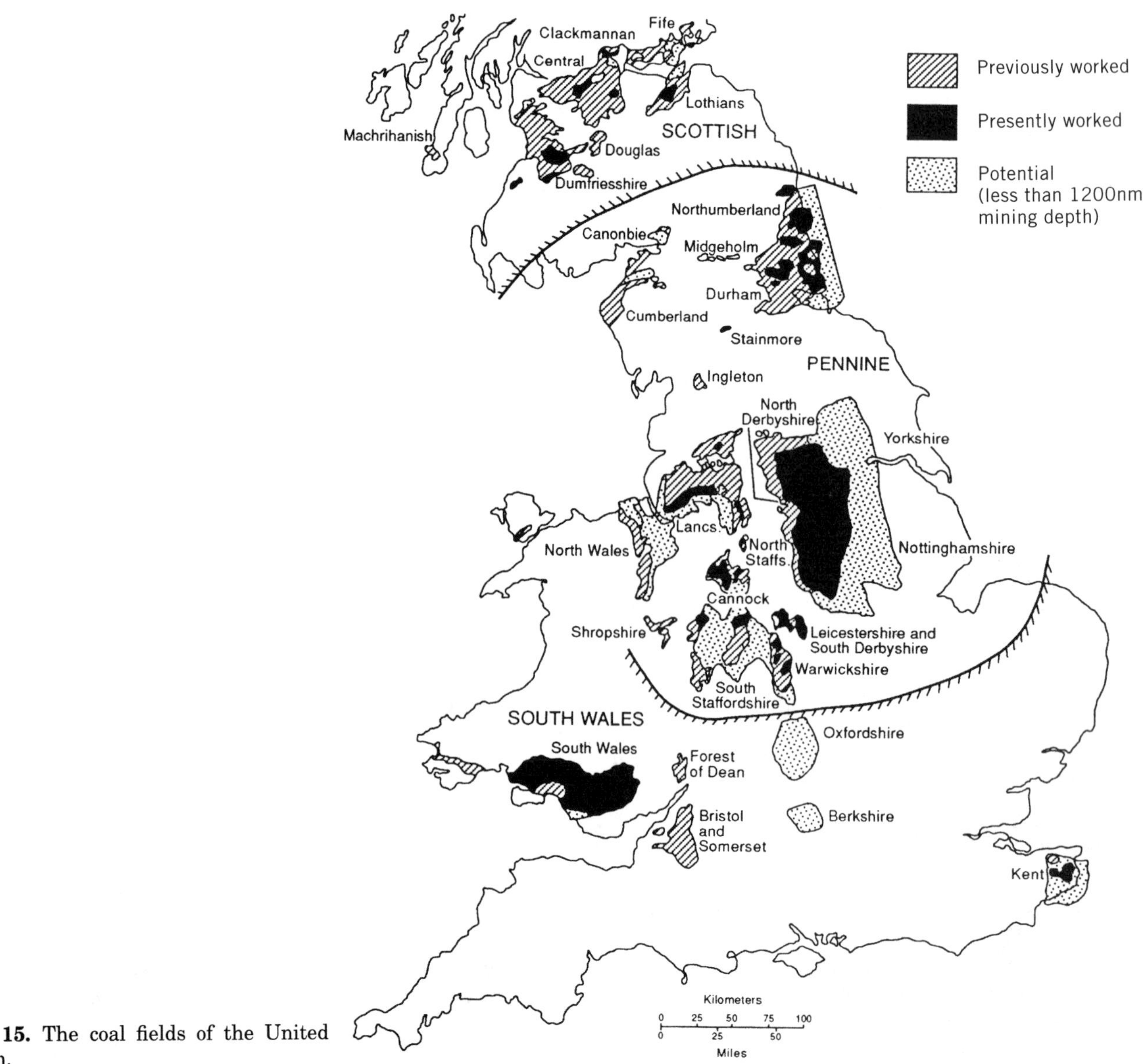

Figure 15. The coal fields of the United Kingdom.

age and were deposited in alluvial plain settings. Usually the coals are thin (<1 m) and of variable ash and sulfur values and range in rank from low to high volatile bituminous coal. Total reserves are likely to be between 3 and 5 Gt (77). The principal coalfields have been structurally separated and many are on the verge of closure.

In the Northumberland and Durham coalfields most of the remaining reserves are under the North Sea and in recent years there have been extensive mine closures. The Yorkshire coalfield (part of the East Pennine area) has seen the most recent development, in particular the Barnsely Seam which reaches up to 2-m thick in the Selby area. This area of 285 km^2 has total reserves of 2 Gt and the Barnsely Seam alone 600 Mt (77). The coals are mainly high to medium volatile coals with moderate to low ash and sulphur values. The Nottinghamshire coalfield in the English Midlands is also being run down but a few deep mines are likely to remain. In the Warwickshire coalfield the Warwickshire thick coal reaches up to 8 m in thickness. Over the next five years Britain's only remaining deep mines are likely to be in the Pennine Basin coalfields. Operating reserves in the East Pennine coalfields are estimated to be 2 Gt (9).

South Wales Basin. The South Wales coalfield exposes Coal Measures of Westphalian age and is over 100 km long but structurally complex (76). There are no current estimates of reserves but 3 Gt have already been extracted and following recent and planned mine closures only a small amount of opencast mining is likely in the future.

Lough Neigh Basin, Northern Ireland. Tertiary lignites occur in Northern Ireland at several localities with a total estimated resource of 1 Gt (78). The Crumlin deposit, 21 km west of Belfast, have a total resource of more than 550 Mt (with 40% on shore and 60% under Lough Neigh) and 100 Mt of minable reserves. The Upper, Lower and Basal Seams are 15–63-m thick with low sulfur and low ash. At Ballymony there is a proven reserve of 630 Mt with 400 Mt extractable by opencast methods. It has a sulfur content of 0.15% (78).

Germany

Both the former East and West Germany contained significant quantities of lignites, bituminous coals and anthracites. Proved recoverable reserves in the former Federal Republic (west) are 23.919 Gt with 60% coking but all underground mined (73). Brown coals and lignites in the Federal Republic total 35.150 Gt. All surface minable and proved reserves in the former GDR (east) are 21 Gt (73). The bituminous coals and anthracites are of Carboniferous age with several basins including Aachen, Ruhr, and Saar all belonging to the tropical Upper Carboniferous Euramerican coal province (Fig. 16). These paralic coal basins contain marine horizons but these are restricted in the Saar Basin (80).

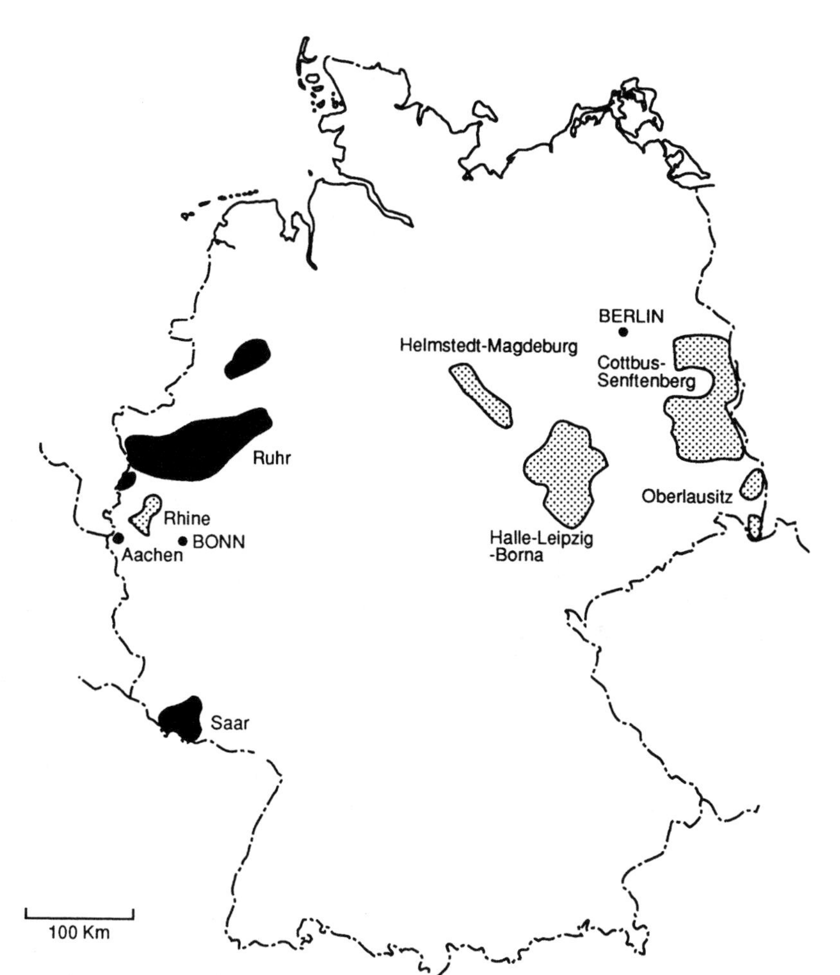

Figure 16. The coal fields of Germany ■, Bituminous coal and anthracite; ▩ brown coal.

Ruhr Basin. The Ruhr Basin was deposited within the Subvariscan foredeep and the coal-bearing strata has been folded and faulted (79). The Upper Carboniferous productive coal measures are 3000-m thick with 300 coal seams (9). In the Ruhr Basin (5000 km^2) several seams occur 0.5–3-m thick of low ash, low sulfur high volatile bituminous coals which have good coking properties (2,9). A reserve of 1.1 Gt is only available for deep mining. In situ resources may be as high as 425 Gt. Recent estimates (72) indicate 2.03 Gt including anthracite with an average analysis with 8.5 volatile matter, 11.1 ash, 0.9 S, and 28.2 GJ/t heating value; and bituminous coals with 22–36 volatile matter, 7–14 ash, 0.7–1.4 S, and 26–30 GJ/t.

Rhine Basin. There are several principal lignite basins in Germany making it the largest producer of brown coal in the world. In the Lower Rhine embayment a main coal basin occurs between Aachen and Cologne (Köln) (Fig. 17). Here the single main seam reaches up to 100 m although this thins both to the north and west (81). Peat was deposited under sub-tropical coastal conditions through the Miocene where subsidence was controlled by both Palaeozoic and Tertiary Faults. Synsedimentary faulting occurred with big thickness changes across the faults. In addition, there are many post-depositional faults. The coals are, however, gently dipping and are low in ash and sulfur. Whilst there is a total resource of 55

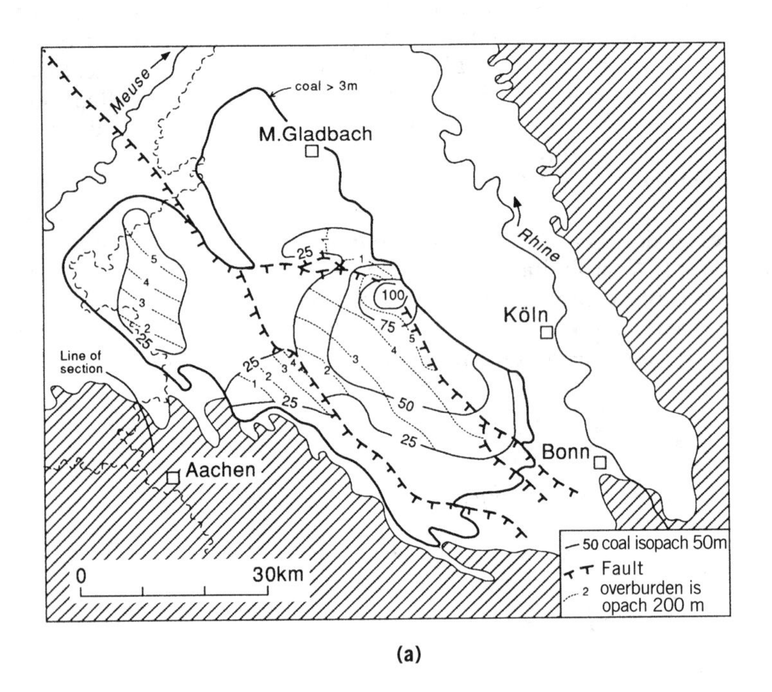

(a)

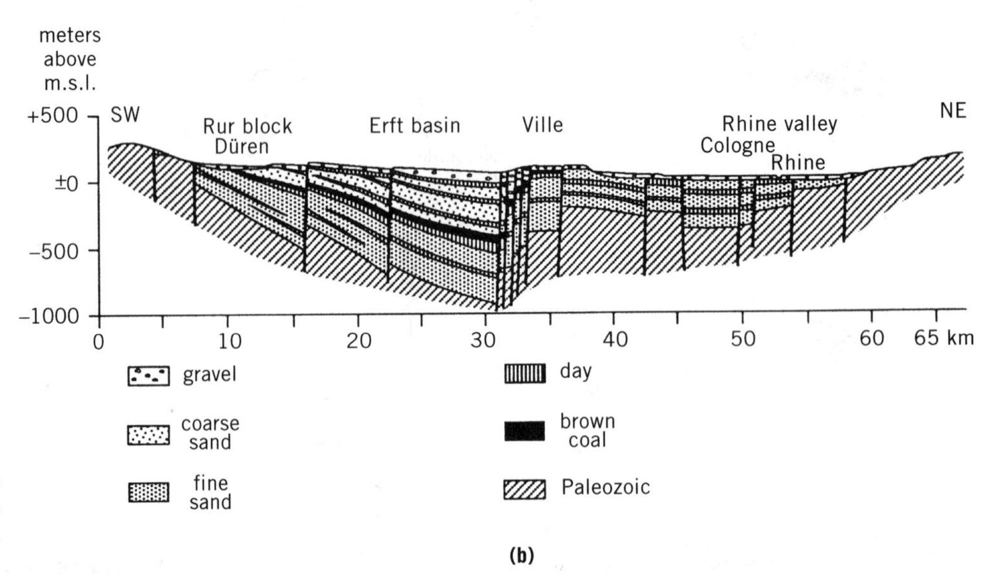

(b)

Figure 17. (**a**) The geology of the Rhine brown coals (81). (**b**) Cross section through the Rhine brown coals (83).

Gt (81), accessible reserves are around 12 Gt all being surface minable. Other estimates (72) indicate 17 Gt of proven reserves. The coal is a low rank lignite, with generally low ash and sulfur (82,83).

Halle–Leipzig–Borna Basin. Proven recoverable reserves of brown coal (mainly of Oligocene age) in the former GDR (east) are 21 Gt (84) but only 10 Gt may be economically minable (84). In the Halle-Leipzig-Borna Basin (3400 km^2) there are 5 seams ranging in thickness from 6–100 m of low ash, low sulfur, high volatile lignites (2). Many of these lignites have proven to be useful for the chemical industry and for conversion processes having a tar yield of up to 30%. Of the 865 Mt accessible (84), some 50% are available for surface mining. Other estimates indicate 9 Gt of proven reserves (72).

Oberlausitz Basin. In the Oberlausitz Basin (1500 km^2) there are two seams (3–22-m thick) of high-volatile low-ash, low-sulfur coal which can be used for coking. Of the accessible reserves of 5 Gt 80% is available for surface mining (2). Recent estimates suggest 13 Gt of recoverable reserves (83).

Cottbus Basin. In the Cottbus Basin of Brandenburg/Saxony there are 13 Gt of proven reserves with seams 3–12-m thick with average values of 15% Ash, 1.5% S, and 8.4 GJ/t (72) but these do not lie close to the surface.

Poland

Poland has significant reserves both of black bituminous coal and anthracite (28.7 Gt) and of lignite (11.67 Gt) of mainly Carboniferous and Tertiary age (73,85).

There are several large Carboniferous basins which were all formed as part of the tropical Euramerican coal province. The three main coal basins are those of Upper Silesia, Lower Silesia, and Lublin (Fig. 18**a**).

Upper Silesia Basin. The Upper Silesia Basin (5000 km^2, 1000 km^2 in the Czech Republic) contains numerous seams up to 4.5-m thick, but averaging between 1.5–2 m, of high volatile bituminous coal with low ash and low sulfur (85), with an accessible reserve of 126 Gt (84). The Upper Carboniferous sequence is 8500-m thick (9) and typical of the Euramerican coal province. Several hundred coal seams, occur with some up to 28-m thick (9). Other recent estimates put the total reserve as 69.74 Gt and with typical analyses of 10.5–35.8% volatile matter, 1.0–10.8% ash, 0.8–1.85% S, and 20.9–33.7 MJ/kg (9).

Lower Silesia Basin. In contrast, the Lower Silesia (550 km^2) Basin is small and the coals are thinner and tectonized. Although the remaining reserves (550 Mt) are deep they include important coking coals (2).

Lublin Basin. The Lublin coalfield (9,000 km^2) is one which has only recently been discovered and it is a large area with large potential reserves estimated at 10 Gt (2). Within the developed area there is 400 Mt of coal (84). The coals which are bituminous with low ash and sulfur contents have strong coking properties. They are less structurally disturbed than those in the Silesian basins and will eventually supplant these basins as the principal source of the country's hard coal (2).

Extensive Tertiary brown coals occur in the Central and southwest regions of Poland. Proven recoverable reserves total 117 Gt with moderate to low ash and low sulfur (73).

Belchatow Basin. In the Belchatow basin (33 km^2) one seam, 50–70-m thick occurs at shallow depth (2). This Miocene coal is a high volatile lignite with a high moisture content and all the estimated reserves of 2 Gt are accessible to surface mining (2). A typical analysis shows 23.6% volatile matter, 7.3% ash, and 2.0% S (78).

In the Konin Basin (1100 km^2) the lignite again has a high ash and moisture content, is 6–20-m thick and all 1 Gt of estimated reserves are available for surface mining (2). In addition, in southwest Poland in the Lower Lansitz and Turoscrow regions, seams of similar quality occur up to 60-m thick.

France

Of the principal western European countries with a long industrial record France has relatively few coal reserves and indeed its coal industry has been in decline for several years. It has 213 Mt reserves of Carboniferous bituminous coal and anthracite (73) and small reserves of about 30 Mt of Tertiary lignite in Provence (2) (Fig. 18**b**).

The Carboniferous coals in France were deposited in two distinctive environments: widespread coals deposited in a marine-influenced paralic basin in the north and more local, but thicker, coals deposited in small limnic rift basins in the centre and south of the country.

In the north of France there are two main coalfields. The Lorainne coalfield (1280 km^2) has several seams of Westphalian (Upper Carboniferous) age, 1.5–2.5-m thick of high volatile bituminous coal, some of which is coking, but which are tectonized and mined by underground methods. The total reserve is about 330 Mt (2). Seams typically have 33–40% volatile matter, 2–17% ash, 0.6–0.85 S, and 33–35 GJ/t (72).

A similar sequence of coals is known in the basin of Nord et Par de Calais (1000 km^2) but here, the coals are thinner, 0.6–1.5 m, and total reserves are only 30 Mt (2). These have moderate to high ash and sulfur values (73) but this area ceased production in 1990.

The small limnic basins contain minor reserves and many mines are now closed. Some mines still are working including at Monceau-les-Mines where low sulfur, low ash bituminous coal is mined.

Spain

Coals are found in several basins in Spain including Carboniferous bituminous coals and Cretaceous and Tertiary lignites (Fig. 18**c**). Total proved recoverable reserves of bituminous coal are 379 Mt, 155 Mt of sub-bituminous coal and 236 Mt of lignite (73). All of these tend to be high in ash and sulfur (73).

During the Carboniferous extensive strike-slip faulting generated several basins, particularly in the north where coals were deposited. The thickest coals are of Cantabrian

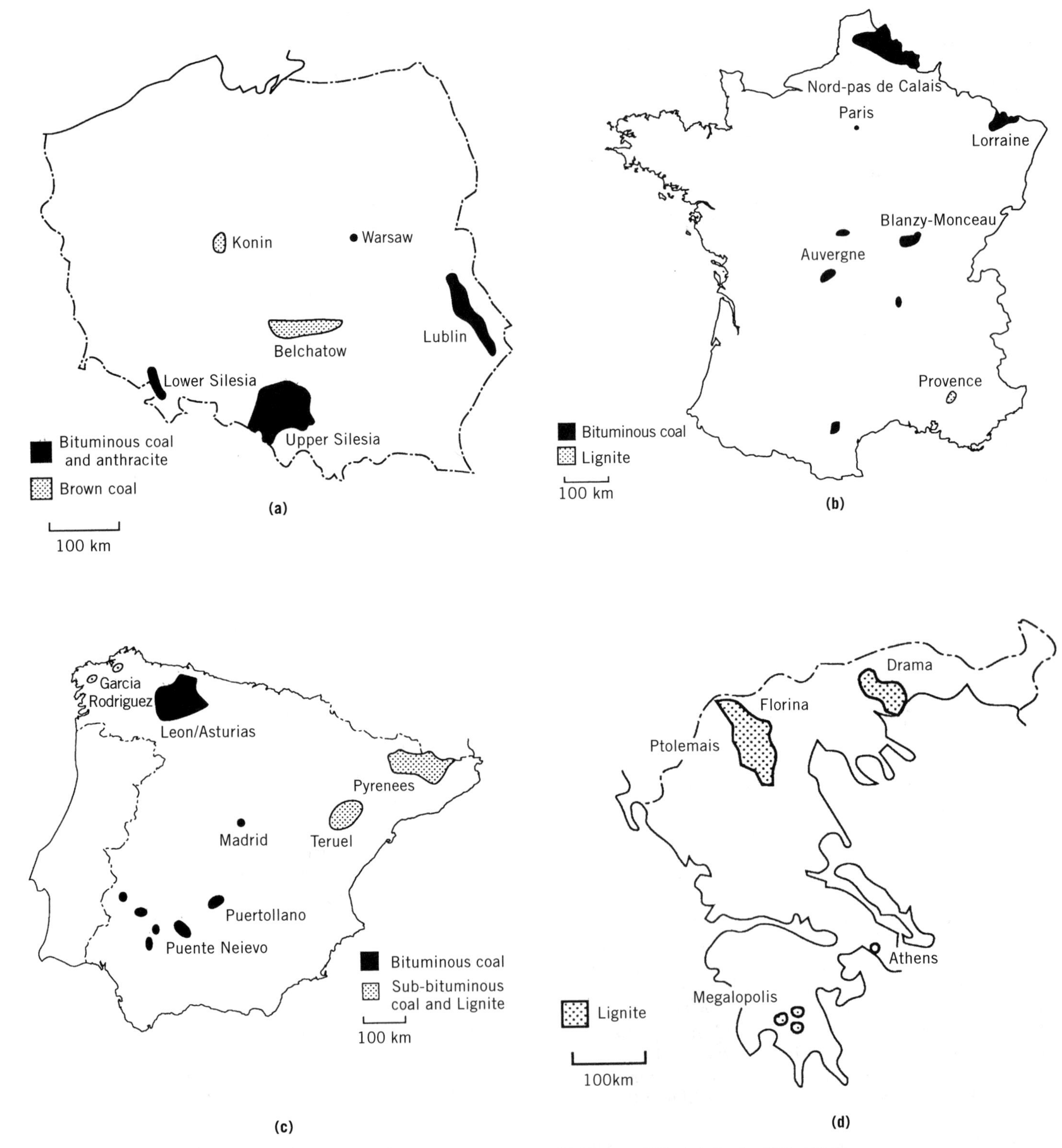

Figure 18. The coal fields of (**a**) Poland (84); (**b**) France (72); (**c**) Spain (72); (**d**) Greece (72).

(Upper Carboniferous) age. The principal basin in the north is in the Leon region (1400 km^2) where there are seams up to 1.5-m thick of low sulfur low volatile bituminous coals, some useable as coking coal, with reserves of 345 Mt (2). Recent estimates but recoverable reserves at 1.35 Gt with anthracite analyses typically 5.8–7.0% volatile matter, 28–34% ash, and 20–23 GJ/t and bituminous analyses of 12–25 volatile matter, 32–46% ash, 0.7–1.6% S, and 17–22 GJ/t (72). In the south of Spain, south west of Puertallono, Carboniferous low volatile bituminous coals are found (2).

Subbituminous coals and lignites occur in several basins. Cretaceous lignites (870 Mt) are known from the Teruel area (4455 km^2) and recoverable reserves have been estimated at 225 Mt with typical analyses of volatile matter 20–26%, ash 28–49%, S 5.4–6.7%, and HV 12–15 GJ/t. In the Pyrenees Basin there are proved reserves of 108 Mt with typical analyses of volatile matter 31%, ash 42%, and HV of 13–14 GJ/t (72). Subbituminous coals are known from the Garcia Rodriguez Basin (300 km^2) (2) which includes the area around La Coruna (Puentes and Meirama) with 225 Mt of recoverable reserves with typical analyses of volatile matter 37–42%, ash 30–39%, S 2.5%, and HV 8.6–9.6 GJ/t (72).

Greece

Coals in Greece are restricted to small Tertiary fault bounded basins in the north of the country where thick lignites occur (Fig. 18**d** 30). There is a total proved recoverable reserve of 3 Gt of lignite (73). Other estimates put this higher at 6.172 Gt (72). At Florina-Amyndaeon (24 km^2) lignites up to 40-m thick occur with a reserve of 484 Mt (72). At Ptolemais (72 km^2) seams of low ash, low sulfur, high volatile lignite occur with a reserve of 2.195 Gt (72). At Megalopolis there are reserves of 466 Mt (72).

Turkey

Coals in Turkey comprise Carboniferous bituminous coals and Tertiary lignites. The Carboniferous sequence bears many similarities with other European basins. Total bituminous coal reserves are estimated at 162 Mt (2). In the Zonguldak coalfield (500 km^2) in the north coals are 0.7–10-m thick and are low ash, low sulfur bituminous coals which are used for coking. There is only 100 Mt of reserves. High gas levels are reported which may be of significance for coal-bed methane exploitation in the future.

Tertiary coals are more widespread. Resources have been indicated at 5 Gt with Recoverable Reserves of 4.76 Gt in 261 deposits (78). The most significant basin is that of Elbistan (120 km^2) (86) (Fig. 19). Whilst within the basin there are rocks of late Cretaceous age to Quaternary the coal-bearing strata appears to be restricted to an unnamed formation of possible Paleocene age but neither data on the detailed stratigraphy nor age is yet available. Where there is a single lignite seam it is 84-m thick, high ash and high sulfur, with reserves of 3 Gt are available for surface mining (2). Recent estimates are 580 Mt of recoverable reserves (78). The lignite has a high volatile content of 60–70% and a huminite reflectance of 0.27% (85). Recent estimates but the proved recoverable reserves of lignite as high as 5.929 Gt (73).

Czech Republic

Most of the coal in the former Czechoslovakia occurs in the Czech Republic (Fig. 20**a**). Upper Carboniferous coals occur in the continuation of the Upper Silesian Basin (1000 km^2) called Ostrava-Karvina Basin. There are over 250 seams 0.7–15-m thick with typical analyses of volatile matter 8.0–42.0%, ash 1.0–30.0%, S 1.0–2.0, and 21.7–30.1 MJ/kg with a total reserve of 5.43 Gt (9). Most coals in the country are predominantly Tertiary low rank coals with high ash and sulfur contents. There are 1.87 Gt of proved reserves of bituminous coal and 8.85 Gt recoverable reserves of lignite (73). The lignites in the Bohemian Basin (around Most) are of Miocene age. This coalfield has an area of 850 km^2 with three seams ranging in thickness from 10–40 m and is subbituminous, low ash and sulfur (84).

OTHER EAST EUROPEAN COUNTRIES

Small coal deposits are also known in several East European countries such as Hungary, Romania, Bulgaria and the former Yugoslavia. Details are documented in Ref. 84.

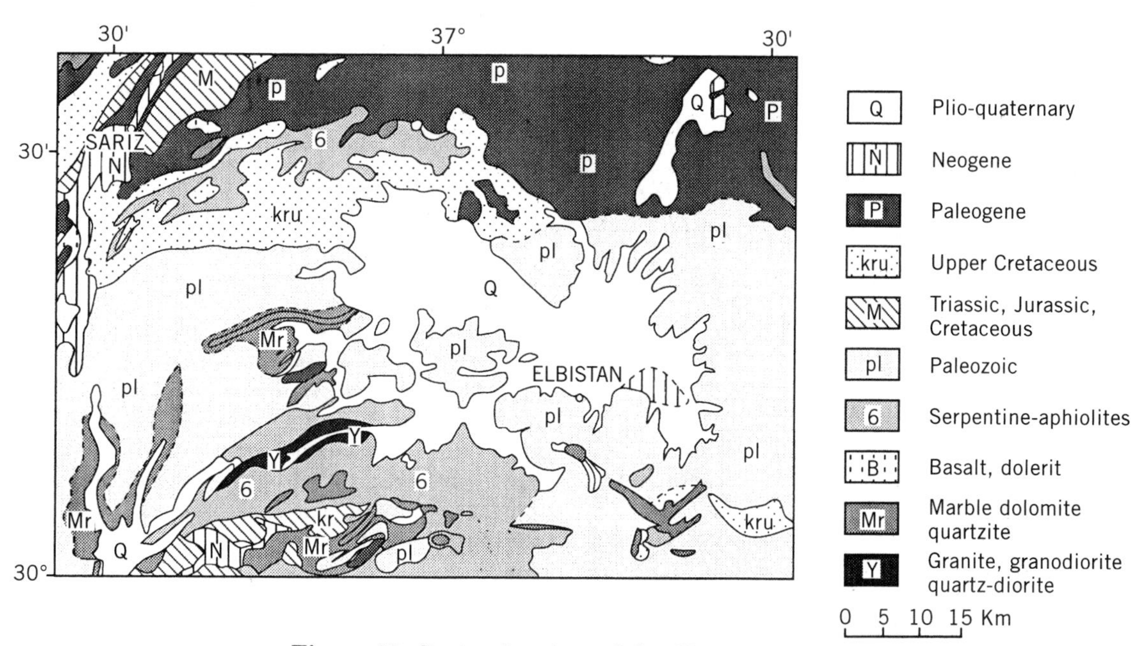

Figure 19. Regional geology of the Elbistan area (86).

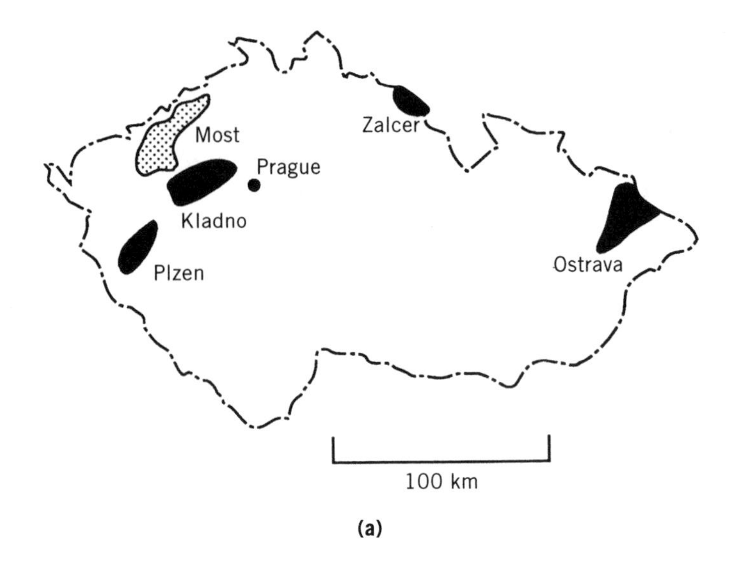

(a)

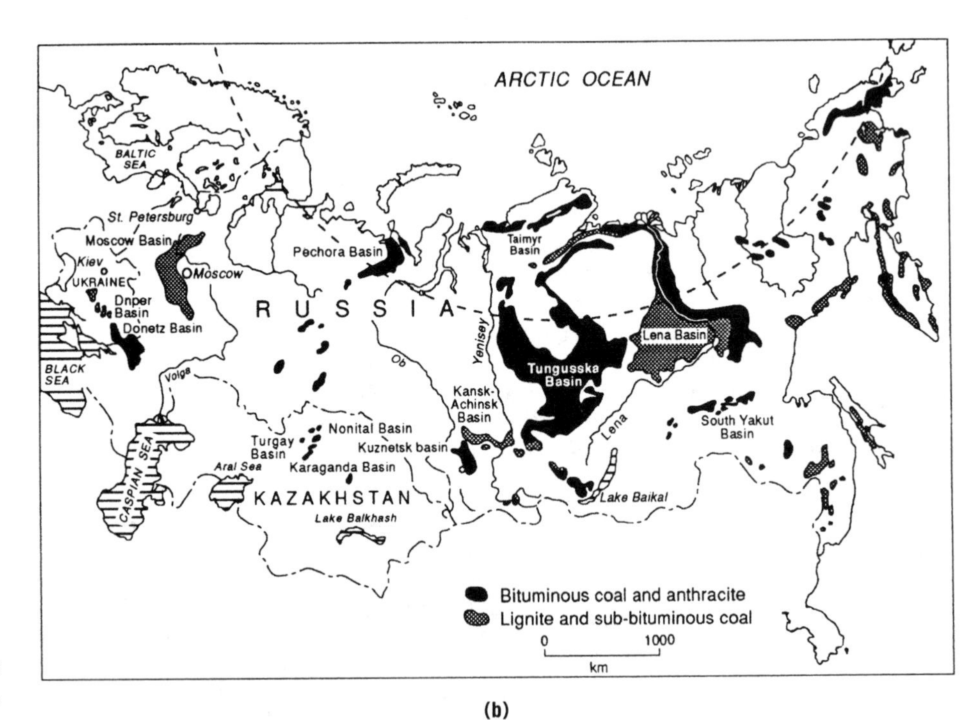

(b)

Figure 20. The coalfields of (**a**) the Czech Republic (84); (**b**) Russia, Ukraine, and Kazakhstan (85).

Ukraine

The Ukraine (part of the former Soviet Union) possesses two major coalfields, one of bituminous coal, the Donetz Basin, and one of lignite, the Dnyepr Basin (Fig. 20**b**).

Donetz Basin. The Donetz basin (4100 km^2) is an Upper Carboniferous paralic basin where coals and marine limestones are interbedded (89). This basin is part of the tropical Euramerican belt and its coal has much in common with those other Carboniferous basins of Europe and North America. The coal-bearing sequence is overlain by Permian limestones and evaporites and is folded and faulted. Within the Upper Carboniferous sequence there are more than 100 seams ranging in thickness from 0.45–2.3 m (2), but most being 0.6–1.8 m (89). The accessible coal is 34 Gt but all is only extractable by underground methods with an average mining depth of 566 m (2). The coal ranges in rank from anthracite to low volatile and good coking bituminous coals but as might be expected with the interdigitation of marine strata these coals have a variable ash content and high sulfur content. Mining conditions are worsening with any new mines likely to be deep (88).

Dnyepr Basin. Palaeogene and Neogene brown coals are found in this basin. The beds belong to the Buchakian Stage and are lignites and subbituminous coals. The 1–

3 seams reach 20 m and their total resources have been calculated at 33 Gt (90).

Kazakhstan

Kazakhstan (part of the former Soviet Union) possesses two principal coalfields, both of Carboniferous age, one of bituminous coal, the Karaganda Basin, and one of lignite and subbituminous coal, the Ekibastuz Basin.

Karaganda Basin. The Karaganda Basin is in the northern part of Kazakhstan (1500 km^2). The coal-bearing succession is 5500-m thick (90). The Carboniferous coals are generally high volatile bituminous coals with good coking properties with a reserve of 7.4 Gt (2). The coals are generally 0.6–8 m thick with a wide range of ash and sulfur values. There are in addition some Jurassic coals. The coals are generally shallow and horizontal and 50% are potentially surface minable.

Ekibastuz Basin. The Ekibastuz Basin (170 km^2), to the north-west of the Karaganda basin, also in the north of Kazakhstan. It is also of Carboniferous age but in contrast to the Karaganda basin the coals are of lower rank being subbituminous with high ash but low sulfur. There are six seams which range in thickness from 6.5–85 m thick and are shallow and gently dipping and most of the 3.4 Gt reserve is near the surface (2).

Russia

The total Russian coal reserves are difficult to estimate as until now all figures published have been for the whole Soviet Union. In 1992 the total reserves for the former Soviet Union were 104 Gt of anthracite and bituminous coal and 137 Gt of lignite. Russia has vast reserves which range in age from the Carboniferous to the Tertiary but equally many of the largest basins are in very remote and inhospitable areas which together with problems of transport infrastructure makes their exploitation difficult.

Moscow Basin. The Moscow Basin, yielding Carboniferous coals in the Padmoskovnyy coalfield (8000 km^2) is unusually a low rank field. There are three main seams up to 3-m thick of high ash, high sulfur subbituminous coal with a reserve of 5 Gt (2). Although the coals are shallow, they are faulted and folded and only 10% may be surface mined and the coalfield has little potential of further development, and production has been falling steadily.

Pechora (Pechorskiy) Basin. The Pechora (Pechorskiy) Basin (1400 km^2) is northeast of Moscow in the Arctic Circle. Permian coals occur here and up to 30 seams are present. The coals are up to 4 m in thickness and are generally subbituminous to high volatile bituminous and some are semicoking and have high ash and low sulfur content. Although there is an estimated reserve of 7 Gt, 85% of the basin experiences deep permafrost, making mining difficult (2).

Kuznetsk Basin. The Kuznetsk Basin (335 $\times$ 110 km, totaling 26,000 km^2 with a worked area of 1000 km^2) is the second largest in Russia in terms of total available coal (41.5 Gt) (2). Other recent estimates indicate reserves of 63 Gt (88). The basin in central south Russia adjoins the western margin of Altai-Sayan mountain range and lies in a shallow depression. To the northwest is the west Siberian plain. The basin exposes coals of varying age including the Carboniferous Lower Balakhonskaya Series (91), the Permian Upper Balakhonskaya Series (91) and Kolchunginskaya Series (91), and the Jurassic Tarbaganskaya Series (91) with over 100 seams ranging in thickness from 1.3–20-m thick. The coals range in rank from subbituminous to anthracites and are generally low ash and sulfur (2). The coals are often steeply dipping and faulted and hence mining may be difficult with 30% of the reserves being potentially mined by surface methods.

Kansk-Achinsk Basin. The Kansk-Achinsk Basin (3000 km^2) occurs just to the north of the Kuznetsk basin in central South Russia. This basin has good development potential and exposes 25 seams, up to 100 m in thickness of Jurassic age. Those coals near the surface are high-ash low-sulfur lignites and of the estimates reserves of 67 Gt much is shallow and gently dipping, most being accessible to surface mining technique (2). At depth there is also some bituminous coal. The Tungusska and Lena Basins occupy much of Central Russia and include the entire range of coals both in terms of age and rank. The very hostile nature of this region makes exploitation of these basins unlikely in the foreseeable future and hence are not dealt with further here.

South Yakutsk Basin. The South Yakutsk Basin (1200 km^2) is in the extreme southeast of asiatic Russia. The coals are of Jurassic age but are thin, generally less than 1.5 m thick. There are many gently dipping seams of high ash, low sulfur, high volatile bituminous coal with good coking properties (2). Of the 1.2 Gt of accessible reserves some 50% is accessible to surface mining. The Neryungri deposit has a coal reserve of 450 Mt (288 Mt coking coal and 162 Mt steam coal) with an average seam thickness of 25.2 m (88).

EAST ASIA

North Korea

Coals are known in both onshore and offshore basins in North Korea (Fig. 21). They include seams of late Carboniferous/?Permian age, Jurassic, and Tertiary age. Some of these may also generate oil in the offshore West Korea Bay Basin (93). The total reserves of coal proved in place are 2 Gt and range from anthracites to lignites. WEC data (73) indicates 300 Mt of bituminous coal and 300 Mt of lignite proved.

Palaeozoic coals were deposited on the Sino-Korean platform which was part of the Eurasian continent. Several phases of burial and tectonism have affected the coals so that all are anthracites (92). There are two main coalfields with significant economic potential: The Pyongyang-North Pyongyang coalfield and the Kowan-Muchon coalfield. Each of these have probable exploitable reserves of 300 Mt (2).

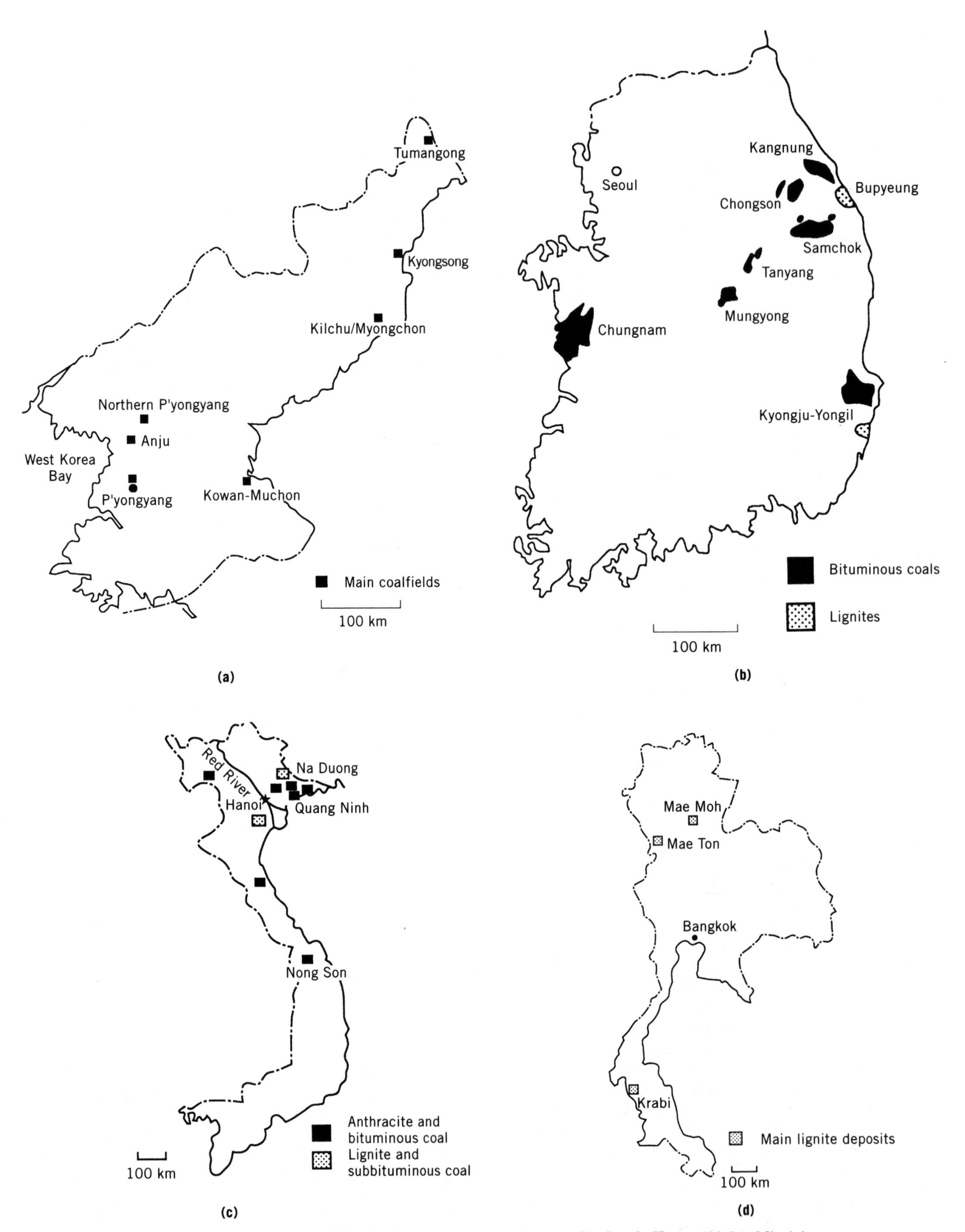

Figure 21. The coalfields of (**a**) North Korea (92,94); (**b**) South Korea (92,94–96); (**c**) Vietnam (97); (**d**) Thailand (94).

Pyongyang–North Pyongyang. This coalfield of 1000 km^2 has eight economic seams which vary in thickness from 1–15 m. The coals are low volatile anthracites with low ash and low sulfur contents. Currently the exploration and exploitation of these coals is progressing northward to the North Pyongyang field (2).

Kowan-Muchon. This coalfield is smaller with an area of 120 km^2. There are three main coals up to 6 m thick which are highly tectonized. They are low volatile anthracites with low ash and sulfur and are only mined underground (1).

Mesozoic coals are mainly of Jurassic age onshore North Korea and are anthracites. Offshore, however, Jurassic coal-bearing strata is known from the West Korea Bay Basin where it is of bituminous rank and may generate some oil (93).

Tertiary coals are more extensive in North Korea where they comprise subbituminous coals and lignites of Eocene, Oligocene, and Miocene age. There are several basins formed during several rifting phases which underwent rapid subsidence in a warm humid climate which favored the deposition of thick peats, some of which may have been formed in shallow lakes with significant algal input. Such sapropelic, oil-prone coals have been used at Tumangong for conversion to liquid fuel and in the production of chemicals. These small rift-basins appear to be controlled by NE–SW trending fold belts. The Tertiary coalfields include Kyongsong, Tumangong, and the larger fields of Kilchu/Myongchon and Anju where Oligocene coals are mined (92). Recent estimates indicate a reserve of 10 Gt of bituminous coal in the Anju Basin (93).

South Korea

As in the North, coals of late Palaeozoic, Jurassic, and Tertiary age occur including lignites to anthracites (Fig. 21**b**). Recent estimates indicate 158 Mt of recoverable reserves of bituminous coals and proved reserves of 238 Mt (73). However, other estimates indicate that anthracites have a total reserve of 1.6 Gt and a recoverable reserve of 754.3 Mt (95) and a proved reserve of 245 Mt (96). The anthracites are Carboniferous in age with low sulfur and seams 1–2-m thick but tectonically disturbed. The principal fields include Samchok (Samcheog) (26.7 Mt recoverable reserves), Chongson (Jeongseon) (45.2 Mt), Kangnung (Gangneung) (20.9 Mt), Danyang (15.8 Mt), and Mungyong (Munkyeong) (27.3 Mt) (93). Small amounts of Jurassic anthracites are known from the Mungyong and the Chunsan (Chungnam) coalfields (2).

Tertiary lignites are known in small areas bordering the south east coast.

Vietnam

The total amount of coal in Vietnam is a matter of conjecture but estimates of proven reserves range from 1 to 20 Gt with proven reserves of only 135 to 300 mt (94). Coals from Vietnam are mainly low volatile bituminous to anthracites mainly of Triassic age but also including Permo-Carboniferous which occur in a broad belt north of Hanoi (2,97) (Fig. 21**c**). Anthracites of Upper Triassic age in the Quang Ninh region have a total reserve of 6.5 Gt and a minable reserve of 557 Mt both opencast (199 Mt) and underground (358 Mt) (97). The main basins include Hon Gai, Cam Pha and Uong Bi, east of Haiphong in Quang Ninh Province. There are four principal coal basins north of Hanoi extending 20 km: Nui Hong, Phan Me, Bo Hai and Hon Yai (97).

In addition there are some medium volatile bituminous coals with coking properties (eg, Nong Son). Lignite (Neogene) occurs in a number of basins in the north of the country. This lignite has 12–15% ash but has a potential total reserve in the Lower Red River area of 200 Gt (97). However there is only 19 Mt mineable reserves in the Na Duong Coalfield (97).

Thailand

Thailand is relatively poor in coal resources but has some semi-anthracite, bituminous coals, and lignites (Fig. 21**d**). Extensive petroleum exploration has shown the more widespread occurrence of Tertiary lignites in rift basins. It has been calculated that the total measured reserves of all coal, but predominantly lignite, is 1.271 Gt which together with an indicated reserve of 1.076 Gt gives a total of 2.347 Gt (99). Other estimates give lower Figures such as the WEC (73) which quotes a proved amount of lignite in place as 1.648 Gt.

Small Mesozoic deposits of low sulfur semi-anthracites occur in the North East at Na Duang.

The Tertiary coals occur in several small basins in the North such as in Mae Moh, Mae Tun (Mae Than) (where there are high volatile bituminous coals), and Vaeng Haeng (Wiang Haeng) where there are 93 Mt of measured reserves (99). The Mae Moh basin is the most extensive where seams belonging to the Na Khaem Formation of Miocene age reach 30 m in thickness and range from low sulfur lignites to high volatile bituminous coals. The basin is a NE-SW trending graben 16 km long and 6.5 km wide (98). The measured reserves in this basin are 820 Mt together with 670 Mt of indicated reserves (99).

In addition Tertiary basins occur around Bangkok at Nong Ya Plong (Khlong Thom) where seams 2–12 m of high volatile bituminous coals are known and in the S.W. at Krabi. Here there are 83.6 Mt of measured reserves (99). Other estimates put this as only 15 mt (94). In addition a reappraisal of Recoverable Reserves in Saba Yoi near Songkhla estimates are given of 615 Mt of high-ash, high-sulfur coals (94).

Taiwan

The coal reserves of Taiwan are relatively small around 100 Mt (2) include a range from anthracites through to sub-bituminous rank, mainly of Tertiary age in the north and central provinces. WEC (73), however, indicates reserve of 100 Mt of bituminous coal and 100 Mt of subbituminous coal.

In the northern province low-ash and low-sulfur high volatile bituminous and subbituminous coals occur but these are highly tectonized and suffer from the effects of igneous intrusions. In Chilung, in the north, semi-anthracites are mined. In addition, low-sulfur, low-volatile bituminous coals, some with good coking properties are mined at Hsinchu, Nonchuang, Schangchi, and Mushan where there are 4 seams more than 1 m thick, all only obtainable by underground mining and future prospects are not favourable.

Japan

Japan is a main coal importer, but although limited, it has over 8.6 Gt reserves, 3.2 Gt mineable (94), (which includes 17 Mt of lignite) ranging in age from Permian to Tertiary (Fig. 22**a**). Most Paleozoic and Mesozoic coals are unimportant (the Ube coalfield in western Honshu contains anthracite) but the fact that Japan is within a major tectonic and volcanic belt has meant that extensive heat flows have created a wide range of coal ranks from anthracites to lignites (100). In 1989 9.6 Mt of coal were produced from 20 mines (101).

On Hokkaido the Ishikari coalfield of over 2300 km^2 yields high-ash, low-sulfur, strongly-caking bituminous coal in seams up to 6-m thick of Paleogene age (100) with a total reserve of 445 Mt (94). The smaller Kushiro coalfield (300 km^2) contains 128 Mt of non-coking bituminous and subbituminous coal (94).

On Kyushu, the Miike coalfield (500 sq km) is mainly offshore and although structurally disturbed, contains three main seams 1.6–2.7-m thick with a total reserve of 116 Mt (2,94). The coal is a good coking bituminous coal. In this coalfield there are 300 m of coal-bearing strata of the Nanaura and Kattachi Formations of Palaeogene age with up to 10 seams in 4 main areas (100). The Chikuho coalfield is similar, although larger, 1700 km^2, with 20 seams, 1–3 m thick of good bituminous coking coal with a reserve of 147 Mt (2,94).

On Honshu the Joban coalfield (300 km^2) is mainly offshore with 3 seams, up to 3-m thick of high ash, high sulfur coal with only 6 Mt of reserves. Lignites also occur in the Mogami and Miyagi coalfields, with seams around 1-m thick (78).

China

Calculations of Chinese coal reserves has been problematical but have almost certainly been underestimated. China has coals of every Phanerozoic Period and of every rank (Fig. 22**b**). Estimates by B. P. in 1992 give the reserves of bituminous coal as 62.2 Gt and lignite as 52.3 Gt. Recent estimates (102) indicate that in total known reserves are about 103.1 Gt and the predicted resources are 380 Gt. The literature on Chinese coals and basins is vast and a most useful guide is Ref. 103. China is divided into five coal accumulation regions (excluding Taiwan, dealt elsewhere) and the country is divided into 32 coal administrative regions. There are over 320 coal basins in China although only a few basins can be considered here. China represents one of the most important areas for potential development.

Within the North–East Coal Accumulation region most of the coal basins are of Jurassic or Tertiary age (105). Early and Middle Coal Measures are found extensively in Inner Mongolia as are coals of late Jurassic and early Cretaceous age. Tertiary basins are found in several small basins. In the North China area Carboniferous coal basins are found in Shanxi and Permian coals are widespread. Major Jurassic basins include several areas, especially in Shaanxi where the Ordos basin has 18 Gt of coal reserves of various ages, especially Jurassic. Tertiary coal basins are widespread in Shandong, Liaoning, and other areas. In the South China accumulation zone there are Permian coals across the region as well as a few deposits of Triassic age such as in Central Jiangxi and on the Sichuan-Yunnan border area (104). Tertiary coals are widespread in Yunnan and Guangdong. In the northwest accumulation area Jurassic coals are found in Gansu and Ningxi as well as in the far north-west area of Urumchi. There are few deposits of principal interest in the Tibet accumulation area.

Coal rank is very variable in China from lignite (8%) to bituminous coal (83%) and anthracite (9%). Coking coal makes up to 25% of the total resources mainly distributed in eastern Heilongiang, Hebei, Shanxi, Ninxia, Gansu, Xinjiang, Guizhou and on the Jiangsu-Shandong-Henan-Anhui border area.

Hebei. This region contains coals of Carboniferous, Permian, and Jurassic age in two main basins: Kailuan and Fengfeng. The coals range in thickness from 0.5–8-m thick and include low-ash, low-sulphur low volatile bituminous coals and anthracites. In the Kailuan basin the coals are of Permian age and those in the Fengfeng basin are both Carboniferous and Permian (9).

Shanxi. This region is one of China's principal mining regions producing more than 200 Mt/a. The main coals are of late Carboniferous age in a large basin which occupies much of the southern part of the province with the mining center of Taiyuan. The other area most recently developed is in the north at Datong where there are late Palaeozoic and Jurassic coals (Fig. 23).

Important Carboniferous basins include Xishan, Yangquan Jincheng, and Pingshuo. The main coal basins south of Taiyuan cover an area 550 × 300 km with Ordovician as the local basement. The basins are part of the north China plate (106) and late Carboniferous paralic coal-bearing strata are overlain by continental red beds. Tectonic movements have split this larger basin into several coalfields. These basins have a reserve of 200 Gt (102). Anthracites account for 25% of the total. The coals vary in thickness, up to 8 m. The coals are generally 15–16% ash, 2–3% sulfur (although locally can be much higher) and 10–20% volatile matter (102). In the south of the basin the main seam is the No 3, a low sulfur coal up to 6-m thick.

In Datong the coals are medium to high volatile bituminous, low sulfur and high ash. The Permian coals in this basin occur in two to ten seams with a cumulative thickness of 10 to 40 m (106). In the Antaibo district the lowest coals average 24 m with ash 21% and sulfur 2% (106,107). The highest coals in this sequence are low-sulfur but high in ash.

Inner Mongolia—Shaanxi. This region contains coals of Carboniferous, Permian and Jurassic age and include two important basins, the Huolinhe and Dongsheng-Shenmu basins. The coal in the Shenmu basin is medium volatile, low ash and low sulfur (9). These coals are of Jurassic age.

Liaoning. This region contains the Mesozoic and Tertiary Coal basins of Fushun, Fuxin and Tiefa. About 60% of Chinese coals were deposited in the Jurassic.

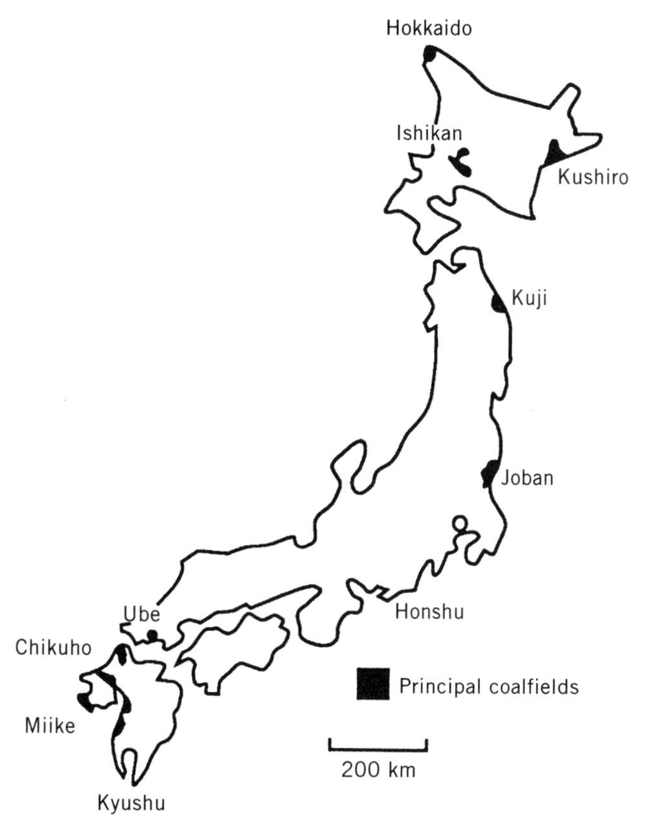

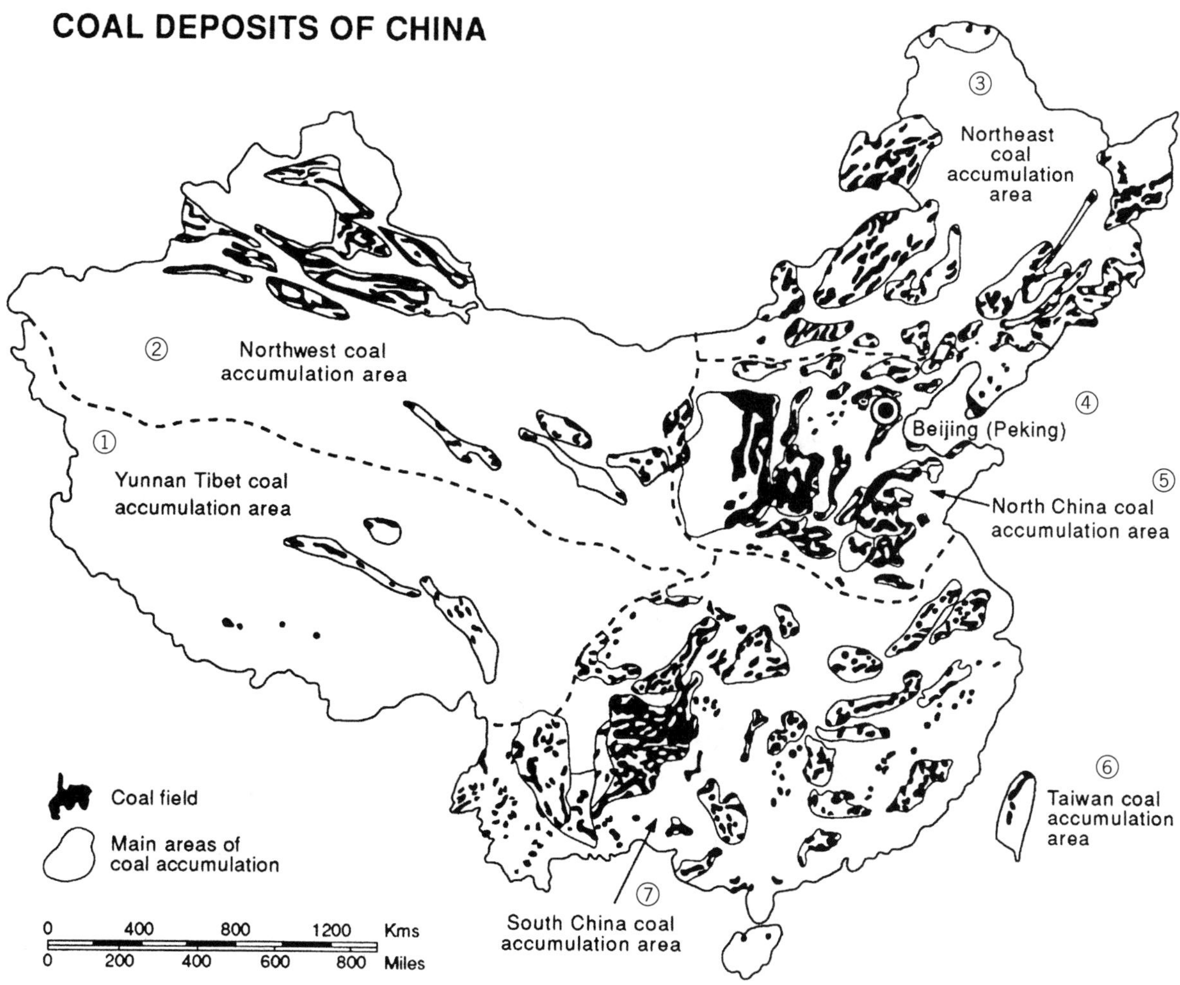

Figure 22. The coalfields of (**a**) Japan (100); (**b**) China (102).

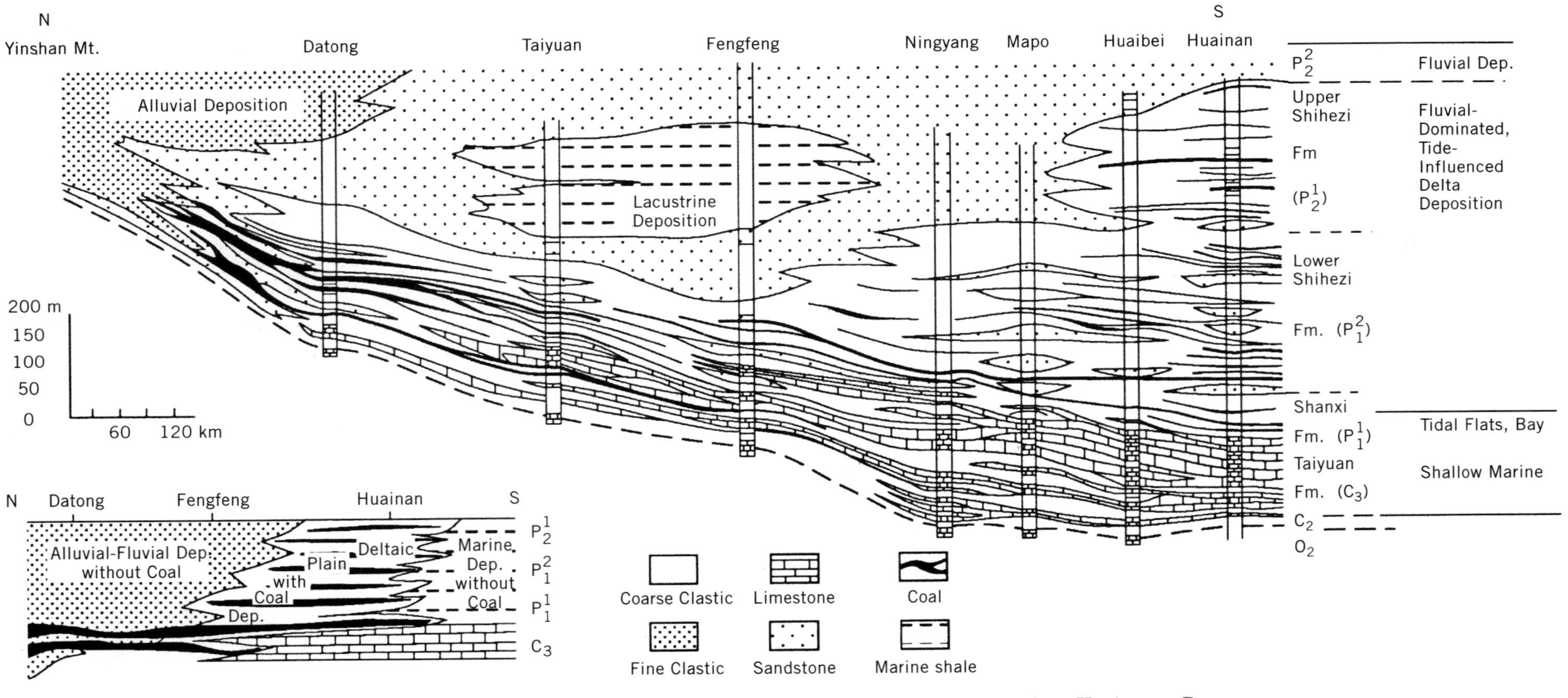

Figure 23. The Carboniferous-Permian coal-bearing systems from Huainan to Datong (106).

The Fuxin Basin in northeast China has been intensively studied. The rift basin is elongate with a length: width ratio of 20 : 3 and covers an area of 2300 km^2. The initiation of the basin occurred with the eruption of basaltic and andesitic volcanics. The margins of the basin are faulted against Precambrian basement. Towards the basin center there are thick lacustrine facies (Shahai Formation) and there are some coals 5 m thick at the top of the Formation. The basin fill is over 4000-m thick. The upper part of the sequence, particularly towards the basin center is dominated by a coal 125-m thick (108). The total coal reserves in this basin are about 2 Gt and available for both opencast and deep mining. The coal ranges from subbituminous to medium volatile bituminous coal and is moderate to low ash and sulfur.

The Fushun Basin, part of the Liaoning Province, is essentially a Tertiary Basin 40 km long and 4 km wide. The basin overlies a basement of Precambrian gneiss and probably began to form in the late Mesozoic. The basin probably represents a small extensional basin. The oldest rocks are of Lower Cretaceous age, 0–25-m thick, belonging to the Longfengkan Formation and comprise grey-green and white sandstones. The Tertiary sequence rests unconformably upon the Cretaceous and a basalt occurs at the base. This is part of the Laohutai Formation which comprises three zones, 50–130 m in the lower part, a middle part 50–125 m with a coal seam (B) 0–25 m (average 7 m) and an upper tuffaceous part 0–120 m also with one coal (A), which is impersistant, 0–10-m thick (the Lizigou Formation of Johnson, 109). The main coal bearing sequence belongs to the Guchengzi Formation which may be interpreted as a complex fluvio-lacustrine sequence formed with a small pull-apart basin and the main coal is 35–120-m thick. This coal splits towards the west of the basin. The overlying strata of the Jijuntun Formation comprises oil shales 80–158 m thick which are mined and processed for oil.

Within the basin the coal reserves are calculated at 900 Mt. The coals are Eocene subbituminous coals with an ash content of 13–15%, a sulfur content of 0.06% and a heating value of 31.4 MJ/kg. Within the coal sequence is a cannel coal, 0–15-m thick, which contains conspicuous amber. Currently there is only one major opencast mine in the area which currently produces 3.5 Mt/a. At its present production level, the pit has still 25 years of life. Two opencast mines are also in the basin.

Heilongjiang. Within this province is the Hegang (Hokang) Jurassic coal basin (500 km^2) which is currently being mined for bituminous coal. The coal seams range in thickness from 2–20-m thick and they are gently dipping and near the surface so that of the 500 Mt reserves (2) some 20% is suitable for surface mining. The coal is low ash, low-sulfur, high-volatile bituminous coal.

The Shuangyashan coal basin contains coals of Carboniferous and Permian age. The Jixi coal basin also contains coals of Carboniferous and Permian age.

Shandong. This region contains coals mainly for Carboniferous and Permian age in Yomzhou. These coals are low-ash, low-sulfur bituminous coals (2,9).

Jiangsu. This region contains many small coal basins including that of Xuzhou (Hsuchon). This basin in the north of the province yields bituminous coals with 4 seams totaling 2.5 m in thickness (2).

Henan. This region contains Permian coal basins including that of Pingdingshan (9) as well as Jurassic basins such as that of Yima (98) where seams are up to 20 m thick, low rank (0.5% R; volatile matter 38–42%) and hence is a *subbituminous* high volatile bituminous coal (110).

Anhui. This province contains two major coal basins of Permian age: the Huainan and Huaibei basins.

The Huainan Basin contains numerous seams ranging in thickness from 1 to 7 m thick. These coals are low-ash, low-sulfur bituminous coals and 40–70% are potential coking coals with over 3 Gt of reserves (2). Igneous intrusions have increased the rank of the coal in the Huaibei basin.

Guizhou. This region contains the important Permian coal basin of Lupanshui (9) which contains bituminous coals.

AUSTRALIA

Although Australia is a big exporter of coal, most of the coal is taken from Permian deposits in the eastern states of New South Wales, the long-time principal producing state of thermal and coking coals, and Queensland, the state which has had the most rapid and greatest increase in production in the past 20 years. Of the other states, Victoria is a very large producer of soft brown coal for domestic markets and both Western Australia and South Australia produce significant volumes of coal. Historically, Tasmania has not been a large coal producer.

By world standards, particularly compared to China, United States, and USSR who together produce 80% of world coal, Australia is not a large coal producer with only 3.8% of world production (111). However, as an exporter, Australia is the world's largest with almost half total production shipped overseas.

New South Wales

The main coal resources of New South Wales are located in the Sydney-Gunnedah Basin which extends from the central coastal region of Sydney, Wollongong and Newcastle, northwest to the New South Wales-Queensland border (Fig. 24**a**). A total of almost 90 Gt of coal have been identified (112) but only 20 Gt are recoverable because of quality constraints, depth to the seams and sterilization as a result of urban development, national parks, and water storage areas.

Sydney–Gunnedah Basin. The Sydney–Gunnedah Basin has been divided into five producing areas: Western Coalfield, Southern Coalfield, Newcastle Coalfield, Hunter Coalfield and Gunnedah Coalfield. Resources for these coalfields are given in Table 8.

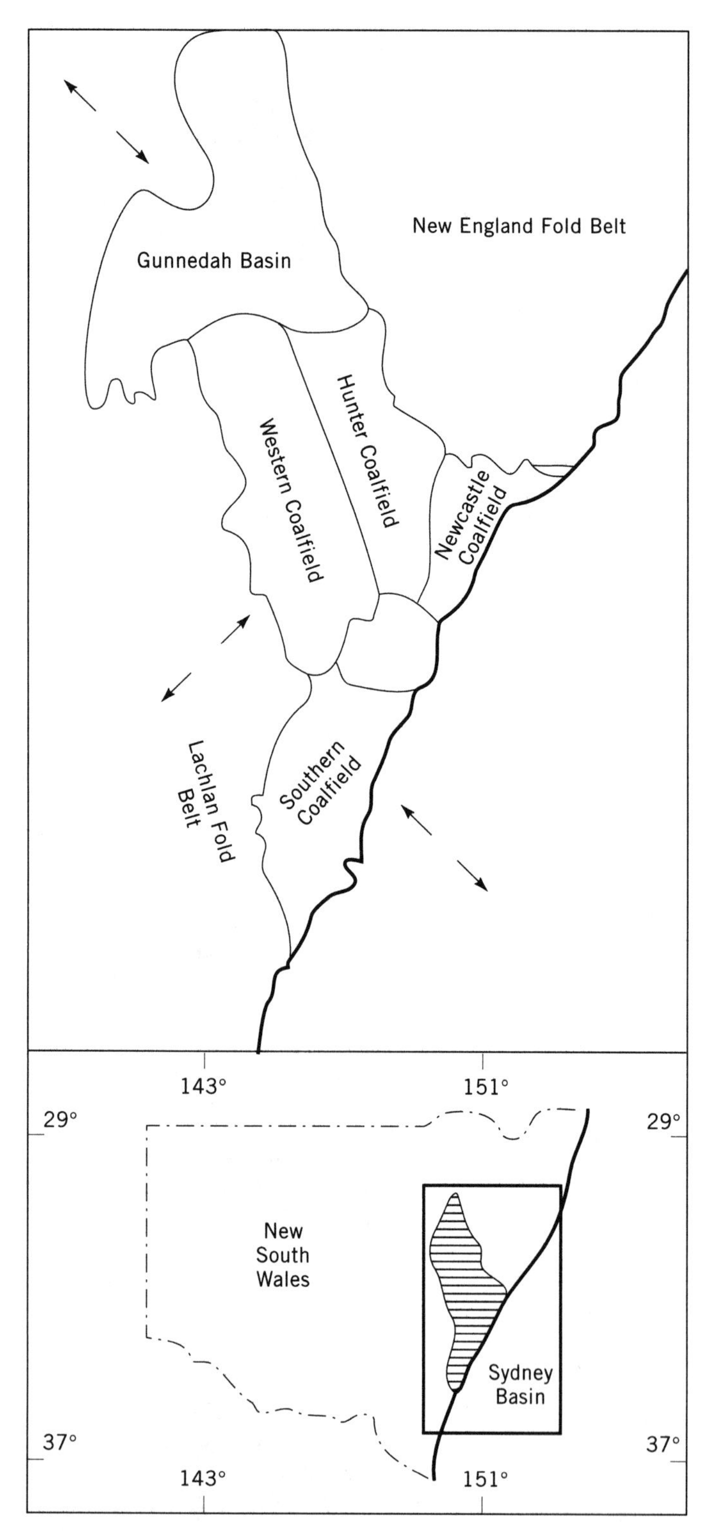

Figure 24. Location of the Sydney and Gunnedah basins.

Western Coalfield. The coal resources of the Western Coalfield are within the Permian Illawarra Coal Measures. In the southern part of the coalfield (Lithgow area and surrounds), the resources are exploited mostly from the Lithgow and Katoomba seams (Fig. 24**b**) which contain high specific energy thermal coals that are accessed by drift entry and, to a lesser extent, open pit mining; a small quantity of coal is taken from the Irondale and Lidsdale seams; Middle River seam has significant resources but little mining of this seam is carried out. Most coal from both the Lithgow and Katoomba seams is medium ash, low to moderate sulfur, low phosphorus, export quality thermal coal (Table 9). The high ash Lithgow seam coal is used for domestic power generation and cement manufacture.

Ten mines are presently working; annual productions range from 0.2 to 2.4 Mt using mostly longwalls augmented with continuous miners where appropriate. Domestic markets are Wallerawang and Mt Piper power stations which are located within 20 km of Lithgow. Export markets include Japan, South Korea, Hong Kong, Malaysia, and Europe. Total recoverable resources are estimated to be 500 Mt of coal.

In the Ulan area of the northern part of the Western Coalfield, the resources are in the Ulan seam of the Illawarra Coal Measures. Coals are typical medium to high ash medium to high volatile bituminous coals that are mined selectively by underground, drift entry (working thickness of 3 m) and washed to produce a low to medium export coal (South Korea, Japan, Taiwan, Hong Kong, Europe) or by open pit methods (working thickness of 12 m) for domestic power generation at Eraring power station in the Hunter Valley. Resources are estimated to be 810 Mt of raw coal with a potential production capacity is 8.4 Mt (547 Mt recoverable) coal per annum although 1991–1992 production was 7.89 Mt of raw coal giving 6 Mt of saleable coal.

In addition to the working mines, 5 proposed mines, with resources of 269 Mt, have been planned.

Southern Coalfield. The Southern Coalfield is noted for its premium quality coking coal which is exploited by underground methods from depths of up to 600 m. Mining is mostly from the Bulli seam of the Illawarra Coal Measures (Fig. 24) with lesser amounts from the Wongawilli seam. In the past, the Balgownie, Tongarra, and Woonona seams have been mined.

Mining occurs in three main areas of the coalfield: Illawarra area originally centered on Wollongong but now extending westwards to deeper coal; the Burragorang Valley (western part of the coalfield); and a smaller but significant area at Berrima. Historically the Southern Coalfield has been the only producer of hard coking coal in New South Wales with most of this production used locally in steel making at Port Kembla (Wollongong) with increased export sales over the past 20 years. In more recent years coking coals have been exported and an increasing tonnage, derived from the middlings from the washeries, is exported as steaming coal. The domestic market for Southern Coalfield steaming coal is almost entirely restricted to the local area.

At Berrima, the Wongawilli seam coal is used almost exclusively for the manufacture of cement although long term plans will utilize the coal as export quality steaming coal either as an unblended product or as a blended product with the Western Coalfield coal. Berrima colliery produces 220 kt/a and this is used exclusively for the manufacture of cement. Measured in situ resource is 93 Mt (51 Mt marketable) and with an indicated resource of 40 Mt.

Table 8. Coal Resources of New South Wales[a]

Coalfield	Method	Resources: Measured-Indicated, Mt	Inferred, Mt	Total, Mt
Western	Open cut	737	400	1,137
	Underground	1,893	1,310	3,203
	Subtotal	2,630	1,710	4,340
Hunter	Open cut	10,947	470	11,417
	Underground	8,071	3,160	11,231
	Subtotal	19,018	3,630	22,648
Newcastle	Open cut	209	160	369
	Underground	4,472	3,150	7,622
	Subtotal	4,681	3,310	7,991
Southern	Open cut	4,346	12,701	17,047
	Underground	4,346	12,701	17,047
	Subtotal			
Gunnedah	Open cut	540	100	640
	Underground	1,606	29,510	31,116
	Subtotal	2,150	29,610	31,760
Gloucester	Open cut	104	30	134
	Underground	39	30	69
	Subtotal	143	60	203
Oaklands	Open cut	1,388		1,388
	Underground		3,260	3,260
	Subtotal	1,388	3,260	4,648
	Total	*34,356*	*54,281*	*88,637*

[a] Ref. 112.

Table 9. Selected Data for New South Wales Coal[9]

Property	Southern Coalfield 1	Southern Coalfield 2	Western Coalfield 1	Western Coalfield 2	Hunter Coalfield 1	Hunter Coalfield 2	Hunter Coalfield 3	Newcastle Coalfield 1	Newcastle Coalfield 2	Newcastle Coalfield 3	Gunnedah Coalfield 1	Gunnedah Coalfield 2
Moisture, % ar	7.5	9.0	9.0	8.0	8.6	10.0	9.0	8.0	8.0	7.0	9.0	9.5
Moisture, % ad	1.0	1.2	2.6	3.2	2.6	3.0	3.9	2.5	2.5	2.6	4.0	3.5
Ash, %	9.7	16.5	13.4	19.4	8.0	12.8	15.1	8.0	15.0	20.5	5.5	9.1
Volatile matter, %	24.1	22.8	29.7	26.7	35.5	33.0	33.7	38.0	30.0	27	37.0	33.6
Phosphorus, %	0.04	0.04	0.6	0.55	0.57	0.6	0.74	0.8	0.4	0.4	0.38	0.4
Specific Energy kcal/kg ad		6,885	6,868	6,526	7,172	6,737	6,750	7,533	6,700	6,112	7,530	7,088
MJ/kg, ad		28.9	28.8	27.4	30.1	28.5	28.1	32.0	28	25	31.5	29.8
CSN	6.4	3	1.0	1.0	4.7	0.7–0.5	5.5	2.0	1.0	4.75	1.6	
Max fluidity, ddm	1,300				280			1,500			1,000	
HGI	75	66	45–48	40	51	50	48.3	41.8	50	50	43	50
AFT, °C, Deform.		1,477	1,400	1,380	1,460	1,342	1,400	1,200	1,450	1,450	1,450	1,412
Flow		1,557	1,500	1,600+	1,504	1,484	1,500	1,520	1,560	1,580	1,515	1,500

The Southern Coalfield enterprises have been severely affected by the recession of the late 1980s and early 1990s and the subsequent downturn in the economy has resulted in the closure of several ventures. Two proposed mines are in the planning stage but these are unlikely to come into production unless there is a dramatic increase in export demand. Domestic consumption is unlikely to increase in the foreseeable future and may in fact decline.

In the Illawarra area, ten mines are currently operating and total raw coal production is 15 Mt per annum (saleable production of 11 Mt) of which 7 Mt is contributed by four mines which exclusively service the Port Kembla steel works. Coal for the steel works is washed to 9% ash and has volatile matter of 21–24%. Total recoverable resource of the Illawarra area is estimated to be 700 Mt. Export coal is either transported by rail or road to the Port Kembla coal loader from where thermal coal is exported to Japan, South Korea, Philippines, Brazil, Hong Kong, France, Taiwan, United Kingdom, other Far East, Europe and Southeast Asian countries and coking coal to Brazil, Japan, other Asian and European countries.

Four mines are currently operating in the Burragorang Valley giving a total production of 4 Mt/a (3 Mt saleable) from a total resource of 43 Mt. Coking coal is exported South Korea, Brazil, Japan and Taiwan; the remainder is exported as thermal coal to Japan, France and other countries in Southeast Asia and Europe.

Two new collieries are planned for the southwestern part of the coalfield and potential resources are given as 265 Mt.

Newcastle Coalfield. The Newcastle Coalfield has three Permian coal-bearing sequences: Greta Coal Measures, Tomago Coal Measures and Newcastle Coal Measures. Total resources are approximately 8 Gt of soft coking and steaming coal. The coalfield is divided into four areas; the western Cessnock area, northern Maitland area, central Teralba area and southern Wyong area.

The western area produces high quality export coking coal for Europe from one colliery using underground longwall methods to exploit the 3.5 m Greta seam (Greta Coal Measures) which is approximately 500 m deep. The coal is a high volatile bituminous coal with a high specific energy and fluidity but low ash. Annual production is 2.2 Mt raw coal (1.87 Mt saleable coal) from a total resource of 53 Mt. The coal is railed to the Newcastle coal loader.

Five seams of the Tomago Coal Measures are exploited in the northern area from open pit mines. Product mix comprises low ash soft coking coal and a medium ash thermal coal, both of which are exported. Annual production is 1.8 Mt of raw coal (1.0 Mt saleable coal) from a resource of 55 Mt. The coal is shipped to the Newcastle coal loader and exported to Japan (coking coal) and Europe, Japan, and Taiwan (thermal coal).

The Dudley and Borehole seams are the main producers of the Newcastle Coal Measures in the central part of the coalfield, producing a low ash, medium fluidity soft coking coal for domestic use in the steelworks at Newcastle and export to Japan, Pakistan and other Southeast Asian countries from the Newcastle coal loader. Medium ash thermal coal is also produced for the domestic market (Eraring power station) and exported to Pakistan and Southeast Asia. Total resource is 280 Mt and annual production from the collieries ranges from 1.75 Mt to 0.65 Mt of raw coal.

From the southern area of the Newcastle Coalfield, coal is taken from the Wallarah, Fassifern and Great Northern seams. Saleable coal is mostly a medium to high ash thermal coal for use in local power stations and for export to Japan. Annual production rates from the collieries range from 0.4 Mt to 1.7 Mt of raw coal.

Hunter Coalfield. Hunter Coalfield is the largest producer in New South Wales with coal exploited from three units, Greta Coal Measures, Wittingham Coal Measures and Wollombi Coal Measures (Fig. 25). At least 50 seams are worked to produce soft coking and thermal coal from a resource totaling 22.648 Gt. The shallow depths to the coal results in large scale, multiseam open pit extraction and large production rates. Geographically, the coalfield is divided into three areas, southern area utilizing the Wittingham Coal Measures, central area utilizing the Wittingham Coal Measures and a northern area utilizing the Greta Coal Measures.

Forty seams produce soft coking coal (low ash, medium fluidity and high volatile content) and a thermal coal (low to moderate ash, low sulfur and comparatively high grindability index) for export from the Wittingham Coal Measures in the southern area. A higher ash thermal coal is also produced for domestic consumption.

A similar range of coals is produced from 8 seams of the Wittingham Coal Measures in the central region of the coalfield. Several new projects are in the planning stage but a marked improvement in the economic climate is needed before significant production can commence. The proposed projects will produce low to very low ash, high energy, low sulfur thermal coals for export.

Five shallow seams of the Greta Coal Measures produce coal in the northern area. Products are low to medium ash thermal coals for the export market.

The recoverable reserve for the Hunter Coalfield are estimated to be 4.56 Gt; annual productions range from 5.7 Mt to 0.5 Mt/a.

Gunnedah Coalfield. Gunnedah Basin has total underground resources of more than 31.116 Gt at depths of less than 300 m and open pit resources of 644 Mt. However, current production is very small, amounting to a little more than 1.5 Mt/a from two open pit and one underground mine using bord and pillar techniques. Total resource for the operating pits is 212 Mt but one colliery has 198 Mt of the total. Two proposed mines have resources of 897 Mt. Most saleable coal is currently railed 320 km to Newcastle and exported to Japan (thermal and coking coal), South Korea (thermal coal) and Taiwan (coking coal).

Gunnedah Coalfield is divided into two areas by the north-south trending Boggabri ridge which is an upward inflection of the basement (Fig. 24). The eastern part of the coalfield contains substantial resources of low ash, high energy, high volatile, low phosphorus thermal coal with a small resource of high volatile, high fluidity soft coking coal.

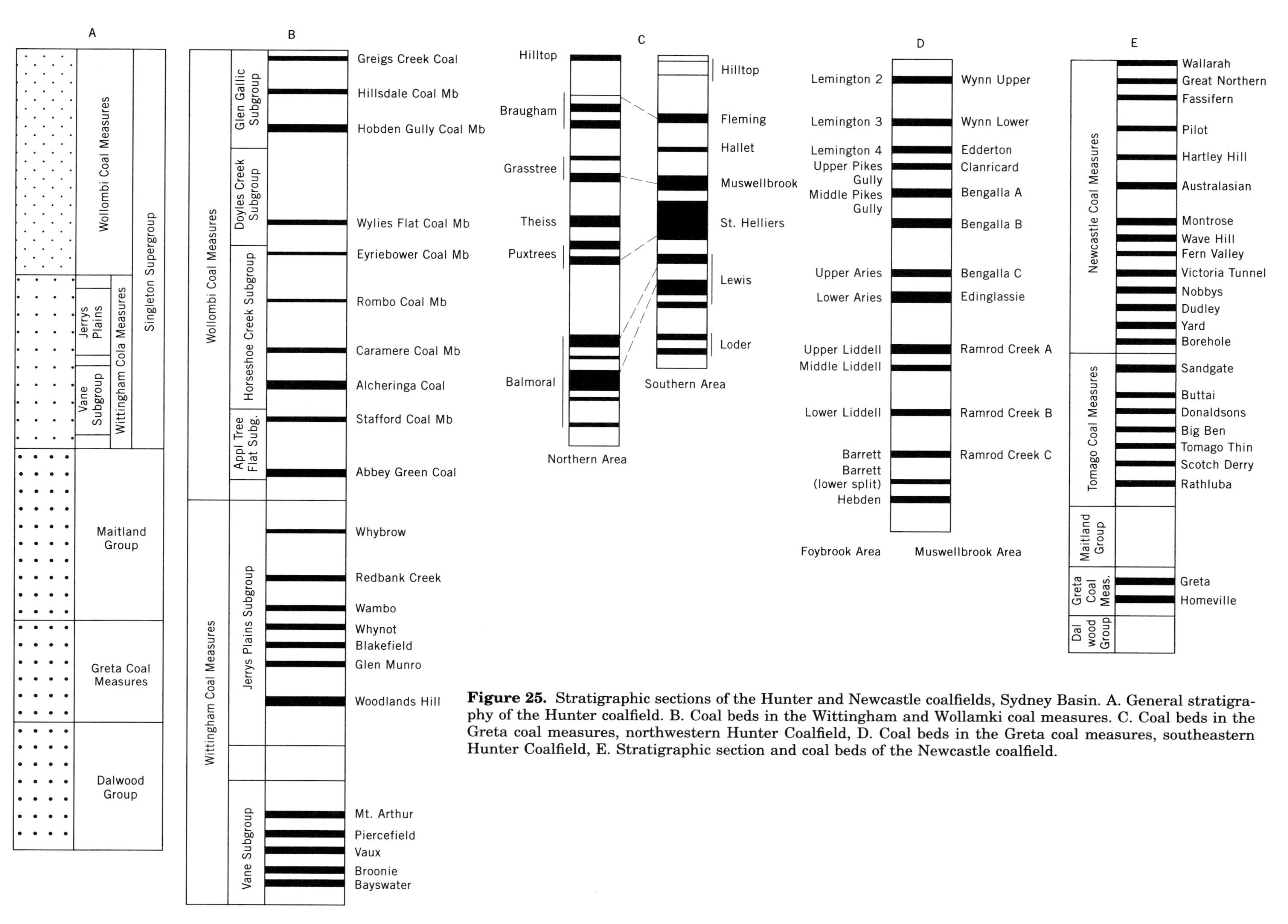

Figure 25. Stratigraphic sections of the Hunter and Newcastle coalfields, Sydney Basin. A. General stratigraphy of the Hunter coalfield. B. Coal beds in the Wittingham and Wollamki coal measures. C. Coal beds in the Greta coal measures, northwestern Hunter Coalfield, D. Coal beds in the Greta coal measures, southeastern Hunter Coalfield, E. Stratigraphic section and coal beds of the Newcastle coalfield.

The western area of the coalfield is much larger and contains very large underground resources in the Hoskissons seam. The upper part of the seam comprises high and medium ash thermal coals; the lower part of the seam contains a soft coking coal that exhibits considerable lateral variability with respect to ash and phosphorus. The southeastern part of the basin contains small tonnages of low ash, low phosphorus soft coking coal suitable for blending. Large tonnages of low to medium ash thermal coal are found elsewhere in the western part of the coalfield.

Gloucester Basin. Gloucester Basin (also known as the Stroud-Gloucester Basin) is a small 10 × 40 km basin, 80 km north of Newcastle. The sequence contains the Gloucester Coal Measures which contain approximately 50 seams many of which are lenticular. Only 3 seams are considered to be economic. Estimated resources are given as 203 Mt.

The lenticular nature of the seams as well as the structural complexity of the basin will probably limit production to open pit mining methods but production is unlikely to commence in the foreseeable future.

Oaklands Basin. Oaklands Basin is located in southern New South Wales near the New South Wales-Victoria border. This small basin contains the Corabin Coal Measures which contains two seams of which the Lanes Shaft seam is considered to be the most prospective. The seam contains moderate to high ash, high moisture subbituminous coal which is amenable to open pit mining. The resource is too far from main shipping ports to have significant export potential; domestic use is likely to be restricted to power generation but this is unlikely as coal with higher specific energy and much greater resources are found in the present coal-producing areas.

Coal properties for the Oaklands coal are as follows:

Moisture, % ar	28.0
Moisture, %, ad	11–17
Ash, %	7–19
Volatile matter, %	23–44
Phosphorus, %	0.002
Specific energy (kcal/kg)	4180
(MJ/kg)	17.5
HGI	100
AFT, °C Deform.	1390
Flow	1540

Queensland

Queensland, with its vast resources of thermal and coking coals, is Australia's largest coal producer with an annual production of more than 84.08 Mt saleable coal of which 69.66 Mt, or 82.5% of production, is exported to 42 countries worldwide. Export coal is transported to five ports (Table 10) by five main rail lines which have weekly haulage loads of up to 526 kt per week. Coking coal exports total 44.9 Mt compared to 24.8 Mt of thermal coal with the division of coking versus thermal coals placed at a Free Swelling Index (FSI) of 4. Japan was the largest purchaser with 31 Mt followed by India (5.5 Mt), Korea (4.8 Mt) and France (3.1 Mt).

Of the 14.9 Mt used domestically, 12.4 Mt is used for electricity generation in state-owned power stations; nonferrous metal and cement making industries are the next big users.

Geographically, Queensland is divided into three mining districts, Northern, Central and Southern Districts with the Central District the largest coal producer (32.1 Mt) followed by the Northern District (28.2 Mt). Underground mining is small compared to open pit mines producing only 7.6 Mt compared to 76.5 Mt.

The main producing coalfields are in the Permian Bowen Basin; two other significant coalfields, Callide and Ipswich which are of Triassic age. Jurassic coalfields have resources second only to the Permian coalfields but little mining has been undertaken in these fields because of the quality and better properties of the Permian coals.

Permian Coalfields. ***Bowen Basin.*** The Bowen Basin Coalfield, Australia's largest producer of export coal and

Table 10. Coal Export Facilities

	1991–1992 Throughput, Mt	Maximum Loading Rate, t	Vessel Size, t	Stockpile Capacity, Mt
Abbot Point	5.9	4,600	35,000 to 165,000	0.75
Brisbane	3.5	3,300	10,000 to 90,000	0.28
Dalrymple Bay	18.4	7,200	200,000	1.8
Gladstone				
Auckland Point[a]	0.05	1,600	65,000	0.3
Barney Point	1.9	2,000	90,000	0.4
Clinton	17.6	4,000	220,000	2.7
Hay Point	22.6	4,000	200,000	2.5

[a] Ceased operating 1991.

home of the world's largest coking coal mine, is a 600 by 250 km triangular-shaped basin in central eastern Queensland (Fig. 26). Most of the dramatic escalation of coal production in Queensland has been from open pit mines in the Bowen Basin (Table 11). Of the total Queensland production up to 1950, only 1.94 Mt or 3.2% of the total had been from open pit methods (113). Of this, export amounted to small spot sales. For example, during 1951–1957 only 70 kt of coal were exported. However, by 1964–1965 1 Mt was being exported and by 1973, 16 Mt or 80.8% of production was exported (114).

Seams in the Bowen Basin exhibit big variations in rank and quality. In the eastern part of the basin, the rank ranges from low volatile to semi-anthracite whereas as in the central part, the rank is medium to high volatile bituminous. Along the western edge of the basin, the rank decreases and coals are generally low ash thermal coals.

Coal seams occur at a number of stratigraphic levels but only four, informally termed Groups I to IV, are considered to be presently or potentially economic. Geographically, the mines in the Bowen Basin are grouped into northern, central and southern Bowen Basin mines.

Group I Coal Measures. Group I contains the oldest coal-bearing units, the Reids Dome Beds which are up to 30-m thick in the Denison Trough in the southwestern part of the basin but thin considerably in the north where the name Lizzie Creek Volcanics is used. The coals are not presently mined. Shallow reserves of good quality thermal coals are found near Emerald. The coals are higher in exinite (6–11%) than most Bowen Basin coals and have 37–40% volatile matter, and 0.77% R_{max}.

Group II Coal Measures. Group II coal measures include the Collinsville Coal Measures in the north and the Blair Athol Coal Measures in the south. The Collinsville Coal Measures lie in an asymmetrical syncline and contains nine persistent seams (5 minable) with the basal Blake seam up to 20-m thick and averaging 7.5 m. Rank of the coal is variable with several areas of elevated rank as a result of intrusions. Typical analyses for Collinsville coal give 14–20% ash, 19–20% volatile matter, and 1.1% R_{max}. Export coking coal and domestic and export thermal coals are produced. Mining of the medium to low volatile bituminous coal is by open pit and underground methods with total resources given as 232 Mt (measured and indicated) comprising 25 Mt underground and 81 Mt open pit coking coal and 28 Mt underground and 98 Mt open pit thermal coal. Current annual production is 2.3 Mt with sales of coking coal to Japan and India and thermal coal for domestic use.

Blair Athol Coal Measures has four seams of medium volatile bituminous coal that are mined by open pit methods (resources of 265 Mt measured and indicated) with average thicknesses ranging from 1 to 29 m. Analyses show 4 to 16% ash, 16 to 34% volatile matter, and 0.7% R_{max}. Current production is 7.5 Mt/a with sales of thermal coal to Japan, Indonesia, Denmark, Hong Kong, and domestic users.

Group III Coal Measures. Group III Coal Measures were deposited on the relatively stable Collinsville Shelf and Comet Platform. The seams are best developed along the western margin of the basin. In the Northern Bowen Basin geographic region the coal is found in the Moranbah Coal Measures which is a transitional, laterally equivalent unit of the German Creek Formation in which the coal is found in the central Bowen Basin region. The coal measures are generally 200 to 400 m thick but have a maximum thickness of 760 m in the east.

Two open pit mines currently produce 9.6 Mt per annum of coking coal from three seams, ranging in thickness from 6 to 8 m, in the Moranbah Coal Measures. Total resources are given as 1.663 Gt of coking coal, of which 30 Mt is underground (not mined at present) and 1.362 Gt is open pit. Principal markets for the coal are Japan, India, South Korea, Argentina, Taiwan, Brazil, Europe and Australia. Analyses of the coal show 8–10% ash, 24–25% volatile matter, and 1.1–1.2% R_{max}.

In the Central Bowen Basin, the German Creek Formation has up to nine seams but in the six mines where mining occurs, 1 to 6 seams are mined. Average seam thickness ranges from <1 to 10 m. Total annual production is in excess of >30 Mt per annum with total resources of 3.42 Gt measured and indicated open cut. Analyses of the low to medium volatile bituminous coking coals, which are exported to Japan, Korea, Taiwan, Brazil, Iran, China, India, Argentina, Turkey and Europe show 8–10% ash, 17–32% volatile matter, and 0.95–1.65% R_{max}.

Group IV Coal Measures. Group IV coals are the most diverse group of coals in terms of quality and are the most widely distributed; in contrast, they have relatively constant, very low sulfur. The seams are part of the Rangal Coal Measures (northern and central areas) and the stratigraphic equivalent Baralaba Coal Measures (southeast); equivalent coals are found in the Bandanna Forma-

Table 11. Aspects of Selected Bowen Basin Mines[a]

Mine	Operating Strike Length, km	Annual Overburden Removal, Mm^3	Annual Fuel Consumption, ML	Annual Production, Mt
Blackwater	22	43	13	4.0
Peak Downs	30	79	17.2	5.5
Goonyella-Riverside	30	120	38	6.0
Saraji	24	71	17	5.0
Norwich Park	11.1	66.5	11.5	4.5
Moura	26	35	8.7	3.5
Gregory	11	27	5.5	3.6

[a] Ref. 115.

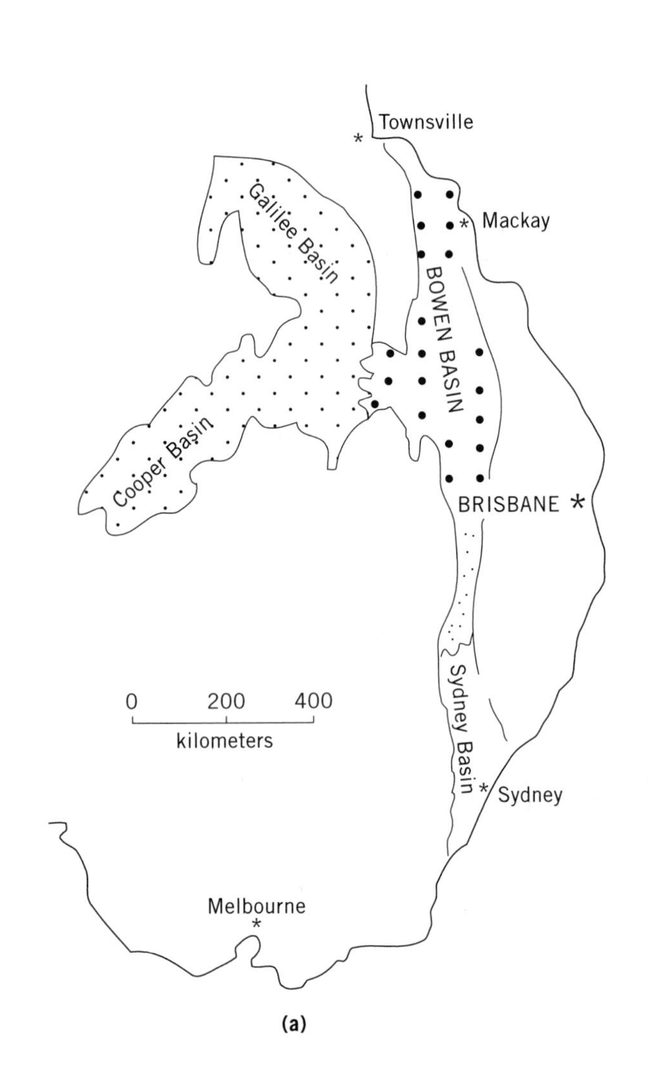

(a)

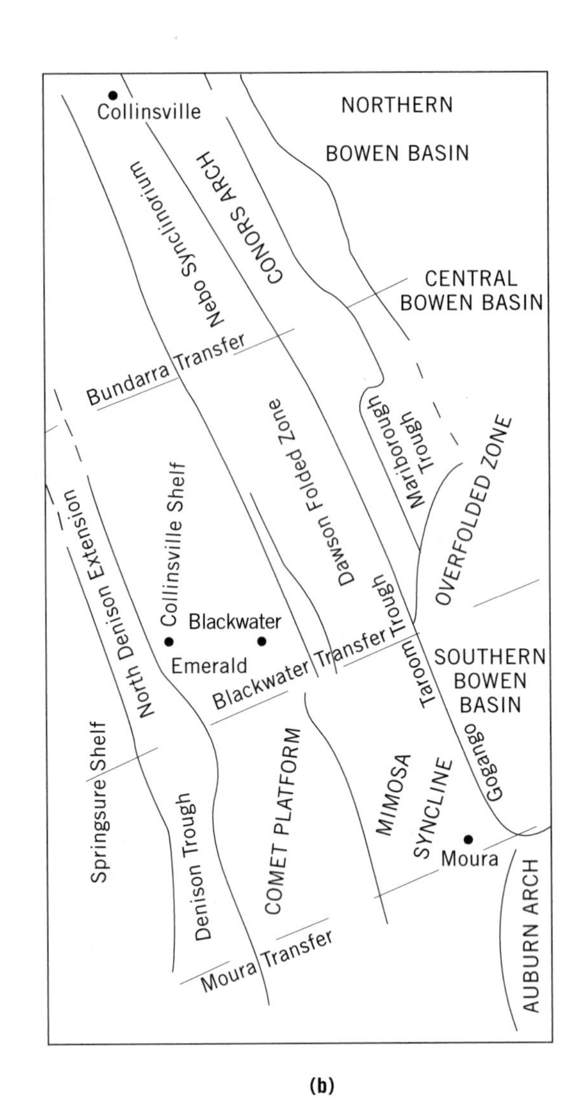

(b)

SOUTHEAST	SOUTHWEST	CENTRAL		NORTHERN	COAL GROUP
Rewan Group	Rewan Group	Rewan Group		Rewan Group	
Beralaba CM	Bandanna Fm	Rangal CM		Rangal CM	IV
Gyranda Fm	Black Alley Shale	Burngrove CM		Fort Cooper CM	IIIA
Flat Top Fm	Peawaddy Fm	Fair Hill Fm			
		Macmillan Fm		Moranbah CM	III
Barfield Fm		German Creek Cm			II
Mt Ox Subg.		West	East	Blenheim Fm	
		BA	Back Creek Fm	CV Gebbie Fm	
BuffelFm	Cattle Creek Fm			Tiverton Fm	
Camboon Volcanics	Reids Dome Beds		Carmila Beds	Lizzie Creek Volcanics	I

(c)

Figure 26. Bowen Basin, Queensland. (**a**) Location; (**b**) structural units; (**c**) stratigraphy.

tion of the Denison Trough. These coals are of major economic importance and are used as thermal and coking coals with sales to domestic (principally power generation) and export to Japan, India, Israel, Pakistan, Taiwan, The Netherlands, Turkey, other Southeast Asian countries and Europe. The coal is mined in 7 open pit mines and 3 underground mines with total annual production of >22 Mt from total resources of 4.98 Gt of which 1.739 Gt is underground and 429 Mt open pit thermal coal, 2.668 Gt underground and 570 Mt open pit coking coal.

Analyses of the low to high volatile bituminous coals show 7–16% ash, 8–32% volatile matter, and 0.6–2.6% R_{max}.

Galilee and Cooper Basins. The Galilee Basin, which is contiguous with the Bowen Basin but separated by the Springsure Shelf, contains >2 Gt of identified subhydrous, noncoking coal but little mining has taken place because of the remote geographic location and relatively unfavorable quality. The main potential is in seams which have been correlated with the Group IV coals of the Bowen Basin; other deeper seams in the Aramac Coal Measures have been intersected during petroleum exploration.

Coal resources of the Cooper Basin are estimated to be >1 Gt but mining has not been undertaken. Coal seams aggregating 25 m have been intersected in the Patchawarra Formation, a unit regarded as being of the same age as the Group I coal measures of the Bowen Basin. Other seams, aggregating >9 m have been intersected in the Toolachee Formation which are considered to be a correlative of the Bowen Basin Group IV coal measures.

Coal has also been intersected in the Mount Mulligan Basin but little resource determination has been carried out.

Triassic Coal. Compared to the Permian coal measures, there are few Triassic coal measures with only the Ipswich and Callide Coalfields being of significant importance.

Callide Basin. The Callide Basin is located in the central coastal area of Queensland and contains the Callide Coal Measures which contains four persistent seams. Three seams are mined in the Callide mine but only the Callide seam, the most economically important seam, is mined at Boundary Hill Mine. The Callide seam has a maximum thickness of 26 m but averages 16 to 18 m in most mining faces. The seam has two or three big splits. Total production from open pit mining in the Callide Basin is 4 Mt/a with total resources of >217 Mt open pit and 553 Mt underground (measured and indicated).

Callide coal is a medium ash, low sulfur high volatile bituminous coal which is mostly used for power generation in the domestic market; a small quantity is used in alumina smelting. Typical analyses show 10–12% moisture, 12–17% ash, 25% volatile matter, and a specific energy of 22 MJ/kg.

Ipswich Basin. Coal from the Ipswich Basin, located on the southern outskirts of Brisbane, has been an important feedstock for power generation for many years and is also exported. Coal seams occur in the Tivoli and Blackstone Formations of the Brassal Subgroup of the Ipswich Coal Measures. More than 20 seams have been worked at various times during the history of the coalfield. Seams are typically banded with many bands lenticular resulting bord and pillar underground methods although two open pit mines are operating. Annual production is 1.2 Mt from a total resources of 6 Mt open pit and 571 Mt underground (measured and indicated).

Tarong Basin. The Tarong Basin is a small fault-bounded basin 190 km northwest of Brisbane and contains the Tarong Beds. This unit has 6 seams of which 3 produce 5.1 Mt of coal per year by open pit methods. The seams, which have been correlated with the Tivoli Formation of the Ipswich coalfield, are up to 24 m thick and total resources are given as 431 Mt (measured and indicated).

Tarong coal is a high volatile bituminous coal which assays at 28% ash, 27–28% volatile matter, specific energy of 19.4 MJ/kg, and 0.6% R_{max}. Use of the coal is exclusively in a nearby power station.

Jurassic Basins. Jurassic coal measures are widely distributed in Queensland and have identified resources second only to the Permian coal measures. Apart from the Moreton Basin, mining has not been undertaken in most basins because of the lesser quality and inaccessibility of the coals compared to the Permian coals.

The Moreton Basin contains the Walloon Coal Measures which have been extensively mined in the Rosewood-Walloon Coalfield by underground methods in the past. Present production is 2.5 Mt/a from open pit and underground mines. Fifty four thin seams have been identified and up to 34 seams have been worked in open pits; only 6 seams, the thickest 6 m, have been worked by underground methods. Total resource for the Moreton Basin is 2.166 Gt open pit and 7 Mt underground (measured and indicated); current minable resource is given as 88 Mt open pit and 7 Mt underground.

The coal is a high volatile bituminous thermal coal with of 0.6% R_{max}. Assays show 13–18% ash, 35–42% volatile matter, and a specific energy of 27–28 MJ/kg. Domestic and export markets are supplied.

Walloon Coal Measures are also found in the Surat Basin where 2.65 Gt of resource has been identified. Mining has not and is unlikely to occur.

The Eromanga Basin and Mulgildie Basins also contain 2.802 Gt and 110 Mt resources in the Injune Creek Group and Mulgildie Coal Measures respectively. Mining occurred in the Mulgildie Basin in the past but has now ceased.

Cretaceous Basins. Noncoking coals are found in the Styx Basin (4 Mt resource), Maryborough Basin and Laura Basin (157 Mt; all underground measured and indicated category), and Cretaceous units of the Eromanga Basin but mining has not taken place.

South Australia

Of the vast widely distributed resources of coal in South Australia, ranging in age from Permian to Tertiary, only the Jurassic Leigh Creek deposit has been exploited; all production is transported by rail to Port Augusta for domestic power generation. Mostly the coals are of low rank and poor quality commonly with very high impurities and suitable only for thermal power generation. Coking coal for the small smelting operations at Whyalla and Port Pirie are imported by ship from the eastern states and thermal coals are also imported for some of the outlying towns.

Coal was first mined in South Australia in 1943 as a result of difficulties experienced in bringing coal from New South Wales to South Australia; it is now a significant part of the state's energy resource and will play an increasing role within the next few decades with consumption likely to grow as a result of new generating capacity that is required by the year 2005.

The substantial resources of Permian coals of the Arckaringa Basin (Table 12) in the central north of the state, are located in six deposits which lie below flat-lying, poorly-consolidated, water-saturated Jurassic and Cretaceous sandstone and mudstone of the Eromanga Basin. The coal occurs in the upper part of the Early Permian Mount Toondina Formation and is interbedded with siltstone, carbonaceous mudstone and minor sandstone.

The Jurassic Lock deposit has resources of 230 Mt of low grade (approximately 25% ash) subbituminous coal generally 35 to 230 m below the surface but commonly less than 130 m. The deposit is 2 to 4-km wide and 15-km long with a cumulative coal thickness of up to 17 m. Poor coal quality, seam variability and groundwater aquifers above and below the seams make this deposit unattractive.

South Australia has vast resources of low rank brown coal of Tertiary age with high moisture, sulfur, chlorine, and sodium, which commonly causes slagging, fouling and corrosion of boilers. Many of the deposits are shallow but poor quality and problems associated with mining make these coals unattractive.

Leigh Creek. Leigh Creek (Fig. 27**a**) is South Australia's only commercial coal project and has an annual production of 2.6 to 3 Mt. Exploitation is by open pit methods using draglines to remove 17 Mm^3 (37 Mt) of overburden annually at a strip ratio of 5.9 cubic meters of overburden per ton of coal. Overburden comprises consolidated rock and thus removal of the waste necessitates drilling and blasting. During early operation after the mine was opened in 1943, dips of the coal remained less than 5° but in more recent operations, dips have been increasing and existing mining practices will see exploitation of 70 to 100 Mt of coal up to depths of 150 to 200 m. To extend the life of the mine, it is planned to introduce terrace or haulback mining to expose the coal and use the overburden as fill in previously mined areas.

The coal is entirely used for domestic power generation at the Port Augusta power station, 225 km south of Leigh Creek.

The deposit comprises four small discrete basins:

1. Lobe A (Copley Basin) has a resource of 11 Mt but not mined as yet.
2. Lobe B (Telford Basin) is the largest of the basins and contains 500 Mt in three seams (Upper Series and Lower Series with up to 8 m of coal and the Main Series with coal 6 to 18-m thick) interbedded

Table 12. Coal Tonnages and Selected Properties of South Australian Coals

			Tonnage, $\times 10^6$		Proximate Analysis					
Deposit		Rank	Measured	Inferred	Moisture	Ash	Volatile	Fixed Carbon	Specific Energy	Sulfur, %
Archaringa Coalfield	Wintinna		1870	2270	38	6	23	33	17.5	1.1
	East	Sub-bit	975	1200	38	5	23	34	17.6	0.5
	Wintinna		940	2690	37	9	23	32	16.1	1.8
	Murloocoppie Westfield		313	470	38	6	22	34	16.7	2.7
Weedina		Sub-bit	1200	6000	38	8	22	32	16.5	0.5
Lake Philipson		Sub-bit		5000	35	11	25	29	16.2	0.7
Leigh Creek	Copley Basin	Sub-bit	11		37	12	20	31	15.2	2.9
	North Field		12		29	22	22	27	13.3	3.0
	Telford Basin		150	350	31	13	21	35	15.2	0.5
Lock		Sub-bit	320		26	23	30	21	14.6	0.4
Northern St Vincent Basin	Bowmans		1250	350	56	6	21	17	10.6	2.6
	Clinton		340	440	53	9	18	20	9.4	1.9
	Beaufort	Brown	255	45						
	Whitwarta		145	185	55	12	19	14	9.4	2.6
	Lochiel		625		61	6	19	14	9.1	1.1
Kingston		Brown	985		53	7	22	18	10.6	1.5
Anna		Brown	58		54	11	21	14	9.9	1.8
Sedan		Brown	184		58	9	19	14	9.4	2.3
Moorlands		Brown	32		55	9	18	18	9.9	1.8

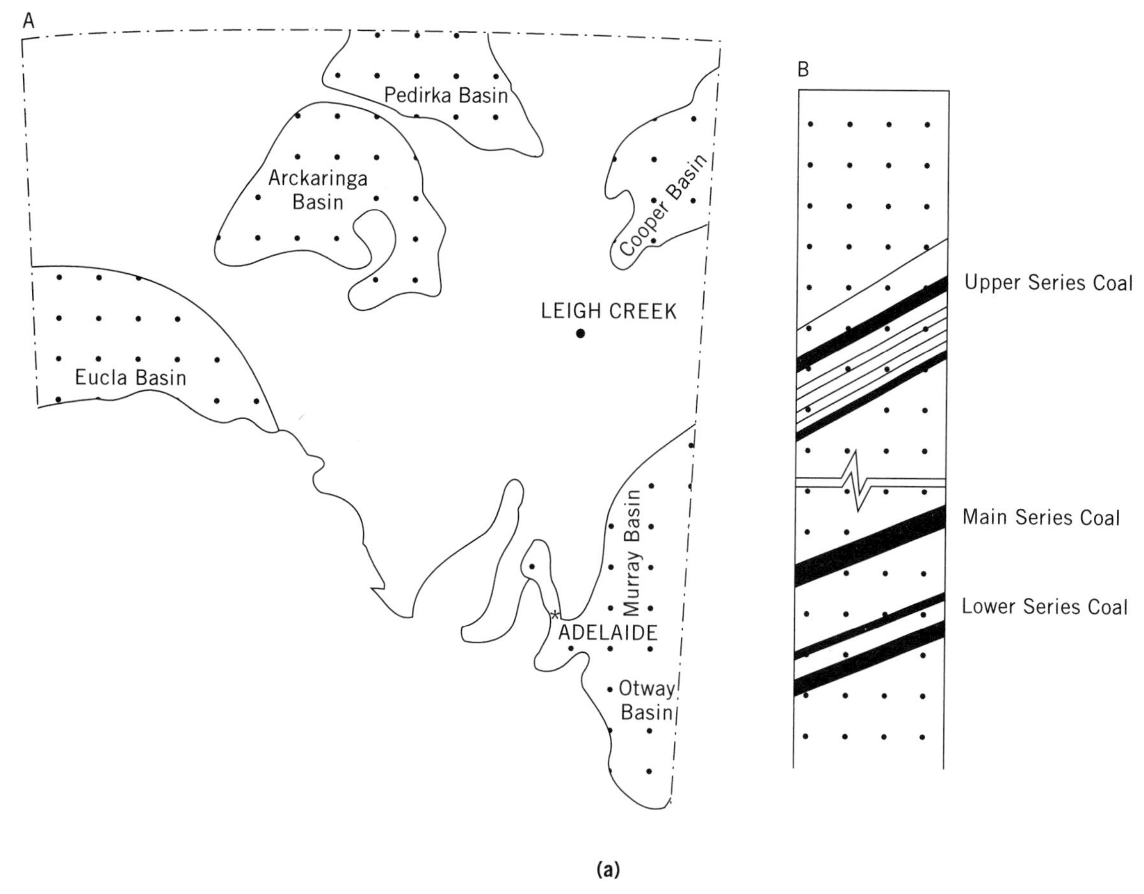

Figure 27. Coalfields of (**a**) South Australia: A. Location of basins: B. Stratigraphy of the Leigh Creek deposit. (**b**) Tasmania: A. Location of the Tasmania Basin; B. Subdivision of the Parmeener Group; C. Stratigraphic section, Fingal Tier. (**c**) Victoria coalfields: A. Location of main coal areas; B. Sketch map of the Gippsland Basin; C. Stratigraphy of the Latrobe Valley area.

with mudstone and siltstone; coal from the two top seams is the only production at present.

3. Lobes C (North Field) has one seam 1.6 to 16 m thick; 8.4 Mt of the 20 Mt resource have been mined.
4. Lobe D (North Field) has two seams with the upper one containing a lower seam 6 to 8 m thick separated from an upper 9 m seam by 10 m of grey shale; the total resource of 20 million tonnes had been mined out by 1976.

Leigh Creek coal is a low grade (28% ash) subbituminous with 27% moisture, 26–27% volatile matter, and 0.7% sulfur. Specific energy is 18.8 MJ/kg (dry basis).

Tasmania

Coal measures are found in two Permian intervals and one Triassic stratigraphic interval in the Tasmania Basin (Fig. 27**b**). Early Permian coal occurs in the Mersey and Preolenna Coal Measures and Late Permian coal occurs in the Cygnet and Adventure Bay Coal Measures of the Parmeener Supergroup which is found in the northern part of the island state. Production from these coals commenced in the latter half of the nineteenth century but ceased in the 1960s. Present resources are unknown but are thought to be small. The coals have very high sulfur (up to 6%) and further exploitation is unlikely.

Largest coal resources are in the Triassic coalfields in the northeast, east and southeast of the state with the most important areas being the Fingal-Mt Nicholas-Dalmayne Coalfields with resources of 250 Mt, 50 Mt inferred, and 160 Mt (measured + indicated). Both the Fingal and Dalmayne Coalfields also have large inferred resources (116). Other measured + indicated resources in Triassic coalfields amount to 70 Mt.

Eight main seams, designated A to H, have been recognized in the Triassic coal measures. Although faulting is common and splitting occurs, the seams can be correlated across the Fingal area. Seams of most economic importance are the Duncan and East Fingal seams. The Duncan seam is the most extensively worked seam and varies from 2 to 3 m with minor claystone and mudstone partings.

The coals are high ash coals (commonly 18–22% but up to 30%), low sulfur (<0.5%), low moisture (3–5%), and medium volatile (26–30%) with low calorific value (23–28 MJ/kg) (117). Only three collieries operate in Tasmania and total annual production is approximately 0.6 Mt with 0.4 Mt saleable. Mining is by underground bord (room)

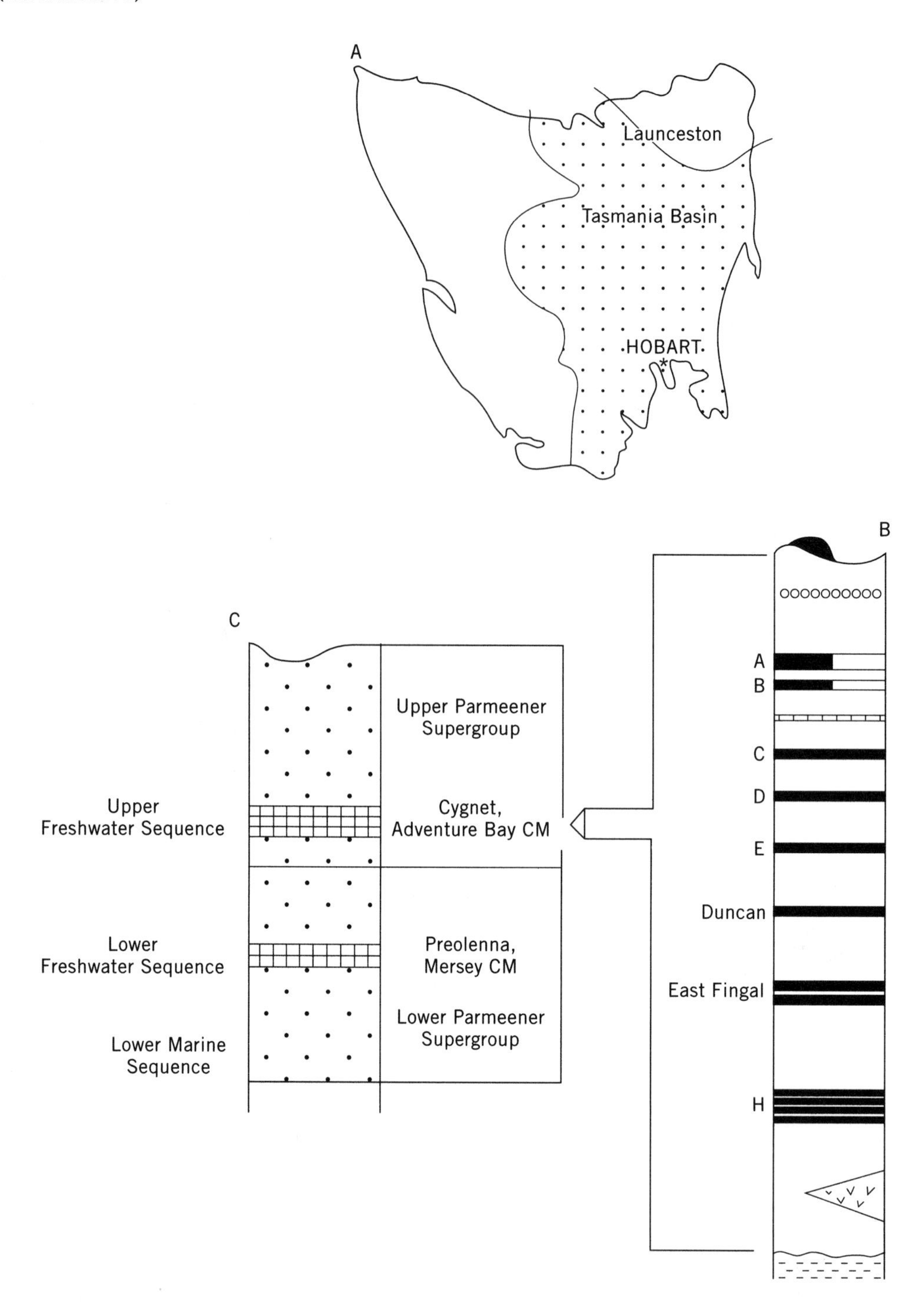

Figure 27. *Continued.*

and pillar methods although longwall mining has been attempted at some mines in the past. The coal is a steaming coal exclusively marketed domestically for use in boilers, cement manufacture and paper mills.

Victoria

Victoria contains Australia's largest resources of Tertiary brown coal. Mining of Cretaceous black coal in several state-owned mines was also carried in the Wonthaggi Coalfield in the past. Mining in the Wonthaggi Coalfield ceased in 1970 (118) after a total production of 22.8 Mt of coal. Drilling programs during the lives of the mines indicated that up to 9 Mt of coal probably occurs adjacent to the mined areas but future exploitation is unlikely. Typical coal properties for Wonthaggi coal include 6–11% moisture, 6–11% ash, 30–35% volatile matter, and a specific energy of 26–29 MJ/kg (dry basis). Seams were a maximum thickness of 3 m and were worked to depths of 440 m.

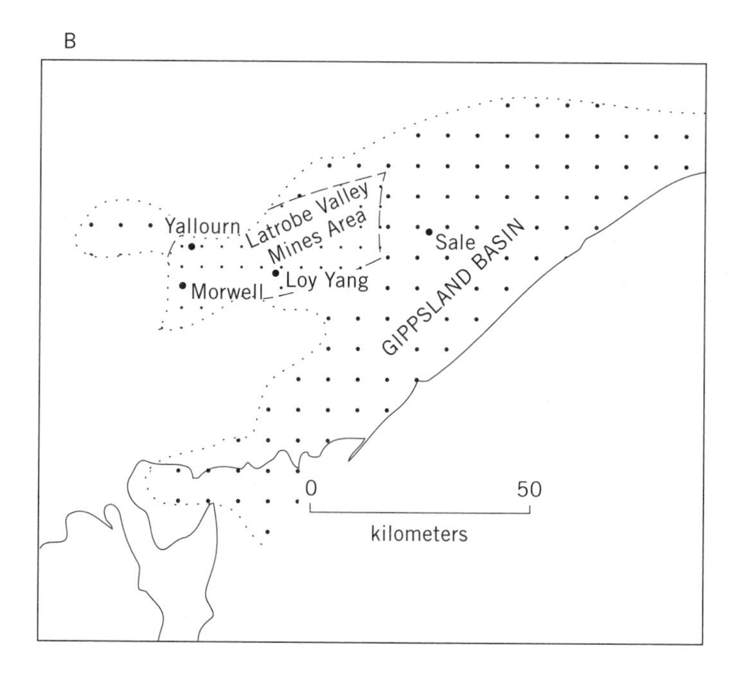

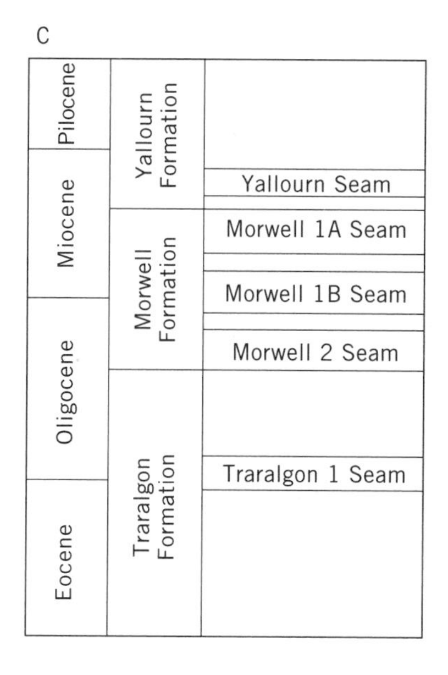

(c)

Figure 27. *Continued.*

Tertiary Coals. ***Gippsland Basin.*** Victoria has at least 98 Gt of brown coal of which 65 Gt has been projected for future mining in several coalfields with the largest being the Latrobe Valley of the Gippsland Basin (Fig. 27**c**). Other brown coal deposits are mined at Anglesea, Bacchus Marsh, Altona and a large deposit has been delineated at Gelliondale in the southeast Gippsland Basin, approximately 50 km south of the Latrobe Coalfield. Gelliondale resources include 420 Mt measured open pit reserves, with coal to overburden ratio varying between 4.7:1 and 2.9:1, in the central part of the coalfield; an additional 500 Mt of indicated resource has been delineated.

Drilling in the Gelliondale Coalfield has shown a single seam with an average thickness of 40 m but ranging between a few meters and 134 m. The coal is overlain by carbonaceous shale and sandy claystone. The coalfield has relatively simple structure and there appears to be little evidence of significant faulting (118).

Proximate analysis of the coals shows: Moisture, 67%; Volatile matter, 16%; Fixed Carbon, 15%; Ash, 2.5%; Sulfur, 1.1%; and Calorific value, 25 MJ/kg (dry basis).

Latrobe Valley Coalfield. The Latrobe Valley Coalfield lies within the Latrobe Valley Depression which is the onshore extension of the Gippsland Basin. The 770 m Tertiary coal-bearing sequence comprises claystone, coal and semi-consolidated to unconsolidated silt, sand and gravel. The total resource of the Latrobe Valley is estimated to be at least 65 Gt with a further 43 Gt indicated. Of the proven resources, mining reserves have been measured at 36 Gt with less than 30 m overburden above the uppermost seam, of which 12 Gt is deemed to be economic given present day costs.

Mining is by open pit methods using electrically-operated dredges for both overburden and coal winning. Belt conveyors transport the coal to nearby power station storage bunkers and overburden to disposal areas. Production is used almost exclusively for thermal power generation with a small quantity sold to other domestic users or converted to briquette in an onsite installation. Present production is from three large mines; annual production figures are Loy Yang, 17.5 Mt; Morwell, 15 Mt; and Yallourn, 17 Mt. Production at Loy Yang is likely to double to 32 Mt/a when the second Loy Yang 2000 MW power station is built.

The coal measures are part of the Latrobe Valley Group which is divided into three formations, each containing at least one coal seam. The seam are extremely thick ranging from a minimum thickness of 40 m to a maximum of 130 m. Seam thickness is related partly to penecontemporaneous basement structural movement (folding and faulting) and partly to localised postdepositional erosion.

The Latrobe Valley coal is a low ash, high moisture, soft brown coal. Representative properties are given in Table 13.

Anglesea Coalfield. The Anglesea Coalfield is located south of Melbourne near the provincial city of Geelong. The coal bearing sequence is 140 m thick and comprises a 20 to 45 m upper seam and three to six thinner seams of variable thickness but with an aggregate thickness of up to 30 m; all seams thin to the east. At the margins of the coalfield the seams dip at 2 to 5° but are subhorizontal near the center. Anglesea coal is a soft brown coal that is harder and drier than the Latrobe Valley coal. Average analyses show 44% moisture, 4% ash (dry), 48% (dry) volatile matter, 26.5 MJ/kg (db) specific energy, and 3–4% sulfur.

Mining is by open pit methods; reserves are given as 160 Mt at a depth of <150 m of which 70 Mt is in the upper seam. Annual production is approximately 1 Mt per annum, all of which is used in local alumina smelting.

Bacchus Marsh Coalfield. Soft brown coal of similar quality to that of the Latrobe Valley coal is mined by open pit methods at Bacchus Marsh. Reserves are estimated to be 100 Mt although 25 Mt is under valuable irrigated land and may not be exploited. Average analyses show 59% moisture, 2.6% ash (dry), 20% (dry) volatile matter, 26 MJ/kg (dry) specific energy, and 2.8% sulfur. Production is for local industry including a paper mill (118).

Western Australia

Although coal of Permian age is found in the Canning, Collie and Perth Basins, it is mined only in the Collie Basin where both underground and open pit operations are used to exploit the coal.

Canning Basin is a large cratonic basin covering 430,000 km^2 on shore of Western Australia and 165,000 km^2 offshore. Widely distributed coal seams have been intersected in drill holes in several of the Permian units but none are commercial.

Collie Basin. Collie Basin is a small structurally-complex basin, comprising three sub-basins separated by a basement ridge and several faults (Fig. 28), surrounded by Precambrian crystalline basement. Coal is mined from the Collie Group of the Cardiff Sub-basin (containing up to 1500 m of Permian sequence) and Muja Sub-basin (containing 800 m of Permian sequence) but not the Schott Sub-basin (600 m of Permian sequence). Total resources have been estimated to be 1.915 Gt (120) of which approximately 350 Mt have been detailed sufficiently to be considered extractable. Total production from both sub-basins in 1991–1992 was 5.5 Mt (121).

Cardiff Sub-basin, with resources of 1.150 Gt (362 Mt measured), contains three coal-bearing units although coal is only taken from the upper two units. Approximately 30 seams have been intersected but many are thin and uneconomic. The thicker coal seams are relatively uniform laterally and with respect to chemical and physical properties. Historically, underground mining has been restricted because the sequence contains many aquifers with the deeper seams commonly bound by aquifers. Present underground operations work the 3.5 m Wyvern seam and the 2.2 m Collieburn seam, both part of the Collieburn Member. Open pit operations principally mine the shallower 2.4 m Neath seam and the 3.9 m Cardiff seam of the Cardiff Member with some mining of the 2.1 m Ewington No 2, 4.8 m Moira Upper, 2.3 m Stockton Upper, and 3.8 m Wallsend seams. Sales are mostly on long term contracts to the state energy authority, alumina plant and other domestic users. Increased production is

Table 13. Coal Quality Properties, Latrobe Valley Brown Coals[a]

	Yallourn	Morwell	Yallourn	Loy Yang	Gormandale	Gooungoolun
Seam	Yallourn	Morwell 1	North Latrobe	Morwell 1B	Traralgon	Traralgon
Moisture, %	66.8	60.9	52.9	62.5	56.0	53.1
Ash, % db[b]	1.8	3.2	4.8	1.5	2.5	2.6
Volatile matter, % db[b]	51.7	49.8	48.8	51.3	52.2	47.4
Fixed carbon, % db[b]	65.9	67.1	65.7	68.3	66.1	68.2
Hydrogen % db[b]	4.6	4.8	4.6	4.8	4.8	4.8
Specific energy, ash free MJ/kg	25.8	27.3	26.7	27.0	26.4	29.2

[a] Ref. 119.
[b] db-dry basis.

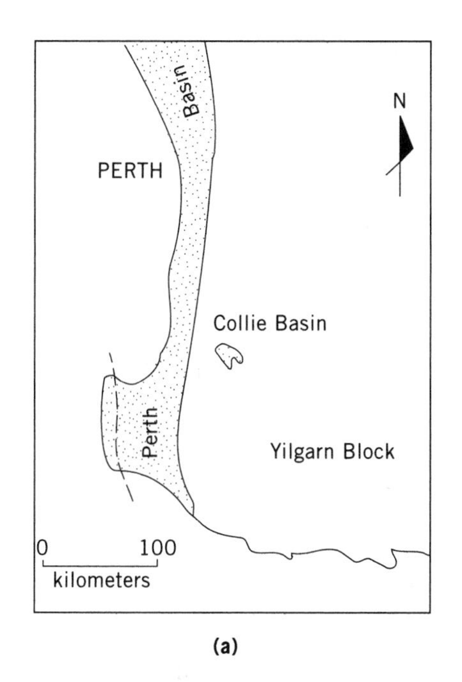

(a)

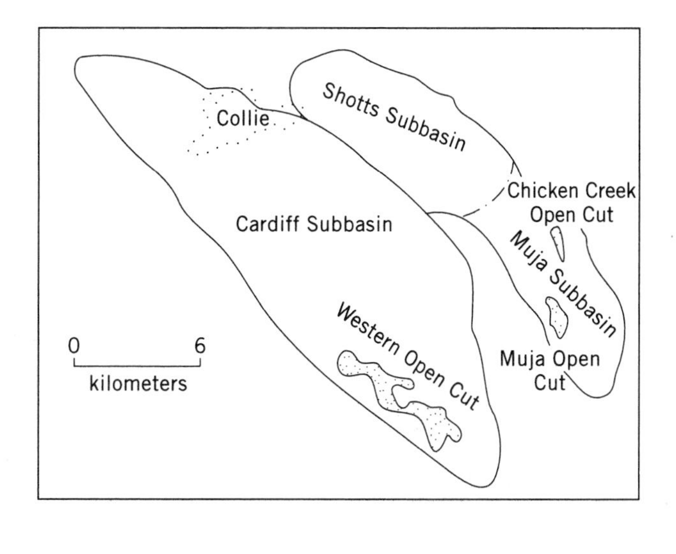

(b)

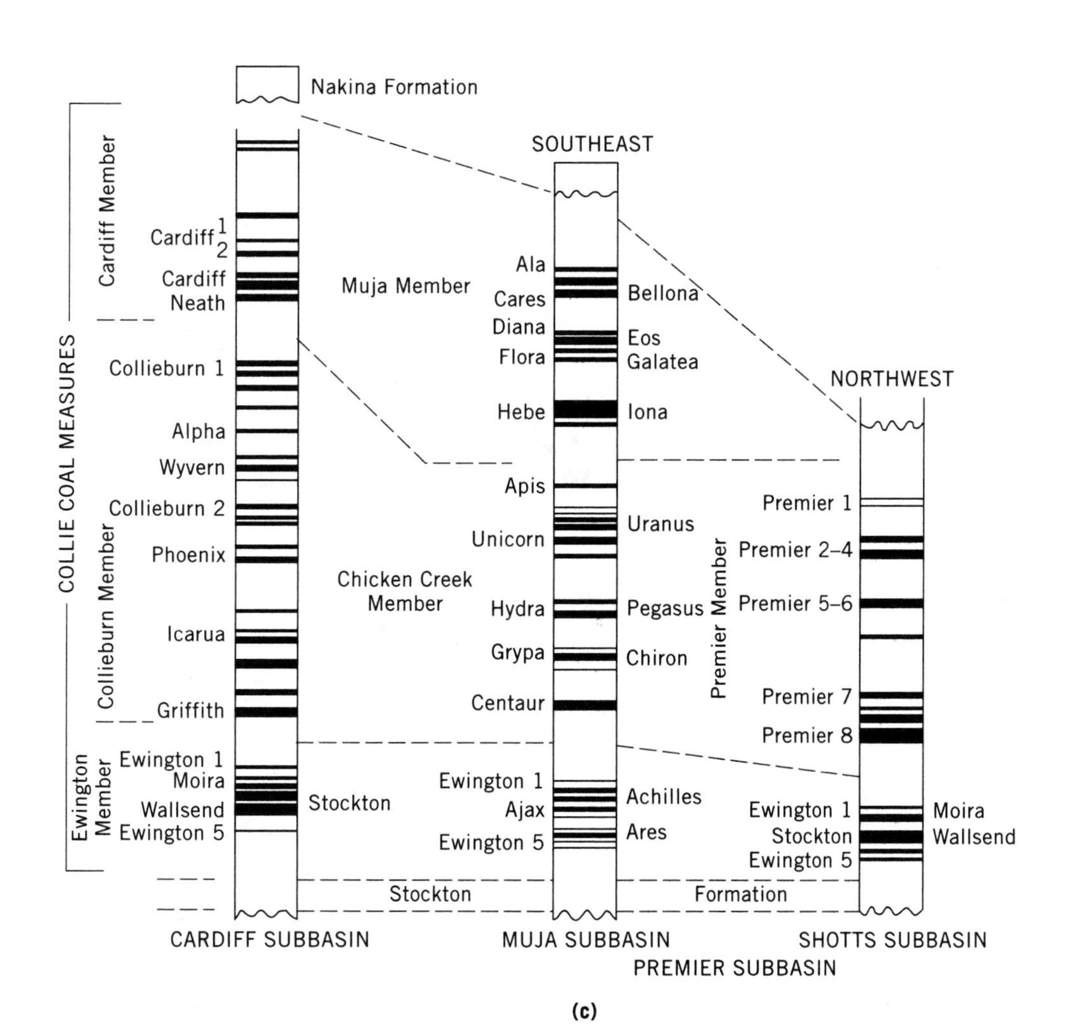

(c)

Figure 28. Coal basins of Western Australia. (**a**) Location of Permian coal-bearing basins; (**b**) Map of the Collie basin; (**c**) Stratigraphy of the Collie basin.

expected in the near future with additional collieries coming on stream as the underground operations decrease output.

Coal in worked by open pit methods in two mines in the Muja Member of the Muja Sub-basin. The largest mine, Muja, is a multiple seam operation in which the Hebe seam, up to 12 m thick, is the most important producer in a sequence of 9 minable seams aggregating 28 m of coal (120). A smaller operation, Chicken Creek Mine, takes coal from four seams, the most important being the Centaur seam which is approximately 3 m thick. Present resources of the sub-basin are given as 63 Mt measured minable. All sales are to the State Energy Commission with contracts up to the year 2010. Plans are to develop to a depth of 230 m.

Collie coal is a low ash, medium volatile bituminous, high moisture thermal coal with poor coking properties. The better quality coal occurs in the lower seams of the sequences and there is a slight deterioration in quality towards the southeast. Typical properties (120) are

Moisture %	26
Ash, %	3–7
Volatile matter, %	26–28
Fixed carbon, %	39–42
Specific energy (MJ/kg)	19–21
Sulfur, %	0.5–0.8

The coal is used mostly for domestic power generation (80% of production) located close to the coalfield. Other markets include sales for use in cement manufacture, an alumina smelter, synthetic rutile production and small sales to local domestic users. A government decision to have a privately-owned, coal-fired power generation plant as the next stage in Western Australia's energy scheme will ensure long term markets for the coalfield.

Other Coal Basins. Perth Basin contains 2.6 km of Permian sedimentary rocks including the 123-m thick Irwin Coal Measures and the Sue Coal measures which has a maximum thickness of more than 1.8 km. The Irwin Coal Measures is a sequence of very fine-grained sandstone, conglomerate, siltstone, carbonaceous claystone and intercalated beds and lenses of subbituminous coal. Although seams are up to 2 m thick, the depth to the seams and coal quality indicate that these coals are unlikely to be exploited.

The Sue Coal Measures contains sub-bituminous coal in seams up to 5.5 m thick and at depths as shallow as 111 m. Quality is similar to that of the coal in the Collie Basin. However, knowledge of the coal is thought not to be sufficiently detailed to permit exploitation.

Tertiary brown coal has been found in several basins including the Eucla Basin which is located in the south east of the state. The coal is of low rank and has a very high ash. Although at relatively shallow depth, these coals are unlikely to be exploited.

NEW ZEALAND

New Zealand is a relatively small producer of coal by world standards with most of its production used for domestic purposes and only small quantities exported. Most of the known coal resources (82% and most in the South Island) are of brown coal rank with approximately 14% subbituminous and 4% bituminous. Many areas of New Zealand have not been subjected to detailed feasibility studies and recovery rates are extremely variable ranging from as low as 18% in some underground mines to 95% in some open pit mines; typical underground recovery rates are approximately 30% (122).

Total coal resources in New Zealand are 15.7 Gt of inground coal of which 13.1 Gt are in the South Island and 2.6 Gt in the North Island. However the disparity in rank and quality of coal in the North and South Island is such that slightly more than 50% of production is from the North Island. Total recoverable resources are estimated to be 8.6 Gt of which 1.9 Gt is classed as measured but only 117 Mt is classed in the reserve category. Recoverable coal of 7.6 Gt is on the South Island and of this only 1.6 Gt is classed as measured; only 36 Mt is classed as reserves (122).

Most New Zealand coal production is of low to medium ash and low sulfur coals with approximately 35% of production in the low ash-low sulfur category where low ash is defined as <5% ash and high coal as >10% ash. Most presently operating mines have production restricted to one seam and commonly the central portion of the seam has <1% ash with higher ash in the floor and roof sections.

Peat occurs widely in New Zealand and is used for heat production. Resources and production are not currently available. Chatham Island peats have a high wax content and resources are large.

The coalfields of New Zealand have been divided into several geographic areas (Fig. 29) and commonly in some coalfields, seams are found in different stratigraphic units. Southland and Otago have the largest brown coal resources with Newvale Mine in the Waimumu Coalfield the largest brown coal producer (approximately 100 kt/a). The Huntly, Rotowaro and Maramarua Coalfields in the Waikato Region are the main sub-bituminous coal producers (1.3 Mt/a combined production) and the Stockton-Webb open pit mine in the Buller Coalfield is the largest bituminous producer (approximately 300 kt/a) most of which is exported (122).

Production on a region by region basis, and based on 1986 figures, is given in Table 14.

Main uses of New Zealand coal are: Export: 15% and Domestic: steel production, 21%; primary processing, 21%; electricity generation, 13%; cement manufacture, 8%; institutions, 8%; domestic uses, 6%; other industries, 8%.

North Island

The North Island has relatively few resources compared to the South Island but production is greater. Three coal regions have been identified, the Northland, Waikato and Taranaki Coal Regions. Production is mostly from the Waikato and Taranaki Coal Regions.

Northland Coal Region. Five coalfields are located in the Northland Coal Region which is the northernmost coal re-

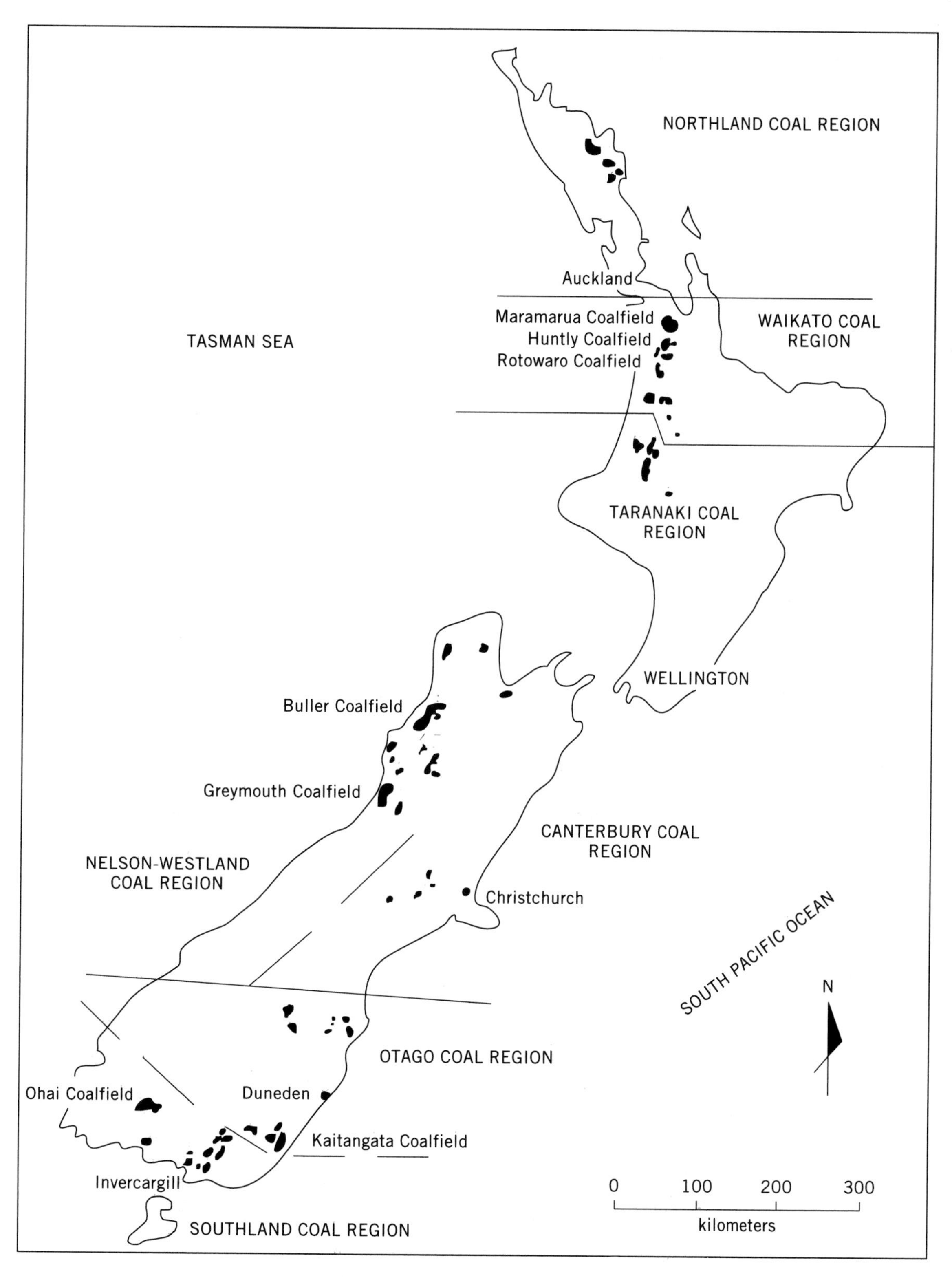

Figure 29. Coalfields of New Zealand.

gion of New Zealand. Coal seams occur in the Late Eocene Kamo Coal Measures at Kawakawa, Hikurangi and Kamo, in a structurally complex area at Avoca, and in a fault angle depression at Kiripaka. Underground mining commenced in the 1860s but virtually finished by the 1920s at Kawakawa and Kiripaka. It continued at Kamo until 1955 with a small amount of open pit mining after the underground mines closed. Total production was estimated to be 7.5 Mt.

Several small scattered coalfields were explored during the 1981–1982 but significant resources were not found.

Peat and brown coal deposits at North Kaipara and Aupouri contain significant resources but are not considered to be economic at this stage.

Table 14. Production for New Zealand Coal Region[a]

Region	Mine Type	Production, kt	Total	Resources			
				Measured	Indicated	Inferred, Mt	Total
Waikato	Underground	420		64.9	6.6	0.5	71.9
	Open pit	906	1324	16.2	0.2	0.1	16.5
Taranaki	Underground	10.2					0.4
	Open pit	23.4	33.6	0.2	0.2		
North Island							
	Underground	427					
	Open pit	930	1357				
Nelson-Westland	Underground	282		3.5	7.6	2.4	13.5
	Open pit	459	741	23.8	3.6	1.5	28.9
Canterbury	Underground					0.4	0.4
	Open pit	3.8	3.8	<0.1		0.3	0.3
Otago	Underground				<0.1		<0.1
	Open pit	80	80			4.2	4.2
Southland	Underground	170		2.5	4.5	4.0	10.5
	Open pit	166	336	6.4	4.0	12.1	22.6
South Island							
	Underground	451	1161				
	Open pit	708					
New Zealand							
	Underground	879					
	Open pit	1639	2518				

[a] Ref. 122.

Waikato Coal Region. The Waikato Coal Region is the largest coal producing coal region in the North Island and produces almost 1.5 Mt of coal annually, more than half the national output, and has total resources of 88.4 Mt. The Huntly and Rotowaro Coalfields are the largest producers, currently supplying both of New Zealand's coal-fired power stations; Huntly West mine is dedicated to Huntly power station and Kopuku (Maramarua) is dedicated to Meremere power station. Other mines also supply the power stations. Expected expanded steel production and increased electricity demand could see increases in the future, and especially in the long term, if natural gas now being used is depleted in 30 years.

The economic seams are part of the fluvial to fluviolacustrine Waikato Coal Measures which are of Late Eocene age. The coals are mainly of subbituminous rank. Inferred recoverable resources are 860 Mt.

Taranaki Coal Region. The Taranaki Coal Region produces approximately 35 kt/a of subbituminous coal, mainly for industrial and domestic use including several single point consumers such as dairy factories. Future uses are the Aukland steel industry and power station feedstock. The Mokau and Waitawhena are the largest producers. Inferred recoverable resources are 173 Mt.

The seams are part of the marginal marine Maryville Coal Measures which are between the marine and marginal marine Lower and Upper Mokau Sandstone.

South Island

The South Island has been divided into four geographic coal regions which produce a total of 1.16 Mt.

Nelson-Westland Coal Region. The Nelson-Westland Coal Region extends from the northwestern tip of the South Island along the western coast and is the largest producer of the four South Island regions with a total production of 74 kt/a. This region is one of the main producers of coking coal which was formerly used for domestic town gas and coke but is now exported for metallurgical uses. Domestic uses include the Westport cement plant and other industrial and domestic uses. The coals are mostly high to medium volatile bituminous coals with some subbituminous coal and are characterized by high specific energies.

Coal is found in three stratigraphic intervals:

1. Miocene Rotokohu Coal Measures: basal fluvial to marginal marine coal measures with some lenticular seams.
2. Eocene Brunner Coal Measures; basal fluvial to marginal marine coal measures with lenticular to extensive seams.
3. Upper Cretaceous-Paleocene Paparoa, Pakawau and Tauperikaka Coal Measures; fluvial to fluviolacustrine intermontane coal measures, commonly in rift basin with seams of variable continuity. The Paparoa coals are generally of higher rank than other Cretaceous coals (123).

The biggest coalfield is Buller Coalfield which produces >450 kt/a from the Brunner Coal Measures. Other significant producers are Inangahua Coalfield (40 kt from Rotokahu and Brunner Coal Measures), Reefton Coalfield (100 kt from Brunner Coal Measures), Garvey Creek (50 kt from Brunner Coal Measures) and Greymouth (75 kt from Paparoa and Brunner Coal Measures). Other coalfields with minor production are Collingwood, Charleston, Flat Creek and South Westland.

Canterbury Coal Region. Canterbury Coal Region is a minor producer with only a few thousand tonnes per year all of which is used by local domestic and industrial consumers. Coals are mostly subbituminous rank with small resources of semianthracite.

The seams occur at the base of transgressive fluvial to marginal marine sequences and seams are typically lenticular. Mt Somers is the main coalfield. Other coalfields are Avoca, Acheron and Malvern Hills.

Otago Coal Region. Otago Coal region produces approximately 80 kt of brown to subbituminous coal annually from the Kaitangata Coalfield (55 kt) and Central Otago Coalfield (10 kt). Inferred recoverable resources are estimated to be 1.086 Gt. Consumers are mostly local industrial and domestic users. Possible future uses are liquid fuel conversion.

Coals of Kaitangata Coalfield occur in the Upper Cretaceous to Paleocene Taratu Formation; multiple seams occur in laterally variable cyclic sequences of fluviolacustrine origin with marginal marine influence near the top. Coals of the Central Otago Coalfield occur in the Miocene Manuherikia Group which is a fluviolacustrine unit below a regressive sequence.

Southland Coal Region. Southland Coal Region is the second significant producer with a total annual production of 340 kt from the Ohai Coalfield (166 kt) and Main Eastern Southland Coalfield (184 kt). Total, inferred recoverable resources are estimated to be 6.051 Gt. Subbituminous to high volatile bituminous Ohai coal has very low sulfur and ash and high specific energy making it economic even if transported as far afield as Christchurch for industrial and domestic use. Eastern coal is of brown coal rank is used locally for paper manufacture and other industrial and domestic uses. Future uses may include liquid fuel conversion.

The seams of Ohai Coalfield are part of the Upper Cretaceous, fluvial to fluviolacustrine Morley Coal Measures which were deposited in rift basins which are unconformably overlain by the Eocene Beaumont Coal Measures.

Drilling in the offshore Great South Basin intersected 2000 m of coal measures with 194 of coal seams with an aggregate thickness of 500 m (124). The coal are of high volatile bituminous rank with sulfur content increasing towards the top of the seams.

INDONESIA

Estimates of the total coal resources of Indonesia have increased rapidly over the past five years changing from 29 Gt (125) of which 12.96 Gt was regarded as demonstrated reserves with 2.39 Gt considered proven increasing to 36 Gt (Table 15). Although coal deposits are widely distributed throughout the archipelago except in Nusatenggara, Maluka and northern Sulewesi, economic deposits are mostly in the Miocene to Eocene strata in Sumatera, Kalimantan and Java and Pliocene to Pleistocene strata in Sumatera (Fig. 30). Indonesia's coals span the spectrum of ranks with at least 14 Mt of anthracite, 760.2 Mt of bituminous coal, 4.4 Gt of sub-bituminous coal and 18 Gt of brown coal (126).

Coal has been known in Indonesia for many centuries but mining only commenced in 1849 in Kalimantan with exports of coal not commencing until the 1930s when coal was sent to neighbouring countries. Peak production of 2.0 Mt was reached in 1941, mostly export, followed by a significant decline after World War II. Following a Presidential Instruction in September, 1976, coal exploration

Table 15. Indonesian Coal Resources, May 1993[a]

Island	Region	Measured	Resources Indicated	Inferred	Hypothetical	Total
Sumatera	Total	2,887.9	11,165.5	2,279.9	8,343.0	24,676.2
	North		1,272.0	2.0	433.0	1,707.0
	Central	717.8	2,370.9	58.0	1,019.0	4,165.8
	South	2,143.0	7,505.5	2,204.0	6,891.00	18,743.5
	Bengkulu	27.0	17.0	15.9		60.0
Kalimantan	Total	1,985.6	1,494.3	3,789.5	4,231.2	11,500.6
	South	1,112.7	668.2	1,848.3		3,629.2
	West	1.8	68.9	211.5	1,837.9	2,120.1
	Central				436.3	436.3
	East	871.1	757.1	1,729.8	1,957.0	5,315.0
Java		12.1	28.6		20.0	60.7
Sulewesi		5.8	12.2	6.6		24.3
Irian Jaya			79.5	3.6		83.1
Total		*4,891.1*	*12,780.0*	*6,079.6*	*12,594.2*	*36,345.5*
Production 1965–1993						

[a] Data from Direktorat Batubara, Indonesia.

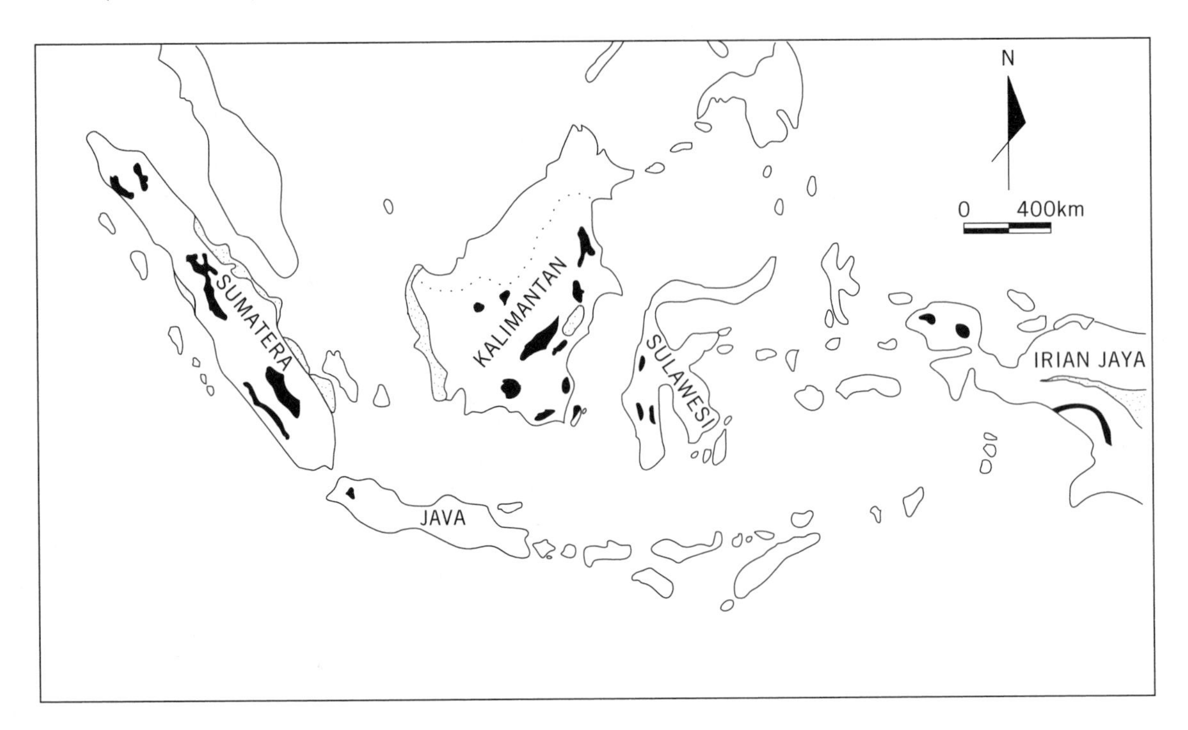

Figure 30. Main coal basins of Indonesia. ■, Coal; ▩ peat.

greatly increased especially in the early 1980s. By 1989, 4.6×10^6 hectares of land in Sumatera and Kalimantan had been covered by exploration programs under the supervision of Perum Tambang Batubara, the state-owned coal enterprise (125). Coincidental with the exploration programs was increased exploitation both on the domestic scene (from a base of 240 kt in 1983, mostly for electricity generation, cement manufacture, railways, metal processing and steam generation) and especially exporting of coal as an earner of foreign exchange. Coal production again increased in the 1980s reaching and 5.1 Mt in 1988 (1.1 Mt exported).

The dramatic increase in production continued with the development of large mines in Sumatera and Kalimantan in 1990–1992. Production in 1992 was 23.62 Mt with 14.1 Mt from Kalimantan and 9.5 Mt from Sumatera and Java (127). Sumatera produced more than 5 Mt for the giant Suralaya power station in 1991 (126). With additional mines coming onstream in Kalimantan in the next few years, exports are expected to steadily increase. Projected production for 2000 is 50 Mt/a with 36 Mt from Kalimantan.

Principal importers of Indonesian coal have been Malaysia, traditionally the most consistent purchaser of Indonesian coal, Thailand, Singapore, Bangladesh and the Korea. The Japanese market is increasing from a base level of 320 kt of coking coal and 51 kt of steaming coal in 1986. Indonesia is strategically placed to further increase its share of Asian coal imports in the future especially as some countries decided to diversify coal supplies in order to have assured supplies. Current Asian consumption of coal is 233 Mt/a and this is expected to increase to 600 Mt/a by 2000. As an example, Japan's long term predictions are to increase coal consumption from 113.6 Mt in 1989 to 142 Mt in year 2000 with the figure remaining steady until 2010.

Indonesian coal production represents only 0.2% of world production but in South East Asia, Indonesia is the largest coal producer. Production growth rate for Indonesia, 14.1%, are much higher than for neighboring countries such as Australia (6.8%) and China (1.2%).

Increases in domestic consumption are linked to the construction of new coal-fired power stations and expansion of cement manufacture which produced 17.4 Mt of cement in 1988. Increases in coal consumption due to small industry and household needs is restricted because of the need for a much improved transportation system. Additional markets could be forthcoming if gasification and coal liquefaction become significant energy sources. Low labor costs and the active encouragement of foreign capital and technology, thus reducing domestic risk in new ventures, are further incentives for an optimistic Indonesian coal industry in the 1990s and beyond. Domestic consumption in 1992 was 5.3 Mt for power generation, 2.6 Mt for cement manufacture and 400 kt for other uses; total consumption was 8.3 Mt. Expected consumption in 2000 is projected to be between 15.4 and 28.9 Mt for electricity generation, 3.2 Mt for cement manufacture, and 7.4 Mt for other uses; total consumption is projected to be 39.5 Mt.

In summary, medium to low grade coals from South Sumatera will be reserved for domestic use which is expected to increase dramatically whereas the medium (low-ash, low-sulfur) and high grade coals from West Sumatera and East Kalimantan will serve export needs.

Indonesia has large deposits of peat totalling 16.1 million hectares concentrated mostly in eastern Sumatera (60%), Kalimantan (39%) and Irian Jaya. Thickness are generally approximately 2 m although one deposit has a recorded thickness of 16 m. Calorific values are approximately 20 MJ/kg (dry equivalent) (127). Most consumption is restricted to local household uses.

Sumatera

Sumatera has the largest coal resources in Indonesia with measured reserves of 1.23 Gt, 42.0 Gt indicated/inferred, and 18.3 Gt hypothetical (125).

North Sumatera: Meulaboh Coalfield. Total resources of the Central Sumatera region is given as 1.27 Gt indicated/inferred and 428 Gt hypothetical. Coal exploration in the Meuloboh Coalfield, located in the West Aceh Basin, commenced in 1984 with a preliminary survey by the Coal and Peat Division of the Directorate of Mineral Resources (Indonesia) with the Pliocene coals especially targeted. Resources of at least 500 Mt of Pliocene coal (brown coal rank) were identified with additional resources of Neogene coal.

Analyses of the Pliocene coal gave 5–12% moisture, 35–45% volatile matter, 2–5% ash, and 40–48% fixed carbon. Sulfur was <1.7% and calorific value ranged between 20.9 to 28.4 MJ/kg. Petrographic analyses gave 70 to 90% vitrinite, 8–20% liptinite, and 1–5% inertinite, with vitrinite reflectance of 0.45–0.7%.

Analyses of the Neogene coal indicated higher moisture, volatile matter, ash, and liptinite content than the Pliocene coal but lower sulfur and vitrinite reflectance.

Utilization of North Sumatera coals is likely to be for domestic use, especially power generation.

West Sumatera: Ombilin Coalfield. Ombilin Coalfield is located in the structurally controlled Ombilin Basin of northwest Sumatera. Total resources are given as 181.6 Mt measured and 27.4 Mt indicated/inferred.

Mining commenced on a very small scale in 1892 after initial exploration by the Dutch in 1868. Data were sparse and the Perum Tambang Batubara embarked on a major exploration program, including extensive geophysical surveys, and completing a 400 hole drilling program totalling 61.3 km. Underground mines are operated in the Sawah Rasau area and open pit mines in the Tanah Hitam area. The large Ombilin project is a three-stage project around the Ombilin mine and envisages a fully mechanized longwall mine as well as open pit mines.

Coal is found in the Sawahlunto Formation which can be divided into a lower Tandikat Member which has very thin coal and carbonaceous shales designated D seam and an upper unnamed member which contains A seam (averaging 2 m) B seam (averaging 0.6 m) and C seam (averaging 6 m). Ombilin coal is closely cleated with a conchoidal fracture. All coals are vitrinite rich (80–85%) with 5–8% liptinite and 3–4% inertinite and are high to medium volatile bituminous coals with vitrinite reflectances of 0.7–0.8%. Proximate analyses given 2–4% moisture, 25–43% volatile matter, 30–55% fixed carbon, and 2–30% ash. Sulfur is variable, ranging from 0.5–3%, and calorific values are 20.9–29.3 MJ/kg. Washability is reasonably good but the high vitrinite content imparts poor coking qualities but good thermal properties.

Central Sumatera Basin: Sinamar, Mampun Pandan and Cerenti Coalfields. Total resources of the Central Sumatera region is given as 524 Mt measured, 1.3 Gt indicted/inferred, and 700 Mt hypothetical.

Limited exploration has been carried out in the Sinamar and Mampun Pandan areas of Central Sumatera by the Indonesian Government in collaboration with a private company PT Rio Tinto Betlehem Indonesia. Resources of 210 Mt (indicated/inferred) with 109 Mt at Sinamar (70 Mt recoverable) with an average depth of 75 m, seam thickness of 11.5 m and over an area of 7 km^2. Mampun Pandan Coalfield has 53 Mt at an average depth 51 to 56 m and seam thicknesses ranging between 5.4 and 11 m.

Coal from these coalfields analyzed at 4 to 8% moisture, 35 to 42% volatile matter, 30 to 40% fixed carbon, 3 to 15% ash, and sulfur less than 1%. Calorific values ranged between 21.8 and 32.2 MJ/kg. Vitrinite reflectance ranged between 0.5 and 0.75% with 80 to 95% vitrinite, 1 to 15% liptinite, and <1% inertinite. Likely use is domestic electricity generation and general combustion.

Exploration in the Cerenti Coalfield did not commence until 1988 when a 52 hole drilling program located coal in the Kerinci Formation at depths of less than 120 m. Seams were 1.6 to 14 m thick, averaging 12.5 m. Resources were calculated to be 1.4 Mt (indicated/inferred) with 266 Mt nominated as open pit reserves with an in situ stripping ratio of 1:4.4. Petrographic analyses were similar to those for other Central Sumatera coals but vitrinite reflectance was found to be 0.2–0.45% (brown coal rank). Analyses gave 18–25% moisture, 30–39% volatile matter, 31–39% fixed carbon, 3–95 ash, and <2% sulfur. Calorific values ranged from 18.4 to 20.3 MJ/kg.

South Sumatera: Banko Barat and Bukit Assam Coalfields. Total resources of South Sumatera are estimated to be 1.8 Gt measured, 3.8 Gt indicated/inferred, and 13.2 Gt hypothetical. Most of this coal is in the South Sumatera Basin.

Coal seams are found in the Muara Einim Formation, which ranges in thickness from 450 to 750 m, near the middle of Palembung Formation in the South Sumatera Basin. Six principal seams (Fig. 31) and several thin discontinuous to semi-continuous seams with an aggregate of 60 to 90 m have been intersected in drilling programs. The Muara Einim Formation is divided into an upper and lower unit with two coal-bearing units, each of which is further subdivided into two coal-bearing sequences numbered M1, M2 (bottom unit) and M2 and M3 (top unit).

M2 contains the minable coal and three seams are recognized although the two top seams Mangus and Suban seams both have an upper and lower split. The five seams are generally designated as A1 and A2 which are equivalent to upper and lower Mangus seam; B1 and B2, equivalent to upper and lower Suban seam; C, equivalent to Petai seam.

The two main coalfields are Bukit Assam and Banko Barat. In the Bukit Assam Coalfield the seams are 4 to 12 m thick. Total resources for the Bukit Assam Coalfield are given as 426 Gt indicated/inferred. Bukit Assam is an integrated project designed to provide coal for the Suralaya power station which cost US$800 million and comprises four power generation units (Suralaya I, II, III and IV) will consume 4.2 to 4.4 Mt of coal when fully opera-

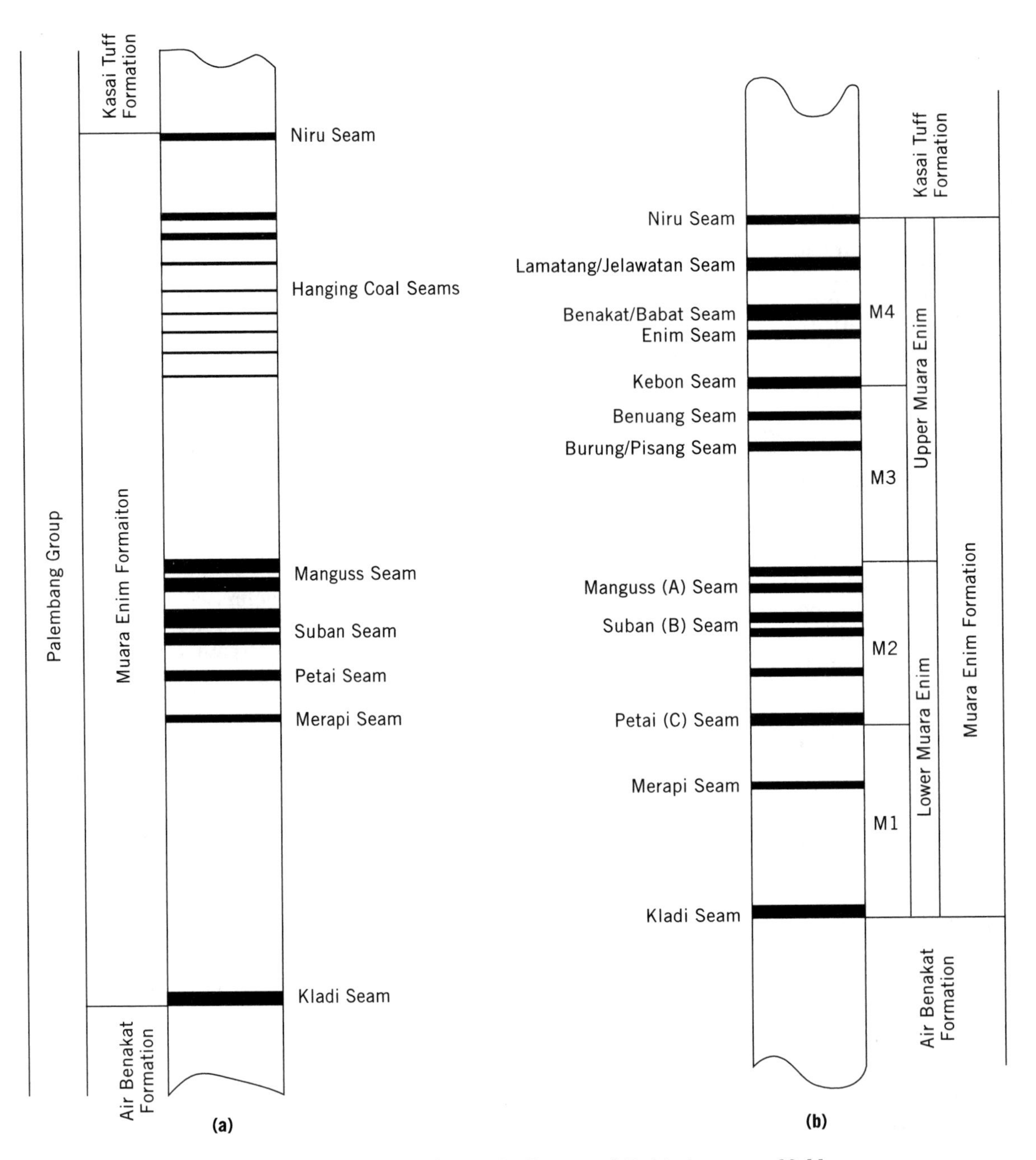

Figure 31. Coal beds of the Banko Barat and Bukit Assam coalfields.

tional. The limiting factor on this project is the railway transport system which is designed to carry 200,000 kt monthly.

In the Banko Barat Coalfield A1 seam splits laterally towards the south and grades into carbonaceous shale in the north areas. A2, B1 and B2 seams are relatively uniform throughout the area although the B seams merge towards the southeast. Seam C also splits into C1 and C2 seams. Resources for the Banko Barat Coalfield are given as 965 Mt indicated/inferred.

Banko Barat coals analyzed at 10–20% moisture, 40–45% volatile matter, 40–50% fixed carbon, 4 to 15% ash, <2% sulfur, and have a specific energy of 25.1 to 29.3 MJ/kg. Vitrinite comprises 80 to 90% of the coal and reflectance is 0.30 to 0.55%.

Sumatera coal is exploited by state-owned Perum Tambang Batubara with the exception of several small mines in the province of Bengkulu which are operated by small private companies such as PT Bukit Sunur and PT Danau Mas Hitam. Total production from these mines was 1.1 Mt in 1988. It is reported that several other small private companies were preparing for similar operations in Bengkulu (128).

PT Allied Indo Coal, a joint venture between two Australian and one Indonesian company, is the largest private contractor in Sumatera with development and production rights in the Parambahan area of Ombilin. Production commenced in 1987 with 53 kt/a which increased to 444 kt/a in 1988. Current production is significantly higher.

Java

Java is a relatively minor producer of coal although resources are estimated to be 30.3 Mt in West Java and 3.8 Mt in Central Java (indicated/inferred status).

The most important coalfield in West Java is the Bayah Coalfield which is part of a Paleogene basin located in south Banten. The seams are part of the Eocene Bayah Coal Group which up to 1500 m thick structurally complex characterized by complex folds and severe faulting, especially in the Cimandiri area. At least 14 seams have been intersected but rarely exceed 1 m, with the thickest 2 m. Several of the seams are lensoidal. The most prospective seams are seam A in the Western area (0.8 m thick and resources of 1.6 Mt) and seams A, B and C in the Eastern area (0.7 m and 1.68 Mt, 1.3 m and 3.5 Mt, and 0.8 m and 0.14 Mt respectively).

Sulawesi

Sulawesi has total resources of 80 Mt (indicated and inferred status) in four basins, of which the Gelumpang Basin with 78 Mt is the most important.

Irian Jaya

Tertiary coals were deposited in Pliocene regressive sequences. Little exploitation of the coals is undertaken. Total resources for Irian Jaya are 4 Mt (indicated and inferred status) which are found in Bintuni and Salawati Basins.

In the Salawati Basin, a 1 m seam in the Pliocene Klasaman Formation, with an overburden of 10 m, crops out on Salawati Island; resources are estimated to be 174 kt. A second seam, possibly 4 m thick, crops on the eastern part of the island.

In the Bituni Basin, coals are found in the Steenkool Formation at Horna, near the northeastern edge of the basin. Seams are approximately 1 to 2.5 m thick and has properties very similar to those of coals from the Ombilin Basin of Sumatera.

Kalimantan

Kalimantan is the second largest producer of Indonesian coal, most of which is taken from the eastern half of the island (Fig. 32). The coal is exported because of its low sulfur, low ash, and ease of mining. Whereas most production in Sumatera is from state-owned mines, exploitation of Kalimantan coals has relied on large private, commonly foreign multinational, capital under the supervision of Perum Tambang Batubara. Total resources of Kalimantan are 8.54 Gt or approximately 26% of the total Indonesia resources including 1.99 Gt (23.3%) measured and 6.55 Gt (76.7%) indicated/inferred.

Largest resources are in the Kutei Basin (55.6% of total resources) followed by Barito and Asam Asam Basin (41.6%) and Tarakan Basin (2.8%). The figure for the Tarakan Basin underestimates the resources because deep seams have not been included. Oil companies have intersected additional seams that have not been included.

Coal types include high-calorific value Eocene sub-bituminous coals to low-calorific value, high moisture Miocene brown coal. Most coals are low ash with <1% sulfur and include some exceptionally low sulfur coals of <0.1% S.

Development of Kalimantan coalfields was expedited in the late 1980s after forecasts indicated that Sumatera production was unlikely to supply short term domestic needs. Negotiations with the foreign companies began in 1978 and three contracts were signed in 1981 with field work commencing in 1982. By 1989, nine contractors had been licensed for exploration and, or leading to, production.

Kutei Basin. Kutei Basin is the largest and deepest Tertiary basin in East Kalimantan and extends offshore to the east and is separated from the Tarakan Basin to the west by the Mangkalihat Arch. Coal seams in the Kutei Basin are in the Early Miocene Pemaluen and Pulubalang Formations and the Miocene-Pliocene Balikpapan and Kampungbaru Formations which have been folded into NNE trending anticlines and synclines south of the Mahakam Delta. Up to 40 seams have been intersected with thicknesses ranging from a few centimetres to 4 m. Dips of the seams range from 5° to 20°.

Three private companies have rights for coal exploitation in Kutei Basin: Pt Kaltim Prima, PT Tanito Harum and PT Multi Harapan Utama. PT Kaltim Prima has extensive resources in the Kutei Basin-Pinang, Bengalon and West Malawan areas in northern Kutei Basin (Table 16) and Santan and Sapari areas, near the Mahakim Delta in southern Kutei Basin. Agreements for the large Sangatta Mine were signed in 1988 and production commenced in 1991. Production in 1992 was 7.4 Mt (31.3% of total Indonesian production); capacity of the mine is at least 13 Mt/a.

Seams are found in the Miocene Pulualang and Balikpapan Formations in the Sangatta area, especially on the Pinang Dome where syncline and faults are thought to have controlled deposition. Four main seams are Prima, Sangatta, Pinang and Kedapat seams (Fig. 33), all of which exhibit lateral thickness variations; Sangatta seam is the most persistent. Seam thickness ranges from 2.5 to 9 m (averaging 5 m) and dips are 5 to 15°. Typical analyses for the Sangatta coal are 4 to 12% moisture, 30 to 40% volatile matter, 30 to 45% fixed carbon, 2 to 5% ash, 0.2 to 1.5% sulfur, and 25.1 to 29.3 MJ/kg (126).

Coal is also found near the Mahakam Delta in the Sigihan, Busang, Loa Tebu, Perjiwa and Loa Kulu coalfields; resources are estimated to be 50 Mt. As many as 40 seams of >0.5 m thickness per kilometer of section have been intersected. Typical analyses for Mahakam Delta (Loa Kulu) coal are 10–15% moisture, 30–40% volatile matter, 3–5% ash, 28–46% fixed carbon, 0.1–3.3% sulfur, and 17.8–26.4 MJ/kg specific energy (126). Coal is vitrinite-rich subbituminous coal with vitrinite reflectance ranging between 0.5 and 0.6%.

Pasir Basin. Coal-bearing sequences in the Pasir Basin are known to occur in the Eocene Tanjung Formation and the Miocene Warukin Formation on Pulau Laut and Sebuku Islands as well as mainland Kalimantan. PT Utah Indonesia commenced exploration in the Pasir Basin in 1981 and identified four seams (Muru, Bekoso, Kendilo, and Lempusu seams, from 1 to 8-m thick and dipping at

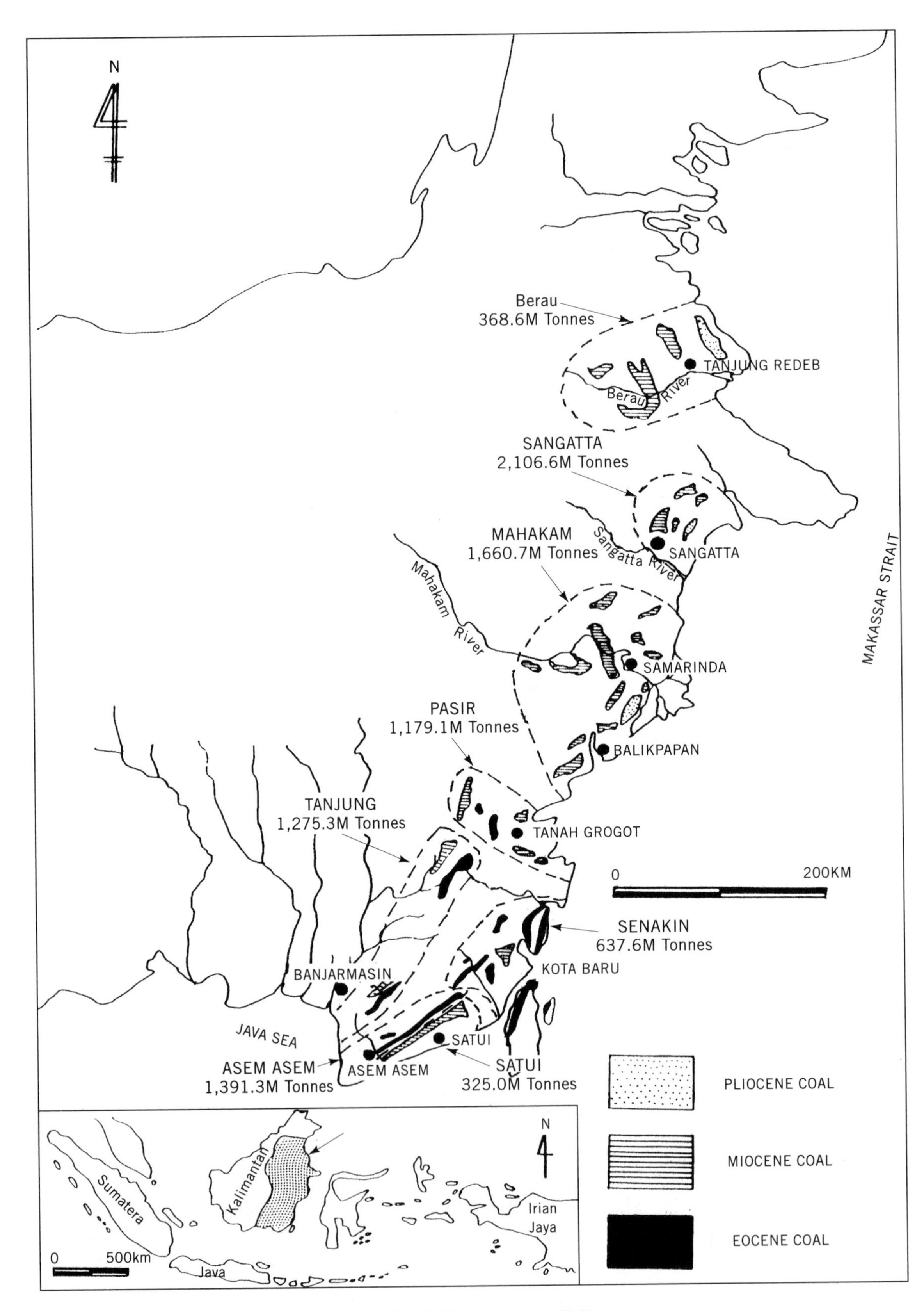

Figure 32. Coalfields of eastern Kalimantan.

Table 16. Resources of Selected Indonesian Basins

Region	Basin	Measured	Coal Resources, ×10^6 ton Indicated	Inferred	Total
Sumatera					
North	West Aceh		500	760	1260
Central	Cerenti		1400		1400
West	Ombilin		160	24	184
	Jambi Simamar		350		350
South	Bukit Asam		333	93	426
	Banko		456	510	966
	Musirawas		500		500
Kalimantan					
North	Tarakan–Berau	55	183		238
West	Ketangau		300		300
	Melawi		125		125
Central	Kutai–Sangatta	376	1730[a]		2107
	–Mahakam	79	1380[a]		1459
	Pasir–Pasir	548	632[a]		1179
South	Barito–Tanjung	300	975[a]		1275
	–Senakin/Satui	633	1648[a]		2281
Java					
West	Bayah, Ciman diri, Bojong manik		0.4	3.0	3.4
Sulewesi South	Malawa, Patuku Todongkurah		11.3		11.3

[a] Refs. 126, 127.

15 to 20°) in the Eocene Kuaro Formation within a narrow belt towards the northern end of the Meratus Mountains. Economic interest is restricted to the Kendilo seam in the Bindu (approximately 6.9 m), Betitit (5.7 m) and Petangis (4.4 m) areas. A total of 67 Mt of measures reserves (<100 m depth) have been identified in the three areas.

Coals are subbituminous to high volatile bituminous rank with 4–16% moisture, 34–50% volatile matter, 34–47% fixed carbon, 1–28% ash, 0.1–5.4% sulfur, and heating values of 20.5–28.1 MJ/kg. Vitrinite reflectance ranges from 0.5–0.65%.

Barito Basin. The Barito Basin is in the southeast corner of Kalimantan and contains coals in Eocene coal measures and in Miocene and Pliocene coal measures. The Eocene coals are in the Tanjung Formation, a 615 m sequence of shale, mudstone, sandstone and limestone which contains seams 4–9-m thick and averaging 6 m; seams dip at 5 to 15°. Splitting is a common feature especially in the Senakin, Sepapah, and Sangsang areas and less so in the Satui area.

Eocene Barito Basin coals are exploited by several private companies, including PT Arutmin Indonesia and PT Chung Hua Overseas Mining Development who have interests in bituminous to high volatile bituminous coal in the Tanjung Formation where the seams are up to 6 m thick. Typical analyses are 2–5% moisture, 38–45% volatile matter, 30–45% fixed carbon, 2–15% ash, and <1% sulfur. Vitrinite reflectance is 0.5–0.65%.

Miocene to Pliocene coals occur in the Warakin Formation especially in the Sarongga and Asam Asam areas. Seams are 1–40-m thick and especially well developed in the Tanjung area where the coal-bearing sequence is 400–450-m thick. Eleven prospective coal seams have been identified in the Tutpan, Wara and Paringin areas. PT Adaro Indonesia has interests in the Warukin coals in Tutupan, Wara and Paringin areas where seams are up to 13-m thick and extend over a strike length of 10–16 km. The coals have 4–15% moisture, 37–40% volatile matter, 30–40% fixed carbon, 2–10% ash, sulfur at <1%, and calorific values are 17.6 to 28.5 MJ/kg. Vitrinite reflectance is 0.3 to 0.45%.

Tarakan Basin. Economic seams occur in the Late Miocene Berau Formation of the Tarakan Basin which is the northernmost coal basin in East Kalimantan. At least 30 seams, ranging in thickness from a few centimeters to 12 m, have been intersected. Best resources are in the Kelai, Lati and Binungan areas where seams dip at 5 to 28°.

Coal is exploited from the Lati area by the private company PT Berau Coal who signed an agreement in 1983. Four seams, ranging in thickness from 1 to 4 m are exploited. Coals are subbituminous with 11 to 18% moisture, 2 to 11% ash, 33 to 44% volatile matter, 42 to 46% fixed carbon, <2% sulfur, and calorific values of 23.9 to 24.7 MJ/kg.

West Kalimantan. The resources of coal in East Kalimantan are so large that little exploration has been carried out in the western two thirds of Kalimantan. Subbituminous, vitrinite-rich coals are known from the Marakai and Nanga Tabidah areas. These coals have specific energies of 23.0–32.7 MJ/kg and would be suitable as thermal coals if exploited.

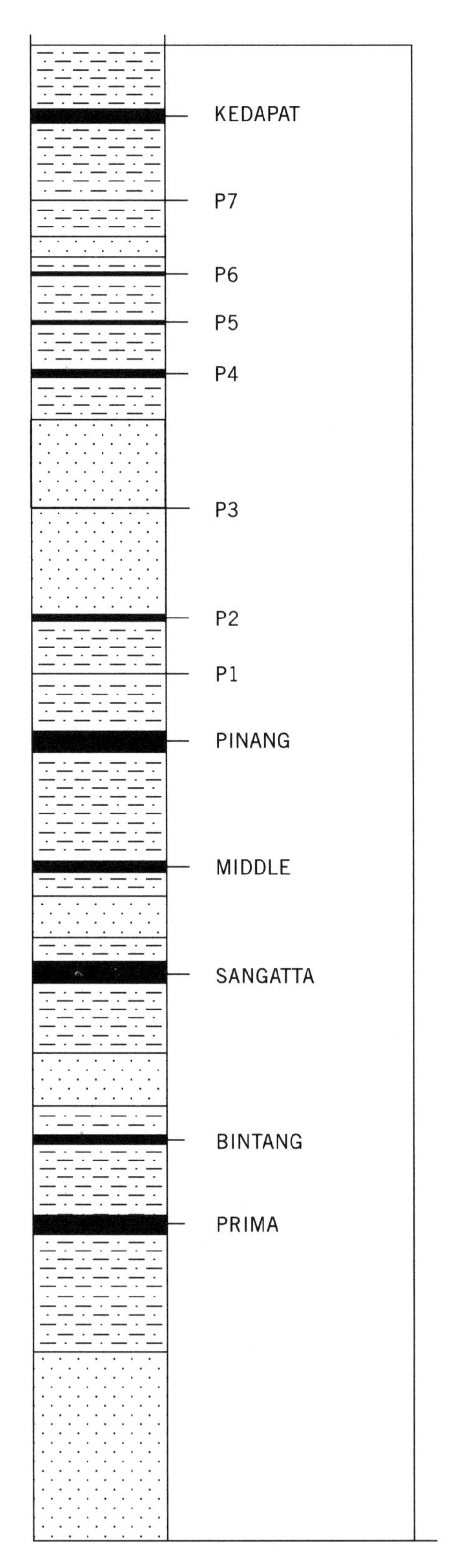

Figure 33. Coal beds of the Sangatta coalfield.

INDIA

India has a long history of coal production covering at least 200 years although systematic exploration for coal did not commence until much later. Surveying of the coalfields began in 1837 and most of the important coalfields had been geologically mapped by 1876. Drilling for proving of coal resources was as early as 1847 when the Bengal Coal Company Limited drilled its first hole at Kalipur, near Durgapur (West Bengal). The Geological Survey of India (GSI) was actively associated with drilling in the latter half of the nineteenth century and drilling became a routine part of exploration during this period.

During the 1950s and 1960s, the Indian Bureau of Mines (IBM) undertook extensive systematic exploration as part of the 'Second Five Year Plan' (1955–1956 to 1960–1961) which emphasized exploitation of India's coal resources. At the end of this phase, IBM has 49 drill rigs actively drilling in nine coalfields. Adding drill rigs from other sources, a total of 100 drill rigs were engaged in exploration.

During the "Third Plan," three government organizations, GSI, IBM and the National Coal Development Organization (NCDO) were active in coal resource programs but by 1964, NCDO was the only participant, having taken over full responsibility for coal resources.

In the early 1970s, coal mines were nationalized and brought under the control of the Central Mine Planning & Design Institute (CMPDI) in January 1974. With the subsequent formation of Coal India Limited in 1975, the entire responsibility for coal exploration was delegated to the Exploration Division of the newly established Central Mine Planning & Design Institute Limited (CEMPDIL).

Coal Resources

Coal resources of India have been calculated for seams greater than 1.2 m thick and are divided into Proven Reserves (those based on geological data derived from outcrop, trenches, mine faces, and drill holes not exceeding 200 m apart), Indicated Reserves (those based on drill hole data not exceeding 1000 m apart or up to 2000 m where continuity is known), and Inferred Reserves (where drill hole data are based on distances of greater than 1000 m or 2000 m where continuity is known). A further subdivision is based on depth (Table 17).

Ninety percent of India's major workable coal resources are of Permian age and are known as Gondwana coals (Fig. 34). Nearly 50 coalfields, ranging size from a few square kilometers to greater than 1500 km^2, have been mapped. Most occur in peninsular India in the quadrant bounded by 78°E longitude and 24°N latitude. Within the basins the number of coal seams ranges from as few as two or three up to 44 with total minable intervals are 0.5 to 160 m (Singrauli) with the rank of coals varying from sub-bituminous to bituminous rank; ash contents are generally high (20 to 35%) and phosphorus contents are low (<0.01%).

Several Mesozoic coals have been reported in Gujarat state but are not of economic significance.

The Gondwana coals are broadly divided into coking and noncoking coals where noncoking coals do not fit any of the categories given in Table 18.

Approximately 25% of the total resources are amenable to open pit mining and 55% of current production (more than 220 Mt in 1991) is taken from these resources. Surface mining conditions are generally good with stripping

Table 17. Bituminous Coal Resources (Mt) of India

Depth, m	Reserves: Proven	Indicated	Inferred	Total
0–600	54,257	71,981	41,488	167,726
600–1200	2,037	9,396	6,886	19,319
Total	*56,294*	*81,377*	*48,374*	*186,045*

ratios of approximately 4:1. Problems encountered with open pit mining include pit fires initiated by spontaneous combustion (an estimated 40 Mt of coal has been destroyed by fire) and flooding during the monsoonal season. Extraction rates are in the range of 1.5 to 6.5 ton per manshift which reflects both low levels of mechanization and poor training.

Mining

Underground mining is generally by room and pillar methods with an increase in longwall methods modified and adopted for thick, steeply dipping seams. Production rates generally low by international standards being limited by manual loading methods although extraction is generally mechanized; production rates of 0.4 to 0.8 tons per manshift are common.

Mining has been quite seriously affected by labor disputes and power shortages in the past and these tend to exacerbate the low production rates. Approximately 700,000 people are employed in the industry. Production cost range from up to US$6 for open pit mines to more than $25 per tonne for less efficient underground mines. Average per tonne cost is given as $17 (9). Cost are influ-

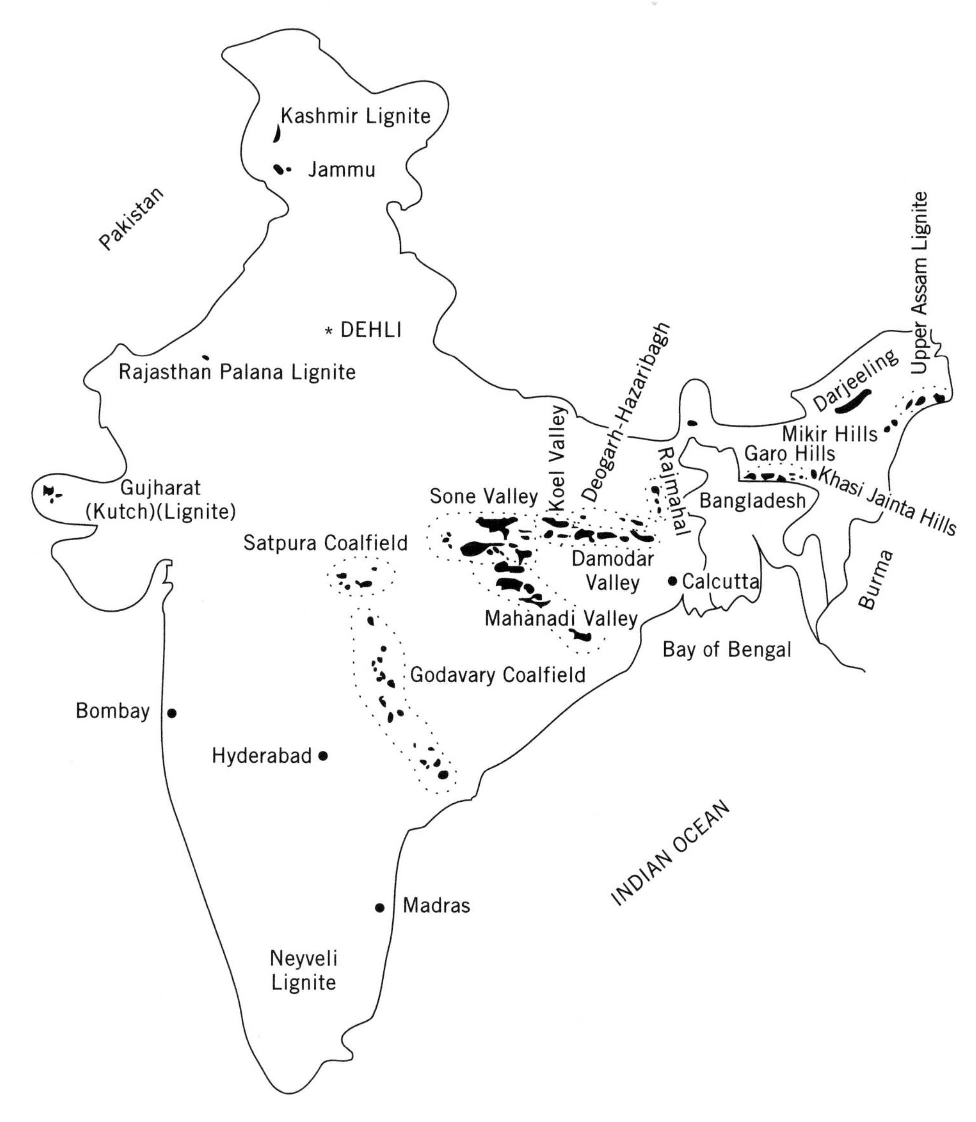

Figure 34. Coalfields of India.

Table 18. Properties of Indian Coking Coals

Coal	Moisture, %	Volatile Matter, %	Carbon, %	Hydrogen, %	Calorific Value, MJ/kg	Reflectance, %
Prime coking	<2	22–32	88–91	4.8–53	36.4–37.1	1.0–1.4
Medium coking						
High volatile	1.0–2.0	32–37	86–88	4.7–4.9	32.7–35.6	0.85–0.95
Low volatile	<1	20–22	90–91.5	4.7–4.9	36.4–36.8	1.4–1.55
Semi-coking	2.0–4.0	37–44	83–85	5.4–5.8	34.3–35.6	0.75–0.85
Weakly coking						
High volatile	3.0–5.0	38–46	83–84	5.1–5.6	34.3–35.2	
Low volatile	1.0	18–20				1.55–1.8

enced by degree of mechanization, geological complexity and labor.

Markets

India is generally regarded as a third world country and can be expected to attempt to increase coal production up to and well into the next century. Coal is already the largest source of energy in India, accounting for over 69% of the domestic energy resources. Most production at present is for the domestic market and is transported throughout the country by rail which accounts for over 90% of all coal movement. Road transport accounts for approximately 10% with coastal shipping moving small amounts, mostly from the northeast of the country to power stations in the south. India's rail network is extensive with over 62,000 km of track (10% electrified (9)) but mostly predates independence in 1948. Many of the facilities and equipment, such as rolling stock and locomotives, require modernization and therefore limits load carrying capacity. Current average haul distance is 750 km but this will decrease as new mines are opened south of existing mines in Bihar and West Bengal.

Lack of export coal has resulted in below average to poor port facilities. The ports of Haldia and Paradip are capable of handling ships with capacities of 60,000 to 90,000 ton (dwt) (9). While upgrading of some ports is expected, the high ash nature of the Indian coals is unlikely to permit a significant increase in export coal by next century.

Increased production of thermal coals can be expected. Earlier estimates indicated a doubling of production to 410 Mt (approximately 360 Mt of thermal coal and 50 Mt of coking coal) by the year 2000, especially if government plans to substantially develop industries. Recent adjusted predictions are for a maximum production of 300 to 325 Mt/a by the year 2000. These production rates may be less than the requirement for industry and India could become a principal importer of coal.

Most present production is from the Bihar and West Bengal coalfields. Most potential for increased production is likely from coalfields further to the south; Talcher, Korba and Singrauli Coalfields are good examples. Surface mines with a potential of 85 million ton per annum capacity are under development and await government approval (9).

Geology

The Permian Gondwana Coalfields are located in three main belts, the Koel-Damodar Valley Basin of East India, the Son-Mahanadi Valley basin of east central India and the Pranhita-Godavari and Satpura (Pench-Kjanban) Valley Basins of south central and Central India. The coal-bearing sequences, known as the Gondwanas, in the peninsular basins fill graben or half graben with edges of the basins commonly bounded by high angle normal faults, many of which are *en echelon.* Many of the seams are transected by smaller faults of several meters throw, causing discontinuity of seams where these faults occur. The absence of Upper Gondwana strata in the Jharia Coalfield (eastern Damodar Valley Basin) and Daltonganj (west Damodar Valley Basin) and their occurrence in adjoining basins such as Raniganj, Bokaro, Karanpura and Hutar can be best explained as a result of faulting. Elsewhere however, such as Mahanadi Basin, the absence of upper Gondwana strata is attributed to uplift and erosion.

The coal measures sequences generally dip at angles ranging from subhorizontal to 20°.

Ultrabasic (mica-peridotite and lamprophyre) and basic (dolerite and basalt) dykes and sills intruded the sequences and in the Damodar Valleys coalfields such as Raniganj Jharia, West Bokaro and East Bokaro, significant reserves, estimated to be several hundred million tonnes, have been sintered.

The Gondwana sequences comprise 5 to 6 km of subhorizontal, shallow-water fluviatile and lacustrine sediments, known as the Gondwana Supergroup. In some literature, the sequences are divided into the Lower Gondwana (Permian) and Upper Gondwana (Triassic-Lower Cretaceous) (Table 19).

Economic coal is found in three units; the Karharbari Formation (or Basal Coal Measures), the Barakar Formation (Lower Coal Measures) and the Raniganj Formation (Upper Coal Measures).

Karharbari Formation. The Karharbari Formation varies from 65 to 120+ m in the Damodar Valley Basin and up to 280 m in the Son-Mahanadi Valley Basin (130). The unit comprises mostly coarse-grained sandstone, conglomerate and several coal seams which range from 1.2 to 9 m thick. Splitting is common and ubiquitous in the peninsular region. Coal is better developed outside the Damodar

Table 19. Stratigraphic Correlation of Selected Permian Basins

	Unit	Koel-Damodar Valley	Son-Mahanadi Valley	Godvari Valley	Satpura
Upper Gondwana	Post Mahadevas		Jabalpur	Chikilia Fm Kota	Jabalpur
	Mahadevas	Mahadeva Fm	Tikki Fm	Dharmaram Fm Maleri Fm	Mahadeva Fm
	Panchet	Panchet	Dalgaon Bed	Yerrapalli Parsora Bed	Mangil
Lower Gondwana	Raniganj Fm	Raniganj Fm	Kamthi Fm	Kamthi Fm	Kamthi Fm Fijori Fm
	Barren Measures	Barren Measures/ Ironstone shale Fm	Shale Barren Measure Fm	Barren Measures	Motur Fm
	Barakars	Barakar Fm	Barakar Fm	Barakar Fm	Barakar Fm
	Karharbaris	Karharbari Fm	Karharbari Fm	Karharbari Fm	Karharbari Fm
	Talchirs	Talchir Fm	Talchir Fm	Talchir Fm	Talchir Fm

Valley Basin but this basin has the best coal to noncoal ratio (Table 20).

Barakar Formation. The Barakar Formation has the largest resources of coal including both steaming and coking varieties. Clastic unit comprise sandstone, conglomerate, siltstone, shale, carbonaceous shale, and claystone. Seam number is variable ranging from three in the Jayanti Basin to more than 30 in the Damodar Basin. The Barakar Formation has been conveniently divided into three subunits although not all three are necessarily found in any one basin.

The lower Barakar Formation is composed of coarse-grained sandstone, with subordinate shale and siltstone, and varies from 220 to 380 m in the Damodar Valley Basin, 290 to 470 m in the Son-Mahanadi Valley Basin, and 60 to 126 m in the Satpura Basin. Seams are thick (Table 20), but are high ash.

The middle Barakar Formation is best developed in the Damodar Valley Basin although not everywhere. Clastic beds tend to be finer grained than the lower Barakar Formation and coal seams are relatively thick attaining a maximum of 24 m. Thickness of the unit ranges from 150 to 370 m in the Damodar Valley Basin.

Coals of the lower Barakar Formation are reasonably good coking coals and are the source of most of the coking coal for steel manufacture in the Jharia Coalfield.

The upper Barakar Formation is also mainly developed in the Damodar Valley Basin. Clastic rocks are predominantly finer grained than the middle Barakar Formation. Coals are generally coking coals.

Raniganj Formation. The Raniganj Formation is the thickest subunit attaining a thickness of 450 m (Jharia Coalfield) to 1035 m (Raniganj Coalfield). The clastic rocks comprise fine grained sandstone and shale with seams of variable thickness. Fewer lenses of conglomerate are found in the Raniganj Formation compared to the lower units. Coals are best developed in the Damodar Valley Basin, such as the Raniganj and Jharia Coalfields. The number of coal seams varies from 14 in the Jharia Coalfield to more than 20 in the Raniganj Coalfield. The seams are moderately thick (up to 20 m) and supply most of the highest quality coking coals mined in India. In addition, the Raniganj Coalfield has semi-coking to weakly

Table 20. Features of the Karharbari and Lower Barakar Formation in Selected Basins[a]

Basin	Unit Thickness, m	Seam Thickness, m	Coal to Noncoal
Karharbari			
Koel-Damodar Valley	65–125	1.2–6.0	1:4–1:25
Son Mahanadi	206–280	1.2–7.5	1:10–1:26
Godvari-Pranhita	78–224	1.4–7.5	1:6–1:25
Satpura	146–229	1.3–1.5	1:48–1:59
Lower Barakar			
Koel-Damodar	22–379	1.5–32.0	1:2.5–1:4
Son Mahanadi	294–470	2.0–42.0	1:2–1:10
Godavari-Pranhita	92–225	2.0–10.0	1:7–1:18
Satpura	60–126	1.5–3.0	1:17–1:21

coking coals and the Jharia Coalfield has medium coking coals.

West Bengal. Raniganj Coalfield, one of the West Bengal Damodar Valley coalfields located 200 km north of Calcutta, is the largest coalfield in India with resources of 30.147 Gt. Seven persistent seams, the three most important seams are known locally as the Chanch, Laikdih and Salanpur seams, are found in the 600 m thick Barakar Formation. In the northwestern part of the basin however, up to 14 seams more than 1 m thick have been intersected, the thickest being 24 m. The 1 km thick Raniganj Formation contains 10 seams persistent seams with the three most important known locally as the Jambad, Dishergarh and Sanctoria seams with the thickest 18 m. The Barakar Formation is separated from the Raniganj Formation by 400 to 600 m of clastic interburden.

Coals are medium coking, blendable, and noncoking coal with the latter suitable for power generation as raw coal.

Raniganj Coalfield is bounded to the south by major east–west faults but to the north the coalfield unconformably overlies metamorphosed basement rocks. The coal measures dip at angels of up to 15° to the south and southwest. Faulting is common with the dominant trends northwest–southeast and northeast–southwest; intrabasinal and longitudinal faults have also been documented.

Typical analyses for Barakar coals are 0.8–2.0% moisture, 15–35% ash, 0.5–0.8% sulfur, <0.20% phosphorus, 25–36% volatile matter, and 34.8 to 36.8 MJ/kg specific energy. Typical analyses for Raniganj coals are 2.5–11% moisture, 13–25% ash, 0.5–0.7% sulfur, <0.15% phosphorus, 39–44% volatile matter, and 31.9–35.4 MJ/kg specific energy. Maceral analyses for medium coking coal are typically 45–65% vitrinite, 3–7% liptinite, 30–50% inertinite, and vitrinite reflectance of 0.85–0.95%. For semicoking coals analyses are 75–85% vitrinite, 4–10% liptinite, 10–20% inertinite, and vitrinite reflectance of 0.75–0.8%.

Other coalfields include the Barjora and Darjeeling Coalfields which have combined resources of 85 million tonnes; none have been developed and use of the coal will be power generation if used although Darjeeling Coalfields has thin seams of semi-anthracite.

Bihar Coalfields. Jharia Coalfield, in the Dhanbad area, is the largest coalfield in Bihar with resources of 19.417 Gt. It is the sole supplier of prime coking coal for metallurgical purposes; other uses of non-coking coal are steam raising and other industries; washery middlings are used for power generation.

Jharia Basin is an east–west trending basin now comprising downfaulted remnants of a previously much larger basin. The southern edges of the basin is faulted against Precambrian basement. The Barakar Formation is 600 to 1000-m thick and contains up to 25 extensive seams with splitting occurring in some areas; seam also merge to give extremely thick intersections of coal. Fifteen seams are known in the Raniganj Formation; locally important seams are Lohapiti, Bhurungiya and the Top, Middle and Bottom Mahuda seams.

Typical analyses for Barakar coals are 0.6–2.0% moisture, 15–35% ash, 0.5–0.8% sulfur, <0.30% phosphorus, 17–35% volatile matter, and 35.3–37.2 MJ/kg specific energy. Typical analyses for Raniganj coals are 1.5–2.2% moisture, 20–25% ash, 0.5–0.7% sulfur, <0.3% phosphorus, 36–40% volatile matter, and 35.3–35.8 MJ/kg specific energy. Maceral analyses for prime coking coal are typically 35–70% vitrinite, 0–2.5% liptinite, 30–65% inertinite, and vitrinite reflectance of 0.9–1.3%. For medium coking coals analyses are 65–85% vitrinite, 2.5–10% liptinite, 10–30% inertinite, and vitrinite reflectance of 0.75–0.85%.

Other coalfields in Bihar include East Bokara, West Bokara, Giridih, North Karanpura, South Karanpura, Ramgarh, Auranga, Hutar, Daltonganj, Deogarh, and the Rajmahal Group which includes Hura, Chuperbhita, Pachhwara, Mahuagarhi, and Brahmani Coalfields. Total resources of these areas are 33.3 Gt with the largest resource in North Karanpura. Coal is mined in the East and West Bokara and Giridih Coalfields. Most coals are suitable only for power generation and industrial use.

Madhya Pradesh Coalfields. Madhya Pradesh area has 19 coalfields most of which contain noncoking coals suitable for power generation of other local domestic industries. Total resources are 37.356 Gt. The number of seams in each field varies considerably as does the seam thickness.

Orissa Coalfields. Talchir and Ib-River Coalfields are second only to Raniganj Coalfield in size and have combined resources of 45.19 Gt. Talchir is the more important of the two fields and contains 9 seams in the Karharbari and Barakar Formations with seam thicknesses ranging from 1 to 50 m.

Maharashtra Coalfields. Four small coalfields with total resources of 9.685 Gt contain noncoking coals suitable for domestic use. Exploitation has been minimal until now.

Andhra Pradesh. The Godavari Valley Coalfield contains 10.81 Gt of non-coking coal suitable for domestic use. Ten seams, ranging in thickness from 1 to 18.3 m have been intersected. Not all seams can be exploited.

Tertiary Coalfields

Subbituminous to high volatile bituminous coal of Tertiary age are found in the Arunachel Pradesh, Assam Nagaland and Maghalaya Coalfields. Total resources are 893 Mt in seams varying from less than 1 m to 33 m. Coal in the Mahum Coalfield in the Assam area have low ash and are suitable for industrial use and coke making if the sulfur content was lower.

Typical analyses are 2–16% moisture, 5–35% ash, 2.0–8% sulfur, <0.01% phosphorus, 13–50% volatile matter, and 30.4 to 36.5 MJ/kg specific energy. Maceral analyses are typically 60–90% vitrinite, 5–25% liptinite, 5–20% inertinite, and vitrinite reflectance of 0.45 to 1.85%.

Brown Coal Resources

Approximately 20 Eocene to Oligocene coalfields have been identified but workable coal is restricted to Kashmir, Rajasthan, Gujarat and Temil Nadu with the most significant deposit being the South Arcot Lignite deposit (Neyveli) of Temil Nadu. Resources of Tertiary coal, 3.542 Gt of which 3.3 Gt are in the Tamil Nadu Coalfield, constitute approximately 1% of total resources. Presently, the Temil Nadu coal is used for power generation and other domestic industries.

SOUTH AFRICA

Coal was first discovered in Natal and Northwest Province and subsequently mined in the Molteno-Indwe Coalfield of North West Province from where it was transported by oxen-drawn wagons to supply the energy needs of diamond mining in the Kimberley region (131). Large scale mining commenced, in the East Rand area, soon after the discovery of diamonds with significant mining in the Witbank, Middleburg, and Vereeniging Coalfields.

A massive increase in coal production occurred after World War II, coinciding with rapid expansion of the South African economy. Production was further stimulated in the 1970s when the energy crisis was experienced. This rapid growth is reflected in the production figures which increased from a total of 34.2 Mt in 1957 to almost double this, 63.0 Mt in 1974.

Resources

The most recent available resources and production figures for South Africa are those given by the Commission of Inquiry into the Coal Resources of the Republic of South Africa in 1975. Minable *in situ* resources were given as 81.275 Gt of which 24.89 Gt was classified as high-grade steaming coal and 3.45 Mt was cited as washed metallurgical coal. Anthracite resources were estimated to 745 Mt. Of the total resources, excluding anthracite, 31% of the resources, or 24.91 Mt was minable by underground methods.

Utilization

South Africa is an important producer and exporter of coal, being the fifth largest producer, 183 Mt in 1990, and the third largest exporter, 47 Mt in 1990 (132). This is a dramatic rise since the mid-1970s when total production was 65 Mt.

Domestically, coal has the dominant role in South Africa's economy with more than 80% of total primary energy requirements sourced from coal, possibly the highest proportion for any industrialized country. Wadley (132) predicted several trends for coal consumption:

1. Declining use of coal as a direct source of energy by small consumers with replacement by coal-sourced electricity.
2. The current plateau, following the historically rapid growth, in synfuel consumption to remain.
3. Continuing steady increase in consumption by the broad-based metallurgical industry.
4. Replacement of other domestic direct-coal consumption by gas and electricity which are more favored because of the stringent environmental antipollution legislation.

Two significant factors which are likely to accelerate the decline in direct coal consumption are increasing transport costs and diminishing reserves of high grade coal.

The State-owned Electricity Supply Commission (Escom) and the South African Coal, Oil and Gas Corporation Limited (Sasol) account for 80% of domestic consumption with growth likely to be restricted to less than 2% per annum, a figure that is likely to be influenced by the state of the economy.

Escom policy has been to locate new power stations in the Eastern Transvaal Coalfields where water is in reasonable supply. Local authorities use 11 Mt for electricity generation. Escom policy of locating new thermal power stations in the coalfields has resulted in Eastern Transvaal becoming the dominant electricity generation region. Fly ash release is reduced by the use of cyclones and electrostatic precipitators whereas the low sulfur content reduces the risk of sulfur dioxide pollution.

Metallurgical consumption of coal, for both steel and ferroalloys, is currently 11 Mt/a (8% of total production). Coke manufacture has increased moderately over the past few years and further growth will be influenced by international metal prices and the availability of suitable moderate to high grade feedstock. Most coking coals are low- to medium-quality coking coals by world standards and blending is necessary for some coals.

Typical analysis of coking coal is volatile matter, 20–30%; ash, 11.5–13.5%; and sulfur, 0.8–1.75%.

Utilization of South African coal for the manufacture of oil, gas and petrochemicals commenced in 1950 with the establishment of Sasol. The oil from coal process is based on two adaptations of the Fischer-Tropsch synthesis, which catalytically combines carbon monoxide and hydrogen. For light hydrocarbon fractions, ranging from liquefied petroleum gas to diesel oil, the moving-catalyst synthesis is employed whereas for heavier hydrocarbon fractions, such as naphtha and waxes, a stationary-catalyst synthesis is used. Initially the coal is crushed to a 5 to 50-mm grain size and pressure gasified in steam and oxygen. The coal used in the Sasol process is regarded as poor quality coal with a high ash content. Typical analysis, on a dry basis, is ash, 34.0%; carbon, 51.7%; hydrogen, 2.3%; oxygen, 10.2%; sulfur, 0.6%; and nitrogen, 1.2%.

A critical factor is the sulfur content. The catalysts used are iron-based catalysts which are susceptible to deactivation by sulfur.

Sasol uses 4 Mt/a of coal (1975 figures) of which 50% is used in the Sasol process and 50% is used for steam and electricity generation. A wholly owned subsidiary of Sasol, Gascor, also produces town gas.

Other uses of coal are coke manufacture, fertilizer, and explosives production, all of which use coal as a fuel and as a raw product.

Future predictions (132) are for increasing pressures on clean coal technologies that can use presently discarded coals, in an environmentally acceptable manner and at a cost comparable with alternative such as electricity and gas. Incentives for the development of these technologies will be the increased market prices of by Escom and Sasol products resulting from major investments to accommodate pollution control facilities under the recent legislation.

Geology

Coal resources of the African continent are restricted to the Permo-Triassic Karoo Supergroup which is found in South Africa, Botswana, Zimbabwe, Zambia, South West Africa, and Mozambique (Fig. 35). In South Africa, the Karoo Supergroup is found in the central and eastern parts of the Cape Province, the southern part of the Orange Free State, Lesotho, Transvaal and Natal.

The Karoo Supergroup is divided into four stratigraphic units: Dwyka Formation, Ecca Group, Beaufort Group and Stormberg Group. The Dwyka Formation, the oldest unit, lies unconformably on the pre-Karoo basement and is up 900 m thick. The uppermost unit of the Dwyka Formation is a distinctive dark carbonaceous

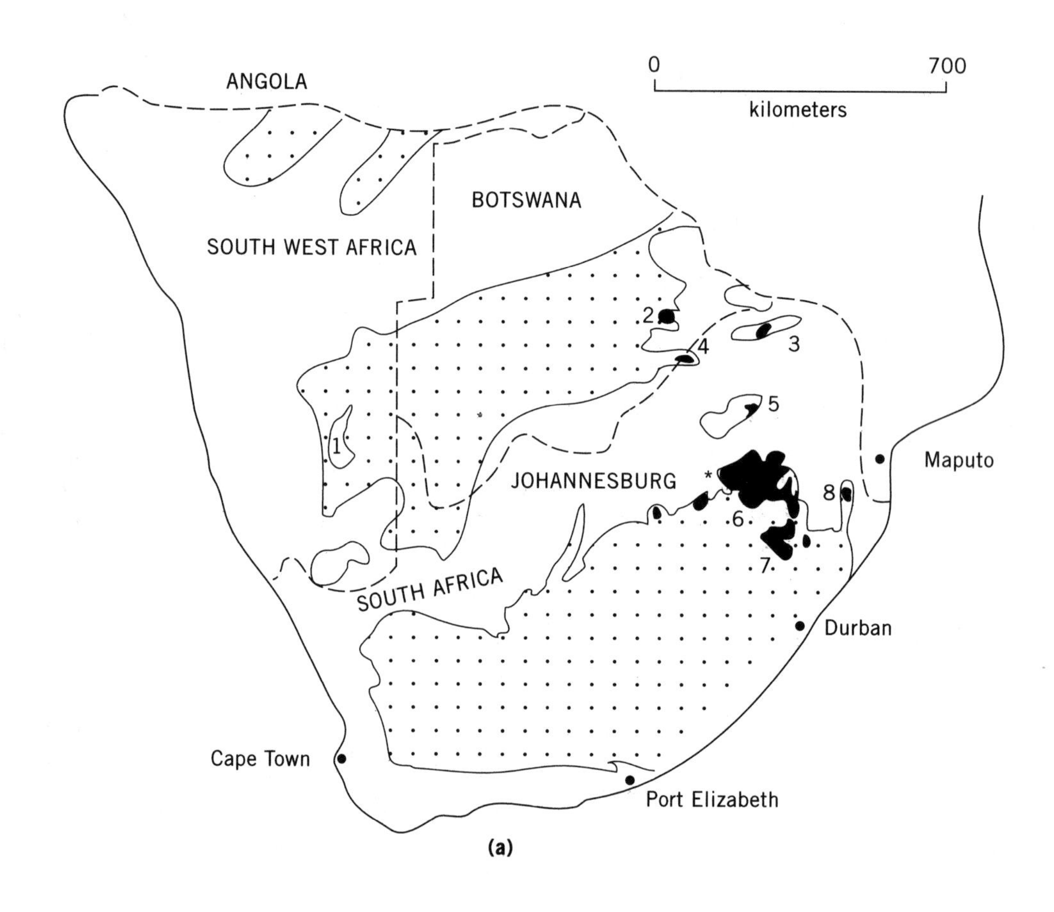

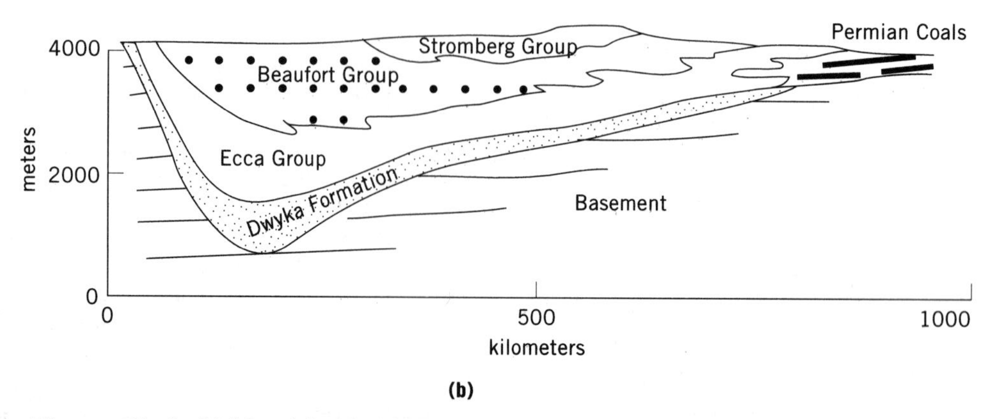

Figure 35. Coalfields of South Africa. (**a**) Location of the basins in South Africa; (**b**) stratigraphy of the Permian sequence. 1, Southwest Africa; 2, Botswana; 3, Limpopo-Southpansberg; 4, Watersberg; 5, Springbok flats; 6, Transvaal; 7, Natal; 8, Swaziland. ⊡ extent of the Ecca group.

shale that weathers to a distinctive white color, hence the name 'white band'. Ref. 133 correlates the white band with the Irati Formation of Brazil and indicated that it may have potential as an oil shale.

The Ecca Group, the main coal-bearing unit, has been divided into three units: lower Ecca, up to 1200 m thick; middle Ecca, up to 1200-m thick; and upper Ecca, up to 1450-m thick.

The Beaufort Group overlies the Ecca Group and contains very thin, poorly developed seams with coal of semianthracitic character with a high ash and low volatile matter. These coals have been identified in the Orange Free State and Natal but are not generally economic although reference has been made two 5.5 and 12 m seams in the Somkele region.

The youngest unit is the Stormberg group which has been divided into four units. The lower unit, the Molteno Formation of Upper Triassic age contains shaly and vitrinite-rich coals. In North West Province, the Indwe, Guba and Molteno seams have been recognized. Coal resources have been estimated at 24 Mt, but numerous dolerite intrusions and numerous shale partings restrict large scale exploitation of these coals.

Ecca Group

Coal is found in the three units of the Ecca Formation which covers an extensive area of the Karoo Basin. Most production is from the Middle Ecca Group although potentially exploitable coals are found in the Upper Ecca Group. Several geographical coal provinces are recognized (131); within each province only one of the three Ecca formations generally containing significant coal resources. The provinces are

- Transvaal Upper Ecca Stage Coal Province
- Witbank-Belfast-Standerton Middle Ecca Stage Coal Province
- Eastern Transvaal-Natal Middle Ecca Stage Coal Province
- Orange Free State Middle Ecca Stage Coal Province
- Sasolberg-Vereeniging-Balfour Middle Ecca Stage Coal Province
- Detached Middle Ecca Stage coal areas.

There is no apparent relationship between depth of burial and coal rank (134). In the northern part of the Karoo Basin, rank increases progressively eastwards from subbituminous near Odendaalsrus in the west, through bituminous rank near Standerton to anthracite in the Vryheid-Lebombo area. Rank is very low in the Orange Free State, Springbok Flats, Waterberg and western Soutpansberg Coalfields: This is attributable to conditions during burial. Dolerite intrusions probably influenced rank in the eastern part of the Karoo Basin.

The most important coalfields are in the eastern Transvaal and Natal regions which contain more than 50% of South Africa's minable *in situ* coal resources. Extensive mining has occurred for many years.

Upper Ecca Stage Coal Province. Coals seam of significant thickness are found in the Waterberg, Springbok Flats, Limpopo and Soutpansberg Coalfields.

Coals in the Watersberg Coalfield are well developed in the Upper and Middle Ecca stages. The Upper Ecca stage averages 60 m and contains seven seams of bright coal. Approximately 36% of the *in situ* reserves are minable by open pit methods. Beneficiation produces a low ash product which can be used as a blend coking coal (10% ash), a petrochemical feedstock (15% ash) and an activated carbon feedstock (6% ash). Open pit reserves are given as 24 Gt with deeper coal resources given as 42 Gt (135).

The Middle Ecca stage averages 55 m and comprises interbedded sandstone, carbonaceous shale and dull coal seams ranging between 1.5 and 8-m thick. Lateral facies changes imparts a significant variation in seam thickness and coal quality. Approximately 10 Gt of coal is minable by open pit methods with a similar amount deeper. Most coal is suitable for electricity generation.

Middle Ecca Stage Coal Province. Five minable seams are found in the Transvaal Coalfield (136) with three, termed the No. 2, No. 4, and No. 5 seams being the most prospective. The No. 2 and No. 4 seams have significant seam splitting which is thought to be related to braided streams transgressing and eroding the peat swamps after deposition. In the eastern part of the Transvaal Coalfield, lenticular torbanite deposits, up to 1.2 m thick, occur in the No. 5 seam.

The Natal Coalfield contains up to 8 coal seams although only three are currently exploitable (137). The lowermost, the No. 1 seam, is not always developed and the topmost seam, the No. 5, commonly has been eroded. The No. 2 seam is the most important seam and locally splits with the lower split known as the No. 2A. The No. 2 coal is medium to low ash (10 to 15%) and has 20 to 25% volatiles.

COAL MINING AND PREPARATION

Coal mines fall into two general classifications: surface and underground. Production of coal by type of mining for the U.S. is shown in Figure 36.

Surface Mining

Surface mining techniques are used when the coal is present near the surface and the overburden is thin enough. These techniques include contour mining, strip mining, and auger mining.

Contour mining is used in hilly countryside areas where the slope of the surface will permit only a narrow bench to cut around the side of a hill; the excavation is backfilled immediately after the removal of coal. It is the only method that can be used on slopes of 15 degrees or higher.

Strip mining is used in flat or gently rolling lands of the Midwest and West where large and efficient equipment can be used. In this technique, the coal is exposed by removing the overlying strata, or overburden. Blast holes are drilled, and explosives are loaded into these holes to shatter the rock cover; earth-moving equipment is used to remove the soil and the shattered rock. The coal then is collected with power shovels or other coal-digging machines and loaded directly into trucks.

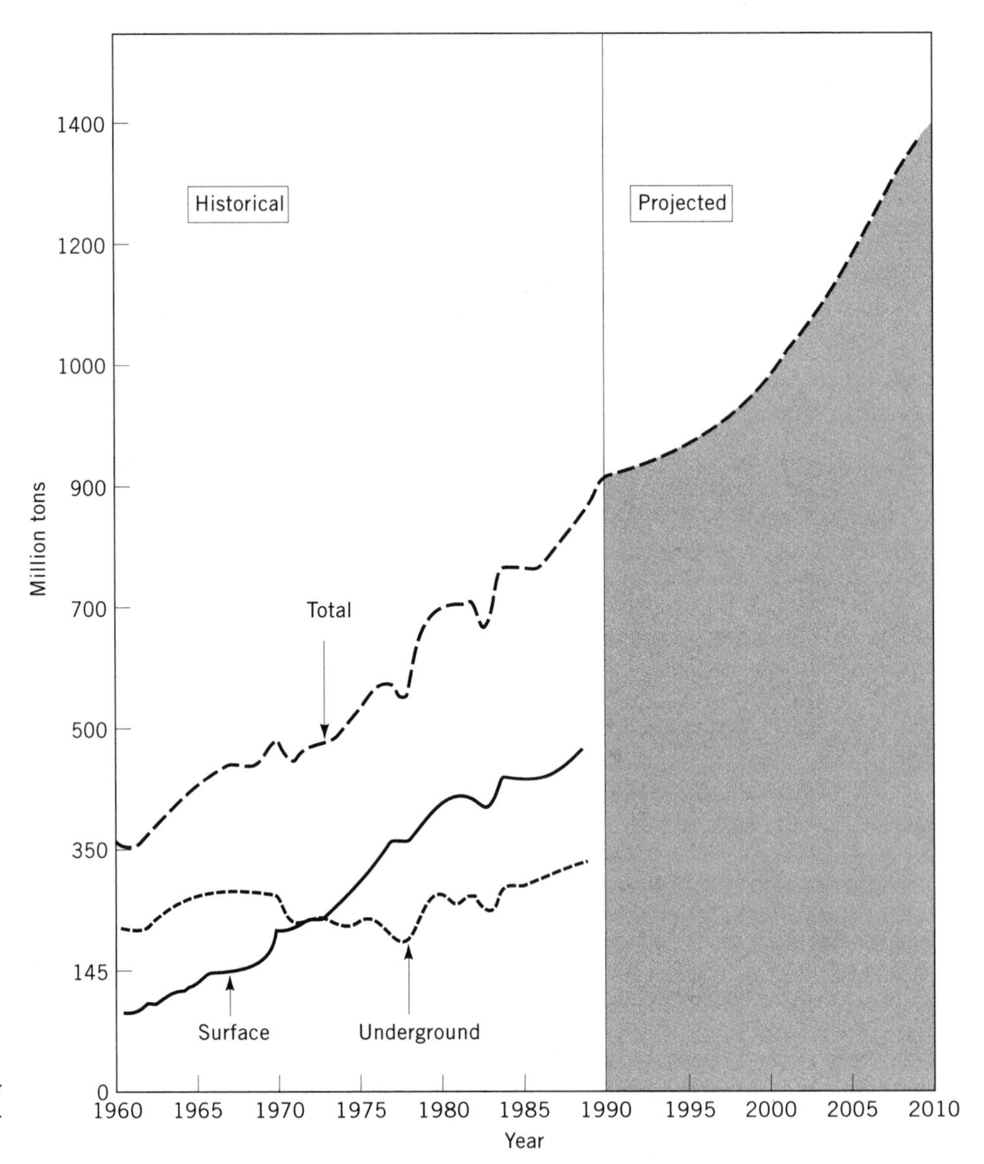

Figure 36. U.S. coal production by mining method (U.S. Energy Information Administration).

Auger mining is a supplementary method used to reach coal in stripped areas where the overburden has become too thick to be removed economically. Large augers are operated from the floor of the surface mines and bore horizontally into the coal face to produce some reserves not otherwise minable.

For a coal deposit where the seam is near the top of a hill, the entire hilltop can be removed to expose the coal.

Underground Mining

Underground mining techniques are somewhat more labor-intensive than surface mining and are used to remove coal located below too much overburden for surface mining; but here, too, machines are used in most instances to dig, load and haul the coal. Access to the coal seam is through a drift (horizontal passage), a slope, or a shaft (Fig. 37), depending on the location of the coal seam.

A *drift mine* is one that enters a coal seam exposed at the surface on the side of a hill or mountain. The mine follows the coal horizontally.

A *slope mine* is one where an inclined tunnel is driven through the rock to the coal, with the mined coal removed by conveyors or truck haulage.

A *shaft mine* is one where a vertical shaft is dug through the rock to reach the coal, which may be at great depth below the surface. The coal then is mined by horizontal entries into the seam, with the resulting coal hoisted to the surface through the vertical shaft.

Two general mining systems are used in underground mining, namely, room-and-pillar and longwall.

Room-and-pillar mining is an open-stopping method where mining progresses in a nearly horizontal or low-angle direction by opening multiple stopes or rooms, leaving solid material to act as pillars to support the vertical load. This system recovers about 50% of the coal and leaves an area much like a checkerboard. It is used in areas where the overlying rock or "roof" has geologic characteristics that offer the possibility of good roof support. This system was used in the old mines where the coal was hand-dug. Two current methods for extracting the coal from the seam are the conventional method, where the

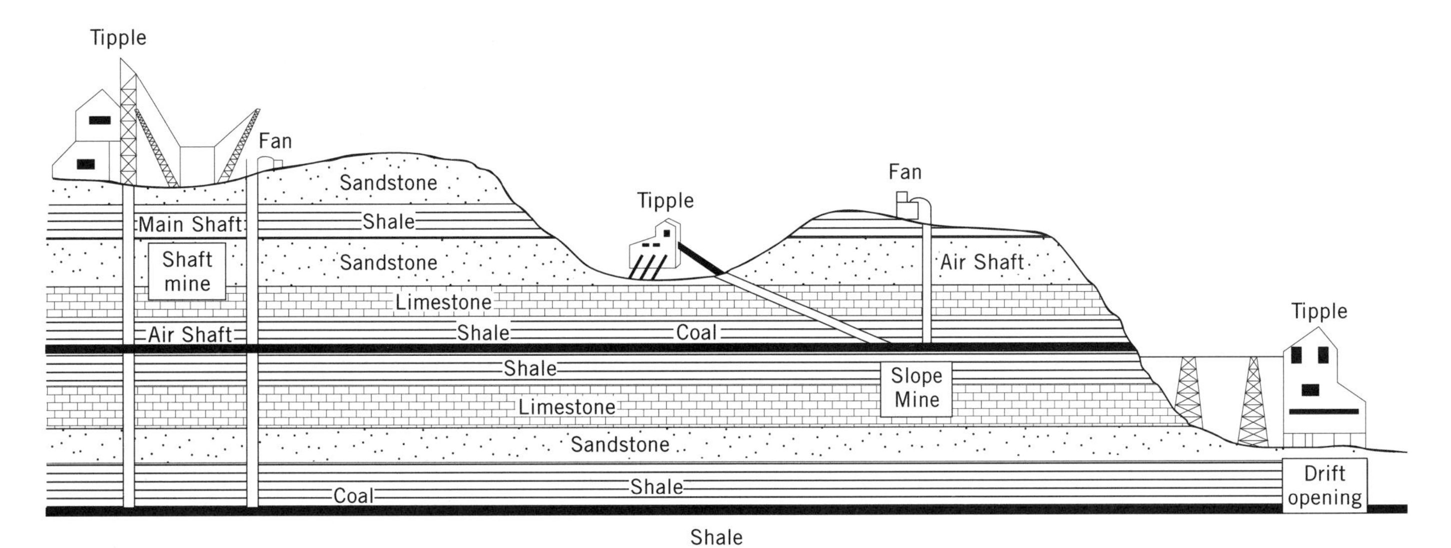

Figure 37. Three types of entrances to underground mines; shaft, slope, and drift (U.S. Bureau of Mines).

coal is undercut and blasted free (Fig. 38), and the continuous method, where a machine moves along the coal face to extract the coal instead of blasting it loose (Fig. 39). Roof control is the major problem of the room-and-pillar method of mining.

Longwall mining uses a machine which is pulled back and forth across the face of the coal seam in larger rooms.

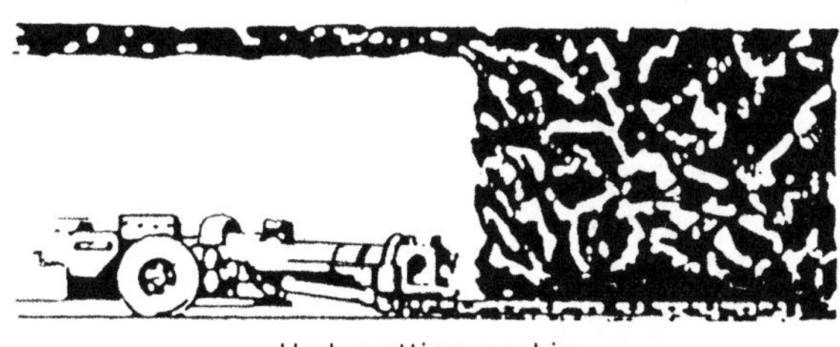

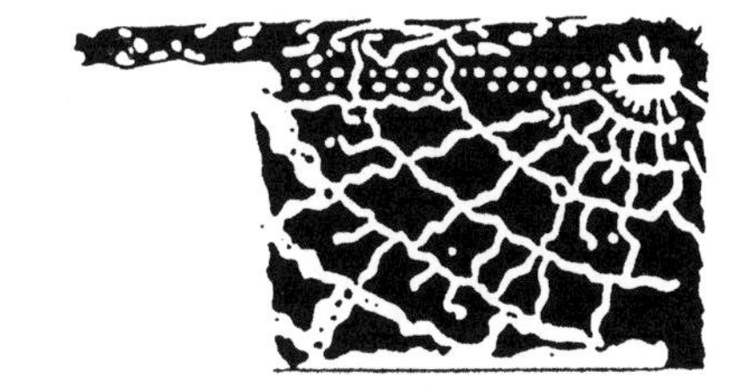

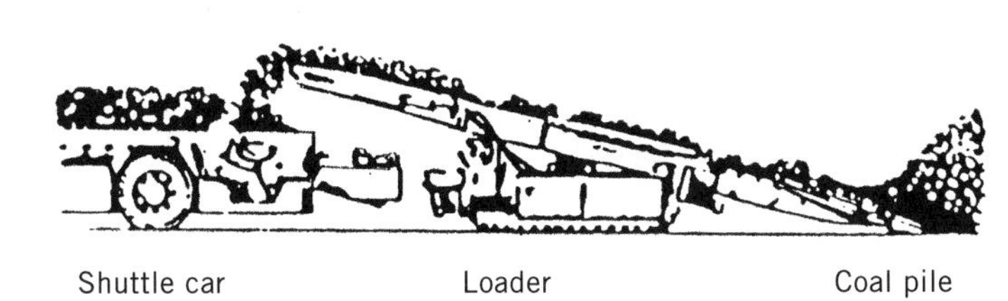

Figure 38. Basic steps in conventional mining (U.S. Bureau of Mines).

It is not a major mining method in the U.S. at this time. Coal recovery using this method is greater than in room-and-pillar mining and can be used where roof conditions are fair to poor. Strong roof rock, however, can be a problem. The seam should be over 42 inches (1.07 m) thick to accommodate this type of operation and the large coal cutter or plow that is used. A large reserve is necessary.

A modification of this method, using a continuous mining machine on faces up to 150 feet (45.7 m) long, is known as *shortwall mining.* It uses the room support system of self-advancing chocks developed for longwall operations.

Coal Preparation

"Coal Preparation or Beneficiation" is a term applied to upgrading of coal to make it suitable for a particular use. It includes blending and homogenization, size reduction and beneficiation or cleaning. It is this last aspect and the degree to which it is required, which most significantly governs the cost of coal preparation. Figure 40 shows levels of cleaning in terms of broad categories.

Only about one-third of the 3.3 Gt of coal produced every year is at present cleaned by breaking, crushing, screening, wet, and dry concentrating processes. This section will briefly describe the processes utilized for coal cleaning.

Coal is generally cleaned by physical methods to remove the mineral matter consisting of rock, slate, pyrite, and other impurities. In general, processes for removal of mineral matter utilize either the differences in density or surface chemical properties of coal and mineral matter. For convenience, coal cleaning can be divided into three parts: coarse-coal cleaning, medium-coal cleaning, and fine-coal cleaning. Figure 41 shows various coal cleaning methods and corresponding size ranges of coal treated.

The first step in most of the coal cleaning operations is size reduction, the main objective being to liberate mineral matter from coal. Size reduction equipment for coal application ranges from heavy-duty crushers and breakers, capable of crushing lumps of up to a meter cube in

Figure 39. Continuous mining machine (U.S. Bureau of Mines).

size, to coal pulverization equipment capable of milling coal to a fine powder. The hardness and the softness of the coal is defined in terms of Hardgrove Grindability Index (HGI). A high (>80) HGI indicates soft coal, easy to grind, and a low (~35) HGI indicates difficult-to-grind coal.

After the coal is crushed it is generally screened to separate the raw coal into various sizes for cleaning operations. Screening of coal above 12-mm size is usually is carried out dry. Double-deck vibrating screens mounted at 17–20° are commonly used for this purpose. For sizing below 12 mm, wet screening is done, using either high frequency vibrating screens or the Sieve Bend. Details on size reduction can be found in various books on coal preparation (138,139).

The specific gravity of coal ranges from 1.23–1.72 depending on rank, moisture, and ash content. The specific gravity of coal increases with rank. Shale, clay, and sandstone have a specific gravity of about 2.6 and mineral pyrite (FeS_2) has a specific gravity of about 5.0. These differences in specific gravities of coal and mineral matter suggest that their separation could be achieved by using a liquid with density in between coal and mineral matter. To determine the preparation method and to identify the best specific gravity for efficient separation of coal and mineral matter "washability" studies (or float-sink analysis) are conducted in the laboratory on a representative sample of coal. Coal is ground to a certain size and is mixed with organic liquids of various specific gravities starting from 1.30 through 1.70. The float and sink fraction of each specific gravity is separated, dried, weighed, and analyzed for ash content. The data is mathematically combined on a weighted basis into cumulative float and cumulative sink and used to develop the washability curves that are characteristic for the coal. A typical washability curve is shown in Figure 42. The scale value of near-gravity material which is determined from the washability curve is listed below. A 0 to 7% near gravity material indicates easy to clean coal, and above 25% represents impossible to clean coal. From Figure 42, it can be seen that if a 5.8% ash clean coal is desired, then cleaning should be performed at a specific gravity of 1.48 and recovery of coal would be about 85.5%. The near specific gravity material will be about 10%, which means it is moderately difficult to clean coal. Scale of value of near-gravity material:

Quantity within +0.10 sp gr Range, %	Separation Problem
0–7	Simple
7–10	Moderately difficult
10–15	Difficult
15–20	Very difficult
20–25	Exceedingly difficult
Above 25	Formidable

Coarse-Coal Cleaning

Jigging and dense-medium separation are generally used for coarse coal cleaning.

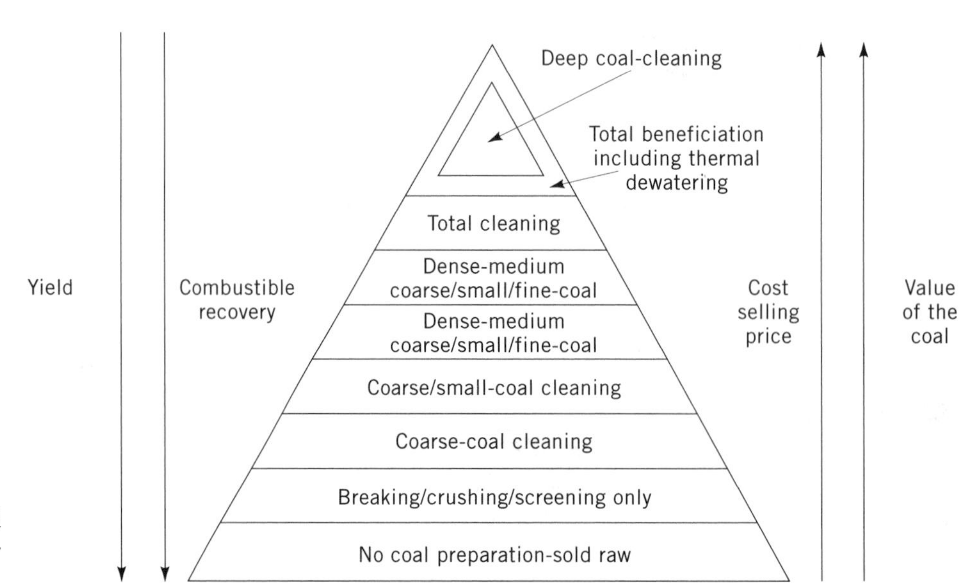

Figure 40. Different levels of coal cleaning and their effect on coal recovery and economics.

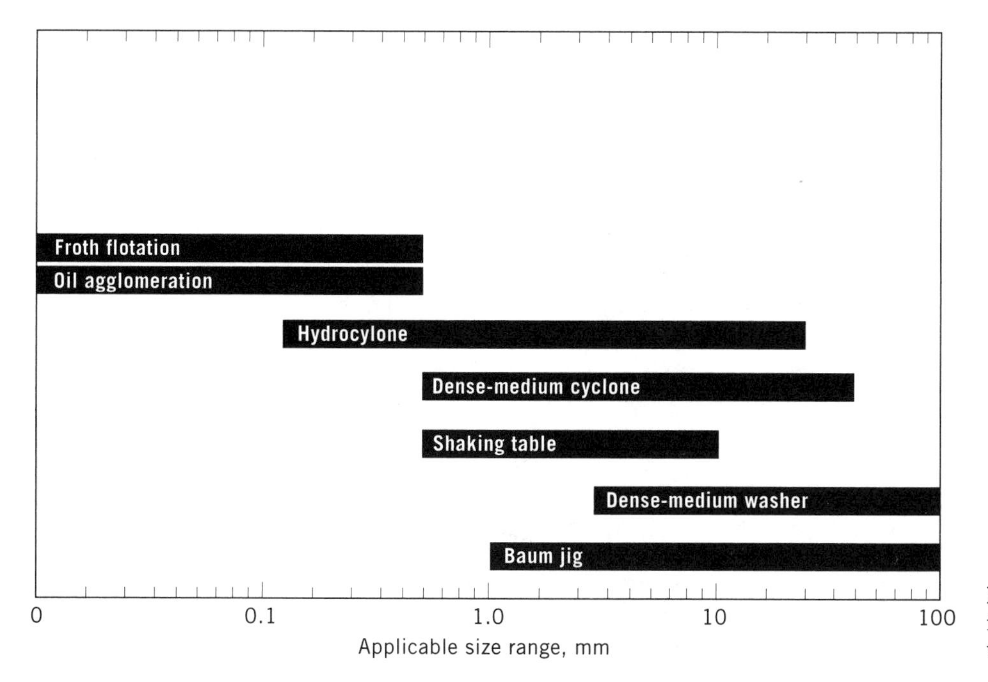

Figure 41. Coal cleaning equipment in common use in the coal industry with respect to the coal size cleaned.

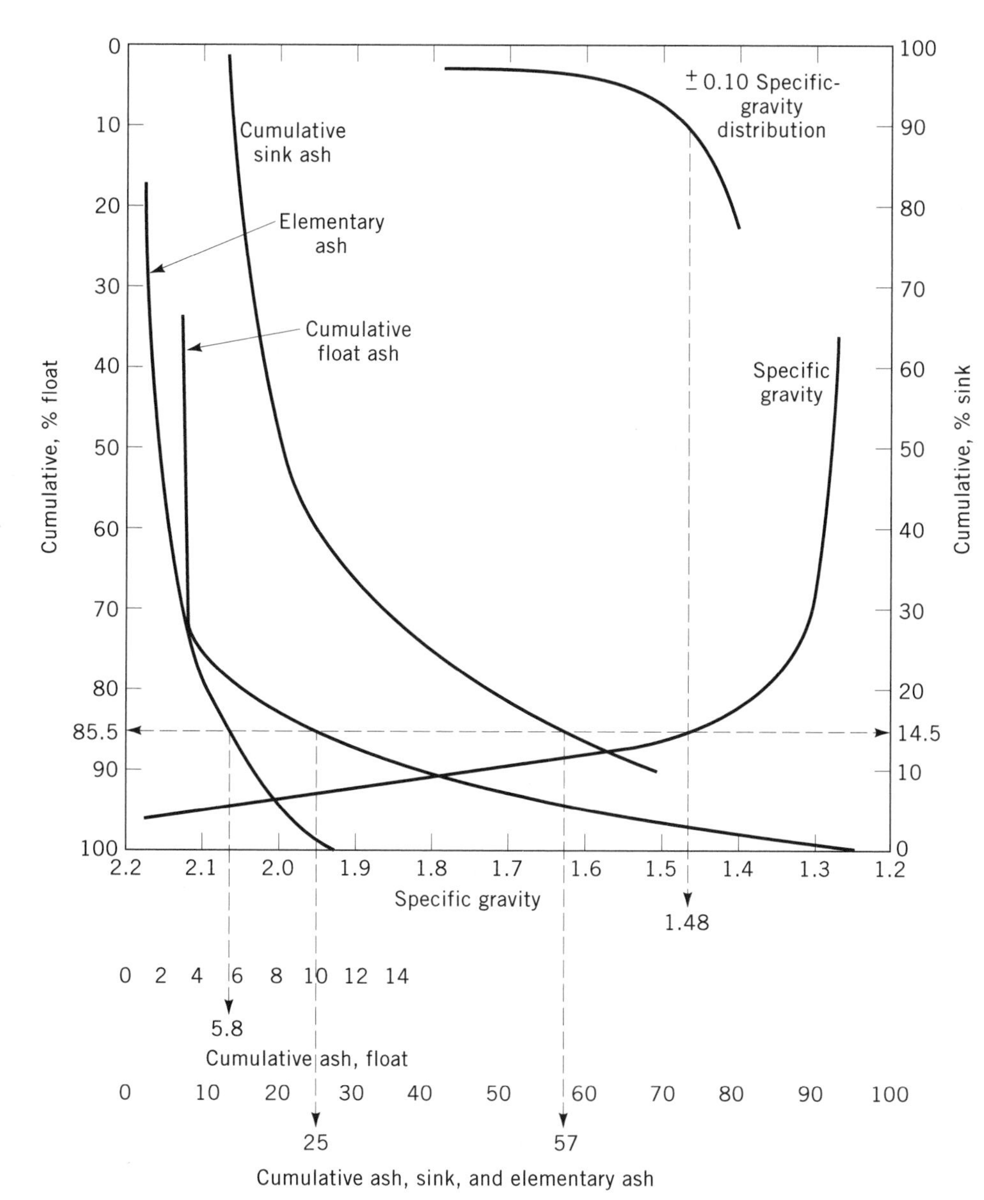

Figure 42. Washability curves showing specific gravity (yield), cumulative % ash float and ash sink, elementary ash, and ±0.10% specific gravity distribution curve.

Jigging. Jigging is among the oldest of mineral processing methods and one of the first adopted for coal cleaning. The separation of coal from mineral matter is accomplished via fluidized bed created by pulsing a column of water which produces a stratifying effect on the raw coal. Water is pulsated through the perforated bottom of an open top rectangular box. The lighter coal particles rise to the top, overflow the end of the jig and are removed as clean product. The denser mineral matter settles on a screen plate and is removed as refuse. Most jigs are of the Baum type (Fig. 43), in which the pulsations are generated by air pressure.

Dense-Medium Washing. Dense-medium separations include processes which clean raw coal by immersing it in a fluid medium having a density intermediate between clean coal and reject. Most dense medium washers use a suspension of fine magnetite in water to achieve the desired density. The process is very effective in providing a sharp separation and is relatively low in capital and operating costs.

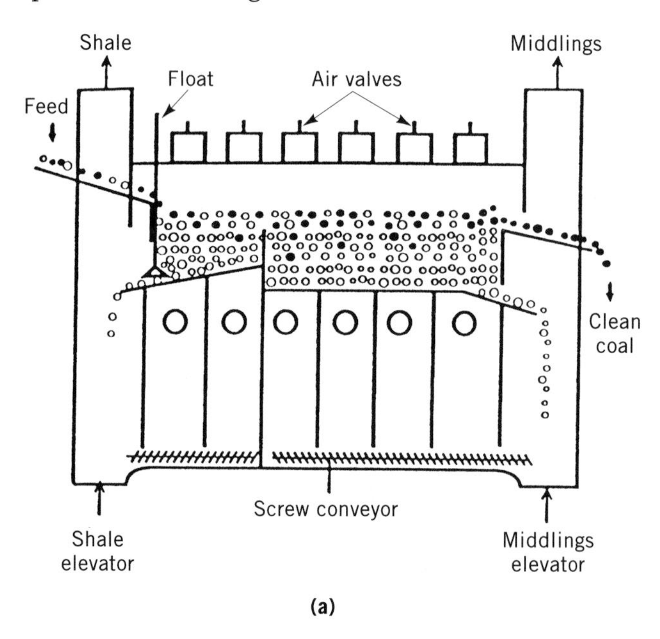

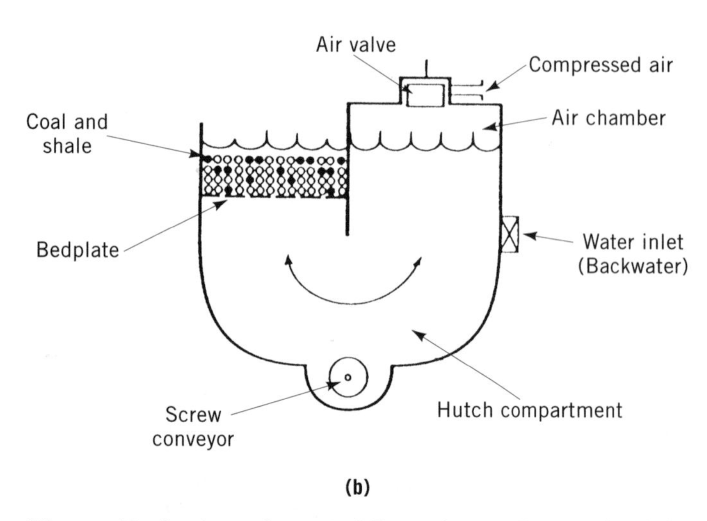

Figure 43. Sections of a typical Baum jig. (**a**) Longitudinal (**b**) Cross section.

Medium Coal Cleaning

Medium-size coal includes coal ranging in size from 3/4 in. (1.9 cm) to 28 mesh (0.5 mm). The principal techniques utilized are the wet concentrating tables, the dense-medium cyclone, the hydrocyclone and spirals. All of these equipment types are widely used in the coal industry.

Wet Concentrating Table. This is also known as a shaking table and works much like the classic miner's pan. The shape of the table is usually a parallelogram, and the deck of the table is at slope with rubber riffles on it. The upper side of the deck contains a feed box and a wash water box. The deck is vibrated longitudinally with a slow forward stroke and a rapid return. Normal decks are 5 m long along the longest edge and 2.5 m wide at right angles to the length. Such a machine can process up to 12 to 15 tons of coal per hour.

Dense-Medium Cyclone. The essential features of a cyclone are shown in Figure 44. A mixture of raw coal and dense medium (magnetite suspension) enters the cyclone tangentially near the top, producing free-vortex flow. The refuse moves along the wall and is discharged through the underflow orifice. The clean coal moves radially toward the cyclone axis, passes through the vortex finder to the overflow chamber and is discharged from a tangential outlet.

Cyclones are available in various diameters; a 500-mm diameter unit can process up to 50 tons of coal per hour, whereas one of 750-mm diameter can process up to 120 tons per hour.

Some other forms of the heavy-medium cyclone are the Vorsyl Separator, the Larcodems, the Dynawhirlpool, and the Tri-flo Separator. Each of these units utilizes cycloning principles.

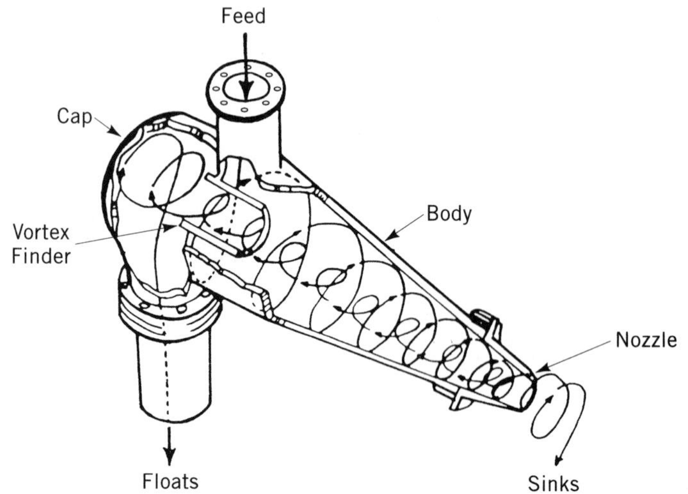

Figure 44. Heavy-medium cyclone separator.

Hydrocyclone. When only water is used in a cyclone for cleaning coal, it is called a hydrocyclone. Hydrocycloning has been applied to process coal finer than 0.5 mm. Its design differs from that of the dense-medium cyclone by having a greater cone angle and a longer vortex finder.

Spiral Separator. The spiral separator is usually 8 to 10 ft (2.44 to 3.05 m) in height and consists of a trough which goes downward in a spiral. The coal slurry is fed in at the top and as it follows the spiral down, centrifugal force separates the coal from the denser mineral matter. A splitter box at the bottom of the spiral can be adjusted to obtain the desired separation.

Fine Coal Cleaning

Coal below 0.5 mm size is classified as fine coal. This size coal is generally processed by surface chemical methods such as froth flotation. In the froth flotation process, the fine coal slurry, to which a small amount of flotation reagents such as fuel oil and a short chain alcohol (methyl isobutyl carbinol, MIBC) are added, is processed through a flotation cell. In the cell, fine bubbles are generated using either forced air or suction. The coal, being hydrophobic, attaches to the air bubbles and rises to the top where it is removed as froth. The refuse, being hydrophilic, remains in the water and is removed from the bottom. The process is very effective in recovering high-grade coal at moderate cost. A simplified diagram of a froth flotation cell is given in Figure 45.

The reagents used in froth flotation of coal are classified as collectors, frothers, and modifying agents. Collectors (kerosene or No. 2 fuel oil) provide a thin coating of oil on the coal and promote attachment of air bubbles to it. Frother (methyl isobutyl carbinol) or polypropylene glycols promotes a stable froth. Modifying agents have several functions such as depressing unwanted material, altering the surface of the particles to aid attachment of air bubbles, and providing acidity or alkalinity to the flotation pulp.

Recently, an advanced froth flotation technique called column flotation has been shown to be superior in obtaining a low-ash clean coal with high recovery of combustibles. It has been proven to be effective in recovering ultra-fine (minus 74 microns) coal (140,141). A schematic diagram of the column flotation process is shown in Figure 46. The basic difference between the conventional and the column flotation system is in the design, which provides quiescent flotation conditions in the column, thus effectively recovering fine coal. Also, gentle washing of froth at the top of the column removes entrained impurities from the froth, providing a low-ash clean coal product. The columns also have been effective in removing high amounts of liberated pyritic sulfur. In 1989 the University of Kentucky Center for Applied Energy Research designed four 2.4 m diameter columns which were installed at the Powell Mountain Coal Company, St.

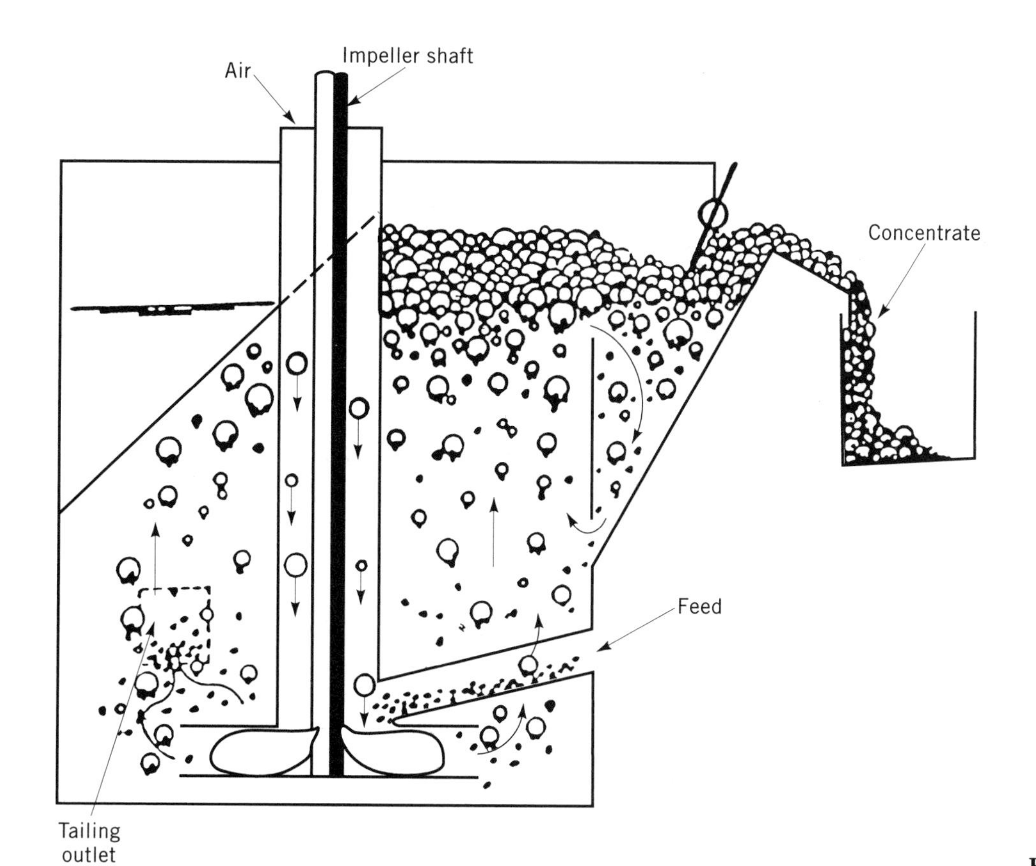

Figure 45. Froth flotation cell.

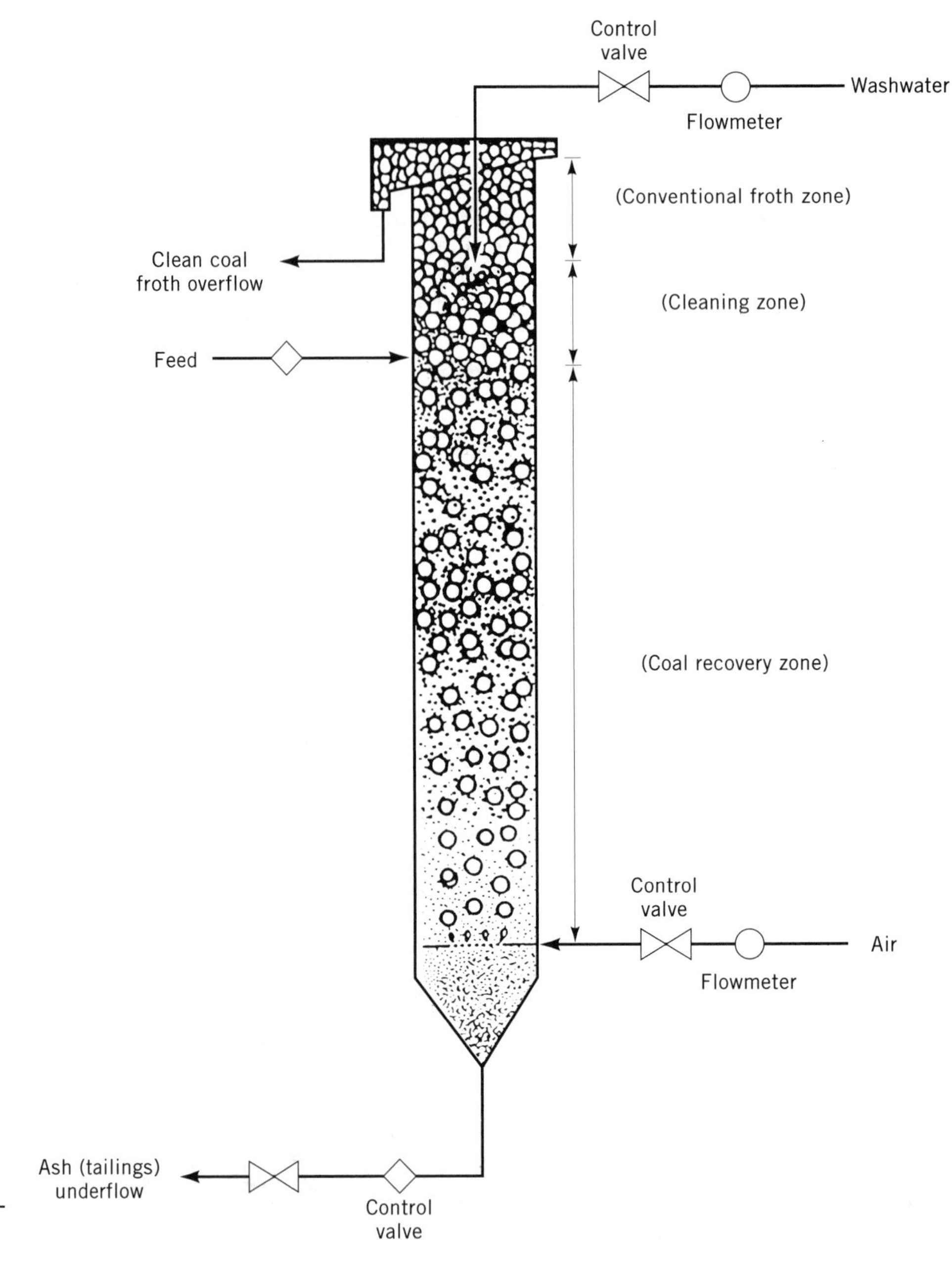

Figure 46. Schematic diagram of column flotation.

Charles, Virginia, for recovery of fine coal from refuse streams.

Recently, an air-sparged cyclone developed at the University of Utah was tested on a pilot scale. In this technique, flotation is performed inside a cyclone that has porous walls through which air is injected.

Other Processes

An oil agglomeration process which utilizes oil or a hydrocarbon to agglomerate coal leaving mineral matter in aqueous suspension has been tested on pilot and commercial scale. The Otisca process, developed by Otisca Industries, Syracuse, New York, utilizes pentane as an agglomerant, which is recovered from clean coal and recycled back in the process. The National Research Council of Canada (NRCC) process utilizes fuel oil as an agglomerant. Another process, the high gradient magnetic and electrostatic cleaning process, utilizes differences in magnetic and electrical charge properties of the mineral matter present on coal. Both of these processes have achieved limited success.

Chemical cleaning processes utilize alkali or acid to leach the impurities present in the coal. TRW's Gravimelt process, which has been tested on a pilot scale, utilizes molten caustic to leach mineral matter. The biggest drawback of chemical cleaning process is the economics.

The chemical methods are not effective in removing major amounts of organic sulfur. Thus, removal of total sulfur to produce a clean burning fuel is very difficult. It is envisioned that the optimum desulfurization of coal will include physical, chemical, and microbial treatment. The raw coal will be ground fine to liberate most of the pyritic sulfur and then processed using the column flotation process to remove most of the liberated mineral matter. Then the clean coal will be treated chemically and finally by microbes to remove organic sulfur from the coal.

Fine Coal Drying. One of the biggest problems in processing fine-size coal is dewatering and drying. Conventional mechanical dewatering devices are effective in dewatering coarse and medium-size coal. For dewatering of fine-size coal, especially if a substantial amount is in the 74 micron range, the conventional techniques provide a dewatered product containing about 30% moisture (142).

Filtration devices that utilize high pressure forces, such as Andritz Hyperbaric and Ama's KDF filters, have the capability of reducing moisture in the filter cake to about 20%. However, the capital and operating costs are high for these devices. Other newly developed techniques which have been tested on a pilot scale include an ultra-high *g* centrifuge, which generates forces up to 4,000 *g* in the centrifuge, and an electro-acoustic technique which utilizes the synergistic effect of electric, ultrasonic, mechanical, and surface chemical forces to remove moisture from the cake.

Thermal drying is generally practiced on coarse and medium-sized coal. Fluidized bed (direct-heating) dryers are most popular in industry. Thermal drying of fine coal is generally avoided due to explosion possibilities and environmental problems.

Re-constitution of fine coal to form pellets and briquettes to improve handling of fines has not been practiced on large scale due to economic reasons. Research efforts are in progress to identify an economical binder, which can provide a low cost pellet or briquette of fine coal.

BIBLIOGRAPHY

1. H. H. Damberger, *Concise Encyclopedia of Mineral Resources,* Oxford, 1989, pp. 47–55.
2. International Energy Agency, *Concise Guide to World Coalfields,* IEA, London, 1983.
3. G. H. Wood, Jr., and co-workers, *Coal Resource Classification System of the U.S. Geological Survey,* U.S. Geological Survey Circular 891, 1983.
4. *International Energy Annual 1990,* Energy Information Administration, U.S. Department of Energy, 1992.
5. H. Flores-Williams, "Chilean, Argentine, and Bolivian Coals," in F. E. Kottlowski and co-workers, *Coal Resources in the Americas,* Geological Society of America Special Paper 179, 1978.
6. F. A. J. Bergmann, "Coal Resources of the Argentine Republic," in F. E. Kottlowski and co-workers, *Coal Resources in the Americas,* Geological Society of America Special Paper 179, 1978.
7. W. W. Olive, "Coal Deposits of Latin America," in F. E. Kottlowski and co-workers, *Coal Resources in the Americas,* Geological Society of America Special Paper 179, 1978.
8. J. Nahuys and S. Piatnicki, *Commun. Serv. Geol. Portugal* **70,** 175–204 (1984).
9. S. Walker, *Geology and Economics of Major World Coalfields,* International Energy Agency, London, 1993.
10. R. da C. Lopes, *Estudos Tecnológicos* **13** (31), 91–112 (1990).
11. M. Marques-Toigo and Z. C. Corrêa da Silva, *Commun. Serv. Geol. Portugal* **70,** 151–160 (1984).
12. Z. Corrêa da Silva, *An. Acad. Brasil. Ciênc.* **44,** 321–331 (1972).
13. E. R. Machado, "Gondwana Coal in Southern Brazil," *Gondwana Geology,* Int. Union Geol. Sci, 1975.
14. E. R. Machado, "Coal in Brazil," in F. E. Kottlowski and co-workers, *Coal Resources in the Americas,* Geological Society of America Special Paper 179, 1978.
15. B. Alpern and J. Nahuys, *Carvão da Bacia de Morungava-Estudo de Seis Sondagens,* Fundação de Ciência e Tecnologia, Porto Alegre, RS, Brazil, 1985.
16. M. Wolf, *Zbl. Geol. Paläont. Tiel I* **3/4,** 591–602 (1983).
17. D. Suescún-Gomez, "Coal deposits in Colombia," in F. E. Kottlowski and co-workers, *Coal Resources in the Americas,* Geological Society of America Special Paper 179, 1978.
18. R. Duran and co-workers, "Evaluación de Reservas de Carbon en Siete Zonas de Colombia," *Geologicas Especiales de Ingeominas* **6,** 1981.
19. T. Bar, *Biuletyn Instytutu Geologicznego* **338,** 71–80 (1982).
20. L. U. Vasquez, "Venezuelan Coal: new projects, new perspectives," in A. H. Pelofsky, *Coal Trading Worldwide: volume IV,* AER Enterprises, 1991.
21. P. Vetter, *Bull. Centres Rech. Explor.-Prod. Elf-Aquitaine* **5,** 291–318 (1981).
22. O. H. Bohnenberger and G. Dengo, "Coal resources in Central America," in F. E. Kottlowski and co-workers, *Coal Resources in the Americas,* Geological Society of America Special Paper 179, 1978.
23. G. H. Wood, Jr., and W. V. Bour, III, *Coal Map of North America,* US Geological Survey Special Geologic Map, 1988, 44 p.
24. JICA, *Prefeasibility study report for the Baja Talamanca coal development project in the Republic of Costa Rica,* Japan International Cooperation Agency, 1983.
25. K. Bolanos, E. R. Landis, S. B. Roberts, and J. N. Weaver, *Coal exploration stage I, Uatsi Project, Baja Talamanca, Costa Rica results and recommendations,* U.S. Geological Survey Open-file report 86-121, 1986.
26. E. R. Landis and J. N. Weaver, *Coal Exploration in Costa Rica: A project assessment,* US Geological Survey Open-file report 89-426, 1989.
27. J. Ojeda-Rivera, "Main coal regions of Mexico," in F. E. Kottlowski and co-workers, *Coal Resources in the Americas,* Geological Society of America Special Paper 179, 1978.
28. L. Toron and S. Cortes-Obregon, *Mining Eng.* **6,** 505–509 (1954).
29. E. T. Dumble, *Geol. Soc. America Bull.* **11,** 10–14 (1900).
30. E. J. Schmitz, *Trans. AIME* **13,** 388–405 (1885).
31. E. G. Tuttle, *Eng. and Mining J.* **58,** 390–392 (1894).
32. E. Ludlow, *Trans. AIME* **32,** 140–156 (1902).
33. J. G. Aguilera, *Eng. and Mining J.* **88,** 730–733 (1909).
34. *Keystone Coal Industry Manual,* Maclean Hunter Publications, Chicago, Ill., 1992.
35. G. G. Smith, *Coal Resources of Canada,* Geological Survey of Canada Paper 89-4, 1989.

36. J. R. Nurkowski, *Amer. Assoc. Petrol. Geol. Bull.* **68,** 285–295 (1984).
37. P. A. Hacquebard and A. R. Cameron, *Int. Jour. Coal Geology* **13,** 207–260 (1989).
38. W. Kalkreuth, W. Langenberg, and M. McMechan, *Int. Jour. Coal Geology* **13,** 261–302 (1989).
39. R. M. Bustin and I. Moffat, *Int. Jour. Coal Geology* **13,** 303–326 (1989).
40. W. Langenberg, W. Kalkreuth, and R. Dawson, *Geologie en Minjbouw* **68,** 241–252 (1989).
41. P. A. Hacquebard, *CIM Bull.* **79,** 67–78 (1986).
42. A. P. Sanada, *Coal* **97,** 48–52 (April 1992).
43. H. H. Damberger, "Coalification in North American coal fields," in H. J. Gluskoter and co-workers, *The Geology of North America, vol. P-2, Economic Geology, U.S.,* Geological Society of America, 1991.
44. A. C. Donaldson and C. Eble, "Pennsylvanian Coals of Central and Eastern United States," in H. J. Gluskoter and co-workers, *The Geology of North America, vol. P-2, Economic Geology, U.S.,* Geological Society of America, 1991.
45. T. L. Phillips and A. T. Cross, "Paleobotany and Paleoecology of Coal," in H. J. Gluskoter and co-workers, *The Geology of North America, vol. P-2, Economic Geology, U.S.,* Geological Society of America, 1991.
46. J. C. Hower, J. R. Levine, J. W. Skehan, SJ, E. J. Daniels, S. E. Lewis, A. Davis, R. J. Gray, and S. P. Altaner, "Appalachian anthracites," *Org. Geochem.* **20** (1993).
47. G. H. Wood, Jr., T. M. Kehn, and J. R. Eggleston, "Depositional and Structural History of the Pennsylvania Anthracite Region," in P. C. Lyons and C. L. Rice, *Paleoenvironmental and Tectonic Controls in Coal-forming Basins of the United States,* Geological Society of America Special Paper 210, 1986.
48. J. C. Ferm and P. J. Muthig, *A Study of the United States Coal Resources,* U.S. Department of Energy, DOE/ET-12548-15, 1982.
49. J. R. Eggleston, M. D. Carter, and J. C. Cobb, *U.S. Geol. Survey Circular* **1055,** 15 (1990).
50. J. C. Hower and A. Davis, *Geol. Soc. Amer. Bull.* **92** (1) 350–366 (1981).
51. J. R. Levine and A. Davis, *Geol. Soc. Amer. Bull.* **95,** 100–108 (1984).
52. D. R. Chesnut, *Stratigraphic analysis of the Carboniferous rocks of the Central Appalachian Basin,* Ph. D. thesis, The University of Kentucky, Lexington, 1988.
53. S. E. Lewis and J. C. Hower, *J. of Geol.* **98,** 927–942 (1990).
54. J. C. Hower and S. M. Rimmer, *Org. Geochem.* **17,** 161–173 (1991).
55. J. R. Levine and W. R. Telle, "Coal rank patterns in the Cahaba coal field and surrounding areas, and their significance," in W. A. Thomas and W. E. Osborne, *Mississippian–Pennsylvanian Tectonic History of the Cahaba Synclinorium,* Alabama Geol. Soc., 28th ann. field trip, 1991.
56. S. F. Greb, D. A. Williams, and A. D. Williamson, *Geology and Stratigraphy of the Western Kentucky Coal Field,* Kentucky Geological Survey, Series XI, Bull. 2, 1992.
57. G. D. Ells, *The Mississippian and Pennsylvanian (Carboniferous) Systems in the United States–Michigan,* US Geological Survey Professional Paper 1110-J, 1979.
58. D. A. Williams, C. T. Helfrich, J. C. Hower, F. L. Fiene, A. E. Bland, and D. W. Koppenaal, *The Amos and Foster Coals–Low-ash and Low-sulfur Coals of Western Kentucky,* Kentucky Geological Survey, Series XI, Information Circular 8, 1990.
59. T. J. Evans, *Bituminous Coal in Texas,* Texas Bureau of Economic Geology, Handbook 4, 1974.
60. P. K. Mukhopadhyay, *Organic petrography and organic geochemistry of Texas Tertiary coals in relation to depositional environment and hydrocarbon generation,* Texas Bureau of Economic Geology, Report of Investigations 188, 1989.
61. J. A. Luppens, "Exploration for Gulf Coast lignite deposits: Their Distribution, quality, and reserves," in J. E. Johnston, *Lignite of the Gulf Coastal Plain, Louisiana and Texas,* Geological Society of America field trip guidebook, 1982.
62. R. M. Flores and T. A. Cross, "Cretaceous and Tertiary Coals of the Rocky Mountains and Great Plains Regions," in H. J. Gluskoter and Others, *The Geology of North America, vol. P-2, Economic Geology, U.S.,* Geological Society of America, 1991.
63. F. J. Rich, R. K. Dorsett, and C. R. Chapman, "Petrography, Sedimentology, and Geochemistry of the Wyodak Coal, Wyodak, Wyoming," *Wyoming Geological Association Guidebook,* 39th field conf., 1988.
64. P. D. Warwick and R. W. Stanton, *Org. Geochem.* **12,** 389–399 (1988).
65. P. D. Warwick, *Depositional Environments and Petrology of the Felix Coal Interval (Eocene), Powder River basin, Wyoming,* Ph. D. thesis, The University of Kentucky, Lexington, 1985.
66. F. W. Campbell and G. H. Roybal, "Characterization of New Mexico coals, Menefee and Crevasse Canyon Formations," in G. H. Roybal and co-workers, *Coal Deposits and Facies Changes Along the Southwestern Margin of the Late Cretaceous Seaway, West-central New Mexico,* New Mexico Bur. Mines & Mineral Resources Bull. 121, 1987.
67. A. T. Cross, "Coals of far-western United States," in H. J. Gluskoter and co-workers, *The Geology of North America, vol. P-2, Economic Geology, U.S.,* Geological Society of America, 1991.
68. R. Choate and C. A. Johnson, "Geologic Overview, Coal and Coalbed Methane Resources of the Western Washington Region," TRW Energy Systems Group, Lakewood, Colorado.
69. R. D. Merritt, "Paleoenvironmental and Tectonic Controls in Major Coal Basins of Alaska," in P. C. Lyons and C. L. Rice, *Paleoenvironmental and Tectonic Controls in Coal-forming Basins of the United States,* Geological Society of America Special Paper 210, 1986.
70. G. D. Stricker, "Economic Alaskan coal deposits," in H. J. Gluskoter and co-workers, *The Geology of North America, vol. P-2, Economic Geology, U.S.,* Geological Society of America, 1991.
71. P. D. Rao and J. E. Smith, *Petrology of Cretaceous Coals from Northern Alaska,* U.S. Department of Energy, DE-FG22-80 PC 30237, MIRL report no. 64, 1983.
72. M. Daniel and E. Jamieson, *Coal Production Prospects in the European Community.* IEACR, 48, 1992, 78 pp.
73. World Energy Conference, *Survey of Energy Resources,* WEC, London, 1989, 190 pp.
74. G. Y. Craig, ed., *Geology of Scotland,* 3rd Ed, Geological Society of London, 1991.
75. A. Trueman, ed., *The Coalfields of Great Britain,* Edward Arnold, London, 1954.
76. C. R. Ward, ed., *Coal Geology and Coal Technology,* Blackwell Scientific Publications, Oxford, 1984, 345 pp.
77. P. McL. D. Duff, and A. J. Smith, eds., *Geology of England and Wales,* Geological Society of London, 1992.
78. G. R. Crouch, *Lignite Resources and Characteristics,* IEA Coal Research 13, 1988, 102 pp.

79. M. Teichmüller, in J. Brooks, J. C. Goff, and B. Van Hoorn, eds., *Habitat of Palaeozoic Gas in NW Europe,* Geological Society Special Publication 23, 1986, pp. 101–112.
80. K. Strehlau, *Int. J. of Coal Geol.* **15,** 245–292 (1990).
81. J. Gliese, and H. Hager, *Geologie en Mijnbouw* **52,** 517–525 (1978).
82. Rheinbraun Braunkohlenwerke, *Focuson Rheinbraun.* Rheinbraun Braunkohlenwerke A G (Rheinbraum) 1985, 46 pp.
83. M. Wolf and J. Tayler, *Geology, Petrography and Geochemistry of the Lignite Deposit of the Lower Rhine Embayment,* Kernforschungsanlage Jülich Gmbtl., 1985, 26 pp.
84. G. Couch, M. Hessling, A.-K. Hjalmarsson, E. Jamieson, and T. Jones, *Coal Prospects in Eastern Europe,* IEA Coal Research **31,** 1990, 117 pp.
85. G. Doyle, *Prospects for Polish and Soviet Coal Exports,* IEA Coal Research, **16,** 70 pp., 1989.
86. A. R. Dogru, A. F. Gaines, I. S. Gökcen, S. L. Gökcen, and M. Wolf, *Int. J. of Coal Geol.* **10,** 193–201 (1988).
87. J. Pesek, and J. Vozar, eds., *Coal-bearing Formations of Czechoslovakia,* Dionyz Stur Inst. of Geol., Bratislava, 1988, 401 pp.
88. H. Sasaguri, *Coal in Asia-Pacific,* **3**(2), 8–17 (1991).
89. D. E. Aizenverg, ed., *Field Excursion Guidebook for the Donetz Basin,* VII. Int. Congr. Carb. Strat. Geol., Moscow, 1975, 360 pp.
90. D. V. Nalivkin, *Geology of the U.S.S.R.* Oliver and Boyd, Edinburgh, 1973, 855 pp.
91. V. E. Evtushenko, *Field Excursion guidebook for the Kuznetsk Basin,* VIII Int. Cong. Carb. Strat. Geol., Moscow, 1975, 172 pp.
92. Dai-Sung, Lee, *Geology of Korea,* Lyohak-Sa Publishing Co., 1988, 514 pp.
93. M. S. Massoud, S. D. Killops, A. C. Scott, and D. P. Mattey, *J. of Petroleum Geol.* **14,** 365–386 (1991).
94. G. Doyle, *Coal supply prospects in Asia/Pacific region,* IEA Coal Research 26, 1990, 84pp.
95. Joo Heon Park, *Coal in Asia Pacific,* **3,** (3), 78–85 (1991).
96. Ministry of Energy and Resources, *Yearbook of Coal Statistics,* Korea Coal Industry Promotion Board, 1991, 283 pp.
97. Hiroaki Hirasawa, *Coal in Asia-Pacific,* **4**(1) 4–14 (1992).
98. C. R. Ward, *Int. J. of Coal Geol.* **17,** 69–93 (1991).
99. P. Premmani, *Coal in Asia-Pacific* **2**(3) 2–8 (1990).
100. R. Takahashi, and A. Aihara, *Int. J. of Coal Geol.* (1989).
101. M. Inone, *Coal in Asia-Pacific* **3**(1), 1–31 (1991).
102. A. C. Scott and Mao Bangzhuo, *Geol. Today* **9,** 14–18 (1993).
103. G. Doyle, *China's Potential in International Coal Trade,* IEA Coal Research 2, 1987, 94 pp.
104. Yang Zunyi, Cheng Yugi and Wang Hongzhen, *The Geology of China,* Monograph on Geology and Geophysics 3, Oxford University Press, 1987, 303 pp.
105. X. Zhu, *Chinese Sedimentary Basins.* Elsevier, 1989, 238 pp.
106. Guanghua Liu, *Int. J. of Coal Geol.* **16,** 73–117 (1990).
107. H. Huai, A. F. Gaines, and A. C. Scott, *Introduction to the Petrology and Infrared Spectra of Shanxi Coals, P.R. China,* in preparation.
108. Li Sitian, Li Baofang, Yong Shigong, Huang Jiafu, and Li Zhen, "Sedimentation and tectonic evolution of Late Mesozoic faulted coal basins in north-eastern China." *Spec. Publ. Int. Ass. Sediment.* **7,** 387–406, 1984.
109. E. A. Johnson, *Int. J. of Coal Geol.* **14,** 217–236 (1990).
110. Fen Miao, Lijun Qian and Xiuyi Zhang, *Int. J. of Coal Geol.* **12,** 733–765 (1989).
111. B. Lee, "Market trends and implication Bowen Basin coals," *Proceedings Bowen Basin Symposium 1990,* 54–73.
112. New South Wales Department of Mineral Resources, *1993 New South Wales Coal Industry Profile,* 190 pp.
113. Australasian Institute of Mining and Metallurgy, *Coal in Australia,* 1953, 832 pp.
114. D. King, "History of exploration and Development," in D. M. Traves, and D. King, eds., *Economic Geology of Australia and Papua New Guinea, 2,* Coal, Australasian Institute of Mining and Metallurgy, Monograph Series, **6,** 5–24, 1975.
115. *The Miner,* 34–42, (Oct., 1992).
116. C. A. Bacon, "The Coal Resources of Tasmania," *Bulletin Tasmania Geological Survey,* **64,** 1991, 170 pp.
117. A. J. Noldart, "Triassic coal in Tasmania," in D. M. Traves and D. King, eds., *Economic Geology of Australia and Papua New Guinea, 2, Coal,* Australasian Institute of Mining and Metallurgy, Monograph Series, **6,** 264–265, 300–301, 1975.
118. J. L. Knight, "Wonthaggi and Other Cretaceous Black Coal-fields, Victoria," in D. M. Traves, and D. King, eds., *Economic Geology of Australia and Papua New Guinea, 2, Coal,* Australasian Institute of Mining and Metallurgy, Monograph Series, **6,** 272–277, 353–355, 356–358, 1975.
119. R. J. Gaulton, *Latrobe Valley coal deposits notes on the geology,* State Electricity of Victoria, Geoengineering Division, 1990, 11 pp. (unpublished).
120. J. H. Lord, Collie and Wilga Basins, in D. M. Traves, and D. King, eds., *Economic Geology of Australia and Papua New Guinea, 2, Coal,* Australasian Institute of Mining and Metallurgy, Monograph Series, **6,** 272–277, 1975.
121. Department of Mines Western Australia, *Annual Review 1991–1992,* 1992, 156 pp.
122. J. F. Anckhorn, M. P. Cave, J. Kenny, D. R. Lavill, and G. Carr, "The Coal Resources of New Zealand," *Coal Geology Report, 4,* Ministry of Energy, New Zealand, 1988, 43 pp.
123. J. Newman, and N. A. Newman, *Tectonic and paleoenvironmental controls on the distribution and properties of Upper Cretaceous Coals on the West Coast of the South Island, New Zealand,* Geological Society America Special Paper, 267, 1992, pp. 347–368.
124. A. M. Sherwood, J. K. Lindquist, J. Newman, and R. Sykes, *Depositional controls on Cretaceous coals and coal measures in New Zealand,* Geological Society America Special Paper, 267, 1992, pp. 325–346.
125. Soehandojo, *Geologi Indonesia,* **12,** 279–325 (1989).
126. Hadiyanto and M. M. Faiz, *Indonesia's Coal Export Potential and its Prospects,* Second Asia/Pacific Mining Conference and Exhibition, March 14–17, 1990, Jakarta, 1990, pp. 648–663.
127. B. Daulay, *Petrology of Some Indonesian and Australian Tertiary Coals,* PhD Thesis, University of Wollongong, 1985, 265 pp.
128. A. Prijono, *Geologi Indonesia,* **12,** 252–278 (1989).
129. H. K. Mishra, *A Comparative Study of the Petrology of Permian Coals of India and Western Australia,* PhD Thesis, University of Wollongong, 1986, 610 pp.
130. P. K. Gosh, and A. Basu, *Stratigraphic Positions of the Kharharbari in the Lower Gondwanas of India;* in A. J. Amos, ed., Gondwana Stratigraphy, Paris, 1968, pp. 537–550.
131. F. S. J. de Jager, "Coal," in C. B. Coetzee, *Mineral Resources of the Republic of South Africa. Handbook,* Geol. Survey, South Africa, Vol. 7, 1976, pp. 289–330.

132. R. G. Wadley, "A View on the South African Domestic Coal Market," *Abstracts Conference on South Africa's Coal Resources,* Witbank, Transvaal Branch Geological Society of South Africa, 6–9 November, 1991.
133. A. M. Anderson, and I. R. McLachlan, "The Oil-shale Potential of the Early Permian White Band Formation in Southern Africa," in A. M. Anderson, and W. J. van Biljon, *Some Sedimentary Basins an Associated Ore Deposits of South Africa,* Geological Society of South Africa Special Publication, Vol. 6, 1979, pp. 83–89.
134. E. P. Saggerson, "Distribution of Rank in South Africa," *Abstracts Conference on South Africa's Coal Resources, Witbank, Eastern Transvaal Branch Geological Society of South Africa, 6–9 Nov., 1991.*
135. J. C. Dreyer, "Waterberg Coalfield," *Abstracts Conference on South Africa's Coal Resources,* Witbank, Eastern Transvaal Branch Geological Society of South Africa, 6–9 Nov. 1991.
136. A. B. Cadle, B. Cairncross, and M. F. Winter, "The Transvaal Coalfield," *Abstracts Conference on South Africa's Coal Resources,* Witbank, Eastern Transvaal Branch Geological Society of South Africa, 6–9 Nov., 1991.
137. A. B. Cadle, B. Cairncross, A. D. M. Christie, and D. L. Roberts, "The Permo-Triassic coal-bearing deposits of the Karoo Basin South Africa," *Abstracts Conference on South Africa's Coal Resources,* Witbank, Eastern Transvaal Branch Geological Society of South Africa, 6–9 Nov., 1991.
138. J. W. Leonard, ed., *Coal Preparation,* 5th ed., American Institute of Mining, Metallurgical and Petroleum Engineers, Inc., Littleton, Colorado, 1992.
139. D. G. Osborne, *Coal Preparation Technology,* Graham and Trotman, 1989.
140. R. P. Killmeyer, R. E. Hucko, and P. S. Jacobsen, "Interlaboratory Comparison of Advanced Froth Flotation Processes," Preprint No. 89-137, Society of Mining Engineers, Inc., Littleton, Colorado, 1989.
141. B. K. Parekh, J. G. Groppo, W. F. Stotts, and A. E. Bland, "Recovery of Fine Coal from Preparation Plant Refuse Using Column Flotation," K.V.S. Sastry, ed., Column Flotation '88, Society of Mining Engineers, Inc., Littleton, Colorado, 1988.
142. B. K. Parekh, and A. E. Bland, "Fine Coal and Refuse Dewatering–Present State and Future Consideration," B. M. Moudgil and B. J. Scheiner, eds., *Flocculation and Dewatering,* Engineering Foundation, New York, 1989.

COAL QUALITY

RONALD W. STANTON
ROBERT B. FINKELMAN
U.S. Geological Survey
Reston, Virginia

Coal is a heterogeneous and complex material having a wide range of physical and chemical properties. These properties play a key role in determining if a coal deposit has commercial value and, if so, how that coal should be best used. Coal quality can be defined in the following way:

Coal Quality, n–compositional and physical characteristics or properties of coal that enhance, restrict, or limit its use as a fuel, source of fuels, products or byproducts, or a raw material in the manufacturing of products.

DEFINITION OF COAL

Coal, as defined by the American Society for Testing and Materials (1): "is a brown to black combustible sedimentary rock (in the geological sense) composed principally of consolidated and chemically altered plant remains." These plant remains had been originally composed of compounds rich in carbon, hydrogen, and oxygen–elements that are the primary contributors to the energy potential of coal. Coal occurs as rock layers or beds associated with other sedimentary rock units such as shales and sandstones.

Coal consists of various proportions of complex organic compounds, inorganic compounds (mainly minerals), liquids, and gases. From the aspect of a fuel, coal is a combustible material that, after burning, leaves a noncombustible residue–ash. By definition, coal yields less than 50% ash material upon combustion (2). The ash consists primarily of inorganic compounds, principally minerals. In its natural state, coal contains volatile materials (gases and petroleum-like oils) that, during combustion, can add to the energy output of the fuel. Coal, having the appropriate quality characteristics may be converted to liquid or gaseous fuels and can also be a source of natural gas (methane).

Coal is not only a source of energy but also has been used in the manufacture of drilling fluids and briquettes and as a source of hundreds of different chemicals used in the preparation of medicines, dyes, paints, and perfumes (3). Coal is also the source of coke used in the smelting of iron.

For maximum efficiency or profit, each usage requires coal having specific characteristics or properties. In short, the use of coal is dependent on its quality.

CHARACTERISTICS

Coal may be considered as a rock primarily composed of microscopically identifiable altered plant remains and minerals. The altered plant remains can be observed in coal as macerals. Macerals are microscopic constituents that are classified into three groups: vitrinite, which is derived from coalified plant tissues and gels; liptinite, which is formed from hydrogen-rich protective coatings on plants and plant reproductive organs and resinous cell fillings; and inertinite, which results from the oxidation of vitrinite and liptinite precursors. These macerals retain certain characteristics diagnostic of their plant origin; however, their original chemical compositions have changed during burial. Minerals, like macerals, are classified by their characteristics as observed under a microscope, but minerals, unlike macerals, have defined chemical compositions and are crystalline. Water and liquid hydrocarbons also are found in coal. The water that is an integral part of the coal bed is referred to as inherent moisture. Some liquid hydrocarbons result from natural fermentation of plant material in the early coal-forming stages or are released from thermal cracking of hydrogen-rich compounds during coalification. Gases, primarily methane and carbon dioxide, also are commonly trapped

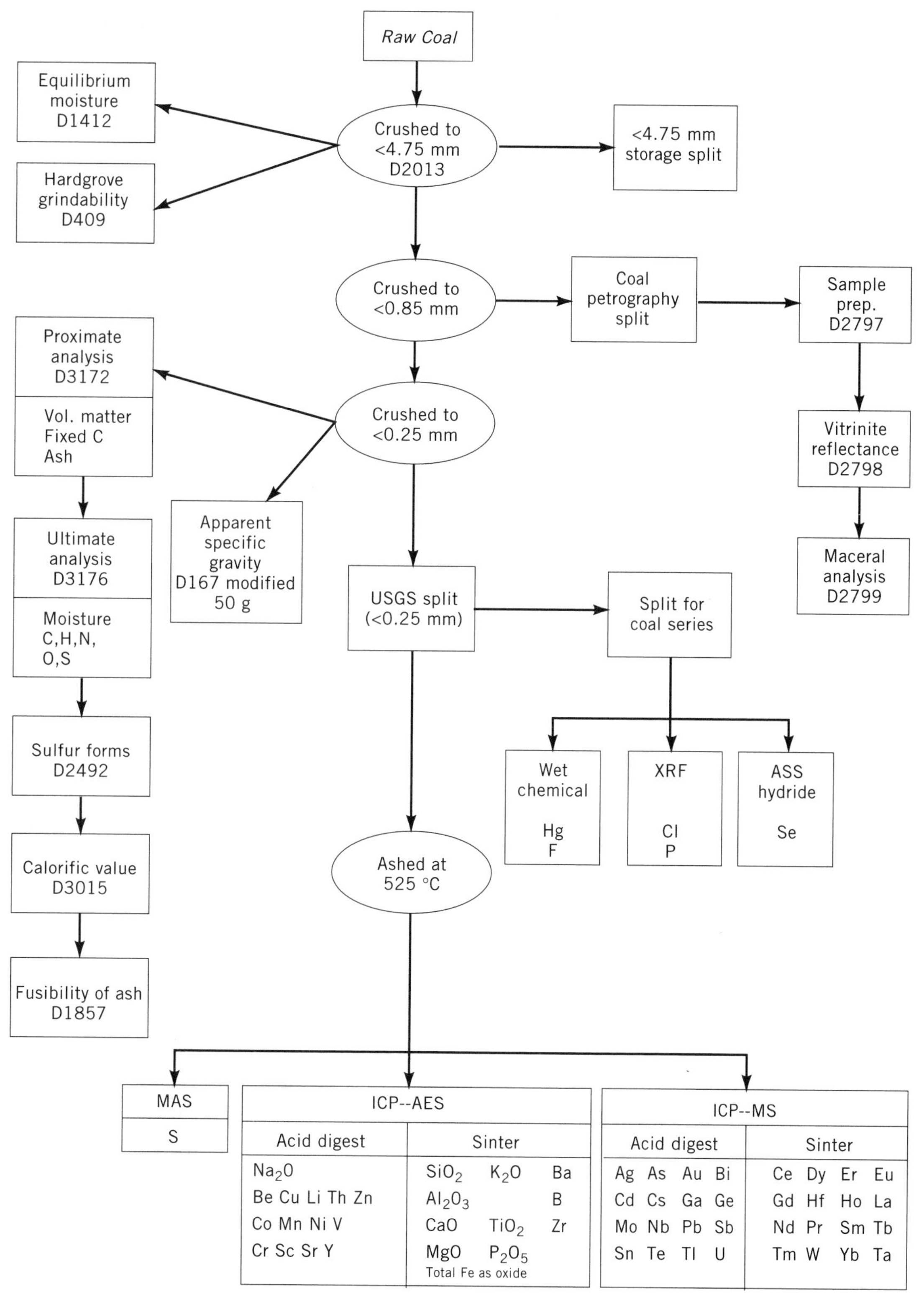

Figure 1. Flow diagram of procedures used for the analysis of coal samples collected by U.S. Geological Survey

in coal beds after being released from plant matter and macerals during biogenic degradation and by thermal cracking of hydrocarbons during coalification.

If coal is considered as a chemical substance, the major chemical elements that compose the macerals are (in order of abundance) carbon, oxygen, hydrogen, nitrogen, and sulfur. Silicon, aluminum, iron, sodium, potassium, calcium, magnesium, and sulfur are the principal chemical elements in the minerals. Both minerals and macerals contain minor to trace amounts of many other elements. The following discussions will demonstrate how these elements profoundly affect coal use.

COAL PROPERTIES

Physical characteristics of coal include such properties as the free swelling index, which is a measure of the agglutinating or agglomerating characteristics of coal (the ability of powdered coal to become plastic upon heating and to form hardened masses upon cooling); bulk density; strength; grindability or resistance to grinding; and degree of cleating (natural fracture). Table 1 contains a list of physical and chemical properties of coal and the analytical tests used to determine them. The standard analyses (1) used to characterize coal composition include: Proximate analysis (moisture, ash, volatile matter, and fixed carbon by difference); Ultimate analysis (moisture, carbon, hydrogen, nitrogen, sulfur, oxygen [indirectly], and ash); forms of sulfur (pyritic, organic, sulfate); calorific value; and fusibility of coal ash (Fig. 1).

The organic petrographic components (macerals) can be estimated petrographically, and their influence on the coal composition can be estimated from bulk chemical data (Table 1). Because macerals differ in their chemical composition, they have different optical properties that can be measured microscopically. The degree of light reflectance of the maceral is the common measure that relates to chemical composition (Table 1). Reflectance of a specific maceral, vitrinite, is the standard used to determine the thermal maturity of coal. Classification of coal by rank, which is based on calorific values and fixed carbon contents (Table 1), is related also to the degree of thermal maturity of the coal.

The trace and minor element contents of coal and coal ash are determined by analytical techniques listed in Table 2. These methods include atomic emission spectroscopy and mass spectroscopy of inductively coupled plasma, neutron activation analysis, atomic absorption spectroscopy, and X-ray fluorescence. These analytical procedures, as applied to coal, are described in ref. 4.

EXAMPLES OF HOW COAL QUALITY AFFECTS COAL USE

Coal utilization can be enhanced by specific properties. For example, a high pyrite content in synthetic fuels is best for catalysis, and a high $CaCO_3$ content in combustion promotes self-cleaning of emissions. High contents of pyrite or potentially hazardous elements are examples of

Table 1. Analytical Tests Used to Characterize the Metallurgical or Steam Coal Quality of Drill Core Samples

Test	Metallurgical Quality	Steam Quality	Appropriate ASTM Standard[a]
Visual description	X	X	none
Chemical			
Proximate	X	X	D3172
Moisture[b]	X	X	D3173
Volatile matter[b]	X	X	D3175
Fixed carbon[b]	X	X	D3175
Ash[b]	X	X	D3174
Equilibrium moisture[c]		X	D1412
Sulfur[b]	X	X	D3177
Ultimate	X	X	D3176
Carbon	X	X	D3178
Hydrogen	X	X	D3178
Nitrogen	X	X	D3179
Oxygen	X	X	D3176
Sulfur	X	X	D3177
Forms of sulfur	X	X	D2492
Chlorine	X	X	D2361, 4208
Ash composition	X	X	D3682
Ash fusion[a]	X	X	D1857
Trace elements		X	D3683
Calorific value[a]	X	X	D3286, 2015
Rheological and Physical			
Gieseler[a]	X[d]		D2639
Dilatation tests (Audibert, Ruhr)	X[d]		none
Free-swelling index[a]	X[d]		D720
Hardgrove index[a]	X	X	D409
Petrographic			
Maceral analysis[a]	X[d]		D2797, 2799
Vitrinite reflectance[a]	X[d]	X	D2797, 2798

[a] Ref. 1.
[b] Minimum test required to properly characterize the coal.
[c] Especially important for drill core samples from low-rank coal.
[d] Bituminous coals only.

Table 2. Analytical Techniques Used to Obtain Coal Quality Information

Technique	Coal Quality Information
Optical emission spectroscopy	Semiquantitative elemental analysis of coal ash
Spark source mass spectroscopy	Semiquantitative elemental analysis of coal ash
Instrumental neutron activation analysis	Quantitative multielement analysis of coal or coal ash
X-ray fluorescence	Quantitative elemental analysis of coal or coal ash
Inductively coupled plasma spectroscopy	Quantitative multielement analysis of coal ash
Atomic absorption spectroscopy	Quantitative elemental analysis of coal ash
Electron microprobe analysis	Elemental analysis of micrometer-size areas
X-ray diffraction	Mineralogical analysis of coal ash or individual particles
Scanning electron microscopy	Morphological and elemental analysis of small areas
Infrared spectroscopy	Molecular characterization of coal ash
Thermometric techniques	Chemical characterization of coal ash
Optical microscopy	Identification of minerals in coal or coal ash
Ion microprobe mass analysis	Isotopic characterization of minerals or macerals
Extended x-ray analysis fine structure	Chemical environment of an atom

coal quality characteristics that can restrict the use of coal because of their environmental effects. A low free-swelling index is undesirable for coking coal applications.

Such restrictions can require special engineering designs to accommodate their use in combustion facilities. For example, coals having relatively high sulfur contents can be limited to use in combustion units that have scrubbers or to use in fluidized bed combustors. However, high sulfur coal may also be the most ideal feedstock for synthetic fuels because it can produce higher yields. Coal of specified quality can be used directly as a fuel or indirectly as a source of other fuels or products (organic chemicals) used in manufacturing.

Metallurgical Coke

Historically, one of the more important properties of coal has been its ability to produce metallurgical coke, a porous, carbon-rich solid used for the reduction of iron ores to metallic iron in blast furnaces (5). It has been stated that more research effort has been devoted to optimize coke-making than was devoted to any other use of coal (5).

The properties of coal that influence its use as a source of coke for use in metallurgical applications include coal rank, maceral composition, ash yield and ash composition, sulfur content, and bulk density. These properties, in varying proportions, will yield cokes of different types and strengths. Blending different coals may result in coal quality properties that will produce cokes of specific strength to support the iron ore in the blast furnace.

Traditionally the organic constituents have been the primary influence on the utilization of coal. Changes in the way coal is used, changes in coal technology, and an increased awareness of environmental issues have focused attention on the inorganic constituents in coal. The inorganic constituents in coal consist of discrete minerals and inorganic elements that are associated with both the minerals and the organic components. Table 3 contains a list of some of the many ways in which the inorganic constituents affect coal use. The following discussion amplifies some of the more widespread and serious problems. Specific inorganic constituents that contribute to technological problems are listed in Table 4.

Fouling

One of the most costly problems of coal utilization is the buildup of sintered ash deposits on the heat exchange surfaces of coal-fired boilers (6). These deposits not only drastically reduce the efficiency of the boiler but also promote corrosion and erosion (16). The size and strength of these deposits are dependent on the configuration of the boiler, the operating conditions, and the chemical composition of the coal being combusted.

There have been numerous attempts (17), using chemical data, to develop empirical formulas, models, or indices for predicting the tendency of a coal to form sintered ash deposits. Most of these attempts have focused generally on one, but no more than a few, compositional variables to account for deposit formation. Sodium is commonly cited as the primary cause of fouling in coal-fired boilers. However, single-element indices have met with limited application and success because the formation of sintered ash deposits is the result of the interaction of many inorganic constituents, not just one or a few. Moreover, none of the indices take into account the elements' modes of occurrence, a factor that will affect the role of the ele-

Table 3. Effects of the Inorganic Constituents on Coal Utilization and Evaluation

Effect	Reference
Boiler fouling (sodium)	6
Boiler slagging (iron)	6
Corrosion (chlorine, fluorine)	6
Erosion of combustors (quartz)	6
Catalysis of conversion processes	7
Poisons methanation catalysis	8
Abrasion of mining and grinding equipment (quartz)	9
Affects washability of coal	10
Lessens tendency of coal to form dust	10
Contributes to pneumoconiosis	10
Affects oxidizability of coal	10
Affects tendency of coal to combust spontaneously	10
Affects calorific value of coal	10
Affects coke strength	8
Affects accuracy of ultimate analysis	11
Useful in bed identification and correlation	10,12
Contributes to environmental pollution	13
Potential economic resource	14
Contains information on environment of deposition, diagenesis, source material	15

Table 4. Inorganic Constituents that Cause Technological Problems

Inorganic Constituent	Problem
Sodium	Fouling
Iron	Slagging
Chlorine	Corrosion
Fluorine	Corrosion
Boron	Corrosion
Quartz	Erosion/abrasion
Silicon/aluminium	Erosion/abrasion

ments in the fouling process. Recent research (18) indicates that ash yield and the concentrations of sodium, calcium, and magnesium were the most important factors influencing the fouling behavior of low-rank coal.

Slagging

Slag is defined as a molten ash deposited on boiler walls in zones subjected to radiant heat transfer (19). Minerals having low melting temperatures tend to stick to the boiler walls and accumulate until massive fused deposits are formed. This process is termed slagging.

To design an efficient boiler, the manufacturer needs to know the slagging characteristics of the coal ash produced. The iron content of the coal ash is of major importance in influencing the ash fusion temperature and the slagging potential of a coal (19). Iron sulfides (primarily pyrite) tend to promote slagging more than do other iron-bearing minerals. The clay mineral, illite, also may play a role in the slagging process by initiating the deposit (20).

A number of empirical tests and indices have been used to predict slagging potential (17). Nevertheless, it is difficult to predict slagging problems in pulverized-coal fired boilers based on laboratory determined properties (21).

Corrosion

Corrosion is the chemical decay of metal in fossil-fuel-fired boilers. It is enhanced by nonsilicate impurities in coal that volatilize and collect on the reactor walls (19). Low-temperature corrosion can be caused by condensation of sulfuric acid. High-temperature corrosion is caused by oxidation, sulfidation, or chloridation of the metal. Apparently, sulfated alkali elements (sodium, potassium) play an important role in the corrosion process. High levels of chlorine and, to a lesser degree, fluorine also have been cited as contributing to corrosion. Coals having chlorine values in excess of 1000 parts per million (ppm) should be considered as potentially corrosive. Coals having chlorine values in excess of 4000 ppm are almost certainly corrosive. Some trace elements (arsenic, lead, and perhaps zinc) also may enhance corrosion under certain conditions. In addition, boron also has contributed to metal corrosion (19).

Erosion and Abrasion

Erosion and abrasion are the physical processes that, like corrosion, lead to wear (that is, loss of material). Erosion and abrasion cause wear not only in boilers but also in mining and grinding equipment.

Erosion and abrasion are primarily due to the presence of abrasive minerals in coal such as quartz and pyrite (22). Particle size and shape are other factors that affect erosion and abrasion.

In the absence of mineralogical data, coal technologists have used the silicon/aluminum ratio as an indicator of abrasiveness. This index is based on the assumption that the higher the ratio the greater the quartz content. Coals having silicon/aluminum ratios greater than 3 are considered to be abrasive.

The Role of Inorganic Constituents in Other Technological Processes

The inorganic constituents in coal can affect coal use in several other technological processes. These influences will be only briefly mentioned here because no clear guidelines exist for assessing the impact of the constituents.

After coal combustion, particulate matter is commonly removed by passing the gas stream through electrostatic precipitators. Sodium in the coal ash has a major effect on the efficiency of these precipitators (17). In contrast to fouling, higher sodium levels are preferable because sodium reduces the resistivity of the ash, thus enhancing precipitator efficiency.

The inorganic constituents in coal can behave as catalysts or poisons for a variety of coal conversion processes. For example, the phosphorus and vanadium in coal are undesirable in the coke-forming process. Other technological effects of the inorganic constituents are listed in Table 4.

Environmental Impact

There are a variety of ways in which coal mining, coal use, and coal waste-product disposal can negatively affect the environment. For example, ambient air quality is degraded by emissions of combustion gases, smoke, and dust. Ground- and surface-water quality can be degraded by acid generation, release of trace elements, and by the influx of particulate matter.

Recent U.S. legislation, the 1990 Amendments to the Clean Air Act, (23) cites twelve elements (excluding radionuclides) in the Hazardous Air Pollutant List (phosphorus, antimony, arsenic, beryllium, cadmium, chromium, cobalt, lead, manganese, mercury, nickel, and selenium). Combustion of coal is a potentially significant source for several of these elements. The legislation requires that the U.S. Environmental Protection Agency (U.S. EPA) gather information, including coal quality data, to determine if regulations are necessary to reduce emissions of these elements from coal combustion and other sources.

Many of the same elements (arsenic, barium, cadmium, chromium, lead, mercury, selenium, and silver) appear in the U.S. EPA's list of potential pollutants in drinking water. Standards setting maximum concentrations for these elements have been published by the U.S. EPA (24).

The U.S. EPA (25) has found that most coal wastes are not hazardous as defined by the Resources Conservation and Recovery Act. The report notes, however, that some ground-water contamination has occurred in the vicinity of waste disposal sites. At one site in Virginia, nearby drinking-water wells were contaminated with vanadium and selenium from coal combustion wastes (25).

Many state regulatory agencies now recognize the potentially harmful effects of trace elements released during coal mining and utilization. Most western U.S. coal-mining States now require baseline information on several trace elements, especially selenium, boron, and molybdenum (26). The release of selenium from coal combustion wastes has caused extensive fish kills at two sites in North Carolina and one in Texas (27). New Mexico and Colorado now require data on 10 to 20 trace elements (26).

Ref. 28 provides detailed discussions on the potential toxicity to plants and animals of arsenic, boron, cadmium, chromium, copper, lead, manganese, mercury, molybdenum, nickel, selenium, and zinc. Brief discussions also are presented on the environmental impact of antimony, barium, beryllium, cobalt, fluorine, silver, thallium, tin, titanium, and vanadium.

Most chemical elements present in exceptionally high concentrations pose potential threats to the environment. It is difficult, however, to determine the threshold level at which the threat becomes a reality. This difficulty arises because the threshold level varies with each element, the chemical form of the element, the technological process in which the coal is being used, the local hydrologic and climatic regimes, and each species of organism.

Selenium is one of the principal elements of concern for environmental regulatory agencies and is included in all legislation aimed at mitigating trace element pollution from coal. Although selenium is an essential micronutrient for humans, small amounts can be potentially toxic to livestock. In the Powder River Basin, coals and associated rocks generally contain less than 1 ppm selenium. Nevertheless, selenium released during coal mining has been sufficient to cause the selenium concentration in the ground water to exceed the State's livestock drinking water standard (29).

Economic Potential of the Inorganic Constituents

Some inorganic constituents in coal have economic by-product potential. High contents of some elements have been reported from coal (Table 5). Historically, mineral resources have been extracted from coal (see ref. 14 for citations). More recently, there has been only sporadic activity in this area. For example, in the 1960s, there was limited production of uranium from some North and South Dakota lignites (30,31). Other recent efforts at by-product recovery from coal are described in refs. 32 and 33. The elements having the greatest potential for economic extraction include gold, silver, and the platinum group elements.

Several researchers (14,33,34) have suggested that the association of minerals and coal can be used as an exploration tool. The high element contents in coal may not be economical to extract but may be indicative of nearby mineralization. The chemical composition of coal may thus be used as a geochemical prospecting tool to locate economic mineral deposits.

Table 5. Examples of High Element Contents in Coal

Element	Content	Location
Gold	3 ppm[a]	Wyoming
Silver	66 ppm[b]	Texas
Arsenic	2,000 ppm	Alabama
Selenium	90,000 ppm	China
Zinc	51,000 ppm	Missouri

[a] At $400 per ounce for gold this concentration is worth about $40 per ton of coal.
[b] At $6 per ounce for silver this concentration is worth about $13 per ton of coal.

MODES OF OCCURRENCE

Information on variations in element contents, however useful, provides only a partial picture of the potential behavior of the trace elements in coal. A better picture can be brought into focus if we know the modes of occurrence of the elements in coal. The mode of occurrence of an element, as well as its concentration, determines technological behavior, environmental impact, and economic potential and reveals its geologic significance.

The classic example is sulfur. It has long been known that sulfur in coal occurs in sulfides, in sulfates, and in organic association (35). Each of the forms of sulfur responds differently to physical, chemical, and biological processes. Even this classification is an oversimplification. Elemental sulfur (36), at least 30 different sulfide minerals (37), and a complex suite of organic sulfur compounds (38) have all been identified in coal.

Many other elements exist in coal in a variety of forms. For example, selenium in Powder River Basin coal occurred in selenium-bearing pyrite, organically bound selenium, selenium-bearing sphalerite, lead selenide, water soluble selenium, and ion-exchangeable selenium (29).

There has been considerable work to determine the modes of occurrence of the elements in coal. Table 6 contains a list of the more likely modes of occurrence for selected elements in coal. Most efforts to determine the modes of occurrence of elements have been qualitative. Future efforts need to focus on ways to quantify the results.

One practical consequence of knowing the modes of occurrence of the elements in coal is the ability to anticipate their fate during coal combustion. During the combustion process, elements can be released into the environment (with the gases emitted from the smoke stack) or they may concentrate in the bottom ash, boiler slag, or fly ash. Anticipating the behavior of the elements during combustion should lead to more efficient methods of capture, thus preventing the elements from polluting the environment. Similarly, knowing the modes of occurrence of the elements may lead to more effective methods to physically or chemically remove these elements during coal cleaning.

Determining the modes of occurrence of the elements in coal is still more of an art than a science. The tools (for example, scanning electron microscopy, x-ray diffraction) and procedures (selective leaching and heating) used in this endeavor are described in refs. 40–42.

Table 6. Probable Modes of Occurrence of Selected Elements in Coal[a]

Element	Modes of Occurrence
Aluminium	Clays, feldspars, perhaps some organic association
Antimony	Accessory sulfide, some organic association
Arsenic	Solid solution in pyrite
Barium	Barite, crandallite, organic association in low-rank coal
Beryllium	Organic association, clay
Bismuth	Accessory sulfide
Boron	Organic association, illite
Bromine	Organic association
Cadmium	Sphalerite
Calcium	Calcite, organic assoc., sulfates, phosphates, silicates
Cesium	Clays, feldspar, mica
Chlorine	Organic association
Chromium	Clays (?)
Cobalt	Accessory sulfides, pyrite
Copper	Chalcopyrite
Fluorine	Perhaps apatite, clays, mica, amphiboles
Gallium	Clays, organics, sulfides
Germanium	Organic association
Gold	Native gold
Hafnium	Zircon
Indium	Probably sulfides
Iodine	Probably organic association
Iron	Pyrite, siderite, sulfates, oxides, some organic assoc.
Lead	Galena, PbSe
Lithium	Clays
Magnesium	Clays
Manganese	Siderite, calcite
Mercury	Solid solution with pyrite
Molybdenum	Unclear; perhaps with sulfides or organics
Nickel	Unclear; perhaps sulfides, organics, or clay
Niobium	Oxides
Phosphorus	Phosphates
Platinum	Native alloys, perhaps some organic association
Rare-earths	Phosphates, some organic association
Rubidium	Probably illite
Scandium	Unclear; clays, phosphates, or organics
Selenium	Organic association, pyrite, PbSe
Silicon	Quartz, clays, silicates
Silver	Perhaps silver sulfides
Sodium	Organic association, clays, zeolites, silicates
Strontium	Carbonates, phosphates, organic association
Tantalum	Oxides
Tellurium	Unclear
Thorium	Rare-earth phosphates
Tin	Inorganic: tin oxides or sulfides
Titanium	Oxides, clays, some organic association
Tungsten	Oxides, organic association
Uranium	Organic association, zircon
Vanadium	Clays, perhaps some organic association
Yttrium	Rare-earth phosphates
Zinc	Sphalerite
Zirconium	Zircon

[a] Modified from ref. 39.

BIBLIOGRAPHY

1. ASTM (American Society for Testing and Materials), *1992 Annual Book of Standards,* Vol. 5, 1992, p. 506.
2. G. H. Wood, Jr., T. M. Kehn, M. D. Carter, and W. C. Culbertson, *U.S. Geol. Surv. Circ.* **891,** 65 (1983).
3. H. H. Schobert, *Coal: The Energy Source of the Past and Future,* American Chemical Society, Washington, D.C., 1987, p. 298.
4. D. W. Golightly, and F. O. Simon, eds., *U.S. Geol. Surv. Bull.* **1823,** 7–14 (1989).
5. R. C. Neavel, in M. A. Elliott, ed., *Chemistry of Coal Utilization,* 2nd Suppl. Vol., John Wiley & Sons, Inc., New York, 1981, pp. 91–158.
6. W. T. Reid, in M. A. Elliot, ed., *Chemistry of Coal Utilization,* 2nd Suppl. Vol., John Wiley & Sons, Inc., New York, 1981, pp. 1389–1445.
7. J. A. Guin, A. R. Tarrier, J. M. Lee, L. Lo, and C. W. Curtis, *Indust. Eng. Chem. Proc. Design Dev.* **18,** 371–376 (1979).
8. R. G. Jenkins, and P. L. Walker, Jr., in C. Karr, Jr., ed., *Analytical Methods for Coal and Coal Products,* Vol. 2, Academic Press Inc., New York, 1978, pp. 265–292.
9. T. G. Callcott, and G. B. Smith, in M. A. Elliot, ed., *Chemistry of Coal Utilization,* 2nd Suppl. Vol., John Wiley & Sons, Inc., New York, 1981, pp. 285–315.
10. R. M. S. Falcon, *Minerals Sci. and Eng.,* **10,** 28–52 (1978).
11. P. H. Given, and R. F. Yarzab, in C. Karr, ed., *Analytical Methods for Coal and Coal Products,* Vol. 2, Academic Press Inc., New York, 1978, pp. 2–41.
12. V. Bouska, *Geochemistry of Coal,* Elsevier, Amsterdam, The Netherlands, 1981, p. 284.
13. U.S. National Committee for Geochemistry, *Trace Element Geochemistry of Coal Resource Development Related to Environmental Quality and Health,* National Academy Press, Washington D.C., 1980, p. 153.
14. R. B. Finkelman, and Brown, Jr., in D. C. Peters, ed., *Geology in Coal Resource Utilization,* TechBooks, Fairfax, Va., 1991, pp. 471–481.
15. R. B. Finkelman, *U.S. Geol. Surv. Open-file Report 81-99,* 1981, p. 322.
16. F. I. Honea, G. G. Montgomery, and M. L. Jones, in W. R. Kube, E. A. Sondreal, and D. M. White, eds., *Technology and Use of Lignites,* Vol. 1, *Proceedings Eleventh Biennial Lignite Symposium,* Grand Forks Energy Technology Center IC-82/1, 1982, pp. 504–545.
17. G. E. Vaninetti, and F. Busch, *J. Coal Qual.* **1,** 22–31 (1982).
18. R. B. Finkelman, and F. T. Dulong, *U.S. Geol. Surv. Open-file Report 89-208,* 1989, p. 155.
19. E. Raask, *Mineral Impurities in Coal Combustion,* Hemisphere Pub. Corp., Washington D.C., 1985, p. 484.
20. R. W. Bryers, in K. S. Vorres, ed., *Mineral Matter and Ash in Coal, Am. Chem. Soc. Symp. Ser. 301,* ACS, Washington, D.C., 1986, pp. 353–374.
21. J. F. Cudmore, in C. R. Ward, ed., *Coal Geology and Coal Technology,* Blackwell Scientific Pub., London, 1984, pp. 113–150.
22. E. Raask, *Erosion Wear in Coal Utilization,* Hemisphere Pub. Corp., Washington D.C., 1988, p. 621.
23. U.S. Statutes at Large, Provisions for attainment and maintenance of national ambient air quality standards, *Public Law 101-549,* 101st Congress, 2nd Session, Vol. 104, part 4, 1990, pp. 2353–3358.

24. U.S. EPA, *Quality Criteria for Water,* U.S. Environmental Protection Agency, Washington, D.C., 1976.
25. U.S. EPA, *Wastes from the Combustion of Coal by Electric Utility Power Plants,* U.S. Environmental Protection Agency Report to Congress, 1987.
26. D. Y. Boon, F. F. Monshower, and S. E. Fisher, in *Billings Symposium on Surface Mining and Reclamation in the Great Plains,* Am. Soc. Surf. Min. & Recl. Reclamation Research Unit Report No. 8704, A-1-1 to A-1-18.
27. M. Shepard, *EPRI J.* 17–21 (1987).
28. D. C. Adriano, *Trace Elements in the Terrestrial Environment,* Springer-Verlag, New York, 1986, p. 533.
29. G. B. Dreher, and R. B. Finkelman, *Envirn. Geol. Water Sci.* **19,** 155–167 (1992).
30. E. A. Noble, in *Mineral and Water Resources of North Dakota, N.D. Geol. Surv. Bull.* **63,** 80–83 (1973).
31. R. W. Schnable, in *Mineral and Water Resources of South Dakota,* Report prepared by the U.S. Geol. Surv. for the Committee on Interior and Insular Affairs, U.S. Senate, 1975, pp. 172–176.
32. G. Burnet, *Energy* **11,** 1363–1375 (1986).
33. J. C. Cobb, J. M. Masters, C. G. Treworgy, and Helfinstine, Ill., *Mineral Notes* **71,** 11 (1979).
34. J. R. Hatch, M. J. Avcin, W. K. Wedge, and L. L. Brady, *U.S. Geol. Surv. Open-File Rept 76-796,* 1976, p. 26.
35. E. S. Moore, *Coal,* John Wiley & Sons, Inc., New York, p. 473.
36. J. E. Duran, S. R. Mahasay, and L. M. Stock, *Fuel,* **65,** 1167–1168 (1986).
37. R. B. Finkelman, in A. T. Cross, ed., "Neuvieme Congress International de Stratigraphic et de Geologic du Carbonifere," *Compte Rendu.* **4,** 407–412 (1985).
38. D. J. Casagrande, in A. C. Scott, ed., *Coal and Coal-bearing Strata: Recent Advances, Geol. Soc. Spec. Pub. No. 32,* 1987, pp. 87–105.
39. R. B. Finkelman, in R. H. Filby, B. S. Carpenter, and R. C. Ragaini, eds., *Atomic and Nuclear Methods in Fossil Energy Research,* Plenum, New York, 1982, pp. 141–149.
40. R. B. Finkelman and H. J. Gluskoter, in R. W. Breyers, ed., *Fouling and Slagging Resulting From Impurities in Combustion Gases,* Engineering Foundation, New York, 1983, pp. 299–318.
41. R. B. Finkelman, C. A. Palmer, M. R. Krasnow, P. J. Aruscavage, G. A. Sellers, and F. T. Dulong, *Energy Fuels,* **4,** 755–767 (1990).
42. D. J. Swaine, *Trace Elements in Coal: Comparison of Australian with Other Coal,* Butterworths, London, 1990, p. 278.

COAL–SPONTANEOUS COMBUSTION AND TRANSPORTATION

WILLIAM BERRY
Co-Ag Consultants
Pittsburgh, Pennsylvania

The susceptibility of coals to spontaneous combustion during storage in stockpiles and during transport in barges and/or ocean-going vessels is a function of many different but interrelated coal quality and physical factors. Depending on the conditions and time involved in storage and transport, one or more of these factors may take-on added significance. Partly, for this reason, predicting the behavior of coal in storage and transport with regard to its susceptibility to spontaneous combustion is difficult.

Coal is an organic substance and will oxidize when exposed to the atmospheric oxygen. Such oxidation is also exothermic and if the heat generated by the oxidation is not dissipated to the atmosphere or its surrounding container the coal will self-heat to combustion. The process of self-oxidation is topochemical by nature so that the rate of heating is dependent on the surface area of the coals and its size consist. The smaller the size-consist, the greater is the propensity to self-heat. Another physical property of coal is its propensity to self-heat when exposed to excess moisture from any source. A reaction known as the "heat of wetting" takes place when a coal is wet and then starts to dry out. This reaction is also exothermic and can enhance the building of a self-heat sump, which can lead to self-ignition.

Coal, in itself, has a wide propensity to self-heat, depending entirely on the rank or maturation of the coal. To understand this differing propensity, it is necessary to discuss the coal types of the world. The Northern Hemisphere of America contains every type of coal from the currently forming peats of the Everglades of Florida to the meta-anthracites of Rhode Island. Coal is a complex mixture of organic debris that with geologic time and pressure, due to burial and tectonic earth adjustments, forms what we know as coal. Being of organic origin, the material is made up of a series of complex hydrocarbons. These long chain hydrocarbons are most complex and unstable in the peat stage and as the metamorphism progresses the coal through the rank stages, these hydrocarbons break down to a simpler chemical make-up and become more stable.

Thus, peat is the most susceptible to radical long chain hydrocarbon breakdown and as metamorphism progresses through coal rank, these chemical bonds become less complex and more stable as the coal progresses: from peat to lignite to sub-bituminous, to high volatile, medium volatile and low volatile bituminous, to anthracite, with the end point being a simple carbon if the process continues that far. Relative to the propensity of coals to spontaneous combustion, the same pattern holds. Peat, being of the most complex structure, is also the most susceptible to spontaneous combustion.

Anyone who has lived or worked close to the Everglades of Florida knows the area as the "sea of grass." Should the water level of the Everglades drop or become lowered for any reason, the "sea of grass" becomes the "sea of fire" as the peat becomes exposed to the air. Some of our western lignites will self-ignite on the mine faces as the coal is being mined. The western and some of the interior basin coals require special handling to control their tendency toward self-ignition. The high volatile coals adjacent to and near the inland waterway system are generally more susceptible to spontaneous combustion than those of the Eastern Appalachian Basin. As the coal evolves to the medium volatile through low volatile coals to anthracite, their propensity to spontaneous combustion becomes less and less.

THE PHENOMENA OF SPONTANEOUS COMBUSTION OF COAL

Spontaneous combustion of coal results from the reactions between carbon and oxygen (oxidation) under conditions in which the rate of heat generated by complex chemical reactions exceeds the rate of heat dissipated. Unless the heat produced from this reaction is dissipated or the contact between the carbon in coal and oxygen ceases, the coal will continue to oxidize and increase in temperature. The most widely accepted mechanism for explaining the spontaneous combustion of coal involves the process by which coal-oxygen complexes form at low temperatures. The formation and decomposition of coal-oxygen complexes are exothermic reactions. The oxidation of coal begins when a hydrogen bond is broken, producing an active carbon site. This reaction occurs gradually in coal whenever new surfaces are formed through breakage. The newly available carbon site, produced via breakage, attaches itself to oxygen in the air and generates heat. The new carbon-oxygen radical formed, subsequently extracts hydrogen from hydrogen-carbon complexes on the coal surface, producing secondary active carbon sites. These carbon sites subsequently combine with oxygen in air to generate additional heat. The aforementioned process continues until oxygen and/or reactive surfaces are depleted. It has been proposed that up to 82°C, the carbon in the coal reacts with oxygen to form low temperature coal-oxygen complexes. At temperatures above 82°C, these coal-oxygen complexes rapidly decompose. At temperatures above 149°C, a second set of high temperature coal-oxygen complexes form, which upon decomposition at ~232°C (~450°F) ignite the coal. Twice as much heat is evolved upon decomposition of the oxygen complexes than on their formation. Thus, it is imperative that no portion of a coal supply in storage or in transport be allowed to exceed 82–88°C. Spontaneous combustion has been known to occur within three days after the temperature reaches 77°C. The increased heat rise of the coal, due to the decomposition of these low temperature oxygen complexes and the subsequent formation of high temperature oxygen complexes once the coal has reached a temperature of approximately 149°C, may generate enough heat in itself to raise the temperature of the coal above its ignition points. From field investigations, we know that the oxidation of coal proceeds at a slow uniform rate until the temperature of the coal reaches ~49°C. Above this temperature and up to 180–185°C, the rate of heat rise more than doubles. If the coal sustains a temperature above the 85°C level, CO_2 and water vapor begin to evolve and the noxious odor of hydrocarbons or sulfur compounds may become prevalent. Since CO_2 is one-and-one half times heavier than air, it may migrate as far down in the holds of ships and/or storage piles as the heat front will allow it to go and, subsequently, displace oxygen upward to combine with lower temperature coal above the hot zone, thus forming additional low-temperature coal-oxygen complexes. Once CO_2 has been detected in significant amounts, an ignition has taken place somewhere in the vessel or stockpile.

Consequently, the significant number to prevent the coal from reaching ignition during storage or ocean-transportation is 82°C. Whether coal is stacked or loaded at 41 or 55°C, if a temperature of 82°C is reached in the coal, the chances of ignition increase dramatically in favor of spontaneous combustion. In fact, many reports in the literature indicate that entire masses of coal can move so quickly between 60 and 82°C, that remedial action should be taken before the 82°C temperature point is reached. Although it is fair to assume that the closer the cargo is to the 82°C temperature upon loading, the sooner it may reach 82°C; the heat rise in a body of coal is very dependent upon its quality. It is here where sufficient characterization of coals becomes crucial. Prevention of the problem becomes one of proper selection, blending and treating of coals and one of monitoring the temperature of coals before and after loading and during their transport to the marketplace.

Coal Characterization

The factors that must be considered as significant when characterizing coals relative to their susceptibility to spontaneous combustion include:

1. Coal rank, as determined by average vitrinoid reflectance, ultimate oxygen and moist mineral matter free, Btu;
2. Coal type as determined through petrographic composition analysis;
3. Coal grade, primarily the amount and form of pyritic sulfur present as determined from forms of sulfur and petrographic examination. The amount and type of ash present has also been reported as an important factor to be considered;
4. Coal size;
5. Coal moisture–both equilibrium and surface;
6. Coal grindability–susceptibility of coal to breakage upon handling;
7. Coal bulk density in storage and/or transport;
8. Initial degree of weathering of coal.

The factors which are unrelated to coal quality, but which must be considered in assessing the susceptibility of a given product to spontaneous combustion include:

1. Type of mining employed to recover coal (size);
2. Type of coal product being shipped–raw or washed (ash and/or mineral content);
3. How freshly mined coal is and its current depth of burial (in situ oxidation);
4. Weather conditions prevailing during the storage and/or transport of coal (eg, humidity, temperature, rain);
5. Length of time since coal has been mined and has been in storage and/or in transport;
6. Maximum temperature coal has reached in storage or in transport;
7. Whether coal blending has been implemented to produce a homogeneous product from a variety of different coal qualities.

In the case of assessing the susceptibility of coals to spontaneous combustion, coal rank is certainly one of the more important factors, but may or may not be the most important. The lower rank coals are generally more susceptible to spontaneous combustion due to their higher reactivities, lower ignition temperatures, higher internal pore surface areas and greater tendency to slack upon exposure to changes in temperature and access to air. The lower rank, higher reactivity coals also contain higher starting oxygen contents, exposing their active carbon sites to oxidation more quickly. High internal pore surface area helps increase the moisture-holding capacity of the coal, which promotes more heat of wetting reactions and decrepitation of lump coal into fines, providing the coal is exposed to repetitive wetting and drying action. The evaporation of moisture is endothermic, while the condensing of moisture or the wetting of coal is exothermic. The exothermic reaction is most prevalent when coal moisture is depressed below its equilibrium level and dry coal surfaces come in contact with external moisture sources such as rain, water in barges and/or ship holds. The heat liberated when a solid is wetted is proportional to the wetted areas. For bone-dry lignite, the heat of wetting 105 J (25 cal/g), which is sufficient to raise the temperature of the coal plus the water used to wet the lignite 27°C. For bone-dry Eastern Interior Region Basin coals, the temperature rise would be only −7°C and at an equilibrium moisture content of 8–10%, the heat of the wetting effect would be negligible. Thus, if wet coal of any rank (excluding lignite) is prevented from drying-out in storage and/or transport, the heat of wetting effects are of minor consideration; however, as hot spots develop in a cargo when in storage and/or transport, and water and/or steam begins to migrate in the cargo, the likelihood of heat from the wetting effects contributing to the further heat rise of cargo increases. However, excessive moisture can also inhibit oxidation by protecting the internal pore surfaces from oxygen attack.

The level of moisture content is also important when estimating thermal diffusivity of coal in storage and in transport. Thermal diffusivity in coal is low since coal is a very poor conductor. With the absence of convection currents, hot spots in a storage pile or cargo hold are confined to very localized areas. Thermal diffusivity, which is the ratio of conductivity to specific heat, will decrease with lower bulk densities, lower rank, higher moisture and finer size consist coal.

When coals of similar rank are compared, duller coals are often found to be less susceptible to spontaneous combustion than Vitrain-rich coals which are more friable and thus, more readily oxidizable. When Vitrain oxidizes, a network of fissures develop, producing a higher surface area for reaction with oxygen. However, some researchers have found that high Exinite coals by virtue of their high internal surface areas, are more oxidizable. Dull coals, however, are more resistant to the crack and fissure development exhibited by the Vitrain. Vitrain is also a very friable component in itself, thus contributing to fines production. Porous Fusain, the most friable of all the coal entities, is thought to have a high oxygen-holding capacity and when it comes in direct contact with Vitrain during storage and/or transport, is believed to accelerate spontaneous combustion.

When coals of similar rank and type are compared, the more weathered the coal is, the more susceptible it will be to further oxidation and spontaneous ignition. Weathered coal is also hydroscopic and generally smaller in size, thus, contributing to more rapid heat rise.

As the size of coal decreases down to 0.3 mm, its surface area increases and the greater is its susceptibility to ignition. The rate of increased reaction with size reduction has been determined to be proportional to the cube root of the exterior surface area. Below 0.3 mm, the oxidation process takes place in the entire volume of the grain and, consequently, the rank and oxidation become dependent on internal pore surface area as well. Finer size-consist coals also retain more surface moisture and thus, the combined effect of high moisture and fine size decrease bulk density to its lowest possible level providing interpore spacing in which entrapped oxygen can react readily with active carbon sites.

The more freshly mined the coal, the more readily the absorption of oxygen on active carbon sites occur. Thus, the chances are less that freshly mined coal will be able to stabilize itself if critically high temperatures are achieved in storage and/or transport.

Raw coals may or may not be more susceptible to spontaneous combustion than their washed counterparts (similar rank and type), depending on the relative amount of ash and fines present in the raw versus the washed product. Of course, the method of mining employed will affect the percentage of fine coal produced, with continuous and long-wall mined coal being finer in size-consist than conventional and strip-mined coal. Regardless of what mining practice is being employed, raw coal will contain more ultra-fines than washed coal due to their loss in the process of cleaning. Counter-balancing this effect is the presence of extraneous ash in the raw coal which dilutes the coal fraction and thus, provides less carbon surface for oxygen to react with. However, washed coals are more homogeneously blended and contain a higher surface moisture, making the product less likely to dry out and be rewet in transport and thus avoiding the heat of wetting effect previously described.

It has been known for a long time that the form of pyritic sulfur present in coal affects the rate of formation of acid mine waters. Pyritic sulfur occurs in coals as distinct euhedral crystals, forming crystal colonies known as framboids, and in a variety of coarser forms found deposited in the cleat and bedding of the coal seams. It is the finer-size pyrite that is most likely to react. The framboidal pyrite appears to alter more readily, while the euhedral is next most reactive followed by the coarse pyrite which is most resistant to alteration. The oxidation of pyrite and the subsequent formation of sulfates and/or sulfuric acid, in the presence of water, are both net heat-producing reactions which contribute to increased coal temperatures. Reaction between oxygen, water and pyrite also causes swelling around the coal pyrite interfaces, resulting in the formation of additional fine coal surfaces for reaction.

Finally, the weather conditions prevailing during the time of storage and loading of a cargo are of importance. Unfortunately, there is little that can be done to prevent an exacerbation of the problem caused by weather conditions. High humidity minimizes the dissipation of heat from a storage pile and has been proven to accelerate oxidation reactions. Oxidizing agents present in rain water, noticeably peroxides, ozone, nitrous oxides and sulfuric acid, can also contribute to an increase in the susceptibility of coal to spontaneous combustion. One stop-gap measure is to stop loading during periods of rain downpours. High surface moisture in combination with coarse, high sulfur coal has several times created galvanic battery attack in the holds of vessels not protected with a bituminous vehicle paint. Having to contend with temperature changes year-round poses serious problems in the areas of: (1) minimum temperature standards accepted for loading, (2) the frequency with which stockpiles must be worked to extend the effective storage life of the coal, (3) the rate of heatup of different quality coals.

The following list is the current suggested procedures taken directly from the International Marine Organization "IMO Code of Safe Practice for Solid Bulk Cargoes," 1991 Edition (5). This material is reproduced with the permission of the International Marine Organization.

Coal Properties and Characteristics

1. Coals may emit methane, a flammable gas. A methane/air mixture containing between 5–16% methane constitutes an explosive atmosphere which can be ignited by sparks or naked flame, eg, electrical or frictional sparks, a match or lighted cigarette. Methane is lighter than air and may, therefore, accumulate in the upper region of the cargo space or other enclosed spaces. If the cargo space boundaries are not tight, methane can seep through into spaces adjacent to the cargo space.
2. Coals may be subject to oxidation leading to depletion of oxygen and an increase in carbon dioxide in the cargo space.
3. Some coals may be liable to self-heating that could lead to spontaneous combustion in the cargo space. Flammable and toxic gases, including carbon monoxide, may be produced. Carbon monoxide is an odorless gas, slightly lighter than air, and has flammable limits in air of 12–75% by volume. It is toxic by inhalation with an affinity for blood hemoglobin over 200 times that of oxygen.
4. Some coals may be liable to react with water and produce acids which may cause corrosion. Flammable and toxic gases, including hydrogen, may be produced. Hydrogen is an odorless gas, much lighter than air, and has flammable limits in air of 4–75% by volume.

General Requirements for All Coals

1. Prior to loading, the shipper or his appointed agent should provide, in writing to the master, the characteristics of the cargo and the recommended safe-handling procedures for loading and transport of the cargo. As a minimum, the cargo's contract specifications for moisture content, sulfur content and size should be stated and especially for whether the cargo may be liable to emit methane or self-heat.
2. The master should be satisfied that he has received such information prior to accepting the cargo. If the shipper has advised that the cargo is liable to emit methane or self-heat, the master should additionally refer to the "Special Precautions."
3. Before and during loading, and while the material remains on board, the master should observe the following:
   - All cargo spaces and bilge wells should be clean and dry. Any residue of waste material or previous cargo should be removed, including removable cargo battens, before loading.
   - All electrical cables and components situated in cargo spaces and adjacent spaces should be free from defects. Such cables and electrical components should be safe for use in an explosive atmosphere or positively isolated.
   - The ship should carry on board appropriate instruments for measuring the following without requiring entry into the cargo space:
     a. concentration of methane in the atmosphere;
     b. concentration of oxygen in the atmosphere;
     c. concentration of carbon monoxide in the atmosphere;
     d. pH value of cargo hold bilge samples; and
     e. temperature of the cargo in the range 0–100°C.

   These instruments should be regularly serviced and calibrated. Ship personnel should be trained in the use of such instruments.
4. The ship should carry on-board and self-contained breathing apparatus required by SOLAS regulation 11-2/17. The self-contained breathing apparatus should be worn only by personnel trained in its use.
5. Smoking and the use of naked flames should not be permitted in the cargo areas and adjacent spaces and appropriate warning notices should be posted in conspicuous places. Burning, cutting, chipping, welding or other sources of ignition should not be permitted in the vicinity of cargo spaces or in other adjacent spaces, unless the space has been properly ventilated and the methane gas measurements indicate it is safe to do so.
6. The master should ensure that the coal cargo is not stowed adjacent to hot areas.
7. Prior to departure the master should be satisfied that the surface of the material has been trimmed reasonably level to the boundaries of the cargo space to avoid the formation of gas pockets and to prevent air from permeating the body of the coal. Casings leading into the cargo space should be adequately sealed. The shipper should ensure that the master receives the necessary co-operation from the loading terminal.
8. The atmosphere in the space above the cargo in each cargo space should be regularly monitored for the presence of methane, oxygen and carbon monoxide. Records of these readings should be maintained. The frequency of the testing should depend

upon the information provided by the shipper and the information obtained through the analysis of the atmosphere in the cargo space. Means should be provided for testing the atmosphere in the space above the cargo without opening the hatch covers. Where such monitoring indicates the presence of methane, a rise in temperature or the presence of carbon monoxide, the master should refer to the relevant "Special Precautions."

9. The master should ensure as far as possible that any gases which may be emitted from the materials do not accumulate in adjacent enclosed spaces.
10. The master should ensure that enclosed working spaces such as storerooms, carpenter's shop, passage ways, tunnels, etc, are regularly monitored for the presence of methane, oxygen and carbon monoxide. Such spaces should be adequately ventilated.
11. Regular hold bilge testing should be systematically carried out. If the pH monitoring indicates that a corrosion risk exists, the master should ensure that all hold bilges are kept dry during the voyage in order to avoid possible accumulation of acids on tank tops and in the bilge system.
12. If the behavior of the cargo during the voyage differs from that specified in the cargo declaration, the master should report such differences to the shipper. Such reports will enable the shipper to maintain records on the behavior of the coal cargoes, so that the information provided to the master can be reviewed in the light of transport experience.
13. The Administration may approve alternative requirements to those recommended in this schedule.

Special Precautions

Coal Emitting Methane. If the shipper has advised that the cargo is liable to emit methane or analysis of the atmosphere in the cargo space indicates the presence of methane, the following additional precautions should be taken:

1. Adequate surface ventilation should be maintained. On no account should air be directed into the body of the coal as air could promote self-heating.
2. Care should be taken to vent any accumulated gases prior to removal of the hatch covers or other openings for any reason, including unloading. Cargo hatches and other openings should be opened carefully to avoid creating sparks. Smoking and the use of naked flame should be prohibited.
3. Personnel should not be permitted to enter the cargo space or enclosed adjacent spaces unless the space has been ventilated and the atmosphere tested and found to be gas-free and to have sufficient oxygen to support life. If this procedure is not possible, emergency entry into the space should be undertaken only by trained personnel wearing self-contained breathing apparatus, under the supervision of a responsible officer. In addition, special precautions to ensure that no source of ignition is carried into the space, should be observed.
4. The master should ensure that enclosed working spaces, such as storerooms, carpenter's shops, passage ways, tunnels, etc, are regularly monitored for the presence of methane. Such spaces should be adequately ventilated and, in the case of mechanical ventilation, only equipment safe for use in an explosive atmosphere should be used. Testing is especially important prior to permitting personnel to enter such spaces or energizing equipment within those spaces."

Self-heating Coals

1. If the shipper has advised that the cargo is liable to self-heat, the master may wish to seek confirmation that the precautions intended to be taken and the procedures intended for monitoring the cargo during the voyage are adequate.
2. If the cargo is liable to self-heat or analysis of the atmosphere in the cargo space indicates an increasing concentration of carbon monoxide or that the temperature of the cargo is rising rapidly, then the following additional precautions should be taken:
   a. The hatches should be closed immediately after completion of loading in each cargo space. The hatch covers can also be additionally sealed with a suitable sealing tape. Surface ventilation should be limited to the extent necessary to remove gases which may have accumulated. Forced ventilation should not be used. On no account should air be directed into the body of the coal as air could promote self-heating.
   b. Personnel should not be allowed to enter the cargo space, unless they are wearing self-contained breathing apparatus and access is critical to the safety of the ship or safety of life. The self-contained breathing apparatus should be worn only by personnel trained in its use.
   c. When required by the competent authority, the temperature of the cargo in each cargo space should be measured at regular time intervals to detect self-heating.
   d. If the temperature of the cargo exceeds 55°C, and the temperature or the carbon monoxide level is increasing rapidly, a potential fire situation may be developing. The cargo space should be completely closed down and all ventilation ceased. The master should seek expert advice immediately and should consider making for the nearest suitable port of refuge. Water should not be used for cooling the material or fighting coal cargo fires at sea, but may be used for cooling the boundaries of the cargo space.

BIBLIOGRAPHY

1. W. F. Berry, and J. S. Goscinski, "Discussion of the New Orleans Hot Coal Problems and Suggestion for the Elimination

of the Problem," Presented at *Lloyds National Maritime Show Program,* Baltimore, Maryland, May 9–11, 1982.

2. W. F. Berry, and J. S. Goscinski, "Hot Coal, Causes and Remedies," Bulk Systems (Oct., 1992).
3. W. F. Berry, and D. H. Tessmer, "The Means to The End of the Hot Coal Problem in Ocean Transport," produced for *Scottish Ship Management,* Edinborough, December, 1982.
4. W. F. Berry, and J. H. Goscinski, "Heating of Coal in Transit–Problems, Causes and Solutions," Seaways (Jan., 1983).
5. "IMO Code of Safe Practices for Solid Bulk Cargo," International Marine Organization, 1991.
6. R. Mason, "Hot Coal: The Problems and Precautions," C. Q. (Fall, 1982).

COAL CLASSIFICATIONS: PAST AND PRESENT

M. J. Lemos de Sousa
H. J. Pinheiro
University of Porto
Porto, Portugal

P. C. Lyons
U. S. Geological Survey
Reston, Virginia

R. A. Durie
CSIRO
North Ryde NSW, Australia

Coal has been known as a useful commodity since the time of the Greeks when it was used as a heat source by blacksmiths during the time of Theophratus (372–287 B.C.), as reported by Thiessen (1). The use of coal by the Chinese is also reported by Marco Polo in his 1280 chronicles (2). The Vikings may have dug anthracitic coal along the coast of Rhode Island because anthracite similar to that of the Narragansett basin of Rhode Island and Massachusetts was found in 1930 by Danish archeologists on the eleventh-century farmstead of Torstein, son of Einik the Red, on the southwest coast of Greenland (3, p. 6). In Europe the earliest references are dated 1113 (*Annals from the Rolduc Abbey,* Heerlen, Holland) and refer to *"terra nigra ad focum faciendum"* being used in Rolduc. Manuscript number 162, "Renier de St. Jacques," refers to coal being used in Liège, Belgium, as far back as 1195 (4). Although coal came into common use during the seventeenth century in England, reports of its use in that country date back to the ninth century (1). In North America, coal was discovered in Nova Scotia in 1672 and was mined there in 1720 (5). Coal mining began in the eastern United States during the 1760s and was widespread in eastern North America by the early 1800s. British soldiers were known to have used coal at Newport, Rhode Island, as a heat source during the American Revolution (1775–1783).

It was a natural outgrowth of interest in and utilization of coal that there was a desire to understand its origin and to classify it like other rocks. The history of coal classifications up to 1945 is treated in detail by (6). The organic origin of coal was recognized in the latter part of the sixteenth century when B. Klein proposed in 1592 that coal originated from wood (7). The first attempt at the systematization of caustobioliths (6) is attributed to A. Libavius in 1599, but Bertrand published in 1763 J. Stockar's scientific classification (8). Beroldingin in 1778 and 1792 was the first to suggest that coal was derived from peat (1), which is the modern interpretation. However, the peat-to-coal theory languished for 40 years but was revived by Witham (9), who examined coal with a microscope and discovered it contained woody components.

By the late nineteenth century, it was widely accepted that coal formed from peat and this scientific activity led to an interest in the classification of coal (6). Although coal systematics dates to 1837 when Regnault (10) attempted a systematic classification of coal, it was Grüner (11) who first applied scientific criteria to Regnault's work. Around the same time, Frazer (12) and Lesley (13) published classifications of coal, and their work was followed by the coal classification of Campbell (14), a review of coal classifications by McCallum (15), and Collier's (16) classification of low-grade coal. Grüner's classification (coauthored with Bousquet) was published in 1911 (17). Raylston (18) graphed the calorific and ultimate analyses of 3000 government analyses and was able to distinguish classes of coal using the system of the U.S. Geological Survey (6). Parr's (19) classification of coal led to the American Society for Testing and Materials (ASTM) classification of coal by rank in 1934. Further details on the history of coal systematics can be found in Thiessen and Francis (20), Thiessen (21), Seyler (22), Francis (23), van Krevelen (2,24,25), Alpern (26,27), Rose (6), Todd (28), Lemos de Sousa (29), Falcon (30), Carpenter (31), and Alpern and Lemos de Sousa (32).

Because of the wide amount of information on the subject of coal classification, this article will focus on the latest approaches to and concepts in coal classification. Main classification systems will be discussed, and these will encompass examples of both national [the Standards Association of Australia coal classification (Ref. 33; AS 2096), the ASTM classification of coal by rank (Ref. 34; ASTM D 388-92a), and other systems will be briefly mentioned] and the international systems [the Economic Commission for Europe of the United Nations (UN-ECE) coal classification].

Prior to dealing with the issue of classification *stricto sensu,* it is essential to clarify certain concepts (other than classification itself) that are linked to modern concepts related to coal systematics. Considering Evaluation as the determination of properties, characteristics, and behavior of a product related to its end use, it is obvious that both classification and specification distinctly contribute to it. However, although specification refers precisely to parameters not necessarily included in a classification, each one is applicable to a specific aspect or to a requirement for a particular use, and Classification conforms to the principles of systematics as referred to above.

On the whole, the concepts of evaluation, classification, and codification are linked to Quality, which may be defined as the totality of characteristics of a product that meet certain needs. These characteristics are verified by means of a qualification process, immediately implying the implementation of a quality system through standardization (ie, processes of formulating, issuing and implementing standards aimed at the achievement of the optimum degree of order in a given context) and Certification (ie, procedure by which a third party gives written assurance that a product conforms to specified requirements (35,36).

Further, with the view of facilitating both the proper ordering of subjects and the progress in the complex field of coal systematics, it has recently become necessary to distinguish *scientific/genetic* classifications from *technical/commercial* ones (Table 1). Both have different fields of application, objectives, and purposes (26,27,29,30). Table 1 compares the main differences between the two types of systematics, but it should be emphasized that in genetic codifications coal is regarded as a geological entity (coal seam), whereas in the technological/commercial codifications coal is considered as an industrial product already beneficiated (from one single seam or in the form of a blend or mixture) for the purpose of being commercialized and utilized using different technological processes. However, it is important to bear in mind that scientific/genetic classifications cannot be regarded as purely theoretical models without any immediate practical application because they may contribute favorably, and primarily, to the calculation of resources and reserves on a common basis.

As far as technical/commercial systems and their immediate practical implications are concerned, the different established categories are expressed, in most cases, through a numerical code linked to chosen parameters and without concern for hierarchy; they serve only practical purposes. For this reason, the majority of technical/commercial systems correspond to what is currently referred to as codifications or codings, thus presently leaving, in a practical sense, the simplified designation of classifications to the scientific/genetic systems in which the different classes or groups are identified through the use of names rather than codes.

Attempts at elaborating internationally acceptable systematizations has, apart from individual contributions from national specialists (eg, B. Alpern from France, M. R. Campbell from the U.S., W. Francis, S. W. Parr from the U.S., and C. A. Seyler from England), also deserved the attention of international organizations such as the Coal Committee/Working Party on Coal of the UNECE in Geneva, the International Committee for Coal and Organic Petrology (ICCP), and the International Organization for Standardization (ISO). In fact, the enormous diversity in existing systematics, on one hand, and the clearly distinct purposes required, on the other, have led to a very confusing situation. This has raised the need to elaborate systematizations that will be of international scope and acceptance because it is clearly impossible to harmonize the different national systems (24,25). See Table 2 for a comparison of the different classifications for medium- and high-rank coals. The term rank describes the stage of carbonification attained by a given coal (7,37).

Table 1. Coal Codification and Classification

Technical/Commercial (Codifications)	Scientific/Genetic (Classifications)
Application	
1. Beneficiated products (crushed, washed, sized)	1. Coal as an organic sedimentary rock (maceral and mineral matter)
2. Seams ex situ and blends	2. Seams in situ
Means	
1. Artificial system	1. Natural system
2. Numerical code	2. Words (descriptive)
3. Unlimited parameters, not in hierarchy	3. Limited parameters, always in a hierarchy
Aims and Objectives	
Commercial: export/import	Scientific/geological purposes
Detailed characterization	Calculate reserves and resources
Evaluate ex situ coals	Evaluate in situ coals
	Teaching (specialists and preliminary)
	Tentative to correlate different national classification systems

Source: Modified after Ref. 27.

NATIONAL SYSTEMS

Since the industrial Revolution of the nineteenth century, and particularly since World War II, remarkable development has taken place in countries with coal-related industries. The major coal-producing nations developed classification systems for solid fossil fuels, but these were essentially elaborated for the purpose of trade and utilization (coal and/or blends) and are based on knowledge gained on their own coals. Although it is rather difficult to correlate between different national systems, several classification systems gained international recognition due to the degree of importance of the respective countries in terms of coal utilization and trade. As examples, the following systems are quoted: Australia (AS 2096-77) (33), France (Ref. 38; NF M 10-001), People's Republic of China (Ref. 39; GB 5751-86), United Kingdom (Ref. 40; BS 3323-78; refer to Table 3), the United States (ASTM D 388-92a) (34), and the former USSR (Ref. 41; GOST 25543-88). Consequently, and due to their level of importance in terms of modern-day systems, the American and Australian systems will be here described as examples.

The ASTM Classification of Coals by Rank

The American Society for Testing and Materials was founded in 1898. It is a scientific and technological organization for "the development of standards on characteristics and performance of materials, products, systems, and services; and the promotion of related knowledge" (43). In 1991, the Society consisted of 132 main technical committees with 2067 subcommittees and has a membership of 32,800. More than half of the members serve as technical experts on committees. The society publishes the *Annual Book of ASTM Standards,* consisting of 68 volumes representing 16 sections. Section 5–Petroleum Products, Lubricants and Fossil Fuels–contains volume 05.05, which deals with gaseous fuels, coal, and coke. "Standard Classification of Coals by Rank" is ASTM D 388-92a; the number following the hyphen indicates the year of last revision.

The ASTM classification of coals by rank was first issued as a tentative standard D 388-34T in 1934 (43). It was established as an adopted standard D 388-37 in 1937 (44), and was revised in 1938 and most recently in 1964,

Table 2. Classes of Old (1956) UN-ECE International Codification System for Medium- and High-Rank Coals (Hard Coals) Compared with "Classes" of the National Systems

International System			National Systems							
Class No.	Volatile Matter Content	Calorific Value[a]	Belgium	Germany	France	Italy	Netherlands	Poland	United Kingdom	United States
0	0–3					Antraciti speciali		Meta-antracyt		Meta-anthracite
1A	3–6.5				Anthracite	Antraciti		Antracyt		Anthracite
						comuni	Antraciet		Anthracite	
			Maigre	Anthrazit						
1B	6.5–10							Pólantracyt		
					Maigre					Semi
						Carboni				anthracite
						magri	Mager	Chudy	Dry steam	
2	10–14		¼ gras	Magerkohle						
			½ gras		Demi-gras					
										Low-
						Carboni semi		Pólkoksowy	Coking steam	volatile bituminous
3	14–20		¾ gras	Esskohle		grassi	Esskool	Metakoksowy		
					Gras	Carboni grassi			Medium	Medium-
4	20–28			Fettkohle		corta fiamma		Ortokoksowy	volatile	volatile bituminous
					à courte		Vetkool		coking	
					flamme					
5	28–33		Gras			Carboni grassi				High-volatile
				Gaskohle		media fiamma		Gazowo		bituminous A
6	>33	8450–7750			Gras	Carboni	Gaskool	koksowy		
	(32–40)				proprement	da gas				
7	>33	7750–7200			dit			Gazowy		High-volatile
	(32–34)					Carboni	Gasvlamkool		High	bituminous B
				Gasflamm-	Flambant	grassi da vapore			volatile	
8	>33				gras			Gazowo-		
	(34–36)	7200–6100		Kohle				plomienny		High-volatile
	>33					Carboni secchi	Vlamkool			bituminous C
9	(36–48)	<6100			Flambant sec			Plomienny		Sub-bituminous

Source: After Ref. 24, Table II.5, p. 27.
[a] Calculated to standard moisture content.

1966, 1977, 1982, 1988, and twice in 1992 (ASTM D 388-92a). The basis of the classification is fixed carbon and calorific value (expressed in Btu/lb) on a mineral-matter free basis (mmf). The higher rank coals are classified on the basis of fixed carbon, on a dry, mineral-matter-free basis. Lower rank coals are classified on the basis of calorific value on a moist, mineral-matter-basis. Some related ASTM standards deal with volatile matter in coal and coke (45) (D 3175), ultimate analysis of coal and coke (46) (D 3176), sulfur in ash (47) (D 1757), fusibility of coal and coke ash (48) (D 1857), preparing coal samples for analysis (49) (D 2013), forms of sulfur in coal (50) (D 2492), total moisture in coal (51) (D 3302), total sulfur in coal and coke (52) (D 3177), free swelling index (53) (D 720), equilibrium moisture (54) (D 1412), and microscopical analysis and determination (D 2797) (55) is the method for preparation of coal samples for reflected-light analysis, D 2798 (56) relates to the method for the determination of reflectance, and D 2799 (57) deals with methods for the determination of the volume percent of the organic components (macerals) and inorganic components (minerals) of coal. Standard definitions of terms relating to coal and coke (58) (D 121) is a useful guide to terms used in the ASTM standards procedures for the chemical and physical analysis of coal.

Parr's Classification of Coal. Parr (19) published a bibliography relating to the classification of coal beginning with Frazer's (12) "Classification of Coals." Other early classifications are Campbell's (14), Parr's (59), and Bement's (60) classifications. Parr (19) credits Regnault (10) in France as the first chemist to make ultimate analyses of coal, Grüner (11) as using ultimate analyses to classify coal, and Wedding in Germany for stressing the usefulness of volatile matter in coal classifications. Seyler (61) also used elemental data in his coal classification system.

According to Parr (19), however, it was Johnson (62) in 1844 who first used the ratio of fixed carbon to volatile matter ("Fuel-Ratio") to organize coals by "evaporative power." It was Frazer (12) who first recognized the significance of the fuel ratio and used it to classify coals. Only after the formation of the U.S. Bureau of Mines in 1910 was sufficient attention given to coal investigation and characterization, as indicated by the large number of bibliographic items in Parr's (19) compilation of the methods of classifying coal (1910–1927).

Parr's (19) classification stressed the fundamental importance of volatile matter and calorific value, that formed the basis of ASTM D 388-34T, "Classification of Coals by Rank." Inherent in Parr's classification is the concept of unit coal (63), which is the pure coal substance, not including "extraneous or adventitious material." Thus, corrections had to be made for the noncoal substances, such as clays, carbonates, and pyrite, that were mixed with the pure coal. These corrections were made by formulas that were referred to as the Parr formulas in the ASTM D 388-34T classification; values for moisture, ash, and sulfur contents are used to calculate fixed carbon to a dry, mmf basis, and calorific value to a moist, mmf basis. Thus, unit Btu value and unit volatile matter of the pure coal substance are determined using these formulas,

Table 3. National Coal Board/British Coal Codification and Classification for Medium- and High-Rank Coals (BS 3323)[a]

Coal Rank Code Main classes	Class	Subclass	Volatile Matter (%, mass, dmmf)	Gray-King Coke Type[b]	General Description
100	—	—	Under 9.1	A	Anthracites
	101 (3)	—	Under 6.1	A	
	102 (3)	—	6.1–9	A	
200	—	—	9.1–19.5	A–G8	Low-volatile steam coals
	201	—	9.1–13.5	A–C	Dry steam coals
	—	201a	9.1–11.5	A–B	
	—	201b	11.6–13.5	B–C	
	202	—	13.6–15.0	B–G	Coking steam coals
	203	—	15.1–17.0	E–G4	
	204	—	17.1–19.5	G1–G8	
300	—	—	19.6–32.0	A–G9 and over	Medium-volatile coals
	301	—	19.6–32.0	G4 and over	Prime coking coals
	—	301a	19.6–27.5	G4 and over	
	—	301b	27.6–32.0	G4 and over	
	302	—	19.6–32.0	G–G3	Medium-volatile, medium-caking, or weakly caking coals
	303	—	19.6–32.0	A–F	Medium-volatile, weakly caking to noncaking coals
400–900	—		Over 32.0	A–G9 and over	High-volatile coals
400	—		Over 32.0	G9 and over	High-volatile, very strongly caking coals
	401		32.1–36.0	G9 and over	
	402		Over 36.0	G9 and over	
500	—		Over 32.0	G5–G8	High-volatile, strongly caking coals
	501		32.1–36.0	G5–G8	
	502		Over 36.0	G5–G8	
600			Over 32.0	G1–G4	High-volatile, medium-caking coals
	601		32.1–36.0	G1–G4	
	602		Over 36.0	G1–G4	
700			Over 32.0	E–G	High-volatile, weakly caking coals
	701		32.1–36.0	E–G	
	702		Over 36.0	E–G	
800			Over 32.0	C–D	High-volatile, very weak caking coals
	801		32.1–36.0	C–D	
	802		Over 36.0	C–D	
900			Over 32.0	A–B	High-volatile, noncaking coals
	901		32.1–36.0	A–B	
	902		Over 36.0	A–B	

Note: Coals that have been affected by igneous intrusions, ie, "heat-affected" coals, occurs mainly in classes 100, 200, and 300 and when recognized should be distinguished by adding the suffix H to the coal rank code, eg, 102H, 201bH. Coals that have been oxidized by weathering may occur in any class and when recognized should be distinguished by adding the suffix W to the coal rank code, eg, 801W. From ref. 40.

[a] Coal with an ash content at over 10% shall be cleaned before analysis for classification to give a maximum yield of coal with an ash content of 10% or less.

[b] Coals with volatile matter of under 19.6% are classified by using the parameter volatile matter alone; the Gray-King coke types quoted for these coals indicate the general ranges found in practice and are not criteria for classification.

[c] In order to divide Anthracite into two classes, it is sometimes convenient to use the Hydrogen content of 3.35% (dmmf), instead of a Volatile Matter of 6.1% as the limiting criterion. In the original Coal Survey rank coding system the Antracites were divided into four classes then designated 101, 102, 103 and 104. Although the present division into two classes satisfies most requirements, it may sometimes be necessary to recognize more than two classes.

which are still in use today. These two values were used by Parr as the basis of his classification into seven coal types [not including "cannels"]: Type 1: anthracite; Type 2: semianthracite; Type 3: bituminous A; Type 4: bituminous B; Type 5: bituminous C; Type 6: bituminous C (subbituminous or black lignite); Type 7: lignite.

Parr makes it clear that this is a classification by "rank." Cannel coals, an eighth "type" of coal in Parr's (19) classification, are treated as a special type of coal because of their high calorific value and high volatile matter content. The term cannel coal is used for sapropelic coal containing predominantly spores, in contrast to sapropelic coal containing predominantly algae, which is termed boghead coal (7).

Comparison of these "types" with ASTM's 1934 coal classification (D 388-34T) to be discussed in a following section, makes it very apparent that this coal classification by rank was derived from Parr's (19) coal classification with major additions and changes of names and limits of the coal types. Parr (19) made comparisons with Seyler's (61) classification, which is a closely related coal classification based on carbon, hydrogen, and oxygen values. However, coal names or divisions are vastly different in the two classifications. For a complete discussion of the History of Classification up to the 1940's reference should be made to Rose (6).

Classification of Coals by Grade. The ASTM tentative specifications for classification of coals by grade were issued in 1934 as tentative standard D 389-34T (64), which was adopted as ASTM D 389-37 (65) (Table 4). This standard deals with the classification of coals by calorific value, ash and sulfur contents, and ash-softening temperature and is largely of historical value because it was discontinued sometime after 1945 (6) because it was never applied in common use. However, it is treated here because it is a typical example of a codification system and shows some relationship to the parameters used in the 1988 UN-ECE International Codification System for Medium- and High-Rank Coals (66) (refer to the international systems later in this article). However, in the latter system, the values cannot be inferred from the codes without a key. In the ASTM classification of coals by grade, each coal classified was given a four-part designation reflecting, respectively, calorific value, ash (A), ash-softening temperature (F), and sulfur content (S). Thus, the designation 132-A8-F24-S1.6 indicates a calorific value of 30.7; MJ/kg (13,200 Btu/lb), an ash content of 6.1–8.0 wt %, an ash-softening temperature of 1315–1421°C (2400–2590°F), and a sulfur content of 1.4–1.6 wt % (see Table 4).

The grade designation could be combined with the rank designation, as shown in the following example: (62–115) 132-A8-F24-S1.6. The numbers in parentheses reflect the rank and the numbers following the parentheses reflect the grade, as shown in the preceding paragraph. Thus, 62 indicates a fixed carbon of 61.5–62.4 [dry, mineral-matter-free (mmf) basis] rounded to the whole number. The number 146 indicates the calorific value rounded to the nearest hundreds of Btu, in this case 15.4 MJ (14,600 Btu), corresponding to a rank of high volatile A bituminous coal (see Table 5).

ASTM D 388-92a Coal Classification by Rank. Since the tentative standard for classification of coals by rank (D 388) was first adopted in 1934 for U.S. and Canadian, there has been little change to the fundamental basis for the classification, the classes and groups, and their limits (Table 5). However, some minor changes were made to the groups and their limits. In the 1934 tentative standard "brown coal" was used as a group instead of lignite B, as in the present ASTM classification (D 388-92a) (34) and the latter was limited to coals with calorific values of less than 14.6 MJ/kg (6300 Btu/lb) (moist, mmf), instead of less than 19.3 MJ/kg (8300 Btu/lb), as in the 1934 standard. Also, the limits of subbituminous A coal were originally greater than 25.6 MJ/kg (>11,000 Btu/lb) and less than 30.2 MJ/kg (<13,000 Btu/lb) (moist, mmf) and subbituminous B coal less than 25.6 MJ/kg (<11,000 Btu) (moist, mmf) rather than at 24.4–26.7 MJ/kg (10,500–11,500 Btu/lb) and 22.1–24.4 MJ/kg (9500–10,500 Btu/lb), respectively, as in the ASTM D 388-92a classification (Table 5). For further details refer to Appendix 1 of standard D 388-92a. The limits of high volatile B and A bituminous coals remained unchanged since 1934. For higher rank coals, the groups and their limits also remained the same except that a fixed carbon value (dry, mmf) of 77% was changed to 78% as a basis of separating medium-volatile and low-volatile bituminous coals. The 1988 and 1992 revisions (D 388-92a) of this classification exclude inertinite-rich and liptinite-rich coals from this classification, ie, it only applies to vitrinite-rich coals that are generally typical of North American coals. Also, D 388-92a includes a correlation (n = 807) of volatile matter with the mean maximum reflectance of vitrinite for bituminous coals.

It is important to note that only coal analyses made in accordance with ASTM standard methods of analysis can be used in the ASTM classification of coal by rank. Excluded are analyses published before 1913, samples that have been altered by exposure, and fixed carbon results from U.S. Geological Survey and U.S. Bureau of Mines analyses designated Laboratory Numbers 5147 and 9120 inclusive. The latter analyses give high results for fixed carbon.

In the ASTM classification of coal by rank, the "Parr formulas" are used to calculate the dry, mineral-matter-free (d, mmf) fixed carbon and volatile matter and also

Table 4. Symbols for Grading Coal According to Ash, Softening Temperature of Ash, and Sulfur

Ash[a]		Softening Temperature of Ash[b]		Sulfur[a]	
Symbol	Percent Inclusive	Symbol	Inclusive °F	Symbol	Percent Inclusive
A4	0.0–4.0	F28	2800 and higher	S0.7	0.0–0.7
A6	4.1–6.0	F26	2600–2790	S1.0	0.8–1.0
A8	6.1–8.0	F24	2400–2590	S1.3	1.1–1.3
A10	8.1–10.0	F22	2200–2390	S1.6	1.4–1.6
A12	10.1–12.0	F20	2000–2190	S2.0	1.7–2.0
A14	12.1–14.0	F20 minus	<2000	S3.0	2.1–3.0
A16	14.1–16.0			S5.0	3.1–5.0
A18	16.1–18.0			S5.0 plus	5.1 and higher
A20	18.1–20.0				
A20 plus	20.1 and higher				

Note: Analyses expressed on the basis of coal as sampled (ASTM D 389-34T (64)).

[a] Ash and sulfur shall be reported to the nearest 0.1% by dropping the second decimal figure when it is 0.01–0.04 inclusive and by increasing the percentage by 0.1% when the second decimal figure is 0.05–0.09 inclusive. For example, 4.85–4.94% inclusive, shall be considered to be 4.9%.

[b] Ash-softening temperatures shall be reported to the nearest 10°F. For example 2635–2644°F inclusive shall be considered to be 2640°F.

the moist, mineral-matter-free (m, mmf) calorific value. These formulas, virtually unchanged since they were implemented by D 388-34T in 1934, estimate a value for mineral matter percentage calculated from ash and sulfur percentages.

"Apparent rank" may be used for determinations based on tipple or shipment samples or samples taken under unspecified conditions. Thus, the concept of apparent rank implies that a standard rank determination cannot be made due to bed-sampling problems.

With reference to Table 5, four classes of coal are recognized in the ASTM Classification of Coals by Rank. From lower to higher rank, they are lignitic, subbituminous, bituminous, and anthracitic. Each class is subdivided into two to five groups that are differentiated on the basis of moist, mineral-matter-free calorific value (Btu/lb); mineral-matter-free fixed carbon or volatile matter, or both in the case of high-volatile A bituminous coal. It is important to realize that the sum of volatile matter and fixed carbon, both calculated on dry, mmf basis, equals 100 wt %, so that either parameter can be derived from the other one. For example a fixed carbon of 88 wt % indicates a volatile matter content of 12 wt %. Thus the rank series starting upward from high-volatile A bituminous coal can be thought as a series of increasing fixed carbon values or of decreasing volatile matter values. As coalification proceeds due to heating as a consequence of the burial of peat, volatile matter (eg, CO_2, SO_2, and CH_4) is progressively driven off and the fixed carbon is progressively increased. The fixed carbon is the solid residue other than ash which is left when the volatile matter is driven off. Thus, fixed carbon is the difference between the volatile matter content and 100%: "It is made up principally of carbon, but may also contain appreciable amounts of sulfur, hydrogen, nitrogen and oxygen" (ASTM D 121). Consequently, fixed carbon is quite different from elemental carbon, which represents the total amount of carbon in a substance. For coals with calorific values over 32.5 MJ/kg 14,000 Btu/lb (m, mmf) they are differentiated by fixed carbon (or volatile matter) and agglomerating or nonagglomerating character. Subject to the ash content (determined from the residue remaining in the crucible after the volatile matter determination) and other secondary extraneous factors such as weathering, oxidation, and thermal effects, bituminous coals are commonly agglomerating (ie, they form a coke button when heated), whereas higher rank coals (ie, anthracitic coals) are nonagglomerating and do not have coking potential. Standard D 388 allows for coals having a free swelling index equal to or greater than one to be classified as agglomerating. Also, lignitic and subbituminous coals are nonagglomerating.

Australian Classification

Whereas the ASTM Classification of Coals by Rank dates back to 1934, the first Classification System for Australian Hard Coal was issued in 1969 by the Standards Asso-

Table 5. ASTM Classification of Coals by Rank (ASTM D 388-92a)[a]

	Fixed Carbon Limits, % (dmmf)		Volatile Matter Limits, % (dmmf)		Gross Calorific Value Limits (m, mmf)[b]			
					Btu/lb		MJ/kg[c]	
Class/Group	Equal or Greater Than	Less Than	Greater Than	Equal or Less Than	Equal or Greater Than	Less Than	Equal or Greater Than	Less Than
I. Anthracitic[d]								
1. Meta-anthracite	98	—	—	2				
2. Anthracite	92	98	2	8				
3. Semianthracite (4)	86	92	8	14				
II. Bituminous coal[e]								
1. Low-volatile	78	86	14	22				
2. Medium-volatile	69	78	22	31				
3. High-volatile A	—	69	31	—	14,000[f]	—	32.6	
4. High-volatile B	—	—	—	—	13,000[f]	14,000	30.2	32.6
5. High-volatile C	—	—	—	—	11,500	13,000	26.7	30.2
					10,500[g]	11,500[g]	24.4[g]	26.7[g]
III. Subbituminous coal[d]								
1. A	—	—	—	—	10,500	11,500	24.4	26.7
2. B	—	—	—	—	9,500	10,500	22.1	24.4
3. C	—	—	—	—	8,300	9,500	19.3	22.1
IV. Lignitic[d]								
1. A	—	—	—	—	6,300	8,300	14.7	19.3
2. B	—	—	—	—	—	6,300	—	14.7

[a] These classification applies only to coals mainly composed of vitrinite in sector 1 of ref. 34.
[b] Moist refers to coal containing its natural inherent moisture but not including visible water on the surface of the coal.
[c] Megajoules per kilogram. To convert British thermal units per pound to megajoules per kilogram, multiply by 0.002326.
[d] Nonagglomerating.
[e] If agglomerating, classify in low-volatile group of the bituminous class.
[f] Commonly agglomerating except as indicated.
[g] Coals having 69% or more fixed carbon on the dry, mineral-matter-free basis shall be classified according to fixed carbon, regardless of gross calorific value.

ciation of Australia as Australian standard K 184-1969. This classification was "primarily intended for commercial and industrial use and is based on the grouping of coals according to properties that can be readily determined in most laboratories in Australia that deal with coal analysis." For the purposes of this classification "hard coal" was defined as "coal having or exceeding a gross calorific value of 6470 kcal/kg on a dry ash-free (daf) basis." Development of the classification was guided by the International Classification of Hard Coals by Type, which was issued by the UN-ECE in 1956 (67). Unlike the latter, however, its application was not limited to coals with less than 10% ash. The classification was based on two parameters: volatile matter, expressed on a d, mmf basis (six class categories) over the range 10–33% and, for the lower rank hard coals, where the volatile matter exceeded 33%, the gross calorific value (four class categories covering the range 8080–6790 kcal). Within these 10 coded class categories the coals were grouped according to crucible swelling number (CSN) into five-coded groups covering CSN 0–9 (Table 6). There was a further grouping according to Gray King coke type (GKCT) into six coded categories to cover the GKCT range A–G9. In recognition that Australian coals have a higher mineral matter content than coals in Europe and North America, the ash yield was introduced as a fourth parameter, designated "ash number," with nine categories covering ash to >32% [dry basis, (db)]. In 1977, AS K 184 was replaced by AS 2096 (33) with little change, primarily to convert to SI units.

In 1987 the Standards Association of Australia issued an extensively revised and extended version of AS 2096 (112), entitled "Classification and Coding Systems for Australian Coals." This standard was extended to cover coals of all ranks and describes the classification of Australian coals into two rank classes. It also sets out systems for coding coals within each class and provides quantitative definitions for traditional names. The stated purposes of the classification are to assist in estimating reserves and for determining the applicability of standard methods for analysis and testing.

Table 6. Coding Systems of Australian Standard AS 2096

Maximum Reflectance of Vitrinite or Vitrinite Precursor (R_v, max) – AS 2486	
Code (R), %	
	Lower Rank Coals
0	From 0.00 to 0.09 inclusive
1	From 0.10 to 0.19 inclusive
2	From 0.20 to 0.29 inclusive
3	From 0.30 to 0.39 inclusive
4	From 0.40 to 0.49 inclusive
5	From 0.50 to 0.59 inclusive
	Higher Rank Coals
03	From 0.30 to 0.39 inclusive
04	From 0.40 to 0.49 inclusive
05	From 0.50 to 0.59 inclusive
...	...
20	From 2.00 to 2.09 inclusive
Gross Specific Energy (Gross Calorific Value), AS 1038.5.1 or AS 1038.5.2	
Code (E), MJ/kg (daf)	
15	From 15.00 to 15.98 inclusive
16	From 16.00 to 16.98 inclusive
17	From 17.00 to 17.98 inclusive
...	...
36	From 36.00 to 36.98 inclusive
...	...
Volatile Matter, AS 1038.3 (higher rank), AS 2434.2 (lower rank)	
Code (V), % (daf)	
08	From 8.0 to 8.9 inclusive
09	From 9.0 to 9.9 inclusive
10	From 10.0 to 10.9 inclusive
...	...
57	From 57.0 to 57.9 inclusive
...	...
Crucible Swelling Number, AS 1038.12.1	
Code (C), Numbers	
0	0 or $\frac{1}{2}$
1	1 or $1\frac{1}{2}$
2	2 or $2\frac{1}{2}$
3	3 or $3\frac{1}{2}$
4	4 or $4\frac{1}{2}$
5	5 or $5\frac{1}{2}$
6	6 or $6\frac{1}{2}$
7	7 or $7\frac{1}{2}$
8	8 or $8\frac{1}{2}$
9	9
Bed Moisture, AS 2434.1	
Code (M), % (as)	
20	From 20.0 to 20.9 inclusive
21	From 21.0 to 21.9 inclusive
22	From 22.0 to 22.9 inclusive
...	...
65	From 65.0 to 65.9 inclusive
...	...
Ash, AS 1038.3	
Code (A), %	
00	From 0.0 to 0.9 inclusive
01	From 1.0 to 1.9 inclusive
02	From 2.0 to 2.9 inclusive
...	...
30	From 30.0 to 30.9 inclusive
...	...
Total Sulfur, AS 1038.6.3 (higher rank), AS 2434.6.1 (lower rank)	
Code (S), %	
00	From 0.0 to 0.9 inclusive
01	From 1.0 to 1.9 inclusive
02	From 2.0 to 2.9 inclusive
...	...
32	From 3.2 to 3.29 inclusive

Classification. The moisture basis in (m, af) refers to moisture-holding capacity carried out according to AS 2434.3 (68) for lower rank coals and to AS 1038.17 (69) for higher rank coals. In this standard coals are classified into higher and lower rank on the basis of the gross specific energy as follows:

1. **Higher Rank Coals.** Coal is classified as higher rank if both the following criteria are satisfied:

(a) Gross specific energy 27.00 MJ/kg, dry ash-free basis (daf) or greater.
(b) Gross specific energy 21.00 MJ/kg, moist, ash-free basis (m, af) or greater.

2. **Lower Rank Coals.** Coal is classified as lower rank if *either* of the following criteria are satisfied:
(a) Gross specific energy (daf) is less than 27.00 MJ/kg.
(b) Gross specific energy (m, af) is less than 21.00 MJ/kg.

With regard to the frequent use of rank-related traditional names for a coal (eg, as in the ASTM D 388 classification), these are defined in order of decreasing rank as follows:

(a) Anthracite. Coal having volatile matter less than 8.0% (daf).
(b) Semianthracite. Coal having volatile matter within the range 8.0–13.9% daf.
(c) Bituminous Coal. Coal having volatile matter >14.0% (daf) and gross specific energy >26.50 MJ/kg (m, af), >24.00 MJ/kg (m, af) provided CSN >1.
(d) Subbituminous Coal. Coal having a gross specific energy in the range 19.00–23.98 MJ/kg (m, af), or to 26.48 MJ/kg (m, af) provided CSN $\geq 0-\frac{1}{2}$.
(e) Brown Coal. Coal having a gross specific energy less than 19.00 MJ/kg (m, af) (4540 kcal/kg; 8170 Btu/lb).

The demarcation boundaries specified are similar but not identical to those in ASTM D 388-92a. However, lignite and not brown coal is used in the ASTM classification.

Coding System. A coding system is included to assist in the initial evaluation of coals for a specific purpose with the caution that the final selection requires a complete analysis to determine all relevant properties.

Six parameters are used for both the higher and the lower rank coals. The first three parameters–mean maximum reflectance of vitrinite (or vitrinite precursor in the case of lower rank coals), gross specific energy, and volatile matter–are common to both higher and lower ranks. The fourth parameter is the CSN for higher rank coals and bed moisture for lower rank coals. The fifth and sixth parameters, ash and total sulfur contents, are common to both higher and lower ranks. In the coding system, all the parameters are represented by a two-digit code except for the CSN for the higher rank and mean maximum reflectance of vitrinite for the lower rank coals, which are each represented by a single-digit code. Thus the complete code for any coal contains 11 digits with spaces between the digit code for each parameter. The fixed order of mention of the six parameters permits the simple use of two six-letter acronyms, based on an obvious letter for each parameter: REVCAS for the higher rank coals and REVMAS for the lower rank coals, where:

R Refers to mean maximum reflectance of vitrinite ($\overline{R}_v$, max; see AS 2486-1989) (70)
E Refers to the gross specific energy (daf) (see AS 1038.5.1, -5.2) (71,72).
V Refers to the Volatile matter (daf) (see AS 1038.3, 2434.2) (73,74).
C Refers to the CSN (higher rank only) (see AS 1038.12.1) (75)
M Refers to the bed moisture (lower rank only) (see AS 2434.1) (76).
A Refers to ash (db) (see AS 1038.3) (73).
S Refers to total sulfur (db) (see AS 1034.6.1, 1038.6.3) (77,78).

Standard procedures are available for the determination of all these parameters. As mentioned in the standards referred to in Table 6, calculation to different bases are done according to AS 1038.16 (79). Determination of the bed moisture content of the lower rank coals requires special sampling and analysis procedures covered by AS 2434.1 (76). This is due to the gel structure of these high-moisture coals, which collapses irreversibly as moisture is lost in the removal of coal from the seam.

To maximize the value of REVCAS and REVMAS codes six important properties in each of the two rank classes of coal are coded in narrow ranges. The details of the units, codes, and total range covered by each parameter are summarized in Table 6.

In the Australian standard AS 2096, provision is made for the use of a partial code where the data on some property parameters are not available. In such instances an X or XX is used in the code for the appropriate parameter. For example, in lower rank coals the mean maximum reflectance of the vitrinite is not usually determined; thus a code number of X 24 49 59 03 03 obviously refers to a lower rank coal with gross specific energy between 24.00 and 24.98 MJ/kg (daf), volatile matter between 49.0 and 49.9% (daf), bed moisture between 59.0 and 59.9%, ash between 3.0 and 3.9% (db), and sulfur between 0.30 and 0.39% (db). Thus, there is a parallel between this codification system and ASTM D 389-37 (65).

INTERNATIONAL SYSTEMS

As a consequence of the difficulties in correlating the different national systems, as shown in Table 2, a first attempt to formalize a systematization of coal for international purposes was elaborated and published in 1956 by the UN-ECE (67) within the scope of hard coals. Hard coals were those with a gross calorific value over 5700 kcal/kg (m, af). The present-day concept of hard coals has been improved. In 1957, as follow-up, the UN-ECE issued a classification of brown coals (80). This classification was later adopted, with modifications, as ISO 2950 (81) of the International Organization for Standardization (Table 7).

Both these international systems were elaborated with final practical intention in mind, both commercial and technological, and were based exclusively on chemical parameters and/or physical–chemical tests. Furthermore, in both systems coals were characterized by numerical codes and, thus, can also be considered as codifications. Both systems are now outdated, having been overtaken by advances in coal utilization technologies and in the world trade of solid fuels. Further, both cited systems are based on Northern Hemisphere coals and therefore do not correspond to all coals presently known and utilized worldwide, particularly regarding hard coals.

In recognition of the obsolescence of the systems referred to above, the Coal Committee/Working Party on Coal of the UN-ECE decided, in the late 1970s, to reapproach the subject by a group of experts representing different countries. At this point, and as far as representativeness of opinions expressed during work discussions are concerned, it should be emphasized that according to the UN-ECE Terms of Reference, other countries and international organizations are entitled to be represented at the meetings, apart from the ECE members themselves. In fact, several relevant countries, other than member countries, have systematically participated in UN-ECE meetings on coal classification/codification, including Australia, China, India, and Japan. Their opinions are considered and taken into account at the same level as the UN-ECE member countries.

Similarly, official representation from several international organizations such as the former Council for Mutual Economic Assistance (CMEA), the European Union (EU), the European Coal and Steel Community (ECSC), the International Committee for Coal and Organic Petrology (ICCP), and the International Energy Agency (IEA-OECD) and ISO have also participated in these meetings.

International Codification System for Medium- and High-Rank Coals

The International Codification System for Medium- and High-Rank Coals (66), aimed at replacing the 1956 system, evolved from a proposal submitted by the government of the former Federal Republic of Germany. This system consists of *basic* and *supplementary* parameters; the former are used in the codification itself, whereas the latter should be considered as a specifications list from which particular analyses may be drawn in accordance with special uses. There are eight basic parameters represented by 14 digits (Tables 8 and 9 and Fig. 1). The supplementary parameters are numerous, some of which are listed in Table 10. Each parameter is determined in accordance with the relevant standards considering the ISO system whenever possible. The application of this codification system was monitored by the Transport Committee of the Economic Commission for Europe (United Nations) during a fixed period that terminated in 1993.

It should be noted that several countries such as the former Republic of Czechoslovakia (82), France (38), Portugal (83), and Turkey (84) have already adopted as national standards the ECE-UN International Codification System for Medium- and High-Rank Coals (66).

ECE International Classification of Seam Coals

In the 1980s, upon recognizing that there should be a classification for coal as a rock and independently from any codification system, the UN-ECE directed its efforts into preparing the so-called International Classification of Seam Coals. This became the first attempt to create an international classification of coals that would contribute to the characterization of coal deposits, but not for commercial or technological purposes.

In modern terms, the systematization of coal as a rock is based on its *three fundamental characteristics* which, according to present-day knowledge, permit to define it unequivocally as such. These characteristics must always be considered *jointly* and *never separately from each other: rank* (or degree of coalification), *petrographic composition* (or "type"), and *grade* (degree of impurities).

The present UN-ECE classification system has resulted from a total number of four ad hoc meetings (1988, 1989, 1991, 1992) and one Task Force (1990) consisting of a group of specialists (called Group of Experts) representing all interested countries and international organizations.

The Group of Experts accepted the French government proposal as the one to be used as the basis for discussion. When presented, this proposal was considered as a "subclassification" due to the nomenclature in use at the time by the UN-ECE. This proposal was taken from those formally presented to the ECE-UN for consideration. The French project basically coincides with the system developed and published by B. Alpern, presented for the first time as an informal document from the Centre d'Etudes et Recherches des Charbonnages de France dated April 12, 1977. Subsequently, the system was published in its original version (26,27) and underwent successive improvements (32,85). A computerized version demonstrating its application to all coals was also published (86). Suffice to say that the application of the system was also demonstrated for "Gondwana-type" coals by Navale and Misra (87) for Indian coals, by Alpern and Nahuys (88) for Brazilian coals, and by Falcon (30) for Southern African coals.

A general agreement was reached and approved by the UN-ECE Working Party on Coal at its second meeting in 1992 (896) and confirmed with minor improvements at the third Working Party Meeting in 1993. During this meeting one of the decisions taken was to request the UN-ECE Meeting of Experts on Research, Management and Transition in the Coal Industry to monitor the practical

Table 7. ISO Code Numbers for Low-Rank Coals[a]

Group Parameter: Tar Yield (%, m, daf), ISO 647[b]	Group Number	Code Numbers					
		Class 1	Class 2	Class 3	Class 4	Class 5	Class 6
>25	4	14	24	34	44	54	64
>20 to 25	3	13	23	33	43	53	63
>15 to 20	2	12	22	32	42	52	62
>10 to 15	1	11	21	31	41	51	61
<10	0	10	20	30	40	50	60
Class parameter: total moisture content of run-of-mine coal (%, af), ISO 1015 or 5068 (96 or 98)		<20	>20 to 30	>30 to 40	>40 to 50	>50 to 60	>60 to 70

[a] After ISO 2950 (81).
[b] Ref. 119. Calculations in different basis as in ISO 1170, 20.

Table 8. UN-ECE International Codification System for Medium- and High-Rank Coals (1988): Basic Parameters and Codification[a]

Vitrinite Random Reflectance (%), ISO 7404-5 (94)				Maceral Group Composition[b] (volume %, mmf), ISO 7404-3 (111)				Crucible Swelling Number, ISO 501		Volatile Matter Content (%, mass, daf) ISO 562[b]		Ash Content (%, mass, db), ISO 1171[b] (114)		Total Sulfur Content (%, mass, db), ISO 334 or ISO 351[c]		Gross Calorific Value (MJ/kg, daf), ISO 1928[b] (95)	
1:2		1:2		4		5		6		7:8		9:10		11:12		13:14	
02	0.20–0.29	33	3.30–3.39	0	0–<10	0	exempted	0	0–½	48	≥48	00	0 to <1	00	0.0 to <0.1	21	<22
03	0.30–0.39	34	3.40–3.49	1	10–<20	1	0–<5	1	1–1½	46	46 to <48	01	1 to <2	01	0.1 to <0.2	22	22 to <23
04	0.40–0.49	35	3.50–3.59	2	20–<30	2	5–<10	2	2–2½	44	44 to <46	02	2 to <3	02	0.2 to <0.3	23	23 to <24
05	0.50–0.59	36	3.60–3.69	3	30–<40	3	10–<15	3	3–3½	42	42 to <44	03	3 to <4	03	0.3 to <0.4	24	24 to <25
06	0.60–0.69	37	3.70–3.79	4	40–<50	4	15–<20	4	4–4½	40	40 to <42	04	4 to <5	04	0.4 to <0.5	25	25 to <26
07	0.70–0.79	38	3.80–3.89	5	50–<60	5	20–<25	5	5–5½	38	38 to <40	04	5 to <6	05	0.5 to <0.6	26	26 to <27
08	0.80–0.89	39	3.90–3.99	6	60–<70	6	25–<30	6	6–6½	36	36 to <38	06	6 to <7	06	0.6 to <0.7	27	27 to <28
09	0.90–0.99	40	4.00–4.09	7	70–<80	7	30–<35	7	7–7½	34	34 to <36	07	7 to <8	07	0.7 to <0.8	28	28 to <29
10	1.00–1.09	41	4.10–4.19	8	80–<90	8	35–<40	8	8–8½	32	32 to <34	08	8 to <9	08	0.8 to <0.9	29	29 to <30
11	1.10–1.19	42	4.20–4.29	9	≥90	9	≥40	9	9–9½	30	30 to <32	09	9 to <10	09	0.9 to <1.0	30	30 to <31
12	1.20–1.29	43	4.30–4.39							28	28 to <30	10	10 to <11	10	1.0 to <1.1	31	31 to <32
13	1.30–1.39	44	4.40–4.49							26	26 to <28	11	11 to <12	11	1.1 to <1.2	32	32 to <33
14	1.40–1.49	45	4.50–4.59							24	24 to <26	12	12 to <13	12	1.2 to <1.3	33	33 to <34
15	1.50–1.59	46	4.60–4.69							22	22 to <24	13	13 to <14	13	1.3 to <1.4	34	34 to <35
16	1.60–1.69	47	4.70–4.79							20	20 to <22	14	14 to <15	14	1.4 to <1.5	35	35 to <36
17	1.70–1.79	48	4.80–4.89							18	18 to <20	15	15 to <16	15	1.5 to <1.6	36	36 to <37
18	1.80–1.89	49	4.90–4.99							16	16 to <18	16	16 to <17	16	1.6 to <1.7	37	37 to <38
19	1.90–1.99	50	>5.00							14	14 to <16	17	17 to <18	17	1.7 to <1.8	38	38 to <39
20	2.00–2.09									12	12 to <14	18	18 to <19	18	1.8 to <1.9	39	≥39
21	2.10–2.19									10	10 to <12	19	19 to <20	19	1.9 to <2.0		
22	2.20–2.29									09	09 to <10	20	20 to <21	20	2.0 to <2.1		
23	2.30–2.39									08	08 to <9			21	2.1 to <2.2		
24	2.40–2.49									07	07 to <8			22	2.2 to <2.3		
25	2.50–2.59									06	06 to <7			23	2.3 to <2.4		
26	2.60–2.69									05	05 to <6			24	2.4 to <2.5		
27	2.70–2.79									04	04 to <5			25	2.5 to <2.6		
28	2.80–2.89									03	03 to <4			26	2.6 to <2.7		
29	2.90–2.99									02	02 to <3			27	2.7 to <2.8		
30	3.00–3.09									01	01 to <2			28	2.8 to <2.9		
31	3.10–3.19													29	2.9 to <3.0		
32	3.20–3.29													30	3.0 to <3.1		

[a] *Source:* From Ref. 45, Characteristics of Reflectogram 1507404-5, see Table 9, Fig. 1.

[b] 4, Inertinite; 5, liptinite. In codification of coals with an ash content >21% (db), order of code numbers is continued in the way shown above, ie, code number 24 indicates an ash content from 2.4 to <25%.

[c] Calculations in different basis as on ISO 1170. In codification of coals with total sulfur content >3.1% (db), order of code numbers is continued in the way shown above, ie, code number 46 indicates sulfur content from 4.6 to 4.7%.

application of the classification system for a period of 2 years commencing in 1994. It was also decided to invite representatives of governments to test the system in their coal industry and to submit comments on the applicability of the system to the assessment of coal deposits.

It is evident that an accurate assessment of coal deposits requires a clearly defined sampling technique. Such a standard is currently being considered by the ISO, and until such an international sampling standard is developed and agreed upon, it is suggested that sampling be based on existing national standards and that the standard used be identified in conjunction with the classification analysis. This will permit any user of the analytical data to independently evaluate the implications of those data for his or her own purposes. As an example, Australian and U.S. standards may be relied upon (AS 2519 and 2617; ASTM D 4596 and D 5192).

Figures 2 and 3 illustrate the latest version of the UN-ECE classification. The following points summarize the main consensus reached to date:

Rank. The lower and upper rank limits considered in the classification are as follows:

- Lower limit (Peat/Ortho-Lignite boundary): <75% moisture content (as received), moisture being bed moisture.
- Upper limit (Meta-Anthracite/Semi-Graphite boundary): <0.8% hydrogen content (daf).

Between the indicated limits, the following *main divisions* are delineated: Lignite, Subbituminous, Bituminous, and Anthracite or, alternatively, Low-, Medium-, and High-Rank divisions. In both cases *subdivisions* are considered as in Figures 2 and 3.

In the absence of a known parameter reliable enough to cover the full range of rank, the boundary limits for the main divisions and subdivisions are based on *mean random vitrinite reflectance percent* ($-\overline{R}r\%$) and *gross calorific value* (GCV) in megajoules per kilogram. Analytical methods for these parameters shall conform with ISO 7404-5 (94) and ISO 1928 (95), respectively.

The *first step* to be taken when classifying a coal is concerned with the concept of *general classification* as included in the International Codification System for Medium- and High-Rank Coals (45). The objective is to distinguish between *low-rank coals* and *higher rank coals*

Table 9. UN-ECE International Codification System for Medium- and High-Rank Coals (1988): Reflectogram Characteristics and Corresponding Code[a]

Code	Standard Deviation(s)	Meaning
0	≤0.1, no gap	Seam coal
1	>0.1, ≤0.2, no gap	Simple blend
2	>0.2, no gap	Complex blend[b]
3	1 gap	Blend with 1 gap
4	2 gaps	Blend with 2 gaps
5	>2 gaps	Blend with more than 2 gaps

[a] From Ref. 66. See also Fig. 1.
[b] A reflectogram as characterized by code number 2 can also result from a high-rank seam coal.

(hard coals) by the following way:

- Low-rank coals are coals with GCV (m, af) < 24 MJ/kg and $\overline{Rr}$ < 0.6%;
- Higher rank coals (or hard coals), ie, all medium- and high-rank coals, are coals with GCV (m, af) ≥ 24 MJ/kg or GCV (m, af) < 24 MJ/kg provided that $\overline{Rr}$ ≥ 0.6%.

For general classification, the *moist basis* utilized to recalculate the GCV to a moist, ash free basis (m, af) is *moisture-holding capacity* determined according to ISO 1018 (97).

The *second step* involves a refinement of the first in order to place a coal in rank divisions and subdivisions. The latter are as in Figures 2 and 3.

In order to avoid misinterpretations, it was decided that:

- If $\overline{Rr}$ ≥ 0.6%, coals must be classified according to mean random vitrinite reflectance percent.
- If $\overline{Rr}$ < 0.6%, coals must be classified according to GCV (m, af) in megajoules per kilogram.

For the low-rank coals, ie, $\overline{Rr}$ < 0.6% and GCV (m, af) < 24 MJ/kg, where the detailed classification is based on the GCV, this parameter is expressed on a moist, ash-free basis using *bed moisture* (ie, total moisture determined according to ISO 1015 or 5068) (96 or 98).

As stated by the UN-ECE Group of Experts on Coal Classification, the reason for using bed moisture instead of moisture-holding capacity to classify low-rank coals is that, in this range, moisture-holding capacity determined according to ISO 1018 (97) does not give reproducible results.

Furthermore, the reasons for establishing the rank subdivision boundary limits shown in Figures 2 and 3 are the following:

1. In low rank coals: statistical data published in Alpern and co-workers (85) based on Hacquebard (99), Ode and Gibson (100), Parks (101), and Parks and O'Donnell (102).
2. In medium- and high-rank coals:
   (a) $\overline{Rr}$ = 0.6%. This boundary coincides with reflectance value, which, when joined with GCV, establishes the limit between low- and higher rank coals (hard coals) in the general classification included in the International Codification System for Medium- and High-Rank Coals (66).
   (b) $\overline{Rr}$ = 1.0%. This boundary corresponds to the maximum of fluorescence intensity of vitrinite.
   (c) $\overline{Rr}$ = 1.4%. This boundary corresponds to the transition from mosaic to fibrous coke textures.
   (d) $\overline{Rr}$ = 2.0%. This boundary corresponds to an average value, easy to use, corresponding to the end of vitrinite swelling properties. Moreover, this limit also corresponds to the boundary between medium- and high-rank coals (ie, between bituminous and anthracite).
   (e) $\overline{Rr}$ = 3.0%. This boundary corresponds to a change in anthracite ignition properties, becoming comparatively more difficult to ignite over this limit.
   (f) $\overline{Rr}$ = 4.0%. This boundary corresponds to the limit between anthracite and meta-anthracite based on illite crystallinity (103).

The Problem of Moisture. As previously stated, and in accordance with the UN-ECE Group of Experts on Coal Classification, the reason for using bed moisture instead of moisture-holding capacity as the basis to calculate GCV (m, af), used to classify low-rank coals, is that in this range moisture-holding capacity does not give reliable values and, thus, is not reproducible, which needs further explanation.

The fact that experimental tests have proved that the reproducibility of moisture-holding capacity is nonexistent for low-rank coals probably results from using ISO 1018, which is only valid for medium- and high-rank coals (hard coals), as well as the United Kingdom standard BS 1016, Part 21. The French standard NF M 03-034, although applicable to all coals, is inspired, albeit with modifications, on the corresponding ISO standard as far as hard coals are concerned.

Experimental tests have shown that the reproducibility limits of Moisture-Holding Capacity (ISO 1018) are large for Low-Rank coals; in fact, ISO 1018 Standard and United Kingdom BS 1016; Part 21 Standard (104) **are only applicable to Medium- and High-Rank coals (Hard coals).** The French NF M 03-034 Standard (105), although stated as applicable to all coals, was based on the corresponding ISO standard for Hard coals.

However, for Low-Rank coals even though reproducibility limits are not established, the method described in ISO 1018 (97) Standard continues to be widely used.

In studying the problem, it is possible to conclude that difficulties may primarily related to different sample conditioning time and particle size considered in different standards. Also, a fundamental problem results from the fact that high-moisture low-rank coals have a gel structure that collapses irreversibly as the moisture is shed after the coal is removed from the seam. The rate conditions under which this moisture is eliminated affects the structure of the coal in different ways which, in turn, results in variable Moisture Holding Capacity values.

An added problem results from the use of different concepts related to moisture in coal, particularly to bed moisture. A good example of this statement is found in Ref. 106 (Table 2.3.1, p. 55), in which ISO, ASTM, and BS nomenclatures are briefly compared. Useful appraisals on moisture in coals have been previously dealt with by Allardice and Evans (107) and Winegartner (108).

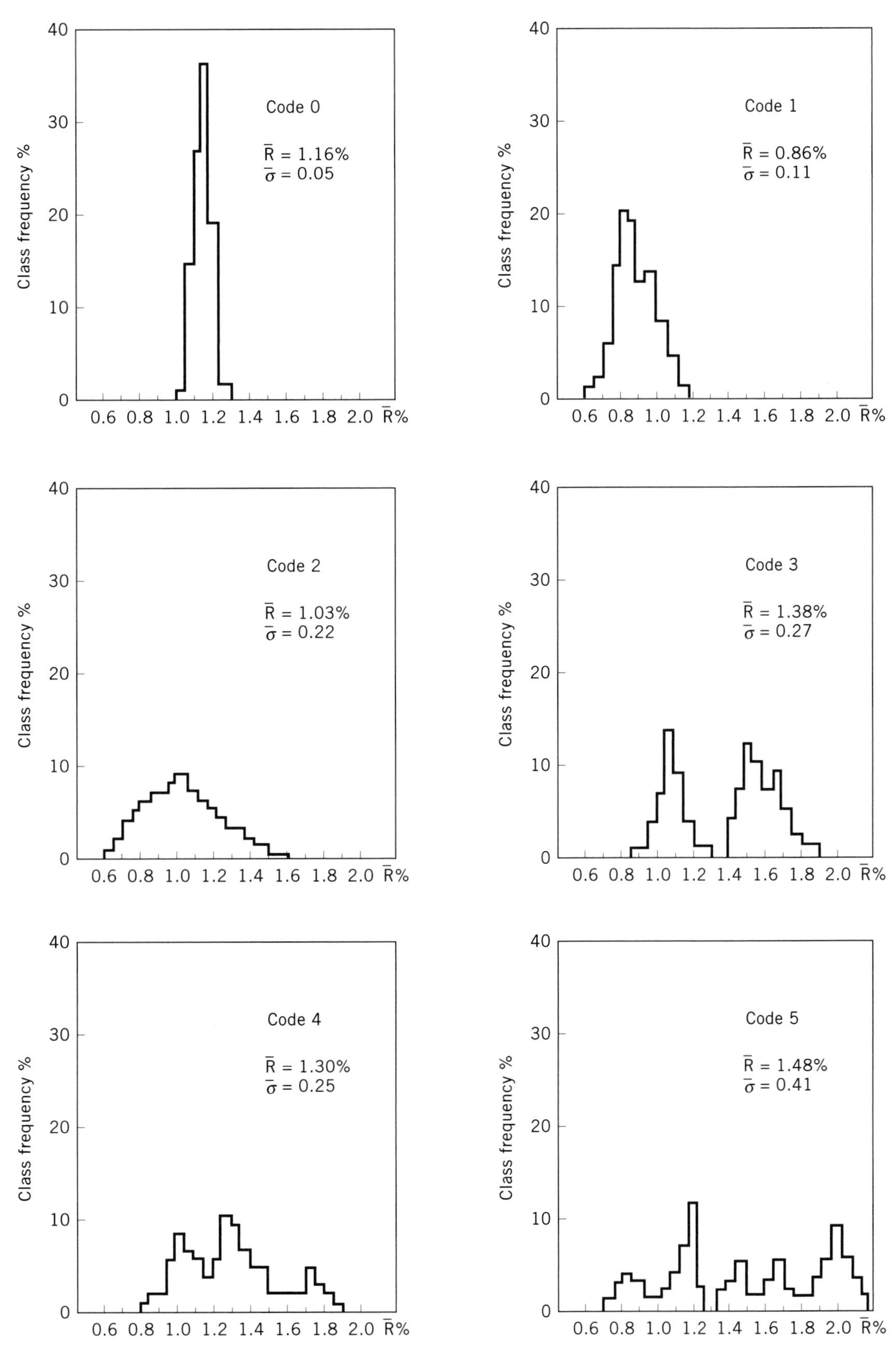

Figure 1. UN-ECE International Codification System for Medium- and High-Rank coals (66). Reflectogram characteristics and corresponding code (see also Table 9).

Table 10. UN-ECE International Codification System for Medium- and High-Rank Coals (1988): List of Supplementary Parameters[a]

Parameters		Standards to be Used	
		ISO[c]	Others[c]
Carbon	%, mass	609 or 625	
Chlorine	%, mass	352	
Composition of ash	%, mass	—	DIN 51729; BS 1016, Part 14; ASTM D 2795
Dilatometric properties	—	349	DIN 51739
Fusibility of ash	°C	540	—
G value	—	—	DIN 51739
Gray-King coke test	—	502	
Hardgrove grindability test	—	5074	
Hydrogen	%, mass	609 or 625	
Mineral matter	%, mass	602	
Moisture-holding capacity	%, mass	1018 (97)	
Net calorific value	MJ/kg	1928 (95)	
Nitrogen	%, mass	333	
Oxygen	%, mass	1994[b]	
Phosphorus	%, mass	622	
Plastometric properties	—	—	ASTM D 2639, GOST 1186
Resinite content	%, volume	7404-3 (111)	
Roga test	—	335	
Size distribution	—	1953	
Sulfur			
Pyritic	%, mass	157	
Sulfate	%, mass	157	
Total moisture	%, mass	589 (109)	
Vitrinite content	%, volume	7404-3 (111)	
Washability test and curve	—	923	

[a] From Ref. 66.
[b] In general, the oxygen content is calculated as 100 − (C + H + S + N)%, daf. In special cases oxygen is determined in accordance with ISO 1994.
[c] Numbers in parens refer to references.

With the exception of the U.S. and Canada, the concept of **Bed Moisture** is taken to be equivalent to **Total Moisture,** as determined by ISO 589 Standard (109) for Medium- and High-Rank coals and by ISO 1015 or 5068 Standards (96,98) for Low-Rank coals, or equivalent national standards. On the other hand, for users of ASTM D 388 Standard (U.S.A., Canada, and Mexico; see also Refs. 58,34), **Bed Moisture** is equivalent to **Equilibrium Moisture** at 96 to 97 per cent relative humidity and 30°C, ie, **Moisture-Holding Capacity,** determined according to ASTM D 1412 Standard, provided that sampling procedures, as stated in ASTM D 388 Standard, are followed.

In Low-Rank coals, Moisture-Holding Capacity (af) is generally lower than Total Moisture (af) eg, Allardice and Evans 1978 (107), Australian AS 2096 Standard (33), and Alpern and co-workers 1989 (85); becoming correspondingly closer with increasing rank and reach practically the same values from about 22% (af) which coincides with the boundary between Low-Rank and Medium-Rank coals. This fact is a supplementary reason for revising the ISO 1018 Standard (97).

All considerations pertaining to moisture have been made with the sole objective of providing solutions to classification problems, and, although moisture has direct implications for coal trade and utilization, these aspects have not been explicitly considered. The problem related to moisture in coal and implications with regards to analyses are further detailed in Lemos de Sousa and Pinheiro (110).

Petrographic Composition. Petrographic composition is expressed in terms of **Maceral Group Analysis (Vitrinite-, Liptinite-, Inertinite- group macerals, mmf, volume percent)** carried out according to ISO 7404-3 Standard (111) and supplemented, where possible, by visual characterization in terms of lithotypes, namely **Bright, Intermediate,** and **Dull coals** (89). The concepts of banded (mainly humic) and nonbanded (mainly sapropelic) have also been taken into account and relate to the megascopic description of coals and coal seams (see ASTM D 2796 Standard; Ref. 113). However, the latter definitions are broad because there are humic coals that can exhibit non-banded characteristics (103).

Grade. The boundary limits for grade, as shown in Figures 2 and 3, are based on **Ash content (dry basis) weight percent** as determined according to ISO 1171 Standard (114).

For **Sapropelic Coals** the upper grade limit is 50% Ash content (db), thus excluding Oil Shales (see Fig. 3). For **Humic Coals** and including carbonaceous rock the upper grade limit is 80% Ash content (db) above which rocks will not be covered by the classification. A subdivi-

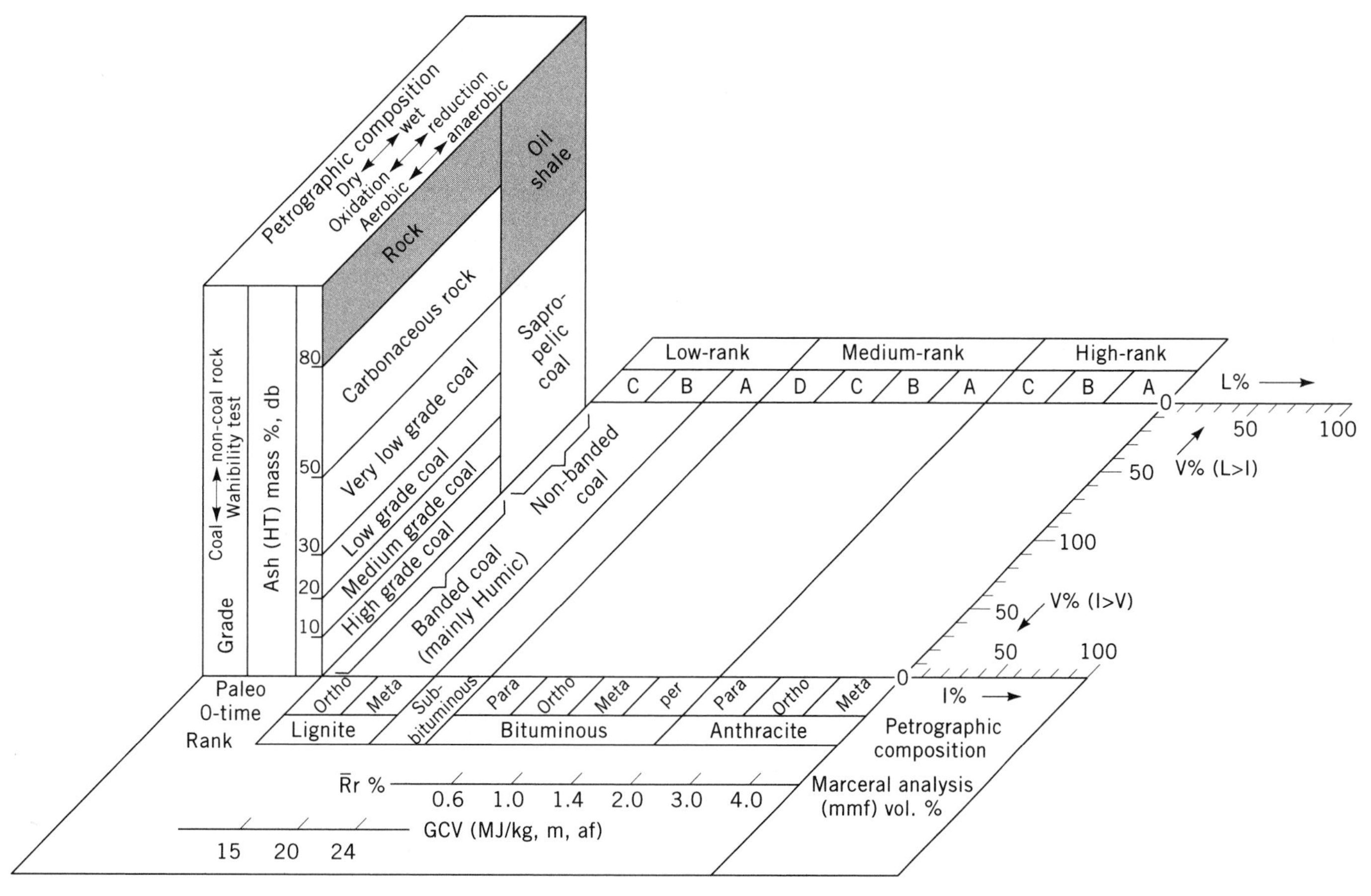

Figure 2. General version of the UN-ECE Classification of Seam Coals (111).

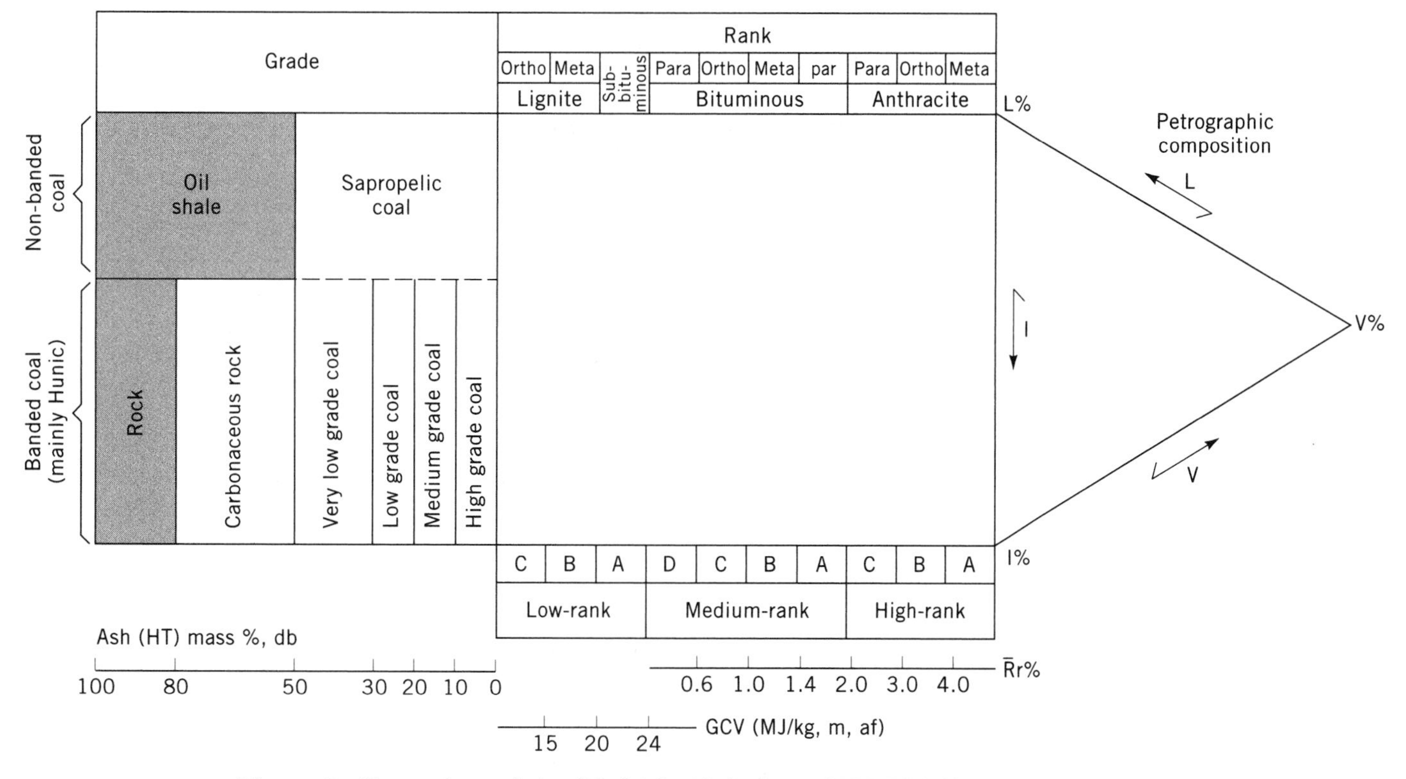

Figure 3. Alternative and simplified (planified) chart of UN-ECE Classification of Seam

sion termed **Carbonaceous Rock** with an Ash content (db) greater than 50% and less than 80% (see Figs. 3 and 4) is used for noncoal with a high percentage of humic organic matter.

The term sapropelic coal designates coal of which the original plant material was more or less transformed by putrefaction (fermentation under reducing conditions (103). Macroscopically, in comparison with most humic coals, sapropelic coal is more or less unstratified, and shows little or no development of cleat fractures. The surface is dull or has a faintly greasy lustre. It breaks conchoidally (7). It should be noted that the most well known sapropelic coals are the cannel and boghead coals, both with liptinite as the dominant maceral group constituent. However, several other sapropelic coals exist where liptinite is not necessarily the dominant constituent, but instead they are rich in vitrinite (saprovitrinite). Such a fact should not be surprising because the concept of "sapropelic" is linked to facies and not to petrographic composition. Therefore, it is impossible to define sapropelic coals on the basis of petrographic composition, least of all, by using liptinite-group content.

The term humic coal designates coal in which the original organic matter underwent change chiefly by humifaction (formation of humic acids under oxidizing conditions (103). In contrast to the sapropelic coals, the humic coals are stratified (7).

A simple washability test is suggested to indicate the amount of finely disseminated mineral matter in the coal. Such information is an additional indicator of facies in relation with grade. For this purpose a representative sample is reduced in size to minus 6 mm, and the recovery and ash content of the fraction that floats at the specific gravity of 1.60 is determined. Other test procedures may rely as far as possible on provisions of the international standard ISO 7936. Bearing in mind that this classification system is not to be used for commercial or technological purposes, the general operating procedures and test methods given by AS 1661, ASTM D 4371, BS 7067, and ISO 7936 (115) are suitable, but the complete size ranges given in the same standards are not to be applied for this particular purpose.

Acknowledgments

We thank Connie S. Gilbert of the U.S. Geological Survey for typing part of this manuscript, Manuela Tavares for compiling the bibliographical references, and M. Manuela Marques for support and advice on graphic design.

BIBLIOGRAPHY

1. R. Thiessen, *Canad. Min. J.* **42,** 64–68 (1921).
2. D. W. van Krevelen, in I. A. Breger ed., *Organic Geochemistry,* vol. 16, Pergamon, Oxford, 1963, pp. 183–247.
3. H. B. Chase, "Coal Exploration and Mining Portsmouth, Newport County, Rhode Island," on deposit at U.S. Geological Survey Library, Reston, Va., 1992.
4. W. J. Jongmans, in *C.R. Congr. avanc. Et. Stratigr. Carbonif.,* W. J. Jongmans, ed., Section Minière de la Geoligish-Mijnbouwkundig Genootschap voor Nederland en Koloniën, Heerlen, Netherlands, 1927, pp. I–L111.
5. H. B. Chase, "Coal Mines and Prospects in New England," on deposit at the U.S. Geological Survey Library, Reston, Va., 1992.
6. H. J. Rose, Classification of Coal, in, H. M. Lowry, ed., *Chemistry of Coal Utilization,* John Wiley & Sons, 1945, pp. 25–85.
7. *International Handbook of Coal Petrography,* 2nd ed., International Committee for Coal Petrology (ICCP), Centre National de la Recherche Scientifique, Academy of Sciences of the USSR, Paris, Moscow, 1963.
8. S. I. Tomkeieff, *Coals and Bitumens and Related Fossil Carbonaceous Substances. Nomenclature and Classification,* Pergamon. London, 1954.
9. H. T. M. Witham, "Observations on Fossil Vegetables; Accompanied by Representations of Their Internal Structure, as Seen Through The Microscope," W. Blackwood, T. Cadell, London, 1831.
10. V. Regnault, *Ann. Mines,* 3rd Sér., **xx,** 161–240 (1837).
11. L. Grüner, *Ann. Min.,* 7th Sér., Mémoires, **4,** 169–207 (1874).
12. P. Frazer, Jr., *Trans. Amer. Inst. Min. Met. Eng.* **6,** 430–451 (1877).
13. J. P. Lesley, *Mining Magazine,* 1879, pp. 144–157.
14. M. R. Campbell, *Amer. Inst. Min. Met. Eng., Bull.* **5,** 1033–1049 (1905).
15. A. L. McCallum, *Mining Soc. Nova Scotia J.* **12,** 113–116 (1908).
16. A. J. Collier, "Classification of Low Grade Coal," discussion of paper by M. R. Campbell, *Econ. Geol.* **4,** 262–264 (1909).
17. L. Grüner and G. Bousquet, *Atlas général des Houillères,* Comité des Houillères de France, Paris, 1911.
18. O. C. Raylston, *U.S. Bureau of Mines Technical Paper,* **93,** 41 (1915).
19. S. W. Parr, *Univ. Illinois, Eng. Exper. St. Bull.* **180,** 1928.
20. R. Thiessen and W. Francis, "Terminology in Coal Research," *U.S. Bureau of Mines Technical Paper,* **446,** 1929.
21. R. Thiessen, *Trans. Amer. Inst. Min. Met. Eng.,* Coal Division. 419–437 (1930).
22. C. A. Seyler, *Proc. S. Wales Inst. Eng.* **53**(4), 254–327 (1938).
23. W. Francis, *Coal, Its Formation and Composition,* 2nd ed., Edward Arnold, London, 1961.
24. D. W. van Krevelen, *Coal. Typology–Chemistry–Physics–Constitution,* Elsevier, Amsterdam, 1961.
25. D. W. van Krevelen, *Coal. Typology–Physics–Chemistry–Constitution,* 3rd ed., Elsevier, Amsterdam, 1993.
26. B. Alpern, *Publ. tech.* **3,** 195–210 (1979).
27. B. Alpern, *Bull. Centres Rech. Explor.-Prod. Elf-Aquitaine* **5,** 271–290 (1981).
28. A. H. J. Todd, *Lexicon of Terms Relating to the Assessment and Classification of Coal Resources,* Graham & Trotman, London, 1982.
29. M. J. Lemos De Sousa, *Bol. Min.* **22,** 1–52 (1985).
30. R. M. S. Falcon, in C. R. Anhaeusser and S. Maske, eds., *Mineral Deposits of Southern Africa,* vol. 2, Geological Society of South Africa, Johannesburg, 1986, pp. 1899–1921.
31. A. M. Carpenter, *Coal Classification,* IEA Coal Research, London, 1988.
32. B. Alpern and M. J. Lemos De Sousa, in *11th C.R. Congr. Internat. Stratigr. Géol. Carbonif.,* J. Yugan and L. Chun, (eds.), Nanjing University Press, Nanjing, 1987, pp. 157–168.
33. *AS 2096-1977, Classification system for Australian Hard Coals, Standards Australia,* North Sydney, New South Wales, 1977.

34. *ASTM D 388-92a, Standard Classification of Coals by Rank.* in *1993 Annual Book of ASTM Standards, Section 5, Petroleum Products, Lubricants, and Fossil Fuels, Vol. 05.05, Gaseous Fuels; Coal and Coke* ASTM, Philadelphia, Pa., 1993, pp. 199–202.
35. *ISO / IEC Guide 2: 1991—General terms and their definitions concerning standardization and related activities.* International Organization for Standardization (ISO), Geneva, 1991, 66 pp.
36. *ISO 8402: 1986—Quality—Vocabulary,* ISO, Geneva, 1986, 12 pp.
37. Standards Association of Australia, North Sydney, Wales, 1959–1991 (various standards in the classification of coal).
38. NF M 10-001, 1990. Combustibles solides. Système International de codification des charbons de rang moyen et de rang supérieur, Association Française de Normalisation (AFNOR), Paris, 1990, 25 pp.
39. *GB 5751-86—Classification of Chinese Coals,* National Standards of PRC, 10 pp.
40. *BS 3323: 1978—Glossary of Coal Terms,* British Standards Institution. London, 1978, 10 pp.
41. *GOST 25543-88—Brown coals, hard coals and anthracites. Classification according to genetic and technological parameters,* 19 pp.
42. *Annual Book of ASTM Standards, 1991, Section 5, Petroleum Products, Lubricants, and Fossil Fuels, Vol. 05.05, Gaseous Fuels; Coal and Coke.* ASTM, Philadelphia, Pa., 574 pp.
43. *ASTM D 388-34T—Standard Classification of Coals by Rank,* in *Book of ASTM Standards,* ASTM, Philadelphia, Pa., 1934, 1257 pp.
44. *ASTM D 388-37, Standard Classification of Coals by Rank,* in *Book of ASTM Standards,* B.S. Suppl. ASTM, Philadelphia, Pa., 1937, p. 144.
45. ASTM D 3175-89a, Standard Test Method for Volatile Matter in the Analysis Sample of Coal and Coke, in *1992 Annual Book of ASTM Standards, Section 5, Petroleum Products, Lubricants, and Fossil Fuels, Vol. 05.05, Gaseous Fuels; Coal and Coke,* ASTM, Philadelphia, Pa., 1992, pp. 321–323.
46. *ASTM D 3176-89, Standard Practice for Ultimate Analysis of Coal and Coke* in Ref. 45, pp. 324–326.
47. *ASTM D 1757-91, Standard Test Method for Sulfur in Ash from Coal and Coke,* in Ref. 45, pp. 223–226.
48. *ASTM D 1857-87, Standard Test Method for Fusibility of Coal and Coke Ash,* in Ref. 45, pp. 227–230.
49. *ASTM D 2013-86, Standard Method of Preparing Coal Samples for Analysis,* in Ref. 45, pp. 239–250.
50. *ASTM D 2492-90, Standard Test Method for Forms of Sulfur in Coal,* in Ref. 45, pp. 279–284.
51. *ASTM D 3302-91, Standard Test Method for Total Moisture in Coal,* in Ref. 45, pp. 352–358.
52. *ASTM D 3177-89, Standard Test Methods for Total Sulfur in the Analysis Sample of Coal and Coke,* in Ref. 45, pp. 327–330.
53. *ASTM D 720-91, Standard Test Method for Free-Swelling Index of Coal,* in Ref. 45, pp. 212–216.
54. *ASTM D 1412-89, Standard Test Method for Equilibrium Moisture of Coal at 96 to 97 Percent Relative Humidity and 30°C,* in Ref. 45, pp. 217–219.
55. *ASTM D 2797-85 (1990), Standard Practice for Preparing Coal Samples for Microscopical Analysis by Reflected Light,* in Ref. 45, pp. 299–302.
56. *ASTM D 2798-91, Standard Test Method for Microscopical Determination of the Reflectance of Vitrinite in a Polished Specimen of Coal,* in Ref. 45, pp. 303–306.
57. *ASTM D 2799-92, Standard Test Method for Microscopical Determination of Volume Percent of Physical Components of Coal,* in Ref. 45, pp. 307–308.
58. ASTM D 121-91, *Standard Terminology of Coal and Coke,* in Ref. 45, pp. 163–173.
59. S. W. Parr, *J. Amer. Chem. Soc.* **28,** 1425 (1906).
60. A. Bement, *Trans. Amer. Inst. Min. Eng.* **39,** 800–805 (1909).
61. C. A. Seyler, *Fuel* **3,** 15–26; 41–49; 79–83 (1924).
62. W. R. Johnson, *Philadelphia Acad. Nat. Sci., Proc. II,* 202–206 (1844–1845).
63. S. W. Parr and W. F. Wheeler, *Univ. Illinois, Eng. Exper. St. Bull.* **37,** (1909).
64. *ASTM D 389-34T, Standard Classification of Coals by Grade,* in *Book of ASTM Standards,* ASTM, Philadelphia, Pa., 1934 Proc., v. 34, I., p.841.
65. *ASTM D 389-37, Standard Classification of Coals by Grade,* in *Book of ASTM Standards,* 1937. Suppl p. 144. ASTM, Philadelphia, Pa., 1937, Suppl, p. 144.
66. Economic Commission for Europe, United Nations (Geneva), "International Codification System for Medium and High Rank Coals," (Document ECE/COAL/115), United Nations, New York, 1988.
67. United Nations, "International Classification of Hard Coals by Type," United Nations, Geneva, 1956.
68. *AS 2434.3—1984, Methods for the analysis and testing of lower rank coal and its chars. Determination of the moisture holding capacity of lower rank coals,* Standards Australia, North Sydney, New South Wales, 1984.
69. *AS 1038.17, 1989, Methods for the analysis and testing of coal and coke, Determination of moisture-holding capacity (equilibrium moisture) of higher rank coal, Standards Australia,* 1989.
70. *AS 2486, 1989, Methods for microscopical determination of the reflectance of coal macerals, Standards Australia,* 1989.
71. *AS 1038.5.1, 1988, Methods for the analysis and testing of coal and coke, Gross specific energy of coal and coke, Adiabatic calorimeters. Standards Australia,* 1988.
72. *AS 1038.5.2, 1989, Methods for the analysis and testing of coal and coke, Gross specific energy of coal and coke, Automatic isothermal-type calorimeters, Standards Australia,* 1989.
73. *AS 1038.3, 1989, Methods for the analysis and testing of coal and coke, Proximate analysis of higher rank coal. Standards Australia,* 1989.
74. *AS 2434.2, 1983, Methods for the analysis and testing of lower rank coal and its chars, Determination of the volatile matter in lower rank coal, Standards Australia,* 1983.
75. *AS 1038.12.1, 1984, Methods for the analysis and testing of coal and coke. Determination of crucible swelling number of coal, Standards Australia,* 1984.
76. *AS 2434.1, 1991, Methods for the analysis and testing of lower rank coal and its chars. Determination of the total moisture content of lower rank coal, Standards Australia,* 1991.
77. *AS 2434.6.1, 1986, Methods for the analysis and testing of lower rank coal and its chars. Ultimate analysis of lower rank coal, Standards Australia,* 1986.
78. *AS 1038.6.3, 1986, Methods for the analysis and testing of coal and coke, Ultimate analysis of higher rank coal. Determination of total sulphur, Standards Australia,* 1986.
79. *AS 1038.16, 1986, Methods for the analysis and testing of coal and coke,* Acceptance and reporting of results, *Standards Australia,* 1986.
80. Economic Commission for Europe, "Mining and Upgrading of Brown Coals in Europe. Developments and Prospects,"

(Document E/ECE297-E/ECE/COAL/124), United Nations (Geneva), 1957.

81. *ISO 2950, 1974, Brown coals and lignites, Classification by types on the basis of total moisture content and tar yield,* 2 pp. ISO, Geneva, 1974, 2 pp.
82. Republic of Chechoslovakia "Kodifikacní Systém Pro Uhlí stredního a Vysokého Prouhelnení," CSN 4413, Praha, in press.
83. Portugal 1992, "Combustíveis sólidos. Carvões de grau médio e superior. Codificação internacional CEE-Nações Unidas, NP 3420 (Solid Fuels. Medium- and High-Rank Coals. International codification ECE-United Nations), 2nd ed., Instituto Português da Qualidade, Lisboa, 1992.
84. Turkey, "Orta ve Yüksek Rankli Kömürler Için Uluslararasi Sayisal Berlirleme (Kodlama) Sistemi."
85. B. Alpern, M. J. Lemos De Sousa, and D. Flores, "A Progress Report on the Alpern Coal Classification," *Internat. J. Coal Geol.* **13,** 1–19 (1989).
86. B. Alpern and co-workers, *Publ. Mus. Labor. miner. geol. Fac. Ciênc. Porto* **1,** 5–31 (1988).
87. G. K. B. Navale and B. K. Misra, *Geophytology* **13,** 214–218 (1983).
88. B. Alpern and J. Nahuys, *Carvão da Bacia de Morungava. Estudo de seis sondagens.* Fundação de Ciência e Tecnologia-CIENTEC, Porto Alegre, 1985.
89. Economic Commission for Europe of the United Nations, International Classification of Seam Coals, ENERGY/WP.1/R.22 ECE-UN document, January 6, 1993 (transmitted by the French Government).
90. *AS 2519, 1982, Guide to the evaluation of hard coal deposits using borehold techniques, Standards Australia.* 1982.
91. *AS 2617, 1983, Guide for the taking of samples from hard coal seams in situ. Standards Australia,* 1983.
92. *ASTM D4596-86, Standard Practice for Collection of Channel Samples of Coal in the Mine,* in Ref. 45, pp. 407–408.
93. ASTM: D5192-91, *Standard Practice for Collection of Coal Samples from Core,* in Ref. 45, pp. 465–468.
94. ISO 7404-5, 1984, *Methods for the petrographic analysis of bituminous coal and anthracite, Part 5; Methods of determining microscopically the reflectance of vitrinite,* ISO, Geneva, 1984, 11 pp.
95. *ISO 1928,* 1976, *Solid mineral fuels, Determination of gross calorific value by the calorimeter bomb method, and calculation of net calorific value,* ISO, Geneva, 1976, 14 pp.
96. *ISO 1015, 1975, Brown coals and lignites, Determination of moisture content, Direct volumetric method.* ISO, Geneva, 1975, 3 pp.
97. *ISO 1018, 1975, Hard coal—Determination of the moisture-holding capacity,* ISO, Geneva, 1975, 6 pp.
98. *ISO 5068, 1983, Brown coals and lignites, Determination of moisture content, Indirect gravimetric method.* ISO, Geneva, 1983, 5 pp.
99. P. A. Hacquebard, *Current Res., Part A, Geol. Surv. Can. Pap.* **84-1A,** 11–15 (1984).
100. V. H. Ode and F. H. Gibson, *U.S. Bur. Mines, Rep. Invest.,* **5695,** 20, 1960.
101. B. C. Parks, *Econ. Geol.* **46,** 23–50 (1951).
102. B. C. Parks and H. J. O'Donnell, *U.S. Bur. Mines, Bull.,* **550,** 1956.
103. E. Stach, M.-Th. Mackowsky, M. Teichmüller, G. H. Taylor, D. Chandra, and R. Teichmüller, *Stach's Textbook of Coal Petrology,* 3rd ed., Gebrüder Borntraeger, Berlin, 1982.
104. BS 1016: Part 21, 1987–*Methods for Analysis and Testing of Coal and Coke, Part 21, Determination of moisture-holding capacity of hard coal.* British Standards Institution, London, 1987, 12 pp.
105. *NF M 03-034, 1968, Determination de la capacité de retention d'humidité des combustibles solides.* AFNOR, Paris, 1968, 5 pp.
106. J. F. Unsworth, D. J. Barratt, and P. T. Roberts, *Coal Quality and Combustion Performance. An International Perspective,* Elsevier, Amsterdam, 1991.
107. D. J. Allardice and G. G. Evans, in *Analytical Methods for Coal and Coal Products,* vol. 1, C. Karr Jr., (ed.), Academic, New York, pp. 247–262, 1978.
108. E. C. Winegartner, in *Coal Technology Europe '81, Cologne, W. Germany, 1981,* Exxon Research and Development Division. Baytown, Texas, 1981, pp. 133–141.
109. *ISO 589, 1981, Hard coal—Determination of total moisture,* ISO, Geneva, 1981, 6 pp.
110. M. J. Lemos de Sousa, and H. J. Pinheiro, "Coal Classification, Basic fundamental concepts and the state of the existing international systems; to be published in *The Journal of Coal Quality.*
111. *ISO 7404-3, 1984, Methods for the petrographic analysis of bituminous coal and anthracite—Part 3: Methods of determining maceral group composition,* ISO, Geneva, 1984, 4 pp.
112. AS 2096, 1987, *Classification and coding systems for Australian coals, Standards Australia,* 1987.
113. ASTM D 2796-88, *Standard Definitions of Terms Relating to Megascopic Description of Coal and Coal Seams and Microscopical Description and Analysis of Coal,* in Ref. 45, pp. 296–290.
114. *ISO 1171, 1981, Solid mineral fuels—Determination of ash,* ISO, Geneva, 1981, 2 pp.
115. ISO 7936,1992, *Hard coal—Determination and presentation of float and sink characteristics,* General directions *for apparatus and procedures,* ISO, Geneva, 1992.

COAL COKING

RALPH GRAY
Monroeville, Pennsylvania

HISTORY OF COKE MAKING

The art of coke making is ancient. It is possible that coal and even coke were used in the People's Republic of China as early as 1500 BC. The earliest reference to coking is that of Theophrastus, a pupil of Aristotle, who in the history of stones (371 BC) noted that some stones (Lipara stone) empty themselves when burning and become like pumice (1).

A relatively modern mention of carbonization was made in 1584 by Julius, Duke of Braunschweig-Luneberg, in Germany. In 1590, John Thornbarough, Dean of York, was issued a British patent to "purify pit coal and free it of its offensive smell" by carbonization. In the following years, a number of British patents were issued concerning the use of coal or coke in iron making. At that time, coal was charred in the same way as wood was in charcoal manufacture. Derby, United Kingdom, was the early center for making cokes, which were used in drying malt. Finney and Mitchell provide an historical account of coke making from the charcoal pile to the by-product oven (2).

However, the usefulness of coke for iron smelting was not developed until the early 1700s. The modern coking industry traces its origin to these 18th-century developments in the iron and steel industry of the United Kingdom. Early iron smelting used charcoal as a fuel to produce sponge iron, which resulted in exploitation of timberlands in Europe and the United Kingdom for charcoal production. The resulting devastation on the forests led to the passage of a British Act in 1558, prohibiting the felling of timber for iron smelting. Attempts to replace charcoal in whole or part with coal and/or coke were then made. Dud Dudley is often credited with smelting iron in a blast furnace in 1619. However, it was not until 1709 that Abraham Darby made 200 tons of coke at Coalbrookdale and also produced the first commercial iron to be successfully smelted by coke (3).

Charcoal Making

The techniques used to make charcoal were adopted by the early coke makers (4). The charcoal mound or meiler consisted of carefully arranged logs, stacked to leave a type of flue at the center. The entire pile was then covered with materials such as turf, dirt, wet leaves, straw, and charcoal fines, and ignited by lighting kindling that had been pushed into the center. The center hold was sealed, and vent holes were made around the circumference. A typical charcoal pile is shown in Figure 1.

In the United States, the collier and one or two helpers "coaled" together to form the charcoal pit, which was clear, level, and 9.1–12.2 m (30–40 ft) in diameter. Two sizes (diameters) of wood, called lapwood and billets, were used to construct the pit or pile. The first tier of stacked wood was called the foot, followed by the waist, and topped by the shoulder or head. A pit held about 98 stones (30 cords) of seasoned wood, which was stacked around a chimney that circled a center pole or fagan and that was filled with kindling for lighting. The pile was covered with charcoal dust from previous runs and leaves. Firing and tending took two or more weeks, at which time the pile "came to foot" or charred to the bottom. The pile was then dug out and hauled to the iron furnace. Wood from an 0.04 hm^2 (1.0 acre) of forest supplied about 6.35 t (7 tons) of pig iron.

Coke Making

When commercial coke making was first practiced, the collier, who set up a rounded heap of coal for coking, followed the same procedures used in charcoal making. Initially, the larger size lumps of coal were used to construct a central flue in the pile (see Fig. 2). Kindling wood was also built into the pile to support the flues. Later, vertical posts were used, and, in 1768, John Wilkinson introduced a permanent brick flue around which the pile was built.

In the United States, coke making generally occurred in mounds or rows, where the coal to be coked was arranged in heaps or pits with longitudinal, transverse, and vertical flues in which wood was distributed for igniting (2). When the fire spread through the mass, the pile was banked with fine dust. The operation lasted for 5–8 days, and the coke yield varied from 33 to 60%.

Beehive Era

In the pre-beehive oven era coke making occurs in piles, rows, and heaps without a surrounding structure. The beehive oven was introduced in 1620 and used until relatively recently (5). The oven, shaped like a beehive, proved to be a major advance in coke making. The first U.S. beehives were probably built in 1833 in Connellsville, Pennsylvania, by L. L. Norton. The ovens were built in groups of 50 or more, called a battery, and were often built with their backs against a hillside or back-to-back. The oven diameter was commonly 3.66 m (12 ft) with a height to the top of the dome of 2.13 m (7 ft). A cross-section of a beehive oven is shown in Figure 3.

The coal was charged through an iron-ringed hole, called the trunnel head, at the top of the oven. The charged coal was leveled through a door in the front of

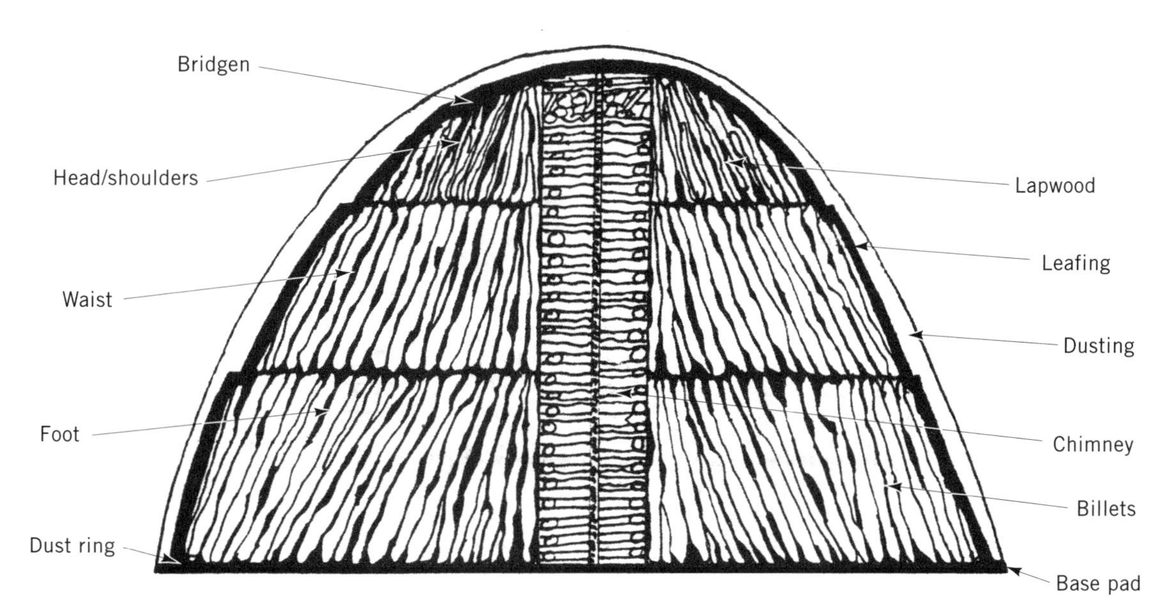

Figure 1. Cross-section of a wood pile used in charcoal making.

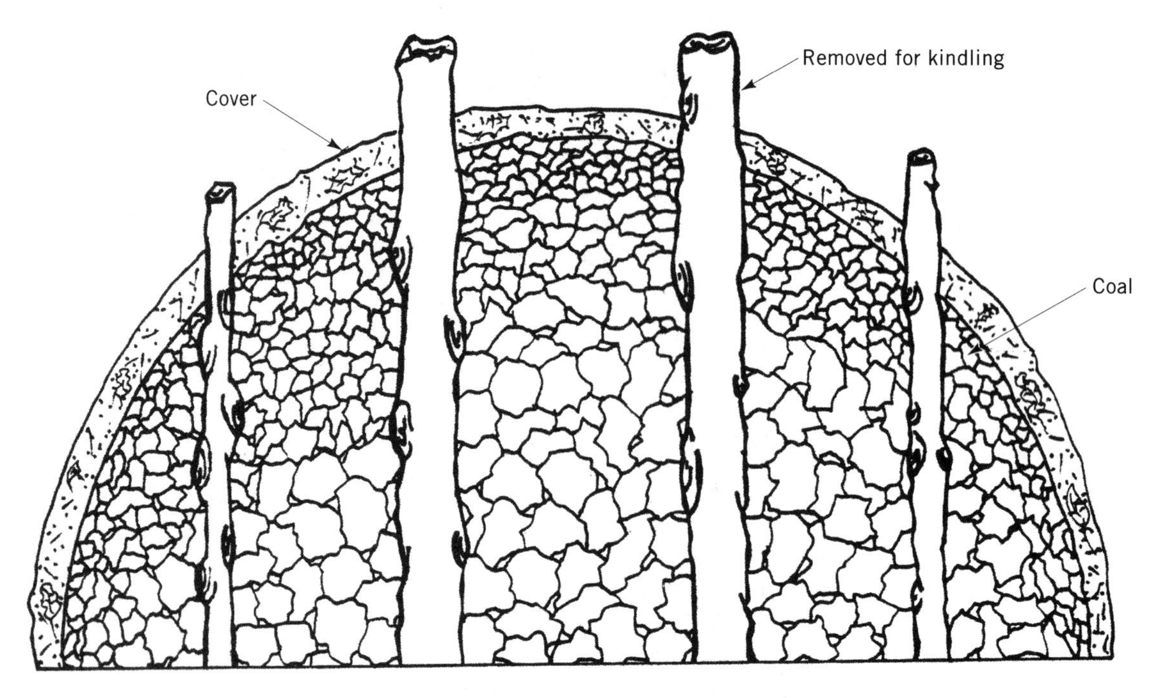

Figure 2. Cross-section of a coal pile used in coke making.

the oven to form a bed that was about 61–76 cm (24–30 in) thick. The hot oven floor ignited the coal, and kindling was added as needed. The oven door was partially and loosely bricked, and luted with wet mud and clay to seal off air when necessary (based on the volume and color of the smoke from the trunnel head). Tars that rose in the bed coked on top of the charge, providing the coke with a silver appearance. The disappearance of smoke indicated completion of the coking process, and the trunnel head was dampered if the coke was not to be drawn immedi-

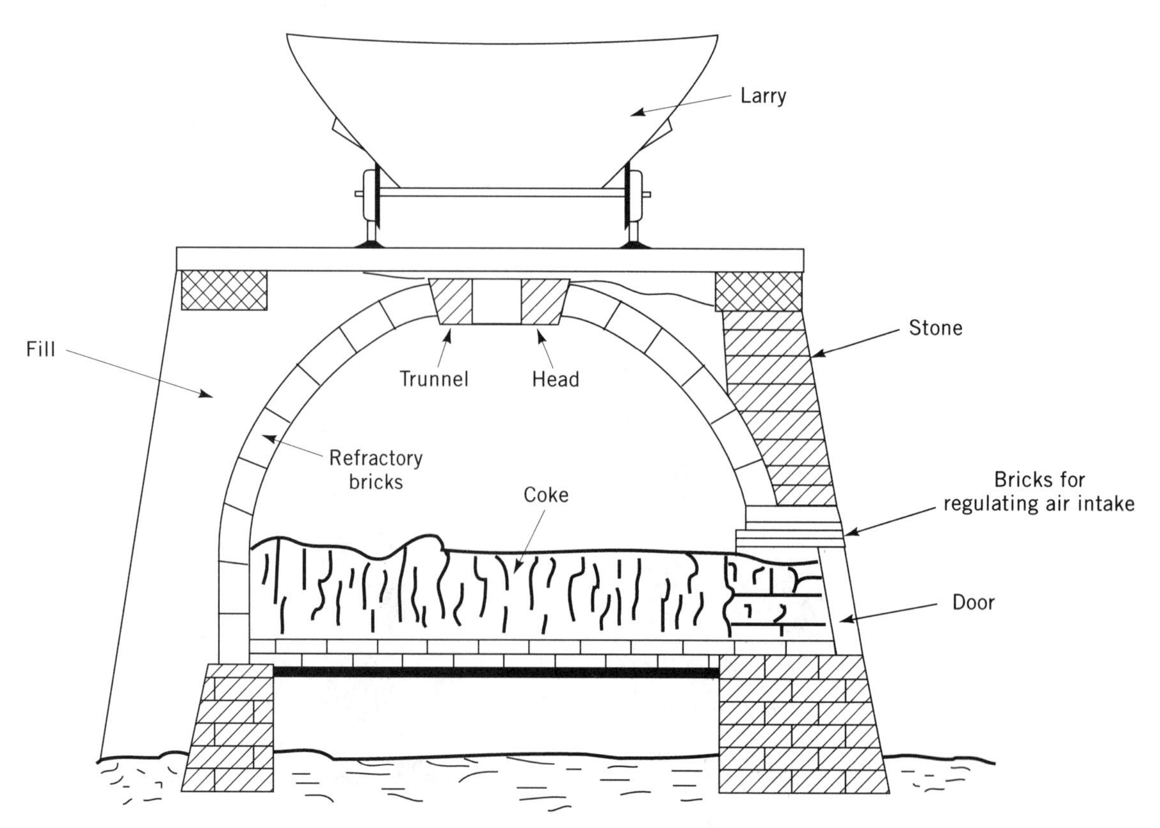

Figure 3. Cross-section of a beehive coke oven.

ately. Finally, the coke was cooled by water quenching in the oven. A coke puller then used a long-handled iron scraper to break and rake the coke from the oven.

A coal charge consisted of about 4.54–7.26 t (5–8 tons) and required about 72 hours to coke. In an efficient operation, a 5.85-t (6.5-ton) charge of coal would yield about 3.98 t (4.5 tons) of coke. All of the by-products were burned in coking, as well as some of the coal.

In the United States, beehive oven construction grew in the latter half of the 19th century. There were 57,400 ovens by 1900, reaching their peak of 100,000 by 1909 (6). By 1930, the number had dwindled to about 20,000 due to the rapid expansion in the construction of by-product ovens. However, the 1930s Depression halted oven construction and limited production in existing plants. The last U.S. beehive ovens in operation are at Alverton, Pennsylvania, where they are charged and pushed once a year as part of a coal and coke heritage festival.

The Anthracite Furnace in the United States

In the United Kingdom and Europe, the transition in blast furnace fuel went from charcoal to coke from coal. In the United States, the transition was somewhat different, particularly in Eastern Pennsylvania (7).

Because the original U.S. colonies had access to cheap British coal, domestic coal development moved slowly. Even after achieving its independence, the United States continued to import coal until President Jefferson embargoed British coal in 1807; the subsequent war with England five years later completely cut off this coal source and led to the rapid development of U.S. coals, particularly the abundant Pennsylvania anthracites. As the anthracite supply increased and its cost decreased, it became a replacement for charcoal in the domestic blast furnaces. After the hot blast furnace was developed in 1830, anthracite coal became the principal fuel in Eastern Pennsylvania (8). A typical anthracite furnace is shown in Figure 4.

Charcoal continued, however, as the principal fuel west of the Allegheny Mountains until the railroad from Philadelphia reached Pittsburgh in 1852, providing Western Pennsylvania with cheaper Eastern iron. Until then, the iron and blast furnace fuel sources relied on plantations. In order to become more competitive as iron and steel producers, Western Pennsylvania and adjacent areas began a conversion to coke, from bituminous coals, for the blast furnaces. The independent operations began integration, resulting in the modern industrial trend of larger furnaces and integrated operations (8).

The Connellsville, Pennsylvania area possessed an abundance of excellent coking bituminous coal and became a center for beehive coke production. It took only about 0.002 hm^2 (0.5 acre) of a 1.8-m (6-ft) coal seam to fuel a furnace that previously required 80–200 hm^2 (2,000–5,000 acres) of forest to supply the charcoal. Thus, the iron and steel industry soon concentrated on supplies of coal and water.

Use of anthracite coal persisted as a blast furnace fuel until around 1890. Coke was cheaper and proved to be a better fuel, but it required a larger furnace. Anthracite coal is denser than porous coke but is less responsive to increased air blowing. As older blast furnaces were phased-out, the newer and larger furnaces were consequently designed to use coke as a fuel.

By-Product Coking Processes

Beehive coking is a nonrecovery process; not only is part of the coal burned, but the gases and tars are generally wasted (9). As early as 1681, a British patent was granted to Becker and Serle for their method of making pitch from coal. The first by-product plant was built in 1766 by Staudt at Sulzbach near Saarbrucker, to remove sulfur from coal so the coke might be used for iron making. A number of individuals have claimed to be the first to produce gas from coal for lighting. In reality, modern by-product ovens are the result of the evolution and inventions that occurred over a century by many workers.

Coke came into its own when it was found to be an ideal blast furnace fuel for iron and steel making. By-product ovens and their recovered chemicals (tar, gas, and ammonia products) became more valued when French coke oven operators at St. Denis, France, piped ammonia gases from coke ovens through tanks of dilute sulfuric acid to produce ammonium sulfate fertilizer. As early as 1841, Gustus von Liebig speculated that rocks and humus store chemicals that plants use as food. In 1847, a British scientist discovered that nitrate of soda and sulphate and chloride of ammonia had a marked effect on plant growth. Thus, fertilizer from coal by-products was exactly what was needed to restore overworked farmland. In Europe, coal-based chemicals and fertilizers may even have held more importance than by-product coke due to the exhausted farmlands. Coal-derived fertilizers were used to replace the lost nitrogen in the soils.

The first U.S. by-product coking plant was built in Syracuse, New York, for the production of soda ash by the Solvey process. By-product ovens continued to be built in the United States as the beehive industry began a decline and as the need for coal-derived dyes and explosives grew as a result of World War I. Utilities' need for gas was a driving force for new ovens beginning in the 1920s. However, the 1930s Depression halted oven expansion, which did not recommence until the 1940s with the onset of World War II. By 1959, there were 15,993 by-product ovens in the United States, with a capacity of 81,447,000 net tons of coke, and only 5,148 beehives with a coke capacity of 4,368,800 net tons.

By-product ovens are elongated rectangular chambers lined with refractory silica brick. The ovens are commonly 8–16 m (26–52 ft) long and 4–8 m (13–26 ft) high, with widths of 35–50 cm (14–20 in). A typical by-product battery is shown in Figure 5. Often, over 100 ovens are arranged side-by-side in a row to form a battery.

Each oven holds about 13.6–27.2 t (15–30 tons) of coal per charge. Each oven is separated from its neighbor by walls, with flues in which gas is burned for heating the coke oven chambers. Coke oven gas with about 55.92–59.08 kJ (530–560 Btu), or blast furnace gas with 29.54 kJ (280 Btu), is used for heating to temperatures of 1,038–1,093°C. The entire mass, and particularly the top, is insulated. Below the ovens, the regenerator chambers recover waste heat for preheating the flue gas used to fire

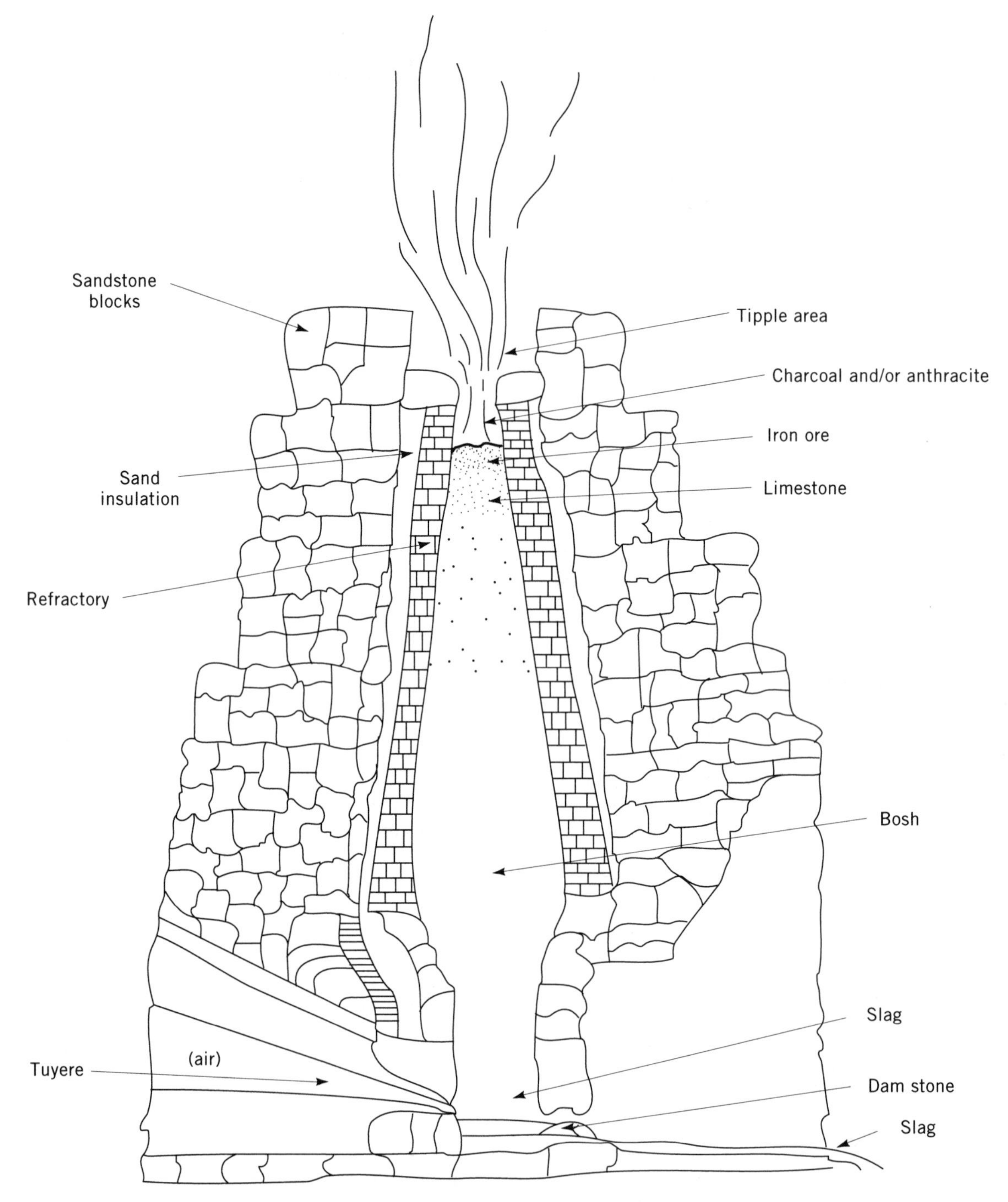

Figure 4. Cross-section of a stone blast furnace that used charcoal and/or anthracite as fuel.

the ovens. The entire pile of bricks sits on a concrete foundation or pad. The mass of oven bricks and other structural elements are connected with steel tie rods that run the width and length of the battery and that tie into vertical I-beams (buckstays) at the ends of the battery.

Ovens are closed on both ends with removable doors, sealed to prevent leakage. Each oven chamber has 2–4 lidded openings on top, called charging holes. Heavy cast iron lids seal the holes after coal is charged.

The coal for coking is stored in a bunker or bin at the end of the battery. Coal from the bunker is discharged into a larry car, which travels on rails over the length of the battery. The larry car receives a coal charge from the bunker; each car has a number of hoppers which corresponds to the number of charge holes per oven. Once stationed over an opened charge hole, a sleeve is dropped to seal each hole, and coal is dropped into the oven. Coal is charged from the larry in stages to minimize the leveling of coal inside the chamber. The hole covers are replaced and sealed with louting once the oven is charged.

The pusher machine that removes the coke from the ovens has a bar that is about the length of the oven. This

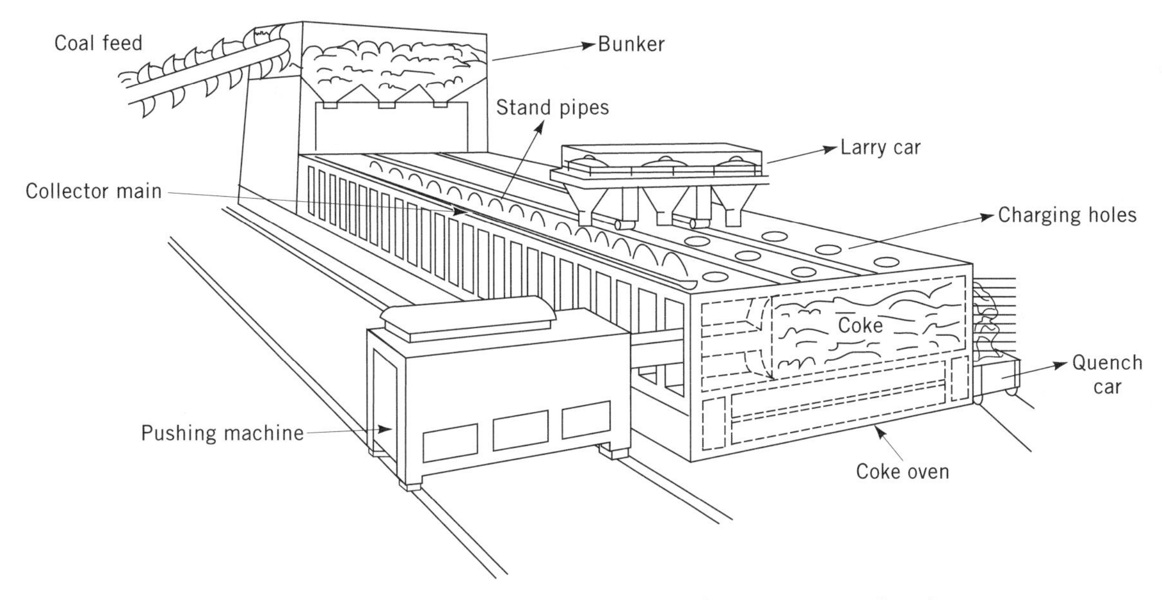

Figure 5. A coke battery with auxiliary equipment and cutaway view of a coke oven.

bar is used to level each coal charge through a chuck door above the oven door. The tunnel formed at the top of a charge allows the gases and vapors to pass from the oven and up an outlet or standpipe at one or both ends of the oven. Standpipe valves are left open during charging; a jet of steam is forced into the standpipe to create a vacuum to suck in gas and coal dust from the oven into the collecting main. Staged charging and steam aspiration help to prevent coal and gas leakage to the atmosphere while charging.

Coking takes about 12–20 hours for blast furnace coke and may exceed 30 hours for foundry coke. At the end of the coking cycle, an extra period is usually allowed for soaking to ensure that coking is complete and that tarry vapors will not escape when the coke is removed from the oven. To remove the coke from the oven, a pusher machine with a ram is stationed before the oven to be pushed. The oven is disconnected from the by-product collecting main, and the doors are removed from the oven. The ram pushes the coke out of the oven and into a waiting railroad car (a quench car), which is hooded or otherwise equipped to prevent escape of polluting emissions. The quench car transports the coke to a tower, where water is used to cool the coke. Some dry quenching is now used, and heat is recovered in the process. The partially cooled coke is transported to an inclined brick-paved area called a wharf, where it is dumped to complete cooling. After cooling, it is then transferred to a belt for transport to a screening station and/or a pulverizer for sizing prior to shipment and utilization.

By-Product Recovery of Chemicals

The volatiles from the coal, which are at about 593–704°C, pass out the top end(s) of the ovens through vertical standpipes, where a water spray lowers the temperature to about 177°C. The vapors condense and pass into a collecting main that runs the length of the battery (10).

The cooling water, or flushing liquor, and tars are transported to a settling tank or decanter for separation. The gas is cooled to about 93°C to remove additional tar and ammonia liquor. This primary cooler tar may be processed separately or mixed with the flushing liquor tar. Accumulated water on the top of the tar in the decanter is removed and reused, and heavy solids that settle from the tar are removed as sludge. A tar decanter is shown in Figure 6.

Some tars are used as fuel, and some are refined by fractional distillation to obtain creosote, naphthalene, phenol, creosols, creosylic acids, pyridine bases, and a semisolid residue called pitch. There are more than 300 chemical substances that can be identified in tar, but only 6–10 are commercial products.

The ammonia liquor is treated with steam and lime, added to the ammonia-rich gas from the tar extractor, and then passed through a saturator. Ammonia is combined with sulfuric acid to form crystalline ammonium sulfate $(NH_3)2SO_4$.

Gas, from which the tar and ammonia have been removed, is cooled and treated with absorbing oil to remove light oil. The gas, rich in hydrogen, methane, ethane, and

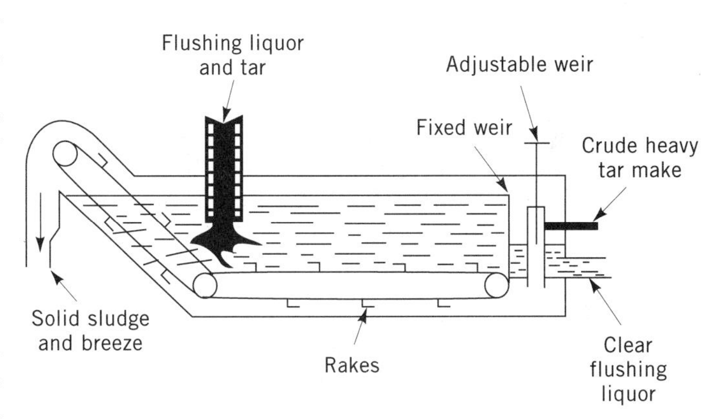

Figure 6. Cross-section of a tar decanter.

ethylene, is generally used as fuel. Steam distillation separates light oil, which is washed with sulfuric acid and fractionally distilled to produce benzene, toluene, xylene, and crude solvent naphthas (11).

Each ton of coal yields about 662 kg (1,460 lb) of coke, which is about 73% of the weight of dry coal charged. In addition, the following by-products are recovered:

Tar	30.3–37.8 (8–10 gal)
Ammonium sulfate	(9.5–10.0 kg (21–22 lbs)
Gas	283.2–297.5 m^3 (10,000–10,500 ft^3)
Benzene	6.62 l (7 qt)
Toluene	0.95 l (1 qt)
Xylene	0.55 l (1 pt)
Naphthalene	0.45 kg (1 lb)

Coal Preheating

Coal preheating is an unconventional addition to conventional by-product coke making. In preheating, the coal is heated to about 150–250°C to drive off moisture and to precondition the coal. About 25% of the heat required for coking is for drying and for initial coal heating. Since the preheated coal enters the oven at a higher temperature than conventionally charged coal, the thermal gradient is lower, the packing densities higher, and the coking time shorter. This results in productivity increases of 25% (12,13).

In the Coaltek process, the wet coal is introduced into a fluidized bed preheater in contact with 800°C inert gas and heated to 250°C. The coal is ground to less than 3 mm, and the entrained gas is separated before the coal enters a loading bin from which it is injected by a pipeline system into the coke ovens (12).

In the Precarbon process, the coal is heated in Rosyn Buttner heaters. The wet coal is first dried to about 2% moisture in 280°C exhaust gas. The coal at a temperature of about 80°C is separated from the gases and heated in a second stage to about 200°C with hot gases at 550°C. After separation from the gases, it is conveyed to a loading bin and then transported to the coke ovens by a Redler conveyor (12).

The Simcar process is similar to Precarbon, except that the second preheating stage raises the coal temperature to 180°C, and coal hoppers are used as an intermediate step followed by metering bins and modified larry car charging (12).

Preheated fine coal must be conveyed and charged in a safe manner. This is accomplished by combining it with an inert gas in an enclosed system, secured to prevent explosions. Coal can segregate in pipeline charging, where it is blown down the length of an oven. Dry coal is very dusty; gas evolution and dust losses on charging can be as much as 1.5% of the coal charged. The dust contaminates the tar, clogging standpipes and overloading tar decanters. Charge density tends to be lower in pipeline charging than in gravity charging; partial fluidization in charging can reduce bulk density and coke oven productivity.

Preheating of coal thickens the plastic layer in which coal is converted to coke. The gases that normally leave on the wall side of the plastic layer tend to permeate the plastic layer on the coal or tar side, increasing the fluid characteristics of the coal. Preheating can improve the coking characteristics of poor coking, low rank coals, but does not improve the coking ability of good coking coals.

Preheating is generally not considered to be a solution to the problems of by-product coke making, and its future use is limited, particularly due to the high capital investment cost required for its use.

Coal Stamping

Coal stamping densifies the coal charge (14). In the stamping process, a chamber is filled with coal while simultaneously stamping the coal to achieve a packing density of as high as 1000 kg/m^3, compared to 750–800 kg/m^3 for conventionally charged coal. The charge is introduced into the oven on a false bottom that travels the length of the oven.

Stamping plants are concentrated in areas producing marginal coking coals, such as Lorraine, France. Stamping brings the poorly swelling and low fluidity coal particles into closer contact, where they tend to fuse better due to the density of the confinement. Stamping has little future potential except where marginal coking coals are available at low cost, as they can be upgraded by this technology to produce acceptable coke (14).

MANUFACTURE OF IRON AND STEEL

The development of coke making as an industry largely depended on the growth of the iron and steel industry. The earliest blast furnaces used charcoal, which was replaced by coal and nonrecovery types of coke such as beehive coke. Eventually, by-product oven coke became the dominant fuel for the iron blast furnace (15).

Since about 1300 AD, iron has been produced in vertical furnaces supplied with a blast. The ore was reduced to a nonmalleable iron–carbon alloy, called pig iron, in a molten form with 2.5–5% carbon. The molten form was used to burn out the carbon to produce steel. This is called an "indirect" process for producing steel, as an additional step after the blast furnace is required. The iron produced was relatively soft, malleable wrought iron, similar to the iron produced directly from the ore.

In 1709, Abraham Darby successfully used coke to replace charcoal in the blast furnace. The hot blast was developed in 1829, which greatly improved blast furnace production. The hot blast involves preheating the air, which is blown into the blast furnace through the tuyeres. Oxygen steel making has now replaced most of the early processes for steel production. Direct iron making processes that do not require coke are being both used and developed (5).

The modern blast furnace is a refractory-lined cylindrical shaped vertical steel shaft, at least 30.5 m (100 ft) high, and consists of four parts: the top, stack, bosh, and hearth (see Fig. 7). The raw materials, consisting of ore, pellets, or sinter, together with limestone and coke, enter the top of the furnace by a double-bell arrangement that allows charging from hopper or scale cars without excess gas escape (3).

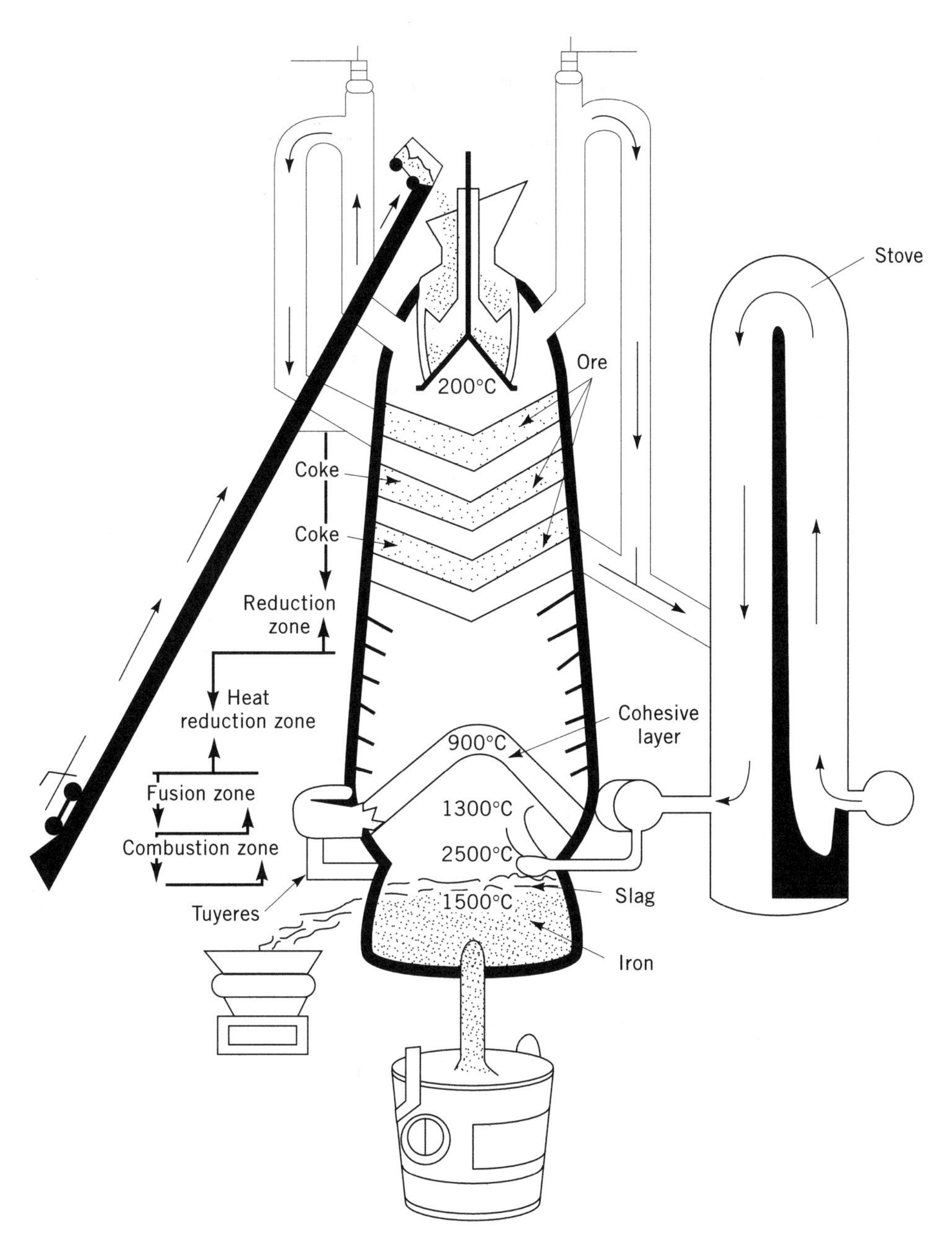

Figure 7. Cross-section of an iron blast furnace.

The raw materials move in a plug fashion counter to the heat source and are dried, preheated, reduced, and melted before the iron and slag products leave or are tapped from the furnace. Hot air from stoves enters the furnace at the tuyeres and combusts the coke, which furnishes the heat, reducing gases, and support for the other raw materials in the blast furnace (12).

The furnace removes moisture from the charged materials; reduces oxides of iron; calcines the flux stone; melts the slag and iron; reduces other oxides such as manganese, silicon, and phosphorus; and removes sulfur from the iron. Each ton of pig iron requires about 1.4 t (1.6 tons) of ore, 0.59 t (0.65 tons) of coke, 0.18 t (0.2 tons) of limestone, about 0.04 t (0.05 tons) of miscellaneous iron and steel scrap, and 3.6–4 t (4–4.5 tons) of air. The furnace yields about 5.4 t (6 tons) of gases and about 318 kg (700 lbs) of slag per ton of pig iron (3). The production of blast furnace iron from iron ores will continue as a basic source of iron for steel making, although direct iron making processes will become increasingly important in the next 10 years. In addition, the production of steel from recycled scrap will increase at the expense of blast furnace iron.

Blast Furnace Coke

Coke is the burden support, fuel, and reductant in the blast furnace (16). In conventional operation, the raw materials that enter the blast furnace are charged in layers. Coke has the lowest apparent density of all the materials charged and therefore occupies the greatest volume. The raw materials move by plug flow, with the coke layers forming the transport on which other raw materials advance. Coke is the only support solid in the bosh and hearth areas. Coke remains solid through its descent in the blast furnace and is consumed by combustion at the tuyeres, as shown in Figure 7. In addition, the coke layers are the highly permeable areas for hot gas movement.

Physically, coke must be coarse and porous to supply permeability. It must also have the strength to support the burden, and it must not degrade mechanically, thereby overloading the dust-collecting system. Further, coke must maintain its physical integrity and resist chemical attack, which can partially consume and weaken the coke by reactions in the stack. Weakened coke cannot provide support in the bosh, hearth, and raceway areas of the blast furnace (16).

Coke shape is also important, as it affects slag and iron drainage in the bosh. It must be uniformly low in moisture, as moisture takes heat (kilocalories) away from the process for its removal. Coke ash (the residue from coal minerals) must be low, as coke is consumed in melting the ash so it can be removed as slag (17). The sulfur content of coke must be low, as sulfur makes iron brittle if it remains. Sulfur is usually controlled by the flux, but coke is required to melt the flux. The flux is formed from the limestone that melts to produce slag, which then picks up sulfur and other removable impurities. External desulfurization is also practiced (18).

Phosphorous must be low in coke, as it is not removed in the blast furnace and its presence in metals adversely impacts properties. Alkalies increase coke reactivity and should be controlled, as they recycle in the blast furnace, attack carbon, and create scales that reduce the working volume of a blast furnace. The coke properties that affect the economics of blast furnace operations have been extensively studied and evaluated (19).

COKING REACTIONS

Coking coals decompose when heated to form gas, liquids, and solids. The devolatilized coal residue is coke, whereas the gas and liquids are by-products.

When coking coal is heated, the reactions involved occur within individual coal particles, between reacting particles, and with decomposition products. Heating coal to about 100°C expels moisture. Continued heating increases the internal energy of the aliphatic side chains, which at about 300–375°C form a metaplast, a pyrolysis product, that helps plasticize the coal macerals. The plasticized coal forms a froth or foam due to the expansion and massaging action of gases on the softened mass.

As the metaplast and plasticized semicoke continue to develop, the mass becomes semisolid at about 500°C. A mosaic (crystalline) coke structure generally develops at about 425°C in the coke walls (see Fig. 8); at temperatures above 700°C, the coke contracts. Carbonization is completed at about 1,000°C (20). Under confinement, the coal particles that are being transformed to coke fuse together by a foaming action (21). The amount and kinds of metaplast that is formed by pyrolysis determines a coal's coking potential (22).

Only a limited group of coals, as defined by their rank, type, and grade, are used for coking. Whereas bituminous rank coals are used commercially to produce coke, some anthracites are used as filler and antifissurants in foundry coke. Rank refers to the maturity of coal; lower rank coals have excessive oxygen content and produce unstable metaplast, which decomposes before the temperature of coal plasticity is reached. These coals sinter but do not coke. Coals of higher rank than bituminous produce insufficient metaplast to produce well-fused coke.

In general, high volatile A rank bituminous coals become fluid when heated and produce coke. However, the product is commonly porous and fissured, as the coals have low coke yields and high gas and tar yields. Therefore, they shrink considerably in producing a fissured coke. Medium volatile bituminous coals are mature coking coals that tend to produce strong cokes with low reactivity to CO_2. Low volatile bituminous coals produce strong coke but are generally not carbonized alone in by-product ovens, as they generate excessive wall pressures during carbonization (22).

Coals contain various amounts of moisture, macerals, and minerals, and mature coals in their natural state contain gas. Macerals are considered to be the organic equivalents of minerals in other rocks; the amount and kinds of macerals determine coal type. The main maceral groups are vitrinite, liptinite, and inertinite. Vitrinite is the principal source of coke bond carbon in the production of coke, whereas liptinite has the highest hydrogen content (present to a significant degree only in high volatile coals) and contributes more to plasticity and gas and tar yields. Inertinite, which has the highest carbon content, does not become plastic when heated; it acts as filler to thicken coke walls. There is an optimum amount of reactives (vitrinite, liptinite, and some semi-inertinite) and inerts that will produce the best coke strength for each coal rank based on vitrinite reflectance. The size of the inerts is also important; the properties of the binder phase which incorporates the inerts is rank-dependent.

Generally, coal grade refers to purity in terms of the ash and sulfur levels, both of which have a deleterious effect on the blast furnace operation. The presence of sulfur can also embrittle the iron produced.

COKING THEORIES

Early experiments demonstrated that some coals produced coke, and others did not. Low and high rank coals did not coke; middle rank bituminous coals did but to varying degrees. Many attempted to determine what makes coal coke. The following describes some of the efforts to explain why only some coals coke.

Solvent Extraction Theory

Experiments conducted as early as 1860 extracted materials from coal that were believed to be responsible for coking. However, a correlation between coking properties and extraction yield was never completely demonstrated. The extraction yield is a variable that depends greatly on the extraction conditions. The "coking principle" idea or solvent extraction theory has recently emerged in a new form (23).

Transient Fusion Theory

This theory suggests that coal may be likened to $KClO_3$, which melts when heated to form an unstable liquid and then decomposes to form a solid as one of the decomposi-

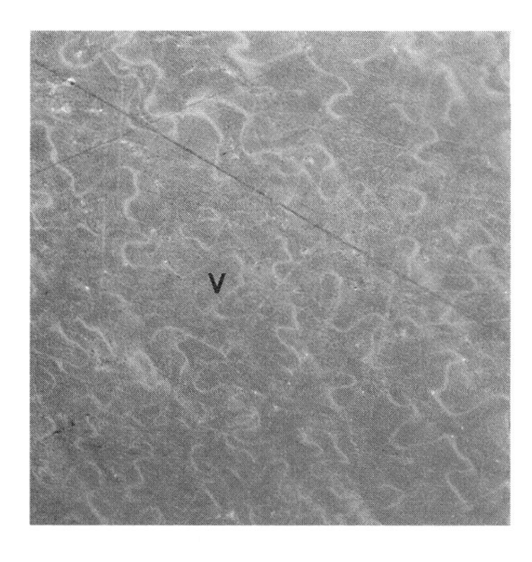

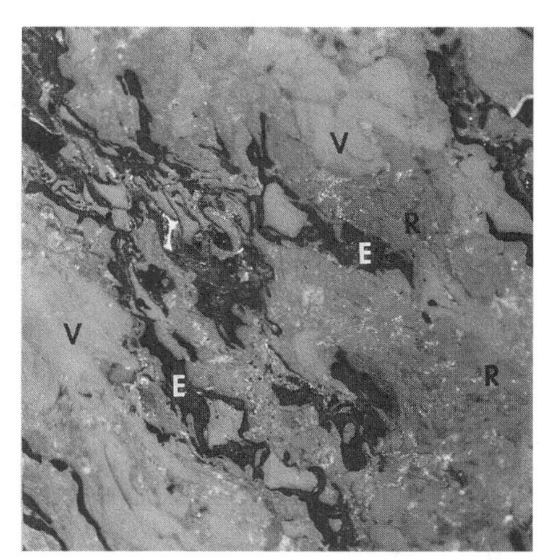

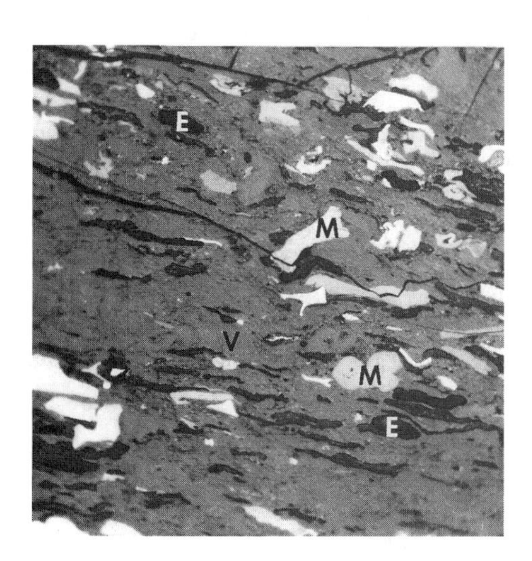

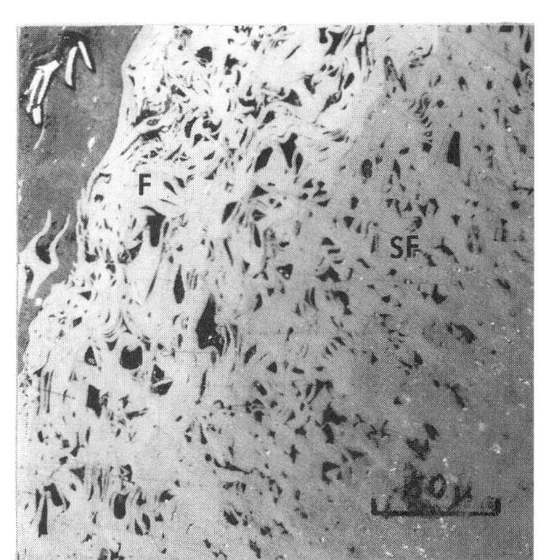

Figure 8. Photomicrographs of coke microtextures (carbon forms) showing isotropic carbon from marginal coking high volatile coal; pinpoint anisotropism from good coking high volatile coal; ribbon anisotropism from low volatile coal and microtextures from a blend of the three coals.

tion products (23). The theory is usually considered to be unsatisfactory.

Precursors of Metaplast Theory

This theory proposes that coal is an organosol consisting of an oily dispersion medium (oil) and a dispersed phase (micelle) (24). The micelle consists of a protective bitumens coating and the micelle nucleus, which is humins. According to Kreulen, the agglutination of inert material by a binder is a process shared by the whole system; this idea has been restated and refined (23,25).

Thermobitumen Theory

This early idea proposes that the plastic behavior of heated coal is due mainly to the formation of liquid primary products of pyrolysis (23).

Metaplast Theory

Pyrolitic decomposition of coal has been divided into three reaction pathways, based on thermobalance studies (25).

1. Coal complex	→	Slow reaction disintegration	→	Primary decomposition product (metaplast) and residue
2. Primary decomposition product (metaplast)	→	Rapid reaction	→	Decomposition product (semicoke) plus gas and tar
3. Residue (semicoke)		———————	→	coke and secondary gas

Numerous researchers have restated these ideas and have added to and modified the concepts. However, the tenet that remains is that coal itself does not contain a coking material, but that some coals when heated produce a decomposition product (metaplast) that cokes or reacts with other materials to render them coking.

Liquid Crystal Theory

When polymeric coal substances are heated, a metaplast is formed *in vitrinite;* this peptizes the reactive coal macerals, producing a plastic state that passes through a liquid crystal state (mesophase) and produces an ordered mosaic structure upon solidification.

Liquefaction Theory

Coals' plastic capability is compared to a liquefaction process (26), where the availability of donor hydrogen determines the thermal stability of plastic transformation products during carbonization.

PETROGRAPHIC CONCEPTS

Coal petrographers separate bituminous coals into reactive and inert materials. The reactive macerals, dominated by vitrinite, produce plastic materials on pyrolysis. The inerts act as filler in the coking process. The idea of a metaplast or thermobitumen and its liquid crystal (mesophase) transformation to a solid coke applies only to the reactive coal macerals. The properties of the final metaplast vary with the rank of the coal being carbonized (27).

Microscopic observations of coal pyrolysis products indicate that most coking coals pass through a plastic state in which optically anisotropic structures develop and, upon solidification, produce an anisotropic mosaic structure. The rank of the coal largely controls the anisotropic domain size and degree of anisotropism. Coke petrographic techniques utilize the determination of isotropic and various anisotropic textures to relate coke carbon forms to the ranks of coals that produced the mosaic textures. The textures from each coal rank remain identifiable after carbonization; only particle boundaries have mixed textures. Coke textures and structures relate to the strength and reactivity of cokes (28).

PROPERTIES OF COKING COAL

Coals are characterized in terms of rank, grade, and type (29). Rank, the most important characteristic, refers to the coal's maturity in the series from lignite to anthracite. Grade generally indicates the purity of coal relative to parameters such as ash and sulfur. Type describes the amounts, kinds, and state of preservation or division of the organic plant materials that formed the coal.

The ASTM D388 terms used in the classification of coal by rank are shown as follows (30):

Class	Group
I. Anthracitic	1. Meta-anthracite
	2. Anthracite
	3. Semianthracite
II. Bituminous	1. Low volatile
	2. Medium volatile
	3. High volatile A
	4. High volatile B
	5. High volatile C
III. Subbituminous	1. Subbituminous A
	2. Subbituminous B
	3. Subbituminous C
IV. Lignitic	1. Lignite A
	2. Lignite B

Coal Rank

Volatile matter, which measures the weight loss on heating coal and decreases as the rank increases, is an accepted rank indicator. Reflectance of vitrinite is also a commonly used and accurate rank parameter for coking coals. Only coals whose rank is classified as bituminous and which are subclassified between high volatile B and low volatile are considered to be coking coals. High volatile C bituminous coals are generally noncoking, whereas high volatile B coals are marginal coking, and high volatile A coals are coking. Medium volatile coals are the strongest coking. Low volatile coals may be good coking coals, but they can exert excessive pressures on coke oven walls when used alone.

Coal Grade

Coking coals are divided into grades by parameters such as ash, sulfur, alkali, chloride, and phosphorous content (31). Almost all coals have some mineral matter content that will produce ash when the coal is burned. In addition, coals commonly contain organic and pyritic sulfur. Ash-forming materials must be melted and removed from the blast furnace, and the sulfur must be fluxed out to reduce its pickup by the metal. Because these processes require heat, additional coke carbon, beyond that which is required for smelting of the iron, will be necessary.

Coal Type

Coals consist of macerals, which are analogous to minerals in rocks (32). Some of these macerals are thermally reactive and melt during coking to form a binder, whereas others are essentially inert and act as filler to thicken coke walls and to reduce porosity. The concepts of coal maturity and reactive and inert macerals are used in techniques to predict coke strength from microscopic analysis of coal (27,33,34).

The optically discernible (in reflected light) organic macerals in the coal are vitrinite, liptinite (exinite), resinite, semifusinite, fusinite, and micrinite. The inorganic materials are collectively referred to as mineral matter and consist of mineral species that can be identified optically or by subsequent chemical analysis techniques. The various macerals and common minerals in the coal are illustrated in the photomicrographs in Figure 9.

Vitrinites, which are moderate in reflectance, are primarily derived from coalified wood materials and form the principal maceral in coals.

Liptinites are the lowest reflecting maceral and represent the remains of plant cuticle (external plant surface, such as waxy leaf coatings), spores, and/or pollen. They are relatively high in hydrogen content.

Resinites are lower than vitrinites in reflectance and are mostly derived from plant-cell fillings and are commonly rounded to oval.

Fusinites are highly reflecting charcoal-like materials that exhibit open cellular structure and are the product of partial oxidation of wood substances. They are high in carbon content.

Semifusinites are materials that are transitional between vitrinites and fusinites in reflectance and appearance.

Micrinites are materials that are similar in reflectance to the fusinites and semifusinites, but do not exhibit cellular structure and are generally finer in size.

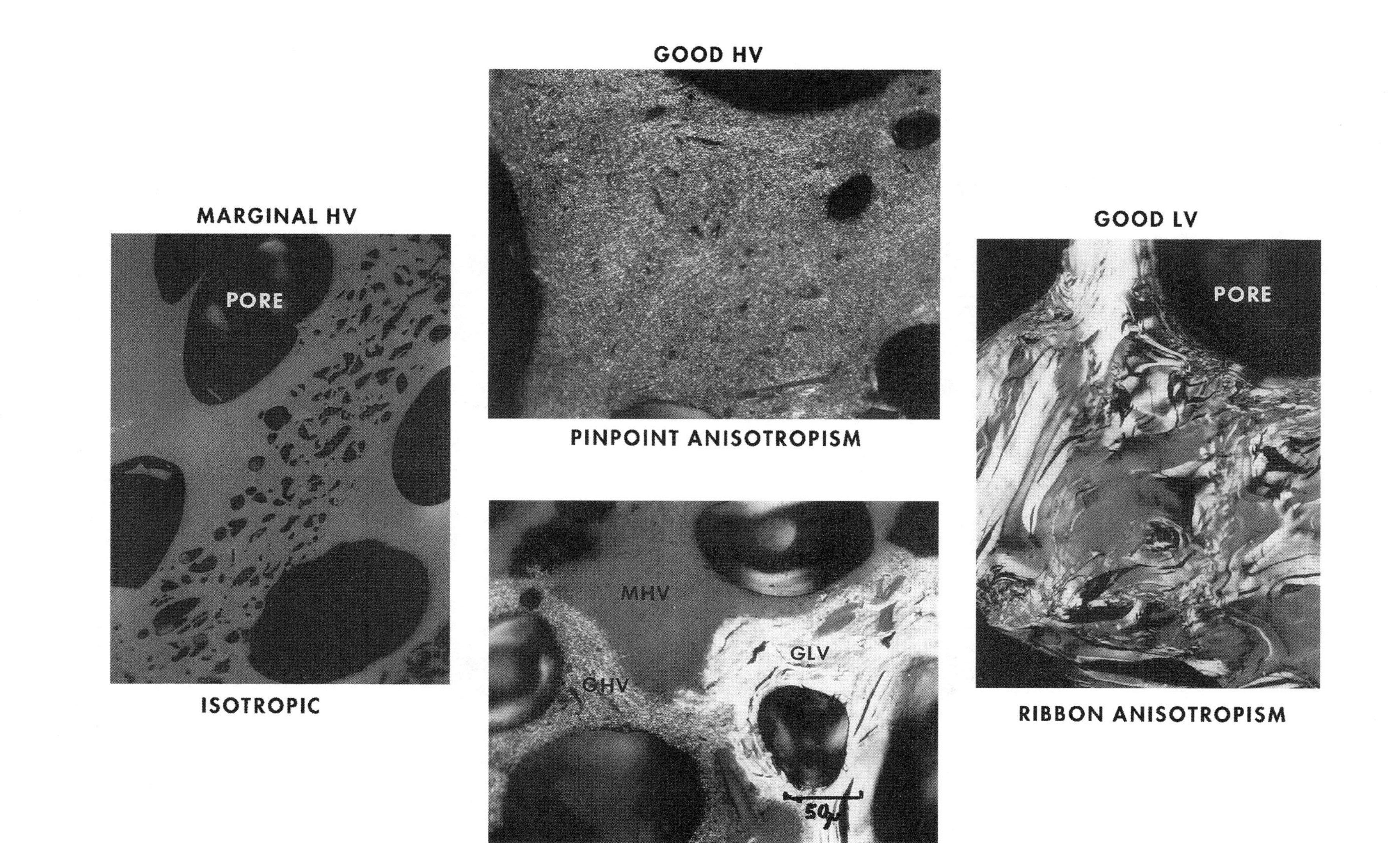

Figure 9. Photomicrographs of macerals in a high volatile bituminous coal: Vitrinite (V), Resinite (R), Exinite (E) or Liptinite, Micrinite (M), Semifusinite (SF), and Fusinite (F).

Mineral matter in most coal is primarily clay, iron pyrite, quartz, and carbonate materials.

Reactives

The vitrinites, resinites, liptinites or exinites, and about one-third to one-half of the semifusinites in bituminous coking coals, soften when heated and form a porous continuous bond structure in coke. They are collectively referred to as the reactive macerals. The proportion of reactive macerals must be high in order for the coal to have good coking properties.

Inerts

The inerts are fusinite, part of the semifusinite, micrinite, and mineral matter. They act as filler in the coking process and when fine-sized and not in excess, they thicken and strengthen coke walls. Coarse and excessive inerts result in a fissured and/or poorly bonded, weak coke structure.

Coke stability or strength can be predicted from petrographic data (27). The underlying principal of the approach is that a given coal or blend should have an optimum reactives–inerts ratio for a specific vitrinoid type to achieve highest coke strength. This idea is expressed as the Composition Balance Index (CBI), where 1 is optimum, less than 1 is inert-deficient, and more than 1 has an excess of inerts. In addition, vitrinoids produce coke carbons that differ in strength for each vitrinoid type. (Different carbon forms from high, medium, and low volatile coals are shown in Figure 8.) This idea is expressed as a Strength Index (SI) or Rank Index (RI), which generally increases with rank up to but not through low volatile coals.

The relationship between the CBI, SI or RI, and coke stability (strength) is shown in Figure 10.

COAL QUALITY PARAMETERS

As discussed previously, bituminous coals are coking coals. Lower rank coals produce a thermobitumen that is unstable, so that the coals decompose during heating; higher rank coals do not soften sufficiently. In addition, only certain bituminous coals qualify for producing coke, as they must meet purity standards based on ash and sulfur (31).

Coals must meet rigid specifications to produce good quality cokes. Table 1 lists chemical and physical properties of coal or blends required to produce commercial coke. Seldom does a single coal fulfill all of the requirements to produce good or acceptable coke, thus the blending of coals to meet specifications is commonly practiced.

The criteria for selecting coals for by-product coke making are well-known. Only bituminous coals coke, and all ranks of bituminous coals ranging from high to low volatile are used. Generally, 70–80% high volatile bituminous coal is blended with 20–30% medium and/or low volatile bituminous coals. High volatile coals tend to produce weak cokes in terms of tumbler stability, whereas low volatile coals tend to produce damaging coke oven wall pressures. Medium volatile coals are the best coking but are limited by supply. More of the better coking medium and low volatile coals can be used in nonrecovery ovens as the charge is not confined.

Reserves of coking coals also impose a restriction on their use (31,35,36). About 82% of all of the U.S. reserves of coking coals are high volatile bituminous (10% are medium and 8% are low volatile coking coals). In addition,

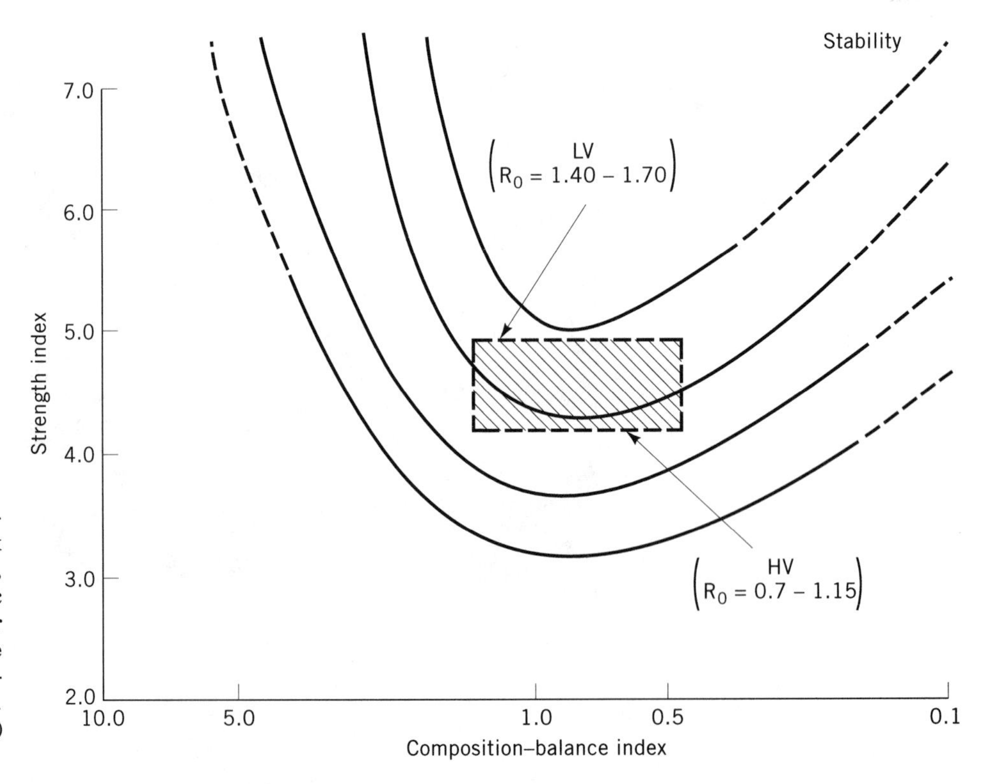

Figure 10. Graph showing the relation of the Strength or Rank Index and the Composition–Balance Index, as calculated from coal petrographic data to the predicted coke stability or strength. The position of low volatile (LV) coals and their vitrinite reflectance (R_o), where R_o = 1.40–1.70. The position of high volatile (HV) coals, where R_o = 0.7–1.15.

Table 1. Characteristics of Coking Coal or Coal Blend

	Good	Acceptable
Volatile matter, %	27	30
Fixed carbon, %	67	62
Ash, %	<6.0	~8.0
Sulfur, %	<0.7	~1.0
Potassium and sodium oxides, % of ash	<1.0	<3.0
Phosphorus, %	<0.01	<0.03
Coal reactives	75	70–80
Coal inerts	25	20–30
Vitrinoid reflectance	1.18	1.10
Composition balance index	~1	0.8–1.5
Rank index or strength index	>4.5	~4
Ash fusion temperature, °C	>1,371	1,260
Free-swelling index	8–9	>7
Gieseler maximum fluidity	>1000	600–10,000
Hardgrove grindability index	80 > 120	>60

sulfur restrictions on steam coals has created competition for low sulfur coking coals.

Blast furnace coke must be chemically low in ash and sulfur, physically strong, relatively coarse, and uniform in size. Table 2 lists the coke quality characteristics. Chemistry assures that the coke will function properly as a fuel and reductant in the blast furnace. Coke strength determines the extent to which coke functions to support the burden and to supply permeability. Cokes have recently been required to meet reactivity specifications (37); increased reactivity to CO_2 at elevated temperatures indicates that more coke will be consumed in the solution loss reaction, which increases the blast furnace coke rate and results in poor blast furnace performance.

The rank, type, and grade of coals used to produce the coke all impact coke reactivity. In general, thoroughly carbonized cokes are lower in reactivity than cokes with higher volatile matter content. Cokes from medium volatile coals are generally less reactive than those from high and low volatile coals. Cokes from low ash coals are generally less reactive than those from higher ash coke. Cokes with ashes that are high in alkalies and iron tend to have higher reactivities.

Table 2. Desirable Chemical and Physical Properties of Blast Furnace Coke

Chemical quality	
Moisture, wt %	5–7
Volatile matter, wt % (dry)	1.0 (max.)
Ash, wt % (dry)	8
Sulfur, wt % (dry)	0.7
Total alkali oxides, wt % (dry)	0.20
Phosphorus, wt % (dry)	0.15
Physical Properties	
Size, mm	50.8–177.8
Stability	60
Hardness	67
Micum 40	80
Micum 10	10
JIS $\frac{30}{15}$	92
Coke reactivity (CRI)[a]	32
Strength after reaction (CSR)[a]	58

[a] Nippon steel test.

Coking Parameters

Only coals that qualify on the basis of rank, chemistry, and type can be used to produce commercially acceptable coke (31). The selection and blending of the proper coals account for about 80% of the coke's properties. In addition to selecting the proper coals for blending, the coals must be processed properly to achieve the highest quality coke. Coal size, bulk density, and heating rate must be controlled (12).

Coke quality is known to be influenced by the distribution of components (macerals, maceral assemblages, and minerals) in the coking coal blend. The more homogeneous the component mixing, the more uniform the coke produced.

Coals are generally crushed to at least 80% minus 3.2 m (1/8 in) and preferably 90% minus 3.2 mm (1/8 in) for blending (38). The duller (higher inert) coal fractions should be ground extremely fine to assure incorporation in the coke structure.

As the coal size is reduced, a loss in bulk density occurs. The coke strength generally increases as the bulk density increases. The lower the rank of the coal, the greater this effect. Oil additions of about 1.1–2.2 l (2–4 pt) per ton of coal are used to control the bulk density. The oil reduces the friction between particles and helps to facilitate packing.

The rate at which the coal is heated is important in coking. In general, the marginal coking, low fluidity, high volatile coals produce stronger cokes when heated faster, whereas strongly coking medium volatile rank bituminous coals produce larger and stronger cokes when heated slowly.

For any given coal blend, the size, bulk density, and heating rate must be optimized, as these factors contribute about 20% to the coke's physical properties.

A wide variety of tests have been devised to sort coals for specific uses. The most stringent requirements are placed on coking coals. Typical tests for metallurgical coals are shown as follows (39):

Typical Tests for Metallurgical Coals

1. Proximate and sulfur analysis
2. Rheological tests
   a. Free-Swelling Index
   b. Gieseler plastometer
   c. Dilatometer
   d. Sole-heated oven
3. Petrographic analysis
4. Oxidation tests
5. Hardgrove grindability
6. Ash fusion and ash mineral analysis
7. Small-scale coke oven tests
   a. (13.6-kg) (30-lb) test oven
   b. Sole-heated oven

8. Optional tests
   a. Ultimate analysis
   b. Sulfur forms
   c. Calorific value
   d. Trace elements
   e. Gray-King Assay
   f. Roga Index
9. Large-scale pilot-oven tests for:
   a. Coke for ASTM stability and hardness testing
   b. Coking pressure
   c. Coke for reactivity testing

A variety of tests are performed on the coke from coal to determine its strength or resistance to breaking and abrasion, as well as true and apparent specific gravity and reactivity to CO_2. In addition, structure and texture analysis can be used to measure pore and wall size and carbon anisotropism (28, 39). Some typical tests of properties of coke are listed below.

Tumbler test	10 kg (22 lb) of coke sized to either 100% 5.1 × 7.6 cm (3 × 2 in) or 50% 5.1 × 3.2 cm (2 × $1\frac{1}{2}$ in) is tumbled for 1,400 revolutions in an ASTM tumbler drum; the tumbled sample is screened and percent plus 2.54 cm (1 in) is reported as the stability factor; plus 0.63 cm ($\frac{1}{4}$ in) is the hardness factor (40).
Stability factor	A relative measure of resistance to impact (abrasion contributes slightly) and a measure of the fractures and fissures in the coke; above 55 is acceptable, and 62 is good.
Hardness factor	A measure of the abrasion resistance of the coke in the ASTM tumbler drum; values range from high 60s to low 70s.
Apparent specific gravity	Weight of the coke including the pore structure; commonly ranges from 0.8 to 1.0 g/cc.
True specific gravity	Weight of coke without pores compared to the weight per unit volume of water; commonly ranges between 1.85 and 2.00 g/cc.
Structure analysis	Microscopic measurement of coke pore and wall sizes.
Texture analysis	Microscopic determination of all isotropic and anisotropic carbon forms in coke.
Reactivity, Bethlehem method	50 g of 18 × 40 mesh dry coke are reacted at 996°C in an excess of CO_2 for two hours; the weight loss (%) adjusted for coke volatile matter is reported as reactivity (%).
Reactivity, Japanese method	200 g of 21 × 19 mm ($\frac{3}{4} \times \frac{7}{8}$ in) coke are reacted at 1,100°C in CO_2 for two hours; coke reactivity (CRI) is calculated from weight loss; coke strength after reaction (CSR) is determined by the amount of plus 10-mm ($\frac{3}{8}$-in) coke remaining after tumbling the reacted coke for 600 revolutions in an I-Drum tester.

The ASTM stability has been correlated with Japanese Industrial Coke Drum Indices (41) and Micum Drum Indices (42).

FOUNDRY COKE

Blast furnace and foundry coke differ in their physical property requirements and methods of production. Foundry coke is coarser, denser, and lower in reactivity than blast furnace coke. Foundry coke should have good impact strength and sufficient resistance to abrasion to minimize attrition upon handling.

Foundry coke is used in cupola-type furnaces for melting and superheating iron. The coke supplies the process energy as well as the carbon for liquid iron absorption. It also helps to reduce silicon and manganese loss in the cupola (13). Coke properties strongly influence cupola back pressure, slag volume, and basicity.

The cupola is a vertical cylindrical shaft furnace in which pig iron, scrap iron, scrap steel, and DRI (direct reduced iron) are melted (14). The molten iron is cast for uses such as engine blocks, brake drums, clutch plates, crankshafts, farm machinery, miscellaneous machinery, pipe, stove plates, grills, railroad car wheels, plow shares, iron bearings, and crusher plates.

Coke properties are achieved by proper selection and blending of coals, and proper processing or carbonization (14). Ideally, the medium volatile bituminous coals are best suited for foundry coke production as they tend to make dense, strong coke of low reactivity, with the least heat required for conversion to coke. Unfortunately, reserves of medium volatile coal are limited. Coke from high volatile coals, which are the most abundant coking coals, tend to be more porous and fissured due to shrinkage, whereas coke from low volatile coals, which are also not abundant, tend to be more abradable and may be more reactive to CO_2.

Commonly about 30–60% of foundry coke blends are medium volatile coals, and the balance is comprised of a mixture of high and low volatile coals and of about 8–16% antifissurants, such as anthracite, coke fines (breeze), and petroleum coke.

The blend should have a volatile matter of about 24%, a vitrinite reflectance of about 1.30–1.40%, 30–40% inerts, and a Gieseler fluidity of about 100–1,000 dial divisions per minute. The Gieseler plastometer test is a rheological test that measures the plastic properties of heated coal. The Free-Swelling Index should be about 6–8. The ash, sulfur, alkali, and phosphorous of the coal blend should be low to help in the production of high fixed car-

Table 3. Summary of Trends in Western World Coke Making Capacity[a]

Year	Metric Tons[b]		
	1988	1990	1995
EEC	52.4	55.577	52.608
Western Europe	9.6	11.056	10.863
North America	33.7	35.661	34.409
Latin America	9.9	13.978	16.317
Africa	6.1	8.700	8.690
Asia	74.0	72.300	75.568
Australia	3.9	4.682	5.443
Total Western world	189.600	201.954	203.898

[a] Ref. 44.
[b] In millions.

bon coke, which ensures adequate carbon absorption in the cupola.

The pulverization level should be about 90% minus 3.2 mm (1/8 in) or more. The heating rate should be about 17.5 mm (0.7 in) per hour to decrease the thermal gradient and to prevent fissuring, which decreases coke size. The bulk density of the coal charge should be high.

COKE MAKING CAPACITY AND USE

Use

The steel and ferrous foundry industries account for about 90% of the world coke used; residential heating is the next largest consumer. Much of this latter use is in Western and Eastern Europe, where the coke is considered a smokeless fuel and is preferred over coal. Ferro–alloy production is the next largest user, followed by lead and zinc smelting. Coke is also used in the production of items such as cement, calcium and silicon carbide, aluminum oxide, elemental phosphorous, slag and rock wool, beet sugar, and other commodities.

There does not appear to be any new significant uses for coke. New demands for coke in the industrial nations will probably be low due to the projected slow population growth (43).

Capacity

A survey of Western world coke making capacity in 1989 estimated a capacity of 207 million 2.07×10^8 t per year (44). This is a decrease of 1.36×10^7 t per year compared to 1985, and a decrease of 4.66×10^7 t per year compared to 1982. Nearly 2.6×10^7 t per year of capacity closed in the last 4 years; most of the reductions were in the United States (6.0×10^6 t per year), Germany (6.6×10^6 t per year), and Japan (3.8×10^6 t per year). Most planned or recent new facilities are in Brazil, India, and Korea; Germany is the only industrialized country that is implementing major modernization programs.

The Western world production of coke oven coke from 1970 through 1988 is shown in Table 3; the U.S. coke supply and demand are shown in Table 4. Coke plants consumed only 5% of the coal produced in the United States in 1989.

Coke consumption has declined over the past 10–20 years due to

1. Decreased steel consumption
2. Reductions in coke rates in blast furnaces from 544.3–635.0 kg (1,200–1,400 lb) per 0.9 t (1 short ton) of iron to 408.2–430.9 kg (900–950 lb).
3. Increased raw-to-finished product yield

Worldwide, about 400 coke batteries produced 3.5×10^8 t of coke, compared to 4.5×10^8 t of coal produced in 1987–1988. About 4.5×10^8 t of raw steel were produced from 8.0×10^8 t of iron ore with this coke.

Coke production in North America and Western Europe has decreased about 45% and 30%, respectively, since 1975; Asia and Latin America are generally increasing production. Part of the decline in North American coke production is due to decreased pig iron production, whereas part is due to improvements in coke and other burden materials, and more efficient blast furnace operations. Fuel injection into a blast furnace is currently not being used extensively in the United States; however, its use will replace large quantities of blast furnace coke in the near future.

ENVIRONMENTAL ASPECTS OF COKE MAKING

Coke batteries are among the largest industrial production units that operate in an open environment. Millions of tons of raw materials are stored, handled, and processed in this environment to produce millions of tons of products that range from solid coke to liquids and gases. Most of the emissions in a coke operation relate to charg-

Table 4. U.S. Coke Supply and Demands, 1980–1990[a,b]

Year	U.S. Coke Production	Imports	Total Supply	Exports	Apparent Consumption
1980	41,850	598	42,449	1,879	37,441
1982	25,506	109	25,615	901	23,384
1984	27,725	528	28,253	948	27,124
1986	23,808	298	23,469	911	22,998
1988	29,397	2,439	31,835	992	30,373
1990[c]	26,989	998	27,987	454	27,442

[a] Ref. 45.
[b] In thousand t.
[c] NCA estimate.

ing coals to the ovens and pushing the cokes from the ovens.

Coke is the principal blast furnace fuel for the integrated steel companies. It remains a critical area in the steel industry for meeting current and proposed emission control regulations.

The coke industry has spent about $6 billion over the past 20 years on environmental controls. About 90% of all emissions are captured, with the possibility of reaching a capture rate of 96–97% with additions of new and available technologies.

The recently enacted Clear Air Act Amendments of 1990, Title III (CAAA), which concerns air toxics, specify U.S. Environmental Protection Agency (EPA) regulations to be placed in effect no later than December 31, 1992, with a compliance date set for January 1, 1998. It is questionable that the coke industry can meet the compliance goals with available technology. Additional major investment in the nonproductive portion of coke making will impact the United States' ability to compete in the world market.

The EPA's cost estimate of complying with the air-toxic provisions of the CAAA is $0.36–5 billion (46). To meet the Maximum Available Control Technology (MACT) standard of 8% leaking doors, it may be necessary to replace doors and jambs on nearly all existing coke ovens. The industry cannot even meet the 3% leaking doors standard (5% for tall ovens) as a condition for receiving an extension. To replace the 95 existing batteries, most of which are old, the capital investment required would be $200–300 million for each new battery built.

To reduce charging emissions, stage charging and steam aspiration have been widely used. In stage charging, the coal is dropped from the charging hoppers into the ovens in a sequence which minimizes the leveling requirements and charging time. While charging, steam is injected in the standpipes to create a suction, to draw the gas and solid particulates into the by-product mains where they can be controlled. Considerable work has been done, and is continuing, on oven door seals and sealing of charging hole lids.

Those companies committed to blast furnace technology will be in the coke business for at least the next 30 years. However, 40% of current U.S. coke making facilities are over 30 years-old and will not meet air-toxic regulations without significant investment. Coke can be imported, but the cost will rise significantly as the level of dependence increases (46).

ALTERNATIVE COKE MAKING PROCESSES

For many years, there has been a widespread interest in alternative coking processes to replace or supplement by-product coking (12). The current growing interest in alternative coking methods is a result of coking emission problems. Some interest also stems from coke oven investment costs and the availability of coking coals.

Formed Coke

Few industries have committed to formed coke processes, in spite of the long historical interest and a large number of proposed processes. Most of the proposed processes use noncoking coals or coal chars with various binders, such as coking coal, tar, or pitch. Most processes use a heating or devolatilization step followed by agglomeration and carbonization. Some of the better known processes are: FMC, BFL, HBNPC, DKS, CCC, Koppelman, ANCIT, AUSCOKE, and CTC (12,47).

The FMC process has been operating for many years in Kemmerer, Wyoming, to produce coke for phosphate furnaces. In this process, 0.75-mm subbituminous noncoking coal is dried in a fluid bed and subsequently heated in two other fluid beds to a maximum temperature of about 900°C to produce a char. The char is then cooled in a fluid bed, briquetted with about 13% of tar binder at about 230°C, and then carbonized at 850–900°C. FMC coke has been tested in blast furnaces but can only replace a portion of the coke burden.

In the BFL process, noncoking coals are first treated at 750°C in a Lurgi-Ruhr reactor to produce a char, which is mixed with coking coal as a binder. The mix is briquetted at about 450°C; the green briquettes can be carbonized.

The French HBNPC process utilizes noncoking low volatile bituminous coals with a minor amount of coking coals for binder. The briquettes are preheated, coked, and cooled in a vertical shaft furnace.

In the DKS process, which was jointly developed by Didier and Sumitomo, a noncoking coal with occasional coke breeze and tar pitch binders are briquetted and coked in ovens of essentially conventional design.

In the CCC process, developed by Conoco and a group of steel companies, a preheated coal (450°C) is mixed with char and recycled coke breeze, and pelletized. The pellets are subsequently carbonized.

In the Koppelman process, lignites and similar carbonaceous materials, including wood, are crushed and slurried with recycled water and then injected into a reactor where pressure and temperature reduce moisture and densify the solids while yielding by-products. A solidified product is discharged from the reactor through lock hoppers into cooling vessels.

The ANCIT process was originally developed by Dutch State Mines. In this process, noncoking low volatile coals or devolatilized high volatile coals are flash-dried and carbonized in a cyclone at 600°C to produce a char. The hot char and coking coal, which is preheated to 460–520°C, are blended and briquetted, then cooled.

The AUSCOKE process was developed in Australia by Broken Hill, PTY. In this process, coal is crushed to at least 6 mm and dried, then mixed with tar and pitches. The blend is briquetted, and the briquettes are hardened in a fluidized bed of sand, then coked in a shaft furnace.

The CTC process utilizes a twin-screw reactor to heat and mix a variety of coals, chars, and binders to produce a plasticized material that can be formed by a variety of procedures before it is coked. The twin-screw reactor will process fluid coking coals.

Processes for forming coal–char and binders are in various stages of development, and it is only a matter of time, political, and/or economic pressure before formed cokes will be commercialized on a large scale.

The nondependency on good coking and costly coals adds to the attraction of form coking. Of greater importance is that form coking is likely to be easier to control from an environmental standpoint. Coal desulfurization (as in the "clean coke process") in conjunction with form coking production would be a major attraction.

Nonrecovery Coke Ovens

Nonrecovery coke ovens dominated the coke industry for many years, until the need for by-products caused the trend toward recovery ovens. Currently, the environmental restrictions and cost of by-product ovens renders nonrecovery coke making even more attractive. There are nonrecovery coke operations run by Sun in Virginia, and PACTI (a demonstration plant) in Mexico (48).

Nonrecovery ovens are constructed of fireclay refractories, rather than the more expensive silica brick used in by-product ovens, and fewer brick forms are used. The gases generated in the process are burned, and the heat is used to heat sole flues in the base of the ovens. Excess gas is scrubbed to reduce sulfur and is used for power cogeneration. High pressure-generating low volatile coals can be coked alone in nonrecovery ovens. However, whereas door emissions from nonrecovery ovens are reduced, leakage to the charge is a problem.

The main issues associated with nonrecovery ovens are

1. Coke quality–chemistry, strength, size, and reactivity
2. Stack combustion emissions cleanup
3. Process controls, such as heating rate, and bulk density
4. Waste heat recovery
5. Excess gas desulfurization and use in power cogeneration
6. Land requirements

Nonrecovery ovens may have a significant future role in supplying the coke needs on a short-term intermittent basis. The ovens are less costly to build and can be taken in and out of service more easily than conventional by-product ovens.

Jumbo-Slot Ovens

A 100-t jumbo oven is currently being investigated in Germany (49,50). Such a large oven minimizes the number of charges and pushes, which are major emission producers, and the sealing surfaces required relative to the volume of coke produced is small. The operation has the added advantage of being similar to conventional by-product coking.

In the past, the coke oven volumes were increased by making the ovens longer and taller (51). However, the leveling and pushing equipment are limiting factors as the length of the ovens is increased; uniform heating is also a limiting factor in taller ovens. Jumbo ovens, in addition to added length and height, are also wider, requiring special coal selection to produce satisfactory coke.

As a rule, the densest and strongest coke in an oven is produced on the wall side of the coke fingers, whereas spongy coke is most common on the inner or tar end of coke fingers. Thus, the amount of porous and weak coke is likely to increase as the oven width increases. In addition, the floor and walls of jumbo ovens will require good durability to handle 100-t charges.

New Iron Making Processes

As most coke is used in the iron blast furnace, any process that eliminates the blast furnace will reduce the demand for coke.

Direct Reduction. There are considerable efforts worldwide to develop processes for direct reduction of iron ore (52). Direct reduction processes are attractive as they reduce capital investment and offer the potential of producing cheaper steel. Direct reduced iron in conjunction with the electric furnace offers a cheaper route to steel making than the blast furnace with basic oxygen steel making.

The American Iron and Steel Institute (AISI), a collective steel industry effort in the United States, is conducting research on direct reduction of iron ore (53). A variety of reactors to reduce iron ore are being considered, such as the shaft with a fixed or moving bed, a fluidized bed, and a rotary kiln. Most of the processes commercialized to date use cheap natural gas, which is not available in the United States. Current U.S. research thus focuses on the use of coal and other solid fuels; gasification of coal to produce reducing gas is a possibility.

Direct reduction of iron ore has been used near the ore source where the fuel source and labor costs are low. The 1990 capacity for direct reduction of steel making grade iron ore was 29.37 t/yr capacity with 17.88-t production.

Fuel Injection. Use of auxiliary fuel injection through the tuyeres of a blast furnace to replace some coke is a fairly dated process. The fuels can be gas, liquid, or solid (54).

Oil injection has been used mostly when oil prices were low. (Tars and pitches have been substituted for oil.) The quantity that can be used is limited by the amount that can be burned within the raceway, the active area in front of the blast furnace tuyere, without producing carbon black. At least 130 kg/t of hot metal can be used, but some systems allow use of greater quantities. Oil reacts endothermically with oxygen and lowers the combustion temperature; additional oxygen can be used to enrich the blast.

Gas injection is used in many industrialized countries; however, availability and cost are limiting factors. Gas decreases flame temperature, although special systems have been proposed to increase the rate of natural gas injection. In addition to tuyere injection, gas has also been injected into the upper bosh of the blast furnace to replace coke (55).

The concept of coal injection in the blast furnace goes back to the past century, but it was not until 1966 that Armco Steel installed the first system on their Bellefonte furnace. Whereas coal injection does not save energy, it does reduce coke use. As coke production decreases in the United States and as the age of the coke ovens is critical with no new facilities, coal injection has become attractive.

In a coal-injection system, the coal must be finely ground to about 80–200% mesh (74 microns), and stored and conveyed under pressure in a controlled atmosphere. Thus, coal injection in the blast furnace is more expensive to install than an oil- or gas-injection system. The future of coal injection will be determined by its total cost relative to that of coke (56).

High coal-injection rates appear possible, and coke rates as low as 250 kg/t of hot metal, and eventually below 200 kg/t of hot metal, may be achieved. Pulverized coal-injection use in Japan has steadily increased, reducing oil consumption and stabilizing the operation of big blast furnaces. The goal is 200 kg/t of hot metal by 1994 at Chiba Works, the largest blast furnace of Kawasaki Steel.

Because coke is the only material in the blast furnace that remains solid from top to bottom, a certain minimum amount of coke will be required to furnish the burden support. High coal injection may decrease the size of the coke grid below the cohesive zone. The coke in this grid will rise rapidly as coal-injection rates increase and will be exposed to additional thermal, mechanical, and erosion forces (55). This will influence the properties of the coke that will be required.

European ironmakers have targeted 181 kg/t of hot metal of coal injection and a complimentary coke rate of 254–272 kg/t of hot metal.

At least half of the pig iron produced worldwide uses some type of fuel injection, which may account for over 10% of the energy consumed in blast furnaces.

In the future, changes in the steel industry will result in a shrinking coke market, and pressures from environmental regulations will result in more imported coke unless new coking processes are developed domestically.

BIBLIOGRAPHY

1. H. H. Schobert, *Coal, The Energy Source of the Past and Future,* American Chemical Society, Washington, D.C., 1987.
2. C. S. Finney and J. Mitchell, "History of the Coking Industry in the United States," *Journal of Metals* (Apr.–Aug., 1961); also published as "The Coke Oven" in *History of Iron and Steel Making in the United States,* American Institute of Mining, Metallurgical and Petroleum Engineers, 1961, pp. 29–56.
3. E. V. Bennett, ed., *The Making of Steel,* 2nd ed., American Iron and Steel Institute, New York, 1964.
4. J. Kemper, III, *American Charcoal Making–In the Era of the Cold-Blast Furnace,* Eastern National Park and Monument Association, National Park Service, U.S. Department of the Interior, 1987, pp. 1–25.
5. W. T. Lankford, Jr., N. L. Samways, R. F. Draven, and H. E. McGannon, eds., *The Making, Shaping and Treating of Steel,* 10th ed., Association of Iron and Steel Engineers, 1985, pp. 149–156.
6. J. K. Gates, "The Beehive Coke Years–A Pictorial History of Those Times," John K. Gates, Uniontown, Penn., 1990, pp. 1–128.
7. D. J. Cuff, W. J. Young, E. K. Muller, W. Zelinsky, and R. F. Abler, eds., *The Atlas of Pennsylvania,* Temple University Press, Philadelphia, 1989, pp. 38–41, 101–105.
8. P. Temin, *Iron and Steel in Nineteenth Century America: An Economic Inquiry,* MIT Press, Cambridge, 1964, pp. 1–299.
9. E. T. Sheridan, *Supply and Demand for United States Coking Coals and Metallurgical Coke,* Bureau of Mines Special Publication, U.S. Department of the Interior, 1976, pp. 1–23.
10. R. J. Wilson and J. H. Wells, *Coal, Coke and Coal Chemicals,* McGraw-Hill Inc., New York, 1950.
11. H. G. Franch and J. W. Sladethofer, *Industrial Aromatic Chemistry,* Springer-Verlag, 1988, pp. 1–485.
12. W. Eisenhut, "High Temperature Carbonization," ch. 14 in M. A. Elliott, ed., *Chemistry of Coal Utilization,* John Wiley & Sons, Inc., New York, 1981, pp. 847–917.
13. K. Beck, "Blast Furnace Coke Production: The Coking Capacity Determines the Special Process Technology," *Erdol and Kohle* **40** (3), 129–131 (1987).
14. J. Boland and co-workers, *CRM Report, 1977;* also *Stahl and Eisin* **97** (26) (Dec. 1977).
15. R. W. Bowman, "Blast Furnace Practice I," *Iron and Steelmaking* (Feb. 1981).
16. J. C. Clendenin, "Coal and Coke," chapter VIII in *Blast Furnace Theory and Practice,* American Institute of Mining Engineers, 1969, pp. 327–436.
17. A. A. Agroskin, "Chemistry and Technology of Coal," translated from Russian, U.S. Department of the Interior and the National Science Foundation, Washington, D.C., 1966.
18. J. C. Agarwal and co-workers, "Economic Benefits of Hot Metal Desulfurization to the Iron and Steelmaking Processes." *Proceeding of the 40th Ironmaking Conference of AIME,* Toronto Ontario, 1981, pp. 146–170.
19. R. V. Flint, "Effect of Burden Materials and Practices on Blast Furnace Coke Rate," *Regional Technical Meeting of the American Iron and Steel Institute,* 1961, pp. 1–36.
20. R. J. Gray, "Theory of Carbonization of Coal," *First International Meeting on Coal and Coke Applied to Ironmaking,* Rio De Janeiro, Brazil, 1987, pp. 551–557.
21. K. Nishioka and S. Yoshida, "Investigation of Bonding Mechanism of Coal Particles and Generating Mechanism of Coke Strength During Carbonization," *Transactions of ISIJ* **23,** 475–481 (1983).
22. R. J. Gray and P. E. Champagne, "Petrographic Characteristics Impacting the Coal to Coke Transformation," *Proceedings of the AIME Ironmaking Conference,* Toronto, Canada, 1988, Vol. 47, pp. 313–324.
23. N. Y. Kirov and J. N. Stephens, *Physical Aspects of Coal Carbonization,* Sydney Kingsway Printers PTY, Ltd., 1967, pp. 1–25.
24. J. W. Kruelen, "A Laboratory Study of the Formation and Structure of Coke," *Fuel* **6,** 171m (1927).
25. D. W. Von Krevelen and J. Schuyer, *Coal Science,* Elsevier, New York, 1957, pp. 286–311.
26. R. C. Neavel, "Coal Plasticity Mechanism: Inference from Liquefaction Studies," *Proceedings of the Coal Agglomeration and Conversion Symposium,* West Virginia Geological Survey and the Coal Research Bureau, 1976, pp. 121–133.
27. N. Schapiro, R. J. Gray, and G. R. Eusner, "Recent Developments in Coal Petrography," *Proceedings of the Blast Furnace, Coke Oven, and Raw Materials Committee,* 1961, Vol. 20, pp. 89–112.
28. R. J. Gray and K. F. DeVanney, "Coke Carbon Forms: Microscopic Classification and Industrial Applications," *International Journal of Coal Geology* **6,** 277–297 (1986).
29. A. M. Carpenter, "Coal Classification," *IEA CR/12,* IEA Coal Research, London, 1988, p. 9.
30. ASTM D388-912, "Standard Classification of Coals by Rank," *1991 Annual Book of ASTM Standards,* Section 5, Vol. 05.05, pp. 202–204.
31. R. J. Gray, J. S. Goscinski, and R. W. Schoenberger, "Selec-

tions of Coals for Coke Making," *Proceedings of the Second International Coal Utilization Conference and Exhibition,* Houston, Tex., 1970, Vol. 4, pp. 129–164.
32. W. Spackman, "The Maceral Concept and the Study of Modern Environments as a Means of Understanding the Nature of Coal," *Trans. N.Y. Academy of Sciences, Series II* **20,** 411–423 (1958).
33. N. Schapiro and R. J. Gray, "Petrographic Classification Applicable to Coals of All Ranks," *Proceedings of the Illinois Mining Institute,* 1970, pp. 83–97.
34. E. Stach, M. T. Mackowsky, M. Teichmüller, G. H. Taylor, D. Chandru, and R. Teichmüller, *Stach's Textbook of Coal Petrology,* Gebrüder Borntraeger, Berlin, Stuttgart, 1982, pp. 87–218.
35. P. Averitt, "Coal Resources of the United States," *U.S. Geological Survey Bulletin* **1412** (1974).
36. "Demonstrated Reserve Base of Coal in the United States on January 1, 1979, *DOE/EIA-0280 (79) UC-88,* U.S. Department of Energy, Apr. 16, 1981, pp. i–iii, 1–121.
37. R. J. Gray and J. S. Goscinski, eds., *Coke Reactivity and Its Effect on Blast Furnace Operation,* Iron and Steel Society, Inc., 1990, pp. 1–158.
38. M. Perch, "Solid Products of Pyrolysis," chapter 15 in M. A. Elliott, ed., *Chemistry of Coal Utilization,* John Wiley & Sons, Inc., New York, 1981, pp. 919–982.
39. R. Loison, F. Pierre, and A. Boyer, *Coke Quality and Production,* (THYSSEN) Butterworth, 1989, pp. 189–192.
40. ASTM D3402-81, "Standard Method of Tumbler Test of Coke," *1991 Annual Book of ASTM Standards,* Section 5, Vol. 05.05, pp. iii–XVI, 367–368, 1–500.
41. J. F. Gransden, W. R. Leeder, and J. C. Botham, "Correlation of ASTM Stability with Industrial Coke Drum Indices–Update," *Report ERP/ERL 78-89(OP), CANMET,* Department of Energy and Resources, Ottawa, Canada, Nov. 1978.
42. "Correlation of Micum Coke Drum Indices," *Document ISO/TC 27/SC 3N91,* International Organization for Standardization, Technical Committee 27 (Solid Mineral Fuels), Subcommittee 3 (Coke) Meeting, London, Sept. 1978.
43. W. Leontief, *The Future of the World Economy,* Oxford University Press, New York, 1977.
44. *Western World Cokemaking Capacity,* International Iron and Steel Institute, Committee on Raw Materials, Brussels, 1989, pp. 1–67.
45. *Facts About Coal, 1991,* National Coal Association, Washington, D.C., 1991, p. 58.
46. Burns & Roe Service Corp., *Coke Oven Emission Control R&D Workshop,* prepared for the U.S. Department of Energy, Pittsburgh Energy Technology Center, Sept. 1991.
47. Battelle Institute eV, ed., *The Production of Formed-Coke and Its Future Impact on Conventional Coke-Making and Blast Furnace Operations in the World,* Batele-Institut, Frankfurt, 1978.
48. J. P. Ciarimboli and R. Sturgulewski, "Recent Technical Developments in PACTI Non-Recovery Cokemaking Systems," *Ironmaking Conference Proceedings,* Vol. 51, 1992, pp. 601–602.
49. H. Bertling, W. Rohde, and K. Wessiepe, "Cokemaking in a Jumbo Coking Reactor," *Ironmaking Conference Proceedings,* Vol. 51, 1992, pp. 601–606.
50. G. Nashan, "Economic Fundamentals for a New Cokemaking System and a New Process Structure of a Coke Plant–Jumbo Coking Reactor," *Ironmaking Conference Proceedings,* Vol. 51, 1992, pp. 349–358.
51. K. G. Beck, "Blast Furnace Coke Production: The Coking Capacity Determines the Special Process Technology," *Erdol Kohle* **40,** 129–131 (1987).
52. T. Fukuyoma, *International Symposium of Future Ironmaking Processes,* Hamilton, June 1990, pp. 1–6.
53. "Progress Report: New Steel Technologies," Pentan Publication, Oct. 1992, pp. 25–34.
54. J. C. Agarwal and co-workers, "A Model for Economic Comparison of Natural Gas, Oil and Coal Injection in the Blast Furnace," *Ironmaking Conference Proceedings,* Vol. 51, 1992, pp. 609–615.
55. A. Poos and N. Ponghis, "The Potential and Problems of High Coal Injection Rates," *Iron and Steelmaking I&SM,* 35–37 (Oct. 1990).
56. I. F. Carmichael, "An Introduction to Blast Furnace Coal Injection," *Iron and Steelmaking I&SM,* 67–73 (Mar. 1992).

COAL COMBUSTION

L. Douglas Smoot
Brigham Young University
Provo, Utah

GENERAL DESCRIPTION

History

Coal combustion is the burning of coal by oxygen, usually in air, for the purpose of generating heat. Direct combustion of coal has been identified in some of the earliest recorded histories. The Chinese used coal as early as 1000 BC, and the Greeks and Romans made use of coal before 200 BC (1). By AD 1215, trade in coal was started in England (2). A dramatic increase in the use of coal occurred in the late 1700s and early 1800s with large-scale use in iron smelting and coal gas production for illumination purposes. As the coal gas industry developed, the use of gas for heating purposes also developed. The use of producer gas increased until the early part of the 20th century, when it was largely replaced by cheap natural gas (1).

Processes for coal combustion progressed from burning in open fireplaces to burning in fuel beds in small household furnaces, which led to large industrial furnaces and then to large, pulverized-coal furnaces such as utility boilers. Pioneering uses of coal (eg, coke, coal tars, and gasification) have continued to be advanced since the late 16th century. The fixed-bed stoker system was invented in 1822, the firing of pulverized coal occurred in 1831, and fluidized beds were invented in 1931. The emergence of electric lighting opened the door to large-scale coal use as a fuel for the production of steam for power generation. These utility boilers were first attempted in 1876 but were not successful in the United States until 1917 when furnace design was improved (3).

Availability and Uses

Specific purposes for uses of direct coal combustion include (*1*) power generation, (*2*) industrial steam and heat production, (*3*) firing of kilns (cement, brick, etc), (*4*) coking for steel processing, and (*5*) space heating and domestic consumption. The various forms in which the coal is used also differ substantially. Coal particle diameters in different processes vary from micrometer size through centimeter size. Use is made of virginal coal, char, and coke. Pulverized coal is also slurried with water or other carriers, both for transporting and for direct combustion (4).

Coal is the world's most abundant fossil fuel. Recoverable world coal resources are estimated to be in excess of 0.9 trillion t (5). World coal production increased from 3.8 billion t in 1980 to 4.7 billion t in 1988 (see COAL AVAILABILITY, MINING, AND PREPARATION). At this level of production, the recoverable world coal reserves should last for about 200 yr. Coal is also the most abundant fuel in the United States, where coal resources represent 90% of all known fossil energy resources (6). Recoverable reserves are estimated to be 264 billion t, with total resources far in excess of this amount. The U.S. coal production was 890 million t in 1989. At this production level, the U.S. recoverable coal reserves should last almost 300 yr (7). Coal represents a smaller fraction of the U.S. energy consumption than of its energy production, because of heavy reliance on imported oil for transportation. In 1989, coal accounted for 23% of total energy consumption, and its share is expected to increase in the future (7). Electric utilities are the largest coal-consuming sector by far and account for most of the growth in coal consumption. In 1989, 86% of coal consumption went into generating electricity, and about 55% of the electricity was produced from coal (6).

Most of the coal presently being consumed is by direct combustion of finely pulverized coal in large-scale utility furnaces for generation of electric power, and this is likely to remain the way through the end of this century. However, many other processes for the conversion of coal into other products or for the direct combustion of coal are being developed and demonstrated, including various coal combustion and gasification processes (see COAL GASIFICATION), magnetohydrodynamic generators, and fuel cells.

Increasing the use of coal presents many technical problems, particularly in protecting the environment while increasing combustion efficiency. To solve these problems and increase the use of coal, many countries in the world are supporting research and development of clean coal technologies. It is imperative for new coal technologies to reach the market in a timely manner with minimal environmental impact, and at a competitive cost. It was against this background that the International Energy Agency ministers decided in 1985 that the first area of emphasis for international collaboration in energy research, development, and demonstration should be the clean use of coal (8).

A key goal of the U.S. National Energy Strategy is to maintain coal as competitive and to establish it as clean fuel (6). The resulting U.S. Clean Coal Technology (CCT) Demonstration Program is a major effort in this direction. A multiyear effort consisting of five separate solicitations is under way. As a result of four solicitations through September 1991, the CCT program currently comprises 42 demonstration projects. Of these, 11 are in advanced electric power generation systems, 21 in high performance pollution control devices, 6 in coal processing for clean fuels, and 4 in industrial applications (9).

Coal Processes and Burning Times

Classification. Various processes that are used directly to combust or to gasify coal can be classified in several different ways. This section presents a discussion of the general features of these practical flames, together with various classification methods. A brief description of various direct coal combustion processes is also included.

Practical coal combustion systems can be classified in several ways, including the following:

Classification	Examples
Flow type	Well stirred
	Plug flow
	Recirculating
Process type	Fixed bed
	Moving bed
	Fluidized bed
	Entrained bed (ie, suspension firing
Flame type	Premixed
	Diffusion
Coal Particle Size	Large
	Intermediate
	Small
Mathematical model complexity	Zero-dimensional
	One-dimensional
	Multidimensional

Classifications based on process type and particle size are among the most common. Various types of coal process equipment have been devised to combust coal. In fixed beds, the coal particle size is in the range of 2 to 3 cm in diameter. The particles are packed tightly into a vessel, and the combusting air flows through the bed. Coal reaction rates in fixed beds of large particles are slow, in the range of many minutes to over an hour (Table 1).

When the particles are ground smaller in the range of 2 to 5 mm, and the velocity of air flowing up from the bottom of the vessel is increased, the particles are lifted and said to be fluidized. Particles are in vigorous motion in a fluidized bed and particles can react more quickly even at lower temperatures. Times required to consume particles in fluidized beds are typically 10 to 500 s, as shown in Table 1. Coal particles can also be ground even more finely, to a face-powder consistency where the particle sizes are in the range of 10 to 100 μm and barely visible individually by the naked eye. These pulverized coal particles can be blown into large furnaces and burned in a matter of seconds at high temperatures.

Table 2 summarizes these classification groups and illustrates typical relationships among the various classifications. For example, a fluidized-bed combustor may be approximated as a well-stirred reactor (zero-dimensional) and may be assumed to be well-mixed and kinetically controlled. An entrained-flow combustor may be considered as a plug-flow reactor (one-dimensional), with solid kinetics and heat transfer controlling the reaction rate.

Table 1. Classification of Practical Coal Flames by Process Type[a]

Process Type	Fixed or Moving Bed	Fluidized Bed	Suspended Bed
Coal size, μm	10,000–50,000[b]	1,500–6,000[b]	1–100[b]
Bed porosity, %	low	95–99[b]	very high
Operating temperature, K	<2,000	1,000–1,400[b]	1,900–2,000[c]
Residence time, s	500–5,000	10–500	<1
Coal feed rate, kg/h	up to 40,000[c]	up to 40,000[b]	up to 120,000[c]
Distinguishing features	countercurrent or cross-flow	particle–particle interaction important, much of heat transfer due to conduction	very small particles, high rates
Advantages	established technology, low grinding, simple	low SO_x and NO_x pollutants, less slagging, less corrosion, low grade fuel possible[b]	high efficiency, large-scale possibilities, high capacity
Disadvantages	emissions, especially particulates,[b] less efficient than other methods	new technology, feeding fuel[b] tube erosion	high NO_x, fly ash, pulverizing expensive
Key operational variables	fuel feed rate, air-flow rate	bed temperature, pressure, fluidizing velocity	air : fuel ratio
Commercial operations	stokers	industrial boilers	pulverizing coal furnaces and boilers

[a] From Ref. 4.
[b] Data from Ref. 1.
[c] Data from Ref. 10.

Practical Process Characteristics. Direct coal combustion processes are most frequently classified according to bed type: fixed or moving beds (large particles), fluidized beds (intermediate-size particles), and suspended or entrained beds (small particles). Coal stokers are an example of the first; fluidized-bed combustors, the second; and pulverized coal furnaces, the third. It is clear that particle size is a major distinguishing characteristic among these types of processes, with the coal size varying from about 10 μm to 5 cm, depending on process type. Flame temperature also varies with the fluidized bed operating at uniquely lower temperatures (1100 to 1200 K) compared with fixed and entrained-bed systems. These predominant differences have a marked impact on residence times of the coal particle in the bed, which vary from less than a second to more than an hour. This observation is well illustrated by measurements and predictions shown in Figure 1 (11), where burning times for particles from 100 to 5000 μm vary from 100 ms to 500 s. Table 3 summarizes some of the characteristics of these coal processes, including extent of use, scale size, and the coal particle sizes employed.

SCIENCE OF COAL COMBUSTION

During combustion of coal particles in these various types of equipment, several reaction processes take place (Fig. 2) (17). These processes include

1. Heating up of the coal particle.
2. Loss of coal moisture by evaporation.
3. Loss of coal volatile matter, referred to as devolatilization (18), as the organic structure of the coal breaks up.

Table 2. Classification of Practical Coal Flames by Flow Type[a]

Flow Type	Process Application	Typical Flame Type and Control	Particle Size
Perfectly stirred reactor	fluidized bed	well-mixed solids, kinetically controlled	intermediate, 500–1000 μm
Plug flow	moving beds, steady fixed beds, shale retort	solid reaction and heat-transfer control	large, 1–5 cm
Plug flow	pulverized coal furnace, entrained gasifier	mixing specified, solid reaction and heat-transfer control	small, 10–100 μm
Plug flow	coal mine explosions, coal process explosions, flame ignition and stability	premixed flame, kinetic and diffusion control	small, 10–100 μm
Recirculating flow	power generators, entrained gasifiers, industrial furnaces	diffusion flames complex control (mixing, gas kinetics, solid kinetics)	small, 10–100 μm

[a] From Ref. 4.

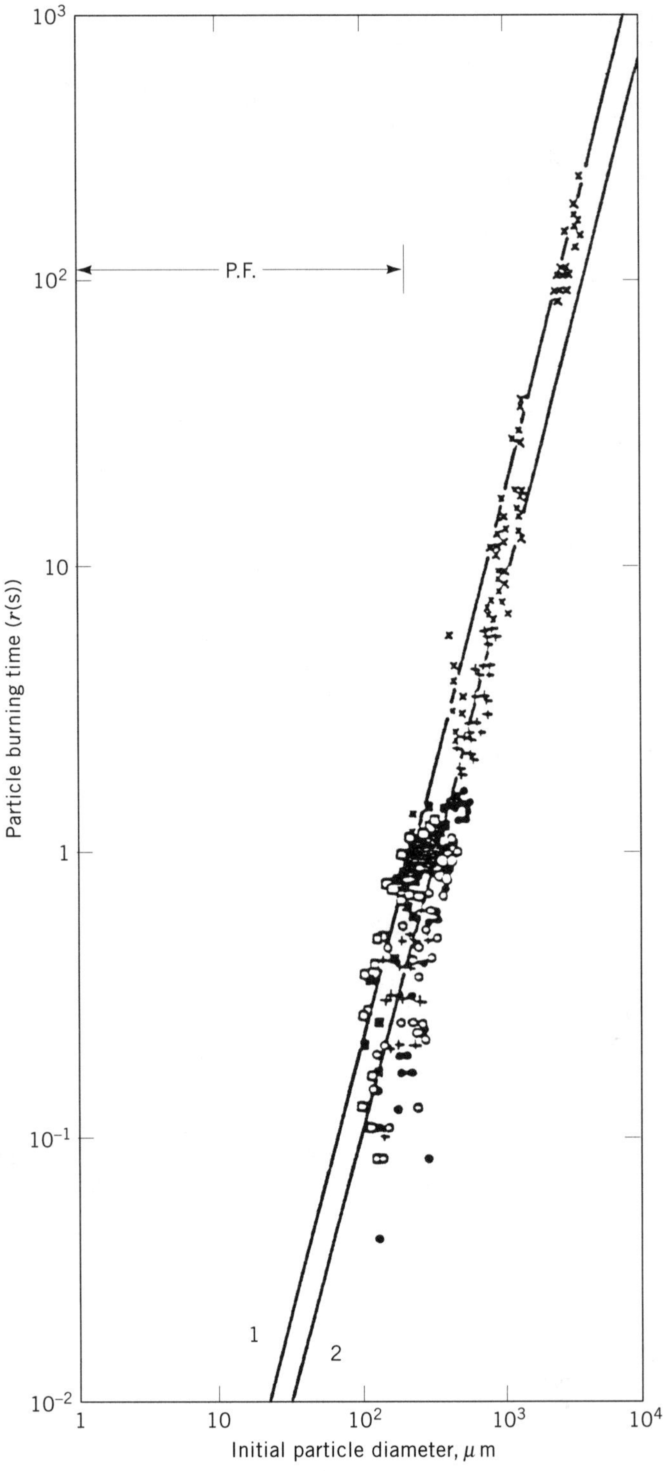

Figure 1. Comparison of theoretical and experimental burning times in air for various particle sizes. ○, gas coal coke (1.013×10^5 Pa); △, anthracite (1.013×10^5 Pa); □, electrode carbon (1.013×10^5 Pa); ●, gas coal coke (3.039×10^5 Pa); ▲, anthracite (3.039×10^5 Pa); ■, electrode carbon (3.039×10^5 Pa); ⦶, gas coal coke (4.559×10^5 Pa); ⧍, anthracite (4.559×10^5 Pa); ⧄, electrode carbon (4.559×10^5 Pa); +, lignite char (1.013×10^5 Pa); ×, bituminous coal char (1.013×10^5 Pa); ◇, graphite (1.013×10^5 Pa). Bituminous coal char calculated from given particle weight, assuming cubic particle of density 1 g/cm³. Graphite was cloud of particles burning in 160% excess oxygen. From Ref. 11.

4. Ignition and burning of the volatile matter with oxygen, usually in air.
5. Ignition of the residual carbonaceous material left after devolatilization, referred to as char.
6. Combustion of the char by oxygen, usually on the char surface, and termed heterogeneous combustion.
7. Breaking up and consumption of the residual char in the last phases of burnout.

These processes can occur at various rates and in sequence or at the same time, depending on coal size, heating rate, temperature, pressure, oxygen concentration, and coal structure. These various fundamental processes are discussed briefly here (see also COAL GASIFICATION; COMBUSTION MODELING).

Coal Composition and Structure

Organic Structure. The chemical properties of coal, including details of proximate and ultimate analyses, plastic properties of coal, coal hydrogenation, solvent extraction of coal components, properties of minerals, and coal chemical structure are discussed elsewhere. A brief treatment related to coal combustion follows.

A recent model for the chemical organic structure of coal is illustrated in Figure 3, based on information from infrared measurements, nuclear magnetic resonance, ultimate and proximate analyses, and pyrolysis data (19). Mineral matter can exist as occlusions within the base organic structure. This is not the specific structure of any particular coal but a model for interpreting the correlating coal conversion data. Infrared (Fourier transform) measurements provide quantitative concentrations for the following constituents in the raw coal or its components: hydroxyl, aliphatic or hydroaromatic hydrogen, aromatic hydrogen, aliphatic carbon, and aromatic carbon. From this and other information, the infrared spectrum of the parent coal or its constituents can be synthesized from the basic organic components (19).

Presently, methods for design and analysis of existing coal conversion processes do not make significant use of this kind of information on the structure of coals. However, with increased interest in the use of coal in the United States, more attention is being given on the relationship of coal structure to coal reaction and conversion processes (20).

Inorganic Structure. Through deposition, inclusion, and coal formation processes, inorganic matter is incorporated into the coal structure as cations and discrete mineral grains (21,22) (Fig. 4). The primary mineral groups that are found in all coals consist of clay minerals, carbonates, sulfides, oxides, and quartz. The specific types of inorganic components present depend on the rank of the coal and the environment in which the coal was formed (22). Inorganic mineral matter present in coal is of substantial concern in the combustion behavior of coal and to environmental and pollution; deposition, slagging, and corrosion; and catalytic effects (23,24). The fate of the inorganic con-

Table 3. Summary of Selected Coal Combustion Processes[a]

Process Type	Description	Coal Use in the United States (% total)	Commercial Use	Scale Size (t/day)	Coal Types	Coal Size (mm)
Power station	commercial electrical	78[b]–80[c]				
Pulverized	rapid burning of finely grained coal		common	907–9,072[d]	all	0.001–0.100[d]
Fluidized bed[e]	well-stirred combustion		pilot	1,1814–7,258[d]	all	1.5–6.0[d]
Stoker	mechanically fed fixed bed		small	91[d]	Noncaking	10–50[d]
MHD[f]	combustion energy capture by magnetic fields		laboratory	726–3,269[g]		
Coal–oil mixture	burning coal–oil mixtures in oil furnaces		demonstration			
Industrial	industrial plant power	8–11				
Heat–Steam						
Pulverized			small	0.9–91		
Fluidized bed			pilot	0.9–91	all	
Stoker			common	0.9–91		
Coal–oil mixture			demonstration	0.9–91		
Domestic–commercial	hand-stoked space heating	1		0.0045–0.045	Noncaking	30–100
Transportation	fuel for railroads	0.01–0.02			Noncaking	

[a] From Ref. 12.
[b] Data from Ref. 13.
[c] Data from Ref. 14.
[d] Data from Ref. 15.
[e] Adaptable, exhibits high heat transfer in bed as well as low level pollutant products.
[f] Magnetohydrodynamic
[g] Data from Ref. 16.

stituents during coal preparation and combustion largely depends on how the inorganic compounds are associated in the coal (25).

The methods used to determine the association of inorganic components in coals have evolved substantially over the past 80 yr. Standard methods typically involve concentration of inorganic components for chemical analysis by ashing or gravity-separation techniques. Recently, more advanced methods such as computer-controlled scanning electron microscopy (ccsem) and chemical fractionation are being used to determine more quantitatively the abundance, size, and association of inorganic components in coals. Example results of ccsem analysis of mineral matter from several coals are shown in Table 4 (22). A more detailed discussion on coal mineral matter can be found elsewhere (see COMBUSTION ASH).

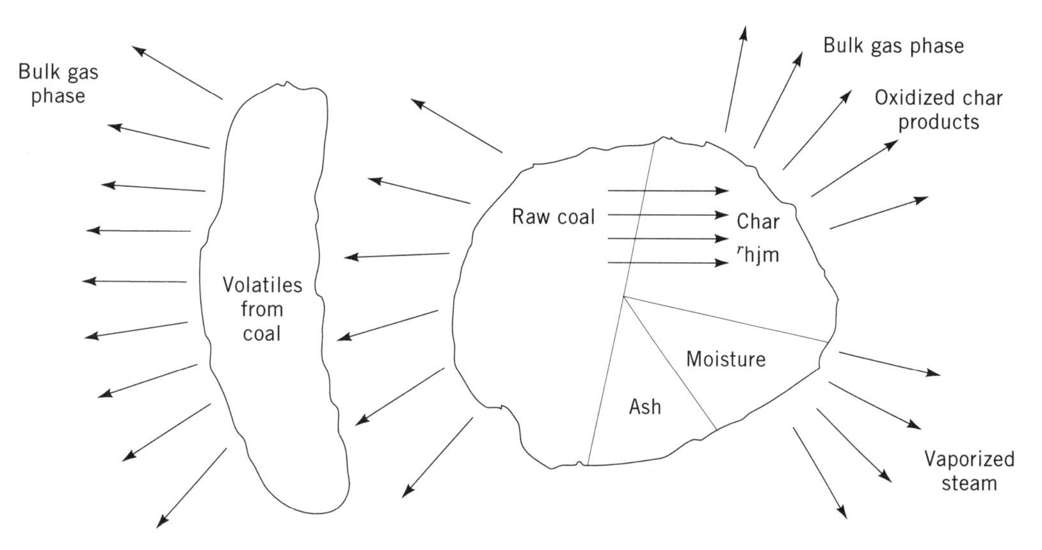

Figure 2. Schematic of burning coal particle. From Ref. 17.

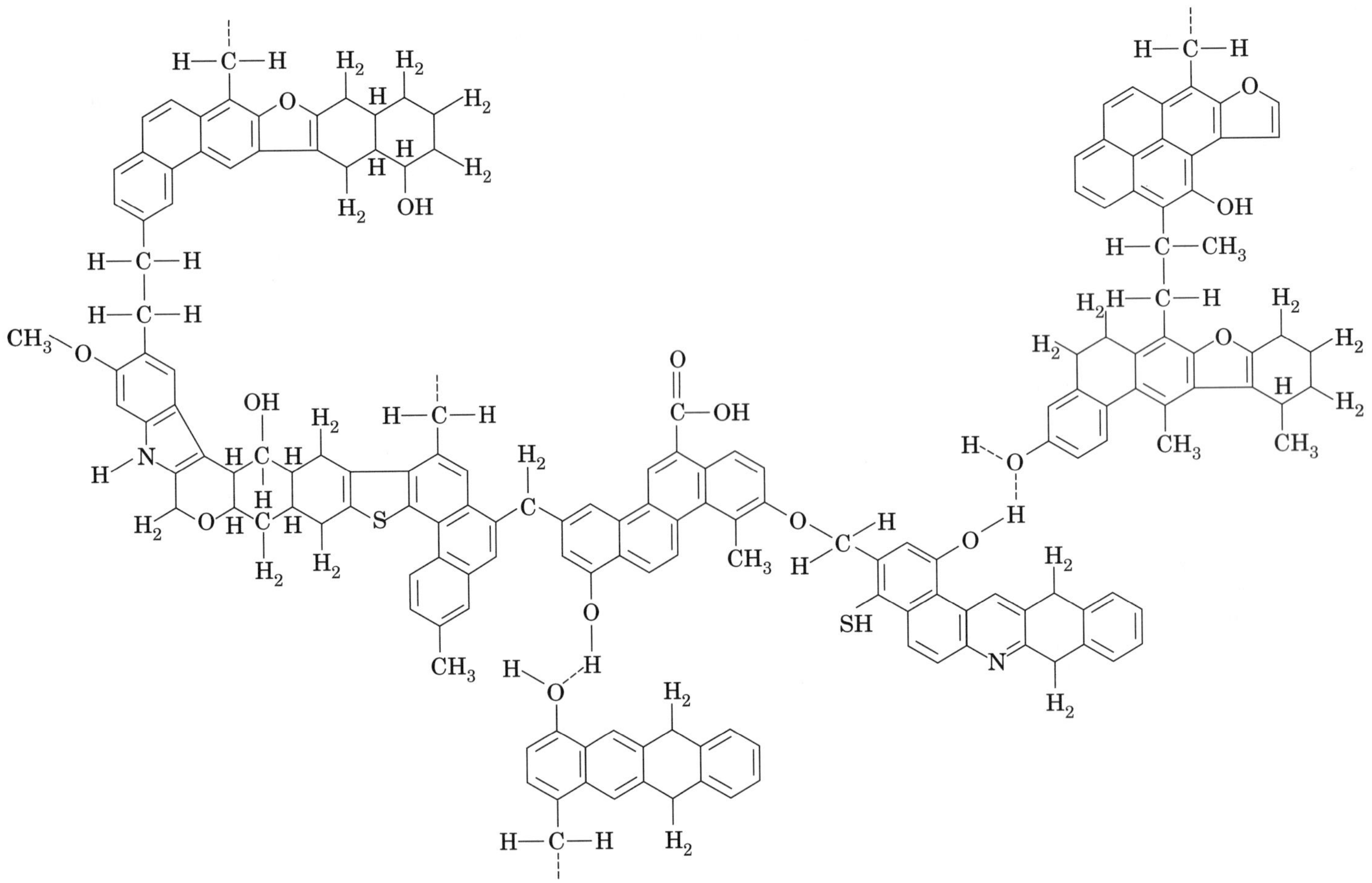

Figure 3. Representation of the structure of a coal molecule. From Ref. 19.

Coal Ignition

For combustion to occur, a coal particle must first ignite. Ignition characteristics of coal strongly depend on the way the coal particles are arranged or configured. Three arrangements can be identified as (*1*) single coal particles, (*2*) coal piles or layers, and (*3*) coal clouds. For single particles, no interaction among particles occurs. This arrangement provides basic data and may provide practical insight in dilute coal flames. Coal piles or layers occur in moving beds, fixed beds, coal storage piles, and dust lay-

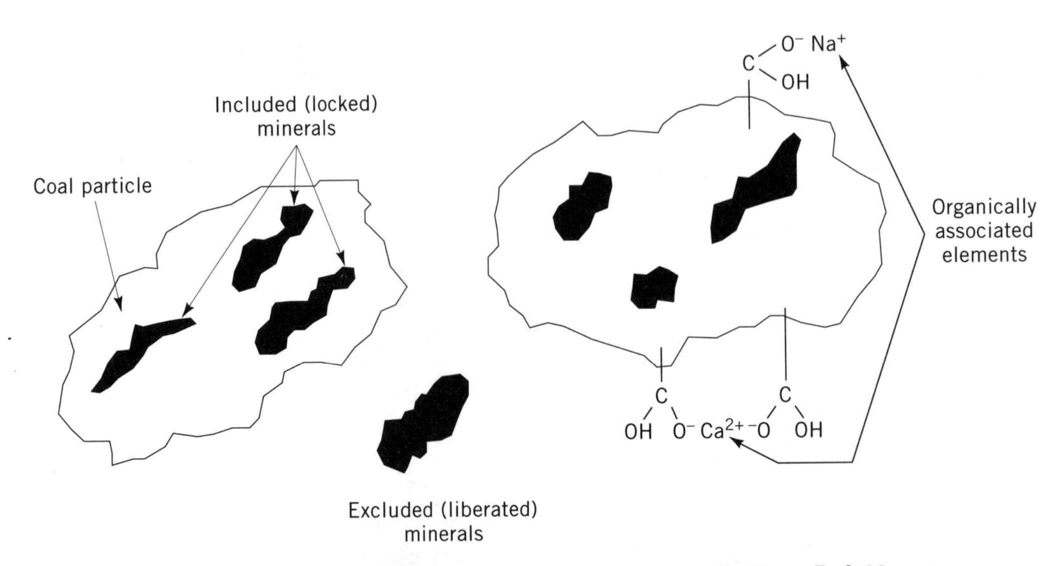

Figure 4. Coal and associated inorganic compounds. From Ref. 22.

Table 4. Quantitative Mineral Content Data for Selected Coals by CCSEM Analysis[a,b]

	Upper Freeport (MVB)		Wyodak (SubC)		Illinois No. 6 (HVCB)		Pittsburgh (HVAB)		Pocahontas No. 3 (LVB)		Blind Canyon (HVBB)	
Mineral	Total wt % (mineral)	Total wt % (coal)	Total wt % (mineral)	Total wt % (coal)	Total wt % (mineral)	Total wt % (coal)	Total wt % (mineral)	Total wt % (coal)	Total wt % (mineral)	Total wt % (coal)	Total wt % (mineral)	Total wt % (coal)
Quartz	15.65	2.24	14.13	0.94	10.91	1.91	14.74	1.50	7.86	0.43	14.91	0.91
Iron oxide	1.52	0.22	1.55	0.10	0.30	0.05	0.67	0.07	22.57	1.22	4.03	0.25
Aluminoscilicate	12.05	1.72	15.92	1.05	9.03	1.58	25.42	2.59	16.81	0.91	22.45	1.37
Ca-Aluminoscilicate	0.22	0.03	3.41	0.23	0.04	0.01	0.22	0.02	0.40	0.02	0.29	0.02
Fe-Aluminoscilicate	2.36	0.34	0.95	0.06	2.61	0.46	1.24	0.13	11.52	0.62	1.34	0.08
K-Aluminoscilicate	31.79	4.54	17.92	1.19	17.21	3.01	25.01	2.55	5.46	0.29	16.50	1.01
Ankerite	0.00	0.00	0.00	0.00	0.00	0.00	0.07	0.01	1.39	0.08	0.78	0.05
Pyrite	29.15	4.17	37.54	2.49	40.18	7.02	25.53	2.60	7.12	0.39	14.41	0.88
Gypsum	0.68	0.10	0.34	0.02	9.56	1.67	0.26	0.03	0.56	0.03	3.16	0.19
Barite	0.00	0.00	0.10	0.01	0.00	0.00	0.00	0.00	0.07	0.00	0.23	0.01
Gypsum–barite	0.00	0.00	1.10	0.07	0.00	0.00	0.00	0.00	0.09	0.01	0.00	0.00
Apatite	0.00	0.00	0.00	0.00	0.00	0.00	0.00	0.00	0.31	0.02	0.00	0.00
Ca-Silicate	0.00	0.00	0.02	0.00	0.00	0.00	0.00	0.00	0.81	0.04	0.20	0.01
Alumina	0.00	0.00	0.00	0.00	0.00	0.00	0.00	0.00	6.29	0.34	0.18	0.01
Aluminosil–gypsum	0.16	0.02	0.59	0.04	0.15	0.03	0.09	0.01	0.03	0.00	0.17	0.01
Ca-Aluminate	0.00	0.00	0.04	0.00	0.00	0.00	0.00	0.00	0.00	0.00	0.00	0.00
Calcite	2.98	0.43	0.31	0.02	5.29	0.93	0.82	0.08	12.21	0.66	7.76	0.47
Rutile	0.33	0.05	1.01	0.07	0.17	0.03	0.00	0.00	0.59	0.03	0.07	0.00
Dolomite	0.00	0.00	0.00	0.00	0.00	0.00	0.00	0.00	0.00	0.00	0.21	0.01
Ca-rich	0.00	0.00	0.00	0.00	0.00	0.00	0.07	0.01	0.12	0.01	1.22	0.07
Si-rich	0.17	0.02	1.05	0.07	0.48	0.08	0.39	0.04	2.70	0.15	5.87	0.36
Periclase	0.00	0.00	0.00	0.00	0.00	0.00	0.00	0.00	0.00	0.00	0.08	0.01
Unknown	2.96	0.42	4.00	0.26	4.08	0.71	5.49	0.56	3.09	0.17	6.14	0.37

[a] From Ref. 22.
[b] Mineral content in a particle-size distribution is available (26).

ers. Coal clouds exist in pulverized coal combustors, furnaces, entrained gasifiers, coal mine dust explosions, and in fluidized-bed systems (more dense in the latter case). Coal ignition characteristics in these various configurations differ substantially.

Results of various typical ignition or onset temperatures are summarized in Table 5 (27,30). Coal onset temperatures vary from that of coal pile self-heating, which occurs at near ambient temperatures, to single-particle ignition temperatures as high as 1200 K. The onset temperatures are functions of coal type, size, and condition as well as experimental method and environment. These representative values emphasize the wide variability in onset temperatures. Chemical reactions proceed at some rate over a wide range of temperatures. It takes several hours at low temperature for a coal pile to ignite, whereas a coal dust cloud can be ignited in milliseconds at high temperature.

No single definition of ignition for coal seems appropriate. In general terms, ignition can be described as a process of achieving a continuing reaction of fuel and oxidizer. Ignition is most often identified by a visible flame. However, reactions can proceed slowly at low temperatures, without a visible flame. Ignition is sometimes said to occur when the rate of heat generation of a volume of combustibles exceeds the rate of heat loss. Ignition is more complex in condensed phases and particularly in heterogeneous solids such as coal in which particles can react slowly by oxygen attack on the surface, leading to spontaneous ignition.

In either inert or reactive hot gases, the coal will react internally, softening and devolatilizing as the particle increases in temperature. In the process, gases and tars are released. Although no flame is visible (and this process is not called ignition), the onset of this thermal decomposition is an ignitionlike process. These off-gases and tars can also be ignited in a surrounding oxidizer. This ignition process may involve only gases or also the tars. The remaining char can be ignited in oxygen by surface reaction processes. The reactants could also be CO_2, H_2O, or H_2, and a visible flame may not result, yet coal reactions continue.

Ignition is often characterized by the time required to achieve a certain temperature, rate of temperature increase, a visible flame, or a certain consumption of fuel for a specified set of conditions. However, no unique ignition time exists either. Potentially important variables that influence coal ignition temperature and time include coal type, system pressure, volatiles content, gas composition, particle size, coal moisture content, size distribution, residence time, gas temperature, coal concentration or quantity, surface temperature, gas velocity, mineral matter percentage, and coal aging since grinding. Variables that dominate the ignition process depend strongly on the configuration of the coal particles, as do the range of ignition temperatures and times (4,31).

Table 5. Typical Ignition (or Onset) Temperatures for Various Coal Reaction Processes[a]

Coal Process	Time	Typical Ignition Temperature Range (K)	Experimental Methods	Reference
Coal pile self-heating (ambient conditions)	hours	303–378	adiabatic calorimeter (air flow 50 cm^3/min)	27
Coal pile auto ignition[b]	minutes	443–500	samples in isothermal electrically preheated cylindrical (8.0 cm ID) oven	28
Coal pile thermal ignition[c]	minutes	473–773	coal layer on isothermal hot plate	29
Coal devolatilization	milliseconds	573–773	several techniques used with various heating rates and residence times	30
Pulverized coal cloud ignition	seconds	673–1073	coal dust injected into isothermal oven that is open at bottom	29
Single-particle ignition	seconds	1070–1173	few particles on tip of platinum wire in isothermal lab furnace	11

[a] From Ref. 4.
[b] Coal pile auto ignition occurs when the coal sample is placed in an environment at some temperature below the ignition temperature, and the coal self-heats to the ignition temperature.
[c] Coal pile thermal ignition occurs when the coal is placed in an environment at some temperature above the ignition temperature, and the coal is externally heated to the ignition temperature.

Coal Devolatilization

Coal devolatilization occurs as the raw coal is heated in an inert or oxidizing environment; the particle may soften (become plastic) and undergo internal transformation. Moisture present in the coal will evolve early as the temperature rises. As the temperature continues to increase, gases and heavy tarry substances are emitted. The extent of this "pyrolysis" can vary from a small proportion up to 80% of the total particle weight and can take place in a few milliseconds or several minutes, depending on coal size and type and on temperature. These released volatile substances can account for up to 50% of the heating value of the coal (32). The devolatilization of large coal particles is significantly different from pulverized coal because of the dominant influence of heat and mass transfer effects. Furthermore, coals of different rank exhibit wide variations in coal devolatilization behavior. The ultimate volatiles yields have been shown to be similar at about 50 wt % for coals through the high volatile bituminous rank, then diminish in coal of higher rank. However, the proportions of gases and tars vary widely, with gases dominating the yields of low rank coals (Fig. 5) (20). Low rank coals exhibit high gas yields and low tar yields, high volatile bituminous coals exhibit high tar yields and moderate gas yields, and high rank coals exhibit moderate or low tar yields and low gas yields (33).

Coals of medium rank exhibit a plastic or fluid behavior when heated. This fluidity normally occurs in coals with 81–92% carbon (daf) but depends on oxygen and hydrogen content as well as heating rate. Plastic properties become more pronounced at higher heating rates up to a point; if the heating rates become too high, coals cannot fluidize because temperatures required for certain cross-linking reactions to occur are reached before the coal structure can relax and fluidize. As coal is heated and becomes fluid, the pores fuse. The subsequent formation of light gas and tar vapor-filled bubbles results in coal particle swelling. Volatiles are transported via bubble formation. However, low rank and high rank coals usually exhibit little fluidity and retain their pore structure during devolatilization. Volatiles are transported by diffusion via pore structures (12).

Five principal devolatilization phases have been identified (34). The first phase is associated with moisture evolution at the boiling point of water. The second phase is associated with the large initial evolution of carbon dioxide and a small amount of tar which begins at about 700 K. The third phase is the evolution of chemically formed water and carbon dioxide in the range from 800 to 1000 K. The fourth phase involves the final evolution of tar, hydrogen, and hydrocarbon gases in the temperature range from 1000 to 1200 K. The fifth phase is the formation of carbon oxides at high temperatures. Structural properties of the coal strongly affect devolatilization behavior. The extent of devolatilization depends on such fac-

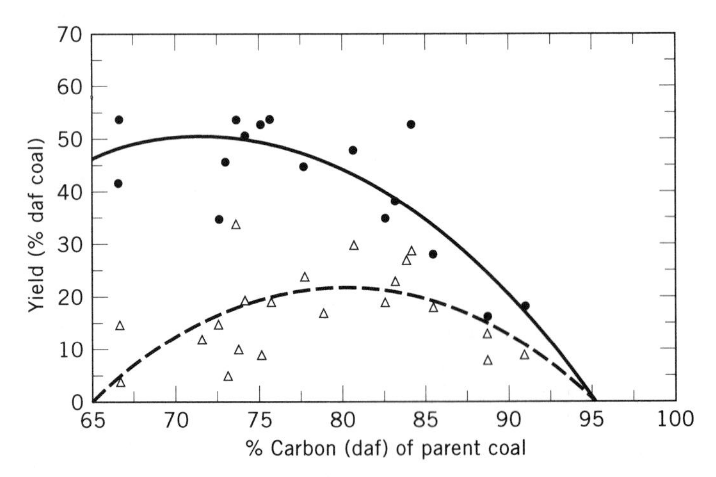

Figure 5. Measured tar and total volatiles yields for a wide range of coals, using carbon content to illustrate coal rank. Solid and dashed lines represent correlation of the total volatiles and tar yields, respectively. From Ref. 20.

tors as peak temperature, heating rate, pressure, and particle size (35,36). However, due to the complexity of coal, a consensus of the mechanisms of devolatilization is only beginning to emerge (20,37).

Coal Volatiles Ignition and Combustion

A variety of products is produced during devolatilization of coal, including tars and hydrocarbon liquids, hydrocarbon gases, CO_2, CO, H_2, H_2O, and HCN (38). These products react with oxygen in the vicinity of the char particles, increasing temperature and depleting oxygen. This complex reaction process is important to the control of nitrogen oxides, formation of soot, stability of coal flames, and ignition of char (39). Some aspects of this include the following:

1. Release of volatiles from coal or char.
2. Condensation and repolymerization of tars in char pores.
3. Evolution of a hydrocarbon cloud through small pores from the moving particle.
4. Cracking of the hydrocarbons to smaller hydrocarbon fragments, with local production of soot.
5. Condensation of gaseous hydrocarbons and agglomeration of sooty particles.
6. Macromixing of the devolatilizing coal particles with oxygen.
7. Micromixing of the volatiles cloud and the oxygen.
8. Oxidation of the gaseous species to combustion products.
9. Production of nitrogen oxides and sulfur oxides by reaction of devolatilized products and oxygen.
10. Heat transfer from the reacting fluids to the char particles.

At high combustion temperatures, up to 70% or more of the reactive coal mass is consumed through this process.

Volatiles combustion in practical systems is complicated by turbulent mixing of fuel and oxidizer, soot formation and radiation, and near-burner fluid dynamics. Such systems usually exhibit volatiles cloud combustion, rather than single particle combustion, making analysis difficult (20,40). One study determined that the combustion rates of the volatiles from coals of different rank do not vary markedly. This was attributed to the documented, similar combustion rates of the common hydrocarbons involved (41). Discussions of the combustion of coal volatiles are available (39,42).

Char Ignition and Oxidation

The residual mass from the devolatilization process (enriched in carbon and depleted in oxygen and hydrogen and still containing some nitrogen, sulfur, and most of the mineral matter) is referred to as char. Char particles are often spherical, have many cracks or holes made by escaping gases, may have swelled to a larger size, and can be porous. The nature of the char depends on the original coal type and size and also on the conditions of pyrolysis. The residual char particles can be oxidized or burned away by direct contact with oxygen at sufficiently high temperature. This reaction of the char and oxygen is thought to occur on the surface of the char, with gaseous oxygen diffusing to and into the particle, adsorbing, and then reacting. This heterogeneous process is often much slower than the devolatilization process, requiring seconds for small particles to several minutes or more for larger particles. These rates vary with coal type, temperature, pressure, char characteristics (size and surface area), and oxidizer concentration. Other reactants, including steam, CO_2, and H_2, will also react with char, but rates with these reactants are considerably slower than for oxygen (39,43). The two processes (ie, devolatilization and char oxidation) may take place simultaneously, especially at high heating rates. If devolatilization takes place in an oxidizing environment (eg, air), then the fuel-rich gaseous and tarry products react further in the gas phase to produce higher temperatures in the vicinity of the coal particles.

The time required for consumption of a char particle is a significant part of the coal reaction process; it can range from a few milliseconds to more than an hour. Heterogeneous reactions are complicated by

1. Coal structural variations.
2. Diffusion of reactants.
3. Reaction by various reactants (O_2, H_2O, CO_2, and H_2).
4. Particle size effects.
5. Pore diffusion.
6. Char mineral content.
7. Changes in surface area.
8. Fracturing of the char.
9. Variations with temperature and pressure.
10. Moisture content of the virgin coal.

The main process in coal combustion is heterogeneous chemical reaction of carbon from coal with oxygen from air to produce carbon dioxide and carbon monoxide.

$$C + O_2 \rightarrow CO_2 \quad \Delta H^\circ_R = 1110.5 \text{ mJ/Kmole} \tag{1}$$

$$C + \tfrac{1}{2} O_2 \rightarrow CO \quad \Delta H^\circ_R = 790.4 \text{ mJ/Kmole} \tag{2}$$

These reactions are highly exothermic (ie, heat releasing) and are the principal sources of the production of heat from coal burning.

The rate-limiting step in the oxidation of char can be chemical (adsorption of the reactant, reaction, and desorption of products) or gaseous diffusion (bulk phase or pore diffusion of reactants or products). The existence of different temperature zones or regimes that determine the controlling resistance have been postulated (44,45). In zone I, which occurs at low temperatures or for small particles, chemical reaction of carbon and oxygen is the rate-determining step. Zone II is characterized by control as a result of both chemical reaction and pore diffusion. Zone III, which occurs at high temperatures, is characterized by mass-transfer limitations outside the particle. Figure 6 illustrates these zones graphically and shows the theoretical dependence of the reaction rate on particle diame-

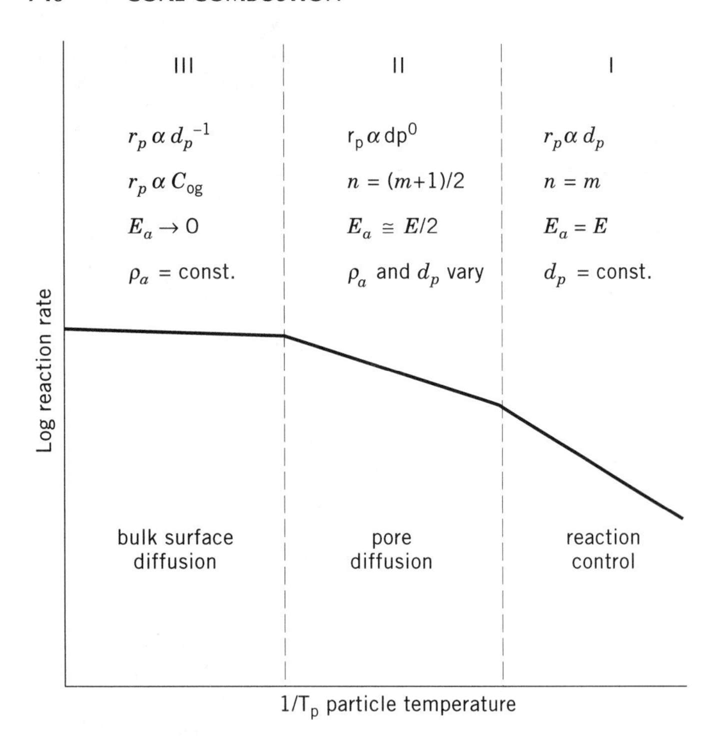

Figure 6. Rate-controlling regimes for heterogeneous char oxidation. From Refs. 44 and 46.

ter and oxidizer concentration (44,46). Kinetic data must be interpreted in light of the conditions under which the data are obtained. To obtain intrinsic rate data, various schemes have been tried to minimize the effects of diffusion, including efforts to conduct tests at low pressure or low temperature, microsampling of gases close to the solid surface, use of high velocity gas streams to decrease the boundary layer thickness, and use of small particles and less active oxidizers (42,43).

The rates of carbon oxidation by steam and carbon dioxide are of the same order of magnitude and are much lower than the reaction of carbon with oxygen, whereas the hydrogenation reaction (ie, H_2 + char) is several orders of magnitude slower than the steam–char and the CO_2–char reactions. Relative rates of the reactions have been estimated and confirmed at 1073 K and 10 kPa as being about 3×10^{-3} for H_2, 1 for CO_2, 3 for H_2O, and 10^5 for O_2 (43,44). At higher temperatures, the differences in rates will not be as extreme, but rates for H_2, CO_2, and H_2O will still remain relatively unimportant as long as comparable concentrations of oxygen are present. Often, only the heterogeneous carbon oxidation reaction with oxygen needs to be considered for combustors (47).

In examining rate data from various sources, it is apparent that temperature and reactant concentration are not the only variables that influence reaction rates. The composition and structure of the coal, its pretreatment, and its thermal history are also important. Some experiments have used pure forms of carbon, while others have used char (48). In general, it has been found that the more pure forms of carbon are less reactive than the chars. Also chars from the lower rank coals are more reactive than those from higher rank coals. Differences in reactivity at different heating rates and temperatures have been noted in chars formed from the same parent coal, but some of the differences in reactivity may be due to additional volatile reactions from the low temperature chars, which were not totally pyrolyzed (20).

The internal and external surface areas of the char are also important. Under some conditions, reactant gases diffuse into the particles before reacting. An observed increase in reaction rate with increasing burnoff, followed by a subsequent decline in rate, has been explained in terms of expansion of the pores, which exposes more and more internal area (49). The surface area continues to increase up to a certain point, then due to the coalescence of pores and thermal annealing the area and reaction rate decrease. Another factor that may affect the oxidation rates is the presence of catalytic impurities in the char (4,48).

Ash Formation and Deposition

As the volatiles and char portions of the coal are consumed during combustion, the remaining solid particles become more and more concentrated in inorganic material. Upon completion of combustion, residual, finely divided, inorganic material called fly ash remains. Deposition, fouling, and slagging occur when ash accumulates on heat transfer surfaces within the furnace. Deposits that form in the flame zone of a furnace are called slag deposits. Deposits that form on surfaces removed from the flame zone, such as in the convective pass of a utility boiler, are called fouling deposits. Slag deposits are exposed to radiation from the flame and are usually highly molten. Fouling deposits contain low levels of molten material that bind solid particles together (22).

The accumulation of ash particles depends on the ability of the inorganic material to arrive at the heat-transfer surface and to form strong bonds with the surface. The extent of ash-related problems depends on the quantity and association of inorganic constituents in the coal, the combustion conditions, and the system geometry. Once a strongly bonded surface has accumulated, the temperature of the surface increases, leading to more efficient collection of impacting ash particles. Both fouling and slagging deposits are significant to combustion systems because they can be associated with severe reduction of heat transfer from the flame and hot gases to steam tubes, reducing the efficiency of utility boilers (22).

Pollutants from Coal Combustion

Pollutants. By far the most striking problem associated with human consumption of coal is the control of air pollutants. The effects of air pollutants from localized regions impact not only extended geographical areas but the global atmosphere as well (50). The inexorable trend for increasing power demand, both by industrially established countries and developing nations, will inevitably lead to increased production and use of fossil fuels. In the wake of increasing petroleum costs, the substitution of coal for relatively cleaner oil and natural gas for power-generating and industrial boilers could continue to accentuate the air pollution problem.

Concern for the quality of the atmosphere is not a recent phenomenon but can be dated to several episodes when coal began to replace wood as the primary source of domestic heating and industrial uses. For example, the dangerous and fatal contamination of coal smoke was noted in the 17th century (51). American steel-producing and manufacturing communities also experienced deterioration of their air quality as a result of the Industrial Revolution. Despite these and other incidents, it was not until the mid-20th century that public support and government resolve were strong enough to regulate air pollutant emissions from stationary combustors.

The combustion of coal produces both primary and secondary pollutants. Primary pollutants include all species in the combustor exhaust gases that are considered contaminants to the environment. The major primary pollutants include CO, hydrocarbon fragments, sulfur-containing compounds (SO_x), nitrogen oxides (NO_x), particulate material, and various trace metals. Even CO_2 reached pollutant status with increasing concern for global atmospheric warming, called the greenhouse effect. Secondary pollutants are defined as environmentally detrimental species that are formed in the atmosphere from precursor combustion emissions. The list of secondary pollutants includes particulate matter and aerosols that accumulate in the size range of 0.1 to 10 μm dia, NO_2, O_3, other photochemical oxidants, and acid vapors. The connection between combustion-generated pollutants and airborne toxins, acid rain, visibility degradation, the greenhouse effect, and stratospheric ozone depletion is well established (see ACID RAIN; NO_x; STACK GAS CLEANUP). The detrimental impact of these contaminants on ecosystems in the biosphere and stratosphere has been the impetus of stricter standards around the world.

Studies have shown that anthropogenic emissions of sulfur-containing pollutants decreased in the United States during the 1970s and early 1980s, while nitrogen oxides continued to increase at a comparable level. In 1985, stationary fuel combustion in electric utilities and industry accounted for approximately 50% of all NO_x emissions and more than 90% of all SO_x emissions. The increase in NO_x emissions can be correlated with the construction of many coal-fired power-generating facilities throughout the U.S. Midwest and West. Coal naturally contains various percentages of sulfur and nitrogen, depending on its primordial origin. Typically, the nitrogen content ranges from 0.5 to 1.5 wt %, while sulfur typically varies from 0.5 to 10 wt %, 1 to 2 wt % being more common.

The disproportionate increase in NO_x emissions occurred as a result of the installation of flue gas desulfurization equipment, while less emphasis was placed on the control of nitrogen oxides. In part, nitrogen oxide technologies were not required on new combustion facilities until the mid-1970s. Strict regulations now require that existing stationary combustion sources be retrofitted with such technologies as advanced low NO_x burners, while new combustion sources are required to implement state-of-the-art or best available NO_x abatement strategies. The current goal is to achieve the lowest emission rate possible on a case-by-case basis.

Health and Ecological Effects of Combustion-generated Pollutants. The emissions of nitrogen and sulfur oxide pollutants negatively impact the environment in a number of ways (see ACID RAIN; NO_x). Exposure to concentrations ranging from 0.01 ppm for a 1-yr period to only a few parts per million over 30 s will affect the health of people, particularly through the respiratory and cardiovascular systems (52,53). The formation of sulfate and nitrate particles by gas-to-particulate conversion in the atmosphere results in small diameter (<2.5 μm) particles that fall in the respirable size range. These particles can reach the alveolar pulmonary processes, which are not coated with a protective mucous layer, where gas exchange occurs. There, the particle clearance time is greater than in the upper respiratory tract; hence the potential for adverse health effects is increased by breathing smaller particles (54). Many of the photochemical oxidants associated with NO_x and SO_x pollutants are also implicated as serious eye irritants.

Nitrogen and sulfur oxides are the key causes of acid rain. The increased acidity of rain, snow, and lakes throughout Central Europe and the U.S. Northeast appears to be directly related to the emissions of sulfur and nitrogen oxides from downwind regions where power utilities are located. The detrimental effects of acid rain deposition on the ecosystem in these regions is well documented (55). Acid rain also is responsible for rapid deterioration of historic buildings and monuments. A comprehensive treatment of acid rain formation and deposition has been given (56). The full extent of the contribution of nitrogen oxides to acid rain is not yet fully understood, although there is increasing evidence that it predominates in the U.S. West and may be responsible for as much as 30% in all industrialized areas (57).

Recent studies also have implicated nitrogen oxides and sulfur oxides in the decline in the growth and vitality of forests in Central Europe and North America that could not be attributed to natural selection and climatic perturbations (57,58) (see ACID RAIN). One hypothesis is that acidic mists, perhaps in combination with atmospheric oxidants, damage cell membranes in leaves and needles, thus disrupting photosynthesis. Another theory is that trees do not properly harden for winter and consequently suffer frost and desiccation damage. This results because the trees receive an excess amount of nitrogen at the end of the growing season through acid deposition to leaf and needle surfaces. Although such theories lack substantial evidence, most investigators believe that multiple stresses may be attributed directly or indirectly to pollutants from combustion.

Nitrogen oxides can lead to an increased production of ozone (O_3) in the troposphere (59). Ozone can harm vegetation and structural materials. The mechanism of unnatural O_3 production is complicated and requires the presence of hydroxyl radicals and hydrocarbons. When these constituents are available in the atmosphere, NO is partially converted to NO_2 and photochemical compounds, such as peroxyacetyl nitrate, without being oxidized by O_3. Solar radiation then decomposes the NO_2 to NO, releasing an oxygen atom which further reacts with molecular oxygen to produce abnormally high concentrations of O_3.

Nitrogen oxides also have a possible connection with global atmospheric warming and with stratospheric ozone depletion. Nitrous oxide (N_2O), like carbon dioxide, water, and methane, traps earth-emitted infrared or heat energy and also leads to the chemical destruction of stratospheric ozone (60). A single N_2O molecule has a greenhouse warming potential of 250 CO_2 molecules. Nitrous oxide has an atmospheric lifetime of 150 yr and is currently increasing in concentration at a rate of 0.2–0.3%/yr.

The principal constituent of fossil fuels, including all ranks of coal, is carbon. Incomplete combustion results in unburnt hydrocarbon fragments and CO emissions; both are regulated by most developed nations. Although CO emission from electric power utilities contribute to less than 1% of the total anthropogenic emissions, they can have a significant impact on local CO concentrations. Carbon monoxide appears to have no detrimental effects on material surfaces, although it can have a harmful effect on higher life forms exposed to a few hundred parts per million for short periods of time. Hemoglobin has an affinity for CO that is approximately 210 times its affinity for oxygen. There is also some evidence that CO is chemically active in smog formation.

Unburned hydrocarbon emissions also can have adverse effects on the health of humans and plants. Certain classes of hydrocarbon emissions have been identified as carcinogenic to humans. These carcinogens include polynuclear aromatic hydrocarbons contained in the soot and tars emitted from coal and oil-fired combustors. Lighter hydrocarbons classified as nonmethane hydrocarbons (NMHC) are also key precursors to photochemical oxidants in the atmosphere. While anthropogenic emissions of NMHC are only half the natural emission level of 65 million t/yr in the United States (56), their detrimental impact in urban areas is well established. A comprehensive review on the mechanisms of formation and the chemical basis of control strategy options for oxidant and gaseous airborne toxic chemicals is available (56).

Particulate matter composed of soot or condensed hydrocarbons also contributes to visibility degradation. Studies reveal that elemental carbon contributes from 33 to 50% of the total aerosol concentrations in the atmosphere downwind of stationary combustion sources (61,62). The remaining fraction is nitrate and sulfate aerosols. However, the carbon-bearing aerosols account for all (>95%) of the light absorption.

Until recently, neither H_2O nor CO_2 from combustion had been viewed as being detrimental to the atmosphere, because both exist in appreciable quantities in the atmosphere and are essential to sustain plant life and marine organisms. Now, a growing concern is that combustion-source CO_2 emissions in conjunction with global deforestation may have affected the observed global atmospheric warming trend (63). CO_2 concentrations have steadily climbed from 280 ppm a century ago to more than 350 ppm in the 1990s.

Since the mid-1800s, average global temperatures have risen 0.5 K. Natural gas creates less CO_2 per unit of energy produced, because it has a higher hydrogen:carbon ratio per Joule of chemical potential than either coal or fuel oil. The proposed connection between CO_2 and global warming has prompted the United States, as a part of its national energy strategy, to aim to reduce CO_2 emissions to 35% below 1987 levels by the year 2010 (64).

Coal also contains variable amounts of mineral matter, depending on its age and origin. Most naturally occurring elements can be found in coal, including trace amounts of some radionuclides. Silica, iron, and sulfur are the major mineral matter constituents, although alkaline and alkaline-earth metals and halides are also common. Mercury, nickel, arsenic, and fluorine are four toxic air pollutants that are liberated during coal combustion (65).

At high temperatures, the mineral matter in coal becomes molten and is partially or completely vaporized. Cooling in the aft-combustion section results in fly ash. Fortunately, technology has provided effective methods for removing and disposing of fly ash. Other chemicals additives that are used for makeup water or other purposes may also be emitted with the combustion exhaust. The fraction of material that is not removed from the flue gases is emitted into the atmosphere as particulate material (66). The fly-ash particulates may be removed in mechanical collectors, fabric filters, wet scrubbers, or electrostatic precipitators (ESP). The fabric filters and the electrostatic precipitators are the leading technologies, the electrostatic precipitator being the most commonly used today. In the electrostatic precipitator, the fly-ash particulates from flue gas are electrically charged, driven to collecting electrode plates by an electrical field, and then removed and disposed of.

NO_x Formation and Reduction. A basic understanding of nitrogen oxide formation during the combustion of fossil fuels has been achieved by years of research. Most of the nitrogen oxides emitted to the atmosphere by combustion systems is in the form of nitric oxide (NO), with smaller fractions appearing as nitrogen dioxide (NO_2) and nitrous oxide (N_2O). Other nitrogenous pollutants include ammonia, hydrogen cyanide, and amine compounds.

Nitric oxide is formed in flames predominantly by three mechanisms: thermal NO_x (the fixation of molecular nitrogen by oxygen atoms produced at high temperatures), fuel NO_x (the oxidation of nitrogen contained in the fuel during the combustion process), and prompt NO_x (the attack of a hydrocarbon free radical on molecular nitrogen producing NO_x precursors). N_2O can be formed by a number of paths in gaseous and coal-laden reactors and is more prevalent at lower temperatures. N_2O levels in coal combustors are generally less than 5 ppm (except for fluidized-bed systems) and are not related to the nitrogen oxide levels in the flue gas (60,67). Experimental studies show that NO_2 is significant only in combustion sources such as gas turbines (68). NO_2 often exists as a transient species that is quickly destroyed in the postflame region (69). Neither N_2O nor NO_2 reaches high concentrations in coal flames or in the unquenched effluent of most gas combustors. Hence, NO is normally the most significant nitrogen oxide species emitted from practical combustors.

It is generally agreed that the reaction network for thermal NO is described by the modified Zel'dovich mechanism (70).

$$N_2 + O \leftrightarrow NO + N \tag{3}$$

$$N + O_2 \leftrightarrow NO + O \tag{4}$$

$$N + OH \leftrightarrow NO + H \tag{5}$$

The first step is rate limiting and requires high temperatures to be effective due to the high activation energy barrier for reaction 3 (314 kJ mol^{-1}); hence, the designation *thermal* NO. The third step is important in fuel-rich environments (71).

Prompt NO occurs by the collision and fast reaction of hydrocarbons with molecular nitrogen in fuel-rich flames. This mechanism accounts for rates of NO formation in the primary zone of the flame that are much greater than the expected rates of formation predicted by the thermal NO mechanism alone. This mechanism is much more significant in fuel-rich hydrocarbon flames. Two such reactions are believed to be particularly significant:

$$CH + N_2 \rightarrow HCN + N \quad (6)$$

$$C + N_2 \rightarrow CN + N \quad (7)$$

The cyanide species is further oxidized to NO (69).

Fuel NO typically accounts for 75–95% of the total NO_x accumulation in coal flames (66). Nitrogen is primarily evolved from tars as HCN and from amines as NH_3 (72,73). The distribution of coal nitrogen between tar and light gases has been found to be similar to the distribution of coal mass between the same volatile phases. The conversion of fuel nitrogen to HCN is independent of the chemical nature of the initial fuel nitrogen and is not rate limiting (74,75). Once the fuel nitrogen has been converted to HCN, it rapidly decays to NH_i (i = 0,1,2,3), which reacts to form NO and N_2 as illustrated by the global reaction scheme:

$$\text{fuel—N} \longrightarrow HCN \longrightarrow NH_i \begin{cases} \xrightarrow{NO} N_2 \\ \xrightarrow{O_x} NO \end{cases} \quad (8)$$

Although concepts for the dominant network steps have been explored by a number of investigators, an exact determination of the complete mechanism is presently not available for generalized fuel types.

In fuel-rich combustion systems, NO can also be reduced by hydrocarbon radicals leading to the formation of HCN (76). Based on such observations, a NO_x abatement technology, termed *reburning,* has been developed by injecting and burning secondary fuel in the postburner flame to convert NO back to HCN. It is generally accepted that light hydrocarbon gases are the most effective staging fuels, although even pulverized coal has been shown to work (77).

In a typical coal flame, an appreciable amount of the coal nitrogen is retained by the char residue. Under certain circumstances, as much as half of the fuel nitrogen remains in the coal after devolatilization (78,79). Char nitrogen undergoes oxidation along with carbon as the char is burned out. NO formation during char reactions often accounts for 20–30% or more of the total NO formed (78,80).

Heterogeneous destruction of gas-phase NO has been determined to be significant during fuel-lean char burnout, but homogeneous decay reactions appear to dominate NO removal in fuel-rich zones of coal combustion (79,81). Soot formation and subsequent conversion of soot-nitrogen to nitrogen pollutants represents yet another way to produce NO_x from coal burning. The mechanism of NO destruction by soot particles is also well documented (82), with results indicating that NO removal rates on carbon black particles are much greater than heterogeneous destruction of NO by char.

Sulfur Oxides Formation and Capture. Sulfur oxides are formed in stationary combustors from the sulfur entering with the coal. The sulfur content of coal typically varies from 0.5 to 10 wt %, depending on rank and the environmental conditions during the coal formation process (see ACID RAIN; COAL QUALITY). Sulfur oxide emissions can be reduced substantially by simply converting from high to low sulfur coal. For certain coals, the sulfur content can be reduced by cleaning the coal before burning it.

There are three forms of sulfur in coal (83): (*1*) organic sulfur such as thiophene (heteroaromatic compounds), sulfides, and thiols; (*2*) pyritic sulfur (FeS_2); and (*3*) sulfates such as calcium or iron salts. Both organic and inorganic sulfur phases undergo significant chemical transformations during coal devolatilization and combustion. Sulfur-containing pollutants formed by burning fossil fuels include SO_2, SO_3, H_2S, COS, and CS_2. Under normal boiler operating conditions, with overall excess oxygen being used, virtually all of the sulfur is oxidized to SO_2 and small quantities of SO_3 (84,85). The reactions of sulfur species in coal flames are numerous and complex. The simplified reaction scheme in Figure 7 has been suggested as the basis of a model to predict sulfur formation (66).

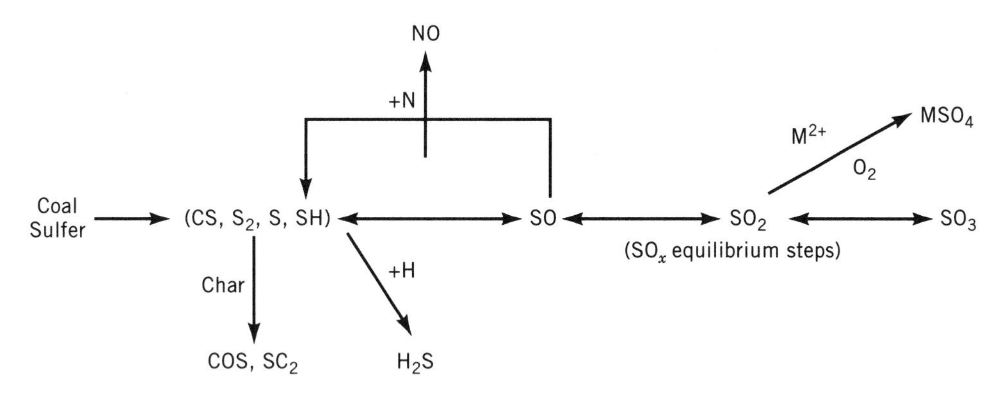

Figure 7. Simplified reaction mechanism for the conversion of coal sulfur to sulfur-containing pollutants. From Ref. 66.

The limestone calcines to form the calcium oxide and to develop pathways called pores in the particle:

$$CaCO_3 \rightarrow CaO + CO_2 \quad (9)$$

The sulfur dioxide penetrates the pores and reacts with the calcium oxide to form the calcium sulfate

$$CaO + SO_2 + \tfrac{1}{2} O_2 \rightarrow CaSO_4 \quad (10)$$

that can be removed with the ash. At high pressure, limestone may not calcine because the carbon dioxide partial pressure is greater than the equilibrium dissociation pressure (75). Dolomite ($CaCO_3 \cdot MgCO_3$) and hydrated lime $Ca(OH)_2$ are also used as sorbents.

Other Pollutants. Coal contains significant amounts of many trace elements. Most trace elements are associated with the mineral portion of the coal, although they are also chemically and physically bound with the organic material. Trace element concentrations vary significantly among coals and even between coals from the same seam (86). During coal combustion, these trace elements become partitioned in effluent streams, depending on their chemical behavior. Classification of trace element emissions have been partitioned into three general groups (87):

Group 1. Elements concentrated in coarse residues, such as bottom ash or slag, or partitioned between coarse residues and particulates that are trapped by particulate control systems, eg, Ba, Ce, Cs, Mg, Mn, and Th.

Group 2. Elements concentrated in particulates compared with coarse slag or ash and tend to be enriched in fine particulate material that may escape particle control systems, eg, As, Cd, Cu, Pb, Se, and Zn.

Group 3. Elements that volatilize easily during combustion, are concentrated in the gas phase, and deplete in solid phases, eg, Br, Hg, and I.

Global emission levels of selected trace elements from coal combustion are shown in Table 6. Emission of elements found in Group 1 is easily controlled by standard particle handling systems. Because Group 2 elements are concentrated in fine particulate material, the release of these elements depends largely on the efficiency of the gas-cleaning system. The emission of volatile trace elements, such as those in Group 3, can be controlled by combustion conditions. For example, trace element release may be lower from fluidized-bed systems than from entrained-flow systems because of the lower temperatures in fluid beds. Furthermore, wet scrubbers used for flue gas desulfurization allow volatile elements to condense and, therefore, provide an effective method of reducing emissions of certain trace elements.

The environmental effects of trace elements released from coal combustion are difficult to determine because plants and animals are exposed to a wide variety of chemicals, many of which occur naturally. Trace elements that are of particular concern from coal combustion are As, Cd, Hg, and Pb. With the exception of Hg, few adverse ecological effects from current atmospheric emissions of trace elements from coal combustion have been identified (86).

Table 6. Estimated Global Trace Element Emissions From Coal Combustion[a]

	Emission Source	
Element	Electric Utilities (μg/MJ)	Industrial and Domestic (g/t)
As	15–100	0.2–2.1
Cd	5–25	0.1–0.5
Cr	80–500	1.7–12
Cu	60–200	1.4–5
Hg	10–35	0.5–3
Mn	70–450	1.5–12.0
Mo	15–150	0.4–2.5
Ni	90–600	2.0–15.0
Pb	50–300	1.0–10.0
Sb	10–50	0.2–1.5
Se	7–50	0.8–2.0
Sn	10–50	0.1–1.0
Tl	10–40	0.5–1.0
V	20–300	1.0–10.0
Zn	70–500	1.5–12.0

[a] From Ref. 88.

PROCESSES AND TECHNOLOGIES

Fixed Beds

The Process. Combustion of coal in fixed beds (eg, stokers) is the oldest method of coal use, and it used to be the most common. In the past few decades, stokers have lost part of their traditional market to entrained and fluidized beds. In the lower range of capacities, the market had been lost to gas years ago. The basic aspects of fixed-bed combustion are illustrated in Figure 8 (89). The run-of-mine coal is crushed typically to 95% less than 3.2 cm and 20–60% less than 0.6 cm (90). The crushed coal is fed pushed, dropped, or thrown on a slowly moving bed of coal particles on a grate. The raw coal is heated, dried, devolatilized, and burned, leaving ash. The layer of ash protects the grate from excessive heat. The primary air flows upward through the grate and through the bed of coal particles. It is heated in the ash zone and reacts with the char in the oxidation zone. The char burns to carbon monoxide and carbon dioxide but, in the presence of oxygen, most of the carbon monoxide is converted to carbon dioxide. Most of the oxygen is consumed in the oxidation zone. In the reduction zone, a part of the carbon dioxide reduces to carbon monoxide. More combustible gases are added in the devolatilization zone. The combustible gases from the reduction and the devolatilization zones must be burned in the space above the bed. The burning is facilitated by introducing secondary or overfire air. The fixed-bed processes may be classified according to the flow patterns of coal and air as countercurrent, cross-current, and concurrent processes. Chemical and physical processes in the fixed beds are similar to those in entrained and fluidized beds. However, fixed beds are distinguished by the flow of

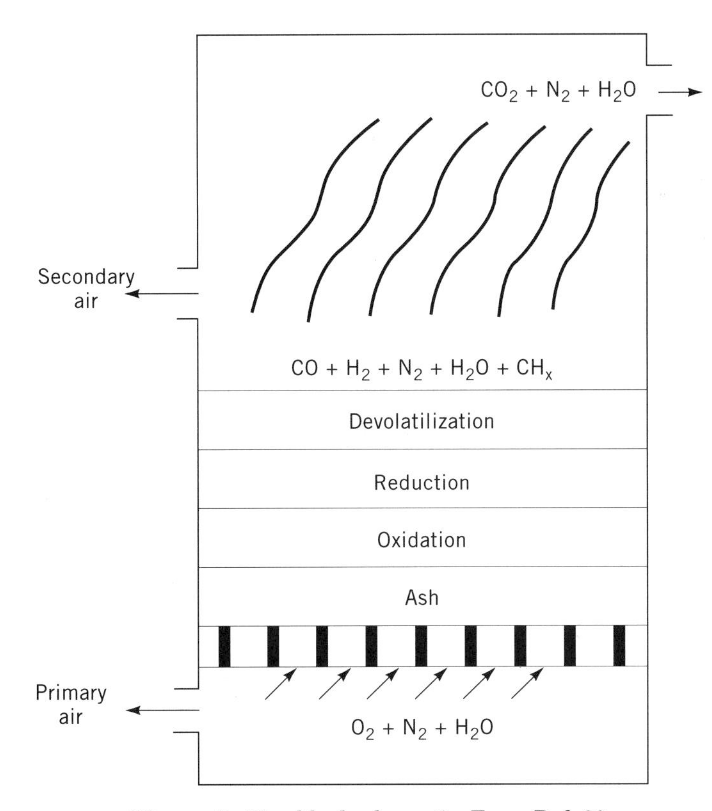

Figure 8. Fixed-bed schematic. From Ref. 89.

solids being almost independent from the flow of gases, and there is increased dependence of the reaction rates on oxygen diffusion.

Technologies. Fixed-bed combustion technologies are well-established technologies, and a great variety of stokers is in use today. Stokers are commonly classified by the way in which coal is fed onto the grate as spreaders, overfeed, and underfeed stokers (91). These categories are, in general, equivalent to countercurrent, cross-current and concurrent processes, respectively. Within each category, the stokers differ according to the way in which the grate handles the ash.

Spreader stokers are the most commonly used type of stokers. A traveling grate, spreader stoker is shown in Figure 9 (92). The coal is thrown and spread over the entire grate surface by mechanical feeders. There is some burning of the suspended coal fines above the bed. It is this suspension burning, coupled with a thin coal bed, that allows fast response to load changes. The smaller spreader stokers use the dumping grates, while the larger ones use the continuous discharge grates. Traveling grates (Fig. 9) are more common than reciprocating and vibrating grates. The spreader stoker can fire a wide range of coals. It has high availability, simplicity of operation, and high operating efficiency, but it also has high fly-ash carryover and high fly-ash combustible heat loss (91).

A traveling grate overfeed stoker is shown in Figure 10 (93). The coal is fed onto the grate from a coal hopper. The coal depth is adjusted by a coal gate. The coal is burned as it passes slowly through the furnace. Ash is discharged continuously into an ash pit. The overfeed stokers use the chain grate, the traveling grate, and the water-cooled vibrating grate. The overfeed stokers are characterized by low fly-ash carryover. They burn most coals, but high coking coals can be a problem. Their response time is longer than that of the spreader stokers. Overfeed stokers require larger grate sizes than spreader stokers of the same capacities (91).

A multiple retort, underfeed stoker is shown in Figure 11 (93). The coal is introduced through long retorts below the level of air tuyeres. Thus the raw coal is at the bottom, the ash moves away from the retort at the top, and combustion takes place in between. The smallest underfeed stokers use the single and the double retort. The coal is fed by a screw or a ram. The ash is usually discharged with the side-dumping grates. The larger underfeed stokers are the multiple retorts inclined at an angle of 25° to 30° to aid the flow of coal and ash. The ash is discharged either intermittently or continuously. The underfeed stokers operate with thick coal beds, causing a high thermal inertia and a slow response to load changes. They have trouble burning high coking coals, low ash bituminous coals, and loose ash subbituminous coals because of the grate overheating. On the other hand, the underfeed stokers have a clean smokeless combustion and low fly-ash carryover (91). The smokeless combustion comes from feeding the coal under the combustion zone. The volatiles escape from the raw coal, flow upward through the combustion zone, and burn almost completely while passing through the zone.

Developments. The most recent major fixed-bed development effort was funded by the American Boiler Manufacturers Association (ABMA), the U.S. Department of Energy (DOE), and the U.S. Environmental Protection Agency (EPA) between 1977 and 1979 (91,94). The main objective of this extensive test program was to reduce emissions and improve efficiency of stoker boilers. The measurements were performed on 18 stoker boilers, 6 spreader stokers, 7 overfeed stokers, and 5 underfeed

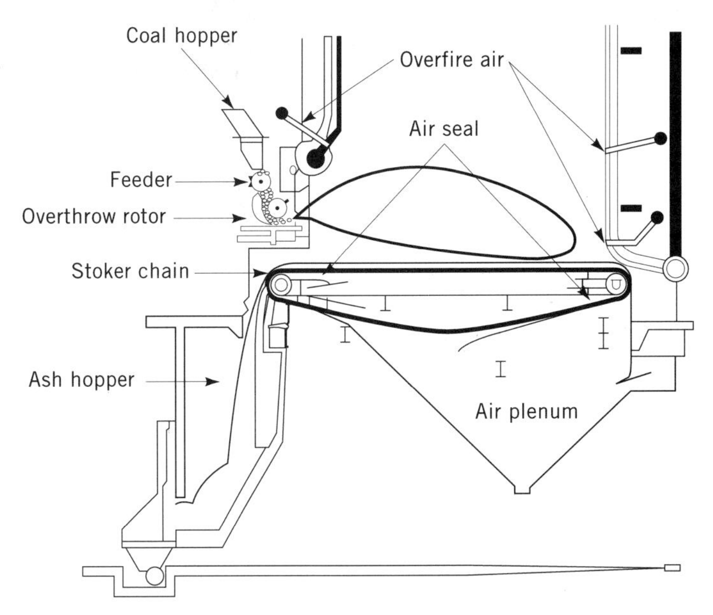

Figure 9. Traveling grate spreader stoker. From Ref. 92.

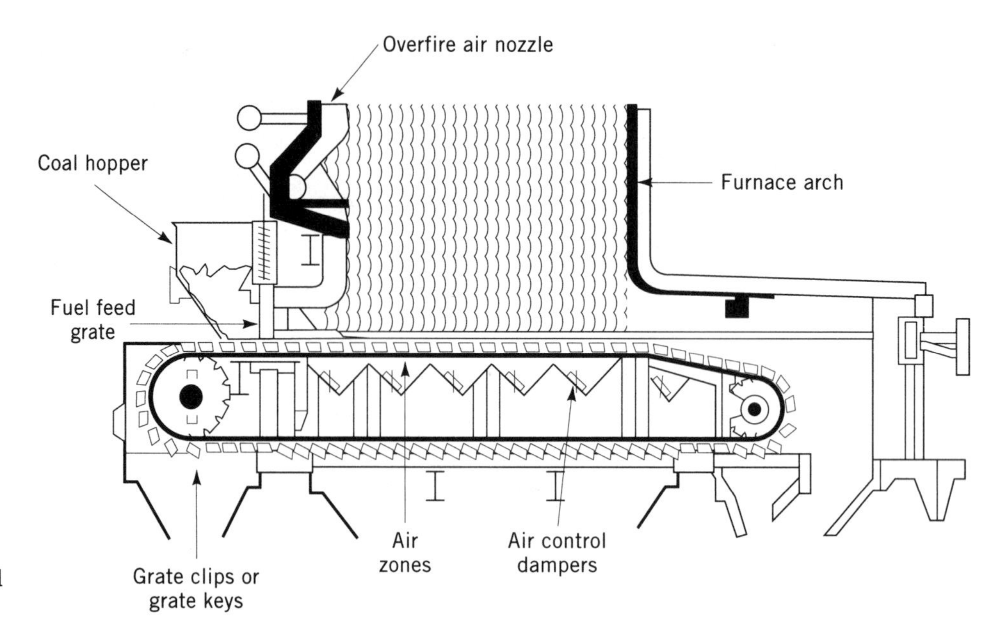

Figure 10. Traveling grate overfeed stoker. From Ref. 93.

stokers. The spreader stokers all had traveling grates. Of the overfeed stokers, 1 had a vibrating grate; 3, a traveling grate; and 1, a chain grate. A total of 4 of the underfeed stokers used multiple retorts, and 1 had a single retort. The operating variables included heat release rate, excess air, overfire air, fly-ash reinjection, and coal properties. The measurements included both uncontrolled and controlled particulate loading, nitrogen oxides, sulfur oxides, oxygen, carbon dioxide, carbon monoxide, unburned hydrocarbons, combustibles in the fly ash and bottom ash, particle size distribution, and boiler efficiency. The particulate loading was found largely to depend on stoker type and degree of fly-ash reinjection. It increased with the heat-release rate, but in many cases, it could be controlled with the overfire air. The nitric oxide increased with the excess air and the grate heat-release rate. The overfire air, as it existed in the boilers at that time, did not affect the nitric oxide. The results also addressed other relationships between the operating variables and the measured emissions and efficiency. Based on these tests, the guidelines for clean and efficient operation of stoker boilers were prepared (91).

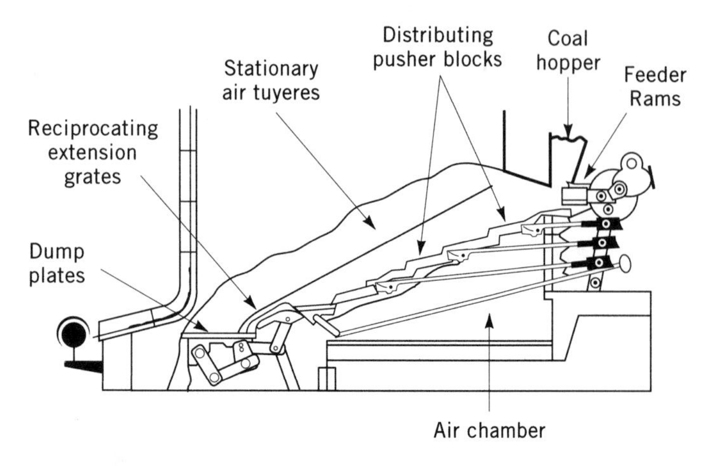

Figure 11. Multiple retort underfeed stoker. From Ref. 93.

The improved overfire air systems for nitric oxide reduction have been discussed (95). Overfire air, limestone addition, and flue gas recirculation for the reduction of nitrogen and sulfur oxides also have been discussed (96). Two flue gas recirculation systems for stokers have been described (97).

Entrained Beds

The Process. The run-of-mine coal is crushed, dried, and pulverized to fine powder in a crusher and a pulverizer or mill. Typically, 70–80% of the pulverized coal particles pass through a 200-mesh sieve of 74-μm apertures. The pulverized coal is transported with a small part of the total air needed for combustion, called primary air, to a burner and into a furnace. A greater part of the total air, the secondary air, is heated and introduced through the burner ports into the furnace to ensure complete combustion. The combustion rate of the pulverized coal is high because of the small particle size. The pulverized coal particles are entrained by the combustion air and transported in suspension through the furnace. They devolatilize, ignite, and burn, leaving ash. The residence time of the pulverized coal particles in the furnace is typically 1–2 s, which usually is sufficient for nearly complete combustion. However, some of the largest particles (400–500 μm) do not always burn out (98). A large proportion of the ash leaves the furnace still entrained as fly ash and must be removed in an electrostatic precipitator. Some ash deposits on the tubes in the furnace and in the flue gas passages cause slagging and fouling. The remaining ash falls to the bottom of the furnace and is removed through an ash hopper.

To dry coal and increase the furnace temperature, the secondary combustion air is often preheated in an air heater. Sometimes, for reactive, high moisture coals, a part of the hot gases is removed after the furnace and used, instead of the primary air, to dry and transport the pulverized coal. The principal advantages of pulverized coal combustion are high reliability, full automation, adaptability to all coals ranks, and excellent capacity for increasing unit size. The main disadvantages are high energy consumption for pulverizing coal, high particulate emissions, and SO_2 and NO_x emissions.

Technologies. Many pulverized coal combustion technologies are available today. Most components are common among pulverized coal combustion technologies, and some are described such as crushers, pulverizers, transport systems, burners, furnaces, and pollution-control systems.

Several types of crushers are commonly in use: Bradford (rotary) breakers, single- and double-roll crushers, and hammer mills (10,90,92). The Bradford breaker, for example, breaks the coal by the force of gravity. It consists of a large, slowly rotating cylinder typically 2.7–3.7 m dia and 6.1–9.1 m long, rotating at approximately 20 rpm. The cylinder is made of perforated steel plates and radial shelves. The size of the perforations determines the size of the crushed coal. Typically, these openings are 3.2–3.8 cm. The coal is fed at one end of the cylinder and carried upward on the radial lifting shelves until it drops and breaks. The production of fines is small. The coal, broken to the size of the perforations, passes to a hopper below. Larger refuse and tramp iron are discharged at the end of the cylinder. The Bradford breaker is probably the most commonly used crusher for large capacities (Fig. 12) (10,99).

A pulverizer or mill grinds, dries, classifies, and delivers the coal to a transport system. Pulverizers can be classified as low, medium, and high speed machines. The low speed pulverizers include the ball or tube mills and the large roll-and-race and ball-and-race pulverizers used in utility plants. The medium speed roll-and-race and ball-and-race pulverizers are used mostly in industrial and in some utility plants. The high speed impact or hammer mills and the high speed attrition mills are used in both industrial and utility plants. All of these machines grind by impact, attrition, compression, or a combination of two or all of these forces.

The medium speed roll-and-race and ball-and-race pulverizers are the most commonly used pulverizers (10). The grinding takes place by crushing and attrition of the coal between two surfaces, one rolling over the other. The rolling elements may be balls or rolls rolling over a race or a ring. In one class, the roller assemblies are driven and the ring is stationary. In another, which is used more often, the ring rotates and the roller assembly is fixed. A roll-and-race pulverizer, also called the bowl mill, is illustrated in Figure 13 (10).

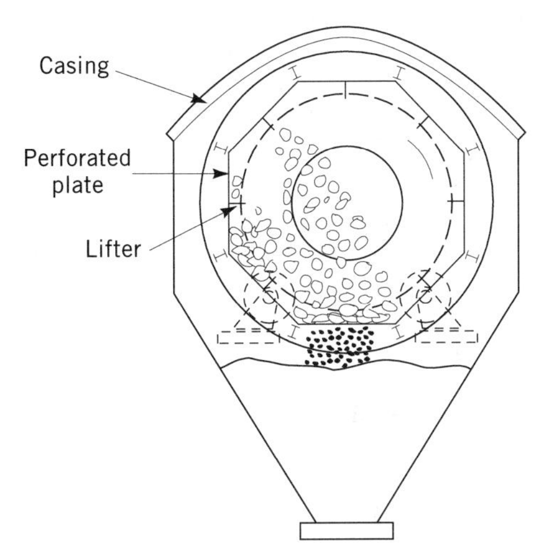

Figure 12. Bradford breaker. From Ref. 99.

The high speed impact or hammer mills are used commonly in Europe for lower rank coals of high moisture content (up to 50%) (92). The hot gases from the furnace are recirculated into the pulverizer to dry the coal. The fineness of the pulverized coal is much less than 70–80% through a 200-mesh sieve (74-μm openings), typical in American practice.

There are two main types of the pulverized coal transport systems: the storage system and the direct system. In the storage system, the pulverized coal is separated from the transport air or gas and stored in a bin. The pulverized coal is discharged from the bin as needed and transported by the primary air to a burner. The storage system has been largely superseded by the direct system. In the direct system, the coal is transported to the burner directly from the pulverizer. This system is simpler and safer. The quantity of the pulverized coal in the system is smaller and the fire hazard is reduced.

Pulverized coal burners are of two main types: the swirl burner and the jet burner. Other types are variations of these two basic designs. In the swirl burner, the coal and primary air as well as the secondary air are introduced into the furnace with a strong, swirling rotation. The swirling motion of the air and the appropriately shaped burner throat give rise to a low pressure region close to the burner throat. The hot combustion products recirculate back toward this low pressure region and provide the ignition energy needed for stable combustion. The recirculation zone can extend several throat diameters into the furnace; ignition takes place mostly within this zone. Each burner is largely independent and has its own flame envelope. The degree of interaction between the burners is limited and depends on the burner and furnace configurations (10). A swirl burner for wall firing is shown in Figure 14 (10).

In the jet burner, the coal and primary air, as well as the secondary air, are introduced into the furnace as jets with no rotation. The jet burners are often placed at the corners of the furnace so that the coal and air jets are tangential to an imaginary vertical cylinder at the center of the furnace. This gives rise to a rotation of the large mass of gases in the center of the furnace, referred to as the fireball. The jet burners operate together and have a single flame envelope (10). A jet burner for corner firing is illustrated in Figure 15 (10).

The pulverized, coal-fired furnaces are usually classified according to the method of ash removal into dry-ash and molten-ash furnaces. The dry-ash furnaces also are called dry-bottom furnaces, and the molten-ash furnaces also are called wet-bottom or slagging-bottom furnaces (10,100). The dry-ash furnaces are much more common in the United States than the molten-ash furnaces. They are simpler, more flexible, and more reliable. However, they also are larger for the same capacity, and most of the ash leaves the furnace still entrained as fly-ash, which must

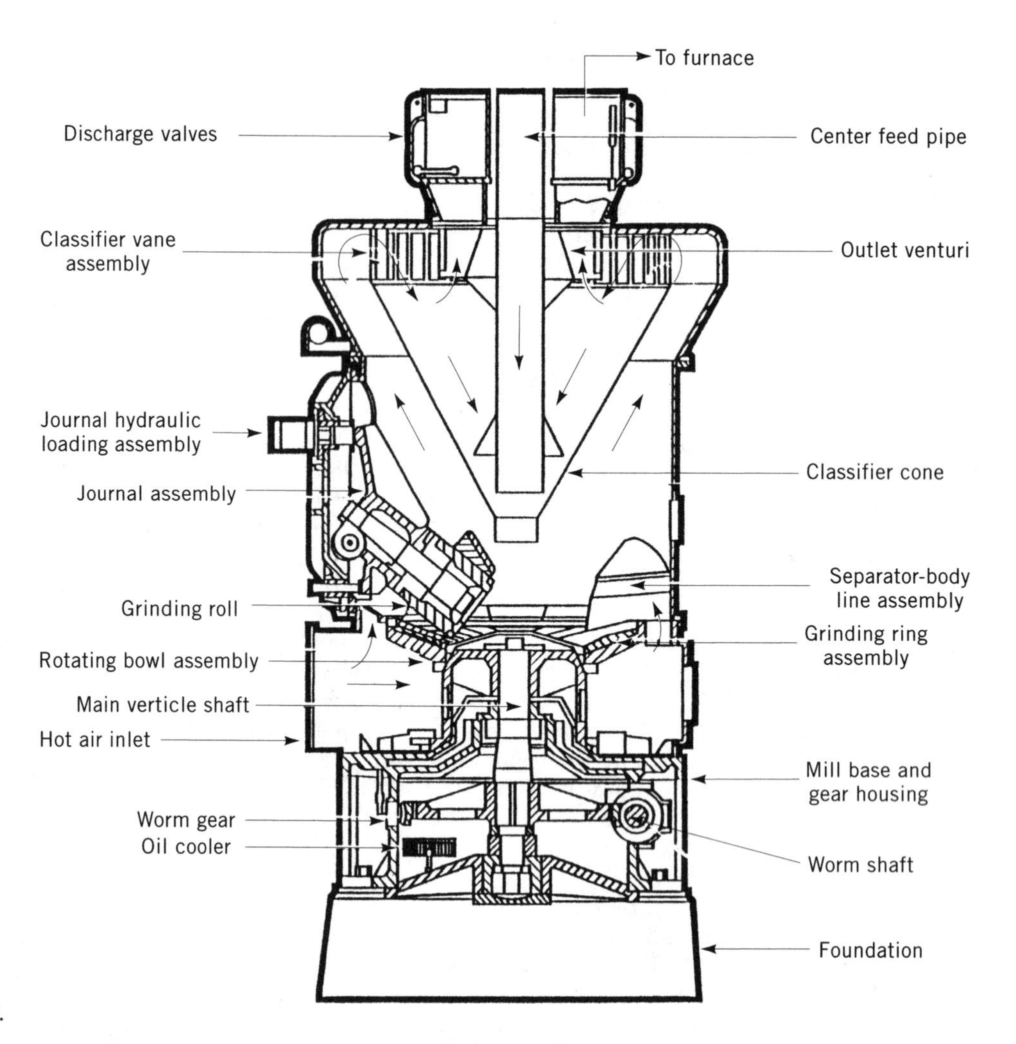

Figure 13. Bowl mill. From Ref. 10.

be removed in the electrostatic precipitator or baghouse filter.

Figure 16 (10) shows a 256-MW subcritical, drum-type boiler for burning pulverized subbituminous coal (10). The jet burners are located at five elevations in the four corners of the furnace for tangential firing. Most boilers installed during the 1980s in the United States were of the 18-MPa subcritical drum type. These boilers will probably continue to dominate the utility market in the 1990s, with expected unit sizes being at least 300 MW. The supercritical, once-through boilers will challenge the drum boilers only in large sizes: 800 MW and above (99).

The molten-ash furnaces are used more extensively in Europe (100). The molten-ash (slag) flowing from the furnace is quenched by water and reduced to a course, granular solid. The furnace ash retention is high; more than 80% is retained in some designs. The heat release rates also are high and result in smaller furnaces. The cyclone furnaces, even though based on the molten-ash removal, are usually not classified as entrained-bed furnaces. In the cyclone furnace, a large portion of the coal is burned on the surface of a moving slag layer.

Developments. Many new pulverized coal combustion technologies are being developed. Most are classified as the clean coal technologies, concerned with controlling and reducing pollution. Some of the technologies being developed in Europe are listed in Table 7. The pulverized coal combustion technologies being developed and demonstrated in the United States under the Clean Coal Technologies Demonstration Program are presented in Table 8.

A demonstration project at the Ohio Edison's Edgewater Station is representative of these developments (101) (Fig. 17). The project objective is to demonstrate, with a variety of coals and sorbents, a limestone injection, multistage burner (LIMB) process for simultaneous control of SO_x and NO_x. The process is expected to achieve up to 60% reduction in both NO_x and SO_x. In the process, the dry limestone sorbent is injected into the furnace at a point above the low NO_x burners. The burners reduce NO_x formation by staging combustion. The sorbent travels through the boiler and is removed along with the fly ash in the existing electrostatic precipitator (ESP). Humidification of the flue gas before it enters the ESP is necessary to maintain normal operation and to enhance SO_2 removal. In addition, by using the duct injection of the dry lime sorbent upstream of the humidifier and the precipitator, up to 80% SO_x reduction is expected. A chemical additive is added in the humidification water to improve SO_2 absorption. The project is conducted on a commercial, wall-fired, 105 MWe boiler.

A clean coal technology of considerable interest is coal slurry technology. Interest in the use of coal slurries, and

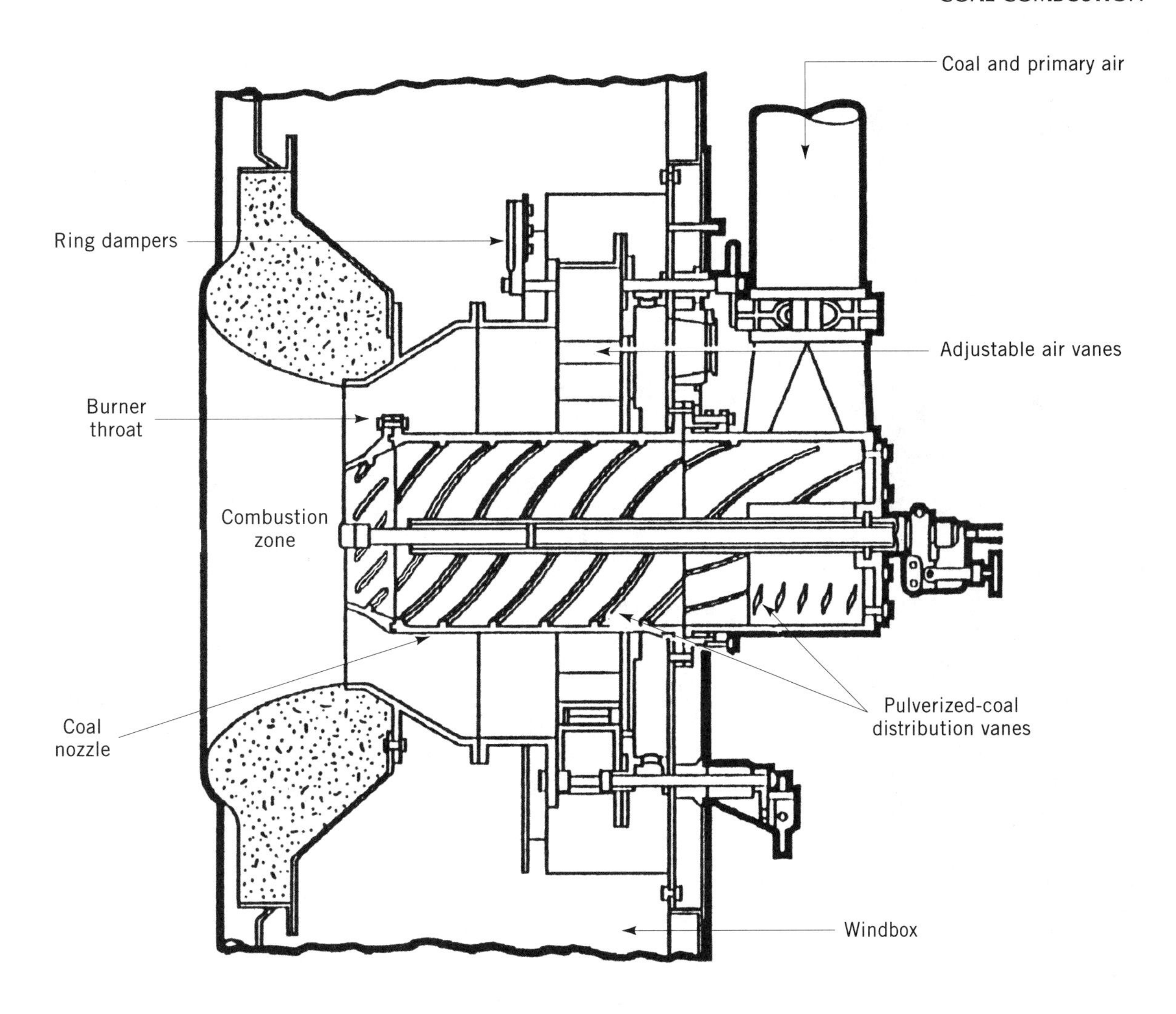

Figure 14. Swirl burner. From Ref. 10.

particularly coal–water mixtures (CWM), centers largely around use of coal slurries as a replacement fuel for existing oil-fired boilers. In addition, CWM may also be useful in ignition of pulverized coal in existing furnace systems and as a substitute fuel in gas turbines. Research results for coal–oil mixtures (COM) have shown little economic incentive as an oil-replacement fuel. Furthermore, full-scale utility boiler test results have uncovered unexpected problems in handling of ash fouling with the use of coal–oil slurries.

Efforts to develop coal–water mixtures have been more encouraging. It has been shown that for a specific 200-MWe oil-fired boiler CWM may have some economic incentive (102). It has been noted that the critical technical issues include development and integration of advanced coal-cleaning methods for up-front reduction of sulfur and ash levels and demonstration of high combustion efficiency, reliability, and performance in a utility boiler. The commercial use of coal slurries depends on higher oil prices.

Fluidized Beds

The Process. Combustion of coal in fluidized beds is becoming increasingly common. The atmospheric fluidized-bed combustion (afbc) technology has been commercialized in the last two decades; pressurized fluidized-bed combustion (pfbc) is in the demonstration phase. Today, the fluidized-bed boilers compete with the stoker boilers in small sizes and with the pulverized coal fired boilers in large sizes (103).

A fluidized bed consists of a bed of particles set in vigorous, turbulent motion by the combustion air blowing upward through the bed. The particles are mostly inert materials such as coal ash or sand, or sulfur sorbents such as limestone or dolomite. Unburned or partially burned coal particles make up only around 1% of the bed mass. At velocities greater than the minimum fluidizing velocity, the bed is fluidized and the air flows through the bed in bubbles, thus the bubbling fluidized bed (BFB). At velocities approaching or greater than the freefall velocity of the particles, the particles become entrained in the air and are carried out of the furnace. The entrained particles are separated from the combustion gases in a cyclone and circulated back to the bed, giving rise to the circulating fluidized bed (CFB) (104).

The basic aspects of the atmospheric bubbling fluidized bed are illustrated in Figure 18 (105). The run-of-mine coal is crushed to less than 3.2 cm for overbed feed systems. For underbed feed systems, the coal is crushed to

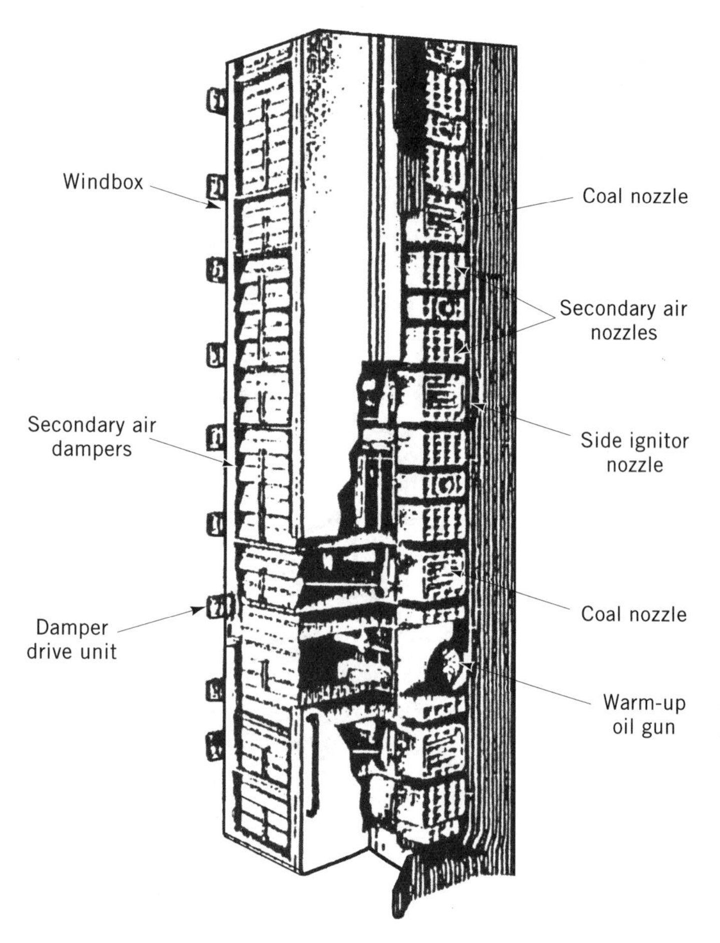

Figure 15. Jet burner. From Ref. 10.

less than 0.6 to 2.5 cm and dried to less than 6% surface moisture for pneumatic conveying (105). If the run-of-mine limestone is used, then it must also be crushed. The crushed coal and limestone are fed to the fluidized bed of coal ash and limestone particles. To start combustion, the bed is preheated to 700–811 K (800–1000°F), depending on the particular coal properties. The coal particles introduced in the fluidized bed heat, dry, devolatilize, ignite, and burn, leaving ash. The residence time of the coal particles in the bed is typically around 1 min, which is usually sufficient for 80–90% burnout. The temperature of the bed is increased to the operating temperature, usually 1120 K (1550°F), and maintained by removing 40–45% of the heat input with an in-bed heat exchanger. The limestone particles heat and calcine in the bed, and the calcium oxide then reacts with sulfur dioxide to form calcium sulfate. Due to intensive mixing, the heat-transfer rates in the bed are high and the temperature of the bed is near uniform. For the same reason, the heat-transfer rate between the bed and the in-bed heat exchanger is high in spite of the low bed temperature. The low bed temperature prevents formation of the thermal nitric oxides and helps reduce the fuel nitric oxides, but N_2O is high. It also helps reduce slagging and fouling.

Some coal, coal ash, and limestone particles are thrown out of the bed in a freeboard zone. The freeboard zone provides additional space and time to complete coal combustion, increase sulfur dioxide capture, and destroy a portion of the nitrogen oxides released in the bed. An additional 10–20% burnout is achieved. The large particles decelerate and return to the bed; the small particles are entrained and must be removed in a cyclone. The unburned carbon and the unreacted limestone contents in the separated particles are usually high and justify recycling to the bed. The remaining fly ash is removed in an electrostatic precipitator or a fabric bag filter.

The atmospheric circulating fluidized beds, developed more recently, are becoming increasingly important for the large units. The circulating fluidized bed is similar to the bubbling bed; the differences stem from smaller coal and limestone particle sizes and higher gas velocities. The bed fills the entire furnace volume although most of the mass is still in the lowest third of the bed. A large portion of the solids is carried out of the furnace, separated in the cyclone, and recirculated to the furnace. Thus combustion takes place throughout the furnace as well as in the cyclone.

Chemical and physical processes in the fluidized beds are similar to those in entrained beds. The main distinguishing features of fluidized beds are complex flow of solids, desulfurization in the bed, and the dependence of the reaction rates on diffusion. At atmospheric pressure, limestone is the preferred sorbent. The principal advantages of fluidized bed combustion are

1. The substantial reduction of SO_x during combustion, eliminating the need for postcombustion SO_x control processes; NO_x is also somewhat lower.
2. The high heat transfer rates in the bed, resulting in compact furnaces and heat transfer exchangers.
3. The low combustion temperature, reducing slagging, fouling, and the dependence on coal-ash properties.

The fluidized bed boilers can burn a wide variety of coals, including high sulfur coal, cleanly and efficiently. However, there are also some disadvantages (103):

1. The increased erosion of components handling solid because of the high solids loading.
2. The refractory in the circulating fluidized beds.
3. Increased quantities of the solid waste because of the sorbent.
4. The increased emission of the nitrous oxide with respect to pulverized coal and stoker firing.
5. Unproven reliability of the large units.

Technologies. Fluidized-bed combustion technologies are still being intensively developed. The atmospheric fluidized-bed technology has been demonstrated for utility boilers in units up to 150 MW (103). However, utility boilers are often in much larger units of 300–400 MW and sometimes even in units up to or exceeding 800 MW (99). For these larger units, the pulverized coal combustion technology is still the technology of choice. The pressurized fluidized-bed technology for utility-size boilers is presently being demonstrated.

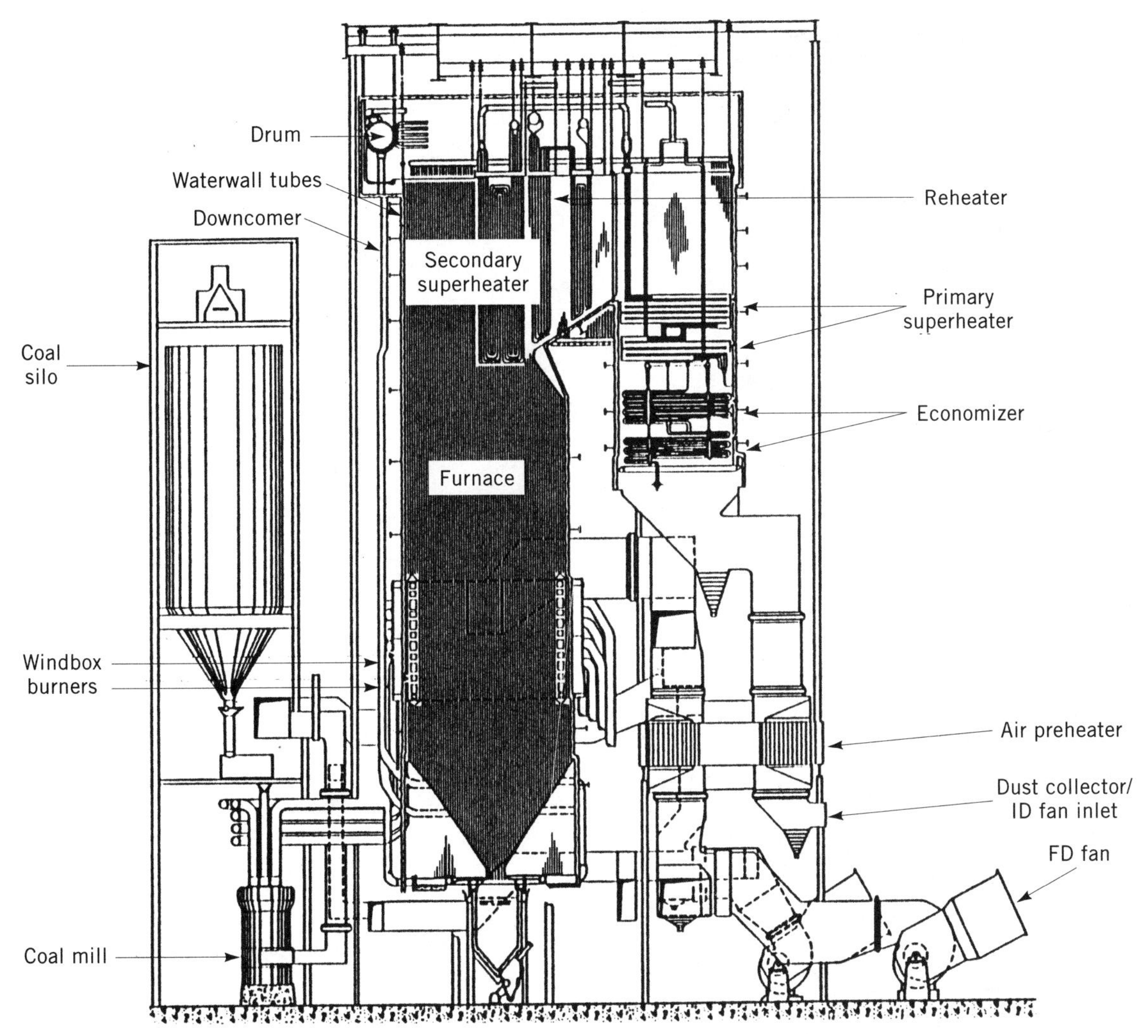

Figure 16. Tangentially fired boiler. From Ref. 10.

An example of the bubbling atmospheric fluidized-bed combustion technology is the Northern States Power Company's Black Dog Unit 2 (105). The plant burned a blend of the Illinois No. 6, high volatile, bituminous coal and a low sulfur, western subbituminous coal to meet the emission limits for sulfur. The switch from the original design coal (the Illinois No. 6) to the western coal blend resulted in increasing furnace deposits and derating from 100 MW to 85 MW. The unit was retrofitted with a bubbling afbc boiler to increase the rating with the western coal to 130 MW while preserving low emissions. The design emission limits were 17.2 mg/MJ for particulates and 516 mg/MJ for SO_2 (Fig. 19) (105).

A system was added for storage, crushing, transporting, and feeding the limestone. The fluidized bed area is divided into the center cell and two side cells, separated by water-cooled walls. The independent operation of cells provides turndown to 25% of the maximum rated capacity. The center cell has six overbed spreader–feeders, two limestone ports, and 12 recycle ports. The steam side includes inbed superheaters. The convection pass is similar to the convection passes for the pulverized coal boilers. Multicyclones collect the larger particles of fly ash containing the unburned carbon and the unreacted limestone. The collected material is recycled to the fluidized bed. The existing electrostatic precipitator was upgraded to accommodate the greater capacity.

The Black Dog Unit 2, retrofitted with the bubbling afbc boiler, was placed in commercial operation in 1986. The plant performance was rated in terms of availability, boiler operating statistics, and on-line startups. The availability for a test period in 1988, including outages to prepare for the boiler performance tests, was 58.3%. The boiler performance tests indicated that efficiency at full load was 86.8%. The cold starts required approximately 8 h; the hot restarts required around 1 h.

One of the important features of bubbling, fluidized beds is the feeding method. There are two in use:

1. The overbed feeders (spreader stokers) feed the coal and the limestone over the bed surface; they are simpler but may result in lower coal and limestone use efficiency.
2. The underbed feeders feed the coal and the limestone through many feeder pipes into the bottom of the bed; the current practice requires a coal feed point for every 0.9–1.9 m^2 of the bed area; they are

Table 7. European Research and Development Technologies in the Clean Use of Coal

Precombustion	During Combustion	Postcombustion	Other and Waste Disposal
	Denmark		
Coal characterization Transport and handling Coal cleaning	low NO_x burners	catalytic de-NO_x FGD and cooling for district heating	transport and handling FGD byproduct utilization fly-ash use
	Belgium		
In situ gasification Hydrogen pyrolysis Ash removal Gasifier development	AFBC[b] (0.6–5.0 MW size for heating systems) burning tailings		high-temperature dust collectors
	Germany		
In situ gasification Gasification Liquefaction Coal preparation	PFBC[c] IGCC[d]	catalytic de-NO_x noncatalytic de-NO_x electron beam dry scrubber for de-NO_x and SO_2 absorption processes	land restoration after mining and coal preparation FGD[e] byproduct use
	Sweden		
Coal characterization Physical coal cleaning Chemical coal cleaning	sorbent injection AFBC PFBC PCFBC[f] coal–liquid mixtures	wet–dry FGD catalytic de-NO_x heavy metal removal dust removal	FGD byproduct use
	Turkey		
Smokeless fuels Gasification and liquefaction	FBC lignites	FGD of lignites	solid waste conversion to byproducts
	United Kingdom		
Smokeless fuels Coal characteristics Gasification for utility and industry Fixed-bed slagging gasification Coal washing Coal liquefaction Coal–water mixtures	AFBC (bubbling and circulating) for industrial use PFBC for utility uses burner modification staged combustion IGCC	prescrubbers for HCl dry lime injection FGD large-scale ESP[g] hot gas cleaning combined cycle studies	comparative economic analysis of clean use technology option FGD byproduct use

[a] From Ref. 9.
[b] Atmospheric fluidized-bed combustion.
[c] Pressurized fluidized-bed combustion.
[d] Integrated gasification combined cycle.
[e] Flue gas desulfurization.
[f] Pressurized circulating fluidized-bed combustion.
[g] Electrostatic precipitator.

more complex and costly but yield better coal and limestone utilization efficiency.

The circulating afbc boilers are similar in many respects to the bubbling afbc boilers. Some of the important differences are listed below:

1. The circulating afbc boilers are taller and have smaller cross-sections because of higher gas velocities.
2. The circulating afbc boilers usually do not include the in-bed heat exchangers because of potential high velocity erosion.
3. The cyclone in the circulating afbc boilers is located before the convection pass to protect the convection surfaces from erosion due to high solids loading and velocity; the cyclone thus operates at high temperatures.
4. The feeding system in the circulating afbc boilers, because of enhanced solids mixing, requires an order of magnitude fewer feed point.

Table 8. Clean Coal Technology Demonstration Technologies[a]

Commercially Available	Recently Developed	In Various States of Demonstration[b]
	Precombustion	
Conventional physical coal preparation Low pressure gasification processes	improved coal gasification for the production of clean fuel gas	advanced physical coal cleaning
	During Combustion	
Modern stoker-fired and pulverized coal-fired boilers for industrial and utility applications Shallow-bed atmospheric fluidized-bed combustion (AFBC) for small industrial applications Combustion control to reduce the generation of oxides of nitrogen (NO_x)	deep bubbling bed AFBC for industrial use circulating bed AFBC for industrial use low NO_x burners for new utility and industrial boilers	coal liquid mixtures advanced coal gasification AFBC for utility use pressurized fluidized-bed (PFBC) for utility use low NO_x burners for retrofit furnace sorbent injection for the removal of sulfur dioxide
	Postcombustion	
Electrostatic precipitators for utility and large industrial use Fabric filters for utility and industrial use Wet lime and limestone flue gas desulfurization (FGD) for utility and large industrial use FGD using sodium alkali sorbent	spray dryer FGD for utility and large industrial applications regenerable FGD with byproduct recovery for utility applications selective catalytic removal of NO_x FGD using ammonia sorbent	furnace sorbent injection for the removal of sulfur dioxide dry sorbent methods of SO_2 and NO_x removal direct injection of alkaline sorbents

[a] From Ref. 9.
[b] There are many other technologies, such as chemical coal cleaning, that are in various stages of laboratory development.

An example of the circulating atmospheric fluidized-bed combustion technology is the Colorado-Ute Electric Association's Nucla Power Station (105). The station used to burn a local coal in three stoker-fired boilers. It was repowered with a circulating afbc boiler to produce 110 MW at low emission levels. The design emission limits were 12.9 mg/MJ for particulates, 172 mg/MJ for SO_2, and 215 mg/KJ for NO_x. The boiler is shown in Figure 20 (105).

The furnace consists of two combustion chambers. The coal is fed to each combustion chamber through three ports. The limestone is injected at four points in each chamber. NO_x is reduced by a two-stage combustion that uses primary and secondary air supplied to combustion chambers. The hot gases and particulates are separated in two cyclones, and particulates are returned to the combustion chambers. The hot gases and remaining fly ash flow through the convection pass. The convection pass is similar to the convection passes for the pulverized coal boilers. The fly ash is collected in four fabric-bag houses.

The Nucla Power Station repowered with the circulating afbc boiler was brought to full load in 1988. The results of acceptance testing indicated boiler efficiency of 88.5% and sulfur retention of 70% at a Ca:S ratio of 1.73. The particulates were under 5% opacity. The NO_x emission was at ~50% of the design limit of 129 mg/MJ. The SO_2 emission was maintained at the compliance level of 172 mg/MJ.

One of the important design features of the circulating fluidized beds is the location of heat transfer surfaces. There are two approaches: (*1*) the external heat exchanger for the solids returning from the hot cyclone and (*2*) the heat exchangers in the upper part of the furnace. The first one is more complex; the second is simpler, but it exposes the high temperature heat exchanger tubes to a strongly erosive environment.

Developments. Fluidized-bed combustion technologies are being developed in many countries of the world. These technologies are attractive for developed countries because of their ability to control and reduce pollution during combustion and for developing countries because of their ability to burn a wide variety of low quality fuels effectively in smaller units. Some of the technologies being developed in Europe and the United States are listed in Tables 7 and 8.

The Tidd PFBC Demonstration Project at the Ohio Power Company's Tidd Plant was described in detail as a representative of these developments (101) (Fig. 21). The project objectives are to demonstrate pfbc at a 70-MW scale, to verify pfbc in a combined cycle application, to achieve more than 90% SO_2 removal and less than 86 mg/MJ NO_x, and to attain an overall power plant efficiency of 38% using the existing steam system.

The pressurized bubbling fluidized bed is operating at 1.2 MPa; the combustion air is supplied by the gas turbine compressor. The coal–water paste is fed to the bed of coal ash and dolomite sorbent. The hot combustion gases and entrained particles pass through the cyclones to remove 98% of the particles. The hot gases are then expanded in the gas turbine to generate 16 MW of electricity. The gases exiting the gas turbine are further

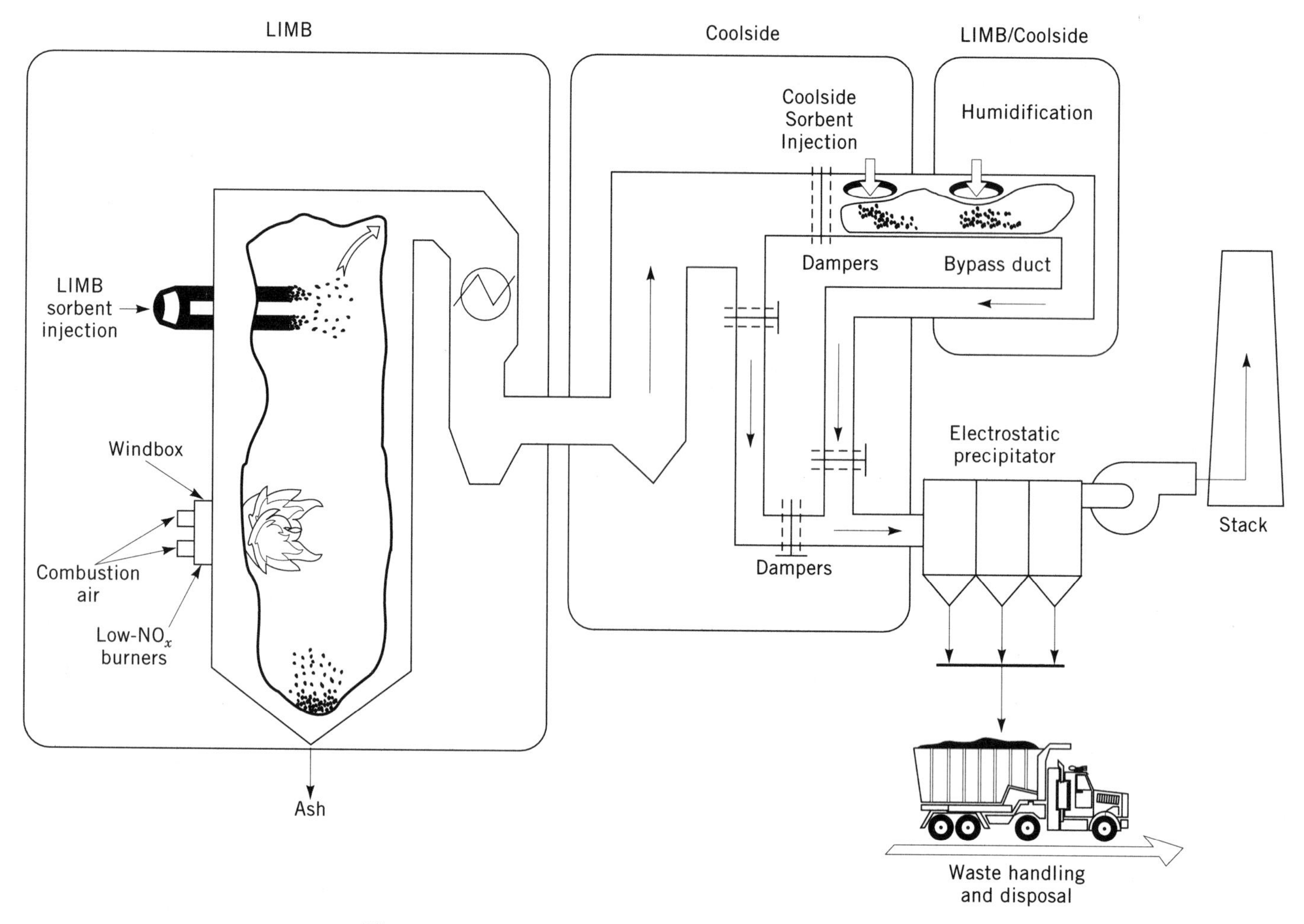

Figure 17. LIMB demonstration project. From Ref. 101.

cooled in the waste heat exchanger, which heats the boiler feed water. The gases are cleaned in the electrostatic precipitator before being discharged. The feed water is converted to superheated steam in the pfbc boiler. The steam passes through the steam turbine to produce an additional 58 MW of electricity. The 3-yr operation and testing phase started in 1991.

Magnetohydrodynamic Generators

The Process. A magnetohydrodynamic (MHD) generator works in a way similar to a conventional electric generator. A conducting gas, instead of a rotating solid conductor, flows through a magnetic field, generating electricity (106,107). The conducting gas is provided by seeding high temperature combustion products with easily ionizable potassium. Temperatures of around 2700 K are required to produce adequate conductivity. The MHD generator produces direct current.

The main advantage of the MHD generator is very high conversion efficiency. A typical commercial MHD generator is expected to achieve efficiency in the range of 80–90%. The main disadvantages are very high operating temperature and related materials problems. An advantage of the MHD generator is an inherent SO_2 control system. The potassium, added to improve conductivity, also reacts with sulfur. The resulting potassium sulfate, which condenses as a solid, is collected in an electrostatic precipitator or a bag house. Almost 100% SO_2 can be removed at 50% excess potassium (106,108,109). With the MHD combustor operating fuel rich, NO_x control can be accomplished in a heat-recovery furnace. It is expected that NO_x emissions can be kept below 17% of the current U.S. New Source Performance Standards (106,108,109).

Technologies and Developments. The MHD generator can be used only when the gas temperature is high enough to maintain thermal ionization ie, higher than around 2300 K (107,108). Thus the MHD generator can be used as a topping cycle in a combined-cycle power plant. The remainder of the energy in the hot gases will be used in a bottoming cycle. U.S. development has centered around a boiler–steam turbine bottoming cycle. The topping cycle components are a high temperature coal combustor, a nozzle for near sonic expansion on the combustion gases through a magnetic channel, a superconducting magnet, a diffuser after the magnetic channel, a regeneration plant to recover the potassium seed and return it to the combustor, and a power-conditioning system to collect power from the electrodes and to convert direct

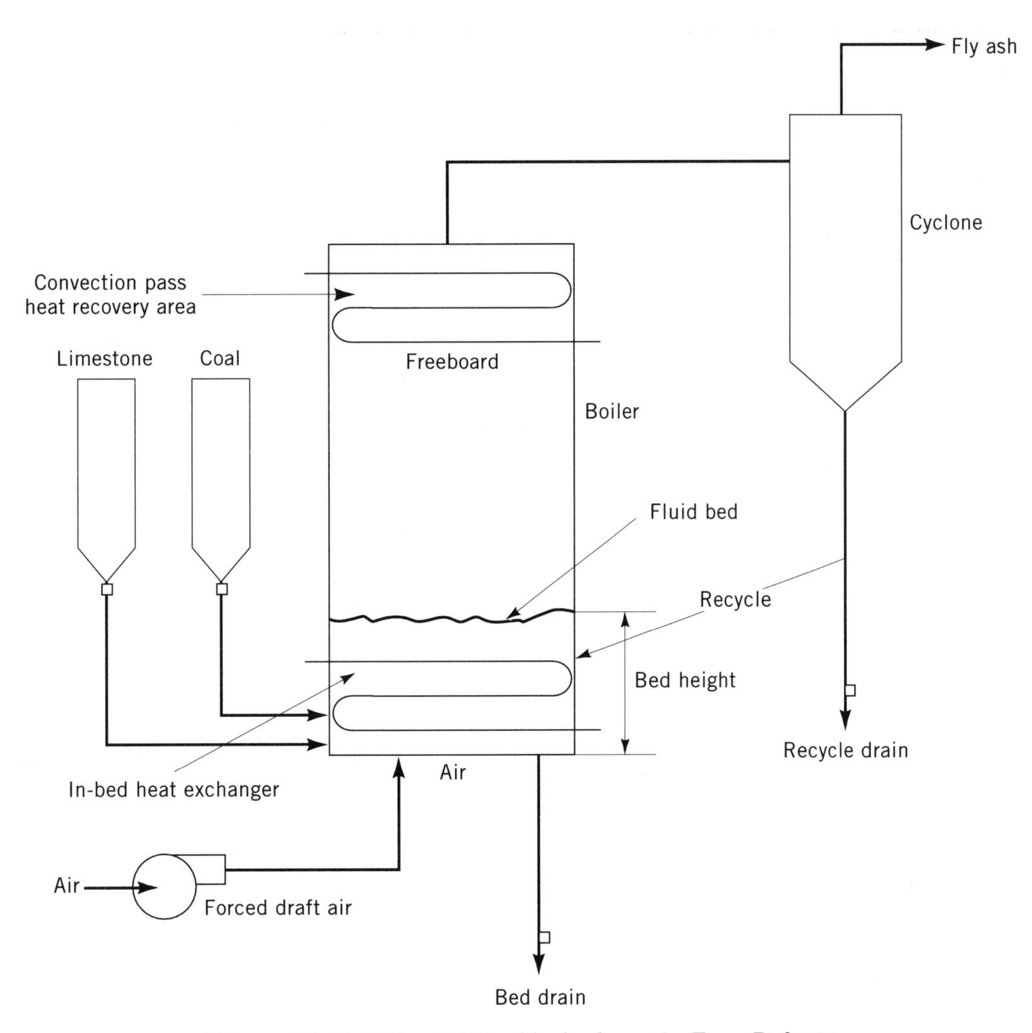

Figure 18. Bubbling fluidized-bed schematic. From Ref. 105.

current into alternating current. The bottoming cycle components are conventional but will operate at substantially higher temperatures and with the combustion gas chemistry changed by the potassium seed. The MHD combined cycle power plants are expected to achieve fuel-to-busbar efficiency of around 50% in the first commercial versions and up to 60% in the advanced versions.

Much of the MHD technology developed in the United States has taken place at two facilities: the 20 MW_t Coal Fired Flow Facility (CFFF) in Tullahoma, Tennessee, and the 50 MW_t Component Development and Integration Facility (CDIF) in Butte, Montana, both funded by the Department of Energy. The CFFF team focuses on bottoming cycle components, while the work at the CDIF concentrates on topping cycle development (110). The current, proof-of-concept test program is scheduled to be completed by the end of 1993. Two conceptual designs for retrofitting an MHD topping cycle to an existing coal-fired power plant have been prepared. It is expected that one of these designs will go into construction, possibly under the Clean Coal Technology Demonstration Program (106). The MHD technology development in Russia has been extensive. It is reported that a 500-MW MHD power plant, fired with natural gas, has been under construction since the late 1980s (108).

Fuel Cells

The Process. A fuel cell, like a battery, converts chemical energy directly into electrical energy. Electrodes, in contact with an electrolyte, are sites of electrochemical reactions releasing or absorbing electrons. In the battery, the electrodes provide fuel for these reactions and thus are eventually depleted. In the fuel cell, the electrodes are catalysts for the reactions of a hydrogen-rich fuel with an oxidant. The fuel and the oxidant are fed continuously and separately to the anode and cathode where electrochemical reactions take place. The electrodes are separated by electrolyte that serves as a transport medium for the ions. By physically separating the oxidation of the fuel at the anode from the reduction of the oxidant at the cathode, the electrochemical reactions are set to produce electricity directly (111,112). Figure 22 (9) illustrates these processes for a typical fuel cell. The oxidation of the fuel at the anode releases electrons, while the reduction of the oxidant at the cathode absorbs electrons, creating a po-

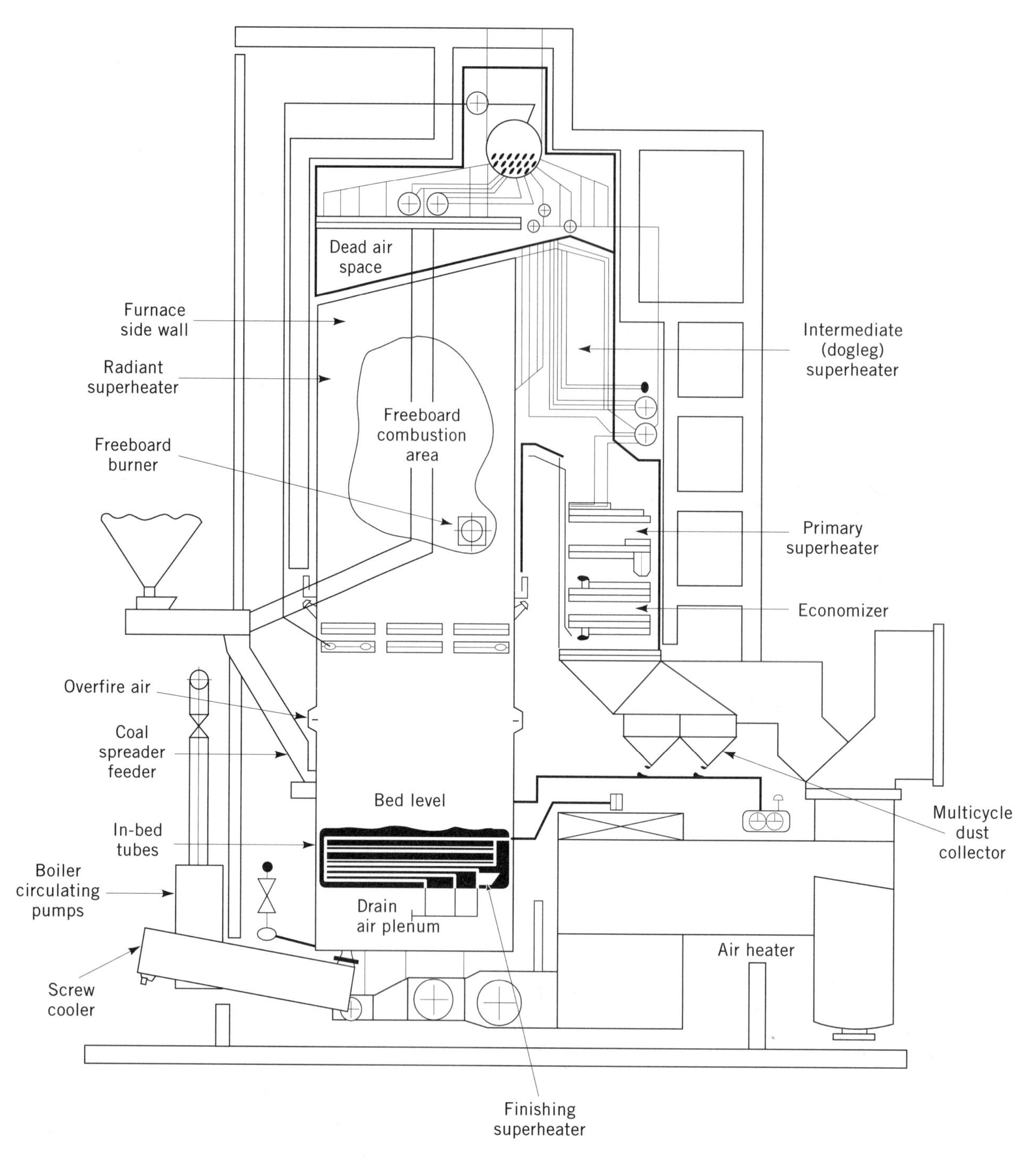

Figure 19. Black Dog bubbling AFBC boiler. From Ref. 105.

tential difference between the electrodes. In the electrolyte, the ions flow from one electrode to the other, completing the electric circuit (112). The fuel cell generates direct current, which must be converted into alternating current.

The advantages of the fuel cell stem from the direct conversion of chemical energy into electrical energy (113). The fuel cell efficiency is not limited by the Carnot cycle and may eventually reach 60%. Pollutant emissions, in the absence of combustion, are extremely low. The fuel cell potentially offers the highest efficiency and the lowest emissions of any coal-based process (111). However, there are also some disadvantages (113):

1. The fuel cell needs a clean, hydrogen-rich fuel; thus the coal must be gasified, the coal gas reformed by steam into a hydrogen-rich gas, and the hydrogen-rich gas treated to remove impurities, which may be expensive.
2. Some components of the fuel cell are made of rare or expensive materials such as platinum, nickel, strontium-doped lanthanum manganite, calcia-stabilized zirconia, yttria-stabilized zirconia, and magnesium-doped lanthanum chromite.

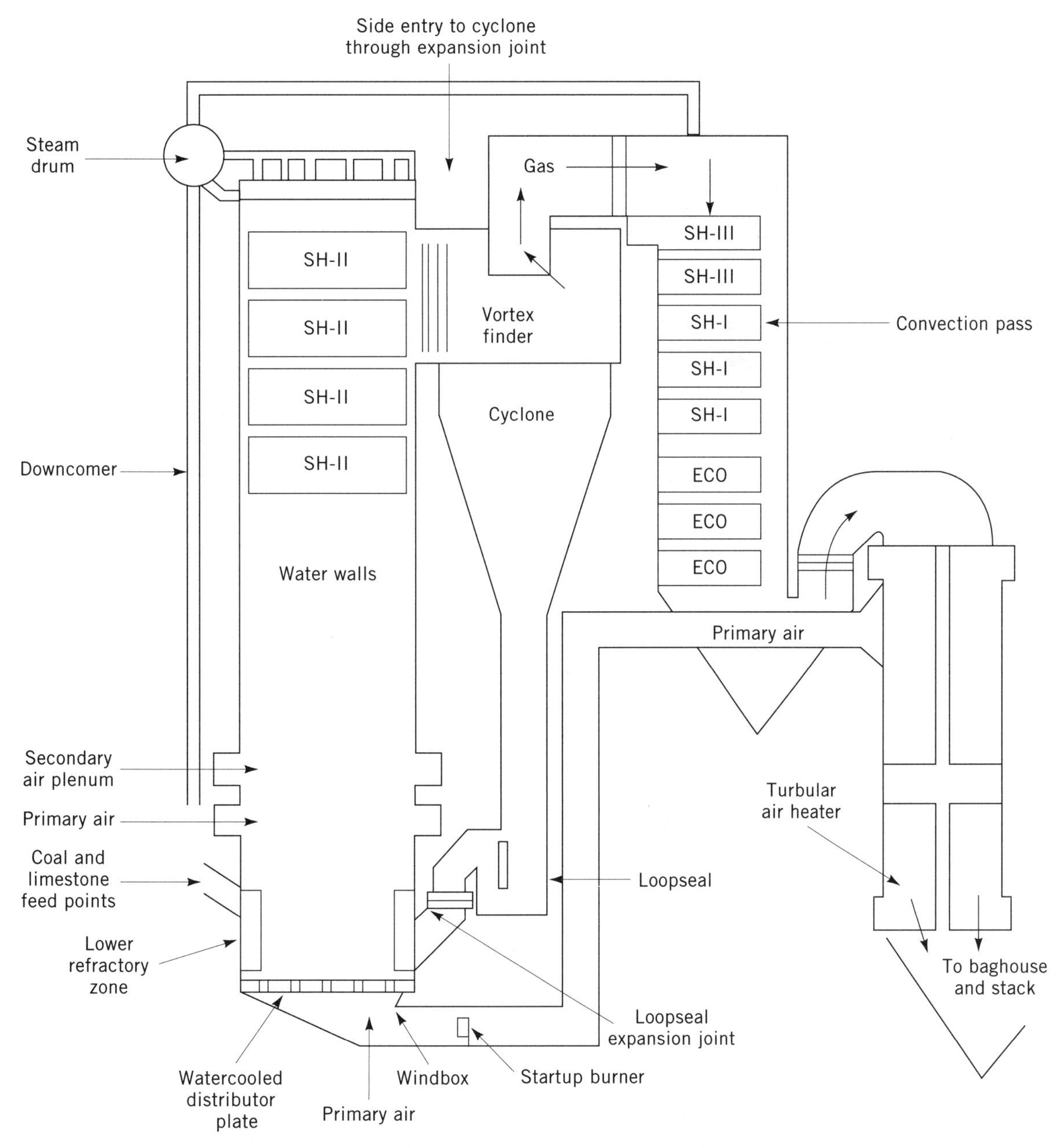

Figure 20. Nucla circulating AFBC boiler. From Ref. 105.

3. The materials are not only expensive but also susceptible to premature failures with respect to the fuel cell operating life (9).

Technologies and Developments. Three types of fuel cells are being developed for utility, industrial, and commercial applications. They are classified by the type of electrolyte as phosphoric acid (PA), molten carbonate (MC), and solid oxide (SO) fuel cells. Phosphoric acid fuel cells (PAFC) and polymer electrolyte fuel cells (PEFC) are also being developed for transportation uses. The operating temperatures are around 480 K for PAFC, 920 K for MCFC, and 1260 K for SOFC (114).

Fuel cell technologies are being developed in many countries, but the research-and-development programs in the United States and Japan seem to be the largest (115). In the United States, PAFC systems for electric utilities are on the verge of commercialization (111,116). PAFC systems, in sizes ranging from 10 kW to 4.5 MW, were demonstrated in the mid-1980s. An 11-MW PAFC power plant is planned for demonstration in the early 1990s. MCFC systems are currently receiving the most attention. A 100-kW MCFC unit will go on-line in the early 1990s and 2-MW MCFC systems will be demonstrated in the mid-1990s (111,117). There are three SOFC configurations now under development: tubular, monolithic, and planar. For the tubular configuration, units up to 5 kW have been tested (116).

In Japan, PAFC systems are also in the final stages of development (118,119); 1-MW PAFC units for the central utility power plants were demonstrated in the late 1980s. Testing of 200-kW PAFC units for on-site power plants

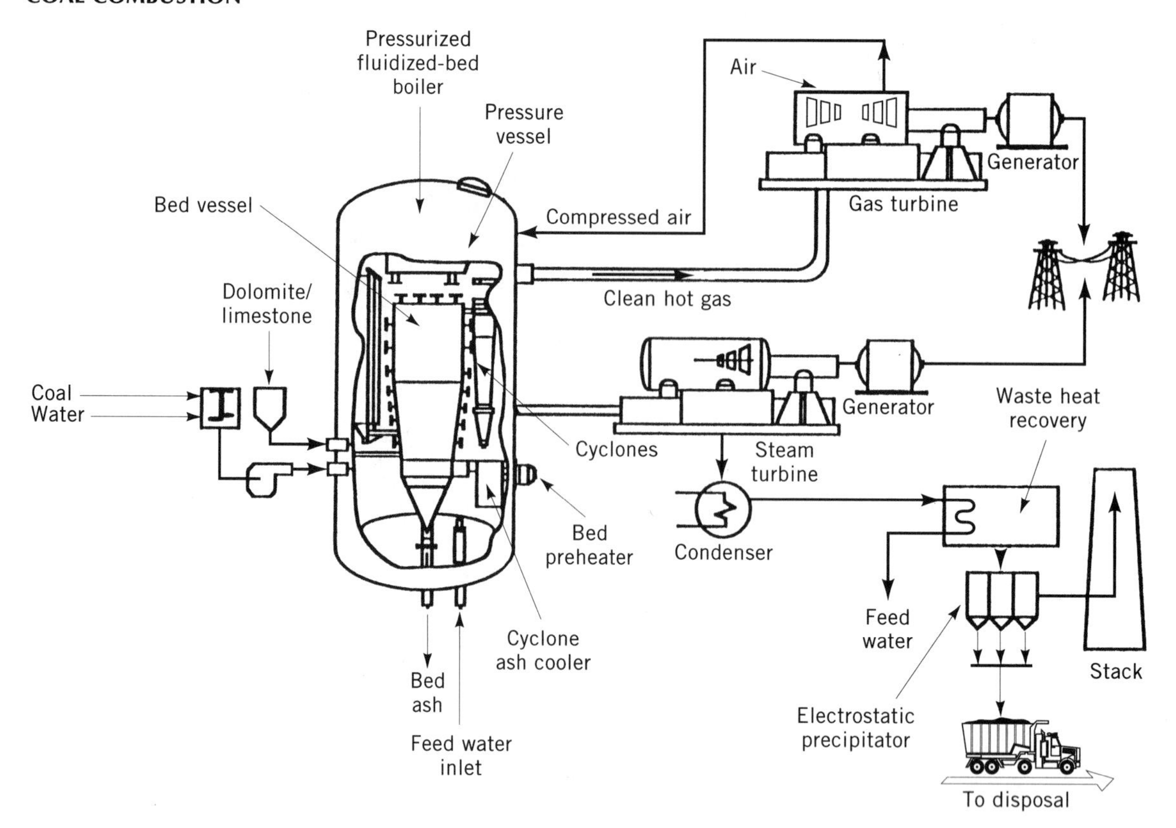

Figure 21. Tidd PFBC demonstration project. From Ref. 101.

and commercial applications is in progress. MCFC stacks of about 10 kW were demonstrated in the mid-1980s and 100-kW stacks are being developed for a 1-MW power plant scheduled for demonstration in the mid-1990s. SOFC stacks of about 100 W range are planned for testing in the early 1990s (9).

ENVIRONMENTAL CONTROL TECHNOLOGIES

Understanding the formation mechanisms and control methods of SO_x and NO_x pollutants is key to environmental concerns about coal combustion. These sulfur and nitrogen species, when emitted into the atmosphere, can cause acid rain. Burning of coal is a major source of these pollutants. Most coals contain about 1% nitrogen by weight, a part of which ends up as NO_x when combusted. Even low sulfur coals contain up to 1% sulfur, which forms SO_x when burned. The control of CO and hydrocarbon emissions is achieved by increasing the efficiency of combustion and, for the most part, has already been accomplished through boiler design and operating experience. The control of particulate matter has been successfully achieved using electrostatic precipitators and fabric filter bag houses (see PARTICULATE CONTROL) and thus is not often a major issue in pollution control research at this time. However, the impact of SO_2 and NO_x abatement technologies can have an adverse impact on the combustion efficiency and must be considered when applying the control applications. Control of trace metals is increasing in importance.

Nitrogen Oxides

Strategies. To comply with regulations for nitrogen oxides emissions, various abatement strategies have been developed. These strategies can be broken down into three categories: (*1*) modification of the combustion configuration, (*2*) injection of reduction agents into the flue gases, and (*3*) treatment of the flue gas by postcombustion denitrification processes. A compilation and discussion of current stationary combustion NO_x control systems, in-

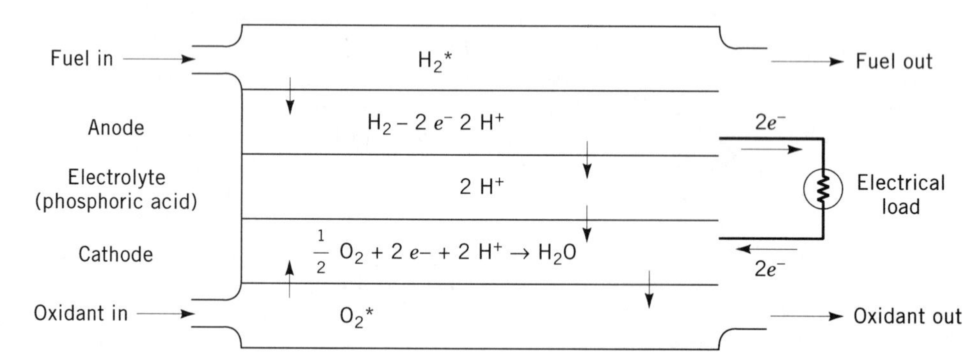

Figure 22. Fuel cell processes. From Ref. 9.

cluding those available by commercial manufacturers and those presently being developed by private and government research organizations, is available (97,120). A determination of the most effective and least cost abatement technology depends on specific boiler firing conditions and the mandated emission limits. A combination of technologies may be needed. Figure 23 (66) shows a composite drawing of an industrial, coal-fired boiler and illustrates the point of application of the various abatement techniques. A summary of the achievable emission rates for the various NO_x control technologies is presented in Table 9.

It is obvious that the selection of the pollution-control system must also include the impact on boiler efficiency because changes can affect the fuel oxidation process. The large variety of combustion configurations and different fuel types requires that the abatement strategy be applied on a case-by-case basis. Figure 24 illustrates potential NO_x reduction for selected technologies applied to pulverized coal furnaces (66).

Combustion Modification Approaches. The formation of nitrogen oxides strongly depends on temperature conditions and oxygen availability; both are influenced by the combustor design and firing configuration. Conventional boiler designs are classified as wall-fired, tangentially fired, cyclone, and stoker-fired boilers. Wall-fired boilers consist of rows of burner nozzles through which air and/or fuel are injected. Cell units employ a different type of wall-fired design with the burners arranged in clusters along the boiler wall. In a tangentially fired boiler, the burners are located in the corners and directed off-center to produce rotation of the bulk combustion zone gas. This ensures a well-mixed condition throughout the boiler. Cyclone boilers are equipped with one or more smaller combustion chambers (cyclones) that burn the fuel at high turbulence and temperatures (1900–2150 K). They are designed to reduce fuel preparation costs (only coal crushing is required), furnace size, and fly ash in the flue gas. Stoker-fired boilers are not generally used for power generation but are used for heating and steam production by industry.

The chief objective of boiler operations is to achieve maximum conversion of the fuel to product to gain the highest possible thermodynamic efficiency. This objective can be accomplished by thoroughly mixing the oxidizing air and fuel and by increasing the excess air. In part, well-mixed conditions are achieved by the air that is used to pneumatically transport classified coal. Preheating the oxidizing air also is used in conventional designs to improve the thermal efficiency.

Unfortunately, the well-mixed high temperature and high excess air conditions that promote efficient oxidation of the fuel constituents also favor the production of nitric oxide (NO). The relative inertness and insolubility of nitric oxide makes flue gas treatment more difficult than sulfur oxide removal. Consequently, it is often easier to intervene right at the point of fuel combustion to avoid the formation of nitrogen oxides and their precursors. Modifications to conventional combustion configuration

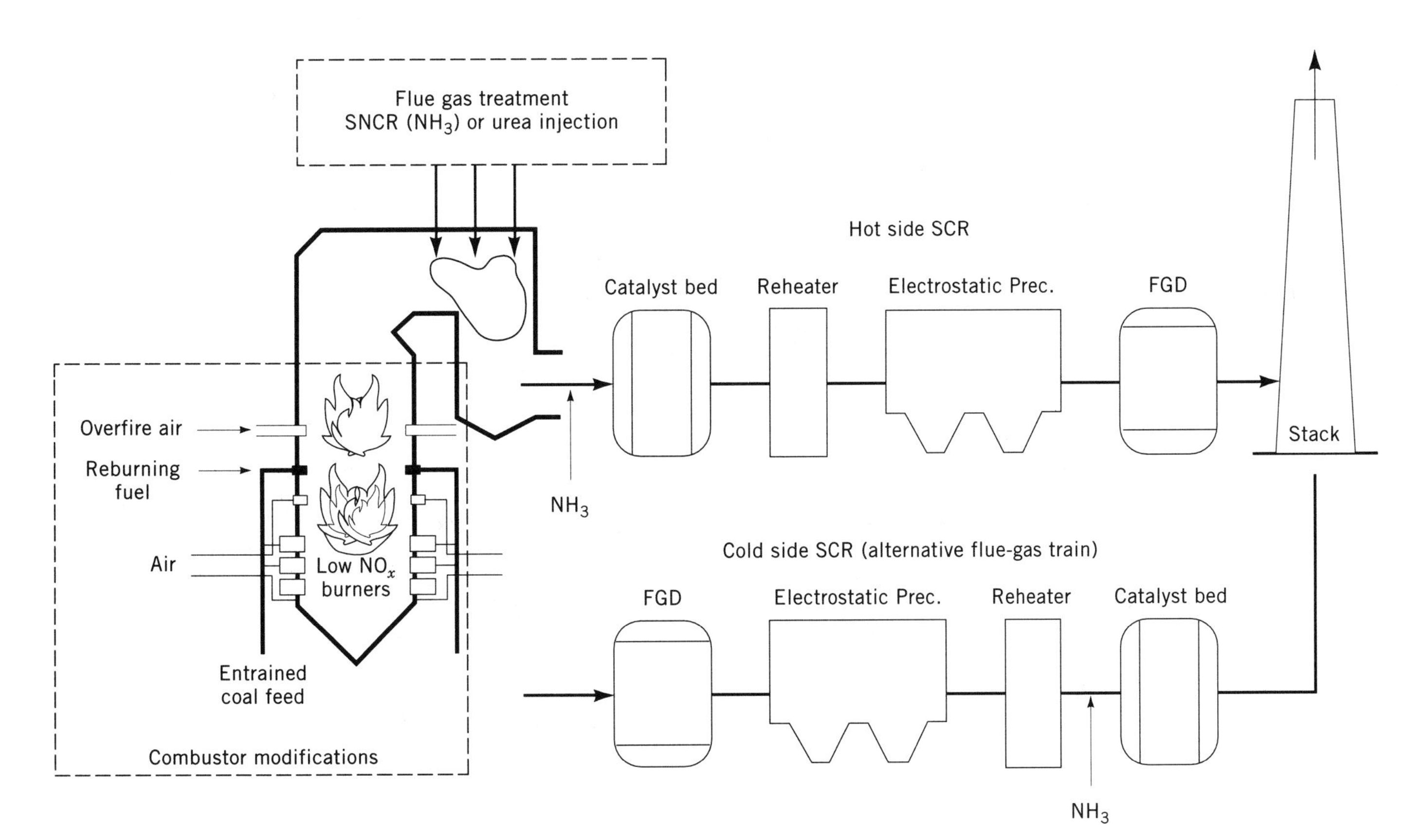

Figure 23. NO_x abatement control technologies and their point of application. From Ref. 66.

Table 9. Achievable Emissions Reduction for Selected NO_x Control Technologies for Coal Combustion[a]

Abatement Technology	Potential NO_x Reduction (%)	Advantages	Disadvantages
Low excess air	0–15	no significant capital cost, increased boiler efficiency	oxygen trim system requires maintenance
Flue gas recirculation	0–5	retrofits are usually possible, can easily be combined with other technologies, effective in reducing thermal NO	requires high temperature duct work and fans, higher capital cost than staged air or staged fuel to achieve equal or greater emissions reductions
Air staging	10–50 30 (typical)	low to moderate operating costs, can be adapted to existing boilers by taking selected burners out of service	may not be applicable to package boilers, access for secondary air ports may not be available
Fuel staging or low NO_x burners	30–75 50 (typical)	low capital costs, retrofits normally possible, low operating and maintenance costs	some increase in static pressure, may require package boilers to operate at a lower rate
Low NO_x burners and air staging or fuel staging	25–80 70 (typical)	highest reduction efficiency without resorting to flue-gas treatment technologies, retrofit of existing combustors normally possible	may require boiler to be taken off-line for an extended period during retrofit, may require package boilers to operate at reduced rates
Ammonia injection	40–90 75 (typical)	lower capital costs than SCR	potential formation of fine particulates, potential NH_3 emissions, moderate to high costs, sensitive to temperature, catalyst poisoning possible
Urea of cyanuric acid injection	30–70	achieves lowest emission rate, can be retrofitted to most combustors	highest capital cost, potential for NH_3 emissions, increased pressure drop, potential for catalyst poisoning
Selective catalytic reduction (SCR)	60–95 90 (typical)		
Combined NO_x–SO_x	50–95 (NO_x) 50–95 (SO_x)	joint removal of NO_x and SO_x, often a salable byproduct is produced	high capital cost, most technologies only in development stage, solid or liquid waste disposal may be required

[a] From Ref. 66.

hardware can reduce nitrogen oxide emissions from 50 to 80%. They are the simplest and least expensive alternatives to apply. Modifications recognized as being technically viable in reducing nitrogen oxide emissions include low excess air, flue gas recirculation, staged air combustion, and staged fuel combustion (commonly called reburning).

Operating with lower excess air is achieved without any retrofit modifications and may reduce nitrogen oxide emissions up to 15% in some, but not all, combustors. Lowering the excess air often improves the thermal efficiency because higher excess air, while serving to reduce hydrocarbon emissions, also acts as a thermal diluent. The trade-offs in combustion efficiency and nitrogen oxide reductions must be monitored to achieve the best air : fuel ratio. Already most boilers, including pre-NSPS boilers, have implemented low excess air practices.

The purpose of flue gas recirculation is twofold: to dilute the inlet oxygen concentration and to lower the combustion zone temperature. Flue gas recirculation primarily affects thermal NO but has little influence on fuel NO. Furthermore, the high cost of equipment and operation (recirculation blowers) makes this modification generally unattractive. In fact, few facilities have retrofitted or designed boilers with flue-gas recirculation capability.

Air staging and fuel staging are effective in controlling fuel nitrogen oxide emissions. In concept, air and fuel staging can be accomplished by advanced burner design (aerodynamic staging) or by externally injecting the air or fuel at optimal locations throughout the furnace. Overfire air has been successfully implemented in wall-fired boilers to achieve nitrogen oxide reduction of up to 30%. Typically, 25% of the air is redirected from the primary combustion zone to an elevation above the fuel burners (Fig. 25) (59). Overfire air has the advantage of reducing nitrogen oxides without sacrificing boiler efficiency; however, thermal NO formation may result in the region of overfire air. Most applications of overfire air have been made to new facilities, although retrofitted applications are proposed for a number of existing facilities.

Air staging is accomplished in tangentially fired boilers by alternating fuel-rich and fuel-lean zones at different levels in the combustion zone. Another technique is to divert part of the combustion air along the boiler wall, thus establishing a concentric envelop of air rotating around a fuel-rich core, as in Figure 25. Staged fuel operation (re-

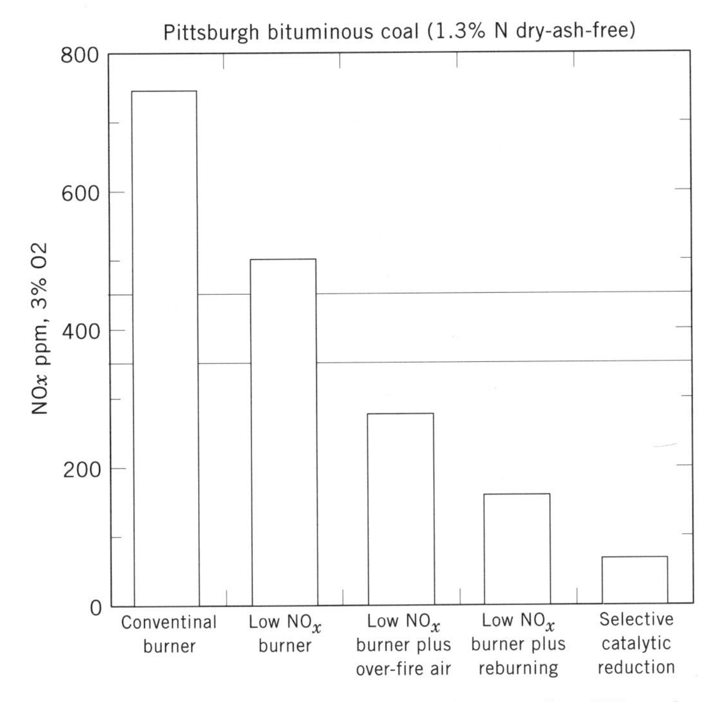

Figure 24. Potential reduction of NO_x by some low NO_x technologies for pulverized coal combustors. From Ref. 66.

burning) involves injecting fuel into more than one combustion zone in the boiler. The objective is to inject part of the fuel with the bulk of the combustion air in the primary combustion zone. This produces a fuel-lean zone that reduces the formation of thermal NO by lowering the peak temperature; however, the potential for formation of fuel NO may be increased. In the secondary combustion zone where reburning fuel is added, the overall air:fuel ratio is fuel-rich. Fuel fragments are produced that react with NO, producing HCN. Subsequently, HCN favors reduction to N_2 under the prevailing fuel-rich conditions. Final air addition is then employed to burn out the unoxidized hydrocarbons at lower temperatures that do not favor thermal NO formation.

Fuel staging is particularly effective for abating fuel NO formation in oil and coal-fired boilers. The most effective reburning fuels are volatile, low nitrogen containing fuel oils and natural gas, although coal has been successfully applied in some pilot-scale tests. A 40–60% reduction in nitrogen oxide emissions can be achieved by reburning. One combustor that is a prime candidate for fuel staging is the cyclone combustor, because most modifications to the cyclone chambers are counterproductive.

Advanced low NO_x burners have been developed and tested by a number of boiler manufacturers and utilities (97). The general purpose of low NO_x burners is to aerodynamically stage the air or fuel. Variation in the momentum of the fuel and air streams, in the swirl of the streams, and in the angle of injection of the streams, affects the mixing and combustion pattern and retards the formation of both thermal and fuel NO. The majority of nitrogen oxides formed in coal flames occurs as the fuel nitrogen, which is released with the volatiles or contained in the char, is oxidized. Fortunately, the chemistry controlling nitrogen oxide formation is slower than the fuel oxidation rates. Thus the aim of low NO_x burners is to establish a fuel-rich zone of low temperature where the volatiles are oxidized but not the fuel nitrogen constituents. This zone is followed by gradual mixing of air into the outer fringes of the flame to burn out the char and soot. Figure 26 (121) illustrates the basic features of a low NO_x burner that is incorporated with some variations in commercially available burners.

Injection of Chemical Reductants. Under the right conditions, nitrogen oxides can be converted to nitrogen and water by injecting chemical reductants into the combustion flue gases. This technique involves selective noncatalytic homogeneous reactions with the decomposition products of the injected compounds. Ammonia is used in a well-established commercial process pioneered in the early 1970s (122). Other compounds that have been successfully tested include cyanuric acid, ($(HOCN)_3$), hydrazine hydrate (N_2H_4) solutions, urea ($CO(NH_2)_2$), and methanol (123,124).

Factors that determine the effectiveness of nitrogen oxide removal by reducing agents include the temperature of the flue gas, injection rate and mixing efficiency, and the composition of the flue gas. Injection jets are designed to achieve rapid and thorough dispersion of the reductant at the optimum temperature. Flue gas radicals initiate decomposition of the chemicals to nitrogen-containing intermediates that then selectively react with NO to form N_2. Thus the presence of moisture in the exhaust gas, or certain additives such as hydrogen and hydrogen peroxide, may enhance the initiation of reduction reactions. Normally, there is a narrow temperature range that is conducive to the reduction reactions. At excessive temperatures, the nitrogen intermediates will be oxidized and produce additional nitrogen oxides. At lower temperatures, undesired amounts of the injectant can slip through the process and be emitted with the nitrogen oxides. In some cases, the unreacted injectant can react with the nitrogen oxides to form undesirable particulate matter (eg, NH_4NO_3 and $(NH_4)_2SO_4$ aerosols).

Research efforts have succeeded in elucidating the chemical kinetics of ammonia decomposition and reaction with nitric oxide to the point at which the process is virtually predictable (122). This knowledge, coupled with practical experience in a variety of commercial combustors, has allowed nitrogen oxide reduction efficiencies of around 75% to be obtained. Some retrofitted applications have achieved in excess of 90% removal of NO. The reaction sequence for NO_x reduction by ammonia is summarized in Figure 27 (122). The principal nitrogen intermediate, NH_2, either reduces NO or is competitively converted to NO. The NH_2 + O reaction has a high activation energy; thus higher temperatures are required for this reaction pathway to be favorable. When the temperature is too low, the reaction of NH_3 with oxygen atoms and hydroxyl radicals is slow, allowing NH_3 to slip. The optimum temperature window for selective reduction of NO ranges from 1250 to 1400 K.

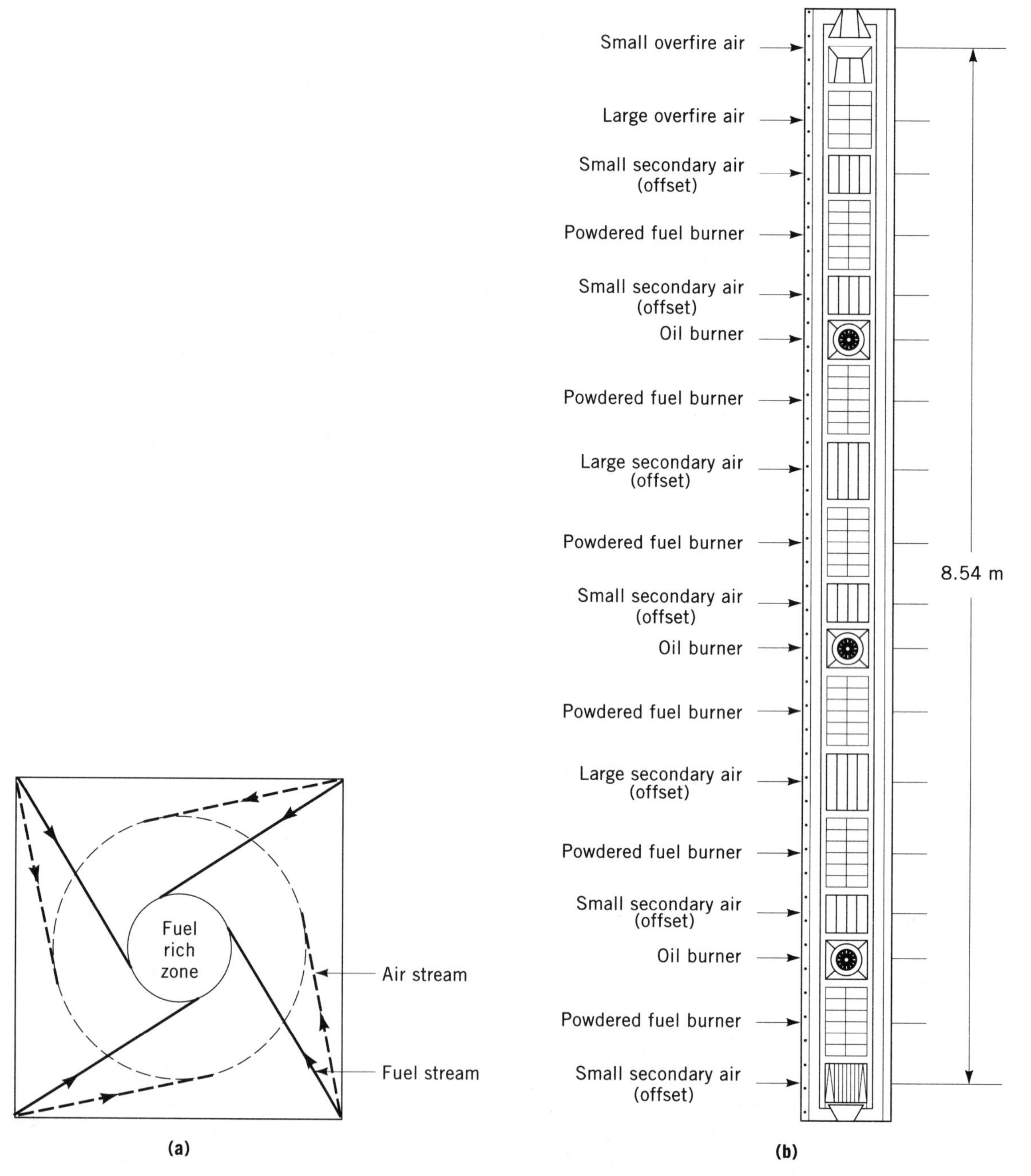

Figure 25. Arrangement of burners and fluid motion of a low NO_x concentric firing system. From Ref. 59.

Ammonia reduction of nitric oxide is effective only when excess oxygen is available to produce oxygen atoms. In fact, the process can be inhibited by increasing the NH_3 concentration above the O_2 concentration or by increasing the moisture content of the flue gas. The presence of moisture can inhibit NO reduction at low temperatures by competing for the oxygen atom. At higher temperatures, moisture competes with the oxidation of NH_2 and inhibits NO formation. The combination of effects results in an increase of the optimum temperature. Additives such as hydrogen and hydrogen peroxide tend to lower the optimum temperature window.

The selection of cyanuric acid as a chemical reductant resulted from the recognition that HNCO could be involved in a series of reactions that would remove NO (125). When cyanuric acid is injected into the exhaust stream, it rapidly decomposes to form isocyanic acid (HNCO), which further decomposes in high temperatures to NH, NH_2, and N_2O. At lower temperatures, surface reactions are required to decompose isocyanic acid. The optimum temperature window ranges from 1000 to 1200 K (124). The detailed chemistry of urea with NO is not completely understood, but the reaction path depends on the thermal decomposition products. It is suggested that urea

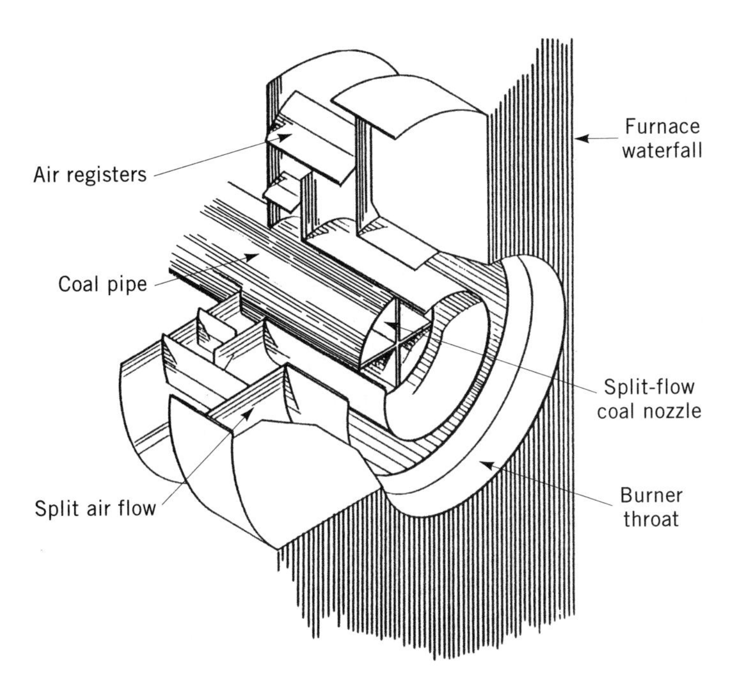

Figure 26. Schematic diagram of a basic low NO_x burner. From Ref. 121.

might decompose into NH_3 and HNCO and thus follow either or both the ammonia or isocyanic acid reaction paths.

One major disadvantage of flue-gas treatment by chemical reductants is the potential formation of byproduct nitrous oxide (N_2O). In view of the increased concern over N_2O emissions, this is an undesirable byproduct. An experimental and theoretical study recently conducted concluded that all chemical injectants, including NH_3, produce N_2O as a byproduct, depending on the amount of reductant used and the conditions at the point of injection (126). Ammonia produced the least concentrations of N_2O, while cyanuric acid produced the highest concentrations. Under some conditions, a molecule of N_2O may be formed for every molecule of NO removed by cyanuric acid (69). Nitrous oxide emissions due to urea injection were found to be between the levels produced by NH_3 and cyanuric acid.

Catalytic Reduction of Nitrogen Oxides. The most effective technology for removing nitrogen oxides from combustion gases is selective catalytic reduction (scr). Selective catalytic reduction involves introduction of ammonia before the catalyst bed (Fig. 23). The catalyst can be placed in different positions in the flue-gas flow as long as the gas temperature is appropriate for the type of catalyst used. Practical applications in Japan and Germany have obtained removal efficiencies approaching 90% (127,128). Nonselective catalytic reduction, in which hydrogen or hydrocarbons in the flue gas act as reducing agents, has also been researched but has not achieved the same level of success as scr.

The basic scr catalyst types that have been researched and are commercially available have been discussed in detail (97,128). These include noble metals (Pt, Pd, Ru/AL_2O_3, and Pt/Al_2O_3), metal oxides (Fe_2O_3/Cr_2O_3, V_2O_5/TiO_2, $V_2O_5/MoO_3/WO_3/Al_2O_3$) and zeolite (synthetic mordenites). Each catalyst exhibits a unique optimum operating temperature and $NH_3{:}NO_x$ ratio. The optimum operating temperature ranges from 500 to 700 K. At low temperatures, the reaction efficiency drops off, while the catalyst is prone to sintering at excessive temperatures. Typically, the catalyst bed will be placed upstream of the elctrostatic precipitator or bag house to take advantage of the hot flue gas. However, a problem exists for fouling and plugging in the catalyst bed by fly-ash particulates. Flue-gas products such as SO_3 can also combine with ammonia to form corrosive ammonium bisulfate $(NH_4)_2SO_4$.

Most metal-based catalysts are subject to a decrease in activity during operation. Catalyst poisons that deactivate the reaction sites include alkaline and alkaline-earth oxides or arsenic compounds. Sulfur oxides can also deactivate certain metal oxide catalysts and can reversibly inhibit noble metal catalysts. A catalyst activity loss of 20%/yr is considered normal for most combustion cleanup applications.

Selective catalytic reduction requires the highest capital and operating costs, despite many recent technological advances in catalyst support geometries and ammonia injection control. However, because of more stringent standards, scr may become the most viable technology for nitrogen oxide removal.

Sulfur Pollutants

Alternatives. Unlike nitrogen oxide pollutants, there are no benign gaseous species to which sulfur-containing pollutants can be converted. Limestone slurry scrubbers, although costly, have been effective in removing more than 90% of the sulfur oxides from the flue gas. A technical description on flue gas desulfurization systems has

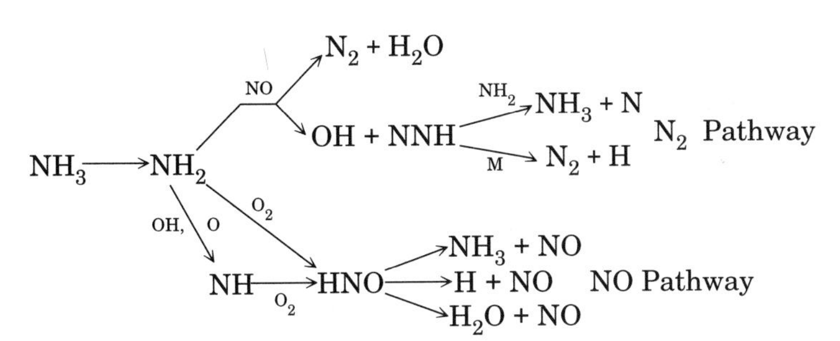

Figure 27. Mechanism of nitric oxide reduction in the flue gas by ammonia injection. From Ref. 122.

been published (129). Advanced combustion configurations, including the slagging combustor, fluidized-bed combustors, and the integrated gasification combined cycle (see COAL GASIFICATION) can also achieve high sulfur pollutant reduction levels, while intermediate stream desulfurization of effluents is a promising technology for gasification processes. However, to reduce current sulfur oxide emissions, control technologies that can be retrofitted to existing facilities are being developed. The injection of sorbents directly into the postcombustion chamber is one option that can reduce SO_2 emissions from new and existing combustor configurations.

During the previous two decades, several desulfurization alternatives have been investigated in the laboratory and on a pilot plant scale. A few commercial applications have been demonstrated. New generation desulfurization can generally be categorized into four groups: (*1*) precombustion coal cleaning, (*2*) advanced combustion configurations, (*3*) *in situ* sulfur reduction, and (*4*) postcombustion cleaning. A fundamental discussion of many of the alternatives is available (130).

Precombustion Coal Cleaning to Remove Sulfur and Minerals. Environmental constraints, together with development of coal-based fuels for oil substitution, have increased interest in the cleaning of coal (131,132). Reduction in mineral and sulfur content of coals is of particular concern.

A certain amount of the sulfur can be removed from the coal before combustion by physically, chemically, or biologically cleaning the coal. Most coal receives some cleaning before it is burned. Physical cleaning typically separates inert matter from the coal by relying on differences in density or surface properties; however, it is not possible to remove physically the organically bound sulfur. Physical cleaning can remove 30–50% of the pyritic sulfur, which is 10–30% of the total sulfur. Advanced coal-cleaning methods are expected to remove even a greater amount of pyritic sulfur. Through chemical and biological cleaning, some of the sulfur that is chemically bound to the coal can also be removed. Chemical or biological cleaning is capable of removing 90% of the total organic and inorganic sulfur, but it is expensive. The predominate method of chemical coal cleaning is to leach both the organic sulfur and mineral matter from the coal using a hot sodium-based or potassium-based caustic. Biological cleaning involves exposing the coal to bacteria or fungi that have an affinity for sulfur. Often, sulfur-digesting enzymes are injected into the coal to stimulate sulfur consumption.

Ultrafine grinding (micronization) of the coal to an average particle diameter of about 10 μm separates much of the mineral matter from the organic coal. Subsequent separation of the coal and mineral matter by differences in specific gravity have led to coals with mineral matter levels of less than 1% and with sulfur levels about 50% the original value. Reduction in sulfur level by physical methods depends strongly on the relative proportions of sulfur in the mineral matter (pyritic and sulfates) and that bound in the organic structure. Work is continuing on several physical and chemical coal-cleaning processes (131,132) in an effort to demonstrate economically viable processes (4).

In Situ Sulfur Capture. Sulfur pollutants can be efficiently removed during the combustion process or in the flue-gas ducts by injecting sorbents into the postcombustion gases or by adding sorbents directly to the furnace or combustor. A variety of sorbents will absorb and react with sulfur pollutants, including calcium and magnesium oxides and metal oxides of zinc, iron, and titanium. The calcium content of coal ash provides some natural sulfur capture capability, which is limited by the Ca:S ratio of the coal (133), and the addition of other sorbents is required for substantial *in situ* sulfur capture. Calcium-based sorbents are particularly attractive due to their low cost and the inertness of the calcium sulfate or calcium sulfide byproduct. The reacted sorbents are typically collected with the fly ash by the bag house or electrostatic precipitator.

Sulfur oxide reduction by sorbents has been demonstrated in a number of pilot-scale facilities. Calcium-based sorbent–SO_2 capture has been characterized in a 380-kW, two-burner level, tangentially fired pilot-scale test (134). At the nominal Ca:S ratio of 2.1, the overall calcium use was 24%. Limestone injection tests conducted in a 440-kW pilot-scale combustor achieved SO_2 reductions of 40–60% for a range of test parameters. Conversions in excess of 50% appear possible for hydrated lime (135).

Fluidized-bed combustors commonly use crushed limestone for bed material. The particles, introduced into the bed, heat and calcine. The calcium oxide then reacts with sulfur dioxide to form calcium sulfate. Some coal, coal ash, and limestone particles are thrown out of the bed in a freeboard zone. The freeboard zone provides additional space and time to increase sulfur dioxide capture. The substantial reduction of SO_x during combustion (about 90%) reduces the need for postcombustion SO_x control processes (66).

Flue Gas Desulfurization. The most widely used technology for SO_x control is flue gas desulfurization (fgd). It is a highly efficient but expensive technology. More than 100 fgd technologies are commercially available or under development throughout the world (136). They are usually classified as wet throwaway, dry throwaway, and regenerative fgd. By far the most common fgd technology is the wet throwaway, or wet scrubbing. The largest user of fgd in terms of installed capacity is the United States, with more than 80,000 MW, followed by Japan, with about 45,000 MW, and Germany, with about 40,000 MW (136). The fgd systems are required on all new coal-fired power plants in the United States. The first wet scrubbers used lime as a sorbent and produced waste that required disposal. Most new wet scrubbers use limestone and produce gypsum, which can be marketed. In the wet scrubber, the flue gas passes through an absorber tower at low velocity while being sprayed by the limestone slurry.

Combined NO_x–SO_x Combustion Gas Treatment Technologies

Numerous processes also have been proposed in recent years for combined denitrification and desulfurization of combustion flue gases. Some are commercially available and have been implemented by full-scale combustion facilities (97,137). Most of the NO_x–SO_x removal processes are considered expensive and may only be considered as more stringent standards become mandatory for new and existing facilities. Combined NO_x–SO_x flue-gas treatment processes are generally classified as wet or dry. Wet processes normally use absorption with complexation, reduction, or oxidation chemistry; the dry processes include catalytic decomposition, absorption by activated carbon beds, spray drying, and electron-beam irradiation.

One innovative, NO_x–SO_x postcombustion concept involves streamer corona discharge to produce radical species in the flue gas. The chemistry proposed by various researchers is based on the interaction of electrons with molecular species, giving rise to atomic oxygen, nitrogen, and amine radicals. In the presence of ammonia and water, the radicals lead to the reduction of nitrogen oxides and SO_2, proceeding through reactions included in the ammonia–NO_x reduction mechanism and other ionic and free-radical reactions. However, in the absence of ammonia and water, NO_x emissions can be further exacerbated. The reactions leading to removal of SO_2 and NO by streamer corona discharge have been discussed in detail (138). Another scheme proposes denitrification of the flue gas using regenerable mixed oxide sorbents (eg, Na_2O, Co_3O_4, and Fe_2O_3). This process involves the sorption of nitrogen oxides onto porous sorbents that can be thermally regenerated on saturation. The concentrated gas stream is then treated by standard catalytic reduction methods. An in-depth description on the removal of nitric oxide from flue gas using mixed oxide sorbents has been published (139).

Demonstrations

Practical NO_x control experience has been obtained in pilot-scale and utility boilers over the past score of years. The majority of this experience has been reported in the *Proceedings of the Joint Symposium on Stationary Combustion NO_x Control* sponsored by the EPA and EPRI since 1979. Researchers from the international community have discussed the implementation of combustion modifications, injection of chemical reductants, and selective catalytic reduction. The bulk of scr experience has been provided by Japan. Other references include publications by the International Energy Agency (IEA), which is currently sponsored by 14 nations (Austria, Australia, Belgium, Canada, Denmark, Germany, Finland, Italy, Japan, The Netherlands, Spain, Sweden, the UK, and the United States). Most of these countries have strong governmental support for NO_x and SO_x abatement demonstration projects.

The U.S. CCT program has a goal to furnish showcase examples of advanced, more efficient, and environmentally viable coal-utilization alternatives. CCT demonstration projects include NO_x and SO_x control concepts that are applicable to conventional coal-fired power plants (eg, flue-gas desulfurization, limestone injection, gas reburning, and coal cleaning) and also nontraditional methods of coal combustion (eg, pressurized fluidized-bed combustion and integrated gasification-combined gas cycle (IGCC)).

Several CCT projects are being carried out to demonstrate sorbent capture of sulfur in entrained-flow and fluidized-bed combustors and also sorbent injection into the flue-gas duct (8). One project involves joint gas reburning and sorbent injection in a 71-MW, tangentially fired boiler burning a 3 wt % sulfur Illinois coal (140). The upper limit to SO_2 reduction using calcium-based sorbents appears to be 50–60%. With popular technology termed limestone injection multistage burner (LIMB) combined with lime injection in the flue-gas duct, reduction of 50–80% SO_2 is projected for a 104-MW wall-fired boiler. Other sorbent–SO_2-capture demonstration projects include injection into an integrated gasification combined cycle, a slagging combustor, and fluidized-bed combustors (141). The latter application is enhanced when the fluidized bed is pressurized (66).

Table 10. Technical Issues and Problems Arising from the Expanding Use of Coal[a]

Problems Area	General Description
Environmental concerns	need for control of emissions from coal processes, including oxides of nitrogen, oxides of sulfur, carbon dioxide, fine particulates, soot, heavy hydrocarbons, and trace elements
Safety and explosions	need for methods to control explosions and fires in mines, utility and industrial boilers, and in coal storage facilities
Combustion and conversion efficiency	need to maintain high combustion efficiency as process variables change for pollutant control, coal variability, etc.
Ash and slag	need to control ash and slag deposits and removal, and associated maintenance of surfaces, corrosion, etc, particularly for coals with high ash levels
Process design and optimization	need for development and verification of more efficient methods for designing, scaling, and optimizing coal conversion processes
Basic process data	need for relating coal characteristics (composition, structure, etc) to conversion and combustion characteristics, including radioactive properties, pyrolysis rates, and char oxidation rates
Coal oil–water mixtures	for retrofitting existing oil-fired furnaces and for direct coal ignition systems; needs for coal preparation, COM–CWM slurry characteristics and full-scale boiler tests

[a] From Ref. 4.

FUTURE OF COAL

Coal will most certainly play a significant role in meeting future energy needs. It is the most abundant fossil reserve. A total of 70% of the world's known fossil reserves, and 94% of the U.S. reserves are coal (142). It is estimated that there are about 200 yr of known world coal reserves using existing technologies at present consumption rates (143). Known recoverable reserves of U.S. coal could satisfy domestic demands at current consumption rates for nearly 300 yr (101). About 90% of the world's energy in 1988 was supplied by fossil fuels (143); about 30% of the total was coal. It is also projected that the use of fossil energy will increase by an estimated 43% over the next 15 yr (142).

Significant progress has been made in the past few decades to reduce harmful emissions related to the use of coal, specifically regarding SO_x, NO_x and particulate matter. However, much remains to be accomplished to ensure clean and efficient future use of coal (Table 10). Destruction of the ozone layer is an important question. N_2O has a long half-life compared with NO. It is produced in high percentages in certain combustors such as fluidized beds, which operate at lower temperatures. Global warming is an important technical issue with political consequences. Control of minerals in conversion of low quality fossil fuels remains a challenge. Variable quality feedstocks, increasingly tight environmental regulations, emissions of toxins and trace metals, acid rain control, and clean incineration of wastes add to research needs. A lifetime of important research is still required to contribute to the solution of these problems in areas of energy generation on a worldwide scale.

Acknowledgment

Craig Eatough, postdoctoral associate in the Advanced Combustion Engineering Research Center, has contributed substantially to this section. Predrag Radulovic's and Richard Boardman's recently published work on coal combustion technologies and pollutant emissions and control, respectively, provided important contributions to this article. Support of the Advanced Combustion Engineering Research Center at Brigham Young University in preparing this manuscript is appreciated.

BIBLIOGRAPHY

1. M. A. Elliott, *Chemistry of Coal Utilization,* Vol. 2, John Wiley & Sons, Inc., New York, 1981.
2. J. Tonge, *Coal,* Archibald Constable and Co, Ltd., London, 1907.
3. A. C. Fieldner and W. E. Rice, *Research and Progress in the Production and Use of Coal,* National Resources Planning Board, Washington, D.C., Oct. 1941.
4. L. D. Smoot, in W. Bartok and A. F. Sarofim, eds., *Fossil Fuel Combustion: A Science Source Book,* John Wiley & Sons, Inc., New York, 1991.
5. U.S. Department of Energy, *International Energy Annual 1988, Report* DOE/EIA-0219(88), Energy Information Administration, Washington, D.C., 1989.
6. U.S. Department of Energy, *"Coal," National Energy Strategy: Powerful Ideas for America, 1991–92,* Report DOE/S-0082P, U.S. Government Printing Office, Washington, D.C., 1991.
7. U.S. Department of Energy, *Annual Energy Outlook 1991 with Projections to 2010,* Report DOE/EIA-0383 (91), Energy Information Administration, Washington, D.C., 1991.
8. U.S. Department of Energy, *Clean Coal Technology Demonstration Program: Program Update 1990 (as of December 31, 1990),* Report DOE/FE-0219P, Washington, D.C., 1991.
9. P. T. Radulovic and L. D. Smoot, in L. D. Smoot, ed., *Fundamentals of Coal Combustion: For Clean and Efficient Use,* Elsevier, New York, 1993.
10. J. G. Singer, ed., *Combustion: Fossil Power Systems,* 3rd ed., Combustion Engineering, Windsor, Conn., 1981.
11. R. H. Essenhigh in Ref. 1, p. 1153.
12. L. D. Smoot, ed., *Fundamentals of Coal Combustion: for Clean and Efficient Use,* Elsevier, New York, 1993.
13. C. L. Wilson, *Coal-Bridge to the Future: Report of the World Coal Study,* Ballinger, Cambridge, Mass., 1980.
14. M. B. McNair, *Energy Data Report: Coal Distribution January–June 1980,* U.S. Department of Energy Report DOE/EIA-0125(80/2Q), Energy Information Administration, Washington, D.C., 1980.
15. R. H. Essenhigh in C. Y. Wen and E. S. Lee, eds., *Coal Conversion Technology,* Addison-Wesley Publishing Co., Inc., Reading, Mass., 1979, pp. 171–312.
16. R. G. Schwieger, *Power* **123,** S.1–S.24 (1979).
17. L. D. Smoot and D. T. Pratt, *Pulverized-Coal Combustion and Gasification,* Plenum Press, New York, 1979.
18. D. B. Anthony and J. B. Howard, *AIChE J.* **22,** 625–656 (1976).
19. P. R. Solomon, *Characterization of Coal and Coal Thermal Decomposition,* Advanced Fuel Research, Inc., East Hartford, Conn., 1980.
20. K. L. Smith, L. D. Smoot, T. H. Fletcher, and R. J. Pugmire, *The Structure and Reaction Processes of Coal,* Plenum Press, New York, 1994.
21. E. Stach, M.-Th. Mackowski, M. Teichmuller, G. H. Taylor, D. Chandra, and R. Teichmuller, *Stach's Textbook of Coal Petrology,* Gebruder Borntraeger, Berlin, 1982.
22. S. A. Benson, M. L. Jones, and J. N. Harb, in Ref. 12, Chapt. 4.
23. J. J. Helble, S. Srinivasachar, and A. A. Boni, *Prog. Energy Comb. Sci.* **16,** 267–279 (1990).
24. C. J. Zygarlicke, M. L. Jones, S. A. Benson, and W. H. Puffe, *Prog. Energy Comb. Sci.* **19,** 195–204 (1990).
25. M. L. Jones, D. P. Kalmanovitch, E. N. Steadman, C. J. Zygarlicke, and S. A. Benson, in H. L. C. Meuzelaar, ed., *Advances in Coal Spectroscopy,* Plenum Press, New York, 1992, pp. 1–28.
26. C. J. Zygarlicke, D. L. Toman, and S. A. Benson, *Am. Chem. Soc. Div. Fuel Chem.* **35**(3), 621–636 (1990).
27. J. M. Kuchta, V. R. Rowe, and D. S. Burgess, *Spontaneous Combustion Susceptability of U.S. Coals,* RI-8474, U.S. Bureau of Mines, 1980.
28. M. Hertzberg, C. D. Litton, and R. Garloff, *Studies of Incipient Combustion and Its Detection,* RI-8206, U.S. Bureau of Mines, 1977.
29. J. Nagy, J. G. Dorsett, and A. R. Coper, *Explosibility of Carbonaceous Dusts,* RI-6597, U.S. Bureau of Mines, 1965.
30. P. R. Solomon and M. B. Colket in *Symposium (International) on Combustion,* Vol. 17, Combustion Institute, Pittsburgh, Pa., 1979, pp. 131–143.

31. R. H. Essenhigh, M. K. Misra, and D. W. Shaw, paper presented at the International Specialists' Meeting on Solid Fuel Utilization, Portuguese Section, the Combustion Institute, Lisbon, 1987.
32. S. C. Saxena, *Prog. Energy Combust. Sci* **16,** 55–94 (1990).
33. W. C. Xu and A. Tomita, *Fuel* **66,** 632–636 (1987).
34. E. M. Suuberg, W. A. Peters, and J. B. Howard in *Proceedings of the 17th Symposium (International) on Combustion,* Combustion Institute, Pittsburgh, Pa., 1978, pp. 117–130.
35. P. R. Solomon, M. A. Serio, and E. M. Suuberg, *Prog. Energy Combust. Sci.* **18,** 133–220 (1992).
36. M. R. Khan, *Fuel* **68,** 1522–1531 (1989).
37. P. R. Solomon, J. M. Beér, and J. P. Longwell, *Energy,* **12,** 837–862 (1987).
38. D. McNeil in Ref. 1, Chapt. 17.
39. L. D. Smoot and P. J. Smith, *Coal Combustion and Gasification,* Plenum Press, New York, 1985.
40. D. Marlow, S. Niksa, and C. H. Kruger, in *Proceedings of the 24th Symposium (International) on Combustion,* Combustion Institute, Pittsburgh, Pa., 1992, pp. 1251–1258.
41. D. W. Shaw, X. Zhu, M. K. Misra, and R. H. Essenhigh, in *Proceedings of the 23rd Symposium (International) on Combustion,* Combustion Institute, Pittsburgh, Pa., 1990, pp. 1155–1162.
42. M. A. Field, D. W. Gill, B. B. Morgan, and P. G. W. Hawksley, *Br. Coal Util. Res. Assoc. Monogr. Bull.* **31,** 285 (1967).
43. D. J. Harris and I. W. Smith, in Ref. 41, pp. 1185–1190.
44. P. L. Walker Jr., F. Rusinko Jr., and L. G. Austin, *Adv. Catal.* **11,** 135 (1959).
45. D. Gray, J. G. Cogoli, and R. H. Essenhigh, *Adv. Chem. Ser.* **131**(6), 72–91 (1974).
46. I. W. Smith, paper presented at the Conference on Coal Combustion Technology and Emission Control, California Institute of Technology, Pasadena, Feb. 1979.
47. M. A. Field, *Br. Coal Util. Res. Assoc. Monogr. Bull.* **28,** 61 (1964).
48. R. H. Hurt and R. E. Mitchell in Ref. 40.
49. W. J. Thomas, *Carbon* **3,** 435 (1977).
50. C. W. Garrett, *Prog. Energy Combust. Sci.* **18,** 369–407 (1992).
51. J. P. Lodge Jr., *Selections of the Smoake of London, Two Prophesies,* Maxwell Reprint Co., Elmsford, N.Y., 1969.
52. J. Q. Koenig, W. E. Pierson, M. Horike, and R. Frank, *Arch. Environ. Health* **37,** 5–9 (1982).
53. G. L. Love, L. Shu-ping, and C. M. Shy, *Arch. Environ. Health* **37,** 75–80 (1982).
54. R. F. Phalen, *Inhalation Studies: Foundations and Techniques,* CRC Press, Inc., Boca Raton, Fla., 1984.
55. J. J. Kraushaar and R. A. Ristinsen, *Energy and Problems of a Technical Society,* John Wiley & Sons, Inc., New York, 1984.
56. B. J. Finlayson-Pitts and J. N. Pitts Jr., *Atmospheric Chemistry: Fundamentals and Experimental Techniques,* John Wiley & Sons, Inc., New York, 1986.
57. C. Hakkarinen, paper presented at the EPRI/EPA Joint Symposium on Stationary Combustion NO_x Control, San Francisco, Calif., 1989.
58. R. I. Bruck, paper presented at the EPRI/EPA Joint Symposium on Stationary Combustion NO_x Control, New Orleans, La., 1987.
59. J. Redman, *Chem. Eng.,* **458,** 35 (Mar. 1989).
60. A. N. Hayhurst and A. D. Lawrence, *Prog. Energy Combust. Sci.* **18,** 529–552 (1992).
61. S. M. Japer, W. W. Brachaczek, R. A. Gorse Jr., J. M. Norbeck, W. R. Pierson, *Atmos. Enviorn.* **20,** 128–1289 (1986).
62. L. W. Richards, R. W. Bergstrom Jr., and T. P. Ackerman, *Atmos. Environ.* **20,** 387–396 (1986).
63. S. W. Matthews and J. A. Sugar, *Natl. Geographic Mag.,* **178,** 66 (1990).
64. Office of Technology Assessment, *Changing by Degrees: Steps to Reduce Greenhouse Gases,* OTA Brief Report, U.S. Congress, Washington, D.C., Feb. 1991.
65. D. Boutacoff, *EPRI J.,* 13 (Mar. 1991).
66. R. D. Boardman and L. D. Smoot, in Ref. 12, Chapt. 6.
67. L. J. Muzio, M. E. Teague, T. A. Montgomery, G. S. Samuelsen, L. C. Kramlich, and R. K. Lyon, paper presented at the EPRI/EPA Joint Symposium on Stationary Combustion NO_x Control, San Francisco, 1989.
68. G. M. Johnson and M. Y. Smith, *Combust. Sci. Technol.* **19,** 67–70 (1978).
69. J. A. Miller and C. T. Bowman, *Progr. Energy Combust. Sci.* **15,** 287–338 (1989).
70. Y. B. Zel'dovich, P. Y. Sadovnikov, and D. A. Frank-Kamentskii, *Oxidation of Nitrogen in Combustion,* Academy of Sciences of the USSR, 1947.
71. G. A. Lavoie, J. B. Heywood, and J. C. Keck, *Combust. Sci. Technol.* **1,** 313 (1970).
72. C. Morley in *Proceedings of the 18th Symposium (International) on Combustion,* Combustion Institute, Pittsburgh, Pa., 1981, pp. 23–32.
73. A. E. Axworthy, G. R. Schneider, and V. H. Dayan, *Chemical Reactions in the Conversion of Fuel Nitrogen to NO_x,* EPA 600/2-76-039, NTIS, Springfield, Va., 1976.
74. C. P. Fenimore, *Combust. Flame* **26,** 249 (1976).
75. R. A. Altenkirch, R. E. Peck, and S. L. Chen, *Combust. Sci. Technol.* **20,** 49–58 (1979).
76. J. W. Glass and J. O. L. Wendt, DOE/ET/15184-1152, 1981.
77. K. J. Knill and M. E. Morgan, paper presented at the EPRI/EPA Joint Symposium on Stationary Combustion NO_x Control, San Francisco, 1989.
78. D. W. Pershing and J. O. L. Wendt, *Ind. Eng. Chem. Process Des. Dev.* **18,** 1 (1979).
79. J. W. Glass and J. O. L. Wendt, in *Proceedings of the 19th Symposium (International) on Combustion,* Combustion Institute, Pittsburgh, Pa., 1982, pp. 1243–1251.
80. M. P. Heap, T. M. Lowes, R. Walmsley, H. Bartelds, and P. LeVaguerese, *Burner Criteria from NO_x in Pulverized Coal Flames,* EPA-600/2-76-061a, 1976.
81. A. C. Bose, K. M. Dannecker, and J. O. L. Wendt, *Energy Fuels* **2,** 301–308 (1988).
82. M. T. Cheng, M. J. Kirch, and T. W. Lester, *Combust. Flame* **77,** 213–217 (1989).
83. J. Speight, *The Chemistry and Technology of Coal,* Marcel Dekker, Inc., New York, 1983.
84. J. C. Kramlich, P. C. Malte, and W. L. Grosshandler, in Ref. 72, pp. 151–161.
85. S. D. Zaugg, A. U. Blackham, P. O. Hedman, and L. D. Smoot, *Fuel,* **68,** 346–353 (1989).
86. L. B. Clarke and L. L. Sloss, *Trace Elements—Emissions from Coal Combustion and Gasification,* IEACR/49, IEA Coal Research, London, July 1992.
87. L. B. Clarke, *Management of By-Products from IGCC Power Generation,* IEARC/38, IEA Coal Research, London, May 1991.

88. J. O. Nriagu and J. M. Pacyna, *Nature* **333,** 134–139 (1988).
89. *Power* **118,** S25–S48 (1987).
90. T. C. Elliott, ed., *Standard Handbook of Powerplant Engineering,* McGraw-Hill Book Co., Inc., New York, 1989.
91. P. L. Langsjoen, *A Guide to Clean and Efficient Operation to Coal-Stoker-Fired Boilers,* DOE, Rept No. EPA-600/8-81-016, Report for the American Boiler Manufacturers Association, U.S. Department of Energy and U.S. Environmental Protection Agency, KVB, Inc., Minneapolis, Minn., 1981.
92. *Steam: Its Generation and Use,* Babcock & Wilcox, New York, 1972.
93. R. A. Wills, N. H. Johnson, and H. L. Knox, in Ref. 90, pp. 1.183–1.206.
94. U.S. Department of Energy, *Emissions and Efficiency Performance of Industrial Coal Stoker Fired Boilers: Data Supplement,* U.S. EPA, American Boiler Manufacturers, Arlington, Va., 1981.
95. R. A. Lisauskas and C. E. McHale, paper presented at the Joint Symposium on Stationary Combustion NO_x Control, New Orleans, La., 1987.
96. M. Angleys, *VDI Berichte* **622,** 95–121 (1987).
97. A.-K. Hjalmarsson and H. N. Soud, *Systems for Controlling NO_x from Coal Combustion,* IEA Coal Research, London, 1990.
98. J. M. Beér, J. Chomiak, and L. D. Smoot, *A Review Prog. Energy Combust. Sci.* **10,** 177–208 (1984).
99. *Power* **132,** B1–B121 (1988).
100. F. J. Ceely and E. L. Daman in Ref. 1, pp. 1313–1387.
101. U.S. Department of Energy, *Clean Coal Technology Demonstration Program: Program Update 1991 (as of December 31, 1991),* Report DOE/FE-0247P, U.S. Government Printing Office, Washington, D.C., 1992.
102. D. L. deLesdernier, S. A. Johnson, and V. S. Engleman in *Proceedings of the Fourth International Symposium on Coal Slurry Combustion,* Vol. 2, Pittsburgh Energy Technology Center, Department of Energy, Pittsburgh, Pa., 1982.
103. J. Makansi, *Power* **135,** 12–33 (1991).
104. D. Merrick, *Coal Combustion and Conversion Technology,* Elsevier, New York, 1984.
105. *The Current State of Atmospheric Fluidized-Bed Combustion Technology,* World Bank Technical Paper No. 107, Electric Power Research Institute, County Department 1, and Technical Department, Europe, Middle East, and North Africa Region, The World Bank, Washington, D.C., 1989.
106. J. N. Chapman and N. R. Johanson, *Mech. Eng.,* **113,** 64–68 (1991).
107. G. R. Seikel in B. R. Cooper and W. A. Ellingson, eds., *The Science and Technology of Coal and Coal Utilization,* Plenum Press, New York, 1984.
108. J. Makansi, *Power* **135,** 61–64 (1991).
109. R. C. Attig, J. N. Chapman, A. C. Sheth, and Y. C. L. S. Wu, *CHEMTECH* **18,** 694–702 (1988).
110. R. A. Carabotta, H. F. Chambers, and W. R. Owens, *IEEE Trans. Energy Conversion* **EC-2,** 549–555 (1987).
111. J. Douglas, *EPRI J.* **15,** 4–11 (1990).
112. M. Warshay in Ref. 107.
113. J. Makansi, *Power* **133,** 86–87 (1989).
114. V. Minkov, K. Krumpelt, E. Daniels, and J. G. Ashbury, *Power Eng.* 35–39 (July 1988).
115. J. H. Hirschenhofer in P. A. Nelson, W. W. Schertz, and R. H. Till, eds., *25th Intersociety Energy Conversion Engineering Conference Proceedings,* Vol. 3, American Institute of Chemical Engineers, New York, 1990, pp. 176–184.
116. G. Hagey in Ref. 115, pp. 159–164.
117. R. H. Goldstein in Ref. 115, pp. 170–175.
118. K. Shibuya, Y. Sakai, and K. Takeda in *23rd Intersociety Energy Conversion Engineering Conference Proceedings,* American Society of Mechanical Engineers, New York, 1988, pp. 221–226.
119. Y. Yamazaki, H. Yamazaki, and A. Fukutome in Ref. 115, pp. 165–169.
120. A.-K. Hjalmarsson, *NO_x Control Technologies for Coal Combustion,* IEACR/24, IEA Coal Research, London, Dec. 1989.
121. T. Moore, *EPRI J.* **26,** 26 (Nov. 1984).
122. R. K. Lyon, *Environ. Sci. Technol.* **21,** 231–235 (1987).
123. R. A. Perry and D. I. Siebers, *Nature* **324,** 657–658 (1986).
124. F. A. Caton and D. I. Siebers, *Combust. Sci. Technol.* **65,** 227 (1989).
125. J. A. Miller and G. A. Fisk, *C&E News* **22** (Aug. 1987).
126. L. J. Muzio, T. A. Montgomery, G. C. Quartucy, J. A. Cole, and J. C. Kramlich, paper presented at the EPRI/EPA Joint Symposium on Stationary Combustion NO_x Control, Washington, D.C., 1991.
127. K. M. Bently and Jelinek, *TAPPI J.* **72,** 123–130 (Apr. 1989).
128. F. P. Boer, L. L. Hegedus, T. R. Gouker, and K. P. Zak, *Chem. Tech.* **20,** 312 (1990).
129. J. C. Klingspor and D. Cope, *FGD Handbook: Flue Gas Desulphurization Systems,* ICEAS/B5, IEA Coal Research, London, May 1987.
130. Y. A. Attia, ed., *Proceedings of the First International Conference on Processing and Utilization of High Sulfur Coals,* Columbus, Ohio, 1985.
131. S. P. N. Singh, J. C. Moyers, and K. R. Carr, *Coal Beneficiation—The Cinderella Technology,* Coal Combustion Applications Working Group, Oak Ridge National Laboratory, Oak Ridge, Tenn., 1982.
132. Y. Liu, ed., *Physical Cleaning of Coal,* Marcel Dekker, Inc., New York, 1982.
133. M. Angleys and P. Lucat, paper presented at the 2nd International Symposium on Coal Combustion Science and Technology, Beijing, China, 1991.
134. W. Bartok, B. A. Folsom, T. M. Somer, J. C. Opatrny, E. Mecchia, R. T. Keen, T. J. May, and M. S. Krueger, paper presented at the Symposium on Stationary Combustion NO_x Control, Washington, D.C., 1991.
135. W. Bartok, B. A. Folsom, and F. R. Kurzynske, paper presented at the EPRI/EPA Joint Symposium on Stationary Combustion NO_x Control, New Orleans, La., 1987.
136. D. Thompson, T. D. Brown, and J. M. Beér in *Proceedings of the 14th Symposium (International) on Combustion,* Combustion Institute, Pittsburgh, Pa., 1973, pp. 787–799.
137. A. A. Siddigi and J. W. Tenini, *Hydrocarbon Proc* **60,** 115–124 (Oct. 1981).
138. J. S. Chang, *J. Aerosol Sci.* **20** (1989).
139. J. Gorst, D. D. Do., and N. J. Desai, *Fuel Process Technol.* **22,** 2 (1989).
140. R. J. Martin, J. T. Kelley, S. Ohmine, and E. K. Chu, *J. Eng. Gas Turbines Power* **110,** 111–116 (1988).
141. J. L. Vernon and H. N. Soud, *FGD Installations on Coal-Fired Plants,* IEACR/22, IEA Coal Research, London, Apr. 1990.
142. J. Siegel, *Energy World* **196,** i–viii (Mar. 1992).
143. W. Fulkerson, R. R. Judkins, and M. K. Sanghvi in *Energy for Planet Earth, Scientific American,* W. H. Freeman and Co., New York, 1991, pp. 83–94.